Encyclopedia of BIOETHICS

VOLUMES 1 AND 2
Complete and Unabridged

Associate Editors

K. Danner Clouser. *Professor of Humanities, College of Medicine, The Pennsylvania State University, Hershey*

H. Tristram Engelhardt, Jr. *Rosemary Kennedy Professor of the Philosophy of Medicine, Center for Bioethics, Kennedy Institute of Ethics, Georgetown University, Washington, D.C.*

John C. Fletcher. *Assistant for Bioethics, Office of the Director, Clinical Center, National Institutes of Health, Bethesda, Maryland*

Stanley Hauerwas. *Associate Professor of Theology, University of Notre Dame, Notre Dame, Indiana*

Albert R. Jonsen. *Associate Professor of Bioethics, School of Medicine, University of California, San Francisco*

Robert Neville. *Professor of Religious Studies and Philosophy, The State University of New York at Stony Brook*

Robert M. Veatch. *Senior Associate, The Hastings Center, Institute of Society, Ethics, and the Life Sciences, Hastings-on-Hudson, New York*

LeRoy Walters. *Director, Center for Bioethics, Kennedy Institute of Ethics, Georgetown University, Washington, D.C.*

Assistant Editors

James J. Doyle
David E. Ost

Managing Editor

Sandra M. Hass

Bibliographer

William Bruce Pitt

Assistant Bibliographer

Deborah Glassman

Consulting Editor

Philip P. Wiener. *Professor Emeritus of Philosophy, Temple University, Philadelphia*

Encyclopedia of BIOETHICS

WARREN T. REICH, *Editor in Chief*
GEORGETOWN UNIVERSITY

VOLUME 1

THE FREE PRESS
A Division of Macmillan Publishing Co., Inc.
NEW YORK

Collier Macmillan Publishers
LONDON

Copyright © 1978 by Georgetown University

All rights reserved. No part of this book may be reproduced or transmitted in any form or by any means, electronic or mechanical, including photocopying, recording, or by any information storage and retrieval system, without permission in writing from the Publisher.

THE FREE PRESS
A Division of Macmillan Publishing Co., Inc.
866 Third Avenue, New York, N.Y. 10022

Collier Macmillan Canada, Ltd.

Library of Congress Catalog Card Number: 78–8821

Printed in the United States of America

printing number

    4 5 6 7 8 9 10

Library of Congress Cataloging in Publication Data
Main entry under title:

Encyclopedia of bioethics.

    Includes bibliographies and index.
    1. Bioethics–Dictionaries. 2. Medical ethics–
Dictionaries. I. Reich, Warren T. [DNLM: 1. Bio-
ethics–Encyclopedias. QH302.5 E56]
QH332.E52   174′.2   78-8821
ISBN 0-02-926060-4
ISBN 0-02-925910-X (this edition)

The *Encyclopedia of Bioethics* project was made possible by a grant from the National Endowment for the Humanities. In funding the project the Endowment matched gifts from non-federal sources including The Joseph P. Kennedy, Jr. Foundation, the Raskob Foundation, The Commonwealth Fund, The Loyola Foundation, Inc., and The David J. Greene Foundation, Inc. The views expressed in this encyclopedia are not necessarily those of the Endowment or of these foundations.

Complete and unabridged edition   1982

Contents

EDITORIAL ADVISORY BOARD	*vii*
PREFACE	*xi*
INTRODUCTION	*xv*
CONTRIBUTORS	*xxiii*
SPECIAL REVIEWERS	*xxxvii*

Encyclopedia of Bioethics

APPENDIX: Codes and Statements Related to Medical Ethics	*1721*
ALPHABETICAL LIST OF ARTICLES	*1817*
SYSTEMATIC CLASSIFICATION OF ARTICLES	*1825*
ADDITIONAL RESOURCES IN BIOETHICS	*1839*
INDEX	*1845*

Editorial Advisory Board

Henry K. Beecher
    Henry Isaiah Dorr Professor of Research in Anaesthesia
    Harvard Medical School, Boston

Jean Bernard
    Director, Hematology Research Institute
    University of Paris

Gunnar Biörck
    Professor of Medicine
    Karolinska Institute, Stockholm

Franz Böckle
    Director, Faculty of Theology
    University of Bonn

Sissela Bok
    Lecturer in Medical Ethics
    Harvard Medical School, Boston

Jacob Bronowski
    Director, Council for Biology in Human Affairs
    The Salk Institute, San Diego

Daniel Callahan
    Director, The Hastings Center
    Institute of Society, Ethics and the Life Sciences, Hastings-on-Hudson, New York

Eric J. Cassell
    Clinical Professor of Public Health
    Cornell University Medical College, New York

Robert E. Cooke
    President
    Medical College of Pennsylvania, Philadelphia

Theodore Cooper
    Provost and Dean
    Cornell University Medical College, New York

Paul B. Cornely
    Senior Health Consultant
    System Sciences, Inc., Bethesda, Maryland
    Past President, American Public Health Association

Charles E. Curran
    Professor of Moral Theology
    The Catholic University of America, Washington, D.C.

William J. Curran
    Lee Professor of Legal Medicine
    Medical School and School of Public Health, Harvard University, Boston

David Daube
    Professor of Law
    University of California, Berkeley

Panagiotes Ch. Demetropoulos
    Professor of Ethics and Christian Sociology
    University of Thessalonike

EDITORIAL ADVISORY BOARD

Arthur J. Dyck
:   Mary B. Saltonstall Professor of Population Ethics
    Harvard University, Cambridge

Paul Edwards
:   Professor of Philosophy
    Brooklyn College of the City University of New York

Claudine Escoffier-Lambiotte
:   Science Editor
    Le Monde, Paris

Emil L. Fackenheim
:   Professor of Philosophy
    University of Toronto

Joel Feinberg
:   Professor of Philosophy
    University of Arizona, Tucson

David M. Feldman
:   Rabbi
    Bay Ridge Jewish Center, Brooklyn, New York

Renée C. Fox
:   Professor and Chairman of Sociology and Professor, Departments of Medicine and Psychiatry
    University of Pennsylvania, Philadelphia

Charles Frankel
:   President
    National Humanities Center, Research Triangle Park, North Carolina

Eliot Freidson
:   Professor of Sociology
    New York University, New York

Paul A. Freund
:   University Professor Emeritus of Law
    Harvard Law School, Cambridge

Josef Fuchs
:   Professor of Moral Theology
    Pontifical Gregorian University, Rome

Willard Gaylin
:   President, The Hastings Center
    Institute of Society, Ethics and the Life Sciences, Hastings-on-Hudson, New York

Otto E. Guttentag
:   Samuel Hahnemann Professor Emeritus of Medical Philosophy
    University of California, San Francisco

Bernard Häring
:   Professor of Moral Theology
    Academia Alfonsiana, Lateran University, Rome

André E. Hellegers
:   Director, Kennedy Institute of Ethics
    Georgetown University, Washington, D.C.

Louis M. Hellman
:   Director of Health Information
    Population Reference Bureau, Inc., Washington, D.C.

Milton Himmelfarb
:   Director of Information and Research Services
    American Jewish Committee, New York

Immanuel Jakobovits
:   Chief Rabbi of the United Hebrew Congregations of the British Commonwealth of Nations, London

Hans Jonas
:   Alvin Johnson Professor Emeritus of Philosophy
    New School for Social Research, New York

Jay Katz
:   Professor (Adjunct) of Law and Psychiatry
    Yale Law School, New Haven

Lester S. King
:   Contributing Editor
    Journal of the American Medical Association, Chicago

Thomas J. King
:   Director, Division of Cancer Research Resources and Centers
    National Cancer Institute, Bethesda, Maryland

Elisabeth Kübler-Ross
:   Psychiatrist
    Flossmoor, Illinois

Chauncey D. Leake
:   Senior Lecturer in Pharmacology and in the History and Philosophy of the Health Professions
    University of California, San Francisco

Philip R. Lee
:   Professor of Social Medicine and Director, Health Policy Program, School of Medicine
    University of California, San Francisco

EDITORIAL ADVISORY BOARD

Jérôme Lejeune
    Professor of Genetics
    University of Paris

Knud E. Løgstrup
    Professor of Theology
    University of Aarhus, Denmark

Richard A. McCormick
    Rose F. Kennedy Professor of Christian Ethics
    Center for Bioethics, Kennedy Institute of Ethics, Georgetown University, Washington, D.C.

Margaret Mead
    Curator Emeritus of Ethnology
    The American Museum of Natural History, New York

Jürgen Moltmann
    Professor of Theology
    University of Tübingen

Robert F. Murray, Jr.
    Professor of Genetics and Human Genetics and Professor of Pediatrics and Medicine
    Howard University, Washington, D.C.

John T. Noonan, Jr.
    Professor of Law
    University of California, Berkeley

Teizo Ogawa
    Professor of Medical History
    Juntendo University, Tokyo

Edmund D. Pellegrino
    Professor of Medicine
    Yale University
    President
    Yale–New Haven Medical Center, New Haven

Paul Ramsey
    Harrington Spear Paine Professor of Religion
    Princeton University

P. J. Roscam Abbing
    Professor of Theology
    University of Groningen, The Netherlands

Norman St. John-Stevas
    Member of Parliament
    House of Commons, London

George E. Schreiner
    Professor of Medicine and Director, Nephrology Division
    Georgetown University Medical Center, Washington, D.C.

Seymour Siegel
    Ralph Simon Professor of Theology
    Jewish Theological Seminary of America, New York

David L. Sills
    Executive Associate
    Social Science Research Council, New York

Irene Taeuber
    Senior Research Demographer
    Princeton University

A. G. M. van Melsen
    Professor of Philosophy
    University of Nijmegen, The Netherlands

Conrad H. Waddington
    Professor of Genetics
    University of Edinburgh, Scotland

William A. Wallace
    Professor of Philosophy and History
    The Catholic University of America, Washington, D.C.

Maurice H. F. Wilkins
    Professor of Biophysics
    King's College, University of London

Preface

When a special encyclopedia—one aiming to be comprehensive in a particular discipline—is the first in its field, its appearance has the potential for marking an important stage in the development of one aspect of human knowledge. Yet its production invariably is attended by unique problems that can be surmounted only by a combination of loyal support, careful planning, and capable personnel. The *Encyclopedia of Bioethics* has been no exception to this rule.

This encyclopedia could not have been developed had it not been for the intellectual vision and continual labors of André E. Hellegers, first director of The Joseph and Rose Kennedy Institute of Ethics at Georgetown University. His constant supportive efforts made possible the launching and sustaining of this project.

Georgetown University provided a stimulating and supportive academic atmosphere in addition to financial assistance for carrying out this project. We are particularly indebted to two Presidents of the University, Robert J. Henle and Timothy S. Healy, for their faith in this project.

The National Endowment for the Humanities, a federal grant-making agency, provided the principal funding for the project by matching gifts from private sources, including The Joseph P. Kennedy, Jr. Foundation, the Raskob Foundation, The Commonwealth Fund, The Loyola Foundation, Inc., and The David J. Greene Foundation, Inc. Ronald Berman, Chairman of the National Endowment for the Humanities during most of this project, and Robert Kingston, Deputy Chairman, showed a strong interest in the project at its inception and gave their consistent support through its final stages. The encyclopedia was completed while Joseph Duffey was Chairman of the Endowment. The *Encyclopedia of Bioethics* is a project of the Endowment's Program of Science, Technology and Human Values. We are grateful to Richard Hedrich, coordinator of that program, for his continuous encouragement and assistance.

The organizations that funded the encyclopedia project, and Georgetown University, were firmly committed to the principle that the editors and contributors would present all views—in an interdisciplinary, intercultural, and international context.

As Senior Research Scholar at The Kennedy Institute during the first year of its existence (1971–1972), I conceived the idea of developing a reference work that would gather together what was known about the scientific state of the art and the full range of ethical views and policy options in matters dealing with the life sciences. After extensive consultation with experts in the organization of knowledge, in the fields bearing on bioethics, in editing, and in publishing, I developed the basic plan for the project and selected the Editorial Advisory Board, the Associate Editors, and the editorial staff. As Editor in Chief, I directed the various phases of the work, commissioned the articles, maintained direct contact with the contributors, and determined the acceptability of the articles and of all other materials in light of the purpose, scope, and desired level of quality of the work.

In the fall of 1972, approximately twenty-two members of the Editorial Advisory Board assembled for two meetings held at Georgetown University to test the feasibility of this encyclopedia and to recommend topics, organizational ideas, and personnel. The sixty members of the Editorial Advisory Board, listed elsewhere in these introductory pages, are from eleven countries and represent all the disciplines converging on bioethics. They reviewed several drafts of the table of contents, made suggestions of topics and contributors, and responded to specific editorial questions that arose in the course of the work. Some of them also reviewed manuscripts. It is a pleasure to thank this distinguished international and interdisciplinary Editorial Advisory Board for their advice and encouragement.

Eight Associate Editors, who played a crucial role in designing and evaluating the contents of the work, were selected on the basis of their outstanding knowledge of the field of bioethics, the diversity of their intellectual and methodological backgrounds, and the special areas in which each of them had excelled. They contributed to this project a level of expertise and dedication that could scarcely have been matched by any other group.

Each Associate Editor submitted descriptions of the intended scope of the articles in a specific area, prepared a list of contributors for each article, judged the quality of the manuscripts, and decided on their acceptability. We integrated the entire plan at several meetings and constantly adapted it as the articles were being prepared.

The Associate Editors and I, working collaboratively, exercised full autonomy in setting editorial policies, selecting contributors, and approving the contents of the work.

A complete editorial staff was hired specifically for this project at The Kennedy Institute. This group included the assistant editors, managing editor, bibliographers, and editorial secretaries. In addition, we hired research assistants, bibliographic assistants, copy editors, proofreaders, an assistant managing editor, computer specialists, and a clerical staff to perform specific tasks as needs arose.

Sandra Hass, the encyclopedia's managing editor, did outstanding work in directing the editorial staff and coordinating contact with the contributors, advisors, reviewers, and publisher. We are indebted to her for her loyalty and her persistence in setting and maintaining high standards of excellence for stylistic editing of the work.

The consulting editor, Philip Wiener, editor of the *Dictionary of the History of Ideas*, gave extensive advice and also made editorial suggestions regarding the articles.

The assistant editors rendered an indispensable service, especially in correcting and evaluating manuscripts at the revision stage. Many phases of the project could not have been brought to successful completion without the competent help of the research and editorial staffs; I would like to thank especially Cheryl Calhoun, Sara Hannum Chase, Ronald S. Gass, Nina Shafran, Mary Ellen Timbol, and Susan White.

The bibliographies are a major component of this encyclopedia. The task of verifying and copy editing all bibliographic listings, as well as occasionally suggesting additional bibliographic items, was carried out with meticulous and professional attention by the bibliographer, William Pitt. He was assisted at various times by a total of twenty-seven persons.

Credit for the quality of the articles must go principally to the contributors. They accommodated themselves to the integrated plan, the scholarly standards, the succinct yet comprehensive style, and the demanding schedule of the encyclopedia. Edited manuscripts were returned to the contributors for their correction and approval; a final copy of the text was mailed to them in the form of printed galleys.

The following standards guided us in the selection of contributors: that each should have highly respected qualifications to write on the topic as well as the ability to write well, and that they should include diverse intellectual backgrounds and appropriate geographical and institutional representation. The names and institutional affiliations of all contributors are listed in the front matter of this encyclopedia, together with the titles of their articles. We are especially grateful to those contributors who prepared excellent articles on extremely short notice as we approached the end of the project.

All manuscripts were thoroughly reviewed for appropriateness of scope, accuracy of content, and clarity of style by at least four editors. In addition, we called on special reviewers to evaluate manuscripts that included material beyond the special expertise of the editors. These special reviewers, whose names are listed elsewhere in the front matter of this encyclopedia, made an important contribution to the accuracy of this work.

My colleagues, the scholars of The Kennedy Institute, have been most generous with their time and advice, and have been a constant and indispensable resource for the development of this project. They include these philosophers, religious ethicists, biologists, physicians, demographers, economists, and lawyers, who have been permanent and visiting scholars of the Institute: Francisco Abel, Robert C. Baumiller, Tom L. Beauchamp, Roy Branson, Frederick Carney, Kenneth Casebeer, James F. Childress, John Connery, Charles E. Curran, H. Tristram Engelhardt, Jr., Murray Gendell, Bernard Häring, Stanley Hauerwas, André E. Hellegers, James Jones, Leon R. Kass, Patricia A. King, Karen Lebacqz, Sid Z. Leiman, Richard A. McCormick, Thomas Merrick, Gene Outka, Seymour Perlin, Ralph B. Potter, Jeanne Clare Ridley, Bruno Schüller, Henry S. Shryock, Joan Sieber, Seymour Siegel, David Smith, Conrad Taeuber, and LeRoy Walters.

In the early planning stages of the encyclopedia project, the editors of several scholarly journals agreed to serve as advisory editors. They were asked to recommend contributors and reviewers and to comment on the topics to be included in the work. Several of them also served as critical reviewers of manuscripts. They are Philip H. Abelson, editor of *Science;* Walter J. Burghardt, editor of *Theological Studies;* Marshall Cohen, editor of *Philosophy & Public Affairs;* T. George Harris, editor in chief of *Psychology Today;* Franz J. Ingelfinger, editor of *The New England Journal of Medicine;* Paul Kurtz, editor of *The Humanist;* Peter J. M. McEwan, editor in chief of *Ethics in Science & Medicine;* Robert H. Moser, editor of *The Journal of the American Medical Association;* Nicholas Rescher, editor of *American Philosophical Quarterly;* Wolfgang Schweitzer, editor of *Zeitschrift für Evangelische Ethik;* and Edward F. Shotter, member of the Editorial Board for the *Journal of Medical Ethics.*

Preparation of the bibliographies, which embrace a wide variety of disciplines and cultures, could not have been accomplished without the resources and personnel of many libraries. We wish to thank the following libraries of Georgetown University: The Center for Bioethics Library, Kennedy Institute of Ethics; Lauinger Library—its Reference, Serials, Government Documents, Acquisitions, Special Collections, Cataloging, and Circulation staffs; Woodstock Theological Center Library; Dahlgren Medical Library; Bloomer Science Library; the Library of the Center for Population Research; and the Dennis Library of the Georgetown University Law Center.

We are also indebted to: The Library of Congress, especially its Reference staff, its Law Library, and its Orientalia Division; the National Library of Medicine, its Reference staff, and its Rare Book staff; the National Institutes of Health Library; the Theology, Philosophy, and Canon Law Library of the Catholic University of America; The George Washington University Library and Burns Law Library; the American University Library; Wesley Theological

Seminary Library; the Howard University Medical-Dental Library; the University of Maryland libraries; and the State Library of Maryland.

There were many university, college, public, and private special libraries in the District of Columbia area that helped us, as well as those librarians who responded to telephone inquiries throughout the United States and Canada. Special reference help came from the United Nations and associated international organizations, as well as from various embassies, notably those of Canada, the United Kingdom, the Federal Republic of Germany, and the Peoples' Republic of China.

Many individuals, too numerous to mention, assisted in the completion of this encyclopedia. We thank all of them, especially David Eggenberger, David Sills, and John Bradley for their advice and encouragement; Dorothy Sawicki for her careful preparation of manuscripts for the printer, her supervision of the proofreading, and her correction of the galley proofs; George Rowland of The Free Press, who served as the publisher's managing editor, for his extraordinary cooperation and steadfast concern for the quality and appearance of this work; William E. May for his astute evaluation and speedy editing of manuscripts in the crucial final stage of the project; Doris Goldstein, Kennedy Institute librarian, for her cooperation in obtaining information and for preparing Additional Resources in Bioethics, which appears at the end of this encyclopedia; David Webster for his helpful legal advice; and the following for their advice on topics and contributors: James Luther Adams, Harry Gordon, James Gustafson, R. M. Hare, Joseph Kitagawa, Frank Reynolds, Daniel Robinson, Lloyd Stevenson, and Owsei Temkin.

A special acknowledgment must be given to The Joseph P. Kennedy, Jr. Foundation for contributing generously to the support of this project and for establishing The Joseph and Rose Kennedy Institute of Ethics, a research center at Georgetown University. This encyclopedia project could not have been completed without the scholarly resources of The Kennedy Institute. It was The Kennedy Foundation's deep interest in the well-being of the mentally retarded and of children—groups so vulnerable to neglect and exploitation—that led the Foundation to create The Kennedy Institute and to support a variety of research and teaching endeavors in the field of bioethics.

Finally, a special expression of gratitude goes to Sargent Shriver, Chairman of The Kennedy Institute's International Advisory Board. He has worked with singular dedication for the support of this project and for the advancement of the discipline of ethics, both basic and applied, in the biomedical and other professions.

WARREN T. REICH
Senior Research Scholar, Center for Bioethics,
The Joseph and Rose Kennedy Institute of Ethics
Associate Professor of Bioethics, School of Medicine
Georgetown University
Washington, D.C.

Introduction

The *Encyclopedia of Bioethics* is the first encyclopedia in its field. Its purpose is to synthesize, analyze, and compare the positions taken on the problems of bioethics, in the past as well as in the present, to indicate which issues require further examination, and to point to anticipated developments in the ethics of the life sciences and health care.

The emergence of bioethics as a field of study is a contemporary phenomenon traceable to several causes. First, the issues of bioethics have captured the contemporary mind because they represent major conflicts in the area of technology and basic human values, those dealing with life, death, and health. Although many bioethical issues have been discussed since ancient times, the introduction of modern biomedical technologies, especially since the 1950s, has intensified some age-old questions and has given rise to perplexing new problems—the prolongation of life, euthanasia, prenatal diagnosis and abortion, human experimentation, genetic interventions and reproductive technologies, behavior control and psychosurgery, the definition of death, the right to privacy, allocation of scarce health resources, and dilemmas in the maintenance of environmental health.

Second, there is an intense and widespread interest in bioethics because it offers a stimulating intellectual and moral challenge. In contrast to earlier eras, when ethical world views were held in common and offered a certain security for dealing with moral dilemmas, today the very tools for coping with these problems are themselves subject of considerable controversy. There is an uncertainty about moral values, ethical principles, and their priorities; the contemporary world is experiencing a philosophical upheaval; and many systems of theological thought are questioning traditional assumptions in religious ethics. Bioethics has already had a significant intellectual impact, for it has precipitated a reexamination of basic moral values and methods of applying them to practical ethical questions.

Third, the rapid growth of the field of bioethics has been facilitated by the openness to multidisciplinary work that characterizes many scholars and academic institutions today, especially in matters dealing with personal and social aspects of human behavior.

Bioethics, like other special fields of learning, has manifested the "explosion of knowledge" characteristic of our era. Courses of study in bioethics have proliferated in universities, professional schools, colleges, and high schools; a number of scholarly journals specially devoted to the field have begun publication; hundreds of journals in disciplines bearing directly on bioethical concerns—such as medicine, law, psychology, philosophy, religious studies, biology, and political sciences—have dramatically increased the amount of space they offer to bioethics articles; monographs and anthologies have multiplied; and research institutes have been established.

The Editors and Editorial Advisory Board of this work were convinced that a reference work was needed that would gather together, as comprehensively yet as succinctly as possible, what is known about the questions of bioethics, the data from which they arise, a full range of responses to them, and the principles and values that account for those responses.

We believe that the articles and accompanying bibliographies in the *Encyclopedia of Bioethics* will be helpful to those who have been seeking a reliable and comprehensive treatment of topics in bioethics. Our prospective readers fall into four groups:

- teachers and their students—in departments or programs of philosophy, religion, biology, medicine, social sciences, psychology, law, health sciences, and other disciplines—who are increasingly involved in education, research, and writing in the field
- nonacademic scholars, journalists, and other writers whose important contributions to bioethics would be greatly strengthened by a reference work in this field
- health professionals who would draw on such a work to inform and enrich their professional lives, e.g., physicians, health administrators, nurses, psychologists, social workers, lawyers, and counselors
- those involved in the formulation and execution of public policy bearing on the life sciences and health care, such as legislators, judges, public officials, health planners, and members of research committees

This encyclopedia was designed, then, not only to meet educational needs, but to serve as a resource for professional decision making, for the development of public policies, and for further research.

Although it is unusual, perhaps unprecedented, for a special encyclopedia to be produced almost simultaneously with the emergence of its field, several reasons persuaded us that this work was not premature. First, many of the issues and methodological roots of bioethics are not new; they were waiting to be gleaned from centuries of literature in the fields of philosophy, medical ethics, history of medicine, and other fields. Second, the plethora of contemporary literature badly needed systematization: Approximately 1,500 significant items are published annually in the English language alone. Third, an encyclopedia could assist the formation of this field of study.

In a field as fast-growing as bioethics, one difficulty is that the encyclopedia might be outdated too quickly. This problem is not unique to bioethics: Most active fields of learning are subject to constant change and development, with the result that most special encyclopedias, and even general encyclopedias, are faced with the need to adapt to a rapid growth in knowledge. An encyclopedia cannot be a vehicle for definitive statements in fixed categories of thought; it must be a tool that reflects and facilitates the development of ideas. The purpose of this encyclopedia, then, was not to freeze knowledge, but to summarize and analyze the historical and current state of knowledge in bioethics in such a way as to facilitate learning what is known in this area and to stimulate further thought and research. Furthermore, we have attempted to minimize the problem of obsolescence by asking the contributors to emphasize major positions and schemes of analysis. Supplements to or revisions of this reference work can serve future needs by reporting new scientific data and summarizing further developments in the field of bioethics.

SCOPE OF THE ENCYCLOPEDIA

The range of subject matter within this encyclopedia was chosen to make the work reasonably self-contained. The scope of topics, discussed in more detail below, can be viewed at several levels: (I) the range of *concrete ethical and legal problems* included within bioethics; (II) the basic *concepts* that clarify bioethical issues and the *principles* appealed to as guides for human behavior in this area; (III) the various *ethical theories* that account for how one knows human values and justifies the norms that should guide human conduct; (IV) the *religious traditions*, which also account for what is good and bad, right and wrong in bioethical matters; (V) *historical perspectives*, particularly in the traditional area of medical ethics which deals specifically with the physician–patient relationship; and (VI) information about *disciplines bearing on bioethics*.

The following outline offers an overview of the contents of the encyclopedia under the six headings mentioned above.

I. Concrete Ethical and Legal Problems. This category of entries deals with specific normative questions—i.e., value questions requiring an answer as to what ought or ought not to be done. These questions may arise, for example, in reference to specific kinds of actions (such as sterilization of the retarded), value aspects of relationships (e.g., What sort of confidentiality should characterize the psychiatrist–patient relationship?), and public policy options and legal structures (e.g., What policy should govern the cessation of medical treatment of the incompetent elderly?). More than any other single factor, the range of these issues defines the scope of bioethics. The following headings represent the broad spectrum of concrete ethical problems in bioethics; under each heading some examples are given representing article titles in the encyclopedia.

The Therapeutic Relationship—e.g., truth-telling, informed consent, confidentiality, medical malpractice, and the patients' rights movement. We have also singled out some dominant health professions as titles of separate articles, trusting that the issues they encounter would serve as models for discussing the ethical problems of the innumerable health professions and specialties. Some of the articles treating individual professions include medicine, nursing, dentistry, surgery, pharmacy, medical social work, public health, and social medicine. The mental health professions are included topically, that is, in articles dealing with such topics as mental health therapies.

Codes of Professional Ethics—articles dealing with the history and ethical analysis of codes of medical ethics. The encyclopedia also includes an Appendix, which is a selective annotated collection of the texts of the most important codes from the earliest to contemporary times. The Appendix is organized under four headings: I. GENERAL CODES FOR THE PRACTICE OF MEDICINE; II. DIRECTIVES FOR HUMAN EXPERIMENTATION; III. PATIENTS' BILLS OF RIGHTS; and IV. CODES OF SPECIALTY HEALTH-CARE ASSOCIATIONS. Highlights of most of these codes are synthesized and analyzed in the Appendix's introductory article on CODES OF THE HEALTH-CARE PROFESSIONS.

Health Care—including such topics as humanization and dehumanization of health care, right to health-care services, rationing of medical treatment, international health, and health policy.

Sociopolitical Problems in Biomedicine—for example, torture and the health professional, biomedical science and war, and racism.

Biomedical and Behavioral Research—including articles on the social and professional control of human experimentation, informed consent in human research, research on children, and research on the mentally handicapped.

Mental Health and Behavioral Issues—embracing such questions as drug use, abuse, and dependence; self-realization therapies; behavior control; psychosurgery; and institutionalization.

Sexuality, Contraception, Sterilization, and Abortion—embracing such topics as sex therapy and sex research; sexual ethics; abortion: historical and contemporary perspectives; and sterilization: ethical and legal aspects.

Genetics—including genetic screening, genetic counseling, gene therapy, eugenics, and genetic aspects of human behavior.

Reproductive Technologies—for example, artificial insemination, in vitro fertilization, sex selection, and asexual human reproduction ("cloning").

Organ and Tissue Transplantation and Artificial Organs—blood transfusion, organ transplantation, heart transplantation, and kidney dialysis and transplantation.

Death and Dying—including such topics as definition and determination of death, euthanasia and sustaining life, suicide, and ethical perspectives on the care of infants.

Population—e.g., history of population policies; compulsory population control programs; and genetic implications of population control.

Environment—e.g., environmental health and human disease, and the problem of growth.

II. **Basic Concepts and Principles.** Also discussed in this encyclopedia are underlying concepts that clarify or give shape to ethical attitudes and positions in bioethics, e.g., health and disease, pain and suffering, life, death in Eastern and Western thought, and paternalism. Principles, i.e., norms of right conduct, are listed under such titles as justice, rights, double effect, and obligations to future generations.

III. **Ethical Theories.** Systematic philosophical methods of viewing the moral life and of accounting for what is good and right in human conduct are explained, for example, in the entry titled ETHICS. This entry—which is intended to be a virtual primer in ethical theory with examples taken from bioethics—contains twelve articles dealing with a broad range of ethical theories, beginning with the introductory article on THE TASK OF ETHICS, and including such titles as UTILITARIANISM, DEONTOLOGICAL THEORIES, RULES AND PRINCIPLES, and THEOLOGICAL ETHICS. The reader would do well to relate these articles and the entries on religious traditions to the entry on BIOETHICS to grasp the ethical underpinnings of the field.

IV. **Religious Traditions.** The great traditions of religious ethics embody beliefs and principles that are applicable to problems in bioethics. The implications of these religious traditions for bioethics have been synthesized —several of them for the first time—in this encyclopedia. They include Buddhism, Confucianism, Eastern Orthodox Christianity, Hinduism, Islam, Judaism, Protestantism, Roman Catholicism, and Taoism. Additionally, some articles deal expressly with the ethical approaches taken by various religious groups to specific issues in bioethics. These are found either as separate titles—e.g., Islamic perspectives on population ethics— or combined with nonreligious views, as in the article on the ethical aspects of sterilization.

V. **Historical Perspectives.** A unique feature of this encyclopedia is the comprehensive 97,000-word entry on the history of medical ethics. Its twenty-nine articles, designed to reward a reading either separately or in sequence, are divided into four sections: PRIMITIVE SOCIETIES; NEAR AND MIDDLE EAST AND AFRICA; SOUTH AND EAST ASIA; and EUROPE AND THE AMERICAS. Each section contains articles in chronological sequence, with such titles as: ANCIENT NEAR EAST; PREREPUBLICAN CHINA; MEDIEVAL EUROPE: FOURTH TO SIXTEENTH CENTURY; NORTH AMERICA: SEVENTEENTH TO NINETEENTH CENTURY; and EASTERN EUROPE IN THE TWENTIETH CENTURY. Other historical articles are situated topically within the encyclopedia, such as those on the history of the therapeutic relationship and the history of human experimentation.

VI. **Disciplines Bearing on Bioethics.** Some entries survey an entire discipline, presenting in succinct fashion an overview of the discipline and its bearing on bioethics. In this category are entries on the philosophy of biology; the anthropology of medicine; the philosophy of medicine; the sociology of medicine; and the sociology of science.

DEFINING THE FIELD OF BIOETHICS

Bioethics is a composite term derived from the Greek words *bios* (life) and *ēthikē* (ethics). It can be defined as the systematic study of human conduct in the area of the life sciences and health care, insofar as this conduct is examined in the light of moral values and principles.

Bioethics is an area of interdisciplinary studies whose focus depends on the kinds of issues it examines and the nature of ethical inquiry. The scope of the field has been outlined earlier in this Introduction under "Scope of the Encyclopedia," especially in the thirteen areas of "Concrete Ethical and Legal Problems." Further explanation may help define more fully the scope of bioethics as presupposed in this encyclopedia and in the interdisciplinary approach we have taken.

Bioethics encompasses medical ethics and extends beyond it. Medical ethics in its traditional meaning deals with value-related problems that arise in the physician–patient relationship. Bioethics is more inclusive in four significant respects:

- It embraces the value-related problems that arise in all health professions, including the "allied" health professions, mental health professions, and so forth.
- It extends to biomedical and behavioral research, whether or not that research has a direct bearing on therapy.
- It includes a broad range of social issues, such as those associated with public health, occupational health, international health, and the ethics of population control.
- It extends beyond human life and health to embrace issues involving animal and plant life, e.g., in the topics dealing with animal experimentation and competing environmental claims.

These four main areas can be accounted for as the cardinal issues of bioethics by two key terms in our definition of bioethics: the *life sciences* and *health care*. Both of these terms have acquired broad connotations in the contemporary world. Health care cannot be restricted to physical health nor confined to the medical profession. A life science may also be defined somewhat broadly as "a branch of science (as biology, medicine, anthropology, or sociology) that deals with living organisms and life processes" (*Webster's New Collegiate Dictionary*, 1973).

The appropriateness of including a broad range of health professions within bioethics is obvious simply because health-care services are rendered by so many professions. Both biomedical and behavioral research can be viewed either as activities of the life sciences or as endeavors inextricably linked with health care or both. Similarly, a broad range of social issues is associated with health care; the competing interests of human, animal, and plant life are deeply enmeshed in the life sciences; and demographic concerns are inseparable from health care and the life sciences.

For example, environmental health and occupational health raise ethical issues at least as important as those encountered on an individual basis in the clinic; and the biological disciplines, so central to the life sciences, embrace not only molecular and cell biology, but organismic biology (including ecology, eugenics, and genetics) and population biology (which has an integral relationship with population ecology, population genetics, and sociology). Furthermore, even the perennial problems of medical ethics, such as the ethics of sterilization and abortion, cannot be adequately discussed except in the context of demographic and environmental perspectives.

Certainly not all scholars would agree on the precise scope of bioethics. The explanation offered here gives a rationale for the broad perspective on the scope of bioethics that is presumed in the structure of this encyclopedia.

Bioethics is further defined by the nature of ethical inquiry and the competence presumed for it.

Contributions are made to bioethics by philosophers, biologists, physicians, theologians, psychologists, lawyers, various health professionals, social scientists, and many others. Some have claimed that the right ethical answers can be known directly from the implications of the biomedical sciences themselves, or simply from sociological studies of human behavior in the area of biomedicine. These claims represent particular theories of moral knowledge; but it is the aim of this encyclopedia to present all significant theories and viewpoints.

The assumption underlying this work is that all the contributing disciplines are concerned with an evaluative endeavor and that each discipline brings a different competence and methodology to such questions as these: What should be done (e.g., in supporting human life)? Who should decide? What policies should guide society in these areas? In assembling the encyclopedia, we have been guided by the special competence of each discipline.

Leading scientists were invited to prepare articles about the biomedical aspects of topics such as sterilization. These scientists were asked to present accurate scientific data on the "state of the art" to the extent necessary for discussing the ethical questions and to give a clear statement of the ethical issues arising in research or practice. The biomedical sciences make further contributions to the delineation of ethical issues by offering conceptual schemes, explaining the consequences of therapeutic actions, and adopting certain notions of professional responsibility.

Articles dealing with social or sociological aspects of bioethical questions explain the behavior of individuals and the policies that guide them, as well as some of the factors that account for behavior and preferred public policies; but the social science contributors were not asked to analyze the ultimate ethical question—what that behavior or policy ought to be. Again, the conceptual schemes offered by the sociologists are helpful in understanding the ethical issues and options, e.g., in the area of human experimentation.

Articles on legal aspects of bioethical questions—for example, the legal aspects of reproductive technologies or of informed consent in the therapeutic relationship—do not offer a mere history of statutory law in the area but are a systematic presentation of the legal principles employed, illustrated by landmark cases. Our assumption has been that the law supplies concepts and principles of action that are useful in arriving at ethical conclusions and in developing better public policies, but that the law itself should be examined from an ethical perspective to test its suitability as public policy.

Articles dealing explicitly and systematically with the ethical aspects of a topic—i.e., with a summary of ethical viewpoints and an explanation of their justifications—were usually prepared by leading ethicists: philosophers and theologians whose special area of expertise is ethics, many of whom have specialized even further in bioethics. They were asked to summarize a variety of ethical views in terms of values and goods, rights and responsibilities, duties and obligations. When these articles were prepared by someone from a discipline other than ethics, such as medicine or law, the contributor was selected particularly because of his or her special competence in ethics.

In some cases, rather than offer two articles, for example, on the medical and ethical aspects of a topic, we achieved a multidisciplinary approach to the subject by arranging for coauthorship or by enlisting a contributor with competence in more than one field.

INTRODUCTION xxi

SPECIAL FEATURES OF THE ENCYCLOPEDIA

The international and interdisciplinary character of the encyclopedia is evident from the fact that the 285 contributors are from fifteen countries and represent many fields of learning including ethics, biology, medicine, the allied health professions, psychology, sexology, philosophy, religion, sociology, anthropology, law, political science, demography, and history.

Contributors were asked to make all topical presentations as international as possible. While they could not be expected to be equally competent in the mores, laws, and ethical systems found in every part of the world, and while some emphasis is placed understandably on the country of origin of this encyclopedia, every effort has been made—both in stating issues and in examples used—to give the work a pronounced international dimension. Articles in the entry MEDICAL ETHICS, HISTORY OF offer an explicit treatment of bioethical themes as they are perceived in many parts of the world.

All of the entries consist of original, signed articles. Contributors were asked to present significant ethical viewpoints from a variety of perspectives, rather than to settle differences of opinion arbitrarily or articulate a purely personal view. They were also asked to compare and contrast various methodologies or approaches, so as to encourage ethical reflection on the part of the reader. A few contributors were asked to write creative essays in areas that have been notably neglected in bioethical studies. A number of the essays will undoubtedly be unique contributions to the literature of bioethics.

All encyclopedias must face the problem of bias, which affects even the most distinguished scholars in all fields of learning. In an encyclopedia dealing exclusively with ethics and values, the presumption is that biases and ideologies might insinuate themselves even more easily. The contributors were asked to give fair representation to a variety of ethical views on a given topic. A number of well-written articles were rejected because they did not meet this requirement. Frequently the perspective of the individual author cannot be entirely concealed: If some ethical, philosophical, or religious perspectives are not fully acknowledged in a given article, we believe this is counterbalanced by the breadth of perspectives offered in related entries.

All articles are directed to the educated person who may or may not be a specialist on the topic. In some instances prior knowledge of the subject will prove helpful, but every effort has been made to offer a clear and direct exposition of the material for the uninitiated reader and to avoid the use of undefined technical terms.

There is a total of 315 articles in the encyclopedia: 314 articles are included in the body of the work and an additional article introduces the Appendix. The articles range in length from 800 to 12,000 words, with an average length of 3,400 words.

Entries are arranged alphabetically. To achieve a systematic and multidisciplinary coherence among related topics, some *entries* are composed of several *articles*. For example:

> REPRODUCTIVE TECHNOLOGIES
> I. SEX SELECTION
> II. ARTIFICIAL INSEMINATION
> III. SPERM AND ZYGOTE BANKING
> IV. IN VITRO FERTILIZATION
> V. ASEXUAL HUMAN REPRODUCTION
> VI. ETHICAL ISSUES
> VII. LEGAL ASPECTS

In the above entry, the ethical principles that can be applied in the first five articles are discussed in depth only in the sixth article. We have followed this pattern on occasion throughout the encyclopedia: One or more articles on a given topic explain the scientific state of the art and present the ethical issues associated with the use of certain technologies or procedures, and a subsequent article in the same entry systematically discusses the relevant ethical principles.

Some entries name in their titles particular populations—e.g., CHILDREN AND BIOMEDICINE, ADOLESCENTS, WOMEN AND BIOMEDICINE, and AGING AND THE AGED. These articles contain discussions of ethical issues concerning both the therapeutic relationship and human research with these populations.

The encyclopedia contains no biographical articles, because the scholars who have made outstanding contributions to bioethics are identified with a great variety of disciplines, and their biographies are generally available in biographical dictionaries and other encyclopedias. Our contributors were asked simply to mention the names and contributions of those figures who have exerted a significant influence on the ethical principles pertinent to the life sciences and health care.

This encyclopedia appears at a time when many writers and publishers are attempting to avoid noninclusive ("sexist") language. The editors of this encyclopedia have urged the contributors to accept inclusive language wherever possible, but also made efforts to avoid the sometimes awkward and inaccurate alternatives to traditional usage.

An important resource of the encyclopedia is the bibliographies following each article. These select bibliographies were prepared by the contributors and were verified, edited, and occasionally supplemented by the encyclopedia's bibliographic staff. The bibliographic items refer the reader to sources used by the contributor, some of which have been cited in the text in parentheses, e.g., (Smith, 1972, p. 50); to further readings; and to works containing additional bibliographies.

Use of this encyclopedia is facilitated by a carefully worked out system of cross-references that lead the reader to other entries or to parts of the Appendix that offer further information relevant to the article at hand. The principal cross-referencing device is found in a paragraph following each article. Here the reader is referred to other articles, which are listed in categories of descending importance as regards their relevance to the present article. Occasionally the cross-references appear in an introductory paragraph preceding an entry, and some are located within the article itself.

In cross-references of every type we have used a uniform style to help the reader differentiate between the titles of entries and articles. Entry titles use large capitals for the first letter of each main word; article titles use all small capitals—e.g.:

> [For further discussion of topics mentioned in this article, see the entry HEALTH AND DISEASE, article on HISTORY OF THE CONCEPTS.]

The alphabetical arrangement of the entries and the various cross-referencing mechanisms make the entire body of the encyclopedia self-indexing. Nonetheless, an indispensable tool for the use of this encyclopedia is its thorough index: Many specific topics and names can be found only by making use of the index.

Further research aids, found in the last volume, include an "Alphabetical List of Articles" with contributors, a "Systematic Classification of Articles," and a list of "Additional Resources in Bioethics."

WARREN T. REICH
Editor in Chief

Contributors

Abdul-Rauf, Muhammad
> Director, The Islamic Center (Washington, D.C.)
> MEDICAL ETHICS, HISTORY OF, section on NEAR AND MIDDLE EAST AND AFRICA, article on CONTEMPORARY MUSLIM PERSPECTIVE

Allen, David Franklyn
> Associate Professor of Psychiatry and Religion, The Divinity School, Yale University
> MENTAL HEALTH SERVICES, article on EVALUATION OF MENTAL HEALTH PROGRAMS

Amundsen, Darrel W.
> Associate Professor of Classics, Western Washington University
> MEDICAL ETHICS, HISTORY OF, section on NEAR AND MIDDLE EAST AND AFRICA, article on ANCIENT NEAR EAST; section on EUROPE AND THE AMERICAS, articles on ANCIENT GREECE AND ROME and MEDIEVAL EUROPE: FOURTH TO SIXTEENTH CENTURY

Anderson, Odin W.
> Professor and Director, Center for Health Administration Studies, University of Chicago
> HEALTH POLICY, article on HEALTH POLICY IN INTERNATIONAL PERSPECTIVE

Annas, George J.
> Associate Professor of Law and Medicine, School of Medicine, Boston University
> PATIENTS' RIGHTS MOVEMENT

Asper, Samuel P.
> Professor of Internal Medicine and Dean, School of Medicine, American University of Beirut
> MEDICAL ETHICS, HISTORY OF, section on NEAR AND MIDDLE EAST AND AFRICA, article on CONTEMPORARY ARAB WORLD

Aufhauser, Marcia Cavell
> Associate Professor of Philosophy, State University of New York, College at Purchase
> SEXUAL DEVELOPMENT

Back, Kurt W.
> James B. Duke Professor and Chairman, Department of Sociology, Duke University
> SELF-REALIZATION THERAPIES

Baier, Kurt
> Professor of Philosophy, University of Pittsburgh
> ETHICS, articles on DEONTOLOGICAL THEORIES and TELEOLOGICAL THEORIES

Bailey, Lloyd
> Associate Professor of Old Testament, Duke University
> DEATH, article on WESTERN RELIGIOUS THOUGHT: DEATH IN BIBLICAL THOUGHT

Bajema, Carl Jay
> Professor of Biology, Grand Valley State Colleges (Allendale, Michigan)
> POPULATION POLICY PROPOSALS, article on GENETIC IMPLICATIONS OF POPULATION CONTROL

Baker, Robert
> Assistant Professor of Philosophy, Union College (Schenectady, New York)
> MENTAL ILLNESS, article on CONCEPTIONS OF MENTAL ILLNESS; VIOLENCE AND THERAPY

Barbour, Ian G.
> Professor of Religion and Professor of Physics, Carleton College (Northfield, Minnesota)
> ENVIRONMENT AND MAN, article on WESTERN THOUGHT

Basham, A. L.
> Professor and Head, Department of Asian Civilizations, Australian National University
> HINDUISM

Bassett, William W.
> Professor of Law, University of San Francisco
> EUGENICS AND RELIGIOUS LAW, article on CHRISTIAN RELIGIOUS LAWS

Bayles, Michael D.
: *Professor of Philosophy, University of Kentucky*
POPULATION POLICY PROPOSALS, article on POPULATION DISTRIBUTION

Beauchamp, Tom L.
: *Associate Professor, Department of Philosophy and Senior Research Scholar, Center for Bioethics, Kennedy Institute of Ethics, Georgetown University*
PATERNALISM

Bleich, J. David
: *Professor of Talmud, Yeshiva University*
POPULATION ETHICS: RELIGIOUS TRADITIONS, article on JEWISH PERSPECTIVES

Block, Ned
: *Associate Professor of Philosophy, Massachusetts Institute of Technology*
REDUCTIONISM, article on PHILOSOPHICAL ANALYSIS

Blomquist, Clarence
: *Department of Psychiatry, Karolinska Hospital (Stockholm)*
MEDICAL ETHICS, HISTORY OF, section on EUROPE AND THE AMERICAS, article on WESTERN EUROPE IN THE TWENTIETH CENTURY

Bloom, Samuel W.
: *Professor of Sociology and Community Medicine, City University of New York*
THERAPEUTIC RELATIONSHIP, article on SOCIOHISTORICAL PERSPECTIVES

Bok, Sissela
: *Lecturer in Medical Ethics, School of Medicine, Harvard University*
DEATH AND DYING: EUTHANASIA AND SUSTAINING LIFE, article on ETHICAL VIEWS; TRUTH-TELLING, article on ETHICAL ASPECTS

Bole, Thomas J., III
: *Assistant Professor of Philosophy, Auburn University*
OBLIGATION AND SUPEREROGATION

Bowker, John W.
: *Professor of Religious Studies, University of Lancaster (England)*
PAIN AND SUFFERING, article on RELIGIOUS PERSPECTIVES

Branson, Roy
: *Senior Research Scholar, Center for Bioethics, Kennedy Institute of Ethics, Georgetown University*
HEALTH CARE, article on THEORIES OF JUSTICE AND HEALTH CARE; PRISONERS, article on PRISONER EXPERIMENTATION

Brieger, Gert H.
: *Professor and Chairman, Department of the History of the Health Sciences, University of California, San Francisco*
HUMAN EXPERIMENTATION, article on HISTORY

Brody, Baruch A.
: *Professor and Chairman, Department of Philosophy, Rice University*
LAW AND MORALITY

Brody, Eugene B.
: *Professor of Psychiatry and Human Behavior, University of Maryland*
MENTAL HEALTH SERVICES, article on SOCIAL INSTITUTIONS OF MENTAL HEALTH

Brown, Peter G.
: *Director, Center for Philosophy and Public Policy, University of Maryland*
POPULATION ETHICS: ELEMENTS OF THE FIELD, article on ETHICAL PERSPECTIVES ON POPULATION

Bryant, John H.
: *Professor and Director, School of Public Health, Columbia University*
POVERTY AND HEALTH, article on POVERTY AND HEALTH IN INTERNATIONAL PERSPECTIVE

Bürgel, J. C.
: *Professor of Oriental Studies, University of Bern*
ISLAM

Burns, Chester R.
: *Rockwell Associate Professor of Medical History and Associate Director, Institute for the Medical Humanities, University of Texas Medical Branch*
DENTISTRY, article on PROFESSIONAL CODES IN AMERICAN DENTISTRY; MEDICAL ETHICS, HISTORY OF, section on EUROPE AND THE AMERICAS, article on NORTH AMERICA: SEVENTEENTH TO NINETEENTH CENTURY

Burrow, James G.
: *Professor of History, Abilene Christian University*
MEDICAL PROFESSION, article on ORGANIZED MEDICINE

Burt, Robert A.
: *Professor of Law, Yale University*
INFORMED CONSENT IN MENTAL HEALTH

Caplan, Arthur L.
: *Instructor in the History and Philosophy of Medicine, School of Public Health, Columbia University; Associate for the Humanities, The Hastings Center, Institute of Society, Ethics and the Life Sciences (Hastings-on-Hudson, New York)*
GENETIC ASPECTS OF HUMAN BEHAVIOR, article on PHILOSOPHICAL AND ETHICAL ISSUES

Capron, Alexander Morgan
: *Professor of Law and Professor of Human Genetics, University of Pennsylvania*
DEATH, DEFINITION AND DETERMINATION OF, article on LEGAL ASPECTS OF PRONOUNCING DEATH; HUMAN EXPERIMENTATION, article on BASIC ISSUES; RIGHT TO REFUSE MEDICAL CARE

Carney, Frederick S.
: *Professor of Ethics, Southern Methodist University*
ETHICS, article on THEOLOGICAL ETHICS

Casebeer, Kenneth M.
Assistant Professor of Law, University of Miami
TECHNOLOGY, article on TECHNOLOGY AND THE LAW

Cassell, Eric J.
Clinical Professor of Public Health, Medical College, Cornell University
INFORMED CONSENT IN THE THERAPEUTIC RELATIONSHIP, article on CLINICAL ASPECTS; THERAPEUTIC RELATIONSHIP, article on CONTEMPORARY MEDICAL PERSPECTIVE

Childress, James F.
Joseph P. Kennedy, Sr. Professor of Christian Ethics, Center for Bioethics, Kennedy Institute of Ethics, Georgetown University
RATIONING OF MEDICAL TREATMENT; RISK

Christiansen, Drew
Research Associate, Woodstock Theological Center, Georgetown University
AGING AND THE AGED, article on ETHICAL IMPLICATIONS IN AGING; ENVIRONMENTAL ETHICS, article on THE PROBLEM OF GROWTH

Clouser, K. Danner
Professor of Humanities, College of Medicine, The Pennsylvania State University
BIOETHICS

Cohen, Sidney
Clinical Professor of Psychiatry, Neuropsychiatric Institute, University of California, Los Angeles
DRUG USE, article on DRUG USE FOR PLEASURE AND TRANSCENDENT EXPERIENCE

Cole, Jonathan R.
Professor of Sociology, Columbia University
SCIENCE, SOCIOLOGY OF

Connery, John R.
Professor of Moral Theology, Loyola University (Chicago)
ABORTION, article on ROMAN CATHOLIC PERSPECTIVES

Cooke, Robert E.
President, Medical College of Pennsylvania
MENTALLY HANDICAPPED

Copenhaver, Brian P.
Director, Honors Program, Western Washington University
DEATH, article on WESTERN RELIGIOUS THOUGHT: ARS MORIENDI

Coulter, Harris L.
Washington, D.C.
DRUG INDUSTRY AND MEDICINE

Cowen, David L.
Professor Emeritus of History, Rutgers, The State University of New Jersey
PHARMACY

Creson, D. L.
Chief, Division of Community and Social Psychiatry, University of Texas Medical Branch
HOMOSEXUALITY, article on CLINICAL AND BEHAVIORAL ASPECTS

Curran, Charles E.
Professor of Moral Theology, The Catholic University of America
ABORTION, article on CONTEMPORARY DEBATE IN PHILOSOPHICAL AND RELIGIOUS ETHICS; ROMAN CATHOLICISM

D'Arcy, Eric
Reader in Philosophy, University of Melbourne (Victoria, Australia)
NATURAL LAW

DeBakey, Lois
Professor of Scientific Communication, Baylor College of Medicine and Tulane University School of Medicine
COMMUNICATION, BIOMEDICAL, articles on MEDIA AND MEDICINE and SCIENTIFIC PUBLISHING

DeBakey, Selma
Professor of Scientific Communication, Baylor College of Medicine
COMMUNICATION, BIOMEDICAL, article on MEDIA AND MEDICINE

De Graeve, Frank
Professor of Comparative Religion, University of Louvain
HEALTH AND DISEASE, article on RELIGIOUS CONCEPTS

Derr, Thomas Sieger
Professor of Religion, Smith College (Northampton, Massachusetts)
RELIGIOUS DIRECTIVES IN MEDICAL ETHICS, article on PROTESTANT STATEMENTS

Draguns, Juris G.
Professor of Psychology, The Pennsylvania State University
BEHAVIORISM, article on HISTORY OF BEHAVIORAL PSYCHOLOGY

Dubos, René
Professor Emeritus of Environmental Biomedicine, The Rockefeller University
GENETIC CONSTITUTION AND ENVIRONMENTAL CONDITIONING

Dukeminier, Jesse
Professor of Law, University of California, Los Angeles
ORGAN DONATION, article on LEGAL ASPECTS

Dummett, Clifton O.
Professor of Dentistry, University of Southern California
DENTISTRY, article on ETHICAL ISSUES IN DENTISTRY

CONTRIBUTORS

Duncombe, David C.
: Assistant Professor of Pastoral Theology, Lecturer in Medicine, and Chaplain, School of Medicine, Yale University
PASTORAL MINISTRY

Dyck, Arthur J.
: Mary B. Saltonstall Professor of Population Ethics, Harvard University
POPULATION ETHICS: ELEMENTS OF THE FIELD, article on DEFINITION OF POPULATION ETHICS

Edel, Abraham
: Research Professor of Philosophy, University of Pennsylvania
SCIENCE: ETHICAL IMPLICATIONS

Ehrman, Lee
: Professor of Biology, State University of New York, College at Purchase
EUPHENICS

Emmott, Carol
: Special Assistant to the Deputy Assistant Secretary of Legislation/Health, Department of Health, Education, and Welfare (Washington, D.C.)
HEALTH CARE, article on HEALTH-CARE SYSTEM

Engelhardt, H. Tristram, Jr.
: Rosemary Kennedy Professor of the Philosophy of Medicine, Center for Bioethics, Kennedy Institute of Ethics, Georgetown University
HEALTH AND DISEASE, article on PHILOSOPHICAL PERSPECTIVES; MEDICINE, PHILOSOPHY OF

Epstein, Samuel S.
: Professor of Environmental and Occupational Medicine, School of Public Health, University of Illinois (Chicago)
ENVIRONMENTAL ETHICS, article on ENVIRONMENTAL HEALTH AND HUMAN DISEASE

Erde, Edmund L.
: Assistant Professor of the Philosophy of Medicine, University of Texas Medical Branch
FREE WILL AND DETERMINISM; MEDICINE, PHILOSOPHY OF

Eshraghi, Rahmatollah
: Chairman, Department of Community Medicine, Ferdowsi University (Meshed, Iran)
MEDICAL ETHICS, HISTORY OF, section on NEAR AND MIDDLE EAST AND AFRICA, article on PERSIA

Faramelli, Norman J.
: Chief of Environmental Management, Massachusetts Port Authority (Boston)
ENVIRONMENTAL ETHICS, article on QUESTIONS OF SOCIAL JUSTICE

Farley, Margaret A.
: Associate Professor of Ethics, The Divinity School, Yale University
SEXUAL ETHICS

Feinberg, Joel
: Professor of Philosophy, University of Arizona
BEHAVIOR CONTROL, article on FREEDOM AND BEHAVIOR CONTROL; JUSTICE; RIGHTS, article on SYSTEMATIC ANALYSIS

Feldman, David M.
: Rabbi, Bay Ridge Jewish Center (Brooklyn, New York)
ABORTION, article on JEWISH PERSPECTIVES; EUGENICS AND RELIGIOUS LAW, article on JEWISH RELIGIOUS LAWS

Finnis, J. M.
: Fellow and Praelector in Jurisprudence, University College, Oxford
ABORTION, article on LEGAL ASPECTS

Fletcher, John C.
: Assistant for Bioethics, Office of the Director, Clinical Center, National Institutes of Health (Bethesda, Maryland)
PRENATAL DIAGNOSIS, article on ETHICAL ISSUES

Fletcher, Joseph
: Professor of Medical Ethics, School of Medicine, University of Virginia
ETHICS, article on SITUATION ETHICS

Flew, Antony G. N.
: Professor of Philosophy, University of Reading (England)
EVOLUTION

Foot, Philippa
: Professor of Philosophy, University of California, Los Angeles; Senior Research Fellow, Somerville College, Oxford University
ETHICS, article on MORAL REASONING

Fost, Norman C.
: Associate Professor of Pediatrics, University of Wisconsin, Madison
CHILDREN AND BIOMEDICINE

Fox, Renée C.
: Professor and Chairman, Department of Sociology and Professor, Departments of Medicine and Psychiatry, University of Pennsylvania
KIDNEY DIALYSIS AND TRANSPLANTATION; ORGAN TRANSPLANTATION, article on SOCIOCULTURAL ASPECTS

Frankel, Mark S.
: Assistant Professor of Political Science, Wayne State University
HUMAN EXPERIMENTATION, article on SOCIAL AND PROFESSIONAL CONTROL; REPRODUCTIVE TECHNOLOGIES, articles on ARTIFICIAL INSEMINATION and SPERM AND ZYGOTE BANKING

Freeman, John M.
: Associate Professor of Neurology and Pediatrics, Johns Hopkins University; Director, Birth Defects Treatment Center, John F. Kennedy Institute (Baltimore)
INFANTS, article on MEDICAL ASPECTS AND ETHICAL DILEMMAS

French, Richard D.
: *Associate Professor, Faculty of Management, McGill University*
ANIMAL EXPERIMENTATION, *article on* HISTORICAL ASPECTS

Fried, Charles
: *Professor of Law, Harvard University*
HUMAN EXPERIMENTATION, *article on* PHILOSOPHICAL ASPECTS

Friedman, Jane M.
: *Professor of Law, Wayne State University*
STERILIZATION, *article on* LEGAL ASPECTS

Fruchtbaum, Harold
: *Associate Professor of the History and Philosophy of Public Health, Columbia University*
PUBLIC HEALTH

Garland, Michael J.
: *Bioethics Program Representative, School of Medicine, University of California, San Francisco*
BLOOD TRANSFUSION

Gass, Ronald S.
: *Research Assistant, National Commission for the Protection of Human Subjects of Biomedical and Behavioral Research (Bethesda, Maryland)*
APPENDIX: INTRODUCTION, *article on* CODES OF THE HEALTH-CARE PROFESSIONS

Gatch, Milton McC.
: *Professor of English, University of Missouri, Columbia*
DEATH, *article on* WESTERN RELIGIOUS THOUGHT: POST-BIBLICAL CHRISTIAN THOUGHT

Gerson, Elihu M.
: *Lecturer in Psychiatry, University of California, San Francisco*
CHRONIC CARE

Gert, Bernard
: *Professor of Philosophy, Dartmouth College (Hanover, New Hampshire)*
ETHICS, *article on* OBJECTIVISM IN ETHICS

Gettner, Alan
: *Former Assistant Professor of Philosophy, State University of New York, College at Purchase*
MENTAL HEALTH, *article on* MENTAL HEALTH IN COMPETITION WITH OTHER VALUES

Girardot, Norman J.
: *Assistant Professor of Theology, University of Notre Dame*
TAOISM

Golding, Martin P.
: *Professor of Philosophy, Duke University*
FUTURE GENERATIONS, OBLIGATIONS TO; POPULATION POLICY PROPOSALS, *article on* POPULATION DISTRIBUTION

Gordy, Michael
: *Assistant Professor of Humanities, University of Houston*
SOCIALITY

Gorovitz, Samuel
: *Professor and Chairman, Department of Philosophy, University of Maryland*
HEALTH AS AN OBLIGATION

Gottesman, Irving I.
: *Professor and Director, Behavioral Genetics Center, University of Minnesota*
GENETIC ASPECTS OF HUMAN BEHAVIOR, *articles on* STATE OF THE ART *and* GENETICS AND MENTAL DISORDERS

Gray, Bradford H.
: *Professional Associate, Institute of Medicine, National Academy of Sciences (Washington, D.C.)*
INFORMED CONSENT IN HUMAN RESEARCH, *article on* SOCIAL ASPECTS

Greenawalt, Kent
: *Professor of Law, Columbia University*
PRIVACY

Gresson, Aaron D.
: *Assistant Professor of Humanistic and Behavioral Studies, Boston University*
RACISM, *article on* RACISM AND MENTAL HEALTH

Grote, Jim
: *Graduate Student, University of Louisville*
TECHNOLOGY, *article on* PHILOSOPHY OF TECHNOLOGY

Gruman, Gerald J.
: *Silver Spring, Maryland*
DEATH AND DYING: EUTHANASIA AND SUSTAINING LIFE, *article on* HISTORICAL PERSPECTIVES

Gutmann, James
: *Professor Emeritus of Philosophy, Columbia University*
DEATH, *article on* WESTERN PHILOSOPHICAL THOUGHT

Haddad, Fuad Sami
: *Clinical Professor of Surgery, American University of Beirut*
MEDICAL ETHICS, HISTORY OF, *section on* NEAR AND MIDDLE EAST AND AFRICA, *article on* CONTEMPORARY ARAB WORLD

Hammer, A. G.
: *Professor of Psychology, MacQuarie University (North Ryde, Australia); Research Associate, The Institute of the Pennsylvania Hospital (Philadelphia)*
HYPNOSIS

CONTRIBUTORS

Harakas, Stanley S.
Dean and Professor of Christian Ethics, Holy Cross Greek Orthodox School of Theology (Brookline, Massachusetts)
EASTERN ORTHODOX CHRISTIANITY; POPULATION ETHICS: RELIGIOUS TRADITIONS, article on EASTERN ORTHODOX CHRISTIAN PERSPECTIVES

Hare, Peter H.
Professor of Philosophy, State University of New York at Buffalo
CIVIL DISOBEDIENCE IN HEALTH SERVICES

Hare, R. M.
White's Professor of Moral Philosophy, Oxford University and Fellow of Corpus Christi College, Oxford
ETHICS, articles on UTILITARIANISM and NON-DESCRIPTIVISM

Häring, Bernard
Professor of Moral Theology, Academia Alfonsiana, Lateran University (Rome)
RELIGIOUS DIRECTIVES IN MEDICAL ETHICS, article on ROMAN CATHOLIC DIRECTIVES

Hartt, Julian N.
William Kenan, Jr. Professor of Religious Studies, University of Virginia
MAN, IMAGES OF

Hauck, George H.
Lecturer in Law, University of California, Berkeley
MEDICAL MALPRACTICE

Hauerwas, Stanley
Associate Professor of Theology, University of Notre Dame
CARE

Havighurst, Clark C.
Professor of Law, Duke University
ADVERTISING BY MEDICAL PROFESSIONALS

Hayflick, Leonard
Senior Research Cell Biologist, Bruce Lyon Memorial Research Laboratory (Oakland, California)
AGING AND THE AGED, article on THEORIES OF AGING AND ANTI-AGING TECHNIQUES

Heelan, Patrick A.
Professor of Philosophy, Dean of Arts and Sciences, and Vice President for Liberal Studies, State University of New York at Stony Brook
PURPOSE IN THE UNIVERSE

Hehir, J. Bryan
Office of International Justice and Peace, United States Catholic Conference (Washington, D.C.)
POPULATION ETHICS: RELIGIOUS TRADITIONS, article on ROMAN CATHOLIC PERSPECTIVES

Hellegers, André E.
Director, Kennedy Institute of Ethics, Georgetown University
ABORTION, article on MEDICAL ASPECTS; FETAL RESEARCH

Hellman, Louis M.
Director of Health Information, Population Reference Bureau, Inc. (Washington, D.C.)
STERILIZATION, article on MEDICAL ASPECTS

Hemphill, John Michael
Department of Neurology, Johns Hopkins Hospital (Baltimore)
INFANTS, article on MEDICAL ASPECTS AND ETHICAL DILEMMAS

Henriot, Peter J.
Staff Associate, Center of Concern (Washington, D.C.)
FOOD POLICY

High, Dallas M.
Professor and Chairman, Department of Philosophy, University of Kentucky
DEATH, DEFINITION AND DETERMINATION OF, article on PHILOSOPHICAL AND THEOLOGICAL FOUNDATIONS

Himmelsbach, Kathryn K.
Chief, Cancer Social Work Section, Clinical Center, National Institutes of Health (Bethesda, Maryland)
MEDICAL SOCIAL WORK

Hook, Ernest B.
Chief, Epidemiology and Human Ecology, New York State Birth Defects Institute (Albany, New York); Associate Professor of Pediatrics, Albany Medical College, Union University
GENETIC ASPECTS OF HUMAN BEHAVIOR, article on MALES WITH SEX CHROMOSOME ABNORMALITIES (XYY AND XXY GENOTYPES)

Hovde, Christian A.
Chairman, Department of Religion and Health, Rush-Presbyterian-St. Luke's Medical Center (Chicago)
CADAVERS, article on GENERAL ETHICAL CONCERNS

Howard, Jan
Research Sociologist and Lecturer, School of Medicine, University of California, San Francisco
HEALTH CARE, article on HUMANIZATION AND DEHUMANIZATION OF HEALTH CARE

Howard, Richard J.
Assistant Professor of Surgery, University of Minnesota
ORGAN TRANSPLANTATION, article on MEDICAL PERSPECTIVE

Izen, Judith E.
Research Analyst, Psychiatry Service, Massachusetts General Hospital (Boston)
PSYCHOPHARMACOLOGY

Jaggi, O. P.
Head, Department of Clinical Research, University of Delhi
MEDICAL ETHICS, HISTORY OF, section on SOUTH AND EAST ASIA, article on INDIA

Jakobovits, Immanuel
: Chief Rabbi of the United Hebrew Congregations of the British Commonwealth of Nations (London)
: JUDAISM

Jameton, Andrew L.
: Postdoctoral Fellow, Bioethics Program, School of Medicine, University of California, San Francisco
: MEDICAL ETHICS, HISTORY OF, section on EUROPE AND THE AMERICAS, article on NORTH AMERICA IN THE TWENTIETH CENTURY; ORGAN DONATION, article on ETHICAL ISSUES

Johnson, James T.
: Associate Professor and Chairman, Department of Religon, Douglass College of Rutgers, The State University of New Jersey
: PROTESTANTISM, article on HISTORY OF PROTESTANT MEDICAL ETHICS

Jones, James H.
: National Endowment for the Humanities (Washington, D.C.)
: RACISM, article on RACISM AND MEDICINE

Jonsen, Albert R.
: Associate Professor of Bioethics, School of Medicine, University of California, San Francisco
: HEALTH CARE, article on RIGHT TO HEALTH-CARE SERVICES; LIFE-SUPPORT SYSTEMS; MEDICAL ETHICS, HISTORY OF, section on EUROPE AND THE AMERICAS, articles on WESTERN EUROPE IN THE SEVENTEENTH CENTURY and NORTH AMERICA IN THE TWENTIETH CENTURY

Kalish, Richard A.
: Professor of Behavioral Sciences, Graduate Theological Union (Berkeley, California)
: DEATH, ATTITUDES TOWARD

Kanoti, George A.
: Professor of Religious Studies, John Carroll University
: HOMOSEXUALITY, article on ETHICAL ASPECTS

Kao, Frederick F.
: Professor of Physiology, Downstate Medical Center, State University of New York
: MEDICAL ETHICS, HISTORY OF, section on SOUTH AND EAST ASIA, article on CONTEMPORARY CHINA

Kao, John J.
: Clinical Fellow, Department of Psychiatry, Harvard University
: MEDICAL ETHICS, HISTORY OF, section on SOUTH AND EAST ASIA, article on CONTEMPORARY CHINA

Katz, Jay
: Professor (Adjunct) of Law and Psychiatry, Law School, Yale University
: INFORMED CONSENT IN THE THERAPEUTIC RELATIONSHIP, article on LEGAL AND ETHICAL ASPECTS

Kaufman, Martin
: Professor of History, Westfield State College (Massachusetts)
: ORTHODOXY IN MEDICINE

Keller, Mark
: Professor Emeritus, Center of Alcohol Studies, Rutgers, The State University of New Jersey
: ALCOHOL, USE OF

Kelman, Herbert C.
: Richard Clarke Cabot Professor of Social Ethics, Harvard University
: RESEARCH, BEHAVIORAL

Kitagawa, Joseph M.
: Professor of History of Religions and Dean, The Divinity School, University of Chicago
: MEDICAL ETHICS, HISTORY OF, section on SOUTH AND EAST ASIA, article on JAPAN THROUGH THE NINETEENTH CENTURY

Kittrie, Nicholas N.
: Director, Institute for Advanced Studies in Justice, and Professor and Dean, School of Law, The American University
: MENTAL ILLNESS, article on LABELING IN MENTAL ILLNESS: LEGAL ASPECTS

Klerman, Gerald L.
: Professor of Psychiatry, Harvard University
: PSYCHOPHARMACOLOGY

Konold, Donald
: Professor of History, Arkansas State University
: CODES OF MEDICAL ETHICS, article on HISTORY

Kosnik, Anthony R.
: Professor of Moral Theology and Dean, SS. Cyril and Methodius Seminary (Orchard Lake, Michigan)
: HOMOSEXUALITY, article on ETHICAL ASPECTS

Ladd, John
: Professor of Philosophy, Brown University
: ETHICS, article on THE TASK OF ETHICS

Laín Entralgo, Pedro
: The Arnaldo de Villanova Institute (Madrid)
: THERAPEUTIC RELATIONSHIP, article on HISTORY OF THE RELATIONSHIP

Landy, David
: Professor of Anthropology, University of Massachusetts at Boston
: DEATH, article on ANTHROPOLOGICAL PERSPECTIVE

Lappé, Marc
: Chief, Office of Health Law and Values, California State Department of Health
: EUGENICS, article on ETHICAL ISSUES

Largey, Gale
: Associate Professor of Sociology and Anthropology, Mansfield State College (Pennsylvania)
: REPRODUCTIVE TECHNOLOGIES, article on SEX SELECTION

Lebacqz, Karen
Associate Professor of Christian Ethics, Pacific School of Religion (Berkeley, California)
INFORMED CONSENT IN HUMAN RESEARCH, article on ETHICAL AND LEGAL ASPECTS; STERILIZATION, article on ETHICAL ASPECTS

Lee, Philip R.
Professor of Social Medicine and Director, Health Policy Program, University of California, San Francisco
HEALTH CARE, article on HEALTH-CARE SYSTEM

Leighton, Dorothea C.
Former Professor and Chairman, Department of Mental Health, University of North Carolina
MEDICINE, ANTHROPOLOGY OF

León C., Augusto
Professor of Medicine, Central University of Venezuela
MEDICAL ETHICS, HISTORY OF, section on EUROPE AND THE AMERICAS, article on LATIN AMERICA IN THE TWENTIETH CENTURY

Levine, Robert J.
Professor of Medicine and Lecturer in Pharmacology, Yale University
INFORMED CONSENT IN HUMAN RESEARCH, article on ETHICAL AND LEGAL ASPECTS; RESEARCH, BIOMEDICAL

Lieb, Irwin C.
Professor of Philosophy, Dean of Graduate Studies, and Vice-President, The University of Texas at Austin
PRAGMATISM

Lister, George
Assistant Professor in Residence, Departments of Pediatrics and Anesthesia, University of California, San Francisco
LIFE-SUPPORT SYSTEMS

Lloyd, Charles W.
Co-Director, Center for Study of Human Sexual Behavior, Professor of Clinical Psychiatry, and Professor of Obstetrics/Gynecology, University of Pittsburgh
SEX THERAPY AND SEX RESEARCH, article on SCIENTIFIC AND CLINICAL PERSPECTIVES

Lloyd, Kenneth E.
Professor and Chair, Department of Psychology, Drake University
BEHAVIORAL THERAPIES

Lloyd, Margaret E.
Associate Professor of Psychology, Drake University
BEHAVIORAL THERAPIES

Loehlin, John C.
Professor of Psychology, The University of Texas at Austin
GENETIC ASPECTS OF HUMAN BEHAVIOR, article on RACE DIFFERENCES IN INTELLIGENCE

London, Perry
Professor of Psychology and Psychiatry, University of Southern California
SEXUAL BEHAVIOR

Louisell, David W.
Elizabeth Josselyn Boalt Professor of Law, University of California, Berkeley
MEDICAL MALPRACTICE

Ludmerer, Kenneth M.
Instructor of Medicine, Washington University (St. Louis)
EUGENICS, article on HISTORY

Lynch, Abbyann
Associate Professor of Philosophy, St. Michael's College, University of Toronto
MEDICAL ETHICS, HISTORY OF, section on EUROPE AND THE AMERICAS, article on NORTH AMERICA IN THE TWENTIETH CENTURY

MacIntyre, Alasdair
University Professor of Philosophy and Political Science, Boston University
BEHAVIORISM, article on PHILOSOPHICAL ANALYSIS

McCarthy, Charles R.
Chief, Legislative Development Branch, Office of the Director, National Institutes of Health (Bethesda, Maryland)
RESEARCH POLICY, BIOMEDICAL

McCormick, Richard A.
Rose F. Kennedy Professor of Christian Ethics, Center for Bioethics, Kennedy Institute of Ethics, Georgetown University
ORGAN TRANSPLANTATION, article on ETHICAL PRINCIPLES; REPRODUCTIVE TECHNOLOGIES, article on ETHICAL ISSUES

McCullough, Laurence B.
Assistant Professor of the Philosophy of Medicine and Head, Department of Humanities in Medicine, College of Medicine, Texas A & M University
MEDICAL ETHICS, HISTORY OF, section on EUROPE AND THE AMERICAS, articles on INTRODUCTION TO THE MODERN PERIOD IN EUROPE AND THE AMERICAS, BRITAIN AND THE UNITED STATES IN THE EIGHTEENTH CENTURY, and INTRODUCTION TO THE CONTEMPORARY PERIOD IN EUROPE AND THE AMERICAS

Macklin, Ruth
Associate for Behavioral Studies, The Hastings Center, Institute of Society, Ethics and the Life Sciences (Hastings-on-Hudson, New York)
REDUCTIONISM, article on ETHICAL IMPLICATIONS OF PSYCHOPHYSICAL REDUCTIONISM; RIGHTS, article on RIGHTS IN BIOETHICS; SEX THERAPY AND SEX RESEARCH, article on ETHICAL PERSPECTIVES

Madden, Edward H.
Professor of Philosophy, State University of New York at Buffalo
CIVIL DISOBEDIENCE IN HEALTH SERVICES

Mahoney, Maurice J.
: Associate Professor of Human Genetics and Pediatrics, Yale University
: FETAL–MATERNAL RELATIONSHIP

Marshall, Carol Paul
: Former Coordinator for Program Planning and Special Services, Central Flushing–Upper Queens Medical Group (New York)
: POVERTY AND HEALTH, article on POVERTY AND HEALTH IN THE UNITED STATES

Marshall, Carter L.
: Professor of Preventive Medicine, College of Medicine and Dentistry of New Jersey (Newark)
: POVERTY AND HEALTH, article on POVERTY AND HEALTH IN THE UNITED STATES

Mastroianni, Luigi, Jr.
: Professor and Chairman, Department of Obstetrics and Gynecology, Hospital of the University of Pennsylvania
: REPRODUCTIVE TECHNOLOGIES, article on IN VITRO FERTILIZATION

May, William E.
: Associate Professor of Moral Theology, The Catholic University of America
: DOUBLE EFFECT

Mechanic, David
: John Bascom Professor of Sociology, University of Wisconsin, Madison
: MEDICINE, SOCIOLOGY OF; THERAPEUTIC RELATIONSHIP, article on CONTEMPORARY SOCIOLOGICAL ANALYSIS

Meister, Joel
: Assistant Professor of Sociology, Amherst College (Massachusetts)
: ELECTRICAL STIMULATION OF THE BRAIN

Michels, Robert
: Barklie McKee Henry Professor and Chairman, Department of Psychiatry, Cornell University
: MENTAL ILLNESS, article on DIAGNOSIS OF MENTAL ILLNESS

Milunsky, Aubrey
: Director, Genetics Division, Eunice Kennedy Shriver Center (Waltham, Massachusetts); Assistant Professor of Pediatrics, Harvard University
: PRENATAL DIAGNOSIS, article on CLINICAL ASPECTS

Mintz, Beatrice
: Senior Member, The Institute for Cancer Research (Philadelphia, Pennsylvania)
: GENE THERAPY, article on PRODUCTION OF FOUR-PARENT INDIVIDUALS

Missett, James R.
: Chief Resident in Psychiatry, Stanford University
: HEALTH, INTERNATIONAL; MASS HEALTH SCREENING

Mitcham, Carl
: Instructor of Philosophy and Psychology, St. Catharine College (Kentucky)
: TECHNOLOGY, article on PHILOSOPHY OF TECHNOLOGY

Molinari, Gaetano F.
: Professor and Chairman, Department of Neurology, The George Washington University
: DEATH, DEFINITION AND DETERMINATION OF, article on CRITERIA FOR DEATH

Moore, Harold F.
: Assistant Professor of Philosophy, University of Notre Dame
: ACTING AND REFRAINING

Morawetz, Thomas H.
: Associate Professor of Law and Adjunct Professor of Philosophy, University of Connecticut
: DYNAMIC THERAPIES

Murphy, Edmond A.
: Professor of Medicine, Johns Hopkins University
: DECISION MAKING, MEDICAL

Murray, Robert F., Jr.
: Professor of Genetics and Human Genetics and Professor of Pediatrics and Medicine, Howard University
: GENETIC DIAGNOSIS AND COUNSELING, articles on GENETIC DIAGNOSIS and GENETIC COUNSELING

Musallam, Basim F.
: Assistant Professor of History, University of Pennsylvania; Research Associate in Population Studies, Harvard University
: POPULATION ETHICS: RELIGIOUS TRADITIONS, article on ISLAMIC PERSPECTIVES

Nadelson, Carol C.
: Associate Professor of Psychiatry, Harvard University
: WOMEN AND BIOMEDICINE, articles on WOMEN AS PATIENTS AND EXPERIMENTAL SUBJECTS and WOMEN AS HEALTH PROFESSIONALS

Nagle, James J.
: Associate Professor of Zoology, Drew University
: EUPHENICS

Najarian, John S.
: Professor and Chairman, Department of Surgery, University of Minnesota
: ORGAN TRANSPLANTATION, article on MEDICAL PERSPECTIVE

Nakamura, Hajime
: Director, The Eastern Institute, Inc. (Tokyo); Professor Emeritus, University of Tokyo
: BUDDHISM; ENVIRONMENT AND MAN, article on EASTERN THOUGHT

Nelson, James B.
> Professor of Christian Ethics, United Theological Seminary of the Twin Cities (New Brighton, Minnesota)
> ABORTION, article on PROTESTANT PERSPECTIVES

Neufeld, Elizabeth F.
> Chief, Section on Human Biochemical Genetics, National Institute of Arthritis, Metabolism and Digestive Diseases, National Institutes of Health (Bethesda, Maryland)
> GENE THERAPY, article on ENZYME REPLACEMENT

Neugarten, Bernice L.
> Professor of Human Development, The University of Chicago
> AGING AND THE AGED, article on SOCIAL IMPLICATIONS OF AGING

Neville, Robert
> Professor of Religious Studies and Philosophy, State University of New York at Stony Brook
> BEHAVIOR CONTROL, article on ETHICAL ANALYSIS; DRUG USE, article on DRUG USE, ABUSE, AND DEPENDENCE; PSYCHOSURGERY

Newman, Lucile F.
> Assistant Professor of Community Health and Anthropology, Brown University
> MEDICAL ETHICS, HISTORY OF, article on PRIMITIVE SOCIETIES

Ninomiya, Rikuo
> Director, Ninomiya-Naika Clinic (Tokyo)
> MEDICAL ETHICS, HISTORY OF, section on SOUTH AND EAST ASIA, article on CONTEMPORARY JAPAN: MEDICAL ETHICS AND LEGAL MEDICINE

Noonan, John T., Jr.
> Professor of Law, University of California, Berkeley
> CONTRACEPTION

Notman, Malkah T.
> Associate Clinical Professor of Psychiatry, Harvard University
> WOMEN AND BIOMEDICINE, articles on WOMEN AS PATIENTS AND EXPERIMENTAL SUBJECTS and WOMEN AS HEALTH PROFESSIONALS

Offer, Daniel
> Professor of Psychiatry, The University of Chicago; Chairman, Department of Psychiatry, Michael Reese Hospital and Medical Center (Chicago)
> ADOLESCENTS

Offer, Judith B.
> Research Associate in Psychiatry, Michael Reese Hospital and Medical Center (Chicago)
> ADOLESCENTS

Ost, David E.
> Assistant Editor, Encyclopedia of Bioethics
> INFANTS, articles on ETHICAL PERSPECTIVES ON THE CARE OF INFANTS and PUBLIC POLICY AND PROCEDURAL QUESTIONS

Page, Benjamin B.
> Assistant Professor of Philosophy and Health Services Administration, Quinnipiac College (Hamden, Connecticut)
> MEDICAL ETHICS, HISTORY OF, section on EUROPE AND THE AMERICAS, article on EASTERN EUROPE IN THE TWENTIETH CENTURY

Panuska, J. A.
> Former Professor of Biology, Georgetown University
> CRYONICS

Parsons, Talcott
> Professor Emeritus of Sociology, Harvard University
> DEATH, article on DEATH IN THE WESTERN WORLD; HEALTH AND DISEASE, article on A SOCIOLOGICAL AND ACTION PERSPECTIVE

Pellegrino, Edmund D.
> Professor of Medicine, Yale University; President, Yale–New Haven Medical Center
> MEDICAL EDUCATION

Perlin, Seymour
> Professor of Psychiatry and Behavioral Sciences, The George Washington University; Senior Research Scholar, Center for Bioethics, Kennedy Institute of Ethics, Georgetown University
> SUICIDE

Pernick, Martin S.
> Instructor of Social History, College of Medicine, The Pennsylvania State University
> MEDICAL PROFESSION, article on MEDICAL PROFESSIONALISM

Petersen, William
> Robert Lazarus Professor of Social Demography, Ohio State University
> POPULATION ETHICS: ELEMENTS OF THE FIELD, article on HISTORY OF POPULATION THEORIES; POPULATION POLICY PROPOSALS, article on DIFFERENTIAL GROWTH RATE AND POPULATION POLICIES: DEMOGRAPHIC PERSPECTIVES

Peterson, M. Jeanne
> Assistant Professor of History, Indiana University
> MEDICAL ETHICS, HISTORY OF, section on EUROPE AND THE AMERICAS, article on BRITAIN IN THE NINETEENTH CENTURY

Poste, George
> Professor of Experimental Pathology, Roswell Park Memorial Institute (Buffalo, New York)
> GENE THERAPY, article on CELL FUSION AND HYBRIDIZATION

Potter, Ralph B.
> Professor of Social Ethics, The Divinity School, and Member, Center for Population Studies, Harvard University
> POPULATION ETHICS: ELEMENTS OF THE FIELD, article on NORMATIVE ASPECTS OF POPULATION POLICY

Powers, Charles W.
  Cummins Engine Company, Inc. (Columbus, Indiana); Affiliate Professor of Social Ethics, Christian Theological Seminary (Indianapolis)
  ENVIRONMENTAL ETHICS, article on QUESTIONS OF SOCIAL JUSTICE

Powledge, Tabitha M.
  Research Associate, The Hastings Center, Institute of Society, Ethics and the Life Sciences (Hastings-on-Hudson, New York)
  GENETIC SCREENING

Rao, K. L. Seshagiri
  Professor of Religious Studies, University of Virginia
  POPULATION ETHICS: RELIGIOUS TRADITIONS, article on A HINDU PERSPECTIVE

Redlich, F. C.
  Professor of Psychiatry, University of California, Los Angeles
  MEDICAL ETHICS UNDER NATIONAL SOCIALISM

Reich, Warren T.
  Senior Research Scholar, Center for Bioethics, Kennedy Institute of Ethics, and Associate Professor of Bioethics, School of Medicine, Georgetown University
  INFANTS, articles on ETHICAL PERSPECTIVES ON THE CARE OF INFANTS and PUBLIC POLICY AND PROCEDURAL QUESTIONS; LIFE, article on QUALITY OF LIFE

Restak, Richard M.
  Instructor in Neurology, Georgetown University
  ELECTROCONVULSIVE THERAPY

Reynolds, Frank E.
  Associate Professor of History of Religions and Buddhist Studies, The Divinity School, The University of Chicago
  DEATH, article on EASTERN THOUGHT

Richardson, William J.
  Professor of Philosophy, Fordham University; Director of Research, Austen Riggs Center (Stockbridge, Massachusetts)
  MENTAL HEALTH, article on RELIGION AND MENTAL HEALTH

Riesenfeld, Stefan A.
  Professor of Law, University of California, Berkeley, and University of California, San Francisco
  HEALTH INSURANCE

Risse, Guenter B.
  Professor of the History of Medicine, University of Wisconsin, Madison
  HEALTH AND DISEASE, article on HISTORY OF THE CONCEPTS; MEDICAL ETHICS, HISTORY OF, section on EUROPE AND THE AMERICAS, article on CENTRAL EUROPE IN THE NINETEENTH CENTURY

Robertson, John A.
  Associate Professor of Law, University of Wisconsin, Madison
  REPRODUCTIVE TECHNOLOGIES, article on LEGAL ASPECTS

Robinson, Daniel N.
  Professor of Psychology, Georgetown University
  PAIN AND SUFFERING, article on PSYCHOBIOLOGICAL PRINCIPLES

Roblin, Richard O.
  Basic Research Program, Frederick Cancer Research Center (Frederick, Maryland)
  GENE THERAPY, articles on GENE THERAPY VIA TRANSFORMATION and GENE THERAPY VIA TRANSDUCTION

Rosner, Fred
  Associate Professor of Medicine, State University of New York at Stony Brook; Director, Division of Hematology, Queens Hospital Center (Jamaica, New York)
  RELIGIOUS DIRECTIVES IN MEDICAL ETHICS, article on JEWISH CODES AND GUIDELINES

Russell, Paul S.
  John Homans Professor of Surgery, Harvard University; Chief, Transplantation Unit, Massachusetts General Hospital (Boston)
  SURGERY

Sagan, Leonard A.
  Palo Alto Medical Clinic (Palo Alto, California)
  PRISONERS, articles on MEDICAL CARE OF PRISONERS and TORTURE AND THE HEALTH PROFESSIONAL

Sai, Frederick T.
  Assistant Secretary General, International Planned Parenthood Federation (London); Former Director of Medical Services (Ghana)
  MEDICAL ETHICS, HISTORY OF, section on NEAR AND MIDDLE EAST AND AFRICA, article on SUB-SAHARAN AFRICA

Sanders, Judith Rose
  Co-Director, Health Education Resources (Brooklyn, New York)
  SEXUAL IDENTITY

Savodnik, Irwin
  Director of Medical Student Education, University of Pittsburgh
  BIOLOGY, PHILOSOPHY OF

Schumacher, Sallie
  Co-Director, Center for Study of Human Sexual Behavior and Associate Professor of Clinical Psychiatry, University of Pittsburgh
  SEX THERAPY AND SEX RESEARCH, article on SCIENTIFIC AND CLINICAL PERSPECTIVES

Schumaker, Miliard
    Associate Professor of Religion and Ethics, Queen's Theological College (Kingston, Ontario)
    OBLIGATION AND SUPEREROGATION

Shaffer, Jerome A.
    Professor and Head, Department of Philosophy, University of Connecticut
    PAIN AND SUFFERING, article on PHILOSOPHICAL PERSPECTIVES

Shaw, Margery W.
    Director, Medical Genetics Center, The University of Texas Health Science Center
    GENETICS AND THE LAW

Sherlock, Richard K.
    Instructor of Philosophy and Religion, Northeastern University
    POPULATION POLICY PROPOSALS, article on DIFFERENTIAL GROWTH RATE AND POPULATION POLICIES: ETHICAL ANALYSIS

Shinn, Roger L.
    Reinhold Niebuhr Professor of Social Ethics, Union Theological Seminary (New York)
    GENE THERAPY, article on ETHICAL ISSUES

Shoemaker, Sydney
    Professor of Philosophy, Cornell University
    MIND–BODY PROBLEM

Sidel, Victor W.
    Professor of Community Health, Albert Einstein College of Medicine; Chairman, Department of Social Medicine, Montefiore Hospital and Medical Center (Bronx, New York)
    WARFARE, article on BIOMEDICAL SCIENCE AND WAR

Sidel, Mark
    Student, Princeton University
    WARFARE, article on BIOMEDICAL SCIENCE AND WAR

Siegel, Seymour
    Ralph Simon Professor of Theology, Jewish Theological Seminary of America (New York)
    DEATH, article on WESTERN RELIGIOUS THOUGHT: POST-BIBLICAL JEWISH TRADITION; MEDICAL ETHICS, HISTORY OF, section on NEAR AND MIDDLE EAST AND AFRICA, article on CONTEMPORARY ISRAEL

Silver, George A.
    Professor of Public Health, School of Medicine, Yale University
    SOCIAL MEDICINE

Singer, Peter
    Professor of Philosophy, Monash University (Melbourne, Australia)
    ANIMAL EXPERIMENTATION, article on PHILOSOPHICAL PERSPECTIVES; LIFE, article on VALUE OF LIFE

Sinsheimer, Robert L.
    Professor of Biology and Chancellor, University of California, Santa Cruz
    REPRODUCTIVE TECHNOLOGIES, article on ASEXUAL HUMAN REPRODUCTION

Smith, David H.
    Associate Professor and Chairman, Department of Religious Studies, and Director, Medicine and the Public Project, Indiana University
    SUICIDE

Smith, Harmon L.
    Professor of Moral Theology and Professor of Community Health Services, Duke University
    HEART TRANSPLANTATION

Solomon, Wm. David
    Milbank Research Fellow, Department of Philosophy, Boston University
    ETHICS, article on RULES AND PRINCIPLES

Solomon, Robert C.
    Professor of Philosophy, University of Texas at Austin
    SEXUAL IDENTITY

Spengler, Joseph J.
    James B. Duke Professor Emeritus of Economics, Duke University
    POPULATION ETHICS: ELEMENTS OF THE FIELD, article on HISTORY OF POPULATION POLICIES

Stanley, Teresa
    Kennedy Nursing Faculty Fellow in Bioethics; Director, Division of Nursing, Incarnate Word College (San Antonio, Texas)
    NURSING

Strauss, Anselm
    Professor of Sociology, University of California, San Francisco
    CHRONIC CARE

Strickland, Stephen P.
    Vice President, Aspen Institute for Humanistic Studies (New York)
    HEALTH POLICY, article on EVOLUTION OF HEALTH POLICY

Sullivan, William M.
    Assistant Professor of Philosophy, Allentown College of St. Francis de Sales (Center Valley, Pennsylvania)
    MENTAL HEALTH THERAPIES

Swazey, Judith P.
    Professor of Socio-Medical Sciences, School of Medicine, Boston University
    KIDNEY DIALYSIS AND TRANSPLANTATION; SURGERY

Takemi, Taro
    President, The Japan Medical Association (Tokyo)
    MEDICAL ETHICS, HISTORY OF, section on SOUTH AND EAST ASIA, article on TRADITIONAL PROFESSIONAL ETHICS IN JAPANESE MEDICINE

Taylor, Carl E.
Professor and Chairman, Department of International Health, Johns Hopkins University
HEALTH, INTERNATIONAL; MASS HEALTH SCREENING

Teitelbaum, Michael S.
Fellow of Nuffield College, Oxford, and University Lecturer in Demography, Oxford University
POPULATION ETHICS: ELEMENTS OF THE FIELD, article on THE POPULATION PROBLEM IN DEMOGRAPHIC PERSPECTIVE

Thielicke, Helmut
Professor of Systematic Theology, University of Hamburg
MIRACLE AND FAITH HEALING, article on THEOLOGICAL PERSPECTIVE

Tooley, Michael
Research Fellow in Philosophy, Research School of Social Sciences, Australian National University
INFANTS, article on INFANTICIDE: A PHILOSOPHICAL PERSPECTIVE

Trainin, Isaac N.
Director, Federation of Jewish Philanthropies of New York
RELIGIOUS DIRECTIVES IN MEDICAL ETHICS, article on JEWISH CODES AND GUIDELINES

Trautmann, Joanne
Associate Professor of Humanities and English, College of Medicine, The Pennsylvania State University
MEDICAL ETHICS IN LITERATURE

Tribe, Laurence H.
Professor of Law, Harvard University
TECHNOLOGY, article on TECHNOLOGY AND THE LAW

Unschuld, Paul U.
Lecturer in Chinese, University of Marburg; Visiting Assistant Professor in Behavioral Sciences, Johns Hopkins University
CONFUCIANISM; MEDICAL ETHICS, HISTORY OF, section on SOUTH AND EAST ASIA, articles on GENERAL HISTORICAL SURVEY and PREREPUBLICAN CHINA

Vanderpool, Harold Y.
Assistant Professor of the History of Medicine, University of Texas Medical Branch
MIRACLE AND FAITH HEALING, article on CONCEPTUAL AND HISTORICAL PERSPECTIVES; PROTESTANTISM, article on DOMINANT HEALTH CONCERNS IN PROTESTANTISM

van Melsen, A. G. M.
Professor of Philosophy, University of Nijmegen
PERSON

Vastyan, E. A.
Professor and Chairman, Department of Humanities, College of Medicine, The Pennsylvania State University
WARFARE, article on MEDICINE AND WAR

Veatch, Robert M.
Senior Associate, The Hastings Center, Institute of Society, Ethics and the Life Sciences (Hastings-on-Hudson, New York)
CODES OF MEDICAL ETHICS, article on ETHICAL ANALYSIS; DEATH AND DYING: EUTHANASIA AND SUSTAINING LIFE, article on PROFESSIONAL AND PUBLIC POLICIES; MEDICAL ETHICS EDUCATION; POPULATION POLICY PROPOSALS, article on GOVERNMENTAL INCENTIVES; TRUTH-TELLING, article on ATTITUDES

Viederman, Stephen
Head, Social and Demographic Research Unit, United Nations Fund for Population Activities
POPULATION POLICY PROPOSALS, article on POPULATION EDUCATION

Walters, LeRoy
Director, Center for Bioethics, Kennedy Institute of Ethics, Georgetown University
TECHNOLOGY, article on TECHNOLOGY ASSESSMENT

Warwick, Donald P.
Institute Fellow, Harvard Institute for International Development
POPULATION POLICY PROPOSALS, articles on CONTEMPORARY INTERNATIONAL ISSUES and SOCIAL CHANGE PROPOSALS

Weber, Leonard J.
Associate Professor of Religious Studies, Mercy College of Detroit
SMOKING

Weiner, Dora B.
Associate Professor of History, Manhattanville College (Purchase, New York)
MEDICAL ETHICS, HISTORY OF, section on EUROPE AND THE AMERICAS, article on FRANCE IN THE NINETEENTH CENTURY

Wellman, Carl
Professor of Philosophy, Washington University (St. Louis)
ETHICS, articles on NATURALISM and RELATIVISM

Wexler, David B.
Professor of Law, University of Arizona
INSTITUTIONALIZATION

Williams, Kenneth J.
President, K. J. Williams and Associates, Inc., Hospital Consultants (Napa, California); Associate Professor of Hospital and Health Care Administration, St. Louis University
HOSPITALS

Winslade, William J.
> Lecturer in Law, Lecturer in Psychiatry, and Co-Director, Program in Medicine, Law, and Human Values, University of California, Los Angeles
> CONFIDENTIALITY

Wolf, C. P.
> Associate Professor, Environmental Psychology Program, City University of New York
> ENVIRONMENTAL ETHICS, article on THE PROBLEM OF GROWTH

Wolstenholme, Gordon
> Director, The Ciba Foundation; President, The Royal Society of Medicine (London)
> MEDICAL ETHICS, HISTORY OF, section on EUROPE AND THE AMERICAS, article on BRITAIN IN THE TWENTIETH CENTURY

Wurzburger, Walter S.
> Rabbi, Congregation Shaaray Tefila (Far Rockaway, New York)
> CADAVERS, article on JEWISH PERSPECTIVES

Yates, Wilson
> Professor of Church and Society, United Theological Seminary of the Twin Cities (New Brighton, Minnesota)
> POPULATION ETHICS: RELIGIOUS TRADITIONS, article on PROTESTANT PERSPECTIVES; POPULATION POLICY PROPOSALS, article on COMPULSORY POPULATION CONTROL PROGRAMS

Young, Ernlé W. D.
> Lecturer in Medical Ethics and Chaplain, The Stanford University Medical Center
> AGING AND THE AGED, article on HEALTH CARE AND RESEARCH IN THE AGED

Zaner, Richard M.
> Easterwood Professor of Philosophy, Southern Methodist University
> EMBODIMENT

Special Reviewers

To assure accuracy of content, articles in this encyclopedia were subject to review not only by the editors but by specialists in the numerous fields of learning discussed in the articles. More than 330 special reviewers were called upon to examine manuscripts for accuracy and comprehensiveness, and, on occasion, to recommend alternate authors and reviewers. These reviewers represent fields as diversified as surgery, Islamic studies, pediatrics, philosophy, environmental sciences, theology, psychiatry, philosophy, law, public health, anthropology of medicine, geriatrics, policy studies, genetics, history, psychology, demography, and health-care administration. We are indebted to the following scholars, from a number of different countries, who drew on their specialized knowledge to offer detailed assessments of this work:

Francisco Abel
Kenneth S. Abraham
Ruth D. Abrams
David C. Abramson
Leo Alexander
David Franklyn Allen
Gordon Allen
Darrel W. Amundsen
Odin W. Anderson
W. French Anderson
Paul H. Andreini
Judith Areen
Gordon Avery
Stephen Axelrad
Lloyd Bailey
Mitchell B. Balter
A. L. Basham
Donald Bates
Robert C. Baumiller
Tom L. Beauchamp
Alex Berman
Lyle W. Bivens
Peter Black
J. David Bleich
Ned Block
Samuel W. Bloom
Leon Bouvier

Joseph V. Brady
Gerard M. Brannon
Roy Branson
Lester Breslow
Harvey Brooks
Peter G. Brown
Walter Brueggemann
Mario Bunge
John Bunker
J. C. Bürgel
John C. Burnham
Chester R. Burns
Robert N. Butler
Jerome Bylebyl
Maureen Canick
Norman Cantor
P. V. Cardon
Frederick S. Carney
Charles Carroll
Kenneth M. Casebeer
Eric J. Cassell
Wing-tsit Chan
Thomas Chalmers
Satya N. Chatterjee
James F. Childress
Drew Christiansen
Ira Cisin

G. Mary Clinton
Mercedes Concepción
John R. Connery
Demetrios J.
   Constantelos
Robert E. Cooke
Harris L. Coulter
Austin B. Creel
D. L. Creson
Ralph C. Croizier
Barbara J. Culliton
Joseph Curl
Charles E. Curran
David Daube
Henry P. David
Kingsley Davis
Gerald Davison
Jean Robert Debray
Panagiotes Ch.
   Demetropoulos
Thomas Sieger Derr
Pat Diamond
Cornelia C. Dimmitt
Dennis J. Doherty
Raymond S. Duff
Clifton O. Dummett
David C. Duncombe

SPECIAL REVIEWERS

Arthur J. Dyck
Abraham Edel
Charles Edwards
R. G. Edwards
Barbara Ehrenreich
Anne H. Ehrlich
Paul R. Ehrlich
Bruce J. Ennis
L. Erlenmeyer-Kimling
Horacio Fabrega, Jr.
Richard M. Fagley
Margaret A. Farley
David M. Feldman
Mark G. Field
Max Fink
Norman J. Finkel
George Foster
Willard O. Foster, Jr.
Renée C. Fox
Donald S. Fredrickson
Daniel Freedman
John M. Freeman
Eliot Freidson
Paul A. Freund
Charles Fried
Jane M. Friedman
Arthur W. Galston
Edwin S. Gaustad
Willard Gaylin
Jack Geiger
Gerald L. Geison
Murray Gendell
Park S. Gerald
Norman J. Girardot
Barney G. Glaser
Bentley Glass
Holly Goldman
Byron J. Good
Irving I. Gottesman
Denis Goulet
Harold Green
Daniel Greenberg
Irving Greenberg
Otto E. Guttentag
John Whitney Hall
Bernard Häring
Bernice Catherine
  Harper
George T. Harrell
Harry Harris
Leston Havens
Robert Hayden
Leonard Hayflick
André E. Hellegers
Louis M. Hellman
Swailem S. Hennein
Richard J. Herrnstein
Ernest R. Hilgard
William J. Hill
Milton Himmelfarb
Kathryn K.
  Himmelsbach
P. Browning Hoffman
Frederick H. Holck
Edwin Holman
Ernest B. Hook
Rollin D. Hotchkiss
Lee Howard
Charles A. Hufnagel
Everett C. Hughes
David Lee Hull
Robert R. Huntley
E. P. Hutchinson
Warren F. Ilchman
Franz J. Ingelfinger
Charles Issawi
Immanuel Jakobovits
James T. Johnson
Renée Johnson
James H. Jones
Stephen Joseph
Michael Kaback
Richard A. Kalish
George A. Kanoti
John Karefa-Smart
Leon R. Kass
Jay Katz
Seymour S. Kety
Lester S. King
Dudley Kirk
Joseph M. Kitagawa
Nicholas N. Kittrie
Gerald L. Klerman
Martin Koen
Robert C. Kolodny
Donald Konold
Julius Korein
Samuel L. Kountz
Fridolf Kudlien
Stephan Kuttner
Marc Lappé
Donald V. Lassiter
Jérôme Lejeune
Charles Leslie
Charles W. Lloyd
George V. Lobo
Margaret Lock
John C. Loehlin
David W. Louisell
Paul Lowinger
Harold S. Luft
Edward Lurie
Abbyann Lynch
Arthur McCormack
Richard A. McCormick
James M. McCullough
Laurence B.
  McCullough
Thomas K. McElhinney
Peter J. M. McEwan
Jim McIntosh
Ronald W. McNeur
Francis MacNutt
Edward H. Madden
Jose Mainetti
David G. Mandelbaum
Herbert Joseph Manz
Vernon Mark
J. Donald Mashburn
Luigi Mastroianni, Jr.
William Maxted
Rollo May
William E. May
Jean Mayer
Ernst Mayr
David Mechanic
Zhores Medvedev
Thomas W. Merrick
John Meyendorff
David W. Meyers
Robert Michels
John Giles Milhaven
Aubrey Milunsky
Robert Mnookin
N. J. Modi
Gaetano F. Molinari
John W. Money
Richard T. Morris
Jerome A. Motto
Arno Motulsky
K. E. Moyer
Marjorie Smith Mueller
Charles W. Murdock
Edmond A. Murphy
Robert F. Murray, Jr.
Basim F. Musallam
Selma J. Mushkin
Henry L. Nadler
Ann Neale
Jacob Needleman
John R. Neil
D. R. Newth
Peter H. Niebyl
John T. Noonan, Jr.
Richard Norris
Frank W. Notestein
Thomas W. Ogletree
Marian Osterweis
Gene Outka
Larry I. Palmer
John Passmore
Edmund D. Pellegrino
Seymour Perlin
Martin S. Pernick
Josef Pieper
Harriet F. Pilpel
Robert Plotkin
Karl H. Pribram
David Rabin
Thomas A. Rakowski
Paul Ramsey
F. C. Redlich
James Reed
Nicholas Rescher
Frank E. Reynolds
Jeanne Clare Ridley
John A. Robertson
Daniel N. Robinson
Jonas Robitscher
Richard O. Roblin
George Rosen
Albert Rosenfeld

Fred Rosner
John Ross
Gottfried Roth
Heinrich Schipperges
George A. Schrader
John J. Schruefer
Bruno Schüller
Stuart O. Schweitzer
Nathan A. Scott, Jr.
William B. Scoville
Belding H. Scribner
John R. Searle
Peter Sedgwick
Irving J. Selikoff
David A. Shephard
Richard Sherins
N. E. Shumway
Ronald K. Siegel
Seymour Siegel
David L. Sills
Milton Silverman
Nathan Sivin
Florence Slepian

David H. Smith
Harmon L. Smith
John E. Smith
Robert Solow
J. Spaander
Roger W. Sperry
Stuart F. Spicker
Robert Spitzer
Frank G. Standaert
Teresa Stanley
John F. Stapleton
Robert L. Stivers
Elizabeth C. Stobo
Robert Stoller
Stephen P. Strickland
Wadi N. Suki
Conrad Taeuber
Michael S. Teitelbaum
Bernard Towers
David Tracy
Isaac N. Trainin
K. N. Udupa
Paul U. Unschuld

William M. Upholt
Joseph van Arendonk
Harold Y. Vanderpool
E. A. Vastyan
Herbert G. Vaughan, Jr.
Ilza Veith
John C. Wakefield
Richard Wasserstrom
Claude E. Welch
J. Philip Welch
Charles C. West
Robert D. Wheelock
Kerr L. White
George Hunston
 Williams
Kenneth J. Williams
John A. Wilson
John H. Wright
Ernlé W. D. Young
James H. Young
K. C. Zachariah
Richard Zeckhauser
Harriet Zuckerman

ABORTION

The article on MEDICAL ASPECTS *explains the physiological and medical data relevant to the ethical issues on abortion. The next three articles,* JEWISH PERSPECTIVES, ROMAN CATHOLIC PERSPECTIVES, *and* PROTESTANT PERSPECTIVES, *represent the consistent ethical methodologies in the dominant Western religious traditions through the mid-twentieth century. The fifth article,* CONTEMPORARY DEBATE IN PHILOSOPHICAL AND RELIGIOUS ETHICS, *continues the discussion of the religious traditions to the present and then compares them with the same points made in contemporary philosophical writings. The final article,* LEGAL ASPECTS, *offers a framework within which laws on abortion can be understood in an international perspective.*

I. MEDICAL ASPECTS	André E. Hellegers
II. JEWISH PERSPECTIVES	David M. Feldman
III. ROMAN CATHOLIC PERSPECTIVES	John R. Connery
IV. PROTESTANT PERSPECTIVES	James B. Nelson
V. CONTEMPORARY DEBATE IN PHILOSOPHICAL AND RELIGIOUS ETHICS	Charles E. Curran
VI. LEGAL ASPECTS	J. M. Finnis

I
MEDICAL ASPECTS

Medical context of abortion

The place occupied by the fetus in the thinking of the medical profession and the theological and philosophical professions has evolved along with the specialty of obstetrics or midwifery.

Thinking about the fetus as a human entity, separate from the mother, has a long tradition in both ethics and medicine. For ethics the subject of abortion has been discussed since pre-Christian days (Connery). For obstetrics, the concept of the fetus as an individual entity at first presented a philosophical problem of how it obtained its "vital spirit," later to be known as oxygen. The debate raged from the days of Greek philosophers until its final resolution when a separate fetal circulation was first described by William Harvey in 1651. Refinements in assessing fetal metabolism date largely to the twentieth century, when fetal physiology and neonatology emerged as distinct specialties.

The stimulus for a refined examination of the fetus in medicine and ethics followed a distinct drawn-out progression of events. At all times in medical practice, and mostly also in ethics, the mother has been considered more important than the fetus. In medicine the induction of death in the fetus to save maternal life has had a long tradition, ranging from the performance of embryotomy for cephalopelvic disproportion to the removal of pregnancies implanted in sites other than the uterus (ectopic pregnancies). For many decades also abortions were performed for a variety of maternal somatic diseases, especially of the lungs, heart, and kidneys, in which it was thought appropriate to perform abortions for the immediate relief of the perceived strains of pregnancy on the maternal renal, pulmonary, and cardiovascular systems. Only when medical institutions that refused to do such abortions on religious grounds adduced data to show that their maternal outcomes were no worse did a shift in thinking occur. It came to be understood that to perform surgical procedures, including abortions, on critically ill women was likely to induce death rather than to prevent it. A philosophy of first treating a critically ill woman medically, until she was fit enough to undergo surgery or abortion, evolved. Inevitably when she became fit enough to undergo surgery, the question arose why she was not sufficiently fit to complete the pregnancy, at least until the fetus was viable. What occurred, then, was a progressive questioning of the practice of deeming abortion necessary to save maternal life.

The questioning took place concurrently with considerable progress in the techniques of epidemiology and biostatistics. Those techniques permitted great refinements in assessing the long-term effects of acting or not acting obstetrically. Studies then became possible not just on the effects of abortion for the immediate saving of maternal life, but on the effects of aborting, or not aborting, on the subsequent life expectancy of women as well. As a result, the issue changed from one of aborting instantaneously to save maternal life to a matter of aborting to preserve an average life span. The new obstetrical question involved abortion for the maintenance of health in the long run, rather than for saving life in the short run.

Toward the middle of the twentieth century many epidemiological studies began to appear that related the quality of fetal outcome to the status of mothers in pregnancy. A key study performed in Australia related the incidence of congenital anomalies in infants to the occurrence of rubella, or German measles, in the first trimester of pregnancy (Gregg). A question arose as to whether abortions should be performed to prevent the birth of children who, by virtue of congenital anomalies, would be a handicap to their parents and at risk of suffering themselves. Thus, the question became whether abortions should be performed for "fetal indications." Proponents of the notion stressed the suffering, for both infant and parents. Antagonists stressed not only the value of any fetal life, but also the fact that, within the limits of the knowledge then available, several normal fetuses had to be aborted to ensure the abortion of the abnormal ones. With development of techniques for culturing amniotic fluid cells it became possible to diagnose fetal defects directly and to abort only defective fetuses, unless the diagnosis was inaccurate. Abortions performed acutely to save maternal life therefore decreased as abortions to prolong the life span or to avoid the birth of anomalous infants increased.

These trends in abortion practice were influenced not just by the availability of prognostic and diagnostic techniques, such as epidemiology and amniotic fluid analysis, but also by progress in obstetrical techniques of other kinds. Committees for analyzing each maternal death were established, and corrective measures and educational programs were instituted until, by 1974, total deaths associated with the pregnancy process had decreased to no more than a few hundred per year for the entire United States. As the obstetrical task of making pregnancy safe for women became almost fully successful, stress was increasingly laid on making it safe for fetuses as well. It became realized that many newborn infants who had not died had, nevertheless, been damaged. Prematurity and fetal distress in the labor process were discovered to be particularly productive of subsequent damage, especially to the brain. Progress in anesthesiology, in Caesarean section techniques, in blood transfusion, and in antibiotic therapy permitted increasing opportunities to perform Caesarean sections with relative impunity. Progressive decrease in perinatal and infant mortality also meant that women needed to undergo fewer pregnancies before desired family size was achieved. As a consequence, Caesarean section scars would be less frequently subjected to risk of rupture in subsequent pregnancies, another factor that fostered an earlier resorting to Caesarean section. The same relative certainty that almost all products of birth could be made to survive led to an increasing concern for the avoidance of infant morbidity as well as infant mortality. The focus of obstetrics and allied sciences, therefore, gradually changed from mothers to fetuses and to their quality.

As data became available to suggest that wanted babies fared better than unwanted ones, the stress on family planning increased. An ethos arose that proclaimed that parents should have the number of children they desired and at the time they desired. Thus the procedure of abortion, which originally was linked to the preservation of maternal life and health, gradually changed to become a backstop to, or an alternative to, other methods of family planning.

Definition of abortion

In medical parlance, abortion is defined as termination of pregnancy, spontaneously or by induction, prior to viability. Thereafter termination of pregnancy is called delivery. To the lay public the term spontaneous abortion is usually referred to as miscarriage. Induced abortions have been classified as therapeutic and nontherapeutic in the past, the terms often being used synonymously with legal and illegal abortion respectively. The synonymous use of the terms stemmed from the fact that licensed physicians, operating in hospitals, were not prosecuted for performing abortions where a threat to the life of the mother existed.

An induced termination of pregnancy prior to viability is not always regarded as an abortion in common parlance. For example, salpingectomy (removal of a fallopian tube) for ectopic pregnancy (pregnancy implanted in the tube) is a common, indeed a universal, practice, even in Catholic hospitals, where the operation is justified by the "principle of double effect" as an "indirect" abortion. Regardless of what the procedure is called in common parlance or in ethics, it conforms to the medical definition of induced abortion. Similarly, most diagnostic uterine curettages are performed in the late stages of the menstrual cycle, when a fertilized ovum may well be present in the tube, unless patients have been told not to have unprotected intercourse in that menstrual cycle. Such advice is rarely given, leading to the loss of thousands of fertilized eggs annually. Thus, the curettage will either not be regarded as an abortion at all or will be judged as "indirect" abortion, providing the curettage is justified on diagnostic, even though not directly therapeutic, grounds. Similarly again, it is often held, though far from settled, that intrauterine devices act by preventing implantation of fertilized eggs. Many disapprove of such "abortifacient" action, yet would not hesitate to do a curettage in the postovulatory (and hence potentially postfertilization) stages of the menstrual cycle. These are some of the vagaries in debates on what constitutes abortion.

Abortion procedures

Where there is an intention to procure an abortion, several procedures are available. The procedures are chosen largely on the basis of uterine size.

First trimester. In the first trimester of pregnancy, the uterine contents are of such size that they are easily evacuated through the cervix (mouth) of the uterus (womb). That is done by dilating the cervix and then emptying the uterus, either by strong suction ("suction curettage") into a bottle, or by scraping the uterine contents from its wall with a spoon-shaped instrument called a curette ("D & C" or dilation and curettage). In general, less uterine dilation is required for suction curettage than for a D & C, because the tubing through which the uterine contents are removed by suction is thinner than a curette. Also, since the pregnant uterus is very soft in consistency, the relatively soft tubing used in suction curettage presents fewer hazards of perforation of the uterus than the steel curette. Thus, suction curettage has become the commonest, safest, and fastest method of performing abortions in the first thirteen or so weeks of pregnancy.

Second trimester. Early in the second trimester uterine contents become of such size that rapid evacuation through the cervix becomes dangerous. Methods of uterine evacuation are therefore used, which increasingly resemble full-term delivery, surgically or medically.

Surgical. Surgical evacuation of the uterus prior to viability of the fetus is called hysterotomy (incision of the uterus) and is a miniature Caesarean section. Either the fetus is delivered first, the umbilical cord tied and cut, and then the placenta delivered, or the placenta is separated from the uterine wall first, thus delivering the placenta and fetus as one entity. The former procedure gives the fetus, if viable, a better chance of survival, for it can be better and earlier oxygenated if it is not retained *in utero* with its oxygen supply via the placenta cut off. Removal of the placenta and fetus simultaneously, however, gives the uterus a better chance of contracting down and stopping bleeding from the placental implantation site.

Medical. Medical evacuation of the uterus seeks to emulate the natural mechanism of labor and delivery at term by initiating uterine contractions, thereby dilating the uterine cervix and expelling the uterine contents. The initiation of contractions is induced by injecting into the amniotic fluid a concentrated solution of salt or twenty-five to forty mg. of a hormone called prostaglandin. The former is thought to act by destroying placental and fetal actions that tend to prevent the uterus from contracting, while the latter is thought to establish uterine contractions directly. Thus, the former tends to kill the fetus directly, while the latter does not. Upon occasion, both agents' action is augmented by the intravenous infusion into the maternal circulation of a pituitary gland extract called pitocin. Pitocin causes uterine contractions in its own right.

Sometimes, also, physicians will insert into the cervical canal laminar or tentlike structures that, by absorbing fluid, will enlarge and thus help dilate the cervix. Additional use of pitocin and laminaria is designed to speed up the time from the induction of abortion to its completion and also increases the percentage of abortions factually achieved by saline or prostaglandin. The disadvantage of added pitocin is that it may overstimulate uterine contractions, with the occasional occurrence of a uterine rupture.

Ethical considerations

It is a recognized fact that physicians often misjudge the existing duration of a pregnancy. In large part, this is because twenty percent of pregnant women bleed at some time during early pregnancy, often at the time of an expected menstrual period. Such bleeding episodes may be mistaken for a regular or slightly abnormal menstrual period. One such bleeding episode may cause the underestimation of gestational age by a month, and two episodes by even more. Occasionally, therefore, abortion procedures may produce live fetuses that approximate viability; in fact, some fetuses have survived abortion by these methods. At issue, then, ethically, is whether second-trimester abortions should be done only by those methods—hysterotomy with immediate fetal delivery, or prostaglandins—which constitute the least direct attacks upon the possibility of fetal survival, in contrast to saline injections. In part, the answer depends upon whether abortion is seen as a physical separation of mother and fetus, two parties who have a conflict of interest, or whether it is seen as a procedure that guarantees for a mother the destruction of her fetus. The American College of Obstetricians and Gynecologists, in a policy statement issued in 1976, takes the view that abortion is a process of separating conflicting parties and does not primarily aim at destruction of the fetus directly. It therefore implies that interruptions of pregnancy should, in the choice of methodology, maximize survival chances of the fetus, where this can be done without increasing the risk of the procedure to the mother. Courts of law have not expressed an opinion on whether abortion is a process of separating mother and fetus, or whether it may guarantee death of the fetus, whether necessary or not.

Complications

In general induced abortion, like any surgical or medical procedure, carries some risks. In general also, they are less frequent and of less severity than childbirth. A representation of those risks in terms of complication rates is contained in the following table. They refer to complications immediately incident to the various procedures. Still at issue are long-term sequellae of abortion, both physically and psychologically.

Complication Rates per 1000 Abortions by Type and Method of Termination, New York City

July 1, 1970–December 31, 1971

TYPE OF COMPLICATION	TOTAL	DILATION AND CURETTAGE	SUCTION	SALINE	HYSTEROTOMY
Hemorrhage	1.1	1.0	0.5	4.0	3.7
Infection	1.5	0.9	0.8	5.4	11.0
Perforated uterus	1.3	1.9	1.3	0.2	5.1
Anesthesia	0.1	0.1	0.1	0.1	0.7
Shock	*	*	*	0.2	0.7
Retained tissue	2.3	0.5	0.5	14.7	1.5
Failure	0.3	*	*	2.4	—
Lacerated cervix	0.2	0.3	0.2	0.1	0.7
Other	0.4	0.3	0.3	1.0	4.4
Unspecified	0.1	0.1	*	0.2	—
Total	7.3	5.1	3.7	28.3	27.8

* Less than 0.05.
Source: J. Pakter and F. G. Nelson, "Effects of a Liberalized Abortion Law in New York City," *Mt. Sinai Journal of Medicine* 39 (1972): 535. Reprinted by permission of the publisher and the authors.

At the physical level, the main unresolved question is the effect of aborting one pregnancy on the outcome of subsequent pregnancies, particularly in teenagers (Russell). The problem seems to revolve around the degree of damage done to the uterine cervix in artificial dilation or by natural dilation occurring at an early age, through either abortion or childbirth. At the psychological level, the problem revolves around how the human mind views its own fetuses.

Finally, it should be realized that any medical procedure is limited by the total biological knowledge available at the time of its development. At all times medical science seeks to make its interventions effective with the least trauma. The psychological degree of trauma induced depends in large part upon the body image that the individual has of himself or herself. Because abortion is a procedure following upon one result of sexual activity, its effects depend, in part, upon the significance that may, from time to time, be attached to that sexual activity.

ANDRÉ E. HELLEGERS

[*While all the articles in this entry are relevant, see especially the articles* CONTEMPORARY DEBATE IN PHILOSOPHICAL AND RELIGIOUS ETHICS *and* LEGAL ASPECTS. *For further discussion of topics mentioned in this article, see the entries:* CONTRACEPTION; DOUBLE EFFECT; FETAL–MATERNAL RELATIONSHIP; GENETIC DIAGNOSIS AND COUNSELING; INFANTS; LIFE; PRENATAL DIAGNOSIS; *and* RISK. *See also:* EMBODIMENT; PERSON; POPULATION ETHICS: ELEMENTS OF THE FIELD; POPULATION ETHICS: RELIGIOUS TRADITIONS; *and* POPULATION POLICY PROPOSALS.]

BIBLIOGRAPHY

American College of Obstetricians and Gynecologists, Executive Board. "Some Ethical Considerations in Abortion." Chicago: 1975. Xeroxed.

CONNERY, JOHN. *Abortion: The Development of the Roman Catholic Perspective.* Chicago: Loyola University Press, 1977.

GREGG, N. MCALISTER. "Congenital Cataract Following German Measles in the Mother." *Transactions of the Ophthalmological Society of Australia* 3 (1941): 35–46.

HARVEY, WILLIAM. *Anatomical Exercitations Concerning the Generation of Living Creatures: To Which Are Added Particular Discourses of Births, and of Conceptions, etc.* London: Printed by James Young for Octavian Pulleyn, 1653.

PAKTER, JEAN, and NELSON, FRIEDA G. "Effects of a Liberalized Abortion Law in New York City." *Mt. Sinai Journal of Medicine* 39 (1972): 535–543.

RUSSELL, J. K. "Sexual Activity and Its Consequences in the Teenager." *Clinics in Obstetrics and Gynaecology* 1 (1974): 683–698.

II
JEWISH PERSPECTIVES

The abortion question in talmudic law begins for some with an examination of the fetus's legal status. For this the Talmud has a phrase, *ubar yerekh imo*, a counterpart of the Latin *pars viscerum matris*. The fetus, that is, is deemed a "part of its mother" rather than an independent entity. Of course, this designation says nothing about the right of abortion; the term is found only in more theoretical contexts. It defines ownership, for example, in the case of an embryo found in a purchased animal, that, as intrinsic to its mother's body, it belongs to the buyer. Moreover, in the religious conversion of a pregnant woman, her unborn child is automatically included and requires no added ceremony. Nor does the fetus have power of acquisition: Gifts or transactions made on its behalf, except by its father, are not binding; it inherits from its father only, in a natural rather than a transactional manner.

Germane as such information might seem to the question of abortion, it tells us little more than, in the words of a modern writer on Roman and Jewish law, Y. K. Miklishanski, that in both systems the fetus has no "juridical personality" of its own. The morality of abortion is a function, rather, of the legal attitude to feticide as distinguished from homicide or infanticide. The law of homicide in the Torah, in one of its several formulations (Exod. 21:12), reads: "*Makkeh ish* . . ." "He who smites a man . . ." Does this include *any* "man," say, a day-old child? Yes, says the Talmud, citing another text (Lev. 24:17): *ki yakkeh kol nefesh adam*, "if one smite *any nefesh adam*"—literally, any human person. The "any" is understood to *include* the day-old child, but the "*nefesh adam*" is taken to *exclude* the fetus in the womb, for the fetus in the womb is *lav nefesh hu* (not a person) until he is born. In the words of Rashi, the classic commentator on the Bible and Talmud, only when the fetus "comes into the world" is it a "person."

The basis, then, for denying capital-crime status to feticide in Jewish law is scriptural. Alongside the *nefesh adam* text is another basic one, in Exodus 21:22, which provides: "If men strive, and wound a pregnant woman so that her fruit be expelled, but no harm befall [her], then shall he be fined as her husband shall assess, and the matter placed before the judges. But if harm befall [her], then shalt thou give life for life."

The Talmud makes this latter verse's teaching explicit: Only monetary compensation is exacted of him who causes a woman to miscarry. And though the abortion spoken of in this biblical passage is accidental, the verse is still a source for the teaching that feticide is not a capital crime (since even accidental homicide could not be expiated by monetary fine).

This important passage in Exodus has an alternate version in the Septuagint, the Greek translation of the Bible produced in Alexandria in the third pre-Christian century. One word change there yields an entirely different statute on miscarriage. Prof. Viktor Aptowitzer's essays analyze the disputed passage; the school of thought it represents he calls the Alexandrian school, as opposed to the Palestinian—that is, the talmudic—view set forth above. The word in question is *ason*, rendered above as "harm," hence, "if [there be] no harm [i.e., death, to the mother], then shall he be fined. . . ." The Greek renders the word *ason* as "form," yielding something like: "if [there be] form, then shalt thou give life for life." The "life for life" clause is thus applied to the fetus instead of the mother, and a distinction is made—as Augustine will formulate it—between *embryo informatus* and *embryo formatus*, a fetus not yet "formed" and one already "formed"; for the latter, the text so rendered prescribes the death penalty.

Among the Christian Church Fathers, the consequent doctrine of feticide as murder was preached by Tertullian (160?–230?), who accepted the Septuagint, and by Jerome (340?–420), who did not. Jerome's classic Bible translation renders this Exodus passage according to the Hebrew text accepted in the Church. The Didache, a handbook of basic Christianity for the instruction of converts from paganism, follows the Alexandrian teaching and specifies abortion as a capital crime. Closer to the main body of the Jewish community, we find the doctrine accepted by the Samaritans and Karaites and, more important, by Philo, the popular first-century philosopher of Alexandria. On the other hand, Philo's younger contemporary, Josephus, bears witness to the Palestinian (halakhic) tradition. Aside from its textual warrant, the latter is the more authentic, according to Aptowitzer, while the other is a later tendency, "which, in addition, is not genuinely Jewish but must have originated in Alexandria under Egyptian-Greek influence" (Aptowitzer, p. 88).

In the rabbinic tradition, then, abortion remains a noncapital crime at worst. But a curious factor further complicates the question of the criminality of the act. This is the circumstance that one more biblical text, this one in Genesis and hence "before Sinai" and part of the Laws of the "Sons of Noah," served as the source for the teaching that feticide is indeed a capital crime—for non-Jews. Genesis 9:6 reads, "He who sheds the blood of man, through man [i.e., through the human court of law] shall *his* blood be shed." Since the Hebrew (*shofekh dam ha'adam, ba'adam* . . .) allows for a translation of "man, in man" as well as "man, through man," the Talmud records the exposition of Rabbi Ishmael: "What is this 'man in man'? It refers to the fetus in its mother's womb." The locus of this text in Genesis—standing as it does without the qualifying balance of the Exodus (Sinaitic) passage—made feticide a capital crime for non-Jews (i.e., those not heir to the Sinaitic covenant) in Jewish law. Some modern scholars hold this exposition to be more sociological than textually inherent, representing a reaction against abuses among the heathen. In view of rampant abortion and infanticide, they claim, Rabbi Ishmael "expounded" the above exegesis in the Genesis text in order to render judgment against the Romans.

Regardless of its rationale, the doctrine remains part of Jewish law, as Maimonides systematically codifies it: "A 'Son of Noah' who killed a person, even a fetus in its mother's womb, is capitally liable. . . . (The Jewish court is obliged to provide judges for the resident alien to adjudicate for them in accordance with these laws [of the Sons of Noah] so that society not corrupt itself. The judges may come either from their midst or from the Israelites.)" Lifesaving therapeutic abortion is not, of course, included in this Noahide restriction. Nor is an abortion during the first forty days of pregnancy included, according to some. The implications of this anomaly of a different law for "Sons of Noah" were dealt with in a Responsum of the eighteenth century: "It is not to be supposed that the Torah would consider the embryo as a person [*nefesh*] for them [Sons of Noah] but not a person for us. The fetus is not a person for them either; the Torah merely was more severe in its practical ruling in their [the Noahides'] regard. Hence, therapeutic abortion would be permissible to them, too."

In the rabbinic system, then, abortion is not murder. Nor is it *more* than murder, as would be the case if "ensoulment" were at issue. Talmudic discussions speak of the moment—conception, birth, postbirth, etc.—at which the soul joins the body. This is seen to be irrelevant to the abortion question, because the soul is immortal no matter when it enters or leaves the body. And, more important than being immortal, it is a pure soul, free of the taint of "original sin." In the sixth century, St. Fulgentius asserted that "original sin" is inherited by the soul of the fetus at conception. This made baptism *in utero* necessary in cases of miscarriage, and also made abortion worse than murder, in the sense that the fetus was being "killed in this world and the next." Judaism has no concept of "original sin" of this kind; in the words of the Talmud and Daily Prayer Book, "My God, the soul with which Thou hast endowed me is pure."

Murder (of the innocent) would be forbidden even to save life. But with abortion removed from the category of murder, therapeutic abortion becomes permissible and, in fact, mandated. The Mishnah sets forth the basic talmudic law in this regard: "If a woman has [life-threatening] difficulty in childbirth, the embryo within her should be dismembered limb by limb, because her life takes precedence over its life. Once its head (or its greater part) has emerged, it may not be touched, for we do not set aside one life for another" (Mishnah, Oholot 7:6).

In analyzing such provisions the Talmud suggested that the reason could well be that the fetus is in the category of an "aggressor"; that is, its life is forfeit under the law that permits killing a "pursuer" in order to save the intended victim. The Talmud, however, dismisses this reasoning, since the fetus is an innocent being, and since one cannot know "who is pursuing whom"; the pursuit must therefore be deemed an "act of God," and this factor does not apply. Yet, in his great Law Code, Maimonides reintroduced the term "aggressor"—his commentators explaining his qualified use of the concept. He formulates the Mishnaic provision as follows:

This, too, is a [Negative] Commandment: Not to take pity on the life of a pursuer. Therefore the Sages ruled that when a woman has difficulty in giving birth, one may dismember the child in her womb—either with drugs or by surgery—because it is like a pursuer seeking to kill her. Once its head has emerged, it may not be touched, for we do not set aside one life for another; this is the natural course of the world [Mishnah Torah, Laws of Murder and Preservation of Life, 1:9].

Some commentators to Maimonides's Code suggest that although abortion is not technically murder it is still so grave an act that Maimonides enlisted the aggressor argument to buttress the existing permission for abortion; its justification is that the fetus is at least *like* an aggressor.

The subsequent rabbinic tradition seems to align itself either to the right, in the direction of Maimonides; or to the left, in the direction of Rashi, above. The first approach can be identified especially with the late Chief Rabbi of Israel, Issar Unterman, who sees any abortion as "akin to homicide" and therefore allowable only in cases of corresponding gravity, such as saving the life of the mother. This approach then builds *down* from that strict position to embrace a broader interpretation of lifesaving situations, which include a threat to her health, for example, as well as a threat to her life. The second approach, associated with another former Chief Rabbi of Israel, Ben Zion Uziel, and others, assumes that no real prohibition against abortion exists—other than antiprocreational—and builds *up* from that lenient position to safeguard against indiscriminate abortion. This includes the example of Rabbi Yair Bachrach in the seventeenth century, whose classic Responsum saw no legal bar to abortion, in the case before him, but would not permit it on other grounds. The case was one of a pregnancy conceived in adultery; the woman, in "deep remorse," wanted to destroy the fruit of her sin. The author concludes by refusing to sanction the abortion, not on legal grounds but on sociological ones, as a safeguard against immorality. Other authorities disagreed on this point, reaffirming the legal sanction of abortion for the woman's welfare, whether life or health, or even avoidance of "great pain."

The criterion in both approaches remains maternal rather than fetal. A principle in Jewish law is *tza'ar gufah kadim*, i.e., that avoidance or prevention of "her pain should be the first consideration." The mother's welfare is primary, and hence maternal indications, rather than fetal— or even the husband's wishes—are determinative. Rabbinic rulings on abortion are thus amenable to the following generalization: If a possibility or probability exists that a child may be born defective, and the mother seeks abortion *on the grounds of* pity for a child whose life would be less than normal, the rabbi would decline permission. If, however, an abortion for that same potentially deformed child were sought on the grounds that the possibility is causing severe anguish to the mother, permission would be granted. The fetus is unknown, future, potential, part of the "secrets of God"; the mother is known, present, alive, and asking for compassion.

One rabbinic authority, writing in Rumania in 1940, responded to the case of an epileptic mother who wanted to interrupt her pregnancy for fear that her child too would be epileptic. He first discusses the question of epilepsy itself, then writes:

For fear of possible, remote danger to a future child that, maybe, God forbid, he will know sickness—how can it occur to anyone to actively kill him because of such a possible doubt? This seems to me very much like the laws of Lycurgus, King of Sparta, according to which every blemished child was to be killed. . . . Permission for abortion is to be granted only because of fear of mental anguish for the *mother*. But for fear of what might be the child's lot—"the secrets of God are none of your business" [Feldman, p. 292].

In the current Tay-Sachs genetic screening controversy, rabbinic authorities recommend screening before rather than during the pregnancy. This is because the alternative would be resort to amniocentesis after the first trimester of pregnancy, with possible abortion on the basis of its results. This abortion for fetal rather than maternal indications would not ordinarily be sanctioned by Jewish law. True, rabbinic opinion permitting abortion for fetal reasons alone is not altogether lacking, but the normative rabbinic view is to permit it for maternal indications only. Yet the one can blend into the other: Fetal risk can mean mental anguish on the part of the mother, so that the fetal indication becomes a maternal one. The woman's welfare is thus the key to warrant for abortion.

Implicit in the Mishnah above is the teaching that the rights of the fetus are secondary to the rights of the mother all the way up until the moment of birth. This principle is obscured by the current phrase "right to life." In the context of abortion questions, the issue is not the right to life, which is very clear in Jewish law, but the right to be born, which is not as clear. The right to be born is relative; the right to life, of existing persons, is absolute. "Life" may begin before birth, but it is not the life of a human person; animal life, plant life, or even prehuman life is not the same as human life. Rabbinic law has determined that human life begins with birth. This is neither a medical nor a court judgment, but a metaphysical one. In the Jewish system, human life in this sense begins with birth. Of course, potential life already partakes of the potential sacredness of actual life, since the latter can have its inception only through the former.

Another slogan-like phrase is dealt with in the same Mishnah, wherein it is ruled that "once the fetus has emerged from the womb, it cannot be touched" even to save the life of the mother, "for we cannot set aside one life for another." The "quality of life" slogan or concept is thus inadmissible. The life of the mother has more "quality"; she is adult and has husband, children, and associations. Still, the *sanctity*-of-life principle means that life is sacred regardless of differences in quality: mother and newborn babe are equal from the moment of birth.

Talmudic statements do use the term "murder," in a figurative sense, of course, to describe even the neglect to conceive. Procreation is a positive mitzvah (commandment), and he who fails to fulfill this mitzvah is called "guilty of bloodshed." Much of the pronatalist attitude of Judaism helps account for its abhorrence of casual abortion. There may be legal sanction for abortion where necessary, but the attitude remains one of hesitation before the sanctity of life and a pronatalist reverence for potential life.

Accordingly, abortion for "population control" is repugnant to the Jewish mind. Abortion for economic reasons is also not admissible. Taking precaution by abortion or birth control against physical threat remains a mitzvah, but never to forestall financial hardship. Material considerations are improper in this connection. In the Jewish community today, with a conscious or unconscious drive to replenish ranks decimated by the Holocaust, contemporary rabbis invoke not the more lenient but the more stringent Responsa of the earlier authorities. The more permissive decisions, they point out, were in any case rendered against the background of far greater instinctive hesitation to resort to abortion. Against today's background of more casual abortion, rabbis are moving closer to the position associated with Maimonides and Unterman, allowing abortion only for the gravest of reasons.

DAVID M. FELDMAN

[*While all the articles in this entry are relevant, see especially the articles* ROMAN CATHOLIC PERSPECTIVES, CONTEMPORARY DEBATE IN PHILOSOPHICAL AND RELIGIOUS ETHICS, *and* LEGAL ASPECTS. *For further discussion of topics mentioned in this article, see the entries:* CONTRACEPTION; GENETIC DIAGNOSIS AND COUNSELING; INFANTS; LIFE; POPULATION ETHICS: RELIGIOUS TRADITIONS, *article on* JEWISH PERSPECTIVES; PRENATAL DIAG-

NOSIS; *and* RIGHTS. *Also directly related are the entries* JUDAISM; MEDICAL ETHICS, HISTORY OF, *section on* NEAR AND MIDDLE EAST AND AFRICA, *article on* CONTEMPORARY ISRAEL; *and* RELIGIOUS DIRECTIVES IN MEDICAL ETHICS, *article on* JEWISH CODES AND GUIDELINES. *See also:* PERSON; *and* POPULATION POLICY PROPOSALS, *article on* DIFFERENTIAL GROWTH RATE AND POPULATION POLICIES: ETHICAL ANALYSIS.]

BIBLIOGRAPHY

APTOWITZER, V. "Observations on the Criminal Law of the Jews." *Jewish Quarterly Review* 15 (1924): 55–118.

BLEICH, DAVID. "Abortion in Halakhic Literature." *Tradition* 10 (1968): 72–120.

FELDMAN, DAVID MICHAEL. *Birth Control in Jewish Law: Marital Relations, Contraception and Abortion as Set Forth in the Classic Texts of Jewish Law.* New York: New York University Press, 1968. Schocken Paperback, 1974.

JAKOBOVITS, IMMANUEL. *Jewish Medical Ethics: A Comparative and Historical Study of the Jewish Religious Attitude to Medicine and Its Practice.* New York: Bloch Publishing Co., 1962. Rev. ed. 1975.

KLEIN, ISAAC. "Abortion" (1959). *Responsa and Halakhic Studies.* New York: Ktav Publishing House, 1975, chap. 4, pp. 27–33.

ROSNER, FRED. *Modern Medicine and Jewish Law.* Studies in Torah Judaism, vol. 13. Edited by Leon D. Stitskin. New York: Yeshiva University, Department of Special Publications, 1972.

III
ROMAN CATHOLIC PERSPECTIVES

Scriptural influence

An understanding of the Roman Catholic perspective on abortion, if it is to be adequate, will call for an acquaintance with the Hebrew tradition that preceded the Christian era. One would expect to discover the source, or at least the expression, of this tradition in the Torah, the first five books of the Bible, which contains the fundamental law of the Jews. Unfortunately, an examination of this part of the Bible yields only one text that deals explicitly with abortion. It comes from Exodus (21:22–25), more specifically from that part of Exodus known as the Book of the Covenant, and imposes a penalty on a person guilty of what seems to be an accidental abortion. According to the law, if two men are fighting and one happens to strike a pregnant woman, causing an abortion, he will have to pay a fine. But if injury occurs (to the woman), he will have to pay in kind according to the *lex talionis*, that is, "an eye for an eye," etc. This has generally been interpreted to mean that the fetus was not considered a human being. If it were, the *lex talionis* would apply and the penalty would be "a life for a life," which would be the penalty if the mother herself died.

Curiously enough, the Septuagint (Greek) version of the above text, made in the early third century B.C., is quite different. The case presented is the same, but the distinction made is not between an abortion and damage to the mother, but between the formed and unformed fetus. If the fetus is unformed, the one guilty will have to pay a fine; if the fetus is formed, the *lex talionis* will apply and he will have to pay a life for life. This statement of the law is quite different from that of the Hebrew text and coincides more with Greek, particularly Aristotelian, thought. According to Aristotle the human soul was infused into the fetus when it took on the human form and figure. This occurred forty days after conception in the male, and ninety days in the female. Before that time it did not have a human soul. The Septuagint makes no distinction between male and female but seems to consider the fetus a human being when it is formed and applies the *lex talionis* to the formed fetus just as the Hebrew text applied it to the mother. Since the lingua franca of the Roman world at the beginning of the Christian era was Greek, it was the Septuagint text that had the greater impact on the Christian tradition. The Vulgate of St. Jerome (Latin version of the Bible, A.D. 389?) followed the Hebrew text, but there is no evidence that the Roman Church ever accepted the view that the fetus is not a human being until it is born.

The Hebrew Bible says nothing about a woman causing an abortion on herself. It might be tempting to conclude from the rabbinic tradition, which considered the fetus a part of the mother, to a liberal attitude toward self-induced abortion. But this would be quite contrary to the entire Jewish tradition of respect for fertility. The only case mentioned in early (pre-Christian or early Christian) Jewish tradition in which it was permitted to sacrifice the fetus was to save the mother. Dismemberment of the fetus (at term) rather than abortion was involved in this case, but the principle was the same. Since the fetus was not considered a human being in the full sense, it was permissible to sacrifice it to save a human being.

There is another text (Gen. 9:6) that Jewish tradition applied to abortion, at least among the Romans. According to this text abortion would be considered a capital crime. No explicit use of this text was made in the Christian tradition but some of the early Fathers condemned abortion as murder without any distinction between the formed and the unformed fetus.

The New Testament says nothing explicit about abortion, but the incompatibility of abortion, as well as infanticide, with the Christian message, came through quite clearly. An explicit condemnation of both is found in the *Didache* and the *Epistle of Barnabas,* first- or early second-century writings. Although these were documents of the early Church, the part that contains these condemnations, known as the *Duae Viae,* seemed to belong originally to the Jewish community and was used to instruct converts to Judaism. If this is correct, these condemnations are as representative of the moral traditions of the Jews of the pre-Christian era as they are of the Christians themselves. Such explicit condemnations were necessary because of the tolerant attitude of the Roman world of the time toward both abortion and infanticide. The apologists of the early Church were able to point to their attitude toward these two practices to distinguish the Christians from the pagans of the times.

Foundations: status of the fetus

Later Christians were not always of the same caliber as the early Christians, but the Church in no way relaxed its attitude toward these practices. When at the beginning of the fourth century the first Church legislation appeared in connection with penitential discipline (reconciliation of sinners), a woman who committed abortion after adultery was not allowed back into the Church for the rest of her life. This legislation, which came from the Council of Elvira, was modified by a council held at Ancyra in A.D. 314 to allow reconciliation after ten years, but even this penance would be considered very severe by modern standards.

Early Church law regarding penance made no distinction between the formed and unformed fetus, but this distinction, already used by some of the Fathers, was introduced into the practice of private penance in the late seventh century. Although this practice differed considerably from the public or canonical penance mentioned above, it was basically the same practice of reconciling sinners to the Church. Formation of the fetus was significant because it was thought that the human soul was infused at that time. Abortion of the formed fetus was consequently considered homicide. But the distinction was not accepted universally on the level of Church law until after the decree of Gratian in 1140. Official recognition of it came in the decree *Sicut ex* of Innocent III in 1211. The decree dealt with an irregularity, an impediment to the reception or exercise of clerical orders, which was incurred, among other reasons, for homicide. According to the decree the irregularity would not be incurred for abortion unless the fetus was animated. Since the time of animation was identified with formation, the decree implied that only abortion of the formed fetus was considered homicide, at least in reference to irregularities. Confusion arose, however, from a parallel tradition that extended the notion of homicide not only to the abortion of the unformed fetus but to sterilization as well. Nonetheless, the distinction was generally accepted during the late Middle Ages, and only abortion of the formed fetus was considered homicide even in reference to sacramental penances. Forty and ninety days were also accepted as the time of animation for the male and female fetus respectively. Aristotle claimed to base this time difference on evidence derived from aborted fetuses, but it may have been based more on false concepts of embryology.

Although the distinction between the animated and unanimated fetus was used as a canonical dividing line, it was never used during the early centuries of Christianity as a moral dividing line. Abortion was condemned as wrong, whether the fetus was animated or not. In the early fourteenth century, however, the Dominican theologian John of Naples introduced an exception into the condemnation. If the fetus was unanimated, he argued, it would be permissible to abort it to save the life of the mother when threatened by disease. This exception was accepted by another Dominican theologian, Antoninus, Archbishop of Florence (1389–1459), and subsequently by several other theologians. The original argument of John of Naples was that, since the unanimated fetus was not yet a human being, it would be permissible to sacrifice it, if necessary, to save a human being. Later theologians, particularly Thomas Sanchez (1550–1610), used the argument of self-defense against unjust aggression (considering the fetus an unjust aggressor in these circumstances) or the principle of totality (looking upon the fetus as a part of the mother). These arguments would give rise to difficulties at a later date.

In the early Church the penance for serious sins was exclusion from communion with the faithful for a determined period of time. By the end of the first millennium, however, sinners were reconciled to the Church immediately upon confessing their sins and before performing whatever penance was imposed on them. Excommunication from the Church then began to appear as a special penalty for certain sins. Excommunication for the sin of abortion was first imposed in the thirteenth century, but only on a local level. The penalty was extended to the whole Church in 1588 by Sixtus V. What is interesting about the decree (called *Effraenatam* from the introductory word) in which the penalty was legislated was that it imposed the penalty not only on the abortion of the formed fetus but on the abortion of the unformed fetus and sterilization as well. In doing this, it was going beyond the practice of the Church in dealing with irregularities. As already mentioned, from the time of Innocent III (1211), an irregularity was attached only to the abortion of the formed or animated fetus. Sixtus V (1588) was following more in the footsteps of the *Si aliquis*, a canon that originated with Regino of Prüm (d. 905) but was made official in the Decretals of Gregory IX (1234). This canon gave expression to the parallel tradition noted above, which classified all of these acts as homicide. The reason behind the rigid position taken by Sixtus V may have been the revival of pagan practices that accompanied the Renaissance in Europe. The legislation was soon modified, however, and in 1591 the excommunication was limited to the abortion of the animated fetus. The estimate of the time of animation was still the one generally accepted, that is, forty to ninety days. This legislation remained in effect until 1869, when Pius IX extended the excommunication to all abortion. By this time theories of delayed animation had given way to the proponents of early animation.

The question of intentionality

For centuries theologians distinguished between intentional and accidental abortion. They discussed at length the circumstances under which an abortion that was not intentional would be imputed to the person performing the act from which it followed. Initially, the person performing the act was not considered responsible for the abortion if what he was doing was necessary. Later the person was relieved of responsibility as long as what he was doing was licit. Generally speaking, in these cases the agent did not foresee the danger of abortion, but the question eventually arose about the responsibility of a person who would foresee such danger. At the end of the sixteenth century a Spanish Franciscan, Anthony of Cordova, set down the principle that, if the procedure was necessary to save the life of the mother and if its nature aimed principally at this goal, it would be permissible, even if it carried a danger of abortion. In this category he placed such remedies as bloodletting, baths, analgesics, and cathartics. But if the procedure in question was aimed principally at causing the abortion, e.g., striking the mother, giving her a poisonous drug, etc., it would not be permissible. This distinction met with immediate and universal acceptance by theologians of the time and has since that time been generally accepted. It has been formalized in the terms *direct* and *indirect* abortion.

Modern controversies: fetal and maternal factors

The exception made by John of Naples regarding the abortion of the unanimated fetus to save the life of the mother did not receive such universal acceptance as indirect abortion. Many important theologians of the seventeenth and eighteenth century argued against his exception: They refused to consider the fetus either an unjust aggressor or part of the mother. If the unanimated fetus could be considered an unjust aggressor, there might be more reason to regard the animated fetus as an aggressor and argue for abortion even of the animated fetus to save the life of the mother. Or if the fetus could be sacrificed as a part of the mother to save her from some intrinsic threat, it could also be sacrificed to save her from an extrinsic threat as well. But the authors who used these arguments rejected their extension to other cases. There were a few theologians during the seventeenth century who wanted to permit the abortion of the unanimated fetus to save the life or reputation of the mother from an extrinsic threat, e.g., from an angry father or husband, but the more important theologians of the period rejected this opinion. It was also condemned in 1679 by Innocent XI.

Another development that would influence the abortion debate began in the early seventeenth century. As mentioned earlier, theories of delayed animation, particularly that of Aristotle, were generally accepted throughout most of the Christian era. In the early seventeenth century a Belgian physician, Thomas Fienus (De Feynes), wrote a treatise on the formation of the fetus in which he defended the thesis that the human soul was infused into the fetus at least by the third day after conception. His basic argument is that since formation begins shortly after conception, some formative principle must be present. This can only be the human soul, unless one wants to admit, as Aristotle does, a succession of souls. Fienus argues at length that postulating a succession of souls leads to ridiculous conclusions. He concludes therefore that the human soul is infused within three days after conception. Another physician of the time, Paolo Zacchia, also came to the conclusion of immediate animation.

Oddly enough, about the same time a Prague physician, Ioannes Marcus, was arguing in favor of the position taken in Roman law, and perhaps in the Jewish Bible, that the fetus was part of the mother until birth and did not become a human being until that time. But this opinion gained no adherents among the theologians and was condemned by Innocent XI at the same time he condemned the opinion allowing abortion to save the mother's life from extrinsic threat, or even to save her reputation. The revival of this old opinion regarding the status of the fetus may well have been the product of Renaissance interest in the ideas of the Classic Age.

The opinion of Fienus and Zacchia on immediate animation, although not rejected by the Church, did run into considerable opposition from theologians. It was objected to on three specific grounds: It was contrary to the Scriptures, to the universal opinion of theologians, and to the practice of the Church. It might be surprising that an opinion that had so much authority against it could survive. But the fact is that it did take hold and gradually replaced theories of delayed animation. As might be expected, it was put into practice first in baptismal discipline. Since the benefit of any doubt was to be given to the fetus, the Church on the strength of this opinion allowed the conditional baptism of fetuses, no matter how early they were aborted, as long as they showed signs of life, or at least as long as certain death had not set in. Again, as might be expected in the area of penal law, where the benefit of any doubt is given to the culprit, the opinion was put in practice last in penal legislation regarding abortion. It was not until 1869 that excommunication was attached to all abortions without distinction. It is important, however, to distinguish here between Church practice or discipline and Church teaching. The Church has made no positive teaching statement regarding the time of infusion of the human soul, and this is true of Vatican II as well as of earlier documents. Some have seen in the dogma of the Immaculate Conception an affirmation of immediate infusion of the human soul, but no such affirmation is implied in this teaching. The only opinion the Church has condemned is that of Ioannes Marcus, that the human soul is not infused until birth.

With the common acceptance of early animation the whole question regarding the abortion of the unanimated fetus to save the life of the mother lost its practical significance. Up to the middle of the nineteenth century theologians were almost unanimously opposed to the abortion of the fetus, once it was animated, even to save the mother's life. But in the middle of the nineteenth century a few theologians, particularly Ballerini, suggested that a medical abortion would be permitted to save the life of the mother. They argued that, since such an abortion did not involve an attack on the fetus, it would not be direct killing. It would involve moving the fetus from a place where it could not survive to another place where it would also die. The death of the fetus would result not from the operation itself but from the inability of the fetus to survive in a different medium.

There were also some theologians in the second half of the nineteenth century who argued in favor of craniotomy to save the life of the mother. They offered several arguments to justify their position, but their main defense seemed to be that the fetus in these circumstances was an unjust aggressor. When the Sacred Penitentiary, a Roman tribunal that among other functions handles cases of conscience, was first presented with a craniotomy case, its response was simply that the petitioners should consult the approved authors for their answer. It was probably the ambiguity of this response that gave rise to the controversy that ensued. The controversy was resolved finally by a response from the Holy Office, the Roman Congregation, which is entrusted with the protection of faith and morals. According to the response the opinion that craniotomy was permissible in this case could not be safely taught. In further responses the Holy Office condemned any kind of direct attack on the fetus, and even medical (therapeutic) abortion. In the response dealing with medical abortion it was refusing to accept the distinction Ballerini and others were trying to make between this procedure and those that would clearly involve a direct attack on the fetus, such as craniotomy.

In the first half of the twentieth century there was general acceptance by Catholic theologians of the condemnations of craniotomy and medical (therapeutic) abortion. A few still wanted to make exception for the case where without these procedures both the mother and the child would die—even after the 1930 encyclical *Casti Connubii*, which condemned direct abortion even in these instances—but they found little support among their confreres. As medical science learned how to handle complications of pregnancy and delivery, the issue of craniotomy and medical abortion was reduced largely to the realm of speculation.

The distinction between direct and indirect abortion would still give rise to difficulties in judging individual cases. A controversy arose in the 1930s over the removal of a cancerous but pregnant uterus. At an even earlier date a similar controversy began regarding ectopic pregnancies. Eventually a consensus developed in favor of considering abortion indirect in these instances—and therefore permissible to save the mother's life—because the direct intent would be the removal of a pathological, though pregnant, uterus or tube.

Conclusion

The abortion debate of the 1960s and 1970s, which is the subject of a separate article in this entry, makes clear that theologians have discussed and adopted a greatly expanded number of positions on the two historically significant focal points—namely, as regards both the value of fetal life and the resolution of conflict cases. Still, the Church's official teaching retains considerable continuity. The Second Vatican Council (1965) summarized in two brief sentences the commitment of the Church to the welfare of the fetus, condemning abortion as an unspeakable crime and asking that the fetus be given the greatest of care right from the moment of conception. This statement was in strict continuity not only with the remote past but also with more recent pronouncements by Pius XI, Pius XII, John XXIII, and Paul VI. These Pontiffs confirmed the traditional teaching in statements too numerous to mention.

JOHN R. CONNERY

[*While all the articles in this entry are relevant, see especially the articles* JEWISH PERSPECTIVES, *and* CONTEMPORARY DEBATE IN PHILOSOPHICAL AND RELIGIOUS ETHICS. *Other relevant material may be found under* DOUBLE EFFECT; FETAL-MATERNAL RELATIONSHIP; POPULATION ETHICS: RELIGIOUS TRADITIONS, *article on* ROMAN CATHOLIC PERSPECTIVES; RELIGIOUS DIRECTIVES IN MEDICAL ETHICS, *article on* ROMAN CATHOLIC DIRECTIVES; ROMAN CATHOLICISM. *For further discussion of topics mentioned in this article, see the entries:* INFANTS, *article on* INFANTICIDE: A PHILOSOPHICAL PERSPECTIVE; PERSON; POPULATION ETHICS: RELIGIOUS TRADITIONS, *article on* ROMAN CATHOLIC PERSPECTIVES; STERILIZATION, *article on* ETHICAL ASPECTS. *See also:* MEDICAL ETHICS, HISTORY OF, *section on* EUROPE AND THE AMERICAS, *article on* MEDIEVAL EUROPE: FOURTH TO SIXTEENTH CENTURY. *See* APPENDIX, SECTION I, ETHICAL AND RELIGIOUS DIRECTIVES FOR CATHOLIC HEALTH FACILITIES.]

BIBLIOGRAPHY

BEUGNET, A. "Avortement." *Dictionnaire de théologie catholique.* Paris: Librairie Letouzey & Ané, 1903, vol. 1, cols. 2644–2652.

BOUSCAREN, TIMOTHY LINCOLN. *Ethics of Ectopic Operations.* 2d ed., rev. Milwaukee: Bruce Publishing Co., 1944.

CANGIAMILA, FRANCISCO EMANUELLO. *Sacra embryologia* (1745). Venice: 1763, pp. 1–43.

ESCHBACH, ALPHONS. *Disputationes physiologico-theologicae tum medicis chirurgis tum theologis et canonistis utiles.* Rev. ed. Rome: Desclee, Lefebvre & Sacii, 1901.

FELDMAN, DAVID MICHAEL. *Birth Control in Jewish Law: Marital Relations, Contraception, and Abortion as Set Forth in the Classic Texts of Jewish Law.* New York: New York University Press, 1968, pp. 251–297.

FIENUS [FEYENS], THOMAS. *De formatrice foetus liber in quo ostenditur animam rationalem infundi tertia die.* Antwerp: Gulielmum á Tongris, 1620.

GRISEZ, GERMAIN G. *Abortion: The Myths, the Realities and the Arguments.* New York: Corpus Books, 1970, pp. 117–185.

HUSER, ROGER JOHN. *The Crime of Abortion in Canon Law: An Historical Synopsis and Commentary.* Catholic University of America, Canon Law Studies, no. 162. Washington: 1942.

NOONAN, JOHN T., JR. *The Morality of Abortion: Legal and Historical Perspectives.* Cambridge: Harvard University Press, 1970, pp. 1–59.

PALAZZINI, GIUSEPPE. *Ius fetus ad vitam eiusque tutela in fontibus ac doctrina canonica usque ad saeculum XVI.* Urbania: Bramantes, 1943.

VISCOSI, DANIELE. *L'Embryotomia nei suoi rapporti colla morale cattolica.* Naples: Tipografia & Libr. di Andrea & Salv. Festa, 1879.

ZACCHIA, PAOLO. *Questiones medico-legales* (1621). Lyons: 1701, pp. 51–53, 686–707.

IV

PROTESTANT PERSPECTIVES

Early Protestant perspectives

Early Protestant abortion attitudes show considerable continuity with those of the pre-sixteenth-century Church. The major reformers—Martin Luther (1483–1546), Philip Melanchthon (1497–1560), and John Calvin (1509–1564)—were at least as conservative as their Roman Catholic counterparts on the issues of ensoulment and the gravity of abortion. Indeed, some historians (e.g., Williams, pp. 13, 41) believe that they indirectly but significantly contributed to the present papal position on the subject.

The reformers insisted upon the full humanity of the fetus from the time of conception. Their insistence arose, however, less from attention to the abortion issue itself than from their concern about the doctrines of original sin and predestination. Full humanity of the conceptus was believed necessary if the mind and spirit as well as the body of nascent life were to be involved in the consequences of the human Fall.

Luther reclaimed the traducian position of the early Church Fathers in regard to ensoulment. This position held that the soul of the fetus along with its body was inherited from its parents. However, influenced by the embryology of Aristotle and Galen, Luther assumed that the semen alone contained life and that the woman provided only nourishment for that life. Hence, the father (through the power of God) was the source of the fetal soul. Melanchthon's theory of ensoulment was creationist, the belief that God directly creates the soul of each nascent life. Both of these reformers, however, insisted upon the full humanity of the conceptus regardless of its gestational stage.

With Melanchthon, Calvin was creationist concerning ensoulment. With Luther, Calvin was concerned about the depth of human need and the radical nature of God's grace in the face of original sin. His particularly strong doctrine of predestination gave additional force to the contention that the fetus from its earliest stage was already homo, fully a person, for this life was believed to be primordially destined to be saved or damned.

The major reformers, then, were rigorously opposed to abortion at any stage of pregnancy. Moreover, they had significantly enhanced the fetal status for reasons more basically doctrinal than for ethical reasons against abortion. Regarding fetal status, they were more conservative than the sixteenth-century Roman Catholic Church, which still maintained the Septuagint's distinction between the "unformed" and the "formed" fetus, and with it a consequent distinction in the gravity of abortion, depending upon its timing. On the other hand, the Reformation's strong affirmations of justification by grace and forgiveness of sin were to provide many later Protestants with theological perspectives and religious resources for dealing with the potential abortion as a situation of ambiguity, compromise, and value conflict.

In seventeenth-century England, both Anglicans and Puritans continued the opposition to abortion, but with the distinction between the unformed and the formed fetus. At the same time, however, the English Puritans initiated changes in the traditional understanding of marriage. Companionship was now seen as marriage's primary end, to which the ends of procreation and the restraint of lust were believed subordinate. This companionate view gradually became the dominant Protestant interpretation, and with it came the groundwork for justifiable abortions in cases of mortal conflict between mother and fetus. Indeed, by the late eighteenth century, English physicians were commonly urging and performing abortions to save threatened mothers.

Eighteenth and nineteenth centuries

American Protestantism manifested at least three major developments relevant to the abortion issue during the eighteenth and nineteenth centuries. One was a loss of interest in the human status of the fetus. Puritan predestinationists pressed Calvin's position on salvation but with different results. Convinced that the individual's eternal destiny was fixed well before conception and gestation, the Puritans saw little to be gained from speculation about fetal ontology. Protestants generally lost touch with the earlier traditions on abortion and were inclined to appeal directly to Scripture, where the early Jewish view and most subsequent texts supported full humanity only at birth. The pietist and revivalist groups of this period were disinclined to reflect about fetal humanity, for they were principally concerned in adult conversion. While they continued the typical Protestant opposition to abortion, their emphasis was often more restricted and practical: Abortion was wrong because it was used, frequently at least, as a coverup for sexual sins.

A strong profertility norm emerged in large segments of nineteenth-century American Protestantism, further buttressing antiabortion sentiments. Women must bear children if they were to be fulfilled as women. Furthermore, as God's New Israel, America had a crucial destiny in the world's salvation, and population growth was desirable to that end. Abortion thus was not simply an individual sin but also a social evil which denied society its needed citizens, economic producers, and defenders.

A significant countervailing influence on the abortion issue also emerged in nineteenth-century American Protestantism, however: the rise of social idealism and the Social Gospel movement. Those trends were less concerned about abortion than about the more visibly disadvantaged. But their general support for the oppressed and their particular concern for women's dignity, expressed in the suffrage movement, laid important foundations for later theories of both justifiable abortion and abortion-on-request.

Twentieth-century influences

During the first half of the twentieth century, while official American Protestant opposition to abortion largely continued, the attitudes of numerous Protestant individuals were gradually being altered by several factors (Potter, pp. 88–89; Yates, pp. 63–78). Under the impact of secularization, certainties that once had undergirded theological opposition to abortion became less certain. If to an earlier Protestant the presence of each fetus in the womb was an expression of God's mysterious providence, to a later Protestant the fetal presence was explained by natural processes, including human mistakes as well as human planning.

Further, the mid-century abortion reform movement drew upon themes of great importance to the Protestant tradition itself, particularly human self-determination and the obligation to exercise rational control over nature. While the latter-day view of self-determination may have been a highly individualistic and secularized version of the Reformation's image of Christian liberty, it was nevertheless inexplicable apart from that historical connection.

A third factor was the waning of the vision of a Protestant America. The failure of the Protestant-inspired Prohibition law was a reminder that religious pluralism had replaced Protestant hegemony. America was no longer a society in which moral convictions that arose from particular Protestant theologies could be sanctioned by legislation. Thus, the distinction between a sin and a crime gained currency: What a particular religious group deemed sinful on its own doctrinal grounds ought not to be socially legislated as criminal without a much broader moral consensus on the issue. This distinction was further to erode Protestant support for anti-abortion legislation.

Patterns in contemporary Protestant thought

In their theological reasoning, Protestants tend to take the church's tradition less seriously than they take the Bible. However, those Protestants who look to Scripture for specific moral instruction rather than for more general ethical principles discover an absence of biblical texts specifically forbidding or permitting induced abortions. Indeed, the Old Testament does not typically view the fetus as fully human life. Exodus 21:22–25, for example, simply demands monetary compensation from the guilty party who causes a woman's miscarriage. Only if the woman herself dies does the charge become homicide and the offender sentenced to death. Nevertheless, there are significant themes in both Old and New Testaments upon which Protestants typically draw for guidance in the abortion question. Life is a gift from God, and human reproduction is a process of sharing in God's creative powers (Psalm 139:13). Human beings are created not simply for comfort and pleasure but more basically for fulfillment, which often involves sacrifice for others (John 15:13). Life and death belong to God's providence, and there is no human right to extinguish life apart from persuasive evidence that such action expresses God's will in a tragic situation (Phil. 1:21–24). However general these biblical principles are, many Protestants insist that even they must be understood within the framework provided by the central ethical norm of Scripture: the love commandment, which includes the dimensions of justice and mercy.

The major historical positions and influences earlier mentioned are present in the contemporary configurations of Protestant opinion on abortion. The perspectives range along a continuum from antiabortion to abortion-on-request, with perhaps the large number of ethicists and Protestant groups affirming the justifiable-but-tragic abortion in certain situations of value conflict.

The antiabortion position is voiced by Dietrich Bonhoeffer, Paul Ramsey, Helmut Thielicke, and, with qualification, Karl Barth. Regarding the ensoulment issue, these ethicists maintain that it is not necessary to describe the fetus as a person. Rather, it is crucial that the fetus be recognized as a human individual; it is en route to becoming personal, but at all times it possesses the full sanctity of human life. Thus, two arguments are interwoven. First, biological data, especially at the stage of segmentation, establish human individuality. Second, Christian faith insists that the dignity of human life is not founded upon the individual's utility but rather is an alien dignity conferred by God and not admitting relative degrees of worth. While additional antiabortion arguments are typically added to the two above, the central issue is held to be the inviolable right to life of the fetus as a human individual, and not the weighing of competing values. Antiabortion Protestants, however, usually incorporate some version of the traditional Roman Catholic "principle of double effect" to argue that in certain extreme situations (threat to the mother's life) abortion is permitted because it is not the direct and willful taking of innocent life but rather the indirect result of saving life.

Abortion-on-request is a more recent and less fully developed position in the literature of Protestant ethics, although it has its articulate defenders. Proponents of this position argue that only wanted babies should be born. They typically maintain that permissive legal systems better protect women's lives and health, minimize social discrimination, and protect the autonomy of the medical profession. However, like the antiabortion position, this stance finally elevates one human right to supremacy. In this instance it is the woman's right to control over her own body and its reproductive processes, against those double standards that would permit others (usually males) to make choices on her behalf. "[M]aybe an unexpressed and powerful objection is that responsible abortion, especially upon request, may give women the final control over reproduction" (Fletcher, p. 27). The earlier Protestant emphases upon self-determination, rational control of nature, equality, social justice, and the dignity of women enter strongly into this position. The real question, proponents argue, is not so much how we can justify abortion, but rather how we can justify compulsory pregnancy.

The third general approach, sometimes called the justifiable abortion, lies somewhere between the first two, affirming arguments of each but differing from both. Rights of both fetus and woman are to be highly valued, but in theory and in practice neither ought to be absolutized. The elevation of one single right, it is argued, removes the inherent moral ambiguity of abortion decisions and oversimplifies other relevant moral factors. Instead of absolute rights, it is better to speak of competing rights and values. Together with the woman's right to self-determination and the value of human fetal life, a complex range of relational, social, medical, economic, and psychological values must be weighed, inasmuch as God cares for all of these dimensions. Thus, each problem pregnancy has its own uniqueness, its own moral tragedy, and its possible alternatives. Christians must rely upon God's grace and forgiveness as they make these ambiguous decisions (Gustafson).

A number of issues related to public policy on abortion have concerned Protestant ethicists of all persuasions. Does a legally permissive policy tend to cheapen or enhance human dignity and the value of human life generally, and how can the evidence be assessed? Is abortion-on-request a disguised, genocidal attack on racial minorities, or does it enhance racial equality? What should be the relation of abortion policies to the population problem? What abortion policies best protect the freedoms of religion and conscience that belong both to those with problem pregnancies and to medical personnel?

In the decades of the 1950s and 1960s several American Protestant denominations and ecumenical groups publicly affirmed positions on the abortion issue. Since the 1973 U.S. Supreme Court rulings (*Roe* v. *Wade*, *Doe* v. *Bolton*) additional Protestant groups have issued policy statements. Most of these pronouncements affirm the recognition of competing values and the possibility of the justifiable abortion as the lesser of the evils in many situations. Some denominations, however, have remained officially silent, encouraging their members to exercise prayerful freedom of conscience in such decisions.

JAMES B. NELSON

[*While all the articles in this entry are relevant, see especially the articles* ROMAN CATHOLIC PERSPECTIVES, CONTEMPORARY DEBATE IN PHILOSOPHICAL AND RELIGIOUS ETHICS, *and* LEGAL ASPECTS. *For further discussion of topics mentioned in this article, see the entries:* DOUBLE EFFECT; PERSON; POPULATION ETHICS: ELEMENTS OF THE FIELD, *article on* HISTORY OF POPULATION THEORIES; POPULATION ETHICS: RELIGIOUS TRADITIONS, *article on* PROTESTANT PERSPECTIVES; PROTESTANTISM; *and* RELIGIOUS DIRECTIVES IN MEDICAL ETHICS, *article on* PROTESTANT STATEMENTS. *See also:* CIVIL DISOBEDIENCE IN HEALTH SERVICES; FETAL–MATERNAL RELATIONSHIP; LIFE; RIGHTS, *article on* RIGHTS IN BIOETHICS.]

BIBLIOGRAPHY

CALVIN, JOHN. *Institutes of the Christian Religion.* 2 vols. Translated by Henry Beveridge. Grand Rapids, Mich.: Wm. B. Eerdmans Publishing Co., 1957, vol. 1.

Doe v. Bolton. 410 U.S. 179. 35 L. Ed. 2d. 201. 93 S. Ct. 739 (1973).

FLETCHER, JOSEPH F. "A Protestant Minister's View." *Abortion in a Changing World.* 2 vols. Edited by Robert E. Hall. The Proceedings of an International Conference Convened in Hot Springs, Virginia, November 17–20, 1968, by the Association for the Study of Abortion. New York: Columbia University Press, 1970, vol. 1, pp. 25–29.

GUSTAFSON, JAMES M. "A Protestant Ethical Approach." *The Morality of Abortion: Legal and Historical Perspectives.* Edited by John Thomas Noonan. Cambridge: Harvard University Press, 1970, pp. 101–122.

LUTHER, MARTIN. *Luther's Works.* Edited by Jaroslav Pelikan. American ed. St. Louis: Concordia Publishing House, 1955–.

POTTER, RALPH B., JR. "The Abortion Debate." *Updating Life and Death: Essays in Ethics and Medicine.* Edited by Donald R. Cutler. Boston: Beacon Press, 1968, pp. 85–134.

Roe v. Wade. 410 U.S. 113. 35 L. Ed. 2d. 147. 93 S. Ct. 705 (1973).

SPITZER, WALTER O., and SAYLOR, CARLYLE L., eds. *Birth Control and the Christian: A Protestant Symposium on the Control of Human Reproduction.* Wheaton, Ill.: Tyndale House Publishers; London: Coverdale House Publishers, 1969.

WILLIAMS, GEORGE HUNTSTON. "Religious Residues and Presuppositions in the American Debate on Abortion." *Theological Studies* 31 (1970): 10–75.

YATES, WILSON. *Family Planning on a Crowded Planet.* Minneapolis: Augsburg Publishing House, 1971.

V

CONTEMPORARY DEBATE IN PHILOSOPHICAL AND RELIGIOUS ETHICS

Since approximately 1950, there has been an ongoing debate about abortion as the availability and public acceptance of abortion have increased considerably in most societies. Many religions consider abortion to be a significant moral issue, and religious ethicists have long been interested in the question. Philosophical ethicists generally have not given much consideration to abortion until the contemporary period. There is no agreement about the morality of abortion among either philosophical or religious ethicists, but religious ethicists as well as philosophical ethicists often appeal to rational arguments. This article will summarize and analyze the contemporary debate about abortion in philosophical and religious ethics.

Various reasons or indications have been proposed for justifying abortion: to protect the life of the mother, to safeguard the physical and mental health of the mother, to act as a remedy against injustices due to rape or incest, to prevent the birth of defective children, to vindicate the right of the woman to determine her own reproductive capacities and to have control over her body, to protect the reputation of the woman, and to alleviate economic, sociological, or demographic problems. These reasons all have some validity, but they must be seen in a larger context including the zygote-embryo-fetus and its value or rights.

Most philosophers and theologians recognize that the heart of the abortion problem is the difficult question of the beginning of human life, since human beings are treated differently from all other beings. A terminological problem exists in even framing the question, for some authors make a distinction between human life and personal life, between human being and human person, between biological existence and fully human existence. To avoid confusion the term "truly human being" or "truly human life" will be used and understood as that human life deserving the value, rights, and protection due the human person as such.

When does truly human life begin?

Opinions about the beginning of truly human life may be conveniently categorized in two ways: (1) in terms of the conclusions about the beginning of truly human life, which range from the moment of conception to various stages in fetal development, viability, birth, and for some time after birth; or (2) in terms of the criteria employed for determining the beginning of truly human life. The following types of criteria best categorize and summarize the abundant literature on the subject: (1) individual-biological, (2) relational, (3) multiple, and (4) the conferral of rights by society. There is no absolute correlation between the criterion employed and the conclusion about the beginning of truly human life; but the following tendencies exist: the individual-biological criterion tends to place the beginning of human life at conception or at early stages of development; the relational and conferral of rights criteria usually accept a rather late point in development for the beginning of truly human life; the multiple approach often results in an intermediary position.

One preliminary point is most important. Biological, genetic, or scientific data alone will not be able to solve the problem of when truly human life begins. The ultimate judgment remains truly a philosophical or human judgment, which gives meaning and interpretation to the biological and other data involved. Such a conclusion is based on the recognition that human existence involves more than just the biological and genetic and cannot be simply identified with just one aspect. However, it must also be pointed out that at times the human and the physical or biological are inseparable. Our personal existence as human beings is intimately connected with our physical-biological existence. Human beings do not exist without or apart from their bodies. Human personal death is described in accord with individual-biological categories—the cessation of brain activity accompanied by other signs, according to increasingly accepted criteria of human death.

Individual-biological criterion. The individual-biological approach generally sees the criterion of truly human existence in terms of some physical, biological, or genetic aspects of the individual. Exponents of the theory placing human life at an early stage based on the individual-biological criterion usually begin with the assumption that all humans accept the born child as a truly human being. The first serious question is raised about birth as the beginning of human life. But the fetus one day before birth and the child one day after birth are not that significantly or qualitatively different in any respect. Even outside the womb the newborn child is not independent but remains greatly dependent on the mother and others. Birth in fact does not really tell much about the individual as such but only where the individual is—either outside the womb or still inside the womb. Viability has often been proposed as the beginning of truly human life, but from the perspective of the individual-biological criterion viability again indicates more about where the fetus can live than what it is. The fetus immediately before viability is not that qualitatively different from the viable fetus. In addition viability is a very inexact criterion because it is intimately connected with rapidly changing medical and scientific advances. In the future it might very well be possible for the fetus to live in an artificial womb or even with an artificial placenta from a very early stage in fetal development.

There are two other stages in the development of the fetus that have been pointed out as marking the beginning of human life. Joseph Donceel has based a delayed hominization or delayed animation theory on the Thomistic concept of hylomorphism (viz., there is a complementarity between the material and formal aspects of being). The form, which in this case is the soul, is received only into matter capable of receiving it. Thomas Aquinas and many other medieval theologians and philosophers accepted a theory of delayed animation. According to Donceel, the insistence on immediate animation in Catholic circles in the last few centuries was based on the erroneous biological theory of preformation and was influenced by a Cartesian dualism. The unity of the human person demands that the bodily or material element be more highly organized in order to receive the truly human form or soul. The fertilized ovum, the morula, the blastula, and the early embryo cannot be animated by an intellectual soul. Donceel concludes that the least that must be present before admitting a human soul is the availability of these organs: senses, the nervous system, the brain, and especially the cortex. Since these organs are not ready during early pregnancy, there is no human person present until several weeks have elapsed. Others might want to insist on the incipient presence of the major organs as the determining point for the existence of truly human life.

Some (e.g., Grisez, p. 283) reply that Thomas Aquinas's own position was determined more by poor biological knowledge according to which the seed was the primary active element in human generation, the ovum was not known, and the seed had to die before new life could come into existence in the womb of the mother. Whether Aquinas's position is based on faulty biological knowledge which is corrected in the light of modern biology and genetics or is based on his philosophical theory will continue to be debated. However, it seems difficult to maintain that such early development of these organs constitutes the qualitative difference between truly human life and no truly human life. Self-awareness and reason are not actually present at this time, and much more growth and development are necessary. The appearance of these organs appears to mark just another stage in the ongoing development—a stage directed by an inner teleology already present in the embryo or fetus.

A somewhat similar theory (Ruff, and proposed but not accepted in practice by Häring, pp. 81–85) places the beginning of human life at the time of the formation of the cortex of the brain. Electrical brain activity is detectable at the eighth week. Personal life is characterized by consciousness and self-reflection which have their indispensable substratum in the cerebral cortex. Many accept brain death—a flat EEG—together with other signs of no responses as a criterion of death. Is it not logical to insist on the same criterion for determining when human life begins?

Within the individual-biological criterion the early formation of the cortex seems to be just another step in the ongoing development and does not constitute a qualitatively different threshold distinguishing truly human life from nonlife. There is validity in the plea for consistency in dealing with the beginning and end of life, but consistency does not necessarily call for the same material test at the beginning as at the end of life. The test really measures whether there is any immediate potentiality for spontaneous life functioning. At the beginning of life this can be present before electrical brain activity can be detected and measured.

Many proponents of the individual-biological criterion emphasize a continual, progressive development of the fetus without any obviously qualitatively different thresholds, concluding that the only logical place to mark the beginning of truly human life is conception. John T. Noonan (pp. 51–59) has accepted as the conclusion of his analysis of the Catholic tradition that whatever is conceived of human beings is human and buttresses his conclusion with three different reasons: the presence of characteristics in the embryo and adult which are similar, an argument based on probabilities which indicates the great difference between the sperm or ovum and the fertilized ovum or zygote, and a critique of alternative proposals. Modern genetics, according to Noonan, supports this position by showing that the zygote is a dynamic blueprint which if it receives proper nourishment and suitable environment grows and develops.

Germain Grisez (pp. 273–287) develops a two-stage argument for his opinion that truly human beings come into existence at conception. The first stage is factual. The fertilized ovum is a single thing derived from two sources. The facts of genetics indicate that a new being comes into existence with conception, for the fertilized ovum is continuous with that which develops from it, while the sperm and ovum which constitute the new being are themselves two different realities continuous with the duality of the two parents. This factual question leads to the moral question: Is this new individual to be regarded as a person? Anyone who recognizes the importance of the body and does not adopt a dualistic understanding of the person must acknowledge that development is a continual process from the time of fertilization so that the conceptus must be considered a truly human being.

Lately, some authors (for example, Chiavacci, Curran, and Ramsey) have proposed on the basis of the individual-biological criterion that an individual is not present until a certain development has occurred which is usually placed somewhere between the second and third week after conception and does not occur before nidation or implantation (*Aborto*, pp. 254–274; Curran, pp. 179–193; Noonan, pp. 60–100). Biological information heavily influences this judgment, but the ultimate reason rests on the recognition that individuality, which is a most fundamental characteristic of the truly human being, is not achieved before this time. Twinning and recombination can occur before this time. Before this time there is no organizer that directs the differentiation of the pluripotential cells so that without this organizer hominization cannot occur. Also this theory contends that the large number (perhaps as many as fifty percent) of fertilized ova that are spontaneously aborted without the mother's being aware of having conceived are not truly human beings. Rebuttal arguments downplay the fact of twinning and recombination, which are comparatively rare and can be explained in other ways, such as the direct infusion of a new soul in the case of twinning.

The reasons proposed thus far have been of a rational nature and not based on any particular religious or theological perspective. Paul Ramsey invokes the religious understanding that the value and sanctity of human life are not something intrinsic in the human being but result from the gift of God (1968, pp. 71–78). The sanctity of life is an alien dignity conferred by God so that even a small zygote can have that dignity and value. Interestingly, Roman Catholic ethics, which has often insisted on an early beginning of human life, usually appeals not to the alien dignity of human life but to its intrinsic dignity. God does bestow the gift of life, but the presence of human life can and should be proved by rational arguments.

The following arguments have been raised against the individual-biological criterion, especially the conclusion—often derived from this criterion—that truly human life has an early beginning. It goes against the common experience, which does not consider the embryo or early fetus to be a human being; it absolutizes the biological and genetic, not giving enough importance to broader understandings of the human; it fails to recognize that in addition to genetic factors environmental aspects are also necessary for human growth; it overemphasizes potentiality and does not give enough importance to development. Counterarguments can be proposed. Common human experience must always be subject to reflection, scrutiny, and critique. Granted that the human is more than just biological, at times the human and the biological are inseparable. The only human beings we know in this world are human beings in their physical and biological nature. Although environmental factors are important for human development, these factors also continue to be of great importance after birth. Potentiality based on something intrinsic in the being itself is a better criterion than a developmental approach that could open the door to differing value attributed to different human lives depending on their developed potential.

Relational criterion. A relational criterion for the beginning of human life generally sees truly human life coming into existence late in the development of the fetus—usually at the time of viability or even birth. The starting point of the relational school emphasizes the common experience, which rejects the notion that the embryo or early fetus is a truly human being and insists that the concept "truly human being" cannot be applicable by virtue simply of the genetic or biological features of a living thing. In fact it is futile to look for a biological moment when truly human life begins even if it were possible to determine such a moment. After brain waves or other acceptable criteria of human death are no longer present, lower biological life can continue to exist in the body even if truly human life is ended. These approaches frequently distinguish between human life and human person, between biological life and personal life, between human life and humanized life.

According to some French Catholic moral theologians, e.g., Pohier, Ribes, Roqueplo (*Avortement et respect de la vie humaine*), fully personal and humanized life must be seen in terms of relationships. An older objectivist epistemology sees the fruit of conception as a human being in itself, but contemporary epistemology insists on a more relational understanding, especially the relationship of the conceptus to the parents. There are economic, psychological, cultural, and even faith aspects of human life in addition to the biological. The mere fact of biological procreation does not constitute a truly human personal life, especially if the parents were not intending such a result and were trying to prevent it. The fetus must be accepted by the parents and also to some extent by the society into which it will be born.

One of the arguments frequently raised against the relational criterion is that the employment of such a criterion logically leads to infanticide and the killing of adult human beings who are no longer capable of entering into such human relationships. Some authors, such as Michael Tooley and Joseph Fletcher, are willing to admit infanticide and/or the killing of some biological adults who do not fulfill the criterion of the human, but most defenders of abortion based on a relational criterion of the beginning of human life do not want to admit such killing and make a distinct effort to show that acceptance of abortion does not logically entail acceptance of these other forms of killing.

It seems that truly human relationships cannot exist until after birth because full human relationships require a reciprocity of giving and taking in a human way that cannot be had until both partners have the requisite self-consciousness to enter into such a relationship. In a true sense fully personal life is not present at birth or even immediately thereafter. The infant does not possess the characteristics of rationality or self-consciousness that are generally proposed as definitive of the human person. Lederberg maintains that the infant becomes a human person only after the first year or so of life.

H. T. Engelhardt has addressed this particular question and proposed a second concept of person in addition to the strict concept used to identify actually self-conscious moral agents—a social category of person based on a social role or function which claims the intrinsic value of actual persons. The newborn infant, although not yet a full person, has such a potential personhood that it can play the role or function of a person within family and society. The mother–child relationship is a social relationship which depends upon the mother's agency and involves the child's being treated as if it were a person. The mother–child relationship is a social and willed structure of interaction and a cultural enterprise as contrasted with the mother–fetus relationship, which is a biological and imposed structure and a physiological enterprise. The biological reality of viability now becomes understood in broader social terms as indicating when the social role of a person can begin. This theory is thus based on two concepts of human life—biological and personal—and on a strict and social concept of person but insists that the social concept of person has exactly the same consequences as the strict concept of person.

Does the relational criterion avoid the danger of exposing infants and other adult human beings to being judged less than fully human persons and therefore killed or sacrificed for others? As a matter of fact it can be pointed out that many people, countries, and cultures which accept abortion do not endorse any form of infanticide or killing of the weak, the retarded and the defenseless. In theory most authors proposing a relational criterion oppose any extension to those who are already born, but there remains a grave doubt about the logic involved in such a criterion. Truly human relationships are not present immediately after birth, and often adult human beings seem to fall below the line of full human relationships. If one admits a distinction between a strict and a social concept of a person, it becomes very easy to say that a strict concept of a person results in more rights or more values than a social concept of a person, thus denying the basic realization that all persons have equal rights and equal values in terms of the fundamental question of life itself.

A multiple criterion. A third type of criterion tries to avoid some of the problems associated with both the individual-biological and the relational by calling for a criterion based on multiple indications. Daniel Callahan, in a well-documented monograph on abortion, argues for a consensus approach to the question which will do justice to all the values involved, and proposes a criterion for the beginning of human life based on biological, psychological, and cultural factors. Callahan (pp. 349–404, 493–501) maintains that even a zygote is individual human life, but full value should not be assigned at once to the life thus begun. Interestingly, he never explicitly develops or describes in great detail when truly human life which is to be valued as such does occur, but he seems to indicate the existence of brain activity as the distinguishing mark. Callahan's criterion enables him to do three things: to allow for the possibility of decisions in which the woman would want an abortion; to avoid the unilateral approach of the biological; and to avoid some of the dangers of the relational criterion, which if used alone could endanger the rights of the newborn and other marginal people in society.

There are many aspects of Callahan's mediating approach which are appealing, but in the last analysis doubts remain, perhaps increased somewhat by his apparent reluctance to discuss in greater detail precisely when human life is to be fully valued as such. The danger in an approach calling for the threefold criterion of biological, psychological, and cultural considerations is that the psychological and cultural aspects are really not all that present at the time of birth and might also cease to exist for some adults after birth. Callahan himself implies a comparatively early point at which full value is achieved (brain activity), and one could really question if enough cultural and psychological development have truly taken place to mark this as the beginning of fully valued human life. Any developmental theory also is going to open the door to the recognition that such a criterion can be employed even after birth to distinguish the values involved in different human lives depending upon their development in accord with psychological and cultural aspects.

Conferred rights criterion. Metaphysical considerations, the recognition that human beings must ultimately make the decision about when human life begins, and the difficulties involved with other criteria are some of the reasons proposed for a fourth type of criterion for the beginning of human life—the social conferral of rights. In the earlier years of the abortion debate after 1950, Glanville Williams and Garrett Hardin advocated such approaches. Williams contends that life is a continuum and the fundamental problem is to discover when human life begins. The basic criterion becomes what answer will have the best social consequences. Williams leans to viability as the dividing line. Hardin also acknowledges that whether or not the fetus is human is a matter of definition and not of fact.

More nuanced philosophical understandings of the same type of criterion have been advanced. R. B. Brandt proposes a contractual understanding of morality based on the criterion that what is right is fixed by what would be permitted in a moral code which rational, impartial persons, with no personal gain, living now or in the future, would prefer. The important condition emphasizes that these individuals would be totally impartial and would in no way know how they themselves as individuals might profit or suffer from their decision. It is prima facie wrong to kill a being that wants to live, Brandt contends, but the fetus is not sentient and cannot choose. Even if these people were altruistic, the fetus would not be an object of their sympathy. Nor can a human person ever truly put itself in the place of a fetus, for the precise reason that the human person is not a fetus. On this basis, Brandt concludes that abortion is not prima facie wrong.

R. M. Green proposed a similar approach of a rational theory of rights as a way of bypassing the thorny and seemingly insoluble problem of when human life begins. Understanding morality as a noncoercive means of settling disputes, Green identifies a class of agents who must have fundamentally equal rights—all those agents with whom we can possibly come into dispute, who display an elementary rational ability and who are capable of understanding and respecting moral rules. Such a broad understanding includes almost all human beings; but there are some who do not have these capacities so that rights in these cases are conferred at the behest of rational agents who are concerned about behavior with respect to third parties that have ultimate effects on their own lives. A threefold criterion is proposed for conferring rights on others: on the basis of the effect on our capacity for sympathy, on the effect on the possible interests of particular agents, and finally on the effect on the character or moral worth of rational agents generally. This threefold criterion would confer rights on the newborn but not on the eight-week-old fetus, although they would place some value on that life. Sissela Bok proposed somewhat similar standards as the reasons for protecting human life, most of which are absent in the case of prenatal life but some of which are clearly present with respect to newborn life, although not as fully as in the case of adult human beings.

Various reasons have been proposed against the criterion of conferred rights. Biological data are given little or no importance, even though such data can never be used by themselves to determine when truly human life begins. It seems one cannot avoid the question of when human life begins, for our sympathy and our concern depend primarily on what we believe the fetus to be. Proponents of this criterion are very conscious of the charge of arbitrariness which might allow many people to be excluded from human rights. One way to avoid the charge is to include as many as possible within the class of moral agents who confer the rights, thus including all races, colors, and creeds. Another approach appeals to the neutral observer who behind the "veil of ignorance" does not know what one's own position is. However, if agents put themselves in the place of the fetus, which all of them were at one time, it seems that they would be quite prone to protect the life of the fetus. One cannot exclude the rational agent from taking that position unless one first decides that the fetus is not a person. There always remains the philosophical problem that human rights must exist prior to any conferral of rights by the state or by individuals representing society.

Value and rights of the fetus

The question of the value and rights of the fetus is intimately connected with the question of when human life begins. For those who see the fetus as a truly human being, the fetus has all the rights and values of every other human being.

An opposite approach is taken by those who view the embryo or fetus as only tissue in the mother and give the mother the right to control her body as her personal property. If women are to be truly free and liberated, they must ultimately have full freedom to control their own reproductive capacities. The woman, according to Wahlberg, cannot truly experience the fetus as anything but a part of her own flesh. There can be no doubt that in practice emphasis on the right of the woman to control her own body has greatly contributed to a wider acceptance of abortion, but in the ethical literature the women's rights reasons have not been developed in great depth. This position seems to ignore the rights of the fetus. Although biological data alone cannot determine the existence of a truly human being, such data prove that the fetus is a biological individual and not merely maternal tissue.

The developmental school, as exemplified in Callahan (pp. 493–501) and others, views the fetus from conception as individual human life, but development counts in assigning value to the fetus. On the basis of this, there should be a strong bias against abortion; but the body life of the conceptus is not of the same value as the person life of the pregnant woman, so that many considerations would justify abortion. Relational approaches could give some value to the fetus as biological life, but fully human life becomes present only when the requisite relationship is present. Since the relational criterion does not generally give that much importance to biological life, the fetus before the time of becoming fully human would seem not to be of great overriding value. The conferred rights approach could also assign different meaning and value to the fetus; but generally speaking, since such an approach gives little significance to the biological aspects, the fetus does not have that much value in such a generic approach.

R. M. Hare in dealing with the issue of abortion wants to avoid both rights talk and the question of when human life begins. The proper question is whether there is anything about the fetus that says we should not kill it. The primary moral question results from the fact that the fetus has the potentiality of becoming a person in the full, ordinary sense of that word. The potentiality principle in the case of the fetus asserts a presumption against the permissibility of abortion, which is fairly easily rebutted if there are good indications, such as the interests of the mother. The above serve as illustrations of the various ways in which the value or rights of the fetus have been understood in the contemporary debate about abortion.

Solution of conflict situations. Various reasons have been proposed to justify abortion: medical, psychological, sociological, economic, humanitarian, fetal, feminist, and demographic. The moral decision involves a proper balancing of the reasons or indications proposed in relationship to the rights, value, or importance of the fetus. The position which regards the fetus as just tissue in the mother would then accept practically any reason the woman proposed based on her right to the control of her body and reproductive capacities. The vast majority of commentators would see abortion as somewhat different from artificial contraception, but reasons of family planning or population control would be a sufficient reason to justify abortion for some. The balancing and weighing of the various reasons ultimately depend upon the value which one attributes to the fetus. The greater the value given to the fetus, the greater the reason needed to justify abortion. Generally speaking, the solution to such conflicts is based on the proportion existing between the value given to the fetus and the indications or reasons proposed for the abortion.

The question most discussed in the literature is the solution to conflict situations if the fetus is looked upon as a truly human being with all the rights of human beings. This is the problem that has been traditionally discussed within Christian ethics and especially within Roman Catholic ethics down to the present. Recently some philosophers have given attention to the same question, often with the intention of showing that the abortion question is not morally closed once one accepts or asserts, without necessarily conceding, that the fetus has the rights of a human being.

Roman Catholic theology in its historical tradition has acknowledged that conflict situations arise in which human lives can be taken; e.g., war, killing in self-defense, capital punishment. In the question of abortion the hierarchical magisterium in the Roman Catholic Church taught that the fetus cannot be considered an unjust aggressor, since the fetus in the womb is just doing what is necessary to preserve its own life—hence the mother cannot kill the fetus in self-defense. Conflict situations involving abortion were dealt with in terms of the principle of the double effect which justified action producing evil effects as well as good effects if the following four conditions were fulfilled: (1) The action itself is good or indifferent; (2) the good effect and not the evil effect is the one sincerely intended by the agent; (3) the good effect is not produced by means of the evil effect (If the evil effect is not at least equally immediate causally with the good effect, then it becomes a means to the good effect and intended as such.); and (4) there is a proportionate reason for permitting the foreseen evil effect to occur.

The decisive criterion is the third—the causality of the act cannot be such that the good effect comes about by means of the evil effect. When applied to the question of abortion, this means that abortion cannot be the means by which the good effect is accomplished—for example, one cannot abort to save the life of the mother which might be endangered because her heart cannot take the pregnancy. The two most famous illustrations of indirect abortion are the removal of the cancerous uterus when the woman is pregnant and the removal of a fallopian tube or part of it that is threatened with rupture from an ectopic pregnancy in the tube.

The accepted Catholic teaching allows less conflict situations for the fetus in the womb than for human life outside the womb because outside the womb the possibility of killing in defense against an unjust aggressor is recognized. One can argue that the older teaching on unjust aggression was not referring to the subjective will of the aggressor nor even to an objectively unjust act, but to the reality of a conflict situation. In this perspective it is easier to see that the right of defense even to the killing of the fetus might be extended to the mother, for truly conflict situations can arise even though the fetus is just attempting to provide what is necessary for its own growth and continued existence.

The theory of double effect has come in for closer scrutiny in the last few years (McCormick; Schüller). Some Catholic scholars continue to uphold it, but often contemporary Catholic authors reject the theory, especially its third condition based on the causal structure of the act itself. Historical scholarship, with some dissenting voices, indicates that the third condition for the double effect was not found in Thomas Aquinas. There is also a strong tendency among Catholic theologians to reject the identification of the moral aspect of the act with the causal and physical structure of the act. In the abortion case where there is some conflict existing between the fetus and the mother, a good number of contemporary Catholic moralists would agree with the conclusion that the physical causality of the act has no moral significance. The human values involved must be carefully considered and weighed. In my judgment abortion can be justified for preserving the life of the mother and for other important values commensurate with life even though the action aims at abortion as a means to an end. The difficult problem concerns what values if any are commensurate with life. In the Catholic tradition, killing an aggressor was justified not only in defense of one's life but even in defense of earthly goods of great value. Obviously there is the danger here of opening the door for reasons that are truly not of grave seriousness, but within the Catholic tradition other values have been proposed as commensurate in some situations with the value of human life.

Protestant authors generally do not accept the principle of double effect and usually employ some type of proportionate reasoning to solve conflict situations involving abortion. Paul Ramsey, an American Protestant ethicist, originally praised the double effect theory, but later he abandoned the all-important third condition of the double effect (1968, p. 78; 1973, pp. 210–226). Some philosophers have lately discussed the principle of double effect in general and have often rejected the third condition, while calling for distinctions between positive and negative obligations, or between foreseen and unforeseen consequences. But even in philosophical circles there are some defenders of the principle of double effect (e.g., Finnis).

Judith Jarvis Thompson sparked an interesting exchange by proposing that, granted only for the sake of argument that the fetus is a truly human being, it does not follow that abortion is always wrong. She draws an analogy with a person who unknowingly and unwillingly is plugged into a famous violinist who has a fatal kidney ailment that can be overcome only by using the kidneys of the other person. Does the person in this case have the right to unplug the violinist even though it will bring about the death of the famous musician? A minimally decent Samaritan or the normal human being is not required to make such a huge sacrifice as being plugged into the violinist for nine months in order to sustain the life of another who has no right to demand this.

Thompson has raised the very important question of conflict situations, but her analogy with the person plugged into the violinist for nine months decidedly limps. It seems that only in the case of rape might one hope to find a comparison between the two cases. A woman has to assume responsibility for her previous actions, and likewise there is a significant difference between an obligation to save the life of another and an obligation not to take the life of another.

There is another possible approach in conflict situations involving abortion which to some extent is also acceptable within the traditional approach of Roman Catholic theology. James Gustafson (in Noonan, pp. 119–122), a Protestant ethicist, accepts the decision of a woman to have an abortion after she has been brutally raped by her former husband and acknowledges some financial and psychological problems in having the child. Bernard Häring (pp. 112–115) and even Gustafson himself recognize that from the viewpoint of Roman Catholic pastoral theology on the basis of an expanded notion of invincible ignorance, which prudently recognizes that one cannot ask another to do what is not possible, a counselor can accept the decision of the woman. Others are continuing to debate whether there are any real differences in the solutions proposed by Gustafson and Häring.

Survey of the contemporary debate

Protestant church teaching and Protestant ethicists until 1950 generally condemned abortion, but since that time many Protestant churches and groups have come out in favor of abortion in certain circumstances and rejected the concept that the fetus from the moment of conception is a truly human being. However, it would be wrong to say that all Protestants are in favor of abortion.

In Germany in 1973 a conference of German Bishops (Catholic) and of the Evangelical Church (Protestant) issued a common statement on abortion which insisted that all decisions that involved human life can be oriented only to the service of life. Generally speaking, more evangelical and conservative Protestant groups strenuously oppose abortion. Also a large number of respected Protestant ethicists in the United States oppose abortion although many others do not consider the conceptus from the time of fertilization to be a truly human being.

Developments have also occurred within Roman Catholicism. The hierarchical magisterium has continued to proclaim the accepted Catholic teaching. The Second Vatican Council's *Pastoral Constitution on the Church in the Modern World* (n. 51) taught that from the moment of its conception life must be regarded with the greatest care and that abortion and infanticide are unspeakable crimes. Pope Paul VI in his 1968 encyclical *Humanae Vitae* (n. 14) declared once again that both the direct interruption of the generative process already begun and every directly willed and procured abortion, even if for therapeutic reasons, are to be absolutely excluded as licit means of regulating birth. The Vatican's Sacred Congregation for the Doctrine of the Faith issued a Declaration on Procured Abortion, 18 November 1974, which reiterated the traditional teaching, and this has also been repeated by national conferences of Catholic bishops throughout the world.

The majority of Roman Catholic theologians still seem to uphold the traditional teaching that from the moment of conception the conceptus is to be treated as a truly human being. There is however some dissent on this issue, most publicized in France, where a small respected group of Catholic theologians has been proposing a relational criterion for determining when human life begins. While an increasing number of Catholic scholars claim that dissent on when truly human life begins is legitimate and justified (since, in this matter, it is impossible to arrive at a certitude that excludes the possibility of error), still the vast majority of Roman Catholic authors accept the traditional teaching or disagree only to the extent that the beginning of truly human life is placed in the second or third week after conception. A comparatively large number of Catholic theologians from all countries reject the traditional use of "double effect" to solve conflict situations involving the fetus.

Differences continue to exist within Judaism. The only authoritative text on therapeutic abortion in the Talmud recognizes that the fetus becomes truly human only if the greater part of it is already born. The controversy obviously centers on the value and importance of the fetus. All authors in the orthodox and conservative cultural tradition (e.g., Novak) accept the therapeutic abortion necessary for the life of the mother and, according to some, also for the mental health of the mother; but there is always a very strong emphasis in favor of the life of the fetus. Within the reformed branch of Judaism there is a greater willingness to attribute less value to the fetus. It is indistinguishable from the mother and may be destroyed for the mother's sake, just as a person may decide to sacrifice a bodily limb to cure a worse malady (Margolies in Hall, vol. 1, pp. 30–33).

One of the characteristics of the contemporary debate has been the interest shown by professional philosophers in the question of abortion. Until very recently, philosophers tended to neglect content questions such as abortion, but changes in philosophy and the strong interest within society on this question have changed this. The writings of the philosophers, while not numerous to date, have frequently added precision of thought to the contemporary debate. In general there is no doubt that in the contemporary debate since 1950 many voices have been raised in favor of the position that the fetus is not a truly human being from the moment of conception and therefore abortion is morally acceptable for various reasons. Such sentiments obviously have had an effect on public opinion and on changing abortion laws in a number of countries, even though there is not a strict correlation between questions of law and questions of morality. However, it is interesting to note in the last few years some former advocates of abortion have expressed either a change of mind or doubt about the morality of abortion, at least to the extent that they want to caution against a too ready acceptance of abortion. The general public, philosophers, and theologians continue to be divided on this issue.

CHARLES E. CURRAN

[*While all the articles in this entry are relevant, see especially the articles* JEWISH PERSPECTIVES, ROMAN CATHOLIC PERSPECTIVES, *and* PROTESTANT

PERSPECTIVES. *Also directly related are the entries* DOUBLE EFFECT; LIFE; PERSON; *and* PRENATAL DIAGNOSIS, *article on* ETHICAL ISSUES. *Other relevant material may be found under* FETAL–MATERNAL RELATIONSHIP; INFANTS, *article on* INFANTICIDE: A PHILOSOPHICAL PERSPECTIVE; RIGHTS; SEXUAL ETHICS; *and* WOMEN AND BIOMEDICINE, *article on* WOMEN AS PATIENTS AND EXPERIMENTAL SUBJECTS. *For discussion of related ideas, see the entries:* CONTRACEPTION; OBLIGATION AND SUPEREROGATION; STERILIZATION, *article on* ETHICAL ASPECTS.]

BIBLIOGRAPHY

Aborto questione aperta: Le Posizioni dei moralisti italiani. Introduction by Ambrogio Valsecchi. L'opinione religiosa in Italia, no. 1. Turin: Gribaudi, 1973.

Avortement et respect de la vie humaine. Colloque du Centre catholique des médecins français (commission conjugale). Paris: Éditions du Seuil, 1972.

BOK, SISSELA. "Ethical Problems of Abortion." *Hastings Center Studies* 2, no. 1 (1974), pp. 33–52.

BRANDT, RICHARD B. "The Morality of Abortion." *Monist* 56 (1972): 503–526.

CALLAHAN, DANIEL. *Abortion: Law, Choice and Morality.* New York: Macmillan, 1970.

CLOUSER, K. DANNER, and ZUCKER, ARTHUR, comps. *Abortion and Euthanasia: An Annotated Bibliography.* Philadelphia: Society for Health & Human Values, 1974.

CURRAN, CHARLES E. "Abortion: Its Legal and Moral Aspects in Catholic Theology." *New Perspectives in Moral Theology.* Notre Dame, Ind.: Fides, 1974, pp. 163–193.

DONCEEL, JOSEPH F. "Immediate Animation and Delayed Hominization." *Theological Studies* 31 (1970): 76–105.

ENGELHARDT, H. TRISTRAM, JR. "The Ontology of Abortion." *Ethics* 84 (1974): 217–234.

FEINBERG, JOEL, ed. *The Problem of Abortion.* Belmont, Calif.: Wadsworth Publishing Co., 1973.

FINNIS, JOHN. "The Rights and Wrongs of Abortion: A Reply to Judith Thompson." *Philosophy and Public Affairs* 2 (1973): 117–144.

FLETCHER, JOSEPH. "Medicine and the Nature of Man." *The Teaching of Medical Ethics.* Edited by Robert M. Veatch, Willard Gaylin, and Councilman Morgan. Hasting-on-Hudson, N.Y.: Institute of Society, Ethics, and the Life Sciences, 1973, pp. 47–58.

FLOYD, MARY K. *Abortion Bibliography for 1973.* Troy, N.Y.: Whitson Publishing Co., 1974. Similar volumes were published for 1970, 1971, and 1972.

GREEN, RONALD M. "Conferred Rights and the Fetus." *Journal of Religious Ethics* 2 (1974): 55–75.

GRISEZ, GERMAIN G. *Abortion: The Myths, the Realities, and the Arguments.* New York: Corpus Books, 1970.

GRÜNDEL, JOHANNES. *Abtreibung—Pro und Contra.* Würzburg: Echter, 1971.

HALL, ROBERT E., ed. *Abortion in a Changing World.* 2 vols. New York: Columbia University Press, 1970.

HARDIN, GARRETT. "Abortion—Or Compulsory Pregnancy?" *Journal of Marriage and the Family* 30 (1968): 246–251.

HARE, R. M. "Abortion and the Golden Rule." *Philosophy and Public Affairs* 4 (1975): 201–222.

HÄRING, BERNARD. *Medical Ethics.* Notre Dame, Ind.: Fides, 1973, pp. 75–119.

HAUERWAS, STANLEY. *Vision and Virtue: Essays in Christian Ethical Reflection.* Notre Dame, Ind.: Fides, 1974, pp. 127–165.

LADER, LAWRENCE. *Abortion.* Boston: Beacon Press, 1966.

LEDERBERG, JOSHUA. "A Geneticist Looks at Contraception and Abortion." *Annals of Internal Medicine* 67, supp. 7 (1967): 25–27.

McCORMICK, RICHARD A. *Ambiguity in Moral Choice.* The 1973 Père Marquette Theology Lecture. Milwaukee: Marquette University Press, 1973.

NOONAN, JOHN T., JR. *The Morality of Abortion: Legal and Historical Perspectives.* Cambridge: Harvard University Press, 1970.

NOVAK, DAVID. "A Jewish View of Abortion." *Law and Theology in Judaism.* New York: Ktav Publishing House, 1974, pp. 114–124.

RAMSEY, PAUL. "Abortion: A Review Article." *Thomist* 37 (1973): 174–226.

———. "The Morality of Abortion." *Life or Death: Ethics and Options.* By Edward Shils, Norman St. John-Stevas, Paul Ramsey, P. B. Medawar, Henry K. Beecher, and Abraham Kaplan. Introduction by Daniel H. Labby. Seattle: University of Washington Press, 1968, pp. 60–93.

RUFF, WILFRIED. "Individualität und Personalität im embryonalen Werden." *Theologie und Philosophie* 45 (1970): 24–59.

SCHÜLLER, BRUNO. *Die Begründung sittlicher Urteile: Typen ethischer Argumentation in der katholischen Moraltheologie.* Düsseldorf: Patmos, 1973.

THOMPSON, JUDITH JARVIS. "A Defense of Abortion." *Philosophy and Public Affairs* 1 (1971): 47–66.

TOOLEY, MICHAEL. "Abortion and Infanticide." *Philosophy and Public Affairs* 1 (1972): 37–65. Edited reprint. "A Defense of Abortion and Infanticide." Feinberg, *The Problem of Abortion,* pp. 51–91.

WAHLBERG, RACHEL CONRAD. "The Woman and the Fetus: 'One Flesh'?" *Christian Century* 88 (1971): 1045–1048.

WILLIAMS, GLANVILLE. "Euthanasia and Abortion." *University of Colorado Law Review* 38 (1965–1966): 178–201.

VI
LEGAL ASPECTS

In all developed legal systems, abortion has been and is more or less explicitly regulated by law. This can hardly fail to continue to be so, for reasons that emerge from an analysis of recent legal developments in the United States.

The specific legal norms in relation to abortion vary widely from country to country, and from one time to another. But three basic models can be discerned. Most actual schemes approximate to one or another of these three models, though of course some actual schemes are compromises between diverse models, and others are in transition from one model to another, so that their practical working differs from their formal legal framework. The characterization of these models will refer not only to the formal normative structure of prohibitions, permissions, and authorizations but also to the values or objectives to be realized or protected by way of that structure. The scope for variation is as wide as it actually is precisely because the functional objectives of a scheme, its formal normative structure, and the procedures for operating and policing that structure are three more or less independent variables, all of which must be taken into account in explaining and classifying any particular scheme.

In the first of the three basic models all abortions are legally prohibited, either absolutely or subject to an exception where continuation of the pregnancy would threaten the life of the pregnant woman. The characteristic objective of this type of scheme is to protect the interests of the unborn child, either for its own sake (as a matter of justice and respect for rights) or for the sake of the social interest in the maintenance or increase of the population, or sometimes (as in France in the first half of this century) for both reasons. In the past those two objectives were often supplemented by a concern to protect the pregnant woman from the medical risks involved in attempts to terminate pregnancy.

The second model is that in which abortion is permitted only with the prior authorization of independent officials, whose grant or refusal of permission is guided by standards relating to defined categories of medical, psychomedical, or quasi-medical conditions. Schemes with this type of formal normative structure are often in fact compromises or transitional phases between the first and the third models, that is to say, devices for relaxing the first scheme (in the interests, e.g., of women's freedom) or for tightening up the third scheme (in the interests, e.g., of women's health, or of stemming a decline in population or in birthrate). But the second model can have the distinctive goal of empowering the medical and psychiatric professions to apply representative standards of judgment—in a matter that deeply affects these professions—in order to maintain professional unity and standards.

In the third model all abortions are legally permitted if performed by medically qualified persons. The characteristic objectives are to give effect to the rights of the woman over her own body and to eliminate unskilled abortions with their attendant medical hazards.

First model: restriction for the sake of the child

Although the matter has been disputed, the better view appears to be that the law of England always adopted the first model. Abortion after quickening (about the sixteenth week of pregnancy) was, according to Chief Justice Coke (c. 1630), a "great misprision" (and he defines "misprision" as a high offense, under, but "nearly bordering" on, the "degree of capital" offense). This position was reaffirmed in judgments or scholarly expositions by all the great common-law writers—Hale, Hawkins, Blackstone. In the nineteenth century many American courts affirmed that in the common law of England, and of the American colonies and states, abortion after quickening was a serious (though not felonious) offense, and courts in two states ruled that the offense extended to abortion at all stages of pregnancy. However, it is also true that English criminal law as enforced in the common-law courts was, as in many other important matters, encumbered by procedural, evidential, and doctrinal rigidities. Prosecution directly for the offense itself was difficult if not impossible in those courts and was left to the ecclesiastical courts, whose law was accepted by common-law lawyers as an important part of the law of the land. But the common law's characterization of (postquickening) abortion as a serious offense appears quite clearly from the common-law rule that to cause accidentally the death of the mother in an effort to abort her was murder. The opinion expressed by the United States Supreme Court in 1973, that "it now appear[s] doubtful that abortion was ever firmly established as a common-law crime even with respect to the destruction of a quick fetus" gives a rather misleading impression of the state of scholarship on this matter, and the Court's conclusion that "at common law . . . a woman enjoyed a substantially broader right to terminate a pregnancy than she [did] in most States [in 1972]" would have been rejected as equivocal or mistaken by any common-law judge or scholar between the beginning of the seventeenth and the end of the nineteenth centuries, at least.

The decadence of the English ecclesiastical courts and the rigidities of the common law gave rise to statutory reform by the Parliament of Great Britain in 1803 (a time of extensive reform of the inadequate common law of crimes). The statute cut through the evidential encumbrances of the common law by penalizing (as felony) all attempts to procure abortion. All abortions were criminal, but postquickening ones were punishable more severely than prequickening ones; in 1837 this distinction disappeared from English law altogether. In 1861 the statutory abortion law of the United Kingdom reached the form it retained until 1967. Any act intended to cause abortion, whether induced by the woman herself (if she were pregnant) or by others (whether or not she was in fact pregnant) was a felony.

The first American legislation against abortion was passed in Connecticut in 1821. Many early American statutes did not penalize abortion before quickening. But in 1827 (Illinois) that restriction began to disappear; by 1860 twenty American states or territories, including Connecticut, had statutes against abortion at all stages of pregnancy. By 1868 thirty-six states or territories had antiabortion statutes. In 1965 all fifty states prohibited abortion and attempts to abort, at all stages of pregnancy. In forty-six states and the District of Columbia the relevant statute explicitly permitted abortion to save the mother's life, while in two of the other four states (as in England) judicial interpretation had recognized a similar exception to the statutory prohibition. A survey both of the legislative history and of accompanying judicial and professional statements leaves little room for the opinion, favored by the Supreme Court in 1973, that the primary intention of the law prevailing in the United Kingdom and the United States up to 1965 was to protect the pregnant woman against the medical risks that were inherent in abortion until the advent of modern medical techniques. On the contrary, the law's primary intention was plainly to prohibit what was considered a wrongful attack on human life in the womb.

Developments in the United States and the United Kingdom after 1965 are discussed below. In the early 1970s, schemes on the first model were legally in force in more than sixty countries, including more than twenty where the formal legal norms absolutely prohibited all abortions, even when performed to save the life of the mother. In some other jurisdictions—such as Belgium, most of the states of Australia, and to some extent New Zealand—judicial interpretation or prosecutorial policy had effected a substantial move toward the third model by permitting abortion when performed by a medical practitioner in order to avert any substantial risk (in the view of that practitioner) to the physical or mental health of the pregnant woman, or to prevent the birth of a handicapped child.

Second model: restriction for the sake of uniform medical practice

The earliest instance of a scheme of legal regulation on the second model—in which each abortion requires (save in emergency) the prior authorization of some official referee(s)—is afforded by the German law enacted shortly after the Nazi accession to power in 1933 and replaced after the Second World War by a law on the first model. Other early instances are the laws and administrative regulations in force in Denmark from 1937 to 1973 and in Sweden from 1939 to 1974. The criteria for authorization by the relevant committee or board in these Scandinavian countries included the principle that, in assessing how giving birth to and caring for the expected child would affect the mental or physical health of the woman, her actual and prospective living conditions are to be taken into account. The Swedish criteria were always a little more restrictive than the Danish, but in both countries the law and administrative practice were several times relaxed to allow wider medical, eugenic, sociomedical, or social "indications." Countries that adopted some form of the second model more recently include Turkey, Brazil, and Singapore.

Between 1967 and early 1973 a dozen jurisdictions in the United States adopted abortion laws more or less patterned on the model legislation suggested in 1962 by the American Law Institute, an unofficial voluntary association of lawyers. This permitted abortion when performed by a licensed physician who believed that there was a substantial risk that continuance of the pregnancy would gravely impair the physical or mental health of the mother, that the child would be born with grave physical or mental defect, or that the pregnancy resulted from rape, incest, or other felonious intercourse. However, although a few of the states amending their laws in this direction adopted somewhat looser categories of permitted circumstances, those few and almost all the others of the dozen reforming jurisdictions sought to strengthen the institutionalization of abortion practice by stipulating that an abortion would be lawful only if performed (in the permitted circumstances) in an accredited hospital after approval by a committee established in the hospital for that purpose. This institutionalization of an evolving consensus of professional medical judgment was stronger than in the Scandinavian countries, in that the hospital committees were to be composed entirely of medical practitioners, but also weaker, in that the hospital committees were subject to no central appointment, direction, or control. A scheme of this kind was adopted by legislation in Canada in 1969.

In retrospect it is difficult to see the Scandinavian and American experiments with prior authorization as anything other than unstable compromises adopted along a society's path from the first to the third model. But it would be wrong to deny that many people genuinely believed that abortion was a matter that society should leave to medical judgment—not individual, but *representative* medical judgment. This belief was rooted in the assumption that the medical profession had an ethic, autonomous and distinct from society's general laws and mores, which, when combined with medical knowledge and technical expertise, would generate a code of practice that society would do well to follow. Traces of this assumption can still be found in the decisive role accorded to "the physician" in the judgments of the United States Supreme Court in 1973—a role that on any other assumption is difficult to reconcile with the Court's basic reliance on the *woman's* constitutional right of personal privacy.

In Eastern European countries such as Hungary, Rumania, and Bulgaria, abortion is lawful only if approved by a state board. Here the objective is to free women for employment while at the same time giving effect to the state's population policy. The general policy has usually been to permit abortion in almost every case, at least up to the sixteenth week of pregnancy. But from time to time a government's concern about falling population growth, or about the medical side effects of repeated abortions, has inspired a severe restriction on the number of authorizations of abortion (e.g., in Rumania in 1966).

Third model: general availability for the sake of women's freedom

The earliest instance, in the modern world, of a scheme on the third model is afforded by the decree of the Soviet Union, in November 1920, that made lawful any abortion performed by a physician in a state hospital. The expressed objective of the decree was to protect women from unskilled abortionists, to whom they resorted because of "moral survivals of the past" (presumably shame at illegitimate pregnancy) and "the difficult economic conditions of the present." But the decree was also part of a general program of women's liberation and sexual equality in work, education, and marriage. In June 1936 the decree was replaced by a law that prohibited abortion except where pregnancy threatened the life or seriously threatened the health of the woman, or when a serious disease of the parents could be inherited. The shift back to a form of the first model seems to have been motivated by concern primarily about declining birthrates, and secondarily about medical side effects and increasingly irresponsible attitudes toward marriage and childbearing. But in November 1955 the decree of 1936 was repealed, for the express reasons of reducing the harm done by unskilled abortionists and giving women "the possibility of deciding by themselves the question of motherhood."

Japan adopted, in 1948, a scheme on the second model but in 1952 amended it to a scheme on the third model, not by extending the grounds for abortion specified in the Eugenic Protection Law but simply by removing the control of district Eugenic Committees over individual physicians' decisions. The primary objective appears to have been population control; a secondary objective was reduction in unskilled abortions.

The Abortion Act 1967, which came into effect in Great Britain in April 1968 and has been copied in Hong Kong, Zambia, and South Australia, amounts in practice to a scheme of the third type: In practice abortion is freely and openly available, at least to anyone who can pay the fees charged by licensed private institutions. Indeed, even in theory the law is on the third model (though this seems not to have been the intention of many of the legislators who supported it), for it permits an abortion where any two medical practitioners are of the opinion, formed in good faith, that continuance of pregnancy "would involve risk to the life of the pregnant woman, or of injury to the physical or mental health of the pregnant woman or any existing children of her family, *greater than if the pregnancy were terminated.*" The last words are here emphasized because they mean that the legal justification of abortion depends on nothing more than the medical evaluation that controls any surgical procedure: The risk of performing it must be less than the risk of omitting it, but neither risk need be in any way serious. Official analysis of the medical data, supplied by medical practitioners to the relevant government department, suggests that in 1973, in the nongovernment institutions in which two-thirds of all notified abortions were performed, ninety-nine percent of the abortions were performed primarily or exclusively for reasons of "risk to mental health" of the mother; in National Health Service hospitals the corresponding proportion was over eighty percent. Abortion is also lawful where two physicians believe that "there is substantial risk that if the child were born it would suffer from such physical or mental abnormalities as to be seriously handicapped." Responsible medical opinion in Great Britain considers that a risk of one in ten is certainly "substantial."

Since 22 January 1973, when the Supreme Court decided *Roe* v. *Wade* and *Doe* v. *Bolton*, the law throughout the United States has been on the third model. In those cases, the Court by majority of seven to two decided that the constitutional right to privacy, protected by the due process clause of the Fourteenth Amendment to the Constitution, entails that (1) no law can restrict the right of a woman to be aborted by a physician during the first three months (trimester) of her pregnancy; (2) during the second trimester the abortion procedure may be regulated by law only to the extent that the regulation reasonably relates to the preservation and protection of maternal health; (3) at the point at which the fetus becomes viable (not before the beginning of the third trimester) a law may prohibit abortion but only subject to an exception permitting abortion wherever necessary to protect the woman's life or health (including any aspects of her physical or mental health); (4) no law may require that all abortions be performed in accredited hospitals (where accreditation has no particularized concern with abortion as a medical or surgical procedure), or that abortions be approved by a hospital committee, or by a second medical opinion, or that abortions be performed only on women resident in the state concerned. The effect of these decisions was to invalidate the laws regulating abortion in every state, except perhaps the already very permissive laws adopted in 1969 and 1970 in New York, Alaska, Hawaii, and Washington.

The theory underlying the Supreme Court's division of pregnancy into three trimesters of differing constitutional status is that during the first trimester the mortality rates for women undergoing abortion "appear to be as low as or lower than the rates for normal childbirth"; as one moves into the second trimester the health risks of abortion begin (on the average) to exceed those of childbirth; and as the third trimester begins the fetus is becoming viable and a state's interest in protecting it, *should the state legislature choose to do so*, overrides the woman's right to privacy in deciding for or against maternity.

Within two years of the Supreme Court's decisions, cases in lower federal courts appeared to have established the following implications or consequences: (1) During the first trimester abortions are exempt from any legal or licensing requirements that do not apply to medical facilities generally, and from any regulations other than those applicable to all other relatively minor operations. (2) Advertisements of medical facilities for first-trimester abortions may not be prohibited. (3) No law may require the consent of the husband or of the parent or guardian of a pregnant, unmarried minor (even where the motive for requiring consent of the parent or guardian is to protect the health of the minor—at least during the first trimester). (4) During the second trimester the decision about abortion must remain absolutely with the woman and her physician; only procedures for carrying out the operation may be regulated, and these only where necessary to protect the woman's health. (5) A state law may not truncate the woman's constitutional rights by declaring that a fetus is to be deemed viable at twenty weeks, or by requiring medical personnel to seek to preserve the life of an aborted fetus between the twentieth and the twenty-fourth week. (6) It is unconstitutional to restrict third trimester abortions to cases where the woman's *physical* health is threatened. (7) Any denial of public assistance funds for abortion is unconstitutional. (8) Public hospital facilities must be made available for abortions, and it is unconstitutional for a state to enact a conscience clause specifying that "no person or hospital or institution shall be coerced, held liable or discriminated against in any manner because of a refusal to perform, accommodate, assist or submit to an abortion for any reason"; for such a clause tends to limit the availability of hospital facilities for abortion. (9) Where a state fails to enact a constitutionally valid abortion statute, there is no restriction on the availability of abortion, performed by a physician, at any time during pregnancy. (10) Subject to the foregoing nine points, there is no legal right to abortion on demand.

Not all the foregoing points, perhaps, are established with equal certainty. In any event, a large number of legal questions remain to be resolved. The following are examples of these questions, most of which would need to be the subject of explicit legal decision (i.e., regulation by law) however widely abortion is permitted. (1) What are the implications of *Roe* v. *Wade* in relation to rights of the unborn child, other than its now legally unrecognized right not to be aborted? There is a considerable body of law granting rights to sue to persons who were tortiously injured before birth, and this law is in many jurisdictions expressly founded on the theory that a human life begins at conception. In *Roe* v. *Wade* the Supreme Court recognized without approving or disapproving that body of law but did declare that "the judiciary, at this point in the development of man's knowledge, is not in a position to speculate as to the answer" to "the difficult question of when life begins." (2) Does the woman's right to privacy really entail that a state is powerless to require medical personnel to make reasonable efforts to save fetuses that are aborted before the end of the second trimester (cf. point [5] above)? In general, is the Supreme Court's theory of viability open or impervious to future advances in medical techniques (artificial wombs, etc.)? In connection with these questions, note that several states require abortion procedures be used which will preserve the life of a viable fetus, and several also provide that a fetus surviving an abortion may be made a ward of the state. (3) Is it lawful for a woman to self-induce an abortion during the first trimester? On the one hand, the Supreme Court said that a state may proscribe any abortion by a person who is not a physician. On the other hand, the Court's basic theory is of the woman's right to privacy, and moreover the Court takes for granted "new medical techniques such as menstrual extraction [and] the 'morning-after' pill," which might well be used without medical supervision (the Court's reference to these new techniques was, oddly, as posing a "substantial problem for precise definition of [the] view" that human life begins at conception). (4) Can a state prohibit all, or some, or any forms of experimentation on fetuses before, during, or after abortion? Is a fetus to be treated as a part of the pregnant woman which she can dispose of at will, at least during the first two trimesters? (5) What is the legal liability of a physician who fails to advise, or to perform, an abortion? (6) What forms of conscientious objection are available to physicians, nurses, hospitals, or other institutions unwilling to perform any or some abortions (cf. point [8] above)? (7) Does the father of an unborn child, even during the third trimester, altogether lack any rights that might prevail against the mother's constitutionally recognized rights? (8) What forms of assistance is a state or federal agency bound to afford to a woman who wants an abortion? (9) Could a state validly *require* a woman to undergo an abortion, whether to prevent the birth of a handicapped child, to protect the health of the woman, or to advance a state policy of population control? The question can arise because in 1973 the Supreme Court refused to consider that another human life (that of the unborn) is involved, and appeared to accept the 1927 decision of the Supreme Court upholding compulsory sterilization of imbeciles afflicted with hereditary forms of insanity or imbecility.

During the early 1970s a number of countries adopted laws approximating the third model. The theory that during the first trimester abortion ought to be available without any restrictions gained popularity; it was adopted in East Germany in 1972, Denmark in 1973, Sweden in 1974, and France in 1975. "Viability," the point at which, in most of these schemes, the interests of the woman cease to be accorded overriding weight, is variously fixed at twenty weeks (Sweden, India), twenty-four weeks (Singapore), and twenty-eight weeks (in the United Kingdom, in practical effect). The French law of 17 January 1975 is representative of much international opinion in the wake of the judicial transformation of United States law in 1973. The French enactment begins with a declaration that the law guarantees respect for every human being from the beginning of life, and that this principle is to be departed from only in case of necessity and according to the conditions defined by the present enactment. But the law authorizes any pregnant woman whose condition puts her in a situation of hardship to request a physician for abortion. Provided that the abortion is performed before the end of the tenth week of pregnancy, there are no preconditions other than: (1) The physician must inform the woman of the medical risks to herself and her future pregnancies, and give her an official guide to the forms of assistance available to all families, mothers, and children, and to relevant social service organizations; (2) the woman must consult one of the listed social services; (3) if the woman still wishes to proceed to an abortion she must confirm this in writing, not earlier than one week after her first request; (4) the abortion must be performed by a physician in a public or recognized private hospital and must be reported to the regional health authorities; (5) the hospital must provide the aborted woman with birth control information. But the law also removes, for a period of five years, all penalties against a doctor who performs an abortion, in a public or recognized private hospital, on any woman before the end of the tenth week. After that stage of pregnancy, right up to the moment of birth, abortion is lawful if two physicians, one of them from an official list, certify that continuation of pregnancy would put the woman's health gravely in peril, or that there is a strong possibility that the child would suffer from an incurable condition. The law concludes with the declaratory statement that in no case should termination of pregnancy be used as a means of birth control.

J. M. FINNIS

[*While all the articles in this entry are relevant, see especially the articles* MEDICAL ASPECTS, *and* CONTEMPORARY DEBATE IN PHILOSOPHICAL AND RELIGIOUS ETHICS. *For further discussion of topics mentioned in this article, see the entries:* CONTRACEPTION; FETAL–MATERNAL RELATIONSHIP; FETAL RESEARCH; GENETIC DIAGNOSIS AND COUNSELING; INFANTS; LIFE; POPULATION POLICY PROPOSALS; PRENATAL DIAGNOSIS; PRIVACY; *and* RIGHTS. *Other relevant material may be found under* LAW AND MORALITY.]

BIBLIOGRAPHY

BYRN, ROBERT M. "An American Tragedy: The Supreme Court on Abortion." *Fordham Law Review* 41 (1973): 807–862.
CALLAHAN, DANIEL J. *Abortion: Law, Choice, and Morality.* New York: Macmillan Co., 1970. Chapters 5, 6, and 7 survey "restrictive," "moderate," and "permissive" codes in various countries.
DELGADO, RICHARD, and KEYES, JUDITH DROZ. "Parental Preferences and Selective Abortion: A Commentary on Roe v. Wade, Doe v. Bolton, and the Shape of Things to Come." *Washington University Law Quarterly* 1974 (1974): 203–226.
Doe v. Bolton. 410 U.S. 179. 35 L. Ed. 2d. 201. 93 S. Ct. 739 (1973).
ELY, JOHN HART. "The Wages of Crying Wolf: A Comment on Roe v. Wade." *Yale Law Journal* 82 (1973): 920–949. Perhaps the most damaging critique of the Supreme Court's decision, by a constitutional lawyer not opposed to abortion.
EPSTEIN, RICHARD A. "Substantive Due Process by Any Other Name: The Abortion Cases." *1973: The Supreme Court Review.* Edited by Philip B. Kurland. Chicago: University of Chicago Press, 1974, pp. 159–185.
FINNIS, JOHN M. "Three Schemes of Regulation." *The Morality of Abortion: Legal and Historical Perspectives.* Edited by John T. Noonan, Jr. Cambridge: Harvard University Press, 1970, pp. 172–219.
GRISEZ, GERMAIN G. *Abortion: The Myths, the Realities and the Arguments.* New York: Corpus Books, 1970. Chapters 5 and 8 contain the best available discussion of the legal questions discussed in this article.
MALEDON, WILLIAM J. "The Law and the Unborn Child: The Legal and Logical Inconsistencies." *Notre Dame Lawyer* 46 (1971): 349–372.
MEANS, CYRIL C., JR. "The Law of New York Concerning Abortion and the Status of the Foetus, 1664–1968: A Case of Cessation of Constitutionality." *New York Law Forum* 14 (1968): 411–515.
———. "The Phoenix of Abortional Freedom: Is a Penumbral or Ninth-Amendment Right About to Arise from the Nineteenth-Century Legislative Ashes of a Fourteenth-Century Common-Law Liberty?" *New York Law Forum* 17 (1971): 335–410. Means's views on the history of the common law, which differ in his two articles, appear to have considerably influenced the Supreme Court in 1973. For a preliminary reply, see Byrn, "An American Tragedy."
QUAY, EUGENE. "Justifiable Abortion—Medical and Legal Foundations." *Georgetown Law Journal* 49 (1960): 173–256, 395–538. Appendix I reproduces all the enactments on abortion in the United States and its territories, as of 1960.
REBACK, GARY L. "Fetal Experimentation: Moral, Legal, and Medical Implications." *Stanford Law Review* 26 (1974): 1191–1207. Analysis in the light of *Roe v. Wade* and its legislative aftermath.
Roe v. Wade. 410 U.S. 113. 35 L. Ed. 2d, 147. 93 S. Ct. 705 (1973).
VEITCH, EDWARD, and TRACEY, R. R. S. "Abortion in the Common Law World." *American Journal of Comparative Law* 22 (1974): 652–696.
World Health Organization. *Abortion Laws; A Survey of Current World Legislation.* Geneva: World Health Organization; London: Her Majesty's Stationary Office, 1971. Surveys legislation up to 1970. Original appearance. *International Digest of Health Legislation* 21 (1970): 437–512.

ACTING AND REFRAINING

It is often claimed that there is a radical difference, from the moral point of view, between making something happen and merely letting something happen. It has been argued that there are instances where taking overt direct action to bring some state of affairs about would be wrong, while merely refraining from preventing the same result is at least sometimes permissible. Important moral consequences are based on the general distinction that directly killing someone is prohibited, while letting someone die, at least in some circumstances, is permissible. Claims like this have direct bearing on medical practice in general and on issues like abortion and euthanasia in particular. The aim of this essay is to trace the outlines of the development of the distinction between acting and refraining, with special attention to two particular issues: the conceptual basis of the distinction and the moral significance of the distinction.

One might begin by considering the difference among (1) shooting someone with the intention of killing him, (2) letting someone die and intending that he die, (3) letting someone die but not intending that he die, and (4) killing someone, but not intending to kill him (e.g., using a high dose of narcotics to ease the pain of a terminal cancer patient which may also depress the respiration of the patient and actively hasten death). Granted that each case involves a death, the issue is whether there is any significant *moral* difference exemplified by these four cases. Prima facie, (1) seems to be morally unacceptable, (2) perhaps less clear. For present purposes (3) and (4) are more significant, for many philosophers and theologians argue that there are decisive moral differences between (1) on the one hand and (3) and (4) on the other. But what could this difference be?

In (1), the model of a prima facie unacceptable act, there is a direct, willing, causal assault on someone's life. In (3) and (4), some or all these features are missing. In (3), there is no overt causal intervention on another's life, rather a mere letting something occur or a refraining from preventing; nor in (3) is the effect an intended one. In (4), although there is causal intervention, the claim is made that the causal intervention bearing on death does not fall within the range of what is intended by the agent, and therefore cannot be analyzed as (1) is analyzed. The morally relevant difference then between (1) and (3) and (4) could be defended by appealing to two different claims: (a) there is a moral difference between "acting" and "refraining," or (b) there is a decisive difference between the consequences of our "actings" and "refrainings" that we directly intend to happen and those consequences that we merely foresee, but don't intend. If both (a) and (b) are indefensible, then it is not clear how moral differences between (1) and (3) and (4) could be established.

Let us begin the discussion with an analysis of (b), for it provides the conceptual link between (3) and (4), both of which presuppose that there is a distinction between what we foresee and what we intend, or between what we "obliquely intend" and "directly intend." The best defense of such a distinction is to be found in the tradition of Catholic moral theology and its development of the so-called principle of double effect.

The principle of double effect in Catholic moral theology

Directly intending and causing the death of an innocent individual is prima facie forbidden in the tradition of Catholic moral theology. Yet the very same tradition permits a physician to "pyramid" painkilling drugs that will have the net effect of shortening the life of the patient. Furthermore, the performance of a hysterectomy on a pregnant woman in the case of uterine cancer is also permitted.

Critics of the tradition, like Glanville Williams, argue that exceptions like these are simply inconsistent and cast serious doubt on the viability of the tradition of moral theology (Williams). Defenders of such exceptions, on the other hand, argue that it is of the utmost moral significance that one individual must always refrain from taking direct action against the life of another. Clearly some principle that will maintain the general prohibition against the taking of life and that can make such exceptions intelligible is needed. Why is it morally justifiable that the doctor refrain from taking direct action against a life, while it is permissible in some circumstances to hasten the moment of death? The principle of double effect (PDE) has been employed to answer questions like these.

Thomas Aquinas is probably the first theologian who articulated both the principle and the general conditions for its use. Aquinas formulated the principle to deal specifically with the issue of whether it is ever morally justifiable to take the life of another in self-defense. Aquinas answered affirmatively, on the general grounds that sometimes one action had more than one consequence, and that some of these further consequences might themselves be evil. Some scholars believe that the four conditions later associated with the principle of double effect were at least implied in Aquinas's argument. Those conditions, according to which an action that results in undesirable consequences was judged permissible, were the following: (1) the act itself must be morally good or indifferent; (2) only the good consequences of the action must be intended; (3) the good effect must not be produced by means of the evil effect; and (4) there must be some grave reason for permitting the evil (Aquinas). (1) and (4) are certainly acceptable principles. The difficulty in interpreting the principle seems to hang on the intelligibility of (2) and (3). Aquinas is relying on two things in accepting both constraints: with regard to (2) he is appealing to the difference between foresight and intention; with respect to (3) he is relying on a principle that had substantial theological and moral weight, "Never do evil so that good might result." Both constraints have in common a plausible moral principle: a moral agent should never intentionally do something that is evil. PDE, in its original formulation, is neither bizarre nor mysterious; whether it can carry the heavy conceptual and moral weight it is made to bear by the Catholic tradition is an issue on which we can reserve judgment temporarily.

The development of the principle from Aquinas bears witness to the ambiguity of (2) and (3) and to the difficulty of applying the principle in a plausible way. Cajetan (1468–1534), for example, in his commentary on Aquinas's *Summa*, was the first to apply the principle explicitly to the killing of the innocent (Mangan). The casuists Sanchez (1550–1610) and Lessius (1554–1623), among others, applied the principle in great detail to the problem of abortion (Grisez, chap. 4). And yet Aquinas does not do so, even though he was not unaware of its relevance (Noonan). Whether or not there is an entirely consistent application of the principle to be found in Aquinas is not clear.

What is clear is that the casuists of the sixteenth and seventeenth centuries did extend the principle to the taking of innocent life. These discussions centered on issues like the just war theory and the conditions under which abortion might be legitimate. A second key transition in the application of the principle emerges here: the introduction of the distinction between direct and indirect means. In general, casuists like Sanchez and Lessius condemned any means of abortion that tended to kill the fetus directly, while under certain conditions means that did not "tend directly" toward the taking of life were permissible. Much the same position was endorsed by St. Alphonsus Ligouri (1696–1787) (Mangan; Noonan). The moral significance of the direct–indirect distinction was debated even at that point in history: Gabriel Vasquez, a Jesuit casuist of the seventeenth century, argued that any abortion that resulted from any overt positive act, even one that employed indirect means, was immoral (Grisez). Most moral theologians, however, have accepted the significance of the direct–indirect distinction. In summation, there seem to have been three roughly distinct periods involved in the development of the principle: its formulation by Aquinas, the extension of the principle to moral problems other than self-defense, and the interpretation of the principle in terms of the direct–indirect distinction.

The modern period has seen almost no serious conceptual development of the principle, but is marked by debate over the correct scope and application of the principle. The principle itself has been accepted by many moral theologians and implicitly endorsed by two popes, Pius XI and Pius XII, but precisely which actions fall under the principle is not yet clear. There seems to be some consensus on its application to the issue of painkilling drugs that hasten death, so-called indirect euthanasia, and there also seems to be consensus on the issue of abortion in ectopic pregnancy and in the case of cancer of the uterus. The application of the principle to the moral issues of modern warfare is a matter of some contention. Contrary to Williams, then, the consistency of PDE over a period of roughly seven centuries is remarkable; the historical difficulties with PDE seem to lie not so much with the principle as with how it is to be applied.

In the contemporary period, the principle of double effect is subject to three sorts of attack: (1) it generates ridiculous consequences; (2) it rests on an untenable distinction between intention and foresight, and (3) the principle is itself morally insignificant. The first two criticisms can be coped with, while the third is more troublesome.

Duff has generated an attack on PDE along the lines suggested by (1). He appeals to the Dudley and Stephens case. Lost at sea on a raft, Dudley and Stephens ate their cabin boy in order to survive. Couldn't they argue, if PDE is an acceptable principle, that they in no way intended the foreseen death of the boy, and hence that their action was justifiable? A second supposed counterexample is a medical one: a doctor removes the heart, lungs, and liver of one patient in order to save the lives of three others. Is it not possible to argue that the death of the first patient was foreseen but unintended and that such a procedure is morally justifiable? Surely the number of such cases can be multiplied indefinitely, and, as Duff argues, if PDE can be used to justify such actions then it is an untenable moral principle.

These alleged counterexamples put little pressure on PDE, however, for they involve misapplication of the principle. In neither case is there one action with good and bad results: the killing and eating in the first case and the killing and the transplantation in the second are distinct and separable actions in both cases. Therefore, the only rational way to regard the killing in both instances is as a means to an end and consequently as falling within the range of the agent's intention. Hence the principle could not be employed to generate this sort of counterexample. It can be said with some confidence that PDE, in general, will not have the outrageous results it is alleged to have by this sort of counterexample. For PDE is not an act-utilitarian principle whereby any particular means might in principle be used to bring about some independent result. To interpret it as such would be to ignore all the constraints Aquinas placed on PDE.

A second distinct attack is the claim that PDE is unacceptable because the distinction between intention and foresight it rests on cannot be defended. While any number of philosophers have attacked the distinction, the claims made by Bentham and Aune are typical. The basic line of argument may be reconstructed as follows: If a person (P) is (a) doing an action A; (b) knows that he is doing A; (c) intends to do A; (d) knows that doing A will bring about event B; and (e) knows that B will not occur unless he does A, then it cannot be claimed that he refrained from B; rather the agent acted and intended in such a way that B obtained. Bentham calls this an instance of "oblique intention." Aune argues similarly that it is only by taking a too narrow view of practical reason that we come to make such a distinction in the first place. Clearly, if this line of counterargument succeeds, PDE must be rejected.

Such an attack could be forestalled, however, if there were at least a general way to indicate why it is conceptually important not to collapse the difference between intention and foresight, and this can be done. We need to consider whether there is a difference between (f) P does A, intending that B will result; and (g) P does A, knowing that B will result.

The difference can in general be marked by discovering the reasons that P has for doing A. In the first instance, one might say that P does A because it will bring about B. In the latter case, the occurrence of B is in *no way* a *reason for* performing A. The agent might do A and be entirely indifferent to the occurrence of B; he might do A in spite of the fact that it will bring about B because on balance he might think that A is the best action to perform, whatever the consequences of his doing so. Insofar as it is possible to understand way an agent is doing something, intention and foresight should be distinguished. It may, of course, not always be possible to specify what the reasons are for P's performing any given action; perhaps the reasons are unconscious, perhaps they are indeterminate. Insofar, however, as it is important to understand the behavior of an individual by understanding his reasons for performing or refraining from any given act, intention should be distinguished from foresight.

Even if one is able in a particular instance to distinguish intention from foresight, however, the moral dimension of the problem remains open. For the principle of double effect to be of any moral significance the following principle seems to be presupposed, and is explicitly held by at least one defender of PDE (Geddes): *One is not morally responsible for the foreseen but unintended consequences of an action.* Prima facie, however, it seems that this principle is false, for we are in general responsible for the foreseen consequences of our action. That a given action was unintended might be a good excuse for what we did or permitted, but that hardly renders us free from our responsibility to account for what we did or permitted. Consider the example of the shopkeeper who sells poisons knowing full well the consequences of such an act (Foot). Even if this case could be made out that the merchant in no way intended anyone's death, this would hardly constitute an adequate moral justification for such an action. And it is just here that PDE seems to flounder. For what seems to be of paramount moral interest is the character of the action being performed or refrained from and not the narrower issue of whether or not the consequence was merely foreseen and not intended. At best the moral theologian could argue that it is only *sometimes* morally justifiable to appeal to PDE, but this has yet to be done, and it is not clear that it could be done in a way that would render clear and concrete guides to action.

"Killing" and "letting die": contemporary debate

Even though PDE is an intrinsically interesting and historically influential principle, its development and debate have precipitated an even more important moral issue, especially for medical practice. Any defense of PDE presupposes that there is an important moral difference between directly acting to bring something about and merely letting (allowing, permitting) something come about; it is just the prohibition against direct, intentional killing, for example, that makes PDE such an important principle for the justifiable taking of life.

Yet it is just the assumption that there is a morally significant difference between "acting" and "refraining" that has been the center of much recent debate. In particular, the debate has centered on the difference between "killing" and "letting die." A number of philosophers and theologians have defended the distinction in various ways (Anscombe; Casey; Dinello; Ramsey); others have argued that no special moral significance is to be attributed to such a distinction (Bennett; Tooley). In examining the issue, I shall argue that the distinction is not, in general, of decisive moral significance; there is, however, one way of interpreting the distinction that makes it more credible.

Consider the following four examples: (1) directly killing someone; (2) refusing to operate on a (retarded) child; (3) suspending treatment for a terminal disease; (4) not using all the life-support measures available for a terminally ill patient. In terms of currently accepted law and medical practice in most countries, the only item in the series that is categorically forbidden is (1). There may be any number of ways of explaining the different stances taken with respect to these particular actions, including historical and psychological reasons. If the difference is to be of any substantial *moral* weight, though, there should be some significant *moral* difference between (1) and the other items on the list.

Now the central difference between (1), on the one hand, and (2), (3), and (4), on the other, is that (1) involves the notion of making something happen while (2), (3), and (4) do not. This in turn suggests that, if there is a significant moral difference between the first instance and the others, there ought to be a difference between causing something to happen and merely letting something happen. Now surely the decision to let something happen is going to have consequences, but these consequences are nonetheless not, in general, going to be *causal* consequences of actions initiated by the agent. Hence, it is reasonable to assume that, if there is a significant moral difference between (1) and (2), (3), and (4), it has something to do with the idea that (1) involves "killing" while (2), (3), and (4) all seem to involve the notion of "letting die." And this seems to presuppose, in turn, that there is a moral difference between initiating a chain of events and forbearing from preventing a chain of events that has already been initiated.

This move has some plausibility, but how is it to be interpreted? One way to mark this difference might be to contrast "action" and "inaction" and argue that moral responsibility, properly analyzed, attaches only to instances of positive action. If "refraining" is interpreted as "not acting," however, then this distinction is of little moral interest because it surely cannot be sustained. Refraining from something might often, though not always, involve *not moving*, but it surely cannot be analyzed as a species of inaction. This is perhaps best seen by comparing refraining with inadvertence. My behavior is inadvertent if I fail to pay attention to something or if I do something unintentionally. But it would not make sense to say that I refrained from something unintentionally or "by accident." The conceptual feature of "conscious attention" precludes analyzing "refraining" as a kind of inaction. Just the opposite is true, for refraining is not only an action but a special kind of action, one that involves reference to a conscious intention. Since we are prima facie responsible for the consequences of our actions one might argue, then, we are prima facie responsible for the consequences of our forbearing.

One might conclude on the basis of considerations like these that there is no significant moral difference between acting and refraining. This is essentially the position taken by Jonathan Bennett and Michael Tooley. Although they defend the position differently they both seem to agree that if the *only* difference between two courses of action is the difference between acting and refraining, then there is no significant *moral* difference between the two courses of action. This suggests that there is no significant moral difference between "killing" and "letting die," which in turn suggests that there is no defensible *moral* basis for analyzing (2), (3), and (4) as issues that are morally different from (1).

The conclusion is tempting for several reasons. The discussion of "refraining" seems to support strongly the claim that there is no intrinsic conceptual feature possessed by instances of forbearance that could underwrite the derived distinction. And surely it is the sheerest form of self-deception to assume that not feeding a child is somehow more "morally responsible" than "directly killing" it.

If we bracket (2), however, which it would be very difficult to defend, and restrict the further discussion to the killing involved in (1) and the letting die involved in (3) and (4), there is something to be said in defense of current medical practice. In particular, the difference between (1), on the one hand, and (3) and (4), on the other, bears directly on the justifiability of euthanasia.

It is typically argued by those in favor of euthanasia that there is a "right to die." The extreme thesis that all rights can be analyzed into obligations does not seem to be either a logical or a moral truth. However, there are a significant number of rights, typically those which depend on the performance or nonperformance of a particular action, which do underwrite obligations. In these performance situations, usually underwritten by either contract or positive law, for A to have a right means A's ability to *claim* something from B, to have something to press against him that, in justice, he owes to A. To assert this sort of right then is to assert something about the prima facie obligations of others.

The notion of "right to die" interpreted as "right to euthanasia" clearly falls under the classification of claims *to* a certain kind of performance. But if this is so then the coherence of "right to die" depends on the following assumption: if an individual has a right to euthanasia, then someone is obliged to kill him. This assumption, however, is extremely implausible. For it is not clear what it would mean to say that one individual is morally obliged to kill another. There are all sorts of reasons one might bring to bear on why it is merciful or perhaps kind to end someone's life. One might even argue that a doctor ought to, when requested by a patient, end the patient's suffering. Yet in each instance the doctor could deny that he is *obliged* to kill the patient, that is, he is not depriving the patient in justice of anything that is due to him. But if this is so then the concept of "right to die" loses it hold.

Some people might find this argument counterintuitive, that is, they might argue that from "A ought to do X for B" it follows, as a logical or moral truth, that B has a right against A. There is an important reason for not accepting any such claim. Basically we should keep distinct the issue of having good or even compelling reasons for doing something (i.e., "X ought to be done") from the issue of whether or not any individual has a right to X. The coherence of the latter notion, unlike the former, is contingent on the ability to assign a reasonable obligation, and it is just this that seems to be lacking with respect to the euthanasia issue.

The notion of "right to X" then is used to underwrite a rational demand that something be done to a specific individual. Insofar as it is possible to defeat the latter demand, the claim to possess a right is cast in doubt. The instances that involve "letting die," on the other hand, place no such extraordinary demand on a doctor or any other person and hence are not contingent on the existence of such rights. There is, then, at least one interesting conceptual difference between acting and refraining and hence at least one prima facie reason for analyzing the issues in different ways. Hence, a doctor can reject a patient's claim to have a "right to die" if this means that the doctor is *obliged* to kill him. It is precisely because he is under no such obligation that there is no such right.

The assignment of obligations, of course, does not yet *settle* anything significant about the moral issues, but rather throws into sharp relief what is often really at issue in such discussions. We tend to assign obligations in terms of what we normally expect from one another, and such expectations often take the form of fairly determinate rules. A doctor who refrains from giving insulin to a diabetic is doing something wrong because of the reasonable expectations and assumptions of the doctor–patient relationship. A doctor who forbears from using all the available measures on a hopeless terminal case is not breaking any obligation because usually no one expects him to make use of them. What such observations show (as the discussion of PDE also shows) is that the issue that should play a central role in this moral discussion is not so much the narrow one of acting or refraining, but rather the rules that will define an appropriate and justifiable set of expectations. Since we are not, in the normal course of events, usually called on or required to account for the things that we let happen, then the claim that we merely "let something happen" is often a perfectly good and relevant moral reason for justifying our behavior. However, it is just as clear (for example, as articulated in child neglect laws) that the appeal to "letting something happen" is contingent for its moral justification on our understanding of what is and is not required or expected of one individual toward another.

Substantial moral progress on these questions, then, demands a careful and detailed account of what individuals *ought* to *expect* from one another, of what ought to be required from both sides of the doctor–patient relationship. What is the nature of "care," what limits does it have, what limits should it have? In short, complete and adequate understanding of the issues raised by the "acting–refraining" problem, as with virtually every other problem in medical ethics, presupposes an account of a humane, reasonable, and morally acceptable set of assumptions about medical care and treatment. If the debate on these issues is to continue, it should focus on the articulation of a reasonable set of presuppositions and not on the narrow issue of "acting" and "refraining."

HAROLD F. MOORE

[*Directly related is the entry* DOUBLE EFFECT. *This article will find application in the entries* ABORTION; *and* DEATH AND DYING: EUTHANASIA AND SUSTAINING LIFE, *article on* ETHICAL VIEWS. *For further discussion of topics mentioned in this article, see the entries:* CARE; RIGHTS; *and* THERAPEUTIC RELATIONSHIP. *Other relevant material may be found under* ETHICS, *article on* UTILITARIANISM; INFANTS, *articles on* ETHICAL PERSPECTIVES ON THE CARE OF INFANTS, *and* INFANTICIDE: A PHILOSOPHICAL PERSPECTIVE; *and* LIFE, *article on* VALUE OF LIFE.]

BIBLIOGRAPHY

ANSCOMBE, G. E. M. "Modern Moral Philosophy." *Philosophy* 33 (1958): 1–20.
AUNE, BRUCE. "Intention and Foresight." *Journal of Philosophy* 60 (1963): 652–654.
BENNETT, JONATHAN. "Whatever the Consequences." *Analysis* 26 (1966): 83–102.
BENTHAM, JEREMY. *Introduction to the Principles of Morals and Legislation* (1780). New York: Humanities Press, 1970.
CASEY, JOHN. "Actions and Consequences." *Morality and Moral Reasoning.* Edited by John Casey. London: Methuen, 1971, pp. 155–205.
DINELLO, DANIEL. "On Killing and Letting Die." *Analysis* 31 (1971): 84–86.
DUFF, R. A. "Intentionally Killing the Innocent." *Analysis* 34 (1973): 16–19.
FOOT, PHILIPPA. "Abortion and the Doctrine of Double Effect." *Moral Problems: A Collection of Political Essays.* Edited by James Rachels. New York: Harper & Row, 1971, pp. 28–41.
GEDDES, LEONARD. "On the Intrinsic Wrongness of Killing Innocent People." *Analysis* 33 (1973): 93–97.
GRISEZ, GERMAIN G. *Abortion: The Myths, the Realities, and the Arguments.* New York: Corpus Books, 1970.
HUGHES, GERARD J. "Killing and Letting Die." *Month* 2d n.s., 8 (1975): 42–45.
KNAUER, PETER. "The Hermeneutic Function of the Principle of the Double Effect." *Natural Law Forum* 12 (1967): 132–162.
MAGUIRE, DANIEL C. *Death by Choice.* Garden City, N.Y.: Doubleday & Co., 1974.
MANGAN, JOSEPH T. "An Historical Analysis of the Principle of Double Effect." *Theological Studies* 10 (1949): 41–61.
NOONAN, JOHN T., JR. "An Almost Absolute Value in History." *The Morality of Abortion: Legal and Historical Perspectives.* Edited by John T. Noonan, Jr. Cambridge: Harvard University Press, 1970, pp. 1–59.
RACHELS, JAMES. "Active and Passive Euthanasia." *New England Journal of Medicine* 292 (1975): 78–80.
RAMSEY, PAUL. *The Patient as Person.* New Haven: Yale University Press, 1970.
THOMAS AQUINAS. *Summa Theologiae*, II–II 64,6 and 64,7.
TOOLEY, MICHAEL. "Abortion and Infanticide." *Philosophy and Public Affairs* 2 (1972): 37–65.
WILLIAMS, GLANVILLE. *The Sanctity of Life and the Criminal Law.* New York: Knopf, 1957.

ADOLESCENTS

Adolescents (ages thirteen to eighteen) have been regarded as minors who need the consent of parents for most forms of medical treatment. This categorization of adolescents together with other minors is, however, controversial because the issues regarding the necessity of parental consent for the medical treatment of adolescents are different from those for children. Attention is rarely paid to the rights of adolescents vis-à-vis their parents, their physicians, and society (e.g., Katz).

This article is concerned with areas where ethical conflicts arise concerning the accepted cultural norm, the parents' wishes, the physicians' recommendations, and the adolescents' wishes and needs. Further, guidelines for adolescents within treatment and research programs are given. These are similar to those utilized when an adult is the patient or subject. The guidelines are presented in order to underscore their importance in the ethical treatment of each adolescent as an individual with rights for self-determination and needs for guidance.

Conflicts between adolescents' rights and the need for parental consent occur in connection with providing contraceptive devices, abortion, medical treatment for chronic drug or alcohol usage, and psychiatric treatment. These are the areas where adult ethical judgments are most widely divided. Treatments relating to sexuality are particularly frequent both because of cultural taboos (along with implicit encouragements for sexual experimentation) and because sexual maturity represents an area in which adolescents' capacities unquestionably differentiate them from younger minors. At times the need for parental consent is an issue because of parental disapproval of the treatment which the adolescent is requesting; but often it is the adolescent who wants to keep the medical treatment confidential because of fear or shame related to confiding in the parent.

The American Academy of Pediatrics Model Act of 1973, providing for consent of minors for health services, recommends that minors should be allowed to give consent for health services when they are "pregnant or afflicted with any reportable communicable diseases including venereal diseases, or drug and substance abuse including alcohol and nicotine. This self consent only applies to the prevention, diagnosis, and treatment of those conditions." Taken in its broadest context, the Model Act may be read as an advocacy of adolescent self-determination on the use of contraceptives and on the procedure of abortion, as one relates to "prevention" and the other, at least tangentially, to either "prevention" or "treatment" of pregnancy.

There are legal guidelines for certain medical treatments that are generally accepted. The age of the patient is not considered for some treatments. If the patient is below eighteen, parental consent is automatically required for other treatments. As an example of the former, in most states adolescents can obtain treatment for venereal disease without the knowledge and consent of their parents. The therapeutic contract is made between the physician and the adolescent with society's sanction and legal protection. Similarly, any treatments deemed to be emergency procedures do not require parental consent for any minors, be they adolescents or children. On the opposite end of the spectrum are other areas of medical treatment for which the consent of the parent is not considered an issue of controversy but accepted as necessary. As a seldom disputed example, if an adolescent wants to have plastic surgery for cosmetic reasons, almost all physicians would require consent from the parents or legal guardians. In relation to the need for parental consent, age or psychological variables such as levels of maturity become issues of ethical concern when the treatment requested is not seen to have a fairly clearly definable physical necessity.

Relationship between age and psychological maturity

Legislators and citizens are continuously grappling with the extent to which the state serves as a protector by legislating "rights" through arbitrary points such as age. Except in unusual cases where parents are deemed incompetent, parents assume the responsibility for their children under the age of thirteen. The law ought to differentiate the youth who attains biological and possibly cognitive maturity during adolescence (Inhelder and Piaget), both from the child whose capacities are less developed and from the adult who has had more years of experience. Adolescents' rights, however, have been minimized, or simply not considered apart from those of minors in general. The legal profession and society at large have, because of a lack of better criteria, used the arbitrary age of eighteen as representing maturity. One reason suggested for the failure to regard adolescents as different from younger children has been a fear that the result would be the deterioration of the family unit (Emson).

Within the period of adolescence, the mere age of an individual may not reflect adequately the level of psychological and physiological maturity. Recent research in a small subsample of adolescents has demonstrated that certain adolescents develop cognitively and emotionally at age fifteen far beyond their peers at age eighteen (Hatcher). Hatcher's study of a homogeneous population of thirteen unwed pregnant adolescents suggested a strong developmental determinant in how the young women reacted to the pregnancies and abortions. From an analysis of interviews and projective test material, three clinical and statistical patterns of adjusting to the experience emerged. The ability to cope with both the pregnancy and subsequent abortion related to the maturity level of the adolescent girls, whereas coping capacities did not correlate with age of the adolescent.

Maturity as a measure of self-determination?

Even if the intent were to correct for the psychological differences among peers of the same age, methodological complications interfere. How confident are we of the reliability and validity of psychological measurements? Within the behavioral and social sciences, when reliable and valid measurements of maturity are attainable, further questions have to be asked: For whom is the treatment, under what circumstances, and what are the personal values of the people involved? These determinations involve definitions or conceptualizations of normality, health, and maturity that can relate the type of treatment sought to the context in which the adolescent is functioning. Is a consensual understanding of what is normal or healthy for a particular adolescent obtainable?

Normality has been described as having four perspectives: (1) *Normality as health* is the traditional medical approach to health and illness; behavior is within normal limits when no gross psychopathology is present. (2) *Normality as utopia* conceives of normality as a harmonious and optimal blending of diverse elements of the mental apparatus that culminates in optimal functioning. (3) *Normality as average* is based on a mathematical principle of the bell-shaped curve; the middle range is normal and both extremes are deviant. (4) *Normality as transactional systems* defines normality as an integration of bio-psychosocial variables over a period of time (Offer and Sabshin). Definitions of normality or health will differ according to the perspective utilized for the definition, the particular cultural contexts, and the psychology of the individual adolescent within the sociological and cultural context (ibid.).

The range in conceptualizations of normality makes elimination of age-related restrictions problematic. However, our inability clearly to define measurable criteria of maturity, to determine a stage where a person is cognitively and emotionally able to make relatively sound decisions concerning his or her physical and emotional welfare, does not mean that we have to maintain the legal status quo. The difficulties in replacing age-related laws requiring parental consent do not mean that these laws should be retained without qualification.

What is the effect of a request for parental consent? Is the adolescent being protected by the request for parental consent or is this request a denial of one's rights as a person separate from one's parents? Does abdicating the need for parental consent give too much power to the physician vis-à-vis the relatively inexperienced adolescent, and, alternatively, is it justifiable to assume that the parent provides reasonable protection?

Judgment by the physician on the "mature minor"

Particularly in the areas which we have mentioned above, the physician needs to evaluate the adolescent's capacity to make reasonable decisions concerning his or her future. If the physician believes that the adolescent is capable of making sound decisions, the patient, at the very least, ought to be classified under the category of "mature minor" (American Academy of Pediatrics, 1974). Similarly, the American Academy of Pediatrics Model Act of 1973 recommends that the minor assume responsibility for medical treatment when he or she seems to the physician to "be capable of making rational decisions."

Recognizing the maturity of the adolescent patient renders treatment of the adolescent subject to the same potential judgmental biases as are present in the treatment of the adult patient. If the physician feels that the adult patient or research subject cannot grasp the significance of the situation, the physician consults relatives (or in extreme situations, the courts) to act on behalf of the best interests of the patient. This same procedure can be followed for the adolescent patient. If the physician has serious doubts about the appropriateness of treatment of an adolescent, the physician ought to consult colleagues, parents, and in some cases the courts. An adolescent should be encouraged to proceed with parental knowledge and consent but not required to do so unless, in the physician's opinion, the individual case necessitates parental consent. Thus, the physician's judgment is substituted for state or federal laws, in which individual personality assessments are not taken into consideration.

There are no scientifically based criteria to aid the physician in deciding the level and extent of maturity. It is frequently assumed that the physician will automatically know when a minor is mature enough to make the "rational decision." The further question to be asked is, How often are individuals of any age who are coming for treatment in a position to be able to use their capacities for making rational and abstract judgments when their own bodies or minds are the "patients"?

Ideally, of course, "rational decision" should be reached by all parties: the patient, the family, the physician, and the society. Where differences arise concerning the nature of rationality, the weight ought to go to the adolescent patient's decision, with the agreement of a qualified physician. The physician, in turn, should examine the decision in the light of recommendations from his or her medical specialty.

What criteria do physicians use to arrive at these decisions? At times, factors other than the welfare of the patient are likely to be involved in the physician's decision to proceed without the consent of the adolescent's parents. Physicians, for example, might be influenced by their own inability to communicate effectively with parents or by an identification with some young patients and their overt rebellion against parental authority. On the other hand, there may be a psychological need on the part of the physician, in order to feel justified in his decision, to share all his knowledge concerning the adolescent patient with the parents. In the latter case, if this might be to the detriment of the patient's well-being, an ethical problem arises.

The physician has an obligation to adolescents not to abuse their confidence; yet the physician should respond when the adolescent's statement seems to contain an indirect plea to the physician to intervene and provide the patient with familial or legal supports which the adolescent has been emotionally unable to request directly. These are the cases where the physician judges that the patient wants guidance. Then, the doctor ought to require some sort of adult consent. This would come from the parents or, in special circumstances, from a "patient advocate." The patient should be refused treatment when the doctor believes that the greater harm would come from performing the treatment without parental or other guidance than from the neglect of the treatment requested.

In some cases, physicians' ethical beliefs prevent them from complying with adolescents' medical requests; this is most prevalent in cases involving requests for abortions or contraceptive pills. Here the physician is responding neither to the maturity level of the patient nor to the legal status of the treatment. Under these circumstances, physicians should discuss their beliefs with the adolescent patients. When the physician's ethical beliefs conflict with those of the adolescent, the patient should be referred to a medical colleague who does not share the physician's ethical beliefs and who can act in accordance with both good medical judgment and the rights of the patient involved.

The adolescent in psychotherapy

Under what conditions should an adolescent be allowed to obtain psychiatric treatment without the knowledge and consent of parents or legal guardians? Is it only when a medical emergency is involved that endangers the young person's life, such as a suicidal patient?

The high school student can consult his or her high school counselor, who is paid by the Board of Education. The student sees the counselor once or twice a week for therapy or counseling. These sessions are not labeled as psychiatric treatment and, therefore, there is no need to obtain permission from the parents. Most of these counselors do practice psychodynamically or behaviorally oriented psychotherapy. But because they are not psychiatrists, they are not, by definition, delivering medical care, and hence no consent is needed. Is there really a significant difference between the adolescent choosing therapy in the high school and the teenager going to a mental health clinic—a difference significant enough to make parental consent necessary for the one treatment and not for the other?

The adolescent may go to a medical clinic seeking psychiatric treatment for relief of depression. The depression may not be severe enough to endanger the teenager's life with suicide, yet there is much discomfort and agitation. The parents of this adolescent may be unwilling to let their child undergo psychiatric treatment because of a disbelief in the validity of psychiatric treatment, an unwillingness to accept the fact that they, the parents, cannot help their own child, or a resentment against the child, feeling that he or she is not "trying" hard enough to function adequately. Here the adolescent's rights conflict with parental judgments. Is it for the society to legislate or for the physician to decide based on his or her knowledge of the case? If the latter, there would be a bias in favor of treatment, as this is the tool the physician has been trained to value and provide. Yet the laws in most states are explicit in stating that the parents have the right to decide whether their child will enter psychotherapy or not (Shlensky). This legal bias seems to collide with the rights of adolescents and generally to disregard their ability to seek help in situations where parents or parental decisions may be irrational. Thus, the parents' irrationality may be legally sanctioned while the rights of the adolescent are ignored.

The right of adolescents to refuse therapy is less clear cut. They can refuse therapy and can make the therapist's role meaningless if they choose not to cooperate. Most psychiatrists know that they cannot effectively treat an adolescent without his or her consent. However, parents, physicians, and courts together can require psychotherapy in the face of the adolescent's refusal when the adolescent is adjudged to be incapable of recognizing his or her needs. Thus, the adolescent's right to refuse psychotherapy is a right which, under certain circumstances, can be countermanded.

Confidentiality

Once treatment has begun, the physician should have absolute protection under the law regarding confidentiality. After proper consent has been given, the patient and the physician enter into a mutual contract of confidentiality. Medical records should not be released to a third party without the written consent of the adolescent patient (Hofmann). This means that the physician's obligation is to the adolescent patients, and only when patients agree should physicians communicate with families, insurance investigators, or educational institutions.

Informed consent within treatment and research programs

In any treatment, objective information must be provided in a manner understandable to both adolescent and parents, where parents are involved. Where a patient is a control for a research project, knowledge of an involvement as a control and of the research as a whole must be provided. Patients should have the knowledge that enables them to accept or reject the treatment program in which they are involved. In areas of insufficient knowledge, patients should be informed of the parameters of what is known and what is hypothesized.

In situations where current theory and practice provide any indication that one treatment is preferable to another, the more generally accepted treatment should be given to the patient unless there is truly informed consent for being a part of an experimental program, where the results have not been demonstrated.

The word "demonstrated" is undoubtedly open to a sliding scale of truths. We must continue our research while treating patients according to our most educated guesses. How best to protect the interest and well-being of the individual yet not stagnate or eliminate research has been a burning issue for some years (Duffy). The American Medical Association has set relatively strict ethical guidelines for clinical investigation (American Medical Association, pp. 10–12). One of the crucial requirements is for the adolescent patient to be informed of the limitations and potential dangers of research undertaken. Consent for participation is meaningless unless it is informed consent.

As psychiatric researchers we suggest the use of empathy, of putting oneself in the other's place, when deciding how much should be told. A study of doctors at Michael Reese Hospital in Chicago reported that ninety percent of doctors did not believe in telling their patients that they had cancer. However, sixty percent of doctors said that they themselves would, in a similar situation, prefer to be told the truth (Oken). A conclusion might be that doctors are not identifying with their patient population but rather consider themselves to be in a class by themselves.

In research on normal adolescent subjects, who have not approached the physician with a request for treatment, knowledge of the use of the data must form a part of a "research alliance" between adolescent and physician. The physician should assess the psychological significance of the relationship with the subject and take precautions against creating situations in which the subject's coping may be threatened without adequate supports applied. When a normal subject in a research project requests aid, the role of the physician–researcher must shift to provide the necessary humane aid or referrals. In cases where requests have not been made but the physician sees a dangerous situation developing, he or she owes it to the subject to intervene. Above all, however, the researchers working with a normal adolescent population must be careful not to intervene too much, intervention being the role to which the physician is more accustomed. Medical researchers must be aware of personal value judgments and must try to empathize with the subject, in this case the adolescent, without influencing the subject where no psychological issue of danger is involved. The physician should not be engaged in sociological engineering unless that has been the stated purpose of the research, and informed consent has been obtained.

In our study of normal male adolescents (Offer and Offer) we followed the guidelines stated above. The investigation began when the subjects were fourteen and continued until the young men were twenty-two. During the selection process, schools, parents, and the high school freshmen were informed of the purposes and extent of the study, and their consents were obtained. We developed the concept of a research alliance, contrasted to a therapeutic alliance. We wanted to gain the subjects' trust and cooperation but were not providing the usual contracts for psychiatric treatment. We needed to be careful not to seduce the subjects into cooperation by allowing them to assume that, since the interviewers were psychiatrists, there was an implicit agreement to treat them. In addition, the psychiatrists had to change their usual role orientation. This was done through the timing of interviews. Throughout the high school years we saw the young men once every four months in order to encourage continued long-term involvement of the teenagers; we believed that this interval was infrequent enough to avoid interfering with their own modes of functioning. The interviews were semi-structured, and care was taken not to delve too deeply when resistance became manifest. In psychiatric research, the resistance would be noted; in psychiatric therapy, it might have been challenged.

Where are the physicians' values and the values of the parents most likely to collide? In our research on normal adolescents, we found the adult population to take most offense when questions about sexual behavior were asked. Here, it was our value judgment that the adolescents would not become sexually active just because they were questioned about their sexuality. Further, we believed that our inquiries would, and did, provide relief to the subject in an area where discussion had often been taboo. This is an example of the predominance of the researcher's value judgments, which are operative within any study. They can, and should, be checked through peer and institutional review in order to ensure that the researchers do not act in a way that is considered to be professionally deviant. Where general societal dictates are contradictory, the researchers' point of view should be ethically acceptable within their field of specialization. What is most necessary is for researchers to be aware of influences upon their subjects and to understand their own biases while logically defending the tactics being utilized to the satisfaction of a professionally accredited medical society.

The researcher and the clinician

Need there be a conflict between researcher and clinician? Within medicine, and certainly within patient care, the researcher and the clinician are often the same individual. The major difference between the labels "researcher" and "clinician" is the systematized way in which data are used, observed, and communicated; possibly, there is also a difference in selectivity of populations treated. These differences should not affect quality of treatment; they are pertinent to use of data rather than to quality of care. Dangerous situations may arise when the researchers disregard the informed consent strictures mentioned above. Advantages for the patient within the research setting must also be noted. These may be provided through the sophistication of knowledge within a highly specialized area that researchers may have or through possible benefits arising from experimental situations. Here we speak of the individual adolescent patient. Clearly, for the society research must be continued in order to advance clinical care. Without it, we can only repeat the same mistakes. In this section attention has not been directed to the need for research, which ought to be self-evident. Our consideration is for the rights of the adolescent patient within the research program, the research itself being a necessity for the advancement of knowledge in any scientific field (Offer, Freedman, and Offer).

Conclusion

Adolescence should be understood as a separate stage of human development with its own tasks and solutions, its specific problems, and moral and legal considerations that differentiate it from both childhood and adulthood.

Decisions on the appropriate response to an adolescent requesting medical care ought to be flexible enough to take into consideration variances pertaining to the individual's level of maturity and the nature of the treatment requested. The need for parental consent for adolescent medical treatment may serve as a detriment to the adolescent, rather than as a protection for the youth's health. Although many adolescents do not attain cognitive and emotional maturity, this does not differentiate them from much of the adult community. Maturity cannot be defined by age alone. The concepts of "mature minor" and "rational decision" are particularly applicable to the adolescent years. In all research and treatment cases the right of the adolescent to give informed consent must be respected. A plan has been entered for the treatment of the psyche as well as the soma of the adolescent patient. The ability to put oneself in the place of the adolescent and of the adolescent's parents might well be the most important determinant in the physician's ethical–medical judgments.

DANIEL OFFER AND JUDITH B. OFFER

[*For further discussion of topics mentioned in this article, see the entries:* ABORTION; CONFIDENTIALITY; CONTRACEPTION; INFORMED CONSENT IN HUMAN RESEARCH; *and* INFORMED CONSENT IN THE THERAPEUTIC RELATIONSHIP. *Also directly related are the entries* CHILDREN AND BIOMEDICINE; PATERNALISM; PERSON; RIGHT TO REFUSE MEDICAL CARE; RIGHTS; SEXUAL DEVELOPMENT; THERAPEUTIC RELATIONSHIP; *and* TRUTH-TELLING. See APPENDIX, SECTION III, PEDIATRIC BILL OF RIGHTS.]

BIBLIOGRAPHY

American Academy of Pediatrics, Committee on Youth. "The Implications of Minor's Consent Legislation for Adolescent Health Care: A Commentary." *Pediatrics* 54 (1974): 481–485.

———. "A Model Act Providing for Consent of Minors for Health Services." *Pediatrics* 51 (1973): 293–296.

American Medical Association, Judicial Council. *Opinions and Reports of the Judicial Council.* Chicago: 1971.

DUFFY, JOHN C. "Research with Children: The Rights of Children." *Child Psychiatry and Human Development* 4 (1973): 67–70.

EMSON, H. E. "The Age of Consent." *Canadian Medical Association Journal* 109 (1973): 687–688.

HATCHER, SHERRY L. M. "The Adolescent Experience of Pregnancy and Abortion: A Developmental Analysis." *Journal of Youth and Adolescence* 2 (1973): 53–102.

HOFMANN, ADELE D. "Confidentiality and the Health Care Records of Children and Youth." *Psychiatric Opinion* 12, no. 1 (1975), pp. 20–28.

INHELDER, BÄRBEL, and PIAGET, JEAN. *The Growth of Logical Thinking from Childhood to Adolescence.* New York: Basic Books, 1958.

KATZ, JAY, with the assistance of CAPRON, ALEXANDER, and GLASS, ELEANOR S. *Experimentation with Human Beings.* New York: Russell Sage Foundation, 1972.

OFFER, DANIEL; FREEDMAN, DANIEL X.; and OFFER, JUDITH L. "The Psychiatrist as Researcher." *Modern Psychiatry and Clinical Research.* Edited by Daniel Offer and Daniel X. Freedman. New York: Basic Books, 1972, pp. 208–233.

———, and OFFER, JUDITH L. *From Teenager to Young Manhood: A Psychological Study.* New York: Basic Books, 1975.

———, and SABSHIN, MELVIN. *Normality: Theoretical and Clinical Concepts of Mental Health.* Rev. ed. New York: Basic Books, 1974.

OKEN, D. "What to Tell Cancer Patients—A Study of Medical Attitudes." *Journal of the American Medical Association* 175 (1961): 1120–1128.

SHLENSKY, RONALD. "Minors' Rights to Psychiatric Outpatient Treatment without Parental Consent in Illinois." *Illinois Bar Journal* 61 (1973): 650–652.

ADVERTISING BY MEDICAL PROFESSIONALS

Advertising by medical professionals is adjudged to be generally unethical in all countries where physicians have established their professional identity. The International Code of Medical Ethics, adopted by the World Medical Association in October 1949, declares unethical "any self-advertisement, except such as is expressly authorized by the national code of medical ethics" (World Medical Association).

Professional codes of ethics typically adopt broad definitions of prohibited advertising. Thus, for example, the British General Medical Council defines "the professional offence of advertising . . . [as arising] from the publication (in any form) of matter commending or drawing attention to the professional skill, knowledge, services, or qualifications of one or more doctors, when the doctor or doctors concerned have instigated or sanctioned such publication primarily or to a substantial extent for the purpose of obtaining patients or otherwise promoting their own professional advantage or financial benefit" (British Medical Association, p. 52). Professional organizations typically attempt to justify such strictures on public interest grounds. Thus, the British Medical Association expresses the conviction that "in taking up the attitude of determined opposition to undesirable methods of publicity it is acting in the best interests of the public as well as of the medical profession" (ibid., p. 42).

Restrictions in the United States

Advertising of physicians' services in the United States has long been restricted by the Principles of Medical Ethics of the American Medical Association (AMA). Formulation of the ethical principle has varied over time, but since 1957 it has read simply, "He [the physician] should not solicit patients." State and local medical societies have maintained similar restrictions and are recognized by the AMA as having the responsibility for giving specific content to the general principles to reflect "local custom" and "local ideals." Though long accepted and actively enforced, the professional restriction against advertising has recently come under attack in the United States and may be subject to significant curtailment under emerging legal principles.

On numerous occasions, the AMA Judicial Council has expressed its views on the meaning and purpose of the AMA rule against solicitation of patients: "The practice of medicine should not be commercialized or treated as a commodity in trade. Respecting the dignity of their calling, physicians should resort only to the most limited use of advertising and then only to the extent necessary to serve the common good and improve the health of mankind" (American Medical Association, p. 27). "Self-laudations defy the traditions and lower the moral standard of the medical profession; they are an infraction of good taste and are disapproved" (ibid., p. 23). The Judicial Council defines *advertising* as "making information or intention known to the public" and contrasts it with *solicitation,* which involves "the attempt to obtain patients by persuasion or influence" through the use of "testimonials," creating "unjustified expectations," implying superior skills, or deceiving the consumer ("AMA Clarifies Policy"). This fuller delineation of "solicitation" was provided in a 1976 statement issued following the filing of an antitrust action against the AMA's restrictions by the Federal Trade Commission. Though offered as a clarification, this statement may have liberalized the AMA's earlier position.

With respect to the scope of the ethical constraint, the Judicial Council concludes that "the public is entitled to know the name of physicians, the type of their practices, the location of their offices, their office hours," and the like (American Medical Association, p. 24). Increased specificity is supplied in the 1976 statement, which allows the giving of some fee information and "biographical and other relevant data." This seeming liberalization is qualified, however, by the use of such unspecific terms as "relevant data" and the requirement that information not be "misleading or deceptive," formulations that would allow professional societies considerable leeway to challenge individual doctors' attempts to convey information to the public.

Physicians "may furnish this information through the accepted local media of advertising and communication . . . [such as] dignified announcements, telephone directory listings and reputable directories" ("AMA Clarifies Policy"). Newspapers are not mentioned, though "the particular use to be made of any medium of communication and the extent of that use are . . . matters to be determined according to local ideals" (American Medical Association, p. 24).

Enforcement of the restriction on advertising may be by the local medical society itself, through means ranging from informal warnings to termination of society membership. In some thirty-four states, however, a statutory prohibition of advertising by physicians allows the state board of medical examiners also to police excesses. Where the law prohibits merely "unprofessional conduct" or the like, the state board will frequently define such conduct in accordance with ethical codes of state or local medical societies. Normally the threat of being charged with unethical conduct is an effective enough deterrent to physicians, although stronger punishments are occasionally imposed.

One area of frequent difficulty has been how to regard publicity given to a physician for either medical or nonmedical accomplishments. The AMA states that a physician's name may appear in "the reporting of proper news" and in connection with civic projects, but that he may not initiate or encourage its use in a news story (ibid., p. 29). Televised interviews with physicians in various capacities are now common and present recurrent problems, especially where the interview appears to boost the physician's book on a medical topic or his style of medical practice.

Justifications for restrictions on medical advertising

The historical justification for the ban on advertising was the perceived need to distinguish quackery from ethical professionalism. The 1920 Code of Ethics of the Massachusetts Medical Society stated, "The distinction between legitimate medicine and quackery should be clearly maintained. Physicians should not advertise their methods of practice" (Massachusetts Medical Society). In the years prior to 1920, when medical science had relatively little to offer to patients in comparison with today, the profession sought to enhance its reputation by substituting cooperation for competition, repressing extramural criticism of individual physicians, and cultivating primary consumer reliance on the profession's image and the individual practitioner's general reputation rather than on his advertised specific qualifications. The Massachusetts code of 1920 provided that "a spirit of competition considered honorable in purely business transactions cannot exist among physicians without diminishing their usefulness and lowering the dignity and standing of the profession" (ibid.). In contrast, the AMA Judicial Council says, "Freedom of choice of physician and free competition among physicians are prerequisites of optimal medical care" ("AMA Clarifies Policy").

Although the policy against advertising originated as part of the early strategy of raising the practice of medicine to "professional" status, the asserted justifications for a restrictive policy have not changed much today. Even though licensure laws have long since removed most unqualified practitioners and the profession has gained the power to remove any remaining ones, the AMA Judicial Council still contends that the advertising ban "protects the public from the advertiser and salesman of medical care by establishing an easily discernible . . . distinction between him and the ethical physician" (American Medical Association, p. 23). Though now conceding that competition has a role, the Judicial Council's position is still that "[s]ome competitive practices accepted in ordinary commercial and industrial enterprises . . . are inappropriate among physicians" ("AMA Clarifies Policy"). The extent of the difference between professional and commercial standards is likely to be established ultimately by the courts, which have frequently conceded that such a difference exists.

While the profession's concerns are largely unchanged, modern conditions may put advertising in a somewhat different light. For example, the technical advances in medicine, in addition to increasing the patient's stake in the physician's skill, may have strengthened the argument that the consumer cannot make the necessary judgments concerning quality of care and should therefore not be trusted with possibly misleading information. Price information in particular may be regarded as difficult to evaluate and misleading in view of the widely variable content and quality of the services rendered. To compensate for the consumer's presumed inadequacy, the medical profession has purported to accept primary responsibility for maintaining the quality of care through peer oversight of practitioners. Moreover, in lieu of expecting consumers to find attentive care through the trial-and-error process of shopping, ethical standards are supposed to assure the consumer of the physician's commitment and concern. The profession is regularly criticized, however, for failing to police the quality of care effectively and for not sharing information with patients in the course of treatment in accordance with the legal and ethical requirements of informed consent. The ethical limitation on advertising, in denying consumers quality-related information and the opportunity to exercise judgment, may thus be seen as part of the larger ethical issue of professional dominance.

The repression of advertising can also be seen as an important aspect of the broad policy issue in medical care of the role and reliability of consumer choice and market forces on the one hand versus efficacy of regulatory and professional cost and quality controls on the other. Inflation in the mid-1970s caused heightened concern that regulatory restraints of all kinds, including restrictions on professional advertising, tend to increase prices, a fear confirmed in a widely cited study of state restrictions on opticians' advertising of eyeglass prices (Benham, p. 344). If primary reliance is to be placed on regulation and controls to guide the health industry's performance, then advertising's utility will be limited, but questions currently being raised about advertising restrictions may reflect distrust of these mechanisms.

Impact on institutional providers

In recent years some important changes have taken place in the organization of medical care delivery and in the nature of financing mechanisms. Advertising has been important in informing consumers of some of these developments. The AMA's 1953 principles held that "solicitation of patients, directly or indirectly, by a physician, by groups of physicians or by institutions or organizations, is unethical." After the formulation of the basic principle was shortened in 1957, the AMA restated the foregoing language as part of its textual explication, omitting the reference to "institutions and organizations." This omission suggested that the new principles were addressed to medical practitioners alone and were not to govern institutional behavior, but group practices were advised that "no physician-member of a clinic may permit the clinic to do that which he may not do" (American Medical Association, p. 3). Advertising by prepaid group practices and—to use the newer, broader term—health maintenance organizations (HMOs), has been actively restricted by some medical societies under this principle. In some areas flexibility has prevailed, but HMOs have generally confined themselves to advertising only the availability of their services and perhaps the monthly rate, eschewing mention of their physicians' qualifications.

Some states have enacted legislation clarifying HMOs' right to advertise. California's Knox–Keene Health Care Service Act of 1975, following earlier legislation, imposes some controls while generally permitting the practice. Most important, the federal Health Maintenance Organization Act of 1973 explicitly declared inoperative with respect to certain HMOs those state laws which would prevent such HMOs from "soliciting members through advertising its services, charges, or other nonprofessional aspects of its operation." Significantly, however, it specifies that it "does not authorize any advertising which identifies, refers to or makes any qualitative judgment concerning, any health professional."

Legal issues

Until 1975 restrictions on advertising appeared to be immune from legal attack. "Free speech" arguments against statutory or regulatory restrictions appeared to be foreclosed by a 1942 Supreme Court ruling that suggested that advertising, as "commercial speech," was outside the scope of First Amendment protection (*Valentine* v. *Chrestensen*). Restrictions imposed through professional societies' ethical codes also appeared valid because the antitrust laws, prohibiting anticompetitive agreements, had never been held to apply to the so-called learned professions.

Both of these immunizing doctrines were subsequently discredited by the Supreme Court. In 1975 the antitrust laws were held to apply to the legal profession (*Goldfarb* v. *Virginia State Bar*), and shortly thereafter the Federal Trade Commission, following this lead, instituted an action against the AMA and professional associations in Connecticut, charging that their efforts to prevent their members from "soliciting business, by advertising or otherwise," violated federal law (Federal Trade Commission, 1975, p. 2). In a 1976 case the Supreme Court held unconstitutional a Virginia law prohibiting the advertising of prescription drug prices, stressing the consumer's First Amendment right to receive useful information through advertising; the Court discounted the state's argument that advertising was antithetical to high professional standards in pharmacy (*Virginia State Board of Pharmacy* v. *Virginia Citizens' Consumer Council*). In 1977, the Court extended this analysis to the legal profession, holding in a five-to-four decision that lawyers cannot be prohibited from using newspaper advertisements to disclose the price of routine services (*Bates* v. *State Bar of Arizona*).

The meaning of these developments for the advertising of medical services is not clear. One lower court has recognized consumers' right not to be denied the basic facts to be included in a physician directory prepared by a health planning agency (*Health Systems Agency of Northern Virginia* v. *Virginia State Board of Medicine*). More commercial types of advertising may still be subject to curtailment, however, under arguments different from those advanced against lawyers' advertisements in the *Bates* case. For example, a nontraditional but possibly convincing argument might focus on the danger of inducing overutilization of medical services by beneficiaries of insurance plans or governmental programs. Finally, advertising that in any way touts the quality of professional services is likely to be regarded with particular suspicion.

Issues about professional advertising that were long dormant seem now to be open for public debate and ultimate judicial resolution. The renewal of interest in this subject caused an apparent liberalization of the AMA's position in 1976 and changed the focus from whether doctors should be permitted to advertise to how they should be permitted to do so. New guidelines, developed not by the profession unilaterally but under public pressure and with some public input, appear to be in the offing.

CLARK C. HAVIGHURST

[*For the proscriptions against advertising medical and other health-care professions, see:* CODES OF MEDICAL ETHICS, *article on* ETHICAL ANALYSIS; DENTISTRY, *article on* ETHICAL ISSUES IN DENTISTRY; PHARMACY; APPENDIX, INTRODUCTION, *article on* CODES OF THE HEALTH-CARE PROFESSIONS. *For the text of such codes, see* APPENDIX, SECTION I: GENERAL CODES FOR THE PRACTICE OF MEDICINE; *and* SECTION IV: CODES OF SPECIALTY HEALTH-CARE ASSOCIATIONS. *See also:* COMMUNICATION, BIOMEDICAL; MEDICAL PROFESSION; *and* ORTHODOXY IN MEDICINE. *For a discussion of advertising of drugs by pharmaceutical companies, see:* DRUG INDUSTRY AND MEDICINE.]

BIBLIOGRAPHY

"AMA Clarifies Policy on Advertising by MDs." *American Medical News*, 19 April 1976, p. 9.
American Medical Association, Judicial Council. *Opinions and Reports of the Judicial Council, Including the Principles of Medical Ethics and Rules of the Judicial Council*. Chicago: American Medical Association, 1971, pp. 22–29.
"The American Medical Association: Power, Purpose, and Politics in Organized Medicine." *Yale Law Journal* 63 (1954): 937–1022.
Bates v. State Bar of Arizona. 45 U.S.L.W. 4895 (1977). U.S. Sup. Ct. Dkt. No. 76-316, argued 18 January 1977.
BENHAM, LEE. "The Effect of Advertising on the Price of Eyeglasses." *Journal of Law and Economics* 15 (1972): 337–352.
British Medical Association. *Medical Ethics*. London: 1974.
Federal Trade Commission, Complaint. *In Re American Medical Association, Connecticut State Medical Society, and New Haven Medical Association*. Docket no. 9064, filed 19 December 1975. Available on request from the Federal Trade Commission.
Goldfarb v. Virginia State Bar. 421 U.S. 773. 44 L.Ed.2d 572. 92 S. Ct. 2004 (1975).
Health Systems Agency of Northern Virginia v. Virginia State Board of Medicine. 424 F. Supp. 2d 267 (1976).
KESSEL, REUBEN A. "Price Discrimination in Medicine." *Journal of Law and Economics* 1 (1958): 20–53.
"Knox–Keene Health Care Service Plan Act of 1975." *California Health and Safety Code*. § 1340–1395 (Supp. 1975).

KONOLD, DONALD. *A History of American Medical Ethics 1847–1912.* Madison: State Historical Society of Wisconsin, 1962.

Massachusetts Medical Society. "Code of Ethics of the Massachusetts Medical Society." *Massachusetts Medical Society: Its Services and Functions.* Boston: 1961, p. 48. Adopted 9 June 1920.

"Restrictive State Laws and Practices." "Health Maintenance Organization Act of 1973." Pub. L. 93–222 Sec. 1311 § 2(b); 87 Stat. 931; 42 U.S.C. § 300e–10(b) (1974).

Semler v. Oregon State Board of Dental Examiners. 294 U.S. 608. 79 L.Ed. 1086 (1935).

Valentine v. Chrestensen. 316 U.S. 52. 86 L.Ed. 1262 (1942).

Virginia State Board of Pharmacy v. Virginia Citizens Consumer Council. 425 U.S. 748. 48 L. Ed. 2d 346. 96 S. Ct. 1817 (1976).

World Medical Association. *International Code of Medical Ethics.* London: Third General Assembly of The World Medical Association, 1949. Printed single sheet. Text in Appendix. Also in *World Medical Association Bulletin* 1 (1949): 109, and in *World Medical Journal* 3 (1956), Supplement, p. 12.

ADVERTISING OF DRUGS

See PHARMACY; COMMUNICATION, BIOMEDICAL, *article on* MEDIA AND MEDICINE; DRUG INDUSTRY AND MEDICINE.

AFRICA

See MEDICAL ETHICS, HISTORY OF, *section on* NEAR AND MIDDLE EAST AND AFRICA, *article on* SUB-SAHARAN AFRICA.

AGING AND THE AGED

I. THEORIES OF AGING AND ANTI-AGING TECHNIQUES   *Leonard Hayflick*
II. SOCIAL IMPLICATIONS OF AGING   *Bernice L. Neugarten*
III. ETHICAL IMPLICATIONS IN AGING   *Drew Christiansen*
IV. HEALTH CARE AND RESEARCH IN THE AGED   *Ernlé W. D. Young*

I
THEORIES OF AGING AND ANTI-AGING TECHNIQUES

Theories of aging

Probably no other area of biological inquiry is susceptible to so many theories as is the science of gerontology. This occurs not only because of a lack of sufficient fundamental data but also because manifestations of biological changes with time affect almost all biological systems from the molecular level up to that of the whole organism. It is therefore easy to construct a theory of aging based on a biological decrement (or decline) that may be observed to occur in time in any system at the level of the molecule, cell, tissue, organ, or whole animal. The significant question will always be: Is the change observed a direct cause of aging or is it the result of changes that may be occurring at a more fundamental level?

If the modern notions of biological development are rooted in signals originating from information-containing molecules, then it would seem reasonable to attribute postdevelopmental changes to a similar system of signals occurring at the molecular level. The following example might be illustrative: One current theory of aging rests on decrements occurring with time in the immune system such that individuals are less able to deal with infections or antibodies "mistakenly" produced against their own cells. Either or both of these events would lead to a reduced ability to deal with environmental insults, and thus an increasing force of mortality would result. That such events might occur cannot be wholly denied, but this logic would suggest that, conversely, "maturation" of the immune system in early life controls development of the animal. Conventional wisdom does not accept this as a likely possibility.

If age changes are caused by fundamental changes in information-containing molecules that result, in chain-reaction fashion, in physiological decrements at higher organizational levels—that is, cell, tissue, organ, whole animal —then the most tenable theories of aging are those based on molecular events occurring in the genetic apparatus.

Error theories. Orgel developed a model for biological aging that is dependent upon a decrease in the fidelity of protein synthesis. In this way an accumulation of errors is thought to occur in enzyme synthesizing systems such that faulty proteins would be specified leading to decrements in cell function. Evidence both for and against this hypothesis has been obtained, although there is currently (1976) a greater preponderance of negative data.

Redundant message theory. Although partly based on error accumulation, Medvedev suggests an interesting alternative hypothesis (Medvedev). His conceptions are based on the fact that within each cell the genetic message is written in a highly redundant fashion. That is, each bit of information can be specified by a number of genes, only one of which functions at any given time. As errors accumulate in the expressed gene, it is repressed, and a second gene providing the same information becomes expressed, once again maintaining the fidelity of protein synthesis. This cycle can occur repeatedly as reserve genes become expressed after functioning genes accumulate errors and become repressed. The well-recognized differences in the life spans of animal species are conjectured to be a manifestation of the degree of gene repetition. Engineers are very familiar with the principle whereby the greater the number of "back-up" systems available for an essential function, the greater the likelihood that the total machine will not fail. A repeated information system simply has a greater chance of preserving intact the integrity of any complex system, although with time ultimate failure is inescapable.

The theory that aging is caused by gene mutations has several advocates (Burnet; Curtis; Szilard), yet the failure of this theory to explain the quantitative aspects of normal and radiation-accelerated aging has been repeatedly observed (Alexander; Clark; Strehler, 1959).

Molecular cross-linking. This thesis, championed by von Hahn, is based on several a priori assumptions, the most important of which is that there exists a universal physiological aging process due to intrinsic causes that is deleterious to the cell and acts progressively with chronological age (von Hahn). These are the essential criteria characterizing biological aging that have been proposed by Strehler (1962) and appear to be generally valid.

The central event of aging, at whatever level of biological complexity, seems to be a progressive diminution of adaptation to stress and of the capacity of the system to maintain the homeostatic equilibrium characteristic of the adult animal at the peak of growth and full development (Comfort).

Von Hahn suggests that information-containing molecules, specifically DNA, can accumulate linkages between adjacent portions of the folded molecule. Such binding will prevent the maintenance of fidelity of daughter molecules produced by this self-duplicating molecule, or it will interfere with the transcription of essential genetic information. Evidence has been reported that points to the occurrence of an age-related increase in the cross-linking of the DNA double helix, dependent on the presence and degree of binding of certain associated proteins. Since DNA strand separation is an essential step in transcription of genetic information, any process that blocks this step will lead to a loss of genetic information within the cell. An accumulation of these events, it is thought, could lead to functional decrements resulting in age changes.

The increased cross-linking with time of the protein collagen is also thought to provide a mechanism for age changes. Collagen constitutes twenty-five to thirty percent of the total body protein and is found in extracellular spaces. Cross-linking of this molecule increases with time and is thought to interfere increasingly with the flow of nutrients and waste products between cells, resulting in age-associated physiological decrements. The increasing flaccidity of the skin by which wrinkling occurs, and the decrease in elasticity of the lung and similar alterations in the vascular system over time are attributed to greater cross-linking in collagen molecules (Kohn).

The physiological changes that can be attributed to cross-linked molecules are analogous to similar changes that occur in a crowd of people who slowly join hands in random fashion until all are so linked. Individual members thus lose their mobility, and the flow of objects between them is slowly reduced. Although cross-linking of molecules undoubtedly occurs, its role as a direct cause of aging is still uncertain.

Programmed genetic events. A final proposition offered to explain age changes at the genetic level is based upon a continuation of those genetic events that result in the development and maturation of any animal. This notion assumes that the switching on and switching off of genes during developmental processes also determines age changes. That is, age changes, like developmental changes, are "programmed" into the original pool of genetic information and are "played out" in an orderly sequence just as developmental changes are. The graying of the hair is not generally thought of as a disease associated with the passage of time. It is regarded as a highly predictable event that occurs later in life after the genetic expression of myriad other programmed developmental events that occur in orderly sequences.

This example may be analogous to attributing the cause of aging to a similar series of orderly programmed genetic events that shut down or slow down essential physiological phenomena after postreproductive age is reached. The programming may be the result of specific gene determinants that, like the end of a tape recording, simply trigger a sequence of events to shut the machine down. Alternatively, the universality of aging might be attributed to functional failures arising from the random accumulation of "noise" in some vulnerable parts of the system, which ultimately interfere with optimum function and produce all of the well-known physiological decrements.

"**Mean time to failure.**" If the "noise" is randomly accumulated, then why do members of each species appear to age at specific, highly predictable times? We may call the span of time during which "noise" accumulates and becomes manifest in some functional decrement the "mean time to failure." This concept is applicable to the deterioration of mechanical as well as biological systems and can be illustrated by considering the "mean time to failure" of, for example, automobiles. The "mean time to failure" of the average machine may be five to six years, which may vary as a function of the competence of repair processes. Barring total replacement of all vital elements, deterioration is inevitable. Similarly, failure of cell function may occur at predictable times that are dependent upon the fidelity of the synthesizing machinery and the degree of perfection of cellular repair systems. Since biological systems do not appear to function perfectly and indefinitely, we are led to the conclusion that the ultimate death of a cell or loss of function is genetically programmed and has a "mean time to failure." The "mean time to failure" may be applicable to a single cell, tissue, organ, or the intact animal itself. It is proposed that the genetic apparatus simply runs out of accurate programmed information that might result in different "mean times to failure" for all of the dependent biological systems. The existence of different life spans for all species may be the reflection of more perfect repair systems in those animals of greater longevity.

Finally, one must consider the two-cell lineages that seem to have escaped from the inevitability of aging or death. These are the continuity of the germ cells (precursors of egg and sperm cells) and continuously reproducing cancer cell populations. It may be possible to explain the immortality of cancer cell populations by suggesting that genetic information is exchanged between these cells in the same way that the genetic cards are reshuffled when egg and sperm fuse. Thus exchange of genetic information may serve to reprogram or reset a more perfect biological clock.

Anti-aging techniques

It is generally believed that the approximately maximum human life span of about 100 years has not changed since recorded history, but what has changed is the larger number of people who approach this apparent limit. Medical achievements have simply resulted in the fact that more people are reaching the limit of what appears to be a fixed life span. Deaths in the early years are becoming progressively less frequent, so that in many privileged countries one can now reasonably expect to become quite old, which is a very new phenomenon.

In a world where the leading current causes of death are resolved, the elimination of cardiovascular diseases, stroke, and cancer in, for example, the United States, would yield a net increase of about eighteen years in life expectation at birth and only slightly less at age sixty-five. This figure is almost identical to the net increase in life expectation at birth achieved in the United States from 1900 to 1950. The increase in life expectation at birth during the first half of this century resulted mainly from the resolution of deaths from infectious diseases occurring before the age of sixty-five. The gain in life expectation in the United States at ages sixty-five and seventy-five from 1900 to 1950 was, respectively, only 1.9 and 1.3 years.

If there is reason to believe that deaths from cardiovascular diseases, stroke, and cancer will be eliminated in the next fifty years, then life expectation will show an increase by A.D. 2025 as profound as that which has occurred since 1900. After that spectacular accomplishment, the leading cause of death will be accidents, which because of their random nature are not likely to yield to total elimination. Thus the social, psychological, political, and economic impacts of resolving the leading causes of death on life expectation in the next fifty years can be reasonably assessed by studying like changes that have occurred in the first seventy-five years of this century, when a similar increase in life expectation occurred.

Let us also consider a world in which all causes of death resulting from disease and accidents are totally eliminated. The theoretical effect on human longevity would be to find all citizens living out their lives, free of the fear of premature death, but with the likely fate that on the eve of their one-hundredth birthday they would suddenly die.

This situation will continue to evolve because

biomedical research has confined its efforts almost exclusively to causes of death associated with what have traditionally been thought of as diseases. Scant attention has been paid to the underlying causes of biological aging that are not disease-associated, which, in clock-like fashion, dictate for each species a specific maximum life span. To be sure, the physiological decrements that occur in advancing years increase vulnerability to disease, but unless more attention is paid to the fundamental non-disease-related biological causes of aging, then our inexorable destiny will be death on or about our one-hundredth birthday.

Prospects for increasing human longevity. As a consequence of these considerations there are two general ways in which biomedical research can be expected to extend human longevity in the next fifty years. The first is to reduce or eliminate the major particular causes of death, such as cardiovascular diseases, stroke, and cancer. The results of ameliorating minor diseases will be minimal. For example, if tuberculosis were completely eliminated in the United States, there would be a mere 0.1 year gain in life expectation at birth. Thus it could be argued that if an increase in life expectation becomes the main goal of biomedical research, most research should be directed toward the elimination of the major causes of death. This position, although less than humane (because it would reduce current research investments proportionally to diseases other than heart disease and stroke), and not likely to attract many adherents, is nonetheless the most logical conclusion to be reached.

The second way in which biomedical research can deal with human longevity is to address itself specifically to the underlying non-disease-related fundamental biological changes that occur with time. These are not diseases but are the basic biological changes that result in those physiological decrements characteristic of aging and upon which are often superimposed an increasing vulnerability to disease. Such an approach, then, does not directly concern itself with efforts to increase human life expectation but rather to extend what appears to be a fixed human life span.

Some immediate possibilities. Of all the areas of human health care, there is probably no other arena of greater quackery than the legion of alleged life-extending nostrums, diets, exercises, injections, and other procedures foisted on a gullible public. Most gerontologists agree that none of these claims has been authenticated by following sound principles of scientific proof. There is, in the judgment of some gerontologists, at least one comparatively innocuous way in which the human life span probably might be extended significantly. The method is based on classic studies made in the 1930s and since confirmed in many laboratories for a number of animal species including rats, in which it was first described (McCay, Pope, and Lunsford). The method is simply to reduce the caloric intake to such a level that undernutrition but not malnutrition occurs. This is done by providing an animal with a diet sufficient in all necessary nutrients but very low in calories. Longevity can then be increased by as much as fifty percent. The effects are most pronounced if caloric restriction diets are initiated when animals are very young. This results in a stretching out of the developmental stages; infancy, puberty, maturity, adulthood and aging simply occur at later than usual points in time, so that the total life span is increased.

On the assumption that undernutrition in man would yield similar results, it is of interest to observe that, in the forty years since this has been known, no human has consciously chosen to do it, even the biologists who know the data best. Considering the number of nostrums and treatments that have been foisted on a gullible public as anti-aging regimens, the lack of interest in underfeeding is, upon superficial consideration, remarkable. On the supposition that the method is widely known, that it works, and that it is not dangerous, the main conclusion that can be drawn from the notable lack of interest in it is that for most people the quality of life is more important than its quantity.

If this is so, then an important lesson can be learned. Any method that might increase human longevity could be unacceptable if it affects the enjoyment of life.

To sleep, perchance to dream. There remains yet another method for increasing life expectancy that bears consideration. Although it will not result in an extension of life on an absolute time scale, it is interesting to consider a form of increased longevity based on the self-evident supposition that life can be lived only when individuals are both physically and mentally active. Since, for most individuals, sleep consumes nearly one-third of our lives, a reduction in the time spent sleeping should result in an increase in productive occupation and the enjoyment of life, that is, if sleep itself is not considered to be either productive or enjoyable. Sleep researchers tell us that no detectable negative effect on health has been observed in those individuals who have learned how to make a modest reduction in the length of time usually spent asleep. The impact of this change would be profound, for if we were to reduce by one half hour the average of eight hours spent sleeping, the net effect on "life-extension" would be an increase of more than two years. This increase in "life expectancy" is equivalent to living in a society where cancer deaths have been totally eliminated.

Instead of any biological advances resulting in the prolongation of human life, this scenario could result from a social movement of an organization of "Awakists," whose members will have learned to do with less sleep in order to gain additional time and, it is to be hoped, spend it productively.

Reduced metabolic rates. Another similar approach to prolongevity would be to mimic in man the reduced metabolism of those animals that pass the winter in hibernation or the summer in estivation. The value of this method presupposes that a slowing of metabolic rates for prolonged periods of time could, in the long run, extend the life span.

Although no compelling evidence exists that a reduced metabolic rate might extend the human life span, this approach has the merit of likely feasibility since the state could probably be induced by drugs or by cold treatment, producing a controlled inhibition of the sympathetic nervous system and a twilight sleep resembling narcosis. Assuming that this were feasible, significant questions of the ethics and societal consequences of such an action must be considered. If it were possible to reduce the metabolic rate in man over a span of, say, ten years, with the result that five more years would be added to the life span, would this be useful? What ten-year period of hibernation would a subject choose in order to effect a gain in years at some future time? One might choose to arrest oneself in a sleep-like trance for, say, ten years, beginning at the age of forty, with instructions to be awakened in a decade so as to enjoy ten more years with one's grandchildren as young adults rather than as infants. The potential societal dislocations resulting from such a scenario are awesome. Family councils might have to be held in order to determine who should be allowed to hibernate, when, and for how long, so as to avoid Father's being awakened at a biological age of forty only to celebrate Mother's eightieth birthday. Other variations on these kinds of asynchronous scenarios can lead to terrible consequences, although others might lead to amusing solutions of current problems. Consider those societies with negative attitudes toward May–December marriages. The older partner might hibernate for a few decades in order to allow the younger partner to either grow up or catch up.

Tampering with our biological clocks

If the control of aging is dependent upon an understanding of basic biological processes, one profoundly important question arises: How desirable is it to gain this understanding and to be able to manipulate our biological clocks? The answer is not simple. The fact that it must be asked is further evidence of the distinction that must be made between disease-oriented biomedical research and gerontological research. Who asks: What are the goals of cancer research, or what are the goals of cardiovascular or stroke research? The answers are so obvious as to preclude asking the question. But the goals of gerontological research are quite a different matter, because we are not certain whether the "resolution" of the physiological decrements of old age will indeed benefit the individual or society as a whole. Take, for example, the possibility that research into the biology of aging might result in the total elimination of all age-related physiological decrements. If this were achieved and no control were established over the biological clock itself, the result might be a society whose members would live full, physically vigorous, youthful lives until their one-hundredth birthday—at which time they would die.

If, on the other hand, we were to learn how to tamper with our biological clocks, with what goal in mind would one choose to reset his clock? Surely one wouldn't choose to spend ten additional years suffering from the infirmities of old age; yet that might, initially, be the only way to intervene. Is society prepared to cope with individuals whose only choice might be between naturally occurring death and ten or more years spent with the vicissitudes of old age? We can hardly deal with a mean maximum life span of, say, 100 years, not to mention the further social, economic, and political dislocations that might occur if we add a decade to this figure.

Aside from this possibility, it is also worth considering the prospect of clock-tampering in which the choice would be to spend more years at a particular stage of our lives than we now do. The clock might be stalled for ten years at, for example, a chronological age of twenty. Is this desirable? Each of us, after pondering this provocative question, would be likely to agree that the time at which we would like our biological clocks arrested should correspond to those years in which maximum life-satisfaction and productivity occurred. Yet if we were forced to make such a decision, it would probably have to be made prospectively. Even more complex is the question of when in the human life span individuals are most productive. An interesting and exhaustive study of this question was made in 1953 by Lehman, whose data showed that, depending upon the particular area of human endeavor, the time of maximum productivity can occur throughout the human life span. Thus clock-tampering becomes a game that very few of us are capable of playing.

In spite of the apparent dilemma in stating goals for gerontological research, there is one goal that appears to be wholly desirable and even attainable as a short-range objective. That would simply be to reduce the physiological decrements associated with biological aging so that vigorous, productive, nondependent lives would be led up until the mean maximum life span of, say, 100 years. Implicit in this notion is that the quality of life is more important than its quantity.

If tampering with our biological clocks ever becomes a reality, it would be tragic if such clock-tampering, instead of extending our most vigorous and productive years, served only to extend the years of our infirmities.

LEONARD HAYFLICK

[*Directly related are the other articles in this entry:* SOCIAL IMPLICATIONS OF AGING, ETHICAL IMPLICATIONS IN AGING, *and* HEALTH CARE AND RESEARCH IN THE AGED. *For discussion of related ideas, see:* LIFE, *article on* QUALITY OF LIFE; *and* POPULATION ETHICS: ELEMENTS OF THE FIELD, *article on* THE POPULATION PROBLEM IN DEMOGRAPHIC PERSPECTIVE.]

BIBLIOGRAPHY

Advances in Gerontological Research. Edited by Bernard L. Strehler. New York: Academic Press, 1964–. Irregular.

ALEXANDER, PETER. "Is There a Relationship between Aging, the Shortening of Life-Span by Radiation and the Induction of Somatic Mutations?" *Perspectives in Experimental Gerontology: A Festschrift for Doctor F. Verzár.* Edited by Nathan Wetheril Shock with F. Bourliere, H. von Hahn, and D. Schlettwein-Gsell. Springfield, Ill.: Charles C. Thomas, 1966, pp. 266–279.

BURNET, F. M. "A Genetic Interpretation of Ageing." *Lancet* 2 (1973): 480–483.

CLARK, A. M. "Genetic Factors Associated with Aging." *Advances in Gerontological Research* 1 (1964): 207–255.

COMFORT, ALEX. *Ageing: The Biology of Senescence.* Rev. ed. New York: Holt, Rinehart & Winston, 1964.

CURTIS, H. J. "The Possibility of Increased Longevity by the Control of Mutations." *Perspectives in Experimental Gerontology: A Festschrift for Doctor F. Verzár.* Edited by Nathan Wetheril Shock with F. Bourliere, H. von Hahn, and D. Schlettwein-Gsell. Springfield, Ill.: Charles C. Thomas, 1966, pp. 257–265.

FINCH, CALEB, and HAYFLICK, LEONARD, eds. *The Handbook of the Biology of Aging.* The Handbooks of Aging, vol. 3. Edited by James Birren. New York: Van Nostrand Reinhold, 1977.

GRUMAN, GERALD J. *A History of Ideas about the Prolongation of Life.* Transactions of the American Philosophical Society, n.s. vol. 56, pt. 9. Philadelphia: 1966.

HAYFLICK, LEONARD. "The Biology of Human Aging." *American Journal of the Medical Sciences* 265 (1973): 432–445.

———. "The Strategy of Senescence." *Gerontologist* 14 (1974): 37–45.

KOHN, ROBERT R. *Principles of Mammalian Aging.* Foundations of Developmental Biology Series. Edited by Clement L. Markert. Englewood Cliffs, N.J.: Prentice-Hall, 1971.

LEHMAN, HARVEY CHRISTIAN. *Age and Achievement.* Memoirs of the American Philosophical Society, vol. 33. Princeton: Princeton University Press, 1953.

MCCAY, CLIVE M.; POPE, FRANK; and LUNSFORD, WANDA. "Experimental Prolongation of the Life Span." *Bulletin of the New York Academy of Medicine* 32 (1956): 91–101.

MEDVEDEV, ZHORES A. "Repetition of Molecular-Genetic Information as a Possible Factor in Evolutionary Changes of Life Span." *Experimental Gerontology* 7 (1972): 227–238.

ORGEL, L. E. "Aging of Clones of Mammalian Cells." *Nature* 243 (1973): 441–445.

STREHLER, BERNARD L. "Origin and Comparison of the Effects of Time and High-Energy Radiations on Living Systems." *Quarterly Review of Biology* 34 (1959): 117–142.

———. *Time, Cells and Aging.* New York: Academic Press, 1962.

SZILARD, LEO. "On the Nature of the Aging Process." *Proceedings of the National Academy of Sciences of the United States of America* 45 (1959): 30–45.

United States Bureau of the Census. *Some Demographic Aspects of Aging in the United States.* Current Population Reports, ser. P-23, no. 43. Washington: Government Printing Office, 1973.

VON HAHN, H. P. "The Regulation of Protein Synthesis in the Aging Cell." *Experimental Gerontology* 5 (1970): 323–334.

II
SOCIAL IMPLICATIONS OF AGING

Sociologists do not deal directly with ethical principles or with the determination of the values that ought to be respected in society. They can, however, supply data and interpretation of social trends that are important in understanding the complexity of ethical issues. This article is intended to highlight some of the implications of rapid social change as they have given rise to new ethical issues regarding aging and the aged.

Long life expectancy is a decidedly modern achievement. At the turn of this century, four of every hundred persons in the United States were sixty-five or older. Now it is ten of every hundred. Although the proportion of older people is not expected to rise a great deal more in the next few decades, their absolute numbers will grow rapidly. By the year 2000 the number is expected to surpass thirty million, and by the year 2020, when persons born in the "baby boom" of the 1950s and 1960s reach old age, the number is expected to surpass forty-two million.

These projections are relatively safe, because everybody who will be old by the year 2020 is already alive, and the numbers therefore depend only upon mortality rates. The projections are made by the U.S. Bureau of the Census, based on only slight declines in death rates to the year 2000, and they assume no new basic biological or medical discoveries.

The same general picture is true of other countries of the world. There are a dozen or more industrialized countries in which the proportions of the old are at present higher than in the United States, and although each nation will have a somewhat different age distribution over time, the world trend is toward "aging societies" in which the proportion of the old is increasing in comparison to the proportion of children and adolescents. That is true because populations, like individuals, age as the benefits of industrialization, rising standards of living, and modern medical research lead to longer life expectancy. In the more economically developed regions of the world, it is anticipated that between the years 1970 and 2000 there will be a 50 percent increase in the numbers of persons sixty and over, and, more strikingly, in the less developed regions of the world, a 158 percent increase (United Nations Report).

Changing relations between age groups

The increasing number of older persons is not itself a social problem. On the contrary, nations prize longevity and regard it as an outstanding accomplishment when, as in the United States, the majority of citizens live to old age. Average life expectancy, or average age at death, is often regarded as the single most important index of the overall quality of life of a society. The social problems arise in adapting social and economic institutions to the needs of older people at the same time that the needs of younger people are being met.

The relative numbers of young, middle-aged, and old affect every aspect of life, including the relations among age groups. In most societies in most periods of history, an equilibrium becomes established whereby all age groups receive a stable share of goods and services that is regarded as equitable. However, in countries where the appearance of large numbers of older persons has been relatively sudden, as in the United States within the past seventy-five years, social dislocations occur, because such societies have been unprepared by prevailing value systems to meet the newly emerging needs.

In many countries the aging society has brought with it varying proportions of older persons who suffer from poverty, preventable illness, and social isolation. Those persons, who can be called the needy aged, create acute problems in the field of social and health care. But broader issues stem from the needs of all individuals, young and old, to adjust to the new rhythms of life that come with increased longevity: multigenerational families, retirement, increased leisure, changing health status, and new opportunities and new problems of adaptation that accompany a long life. For the society at large innumerable policy questions arise as the whole social fabric accommodates itself to the changing age distribution.

In oversimplified terms the contemporary American society faces two different sets of issues. On the one hand, there are increasing numbers of the "young-old": persons in their late fifties, sixties, and early seventies who are relatively healthy and vigorous, a large number of whom are retired, who seek meaningful ways to use their time, either in self-fulfillment or in community participation, and who represent a great resource of talent for society. On the other hand, there are even more dramatic increases in numbers who might be called the "old-old": persons in their mid-seventies, eighties, and nineties, an increasing minority of whom remain vigorous and active, but a majority of whom need a wide range of supportive and restorative health services and social services (Neugarten).

For various historical reasons, the old-old of the 1970s in the United States represent a disproportionately disadvantaged group. Many were immigrants who had little formal education; many worked most of their lives at low-skill occupations; many lost their occupational moorings during the Great Depression and did not recoup in the period of prosperity ushered in by the Second World War, nor did they build up sizable equities under the social security program as it developed in the 1940s and 1950s.

In succeeding decades more and more older people will have been native-born, will have grown up in urban areas, will have had the advantages of regular medical care during their lifetimes, will have had high school and college educations, and will take for granted pension programs, social security, and government service programs of all types. Thus persons growing old in the future will have very different characteristics, different values and expectations, from those who are currently old.

The relations among age groups are also influenced by changing perceptions of the life cycle and the periods of life. Age groups have become increasingly differentiated over time. It was not until the seventeenth century that childhood became a discernible period of life with its special needs and characteristics (Ariès). Adolescence, socially defined, can be viewed as an invention of the late nineteenth century, and "youth" an invention of the twentieth century (Coleman et al.). In the past few decades middle age has become a newly delineated stage in the life cycle. Persons no longer move abruptly from adulthood—the period of full commitment to work and family responsibilities—into old age, but they move instead through a relatively long interval of middle age when family responsibilities are diminished but work continues, even though specific work roles may change—for example, women reentering the labor market in their forties and fifties—and when physical vigor remains high. Still another meaningful division is now appearing between the young-old and the old-old.

The young-old

Although age is not, in itself, the distinguishing characteristic, the young-old are drawn mostly from those who are fifty-five to seventy-five. Retirement is the primary event that distinguishes the young-old from the middle-aged. Granted that it is arbitrary to use a single life event as the criterion, retirement is nevertheless a meaningful marker, just as the departure of children from the home is a useful marker with regard to middle age.

This fifty-five to seventy-five grouping is not one to which we are accustomed. Age fifty-five is becoming a meaningful lower age limit because of the lowering age of retirement. Many workers are now voluntarily retiring just as soon as they can live comfortably on their retirement incomes. The 1970 U.S. census indicates that only eighty-one percent of fifty-five- to sixty-four-year-olds were in the labor force, as compared to ninety-two percent of the next younger group.

Whether this trend toward earlier retirement will continue depends upon rates of economic and technological growth, the number of young workers, the number of women workers, increases in part-time work opportunities, the development of share-the-work programs, and so on. Most observers predict, however, that the downward trend will continue over the next two or three decades (Jaffe). By and large, then, the young-old will increasingly become a retired group. It is already a relatively healthy group. It is estimated that, while a large number have one or more chronic illnesses, only twenty to twenty-five percent of persons aged fifty-five to seventy-five need to limit their major activities for reasons of health.

Because of the enormous diversity of lifestyles among the young-old, meaningful differences are obscured in aggregated data. Nevertheless, the gross characteristics of this age group are worth noting. At present the young-old constitute more than fifteen percent of the U.S. population, as compared with the old-old, who constitute less than four percent.

Because mortality rates for women are lower, women considerably outnumber men, and this imbalance is expected to grow even larger. Nevertheless, because most men marry women somewhat younger than themselves, the young-old as a total group are more like their younger than their older counterparts with regard to family relationships. About eighty percent of the men and well over half the women were married in 1970 and living with their spouses. (The rates of widowhood are very different for the two sexes; about seven percent of the men, but more than thirty percent of the women, were widowed.) By far the most common pattern is husband-wife families who live in their own households—some seventy percent own their own homes—and only a small minority who move from one house to another within a five-year interval.

The young-old see their children frequently,

and some seventy-five to eighty-five percent of those who have children live within thirty minutes of the nearest child (Shanas et al.). They expect that when they grow to advanced old age and can no longer manage for themselves their children will come to their aid—not financially, for the government is looked to as the expected source of financial and medical assistance, but emotionally. As a number of studies have shown, these expectations are usually met. Various forms of assistance and services are being exchanged across generational lines, ties of affection and obligation are strong, and the family continues to be an important part of daily life.

It is noteworthy that a large proportion of young-old have a living parent. The estimate in 1972 was one of every three sixty-year-old Americans; and this proportion will increase, because the numbers of old-old are growing faster than the numbers of young-old.

The economic status of the young-old is not easily summarized. Income from work is a major factor in economic status, and for many persons income drops precipitously upon retirement, despite public and private pensions. If present trends continue, the adjustment to lower incomes may be timed closer to age fifty-five than to age sixty-five. Nor is the economic status of the young-old easily described compared to other age groups, for satisfactory data are available only with regard to current money income, and money income is only part of total economic resources. For instance there is Medicare, the value of rent to homeowners, tax adjustments for those sixty-five and over, intrafamily transfers, and other assets that create differences between age groups. Overall, the majority of the young-old in the United States are currently neither poor nor "near-poor," and most likely the trend toward improved economic status for older people will continue, given the increases in social security and the growing number of public programs directed at protecting the financial security and general well-being of older people. Other programs, such as national health insurance, already provided in other Western countries, and improvements in private pension and profit-sharing plans, can be anticipated. The likely outcome is that not only will the future young-old be more financially secure than their predecessors, but their economic position will continue to improve when compared to younger adults. The ethical question nevertheless arises: What proportion of the gross national product should go into private and public pension systems?

The young-old are already much better educated than the old-old, but the more important fact is that successive cohorts will be in a less disadvantaged position in comparison to the young. So marked are the gains in educational level that by 1980 the average fifty-five-year-old in the United States will be a high school graduate; by 1990 this will be true of all the fifty-five-to-seventy-five group. Furthermore, with the growth in adult education programs, it can be anticipated that the educational differences that now exist between young, middle-aged, and young-old will be much reduced (Carnegie Commission).

With regard to political participation there is no evidence that age blocks are forming or that a politics of age is developing in the United States (Hudson and Binstock). A quite different picture emerges, however, when general political participation and voting patterns are examined. When national data are corrected for income and education, overall political participation is highest for the age group fifty-one to sixty-five. It falls off only a little for persons over sixty-five (Verba and Nie). Thus, the young-old are disproportionately influential in the electorate.

What do the young-old want?

A vigorous and educated young-old group can be expected to develop a variety of new needs with regard to the meaningful use of time. They are likely to want a wide range of options and opportunities for self-enhancement and for community participation—in general, for what might be called an age-irrelevant society in which arbitrary constraints based on chronological age are removed.

The young-old want a wide range of choices with regard to work. Some opt for early retirement; some want to continue working beyond age sixty-five; some want to undertake new work careers at one or more times after age forty. Even more in the future than in the present, they are likely to encourage economic policies that hasten the separation between income and work and that move toward the goal of providing retirees with sufficient economic resources to maintain their preretirement living standards.

We are already seeing a wider range of life patterns with regard to the related areas of work, education, and leisure. More middle-aged and older people are returning to education, some because of obsolescence of work skills and others for recreation or self-fulfillment. Plans are now going forward in various parts of the United States to create intergenerational educational campuses.

While age-segregated communities are on the increase, they now accommodate only a small percentage of the older population. Their growth will depend upon the extent to which the young-old will be provided opportunities for meaningful community participation in their present locations.

The vast majority of the young-old will continue to live as married couples, but the large number of widows and the increased number of single and divorced persons among them will probably lead to the formation of more group households composed of nonfamily members. At the same time, many will want housing arrangements that make it possible to maintain an aged parent at home. Family interactions of other types may also increase rather than decrease. Contrary to the concerns often expressed over the "vanishing" family, some observers now predict that, as the more instrumental aspects of life—education, income maintenance, health services—are shifted to other social institutions, the family may become more rather than less important with regard to the expressive aspects of life—that is, in providing lasting emotional ties, a sense of identity and self-worth, and respect for the worth of others.

Overall, as the young-old articulate their needs and desires, the emphasis is likely to be upon improving the quality of life and upon increasing the choices of life-styles.

The old-old

The number of those over seventy-five is growing at a more rapid rate than the young-old. Their claims upon the society are at least as morally compelling as those of other age groups.

The prevailing assumptions are that life expectancy will continue to increase, although slowly, and that in general older persons will have improved health in the future. But this says little regarding the period of disability that can be expected to occur for most people in the very last phase of life. There is little basis at present for predicting that this period will become shorter by the year 2000. Most persons would probably regard it as their goal to maintain physical and psychological intactness through their eighth or ninth decade, then die relatively quickly without a long period of terminal illness. (An example is Picasso, who reputedly maintained his full physical, mental, and creative abilities to the day before his death.) But the prospect may be the opposite: With increasing age, people may have a drawn-out period of physical and psychological debility, with medical services that can keep them alive but neither healthy nor happy. The latter is the specter that haunts most persons when they think of their own old age, and it is the specter that haunts medical and social planners when they ponder the implications of the aging society. For the latter, the question arises, for example, whether ours is an equitable distribution, with thirty percent of all health services now going to the ten percent who are over sixty-five. Is this too much or too little?

Whatever the uncertainties regarding the period of disability, the needs of the old-old for meaningful ways of spending time, for special housing, and for transportation will depend in large measure upon health status. The majority will probably live independently in the future, as now, but they will need both supportive social services and special features in the physical environment to enable them to function as fully as possible. The old-old require not only health services aimed at slowing physical and mental deterioration, but social services designed to prevent unnecessary decline in the individual's feelings of self-worth and dignity. Can opportunities for social interaction and social contributions be provided?

Some of the old-old will require nursing home care; others will require new forms of home health services. Will a larger share of the public budget go to meeting these needs of the old-old?

For persons who are terminally ill or incapacitated, the problems are likely to multiply with regard to providing maximum social supports, the highest possible levels of care and comfort, the assurance of dignified death, and, at the same time, an increasing element of choice for the individual himself or for members of his family in deciding how and when his life shall end. New ethical issues will arise over ways of achieving a "best death" for each individual.

Other social and ethical questions will become more pressing: What will be the principal deleterious and the principal beneficial effects of increased numbers of older persons on the society at large? Will vigorous older persons wish to employ their newly won leisure to become important agents of social change? Will they create an attractive image of aging, thus allaying the fears of the young about growing old? Will they be provided with the necessary resources? Or will their increased numbers so aggravate the problems of income maintenance, housing, health care, and social services that their situation will become generally worse rather than better?

In considering what proportion of the society's resources should go to the young, to adults, and to the aged, the question arises whether age itself constitutes a meaningful basis for categorizing people. Should public programs be addressed to the needs of persons who are ill or poor, but not to persons who happen to be old? Just as we have tried to erase distinctions based on sex or on race, should we try to erase distinctions based on age?

Finally, an additional set of value issues arises with regard to the aged: to what extent society should follow a cost-benefit analysis, with benefits to older people weighed in terms of their past or potential productivity; whether the analysis can be based on the very fact that the old are equal members of the society; and whether societies will (or should) develop higher forms of culture where the aged are cherished because their very presence symbolizes man's most humane values.

BERNICE L. NEUGARTEN

[*Most directly related to this article are its companion articles in this entry:* THEORIES OF AGING AND ANTI-AGING TECHNIQUES, ETHICAL IMPLICATIONS IN AGING, *and* HEALTH CARE AND RESEARCH IN THE AGED. *For discussion of related ideas, see the entries:* ADOLESCENTS; *and* CHILDREN AND BIOMEDICINE. *See also:* DEATH; *and* RATIONING OF MEDICAL TREATMENT.]

BIBLIOGRAPHY

ARIÈS, PHILLIPE. *Centuries of Childhood: A Social History of Family Life.* Translated by Robert Baldick. New York: Alfred A. Knopf, 1962.
Carnegie Commission on Higher Education. *Toward a Learning Society: Alternative Channels to Life, Work, and Service: A Report and Recommendations by the Carnegie Commission on Higher Education.* New York: McGraw-Hill, 1973.
COLEMAN, JAMES S.; BREMMER, ROBERT H.; CLARK, BURTON R.; DAVIS, JOHN B.; EICHON, DOROTHY H.; GRILICHES, ZVI; KETT, JOSEPH F.; RYDER, NORMAN B.; DOERING, ZAHAVA BLUM; and MAYS, JOHN M. *Youth: Transition to Adulthood—Report on Youth of the President's Science Advisory Committee.* Chicago: University of Chicago Press, 1974.
HUDSON, ROBERT B., and BINSTOCK, ROBERT H. "Political Systems and Aging." *Handbook of Aging and the Social Sciences.* Edited by Robert H. Binstock and Ethel Shanas. Handbooks of Aging Series, vol. 1. Edited by James E. Birren. New York: Van Nostrand Reinhold Co., 1976, chap. 15, pp. 369–400.
JAFFE, A. J. "Has the Retreat from the Labor Force Halted? A Note on Retirement of Men, 1930–1970." *Industrial Gerontology,* no. 9 (1971), pp. 1–12.
NEUGARTEN, BERNICE L. "Age Groups in American Society and the Rise of the Young-Old." *Annals of the American Academy of Political and Social Sciences* 415 (1974): 187–198.
SHANAS, ETHEL; TOWNSEND, PETER; WEDDERBURN, DOROTHY; FRIIS, HENNING; MILHØJ, POUL; and STEHOUWER, JAN. *Old People in Three Industrial Societies.* New York: Atherton Press, 1968.
United Nations, General Assembly. *Question of the Elderly and the Aged: Conditions, Needs and Services, and Suggested Guidelines for National Policies and International Action: Report of the Secretary General.* A/9126. New York: 1973.
VERBA, SIDNEY, and NIE, NORMAN H. *Participation in America: Political Democracy and Social Equality.* New York: Harper & Row, 1972.

III
ETHICAL IMPLICATIONS IN AGING

Ethical issues related to aging and old age—problems pertaining to the qualities of life of the elderly, the use of anti-aging technologies, etc.—are ultimately influenced by values (or disvalues) associated with longevity and aging.

Values and longevity: a history

Throughout the course of Western civilization complaints over the pains and disabilities of old age mix with exhortations to virtuous resignation and praise for old people who lead lives of strength and dignity. Generally, elderly men and women are given the most esteem when they continue to carry on the roles of adult life. Conversely, the loss of capacity through old age is often the occasion of ridicule and complaint.

In premodern Europe, four principal cultural traditions on aging are to be found: (1) The early Greeks regarded a happy old age as a fortuitous blessing. (2) With the spread of philosophical theories of virtue, the circumstances of old age became less important than one's attitude and bearing toward them. (3) The Hebrew tradition took note of the religious fidelity of old people and showed greater appreciation of the aged couple than Greco-Roman culture. Early Christianity, moreover, established a special status for older women as "widows." Medieval monasticism developed the ideal of old age as a preparation for death but also carried on the late classical ideal of leisured retirement. (4) In the later Middle Ages and the Renaissance old women lost the protection of the Church through the rise of bourgeois mores, and with the Reformation the work ethic eclipsed leisured retirement as a life task of old people.

Classical Greece. Solon summarized an early Greek view of life in a terse sentence, "Man is accident." A happy old age was not something to be expected (Herodotus), and especially not something to be planned for. Life was an aesthetic whole, which misfortune could mar irreparably. The tragedians reminded their audiences of the sober dictum, "Call no one happy until he dies." Not length of life but its quality counted. Stories related by Herodotus in *The Histories* show that in Solon's time a happy old age consisted in favorable circumstances, success in multiple roles, and an honorable death. Fortune, however, not human wit, was regarded as the decisive factor for a happy old age.

Sophocles' *Oedipus at Colonus* is the first major literary work with an old person as hero. Sophocles wrote the play when he was past ninety, and legend says that he presented it for the first time in court as a defense against the charge of senility. In terms of the development of moral stances toward aging, the most important theme of the play is the coincidence of heroic choice and resignation in the old Oedipus. Willing consent to the sufferings of age marks a departure from the activist naturalism of the lyric age toward the ideal of spiritual appropriation of old age, which would dominate the classical period.

With Socrates' account of virtue in Plato's early dialogues the new attitude has fully emerged. In Aristotle's discussion of happiness in the *Nichomachean Ethics*, a generation later, virtue rather than fortune appears as the key to a successful life. The philosopher acknowledges that men can show nobility even in misfortune, not out of insensitivity to pain but through greatness of soul.

Rome. Virtue is a key idea in Roman views of old age too. Cicero's *De Senectute (Old Age)*, the fullest treatment of the topic from antiquity, notes that responses to old age generally reflected life-long dispositions. People who never showed self-control found their last years troublesome. Those who practiced virtue found no evil in aging, because they viewed it as a natural process.

Cicero also informs us about the sociology of aging among the upper classes of his day. Continuance in public service was highly admired, and control over one's family nearly an imperative. Cicero himself retired for political reasons, and he describes with pleasure the enjoyments of literature and gardening. His description of those avocations served as a model of leisured retirement (*otium*) for centuries and was an important influence on the monasteries and convents of the early Middle Ages.

He also records Stoic attitudes toward health and death. To resist old age and compensate for its defects is a duty. Aging is to be combated as one would a disease, with moderate exercise and a spare diet. His readers were admonished to take no thought of death; when it was time to die, they should not vainly cling to life. The Stoic value of unimpassioned judgment gave old age a prized status.

Hebrew and Christian values. The Jewish and Christian traditions show even greater tolerance for the difficulties of the last stages of life. Old people who have been faithful to God, the Law, and the coming Messiah are presented as models of devotion. Stories about the afflictions of age stress constancy in adversity as a basic disposition. Care of parents was valued (Exod. 21:15, 21:17; Lev. 20:91; Deut. 27:16; cf. Matt. 15:3–7); yet the usual form of the commandment to "honor thy father and thy mother" indicated a broader, more inclusive ideal than simple support (Exod. 19:3). It enjoins an attitude of reverence and affection akin to worship, and the promise of long life for those who observe it made it a solemn covenantal obligation.

From an early date, the Christian Church assigned a position of special esteem to widows. Women sixty years and older without family to support them could give themselves to a life of prayer and good works in return for maintenance by the early Christian Church (1 Tim. 5:3–16). Widowhood was a significant and flexible social invention: By the second century, it included both genuine widows who carried out works of mercy and unmarried women living a secluded life of prayer; and by the twelfth century, women who had been released from their marriage vows, e.g., to found convents, could be admitted to formal widowhood (*Monumenta de viduis . . .*). But the bourgeois customs associated with the rise of a new middle class (Harksen) and the revival of Roman law and of Greco-Roman family styles during the Renaissance (Sachs) contributed to the decline of widowhood as a distinctive status by reasserting the domestic duties of women and their subordination to their husbands.

In late antiquity and the Middle Ages, Christianity developed still another role for old age: preparation for death. Death after the model of Jesus was a religious ideal. With the passing of persecution, the enthusiasm that had been attached to martyrdom was transferred to monasticism and the ascetic life generally. Lives of saints from Anthony of Egypt in the fourth century to Aelred (or Ethelred) of Rievaulx in Yorkshire in the twelfth century record highly active old people who retire to a life of prayer and asceticism in their very last years in anticipation of death. The later manuals on the art of dying were popularizations of a way of life that first arose in the monasteries.

Transition to modernity. Through the late Middle Ages religious and humanistic ideals of retirement mix easily with one another. The iconography of the period pictures the old man as a scholar, indicating the social role that at least some elderly persons were expected to play. With the advent of Protestantism, and Puritanism in particular, this ideal fades in favor of continued participation in the roles of adult life. The activism never wholly absent from Western ideals of aging reasserts itself over more contemplative models of old age (Demos).

Moral confusion in modern societies. Values are intimately related to social structure. The most relevant fact about modernity, first in the West but increasingly in all industrialized nations, is the dominance of economic over other social factors. The debate between segregationists and integrationists among social gerontologists reflects a basic disagreement about the effects of economic structures and values on old age (Shanas et al.). The segregationists stress the disengagement of the elderly from their primary roles, especially employment, while the integrationists emphasize continued participation in a wide range of social involvements, generally outside the market place. Because of the dominance of job-related roles, the first group downplays nonpublic, personal pursuits and has little positive to say about old age and retirement. Looking to the family, the neighborhood, and voluntary associations, the second group sees the possibility of a shift of roles and goals in old age with an increase in the intensity of interpersonal relations and the expansion of personal pursuits. Both groups represent one face of aging in modern societies, but their differences expose the moral ambivalance of old age in modern times. The work ethic makes adjustment to nonwork roles problematic. Workers are unused to managing large periods of leisure time. Strong career identification builds barriers against fostering personal goals, and the nuclear family and antinatalist trends make it less easy to enjoy being a grandparent.

Ethical dilemmas with respect to aging concern maintaining a balance between work and leisure, freedom and dependence, economic enterprise and interpersonal life, health care and the acceptance of impairment.

Work and leisure. In advanced industrial societies, the balance between work and leisure is again shifting. Social security, retirement plans, associations of old people, and other factors open possibilities of rich retirement experiences even in nations where the work ethic is strongest. Both the place of work in contemporary notions of worth and identity and the difficulties of many in facing retirement suggest that adjustment to new roles may be eased by extended or part-time employment, or by gradual shortening of the workweek. A less problematic old age, however, will demand changes in social attitudes: appreciation of leisure, the acceptance of personal goals and private time, belief in the potentialities of old persons, and encouragement to participate in cultural and religious activities. Psychologists have noted life review and integration as tasks of old age, and sociologists have suggested the resumption of interrupted interests (Markson; Erikson, 1968, 1976; Havighurst and Albrecht). But these are yet to become accepted and supported patterns of aging.

Freedom and dependence. A peculiar difficulty of aging in many societies, and Western societies particularly, is the assumption that freedom means independence and mastery. This conviction makes it exceedingly difficult to accept the losses and dependence that accompany aging. The same attitude causes strain for people who are faced with the subtle changes of more intimate personal relations—for example, extended daily contact with spouse and relatives—after years of involvement in the impersonal activity of economic life. Conflicts arise because cooperation at an interpersonal level and acceptance of diminution are alien attitudes in modern culture where unimpeded freedom is the paramount value. Affirmation of interdependence with others and of resignation to necessity, therefore, is a constituent of an adequate philosophy of human freedom forced on our attention by the losses of old age.

Interpersonal life. The return to active social relations in old age is also an unsettled area. Despite myths of abandonment and segregation, the vast majority of old people in the United States are in contact with their families and receive support from them in time of need (Shanas et al.). Still, the transition to a more intense social life is not easy. In old age, women adjust more readily than men, because they are usually responsible for maintaining familial and kin ties. Accordingly, male retirement sometimes suffers from reverse sexual discrimination and the stereotyping of social activity as a feminine role. Another severe strain on the elderly in many places is the denigration of family values. Although in Eastern Europe and elsewhere grandparents seem to have a respected role, grandparenting has lost some of its important functions in the United States, where peer relations have emerged as a substitute involvement.

Attitudes toward sickness. Attitudes toward sickness and impairment are related to independence as a dominant value. Fears of dependency due to chronic disability are common, particularly in the United States. Materialism and the cult of youth reinforce preoccupation with bodily health and aggravate feelings of chagrin and indignation when they accompany age (ibid.). An important area of concern, therefore, is the set of values one brings to aging. It is relevant to consider whether any set of values is appropriate to only one or two stages of life and not to others, whether it engenders the moral strength needed in adversity, and whether it overcomes preoccupation with the body.

Against the background of the foregoing values and value conflicts associated with aging, an ethical analysis will be offered of two sets of problems that can be properly understood only in conjunction with each other: first, problems related to the general condition and care of the aged, and second, issues raised by the use of anti-aging and other life-prolonging techniques.

Ethical analysis of problems of the aged

When old people are a marginal group, as they have been until recently in industrial countries, they tend to be penalized like any marginal group. Social losses are heaped onto the natural losses of age. The most general rule of geriatric ethics is that losses should not be compounded (Christiansen). Principles that apply to others in the wider society should apply equally to old people. They include principles of justice, rules against discrimination, and codes of medical ethics. There are at least three areas, however, in which general ethical principles can be given specification with respect to common problems of aging: dependency, financial and moral support, and medical care.

Autonomy. General aversion to dependency and to the real but often unexpected burdens it creates frequently leads to undue loss of dignity by the aged, as measured in terms of their actual need. In such cases the general moral rule attempts to ensure that dependency does not turn into humiliation. Respect for the autonomy of the elderly remains an important factor in their ability to age with full dignity. In the family situation, the ability of old people to direct their own affairs is a matter of some importance (ibid.). Family members are helpers; preempting responsibilities older members can still fulfill is a serious infringement of the rights of the elderly. Conversely, retaining powers of decision, especially over health care, enhances their dignity. In many cases at least, aging is a gradual process with an incremental scale of loss. It helps, therefore, to adjust the withdrawal of responsibility to the degrees of impairment. At every stage of senescence the moral optimum is attained when the aged themselves and their families foster that range of personal autonomy which remains to them after their other losses.

Familial justice. With respect to support of the elderly, equitable sharing of the burden among relatives is a grave but seldom treated issue. Psychologically and sociologically it is understandable that care for aged parents usually falls on one family member, generally a daughter. But that burden can seem excessive, especially where self-fulfillment is a value and women have access to professional careers. Graduated retirement might assist in alleviating financial problems for at least some period of time. When the support does fall to adult children, however, the monetary burden needs to be shared more equitably and compensation given for service rendered. One solution might be the encouragement of a temporary career, discontinuous with other life careers, which would have as its purpose the care of the elderly infirm, similar to a career in the care of children. Yet if such a career is to have social approval and ease the burden of care, it should be accorded suitable tax preferences. Similarly, temporary care institutions would allow necessary time for rest, illness, and recreation for the children, so that uninterrupted care does not become oppressive. It should be noted, too, that the needed support is not simply monetary. Since older people have more personal time, there is a need for social exchange that cannot be satisfied by one or two close relatives.

Medical care. Finally, the principle that losses ought not to be compounded in reference to the aged may be applied to the medical context.

First, health policy should not overlook basic services. Old people need simple things: eyeglasses, false teeth, comfortable shoes, and foot care. The structure of technological medicine, however, often prevents the serving of these needs. Similarly, rehabilitation and quality home care would seem to merit priority over research and development of technology for treating catastrophic illness. Second, since communication of health information is critical to the disposition of the lives of old people themselves, reports on their condition and prognosis ordinarily ought to be made to them fully and without mediation. Family members are helpers and as such are not entitled prima facie to privileged access or control of health information. Third, instead of the elderly infirm finding themselves in the sick role for an indefinite period, health professionals should encourage them to resume as much responsibility as their condition allows. Fourth, in view of the personal crises brought on by impairment in old age, professionals should regard the psychological and religious growth of the aged as an important aspect of care.

Anti-aging techniques: ethical assessment

The quest for immortality is as old as Western culture, but so also is the acceptance of mortality as the highest wisdom. With the dawn of scientific knowledge, Francis Bacon proposed that prolongation of life could be the highest gift of scientific medicine (Bacon). During the last hundred years, the curbing of infectious diseases has extended life expectancy by many years. More recently, organ transplants, renal dialysis, heart-lung machines, and other advanced techniques have given extended life to many, and developments in molecular biology offer the promise of controlling at least some aspects of the aging process itself. Forecasts as of 1975 suggest that adequately funded research on such methods could increase life expectancy once more from fifteen to thirty years (Segerberg).

Moral assessment of anti-aging techniques—which are but one issue of an emerging geriatric ethics—includes the following considerations: (1) the use of various kinds of anti-aging techniques and their place in overall health priorities; (2) their social impact, and (3) their moral and anthropological effects.

Use of anti-aging techniques. Anti-aging techniques include everything from intelligent nutrition and physical fitness programs to advanced prosthetic devices, new treatments for catastrophic diseases such as heart disease and stroke, and yet undeveloped but plausible techniques to inhibit cytolysis, the breakdown in cellular life, which comes with age.

Little or no objection can be raised to the simple techniques of improved nutrition and exercise programs. Indeed many moral traditions view such prudent health measures as a moral imperative. Because these practices would improve the lives of those already living at relatively low cost, there is reason to give them greater weight than more exotic techniques that may reach only select populations at some future time. In addition, such limited techniques tend to involve people in their own care, reduce dependence on experts, and circumvent large expenditures on technologies that are only marginally useful (Illich, 1975).

The development of prosthetic devices is a more ambiguous matter. Medicine's capability of preserving life even though conscious, meaningful existence has ceased has created situations in which neither patient nor family is permitted to accept the reality of death and the burden of moral responsibility. Techniques like renal dialysis raise very difficult questions of just allocation of scarce resources and moral-psychological problems about man–machine synergies. Again, there are important issues of complex dependence on societal resources with related issues of cost–benefit analysis (Arrow; Rice and Cooper). Simple devices, such as pacemakers, offer the least difficulty in that they involve only limited institutional commitments and allow patients to function on their own with only intermittent hospitalization.

Evaluation of advances in cellular biology is difficult. Knowledge of the aging process is rapidly increasing, but specific techniques to counteract the ill effects of aging are still to be developed. In general, proposals for research and development of anti-aging techniques should weigh: (1) the degree of availability that will be feasible with a given technique; (2) the impact of the techniques on the just allocation of all biomedical resources and on the quality of overall health care; (3) the degree of dependence on experts, institutions, and technologies; and (4) the utility of the invention in relation to the general health and full life of the patient.

A variety of life-prolonging technologies is already in use and under development. In some areas, such as cancer and cardiovascular ailments, public commitment is substantial. Should applied research aimed specifically at inhibiting the aging process itself be added to existing programs to fight the major age-related diseases? To answer this question, choices must be made about the lines of research to be adopted, about the structure of a desirable pattern of health care, and about the allocation of resources. Basic research on aging will be important. But should higher priority be given to study and relief of infirmities like arthritis that now afflict old people? What is the proper balance between resources employed for rehabilitation now and those applied to research that will benefit future generations? Is there a just allocation of resources when the poor lack basic nutrition and health services and at the same time sophisticated medical technologies are developed at public expense? Will a majority of the population benefit from these new technologies, or will extended life be a boon only to an elite? Will life-extending technologies lead to a fuller, healthier life for the elderly, or only to prolonged illness and suffering? Answers to such questions will disclose important differences in value judgments concerning the weight given the present generation over the future, the drive for scientific breakthroughs vis-à-vis the demand for the relief of suffering, the relative power of technological versus human imperatives, and the balance between health maintenance and crisis medicine as ideals of health care.

Social impact. Prolongation of life carries with it the potential for massive economic, social, political, and psychological dislocations. Life-extension at the far end of the life cycle could place an immense financial burden on the middle generation. Opportunity for retirement might be reduced simply to offset the cost of subsidizing the elderly. The demographic shifts resulting from anti-aging policies could also lead to political upheavals. The elderly, with larger numbers and resources, might press for greater services, and younger people might rebel at having to support a large, unproductive class of people. One solution would be to focus on the middle years as the phase of life that would be prolonged. But this too would have its difficulties, such as providing employment for the increased pool of workers.

The psychological impact of life extension must also be assessed. Some techniques, such as low-temperature hibernation, would result in gaps in an individual's life experience and in disconnections in relations with others. These side effects seem to provide occasion for identity conflicts and social tensions. Furthermore, when compared with old people in traditional societies, the aged in modern industrial countries already live in a relatively anomic social situation. Meaningful, socially sanctioned ideals and roles are not available for them (Callahan and Christiansen). Without consensus on the values of old age, life extension offers potential for great personal disorientation among the elderly.

Life extension has serious effects on demographic policy too. If stabilized population size is a matter of policy, life extension would have to be compensated for by further curtailment of births. Accordingly life tasks related to child care and the family would be reduced. Other forms of mutual support and bonding would arise and would be of special significance in providing companionship and care for the elderly infirm. But with fewer children proportionally in the population, society's ability to renew itself generally in ideals and ideas might be jeopardized. This is a cost that must be weighed in terms of an overall demographic policy. Where the threat of "overpopulation" has not yielded a coherent government population policy, the imbalances threatened by wide use of anti-aging techniques may well force the development of such plans.

Moral and anthropological effects. Scientific immortality is not in prospect, but significant delays in aging and dying are near. Such human power over death poses substantive questions about morality and human nature. New questions arise about the termination of human life. Will people be permitted to choose a natural death? Will their choices be classified as suicides? At what point will cessation of life-prolonging treatment be allowed? Will society set upper limits to the extension of life, and by what criteria? In short, anti-aging technology will force redefinition of suicide and euthanasia.

Extended life may also bring substantive changes in the structure of moral life. A decline in fertility may mean a decrease in intergenerational responsibility, the paradigm case of altruistic behavior and moral obligation. Will prolonged life lead to an increase in the excesses of individualism, indifference to the weak and powerless, increased competitiveness, and a consequent increase in legal control of interpersonal relations? What areas will remain for moral responsibility in the traditional sense?

Summary approaches

Two approaches to humanism conflict in weighing the prospects for life-prolonging technologies. One attitude might be called neonaturalism, because it regards old age and death as natural events that are to be freely appropriated by the aging and the dying (Kübler-Ross). Others argue for natural limits to technological innovation on the grounds that those satisfactions which can be legitimately expected from technology must be distinguished from those that are unrealistic in light of the human condition (Callahan), such as immortality. The second approach, technological humanism, insists to the contrary that the control of death is a desirable goal, and that there is no warrant for depriving people of realizing the ancient longing for immortality (Harrington).

The debate over natural death and the prolongation of life concerns the desirability of certain sets of virtues and alternative views of human nature, especially of freedom. For the humanism that sees death as natural, old age places activism and individualism proper to the other phases of the life cycle in perspective. Trust, sociality, and contemplation reassert themselves over the drives for mastery and aggressive self-fulfillment. The passive virtues are necessary for coping with the limits of the human condition.

For technological humanism, however, restraint and resignation are values only when there are no alternatives. The role of technology is to push back the limits of human life and increase the opportunities for growth and enrichment. In the new world of technological immortality, passive virtues, such as contemplation and moderation, are merely optional matters of life-style. Technological developments give people a choice. They can continue their pursuit of self-fulfillment or not, as they choose. On this view, anti-aging techniques would increase the freedom of individuals to realize their own desires. For this group, then, human nature is defined by freedom as a power of self-fulfillment, a power greatly and rightly enhanced by technology. For neonaturalism, however, humanity is differently defined. It consists in the coincidence of freedom with natural and social limitations (Ricoeur). As a result, individual fulfillment is realized in a moral universe that sets important limits to one's desires and power (Callahan).

DREW CHRISTIANSEN

[Most directly related to this article are its companion articles in this entry: THEORIES OF AGING AND ANTI-AGING TECHNIQUES, SOCIAL IMPLICATIONS OF AGING, and HEALTH CARE AND RESEARCH IN THE AGED. For further discussion of topics mentioned in this article, see the entries: KIDNEY DIALYSIS AND TRANSPLANTATION; LIFE-SUPPORT SYSTEMS; and ORGAN TRANSPLANTATION. Other relevant material may be found under DEATH; HEALTH AND DISEASE; JUSTICE; and RATIONING OF MEDICAL TREATMENT. See also: DEATH AND DYING: EUTHANASIA AND SUSTAINING LIFE; POPULATION ETHICS: ELEMENTS OF THE FIELD, article on THE POPULATION PROBLEM IN DEMOGRAPHIC PERSPECTIVE; POVERTY AND HEALTH, article on POVERTY AND HEALTH IN THE UNITED STATES; RISK; and SUICIDE.]

BIBLIOGRAPHY

ARROW, KENNETH J. "Government Decision Making and the Preciousness of Life." With commentaries by Guido Calabresi and Edmund D. Pellegrino. Tancredi, *Ethics of Health Care*, pp. 33–64.

BACON, FRANCIS. *Historia Naturalis: Historia Vitae et Mortis* (1623). "Historia Vitae et Mortis." *The Works of Francis Bacon*. 15 vols. Edited by James Speddings, Robert Leslie Ellis, and Douglas Denon Heath. Vol. 2: *Philosophical Works, Vol. II*. London: Longman & Co., 1887–1901, pp. 89–226. Translation. "The History of Life and Death, or the Second Title in Natural and Experimental History for the Foundation of Philosophy: Being the Third Part of the Instauratio Magna." *The Works of Francis Bacon*, Vol. 5: *Translations of the Philosophical Works, Vol. II*, pp. 215–335.

BLAU, ZENA SMITH. *Old Age in a Changing Society*. New York: New Viewpoints, 1973.

CALLAHAN, DANIEL. *The Tyranny of Survival*. New York: Macmillan Co., 1973.

CALLAHAN, SIDNEY, and CHRISTIANSEN, DREW. "Ideal Old Age." *Soundings* 57 (1974): 1–16. Special Issue: Leisure, Retirement and Aging.

CHRISTIANSEN, DREW. "Dignity in Aging." *Hastings Center Report* 4, no. 1 (1974), pp. 6–8.

CICERO, MARCUS TULLIUS. *De Senectute, De Amicitia, De Divinatione*. Translated by William Armistead Falconer. Loeb Classical Library. Cambridge: Harvard University Press, 1938, pp. 1–99.

COWGILL, DONALD O., and HOLMES, LOWELL D., eds. *Aging and Modernization*. New York: Appleton-Century-Crofts, 1972.

CUMMING, ELAINE, and HENRY, WILLIAM E. *Growing Old: The Process of Disengagement*. New York: Basic Books, 1961.

DEMOS, JOHN. *A Little Commonwealth: Family Life in Plymouth Colony*. New York: Oxford University Press, 1970.

ERIKSON, ERIK H. "The Life Cycle: Epigenesis of Identity." *Identity: Youth and Crisis*. New York: W. W. Norton & Co., 1968, chap. 3, pp. 91–141.

———. "Reflections on Dr. Borg's Life Cycle." *Daedalus* 105, no. 2 (1976), pp. 1–28.

HARKSEN, SIBYLLE. *Women in the Middle Ages*. Translated by Marianne Herzfeld. The Image of Woman. New York: Abner Schram, Universe Books, 1975.

HARRINGTON, ALAN. *The Immortalist: An Approach to the Engineering of Man's Divinity*. New York: Random House, 1969. Paperback ed. New York: Avon, Discus Books, 1970.

HAVIGHURST, ROBERT J., and ALBRECHT, RUTH. *Older People*. New York: Longmans, Green, 1953.

HERODOTUS. *Herodotus.* Translation by A. D. Godley. Loeb Classical Library. New York: Putnam, 1931, vol. 1, bk. 1, nos. 30–32, pp. 32–41.
ILLICH, IVAN D. *Medical Nemesis: The Expropriation of Health.* London: Calder & Boyars, 1975.
———. "The Political Uses of Natural Death." *Hastings Center Studies* 2, no. 1 (1974), pp. 3–20.
KÜBLER-ROSS, ELIZABETH, ed. *Death: The Final Stage of Growth.* Human Development Books: A Series in Applied Behavioral Science. Englewood Cliffs, N.J.: Prentice-Hall, 1975.
KAPLAN, JEROME. "In Search of Policies for Care of the Aged." With commentary by Sissela Bok. Tancredi, *The Ethics of Health Care*, pp. 281–313.
MARKSON, ELIZABETH WARREN. "Readjustment to Time in Old Age: A Life Cycle Approach." *Psychiatry* 36 (1973): 37–48.
Monumenta de viduis, diaconissis virginibusque tractantia. Edited by Josephine Mayer. Florilegium Patristicum tam veteris quam medii aevi auctores complectens, vol. 42. Bonn: Peter Hanstein, 1937.
RICE, DOROTHY P., and COOPER, BARBARA S. "The Economic Value of Human Life." *American Journal of Public Health and the Nation's Health* 57 (1967): 1954–1966.
RICOEUR, PAUL. *Freedom and Nature: The Voluntary and the Involuntary.* Translated, with an introduction, by Erazim V. Kohák. Evanston, Ill.: Northwestern University Press, 1966.
ROSENFELD, ALBERT. *Prolongevity: A Report on the Revolutionary Scientific Discoveries Now Being Made about Aging and Dying, and Their Promise of an Extended Lifespan, without Old Age.* A Borzoi Book. New York: Alfred A. Knopf, 1976.
SACHS, HANNELORE. *The Renaissance Woman.* Edited by Bernhardus Geyer and Johannes Zellinger. Translated by Marianne Herzfeld. Revised by D. T. Rice. The Image of Woman. New York: McGraw-Hill, 1971.
SEGERBERG, OSBORN, JR. *The Immortality Factor.* New York: Dutton, 1974.
SHANAS, ETHEL, and STREIB, GORDON F., eds. *Social Structure and the Family: Generational Relations.* Englewood Cliffs, N.J.: Prentice-Hall, 1965.
———; TOWNSEND, PETER; WEDDERBURN, DOROTHY; FRIIS, HENNING; MILHØJ, POUL; and STEHOUWER, JAN. *Old People in Three Industrial Societies.* New York: Atherton Press, 1968.
TANCREDI, LAURENCE R., ed. *Ethics of Health Care.* Conference on Health Care and Changing Values, Institute of Medicine, 1973. Washington: Academy of Sciences, 1974.

IV
HEALTH CARE AND RESEARCH IN THE AGED

This article will deal with (1) the provision of health care for the aged, with special consideration of medical and hospital services, nursing home care, and mental health care in the United States, and (2) experimentation involving the aged.

Health care for the aged

Medical and hospital services. The majority of older Americans below the age of seventy-five are independent. Among those not in this majority, and among those seventy-five years of age and older, intractable problems of ill health are prevalent, straining to the limit the resources of the health-care delivery system, and exposing its deficiencies. In the United States the sick aged (bedfast and housebound) number more than three million, or between twelve and fourteen percent of those over sixty-five (Anderson, pp. 89 f.). Elderly patients remain in hospital twice as long and require hospitalization twice as often as those under sixty-five. They account for one-fourth of the nation's health expenditures, are the prime users of long-term care facilities, and consume twenty-five percent of all drugs (Butler, p. 174). Unfortunately, the emphasis in American society has been on "high medical technology" designed to meet acute, episodic needs, rather than on adequate care for the chronically ill. Yet, as one commentator observes, "while the aged have need for acute medical care, their major requirement is in the continuum of services for the chronically disabled that will enable them to function optimally" (Brody, p. 414).

Our ability to respond to this principal requirement by managing the chronic ailments associated with aging is disastrously poor. This is evident in a widespread apathy toward the old within the medical profession. House-call services are rare. Emergency room treatment is typically depersonalizing. Doctors' fees are beyond the means of many (Butler, pp. 183 f.). Medicare, with its array of deductibles, coinsurances, and rising premiums, meets less than forty percent of the costs of health care (Corman, pp. 83 ff.). As the 1976 report by the Congressional Subcommittee on Health and Long-term Care of the Select Committee on Aging points out, this situation is exacerbated by two underlying factors. One is that "both government and carriers of health insurance accept as given a tightly defined medical model as the premise for defining benefits and payments" (*New Perspectives*, p. 2). This medical model "has severe limitations which hamper its application in designing benefits for chronic illness, geriatrics and long-term care services" (ibid.). The other factor is that the scope of and eligibility for health-care services are normatively designated in terms of levels. "The levels of care needed by the users and the level of services provided by classes of providers not only have complicated and limited the provision of patient care, but have also created an administratively cumbersome mechanism for monitoring the quality and quantity of services" (ibid.).

Together, those two factors contribute to the following problems: delayed entry into the health-care system because of constraints on early detection and treatment; lack of incentives for optimum use of health resources and cost-efficient use of home health services; increased dependency upon institutionalization for persons whose needs are for maintenance and rehabilitation; perpetuation of minimum standards for providers, which discourage the provision of individualized care; overemphasis on quantitative factors (such as size of physical facilities) rather than a qualitative concern; and the neglect of the needs of a sizable patient population whose requirement for care does not conform to the artificial structure of the reimbursement systems. In practice this means, inter alia, that the elderly encounter obstacles when they require hospitalization. Acute brain syndromes are frequently overlooked in emergency rooms; chronic brain syndromes are commonly dismissed as "senility." Older patients are treated symptomatically and released, or transferred from voluntary to overburdened and underfinanced community hospitals. Nor have Medicare and Medicaid succeeded in eradicating a double standard in medicine, whereby younger and less impecunious patients receive better hospital care than those both aged and poor.

This generally dismal situation calls for a review of methods of medical and hospital practice, fee schedules, reimbursement formulae, and incentive grants, and of the universal applicability of the medical model. The moral imperative is to develop a more adequate system of medical and hospital services to the aged. The ethical principles of respect for persons and distributive justice might serve as criteria for doing this (Jonsen, pp. 97 ff., 100 ff.). These norms underlie many of the recommendations made by the 1976 congressional subcommittee in the report, among the more significant and far-reaching of which are the following: a system of community long-term care centers to coordinate the provision of health services for older Americans, allowing to the elderly high-quality care in the setting of their choice; a revision of the congressional committee structure to provide better coordination of legislation affecting the aged; and legislation to provide for the gathering and disseminating of information concerning the various public and private agencies delivering home health care and correlative services to the elderly (*New Perspectives*).

Nursing homes. Nonprofit and government homes, providing residential and personal care rather than nursing services, accommodate twenty percent of those older persons who are in institutions. The remaining eighty percent are in commercial or proprietary nursing homes, which now constitute a $9 billion industry (ibid., p. x). Despite exceptions, it is in this latter category that gross abuses occur in the wake of a conflict between profit and service. This has led one commentator necessarily to conclude that "a nursing home is a facility that has few or no nurses and can hardly qualify as a home" (Butler, p. 263). Another investigator and her collaborators document Butler's view in devastating terms (Townsend). Of the persons residing in nursing homes, seventy percent are female; fifty percent either have no living relatives or contact with even distant relatives; the average age is seventy-eight; ninety-six percent are white—because of the shorter life expectancy of minorities and widespread discrimination against them; sixty to eighty percent are poor; eighty-five percent die in nursing homes; almost all have more than one physical ailment; about thirty-two percent have either serious hearing defects or visual disabilities (Butler, p. 267). Considering that in 1969 there were approximately 815,000 residents sixty-five years of age and over in 18,000 nursing and personal care homes (Anderson, p. 91), and that those figures have now increased appreciably, one can see that victimization of the elderly is occurring on a massive scale. Federal programs have not only contributed to making nursing homes profitable to the owners; they have also failed to enforce with any rigor the comprehensive national standards. There is a dearth of audits of nursing homes across the nation (*New Perspectives*, p. 29). Many government surveys, such as the Office of Nursing Home Affairs' 1975 Facility Improvement Survey (United States), and many of the state fiscal audits of nursing homes are confidential, ensuring the anonymity of the institutions being surveyed. In many cases this policy does not serve the best interests of the public: Disclosure of information "may encourage action that will result in correction of deficiencies and improvement of life safety conditions for residents" (*New Perspectives*, p. 31).

The use of nursing homes as extended-care facilities providing skilled nursing has been hampered by the arbitrary, inflexible, and inconsistent nature of social security guidelines in the United States, and by the fact that the most obdurate conditions do not fall into the "acceptable" category or, if they do, do so for a limited time only. This has led one author to assert that

"Medicare essentially ends when long-term helplessness and need for surveillance starts" (Anderson, p. 90). Aggravating this bleak state of affairs is the propensity of associations of homes for the aged to seek to perpetuate the status quo (Butler, pp. 285 ff.).

The profound deficiencies in nursing homes and Medicare-approved homes thwart the ideal of adequate medical care for the elderly. Reforms are long overdue. Several eminently reasonable proposals have been made in this regard (ibid., pp. 295–299). A massive enlistment of public support for radical change is both warranted and wanting. One of the more acute moral questions concerns the propriety of placing home health services, a logical alternative to nursing homes, under the jurisdiction of the Office of Nursing Home Affairs. Since "the nursing home industry and the home health industry are, at least theoretically, competing for the same market" (*New Perspectives*, p. 8), present policy seems clearly to involve a substantial conflict of interest.

Mental health care. Given the conditions described briefly above, and similar formidable difficulties encountered in respect to housing, transportation, the cost of living, family and societal segregation and consequent isolation, and employment, and consequent upon the deterioration of physical functions, it is not surprising that those sixty-five and over are the most susceptible to mental illness, with males the most prone to committing suicide (ibid., p. 37).

There is a disproportionately high number of elderly in mental hospitals: the total population over 65 in mental hospitals ranges from 28 to 40 percent although the elderly comprise only about 10 percent of the total population. And while there are only about 5 percent of the total population in all forms of institutions at any one point in time, almost 1 in 4 elderly persons will be institutionalized at some point in their later years [ibid., p. 34].

In spite of this situation, substantial increases of funds, personnel, and facilities for matching the mental needs of the aged have not been forthcoming.

Psychiatry as a whole "shares a sense of futility and therapeutic nihilism about old age" (Butler, p. 231). The label "senility" is often indiscriminately applied as a means of evading thorough diagnosis and treatment. Drugs are often used to pacify, control, and treat symptomatically aged persons with remediable disorders (ibid., pp. 235 f.). Federally funded community health centers provide deficient care because of budgetary constraints and a lack of trained staff. Private mental hospitals, affording the best care, are beyond the means of most of the elderly. State and county hospitals thus become their chief resource, but most are dubiously effective in respect to basic services, trained psychiatrists, and their environment. Transfer out of these institutions to boarding homes is often precipitous and more in the interest of cutting costs than of benefiting mental patients (ibid., pp. 225–259). On the other hand, "it is clear that many elderly persons are in mental hospitals, not because of severe mental impairment, but because there is no other place for them to go" (*New Perspectives*, p. 34). The President's Task Force on Aging in 1970 expressed concern about the use of state mental hospitals as custodial facilities for the elderly who are not in need of psychiatric care simply because of the absence of alternative living arrangements and psychological support.

Ethical considerations. This unexaggerated delineation of shortcomings in the provision of medical and hospital services, nursing homes, and mental health care for the aged gives weight to the moral arguments advanced by Jonsen and others. Not only is the form of care in the United States, being predominantly acute rather than chronic, unsuited to the elderly; it also compromises their individuality and diversity. Instead of respecting and enhancing their autonomy, it contributes to their deterioration and dependency and, in so doing, compromises their constitutional right to liberty. In addition, it raises sharply the question of allocating limited resources. The principle of distributive justice, of which John Rawls is an outstanding contemporary exponent, gives grounds for contesting productivity as a preferred societal value, for challenging our cultural equation of age with obsolescence and ordering our priorities accordingly, for claiming that the aged represent the acme of our human vulnerability to physical and psychological insult, and for urging that we should not invoke "lifeboat ethics" until we have exhausted other, less desperate alternatives. But distributive justice is a philosophical concept beyond the immediate horizons of those engaged in politics. "It is largely impotent unless it attracts the force of constituencies' interests and social urgency" (Jonsen, p. 103). Therefore, it is not merely hortatory but strategically necessary for Jonsen to pose the moral question: "Who will raise the cry of injustice that the sick aged cannot utter? How can their injustices outweigh the competing injustices of other more vocal groups, not to say the competing claims of the already powerful?" (Ibid., p. 103.)

Experimentation involving the aged

Not only is research into the processes and experiences of aging requisite as a means to providing better care, but it promises benefits to society as a whole, all of whose members are ineluctably growing older. In order to extend and improve the life of the general population, as well as of those now aged, we need to know more about the multifarious ramifications of aging. At present, disproportionately small amounts of time and money are spent on such work. Much research can and is being done in vitro and with animal models. Moreover, many gerontological studies do not place subjects at risk. However, because it is not possible simply to extrapolate from animal or in vitro findings to human beings, and because geriatric research involves risks of harm as well as potential benefits, experimentation with elderly persons is fraught with moral dilemmas. Of these, three of the more important deserve mention.

Therapeutic and nontherapeutic research involving the aged. The Declaration of Helsinki, adopted in June 1964 by the World Medical Association and endorsed by the major American investigative and medical associations, makes a clear and crucial distinction between therapeutic research, in which investigation occurs coincidentally with the provision of care, and nontherapeutic research, in which the element of care for the individual is absent. The full text of the Declaration will be found in the Appendix to this encyclopedia. Sections II and III are relevant to this discussion.

I have already alluded to the need for investigation with this segment of the total population, but what of its ethical propriety? Therapeutic research raises fewer objections in this regard than experimentation of a nontherapeutic nature. Provided that the potential benefits of the therapy are greater than or equal to those of alternative modalities of treatment, and provided that the informed consent of the patient-subject is obtained, there is little to impugn the moral acceptability of therapeutic research. However, nontherapeutic research with the elderly as subjects is a more controversial matter. Such research involves an inescapable conflict of values: the potential benefits for aged persons, present and future, as a group versus respect for the dignity, security, and well-being of the individual.

The aged as a captive population. Should the aged be regarded as a controlled or "captive" population and, as such, be exempt from research, or at least receive special consideration? A case might be made for outright exemption. Most aged persons are deprived, not only economically, but because of social, cultural, administrative, and political exigencies as well; they are thus more controlled than the general population. The elderly are institutionalized more than any other group and, as "captives," are vulnerable to studies of questionable ethicality. Moreover, those suffering from diagnosed organic brain syndromes might be regarded as among the mentally incompetent, commonly excluded from research. These factors, however, seem to warrant special consideration for the aged as research subjects, rather than exemption. To exclude all because some have an impaired capacity to comprehend or by virtue of their external circumstances cannot make autonomous decisions would be to stereotype and degrade many. Recognizing that the aged constitute a heterogeneous group should cause us instead to urge extreme caution in their involvement in research. Jonas's principle of a descending order of permissibility ought to be applied with particular care: "The poorer in knowledge, motivation, and freedom of decision . . . the more sparingly and indeed reluctantly should the reservoir be used" (Jonas, p. 20).

Informed consent and the aged. If it is discriminatory to exempt all elderly persons from research because some are incapable of comprehending information about the risks and benefits of, and alternatives to, contemplated studies, or incapable of consenting freely to participate, the onus should be on investigators to distinguish between those who are competent and uncoerced and those who are not, and to apply Jonas's principle with discretion and restraint. This presupposes that researchers will establish genuine interpersonal relationships with those to be involved in experimentation, in which their abilities or disabilities to comprehend and consent can be assessed and in which, if they are judged competent, their collaboration can be enlisted without constraint. Consideration should also be given to the establishment of proxy or surrogate consent mechanisms as a means for providing additional protection to elderly research subjects.

Conclusion

While it is true that the aged have special needs, it is not clear that special ethical principles are needed to respond to the moral dilemmas raised by health care and biomedical research in the aged. Biomedical policies and practices affecting the aged should be scrutinized in light of an ethic of caring, of distributive justice, and of respect for the autonomy of the individual in a context of reasonable advances in scientific knowledge.

ERNLÉ W. D. YOUNG

[In addition to the foregoing three articles on aging in this entry, the following are directly related. HEALTH CARE; CHRONIC CARE; and POVERTY AND HEALTH. See also HOSPITALS; HUMAN EXPERIMENTATION, articles on BASIC ISSUES, PHILOSOPHICAL ASPECTS, and SOCIAL AND PROFESSIONAL CONTROL; INFORMED CONSENT IN HUMAN RESEARCH; INSTITUTIONALIZATION; JUSTICE; and PATIENTS' RIGHTS MOVEMENT.]

BIBLIOGRAPHY

ANDERSON, ODIN W. "Reflections on the Sick Aged and the Helping Systems." Neugarten, Social Policy, pp. 89–96.

BERNSTEIN, JOEL E. "Medical Experimentation in the Elderly." Journal of the American Geriatrics Society 23 (1975): 327–329.

BEAUVOIR, SIMONE DE. The Coming of Age. Translated by Patrick O'Brian. New York: Warner Paperback Library, 1973.

BRODY, STANLEY J. "Comprehensive Health Care for the Elderly: An Analysis." Gerontologist 13 (1973): 412–418.

BUTLER, ROBERT N. Why Survive? Being Old in America. New York: Harper & Row, 1975.

Commission on Chronic Illness. Chronic Illness in the United States. 4 vols. Vol. 2: Care of the Long-Term Patient. Cambridge: Commonwealth Fund, Harvard University Press, 1956.

CORMAN, JAMES C. "Health Services for the Elderly." Neugarten, Social Policy, pp. 81–88.

COWDRY, EDMUND VINCENT, and STEINBERG, FRANZ U. The Care of the Geriatric Patient. 4th ed. St. Louis: Mosby Co., 1971.

Group for the Advancement of Psychiatry, Committee on Aging. "The Aged and Community Mental Health: A Guide to Program Development." Group for the Advancement of Psychiatry. Report 8 (1971): 1–96. Report no. 81.

——. "Toward a Public Policy on Mental Health Care of the Elderly." Group for the Advancement of Psychiatry. Report 7 (1970): 651–700. Report no. 79.

JONAS, HANS. "Philosophical Reflections on Experimenting with Human Subjects." Experimentation with Human Subjects. Edited by Paul A. Freund. New York: George Braziller, 1969, pp. 1–31.

JONSEN, ALBERT R. "Principles for an Ethics of Health Services." Neugarten, Social Policy, pp. 97–104.

KATZ, JAY, ed. Experimentation with Human Beings: The Authority of the Investigator, Subject, Professions, and State in the Human Experimentation Process. New York: Russell Sage Foundation, 1972.

LASAGNA, LOUIS. The Conflict of Interest between Physician as Therapist and as Experimenter. Philadelphia: Society for Health and Human Values, 1975. Originally prepared for the Committee on Human Experimentation of the Society for Health and Human Values, 1971.

MENDELSON, MARY ADELAIDE. Tender Loving Greed: How the Incredibly Lucrative Nursing Home "Industry" Is Exploiting America's Old People and Defrauding Us All. New York: Alfred A. Knopf, 1974.

National Academy of Sciences. Experiments and Research with Humans: Values in Conflict. Academy Forum, 3d, 1975. Washington: 1975.

NEUGARTEN, BERNICE L., and HAVIGHURST, ROBERT J., eds. Social Policy, Social Ethics, and the Aging Society. NSF/RA 76-000947 Washington: Government Printing Office, 1976. Report prepared by Committee on Human Development, University of Chicago, for the National Science Foundation, RANN-Research Applications Directorate.

New Perspectives in Health Care for Older Americans: Recommendations and Policy Directions of the Subcommittee on Health and Long-term Care: Report, Together with Additional and Supplemental Views. Subcommittee on Health and Long-term Care, Select Committee on Aging, House of Representatives. 94th Cong., 2d sess. Washington: Government Printing Office, 1976.

Older Americans Act of 1965. Pub. L. No. 89–73. 79 Stat. 218 (1965).

President's Task Force on the Aging. Toward a Brighter Future for the Elderly: A Report. Washington: Government Printing Office, 1970.

RAWLS, JOHN. A Theory of Justice. Cambridge: Belknap Press, Harvard University Press, 1971.

SLOATE, NATHAN. "Old Age." A Concise Handbook of Community Psychiatry and Community Mental Health. Edited by Leopold Bellak. New York: Grune & Stratton, 1974, pp. 91–104.

TOWNSEND, CLAIRE, project director. Old Age: The Last Segregation. New York: Grossman Publishers, 1971.

United States, Office of Nursing Home Affairs. Long-Term Care Facility Improvement Study: Introductory Report. DHEW Publication no. (OS) 76-50021. Washington: Government Printing Office, 1975.

WOODRUFF, DIANA S., and BIRREN, JAMES E., eds. Aging: Scientific Perspectives and Social Issues. New York: D. Van Nostrand Co., 1975. Project of the Ethel Percy Andrus Gerontology Center, University of Southern California.

ALCOHOL, USE OF

Three behaviors with alcoholic beverages pose ethical issues: drinking, drunkenness (getting drunk), and alcoholism. Numerous secondary ethical issues derive from attitudes toward and responses to these behaviors by individuals and by agencies of society.

Drinking

Many people see drinking itself as an ethical issue. The classical antialcohol (temperance) movement perceived that biological as well as social harms result from intake of alcohol and drew its moral conclusions from that fact. It held, first, that harm to the self was inherently wrong, and second, that even those who drank most moderately, so that self-harm could not be presumed, were nevertheless practicing an inessential indulgence that could harm others and were thus under an ethical imperative to desist. The harm to others was seen as in the nature of setting an example that the "weak brother" might follow to a harmful degree ("Wesley").

The morality of drinking and the belief that any drinking of alcoholic beverages may be immoral or unethical are subjects still discussed by many. Some old religions and some comparatively new religious developments forbid alcohol: Buddhism, Brahmanism, Islam, and, among Christian groups, the Baptists, Christian Scientists, Mormons, Quakers, Seventh-Day Adventists, and Witnesses. The Methodists only relatively recently modified the doctrine that made drinking a sin. Thus, for multitudes of religious people drinking itself is the ethical issue, and the alcohol industry is tainted with immorality. The ethical problems connected with alcoholism, then, are derivative. In addition, some churches that do not require abstinence advocate it as morally desirable (Cherrington).

Religions and churches were not alone in perceiving a moral issue in drinking. Social reformers, often religiously inspired (for example, Frances Willard, founder of the Women's Christian Temperance Union), often proclaimed the harm and recognized no compensatory virtue in alcohol. In addition, some nonreligious and even antireligious reformers also opposed drinking and the drink trade as social evils, ethically reprehensible. These latter included a portion of the nineteenth- and early twentieth-century socialist and labor leadership. And while in the United States labor generally opposed the imposition of national prohibition, seeing it as representing exploitative interests, in several European countries labor and socialist support helped to impose prohibition or stringent legal controls on the availability of alcohol, which was seen as compounding the workers' economic disadvantagement.

On the other side, against the outlook of the alcohol-rejecting groups, were the views of those who held drink to be either neutral or good—"The Good Creature of God"—to be used, in moderation, for man's benefit. They had reason to think that in moderate drinking there was not only no harm but positive benefit. The possibility of harmful misuse was recognized, but that was true of many things on earth and did not demand abstinence by the majority who did not misuse.

The question whether a primary ethical issue exists apparently hinges on whether any drinking, even in unchallengeable moderation, can be injurious to health. At present the consensus among health and biological authorities is that moderate drinking—here defined arbitrarily as small amounts of alcohol taken occasionally, for example, within the famed Anstie's (1864) limit of 1 1/2 ounces of absolute alcohol per day—causes no demonstrable injury to health. Whether the moderate drinker sets a seductive example, and whether this constitutes an ethical obligation to refrain entirely, are questions not answerable by present scientific knowledge. The problem is not the same as with moderate smoking, where it is argued that the smoker (at least in public places) pollutes the air that others must breathe. But it is a problem that may confront many potentially example-setting behaviors.

Whatever applies to the ethical problems connected with drinking is relevant also to engaging in the drink trade. At a time when alcohol—the substance itself, personified as the Demon Rum—was seen as the enemy of the good life, the Methodist Church in America, for example, doctrinally held traffic in alcohol to be sinful. Yet the Mormon Church, which forbids drinking, holds it no sin to sell alcohol to gentiles who want to drink.

Universal conflict thus exists about the rightness of drinking and of engaging in the drink trade. In the United States some 95 million adults and an additional nineteen million persons aged over thirteen but under twenty-one years are drinkers. Nearly one-third of the adults abstain, however (Keller and Gurioli). In other English-speaking countries, and in the Germanic, Romance, and Slavic lands, even larger majorities are drinkers. Thus it seems that the vast multitudes in these countries recognize no inherent wrong in drinking itself and consequently, it seems reasonable to presume, in the trade.

Drunkenness

Almost universally regarded as wrong is overdrinking—variously expressed as excessive, immoderate, heavy, irresponsible drinking—and, especially, getting drunk. A distinction is here made between drunkenness and alcoholism. People who do not have the disease of alcoholism (as defined below) may on occasion drink enough to achieve an alcoholic "high"—marked by some functional impairment—or even enough to get drunk, for a variety of reasons or purposes.

Although most people apparently regard drinking for the explicit purpose of getting drunk to be wrong on the grounds that it impairs a person's self control and poses a risk to the well-being of others, there are some who believe that even this type of behavior can be justified on occasions. Thus, for example, some condone fiestal drinking, often to drunkenness sometimes lasting several days, practiced by whole communities (Washburne). Such may be the overdrinking required, by consensus, of the young man being initiated into a college fraternity, or expected by the friends of the bridegroom at his "bachelor party." Thus even drunkenness evokes equivocal responses, for in the eyes of some it too has its season.

Despite this benign view condoning occasional drunkenness, it is important to recognize that the intake of alcohol in beverages (as beer, wine, distilled spirits) beyond some safe quantity (or combination of quantity and frequency) can certainly injure the human biological organism. It can cause or contribute to the development of many physical and behavioral or mental disorders. On this ground, excessive drinking (here defined as that which is sufficient to cause drunkenness or other psychic or physical disorder) is condemned by ethicists who believe that self-indulgence which leads to self-injury is wrong. This position is the same as the moral teaching of the religions that do not forbid moderate drinking.

Alcoholism

Drinking and alcoholism. If drinking has been a source of conflict in the realm of ethics—churches and citizens seeing it as both right and wrong—alcoholism has been even more divisive of attitudes.

From ancient times, drinking has been connected with excess and derived troubles (Keller, 1974, pp. 442–443). Drunkenness seems to have occurred early, and for the most part most people did not distinguish between drunkenness and alcoholism as it is understood today—that is, a psychobiological disease (a drug addiction or drug dependency) marked by an inconsistent helplessness of the victim to refrain from re-

sorting to alcohol, and inconsistent helplessness to refrain from drunkenness (Keller and McCormick, p. 12). It was therefore natural to condemn all drunkenness as wrongdoing. Yet an insightful first-century philosopher already made the distinction between drunkenness and addiction (Seneca, Epistle lxxxiii), and a long line of thinkers, including physicians, agree with him to this day.

Alcoholism as a disease. Alcoholism, conceptualized as alcohol addiction (Keller and McCormick, p. 14), or alcohol dependence, or the alcohol-dependence syndrome (Edwards et al., p. 1364), is regarded as a disease by most authorities; it is so classified in the nomenclatures of the authoritative American and international medical organizations (United States; National Conference; American Psychiatric Association; World Health Organization). The disease of alcoholism, so conceptualized, is thus to be distinguished from drinking as such and even from drunkenness. Of course, if drinking as such caused alcoholism, then it would come under the ethical ban of self-injurious behavior. But most authorities regard as simplistic the idea that drinking or alcohol alone is the cause of alcoholism. They hold that, while there is no alcoholism without drinking, the fact that the vast majority of drinkers—perhaps ninety-five percent in the United States (Keller and Gurioli) —do not develop alcoholism indicates that another cause must operate to induce addiction, and that the uncontrollable alcohol ingestion by alcoholics is not the same thing as the normal personal and social behavior of drinking (Bacon).

From the viewpoint of those who demand abstinence (and abolition of the alcohol trade) on account of the weak brother, the fact that most drinkers do not develop alcoholism is irrelevant. No one would develop alcoholism if alcohol were banished utterly. The problem, however, is not so simple. What would happen to the people who develop alcoholism if there were no alcohol? Would they escape trouble and harm? What does happen to such people in societies that do not have alcohol? There is evidence that some people, especially women, by way of "choice of symptoms," develop affective disorders rather than alcoholism (Winokur et al., p. 110), and this is suggested as one explanation of the higher frequency of alcoholism among men in contrast to the higher frequency of affective disorders among women. There are also indications in the medical literature that for some people alcoholism is a disorder that enables them to avert other psychic traumas that are more serious (Bird; Smilde). If, then, the abolition of alcohol would only prevent alcoholism but not alternative illness in susceptible people, the argument for an ethical imperative to abolish alcohol in order to prevent alcoholism loses its force.

The moral issue. If alcoholism is illness, can it still be immoral? Courts have sometimes struggled with the problem of self-inflicted disease—after all, alcoholics voluntarily "put an enemy in their mouths to steal away their brain" (Shakespeare, *Othello*, II, 3, 293) and their good health. A Catholic theologian, John Ford, has put the ethical issue well: The alcoholic may have been at fault, guilty of the sin of gluttony (by drunkenness), when he was still able to choose whether or not to drink, and how much. But once he has become addicted, once he has passed a line that marks him off as diseased and by reason of his disease no longer always has freedom with respect to alcohol—if under the compulsion of his disease he now gets drunk—he is guiltless. In recent years this understanding has been written into many state laws explicitly defining alcoholism as a disease and has been confirmed in decisions of high courts (Ford).

But this view is not accepted universally, not even by physicians. A surviving body of opinion holds that alcoholism is not a disease, an opinion apparently based on the belief that alcoholics are able to control their behavior but fail to make an adequate effort to do so (Todd; Schmidhofer). Some of this opinion appears to be based on misunderstanding of the conception of "loss of control over drinking." If "loss of control" is understood, according to one definition, to mean inability to stop drinking once it is started, then, the argument goes, the alcoholic is blameworthy for taking the first drink. But the more sophisticated conception of "loss of control" emphasizes the fundamental incapacity or disablement of the addict, at times, to choose whether to drink—to resist an overwhelming inner drive (Keller, 1972; Glatt; Ludwig and Wikler), possibly the effect of biodynamic changes in central neurons (Kesner, Priano, and DeWitt; Gross).

At one stage of alcoholism, at least, the behavior of alcoholics in resorting to alcohol, in drinking to and beyond drunkenness, seems surely a consequence as well as a manifestation of disease. At this point the alcoholic seems not to be free in his choices. This occurs in the course of a bout—whether or not its initiation is seen as blameworthy—when the alcoholic, pausing in alcohol intake, experiences the beginning of the alcohol withdrawal syndrome. The symptoms, unrelieved, may become intolerably painful, even threatening to sanity and life. The desperate resort to alcohol then, which may be the only reliable and effective medicine available to the alcoholic, appears to be beyond censure.

Of course, those who recently have challenged all belief in mental or behavioral illness reject utterly a conception of alcoholism as disease (Szasz). They are followed by a new trend in social science and behavioral psychology to reject what is perceived as "the medical model" of behavioral disorders. It is not yet clear where this trend leads with respect to the ethical issues of alcoholism. Those who refuse to grant that alcoholism is a disease would leave alcoholics alone: Certainly they would be against any compulsory treatment. Others, in particular the psychologists who believe that the behavior is environmentally induced, advocate behavioral therapies.

Some ethicists would question the ethics of leaving alcoholics alone. As against the idea that it is wrong to interfere with nonconformist behavior, they question the morality of not helping sick people but leaving them to inevitable disaster. The fate of alcoholics who recover is extremely different from that of those who suffer the severe and fatal consequences of unrelieved alcoholism. Is "leave them alone" a humane attitude?

Problems of response to the alcoholic

Renewal of the historical disagreement over whether alcoholism is a disease has raised questions about the ethical response to alcoholics and other drunken persons by various professions, by numerous social institutions, by related and unrelated individuals. Foremost among the concerned professions is medicine. The accusation has often been leveled that physicians try to avoid treating alcoholics, because alcoholics are difficult persons, unresponsive to sound medical advice (especially "you must stop drinking"), usually incurable, and unreliable in paying their bills. Physicians have been accused of being motivated to assume a moralistic view of alcoholism as an excuse for avoiding alcoholics. Nurses and other health professionals, as well as social workers, have been depicted similarly. To the extent that these accusations may be true of a part of the personnel of the health professions, a serious breach of professional ethics would prevail.

Among the social institutions, hospitals and clinics have been accused of refusing admission to alcoholics, sometimes even when they are acutely ill, if "drunkenness" is obvious. In 1956 the American Medical Association through its officers found it necessary to urge hospitals to admit alcoholics like other sick people ("Reports of Officers"), and subsequently the American Hospital Association echoed this sentiment.

Both public and private social service agencies have at times by policy refused to help alcoholics and families in which the perceived cause of distress was the drunkenness of the man of the house. Most social caseworkers in the field considered this policy to be wrong and frequently countered it by benignly (if unethically) neglecting to record the alcoholism in their reports to the agencies. Even jails, one social institution to which at least lower-class alcoholics are referred with special frequency, have been accused of discrimination against alcoholics by confining them in the infamous "drunk tanks" and failing to provide urgently needed medical care to the severely ill, sometimes with resulting fatalities. Jails and mental hospitals have been accused of seeking alcoholics as inmates so as to exploit them for housekeeping tasks. Clearly failures of normal humane treatment by health and social personnel and institutions are related to beliefs about drunkenness as a form of wrongdoing, deserving only punishment, and ignorance about or disbelief in the reality of alcoholism as a disease compelling the drunkenness of the alcoholic.

The belief that his repetitive ingestion of harmful amounts of alcoholic beverage is willful drunkenness, which the alcoholic could stop by sheer will power, is at the root of the discriminatory treatment of alcoholics. In the eyes of those who believe that alcoholism is a disease that most alcoholics are unable to overcome without therapeutic help, the result is unjust, inhumane, and unethical treatment of the alcoholic. The conflict of beliefs, whether it is the behavior of the alcoholics which is immoral, or the behavior of those who deny them help, remains an open challenge in ethics.

Experimentation and therapeutics

Biological and psychological interest in effects of alcohol on the organism and on behavior, and in the causes and treatments of alcoholic diseases and of alcoholism itself, evokes the same ethical issues with respect to experimentation and therapeutic approaches as interest in other biological–psychological functions and disorders. In addition, alcoholism presents some unique ethical problems.

A serious question seems to arise with informed and free consent, but there is really no difficulty in answering that question. Alcoholics may have a mental illness, one of the special alcoholic disorders, or coincidentally a psychosis such as schizophrenia. In such cases their capacity to decide and consent is presumably impaired. But otherwise, alcoholism is not in the class of psychic illnesses that render a person mentally incompetent, and alcoholics are capable of giving informed consent. In this connection it needs to be borne in mind that they are often found in mental hospitals only because that is where some authorities have found it convenient to situate treatment and rehabilitative programs for alcoholics.

A more serious issue arises when the experiment involves giving people large amounts of alcohol repeatedly and over prolonged periods, with the possibility of inducing addiction. With nonaddicted subjects this would presumably be ethically impermissible, regardless of informed consent, since here it is not a question of risking a curable disorder but of acquiring a catastrophic chronic disease that at present is rarely curable. Such experiments were conducted in the past, the subjects being either alcohol addicts or addicted to other drugs, with resulting induction of withdrawal symptoms (interpreted by the experimenters as indicative of physical addiction). It is unlikely that such an experiment would be undertaken nowadays in human subjects who are not already specifically alcohol addicts. But in the case of alcohol addicts who are going to ingest all the alcohol they can get and hold in any event, experimenters argue that to provide them with the alcohol under ideal, controlled, metabolic-ward conditions, with twenty-four-hour-a-day nursing supervision, medical care, and nutritional protection, is to their advantage rather than harmful.

It had been generally accepted throughout the world that the aim in treating alcoholism was not cure, which would allow normal drinking, but remission through total abstinence. A number of behavioral psychologists have reported experiments, with some claims of success, in treating alcoholic patients so that they could resume moderate drinking. But some have questioned such attempts and claims. They note that success in the relief of alcoholism has been almost exclusively based on the work of Alcoholics Anonymous, a self-help fellowship whose program is firmly rooted in the abstinence ideology, and on treatments by professionals who likewise have helped alcoholics to avoid any drinking. Further, they hold that remission into moderated drinking is classically temporary; only very rarely have reliable cases been reported of alcoholics who returned to normal drinking for several years. The treatment of alcoholics with a controlled-moderate drinking goal has therefore been challenged as both impractical and unethical.

The psychologists who have experimented with behavioral techniques to moderate the uncontrolled alcohol intake of alcoholics believe that the idea of hopeless, permanent incurability of alcoholism is illogical, and that past failure to cure it was due to failure to apply appropriate techniques. They consider it a necessary enterprise, from the viewpoint of public health, and of a majority of alcoholics who cannot achieve permanent total abstinence, or will not try it, to seek ways of achieving controlled drinking by alcoholics, or even moderated drinking that would at least reduce the harm.

MARK KELLER

[*Directly related are the entries* BEHAVIOR CONTROL; DRUG USE; HEALTH AND DISEASE; *and* HEALTH AS AN OBLIGATION. *See also:* BEHAVIORAL THERAPIES; HUMAN EXPERIMENTATION; INFORMED CONSENT IN THE THERAPEUTIC RELATIONSHIP; *and* PUBLIC HEALTH.]

BIBLIOGRAPHY

American Psychiatric Association, Committee on Nomenclature and Statistics. *Diagnostic and Statistical Manual: Mental Disorders with Special Supplement on Plans for Revision.* Washington: 1965. 2d ed. 1968.

ANSTIE, FRANCIS EDMUND. *Stimulants and Narcotics, Their Mutual Relations: With Special Researches on the Action of Alcohol, Aether, and Chloroform on the Vital Organism.* London: Macmillan & Co., 1864. Sets a limit of 1.5 ounces of absolute alcohol per day, which was newly confirmed in United States, Department of Health, Education, and Welfare, Public Health Service, National Institute on Alcohol Abuse and Alcoholism. *Second Special Report to the U.S. Congress on Alcohol and Health from the Secretary of Health, Education, and Welfare: New Knowledge.* Morris E. Chafetz, chairman of the Task Force. Edited by Mark Keller. DHEW Publication no. (ADM) 75-212. Washington: Government Printing Office, 1974.

BACON, SELDEN D. "Alcoholics Do Not Drink." *Annals of the American Academy of Political and Social Science* 315 (1958): 55–64.

BIRD, BRIAN. "One Aspect of Causation in Alcoholism." *Quarterly Journal of Studies on Alcohol* 9 (1949): 532–543. This wise psychiatrist wrote, "If all the help you have to offer an alcoholic is to stop him from drinking, you might as well leave him alone."

BOWMAN, KARL M., and JELLINEK, E. MORTON. "Alcoholic Mental Disorders." *Quarterly Journal of Studies on Alcohol* 2 (1941): 312–390. On "ethical degeneration" in alcoholism these reviewers quote from Eugen Bleuler's *Lehrbuch der Psychiatrie.* 6th ed. Berlin: J. Springer, 1937.

CHERRINGTON, ERNEST HURST, ed. *Standard Encyclopedia of the Alcohol Problem.* 6 vols. Westerville, Ohio: American Issue Publishing Co., 1924–1930. Details the history of the world temperance movement, the tenets of the churches, and the work of individual adherents and reformers.

EDWARDS, GRIFFITH; GROSS, MILTON M.; KELLER, MARK; and MOSER, JOY, eds. "Alcohol-Related Problems in the Disability Perspective: A Summary of the Consensus of the WHO Group of Investigators on Criteria for Identifying and Classifying Disabilities Related to Alcohol Consumption." *Journal of Studies on Alcohol* 37 (1976): 1360–1382.

FORD, JOHN C. *Depth Psychology, Morality, and Alcoholism.* Weston, Mass.: Weston College, 1951.

GLATT, MAX M. "Loss of Control: Extensive Interdisciplinary Borderland, Not a Sharp Pharmacological Borderline." *Alcoholism: A Medical Profile.* Edited by Neil Kessel, Ann Hawker, and Herbert Chalke. London: B. Edsall & Co., 1974, pp. 122–132.

GROSS, MILTON M. "The Psychobiological Contributions to the Alcohol Dependence Syndrome: A Selective Review of Recent Research." *Alcohol-Related Disabilities.* Edited by Griffith Edwards, Milton M. Gross, Mark Keller, Joy Moser, and Robin Room. WHO Offset Publication, no. 32. Geneva: 1977, pp. 107–131.

KELLER, MARK. "Alcohol Consumption." *The New Encyclopaedia Britannica.* 15th ed. 30 vols. Chicago: Encyclopaedia Britannica, 1974, vol. 1, pp. 437–450.

———. "On the Loss-of-Control Phenomenon in Alcoholism." *British Journal of Addictions* 67 (1972): 153–166.

———, and GURIOLI, CAROL. *Statistics on Consumption of Alcohol and on Alcoholism.* New Brunswick, N.J.: Rutgers Center of Alcohol Studies, 1976. Basic numerical data on drinking and alcoholism.

———, and McCORMICK, MAIRI. *A Dictionary of Words about Alcohol.* New Brunswick, N.J.: Rutgers Center of Alcohol Studies, 1968.

KESNER, RAYMOND P.; PRIANO, D. J.; and DeWITT, J. R. "Time-Dependent Disruption of Morphine Tolerance by Electroconvulsive Shock and Frontal Cortical Stimulation." *Science* 194 (1976): 1079–1081.

LUDWIG, ARNOLD M., and WIKLER, ABRAHAM. "'Craving' and Relapse to Drink." *Quarterly Journal of Studies on Alcohol* 35 (1974): 108–130.

National Conference on Medical Nomenclature. *Standard Nomenclature of Disease and Operations.* 5th ed. Edited by Edward T. Thompson. New York: McGraw-Hill, 1961.

"Reports of Officers: Hospitalization of Patients with Alcoholism." *Journal of the American Medical Association* 162 (1956): 750.

SCHMIDHOFER, ERNST. "Alcoholism is NOT a Disease." *Maryland State Medical Journal* 18, no. 3 (1969), pp. 59–64. Example of the contemporary opposition to the concept of alcoholism as disease. See also Todd.

SMILDE, J. "Risks and Unexpected Reactions in Disulfiram Therapy of Alcoholism." *Quarterly Journal of Studies on Alcohol* 24 (1963): 489–494. Reports on alcoholics who, deprived of alcohol by pharmacological intervention and lacking other support, committed suicide.

SZASZ, THOMAS STEPHEN. *The Myth of Mental Illness: Foundations of a Theory of Personal Conduct.* New York: Hoeber-Harper, 1961.

TODD, JOHN EDWARD. *Drunkenness a Vice, Not a Disease.* Hartford, Conn.: Case, Lockwood & Brainard, 1882. Example of the historic, extremely moralistic view See also Schmidhofer.

United States, Public Health Service. *Manual for Coding Causes of Illness According to a Diagnosis Code for Tabulating Morbidity Statistics.* Miscellaneous Publication, no. 32. Washington: Government Printing Office, 1944.

WASHBURNE, CHANDLER. *Primitive Drinking: A Study of the Uses and Functions of Alcohol in Preliterate Societies.* New York: College & University Press, 1961. Describes the survival and contemporary forms of drinking practices of early social groups.

"Wesley, John." Cherrington, *Standard Encyclopedia of the Alcohol Problem*, vol. 6, p. 2816.

WINOKUR, GEORGE; REICH, THEODORE; RIMMER, JOHN; and PITTS, FERRIS N., JR. "Alcoholism. III. Diagnosis and Familial Psychiatric Illness in 259 Alcoholic Probands." *Archives of General Psychiatry* 23 (1970): 104–111.

World Health Organization. *Manual of the International Statistical Classification of Diseases, Injuries, and Causes of Death: Based on the Recommendations of the Eighth Revision Conference, 1965, and Adopted by the Nineteenth World Health Assembly.* 2 vols. Geneva: 1967–1969.

ALLOCATION OF MEDICAL TREATMENT

See RATIONING OF MEDICAL TREATMENT; HEALTH CARE, *articles on* RIGHT TO HEALTH-CARE SERVICES *and* JUSTICE AND HEALTH CARE; JUSTICE.

ALLOCATION OF SCARCE RESOURCES

See RATIONING OF MEDICAL TREATMENT; JUSTICE; KIDNEY DIALYSIS AND TRANSPLANTATION; HEART TRANSPLANTATION.

ALLOWING TO DIE

See DEATH AND DYING: EUTHANASIA AND SUSTAINING LIFE; RIGHT TO REFUSE MEDICAL CARE; ACTING AND REFRAINING.

AMNIOCENTESIS

See PRENATAL DIAGNOSIS.

ANIMAL EXPERIMENTATION

I. HISTORICAL ASPECTS Richard D. French
II. PHILOSOPHICAL PERSPECTIVES Peter Singer

I
HISTORICAL ASPECTS

Man's use of animals for experimental purposes—to understand form, function, and their relationship in the animal and the human body—dates from pre-Christian times. Its nature and frequency have varied with the intellectual predispositions and technical capacities of biomedical scholars and have been strongly affected by the religious, philosophical, and political atmosphere in which those scholars have practiced. This article describes: (1) the rise of animal experimentation as a method in biological and medical research, (2) its institutionalization in the latter part of the nineteenth century, (3) the emergence of opposition to animal experimentation, the modern antivivisection movement, and (4) the current situation in animal experimentation and recent developments in antivivisectionism.

Origins of animal experimentation

The dissection of the dead animal body, incidental to the activities of hunters and cooks since the dawn of human existence, was carried out from motives of curiosity by the Greeks as early as five hundred years before Christ (Singer). The objective was essentially improved understanding of the internal structure of the body. When living animals were first used—probably by the Alexandrian physicians Herophilus and Erisistratus in the third century B.C.—the approach was similarly anatomical and morphological rather than physiological in nature. Death was understood to cause physical changes in internal organs such that observation in vivo rather than by dissection provided a more accurate sense of the shape, texture, color, interconnections, and other characteristics of the organs. From such observations, analogical reasoning based on the similarity between the viscera and everyday objects guided speculation as to the functions of the former.

Our knowledge of the medical researches at Alexandria derives from the works of later writers, especially the Roman physician Galen of Pergamum (A.D. 129–199), who codified the medical writings of the ancients. At least as important, his own researches dominated the biomedical theory that emerged in the rediscovery of the classical authors after the Dark Ages. Galen was a superb operator who dissected a variety of animals. His *De anatomicis administrationibus* [On anatomical procedures] described for the first time the methods and instruments used in specific experiments on living animals. This work recorded his observations on the effect of section at various points of the spinal cord in the living animal—a state of knowledge only modestly increased in the seventeen hundred years that followed.

The Dark Ages saw little development in what we recognize as the Western medical research tradition. Dissection of the human body —for purposes of postmortem and illustration of the classical and Arabic authorities rather than for research per se—seems to have emerged in the Northern Italian universities toward the end of the thirteenth century. By the fifteenth century it had become accepted in the leading medical schools, though hardly common and not without having attracted the attention of the ecclesiastical authorities from time to time.

Andreas Vesalius (1514–1564) created the first modern anatomy by systematic dissection, direct observation, and precise illustration. He experimented on living animals, as did a number of his contemporaries (Schiller, 1967). Francis Bacon (1561–1626), prophet of the scientific revolution, argued the value of animal experiment for scientific knowledge of the human subject in his *De augmentis scientarum* [The advancement of learning] in 1623.

> Wherefore that utility may be considered as well as humanity, the anatomy of the living subject is not to be relinquished altogether . . . since it may be well discharged by the dissection of beasts alive, which, notwithstanding the dissimilitude of their parts to human, may, with the help of a little judgment, sufficiently satisfy this inquiry [Bacon].

The seventeenth century saw an enormous outburst of scientific activity—not least in the biomedical sciences, where new instruments and techniques were developed for experimentation on living animals. William Harvey's epochal demonstration of the circulation of the blood in 1628 was only the most important of a series of discoveries made by a combination of dissection and animal experimentation during the seventeenth and eighteenth centuries (Schiller, 1967). Harvey's example inspired a host of medical men and natural philosophers to imitate his logical exposition and brilliantly conceived and executed experiments. For the first time the practice of experimentation upon living animals impinged on the consciousness of the educated elite in general.

In the 1820s and 1830s extensive animal experiments by François Magendie (1783–1855) finally destroyed the anatomically based notions of structure–function relationships in favor of the concept of a function as the product of several organs. No longer were elaborate physiological systems to be developed from speculation and analogy applied to the structures of the body. Rather, physiological ideas would have to be developed empirically, by systematic operations on living animals and observation of the results (Schiller, 1967, 1973). The introduction of anesthesia in 1847 was at least as important for experimental medicine as it was for surgery, for it made feasible and humane a great many experiments previously impossible because of the pain and trauma they inflicted on the experimental subject (Cranefield).

Claude Bernard's classic, *Introduction à l'étude de la médecine expérimentale*, published in the 1860s, embodied the full technical sophistication and philosophical rationale for the new physiology. During the same mid-century period, the spread of the critical research tradition from biblical and classical studies through other disciplines in the highly competitive German university system affected the medical sciences just as it affected all of the natural sciences. The establishment of new university positions required aspirants to establish competence by virtue of successful experimental research. Laboratories in physiology spun off research traditions in the sciences of pathology, pharmacology, bacteriology, and the like, all of them invigorated by the practice of animal experimentation.

Institutionalization of animal experimentation

The new methodology involved the control of variables affecting a physiological system, isolating one variable, that one to be affected by experimental intervention—be it by ablation or extirpation, pharmaceutic agent, or introduction of an infectious agent—in order to observe the precise impact of the intervention. Propagandists since at least the seventeenth century had been promising great benefits to the clinical practice of medicine from the experimental approach, but it was not until the last decades of the nineteenth century that support among the great bulk of the medical profession and the university and bureaucratic authorities was such that resources were available for experimental medicine, and professional standards began to demand systematic knowledge of it. The newly recognized and reorganized medical professions of Northern Europe and America successively took up the standard of experimental medicine not only because they were convinced of its value but because they could use it as a kind of acid test for professionalism with which to exclude the unorthodox and isolate the older and more traditional members of the profession.

Spectacular medical advances attracted enormous public attention and acclaim. The emergence of the science of immunology in the 1880s was probably the most important single accomplishment to demonstrate in convincing fashion not only the intellectual validity but the practical benefit to the public of the experimental approach. There could be no denying that the discoveries involved were based on experiments on living animals. The consequences for medical practice eventually affected the lives of millions of people and received unequivocal endorsement from the medical establishment and governments. Similar advances in chemotherapy, surgery, and preventive medicine, among many others, stimulated a commitment to experimental medicine in virtually every developed nation by the third decade of the twentieth century.

In one hundred years, experimental research had emerged from the dingy back rooms of clinic and lecture theater and the scarcely read pages of arcane journals to recast the fundamentals of medical practice on the basis of laboratory research with living animals and of new techniques with vastly improved instrumentation. The power and prestige gained for the medical profession by the demonstrable public benefits of these research advances in turn made of it, for good or ill, the prototype of the modern profession in terms of independence and self-governance, state recognition, control over admission, and sole guardianship of its intellectual and political legacy.

Opposition to animal experimentation

Emergence of antivivisection movement. Early discussion of the ethics of vivisection was largely confined to its use on humans, a practice that was occasionally ascribed to the anatomists of antiquity, and more frequently to those of the sixteenth century. At least as far as the Renaissance anatomists were concerned, it is doubtful whether there was much validity to the charge, but they were anxious to avoid even the appearance of experimentation on humans, and this concern may have had some inhibiting effect on the resort to animal experimentation.

Opposition to the use of living animals in experimental procedures paralleled the spread of the practice following William Harvey's demonstration of the circulation of the blood in the early seventeenth century. One hundred years later, the experiments of the Reverend Stephen Hales and others inspired disapproving comments from literati like Alexander Pope, Joseph Addison, and Samuel Johnson. François Magendie and his French successors were likewise attacked on grounds of cruelty, but it was only when the research tradition of experimental medicine was imported from France and Germany to Great Britain—the birthplace of anticruelty movements of all kinds—that an organized movement against experiments on living animals emerged.

Led by Frances Power Cobbe (1822–1904), a skillful propagandist, the British movement created in the mid-1870s substantial public concern over vivisection—a term used to denote all animal experimentation, including the administration of drugs or bacteria where the literal implication of surgical incision is not present. A Royal Commission called by Disraeli's government in 1876 found no specific evidence of abuse by British experimenters but recommended regulation of the practice. After considerable political maneuvering, the Cruelty to Animals Act of 1876 was passed (French). The act, in force in Great Britain to the present day, required the registration of places where experiments on living animals were to take place, the licensure of experimenters, certification for particular kinds of experiments, and meticulous record-keeping and reporting on the part of licensees. British antivivisectionists have never been happy with the act or its administration, and, in part because of their pressure, subsequent government inquiries reported on the issue in 1913 and 1965.

As with the humane movement in general, other countries took up the style and philosophy of the activists who created the British antivivisection agitation. The campaign spread through Northern Europe (Bretschneider), to America (Dennis) and to the other English-speaking countries. Experiments on animals are illegal in Liechtenstein and are regulated in Austria, Denmark, Norway, Sweden, Poland, the Bahamas, Jamaica, Argentina, and parts of Switzerland and Australia (Lapage). Anticruelty statutes exist in most countries, and certain of these also have statutes covering the provenance and care of experimental animals. Historically, the Western anticruelty and antivivisection movements have had greater success in Protestant than in Catholic countries (Passmore; Stevenson, 1956).

The campaign against the use of animals for experiments stimulated the political development of the scientific and medical communities (French) and resulted in the establishment of organizations such as the Association for the Advancement of Medicine by Research and its successor, the Research Defence Society, in Great Britain and the National Society for Medical Research in the United States.

Antivivisection arguments. Antivivisectionists have attacked experimental medicine from two basic lines of argument. First, they have argued from the immorality of the method: In their view, whatever the medical or scientific benefits of scientific experiments, such benefits are too dearly bought at the price of complicity in a brutal and degrading practice. This line of argument has tended to coincide with the view that abolition of all experiments on animals—including painless ones—is the only legitimate objective. This approach has sought intellectual support in religion and philosophy (Passmore; Stevenson, 1956).

A second line of argument has been from the inutility of the results of the method. Antivivisectionists, both medical and lay, have ransacked the technical literature to deny the scientific validity of conclusions drawn, to deny the medical value of knowledge gained, or to contend that methods other than those using living animals, such as dissection or clinical research, would have resulted in the same conclusions. Regular themes in such critiques were that the trauma resulting from experimental incursion into the subject organism invalidated any conclusions that could be drawn, or that species error—arising from the difference between the experimental animal and the human patient—provided an ipso facto negation of any practical clinical benefit. As medical research became increasingly inaccessible to the uninitiated, this style of polemic became relatively rare; still, it has not fallen into complete disuse, especially with respect to the use of animals in experimental psychology. The argument from the inutility of the method has appealed both to total abolitionists and to those who seek more stringent regulation of experiments on animals.

On the whole, medical scientists have preferred to ignore antivivisectionist attacks, except when they manifest themselves in legislative initiatives. A small literature of experimental apologetics does, however, exist (e.g., Lapage; Visscher; White).

Antivivisection is a complex phenomenon: varied in form and impact by time and national context; rent by internal divisions and subject to that classical affliction of voluntary movements, the iron law of oligarchy; motivated at various times by myriad social, political, and psychological forces, from antiscientism to evangelicalism to feminism to simple love of animals. In the end its failure to achieve anything resembling its ultimate objectives would seem to depend upon broad public confidence in the medical profession and in the profession's claim that animal experimentation conduces to human health (National Opinion Research Center).

The current situation

The use of living animals for purposes of research in the biosciences, experimental medicine, and experimental psychology, as well as in medical technology—the production of biological extracts and the testing and standardizing of extracts, of drugs, of consumer products, of clinical samples, of water, and of food—is well entrenched in virtually all developed countries. The subjects used number in the millions per year, though precise statistics are available only for Great Britain. The main constraining factor in the growth of animal experimentation has historically been, and remains, the amount of resources available to the research community. The objective of total abolition of such experiments appears more distant today than when first fully articulated in the nineteenth century.

Two major responses to this situation may be detected among those concerned with the promotion of the anticruelty and antivivisection issues. The first is the attempt by intellectuals to construct a defensible philosophical underpinning for the movement, in the recognition that the attack on the medical and scientific utility of experimental medicine has failed, and that ethics rather than epistemology provides the only route of potential advance for the issue.

The second is the attempt by moderate antivivisectionists to promote the use of tissue culture and other nonvivisectional means of biomedical investigation by funding and publication of research developing or using such methods (Fund for the Replacement of Animals in Medical Experiments).

RICHARD D. FRENCH

[*Directly related is the other article in this entry:* PHILOSOPHICAL PERSPECTIVES. *Other relevant material may be found under* LIFE, *article on* VALUE OF LIFE; *and* RESEARCH, BIOMEDICAL. *For discussion of related ideas, see:* HUMAN EXPERIMENTATION. *See also:* RESEARCH, BEHAVIORAL; *and* RESEARCH POLICY, BIOMEDICAL.]

BIBLIOGRAPHY

BACON, FRANCIS. *De augmentis scientarum* (1623). Translated as "Of the Dignity and Advancement of Learning." *The Works of Francis Bacon.* Edited by James Spedding, Robert Leslie Ellis, and Douglas Denon Heath. Vol. 4: *Translations of the Philosophical Works, Vol. I.* London: Longman & Co., 1860, bk. 4, chap. 2, p. 386. Bacon says much the same thing in an earlier English work *The Twoo Bookes of Francis Bacon of the Proficience and Advancement of Learning: Divine and Humane* (1605). *The Works of Francis Bacon*, vol. 3, bk. 2, p. 374.

BRETSCHNEIDER, HUBERT. *Der Streit um die Vivisektion im 19. Jahrhundert: Verlauf, Argumente, Ergebnisse.* Medizin in Geschichte und Kultur, vol. 2. Stuttgart: G. Fischer, 1962.

CRANEFIELD, PAUL F. *The Way In and the Way Out: François Magendie, Charles Bell and the Roots of the Spinal Nerves: With a Facsimile of Charles Bell's Annotated Copy of His Idea of a New Anatomy of the Brain.* Mount Kisco, N.Y.: Futura, 1974.

DENNIS, CLARENCE. "America's Littlewood Crisis: The Sentimental Threat to Animal Research." *Surgery* 60 (1966): 827–839.

FRENCH, RICHARD D. *Antivivisection and Medical Science in Victorian Society.* Princeton: Princeton University Press, 1975.

Fund for the Replacement of Animals in Medical Experiments (FRAME). *Is the Laboratory Animal Obsolete?* London: 1970.

Great Britain, Home Department, Departmental Committee on Experiments on Animals. *Report of the Departmental Committee on Experiments on Animals Presented to Parliament by the Secretary of State for the Home Department by Command of Her Majesty, April, 1965.* Cmnd. 2641. London: Her Majesty's Stationery Office, 1965.

LAPAGE, GEOFFREY. *Achievement: Some Contributions of Animal Experiment to the Conquest of Disease.* Cambridge: W. Heffer, 1960.

National Opinion Research Center. *Animal Experimentation: A Survey of Information, Interest, and Opinion on the Question among the General Public, High School Teachers and Practicing Physicians.* Report no. 39. Chicago: 1949.

PASSMORE, JOHN. "The Treatment of Animals." *Journal of the History of Ideas* 36 (1975): 195–218.

SCHILLER, JOSEPH. "Claude Bernard and Vivisection." *Journal of the History of Medicine* 22 (1967): 246–260.

———. "The Genesis and Structure of Claude Bernard's Experimental Method." *Foundations of Scientific Method: The Nineteenth Century.* Edited by Ronald N. Gieve and Richard S. Westfall. Bloomington: Indiana University Press, 1973, pp. 133–160.

SINGER, CHARLES JOSEPH. *A Short History of Anatomy from the Greeks to Harvey.* 2d ed. New York: Dover Publications, 1957.

STEVENSON, LLOYD G. "Anatomical Reasoning in Physiological Thought." *The Historical Development of Physiological Thought: A Symposium Held at the State University of New York Downstate Medical Center.* Edited by Chandler McC. Brooks and Paul F. Cranefield. New York: Hafner Publishing Co., 1959, pp. 27–38.

———. "Religious Elements in the Background of the British Anti-Vivisection Movement." *Yale Journal of Biology and Medicine* 29 (1956): 125–157.

———. "Science down the Drain." *Bulletin of the History of Medicine* 29 (1955): 1–26.

VISSCHER, MAURICE B. "Medical Research and Ethics." *Journal of the American Medical Association* 199 (1967): 631–636.

WHITE, ROBERT J. "Antivivisection: The Reluctant Hydra." *American Scholar* 40 (1971): 503–507.

II

PHILOSOPHICAL PERSPECTIVES

Although the practice of conducting scientific experiments on living animals, vivisection, goes back at least to Galen (A.D. 130?–200?), the modern period of experimentation stems from the seventeenth century, when scientific inquiries were beginning to be made in many fields. This was the period of the philosopher and scientist René Descartes (1596–1650). For Descartes and the physiologists who declared themselves his followers, cutting open a fully conscious animal posed no ethical problem, since, Descartes said, animals are mere machines, more complex than clocks but no more capable of feeling pain.

If this convenient view of the nature of nonhuman animals is rejected, however, a serious ethical problem about experimenting on animals does arise, because the infliction of suffering and death on an animal seems, in itself, to be an evil. On the other hand, supporters of vivisection argue that such experiments provide great benefits for humans. Animal experimentation, therefore, raises the issue of whether the end justifies the means (an issue also raised by experiments on humans) and in addition forces us to consider what place nonhuman animals have in our ethical deliberations.

The nature and extent of experiments on animals

The number of experiments performed on animals has increased remarkably in the last hundred years. The present extent of experimentation, worldwide, is impossible to ascertain with accuracy, since few countries compile the necessary statistics. In the United Kingdom, according to an annual statement published by the Home Office, more than five million experiments "calculated to cause pain" are performed on live vertebrate animals every year. This figure is low compared to that of the United States, where the number of animals used yearly has been reliably estimated as being in excess of sixty million (Singer). Other countries using large numbers of animals include Russia, Japan, West Germany, and France. The worldwide tendency of smaller nations to follow Western scientific techniques has meant that there are now very few nations in which animal experiments are not being performed.

The animals most often used are mice and rats, but dogs, cats, and monkeys are also used in large numbers. It is commonly assumed that animals are experimented upon only for important medical research, but closer scrutiny reveals that only a minority of experiments can be classified as "medical" at all (Ryder). Many of the most painful experiments are carried out by psychologists and are intended to test theories about learning, punishment, maternal deprivation, and so on. Millions of animals are used to test foodstuffs, pesticides, industrial products, weapons, and even nonessential items like cosmetics, shampoos, and food-coloring agents.

Many of these experiments involve severe and lasting pain for the animals. To test the safety of a foodstuff or cosmetic, the substance is fed in concentrated doses to a group of animals, until a level is found at which half of the sample dies. This means that most of the animals become very sick before some die and others pull through. The dose at which half of the sample dies is supposed to give an indication of the toxicity of the substance, but, since different species have different tolerances, it is at best a very rough guide to the safety of the product for humans (thalidomide, for instance, was tested on several species of animals before being released to humans, and no deformities were found) (Ryder).

Psychologists' experiments about punishment or learning may involve hundreds of severe and inescapable electric shocks. Some psychologists have made monkeys permanently neurotic by rearing them in total isolation (these experiments have been repeated, with minor variations, over and over again). In experiments on stress, monkeys have been locked into iron chairs for more than a year and made to perform tasks in order to avoid electric shock. To study the effects of heatstroke, medical researchers have slowly heated fully conscious dogs to death. Further examples, drawn from recent scientific journals, may be found in Ryder and in Singer.

On the other hand, it is true that many experiments involve little or no suffering for the animals involved. This may be because the experiment is of a harmless nature (such as running a rat through a maze) or because the animal is totally anesthetized during the operation and killed painlessly afterward.

Moreover, some animal experimentation has been of considerable benefit to humans. In such areas as the identification of necessary vitamins and minerals and the development of new surgical techniques and new drugs, discoveries have been made through animal experimentation that could not have been made, or would have been much more difficult to make, had animals not been used. Diabetes is often cited as an example of a disease that has lost its terror through a cure first developed on animals ("Vivisection-Vivistudy").

Legislation

Laws governing experiments on animals vary from country to country, but in no country does the law prohibit outright painful experiments or require that the experiment be of sufficient importance to outweigh the pain inflicted.

The first law specifically regulating experiments was the British Cruelty to Animals Act of 1876. This law, which has never been amended, requires the use of an anesthetic, except where "insensibility cannot be produced without necessarily frustrating the object of such experiments"; but neither the statute itself nor the officials who administer it make any attempt to assess whether the object of the experiment is itself worth the pain caused. Clearly, a psychologist wishing to test the effects of electric shock on the behavior of dogs cannot anesthetize the animals without frustrating the object of his experiment. Even in toxicity tests of cosmetics anesthetics are not used, because it is thought that they might distort the result of the test. Therefore such experiments are permitted in Britain.

In the United States, the Animal Welfare Act of 1970 sets standards for the housing, transportation, and handling of animals; but the Act does not control the nature of the experiments performed, except to the extent of requiring research facilities to lodge a report stating that, when painful experiments were performed without the use of pain-relieving drugs, this was necessary to achieve the objectives of the research project. Again, no attempt is made to assess the importance of these objectives, and in fact one section of the law specifically disavows any intention of interfering with the design or performance of research or experimentation. Moreover, since the law is a federal one, facilities not receiving federal funds and not involved in interstate commerce do not have to comply.

A 1972 West German law requires the use of alternatives to experiments on animals whenever possible. Amendments to the same effect have been proposed in Britain, Denmark, and Holland, but it remains to be seen how effective the West German provision will be. Much depends on what is deemed to be a "possible alternative." Many countries, including France, Spain, Brazil, and Japan still have no legislation regulating experiments on animals.

The case for experimenting on animals

The simplest argument for the permissibility of experiments on animals is the Cartesian one: animals do not suffer, and so there is nothing wrong with experimenting upon them. But both common sense and the great majority of experts agree that mammals and probably other vertebrate animals, at least, are capable of suffering both physical pain and some kinds of emotional distress, such as fear.

If animals do suffer, how is their suffering to be justified? The usual justification offered is that the suffering of animals is outweighed by the benefits, for humans, of the discoveries made by the use of animals. Sometimes, however, it is said that the goal of increasing our understanding of the universe is sufficient justification.

Behind these justifications may lie one of a variety of philosophical positions. For instance, it may be said that, as related in Genesis, God has given man "dominion" over the other animals to use as man pleases. Combined with other theological notions, such as the idea that man, alone of all animals, has an immortal soul, this idea has been influential throughout the Christian world. But it can also be turned the other way: as long ago as 1713, Alexander Pope argued against cruel experiments on the grounds that man's dominion requires him to play the role of the good shepherd, caring for his flock (Turner, p. 48).

It has also been said, by writers as diverse as St. Thomas Aquinas, Immanuel Kant, and D. G. Ritchie, that animals are not "ends in themselves" or that they have no rights (Passmore; Regan and Singer). In support of this it is alleged that the status of a being that is an "end in itself," or has rights, belongs only to a being that is rational, capable of autonomous action, or a moral agent. Whereas humans satisfy this requirement, animals, it is alleged, do not. A difficulty with this argument is that *some* humans come no closer to satisfying the requirement than the animals experimented upon. Mentally retarded human beings, for instance, may be no more rational than a dog; yet we do not consider that they are entirely devoid of rights. Other writers have denied that rationality, autonomy, or moral agency is required before we can grant that a being has rights (Feinberg). Still others have taken the approach that we have obligations not to inflict suffering, irrespective of whether we can meaningfully attribute rights to the being in question (Singer).

A utilitarian case for animal experimentation is based on the idea that more suffering is alleviated by it than is caused by it. The classical utilitarian writers, however, all accepted that a utilitarian must take *all* suffering—human and animal—into consideration, and this makes the factual claim that animal experimentation relieves more suffering than it causes more difficult to defend. Nevertheless, there are probably some experiments—those that do not involve much suffering for the animals and promise major benefits for humans or animals—that can be defended on this ground.

Finally, defenders of experimentation often accuse their opponents of inconsistency in objecting to the deaths of animals in laboratories while continuing to participate in the practice of rearing and killing animals for food. This argument holds no validity for antivivisectionists who are also vegetarians. Even when directed against someone who does eat meat, it hardly amounts to a positive defense of experimentation.

The case against experimenting on animals

Opponents of experiments on animals tend to divide into two groups: absolute abolitionists and reformers. Absolute abolitionists usually rely on the principle that the end does not justify the means. To inflict pain and death on an innocent being is, they maintain, always wrong. They point out that we do not think the possibility of advancing scientific knowledge justifies us in taking healthy human beings and inflicting painful deaths upon them; similarly, they say, the infliction of suffering on animals cannot be justified by reference to future benefits for human or other animals.

The weakness of the absolutist position is that, when the end is sufficiently important, we do sometimes think otherwise unacceptable means are justifiable if there is no other way of achieving the end. We do not like invasions of privacy, but we countenance telephone taps on suspected criminals. Similarly, if the prospects of finding a cure for cancer depended upon a single experiment, should we have any doubt about its justification?

Reformers usually take a more utilitarian line. They concede that some experiments may be justifiable but contend that most are not, because the experiments bring certain suffering and death to animals with no likelihood of significant benefits. In reply to the general argument that experiments on animals benefit humans, the reformers demand that any such benefits be sufficient to offset the costs to the animal subjects; they urge that every experiment come under close prior scrutiny to determine if the benefits are likely to outweigh the costs. Were this done, they maintain, only a small fraction of the experiments now performed would be seen to be justifiable.

Reformers claim (like Ryder and Singer) that alternative methods, not involving animals, could replace many of the experiments now being carried out on animals. Techniques using tissue cultures, for instance, have already replaced animals in the production of certain vaccines, and opponents of animal experimentation suggest that other alternative methods would be developed more rapidly if they were to receive government support (Ryder, chap. 12).

Although the absolutists and reformers disagree in important respects, they are united in seeking to narrow the ethical gulf that now separates humans from other animals in our conventional morality. This may be the most philosophically interesting question raised by the vivisection controversy, and it has implications that go beyond experimentation to our treatment of animals in general.

The moral status of animals

Is there any ethical justification for the sharp distinction we now make between our treatment of members of our own species and members of other species? Although it is commonly said that humans are superior to other animals in various respects, or that humans are "persons," while animals are not, both Ryder and Singer have pointed out that certain categories of human beings—infants and mentally retarded humans—actually fall below some adult dogs, cats, pigs, or chimpanzees on any test of intelligence, awareness, self-consciousness, moral personality, capacity to communicate, or any other capacity that might be thought to mark humans as superior to other animals. Yet we do not think it legitimate to experiment on these less fortunate humans in the ways in which we experiment on animals. Ryder and Singer claim that our respect for the interests of these humans, and our neglect of the interests of members of other species with equal or superior capacities, is mere "speciesism"—a prejudice in favor of "our own kind" that is analogous to, and no more justifiable than, racism.

Certainly it does seem that those supporters of experimentation who have cited the benefits the experiments may bring to humans would need to explain whether this argument also justifies experiments on mentally retarded humans, and, if not, why not. The fact that a being is not a member of our own species does not, in itself, seem to be a sufficient reason for experimenting upon it, if we would refuse to perform a similar experiment upon a member of our own species with similar potentialities.

Defenders of vivisection have had surprisingly little to say about their reasons for disregarding or discounting the interests of nonhuman animals. R. J. White, who has himself carried out experiments in which the heads of monkeys are kept alive and conscious after being severed from their bodies, is perhaps representative of many experimenters when he writes that "the inclusion of lower animals in our ethical system is philosophically meaningless" (White, p. 507). But White does not explain why the clear proposal of utilitarian writers from Jeremy Bentham onward that pain, as such, is an evil, whatever the species of the being that suffers it, is devoid of meaning. It may sometimes be difficult to compare the suffering of a human and, say, a dog; but if rough comparisons can be made, surely the mere fact that the dog is a "lower animal" is no reason to give less weight to its suffering.

Seen in this light, the argument that restricting experiments on animals interferes with scientific freedom and medical progress also appears less conclusive. We do not grant scientists the freedom to experiment at will on humans, although such experiments might advance medical knowledge. It would seem, therefore, to be incumbent upon the defenders of experiments on animals to show that there is a relevant difference between humans and other animals that justifies experiments on the latter, but not the former; but this is a question to which the experimenters have not addressed themselves.

Conclusion

While there has been considerable controversy over the ethics of experiments on humans, there has been little serious discussion of the morality of the far more numerous experiments on animals in recent years. Antivivisectionists have, by and large, been regarded as oversentimental animal lovers or as eccentrics. In the first half of the 1970s, however, three books have appeared containing criticisms of animal experimentation based on carefully reasoned ethical considerations. It would not be surprising if these books provoke a more serious consideration of the entire issue.

PETER SINGER

[*For further discussion of topics mentioned in this article, see the entries:* PAIN AND SUFFERING; PERSON; *and* RIGHTS. *Also directly related are* LIFE, *article on* VALUE OF LIFE; *and* RESEARCH, BIOMEDICAL. *For discussion of related ideas see* HUMAN EXPERIMENTATION. *See also* APPENDIX, SECTION IV, AMERICAN PSYCHOLOGICAL ASSOCIATION.]

BIBLIOGRAPHY

FEINBERG, JOEL. "The Rights of Animals and Unborn Generations." *Philosophy and Environmental Crisis.* Edited by William T. Blackstone. Athens: University of Georgia Press, 1974. Argues that animals can possess rights.

GALTON, LAWRENCE. "Pain is Cruel, But Disease is Cruel Too." *New York Times Magazine,* 26 February 1967, pp. 30–31. A sympathetic view of animal experimentation.

GODLOVITCH, STANLEY; GODLOVITCH, ROSLIND; and HARRIS, JOHN, eds. *Animals, Men and Morals.* New York: Taplinger, 1972. A collection of essays by philosophers and others on our treatment of animals. Two essays by Richard Ryder and Brigid Brophy are concerned specifically with experimentation.

PASSMORE, JOHN. "The Treatment of Animals." *Journal of the History of Ideas* 36 (1975): 195–218. Treats the historical ethical issues in cruelty to animals.

REGAN, THOMAS, and SINGER, PETER, eds. *Animal Rights and Human Obligations.* Englewood Cliffs, N.J.: Prentice-Hall, 1976. An anthology of writings, ancient and modern, on ethical aspects of our relations with animals.

Report of the Departmental Committee on Experiments on Animals. Sir Sydney Littlewood, Chairman. London: Home Office, Her Majesty's Stationery Office, 1965. The result of an official inquiry into animal experiments in Britain; popularly known as the "Littlewood Report," after the committee chairman.

RYDER, RICHARD DUDLEY. *Victims of Science: The Use of Animals in Research.* London: Davis-Poynter, 1975. The most comprehensive account yet published of experiments on animals, the legislation governing experiments, and the case for reform, with principal focus on Britain.

SINGER, PETER. *Animal Liberation.* New York: Random House, New York Review of Books, 1975. Argues for a radical revision of our attitudes to animals; contains a long chapter discussing experiments, including descriptions of experiments in America and Britain.

TURNER, ERNEST S. *All Heaven in a Rage.* London: Michael Joseph, 1964. A historical account of the growth of compassion for animals, with sections on the beginning of vivisection and the antivivisection movement.

"Vivisection–Vivistudy: The Facts and the Benefits to Animal and Human Health." *American Journal of Public Health* 57 (1967): 1597–1626. In four separate papers presented at a symposium, medical and veterinary specialists discuss some of the benefits of animal experimentation for human and animal health.

VYVYAN, JOHN. *In Pity and in Anger.* London: Michael Joseph, 1969. A historical account of experiments on animals and the opposition to such experimentation.

———. *The Dark Face of Science.* London: Michael Joseph, 1971. A continuation of the above.

WESTACOTT, E. *A Century of Vivisection and Anti-Vivisection.* Ashingdon, England: C. W. Daniel & Co., 1949. Account of vivisection controversy from the beginnings to the time of publication, including a full account of the evidence presented by both sides to the British Royal Commission on Vivisection of 1875 and 1906.

WHITE, ROBERT J. "Anti-Vivisection: The Reluctant Hydra." *American Scholar* 40 (1971): 503–512. One of the few articles in which an experimenter defends his right to use animals.

ANTHROPOLOGY OF MEDICINE
See MEDICINE, ANTHROPOLOGY OF.

ANTIVIVISECTION
See ANIMAL EXPERIMENTATION.

ARAB NATIONS
See MEDICAL ETHICS, HISTORY OF, *section on* NEAR AND MIDDLE EAST AND AFRICA, *articles on*

CONTEMPORARY ARAB WORLD *and* CONTEMPORARY MUSLIM PERSPECTIVE. *See also* ISLAM; POPULATION ETHICS: RELIGIOUS TRADITIONS, *article on* ISLAMIC PERSPECTIVES.

ARTIFICIAL INSEMINATION
See REPRODUCTIVE TECHNOLOGIES, *articles on* ARTIFICIAL INSEMINATION, ETHICAL ISSUES, *and* LEGAL ASPECTS.

ARTIFICIAL ORGANS
See LIFE-SUPPORT SYSTEMS; ORGAN TRANSPLANTATION; KIDNEY DIALYSIS AND TRANSPLANTATION.

ASEXUAL HUMAN REPRODUCTION
See REPRODUCTIVE TECHNOLOGIES, *articles on* ASEXUAL HUMAN REPRODUCTION, ETHICAL ISSUES, *and* LEGAL ASPECTS.

ASIA
See MEDICAL ETHICS, HISTORY OF, *sections on* NEAR AND MIDDLE EAST AND AFRICA *and* SOUTH AND EAST ASIA.

AUSTRIA
See MEDICAL ETHICS, HISTORY OF, *section on* EUROPE AND THE AMERICAS, *article on* WESTERN EUROPE IN THE TWENTIETH CENTURY.

AUTOPSY
See CADAVERS.

B

BEHAVIOR CONTROL

This encyclopedia includes three entries on behavior. The first, BEHAVIOR CONTROL, *is divided into two articles:* ETHICAL ANALYSIS, *which gives a general definition of behavior control, and* FREEDOM AND BEHAVIOR CONTROL, *which describes the ways in which behavior control both enhances and threatens different kinds of freedom. This entry is related to, but distinct from, the entry on* BEHAVIORAL THERAPIES, *which raises the ethical issues created by the practices of behavioristic therapies, such as the freedom of the therapist–client relationship and questions concerning the mechanistic approach to training and curing. The final entry,* BEHAVIORISM, *offers a set of theories that are relevant for understanding the conceptual background of the other two entries. It accomplishes this in two articles—one that traces the history of the various schools of behavioral psychology and the issues of ethical interest under the title* HISTORY OF BEHAVIORAL PSYCHOLOGY, *and the other,* PHILOSOPHICAL ANALYSIS, *which discusses the philosophical and ethical implications of behavioristic schools of thought.*

I. ETHICAL ANALYSIS   *Robert Neville*
II. FREEDOM AND BEHAVIOR CONTROL
                                           Joel Feinberg

I
ETHICAL ANALYSIS

Behavior control in its most general sense is "the ability to get someone to do one's bidding" (London, p. 3). Ethical issues emerge concerning behavior control when questions are asked about just who has the ability to control, whose bidding is served, and whose behavior is controlled. What values should shape the controlled behavior? Who determines and approves these values? What is the relation between valuable behavior and freedom?

Innovations in behavior control

Two general classes of innovations give behavior control its currency as a special problem in bioethics. First and most obvious is the invention of technologies for control. For example, behavior can be intentionally modified by direct manipulation of the brain, as in psychosurgery, electrical stimulation of the brain (ESB), or the infusion of chemicals through cannulas or hollow tubes leading to specific organs of the brain. Other behavioral effects can be obtained by indirectly manipulating the brain through drugs. Although the brain mediates all behavior, some new technologies such as those associated with behaviorism focus instead on the manipulation of environmental stimuli and reinforcements. Yet other new technologies, for instance those employed by various forms of dynamic psychotherapies, manipulate the affective and cognitive symbols structuring behavior. Even the technologies that have no new hardware, such as hypnosis and psychoanalysis, have new techniques, e.g., the diagnostic use of word association and the interpretation of dreams and slips, that systematically do what was usually only coincidentally done before the nineteenth century.

The second class of innovations augmenting behavior control consists of theoretical advances in the understanding of human behavior. As quite ordinary processes of behavior are understood theoretically, it becomes possible to control them more. For instance, as Perry London has shown, understanding how people deal with information makes it possible partially to control their responses by manipulating the information (London, chaps. 3 and 4).

There may be nothing new in the methods of controlling information, but there is much that is new in the use of those methods of control. Furthermore, new media of information, such as television, which were developed for reasons having little to do with behavior control, can now be used for that purpose with some effect (or so the advertising agencies tell their clients). Another example of control made possible by new knowledge is that in "total institutions." Erving Goffman has described some of the ways by which total environments such as prisons or mental hospitals determine behavior simply by their institutional structures, over and above specific therapies within the institution. Now that these are somewhat understood, the institutions can be deliberately arranged so as to produce effects intentionally. Early childhood education affords yet another example of an area where increased knowledge leads to greater possibilities of control. Observation of such unconscious communication between parent and child as eye contact or method of touch has led to increased understanding of the conditions for enhanced alertness and capacity for affection. Obstetric wards and play schools can now be ordered to certain behavioral goals.

For the most part the new technologies and the new uses of scientific knowledge for behavior control have been developed for benevolent purposes. As is usual with technologies, however, they can also be used malevolently, and they can cause great harm even with the best of intentions.

But there is an even greater dimension to the problem. Every society civilizes to some degree the powers functioning within it; that is, socially effective powers are put in contexts that define their acceptability, that set limits to their use, and that provide countervailing forces where the powers transcend proper limits. The older forms of behavior control, such as they are, have been more or less civilized; most societies have moral customs as well as legal traditions that civilize the power of parents over children, the use of weapons, and the application of the criminal sanction, to name only a few traditional behavior controls. But many of the newer forms of behavior control have not been civilized. At best, people reason from uncertain analogies, for instance when control of television news is considered as a First Amendment problem of free speech. The civilizing structures that ordered practitioners when their powers were not very effective may not do the same when their powers are great. Physicians treating the mind, for example, were civilized by the Hippocratic Oath when they could in fact do little and merely avoided doing harm; but more is needed now, when tranquilizers and electroshock are at their disposal. The general problem of behavior control is to understand its forms, assess its cumulative effects, and civilize its powers.

Dimensions of behavior control

Spelling out some elementary value dimensions of situations in which behavior control is possible is a first step toward understanding the importance of behavior control. Ethical judgments must balance the values involved in these many situations.

Of the many aspects of behavior that can be controlled overt public behavior comes first to mind. But there are also more internal senses of behavior. Thinking is a kind of behavior and it can be controlled sometimes through the manipulation of information. Emotion and affect are also inner sorts of behavior and are subject to control by drugs, psychoanalysis, or good or bad news. Perhaps even more inward are moods such as depression or feeling "high"; these are similar to but not identical with moods such as alertness, dullness, and feelings of vitality that are closely allied with physiological states and that also can be technologically modified.

Clearly one of the most morally weighty problems of behavior control is the theoretical one of conceiving the relations among these senses of behavior. For different technologies affect behavior at different points. Any harmonious full human action rises from the base of a person's mood and affect, involves deliberate thought reflecting emotional valuations, and makes its mark in a public environment in proximate and remote ways. It can be analyzed from inside to outside by terms meaningful within certain physical systems, and it can also be analyzed by terms taking their meaning from the realm of human interests and symbols. In common experience these all ordinarily interact.

Overt behavior reflects inner states and intentions, which in turn are responsive to overt stimuli and rewards. Bad news turns the bowels to water, and an undigested bit of beef is a plausible explanation of Marley's ghost. The ancient philosophical problem of the relation between mind and body is of utmost practical urgency if the connections between the various senses of behavior that might be subject to control are to be conceived coherently.

Behavior control, at least of the sort important to bioethics, is a technological ability. A technology is an organized, socially learnable way by which people can attain an effect. Behavior can be determined in part by factors other than behavioral technology—for instance, by conscious free choice, according to some thinkers, or by a chance concatenation of events. The determination of behavior by a technology requires a *method*, an organized way of attaining the behavioral effect. Hardware is a common connotation of technology in the twentieth century, particularly instruments using electricity. Psychosurgery, electroshock, and television are instances of hardware behavioral technologies. But many new technologies of behavior control are ways of organizing the preexisting social and physical environment, such as the talk in psychoanalysis or the manipulation of a total institution. The novelty in a method without hard tools is not the appearance of a new device but rather a new organization of the determinants of a person's behavior, for instance, conversation with the person so as to bring about the effect with some predictable regularity.

It is sometimes thought that behavior control implies someone who wants to control, that there is someone responsible who is potentially morally responsible. But sometimes no one intends to control, as when the behavioral effects of a drug are unknown or when the determinants of a total institution's structure are too diverse. The ethical dimensions of behavior control, however, arise because someone's behavior is controlled, not because someone actually controls it. If the control is technological, then perhaps someone wields the technology and in this sense, of course, is a controller. But even if the controller does not intend what he does, the control is a moral problem because of its effect upon the controlled party. Perhaps because of that effect the control should itself be brought under control or undertaken with greater deliberate effort. Even when deliberate, the use of a behavior control technology in a given case is not necessarily moral; the deliberation may involve bad judgment, as when it mistakes the cost or serves bad purposes, or the use may be in willful contradiction to the results of deliberation, as when the controller knows he is doing wrong.

The use of a technology, while intentional, may be *unconscious*. One does not have to be a Freudian to accept the notion that people systematically pursue intentions of which they are not aware, or of which their awareness is the peculiar sort they themselves cannot admit. Hardly anyone consciously intends to drive another person crazy, for instance, yet many act so as to do so, and in methodical ways such as establishing pathological dependencies or creating double binds. Unconscious employment of behavior control technology is not always a bad thing; parents, for example, should shape their children's behavior unconsciously as well as consciously, for otherwise all parties would go mad.

Whether the use of a technology is intentional does not of itself, then, determine either morality or responsibility. The morality is determined by a cumulative assessment of the worth of all the effects and of the legitimacy of the agent employing controls relative to the party controlled. Responsibility is determined by an even more complex assessment of causal determinations, of authority, and of rights. Nevertheless, the intentionality of a use of a behavior control technology is frequently an important factor in composing a picture of the relevant factors to be assessed.

Controlled and free behavior

The first impression most people have of the ethical dimension of behavior control is that it conflicts with freedom. On second thought, this is seen to be an oversimplification. Some forms of control provide greater freedom, particularly those aiding self-control; others, such as some of the forms of social control, aid in establishing contexts where greater freedom is possible. Behavior control may, however, run into direct conflict with freedom when it takes a technological form that may enslave the controlled person to the will of others; this is discussed below in connection with the role-systems of behavior control. It may also contradict freedom in a subtler sense, namely, when it leads to behaviors or experiences that themselves are not free, that exclude freer ways of behaving and experiencing. The controlled person who lives in a vegetative state or acts like an automaton, or even who is led into an extreme dependency on the controller, is made unfree by the control technology, however many other benefits may be achieved. This distinction between controlled and free behavior is our concern here.

Controlled behavior differs from free behavior, according to many thinkers, in that the latter is motivated at some relevant stage by the appeal of the behavior itself. The general line of argument is that unfree behavior is performed without the person's motives being relevant to the worth of the behavior; either he is deceived about its worth, is motivated by someone's threat, or is acting mechanically without a motive. Of course, some thinkers object to the concept of free behavior. They believe that all behavior is learned in a sense that allows in principle for the learning process to be controlled. Whether or not responding to the appeal of something is a form of learning in the requisite sense is a complicated question not to be dealt with here. But it is possible to describe various senses in which the appeal of behaviors can be responded to, senses that constitute aspects of what some, though not all, thinkers call free behavior.

At one extreme is a rather unreflective, immediate appeal, such as a baby's grab at something bright, new, and moving. Immediate appeal frequently determines such behavior as eye movements, body posture, scratching, laughing, and, in general, one's comportment to the aesthetic aspects of the physical and social environment; the train of associations in one's inner psychological environment may also at times approach the immediate extreme.

Moving away from this extreme, there is a class of behaviors whose appeal includes the fact that they would satisfy intentions. That is, the behaviors are viewed symbolically in reference to the agent's purposes and intentions. It makes sense to say of these behaviors that people engage in them because they "mean" such and so to the people. Of course, there is a wide range of symbolic functions in the appeal of projected behavior, from the most rudimentary to the very complicated, from the confused and inarticulate to the highly refined. Perhaps even the unreflective and immediate behaviors discussed previously have a symbolic element, although that is not the important part of their appeal.

Yet another complicating dimension to the appeal of behaviors, over and above their symbolic promise of fulfilling intentions, may be their worth as self-chosen. That is, some aspect of their appeal may consist in the fact the chooser constitutes himself to be the chooser of the behaviors in the acts of choosing. For this dimension of appeal to be real and not merely illusion, as some people believe, it must be possible for the chooser to have chosen something else; his choice could not have been completely necessitated by antecedent factors, however influential they may have been. Whatever other aspects of appeal it has, a freely chosen behavior has a special appeal bestowed upon it by the chooser in adopting it as his choice. Whether this is real, of course, is the old problem of free choice, and some people believe this kind of appeal is necessarily illusory.

A further dimension of behavioral appeal can roughly be called "good reasons." This is the appeal that deliberation uncovers in a prospective behavior. Again, the forms of this appeal vary from the rudimentary to the sophisticated. Some behaviors have appeal stemming from a crude felt comparative evaluation with alternatives. Others derive from careful moral analysis, reflective of ethical theory and of scientific study of the contours of causation. Sophisticated deliberation includes evaluation of the behavior in reference to the identity of the performer as well as in reference to its particular world context.

More complex yet are the dimensions of appeal that stem from the performer's developing character. An action that may or may not be appealing in its own right has an added appeal when it expresses the ongoing character of the actor as a person. This dimension is connected with the fact that the element of the environment most frequently and seriously affected by a person's behavior is his own character. Related to this is the appeal a behavior has resulting from its being a continuation of the work of the agent in the past, whatever its own particular merits. Insofar as an action connects with a person's personality and past work, it has a special appeal because of its place in his career.

The far end of the spectrum of appeal is reached in those behaviors that are part of the tissue of creative life, where aesthetic responsiveness, intention, choice, deliberation, and connection with self and world have been internalized to the behavioral process. A creative person seems to come full circle to immediacy again with actions that seem to others and himself to be as spontaneous as the baby's reach yet to incorporate all the disciplines of maturity.

Behavior is itself free and perhaps responsible, many people believe, when it arises out of one or several of the senses of appeal mentioned above (and perhaps other senses). A person behaves freely upon presentation of the relevant appeal if his body and character have sufficient conditions to make that response possible. The appeal alone does not cause the behavior, but only in conjunction with the state of the person. The state of a person's actual conditions, however, may not allow him to behave according to the relevant appeal. This discordance of appeal and response provides much of the ambiguous ethical quality of behavior control technology.

On one hand, behavior control is inimical to free behavior when it affects the body or character so that there are insufficient conditions for response. For instance, drugs or surgery can stupify, psychotherapy can create enslaving dependencies, information control can mislead people about the true worth of things. On the other hand, behavior control is among the means people employ to make themselves free to respond to what appeals to them. A person drinks coffee (for its caffeine) to become alert enough to do his work when the appeal of the work is insufficient to overcome his groggy mind. A person goes to a psychologist or psychiatrist for behavioral help in overcoming his actual conditions which prevent him from doing what appeals to him.

Most of the biological technologies of behavioral control have been developed to give people more freedom, particularly psychological freedom. Some of these have effects that themselves make people's behavior less free. The ethical question is whether the benefits deriving from the use of those technologies outweigh the deficits, case by case. Freedom is not the only advantage to be gained by behavior control, nor the only loss that might be suffered.

The role-systems of behavior control

The technologies of behavior control do not merely happen to people. They are employed in the context of various social systems, and serious ethical questions arise from the fact that those systems distribute the power people have over each other and themselves.

Who desires a behavior to be controlled? Who exercises the control? Over whom? In whose interests? By what right? There are three broad classes of role systems in regard to motive: (1) self-control or control of oneself for one's own interest; (2) social control or control of individuals by a group of which they are members for the good of the group; and (3) other-control or the control of one party (a person or group) by another party in the interest of the latter. In all these role-systems, technical experts such as physicians, penologists, or advertising agencies may be employed as agents of the various parties; confusions may arise as to just whose agent the expert is. The following discussion of the three main roles intends only to point out controversial areas of value questions. Readers are referred to separate articles on particular means of behavior control for problems concerning their functions in the various roles.

Self-control. Uses of technology to control oneself seem to be the simplest to evaluate. Assuming everything is aboveboard and the technologies are employed competently, it would seem that this use of behavior control is in fact an enabler or extension of a person's own freedom. The behavior control technician represents the person to be controlled, acting where the person cannot act for himself. But the issue is more complex.

Consider behavior control of self aimed principally at proximate pleasure, for instance, through the use of alcohol or psychedelic drugs. Does this pursuit of pleasure detract from a person's social responsibilities? How is pleasure-seeking to be regarded as a part of human life? What are the costs of the pleasure to the person? To others? In the long or short run? Whose concern is it to see that these questions are answered? Are there rights of society to limit a person's pursuit of pleasure? What are the rights of neighbors in this regard? These questions have a special twist when the pleasure is sought through technological means that the society sanctions.

Most self-control behavior technology aims not at immediate pleasure but at making possible various activities; the activities may be pleasurable or worthwhile in themselves or they may be instrumentally valuable. Some behavior control seeks to enhance activities beyond their normal range; amphetamines, psychoanalysis, self-realization therapies, and the methods of athletic trainers do this. Special value questions here have to do with the luxury of spending scarce resources on enhancement, particularly those which in medicine are developed and earmarked for therapeutic purposes, that is, for bringing people up to normal. (Psychiatrists who practice psychoanalysis on essentially normal people are frequently criticized in this regard.)

There are many ways in which people can be "less than normal," but relative to behavior control they fall into three principal classes.

Bad behavior. The first class includes those people whose below-normal behavior is "bad," and who engage in it out of their own choice and responsibility, i.e., the *immoral* people. (Some thinkers argue that if determinism is true there are no people in this class, and that all behavior is either pathological or the result of ill-education.) Although behavior control techniques can be employed to "normalize" bad behavior, the roles do not make this a case of self-control because the self chooses something different; rather it is a matter of *social control* (to be treated in the next section). The second and third classes of deviation downward from the normal are pathology and ill-education respectively.

Pathology. Behavior control is called "therapy" when the behavior it alters is thought of as pathological. Although the therapist is acting as the patient's agent in the therapeutic relationship, he frequently takes a larger role in defining the patient's interest than agents would in other self-control situations. In therapy a patient is supposed to approve the appealing normal behavior and employ the therapist as his agent; the roles are those of the self control model. But if the patient does *not* want to get well, he is either viewed as immoral—wherein the sanctions of social control are employed to make him take his medicine (there are legal procedures for setting aside the requirements of informed consent)—or as "sick in the head" so that the diseased organ is the one that would have to give consent. It is crucial to note that in the last case, where the disease includes not wanting to get well (in severe depression, for example) the therapeutic relationship is reinterpreted. Either it is shifted from the self-control system to that of social control, where the person is treated for the good of the group, or the concept of a proxy is used to provide an analogy with self-control. Some of the moral problems of involuntary commitment and treatment arise because therapists are thought to carry many powers that would be justified in a straightforward self-control role-system (because the patient himself wants them) over into a social control system in which those powers are not justified, or into a mere analogy with self-control.

Protection of personal integrity in a self-control system rests on some form of consent; in the medical therapeutic situation this has been codified in some countries into laws of informed consent. But consent is what cannot be given if the patient's below-normal character includes not wanting to be normal. If forcibly treated under the role-system of social control, then his integrity must be protected by the due process justifying actions within that role-system, principally those laws legitimating authorities that act with the authority of the society; these laws are not easily applied to therapeutic situations.

Ill-education. It is very difficult to define the difference between pathology and ill-education, but the difference is crucial to the authority of medical psychotherapists. By ill-education is meant a range of below-normal states running from bad socialization, deficient early experience, and lack of normal skills to lack of normal character development and lack of normal knowledge. In contrast to a pathology, for which the patient is not held responsible, a matter of ill-education is more complex regarding responsibility. An infant or young child is not held responsible for its education; as it matures it is held more and more responsible, until finally there is responsibility for having the behavioral abilities normal for the adult group. Furthermore, in many cases an adult is held responsible for remedial work to correct a deficiency for which someone else was responsible when he was a child.

Educators who wield behavioral technologies have a different kind of authority from that usually accorded therapists. Because most early education is not planned but simply results from the sociophysical environment, its authority is noticed only in the breach—for instance, when someone suggests a radical alteration in the family structure. Where authority is deliberately vested in an educator for planned education at an early stage in a person's life, for instance, in a schoolteacher or a government regulating the paideic (*paideia*—a harmonious physical, mental, and cultural development) effects of a community, it is done so usually on a social control model, e.g., "children must be educated because democracy requires an educated citizenry." But the justification of democracy's intent to legitimize education as a social demand is that people should assume control over their own education within the limits of certain social demands. As children grow, educators cease to derive their authority from the social control system and begin to derive it as agents of the student in a self-control system.

Both mental health experts and educators may distort the self-control model in yet another way. By virtue of their expertise they help the person define what his goals are, in accord with which he will be controlled. In varying degrees, psychology has defined mental health for our culture, and therapists make this specific for their patients. (Some clinical psychologists deal with clients only on a contract basis in which the client sets the goals; but even in these rare cases the behavioral technology shapes the goals it is supposed to serve.) Similarly in education, the educator cannot tell the student what he will learn in advance without giving away the content. In both the mental health field and in education, the therapeutic agent not only performs the will of the controlled person, but helps shape it. Self-control is partially a myth as a model for these situations, although the usual protection from abuse—consent—derives from that model.

Besides questions of efficiency and the right of a person to determine his own behavior (and get help in behaving that way), the basic value questions in the self-control role-system are the extent to which the behavior agent should set the goals for the sake of which the behavior is controlled, and the extent to which his authority to alter behavior derives from the person controlled or from the society.

Social control. The social control model is that in which an individual or group is controlled in the alleged interest of the larger society. Insofar as the controlled person is a member of the society he has interests that are served by the very process that controls his more personal interests; so the professional thief has his own property protected by the laws that make his profession a difficult one. But the society would not allege that its control is for the benefit in particular of the controlled person; it is not therapy or education.

The general moral problems connected with social control are, of course, those of politics. What rights does a society have as against its members, and vice versa? How does a society authorize particular agents to express and serve its interests? What are the procedures for challenge by dissidents? What are the protections of due process?

Medicine, of reasonable necessity, is sometimes practiced in institutions designed for or greatly facilitating social control, for instance, prisons, the military, or mental hospitals. Over and above the problem of misplaced therapeutic authority discussed above, there are other problems regarding medicine in these institutions.

One problem is the abuse of the therapeutic ideal by smuggling in the goals of the institution as if they were therapeutic. For instance, a prison or mental hospital may have a special interest in inmate docility, and treatments of disease may be chosen for their special sedative side effects. A soldier in the military may be given short-range treatments that return him quickly to action even though his alternatives are therapeutically far better in the long run. This is not to say that the social interest in docility or a useful fighter is unwarranted; but the decision that it is warranted, and the authority to control the behavior, ought to come through the due process of social control, not merely reflecting the private social views of the treating physician or institution.

A second problem is that some of the new medical technologies of behavior control, such as psychotropic drugs, psychosurgery, electroshock and ESB, as well as behavior conditioning techniques, can be used as adjuncts or alternatives to more traditional technologies of social behavior control such as prisons, fines, and so forth. Given the unhappy state of prisons, for instance, behavior control technologies are frequently suggested as alternatives to incarceration. But penology is confused enough as it is, vacillating among goals of punishment, social protection, and rehabilitation. Psychosurgery or mandatory tranquilization offers similar ambiguities regarding punishment, protection, and rehabilitation, but without the tradition of legal remedies and personal integrity possible (if not likely) in prisons. The concern to civilize these technologies is of utmost importance to the physicians who are required to use them not as healers but as agents of the society.

The education of newborn and older infants and young children is one of the areas where new understanding is making possible new control. Long before the Greek conception of *paideia*, societies understood that babies become people because of social conditions; but only recently has there been much understanding of how and why this works. José Delgado has pointed out that behavior control technologies connected with education now require asking, "What do we want man to be?" Whatever answers are given, the existence of the new technologies, and those in prospect, means that education will not be what it was before, innocent as that was, even if no decision is made by anyone.

Other-control. Other-control is the situation in which one party is controlled by another in the interest of the controlling party. The evil genius hypnotizing politicians or innocent maidens to do his will is a familiar bogeyman. Yet hardly any of the new biomedical technologies involving hardware has added a new dimension to this role, save when an agent in the self-control or social control systems begins to act on his own account.

New psychological knowledge is a different story, however. Motivational research provides sellers with powers to increase people's appetites for their products. Government policy reflects, in part, anticipations of psychological effect. Private citizens try to control one another with techniques learned from psychology. The principal ethical dimension of the new technologies used in this context is duplicity. In the old techniques, for instance bribery, the taker of the bribe, whose behavior was modified accordingly, knew what was going on. Even flattery was an obvious technique to observers. But a person now can be moved to buy a car or elect a President by manipulation of anxieties he does not even know he has.

The development of new powers with which one party can control another should not lead to the assumption that such control is always wrong. People generally approve certain kinds of such control in defense of property, for instance, or one's living environment, or in the management of certain social contexts. Western societies generally have legitimated certain forms of other-control by one or both of two protections of the controlled individual. One, derivative from the self-control system, suggests that other-control is legitimate if the controlled person consents to a situation in which he knows he might be controlled—let the buyer beware. The other, derivative from the social control systems, legitimates other-control if both the controlled and the controlling parties occupy social roles that the society approves for the control, for instance children and parents in a family.

Right of access

To say that behavior control provides new powers that must be civilized states only half the problem, and it casts an ambience of danger over the control enterprise. The other half of the problem is to provide access to the new powers for those who might value or profit from them. If a "happiness pill" were invented with more acceptable side effects than heroin or LSD, why should the unhappy be deprived of it? If there are treatments for psychopathology, how can the sick get them? If there are means to educate, why should anyone be deprived? Of course nearly all the issues discussed in previous paragraphs constitute partial answers to those "Why" questions. There is nevertheless a powerful argument that the scruples about moral civilization that have characterized much of the behavior control debate (and are represented above) in fact are the expression of a favored intellectual and social class that does not need behavior control. There are people who do need it.

Why is it that the public discussion so frequently tips the scales in favor of considering dangers of technology rather than its promises? Some would argue that it is because there is a natural momentum to the development of technology and that moral intervention most frequently takes the form of a brake. Others would argue that human life is naturally untechnological and that new technologies must prove themselves before wide employment is permissible. Whatever the reasons, clearly a formulation of rights of access to new technologies is an essential counterbalance to rights of protection from their abuse.

ROBERT NEVILLE

[*Directly related to this article is the companion article on* FREEDOM AND BEHAVIOR CONTROL. *This article will find application in the entries* BEHAVIORAL THERAPIES; DRUG USE; DYNAMIC THERAPIES; ELECTRICAL STIMULATION OF THE BRAIN; ELECTROCONVULSIVE THERAPY; HYPNOSIS; PSYCHOSURGERY; PSYCHOPHARMACOLOGY; *and* SELF-REALIZATION THERAPIES. *For discussion of related ideas see the entries:* ALCOHOL, USE OF; BEHAVIORISM; GENETIC ASPECTS OF HUMAN BEHAVIOR; INSTITUTIONALIZATION; MENTAL HEALTH SERVICES, *article on* SOCIAL INSTITUTIONS OF MENTAL HEALTH; POPULATION POLICY PROPOSALS, *articles on* GOVERNMENTAL INCENTIVES, SOCIAL CHANGE PROPOSALS, COMPULSORY POPULATION CONTROL PROGRAMS, *and* GENETIC IMPLICATIONS OF POPULATION CONTROL; PRISON-

ERS; PSYCHOPHARMACOLOGY; SEX THERAPY AND SEX RESEARCH; and VIOLENCE AND THERAPY. *See also:* COMMUNICATION, BIOMEDICAL, article on MEDIA AND MEDICINE; FREE WILL AND DETERMINISM; INFORMED CONSENT IN THE THERAPEUTIC RELATIONSHIP; and MENTAL HEALTH THERAPIES.]

BIBLIOGRAPHY

BANDURA, ALBERT. *Principles of Behavior Modification.* New York: Holt, Rinehart & Winston, 1969.
"Behavior Control in Prisons." *Hastings Center Report* 5, no. 1 (1975), pp. 16–48.
BERTALANFFY, LUDWIG VON. *Robots, Men and Minds: Psychology in the Modern World.* New York: George Braziller, 1967.
BIRDWHISTELL, RAY L. *Kinesics and Context: Essays on Body Motion Communication.* Philadelphia: University of Pennsylvania Press, 1970.
COMSTOCK, GEORGE A.; MURRAY, JOHN P.; and RUBENSTEIN, ELI A., eds. *Television and Social Behavior.* 5 vols. Rockville, Md.: National Institute of Mental Health, 1972.
DELGADO, JOSÉ. *Physical Control of the Mind.* New York: Harper & Row, 1969.
GAYLIN, WILLARD; MEISTER, JOEL; and NEVILLE, ROBERT, eds. *Operating on the Mind.* New York: Basic Books, 1975.
GOFFMAN, ERVING. *Asylums.* Garden City, N.Y.: Doubleday, 1961.
KITTRIE, NICHOLAS. *The Right to be Different: Deviance and Enforced Therapy.* Baltimore: Johns Hopkins Press, 1971.
KLERMAN, GERALD L. "Psychotropic Hedonism vs. Pharmacological Calvinism." *Hastings Center Report* 2, no. 4 (1972), pp. 1–3.
KOESTLER, ARTHUR. *The Ghost in the Machine.* New York: Macmillan Co., 1967.
LONDON, PERRY. *Behavior Control.* New York: Harper & Row, 1969.
NEVILLE, ROBERT. *The Cosmology of Freedom.* New Haven: Yale University Press, 1974.
———. "The Limits of Freedom and Technologies of Behavior Control." *Human Context* 4 (1972): 433–446.
SCHWITZGEBEL, RALPH K. *Development and Legal Regulation of Coercive Behavior Modification Techniques with Offenders.* U.S. Public Health Service Publication no. 2067. Chevy Chase, Md.: National Institute of Mental Health, Center for Studies of Crime and Delinquency, 1971.
SIEGLER, MIRIAM, and OSMOND, HUMPHRY. *Models of Madness, Models of Medicine.* New York: Macmillan Publishing Co., 1974.
SKINNER, B. F. *Walden Two.* New York: Macmillan Co., 1948.
"Symposium: Psychosurgery." *Boston University Law Review* 54 (1974): 215–353.
SZASZ, THOMAS. "The Ethics of Addiction." *American Journal of Psychiatry* 128 (1971): 541–546.
ULRICH, ROGER; STACHNIK, THOMAS; and MABRY, JOHN. *Control of Human Behavior.* 2 vols. Glenview, Ill.: Scott, Foresman & Co., 1966, 1970.
VEATCH, ROBERT M. "Drugs and Competing Drug Ethics." *Hastings Center Studies* 2, no. 1 (1974), pp. 68–80.

II

FREEDOM AND BEHAVIOR CONTROL

There have been stubborn disagreements among theorists about whether behavior control can be compatible with freedom, and whether, in cases of conflict between the two, the loss of freedom can be a reasonable price to pay for the ends gained by techniques of human manipulation. The resolution of these controversies presupposes a common understanding of the concept of freedom itself, an accord that is difficult to achieve, partly because the uses of the word "free" are many and various. For example, we make both singular judgments to the effect that a particular person is free in a particular respect, and "on-balance judgments" to the effect that a given person or class of persons is free "on the whole, all things considered." Persons are sometimes said to be free from irritations, emotions, and other disliked states or circumstances, and (perhaps more commonly) free from constraints on their behavior. They are said to be free to act or free to choose, to enjoy or lack "freedom of the will" or the status of free subject or citizen. None of these usages is free of obscurities and ambiguities, and these in turn have generated conflicting philosophical accounts of the nature, limits, and value of freedom.

Free action

One leading controversy concerns the analysis of "freedom from constraint" and the relation between constraints and desires. The ancient Stoics and Epicureans claimed to be speaking for common sense when they maintained that freedom (from constraint) consists simply in the ability to do what one wants most at the time to do. An advantage of this simple account is that it provides an explanation of why we consider freedom to be a valuable thing: when one is prevented from doing what one wishes to do (constrained) or forced to do what one does not wish to do (compelled), one's desires are frustrated, and frustration is usually a very unhappy experience. Since one is rarely powerful enough to do everything one desires, and greater power is hard to come by, the Stoics and Epicureans recommended that persons modify or extinguish their desires in order to minimize the chances of frustration, thereby achieving both contentment and freedom. "Demand not that events happen as you wish but wish them to happen as they do happen, and you will get on well" (Epictetus, *The Enchiridion*, VIII).

If there is only one thing that authorities or circumstances permit a person to do at a given time, but that thing happens to be the thing he wants most to do at that time, then he can do what he wishes and hence is free, according to the Stoic–Epicurean conception. Therein lies the weakness of that conception according to its critics, for to say that circumstances permit only one action (whether desired or not is immaterial) is precisely to say that one is compelled to perform that action, and compulsion is the very opposite of freedom. One can act quite willingly under compulsion, but the willingness does not change the fact of compulsion. Freedom to act, in this second conception, requires not (or not only) that one can do what one wishes to do, but that one can do many things other than what one wishes to do. In fact, the more things a person can do—the greater his options, the more alternatives kept open for him—the freer he is, no matter what his desires happen to be. The great advantage of this account is that in no circumstances is it committed to the identification of freedom and compulsion. Its main difficulty is to explain why freedom is as valuable as most people think, how there can be value in being able to do things one does not want to do. Some writers base the value of open alternatives on the security they offer to persons whose desires might change, for all they know, in the future. Others find an intrinsic value in having options to choose among even when most of the alternatives will not seem eligible for their choice. In any case, writers in this camp insist, freedom is only one value among many. It may well conflict with contentment in some circumstances and may cause confusion, unwise decisions, and frustration. It is no aid to clarity to identify freedom with other goods or to deny that it can ever conflict with other things of value.

Another controversy concerns the sorts of things that can properly be called "constraints" in the analysis of freedom. Some would identify lack of freedom to act with inability to act, whatever the nature or source of the inability, whether it be policemen's guns and locked prison gates; poverty, sickness, or ignorance; private emotions, repressions, or scruples; laws of nature (we are not "free" to high-jump twenty feet); or laws of logic (not even God is "free" to be and not be at the same time).

On the other side, it is said that freedom is not simply the absence of *any* factor that might impede a possible action, but rather the absence of one specific kind of constraint, namely, *coercion* by human beings and its indirect inhibitory consequences. Coercion includes direct forcing, for example pushing and arm twisting, as well as threats of punishment or other evils, whether by legal authorities or by private persons. Insofar as such states as poverty, sickness, and ignorance are the consequences of coercive rules and practices, they too count as constraints on freedom.

It will not do simply to identify the lack of freedom to act with any inability to act whatever its source, for there are many things persons are unable to do even though it would be natural to say they are free to do those things. I cannot compose symphonies, speak Russian, or shoot par on a golf course because I lack the requisite talent or knowledge, not because I lack freedom. Moreover, if the source of constraints is left completely unrestricted in our account of freedom, the notion of perfect freedom will be an empty and unapproachable ideal. No matter how much our freedom is enlarged, there will still be an infinite number of things we are not free to do. Perfect freedom on the alternative account consists in the largest number of open options consistent with reasonable rules and social order, an ideal that is in principle achievable.

The free person

Sometimes people are said to be free not from some specific constraint, not to do or omit doing some specific action, but simply free *tout court*. This can mean two things of quite different kinds. The judgment that a person is free "on balance" may be a rough generalization from the various singular judgments that can be made about his freedom to do or omit doing various actions. In that case it means that he is generally free from constraints to do a great many things, that he has open alternatives among which to choose in respect to a great many areas of life. A person is free, in this sense, to the extent that he has *open options*.

The second thing that can be meant when persons are described as free *tout court* is that they are autonomous, self-governing, or independent, that they have a "sovereignty" over their own affairs. Autonomy or self-government, on the one hand, and freedom from constraint (open options), on the other, are quite distinct concepts of freedom, not reducible one to the other. Isaiah Berlin, following established usage, labels the conception of freedom as autonomy "positive freedom" and the conception of freedom as absence of constraint (open options) "negative freedom." "The answer to the question 'Who governs me?' is logically distinct from the question, 'How far does government interfere with me?' It is in this difference that the great contrast between the two concepts of negative and positive liberty, in the end, consists" (Berlin, p. 130).

Applying the political metaphor of self-government to the individual person requires a clear conception of the "boundaries" of the self that is the subject of freedom. The autonomous person is not only independent of the control of external selves, he must be firmly in control of the self that is his own. This in turn has suggested to writers from Plato on that the self must be divided into a "ruling part" and a lower nature that is subject to its authority. The language of autonomy seems to require a conception of an "inner core" self, consisting of reason, conscience, or those values that are highest in a personal hierarchy and with which a person most intimately and securely identifies himself. This inner core self must be sufficiently narrow to be contrasted with other internal elements of the wider self, which it is said to govern, such as specific desires, purposes, and affections.

Many writers have identified the "ruling part" of the self with reason (Plato; Kant). David Hume, on the other hand, in a characteristic hyperbole, argued that "reason is, and ought only to be, the slave of the passions" (Hume, p. 415). If Hume had spoken of a servant instead of a slave and endorsed a democratic conception of authorities as "public servants," his metaphor might not have been far from the mark, for then it would have permitted us to derive the authority of practical reason to regulate particular aims and desires from the interests of those particular "passions" themselves. So conceived, reason is like a traffic cop directing cars to stop and go in an orderly fashion so that they might get to their diverse destinations all the more efficiently, without traffic jams and collisions. The person whose desires lack order and structure and obey no internal regulator will be torn this way and that and fragmented hopelessly. Such a person fails to be autonomous not because of outside government but because of his failure to govern himself. He will also fail to be free on balance from constraint insofar as his constituent desires constrain one another in internal jams and collisions. At its worst, such a condition approaches that which Emile Durkheim called "anomie," one of the leading causes of suicide (Durkheim).

Respect for a person's autonomy is respect for his unfettered voluntary choice as the sole rightful determinant of his actions except where the interests of others need protection from him. Whenever a person is compelled to act or refrain from acting on the ground that he must be protected from his own folly, his autonomy is infringed. When the state, for example, refuses to permit a citizen to do something he has chosen to do on the sole ground that his choice seems contrary to his own welfare, that citizen's autonomy is violated. If we assume that "on-balance freedom" is a constituent of a person's welfare, and we therefore refuse to permit him to trade some of his "open options" in exchange for benefits of another kind, then we have violated his autonomy for the sake of his "on-balance freedom" in the long run. Even if a person is prevented, on wholly paternalistic grounds, from selling himself into slavery in exchange, say, for a million dollars to be delivered in advance to a charity or favored cause, his voluntary choice is overruled for the sake of his future freedoms. This example shows vividly that personal autonomy and freedom as "open options" are distinct values quite capable of conflict. A perfectly autonomous person would have the "power of voluntarily disposing of his own lot in life" (Mill, p. 125), even if that involved forfeiture of most of his de facto liberty in the future.

Free will

It is possible for a person to have a great deal of freedom of action and yet not be a free person on balance. Such a person would be largely free of external coercion and thus have ample capacity to do whatever he might come to choose, but in respect to one or more large categories of action he would lack the ability to choose or not choose as he might come to wish. In that case many of his options to choose (or to will, or even to desire) are closed by internal psychological conditions even though there are no external constraints to his conduct. His freedom of action is unimpaired but he falls short of freedom on balance because, for a large range of possible choices, he lacks free will.

The clearest difference between the person who has free will with respect to a given area of choice and the man who suffers from an impaired psychological capacity (to choose, will, intend, decide, or desire) is that the former's choice would be different were he presented with what he himself acknowledged to be decisively good reasons for choosing differently, whereas the latter's intention remains fixed in the presence of excellent reasons for choosing differently. Because this crucial difference is a matter of degree, it follows that free will is a matter of degree too. Even the most confirmed alcoholic will probably choose to decline another drink if he is led to believe that it will cause his painful death, though the beliefs that it will merely make him an unsafe driver or an incompetent husband, which would move the will of a freer man, leave his intention unaltered. When a person's choices are unalterable by acknowledged good reasons for changing them (or by reasons that would be so acknowledged if the person were in a rational state), then he suffers from the sort of impaired psychological capacity that diminishes his free will. In that case his intentions can be altered, if at all, only by severe threats or "by some form of manipulation such as behavior therapy or drugs" (Glover, p. 98).

Part of the reason why "freedom to will" is an intelligible notion referring to something of great value is that persons are capable of desiring that their desires, preferences, choices, and willings be other than what they are. It is precisely this "capacity for reflective self-evaluation" that renders human beings *persons*, in contrast to the lower animals, which may have wills and purposes but no "desires of the second order" (Frankfurt, p. 7). Even though a drug addict may desire to choose not to take another dose (he may hate his addiction), he finds himself desiring, and then choosing, to take it anyway. If his addiction is complete, the option to choose to abstain, in accordance with his second-order desire, is not open to him. He cannot desire as he wants to desire, and he cannot choose as he wishes to choose. On the other hand, the person whose will is free in respect to drug-taking can choose to take drugs or choose to abstain, and if he prefers to abstain that is what he will choose to do.

There is no limit in principle to the number of higher-order desires a person might form, though in fact persons rarely have occasions to form desires above the third order (a desire to desire to will to do . . .). Typically, however, persons "identify decisively" with desires of the second order and equate their freedom with the capacity of the inner-core self so defined to will or not will in accordance with its preferences. Free will, so conceived, is valuable to its possessor because it affords him the satisfaction of knowing that "his will is his own . . . as opposed to being estranged from one's self or being a helpless or passive bystander to the forces that move him" (ibid., p. 18).

The effects of behavior control on freedom

Primarily because it is a form of manipulation, behavior control raises philosophical perplexities about the nature and value of freedom. Manipulation is one of three ways of influencing the behavior of others and can best be understood in contrast to its alternatives. Manipulation, or "dextrous management" of another person, can be distinguished both from rational persuasion and coercion, though it has more significant similarities to the latter. We induce others to behave as we wish by rational persuasion when we offer them advice, issue reminders, present new facts, invoke common principles, in short, when we present good reasons for behaving in one way rather than another. Our advice becomes effective when the other party adopts our reasons as his own and acts accordingly. In that case our persuasion was both rational and noncoercive; it did not by force close the other's option but only gave him a reason to exercise his option in a certain way. Right up to the moment of his decision, it was up to him alone whether to decide one way or the other.

Coercion is a blunter and more visible way of influencing another person's behavior. We can close one of his options by force (twisting his arm, dragging, pushing, clubbing), by expropriating the necessary means to act (money, tools, weapons), by creating physical barriers (walls, moats, stakes), or by intimidation (threats of death, mayhem, or legal punishment). Typically, of course, coercion is employed to close one or more of a person's options and thereby diminish his freedom of action. But coercion can also be used to open up some of a person's options (against his will) in which case the person is, in Rousseau's phrase, "forced to be free" (Rousseau, p. 64). Rousseau meant that a person's misinformed choices and irrational desires could be put down by external force in support of his true inner-core self, the rational will that rightly governs him. To be forced to do what one's rational will would choose to do if liberated, he argued, is to be forced into true freedom for one's real self.

Isaiah Berlin, on the other hand, speaks for many twentieth-century philosophers when he characterizes this conception as a "monstrous impersonation, which consists in equating what X would choose if he were something he is not, or at least not yet, with what X actually seeks and chooses" (Berlin, p. 133).

There are, however, less paradoxical applications of the idea of forcing someone to be freer on balance than he is or than he wants to be. It is possible to have the area of one's freedom to act enlarged by force, and, when this happens, some of one's options are closed by a coercive act that at the same time causes many more options, or options of greater importance, to open. Thus a person might be dragged struggling and kicking over the border from a cruel police state into a liberal democracy. He may have resisted out of ignorance or because he genuinely preferred tyranny to freedom. Not every one appreciates having open options and the difficult burden of having always to choose for oneself what one shall do. But, whatever a person's motives for resisting the expansion of his options, the condition so described is one of greater freedom of action. It is important to note, however, that, benign as our motives might be, insofar as we force a person against his will into a condition he did not choose, we undermine his status as a person in rightful control of his own life. We may be right when we tell him that greater freedom of action is for his own good in the long run, but we nevertheless violate his autonomy if we force our conception of his own good upon him.

Manipulation of another person characteristically works directly upon his beliefs, motives, and psychological capacities and effects change neither by means of rational persuasion nor by means of external hindrances, prods, and deprivations. Some forms of manipulation, for example extortion and aversion therapy, employ threats or something very like threats and are thus properly located near the vague boundary between coercion and manipulation. Certain forms of psychotherapy that bring the patient greater insight into his own motives through such techniques as self-revelation, free association, and suggestion are very near the equally vague border between manipulation and rational persuasion. One more clearly defined category of manipulation includes the many modes of nonrational persuasion: stimulation of stereotyped emotional responses; propaganda; posthypnotic suggestion; subliminal advertising; salesmanship that subtly associates a product with a social status or a desired state (for example, sexual attractiveness) with which it has no actual connection; and indoctrination by appeal to authority, emphatic reiteration, even the "hypnopaedic sleep teaching" of *Brave New World*. Another category includes the various techniques of behavior control: psychosurgery, electric stimulation of the brain, psychotropic drugs, and aversive conditioning. These techniques can be employed either to close or to open a person's options, either with or without his voluntary consent. Each of these four combinations has its own effect on freedom.

Where manipulative techniques are used to open a person's options with his voluntary consent, there is an enlargement of freedom and no violation of autonomy; hence, this is the least troublesome category. For example, suppose a person so suffers from claustrophobia that he cannot bring himself to enter any small confined place, not even the closet where his guests have left their coats. Suppose a therapist proposes a series of treatments designed gradually to increase his tolerance through conditioning and hypnosis. The patient consents to the treatments, fully understanding the theories on which they are based and the risks they might involve. As a consequence his phobia is destroyed and he now has one kind of option he did not have before, namely to choose to enter or not to enter confined places as he pleases. Indirectly, many other options are also opened up. It is now open to him, for example, to enter elevators, to work in photographic dark rooms, or to become a cloakroom attendant.

Another patient might consent in a fully voluntary way to manipulative behavior control designed to close one of his options and thereby open up many others. An alcoholic, for example, may voluntarily take a drug that will make him violently ill if he should ingest as little as a half ounce of alcohol in a two-week period, thus effectively shutting off his option to accept drinks. As a consequence, many other options far more valuable to him are now opened, and he can function in the world once more. He was "manipulated into his freedom," to be sure, but his own consent to the treatment made him a party to it, so that it became a form of self-manipulation as morally innocuous as setting an alarm clock before retiring and the many other "tricks" free men play on themselves.

In other cases, persons freely consent to manipulative treatment designed to close some of their options where there is either a certainty or a high risk that this loss of freedom will *not* be compensated by a greater gain in freedom on balance. A person who suffers from uncontrollable rages or chronic suicidal depression may well elect to undergo psychosurgery knowing full well that the technique is inexact and may not have its intended effects, and also that there is a substantial risk that the treatment may restore him to tranquility at the cost of his becoming a passively docile vegetable, unable to initiate any projects or enterprises of his own. The risks may seem reasonable to him and unreasonable to those with authority over him, or he might actually have a considered preference for vegetative docility over his present intolerable condition, a preference that seems reasonable to him but unreasonable to those who have authority over him. The question these cases pose for social policy is a troublesome one: should such persons be permitted to consent to treatment that is likely to diminish their freedom by restricting their future options irrevocably, for the sake of a good that they have come to value more than their freedom? On the assumption that consent *can* be fully voluntary in such circumstances, it would be to respect an individual's own choice to permit him his dangerous manipulative treatment, and it would violate his autonomy to deny it for what *we* take to be his "own good." An autonomous being has the right to make even unreasonable decisions determining his own lot in life, providing only that his decisions are genuinely voluntary (hence truly his own), and do not injure or limit the freedom of others.

The most odious kind of case is the manipulation of a person without his consent (even without his knowledge) in order to close many of his options to choose as he pleases in the future. Patients or prisoners (the two are easily confused in totalitarian countries) can be drugged, put under total anesthesia, and then made to undergo lobotomies or other surgical manipulations or mutilations of the brain. Psychotropic drugs used in small quantities and electric stimulation of the brain for shorter periods have less severe effects and are revocable, but when imposed on a person without his consent or, worse, without his knowledge, they are hardly distinguishable on moral grounds from assault and battery. The situation is not affected morally by the motives of the controller. Whether he be kindly and benevolent or malicious and cynical, the effect of his actions is drastically to diminish freedom of choice and to violate personal autonomy. The case is complicated morally, however, when the patient, having lost his competence to govern himself and his capacity, therefore, to grant his voluntary consent, has no autonomy left to violate. In such cases, a person can be made no worse in respect to freedom and autonomy than he already is, and behavior control may reduce his pain and anxiety and promote the convenience of those who must govern him. Respect for personal autonomy, however, requires that the benefit of every doubt be given to the patient, that every effort be made to improve his lot without irrevocable destruction of his capacity to govern himself.

The final kind of case is perhaps the most troublesome. That is where manipulative techniques are used to open a person's options and thus increase his freedom on balance, but without his consent. Here indeed a person is "manipulated into freedom," not with his own connivance, but without his knowledge, and perhaps even against his will. Being involuntarily suffered, such treatment is necessarily a violation of a person's right to be his own master and to make the choices himself that vitally affect his future. It therefore abridges his freedom in one sense (autonomy), while expanding it in another (by opening options). Where interests of third parties are not involved, every person's moral right to govern himself surely outweighs the "right" of benevolent intermeddlers to manipulate him for his own advantage, whether that advantage consist of health, wealth, contentment, or freedom. At the very most, a man's options to choose can be opened for him by manipulation to which he has not consented, only to protect others from harm at his hands.

The value of freedom

It is worth reemphasizing that freedom is only one of the good things in life, for many writers have claimed that any putative freedom that can conflict with such other undoubted goods as contentment and fulfillment of desire cannot be "true freedom" at all. As one among many values, freedom must sometimes be sacrificed or compromised in cases of unavoidable conflict. B. F. Skinner, in his provocative book *Beyond Freedom and Dignity*, has argued that freedom (and dignity too) are vastly overrated and that the value placed on them by misguided humanitarians has been one of the major obstacles to human progress. Our physical technology, he maintains, has created as many problems for mankind as it has solved, and what is needed now is a highly developed "technology of behavior." Birth control devices, for example, will not solve the population problem unless human beings in overcrowded countries can be induced to use them; pollution-control devices and recycling techniques will not protect the environment unless industrialists and consumers change their attitudes toward them; policemen and prison officials will fight a losing battle against crime until the motives, characters, and opportunities of the criminals themselves are changed. What is required to solve these problems, in short, is more effective control of human behavior. The partisans of freedom and dignity, Skinner argues, consider all kinds of control by some persons over others exploitative and inconsistent with freedom. Skinner, on the other hand, considers only certain forms of coercive control objectionable; he would minimize or eliminate if possible the use of "aversive stimuli" to control others, that is, harmful or disagreeable reinforcers that human organisms naturally "turn away from." Nonaversive controls, however, are morally unobjectionable according to Skinner, even though incompatible with freedom and dignity as those terms are commonly interpreted.

Skinner is perhaps most persuasive when he applies his thesis to the problem of criminal punishment. The current system leaves people "free" to decide whether to obey the law or risk punishment; therefore they get credit for their law abidingness and blame and punishment for exercising their freedom of choice in the wrong way. In either case, they can take responsibility for what they do and thus maintain their dignity. People who are unable to choose to behave badly cannot take credit for their goodness, and indeed, in a world in which everyone behaved well because he had no option of behaving badly, there would be no goodness of character to take credit for. On the other hand, in such a world there would be no crime. And no victims of crime. The problem, writes Skinner, "is not to induce people to be good but to behave well" (p. 70). This result can be achieved most economically by creating circumstances in which crime is not likely to occur or in which it can have no point or profit; in providing acceptable outlets for aggressive energies and other means of sublimation; or finally, if all else fails, resorting to "hormones . . . to change sexual behavior, surgery (as in lobotomy) to control violence, tranquilizers to control aggression" (ibid., p. 68).

There is considerable plausibility in Skinner's contention that freedom and dignity are often overvalued to the detriment of social progress. Gun-control legislation, which would make certain kinds of firearms inaccessible to most people, has been opposed, for example, on the grounds that the credible threat of severe punishment for misuse of weapons would be almost as effective in preventing crime, and more consistent with human freedom. A similar principle is involved when the construction of highway overpasses as a way of obviating collisions at dangerous intersections is rejected in favor of heavier penalties for negligent and reckless driving. The latter policy would leave people "free" to behave or misbehave and punish those autonomous agents who exercise their freedom in the wrong way, as if that would be compensation for the lives and limbs of their victims. On the other hand, few protested when exact change became mandatory on public buses in New York and Washington, so that bus drivers no longer carried money to attract hold-up men. Almost overnight, bus robberies ceased once and for all, and the need for deterrent punishment vanished. These are examples of "engineering the environment" to destroy the occasions for criminal conduct, at the cost (obviously minor) of closing options to choose antisocial conduct and depriving good people of credit and bad people of blame. If all occasions for crime could be thus eliminated, the elaborate ritual complex of sin, punishment, and remorse would vanish as well, a consequence Skinner would not regret.

Similarly, some of the forms of *human* engineering, to which Skinner would resort in the end, if reliable, might seem to be good bargains even at the expense of freedom and dignity. If conditioning or psychosurgery could prevent "vicious criminals" from ever choosing to commit wantonly violent attacks on others, without at the same time increasing their own vulnerability to such attacks (unlike the crude techniques depicted in Anthony Burgess's novel, *A Clockwork Orange*) it would be difficult to appreciate the objections to their use, especially when consented to by the criminal.

Even if the prisoner's consent is not fully voluntary (since his alternative is continued incarceration) or if the operation is performed or the treatment imposed against his will, the invasion of his autonomy would probably seem to Skinner (though many or most writers disagree) to be a more economical and humane method of protecting his potential victims than a continuation of his punitive confinement.

But even if that assessment of the relative weights of efficiency and humanity against autonomy in an extreme case is correct, it does not imply what Skinner seems to say, that personal autonomy, freedom of action, and freedom of choice are only minor or trivial values. To appreciate this point one need only consider Aldous Huxley's imaginary society in *Brave New World*, where most wants are gratified instantly, anxiety and depression are eliminated, and crime and violence are unknown. All the people (with the exception of a few aberrant dissidents who are sent into a very pleasant exile) are perfectly content, even though many of the values that we readers, looking in from the outside, regard as essential to a good life, are missing. In some respect, freedom of action in *Brave New World* is enlarged: ". . . you're so conditioned that you can't help doing what you ought to do. And what you ought to do is on the whole so pleasant, so many of the natural impulses are allowed free play, that there really aren't any temptations to resist" (Huxley, p. 185). But on balance freedom to act is severely diminished, for persons are not permitted to travel where they will, read whatever they might choose, express any of a wide range of opinions, and so on. The closed options, of course, do not trouble or frustrate the happy citizens of *Brave New World* in the slightest, for they do not want to do, and could not choose to do, anything but what they can do. Since they are unable to will in accordance with any "second-order desires" other than the ones they have been conditioned to have, they lack open options to choose, or "free will" in the sense discussed above. But that doesn't trouble or frustrate them either, because they *can* will as they want to will, even though they *cannot* will other than as they do. Nor do they enjoy personal autonomy, the rightful power to make on their own the decisions that will determine their own lot in life. Yet the citizens of *Brave New World* are content, even without much freedom. Insofar as we readers find this picture both convincing and repugnant, Huxley's didactic novel proves to us that we do not value freedom merely as a means to contentment, but attribute to it an independent value of its own.

Deriving and explaining the distinctive value of freedom is one of the great unfinished tasks of philosophy. If the one thing we all value as an end in itself is happiness, and if we can get happiness without freedom, how in that case can freedom be of any further value to those who lack it? *Brave New World* provides some suggestive clues. The contented, brain-washed automata in that fictitious land, being incapable of concern or affection for one another, or original thought, or high ambition, are so alien that we hardly recognize them as members of our own species. It may be a truism that all men desire happiness, but the happiness they desire is a recognizably *human* happiness, and the contentment that does not presuppose freedom fails to satisfy that description.

JOEL FEINBERG

[*Directly related to this article is the companion article,* ETHICAL ANALYSIS. *This article will find application in the entries* BEHAVIORAL THERAPIES; DRUG USE; DYNAMIC THERAPIES; ELECTRICAL STIMULATION OF THE BRAIN; HYPNOSIS; ELECTROCONVULSIVE THERAPY; PSYCHOSURGERY; PSYCHOPHARMACOLOGY; *and* SELF-REALIZATION THERAPIES. *For further discussion of topics mentioned in this article, see the entries:* ALCOHOL, USE OF; ENVIRONMENTAL ETHICS, *articles on* QUESTIONS OF SOCIAL JUSTICE, *and* THE PROBLEM OF GROWTH; FREE WILL AND DETERMINISM; PATERNALISM; PERSON; PRISONERS; SUICIDE; *and* TECHNOLOGY. *Also directly related are the entries* BEHAVIORISM; GENETIC ASPECTS OF HUMAN BEHAVIOR; GENETIC CONSTITUTION AND ENVIRONMENTAL CONDITIONING; INSTITUTIONALIZATION; *and* SEXUAL BEHAVIOR. *See also:* MENTAL HEALTH THERAPIES; RESEARCH, BEHAVIORAL; TRUTH-TELLING; *and* VIOLENCE AND THERAPY.]

BIBLIOGRAPHY

BENN, STANLEY I., and WEINSTEIN, WILLIAM L. "Being Free to Act, and Being a Free Man." *Mind* 80 (1971): 194–211.

BERLIN, SIR ISAIAH. *Four Essays on Liberty.* London: Oxford University Press, 1969.

BURGESS, ANTHONY. *A Clockwork Orange.* New York: W. W. Norton, 1963.

DURKHEIM, EMILE. *Suicide.* Translated by John A. Spaulding and George Simpson. New York: Free Press, 1951, pp. 241–276.

EPICTETUS. *The Enchiridion* (A.D. 138). Translated by Thomas W. Higginson. New York: Liberal Arts Press, 1948.

FEINBERG, JOEL. "The Idea of a Free Man." *Educational Judgments.* Edited by James F. Doyle. London: Routledge & Kegan Paul, 1973, pp. 143–169.

FRANKFURT, HARRY G. "Freedom of the Will and the Concept of a Person." *Journal of Philosophy* 68 (1971): 5–20.

GLOVER, JONATHAN. *Responsibility.* International Library of Philosophy and Scientific Method. London: Routledge & Kegan Paul; New York: Humanities Press, 1970.

HUME, DAVID. *A Treatise of Human Nature* (1739–1740). Oxford: Clarendon Press, 1888 and 1941.

Huxley, Aldous. *Brave New World* (1932). Harmondsworth, England: Penguin Books, 1955.
Kant, Immanuel. *Fundamental Principles of the Metaphysic of Morals* (1785). Translated by Thomas K. Abbott. New York: Liberal Arts Press, 1949.
Mill, John Stuart. *On Liberty* (1859). New York: Liberal Arts Press, 1956.
Rousseau, Jean-Jacques. *The Social Contract* (1762). Translated by Maurice Cranston. Harmondsworth, England: Penguin Books, 1968.
Skinner, B. F. *Beyond Freedom and Dignity.* New York: Alfred A. Knopf, 1971.

BEHAVIOR MODIFICATION

See Behavior Control; Behaviorism, *article on* History of Behavioral Psychology; Behavioral Therapies; Research, Behavioral.

BEHAVIORAL GENETICS

See Genetic Aspects of Human Behavior.

BEHAVIORAL RESEARCH

See Research, Behavioral.

BEHAVIORAL THERAPIES

This entry is related to, but distinct from, the previous entry on Behavior Control. *This entry raises the ethical issues created by the practices of behavioristic therapies, such as the freedom of the therapist–client relationship and questions concerning the mechanistic approach to training and curing.*

People who make decisions about the behavior of others—parents, educators, legislators, judges—rarely refer to the psychological laboratory for verification of their notions of human behavior. Intuition, common sense, and nonscientific writings may seem to offer more immediate and relevant answers to human problems. However, a science of human behavior has recently become available. Principles developed largely in experimental animal learning laboratories have been utilized by behavioral psychologists and therapists to modify socially relevant human behaviors.

Two areas of animal learning have contributed to the behavioral therapies: classical or Pavlovian conditioning and operant conditioning. Pavlovian conditioning presents the following model. When food (an unconditioned stimulus) is given to a dog, it salivates (an unconditioned response); after food has been repeatedly presented following the ringing of a bell (a conditioned stimulus) the dog will salivate (a conditioned response) when only the bell is rung. The conditioned response will eventually cease to occur (extinguish) if the conditioned stimulus is no longer paired with the unconditioned stimulus. Classical conditioning affects only the performance or nonperformance of innate behaviors. Perhaps the most well-known application of these principles is desensitization by reciprocal inhibition, a technique used primarily with phobic individuals. Aversion therapy for problems of alcoholism or sexual deviation is also based on a classical conditioning model. The term "behavior therapy" is often used to describe therapeutic techniques at least loosely based on classical conditioning (Franks).

Operant conditioning emphasizes the relationship between a response and environmental events that closely follow it in time. Consequent environmental events that result in an increase in response frequency are classified as reinforcing events. Those resulting in a decrease in response frequency are classified as punishing events. If a dog is given a piece of popcorn every time it whines and the rate of whining increases, the popcorn would be called a reinforcer. If the popcorn were later withheld, the rate of whining would decrease (extinguish). Operant principles have been used both to increase the frequency of some behaviors such as positive statements in job interviews, studying in college, or cleaning one's room, and to decrease the frequency of other behaviors such as hallucinatory speaking, littering, or overeating. The special advantage of operant conditioning is that behaviors never previously emitted can be taught by reinforcing each component approximation of the target behavior. The term "behavior analysis" is often given to therapeutic procedures based on operant conditioning.

The early methodological behaviorists stressed that all human learning could be understood in terms of observable environmental events and observable responses (Craighead, Kazdin, and Mahoney). Now behavior therapists in social learning (e.g., Bandura, 1974) argue that much behavior is a response to environmental stimuli mediated through the intervention of thoughts that predict consequences; the behavior is partially a reaction to those predictions. Behavior analysts, on the other hand, argue that thoughts and feelings are observable responses even though they are observable only to the person who experiences them. Thoughts and feelings, like other response classes, may be modified by altering their consequences (Skinner, 1953). Both behavior therapists and behavior analysts deal with cognitive processes (Bandura, 1969, pp. 564 ff.; Craighead, Kazdin, and Mahoney, pp. 1–34; Rimm and Masters, pp. 416 ff.). This fact is not widely understood by nonbehaviorists, who may consequently tend to label behavior therapies as mechanistic.

Behavior therapy as a generic term is frequently used to encompass all therapeutic techniques based on either the classical or the operant animal learning literature. The techniques may differ in the extent to which they are theoretical rather than empirical, but all aspire to an empirical basis. Behavior therapy is part of a broad spectrum of treatment strategies within psychology and psychiatry whose emphases vary widely, e.g., existential therapy, gestalt therapy, client-centered therapy, and psychoanalysis (Craighead, Kazdin, and Mahoney, pp. 93 ff.).

Historical review

The principles of behavior therapy have been used unsystematically since the beginning of time, but their systematic application is relatively new. Although the first systematic applications of learning principles to human behavior occurred in the 1920s, psychotherapeutic procedures derived from the medical model, which located the cause of behavior inside the individual (e.g., in a conflict between the id and ego), rather than in environmental events, predominated. In the 1950s three events seem to have occasioned the rapid development of behavioral treatment procedures: (1) a questioning of the effectiveness of psychotherapeutic techniques based on the medical model (Eysenck); (2) the development of systematic desensitization procedures (Wolpe); and (3) the analysis of social problems in terms of the principles of operant conditioning (Skinner, 1953). Initially, behavior analysts worked with populations that had largely been ignored by traditional psychotherapists, those labeled retarded or psychotic. Substantial success was reported both with individual programming (e.g., language) and group programming (e.g., token economies) with those populations. Behavior therapists at first worked largely with client populations previously handled by traditional psychotherapists. Later, therapists in both areas expanded their treatment efforts to other populations and problems with a consequent blurring of the distinctions between the two. Recent behavioral textbooks list (1) the following problem areas: anxiety, depression, sexual deviation, social skills, alcoholism and drug abuse, obesity, smoking, and marital problems; (2) the following settings: classrooms, mental hospitals, medical hospitals, institutions for the retarded, prisons, and experimental communities; and (3) the following procedures: desensitization, modeling, assertive training, contingency management, self-control, extinction, positive control, aversive control, and cognitive methods (Craighead, Kazdin, and Mahoney; Rimm and Masters; Bandura, 1969).

Controversial issues related to behavior therapy

Mechanism. The attention to behavior rather than personality and intrapsychic processes may appear mechanistic or at least nonhumanistic. Behavioral psychologists do not find it useful to characterize a client as schizophrenic, delinquent, or defensive. Instead they describe salient high- or low-frequency responses emitted by the client, which were reinforced or punished by the environment. Similarly the behaviorally trained teacher or parent will not characterize a child as good or bad but will indicate that particular responses were well executed or not.

The distinction between reinforcing and punishing a response rather than a person is important. When reinforcement occurs a response increases in frequency. Procedures for reinforcing a whole person do not exist. Punishment is operating if a response followed immediately by an environmental event decreases in frequency. A mother may turn off the television for five minutes (time-out from television) every time her children bicker. If bickering decreases she has punished that response. Procedures for punishing a whole person do not exist.

Misapprehensions. Retributive punishment is directed toward the whole person. It would not be defined as punishment by behaviorists since in retributive punishment the punishing event is not necessarily contingent on a response (but may be contingent on having inadequate legal defense) and is rarely documented to decrease the frequency of the response (as attested by high recidivism rates).

The term "punishment" in the context of the judicial system is not synonymous with the same term in a behavioral context. Nonbehaviorists have tended to assume mistakenly that prison terms, solitary confinement, and other isolation procedures in prisons and hospitals for the retarded and mentally disturbed were always behavioral procedures. Thinkers who believe that people always should be treated as persons may in fact defend retributive punishment for its "human" orientation, arguing that behavioral punishment "dehumanizes" people by recognizing them only as sets of conditioned responses. Their opponents may argue that isolated conditioned responses are the only aspects of a person that society has a right to change because only such isolated behaviors can be examined and judged in a socially equitable way; society, they may argue, has no right to a whole person. Behavioral procedures have also been erroneously associated with various surgical techniques and with the use of psychoactive drugs. To be categorized as behavioral, a procedure must be based on the laws of learning as they have been discovered in the basic science laboratory (National Association).

Although there is nothing new in the attempts of one or more persons to change the behaviors of other persons, deliberate planning of behavior change based on established principles of behavior can seem ominous. If such planning also includes the introduction of extraneous or arbitrary reinforcing stimuli such as giving candies to schoolchildren contingent upon the completion of assignments, some long-held values about learning for its own sake seem violated. Nonbehaviorists may think that natural reinforcers ought to be sufficient. The behaviorist would argue that if certain appropriate responses are lacking in a person then whatever natural reinforcers may be available have clearly been ineffective in establishing the target responses. Additional, arbitrary reinforcers may be temporarily needed to strengthen responding so that the individual can begin to come into contact with any natural consequences that may exist in the situation. All too often society's natural consequences are the unpleasant things that happen to persons who do not exhibit the appropriate behaviors.

Problems created by treatment effectiveness. It is important to remember that the classification of environmental events is based on the direction of the behavior change. There are no events that are always reinforcing or always punishing. If a priori definitions of punishers and reinforcers are used, behavior change may be in unexpected directions. A teacher may think that reprimanding a student for breaking classroom rules is punishing. If the student continues to break the rules, an alternative explanation, locating the cause of rule-breaking in the moral character of the student instead of the consequent environmental events, may be made. The more appropriate analysis would be that teacher attention (even though delivered in what would be subjectively perceived as a negative manner) was functioning as a reinforcer, thereby increasing the frequency of the behavior it followed.

It is incorrect to suggest that a behavioral principle did not work in the above instance. Rather, the situation was analyzed incorrectly and a nonpunishing event was chosen to follow upon the behavior. If an appropriate analysis is made and if a demonstrated consequent event is available the probability of a change in behavior is very high. Furthermore, that change would be publicly demonstrable. A behaviorist insists upon replicate intervention strategies as well as direct and quantifiable observations of the behavior before, during, and after treatment.

The documented success of behavior therapy and the potential associated with that success have occasioned ethical concern about the choice of particular target responses as well as about who will make those choices. The concern becomes especially evident when the values of the client, the client's agent, the therapist, and society at large are not consonant. For example, early behavioral interventions in elementary classrooms dealt with the reduction of disruptive behavior (Winett and Winkler). Decreasing the disruptive behavior was not always accompanied by an increase in on-task behavior, and therefore not consonant with the values of all concerned. Subsequent behavioral projects in elementary classrooms have targeted responses such as task completion rather than disruptive behavior. A similar but unresolved problem lies in the selection of the target response when the overall problem relates to the sexual identity of the client.

While these two areas have received considerable attention, almost every therapeutic intervention involves problems related to the choice of the target response and its ultimate frequency of occurrence. How many chores should a teenager be expected to do around the house, how much eye contact should a nonassertive person develop, how many pleasant statements should a husband make to his wife, how many laps in the exercise yard should the prisoner run?

These issues in behavior therapy are particularly appropriate if the client is a resident of an institution for the retarded, for the psychotic, or for prisoners. Such clients are often less knowledgeable about available treatment choices, and target behaviors are often defined by someone other than the client (ward aides object to incontinence or self-destructive behavior). The behaviorists employed by institutions may be reinforced for producing client behavior that furthers institutional goals rather than client interests (Stolz, Wienckowski, and Brown).

Behavior therapies and the institutionalized

The U.S. courts abandoned their hands-off policy toward the care and treatment of the institutionalized in the mid-1960s. While changes in the behavior of residents as a result of token economies and other behavioral procedures were well documented, it became clear that abuses had also occurred (Wexler), sometimes as a result of institutional practices unrelated to treatment and sometimes as a result of behavioral or other treatment procedures ("Viewpoints"; Budd and Baer; National Association, pp. 1–6; Martin; Wexler). Judicial decisions that have affected the activities of behavioral psychologists in institutions are discussed below.

The treatment procedures selected for any institutionalized patient must be the least restrictive of the treatment alternatives available, e.g., *Covington* v. *Harris* (Budd and Baer); for example, giving a client tokens for social interactions may be more restrictive than asking him to self-monitor those behaviors (both behavioral procedures), but less restrictive than electroconvulsive therapy (a nonbehavioral procedure). It is unclear if the courts would declare a more restrictive treatment that was quickly successful to be less or more restrictive than a less restrictive treatment that took longer.

The courts have also proscribed the use of some aversive procedures, e.g., nausea-induction in *Knecht* v. *Gillman* ("Viewpoints," p. 63), physical restraint in *Wyatt* v. *Stickney* and seclusion in *Inmates of Boy's Training School* v. *Affleck* and *Morales* v. *Turman* (Budd and Baer). Such procedures had in the past been used by both behavioral and nonbehavioral psychologists. When less restrictive aversive procedures are unavoidable, e.g., time-out, their use may be monitored by a human rights review board composed of institutional employees, parents, lawyers, community residents, and behavior analysts.

Prior notice must be given to the client before a treatment program is begun and an opportunity for a hearing must be made available. The risks, benefits, and other consequences must be explained (Stolz in Wood, p. 246; "Viewpoints," pp. 91–92; Budd and Baer, pp. 229–233) so that the resident can make an informed consent. The possibility of an institutionalized person giving consent independent of coercion may be questioned, e.g., *Kaimowitz* v. *Michigan Department of Mental Health* ("Viewpoints," p. 59), especially if there is a likelihood that release from the institution is in any way contingent upon participation in the treatment. Behavioral analyses of responses and contingent consequences involved in consent and coercion have introduced the notion of degrees of consent and coercion (ibid., p. 81) in terms of sets of alternative responses or choices available to the client (Goldiamond, pp. 54–62). The legal and behavioral aspects of consent involve many conflicting assumptions about the fundamental variables controlling behavior. Traditionally, freedom has implied absence of aversive control by powerful persons. The behaviorist's stress on the importance of immediate positive consequences in controlling responses seems to violate some long-held ideas of freedom, e.g., being able to consent freely. A number of alternative choices may be planned in an apparently free manner yet the pattern of choices may be readily related to prior positive consequences for particular choices (ibid.). The present notion that freedom exists if aversive controls are absent may change with understanding of the principles of positive reinforcement.

Another issue for behavior analysis in institutions is the relationship between noncontingent reinforcement and the notion of absolute rights. The courts have made a distinction between absolute rights and contingent rights, i.e., those privileges which shall be available freely and those which may be made available only following some specific response. The list of absolute rights includes, in part, visits and telephone calls, meals in dining rooms, wearing of personal clothing, adequate physical space, and comfortable beds, e.g., *Wyatt* v. *Stickney* (Budd and Baer). "The crux of the problem, from the point of view of behavior modification, is that the items and activities that are emerging as absolute rights are the very same items and activities that the behavioral psychologists would employ as reinforcers—that is, as 'contingent rights'" (Wexler, p. 11); e.g., a patient might have been required to brush his hair and wash his face before being allowed into the dining room for breakfast. If self-grooming is to remain a target behavior, then some other reinforcer must be found. Absolute rights may not be entirely eliminated as reinforcers, since they can vary in frequency, duration, or kind. The duration of visiting hours, the number of telephone calls per day, or being served a preferred softboiled egg instead of the ordinary hardboiled egg could conceivably be made contingent. Absolute rights do not exhaust the possibilities for reinforcers, but alternatives are often either beyond the financial means of institutional budgets or not functionally reinforcing stimuli for institutionalized persons, e.g., the arts, theater, and travel.

The problem goes beyond that stated by Wexler as quoted above. If a reinforcing event is presented without regard to what the client is doing, whichever response happens to be occurring just before the reinforcer is presented is likely to be increased. A reinforcer will have an effect on a response whether or not anyone explicitly intended an effect. Accidental contingencies are as effective as planned contingencies. Meals as absolute rights in institutions may simply increase the frequency of sitting and staring when they could serve to increase more adaptive behaviors. The guarantee of absolute rights guarantees an ignorance of which response(s) will be reinforced and eliminates the requirement for institutionalized persons to do anything in order to receive food, medical care, and other needs. Although society may ultimately decide this requirement is wrong, that policy sharply differentiates the institutionalized from the noninstitutionalized persons who must usually work consistently to earn absolute rights for themselves.

The recipients of bountiful help are rather in the position of those who live in a benign climate or possess great wealth. They are not strongly deprived or aversively stimulated and, hence, not subject to certain kinds of reinforcement. . . . But such people do not simply do nothing; instead, they come under the control of lesser reinforcers [Skinner, 1975, p. 626].

The "lesser" reinforcers associated with leisure are less likely to be art, music, literature, and science than sweets, alcohol, tobacco, or television. In an institution, alternative reinforcers are not available. The residents have nothing to do. "They may be reinforced for mild troublemaking, and if possible they escape, but otherwise we say their behavior tends to be marked by boredom, abulia . . . and apathy" (ibid., p. 630). Helping someone has traditionally meant giving them free reinforcers. A greater help may be in providing simple contingencies between useful responses and available reinforcers.

The courts have recognized that "the mere characterization of an act as 'treatment' does not insure that the client will benefit," e.g., *Knecht v. Gillman* ("Viewpoints," p. 64). Time-out from reinforcement has not been prohibited by the courts although seclusion has (Budd and Baer). This important distinction by the courts between procedures that have been demonstrated to change behavior and those that have not suggests that the courts may soon specify that the right to treatment implies the right to effective treatment. The effectiveness of treatment procedures may be demonstrated in a number of ways. Quantifiable changes in client behavior may be made public. Questionnaires that solicit the satisfaction or dissatisfaction of persons who eventually will judge the effectiveness of the treatment (parents, teachers, judges, clients) provide important information to the therapist (Wood, pp. 131–157). Public specification of target dates for accomplishing certain treatment goals further protect clients.

Implications beyond the institutions

The decisions reached by the courts regarding behavior analysis and the institutionalized have implications for the entire society. Nonbehavioral views of human nature have held that entities such as a free will or rational mind direct behavior. Our laws were designed to control the behavior of persons possessing such entities. The model of behavior implicit in our legal system and the principles stated explicitly in the analysis of behavior are at odds. An example of a resulting problem is in the issue of the right to privacy in one's own home. Parents are permitted to raise their children in almost any way they see fit. However, it is now known that attention of any sort is a powerful reinforcing stimulus for many young children. It is also known that parents and older siblings are likely to attend to children when they misbehave and to ignore them when they behave appropriately. These child-raising procedures increase the frequency of the very responses parents wish to eliminate. This paradigm is so common that most behavioral interventions involve little more than shifting the usual consequences for desired responses to undesired responses and vice versa. It may be unethical to permit parents to raise their own children without providing parent-training programs that make the consequences of the above situation explicit.

Another ethical issue becomes immediately apparent. If parents were trained to teach the responses they desired in their children, how many parents would find it more reinforcing to teach responses that would be in the long run more useful for their children, and how many would find it more reinforcing to produce responses that were in the short run more useful to themselves? That at least some parents would prefer to reinforce rapid obedience and long periods of quiet at the expense of discussion and exploration seems reasonably high (Winett and Winkler). The rights of children to treatment (in this case the opportunity to learn the more useful behaviors) seem to conflict with the rights of the parents to privacy in child-raising (Martin, pp. 134–136).

The continued success of behavioral treatment strategies has encouraged behaviorists to think beyond the treatment of the individual to that of the entire social system in which individuals would collectively plan and arrange consequences for responses in every aspect of social living. Currently accepted social customs involved in childrearing, religious beliefs, economic and political theory, and rational philosophy would inevitably be seriously questioned. Whatever subsequent changes in behavior patterns might eventuate would be based upon an objective analysis of the relationships among response frequencies and environmental consequences.

KENNETH E. LLOYD AND
MARGARET E. LLOYD

[*For further discussion of topics mentioned in this article, see the entries:* BEHAVIOR CONTROL; *and* BEHAVIORISM. *Other relevant material may be found under* INFORMED CONSENT IN MENTAL HEALTH; INFORMED CONSENT IN THE THERAPEUTIC RELATIONSHIP; INSTITUTIONALIZATION; MENTAL HEALTH THERAPIES; PRISONERS; PRIVACY; *and* RESEARCH, BEHAVIORAL.]

BIBLIOGRAPHY

BANDURA, ALBERT. "Behavior Therapy and the Models of Man." *American Psychologist* 29 (1974): 859–869.
———. *Principles of Behavior Modification*. New York: Holt, Rinehart & Winston, 1969.
BUDD, KAREN S., and BAER, DONALD M. "Behavior Modification and the Law: Implications of Recent Judicial Decisions." *Journal of Psychiatry and Law* 4 (1976): 171–244.
CRAIGHEAD, W. EDWARD; KAZDIN, ALAN E.; and MAHONEY, MICHAEL J. *Behavior Modification: Principles, Issues, and Applications*. Boston: Houghton Mifflin, 1976.
EYSENCK, H. J. "The Effects of Psychotherapy: An Evaluation." *Journal of Consulting Psychology* 16 (1952): 319–324.
FRANKS, CYRIL M., ed. *Behavior Therapy: Appraisal and Status*. McGraw-Hill Series in Psychology. New York: 1969.
GOLDIAMOND, ISRAEL. "Toward a Constructional Approach to Social Problems: Ethical and Constitutional Issues Raised by Applied Behavioral Analysis." *Behaviorism* 2 (1974): 1–84.
MARTIN, REED. *Legal Challenges to Behavior Modification: Trends in Schools, Corrections, and Mental Health*. Champaign, Ill.: Research Press, 1975.
National Association for Retarded Children, Research Advisory Committee, Joint Task Force. *Guidelines for the Use of Behavioral Procedures in State Programs for Retarded Persons*. Assembled by the Florida Division of Retardation and the Department of Psychology of the Florida State University. Monograph no. 1. 1975. 73 pp. Pamphlet available from the National Association for Retarded Children, 2709 Ave. E East. Arlington, Texas, 76011.
RIMM, DAVID C., and MASTERS, JOHN C. *Behavior Therapy: Techniques and Empirical Findings*. New York: Academic Press, 1974.
SKINNER, B. F. "The Ethics of Helping People." *Criminal Law Bulletin* 11 (1975): 623–636.
———. *Science and Human Behavior*. New York: Macmillan Co., 1953.
STOLZ, STEPHANIE B.; WIENCKOWSKI, LOUIS A.; and BROWN, BERTRAM S. "Behavior Modification: A Perspective on Critical Issues." *American Psychologist* 30 (1975): 1027–1048.
"Viewpoints on Behavioral Issues in Closed Institutions." *Arizona Law Review* 17 (1975): 1–143. Special issue.
WEXLER, DAVID B. "Token and Taboo: Behavior Modification, Token Economies, and the Law." *California Law Review* 61 (1973): 81–109. Reprint. *Behaviorism* 1, no. 2 (1973), pp. 1–24.
WINETT, RICHARD A., and WINKLER, ROBIN C. "Current Behavior Modification in the Classroom: Be Still, Be Quiet, Be Docile." *Journal of Applied Behavior Analysis* 5 (1972): 499–504.
WOLPE, JOSEPH. *Psychotherapy by Reciprocal Inhibition*. Stanford: Stanford University Press, 1958.
WOOD, W. SCOTT, ed. *Issues in Evaluating Behavior Modification*. Proceedings, First Drake Conference on Professional Issues in Behavior Analysis, 1974. Champaign, Ill.: Research Press, 1975, pp. 1–264.
Wyatt v. Stickney. 325 F. Supp. 781 (M.D. Ala. 1971). 334 F. Supp. 1341 (M.D. Ala. 1971). 344 F. Supp. 373 (M.D. Ala. 1972). 344 F. Supp. 387 (M.D. Ala. 1972). Appeal docketed sub nom. Wyatt v. Aderholt. 503 F. 2d 1305 (5th Cir. 1974).

BEHAVIORISM

This entry offers a set of theories that are relevant for understanding the conceptual background of the previous two entries. It accomplishes this in two articles—one that traces the history of the various schools of behavioral psychology and the issues of ethical interest under the title HISTORY OF BEHAVIORAL PSYCHOLOGY, *and the other,* PHILOSOPHICAL ANALYSIS, *which discusses the philosophical and ethical implications of behavioristic schools of thought.*

I. HISTORY OF BEHAVIORAL PSYCHOLOGY
    Juris G. Draguns
II. PHILOSOPHICAL ANALYSIS   *Alasdair MacIntyre*

I
HISTORY OF BEHAVIORAL PSYCHOLOGY

Origins of behaviorism

Unlike many other events in the history of science, the origin of behaviorism can be readily pinpointed. In 1913 John B. Watson published an article, "Psychology as the Behaviorist Views It," in which the tenets, methods, and objectives of behaviorism were presented. In that article and in several later publications, Watson introduced a new theoretical framework in psychology that stood in contrast to the then prevailing conceptualizations in method, substance, and objective. In method, Watson proposed a sharp break from introspection or disciplined self-observation as the basic avenue of data-gathering in psychology in favor of recording the overtly observable behavior of organisms. In substance, he redefined the subject matter of psychology from the study of the contents of the mind to the investigation of behavior. In reference to objectives, he stressed prediction and control of behavior over its explanation and understanding for its own sake.

The impact of this triple shift in emphasis was to redirect the efforts of psychologists from the search for the general laws that govern the functioning of the average adult mind to the study of subjects and organisms that could not sidetrack the investigator by their spontaneous or elicited introspective reports, i.e., animals and young children. Research with animals, especially white rats, came to occupy a prominent place in behavioral psychology, on the assumption that the principles underlying human behavior could be discovered by experimenting with animal subjects. Such differences as existed between animals and humans were thought to be quantitative and not qualitative. Thus, animals could be observed and the observations extended to humans. The research strategy inaugurated by Watson and followed by many prominent behaviorists to the present day involved the investigation of simple responses in order to proceed from these studies to more complex behavior. Watson was centrally concerned with the conditions that caused behavior to change, i.e., with learning, and saw human as well as animal activity in a constant flux, ever subject to modification by the environment. His belief in the malleability of human as well as animal behavior was pronounced and is well expressed in the following quotation:

Give me a dozen healthy infants, well-formed, and my own personal world to bring them up in and I'll guarantee to take any one at random and train him to become any type of specialist I might select—doctor, artist, merchant-chief and, yes, even beggarman and thief, regardless of his talents, penchants, tendencies, abilities, vocations, and race of his ancestors [Watson, 1930, p. 82].

Anyone could, in short, become anything, and the task of behavioristic psychology was to uncover and specify the conditions under which this would be possible—not by means of the utopian experiment proposed in the passage above, but by a laborious program of systematic experiments, for the most part with animals.

Antecedents of behaviorism

Although behavioral psychology represented a radical break with the concepts and operations of psychology in the early twentieth century, its advent did not occur in a conceptual void. Two influences in particular deserve to be pointed out. Edward L. Thorndike's experimental studies on learning in cats by means of trial and error and his formulation of the law of effect represented an important antecedent to behaviorism. The law of effect postulated that responses followed by satisfaction and pleasure tend to be repeated; responses followed by discomfort are less likely to occur (Thorndike).

Another important development took place in Russia, where I. P. Pavlov demonstrated a type of learning that came to be known as classical or Pavlovian conditioning. In Pavlov's laboratory, animals came to respond to a previously neutral stimulus, such as the sound of a bell or metronome, with behavior formerly elicited by the presentation of meat powder, i.e., by salivation. In Pavlov's terminology, the buzzer was the conditioned stimulus and salivation in response to it, the conditioned reflex; meat powder served as the unconditioned stimulus (Pavlov). These experiments and similar, although independently conceived studies by Pavlov's countryman, Bechterev (1913), provided, together with Thorndike's work, an empirical foundation for the program of investigations on learning that Watson envisaged.

Comprehensive learning theories

Watson participated only in the early stages of the investigation. It fell to others to propose and develop comprehensive theories of learning. Notable in this respect were the contributions of Hull, Tolman, and Guthrie.

Hull's effort was an ambitious attempt to develop a hypothetico-deductive theory of learning composed of a set of assumptions and theorems. He proceeded from observations of rote learning in animal species, particularly rats, and extended them to the entire range of human behavior. By the time of his death his effort was not yet complete, but the basic principles on which the theoretical structure was to rest were enunciated. In a simplified form, Hull viewed all behavior as a function of habit and drive, which stood in a multiplicative relationship. It follows then that both habit and drive are necessary for a specific response to occur. All behavior is thought to be at the service of drive reduction; the organism forever tends toward a tensionless state (Hull, 1943, 1952).

Hull's great opponent was Tolman, who stressed the purposeful character of learning even in subhuman species. The rat's behavior in a maze was determined by hypotheses or a cognitive map. Proceeding from this view, a distinction was drawn between latent learning, as accumulation of experience, and its manifestation in overt behavior (Tolman). This theme reverberates in a number of recent formulations concerning the nature of learning.

In a different way, Guthrie drew a sharp distinction between learning and performance. All learning, he held, occurs in one trial, but many trials are necessary to tie the various learned responses together and to cause them to occur in the course of one act (Guthrie).

These global explanations of learning attracted both adherents and controversy, but fascination with such explanatory structures dwindled after their originators' deaths. At the time of this writing (1976), there are few psychologists who identify themselves as followers of Hull, Tolman, or Guthrie, even though specific features of their contributions continue to inspire theoretical elaboration and empirical research.

Recent developments

From the vantage point of the 1970s it can be said that behaviorism as a current of conceptualization has been tremendously successful. With the exception of a few dissenting voices (Miller; Vexliard), there is broad agreement in current psychology textbooks and other writings that psychology is the science of behavior. The concept of behavior, however, has been broadened beyond the glandular secretions and muscular contractions to which Watson restricted this term. Behavior has come to encompass speech, thought, feeling, and perception, and the range of psychologists who call themselves behaviorists has greatly expanded. Thus, there are cognitive behaviorists (e.g., Boneau; Mahoney). Strictures against the study of private internal states have been greatly loosened. Proposals for reintroduction of introspection as a method of psychology have been heard (Radford), even though there are few, if any, psychologists who would return it to the prominent place that it occupied before Watson.

Throughout all of these developments, the central figure in the mainstream of behaviorism has been B. F. Skinner. In marked contrast to Hull, he has eschewed construction of formal theories in order to concentrate on the specification of the conditions necessary for the occurrence of a response. His approach has been empirical, inductive, and minimally burdened with inference. His focus has been on the study of the operant response by means of which the organism alters, or acts upon, its environment. He has held throughout four decades that the contingencies for such responses can be found in the external environment and that operant responses are maintained and shaped by externally mediated reinforcement. The control and prediction of a response has been Skinner's central concern; any reference to internal states, subjective or physiological, is unnecessary to that end. Like Watson, Skinner is articulately deterministic. Since all behavior is under the control of its consequences, Skinner's focus in his more speculative writings (1948, 1972) has been on developing more rational and efficient ways of controlling behavior. This has taken him into the area of social design and the formulations of policies that would control behavior in the economic, political, and social areas following principles developed in his laboratory.

Social impact of behaviorism

The last two decades have been characterized by the expansion of behaviorism beyond the confines of laboratory research. Prominent among developments in that direction is behavior modification, which represents the application of behavioral principles, notably of classical conditioning and operant learning, to the changing of maladaptive or disturbed behavior (Bandura; Eysenck; Goldfried and Davison; Wolpe). These developments have had highly beneficial and promising results (Mahoney, Kazdin, and Lesswing), although there continues to be controversy concerning the scope of applications of the techniques and their relationships to their original laboratory models of learning (London, 1972). The techniques of behavior modification have expanded to include modeling or imitation (Bandura) as well as self-observation and self-control (Goldfried and Merbaum; Mahoney). The methods of behavior modification have been applied outside of hospital and clinic settings in schools (Klein, Hapkiewicz, and Roden; Harris; O'Leary and Drabman), correctional institutions (Stolz, Wienckowski, and Brown), and a variety of other real-life environments (Tharp and Wetzel).

The effectiveness of behavior modification techniques has sparked criticisms and apprehensions, many of which had been anticipated in earlier critical reactions to behaviorism and its major exponents (Koestler; Rogers and Skinner; Bertalanffy). Since behavior modification involves the exercise of power and control, the old Platonic question has been raised: Who will control the controllers? The answer the behaviorists provide is that the behavior modifier is not to be thought of as an external agent, unilaterally impinging upon the objects of his techniques. Rather, as Skinner (1972) and others (Goldfried and Davison) have asserted, he who reinforces is reinforced by the recipients of reinforcement. It has been demonstrated that experimental control produces countercontrol by the subjects of the experiment (Ulrich).

Other objections to behavioral techniques have concerned the use of painful and aversive means to modify behavior (Goldfried and Davison; Stolz, Wienckowski, and Brown) and the application of behavior modification techniques to involuntary subjects in correctional institutions (Stolz, Wienckowski, and Brown), schools (ibid.), and other settings. The consensus among recent writers on this subject (London, 1969; Mahoney, Kazdin, and Lesswing; Stolz, Wienckowski, and Brown) is that safeguards must be carefully applied and that the interests of the individual whose behavior is being modified have priority over institutional or social preferences. Finally, questions have been raised concerning the ethical propriety of modifying such deviant behavior as homosexuality, especially in the absence of techniques for helping homosexuals develop and maintain their homosexual potential (Goldfried and Davison). While most practitioners of behavior modification would agree that these techniques can be legitimately applied only to consenting clients and only with a minimum of discomfort or pain, this does not exhaust the subtlety and complexity of ethical issues that have been raised. Some of them remain unresolved at the time of this writing.

JURIS G. DRAGUNS

[*Directly related are the entries* BEHAVIOR CONTROL; BEHAVIORAL THERAPIES; *and* RESEARCH, BEHAVIORAL. *Also directly related is the other article in this entry,* PHILOSOPHICAL ANALYSIS.]

BIBLIOGRAPHY

BANDURA, ALBERT. *Principles of Behavior Modification.* New York: Holt, Rinehart & Winston, 1969.

BECHTEREV, VLADIMIR MICHAILOVITCH. *General Principles of Human Reflexology* 4th ed. (1928). Translated by Emma Murphy and William Murphy. New York: International Publishers, 1932. Reprint. Classics in Psychology. Edited by Howard Gardner and Judith Kreiger Gardner. New York: Arno Press, 1973.

BERTALANFFY, LUDWIG VON. *Robots, Men and Minds: Psychology in the Modern World.* New York: G. Braziller, 1967.

BONEAU, C. ALAN. "Paradigm Regained? Cognitive Behaviorism Restated." *American Psychologist* 29 (1974): 297–309.

CHAPLIN, JAMES PATRICK, and KRAWIEC, THEOPHILE STANLEY. *Systems and Theories of Psychology.* 3d ed. New York: Holt, Rinehart & Winston, 1974.

EYSENCK, HANS JURGEN, ed. *Behavior Therapy and the Neuroses: Readings in Modern Methods of Treatment Derived from Learning Theory.* Oxford: Pergamon Press, 1960.

GOLDFRIED, MARVIN R., and DAVISON, G. C. *Clinical Behavior Therapy.* New York: Holt, Rinehart & Winston, 1976.

———, and MERBAUM, MICHAEL, eds. *Behavior Change through Self-Control.* New York: Holt, Rinehart & Winston, 1973.

GUTHRIE, EDWIN RAY. *The Psychology of Learning.* Rev. ed. Harper Psychology series. New York: Harper, 1952. Reprint. Gloucester, Mass.: Peter Smith, 1960.

HARRIS, MARY B., ed. *Classroom Uses of Behavior Modification.* Columbus, Ohio: Merrill, 1972.

HULL, CLARK LEONARD. *A Behavior System: An Introduction to Behavior Theory Concerning the Individual Organism.* New Haven: Yale University Press, 1952. Science edition. New York: John Wiley & Sons, 1964.

———. *Principles of Behavior: An Introduction to Behavior Theory.* New York: Appleton-Century-Crofts, 1943.

KLEIN, ROGER D.; HAPKIEWICZ, WALTER G.; and RODEN, AUBREY H., eds. *Behavior Modification in Educational Settings.* Springfield, Ill.: Charles C Thomas, 1973.

Koestler, Arthur. *The Ghost in the Machine.* New York: Macmillan Co., 1968.
London, Perry. *Behavior Control.* New York: Harper & Row, 1969.
———. "The End of Ideology in Behavior Modification." *American Psychologist* 27 (1972): 913–920.
Mahoney, Michael J. *Cognition and Behavior Modification.* Cambridge, Mass.: Ballinger, 1974.
———; Kazdin, Alan E.; and Lesswing, Norman J. "Behavior Modification: Illusion or Deliverance?" *Annual Review of Behavioral Therapy: Theory and Practice, Vol. 2.* Edited by Cyril M. Franks and Terence Wilson. New York: Brunner-Mazel, 1974, pp. 11–40.
Miller, George A. *Psychology: The Science of Mental Life.* New York: Harper & Row, 1962.
O'Leary, K. Daniel, and Drabman, Ronald. "Token Reinforcement Programs in the Classroom: A Review." *Psychological Bulletin* 75 (1971): 379–398.
Pavlov, I. P. *Conditioned Reflexes: An Investigation of the Physiological Activity of the Cerebral Cortex.* Translated and edited by G. V. Anrep. London: Oxford University Press, 1927.
Rachlin, Howard. *Introduction to Modern Behaviorism.* San Francisco: W. H. Freeman & Co., 1970.
Radford, John. "Reflections on Introspection." *American Psychologist* 29 (1974): 245–261.
Rogers, Carl R., and Skinner, B. F. "Some Issues Concerning the Control of Human Behavior: A Symposium." *Science* 124 (1956): 1057–1066.
Skinner, B. F. *About Behaviorism.* New York: Alfred A. Knopf, 1974.
———. *The Behavior of Organisms: An Experimental Analysis.* The Century Psychology Series. Edited by R. M. Elliott. New York: Appleton-Century Co., 1938.
———. *Beyond Freedom and Dignity.* New York: Alfred A. Knopf, 1972.
———. *Contingencies of Reinforcement: A Theoretical Analysis.* The Century Psychology Series. New York: Appleton-Century-Crofts, 1968.
———. *Science and Human Behavior.* New York: Free Press; London: Collier-Macmillan, 1953.
———. *Walden Two.* New York: Macmillan Co., 1948.
Stolz, Stephanie B.; Wienckowski, Louis A.; and Brown, Bertram S. "Behavior Modification: A Perspective on Critical Issues." *American Psychologist* 30 (1975): 1027–1048.
Tharp, Ronard G., and Wetzel, Ralph J. *Behavior Modification in the Natural Environment.* New York: Academic Press, 1969.
Thorndike, Edward L. "Animal Intelligence." *Psychological Review, Monograph Supplements* 2, no. 4 (Whole no. 8) (1898), pp. 1–109.
Tolman, Edward Chace. *Purposive Behavior in Animals and Man.* The Century Psychology Series. Edited by Richard M. Elliott. New York: Century Co., 1932.
Ulrich, Roger. "Behavior Control and Public Concern." *Psychological Record* 17 (1967): 229–334.
Vexliard, Alexandre. *La Psychologie n'est pas une science du comportement.* Ankara: Ankara University, 1967.
Watson, John Broadus. *Behavior: An Introduction to Comparative Psychology.* New York: Henry Holt & Co., 1914.
———. *Behaviorism.* 2d ed. New York: W. W. Norton & Co., 1930. 1st ed. 1925.
———. "Psychology as the Behaviorist Views It." *Psychological Review* 20 (1913): 158–177.
———. *Psychology from the Standpoint of a Behaviorist.* Philadelphia: J. B. Lippincott & Co., 1919.
Wolpe, Joseph. *Psychotherapy by Reciprocal Inhibition.* Stanford: Stanford University Press, 1958.

II
PHILOSOPHICAL ANALYSIS

Behaviorism is often thought to be a doctrine on the truth or falsity of which a great deal is at stake for ethics generally and for bioethics in particular. For the account that we give of such concepts as those of a person, of the relationship between a person and his actions, of a motive, and so on, will all clearly be affected by our beliefs about the success or failure of behaviorist claims. Yet there is an initial difficulty.

Behaviorism is not one single doctrine, but a cluster of doctrines, not all mutually compatible, which derive from a single project, that of understanding mental activities, states, events, and processes as a species of physical behavior. The arguments in favor of behaviorism are often, although not always, negative in form. Their proponents wish to deny some particular thesis about the mental and sometimes seem to take this denial as providing positive support for behaviorism. Yet what is denied is at least as various as what is asserted: Descartes's dualism, James's stream of consciousness, and the introspective techniques of Wundt and Titchener have all been targets for behaviorist attack. But such doctrines clearly do not all stand or fall together. Debates over behaviorism have therefore been of very different kinds, and we need to discriminate a number of separate, if related, debates in each of which a distinctive behaviorist thesis has been asserted. In all these debates considerations quite independent of the arguments for or against behaviorism have intruded and behaviorism has therefore acquired a set of associated beliefs, which have been its allies historically, but the strength of whose connection with behaviorism is not always clear.

Overt behavior

First it has been contended by methodological behaviorists that only overt behavior is available for investigation by the methods of natural science. The view of natural science presupposed is one according to which experimental evidence enables us to link by means of a lawlike generalization some type of antecedent observable state of affairs with some type of consequent observable state of affairs. J. B. Watson, who introduced this version of behaviorism into psychology, linked his general thesis to a particular view of learning according to which a learned response is always and only produced by a stimulus with which it has become associated through conditioning. The conditioned reflex was taken as the paradigm for the explanation of human behavior. In the course of Watson's own intellectual development two further tendencies became associated with his behaviorism: One was an emphasis on peripheral nervous mechanisms rather than on the brain in the explanation of behavior, and the other was an extreme version of the view that environment, and not heredity, molds behavior.

In the writings of some of Watson's successors, most notably Clark L. Hull, the view taken of the canons of natural science is self-consciously derived from the philosophy of science of the Vienna Circle. Edward C. Tolman had developed Watson's approach so that the task of the psychological theorist became that of specifying the intervening variables by which stimulus and response were linked. Hull attempted to set out a completely general theory of this kind in hypothetico-deductive form. The influence of P. W. Bridgman's operationalism on this project was crucial. No concept is to be admitted into Hull's theory which cannot be specified in terms of experimental manipulation. But this attempt at methodological rigor accompanies a curiously speculative bent. Hull's central thesis—derived from Thorndike's Law of Effect—was that, if a response to some stimulus is followed by a reduction in some biological drive, the tendency for that stimulus to produce that response will be strengthened. The result is that, just as Watson was self-indulgent in allowing himself to invoke speculative peripheral mechanisms, so Hull is self-indulgent in invoking and indeed inventing not only peripheral mechanisms, but also whatever biological drives are required by his theory.

There are two other central weaknesses in Hull's approach. One derives from his attempt to generalize from very limited experimental evidence; the other is carried over from the weaknesses of the positivist philosophy, which dominated his view of science. The positivist ban on the introduction of unobservables into scientific theories was at one time based on a thesis about meaning formulated by Friedrich Waismann in 1930: The meaning of a statement *is* the method of its verification. When it became clear that it was quite implausible to suppose that statements about electrons or about heat were equivalent in meaning to statements about trails in Wilson cloud chambers or about mercury in thermometers, the whole relationship of evidence to meaning was put in question, and no cogent philosophical account of that relationship from which a clear ban on unobservable entities can be derived has ever been elaborated since. Moreover, such studies as those which experimental psychologists have made of afterimages depend on reports of private experience without violating any of the conditions of a valid experiment, as that is understood by most contemporary philosophers of science. Operationalism is nowadays a doctrine usually held by psychologists and not by philosophers.

Yet these considerations are not conclusive against Hull's type of theory; but its merits, if any, must now be derived from its explanatory power and not from its logical form. One merit was certainly claimed for it. Hull seemed able to give quantitative formulation to increases in strength of association, and so seemed able to produce a number of hypotheses precise enough to be falsified or corrected. But nothing in his work in fact passes beyond the speculative and the programmatic; and in any case this feature of Hull's work has no necessary connection with and provides no particular support for his behaviorism.

There is therefore no good reason to accept the behaviorist thesis when it is formulated as the methodological contention that only overt behavior can be studied by scientific methods. The same is true of the stronger metaphysical thesis that a belief in mental activities, states, events, and processes is a belief in a set of mythological nonexistent items. Watson sometimes allowed his rhetoric to betray him into what appeared to be this position, and he was in fact rash enough to deny the occurrence of images. But no psychological behaviorist since Watson has ever seriously held this position, although it has been imputed to some behaviorists by their critics.

Inner states or events

The second important debate is a philosophical one. It has been argued by Gilbert Ryle that those expressions in ordinary language which have been construed as naming inner mental states or events do in fact name patterns of behavior or dispositions to behave in particular ways. Ryle does argue that some expressions taken to name inner occurrences name nothing at all—"volition" is a prime example. But such expressions are on his account barren technical terms invented by confused theorists; they are not part of the ordinary language of transactions between agents. The target of Ryle's attack is Cartesian dualism, what he calls the myth of the ghost in the machine. Ryle does not argue in *The Concept of Mind* that there are no inner events or states. But he says, for example, that it is logically improper to speak of observing or witnessing sensations or feelings, although we certainly do have twinges and pains, chills of disquiet, and tugs of commiseration. However, in experiencing these an agent never merely reports a private occurrence of an event only discernible by him; there is no privileged access to his own consciousness by each agent, and if by "consciousness" we were to mean that to which we have privileged access, as some philosophers have claimed, then there is no such thing as consciousness. It was this denial which led to Ryle being sometimes misread as simply denying the occurrence of inner events or states.

Ryle's philosophical behaviorism is less extreme than that propounded by B. A. Farrell. Farrell takes it that words such as "consciousness" and "experience" have been displaced from their traditional role in our language by the advances of experimental psychology. Experimental psychology teaches us that everything that needs to be said or can be said about color discrimination, for example, can be expressed in terms of behavior and dispositions to behave. Ordinary language has not yet reflected this advance, but in Farrell's view if Western societies assimilate what psychology has to teach it is possible that the notion of "experience" will come to be treated as embodying a delusion.

One of the defects in Ryle's account is that it seems to make the philosophical views presupposed by ordinary language normative for philosophy and immune from scientific criticism of the type that Farrell envisages. One of the defects of Farrell's account is that he does not seem to notice that the experimental psychology which he cites is not in fact a philosophically neutral, scientifically authoritative source. For the experimental psychology on which he relied was itself the product of a particular philosophical standpoint and it was this standpoint and not the experimental findings of the psychologists which entailed behaviorism. Since restrictions were placed by behaviorists on counting as a valid part of any experiment any feature of it which did not conform to behaviorist canons, it was scarcely surprising and not very impressive that reports of such experiments revealed nothing contrary to behaviorism. Farrell allows that in using the reports of experimental human subjects as evidence—as many psychologists do—we are going beyond what behaviorism required, but he insists that we always could dispense with such reports. We use them merely as a matter of convenience.

Accounting for human actions

The most important criticism of the philosophical grounds for behaviorism advanced so far concerns its account of human actions. For a behaviorist of Hull's kind an action is in the end no more than a highly complex set of bodily movements. Hull did not deny the reality of purposive acts, or wish to exclude the use of such concepts as those of intelligence, insight, or intention. He aspired rather to justify the use of such concepts by deducing them from what he took to be more elementary objective primary principles. But he certainly believed that ultimately statements about intentional action could be analyzed into statements about overt behavior. It has been the contention of a number of philosophers that this is not possible, that statements about actions are in no way reducible to statements about bodily movements. It is certainly the case that the truth conditions of statements about actions differ from the truth conditions of the corresponding statements about bodily movements. "He paid his debts" can be true if any one out of the following set of statements is true: "He handed over various pieces of metal and paper," "He wrote his name on a certain document," "He said certain words under certain conditions." That is to say, one and the same action may be given effect in a number of very different bodily movements. Equally one and the same set of bodily movements may be the bearer of a number of different actions: That "He wrote his name" is true can satisfy a truth condition necessary for "He signed a check," "He applied for a passport" or "He tested his fountain pen." It follows that statements about human action are never logically equivalent to statements about observable human movement.

To identify an action we must be able to say what intention was expressed in the behavior in question and not merely what behavior was enacted. The issue over behaviorism then becomes the question of whether we can or cannot give an adequate behaviorist account of the concept of intention, or perhaps whether we can find an alternative way to specify the differences between different actions embodied in similar or the same sets of bodily movements which will satisfy behaviorist criteria, so that all references to intentionality may be eliminated from our discourse. This behaviorist program is characteristically defined as an attempt to replace intentional sentences such as "He applied for a passport" and "He tested his fountain pen" by conjunctions of nonintentional sentences which conjoin categorical statements about an agent's behavior with hypothetical statements about how such an agent would behave in a range of other situations. It is by reference to these sets of hypotheticals that the behaviorist hopes to match the discriminations that are ordinarily made by references to intention.

About this program two remarks need to be made. The first is that even its achievement would not be enough to show that nothing important in the meaning of intentional sentences had been omitted in their replacement by nonintentional sentences. The second is that this program has never been carried through satisfactorily even for one single example by any of the philosophers who have propounded it. A crucial argument against the very possibility of carrying through such a program is that the relevant sets of hypotheticals could never be adequately specified, except by ending the list of hypotheticals with some such phrase as *"and so on"*; and that the only way to delimit the class of hypotheticals intended would be by reference to that very intentional sentence which the program aspired to eliminate. But this argument has never yet been set out by any critic of behaviorism in a form sufficiently cogent to convince behaviorists.

B. F. Skinner

Unlike Ryle, most contemporary behaviorists agree with their critics in allowing that ordinary language, at least in its most plausible interpretation, is antibehaviorist. The conclusion drawn by B. F. Skinner is that ordinary language is therefore almost irremediably superstitious. Skinner has no wish to deny either that behaviorism cannot represent itself as giving the meaning of what we ordinarily say about mind or that there are indeed inner states, events, and processes. What he contends is that no reference to the inner has any place in a scientific explanation of the production of behavior. What Skinner offers positively is a program for a science of overt behavior, a program whose achievement would make the use of our ordinary sentences about actions intellectually unjustifiable. If Skinner is right, the whole philosophical debate is largely beside the point.

Skinner's position differs in two important respects from that of his psychological predecessors. First, the model of the conditioned reflex is completely abandoned. Skinner does not envisage each piece of behavior by an organism as a response to a stimulus. Instead he sees a great deal of spontaneous behavior by organisms, behavior that is gradually modified insofar as it produces either agreeable or disagreeable consequences. This process of *operant conditioning* by positive and negative reinforcements shapes behavior in such a way that complex patterns are built up out of simple movements. Some schedules of reinforcement are more effective than others; positive reinforcement is generally more effective than negative. Because all behavior is the outcome either of biological factors or of the contingencies of the environment or of both, the study of the inner, of mental states, is usually completely irrelevant to the causal explanation of behavior (and at best might provide us with suggestive clues to the presence of dispositions to behave in certain ways).

Secondly, Skinner believes that experimental psychology should abjure any reliance on neurophysiological explanations. The *only* factors to be appealed to are changes in the environment and changes in the subsequent behavior of the organism. The only defensible way of studying the causal relationship between these two types of change is an experimental method which does not make use of anything but the observation of overt behavior. Reports of experimental subjects are to be excluded, if possible, as a source of error and observations undisciplined by experimental conditions can only mislead us. Skinner himself invented a new type of experimental environment where variation in environment could be strictly controlled, the so-called Skinner box. His work on pigeons provided a model for subsequent work, which has extended outside the laboratory to human behavior in programmed learning and in producing changes in the behavior of retardates, the mentally ill, and delinquents.

The principal criticisms to be made of Skinner's theory are three. First, it has been argued by Noam Chomsky that when Skinner extends his thesis to the explanation of language-using behavior his position becomes speculative and unfalsifiable. In particular he argues that the use of such expressions as "stimulus," "response," and "reinforcement," when they are detached from their application in narrowly experimental contexts, becomes vague and arbitrary. Moreover, Skinner and Chomsky differ over the place of rules in the production of human behavior. Skinner's position is—and given his behaviorism must be—that rules can play no part in the explanation of behavior; to say that behavior accords with a rule is to speak only of a regularity that has been produced by environmental contingencies, whereas for Chomsky the rules of a grammar play a key part in the explanation of linguistic competence.

Second, Skinner's methodological canons are such that his experiments could produce no evidence that would provide counterexamples to his own fundamental position. Hence Skinner's position can be progressively confirmed, but it cannot be refuted, or rather it can be refuted only by the use of procedures that Skinner disallows. This makes rational debate between those who agree with and those who disagree with Skinner difficult. Where complex behavior outside the laboratory is concerned, that of delinquents for example, there exists no established method for deciding on the truth or falsity of a Skinnerian interpretation, as against that of rival theorists such as Freudians; hence, perhaps, the observable fact that controversy over Skinner's positions is notably sterile.

Third, there is a paradox at the heart of Skinner's behaviorism. One dominant reason for adopting behaviorism in psychology, which Skinner shares with many of his predecessors, has been the belief that only thus can psychology be scientific. We have already seen that a necessary requirement for vindicating any behaviorism is the elimination of intentional language from the description of behavior. But the concept of science invoked by behaviorists is itself an intentional, and indeed normative, concept; to prescribe the methods of science is to invite us to be guided by one type of reason and intention rather than another, to deliberate in certain ways and to frame our purposes in accordance with our deliberations. Without such a concept of science (and to eliminate intentionality from our language would be to abolish that concept) we would have no way of justifying any one set of methodological procedures, including those of behaviorism, rather than another. Skinner thus seems simultaneously committed both to the retention and to the elimination of intentional language. His commitment to science is in conflict with his commitment to behaviorism. Hull avoided this paradox only by adopting the logically untenable position that intentional concepts and statements can somehow be logically derived from nonintentional concepts and statements. No one has yet shown that there is any way of formulating a behaviorism that does not fall victim either to the fallacy in Hull's position or to the paradox in Skinner's.

Conclusion

Both critics and protagonists of each of the behaviorisms so far formulated have often thought that the moral implications of asserting or denying behaviorism are easily spelled out. B. F. Skinner, for example, has suggested that traditional concepts of freedom and dignity ought not to survive in the light of his behaviorist findings. But if Skinner's behaviorism were true and all intentional language were inappropriate, it is difficult to see that there would be any place for *any* moral view, including Skinner's own. For any morality clearly requires an ability to follow some rules and precepts just because they are reasonably held to be the correct rules and precepts; and such an ability cannot be described except in terms of reason-governed intentions and purposes.

Clearly no such radical consequence follows from either Ryle's or Hull's behaviorism. But less immediate consequences of their views might be important for ethics in general and for bioethics in particular. Could any behaviorist analysis yield an adequate account of an agent's responsibility? The answer is that we do not know, simply because no behaviorist analysis has ever been adequately carried through. It is just because the force and grounds of behaviorist contentions themselves still remain unclear that the moral implications of behaviorism also remain unclear.

ALASDAIR MacINTYRE

[*Directly related are the entries* BEHAVIORAL THERAPIES; FREE WILL AND DETERMINISM; MIND–BODY PROBLEM; *and* REDUCTIONISM. *See also:* BEHAVIOR CONTROL; BIOLOGY, PHILOSOPHY OF; EMBODIMENT; GENETIC ASPECTS OF HUMAN BEHAVIOR; *and* GENETIC CONSTITUTION AND ENVIRONMENTAL CONDITIONING.]

BIBLIOGRAPHY

CHOMSKY, NOAM. "A Review of B. F. Skinner's *Verbal Behavior*." *The Structure of Language*. Edited by Jerry A. Fodor and Jerrold J. Katz. Englewood Cliffs, N.J.: Prentice-Hall, 1966, pp. 547–578.

FARRELL, B. A. "Experience." *The Philosophy of Mind*. Edited by Vere Claiborne Chappell. Englewood Cliffs, N.J.: Prentice-Hall, 1962, pp. 23–48.

GOLDMAN, ALVIN I. *A Theory of Human Action*. Englewood Cliffs, N.J.: Prentice-Hall, 1970.

HAMLYN, D. W. "Behavior." *The Philosophy of Mind*. Edited by Vere Claiborne Chappell. Englewood Cliffs, N.J.: Prentice-Hall, 1962, pp. 60–73.

HULL, CLARK L. *Principles of Behavior: An Introduction to Behavior Theory*. New York: Appleton-Century, 1943.

KOCH, SIGMUND. "Psychology and Emerging Conceptions of Knowledge as Unitary." *Behaviorism and Phenomenology*. Edited by T. W. Wann. Chicago: University of Chicago Press, 1964, pp. 1–41.

RYLE, GILBERT. *The Concept of Mind*. London: Hutchinson, 1949.

SKINNER, B. F. *About Behaviorism*. New York: A. Knopf, 1974.

———. *Beyond Freedom and Dignity*. New York: A. Knopf, 1971.

———. *Cumulative Record: A Selection of Papers*. 3d ed. New York: Appleton-Century-Crofts, 1972.

———. *Science and Human Behavior*. New York: Macmillan, 1953.

———. *Verbal Behavior*. New York: Appleton-Century-Crofts, 1957.

TOLMAN, EDWARD CHACE. *Purposive Behavior in Animals and Men*. London, New York: Century, 1932.

WATSON, JOHN BROADUS. "Psychology as the Behaviorist Views It." *Readings in the History of Psychology*. Edited by Wayne Dennis. New York: Appleton-Century-Crofts, 1948, pp. 457–471.

BENEFIT-HARM CALCULUS

See RISK; TECHNOLOGY, *article on* TECHNOLOGY ASSESSMENT.

BEREAVEMENT

See DEATH; DEATH, ATTITUDES TOWARD; PASTORAL MINISTRY.

BIOETHICS

Introduction

An Encyclopedia of Bioethics—or of Psychology or Religion or any special field of knowledge—need not have an entry on the very title of the work itself. The present encyclopedia is an extensive, almost exhaustive, collection of those issues and topics thought to be part of or directly relevant to bioethics. Yet beyond the scope of all these articles perhaps some questions will remain for the inquisitive reader—such questions as: How is bioethics related to ethics? Where does medical ethics fit in? Does bioethics have special problems or special principles? Is bioethics related to scientific facts in a way ordinary ethics is not? These, and related questions, will constitute the focus of this article.

It would be inappropriate at this point to define "bioethics" in the sense of setting limits to what should and should not count as bioethics. Rather this article, in dealing with the questions mentioned, will aim at provoking reflection on the nature of bioethics. Such reflection is not necessary for "doing" bioethics, that is, for understanding or working through issues within bioethics. It is intended for the inquisitive who are seeking a conceptual grasp of the field itself—an overview of what it is up to and of its relations with other, but similar, concerns. This article will be a guided tour through some of the questions, tensions, alternatives, and arguments that surround the concept of bioethics but would not necessarily arise in the course of pursuing issues *within* bioethics.

Fitting the object of our focus into its natural landscape is a helpful first step. The varieties of factors giving rise to concerns labeled "bioethics" have involved two main factors: increased capability and knowledge. Being capable of something (keeping a dying patient alive, discovering fetal characteristics before birth, transplanting body organs, etc.) forces the question, "Ought we to do it?"—a question irrelevant to practice so long as we are not capable of doing it. Increased knowledge, insofar as it discovers unforeseen consequences of our actions (spray cans and the ozone layer, nuclear energy plants and perinatal development, etc.), forces moral decisions we had not anticipated. Moreover, increased capability and knowledge can bring more benefits, which in turn raises the problem of their fair distribution. This mixture is then ignited by a general, culturewide emphasis on individual rights. Additionally, increased knowledge has led to specialization, producing "the expert" into whose hands we have entrusted everything. He alone, we thought, had the necessary competence. Supposedly these were just technical, factual matters that we had relinquished to the expert. The economist, the highway engineer, the energy expert, the diplomat, the financial wizard, the doctor—they all knew best. But we eventually came to realize that *values*, unbeknownst even to the experts, were being unwittingly employed in ostensibly factual equations. The resultant decisions greatly affect our quality of life.

The preceding paragraph is merely a reminder of the context that has given emphasis to a discipline called "bioethics." It is mentioned only to reinforce our seeing bioethics as a natural response to new dilemmas, increased knowledge, and threatened rights—rather than as at bottom a new discovery of basic principles or derivations therefrom. Also, in putting bioethics into perspective, we are reminded that it is part of a general move toward ferreting out values that had been camouflaged by factuality and clarifying rights to which we had been blinded by the glare of the unquestioned "common good." All this ties in with the position that, by and large, will be taken in this article—namely, that bioethics is not a new set of principles or maneuvers, but the same old ethics being applied to a particular realm of concerns.

Some preliminary metaethical points

Given what was just said about the position to be taken in this article, it would seem that one's metaethical views would remain undisturbed. That is, what constitutes one's view of the "same old ethics"—whether it be subjectivism, emotivism, cognitivism, intuitionism, or some other—is simply carried over to the biomedical arena. This article cannot deal with what *ethics* is, but only with what bioethics is in relation to whatever "ethics" is taken to be.

However, it may be both helpful and fair to state with (precarious) brevity the general orientation toward ethics assumed in this article. To that end three aspects of this view should be mentioned: (1) It is an ethics based on rationality; (2) it is primarily an avoidance of causing evil or a prevention of evil rather than a promoting of good; and (3) its basic rules are applicable to all people, at all times and places, equally (Gert). There are few, if any, goods that we can all agree on. There are, however, a small number of evils that all rational people would want to avoid (unless they had a reason not to—such as that it would avoid an even greater evil, or would benefit someone else). Death, pain, or loss of freedom, opportunity, and pleasure would be evils to be avoided. In short, a rational person does not want harm to himself or to others for whom he is concerned, and therefore the rules most likely to be advocated by all rational people are those which proscribe behavior harmful to others (ibid.).

It is clarifying to distinguish the dedication to and achievement of some good from the avoidance of evils. The former may comprise a "philosophy of life" or a "guide to living," but the latter seems to be what has been essential to ethics. What is one person's good might not be another's, and it might be disastrous for the former to believe himself "morally obligated" to promote that good. Some "goods" or "values" that individuals have claimed (such as domination of others, racial purity, etc.) are downright immoral. They could not be labeled as immoral if morality simply meant acting in accord with one's own principles or promoting one's own values. Furthermore, if morality were "promoting good," we could not fulfill the criteria of being moral toward all people at all times equally. So we would, by definition, be "immoral" toward most people most of the time. At most we can only avoid causing evil to all people at all times equally.

This has not been an effort to defend a position, but only to describe vaguely an orientation, in hopes of helping the reader relate to, uncover problems in, or discount various points in, all that follows. The article's aim is to explore the nature of bioethics, to push and probe, to bring issues, angles, and possibilities to light. Basically, the strategy for accomplishing this will be to take a position and then subject it to challenges and counterchallenges. No winner will be declared, but at least most of the points will have been examined. Finally, some overall considerations and relationships will be elaborated.

A first approximation: medical ethics

The issues will be more manageable if we approach "bioethics" by first examining the more limited case of medical ethics. It is a particular instance of the same position that will be taken with respect to bioethics. The basic point can be made and questioned in this more limited arena, and then expanded to the sphere of bioethics. (It should be noted that the distinction between "moral" and "ethical" will not be rigorously adhered to in this article.)

Medical ethics is a special kind of ethics only insofar as it relates to a particular realm of facts and concerns and not because it embodies or appeals to some special moral principles or methodology. It is applied ethics. It consists of the same moral principles and rules that we would appeal to, and argue for, in ordinary circumstances. It is just that in medical ethics these familiar moral rules are being applied to situations peculiar to the medical world. We have only to scratch the surface of medical ethics and we break through to the issues of "standard" ethics as we have always known them. Of course, understanding the facts, distinctions, relationships, and concepts of the world of medicine in order to apply these "ordinary" moral rules is an immense task. But it is precisely that task which primarily comprises medical ethics.

"Do not kill" is a moral rule. It would not be this rule itself that would be challenged or justified within medical ethics. Nor would medical ethics as such be concerned with articulating and defending criteria for exceptions to the rule. Rather medical ethics prepares the ground for the constructions of ordinary ethics. Is withdrawing lifesaving therapy an instance of killing? Is the refusal to initiate a life-support system an instance of killing? Are we killing one when we give the limited lifesaving facilities to another instead? We are not questioning whether we should kill or not; we accept that "ordinary" moral rule. What we are questioning is whether, in these special and difficult situations, our action is appropriately labeled "killing." If we decided these were instances of killing, we would probably take the next step: deliberating whether they were justifiable exceptions to the moral rule, "Do not kill." In considering the nature of exceptions and their justification, we would in effect be back in "ordinary" ethics. But in examining the medical details for distinctions, relationships, implications, and the like in order to apply the criteria for exceptions, we would be doing medical ethics. That is, we are preparing the ground for the "application" of ethics.

We might see the same points with issues less weighty than that of killing. For example, "Do not deceive" is a moral rule of ordinary ethics (or so it might be argued). But what constitutes deceiving in the medical realm? This again calls for a careful understanding of what goes on in that realm and perhaps for subtle distinctions and conceptual analyses in order to determine when deceiving is really taking place. And, again, intimate knowledge of these particulars would be important in determining when exceptions to the rule are justified.

This clean distinction between ethics, on the one hand, and the world of medicine, on the other, smudges a bit upon closer inspection, though not enough to invalidate the basic point. Two factors are responsible for blurring the demarcation. First, in what has been called "preparing the ground" (i.e., classifying, drawing distinctions, sorting out causes and effects, uncovering implicit values in the medical realm) there may very well be an ethics-laden element. How one views ethics—its point, methods, justification, etc.—may influence how one prepares the factual and conceptual ground to receive the moral rules. There could conceivably be a prejudice in arranging the data for moral deliberation—a familiar phenomenon whenever data and theory are wed.

Second, problems arise within medicine as a result of unusual capabilities and circumstances that were not in view as morality was forged. These in turn put pressure on ethics to answer questions for which it was not designed or to refashion itself in order to answer those questions. If refashioning occurred in the face of these special circumstances, would not the result be what is truly medical ethics and not simply ordinary ethics applied to special circumstances? Very likely, but it is not at all clear that such refashioning has been necessary. However, it is a possibility to which the reader should remain alert. The situation seems analogous to that of the implied rights and powers of the Constitution. Do these implications, having been drawn out and articulated to meet special circumstances, really comprise a new constitution, or are they simply applications of the "old" Constitution to new circumstances?

Challenges to this view of medical ethics

Before going on to expand the limited case of medical ethics to the more general instance of bioethics, there are questions that ought to be posed concerning the preceding view of medical ethics.

Proliferation of kinds of ethics. A sort of *reductio ad absurdum* might go like this: If every area where ethics is applied gets its own name, we will have a proliferation of "kinds" of ethics beyond compare. There will be "baker ethics," "banker ethics," "barber ethics," "bartender ethics," "bicycler ethics," and on and on.

But this does not seem to be a good argument against the view presented, simply because what purports to be *absurdum* is not so absurd after all. All the different realms of activity do in fact have ethical concerns. (Indeed, physicians frequently wonder why *they* are singled out for ethical emphasis instead of car mechanics and myriad others!) What sounds silly, of course, is that each would have its own name. That we do not in fact have separate names for the infinite variety of activities where ethics is applicable can be explained in other ways. Most of these activities are more easily understood, present few conceptual difficulties, do not encounter unique situations that strain ordinary ethics, and are not intimately and deeply involved with human life, death, and well-being. Hence they are not highlighted with names of their own. Notice that some instances of naming ethics for particular areas of concern do not strike us as odd at all: "environmental ethics," "business ethics," "legal ethics," "political ethics." The labels both express and reinforce any focus on concerns that hold for us considerable importance, interest, and complexity.

Actually the *reductio ad absurdum* could be turned on the very position it purports to be defending, namely, that medical ethics is a special kind of ethics, with its own principles, rules, and methods. By extension this might suggest that every area of concern and activity has its own ethical principles—butcher, baker, candlestick maker, and everybody else. This would make a farce of ethics, which at the very least should be the same for all of us.

Why should ethics not change? Another critical challenge to the position taken on medical ethics could be stated this way: If new knowledge is so rapidly increasing in medicine (and other fields) how can the ethics *not* change? We now understand matters differently and different situations obtain, so why shouldn't the ethics change?

To respond to this question requires that a distinction be drawn between basic and derived moral rules (Gert, p. 67). Basic moral rules would not be culturally relative; they would be those we would find applicable to all persons at all times and places. (It is not being argued here that there are such, but only that such a distinction is clarifying.) These basic moral rules would have to be general in not referring to particular beliefs, practices, or institutions. For example, "Do not be unfaithful to your spouse" would not be a basic rule, since it counts on the institution of marriage, which may not be a universal phenomenon. However, "Do not cheat" might well be a basic moral rule, independent of particular cultures and its institutions. What actions constitute cheating in a culture will depend upon the beliefs, expectations, and institutions of that culture. What remains universal is "Do not cheat"; what actions are seen as cheating will, of course, vary with understandings, practices, and contexts.

A derived moral rule would be one that was implied by a basic moral rule in conjunction with a particular practice or institution. Thus, "Do not cheat" plus the institution of marriage may yield a derived moral rule such as, "Do not be unfaithful to your spouse." Similarly, "Do not kill" (or "Do not cause pain," or some such) plus the existence of cars and highways may elicit the derived moral rule, "Do not drive while drunk." This could not be a basic moral rule because cars and, perhaps, intoxicating beverages did not exist in every place and time.

The relevance of this for medical ethics is that new knowledge and practices may indeed yield new *derived* moral rules (what some may call "middle principles"), though not necessarily new *basic* moral rules. Suppose that the essence of the basic moral rules were that they proscribed causing harm. What constitutes harm and what procedures and substances are seen as causing harm may change from one discovery to another, so that what one is specifically admonished not to do may vary, though the basic rule "Do no harm" remains unchanged. We now avoid giving extra oxygen to neonates, because we have discovered the bad consequences; we are now beginning to question the advice not to impose discipline on children, because of the results. In all cases the goal remains the same, namely, to avoid causing harm; only our view of the means to that end has changed. ("Avoiding harm" is used throughout this article to suggest the general focus of the ethical theory assumed herein. But it is only very roughly accurate; it is neither necessary nor sufficient for an act's morality.)

The point of Van Rensselaer Potter's *Bioethics* can best be seen within the context of derived rules. Potter's 1971 book must have made one of the first uses of the term "bioethics," making comment on it here appropriate. His use of the term is entirely different from this article's position, though the thesis of his book is quite consistent with what is said here. For him, bioethics is the enterprise of utilizing the biological sciences for improving the quality of life. It also helps us formulate our goals by helping us better understand the nature of man and the world. It is the "science for survival"; it helps write prescriptions for "happy and productive lives" (Potter).

It seems odd to call applied science "ethics." Much of science has been used to advise us on improving the quality of life, but we have not been tempted to call it "ethics." Such science is seen as a means to an end, and both means and ends might be subject to moral appraisal. In the terms of this article, science does help with the formulation of *derived* moral rules. That is, given that we already have general moral rules proscribing harm to one another, science may help fill in that list of actions which will in fact lead to harm.

But Potter also says that we can use science to formulate our goals by better understanding the nature of man and the world. What can this mean? Again, it really seems to be dealing with means. The biosciences might tell us what some or even all persons in fact find deeply satisfying, or that nuclear pollution can cause cancer. But it still does not tell us whether we are obligated to promote another's satisfaction, or have a right to our own satisfaction, or whether the benefits of nuclear power outweigh its lethal detriment. In short, it seems odd to call such an enterprise "ethics," though this in no way demeans the great importance of scientific knowledge for carrying out our purposes and achieving fulfillment.

Traditional meaning of medical ethics. The last challenge is an appeal to what has been traditionally meant by "medical ethics." This has included an extensive range of admonitions varying with culture, time, and type of medicine, and dealing with demeanor, rebates for referrals, decorum in consultations, soliciting another's patient, rumor mongering about other doctors, advertising, and endless other behavioral possibilities within medicine. There is no denying that this is largely what has been known in the past as "medical ethics." But does this traditional view of medical ethics invalidate the view of medical ethics taken in this article? The answer is probably that it partly does and partly does not. That is, traditional medical ethics is such a conglomeration of things that some aspects fall in nicely with the view espoused here, and some do not. But staking claim to the term "medical ethics" is not the goal; understanding is. What follows is an attempt to sort out several strands within traditional medical ethics and to put them in perspective from an ethical point of view. The upshot will be that one strand supports the espoused view, another is consistent with it, and a third is irrelevant to it.

The strand that confirms the view expressed here is that which is attentive to the harm that can be done others. Urging the physicians to keep confidences, to comfort patients when therapy is no longer effective, and not to insult one another might be seen as a translation of general morality into the medical context. These, and many such rules, can be seen as tailoring the general rules of morality, which are binding on us all, to specific situations within medicine. The object, as in general morality, is (very roughly) to avoid causing harm to individuals and to the community at large (e.g., by a general breakdown of trust).

Consistent with the stated view is that strand which announces what the physician's "duty" is; such as, "Consider the well-being of the patient above all else," or "Do not have sex with patients," or even those imperatives guiding referrals or pharmaceutical rebates. A strong case could be made for the rule "Do your duty" as being a basic moral rule, which as such would hold for all people in all times and places (Gert, pp. 121–125). Of course the duties themselves of different people in their various roles (as mother, as neighbor, as airline pilot, as lawyer, etc.) vary immensely, but what remains the same is the responsibility for doing your duty, whatever it is (so long as it is consistent with the other moral rules). In essence the justification for such a rule is that we all, by arbitrary decision or tradition, come to count on each other's doing his duty. We organize our lives around this expectation. Failure to fulfill one's role causes harm to all who had depended on its being performed. So it is a rule that we all would endorse and urge others to follow, inasmuch as noncompliance hurts us all. Again, what the specific duties are is by no means universal; they could easily vary from one community to another. But whatever they are, we all urge that they be done, because it is important for our functioning that we be able to depend on their performance.

A great deal of medical ethics can be seen as an effort to decide and make explicit what these duties are or should be for physicians. Of course, there is nothing universal about the particulars arrived at, and they fluctuate according to different society structures, differing views of medicine, health, and cure, and different capabilities. The important thing is that they be explicit so that society knows what to expect and that they be consistent with the other moral rules. Indeed, many of the delineated duties are best seen simply as rational social arrangements. However lofty and solemn they may sound, they are at bottom a spelling out of "promises," shaped by abilities, needs, traditions, and expectations. They could be quite other than they are and still be good, so long as they made explicit what could be expected by patients. This interpretation fits nicely with our tendency to refer to these matters as medical "etiquette" rather than "morals." That tendency bespeaks our realization that these matters are of a different order from morality. Yet, as these last two paragraphs have attempted to show, these "social arrangements" are the spelling out of something that is itself a basic moral rule, namely, "Do your duty."

The third strand distinguishable in codes of medical ethics seems not to be concerned with ethics, according to the position taken in this article. The strictures, "Keep heads clear and hands steady," "Be humble," and "Give respect and gratitude to teachers" sound like tactical advice. Rather than admonishing one not to cause this or that kind of harm, these sayings are recommending how one's life might be enriched and how one might even promote happiness in others. In the ethical view assumed (rightly or wrongly) in this article, these statements are more a "guide to life" than morality; but it is important to observe that other views of ethics might regard them as pertaining to the essence of morality.

In this category one might also place those admonitions that are really self-serving—which protect the physician and his guild and which promote medicine as a profession. These elements are best regarded as strategical advice. Examples are here being avoided because they are individually open to a variety of interpretations, and this is not the place to argue for one or another interpretation, but only to establish this classification. An instance would be the restriction against advertising: Though it seems primarily self-serving, a case might be made for its protecting the public.

Because "medical ethics" has meant such a variety of things, and because our ethical concerns with medicine often lead us beyond issues of medicine narrowly defined, another term might be useful. "Bioethics" is a good candidate. This term offers the advantage (1) of suggesting a much wider concern than simply rules of behavior for a particular guild, and (2) of connoting more broadly both the ethics and the science–technology out of which the issues grow.

Bioethics

The plan has been to examine medical ethics prior to dealing with "bioethics." One reason for this approach was that the same stance would be taken with respect to both, namely, that both are instances of ordinary morality applied to new areas of concern. Another reason was the expectation that, if medical ethics is understood as a more particularized case of ethical concern, the expansion to the more general case of bioethics will be easier. Furthermore, the variables unique to each are thereby kept separate, facilitating clarity.

The proposition that "medical ethics" is basically just ordinary ethics applied to the realm of medicine may be vague, but is not very surprising. On the other hand, it may seem more surprising to say that bioethics is just ordinary ethics applied to the "bio-realm," yet that is what will be held, if only to force the relevant issues into the open.

"Bio-realm" obviously designates a much more inclusive universe of concerns than does "medical realm." The impetus for ethical reflection in areas connected with the biological sciences seems to have been biology's startling capabilities both actual and projected, and the consequent implications for humankind. Though capabilities in the physical sciences and in technology have had similar disconcerting implications and have generated some ethical and societal reflections, it seems to have been recent biomedical developments that have precipitated the major ethical concerns which we group under "bioethics." The ripeness of the times for moral sensitivities was no doubt a contributing factor; civil rights, environmental issues, and population problems were high in the general consciousness. In any case, "bioethics" became the focal point, and subsequently many science-technology-society issues became its concern. This has resulted in a conglomeration that almost defies characterization. The field is more likely defined de facto in terms of its issues than by any shared essence or scientific perspective. (It should be noted that there is an issue of values-in-science that is not being raised in this article. It concerns whether and how value judgments encroach on science, influencing—according to some—the problem selection, observation, reasoning processes, and drawing of conclusions. These values are as frequently aesthetic, political, and social as ethical, and the matter has a literature of its own that predates the recent emphasis on bioethics.)

Assuming that bioethics is just ordinary ethics applied to a specialized area of concern, it becomes a matter of interest to determine whether problems arise in this special area that *in principle* are not amenable to traditional ethics. Whether or not they do cannot be settled here, but identifying and examining the pressures and tensions can be instructive. What is at issue is whether this new field of bioethics will have to develop new ethical principles (not just "derived rules" as discussed earlier) in order to deal with the problems confronting it. What follows should be seen as challenges to traditional ethics, brought on by new developments in the biological sciences. Their success as challenges will not be decided here, inasmuch as that would require a depth study of traditional ethics far beyond the scope of this article. Furthermore, the challenges themselves are too new to have their full complications and implications drawn and assessed. Nevertheless, what follows should elicit some underlying issues and suggest to the reader still other possible challenges.

Concerning what is "naturally given." The basic point of this challenge can be quickly seen, though it is not at all clear that it can be sustained under careful scrutiny. It is this: The "naturally given" seems, rightly or wrongly, to influence moral considerations, generally by constituting the inviolable starting point, the point of departure for ethical deliberations. Now, if we are able to tamper biologically with the very framework within which ethical deliberations make sense, then it would seem that more basic ethical principles must be formulated to guide us in how and what may be tampered with. But this point needs considerable elaboration. It is helpful to look at this from the perspectives of individual ethics and social ethics.

Individual ethics. In individual ethics, when two people encounter each other, the "natural" makeup of each constitutes the given, the point of departure for moral behavior. The preferences, aversions, anger, temperament, etc., of each must be worked with and around by the other. These characteristics constitute the inviolable sphere of the self and the zone of privacy, and, in effect, litigation takes place among these givens. Each person tries not to harm the other unduly, given the other's temperament, aversions, and such characteristics. But what if these personal characteristics could be changed, either before birth, during childhood, or now? What if we could cause a person to have less of a temper or preferences that would coincide with ours? As it is now, we start with complete persons as given—as completed units that barter, bargain, give and take, fashioning a symbiotic relationship that may restrain or require certain behavior but leaves the basic units or persons intact. However, with new capabilities we could perhaps alter "that" person before we even encounter him as a person.

What would guide our decision as to whether and how we would change this other? Currently, we are guided by rationality, that is, by those principles of behavior we can both agree on, being willing to follow them and have them followed with respect to us. But what criteria could we use if we were involved in changing the very makeup of the other? One answer might be: only with the consent of the other. If so, there is no challenge to standard ethics. Once again it is treating the whole person that confronts us as a given; it is not really altering that given until *after* the encounter, and then with that person's permission.

Changing a person without his consent is certainly wrong, and ordinary ethics would find it so. But we can make our problem tougher by casting doubt on the quality of consent (that the "consenter" is not really informed, that he is not in the right frame of mind, etc.). That is, reasons can of course be found for doing something against someone's expressed desire whether the act would be clearly a harm or a benefit. But even this does not necessarily force us beyond the ken of ordinary ethics, inasmuch as it is similar to problems we have had all along.

Tampering with personality characteristics prior to birth may push the problem further. There would be no ethical difficulty correcting a physical defect known to be painful or disabling to the eventual person. But what would justify changing—or refusing to change—the emotions, attitudes, or capabilities of the eventual person? The issue is no longer one of simply avoiding the causing of evil (a principle we would expect all rational beings to advocate) or even one of preventing evil. Rather this has to do with creating a good, and hence it is beyond the scope of ethics (as assumed in this article), which is essentially an avoidance of evils. Here we would not just be avoiding the causing of an evil to some given; we would be *creating* the given. Whereas we might expect universal agreement on what evils to proscribe, rational persons can and do differ widely on what things are good. Hence the problem: What criteria could be appealed to in the cases of whether and how to tamper with the endowments of eventual persons? On an individual basis there seem to be no guidelines and no possibility of there ever being any. It seems that guidelines could be framed only in terms of social goals. So the temptation would be to allow the "natural lottery" to continue; the randomness seems more fair in the absence of universal agreement on what would constitute a good person.

There may be promise of finding relevant

ethical guidelines embedded in our raising children, which is, in a sense, creating the eventual person. Yet raising children is also largely fashioned by social goals rather than by ethical imperatives. That is, in molding children we go far beyond merely what morality requires. Furthermore, in raising children as distinguished from juggling genes, it is generally assumed that our influence is not irreversible. If so, this would be a morally significant difference between the two.

Social ethics. New capabilities of altering what is "naturally given" might raise problems in the area of social ethics as well. The matter of just distribution may come up in a way it never has before. Perhaps society will no longer be willing to rely on the "natural lottery" of talents, dispositions, capabilities, etc., if these variables could be "evened" up before birth. It might be argued that justice would require an equal distribution of talent if that should become possible. Is some new principle of justice required? Or would it simply call for a decision as to the general social desirability of having so many similar and equal characteristics and talents? Yet, if these characteristics make a difference in the advantages one has in society, justice may well require us to make the characteristics equal. (Of course, the most likely solution would be not to make them all alike, but to stop allowing these characteristics to make a difference in one's obtaining rewards and status in society!)

It may be that this problem is no different from any other problem of justice. That is, we have believed justice required an evening up of talents and opportunities, insofar as this is possible with special training. Perhaps there is no difference in principle between balancing the disparities before and after birth. This does not make figuring the moral policy any easier, but it would indicate that it was the same old problem we have always had, and not something peculiar to new technological capabilities.

However, there are slight differences between the tampering before birth and the compensating endeavors after birth, and perhaps someone could argue that the differences are morally relevant. Before birth the mutation of talents, dispositions, and abilities is "artificial," that is, it is induced. After birth it takes some effort of that person to overcome his shortcomings; opportunity would be equal, but results would vary with the effort put forth, which seems intuitively fairer than "magically" making them all equal *in utero.*

Another difference between the two cases which might be seen as morally relevant is the matter of consent. Initiating change after birth at least might have some form of consent from the person himself or from his duly determined proxy.

Similarly some might see a morally relevant difference in the implicit employment of a standard or norm in the case following birth. Justice has seemed to require that corrective help be available to those who fall short in mental, physical, or dispositional ability. But to do this, what is "normal" in each category must be established. Then the obligation is to bring them up to normal, though there seems to be no obligation to help the already normal to rise above normal (or, at least, the below normal seem to have priority on the resources for improvement). Presumably, with the before-birth case, the magical equalizing injections would maximize each attribute rather than recognize norms. But it is not clear how this could be seen as a morally relevant difference of the before-birth case, such that it presents a challenge different in kind from the case following birth. There may be a question as to whether justice requires equal magical potions (and consequently unequal results) or unequal potions (in order to obtain equal results). But that is a frequent dilemma of justice. Furthermore, the postpartum case is probably *more* difficult, since it involves priorities (e.g., Which should be first, the neediest or the most promising?).

Perhaps the issue of distribution deserves further comment. Does the "new biology" (recent biomedical discoveries and technology, current and envisioned) in any sense challenge the "old" concepts of justice such that new basic principles must be articulated? This might be thought to be so in the case of our current practices of awarding certain benefits according to accomplishments and abilities. If those accomplishments and abilities which justified the awards were *themselves* induced by biologic tampering, then by what criteria would we determine who was to receive the benefits of biologic tampering? In short, if we can "artificially" create those merits in virtue of which certain benefits are bestowed, then how do we decide who should receive this merit-making magic? (Shapiro.)

Yet it is not clear that this is different in principle from the ordinary problems of distributive justice. Education and training given to everyone equally have become the basis for subsequent rewards. Of course, it might be argued that natural ability and effort are what is being rewarded. But even in the case of biologic tampering, effort would be relevant. So in any case it is not obvious that the "new biology" presses on us any problem of just distribution that we have not had in principle all along. Whether to award goods on the basis of effort, worth, need, or desert continues to be a basic question. These operate within the context of social ideology, which no doubt helps determine what gets emphasized. And if biotechnology enabled us to make everyone the same in all admirable respects (courage, ability, endurance, etc.) it would be arguments from boredom or versatility in evolution or some such social ideal that would lead us to distribute the goods differently.

The scope of morality. Another conceivable challenge might be voiced: Do the new developments in biological knowledge and technology expose as inadequate the compass of traditional ethics? Specifically, can traditional ethics handle obligations and rights in borderline cases (fetuses, the severely retarded, the totally senile) or among those living things normally outside the sphere of morality (animals, plants, streams)? The thrust of the challenge is that because of increased knowledge new capabilities and the resultant heightened sensitivities, we now see that rights must exist where traditional ethics had never envisioned them.

This cannot be done with the precision one would like. The concept of rights is a huge area. Furthermore, the variety of views on rights is compatible with a variety of ethical theories. But what can be done here is to sketch the general kind of reflection that might be stimulated by "the new biology" with respect to the inadequacy of the "old morality," and then to sketch a reply, arguing that no changes are necessitated, provided we understand rights in a certain way.

There has been a general raising of consciousness concerning the world around us. The intimate interrelationship of humankind with the environment, the interdependence of all living things, and the cumulative effect of any disruption of this balance are matters keenly impressed on us of late. Much of this results from the explosion of knowledge, particularly in the biological fields. It coincides as well with social and political movements that direct attention to environment, conservation, and appropriate lifestyles. The upshot has been pressure to reconsider the scope of moral rights and obligations, specifically whether they should be extended to include plants and animals. It is not enough, the argument goes, to have laws protecting trees and animals from various abuses; rather, we must think of them as having rights of their own, because this alone will give the necessary edge to their defenses, shifting the burden of proof to the proper contestant. With the prima facie right to noninterference anchored firmly in the living thing itself, the challenger would have to show why he should be allowed to harm it, rather than the "defendant's" having to show why not. In short, we would assume that one could not infringe on a tree or an animal without producing adequate reasons, instead of our assuming that these things may be freely infringed upon unless there are adequate reasons to the contrary.

Sensitivity to the plight of animals is not new, but there is a recent surge of concern that appears to be part of the general bioethics movement. Perhaps it is the great empathy for the voiceless oppressed (fetuses and minorities), perhaps it is increased awareness of all suffering, or perhaps it is the scientific eroding of essential differences between man and some animals, but whatever it is, it is being expressed in an effort to include animals within the moral realm (Regan and Singer).

The move to protect the environment is well known. What is not so well known is the possibility and need that elements of the environment have basic rights. One author argues thoroughly and convincingly that natural objects (such as trees and streams) have rights to life and health (Stone). Without the "discovery" of such rights, the abuse of these natural objects can be stopped only if detrimental effects to a human's health or livelihood can be shown. However, if a natural object's right to life and health were acknowledged, evidence of possible harm simply to it would suffice for enjoining the abuse. Stone argues that no amount of rules or laws protecting the object in question are ever equivalent to acknowledging *the right* of that object to health and life. The right has connotations, force, and presumption never embodied completely in a set of rules.

The preceding considerations, though only roughly sketched, may represent a thrust that will significantly alter traditional ethics in scope, if not in substance. These basic intuitions appear to put a stress on the framework of ethics which may force a change whereby bioethics will be essentially different from ethics. However, there are other ways of understanding the matter that, if correct, would obviate the need for basic revision. What follows is one such possibility. Like all else in this article, it is not a detailed argument, though it is a hint of one. Its role here is to suggest that the situation does not necessarily require a new ethics but that the old, adequately understood, can apply to the developing new circumstances.

The suggestion to be made could be called the concept of "bestowed rights." It builds on what is taken to be the purpose of morality, which in turn establishes the scope of morality. If an essential characteristic of a rational person is that he will avoid harm to himself (unless he has a reason not to), then we would expect that the community of rational persons would work to articulate and advocate a morality. That is, they would attempt to formulate rules and procedures to which they could all subscribe in order to settle their differences and otherwise avoid a life that would be "poor, nasty, brutish and short." Thus the rational person's instigating incentive for formulating and advocating moral rules is the avoiding of harm to himself and to others he cares about. As such, morality is a kind of agreement for mutual benefit, and rationality (avoiding of harm to oneself) is the foundation (Gert). Rationality may not compel one to be moral, but it would compel one to urge that *others* be moral. Furthermore, many things and ideas might motivate one to be moral, other than simply avoiding harm to oneself. It is just that avoidance of harm is the one motivation we can count on all rational persons sharing. This motivation has a purchase on those who can be responsible for their actions and who can realize that they are able to harm others and that others can harm them. All others unable to be gripped by this sanction are outside the community of moral agents.

This however does not mean that all others—the senile, the severely retarded, infants, fetuses, the insane—are without rights, but it does mean that the source of their rights is quite different. This is where the "bestowed right" becomes relevant. Those excluded from the moral community by virtue of their inability to act and plan responsibly and their inability to appreciate the force of the sanction at the base of morality are still significant members of society. The moral community has an interest in protecting them: We ourselves might one day be in their position; we may have an emotional attachment to them; many of us have a "natural sympathy" with all forms of life; we could all become brutalized and consequently suffer if we failed to respect these other lives. Therefore, rights are bestowed on beings outside the community of moral agents. These are extremely important rights, though their foundation is different from those within (Clouser, 1976; Green). This interpretation seems to be more realistic with respect to where *in fact* the sanctions lie and would help arguments for and against rights to focus on the appropriate grounds.

Fetuses, animals, and trees cannot act responsibly or express their interests and desires, nor can they threaten us with retaliation. There is no basis for them to "claim" rights. On the other hand, it would be terribly shortsighted of us not to ascribe rights to them, or bestow rights on them, to avoid long-range harm to ourselves. When neither their interest, nor desire, nor consent can be solicited, what other argument (other than the long-range good of humankind) could ultimately be made for infringing on what would be a basic right of any of these nonrational beings? The point is that no matter what we postulate as to the "basic rights" of these nonrational beings, our ultimate criterion for honoring the rights will be in terms of the long-range good of humankind.

The point of all this has not been the argument itself, but rather to show that bioethics is not necessarily a new ethics: It might still be seen as the old ethics being applied to new circumstances.

Other possible challenges. The matter of obligations to future generations is almost always included in any listing of bioethical issues, and yet it is not clearly a new issue. It is in fact a good example of what is probably the heart of bioethics, namely, an old issue that is exacerbated by new discoveries. For a long time we have been able to affect subsequent generations, but only recently have we been able to do it and realize it so vividly. So it is not new and different in principle, though its resolution may be more urgent. All attempts to deal with it have been attempts to draw out the implications of traditional morality, for example, by pointing out that the nature of obligation is always future oriented, or that the nature of a rule is timeless, or that from a moral perspective (whether that of the "ideal observer," the "rational man," or those "behind the veil of ignorance") all persons who are or ever will be must be regarded as contemporaneous.

In short, bioethics would seem to be the response of traditional ethics to particular stresses and urgencies that have emerged by virtue of new discoveries and technology. Ethics is pressed, not to find new principles or foundations, but to squeeze out all the relevant implications from the ones it already has. This, of course, does not solve every problem nor even necessarily narrow down the options significantly. Even after the ethical criteria are met, many alternatives may remain, thus requiring criteria other than morality for settling on a single line of action (Clouser, 1975).

This general theme is encountered time and again. The hackneyed example of the respirator, or of any life-extending apparatus, makes the same point. With new technology we are forced to make certain decisions with a regularity and urgency we had not faced before. This does not call for new ethical principles, but new applications of the old. Our concern for not harming a person and for honoring his wishes remains the same; we ponder only how to express it in these particular circumstances.

A similar case is the prevalent matter of behavior control. We have always been able to control some behavior, but new techniques make such control easier, quicker, and more certain, and have made moral clarity on the issue immensely more urgent. Again, the basic admonition not to interfere with a person's freedom without his consent still stands. But the new circumstances and capabilities drive us to more precision on what constitutes freedom, autonomy, and consent, in order to apply the basic rule. Sometimes it is clear that interference with freedom leads to even greater freedom (e.g., compulsory education); sometimes it is not clear whether meaningful consent is being given (e.g., Does the person *really* know the consequences and alternatives?); sometimes it is not clear whether the "real" self is speaking (e.g., the person before or after the mood-elevator, the shock treatment, or the hemodialysis). Furthermore, much of the concern (as is fairly typical of bioethical issues) has not been over *whether* to restrict the use of these new technologies, but *how*. That is, it is often clear when it is moral and when it is immoral to use these techniques of behavior control, but the problem comes in formulating a general policy that can make and incorporate the distinctions and subtleties necessary to capitalize on the worthy uses without risking the misuses.

Bioethics, science, and public policy

The position taken in this article has been that the revelations and capabilities mediated by science create an urgency for moral guidance but do not require a new morality, revised in its basic principles. What is required is analysis of the new circumstances and their pivotal concepts, so that the applications and implications of the "old" morality might be clearly seen. Additionally, ethics itself may be pressed to yield clarity and distinctions it had not heretofore manifested. But it would be odd to call this a "new" morality. We are more likely to see it as a further uncovering, exploration, and articulation of the old, simply because we think of morality as applying to all persons at all places and times. Morality is not invented or legislated; at most it is "discovered," that is, it is an unpacking, explication, and articulation of our deepest intuitions about what we ought and ought not to do. Of course new "derived rules" (as explained earlier) will emerge, as certain lines of action are discovered to lead to certain results that conflict with basic moral rules. It is in this role that the empirical sciences become very important. Spelling out the cause-and-effect relationships makes clearer where restraints should be imposed, so as to be in accord with the basic moral rules. The facts of sociology, biology, psychology, medicine, and other sciences are crucial in determining the outcome of various acts and policies, apart from which their morality generally cannot be judged. Analyses of pivotal concepts within these sciences (whether done by philosophers or scientists) are also crucial in determining morality: Has a person's "freedom" in fact been curtailed? Is he appropriately labeled "sick"? Was he acting "freely"? Can he really be said to have been "deceived"? And with this kind of conceptual analysis and figuring of consequences lies most of the work of bioethics. The remainder lies with probing the foundations of ethics for implications that relate more exactly to the novel predicaments characteristic of "the new biology."

Bioethics appears to have no essence that would mark it off. Rather it seems to be individuated by a de facto list of issues, extended and interrelated by "family resemblances." The initial set of dilemmas was no doubt introduced by biological discoveries, real and foretold. But we gradually became familiar with a wider list of issues: genetic engineering (cloning, in vitro fertilization, eugenics, etc.), allocation of limited health resources, obligations to future generations, environmental health and pollution, population control, abortion, euthanasia, behavior control (behavior modification, drugs, psychosurgery), human experimentation, organ transplantation, etc. By no means were all these forced on us by new discoveries; most have been around for a long time. Historical and sociological factors in addition to the moral urgency of new knowledge and capabilities contributed greatly to the coagulation of issues distinguishing itself from ordinary ethics. The result has been a considerable focusing on a set of issues, marked by urgency and importance.

Consequently, a common thread joining all the issues is hard to find. Invariably an article or book dealing with bioethics will mention quickly the "new biology" and the "new technology," and then immediately cite examples, namely, an enumeration of those issues which simply as a matter of fact are labeled "bioethics." Yet the conglomeration is not completely haphazard: Business ethics is not included, nor is economic ethics, nor diplomatic ethics. On the other hand, it is difficult to articulate a common essence that unifies the conglomerate. The core and impetus of the de facto list was no doubt concern over the technology of control of man's body, mind, and quality of life. The more such dramatic possibilities were highlighted and the more we became attuned to the issues, we began seeing similar, albeit less dramatic, instances all around us: Not only does the psychosurgeon alter our person, but so do parents, educators, and social systems; not only does the neonatologist determine quality of life, but so does the Highway Commission. Until we saw the grosser possibilities we were not sensitive to the subtler forces already at work. And thus the list grew. Perhaps an original criterion was that the issue have an obvious biological component, which affects coming into or going out of life, the nature of a person, or the quality of life. This of course directly or indirectly included almost everything. And so it must, to extend from the economics of feeding the world to paternalism in the doctor–patient relationship, from cultural determinations of illness and death to a tree's claim of a right to life! Adding to this the relevant data from law, sociology, psychology, biology, philosophy, theology, and other fields results in a vast network of issues categorized roughly as "bioethics." The aim of it all should not be lost in the vastness: to decide how humankind ought to act in the biomedical realm affecting birth, death, human nature, and the quality of life.

In this context, and as a concluding observation, the matter of public policy should be put in perspective. Much of what is compressed under the umbrella of "bioethics" is really concerned with public policy—that is, with the legislation, policies, and guidelines that should be enacted with respect to all these issues. It is helpful to see this as distinct from ethics per se. Though many views of ethics exist (and this is not the place to thrash out the differences) the assumption of this article has been, very roughly, that ethics concerns those prohibitions all rational persons would urge everyone to obey in an effort to avoid those evils on which they all agree. Specifically, disregarding these prohibitions in order to "promote good" would be regarded as immoral (unless, of course, they met criteria for justifiable exceptions).

However, on a societal level the matter is different. It is not a matter of one individual's deciding on his own how to treat another. Rather, if there is a democratically arrived at consensus that a certain good must be promoted (say, education or welfare), though it means causing some harm (taxation and some loss of freedom), it is morally acceptable. It is at this level that much of what is called "bioethics" is taking place. Whereas individual morality is primarily a system of restraints, society-wide policy is on a level where promotion of goods is a moral option. Here the question becomes "What goods ought to be promoted?" or, contrariwise, "Which goods ought to be restrained (e.g., scientific research)?" Priorities, values, and goods are at center stage, where they must be weighed, balanced, and compared. The sciences and humanities are highly relevant in analyzing and prognosticating and in otherwise assisting a pluralistic society to settle on goods to be promoted. This level and direction of discussion accounts for a large portion of what comes to be called "bioethics," although, on the ethical theory assumed in this article, it is not really ethics so much as value theory and political theory.

But there is a still higher level at which significant bioethical analysis is done. When a society promotes various goods, inevitably some moral rules are broken with respect to some individuals (depriving them of something, a sacrifice necessary to support the "common good"). Which rules are broken with respect to whom becomes an important consideration in determining the justifiability of the goals or goods fixed upon. This determination is the jurisdiction of justice, the ethics pertaining to the societal level. And insofar as this has to do with the distribution of benefits and burdens relating to biomedical issues, it is a crucial part of bioethics.

K. DANNER CLOUSER

[*Directly related are the entries* ACTING AND REFRAINING; ETHICS; FUTURE GENERATIONS, OBLIGATIONS TO; HEALTH CARE; JUSTICE; LIFE; MEDICAL PROFESSION; RATIONING OF MEDICAL TREATMENT; *and* SCIENCE: ETHICAL IMPLICATIONS. *See also:* ANIMAL EXPERIMENTATION; BEHAVIOR CONTROL; DEATH AND DYING: EUTHANASIA AND SUSTAINING LIFE; ENVIRONMENTAL ETHICS; GENETIC CONSTITUTION AND ENVIRONMENTAL CONDITIONING; INFORMED CONSENT IN HUMAN RESEARCH; INFORMED CONSENT IN THE THERAPEUTIC RELATIONSHIP; POPULATION ETHICS: ELEMENTS OF THE FIELD; *and* PSYCHOSURGERY.]

BIBLIOGRAPHY

BLACKSTONE, WILLIAM T., ed. *Philosophy and Environmental Crisis.* Athens: University of Georgia Press, 1971.
BROADIE, ALEXANDER, and PYBUS, ELIZABETH M. "Kant's Treatment of Animals." *Philosophy* 49 (1974): 375–382.
CALLAHAN, DANIEL. "Bioethics as a Discipline." *The Hastings Center Studies* 1, no. 1 (1973), pp. 66–73.
CLOUSER, K. DANNER. "Bioethics: Some Reflections and Exhortations." *Monist* 60, no. 1 (1977), pp. 47–61. This issue is devoted to bioethics.
———. "Medical Ethics: Some Uses, Abuses and Limitations." *New England Journal of Medicine* 293 (1975): 384–387.
———. "Some Things Medical Ethics Is Not." *Journal of the American Medical Association* 223 (1973): 787–789.
———. "What Is Medical Ethics?" *Annals of Internal Medicine* 80 (1974): 657–660.
FEINBERG, JOEL. "The Rights of Animals and Unborn Generations." Blackstone, *Philosophy and Environmental Crisis,* pp. 43–68.
FOX, RENÉE C. "Ethical and Existential Developments in Contemporaneous American Medicine: Their Implications for Culture and Society." *Milbank Memorial Fund Quarterly* 52 (1974): 445–483. Bibliography.
GERT, BERNARD. *The Moral Rules: A New Rational Foundation for Morality.* New York: Harper & Row, 1970.
GREEN, RONALD M. "Conferred Rights and the Fetus." *Journal of Religious Ethics* 2, no. 1 (1974), pp. 55–75.
HARDIN, GARRETT; STORR, ANTHONY; LEISS, WILLIAM; and SHEPARD, PAUL. "Rights: Human and Nonhuman." *The North American Review,* Winter 1974, pp. 14–42. A collection of papers.
KASS, LEON R. "The New Biology: What Price Relieving Man's Estate?" *Science* 174 (1971): 779–788.
KING, LESTER SNOW. "Development of Medical Ethics." *New England Journal of Medicine* 258 (1958): 480–486.
PASSMORE, JOHN. "The Treatment of Animals." *Journal of the History of Ideas* 36 (1975): 195–218.
POTTER, VAN RENSSELAER. *Bioethics: Bridge to the Future.* Prentice-Hall Biological Science Series. Edited by Carl P. Swanson. Englewood Cliffs, N.J.: 1971.
REGAN, TOM, and SINGER, PETER, eds. *Animal Rights and Human Obligations.* Englewood Cliffs, N.J.: Prentice-Hall, 1976.
SHAPIRO, MICHAEL H. "Who Merits Merit? Problems in Distributive Justice and Utility Posed by the New Biology." *Southern California Law Review* 48 (1974): 318–370.
SINGER, PETER. *Animal Liberation: A New Ethics for Our Treatment of Animals.* New York: New York Review, Random House, 1975.
STONE, CHRISTOPHER D. "Should Trees Have Standing?—Toward Legal Rights for Natural Objects." *Southern California Law Review* 45 (1972): 450–501.
VEATCH, ROBERT M. "Medical Ethics: Professional or Universal?" *Harvard Theological Review* 65 (1972): 531–559.

BIOETHICS AS A DISCIPLINE

See MEDICAL ETHICS, HISTORY OF, *section on* EUROPE AND THE AMERICAS, *the six articles on* CONTEMPORARY PERIOD: THE TWENTIETH CENTURY, *especially the article on* NORTH AMERICA IN THE TWENTIETH CENTURY. *See also* MEDICAL ETHICS EDUCATION; *and* BIOETHICS.

BIOHAZARDS

See RISK; RESEARCH POLICY, BIOMEDICAL; GENETICS AND THE LAW; TECHNOLOGY, *article on* TECHNOLOGY AND THE LAW.

BIOLOGICAL WARFARE

See WARFARE, *article on* BIOMEDICAL SCIENCE AND WAR.

BIOLOGY, PHILOSOPHY OF

The philosophy of biology refers to the systematic investigation of biology from the standpoint of its fundamental assumptions, propositions, and theories as well as its methodology. In its wider sense the philosophy of biology includes issues related to speculative metaphysics (e.g., the philosophy of Henri Bergson), evolutionary ethics (as initially proposed by Darwin in *The Descent of Man*), and biologically founded social philosophy (e.g., "the self-reproducing society" of Julian Huxley). Within the narrower sense of the term, the philosophy of biology concerns itself with the logical status of biological theory—in particular with the issue of the reduction of such theory to that of physics and chemistry. This article will concern itself first with the more limited view of the philosophy of biology insofar as it bears on the ethics of the life sciences, medicine, and health care.

The problem of reduction

The question of whether or not biology can be reduced to physics can be viewed in several ways. From the viewpoint of bioethics, the issue of reduction bears significantly on such issues as the sanctity of life or the uniqueness of living things. In controversies dealing with euthanasia, abortion, and genetic engineering, arguments will often cleave along lines separating those who believe that living systems are ultimately complex physical organizations from those who hold that living systems embody certain organizational features not found in the inanimate world and, which, thereby, render them unique. The response to biological reductionism can take numerous forms, including theological. The work of Teilhard de Chardin, for instance, represents an effort to indicate the uniqueness of man by incorporating him into an overall cosmological evolutionary process. At the same time, Teilhard de Chardin holds that science is aptly suited to deal with living phenomena, thereby avoiding the criticism that such a view of man is prima facie nonscientific.

It is clear from the outset that biological systems are markedly different in their behavior from nonbiological or inanimate ones, although such entities as crystals and viruses seem to share certain important characteristics with living systems. On both a macro- and a microscopic level biologists deal with a class of entities that are, for the most part, clearly delineated from the rest of the physical world. To the extent that mitochondria, membranes, and genes are not rocks, stars, and water, biology may be regarded as not reducible to physics.

In a somewhat more interesting sense, biological terminology may be regarded as autonomous with respect to physics in that its theories and propositions are phrased in a vocabulary not derived from the physical sciences. Thus, genetics and evolutionary theory are systematic attempts to order certain phenomena regarding individual "organisms" and multiple "species" in terms not borrowed, derived from, or defined in terms of physical science. Insofar as the initial goals of these two biological theories are concerned, what is contained within other scientific theories has no formal bearing on the integrity of these intrinsically biological points of view. In a restricted sense there seems to be no need to use any terminology or scientific law outside their respective self-defined domains when describing biological events. The question remains, though, as to the ultimate theoretical possibility of stating the terms and propositions of biology in the language of physics. That is, whether or not the science of biology is able to describe its own phenomena by using an indigenously biological, theoretical, and descriptive vocabulary is quite independent of the issue as to whether or not the laws of biology are ultimately statable in terms of those of physics.

Biology can be viewed as a series of empirical or descriptive generalizations having neither the universality of physical laws nor the capacity to be incorporated into a comprehensive theory dealing with its extensive domain. There are two ways out of this minimal characterization of biology. The first is to deny any difference in kind between physical and biological laws. The second way explores critically the theoretical possibility of the reduction of biological to physical theory.

This second point of view is that of organismic biology and general systems theory. Rather than assert that biology is totally reducible to physics either theoretically or methodologically, or that biology should not enjoy the same respect as physics with regard to its status as a science, such men as Ludwig von Bertalanffy and E. S. Russell claim that biology has its own conceptual structure distinct from that of physics. The domain of biology is distinct and different enough to warrant its having unique principles not derivable or transferable from physical science. Thus, they claim that the organization of living systems is hierarchical and so different from inanimate modes of organization that the laws of physics cannot apply adequately in an attempt to explain fully the workings of living things. As von Bertalanffy claims, the organizational principles applying to the ordering of living systems are to be derived from general systems theory and not from physics.

The issue concerning the reducibility of biology to physics and chemistry has both historical and logical dimensions; that is, the arguments for or against reduction can proceed by asserting that history is on one side or the other. Classically, this is the position of the mechanists contra the vitalists (both of whose positions are defined below), who assert that the history of biological research clearly proves that vitalism is a dead issue. History, for the mechanists, seems to obviate any need to provide a formal argument that definitively dismisses vitalism as a viable theory of living systems. On the other hand, history is never definitive with respect to the open texture of science, and there is little doubt that a need exists for a formal demonstration that one or the other position or neither is correct. In this regard certain logical features of vitalism, mechanism, and holism or organismic biology will be considered briefly at this point.

Mechanism, vitalism, and holism

Mechanism. To view a living system "mechanically" may involve more than one sense. There are at least three different meanings to be ascribed to the concept of mechanism insofar as it applies to the nature of biological entities:

1. "Mechanical explanation" refers to the kind of explanation that is based entirely on the laws and theories of the science of mechanics.
2. "Physicochemical" refers to the statements constituting mechanical explanation that utilize the terms and concepts of physics and chemistry.
3. "Causal explanation" in biology states that all propositions describing a living system gain their intelligibility by being subsumed under at least one causal law.

It should be stated that classical mechanics is often touted as more than it is. For instance, it is clear that this branch of physics is insufficient to serve as a foundation for physics as a whole. Similarly, its extension to the realm of biology becomes even more problematical in this light. Also, while mechanics in a more general sense does not in itself constitute an ethical position, it does have ethical implications. This is the view of Jacques Loeb, who, in the period just prior to the First World War, asserted in his book *The Mechanistic Conception of Life* that the "riddle of life" whose solution lies in the principles of mechanics bears directly on the ethical principles upon which our lives ought to be run. The view of life as essentially mechanical influences conceptions of life and death, the autonomy of the individual, and the problem of free will.

Vitalism. Opposed to the mechanistic view is the philosophy of vitalism, which for the most part is largely abandoned in contemporary biological theory, although, historically, vitalism seemed to offer a bridge to ethics (e.g., Bergson's *élan vital*). In its least sophisticated form it asserts that the matter out of which living things are constituted is qualitatively different from inanimate matter. With the development of biochemistry and the occurrence of such events as the synthesis of urea by Wohler in the nineteenth century and, more recently, the unraveling of the alpha helical structure of DNA, this "substantive" view of vitalism is unsupportable. A somewhat more sophisticated view offered by the biologist Hans Driesch is, in essence, an entelechy theory, i.e., a theory of the organism which asserts that living systems operate in accordance with mechanistic principles up to a point and then require the guidance of an immaterial vital force in order for certain processes to occur—in particular, adaptation and repair. As a theory of explanation, the entelechy is an explanatory deus ex machina that is employed when traditional explanatory prowess is exhausted. Clearly, anyone at all sympathetic to the tenets of logical positivism would be unalterably opposed to such a viewpoint since Driesch's entelechy could have no empirical referent whereby propositions containing the term could be either verified or refuted. The vitalist position, then, served as an ideal target for Moritz Schlick, who attacked it quite clearly from the point of view of logical positivism by showing that it violated the verifiability principle upon which that entire philosophical school was built. For Schlick, ethics was reduced to the psychological expression of the emotions.

Driesch, however, felt that his argument for an entelechy also was an argument for the existence of a soul. This latter point is part of a broad metaphysical system Driesch erected around his considerations about the nature of life. Thus, individual responsibility and the personal awareness of a sense of obligation is derived from a suprapersonal wholeness that is continually evolving toward a goal which is unknown to each individual. Driesch's endowment of material man with a soul constitutes an aspect of his metaphysics in which the individual is seen to have an evolving spiritual dimension.

What mechanical explanation is to mechanism, teleology is to vitalism. That is, explanation in terms of end-points, goals, and the like is viewed as being most naturally fitted to the vitalist's position. An important point, however, is that a subscription to a teleological explanatory point of view does not necessarily commit one to the more severe ontological position of vitalism. Hence, while the vitalism–mechanism debate is essentially resolved in favor of the former, its heir is, in part, the debate over whether or not teleological explanation as well as teleological phrases have any place in rigorous scientific discourse and whether they are capable of being or ought to be translated into linguistic forms congruent with mechanical explanation.

Holism. A resolution of the vitalism–mechanism controversy has implications for bioethics in that such a solution necessarily commits itself to a view of the organism (and possibly the self) wherein certain ethical considerations concerning the uniqueness of the living system may be suggested. Such a resolution is offered by the concept of holism. Holism asserts that the organization of biological systems is what distinguishes them from nonliving systems, and hence, in a manner of speaking, the whole of the living system is greater than the sum of its parts. That is, the propositions constituting an explanation of all of the physical and chemical components of the living system cannot, by their very nature, suffice to elaborate the intrinsically biological or living character of the organism. Specific organizing principles, which, as von Bertalanffy holds, derive from general systems theory, are necessary if the living system is to be understood in its uniqueness and distinctness from the rest of the natural world. In particular, the organism is viewed as a certain sort of hierarchical organization whose individuality is a function of such a mode of constituency. Organisms, from this point of view, are multilayered and complex phenomena the explanation of which requires specific principles specifying the character of the hierarchy. This position has considerable appeal to many biologists and is sometimes associated with the not unproblematic and broader metaphysical concept of emergentism, which, by the way, can be viewed as a methodological principle rather than a more speculative metaphysical one. Emergentism holds that the properties of such hierarchically arranged systems cannot be predicted through an elucidation of the parts of that system. For example, the consistency, taste, and color of water is not deducible from an examination of the properties of hydrogen and oxygen alone. The property of water is seen as an emergent one with respect to the properties of its individual constituents. With organisms, then, their properties are a function of their complex organization, and a mechanical explanation of the various constituents of the organism cannot suffice to specify the peculiar emergent traits that are manifest as a result of this organization. An objection to holism, or organismic biology as it is called by some, is that organizational principles are involved in nonliving systems also, so that the concept of hierarchical organization of even a highly complex form does not suffice as a logical principle whereby living and nonliving systems are distinguished from one another.

Just as with mechanism and vitalism, holism too can be seen to have implications for ethics. The organismic biologist bases his argument for biological uniqueness not upon the existence of a vital principle but upon principles of organization to which science is more congenial. Hence it is possible, from the perspective of the holist, to argue for individuality and the uniqueness of life as does Driesch, and at the same time not abandon sound scientific principles. Furthermore, mechanism is not viewed as the only way to be scientific in these matters. Given such a position, issues such as free will and moral responsibility may take on somewhat different perspectives. The idea of free will may be reconcilable with that of the material basis of life since the organizing relations that characterize the living system are such as to render it unique, with the possibility that this uniqueness entails human freedom.

In broad view, then, each point of view concerning reductionism, mechanism, and vitalism carries with it certain ethical considerations. Mechanism denies the existence of a soul, whereas vitalism proposes it. The reductionist may (but not necessarily) argue on the basis of his position that free will is illusory. The vitalist or organismic biologist will argue to the contrary. Hence, such technical considerations in the philosophy of biology bear directly on bioethical considerations, to which we now turn.

Biology and ethics

There are two broad ways in which ethics can intersect with biological science such that philosophically and biologically interesting questions are raised. The first has to do with man's intervention in what have been traditionally regarded as natural processes of which man is, to a greater or lesser extent, a part. The second concerns itself with the possibility of developing explicit ethical or metaethical points of view from various parts of biology, usually those areas related to the theory of evolution. These two areas will be considered in turn.

Intervention in natural processes. For the greater part of his history man has looked upon natural processes, both biological and physical, as happenings over which he had little or no control. Disease, evolutionary trends, mutations, and the like were not regarded as susceptible to the willful or even the inadvertent actions of human beings. Whatever occurred in the natural world, living or nonliving, was—except for that small realm in which man had some influence—regarded as the product of natural, i.e., non-human forces. With the growth of biological methodology and modern medicine (e.g., the germ theory of disease), the vision of man as a largely passive recipient of the whims of nature began to decline. Thus, concerns about man's willful and accidental intervention in the sphere of natural processes has grown enormously in the twentieth century. Along with this concern have come ethical considerations that bear directly on these issues. This instance of ethical reflection constitutes not a new view of ethics or metaethics but rather a new sphere for those modes of ethical inquiry already in vogue.

As examples of these new concerns and the questions of priority that are raised, two illustrations having to do with willful and inadvertent human intrusion into natural processes can be offered here. The first is the question of the moral responsibility of those individuals who are engaged in efforts to synthesize living organisms and quasi-living systems such as viruses. As the emergentists point out, it is virtually impossible to predict the possible pathological potential of any successfully synthesized organism for the human race. It is a real possibility, for instance, that a severely pathogenic virus could be synthesized for which no treatment is known and which could wipe out large portions if not all of the world's population. Is the growth of biological knowledge in this domain worth the risks entailed in this example? On the other hand, one could argue, such research might result in the elimination of most of man's diseases. Thus, the possibility of an epidemic is countered by the possibility of an opposite practical consequence of the research. The legislation of such decisions as to whether or not to pursue a certain line of inquiry in the biological sciences constitutes an area in which ethical considerations can have the profoundest consequences.

In the case of inadvertent intrusion into natural processes, the issue is often raised with respect to ecological considerations. A chemical plant may produce valuable products for industry and society at large but may, at the same time, pollute the river that is adjacent to it. In this case there are at least two considerations to be raised. First, one must ask what priorities are to be held and in what order they will be embraced. If, for example, the public desires a clean river more than a particular chemical, then through any number of social and political processes, the plant may be compelled to close its operations. This case is not unlike the last one stated, in that a question of priorities is raised and specific ethical considerations can be brought to bear on the problem in an effort to resolve it. The second consideration, however, is rather more problematic in that it raises the issue of man's responsibility both to other living things (in this case, the fish and plants in the river) and to other natural entities (the river and its environs). How far does the right to life extend, for instance? What is the value of natural beauty in the hierarchy of values individuals construct for themselves in order to run their lives? Is it morally defensible to extinguish a species of wildlife because of certain human needs or desires? These are questions that biology, as it has grown, raises in the minds of philosophers, politicians, and the general citizenry.

Evolution and ethics. An interesting point of intersection of ethics and biology is seen in the development of specific ethical or metaethical points of view based upon the theory of evolution. The work of C. H. Waddington and Sir Julian Huxley serve as examples in this context. What has been called evolutionary ethics consists in the building of ethical theories or systems from the framework of evolutionary theory. Biologists, more than philosophers, have been drawn to this form of theorizing. They see in it an external source of values for mankind. Evolutionary order can, say the evolutionary ethicists, serve to provide a hierarchy of values, a process for legislating between good and evil, and a theory about the moral roots of human nature. It is certainly an optimistic point of view but one that has not had the enthusiastic endorsement of many philosophers. The main reason is that the later philosophers have traditionally held the opinion that such an approach to ethics involves the "naturalistic fallacy." This fallacy, whose explication lies initially in the work of Hume and which was analyzed in twentieth-century philosophy by G. E. Moore, points out the logically illicit deduction of an evaluative statement, moral or otherwise, from solely descriptive statements about the world. The extraction of value statements from descriptive ones is precisely the naturalistic fallacy of evolutionary ethics, according to its philosophical critics. While there has been some weakening of the role that the naturalistic fallacy has played in some ethical works of the mid-twentieth century, its deeply rooted place in the minds of many moral philosophers disposes many critics to discount those who seek to derive an ethical theory or a theory of ethics from the theory of evolution. Thus, those moral principles based upon the patterns and processes of evolutionary phenomena may have to await more hospitable times when philosophers are not as ill-disposed to such theorizing or else find another base of justification.

A newer approach to the issues of bioethics and evolution is to be found in the work of Edward O. Wilson on sociobiology. Wilson takes it as a major task of his to explain altruism, a trait not readily explainable in terms of traditional evolutionary theory. By changing the focus of concern from that of the organism to that of the gene, Wilson is able to develop his theory of altruism along highly creative lines. From the perspective of bioethics, he is involved in the nature/nurture controversy (siding more with the nature theorists) and issues related to social Darwinism. Whether Wilson's approach to ethics via examination of genetically influenced or determined behaviors is ultimately successful remains to be seen. As a movement in bioethical theory, however, it is the most recent and, perhaps, the most significant one in the latter part of the twentieth century. How successful this approach is to meeting the above-mentioned criticisms leveled against other bioethical theories will, in part, determine the ultimate impact of the point of view it puts forth.

IRWIN SAVODNIK

[Directly related are the entries ENVIRONMENTAL ETHICS, article on QUESTIONS OF SOCIAL JUSTICE; EVOLUTION; MEDICINE, PHILOSOPHY OF; and REDUCTIONISM. For discussion of related ideas, see the entries: ENVIRONMENT AND MAN, article on WESTERN THOUGHT; FREE WILL AND DETERMINISM; and MIND–BODY PROBLEM. Other relevant material may be found under ETHICS, article on NATURALISM; and SCIENCE: ETHICAL IMPLICATIONS.]

BIBLIOGRAPHY

AYALA, FRANCISCO JOSÉ, and DOBZHANSKY, THEODOSIUS, eds. *Studies in the Philosophy of Biology: Reduction and Related Problems*. Berkeley: University of California Press, 1975.

BECKNER, MORTON. *The Biological Way of Thought*. Berkeley: University of California Press, 1968.

BERTALANFFY, LUDWIG VON. *Modern Theories of Development: An Introduction to Theoretical Biology*. Translated by J. H. Woodger. New York: Harper & Row, 1962.

——. *Problems of Life: An Evaluation of Modern Biological Thought*. New York: John Wiley & Sons, 1952.

BLUM, HAROLD FRANCIS. *Time's Arrow and Evolution*. Princeton: Princeton University Press, 1951.

CANFIELD, JOHN V., ed. *Purpose in Nature*. Contemporary Perspectives in Philosophy Series. Englewood Cliffs, N.J.: Prentice-Hall, 1966.

FLEW, ANTONY GARRARD NEWTON. *Evolutionary Ethics*. New Studies in Ethics, vol. 8, no. 6. New York: St. Martin's Press; London: Macmillan & Co., 1967.

GRENE, MARJORIE GLICKSMAN, and MENDELSOHN, EVERETT, eds. *Topics in the Philosophy of Biology*. Boston Studies in the Philosophy of Science, vol. 27. Synthese Library, vol. 84. Boston: D. Reidel, 1975.

HALDANE, J. S. *Mechanism, Life and Personality: An Examination of the Mechanistic Theory of Life and Mind*. London: John Murray, 1921; New York: E. P. Dutton & Co., 1923. Reprint. Westport, Conn.: Greenwood Press, 1973.

HEMPEL, CARL GUSTAV. *Aspects of Scientific Explanation, and Other Essays in the Philosophy of Science*. New York: Free Press; London: Collier-Macmillan, 1965.

HULL, DAVID L. *Philosophy of Biological Science*. Prentice-Hall Foundations of Philosophy Series. Englewood Cliffs, N.J.: 1974.

HUXLEY, JULIAN SORELL. *Evolutionary Ethics*. Romanes Lecture. London: Oxford University Press, 1943.

HUXLEY, THOMAS HENRY. *Evolution and Ethics and Other Essays*. Huxley's Collected Essays, vol. 9. New York: D. Appleton & Co., 1929.

LOEB, JACQUES. *The Mechanistic Conception of Life*. Edited by Donald Fleming. Cambridge: Belknap Press, Harvard University Press, 1964.

MUNSON, RONALD, ed. *Man and Nature: Philosophical Issues in Biology*. A Delta Book. New York: Dell Publishing Co., 1971.

NAGEL, ERNEST. *The Structure of Science: Problems in the Logic of Scientific Explanation*. New York: Harcourt, Brace & World, 1961.

RUSE, MICHAEL. *Philosophy of Biology*. London: Hutchinson, 1973.

SCHAFFNER, KENNETH F. "Theories and Explanations in Biology." *Journal of the History of Biology* 2 (1969): 19–33.

——. "The Watson–Crick Model and Reductionism." *British Journal for the Philosophy of Science* 20 (1969): 325–348.

SCRIVEN, MICHAEL. "Explanation in the Biological Sciences." *Journal of the History of Biology* 2 (1969): 187–198.

SIMON, MICHAEL A. *The Matter of Life: Philosophical Problems of Biology*. New Haven: Yale University Press, 1971.

SIMPSON, GEORGE GAYLORD. *The Meaning of Evolution: A Study of the History of Life and of Its Significance for Man*. New Haven: Yale University Press, 1967.

WILSON, EDWARD OSBORNE. *Sociobiology: The New Synthesis*. Cambridge: Belknap Press, Harvard University Press, 1975.

WOODGER, JOSEPH HENRY. *Biological Principles: A Critical Study*. Reissued, with a new Introduction. International Library of Psychology, Philosophy and Scientific Method. London: Routledge & Kegan Paul; New York: Humanities Press, 1967.

BIOMEDICAL RESEARCH

See RESEARCH, BIOMEDICAL; HUMAN EXPERIMENTATION; ANIMAL EXPERIMENTATION; RESEARCH POLICY, BIOMEDICAL.

BIRTH CONTROL

See CONTRACEPTION.

BIRTH DEFECTS

See FETAL–MATERNAL RELATIONSHIP; GENETIC DIAGNOSIS AND COUNSELING; GENETIC SCREENING; INFANTS; LIFE, *article on* QUALITY OF LIFE; PRENATAL DIAGNOSIS. See also MENTALLY HANDICAPPED.

BLOOD TRANSFUSION

The procedure

The process called blood transfusion is the introduction of whole blood or blood derivatives directly into the body's circulatory system. Transfusion replenishes depleted blood volume as in cases of hemorrhage, burns, injuries to blood vessels, and shock during surgery, or corrects blood disorders as in anemia, leukemia, hemophilia, and immune deficiencies.

Current clinical practice benefits from techniques for separating and storing blood derivatives developed during the past thirty years. Any of the cellular components of blood (white, red, and platelet cells), plasma alone, or various plasma proteins may now be used to respond to specific clinical needs.

Efforts to meet the demand for blood must deal with the constraints of short storage life, safety limits for repeated donations, and exclusion of some potential donors to protect recipients from blood transmissible diseases, notably, serum hepatitis (Mollison).

Ethical problems

Transfusion generates three sets of ethical problems: religio-cultural attitudes toward blood, social organization for securing blood, and maintenance of professional standards.

Attitudes toward blood. Confronted by cultural traditions not deeply affected by the scientific world view, donor recruitment programs must often overcome beliefs about evils resulting from the drawing of blood, e.g., impotence, infertility, general weakness, susceptibility to witchcraft, or guilt or sacrilege (League of Red Cross Societies, p. 8). In such circumstances, transfusion becomes involved in ethical problems of resolving conflicts in cross-cultural values. Issues of power, coercion, independence, and individual and cultural dignity enter into the task of securing needed blood supplies.

Attitudes also affect the receiving of transfusions. Cultural values may oppose the mingling of racially different blood, e.g., in South Africa and the Southern United States (Titmuss, p. 193). Blood service policies are thus drawn into large ethical issues of interracial justice and equality. On religious grounds, Jehovah's Witnesses oppose any transfusion as being a use of blood forbidden by God (Farr, p. 44). Refusal of transfusion for racial or religious reasons raises two sets of problems rooted in personal autonomy and religious liberty: first, refusal by an adult for himself; second, refusal by an adult responsible for a dependent person, such as a child or unconscious spouse.

Social organization of blood supply. Questions of distributive justice and social solidarity enter into the evaluation of blood supply systems, of which there are two basic approaches: voluntary unpaid donors and quid pro quo. In the latter, blood may be exchanged for cash, social rewards, fringe benefits, or security regarding transfusions for self or family (Titmuss, pp. 75 ff.).

If equitable distribution of burdens and benefits is taken as the norm, cash systems appear to be the least just. Here the poor and derelict become the major suppliers of blood, since the motivational appeal addresses their economic situation most directly. Yet they are the least able to afford blood when they need it, because the cost of commercially purchased blood is from five to fifteen times higher than blood supplied through voluntary unpaid donor systems (ibid., pp. 205 ff.).

Other quid pro quo systems appeal especially to the working and lower middle class. The base of appeal is broader than cash systems, the spread of the burden more equitable, and added costs, if any, less than those of cash systems. However, the widest spread of the burden is achieved in voluntary unpaid donor systems. These also generate the least cost in the process of supplying blood and keep the benefit of blood transfusion most widely accessible (ibid., pp. 120 ff.).

Unpaid systems produce blood supplies of higher quality than cash systems. Despite improvement of tests for discovering blood transmissible hepatitis (Mollison, pp. 603 ff.), blood supply systems must rely heavily on donors' testimony that they are not infected. However, in cash systems the major suppliers come from populations noted for high incidence of hepatitis and have a conflict of interest in revealing a fact that would disqualify them from selling blood. Studies indicate that post-transfusion hepatitis rates are much higher when blood from commercial banks is used.

Blood supply systems give rise to the issue of social solidarity, because the giving of blood is an important means of promoting community consciousness and altruism (Titmuss, pp. 237 ff.). In voluntary unpaid systems, blood is literally a gift of oneself for the benefit of unknown neighbors. Giving the gift in the knowledge that there are many other such givers helps create confidence in one's community. Quid pro quo systems counteract these values. They tend to atomize individuals in private worlds of self-interest by closing off this opportunity for the expression of altruism and community consciousness.

Professional standards. The traditional norm of medical ethics, "Do no harm," dictates that, in the handling, processing, and administration of blood and blood products, great care must be taken to avoid injury to recipients resulting from human and computer errors in blood group identification, screening for contaminated blood, cross-matching, labeling, and patient identification. Similar care must be taken to protect donors from excessive giving of blood, infection, and various postdonation hazards (Mollison, pp. 2–10).

The same norm, at the level of social policy, requires concern about commercial methods for supplying blood. Both the International Red Cross and the World Health Organization have recommended voluntary unpaid donor systems as preferable for protection against blood transmissible diseases (Bowley, p. 16).

MICHAEL J. GARLAND

[*Directly related are the entries* HEALTH CARE, *articles on* RIGHT TO HEALTH-CARE SERVICES *and* THEORIES OF JUSTICE AND HEALTH CARE; ORGAN DONATION; ORGAN TRANSPLANTATION, *article on* SOCIOCULTURAL ASPECTS (*for the concept of gift*); RATIONING OF MEDICAL TREATMENT; *and* RIGHT TO REFUSE MEDICAL CARE. *See also:* PROTESTANTISM, *article on* DOMINANT HEALTH CONCERNS IN PROTESTANTISM; *and* RACISM.]

BIBLIOGRAPHY

BOWLEY, C. C.; GOLDSMITH, K. L. G.; and MAYCOCK, W. D'A., eds. *Blood Transfusion: A Guide to the Formation and Operation of a Transfusion Service.* Geneva: World Health Organization, 1971.

FARR, ALFRED D. *God, Blood and Society.* Aberdeen, S. Dak.: Impulse Publications, 1972.

League of Red Cross Societies. *IIIrd Red Cross International Seminar on Blood Transfusion: Blood Donor Motivation and Training of Auxiliary Personnel for Blood Transfusion Centers.* Medico-Social Documentation no. 27. Geneva: League of Red Cross Societies, 1965.

MOLLISON, PATRICK L. *Blood Transfusion in Clinical Medicine.* 5th ed. Oxford: Blackwell Scientific Publications, 1972.

TITMUSS, RICHARD M. *The Gift Relationship: From Human Blood to Social Policy.* New York: Random House, 1971.

BRAIN DEATH

See DEATH, DEFINITION AND DETERMINATION OF, *article on* CRITERIA FOR DEATH.

BRITAIN

See MEDICAL ETHICS, HISTORY OF, *section on* EUROPE AND THE AMERICAS, *articles on* MEDIEVAL EUROPE: FOURTH TO SIXTEENTH CENTURY; INTRODUCTION TO THE MODERN PERIOD IN EUROPE AND THE AMERICAS; BRITAIN AND THE UNITED STATES IN THE EIGHTEENTH CENTURY; BRITAIN IN THE NINETEENTH CENTURY; *and* BRITAIN IN THE TWENTIETH CENTURY.

BUDDHISM

Buddhist teaching: a model on the pattern of pathology

The Buddha endeavored to heal diseases of the mind. He was often called "the Great Physician" or "the King of Physicians." The outline of an important Buddhist teaching called "the Four Noble Truths" was made according to the pattern of pathology (*nidāna*) of ancient India. Why are we afflicted by suffering in our worldly existence? Why are we involved in the round of transmigration? The Buddha's inquiry into the cause of worldly suffering or afflictions proceeded on a practical, psychological level, and he discovered that the real cause of human suffering is ignorant craving or blind desire (*tṛṣṇā*). He then showed human beings that the right and effective way to deliverance from suffering is by the cessation of such craving.

The Buddha is then called a physician metaphorically. Just as a doctor must know the diagnosis of the different kinds of disease (Jolly), and must know their causes (*Vinaya, Mahāvagga* VI, 14, 1–5), and the antidotes and remedies to cure them, and must be able to apply them ("Greater Discourse"), so also the Buddha taught the Four Noble Truths, in order to indicate the range of "suffering," its "origin," its "cessation," and the "way" which leads to its cessation. According to this doctrine, if we can get rid of the cause of suffering, we shall be able to attain deliverance, comparable to the healing of a disease.

The Buddha's general opinion about the problems of ethics is typically expressed in the doctrine of the following "Four Noble Truths" (Jolly):

1. "The Noble Truth concerning suffering or afflictions. Birth, decay, disease, death, union with the unpleasant, separation from the pleasant, and any craving that is unsatisfied are forms of suffering. In brief, suffering springs from instinctive attachment, that is, the conditions of individuality." Very much of this truth is comparable to the diagnosis of diseases.

2. "The Noble Truth concerning the origin of suffering. It is the craving, comparable to thirst, that causes the renewal of the process that is accompanied by sensual delights, and seeks satisfaction, now here, now there, that is to say, the craving for the gratification of the senses, or the craving for existence, or the craving for non-existence." Our craving is so strong and blind that it was compared by the Buddha to thirst. When we are thirsty, we cannot help desiring water, forgetting everything else. In the same way our craving compels us to seek objects. This truth is comparable to knowing the causes of diseases.

3. Next, the healing of diseases should be sought. "The Noble Truth concerning the cessation of suffering. It is the vanishing of afflictions so that no passion remains. It is the giving up, the getting rid of, the emancipation from, the harboring no longer of this craving or thirst." The word "cessation" (*nirodha*) originally and etymologically meant "control." To control this craving thirst is truly the ideal state.

4. Finally, the means for healing should be made clear. "The Noble Truth shows the way that leads to cessation of suffering. It is the Noble Eightfold Path which consists of right views, right aspirations, right speech, right conduct, right mode of livelihood, right effort, right mindfulness, and right concentration."

These eight are comparable to remedies of an illness. The Path pointed out by Gautama is called the Noble Path; the truths enumerated are called the Noble Truths. They permit full elaboration by his followers. All the items of the Noble Fourfold Truths were commented upon in full detail by later expositors.

In Mahāyāna Buddhism medical science (*cikitsā-vidyā*) was one of the five sciences which should be learned, the other four being metaphysics (*adhyātma-vidyā*), logic (*hetu-vidyā*), science of language (*sabda-vidyā*), and technology (*silpakarmasthāna-vidyā*).

Health care

Despite the medical model that can be seen to underlie the Buddha's teaching, early Buddhism discouraged monks from engaging in medical care because at that time medical science was considered a sort of worldly pursuit, classified with magic and sorcery.

"Let a monk not apply himself to practicing [the hymns of] the Āthabbaṇa [the Atharvaveda], to [the interpretation of] sleep and signs, nor to astrology; let not my follower [a Buddhist monk] devote himself to [interpreting] the cry of birds, to causing impregnation, nor to [the art of] medicine [*tikicchā*; literally, 'treatment of diseases']" (*Suttanipāta*). But to help the sick was regarded as a virtue. A scripture of early Buddhism tells the following story. In the days of the Buddha, a sick brother was once neglected by the other members of the monastery. The Buddha washed him and tended him with his own hands, saying afterward to the negligent monks, who would have been eager enough to serve him, "Whosoever would wait upon me, let him wait upon the sick." He claims his oneness with humanity so that service to the sick or the destitute is in reality rendered to himself (*Vinaya, Mahāvagga* VIII, 26).

Subsequently, Buddhist orders established medical institutions. For example, monasteries of early Buddhism had halls for the sick monks. During the reign of King Aśoka (third century B.C.), a man greatly influenced by the compassionate spirit of Buddhism, hospitals were established for sick humans as well as for sick animals. This antedated anything similar in the West and animal clinics (*pinjpols*) are found to this day in Western India (Bhandarkar).

In his Rock Edict II, King Aśoka tells us that in his own dominions as well as in those of neighboring potentates, he established two kinds of medical treatment, one relating to humans and the other to animals. And he further informs us that medical herbs, roots, and fruits, whenever lacking or in short supply, were imported and planted everywhere. After Aśoka's reign hospitals were institutionalized in ancient India.

From the records of the eighteenth century it is quite clear that in two states, Mahārāṣṭra and Gujarāt, kings and chiefs frequently arranged for free medical help to be given to the needy and indigent; as a consequence, the physician was often rewarded with grants of rent-free land in a village, and in some cases the purpose of the grants was expressly stated to be the growing of medicinal herbs on the land.

The care of the sick was recognized as a duty and a meritorious act in all Buddhist countries and is recommended by the example of Buddha himself. The *Mahāvaṃsa*, a chronicle of Ceylon, repeatedly asserts that kings founded hospitals and distributed medicines. The Buddhist king Buddhadāsa of Ceylon (fourth century A.D.) cured patients, appointed doctors and provided them some compensation, established asylums, and wrote a medical work entitled *Sāratthasaṅgaha* (Turnour). A certain doctor named Caraka is said to have been the personal doctor of the Buddhist king Kaniska (second century A.D.).

After Buddhism was introduced into China sometime in the first century A.D., Buddhists engaged in a great many altruistic activities in accordance with the ideal of "Compassion." Ever since Shih-lo (A.D. 329?), of the later Chao period, the inculcation of Buddhist teachings became widespread. The social work of the Buddhist priests was especially noticeable in its medical treatment and relief of the poor. In the Eastern Chin period (A.D. 317–420), Fo-t'u-ch'eng, Fa-k'uang, K'o-lo-chieh, An-hui of Loyang, and Tan-tao of Lo-ching-shan helped people obtain medical attention and care. In the T'ang dynasty (618–907), the system of temple hospitals was established, and institutions for the poor, the sick, and the orphaned were built. Buddhist priests endeavored to build bridges, plant trees, dig wells, and construct rest houses.

Prince Shōtoku (574–622), the founder of Japanese Buddhism, established Shitennōji Temple in 587 in what is now the city of Ōsaka, and this temple was renowned for its creative undertaking of the relief of suffering people. The temple was laid out in four main divisions: Kyōden-in, the great central hall or religious sanctuary proper, used for training in Buddhist discipline and in aesthetic and scholarly pursuits; Hiden-in, a hall where the poor could obtain relief; Ryōbyō-in, a hospital or clinic where the sick could receive treatment without charge; and Seyaku-in, a dispensary where medical herbs were collected, refined, and distributed free of charge. It is not clear whether Prince Shōtoku established an animal hospital there, but judging by the name, the Kyōden-in (which means "institution based on respect for existent beings") was aimed at promoting the happiness and welfare of all living beings, human and animal alike. Moreover, according to the *Nihongi*, Shōtoku and other members of the imperial family, as well as officials of the court, used to set aside fixed days for the purpose of gathering medicinal herbs, and the court is known to have shown special consideration to the needy, whether living alone, destitute, or aged.

Empress Kōmyō (eighth century) engaged in philanthropic activities. Legend has it that in the public bathing room she washed and tended a leprous beggar covered with loathsome sores, who suddenly revealed himself as the Buddha Akṣobhya.

Later, the priest Ninshō (1217–1303) began such social welfare works as caring for the suffering and the sick. He dedicated his whole life to the service of others and was even criticized by his master, Eison: "He overdid benevolence." Although it was a breach of customary discipline to dig ponds or wells or to give medicine and clothing to the sick or to collect money for them, Ninshō never let himself be deterred from doing any of these things.

In Cambodia an inscription of Jayavarman VII (A.D. 1185?) reveals that there were 102 hospitals in his kingdom. He evidently expended much care and money on them. One is led to suppose that Buddhism took a more active part than Brahmanism in such works of charity (Eliot).

Medical science as revealed in Buddhist literature

The Bower manuscripts are the oldest extant works of Buddhist influence. The manuscripts, which were found in a Buddhist cairn in Kashgar (China), were probably written by immigrants from India about A.D. 450. The sixth and seventh texts of the Bower manuscripts often refer to Buddha, Tathāgata, Bhagavat, and synonyms of Buddha (Jolly). The texts describe types of early medical treatment.

In a scripture entitled "Nine Causes of Unexpected Early Death" the reader is warned against overeating, eating wrong foods, and disregarding Buddhist precepts. The ways in which one should console a seriously sick person at death are also explained.

The Chinese Buddhist pilgrim I-tsing (671–695) gives a detailed description of the Indian medicine of his time, including symptoms of bodily illness, rules for giving medicine, medical treatment and the use of herbs, rules for diagnosis, and fasting to cure disease (I-tsing). Contents of medicines are not specified. In ancient India the urine of cows was used as medicine, attributable to the general tendency in India to worship cows, and Buddhists were no exception. In Japan this custom was never followed. Medicines obtainable by killing animals were disliked because of the traditional ideal of not injuring any living beings. But nowadays these medicines are tolerated in Japan because of the persistence of the idea of Shinshu Pure Land Buddhism, which tolerates killing of animals in case of necessity but with the feeling of sinfulness.

Vajrayāna, a school of Mahāyāna, which developed after the fifth century A.D. in India, incorporated various forms of popular beliefs prevalent among common people. Wherever there is any reference to diseases and treatment in Vajrayāna scriptures, the usual course recommended is the recitation of spells (dhāraṇīs).

Ancient China and Japan have left a large number of medical texts which have not yet been explored and it has not as yet been ascertained to what extent they have been influenced by Buddhism.

Mutilation and suicide

In Buddhism mutilation of any sort of the human body was abhorred and discouraged. It was assumed that legal punishment should be limited to imprisonment, confiscation, scolding, and so on; but capital punishment and mutilation should not be inflicted upon criminals. The aim of legal punishment is to save one from evils and to make a man of good character. This ideal was advocated in Buddhist sacred books.

Buddhism does not necessarily prohibit suicide, but according to Buddhist sacred texts suicide is meaningless, for by resorting to suicide one cannot save oneself from the miserable condition of mundane existence (or transmigration); karma (actions) are supposed to accompany a person who has committed suicide even after his death. The shocking cases of self-immolation by Buddhist monks in order to protest against the policy of the ruling class, as took place in Vietnam in recent times, were not based upon any tenet of Buddhist teachings. This kind of action is not encouraged in any Buddhist teaching. No Japanese Buddhist priest committed suicide in order to protest either in wartime or during the American occupation. In China and Japan there were some rare cases in the past when disciplined Buddhist monks committed suicide, immolating themselves by starvation resulting from rigorous spiritual practices. They were respected. Some of them by way of repentence or redemption redeemed their own sins or crimes, even for murders committed while young. But such self-immolation was quite exceptional and has no scriptural approval.

HAJIME NAKAMURA

[*For further discussion of topics mentioned in this article, see the entries:* HOSPITALS; PAIN AND SUFFERING; *and* SUICIDE. *Also directly related are the entries* ENVIRONMENT AND MAN, *article on* EASTERN THOUGHT; MEDICAL ETHICS, HISTORY OF, *section on* SOUTH AND EAST ASIA, *articles on* PREREPUBLICAN CHINA, CONTEMPORARY CHINA, JAPAN THROUGH THE NINETEENTH CENTURY, TRADITIONAL PROFESSIONAL ETHICS IN JAPANESE MEDICINE, *and* CONTEMPORARY JAPAN: MEDICAL ETHICS AND LEGAL MEDICINE. *Compare:* CONFUCIANISM; HINDUISM; *and* TAOISM. *See* APPENDIX, SECTION I, FIVE COMMANDMENTS AND TEN REQUIREMENTS.]

BIBLIOGRAPHY

BAGCHI, PRABODH CHANDRA. "New Materials for the Study of the *Kāśyapasaṃhitā* in Chinese." *Indian Culture* 9 (1942?): 53–64. Cf. *Taishō*, no. 1385.

———. "New Materials for the Study of the *Kumāratantra* of Rāvaṇa." *Indian Culture* 7 (1940?): 269–286. A translation of *Taishō*, no. 1330, also known as *Rāvaṇakumāra-tantra*.

BHANDARKAR, DEVADATTA RAMAKRISHNA. *Asoka.* The Carmichael Lectures, 1923. Calcutta: University of Calcutta, 1925. 3d ed., rev. 1955, pp. 167–168.

The Book of the Discipline (Vinaya-Piṭaka). 6 vols. Vol. 4: (*Mahāvagga*) [The Great Division]. Translated by Isaline Blew Horner. Sacred Books of the Buddhists, vol. 14. London: Luzac & Co., 1951.

Le Canon bouddhique en Chine: Les Traducteurs et les traductions. Translated by Prabodh Chandra Bagchi. 2 vols. Sino-Indica: Publication de l'Université de Calcutta, vols. 1, 4. Paris: P. Geuthner, 1927–1928, vol. 1, pp. 302–303. Cf. *Taishō*, no. 793.

"Channa." *The Book of the Kindred Sayings (Sanyutta-nikāya) or Grouped Suttas.* 5 pts. Pt. 4: *The Saḷāyatana Book (Saḷāyatana Vagga).* Translated by F. L. Woodward. Pali Text Society Translation series, no. 14. London: Routledge & Kegan Paul, 1972, text iv, 55; XXXV, II, 4, sec. 87(4); pp. 30–33. Ways of consoling a seriously ill person on his deathbed. *Saṃyutta* or *Sanyutta* is the Pali recension of *Saṃyukta*.

Eliot, Charles Norton Edgecumbe. *Hinduism and Buddhism: An Historical Sketch.* 3 vols. London: E. Arnold & Co., 1921; Routledge & Kegan Paul, 1954–1957; New York: Barnes & Noble, 1954, vol. 3, p. 124.

"Greater Discourse on the Lion's Roar (Mahāsīhanāda-sutta)." *The Collection of the Middle Length Sayings (Majjhima-Nikāya).* Translated by Isaline Blew Horner. 3 vols. Vol. 1: *The First Fifty Discourses (Mūlapaṇṇāsa).* Pali Text Society Translation series, no. 29. London: Luzac & Co., 1954, chap. 12, pp. 108–110. *Mahāsīhanāda-sutta* I, 82–83. *Majjhima* is the Pali recension of the *Madhyama*.

I-tsing. *A Record of the Buddhist Religion as Practised in India and the Malay Archipelago, A.D. 671–695.* Translated by Junjirō Takakusu. Oxford: Clarendon Press, 1896, pp. 126–140. For the theory of the four humors, see p. 130. The standard romanization for the author's name is I-ching.

Jolly, Julius. *Medicin.* Strassburg: K. J. Trübner, 1901, pp. 14–15.

Sen, Satiranjan. "Two Medical Texts in Chinese Translation." *Visva-Bharati Annals* 1 (1945): 70–75. Cf. *Suśruta, Sūtrasthāna* XIV, 34–35; XXIV, 8.

Suttanipāta, Tuvatakasutta, v. 927. "The Sutta-nipâta: A Collection of Discourses, Being One of the Canonical Books of the Buddhists." Translated from the Pâli by V. Fausböll. *The Sacred Books of the East.* Edited by F. Max Müller. Oxford: Oxford University Press, 1881; Delhi: Motilal Banarsidass, 1965, vol. 10, pt. 2, sec. 14:13, p. 176.

Suśruta, Sūtrasthāna I, 24; I, 37; XXI, 3–4.

Taishō Shinshū Daizōkyo [Taisho Edition of the Tripiṭaka in Chinese]. Edited by Junjirō Takakusu, K. Watanabe, and G. Ono. 100 vols. Tokyo: 1929–. Sometimes called the *Taishō Tripiṭaka.* Some interesting portions on medicine are: *Ekottaragāma-sūtra;* Chinese version, vols. 44, 50, 31; *Taishō,* vol. 2, pp. 776, 811, 670. *Saṃyuktāgama-sutta;* Chinese version, no. 684 and vol. 41, no. 1122; *Taishō,* vol. 2, p. 186, pp. 297c–298b. Also: *Taishō,* vol. 17, pp. 591, 747; vol. 2, p. 883ab. *Rāvaṇakumāra-tantra; Taishō,* no. 1330, see Bagchi, "New Materials for *Kumāratantra.*" *Kaśyapa-ṛṣi-prokta-strīcikitsā-sūtra; Taishō,* no. 1385, see Bagchi, "New Materials for *Kāśyapesaṃhitā.*" *Bhaisajyarāja-sūtra; Fo shuo fo yi wang king; Taishō,* no. 793, see *Canon Bouddhique en Chine.* Also: *Taishō,* nos. 1059, 1060.

Turnour, George, ed. and trans. *The Mahāwanso in Roman Characters with the Translation Subjoined; And an Introductory Essay on Páli Buddhistical Literature: In Two Volumes. Vol. 1 Containing the First Thirty Eight Chapters.* Ceylon: Cotta Church Mission Press, 1837, pp. 243–245. Only one volume published. Translation reprinted, with notes and emendations, and completed by L. C. Wijesiṇha, Colomba: G. J. A. Skeen, 1889. *Mahāwanso* is a varient of *Mahāvaṃsa* or *Mahāvaṇsa,* attributed to Mahanama, 5th century.

Vinaya, Mahāvagga VIII, 26. *Book of Discipline,* vol. 4, pp. 431–434.

Vinaya, Mahāvagga VI, 14, 1–5. *Book of Discipline,* vol. 4, pp. 278–279. See also *Vinaya, Mahāvagga* VIII, 1f, *Book of Discipline,* vol. 4, p. 379f.

Warder, A. K. *Indian Buddhism.* Delhi: Motilal Banarsidass, 1970. Contains a bibliography and guide to texts.

BURIAL
See Cadavers.

CADAVERS

I. GENERAL ETHICAL CONCERNS
    Christian A. Hovde

II. JEWISH PERSPECTIVES
    Walter S. Wurzburger

I
GENERAL ETHICAL CONCERNS

Several questions arise in thinking about the disposal and use of the human cadaver. The following article, after a brief reflection on the significance of the human cadaver, centers on autopsy, the use of cadavers as a source of organic tissue, experimentation on fetal cadavers, and burial and cremation.

The significance of the body

Like other animal bodies the human body is a combination of highly organized cell populations mutually dependent upon each other and, in proper combination, forming something more than the cell populations themselves. Unlike other animal bodies the human body is *human*, that is, it participates in the dignity of the human person. Even after death the human cadaver reminds us that once there was present in our midst a human person, a being like ourselves, sharing in our dignity and value.

For this reason civilized people have always been outraged at cannibalism and at the desecration of the dead. The cadavers of human persons are not, as Homer sings, to be the feast of birds nor are they to be considered as commercial property, a point brought out in a legal decision in Rhode Island in 1872 when the court declared: "There is a duty imposed by the universal feelings of mankind to be discharged by someone towards the dead; a duty, and we may say a right, to protect from violation; and a duty on the part of others to abstain from violation" (*Pierce v. Swan Point Cemetery*, 10 R.I. 227 [1872].

It is not that the human cadaver is something of inherent value, for it will soon decay unless measures are taken to prevent this. Rather it is that the human cadaver is a symbol of the human person, a reminder to us of our life together as a community of persons. As a symbol, the human cadaver demands respect from the living.

Autopsy

Autopsy procedures are routinely used in medical institutions today. For reasons of public health and safety, the bodies of those who have died without benefit of medical aid or in whom the cause of death is not immediately apparent are subjected to autopsy carried out in order to discover the cause of death and to establish where possible the responsible agent of that death.

Those who see no symbolic value in the human cadaver find few problems with autopsy other than the aesthetic ones having to do with mutilation. But for those who believe that the human cadaver still possesses human significance some problems are raised. Even if one grants that the common good of the entire community can be served by an autopsy, one wants to secure that good and at the same time to protect the cadaver from violation, to prevent this symbol from being used inhumanly. Thus those who, for religious or humanistic reasons, regard the human cadaver with respect, while permitting autopsy for valid reasons, insist that tissues thus examined be restored to the cadaver before burial.

Historically the first major impetus for autopsies was provided when Frederick II, emperor of the Holy Roman Empire, instructed physicians studying at Salerno or Naples to spend at least one year in the study of anatomy. The imperial edict was given about A.D. 1240 (Corner; Packard). This was followed by edicts from Pope Sixtus IV, A.D. 1414–1484, and Pope Clement VI, A.D. 1478–1534, permitting the opening of the human body for dissection (Lassek, p. 74). Christian theologians, in commenting on these edicts, expressed the belief that such dissection of the human cadaver could be done with proper respect for the dead, so long as the organs were restored to the body prior to burial.

On an individual level, emotional reactions to the prospect of postmortem desecration or mutilation of one's own body or that of someone cared for very often influences the way in which a person responds to the possibility of autopsy. Most people will acknowledge intellectually that something worthwhile may be discovered through the use of autopsy procedures. But to many, the feeling of guilt over their failure to protect that body during the last hours of life produces a rejection of autopsy in order to protect it after death. Although such feelings can be overcome and often are, the clash produces pain, uncertainty, and conflict of interest. Coupled with the emotional response is an idea that has some support in past history. In earlier years, bodies used for dissection or autopsy were obtained from the ranks of malefactors, disreputable people, paupers, or unclaimed persons (ibid., p. 73). In the United States cadavers have at times been obtained by criminal means through grave robbing and murder, and such desecration of the body contributed to the pejorative attitude of very many people toward autopsy and medical school dissection.

Since medical science cannot afford to be ignorant of the detailed structure and functioning of the human body, especially in the teaching of new physicians, dissection of the body must be undertaken. The Uniform Anatomical Gift Acts enacted in most of the states in the United States have provided a more acceptable and uniform method of providing cadavers or body parts (ibid., p. 254; Luyties, 1969). These acts make it mandatory that an unclaimed body be reported by hospital or morgue authorities to a centrally located office and prepared for use there. When there is need for cadaver material, the body is turned over on the request of the laboratory or medical school. Bodies may also be obtained from a second category of persons. Those who wish to do so may make, prior to their own deaths, an arrangement with the authorities of the central office or society to release their own bodies for this purpose. Similar instructions may be included in their wills. However, responsibility for carrying out those arrangements rests with the surviving relatives and friends, who, if they have not known about and come to terms with the wishes of the deceased person, may object and refuse to comply.

By far the greatest number of cadavers are still obtained from among those in long-term hospitals who die without known relatives or friends to claim the body. In some places where the act has not been legislated or where it is not enforced, there is a tendency for those in the undertaking business to claim the body because of the availability of money for burial expense provided by third-party payers.

A favorable public opinion with respect to this new way of providing cadavers for dissection is created by public relations activities of medical schools and pathological societies in educating the public to the need for bodies in teaching and biomedical research. In most cases, the arrangements can also specify the final disposition of the unused or remaining cadaveric products following the termination of dissection or experimental procedures. These can be returned either to a relative or to some organization for burial, or be cremated followed by disposition of the ashes.

Organ transplants and other uses of the cadaver

While the most common medical use of the cadaver is autopsy, bodies in increasingly large number are used as the source of tissues or organs for transplant into other living bodies to replace damaged and nonfunctioning tissues. Pathological societies and organizations set up to encourage donation and to receive parts of bodies provide these parts to hospitals on call. As in the case of cadaver donation, arrangements can be made either by the donor or by relatives to give specific tissues or organs for specific purposes.

The donation of and use of body tissues for transplantation are accepted more widely and easily than is total body dissection. This is due in part to the good public relations efforts of such organizations as the Kidney Foundation, several groups concerned with eyesight, and the surgeons and medical organizations dealing with cardiac replacement. It is also due to the fact that only small parts of the body are removed such as corneas and lenses of the eye, one or both kidneys, pieces of bone, and skin. The immediate benefit of these removals is easily seen and can be justified on the basis of aiding the recovery of function and/or survival of another, usually genetically close, human being.

The widely acknowledged value of using cadaveric organs as transplant tissue that can enable living persons in need of such tissue to live has caused those, primarily in the Judaic and Christian traditions, who respect the body of the dead as symbolic of the human person and his dignity, to rethink their attitudes toward the use of cadaveric materials. Earlier it was noted that permission for autopsies was given on the supposition that the tissues examined would be restored to the cadaver prior to final interment. This, obviously, cannot be done when such tissues are used for transplant purposes. Yet those who at one and the same time believe that it is right to use the cadaver for such purposes and that the cadaver requires human respect hold that these two attitudes are by no means necessarily incompatible.

In the case of heart transplant and less stringently with kidney transplant, several technical problems occur. The organ must be removed as close to the time of death as possible in order to ensure proper function of the organ after transplant. This has required in the recent past the use of donors whose brain function has been determined to have ceased. This criterion has been used as the determinant of the state of life or death. Automatic breathing and continued heartbeat, parts of the classic signs of vitality, have lost much of their significance, in favor of signs of brain activity and function. This problem is discussed elsewhere in this encyclopedia in greater detail, but it should be pointed out here that the question of when death actually occurs is of great importance to medical, moral, and ethical authorities, not to mention the donor and his relatives.

The cadaver is also used as the base for the study of embalming and preservation of the body after death. The study of the anatomy of the body by artists in the past was extremely important not only for accurate representation of the body but also in the efforts to obtain permission from secular and religious authorities for dissection. Both aspects of this material are discussed in detail by Lassek (pp. 38–104).

In the process of continuing medical education and research, cadavers are often used by physicians and researchers whose concern is the more efficient solution of a surgical problem or the development of devices to visualize internal body parts more clearly.

Research on fetal cadavers

There is increasing need felt by the medical profession for examination and experimentation upon fetal tissue in order to obtain information leading to better care and better understanding of disease processes and normal development of the human being.

One of the major problems associated with use of the fetal cadaver is related to the religious and secular views on the value of the human fetus and the respect due to it. Although abortion practices are widely accepted by public and legal institutions as well as by private persons, there is a commonly accepted principle that respect is due to the human fetus. Consequently, we find today policies that place some restrictions on research with human fetuses. There continues to be ethical debate concerning the relevance of such factors as viability, whether a decision has been taken to abort the fetus, whether the abortion is in fact being performed or has been performed, etc.

Public policy on fetal research in Great Britain follows the guidelines proposed by the "Peel Committee Report" in 1972. The committee recommended that no harmful research be performed on the fetus *in utero* and that research on the previable fetus outside the uterus be conducted only when it promises to provide scientific information that cannot be secured in any other way. Research on the dead fetus or on tissues from dead fetuses was permitted by the committee, provided that applicable provisions of the Anatomy Acts of 1832 and 1871 and the Human Tissue Act of 1961 were observed (Great Britain).

In the United States, national policy on fetal research has been established by two sets of federal regulation. The first set, proposed in November 1973 and published in August 1974, was developed by an interagency group within the Department of Health, Education, and Welfare. This set of regulations prohibited nontherapeutic research on the fetus *in utero* in anticipation of abortion but allowed research on the "abortus," provided that such research did not terminate the heartbeat or respiration of the aborted fetus. Research involving the dead fetus was permitted if conducted "in accordance with any applicable State or local laws regarding autopsy" (U.S., DHEW, 1974, p. 30654).

A second set of regulations for fetal research was promulgated by the Assistant Secretary for Health in 1975; several amendments to the regulations followed in early 1977. These revised guidelines were based upon an intensive examination of the fetal research issue undertaken by the National Commission for the Protection of Human Subjects of Biomedical and Behavioral Research, an interdisciplinary public body established by the Congress and appointed by the Secretary of Health, Education, and Welfare. The National Commission solicited testimony on the medical, ethical, and legal aspects of fetal research prior to publishing a detailed report on the issue (National Commission). The commission's recommendations, accepted and translated into regulations, required equal treatment for all fetuses *in utero*, without respect to the mother's intention to carry the fetus to term or to have it aborted. Some types of research on the living but clearly nonviable fetus during or following induced abortion were permitted, but only on condition that the research would not alter the duration of fetal life. Research on the dead fetus was permitted, provided that it was conducted in accordance with "any applicable State or local laws regarding such activities" (U.S., DHEW, 1975, pp. 33529–33530; U.S., DHEW, 1977, p. 2793).

Burial and cremation

The two most commonly used methods of final disposal of a human cadaver are burial and cremation (Lassek, p. 17). Entombment was seen in China and Egypt as a method of protection of the body for its eventual reuse by the spirit following a period of purification and/or trial. Mummification and embalming were used to preserve the tissues of the body so that they could be activated upon the entry of the spirit.

With the advent of Judaism and Christianity, the indissoluble relationship of body and soul was emphasized. Christianity also taught the resurrection of the body for life eternal.

Burial of the cadaver in the earth provides the most direct and simple way of returning all parts of the body to the substance of the earth and is therefore the method of choice for those who strictly hold these beliefs. Anything that interferes with the dissolution processes of the body should be avoided. For this reason, embalming is ill thought of in Jewish tradition, although it is permitted when the law requires it for transport of the body over long distances. It is also seen as possible mutilation of the body, which is strictly forbidden in Jewish tradition. Christian thought and practice agree with the Judaic thought, although they have avoided the strict interpretation Jews have used.

In the United States Christian burial rites have changed or varied from place to place depending upon technical skills and materials present, and they have also been strongly influenced by the rise of the undertaking business. In primitive or poor areas, embalming, extended viewing of the body, and elaborate coffins and vaults to contain the coffin were not used. In more affluent or technologically sophisticated regions, with the support of professionals in the field, more and more of the above are found. The question being raised about these practices is whether they provide the proper marks of respect for the deceased person consistent with his or her religious belief, or whether in some cases they are designed more for the survivors than the needs of the deceased. There is no question that modern thought concerning mourning and grief supports the need for rites in which the realization of separation and loss can begin and gain expression, but it is also true that many modern mortuary practices appear antithetical to religious concepts of the real significance of the body.

Cremation was used in China for long periods as a normal method of disposal of the cadaver. Its use in the Western world has not been a standard part or an alternative to burial until relatively recently. That may be due in part to the feeling that burning the body essentially damages it and may inflict anguish upon the soul or spirit. Jewish tradition forbids cremation as mutilation, and, according to Lannin, Orthodox Jews are not permitted to use funeral or memorial rites nor may they bury ashes of a cremated person in a Jewish cemetery for this reason (Lamm).

Christian philosophy has followed the same outline with regard to the return of the entire body to the earth, and for many Roman Catholics and conservative Protestants cremation was seen as an interference with the natural process of dissolution of the body providing unity with the earth. That opinion has changed gradually with the changes in total popular opinion as well as with a greater appreciation of the total physical environment, including the earth, in which the body is found.

Cremation has been adopted by some advocates of ecology as being a more acceptable social policy, preserving cemetery lands for other uses and reducing costs in the disposition of the body.

Conclusion

There is a great deal of evidence that all organized groups of people have struggled both with the concrete problem of the disposition of cadaver material and with the meaning of the cadaver to the deceased person and to the group or society. The cadaver has been feared for the possibilities of what it might become as an agent of evil or unsatisfied spirits; it has been respected for what it has been in the past as the physical representation of the person; and it has been treated with wonder for what it represents in the scheme of creation. Primitive man as well as modern man recognized the physical attributes of the cadaver; it has been the nonphysical, and the relationships between the two, that have puzzled and bedeviled them. Perhaps the worst solution to the problem has been the tendency to ignore the cadaver, for that produces a degradation of all life. The best solution may be the attitude that dictates respect and protection from embarrassment for the helpless and the stranger in our midst.

CHRISTIAN A. HOVDE

[*Directly related are the entries* DEATH, DEFINITION AND DETERMINATION OF, *article on* CRITERIA FOR DEATH; FETAL RESEARCH; HEART TRANSPLANTATION; KIDNEY DIALYSIS AND TRANSPLANTATION; ORGAN DONATION; *and* ORGAN TRANSPLANTATION. *For discussion of related ideas, see the entries:* EMBODIMENT; *and* PERSON.]

BIBLIOGRAPHY

CORNER, GEORGE W. "The Rise of Medicine at Salerno in the Twelfth Century." *Annals of Medical History* n.s. 3 (1931): 1–16.

Doe v. Bolton. 410 U.S. 179. 35 L. Ed. 2d 201. 93 S. Ct. 739 (1973).

FRAZER, JAMES GEORGE. *The Golden Bough: A Study in Magic and Religion.* Vol. 1: abr. ed. New York: Macmillan Co., 1951.

GREENWALD, YEKUTIEL YEHUDAH. *Ko bo 'al avelut* [Compendium of laws about mourning]. 2 vols. New York: Feldheim Publishing Co., 5716[1955].

Great Britain, Department of Health and Social Security, Scottish Home and Health Department, Welsh Office. *The Use of Fetuses and Fetal Material for Research: Report of Advisory Group.* London: Her Majesty's Stationery Office, 1972. The Peel Committee report.

HOMER. *Iliad.* Translated by Samuel Butler in *The Great Books of the Western World.* Edited by Robert Maynard Hutchins. Vol 4: *The Iliad of Homer and the Odyssey.* Chicago: Encyclopaedia Britannica, 1952.

JAKOBOVITS, IMMANUEL. *Jewish Medical Ethics: A Comparative and Historical Study of the Jewish Religious Attitude to Medicine and Its Practice.* New York: Bloch Publishing Co., 1962. 2d ed. 1975.

LAMM, MAURICE. *The Jewish Way in Death and Mourning.* New York: Jonathan David Publishers, 1969.

LASSEK, ARTHUR MARVEL. *Human Dissection: Its Drama and Struggle.* Springfield, Ill.: Charles C Thomas, 1958.

LIEBES, YITZCHAK ISAAC. "Be'inyan hashtalat eivarim" [Regarding the transplantation of organs]. *Noam,* vol. 14. Edited by Mosheh Shelomoh Hasher. Jerusalem: Torah Shelemah Institute, 5731[1970], pp. 28–111.

LUYTIES, FREDERIC A. "Suggested Revisions to Clarify the Uncertain Impact of Section Seven of the Uniform Anatomical Gift Act on Determinations of Death." *Arizona Law Review* 11 (1969): 749–769.

MOORE, GEORGE FOOT. *History of Religions.* 1 vol. only. International Theological Library. New York: Charles Scribner's Sons, 1913, vol. 1.

National Commission for the Protection of Human Subjects of Biomedical and Behavioral Research. *Report and Recommendations: Research on the Fetus.* DHEW Publication no. (OS) 76–127. Washington: Department of Health, Education, and Welfare, 1975.

PACKARD, FRANCIS R. "History of the School of Salernum." *The School of Salernum: Regimen Sanitatis Salernitanum.* Edited by John Harrington. New York: Paul B. Hoeber, 1920. Reprint. New York: Augustus M. Kelley, 1970.

Roe v. Wade. 410 U.S. 113. 35 L. Ed. 2d 147. 93 S. Ct. 705 (1973).

SHILOH, AILON, and SELAVAN, IDA COHEN, eds. *Ethnic Groups of America: Their Morbidity, Mortality, and Behavior Disorders.* Vol. 1: *The Jews.* Springfield, Ill.: Charles C Thomas, 1973.

SOPHOCLES. *Antigone.* Translated by Richard C. Jebb in *Great Books of the Western World.* Edited by Robert Maynard Hutchins. Vol. 5: *Aeschylus, Sophocles, Euripides, Aristophanes.* Chicago: Encyclopaedia Britannica, 1952.

United States; Department of Health, Education, and Welfare; Office of the Secretary. "Protection of Human Subjects: Fetuses, Pregnant Women, and In Vitro Fertilization." *Federal Register* 40 (1975): 33526–33551.

———; Department of Health, Education, and Welfare; Office of the Secretary. "Protection of Human Subjects: Proposed Amendments Concerning Fetuses, Pregnant Women, and In Vitro Fertilization." *Federal Register* 42 (1977): 2792–2793.

———; Department of Health, Education, and Welfare; Office of the Secretary. "Protection of Human Subjects: Proposed Policy." *Federal Register* 39 (1974): 30648–30657.

II
JEWISH PERSPECTIVES

Jewish religious thought emphasizes the unique status of human beings as bearers of the divine image. It is this conception which provides the matrix for a variety of regulations governing the treatment of human cadavers.

Sanctity is attributed to the total human personality, not merely to selected elements or aspects. A radical dualism that relegates the body to the realm of imperfection or evil while assigning the soul to the domain of goodness or divinity is totally alien to Judaism. Hence, the physical remains of man, the bearer of the divine image, must not be degraded. A person has only limited rights to his body: One must not dispose of one's body in a manner that violates its dignity.

A human cadaver belongs neither to the heirs nor to society at large. Its inviolability is guarded by stringent prohibitions against deriving any form of benefit from the use of a human corpse. Mutilating or disfiguring a cadaver constitutes a grievous offense, unless circumstances demand it for the enhancement of the dignity of the deceased or for the saving of human life.

Judaism does not merely frown upon outright desecration of the physical remains, but calls for avoidance of even the appearance of insensitivity to the departed. Lest one be guilty of "scoffing at the poor," the Halakhah (Jewish law) prohibits eating or drinking in the presence of a corpse. Even the performance of religious rituals, unless specifically intended to honor or benefit the deceased, would be regarded as a source of embarrassment to the dead who can no longer participate in such activities (Greenwald, p. 35).

Respect for the dead also imposes the obligation to take positive action to protect their dignity. Traditional Jewish law stipulates that the human cadaver must not be left unattended. Interment should take place on the day of death before sunset, unless postponement would significantly enhance respect for the deceased. Burial serves a twofold purpose: (1) removal of a decomposing body eliminates a source of degradation of human dignity; (2) interment is viewed as helpful in attaining atonement for the deceased (Arieli, p. 83). Even explicit instructions of a deceased not to be interred must be disregarded (Greenwald, p. 53). Cremation is strictly forbidden by traditional Jewish law. Embalming is also frowned upon unless necessary to comply with the wish of a deceased to be buried at a particular site.

While Judaism demands respect for a corpse, it does not attribute to it any special sanctity that did not belong to it in life. In fact, separated from the soul, the body is viewed as a source of spiritual impurity. Anyone coming into contact with a human cadaver is disqualified from entering the site of the ancient Temple in Jerusalem or from performing a number of religious rituals until the impurity has been removed by special rites of purification. A Kohen, a member of the priestly tribe and a descendant of the original priest, Aaron, is prohibited from being in the same area or having contact with human cadavers except in the case of the burial of a member of the immediate family.

These restrictions create numerous problems for Orthodox Jews. There are serious questions, for example, whether it is permissible for a member of the priestly tribe to receive cadaver transplants (Liebes, pp. 64–70). Much attention is given to the consideration of the conditions that would permit a physician who is a member of the priestly tribe to attend to a dying patient.

Orthodox Jewish law presents other impediments to medical practice and research. Dissection of cadavers violates the injunction against disfiguring the dead. According to prevailing opinion of orthodox scholars, autopsies can be condoned only when there are indications that the information accruing from them may be of immediate value in saving the life of another individual. Thus postmortem dissections are indicated when an experimental drug or surgical procedure was utilized and the autopsy is likely to shed some light on the merits of the treatment. Similarly, when death was caused by contagious disease or genetic disorder, autopsies are warranted for the purpose of instituting prophylactic treatment or helping with genetic counseling. Most rabbinic authorities also permit postmortem dissections for forensic purposes when mandated by law. But in all cases where autopsies are indicated, they must be limited to the special areas where relevant information may be obtained. Following the examination, all organs must be returned for burial (Tendler, pp. 58–60).

Of late, with the growing utilization of organ transplants, insistence upon the inviolability of the cadaver has created new difficulties. Militating against cadaver transplants are (1) the prohibition against dissection, (2) the prohibition against deriving any benefit from the use of the human corpse, and (3) the obligation to bury the complete remains. Most authorities, however, sanction the removal and use of the cornea when consent has been given by the deceased, since restoration of eyesight is considered a life-saving act, and in addition, the prohibition against the use of external tissue is less stringent than that against the use of organs (Jakobovits, pp. 285–286). Some authorities permit the transplant of organs, provided they are needed for the preservation of life and that explicit permission has been given by the donor or his family (Liebes, p. 62).

WALTER S. WURZBURGER

[*Directly related are the entries on* JUDAISM *and* RELIGIOUS DIRECTIVES IN MEDICAL ETHICS, *article on* JEWISH CODES AND GUIDELINES. *For discussion of related ideas, see the entries:* ORGAN DONATION, *article on* ETHICAL ISSUES; *and* ORGAN TRANSPLANTATION, *article on* ETHICAL PRINCIPLES.]

BIBLIOGRAPHY

ARIELI, YITZCHAK. "Baayat nituchei meitim" [The Problem of Autopsies]. *Noam*, vol. 6. Edited by Mosheh Shelomoh Kasher. Jerusalem: Torah Shelemah Institute, 5723 [1962], pp. 82–103.

FEINSTEIN, MOSHEH [MOSES]. *Igrot Mosheh* [Epistles of Moses]. 5 vols. Vol. 2: *Yoreh deah*. New York: M. Feinstein, 1973.

GREENWALD, YEKUTIEL YEHUDAH. *Kol bo al avelut* [Compendium of Laws about Mourning]. New York: Feldheim Publishing Co., 5716 [1955].

JAKOBOVITS, IMMANUEL. *Jewish Medical Ethics: A Comparative and Historical Study of the Jewish Religious Attitude to Medicine and Its Practice.* New York: Bloch Publishing Co., 1959. New ed., 1975.

LAMM, MAURICE. *The Jewish Way in Death and Mourning*. New York: Jonathan David Co., 1969.

LIEBES, YITZCHAK ISAAC. "Be'inyan hashtalat eivarim" [Regarding the Transplanting of Organs]. *Noam*, vol. 14. Edited by Mosheh Shelomoh Kasher. Jerusalem: Torah Shelemah Institute, 5731 [1970], pp. 28–111.

RABINOVITCH, NACHUM L. "What is the Halakhah for Organ Transplants?" *Tradition* 9, no. 4 (1968), pp. 20–27.

ROSNER, FRED. *Modern Medicine and Jewish Law*. Studies in Toral Judaism, vol. 13. Edited by Leon D. Stitskin. New York: Yeshiva University, Department of Special Publications, 1972.

TENDLER, M. D., ed. *Medical Ethics: A Compendium of Jewish Moral, Ethical and Religious Principles in Medical Practice*. 5th ed. New York: Federation of Jewish Philanthropies, Committee on Religious Affairs, 1975.

CANADA

See MEDICAL ETHICS, HISTORY OF, *section on* EUROPE AND THE AMERICAS, *articles on* NORTH AMERICA: SEVENTEENTH TO NINETEENTH CENTURY; *and* NORTH AMERICA IN THE TWENTIETH CENTURY.

CARE

This entry deals with the value implications of the concept of care as it is used in various biomedical settings and as it affects the physician–patient relationship and the direction of health-related activities.

The ambiguity of "care"

Care and medicine have become closely identified, if not synonymous, in the minds of many. For example, medicine, nursing, and other health-related activities are often referred to as "caring professions." Or "care" is used to indicate that someone is receiving medical aid, e.g.: "She is getting the best care possible." But it is not clear why we associate medicine with care except as we think of both care and medicine as appropriate responses to people in distress.

On analysis, care proves to be an extremely ambiguous notion. We sometimes use "care" to indicate an attitude, feeling, or state of mind about a person or state of circumstance—"I really care for Judy," or "Everyone ought to care for the outdoors." To say that I care for X is, therefore, very similar to saying I like X, except "to care" may denote a stronger intention "to pay particular attention to."

"Care" is also used in a manner that does not involve any attitude, but rather is a correlative to someone's having a certain skill. For example, we say that a mechanic cares for our car not because he likes our car, but because he possesses a technical ability that is necessary to be able to repair a car. He cares for our car because we have established a particular relationship with him, namely, we pay him for his service.

Both senses of care require further specification to determine for what or how one ought to care. "Care" is a context-dependent term, i.e., it carries no particular meaning apart from the context in which it is used. Thus to say that "we ought to care about X or Y" does little more than to remind us that certain kinds of attitudes or skills are appropriate in certain contexts. "Care," like the word "good," is a notion that is incomplete because its significance depends on further specification in relation to particular roles, principles, expectations, or institutions (Kovesi, p. 124).

The reason "care" seems so appropriate to medical and health-related activities is that these activities involve both senses of care. For example, the doctor is expected to care for his patients because he has a responsibility to have a positive attitude toward securing their well-being. But doctors are also expected to have the skills that give them the ability to "take care" of their patients beyond simply "caring" for them. But as I will suggest below, it is by no means clear how these senses of care are interrelated or specified in contemporary medicine.

Moral ambiguity of "medical care"

The identification of medicine with care seems to be based on the assumption that we owe it to someone to pay special attention to them because they are in particular need or trouble. However, when used in this manner the moral force of care is ambiguous. For it is not clear if care as an attitude and care as a skill are both required and, if they are, on what grounds. For example, we normally think that it is a good thing to care for people in need, but we do not generally assume that it is a moral obligation to do so; that someone may know how to repair cars does not mean that he is obligated to repair our car. Do the skills that doctors possess mean that they are morally obligated to care, and if so, why?

Hans Mayeroff has suggested that to care for someone, "I must know many things. I must know, for example, who the other is, what his powers and limitation are" (Mayeroff, p. 13). But in the case of injury or illness we normally assume that the need to maintain basic physical integrity clearly sets the context for the kind of care appropriate. Moreover, because physical need is prerequisite for all other activities we assume that if persons with such need can be helped they should be helped. Thus, while we have no obligation to help some to have a better running car than others, we may feel we should care for them through the office of medicine if it is necessary to maintain their physical existence.

Thus the language of "care" when used in a medical context may assume that, because doctors have the training and skill to care for our physical integrity, they have an obligation to care for those in distress. Care and medicine are closely identified because many of the skills associated with medicine are correlative or even necessary for our most basic human needs. But it does not follow from our assumption that medicine is such a basic form of care that doctors are morally required to provide their service. Such a conclusion must be based on further argument that involves issues that cannot be settled simply by determining whether medicine is or is not a way to "care." Moreover, even if it can be shown that it is a good thing to try to care for someone through the means of medicine, it does not follow that we always ought to do so by these means.

It may well be that we should care for the injured or the ill, but it is by no means clear that medicine offers the best or only way to care for them. Whether the "care" that medicine can provide should be provided will depend on the kind and extent of the medical skill that has been developed. For example, are we required to provide all the "medical care" that is technically possible to someone suffering from kidney disease? Such questions are not meant to deny that medicine may be a basic form of care, but they do make clear that, even if that is the case, whether such care should be provided remains open to decision by doctor and patient.

Paul Ramsey, however, has argued that "care" is not meant to provide a basis for judgment for specific actions in the medical context. Rather, "care" is the "source of all particular obligations and one's court of final appeal for deciding the features of actions and practices that makes what we do right or wrong in any context" (Ramsey, 1976, p. 47). Ramsey suggests that this sense of care generates basic rules of practice that embody the physician's commitment to the "preciousness of life." Indeed, it is Ramsey's contention that such an ethics of care provides the basis for a professional ethic that consists in: "(1) rules constitutive of medical care, e.g. the consent requirement, the prohibition of direct killing, and randomizing life-and-death decisions to insure equality of access to sparse resources, which are always binding; (2) directives to cure and save life which true care sometimes suspends and replaces by comfort and dignity for the dying; (3) balancing situational decisions, such as to operate or not to operate, or to use this or that research protocol" (ibid., p. 51). However, as we shall see below, many doubt if "care" in itself can generate the kind of ethic Ramsey thinks it implies.

Medical care as personal care

The emphasis on care as a morally significant notion for medicine often serves as a way to stress that patients have needs that are other than "strictly medical." It is not enough for physicians to provide the best technological care available; they also have a responsibility to treat the patient as a "whole person" (Menninger). The demand for a more humane practice of medicine is not a call for the physician to be personally concerned about the patient over and above what he or she can do for the patient medically, but rather it is to remind physicians that their responsibility is to treat not a disease or a medical problem but the person who is subject to the disease or injury. In other words, care and compassion for the patient are not just something nice for the physician to have beyond his responsibilities as a physician, but such personal care is an integral component for the practice of good clinical medicine.

To care in this manner means that the physician must have the capacity to share in the pain and anguish of those who seek help from him. It means that the physician must "have some understanding of what sickness means to another person, together with a readiness to help and to see the situation as the patient does. Compassion demands that the physician be so disposed that his every action and word will be rooted in respect for the person he is serving" (Pellegrino, p. 1289). Such care should not be confused with pity, condescension, or paternalism. Rather it is to respect the uniqueness of each patient by helping the patient to make those choices that are best for him or her.

Though most assume the physician should "care" in this compassionate or personal manner, the matter is not quite as straightforward as it seems. For, while few deny the medical importance of empathy for the physician's treatment of the patient, it cannot be forgotten that competent care is equally important. "Which would you rather have—warm, compassionate care to usher you into the next world, or cool scientific care to pull you back to this one?" (Greenberg, pp. 205–206.) It is, of course, hoped that this rhetorical question does not present a genuine alternative, since good medicine should combine both (Cassell, 1975, p. 2). The problem for doctors and patients alike is "how the priceless personal equation can be retained in the face of the constantly expanding arsenal of knowledge and inexorable trend toward mass production in medicine no less than in other fields. For, to be true to his calling, a doctor must always complement his expertise with his understanding and view the patient as a whole person rather than merely the sum of his symptoms" (Greenberg, p. 206).

It is important to note, however, that what it means for anyone to care for the "whole person" remains ambiguous. The problem of the kind of personal care the physician should give the patient is not just occasioned by the increasingly sophisticated technology or the demand of justice to try to extend the physician's skill to as many as possible. But it is not clear that care of the "whole person" simply means to treat patients with empathy or compassion. Indeed, it may well be that to treat someone impersonally is a way of caring, especially if we remember that respect is an important aspect of all care. For, if respect is missing, the physician's concern for the patient, even with the best will, can too easily become paternalistic manipulation. For example, the respect due the patient may perhaps mean that the physician must allow the patient not to choose to be "cared" for medically. Open-heart surgery may help many, but that does not mean that the physician has grounds for urging all his patients to undergo such surgery.

Care and the primacy of the patient's interest

The importance of understanding "care" in terms of respect is the basis of Paul Ramsey's suggestion that "care" is the term that best expresses "the ultimate requirement or standard or warrant binding in all cases upon the helping and healing profession" (Ramsey, 1973, p. 20). That the physician must "care" does not mean just that the patient should be given personal care, but that the physician has a commitment to each individual patient that is not and cannot be overridden by any other consideration. Care or respect for the patient, therefore, carries the substantive commitment that no patient should be cared for as a means for the betterment of others without his or her consent. In the language of normative ethics this means that the requirement of the physician to care for each individual patient is a basic deontological commitment that cannot be overridden by any considerations including teleological ones, e.g., the physician must continue to give care to the aged even though by ignoring them he might be able to serve more patients. Such a commitment sets the primary task of medical ethics, according to Ramsey, which is to reconcile the welfare of the individual with the welfare of mankind when both must be served (Ramsey, 1970, p. xiv).

Even if it is generally accepted that a physician may have an overriding commitment to each patient in his care, medicine and its practitioners face issues that such a commitment does not resolve. Such grounds make problematic the commitment of public health medicine to "care" for a "patient" that may be a city, state, or country. For example, does the commitment to care for the individual patient mean physicians should not recommend everyone be inoculated against an illness because they know that a few will die from the inoculation itself?

The commitment to care for the individual patient is certainly an important and perhaps even crucial commitment for medical ethics. The problem is that such a commitment, in itself, is not sufficient to show how we should care in such complex matters as the just allocation of scarce medical resources (Fried, 1970, pp. 183–207). Nor is it sufficient to determine the ethical guidelines as to how we should think about the use of statistical lives, random clinical trials, and the risks that are inherent in the development and practice of normal medicine. Because of these kinds of problems some argue that the kind of "care" offered by medicine cannot be limited to the needs of the individual patient.

Even if "care" is so limited, many questions remain unanswered about what form such "care" should take. For it is often unclear what it means for the doctor to act or to refrain from acting in the patient's interest. The definition of the patient's "interest" that is important to guide the physician depends on the definition of health, which is itself in need of clarification. "The concept of good health implies a concept of the good life, and the goodness of life includes a large number of other factors besides simply its length" (Fried, 1974, p. 150). But without a more detailed or concrete sense of what those "factors" should be, and what kind of particular responsibility medicine has for sustaining and enhancing them, there can be little consensus about what kind of "care" the physician is obligated to give the patient. Indeed, some have recently argued that there is no readily apparent moral position or philosophy that can provide the moral direction that medicine necessarily requires (MacIntyre, pp. 97–111).

It is often assumed that for a physician to care for a patient means that he should try to cure the patient. To confuse care with cure often results in the cruel abandonment of the dying, as we assume we can no longer care for them if we cannot cure them. Moreover, it may be that some patients are subjected to attempts to cure them in a manner that is antithetical to caring for them. For example, to encourage some patients to undergo surgery that only sustains but does not enhance their lives may be incompatible with care. Ironically the technological power of modern medicine raises the question whether it is not often the best way to care for the patient by refraining from doing what medically we have the skill and power to do. This kind of issue makes many decisions in neonatal care today particularly agonizing (Hauerwas, pp. 222–236).

The problem of the relation between caring and curing is perhaps most clearly, but by no means exclusively, illustrated in relation to how we deal with the dying. For example, even though Ramsey argues that medical care as a moral institution can never act directly to take the life of a terminal patient, there comes a time when to respect life means *only* to care for the dying. This means we must be ready to be with the dying, to comfort them, to assure we will not desert them, but at the same time we will not oppose their death. This means that the requirement to care and to save life is not always to be applied strictly in medical practice. "Care never ceases; yet care, never ceasing, has no duty to do the impossible or useless" (Ramsey, 1973, p. 24; 1970, pp. 124–132).

Ramsey, therefore, seems to indicate that the "care" that is incumbent on doctors does not involve simply their medical skills, but the moral skill to be present with those who are suffering. However, it is not clear why the doctor's role involves such a skill. To be sure, communities should provide someone to "only care for the dying," but there is no reason to think the doctor has any different obligation from those of any of us in this respect. It may be, however, that because of their experience in dealing with the sick and the dying doctors have learned better how to care for them, that is, to be with them, better than most of us.

Care, respect, and truth-telling

If care is not equivalent to cure, then the question of what kind of care is due the patient remains open. While no general notion of "care" can be given to account for every kind of medical context, it is clear that a fundamental respect for the patient is required in all medical care. This is often difficult in the medical context because the patient's helplessness and suffering often seem to require that the independence of the patient be qualified in order to "help" the patient. But the inequity of power in such helping situations should but remind us that care for another must be done in a manner that his integrity is not violated (Mayeroff, p. 17; Nelson, pp. 29–30). To care for another is to help him maintain or establish an independent existence, which means to help him care for something or someone else. It also means to help the person to care for himself and become responsive to his own need to care and be responsible for his life (Mayeroff, pp. 10–11).

Charles Fried, therefore, suggests that, whatever else care may involve, it must provide the conditions for the maintenance of the patient's lucidity, autonomy, fidelity, and humanity (1974, pp. 101–104). Lucidity requires that the patient know all the relevant details about the situation in which he finds himself. Autonomy means that, even if a patient is fully informed but does not wish to undergo the therapy recommended by the physician, he cannot be forced to do so. For a person's autonomy to be respected requires that he be allowed to dispose himself according to the life plan and conception he has chosen.

Fidelity requires that we meet the justified expectations that develop from our dealing with one another. Such expectations are often not articulated, but they are not any less significant because they are implicit. Not to meet such expectations is a form of deceit of which lying is but the most dramatic example. The notion of humanity, while admittedly vague, requires that a person should be treated in a manner that does justice to his particular wishes and desires. A person may have no right to be treated affirmatively, but once we are in a significant relationship, our wants, needs, and vulnerabilities should not be ignored. Simply being treated honestly and with autonomy is not sufficient. We should also be noticed (ibid., p. 103).

Assumed in this account of "care" is the importance of telling the patient the truth about his condition. For to withhold the truth is to fail to respect his status as a moral agent capable of being lucid, autonomous, and faithful. Put positively, to care for a patient means that he is to be treated in a manner that assumes that he or she is capable of acting in a morally responsible way. Concretely this means that the simple fact that we are dying does not release us from being held morally responsible for how we die. Moreover, if the patient is unwilling to know the truth about his condition, that does not mean his family or the physician has the right to withhold the truth. However, the importance of the patient's knowing the truth does not mean that he or she must be told the truth bluntly or without feeling. Rather, part of what it means to be truthful in the context of medical ethics is that the truth be spoken to a patient in a skillful, kind, and caring manner.

Conclusion

Care often appears to be a more important regulative notion for determining the moral basis and direction of health-related activities than is morally justified. Care, however, is a significant notion that reminds us that medicine serves as one of the ways we can help others maintain basic physical and psychological integrity. Moreover, care directs our attention to the concrete patient in need without subjecting him or her to manipulation for the good of others. However, it is important that the care given the patient be based on the respect due each of us, well or ill, for otherwise our attempts to care can lead to sentimental or paternalistic perversions.

STANLEY HAUERWAS

[*For further discussion of topics mentioned in this article, see the entries:* CHRONIC CARE; HEALTH CARE; INFORMED CONSENT IN THE THERAPEUTIC RELATIONSHIP; PATERNALISM; PERSON; RIGHTS; THERAPEUTIC RELATIONSHIP; *and* TRUTH-TELLING. *For discussion of related ideas, see the entries:* ACTING AND REFRAINING; HEALTH AS AN OBLIGATION; HEALTH AND DISEASE; MEDICAL MALPRACTICE; *and* RATIONING OF MEDICAL TREATMENT.]

BIBLIOGRAPHY

CASSELL, ERIC. *The Healer's Art: A New Approach to the Doctor–Patient Relationship.* New York: J. B. Lippincott Co., 1976.

———. "Preliminary Explorations of Thinking in Medicine." *Ethics in Science and Medicine* 2 (1975): 1–12.

FRIED, CHARLES. *An Anatomy of Values: Problems of Personal and Social Choice.* Cambridge: Harvard University Press, 1970.

———. *Medical Experimentation: Personal Integrity and Social Policy.* Clinical Studies, A North-Holland Frontier Series, vol. 5. Edited by A. G. Bearn, D. A. K. Black, and H. H. Hyatt. New York: 1974.

GREENBERG, SELIG. *The Quality of Mercy: A Report on the Critical Condition of Hospital and Medical Care in America.* New York: Atheneum, 1971.

HAUERWAS, STANLEY. "The Demands and Limits of Care: Ethical Reflections on the Moral Dilemma of Neonatal Intensive Care." *American Journal of the Medical Sciences* 269 (1975): 22–236.

KOVESI, JULIUS. *Moral Notions.* Studies in Philosophical Psychology. Edited by R. F. Holland. New York: Humanities Press, 1967.

MACINTYRE, ALASDAIR. "How Virtues Become Vices: Values, Medicine and Social Context." *Evaluation and Explanation in the Biomedical Sciences: Proceedings of the First Trans-Disciplinary Symposium on Philosophy and Medicine, Held at Galveston, May 9–11, 1974.* Edited by H. Tristram Engelhardt, Jr. and Stuart F. Spicker. *Philosophy and Medicine,* vol. 1. Dordrecht-Holland and Boston: D. Reidel Publishing Co., 1975.

MAYEROFF, MILTON. *On Caring*. World Perspectives, vol. 43. Edited by Ruth Nanda Anshen. New York: Harper & Row, 1971.

MENNINGER, W. WALTER. " 'Caring' as Part of Health Care Quality." *Journal of the American Medical Association* 234 (1975): 836–837.

NELSON, JAMES BRUCE. *Human Medicine: Ethical Perspectives on New Medical Issues*. Minneapolis: Augsburg Publishing House, 1973.

PELLEGRINO, EDMUND D. "Educating the Humanist Physician: An Ancient Ideal Reconsidered." *Journal of the American Medical Association* 227 (1974): 1288–1294.

RAMSEY, PAUL. "Conceptual Foundations for an Ethics of Medical Care: A Response." *Ethics and Health Policy*. Edited by Robert M. Veatch and Roy Branson. Cambridge: Ballinger Publishing Co., 1976, pp. 35–55.

———. "The Nature of Medical Ethics." *The Teaching of Medical Ethics: National Conference on the Teaching of Medical Ethics Sponsored by the Institute of Society, Ethics and the Life Sciences and Columbia University, College of Physicians and Surgeons, June 1–3, 1972*. Edited by Robert M. Veatch, Willard Gaylin, and Councilman Morgan. Hastings-on-Hudson, N.Y.: Hastings Center Publication, 1973, pp. 14–28.

———. *The Patient as Person: Explorations in Medical Ethics*. The Hyman Beecher Lectures, 1969. New Haven: Yale University Press, 1970.

CATHOLIC ETHICS
See ROMAN CATHOLICISM.

CATHOLIC HOSPITALS
See RELIGIOUS DIRECTIVES IN MEDICAL ETHICS, *article on* ROMAN CATHOLIC DIRECTIVES; ROMAN CATHOLICISM; HOSPITALS.

CERTIFICATION OF DEATH
See DEATH, DEFINITION AND DETERMINATION OF, *article on* LEGAL ASPECTS OF PRONOUNCING DEATH.

CHAPLAINS
See PASTORAL MINISTRY.

CHEMICAL WARFARE
See WARFARE.

CHILD ABUSE
See CHILDREN AND BIOMEDICINE; *and* INFANTS, *articles on* ETHICAL PERSPECTIVES ON THE CARE OF INFANTS *and* PUBLIC POLICY AND PROCEDURAL QUESTIONS.

CHILDBIRTH
See WOMEN AND BIOMEDICINE, *article on* WOMEN AS PATIENTS AND EXPERIMENTAL SUBJECTS.

CHILDREN AND BIOMEDICINE

Introduction: definitions and scope

The definition of the child is elusive; dictionaries give the following meanings of the term: "unborn or recently born," "a young person," or simply, "a son or daughter of human parents." In medicine childhood is generally defined as the "period between infancy and puberty" (Stedman). In law, a "child of tender age" is any "progeny . . . less than 14 years" (Black), but state statutes vary.

The boundary between fetus and infant is a shifting one, and ethical issues affecting fetuses are of obvious relevance to the well-being of postnatal children. The boundary between childhood and adulthood is still harder to define; it is less distinct biologically and spans a broad range of psychological and social characteristics. Definitions are even more arbitrary at this end of the scale, and, as with the fetal boundary, the definition may have profound implications for the health and well-being of a particular child.

The drawing of lines based strictly on age—whether between childhood and adulthood, embryo and fetus, fetus and child—can be seen as arbitrary, for this criterion implies that an organism passes in an instant from one status to another. Boundaries may be necessary as guides, to warn us when a significant transition may be occurring, but in an individual case it would be advisable to avoid decisions based on such strict status criteria.

A second definitional problem involves the proper scope of "medical-ethical" concerns. A child's right to health cannot be separated from his nutrition, social status, and education. A consideration of medical-ethical issues concerning children cannot be complete if it confines itself to such classic problems as experimentation and transplantation, which directly affect a minute percentage of the population, while ignoring ethical issues affecting the physical and emotional health of many more children. Such problems—including malnutrition, child abuse, custody proceedings, and citizenship rights in general—can only be mentioned in passing. This article will be confined to general considerations in the therapeutic and experimental relationships.

Finally, there are medical-ethical issues—such as the allocation of scarce resources, management of the terminally ill, and the boundary between research and innovative care—that are not peculiar to children but are compounded in the pediatric setting. The central complicating factor for the child is his inability, presumed or real, to speak for himself, leaving him both vulnerable to adult perceptions of his best interest and powerless against competing adult self-interests.

Historical aspects

"The history of childhood is a nightmare from which we have only recently begun to awaken" (DeMause). While DeMause's prose might be considered lurid by some, there is ample documentation of the lowly status of children throughout history. Even Ariès, who argues that the child's plight is becoming worse and not better, asserts that childhood was only "discovered" around the thirteenth century in the Christian era (Ariès).

The documented history of child abuse and infanticide provides dramatic evidence (DeMause; Langer). It is likely that "a very large percentage of the children born prior to the eighteenth century were what would today be termed battered children" (DeMause). It is not until 1690 that one can discover a biography of a child who has not been beaten, and in one Western European country eighty percent of the parents still admit to beating their children, thirty-five percent with canes (ibid.).

The widespread tolerance of infanticide has been easier to document. It has long been used as a means of population control and continues to be accepted in some underdeveloped cultures (Langer). From ancient times until the twentieth century the low status of infants and children made such practices commonplace. In seventeenth-century China thousands of babies were thrown on the streets like refuse each day, and in nineteenth-century London, Paris, and St. Petersburg the public "foundling homes" were admitting up to 5,000 infants annually for a more socially acceptable death sentence. Even where laws existed they were seldom enforced, so that in Disraeli's London infanticide flourished and the "police seemed to think no more of finding a dead child than of finding a dead dog or cat" (ibid.).

By the end of the nineteenth century, in response to exposés in the press and pressure from the medical societies, England had begun to tighten and enforce protective statutes. The early American colonies were apparently more solicitous, although by 1826 the laws of New York would still not allow state intervention with parents who grossly abused their children (Radbill). In 1838 the Pennsylvania Supreme Court invoked the doctrine of *parens patriae* for superseding parental authority. In the mid-twentieth century social acceptance of active killing of normal children had all but disappeared from Western societies, but passive euthanasia of infants with birth defects is still openly practiced in England and the United States (Duff and Campbell; Lorber).

The child and health care

The recent trend in the United States is for increasing state intervention on behalf of the child's health. Statutes on child abuse, now in effect in every state of the United States (Paulsen), encourage or require physicians and others to report suspicions to appropriate authorities, with civil or criminal penalties for failing to do so. In some jurisdictions a child may sue his parent for a negligence tort (*Emery* v. *Emery*), a principle generally denied earlier in this century. The juvenile court system—a well-intentioned attempt to protect children from hazards of the criminal justice system—was, in the mid-1970s, under pressure from judicial decisions requiring that children be given equal status in matters such as right to counsel, cross-examination, and protection from self-incrimination (In re *Gault*).

Concurrent with these legislative and judicial reactions to gross abuse has been a trend for state involvement in positive, preventive measures affecting the health of children. Compulsory immunization and fluoridation have been upheld not only for the protection of the community but as being in the interest of individual children (*Cude* v. *State*; *Mannus* v. *State*). Nearly all U.S. states require screening of newborns for phenylketonuria (National Research Council), a rare and preventable cause of mental retardation, and most states have enacted laws enabling minors to seek medical care without parental consent under specified circumstances, particularly when venereal disease is involved (Hofman and Pilpel).

Some see these trends as an excessive swing of the pendulum. The granting to young adolescents of full rights to medical care may undermine a parent's authority to affect the child's behavior, and may perhaps widen the gulf between parent and child, depriving the latter of a potentially valuable counselor. The winning of such rights, in this view, can be a pyrrhic victory for some children who become unintentionally isolated. In some states a child of *any* age may receive treatment for venereal disease without parental knowledge or consent (Wisconsin Statutes). To illustrate the undesirable consequences, consider an eleven-year-old girl who sought treatment for gonorrhea. She insisted that the treating physician promise not to inform her parents. Because of the reporting requirement, a public health nurse conducted an inquiry and discovered that the child was being repeatedly sexually assaulted after school by a sixteen-year-old neighbor prior to the parents' arrival from work. The nurse felt that the pledge of confidentiality by the doctor bound her to conceal the facts from the parents. It later became apparent that, with modest encouragement, the child was able to discuss this with her parents and enlist their support. While the statute enabling the physician to treat the child did not require him to exclude the parents, a narrow concept of the "rights" of the child may have led the health professionals to become involved in a drama that did not serve the child's best interests.

In the area of child abuse, it is also argued that excessively broad notions of neglect permit or encourage premature or inappropriate state intervention. Abuse is often not defined (Paulsen) and, particularly when extended to include emotional injury, may be used by zealous physicians or social workers to intervene punitively in families where there is either no clear abuse, or where a custody change may be more harmful than an admittedly unsatisfactory home situation (Goldstein, Freud, and Solnit).

Implicit in all deprivations of equal status for children is the notion that it is permissible for an adult to substitute his judgment for that of the child (Dworkin; Schrag), based on the assumptions that the adult has the best interests of the child at heart, and that he or she is better qualified than the child to make decisions in the child's interest. While these assumptions are true enough to gain acceptance as general principles, the exceptions are sufficiently common to arouse concern (Fost; McCollum and Schwartz; Robertson and Fost). This concern over paternalism applies to purely therapeutic as well as experimental interventions. It is not clear, for example, that a parent who advocates passive euthanasia for a newborn with Down's syndrome and intestinal obstruction is primarily seeking relief for the child rather than himself (Gustafson). Nor is it always clear that such parents have any meaningful information to support the contention that death is more advantageous to the child than life. Similarly, a parent who initiates a behavior modification program for his child's bed-wetting may be primarily seeking relief of his own annoyance and might have difficulty establishing that the benefit–risk ratio of the behavior program is more advantageous to the child than continuation of the symptom.

While the limitations of substituted judgment by parents can be argued in the traditional therapeutic setting, it is in the area of experimentation where the validity of such proxy consent has been most extensively discussed. This will be analyzed further in the next section.

Experimentation in children

The many and complex problems of human experimentation are compounded in the pediatric setting by four factors. First, children are not simply small adults, but are biologically different in many ways, so that knowledge acquired from adult subjects often cannot be applied to children without testing them. Moreover, the U.S. Food and Drug Administration requires that drug labeling confine recommendations to adults unless testing on children has been conducted. Second, there are many diseases that occur exclusively in children, so that advances in understanding may be absolutely dependent on the use of children (Capron; Lowe, Alexander, and Mishkin). Third, the hazards of experimentation may be greater in younger age groups. Physical injuries that might be trivial in a mature person (such as radiation of the growing portion of a long bone) can become magnified when occurring early in the developmental process. Events that may have little psychological significance for an adult, such as a short hospitalization or repeated venepunctures, may have profound detrimental effects on a child. Fourth, uncoerced informed consent, the keystone of protection of human subjects, is often unattainable from the minor.

There is broad consensus that therapeutic experimentation—nonstandard interventions that have the explicit primary intent of helping the subject to whom they are being applied—can be practiced in children without major modification of the rules that apply to experimentation in general, so long as consent from a legally authorized representative is given (Curran and Beecher; Great Britain; World Medical Association). This is not to imply that present regulation of adult experimentation is satisfactory, or that parental consent for nonexperimental therapy is without controversy. But it is in the realm of nontherapeutic experimentation where concerns are greatest. Ideally, risk, benefit, and consent should all reside in the same person. When nontherapeutic research is conducted on a child, it is the child who bears the risk, while future persons gain the possible benefits, and a third party, without risk or benefit, gives consent.

Whether or not parents can legally consent to nontherapeutic interventions on their children is unresolved. The Nuremberg Code does not mention children but implies they should be excluded as subjects by stating it is "absolutely essential . . . that the person involved should have legal capacity to give consent." The Declaration of Helsinki allows nontherapeutic studies on children with consent of the legal guardian. Current (1976) regulations of the U.S. Department of Health, Education, and Welfare (DHEW) do not distinguish between the right of the guardian to consent for nontherapeutic versus therapeutic research.

In the United States, statutory law on the subject has been nonexistent until the 1970s, and as of 1975 the specific question had never been explicitly decided in a court. A 1974 federal statute created the National Commission for the Protection of Human Subjects of Biomedical and Behavioral Research, but this body was given a strictly advisory role (Public Law 93-348). The case of *Neilsen* v. *Regents of University of California* raised, for the first time, the explicit question of whether a parent may volunteer a child for nontherapeutic research.

Related cases come from the field of transplantation, arising when physicians and/or parents seek to use a legally incompetent person as kidney donor for a relative. The general response of courts has been to require a finding of benefit for the donor, such as the advantages derived from future experiences with the recipient should he survive, or the avoidance of remorse due to loss of companionship or later guilt from realizing that one had failed to come to the aid of his sibling (Baron, Botsford, and Cole). In instances where the donor was incapable of experiencing such benefit because of severe mental disability or where the social contact with the sibling was not close, permission to allow the transplantation has been denied (*Lausier* v. *Pescinski;* In re *Richardson*). While many of the cases regarding incompetent donors involved intelligent adolescents or retarded adults, the general principle of the legitimacy of proxy consent would seem applicable to children in general, and at least one appellate case did involve a seven-year-old donor (*Hart* v. *Brown*).

In contrast is the famous and much disputed case of *Bonner* v. *Moran,* in which an appellate court ruled that a skin graft taken from a fifteen-year-old boy could not be upheld unless the mother consented. One legal scholar takes the decision to imply that such parental consent would have made the procedure valid, despite the absence of a direct benefit (Curran and Beecher), but the decision as written leaves room for disagreement as to how much it can be extended to nontherapeutic experimentation (Capron).

In Great Britain the Medical Research Council has stated that "parents and guardians of minors cannot give consent on their behalf to any procedures which are of no particular benefit to them and which may carry some risk of harm," although legal precedent or authority for this is obscure (Curran and Beecher).

Proxy consent

As with other ethical issues, the already complex and unresolved problems of consent are compounded when children are involved. The assumption that parents can provide uncoerced informed consent for their children rests at least on the presumption that they can provide it for themselves. There is evidence that educated, competent adults are frequently not adequately informed to give meaningful consent (Fellner and Marshall; Fletcher; Gray), so their ability to speak for others may be questioned. The barriers may not be so much a lack of intelligence or motivation on the part of patient or physician as more complex forces such as the anxiety of illness, the intimidating milieu of the hospital, and a sense of awe, trust, and dependence on the physician—all of which may conspire to make solicitation of consent a ritual wherein few are meaningfully informed (Ingelfinger).

In addition to these general barriers to informed consent, there are aspects peculiar to the parent–child relationship that complicate the matter further. The perceived vulnerability of the child may evoke excessive anxiety, which further clouds judgment. The parent may be acting out unconscious hostile wishes against the child, particularly if he is retarded or deformed. There may be a wish to repay the physician–investigator for prior service by offering the child as a sacrifice.

Even if these obstacles were overcome, the practice of proxy consent rests on other notions under challenge: the claim that a person may *ever* consent to a nontherapeutic intervention on another (Ramsey, 1970, 1976); the assumption that adults can reliably assess what is in a child's best interest; and the faith that they are capable of acting in a person's best interest, even if it can be accurately identified. The observation that adults frequently cannot identify or act in their own best interest—as manifested by decisions ranging from smoking and drinking to marriage and financial investments—undermines these assumptions.

Two common justifications for proxy consent are the *substituted judgment* doctrine, a legal doctrine which suggests that decisions be based on what the person would be likely to do if competent (Robertson), and the ethical notion that a child can be volunteered on the basis of what he *ought* to do (McCormick).

The substituted judgment doctrine is weakened by the inability of an adult to know the child's mind. Put another way, the adult may err by being able only to imagine what a reasonable adult would do if in the situation of the child, thereby fallaciously equating adult values and preferences with those of children. An adult might gladly consent to a venepuncture, a procedure of minimal risk and annoyance, but for a child the same event could be a major psychological trauma. Presumptions of what adults would do may suffer from misperceptions of the empirical situation. One could presume that adults probably would consent to minimally hazardous nontherapeutic procedures, but in many communities they in fact do not participate, unless offered significant inducements. One implication of this might be that a minimum requirement for the use of nonconsenting children in nontherapeutic studies (assuming other objections can be overcome) would include some age-appropriate indirect benefit or reward comparable to the monetary rewards that an adult would receive.

McCormick has argued that children may be used for nontherapeutic studies to which they *ought* to consent, on the grounds that there are some sacrifices all members of the human community ought to make, and it is in their interest to do so. While there might be agreement on what sacrifices people ought to make, many would object to the recruitment of children without their consent, unless adults were also required to participate on the same grounds, unrelated to their consent.

Ramsey has taken the extreme view that proxy consent for nonbeneficial experiments in children are unethical without qualification (Ramsey, 1970). He later modified this by acknowledging that such research might be done so long as one acknowledged that he was "doing wrong for the sake of the public good" (1976). Reluctantly admitting that it might be wrong, in some circumstances, *not* to do the research, he seemed to be pointing up an aspect of all true ethical dilemmas: that they involve conflicts of two important obligations, with the inevitable consequence that something of value will be lost no matter which way the conflict is resolved.

As the child gets older—and intellectual and emotional maturity allow him increasingly to participate in decisions—some have advocated a requirement that the child consent in addition to the parent, even though the child may be legally incompetent (U.S. DHEW, National Institutes of Health). While such a practice sounds appealing, there is some evidence that a principle may be honored at the expense of the child's emotional well-being. One study concluded that children informed of the research nature of their hospitalization experienced overwhelming anxiety, due in part to fantasies aroused by their primitive notions of research (Schwartz).

Rules based on inflexible age boundaries may not serve the needs of individual persons. Many children can and should be included in discussions of their participation in research; many adults are incapable of participating meaningfully. Ideally, decisions would be based on a full consideration of the facts in each case, including the probability of discomfort and risk and the emotional and intellectual capacity for consent in each child.

Requirements for consent serve two functions: protection from unacceptable risks and respect for the autonomy of each individual. It appears that trust in consent as protection from unacceptable risks has been misplaced (Fletcher; Gray) and, in the case of young children, respecting autonomy may be impossible unless nontherapeutic research is avoided entirely. If such studies are to continue, it will be necessary to pay attention to Ramsey's reminder that it is not an unmitigated good, that something of value is being lost, and that it is possible for a subject to be "wronged without being harmed" (Ramsey, 1970).

While there is obviously no consensus on a precise formulation of rules affecting nontherapeutic research on children, the following principles would probably attract broad support: (1) Children should be used only as a last resort, when information cannot be obtained in any other way. (2) Nontherapeutic studies on children should be of minimal risk. (3) The right of the child to withdraw should be respected. (4) Proxy consent, if appropriate at all, should be based on a notion of what a competent and reasonable person would do: Whether parents will actually make such a judgment in the best interest of the child is only a matter of presumption.

NORMAN C. FOST

[*For further discussion of topics mentioned in this article, see the entries:* DEATH AND DYING: EUTHANASIA AND SUSTAINING LIFE, *articles on* ETHICAL VIEWS *and* PROFESSIONAL AND PUBLIC POLICIES; HUMAN EXPERIMENTATION; INFORMED CONSENT IN THE THERAPEUTIC RELATIONSHIP; INFORMED CONSENT IN HUMAN RESEARCH; ORGAN TRANSPLANTATION; PATERNALISM; PATIENTS' RIGHTS MOVEMENT; RATIONING OF MEDICAL TREATMENT; RIGHT TO REFUSE MEDICAL CARE; RIGHTS; *and* RISK. *For discussion of related ideas, see the entries:* ADOLESCENTS; AGING AND THE AGED, *article on* HEALTH CARE AND RESEARCH IN THE AGED; CONFIDENTIALITY; FETAL RESEARCH; *and* INFANTS. *See* APPENDIX, SECTION III, PEDIATRIC BILL OF RIGHTS.]

BIBLIOGRAPHY

"AMA Ethical Guidelines for Clinical Investigation." *Annals of Internal Medicine* 67, no. 3, pt. 2, supp. 7, appendix 5 (1967), pp. 76–77. Adopted by the House of Delegates, American Medical Association, 30 November 1966.

ARIÈS, PHILIPPE. *Centuries of Childhood: A Social History of Family Life.* Translated by Robert Baldick. New York: Vintage Books, 1965. Translation of *L'Enfant et la vie familiale sous l'Ancien Régime.*

BARON, CHARLES H; BOTSFORD, MARGOT; and COLE, GARRICK F. "Live Organ and Tissue Transplants from Minor Donors in Massachusetts." *Boston University Law Review* 55 (1975): 159–193.

BEECHER, HENRY KNOWLES. *Research and the Individual: Human Studies.* Boston: Little, Brown & Co., 1970. Contains an extensive collection of codes.

BLACK, HENRY CAMPBELL. *Black's Law Dictionary.* 4th ed. Minneapolis: West Publishing Co., 1968.

Bonner v. Moran. 126 F. 2d 121 (D.C. Cir. 1941).

CAMPBELL, A. G. M. "Infants, Children, and Informed Consent." *British Medical Journal* 3 (1974): 334–338.

CAPRON, ALEXANDER MORGAN. "Legal Considerations Affecting Clinical Pharmacological Studies in Children." *Clinical Research* 21 (1973): 141–150.

Cude v. State. 237 Ark. 927. 377 S.W. 2d 816 (1964).

CURRAN, WILLIAM J., and BEECHER, HENRY K. "Experimentation in Children." *Journal of the American Medical Association* 210 (1969): 77–83.

DEMAUSE, LLOYD. *The History of Childhood.* New York: Psychohistory Press, 1974.

DUFF, RAYMOND S., and CAMPBELL, A. G. M. "Moral and Ethical Dilemmas in the Special Care Nursery." *New England Journal of Medicine* 289 (1973): 890–894.

DWORKIN, GERALD. "Paternalism." *Morality and the Law.* Edited by Richard A. Wasserstrom. Belmont, Calif.: Wadsworth Publishing Co., 1971, pp. 107–126.

Emery v. Emery. 289 P. 2d 218 (Cal. 1955).

Federal Food, Drug, and Cosmetic Act. 21 U.S.C. sec. 301 (1962).

FELLNER, CARL H., and MARSHALL, JOHN R. "Kidney Donors—The Myth of Informed Consent." *American Journal of Psychiatry* 126 (1970): 1245–1251.

FLETCHER, JOHN. "Realities of Patient Consent to Medical Research." *Hastings Center Studies* 1, no. 1 (1973), pp. 39–49.

FOST, NORMAN C. "Ethical Problems in Pediatrics." *Current Problems in Pediatrics* 6, no. 12 (1976), pp. 1–31.

———. "How Decisions Are Made: A Physician's View." *Decision Making and the Defective Newborn: Proceedings of a Conference on Spina Bifida and Ethics.* Edited by Chester A. Swinyard. Foreword by Robert E. Cooke. Springfield, Ill.: Charles C Thomas, 1978, chap. 13, pp. 220–258. Includes discussion.

In re Gault. 387 U.S. 1. 18 L. ed. 2d 527. 87 S. Ct. 1428 (1967).

GOLDSTEIN, JOSEPH; FREUD, ANNA; and SOLNIT, ALBERT J. *Beyond the Best Interests of the Child.* New York: Free Press, 1973.

GRAY, BRADFORD H. *Human Subjects in Medical Experimentation: A Sociological Study of the Conduct and Regulation of Clinical Research.* Health, Medicine and Society Series. New York: Wiley-Interscience, 1975.

Great Britain, Medical Research Council. "Responsibility in Investigations on Human Subjects: Statement by the Medical Research Council." *Report of the Medical Research Council for the Year 1962–1963.* Cmnd. 2382. London: Her Majesty's Stationery Office, 1964, pp. 21–25.

GUSTAFSON, JAMES M. "Mongolism, Parental Desires, and the Right to Life." *Perspectives in Biology and Medicine* 16 (1973): 529–557.

Hart v. Brown. 29 Conn. Supp. 368. 289 A. 2d 386 (Super. Ct. 1972).

HELFER, RAY E., and KEMPE, C. HENRY, eds. *The Battered Child.* 2d ed. Foreword by Katherine B. Oettinger. Chicago: University of Chicago Press, 1974.

HOFMAN, ADELE D., and PILPEL, HARRIET. "The Legal Rights of Minors." *Pediatric Clinics of North America* 20 (1973): 989–1004.

INGELFINGER, FRANZ J. "Informed (But Uneducated) Consent." *New England Journal of Medicine* 287 (1972): 465–466. Editorial.

Landeros v. Flood. 50 Cal. App. 3d 115. 123 Cal. Rptr. 713 (1st D. 1975).

LANGER, WILLIAM L. "Infanticide: A Historical Survey." *History of Childhood Quarterly* 1 (1974): 353–388.

Lausier v. Pescinski. 27 Wis. 2d 4. 226 N.W. 2d 180 (1975).

LORBER, JOHN. "Selective Treatment of Myelomeningocele: To Treat or Not to Treat?" *Pediatrics* 53 (1974): 307–308. Commentary.
LOWE, CHARLES U.; ALEXANDER, DUANE; and MISHKIN, BARBARA. "Nontherapeutic Research on Children: An Ethical Dilemma." *Journal of Pediatrics* 84 (1974): 468–472.
MCCOLLUM, AUDREY T., and SCHWARTZ, A. HERBERT. "Pediatric Research Hospitalization: Its Meaning to Parents." *Pediatric Research* 3 (1969): 199–204.
MCCORMICK, RICHARD A. "Proxy Consent in the Experimentation Situation." *Perspectives in Biology and Medicine* 18 (1974): 2–20.
Mannus v. State. 240 Ark. 42. 398 S.W. 2d 206 (1966).
National Research Council, Committee for the Study of Inborn Errors of Metabolism. *Genetic Screening: Programs, Principles and Research.* Washington: National Academy of Sciences, 1975.
Nielsen v. Regents of the University of California. Case no. 665-049 (Super. Ct. of Cal., County of San Francisco, as amended 20 December 1973).
"Nuremberg Code, 1946–1949." Beecher, *Research and the Individual*, pp. 227–234.
PAULSEN, MONRAD G. "The Law and Abused Children." Helfer, *The Battered Child*, pp. 153–178.
"Pediatric Bill of Rights: National Association of Children's Hospitals and Related Institutions." *Law, Medicine and Forensic Science.* 2d ed. *1974 Supplement.* Edited by William J. Curran and E. Donald Shapiro. Boston: Little, Brown & Co., 1974, pp. 129–130. Adopted 24 February 1974.
Public Law 93-348. 88 Stat. 348. Title II. Protection of Human Subjects of Biomedical and Behavioral Research (12 July 1974).
RADBILL, SAMUEL X. "A History of Child Abuse and Infanticide." Helfer, *The Battered Child*, pp. 3–21.
RAMSEY, PAUL. "The Enforcement of Morals: Nontherapeutic Research on Children." *Hastings Center Report* 6, no. 4 (1976), pp. 21–30.
———. *The Patient as Person: Explorations in Medical Ethics.* New Haven: Yale University Press, 1970, p. 14.
In re Richardson. 284 So. 2d 185 (La., Ct. of App. 1973).
"The Rights of Children." *Harvard Educational Review* 43 (1973): 479–668; 44 (1974): 6–157. Special issues, pts. 1 and 2.
ROBERTSON, JOHN A. "Organ Donations by Incompetents and the Substituted Judgment Doctrine." *Columbia Law Review* 76 (1976): 48–78.
———, and FOST, NORMAN C. "Passive Euthanasia of Defective Newborns: Legal Considerations." *Journal of Pediatrics* 88 (1976): 883–889.
SCHRAG, FRANCIS. "Rights over Children." *Journal of Value Inquiry* 7 (1973): 96–105.
SCHWARTZ, A. HERBERT. "Children's Concepts of Research Hospitalization." *New England Journal of Medicine* 287 (1972): 589–592.
STEDMAN, THOMAS LATHROP. *Stedman's Medical Dictionary.* 23d ed. Baltimore: Williams & Wilkins Co., 1976.
United States; Department of Health, Education, and Welfare; National Institutes of Health. "Protection of Human Subjects." *Federal Register* 38 (1973): 31738–31749, esp. 31742. Draft additions to proposed regulations.
———; Department of Health, Education, and Welfare; Public Health Service; National Institutes of Health. *The Institutional Guide to DHEW Policy on Protection of Human Subjects.* DHEW Publication no. (NIH) 72-102. Washington: Government Printing Office, 1971.
WILKERSON, ALBERT E., ed. *The Rights of Children: Emergent Concepts in Law and Society.* Philadelphia: Temple University Press, 1973.
Wis. Stat. sec. 143.07. Venereal Disease.
World Medical Association. "Human Experimentation: Code of Ethics of the World Medical Association: Declaration of Helsinki." *British Medical Journal* 2 (1964): 177. Reprint. "Declaration of Helsinki, 1964." Beecher, *Research and the Individual*, pp. 277–279.
WORSFOLD, VICTOR L. "A Philosophical Justification of Children's Rights." *Harvard Educational Review* 44 (1974): 142–157.

CHINA

See MEDICAL ETHICS, HISTORY OF, *section on* SOUTH AND EAST ASIA, *articles on* GENERAL HISTORICAL SURVEY; PREREPUBLICAN CHINA; *and* CONTEMPORARY CHINA. *See also* BUDDHISM; CONFUCIANISM; TAOISM.

CHIROPRACTIC

See MEDICAL PROFESSION, *article on* ORGANIZED MEDICINE; ORTHODOXY IN MEDICINE.
See also APPENDIX, SECTION IV, AMERICAN CHIROPRACTIC ASSOCIATION.

CHRISTIAN SCIENCE

See PROTESTANTISM, *article on* DOMINANT HEALTH CONCERNS IN PROTESTANTISM.

CHRONIC CARE

Chronic illnesses represent an ever increasing percentage of total illness, at least in the more industrialized nations. Approximately half the U.S. population suffers from one or more chronic illnesses. High on the list are "heart conditions," arthritis, rheumatism, and impairments of back and spine (U.S., DHEW, pp. 1–4). Although afflicting many young people, chronic illnesses are highly associated with advancing age and also with low income. The rising rate of chronicity is linked with the continuing advance of medical technology, which not only contributes to the control of infectious and parasitic disease but, paradoxically, has converted formerly terminal medical conditions into debilitating chronic ones. Chronic illnesses, therefore, are now a major class of medical problems faced by patients, relatives, and health personnel. With these medical problems, a variety of ethical and "quality of life" problems are directly associated.

Properties of chronic disease. There are several general properties of chronic disease. First, it is long-term; hence, whether treatments are occasional, frequent, or virtually continuous, characteristically the disease spans many months, even years. Hospitals are oriented toward treating the acute phases of these diseases, while outpatient services are focused on the intermittent monitoring of symptoms, regimens, and disease. Neither type of institution is geared very effectively toward continuity of care or offering counsel concerning effects of the disease and its treatment on the life-styles of patient or family.

Many chronic diseases are also uncertain in prognosis, phasing of the disease, and response to treatment. These uncertainties force patients to reorganize and restrict their lives in an attempt to handle the unpredictable. Again, the issues are as much social and psychological as medical.

Chronic diseases require more focus on "relief," "comfort," and "care," since cure is problematic or impossible. Correspondingly, they involve considerable attention to minimization of symptoms and management of pain and other discomforts, as well as attention to psychological and social problems associated with anxiety, grief, disfigurement, stigma, relative immobility, and enforced or voluntary isolation.

Chronic diseases are often multiple diseases, a single chronic condition often leading to multiple chronic conditions. Persons may also suffer from two or more unrelated diseases—the older they are, or the lower in socio-economic scale, the more likely are multiple afflictions to occur. Side effects of medical treatments can lead to additional chronicity. Multiplication of diseases and symptoms implies a multiplication of management (medical, psychological, social) problems.

Chronic diseases tend to intrude upon the lives of patients and their families. Thus, reduced mobility may force changes in household arrangements, in amount and kind of work. Impairment of hands or loss of energy may result in the slowing down of household duties and even in simple routines like dressing oneself. Prescribed regimens may contribute (sometimes more than the symptoms) to embarrassment before others or to the absorption of time and energy. Work lives may be affected so that, even when the disease is not completely disabling, inroads on skill, energy, and time may interfere drastically with job requirements. Furthermore, employers often refuse to hire or rehire the chronically ill. Some diseases (epilepsy or sickle-cell disease) make it difficult (at least in the United States) for the sufferer to get jobs if his condition is known, and difficult even to obtain requisite health insurance. Another notable effect of chronic illness is its potential for causing some degree of social isolation from family, friends, and community. Reduction of mobility or energy can contribute to isolation; because of changes in appearance either the patient voluntarily withdraws from social contact or others withdraw from him.

Chronic diseases are expensive. The occurrence of crises, the long-time use of drugs, and routine monitoring all contribute toward the cost, quite aside from particular diseases that regularly or occasionally require expensive technologies. The use of ancillary services also adds to the costs. All these costs are shared differently in different countries; in the United States, burdens on patients and relatives can be considerable; continuing expenses plus potential unemployment or underemployment can reduce families to welfare recipients. Unemployment or financial drain, whether total or not, everywhere leads to marital and family disintegration, psychological impairment, damaged identities, and loss of social status.

Care for chronic disease. Proper care for chronic disease requires a variety of ancillary services. Psychological counseling or therapy may be needed to cope with the impact of the disease, symptoms, or regimen. Special educational services may be needed for ill children and their siblings. Physical therapy, occupational counseling and retraining, and even marital counseling may be necessary. (In the United States, social workers can, or could, help to manage the maze of regulations, forms, and agencies involved in financing a patient's care and in reaching specialized services and agencies.) In addition, legal and financial services may be necessary. Most needed, perhaps, is counseling from a sophisticated and compassionate third party who suggests plausible modes for managing the diverse problems that arise during the course of an illness—whether rearranging household spaces and duties, suggesting "easier to handle" types of clothing, planning strategies for managing insensitive health personnel, or obtaining results from bureaucratic health agencies.

A serious barrier to improving the care of the chronically ill is that health personnel tend implicitly to utilize an "acute disease" model, one appropriate for infectious and parasitic disease but hardly appropriate to the management of chronic diseases. The acute-care and disease-oriented model tends toward an emphasis on hospital care with the utilization of a complex medical technology—drugs, surgery, transplants, complicated machinery—and concentrates on the peak periods of the disease and the weeks immediately thereafter. Organizationally, the major rewards are generally garnered by those who are skilled at that kind of acute care. Ambulatory care seems less heroic and less important, and on the whole its practitioners receive fewer monetary and status rewards. Also, the auxiliary services tend to be underused or underdeveloped in light of the range of disease-linked but nonmedical needs of patients and their families. In short, the acute-care, disease-oriented model leads health personnel to underplay the full range of strategies and organizational arrangements that would be required in order to give more humane and more efficient extensive care.

Normalization as an ethical problem. In the largest sense, *the* problem of any chronically ill person is normalization, i.e., how to live as normally as possible despite one's illness. Society in general is not sufficiently aware of the extent of the problem. Few countries, for instance, have taken into account that even streets and public buildings are designed for normal people who have full mobility and energy. Public landscapes provide clear examples of the central issue, namely, that chronicity is much more than a medical problem. It is very much a social problem. It is also an ethical problem, one in which the family and the general public, whether they recognize the fact or not, are deeply implicated.

Unquestionably needed is a more general ethical awareness of how countries are arranged —spatially, physically, socially, financially—for citizens who are relatively free from the disabilities of chronic disease, thus rendering the struggles of the chronically ill to live satisfying lives, as close to normal as possible, poignantly difficult. They become all the more difficult through the predominantly "medical" approach to chronicity. Yet one might plausibly predict that a combination of factors—increasing rates of chronicity, increasing costs of the acute-care model, and an increased sense of the rights of consumers in many countries—will eventually contribute to improving the chances for the chronically ill to live more satisfactorily than is now possible. (Improved medical technology and practice will certainly help alleviate their suffering, but it will also inevitably add more sufferers to the population.) To be unaware of that future and feasible aim would be to compound today's relatively unwitting moral irresponsibility.

ANSELM STRAUSS AND ELIHU M. GERSON

[*Directly related is* HEALTH CARE, *article on* HUMANIZATION AND DEHUMANIZATION OF HEALTH CARE. *Other relevant material may be found under* AGING AND THE AGED, *article on* SOCIAL IMPLICATIONS OF AGING; HEALTH CARE, *articles on* RIGHT TO HEALTH-CARE SERVICES *and* THEORIES OF JUSTICE AND HEALTH CARE. *See also:* CARE; MENTAL HEALTH SERVICES; *and* RATIONING OF MEDICAL TREATMENT.]

BIBLIOGRAPHY

BENOLIEL, JEANNE QUINT. "The Developing Diabetic Identity: A Study of Family Influence." *Communicating Nursing Research: Methodological Issues.* Third WICHE Nursing Research Conference. Edited by Marjorie V. Batey. Boulder, Colo.: Western Interstate Commission for Higher Education, 1970, pp. 14–32.

CALKINS, KATHY. "Shouldering a Burden." *Omega* 3 (1972): 23–36.

DAVIS, MARCELLA Z. *Living with Multiple Sclerosis.* Springfield, Ill.: Charles C. Thomas, 1973.

DEBUSKEY, MATTHEW, ed. *The Chronically Ill Child and His Family.* Springfield, Ill.: Charles C. Thomas, 1970, pp. 196–198.

FAGERHAUGH, SHIZUKO. "Getting Around with Emphysema." *American Journal of Nursing* 73 (1973): 94–99.

FUTTERMAN, EDWARD H.; HOFFMAN, IRWIN; and SABSHIN, MELVIN. "Parental Anticipatory Mourning." *Psychosocial Aspects of Terminal Care.* Edited by Bernard Schoenberg, Arthur C. Carr, David Peretz, and Austin H. Kutscher. New York: Columbia University Press, 1972, pp. 243–272.

GUSSOW, ZACHARY, and TRACY, GEORGE. "Status, Ideology, and Adaption to Stigmatized Illness: A Study of Leprosy." *Human Organization* 27 (1968): 316–325.

LILIENFELD, ABRAHAM, and GIFFORD, ALICE, eds. *Chronic Diseases and Public Health.* Baltimore: Johns Hopkins Press, 1966.

REIF, LAURA. "Cardiacs and Normals: The Social Construction of a Disability." Ph.D. dissertation, Department of Social and Behavioral Science, University of California at San Francisco, 1975.

———. "Managing a Life with Chronic Disease." *American Journal of Nursing* 73 (1973): 261–264.

STRAUSS, ANSELM. *Chronic Disease and the Quality of Life.* St. Louis: C. V. Mosby, 1975.

United States, Department of Health, Education, and Welfare. *Chronic Conditions and Limitations of Activity and Mobility, U.S., July 1965–June 1967.* Vital Health Statistics, Data from the National Health Survey, ser. 10, no. 61. Washington: Public Health Service, Health Resources and Mental Health Administration, 1971.

CIVIL COMMITMENT
See INSTITUTIONALIZATION; MENTAL HEALTH SERVICES, *article on* SOCIAL INSTITUTIONS OF MENTAL HEALTH; INFORMED CONSENT IN MENTAL HEALTH.

CIVIL DISOBEDIENCE IN HEALTH SERVICES

Before showing how the concept of "civil disobedience" can be applied to medicine and health services, it is first necessary to introduce some order into the various casual uses of the term. Then we shall illustrate the occurrence of civil disobedience in health care in the particularly clear cases of birth control, the care of infants with drastic birth defects, Captain Levy's refusal to train members of the Special Forces, and protests by feminist health centers. These cases are merely illustrative; there are many cases of civil disobedience scattered throughout the literature of the various areas of health services. Moreover, in medicine there is much dissent that does not count technically as civil disobedience but which is, as we shall see, similar in certain respects to civil disobedience.

The meaning of civil disobedience

The concept of civil disobedience is extremely rich and diverse, not at all precise and specific. Yet much can be done to analyze and clarify the concept though not formally define it. Civil disobedience, first of all, can occur only within some structure of law, enforced by established governmental authorities, some aspect of which the person engaged in civil disobedience is trying to change. Disobedience in the context of family, church, lodge, business, university, or profession does not count as civil disobedience. Being civilly disobedient, then, consists in publicly announcing defiance of specific laws, policies, or commands of the legal structure that an individual or group believes to be either unjust or unconstitutional or both. To be defiant is not enough; the defiance must be made public since its purpose is to bring an injustice to the attention of the public and government, either for the purpose of stirring their consciences to rectify it, or to pressure them into rectifying it. The notion of civil disobedience requires not only the breaking of a law on moral grounds but also the pointing up of the disobedience for its symbolic and pressure-exerting value.

To give first an example of civil disobedience outside medicine: During the Vietnam War the burning of draft cards at public meetings in the United States was an act of civil disobedience. The men burning their draft cards were disobeying Selective Service law in protest against the government's policy of continuing the war. The law was broken because of moral objections to the war, and the disobedience was intended to bring the immorality of the war to the attention of the public and the government in the hope that government policy would be changed.

Civil disobedience in health services

The area of medicine and health services that appears to have produced the most clear-cut cases of civil disobedience has been that of birth control. In the cases, e.g., of Emma Goldman, Van Kleek Allison, Margaret Sanger, and Ethel Byrne, birth control laws were defied and attention explicitly drawn to the breaking of the law for the purposes of marshaling moral and legal support for their repeal. The classical cases of civil disobedience in the birth control movement are those of Sanger and Byrne. In 1916 these nurses directly challenged Section 1142 of the New York State Statutes, which prohibited the dissemination of birth control information and the distribution of contraceptive devices, by establishing a birth control clinic in the Brownsville area of Brooklyn. Leaflets announcing the fact were widely distributed. An immediate success, the clinic was swamped with working-class women in desperate need of advice. Byrne was found guilty of violating Section 1142 by distributing contraceptive information and devices, and Sanger was found guilty of violating the same law by conducting a clinic for such purposes, as well as for maintaining a "public nuisance." Byrne was sentenced to thirty days in the workhouse on Blackwell's Island and immediately went on a hunger strike. She was forcibly fed by prison officials. Sanger and Byrne enlisted the aid of the *New York World* to keep the attention of the public focused on Byrne's ordeal and on the just cause of birth control. Sanger was surprised in her case that the prosecutor tried so hard to prove her guilty. Of course she had violated the law; that was the point of her case! She was convicted and sentenced to thirty days in the Queens County Penitentiary, where she tried to disseminate birth control information among the inmates.

It would be a mistake to think that civil disobedience stopped after birth control had become a "respectable" movement. The dissemination of birth control information had become uncontroversial as long as the information was given only to married women. In the early 1960s William Baird witnessed the death of a young mother of eight children after an effort to abort herself with a coat hanger. Since she was unmarried she had been ineligible to receive information about birth control. Baird converted an old moving truck into a birth control information dissemination unit and used it in the poverty areas in New York City and vicinity. In 1966 he was arrested in Freehold, New Jersey, while showing a woman a diaphragm and describing its use. As a direct result of Baird's case, the New Jersey law concerning the distribution of birth control information and devices was liberalized. Baird advocated making birth control advice available to *any* unmarried women, even adolescents, and making contraceptives available in supermarkets as well as drug stores. In Massachusetts in 1967 he was convicted of the distribution of contraceptive devices to unauthorized persons. Sentenced to three months in prison, he appealed to the United States Supreme Court, which eventually sustained his plea in 1972. The Court struck down as a denial of "equal protection" to the unmarried the Massachusetts laws that forbade distributing contraceptives to unmarried persons.

The disclosures of R. S. Duff and A. G. M. Campbell in 1973 count as a courageous instance of civil disobedience in special care clinics. According to some authorities, the principle is well established both in law and ethics that physicians must use all "ordinary" means to preserve and prolong life but that the decision to apply "extraordinary" means is a matter of discretion. In the special-care nursery of the Yale–New Haven hospital, fourteen percent of 299 deaths occurred as a direct result of *withholding* treatment to infants with severe birth defects. In the words of Duff and Campbell, "After careful consideration of each of these 43 infants, parents and physicians in a group decision concluded that prognosis for meaningful life was extremely poor or hopeless, and therefore rejected further treatment" (Duff and Campbell, p. 890). Realizing that some people may argue the law has been broken, Duff and Campbell conclude that if what they have done *is* in violation of the law, then "we believe the law should be changed"; and their publication of these case reports was made in the hope that "out of the ensuing dialogue perhaps better choices for patients and families can be made" (ibid., p. 894). Subsequent legal scholarship has made it plain that Duff and Campbell were breaking numerous laws (Robertson).

A third example of civil disobedience in medicine is the refusal by Captain Howard B. Levy, a young dermatologist, to train Special Forces medics for service in Vietnam in 1967. On moral grounds he disobeyed the direct orders of agents of the government, and on many occasions publicly stated his opposition to the war. He spent twenty-six months in prison because he believed that to obey orders would have compelled him to violate canons of medical ethics. The aidmen Levy was ordered to train were not medics in the usual sense; in addition to their medical duties, they were expected to engage in guerrilla warfare. Levy felt that his moral obligation as a physician was to train personnel in the healing arts, and only in the healing arts, but he was directed to train servicemen expected to kill and wound as well as heal. His response was civil disobedience.

A fourth and final example of civil disobedience in medicine is the action taken in 1974 by the Feminist Women's Health Center in Los Angeles. The women there made a moral protest against the action or inaction of various government agencies in supporting or at least condoning the abortion clinic of Harvey Karman. Karman's allegedly substandard medical procedures they regarded as hazardous to the health and lives of women. Their protest took the form of disobedience of laws against theft and the public distribution of what they called a "writ of mandamus," which concluded: "Having confiscated the medical equipment, supplies, and office furniture necessary to the functioning of this experimental laboratory, we order these delegated official agencies to do their duty, to cease their financial aid of Harvey Karman and to stop his activities" (Feminist).

The nature of civil disobedience in medicine can be further clarified by some examples of actions that are clearly *not* civil disobedience, though they may be mistaken as such.

In 1968, when Patrick Cardinal O'Boyle, archbishop of Washington, delivered a sermon urging absolute obedience to Pope Paul VI's encyclical barring contraception, two hundred parishioners walked out of the church in the middle of Mass ("For Birth Control"). Although this walkout was a moral protest against authority, it cannot be considered civil disobedience since it did not involve disobedience of law and defiance of governmental authority.

Neither can the strikes by physicians that have become increasingly common in the 1970s be considered civil disobedience. In general, strikes are not acts of civil disobedience both because no laws are disobeyed and because the primary goal of the strikers is personal gain. Even in cases where laws are disobeyed (e.g., laws against strikes by government employees), strikes are not normally acts of civil disobedience because they lack the crucial element of moral protest. This is not to say that a civilly disobedient strike by physicians is inconceivable. If a group of physicians in Nazi Germany, in moral protest against the callous use of human subjects in experiments, had disobeyed directives from the government to treat patients, such a strike would have been preeminently an instance of civil disobedience. However, a strike by physicians in protest against, say, rates of premiums for malpractice insurance cannot be considered protest on moral grounds and consequently cannot be regarded as civil disobedience.

Dissent in medicine

It would be possible to cite many other cases in the areas already discussed, as well as in other areas, but to multiply such cases and areas would fail to disclose the focus of much current dissent against medicine and health services in the United States. One of the reasons why physicians are not more involved in what can properly be called civil disobedience is that they have considerable power to make authoritative rules that are equivalent to established laws. Civil disobedience is traditionally a method of bringing social change that the relatively powerless must use to change the policies and laws administered by those in power, usually government officials. In the case of medicine, the physicians *are* those with the power. The common practice of physicians establishes, in a sense, the common law of medicine. There are not many laws on the books governing medical practice beyond those governing birth control and abortions. Society has traditionally given physicians great power to practice medicine as they see fit, restricted by very few laws; so *civil* disobedience, as we have defined it, is not called for in most areas in which change, of an extremely controversial sort, is needed in medical practice. However, a great deal of current dissent against the medical establishment in the United States is clearly similar to civil disobedience. Medical practice *is* the law, so to speak, and the parallel to civil disobedience in this case is disobedience against this common law of medical practice, a common law that has become institutionalized in medical organizations that are capable of wielding powerful sanctions against dissent.

The nature of disobedience against this power structure has been manifested in numerous ways. Self-help clinics for women have sprung up across the United States. Another concrete manifestation of disobedience to the power structure is the Health Policy Advisory Center (Health-PAC) in New York City, a research and action collective. Similar groups are active in other cities. Also ghetto organizers and community groups are becoming increasingly interested in the idea of neighborhood clinics. The thrust against both the American Medical Association and the medical schools is toward participatory democracy, where the community of citizens is seen to be active in decisions about where and how health resource funds are to be allocated. Moreover, nursing schools and other paramedical groups are fighting for financial and moral independence of the AMA and the influential medical schools. Finally, the general populace has become increasingly militant against the established medical profession. The legal structure is being used extensively as a way of controlling the common law practices of the medical profession, and the increase in medical malpractice suits in recent years has been enormous.

EDWARD H. MADDEN AND PETER H. HARE

[*This article will find application in the entries* ABORTION, *article on* LEGAL ASPECTS; CONTRACEPTION; INFANTS, *article on* PUBLIC POLICY AND PROCEDURAL QUESTIONS; MEDICAL ETHICS UNDER NATIONAL SOCIALISM; *and* WARFARE. *Other relevant material may be found under* LAW AND MORALITY; MEDICAL PROFESSION; *and* ORTHODOXY IN MEDICINE.]

BIBLIOGRAPHY

American Medical Association, Judicial Council. *Opinions and Reports of the Judicial Council.* E. G. Shelley, Chairman. Chicago: 1971.

BEDAU, HUGO A., ed. *Civil Disobedience: Theory and Practice.* New York: Pegasus, 1969.

Boston Women's Health Book Collective. "Women and Health Care." *Our Bodies, Ourselves: A Book by and for Women.* New York: Simon & Schuster, 1973, pp. 236–276.

Branson, Roy. "The Secularization of American Medicine." *Hastings Center Studies* 1, no. 2 (1973), pp. 17–28.

Cohen, Carl. *Civil Disobedience: Conscience, Tactics, and the Law.* New York: Columbia University Press, 1971.

Dienes, C. Thomas. *Law, Politics, and Birth Control.* Urbana: University of Illinois Press, 1972.

Duff, Raymond, and Campbell, A. G. M. "Moral and Ethical Dilemmas in the Special-Care Nursery." *New England Journal of Medicine* 289 (1973): 890–894.

Feminist Women's Health Center. "Writ of Mandamus." *Monthly Extract: An Irregular Periodical* 3, issue 3 (1974), p. 2.

"For Birth Control, a Mass Walkout." *Life*, 4 October 1968, pp. 30–33.

Langer, Elinor. "The Court-Martial of Captain Levy: Medical Ethics v. Military Law." *Science* 156 (1967): 1346–1350.

Madden, E. H., and Hare, P. H. "Reflections on Civil Disobedience." *Journal of Value Inquiry* 4 (1970): 81–95.

Michaelson, Michael G. "The Coming Medical War." *Readings on Ethical and Social Issues in Biomedicine.* Edited by Richard W. Wertz. Englewood Cliffs, N.J.: Prentice-Hall, 1973, pp. 269–285.

Ribicoff, Abraham. "Medical Malpractice: the Patient vs. the Physician." *Trial, The National Legal Newsmagazine*, February–March 1970, pp. 10–13, 22.

Robertson, John A. "Involuntary Euthanasia of Defective Newborns: A Legal Analysis." *Stanford Law Review* 27 (1975): 213–269.

Sanger, Margaret. *Margaret Sanger: An Autobiography.* New York: W. W. Norton & Co., 1938.

CLINICAL RESEARCH

See Human Experimentation; Informed Consent in Human Research; Research, Biomedical; Research, Behavioral; Informed Consent in the Therapeutic Relationship, *article on* Clinical Aspects; Fetal Research; Children and Biomedicine; Aging and the Aged, *article on* Health Care and Research in the Aged; Prisoners, *article on* Prisoner Experimentation; Women and Biomedicine, *article on* Women as Patients and Experimental Subjects.

CLONING

See Reproductive Technologies, *articles on* Asexual Human Reproduction; Ethical Issues; *and* Legal Aspects; Gene Therapy, *article on* Cell Fusion and Hybridization.

CODES OF MEDICAL ETHICS

I. HISTORY   *Donald Konold*
II. ETHICAL ANALYSIS   *Robert M. Veatch*

I
HISTORY

In the ethics of medicine and health care, ethical standards have been formulated for physicians in their professional duties, for persons conducting medical experiments involving human subjects, and for the various health care professions. While codes of ethics have long been regarded as the classic expression of these directives, various principles and professional rules of conduct have also been stated in the form of prayers, oaths, creeds, institutional directives, and statements of professional organizations. This historical essay is concerned with the principles and rules of conduct for the medical profession as they have been formalized in prayers, oaths, and codes. Medical prayers state a very personal commitment of the physician to his professional duty; oaths publicly pledge the new physician to uphold the recognized responsibilities of his profession; and codes provide more comprehensive standards to guide the practicing physician. Each form of ethical statement implies a moral imperative, either to be accepted by the physician personally or to be enforced by a medical organization upon its members.

Most formal statements of medical ethics have been advanced by physicians themselves. Although governments and religious groups have long prescribed some controls by law and precept on the conduct of medical practitioners, only in the last century have other groups developed ethical standards for the physician–patient relationship.

Medical prayers

The most beautiful and moving expressions of dedication to medical practice are in the form of prayers. Doctors of all ages have composed deeply moving prayers expressing gratitude for divine blessings and asking for divine inspiration in their professional conduct. The most widely acclaimed of these is the Daily Prayer of a Physician, once ascribed to the philosophical Jewish physician Moses Maimonides (1135–1204) but now believed to be the work of the eighteenth-century German Jewish physician Marcus Herz (Rosner, 1967, pp. 451–452). In the manner of most medical prayers, the Daily

Prayer asks for courage, determination and inspiration to enable the physician to develop his skills, meet his responsibilities, and heal his patients. It commits the physician to place his duty to patients above his own concerns and voices high aspirations without reference to the specific issues of medical ethics.

Oaths for physicians

In the ancient world physicians expressed their ethical concepts most often in the form of oaths, which were an integral part of the initiation ceremony for medical apprentices. Like many medical prayers, ancient oaths reflect the physician's belief that success in his profession required that he ally himself with the deity in the treatment of disease. All the ancient oaths beseech the deity to inspire physicians to fulfill their moral obligations, reward those who honor their sacred trust, and punish those who violate it.

One of the oldest of ancient oaths, a medical student's oath taken from the *Charaka Samhita* manuscript of ancient India, contains concepts which had pervaded Indian ethical thought for many centuries before their inclusion in the oath about A.D. 1 (Menon and Haberman, pp. 295–296). Pledging the medical student to live the life of an ascetic and a virtual slave of his preceptor in accordance with Indian custom for apprenticeships, the oath requires a personal sacrifice and commitment to duty from the student comparable to the physician's responsibilities to patients. By the eloquent terms of the oath the student physician is to place the patient's needs above his personal considerations, serving day and night with heart and soul; abstaining from drunkenness, crime, and adultery; and observing professional secrecy scrupulously. In sharp contrast to the medical ethics of the Western world, the Indian oath obliges the physician to deny services to enemies of his ruler, evildoers, unattended women, and those on the point of death. Ancient Indian thought condemned aid to anyone immoral, interference with the process of dying, and any circumstance that might suggest illicit sexual contact. Despite these differences, the oath of the Indian student reveals significant parallels between the medical ethics of India and those of the Western world and suggests a diffusion of ideas, probably from India to the West.

The most honored and enduring medical oath of Western civilization is the Oath of Hippocrates. Despite its renown, its origin is obscure. It is a part of the Hippocratic Collection, which was catalogued and edited by a group of Alexandrian librarians sometime after the fourth century B.C. Copies of these writings available to modern scholars, however, date from the tenth to the fifteenth century A.D. and do not preserve the original text with verbal accuracy. None of the manuscripts of this collection can be positively verified as genuine works of the great physician, and clearly the documents are the products of many contributors, with the earliest predating the latest by at least a century. Recent investigations have revealed substantial evidence that the Oath conforms closely to the teachings of Pythagoreans of the fourth century B.C. and that it was used in their rituals (Edelstein, 1943). It proclaims a more strict morality for physicians than was established by Greek law, Platonic or Aristotelian ethics, or common Greek medical practice.

The Oath of Hippocrates actually consists of two parts, the first serving as a contractual agreement between pupil and teacher and the second constituting an ethical code. The opening sentences pledge the novice physician to become an adopted member of his teacher's family, to help support his teacher and his teacher's children in case of need, and to pass on his instruction to the teacher's children free of charge. Instruction is forbidden by the Oath for any student who will not agree to this stipulation or take the same oath. Since familial bonds between teacher and pupil implied careful selection of those admitted to the family group, the covenant enabled physicians to prevent unworthy persons from entering the profession by this route.

The ethical code contained in the Oath of Hippocrates places restrictions on the medical techniques of the physician and defines his relations with his patient's family. He who takes the Oath agrees to employ dietetic means to promote recovery, to refuse to dispense poisons or abortive remedies, and to leave surgery to craftsmen trained in that art. These measures, the Oath asserts, will protect the patient from unjust treatment and preserve the purity and holiness of the physician and his art. When he visits the home of his patient, the physician is under oath not to commit acts of injustice or mischief, and to abstain from having sexual relations with women and slaves as well as men. Finally he promises to protect the secrecy, not only of professional confidences, but also of those things learned outside the profession that should not be made public.

The paternalism implicit in the Oath's instruction that the physician refuse his patient's requests in some cases and judge what confidences to keep is an important element of its legacy. The Oath's provisions contrast sharply with the standards of actual Greek medical practice, which permitted physicians to abet suicide and infanticide and to perform surgery, including lithotomy (removal of a stone from the urinary bladder). They also set a higher standard for the equal treatment of all social classes than might be expected in Greek society. Even the secrecy requirement, extending to information obtained outside the professional relationship, is exceptionally stringent. Yet these precepts, representing the thought of only a small group of medical practitioners, have outweighed all others in shaping the development of medical ethics in the modern world.

For centuries following the appearance of the Hippocratic Oath, the medical profession showed no inclination to accept it. Hellenistic physicians violated its injunctions without compunction. Then the rise of Christianity produced a new idealism which was generally in agreement with the Hippocratic ethics. Increased attention to the Oath led to modifications bringing it into more perfect harmony with Christian ideological concepts and practices. The earliest of these extant revisions, entitled From the Oath According to Hippocrates Insofar as a Christian May Swear It, substitutes a statement of Christian adoration of God for the references to Greek deities in the original Oath and replaces its covenant with a statement of teaching responsibilities in terms of Christian brotherhood, pledging the physician to teach his art to whoever wants to learn it and without any stipulation (Leake, 1927, pp. 215–218). The injunction against surgery does not appear in this version of the Oath, but no reason is known for this omission, and later Christian versions do contain it. The Oath of Asaf, from the seventh-century *Sefer Asaf* manuscripts of the oldest Hebrew medical work, reveals Hippocratic influences in its injunctions against administering poisons or abortifacient drugs, performing surgery, committing adultery and betraying professional confidences (Rosner and Muntner, pp. 318–319). Like the medieval Christian oaths, it instructs physicians to give special consideration to the poor and needy. The Oath of Hippocrates also appeared in medieval Muslim literature, where the only significant changes replaced references to Greek gods with statements in harmony with Islamic theology. The Oath in its original form was also known to Christian and Muslim scholars, of course, and it may have been taken by physicians practicing in both medieval societies.

Following the transition from medieval to modern Western civilization, the Oath of Hippocrates continued to be a model for ethical pledges by physicians. Medical schools, seeking to commit their students to the pursuit of high ethical ideals, continued a tradition begun in the Middle Ages to incorporate Hippocratic concepts in oaths for their graduates, especially the covenant's requirement for the physician to instruct his teacher's children and the ethical injunctions for secrecy and against administering harmful drugs. Oaths taken by graduates of medical schools of India, most of which have been oriented toward Western medicine since the eighteenth century, have also been modeled after the Hippocratic tradition. During the nineteenth century some medical schools in the United States required their graduates to take the Hippocratic Oath in its original form, and that continued to be a common practice in the twentieth century, despite the fact that many of the Oath's provisions were archaic.

A significant revision of the Oath of Hippocrates appeared in 1948, when the newly organized World Medical Association adopted the Declaration of Geneva (*Declaration*). The Association recommended the new oath to medical schools and practicing physicians in all nations as a pledge which would inspire doctors in the fundamental principles of medical ethics. The Declaration represents an attempt to make the original Oath fully applicable to modern conditions of medical practice and to diverse cultural, religious, and ethnic factions in the world community which the Association represents.

The Declaration of Geneva is a secular oath that contains no reference to religious tenets or loyalties. Its phraseology not only makes the Declaration more acceptable to physicians of diverse religious convictions and affiliations, but also avoids making religion a source of professional discord. By pledging physicians to uphold the honor and traditions of the medical profession, the Declaration offers a basis for professional pride and solidarity. It also appeals for medical unity by eliciting from the Hippocratic covenant obligations for the modern physician to give respect and gratitude to his teachers and to recognize his colleagues as his brothers. The family unit idealized in the original Oath has thereby been expanded to embrace the international profession.

In describing the physician's responsibilities to his patient, the Declaration of Geneva renews the Hippocratic emphasis on service but significantly modifies the Oath's specific provisions. Pledges by the Declaration's physician to make the patient's health his first consideration and to respect professional confidences are simple reiterations of basic principles in the original Oath, although an amendment of 1968 specifies that physicians should honor confidences even after the patient's death. The Oath's surgical restriction, however, is omitted from the Declaration in deference to the major role of surgery in modern medicine, and the Oath's injunction against improper behavior with women and slaves in the patient's household is redirected to the patient himself, and to more relevant social categories. The physician of the Declaration vows not to permit considerations of religion, nationality, race, party politics, or social standing to interfere with his duty to his patient. Obviously, those who conceived and adopted the Declaration found united support for a clearer condemnation of these prejudices than the original Oath provided. In sharp contrast, however, the Declaration's statement of the physician's responsibility regarding suicide, mercy killing, and abortion is carefully obscured in generalities which conceal modern controversy on these matters among physicians and laymen alike. The physician of the Declaration pledges only to maintain respect for human life from the time of conception and not to use his medical knowledge contrary to the laws of humanity.

A government-sponsored medical oath has recently appeared in the Soviet Union, where the Presidium approved the Oath of Soviet Physicians in 1971 ("The Oath"). Modeled after an oath used at the University of Moscow since 1961, the Soviet oath pledges the physician to conduct himself in accordance with Communist principles and to honor his responsibility to the Soviet government. This commitment to political creed and government is unique among medical oaths. The Soviet oath does not neglect other moral obligations, however, for it instructs the physician to honor professional secrets, constantly improve his knowledge and skill, be available always to calls for medical care or advice, and dedicate all his knowledge and strength to his professional activities. Like other recent oaths, the Soviet oath voices virtually the same ideal of humanitarian duty to individual patients that appears in the earliest medical creeds, but it also pledges the physician to serve the interests of society.

Professional codes

Physicians of the modern world have not been content with the spiritual inspiration of prayers and the moral commitments of medical oaths. The large medical institutions of urban society have required complex relationships among medical personnel who demand detailed procedures to prevent embarrassing ethical controversy and disruption of services. Lengthy treatises on medical subjects, which had enlightened physicians on ethical matters since the earliest times, were not easy to cite by paragraph and line and frequently concealed ethical instruction in needless verbiage. By reducing these essays to lists of rules, proponents of professional control produced elaborate ethical codes.

One of the earliest codes of medical ethics appeared in China, where the Oath of Hippocrates has never made a significant impression. Instead, an indigenous Chinese tradition in medical ethics developed from sixth-century origins in Sun Ssu-miao's *The Thousand Golden Prescriptions*. This tradition received clear expression in the Five Commandments and Ten Requirements listed by Ch'en Shih-kung in a seventeenth-century treatise on surgery (Lee, pp. 271–273). Along with much guidance for social intercourse, Ch'en's precepts included several ethical principles of merit. In keeping with a tradition of gratuitous medical service by physicians of independent wealth in China, the precepts instruct physicians to give equal treatment to patients of all ranks, to keep expenses modest, and to treat the poor without charge. Physicians should visit women only in the presence of an attendant, a rule that had become nearly universal by the Middle Ages, and should observe professional secrecy regarding female diseases. They should keep well informed in medicine, equip their offices fully, and prepare remedies of orthodox formulae to avoid disputes. Finally, they should be humble and not insult each other. These instructions continue to characterize Chinese medical ethics in modern times, but they have had little influence elsewhere. Although they bear some resemblance to ethical concepts in Western medicine, there is little evidence of cross fertilization.

In the West, a treatise published in 1803 by Thomas Percival, an eminent physician of Manchester, England, strongly influenced the development of codes of medical ethics (Leake, 1927, pp. 61–210). Originally prepared in 1794 to guide professional conduct relative to hospitals and other medical charities and expanded in 1803 to include physicians in general practice, *Medical Ethics; Or, A Code of Institutes and Precepts Adapted to the Professional Conduct of Physicians and Surgeons* expresses standards of morality and etiquette that were in sharp contrast to the quarrelsome conduct of British practitioners of that era. Percival's treatise places greatest emphasis on the professional relationships of physicians to one another; to hospital personnel, apothecaries, and others engaged in the treatment and care of the sick; and to the law. It advances rules for conduct that would permit physicians to treat patients properly and at the same time uphold the dignity of the profession.

In its advice to physicians to treat patients with compassion, fidelity, and humanity, and to observe secrecy warranted by circumstances, Percival's *Medical Ethics* acknowledges the Hippocratic tradition. It urges physicians to keep their heads clear and hands steady by observing the strictest temperance, to avoid the use of fraudulently advertised medicines, to publicize those remedies found to be effective, to continue to treat and comfort patients during the final stages of a fatal condition, and to consider benevolence and virtue more than wealth and rank in determining fees. It also recommends the separation of physic and surgery wherever there are enough physicians to permit the excellence in performance which accompanies specialization. Unlike the Hippocratic Oath, however, *Medical Ethics* recognizes both arts as honorable branches of the medical profession and subject alike to its ethics.

A principal concern of Percival's *Medical Ethics* is with those aspects of professional conduct governed by etiquette. It offers elaborate procedures for consultation among physicians in difficult cases and for preservation of distinctions of rank in relationships between junior and senior physicians on hospital faculties and in consultations. It cautions physicians to display respect for one another, to avoid criticism of the practice of their colleagues, to conceal professional differences from the public, and not to steal patients from one another. In justifying these procedures, Percival reasoned that criticism of the profession was usually unfounded and always degrading both to the doctors criticized and to the profession.

In most of its provisions Percival's *Medical Ethics* suggests a modified utilitarian philosophy, calling for individual physicians to conduct themselves in a manner that would enhance public respect for the entire medical profession. Although Percival frequently voiced the Christian ideals of duty and sacrifice in his treatise, he recommended that physicians carefully control their responses to those ethical demands. Wealthy physicians should not offer gratuitous treatment to patients who can afford to pay, for that would deprive other physicians of needed means of support. Physicians should expose unprofessional conduct only to the proper medical or legal tribunals, not to the patients. Only in extreme cases of incompetence or neglect should a physician interfere in another's case, and even then he should warn the patient only when the offending doctor ignores his advice. When new methods are demanded by the failure of conventional practice, physicians should proceed only after consultation with peers and only when reason and evidence suggest a favorable result. These provisions and others make it clear that physicians should police their profession against malpractice, but limitations imposed by procedural safeguards contradict the ideal of moral responsibility exercised by the individual physician.

When the American Medical Association was organized in 1847, it adopted a Code of Ethics drawn largely from Percival's *Medical Ethics*. The Code of Ethics made no mention of etiquette for hospital staffs and barely mentioned the relations of physicians with pharmacists and courts of law, but it expanded and elaborated the principles for physicians in private practice, even to include a statement of obligations of patients and the public to physicians. The medical profession in the United States was faced with a crisis in public confidence in 1847. Medical regulations in most states had been repealed with the result that uneducated practitioners and rank imposters had begun to compete with regular physicians for patients. Exponents of the Code of Ethics hoped that the public would cooperate with doctors in establishing standards for medical practice that would prevent the worst abuses of the doctor–patient relationship and re-establish public respect for the medical profession.

In response to the American profession's experience with pretenders, the Code of Ethics contained a variety of restrictions on open competition among physicians. It branded as quacks all medical practitioners who lacked orthodox training, claimed special ability, patented instruments or medicines, used secret remedies, or criticized other practitioners. The requirement of orthodox training made outcasts of doctors who belonged to medical sects such as the Homeopaths, the Eclectics, the Thomsonians, and later the Osteopaths and Chiropractors. Since each sect claimed superior results from its form of treatment, practitioners with sectarian designations were guilty of claiming superior ability as well as handicapped by their incomplete education. Claiming that these offenses resulted from selfishness and efforts to discredit rivals, the Code of Ethics demanded that reputable physicians avoid any action that might appear as an attempt to solicit the patient of another doctor. Although these provisions united the profession against quackery, the prohibition on claims of special ability produced conflict between general practitioners and aspiring specialists. This ethical rule ceased to cause dissension only after the establishment of specialist organizations to certify the credentials of their members and after specialization won sufficient acceptance to permit physicians to restrict practice to their specialties.

The Code of Ethics provided orthodox physicians with one means of exposing those undeserving of confidence. It stated that physicians should not consult professionally with anyone who lacked a license to practice or was not in good professional standing. Since professional standing was determined by local medical societies, this provision had the effect of substituting a collective professional judgment for that of individual physicians and patients. In those cases where the patient insisted on inviting a consultant who was not approved by the local medical organization, the attending physician would have to retire from the case in order to retain his professional standing. While physicians argued that they could not fulfill their obligation to patients by admitting a right for fraudulent practitioners to advise in any capacity, their ethics required that they withdraw and thus give full charge of the case to the allegedly unqualified practitioner. Moreover, the majority of physicians found the consultation restriction a useful means for excluding many qualified doctors from association with the dominant organizations. Before 1870 regular medical societies excluded from membership and forbade consultations with female physicians and Negro physicians and, throughout the latter half of the century, with physicians who adopted a sectarian designation, even when they were certified by licensing boards. Because of mounting criticism, the consultation restriction was eliminated from the Code of Ethics in 1903, but its spirit was revived by a 1924 resolution of the American Medical Association forbidding voluntary association of its members with cultists.

In the twentieth century a number of national governments have incorporated elaborate ethical codes into legal statutes governing the medical profession, to be enforced by an official medical board. The precepts in these codes generally accord with the broader principles of the Percival tradition, but many provisions deal with problems of recent origin and reflect a modern concern for public as well as individual welfare. The new codes reveal an increased concern for the ethics of abortion, which they generally allow only in accordance with legal determination or as a therapeutic measure to save the mother's life. Most of them clearly prohibit advertising, self-serving publicity, and even indirect promotional tactics by physicians. They forbid physicians to pay or receive commissions for referral of patients, to accept pharmaceutical rebates, or to offer their services in group medical care plans under conditions that would reduce their freedom of judgment. They restrict specialists to their specialties, discourage rivalry or acrimony between specialists and general practitioners, and forbid collusion or fee splitting by specialists and general practitioners against the patient's interest. Most modern official government codes also place emphasis on professional secrecy as permitted by law; for example, the regulations for West German physicians cover ethical aspects of the physician's responsibilities for secrecy and disclosure in great detail.

Establishment of the World Medical Association in 1948 encouraged physicians to develop international standards of medical ethics. The new organization adopted an International Code of Medical Ethics in 1949, which attempted to summarize the most important principles of medical ethics ("International Code"). Since 1900, certification laws had reduced the prevalence of unqualified medical practitioners, and scientific advances had increased the effectiveness of trained physicians. By mid-century physicians were directing their attention more to the actual treatment of patients and less to the formality of relations between one another or between doctor and patient. The International Code reflects the new concerns in a shift away from the detailed regulations of the preceding one and a half centuries. In place of elaborate etiquette for consultations and other medical confrontations, it recommends only that physicians behave toward colleagues as they would have colleagues behave toward them, that they call specialists in difficult cases, and that they not entice each other's patients. It warns against the profit motive and prohibits unauthorized advertising, medical care plans that deprive the doctor of professional independence, fee splitting or rebates with or without the patient's knowledge, and refusal to treat emergency cases. It also reminds physicians of their obligation to honor professional secrecy, an obligation which continues after the death of the patient, according to an amendment to the Code adopted in 1968.

The International Code only hints at the ethical problems of abortion and euthanasia by asserting the physician's responsibility to preserve life. It does, however, warn specifically against any action that would weaken the patient's resistance without therapeutic justification. Applicable to the dying patient and experimental subject alike, this standard requires the physician to consider the patient's well-being above all else. The International Code also recognizes the need for adequate testing of innovations by urging great caution in publishing discoveries and therapeutic methods not recognized by the profession.

Using the International Code of Ethics as an example, the American Medical Association reduced its elaborate code to ten Principles of Medical Ethics in 1957 ("Ten Principles"). Most of these principles had been anticipated in the International Code, but there are a few noteworthy exceptions. Reflecting a continuing distrust of sectarian practitioners by regular physicians in the United States, the 1957 Principles warn against professional associations with unscientific practitioners. They also obligate physicians to expose the legal and ethical violations of other doctors. Instead of warning against premature publication of discoveries, the 1957 Principles urge physicians to make their attainments available to patients and colleagues. Finally, while reaffirming the principle of professional confidence, the 1957 Principles authorize physicians to violate this principle as required by law or to advance the welfare of the individual or the community. This provision suggests more discretionary authority for the physician than do the codes of most nations and the World Medical Association, which emphasize the inviolability of professional secrecy.

Scientific advances and changing social standards in recent decades have raised ethical questions in a number of areas that are not adequately covered by existing general codes. In some instances additional guidelines have appeared to help conscientious physicians make ethical decisions. The growing popularity of organ transplantation as a surgical procedure prompted the American Medical Association to issue a supplementary statement in 1968 regarding the ethics of organ transplantation ("Ethical Guidelines"). It cautions physicians to protect the rights of both donor and recipient by informing them fully of the procedures to be used and of reasonable expectations of results, and, in the case of vital organs, to have the death of the donor certified by one or more physicians other than the recipient's attendant. It stresses the need for these procedures to be conducted only by specially qualified physicians, in adequate facilities, after alternatives have been expertly evaluated. In that same year the World Medical Association adopted the Declaration of Sydney, which proposes rather technical precautions in the determination of death of organ donors ("A Statement on Death").

The development of elaborate and costly techniques for sustaining human life in cases where death threatens has also raised questions regarding the responsibility of attending physicians. The American Medical Association attempted an answer in 1973 with a guideline for treatment of terminal illnesses ("The Physician and the Dying Patient"). The Association's statement condemns mercy killing but authorizes the physician to abide by any decision of the patient or his immediate family to discontinue extraordinary efforts to prolong life when evidence of imminent death is irrefutable. Closely related to these developments are the intense efforts of medical scientists to reach agreement on guidelines for clinical research. The accelerated pace of medical research has resulted in the appearance of numerous codes governing experimentation with human subjects.

Biomedical advances and changing social imperatives have raised a number of moral questions for which no widely accepted answers yet exist. The rapidly developing science of fertility control, including legal abortion, contraceptives, and artificial insemination, has provoked stormy moral controversy. Experiments in genetic control present issues on which it may be even more difficult to reach a moral consensus. New chemicals—antibiotics, cancer controls, mood changers—with extensive and often unknown effects beyond their therapeutic purpose, also create moral dilemmas for those who would prescribe them. Finally, the extremely high cost of much contemporary medical therapy limits its availability and creates problems of distribution, pricing, and rationing with grave moral overtones. In view of growing public concern, ethical codes of the future will undoubtedly contain guidelines for physicians who deal with these problems.

Codes from outside the profession

Some recent codes of medical ethics have been sponsored by groups that do not principally represent the medical profession: religious bodies, civil governments, and consumer groups.

Unlike other contemporary world religions, the Catholic Church has, in modern times, promulgated formal codes of medical ethics in various parts of the world, where they have generated a rather widespread social effect. They are not only considered binding on Catholics but also affect non-Catholics who are associated with Catholic health facilities.

To regulate its many hospitals in the United States, the National Conference of Catholic Bishops has adopted its own code, Ethical and Religious Directives for Catholic Health Facilities, which establishes standards of medical practice within the Catholic institutions in conformity with the moral and religious teachings of the Church. Although its statements on the subjects of secrecy, consent, organ transplantation, and terminal cases closely resemble those in other codes, it has more rigorous rules for protection of reproductive functions. It prohibits abortion except as an unintended result of a procedure employed to protect the mother, it prohibits both male and female sterilization except in treatment of a serious pathological condition, and it prohibits artificial insemination. The firm position of the Catholic Church is being subjected to increasing scrutiny from a larger society that lacks a clear moral consensus on these matters. Meanwhile, the Church's restrictions continue to have a significant effect on the nature of medical service available to patients in communities where Catholic health institutions are the only ones. By contrast, the Declaration of Oslo, adopted by the World Medical Association in 1970, expresses a more liberal view of abortion ("Statement on Therapeutic Abortion"). While the Declaration cautions that abortion should be performed only as a therapeutic measure and that physicians need not perform abortions against their personal convictions, it authorizes abortion in the vital interests of the mother, where the law permits. Even this statement falls short of the views of many physicians and the official opinion of the United States Supreme Court (*Roe* v. *Wade*; *Doe* v. *Bolton*, January 22, 1973) that, in the early stages of pregnancy, abortion is the mother's prerogative.

The modern consumer movement has also influenced the ethics of medical practice. As hospitalization has become a major consumer service, consumers have become vocal in demanding the right of patients to minimum standards of care and respect. In 1972 the American Hospital Association responded to consumer pressure and adopted a Patient's Bill of Rights, which pertains primarily to hospitals but involves physicians with several responsibilities to patients ("Statement"). By its provisions the physician is obligated to keep the hospitalized patient informed of diagnosis, treatment, and prognosis, to instruct the patient fully regarding possible consequences and alternatives before obtaining consent for medical procedures, to honor a patient's refusal to consent to treatment to the extent permitted by law, to protect the patient's right to confidentiality and privacy from physicians and staff not involved in his case, and to instruct the patient about his care requirements after discharge. These standards represent a significant departure from the traditional paternalism governing the doctor–patient relationship. The physician, who once was empowered by his ethics to be sole arbiter in deciding how to manage a case and what to tell the patient, would henceforth have to take the patient into the decision-making process and even recognize the patient's superior authority in the process. The Patient's Bill of Rights applies only to patients in hospitals, but it sets a precedent that physicians may eventually be obligated to follow in their own ethical codes.

Conclusions

The difficulties confronting professional leaders who undertake to establish ethical standards on new issues reflect the conflicts in fundamental values inherent in diverse views of medical ethics. The traditional ethics of the professional physician tends to ennoble him and to place great emphasis on the virtue of benevolence and the doctor's need to devote himself to the service of his patients. This tradition honors the individuality of the doctor–patient relationship, professional secrecy, and the physician's duty to promote the patient's welfare. In these and other matters, ethical formulations by physicians have been paternalistic, making the physician the dominant party in determining what action will best advance both the doctor's and the patient's interests. Recent codes prepared by interests outside the medical profession have advanced other philosophical tenets as foundations for medical ethics. Religion's dedication to moral rectitude has challenged the reliance by the medical profession on the benefits of medical care as a justification of ethical standards. Government-sponsored codes have given the general welfare priority over the interests of individual patients in some instances, and consumer groups have demanded more rights for patients to participate in decisions affecting their welfare. This has resulted in mounting ethical confusion as physicians become subject to competing ethical authorities with conflicting standards.

Responsibility for the development of ethical guidelines relative to the physician–patient relationship may be shifting from the physician to society as a whole. Only in those contingencies not anticipated by accepted guidelines does the responsibility for ethical criteria still remain with the individual physician, and medical societies are codifying the ethics of these moral frontiers as rapidly as consensus is being reached. Future success in the use of codes to control medical practice may well depend on an accommodation of the ethical norms of physicians with those of the larger society.

DONALD KONOLD

[*For the text of codes discussed in this article as well as other related codes, see* APPENDIX, *especially* SECTIONS I, III, *and* IV. *Directly related to this article is the companion article in this entry,* ETHICAL ANALYSIS, *as well as the article* CODES OF THE HEALTH-CARE PROFESSIONS *in the* INTRODUCTION *to the* APPENDIX. *A number of articles on the history of medical ethics are directly relevant, especially:* MEDICAL ETHICS, HISTORY OF, *section on* NEAR AND MIDDLE EAST AND AFRICA, *article on* ANCIENT NEAR EAST; *section on* SOUTH AND EAST ASIA, *articles on* INDIA *and* PREREPUBLICAN CHINA; *section on* EUROPE AND THE AMERICAS, *articles on* ANCIENT GREECE AND ROME, BRITAIN AND THE UNITED STATES IN THE EIGHTEENTH CENTURY, BRITAIN IN THE NINETEENTH CENTURY, NORTH AMERICA: SEVENTEENTH TO NINETEENTH CENTURY, *and* NORTH AMERICA IN THE TWENTIETH CENTURY. *Also directly related is the entry* RELIGIOUS DIRECTIVES IN MEDICAL ETHICS. *For further discussion of topics mentioned in this article, see the entries:* ABORTION; ADVERTISING BY MEDICAL PROFESSIONALS; CONFIDENTIALITY; DEATH AND DYING: EUTHANASIA AND SUSTAINING LIFE; DEATH, DEFINITION AND DETERMINATION OF; INFORMED CONSENT IN THE THERAPEUTIC RELATIONSHIP; LIFE; MEDICAL PROFESSION; ORGAN TRANSPLANTATION; ORTHODOXY IN MEDICINE; PATERNALISM; PATIENTS' RIGHTS MOVEMENT; RACISM; STERILIZATION; *and* SURGERY.]

BIBLIOGRAPHY

BARNS, J. W. B. "The Hippocratic Oath, An Early Text." *British Medical Journal* 2 (1964): 567. Analysis of a portion of the Hippocratic Oath found on a fragment of papyrus from the third century suggests the Oath was in use by that time.

BARTON, RICHARD THOMAS. "Sources of Medical Morals." *Journal of the American Medical Association* 193 (1965): 133–138. Barton traces two conflicting deontological traditions in medical ethics from ancient times to the present.

BURNS, CHESTER RAY. "Medical Ethics in the United States before the Civil War." Ph.D. dissertation, Johns Hopkins University, 1969. Discussion of several state and local codes and the American Medical Association code of 1847.

CHAKRAVORTY, RANES. "The Duties and Training of Physicians in Ancient India as Described in the Sushruta Samhita." *Surgery, Gynecology, and Obstetrics* 120 (1965): 1067–1070. Summary of a fifth-century Indian medical text, including a commentary on ethical duties.

Code of Medical Ethics Adopted by the American Medical Association at Philadelphia in May, 1847 and by the New York Academy of Medicine in October, 1847. New York: H. Ludwig & Co., 1848.

Declaration of Geneva, Declaration of Helsinki, Declaration of Sydney, Declaration of Oslo. New York: World Medical Association, n.d.

EDELSTEIN, LUDWIG. "The Hippocratic Oath: Text, Translation and Interpretation." *Bulletin of the History of Medicine,* supplement no. 1 (1943), pp. 1–64. Edelstein advances a highly original hypothesis regarding the origin and meaning of the Oath of Hippocrates.

———. "The Professional Ethics of the Greek Physician." *Bulletin of the History of Medicine* 30 (1956): 391–419. Edelstein traces the humanistic ideal in Greek and Roman medical ethics.

Ethical and Religious Directives for Catholic Health Facilities. Washington: U.S. Catholic Conference, Department of Health Affairs, 1971.

"Ethical Guidelines for Organ Transplantation." *Journal of the American Medical Association* 205 (1968): 341–342.

ETZIONY, M. B. *The Physician's Creed: An Anthology of Medical Prayers, Oaths and Codes of Ethics Writ-*

ten and Recited by Medical Practitioners Through the Ages. Springfield, Ill.: Charles C. Thomas, 1973. Includes editorial comments on the historical and deontological significance of each entry and a useful bibliography for topics of ethical concern.

GARCEAU, OLIVER. "Morals of Medicine: Bibliography." *Annals of the American Academy of Medicine* 363 (1966): 60–69. Comprehensive bibliography includes many references to oaths and codes, and commentaries on them.

"International Code of Medical Ethics." *World Medical Association Bulletin* I (1949): 109–111.

JAKOBOVITS, IMMANUEL. *Jewish Medical Ethics: A Comparative and Historical Study of the Jewish Religious Attitudes to Medicine and Its Practice.* 4th ed. New York: Bloch Publishing Co., 1975. Jakobovits refers to prayers and oaths as well as other documentary expressions of Jewish medical ethics.

JONES, W. H. S. *The Doctor's Oath. An Essay in the History of Medicine.* Cambridge: University Press, 1924. Contains careful analysis of problems of translation, especially helpful to interpreters of the Oath of Hippocrates, and translations of several medieval and modern adaptations of the Oath.

KING, LESTER SNOW. *The Medical World of the Eighteenth Century.* Chicago: University of Chicago Press, 1958. Chapter 8, "The Development of Medical Ethics," discusses developments in medical ethics culminating in Percival's treatise.

KONOLD, DONALD E. *A History of American Medical Ethics 1847–1912.* Madison: State Historical Society of Wisconsin, 1962. Analyzes the Code of Ethics adopted by the American Medical Association in 1847 and traces its revisions down to the present AMA code.

KUDLIEN, FRIDOLF. "Medical Ethics and Popular Ethics in Greece and Rome." *Clio Medica* 5 (1970): 91–121. Explores the traditional and practical influences on medical ethics in the classical world.

LEAKE, CHAUNCEY D. "Theories of Ethics and Medical Practice." *Journal of the American Medical Association* 208 (1969): 842–847. Survey of ethical traditions in medicine provides a background for Leake's discussion of current ethical issues and situational ethics.

———, ed. *Percival's Medical Ethics.* Baltimore: Williams and Wilkins, 1927. Percival's treatise in its entirety, as well as an introductory essay discussing the background and deontological significance of Percival's ethics and a photostat and translation of a medieval Cruciform Oath.

LEE, T'AO. "Medical Ethics in Ancient China." *Bulletin of the History of Medicine* 13 (1943): 268–277. A translation of the five commandments for physicians and summaries of other important documents are included.

LEVEY, MARTIN. "Medical Deontology in Ninth Century Islam." *Journal of the History of Medicine and Allied Sciences* 21 (1966): 358–373. Several translations of Arabic documents are included.

———. *Medical Ethics in Medieval Islam with Reference to al-Ruhawi's "Practical Ethics of the Physician."* Transactions of the American Philosophical Society, New Series 59, pt. 3. Philadelphia: American Philosophical Society, 1967. An interpretive essay introduces a complete translation of al-Ruhawi's treatise.

LEVINE, MAURICE. "The Hippocratic Oath in Modern Dress." *Journal of the Association of American Medical Colleges* 23 (1948): 317–323. To Levine the Hippocratic Oath is essentially a statement of some of the problems of countertransference in medical practice.

MACKINNEY, LOREN C. "Medical Ethics and Etiquette in the Early Middle Ages: The Persistence of Hippocratic Ideals." *Bulletin of the History of Medicine* 26 (1952): 1–31. Manuscripts described reveal an early and widespread diffusion of Hippocratic ethical concepts in medieval Europe.

"The Medical Oath: Empty Ritual or Wellspring of Dedication?" *Spectrum*, Winter 1965–1966, pp. 6–9. Brief survey of medical oaths, illustrated with reproductions of many of them.

MENON, I. A., and HABERMAN, H. F. "The Medical Students' Oath of Ancient India." *Medical History* 14 (1970): 295–299. The oath is taken from the *Charaka Samhita*, a medical text written about A.D. 1. A commentary is included.

MUNTNER, SUSSMAN. "Hebrew Medical Ethics and the Oath of Asaph." *Journal of the American Medical Association* 205 (1968): 912–913. Muntner discusses Asaf's Oath as an integral part of a Hebrew tradition.

"The Oath of Soviet Physicians." *Journal of the American Medical Association* 217 (1971): 834.

OLIVER, JAMES H., and MAAS, PAUL L. "An Ancient Poem on the Duties of a Physician." *Bulletin of the History of Medicine* 7 (1939): 315–323. Inspirational message documents the ideal standards for medical practice in ancient Greece.

"The Physician and the Dying Patient." Report of the Judicial Council, American Medical Association, December 1973.

REDDY, D. V. SUBBA. "Medical Ethics in Ancient India." *Journal of the Indian Medical Association* 37 (1967): 287–288. Summaries of ethical concepts in the *Kasyapa Samhita* and the *Kalyana Karaka*.

ROSNER, FRED. "The Physician's Prayer Attributed to Moses Maimonides." *Bulletin of the History of Medicine* 41 (1967): 440–454. Analysis of the controversy over the prayer's authorship, including a complete English translation.

———, and MUNTNER, SUSSMAN. "The Oath of Asaph." *Annals of Internal Medicine* 63 (1965): 317–320. A complete text of the Oath of Asaph accompanies a comparison of it with the Oath of Hippocrates.

SMITHIES, F. O. "On the Origin and Development of Ethics in Medicine and the Influence of Ethical Formulae upon Medical Practice." *Annals of Clinical Medicine* 3 (1924–1925): 573–603. Concentrates on the relationship between the conduct of physicians and ethical rules.

"Statement on a Patient's Bill of Rights." *Hospitals*, February 1973, p. 41.

"A Statement on Death." *World Medical Journal* 15 (1968): 133–134. Declaration of Sydney.

"Statement on Therapeutic Abortion." *World Medical Journal* 17 (1970): 125. Declaration of Oslo.

"Ten Principles of Medical Ethics." *Journal of the American Medical Association* 164 (1957): 1119–1120.

"Zum deontologischen Gehalt ärztlicher Standesordnungen." *Arzt und Christ* 17 (1st Quarter 1971): 1–41. The ethical regulations of medicine for twelve nations and the World Medical Association, along with a useful bibliographical listing of such regulations for twenty-eight other nations.

II
ETHICAL ANALYSIS

Codes, oaths, and prayers of medical ethics have emerged over the centuries from disparate sources, representing disparate societies, time periods, organizations, and perspectives. It is not surprising that they differ significantly in style and content. This article will examine systematically the ethical content of this divergent collection of documents from the earliest to contemporary times.

Ethical languages

Before looking at substantive differences in ethical content it is important to note the different languages used in various codes of medical ethics. Some of the earliest documents take the form of an oath or pledge. The Hippocratic Oath is sworn in the first person in the name of "Apollo Physician and Asclepius and Hygieia and Panaceia." The strong construction "I will" is used throughout. The Indian Oath of Initiation appearing in the Caraka Samhita and the Oath of Asaph (the oldest Hebrew medical ethical text), on the other hand, are both written in the teacher's words, to be administered to the new initiate. The Caraka Samhita uses the command language "thou shalt," the Oath of Asaph, the imperative. Modern codes, however, have tended to use the command language, secularizing the text, and frequently use the third person. The code written by Thomas Percival at the end of the eighteenth century uses "should" and "ought" constructions. In the first principle of his code, Percival says that hospital physicians and surgeons *"should* study . . . in their deportment, so to unite tenderness with steadiness, and condescension with authority" (Leake). The third-person use of "should" and "ought" was kept in the American Medical Association's (AMA) code in 1847, which was heavily dependent upon Percival. It remains in the twentieth-century AMA codes, the International Code of Medical Ethics of the World Medical Association (WMA) (1949), and many other medical professional organizations. The WMA's Declaration of Geneva (1948, revised 1968) and the Oath of Soviet Physicians (1971) returns to the earlier first-person language where the physician begins, "I solemnly swear—."

In the mid-twentieth century, especially following the Nazi experience, the language of rules becomes the common mode of expression in many professional codes, especially codes or statements dealing with human experimentation. This is seen in the Nuremberg Code (1946), the Declaration of Helsinki (1964, revised 1975), and the International Code of Medical Ethics (1949).

What is striking is the absence from the codes of physicians and most other medical professional groups of any "rights" language. The concept of rights is fundamentally modern. It is in the recent nonprofessional codes or "bills of rights" that rights are claimed, usually on behalf of patients or other nonprofessionals. The Patient's Bill of Rights (1972), the Pediatric Bill of Rights (1974), and the Declaration of General and Specific Rights of the Mentally Retarded (1968) are all documents where rights claims dominate. The Ethical and Religious Directives for Catholic Health Facilities (1971), a document from a religious denomination drafted and issued by the Roman Catholic hierarchy of the United States, mixes the language of rights with that of obligations or duties. Occasional references to rights, sometimes for health-care consumers, but often of health-care providers, occur in other modern documents such as the codes of the American dental, nursing, and chiropractic associations. In rare cases—for example, in the codes of the American Nurses' Association (1950, revised 1976) and the American Psychological Association (1963, revised 1972)—the descriptive sentences are used with clear hortatory implications, e.g., "The nurse provides services with respect for the dignity of man."

The central ethic

Ethical analysis of the codes of medical ethics creates problems. It is unfair to expect them to be fully developed, systematic theories of medical ethics. On the other hand the codes, at least the modern ones, are normally the product of much discussion, debate, and review. These, along with the historical documents that have had lasting significance, can reasonably be expected to reflect the basic ethical views of the organizations that have endorsed them. In fact it might be argued that documents that are the product of practitioners rather than theoreticians reflect even more accurately the ethical stance of the group than do the more systematic efforts at developing theories of medical ethics. For this reason it is valuable to examine the principles that seem to be implied in them.

When one turns to the substance of the codes, especially the physician codes, one can identify what might be called a central ethical obligation, a basic principle that provides the physician with a core moral stance for resolving ethical dilemmas. A striking feature is the presence of contradictions among the codes and the controversial nature of these central ethics.

Modern Western medical ethics has reiterated the central ethic of the Hippocratic Oath into the twentieth century. In spite of the fact that the Hippocratic position was a minority ethic for Greek society, it seems to be the foundation of much modern non-Marxist, physician ethics. The core ethic of the Hippocratic Oath is the physician's pledge to do what he thinks will benefit his patient. This is repeated twice in the Oath. Following the Pythagorean tripartite division of medicine—into dietetics, pharmacology, and surgery—the physician first pledges, "I will apply dietetic measures for the benefit of the sick according to my ability and judgment." Later, with regard to entering the patient's home, he pledges, "I will come for the benefit of the sick, remaining free of all intentional injustice."

The principle that the physician's first obligation is to do what he thinks will benefit the sick person is picked up in the Declaration of Geneva, where the physician swears, "The health of my patient will be my first consideration," and in the WMA's International Code of Medical Ethics, which proclaims that "the doctor owes to his patient complete loyalty and all the resources of his science."

The controversial nature of the "patient-benefiting" ethic, which at first seems so innocuous as to be platitudinous, is seen when it is contrasted with other major ethical positions both among physician groups and in society more generally. The first characteristic of the Hippocratic ethic is that it is individualistic: It concentrates only on benefit to the patient. In contrast, classical utilitarian ethics of the tradition of Bentham, Mill, and G. E. Moore would consider such a narrow focus on consequences for the patient to be ethically unjustified. Only if there were an intermediate empirical claim—a claim that, by having physicians concern themselves only with benefits to their individual patients, the greater good of the greater number will be served in the long run—would a holder of an ethical theory concerned with net benefits in aggregate find the Hippocratic principle acceptable. There is no evidence that the Hippocratic authors or their twentieth-century counterparts had such an indirect utilitarianism in mind. Rather they seem to hold that the physician has a special ethical obligation to benefit his patient, independently of the net consequences for others who are nonpatients. The real test comes in a case where the physician believes that one course will produce the most good in total, but another course will most benefit the patient. A physician who says he must choose the course most beneficial to the patient is faithfully following the Oath and rejecting the utilitarian alternative.

The AMA in its 1957 Principles of Medical Ethics could not accept the Hippocratic individualism. It instructs the AMA physician that "the principle objective of the medical profession is to render service to humanity [not the individual patient]." The tenth principle makes this interpretation unambiguous:

The honored ideals of the medical profession imply that the responsibilities of the physician extend not only to the individual, but also to society where these responsibilities deserve his interest and participation in activities which have the purpose of improving both the health and well-being of the individual and the community.

Here the AMA is closer to the Soviet physicians' oath than to Hippocrates'. The Soviet physician more boldly swears "to work conscientiously wherever the interests of the society will require it" and "to conduct all my actions according to the principles of the Communistic morale, to always keep in mind the high calling of the Soviet physician, and the high responsibility I have to my people and to the Soviet government."

Second, the central ethic of the Hippocratic ethic is paternalistic. The physician is to benefit his patient "according to my ability and judgment." Physicians, according to Percival, should study not only tenderness and steadiness but "condescension and authority, as to inspire the minds of their patients with gratitude, respect, and confidence" (Leake). The AMA and the 1959 British Medical Association (BMA) codes held that medical confidences could be broken if, in the judgment of the physician, it is in the patient's interest for them to be broken. Ludwig Edelstein gives a paternalistic interpretation to the Hippocratic pledge that the physician will not only benefit the sick but also "keep them from harm and injustice." Edelstein, on the question of from whom the physician is to protect the patient, addresses the arguments of some scholars that the patient is to be protected from his family, from friends, or the physician himself. He rejects those arguments, citing an axiom of Pythagorean dietetics that the bodily passions may lead one to inflict harm on himself. He concludes that the Oath means that "the physician must protect his patient from the mischief and injustice which he may inflict upon himself if his diet is not properly chosen" (Edelstein, p. 25).

Finally, one sees the controversy of the Hippocratic patient-benefiting ethic when it is contrasted with other nonconsequentialist ethical theories in which certain principles are taken to be simply inherently right-making or where certain claims are taken to be "inalienable rights." The Kantian ethical tradition, for instance, holds that one can by "pure practical" reason establish certain maxims, which one can act upon and at the same time will that they be universal laws of nature. The deontological or formalist theories of the twentieth century that follow Kant would similarly hold that certain characteristics of actions are right-making—at least when they do not conflict with or are not overridden by other prima facie duties. Such duties as truth-telling, promise keeping, justice, reparation, and gratitude appear on the list of W. D. Ross, one of the leading proponents of such a view. The limitation of morally relevant benefits to those of the patient already has a deontological quality. Holders of such views consider something morally relevant besides consequences, i.e., the one to whom the consequences accrue is also morally relevant. Holders of views in which there are certain characteristics of actions that make them inherently tend toward being right (other things being equal) or holders of the view that certain things like life, liberty, and the pursuit of happiness are "inalienable rights" would have to reject the ethic of doing what one thinks will *benefit* the patient, at least in cases where benefiting the patient will be at the expense of fulfilling prima facie duties or fulfilling basic rights of the patient, such as the right to reasonable information upon which to base consent or giving some nonpatient fair access to needed medical care.

There may be a paradox in the Hippocratic Oath. The physician is to do what he thinks will benefit the patient but is not to give a deadly drug or "use the knife, not even on sufferers from stone." What is the physician to do if he believes that giving a deadly drug or an abortifacient remedy, or using the knife will benefit his patient? Perhaps this apparent contradiction is resolved by the belief of the Pythagorean physician that such actions can never be beneficial to the patient. In that case the Oath simply spells out some rules that guide the physician in deciding what will be beneficial. Alternatively these actions are seen as inherently wrong even if they might be of benefit. If so, then the Hippocratic ethic abandons its consequentialism, at least for these cases.

Specific ethical injunctions

The strictures against abortion, euthanasia, and surgery are examples of specific injunctions that occur from time to time in the codes and oaths of medical and physician ethics. Code-by-code systematic comparison of these injunctions reveals interesting differences. The conflict among the codes on the question of confidentiality is perhaps the most dramatic. It will be taken up first, followed by several other specific injunctions.

1. Confidentiality. The Hippocratic injunction on confidentiality is sometimes taken to forbid breaking medical confidences. The text is really much more ambiguous. It says, "Whatever I may see or hear in the course of the treatment or even outside of the treatment in regard to the life of men, which on no account one must speak abroad, I will keep to myself holding such things shameful to be spoken about." The individual physician, however, is left with the question, just which things which he hears "on no account must be spoken abroad"? Possibly we are to use the "patient-benefiting" criterion for deciding when breaking the confidence is appropriate. That was the explicit principle in the British Medical Association Code in its 1959 version, which said:

It is a practitioner's obligation to observe the rule of professional secrecy by refraining from disclosing voluntarily without the consent of the patient (save with statutory sanction) to any third party information which he has learnt in his professional relationship with the patient. The complications of modern life sometimes create difficulties for the doctor in the application of this principle, and on certain occasions it may be necessary to acquiesce in some modification. Always, however, the overriding consideration must be adoption of a line of conduct that will benefit the patient, or protect his interests.

The World Medical Association's International Code of Medical Ethics (1949) and the Declaration of Geneva (1948, revised 1968) both close any such patient-benefiting loophole in the confidentiality principle. They simply require "absolute secrecy," much as did the ancient Jewish Oath of Asaph. No exception is considered even in a case where the physician has learned that his patient is about to commit mass murder. The Ethical and Religious Directives for Catholic Health Facilities (1971) are almost as blunt. They require that "professional secrecy must be carefully fulfilled not only as regards the information on the patient's charts and records but also as regards confidential matters learned in the exercise of professional duties. Moreover, the charts and records must be duly safeguarded against inspection by those who have no right to see them." The only escape might be in the interpretation of "those who have no right to see them." That qualifier has somewhat the character of the Hippocratic "which on no account one must spread abroad."

Keeping with their more social commitment to the welfare of others as well as the patient, the American Medical Association Principles (1957, revised 1971) (and the American Psychiatric Association's [1973] which are based on them) are quite explicit in providing three exceptions to the general principle of confidentiality:

A physician may not reveal the confidences entrusted to him in the course of medical attendance, or the deficiencies he may observe in the character of his patients, unless he is required to do so by law or unless it becomes necessary in order to protect the welfare of the individual or of the society.

Confidences can be broken not only when the physician thinks it will benefit the patient but also when he thinks it will benefit society or when it is required by law, e.g., informing the police of a bullet wound incurred in a crime. The ethical problem of such broad exceptions, of course, is not only the paternalism of the patient-benefiting exclusion but the potential subordination of the patient's interests and rights to the interests of the society. The Soviet oath, which one might expect to be as explicitly societal in its concerns as the AMA Principles, ambiguously pledges to "keep professional secrets."

The British Medical Association was confronted by a particularly difficult case in which the physician disclosed to the parents of a sixteen-year-old that she was taking birth-control pills. He defended the breaking of the confidence on the grounds that he thought it was for her benefit. This being explicitly permitted by the existing British Medical Association code, the General Medical Council acquitted him of the charge of unprofessional conduct. After that case the BMA in 1971 amended its confidentiality principle to provide the first recognition of the patient's right to confidence in cases where the patient and physician disagree about the proper course. The new code states:

If, in the opinion of the doctor, disclosure of confidential information to a third party seems to be in the best medical interest of the patient, it is the doctor's duty to make every effort to allow the information to be given to the third party, but where the patient refuses, that refusal must be respected.

2. Abortion. On the controversial subject of abortion, groups authoring codes have followed the ethical stance of the group subculture. The Hippocratic Oath follows the Pythagorean prohibition on abortion even though abortion was not generally considered unethical in the broader Greek culture (Edelstein, pp. 9–20). In the Oath of Asaph the early medieval Jewish medical initiate is instructed, "Do not prepare any potion that may cause a woman who has conceived in adultery to miscarry." The 1971 Ethical and Religious Directives for (U.S.) Catholic Health Facilities follow, consciously and precisely, a traditional theological explanation of official Church teaching, devoting six of its forty-three principles to the subject. Directly intended termination of pregnancy before viability is never permitted, nor is the directly intended destruction of a viable fetus. Treatments not intended to terminate a pregnancy but which nonetheless have that effect are permitted provided there is a proportionately serious pathological condition of the mother and the treatments cannot be safely postponed until after the fetus is viable.

When the cultural base of the group writing the code is very broad, the code is predictably less specific about the ethics of abortion. The Declaration of Geneva (1948, revised 1968) says, "I will maintain the utmost respect for human life from the time of conception" without directly prohibiting abortion. The WMA's International Code in its draft, but not in its finally adopted form, stated, "Therapeutic abortion may only be performed if the conscience of the doctors and the national laws permit." The American Nurses' Association (ANA), also representing individuals with a wide variety of viewpoints, also avoids direct comment. In their code revised in 1968 and in effect prior to the 1976 revision they say that "the nurse's respect for the worth and dignity of the individual human being extends throughout the entire life cycle, from *birth* to death" (italics added). The implication may be that fetal life is not included.

A 1966 statement approved by the ANA Board of Directors recognizes "the right of individuals and families to select and use such methods for family planning as are consistent with their own creed and mores," again appealing to individual conscience. Is the combined implication a tolerance of the nurse's participation in abortion? If so, what is the Catholic nurse to do, or what is the nurse who believes in the right of the individual to select methods for family planning to do if she works in a Catholic health facility? These conflicts for individuals who are simultaneously members of more than one group authoring a code need serious attention.

3. Euthanasia. An explicit obligation to preserve life is strikingly absent from the codes of ethics, both professional and public. In the light of a widely held view that the duty, or one of the duties, of the physician is to preserve life one would expect to find this duty emphasized. The only explicit reference with which we are familiar is the weak formulation in the International code (1949), which says that "a doctor must always bear in mind the obligation of preserving life." This obligation to "bear in mind" rather than explicitly attempt to preserve life is a very soft injunction, especially when it is combined with the patient-benefiting principle that an act which could "weaken physical or mental resistance of a human being may be used only in his interest."

Proscribing active killing is much more common in the codes, as might be expected from the general ethical prohibition on active killing even for mercy in many cultures and subcultures. The Hippocratic formula is, "I will neither give a deadly drug to anybody if asked for it, nor will I make a suggestion to this effect." Interpretation of this prohibition is controversial, some taking it to forbid any criminal attempt on a patient's life. That seems less likely, however, than a prohibition against assisting in suicide. While suicide, especially in the face of medical suffering, was not uncommon in ancient society, it was forbidden by the Pythagorean cult. This fact is cited by Edelstein in his defense of the hypothesis that the Hippocratic Oath is a Pythagorean document. "Acts causing another's death" were one of the few things the Indian medical student should not do at his teacher's behest according to the Caraka Samhita. The Oath of Asaph instructs the Jewish medical student to "take heed that you kill not any man with a root decoction." The prohibition against assisting in an act of killing, however, has never been extended to apply to cooperating in withdrawal from treatment. The distinction between active killing and withdrawal of certain treatments is clear in the Ethical and Religious Directives for Catholic Health Facilities, where it is held that "the directly intended termination of any patient's life, even at his own request, is always wrong," and that "euthanasia ('mercy killing') in all its forms is forbidden." The directives go on, however, to say that "failure to supply the ordinary means of preserving life is equivalent to euthanasia. However, neither the physician nor the patient is obliged to use extraordinary means." This also makes clear that "giving a dying person sedatives or analgesics for the alleviation of pain, when such a measure is judged necessary, even though they may deprive the patient of the use of reason, or shorten his life," is not taken to be euthanasia.

The distinction between active killing and omitting treatment by medical professionals is made clearer when the rights language is used, as in A Patient's Bill of Rights (1973). That document proclaims that "the patient has the right to refuse treatment to the extent permitted by law," presumably even if the result will be the death of the patient. It is clear, however, that the authors were not saying, nor did they intend to say, that the patient has the right to demand of the physician that drugs be given that will actively hasten death. Without using rights language the AMA, in its House of Delegates meeting in December 1973, took a similar position stating that, while intentional termination of the life of one human being by another is contrary to that for which the medical profession stands, "the cessation of the employment of extraordinary means to prolong the life of the body when there is irrefutable evidence that biological death is imminent is the decision of the patient and/or his immediate family" ("The Physician and the Dying Patient," report of the Judicial Council of the AMA as adopted by the House of Delegates, 4 December 1973). This statement, of course, does not go as far as the Patient's Bill of Rights in granting the right of the patient to refuse treatment insofar as it is legally permissible. It goes much further, however, than the more paternalistic resolution adopted by the New York State Medical Society House of Delegates in February of that year, which granted "the right to die with dignity with the approval of the family physician."

4. Truth-telling. One of the most conspicuous conflicts between the patient-benefiting principle

and the more deontological ethical theories is over the question of what one ought to tell a dying patient. Many of the professional codes are simply silent, presumably expecting the patient-benefiting principle to apply. The Indian oath of the Caraka Samhita is more explicit, stating, "Even knowing that the patient's span of life has come to its close, it shall not be mentioned by thee there, where if so done, it would cause shock to the patient or to others." The 1847 version of the AMA code instructs: "A physician should not be forward to make a gloomy prognostication . . . but should not fail, on proper occasions, to give to the friends of the patient timely notice of danger, when it really occurs, and even to the patient himself, if absolutely necessary." The violation of confidentiality in communicating to family or friends before informing the patient either is not noticed or is justified on patient-benefiting grounds. The grounding of the violation of any obligation to tell the truth in the patient-benefit principle is traditional in professional physician ethics. The AMA code makes the grounding explicit: "It is, therefore, a sacred duty . . . to avoid all things [that] have a tendency to discourage the patient and to depress his spirits."

Even the authors of A Patient's Bill of Rights seem to yield to the paternalistic patient-benefiting principle when it conflicts with the patient's proclaimed right to know. The Bill states that "the patient has the right to obtain from his physician complete current information concerning his diagnosis, treatment, and prognosis in terms the patient can be reasonably expected to understand." But it then qualifies this by stating that, "when it is not medically advisable to give such information to the patient, the information should be made available to the appropriate person in his behalf." The potential conflict of such an exception with the later-proclaimed right to privacy or the right to receive information necessary to give informed consent are not discussed. In any case even A Patient's Bill of Rights would be found wanting by one fully committed to a more deontological ethic in which truth-telling was at least a prima facie duty.

5. Justice in delivering health care. Many of the codes of physician and other medical ethics have some reference to the duty to deliver health care equitably. The Hippocratic Oath commits the physician to "remain free of all intentional injustice." The Greek term, *adikiē*, which is translated "injustice," has the meaning of "wrongdoing" more generally. The Hippocratic commitment to equal treatment of males and females, free and slave, in abstaining from sexual relations during a medical visit is as close as the Hippocratic text comes to a pledge to equal treatment. The commandments written by Chen Shih-Kung, a seventeenth-century Chinese physician, include a much more explicit commitment that "physicians should be ever ready to respond to any calls of patients, high or low, rich or poor." The twentieth-century Declaration of Geneva holds forth the same ideal: "I will not permit considerations of religion, nationality, race, party politics, or social standing to intervene between my duty and my patient," as does the American Nurses' Association indicative statement of the ideal: "The nurse provides services with respect for the dignity of man, unrestricted by considerations of nationality, race, creed, color, or status." Equality of access seems generally recognized as an ideal. That providing general access to professional services may directly conflict with fulfilling other ethical duties—such as the duty to do what will benefit the (present) patient—has not been explicitly addressed in the codes.

6. Rules, laws, and cases. In the mid-twentieth century a debate emerged in ethical theory between those who, like Joseph Fletcher, hold that what is right must be decided on a case-by-case basis and those who, like John Rawls in philosophy and Paul Ramsey in theology, give a more significant place to rules or law in deciding what is right ethically. It is the nature of codes of ethics that they are general statements of right and wrong kinds of acts—i.e., guidelines or sets of rules. Thus it is not surprising that, especially when dealing with specific issues such as abortion and euthanasia, they are seen as raising the problems of systems of ethics that emphasize rules.

Nevertheless, the overarching centrality of the patient-benefiting principle provides an out for the situationalist. Thus the Hippocratic Oath says with regard to confidentiality that the physician is not to disclose those things that ought not to be disclosed. The American Osteopathic Association code (1965) warns that "no code or set of rules can be framed which will particularize all ethical responsibilities of the physician in the various phases of his professional life." This commitment to the uniqueness of individual cases is typical of the strong situationalism of the medical professions in contrast with other groups and other ethical theories. It contrasts, for instance, with the governmental requirement that informed consent be obtained before a subject is used for research regardless of the uniqueness of the situation. In spite of that situationalism, however, some of the modern codes—especially those that are open to consideration of societal as well as patient interests—emphasize the necessity of obeying the law. The Osteopaths' code specifies that "the physician shall cooperate fully in complying with all laws and regulations pertaining to practice of the healing arts and protection of the public health." The AMA permits breaking confidentiality when required by law.

The ethics of professional relations

Virtually all of the professional codes—in contrast with the lay and public codes or bills of rights—devote major attention to relationships among professionals. The Hippocratic Oath begins with a covenant by which the new physician pledges "to hold him who taught me this art as equal to my parents and to live my life in partnership with him, and if he is in need of money to give him a share of mine, and to regard his offspring as equal to my brothers in male lineage and to teach them this art—if they desire to learn it—without fee and covenant." It includes a pledge to keep secrets much as any initiation ritual into a cult might. The longest of the three sections of the AMA code of 1847 is devoted to "the duties of physicians to each other and to the profession at large." Since many of the codes emerged at a point historically when the profession was separating itself from others claiming to offer treatments and cures, there is often strong language forbidding association with those not properly members of the group. The American Osteopathic Association, for instance, requires that a physician "shall practice in accordance with the body of systematized knowledge related to the healing arts and shall avoid professional association with individuals or organizations which do not practice or conduct organization affairs in accordance with such knowledge."

In terms of the sociology of the professions, it has been suggested that restraints on advertising, rules structuring the referral of patients, instructions on the ways of handling the incompetent member of the profession, or exclusion of those not properly initiated into the profession play important functions in maintaining the professional monopoly. Apart from the sociological functions of the codes, i.e., as rationalizations for protecting professional interests, it is also pertinent to analyze them as sets of ethical obligations.

If, for purposes of ethical analysis, we presume that the function of the sections of the codes dealing with relations among professionals is to spell out what are really perceived as ethical obligations, it should still be clear that appeals are being made to specific types of underlying ethical principles. Three different kinds of ethical arguments may underlie the detailed formulations of professional obligations to other professionals. First, such duties to one's colleagues may be defended on what could be called "universal" grounds. That would be the case if the ethical principles claimed as the foundation of such intraprofessional obligations are principles generally recognized by all persons. For instance, the AMA code of 1847 states detailed rules regarding professional consultation prohibiting "exclusion from fellowship" of duly licensed practitioners, punctuality in visits of physicians when they hold consultations, and secrecy and confidentiality so that the patient will not be aware that any of the consultants did not agree. These standards for consultation, however, are defended on the grounds that "the good of the patient is the sole object in view." Although it is generally not argued, there is a presumption that rational patients should accept this underlying principle. We have seen, of course, that the principle of patient-benefit is quite controversial when put up against competing ethical principles.

A second foundation for intraprofessional duties might be a special ethic for a special group, which, in principle, nonmembers of the group would not be expected to share. The ethic of a profession is in part the ethic of fraternal loyalty, of special obligation to one's adopted brothers. This special professional ethical obligation may be seen as deriving from the professional nexus rather than from some more universal source. It is a special ethic of a special cult.

The ethic of the AMA 1847 code, as of that written by Percival, is an ethic of the gentlemanly class. The ethic is one of dignity and honor among gentlemen. According to the AMA's 1847 code, "there is no profession, from the members of which greater purity of character and a higher standard of moral excellence are required, than the medical." It introduces the discussion of duties of physicians to each other by stating that "each physician on entering the profession, as he becomes thereby entitled to all its privileges and immunities, incurs an obligation to exert his best abilities to maintain its dignity and honor, to exalt its standing, and to extend the bounds of its usefulness." The text goes on to entreat the physician to avoid "all contumelious and sarcastic remarks relative to the faculty, as a body; and while by unwearied diligence, he resorts to every honorable means of enriching the science, he should entertain a due respect for his . . . seniors who have, by their labors, brought it to the elevated condition in which he finds it."

This gentlemanly ethic of honor and purity (the Hippocratic phrase is "purity and holiness") gives rise to special ethical burdens for the medical professional that the layman cannot be expected to grasp. Professional "courtesy" (gratuitous services for practitioners, their wives, and their children) should probably be understood in these terms. "Courtesy" is an ethical expectation for members of the brotherhood.

A third possible foundation conflates the two. It could be that professional duties are defended as being in the public interest (or in some other manner consistent with a more universal ethic), but that only members of the profession can be expected to understand this to be so. Advertising, for instance, could be attacked as it is in the AMA's 1847 code as "derogatory to the dignity of the profession" and necessary to separate the profession from "the ordinary practices of empirics." The authors might well hold that it is really in the public interest that the separation be made, but also concede that only members of the profession could see the necessity of that separation.

If there are special ethical obligations for members of the profession that in principle cannot be recognized from outside the professional group, it follows that there are likely to be conflicts between the profession's formulation of its ethical obligation and the broader public's formulation. The issue is not the existence of different ethical responsibilities attaching to different roles but rather a disagreement between the profession and the broader public over what constitutes the proper behavior of the professional in his specific professional role. Even if a profession agreed that it had a special duty to preserve life or limit advertising, it is still an open question whether the public wants physicians always to act on that norm. If the professional group holds that there is a special professional source of norms, then conflict is predictable.

A specific example of such conflict involves the ethics of advertising. Many professional codes in the manner of the 1847 AMA code prohibit or restrict advertising by members of the profession. The 1957 Principles of Medical Ethics of the AMA claim that "this principle protects the public from the advertiser and salesman of medical care by establishing an easily discernible and generally recognized distinction between him and the ethical physician." While such prohibitions on advertising might be seen as the behavior of a cartel restraining price competition, it is also possible that physicians really believe that they are engaged in a service that must be radically separated from peddling medical services as a commodity. Whether the medical profession sees such advertising as unethical or not, the public may see restraint on advertising as unethical. At stake are not only two different perceptions of ways of maximizing benefits to potential patients, but also two sources of ethical norms—one from within the professional nexus and the other from the broader society.

Conclusion

The codes, oaths, prayers, and bills of rights derive from radically different contexts, representing differing professional groups, public agencies, and private lay organizations such as churches and patients' groups. It is not surprising that radically different ethical conclusions are reached and that they are based on radically different fundamental ethical theories and methods of ethical reasoning.

ROBERT M. VEATCH

[*For the text of the relevant codes, see* APPENDIX, SECTION I: GENERAL CODES FOR THE PRACTICE OF MEDICINE, SECTION III: PATIENTS' BILLS OF RIGHTS, *and* SECTION IV: CODES OF SPECIALTY HEALTH-CARE ASSOCIATIONS. *For further discussion of topics mentioned in this article, see the entries:* ABORTION; ADVERTISING BY MEDICAL PROFESSIONALS; CHILDREN AND BIOMEDICINE; CONFIDENTIALITY; CONTRACEPTION; DEATH AND DYING: EUTHANASIA AND SUSTAINING LIFE, *article on* PROFESSIONAL AND PUBLIC POLICIES; ETHICS, *articles on* RULES AND PRINCIPLES, DEONTOLOGICAL THEORIES, TELEOLOGICAL THEORIES, *and* SITUATION ETHICS; HEALTH CARE, *articles on* RIGHT

TO HEALTH-CARE SERVICES *and* THEORIES OF JUSTICE AND HEALTH CARE; HUMAN EXPERIMENTATION, *articles on* BASIC ISSUES *and* SOCIAL AND PROFESSIONAL CONTROL; INFORMED CONSENT IN HUMAN RESEARCH, *article on* ETHICAL AND LEGAL ASPECTS; INFORMED CONSENT IN THE THERAPEUTIC RELATIONSHIP, *article on* LEGAL AND ETHICAL ASPECTS; MEDICAL PROFESSION; NURSING; PATERNALISM; PATIENTS' RIGHTS MOVEMENT; RIGHT TO REFUSE MEDICAL CARE; SURGERY; *and* TRUTH-TELLING. *See also:* RELIGIOUS DIRECTIVES IN MEDICAL ETHICS.]

BIBLIOGRAPHY

[The bibliography for this article is the same as for the previous article. In the Appendix, the reader will find the texts of codes and additional bibliography of codes and commentaries on codes for ethics of the medical and other health professions.]

CODES OF THE HEALTH-CARE PROFESSIONS

See APPENDIX, INTRODUCTION, *article on* CODES OF THE HEALTH-CARE PROFESSIONS. *See also the texts of numerous codes, oaths, prayers, and statements in the* APPENDIX.

COMMISSION AND OMISSION

See ACTING AND REFRAINING; DEATH AND DYING: EUTHANASIA AND SUSTAINING LIFE; *and* TRUTH-TELLING, *article on* ETHICAL ASPECTS.

COMMUNICATION, BIOMEDICAL

The two articles in this entry, MEDIA AND MEDICINE *and* SCIENTIFIC PUBLISHING, *deal with ethical questions in communicating biomedical information in the mass media and the ethics involved in the publication of scientific findings.*

I. MEDIA AND MEDICINE   Lois DeBakey
                         and Selma DeBakey
II. SCIENTIFIC PUBLISHING   Lois DeBakey

I
MEDIA AND MEDICINE

Historical basis of journalistic ethics

Medicine is a discipline whose direction and consequences affect the public as a whole. Since science writers, through whom the public receives much of its medical information, can have an important impact on human health and social welfare, the question of moral and ethical responsibility for the information disseminated in the mass media assumes importance. Like most other social and ethical concepts, those concerning the responsibilities of the press have changed with time. Journalistic ethics, in fact, is largely a development of the twentieth century, the very existence of a free American press having originally been interpreted as assurance that it would serve the public welfare. In the authoritarian and totalitarian systems, the press is subject to direct political control, and any ethical principles observed would be at the discretion of the government in power. A brief review of ethical developments in general journalism will help place in perspective the present status of ethics in medical reporting.

The concept of freedom of the press originated in the libertarian theory, which held that censorship abridged man's natural right of free expression and that the press operated, ultimately, in the public interest. If man had free access to information and ideas, he could presumably, as a creature of reason, find "truth" himself. Freedom of the press thus carried no guarantee of verity or ethical propriety.

In colonial America, newspapers carried heavy medicinal advertising to subsidize printing costs, and manufacturers of proprietary drugs, knowing that early settlers depended on self-medication, exploited the public press. Henry Jarvis Raymond, the founder of *The New York Times*, wrote such advertisements for a quack for fifty cents apiece (Lee). With increasing technology and urbanization in the 1830s, newspapers acquired a mass audience, and yellow journalism flourished. Scare headlines, fictitious items, and stolen pictures became commonplace; ethics was totally ignored. About mid-nineteenth century, however, there emerged a sense of social responsibility among journalists.

In the first decade of the twentieth century—an era of moral awakening—newspapers made notable ethical advances, with new standards for the editorial, advertising, and circulation departments. Regulations were also imposed by state and national legislation, with more stringent libel laws. The threat of greater governmental intervention prompted the industry to establish, voluntarily, codes of professional behavior. Many newspapers refused medical advertising that promised cures for incurable diseases or contributed to drug addiction. In 1923, the American Society of Newspaper Editors adopted a code of ethics ("Code of Ethics or Canons of Journalism"), signaling a break with the libertarian tradition. Fourteen years later, the radio broadcasters adopted their code ("Standards of Good Practice for Radio Broadcasters of the United States of America"), and fifteen years still later, in 1952, the television industry followed suit ("The Television Code of The National Association of Radio and Television Broadcasters"). Although the mass media thus acknowledged a social responsibility to act in the public interest, ethics remained largely theoretical.

Whereas other professionals with a public trust, such as physicians, lawyers, and teachers, must pass licensing examinations before they can practice, journalists are not required by law to have educational qualifications, certification, or licensure, or to take a Hippocratic-like oath. Anyone can be a journalist who can find someone to publish his material. This phenomenon is paradoxical in view of the fact that those in the mass media reach, and possibly influence, more people than any other profession does. Total journalistic objectivity is, of course, impossible, and many journalists subtly inject subjective opinions in "straight news" reporting or editorialize to some extent when they are presumably merely informing. In a television program, for example, a newsman—a nonphysician—endorsed mammography for the detection of cancer, to the consternation of those physicians who question the safety of this procedure. The newsman's defense was that the medical reporter's responsibility is not merely to transmit information, but to interpret it as well. The reliability of such interpretation, and of any medical guidance derived therefrom, by those without a medical education or a medical license, however, raises ethical questions. Sometimes opinions are conveyed nonverbally: a raised eyebrow, a skeptical frown, or a sardonic smile can often be more telling and more devastating than anything spoken. But even if journalists do not tell the public *what* to think, they do influence, by the very selection of the material disseminated, what the public thinks *about*. The need for the industry to observe ethical criteria is therefore real.

Recognizing this need, the National Association of Science Writers, in June 1960, followed the precedent established by the rest of the news industry by formulating its own code of ethics to ensure the transmission of accurate, truthful, impartial information (Cohn).

An early example of medical reporting in the American public press had obvious socioethical implications. *Publick Occurrences,* identified by some historians as the first newspaper printed in the United States, devoted two paragraphs on 25 September 1690 to "Epidemical Fevers and Agues," a public health problem of serious concern to the colonists, who had experienced the ravages of smallpox (Harris). In the following year (1691), John Dunton established the *Athenian Mercury* in England (Cormick). Questions submitted by readers regarding medicine (primarily melancholy and lunacy) and other subjects were answered by the "authorities" of the Athenian Society, a fictional board of experts. Journalistic ethics was thus hardly a consideration.

With the advent of specialization in the late nineteenth and early twentieth centuries, American scientists began to cultivate a private jargon that only their peers understood. Research became deified, and professionals considered explanations to the uninitiated unworthy of their status. The reluctance of scientists to discuss their work publicly was reinforced by the medical code of ethics prohibiting publicity and by the early emphasis on sensationalism in medical journalism. After World War I, however, the medical profession's attitude toward journalists improved, partly because a corps of professional science writers assumed greater responsibility for accuracy. With the availability of governmental research funds after World War II, scientific experimentation mushroomed, and research became a professional status symbol. About mid-twentieth century, however, an antiscience attitude emerged. Exposés like Rachel Carson's *Silent Spring* and James Watson's *The Double Helix* spurred the demands for social and ethical accountability. The retrenchment of governmental funds forced scientists to go public in an attempt to prove the social relevance of medical research.

Science writing, evolving as it did from a tradition of almost unlimited freedom and invulnerability of the press, largely ignored ethical criteria and evoked varying degrees of public umbrage and distrust. This distrust, coupled with the opposing precepts of confidentiality in medicine and disclosure in journalism, erected a barrier between the two professions that made access to reliable sources of information difficult for early medical reporters. The tendency of zealous scientists to overestimate research findings and their inexperience in explaining scientific activities in lay language also impaired competent medical reportage in the mass media.

Responsible journalism

Medical reporting in the public interest. Ethical journalism is responsible journalism. Since the effects of irresponsible medical reporting extend far beyond personalities to possible impairment of scientific progress, unjustified public hostility toward science, and application of potentially dangerous medical treatment or procedures, professional medical reporters have a responsibility to convey accurate, reliable information to the public. Journalists have helped effect a number of health reforms. Exposure by crusading journalists of the maltreatment of patients in mental hospitals spurred authorities to ameliorate the sordid conditions in psychiatric institutions. When health hazards or crises occur, the mass media can alert the public almost instantaneously. Notifying consumers of contaminated or otherwise dangerous products, alerting a community to the increasing incidence of venereal or other infectious diseases, debunking quackery, and exposing other illicit or unethical practices are all in the public interest.

Television is a powerful educational tool that can also inform the public on controversial medical subjects, including ethically questionable experimentation, abortion, passive euthanasia, birth control, drug addiction, and psychic surgery. In one documentary, for example, scientists debated the dangers of genetic engineering and recombinant DNA (deoxyribonucleic acid), discussed the implications of the methods, agreed to a self-imposed moratorium on certain types of experiments, and formulated a set of guidelines to regulate future research. Providing a balanced, comprehensible discussion of such a complex, but socially important, research activity helps the public understand the prolonged and careful deliberations that precede medical experimentation and the responsibility that the scientific community assumes for its activities.

General ethical criteria. Because lay medical journalism is relatively young, little has been published on the ethics of the discipline. Many journalists readily admit, in fact, that they have never seen a code, and of those who have, few refer to it for guidance. Rather, most follow their own judgment, which must then pass the scrutiny of an editor, producer, or other higher authority before publication. Ethical violations in medical reporting may relate to the subject matter involved, the manner in which information is obtained, or the quality of the story published. Is the subject appropriate for public dissemination? Was the information obtained without duress or bribery from a reliable, reputable source? Is the story clear, accurate, and unbiased? Transgressions may include premature or immature reports in the mass media, invasion of privacy, sensationalism, and misquotations or disclosure of "off-the-record" remarks (DeBakey, 1974). Also to be avoided is conflict of interest, in which the material disseminated is influenced by emoluments, bribery, awards, or other such considerations.

Particular ethical criteria

Attribution. Attribution is requisite in medical news stories, since credibility depends on the authority of the source. Assigning statements or opinions to "reliable" but unnamed sources, as is sometimes done in other forms of journalism, is therefore inappropriate in medical reporting. Credibility is enhanced if the credentials of the writer or speaker are unassailable. Medical information conveyed by a physician is more likely to be reliable than that provided by a journalist who has no firsthand knowledge of the subject but acts merely as a medium of transmission. The information disseminated by nonphysician columnists has, in fact, sometimes been inaccurate and, on occasion, has had to be retracted. Although medical columns have long appeared in newspapers under physicians' names, it was not until the latter half of the twentieth century that physicians began to have regular health programs on radio and television, medical ethics having previously discouraged such activities as a form of advertising. Some still question the ethical propriety, however, of a physician's dispensing advice over the airwaves while remaining in the private practice of medicine.

Verbal and visual content. The medical journalist has a responsibility to convey information in simple, clear language that is comprehensible to the reader and is socially acceptable. In addition, most newspaper, radio, and television executives impose certain ethical restrictions on language used. The latter half of the twentieth century witnessed considerable relaxation of previous standards of propriety and good taste in mass communication. Before 1936, the mass media avoided the word "syphilis," a scheduled appearance of a high public health official having been canceled because he refused to delete that word from his prepared text. The purpose and focus of a story, to a great extent, govern its ethical propriety. The names of the sexual and reproductive organs, so carefully avoided in the past, are now often seen and heard in open educational discussions in the mass media. Until the 1970s, the female breast was not shown exposed on television, but today health education films regularly show how to perform self-examination of this region. The delivery of a baby and various operations, including transplantation and vasectomy, have been shown on television, even though some viewers find the sight of blood disturbing. Permission should, of course, be obtained to photograph any patient for publication.

Ethical problems

Premature publication. The ethical medical reporter obtains his information from unimpeachable sources rather than from an unreliable informer through subterfuge, intimidation, or bribery. He honors mutually approved ground rules and release dates. Since direct financial reward for contributions to medical knowledge is rare, public recognition is a strong motive in medical research, and priority is therefore of critical import. Dissemination of new scientific information in the news media before publication in professional journals has been a subject of ethical debate and a source of contention between the professions of journalism and medicine (DeBakey, 1976). Opponents of this practice point to the false hopes that may be aroused by precipitate reporting of a medical advance that has not yet been fully tested, and some have cited alleged "breakthroughs" and "wonder" drugs that were later proved false. Proponents of prepublication argue that the public has "a right to know," and to know immediately, the results of scientific research, particularly that supported by public funds. In the rare instances in which the public health is at risk, the information should obviously be released immediately. The rush to print, however, is often motivated by competition among the news services rather than eagerness to serve the public weal. The ill effects of premature enthusiasm are well illustrated by the unfortunate experiences with krebiozen, Liefcort, thalidomide, and especially human cardiac transplantation, the circus-like publicity of which hampered sober assessment. Since peer review is intended to safeguard standards of accuracy and validity, those who bypass it to make announcements first in the uncensored mass media may be interested more in self-advancement than in the integrity of science. Some physicians and medical scientists have circumvented critical peer review by publishing books, which are then widely publicized, and favorably reviewed, in the mass media, even though their content may be censured by medical experts as scientifically invalid and even dangerous.

One source of medical news in the public press is the scientific meeting, where physicians and scientists present research results, often before their publication in professional journals. When material is presented at an open meeting, it enters the public domain, but because the oral presentation represents the speaker's intellectual property, the speaker's approval should be obtained before others record or publish it. Prepublication may, in fact, deny the speaker publication of his address in a first-rate professional journal, in which originality is highly valued. For a scientific author to be victimized in this way when he has not, himself, released his report to the press prematurely is unfair. When a speaker's visual aids are copyrighted, permission to reproduce them is, of course, legally required.

When the *Washington Post* published a story on the relation of estrogens to breast cancer three days before the original scientific article appeared in the *New England Journal of Medicine*, F. J. Ingelfinger, the *Journal*'s editor, censured the reporter for "ignoring the convention not to release news of a report scheduled to appear in a scientific journal until the official publication date of that report" (Ingelfinger). An editor's note accompanying the critical Ingelfinger letter in the *Washington Post* contended that the reporter had received an advance copy of the scientific article independently. The question Ingelfinger posed is crucial: "Did [the reporter] have the right to use for himself and his paper material that clearly belonged to others?"

Leaks may have broad implications. If it is rumored that an article soon to be published in a professional journal is favorable or unfavorable to a certain medication, the stock market can be manipulated accordingly. Some journalists have pressured editors of scientific journals to release details of such articles prematurely by threatening to publish the leaked information, accurate or not. Suppression of news may, indeed, lead to publication of misinformation, but threats and duress are incompatible with ethical journalistic conduct.

Sometimes the scientist himself releases news directly to the press. Before addressing the

British Medical Association, a scientist reportedly circulated a press release about three "test-tube babies," at least one of whom, he stated, was developing normally ("The Baby Maker," p. 58). His formal presentation, however, reportedly made no reference to this claim (ibid.). Newspapers throughout the world promptly heralded the first "test-tube babies" as a miracle of modern research, leaving the impression with many that fetuses were actually conceived in test tubes. During the storm that ensued over the "unprofessional manner" of the disclosure, the scientist refused to divulge the identity of his patients, despite an offer of $72,000 from one newspaper and $120,000 from another. Declaring himself "fed up" with the publicity, the scientist announced that he was abandoning the research. Although the bizarre may constitute news, journalists should not, as Irvine Page cautioned, present it as "proved, generalized, and accepted by fully competent critics" (Page, p. 646).

Experienced journalists sometimes recognize overly exuberant claims of medical scientists and exercise skepticism and restraint in reporting such claims. At a press conference on cancer research at the National Academy of Sciences in October 1971, for example, a scientist announced that he had detected, in the milk of women with a family history of breast cancer, virus-like particles similar to milk-borne viruses known to transmit breast tumors (Wade). Women with such a family history, said the scientist, would do well not to nurse their babies. Rather than run a scare story based on an incautious announcement, professional medical reporters posed incisive questions that elicited the proper qualifications and limitations of the study, then wrote circumspect stories indicating that a diagnostic test was not yet generally available and that evidence was inconclusive that the suspected virus causes breast cancer.

News stories sometimes make broad inferences that are unsupported by the data. Extrapolation of animal data to human beings, especially when done by nonprofessionals, is not only risky from a scientific standpoint, but may cause unnecessary panic among the people. Conjecture, when advanced, should be based on adequate knowledge of the subject involved and on sound research data. Debating a medical problem superficially in the mass media before adequate evidence has been accumulated may confuse rather than enlighten the public, which is not generally qualified to weigh the pros and cons of a controversial medical issue. The publicity surrounding the "tolbutamide debate" and the saccharin ban created panic among patients, who look to the medical authorities for guidance based on scientific resolutions, not equivocation.

Falsification of scientific data. The critical need for medical reporters to view announcements of scientific "breakthroughs" skeptically has been further dramatized in recent years by disclosures of falsification of scientific data. Most widely publicized, perhaps, was the painted-mouse incident at a well-known cancer research institution. In 1969, science writers described as a "breakthrough" the researcher's experimental work to prolong the life of skin removed for grafting. Despite consistent reports of experimental success by the researcher in overcoming transplantation rejection, others, including Nobelist Sir Peter Medawar, had been unable to reproduce the results. In 1974, artificial coloring of one of the researcher's experimental mice was disclosed, and the validity of the entire program collapsed. The falsification was ascribed to pressure exerted on the scientist to publicize positive research results. The scientist suspected of painting the mouse left the research field, but others investigating methods of culturing to solve immunologic rejection reaped the negative effects of the scandal. Despite some promising findings, the undue skepticism within the scientific community created difficulties in finding professional journals to publish the data ("Was the Painted-mouse Doctor Right?").

Premature claims and overzealousness. Particularly in cases of dubious or unconfirmed "firsts" and "breakthroughs," the reliable journalist exercises caution, regardless of the prestige of his sources. The need for such skepticism on the part of journalists is substantiated by scientists' occasional recantations. Two respected scientists, after having discovered that an undergraduate assistant had falsified letters of recommendation, published a "potential retraction" of scientific articles that were co-authored by the assistant (Dressler and Potter, "Authors' Statement," "Transfer Factor"). Interestingly, a Nobelist had hailed this work as "one of the biggest things ever to come out of the Harvard labs" ("Researchers Retract Cancer Study Report").

In some instances, the scientist's zeal for his research may create problems. In April 1973, at a press conference during a meeting of the National Academy of Sciences, a well-known scientist announced that he had almost conclusive evidence that herpes viruses cause certain types of human cancer. At a science writers' seminar one year later, however, he disclosed his inability to repeat his earlier findings, and published in a professional scientific journal an official retraction of his earlier data (Sabin).

Invasion of privacy. Among the most vehement charges leveled at the press is the invasion of privacy. Such transgressions to obtain a titillating story serve little purpose beyond satisfying perverse curiosity. Compassion and good taste are not incompatible with good journalism. Good taste, admittedly difficult to define, was likely violated by the headline "Family Refuses to Give Heart" for a story about a teenage youth who had died playing Russian roulette. Consideration for the grief of the family might properly have overridden the journalist's instinct for drama, particularly since the disclosure had little constructive value.

Traditionally inviolate is the physician–patient relationship and the confidentiality of personal information transferred therein, protection of which dates back to the Hippocratic Oath (Perr). Several developments of the twentieth century have threatened the confidentiality of medical information: third-party insurance companies, computers, and the Freedom of Information Act. Because private medical information is now accessible through data banks, the journalist should exercise scrupulous ethical judgment in selecting material for public dissemination. Apprehension that private, intimate details of one's health may find their way into ill-intentioned hands has increased since passage of the Freedom of Information Act. Of special significance is psychiatric information that a patient may not wish divulged. News of the break-in of the office of Daniel Ellsberg's psychiatrist on 3 September 1971 outraged Americans and was vigorously denounced by the press as a flagrant intrusion of the patient's privacy, but the same press doggedly probes for intimate medical information about newsworthy people.

The practice of serializing, in the press, detailed medical information about prominent persons is a relatively recent development. President Cleveland was secretly whisked on board a yacht in New York Harbor to have an operation for cancer, and the full story of Franklin Roosevelt's failing health during his presidency was withheld until long after his death. Newspapers carried detailed accounts, however, of President Eisenhower's ileitis and his heart attacks; of President Kennedy's back ailments and purported Addison's disease; of Lyndon Johnson's operation for kidney stones and his heart attacks; of the emotional problems of political figures and their families; and of the mastectomies performed on the wives of high elected officials, including the President and Vice President of the United States. The ensuing public debate about the merits of radical mastectomy versus lumpectomy and the survival statistics may have frightened and perplexed more people than it edified. Moreover, whether a reporter has a right to divulge personal information about a relative of a "public figure" when the relative seeks privacy may create an ethical dilemma.

The line between the public's "right to know" and sheer medical gossip is not always easy to draw. In the coverage of a prominent senator's radical cystectomy in a weekly news magazine, not only were details given of the precise nature of the incision, but readers were informed of the cost of the hospital room, the frequency with which he would need to empty the plastic bag he would have to wear, and the "virtually inevitable impotence" that would ensue. Some readers wrote letters to the editor to protest the invasion of privacy and to recommend greater decency and respect.

To preserve freedom of the press, our laws make it extremely difficult to collect damages for misleading, false, or deprecatory material published about a person; one must *prove* knowing or reckless falsehood, and that is not easy to do. Nevertheless, some whose medical histories have been made public are beginning to take legal action. The Duchess of Windsor, for example, sued a television channel and a newspaper for using films and a photograph of her being supported by others as she walked in her garden. The suit charged that such photography, representing her as a helpless old woman, was damaging and constituted an invasion of her private life.

Sometimes the health of a public official may be of legitimate concern to the people. During the prolonged illness of a United States Supreme Court Justice, the press unsuccessfully sought information about how the state of his health may have affected the work of the Supreme Court. If an official's health affects public business, the public is entitled to be informed.

Sensationalism. The responsible medical science writer reports the facts, both favorable and unfavorable, and avoids slanting, exaggeration, and sensationalism. The inexperienced reporter, however, cannot always distinguish the natural enthusiasm of the dedicated scientist from the exaggerated claims of the publicity seeker. If a lack of scientific training does not allow the medical reporter to evaluate a startling new discovery, he should try to authenticate the significance of the discovery before he publicizes it as a "breakthrough." One way to enhance the accuracy of a medical story is to ask a recognized authority on the subject to read it—something that many journalists are reluctant to do, even though mistakes in this field can be extremely dangerous. To help ensure accuracy in scientific programs, the British Broadcasting Company employs a committee of scientists to serve in an advisory capacity. Periodic claims of cancer "cures" carried by newspapers and popular magazines, but later proved false, have caused needless anguish to victims of the disease, as well as to their families. It is true that imminent deadlines often preclude such detailed preliminary review, but getting it right is more in the public interest than getting it first. Publicizing of dangerous weight-reduction regimens, which have been responsible for deaths of gullible laymen, certainly might have been delayed until adequately assessed.

Sometimes the dramatic nature of a medical "discovery" invites journalistic excess, as in the case of the first human heart transplantation. Almost a decade later, after a plethora of insipid articles on the heart as the seat of the soul and on other irrelevant, meretricious, and often sordid stories had thoroughly desiccated the subject of its drama, human heart transplantation was discontinued by most heart surgeons as a therapeutic procedure. Christiaan N. Barnard, who was catapulted by the press into the limelight after performing the first human heart transplantation, later condemned the news media for creating dissension within the medical profession, exploiting the public taste for the macabre, and paying relatives or friends for controversial statements (Barnard). Such hysterical medical reporting generally disrupts the normal conduct of research and interferes with rational thinking. The heart transplant frenzy, however, had at least one salutary effect: the public debate regarding its ethical implications led to their serious consideration by the medical, as well as nonmedical, communities and contributed to the development of a new and needed discipline, bioethics.

Science writer Joseph Hixson censured television for valuing the visual impact of news items over balance and good judgment. In October 1974, a New York television channel showed ear-stapling being used for treatment of obesity, alcoholism, and depression without, Hixson argued, presenting the views of opponents of the procedure or giving adequate warning of the dangers involved. Such a "gimmicky piece of arrant quackery," Hixson maintained, violates the basic principles of responsible reporting (Hixson, p. 197).

Sensationalism is often the keystone of exposés, and medical exposés became prolific and profitable in the 1960s and 1970s. Articles and books on medical malfeasance flourished, including Medicare excesses, questionable therapeutic practices, and illegal dispensing of prescription drugs. Some were based on flimsy or inadequate evidence, but writing that is slanted to convey an unsubstantiated negative view, although unethical, may not be illegal. In a nationally televised segment entitled "Ghost Surgery," a reporter elicited from a patient a denial of any foreknowledge that a senior resident physician, not the chief surgeon, was to perform his operation. The chief surgeon insisted, however, that the patient had agreed, in advance, to allow the resident (who had previously performed more than 100 similar hernia operations) to do the operation under his chief's supervision, and he speculated that the patient may have simply been confused on the first postoperative day ("Clearing a TV Ghost"). The purported method of obtaining permission to make the film in this case raises ethical questions. Network representatives were reported to have requested permission to film "the training of a surgeon," not a story about medical ethics. Gaining access for one stated purpose and focusing the film on another, particularly if it has a negative focus, not only has ethical implications, but creates distrust. In this instance, a well-known surgeon's reputation may have been damaged, even though he denied the charge of ghost surgery. Persons whose character is impugned or whose reputations are thus damaged have recourse through laws against libel (malicious written defamation) and slander (oral defamation), but intentional malice is hard to prove.

Misquotations. Garbled quotations, whether direct or indirect, may not only create dissension between the press and the subject quoted, but may even have international implications. A neurologist, protesting the accuracy of a newspaper story about his visit to the Soviet Union, objected to the headline "Soviet Brain Study

Aim Is Behavior Control," which attributed "a sinister connotation and an implication of malicious intent" to his Russian hosts. The physician denied the reporter's implication "that patients are put into hospitals in the U.S.S.R. for the primary purpose of human experimentation," which, he contended, was "not a factual report of what I said" (Fields). Fear of distortion, false emphasis, or unbecoming levity has made some medical scientists acutely wary of the press.

Controversies. An ethical reporter does not create controversy for controversy's sake or falsely attribute intellectual differences among scientists to personal rivalry. Although such stories may sell newspapers, magazines, and books, they serve little constructive purpose. Open discussion of dubious practices, on the other hand, may have beneficial effects. The wide press coverage given to the questionable ethics of the cancer experiments on aged patients at the Jewish Chronic Disease Hospital in New York and of the Tuskegee syphilis study undoubtedly contributed to the formulation of ethical guidelines for human experimentation.

Headlines. Although the journalist is usually blamed for all errors in news stories, the responsibility sometimes lies elsewhere. Among the most common complaints about newspapers is that headlines are misleading, inaccurate, or sensational. The journalist has no more control, however, over headline writers who may cheapen his story with a shocking banner than he has over editors and rewriters who may change his copy. Headlines, designed as they are to attract attention, sometimes take liberties with the truth. Since, however, many people do not read beyond the bold print, accuracy is important. The impact of jarring headlines such as these may be profound: "Studies Say Doctors Harmful to Health," "Harvard Surgeon Reports Starvation in Hospitals," "One Step Nearer to the Super Race," "Research Group Says Milk Can Be Harmful," "Jews, Italians Complain More of Pain," "The Pill Will Kill, Says Doctor." Such headlines hardly suggest a sober treatment of the subject matter. Analogous to the sensational headline is the shocking title for a television program, such as "Don't Get Sick in America," which organized medicine considered biased and designed to frighten more than inform.

Misinformation. Although television is relatively young as a journalistic medium, its power to shape public opinion and policy is widely acknowledged. That power extends beyond informational programs to entertainment and commercials. Television dramas about medicine, by projecting an idealized image of physicians, may create unrealistic expectations in patients. The fictional physician in television serials is totally self-sacrificing and devotes his full attention to a single patient at a time, often extending his ministrations to resolution of the patient's personal problems. Diagnoses and treatment are quick and certain, rarely ambiguous or equivocal as in real life. Conveying such inaccurate impressions is not in the public's best interest. Also questionable is the ethicality of certain television advertisements of over-the-counter drugs, often with testimonials by actors posing as pharmacists or other health professional surrogates. Not only have these been censured as misleading, but the safety and efficacy of some of the products have also been under attack. Some firms, in fact, have been required to withdraw therapeutic claims for products that could not be proved.

Corrections and regulations

Corrections. If each science writer maintained a file on the "major medical advances" he has reported and, after a reasonable interval, arranged follow-up interviews with the scientists who announced them, he would discover that a large proportion of the "advances" fail the test of time. Subsequent emendation of such extravagant stories would clearly show the tentative and fragmentary nature of scientific "truths." Because newspapers serve as part of our historical archives, moreover, setting the public record straight is vital. Corrections, like retractions, rarely appear in the mass media, however, and when they do, they are usually buried in an obscure place or are relegated to letters to the editor.

Feedback. The American mass media, owned by private enterprise and largely autonomous, have a responsibility to allow dissenting opinions to be published. Most newspapers, in a spirit of fairness, have a regular feature of letters to the editor, and broadcasting stations provide time for the expression of contrasting opinions. Radio and television talk shows also permit listeners to call in and voice their opinions on various matters. Permitting such feedback by the public is an ethical responsibility of the free press, for the right of free expression extends beyond the press to the people.

Self-regulation. Self-analysis and self-regulation are beneficial in any profession, and especially in a free press. The British Broadcasting

Company has a three-man commission that acts as an appellate court in evaluation of viewers' complaints about news reporting and editing. Attempts in the United States to establish a press council to investigate complaints against journalists have largely failed, most representatives of the media considering such a body unnecessary and undesirable. Newspeople view themselves as representatives of the people, but since they are self-appointed, not elected, and since they have a vested commercial interest in their product, adherence to a code of ethics is essential. Such a code is especially important for medical reporters, who disseminate information that may affect the health and lives of human beings. The most practical ethical principles that medical reporters can follow are to obtain information from reputable, reliable sources, use common sense and fair play, and treat each story honestly and accurately.

LOIS DeBAKEY AND SELMA DeBAKEY

[Directly related is the other article in this entry, SCIENTIFIC PUBLISHING. Also directly related are the entries CONFIDENTIALITY; PRIVACY; RESEARCH, BEHAVIORAL; and RESEARCH, BIOMEDICAL. For discussion of related ideas, see: ADVERTISING BY MEDICAL PROFESSIONALS; and DRUG INDUSTRY AND MEDICINE. See APPENDIX, SECTION I, MEDICAL ETHICS: STATEMENTS OF POLICY DEFINITIONS AND RULES, BRITISH MEDICAL ASSOCIATION.]

BIBLIOGRAPHY

"The Baby Maker." *Time*, 29 July 1974, pp. 58–59.
BARNARD, CHRISTIAAN N. "Medicine and the Mass Media." *American Journal of Cardiology* 30 (1972): 579–580.
"Clearing a TV Ghost." *Medical World News* 18, no. 6 (1977), p. 11.
"Code of Ethics or Canons of Journalism." Adopted by the American Society of Newspaper Editors at annual meeting in Washington, D.C., 27 April 1923. American Society of Newspaper Editors, Box 551, 1350 Sullivan Trail, Easton, Pa. 18042.
COHN, VICTOR. "Science Writer Ethics." *National Association of Science Writers Newsletter* 8, no. 3 (1960), p. 11.
CORMICK, JEAN. "Medical Advice in Seventeenth-Century Journalism." *Journal of the American Medical Association* 224 (1973): 83–86.
DeBAKEY, LOIS. "Medicine and the Press: Resolving the Conflicts." *Clinical Research* 22 (1974): 214–225.
———. *The Scientific Journal: Editorial Policies and Practices: Guidelines for Editors, Reviewers, and Authors*. St. Louis: C. V. Mosby Co., 1976, chap. 8, pp. 36–39.
DRESSLER, DAVID, and POTTER, HUNTINGTON. "Authors' Statement." *Proceedings of the National Academy of Sciences of the United States of America* 72 (1975): 409.
———. "Transfer Factor: Warning on Uncertainty of Results." *Annals of Internal Medicine* 82 (1975): 279.
FIELDS, WILLIAM S. "Doctor Lodges Protest of Reported Interview." *Houston Chronicle*, 5 September 1971, sec. 3, p. 13. Letter to the editor.
HARRIS, BENJAMIN, pub. *Publick Occurrences Both Forreign and Domestick*, 25 September 1690, pp. 1–2. Printed in Boston by R. Pierce at the London-Coffee-House. Evans 546. *Historical Magazine* (Boston) 1 (1857): 228–231.
HIXSON, JOSEPH. "Epilogue: The Responsibility of the Media." *The Patchwork Mouse*. Garden City, N.Y.: Anchor Press/Doubleday, 1976, pp. 183–197.
INGELFINGER, F. J. "The Case of the Hoover Paper." *New England Journal of Medicine* 295 (1976): 896–897.
LEE, JAMES MELVIN. *History of American Journalism*. Boston: Houghton Mifflin Co., 1917, pp. 225–226.
PAGE, IRVINE H. "Science Writers, Physicians, and the Public—Ménage à Trois." *Annals of Internal Medicine* 73 (1970): 641–647.
PERR, IRWIN N. "Current Trends in Confidentiality and Privileged Communications." *Journal of Legal Medicine* 1, no. 5 (1973), pp. 44–47.
"Researchers Retract Cancer Study Report: Student Forged Letters." *Houston Chronicle*, 17 December 1974, sec. 1, p. 20.
SABIN, ALBERT B. "Herpes Simplex-Genitalis Virus Nonvirion Antigens and Their Implication in Certain Human Cancers: Unconfirmed." *Proceedings of the National Academy of Sciences of the United States of America* 71 (1974): 3248–3252.
"Standards of Good Practice for Radio Broadcasters of the United States of America." Washington: National Association of Broadcasters, 1771 N St. N.W., Washington, D.C. 20036, 1937.
"The Television Code of The National Association of Radio and Television Broadcasters." Washington: National Association of Broadcasters, 1771 N St. N.W., Washington, D.C. 20036, 1952.
WADE, NICHOLAS. "Scientists and the Press: Cancer Scare Story That Wasn't." *Science* 174 (1971): 679–680.
"Was the Painted-mouse Doctor Right?" *Medical World News* 18, no. 8 (1977), pp. 28, 30.

II
SCIENTIFIC PUBLISHING

The history of biomedical communication, and particularly of the biomedical journal, is intimately associated with the ethos of science—the sharing of new knowledge by a discoverer without his risking priority for his original intellectual concepts. That ethos, tacitly understood and widely accepted, is inherent in, and essential to, the very being of science. It implies honesty, mutual trust, and good faith among all those involved in the biocommunication system, including the researcher-author, editor, reviewer, publisher, reader, advertiser, speaker, listener, science reporter, and others. The ethics of biomedical communication remains largely uncodified, and the following discussion will therefore treat primarily practices that have evolved through general agreement.

Establishing priority

The inauguration of the *Journal des Scavans* in France in January 1665, and of *Philosophical Transactions* in England in March of the same year, allowed scientists for the first time to share their observations without fear of losing priority. Thereafter, the secrecy practiced by early scientists began to be replaced by disclosure. When Oldenburg, editor of *Philosophical Transactions* of the Royal Society, agreed to publish several of Robert Boyle's papers, he assured the author that the papers were "now very safe, and will be wthin [sic] this week in print" and that he therefore no longer need fear "a Philosophicall [sic] robber" (Hall and Hall, p. 291). Because of the lag between submission of a manuscript and its publication, some journals still protect authors' priority by publishing the date of receipt or acceptance in a footnote.

Before the era of professional journals, scientists used various means of protecting their intellectual property, including anagrams and sealed letters. Some scientists deposited with learned societies sealed letters, carefully dated, or, in later years, mailed themselves certified letters containing ideas or data, to be kept as proof of priority. Today, scientists establish priority more often through publication, in a conventional journal, of a letter to the editor, brief note, or abstract, which can be published more quickly than a full report, or through publication of a longer communication in a rapid-publication journal.

Despite the presumably free flow of information today, many scientists are still reluctant to disclose their work fully until they have firmly established their priority. The need for free discussion while guarding priority led to the formation of "invisible colleges," which restrict access to information to a select group. The question of ownership of information has become more complicated as its transmission has become easier.

Biocommunication ethics

Because physicians, scientists, and others who use the biomedical communication system assume that the information so stored and retrieved is honest and accurate, and because the application of that information often affects the lives of human beings, a code of biocommunication ethics is important. Some aspects of such ethics are assumed and therefore left to the integrity of individual persons, whereas others are safeguarded by guidelines designed to impose external control. These guidelines have not yet been systematized, however, as have those for the biomedical research that underlies much of the information.

Peer reviewing of manuscripts

One of the ethical guidelines in biomedical communication is validity of material published. To help ensure validity, most journal editors, with the help of expert reviewers in the subject under consideration, screen manuscripts before accepting them for publication. Two opinions are usually adequate for acceptance of a manuscript, but if one of the first two reviewers recommends rejection, most editors consider it appropriate to seek a third opinion. Papers read at professional meetings are sometimes published without benefit of peer review, particularly in society-owned or society-sponsored journals. At the base of the peer review system is the tradition of scientific skepticism, which dates back to Heraclitus and requires scientists to consider, and account for, contrary views. As in the legal system, however, the literary judges are not themselves without serious ethical responsibility (DeBakey and DeBakey, 1975), and the fact that a journal has a peer review system does not eliminate the chance for errors or bias, since reviewers, like everyone else, are subject to human frailties.

The reviewer is expected to maintain confidentiality, that is, to avoid disclosing the substance of the paper or using his privileged knowledge to the detriment of the author. Once a reviewer has certain information, it is difficult for him to suppress any ideas it may prompt for further research, but he should guard against doing anything that will deny the author his rightful priority.

Punctuality in evaluating manuscripts is another responsibility of reviewers. Delayed reviews create problems not only for authors and editors but for readers as well. Prompt publication of worthwhile information allows scientists to use the new knowledge to generate additional information or to prevent costly and time-consuming duplication of research that has already proved fruitless. If a reviewer cannot examine a manuscript within about two weeks, he should return it and so advise the editor.

Influence of authorship on reviewing. It is the scientific merit of an article that editors and reviewers are expected to evaluate—the plausibility, originality, validity, and comprehensibility of the thesis, methods, results, and conclusions —not the professional reputation, stature, nationality, or political views of the author. Many believe, however, that borderline ideas or questionable theses are more likely to be rejected if proposed by obscure scientists than by those of wide repute. Yet removal of authors' names from manuscripts before circulation to referees is not only impractical but will not erase other readily identifiable signs, such as references cited, methods used, and opinions stated.

Anonymity of reviewers. Traditionally, reviewers of manuscripts for scientific journals have been anonymous, serving as advisers to editors, who are the visible "gatekeepers" of scientific periodicals. Since most scientific reviewers serve without remuneration, they should not, it is reasoned, be exposed to importunings, rebuttals, or acrimonious communications from rejected authors. Whether or not reviewers are sometimes prejudiced in evaluating manuscripts, their anonymity has led to suspicion of dark motives in the minds of some rejected authors. Such prominent scientists as Thomas Huxley have openly expressed their distrust (Huxley, p. 141) and have objected to "psycho-political manoeuvres," such as protracted delays and demands for successive revisions, used by the reviewer while he, himself, rushes into print (Wright).

Cloaked in anonymity, some reviewers may reject a paper with such blanket condemnation as "topic inappropriate," "defective methodology," or "data weak," or they may consign the paper to oblivion with such an unsupported verdict as "ludicrous," "asinine," or "totally incompetent research." The author whose paper receives such a judgment has been peremptorily denied an opportunity to present his ideas in the journal of his choice. He may understandably feel that, just as *he* is required to provide supporting evidence for statements in his manuscript, so should editors and reviewers be expected to support their judgments. When denied a forum in scholarly journals, some authors will publish their material in book form and thus bypass peer review.

The critic who privately points out weaknesses in a manuscript is protecting the author against their public disclosure and should therefore be considered his ally rather than his adversary. The reluctance of a reviewer to sign a fully documented evaluation, even if it is negative, is puzzling. Most book reviews are signed today, and even though such critiques are made in public *after* the material has been published, few serious problems have resulted. In nine years as editor of *The New England Journal of Medicine,* Ingelfinger (p. 1372) had "never seen an abrasive or insulting word used by a reviewer who identified himself." Identification of reviewers, by promoting accountability, should foster careful documentation of criticism, which, in turn, will assist authors in improving their manuscripts and will reduce their distrust (DeBakey and DeBakey, 1975, 1976).

Editorial criteria for ethical research

Until relatively recently, editors and reviewers, in evaluating manuscripts for publication, paid little attention to ethical criteria of experimentation (DeBakey, "Ethically Questionable Data"). It was a biomedical communication that jolted the scientific community and aroused its conscience regarding the ethics of human experimentation; an exposé published in 1966 cited numerous examples of ethically questionable practices found among research reports examined (Beecher). The transplantation of the first human heart in 1967 focused further attention on the ethics of research on human subjects. The inhumane practices uncovered during the Nuremberg trials were cited for comparison, and various groups began debating ethical issues in an effort to prevent scientific zeal from infringing human rights. These discussions by scientists, clergymen, jurists, sociologists, philosophers, and others conferred new prominence on the discipline of medical ethics.

Notwithstanding disclaimers, readers usually assume that the entire contents of a journal carry the imprimatur of the editor. Aware of this assumption, most editors today examine the ethical as well as the scientific aspects of submitted material. Since most manuscripts of experimental research emanate from established scientific institutions, some editors rely heavily on mechanisms within these institutions to ensure the ethicality of the research. The Department of Health, Education, and Welfare (DHEW) has published rather stringent guidelines to encourage ethical human and animal experimentation for the many projects it sponsors (Transportation, Sale, and Handling; United States). The guidelines include the establishment of institutional review boards (IRBs) to attest to the ethicality of all experimentation under DHEW grants.

Approval by an institutional review board usually signifies only that the research protocol submitted prior to the experimentation has been adjudged to meet ethical guidelines—not that the research itself has been monitored or certified to have been ethical. The experiments being reported may differ substantially from the proposal approved, since the direction research takes must sometimes be altered during the experimentation. Because the degree of responsibility exercised by some IRBs may be rather perfunctory, some editors may require more than a simple statement of approval by the IRB. The method section of the report, for example, may be required to identify the ethical criteria observed and the mechanism by which they were met.

Ethically questionable data. A 1955 report on human experimentation in the Netherlands advocated that medical editors reject articles based on unethical experiments (Public Health Council). The editor of *The Practitioner* similarly urged editors to deny publication to any manuscript based on ethically improper practices, even if it is an "epoch-making article" (Thomson). In 1960 the editor of *The Lancet* announced his refusal to publish material "wrongly obtained" according to professional ethics (Fox). Several years later the British Medical Research Council recommended rejection of manuscripts describing ethically dubious research (Medical Research Council). The guidelines of the Massachusetts General Hospital of 1970 also called for the suppression of unethically obtained data, in an effort to curb unacceptable practices (*Massachusetts General Hospital Human Studies*). And in the same year an editorial in *The New England Journal of Medicine* exhorted institutions to exercise responsibility in certifying the ethicality of experimentation, to relieve editors of the dilemma regarding publication of unethical material ("Ethics of Ethical Evaluation").

Readers, too, have called for stricter editorial control regarding ethical experimentation. One reader denounced editors who, by publishing certain reports, sanction harmful procedures on patients who cannot possibly benefit from those procedures (Shapiro). Another called upon editors and publishers to demand that manuscripts contain a full explanation of the method of consent used (Blumgart). And Sir John Eccles urged that publication of data derived from unethical experiments be suppressed as a means of discouraging such practices (Eccles).

Some contend that, once research is completed, it is unethical to withhold from publication any useful information so derived—a retroactive justification of means on the basis of the end. Any breaches cannot be corrected at this point, they reason, and requiring the research to be repeated under ethical circumstances, even if possible, would not only be wasteful but would also expose additional subjects to risk. Rejection of unethical data, others say, would only remove experimental improprieties from visibility, not from commission, whereas publication would promote open discussion, help resolve conflicting values, and perhaps even deter further breaches. Still others suggest that ethically questionable data be published, but only if accompanied by an editorial commenting on or questioning the ethicality of the research. In such instances, the challenged author should be permitted to respond to the criticism, they argue, for denial of such an opportunity would suppress controversy and stifle open discussion, both vital to scientific progress.

Multiple publication

Most editors will not knowingly publish material that has already appeared elsewhere, whether in a scientific or a lay periodical. Authors are therefore expected not to submit to one journal essentially the same material that has already appeared elsewhere and not to submit the same manuscript simultaneously to two or more journals in an effort to expedite publication. Although multiple publication may expand the author's bibliography, it ill serves the scientific community. Editors have, in fact, publicly reproached authors whose material they have published without realizing that it had appeared several months earlier in another journal.

Confidentiality of medical data

In case reports, authors have long identified patients by initials as well as hospital number, but some journal editors have abandoned this practice to protect the identity of patients. Some editors also require signed consent forms before they will publish photographs of patients, even when the eyes are blocked out in the illustration.

Confidentiality of the patient's medical history is widely endorsed, but the advent of the computer has increased the risks of disclosure. The threat to privacy posed by easily accessible medical information stored in a computer data bank is of concern to many who envision potential abuses. The illnesses of well-known persons, for example—including politicians—have always been of keen interest to newspeople. Physicians, allied health personnel, and hospital administrators are expected by some patients to guard against unwanted publicity. Except when legally required, medical information about a patient should not generally be released without his specific approval.

Retraction and refutation

Although retraction of published information that later proves to be erroneous would seem to be an intellectual responsibility, scientific errors are rarely disavowed by their authors or by the editors who published them. Repudiation sometimes appears in subsequently published letters or articles, but these are usually written by readers, not the original authors. Since refutation generally brings less recognition than new discoveries, fallacious ideas are often simply ignored and allowed to die painlessly.

Mutual criticism of research and publications is accepted practice among scientists. It is, in fact, considered essential and is the rationale for peer reviewing. The criticism is expected to be unprejudiced and confined to the scientist's work rather than his person. Rarely has the right of an editor to publish critical material been questioned, but an occasional author may resort to the courts over derogatory statements about his work. In 1970 a dental surgeon, Drummond-Jackson, sued the British Medical Association and the authors of a paper published in the association's official organ, the *British Medical Journal*, for defamation. The article was critical of an anesthetic technique introduced by Drummond-Jackson. At the Court of Appeal, Lord Denning reportedly said that "it would be a sorry day if scientists were to be deterred from publishing their findings for fear of libel actions. So long as they refrained from personal attacks, they should be free to criticize the systems and techniques of others. Were it otherwise, no scientific journal would be safe" ("Science in Court"). Fortunately, scientists who disagree with one another usually express their views in professional periodicals, not by bringing libel suits.

Authorship and the by-line

Because a scientist's bibliography is a weighty criterion for academic promotion, admission to prestigious professional societies, and general professional advancement, a mentor may generously include in the by-line of an article the name of a protégé who contributed little or nothing. The names of some research directors have also been routinely added to by-lines, in deference to their seniority. As a result, the by-line no longer necessarily identifies the actual writers, as it once did in the scholarly tradition.

Despite the looseness with which "author" is used today, dictionaries continue to define the term as one who originates or composes a piece of writing. Thus, authorship refers denotatively to the origin of a *literary* production, not to experimentation. By this definition, one would expect the by-line to contain the names of those who contribute substantively to the form, as well as the content, of the report, and to display those names in descending order of magnitude of their contribution to the intellectual concepts in, and the literary exposition of, the report (DeBakey and DeBakey, 1975). Such a policy would eliminate granting authorship to a colleague who merely reads a manuscript and offers suggestions for its improvement, to a superior or subordinate who is unassociated with the project under consideration, or to a technician, resident physician, or anyone else who performs prescribed, remunerated duties. Yet the names of all such persons have appeared in by-lines. This laxity in assigning auctorial credit led Price to propose that editors refuse publication in cases in which team members are given credit because of "participation in the work reported rather than . . . specific responsibility for the publishable contribution" (Price).

Ghostwriting. A relatively recent development is for experimental or clinical researchers to turn over the preparation of the research report to a ghostwriter. But is there a place in the biosciences for the surreptitiousness intrinsic in ghostwriting? By its very designation and definition, ghostwriting involves pretense; one person publicly claims as his own the literary efforts of another (DeBakey, "Honesty in Authorship"). Since form and content, or language and thought, are difficult to separate, the ghostwritten product may also represent, at least in part, the intellectual efforts of the spectral scribe. The accolades derived from such publications may thus belong to the ghostwriter rather than to the auctorial impostor (DeBakey, "Rewriting and the By-line"). Editors often solicit editorials or articles from "experts," and others make awards, on the basis of previous publications. If those publications were ghostwritten, the basis for solicitation or recognition is fraudulent, and in a discipline dedicated to intellectual honesty and the pursuit of truth, fraudulence of any kind is inimical. Since integrity and authenticity are of the essence in the scholarship of biomedical science, phantom authorship is indefensible. If a person will pose as the author of a composition that someone else wrote, can he be relied on to report his data completely honestly? Some have suggested that ghostwriters be acknowledged by a phrase such as "as told to . . ." or "with the literary assistance of . . . ," but the commercial ring of such phraseology seems incongruous in a scholarly journal.

Editorial revision. If ghostwriting of material to be published in a learned scientific journal is unethical, what about extensive editing or rewriting? Is it appropriate for a journal editor or copy editor to rewrite an author's article before publication? How much may an editor properly alter a manuscript? It seems fair to expect the published material to represent principally the conceptual, ratiocinative, and literary efforts of those named in the by-line and insignificantly those of the editor or others (ibid.). Such slight efforts may be directed toward making the manuscript conform to the policies of the journal in citation of references, abbreviations, and similar mechanical details; remedying an occasional grammatical lapse; clarifying an ambiguous phrase; or refining an awkward syntactic structure. To go beyond that by reorganizing a paper, reconstructing tables or graphs, or recasting the entire text is tantamount to ghostwriting and is therefore to be discouraged. Those who wish to publish in scholarly journals have a responsibility to learn to write comprehensibly so as not to require extensive editorial revision of their manuscripts. If a manuscript is so poorly written as to warrant major revision, the reviser risks misinterpreting the original text and thus distorting the author's intended meaning. Even if the author is given an opportunity to approve the revision, he may overlook subtle errors in the rewritten version while focusing on major ideas.

Citation of references

Ethics requires that the ideas and the words of others be acknowledged by citation of their sources. To pad the list of references cited merely to appear scholarly is as improper as to omit essential references because of professional rivalry or opposing views. The ethical author documents statements that have not become general knowledge and that the reader may want to consult to evaluate the thesis being presented. All direct quotations should, of course, be properly acknowledged. Because "unpublished data" and "personal communication" are inaccessible to the reader, they are inappropriate in references cited, although the information and source may be referred to, when appropriate, in the text.

Advertisements

Since advertisements for medical products carry scientific information that can affect the health, welfare, and even the life of human beings, a code of advertising ethics is essential. The United States Food and Drug Administration exercises some control over advertisements for prescription drugs, but it is possible to fulfill the letter of the law while violating the spirit of the law and, as in other spheres, much therefore depends on the integrity of the manufacturer. Since advertising revenue helps support some scientific publications, the editor may be confronted with a conflict of interest in evaluating advertisements. An editor would therefore do well to provide potential advertisers with precise policies governing handling and publication of advertisements, and to adhere to these as much as possible in the selection of promotional material. Manuscripts submitted for publication that raise questions about a product advertised in a journal should not be rejected solely on that basis, but should be subjected to the same criteria for acceptance as all other manuscripts.

The pre-meeting abstract and oral presentation

Abstracts submitted to program committees for consideration for presentation at scientific meetings should contain authentic data obtained from completed, ethical research. Abstracts written before the research is undertaken, or before it is completed, and containing anticipatory results are obviously unethical. Publication of such abstracts by the sponsoring organization carries additional risks for the author who submits fictitious material.

Ratnoff and Ratnoff interpret the reporting, at a meeting, of unjustifiable risks to human experimental subjects as an endorsement by the sponsoring society (Ratnoff and Ratnoff). Society may sometimes have to wait for answers to questions, they contend, until those answers can be obtained ethically. Program committees may well inform scientists who submit ethically questionable reports that the material is unacceptable on ethical grounds. Editors must, however, be cautious about exerting undue policing pressures while guarding against unethical practices.

Conclusion

Despite the lack of a formal code of ethics in biomedical communication, certain tenets that have evolved through the years are now widely accepted in the scientific literary community. These include intellectual honesty; scientific integrity; acknowledgment of the original concepts and contributions of others; respect for the rights, and protection of the privacy, of patients and experimental subjects; unbiased evaluation of the work of colleagues; and recognition and admission of errors. Basic to this ethical code is the placement of fundamental human rights and the advancement of science above any selfish personal goals of the researcher.

LOIS DEBAKEY

[*Directly related is the other article in this entry,* MEDIA AND MEDICINE. *Also directly related is* RESEARCH, BIOMEDICAL. *For further discussion of topics mentioned in this article, see the entries:* CONFIDENTIALITY; HUMAN EXPERIMENTATION, *article on* BASIC ISSUES; INFORMED CONSENT IN HUMAN RESEARCH, *article on* ETHICAL AND LEGAL ASPECTS; MEDICAL ETHICS UNDER NATIONAL SOCIALISM; *and* PRIVACY. *Other relevant material may be found under* ADVERTISING BY MEDICAL PROFESSIONALS; DRUG INDUSTRY AND MEDICINE; HUMAN EXPERIMENTATION, *article on* SOCIAL AND PROFESSIONAL CONTROL; MEDICAL MALPRACTICE; MEDICAL PROFESSION, *article on* MEDICAL PROFESSIONALISM; *and* RESEARCH, BEHAVIORAL.]

BIBLIOGRAPHY

BEECHER, HENRY K. "Ethics and Clinical Research." *New England Journal of Medicine* 274 (1966): 1354–1360.
BLUMGART, HERRMAN L. "The Medical Framework for Viewing the Problem of Human Experimentation." *Daedalus* 98 (1969): 248–274.
DEBAKEY, LOIS. "Ethically Questionable Data: Publish or Reject?" *Clinical Research* 22 (1974): 113–121. Editorial.
———. "Honesty in Authorship." *Surgery* 75 (1974): 802–804. A letter to the editor.
———. "Rewriting and the By-line: Is the Author the Writer?" *Surgery* 75 (1974): 38–48.
———, and DEBAKEY, SELMA. "Ethics and Etiquette in Biomedical Communication." *Perspectives in Biology and Medicine* 18 (1975): 522–540.
———, and DEBAKEY, SELMA. "Impartial, Signed Reviews." *New England Journal of Medicine* 294 (1976): 564. Correspondence.
ECCLES, JOHN. "CIOMS Round Table Conference: Fourth Discussion." *Biomedical Science and the Dilemma of Human Experimentation: Round Table Organized with the Assistance of Unesco and the World Health Organization.* CIOMS Round Tables, no. 1. Edited by Vittorio Fattorusso. Paris: Council for International Organizations of Medical Sciences, 1967, p. 107.
"Ethics of Ethical Evaluation." *New England Journal of Medicine* 282 (1970): 449–450. Editorial.
FOX, T. F. "The Ethics of Clinical Trials." *Medico-Legal Journal* 28 (1960): 132–141.
HALL, A. RUPERT, and HALL, MARIE BOAS, eds. and trans. *The Correspondence of Henry Oldenburg.* Vol. 2: *1663–1665.* Madison: University of Wisconsin Press, 1966.
HUXLEY, LEONARD. *Life and Letters of Thomas Henry Huxley.* 2 vols. London: Macmillan & Co., 1900, vol. 1.
INGELFINGER, F. J. "Charity and Peer Review in Publication." *New England Journal of Medicine* 293 (1975): 1371–1372. Editorial.
Massachusetts General Hospital Human Studies—Guiding Principles and Procedures. 2d ed. Boston: 1970. Cited in *Experimentation with Human Beings: The Authority of the Investigator, Subject, Professions, and State in the Human Experimentation Process.* Jay Katz, with Alexander Morgan Capron, and Eleanor Swift Glass. New York: Russell Sage Foundation, 1972, p. 935.
Medical Research Council. "Responsibility in Investigations on Human Subjects: Statement by Medical Research Council." *British Medical Journal* 2 (1964): 178–180.
PRICE, DEREK J. DE SOLLA. "Ethics of Scientific Publication." *Science* 144 (1964): 655–657.
Public Health Council of the Netherlands. "Human Experimentation." *World Medical Journal* 4 (1957): 299–300.
RATNOFF, OSCAR D., and RATNOFF, MARIAN F. "Ethical Responsibilities in Clinical Investigation." *Perspectives in Biology and Medicine* 11 (1967): 82–90.
"Science in Court." *Nature* 225 (1970): 1179.
SHAPIRO, SAMUEL. "Ethical Aspects of Urethral Manipulation." *New England Journal of Medicine* 286 (1972): 1160. Correspondence.
THOMSON, WILLIAM A. R. "Editorial Responsibility in Relation to Human Experimentation." *World Medical Journal* 2 (1955): 153–154.
Transportation, Sale, and Handling of Dogs, Cats, and Certain Other Animals for Research Purposes. 7 U.S.C. § 2131 et seq. (1970), Pub. L. 89–544, 80 Stat. 350: amended by the Animal Welfare Act of 1970, Pub. L. 91–579, 84 Stat. 1560.
United States, Department of Health, Education, and Welfare; The Secretary. "Protection of Subjects: Technical Amendments." *Federal Register* 40 (1975): 11854–11858.
WRIGHT, R. DOUGLAS. "Truth and Its Keepers." *New Scientist* 45 (1970): 402–404.

CONFIDENTIALITY

Conceptual analysis

Essential elements of confidentiality. Concern about confidentiality begins when a child first experiences a desire to keep secrets as well as to share them. The desire to keep a secret is a manifestation of a developing sense of self as separate from others; the desire to share a secret stems from a need to retain or estabish intimate relationships with others (Ekstein and Caruth). A child will often keep secrets he or she believes to be embarrassing, shameful, or potentially harmful. To be willing to tell secrets to another, a child must have confidence that the other can be trusted. Other persons may, however, fail to keep the secrets entrusted to them. They may not wish to enter into an intimate relationship; they may not be trustworthy; they may believe that other factors outweigh the importance of keeping the secret; they may be tricked or forced into giving away secrets. In addition, unless a child feels that the environment is safe, for example, that no one else is listening in, he or she may be afraid to reveal secrets.

The example of childhood secrets presents in microcosm many of the elements germane to an analysis of confidentiality. It brings out that the keeping and sharing of secrets is linked to a complex set of personal attitudes, beliefs, and expectations that are intertwined with the conduct of other persons.

The adult social practice of designating certain information as confidential, like the behavior of children keeping secrets, has a twofold aim. On the one hand, it seeks to facilitate communication pertaining to intimate or other sensitive matters between persons standing in special relationships to each other. On the other hand, the practice is designed to exclude unauthorized persons from access to such information. Thus, confidentiality is essentially linked to *control* over the disclosure of and access to certain information.

To understand better the nature of confidentiality as a social practice, it is helpful to distinguish three of its aspects: subject matter, special relationships, and procedures. Sometimes attention is focused upon the subject matter—the informational content—which is confidential. For example, the use of the label "confidential" on certain documents is one familiar method to identify sensitive information. At other times the persons who respectively receive, transmit, or have access to such confidential information are the target of concern. Whether such persons stand in a special relationship which includes the authority to disclose or obtain confidential information is often at issue. For example, attorneys and physicians are designated by law as persons eligible to receive and preserve confidential communications in appropriate circumstances. A third important aspect of confidentiality is that of the procedures for protecting or limiting its scope. For example, locked files and soundproof rooms preserve confidentiality, while unregulated computerized records and hidden microphones violate it. These three aspects of confidentiality deal with *what* information is confidential, *who* has control over such information, and *how* confidentiality is affected by communication procedures.

Scope of confidentiality. Two essential aims (facilitating sensitive communication and excluding unauthorized persons) and three important aspects (subject matter, special relationships, and procedures) of confidentiality have been identified. Nevertheless, the boundaries of the concept of confidentiality are unclear because it is so closely tied to a cluster of concepts such as confidence, confession, trust, reliance, respect, security, intimacy, and privacy. Consider the following illustration: I may wish to keep personal information confidential to prevent exposure to ridicule or humiliation. For example, I might not want others to know of my fear of insects or certain sordid facts about my past. But I may wish to discuss my fears with or to confess such facts to a therapist or a friend. Unless I have confidence that I can trust another person to keep my secrets, I may withhold such information. Even if I believe that the person is worthy of respect and can be relied upon not to betray my confidences, I may fear that other persons or the state might seek to force or deceive the other person to obtain such information. A reasonable expectation that my communications to the other person will be protected may be a prerequisite to my disclosure of facts about myself. However, there may be circumstances in which I might choose to relinquish confidentiality or circumstances in which others might feel justified in refusing to respect my desire for confidentiality.

An explication of the many logical and contingent connections among the related concepts just mentioned would reveal both a complex and an overlapping conceptual pattern. It is difficult, therefore, to distinguish sharply and precisely define confidentiality apart from its companion concepts. As a result, one might say that the concept of confidentiality has "blurred edges" (Wittgenstein). For example, privacy and confidentiality are easily confused. Some writers define privacy as the right to control information about oneself (Fried). This suggests that privacy and confidentiality are coextensive. Other writers define privacy in terms of the right to personal autonomy and freedom. Confidentiality, in this view, is only one aspect of the right to privacy.

Just as the concept of privacy is elastic and vague, so also is the concept of confidentiality. One might define confidentiality narrowly in terms of protection of *communications* among persons in certain special relationships. But one might also define confidentiality more broadly, in terms of the right to control *information* about oneself. We are not here concerned with the appropriateness of alternative definitions of privacy and confidentiality. The examples are presented only to bring out that confusion may arise out of conflicting definitions, which imply differences in conceptual scope.

Value of confidentiality. The social practice of maintaining confidentiality provides one means to facilitate, control, and protect communications about intimate or other sensitive information. In this context confidentiality has instrumental value; it is not an end in itself. The measure of instrumental value is determined by the extent to which confidentiality contributes to achieving the desired goal. Confidentiality is more or less valuable depending upon whether other means provide better ways to accomplish the same purposes.

Like any instrument, however, confidentiality can be used for good or evil purposes. If the goal to be achieved with the aid of confidentiality is thought to be desirable—such as promoting love, friendship, or trust—then efforts to preserve confidentiality should be made. But if confidentiality is used, for example, to help perpetrate crimes, it should not be protected. In this respect the value of confidentiality is derived from the worth of the purposes that it serves.

It is uncertain whether confidentiality has intrinsic value as an end in itself. Charles Fried has argued that privacy, identified as a necessary condition for valuable forms of interpersonal intimacy—respect, love, friendship, and trust—has an intrinsic significance. If confidentiality is coextensive with privacy because both are defined in terms of control over information about oneself, then confidentiality would also be intrinsically valuable. However, others have claimed that privacy has intrinsic value because it is the foundation for a moral recognition of one's status as a person (Reiman). Accordingly, one might argue that confidentiality has value only as one means to protect privacy, not as an end in itself. A sense of self may begin with the keeping and sharing of secrets, but the development of personhood requires a capacity to experience intimacy and care for other persons. The control and exchange of confidential information is at best only one limited aspect of privacy.

The value of confidentiality can be brought out in another way: Imagine a world without it. First assume that no information about oneself would be treated as confidential by other persons; there would be no assurance that secrets would be kept. As a result relationships among persons could not be established upon a basis of trust that humiliating or harmful information will not be disclosed. A person would naturally be very reluctant to reveal sensitive information. Relationships among persons would become fragmented and superficial; this would impair, if not undermine, certain forms of intimacy such as marriage or friendship. One can also imagine a world in which secrets are valued and persons desire to be trustworthy, but the environment does not provide an opportunity for confidentiality. In a world such as that depicted by George Orwell's *1984* (updated to include more sophisticated technological monitoring and eavesdropping devices), fear, suspicion, and insecurity would prevail even among persons of goodwill. The context of human interaction would be so threatening that no one would be able to rely on the confidentiality of interpersonal communications. Either a world in which persons do not desire to form confidential relationships or a world in which the forming of such relationships is prevented would preclude the possibility of certain desirable forms of human interaction.

Protection of confidentiality

Persistent demands for increased protection of confidentiality are challenged by equally persistent demands for limitations on confidentiality. Policymakers are frequently faced with a conflict between the need to facilitate confidential communications about sensitive matters and the need for public information to protect the safety and promote the welfare of society. The purpose of this discussion is not to propose how particular competing claims should be evaluated. Instead, it is to examine different methods for protecting confidentiality and consider factors that justify limiting its scope. For convenience reference will be made only to practice in the United States.

Restriction of information. Confidentiality is, or is thought to be, protected by conventional social practices. In American culture many persons have the attitude that all information about one's body, including medical information, is confidential; it may be disclosed to others only for very good reasons. For example, public health laws, designed to prevent unnecessary and preventable suffering, require physicians to report instances of certain communicable diseases and to warn their patients of their condition. Recently questions have arisen about the extent to which genetic information discovered in the course of a physician–patient relationship about a person is confidential. May a person with a genetic disease or defect prevent this information from being disclosed to an employer, or even to members of his or her family? Does a person's fear of loss of livelihood or a desire to avoid feeling shame about an imperfection override the possible harmful consequences to society or family members resulting from their ignorance of a genetic disease or defect? Attitudes and beliefs in this area are uncertain and varied; we lack agreement about what ethical and legal principles should apply to such cases.

The cultural attitude about confidentiality of information concerning one's body partially explains the public outcry in response to reports that the confidentiality of medical records is sometimes inadequately protected. Controversy about medical records is complicated, however, because cultural attitudes come into conflict with professional practices and policies. Medical records typically contain, in addition to factual data, reports and interpretations of patient information made by hospital personnel, institutional data, and notes made by hospital staff concerning treatment. As a result disputes sometimes occur between a hospital's and a patient's claim to ownership of the medical records. The question of ownership then becomes the crux of the issue of control over the disclosure of information contained in the records. A person may wish to prevent or permit disclosure, for example, to an employer or an insurance company, but may be restricted by hospital policy. If control of medical records is vested solely in patients, then hospitals and their staff would be reluctant to enter relevant information that they, as professionals, may wish to keep confidential. And both hospitals and patients may fear disclosure of confidential information to unauthorized or undesirable third parties, including the government, private attorneys, employers, or insurance companies.

A related problem pertaining to restriction of information arises because of the widespread use of computerized files and data banks. Health facilities, insurance companies, and government agencies have compiled complex information systems in order to facilitate the rapid retrieval of relevant information. However, such practices also pose a threat to confidentiality. Most persons are not fully aware of the extent to which personal and private information, including medical information, is vulnerable to exposure. As a result government and other administrative agencies as well as consumer organizations have sought new methods of monitoring access to and use of computerized data.

Virtue of those entrusted. Confidentiality may be adequately protected in some situations by the psychological disposition or the personal virtues of the recipients of confidential information. For example, a family member might refuse to disclose a secret to an outsider because of fidelity to the clan or because of a firm moral commitment not to reveal secrets. However, trust can be misplaced; secrets may be intentionally revealed. Sometimes confidential information is exposed because of carelessness or mistakes. Fraud and deception are also used to obtain confidential information. Thus confidentiality in interpersonal relationships is limited by inherent human weaknesses and evils.

In the context of relationships between professionals and their clients, ethical backing is typically given to the confidentiality of clients' communications by codes of professional responsibility such as the Hippocratic Oath or the American Medical Association Principles of Medical Ethics. However, the degree of protection afforded to confidential information and communications varies among professions and from country to country. These variations result from the way in which conflicts are resolved among other fundamental values as well as conflicts that arise between the claims of confidentiality and the need for public information.

The very existence of ethical codes for protection of confidentiality in the professional–client relationships reflects not only the need for guidance but also the presence of human limitations. It is all too common that professionals publicly discuss clients' confidential communications. Indeed, some believe that codes of professional responsibility have a very limited effect on the conduct of professionals. Even when codes of professional responsibility have legal backing, it is common that the codes are not strictly enforced. Part of the explanation for this is that usually professionals in the same category, such as attorneys, physicians, or psychotherapists, administer the disciplinary proceedings against their own members. As a result, only the most flagrant violations of standards of professional conduct are prosecuted.

Bills of rights. Another form of ethical support for confidentiality comes from a 1972 "Statement on a Patient's Bill of Rights" of the American Hospital Association, which has been adopted by many health facilities ("Statement"). It states in broad terms that "case discussion, consultation, examination, and treatment are confidential and should be conducted discreetly." If such documents do not become merely window dressing and are given operative force, the scope of confidentiality will at least be reinforced, if not enlarged.

Legal protections. In addition to support provided by ethical codes, confidentiality is protected by law in some states which have adopted codes of professional responsibility or established patients' rights that are legally binding on appropriate persons. Further legal recognition of confidentiality is found in standard tort law doctrines, which permit lawsuits for libel, slander, and invasion of privacy for breach of confidential communications.

Another form of protection for some confidential communications in the context of special interpersonal relationships is the statutory law of evidentiary privileges. A "privileged communication" between persons in special relationships —such as attorney–client, physician–patient, or psychotherapist–patient—is one that is protected from disclosure in a legal proceeding even though it would otherwise be relevant evidence. For example, certain communications made in confidence by patients to their physicians relevant to medical treatment may not be disclosed by the physician without the consent of the patient, or his or her legal representative after death, unless the privilege has been waived or certain exceptions apply. The reasoning that underlies the privileged communications doctrine is that disclosure by a client of personal or other sensitive information may be necessary for obtaining adequate professional services. It is important to keep in mind that the holder of the privilege is the client or patient, not the professional. However, in appropriate circumstances the professional may invoke the privilege on behalf of the patient.

A few states have enacted laws which give greater protection to confidential communications disclosed in psychotherapy. Because the revelations that occur in psychotherapy often pertain to matters of great emotional significance, a person undergoing psychotherapy feels especially vulnerable to humiliation or harm. To enable the process of psychotherapy to go forward, laws have been established to assure psychotherapeutic patients that their disclosures will be adequately protected. As we shall see more fully later, even in the psychotherapeutic setting confidentiality is not unlimited.

Confidentiality is given further legal backing as a result of the development and expansion of a limited constitutional right of privacy from governmental intrusion through opinions of the U.S. Supreme Court. In the case of *Roe* v. *Wade* (1973) the Court determined that a woman's right to privacy entitles her, with the aid of her physician, to terminate her pregnancy. In a separate concurring opinion Justice William O. Douglas emphasized that the physician–patient relationship has a conspicuous place within the right of privacy.

The U.S. Congress passed the Privacy Act of 1974, which expressly deals with the right of an individual "to safeguard individual privacy from the misuse of Federal records" and "to provide that individuals be granted access to records concerning them which are maintained by Federal agencies." Individuals are provided with the right to determine the uses made of their personal records, to correct erroneous information, and to prevent unauthorized uses of such information. One can readily see that general concern about access to, accuracy of, and use of government records will have a spillover effect on medical records. Indeed, many state legislatures in the United States favor laws similar to the federal Privacy Act which would deal, among other things, specifically with the problems connected with medical records.

The preceding discussion brings out both ethical and legal issues germane to the conflict between private interest in confidentiality and public interest in obtaining sensitive information. Although the legal literature on privacy is extensive, much less scholarly attention has been given to ethical problems related to confidentiality. There is a need for formulation, clarification, and justification of the ethical principles that can help to resolve ethical conflict and guide social practices. This is especially important in an age dominated by rapid changes in the technology of communication systems, increasing interpersonal interdependence, and expansion of governmental and institutional control over sensitive information. Problems frequently arise, and are not easily resolved, with regard to who should have control over access to and disclosure of personal information as well as sensitive information possessed by government agencies and other institutions.

Limitations on confidentiality

Throughout the previous section emphasis has been placed upon the tensions between pressure to expand the scope of confidentiality and forces that would restrict it. It should also be mentioned that some persons believe that many demands for confidentiality are excessive. For example, some legal scholars dispute the value of privileged communications laws on the grounds that such laws merely obstruct truth in the legal process. Some critics of the U.S. Supreme Court believe that individual rights of privacy have been expanded at the expense of the public need for information and social control.

However the merits of such controversies should be assessed, it is generally agreed that confidentiality is not an inalienable right. Even if a person has a moral or legal right to protection of confidential communications, this right may be waived. For example, a person might consent to the disclosure of confidential information for educational purposes, for economic gain, or personal recognition.

In other contexts, such as litigation, a person who sues for injuries suffered, for instance, in an automobile accident must allow his or her treating physician to disclose confidential information gained in the course of treatment that pertains to the alleged injury. If a choice is made to bring a lawsuit, the need for confidentiality is outweighed by the need for truth in litigation.

More difficult issues emerge when the patient-litigant has been treated by a psychotherapist for alleged emotional injuries. It is sometimes argued that because confidentiality is essential for the process of psychotherapy to be effective, confidentiality here is more critical than in other aspects of medical practice. Nevertheless, the general rule is that confidential communications, even in psychotherapy, may in certain circumstances be overridden by the need for relevant information in litigation.

Confidentiality is also limited by the need for protection of public safety. For example, the confidentiality of communications between attorneys and clients is protected if a client reveals that he or she has committed a crime in the past. But the expression of an intention to commit a crime in the future is not protected. Similar reasoning underlies public health laws. For example, a physician must report incidents of suspected cases of child abuse, even if violation of confidentiality may be required. Psychotherapists are permitted, and in some area of the United States are required, to take steps to protect the public against dangerous patients. This may not be possible without disclosure of confidential information against a patient's wishes. Such problems were confronted by the California Supreme Court in the case of *Tarasoff* v. *Regents of the University of California*, in which it was concluded:

that the public policy favoring protection of the confidential character of patient–psychotherapist communications must yield to the extent to which disclosure is essential to avert danger to others. The protective privilege ends where the public peril begins [*Tarasoff*].

WILLIAM J. WINSLADE

[*Directly related are the entries* PRIVACY; RIGHTS; THERAPEUTIC RELATIONSHIP; *and* TRUTH-TELLING. *This article will find application in the entries* GENETIC DIAGNOSIS AND COUNSELING; GENETIC SCREENING; *and* SEX THERAPY AND SEX RESEARCH, *article on* ETHICAL PERSPECTIVES. *See* APPENDIX, SECTION I, MEDICAL ETHICS: STATEMENTS OF POLICY DEFINITIONS AND RULES (BRITISH MEDICAL ASSOCIATION); SECTION III: PATIENTS' BILLS OF RIGHTS; *and* SECTION IV, *codes of the* AMERICAN NURSES' ASSOCIATION, AMERICAN OSTEOPATHIC ASSOCIATION, *and* AMERICAN PSYCHOLOGICAL ASSOCIATION.]

BIBLIOGRAPHY

DAVIDSON, HENRY A. "Professional Secrecy." *Ethical Issues in Medicine: The Role of the Physician in Today's Society*. Edited by E. Fuller Torrey. Boston: Little, Brown & Co., 1968, chap. 9, pp. 181–194.

EKSTEIN, RUDOLF, and CARUTH, ELAINE. "Keeping Secrets." *Tactics and Techniques in Psychoanalytic Therapy*. Edited by Peter L. Giovachinni. New York: Science House, 1972, chap. 10, pp. 200–215.

FRIED, CHARLES. "Privacy." *Yale Law Journal* 77 (1968): 475–493.

In re Lifschutz. 2 C.3d 415. 85 Cal. Rptr. 829. 467 P.2d 557 (1970).

"The Principles of Medical Ethics with Annotations Especially Applicable to Psychiatry (Section 9)." *American Journal of Psychiatry* 130 (1973): 1063.

The Privacy Act of 1974. Pub. L. No. 93-579. 57 U.S.C. sec. 215 (1970). 88 Stat. 1896.

REIMAN, JEFFREY H. "Privacy, Intimacy, and Personhood." *Philosophy and Public Affairs* 6 (1976): 26–44.

Roe v. Wade. 410 U.S. 113. 35 L. Ed. 2d 147. 93 S.Ct. 705 (1973).

"Statement on a Patient's Bill of Rights." *Hospitals* 47, no. 4 (1973), p. 41.

Tarasoff v. Regents of the University of California. 131 Cal. Rptr. 14, 551 P.2d 334 (1976).

WASSERSTROM, RICHARD. "Legal and Philosophical Foundations of the Right to Privacy." Los Angeles: University of California, Los Angeles, February 1976.

WIGMORE, JOHN HENRY. *Evidence in Trials at Common Law* [Wigmore on Evidence]. 10 vols. Rev. ed. by John T. McNaughton. Boston: Little, Brown & Co., 1961, vol. 8, sec. 2380–2385, pp. 818–849.

WITTGENSTEIN, LUDWIG. *Philosophical Investigations.* 3d ed. Translated by G. E. M. Anscombe. New York: Macmillan Co.; London: Blackwell, 1953.

CONFUCIANISM

The effects of Confucianism on the development of health care and medical ethics in China were manifold; thus, medical ethics was a direct outgrowth of the basic content and tendency of Confucian social theory. Decision makers of the Former Han dynasty (206 B.C.–A.D. 25) chose to recognize Confucianism (rather than any of several other social theories that were offered by philosophical schools during that time) as the most appropriate ideology to stabilize the social pattern that had emerged under their rule and to prevent social crises that might have led to social change.

Ensuing actions on the part of Confucian policymakers consequently were strictly conservative in the sense of "conserving in Chinese society the distribution pattern of resources preferred by the ruling class." Confucianists were aware that accumulation of control over potentially powerful resources by specialized groups might lead to shifts in the pattern of distribution of those resources within society, with social change as the unavoidable consequence. Countless decisions were made to prevent this from happening. Deliberate measures to destroy or erode the power bases of groups that had, indeed, managed to gather control over certain resources during times of administrative weakness were a rather constant theme of the government. Official policy toward salt merchants, financial experts, military leaders, and others reveals these efforts.

Two diametrically opposed alternatives were open to Confucian politicians: either to dilute control over potentially dangerous resources among all the people, or to concentrate all control in the hands of the government, that is, the ruling bureaucracy. The resources watched most carefully included those relating to the knowledge and practice of medicine.

In the following a distinction will be made between primary medical resources and secondary medical resources. The former include medical knowledge and skills, drugs and medical technology, and medical equipment and facilities. Secondary medical resources are defined as rewards of material or nonmaterial kind, i.e., money, gifts, prestige, social power, to be gained through medical practice. Access to secondary resources is generally achieved by a group only after having gained control over existing primary resources. The result is the emergence within society of an influential elite group, with consequences harmful to the continuity of social structure. Policy toward medical practice, therefore, followed the line determined by these premises.

Inherent in Confucianism is the appreciation of life and the desire to keep the body from untimely or unnecessary death. It is difficult, however, to assess all the corollaries of that basically positive attitude toward life. There is some indirect support for the assumption that children with certain congenital malformation were killed in prerepublican China, though few such cases have been recorded. Also, as sources indicate, abortion was frequently induced—a practice that evoked little, if any, concern among Confucian thinkers. Confucian ideology contains several values of bioethical relevance, for instance *jen* (humane benevolence) and *tz'u* (compassion). These, however, were applied at least by some dogmatists within strict socially defined limits. Thus, some Confucians argued that a medical practitioner should not follow every call he received but should only respond to requests for help by those who were suitable for social intercourse.

Mencius (372–289 B.C.), an early Confucian philosopher, commented on the universality of benevolent human nature when he pointed out that anyone who happened to see a child fall into a well would certainly rush to the scene and attempt to rescue it from drowning. Furthermore, Confucianism regarded *hsiao* (filial piety) as one of the key values necessary to maintain social stability. The duty to assist one's parents and other relatives to reach advanced age without unnecessary suffering naturally entailed providing medical care for them. The Confucian tenet stressing the role of the individual layman and his ability to assist his relatives became the focus of political efforts designed to spread control of medical resources over society as a whole in order to make the impact of medical practitioners of marginal importance if not superfluous. Time and again statements and admonitions released by concerned Confucianists asserted that possession of sufficient medical knowledge was necessary to fulfill one's obligations of filial piety.

In spite of this attempt to distribute responsibility for primary health care, social circumstances and political expediency (the nature of which is not always entirely clear) led Confucian decision makers to promulgate a variety of public health measures; the ensuing institutions had to be staffed with at least partially specialized personnel. Thus, ironically, the stimulus for the professionalization of a specifically trained group within Confucian society was a result of Confucian health policy itself. It is a moot point to what extent credit for humanitarian health-policy decisions during the imperial era can be attributed to the voluntary acts of the Confucianists and how much was forced upon them by external events. Still, the mere fact that public health solutions to impending problems were conceived and that attempts were made to put them into practice must be regarded as an achievement of the Confucian culture. Again, today's readers of classical Chinese accounts of public health measures must be cautioned, for such reports do not reveal much about the actual implementation of the programs decreed.

At the beginning of the Confucian era, now slightly more than 2,000 years ago, medical practitioners enjoyed the status of more or less renowned craftsmen. Some of them were itinerant and offered their services to the general populace; others whose skills had created sufficient fame for them would approach the rulers at the many feudal courts existing during that period (Bridgman). "Medical workers" were attached to the military and to administrative posts from Han times on. Somewhat later, better trained experts were required, and it was only natural in Confucian society that the necessary training supplement the standard education of Confucian civil servants, again as a strategy to preserve control over medical resources for the ruling class. During the early seventh century this led to the establishment of institutions for the teaching of medicine outside the imperial court. By the twelfth century, medical training in certain government institutions was restructured and became even more integrated with classical nonmedical knowledge.

Possibly as a response to the opening of such an institution under Buddhist auspices in the fifth century, one finds soon thereafter the first reports of the establishment of a hospital under secular control. The competitive character of public health policies when directed toward followers of Buddhism became even more evident in A.D. 653 when Buddhist (and also Taoist) monks and nuns were prohibited from the practice of medicine, and in A.D. 845 when all Buddhist-controlled hospitals were secularized (Needham, pp. 265, 277, 278).

After a period of weakness during which even Chinese emperors had shown commitment to the Buddhist faith, this kind of loss of control over important primary and secondary medical resources could not be tolerated by a reviving Confucian society.

From the eighth through the twelfth century a mass migration of people from northern regions to southern parts of China occurred as a result of climatic changes in the north. It is estimated that previously ninety-five percent of the Chinese population inhabited the plains of the north, which were more favorable for husbandry and agriculture; during the period of the Southern Sung dynasty (1127–1179), approximately forty-five percent of the total Chinese population lived in the south with its mountainous geography (Kracke, pp. 479–488). The population shift produced unprecedented large cities with often hundreds of thousands and, not infrequently, more than a million people crowded together. Possibly as a consequence of the living conditions in the cities, in which so many people had been uprooted from their former style of life and natural environment, a welfare program was initiated during the eleventh century which entailed the establishment of public dispensaries and state-controlled polyclinics. That system continued in operation at least until the sixteenth century, possibly even until the eighteenth century. The impact of these institutions cannot be regarded as decisive for public health at any time. The numbers of "offices" established were too small for this purpose. Furthermore there were such severe scandals surrounding them that even the Confucian chroniclers could not avoid mentioning them. Not surprisingly, in a slight modification of the official title, the public dispensaries were called by the people "offices for the welfare of the bureaucrats" (Unschuld, 1973, pp. 9–14; 1975, n. 19).

Public health programs were continued even during times of alien rule over China. Reports from the time of the Mongols (Yüan dynasty, 1206–1368), for instance, show an abundance of directions to provide medical assistance to prisoners through "medical workers," the lowest category of governmentally employed practitioners.

The foregoing reference to examples of public health policies in Imperial China may suffice to indicate that it was possible for any medical group to achieve a higher level of professionalization and thus to become socially influential. Orthodox Confucianists, however, observed these developments closely and continuously promoted what might in effect be called an anti-medical-professionalization campaign.

By the twelfth century, three distinguishable main groups were contending for control over medical resources. The first were the free-practicing physicians outside Confucianism. This group included Taoists, Buddhists, and others who practiced medicine to make a living or as a sideline when they were economically independent. Many famous physicians and medical writers belonged to the group. They should be distinguished from those favored by Confucian dogmatists, persons who used their medical abilities only to assist family members, friends in need, or during medical assignments as civil servants.

One outstanding early representative of free-practicing physicians was Sun Ssu-miao (581–682?) who possibly was the first to recognize the need of the free-practicing group for explicit medical ethics to advance in professionalization against the Confucian class. It is not surprising that many of the famous members of this group were offered official positions at the imperial court or elsewhere. This may not only be understood as an attempt by the decision makers to secure the best physicians for their own care but also to alienate them from free-practicing peers who might have gained in status by association with outstanding personalities. The second group is the Confucian class itself, whose interests have been dealt with earlier in this article. Within this group a third one emerged, composed of those Confucianists who because of either an official assignment or personal interest practiced medicine outside of their own families on behalf of clients of all kinds, often accepting money and other material rewards. The third group did not adhere to the famous saying of Confucius (551–479 B.C.): "The accomplished scholar is not a utensil!" Its strategy, carried out to gain acceptance within its own superior group and to compete successfully with the group of the free-practicing physicians, was twofold.

In their explicit medical ethics these people on the one hand had to assure the orthodox Confucianists that their medical practice was in perfect accordance with Confucian values. On the other hand they continuously stressed that "common physicians," as non-Confucian practitioners were called officially, constituted a source of permanent evil practice. In this latter effort the group was supported, of course, by the orthodox Confucianists.

A severe blow against the social acceptance of medicine as more than just an ordinary craft resulted from some sayings of Chu Hsi (1130–1200), the leading Confucian philosopher of his time. In commenting on the ancient classic *Lun yü* ("Analects"), which supposedly comprises sayings by Confucius and his disciples, compiled by the latter, Chu Hsi made some statements regarding medicine that were bound to stigmatize all who wished to practice medicine, outside the domain of their own families, as a means of earning a living.

Confucius had originally stated: "The people in the South have a saying: 'A man without persistency cannot become a sorcerer-physician.' Good!" (Legge, vol. I, p. 272). There were at least two grammatically correct ways to interpret these words. Chu Hsi chose to separate the terms "sorcerer" and "physician" and commented: "Sorcerers communicate with demons and spirits; physicians are entrusted with matters of death and life. If such petty personnel (cannot do without persistency), how much more is this true for others!" (Chu Hsi, 1958, pp. 598–601). Elsewhere in the *Lun yü* the following remark is attributed to a disciple of Confucius: "Petty teachings certainly also contain some aspects which cannot be disregarded. But if you carry them far, you will become soiled. The noble man, therefore, does not deal with them" (Legge, vol. I, pp. 340–341). In this case Chu Hsi wrote in his commentary: " 'Petty teachings' does not mean 'strange (or: heterodox) principles'; 'petty teachings' refers to 'teachings' too, but just to 'petty' ones. These are for instance: agriculture, horticulture, medicine, divination, and all other specialized occupations" (Chu Hsi, 1965, chap. 19, p. 23a).

These two comments exerted an almost continuous influence over most of the remaining centuries of the Confucian era. They provoked many defensive criticisms by medical writers from both the group of free-practicing physicians and the group of Confucian medical practitioners. A well-known physician named Hsü Ch'un-fu, who lived around 1556 and had served as an official at the Imperial Medical Office at court for a while, questioned Chu Hsi's interpretation of Confucius's statement and argued that "sorcerer-physicians" were to be regarded as one single category of people, namely, those who would dance, pray, and offer sacrifices to avert sickness; through this sort of practice they proved not to have any knowledge of medicine and drugs. Therefore, they should not be confused with true medical practitioners.

Later, at the end of the sixteenth century, Lai Fu-yang focused on the meaning of "petty" and argued: "If one directs his medical knowledge only on his own body, this is to be regarded as 'petty'; if one spreads the application (of his medical knowledge) all over mankind, this is not to be regarded as 'petty'!" Here Lai Fu-yang attempted to reverse completely the orthodox Confucian attitude toward medical practice.

Chang Chieh-pin (fl. 1624) is a third and last example of criticism of Chu Hsi's remarks on medicine. Chang Chieh-pin was a noted medical writer who originally was in the group of free practitioners but became a close follower of the ideas of Sung-Confucian medicine, a healing system that integrated the use of drugs into medical theories of the Confucian tradition. The use of drugs had been developed mainly in Taoist traditions. Chang Chieh-pin did not dare or did not find it effective enough to voice his strong criticism of Chu Hsi's comments himself. He placed his thoughts into the mouth of a "strange man," whom he allegedly met somewhere in the wilderness. This "strange man" in conversation with Chang Chieh-pin showed an outburst of anger when the latter told him that, although medicine was to be regarded as a "petty teaching," he was still aware of the necessity to be careful in its application. The stranger then gave him an emphatic lecture on the importance and subtlety of medicine and admonished him to keep it in high esteem. Chang Chieh-pin as a result of this lecture felt extremely ashamed and noted it down so that it might not be forgotten.

Another dimension of the Confucian attitude toward medicine that was expressed in many ethical statements focused on what were called "heterodox" practices and beliefs. Orthodox or officially endorsed medicine consisted of theoretical foundations and actual treatments by means of acupuncture or other techniques, including drugs at a later time. All of these were related to the same cosmogonic theories and philosophical concepts of nature that were also basic to the Confucian social theory.

Effective medical practice outside of those theories and philosophies might have jeopardized the validity of the Confucian paradigms as a whole and could not be tolerated. This is not surprising, for the way of life recommended by Confucian-endorsed medicine as a guarantee for physiological harmony, i.e., health, is the same way of life demanded by the Confucian social theory to maintain social order and stability. It was again Sun Ssu-miao, a leading thinker critical of the Confucian way of life, who recommended demonic medicine as an alternative. The notion inherent in this healing system was that one is constantly subject to attacks by evil demons, which may cause diseases if one doesn't protect oneself carefully enough. The consequences of this theory of disease causation for individual and social behavior were quite different from those put forth by the Confucian social theory.

The resulting ideology toward heterodox knowledge and practices was part of the overall competition between Confucianists and the followers of other social theories. Shamans, Taoists, Buddhists, and often private individuals who had taken up a healing practice on their own by using charms, prayers, or other means that were not in conformity with Confucian medicine, provided the major target for accusations of heterodox practice.

The need to advance over other groups in society competing for control of secondary resources (i.e., the material or immaterial rewards to be gained from practice) is one of the major forces behind medical progress. However, several well-documented historical examples indicate that the pace of medical progress is generally retarded when the control over primary medical resources rests in the hands of a group whose access to secondary resources is not wholly dependent on the use of these primary resources but possesses other means to maintain its control over the people.

Orthodox Confucianists constituted such a group. They attempted to control primary medical resources or to spread them all over society merely to prevent any specialized group from rising to influence and power in society. Their own power was by no means solely dependent on primary medical resources. Therefore, orthodox Confucianists had no interest as such in increasing or improving primary medical resources. In contrast the free-practicing physicians and those Confucianists who practiced medicine outside their own families appear to have been genuinely interested in improving their skills and expanding the available materia medica. The resulting antagonism manifested itself, for example, in the delineation of medical jurisdiction and in repeated admonitions to practitioners not to develop new theories but to stick to the old interpretations. This restrictive policy could not prevent the invention or importation of new primary medical resources, for example, theories, drugs, techniques, and facilities. However, it is impossible to estimate to what extent it may have impeded progress and served to preserve those archaic medical paradigms which dominated Chinese medical literature up to this century.

One of the last Confucianists to resort to this kind of argument was Li Han-chang (1821–1899), elder brother of the famous statesman Li Hung-chang (1823–1901). In the preface to a medical work of a previous century he castigated those who would originate or follow other than the classical theories and condemned those who would try their "unintelligible prescriptions" at the cost of thousands of human lives. It is not unreasonable to assume that these criticisms were meant to discourage those practitioners who were just then experiencing the initial impact of Western surgical and pharmaceutical knowledge. A subsequently increased influx of Western primary resources of many kinds into China brought on dramatic changes in the overall pattern of resource distribution. The millennia-old maxims of Confucian ideology concerning professionalization proved to be infeasible at a time when the amount and the character of primary resources available in the field of medicine and elsewhere seemed to necessitate groups of specialists to control and handle them. The Confucian social system collapsed with the revolution of 1911.

PAUL U. UNSCHULD

[*For further discussion of topics mentioned in this article, see:* MEDICAL ETHICS, HISTORY OF, *section on* SOUTH AND EAST ASIA, *articles on* GENERAL HISTORICAL SURVEY *and* PREREPUBLICAN CHINA. *Also directly related are the entries* BUDDHISM; DEATH, *article on* EASTERN THOUGHT; ENVIRONMENT AND MAN, *article on* EASTERN THOUGHT; *and* TAOISM. *For discussion of related ideas, see the entries:* MEDICAL PROFESSION; PUBLIC HEALTH; *and* RATIONING OF MEDICAL TREATMENT.]

BIBLIOGRAPHY

BRIDGMAN, R. F. "La Médecine dans la Chine antique." *Mélanges chinois et bouddhiques* 10 (1952–1955): 1–213.
CHU HSI, commentator. *Lun yü chi chu* [Commented Analects], by Confucius. 10 vols. Taipei: Chung-hua ts'ung-shu wei-yüan-hui, 1958. Translated by Pyun Yung-tai as Confucius. *Analects.* 2 vols. Seoul: Minjungsugwan, 1960.
———. *Chu-tzu ta ch'üan* [Master Chu's Complete Works]. *Ssu-pu-pei-yao,* vols. 396–406. Taipei: Chung-hua-shu-chü, 1965.
KRACKE, E. A., JR. "Sung Society: Change within Tradition." *Far Eastern Quarterly* 14 (1954–1955): 479–488.
LEGGE, JAMES. *The Chinese Classics.* Hong Kong: Hong Kong University Press, 1960.
NEEDHAM, JOSEPH R. *Clerks and Craftsmen in China and the West.* London: Cambridge University Press, 1970.
UNSCHULD, PAUL U. *Die Praxis des traditionellen chinesischen Heilsystems.* Wiesbaden: F. Steiner Verlag, 1973.
———. *Medizin und Ethik: Sozialkonflikte im China der Kaiserzeit.* Münchener ostasiatische Studien, vol. 11. Wiesbaden: F. Steiner Verlag, 1975. This monograph gives the original sources of all Chinese citations provided in this article.

CONSANGUINITY

See EUGENICS AND RELIGIOUS LAWS.

CONSENT

See INFORMED CONSENT IN HUMAN RESEARCH; INFORMED CONSENT IN MENTAL HEALTH; *and* INFORMED CONSENT IN THE THERAPEUTIC RELATIONSHIP.

CONSEQUENTIALISM

See ETHICS, *articles on* TELEOLOGICAL THEORIES *and* UTILITARIANISM.

CONTRACEPTION

Early methods

Ethical judgment on contraception presupposes the existence of contraceptive technique. Five papyri from the period 1900–1100 B.C. show that Egyptian medicine possessed prescriptions designed to prevent pregnancy (Deines, Grapow, and Westendorf). Crocodile dung figures prominently in two recipes; another stresses acacia tips, coloquintada, and dates. In each case the aim was to block or kill the male semen. Like other formulas of ancient and medieval medicine, those contraceptive compounds were subjected to insufficient experimental control to be established as truly effective; they do reveal an appreciation of the objective of preventing conception and a rational sense of the means that might achieve the objective; their existence implies the ethical acceptance of contraception by the physicians of Egypt.

Among the "Incantations and Ceremonies for Procreation," the Force Brāhamana of the *Brihad-Araṇyaka Upanishad* (seventh century B.C.), is the formula, "With power, with semen, I reclaim the semen from you." It is to be said by the male during intercourse if he does not want the woman to conceive; if it is properly uttered, "she comes to be without seed." By 300 B.C. the Tantras taught that the true yogi would have the self control to have intercourse without emitting semen; Śiva was said to have had such intercourse with Pārvatī for a thousand years, until the gods interrupted them (O'Flaherty, pp. 261–262). But Tantrism was a Hindu heresy, and mainline Hinduism, by teaching that a male descendant was necessary to perform the rites to save one after death, placed an enormous value on procreation.

In the Greco-Roman world, potions are the first and most common contraceptive to be mentioned. A drink of *misy*, apparently a distillate of copper, appears as a contraceptive in *The Nature of Women* 93, a treatise of the fifth-century B.C. Hippocratic school. In the fourth century B.C., Aristotle observed that conception could be impeded by the use of cedar oil, ointment of lead, or frankincense and olive oil on the part of the woman's body where "the seed falls," (*History of Animals* 7.3, 583a). By the first century A.D., writers on pharmacology are familiar with a variety of contraceptive drinks (e.g., Dioscorides, *Materia medica* 1.135; Pliny, *Natural History* 20.51, 142–143) and with use of pessaries like the Egyptians' (e.g., Dioscorides, *Materia medica* 3.36, 136). Both Dioscorides and Pliny also list salves or ointments to be applied to the male genitals to act as spermicides. The classical world was also familiar with physiological ways of avoiding conception. The Hippocratic school had theorized that the period when a woman was most likely to be fertile was just after menstruation. The foremost of Greek gynecologists, Soranos of Ephesus, recommended avoiding this time if conception was to be avoided (Soranos 1.19.61), while he noted that conception was unlikely during menstruation or just before its onset (ibid., 1.10.36). Herodotus reported that Pisistratus, the sixth-century tyrant of Athens, avoided having children by having intercourse with his wife "not according to custom" (*History* 1.61.1); either anal intercourse or coitus interruptus is meant.

The most celebrated report of coitus interruptus is in Genesis 38:8–10, the story of Onan who habitually spilled his seed rather than carry out his obligation to his brother's widow. This single reference to contraceptive technique by the Jews of the Old Testament may be supplemented by references in the *Babylonian Talmud* going back to at least the first century of the Christian era. Coitus interruptus is mentioned in *Yebamoth* 34b and *Niddah* 13a, pessaries in *Yebamoth* 35a and *Niddah* 32, and potions made from roots in *Shabbath* 110a–110b and *Yebamoth* 8.4.

Although references to charms and amulets as contraceptives suggest the failure to establish any single reliable contraceptive, the evidence is ample that the ancient Mediterranean world —Egyptian, Jewish, Mesopotamian, Greek, and Roman—knew enough physiology to try to have sexual intercourse without conception and enough chemistry to attempt to prevent conception. Contraceptive practices were believed to be engaged in by rulers like Pisistratus; by slaves who did not want to bear children for their captors (*Babylonian Talmud, Kethuboth* 37a), by prostitutes who did not want the inconvenience of childbearing (*Yebamoth* 35a), by women concealing adultery (Soranos 1.19.60–61), by women preserving their looks (ibid.), and by women with too many children (Pliny, *Natural History* 29.27.85). What ethical judgment was given upon it?

Ancient ethical thought

Judaism. In Jewish thought, the intense consciousness of Israel as a people, the descendants of Abraham, the elect of God, led to a strong emphasis on the good of progeny and the duty of procreation. From "Increase and multiply" spoken by God in Genesis 1:27–28 to the promise of God in Deuteronomy 7:13, "No man or woman among you shall be childless," there was divine encouragement of childbearing. Sterility was a curse; a big posterity and, by implication, a big family were blessings (Gen. 15:5, Job 42:13, 1 Kings 1:1–28). No law in the Jewish scriptures condemned contraception as such. But its emphasis on fertility disfavored the practice. What Onan did "displeased Yahweh, who killed him" (Gen. 38:10). Onan had disobeyed his father in failing to provide his brother's widow with offspring; his crime was against the family. Although his contraceptive habit was not the explicit ground of Yahweh's judgment, the story suggested that contraceptive intercourse was the way of a wicked man.

Later Jewish thought, citing the punishment of Onan, treated coitus interruptus as a serious sin. One tradition linked the loss of seed to a Messianic theme: The Son of David would not come until all the souls of the unborn were born (*Babylonian Talmud, Niddah* 13a and 13b). Jewish opinion, however, divided on the duty to propagate the race, some rabbis teaching it to be the obligation of both men and women, others only of a man, and some limiting the duty to begetting two boys or a boy and girl (*Yebamoth* 61b, 65b). Associated with the milder views on procreative duty was the recognition by some rabbis of cases where a woman might legitimately use root potions as a contraceptive (e.g., *Babylonian Talmud, Niddah* 45a; *Tosefta, Yebamoth* 8.4).

Greco-Roman world. Greek philosophy and medicine took less interest in the morality of contraception. Nothing similar to the pledge not to give abortifacients appears in the Hippocratic Oath with reference to contraceptives. Aristotle raised no moral objection to the contraceptive technique he reported in the *History of Animals*. Plato, in his *Republic* (372c), portrayed Socrates as sketching an ideal city whose inhabitants do not beget too many children "lest they fall into poverty or war"; the avoidance of conception was implied as a civic obligation.

Conception was seen as controversial in the Greco-Roman world when population was related to the security of the state. By the time of Augustus the Roman government made modest efforts to reward childbearing by the upper class (e.g., Gaius, *Institutes* 1.145, 1.194) and later even celebrated fertility among freedmen (Pliny, *Natural History* 7.13.60). Imperial Roman law attempted to regulate the sale of abortifacient and contraceptive "medicines" (e.g., Justinian, *Digesta* 48.8.3). Effective enforcement, however, probably did not occur.

A second and more powerful current against contraception in the Mediterranean world came from philosophers. The Stoics taught that the purpose of sexual intercourse was procreation. Anything else was unnatural and therefore wrong (Musonius Rufus, *Reliquiae*, sec. 63). The neo-Pythagoreans, imbued with Stoic doctrine, similarly taught that "we have intercourse not for pleasure but for the purpose of procreation" (Ocellus Lucanus, *The Nature of the Universe*, sec. 44). The Essenes, according to Josephus, held the same view (Josephus, *The Jewish War* 2.120, 160). By necessary implication, these views condemned contraception.

Civic responsibility, Stoic natural law, and the authority of the gods were combined by Musonius Rufus, the first-century Roman Stoic, when he argued that the lawgivers had forbidden contraception, and he who practiced it sinned "against his ancestral gods and Zeus, protector of the family" (Musonius Rufus, "Should All Children Born Be Brought Up?", *Reliquiae*). Philo, the Alexandrian intellectual who melded classical philosophy with his Jewish heritage, appropriated the Stoic distinction between marital intercourse for procreation and intercourse for the unlawful purpose of pleasure (Philo, *Joseph* 9.43). A man acted evilly, he taught, if he married a woman known to be sterile. "Those persons," Philo added, "who make an art of quenching the life of the seed as it drops stand condemned as the enemies of nature" (*The Special Laws* 3.36).

Early Christianity. The Christian position on contraception emerged in the first century in Palestine. It represented a confluence of doctrinal elements and a response to the Mediterranean milieu. In an ethic in which love of God and love of neighbor were the great commandments (Matt. 22:40), there was a specific doctrine on sexuality, analytically reducible to five themes: the superiority of virginity (e.g., Luke 1; Luke 20:34–36); the institutional goodness of marriage (Mark 10:7–8; John 2:1–12; Eph. 5:25–33); the sacred character of sexual intercourse (1 Cor. 6:16; 1 Thess. 4:4; 1 Pet. 3:2,7); the goodness of procreation (John 16:21; 1 Tim. 2:15); and the evil of extramarital intercourse and homosexual conduct (Matt. 5:27–28; Rom. 1:24–27). In the context formed by these themes, the Christian community found a balance, permitting and rewarding a way of life in which sexual intercourse was renounced for the higher good of following the personal example of Jesus' virginity, and permitting and rewarding a way of life in which intercourse occurred in marriage. Against the Christian Gnostics, who were found within the community even in its beginnings, it was urged that there were limits to the sexual liberty of the Christian: women would be saved "through motherhood" (1 Tim. 2:15). *Pharmakeia*, i.e., abortifacient and contraceptive drugs, were denounced as sinful by St. Paul (Gal. 5:20) and by the Apocalypse (9:21, 21:8, 22:15).

As Christianity permeated the Mediterranean world, it entered a society in which women were widely regarded as objects of pleasure; prostitution, particularly of slaves, was widespread; divorce and sexual infidelity were common. Infanticide was a frequent phenomenon (e.g., Seneca, *De ira* 1.15; Suetonius, *Gaius Caligula* 5; Tacitus, *Histories* 5.5). The Christian view of contraception may be understood as in part a reaction against these societal tendencies to treat woman as a thing, to destroy the special place of marital love, and to encroach on human life. In the *Didache* or *Teaching of the Twelve Apostles* 5.2, a late first-century document, "killers of offspring, corrupters of the mold of God," were classed with murderers and adulterers, whose life is the "Way of Death." As it was difficult for even the best gynecologists to distinguish contraceptive from abortifacient potions (Soranos 1.19.60–63), so Christian moralists did not distinguish between them.

Not only reaction against society, but reaction against Gnostic trends within the Christian community fortified the Christian exclusion of contraception. The Gnostic Right asked, What was the purpose of procreation? Now that the Messiah had come, now that death was overcome, should not every Christian imitate Christ in a total self-control excluding sexual conduct? (Clement 3.12.90.) The Gnostic Left asked if Christians were not "Lords of the Sabbath," free of any Mosaic commandment including the law against adultery (Clement 2.4.30). To Right and Left, the Christian Center had an answer: Virginity was desirable, but marriage was good. Christ had superseded the law of Moses but had not abolished the law of nature. The formula of the Center adapted the teaching of the Stoics to the needs of the Christian community. The Christian law, Clement of Alexandria wrote, was for "husbands to use their wives moderately and only for the raising up of children" (Clement 3.11.71.4). To have coitus "other than to procreate children," he taught, "is to do injury to nature" (Clement, *Pedagogus* 2.10.-95.3). Responding to internal challenges, rebuking the societal evils of abortion, adultery, and prostitution, the Stoic rule adopted by the Christians excluded contraception.

Manicheanism. In the fourth century A.D., as Christianity became the state religion of the Roman Empire, it was challenged by an underground religion, Manicheanism, whose prophet Mani had been put to death in Persia; his teachings were conveyed by a sacred scripture, and his organization had the shape of a hierarchical church. The central Manichean myth taught that human beings originated in the lustful intercourse of the princes and princesses of darkness after they had devoured the sons of the King of Light. Within humans, particles of light subsisted, which struggled to be free but remained imprisoned as long as man emulated his devilish sires and procreated. This view of the universe in which alienated lights sought to return to the Father of Light and procreation was the worst of sins, appealed to a wide variety of intelligent minds, including that of Augustine, who between the ages of eighteen and twenty-nine was a Manichee. It was a view that did not exclude sexual intercourse but did exclude procreation and favored contraception, including the practice of coitus interruptus (Augustine, *Against Faustus* 22.30), the use of contraceptive potions (Augustine, *Marriage and Concupiscence* 1.115.17), and intercourse at times believed (erroneously) to be sterile (Augustine, *The Morals of the Manichees* 18.65). Manichees may be found in conflict with Zoroastrians in mid-fourth-century Persia and with Catholics in mid-fourth-century Charcar, Mesopotamia (Hegemonius, *Acts of Archelaus* 16), in Bostra, Asia Minor (Titus, *Against the Manichees* 2.33, *Patrologia graeca* 18:1197), in Constantinople (John Chrysostom, Homily 62 on Matthew 17), in the Mediterranean littoral (Epiphanius, *Panarion* 66.26), and in Italy (Ambrose to Chromatius of Aquileia, Letters 50). Manicheanism may be taken as the apogee of the contraceptive mentality prior to the mid-twentieth century.

Augustine. Opposition to the Manichees by the Roman government and by the Catholic Church was intense. Beginning in 320 a series of imperial decrees outlawed them and their practices. The climax of the Catholic Church's opposition came as late as 444, when Pope Leo the Great drove the Manichees from Rome and instructed the bishops of Italy to give them no respite anywhere. But more important than this official persecution was the powerful reinforcement given in Christian ethics to the Stoic rule that marital intercourse must be procreative. For Augustine, this was the most significant difference between the morals of the Manichees and the morals of the Catholic Church, which he had now joined (*Against Faustus* 22.30). The intensity of Augustine's own reaction against Manicheanism, reenacting in the life of one man the reaction of Christianity, was of extraordinary significance for Western ethical thought. The most influential of Western moralists became a convinced enemy of contraception.

In Augustine's analysis, offspring were one of the goods of marriage, and marital intercourse, to be entirely free from sin, must have offspring in view (*The Good of Marriage* 16.18). Mere absence of procreative intent made marital intercourse venially sinful; positive prevention of procreation turns "the bridal chamber into a brothel" (*Against Faustus* 15.7). Capping his anti-Manichean stance, Augustine explained, against Pelagius, that original sin was transmitted in the act of generation, since concupiscence, that is "heat" or "the confusion of lust," always accompanied the act of intercourse (*Marriage and Concupiscence* 1.18.21). Offsetting concupiscence, so as to justify marital intercourse, was the good of offspring. Forging the synthesis in which the doctrine of original sin was tied to the defense of procreative intercourse, Augustine formulated a comprehensive anticontraceptive ethic. Husbands and wives who used "the poisons of sterility" systematically to exclude conception were "not joined in matrimony but in seduction" (ibid., 1.15.17). Reasserting the link of Christianity with the Jewish scriptures, he taught that Onan's sin was contraceptive marital intercourse, "and God killed him for it" (*Adulterous Marriages* 2.12.12). Augustine's synthesis of personal experience and reaction, Old and New Testaments, anti-Manicheanism and anti-Pelagianism, Stoic natural law, and Christian sacramental theory was to constitute the strongest ethical case against any form of contraception.

Islam. By the seventh century the ethical tradition of Islam was dominant in the Middle East and North Africa, where Christianity had once flourished. The Koran has no expressed teaching on contraception, and although Islam inherited the propopulation outlook of the Bible, the major theologians and jurists of Islam did not condemn the contraceptive measures developed by Arabic medicine. By the time ibn Sina (known to the West as Avicenna) wrote his encyclopedic *Canon of Medicine* in eleventh-century Damascus, a wide range of contraceptives was known. In his pharmacopoeia he listed a plant that acts as a spermicide, another that acts as a pessary, another that is a talisman against conception (Avicenna 2.2.163, 277, 495). In book three, under "The Prevention of Pregnancy," suppositories, spermicides, and a single potion were listed. He carefully distinguished contraceptives from abortifacients (3.21.2.12). Like the preventive measures of the Greco-Roman world on which it drew, the Islamic panoply of techniques against conception had a variety reflecting a lack of a single certain method. Apparently because of theological disapproval, coitus interruptus was not mentioned by the medical writers.

Medieval ethical thought

Early medieval Europe. The Celtic and Germanic peoples who came to inhabit Western Europe were familiar with herbal potions designed to prevent conception. Condemnation of these means was a constant theme of the penitential literature developed under monastic auspices from the sixth through the eleventh centuries. Part of the objection to them was their association with pagan magic, but the rationale of opposition also included objection to the killing of the seed and making infecund what God made fertile. The formula owed to Caesarius of Arles, "so many conceptions prevented, so many homicides," became popular. Burchard of Worms noted that the difficulty of feeding the children was not an excuse (*Patrologia Latina* 140:972). The penitentials, picking up the rabbinic and early Christian objection, also vigorously condemned anal and oral intercourse, where of course procreation was impossible. The eighth-century legislation of Theodulphus, bishop of Orléans, included condemnation of coitus interruptus. Implicitly, a standard of natural marital intercourse was assumed. Besides classifying the contraceptive acts as sinful, the penitentials prescribed inquiry to detect them and penances to punish them.

Catharism. Beginning in the early tenth century with Bogomil of Bulgaria, a new form of Gnosticism challenged the value of procreation. Organized as a church, Bogomil's adherents spread to the Eastern Empire's capital, Constantinople, and then westward to Bosnia. By the eleventh century they had appeared in Aquitaine, at Arras, and at Monteforte. A century later they were well established in Southern France and in some of the Northern Italian cities as the Cathar Church. The troubadors of courtly love from Languedoc and Aquitaine were drenched in their ideology of "pure," i.e., nonprocreative, love. Their doctrine opposed sexual intercourse leading to procreation. The devil had given "seed to the children of this world"

(*Un Traité cathare*, ed. C. Thouzellier, p. 96). When a woman conceived, Satan resided there (*Le Traité contre les Bogomils*, eds. Puech and Vaillant, p. 345). A pregnant woman was always to be denied the Cathar sacrament of the *consolamentum* (Borst, *Die Katharer*, p. 181).

To the Orthodox of the East and the Catholics of the West, the Cathars were a reappearance of the Manichees, and they were fought with the scriptural texts favoring procreation which had combated Manicheanism. Between 1140 and 1234, the most creative period of Western Catholic canon law, the Cathars were perceived as a socioreligious threat, challenging the Catholic Church in Southern France and Northern Italy and, like the Manichees, endangering the institution of marriage and the procreation of the species. The extent of the institutional Catholic response indicated the extent of the crisis. Ecumenical councils beginning with the Second Lateran in 1139 condemned the new heretics. The Dominican order was founded to convert them. The Inquisition was used to discover them. The Fourth Lateran Council in 1215 encouraged a holy crusade against them. The Hail Mary, with its key words "Blessed are you among women, and blessed is the fruit of your womb," became a popular prayer, prescribed to counter Cathar sentiment. The canon law, drawing on Augustine in particular, was shaped to include public teaching that condemned contraception and contraceptive marriage (Gratian, *Decretum* C. 32 q. 2 c. 7 and Gregory IX, *Decretals* 5.12.5).

Medieval Europe. Western Europe after 1150 knew the full range of contraceptives set out in Avicenna's *Canon of Medicine*, which, translated into Latin, was the standard textbook of general medicine for the next five hundred years. Actual practice is difficult to ascertain, but in Chaucer's *Canterbury Tales*, three methods of contraception were set out under the Sin of Wrath, including "drynkynge veneouse herbes thurgh which she may not conceive" ("The Parson's Tale," lines 570–580). Peter de Palude noted that a husband might seek to practice coitus interruptus because he had "more children than he can feed" (*On the Sentences* 4.31.3). Catherine of Siena had a vision whose contents suggest that contraceptive acts were common among the married bourgeoisie. The fifteenth-century Franciscan, Cherubino of Siena, in *The Rule of Married Life*, a book for the laity, treated coitus interruptus as a familiar sin. The canons of the Church, the teaching of the theologians, and inquiry in confession were directed against these practices. The contraceptive act was condemned by all articulate ethical opinion as destroying potential life, as frustrating the function of coitus, and as violating a principal purpose of marriage (e.g., Thomas Aquinas, *Evil* 15.2).

Christian doctrine, nonetheless, contained currents that ran in a counter direction. In a famous passage, St. Paul had taught, "Let the husband render to his wife what is due her, and likewise the wife to her husband" (1 Cor. 7:3). As developed by scholastic theologians, the marital debt was a duty, and the satisfaction of it was virtuous even though procreation was not the purpose (Thomas Aquinas, *On the Sentences* 4.32.1.2). Moreover, the Church permitted the sterile and those past the age of childbearing to marry, implicitly conceding that procreative purpose was not necessary for a marriage. Finally, and perhaps most importantly, the theologians taught that "the good of offspring" meant not merely their procreation but their education as rational, spiritual creatures (e.g., William of Auxerre, *Summa* 4, f.287v). The universal acceptance of religious education as a constituent element in the good of offspring was a rejection of quantity and of population for its own sake. With the emphasis on education, Western Christianity carried the possibility of a different crystallization of values in the rule on contraception.

Reformation and post-Reformation thought

Protestantism. No change, however, occurred with either of the two upheavals that shook the doctrinal unity of medieval Christendom. The Protestant leaders were as severe on matrimonial conduct as their Catholic predecessors. Martin Luther, invoking Augustine for his theology, could scarcely have been anti-Augustinian in his morality. His *German Catechism* taught that the purposes of marriage were for a husband and wife "to live together, to be fruitful, to beget children, to nourish them, and to bring them up to the glory of God" (Luther). This was standard scholastic doctrine. John Calvin was more passionate. Coitus interruptus was "doubly monstrous." To commit Onan's act was "to extinguish the hope of the race and to kill before he is born the son who was hoped for" (Calvin). Not only did the new religion offer no alternative or critique of the old on contraception; the religious rivalry that followed led to a new premium on population. In Germany, the

Low Countries, and France, serious shifts in population growth would have affected the religious composition of the country and political control. That the strength of a state lay in its numbers was now stated with Machiavellian dispassionateness by that astute political observer Giovanni Botero in *Greatness of Cities* 3.1.

Rationalism. Analogously, the second upheaval, the spread of rationalism and deism in the seventeenth and eighteenth centuries, led to no alteration in expressed ethical opinion. While the old ethic on usury and on the indissolubility of marriage was under constant criticism, no one publicly advocated contraception as ethical conduct. The absence of challenge to the Christian rule may be explained in part by the reserve that all premodern people have experienced about the mystery of sex; in part by the lack of any institutional group desiring change; in part by the outlet the European colonies afforded for population; and in part by the absence of technological improvements. One new contraceptive device, the condom, did appear in the middle of the seventeenth century, but it was neither cheaply nor efficiently produced, and it won no medical support. Coitus interruptus and potions continued to be the chief contraceptive techniques. This limited technology remained the result and the cause of ethical intransigeance.

China. In non-Western societies contraceptive practice had little impact on population. China, for example, increased its numbers from about 150 million in 1650 to 430 million in 1850 (Ho, pp. 277–278). As the population outstripped economic resources, Hung Liang Chi, "the Chinese Malthus," in 1793 wrote two essays, "Reign of Peace" and "Livelihood," which linked overpopulation to social misery (ibid., pp. 271–272). But the principal means of meeting the problem was infanticide, especially of girls. The scholar Wang Shih-to (1802–1889) did recommend the spread of drugs to sterilize women and also recommended a tax on families with more than two children, the postponement of marriage, the increase of nunneries for women, and the forbidding of remarriage to widows, but he also urged the wholesale infanticide of females (ibid., p. 274). In fact, the killing of baby girls remained a nineteenth-century practice (ibid., pp. 58–61). Milder means of control do not appear to have been widely known.

Modern developments

The first advocates. At the very end of the eighteenth century, Thomas Malthus, "the English Hung," wrote his famous *Essay on the Principle of Population*, positing a geometrical increase in population every twenty-five years if growth were unchecked; but his solution was "moral restraint," i.e., postponement of marriage, not contraception. About the same time, Jeremy Bentham hinted at the use of "sponges" to limit the poor and so reduce the poor rates. In 1806 Jean Baptiste Étienne de Senancour in *De l'amour* recommended "precautions" for women desiring to avoid unpleasant consequences from fornication; but he was unspecific. Francis Place, the first English advocate of particular means—coitus interruptus and sponges—remained anonymous in his advocacy. In the United States the first open advocacy of the means of birth control was undertaken in 1830 by Robert Dale Owens in *Moral Physiology: or a Brief and Plain Treatise on the Population Question*. Owens recommended coitus interruptus as a well-tested French technique.

Practice in France had in fact outstripped theory in the last part of the eighteenth century. A precipitous drop in the French birthrate from 1750 to 1800 reflected its effect, "the most important fact of all [France's] history" (Sauvy, p. 13). Disaffection from Catholic belief, followed by revolutionary emancipation from ecclesiastical control, lay at the root of this shift. The age of birth control had its roots in the age of *l'homme machine*, of a clear-eyed rationalism impatient with the mysterious and convinced that nature was controllable and man perfectible. Yet it was not in obedience to any express injunction of the rationalist philosophers but in response to a silent logic and in rejection of the old spirituality that birth control spread in France.

Christian theology, Catholic and Protestant, remained adamantly opposed to contraception, and medical and governmental opinion remained hostile. Even revolutionaries were unsympathetic: Proudhon in 1858 predicted that "Malthusianism" would depopulate France as it had done the Roman Empire (*De la justice dans la révolution et dans l'église*, I, pp. 348–349). After the debacle of the Franco-Prussian war, analysts of French society and journalists joined in attacking contraception as dooming France as a great power. In England in 1877, the government prosecuted Annie Besant and Charles Bradlagh for distributing Charles Knowlton's American text on contraception, *The Fruits of Philosophy*. In the United States, the federal Comstock law, enacted in 1873, forbade the mailing or importation of contraceptives, and most American states forbade their sale or advertisement (Smith, pp. 275–277).

The acceptance of contraception in the West. By the last quarter of the nineteenth century a small number of persons were convinced that contraception was the answer to miseries attributable to "over-childbearing." A Malthusian League was formed in England in 1878 to propagate this viewpoint. Analogues followed in Germany (1889), France (1898), Bohemia (1901), Spain (1904), Brazil (1905), Belgium (1906), Cuba (1907), Switzerland (1908), Sweden (1911), Italy (1913), and the United States (1913). By the twentieth century the birth control movement was clearly a Western European–American phenomenon (Chachaut, pp. 195–213, 453). International congresses were held in Paris in 1900, in Liège in 1905, in The Hague in 1911, in Dresden in 1912, in London in 1922, and in New York in 1925. Margaret Higgins Sanger, the American apostle of birth control, brought the message to Japan in 1921 and to India in 1936.

New forms of old contraceptive methods were developed in that period of expansion. In 1880 Wilhelm Mensinger developed a diaphragm for use as a pessary. By 1935 some two hundred types of mechanical devices, either condoms or pessaries, were in use in Western societies, and a wide range of chemicals was being employed as spermicides or occlusive agents; contraception was now a substantial business (Himes, p. 321).

In the period 1880–1930, medical opinion shifted from disapproval of contraception to doubt to acceptance. The shift paralleled changes in the learned worlds of science, sociology, and economics, where the humanist arguments of the advocates of birth control won much support. As a girl Margaret Higgins had been moved by the misery and ill health caused by undesired pregnancies among the very poor (Sanger, 1938, pp. 88–90). She also saw contraception as a substitute for "the barbaric methods" of infanticide and abortion (Sanger, 1969, p. 133). The relation between population growth and war was often stressed. The Sixth International Congress in 1925 declared flatly, "Overpopulation produces war." Malthus's predictions were refined and restated. It was estimated that by 1964 the United States would have a population of 214,000,000, which could not be agriculturally supported (East, p. 167). As the solution to overpopulation, as a substitute for abortion, and as an aid to personal happiness, contraception won the support of humanist ethicists in the West.

In Belgium, Ireland, and Spain, legislation of Catholic inspiration banned the sale of contraceptives. In France basically nationalistic legislation of the 1920s was directed to the same end. In the dictatorships of Germany, Italy, and Russia, contraception was discouraged as the nations prepared for war (Noonan, pp. 410–411). But after the Second World War nationalistic opposition to contraception began to disappear in Western countries.

In the United States contraception was widely practiced and, under the leadership of the Planned Parenthood movement, encouraged as "family planning." The last vestiges of secular opposition to contraception disappeared when the United States Supreme Court in *Griswold* v. *Connecticut* declared unconstitutional Connecticut's nineteenth-century statute against the use of contraceptives and when the same court in *Eisenstadt* v. *Baird* declared unconstitutional a Massachusetts law against selling contraceptives to the unmarried. From being illegal, contraception, now seen in the context of ethical concern with overpopulation, had become governmental policy, funded in the United States by such laws as the Family Planning Services and Population Research Act of 1970. On the international level, the Population Council, founded by John D. Rockefeller III in 1952, brought American technology and know-how in contraceptive technique to a variety of Asian, African, and South American nations. The United States moved officially to do the same with the Foreign Assistance Act of 1971.

Asia and Africa. The ethical-political response to contraception in the non-Western world remained mixed. Under the impetus of the American occupation, Japan had become fully committed to contraception with the Eugenic Protection Law of 1948. In other American client-states, such as Taiwan and South Korea, successful family planning programs flourished. Some of the new African states, however, did not value contraception highly, stressing their own relative underpopulation and observing the traditional importance of large families. Other countries like Indonesia gave lip service to the ideal of family planning but did little to implement it. India, with a government dedicated to population control, became a mammoth test case. In that country, "for all but a relative few, a woman's destiny lies in her procreation; the mark of her success as a person is in her bearing thriving children" (Mandlebaum, p. 16). Partly because of Hindu emphasis on the need for a son to perform salvific rites, partly for economic reasons, partly because of conflicts between villages and classes that could be won by manpower, and partly for personal status, large families were prized, and the value of contraceptives was questioned by the people (ibid., pp. 16–33). In competition with these traditional reasons were considerations of personal health and the national interest in curtailing population growth.

In Communist China birth control was officially encouraged in the cities in the mid-1950s, but after the Great Leap Forward, Chairman Mao in May 1958 asserted that China's numbers were a national asset (Aird, p. 699). Following the Cultural Revolution of 1966–1969, however, the government gave much publicity to marital postponement and birth limitation within marriage (Tien, p. 709). Ethical argument within the ruling party attempted the reconciliation of a belief in "planned population increase" with the ideology that "the question of 'overpopulation' does not exist in China" (Chi Lung, quoted in ibid., p. 710).

Modern Judaism. Orthodox Jewish thought had always insisted on the evil of coitus interruptus; it viewed use of a condom as an analogous practice, destructive of the semen and the normal completion of the heterosexual process, e.g., the nineteenth-century *Responsum Shevet Sofer*, n. 2 (Feldman, p. 229). In Europe even the position on the use of contraceptives by women had hardened into opposition with the rulings of Rabbi Meir Posner of eighteenth-century Danzig, Rabbi Akiva Eger of nineteenth-century Poland, and Rabbi Moses Sofer of Pressburg (died 1839) (ibid., pp. 213–218). The permissive tradition as to women, however, had been strongly upheld by Rabbi Solomon Luria (died 1573). In the twentieth century it was reasserted. The diaphragm, analyzed as a means that did not interrupt the coital act, was accepted by rabbinic authorities (ibid., pp. 227–228). When the anovulant pills were developed, they, too, fell within the range of permitted means (ibid., pp. 245–246). Merely cultural Judaism had already accepted the mores of the modern world.

Christian changes. Christian ethicists at first had remained critical. But in 1930 a significant break in the Christian front occurred. The bishops of the Anglican church by a vote of 193 to 67 approved a cautious acceptance of "methods" other than those of sexual abstinence to avoid parenthood ("Resolutions of the Lambeth Conference"). Between that date and 1958 the major Protestant churches all publicly abandoned the absolute prohibition of contraception by married couples. Leading Protestant theologians such as Karl Barth, Jacques Ellul, and Reinhold Niebuhr also accepted contraceptive practice. By 1959, when the World Council of Churches endorsed it, the Protestant consensus in its favor was overwhelming (Fagley, pp. 195–208).

The Orthodox Churches of the East did not alter the rule they had inherited. The Catholic Church remained adamant. The national hierarchies of Belgium (1904), of Germany (1913), of France (1919), and of the United States (1920) published strong pastorals denouncing contraception in their respective countries (Noonan, pp. 419–424). The climax of this episcopal and theological opposition was the encyclical letter *Casti Connubii* issued by Pius XI on 31 December 1930. "Any use whatever of marriage," the Pope declared, "in the exercise of which the act by human effort is deprived of its natural power of procreating life, violates the law of God and nature, and those who do such a thing are stained by a grave and mortal flaw" (Pius XI, 1930).

Catholic debate. In the late 1950s a new advance in technology offered occasion for reconsideration of the Catholic position: Progesterone pills or anovulants producing temporary sterility were developed. The first tentative essay in defense of their morality by Louis Janssens at the University of Louvain was met by a strong rebuke from Pius XII (pp. 735–736). But the theologians continued to probe, and a lay Catholic, John Rock, one of the developers of the pills, wrote *The Time Has Come* (1963) in favor of their legitimacy. John XXIII appointed a commission to investigate.

Within Catholic theology itself, the anti-Augustinian current had expanded from the base which the medieval recognition of non-procreative intercourse had afforded it. Beginning with the fifteenth-century theologian Martin Le Maistre, it had been argued that marital intercourse might without sin be for pleasure. The influential seventeenth-century theologian Tomás Sanchez, found no sin in spouses who intended "only to copulate as spouses" (*The Holy Sacrament of Matrimony* 9.8). By the end of the eighteenth century, the Catholic theological consensus no longer insisted on procreative purpose in marital intercourse; a part of the underpinning of the rule on contraception had been removed. The next step was to see a positive value in nonprocreative intercourse. The step was taken in the nineteenth century by the French Jesuit, Jean Gury, who taught that intercourse might "manifest or promote conjugal affection" (Gury, n. 688). The view became common. In the work of Herbert Doms, *Vom Sinn und Zweck der Ehe* (1935), love was shown to be central to the moral meaning of conjugal coitus. The Second Vatican Council (1963–1965) incorporated this long-matured doctrine, teaching that love "directed from person to person" was "singularly expressed and perfected by the proper work of marriage" (Second Vatican Council, sec. 49). *Casti Connubii* itself had approved marital intercourse at times when the wife would be sterile.

In the theological climate generated by the Council, a majority of the papal commission did in fact recommend the permission of contraceptive intercourse in marriage (Hoyt, pp. 76–77). The leading moral theologian on the commission, Josef Fuchs, was of the majority, and he was supported outside the commission by the equally eminent moralist Bernard Häring. Paul VI, however, decided otherwise, issuing on 25 July 1968 the encyclical letter *Humanae Vitae*, which repeated the old condemnation in familiar terms (Paul VI).

The Catholic world responded to the letter in a divided way. National episcopates such as those of England, Ireland, Italy, and the United States issued statements accepting it without reserve. Other episcopates, notably those of Belgium, Canada, France, Germany, Indonesia, and, above all, the Netherlands treated the condemnation in such a way as to permit conscientious Catholics to practice contraception (Horgan). The discussion did remain open; Paul VI's authoritative but fallible statement of Catholic doctrine had not closed debate; in practice many Catholics adopted a different rule.

Recent issues

By 1970 a new trend was visible in the United States with 2.75 million American couples of childbearing ages choosing to be sterilized, half by vasectomy and half by tubal ligation (Bumpass and Presser, p. 532). Irreversible action to produce sterility raised the question of whether commitment to contraception had not led many young persons to give up prematurely a basic human good.

The issue was also raised as to whether a strong public policy fostering contraception did not also lead to dependence on abortion when contraception failed. For example, in Japan a correlation between the use of contraceptives and the acceptance of abortion was observed (Nizard, p. 1252).

The problem of ancient medicine remained current: It was difficult to discriminate abortifacients from contraceptives. The operating principle of a popular contraceptive, the intrauterine device (IUD), was uncertain: It probably acted by preventing the nidation of a fertilized human ovum. But the solution to this problem depended not so much on ethical reflections as on refinement of the experiments testing the IUD's mode of action.

In the late 1960s it was argued that voluntary contraception, governmentally aided, would not reduce the world birthrate sufficiently, and a variety of coercive government measures were called for, chiefly by U.S. scientists. Among the proposals were Kenneth Boulding's to license the procreation of children; Paul R. Ehrlich's to add sterilants to the water supply and to make population control "the price of food aid" from the United States; and William Shockley's to sterilize all girls temporarily with reversibility allowed only on governmental approval (Berelson, pp. 1–3). These measures were opposed as unethical violations of human freedom by the mainstream of American thought (ibid., pp. 8–9).

The very advocacy of such means raised the suspicion in some quarters that government-aided contraception in the United States had a racist objective (Weisbod, p. 571). In the United States, as in India and in all countries divided into social or ethnic classes, the ethics of contraception remained tied to the ethics of group competition.

Marxist thought did not treat contraception as intrinsically evil but saw an emphasis on family planning programs as a capitalist ruse to divert attention from fundamental economic and political problems. Faithful to this Marxist viewpoint, the head of the Chinese delegation at the 1975 World Population Conference in Bucharest declared that "the primary way of solving the population problem lies in combatting the aggression and plunder of the imperialists, colonialists, and neo-colonialists, and particularly the superpowers, breaking down the unequal international economic relations" (Finkle and Crane, p. 106). At Bucharest, orthodox Marxists joined with Third World nations suspicious of the United States and with the official Vatican delegation to deemphasize contraceptive programs as the principal governmental approach to problems of population. The ethical judgment made was that contraception could not be used as a shortcut to economic and political justice.

Conclusions

Contraceptive technology has presented ethical problems for mankind almost from the beginning of recorded history. As long as the technology was weak, however, no strong pressure existed against ethical judgments unfavorable to its use. With the existence of a panoply of effective techniques, the ethical questions have shifted to the degree of governmental involvement in programs of contraception and the effect of contraception on group, class, or national security.

Contraception, however, is not mere technology in the way that, for example, bathroom plumbing is technology. By definition an act making intercourse infertile, it affects the act still expressive in all cultures of the unity of love, and it affects the transmission of life. Ethical judgment cannot be made upon it without taking into account its impact on a society's sexual morals and valuation of human life, nor without an appreciation of the demographic context, the commitment of parents to the education of the children they conceive, and the value of interpersonal human love. The weighing of these disparate elements has led to one consensus in the past and to another consensus in the present. A moralist sensitive to history will be aware that the balance now struck is not eternal.

JOHN T. NOONAN, JR.

[For further discussion of topics mentioned in this article, see the entries: ABORTION; INFANTS, article on INFANTICIDE: A PHILOSOPHICAL PERSPECTIVE; NATURAL LAW; POPULATION ETHICS: RELIGIOUS TRADITIONS; and POPULATION POLICY PROPOSALS. For discussion of related ideas, see the entries: CIVIL DISOBEDIENCE IN HEALTH SERVICES; SEXUAL ETHICS; and STERILIZATION.]

BIBLIOGRAPHY

[Classical sources are cited in this article with standard numerical divisions; the Oxford Classical Dictionary should be consulted for references to available texts and English translations.]

AIRD, JOHN S. "Review Symposium: China's Population Struggle by H. Yuan Tien." Demography 11 (1974): 695–701. A sharply critical discussion of H. Yuan Tien's China's Population Struggle: Demographic Decisions of the People's Republic, 1949–1969.

AUGUSTINE. "Adulterous Marriages (De incompetentibus nuptiis)." Translated by Charles T. Huegelmeyer. Writings of Saint Augustine. Vol. 15: Treatises on Marriage and Other Subjects, pp. 53–132, especially bk. 2, chap. 12, par. 12, pp. 115–117. "De conjugiis adulternis ad Pollentium." Patrologia Latina, vol. 40, cols. 453–486, esp. cols. 478–479.

———. Against Faustus 15.7. "Reply to Faustus the Manichean." The Works of Aurelius Augustine, Bishop of Hippo: A New Translation. Edited by Marcus Dods. Vol. 5: Writings in Connection with the Manichean Heresy. Translated by Richard Stothert. Edinburgh: T. & T. Clark, 1822, bk. 15, chap. 7, pp. 275–278, esp. p. 276. "Contra Faustum Manichaeum." Patrologia Latina, vol. 42, cols. 309–311, esp. col. 310.

———. "The Good of Marriage (De bono coniugali)." Translated by Charles T. Wilcox. Writings of Saint Augustine. Vol. 15: Treatises on Marriage and Other Subjects, pp. 1–51, esp. chap. 16, par. 18, pp. 32–33. Patrologia Latina, vol. 40, cols. 374–396, esp. col. 386.

———. Marriage and Concupiscence 1.19.21, 1.15.17. "What Is Sinless in the Use of Matrimony? What Is Attended with Venial Sin, and What with Mortal?" and "Thus Sinners Are Born of Righteous Parents, Even as Wild Olives Spring from the Olive." "On Marriage and Concupiscence." A Select Library of the Nicene and Post-Nicene Fathers of the Christian Church. Edited by Philip Schaff. Vol. 5: Saint Augustine: Anti-Pelagian Writings. Translated by Peter Holmes and Robert Ernest Wallis. Revised by Benjamin B. Warfield. New York: Christian Literature Co., 1887, bk. 1, chap. 17 [XV]; chap. 21 [XIX], pp. 270–273. "De nuptiis et concupiscentia." Patrologia Latina, vol. 44, bk. 1, chap. 15, sec. 17; chap. 19, sec. 21; cols. 421–422, 424–425.

———. Writings of Saint Augustine. Vol. 15: Treatises on Marriage and Other Subjects. The Fathers of the Church: A New Translation, vol. 27. Edited by Roy Joseph Deferrari. New York: 1955.

AVICENNA. Canon medicinae. Translated by Gerard of Cremona. Venice: Apud Juntas, 1582. The arsenal of contraceptive prescriptions known to antiquity, the Arabic world, and medieval Europe.

BEHRMAN, S. J.; CORSA, LESLIE, JR.; and FREEDMAN, RONALD, eds. Fertility and Family Planning: A World View. Ann Arbor: University of Michigan Press, 1969.

BERELSON, BERNARD. "Beyond Family Planning." Studies in Family Planning, no. 39 (February 1969), pp. 1–16. Current ethical issues.

BERGUES, HÉLÈNE; ARIÈS, PHILIPPE; HÉLIN, ÉTIENNE; HENRY, LOUIS; RIQUET, R. P. MICHEL; SAUVY, ALFRED; and SUTTER, JEAN. La Prévention de naissances dans la famille: Ses origins dans les temps modernes. Institut national d'études démographiques: Travaux et documents, cahier no. 35. Paris: Presses Universitaires de France, 1959. An historical account of French contraceptive practice.

BUMPASS, LARRY L., and PRESSER, HARRIET B. "Contraceptive Sterilization in the U.S.: 1965 and 1970." Demography 9 (1972): 531–548.

CALVIN, JOHN. Commentarius in Genesim 38.8–10. Translated by John King as Commentaries on the First Book of Moses Called Genesis. 2 vols. Grand Rapids, Mich.: Wm. B. Eerdmans Publishing Co., 1948, vol. 2, chap. 38, v. 8, p. 281.

CHACHAUT, M. *Le Mouvement du "Birth Control" dans les pays Anglo-Saxons.* Bibliothèque de l'Institut de droit comparé de Lyon, sér. centrale, vol. 32. Paris: M. Giard; Lyons: Bosefrères, M. & L. Riou, 1934. A history of the movement.

CHANDRASEKHAR, SRIPATI, ed. *Asia's Population Problems, with a Discussion of Population and Immigration in Australia.* New York: Frederick A. Praeger, 1967. Problems and issues from the perspective of a convinced family planner.

CLEMENT OF ALEXANDRIA. *Stromata* 2.4.30, 3.12.90. Bk. 2 translated as "The Stromata, or Miscellanies." *The Ante-Nicene Fathers: Translations of the Writings of the Fathers down to A.D. 325.* Edited by Alexander Roberts and James Donaldson. Revised by A. Cleveland Coxe. Vol. 2: *Fathers of the Second Century: Hermas, Tatian, Athenagoras, Theophilus, and Clement of Alexandria (Entire).* American reprint. Grand Rapids, Mich.: Wm. B. Eerdmans Publishing Co., 1956, bk. 2, chap. 4, sec. 30, pp. 349–351. Bk. 3 translated by John Ernest Leonard Oulton as "On Marriage: Miscellanies, Book III, The Text." *Alexandrian Christianity: Selected Translations of Clement and Origen.* Introduction and notes by John Ernest Leonard Oulton and Henry Chadwick. Philadelphia: Westminster Press, 1954, pp. 82–83. Oulton and Chadwick do not include book 2. Roberts and Donaldson do not translate book 3 because it is "necessarily offensive to our Christian tastes" (p. 381).

DEINES, HILDEGARD VON; GRAPOW, HERMANN; and WESTENDORF, WOLFHART. *Übersetzung der medizinischen Texte.* 2 vols. Grundriss der Medizin der alten Ägypter, no. 4. Berlin: Akademie-Verlag, 1958.

DOMS, HERBERT. *Vom Sinn und Zweck der Ehe.* Breslau: Ostdeutsche Verlagsanstalt, 1935. Translated by George Sayer as *The Meaning of Marriage.* New York: Sheed & Ward, 1939. Fundamental to modern Catholic views on the purpose of marriage.

EAST, EDWARD M. *Mankind at the Crossroads.* New York: Charles Scribner's Sons, 1923.

Eisenstadt v. Baird. 405 U.S. 438. 31 L. Ed. 2d 349. 92 S. Ct. 1029 (1972).

FAGLEY, RICHARD M. *The Population Explosion and Christian Responsibility.* New York: Oxford University Press, 1960. A review of modern Protestant positions on contraception.

Family Planning Services and Population Research Act of 1970. Pub. L. 91-572. 84 Stat. 1504. 42 U.S.C. sec. 300r.

FELDMAN, DAVID MICHAEL. *Birth Control in Jewish Ethics: Marital Relations, Contraception, and Abortion as Set Forth in the Classic Texts of Jewish Law.* New York: New York University Press, 1968. A comprehensive survey arguing that literary history, not economic-historical factors, has determined the rabbinic positions.

FINKLE, JASON L., and CRANE, BARBARA B. "The Politics of Bucharest: Population, Development, and the New International Economic Order." *Population and Development Review* 1 (1975): 87–114. The current secular debate on contraception in terms of Marxism and international politics.

GLASS, D. V., and REVELLE, ROGER, eds. *Population and Social Change.* London: Edward Arnold; New York: Crane, Russak, 1972. Focused on demographic history, this symposium is a good introduction to the relations between social contexts and contraceptive practice.

Griswold v. Connecticut. 381 U.S. 479. 14 L. Ed. 2d 510. 85 S. Ct. 1678 (1965).

GURY, JEAN. "De Matrimonio" [Marriage]. *Compendium Theologiae Moralis* [Compendium of moral theology] (1846).

HAUSER, PHILIP M., ed. *The Population Dilemma.* 2d ed. Englewood Cliffs, N.J.: Prentice-Hall, 1969. The modern "population problem."

HIMES, NORMAN EDWIN. *Medical History of Contraception.* Foreword by Robert Latou Dickinson. Preface by Alan F. Guttmacher. New York: Gamut Press, 1963. Unreliable as history due to Himes's unfamiliarity with foreign languages; useful on the modern birth control movement.

HO, PING-TI. *Studies in the Population of China, 1368–1953.* Harvard East Asian Studies, vol. 4. Cambridge: Harvard University Press, 1959. Primarily a demographic study, this touches incidentally on means of control.

HORGAN, JOHN, ed. *Humanae Vitae and the Bishops: The Encyclical and the Statements of the National Hierarchies.* Analytic Guide by Austin Flannery. Shannon: Irish University Press, 1972. A complete collection of episcopal responses to the encyclical. Text of Encyclical in English and Latin.

HOYT, ROBERT G., ed. *The Birth Control Debate.* Kansas City: National Catholic Reporter, 1968. The majority and minority reports of the papal commission on contraception.

LEE, LUKE T., and LARSON, ARTHUR, eds. *Population and Law: A Study of the Relations Between Population Problems and Law.* Leiden: A. W. Sijthoff; Durham, N.C.: Rule of Law Press, 1971. A worldwide compendium of statutes on birth control.

LUTHER, MARTIN. "Der grosse Katechismus, 1529." *Luthers Werke in Auswall.* Vol. 4: *Schriften von 1529 bis 1545.* 5th rev. ed. Edited by Otto Clemen. Berlin: Walter de Grunter & Co., 1959, p. 31–34. On the Sixth Commandment.

MACURA, MILOS. "Population Policies in Socialist Countries of Europe." *Population Studies* 28 (1974): 369–379. Eastern European governmental attitudes on contraception.

MANDLEBAUM, DAVID G. *Human Fertility in India. Social Components and Policy Perspectives.* Berkeley: University of California Press, 1974. Practice and attitudes affecting ethical judgment in India.

NIZARD, ALFRED. "Le Japon vingt ans après la loi eugénique." *Population* 25 (1970): 1236–1262. A comprehensive study of population and means of control since 1948.

NOONAN, JOHN T., JR. *Contraception: A History of Its Treatment by the Catholic Theologians and Canonists.* Cambridge: Belknap Press, Harvard University Press, 1965. The history of doctrine on contraception from its pre-Christian roots to the Second Vatican Council.

O'FLAHERTY, WENDY DONIGER. *Asceticism and Eroticism in the Mythology of Śiva.* School of Oriental and African Studies. London: Oxford University Press, 1973.

Patrologiae cursus completus: Series Graeca [*Patrologia Graeca*]. 161 vols. Paris: Apud Garnier Fratres, editores & J.-P. Migne successores, 1857–1894.

Patrologiae cursus completus: Series Latina [*Patrologia Latina*]. 221 vols. Paris: Apud Garnier Fratres, editores & J.-P. Migne successores, 1851–1894. *Supplementum.* Paris: Éditions Garnier frères, 1958–.

PAUL VI. "Humanae Vitae." *Acta Apostolicae Sedis* 60

(1968): 481–503. Translated as "Humanae Vitae (Human Life)." *Catholic Mind,* September 1968, pp. 35–48.

Pius XI. "Casti Connubii." *Acta Apostolicae Sedis* 22 (1930): 539–592, esp. 560. Translated as "On Christian Marriage." *Catholic Mind* 29 (1931): 21–64.

Pius XII. "Essais de solution." *Acta Apostolicae Sedis* 50 (1958): 732–737. Especially pp. 735–736. Text in French.

"Resolutions of the Lambeth Conference, 1930: II. The Life and Witness of the Christian Community: Marriage and Sex." *Report of the Lambeth Conference, 1930: Encyclical Letter from the Bishops with Resolutions and Reports.* London: Society for Promoting Christian Knowledge; New York: Macmillan Co., n.d., resolution 15, pp. 43–44.

Rock, John. *The Time Has Come: A Catholic Doctor's Proposals to End the Battle over Birth Control.* Foreword by Christian A. Herter. New York: Alfred A. Knopf, 1963.

Sanger, Margaret. *Margaret Sanger: An Autobiography.* New York: W. W. Norton & Co., 1938.

———. *My Fight for Birth Control.* New York: Maxwell Reprint, 1969.

Sauvy, Alfred. *La Prévention des naissances.* "Que sais-je?" Le Point des connaissances actuelles, no. 988. Paris: Presses universitaires de France, 1962. 3d ed. 1967.

Second Vatican Council. "Pastoral Constitution on the Church in the Modern World." *The Sixteen Documents of Vatican II and the Instruction on the Liturgy.* Boston: St. Paul Editions, 1962, pp. 511–625; esp. pt. 2, chap. 1, sec. 49, pp. 563–564. *Gaudium et Spes.*

Smith, Peter. "The History and Future of the Legal Battle over Birth Control." *Cornell Law Quarterly* 49 (1963): 275–303. A survey of statutory and case law in the United States.

Soranos [Soranus]. *Peri gynaikeiōn pathōn* [Gynecology] 1.10.36, 1.19.60–63.

Thomas Aquinas. *Evil* 15.2. "Utrum omnis actus luxuriae sit peccatum mortale" [Whether all sex acts are mortal sins]. *Questiones disputatae.* 7th ed. 5 vols. Vol. 2: *De malo.* Rome: Marietti, 1942, pp. 245–250.

Tien, H. Yuan. "Review Symposium: *China's Population Struggle* by H. Yuan Tien: Reply." *Demography* 11 (1974): 708–714. A spirited rejoinder to Aird.

Weisbod, Robert G. "Birth Control and the Black American: A Matter of Genocide?" *Demography* 10 (1973): 571–590.

COST-BENEFIT ANALYSIS

See Risk; Life-Support Systems; Technology, *article on* Technology Assessment; Life, *article on* Value of Life.

CRYOBANKING OF SPERM, OVA, AND EMBRYOS

See Reproductive Technologies, *articles on* Sperm and Zygote Banking, Ethical Issues, *and* Legal Aspects.

CRYONICS

The study of the biological effects of cold is known as cryobiology. Cryonics is a more specific term used to identify the practice of freezing human bodies in the hope of achieving suspended animation. Luyet and Gehenio in 1940 compiled the first comprehensive survey on the survival of organisms at low temperature. In 1964 an increasing volume of laboratory work on low temperature biology precipitated the founding of a new journal, *Cryobiology,* devoted exclusively to the subject. Since the mid-1960s reviews dealing with various aspects of cryobiology have been published (Meryman; Mazur; Popovic and Popovic).

Human beings, like other mammals but unlike most animal species, normally maintain their body temperature within very narrow limits. In a mammal, the decline of the body temperature is accompanied by a corresponding reduction in vital activities. Actual freezing of tissues essentially halts all life processes. States of low temperature above freezing (hypothermia) have been found useful in the preservation of isolated organs and in some surgical procedures. Segregated mammalian cells and some tissues have been frozen to very low temperatures and successfully recovered. The freezing of most cells without previous exposure to cryoprotective solutions results in their death.

The more ambitious goal of maintaining the whole body of a human or of an animal in the frozen state and eventually resuscitating it has not been attained. Nevertheless, this prospect has stimulated the organization of "cryonics" or "life extension" societies, which disseminate information about the freezing of human bodies and aid those who are interested in applying cryonic interment, rather than conventional burial or cremation, to bodies of relatives or friends.

In most cases of human cryonic preservation, the bodies are first perfused with a cryoprotective agent, usually Dimethylsulfoxide (DMSO), temporarily frozen in dry ice, and then permanently stored in liquid nitrogen at $-196°C$. Endowments must be provided so that the liquid nitrogen can be replaced periodically. Ettinger (1964, pp. 170–180) and other proponents of this cryonic interment have been forecasting its widespread application since the mid-1960s. The reason for the freezing of bodies is the anticipation of the possibility of resuscitation at some distant future date when science is greatly advanced and both the pathology that caused the death and the injury resulting from the freezing process itself can be reversed.

Possible methods of cooling

There are at present three cooling methods whereby a human body might be maintained in a reduced metabolic state for a long period: hypothermia, hibernation, and the frozen state.

Hypothermia. Hypothermia involves the cooling of the body to a temperature below normal but above freezing. This can take place when the body temperature regulating mechanisms fail, or whenever the heat production of an individual is too small to compensate for the heat loss resulting from cold exposure. Some men and women have accidentally become deeply hypothermic and have subsequently survived. However, cooling was of relatively short duration. No records of human survival following extended hypothermia (weeks or months) exist. The very low body temperatures required for long-term preservation cause injury even if the period of cooling is brief. Generally in medical usage, regional or differential hypothermia is preferred over whole body cooling. The reasons for limited survival time in deep hypothermia are still not understood, but it seems that inadequate tissue perfusion is responsible for death.

Hibernation. Hibernation, a state of greatly reduced body temperature, occurs naturally in some mammals and birds. The hibernating animal retains the capability of spontaneously rewarming to its normal level of around 37°C by a mechanism of internal heat production without absorbing heat from its environment. Hibernating animals are able to recover fully from week-long periods of profound torpor, a state in which the temperature of the entire body falls to within a few degrees of freezing, heart rate is slowed to a few beats per minute, and metabolism is depressed as much as ninety-nine percent. The physiological depressions are reversible in a hibernator when arousal takes place. It could be useful for research, for surgery, and for future space travel to be able to induce states of hibernation at will in man, but at present it cannot be done. Despite an extensive research effort, the mechanism of hibernation production is still unknown.

At present, therefore, hibernation is not a practical method of producing long-term suspended animation of man, and in any case the aging process would continue.

Freezing. Total body freezing is presumably the best method for long-term preservation of man in a state of suspended animation, if it could be technically accomplished. Cryopreservation of individual human cells and some small or thin pieces of tissue have met with considerable success. Numerous attempts to freeze adult kidneys or hearts have not produced methods allowing for subsequent survival of rewarmed organs on transplantation. The lethal effects of freezing at the mature organ level are not yet controllable, even by the introduction of cryoprotective agents.

Probably because of its complexity, the freezing of whole animals has received relatively little serious attention. The most thorough studies were conducted on golden hamsters by Smith and co-workers (Smith, pp. 199–205). Like most hibernators the hamster has a high resistance to cold injury, but even under these circumstances, full recovery of a completely frozen animal has never occurred. Attempts to "supercool" (i.e., achieve below freezing temperatures without the crystallization associated with freezing) whole mammals, thus avoiding the adverse effects of crystal formation, have also failed to provide a method for long-term preservation. According to Malinin, at present any attempts to preserve humans by freezing must be regarded as based on unfounded hopes. There is essentially no material "prospect of immortality" for those bodies frozen by the cryonic methods now employed.

Moral issues in human freezing

Moral arguments against the use of cryonic interment are based on questions of justice to immediate survivors as well as to future generations, the use of extraordinary means in attempts to extend life even briefly, and religious, sociological, and anthropological considerations. Moral implications vary depending on whether actual current practice or the use of a possible perfected technique is being considered.

Current freezing of bodies. It is argued that if resuscitation from the frozen state appears not to be possible even in the future, there is nothing to be lost by subjecting a body to freezing rather than to burial. This is not necessarily the case, however. False expectations could be encouraged, thereby delaying an acceptance of the reality of a particular death and risking adverse psychological effects on the related family. Further, the freezing and maintenance procedure is costly and would place a large burden on most families. Anxieties could be developed in those who cannot afford the procedure and whose sense of love and loyalty might encourage them to try to use the procedure as something truly desirable.

If the practice ever grew, the cumulative effect on society could be great as resources were diverted to this activity. Perhaps lower age limits would be imposed and only certain types of deaths would justify the procedure. Any exaggerated resource-consuming efforts for individuals would have to be limited by the common good.

A materialistic and, according to some, even a selfish view of human life seems inherent in this procedure, especially if it became widespread, for it would be based on the assumption that the extension of this material life is desired at almost any cost.

The philosophy favoring cryonic interment seems to be contrary to the Christian understanding of the purpose of life and the meaning of death, but Ettinger (1967) does not see this conflict. He understands cryonic interment to be an "urgent spiritual goal," a goal that can serve as an "antidote" to defections from Christianity. He proposes a vision of "supermen" greatly enriched by biological discoveries and advancing toward the material greatness envisioned by the creator himself. Ettinger sees this as an important element in the development of a Christianity that is universal and evolutionary rather than narrow and static.

If freezing of humans were perfected? If perfected, freeze preservation of a human for days or months would present the same moral questions facing any patient and doctor before a significant medical procedure. If the service is limited or the cost is very high, factors such as wishes of the patients, age, likelihood of real utility, demand on resources, etc., would contribute to the final judgment.

If perfected, preservation for extensive periods of years or decades might be defended in circumstances such as long space flights, to await medical developments, or even for purposes of historical continuity of mankind. Each situation would have to be evaluated individually. Widely applied long-term human preservation, however, would confront us with a very different set of moral questions, some of which have already been considered above in the treatment of current freezing methods.

According to Vaux (pp. 96–110), man, as a social being, must learn to die for the sake of the future. Vaux maintains we must learn that death energy is creative. We face population and ecological disaster if we disrupt the evolutionary death now in process without simultaneously disrupting birth. The success of the freezing procedure would not of itself mean that man's actual active biological life would be significantly extended, although his chronological life span would be increased. It could mean that age proportions in our population would be disrupted. Maybe our concentration on health care would be greatly emphasized. Unless there is some limit to birth rate, and death rate is at least maintained in step with this limit, the quality of life on this planet may be radically altered.

In view of the aforesaid problems involved in large scale preservation, someone or some group would have to face crucial choices about who is to be freeze-preserved, when this is to happen, and, perhaps even more important, at what point thawing and resuscitation are to take place. The possibility of population manipulation is as evident here as it would be in cases of genetic engineering. Who would make these choices? On what basis? Obviously, to avoid chaos, economic and social balance would require that humans be reintroduced into society in some planned manner. What would be the basis for this planning?

How would human freedom be affected? Would the right to die be respected? Would the consent of the individual be required before preservation? In the frozen state, where there can be no dialogue with the individual, how are rights negotiated? Is the preserved individual legally dead? What happens to his property, his family relationships? Does anyone own him?

Man's behavior is dependent on the constant interactions between himself and the surrounding objects and actions that impinge upon him. He might hope to be unchanged after a long period of preservation, but his ambience, his historical context, and his associations would be radically disrupted. What would be the effect on his own development? What would be the effect of a long cessation of biological rhythms, so essential for a balanced human existence? What subtle behavioral variation might result in an individual or a population? Long-term preservation would probably produce individuals who are out of step with the historical development of man. And, of course, it is more than likely that the preservation procedures would be at least slightly imperfect and would yield large numbers of either mentally or physically unhealthy individuals. Would society accept a moral responsibility to see that such defective persons are cared for until they die naturally? There are many other unpredictables involved, such as the effect on human motivation, expectancies, and drives, which would result from what is essentially a lack of unity and compactness in the individual life.

It must be emphasized that possible moral ramifications should not necessarily deter scientists from seriously trying to perfect preservation methods. There are clear advantages for man if such methods are available, especially for short-term emergency use. And intermediate successes, such as the successful banking of frozen organs, would have vast benefits. Moral questions become much more complex, however, when the goal of the methodology becomes a frustration of the life–death rhythm of human existence.

J. A. PANUSKA

[*For further discussion of topics mentioned in this article, see the entries:* AGING AND THE AGED, *article on* THEORIES OF AGING AND ANTI-AGING TECHNIQUES; CADAVERS, *article on* GENERAL ETHICAL CONCERNS. *Also directly related are the entries* DEATH, ATTITUDES TOWARD; DEATH, DEFINITION AND DETERMINATION OF; LIFE; *and* LIFE-SUPPORT SYSTEMS. *See also:* FUTURE GENERATIONS, OBLIGATIONS TO; HUMAN EXPERIMENTATION; *and* ORGAN TRANSPLANTATION.]

BIBLIOGRAPHY

BRYANT, CLIFTON D., and SNIZEK, WILLIAM E. "The Cryonics Movement and Frozen Immortality." *Society*, November–December 1973, pp. 56–61.

ETTINGER, ROBERT C. W. *The Prospect of Immortality*. Garden City, N.Y.: Doubleday & Co., 1964.

———. "Cryonics and the Purpose of Life." *Christian Century* 89 (1967): 1250–1253.

LUYET, BASILE JOSEPH, and GEHENIO, P. N. *Life and Death at Low Temperatures*. Normandy, Mo.: Biodynamica, 1940.

MALININ, THEODORE I. "Freezing of Human Bodies." *Journal of the American Medical Association* 221 (1972): 598.

MAZUR, PETER. "Cryobiology: The Freezing of Biological Systems." *Science* 168 (1970): 939–948.

MERYMAN, HAROLD T., ed. *Cryobiology*. New York and London: Academic Press, 1966.

POPOVIC, VOJIN, and POPOVIC, PAVA. *Hypothermia in Biology and in Medicine*. New York: Grune & Stratton, 1974.

SMITH, AUDREY U., ed. *Current Trends in Cryobiology*. New York: Plenum Press, 1970.

VAUX, KENNETH. *Biomedical Ethics*. New York: Harper & Row, 1974.

DEATH

Because views on the ethical issues surrounding death and dying are constantly changing, this entry offers a history of the conceptual and value perspectives on death in different cultures and philosophies as background material for the discussions of the concrete ethical issues presented in the entries on DEATH AND DYING: EUTHANASIA AND SUSTAINING LIFE, *and* DEATH, DEFINITION AND DETERMINATION OF.

  I. ANTHROPOLOGICAL PERSPECTIVE   *David Landy*
 II. EASTERN THOUGHT   *Frank E. Reynolds*
III. WESTERN PHILOSOPHICAL THOUGHT
                                     James Gutmann
IV. WESTERN RELIGIOUS THOUGHT
  1. DEATH IN BIBLICAL THOUGHT
                                     Lloyd Bailey
  2. POST-BIBLICAL JEWISH TRADITION
                                   Seymour Siegel
  3. POST-BIBLICAL CHRISTIAN THOUGHT
                                 Milton McC. Gatch
  4. ARS MORIENDI   *Brian P. Copenhaver*
 V. DEATH IN THE WESTERN WORLD
                                     Talcott Parsons

I
ANTHROPOLOGICAL PERSPECTIVE

Peoples of every known human society value life and deplore its end. Each society has evolved a set of conventions for determining under what circumstances a life may be terminated. There are transcultural variations regarding the extinction of life deliberately or accidentally as well as in the perception of what constitutes natural death. While no society welcomes death, there are conditions under which a death may be intentionally invited and implemented by victim and community; death may become a social obligation. This article is confined to societies termed preindustrial, nonliterate, and non-Western.

Natural death and the desire for immortality

Hartland expresses a long-held view that there is a universal abhorrence of death and a tacit refusal to accept its inevitability (Frazer, 1913, 1927). Myths of many peoples suggest that humanity was originally immortal but was brought down to mortal status by gods angered at some human indisposition: disobedience, deception, betrayal of spirits, or violation of moral and social norms. Hocart ("Death Customs") rejects the notion that preindustrial peoples "are alleged to feel an overwhelming fear of death which prompts measures for self-protection. This is not confirmed by observation." It appears to be not so much horror of the corpse as fear of the symbolic power of spirits as well as love and reverence for the deceased and desire for immortality that motivates human ethics of death (Malinowski). This ambivalence—a reflection of the ambivalence toward the living, whether relative, leader, stranger, or friend—characterizes cults of the dead and ancestor worship and influences forms and functions of mortuary rites (Bendann).

Myths of immortality, often transmitted through tales of sick and old persons through dreams and visions, predicate the notion prevalent in many cultures that the death of a mature adult is not a natural happening but the effect of spiritual retribution, or of sorcery or witchcraft, i.e., human retribution (Frazer, 1913, 1927). Some writers insist that all preindustrial peoples hold the idea of death as unnatural, but the evidence does not support this. Simmons (pp. 217–220) found that, of forty-seven societies in his cross-cultural survey, seventeen regarded all death as unnatural, four believed it to be natural, and twenty-six "partially admitted" the possibility. Thus nearly two-thirds of these societies thought natural death to be possible, seemingly contrary to the more florid ethnological generalizations.

There appears to be a universal tendency to conceive of each person as possessing a vital substance, an élan vital, that animates his behavior and quickens his body. Sometimes the dead will be conceived to have more than a single essence. The Trobriand Islanders divide it into the *kosi* (ghost) and *baloma* (spirit), each with its own characteristics and vicissitudes (Malinowski). Frazer and Tylor felt that the wish for immortality springs from experience (especially of dreams), not intuition, and frequently results in the raising of dead ancestors to godhood in a pantheon that also includes nature spirits. The desire to live indefinitely combined with the actual hazards of life in an unpredictable environment lead easily to the notion that, but for spiritual or human trickery, each mortal would be immortal. Therefore those states that conduce toward death, such as sickness, starvation, accidents and injuries, and other misfortunes, are a consequence of natural, supernatural, and human malevolence.

All peoples attempt to protect themselves from lethal forces and events with often elaborate precautionary systems. Life at its fullest is coterminous with health and happiness (as indigenously defined). But serious illness or injury, states of decrepitude, and disabilities of age are seen as related more to death than to life. When a member of the Murngin of Australia has been "boned" (condemned to death by sorcery) his community classifies him as "half-dead" (Warner, 1958 ed.), a state that Van Gennep and Turner define as "liminal" or "betwixt and between" two conditions while being neither. Since this vulnerable state poses a threat not only to the victim but through ritual contagion to his family and clan, this hunter–gatherer community withdraws physically and psychologically, behaving as though their ill-fated colleague no longer exists. Then they subject him to a mourning ceremony to hasten his journey to the land of the dead. Through transgression of moral norms the victim has endangered his entire group and cannot be allowed to live.

Durkheim (1965) and Hertz (1960) accepted the universality of the human belief in immortality but felt that it derives not from individual psychological need but from "the naive expression of a permanent social need. Indeed society imparts its own character of permanence to the individuals who compose it: because it feels itself immortal and wants to be so, it cannot normally believe that its members, above all those in whom it incarnates itself and with whom it identifies itself, should be fated to die" (Hertz, p. 77). Van Gennep asserted that death was the last in a series of "rites of passage," beginning with birth, that punctuate the life cycle as individuals or cohorts of individuals pass from one status and role-set to another. The implementation of these ceremonies of individual-social transformation occurs in three phases: separation, transition, and incorporation. Various points of potential crisis (birth, adolescence, marriage, death) emphasize one or another phase, but all contain some aspects of each, and frequently include subphases. In separation rituals, survivors mourn their loss, attempting thereby to adapt to bereavement by working out feelings regarding the deceased. In the transition phase the corpse is not yet completely dead (in the social and psychological sense) nor alive, and its ghost hangs about to ensure that appropriate actions have been taken by the living. Rites of incorporation occur finally in which ghost becomes transformed into spirit and a permanent inhabitant of the afterworld, though in touch with mundane affairs and influencing and controlling them through stringent moral and ethical sanctions. In death the rites of passage deal not only with the organic end of the person but with his rebirth in spiritual form (Van Gennep).

Durkheim and Hertz remark upon the fact that in most cultures there is not one but often two or more rituals connected with disposal of the dead. It takes time for the ghost of the deceased to find a comfortable place in the society's collective consciousness, since the familiar image of the dead person seems a long distance from the hazy, exalted image of the worshiped and dreaded ancestors. During this period the group recoils from the shock of bereavement and through ritual finally regains its new equilibrium. Another way to conceive of the process might be as a patterned struggle to reestablish the balance of power between the quick and the dead. Relations between the living society and its spiritual environment are patterned after its internal social relations. Not only psychological balance but a reorganization of the social structure is required.

Homicide and suicide

The first systematic theory of suicide, proposed by Durkheim (1951), had little to say about homicide, and anthropological theories concerning both have since been few (Bohannan). Durkheim suggested that the greater the integration of social groups, the lower the suicide rate, and vice versa. Social institutions are most integrated when they command the greatest loyalty. Persons lying outside the major institutions, or unable to subscribe to their values and demands, are most likely to resort to suicide. Durkheim termed such suicide "anomic." In areas of social life that remain uninstitutionalized (a measure of the "openness" of a society), where pressures to perform in certain ways are not simply inadequate, as in anomic suicide, but nonexistent as social consensuses, resort to suicide is termed by Durkheim "egoistic" and "is a concomitant of a type of social structure, not the result of social pathology." A third form of suicide, "altruistic," occurs in extremely highly integrated institutions in which the individual's adherence to cultural values overrides valuation of his own life. Like egoistic suicide, altruistic suicide is not a criterion of social malfunctioning. A fourth type, vengeance or "samsonic" suicide, which occurs when a person takes his life as a form of revenge against others, was suggested by Jeffreys.

Examples of all four types appear in ethnographic literature. Bohannan suggests that the Durkheim model of suicide could well be applied, with modifications, to homicide. Either action involves the taking of human life and is therefore threatening to any society. Indeed, suicide may be perceived as more threatening, since it symbolizes a person's decision that life in his social groups is not worth living. Suicide frequently stimulates intense emotional responses of alarm, fear, anger, outrage, and even revenge (upon the suicide's property, kin, etc.), as well as sorrow and sympathy.

"Homicide within the society is, under one set of conditions or another, legally prohibited everywhere. Likewise, it is universally recognized as a privilege-right under certain circumstances, either in self-defense against illegal, extreme assault (including sorcery) or as a sanction for certain illegal acts" (Hoebel, p. 286). Since the social subordination of women is nearly universal, it has generally been considered justifiable for husbands in preindustrial societies to kill adulterous wives. In societies where legal institutions are rudimentary, homicide may be used by strong men to achieve power, wives, property, and honor, or to avenge some real or imagined offense. Blood revenge is expected in most Eskimo tribes, except the Copper, Iglulik, and East Greenlanders (ibid., pp. 85–99), though a feuding chain reaction is avoided through certain moral substitutes: institutionalized wrestling, butting, or boxing, frequently as accompaniment to a song duel. Social catharsis may be achieved through vicarious combat. Justice may not be served, since the offender or his representative may vanquish the offended. The victor wins admiration, the loser loses rank. A single homicide may result in a feud, but a second murder by the same person transforms him into a public enemy who is then executed by a public-spirited man with community consent, thus providing another check on excessive slaughter.

Categories of homicide vary with a society's values and social organization. The Tiv of Nigeria practice "ritual" killing (to obtain anatomical items for fetishes), murder of thieves, infanticide, "accidental" killings (usually from poisoned arrows in communal hunts), killings involving sexual jealousy or adultery, and slayings because of illness believed to be caused by witches or sorcerers (Bohannan). Among the Gisu of Uganda, "*all* deaths were considered murder in that responsibility for them was laid at someone's door." However, not all such accusations were followed by revenge since frequently agreement could not be reached on the identification of the culprit (La Fontaine, pp. 94–129).

Deaths of murderer and victim in most cultures are treated somewhat differently. Even in justifiable homicide, frequently the killer must undergo rites of purification since the shedding of blood is a ritually dangerous condition. The victim may also undergo lustration, but in addition the corpse may be disposed of more quickly, perfunctorily, and with less ceremony. Among the LoDagaa (Goody, pp. 112 ff.) only other murderers may handle the corpse. The killer's bow, hung out of reach of children, is broken upon his death. Special obsequies are performed by the eldest killers. Intraclan homicides are also regarded differently from interclan ones. Anyone who has been in contact with a dead or dying killer, especially his wife, must take a special medicine to ward off harmful mystical effects. Just as all societies make a distinction among classes of persons who may be killed, so do they also with respect to categories of non-human animals. Domestic animals are differentiated from undomesticated ones, nontotemic from totemic ones, and animals considered to be in some way connected with evil are contrasted with those considered harmless, as permissible victims (ibid., pp. 116–121).

While homicides in most societies frequently occur between related or acquainted persons, *permissible* homicide generally (though not invariably) seems to coincide roughly with the degree of social and geographical distance—and therefore of difference—between killer and killed. Under conditions of deep-going political and economic change homicide may become a "way of life." In the Mayan Indian township of Teklum frequent recourse to homicide occurs under conditions of sharpened economic competition, changed political conditions (power shifting from indigenous leaders to town president), loss of leadership and esteem by curers (becoming increasingly competitive, losing control over recruitment to the role and respect for their power to vanquish witchcraft), and increased emotional insecurity due to loss of faith in the protective power of ancestral spirits and their earthly mediums (the curers). Traditional social control mechanisms operating through the curers, local officials, and others that served to limit effects of witchcraft and resort to murder to settle disputes and conflicts are critically weakened, and homicide and witchcraft accusations have increased, even though both are outlawed by national Mexican law (Nash).

The relationship between suicide and homicide is often close and complex. Among the Maria aborigines of India similar circumstances provoke homicide in one personality and suicide in another. Disputes over possession of dancing equipment lead one man to murder his betrothed, while another who has his equipment stolen kills himself rather than the thief. One person insulted by a kinsman reacts by killing himself, another by killing the offender. Some persons kill the sorcerer they believe has afflicted them, others take their own lives to break the spell. Both homicide and suicide have a high incidence, and most homicides are against fellow tribesmen, not strangers. Most murderers are men, but the sex ratio of suicides is nearly one to one (Elwin).

Some preindustrial peoples may become preoccupied with actual or threatened suicide, e.g., the Mohave Indians of California, who rarely commit murder. The Mohave myth of the origin of death is at the same time the myth of the origin of suicide. The Mohave define a broad variety of events as suicides, some actual, others symbolic, and all are bound up with Mohave social structure, religion, and mythology: (1) stillbirths, (2) death of an infant after abrupt weaning to nurse an anticipated sibling, (3) death of one or both twins in childbirth or at any time prior to marriage, (4) symbolic suicide involving incestuous marriages between kinspersons, (5) several types of vicarious suicide, (6) funeral suicides, (7) actual suicide (Devereux, 1961, pp. 286–484). The Gisu believe suicide to be evil because it is caused by conflict between men or between men and ancestors. Suicide is contagious, so a nonrelative removes the body for pay and the suicide's evil spirit is appeased through sacrifice. Because it is usually ill feeling among people that causes suicide, though impelled by the ancestors, those closest to the dead person are blamed, and so a suicide must be prevented if possible. The threat of suicide may be used to extort concessions from relatives (La Fontaine).

Various Eskimo tribes also have recourse to suicide because of old age and a feeling that the individual's social usefulness is ended, to spare families the burden of caring for an unproductive member, to terminate a condition of serious illness, from remorse or unrequited love, in the face of starvation, and for other reasons. Generally Eskimo suicide is nonritualized and quite individualistic and variable; but among the St. Lawrence Island Eskimo, patterns are more elaborate, involve more people, and provide notice to family and kin with opportunity to prevent the act. A close relative is usually asked to act as executioner, but if he refuses the individual has to take his own life or ritually appease the spirits who already anticipate his end (Leighton and Hughes).

Balikci believes Netsilik suicide is best explained as Durkheimian "egoistic" (I would say "anomic"), resulting from the isolation and alienation of individuals attempting to deal with contradictions between integrative forces of this culture (extended kin ties, cooperation, dyadic relations) and disintegrative historical and social forces (restriction of kin alignments, jealousies and hatreds born of tight family ties but weak communal ones, unpredictably harsh spiritual beings, lack of relatedness to larger social groupings).

In the period before European contact the Iroquois and Huron Indians held an ambivalent attitude toward suicide: general hostility toward suicide as a cowardly and vengeful act that would exclude the soul from a place in the land of the dead, especially toward males who killed themselves to avoid torture after capture or responsibility for misdeeds, coupled with approval of females who committed suicide out of anger, shame, and revenge against mistreatment or betrayal by spouses or lovers. After conquest and conversion the Iroquois-Huron attitude crystallized into solid opposition toward suicide as a sin, in conformity with Christian dogma. Throughout the centuries after contact "the same motives, the same methods, and similar beliefs concerning the fate of souls prevail," so that enduring suicide patterns are a diagnostic of cultural stability for this long period (Fenton).

Cannibalism and war

Some forms of killing may not always be defined as such, in particular cannibalism and war. Cannibalistic killing may spring from a number of motives both aggressive and affectionate, but it has been widespread in human history and prehistory, and under extreme conditions it has been sanctioned in Western society. Whether cannibalism can be perceived as a stage in cultural evolution, as Sagan suggests, is difficult to say, but he nevertheless concludes that cannibalism is not correlated with level of cultural or technological complexity, and in all places where it was socially and ritually approved, there were always many individuals who were repulsed and abstained.

Vayda (p. 468) notes that "war among nonliterate peoples ranges from the hit-and-run raids and ambuscades of warriors from autonomous local communities of primitive horticulturalists or hunters and gatherers to the military campaigns carried out by the armies of such state-organized societies as the old African kingdoms and the Inca empire of the New World." Warfare, of whatever degree, though it causes death, destruction, and suffering, serves a number of important regulating functions: psychological (reduction of anxiety and aggressiveness), political (settling disputes, rebelling against authority, overcoming tyranny, regulating relations with other groups), economic (control of trade routes, distribution of goods and resources, territorial expansion, taking captives as slaves), religious (taking of captives for sacrifice or ritual cannibalism), and demographic (reduction of population pressure, room for population expansion) (ibid., pp. 468–472). In war killing is morally sanctioned, and death as a warrior is culturally honored, at times even sought.

Not all anthropologists agree with Vayda's broad conception of "primitive" warfare. Some doubt seriously that the patterned feuding, blood revenge, and raids for booty, women, slaves, or scalps in small preindustrial societies represent warfare at all so much as culturally prescribed ways of settling conflicts and dissipating tensions with relatively minimal violence. Such "wars" seem mainly symbolic in character, and, with some exceptions, result in little loss of life. Wars require, these scholars assert, a high level of social, economic, and military development and a level of technology not reached by preindustrial cultures. (For arguments on both sides see Bramson and Goethals; Fried, Harris, and Murphy.) However, the question of the moral and ethical valuation by the participants themselves of such culturally patterned infliction of death on fellow humans is known incidentally and superficially. What does seem universal is that for large-scale killing of other humans their group must be defined by the killers as strange, foreign, evil, dangerous, and/or less than human.

Feticide, infanticide, and the death of children

Most human societies have discovered methods of aborting unwanted pregnancies so that feticide is very nearly universal (Devereux, 1955). Frequently the unwanted offspring is a consequence of a casual or illegitimate procreative encounter, and in the latter case the mother may be severely punished or even killed. Penalties seem less frequent and less severe for the biological father, but where children born out of wedlock are condemned by society they may receive ridicule, physical punishment, and even death should they be permitted to come to term. On the other hand, some societies may welcome and adopt such children, and, as in rural Puerto Rico (Landy), they may grow to adulthood with as much love and care as children born of legal unions. Motives for abortions vary widely among cultures and include marital discord, adultery, jealousy among co-wives, religious and magical demands or omens, survival of older siblings in times of scarcity, rape, incest (as culturally defined), and what Devereux terms "unconscious" factors. In most preindustrial societies in cases of complicated births the decision usually is to destroy the fetus in order to save the mother's life.

Infanticide, the voluntary killing of a neonate by its parents or with their consent, should be distinguished from other forms of child death, including child sacrifice (Hocart, "Infanticide"). Infanticide, so defined, is found among many Eskimo tribes but may be absent in some. A group among whom infanticide has been reported of high frequency is the Netsilik (Balikci; Freeman; Hoebel). Most frequently it is baby girls who are killed, but paradoxically by the time of adulthood the sex ratio becomes balanced again due to the high casualty rate among male hunters in the harsh environment. Balikci (pp. 147–162) concludes that, while infanticide was ecologically adaptive by reducing population size, it was socially maladaptive and "led to an imbalance in the sex ratio, an effort to keep marriageable girls within the kinship unit, and a consequent division of the community into many small, mutually suspicious, unrelated kinship groups." However, Freeman contends that the social motive of female infanticide among the Netsilik is not simply keeping down population numbers in an environment that will not sustain them but the "assertion of male dominance within the household." An adaptive advantage is the preservation of more older persons, for not only are they more productive than infants, but their accumulated knowledge increases information storage and retrieval for the group, thereby improving both its ability to control the environment and its demographic stability (Freeman).

Infanticide not only occurs in harsh environments among hunter-gatherer cultures but has been reported from many parts of the world, including the indigenous societies of Africa and Oceania and the upper classes of China, India, Greece, and Rome (Hocart, "Infanticide") for a variety of social and cultural reasons, only some of which are related to ecology or demography. One or both twins may be destroyed when they are deemed harbingers of evil, and deformed or "weak" infants may be killed in many societies either because they are doomed by religious or social dogma to a life of tragedy or herald misfortune for their families, and/or because their economic liability is too heavy to bear for groups living on the edge of bare subsistence (ibid.).

Some cases of reported infanticide in fact seem to involve children far past infancy and ought to be more accurately classed as child murder. The death of a child from whatever cause is not regarded in many preindustrial or peasant societies with the same seriousness as the death of an adult. In rural Puerto Rico, as in many Latin countries, a small child is considered to lack full intellectual and moral capacity, to be not yet a complete person, and to exist in a state of moral purity, unblemished by the sins and corruption of adult life (Landy). The death of a child is viewed not as a calamity but as a guarantee that the child's soul will migrate directly to the company of angels, and mortuary rites are light-hearted and joyful. Among the Arapaho Indians, however, death of a child was regarded as seriously as that of an adult (Hilger). The Arapaho believed that all persons, including children, continued to live in the land of spirits after death, the only exceptions being suicides or persons who had grievously erred morally and did not become spiritually repentant. The dead child was treated as an adult and buried in the same manner, with parents slashing themselves in grief.

In a society with a high infant death rate, as the LoDagaa, children are wanted and highly valued. Paradoxically, however, when a child dies unweaned, or before the advent of a following sibling, he is assumed not to have achieved a social personality and to be not yet fully human; indeed, in this state the child represents a mystical threat to the living, a still wild being who has ventured on earth to plague its parents. Deaths of such children are greeted with consternation, and in order that they may not be reincarnated to return again and again to earth to haunt the parents they are buried quickly with scant ritual. However, after weaning a son's death is mourned with great display (Goody).

Senilicide and death of the aged

Because of the hazards of life in many subsistence economies, relatively few persons reach old age. Infant and child mortality are high, and life in early or middle adulthood may be ended suddenly and agonizingly by starvation, migration, accident, environmental cataclysm, disease, enemy assault, or sorcery. Death is a frequent and familiar guest. Where elders are respected they may also be feared for their greater knowledge and envied for superior power and privilege. Children may have more to gain than to lose by the death of a parent. The old may be seen as having lived their full measure, and as they become weakened by disease and the infirmities of age they constitute economic handicaps for younger members. In myth as in actuality attitudes toward the aged and toward their final departure carry negative as well as positive emotional valence. Still, the desire to reach mature years persists in all cultures and is reflected in beliefs and lore regarding rejuvenation; and in prayers and rituals, potions taken, counsel sought, and behavioral techniques designed to attain longevity and a place in the afterworld. The Hopi Indians pray frequently toward this goal and believe that the person who works hard and cooperatively with others, obeys his parents, and shuns conflict, trouble, and worry will achieve a ripe age (Kennard). Each individual anticipates a long term of years, but this may be interrupted by misfortune or failing to live according to Hopi ethics. When a person becomes sick and does not recover, his death is laid to his lack of strength or will to live, not to failure of the cure. Old people are expected to die soon, become increasingly disengaged from the social and religious structure, and are treated with indifference; if they persist too long they may be blamed for putting off their own demise.

The Murngin of Australia, on the other hand, venerate age. The society is rigidly age-graded with complex ceremonials marking each passage from one status to another. Women remain in a profane status throughout much of the life cycle, but men acquire degrees of sacredness, and old men are especially inviolable. They are the repositories of tribal lore and wisdom, their counsel respectfully sought, their authority dominating group decisions. The death of an older male is particularly poignant for the band as a whole, and especially his sons. Male elders stand at the apex of family, kinship, and social structure; they embody the cultural ideal and model for socialization of young males (Warner).

Older males in LoDagaa culture maintain dominance and power by clinging to wealth and social positions until they die, ensuring that their sons will work and provide for them. The death of an older man, though celebrated with full ceremony, is not a time of deep sorrow to his sons, who are waiting impatiently to inherit rights in property, women, and offices. A grandparent is provided a special burial in his own courtyard, "for it is he who has successfully established that home by producing two generations of descendants" (Goody). Since a grandparent has had a long and full existence, permanent place in the lineal social structure, and certainty of ancestorhood, "very soon after the beginning of his funeral, his joking partners begin to clown and the young women to dance, and the whole funeral is suffused with a spirit of enjoyment, rather than mourning" (ibid.).

Even societies that revere old age may desire the elderly to get on with the business of dying, so that paradoxically they may be treated with neglect and sometimes cruelty. When a nomadic group is forced to migrate, the elderly and sick are often abandoned, at times almost indifferently, but usually with some show of solicitude. Since these eventualities are traditional and anticipated by all, they are not evaded but usually accepted with equanimity by the concerned participants. Missionaries attempting to save such old persons from death frequently reported that they declined aid and confirmed their inability to be of service and the justice of their fate. Old age and infirmity may become a time of pain and travail. Simmons found that, while advanced age is a time of misery in many societies, actual abandonment is not as widespread, occurring most frequently among nomadic collectors and hunters, less so among herders and fishermen, and least in agricultural societies. Nevertheless, in societies where the elderly are honored and respected, they are more likely to prepare for their own death in a dignified, calm manner, surrounded by families and friends, treated with affection and admiration, and afforded special privileges and favors (Simmons, pp. 241–244).

What are the ecological, social, and cultural concomitants of senilicide? It seems to be positively correlated with severity of climate, residential impermanence, and irregular food supply. With increasing residential stability and food supply, real property rights, centralized authority, and codified laws, the practice tends to occur infrequently or not at all. Sympathy and respect are greatest in those cultures where the necessity seems obvious for group survival and where the elderly volunteer to die with grace and equanimity (Simmons, pp. 239–241).

Hoebel (pp. 76 ff.) attributes patterned slaughter of the elderly mainly to environmental vicissitudes and the need for group survival, but Balikci says that the reasons for senilicide (whether through the Janus-headed alternatives of homicide or suicide) are estrangement of the aged from their social networks and desperation over a now unproductive life that no longer seems worth preservation.

Implications for bioethics

This brief article points up the need to consider the bioethics of death in cross-cultural perspective. Not only are there often striking differences in the ethics of the termination of life, voluntary or involuntary, among the cultures of the world, but anthropology makes it obvious that these differences in values and practices surrounding death in any society can be understood only within the context of that society's ecological and demographic situation, social system, and cultural beliefs. Killing the sick, the infirm, and the aged may be not only an act of mercy but an economic necessity if the group is to survive, although, as I have suggested, a consistent food supply makes this form of euthanasia less probable. Infanticide poses painful ethical problems in industrialized societies but may be imperative for groups living on the edge of subsistence. As I have shown, homicide and suicide become socially sanctioned under certain cultural-historical conditions.

An anthropological view of the bioethics of death suggests that ethical judgments of the attitudes and actions regarding death in other societies require ethnographic and historical knowledge, and even so should be made with extreme caution. It also suggests that such knowledge provides a profound and essential framework in which to cope with the ethical dilemmas that confront our own society.

DAVID LANDY

[*For concepts of death in other societies and cultures, see the other articles in this entry:* EASTERN THOUGHT, WESTERN PHILOSOPHICAL THOUGHT, WESTERN RELIGIOUS THOUGHT, *and* DEATH IN THE WESTERN WORLD. *For further discussion of topics mentioned in this article, see the entries:* AGING AND THE AGED, *articles on* SOCIAL IMPLICATIONS OF AGING *and* ETHICAL IMPLICATIONS IN AGING; *and* SUICIDE. *For discussion of related ideas, see the entries* DEATH AND DYING: EUTHANASIA AND SUSTAINING LIFE, *article on* ETHICAL VIEWS; INFANTS, *articles on* ETHICAL PERSPECTIVES ON THE CARE OF INFANTS, *and* INFANTICIDE: A PHILOSOPHICAL PERSPECTIVE. *See also:* ABORTION; DEATH, ATTITUDES TOWARD; *and* LIFE.]

BIBLIOGRAPHY

BALIKCI, ASEN. *The Netsilik Eskimo.* Garden City, N.Y.: Natural History Press, 1970. Recent study of well-known Eskimo tribe, with thoughtful material on death, homicide, suicide, infanticide, and senilicide.

BENDANN, EFFIE. *Death Customs: An Analytical Study of Burial Rites.* The History of Civilization. Edited by C. K. Ogden. New York: Alfred A. Knopf; London: Kegan Paul, Trench, Trubner & Co., 1930. London: Dawsons of Pall Mall, 1969. Ann Arbor, Mich.: Gryphon Books, 1971. Useful attempt to explain cross-cultural similarities and differences in beliefs and practices regarding death.

BOHANNAN, PAUL, ed. *African Homicide and Suicide.* Princeton: Princeton University Press, 1960. Specifically referred to in this article are Bohannan's "Introduction," "Theories of Homicide and Suicide," and "Homicide Among the Tiv of Central Nigeria." Pioneering anthology of research among tribal groups in Nigeria, Uganda, Kenya, and Tanzania.

BRAMSON, LEON, and GOETHALS, GEORGE W., eds. *War: Studies from Psychology, Sociology, Anthropology.* New York: Basic Books, 1964. Relevant pieces by W. G. Sumner, R. E. Park, B. Malinowski, M. Mead, and J. Schneider.

DEVEREUX, GEORGE. *Mohave Ethnopsychiatry and Suicide: The Psychiatric Knowledge and the Psychic Disturbances of an Indian Tribe.* Bureau of American Ethnology Bulletin no. 175. Washington: Government Printing Office, 1961. Includes multifaceted analysis and rich documentation with case studies of suicide.

———. *A Study of Abortion in Primitive Societies: A Typological, Distributional, and Dynamic Analysis of the Prevention of Birth in 400 Preindustrial Societies.* New York: Julian Press, 1955. Unique, but non-statistical, cross-cultural study of 400 preindustrial societies.

DURKHEIM, EMILE. *The Elementary Forms of the Religious Life: A Study in Religious Sociology.* Translated by Joseph Ward Swain. London: George Allen & Unwin, 1950. New York: Free Press, 1965. Classic monograph purporting to show origins of religious systems reflected in those of Australian Aborigines.

———. *Suicide: A Study in Sociology.* Translated by John A. Spaulding and George Simpson. Glencoe, Ill.: Free Press, 1897, 1930, 1951. Using his own methodological principles, Durkheim analyzes suicide trends and sociological concomitants, and proposes a theory of suicide.

ELWIN, VERRIER. *Maria Murder and Suicide.* Foreword by W. V. Grigson. London: Oxford University Press, 1943, 1950. Rich data, generally well organized, occasionally marred by anachronistic comments on tribal society.

FENTON, WILLIAM NELSON. "Iroquois Suicide: A Study in the Stability of a Culture Pattern." *Anthropological Papers.* Bureau of American Ethnology Bulletin no. 128. Washington: Government Printing Office, 1941, pp. 79–137.

FRAZER, JAMES GEORGE. *The Belief in Immortality and the Worship of the Dead.* 3 vols. London: Macmillan & Co., 1913–1922. Vol. 1 contains Frazer's major ideas on the topic.

———. *Man, God, and Immortality: Thoughts on Human Progress.* New York: Macmillan Co., 1927. Excerpts from Frazer's huge corpus, with much related to death and immortality in preindustrial societies.

FREEMAN, MILTON M. R. "A Social and Ecologic Analysis of Systemic Female Infanticide among the Netsilik Eskimo." *American Anthropologist* 73 (1971): 1011–1018.

FRIED, MORTON; HARRIS, MARVIN; and MURPHY, ROBERT,

eds. *War: The Anthropology of Armed Conflict and Aggression.* Garden City, N.Y.: Natural History Press, 1968. An interesting symposium of anthropologists' views.

GOODY, JOHN RANKINE. *Death, Property and the Ancestors: A Study of the Mortuary Customs of the LoDagaa of West Africa.* Stanford: Stanford University Press; London: Tavistock Publications, 1962. Undoubtedly the finest anthropological analysis and documentation of social-cultural processes and interrelations between death and society.

HARTLAND, EDWIN SIDNEY. "Death and Disposal of the Dead." *Encyclopedia of Religion and Ethics.* 12 vols. Edited by James Hastings. New York: Charles Scribners' Sons, 1912, vol. 4, pp. 411–444. Useful anthropological analysis despite anachronisms.

HERTZ, ROBERT. *Death and the Right Hand.* Translated by Rodney and Claudia Needham. Introduction by E. E. Evans-Pritchard. Glencoe, Ill.: Free Press, 1960. Translated in part from "La Représentation collective de la mort." *Mélange de sociologie religieuse et folklore.* Paris: F. Alcan, 1928. Original appearance. *Contribution à une étude sur la répresentation collective de la mort.* Paris: F. Alcan, n.d. Also, *L'Annee sociologique,* 1907, and *Revue philosophique,* 1909. Paris: Press Universitaires de France. Seminal essay on cultural and social forces underlying multiple funeral practices.

HILGER, M. INEZ. *Arapaho Child Life and Its Cultural Background.* Bureau of American Ethnology Bulletin no. 148. Washington: Government Printing Office, 1952. Includes beliefs and practices concerning death of children.

HOCART, ARTHUR MAURICE. "Death Customs." *Encyclopaedia of the Social Sciences.* Edited by Dewin R. A. Seligman and Alvin Johnson. New York: Macmillan Co., 1937, vol. 5, pp. 21–27.

———. "Infanticide." *Encyclopaedia of the Social Sciences.* Edited by Dewin R. A. Seligman and Alvin Johnson. New York: Macmillan Co., 1937, vol. 8, pp. 27–28.

HOEBEL, EDWARD ADAMSON. *The Law of Primitive Man: A Study in Comparative Legal Dynamics.* Harvard University Press, 1954; New York: Atheneum Publishers, 1973. Contains useful material on cultural-legal implications of homicide, suicide, etc.

"Infanticide." *The New Encyclopaedia Britannica: Micropaedia.* Chicago: Encyclopaedia Britannica, 1974, vol. 5, p. 350.

JEFFREYS, M. D. W. "Samsonic Suicides: Or Suicides of Revenge among Africans." *The Sociology of Suicide: A Selection of Readings.* Edited by Anthony Giddens. London: Frank Cass & Co., 1971, chap. 13, pp. 185–196.

KENNARD, EDWARD A. "Hopi Reactions to Death." *American Anthropologist* 39 (1937): 491–496.

LA FONTAINE, JEAN. "Homicide and Suicide among the Gisu." Bohannan, *African Homicide and Suicide,* pp. 94–129.

LANDY, DAVID. *Tropical Childhood: Cultural Transmission and Learning in a Puerto Rican Village.* Chapel Hill: University of North Carolina Press, 1959. New York: Harper & Row, 1965.

LEIGHTON, ALEXANDER H., and HUGHES, CHARLES C. "Notes on Eskimo Suicide." *The Sociology of Suicide: A Selection of Readings.* Edited by Anthony Giddens. London: Frank Cass & Co., 1971, chap. 13, pp. 158–169.

MALINOWSKI, BRONISLAW. *Magic, Science and Religion and Other Essays.* Introduction by Robert Redfield. Glencoe, Ill.: Free Press, 1948. Garden City, N.Y.: Doubleday & Co., 1954. In addition to title essay, also "Myth in Primitive Psychology" and "Baloma: the Spirits of the Dead," all apposite to the anthropology of death.

NASH, JUNE. "Death as a Way of Life: The Increasing Resort to Homicide in a Maya Indian Community." *American Anthropologist* 69 (1967): 455–470.

SAGAN, ELI. *Cannibalism: Human Aggression and Cultural Form.* Foreword by Robert N. Bellah. New York: Harper & Row, 1974. Cannibalism seen as step in cultural evolution.

SIMMONS, LEO WILLIAM, *The Role of the Aged in Primitive Society.* New Haven: Yale University Press, 1945. New York: Archon Books, 1970. Excellent cross-cultural study using Human Relations Area Files.

TURNER, VICTOR WITTER. *The Ritual Process: Structure and Anti-Structure.* Chicago: Aldine Publishing Co.; London: Routledge & K. Paul, 1969. Turner's ideas on symbolism in ritual with a focus on death and life rituals and development of Van Gennep's notion of liminality in transition states.

TYLOR, EDWARD BURNETT. *Religion in Primitive Culture.* Introduction by Paul Radin. Library of Religion and Culture. New York: Harper & Brothers, 1968. Original appearance. *Primitive Culture: Researches into the Development of Mythology, Philosophy, Religion, Art and Custom.* London: J. Murray, 1871. Reprint. New York: Gordon Press, 1974, vol. 1, chap. 11, pp. 377–453; vol. 2, chaps. 12–19, pp. 1–410. Deals mainly with Tylor's theory of animism as foundation of "primitive" religion with much on attitudes and beliefs regarding death.

VAN GENNEP, ARNOLD. *The Rites of Passage.* Translated by Monika B. Vizedom and Gabrielle L. Caffee. Introduction by Solon T. Kimball. Chicago: University of Chicago Press, 1960. Original appearance. *Les Rites de passage: Étude systematique des rites de la porte et du seuil, de l'hospitalité, de l'adoption, de la grossesse et de l'accouchement, de la naissance, de l'enfance, de la puberté, de l'initation, de l'ordination, du couronnement des fiançailles et du mariage, des funérailles, des saisons, etc.* Paris: É. Nourry, 1909. Classic essay on social and cultural functions of ceremonies of transition, including birth and death.

VAYDA, ANDREW P. "Primitive Warfare." *International Encyclopedia of the Social Sciences.* Edited by David L. Sills. New York: Macmillan Co., Free Press, 1968, vol. 16, pp. 468–472. Concentrates on functions of war, assumes it to be universal.

WARNER, WILLIAM LLOYD. *A Black Civilization: A Study of an Australian Tribe.* New York: Harper & Brothers, 1937. Rev. ed. 1958. Richly documented ethnography of the Murngin tribe, focus on religion and social structure, much on death.

II
EASTERN THOUGHT

The enigmas of dying and death have been a serious preoccupation in the philosophical and religious thought of all peoples, not least the peoples of the "East." In each of the great traditions men have sought to understand and develop appropriate responses to the reality of death as it impinges on the life of each individual human being. They have grappled with the problem of communal continuity in the face of the inevitable intrusions of death. And they have sought to identify the place and function of death in the cosmic order, however that order may be defined.

In Eastern thought, as in human thought more generally, a number of themes emerge from man's encounter with death which cut across all geographical, cultural, and social divisions. So, too, particular kinds of groups in widely different contexts share common understandings of death and its meaning; for example, some emphases are common to the warrior and ascetic elements in the poulation all across classical Asia, whereas other emphases are broadly disseminated among peasant communities. However, for our purposes it will be most appropriate to treat the subject matter in terms of three basic historical traditions—the Sinitic tradition including both its Confucian and Taoist strands, the Vedic–Hindu tradition, and the Buddhist tradition.

Death and Sinitic thought

Many commentators on Chinese thought have noted its distinctively "this-worldly" cast and the emphasis it has traditionally placed on the specifically human mode of existence. Thus it is not surprising that any premature or "unnatural" death was considered to be an aberration; according to the tradition such a death not only prevented the natural fulfillment of the individual involved but also, in so doing, produced an evil, demonic spirit, which posed a continuing threat to the peace and well-being of the community. From the time of the early Chou dynasty (ca. 1050–222 B.C.) we have clear evidence that the avoidance of such a premature death and the attainment of longevity were preeminent concerns. In the supplicatory prayers that have been preserved from this very ancient time, longevity is the boon which is sought after more commonly and more ardently than any other. Moreover, in spite of a deeply embedded popular and to some extent Confucian tendency toward fatalism concerning the timing of death (and many other things as well), the goal of longevity remained a primary focus of attention throughout all of the later periods of Chinese history. For example, in certain of the classical funerary customs of China strongly emotional expressions of mourning were considered to be completely appropriate in cases where the deceased was a younger person; on the other hand, the appropriate response at the funeral of an individual who had attained the age of seventy or more was thought to be celebration of a life that had been lived to the full.

Among some segments of the population in China this continuing concern with the postponement of death and its effects was pushed much farther. In the period of the Shang dynasty (ca. 1500–1050 B.C.) burial remains clearly indicate the presence of the conviction that a thoroughly "this-worldly" mode of existence continued beyond the grave, at least for the elite who could follow the appropriate burial prescriptions. During the period of the early Chou dynasty a rather different extension of the conceptions concerning longevity is reflected in prayers in which men sought to escape from the ravages of old age and to postpone their death into an indefinite future. Moreover the emphases on "the preservation of the body," "long life," and "no death," which were established in these very early times, persisted as important strands throughout the subsequent history of traditional Chinese thought.

Still another dimension of the Chinese concern to avoid or to overcome the implications of death appears in specifically Taoist circles around the fourth century B.C. with the emergence (or possibly the intrusion from India) of the belief that a new, truly immortal mode of being could be attained. At first the attainment of such a radically new mode of being was associated exclusively with certain reclusive sages (*hsien*) who retired to mountain areas, practiced a highly disciplined life, and learned to live in harmony with the processes of nature and ultimately to control them. However, in subsequent centuries the belief that this new, deathless mode of being could be acquired through the transmutation of one's ordinary physical body into an indestructible subtle body became a goal avidly sought by many members of the elite, including a number of important emperors. Thus during the period of the Han dynasty (206 B.C.–A.D. 220) a widespread cult of immortality came into being in which various esoteric sacrifices, drugs, purifying diets, and the like were utilized in an effort to achieve the desired result. But it should also be noted that, as a part of this same process of popularization, the goal of true immortality soon became inextricably fused with the earlier and still persisting concern for the extension of more normal patterns of worldly and social existence.

Despite the strong emphasis the Chinese have

traditionally placed on avoiding or overcoming the implications of death, and despite the reluctance of many of the literati (including, most notably, Confucius himself) to speculate in any detail on the fate of the dead, most Chinese thought has recognized death as the inevitable dénouement of human life and has understood it as a transformation into new though still limited modes of existence. Generally speaking, in classical Chinese thought death has been viewed as the dispersion of the complex of souls (often two—one associated with the yang [male] principle and one with the yin [female] principle—but in some cases more) and body that constitutes each individual human being. Moreover the Chinese have viewed this dispersion as a process through which men are transferred into the role of an ancestor and, at the same time, enter into a postmortem existence conceived in terms of either heavenly enjoyment or the torments of hell. Certainly a man's postmortem condition as an ancestor, along with the resulting extension of the system of family reciprocity and filial piety which have traditionally been at the very core of practically every aspect of Chinese culture from the period of the earliest dynasties up to the twentieth century, represents the more ancient, the more pervasive, and the more comforting of the two conceptions. Nevertheless, it is also true that at least during the long medieval period of Chinese history large segments of the Chinese population were convinced that their death would also result in a postmortem existence in which their virtues and piety would be rewarded or (much more likely) their various sins and transgressions would be punished.

Though, as we have already suggested, most traditional Chinese believed that the souls of the individual maintained an existence after death, many of the more profound thinkers realized that the phenomena of continuation, even if it were true, constituted no final solution to the problem of death. Rather, they recognized that the various souls, like the body with which they had been associated, must sooner or later return to the natural cosmic source from which they had come. As one might expect given the basically positive attitude the Chinese have always taken toward what they have conceived to be the natural cosmic source of life (the Tao), those thinkers who reached such a conclusion seldom evaluated this final dissolution in a totally negative or morbid way. But, on the other hand, there were clearly significant differences in the attitudes taken by various sages and traditions. For the most part the Confucians did not place great emphasis on the process, and when they did consider it they exhibited an attitude characterized more by equanimity than by enthusiasm. On the other hand, in the thought of many of the more philosophically oriented Taoists an appreciation of the process of the individual's return to the Tao was cultivated as the essence of wisdom, and its full realization was equated with the attainment of salvation.

Vedic–Hindu conceptions of death

In India, as in China, any serious discussion of traditional conceptions related to death must take account of the primacy that has consistently been given to the goal of maintaining this-worldly existence. In the Vedas, most of which were composed in the second millennium B.C. and constitute our earliest source of Indian thought, the positive value of vigorous, worldly living is strongly emphasized, and the quest for longevity through prayer and sacrifice is a predominant motif. In fact in several passages a full life of one hundred years is equated with the attainment of immortality. This same basic concern for a full life prior to death is reflected in a very different way in the later *ashrama* ideal of the orthodox Brahmanic tradition according to which each individual (at least each individual included within the true "Aryan" community constituted by the three upper castes of Hindu society) should pass through a complete cycle of adult activity in which a period of education and a period of social responsibility as a householder are followed by two periods of increasingly austere preparations for death. In addition, the mythology, ritual, and devotion of later Hinduism clearly reflect the concern for longevity; for example, in the mythic tradition there are numerous stories that give expression to the belief that old age and death can be averted and youth magically restored through various means including living in a particular holy place, the aid of a saint, an exchange of life with another mortal, or the hearing of an especially sacred text. Nevertheless, in spite of the hope that longevity might be achieved and even that the restoration of youth might occasionally be possible, most Indians have adopted a more fatalistic attitude toward the arbitrariness and unpredictability of death, often associating it with retribution for their own deeds in previous existences (karma). Moreover, throughout the long history of Indian philosophy and religion, conceptions concerning man's destiny after death have played a very significant role.

Already in Vedic times the belief that death involves a transition through which the individual takes on the role of an ancestor was firmly established; and in the later Brahmanic–Hindu tradition this belief came to be expressed in a highly elaborated series of funerary ceremonies through which the success of the transition is assured and the proper reciprocal relationships between the ancestral community and the continuing family unit are established. In addition to this ancestral role, the fate of the deceased has been perceived in a variety of other ways as well. In the earlier Vedas the dominant view seems to have been that, for those who act in accordance with the universal cosmic law (*rta*), death will constitute a transition into a realm not essentially different from the land of the living—a realm in which the individual's bodily existence, social relationships, and worldly enjoyments will be restored. In the period of the Brahmanas (commentaries on the Vedas dated ca. 950–700 B.C.) the conviction developed that through his ritual action a man could actually create for himself the postmortem "body" or "world," which he would inhabit after his death. Beyond this, some of the more profound priestly writers speculated that the various mystical correlations made in the ritual between the elements of a sacrificer's being and the elements of the sacrificial altar which he built, and between those same elements of the sacrificial altar and various aspects of the cosmos, assured the sacrificer that the "immortal" aspects of his being (his eye, his breath, etc.) would survive the decay of the gross elements of his body and rejoin the cosmic elements to which they corresponded (the sun, wind, etc.). In the context of Hinduism proper, beliefs in the possibility of a blissful postmortem existence have also been very much to the fore, but important differences from the earlier Vedic tradition are evident. For example, the great theistic sects have strongly affirmed the possibility of a heavenly afterlife, but they generally held that the attainment of such a goal is dependent not upon man's action (either moral or ritual), but rather on the grace of a preeminent deity such as Vishnu, Śiva or the Goddess. On the more popular level it has commonly been believed that those whose mind is focused on a proper thought at the moment of death will attain a heavenly rebirth; and there has been (and still is) a widespread conviction in India that a heavenly destiny is assured to anyone who dies in an appropriate holy place—notably at Benares—and especially to anyone whose ashes are purified by the sacred Ganges River.

In India, however, more somber conceptions concerning man's destiny appeared at a very early date and ultimately came to dominate much of the Hindu ethos. Even in the Vedas (especially in the fourth or Atharva Veda) there are passages affirming that death may be followed by an afterlife of suffering and torment; in the Brahmanas the possibility of a terrifying "second death" is raised; and in the much later texts of medieval Hinduism a great deal of emphasis is placed on the existence of various hells in which men suffer the results of their earthly transgressions. But it is more important to emphasize the fact that in the period which immediately followed the completion of the Brahmanas a new conception emerged in which death came to be associated with a kind of suffering that posed a more basic threat than any of the hells the Indians were able to conceive. According to this new and extremely influential conception, which came to the fore in texts known as the Upanishads (600 B.C.–?) and was later expressed in different ways in the various orthodox traditions of medieval Hindu philosophy, the core of each man's being is constituted by an indestructible soul or self (*atman*), which, through the power of ignorance and/or the desire for bodily attachment, has become entrapped in an ongoing cycle of death and rebirth. Though the structure of this cycle of transmigration is conceived in such a way that men are able to enjoy limited satisfactions (including the possibility of attaining more advantageous rebirths through proper ritual practice or moral action), existence within it is seen as devoid of any ultimate meaning and is therefore experienced as a hopeless process of indefinitely extended suffering and dying.

During the past 2,500 years, the most authoritative and esteemed sages and thinkers of classical India have viewed death in this context —that is to say, they have viewed it as *the* pervasive fact of the cosmic existence in which men are imprisoned. But these same sages and thinkers have also insisted that death, when it is viewed from a slightly different point of view, may serve as a paradigm for a truly soteriological experience of Release; it is at the point of death, they have noted, that the soul which has been dragged into the cycle of transmigration gains, for a very brief moment, a certain separation from the grosser elements of psyche and body that hold it in bondage. Thus, at this most profound level of Indian thought, death comes to serve as a prefiguration and model for various meditations and disciplines (classical yoga, devotional disciplines, tantra, etc.) through which the ties that bind man's self or soul to cosmic impermanence can be completely broken, and through which the ultimate soteriological goals of immortality and freedom can be finally and definitively attained.

Buddhist conceptions of death

Though Buddhist conceptions of death can never be truly understood apart from the Indian ethos in the last half of the first millennium B.C. within which Buddhism emerged, they are intimately bound up with a religious and philosophical perspective that cannot be interpreted simply in terms of Vedic or Hindu modes of thought. Moreover, in the process of Buddhism's establishment and development in other parts of Asia many of its specifically Indian emphases have been profoundly influenced and even transformed through their interaction with a variety of indigenous traditions.

At the more popular level, Buddhists have shared the common human concern that death be postponed as long as possible. And contrary to some stereotyped views of its strictly "otherworldly" character, Buddhism has always provided a framework within which such a goal has a limited place and has provided a variety of means for its adherents to use to achieve it. In the Buddhist perspective it was generally assumed that the deeds that one had done in one's previous lives were the primary determining factor in establishing not only the basic conditions of one's present life but also its duration (thus generating a kind of "fatalism" that has been observed in the attitudes of many Buddhist peoples); but Buddhist teaching has also affirmed that good and pious deeds done in one's present life can help not only to improve the quality of that life but also to extend its length. In addition, in practically all Buddhist traditions special forms of chanting and other "magical" practices have been developed to ward off various evil influences, including death. In the later Mahayana (Great Vehicle) phases of Buddhist history, beginning shortly after 100 B.C., proper devotion came to be recognized as an efficacious means for the attainment of many boons, including longevity; and in many esoteric Buddhist traditions, which first began to appear about A.D. 350, longevity emerged as one of the many more mundane goals sought through the cultivation of spiritual and magical power. Perhaps the ultimate extension of this Buddhist quest for longevity was reached many centuries later in Japan, when under the influence of the esoteric Shingon school a tradition developed in which a few very determined ascetics strove to attain permanent longevity through a highly elaborate process of self-mummification.

Since very early Buddhist times the great majority of Buddhist faithful have held firmly to the belief that when death comes it leads to rebirth in one of the many heavenly realms that extend above the earth, in a favorable or unfavorable condition in the human world, or in one of the four lower worlds of suffering and woe. For a very few practitioners of the higher forms of meditation this meant rebirth in the highest heavenly realms, which corresponded to the various stages of meditational trance. But for the great majority of men the mode of rebirth was thought to be determined by the balance between a man's good and pious deeds on the one hand, and his evil or impious deeds on the other. Though great emphasis was often placed on the importance of a man's last thought before death in determining the character of his next existence, most Buddhist thinkers have been careful to make clear their view that this thought was itself the result of what the individual had believed and done during his lifetime.

Throughout Buddhist history the idea of rebirth has continued to play a crucial role, but the conceptions related to this idea have undergone a number of significant changes. At a very early point in Buddhist history the original belief that one's own good and evil deeds were the sole factor in determining the quality of one's future existence was compromised by the conception that the fate of a deceased individual might be affected by merit that had been gained through the pious deeds of a living relative and then dedicated to his benefit. Moreover, this view made an important contribution to the development of a very rich tradition of Buddhist funerary practices including various "masses" for the dead, which Chinese Buddhists developed into highly elaborate rituals. In the later, especially Mahayana, traditions devotion to a preeminent Buddha or Bodhisattva (a future Buddha who had postponed his entry into Nirvana in order to work for the salvation of all beings) replaced moral or cultic action as the dominant motif, at least as far as access to a favorable rebirth was concerned. This development was extended even farther in more recent, especially Japanese, schools of Buddhism in which the effective force in attaining a favorable rebirth has come to be seen not as the devotion itself but rather as the infinite compassion of the Buddha or Bodhisattva to whom the devotion is directed. Finally, it is important to note that in these later forms of devotional Buddhism the attainment of a paradisal rebirth in the realm of a great Buddha or Bodhisattva came to be more or less equated with the attainment of Final Release.

It is nevertheless true that most Buddhist thinkers from the time of the Buddha onward have shared the more somber attitudes toward death that characterized the views of the Upanishadic sages and the philosophers of orthodox Hinduism. That is to say, they have viewed death as a constitutive flaw that is endemic to all forms of this-worldly existence, including even those associated with the highest and most paradisal heavens. In fact, various meditational practices have been developed in order to implant this conception deeply into the consciousness of the more serious Buddhist adherents; for example, the contemplation of corpses and their decomposition has been practiced from very early Buddhist times, and contemplations that continually focus attention on the onset of one's own death have been developed by Buddhists who had close associations with the Samurai warrior tradition in Japan. What is more, Buddhists have also shared the Upanishadic and Hindu conception that death may serve as a prefiguration and model of the complete break with worldly existence which can bring Final Rest and Release (Nirvana). However, in this connection it is crucial to emphasize that, unlike other Indian and many other soteriologies, the path for the attainment of Release that the Buddhists have propounded does not depend on belief in any kind of soul or self that has suffered bondage and must regain its state of pristine purity. Quite to the contrary, it involves the conviction that any such belief is a delusion which breeds attachment and therefore results in a continuation of the ongoing cycle of rebirth, suffering, and death; or, to state the matter more positively, it is through a thoroughly appropriated insight into the truth that there is no soul or self of any kind that attachment, suffering, and death may be overcome and Release attained.

In the course of Buddhist history there have been many variations on the themes of man's bondage to death and the nature of Release. In the early tradition suffering and death are seen as the inevitable results of actions motivated by ignorance (i.e., any belief in the existence of a soul or self) and craving or attachment; the conquest of death and the attainment of Nirvana are, on the other hand, equated with the practice of an Aryan or Noble Path, which is a radically distinct mode of action motivated by wisdom (the insight into the self-less character of reality) and compassion. In the Mahayana tradition there is a significant shift in emphasis, and the basic distinction becomes primarily epistemological; through an insight into the truth that all reality is self-less and void one realizes that in the final analysis Nirvana and the cycle of transmigration (including death) are one and the same and cannot be distinguished. When this insight is fully appropriated, death is conquered and true Release is experienced. As this Mahayana perspective became established in East Asia and was gradually adapted to indigenous, particularly Taoist, modes of thought, still further changes occurred. Thus many of the most important Chinese Buddhist thinkers came to insist that the bondage of death could be finally overcome only when men awakened and recognized in themselves the Buddha-nature, which, in this new context, had come to be viewed as the true source of both individual and cosmic life. And finally, in Japan, this process of adaptation was carried to its limit in the works of Saigyo and Basho and other classical poets who attributed man's bondage to suffering and death to his rejection of the natural world. For those Japanese writers the awakening to one's own Buddha-nature and the attainment of a complete harmony with the rhythms and transformations of the natural world were two sides of exactly the same process —and, what is more, both were taken to be absolutely identical with the attainment of Release.

Conclusion

The full range of Eastern thought concerning death has by no means been exhausted in this discussion of certain aspects of the Sinitic, Vedic–Hindu, and Buddhist perspectives. Important and quite distinct traditions that have been touched upon only briefly or not discussed at all include, among others, the Maoist in China, the Ajivaka and Jain in India, the Shinto in Japan, and the Bon Po in Tibet, as well as various forms of Islam and Christianity that are firmly established in several different Asian contexts. However, perhaps enough has been said to suggest something of the diversity, richness, and depth of the thought that has emerged as Eastern men have struggled with the enigmas and paradoxes of death.

Obviously there is no single "Eastern" way of formulating the problematics of death, or of resolving the ethical issues it poses. However, when the various Eastern conceptions are set over against the dominant attitudes that have developed in the modern West two points are worth emphasizing. First, the various Eastern views place a strong emphasis on the fact that meaningful efforts to extend human life expectancy must involve not only the utilization of medical procedures and techniques, but also the development of moral attitudes and styles of life that place men in harmony with the reality and rhythms of the universe in which they live. Second, the various Eastern views (with the exception of certain Taoist traditions which held out the hope that death could be permanently postponed) emphasize the importance of developing and cultivating disciplines, meditations, and insights which have the capacity to rob death of its destructive power and in some cases even to transform it into a positive opportunity for human fulfillment. Perhaps in our modern struggle against death we have now reached a point at which it would be both possible and advisable to take these traditional approaches and conceptions more seriously into account.

FRANK E. REYNOLDS

[*Other relevant material may be found under* PAIN AND SUFFERING, *article on* RELIGIOUS PERSPECTIVES; SUICIDE. *For discussion of related ideas, see:* BUDDHISM; CONFUCIANISM; HINDUISM; *and* TAOISM. *See also:* MEDICAL ETHICS, HISTORY OF, *section on* SOUTH AND EAST ASIA, *articles on* INDIA, PREREPUBLICAN CHINA, *and* CONTEMPORARY CHINA.]

BIBLIOGRAPHY

"Death and Disposal of the Dead." *Encyclopaedia of Religion and Ethics.* Edited by James Hastings. New York: Charles Scribner's Sons; Edinburgh: T. & T. Clark, 1912, vol. 4, pp. 411–511. Those who are interested in the conceptions reflected in the communal practices related to dying and death which have been largely bypassed in our own discussion should begin by consulting, for India and China, the relevant articles under the above general topic. For Buddhist communal practices in various areas, see pp. 485–497 on Japanese customs, and the works by Tambiah and Welch.

FREMANTLE, FRANCESCA, and TRUNGPA, CHÖGYAM, trans. *The Tibetan Book of the Dead: The Liberation through Hearing in the Bardo.* Clear Light Series. Berkeley: Shambhala Publications, 1975. A fascinating primary text which describes in detail the states through which, according to the Tibetan Buddhist conception, every individual passes in his journey from death to rebirth.

HOLCK, FREDERICK H., ed. *Death and Eastern Thought: Understanding Death in Eastern Religions and Philosophies.* Nashville, Tenn.: Abingdon Press, 1974. This collection of essays is very useful for gaining a further orientation to the topic. It contains five essays on various aspects of the Indian tradition (including one that focuses primarily on Buddhism), one dealing with China, and one on Japan in which specific attention is given to certain aspects of Japanese Buddhism. For the books that provide an overall orientation see Reynolds and Waugh, and Lemaître.

JAN YÜN-HUA. "Buddhist Self-Immolation in Medieval China." *History of Religions* 4 (1964): 243–268.

LAMOTTE, ÉTIENNE. "Le Suicide religieux dans le bouddhisme ancien." *Academie royale des sciences, des lettres, et des beauxarts de Belgique: Bulletin de la classe des lettres et des sciences morales et politiques* 5th ser., 51 (1965): 156–168.

LEE, JUNG YOUNG. *Death and Beyond in the Eastern Perspective.* New ed. New York: Gordon & Breach Science Publishers, 1974. An introductory book which should be consulted only with the greatest caution.

LEMAÎTRE, SOLANGE. *Le Mystère de la mort dans les religions d'Asie.* 2d ed. rev. & corr. Paris: Adren Maisonneuve, 1963. An introductory book which covers several major areas in very general terms.

NEEDHAM, JOSEPH. "The Drug of Deathlessness; Macrobiotics and Immortality-Theory in East and West." *Science and Civilization in China.* Vol. 5: *Chemistry and Chemical Technology.* Pt. 2: *Spagyrical Discovery and Invention: Magisteries of Gold and Immortality.* London: Cambridge University Press, 1974, pp. 71–127.

REYNOLDS, FRANK E., and WAUGH, EARLE H. *Religious Encounters with Death: Insights from the History and Anthropology of Religion.* University Park: Pennsylvania State University Press, 1977. This broadly crosscultural collection contains fifteen essays, five of which concern "Eastern" traditions (three on Hinduism, one on Indian Zoroastrianism and one on Japanese Buddhist meditational techniques which emphasize the practice of dying).

SMITH, D. HOWARD. "Chinese Concepts of the Soul." *Numen* 5 (1958): 165–179.

TAMBIAH, S. J. "Death, Mortuary Rites and the Path to Rebirth." *Buddhism and the Spirit Cults in North-East Thailand.* Cambridge Studies in Social Anthropology, no. 2. London: Cambridge University Press, 1970, pp. 179–194.

WELCH, HOLMES. "Rites for the Dead." *The Practice of Chinese Buddhism: 1900–1950.* Harvard East Asian Studies, no. 26. Cambridge: Harvard University Press, 1967, pp. 179–205.

YÜ, YING-SHIH. "Life and Immortality in the Mind of Han China." *Harvard Journal of Asiatic Studies* 25 (1964–1965): 80–122.

III

WESTERN PHILOSOPHICAL THOUGHT

Philosophers, presumably like other humans, have always been aware of death, though the extent of their interest and concern has varied from individual to individual and from age to age. Special problems connected with death have arisen from time to time, though some—such as the question of survival and an afterlife—have persisted in various cultures. Not only individual differences but the emphases and assumptions of established traditions and the cultural atmosphere of which successive philosophical views have been a part have inevitably influenced thoughts about death. Changes in social, political, and economic conditions and in religious, artistic, and scientific outlook have thus affected attitudes toward death.

In Western philosophical thought six such cultural periods can usefully be distinguished: (1) ancient Greek philosophy, comprising (a) the pre-Socratics as background to (b) the classic philosophies of Socrates, Plato, and Aristotle; (2) Hellenistic, that is, Greco-Roman thought, especially the Epicurean, Stoic, and other schools of the post-Aristotelian ancient world and also contributions of Hebrew–Christian traditions that entered the stream of Western philosophy during this period; (3) medieval thought of both (a) the early Middle Ages and (b) from the eleventh to the fifteenth century, including the various Renaissance movements of the period; (4) early modern philosophy of the sixteenth and seventeenth centuries with their social, religious, and scientific revolutions and reformations; (5) philosophy in the eighteenth, nineteenth, and early twentieth centuries, the periods of "Enlightenment" and "Romanticism," of pre-Darwinian and evolutionary science; and (6) recent and contemporary philosophical thought in its positivistic and pragmatic critiques and in its existentialist emphasis on problems associated with death.

Ancient Greek philosophy

The pre-Socratics. In Jacques Choron's *Death and Western Thought*, which provides a partial basis for this article, the author writes in discussing the contemporary German philosopher Martin Heidegger (1899–1976): "Perhaps the new answer to death to which Heidegger is being led is the consoling certitude of the harmony between human existence and the ground of Being, which he believes Western man has not known or felt since the Pre-Socratics" (Choron, p. 240).

Harmony between man and nature (*physis*) was for the pre-Socratics an unquestioned assumption, not as for Heidegger and other moderns the outcome of subtle and sophisticated argument. Moreover the "harmony" assumed by the pre-Socratics was variously expressed in the references and quotations that survive in later writings and constitute the basis of knowledge of their views. Awareness of continuity and change pervades the thought of pre-Socratic scientist-philosophers. Their observation of the natural order, and of human life as part of it, sometimes leads to emphasis on elements that are permanent in the midst of change, sometimes to a sense of universal mutability. It is significant that those Greeks of the sixth and fifth centuries B.C. lived in outlying areas of the Greek world, in Asia Minor and in what today is Italy, though one can only speculate about the influence of that circumstance. What other factors led to various interpretations of continuity and change remain unknown: Parmenides stressed the fundamentally unchanging order of things; Heraclitus, the constant change in everything except the *Logos*, the unchanging law of change. The melancholy that seems to pervade Heraclitus's thought may well have been derived from his pervasive sense of mutability. Tradition contrasts Heraclitean pessimism with the view of Parmenides and also with a later figure, Democritus (born 460? B.C.), whose commitment to atomistic materialism was supposedly connected with a cheerful outlook.

In contrast to these and other observers of the natural order, including Empedocles, whose doctrine of the four elements—earth, air, fire, and water—left a lasting imprint on later theorists, the mathematician-musician Pythagoras (died 497? B.C.) alone presents a comprehensive doctrine with regard to death. Supposedly derived from traditional mystery cults, the Pythagorean view holds that after death souls transmigrate from one animate being to another.

The background of classic Greek philosophy, of which Socrates, Plato, and Aristotle are the principal spokesmen, includes in addition to Homer, Hesiod, and the pre-Socratics the dramatists Aeschylus, Sophocles, Euripides, and Aristophanes, and the historians Herodotus and Thucydides. Herodotus's *History of the Persian Wars* focuses on the glory of the Greek triumph over Persian power and surveys calmly the rise and fall, the life and death of whole civilizations. Herodotus offers a view of death as "no evil," perhaps better than life, indeed "the gods' best gift to man." Thucydides' tragic account of the history of *The Peloponnesian War* provides the immediate background for the drama of Socrates' life and death and the philosophy of Plato's dialogues. It should be noted that in contrast to the pre-Socratic philosophers the three spokesmen of classic Greek thought lived and taught in the center of Greek culture, Athens.

Socrates, Plato, and Aristotle. Though countless scholars have labored to disentangle the specific strands of Socrates' thought from those of Plato, whose dialogues constitute the record of Socratic philosophy, this distinction is insignificant in regard to reflections on death since these are expressed most eloquently in the *Apology*, *Crito*, and *Phaedo*, which relate Plato's account of the trial, imprisonment, and execution of Socrates. Moreover, whether one accepts Plato's praise of Socrates for having brought philosophy down to earth to dwell among men, or agrees with Nietzsche's—and Heidegger's—view that Socrates' teaching disrupted the harmony between man and nature, it is Socrates—Plato's Socrates—who remains the primary philosophical fountainhead of Western ethical thought, including much subsequent thinking about death. This is true of later classical philosophy, of the traditions of Epicureanism as well as Stoicism and other Hellenistic philosophies, and of much of later Greco-Roman, medieval, Renaissance, and modern thinking.

The drama of Socrates' trial and execution is central to Plato's discussion of death though the *Apology*, *Crito*, and *Phaedo* can be supplemented by references in other dialogues, including the *Republic*, *Gorgias*, and *Symposium*. One need not take literally the passage in the *Phaedo*, the account of Socrates' last hours, likening death to the daily life of the philosopher: "Death, the separation of soul and body, is something which a philosopher who dies to the body every day he lives ought to welcome." And in the *Phaedo* he proclaims:

I desire to prove to you that the real philosopher has reason to be of good cheer when he is about to die, and that after death he may hope to obtain the greatest good in the other world. . . . Is death not the separation of soul and body? And to be dead is the completion of this; when the soul exists in herself, and is released from the body and the body is released from the soul what is this but death? [Plato, p. 447.]

It should be repeated that those words are spoken by Socrates as he faces immediate death and that there is no preoccupation with the theme of death in the Platonic dialogues. Moreover, though Plato's philosophy and particularly his theory of Ideas or Forms has been interpreted as separating the world of matter from the realm of spirit, the distinction by no means constitutes a sharp division in Plato's thought. For Plato, as George Santayana has expressed it in his *Life of Reason*, all things ideal have a natural basis and all things natural have ideal fulfillment or possibilities. Similarly, the radical separation of body and soul as presented in the *Phaedo* is by no means as sharp in other Socratic statements and may be read as the expression of Socrates' effort to comfort his friends rather than as a formal doctrine.

Though Aristotle did formulate and criticize the Platonic theory of forms or ideas, his own views of the relation of body and soul and of immortality in the life of reason are essentially Socratic and find many parallels in Plato's dialogues. In early writings attributed to Aristotle, the dialogue *Eudemos* and the *Protrepticos*, there are echoes of a Pythagorean transmigration theory and a view of the body as the soul's "prison." But the major works of Aristotle, notably the *De Anima*, treat the theory of transmigration as a myth. Here Aristotle regards the soul as the entelechy, the "form-giving cause," of the body just as seeing is the entelechy of the eye. For Aristotle, as for the pre-Socratics, the world of nature is of primary interest, and he views it, as they apparently did, as the setting of human life and continuous with human nature. But Aristotle follows the Platonic Socrates in the centrality of his interest in human beings in their social setting.

One notes that Aristotle's father was a physician and that Aristotle's absorption in biological studies may well be the most significant factor in his approach to ethical questions, including problems connected with death. By contrast, Plato's acceptance of mathematical models —not improbably a Pythagorean influence— surely affected his thinking, e.g., his theory of Forms or Ideas. Yet a common element, Socratic in its emphasis, is readily discernible.

Hellenistic thought

Aristotle's death in 322 B.C. has conventionally marked the end of classical Greek philosophy. Post-Aristotelian—Hellenistic as distinguished from Hellenic—thought continued many of the lines of classical philosophy. But even in the traditions of Platonism and Aristotelianism there were marked changes, and this is true of the several schools that claimed to derive from Socrates: Cyrenaics, Cynics, Skeptics and their successors the Epicureans, Stoics, and other Schools of the Greco-Roman world. Such interest as there had been in problems related to death among the pre-Socratics and in the philosophies of Socrates, Plato, and Aristotle had never been the central focus of their thought. But in the Hellenistic age, marked by what Gilbert Murray called a "failure of nerve," the questions of death and dying became a preoccupying and sometimes an obsessive concern.

A few external circumstances may be briefly summarized. The disastrous conclusion of the Peloponnesian War precipitated the decline and death of the Greek city-states. The tragic irony may be noted that it was Aristotle's most famous student, Alexander, who brought to an end the city-state, which made possible, according to Aristotle, a truly good life of free citizens. The transfer of imperial power to Rome brought with it the transfer of surviving elements of Hellenic culture: The Roman pantheon was expanded to accommodate the entire family of Olympian deities, and the closing of the schools of Athens led to their Roman substitutes. Roman power, far more expansive and efficient than the Greek, brought great military achievement; but the succession of wars and conquests—witness Tacitus (A.D. 55–117)—produced diminishing satisfaction. The wheels of fortune turned rapidly; great wealth and poverty coincided. Social and political problems abounded; the ancient gods, as Augustine (A.D. 354–430) later observed, proved unavailing. Ancient pieties sought to fortify themselves by turning to the East, to exotic faiths and mystery cults.

To such a world, Greek in its background, Roman in its laws and organization, the surviving philosophic schools of Epicureanism, Stoicism, and the rest brought their distinctive messages —not philosophy as continued Socratic inquiry, but rather prescriptions for a way of life and a way to face death.

Epicureanism. For Epicurus (341–270 B.C.) philosophy was not a love of wisdom as good in itself but rather as a "remedy for the soul." This distinction is indeed the watershed between Hellenic and Hellenistic philosophers, even though the latter claimed to look back for their inspiration to classical thinkers, notably to Socrates. Epicurus looked back to Democritus too, but knowledge of atomistic physics was no longer desired for itself but rather to give reassurance. Epicurus's remedies were designed to liberate men from fear of the gods, to assure them that death was a relief to be welcomed, that evil could be endured if it could not be avoided in cloistered gardens, and that a relatively good life could be attained in simplicity and seclusion. The bad reputation that Epicureans subsequently acquired is to be explained by their neglect of the duties of active public life and citizenship and their supposed "atheism." It should be noted, however that most Epicureans (e.g., the poet Lucretius, 96?–55 B.C.) retained the Olympiad as models of Epicurean *ataraxia*, life free from worry. Epicurean arguments against immortality were based on Democritean atomism; after death the soul, being material, made of the finest atoms, would quickly be dissipated. There is nothing to fear: No man will attend his own funeral. Life is a feast and we have banqueted; should not the worm as well?

Stoicism. Stoicism, though often contrasted with Epicureanism, shared its essential purposes: to provide a way of life designed to make existence bearable and a system of knowledge primarily valued as a foundation for ethics—in the case of Stoicism an ethics of rigorous duty and civic virtue. For the Greek Stoics (e.g., Zeno, 335–265 B.C.) physics had been basic; for the three great Roman Stoics, Seneca, Epictetus, and Marcus Aurelius, the ethics of *apatheia* (indifference) and *aequanimitas* (peace of mind) is all-important. Though they retain a traditional pantheism, sometimes suggesting an undogmatic monotheism in which Zeus is identified with the universe, these doctrines in Roman Stoicism are very tentative. So too, "living according to nature" is assumed to be equivalent with "living according to reason," which, in turn, interprets the "life of reason" as a life reasonably disposed to accept those evils that cannot be avoided or brought under control as necessary. Thus Epictetus declares in his *Discourses*:

When death appears an evil we must have ready to hand the argument that it is fitting to avoid evils, and death is a necessary thing. What am I to do? Where am I to escape it? ... I cannot escape death: am I not to escape the fear of it? Am I to die in tears and trembling? For trouble of mind springs from this, from wishing for a thing which does not come to pass [Oates, p. 271].

For the Stoics death was an escape from otherwise unavoidable evil—the door was always open—and they held to the conviction (Seneca put it into practice) that suicide was a way out. It is curious how often it is forgotten that the Stoics gave suicide their approval, whereas Schopenhauer, who because of his avowed pessimism is widely though mistakenly believed to have supported suicide, in fact argues against it as an erroneous submission to "Will." It may be noted that the argument for the "open door" has been revived in recent times by Jean-Paul Sartre. For the French existentialist suicide represents a way for man to assert his freedom.

Seneca's prescription for escaping the fear of death by thinking of it constantly is characteristic of Stoic ethics, but Jacques Choron's assertion that Marcus Aurelius's *Meditations* are "meditations on death and almost nothing else" (Choron, p. 73) seems an exaggeration. Death and its inevitability are, indeed, of concern to the emperor-philosopher but there are numerous other interests.

Hebrew–Christian traditions. There is a noble sobriety in the ethical doctrines of the Hellenistic schools, but, in contrast to the Socratic teaching from which they claimed to have been derived, their emphasis is negative—deliverance from evil, escape from an alien world, salvation from sin. Along with Epicureanism and Stoicism, other philosophical systems, some of them revivals of classic formulations such as neo-Pythagoreanism and neo-Platonism, offered themselves as cures for failure of nerve. And along with such philosophies men turned to exotic faiths to supplement or replace their traditional religion. Among these, in competition with Persian Mithraism and, later, Manicheanism, for example, came the promise of salvation through faith in Jesus of Nazareth as preached by Paul of Tarsus. The Old Testament, as well as the New, now entered the mainstream of Western culture, and the Hebrew–Christian ethic came to dominate Western thinking, including thought about death.

There is relatively little concern with death in the Hebrew scriptures. Though later ages stressed Adam's sin as having "brought death into the world and all our woes," the prevailing mood of Hebraism was expressed in the formulation of Koheleth that "there is a time to be born and a time to die" (Ecclesiastes 3:2). Between birth and death there should be a full life, and after death human souls (*rephaim*) may lead a shadowy existence in an afterworld (*Sheol*). Real immortality pertains to the race, rather than to individuals, and to vicarious immortality through one's descendants.

Christianity, though rooted in the Old Testament, reversed these emphases, and the Hebrew–Christian dispensation reflects the preoccupations of the Hellenistic age. Jesus of Nazareth and the Christian promise of divine love supplied a doctrine of salvation from evil—above all from the sins of concupiscence and sexual desire—a doctrine that offered the release of grace from the threat of everlasting hellfire and the hope of eternal blessedness. From the philosophical elements in Patristic theology, from Origen (185?–254?) to Augustine (354–430) to Thomas Aquinas (1225–1274), the natural world was viewed primarily as the scene of human salvation and humankind as essentially sinful, doomed to die and to suffer eternal damnation unless rescued by supernatural Grace.

Medieval thought

This was the framework within which medieval philosophers operated, using tools derived from ancient thinkers. They included Platonic traditions in a limited knowledge of the dialogues and, for scholastic philosophy, the available works of Aristotle, "master of all who know" in Dante's famous epithet. When more complete records came from the Arab world as a result of Muhammadan conquests and penetration of the Iberian peninsula, the teachings of Avicenna (980–1037) and Averroes (1126–1198) added to Western knowledge of ancient philosophy and scientific interests. To this increased understanding there were also contributions by medieval Jewish philosophers such as Abraham Ibn Ezra (1092–1167) and especially Moses Maimonides (1135–1204). These factors helped to modify medieval otherworldliness and twelfth- and thirteenth-century anticipation of the revival of naturalistic philosophy, literature, and art associated with the later Renaissance of the fifteenth century. Though the Renaissance and the sixteenth-century religious Reformation have often been viewed as terminating the Middle Ages, latter-day historians have recognized the earlier shift of emphasis and, on the other hand, the continuity of stress on sin and salvation in various aspects of Reformation thought, notably in the philosophy of Calvin (1509–1564).

Early modern philosophy (16th and 17th centuries)

The waning of the Middle Ages and the dawn of modernity was of course not a sudden shift. The transition was gradual, and elements that characterize modern culture can readily be discerned in late medievalism. The growth of capitalism and of national states, continuing Renaissance naturalism and new sciences took time to develop, and, correspondingly, aspects of a predominantly "medieval" religious culture persist into the seventeenth century and later. But from then on Western thought becomes predominantly secular and scientific.

Two thinkers express this outlook, and each has been called the father of modern philosophy: Francis Bacon (1561–1626) and René Descartes (1596–1650). The very titles of some of Bacon's books indicate his sense of the novelty of his outlook: *Novum Organum*, the *New Method* designed to replace the old methodology, the organon of scholastic Aristotelianism; and the *New Atlantis*, a forecast of cooperative, modern science, which anticipates developments in the investigation of nature until the eighteenth and nineteenth centuries' laboratory science, the practical applications of knowledge as power, the pragmatic mood in philosophy.

In his criticism of the Stoics Bacon echoes the view of his predecessor Montaigne, whose essay on death, "To Philosophize Is to Learn How to Die," concludes: "If we have learned how to live properly and calmly, we will know how to die in the same manner."

As for Bacon, so for Descartes "the sciences have a definitely practical aim, the harnessing of nature to the purposes of man. The will o' the wisp of his life was the conquest of death not only for the soul but also for the body" (Choron, p. 111). Like Bacon, too, Descartes's concern is to extirpate the fear of death, though unlike Bacon his chief argument is the traditional one of assurance of the survival of the soul. Cartesian dualism, the sharp dichotomy between body (*res extensa*) and mind (*res cogitans*) provides ground for Descartes's assurance but leaves the problem of their relationship, the "mind–body problem" of how two radically different kinds of thing can interact, as a continuing problem throughout modern philosophy.

In the context of the continuing preoccupation with the theme of death, a famous dictum of Spinoza (1632–1677) is to be read. He writes: "A free man, that is to say, a man who lives according to the dictates of reason alone, is not led by the fear of death. . . . He thinks, therefore, of nothing less than of death, and his wisdom is a meditation upon life" (Spinoza, bk. IV, prop. LXVII). There is no reason to accept critical interpretations of Spinoza that attempt to psychoanalyze him as himself harboring an extreme fear of death or suspect his argument "that death is by so much the less injurious to us as the clear and distinct knowledge of the mind is the greater" (bk. V, prop. XXVIII, note). This becomes especially cogent in the light of Spinoza's own criticism of simplistic rationalism (e.g., the Stoics) and his conviction that a lesser emotion can be displaced only by one that is more powerful, the most powerful being "the intellectual love of God or Nature" (*Amor intellectualis Dei—Deus sive Natura*).

Another mathematical genius, co-discoverer with Sir Isaac Newton of the calculus, was Gottfried Wilhelm von Leibniz (1646–1716). He reacted against what he took to be the excessive mechanism of dominant Cartesian views and sought to reconcile science and religion. In this context he distinguished mere unconsciousness from "*absolute death* in which all perception would cease. This has confirmed the ill-founded opinion that some souls are destroyed, and the bad ideas of some who call themselves freethinkers and who have disputed the immortality of the soul" (Leibniz). Hence Leibniz argued that no animate being can "entirely perish in what we call death" and that "God will always conserve not only our substance, but also our person." Leibnizian optimism had wide influence in the eighteenth century as expressed in Alexander Pope's couplet seeing in "all discord, harmony not understood; all partial evil, universal good."

Philosophy in the 18th to 20th centuries

There were dissenting voices against the Leibnizian optimism. David Hume attacked the entire doctrine of immortality, and the French philosophes of the Enlightenment (D'Holbach, 1723–1789; Condorcet, 1743–1794; Diderot, 1713–1784) held it to be a "priestly lie." Consequently, "Fear of death is the only true enemy that has to be conquered, and that there is no afterlife makes us free from the power of the priests" (Choron, p. 135). Finally Voltaire in his *Candide* demolished Leibnizian optimism in his caricature of the philosopher "Dr. Pangloss."

In Immanuel Kant, Johann Gottfried Herder, Goethe, and the post-Kantian German philosophers there are fundamentally different views of the problems of death and immortality. Herder held that belief in a future life was universally natural to man. Goethe saw death as "Nature's stratagem to secure more abundant life," and his challenging phrase: *stirb und werde* (die and live anew) was echoed by post-Kantian philosophers of Romanticism. So was Kant's ethical argument for immortality: Though it cannot be established by "pure Reason," the requirements of moral necessity can be validated by "practical Reason" without which "the moral worth of actions, on which alone the worth of a person and even of the world depends in the eyes of supreme wisdom, would not exist at all" (Kant, p. 247). This conviction was followed by the philosophers of the Romantic period (Fichte, 1762–1814; Schelling, 1775–1854; Schleiermacher, 1768–1814) in the phrase of Arthur Schopenhauer (1788–1860): "Death is the true inspiring genius or the muse of philosophy" (Schopenhauer, vol. 3, p. 249).

For Georg Wilhelm Friedrich Hegel, Kant's moral argument had other implications. In Hegel's view "Death has the peculiar effect of uniting the individual with universal matter. The living individual is a particular person; once dead, however, he becomes, through bodily corruption indistinguishable from abstract being" (Hegel, p. 192). Hegel held immortality to be a quality of the living spirit, not an event in the future, and this view guided Hegel's follower, Ludwig Feuerbach. In his *Thoughts on Death*, Feuerbach argued that "Immortal life is the life which exists for its own sake and contains its own aim and purpose in itself—immortal life is the full life, rich in contents" (Choron, p. 191).

Søren Kierkegaard sets himself apart from all this. Attacking Hegel and traditional philosophy in general for concentrating on the "essence" of things as against the "existence" of the specific individual, Kierkegaard and the existentialists who follow his teachings emphasize the values of "immediate experience" in contrast to all metaphysical abstractions. In spite of Kierkegaard's attacks on traditional thought, existentialists have much in common with the attitude of Romanticist philosophers in their emphasis on the values of the individual, and like all exponents of "philosophies of life" they stress the significance of the crucial experiences of human lives, including the experience of death.

Friedrich Wilhelm Nietzsche, though influenced by Kierkegaard as well as by Schopenhauer, rejected the Romantic view of death as "the muse of philosophy" and argued that, though man's mortality is of the greatest importance for any philosophy of life, the act of dying is not. In his view the "will to die" can be countered by the affirmation of life in art and by the heroic acceptance of "eternal recurrence" in the experience of finitude.

Recent and contemporary thought

The individualism of Romanticists and of Kierkegaardian existentialists contrasts with an outlook that "developed during the 19th century and found its expression in Auguste Comte's positivism, an intellectualized form of nationalism." For Comte "society is composed of both the dead and the living. . . . For the dead have gone through the moment of change, and their monuments are the visible sign of the permanence of their city" (Ariès, p. 73).

Comtian sociological philosophy, though scientific and positivistic, was essentially pre-Darwinian in its perspective. Sharing a scientific outlook, the American pragmatists—Charles Sanders Peirce, William James, and John Dewey—were biologically oriented. Thus Peirce held that "death and corruption are mere accidents or secondary phenomena. Among some of the lower organisms, it is a moot point with biologists whether there be anything which ought to be called death" (Fisch, p. 109). With this naturalistic approach and, perhaps, as in the case of the "unalterable optimism of the dentists in an American small town . . . Relax, take it easy, it's nothing," as Jacques Maritain wrote in his *Reflections on America* (p. 91), pragmatists have focused attention on life, not death. However, there are significant reflections on death in William James's view that "life and its negation are beaten up inextricably together. But if life be good, the negation of it must be bad. Yet the two are equally essential facts of existence; and all natural happiness thus seems infected with a contradiction. The breath of the sepulchre surrounds it" (James, p. 139). Of course, James is writing about the attitude of "the sick soul" and from the viewpoint of a sympathetic psychologist.

Other philosophies indicating the importance of post-Darwinian biology include the vitalism of Hans Driesch (1867–1941) and the creative evolutionism of Henri Bergson (1859–1941). Bergson finds in his élan vital, the drive of life throughout reality, a ground for faith in life after death, and Alfred North Whitehead (1861–1947) finds comparable assurance by subsuming human immortality under the inclusive rubric of "the immortality of realized value."

More extensive treatments of the problems of death in late nineteenth- and twentieth-century philosophical thought are in works by Max Scheler and Georg Simmel. Contemporary existentialists devote great concern to the theme of death. Martin Heidegger places death in the very center of his consciousness as the way to "disarm" it and to offset "animal anxiety." Gabriel Marcel uses the despair evoked by the persistent consciousness of death to examine and deepen religious faith, and Jean-Paul Sartre finds in the experiencing of death ground for asserting human freedom.

Though contemporary philosophical analysts are disposed to reject "unanswerable questions" as meaningless, a respected spokesman for this point of view writes about death: "It is a meaningful question, because we can indicate ways in which it could be solved. One method of ascertaining one's own survival would simply consist in dying. It would also be possible to describe certain observations of scientific character that would lead us to accept a definite answer" (Schlick).

Though it is too soon to discern the philosophical implications of recent developments in the biological sciences and in medicine, a new concern for the dying and for human dignity in its final phases may be noted as characteristic of such movements as euthanasia and thanatology. In general it may be asserted that not since the later Middle Ages has there been as much consideration for problems associated with death as is evident at present. Although the anthropologist Geoffrey Gorer could assert in 1965 that there was a "conspiracy of silence" about death as though it were a "pornographic" subject, it is now clear that Philippe Ariès is accurate when he writes, in the early 1970s, that "death is once again becoming something one can talk about" (p. 103).

JAMES GUTMANN

[*Directly related are other articles in this entry:* WESTERN RELIGIOUS THOUGHT *and* DEATH IN THE WESTERN WORLD. *For a discussion of the meaning of death in other cultures, see the other articles in this entry:* ANTHROPOLOGICAL PERSPECTIVE *and* EASTERN THOUGHT. *For discussion of related ideas, see:* DEATH AND DYING: EUTHANASIA AND SUSTAINING LIFE, *article on* HISTORICAL PERSPECTIVES; MIND–BODY PROBLEM; *and* SUICIDE.]

BIBLIOGRAPHY

ARIÈS, PHILIPPE. *Western Attitudes Toward Death: From the Middle Ages to the Present.* Translated by Patricia M. Ranum. Johns Hopkins Symposia in Comparative History. Baltimore: Johns Hopkins University Press, 1974.
ARISTOTLE. *The Basic Works of Aristotle.* Edited by Richard McKeon. New York: Random House, 1941.
BROAD, CHARLIE DUNBAR. *Lectures of Psychical Research: Incorporating the Perrott Lectures Given in Cambridge University in 1959 and 1960.* International Library of Philosophy and Scientific Method. New York: Humanities Press; London: Routledge & Kegan Paul, 1962.
CHORON, JACQUES. *Death and Western Thought.* New York: Collier Books, 1963.
CORNFORD, FRANCIS MACDONALD. *From Religion to Philosophy: A Study in the Origins of Western Speculation.* The Library of Religion and Culture. Edited by A. J. Ayer. New York: Harper & Row, 1957.
DUCASSE, CURT JOHN. *Nature, Mind, and Death.* The Paul Carus Lectures, 8th ser. LaSalle, Ill.: Open Court Publishing Co., 1951.
FEIFEL, HERMAN, ed. *The Meaning of Death.* New York: McGraw-Hill, 1959.
FISCH, MAX HAROLD, ed. *Classic American Philosophers: Peirce, James, Royce, Santayana, Dewey, Whitehead: Selections from Their Writings.* New York: Appleton-Century-Crofts, 1951.
FLEW, ANTONY GARRARD NEWTON, ed. *Body, Mind, and Death: Readings.* New York: Macmillan Co., 1964.
GORER, GEOFFREY. *Death, Grief, and Mourning.* Problems of Philosophy Series. Garden City, N.Y.: Doubleday, 1965. First published in London as *Death, Grief, and Mourning in Contemporary Britain.*
HEGEL, GEORGE WILHELM FRIEDRICH. *Hegel: Selections.* Edited by Jacob Loewenberg. The Modern Student's Library, Philosophy Series. Edited by Ralph B. Perry. New York: Scribner, 1929.
JAMES, WILLIAM. *The Varieties of Religious Experience: A Study in Human Nature: Being the Gifford Lectures on Natural Religion Delivered at Edinburgh in 1901–1902.* New York: Longmans, Green & Co., 1928.
JOLIVET, RÉGIS. *Le Problème de la mort chez M. Heidegger et J.-P. Sartre.* Paris: Éditions de Fontenelle, 1950.
KANT, IMMANUEL. *Critique of Practical Reason: And Other Writings in Moral Philosophy.* Translated and edited by Lewis White Beck. Chicago: University of Chicago Press, 1949.
LAMONT, CORLISS. *The Illusion of Immortality* (1935). 4th ed. Introduction by John Dewey. New York: Frederick Ungar Publishing Co., 1965.
———. *Issues of Immortality: A Study in Implications.* New York: Henry Holt & Co., 1932. Ph.D. dissertation, Columbia University, 1932.
LEIBNIZ, GOTTFRIED WILHELM. *The Monadology and Other Philosophical Writings.* Translated by Robert Latta. London: Oxford University Press, 1898.
MARITAIN, JACQUES. *Reflections on America.* New York: Scribner, 1958.
OATES, WHITNEY JENNINGS, ed. *The Stoic and Epicurean Philosophers: The Complete Extant Writings of Epicurus, Epictetus, Lucretius, Marcus Aurelius.* New York: Random House, Modern Library, 1940.
PLATO. *The Dialogues of Plato.* Translated by Benjamin Jowett. New York: Random House, 1937.
SCHELER, MAX FERDINAND. "Tod und Fortleben." *Gesammelte Werke.* Vol. 10: *Schriften aus dem Nachlass* (1933). Edited by Maria Scheler. Bern: Francke, 1957, pp. 9–52.
SCHLICK, MORITZ. "Unanswerable Questions." *A Modern*

Introduction to Philosophy. The Free Press Textbooks in Philosophy. Edited by Paul Edwards and Arthur Pap. New York: Free Press, 1973, pp. 791–795.

SCHOPENHAUER, ARTHUR. *The World as Will and Idea.* Translated by Richard B. Haldane and John Kemp. 3 vols. New York: Scribner; London: Routledge & Kegan Paul, 1949.

SIMMEL, GEORG. "Zur Metaphysik des Todes." *Logos: Internationale Zeitschrift für Philosophie der Kultur* 1 (1910–1911): 57–70.

SPINOZA, BENEDICT DE. *Ethics: Preceded by On the Improvement of the Understanding.* Edited by James Gutmann. New York: Hafner, 1949. Based on the translation by William Hale White as revised by Amelia H. Stirling.

IV

WESTERN RELIGIOUS THOUGHT

1

DEATH IN BIBLICAL THOUGHT

The Old Testament

In Genesis 2–3, Israelite folk literature preserved and possibly combined two "explanations" (etiologies) of human mortality: (1) A protohuman couple, in primeval time, disobeyed their creator, and the two were placed under the sentence of death ("When you eat from [the tree of knowledge] you are doomed to die," 2:17). Presumably, they were created immortal and might have remained so. (2) The protohuman couple was created mortal. "The Man" (Hebrew: *ha-'adam*, later shortened to a proper name, Adam) was thought to have been fashioned from the soil (*ha-'adamah*) since humans are buried in the ground and their deteriorating bodies seem to blend into it. The image is that of the potter who fashions vessels from ceramic clay. Life results when the creator forces breath into the nostrils, resulting in a "living creature" (2:7). Presumably, mortality was intended, just as it was for all of the other "creatures" who are likewise fashioned from the soil (2:19). The couple is banished from the primeval garden lest they eat from the "tree of life."

The account can also be read as a continuous, integrated story: The couple forfeits access to the "tree of life" by their disobedience in eating from the "tree of knowledge." The result is that they must remain as they were created: mortal. However, later generations, particularly in the early Christian era, would read the unified story to mean that death was a tragic disruption of the Creator's plan. Thus, the second folk explanation will be indicative of Israelite and Jewish thought, while the first will predominate in Christian thought.

The term "death" is used in the Old Testament in at least three senses: (1) for biological cessation, the end of one's historical existence; (2) as a metaphor for those things which detract from life fully lived—illness, persecution, despair, etc.—e.g., God is described as one who "kills and brings to life" (1 Sam. 2:6), with no implication that biological death or resurrection is intended; (3) as an active power in opposition to the created order. This is usually only a literary vestige from the pre-Yahwistic period or from Israel's neighbors, e.g., when one of Job's friends describes deterioration of the body, he uses a formalized idiom: "the first-born of death consumes his limbs" (Job 18:13). He need not mean thereby what a non-Israelite likely would mean: that the god Mot (death), a demonic, autonomous power, had seized the person.

In Israelite anthropology there is nothing corresponding to the Greco-Christian concept of a "soul" (itself a very nebulous term). Humans are a totality, flesh animated by a life force (*nefesh* or *ruach*), which was thought to reside in the blood or in the breath. The departure of the life force results in death. The body, thus weakened, is placed in the tomb, where a subexistence was sometimes associated with it and especially with the bones. Hence, desecration of the remains was considered sacrilege (Amos 2:1). Sometimes a shadowy, weakened remnant of the person was thought to reside in the Underworld (Hebrew: *Sheol;* cf. the "shadows" of the Homeric Hades). For example, a medium claims to have made contact with the deceased prophet Samuel, still recognizable as an old man dressed in a robe (1 Sam. 28). However, such ideas and practices are largely vestiges of an older cult of the dead, which official Yahwism transforms, ignores, or forbids: The potent death-demons of neighboring cults are denied existence; mediums are to be executed (Exod. 22:18); priests are forbidden to participate in rituals for death and burial (Lev. 21); tomb gifts are vestigial (vessels are empty); the dead are reduced to a state approaching nonexistence and hence can scarcely influence the living (Eccles. 9:5–6); no judgment of the dead or expectation of resurrection is affirmed, save at the very end of the Old Testament period.

While there is a variety of responses to biological death (as one might expect from a collection ranging over at least a millennium), the predominant response in old age is one of calm acceptance. Considerable comfort seems to lie behind the oft-repeated refrain, "N died old and full of years" (Gen. 25:8). The fact of human mortality does not engender the belief that life is thereby meaningless. Laments do not focus upon biological death as a theological problem (2 Sam. 1:19–27). The question, "Why should I/we die?" is not asked except in situations where foolish action would lead to premature death (Gen. 47:15).

Occasionally the positive values of mortality are implied: the rebellious protohuman couple, Adam and Eve, barred from "the tree of life" (Gen. 3:22–24); the human life span is shortened in order to limit unacceptable behavior (Gen. 6:1–4); it was necessary for an entire generation to die before Israel could move on toward her ordained destiny (Num. 32:13); Job, overcome with pain and grief, observes that death would be a form of release (Job 3:1–19); the psalmist suggests that human finitude can be an incentive to seek the ultimate meaning of life (Ps. 90:12).

While the Wisdom literature most often is concerned with premature biological death (how to avoid it: "Fools die for lack of sense," Prov. 10:21), Ecclesiastes in particular focuses upon the fact of mortality. Wise or foolish, one fate comes to all, and thus all achievements are negated (Eccles. 2:15–22). Thus, one is forced to concentrate upon the present, which can be seen as a gift from God (Eccles. 3:1–15). Ultimate nihilism is overcome by keeping death within the domain of God's wisdom, even though it be incomprehensible to humans.

In sum, then, what happened after death was not a matter of religious importance, and mortality is calmly accepted as part of the definition of being human (a creature). Attention is focused almost entirely upon this world and its activities: responsible life in a covenant community called to serve God. Anything that interferes with life in this fashion is to be avoided and is metaphorically described as "death." Seemingly, anxiety about mortality has largely been transferred to anxiety about things that interfere with life at its fullest. Protest against biological death, apart from Ecclesiastes, is directed only toward: (1) premature death (Isa. 38:2–3, a response to a death-oracle at age 39); (2) an "evil" death, characterized by violence; and (3) severance of relationship with God ("For [those in] Sheol cannot thank thee, death cannot praise thee," Isa. 38:18). Some scholars think that the desire for unending relationship with God is the bridge that finally leads to assertion of life after death, e.g., in Psalm 73:23–26.

On occasion, attention is focused on, and comfort derived from, those things which will survive the mortal human: one's reputation (Prov. 10:7; Sirach 44:8–9); male offspring to preserve one's memory and property (2 Sam. 18:18; Isa. 56:3–5; such desire may be a vestige of the ancestor cult, wherein the function of the eldest son was to bring offerings to the tomb); the elect, worshiping community Israel (Gen. 48:21); and God, who alone is immortal (Ps. 90; Hab. 1:12, Tiq. Soph.).

The intertestamental literature

In the Old Testament literature, the primary hindrance to Israel's corporate existence was a human tendency toward rebellion (Num. 11–16). The prophets proposed that the reverses of history might serve a didactic purpose, leading to an age of spontaneous obedience (Jer. 31:31–34), or that God would give the people a "new heart" (Ezek. 18:31; 36:26). Because of the failure of such reform to materialize, because of the disintegration of traditional social and economic structures after the fall of Jerusalem (587 B.C.), and because of the harshness of subsequent oppression by Babylonians, Persians, and Greeks, some persons found it impossible any longer to hope for the transformation of society through human initiative or response. Influenced by ancient conflict myths, the apocalyptic school of thought slowly developed, beginning toward the end of the Old Testament period. It proposed that a catastrophic divine intervention would be necessary in order to restore the world to the Creator's design. Death, increasingly viewed as an active power or agent (and later identified with the devil: Wisd. of Sol. 1:13; 2:23–24), would be eliminated from human experience (Isa. 25:8) and the deceased might even be resurrected to enjoy the new age (Dan. 12:1–3). The possibility of a semiconscious interim existence (usually in the Underworld) was then perceived, possibly influenced by contact with the anthropology of the Greek mystery religions: the life force and consciousness are fused into an entity able to survive cessation of bodily functions.

Intense and sustained persecution sometimes produced a positive evaluation of biological death. The paradigm of a martyred mother and seven sons is offered by 2 Macc. (second century B.C.): death, to be followed by resurrection, is much to be preferred to some forms of this-worldly existence. In the Wisdom of Solomon (first century B.C.), the traditional Old Testament stance is reversed: premature death of the righteous indicates that they are worthy of the divine presence (Wisd. of Sol. 3:4–5) and frees them from further suffering (Wisd. of Sol. 4:11). Thus, death need be a source of anxiety only to the wicked.

In the sectarian literature from the ascetic community at Qumran (the Dead Sea Scrolls, first century B.C.), biological death is scarcely an issue. Rather, attention is focused upon the possibility of transition from one this-worldly mode of existence to another—from what is metaphorically described as "death" to "life." Through membership in the Community, with its regimented life-style and esoteric scriptural lore, one can escape the human inclination toward evil, from the assaults of the "angel of darkness," and be in communion with the "heavenly hosts." This present possibility will be augmented by a final war, which will usher in an era of divine favor centered in the city of Jerusalem. (Whether and how the deceased will participate in this new age is still a matter of scholarly debate.)

By the second century B.C., the protohuman couple in the primeval garden was understood not only to have become mortal because of sin (a move in interpretation from etiology number two toward number one), but to have been the means whereby death grasped their descendants: "Woman is the origin of sin, and it is through her that we all die" (Sirach 25:24). This understanding will play a major role in the thought of some New Testament writers.

The New Testament

Although this literature comes from a narrow period of time (little more than a century), it contains a variety of perspectives, definitions, and emphases that cannot easily be harmonized. In general, it builds upon apocalyptic thought and develops death into a larger theological issue than has been the case previously.

The earliest spokesman, Paul (A.D. 50?), concentrates upon biological death as a consequence of sin (Rom. 6:23), unleashed by the disobedience of Adam and Eve (Rom. 5:19). He perceives sin as a power, related to the devil, that has infected the entire cosmos, setting it in opposition to the Creator's design (Rom. 8:22). Death is thus the enemy, the paradigm of all existence, the background of all patterns of possibility: demoralizing, challenging, and negating all human vitality and sense of purpose. It is not merely that the individual is mortal: It is that all existence must be perceived as futile, as death-directed (Rom. 8:18–25). "Life" has been engulfed by "death." However, since death is a "historical" problem rather than a metaphysical one (i.e., "Adam" has been historicized), it is subject to a historical solution. Jesus, perceived as having risen from the dead, illustrates that the resurrection hope of apocalypticism is well founded and that the transition to the new age is imminent. The fact that death has been defeated, that God will shortly vindicate himself, makes it possible to perceive the cosmos in a new light, to see new possibility patterns, to "walk in newness of life" (Rom. 6:4). It is this new perception that now empowers the followers of Jesus to act. They believe that the power of death has been negated, although it temporarily continues to manifest itself biologically. That will cease at the transition to the new age, when all followers of Jesus, living or deceased, will be granted "immortality" (1 Cor. 15:53–54).

In the Synoptic Gospels, biological death is cited primarily as an incentive to prepare for the impending arrival of the new age; the insecurity and shortness of life serve as sanctions for obedience to the appeals of Jesus. Death, especially biological death, is not perceived as a serious theological problem. Little attention is given to the details of life after death.

In the latest Gospel, John, the apocalyptic theme of catastrophic transition to a new age (including the hope for resurrection of the dead), is pushed even farther to the margins of discussion. The quality of one's existence, here and now, becomes the issue. Biological death is not man's fundamental problem; it is how life is to be lived. Hence, when the author speaks of "eternal life," he is not thinking fundamentally of an apocalyptic transition to life without end still lying before the individual. Rather, "death" is a metaphor for a quality of existence that the followers of Jesus are able to transcend: They will "never die" (John 11:26), i.e., not be subject to a mode of existence, even though they will eventually biologically expire. The choice for humans is "eternal life" versus living "death."

Conclusions and implications

The Bible, like any other document, is heavily dependent upon the cultural milieu of which it is a part—a milieu radically different from that of the modern Western reader. This, plus the fact that biomedical technology has thrust upon us problems beyond the imagination of the biblical writers, makes it extremely hazardous to propose simplistic "contemporizations." Nonetheless, the following observations, restrictions, and implications are offered, tentatively, for those whose ultimate points of reference include the biblical tradition.

1. The Bible contains a variety of perspectives, which cannot be harmonized or easily systematized and need not be arranged in a hierarchy of values (such as, "the perspective[s] of Jesus are normative"). Such variety is a reflection of changing historical situations; it is an accurate perception of the ambiguity of reality. It may be a strength rather than a weakness, however frustrating this may be to the ethicist.

2. Death is more than biological cessation. The Bible's metaphorical use of the term corresponds, in part, to such modern perceptions as psychological death, social death, and spiritual death.

3. Life is more than biological functioning. Therefore, one's commitments should be directed to the conditions under which life in community is lived and to the values that make life meaningful (and the Bible has very definite ideas about what such conditions and values are, in social, economic, political, and religious terms). It is precarious to go beyond this and assert, as is sometimes done, that the lack of emphasis in biblical books upon biological life as intrinsically valuable (see, however, paragraph 4 below) would support an ethical stance in which extraordinary life-support systems should be withheld or withdrawn from some categories of dying patients.

4. The emphasis upon biological life as the result of the deity's creative act (Gen. 2:7, Ps. 104:30) and the insistence that even animal life cannot be taken without the enactment of a ritual which acknowledges God's sovereignty over it (Lev. 17:1–16; secular slaughter of an animal is regarded as murder, 17:4) serves as a caution against any simplistic human desire to terminate, directly or indirectly, the life of one who is suffering.

5. Physicians who tend to view death as an "enemy" to be overcome at all costs may feel more kinship with a New Testament stance (death as a "power," which God will soon eradicate) than with that of the Old Testament (death as a natural event).

6. Since the biblical thinkers could not conceive of existence in noncorporeal terms and thus forbade desecration of corpses or even bones, Orthodox Judaism quite logically has usually opposed autopsies and cremation as sacrilege.

7. The idea of life after death, being in some measure culturally and historically conditioned, may only with difficulty become a criterion in ethical decisions. In the biblical text itself, it is not used to devaluate this-worldly existence.

LLOYD BAILEY

[*Directly related are the entries* CADAVERS; EMBODIMENT; *and* JUDAISM.]

BIBLIOGRAPHY

BAILEY, LLOYD R. "Death as a Theological Problem in the Old Testament." *Pastoral Psychology* 22, no. 218 (1971), pp. 20–32.

BRUEGGEMANN, W. "Death, Theology of." *The Interpreter's Dictionary of the Bible: An Illustrated Encyclopedia.* 4 vols. and supp. *Supplementary Volume.* Edited by Keith Crim, Lloyd R. Bailey, Sr., Victor P. Furnish, and Emory S. Bucke. Nashville: Abingdon Press, 1976, pp. 219–222.

KECK, LEANDER E. "New Testament Views of Death." *Perspectives on Death.* Edited by Liston O. Mills. Nashville: Abingdon Press, 1969, chap. 2, pp. 33–98.

LAPP, PAUL W. "If a Man Die, Shall He Live Again?" *Perspective* 9 (1968): 139–156. *Perspective* is a journal of the Pittsburgh Theological Seminary.

2
POST-BIBLICAL JEWISH TRADITION

Rabbinic period

The rabbinic period (c. 200 B.C.E. to 500 C.E.) provided important ideas and practices, which had a crucial impact on the unfolding of the Judaic view of death.

Fate after death. First, there was a definite expression of Jewish dogma concerning the fate of the person after death. The Pharisaic doctrine, which became normative both for Judaism and Christianity, was expressed in the phrase, *tehiat hametim*, the resurrection of the dead. This meant that death was not seen as the final end of existence. It was a prelude to the *olam haba*, the world to come. There is no doctrinal clarity as to the exact nature of the *olam haba*. Some believed that until the day of the resurrection (which would be ushered in by the Messiah) the soul slept waiting. The day of the resurrection would also be a day of judgment when the righteous would be ushered into the *olam haba*. Others believed that the soul enjoyed immortality in a disembodied state in the world to come. During this time there is a preliminary judgment; the wicked being purged in *Gehenna* (hell) and the righteous enjoying the delights of heaven. This would end when the time of *tehiat hametim* arrived, the souls rejoining the reborn bodies to live in eternal bliss. It is important to note that it was the resurrection of the body which was seen as the *final* consummation of salvation. It is also important to note that a religion which was, in general, loath to enunciate specific dogmas, insisted on the belief in the resurrection of the dead as one of the few specific doctrines considered to be indispensable for a believing Jew.

Customs and rituals surrounding death. The second important development during the rabbinic period was the formulation of specific customs and rituals surrounding death. The dying were to be treated with special consideration, even to the extent of withholding the entire truth about their condition so that they might not be thrown into despair. The obligation of *bikkur holim*, the visiting of the sick, especially the dying, was seen as very important. When death seemed to be drawing near, the person was advised to confess, "I acknowledge before Thee, O Lord my God and God of my fathers, that my life and my death are in Thy hands. May it be Thy will to heal me. But if death is my lot, then I accept it from Thy hand with love." Then came a simple confession of sin and a prayer: "May my death be an atonement for whatever sins and errors and wrongdoings I have committed before Thee. In Thy mercy grant me the goodness that is waiting for the righteous and bring me to eternal life." The confession ends with the historic affirmation of Jewish faith: "Hear, O Israel, the Lord our God, the Lord is One."

The dead are to be treated with dignity. Desecration of the body is forbidden, and burial must be immediate; it must be through interment in the soil; the dead are to be buried simply in white shrouds. The funeral is simple; the eulogy is for the honor of the deceased and the comfort of the bereaved. Close relatives observe a prescribed period of mourning, during which time they stay at home to receive friends and neighbors who come to console them, using the formula: "May the Lord comfort you amongst the mourners of Zion and Jerusalem." In later Judaism, these mourning customs were elaborated, the most important aspect being the obligation of the sons to recite the *kaddish*, a prayer in praise of God at every synagogue service for eleven months. This is seen as an expression of the commandment "Honor thy father and thy mother." In another later elaboration, the *kaddish* is recited also on the *yahrzeit*, the anniversary of the day of the death, and there are memorial prayers (the *yizkor*) during the important festivals of the liturgical year. The Judaic rites of burial and mourning have been described as providing an "expression for grief, for strengthening family and community solidarity, honoring God and inculcating the acceptance of His will" (Eckhardt, p. 497).

Ethical rules in death and dying. A third element that was important during the rabbinic period was the formulation of rules of conduct in areas connected with death and dying. We find, for example, a formulation of the definition of death (the cessation of breathing and of the heartbeat); rules concerning the treatment of the fetus threatening the life of its mother (it should be destroyed); and the treatment of the terminally ill. The latter are in no manner to be *actively* deprived of even a moment of life. "The dying person is to be considered as living in all things" (Talmud, Shabbat, 151a). There is indication, however, that it is permissible to remove an "impediment which prevents the soul from leaving the body" (Code of Jewish Law, Yoreh Deah, 339). There is a condemnation of suicide, and persons who "destroy themselves willingly" are not entitled to the dignity of a funeral or mourning. Especially interesting is the stress on the dead as a source of ritual impurity, thus negating pagan notions of the apotheosis of the dead. This is, of course, biblical legislation spelled out in rabbinic codes.

Meaning of death. Fourth, the rabbis speculated as to the reasons why death came into the world. One school of thought held that death was the result of sin. Since there is no righteous man who does not sin, death is universal. Another school of thought believed that death was the natural course of events, reflecting the inevitable cycle of birth, maturity, and decay. The most striking statement expressing this attitude is the comment on Genesis I by the great rabbinic sage, Rabbi Meir, who was a second-century scholar. " 'Behold it was very good'—that refers to death" (Midrash, Bereshet Rabba 9:5). According to both schools, the fear of death is an element of human existence, and its contemplation should guard one from sin. Death for the Sanctification of the Name (martyrdom) was especially praiseworthy.

Post-talmudic period

In the post-talmudic period, two streams of thought expressed the principles of Jewish faith: the mystical school, whose classic work was the *Zohar*, The Book of Splendor, and scholastic philosophy, exemplified by the work of Moses Maimonides (1135–1204).

The mystical school. In the literature of the Jewish mystics there is, of course, an acceptance of the principles of normative Judaism. There is also an acceptance of the idea of the transmigration of soul (*gilgul*) as a form of completion of unfinished earthly tasks or as a form of punishment. In addition, there was the idea of *ibbur* (impregnation) which means the "entry of another soul into a man during his life" (Scholem, p. 348). *Ibbur* of a wicked man into the soul of another man is also called a *dibbuk* (something attached). These unwanted inhabitants could be exorcised only by special formulas. The mystical tendency within Judaism was sensitive to the demonic aspects of existence. This strengthened folk superstitions about the harm that the dead spirits might cause. Customs were added to standard Jewish practices such as covering mirrors in houses of mourning, avoiding cemeteries at special times, and guarding people in special circumstances against possible onslaughts by demons. It was also thought that the dead could intervene on behalf of their relatives and friends. These tendencies were strengthened in areas of Jewish settlement influenced by *hasidism*, a pietistic movement in Eastern Europe beginning in the eighteenth century that based many of its beliefs and customs on mystical teachings. The mystic-hasidic movement also developed extensive and imaginative descriptions of heaven and hell (Scholem).

Scholastic philosophy. The scholastic interpretation of Judaism, whose greatest figure was Moses Maimonides, identified intelligence as the most important part of the soul. It was this part of man which enjoyed eternal bliss, since it was concerned with eternal things, especially the nature of God. This meant that an individual's immortality and his real worth were gauged by the degree to which he had acquired knowledge of eternal truths. Immortality was seen as the soul's vision of God unencumbered by the body. According to Maimonides, the resurrected body would also die, the soul returning to paradise to enjoy the beatific vision of the godhead.

Late medieval times saw the creation of normative codes of Jewish law. The most important of these was that of Moses Maimonides called the *Yad Hachazak* (*Strong Hand*) and that of Rabbi Joseph Caro (1488–1575) entitled the *Shulhan Arukh* ("The Prepared Table"). It is to these works that Jews look for specific guidance as to how to treat the dying, how to relate to the corpse, how to mourn for the dead, obligations for self-sacrifice, etc. These codes together with their commentaries are the authoritative sources for Jewish practice until this day, especially for orthodox and conservative Jews.

Modern period

With the beginning of the modern period, reform movements affected Jewish law, tradition, and theology. There was in these movements little stress on the doctrine of the resurrection of the dead. In some liberal prayer books references to this dogma were removed. The idea of the immortality of the soul was preferred. This was evident in the work of Moses Mendelssohn (1729–1786), the first important Jewish thinker in modern times. The Pittsburgh Platform of the American Reform movement stated: "We reassert the doctrine of Judaism that the soul is immortal, grounding this belief on the divine nature of the human spirit, which forever finds bliss in righteousness and misery in wickedness. We reject as ideas not rooted in Judaism, the beliefs both in bodily resurrection and in *Gehenna* and Eden [hell and paradise] as abodes for everlasting punishment and reward" (Philipson, p. 357). Conservative and Orthodox Judaism retain references to the resurrection of the dead in the liturgy, reserving the right to interpret the doctrine as a symbol for the total salvation in God (both body and soul). The rise of reform Judaism also meant that some of the traditional practices were to be liberalized in the area of autopsies, burial, and mourning customs.

Some of the leading philosophers of contemporary Judaism have discussed the idea of death. Franz Rosensweig (1886–1929) opens his great work, the *Star of Redemption* (1921), with the words: "All cognition of the All originates in death; in the fear of death; and this mortal lives in this fear of death." Abraham Joshua Heschel (1907–1972) sees death as the "ultimate self-dedication to the divine. Death so understood will not be distorted by the craving for immortality, for this act of giving away is reciprocity on man's part for God's gift of life. For the pious man it is a privilege to die" (Riemer, p. 73). Writers such as Will Herberg have been influenced in their view of the doctrine of the resurrection of the dead by the writings of Reinhold Niebuhr. Herberg sees the doctrine as an expression of the basic theological assertion that fulfillment is "not a disembodied soul that has sloughed off the body, but the whole man—body, soul and spirit joined in an indissoluble unity" (Herberg, pp. 229–230).

In contemporary Judaism there is also some concern that Jewish funeral practices have become too elaborate and too expensive. There are calls for a return to the more simple rituals of ancient times.

Summary and conclusions

Several important conclusions emerge from the reflections on death in post-biblical Judaism:

1. There is an emphatic stress on the importance of human life. Therefore, to prolong a human life even for a short while is a good deed. Though death is inevitable, it is seen as an event to be opposed. Therefore, in general, traditional Judaism has been opposed to euthanasia, abortion, and suicide.
2. The dead are to be honored because they once had life. The human body, which houses the immortal soul of man, even when life has ceased, is entitled to be treated with dignity and respect. Therefore, in general, traditional Judaism has been opposed to autopsies for no real purpose, cremation, and displays of the dead.
3. There is emphasis on the reality of death. It is seen as a real event in which the real presence of the dead is removed. There is an emphasis on the communal loss as well as the communal responsibility to help mourners overcome the pain of their loss and to start life again. The dead are really dead. Therefore, they must be mourned and the survivors have to learn to live without them.
4. Yet, there is hope that death is not the end of the human drama. The doctrine of immortality and especially the resurrection of the dead means that humans are responsible for their deeds—a responsibility that will somehow be exacted even after death—but it also means that in God they will find their fulfillment and salvation. The purpose of these doctrines is not to make death easier but to enhance life.

SEYMOUR SIEGEL

[*Directly related are the entries* ABORTION, *article on* JEWISH PERSPECTIVES; CADAVERS, *article on* JEWISH PERSPECTIVES; DEATH AND DYING: EUTHANASIA AND SUSTAINING LIFE, *article on* ETHICAL VIEWS; JUDAISM; *and* SUICIDE. *See* APPENDIX, SECTION I, DAILY PRAYER OF A PHYSICIAN ("PRAYER OF MOSES MAIMONIDES").]

BIBLIOGRAPHY

GENERAL LITERATURE

"Death." *Encyclopaedia Judaica.* 16 vols. New York: Macmillan Co., 1972, vol. 5, pp. 1420–1427.
"Death, Views and Customs Concerning." *The Jewish Encyclopedia.* 12 vols. Edited by Cyrus Adler. New York: Funk & Wagnalls, 1901–1906. New ed. 1925, vol. 4, pp. 482–486.
ECKARDT, A. ROY. "Death in the Judaic and Christian Traditions." *Social Research* 39 (1972): 489–514.
HELLER, ZACHARY I. "The Jewish View of Death: Guidelines for Dying." *Death: The Final Stage of Growth.* Edited by Elisabeth [Elizabeth] Kübler-Ross. Englewood Cliffs, N.J.: Prentice-Hall, 1975, pp. 38–43.
LAMM, MAURICE. *The Jewish Way in Death and Mourning.* New York: J. David, 1969.
RIEMER, JACK, ed. *Jewish Reflections on Death.* Foreword by Elizabeth Kübler-Ross. New York: Schocken Books, 1974.

TALMUDIC PERIOD

BÜCHLER, ADOLF. *Studies in Sin and Atonement in the Rabbinic Literature of the First Century* (1928). Library of Biblical Studies. New York: Ktav Publishing House, 1967.
FINKELSTEIN, LOUIS. *The Pharisees: The Sociological Background of Their Faith.* 3d ed. The Morris Loeb series. Philadelphia: Jewish Publication Society of America, 1962.
SCHECHTER, SOLOMON. *Some Aspects of Rabbinic Theology.* New York: Macmillan, 1909.
———. *Studies in Judaism.* 3 vols. Philadelphia: Jewish Publication Society, 1896–1924, vol. 1.

MEDIEVAL PERIOD

SCHOLEM, GERSHOM. *Kabbalah.* Library of Jewish Knowledge. New York: Quadrangle/ New York Times Book Co., 1974.
TRACHTENBERG, JOSHUA. *Jewish Magic and Superstition: A Study in Folk Religion.* Meridian Books and the Jewish Publication Society, no. 33. Cleveland: World Publishing Co., 1961.
TWERSKY, ISADORE, ed. *A Maimonides Reader.* Library of Jewish Studies. New York: Behrman House, 1972.

MODERN PERIOD

HERBERG, WILL. *Judaism and Modern Man: An Interpretation of Jewish Religion.* Jewish Publication Society series, no. JP10. New York: Meridian Books, 1960.
PHILIPSON, DAVID. *The Reform Movement in Judaism.* Rev. ed. New York: Macmillan Co., 1931.

3
POST-BIBLICAL CHRISTIAN THOUGHT

Historians and theologians have sometimes called the twentieth century the "post-Christian" age, implying that Christian institutions and values have ceased to play a normative role in contemporary civilization. Arguable though this view may be, it is nevertheless true that positions popularly perceived to represent the tradition of Christian teaching have influenced debate in such areas of practical and theoretical bioethical concern as the definition of death, the propriety of life support for the terminally ill, and euthanasia. In fact, Christian theology has had little to say about death as a biological phenomenon. It has, however, been deeply concerned with the problem of the destiny of humanity, the fulfillment of the yearning for perfection of both the individual and the social order—and in this context the theological tradition has had much to say concerning the meaning of individual life and human history. This fulfillment has been discussed in terms transcending historical life. Thus, questions concerning the meaning of death in Christian thought have arisen largely in the context of discussions of (1) a postmortem destiny for the person and (2) the future of the Christian community.

The common modern assumption, both lay and clerical, has been that the immortality of the soul is a fundamental aspect of Christian doctrinal and credal development. Recent scholarship on biblical and post-biblical thought agrees, however, that the matter is very complex. The basic biblical concept is the resurrection of the body at the end of history (the "Last Times"), a doctrine that does not necessarily imply an active afterlife between death and the resurrection or inauguration of the Kingdom of God. Against this is set the doctrine of immortality, which presupposes an active afterlife for the disembodied soul. Some proponents of immortality have seen it as only filling the interim between death and resurrection; others have virtually assimilated resurrection within the concept of immortality. A sketch of the history of Christian eschatology must, therefore, be an account of the developments of these often polarized views. (By "eschatology," theologians mean both the vision of the consummation of history in the Last Times and the group of doctrines treating the destiny of the person from death through the Last Judgment.)

Early Christian thought

In the Judaic tradition, man was viewed as a unity whose death was regarded as a natural and, usually, final phenomenon. By New Testament times two developments had modified this ancient tradition. One was the indigenous development of apocalypticism and the related doctrine of the resurrection of the body, which formed an important element in the teaching of Jesus. This was basically a political—and certainly a corporate—doctrine arising from the yearnings of the people: In the fullness of his time God would gather his people—raising their dry bones—to participate in his fulfilled, perfect reign. The second development arose from contact with Hellenistic religious and philosophical notions of immortality. The Hellenistic notion of immortality was that man's soul is released by death from the body, from entrapment in time and space. The most extreme version of the doctrine was the Neoplatonist teaching that the soul is an emanation of the One and thus is eternal in the senses of both preexistence and continuing existence.

So pervasive and common were the notions of the separability of soul and body and of the body's inferiority to the soul in the Hellenistic world to which the Gentile mission of the primitive Church was sent that it was inevitable some accommodation (linguistic, at least, if not conceptual) would have to be made to the doctrine of the soul and its immortality. Thus the basic traditional Christian definition of death as the separation of body and soul is as old as are investigations in the philosophical vein of the nature of the soul. The soul is often conceived as the divine element in the human and, hence, the object of salvation.

Yet the Church Fathers were reluctant in the extreme to speculate on the nature of the soul's existence following death or even to posit for it a conscious or active existence. For example, Irenaeus of Lyons in the second century speaks of the soul as waiting for the apocalypse of the Last Judgment in an "invisible place" and in an inactive state (*Against Heresies* V, 31). It was sometimes said the martyrs constituted an exceptional class whose souls might not have to await the resurrection to enjoy communion with the godhead. However, the generally conservative view of the postmortem destiny of the individual grounded in Hebrew apocalyptic literature continued to hold sway at the same time as room was made to accommodate the concept of a soul and the custom of prayers for the dead. Gregory of Nyssa, in *On the Soul and the Resurrection* (380?), may exemplify this continuity among the Greek fathers. The soul is the divine or immortal element in man; yet it "exists in the actual atoms" it animates. There is no cause to speculate on the locus of the soul after death, although (metaphorically) it can be said to tend to be drawn toward evil or good according to patterns set during life. Conscious of the writings of Plato and of the platonic tradition, Gregory nevertheless maintained that body and soul cannot fully exist in isolation from each other: The postmortem interim is primarily to be characterized as a period of waiting for their reunion at the resurrection.

The Latin or Western Fathers continue the same kinds of teaching. Augustine's early works (e.g., *Soliloquies*) are often regarded as Neoplatonic exercises on the nature of the soul and its knowledge of God. Yet Augustine broke off the *Soliloquies* before he got to the question of the nature of the soul's postmortem knowledge. His related and contemporary treatise, *On the Immortality of the Soul*, he himself later found uselessly obscure, and it is evasive on this issue. The letter (No. 147) on *The Vision of God* treats of the postresurrection knowledge of the godhead; and the great treatment of the subject is *The City of God*, in which Augustine is primarily concerned with the problem of history and its consummation.

Later theologians often pointed to the *Dialogues* of Gregory the Great (Pope, 590–604) as a locus classicus for the notion that the soul may undergo purgation after death, while it continues its active existence in the interim between the separation of body and soul and their reunion. Gregory deals in this work with the problem of miracles as signs of sanctity and with the nature of the contemplative (or monastic) life; and ultimately his concern is with miracles and visions of the afterlife as signs of sanctity. For Gregory, the souls of the perfect (as, for earlier writers, the souls of the martyrs) enjoy immediate beatification. There are no grounds to suppose, however, that the Pope does more than allow the possibility that divine judgment *may* be meted out in the period between death and the resurrection. Gregory is a great exponent of the biblical doctrine of eschatology; and in the sermons on the Gospels and other works, the emphasis is clearly on the consummation of all things and reunion of the soul and body.

Medieval developments

The same may be said of most theologians of the early Middle Ages (to approximately A.D. 1100). Nevertheless, the possibilities for an active postmortem existence inherent in the doctrine of the immortality of the soul did not go unnoticed. Many of the documents of the Apocryphal New Testament (i.e., writings with some claim to apostolic authority but rejected in the process of formation of the New Testament canon because they were of spurious authorship and questionable orthodoxy) pictured the soul (whether metaphorically or literally is often difficult to say) as having an active existence after death. The *Vision of Paul* is a particularly influential item in this group and is often regarded as an important source for Dante's *Divine Comedy*. From it comes, for example, the tradition that the soul returns periodically to the body to castigate or to praise the body for the state the soul currently endures or enjoys; and from it descends likewise a tradition, in Latin and the Western vernacular languages, of the soul's didactic addresses to the body or of debates between the two.

The state of the question of the afterlife in the later Middle Ages is extremely complex. One possible generalization is that academic or scholastic theology absorbed the notion of an active afterlife for the soul into the doctrine of penance. At the same time, traditional eschatology and its literary topics continued to circulate widely, especially in popular and Benedictine monastic theology. Thus in the fourteenth and fifteenth centuries one continues to find corporate, communal, heroic Christology and eschatology in such more "popular" works as Langland's *Piers Ploughman* and the medieval drama, even while a more individualized eschatology and doctrine of the soul achieved dominance in works emanating from the universities and the mendicant orders. Meanwhile, however, didactic exempla (cautionary and exemplary narratives) of the soul's continued active existence had also permeated popular theology with what often seems to us a gruesome preoccupation with the morbid and with the corruptibility of the body. As Ariès has argued, this phenomenon probably reflects not so much a devaluation of life in the world or of the body as the realization that because life is short each moment of it is urgent in fixing one's eternal destiny and is to be highly valued.

For an appreciation of later developments in both Catholic and Protestant theology, it is necessary to dwell briefly on the development of the doctrine of purgatory within the framework of the teaching about penance and on the implications of this development for the notion of an active afterlife between death and the resurrection of the Last Day. In the social (feudal) metaphor for the work of Christ that became dominant with Anselm's *Cur Deus Homo?* (1097), mankind owed God so enormous a satisfaction or compensatory payment for unlawfulness or sinful breach of the covenantal relationship between creator and creature that only God himself in Christ could pay it. After Christ's atonement (or, for the individual, after baptism), the individual was responsible for new breaches. These penalties could be worked off on the basis of penances imposed by the Church in the penitential discipline. Should the penitent die before completing this task, or should the penance be imposed at confession at the point of death, it was conceived that postmortem purgation was allowed. Hence (in large part) the doctrine of purgatory and the notion of suffering were emphasized in discussion of the period following separation of body and soul; concomitantly, active bliss was posited for those who had no penance to complete or had already completed it; and it was argued the temporal church might grant postmortem mitigation of temporal penalties. These developments contributed materially to the shift in emphasis in Christian thought from the resurrection of the Last Times as inaugurating active future life to the fate of the soul between death and the resurrection. A judgment of the individual soul was said to take place at the instant of its separation from the body.

The great work of Thomas Aquinas illustrates this development. For Thomas, the resurrection and Last Judgment remain the necessary culmination of the history of the race and the individual; but primary interest is centered on the soul, the form of the body, which must, in the light of God's justice, receive merited reward as well as merited punishment. Thus the developments after the reunion of body and soul at the resurrection will confirm a judgment and state of being that essentially date from the moment of death—the only change being that some souls will have completed temporal purgation in this period.

Trends in modern Christian thought

One of the major theses of the Protestant reaction was against the excesses of the penitential doctrine of purgatory. Luther, Calvin, and other Reformers were still late medieval theologians, however; and although purgatory was anathema to them and their successors, increased emphasis on the afterlife of the soul before the resurrection remained characteristic of Protestant as well as of Catholic theology in the sixteenth century. For Calvin, for example, Christ's satisfaction for mankind's sins is absolutely sufficient; thus purgatory is unnecessary. But from the earliest versions of the *Institutes*, the immortality of the soul is "beyond controversy" for Calvin, and immortality seems to have greater force as the image of the postmortem destiny of humans than does resurrection. Thus both Catholic and Protestant theology entered the modern period with the idea of resurrection, the ancient cornerstone of Christian eschatological teaching, retaining only a vestigial or metaphorical place in the theological system. The notion of an active postmortem existence enjoyed by far the greater prominence. (The sociological correlatives of these phenomena are well discussed by Ariès.)

The situation depicted above tended to persist in the modern period. Catholic theology from the Council of Trent (1545–1563) through the neo-Thomist period of the twentieth century followed approximately what was understood to be the position of Thomas Aquinas; and Protestant theology through neo-orthodoxy was based on the outlines laid down by the Reformers. Perhaps the greatest influence on both, but most especially the Protestants, was the rise of empiricism in the seventeenth century. For the empiricist, that which is true is that which is authenticated only by experience; and even revelation is an individual and subjective experience. The apocalyptic eschatology of early and medieval Christianity became an embarrassment, and more and more a spiritualized picture of the disembodied soul in the immediate afterlife took its place. Often the issue of divine judgment was evaded in "liberal" circles.

Not until the twentieth century has a sense of biblical eschatology been regained. At present, historical scholars generally agree that any Christian doctrine of immortality ought to be tempered in the light of the more community- and judgment-oriented eschatology of the earlier centuries, but there has been a general failure to communicate this consensus widely in the lay community. It might further be generalized that those who are more concerned with the state of the individual tend to emphasize immortality in their discussions of eschatology, and those more concerned with the social implications of Christianity tend to find the apocalyptic picture of the redeemed community gathered at the resurrection more meaningful.

Conclusions and observations

Recent studies in the history of Christian eschatology perhaps have not been adequately assimilated by students of Christian ethics, who have increasingly been drawn into debate on bioethical issues. This may be in part because discussion of the dogmatic problem of immortality and resurrection seemingly complicates the issues and in part because issues touching the problem of afterlife (even if viewed only metaphorically) are uncomfortable in the contemporary intellectual milieu. Yet eschatology, the doctrine of hope (together with the doctrine of Christ, which is the basic substance of Christian faith), is the basis and sanction for Christian ethics, the doctrine of love. The tenacity with which common opinion holds to what seems to many in academic theological circles a questionable view of the centrality of immortality and the active afterlife immediately following death is at the core of most difficulties concerning the Christian view of death. In order to address the bioethical implications of Christian ethics, it is necessary first to attempt to reconcile the apparent conflict between immortality, with its emphasis on the destiny of the individual, and resurrection, with its stress on the destiny of the community. The values of both views ought to be weighed carefully.

Regarding such matters as the definition of physical death, there seems little in theological reflection on the subject that helps or hinders in the determination of the moment at which what has been called the separation of body and soul occurs. In discussions of "death with dignity," Christian ethical thought should have suggestions to offer with an eye to reconciling the conflicts between the needs of the individual, the requirements of society, and the traditions of professional medical ethics. Death will always be a fearful event for the dying, their friends, and those who minister to them. There is now a widespread recognition that ethical concerns surrounding death should be faced and discussed rather than evaded, as they have tended to be since the rise of modern medicine and the transfer of care for the dying in urban society from home, church, and family to professional institutions and professional persons. Christian ethics needs to come to terms with Christian eschatology in order to help frame new and helpfully realistic approaches to care for the dying and to the new bioethical issues raised by modern research and applied medicine.

In the past, Christian thought has valued life in the world as the arena in which humanity's yearning and struggle for redemption are worked out. The moment of death thus has been one for assessment or judgment of the individual's contribution to that struggle. At best, Christian theology has reflected on death so as to emphasize the importance of life in the great drama of salvation. Some would say that, at its worst, Christian teaching has seemed to imply that worldly life—in contrast with the ultimate goal of beatific vision—is a status without value from which death is a release. Perhaps the most needed contribution of Christian ethics is to communicate more widely that Christian thought on death is neither so simple nor so "dogmatic" as it is sometimes perceived to be.

MILTON McC. GATCH

[*Directly related are the two preceding articles,* DEATH IN BIBLICAL THOUGHT *and* POST-BIBLICAL JEWISH TRADITION, *and the following article on* ARS MORIENDI. *For discussion of related ideas, see:* PAIN AND SUFFERING, *article on* RELIGIOUS PERSPECTIVES.]

BIBLIOGRAPHY

ARIÈS, PHILIPPE. *Western Attitudes toward Death: From the Middle Ages to the Present.* Translated by Patricia M. Ranum. Johns Hopkins Symposia in Comparative History. Baltimore: Johns Hopkins University Press, 1974.

BOASE, THOMAS SHERRER ROSS. *Death in the Middle Ages: Mortality, Judgement, and Remembrance.* Library of Medieval Civilization. New York: McGraw-Hill, 1972. Collects material from the iconography of medieval churches, manuscripts, and funerary monuments to illustrate both the learned and the popular theological traditions.

CHORON, JACQUES. *Death and Western Thought.* New York: Collier Books, 1963. More philosophical than theological in its focus, but a useful survey.

FROOM, LE ROY EDWIN. *The Prophetic Faith of Our Fathers: The Historical Development of Prophetic Interpretation.* 4 vols. Washington: Review & Herald, 1950–1954. Written from the point of view of the tradition that seeks clues in history and scripture to the approach of the Last Times, but a mine of information on the history of apocalyptic thought.

GATCH, MILTON McC. *Death: Meaning and Mortality in Christian Thought and Contemporary Culture.* New York: Seabury Press, 1969.

———. "Some Theological Reflections on Death from the Early Church through the Reformation." *Perspectives on Death.* Edited by Liston O. Mills. Nashville: Abingdon Press, 1969, chap. 3, pp. 99–136.

RAHNER, KARL. *On the Theology of Death.* Translated by Charles H. Henkey. Quaestiones disputatae, no. 2. New York: Herder & Herder, 1961. 2d ed. Freiburg: Herder; London: Burns & Oates, 1965. Very influential modern Catholic essay.

STENDAHL, KRISTER, ed. *Immortality and Resurrection: Four Essays.* Ingersoll Lectures, Harvard University, 1955–1959. New York: Macmillan, 1965.

4

ARS MORIENDI

Ars moriendi, or *The Art of Dying,* identifies an important subgenre of later medieval conduct literature. Like similar treatises on courtesy, courtship, education, recreation, and warfare, these little manuals on dying were meant to guide the reader's behavior on an occasion of some importance to him. Today, about three hundred manuscripts of the *Ars moriendi* survive along with a hundred or so incunabula, including both block-books and editions printed in movable type. The best evidence indicates that the first woodcut edition appeared by 1465, and the fact that some twenty percent of all surviving block-books are *Artes moriendi* is enough to show the extraordinary importance of these books for the history of early printing. Both Latin and vernacular texts (in seven modern languages, eventually) had appeared in both woodcut and printed form by 1475.

The *Ars moriendi* occurs in a longer and a shorter version. Almost all the manuscripts, the majority of the vernacular texts, and most of the typographic editions are of the longer version, which is divided into six sections: (1) a miscellany of quotations on death from Christian authorities; (2) advice to the dying man (*Moriens*) on overcoming temptations to faithlessness, despair, impatience, pride, and worldliness; (3) catechetical questions whose correct answers lead to salvation; (4) prayers and rules for imitating the dying Christ; (5) advice to persons attending the dying man; (6) prayers to be said by the attendants.

The shorter *Ars moriendi* seems to be an abridged derivative of the longer, which is about triple its size. The smaller work occurs in almost all the woodcut editions. A brief introduction and conclusion summarize the material contained in sections 1, 3, 4, and 5 of the longer version. The body of the shorter version corresponds to section 2 of the longer, but this material is transformed and dramatized as a *psychomachia*, a struggle between good and bad angels for the dying man's soul. *Moriens* must choose either the five vices mentioned in the longer text or their contrary virtues: faith, hope, love, humility, and detachment. In a shorter *Ars moriendi* eleven striking woodcuts in a fifteenth-century Flemish style illustrate the fight against temptation: five cuts for the virtues, five for the vices, and one for the eventual delivery of *Moriens*'s soul to the good angel.

The origins of the *Ars moriendi* are in the compendia of piety and doctrine that appeared when later medieval Church councils took steps to educate the laity in the fundamentals of Christianity. Jean Gerson's *Opus tripartitum* (before 1408) grew out of such concerns, and the third part of this work, titled *De arte moriendi*, is the source of much of the *Ars moriendi*. Other important sources are to be found in the Bible, the Fathers, medieval liturgies, papal and conciliar statements, medieval patristic collections, and later medieval devotional and doctrinal literature. The authorship of the *Ars moriendi* remains in doubt, but there is good reason to locate its composition in southern Germany, at the time of the Council of Constance (1414–1418), and with the Dominican Order.

Both the fifth section of the longer *Ars moriendi* and Gerson's *De arte moriendi* advise those who attend the dying man not to give him false hopes of regaining the health of his body. This is a commentary not only on the state of medicine in the later Middle Ages and on contemporary attitudes toward it but also on the Christian understanding of the death of the body. Since death was a beginning, not an end, there was no reason to take heroic measures to postpone it. Innocent III, whose decretals also found their way into the *Ars moriendi*, warned that none were to administer bodily medicine to *Moriens* until the ills of his soul had been cared for. The antecedents of this attitude are as old as Plato (*Charmides* 156–157) and the Gospels (Matt. 8:5–13; John 5:1–14).

The woodcut that depicts *Moriens*'s being tempted against faith also sets deathbed medicine in a problematic light. Three learned men, apparently physicians, stand in consultation at *Moriens*'s right while a demon points at them and whispers the words *"infernus factus est"* into the dying man's ear. Since the context of this illustration is the temptation against faith, and since other figures in it represent idolatry, suicide, and other errors, some authorities have concluded that the three figures in question are not physicians but heretics. If so, the demon who says "hell has been prepared" would be warning *Moriens* against the sin of heresy— hardly a typical concern for a dying Christian and, in any event, not the sort of advice a demon ought to be giving. In light of what the *Ars moriendi* says elsewhere about corporeal medicine, it is more likely that the demon is enticing *Moriens* into the error of caring for his physical health just at the moment when spiritual health ought to be his chief concern. The demon raises the prospect of damnation in order to frighten *Moriens* away from a good Christian death.

Johan Huizinga and other authorities have mentioned the *Ars moriendi* in the same breath with the *danse macabre*, plague books, grisly funerary art, and other manifestations of an obsession with death in the later Middle Ages, an obsession consequent, perhaps, upon the great plague, which arrived in Europe in 1348 and recurred for several centuries. While the great popularity of the *Ars moriendi* may support such a view, its contents generally do not. True, crowds of demons populate its woodcuts, and its *Moriens* looks like a dying man, but on the whole there is little of the macabre in the advice the *Ars moriendi* gives. Its purpose was not to terrorize but to point the way to a good death. It ended in the expectation of heaven, not with the horrors of hell.

For a society whose members viewed death as a passage to an afterlife, a manual on the art of dying had an eminently practical function. The recent tendency to see the *Ars moriendi* as a feature of a necrophile culture may tell us more about modern anxieties than about medieval attitudes.

BRIAN P. COPENHAVER

[*Other relevant material may be found under* DEATH, ATTITUDES TOWARD.]

BIBLIOGRAPHY

BOASE, THOMAS SHERRER ROSS. *Death in the Middle Ages: Mortality, Judgment, and Remembrance.* Library of Medieval Civilization. London: Thames & Hudson; New York: McGraw-Hill, 1972.

HUIZINGA, JOHAN. *The Waning of the Middle Ages: A Study of the Forms of Life, Thought and Art in France and the Netherlands in the XIVth and XVth Centuries.* Translated by F. Hopman. New York: Longmans, Green, 1949.

O'CONNOR, MARY CATHARINE. *The Art of Dying Well: The Development of the Ars moriendi.* Columbia University Studies in English and Comparative Literature, no. 156. New York: Columbia University Press, 1942. Still the fundamental work.

V

DEATH IN THE WESTERN WORLD

Death has been a "fact of life" for as long as humans can remember, but this in no way puts to rest their struggle to understand its meaning. Some conceptualization, beyond common sense, of a human individual or "person" is necessary in order to understand the problematic of death. Therefore, a few comments on this topic are in order before proceeding to a reflection on some of the more salient features of death as it has been understood in the Western world.

The person and the problematic of death

The human individual has often been viewed in the Western world as a synthesized combination of a living organism and a "personality system" (an older terminology made the person a combination of "body" and "mind" or "soul").

It is in fact no more mystical to conceive of a personality *analytically* distinct from an organism than it is to conceive of a "culture" distinct from the human populations of organisms who are its bearers. The primary criterion of personality, as distinct from organism, is an organization in terms of symbols and their meaningful relations to each other and to persons.

Human individuals, in their organic aspect, come into being through a process of bisexual reproduction. They then go through a more or less well-defined "life course" and eventually die. That human individuals die as organisms is indisputable. If any biological proposition can be regarded as firmly established, it is that the mortality of individual organisms of a sexually reproducing species is completely normal. The death of individuals has indeed a positive survival value for the species.

As Freud said, organic death, while a many-faceted thing, is in one principal aspect the "return to the inorganic state." At this level the human organism is "made up" of inorganic materials but is organized in quite special ways. When that organization breaks down—and there is evidence that this is inevitable by reason of the "aging process"—the constituent elements are no longer part of the living organism but come to be assimilated to the inorganic environment. Still, even within such a perspective on the human individual as an organism, life "goes on." The human individual does not stand alone but is part of an intergenerational chain of indefinite durability, the species. The individual organism dies, but if he or she reproduces the "line" continues into future generations.

But the problematic of human death arises from the fact that the human individual is not only an organism but also a user of symbols, who learns symbolic meanings, communicates with others and with himself through them as media, and regulates his behavior, thought, and feelings in symbolic terms. He is an "actor" or a "personality." The human "actor" clearly is not born in the same sense in which an organism is. The personality or actor comes into being through a gradual and complicated process sometimes termed "socialization."

Furthermore, there is a parallel—in my judgment, something more than a mere "analogy"—between the continuity of the "actor" and that of the organism. Just as there is an intergenerational continuity on the organic side, so is there an intergenerational continuity on the personality or action side of the human individual. An individual personality is "generated" in symbiosis with a growing individual organism and, for all we know, "dies" with that organism. But the individual personality is embedded in transindividual action systems, both social and cultural. Thus the sociocultural "matrix" in which the individual personality is embedded is in an important sense the counterpart of the population–species matrix in which the individual organism is embedded. The individual personality "dies," but the society and cultural system, of which in life he or she was a part, goes on.

But what is it that happens when the personality "dies"? Is the death of a personality to be simply assimilated to the organic paradigm? It would seem that the answer is yes, for just as no personality in the human sense can be conceived as such to develop independently of a living organism, so no human personality can be conceived as such to survive the death of the same organism. Nevertheless, the personality or "actor" certainly influences what happens in the organism—as suicide and all sorts of "psychic" factors in illnesses and deaths bear witness. Thus, although most positivists and materialists would affirm that the death of the personality must be viewed strictly according to the organic paradigm, this answer to the problem of human death has not been accepted by the majority in most human societies and cultures. From such primitive peoples as the Australian aborigines to the most sophisticated of the world religions, beliefs in the existence of an individual "soul" have persisted, conceivably with a capacity both to antedate and to survive the individual organism or body. The persistence of that belief and the factors giving rise to it provide the framework for the problematic of death in the Western world.

Christian orientations toward death

Because the dominant religious influence in the history of the Western world has been that of Christianity, it is appropriate to outline the main Christian patterns of orientation toward death.

There is no doubt of the predominance of a duality of levels in the Christian paradigm of the human condition, the levels of the spiritual and the material, the eternal and the temporal. On the one hand, there is the material-temporal world, of which one religious symbol is the "dust" to which man is said to return at death. On the other hand, there is the spiritual world of "eternal life," which is the location of things divine, not human. The human person stands at the meeting of the two worlds, for he is, like the animals, made of "dust," but he is also, unlike the animals, made in the image of God. This biblical notion of man, when linked to Greek philosophical thought, gave rise to the idea in Catholic Christianity that the divine image was centered in the human "soul," which was conceived as in some sense an "emanation" from the spiritual world of eternal life. Thus arose the notion of the "immortal soul," which could survive the death of the organism, to be rejoined to a "resurrected" body. The hope of the resurrection, rooted in the Easter faith of the Christian community, was from the beginning a part of the Christian faith and provided another dimension behind the teaching on the immortality of the "soul."

The Christian understanding of death as an event in which "life is changed, not taken away" in the words of the traditional requiem hymn, *Dies Irae,* can be interpreted in terms of Marcel Mauss's paradigm of the gift and its reciprocation (Parsons, Fox, and Lidz). Seen in this way the life of the individual is a gift from God, and like other gifts it creates expectations of reciprocation. Living "in the faith" is part of the reciprocation, but, more important to us, dying in the faith completes the cycle. By the doctrine of reciprocation mankind assumes, it may be said, three principal obligations, namely, to "accept" the human condition as ordained by the Divine Will, to live in the faith, and to die in the faith—and to die in the faith means to die with the hope of resurrection. If these conditions are fulfilled, "salvation," life eternal with God, will come about.

This basically was the paradigm of death in Catholic Christianity. Although the Reformation did "collapse" some elements in the Catholic paradigm of dualism between the eternal and the temporal, it did not as such alter fundamentally the meaning of death in societies shaped by the Christian faith. Still, the collapse of the Catholic paradigm did put great pressures on the received doctrine of salvation. The promise of a *personal* "eternal life" became increasingly difficult to accept, and the doctrine of eternal punishment in "hell" proved even more difficult to uphold.

The conception of a "higher" level of reality, a "supernatural" world in which human persons survived after death, did not give way but became more and more difficult to "visualize" (along with the meaning of death as an event in which one gave life back to its Giver and in return was initiated into a new and eternal life) by simple extrapolation from this-worldly experience. In addition to the changes in conceptualization set in motion by the Reformation, the rise of modern science, which by the eighteenth century had produced a philosophy of scientific "materialism," posed an additional challenge to the Christian paradigm of death, manifesting itself primarily in a "monism" of the physical world. There was at that time little "scientific" analysis of the world of action, and there was accordingly a tendency to regard the physical universe as unchanging and hence eternal. Death then was simply the return to the inorganic state, which implied a complete negation of the conception of "eternal life," since the physical, inorganic world was by definition the antithesis of life in any sense.

Contemporary scientific orientations

The subsequent development of science has, however, modified or at least brought into question the monistic and materialistic paradigm generated by the early enthusiasm for a purely positivistic approach. For one thing, beginning in the nineteenth century and continuing into the twentieth the sciences of organic life have matured, thanks largely to placing the conception of evolutionary change at the very center of biological thought. This resulted in the view, which we have already noted, that death is biologically normal for individual members of evolving species.

A second and more recent development has been the maturing of the sciences of *action*. Although these have historical roots in the "humanistic" tradition, they have only recently been differentiated from the humanistic trunk to become generalizing *sciences*, integrating within themselves the same conception of evolutionary change that has become the hallmark of the sciences of life.

The development of the action sciences has given rise, as already noted, to a viable conception of the human person as *analytically* distinct from the organism. At the same time these sciences, by inserting the person into an evolutionary sociocultural matrix analogous to the physico-organic species matrix within which the individual organism is embedded, have been able to create an intellectual framework within which the death of the personality can be understood to be as normal as the death of the organism.

Finally, the concept of evolutionary change has been extended from the fields of the life sciences (concerned with the organism) and of the action sciences (concerned with the person-actor) to include the *whole* of empirical reality, and *at the same time* we have been made aware —principally by the ways in which Einstein's theory of relativity modified the previous assumptions of the absolute empirical "givenness" of physical nature in the Newtonian tradition— of the *relative* character of our human understanding of the human condition.

Thus there is now a serious questioning of absolutes, both in our search for absolutely universal laws of physical nature and in our quest for "metaphysical" absolutes in the philosophical wake of Christian theology.

The Kantian impact and the limits of understanding

The developments in a contemporary scientific understanding of the human condition are both congruent with and in part anticipated and influenced by Kant, whose work during the late eighteenth century was the decisive turning point away from both physical and "metaphysical" absolutism. Kant basically accepted the reality of the physical universe, as it is humanly known, but at the same time he "relativized" our knowledge of it to the categories of the understanding, which were not grounded in our direct "experience" of physical reality but in something "transcending" this. At the same time Kant equally relativized our conceptions of "transcendental" reality, whose existence he by no means denied, to something closer to the human condition. Indeed, it may be suggested that Kant substituted "procedural" conceptions of the "absolute," whether physical or "metaphysical," for "substantive" propositions.

While relativizing our knowledge both of the physical world, including the individual human organism, and of the "metaphysical" world, with its certitude about the immortality of the soul, Kant nonetheless insisted on a transcendental component in human understanding and explicitly included belief in personal immortality in the sense of eternal life.

With respect to the bearing of Kant's thought and its influence through subsequent culture on the problem of the meaning of death, I have already noted that he prepared the way, procedurally, for the development of the action sciences and their ability to account intellectually for the personality or "actor" experienced as one aspect of the human individual without the need to infer, of necessity, the existence of a spiritual soul existentially and not merely analytically distinct from the living organism. The action sciences, in a very real sense, attempt to provide a coherent account of human "subjectivity," much as Kant himself attempted to do in his *Critique of Judgment,* without "collapsing" the difference of levels between the physical and what may be called the "telic" realm.

The framework provided by Kant's thought is indeed congenial to the scientific perspective on the normality of the death of a person, conceived as an actor whose coming into existence is in symbiosis with a growing individual organism and whose individual personality, while continuing into a new generation in the same sociocultural system, can be understood to "die" in symbiosis with the same organism. Nonetheless, if Kant was right in refusing to collapse the boundaries of the human condition into the one vis-à-vis the physical world, the meaning of human individual death can no more be exhausted by that of the involvement of the human individual in a sociocultural system of more comprehensive temporal duration than can the meaning of our sensory experience of empirical reality be exhausted by the "impressions" emanating from that external world, or even the theoretical ordering of those impressions.

If Kant's fundamental position is accepted, then his skepticism about absolutes must apply to both sides of the fundamental dichotomy. Modern biology certainly must be classed as knowledege of the empirical world in his sense, and the same is true of our scientific knowledge of human action. In his famous terminology, there is no demonstrable knowledge of the thing in itself in *any* scientific field.

In empirical terms organic death is completely normal. We have, and acording to Kant we presumably can have, no knowledge of the survival of any organic entity after death except through the processes of organic reproduction, through which the genetic heritage does survive. Kant, however, would equally deny that such survival can be excluded on empirical grounds. This has an obvious bearing on the Christian doctrine of the resurrection of the body. If that is meant in a literal biological sense (though this is by no means universally the way in which Christians understand it), then the inference is clearly that it can never be proved, but it can still be "speculated" about and can be a matter of "faith," even though it cannot be the object of either philosophical or scientific demonstration.

The same seems to hold for the personality-action component of the human individual. Empirically, the action sciences can account for its coming-to-be and its demise without postulating its survival. But neither can they exclude the possibility of such survival. Thus the "eternal life" of the individual soul, although metaphysically unknowable, can, like resurrected bodies, be speculated about and believed in as a matter of "faith."

Thus, included in the victims of Kant's skepticism or relativization is belief in the cognitive *necessity* of belief in the survival of human individuality after death as well as belief in the cognitive *necessity* of belief in the nonsurvival of human individuality after death. Kant's relativization of our knowledge, both empirical and metaphysical, both closed and opened doors. It did, of course, undermine the traditional specificities of received beliefs; but at the same time and for the very same reason it opened the door, by contrast to scientific materialism, not merely to one alternative to received Christian belief but to a multiplicity of them.

This leaves us with the position that the problem of the meaning of death in the Western tradition has, from a position of relative closure defined by the Christian syndrome, been "opened up" in its recent phase. There is above all a new freedom for individuals and sociocultural movements to "try their hands" at innovative definitions and conceptions. At the same time, the "viability" of their innovations is subject to the constraints of the human condition, both empirical and transcendental, noted by Kant.

The problem of the meaning of death in the West is now in what must appear to many to be a strangely unsatisfactory state. It seems to come down to the proposition that the meaning of death is that, in the human condition, it cannot have any "apodictically certain" meaning without abridgement of the essential human freedom of thought, experience, and imagination. Within limits, its meaning, as it is thought about, experienced for the case of others, and anticipated for oneself, must be autonomously interpreted. But this is not pure negativism or nihilism, because such openness is not the same as declaring death, and of course with it individual life, to be "meaningless."

Conclusion

Insofar as it is accessible to cognitive understanding at all, the problem of the meaning of death for individual human beings must be approached in the framework of the human condition as a whole. It must include both the relevant scientific understanding and understanding at philosophical levels, and must attempt to synthesize them. Finally it must, as clearly as possible, recognize and take account of the limits of both our scientific and our philosophical understanding.

If the account provided in the preceding sections is a correct appraisal of the situation in the Western world today, it is not surprising that there is a great deal of bafflement, anxiety, and indeed downright confusion in contemporary attitudes and opinions in this area. Any consensus about the meaning of death in the Western world today seems far off, although the attitude reflected in this article would seem to be the one most firmly established at philosophical levels and the level of rather abstract scientific theory.

A very brief discussion of three empirical points may help, however, to mitigate the impression of extreme abstractness. First, though scientific evidence has established the fact of the inevitability of death with increasing clarity, this does not mean that the *experience* of death by human populations may not change with changing circumstances. Thus, we may distinguish between inevitable death and "adventitious" death, that is, deaths that are "premature" relative to the full life span, and in principle preventable by human action (Parsons and Lidz). Within the last century and a half or so, this latter category of deaths has decreased enormously. The proportion of persons in modern populations over sixty-five has thus increased greatly, as has the expectancy of life at birth. This clearly means that a greatly increased proportion of modern humans approximate to living out a full life course, rather than dying prematurely. Persons living to "a ripe old age" will have experienced an inevitably larger number of deaths of other persons who were important to them. These will be in decreasing number the deaths of persons younger than themselves, notably their own children, and increasingly deaths of their parents and whole ranges of persons of an older generation, such as teachers, senior occupational associates, and many public figures. Quite clearly these demographic changes will have a strong effect on the balance of experience and expectations, of the deaths of significant others, and of anticipation of one's own death.

Second, one of the centrally important aspects of a process of change in orientation of the sort described should be the appearance of signs of the differentiation of attitudes and conceptions with regard to the meaning of the life cycle. There has indeed already been such a process of differentiation, apparently not yet completed, with respect to both ends of the life cycle (Parsons, Fox, and Lidz). With respect to the beginning, of course, this centers on the controversy over the legitimacy of abortion and the beginning of life. And concomitant with this controversy has been an attempt at redefinition of death. So far the most important movement has been to draw a line *within* the organic sector between what has been called "brain death," where irreversible changes have taken place, destroying the functioning of the central nervous system, and what has been called "metabolic death," where, above all, the functions of heartbeat and respiration have ceased. The problem has been highlighted by the capacity of artificial measures to keep a person "alive" for long periods after the irreversible cessation of brain function. The main point of interest here is the connection of brain function with the personality level of individuality. An organism that continues to "live" at only the metabolic level may be said to be dead as an actor or person.

Third, and finally, a few remarks about the significance for our problem of Freud's most mature theoretical statement need to be made. It was printed in the monograph published in English under the title the *Problem of Anxiety*. In this, his last major theoretical work, Freud rather drastically revised his previous views about the nature of anxiety. He focused on the expectation of the loss of an "object." For Freud the relevant meaning of the term "object" was a human individual standing in an emotionally significant relation to the person of reference. To the growing child, of course, his parents became "lost objects" in the nature of the process of growing up, in that their significance for the growing child was inevitably "lost" at later ages. The ultimate loss of a concrete human person as object—of cathexis, Freud said—is the death of that person. To have "grown away" from one's parents is one thing, but to experience their actual deaths is another. Freud's own account of the impact on him of the death of his father is a particularly relevant case in point.

Equally clearly an individual's own death, in anticipation, can be subsumed under the category of object loss, particularly in view of Freud's theory of narcissism, by which he meant the individual's cathexis of his own self as a love object. Anxiety, however, is not the actual experience of object loss, nor is it, according to Feud, the fear of it. It is an anticipatory orientation in which the actor's own emotional security is particularly involved. It is a field of rather free play of fantasy as to what "might" be the consequences of an anticipated or merely possible event.

Given the hypothesis that, in our scientifically oriented civilization, there is widespread acceptance of death—meant as the antithesis of its denial—there is no reason why this should lead to a cessation or even substantial diminution of *anxiety* about death, both that of others and one's own. Indeed, in certain circumstances the levels of anxiety may be expected to increase rather than the reverse. The frequent assertions that our society is characterized by pervasive denial of death may often be interpreted as calling attention to pervasive anxieties about death, which is *not* the same thing. There can be no doubt that in most cases death is, in experience and in anticipation, a traumatic event. Fantasies, in such circumstances, are often characterized by strains of unrealism, but the prevalence of such phenomena does not constitute a distortion of the basic cultural framework within which we moderns orient ourselves to the meaning of death.

Indeed, the preceding illustrations serve to enhance the importance of clarification, at the theoretical and philosophical levels, to which the bulk of this article has been devoted. This is essential if an intelligible approach is to be made to the understanding of such problems as shifts in attitudes toward various age groups in modern society, particularly the older groups, and the relatively sudden eruption of dissatisfaction with the traditional modes of conceptualizing the beginning and the termination of a human life and with allegations about the pervasive denial of death, which is often interpreted as a kind of failure of "intestinal fortitude." However important the recent movements for increasing expression of emotional interests and the like, ours remains a culture to which its cognitive framework is of paramount significance.

TALCOTT PARSONS

[While all the articles in this entry are relevant, see especially the articles WESTERN PHILOSOPHICAL THOUGHT *and* WESTERN RELIGIOUS THOUGHT. *Also directly related is* DEATH, ATTITUDES TOWARD. *For discussion of related ideas, see the entries:* EMBODIMENT; MIND–BODY PROBLEM; ABORTION; *and* DEATH, DEFINITION AND DETERMINATION OF.]

BIBLIOGRAPHY

BELLAH, ROBERT N. "Religious Evolution." *American Sociological Review* 29 (1964): 358–374.

BURKE, KENNETH. *The Rhetoric of Religion: Studies in Logology.* Boston: Beacon Press, 1961. Reprint. Berkeley: University of California Press, 1970.

CHOMSKY, NOAM. *Syntactic Structures.* Janua Linguarum, no. 4. 'S-Gravenhage: Mouton & Co., 1957.

DURKHEIM, ÉMILE. *The Division of Labor in Society* (1893). Translated by George Simpson. Free Press Paperbacks. Glencoe, Ill.: 1933, 1964.

———. *The Elementary Forms of the Religious Life* (1912). Translated by Joseph Ward Swain. London: Allen & Unwin, 1954. Reprint. Glencoe, Ill.: Free Press, 1976.

———. *Sociology and Philosophy* (1924). Enl. ed. Translated by D. F. Pocock. New York: Free Press, 1974.

FREUD, SIGMUND. *Beyond the Pleasure Principle* (1920). Standard Edition, vol. 18, pp. 7–64.

———. *The Ego and the Id* (1923). Standard Edition, vol. 19, pp. 12–66.

———. *The Future of an Illusion* (1927). Standard Edition, vol. 21, pp. 5–56.

———. *Inhibitions, Symptoms and Anxiety* (1926). Standard Edition, vol. 20, pp. 77–175.

———. *The Origins of Psycho-analysis: Letters to Wilhelm Fliess, Drafts and Notes, 1887–1902.* Edited by Marie Bonaparte, Anna Freud, and Ernst Kris. Translation by Eric Mosbacher and James Strachey. Introduction by Ernst Kris. New York: Basic Books, 1954.

———. *The Standard Edition of the Complete Psychological Works of Sigmund Freud.* 24 vols. Edited by James Strachey, Anna Freud, Alix Strachey, and Alan Tyson. London: Hogarth Press, 1955–1974.

HENDERSON, LAWRENCE JOSEPH. *The Fitness of the Environment: An Inquiry into the Biological Significance of the Properties of Matter.* In part delivered as lectures in the Lowell Institute, February, 1913. New York: Macmillan Co., 1913. Paperback ed. Boston: Beacon Press, 1958.

———. *The Order of Nature: An Essay.* Cambridge: Harvard University Press, 1917.

———. *Pareto's General Sociology: A Physiologist's Interpretation.* Cambridge: Harvard University Press, 1935.

KANT, IMMANUEL. *Critique of Judgment* (1790). Translated by James Creed Meredith. Oxford: Clarendon Press, 1964.

———. *Critique of Practical Reason and Other Writings in Moral Philosophy* (1788). Translated and edited by Lewis White Beck. Chicago: University of Chicago Press, 1949.

———. *Critique of Pure Reason* (1781). Translated by Norman Kemp Smith. London: Macmillan & Co., 1929.

LEACH, EDMUND R. *Genesis as Myth, and Other Essays.* Cape Editions, no. 39. London: Jonathan Cape, 1969.

LÉVI-STRAUSS, CLAUDE. *Structural Anthropology.* Translated by Claire Jacobson and Brooke Grundfest Schoepf. New York: Basic Books, 1963.

LOVEJOY, ARTHUR ONCKEN. *The Great Chain of Being: A Study of the History of an Idea.* The William James

Lectures delivered at Harvard University, 1933. Cambridge: Harvard University Press, 1936. Reprint. Harper Torchbooks, 1960.

Mauss, Marcel. *The Gift: Forms and Functions of Exchange in Archaic Societies* (1925). Translated by Ian Cunnison. Glencoe, Ill.: Free Press, 1954.

Nock, Arthur D. *Conversion: The Old and the New in Religion from Alexander the Great to Augustine of Hippo.* London: Oxford University Press, 1933, 1961.

Parsons, Talcott. *Social Systems and the Evolution of Action Theory.* New York: Free Press, 1977.

———; Fox, Renée C.; and Lidz, Victor M. "The 'Gift of Life' and Its Reciprocation." *Social Research* 39 (1972): 367–415. Reprint. *Death in American Experience.* Edited by Arien Mack. New York: Schocken Books, 1973, pp. 1–49.

———, and Lidz, Victor. "Death in American Society." *Essays in Self-Destruction.* Edited by Edwin S. Shneidman. New York: Science House, 1967, chap. 7, pp. 133–170.

Warner, William Lloyd. *The Living and the Dead: A Study of the Symbolic Life of Americans.* Yankee City Series, vol. 5. New Haven: Yale University Press, 1959. Reprint. Westport, Conn.: Greenwood Press, 1975.

Weber, Max. *The Protestant Ethic and the Spirit of Capitalism* (1904). Translated by Talcott Parsons. Foreword by R. H. Tawney. London: G. Allen & Unwin, 1930. Reprint. New York: Charles Scribner's Sons, 1958.

———. *The Sociology of Religion* (1922). Translated by Ephraim Fischoff. Boston: Beacon Press, 1963.

DEATH AND DYING: EUTHANASIA AND SUSTAINING LIFE

The article on HISTORICAL PERSPECTIVES *focuses on historical views of euthanasia and examines the origins of the ethical distinctions in contemporary thought. The second article,* ETHICAL VIEWS, *systematically examines the ethical issues involved in decisions affecting the termination of human life in a biomedical setting and discusses such questions as the right to die, and indirect and direct killing. The final article,* PROFESSIONAL AND PUBLIC POLICIES, *discusses the moral arguments for and against the different options in professional and public policy concerning euthanasia and the prolongation of life.*

I. HISTORICAL PERSPECTIVES   *Gerald J. Gruman*
II. ETHICAL VIEWS   *Sissela Bok*
III. PROFESSIONAL AND PUBLIC POLICIES
                                     Robert M. Veatch

I
HISTORICAL PERSPECTIVES

Introduction

This article will survey in historical context a number of key concepts of the modern era from the Renaissance on, concerning the ethics of sustaining and lengthening life or, contrariwise, hastening death. Such an outline is meant to help elucidate contemporary questions in biomedicine and to provide a useful perspective for the subsequent articles on ethics and on public policy.

The focus of this article on the modern era is not meant to imply that the questions treated here first appeared in the Renaissance but rather that it was then that they became readily relevant to the secular, scientific culture of the twentieth century. Already in prehistoric times, measures had been taken to hasten death, if we may judge from the observed practices of "primitive" cultures. In Graeco-Roman antiquity, there was a generally recognized "freedom to leave" that permitted the sick and despondent to terminate their lives, sometimes with outside help. The combination of tolerance and unconcern that allowed such practices was ended during the rule of Christianity in the medieval West. Although Christian charity brought a heightened responsibility to relieve suffering, the Sixth Commandment (Fifth Commandment in Roman Catholic and Lutheran traditions) seemed to prohibit absolutely the taking of the patient's life.

As to sustaining life, yearnings for extended longevity were characteristic in early folklore and found expression in the religion of ancient Egypt. With the Latin alchemy of the Middle Ages, an ambitious quest for the prolongation of life entered Western culture. On the one hand, man's dominion over nature inspired the thirteenth-century monk Roger Bacon to claim that Christian medicine would surpass pagan science by conquering senescence. On the other hand, medieval Christian thought was pervaded by a disdain for this-worldly considerations compared with the importance of supernatural salvation.

In contemporary discussions of biomedical ethics, euthanasia often is contrasted with the prolongation of life almost as if these aims inherently were opposed. This was not previously the case; the meanings of the terms have changed over the years. Until the seventeenth century, "euthanasia" generally referred to any means for an "easy" death; for example, by leading a temperate life or by cultivating an acceptance of mortality. However, upon entering the domain of medicine, with Francis Bacon's *Advancement of Learning* (1605), "euthanasia" increasingly came to connote specifically measures taken by the physician, including the possibility of hastening death. It is the latter meaning of a more rapid death that has been prevalent in the twentieth century, often in reference to the movement to reform legislation for that purpose.

With the onset of the idea of progress, the vision of the Christian alchemist Roger Bacon became secularized and adapted to the new sciences. The resulting belief I termed "prolongevity" (1956), i.e., the idea that it is possible and desirable to increase significantly the length of healthful, effective life by means of biomedical science. After the Enlightenment, biomedicine in general took on a sense of mission that may be called "meliorism," i.e., the belief that humanity can and should act to make this world better.

A different meaning of prolongation of life, more analogous to life sustaining, is "life support"—contingent measures to keep a patient alive who otherwise shortly would die. This aspect of modern biomedicine, growing out of heroic experiments in resuscitation, has evolved from temporary expediency to become long-term and routine, as in the various methods maintaining comatose patients and cases of renal failure. This development of life support has tended recently to pre-empt the meaning of "prolongation of life" thereby placing it in contradistinction to euthanasia. Moreover, it has tended to displace the popular import of "euthanasia" from active intervention to a passive withholding of "extraordinary" measures.

The evolution of these concepts and issues now will be examined more closely; the article proceeds in chronological order but will take the liberty in each era of drawing inferences directly to the present day (Kastenbaum and Aisenberg, chaps. 8, 9).

Renaissance humanism (1341–1626)

"Temperate" life and "natural" death are the concepts that connect euthanasia and the prolongation of life in Renaissance humanist thought. According to the prevalent physiological theory, each person is born with a certain amount of vital substance; if this is utilized with restraint in the course of a long life, death will be "natural" and benign. But if one's animating principle is consumed by unnatural, inordinate activity or disease, dying is agonized. This classical idea of easy dying was imbued by humanists with the values and rewards of Christian conduct: thus, "natural" connoted a better or higher kind of death. Luigi Cornaro's *Temperate Life* (1558 ff.), frequently consulted into the eighteenth century, featured both an easy ("holy") terminus in advanced years and the prospect of longer life—up to 120 years.

In contrast to Cornaro, Francis Bacon was inspired by the promise of planned experimental research as the key to controlling bodily processes, either to lengthen life or, when indicated, to end it painlessly. He praised prolongevity as the "most noble" purpose of biomedicine, and he also considered "euthanasia," as he termed it, an essential area of medical skill. He inferred that relief of suffering is central in terminal care, and, accordingly, the physician sometimes may hasten death. Although their approaches differ, the intentions of Cornaro and Bacon are similarly in keeping with Christian humanist concepts of man's dignity and well-being: longevity liberated from the infirmities of senescence and dying freed from mortal pain.

The way of Cornaro, with its quietist reliance on nature, is an antecedent of passive euthanasia or "letting die." This standpoint persists in countries where the customs and judicial codes show traditional reverence for natural law. Omission of extraordinary measures continues to summon an imagery of natural death by "exhaustion" that seems preferable to the comparatively "unnatural" procedures of active euthanasia, connoting violent "extinction" and something of the illicit. Until recent decades, many patients did reach a condition of drastic debility where death appeared natural. However, there were many other instances in which nature seemed far from benevolent, and clinical mitigation was inadequate. Moreover, present-day care is inescapably a complex mixture of factors in which the natural, technical, and synthetic components cannot be separated. These considerations bring to mind Bacon's active interventionist aim of using nature's secrets to attain, for the moribund patient, a more *human* death. ("Exhaustion" and "extinction" are Aristotelian terms.)

Mercantilism and the social contract (1498–1714)

During two centuries of warfare, the Baroque ethos transformed Renaissance culture into an instrument of regimentation serving church, state, and the balance of trade. It should be noted that both Francis Bacon and Thomas More not only were humanists but also had official positions in mercantilist governments. The fierce competitiveness of mercantilism was defined by Bacon's maxim that what is gained in one place *must* be lost at another. And the same consciousness of limited resources appeared in More's *Utopia* (1516), which outlined the first organized *system* of euthanasia, in which patients with painful, hopeless diseases are advised by a panel of priests and magistrates to embrace a rapid death either by suicide or by the action of the authorities. Mercantilism favored energetic, mandatory direction of the individual; More's panel of advisers ruled as the "will of God." Humanist self-assertion was replaced by a calculating kind of heroic self-sacrifice that was concisely defined by the famous last words of Sir Philip Sydney (1586): "Thy need is greater than mine."

Amid the assumptions of scarcity economics, there arose a concept we may term "thrift euthanasia," which would permit the community to close off, by one means or another, lives requiring "undue" expenditure of inherently limited resources. The quantification of life-and-death questions was remarkably parallel to the managerial accounting in economic and political matters. The body thus has a fixed sum of vital force: enough, e.g., for "x" number of heartbeats or hours of work. Authorities, including the individual's own conscience, must vigilantly prevent any waste of vital powers (on loan from God and nature) in sexual indulgence, luxury, or even illness. In the contractual imagery of that era, death is a veritable *debt*, payable at three score years and ten, if not sooner. Suicide, however, was denounced as a personal indulgence violating the work ethic, according to which a lifetime of productive labor is equated with saintly suffering that may earn Christian redemption. There is room in heaven only for a "predestined" elite, and, likewise, the allotment of limited medical facilities favored a select few.

Mercantilist regimes would have furthered the prolongation of life only if it had been possible to increase the years of greatest productive vigor. They did not at all want larger numbers of aging persons. The scientific outlook, including statistics, was applied to the population rather to promote orderliness of flow from birth to death and an assured incremental by-product of net social profit. One may note analogous present-day interest in social strategies of selective implementation of birth control and suicide prevention while simultaneously withholding costly therapies, except in patients of unusual social "worth." Such trends are predictable in a century of wartime mobilizations, economic crises, and an apparent depletion of basic resources.

The thrift factor in euthanasia raises difficult problems concerning economics and the public policy aspects of biomedicine. On one hand, modernization creates the possibility of abundance in matters of human value (Fox); on the other, it is customary to submit to the presumed necessity for restrictive slice-of-pie allocations from a limited amount of social wealth. The latter assumption can be said to comport to an "ideology of death" (Marcuse), in that the point at which biomedicine ceases trying to prevent death and to restore health is decided not by inherently biological considerations but by social ones; and much of this bias toward economic scarcity stems from the power struggles of the seventeenth century.

Conflicts about euthanasia and the prolongation of life cannot be reduced to a quarrel between the religious and the laity or to a question of human rights per se. John Locke's *Treatise on Government* (1690), the landmark social-contract document of the liberal secular state, reasons that life not only is a right but also is an inalienable one: It can be neither taken nor given away. Markedly extended longevity did not seem to interest Locke; a physician as well as a philosopher, he always referred to life "preservation," a cautious word. Yet his idea of the natural persistence of the living organism (similar to Newton's concept of momentum) later stimulated Enlightenment theories of prolongevity.

Just as Locke was vigilant against tyrants' usurping of power, so also was he mistrustful that the people would willingly relinquish their God-given and natural prerogatives. Thus he did not condone a freedom to die. Patrick Henry's "liberty or death!" was more in keeping with Renaissance humanism than with Lockean liberalism. Thus, mercantilist tendencies toward euthanasia were blocked by Locke's philosophy in which life preservation, liberty, and the pursuit of property are inseparably interrelated. A possible exception would be a situation that violated integrity of person so grossly that basic contractual responsibilities ended. That would be comparable to a nation at the brink of justified revolution—a highly unusual occurrence.

Something that is "inalienably" invested cannot be disposed of at pleasure, just as the steward or servant of a merchant must guard and augment the wealth of his employer. This stewardship of life and property, to be accumulated rather than enjoyed, initiated hard-working industrial expansion but not greater personal freedom. Partly as a result of this, there is a bias in contemporary society (Arendt) against suicide and voluntary euthanasia, the means by which the individual could shorten his life in order to escape bad conditions; yet, where the individual might wish to lengthen his life, it becomes evident that precedence is given to the ongoing preservation of the governmental regime, the society, or the entire species.

The Enlightenment and the idea of progress (1687–1804)

Enlightenment thinkers, seeking to relax the rigidities of early liberalism, explored two major ethical pathways: hedonism and relativism. Jefferson, exemplifying the former trend, changed Locke's focus on the acquisition of property to a "pursuit of happiness." There were, however, definite limitations in Jefferson's approval of the pleasure principle: Although "cruel and unusual" reprisals against suicide were abolished, self-death nevertheless was to be "pitied" by public opinion. Jefferson thus continued the Lockean supposition that one who takes his own life must be "alienated," a word implying mental imbalance. This interpretation persists today in the assumption that one choosing suicide or euthanasia is temporarily "not himself" (Parsons and Lidz).

It was left to Bentham to affirm an unconditional hedonist rejection of suffering, rephrasing Hamlet: "To be *happy* or not to be at all." Also significant was Hume's philosophy recognizing that different persons and interest groups cannot be kept within Locke's severe formulation but should be permitted to express freely the *relative* truth which each discerns. This open-eyed acknowledgment by Hume of the variety of conditions led him to picture some cases in which the torments of illness justified "courageous escape." He also reasoned that, if mankind legitimately could seek to lengthen life, it follows that similarly life justifiably might also be shortened. Yet Hume and his associates were publicly conservative, and his own composure in terminal illness, in reality eased by potent drugs, was ascribed instead to his unique rationality and superior character. Such ambiguity severely restricted the impact of eighteenth-century libertarian thinkers. They were counterbalanced too by the growing influence of Rousseau's theory of the "general will," a more democratic and romantic version of the social contract. The general will became equated with rule by majority opinion, which frequently turned out to be conventional and unenlightened about life-and-death issues.

Meanwhile, the Enlightenment faith in progress led to the elaboration of a number of ambitious, this-worldly alternatives to Christian salvation from death. There was, for example, intense interest in prolongevity. Benjamin Franklin boldly declared senescence to be not a natural process but a "disease" to be cured, and he predicted that longevity might reach a thousand years or more. The mathematician Condorcet speculated about virtually immortal life, an idea that has reappeared because of the present-day biomedical revolution (Harrington). The seeming lack of concern about euthanasia in Enlightenment biomedicine calls for further scholarly investigation. Present evidence indicates a heroic approach preoccupied with visions of large-scale research and public health programs, possibilities of much-increased life expectancy, and fascinating methods for restoration of persons "apparently" dead. Humane societies were devoted to resuscitation of the drowned and asphyxiated. Also characteristic of progressist medicine was curiosity about suspended animation by freezing or chemical means, a technique that might resolve the plight of the moribund in a completely new way. And there was much anxious apprehension regarding premature burial.

The invention of the now-maligned guillotine, to rid execution of the physical torture then customary, does indicate that research was done at least indirectly relevant to euthanasia. But during the Age of Reason, patients with unbearable pain or long-continued senility usually were considered mentally "alienated" by their extreme situation and not really responsible, thus causing a breakdown in doctor–patient communication. The paucity of eighteenth-century euthanasia debate reflects also the adoption of the reassuring natural-death model by the bourgeoisie. Their life-style aimed at the prudent accumulation of status and wealth during a life sufficiently long gradually to deplete vitality and thus conclude with a relatively easy demise at home; there at the death bed, blessings and property were transferred to the sons. Natural death thus acquired an aura of bourgeois comfort and the sentimental cult of the family. Life insurance was added, and there was a modicum of anticlerical thought to dispel "superstitious" fears. Money and connections made available the services of bedside physicians who applied measures, discreetly vague, for easier, and perhaps speedier, dying.

Positivist mortalism (1800–1906)

In order to see how Enlightenment heroic medicine became transformed en route to the present, it is necessary to examine the philosophy of positivism. The pattern for positivist thought actually was provided by medical practice, in particular the habitual strategem of using death as the pragmatic operative limit of clinical effort. Thus, Bichat in 1800 defined "life," in this common-sense way, as the ensemble of all the functions that *resist* dying. Medicine, he indicated, does not seek to conquer death or to preserve life interminably; the doctor strives only to slow down the tragically inevitable processes of decline. This medical "mortalism" (Foucault) removes death as a final catastrophe and instead spreads dying, as a series of little deaths, along the entire life span from birth on. Bichat, thus, subtly undermined progressist biomedicine.

Adopting Bichat's medical insight, Auguste Comte created a systematic *social* mortalism assuming that civilization inherently is beset by forces of decay and, upon reaching a mature stage, is fated to decline and fall. The worthwhile goal, he stated, is to avoid unprofitable "metaphysical" controversy and instead to utilize known, "positive" facts to stave off such social diseases as war and revolution, thereby preventing *premature* termination of Western civilization. To this purpose, he proposed that a corps of "sociologists" should employ an ensemble of ideas and beliefs, termed "ideology," to mold public opinion continuously to correlate with the everchanging requirements of modern society.

The positivist authorities were to control not only education but the entire process of human development. The life span was divided by Comte into functional age stages: childhood, youth, early adulthood, maturity, and retirement at sixty-three. Death usually would occur by age seventy, and seven years later a solemn judgment would select those whose "sanctified" remains merited interment in a monumental cemetery. Despite his pretentious thoroughness nearly all of Comte's notions were adopted, in one form or another, in Europe and the Americas during the latter nineteenth century. Indeed, the positivist model of expert intervention and guidance at each transitional life crisis still underlies current discussion of the moral role of biomedicine in death and dying.

Positivist-style medicine was effective in reducing the premature death rates of early life; but there is little sign that an aging population was desired or even responsibly foreseen. Comtean "Progress" carefully was balanced with "Order": In generational terms, youth represented innovation whereas the elders were an equally essential conservatism. Both birth control and death control were apprehended as potentially disruptive of smooth social functioning. In this calculated equilibrium, however, culture is not static, for industrial society requires incessant motion and change; but the positivist stream of novelty designedly lacked the profound ethical dimension of Enlightenment advance. Completely contrary to Condorcet's prolongevity, Comte held that death *contributes* to "Progress," a view continued today in the neopositivism of Parsons.

More studies are needed into positivist attitudes about euthanasia. Evidence now at hand indicates there was much less concern about the suffering and the ethical and legal prerogatives of patients than with their duty as citizens to set an inspiring example for the community. The sociologist Émile Durkheim, for example, was hostile to suicide and recommended that the individual's morale and "altruism" be strengthened by increased ties of "solidarity" with colleagues and family. George Eliot, the English writer, reflected positivist ideals in her philosophy of bearing life's pains "without opium." Later however, Eliot protested against the positivist doctrine that "pain is not an evil" (Guyau) and invented the word "meliorism" to emphasize the ethical imperative to better the world and, preeminently, to lessen human anguish. The introduction of anesthesia in the mid-nineteenth century also signified that professional and public attitudes could be favorable to pain prevention, as did William Munk's treatise on "euthanasia" as specialized care to ease the distress of the moribund. Yet positivist constraints are revealed in the unique honors paid to obstetric anesthesia (because it aided the function of motherhood) and in Munk's overly explicit disclaimers of hastening death. Similarly, Ilya Mechnikov's proposals to ameliorate senescence were restricted to so-called premature aging, and he invoked both natural death and an "instinct" of death to keep his project within positivist bounds.

A positivist kind of heroic medicine continues today in attempts to conduct an arduous, defensive "resistance" to death. This differs sharply from Enlightenment heroic medicine, with its aim of decisive life-affirming power over natural forces. The philosophes had been fascinated by prospects for the rapid restoration of effective health, either by resuscitation or by a reversal of senescence; in extremity, they considered suspended animation, with clear provisos for radical cure in the future. They had not wanted painfully to postpone dying, day by day, within a complex equilibrium between vectors of clinical endeavor and inevitable decline. In contrast to Enlightenment medicine, the tenacious positivist maintenance of life is in the social-contract tradition that there is a correct time of death which can be determined by a conflict of contending forces. Such a process has come to be criticized as a "prolongation of dying." Positivist medicine also demands extensive diagnostic and prognostic procedures, aside from curative purposes, in order to provide suitably exact or "positive" knowledge of the changing nature and timing of the fatality. Thus considerable efforts are taken to predict and record the fluctuating parameters of the case, while the medical team makes a suitable show of resistance to "premature" termination.

Revolt and neopositivism (1885 to the present)

In the latter nineteenth century, an intellectual and cultural rebellion began to gain ground against positivism. The most brilliant of the German antipositivists, Nietzsche, was intrigued by the idea that the "right" time for death could be decided not by calculation but by proud, "barbaric," egoistic choice. He scorned the "altruistic" self-denial of positivist ethics, and he praised suicide when "appropriate." However, Nietzsche himself, aware of a core of inner self-destructiveness, instead chose to endure and create despite his long years of painful illness. Confidence in freedom to choose death was undermined further by Freud's depth psychology of conflict and ambivalence. Like Nietzsche, Freud came to reject mortalism and affirmed the forces of life, Eros, against the "death drive," Thanatos.

An even more fundamental antipositivism came from the Russian philosopher Nicholas Fyodorov (1828–1903), who completely rejected the positivist ethos as a shattering betrayal of Christian charity and hope. Fyodorov was unique in his linking of an explicitly Christian vision of salvation with the promise of planned melioristic research. He demanded that science "begin to do" its immense task of actually saving humanity from mortality. The vaunted "progress" of the nineteenth century was derided by Fyodorov as, by and large, a parade of novel consumer products and deadly armaments. He also pointed out the ethical dilemma of generational displacement occurring in modern society in which youth holds superiority merely by arriving at a later, "higher" stage of history. Such a sacrificial succession of generations was being justified by positivists and Darwinists, as in Winwood Reade's *Martyrdom of Man* (1872).

The triumph of Darwinism caused widespread withdrawal of sympathy from the ill and old, who were viewed as inferior and even parasitic beings. According to the important evolutionary theory set forth by August Weismann in 1882, nature programs higher organisms to die at the end of the years of reproduction; this was a severe blow to comforting notions about natural death. Even more ominous was the interpretation being given to Bichat's two-level hypothesis of mortality that contrasted brain death with that of the heart and lungs. It was implied that women, the aged, and the uneducated existed on a lower, less creative "vegetative" plane almost akin to the comatose; therefore, their dying was relatively uncomplicated and unimportant. In 1905 Osler, influenced by Weismann, set the cessation of creativity and the onset of "comparative uselessness" in men at age forty. At almost the same time, a leading German biologist, Ernst Haeckel, recommended that hundreds of thousands of "useless" persons be rapidly poisoned (Gasman).

Meanwhile, a new positivism, with a heroic, romantic bent and an emphasis on national-imperial mission, became the prevalent social ideology. ("Neopositivism" here refers to that ideology, not to logical-positivist philosophy.) Thus, on the eve of the First World War, Durkheim affirmed the primary role in ethics of personal sacrifice on behalf of the community. Another neopositivist theme was the "obligatory-gift" relationship, a modern industrial version of the social-contract theory, that holds each citizen morally accountable for putting his lifetime to socially productive and profitable use. This concept of life as an investment became the basis for the most cogent sociological study of American behavior regarding death and dying (Parsons and Lidz; Parsons, Fox, and Lidz).

Twentieth-century wars and revolutions brought the dire premonitions of Fyodorov to startling reality as entire generations were consigned to the "dustbin" of history. One's political or socioeconomic enemy was seen as a diseased, death-tainted force to be eradicated by the agencies of "real" or authentic life. During the First World War, each side claimed to be the champion of "life against death," and such ideological conflict persisted into the 1930s until it took a strangely twisted form in the Spanish fascist slogan, "Long live death!" With National Socialist racism, a "final solution" (1942 ff.) attempted to promote "racial health" by liquidating allegedly alien "races." This genocide program was titled "euthanasia," and techniques of efficient killing first were developed (1939–1942) in eliminating thousands of patients with chronic disease or mental illness. Such Nazi actions can be cited to exemplify an "opening wedge" argument against euthanasia. For that interpretation, it is necessary first to estimate the extent to which genocide actually represented the continuation of earlier eugenic and public health innovations. Secondly, one must consider the continued existence today of widespread cultural tendencies to evade social problems by victimizing a scapegoat category of people. The first, or social-medical, factor has been debated extensively, but it is the lethal expression of prejudice that is likely to prove the greater danger.

Conclusion

As we have seen, the terms "euthanasia" and "prolongation of life" are not necessarily antithetical. Both aims can be seen generally as expressions of the aspirations of modern culture to control the forces of nature. This article has indicated some of the viewpoints that have considered such newly acquired powers to be conducive to human freedom, dignity, and well-being. On the other hand, it has reviewed attitudes that would tend to bar or to delimit significantly such intervention. But it has not attempted to deal directly with the teachings of organized religions or with the more specialized literature of medicine and law. Instead, it has focused on those turning points in modern history that illuminate present-day secular cultural and intellectual currents as well as the influences of economics and social ideology.

It is questionable if today the young comatose patient is the prototype for the discussion of issues of death and dying. A more probable crucial issue is that of the elderly: a reservoir of relatively defenseless persons, perceived, through bigoted "ageism," as unproductive and pejoratively dependent. In them, modernization has created a population stratum that, in a state of nature or conditions of scarcity economics, "ought" to be dead. Here one confronts the recoil from the unfamiliarity of modernization, with its overturning of traditional age, sex, and generational patterns. To antimodern thinkers, an aging population seems a symptom of socio-cultural decadence, and they typically have urged a revolt of the "young" against the "old"—sometimes arbitrarily defining a particular nation or race as young (Stern; Butler).

The perverse attraction to death that motivated genocide personnel (Alexander, 1949, "Destructive," "Molding") is not uncommon in contemporary society, in which the "necrophilous" personality aligns itself with technology's cold efficiency (Fromm). It may be that human history is veering toward the "death worship" foreseen by Orwell in *1984*—anonymous sacrifice of seemingly expendable individuals to the on-living social unit. When an unconditional mortalism prevails, human personality is stripped of significance and even reality (Borkenau). Thus, it is ethically necessary to challenge both mortalism and antimodernism (see also Veatch, chap. 8).

Genocide can be analyzed as part of a rebellion by desperately alienated masses of people against modernization—a regressive leap into an idealized past. To prevent that temptation, Ernst Nolte, in his profound study of the twentieth-century situation, suggests that human nature be considered something not yet fully formed, but rather engaged in a continuing process of unfolding realization through historical time. In this perspective, modernization is an alienating but necessary "transcending" of our familiar, incompletely moral world. In the open, adaptational dimension of human nature, there is an indispensable role for the therapeutic principle of biomedicine. For that task, mortalism can be replaced by an ethos in which life is the defining absolute and death the variable, "moving target." Discussion of euthanasia and prolongevity may once again come into affiliation with each other in the context of melioristic commitment. The great worth of individual human life is not yet a fact, but it is, for biomedical ethics, an essential goal.

GERALD J. GRUMAN

[*For further discussion of topics mentioned in this article, see the entries:* AGING AND THE AGED; DEATH, *articles on* WESTERN PHILOSOPHICAL THOUGHT *and* DEATH IN THE WESTERN WORLD; *and* SUICIDE. *Directly related are the other articles in this entry:* ETHICAL VIEWS *and* PROFESSIONAL AND PUBLIC POLICIES. *Also directly related are the entries* DEATH, ATTITUDES TOWARD; PAIN AND SUFFERING; *and* RIGHT TO REFUSE MEDICAL CARE. *For discussion of related ideas, see the entries:* CRYONICS; LIFE-SUPPORT SYSTEMS; LIFE; *and* MEDICAL ETHICS UNDER NATIONAL SOCIALism. *Other relevant material may be found under* Ethics, *articles on* teleological theories, utilitarianism, *and* rules and principles; *and* Medical Ethics, History of, *section on* europe and the americas, *article on* medieval europe: fourth to sixteenth century.]

BIBLIOGRAPHY

[Documentation of primary sources, e.g., Bacon, Cornaro, Locke, Condorcet, Bichat, etc., is presented in the Gruman studies cited below.]

Alexander, Leo. "Destructive and Self-Destructive Trends in Criminalized Society: A Study of Totalitarianism." *Journal of Criminal Law and Criminology* 39 (1949): 553–564.

———. "The Molding of Personality under Dictatorship." *Journal of Criminal Law and Criminology* 40 (1949): 3–27.

———. "War Crimes and Their Motivation: The Socio-Psychological Structure of the SS and the Criminalization of a Society." *Journal of Criminal Law and Criminology* 39 (1948): 298–326.

Arendt, Hannah. *The Human Condition.* Charles R. Walgreen Foundation Lectures. Chicago: University of Chicago Press, 1958.

Borkenau, Franz. "The Concept of Death." *Death and Identity.* Edited by Robert Fulton. New York: John Wiley & Sons, 1965, pp. 42–56.

Butler, Robert N. *Why Survive?: Being Old in America.* New York: Harper & Row, 1975.

Foucault, Michel. *The Birth of the Clinic: An Archaeology of Medical Perception.* Translated by A. M. Sheridan Smith. World of Man. New York: Pantheon Books, 1973. Reprint. New York: Vintage Books, 1975.

Fox, Daniel M. *The Discovery of Abundance: Simon N. Patten and the Transformation of Social Theory.* Ithaca, N.Y.: Cornell University Press, 1967.

Fromm, Erich. *The Anatomy of Human Destructiveness.* New York: Holt, Rinehart & Winston, 1973.

Fyodorov, Nicholas. "The Question of Brotherhood or Relatedness, and of the Reasons for the Unbrotherly, Dis-Related, or Unpeaceful State of the World, and of the Means for the Restoration of Relatedness." (1906). Translated by Ashleigh E. Moorhouse and George L. Kline. *Russian Philosophy.* 3 vols. Edited by James M. Edie, James P. Scanlan and Mary-Barbara Zeldin. Chicago: Quadrangle Books, 1969, vol. 3, pp. 16–54. Reprint. *Death As a Speculative Theme in Religious, Scientific, and Social Thought: An Original Anthology.* The Literature of Death and Dying. New York: Arno Press, 1977, pp. 16–54.

Gasman, Daniel. *The Scientific Origins of National Socialism: Social Darwinism in Ernst Haeckel and the German Monist League.* History of Science Library. New York: American Elsevier; London: Macdonald & Co., 1971.

Gruman, Gerald J. "An Historical Introduction to Ideas about Voluntary Euthanasia: With a Bibliographic Guide for Interdisciplinary Studies." *Omega: Journal of Death and Dying* 4 (1973): 87–138.

———. *A History of Ideas about the Prolongation of Life: The Evolution of Prolongevity Hypotheses to 1800.* Transactions of the American Philosophical Society, n.s. vol. 56, pt. 9. Philadelphia: 1966. Reprint. New York: Arno Press, 1977.

———. "Longevity." *Dictionary of the History of Ideas: Studies of Selected Pivotal Ideas.* 5 vols. Edited by Philip P. Wiener. New York: Charles Scribner's Sons, 1973–1974, vol. 3, pp. 89–93.

Guyau, Jean-Marie. *The Non-Religion of the Future: A Sociological Study (1887).* New York: Schocken Books, 1962.

Harrington, Alan. *The Immortalist: An Approach to the Engineering of Man's Divinity.* New York: Random House, 1969.

Kastenbaum, Robert, and Aisenberg, Ruth. *The Psychology of Death.* New York: Springer Publishing Co., 1972.

Marcuse, Herbert. "The Ideology of Death." *The Meaning of Death.* Edited by Herman Feifel. New York: McGraw-Hill, 1959. Paperback ed., 1965, chap. 5, pp. 64–76.

Munk, William. *Euthanasia; or, Medical Treatment in Aid of an Easy Death.* London: Longmans, Green & Co., 1887. Reprint. The Literature of Death and Dying. New York: Arno Press, 1977.

Nolte, Ernst. *Three Faces of Fascism: Action Française, Italian Fascism, National Socialism.* Translated by Leila Vennewitz. New York: Holt, Rinehart & Winston, 1966.

Parsons, Talcott, and Lidz, Victor. "Death in American Society." *Essays in Self-Destruction.* Edited by Edwin S. Shneidman. New York: Science House, 1967, chap. 7, pp. 133–170.

———; Fox, Renée C.; and Lidz, Victor M. "The 'Gift of Life' and Its Reciprocation." *Death in American Experience.* Edited by Arien Mack. New York: Schocken Books, 1973, pp. 1–49.

Stern, Fritz. *The Politics of Cultural Despair: A Study in the Rise of the Germanic Ideology.* Berkeley: University of California Press, 1961. Reprint. Garden City, N.Y.: Doubleday/Anchor, 1965.

Veatch, Robert M. *Death, Dying, and the Biological Revolution: Our Last Quest for Responsibility.* New Haven: Yale University Press, 1976.

II

ETHICAL VIEWS

Introduction

Every society has tried to stave off death and set limits to killing. But differences have arisen at times when living comes to be so painful or oppressive that some desire to die, perhaps even ask to be killed. In such cases of suicide and mercy killing, the prohibition on taking innocent life has collided with the demand for mercy in the face of intolerable suffering.

Arguments based on Judeo-Christian beliefs and/or natural law have traditionally formed the mainstay of the opposition to all such acts. To take an innocent life is, in these traditions, to usurp the right over life and death attributed to God. For some theorists of natural law, such acts violate the most fundamental human end of self-preservation. Others, both within and outside these traditions, have argued, on the contrary, that there are certain predicaments so agonizing that even the drive to self-preservation and the respect for all life are overpowered.

Throughout the discussions of these questions the issue is one of setting limits to what human beings can do to one another and to themselves when life is at stake. Should human beings ever have the right to seek their own death? If so, under what conditions? And should others be allowed to assist them in dying? Can innocent persons ever be rightfully put to death against their own will? What ethical lines against abuses and mistakes should be drawn by religion, by law, and in professional codes?

These problems have often been debated under the heading of "euthanasia," a word deriving from the Greek for "good death." Mercy for the suffering patient who desires to die has been the primary reason given by those who advocate legalizing voluntary euthanasia. But the word has been applied to such different acts, in such diverse circumstances, that much confusion has resulted. In spite of great differences among the views about rightful dying and allowable killing, the debates concerning the standards that ought to prevail have stressed time and again certain themes or aspects of the problems: (1) whether death is actively brought about or comes because support is withheld; (2) whether every conceivable effort should be made to ward off death; (3) what our responsibility is for death which has come through our action though we have not intended it; (4) whether death is voluntary or involuntary; (5) whether the agent brings death to himself as in suicide, or to another. In all these aspects efforts have been made to draw lines, to set procedures, and to make prohibitions that will protect the innocent and the unwilling.

This article attempts a systematic overview of these problems. It considers the different kinds of line-drawing and the protections that have been worked out in the face of human suffering, brutality, and coercion. And it examines the views concerning rights and powers to *decide* whether or not death should be sought by persons for themselves and for others.

Omission and commission

To omit is to leave undone, to fail or to forbear to perform an action that is within one's range of awareness and capability. To commit an act, on the other hand, is to perpetrate or perform it. Both "omission" and "commission" may be disapproved; one can be accused of omitting a lifesaving measure, or of committing a crime or a folly. They are narrower terms than "action" and "inaction," which generally signify the presence or absence of purposeful activity. In the context of what is done or not done to bring about *death*, they point to a basic distinction: that between death which is actively brought about and death which comes through the leaving out or neglecting of life-preserving measures. Thus to shoot someone is to commit an act; but to refuse help to the victim of the shooting is an omission. In discussions of euthanasia, the terms "active" and "passive" are often used to denote the same distinction. Jointly, active and passive euthanasia are then distinguished from a "natural" death.

The distinction between omission and commission is also a necessary one for setting limits to what must be done to preserve life. Legal, moral, and religious traditions have all regarded it as more crucial to specify the harm that people must not do to one another than the helpful acts they must not omit. Otherwise, everyone might be held responsible for all the accidents and deaths in the world from which they could conceivably have protected others. To sleep at night, for instance, rather than to patrol rivers and lakes looking for drowning victims would then be turned into a culpable omission. Clearly, this distinction between omission and commission corresponds to a fundamental limiting of human responsibility. And the failure to draw the distinction leads to fallacies such as that expressed in the loose phrase: "If you are not part of the solution, you are part of the problem."

While the distinction between omission and commission is necessary, however, it is obviously not sufficient. There are circumstances in which it is no defense at all to say that one did not do anything, and should therefore not be considered responsible for an accident or a death. A parent who does not feed an infant, or a doctor who fails to provide standard treatment to a patient in a case where he had a clear duty to act cannot merely argue that such inaction is justifiable *because* it is an omission.

There are circumstances, then, when such heavy responsibility can be assigned for *not* saving or prolonging a life, that the distinction between omission and commission evaporates. These are the circumstances where there is both a *relationship*—such as between parents and children, or in certain cases physicians and their patients—that entails duties to preserve life and to help, and also a chance that the help provided may succeed. In the law, similarly, inaction is held tantamount to action when there is a duty to act and failure to do so.

Intentional acts of killing are thus prohibited in almost all societies, the major exceptions being those of killing in war, in self-defense, and as lawful punishment. Omitting to save or prolong someone's life, on the other hand, has usually been viewed as only comparably wrong where there exists a *duty* to preserve that person's life and a chance of success. To omit help which it is a duty to provide constitutes abandonment.

But choices are not so simple, even in relationships where there are duties to patients and dependents. For it is increasingly possible to provide life support far beyond what many patients wish to receive. The very concept of "help" then comes into question, and a conflict arises as to whether to preserve the lives of such patients can realistically be thought of as "helping" them.

As a result, societies must now determine under what circumstances, even in relationships where there *is* a duty in some sense to take care of another and the ability to prolong life, omission is morally justified. The clearest case is one in which existing duty has ceased because the patient, while manifesting some physiological activities, is no longer alive. Another clear case is that in which the competent patient rejects the life-prolonging treatments available, thus releasing relatives and health personnel from any duty to prolong his or her life.

It is the effort to distinguish *further* criteria for justifying certain omissions that now preoccupies theologians, medical personnel, lawyers, and philosophers. At what cost to the patient in terms of suffering, effort, and exhaustion of resources can efforts to prolong life be carried on? And how are these efforts to be balanced against the neglect of others in need of assistance? Should the likelihood of success of such efforts be a factor in weighing the length to which they should be pushed?

In considering *omission* of action to prolong life under such complex circumstances, the debate has focused on the distinction between ordinary and extraordinary means. The debate concerning *commission*—concerning acts that can and cannot be performed so as to cause death—has centered on the distinction between direct and indirect action. In the next three sections, these two kinds of distinctions will be taken up.

Ordinary and extraordinary means

Physicians have the moral and legal duty to continue appropriate care for patients once they have accepted such a responsibility in the first place. They may not abandon their patients. Yet it is possible for them to continue some kinds of support while not going to every length to prolong lives that are ebbing away, or when the support is useless, unavailable, or unwanted. They need not, for instance, provide resuscitation for a patient who is virtually certain not to survive the effort, even though resuscitation may be indicated for other patients with higher chances of survival.

How, then, can one know where and when it is permissible to omit certain life-supporting efforts, and which ones these should be? If a patient is in a permanent coma, should a respirator be employed? And should a patient near death from both painful cancer and debilitating heart disease be resuscitated? What if this patient develops pneumonia—should antibiotics be administered? Should surgery be undertaken on this patient to correct a condition unrelated to the cancer and the heart condition? And what measures can be omitted from the care of severely malformed newborns?

There is great need, then, for criteria permitting the omission of some life-preserving means at certain times. For this reason, health professionals have looked to the distinction stressed by Catholics, and adopted by many others, between ordinary and extraordinary means. Pope Pius XII said, in a widely cited address to an international congress of anesthesiologists, that "normally one is held to use only ordinary means [for the preservation of life and health]—according to circumstances of persons, places, times, and culture—that is to say, means that do not involve any grave burden for oneself and others" (Pius XII, p. 395).

Certainly, in clear cases, this distinction corresponds to a powerful intuition. We sense that there is a difference between resuscitating a dying man over and over again, on the one hand, and resuscitating someone who has a likelihood of continuing to live, on the other hand. We know that means exist in some hospitals not available in others. But in the many complex cases where one cannot be so certain, the frequent use of the term "ordinary means" has been ambiguous and self-serving. People have tended to say that what they wanted to omit was extraordinary; because of the vagueness of the term, it has been hard to pin down exactly what they meant, and thus to make a coherent argument in support or in contradiction. These terms have both normative and descriptive meanings; if they are not seen as separate, confusion results.

The basic use of the terminology is to signal a distinction between what is binding and what is not binding, between the mandatory and the merely allowable. When we ask "binding for what reason?" several different criteria emerge, often confusedly.

One distinction is between that which is helpful to prolonging life, and called ordinary, and that which merely prolongs the process of dying, and thus is extraordinary, useless, and at the very least not required. This is an important distinction, but it does not speak to all that concerns us about life support. For instance, say that a surgical procedure actually could prolong a life for a period of months, but at an extreme cost of pain for the patient and of resources for society, should it be considered ordinary? Most would follow Pope Pius in considering the gravity of the burden of life-support, thus rendering this first distinction insufficient.

A second distinction, frequently made by physicians, is that between standard treatment and unusual treatment, thought to be extraordinary. In this sense, food and shelter are basic and ordinary, whereas, at the other extreme, rare or experimental or expensive treatment is extraordinary. Again, this distinction speaks to only one factor, but it cannot cover all of what we mean by extraordinary. Take the case of the person dying of cancer in great pain, close to death, who develops pneumonia. Certainly, antibiotics are now part of the standard arsenal of medicine; yet many would consider it cruel and extraordinary to use them to prolong such a patient's care.

A third kind of distinction encompasses these two but stresses the *circumstantial* nature of what is considered ordinary and extraordinary. A procedure may be called extraordinary if there is *any* overwhelming reason why it ought not to be undertaken. Perhaps the procedure will leave the patient in intolerable pain or require resources not easily acquired; perhaps it is morally unacceptable, as when it requires taking an organ from an unwilling donor; or perhaps there is too small a chance of success or survival. According to such a distinction, antibiotics—by now standard or usual treatment in a number of circumstances—might be extraordinary in the sense of not being morally necessary if they have to be flown in from far away in a snowstorm, or if the patient already suffers from a fatal illness and is in great pain.

A further question with respect to the use of the "ordinary–extraordinary" distinction is the following: To whose choices does it apply? Some hold that all that is ordinary is mandatory not only for health professionals but also for patients. Since life is not ours but God's, they argue, we cannot abandon it except in extraordinary circumstances, for ourselves or for others in our care.

A different view holds, on the other hand, that what the patient wants is part of what renders circumstances ordinary or extraordinary. For a Jehovah's Witness, for example, receiving a blood transfusion may be thought immoral, destructive of one's soul, and to be avoided at all costs, however common such a procedure has now become for many patients in our hospitals. For a patient who wishes to die at home, going to the hospital might prolong life considerably yet be considered morally extraordinary.

If the patient's own circumstances and desires are to be taken into account in this view of the two concepts, special difficulties will arise in the increasingly frequent circumstances where patients have lived past the point of being competent to express and enforce their desires. To some extent, this difficulty is being anticipated by the numerous "living wills" now being written in anticipation of just such circumstances. Another effort to cope with such a situation is represented by hospital policies concerning patient treatment in terminal stages of illnesses where continued life prolongation appears increasingly cruel or hopeless.

Stopping versus not starting a procedure

Many health professionals who recognize that a particular life-supporting treatment has become useless are reluctant to discontinue it, especially if the patient's life will ebb away as a result. Even if their patient is permanently unconscious, with no possibility of mental function left, they hesitate to turn off the respirator which keeps the body in partial "life." And they draw a distinction between *ceasing* such support and not starting it in the first place. For a patient similarly unable to regain consciousness, they might have no hesitation at all in deciding not to institute respirator support. But they do hesitate to cease existing support. In part, the hesitation flows from the fact that to stop the respirator or turn it down is itself an act—a commission—and thus surrounded by special precautions where life is at stake. But this need not be the case in all questions of stopping treatment. To cease giving medication, for example, is at times more clearly an omission than to cease respirator support.

Even when one has concluded, therefore, that a procedure may be omitted in a case where it is unquestionably extraordinary, one has to ask whether "omit" here means omit beginning a process, or whether it can also be extended to mean omit continuing it once started. Ramsey holds that there is no difference between the two; that the same moral warrant is required for deciding to stop "extraordinary" lifesaving treatments as for deciding not to begin to use them (Ramsey, p. 121). It is important to stress here the word "extraordinary." For there certainly *is* a great difference between the two types of decisions in most situations of life. Once we undertake some form of help, such as to house an orphan, for example, we have created an obligation not to cease it, greater than before we had stepped in.

But we need to specify what *kind* of cessation would be no different from not starting, even in the realm of extraordinary procedures. Ramsey's equation of the two may be correct only where extraordinary is taken to mean "death-prolonging." If a respirator merely prolongs the process of dying of an organism, with absolutely no hope of reversal, then the moral warrant for turning off the machine is the same as for not instituting such care in the first place. But other forms of extraordinary support *are* more difficult, morally speaking, to stop once they have been started. An extraordinarily expensive treatment, for example, such as the constructing of a special chamber where a single patient is kept free from infection, might not have been undertaken; if undertaken, it might not have been devoted to a particular patient. But to withdraw support from a patient in need once it has been undertaken and so devoted is of a very different moral order. A relationship has been created; expectations have been aroused. And while the withdrawal of support may be justified, such justifications must overcome the claims of the relationship and the expectations; this is not the case *before* the procedure has been undertaken.

Renée Fox and Judith Swazey describe similar difficulties for physicians in discontinuing treatment such as dialysis once they have begun, or deciding not to do successive renal implants on a patient who has rejected previous ones (Fox and Swazey, p. 323). Physicians then feel bound to their terminally ill patients in ways that make it nearly unbearable to cease treatment even where treatment is no longer indicated. There are psychological barriers in such cases which render stopping a procedure quite different from not starting it, quite apart from the moral reasons for and against such acts. From a moral point of view, the refusal of continued treatment by a competent patient may be seen as removing the duty to continue what has been started. What *makes* the treatment "extraordinary," then, is that the patient no longer wants it. Yet in practice it is much harder for a patient to secure the cooperation needed to cease dialysis or to leave the hospital than to refuse dialysis or hospitalization before beginning them. Stopping and not starting extraordinary forms of care, then, are felt differently, and have different moral and psychological warrants in a number of situations. But where a procedure is merely death-prolonging, stopping and not starting are much more alike.

Direct and indirect

Indirect acts are defined as those not directly aimed at or attained. Direct acts, on the contrary, are those with results that are either intended or take effect without intermediate instrumentality. Aiming and attaining—these concern the two separate domains of intention and causality. One or the other, or both together, may be at issue in writings on "indirect killing." Thus Joseph Fletcher (p. 147) calls euthanasia "direct" if life is ended actively, as by swallowing something or pulling out a tube, while "indirect" euthanasia is allowing death to come (ibid.). And Pope Pius XII describes the omission of resuscitation efforts for a dying patient as "never more than an indirect cause of the cessation of life" (Pius XII, p. 397).

A second meaning of "indirect" is one in which death *is* actively caused, but only indirectly. There may be human or mechanical intermediaries to trigger off the actual killing. This can happen at great distance in space and time, as where villages are bombed from planes by military personnel knowing that people will be killed, yet never confronting the victims face to face. Those who kill indirectly in this sense often feel that the indirectness reduces or even removes their moral responsibility for the consequences of their actions. Indirectness, here, serves as a *psychological* buffer; whether or not their *moral* responsibility is in reality reduced, indirect killers sometimes believe that it is.

It is when "direct" and "indirect" are used in a third sense to describe the intention guiding the act which results in death, however, that an especially important moral distinction is illuminated. For while the taking of life is clearly ruled out in most instances, there are certain cases in which action *leads* to death while it is not primarily *intended* to do so. (The same is true of omissions where there is a duty to act.) And in these cases the actions are sometimes seen as more readily justifiable.

Thomas Aquinas made this distinction very clear with respect to self-defense. If a man is attacked and kills his attacker, his intention is to defend himself, even though the effect of his action is also to take a life. This kind of action is characterized by the fact that there are at least two consequences of the action, one good, the other bad, and that the good consequence is the one primarily intended (Thomas Aquinas II-II, 64, 7).

An act of killing, according to this third distinction, is characterized as "direct" or "indirect" according to its goal or purpose. "Indirect euthanasia" would then refer to killing by an action that is primarily intended to relieve suffering (or promote some other good) but is also known to be potentially lethal. By contrast, the term "direct euthanasia" would describe any situation in which the death of the patient is the primary goal. The textbook case of indirect killing is the "merciful overdose"—where fatal doses of narcotics are administered to a terminal patient in unbearable pain. Here the action is characterized as indirect because the real purpose in giving the drugs is held to be to relieve the patient's pain rather than to cause his death. The motive is in fact often a mixed one where a "merciful overdose" is administered. The act is therefore one, often, of both direct and indirect killing; it differs from the clear indirectness described by Aquinas in acts of killing in self-defense.

Clearly, not all actions that do severe damage in the pursuit of some good consequence can be thus defended. To give an overdose of narcotics to save a child from the pain of a curable illness is not excused by the self-professed good intention of the agent. Here we have a case of morally unjustifiable indirect killing. Yet it is equally clear that there are times when we require a moral distinction between two situations in which identical results, equally foreseen, have been achieved with very different intentions. This question has long been debated (in Catholic moral philosophy) in terms of the principle of double effect, which presents four criteria guiding conduct in a dilemma arising from an action with several effects, one bad and at least one good (McCormick, 1973, pp. 1–112). The four criteria are the following: (1) The act itself must be good or morally indifferent; it cannot be morally evil. (2) The evil effect must not be intended, merely allowed. (3) The good effect must flow from the action and not from the bad effect; that is, the bad effect must not be the means to the good. (4) There must be a proportionately good effect, sometimes called a "commensurate reason," to overcome the evil effect —some proportionate reason for allowing the evil to occur.

It is often possible to make a clear distinction between intended wrongs and wrongs foreseen but not intended, using these four criteria (Fried, p. 186). But in difficult or borderline situations, all four become hard to evaluate. The second, which asks that the agent's intention not be aimed at producing the evil effect, then becomes especially elusive. A mere personal claim of innocence of intention on the part of the agent is obviously not sufficient. Yet the knowledge others can have of his intentions is of necessity limited.

At times when people disagree as to whether an act results in any evil effects in the first place, moreover, the distinction is persuasive only for those who on separate grounds believe there is an evil effect, or believe that such evil effects must not be intended for whatever reason. Thus for a woman to take drugs that are needed for a uterine disease but have the foreseeable yet unintended side effect of making her sterile would be permissible in Catholic moral philosophy. It would represent an indirect sterilization (McCormick, 1973, p. 3). But to take the same drug in order to achieve sterility would be a direct sterilization, and as such forbidden. For many others, on the other hand, neither direct nor indirect sterilization requires such special efforts at justification, since they see no "evil effect" in either.

Richard McCormick has suggested that intention be judged in terms of "proportionate reason" (ibid., p. 69). Such proportionate reason is held to exist when (1) the value at stake is at least equal to that sacrificed; (2) there is no other way of salvaging it here and now; and (3) its protection here and now will not undermine it in the long run. Using such a distinction, one can justify certain crisis choices such as that of the truck driver confronted with running into either two men or a busload of children, or that of giving narcotics to still the intolerable pains of a dying person at the risk of speeding the moment of death. But if alternatives do exist, such as narcotics that do not threaten life or a different path for the truck driver, the indirectness of the intention is called into question. Particular care must therefore be taken to assess the attitudes of those potentially involved in an act of euthanasia. Clearly a doctor or nurse could not legitimately be said to have the relief of pain as a primary intention in a case where narcotics are administered in doses known to exceed the quantity required to achieve analgesia, or where the patients show no signs of suffering in the first place (e.g., the comatose patient).

But while the direct–indirect distinction reflects our instinctive sense of the moral significance of right intention, it does not show us how to transcribe this moral sentiment into firm rules of decision making. While there are clear cases of direct and indirect killing, there is a "gray area" between them. Even as we gain insight into current problems, moreover, the issue of indirectness will develop new facets. The justification for the "merciful overdose" depends on the contingency that the most adequate analgesics for many terminal patients happen to have debilitating side effects. If this connection is broken by pharmacological advance or increased sophistication in the use of existing drugs, medicine will have lost this means of coping with the question of how long to try to prolong the lives of terminal patients in great pain. Even without these specific developments, life-support systems are bound to improve in general, so that it will be increasingly relevant to ask whether the length of a patient's life should be compromised for the sake of some other good. Thus the issues surrounding the distinction between direct and indirect euthanasia, particularly the question of the moral significance of right intention, are likely to assume even greater importance with the advance of biomedical technology.

Voluntary and involuntary dying

Every society, then, has set limits to killing. But the more that can be done to *prolong* a life that is ebbing away, the more necessary it becomes also to set limits to avoid assault and battery upon helpless, at times unwilling, victims in the name of prolonging their lives. All the distinctions mentioned up to now represent efforts made through the centuries to separate the permissible from the impermissible in these regards.

There is another indispensable distinction, however; that between voluntary and involuntary dying. It concerns the *attitude* of the person whose life or death is at stake.

Voluntary dying represents a vast spectrum of attitudes, from the peaceful acceptance with which death can be greeted to acts of suicide and euthanasia. It can encompass many choices, such as asking to be kept home from the hospital, ceasing to struggle against disease, or asking medical personnel to assist in suicide or even to perform acts of killing. A voluntary death can be achieved through suicide, through martyrdom or sacrifice for the welfare of others; it can also take place through voluntary euthanasia or mercy killing where what is at issue is the welfare of the victims themselves.

Nonvoluntary dying occurs in all acts of killing of unwilling or nonconsenting victims—either those who expressly oppose dying or those who are unable to express any opinion at all. Infanticide and the killing of unconscious patients, even when undertaken for merciful purposes, differ from voluntary dying since the persons killed have expressed no desire to die. And "involuntary euthanasia" refers to such programs of exterminating the sick and the disabled as were undertaken by the Nazis and others.

Some hold that the distinction between voluntary and involuntary dying is so important that it obviates all the previous ones mentioned. So long as a person *wants* to die, they say, and has a powerful reason for such a desire, it ought not to matter whether death comes by omission or commission, or what means are used, or whether there is direct or indirect intention to take life. Thus, advocates of the legalization of voluntary euthanasia hold that it ought to be possible actively to bring death by poison or other means to a suffering and dying individual requesting such assistance (Williams, 1957).

Religious arguments. Others have maintained their opposition to all kinds of suicide and mercy killing. Chief among them are Christian writers closely connected with the tradition of natural law. Other religions may outlaw these acts summarily or permit them under certain circumstances. But without strong prohibitions, Christianity, with its explicit outlining of the nature of life after death, would run the risk of tempting believers to speed death in order to achieve such a new life. Thus St. Augustine pointed out that if suicide were permissible in order to avoid sin, then it would be the logical course to choose for all those who were fresh from baptism. The "epidemics" of suicides among early Christians show that there have been times when those temptations were very great and help to explain the strong condemnation of suicide by Christian writers such as Augustine and Thomas Aquinas.

The great majority of the religious arguments supporting the condemnations of suicide hold that the individual has no right to desire to end his life (nor therefore to ask another to end it) since the right over life and death belongs exclusively to God, who is held to have given human beings the prescription not to kill. This prescription, however, has never been interpreted as being without exceptions. Even within the religious traditions, some have asked why suicide and voluntary euthanasia could not be exceptions, just as are killing in self-defense and killing in just wars for Christian theologians. Such an exceptional status is nevertheless denied to voluntary dying by most writers in the Christian tradition. To want to end one's life is thought to go against God's right; the right of God over life and death corresponds, they have held, with a duty on the part of men to protect their lives even from themselves. For most religious theorists of natural law, suicide and mercy killing would violate the natural end of self-preservation.

Proposals to legalize voluntary euthanasia. Proposals to legalize euthanasia would permit acts of both omission and commission resulting in death, given certain safeguards (Kamisar; Williams). The requirement that the patient should have *requested or consented expressly* to the act of euthanasia is clearly the most important of these safeguards, a requirement thought by sponsors of such proposals to provide a simple and clearcut distinction between voluntary and nonvoluntary dying. However, such a requirement raises new, and perhaps even more difficult, problems of how to draw a clear line. These problems are always present when consent is involved and are especially grave when a person's life is at issue. Opponents ask, Is there not an important distinction between request and consent in the first place? Might there be a danger of slipping from request to resigned acquiescence? Might there be a risk of someone's asking to die out of a concern for the burden he places on his family? How does one determine whether the request or the consent is not the result of a temporary aberration or of medication-induced confusion or depression? Or what if the desire to die is based upon an unrealistic view of the present situation, or of the prognosis for the future, based on false or insufficient information? And how is the situation altered if the patient requesting to die has dependents? (Cantor.)

In order to meet some of these concerns, a number of additional requirements, beyond mere consent or request, are suggested by those in favor of allowing voluntary euthanasia: The patient must be over the age of twenty-one; the agent must be a physician who should have consulted with another physician or with some specified authority; sometimes a period of time is required between the request for euthanasia and the act itself during which the patient's possible change of mind must be heeded.

Given such safeguards, these advocates argue that it is cruel to prolong intense suffering on the part of someone who is mortally ill and desires to die. Mercy dictates intervention. A secondary argument holds that a person has the *right* to decide whether he should continue to live or not, and that such a decision can be reached after rationally weighing the benefits of continued living against the suffering involved. If a person has such a right, it is held, then it cannot be wrong for him to ask another to help him carry out his desire, nor can it be wrong for another to do so.

Opponents argue first that acts of killing would not, in fact, always truly be merciful for the patients requesting them. There are risks that some patients might die as a result of an error in the prognosis of their disease. Others might die who could have recovered as a result of a new approach to their illness such as that provided when penicillin came into use. Still others might die who did not really wish to die, given the difficulties already mentioned of knowing whether the request for euthanasia is genuine and, even if genuine, is truly in the best interest of the patient. They point to the familiar cases where patients have pleaded to be allowed to die, only to recover with gratitude that the physician did not respond to their plea.

Secondly, these critics argue, even where dying would be more merciful, and even if patients have the right to determine whether they want to continue to live or not, such a right provides the justification for *suicide* and for the *refusal of life-prolonging treatment*, but not for another to engage in an act of killing. "Helping" a person to end his life is seen as fundamentally different from helping him to build a house or to find his way; and the question of whether the killing is helpful or harmful, lawful or prohibited, cannot, therefore, be decided merely by establishing that the victim asked for help.

The third argument stresses the small number of those who would actually be helped by new legislation. There is general agreement that very few of the cases publicly debated under the rubric of euthanasia would fit the requirements suggested in the different proposals. True, many persons suffer, and many are near death, but those who suffer *and* are near death and who are willing and able to ask, in a manner acceptable to courts and to physicians, to be killed, constitute a much smaller group. All acts of euthanasia contemplated by relatives or friends would be ruled out. And the familiar cases of infanticide sometimes referred to as euthanasia would also be ruled out, both because of the absence of consent by the victim and because the victim is under the age of twenty-one.

Most important, those *not competent* to give consent, even if they are over the age of twenty-one, would not fit the requirements. Those who are legally incompetent—such as great numbers of the retarded—have not been held capable of initiating procedures to resist medical treatment. Much less, then, would they be able to request euthanasia. Those who are physically unable to communicate, such as patients in coma, would likewise be excluded from consideration by euthanasia legislation requiring consent, in spite of the fact that much public concern with euthanasia stems from an awareness of the conditions under which their lives are prolonged.

Nevertheless, if these were the only objections to euthanasia, they might well be overcome through careful safeguards. Even if euthanasia does not always represent the most beneficial act for patients, and even if there are possibilities of cures, those few patients who would request euthanasia, it could be argued, ought to have the right to decide whether or not they wish to take such odds and continue to suffer, or choose to die. Even if helping someone to die differs from other forms of help in that it involves destruction, and even if administered euthanasia differs from suicide, given that the act is lawful and the agent willing, those who *want* to perform such acts could be empowered to do so. And even if very few would actually request euthanasia, their small number does not in itself bar providing them with relief. Finally, problems of competence and incompetence could be carefully worked out here, as elsewhere, in medicine and law. All of these factors, then, so long as they are considered purely on an individual basis, fail to make a sufficient case against voluntary euthanasia.

These factors take on new significance in the light of the most serious risks which the acceptance of voluntary euthanasia opens up—risks to social structures, and thus to individuals who have *not* requested euthanasia. These are risks of abuses and errors that might result from a relaxation in the present strong prohibitions against killing. Rules may be misapplied. Practices may be extended to groups of patients beyond the original few who fit the strict requirements. Distinctions may be blurred, so that patients may come to die without having requested euthanasia, perhaps quite against their wishes. The fears of such risks are supported by a concern for the defenselessness on the part of groups such as the newborn or the senile, and by a lack of confidence in the social resistance to harming them.

Suicide. Suicide has often been held to be an exception to the prohibitions universally placed upon taking innocent life. It does not pose the same threats to the social fabric as legalized voluntary euthanasia. There is much less to be feared from a spread of abuses and mistakes. And since suicide is the killing of self, there can be no concern, as in situations where others are the agents, about the brutalization of those professionals asked to participate in taking lives.

In the long debates concerning suicide, it has always been assumed that most suicides should be prevented; the disputed cases have been those where a person seems to be laboring under the compulsion of some painful and inevitable misfortune (Plato IX, 8). Such debates have then focused on the question of the duties owed by persons in respect to their own life. Some have held that, in killing oneself, one need only consider one's own life. Others have held, on the contrary, that duties concerning one's life are owed to God, to the state, to nature, and to one's family, or to some combination of these. Hume argued that, when a man is miserable enough to contemplate suicide, he is not obligated to give society the questionable benefit of living on at the expense of great suffering for himself. Nor could suicide be a transgression of the duty to God since God has already given men the power to alter the course of survival in building dwellings or inoculating children for smallpox (Hume). As for the duty to oneself, Hume argued that no one would kill himself if it were best for him to remain alive, and stated that he intended his arguments for suicide to restore men to their native liberty.

Kant, on the contrary, held that suicide violates man's duty to himself in the most fundamental way and thus is opposed to the supreme principle of all duty. He held that to destroy the subject of morality in one's own person is to root out the existence of morality itself from the world so far as this is in one's own power (Kant).

The growth of the discussion of rights in the last two centuries has brought to the fore the question of whether or not there is a right, more basic than all others, over one's own life. The act of suicide is no longer held to be a crime in many societies. While great efforts are still made to prevent most suicides, some categories of justifiable suicide may come to be set apart, where prevention of the suicidal action may not be undertaken. In such cases, the role of the physician in providing the means for suicide is once again in question, as it was in antiquity.

Conclusion

The distinctions surveyed above represent deeply felt differences. They often succeed in marking off regions of abuse or injustice and in exploring complex dilemmas. But these distinctions are at times very hard to draw. There are borderline regions between omission and commission, between extraordinary and ordinary means, between indirect and direct killing, and between voluntary and involuntary dying. We do not always possess clear natural lines.

Such a realization is sometimes thought to imply that all distinctions are useless, so long as they are not mirrored in nature. But it is crucial to see that, even though a line is not drawn in nature, it may well be needed in practice. Most norms are of this nature. Sometimes the precise location of the line drawn is somewhat arbitrary, as with age limits for driving, yet some line is clearly needed.

All social policy requires the drawing of lines, and those drawn to protect against killing are among the most universal. Prohibitions have to be established and distinctions made even where human affairs are uncertain and hard to classify. Such lines are troubling because they can seem arbitrary or harsh. Yet they are necessary in order to avoid considering afresh each new choice that arises. The most important question remains: Who is to draw the different lines, within what bounds, and with what degree of accountability?

SISSELA BOK

[*For further discussion of topics mentioned in this article, see the entries:* DOUBLE EFFECT; *and* SUICIDE. *Also directly related are the entries* DEATH; DEATH, ATTITUDES TOWARD; *and* DEATH, DEFINITION AND DETERMINATION OF. *For discussion of related ideas, see the entries:* ACTING AND REFRAINING; CARE; HEALTH AS AN OBLIGATION; INFANTS; LIFE-SUPPORT SYSTEMS; *and* RIGHT TO REFUSE MEDICAL CARE.]

BIBLIOGRAPHY

AUGUSTINE. *The City of God* 1.17,19,20,21. Translated by Demetrius B. Zema and Gerald G. Walsh as *Saint Augustine.* Vol. 6: *The City of God, Books I–VII.* Introduction by Étienne Gilson. The Father of the Church: A New Translation, vol. 8. Edited by Roy Joseph Deferrari. New York: 1950, bk. 1, chaps. 17, 19, 20, 21; pp. 46–47, 49–54.

BEHNKE, JOHN, and BOK, SISSELA, eds. *The Dilemmas of Euthanasia.* Garden City, N.Y.: Anchor Press/Doubleday, 1975.

BRIM, ORVILLE G., JR.; FREEMAN, HOWARD E.; LEVINE, SOL, and SCOTCH, NORMAN A., eds. *The Dying Patient.* New York: Russell Sage Foundation, 1970.

CANTOR, NORMAN L. "A Patient's Decision to Decline Life-Saving Medical Treatment: Bodily Integrity versus the Preservation of Life." *Rutgers Law Review* 26 (1973): 228–264.

CRANE, DIANA. *The Sanctity of Social Life: Physicians' Treatment of Critically Ill Patients.* New York: Russell Sage Foundation, 1975.

FLETCHER, JOHN. "Abortion, Euthanasia, and Care of Defective Newborns." *New England Journal of Medicine* 292 (1975): 75–78.

FLETCHER, JOSEPH. "Elective Death." *Ethical Issues in Medicine: The Role of the Physician in Today's Society.* Edited by E. Fuller Torrey. Boston: Little, Brown & Co., 1968, chap. 7, pp. 139–157.

FOOT, PHILIPPA. "The Problem of Abortion and the Doctrine of the Double Effect." *Oxford Review,* no. 5 (1967), pp. 5–15. Reprint. *Moral Problems: A Collection of Philosophical Essays.* 2d ed. Edited by James Rachels. New York: Harper & Row, 1975, pp. 59–70.

FOX, RENÉE C., and SWAZEY, JUDITH P. *The Courage to Fail: A Social View of Organ Transplants and Dialysis.* Chicago: University of Chicago Press, 1974.

FRIED, CHARLES. "Right and Wrong—Preliminary Considerations." *Journal of Legal Studies* 5 (1976): 165–200.

GOUREVITCH, DANIELLE. "Suicide among the Sick in Classical Antiquity." *Bulletin of the History of Medicine* 43 (1969): 501–518.

GRUMAN, GERALD J. "An Historical Introduction to Ideas about Voluntary Euthanasia, with a Bibliographic Survey and Guide for Interdisciplinary Studies." *Omega* 4 (1973): 87–138.

HUME, DAVID. "Of Suicide." *Essays Moral, Political, and Literary.* 2 vols. Edited by T. H. Green and T. H. Grose. The Philosophical Works of David Hume, vols. 3–4. London: Longmans, Green, & Co., 1875, vol. 2, pp. 406–414.

KAMISAR, YALE. "Some Non-religious Views against Proposed 'Mercy-Killing' Legislation." *Minnesota Law Review* 42 (1958): 969–1042.

KANT, IMMANUEL. *Grundlegung zur Metaphysik der Sitten* (1785). Translated and edited by H. J. Paton as *The Moral Law: Kant's Groundwork of the Metaphysic of Morals.* New York: Barnes & Noble, 1967.

KOHL, MARVIN, ed. *Beneficent Euthanasia.* Buffalo, N.Y.: Prometheus Books, 1975.

KÜBLER-ROSS, ELISABETH. *On Death and Dying.* New York: Macmillan Co., 1969.

MCCORMICK, RICHARD A. *Ambiguity in Moral Choice.* The 1973 Père Marquette Theology Lecture. Milwaukee: Marquette University Press, Theology Department, 1973.

———. "To Save or Let Die." *Journal of the American Medical Association* 229 (1974): 172–176.

MAGUIRE, DANIEL C. *Death By Choice.* Garden City, N.Y.: Doubleday, 1974.

PIUS XII. "The Prolongation of Life." *The Pope Speaks* 4 (1958): 393–398.

RACHELS, JAMES. "Active and Passive Euthanasia." *New England Journal of Medicine* 292 (1975): 78–80.

RAMSEY, PAUL. *The Patient as Person: Explorations in Medical Ethics.* The Lyman Beecher Lectures at Yale University, 1969. New Haven: Yale University Press, 1970.

THOMAS AQUINAS. *Summa Theologiae* II-II 64,7; Response. Translated by the Fathers of the English Dominican Province as "Whether It Is Lawful to Kill a Man in Self-defense?" *Summa Theologica: First Complete American Edition.* 3 vols. Vol. 2: *Second Part of the Second Part*, QQ. *1–189 and Third Part,* QQ. *1–90, with Synoptical Charts.* New York: Benziger Brothers, 1947, question 64, article 7, pp. 1471–1472; especially p. 1471.

VEATCH, ROBERT M. *Death, Dying and the Biological Revolution: Our Last Quest for Responsibility.* New Haven: Yale University Press, 1976.

WILLIAMS, GLANVILLE L. *The Sanctity of Life and the Criminal Law.* James S. Carpentier Series, 1956. New York: Alfred A. Knopf, 1957.

III
PROFESSIONAL AND PUBLIC POLICIES

Ethical issues and ethical positions are reflected in explicit philosophical arguments about caring for dying patients, the right to refuse life-saving medical treatment, and the active termination of human life. They are also reflected, at least implicitly, in professional and public policies related to death and dying. Some of these policies deal with the definition of death, the use of cadaver organs for transplantation and other purposes, suicide, and what dying patients should be told. These policies are discussed in the relevant entries elsewhere in this Encyclopedia and are not examined here. Many of the most important and controversial policies deal, however, with the decision to let a dying patient die or to intervene actively to cause death for reasons of mercy. It is such policies that will be examined in this article.

It might be argued on moral grounds that in principle there ought to be no policies regarding letting dying patients die or active intervention to cause death; that such matters should be left in the hands of God or fate. It could be a matter of hospital, professional, or public policy that in all cases everything possible must be done to preserve the life of every dying patient. That policy itself, however, reflects a particular ethical commitment—the commitment to the preservation of life at all costs. While individuals may have held (or thought they have held) such a moral policy position, no society, government, professional, or religious group has even taken such a position. Generally it is recognized that as a matter of policy some things that can be done to preserve life ought not to be done or at least ought not to be required. Different societies and different professional groups have, however, from time to time held different policy positions apparently reflecting different moral positions regarding such decisions.

What follows is a systematic review of policy options focusing first on policies related to professional groups and organizations—that is, policies appropriate for physicians, hospitals, and other professionally organized bodies. Then this article will turn to more public policies: informal policy mechanisms for individual patients as well as laws establishing policies regulating death and dying decision making.

Policies favoring professional decisions

One broad group of policies relating to the care of the dying focuses on the responsibility of the individual professional or group of professionals to use discretion in deciding when it is appropriate to stop care of a dying patient. This discretion is normally exercised in a context of legal limits established by the society. Even contemporary society considers active killing of a terminally ill patient even on grounds of mercy to be an offense (Baughman, Bruha, and Gould, pp. 1203–1204; St. John-Stevas, p. 264). The details of these laws will be discussed later in this article. Such law, moreover, is generally reflected in professional codes and policies.

Policies favoring decision making by physicians. Policies permitting or supporting individual professional decision making focus on the use of professional judgment whether to stop or refuse to initiate treatment of a dying person. Traditionally in Western society this has been seen as part of the physician's prerogative, indeed responsibility. Although occasionally it is suggested that a decision by a physician (at least without court authorization) to let a patient die should and would be treated as a homicide, there has never been a prosecution in the United States for such a decision, and there are no reports available of conviction in other jurisdictions. In fact courts have consistently supported the rights of patients to refuse treatment for any reason whatsoever provided the patient is competent and the treatment is being offered for the patient's own good rather than the good of another (as in public health cases). There have even been decisions supporting such treatment-stopping decisions when they were made by others on behalf of an incompetent patient. Thus, even though the law does not permit a physician to decide to kill for mercy actively, it has generally tolerated reasonable decisions by physicians that treatment should not go on.

This procedure is being endorsed more explicitly in some countries as the debate about the care of the dying becomes more public. For example, the Swiss have had significant public debate about the policy of physician discretion. In 1975 a Swiss physician, Urs Peter Haemmerli, was arrested and accused of murdering by starvation elderly patients at Triemli City Hospital where he was chief of medicine. The patients were described as paralyzed, unconscious persons for whom there was no chance of successful treatment. He was cleared of the charges, the court finding that in nine cases specifically under review "the withholding of nourishment was justified by the fact that it appeared useless to continue feeding a patient whose brain had ceased to function" (Culliton). Although the case could be and has been interpreted as a justification of the policy of permitting physicians to use discretion, there was much confusion over the issues. At some points, arguments were presented that the patients were dead according to brain criteria. At most the Haemmerli decision can be interpreted as justifying the physician's decision in cases where the patient's condition resembled the ones in this case.

In the aftermath of this case the Swiss Academy of Medical Sciences in 1977 issued guidelines for physicians claiming for the physician wide discretion. The Academy favored permitting doctors to discontinue medication as well as technical measures such as respirators, blood transfusions, hemodialysis, and intravenous nourishment when use of these measures "would mean for the dying an unreasonable prolongation of sufferings and if the [patient's] basic condition has taken an irreversible course."

Many physicians in other countries share the view that they should be permitted to use their judgment in such cases. In France, fifty-three percent of physicians surveyed by the French edition of *Medical Tribune* in 1976 said they would consider "omission of an action which hastens the end" when faced with a terminal patient whose suffering seems unbearable (Joly and Neron). Yet it is still an open question whether professional discretion is the most appropriate policy and, indeed, whether it is legal. It could still be illegal to exercise such discretion even if professional medical groups within a country favored such a policy.

The moral objections to this policy of "individual physician" decision making are of two kinds. One group of opponents objects substantively to the position that physicians should ever be placed in a position of deciding not to do something that they could do to save a life. Another argument focuses more on procedure than on substance. A policy of full professional autonomy in deciding when treatment should be stopped could force treatment on patients who do not want it. It could also lead to stopping treatment on a patient who wants the treatment to continue. It is argued that at least for competent patients this professional decision making is an unacceptable policy. The Swiss directive is qualified to the extent that it insists that doctors must "respect the will of the patient" when such wishes can be expressed even if it does not correspond with what is called the medical indications. The House of Delegates of the Medical Society of the State of New York in 1973 approved a statement that appeared to reject the role of the professional as decision maker, yet holds on to the notion insofar as the physician must approve the decision made by the patient and/or his family. This contrasts with the more patient-centered policy adopted by the American Medical Association's House of Delegates later that same year. The AMA made clear that it rejected active killing, but recognized the right of the patient or his family to make treatment-stopping decisions at least in cases where death is imminent.

The arguments against a policy of granting the individual physician or the medical profession as a whole the authority to make such decisions is grounded variously in the fear that the physician may not adequately know the values of the patient, that the physician may hold special professional values (such as the duty to preserve life) that should not be inflicted on the patient, or that patient and familial freedom and integrity require a policy of removing such authority from the professional realm.

Policies using hospital committees. All of these difficulties with the policy of permitting the individual physician to decide when and if treatment should be stopped have led to proposals that a committee of professionals be used to make such decisions (Veatch, 1977). In the United States such a policy received great impetus from the case of Karen Ann Quinlan, where a "hospital ethics committee" was required by the judgment of the Supreme Court of the State of New Jersey to review the case (*In the Matter of Karen Quinlan*).

Several proposals for transferring decision-making authority to committees have been put forward (Teel; Rabkin, Gillerman, and Rice). Some committees actually have begun the process although often in more of an advisory than a final decision-making capacity (Critical Care Committee of Massachusetts General Hospital). The policy of having a committee rather than an individual physician make the decision to stop treatment is put forward for several reasons. It is seen as a way of neutralizing the individual biases that a single physician might possess. It is also thought to be a way of diffusing authority for such decisions.

Yet there are objections to the committee approach as well. Although a committee might in principle eliminate the idiosyncratic biases of the individual physician, if it were made up of physicians, it would in no way eliminate any systematic biases that physicians as a group might have. Furthermore, it would not give authority to individual values, rights, and responsibilities of the patient or his family. It is also argued that, although diffusion of personal responsibility for the decision may be psychologically satisfying for the physician, it is a morally risky procedure. Rather, it is argued by opponents of the committee mechanism, someone should have clear responsibility for such a crucial decision. They also argue that, while the individual physician may have personal knowledge of the patient's beliefs and values about treatment stopping, the more impersonal committee can never have this crucial knowledge except indirectly.

There is debate over the various functions that such a committee might perform. In the *Quinlan* court opinion a "hospital ethics committee" was to determine whether "there is no reasonable possibility of Karen's ever emerging from her present comatose condition to a cognitive, sapient state"; that is, they are to confirm prognosis. They are not given the task of reviewing, approving, or disapproving the decision to stop treatment. One task for such committees, then, is to review prognosis, while leaving to others the decision about what to do about stopping treatment. Other committees such as the Critical Care Committee of the Massachusetts General Hospital see themselves much more as advisers to the individual physician. Still other committees deal with problems of allocating scarce resources (such as deciding who receives a single hemodialysis machine when more than one patient needs it) or of general hospital policy. These functions should be kept quite separate from the use of a committee to decide when and if to stop treatment.

Policies favoring patients, other lay persons, and public bodies

The problems with policies focusing on the decision making of individual medical professionals and professional committees have led to consideration of other types of policies with more direct involvement of patients, other lay people, and the public as a whole. Some policies emphasize the role of personal communication by individuals in anticipation of a day when they may not be able to take an active part in deciding whether treatment should continue. Another type of public policy involves legislation to clarify the rights and responsibilities of various actors in decisions about stopping treatment and about active killing for merciful reasons.

Personal communication policies

Living wills and their rationale. There are a number of responses to the awareness that increasingly decisions must be made about the stopping of medical treatment that prolongs the dying process. One has been the advocacy of informal communication undertaken while a person is of sound mind. Such communication has the objective of indicating the person's wishes regarding terminal care. The most important development has been the preparation of a model document called a "living will." The document, prepared and distributed by the Euthanasia Educational Council, the successor organization to the Euthanasia Society of America, is in the form of an informal letter addressed to "my family, my physician, my lawyer, my clergyman; to any medical facility in whose care

I happen to be; to any individual who may become responsible for my health, welfare, or affairs."

The movement to informal and formal personal expressions of wishes in "living wills" and similar documents has been taken up in many countries in addition to the United States. In Sweden a group called "The Right to Our Death" has promoted such communication as well as legislation. Reportedly thousands of Swedes have signed such documents. In Denmark a group called "My Life Testament" reports similar activity. Similar groups exist in Switzerland and Britain. There is less formal activity in Catholic Europe, such as Italy and France, although groups there are also advocating the "right to die."

The model living will of the Euthanasia Council in the United States includes this request: "If the situation should arise in which there is no reasonable expectation of my recovery from physical or mental disability, I request that I be allowed to die and not be kept alive by artificial or 'heroic' measures." The letter, by itself, is a "request." It probably has no legally binding effect. An earlier version of the document explicity said it was not legally binding.

Other individuals and groups have prepared other versions of such informal communications. The "Christian Affirmation of Life" has since 1974 been distributed by the U.S. Catholic Hospital Association. It includes a more theological introduction, but its operative language is remarkably similar to the Euthanasia Council's document, saying, "I request that, if possible, I be consulted concerning the medical procedures which might be used to prolong my life as death approaches. If I can no longer take part in decisions concerning my own future and there is no reasonable expectation of my recovery from physical or mental disability, I request that no extraordinary means be used to prolong my life."

Objections to these policies. There have been several criticisms of the informal letter devices. They are not, by themselves, legally binding documents, so that if one is worried that he will be treated in a situation where he would not want to be (or would not be treated in a situation where he would want to be) the documents will, at most, provide some guidance; they cannot give a legal guarantee that one's wishes will be followed.

Another criticism is that such documents may be written while one is healthy—perhaps many years before one is critically ill when the document would have to be used. It has been argued that one might change one's mind in the interim. This is said to decrease the reliability of such documents. On the other hand defenders have countered that, while it is true that the writer may have changed his mind, it is more reasonable to presume that the views explicitly set forth in such documents rather than some other views are his current convictions.

Another problem with the letters is that, in part because they are written while one is healthy, they are necessarily very vague. One cannot describe all of the possible conditions under which one may be critically ill. Some physicians have attempted to write much more explicit and specific instructions regarding their own treatment, including such items as the number of minutes without breathing after which resuscitation should not be attempted. It is doubtful than even a well-trained physician could envision all of the specific events that might occur many years in the future. Furthermore one's desire for treatment may depend not only on the medical facts but also on one's life situation. In any case the medical lay person is not able to write instructions with such specificity.

Alternative communications. One alternative that is still in the category of personal communications by individuals prior to the time they are terminally ill is the preparation of a "power of attorney" document that would have as its objective not the specification of conditions under which treatment should or should not be given, but the designation of someone—a spouse, relative, lawyer, clergyman, or friend—as one's agent for the purpose of refusing treatment should the writer become incapable of exercising his own judgment. Some such power of attorney forms include general guidelines to instruct the agent and to make clear to others evaluating the agent's actions what the writer's wishes are. The device leads to decisions made at the critical time by an agent rather than in advance by the writer. It is meant to be legally binding. It provides another alternative for personal expression of wishes about terminal care.

Legislative options. A more formal public policy is also evolving rapidly for the regulation and structuring of decisions concerning death and dying in both case law and legislative proposals. The case law has grown out of court decisions evaluating the right of individual patients or their agents to refuse medical treatment. Several hundred cases in the United States have given substantial clarity to the policy growing on the common law foundation. [This case law is examined in the article RIGHT TO REFUSE MEDICAL CARE.] The legislative options have been receiving increasing attention since about 1970.

Common law countries have traditionally been governed by prohibitions against homicide and suicide, but the precise application to decisions pertaining to the terminally ill has never been clear. The movement to statutory law began in Great Britain in the 1930s with the establishment of the Voluntary Euthanasia Legislation Society. A bill was introduced into the House of Lords in 1936 sponsored by the British society. The bill required that a person sign an application asking to be put to death before the euthanasia could be carried out. The person had to be over twenty-one and suffering from an incurable and fatal disease accompanied by severe pain. Apparently active killing of the one signing such a document was envisioned. The application signed in the presence of two witnesses was to be submitted to a "Euthanasia Referee" appointed by the Minister of Health who was to review the case.

In 1937 a euthanasia bill had been introduced into the Nebraska legislature. About the same time the Euthanasia Society of America was founded by the Reverend Charles Potter. It had first intended to include nonvoluntary euthanasia for eugenic purposes. The legislation proposed for Nebraska had similarly made provision for nonvoluntary killing. A survey of physicians in 1941 indicating resistance to this led the society to limit its activities to voluntary euthanasia. In 1946 a committee including 1,776 physicians worked for legalizing euthanasia in New York State. A bill was introduced into the state legislature in 1947. Reaction to the Nazi experience with compulsory euthanasia for medical as well as genetic reasons combined with the failure of the American movement to distinguish between active killing and omission of treatment led to a diminishing of interest until the late 1960s. At that point increasing use of only partially successful life-extending technologies maintaining individuals in a prolonged debilitated state together with the evolution of an American egalitarian and individual civil rights movement led to a new round of legislative activity. The American legislative proposals rapidly increased in number and variety. Three basic types of legislation appeared. (1) The first type, following the earlier pattern, would make active killing legal—normally upon request. (2) The second is limited to omission of treatment, but would attempt to clarify the policy regarding decision making in the case of incompetent patients. (3) The third type simply clarifies the rights of the individual to refuse medical treatment and to write a document to take effect should the person become incompetent.

Legislation proposing active killing. The legislative activity in the late 1960s began with the introduction of a bill into the British Parliament that would have made it lawful "for a physician to administer euthanasia to a qualified patient who had made a declaration that is for the time being in force" (Downing, pp. 201–206). Even though the law regarding certain stopping and omission of treatment is unclear, in every country in all legal systems active killing even for mercy and at the request of the one killed is an offense (Silving, p. 378). In modern continental European codes of criminal law motive can be a mitigating factor. The Prussian Landrecht in 1794 imposed a penalty, similar to negligent killing, for "killing with intention believed to be good" of a deadly wounded or otherwise dying person (ibid., p. 368). The codes of Wuerttemberg of 1839 and of Thuringia of 1850 and of other German states also imposed reduced penalty when killing took place at the request of a fatally ill person (ibid.). The Norwegian Penal Code of 1902, which is still in effect, has a similar provision, as do the Russian Penal Code of 1903, the Polish Code of 1932, similar codes in the Baltic countries, the Netherlands, Italy, Norway, Germany, Switzerland, Austria, and Spain (ibid.; Meyers). The 1933 Penal Code of Uruguay provides, "The judges are authorized to forgo punishment of a person whose previous life has been honorable where he commits a homicide motivated by compassion, induced by repeated requests of the victim."

In Germany as well as Switzerland mercy killing is classified not as murder but as manslaughter. German policy is unique in being influenced by the experience of the Nazi period. The eugenic idea of the "life not worthy to be lived" and the life that was a useless burden on the community was crystallized in Karl Binding's publication *Die Freigabe der Vernichtung Lebensunwerten Lebens* (permitting the extermination of life not worth living), but the idea is traced to Luther. One interpretation has it that the German mass extermination program can be traced to the attitude toward the nonrehabilitable sick

(Alexander), but others argue that the racist ideology of a pure *Volk* was the critical element behind both the euthanasia program and the "Final Solution" (Dawidowicz, p. 131). It is true that in late 1938 or early 1939 Hitler received a request from the father of a deformed child asking that the child be killed. He turned the request over to Brandt to investigate with authorization to inform the physicians, in Hitler's name, to carry out the euthanasia. In the spring of 1939 the killing of the mentally deficient and physically deformed children was regularized. In July he turned to the murder of the adult insane and rapidly to the "non-Aryans" and racially unfit. The historical debate, however, centers on whether these beginnings actually led to the development of a new moral policy—the first step on a "slippery slope"—or simply served as a convenient starting point for the execution of a policy already implied in Nazi notions of racial health, without which the whole development would not have occurred.

In contrast under Roman, canon, and common law and under English and American statutory penal law consent or request of the victim is irrelevant and the mercy motive is not a guilt-mitigating factor (Meyers, p. 155; Silving, p. 380). In England the Royal Commission on Capital Punishment in 1953 concluded "reluctantly" that voluntary euthanasia could not be taken out of the category of murder (Meyers, p. 146). Thus if there was to be a change it would have to be by statute. "Euthanasia" was defined in the British bill proposed in 1969 as "the painless inducement of death." A qualified patient was "a patient over the age of majority in respect of whom two physicians (one being of consultant status) have certified in writing that the patient appears to them to be suffering from an irremediable condition." An "irremediable condition" was "a serious physical illness or impairment reasonably thought in the patient's case to be incurable and expected to cause him severe distress or render him incapable of rational existence." The declaration made by the patient was one requesting "the administration of euthanasia at a time or in circumstances to be indicated or specified by me or, if it is apparent that I have become incapable of giving directions, at the discretion of the physician in charge of my case."

Similar bills following the British model were introduced into the legislatures of the state of Idaho in 1969 and Oregon in 1973. All of these bills would have permitted active killing. Further they would have required the request of the patient executed in writing. The criticism of the bills has been strong on several grounds. The most direct opposition comes from those who object to active killing on moral grounds. The distinction between active killing and omission of treatment is deeply rooted in the moral conscience. Consistently in opinion polls only a small portion favor legalization of active killing while omission of treatment under specific conditions receives substantially more support. Active killing other than on the request of the patient has been found particularly morally objectionable whether the killing is proposed on humanitarian or eugenic grounds. On the other hand, active killing upon the request of the competent patient, as is proposed in the above draft bills, meets the needs of relatively few patients, because those who are sufficiently competent to execute the document are usually not in a condition in which they would request the action. Especially when omission of treatment in order to let the dying process continue (with pain controlled by adequate medication) is an available alternative in all or virtually all cases, the proposals for a policy of legalization of active killing upon request have received little public support.

Laws clarifying decision-making authority. A second type of legislative proposal was introduced into the Florida state legislature by physician-legislator Walter Sackett in 1970. It would have provided that "any person, with the same formalities as required by law for the execution of a last will and testament, may execute a document directing that he shall have the right to death with dignity and that his life shall not be prolonged beyond the point of meaningful existence." The bill went on, however, to clarify who should make such a decision in the case of a patient incompetent by reason of mental or physical incapacity. The first authority was a spouse or person or persons of first degree of kinship. If there were no relatives, it provided that "death with dignity shall be granted any person if in the opinion of three physicians the prolongation of life is meaningless." The bill, which never passed either house in Florida, was the first model focusing on the decision-making authority of guardians or agents for the incompetent patient. A somewhat similar bill was introduced in the legislature of West Virginia in 1972. There are other variants proposed on this model (Veatch, 1976, pp. 164–203). Some would give the first authority to someone designated by the individual while competent, by making use of a power-of-attorney type of instrument. This would avoid difficulties when an individual does not want the next of kin to make the decision, when he does not want to burden the next of kin, or when there is more than one person of equal degree of kinship. Some variants propose that, in cases where there are no relatives, a court-appointed guardian be used rather than a group of physicians. All of these models presume the existing legal remedy of judicial review whenever there is reason to suspect that the designated guardian is acting foolishly or maliciously.

These proposals avoid the moral objections to active killing. The main opposition has come from those who believe that no one other than the patient should be given the responsibility for making such treatment-stopping decisions or believe that relatives cannot be trusted to make them in the interest of the individual who is not competent. It is also argued that it places an undue burden on the family. The rebuttal emphasizes the presumed authority given parents to make decisions about accepting or rejecting treatment on behalf of their children. It is also argued that there has been little difficulty with such a policy when backed by the right of the court to intervene if necessary. Critical to the defense of the policy is a theory of the rights and responsibilities of the family. Defenders claim that the family is not only in the best position to speak for the interests of the patient, but has a moral obligation to do so. It is part of the nature of being a good family member even if at times the task is burdensome. The importance of the family in protecting the rights and interests of an incompetent member was central to the most important court case in this area, the case of Karen Ann Quinlan. Defenders of this type of legislation argue that someone must make decisions about what constitutes appropriate care of the incompetent dying patient. Those decisions should be based on the religious and ethical values of the patient if the patient had been competent. Or if the patient has never been competent, then the family, functioning within the limits of the law, may speak through the next of kin as the proper presumed guardian for purposes of initiating decisions to refuse treatment.

Legislation clarifying the right to refuse treatment. A third type of bill is much more modest in its intent. It would not make active killing legal even upon request. It would not even clarify the decision-making process for the incompetent patient. It simply affirms the right of the competent patient to refuse medical treatment—a right already existing in U.S. law—and makes legal certain written instructions should the patient become incompetent. One of the legal problems in American jurisdictions has been that, even though it is clear that the competent patient could refuse any treatment whatsoever, it has not been clear that the refusal would remain valid if a patient lapsed into incompetency.

Walter Sackett, having experienced defeat of his previous proposals, introduced such a bill into the Florida legislature in 1973. It provided that any person may at any time execute a document directing that medical treatment designed solely to sustain the life processes be discontinued. Resistance to this bill (defeated by a close vote) came from some who opposed any legislation in this area and others who felt the bill did not go far enough. A similar bill was introduced into the Massachusetts legislature that year and in the next few years similar bills were proposed in at least thirty-nine states.

On 30 September 1976, the governor of the state of California signed the first law providing for what was called a "natural death" (California Health and Safety Code). The law specifies that "any adult person may execute a directive directing the withholding or withdrawal of life-sustaining procedures in a terminal condition." The directive is a statement of a desire that "my life shall not be artificially prolonged." It is limited to very narrow conditions when "I should have an incurable injury, disease, or illness certified to be a terminal condition by two physicians, and where the application of life-sustaining procedures would serve only to artificially prolong the moment of my death and where my physician determines that my death is imminent whether or not life-sustaining procedures are utilized."

One line of objection has come from those who argue that this is an opening wedge for more offensive euthanasia proposals or is, in itself, morally offensive in surrendering the obligation to prolong life as long as possible. Stronger objection has come, however, from those who argue that the bill is either not needed or does not go far enough. Some have argued that in the light of the already existing right to refuse medical treatment the statute adds nothing and in fact, by being limited to certifiably terminal illness, may give the impression that those who are not certifiably terminally ill do not have the right to refuse lifesaving medical treatment. The statute at least makes clear that, if you execute such a document at least fourteen days after becoming certifiably terminally ill, the directive must be followed by the physician, unless revoked, or the physician will be guilty of professional misconduct. This legal authorization of the written document was never clear before the passage of the California bill.

The bill, however, does not remove the fear of those who execute a document or a "living will" before they are certifiably terminally ill. In such a case "the attending physician may give weight to the directive as evidence of the patient's directions regarding the withholding or withdrawal of life-sustaining procedures and may consider other factors. In this case the patient's written instructions are still not binding. In fact, the physician is authorized as never before to make his own decision over and against the expressed wishes of the patient and cannot be criminally or civilly liable for failing to follow the document. Furthermore the family is given only a marginal role of presenting information rather than the key responsibility as advocated by those supporting the type of policy discussed above.

Not only does the document allow substantial physician discretion; it defines terminal illness very narrowly. A terminal condition is defined as one that is incurable "regardless of the application of life-sustaining procedures" and "would, within reasonable medical judgment, produce death, and where the application of life-sustaining procedures serve only to postpone the moment of death of the patient." Thus a patient with severe physical or psychological suffering from hemodialysis is not terminally ill, because if treated he can live indefinitely. Those supporting the moral traditions that recognize this kind of grave burden to justify refusal of treatment (e.g., in the "extraordinary means" ethic) would find this unacceptable. Similar bills have been considered in many states and passed in several. Policy regarding the never competent patient and the patient who lapses into incompetency without executing a document or executes one before being certifiably terminally ill remains vague in all American jurisdictions as does policy on these matters in other countries.

Conclusion

Policy choices must be made by individual patients, their families, physicians, hospitals, and governments. The policies, both professional and public, relating to the decisions about the care of the dying patient have incorporated and must incorporate basic positions on the key ethical issues. The policy choices reflect stances on the value of life, the meaning and importance of freedom, and duties to society and individuals.

ROBERT M. VEATCH

[*Directly related are the entries* ACTING AND REFRAINING; INFANTS, *article on* PUBLIC POLICY AND PROCEDURAL QUESTIONS; *and* RIGHT TO REFUSE MEDICAL CARE. *For discussion of related ideas, see the entries:* CHILDREN AND BIOMEDICINE; DEATH, ATTITUDES TOWARD; INFORMED CONSENT IN THE THERAPEUTIC RELATIONSHIP, *article on* LEGAL AND ETHICAL ASPECTS; LIFE; MEDICAL ETHICS UNDER NATIONAL SOCIALISM; *and* PAIN AND SUFFERING. *See also:* DEATH, *articles on* WESTERN PHILOSOPHICAL THOUGHT *and* DEATH IN THE WESTERN WORLD; DEATH, DEFINITION AND DETERMINATION OF; *and* SUICIDE. *See* APPENDIX, SECTION I, PRINCIPLES OF MEDICAL ETHICS, WITH REPORTS AND STATEMENTS (AMERICAN MEDICAL ASSOCIATION).]

BIBLIOGRAPHY

ALEXANDER, LEO. "Medical Science under Dictatorship." *New England Journal of Medicine* 241 (1949): 39–47.

ARISTOTLE. *Politics.* 7.16 1334B-1336A.

BAUGHMAN, WILLIAM H.; BRUHA, JOHN C.; and GOULD, FRANCIS J. "Euthanasia: Criminal, Tort, Constitutional, and Legislative Questions." *Notre Dame Lawyer* 48 (1973): 1203–1260.

California Health and Safety Code. Sec. 7185. Div. 7, pt. 1, chap 3.9, The Natural Death Act.

Critical Care Committee of the Massachusetts General Hospital. "Optimum Care for Hopelessly Ill Patients." *New England Journal of Medicine* 295 (1976): 362–364.

CULLITON, BARBARA J. "The Haemmerli Affair: Is Passive Euthanasia Murder?" *Science* 190 (1975): 1271–1275.

DAWIDOWICZ, LUCY S. *The War against the Jews: 1933–1945.* New York: Holt, Rinehart & Winston, 1975.

DOWNING, A. B., ed. *Euthanasia and the Right to Death: The Case for Voluntary Euthanasia.* Contemporary Issues series, no. 2. London: Peter Owen, 1974.

In the Matter of Karen Quinlan. 2 vols. Vol. 2: *The Complete Briefs, Oral Arguments, and Opinion in the New Jersey Supreme Court.* Introduction by Daniel N. Robinson. Arlington, Va.: University Publications of America, 1976.

JOLY, JEAN-MICHEL, and NERON, MICHEL. "Fifty-three Percent of French GPs Would Consider Passive Euthanasia." *Medical Tribune,* 19 May 1976, pp. 1, 12.

MEYERS, DAVID W. *The Human Body and the Law: A Medico-Legal Study.* Chicago: Aldine Publishing Co., 1970.

PLATO. *Republic.* 5.6 460A.

RABKIN, MITCHELL T.; GILLERMAN, GERALD; and RICE, NANCY R. "Orders Not to Resuscitate." *New England Journal of Medicine* 295 (1976): 364–366.

ST. JOHN-STEVAS, NORMAN. *Life, Death, and the Law:*

Law and Christian Morals in England and the United States. Bloomington: Indiana University Press, 1961.

SILVING, HELEN. "Euthanasia: A Study in Comparative Criminal Law." *University of Pennsylvania Law Review* 103 (1954): 350–389.

TEEL, KAREN. "The Physician's Dilemma. A Doctor's View: What the Law Should Be." *Baylor Law Review* 27 (1975): 5–9.

VEATCH, ROBERT M. *Death, Dying, and the Biological Revolution: Our Last Quest for Responsibility.* New Haven: Yale University Press, 1976.

———. "Hospital Ethics Committees: Is There a Role? Some Possible Functions . . . and Problems." *Hastings Center Report* 7, no. 3 (1977), pp. 22–25.

WILLIAMS, GLANVILLE. *The Sanctity of Life and the Criminal Law.* James S. Carpentier series, 1956. New York: Knopf, 1957; London: Faber & Faber, 1958.

DEATH, ATTITUDES TOWARD

The simple expression "attitudes toward death" encompasses a complex of interrelated but far from identical concepts. The first term, *attitudes*, is recognized as having cognitive, affective, and behavioral components, so we must consider how people think about death, how people feel about death, and how people behave in regard to death.

Nor is the term *death* a unitary concept. When used in this context, it normally includes the dying process, the moment of death, and the state of being dead. Each of these may be perceived as something imminent or something indefinitely far off in time, as something that will happen to oneself or that will happen to others, as something attended to as an abstract concept or as something personal and real. "Attitudes toward death," then, includes, for example, how I feel about my being dead, what I think about your being in the process of dying, and how I behave in response to a funeral procession.

Attitudes, by definition, are relatively enduring, but they also occur within a context. Thus global statements about either an individual or a culture as being death-denying or fearful of death require careful delineation. An individual might properly be described as death-denying, yet he or she might discuss the latest grotesque homicide with glee. Western culture is generally described as death-denying, yet newspapers are filled with accounts of death; cemeteries and mortuaries are visible to the casual passerby; and an immense number of expressions using the word *death* and its derivatives have crept into Western languages. In order to understand death-related behavior at any given moment, it is necessary to understand the broad social context, the immediate situation, and the life history and present circumstances of the individual affected.

Another concern in understanding the death attitudes of an individual or of a community is the process through which such attitudes are communicated. The attitudes of individuals can be ascertained by asking them how they think or feel, by observing their behavior (how they adopt health practices, drive automobiles, observe religious rituals, participate in life-threatening activities), by asking others who know them, by evaluating their creative output (literature, painting, music), or by some of the more subtle kinds of measures developed by behavioral scientists over the years.

A similar tactic can be used to understand the death attitudes of a community (e.g., society, culture, subculture, people, group). In both instances, we must attend to both the most frequent responses/occurrences and the variation of responses/occurrences. However, whether our focus is the individual or the community, we need to remain alert to the potential for error in the information. Broadly speaking, such error can arise from three sources: distortions and misrepresentations in what various individuals or other sources communicate; ambiguities and incompleteness in the message; and biases and inaccuracies in the interpretation.

Because the likelihood of error arising from these sources is substantial, caution is required in translating the results of any attitude studies into policy, programs, or practice. Feifel and Branscomb have pointed out that responses to death-related stimuli are very different when elicited at the level of conscious awareness from those elicited at levels of less conscious awareness. This makes the research base for policy decisions more problematic, since the decisions are so significant and the research is still so new. A particularly relevant instance is the attitude toward permitting people to die if they want to. Results of one study indicated that about half the respondents agreed that people should be given this prerogative, especially if they are in pain or dying anyway. Of those who did not approve, explanations primarily revolved around the belief that only God has the right to take a life (forty-nine percent) or that there is always hope (thirty-two percent) (Kalish and Reynolds). The ethical issue is not only whether or not to allow a person to die, or even whether or not to accelerate the dying process, but the prior issue of how to evaluate survey data when the implications are literally matters of life and death.

Attitudes of persons facing their own imminent death: stage approaches

The work of Elisabeth Kübler-Ross has undoubtedly had greater influence on both theory and clinical practice than any previous writing. Based on many discussions with terminally ill persons, plus other sources of information, Kübler-Ross described five stages through which dying persons were seen to move: (1) *denial* and isolation, (2) *anger* and resentment, (3) *bargaining* and an attempt to postpone, (4) *depression* and sense of loss, and (5) *acceptance*. Kübler-Ross views this progression as both normal and adaptive, especially when the final stage of acceptance is reached and maintained. Medical caretakers are encouraged to enable their dying patients to advance through the stages, albeit at their own rate, hoping for minimal regression—that is, without returning to denial or anger after a subsequent stage has been attained. Inadequate medical or psychosocial care is often held responsible when a patient who has come to accept his or her coming death returns to a less adaptive stage.

Both the value of the stage approach and its validity have been questioned by others. Concern has been expressed that the stages have become so familiar to many health and social practitioners, as well as to the patients themselves and their families, at least in North America, that there is danger of the stages becoming a self-fulfilling prophecy. It is difficult to ascertain whether the stages are universal, modal, culture-bound, or even adaptive, since no consistent research findings have been reported, and medical clinicians themselves are in disagreement. One summary of the existing literature (Schulz and Alderman) reports little published support for the stages as occurring in the indicated sequence, but the stages appear to provide a useful framework for those who work with terminal patients. Kübler-Ross cautions that the sequence is not immutable, that many patients move back and forth or even maintain two or more stages virtually simultaneously. Kastenbaum reflects the concerns of many workers in the field when he states that "the rapid acceptance of the stage theory of dying has quite outdistanced any attempt to examine the theory empirically or logically . . . and no effort has been made to test out the theory as it continues to become more widely disseminated and applied" (p. 42).

There is little doubt that Kübler-Ross's name is better known than that of anyone else working on the concerns of death and dying or that her stages are the single best-known piece of information extant. Given the lack of empirical validation of her work and the extent to which her stages have been questioned by other knowledgeable persons, an ethical question arises as to the degree to which her writing should function as a basis for policy or practice. The emerging ethical issue is whether her five stages are not merely descriptive but also optimally adaptive and, if the latter, whether policy should encourage practice to intervene with dying persons to increase the probability that they will pass successfully through all five stages in time to die having reached the stage of acceptance. That this kind of intervention is already being practiced is common knowledge; the degree to which it is good or poor practice has not yet been determined.

Attitudes of persons facing their own imminent death: other approaches

Other investigators, using quite different methodologies and asking different questions, have emerged with other kinds of insights. One question that has been studied on several occasions is the desire of dying persons to know of their terminal condition. Through interviews with the dying, evidence has accumulated that not only do they wish to have a reasonable understanding of their prognosis but, given the opportunity, most will spontaneously mention awareness of their own condition, even though relatively few had received official medical diagnosis (Hinton, 1966). Those with dependent children and those in greater physical distress were more likely to indicate awareness.

Weisman adds another dimension to understanding the awareness the dying person has of his or her own condition. "Somewhere between open acknowledgement of death and its utter repudiation is an area of uncertain certainty called *middle knowledge*" (Weisman, p. 65). He continues his warning to avoid attempts to establish firm categories: "Patients seem to know and want to know, yet they often talk as if they did not want to be reminded of what they have been told" (ibid., p. 66).

The ability of dying patients to cope psychologically with their impending death has also been investigated. Although previous neurotic symptoms are apparently not related to effective coping, such factors as being physically comfortable, having good marital relationships, having good interpersonal relationships in general, expressing greater life satisfaction, and maintaining open communication about dying all appear prognostic of more positive attitudes in the face of death (Carey, pp. 435–438; Hinton, 1975, pp. 99–110; Kalish).

Certain qualities appear to enable dying persons to survive longer than statistically predicted by their illness. Those patients who outlive anticipated life expectancy are more likely to have good relationships with others and to maintain a higher level of intimacy with family members and friends, even to the time of death. They were also capable of requesting and receiving considerable support, in terms of both medical care and emotional relationships. These individuals were able to accept the reality of having a serious illness, although they often rejected the idea that death was inevitable (perhaps another self-fulfilling prophecy). Similarly, they were seldom deeply depressed, but they were quite likely to express resentment about various aspects of their treatment and illness (Weisman and Worden, p. 70). Such qualities obviously resemble those described in the previous paragraph.

Given these initial findings, including the suggestion that outliving the life tables may be in part a function of being able to express resentment about both health condition and health care, should health caretakers encourage the development of such characteristics among the dying? Since these qualities would be difficult to develop at the time of dying and would, in most instances, need to have occurred previously, are health caretakers ignoring their responsibilities by not participating in health education and mental health programs that would improve family relationships and the ability of patients to request support? Further, since accepting the reality of the prognosis also appears to be life-extending, what responsibilities do physicians and others have to encourage that kind of understanding? All of this appears to lead us back to a familiar dilemma: When it is learned that nonmedical factors contribute to good or poor health, are physicians who are trained to deal with medical factors also responsible to their patients for improving other aspects of their lives? This dilemma is no less acute for the terminally ill patient than for any other person.

It readily becomes obvious that the dying process and, therefore, attitudes toward it and toward one's own imminent death vary greatly as a function of numerous factors. Glaser and Strauss describe one such factor, the death trajectory, in considerable detail. The cause of death, the pace of decline, and perceptions the patient has of his or her condition all affect death attitudes. A patient dying from lung cancer faces a much different situation from that of a patient who is being rushed to the hospital with a second major coronary; each will undoubtedly differ in attitudes from the patient whose self-inflicted gunshot wounds did not lead to immediate death or the patient whose first stroke has left much of his or her body paralyzed.

Other relevant situational factors include the kind and extent of pain and discomfort and the effectiveness in controlling discomfort; the kind and extent of familial support, which reflects past family relationships; the expectations the patient has of dying, which may be influenced by both cultural and personality factors; the age, and sex, familial roles, extent of unfinished projects, and unresolved tensions in significant human relationships; and location (home, hospital, nursing home, hospice) in which the dying process is occurring.

Attitudes of persons not facing their own imminent death

Innumerable studies have been conducted to determine how various segments of the public feel about death or how they respond to situations in which death is a factor. These have been carried out with psychiatric patients, the elderly, children, adolescents, parents of dying children, physicians, nurses, nursing home personnel, various religious and ethnic groups, the ubiquitous university student, hospital patients, and general respondent populations (Hinton, 1972; Kastenbaum and Aisenberg, pp. 40–246).

Probably the most comprehensive study, based on a quota sampling design, was directed by Kalish and Reynolds. More than four hundred adults in the Los Angeles area were interviewed for one hour each; respondents were about evenly divided among blacks, Japanese Americans, Mexican Americans, and Anglo-Americans; among young, middle-aged, and old; and among men and women. Results showed many significant differences between groups and emphasized the importance of social role as a factor in death attitudes.

A relevant and particularly significant treatise by Becker not only placed fear and denial of death as *the* major determinant of human motivation, but also reinterpreted many Freudian concepts as based on decay, deterioration, and awareness of finitude rather than on Freud's original sex-related considerations (pp. 11–46). Becker posits a "healthy-minded" argument versus a "morbid-minded" argument regarding the fear and denial of death. The former assumes that fear of death is learned and, therefore, can be unlearned or, even better, never learned at all as the result of "healthy" early learning. The latter assumes that the fear of death is natural and is present in everyone, that it is "the basic fear that influences all others, a fear from which no one is immune, no matter how disguised it may be" (p. 15). Optimum learning and social environment can reduce the fear, or terror, but cannot eliminate it; indeed, Becker contends that much of our behavior arises from just such attempts to reduce the fear of death, not the least of which are our attempts to be in control, to attain power, and to relate to transcendent phenomena.

The position one takes regarding the nature and origin of the fear of death thus obviously relates to one's perceptions as to how to reduce this fear or whether to try. For those whom Becker terms "healthy-minded," the attempt to reduce the death fears of others through educational and therapeutic programs might be perceived as virtually a responsibility; for the "morbid-minded," the task called for would have more modest results. Further, if one continues to follow Becker, he would have to deal with the use of rituals and adherence to concepts of transcendence to reduce death fears, while others might espouse more familiar kinds of learning. Thus pedagogy, therapy, and ethics intertwine in the wake of religion and values.

Attitude surveys are not consistent with Becker's assumptions. One review of the literature has shown that, when asked directly, only a small proportion of individuals will admit fear of death (Kalish). Although this could be considered an indication of extensive denial rather than low fear of death, it is more likely that death fear is so complex that direct attitude questions do not effectively probe its meaning for most individuals.

Attitudinal studies of correlates of death fear have also been reviewed, and it was found that high fear is related to high scores on measures of depression, impulsivity, hysteria, manifest anxiety, need for heterosexual activity, need to be succorant, and low need for change. Conflicting results have emanated from studies of religiousness and death fear, with the less religious stating the higher fear of death. However, when curvilinear rather than linear functions were tested, it appeared that the highest death fear was found among persons who were intermediate in religiousness and who were irregular churchgoers, while both the non-religious and the devoutly religious displayed considerably less fear (ibid.).

Differences in responses from the ethnic communities did not follow any particular patterns, but could be interpreted primarily in terms of the ethnology of the community. Thus, of four ethnic groups, Japanese Americans were more likely than others to say they would try very hard to control the way they showed their emotions in public following the death of a loved one; Mexican Americans and black Americans were most likely to say that the death of an infant is less tragic than the death of persons of other ages; white Americans were most likely to have made out a will and least likely to agree that "accidental deaths show the hand of God working among men." When asked directly how they felt about death, Japanese and Mexican Americans were more likely than the others to say they were afraid.

It is important to keep in mind that attitude surveys measure what people say they feel or believe, which presumably correlates with what they actually feel or believe. The common assumption made concerning people's fear of death is that appropriate interventions will help reduce the fear. That this is true is most probable, but the nature of the optimum kinds of interventions is less certain. On the one hand, death educators believe that proper early learning experiences will keep fear of death to a minimum; on the other hand, other experienced persons believe that death fears are maximally reduced through reducing fears of abandonment and loss of control. These are not mutually exclusive, but they most certainly point to differing ameliorative approaches.

Attitudes of professionals

Work-related involvements with the dying and the dead are found in many vocational fields: medicine, nursing, social work, ministry and chaplaincy, cemetery and funeral work, florist industry, police, military, and a variety of others. Much related literature has been critical of the professionals for focusing too much on their specific task and not relating to the feelings of dying persons and their families. This criticism has frequently been directed toward physicians. Thus one study describes its findings as showing "a number of deficiencies and discrepancies in our approach to the terminally ill patient." Such studies emphasize "the patient's desire for complete openness and honesty in discussions regarding diagnosis and prognosis, the physician's reluctance to be that candid, the resident's relative lack of concern for the patient's emotional needs, and the social worker's tendency to minimize the problem" (Mount, Jones, and Patterson, p. 741). Feifel and his colleagues provide additional supportive data showing that a sample of eighty-one physicians exhibited more fear of death than either a group of physically ill persons or a group of healthy normal individuals.

The issue of communication between the dying patient and the health caretakers has been extensively discussed, and the overwhelming weight of opinion seems to support the view that open awareness is, for most dying persons, an optimum setting. Nonetheless, large numbers of physicians have indicated in surveys that they seldom or never inform their patients about their own death (Fitts and Ravdin). With amazing consistency, however, individuals indicate themselves as capable of and desiring the necessary information to understand their prognosis, while being less certain that others are similarly deserving (Kalish and Reynolds). This has been shown equally true for physicians themselves.

A question that is frequently posed is whether individuals who have not fully handled their own feelings about their own death are capable of working effectively with others who are dying. Although no empirical evidence is available, the weight of professional opinion is that people who have great difficulty with their own feelings concerning death will definitely have trouble relating to others who are dying. This does not mean, however, that all vestiges of death fear must be eliminated prior to working with the dying, or that physicians who have not worked through their feelings about death are ineffective in working medically with the dying. Rather, it suggests two possibilities. First, those physicians who prefer avoiding the psychosocial relationship with their dying patients should consider requesting the help of another relevant person to supplement their own functioning; second, physicians and others who have extensive contact with dying persons should develop their own support system, if they lack support through family members, since the professional person's confrontation with death and loss may prove destructive of both his work-related and his personal involvements.

Conclusion

Many issues, not all of which have been touched upon here, arise in any discussion of attitudes toward death. Substantive issues include the effects of religious belief systems on the ability to deal with death and dying, the implications of changing funerary and burial practices, the possible emergence of a new group of professionals trained particularly to work with the dying, the reasons why death has become a significant concern among sociomedical and behavioral scientists and practitioners at this time, and many more. Although a modest body of research and commentary now exists, the data and the insights are just beginning to accumulate.

At the same time, ethical issues are seen as increasingly important. Among the more important ethical issues, aside from those mentioned previously in this article, are the implications of brain death; the right of an individual to opt for euthanasia or suicide; the issue of who should participate in the decision for an abortion; the potential for psychological damage to research participants in death attitude studies; the validity of what people state on a questionnaire about their attitudes toward death and whether this should be perceived as a base for policy; the need to change the health-care financing system to permit new kinds of care for the dying; the development of standards for persons working in one-to-one relationships with the dying; the need for standards for death educators, especially those serving in the public school system; and the impact of these new death education programs on their students.

There appears little reason to doubt that the current interest in death and dying is not merely a fad but has become a fashion. The degree to which the fashion will embed itself into the social fabric of Western and other nations is still unknown.

RICHARD A. KALISH

[*Directly related are the entries* RIGHT TO REFUSE MEDICAL CARE; *and* TRUTH-TELLING. *Other relevant material may be found under* DEATH; DEATH AND DYING: EUTHANASIA AND SUSTAINING LIFE, *articles on* ETHICAL VIEWS *and* PROFESSIONAL AND PUBLIC POLICIES; DEATH, DEFINITION AND DETERMINATION OF; *and* SUICIDE. *See also:* CARE; *and* THERAPEUTIC RELATIONSHIP.]

BIBLIOGRAPHY

BECKER, ERNEST. *The Denial of Death*. New York: Macmillan Publishing Co., Free Press, 1973. An analysis of the pervasive meaning of death for all human activity; written by a social anthropologist, but based more on writing and thinking associated with psychology and philosophy.

CAREY, RAYMOND G. "Emotional Adjustment in Terminal Patients: A Quantitative Approach." *Journal of Counseling Psychology* 21 (1974): 433–439. A study, based on ratings of chaplains, of factors related to emotional adjustment in dying patients.

FEIFEL, HERMAN, and BRANSCOMB, ALLEN B. "Who's Afraid of Death?" *Journal of Abnormal Psychology* 81 (1973): 282–288. Measures death fear in several populations at three awareness levels: conscious, fantasy, and below the level of awareness.

———; HANSON, SUSAN; JONES, ROBERT; and EDWARDS, LAURI. "Physicians Consider Death." *Proceedings of the 75th Annual Convention of the American Psychological Association*. Washington: 1967, vol. 2, pp. 201–202. Reports a segment of a larger study on attitudes of physicians, nurses, critically and terminally ill persons, and healthy persons regarding death and dying.

FITTS, WILLIAM T., and RAVDIN, I. S. "What Philadelphia Physicians Tell Patients with Cancer." *Journal of the American Medical Association* 153 (1953): 901–904. A mail survey of physicians in Philadelphia disclosed that only a minority regularly informed their patients concerning their terminal prognosis.

GLASER, BARNEY G., and STRAUSS, ANSELM L. *Awareness of Dying*. Chicago: Aldine Publishing Co., 1965. A comparison of various awareness contexts, for example, open awareness, mutual pretense, suspicion awareness. In each context, the extent to which the dying person is aware of his prognosis and the significance of his awareness for communication with others is discussed.

———, and STRAUSS, ANSELM L. *Time for Dying*. Chicago: Aldine Publishing Co., 1968. Describes the results of a grounded theory study of dying trajectories, based on extensive observations and interviews.

HINTON, JOHN M. *Dying*. 2d ed. Studies in Social Pathology. Edited by C. M. Carstairs. Baltimore: Penguin Books, 1972. A review of death and dying, with particular attention to the dying patient and the role of the caretaker.

———. "Facing Death." *Journal of Psychosomatic Research* 10 (1966): 22–28. A study based on interviews with terminally ill patients; includes discussion of related studies.

———. "The Influence of Previous Personality on Reactions to Having Terminal Cancer." *Omega* 6 (1975): 95–111. An investigation of mood, satisfaction with care, and life satisfaction of patients with terminal cancer; numerous correlates are statistically evaluated.

KALISH, RICHARD A. "Death and Dying in a Social Context." *Handbook of Aging and the Social Sciences*. Edited by Robert H. Binstock and Ethel Shanas. Handbooks of Aging Series. Edited by James E. Birren. New York: Van Nostrand Reinhold, 1976, pp. 483–503. An overview of death and dying in the later years.

———, and REYNOLDS, DAVID K. *Death and Ethnicity: A Psychocultural Study*. Los Angeles: University of Southern California Press, 1976. A research monograph based on the findings of a three-year study; data are reported across ethnic groups and within ethnic groups.

KASTENBAUM, ROBERT. "Is Death a Life Crisis?: On the Confrontation with Death in Theory and Practice." *Life-Span Developmental Psychology: Normative Life Crises*. Edited by Nancy Datan and Leon H. Ginsberg. The Fourth West Virginia University Conference on Life-Span Developmental Psychology, 1974. New York: Academic Press, 1975, pp. 19–50. An examination of how death is seen as a variable, event, state, analogy, and mystery.

———, and AISENBERG, RUTH. *The Psychology of Death*. New York: Springer Publishing Co., 1972. A comprehensive book discussing the entire range of thought, research, and theory concerning death and dying, with particular emphasis on psychological factors.

KÜBLER-ROSS, ELISABETH. *On Death and Dying*. New York: Macmillan Co., 1969. The now classical description of five stages in the dying process, accompanied by extensive case materials.

MOUNT, BALFOUR M.; JONES, ALLEN; and PATTERSON, ANDREW. "Death and Dying: Attitudes in a Teaching Hospital." *Urology* 4 (1974): 741–748. A survey of staff and patients regarding specific issues concerning the role of health caretakers for the terminally ill. Results showed that members of different health professions perceived their roles and their own effectiveness much differently from members of other health professions or the patients themselves.

SCHULZ, RICHARD, and ALDERMAN, DAVID. "Clinical Research and the Stages of Dying." *Omega* 5 (1974): 137–143. A critical evaluation of both research and commentary concerning various stage theories of dying.

WEISMAN, AVERY D. *On Dying and Denying*. Foreword by Herman Feifel. New York: Behavioral Publications, 1972. A thorough discussion of what dying means and how it takes place, written by a highly experienced psychiatrist.

———, and WORDEN, J. WILLIAM. "Psychosocial Analysis of Cancer Deaths." *Omega* 6 (1975): 61–75. A determination of which psychosocial factors are related to living beyond the statistical prognosis for specific cancer conditions.

DEATH, DEFINITION AND DETERMINATION OF

I. CRITERIA FOR DEATH   *Gaetano F. Molinari*
II. LEGAL ASPECTS OF PRONOUNCING DEATH
    Alexander Morgan Capron
III. PHILOSOPHICAL AND THEOLOGICAL FOUNDATIONS   *Dallas M. High*

I
CRITERIA FOR DEATH

Introduction

Technical advances in medicine now permit maintenance and support of cardiac and respiratory function in man long after massive or even total destruction of the brain. Conversely, in specially equipped operating rooms the biological integrity of the human brain can be protected and maintained for extended, though finite, periods of time, while the function of the heart and lungs is deliberately but reversibly suspended in order to permit successful open-heart surgery. Whereas previously cerebral function and cardiorespiratory function were intimately mutually dependent, modern medical technology permits dissociations among individual vital functions. These dissociations have outstripped the usefulness of a monolithic set of criteria for the pronouncement of death based upon the vitality of a single organ system.

Terminology

Borderline states between life and death have been recognized for many years (Mollaret and Goulon; Bertrand et al.) and a series of often lyrical, sometimes practical neologisms has evolved very rapidly to identify varying degrees of biological integration and survival short of complete cognitive, sentient life. Such terms include coma depassé, coma-vigil, artificial survival, and more recently irreversible coma and persistent vegetative state. Many of these terms have been defined only descriptively; some have been used quite precisely and have been correlated pathologically with massive brain destruction; others correlate with smaller lesions in specific but critical brain centers.

In dealing with current medical developments, the terms cerebral death and brain death have crept into common usage among physicians. While the cerebrum is a part of the brain with distinctive ontogenetic and phylogenetic characteristics, the adjective "cerebral" is often used imprecisely. Terms such as cerebral circulation and cerebral vasculature refer to the blood supply to the entire brain rather than to the cerebrum specifically. Many physicians use the terms "brain death" and "cerebral death" interchangeably, while a few insist on precise distinctions, adding "cortical death" as yet another entity (Korein).

The former group uses the term brain death or cerebral death in contradistinction to cardiac death. The others equate the irreparable destruction of a part of the organ with the "death" of that part. Still others equate certain microscopic characteristics of individual nerve cells with cellular "death." These terms are still more confusing to those outside of medicine, because they imply that there are different kinds of death.

Better terms for the more generic application of "brain death" and "cardiac death" are death as determined by neurological criteria and death as determined by cardiological criteria.

To distinguish among cortical death, cerebral death, and brain death implies that there are degrees of death stratified in accordance with the hierarchical organization of the nervous system. Throughout this treatise, the use of ambiguous, controversial, or confusing terms will be avoided. However, that will not always be possible when referring to the work of some authors.

Classical criteria

The signs of life used in day-to-day medical care, routinely observed, measured, and recorded on hospitalized patients' daily records, are body temperature, pulse, respiratory rate, and blood pressure. While determination of these vital signs is generally left to the nursing staff, their absence may be used as criteria for death only by a physician. Therefore, the pragmatic but traditional criteria for death have been the cessation of heartbeat or cardiac pulse, respiration, blood pressure, followed by a fall in body temperature as determined or confirmed by a physician. The first three are eminently more practical than the last; in fact, the pronouncement or declaration of death is rarely delayed until the body temperature falls.

Despite the ready availability of physiological monitoring of cardiovascular functions, such as electrocardiography, no moral, medical, or legal principles mandate the use of the most sensitive technical device available to determine death under ordinary circumstances. Nonetheless, when such equipment is already in use in operating rooms and in coronary care units, the increased sensitivity afforded the physician is exploited. Death has not usually been declared in the presence of electrical evidence of cardiac function, even though blood pressure, pulse, and respiration have disappeared. It should also be noted that no medical policy or legal precedent specifies a duration of time over which the classical criteria must persist.

Irreversible coma

Coma is defined as a pathological state of depressed consciousness from which the patient cannot be aroused even by painful stimulation. Coma may be caused by a wide variety of disease states that affect the entire brain or merely a strategic part of the brain. When the cause of the coma is known, and that cause is a disease known to be irreparable, a state of irreversible coma exists, a coma from which there is no hope of recovery.

Irreversible coma merely predicts death without return of consciousness; it neither implies that death has occurred nor does it predict precisely when death will occur. Patients in irreversible coma may have portions of the nervous system intact, which permit blood pressure, pulse, respiration, and normal body temperature to persist indefinitely without artificial or mechanical support.

Irreversible destruction of the entire brain

Respiratory movement is initiated by centers within the primitive brain or brain stem; but cardiac muscle has the intrinsic properties of excitability and contractility independent of nervous system control. Hence, the complete destruction of the brain, including loss of spontaneous respiration, may occur while the cardiovascular system continues to function autonomously. Apnea (the absence of spontaneous respiration) has been considered by most medical authors to be a cardinal sign of total brain destruction. Other practical and reliable indices of brain-stem function are the cephalic reflexes, or those reflexes the presence of which require anatomical, physiological, and biochemical integrity of specific brain-stem centers.

Pupil size, pupillary reactions to light, eye movement in response to passive head movement (oculocephalic), eye movement in response to caloric stimulation of the inner ear (vestibulo-ocular), pharyngeal, swallowing, and cough reflexes, in addition to respiration, all are determined by the activity of specific brain-stem centers.

Therefore, the basic criteria for the suspicion of a totally destroyed brain are unresponsive coma, apnea, and absent cephalic reflexes. When these criteria are judged to be permanent, the brain may be considered irreversibly destroyed. However, two reversible causes of coma are also known to produce this constellation of clinical signs, namely, drug intoxication and low body temperature. Allen and his associates have recently analyzed and reported the predictive and discriminative value of the clinical findings in individuals suspected to have irreversibly destroyed brains. In addition to clinical signs, most authors have recommended either objective confirmatory tests, prescribed durations for persistence of the clinical signs, or both, in order to determine if a brain is irreversibly destroyed and to pronounce death on that basis in jurisdictions authorizing such pronouncement (Allen et al.).

Prolonged nonfunctional state

In its 1968 report, the Harvard Committee on Irreversible Coma described the clinical and electroencephalographic characteristics of the nonfunctioning brain. A prerequisite the Committee required was the exclusion of patients with drug intoxication and hypothermia, the reversible conditions known to depress cerebral function, reflex activity, and electrical activity measured by the electroencephalogram (EEG).

In the absence of hypothermia and drug intoxication, the Committee suggested that concomitant unresponsive coma, apnea, absence of cephalic reflexes, absence of spinal reflexes, and an isoelectric electroencephalogram were characteristic of patients with a nonfunctioning brain. Persistence of all these features for twenty-four hours was thought adequate grounds to consider the nonfunctioning state permanent. Although the Committee expressed some doubt about the necessity for the absence of spinal reflexes, that group, in fact, equated brain death with the "permanently nonfunctioning brain."

Clinical experience

Mohandas and Chou recommended far less stringent criteria. Their data suggested that, in patients who have known brain damage of a type known to be irreparable, no spontaneous movement, apnea, and absent brain-stem reflexes persistent for twelve hours, predict an invariably fatal outcome (Mohandas and Chou). While these criteria may predict eventual death with certainty, they rely totally on the responsible physician's diagnostic skills, his knowledge of the natural history of the specific disease diagnosed, and his confidence in the universal applicability of statistical outcome data. Not only do these criteria admit to the possibility of human error at multiple levels, they fail to distinguish the dead from the dying brain. Moreover, in an analysis of 503 prospectively studied brain-death suspects, seventeen patients would have met these criteria even while they had evidence of cerebral activity electroencephalographically (Walker and Molinari).

Cardiovascular collapse

A study undertaken in 1973 in Japan concluded that a diagnosed gross primary brain lesion, deep coma, bilateral dilated pupils with absent pupillary and corneal reflexes (a selection of specific brain-stem reflexes), and an isoelectric EEG characterize moribund patients. Furthermore, the Japanese group observed that a fall in the blood pressure of 40 mm of mercury and persistent low blood pressure for six hours signal the imminence of death (Ueki, Takeuchi, and Katsurada).

These criteria redirect the major focus for the difference between irreversible coma and imminent death from the neurological characteristics of the patients back to classical circulatory system criteria. The progressive and irreversible failure of another of the "vital signs," namely blood pressure, is used as the decisive factor, adequate for termination of artificial support mechanisms.

Scandinavian criteria

At a symposium on this subject, reported in 1972 (Ingvar and Widén), criteria were recommended for use in Scandinavia incorporating the combination of clinical features, cerebral electrical activity, and arrest of the circulation to the brain (regional or local circulatory arrest).

The symposium participants suggested that patients could be dead who had known primary or secondary brain lesions, unresponsive coma, apnea, and absence of all cerebral functions including brain-stem reflexes; a single electroencephalogram showing no biological activity, and radiological evidence of shutdown of the cerebral blood vessels were considered confirmatory objective evidence of a nonviable brain. Furthermore, they postulated that failure to demonstrate perfusion of cerebral arteries, using standard X-ray methods, on two trials twenty-five minutes apart, indicates impairment of circulation to the point of total destruction of the brain.

These criteria are very attractive, because (1) they utilize multiple parameters independently measured and thereby are cross-confirmatory; (2) they eliminate the need to exclude drug intoxication and other metabolic disturbances by biochemical methods, since these etiologies do not grossly affect the cerebral vasculature, and therefore brain circulatory patterns are normal; and (3) they reduce the requirement for persistence of the criteria to the absolute minimum (twenty-five minutes). These criteria therefore equate death by neurological criteria with total brain infarction.

The collaborative study of brain death

A prospective collaborative study of deeply comatose, apneic patients was undertaken in the United States and published in 1975 (Walker). Five hundred and three patients were followed by clinical and electroencephalographic examinations at regular intervals until outcome. Pathological confirmation of the clinical diagnosis of brain death was sought in all fatal cases coming to autopsy.

The findings of that study were somewhat unsettling. First, opinions regarding the histological characteristics of the brains from respirator-dependent patients varied among pathologists, even those specializing in nervous system diseases (Moseley, Molinari, and Walker). Second, in some patients the pathology failed to confirm the clinical diagnosis of total brain destruction (Walker, Diamond, and Moseley). A few patients who were pronounced dead after prolonged periods of coma, apnea, absence of cephalic reflex activity, and electrocerebral silence showed few changes at autopsy of the types reputed to indicate nerve-cell death. Conversely, a few who continued to record biological activity in the EEG until the moment of final cardiac arrest showed postmortem changes consistent with long-standing, advanced, and widespread cellular death in the brain. While reasons for these variations exist, it seems clear that the morphological changes of neuronal death lag behind the physiological and biochemical end points of irreversibility. Therefore, one cannot expect objective confirmation of the clinical diagnosis of death by neurological criteria by the postmortem changes perceptible by light microscopy.

Third, in objective screening of blood samples from deeply comatose patients more instances of drug ingestion (or administration) were uncovered than had been suspected clinically. Even in clear cases of attempted suicide by massive drug overdose, quantitative measurements of the offending drug in the blood were disproportionately low compared to the severity of the clinical syndromes produced. Qualitative identification of drugs alone and in combination was limited by the adequacy of local screening techniques and procedures. The absence of a history of drug ingestion in comatose patients does not exclude the possibility of their presence in the blood. Drugs taken by the patient before a severe accident involving brain injury or, indeed, drugs given therapeutically for convulsions caused by the cerebral disease, may contribute to the overall depression of clinical and electrical indices of nervous system function; consequently the presence of any amount of drugs in the blood introduces doubt as to the irreversibility of the clinical syndrome.

Therefore, it seemed highly desirable to resort to criteria for irreversible destruction of the brain that would be independent of the nature of the disease or cause of the coma (Walker). The criteria recommended for use in Scandinavia seemed to meet this requirement, but in the 503 cases studied in the United States only seventeen X-ray studies of the circulation were performed (Walker and Molinari). The latter fact probably represents the reluctance of American physicians to subject their moribund or desperately ill patients to even slight additional risk. Moreover, neither the equipment required nor respirator-dependent patients themselves are portable, so obtaining X-ray confirmation is a major operational problem. Therefore, these criteria seem impractical.

Indirect assessment of the cerebral circulation is possible, however, by any of a number of parameters. Several ophthalmological methods for indirect assessment of the cerebral circulation have been reported over the years (Pines; Kevorkian; Lobstein, Tempe, and Payeur). Echoencephalography permits ultrasonic detection of brain pulsations at the bedside (Uematsu and Walker). The perfusion of the cerebrovascular bed may be detected using tracer amounts of innocuous radioisotopes and portable equipment (Braunstein et al.). Although none of these techniques has been studied in a large sample of patients, preliminary tests of validity and reliability of the radioisotope technique compared to contrast (X-ray) angiography have been promising (Korein et al.).

Conclusion

This article has traced a spectrum of descriptive criteria for death based upon a variety of clinical observations, often confirmed by one or more of the currently available diagnostic techniques.

It is evident from the criteria described that concepts of death vary among medical authors, ranging from inevitable mortality to documentary evidence of total destruction of the brain.

The determination that death will occur in critically ill or brain-injured patients has never been a diagnostic problem. The prediction of the moment at which death may be pronounced using classical cardiorespiratory criteria is more difficult but becomes simpler as cardiovascular and respiratory system physiology deteriorates. The concept that death may occur while the cardiorespiratory physiology is being artificially sustained is a by-product of modern medical technology. While pronouncement of death based on neurological criteria seems reasonable and practical, proof of the accuracy of that pronouncement is not always possible.

The criteria used in pronouncing death while certain organs continue to show signs of function continue to be enigmatic to physicians and scholars alike. May the destruction of the most complex portion of the organ, the brain, alter the nature of the organism, man? May the total destruction of the organ be equated with the death of the organism? Does the proximity or imminence of death permit a declaration of death? The criteria discussed above presume that one or more of these issues may be decided in the affirmative.

GAETANO F. MOLINARI

[*Directly related are the other articles in this entry:* LEGAL ASPECTS OF PRONOUNCING DEATH *and* PHILOSOPHICAL AND THEOLOGICAL FOUNDATIONS. *See also:* DEATH AND DYING: EUTHANASIA AND SUSTAINING LIFE; DECISION MAKING, MEDICAL; *and* LIFE-SUPPORT SYSTEMS.]

BIBLIOGRAPHY

Ad Hoc Committee of the Harvard Medical School to Examine the Definition of Brain Death. "A Definition of Irreversible Coma." *Journal of the American Medical Association* 205 (1968): 337–340.

ALLEN, NORMAN; BURKHOLDER, JAMES D.; COMISCIONI, JOHN; and MOLINARI, GAETANO F. "Predictive Value of Clinical Criteria in Cerebral Death." *Neurology* 26 (1976): 356–357. Abstract.

BERTRAND, IVAN; LHERMITTE, FRANÇOIS; ANTOINE, BERNARD; and CUCROT, HENRI. "Nécroses massives du système nerveux central dans une survie artificielle." *Revue Neurologique* 101 (1959): 101–115.

BRAUNSTEIN, P.; KOREIN, J.; KRICHEFF, I.; COREY, K.; and Chase, N. "A Simple Bedside Evaluation for Cerebral Blood Flow in the Study of Cerebral Death: A Prospective Study on 34 Deeply Comatose Patients."

American Journal of Roentgenology, Radium Therapy and Nuclear Medicine 18 (1973): 757–767.

INGVAR, DAVID H., and WIDÉN, LENNART. "Hjärndöd: Sammanfattning av ett symposium." [Brain death: abstract of a symposium]. *Läkartidningen* 69 (1972): 3804–3814.

KEVORKIAN, JACK. "Rapid and Accurate Ophthalmoscopic Determination of Circulatory Arrest." *Journal of the American Medical Association* 164 (1957): 1660–1664.

KOREIN, JULIUS. "Neurology and Cerebral Death—Definitions and Differential Diagnosis." *Transactions of the American Neurological Association* 100 (1975): 210–212.

———; BRAUNSTEIN, PHILLIP; KRICHEFF, IRVIN; LIEBERMAN, ABRAHAM; and CHASE, NORMAN. "Radioisotopic Bolus Technique as a Test to Detect Circulatory Deficit Associated with Cerebral Death." *Circulation* 51 (1975): 924–939.

LOBSTEIN, ANDRÉ; TEMPLE, JEAN-DANIEL; and PAYEUR, GEORGES. "La Fluoroscopie rétinienne dans le diagnostic de la mort cérébrale." *Documenta Ophthalmologica* 26 (1969): 349–358.

MOHANDAS, A., and CHOU, SHELLEY N. "Brain Death: A Clinical and Pathological Study." *Journal of Neurosurgery* 35 (1971): 211–218.

MOLLARET, P., and GOULON, M. "Le Coma dépassé." *Revue Neurologique* 101 (1959): 5–15.

MOSELEY, JOHN I.; MOLINARI, GAETANO F.; and WALKER, A. EARL. "Respirator Brain: Report of a Survey and Review of Current Concepts." *Archives of Pathology & Laboratory Medicine* 100 (1976): 61–64.

PINES, N. "The Ophthalmoscopic Evidence of Death." *British Journal of Ophthalmology* 15 (1931): 512–513.

UEKI, K.; TAKEUCHI, K,; and KATSURADA, K. "Clinical Study of Brain Death." Fifth International Congress of Neurological Surgery, Tokyo, 7–13 October 1973. Presentation no. 286.

UEMATSU, S., and WALKER, A. E. "A Method for Recording the Pulsation of the Midline Echo in Clinical Brain Death." *Johns Hopkins Medical Journal* 135 (1974): 383–390.

WALKER, A. EARL. "Cerebral Death." *The Nervous System.* 3 vols. Edited by Donald B. Tower. Vol. 2: *The Clinical Neurosciences.* Edited by Thomas N. Chase. New York: Raven Press, 1975, pp. 75–87.

———; DIAMOND, EARL L.; and MOSELEY, JOHN. "The Neuropathological Findings in Irreversible Coma: A Critique of the 'Respirator Brain'." *Journal of Neuropathology and Experimental Neurology* 34 (1975): 295–323.

———, and MOLINARI, GAETANO F. "Criteria of Cerebral Death." *Transactions of the American Neurological Association* 100 (1975): 29–35.

II
LEGAL ASPECTS OF PRONOUNCING DEATH

The recently developed capability of biomedicine to sustain vital human functions artificially has presented new problems for the public and its legal institutions as well as for medical practitioners. Determining that a person has died is no longer the relatively simple matter of ascertaining that his heart and lungs have stopped functioning. Mechanical respirators, electronic pacemakers, and analeptic and tensive drugs give the appearance of circulation and respiration in what is otherwise a corpse. As lay people became aware of those medical developments in the 1960s, an undeniable need arose for change in the public policy on when and how death could be declared. The great drama of heart transplantation, beginning with the operation performed in Cape Town, South Africa, on 3 December 1967 by Dr. Christiaan Barnard, was all the more astonishing because a heart taken from a woman who had been declared "dead" beat in a living man's chest in place of his own.

Although cardiac transplantation provides the boldest illustration of medicine's new powers, the need for a new way of determining whether death has occurred extends to the far more numerous cases of lengthy, intensive maintenance of moribund patients long after they would have formerly ceased living. In some of those patients, medical intervention can end because it has been successful in permitting recovery to occur, and in others it terminates because the patient's bodily systems have so totally collapsed that circulation and respiration cannot be maintained. But in a significant number, deep in coma from trauma or deterioration, artificial support is successful enough to be continued indefinitely, with no prospect that the patient will ever recover but also no point in sight at which bodily functioning will cease so long as medical care continues.

The response of the medical profession in the 1960s was to develop new criteria, which rely on irreversible changes in the central nervous system, to guide physicians in determining that death had occurred in artificially maintained patients. According to the criteria suggested by an ad hoc committee of the Harvard Medical School and generally followed by American and British physicians, death may be declared when the clinical signs of brain activity—such as response to external stimuli, spontaneous respiration, movements, and reflexes—are absent for twenty-four hours; an isoelectric electroencephalogram confirms the diagnosis (Ad Hoc Committee, pp. 337–338). The Mollaret criteria used in France, like the Harvard criteria, measure indicators of continued brain function, while the methods employed in Austria and Germany search for the preconditions for brain function, such as intracranial blood circulation (Van Till, pp. 139–144). Despite this divergence in reasoning and method, the underlying consensus that emerged in the medical profession was that the total and irreversible absence of brain function is equivalent to the traditional indicators of death. The new medical criteria were not in accord with the common understanding of the lay public nor with the rules embodied in custom and law. Anglo-American common law, for example, requires the total cessation of *all* vital functions ("Death"). Two issues are thus presented: an issue of process—how ought lawmaking bodies to respond to the changes in medical practice and doctrine?—and one that concerns product—what changes should be made in the law?

Process: the authority to frame definitions

Let the physicians decide. A number of routes have been advanced for arriving at a new "definition of death" that would encompass a neurological understanding of the phenomenon. (The commonly employed phrase, "definition of death," is useful shorthand but should not be taken to mean the *explanation* of a fact but rather the *choice* about the significance of certain facts in the task of determining whether, and when, a person has died.) Some commentators have proposed that the task should be left to physicians, because the subject is technical and because the law might freeze the definition prematurely, leading to conflicts with the developments that will inevitably occur in medical techniques (Kennedy, pp. 946–947). Yet the belief that defining death is wholly a medical matter misapprehends the undertaking. At issue is not a biological understanding of the inherent nature of cells or organ systems but a social formulation of humanhood. It is largely through its declaration of the points at which life begins and ends that a society determines who is a full human being, with the resulting rights and responsibilities. So long as the standards being employed by physicians to pronounce death in individual cases are stable and congruent with community opinion, most people are content to leave the entire matter in medical hands. But the underlying extramedical aspects of the definition become visible, as they have recently, when medicine either departs or appears to depart from the common understanding of the concept of death.

Since physicians have no special competence on the philosophical issue of the nature of human beings and no special authority to arrogate the choice among definitions to themselves, their role is properly one of elucidating the significance of various vital signs. Nevertheless, medical statements on death help to frame the relevant issues, and the attention devoted to the subject by the biomedical professions has been a major stimulus to the process of redefinition. The public might decide to reject the medical view if, for example, it concluded that a prognosis based on the new brain-oriented criteria cannot be sufficiently accurate. A new definition will be forthcoming not simply to accommodate biomedical practitioners' wishes, but as a result of perceived social need and of evidence that brain tests for death are as reliable as the traditional heart–lung tests. The public might also reject the medical view that brain-oriented criteria ought to be used even if it were convinced of their reliability in measuring brain function if that public did not share the medical profession's concept of personhood.

Let the courts decide. The medical definitions may take on legal status were the courts to defer to them as the issue arose in litigation. In the United States and other common law countries, law is to be found not only on the statute books but in the rules enunciated by judges as they resolve disputes in individual civil and criminal cases. Faced with a factual situation that does not fit comfortably within the existing legal rules, a court may choose to formulate a new rule in order to reflect current scientific understanding and social viewpoints more accurately.

Nonetheless, there are problems of principle and practicality in placing primary reliance on the courts for a redefinition of death. Like the medical profession, the judiciary may be too narrowly based for the task. While the judiciary, unlike the medical profession, is an organ of the state with recognized authority in public matters, it still has no means for actively involving the public in its decision-making processes. Judge-made law has been most successful in factual settings embedded in well-defined social and economic practices, with the guidance of past decisions and commentary. Courts operate within a limited compass—the facts and contentions of a particular case—and with limited expertise; they have neither the staff nor the authority to investigate or to conduct hearings in order to explore such issues as public opinion or the scientific merits of competing "definitions." Consequently, a judge's decision may be merely a rubberstamping of the opinions expressed by the medical experts who appeared before him. Moreover, testimony in an adversary proceeding is usually restricted to the "two sides" of an issue and may not fairly represent the spectrum of opinion held by authorities in the field.

In the cases in which parties first argued for a redefinition, the courts were unwilling to disturb the existing legal definition. Such deference to precedent is understandable, because everyone needs to be able to rely on predictable legal rules in managing one's affairs and cannot always be overcome by arguments on the merits of a new rule. As recently as 1968 a California appellate tribunal, in a case involving an inheritorship issue, declined to redefine death in terms of brain functioning despite the admittedly anachronistic nature of an exclusively heart–lung definition.

In light of possible judicial adherence to existing but outmoded rules, reliance on the judicial route to a new definition of death is likely to create a considerable period of uncertainty, which is especially unfortunate in an area where private decision makers must act quickly and irrevocably. An ambiguous legal standard endangers the rights, and in some cases the lives, of the participants and fosters public confusion. A physician's choice of one course over another may depend less on his or her view of their relative merits than on his or her willingness to face a court challenge on the issue.

The unfortunate consequences for physicians and patients of the unsettled state of the common law definition of death is illustrated by three American cases. In the first, *Tucker* v. *Lower*, which came to trial in Virginia in 1972, the brother of the man whose heart was taken in an early transplant operation sued the doctors, alleging that the operation was begun before the donor had died. The evidence showed that the donor's pulse, blood pressure, respiration, and other vital signs were normal but that he had been declared dead when the physicians decided these signs resulted solely from medical efforts and not from his own functioning, since his brain had ceased working. The trial judge initially indicated that he would adhere to the traditional definition of death, but he later permitted the jurors to find that death had occurred when the brain ceased functioning irreversibly, and a verdict was returned for the defendants. Since the court did not explain its action, the law was not clarified, and future parties face continued uncertainty.

The other two cases arose in 1974 in California when two transplant operations were performed using hearts removed from the victims of alleged crimes. The defendant in each case attempted to interpose the action of the surgeons in removing the still-beating heart of the victim as a complete defense to the charge of homicide. One trial judge accepted this argument as being compelled by the existing definition of death, but his ruling was reversed on appeal, and both defendants were eventually convicted. This graphic illustration of legal confusion and uncertainty led California to follow a third route to redefining death, the adoption of a statutory definition.

Let the legislatures decide. The legislative process permits the public to play an active role in decision making and allows a wider range of information to enter into the framing of standards for determining death. That is important because basic and perhaps controversial choices among alternative definitions must be made. For example, some commentators have argued for a new concept of death as solely the cessation of the higher faculties of the brain upon which cognition and personality depend (Veatch). Faced with this option, all legislatures have thus far concluded that it is preferable instead simply to restate the existing understanding of death, based on the recognition that, if cardiac and pulmonary functions have ceased, brain functions cannot continue, and if there is no brain activity and respiration has to be maintained artificially, the same state (i.e., death) exists. Because they provide prospective guidance, statutory standards have the additional advantage of dispelling public and professional doubt, thereby reducing both the fear and the likelihood of cases for malpractice or homicide against physicians. The greatest danger of legislation is poor draftsmanship, which can best be addressed in the context of particular examples.

Product: the contours of a definition

The general principles for drafting a statute on death grow out of the objectives that the statute ought reasonably to serve. First, the phenomenon of interest to physicians, legislators, and the public alike is a human being's death, not the "death" of his cells, tissues, or organs. This step will resolve the problem of whether to continue artificial support in only some of the cases of comatose patients. Additional statutory guidance may also be desired by society concerning the cessation of treatment in patients who are alive by brain or heart–lung criteria but for whom further treatment is considered (by the patients or by others) to be pointless or degrading. This question of "when to allow to die" is distinct from "when to declare death."

Second, the merits of a legislative "definition" are judged by whether its purposes are properly defined and how well the legislation meets those purposes. In addition to its cultural and religious importance, society needs a definition of death for a number of decisions having legal consequences; besides terminal medical care or transplantation, those include homicide, damages for the wrongful death of a person, property and wealth transmission, and determination of insurance, taxes, and marital status. From this it can be argued that a single definition is inappropriate because different policy objectives may exist in those different contexts (Dworkin, p. 631). No special purposes requiring separate definitions have been suggested, however, and a single definition of death seems capable of being applied in a wide variety of contexts, as indeed the traditional definition has been.

A different rationale for multiple definitions of death may be drawn from the philosophical debate over the nature of death. Death may be viewed as a process since not all parts of the body cease functioning equally and synchronously (Morison, pp. 695–696). But society, through generally accepted medical and social practices and in its laws, recognizes that a line can and must be drawn between those who are alive and those who are dead (Kass, p. 701). The ability of modern bioscience to extend the functioning of various organ systems may have made knowing which side of the line a patient is on more problematic, but it has not erased the line. The line drawn by society is an arbitrary one in the sense that it results from human choice among a number of possibilities, but it need not be arbitrary in the sense of having no acceptable, articulated rationale.

Having a single definition to be used for many purposes does not, needless to say, preclude reliance on other events besides death as the trigger for some decisions. Most jurisdictions, for instance, make provision for the distribution of property and the termination of marriage after a person has been absent without explanation for a period of years, and some term this a "presumption of death," although a person covered by such a determination could not be treated as a corpse were he actually still alive (Capron, pp. 642–643). Similarly, a status such as "total, irreversible loss of cognitive abilities" might be established as the predicate for distributing the bulk of a terminally ill person's estate; it would not, however, be an alternative definition of death and would need to be surrounded by procedural safeguards not necessary for such a definition.

Another principle of legislative drafting is that the standards must be uniform for all persons. It is, to say the least, unseemly for a person's wealth or potential social utility as an organ donor to affect the way in which the moment of his death is determined.

It is often beneficial for the law to move incrementally, particularly when matters of basic cultural and ethical values are implicated. Thus, what is needed is a modern restatement of the traditional understanding of death that will tie together the accepted cardiopulmonary standard with a new brain-based standard that measures the same phenomenon: the irreversible cessation of the integrated functioning of the three vital systems together.

Finally, in making law in a highly technological area, care is needed that the definition be at once sufficiently precise to determine behavior in the manner desired by the public and yet not so specific that it is tied to the details of contemporary technology. Such flexible precision can be achieved if the definition is confined to the general *standards* by which death is to be determined, for example, the irreversible cessation of spontaneous respiration or the irreversible loss of the ability to respond or communicate. The law can then leave to the continually developing judgment of biomedical practitioners the establishment and application of appropriate *criteria* and specific *tests* for determining that the standards have been met. To illustrate, deep coma, the absence of reflexes, and the lack of spontaneous muscular movements and respiration are among the criteria that have been proposed for determining that one standard—irreversible cessation of spontaneous brain functions—has been met (Ad Hoc Committee, pp. 337–338). Those criteria are implemented in turn by certain tests, such as applying painful stimuli, observing respiration and reaction to light, and performing electroencephalography at specified intervals. As medical understanding and technique develop, new tests and criteria may be established that will provide simpler, less expensive, and even more accurate means of fulfilling the standards set by the law.

Movement toward legal definitions

As one would expect, adoption of a modern definition of death has come about mostly through legislation; the court cases involving the issue have seldom eventuated in appellate opinions providing a new standard of general applicability, and medical practices, although better understood and more widely accepted by the public, still lack explicit legal sanction in most places.

The first statutes on the definition of death, which were addressed primarily to organ transplantation, erred on the side of generality. The British Human Tissue Act 1961, the Danish Act of 1967, and the Uniform Anatomical Gift Act, which was proposed in 1968 and adopted in all American jurisdictions by 1971, leave the standards for death too undefined to be any guide for the decisions of physicians called upon to declare death (Meyers, p. 113). The legislation enacted in France in April 1968, which was also intended to facilitate transplantation, went to the opposite extreme and defined death in clinical terms as occurring when an artificially maintained patient has lesions incompatible with life and has a flat electroencephalographic tracing for at least ten minutes.

Although not free from problems, the legislation adopted in a number of American states since 1970 has steered a middle course between excess generality and undue specificity. Nonetheless, the first American legislation, enacted by the State of Kansas in 1970, suffered in other respects. First, it set forth cardiopulmonary and brain definitions separately and without explaining their relationship. The practitioner is given no guidance on when one or the other is to be applied, and the public is left with the impression that they are two separate phenomena rather than two ways of measuring the same thing. This is especially unfortunate since the legislation was adopted to assist heart transplantation, and a lay person could interpret it as permitting doctors to use a new definition—"brain death"—to declare a person dead earlier than would be possible under existing law. The second problem with the Kansas statute is that it speaks in terms of pronouncing death when further efforts are "hopeless" or appear not to be successful. Such language wrongly implies that legislation defining death extends to passive euthanasia, that is, ceasing treatment on a dying patient in order to permit death to occur when no hope of recovery remains.

A model statute suggested in 1972 provides two interconnected general physiological standards for identifying the two primary manifestations of the phenomenon of death:

A person will be considered dead if in the announced opinion of a physician, based on ordinary standards of medical practice, he has experienced an irreversible cessation of spontaneous respiratory and circulatory functions. In the event that artificial means of support preclude a determination that these functions have ceased, a person will be considered dead if in the announced opinion of a physician, based on ordinary medical practice, he has experienced an irreversible cessation of spontaneous brain functions. Death will have occurred at the time when the relevant functions ceased [Capron and Kass, p. 111].

This proposal, which has since been adopted in several states, applies equally to all persons and not merely to those who are potential organ donors or even those who are hospitalized at the time of death. The legislation gives legal sanction to declaring a person dead based on an absence of all neocortical and lower brain activity without being tied to any particular medical procedures. Thus it would permit use of either the so-called Harvard criteria, which are generally accepted in the United States, or other accepted methods, such as those followed in Europe, which measure blood flow in the brain (Van Till, pp. 143–144). The specific practices prevailing in an area would be a matter of fact in each case, recognizing that more than one set of procedures may reach the necessary degree of certainty to be adopted by physicians.

In 1975 the American Bar Association endorsed similar legislation but defined death solely in neurological terms: "For all legal purposes, a human body with irreversible cessation of total brain function according to usual and customary standards of medical practice, shall be considered dead" (American Bar Association). This proposal has the advantage of removing the last trace of the "two deaths" misconception; yet it ignores the fact that physicians will continue in most cases to employ cardiopulmonary tests, which are easier to perform, more accessible and acceptable to the lay public, and perfectly adequate for determining death in most instances. In recognition of that fact, the California legislature in adopting an ABA-type bill added a clause permitting physicians to continue to employ "usual and customary" standards. Such legislation thereby resurrects the "two deaths" concept while failing to explain the connection between the new and the traditional measures.

Unfulfilled obligation: education for change

As can thus be seen, the movement toward a modern formulation of the bases for pronouncing death has not been completed. In some societies that task may be left to the medical profession, since the problems faced in medical practice have provided the impetus for change. Tradition as well as sound policy suggests, however, that the ground rules for decisions about individual patients should be established by public authorities. Whether the new legal definition emerges from the resolution of court cases or from the legislative process, it will be greatly influenced by opinion from the medical community. Recognition that the standards for determining death are matters of social and not merely professional concern only serves to underline the education of the public on this subject as an important ethical obligation of the profession.

ALEXANDER MORGAN CAPRON

[*Directly related is* DEATH AND DYING: EUTHANASIA AND SUSTAINING LIFE, *article on* PROFESSIONAL AND PUBLIC POLICY. *See also the other article in this entry,* CRITERIA FOR DEATH; *and the entry* ORGAN DONATION.]

BIBLIOGRAPHY

Ad Hoc Committee of the Harvard Medical School to Examine the Definition of Brain Death. "A Definition of Irreversible Coma." *Journal of the American Medical Association* 205 (1968): 337–340. The highly influential statement by Harvard physicians and others on the means for determining an irreversible loss of total brain functioning.

American Bar Association. "Section of Insurance, Negligence and Compensation Law (Report No. 102)." *Summary of Action Taken by the House of Delegates of the American Bar Association, 1975, Midyear Meeting.* Chicago: 1975, p. 19. Reprint. *American Bar Association Journal* 61 (1975): 463–466. Recommends a definition of death as an irreversible cessation of total brain function.

CAPRON, ALEXANDER MORGAN. "The Purpose of Death: A Reply to Professor Dworkin." *Indiana Law Journal* 48 (1973): 640–646. Argues in favor of developing a definition that comports with social reality and can be employed in as many legal settings as it suits.

———, and KASS, LEON R. "A Statutory Definition of the Standards for Determining Human Death: An Appraisal and a Proposal." *University of Pennsylvania Law Review* 121 (1972): 87–118. General discussion of the procedures and objectives for lawmaking on this subject.

"Death." *Black's Law Dictionary.* 4th rev. ed. St. Paul, Minn.: West Publishing Co., 1968, p. 488. Common-law definition of death.

DWORKIN, ROGER B. "Death in Context." *Indiana Law Journal* 48 (1973): 623–639. Contends that the law should be framed in terms of the consequences of cessation of human functioning and not in terms of a single definition of death.

KASS, LEON R. "Death as an Event: A Commentary on Robert Morison." *Science* 173 (1971): 698–702. Refutes Morison's thesis that death does not occur at an identifiable time and explores the social rules that follow from this view.

KENNEDY, IAN MCCOLL. "The Kansas Statute on Death: An Appraisal." *New England Journal of Medicine* 285 (1971): 946–949. Criticizes the Kansas legislation in detail; urges that the defining of death be left in medical hands.

MEYERS, DAVID W. *The Human Body and the Law: A Medico-Legal Study.* Edinburgh: Edinburgh University Press; Chicago: Aldine Publishing Co., 1970. Useful survey of American, British, Scottish, and Continental law on euthanasia, sterilization, etc.

MORISON, ROBERT S. "Death: Process or Event?" *Science* 173 (1971): 694–698. Contends that death is a process, not a single event; examines the ethical implications for physicians.

VAN TILL-D'AULNIS DE BOUROUILL, ADRIENNE. "How Dead Can You Be?" *Medicine, Science and the Law* 15, no. 2 (1975), pp. 133–147. Compares American diagnositic criteria with those used in France, Austria and Germany; also differentiates ceasing artificial maintenance from murder or active euthanasia.

VEATCH, ROBERT M. "The Whole-Brain-Oriented Concept of Death: An Outmoded Philosophical Formulation." *Journal of Thanatology* 3, no. 1 (1975), pp. 13–30. Examines the current concept of brain death from philosophical and medical viewpoints.

III
PHILOSOPHICAL AND THEOLOGICAL FOUNDATIONS

Every definition of death presupposes some form of philosophical and/or theological reflection. It may be explicit by way of careful reasoning and presentation of beliefs about the meaning of life and death (or in conceptual analysis, about the terms "life" and "death"), or it may all be implicit as is a cultural sensibility. Even if a person believes that the definition of death is a purely medical issue, the justification of such a claim must proceed on philosophical or theological grounds. On the other hand, if it is recognized that definitions and meanings of death are valuational, then philosophical or theological considerations, whether lay or professional, are primary. The strong version of this claim argues that a concept of death is never medically, empirically verifiable. Weaker versions of the claim regularly argue that any concept of death involves consideration of a family of ideas, such as, "human," "life," "birth," "soul," and "consciousness," which appeal, in part, to philosophical and theological notions. In bioethics, the philosophical–theological question regarding death is often posed in the following way: What is essential to the nature of man the loss of which would warrant calling an individual dead?

Although some philosophers would claim that attempts to determine the essence of anything are misleading, there is a long tradition in philosophical inquiry and theological doctrine to determine what is significant about man in just that way. Moreover, many contemporary bioethicists believe that, if a sufficiently simple, clear, and satisfactory answer can be given, its relevancy to criteria in determining an individual's death is immediate. For example, the widely known "Report of the Ad Hoc Committee of the Harvard Medical School to Examine the Definition of Brain Death" offers criteria for irreversible coma and tells us nothing of what is essentially significant about the nature of man. Although bioethicists do not argue that the Harvard committee should have specified the latter, many do argue that, unless the essential nature of man is somewhere satisfactorily formulated, no one can fully decide whether a set of criteria is appropriate.

Belief in the human soul

According to one widely popular view in Western religious tradition, man is distinctive by possession of a soul. In its simplest form such a view defines death (and the moment of death) as the departure of the soul from the earthly body. This view usually presupposes that a union of body and soul is essential to the spatiotemporal existence of man. The destruction of this union, whether the soul is affected or not, constitutes death and is very often interpreted as "death of the body." As to what a "soul" is, there is no general agreement. Apart from the popular view, some theologians have put forward rather sophisticated understandings of "soul." For example, Ladislaus Boros in *The Mystery of Death* argues for an interpretation of Thomistic metaphysics which denies that man is a composition of two things. The essence of man is a single whole in which the body is the whole work of the soul, and the body is included in the actuality of soul. The essence of man is a unity, not a duality. Consequently, the separation of the soul and body is not a simple separation of two existent things. Death affects both corporeality (body) and the form (soul), yet the soul as "subsistent form" remains indestructible. Consequently, on this Thomistic interpretation, death is defined as destruction not only of the body and the soul–body unity, but destruction of the soul insofar as it appears in material concretion. The clinical implications of this belief are far from clear. Some believers might be willing to assert that it means nothing more than a definitional specification of the "moment of death," irrespective of the practical problems of determining that moment. Alternatively, the belief may be taken as an argument that opposes efforts to define death in any empirical and quantifiable manner. Other theologians have argued that the word "soul" is best used as an indicator of value and in no way names a spiritual substance or entity (Hick, pp. 1–29). In this case talk of a soul is seen as mythology, which reveals a people's belief in the intrinsic value and worth of a human individual.

Contemporary philosophers generally have not placed much credence in talk of a soul and have, instead, chosen to expose the mistakes of conceiving of the soul (or mind) as a substance. However, Ludwig Wittgenstein (1889–1951) did suggest that talk of a soul is an appropriate way of thinking of a human being. It is to admit of higher life; not to believe that a human being possesses a soul, but to have "an attitude towards a soul" (Wittgenstein, 1953, p. 178e). As such, human life cannot be exhaustively explained as a phenomenon of natural objects.

Process or event?

Some philosophers and physicians argue that the traditional or literary concept of death as a definite event which occurs at a specific moment is mistaken. It is argued that death should be conceived as a gradual process. This view finds support in the empirical knowledge that some parts of the body may disintegrate while others continue indefinitely. More important, it follows directly from a philosophical claim that no system fails all at once or, for that matter, no system can cease altogether and, hence, become nothing. Charles Hartshorne, in following the work of Alfred N. Whitehead, offers just that argument (Hartshorne). According to Hartshorne, death can never mean sheer destruction, including references to individual persons, since the "ego" is not identical with any single configuration of atoms or particles. There are no essential features necessary to the individual. Yet an individual does not "de-become." Death is merely the affixing of a particular quantum of reality. Death is the "last page" of the book, affixed, perhaps, retrospectively and somewhat arbitrarily. The answer to the basic question of death, as Hartshorne sees it—How rich and how complete is the book?—spells the tragedy or lack of tragedy of the quantum.

As an extension of the process view to bioethical questions, Robert Morison argues that the practical judgments concerning death can be solved by use of a cost-benefit analysis concerning the worth or worthlessness of life prolongation, an analysis that is remarkably close to what Hartshorne defines as the basic question of death. Morison holds that the usual bioethical question of "when an individual is dead" is not distinct from but is an instance of the question, "When is a person's life no longer worth prolonging?" Hence, the practical matter of death certification is seen as a value judgment, not a question of fact. According to Morison we need to abandon both efforts to define a moment of death and attempts to establish precise empirical criteria for death determination, since no real border between life and death exists (Morison).

There are, however, several objections to the process view. One objection consists in defending conceptual distinctions either overlooked or discounted by the process proponents. Dying should be distinguished from death, and both of these from aging. The question of when a life is no longer worth prolonging should be distinguished from the question of when in fact a person is dead (Kass, pp. 698–700). Another objection charges that the process view denies the reality of death by making it an empty concept or by using the term as a highly qualified metaphor. Finally, many persons argue that if one can legitimately claim an enduring self-identity (not necessarily substance) from birth till death or an identity of the organism as a whole, then the processive understanding of both the individual and death must be challenged for confusing character, biography, personality, and bodily function with the organism as a whole or the self as an agent.

Brain and consciousness

Focus of attention on a neurological locus of the essential nature of the human being is widespread. It has arisen not only through increased empirical knowledge of the neurological system, especially the brain, but through the continued interest of philosophers in the philosophic-scientific aspects of the mind–body problem, including efforts to understand such concepts as consciousness, remembering, and reasoning. The extension of such interests to a brain-oriented concept of death has arisen chiefly out of increased medical ability to transplant organs, together with increased ability to sustain heart and lung functions artificially. A brain-oriented concept of death might use the following kind of argument: If consciousness, remembering, reasoning, etc., are essential to the nature of man, and if these features are identical with processes of the brain, then the loss of brain processes (brain death) is sufficient warrant to call a person dead, i.e., the essential features of a person are absent.

One widely debated form of philosophical persuasion that would attempt to give some credence, at least by implication, to the above form of argument is "the identity thesis," a hypothesis of the identity of consciousness and brain processes (Rosenthal). Because the thesis generally claims that the confirmation must be the result of scientific neurophysiological investigation, it can claim no higher status than a hypothesis. Basically, the arguments for the thesis take the form of answering objections and attempting to show that the thesis is neither self-contradictory nor self-evidently false. The thesis proposes that consciousness (a) is identical with brain processes (b) in the sense that what is true of (a) is true of (b) without the identity's becoming trivially true as in the case of "bachelors are identical with unmarried males." Statements about (a) and (b), so it is claimed, can have different meanings but identical references. It is not a thesis that consciousness and other mental states are merely associated with or dependent upon brain processes, which everyone would grant, but rather that consciousness and other mental states are factually discernible as brain processes. For the concept of death, this implies that, although the statements "Jones is dead" and "Jones's brain is dead" may have different meanings, they have identical factual references. That is, if brain processes are absent, there can be no other factual reality to death. Loss of consciousness, reason, etc., is nothing more than the diminution of brain processes.

It should be noted that "the identity thesis" has not to date received decisive empirical verification or general acceptance, even though as a hypothesis it has received two decades or more of discussion. However, many neurologists do generally *assume* in their model of explanation that conscious mental processes can be interpreted, without remainder, as material processes. But the neurological model is the philosophical basis of the empirical investigation and as such cannot confirm the identity thesis without circularity. Additionally, it runs the risk of being logically untenable, on the ground that totally different propositions cannot have an identical reference. On the other hand, John C. Eccles, a noted neurophysiologist, claims refutation of the strong version of the identity thesis and has gone so far as to suggest that the neurological model of the brain be amended. Similarly, Sir

Francis Walshe, a neurophysiologist who from time to time has reflected on the philosophical implications of his own enterprise, has argued that neurobiology, dominated by mechanistic and reductionistic ideas, has not provided an adequate account of consciousness and that "identification" efforts have produced a cluster of paradoxes (Walshe).

Efforts to define the death of a person as "death of the brain" encounter other philosophical difficulties as well. It is regularly pointed out that the argument for the identification either begs the question or makes an equivocal claim that conscious life, capacity for reasoning, acting, etc., "reside" in the brain. If the term "reside" is taken to mean "associated with" or "necessarily conditioned," then the argument will not yield the conclusion of identification of so-called brain death with "death of a person" since necessary conditions always leave open the possibility of a reality more comprehensive than the conditions themselves. On the other hand, if "reside" is taken to mean "consciousness is the same as brain processes," then a reckless extension of predicates of the human being to nonhuman beings or objects may be the result. For example, while it is appropriate to say "Jones knows that it is raining" it is a logically dubious personification to say "Jones's brain knows that it is raining" when knowing is accurately taken as a predication of a person (Kenny, pp. 65–74). Likewise, it may well be inadequate to argue from "Jones's brain is dead" to "Jones is dead." By analogy, clearly one would want to avoid argument from "Jones's brain has an undersupply of blood" to "Jones has an undersupply of blood." The difficulty over the "brain death" designation consists of both a reckless extension of predicates and a lack of precision regarding the meaning of brain dysfunction, for surely at death the brain does not cease to exist even though the human being, Jones, does. Just as the human being as a whole thinks, knows, and is conscious, not the brain *in vacuo* as it were, so, too, the human being as a whole dies, and not simply that certain functions cease in a brain. The recent impulse to discover a neurological locus of death coupled with a latent sense for whole entities may well account for the crude and paradoxical nature of the "whole-brain-oriented concept of death." That is to say, a "whole-brain-oriented concept of death" attempts to be more precise about the essential features of man than the heart-lung oriented concept of death, yet by focusing on the destruction of the whole brain it runs the risk of "false positive tests for life"—the same criticism to which the heart-lung oriented concept has been subjected. This has led some bioethicists to argue that the essential nature of man must be specified in terms of more precisely limited anatomical characteristics (Veatch). Nevertheless, caution is still needed. Even if one does distinguish between the whole-brain-oriented concept and the narrower higher-brain-oriented concept, any brain definition of death may well be subject to the philosophical criticism that it is "a curious revenant of the old soul–body dualism" (Jonas, p. 139), an exaggeration of the brain, forming a new dualism of brain and body, just as once the soul was exaggerated in attempting to understand the nature of personal identity.

The logic and ontology of death

Among twentieth-century philosophers, Martin Heidegger and Ludwig Wittgenstein have contributed most notably to discussions of death. Heidegger has made his impact by developing a voluminous existential phenomenology keyed to the subject, especially in *Being and Time*, while Wittgenstein has made impact by a few aphoristic comments about death, especially in *Tractatus Logico-Philosophicus*.

Heidegger's focus on the subject of death is ontological in character; that is to say, he claims that one's awareness of death raises the fundamental questions of what it means for anything to be. The chief feature of Heidegger's work is a theory of the meaning of death which shows that facing one's own personal death concretely offers life a purposefulness and urgency that otherwise would be lacking. The problem of death for Heidegger is not the death of others but clearly "my own death," i.e., death in the first person. Death is unique in the sense that no one else can die my death although it is still the case that not everyone lives his own life. Death is the termination of me and is not just one event among other events or possibilities on the horizon. This means that any phenomenological explanation of death must include more than an investigation of the disintegration of animal structure. Not only does awareness of death provide a frame of reference for one's own authentic existence, but an accurate account of human existence is not "full" or "complete" without proper attention to death. In this sense, death, too, is essential to human nature since it provides the proper perspective for viewing the totality of human existence even if it cannot be something one can experience. Death is the inevitable end of all experience, the end of existence, and the annihilation of one's being.

At the very least, Heidegger's analysis, as applied to bioethical questions, underscores the point that the meaning and definition of death are not essentially experiential issues. Anyone who might claim that he does not yet know what death is because he has not experienced it has only revealed conceptual confusion. More important, perhaps, Heidegger's account provides a forceful account of man's finitude by focusing on meaning as possibility, not actuality, since the question of the ontological actuality of my death cast in any experiential sense is a meaningless question. The parallel questions of what it means to be and what it means not to be are answerable by Heidegger's account not by inquiring into what man's essential nature is but by looking to possible existence, as awareness of death incites authentic existence, and possible nonexistence.

Wittgenstein, on the other hand, has provided some illumination of the logical peculiarities of the concept of death in a few brief remarks. He says, "As in death, too, this world does not change, but ceases. Death is not an event in life. Death is not lived through. . . . Our life has no end in just the way in which our visual field has no limits" (1933, 6.431, 6.4311). Wittgenstein's remarks exhibit quite precisely that death is not a concept characterized as an experience. In this way Wittgenstein's analysis shows the radical force of the concept of death. Although the death of another person may entail certain empirical propositions predicated by observers, there cannot be any empirical predication by the person who is dead. That is, one cannot seriously conjugate the verb "to die" in the past tense, first person singular (Poteat). So far, such a view of death raises serious questions about the appropriate use of such locutions as "Death is a part of life."

What is often overlooked is the peculiar logic of using the term "death" with reference to the world—"as in death . . . the world ceases." At first sight Wittgenstein's claim may appear contrary to all evidence or simply an illicit transition from "my world" to "the world." Not surprisingly, he shares company here with Heidegger and some other existentialists in showing that, in a significant sense, our death cannot be imagined if it is merely a cessation of experience (Van Evra, p. 176), and, as a corollary, he is affirming that in a significant sense only subjects die. The "world" and the use of the expression "the world of objects" is not only parasitic upon subject terms, but also "the world" or "the world of objects" becomes a logical impossibility without subjects. Strictly speaking a world only of objects or a world with only demonstrative references (the, this, that), if imaginable, would be a world in which there is no death. As a result, it makes sense to say that in death the world ceases. To assert that this is an illicit transition from "my world" to "the world" is in fact to entertain an equivocal concept of "world" as well as to assert the impossible priority of a world of objects. Wittgenstein also claims that life has no end in the way that our visual field has no limit. There is no limit to our visual field in the sense that the limit cannot be seen. Death is an analogous limit and is just as inexperienceable and literally inconceivable. To do otherwise is to be within the boundary. But this further means that death is best understood as an "ordering device" analogous to the notion "absolute zero" in thermodynamics and cannot be characterized as an extensional state (Van Evra, pp. 173–175). Such an analysis is opposed to the view that there can be an "awareness" of death as if it were a living recognition (a seeing) of the limit.

Irreducible structure of life

Alongside the strong philosophical warnings about the confusions arising from imagining death to be a thing or an existing entity, there is a strong contemporary, antireductionistic persuasion that the structure of human life itself is not explicable in "thing" language. Rather, an existent being must be seen to have an operating principle as a comprehensive entity that is not simply the sum total of operational principles of parts of the whole. If the "self," "person," or "human being" is not a fully reductive concept, then "grocery lists" of the characteristics of human nature or what is essential to man are at the very least inadequate. These lists have ordinarily included such functions as capacity for experience, rationality, remembering, social interaction, integration of bodily functions, consciousness, and others. Among the most notable alternatives to reductionism is the contribution of Michael Polanyi, a physician, physical chemist, and philosopher. His later work has prompted several advances in philosophical biology.

In several places Polanyi offers arguments and analogies exhibiting the irreducible structure of life (1958, pp. 327–405; 1969). His arguments conclude that an "operational principle" of a comprehensive entity has a reality of its own that is not reducible to lower-level principles. While the constituent particulars (and their laws) set conditions for the successful operation of a comprehensive entity, the conditions can apply to a variety of circumstances. The operational principle of the comprehensive entity provides what Polanyi calls the "boundary control" which forms the frame of relevancy, i.e., enables us to say which of many laws (conditions) are relevant. Thus, a comprehensive entity functions as a system of higher and lower orders, but the relationship of the levels of orders is asymmetrical. For example, certain physiochemical laws continue operation even when a comprehensive entity is smashed, as in the case of a clock. As a result, a comprehensive entity should not only be acknowledged as different from the aggregate of its parts, but the meaning of any part of a fully functioning comprehensive entity must be attended to from the higher operational principle of the entity. That is why it is tempting, but fallacious, to transfer the operation of the comprehensive entity to one or several constituent particulars.

On this view one must give attention to the successively higher levels of operational principles in order to understand and ascribe meaning to life. One set of operational principles always leaves open the possibility of yet another, higher operational principle. Consequently, a description of life and normal functioning requires an antecedent knowledge of the operational features of the comprehensive entity. As a counterpart, recognition of malfunction requires use of the higher order operational principle in order to determine the relevancy of many laws which may apply at a lower level.

Although Polanyi has not applied his findings to the issue of death, there are implications that directly follow (High). Not only is it impossible to give a complete physiochemical topography of a living organism, but acknowledgment of the higher operational principles of a comprehensive entity is paramount. The latter provide the basis for death recognition by way of determining the relevancy of ascertaining that certain functions have failed. It is never a simple matter of saying "This organ has failed; therefore, this person is dead." Indeed, that is why some physicians and bioethicists insist that in death the human being as a whole (as distinguished from whole human being) dies or that it is the individual who dies and not just an organ. In this sense it is quite correct to say that organs do not die, persons do. This means not only that the concept of death must be kept distinct from medical criteria of death, but such criteria must be accorded no higher status than contingent clues, evidence, and justification for acknowledging that an individual, a person, has ceased to exist. Attempts to "locate death" in less than a comprehensive way are at best misleading, and the clues, evidence, and justifications must always be seen as open-ended and incomplete. On such a view, death is the termination of existence of an individual comprehensive being, a person, and not simply a dysfunction of one or more constituent particulars of that individual being, however much the latter may be used as evidence of individual demise or cited as causes of death. Even though certain dysfunctions or changed functions of constituent particulars, e.g., the brain, may be entailed by saying "I will die," the answer to the question of what is essentially lost or absent in death is on this view surprisingly simple and clear: myself.

Conclusion

The alternative candidates for specifying the essence of man and the philosophical/theological debates about conceptual foundations of death are complex and varied. The issues remain open-ended and invite increased attention from bioethicists and the public. However, it is not a logical necessity that the conceptual foundations must have universal agreement before criteria can have validity, even though the degree of confidence that can be placed in criteria will depend on the acceptability of those conceptual foundations.

Two salient features seem to have emerged from the current debates and are likely to have an influence on the continued discussions. First, the primacy of the person or person as a whole has gained a renewed status in understanding death. Whatever else may be said of characteristics or losses of particular functions, the debates constantly remind us that it is the person who dies. Second, a fairly clear distinction has emerged between a definition (or concept) of death and criteria for certifying death. As a result the debates may have inadvertently shown that the so-called traditional definition of death (cessation of circulation and respiration) was not a definition at all, popular assumptions to the contrary. Rather it was and is a set of criteria for determining that an individual has died, however much the criteria may need revising.

Twentieth-century man has not encountered a new death or a need to update death, popular language to the contrary, but has gained, perhaps, a renewed sensitivity to mortality.

DALLAS M. HIGH

[*Directly related is* DEATH, *articles on* EASTERN THOUGHT, WESTERN PHILOSOPHICAL THOUGHT, *and* WESTERN RELIGIOUS THOUGHT. *Other relevant material may be found under* EMBODIMENT; LIFE; MIND–BODY PROBLEM; *and* PERSON. *See also:* BIOLOGY, PHILOSOPHY OF.]

BIBLIOGRAPHY

BOROS, LADISLAUS. *Mysterium mortis: Der Mensch in der letzten Entscheidung.* Otlen und Freiburg im Breigau: Walter-Verlag, 1965. Translated by Gregory Bainbridge as *The Moment of Truth: Mysterium mortis.* London: Burns & Oates, 1965. And as *The Mystery of Death.* New York: Herder & Herder, 1965. Reprint. New York: Seabury Press, 1973.

Death Inside Out: The Hastings Center Report. Edited by Peter Steinfels and Robert M. Veatch. Preface by Daniel Callahan. A Harper Forum Book. New York: Harper & Row, 1975.

ECCLES, JOHN CAREW. *The Understanding of the Brain.* New York: McGraw-Hill Book Co., 1973.

HARTSHORNE, CHARLES. "Time, Death, and Everlasting Life." *The Logic of Perfection, and Other Essays in Neoclassical Metaphysics.* LaSalle, Ill.: Open Court Publishing Co., 1962, pp. 245–262.

HEIDEGGER, MARTIN. *Being and Time* (1927). Translated by John Macquarrie and Edward Robinson. The Library of Philosophy and Theology. London: SCM Press, 1962.

HICK, JOHN. *Biology and the Soul.* Arthur Stanley Eddington Memorial Lectures, no. 25. London: Cambridge University Press, 1972.

HIGH, DALLAS M. "Death: Its Conceptual Elusiveness." *Soundings* 55 (1972): 438–458.

JONAS, HANS. "Against the Stream: Comments on the Definition and Redefinition of Death." *Philosophical Essays: From Ancient Creed to Technological Man.* Englewood Cliffs, N.J.: Prentice-Hall, 1974, pp. 132–140.

KASS, LEON R. "Death as an Event: A Commentary on Robert Morison." *Science* 173 (1971): 698–702. Reprint. *Death Inside Out,* pp. 71–78.

KENNY, ANTHONY J. P. "The Homunculus Fallacy." *Interpretations of Life and Mind: Essays around the Problem of Reduction.* Edited by Marjorie Grene. London: Routledge & Kegan Paul; New York: Humanities Press, 1971, pp. 65–74.

MORISON, ROBERT S. "Death: Process or Event?" *Science* 173 (1971): 694–698. Reprint. *Death Inside Out,* pp. 63–70.

POLANYI, MICHAEL. *Personal Knowledge: Towards a Post-Critical Philosophy.* Chicago: University of Chicago Press, 1958.

———. "The Structure of Consciousness" and "Life's Irreducible Structure." *Knowing and Being: Essays.* Edited by Marjorie Grene. Chicago: University of Chicago Press; London: Routledge & Kegan Paul, 1969, pp. 211–224, 225–239.

POTEAT, WILLIAM H. "'I will die': An Analysis." *Philosophical Quarterly* 9 (1959): 46–58.

ROSENTHAL, DAVID M., ed. *Materialism and the Mind–Body Problem.* Englewood Cliffs, N.J.: Prentice-Hall, 1971.

VAN EVRA, JAMES. "On Death as a Limit." *Analysis* 31 (1971): 170–176.

VEATCH, ROBERT M. "The Whole-Brain-Oriented Concept of Death: An Outmoded Philosophical Formulation." *Journal of Thanatology* 3 (1975): 13–30.

WALSHE, FRANCIS. "Personal Knowledge and Concepts in the Biological Sciences." *Intellect and Hope: Essays in the Thought of Michael Polanyi.* Edited by Thomas A. Langford and William H. Poteat. Durham: Duke University Press, Lilly Endowment Research Program in Christianity and Politics, 1968, pp. 275–314.

WITTGENSTEIN, LUDWIG. *Philosophical Investigations.* Translated by G. E. M. Anscombe. Oxford: Basil Blackwell, 1953.

———. *Tractatus Logico-Philosophicus.* Translated by C. K. Ogden and F. P. Ramsey. Introduction by Bertrand Russell. International Library of Psychology, Philosophy and Scientific Method. London: Kegan Paul, Trench, Trubner & Co., 1933.

DECISION MAKING, MEDICAL

To make medical decisions in an ethical fashion, neither good intentions nor sound knowledge of fact is sufficient. There is no substitute for a careful analysis of the issues. A rational theory has been elaborated for optimizing decisions. In practice it falls short at two points. For most diseases many of the empirical facts are missing—frequencies of diseases and distributions of diagnostic findings in them, for instance. This defect is readily remediable in principle. A more elusive problem is the development of an appropriate theory of values.

This article shall first explain the general characteristics of decisions, paying some attention to those peculiar to medical practice. The formal principles for the well-characterized problem will then be briefly discussed.

General characteristics of medical decisions

Explicitness. An appropriate course of action in medicine is to be chosen by reflection on sound fact and with as clear a discernment as possible of the immediate and long-term consequences. Analysis cannot be subordinated to custom, nor should data be limited to the purported truths of textbooks, however well respected. Embarking on a sequential course of action—i.e., one modified conditionally on the intermediate outcomes—requires some provision for unfavorable outcomes.

Revocability. Some procedures are irrevocable (gastrectomy); some are doubtfully revocable (vasectomy); some are revocable (simple colostomy); and some involve recurrent acts (a regimen of insulin). The onus of decision is evidently greatest where the process in irrevocable.

The critical and the decisive. While the basis for a decision is not necessarily critical, the execution often is. There may be unresolvable doubt as to whether a patient has a perforated peptic ulcer; but the decisive policies—to operate or not—admit of no compromise. Some actions may be graduated—for instance, the amount of insulin given to a patient with a doubtful diabetic coma; others (e.g., tying the femoral artery) do not allow gradations.

Uncertainty. Decisions must commonly be made in the face of uncertainties of various types.

1. Personal ignorance on the part of the physician due to inexperience (in which case consultation is advisable). It may reflect culpable incompetence.

2. Illusory knowledge, based on what "everyone knows," quoted, but without good evidence, in influential textbooks. Policing this area is a corporate responsibility of the medical profession and especially of the academic leaders. Close reading of the literature suggests that the responsibility is not taken so seriously as it should be.

3. A limited corpus of knowledge about diseases only recently recognized or rarely encountered.

4. Incompleteness of data concerning the individual patient due to serious risks involved (e.g., myelography, though highly informative, is not lightly undertaken); or legal constraint (where permission is denied for a test by parents); or a test, though harmless and scientifically sound, that is morally offensive to the patient or doctor (e.g., reading the patient's private diary for evidence of mental disease).

5. An emergency decision that may have to be made without information that would take perilously long to obtain. Also, there may be insufficient time for reflection.

Conflicts. There are two broad kinds of conflicts. Intrapersonal conflicts are those in which the beneficiary is one person. For instance, amputation of a leg for osteogenic sarcoma mutilates the patient but offers him relief of pain and a chance of cure. The principle of doing what is best for the patient may not dispose of the problem because of the incommensurability between length of life, disability, pain, and so forth. But at least the patient can be a party to all aspects of the decision.

Interpersonal conflicts are a different matter. They may call for altruism (e.g., donation of a kidney), where there are dangerous opportunities for insidious moral pressures. Free consent of the patient may not be possible (e.g., commitment of a psychotic patient for the public safety). They may offer nothing at all to the deprived party (e.g., a test may be denied because the laboratories are overloaded). Insufficient thought has been given to relative and absolute absence of constraints. The right to reproduce, which, even among the mentally retarded, is widely respected, cannot be absolute; however, when the density of population is low it may be mistaken for an absolute principle.

Pragmatism. Decisions, by nature pragmatic acts of the will, are not necessarily capricious but can be arrived at by conscientious, rational arbitration of the facts and arguments. Such an imperfect formulation can be ethically acceptable for reasons of urgency when even vacillation and postponement themselves may constitute implicit decisions. There is a real peril of mistaking for an essential principle such imperfect grounds for arbitrary decision. For instance, to prevent exploitation, the law protects the mentally retarded, e.g., in making contracts. To make such laws feasible legislators must arbitrarily define the mentally competent. Such an arbitrary pragmatic decision should not be endowed with an essential character from which a body of ethical theory should be deduced. If we believed in hard and fast categories—e.g., "the normal" and "the abnormal"—criteria for *one* decision would suffice for *many* decisions. A sound test of the validity of such a classification is how far we would be prepared to define it when the area of its application is *indefinite*.

Judgment. The professional person, unlike the scholar, may be forced to make decisions even in the face of insufficient knowledge. The gaps in knowledge are filled in by an indefinable quality, "professional judgment," the need for which is not in dispute. But such judgment—as Alvan Feinstein has so aptly pointed out—is not a substitute for knowledge, for the responsibility to amass knowledge, or for conscientious pursuit of whatever method or process of reasoning the problem may demand (Feinstein). It is not a substitution of sentiment for rationality, nor should it be an ornate rationalization for personal prejudices or a defense for neurotic guilt.

Professional judgment is appropriate in the more elusive evaluations—for reconciling conflicting but not clearly commensurable factors, or in marshaling common sense when all else fails.

Simplicity and practicality. Especially in emergencies, practical considerations may require simplifications. There may be logistical problems: A neurosurgeon may not be available quickly enough to deal with head injuries, or there may be insufficient staff to cope with a massive disaster. Alternatively the defect may lie in the lack of a formed attitude for dealing with foreseeable contingencies. The former class of problems reflects what society can afford. The second class of problems is less easily evaded. While a physician cannot be expected to solve any ethical problem instantaneously and infallibly, he could be prepared for many of them with formal training in the theory of making decisions and with sound prescriptions for the major types of cases he will encounter. That he is not usually so prepared is a lamentable comment on medical curricula. Simplicity and practicality are commonly used as evasions of responsibility to the point of slovenliness. Indeed, the foregoing example illustrates the difference between responsibility as a virtue and as a sentiment. Industrial poisoning with anthracite (which is slow) is not necessarily less important than rapid industrial poisoning with carbon monoxide; but *sentiments* of responsibility are much more easily evoked in the latter case.

Formal theory of decision making

In the past, medical ethics has commonly been cast in syllogistic terms. For practical phenomena, at least, deductive and inductive logic may be recast as special cases of a more general probabilistic decision theory, which should now replace the older methods. George Polya expounds the broad principles of such a theory in logical terms; there are several statistical (Lindley; Ferguson) and even clinical (Murphy) counterparts. A fundamental assumption of the formal theories of medical decision making that presents the most formidable difficulties is that all the penalties—loss of life, pain, disability, embarrassment, shame, loss of money, moral affront, etc., whatever cannot be immediately deduced from brute fact—can be condensed accurately into one mathematical function of the variables on which the decision is to be based. In the same accurate way, the benefits (relief of a disability, beauty, fertility, improved sense of well-being) can also be expressed as a mathematical function of those variables. Benefit may be viewed as negative cost, and a single mathematical formula may be produced, to which we give the technical name *cost function*. It may be positive, in which case treatment does not seem warranted (the cost exceeding the benefit), or negative. If there are several strategies, that which gives the highest negative cost would be the appropriate one.

Much of the attempt to adapt general ethics to medicine has been based on the assumptions of unambiguous diagnoses and fixed costs. Mature consideration suggests the need for a more elaborate structure. There are at least three major components in such a structure.

1. *Probability*. A patient may have six fingers, easily established by examination. Most diagnoses are not of this kind. A common principle is to make the most likely diagnosis and act accordingly. But, as we shall see, this is not the most reasonable strategy.

2. *Nosological characteristics*. Some few conditions are clearly categorical (a traumatic Colles fracture of the wrist, for example). Then treatment and prognosis will be the same almost regardless of the findings, which are important only for diagnosis. The degree, as well as the presence, of the disease may bear on decisions (for instance, coronary disease). In some instances (e.g., "essential hypertension") the findings alone matter: The "diagnosis" does not in itself make any difference to either treatment or prognosis.

3. *Fixity of cost*. In some diseases, such as metastatic (disseminated) cancer, the cost functions are but slightly influenced by the stage of the disease. In certain infectious diseases, the cost of not treating varies according to the current threat to life (e.g., pneumonia), the prospects for preventing irreparable damage (e.g., tuberculosis), and the risks of spread (e.g., smallpox).

Ethical theorists have formulated their ideas on what medical theorists have supposed (often unrealistically) that sound practitioners do. They have thus tended to separate artificially these three major components. For example, arteriosclerosis is a disease; it has a serious prognosis, and therefore treatment is justified. It is not at all clear what it means to say that arteriosclerosis is a disease, because it is difficult to give a watertight definition of a disease. But there is no need to invoke the category "disease" to make rational decisions. The responsible physician has tended to argue broadly that what benefits a patient is warranted, and what does not is unwarranted. The fiction that an ethical decision proceeds in three mutually isolated steps has led to inconsistencies in diagnostic practices. For instance, in the course of one weekend a hospital changed the diagnosis of "sociopathic personality" from a nondisease to a disease because of the changing judicial principles about insanity as a legal defense (*Rosenfield* v. *United States*).

But decision theory would sustain the de facto practice of accepting a "rate of exchange" between uncertainty of diagnosis and cost. For instance, in a disorder of the spinal cord a test for vitamin B_{12} deficiency would be warranted on bare suspicion, because the cost of not treating (in what is a curable disorder) is disastrous. Surgeons demand a less certain diagnosis of appendicitis in a child than in an adult, because peritonitis is so much more devastating a complication in a child.

The treatment of secondary syphilis, once the diagnosis is made, on whatever grounds and with however much uncertainty, will be the same. However, the treatment of high blood pressure varies from inactive vigilance to elaborate and urgent medication: It is not the diagnosis but the floridness of the manifestations that dictates the course of action. Indeed, it matters little what the physician calls the condition or even how he classifies it. Pickering and Platt (to revive a famous controversy) have held diametrically opposite views about the nature of "essential hypertension" but manage cases in much the same way (Murphy).

Ideally we proceed as follows: Where diagnosis is uncertain we define probability distributions of the findings conditional on certain diseases. Particular signs or certain ranges of values may be exceptional in one state, common in another. But all competing diagnoses must be conscientiously considered. An atypical form of a common disease may be a more likely diagnosis than a typical form of a rare disease. The differential diagnoses of a large spleen would differ in emphasis in West Africa and St. Louis, or even from clinic to clinic in the same hospital. Again, diagnosis will be colored by family history. What might be called "tics" in a patient taken at random would be viewed as highly significant early signs if the patient's father had Huntington's chorea. The probabilistic assessment, then, depends on at least three major components: (1) the prior probability of the disorder; (2) the impact of ascertainment (the means by which the patient comes to the attention of the clinician); and (3) the conditional probability of the findings if the patient has the conjectured disease.

These three separate quantities multiplied give the joint probability. If the finding is clearly categorical—for instance the Babinski sign—the result will be a joint probability of two events: first that the sign is present; and second that the patient has, say, multiple sclerosis. If the finding is a measurement on a continuous scale (e.g., a blood sugar level), the result will be a probability density, that is, roughly an index of the relative probability of the result in some defined small neighborhood of the actual observation. (Most of the measurements will be subject to error.)

With this broad background, I shall try to *illustrate* principles. We are torn between realistic examples and uncomplicated explanations. The clinician will object with some truth that "nobody makes decisions on a single datum"; but while an analysis based on many tests would be more plausible, the mathematical manipulation would be correspondingly more complicated.

Intrapersonal and interpersonal decisions

Intrapersonal decisions. The objective is to make the cost as negative as possible. There are two major types of problems.

First, consider cases in which the category to which the patient belongs is known or (what is equivalent) there is only one category. Examples might be arteriosclerotic gangrene of a toe, where the basis for decision is to be the blood flow; and blood pressure level in the absence of diseases—e.g., chronic nephritis or Addison's disease—that perturb it. In both cases the prognoses are related to the measurement. Then the basis for decision is to compare the whole future cost function with treatment (since treatment itself carries some side effects, danger, expense, temporary disability, etc.) with that without treatment. If the former is the greater, there seems to be no place for treatment. Note that in this instance diagnosis does not matter. It is immaterial whether we identify some class of blood pressure levels as "hypertensive."

A special modification may be mentioned. The measurable characteristic may show catastrophic changes in prognosis. In cancer cases the prognosis is markedly different in patients with and those without distant spread, but it scarcely matters how many metastatic lesions are recognized. The cost function with treatment would thus show a sudden leap up as the number of secondaries increases from 0 to 1, since the benefits are drastically curtailed and perhaps completely abolished. A similar case would be the degree of distention of the aorta in the Marfan syndrome, since at some critical point it will rupture, and the surgical mortality abruptly increases when it does.

Second, it may not be known with certainty to which of two or more categories a patient belongs. We must be content with a probabilistic assignment, which will depend on the joint probability functions discussed previously: The problem is one of discrimination. The cost function may itself be a function of the discriminating variable. For instance, red cell sedimentation rate aids in the diagnosis of rheumatism; it is also a prognostic index, and a guide to the efficacy of treatment. In general, then, we have at least four sets of cost functions to deal with: with and without the disease and with and without treatment. If the presence of the disease is in some doubt, the appropriate cost function would be a mixture of the two cost functions according to their relative plausibilities, and likewise for the benefit functions. A rational criterion of action is to minimize the (average) cost function. (This principle would also determine which of several regimens of treatment, all beneficial, should be picked.) The cost functions might also incorporate other factors (e.g., age, sex) that in the particular case bear on prognosis and side effects but not treatment or categorization. In view of the composite origin of the cost function, in making decisions one cannot rationally separate the probabilities and the costs; and an infallible categorization is neither necessary nor, in some cases, even intelligible.

Interpersonal decisions. In some degree the clinician has always been aware of the conflicting demands of more than one person: between the mother and the child at birth, or in distributing his attentions during a physical disaster. But for the most part his main responsibility has been to the individual patient, and the conflicts are intrapersonal. Certain modern procedures—mass screening, routine physical examination, annual examinations, and multiple channel analyzers—have shifted emphasis from the care of the sick to the anticipation of sickness in the symptomless. The profound change in prior probabilities (most of the characteristics in these newer procedures being a priori unperturbed by illness) means that specificity and sensitivity must receive more attention. Screening tests (e.g., listening to the heart) are generally simple, harmless, and cheap, whereas confirmatory tests, such as cardiac catheterization, are often complex, dangerous, and expensive.

So some principles of selection, inevitably fallible, must be used. The patient, and even the personal physician, may find it hard to accept the idea that the patient's interests are being sacrificed to a more deserving case and that what is being optimized is not the patient's interests but the public good. Where the limiting factor is not cost but what is available (professional skill, supplies of a drug, etc.), more or less serious ethical problems arise in subordinating the good of society as a whole to the partisan demands of the patient or vice versa. The only defense for this policy choice would be a political one, which is outside the scope of this discussion.

EDMOND A. MURPHY

[*Directly related is* THERAPEUTIC RELATIONSHIP, *article on* CONTEMPORARY MEDICAL PERSPECTIVE. *For further discussion of topics mentioned in this article, see the entries:* ACTING AND REFRAINING; PRAGMATISM; *and* RISK.]

BIBLIOGRAPHY

FEINSTEIN, ALVAN R. *Clinical Judgment.* Baltimore: Williams & Wilkins, 1967.
FERGUSON, THOMAS SHELBURNE. *Mathematical Statistics: A Decision Theoretic Approach.* Probability and Mathematical Statistics, A Series of Monographs and Textbooks, no. 1. New York: Academic Press, 1967.
In re Rosenfield. 157 F. Supp. 18 (D. D. C. 1957).
LINDLEY, DENNIS VICTOR. *Making Decisions.* New York: Wiley-Interscience, 1971.
MURPHY, EDMOND A. *The Logic of Medicine.* Baltimore: Johns Hopkins University Press, 1976.
POLYA, GEORGE. *Mathematics and Plausible Reasoning.* 2 vols. London: Oxford University Press; Princeton: Princeton University Press, 1954.

DECLARATION OF GENEVA
See APPENDIX, SECTION I, DECLARATION OF GENEVA.

DECLARATION OF HELSINKI
See APPENDIX, SECTION II, DECLARATION OF HELSINKI.

DEFINITION OF DEATH
See DEATH, DEFINITION AND DETERMINATION OF.

DEMOGRAPHY

See POPULATION ETHICS: ELEMENTS OF THE FIELD, article on THE POPULATION PROBLEM IN DEMOGRAPHIC PERSPECTIVE.

DENTISTRY

I. ETHICAL ISSUES IN DENTISTRY
    Clifton O. Dummett
II. PROFESSIONAL CODES IN AMERICAN DENTISTRY
    Chester R. Burns

I
ETHICAL ISSUES IN DENTISTRY

The pursuit of modern dentistry is beset with a number of moral and social issues, which have inundated the profession and affected its daily practice to a considerable degree. Additionally, dramatic changes in the medical profession are having a profound influence on dentistry and have served to alert dentists to the possibilities of similar modifications.

Ethical issues

Among the contemporary ethical and social issues confronting the dental profession are those relating to accountability of dental health professional personnel; high-risk procedures; informed consent for dental and oral health services; peer review and quality of services; dental malpractice; expenditures for comprehensive dental care; dental professional advertising; denturism (unqualified, illegal practice of dentistry in any form on the public); dental auxiliary utilization and interprofessional relations; dental care of the aged and critically ill patients; community dentistry's responsibilities; dental experimentation in human subjects; and ethical emphases in undergraduate, postgraduate, graduate, and continuing dental education.

Oral disease prevention and oral health care for all people, advantaged and disadvantaged, are important responsibilities of community dentistry. Ethical issues are involved when it is recognized that implementation of these objectives requires interprofessional cooperation by all providers of health care. Some subjects have stimulated more attention than others. Extensive publicity has attached to propaganda stressing the need for dentures without emphasis on the equal need for the scientific knowledge that should undergird the technical expertise. This trend in consumerism places dentists in an adversary role in relation to the laboratory technicians whose outspoken proponents advocate direct denture fabrication, bypassing the dentist.

Dentists in general and oral surgeons in particular are sensitive about the question of informed consent—especially as regards the after-effects of anesthesia and X-ray, and clear statements of costs prior to initiating treatment procedure. Interest in the issue of informed consent is due largely to the growth in numbers of patients who have instigated legal action and the sympathetic juries who have made awards on the basis of the absence of informed consent. While the public has an increasing concern about patients' rights, burgeoning costs of malpractice insurance pose similar concerns among physicians and dentists. Peer review—the review of a clinician's services by his professional equals—is becoming an essential component of those dental care delivery systems which are sincerely interested in their accountability to the public for high-quality services. Increasing expenditures for health care have continued to occupy a large share of the public's attention. Consumers of hospital, medical, and dental services complain that health-care costs are rising at twice the rate of the cost of living; thus dentistry is caught up in the problem of justice in health-care delivery. With the professed aim of reducing health-care costs through competition, some states have introduced legislation that would allow various health professionals to advertise. Dental professional advertising has remained a perennial example of unethical practices. Its reevaluation may be one of the outcomes of court declarations with regard to legal and medicopharmaceutical advertising. The oft-stated shortage of health manpower as a primary factor in the health-care "crisis" has revived wide interest in dental auxiliary utilization and has fostered innovations in duty assignments and personnel relations. Even though dental care of the aged and the critically ill has not as yet occupied center stage in the attentions of professional planners, there have been stirrings from senior citizens whose numbers as well as oral health needs have steadily increased. The rights and responsibilities of these persons as well as those of all other consumers of health care have given impetus to renewed demands for a stricter accountability of dental clinicians.

Because the emphases in dental education usually have been oriented toward the acquisition of technical skills, there probably has not been the highest appreciation of the differences that distinguish dentistry as a health profession from the general series of profit-motivated occupations. There is growing recognition among many of today's professional leaders that schools of dentistry in the United States need to review and expand their curricula in the teaching of dental ethics and related health professional topics.

Dentist–patient relationship

The highly specialized knowledge relating to comprehensive treatment of oral diseases has been one reason for patient dependence upon the dental practitioner for the protection of the patient's interests. The quality of protection is directly related to the dentist's maintenance of high professional standards, altruistic ideals, and dedicated service.

The possibilities of patient exploitation have always been plentiful, and there are numerous examples in which improprieties have occurred. Publicity accorded several indiscretions and numerous individual unethical practices have alarmed a trusting public. Public opinion polls have demonstrated increasing lapses in the public's confidence in the occupational and personal integrity of all professionals—physicians and dentists included. As a result, many Americans are accepting greater responsibility for their own oral health and are attempting to maintain individual control over their medical, dental, and hospital care. Such developments have affected traditional dentist–patient relationships in which dental clinicians consistently made the choices of therapy for their patients. There is some evidence that dentists are gradually agreeing to study and accept the viewpoints of patients much more than they did in earlier times.

Dental education

The dental schools' responsibilities for student indoctrination in ethical considerations have remained as clearcut as ever; yet there are indications of a less than desirable effectiveness. In a longitudinal study of professional socialization in which 270 dental students from three dental schools in California participated, Morris and Sherlock reported that professional ethics declined steadily, while cynicism increased, especially in the clinic years.

There is consensus among significant numbers of dental students that their attitudes toward the profession underwent some objectional transformations during their clinical apprenticeships. This is the way it has been for generations of dental students—which suggests that the clinical faculty may also be in need of sensitization to human values in dental care.

In dental schools generally, there are few if any teachers appointed with sole or primary commitments in dental ethics instruction. Most teachers are either dentists who have developed an interest in ethics as it would apply to dentistry, or dentists who have no interest in the subject but nevertheless have been arbitrarily assigned to teach it. It would be unusual to find an ethicist teaching in a dental school.

Prompted by some of the changes in medical education, which include the expansion of the teaching of medical ethics, a few dental educators have been pressing for similar expansion in the teaching of dental ethics. The problems are common to medical and dental schools: The demands of currently established technical courses hinder the recognition, economic support, and ideological consideration of new courses.

Social issues in dental ethics

The American Dental Association's Code of Ethics and the various state dental codes have established laudable guidelines and ethical norms for dentists in their interpersonal relationships, but the dental profession has been reluctant to admit that it had an ethical responsibility in reference to the larger social issues. The quest for human and civil rights in the 1950s and 1960s in the United States compelled all health professions to weigh a professed adherence to ethical principles against widespread practices of discrimination on the basis of race, religion, and national origin. Following the U.S. Supreme Court rulings against segregated education in 1954, dental schools were obliged to reassess their admissions policies, which had denied minorities unhampered access to a dental education. Additionally, the country's social upheaval had an impact on the individual practitioners who were called upon to put dental ethical behavior above regional customs and personal prejudices. Many white practitioners of medicine and dentistry in the United States were guilty of unethical behavior in that they withheld or diluted their services in the delivery of health care to minorities. Moreover, many white dentists resisted admitting minority dentists into local and national dental societies.

In 1958 increased pressures to remove discriminatory barriers prompted the American Dental Association's Board of Trustees to request ADA constituent societies to make a study of the Association's bylaws to see that licensure and adherence to the principles of ethics were the sole essential qualifications for membership in the Association.

Conclusion

Past experiences underscore the unnecessary tensions and confusion created by an unwillingness to face forthrightly the ethical dilemmas in many sensitive areas of dental practice. No less urgent, complex, or far-reaching are contemporary issues that have been opened to public scrutiny. Dentists will have to make some difficult decisions based upon ethical considerations.

Thus academicians, general practitioners, and their representative organizations must make a joint effort to formulate an attainable, ethical base supportive of the delivery of high-quality oral health services. Objective appraisal and careful reorganization of the tenets of various codes must be accomplished within the framework of a morality that recognizes the vagaries of human behavior.

CLIFTON O. DUMMETT

[Other relevant material may be found under ADVERTISING BY MEDICAL PROFESSIONALS; AGING AND THE AGED, article on HEALTH CARE AND RESEARCH IN THE AGED; INFORMED CONSENT IN THE THERAPEUTIC RELATIONSHIP; JUSTICE; MEDICAL ETHICS EDUCATION; and PATIENTS' RIGHTS MOVEMENT. See APPENDIX, SECTION IV, AMERICAN DENTAL ASSOCIATION.]

BIBLIOGRAPHY

"An Act to Amend Section 1680 of and to Repeal Section 651, 651.2, 651.3, 651.4, 2556 and 3129 of the Business and Professions Code Relating to Advertising by Professionals." SB 1974, 1976 sess., State of California. Introduced 23 March 1976 by Senator John F. Dunlap (D.-Napa). Sent to Committee on Business and Professions. Short title: "Price Advertising."

DUMMETT, CLIFTON O. "A Chronology Updated—Recent Events in the Negro's Advancements in Dentistry in the U.S." *Quarterly of the National Dental Association* 21 (1963): 145–201.

———. "Year 2000: Community Dentistry." *Journal of the American Dental Association* 82 (1971): 280–285.

GOLDBERG, ARTHUR J. "Ethics in the Professions." *Journal of the American College of Dentists* 42 (1975): 218–223.

GOLDIAMOND, ISRAEL. "Protection of Human Subjects and Patients: A Social Contingency Analysis of Distinctions Between Research and Practice, and Its Implications." *Behaviorism* 4 (1976): 1–41.

Great Britain, Radioactive Substances Advisory Committee. *Code of Practice for the Protection of Persons Against Ionizing Radiations Arising from Medical and Dental Use.* 3d ed. London: Her Majesty's Stationery Office, 1972.

JONSEN, ALBERT R., and HELLEGERS, ANDRÉ E. "Conceptual Foundations for an Ethics of Medical Care." *Ethics of Health Care: Papers of the Conference on Health Care and Changing Values, November 27–29, 1973.* Edited by Laurence R. Tancredi. Washington: Institute of Medicine, National Academy of Sciences, 1974, pp. 3–20. Reprint. *Ethics and Health Policy.* Edited by Robert M. Veatch and Roy Branson. Cambridge, Mass.: Ballinger Publishing Co., 1976, pp. 17–34.

LAWS, PRISCILLA W. *Medical and Dental X-Rays: A Consumer's Guide to Avoiding Unnecessary Radiation Exposure.* Washington: Public Citizen Health Research Group, 1974.

MORRIS, R. T., and SHERLOCK, B. J. "Decline of Ethics and the Rise of Cynicism in Dental School." *Journal of Health and Social Behavior* 12 (1971): 290–299.

PARISH, JACK R. "Professional Conduct in Dental School and After." *Journal of Dental Education* 32 (1968): 326–329.

VEATCH, ROBERT M. and SOLLITTO, SHARMON. "Medical Ethics Teaching: Report of a National Medical School Survey." *Journal of the American Medical Association* 235 (1976): 1030–1033.

II
PROFESSIONAL CODES IN AMERICAN DENTISTRY

The quest for ethical ideals in American dentistry has been intimately associated with the evolution of professional organizations. By 1859, the year of the founding of the American Dental Association, there were ten local dental societies, two state societies, and four dental colleges. Nineteen years after the American Medical Association adopted a code of ethics, and in spite of opposition expressed by numerous dentists, the American Dental Association adopted a code at its sixth annual meeting in 1866 (American Dental Association, 1866; Burns).

There were many similarities between the code for physicians and the one for dentists. Dentists should exhibit firmness, kindness, and sympathy toward their patients. They should be temperate, make no false promises, and explain their professional procedures. A dentist should be a gentleman in all relationships and never speak disparagingly of a colleague's practices. Physicians and dentists should avoid interprofessional conflicts by recognizing distinctions between their specialized interests. Dentists, like physicians, were expected to honor the public's trust by exposing all quacks.

In 1880 the American Dental Association made adoption of its code a mandatory requirement for membership, as had physicians some twenty-five years previously (American Dental

Association, 1881). As with the physicians' code, the code of dental ethics had no provision for enforcement. By the turn of the twentieth century, only one dentist had lost his membership in the American Dental Association for unprofessional conduct.

In 1897 the Southern and American associations merged to form the National Dental Association, and two years later a revised code was adopted (National Dental Association). The revision contained two significant changes: encouragement of consultations and permission to use cards and newspaper announcements for advertising names and office addresses.

In 1922 the national organization of dentists assumed its original name, the American Dental Association, and adopted a new, substantially different code of ethics. Previous claims about mutual duties of the profession and the public and about interprofessional duties of dentists and physicians were omitted. Statements about the obligations of dentists to patients were summarized in a single exhortation: The dentist should conduct himself "in accordance with the Golden Rule." The remaining sections of the code dealt exclusively with transactions among dentists. A new section in the revised code of 1928 required dentists to report "illegal, corrupt, or dishonest conduct on the part of any member of the dental profession" to the proper authorities. A mechanism for judging and punishing unethical dentists was described in a note appended to the code. The code of 1928 also urged dentists to be good citizens and to conduct themselves as members of a profession "whose prime purpose is service to humanity" (American Dental Association, 1947).

A revision for 1936 included sections about patents, contracts, and group practices (ibid.). The duty to report illegal and unethical conduct was omitted in the revision of 1944, as were the imperatives to be good citizens and serve humanity. The codes of 1928, 1936, and 1944 had an important disclaimer: The ideals of these codes would not "cover the whole field of ethics for the members of the profession." Two very significant sentences were added in the code of 1944: "There are many obligations assumed by those who choose dentistry as their life's work, in addition to those included in the foregoing statements. To know the answers to most questions not presented in this code, we need but to be guided by the Christian rule to do unto others as we would have others do unto us" (ibid.). This conclusion allowed for ethical pluralism and liberal interpretations; but it was also an invitation to oversimplify the ethical conflicts generated by the emerging social and technical complexities of twentieth-century dental practice.

A rewritten code, principles unchanged, was adopted in 1950 (American Dental Association, 1950, 1954). The Judicial Council of the American Dental Association subsequently converted the sections of this code into a set of "principles of ethics," which was approved in October 1955 and revised slightly in November 1958 (American Dental Association, 1958). Traditional ideals of professional loyalty and honor were sustained, but important changes reflected new demands of educators, and allied professionals and paraprofessionals, as well as increasing pressures from both community and governmental groups for more attention to dental health broadly conceived, including fluoridation and preventive practices. A revision of the *Principles* occurred in 1960 and again in 1974 (Conway and Rutledge; American Dental Association, 1975). The Judicial Council of the American Dental Association continues to offer advisory opinions based on their interpretations of these *Principles*.

Repeatedly revising and reassessing their codes, American dentists have discovered norms that would uphold rights for professional self-regulation and provide means for intraprofessional policing. For individual dentists, a set of principles is an elaborate loyalty oath, with disobedience leading to dishonor and disfranchisement. These principles also constitute a bill of professional rights for dental practitioners. American communities may honor these rights and acknowledge the distinctiveness of their associated values if dentists, as groups and individuals, fulfill the obligations associated with the rights.

CHESTER R. BURNS

[*See* APPENDIX, INTRODUCTION, *article on* CODES OF THE HEALTH-CARE PROFESSIONS; *and* SECTION IV, AMERICAN DENTAL ASSOCIATION. *Directly related is the other article in this entry:* ETHICAL ISSUES IN DENTISTRY.]

BIBLIOGRAPHY

American Dental Association. "Code of Dental Ethics." *Transactions of the American Dental Association* 6 (1866): 401–405. Legislative history appears on pp. 228, 234, and 238 of the same volume.

———. "Code of Ethics." *Digest of Official Actions, 1922–1946.* Chicago: 1947, pp. 199–211. Digest containing official actions: Constitution and Administra-

tive By-Laws; Code of Ethics, each with all changes, 1922–1946.
———. "Ethics." *Digest of Official Actions, 1946–1953.* Chicago: 1954, pp. 118–120.
———. *Principles of Ethics,* Chicago: 1958. Pamphlet.
———. "Principles of Ethics." *Journal of the American Dental Association* 90 (1975): 184–191.
———. "Revised Code of Ethics to Be Considered at Atlantic City Meeting." *Journal of the American Dental Association* 40 (1950): 612–615.
———. "Standing Resolutions." *Transactions of the American Dental Association* 21 (1881): 36.
BURNS, CHESTER R. "The Evolution of Professional Ethics in American Dentistry." *Bulletin of the History of Dentistry* 22 (1974): 59–70.
CONWAY, BERNARD J., and RUTLEDGE, C. E. "The Ethics of Our Profession." *Journal of the American Dental Association* 62 (1961): 333–342.
National Dental Association. "Code of Ethics: Report of Committee." *Transactions of the National Dental Association* 3 (1899): 481–483. Legislative history appears on p. 17 of the same volume.
NOYES, EDMUND. *Ethics and Jurisprudence for Dentists.* Chicago: W. B. Conkey Co., 1923.

DEONTOLOGICAL ETHICS
See ETHICS, *article on* DEONTOLOGICAL THEORIES.

DESCRIPTIVIST ETHICS
See ETHICS, *article on* NON-DESCRIPTIVISM.

DETERMINISM
See FREE WILL AND DETERMINISM.

DIALYSIS
See KIDNEY DIALYSIS AND TRANSPLANTATION.

DISCLOSURE
See TRUTH-TELLING; INFORMED CONSENT IN THE THERAPEUTIC RELATIONSHIP; INFORMED CONSENT IN HUMAN RESEARCH; INFORMED CONSENT IN MENTAL HEALTH; PRIVACY; CONFIDENTIALITY.

DISEASE
See HEALTH AND DISEASE; MENTAL ILLNESS.

DOUBLE EFFECT

In an effort to stipulate the conditions under which one may rightfully cause evil, e.g., the death of another person and in particular an innocent person or of oneself, Roman Catholic moral theologians developed what has come to be called "the principle of double effect." Though this principle has been of major concern principally to writers within the Roman Catholic tradition, the moral questions that it raises have been and are of interest to writers in other traditions. They have special relevance to many important problems in bioethics, e.g., abortion, euthanasia, and medical experimentation. This article will (1) provide a preliminary description of the principle, its underlying presuppositions, and its purpose; (2) offer a brief sketch of its historical development and the typical cases to which it has been applied; and (3) discuss some of the major contemporary debates centering on an understanding of the principle.

Description of the principle

As commonly formulated, the principle stipulates that one may rightfully cause evil through an act of choice if four conditions are verified: (1) the act itself, prescinding from the evil caused, is good or at least indifferent; (2) the good effect of the act is what the agent intends directly, only permitting the evil effect; (3) the good effect must not come about by means of the evil effect; and (4) there must be some proportionately grave reason for permitting the evil effect to occur (Mangan, p. 43; Conway, p. 137; Grisez, *Abortion,* p. 329; McCormick, p. 1; Curran, pp. 173–174).

Two of the major presuppositions of the principle around which contemporary discussion centers are that there is a morally significant difference between intending evil and permitting evil (condition 2) and that it is wrong to use an evil means to obtain a good end (condition 3). Condition 3, in short, rejects the view that a good end can justify an evil means, and some of the Roman Catholic defenders of the principle hold that this rejection is grounded in revelation, specifically in Romans 3:8 (Farraher, p. 71; Conway; Kelly, pp. 59–60).

The purpose of the principle is not, as some have recently suggested (Callahan, pp. 428–429), to allay subjective scruples about a "dirty" conscience, but rather, as others (Ramsey, pp. 207–211) have observed, to limit the area of moral ambiguity in "hard" cases by specifying the objective criteria for determining to what extent we can cause evil in the pursuit of good. Whether the principle is in fact capable of achieving this purpose is another question and is related to the debates to be considered later.

Historical development

Scholars agree that the principle was not fully formulated until the middle of the seventeenth century, but there is lively debate over its precise historical development. Some, for example Mangan, argue that the principle was basically, although by no means clearly, expressed in the thirteenth century by Thomas Aquinas in his discussion of killing in self-defense (in the *Summa Theologiae* II–II, 64, 7), and that it was in reference to this teaching that such sixteenth- and seventeenth-century authors as Thomas de Vio Cajetan (1468–1534), Louis Molina (1536–1600), Thomas Sanchez (1550–1610), and, in particular, the Salmanticenses (the Discalced Carmelites of Salamanca), preeminently Domingo de Santa Teresa (1600–1654), made the principle more and more explicit.

Others, inclining to the position of J. Ghoos, maintain that the principle is in no way entailed in Aquinas's teaching on killing in self-defense and that its development was stimulated by an endeavor to bring together two notions that were distinguished by Aquinas, namely, the "indirect voluntary" and the "voluntary in its cause." For Aquinas these were distinct concepts, with the first (*Summa Theologiae* I–II, 6, 3) referring to an effect that came about by the failure to act, and the second (ibid., I–II, 77, 7) referring to an effect that was willed not in itself but in its cause. For Aquinas, in other words, the "indirectly voluntary" was conceptually akin to the modern notion of refraining from acting, and Aquinas clearly held that a person is morally responsible for evils that result from such omissions if that person can and ought to take the action necessary to prevent the evil from occurring. During the sixteenth and seventeenth centuries these two terms gradually became synonymous, having the meaning that "voluntary in its cause" had in Aquinas, and these terms figured prominently in the evolution of the principle of double effect. Its precise formulation, according to this interpretation, was the work of John of St. Thomas (1589–1644) in his *De bonitate et malitia actuum humanorum*.

No matter which historical interpretation is correct, there is an undeniable close conceptual affinity, manifest in contemporary discussions of the principle (Grisez, *Abortion*, pp. 326–329, "Toward a Consistent," pp. 73–79; Ramsey, p. 220; Curran, pp. 174–182), between the reasoning at work in Aquinas's teaching on killing in self-defense and that entailed in the principle of double effect. Aquinas clearly held that a person may rightfully choose to defend himself from attack even if his act of self-defense will result in the death of the assailant, provided that (1) the agent intends only to preserve his own life (a good) and does not intend to kill the assailant (an evil) in order to protect himself, and (2) the force used to repel the attack is proportionate. This measured act of self-defense is justified because the good of life merits protection and one can rightfully protect this good by appropriate acts so long as choosing to do them does not of necessity require setting one's will on the foreseen evil. The reasoning involved in this argument closely parallels that expressed in the first, second, and fourth conditions of the principle of double effect. The difficulty is in reconciling the argument given by Aquinas with the third condition, a subject to which we shall return.

It is noteworthy that in the subsequent historical development of the principle of double effect the majority of Roman Catholic moral theologians did *not* use it to justify killing in self-defense. Although Cajetan did so (Mangan, p. 52), most writers in subsequent centuries, beginning with Molina in the sixteenth and continuing until very recent times, argued that one could rightfully intend to kill an aggressor, even one materially and not formally unjust, in self-defense (ibid., p. 54, n. 25).

During the sixteenth and seventeenth centuries the principle was applied typically to cases involving the loss of semen brought on by sexual desires that unavoidably arose while a person was engaging in otherwise unobjectionable pursuits and to cases involving the causing of scandal. The theologians of the era used the principle to justify, for example, a medical student's research even if it should cause him to ejaculate and to condone a young maiden's stroll through the village even if, modestly attired as she might be according to the customs of the day, she foresaw that her promenade would provoke lustful desires on the part of village rubes (Ghoos, pp. 40–47). It was also applied to cases involving the killing of the innocent, particularly in war, and to exposing oneself to mortal danger for a good cause (Mangan, pp. 54–55).

Eventually the principle was given wider application, particularly during the nineteenth century by Jean Pierre Gury in his often reedited and influential *Compendium Theologiae Moralis* (ibid., pp. 59–60; Connell, p. 1021), until it came to embrace almost the entire field of moral theology. It is invoked today to cope with many problems of crucial importance in bioethics, for instance, abortion, euthanasia, suicide, the hazards of medical experimentation, and the ethics of procreation (contraception and sterilization). The way the principle is commonly understood and applied can be well illustrated by taking the example of abortion. Until the relatively recent past Roman Catholic writers used the principle to justify abortions in ectopic pregnancies and in hysterectomies to remove a cancerous womb. In such instances the intention and the immediate effect of the act is the removal of a pathological organ (the fallopian tubes or uterus) that happens to contain a fetus. The removal of the fetus and its subsequent death are neither directly willed nor directly done (Bouscaren, pp. 147–155). As commonly understood, however, the principle cannot justify procedures involving "direct" abortions, even if such procedures are necessary in order to save the mother from imminent death, not even if it is morally certain that both mother and fetus will die if nothing is done (ibid., pp. 3–16; Grisez, *Abortion*, p. 179). In the latter type of case, it is commonly held, the principle is inapplicable inasmuch as the third condition is not capable of being fulfilled.

Contemporary debates on the principle

There seem to be inconsistencies in the application of the principle, e.g., allowing abortifacient procedures when organs are pathological while refusing them if no pathology exists although death of both mother and fetus is imminent. Partly for this reason but more basically because of serious difficulties in understanding the significance of its requirements, in particular conditions 2 and 3, there has recently been a lively discussion of the principle by many authors interested in bioethical problems.

One group of writers focuses on the second condition: that the evil be permitted and only indirectly intended whereas the good alone is to be directly intended. Williams (pp. 280–291) and Fletcher (1973) argue that this distinction is dishonest at worst and meaningless at best, although central to the reasoning of those who see a vast moral difference between killing a person for merciful reasons and allowing a dying person to die by refusing to continue the use of life-prolonging but extraordinary and morally optional techniques. For Williams and Fletcher the distinction is simply specious and an instance of moral quibbling inasmuch as the result is the same: A person dies. Defenders of this distinction, crucial to condition 2, have responded by pointing out that this criticism is predicated upon a consequentialist or utilitarian understanding of human action and that a significant difference is at stake inasmuch as the result is obtained in radically different ways: In the one instance the person dies from some underlying disease or injury that it is no longer reasonable to combat, whereas in the other instance the person dies from an act truthfully describable as killing (Dyck, pp. 100–105).

Possibly Williams and Fletcher have confused the distinction between directly intending evil and only indirectly intending or permitting evil with the distinction between acting and refraining. The distinctions are not the same. If a person refrains from acting to save another's life when he can and ought to do so (e.g., if he fails to throw a drowning person a life preserver when he can easily do so), he is morally responsible for that person's death. His act of refraining from doing what he can and ought to do is the equivalent of doing the evil deed with direct intent.

With Williams and Fletcher, Aune claims that the distinction between intending and foreseeing, also central to condition 2, is indefensible (Aune). This claim seems unfounded. For instance, a dentist clearly foresees that in extracting an impacted wisdom tooth he will inflict pain on his patient. Yet he is surely not intending the pain that he will inflict. Were he properly intending the infliction of pain, his act could hardly be said to be one of proper dentistry (Grisez, *Abortion*, pp. 327–329). In fact, if a dentist does properly intend pain, his patients would be well advised to go elsewhere.

Somewhat similarly to Aune, Foot, while admitting the validity in theory between intending and permitting/foreseeing, contends that it is quite subsidiary to the distinction between positive and negative duties and argues that it is not reasonable to claim that a person can be held to be not morally responsible for the unintended yet foreseeable effects of his actions (Foot). Thus, should a shopkeeper knowingly sell poisons to someone who he knows will use them to kill others, he cannot avoid responsibility for the subsequent deaths. Although Foot's comments have some validity, they do not substantively affect the principle. The act, unwarrantable in the example given, would not be justifiable on grounds of the principle of double effect insofar as its fourth requirement would not be capable of fulfillment.

Another group of critics (Knauer; Van der Marck; Van der Poel; Janssens) take as their point of departure the third condition of the principle. They argue that in its traditional understanding this third condition has been interpreted too materialistically. They maintain that in instances such as killing in self-defense the evil effect (the death of the assailant) is physically direct and is the means to the good effect; and in making this argument some, for instance Knauer (pp. 138–141) and Janssens (pp. 116–133), explicitly appeal to the teaching of Aquinas on killing in self-defense. Although these authors differ markedly among themselves, they agree in making a crucial distinction between physical (ontic, premoral, or nonmoral) evil and moral evil. They then concur in holding that one can rightfully intend the physical (ontic, premoral, or nonmoral) evil, e.g., death, sterilization, mutilation, so long as a proportionate good or commensurate good will be served.

On this interpretation of condition 3, condition 2 becomes almost superfluous. In fact, some of these authors, e.g., Van der Marck (pp. 56–58) and Van der Poel (pp. 207–210) see no real meaning in the distinction between the directly and indirectly intended. Knauer retains the terminology of direct–indirect, but he so modifies its meaning that, as critics note, the distinction becomes inoperative and its function is supplanted by the notion of the commensurate good (Grisez, *Abortion*, p. 331; McCormick, pp. 11–12).

The general thrust of this critique has been accepted by others (Curran; McCormick) although each objects to specific features in the arguments advanced by Knauer, Van der Poel, and Van der Marck. McCormick believes that their arguments converge with others recently advanced and that they can be summarized by saying that it is morally permissible to intend directly an evil, in its physical (ontic, premoral, or nonmoral) sense, in itself but not for itself. This is true as long as there is a proportionate reason. If there is no proportionate reason, then the evil is intended not only in itself but for itself and thus becomes morally, and not merely physically, evil. McCormick himself admits the validity of the direct–indirect distinction but holds that it is of limited value and simply means that a greater proportionate reason is necessary if one is directly to intend the evil and not merely to permit it or intend it only indirectly (McCormick, pp. 68–106).

A third group of authors, preeminently Grisez and Ramsey, basically accept the principle of double effect. They reject the interpretation given above inasmuch as they believe that it is not only in effect a repudiation of the principle but an acceptance of a consequentialism (cf. also May). With the above group, however, Grisez and Ramsey agree that the third condition of the principle was too restrictively understood in the past. They grant that the evil effected may be direct physically but argue that it is permissible only if it is an unavoidable concomitant or aspect of an act that in itself is immediately targeted on the good (Grisez, *Abortion*, pp. 333–346; "Toward a Consistent," pp. 87–91; Ramsey, pp. 211–222). Although these authors also disagree among themselves, they agree that any evil effected through human acts can be morally justifiable only if it is "indirect," that is, a partial concomitant of an act that is directly intended and targeted on the achievement of the good. In their judgment, a human person makes himself to be morally evil if he chooses to do deeds in which he of necessity makes evil to be the precise end of his will, that is, of his person.

The argument over the meaning of the principle, and in particular of conditions 2 and 3, continues. It centers on some of the most critical issues in ethics and moral theory. Those who accept the principle, even as reinterpreted by the third group of authors discussed, hold that we are very limited in the evil that we may rightfully cause in our pursuit of the good and consequently are inclined to be very much opposed to abortion, artificial insemination, in vitro fertilization, nontherapeutic experimentation on children, etc. On the other hand, those who reject the principle or reinterpret it along the lines of the second group of authors discussed are more ready to accept procedures that give promise of securing proportionately good results despite the harm entailed.

WILLIAM E. MAY

[*Directly related is* ACTING AND REFRAINING. *This article will find application in the entries* ABORTION, *articles on* ROMAN CATHOLIC PERSPECTIVES *and* CONTEMPORARY DEBATE IN PHILOSOPHICAL AND RELIGIOUS ETHICS; CONTRACEPTION; DEATH AND DYING: EUTHANASIA AND SUSTAINING LIFE; HUMAN EXPERIMENTATION; INFANTS, *articles on* ETHICAL PERSPECTIVES ON THE CARE OF INFANTS, *and* INFANTICIDE: A PHILOSOPHICAL PERSPECTIVE; STERILIZATION, *article on* ETHICAL ASPECTS; *and* SUICIDE. *Other relevant material may be found under* ETHICS, *articles on* TELEOLOGICAL THEORIES *and* UTILITARIANISM; RELIGIOUS DIRECTIVES IN MEDICAL ETHICS, *article on* ROMAN CATHOLIC DIRECTIVES; *and* ROMAN CATHOLICISM.]

BIBLIOGRAPHY

AUNE, BRUCE. "Intention and Foresight." *Journal of Philosophy* 63 (1966): 652–654. Abstract of a symposium paper.

BOUSCAREN, TIMOTHY LINCOLN. *The Ethics of Ectopic*

Operations. 2d ed., rev. Milwaukee: Bruce Publishing Co., 1944.
CALLAHAN, DANIEL J. *Abortion: Law, Choice, and Morality.* New York: Macmillan Co., 1970.
CONNELL, F. J. "Double Effect, Principle of." *New Catholic Encyclopedia.* New York: McGraw-Hill Book Co., 1967, vol. 4, pp. 1020–1022.
CONWAY, WILLIAM. "The Act of Two Effects." *Irish Theological Quarterly* 18 (1951): 125–137.
CURRAN, CHARLES E. "The Principle of Double Effect." *On-going Revision: Studies in Moral Theology.* Notre Dame, Ind.: Fides Publishers, 1975, pp. 173–209.
DYCK, ARTHUR J. "An Alternative to the Ethics of Euthanasia." *To Live and to Die: When, Why, and How.* Edited by Robert H. Williams. New York: Springer-Verlag, 1973, pp. 98–112.
FARRAHER, JOSEPH J. "Current Theology: Notes on Moral Theology: Suicide and Moral Principles." *Theological Studies* 24 (1963): 69–79.
FLETCHER, JOSEPH FRANCIS. "Ethics and Euthanasia." *To Live and to Die: When, Why, and How.* Edited by Robert H. Williams. New York: Springer-Verlag, 1973, pp. 113–122.
———. *Morals and Medicine: The Moral Problems of the Patient's Right to Know the Truth, Contraception, Artificial Insemination, Sterilization, Euthanasia.* Foreword by Karl Menninger. Princeton: Princeton University Press, 1954.
FOOT, PHILIPPA. "The Problem of Abortion and the Doctrine of the Double Effect." *Moral Problems: A Collection of Philosophical Essays.* Edited by James Rachels. New York: Harper & Row, 1971, pp. 29–41. 2d ed. 1975, pp. 59–70. Also *Oxford Review* 5 (1967): 5–15.
GHOOS, J. "L'acte à double effet: Étude de théologie positive." *Ephemerides Theologicae Lovanienses* 27 (1951): 30–52.
GRISEZ, GERMAIN GABRIEL. *Abortion: The Myths, the Realities, and the Arguments.* New York: Corpus Books, 1970, pp. 179, 321–346.
———. "Toward a Consistent Natural-Law Ethics of Killing." *American Journal of Jurisprudence* 15 (1970): 64–96.
KELLY, GERALD. "Current Theology: Notes on Moral Theology, 1951: General and Pastoral." *Theological Studies* 13 (1952): 59–63, especially pp. 59–61.
KNAUER, PETER. "The Hermeneutic Function of the Principle of Double Effect." *Natural Law Forum* 12 (1967): 132–162.
JANSSENS, LOUIS. "Ontic Evil and Moral Evil." *Louvain Studies* 4 (1972): 115–156.
MCCORMICK, RICHARD A. *Ambiguity in Moral Choice.* The 1973 Père Marquette Theology Lecture. Milwaukee: Marquette University Press, 1973.
MANGAN, JOSEPH T. "An Historical Analysis of the Principle of Double Effect." *Theological Studies* 10 (1949): 41–61.
MAY, WILLIAM E. "Becoming Human in and through Our Deeds." *Becoming Human: An Invitation to Christian Ethics.* Dayton, Ohio: Pflaum Publishing, 1975, chap. 4, pp. 79–112.
RAMSEY, PAUL. "Abortion: A Review Article." *Thomist* 37 (1973): 174–226.
VAN DER MARCK, WILLIAM H. M. *Toward a Christian Ethics: A Renewal in Moral Theology.* Translated by Denis J. Barrett. Westminster, Md.: Newman Press, 1967.
VAN DER POEL, CORNELIUS J. "The Principle of Double Effect." *Absolutes in Moral Theology?* Edited by Charles E. Curran. Washington: Corpus Books, 1968, pp. 186–210.
WILLIAMS, GRANVILLE LLEWELYN. *The Sanctity of Life and the Criminal Law.* New York: Alfred A. Knopf, 1957.

DRUG INDUSTRY AND MEDICINE

Medical practice in the industrially advanced countries is heavily oriented toward drug treatment, necessitating a more or less intimate tie between physician and drug manufacturer. Despite the potential conflict between the manufacturer's pursuit of profit and the physician's duty to his patient, until the early 1960s the interests of the three parties were considered to be in harmony—industry's desire for new products coinciding with that of the physician and patient for more effective therapy. However, the revelations of the U.S. Senate and House committees chaired by Estes Kefauver, Gaylord Nelson, Edward Kennedy, and others have cast doubt upon this harmony.

Development and use of drugs

Does industry's economic need for new and patentable drugs really promote the health of the public? If not, what can be done about it? The 1969 *Final Report* of the DHEW (U.S. Department of Health, Education, and Welfare) Task Force on Prescription Drugs defined "rational prescribing" as "the appropriate selection of a drug—the right drug for the right person, in the right amounts at the right time" (U.S. DHEW, p. 21). The relations between the drug industry and the medical profession can best be analyzed in terms of whether industry's demand for new products distorts this ideal of "rational prescribing."

Pharmacological education. After acknowledging that the American physician, who prescribes $25,000 worth of drugs every year, often does so in an irrational way, the *Final Report* states that "lack of knowledge and sophistication in the proper use of drugs is perhaps the greatest deficiency of the average physician today" (ibid., p. 23; Council on Economic Priorities, p. 27). Paradoxically, the physician who spends his professional life prescribing drugs knows relatively little about them.

Many authorities have ascribed this deficiency to the limited and inadequate pharmacological education of physicians. The medical school graduate lacks the background to examine critically the claims of industry for its products—contrary to the true aim of medical education and to medical tradition. This happens because initiative in drug development has passed from the medical profession and the medical schools to the pharmaceutical industry, distinguishing modern medicine from all previous eras when the pharmacist was subordinate to the physician. The change began in the 1880s with the introduction of "proprietary remedies"—medicines and medicinal mixtures promoted to the public and the profession for use in specific diseases or conditions (Coulter, pp. 417 ff.). The advertising of "proprietaries" in medical journals was initially considered unethical, undermining the physician's traditional responsibility for knowledge of his instruments of cure, but the profession used "proprietaries" more and more, and the de facto situation was legitimized in 1903 by an appropriate change in the Code of Ethics of the American Medical Association (AMA).

A medical journal in 1905 contrasted the resulting situation with "forty years ago [when] the physician who prescribed a medicine without some knowledge of its composition and effects was considered to have violated the Code of Ethics and regarded by his professional brethren with suspicion. There has been a great change" ("Action of Gases on Bacteria").

The change has meant that today medical schools and physicians simply perform the pharmacological tasks assigned them by the drug industry. Since drug research is often essential to professorial advancement, the drug industry underwriting such research has a measure of control over the pharmacological thinking of physicians and professors who might otherwise offer a counterweight to industry views. The company providing the drug for investigation and the financing often influences the design of the study—which may be used more to advertise the given product than to yield scientific data (Goddard). Some pharmacological research in medical schools is government funded, but its role is often merely one of disproving (at public expense) the claims of some company for a drug upon which it has already earned a profit (University Group Diabetes Program, *Diabetes* XIX [1970], p. 467). The involvement of industry in pharmacological education is hence a source of serious ethical conflict between the profit motive and the health needs of the public.

Informed consent. The competition among drug companies leads to pressure on physicians conducting clinical trials. Henry K. Beecher was one of the first, in 1966, to call attention to the consequence—failure, in many cases, to obtain patient consent to excessively dangerous experiments (Beecher). Some of the more striking examples cited by him were: the withholding of appropriate treatment from 109 servicemen with streptococcal respiratory infections, giving rise to two cases of acute rheumatic fever and one of acute nephritis; the administration of a drug to fifty healthy mentally retarded children and juvenile delinquents to ascertain if the drug caused liver damage, leading to a high incidence of hepatic dysfunction in the group after four weeks of the trial; and the administration, to forty-one randomly selected patients, of a drug known to cause aplastic anemia, so that many developed toxic bone-marrow depression. Beecher gave fifteen other instances and stated that examination of 100 consecutive human studies published in a leading journal indicated that twelve appeared to have been conducted without obtaining patient consent.

The problem of informed consent becomes acute in the case of children, mental defectives, and prisoners (U.S. Congress, Senate, Select Committee on Small Business, pt. 14, pp. 5689–5702). Prisoners are employed extensively by the so-called proof mills often used by drug companies to generate evidence favorable to their product (ibid., pp. 5702 ff.).

Duration of drug trials. The appropriate duration of a drug trial is another ethical issue related to industry's desire to market its products rapidly. Adverse reactions and lack of efficacy may take years to become manifest. The antidiabetic drug Orinase (tolbutamide) had been available since 1956 and was grossing the manufacturer more than $50 million a year when a 1970 study by a group of medical schools demonstrated it to be a cause of cardiovascular disease (University Group Diabetes Program, 1970, 1971). But this became evident only after a patient had taken the drug for three years. The first oral contraceptive was approved for sale in 1957, but only in the late 1960s did physicians become convinced that substances in them caused blood clotting.

Drug companies and their clinical investigators have refused to do follow-up studies of subjects of drug trials when subsequent research has shown such drugs to pose a previously unsuspected threat to health (U.S. General Accounting Office, pp. 37–49).

Reporting and labeling. Another series of ethical issues arises at the reporting stage of the clinical trial. It has been stated that one pharmaceutical company suppressed the report of one of its investigators showing its antidiabetic drug, Diabinese, to cause "side effects" (central nervous system damage, jaundice) in twenty-seven percent of the test subjects (Mintz, 1967, pp. 17–18). The U.S. General Accounting Office (GAO) reported in 1973 that internal controls in drug trials were often inadequate, with physicians involved in the human phase not being alerted in time to unfavorable results of animal studies (U.S. GAO, 1973, pp. 33–36). The U.S. Food and Drug Administration (FDA) has charged a number of drug companies and their officers with concealment and even falsification of experimental data (Silverman and Lee, pp. 64–67).

The same issues arise in connection with the label and package insert for a new medicine. Former FDA Commissioner James Goddard wrote, "This is extremely important to the pharmaceutical companies because this is where the money is. They have to convince the physician that this is the best possible product in its field. FDA, on the other hand, must make sure that the advertising claims remain within the bounds of scientific evidence" (Goddard). One company's package insert for Diabinese began: "Side effects are generally of a transient and non-serious character" (Mintz, 1967, p. 18). In 1971 the U.S. Department of Justice charged a drug company and two of its officers with mislabeling a product used to diagnose for hepatitis—failing to report that it was only thirty-five percent effective ("Abbott Firm Charged"). Still another drug firm, with $70 million in annual sales of its estrogen compound Premarin, used to treat menopause, failed to note on its package insert that the substance had been implicated as a cause of uterine cancer; a company vice-president, moreover, distributed to physicians a letter deriding the possibility of such a relationship ("FDA Criticizes Firm's Letter"; Mintz, " 'Misleading' Drug Promotion Letter?"; Mintz, "Reversing a Stand.")

Advertising and marketing of drugs

Drug industry pressure on physicians and the ethical conflicts associated with this become most acute in the advertising and marketing phase. With 7,000 single entities and 3,000 compounds available on prescription, not to mention an estimated 100,000 additional "over-the-counter" (nonprescription) preparations, a heavy investment in promotion is needed to bring any one of them to the physician's attention. The prescription drug industry spends more than $1 billion a year for promotion—about $5,000 for each of the 200,000 prescribing physicians in the United States (a sum larger than the 1970–1971 teaching budgets of all U.S. medical schools).

Companies manufacturing "over-the-counter" drugs spend another $1 billion or more in advertising. In 1972 $24 million was spent merely to promote Bayer Aspirin, and $11 million to promote Bufferin (Silverman and Lee, pp. 214–215). Since the consumer is rarely able to judge the truth of these claims, he can be easily misled.

About 25,000 "detail men" visit physicians to dispense information on their company's latest product. Two billion free samples are distributed annually to physicians, pharmacists, and medical students. Free color television sets, freezers, and other objects of value are presented to physicians as inducements to prescribe "correctly." All-expense-paid seminars are organized in Bermuda and other exotic locations to inform physicians of the latest drug-industry wonders (ibid., pp. 55–57).

All evidence indicates that these stimuli affect prescribing patterns. It is humanly impossible for the physician to keep abreast of this torrent of products. A professor of pharmacology testified at the Nelson hearings that he could barely keep up with the new offerings in his subspecialty, let alone the whole field (U.S. Congress, Senate, Select Committee on Small Business, pt. 2, p. 461).

The problem of obtaining reliable information on drugs is compounded by the existence of about 5,000 medical journals in the world, publishing upwards of 100,000 articles a year (May). This huge volume of information serves the interests primarily of advertisers and publishers. Companies use them as advertising vehicles, overwhelming the physician not only with the advertisements themselves but also with article after article on the product that is being heavily promoted at the time. Or they confuse legitimate scientific inquiry by placing articles designed to reflect industry views.

In 1970 four of the five trade journals with the largest advertising volume were medical—with *Medical Economics* ($11 million gross) in the lead (Silverman and Lee, p. 56). An authoritative observer estimated that a medical journal such as this earns for its owners a profit of forty percent of the income from advertisements and subscriptions, since no payment is made for contributions (U.S. Congress, Senate, Select Committee on Small Business, pt. 10, p. 3948).

Since profitability depends directly on the volume of advertisements, the commercial pressure on editorial boards is immense. A former editor of the *Journal of the American Medical Association* testified in 1969 that the "lack of ethical standards" within the AMA was a major cause of the misleading and tendentious advertising of drugs (ibid., pt. 14, p. 5723). In 1972 the AMA abolished its Council on Drugs, which had apparently exerted a modest restraining influence (Silverman and Lee, p. 292). A critic reported: "The AMA could no longer keep both an effective Council on Drugs and the support of drug advertisers. One of them had to go" (ibid.).

Overprescribing of drugs

Promotional pressure leads to overprescribing by physicians. A. Dale Console, former medical director of E. R. Squibb and Sons, estimated that only half the U.S. consumer's drug bill is medically justified (U.S. Congress, Senate, Select Committee on Small Business, pt. 11, p. 4491). Louis Lasagna wrote that physicians prescribe antibiotics for sixty percent of their patients with colds, even though these medicines are ineffective against the common cold (Silverman and Lee, p. 290). Harry Dowling, a specialist in infectious diseases and former chairman of the AMA Council on Drugs, has stated that the average patient needs an antibiotic once every five or ten years, while enough is manufactured every year to treat two illnesses of average duration for every man, woman, and child in the United States (Kunin, Tupasi, and Craig, p. 556). Between 1960 and 1970 the production of antibiotics increased by 320 percent. The average hospital patient receives from six to ten different medicines during his stay (Gardner and Cluff, p. 83).

Overprescribing has caused an epidemic of "adverse reactions" and doctor-induced (iatrogenic) disease. From ten to eighteen percent of hospital patients manifest an "adverse reaction," with a five percent case fatality (ibid., pp. 77–78). The cost of treating such iatrogenic illness is estimated at $4.5 billion a year in the United States; and while there is still a need for adequate data, the death rate from such disease is estimated at between 30,000 and 130,000 per year (the death rate of U.S. servicemen during the Second World War was 80,000 per year) (Silverman and Lee, p. 265; U.S. Congress, Senate, Committee on Labor and Public Welfare, pt. 5, pp. 1545–1546).

Significant morbidity and mortality have been associated with the use of tolbutamide in diabetes, chloramphenicol in infectious diseases, isoproternol as an inhalant in asthma (in the United Kingdom), isoniazid in tuberculosis, and potassium-thiazide diuretics (still commonly prescribed in pregnancy) (Davidson, p. 853; Silverman and Lee, pp. 59–61, 249–250). Chloroquine sulphate taken for rheumatoid arthritis has caused blindness, neomycin used for diaper rash and dihydrostreptomycin have been associated with deafness (a recent survey of Americans found a higher than expected incidence of deafness in those born since the onset of the antibiotic age) (Kagan, p. 306). The immunosuppressive drugs used to overcome "rejection" during organ transplants can cause cancer; a number of drugs, including thalidomide, oral contraceptives, and tranquilizers have been linked with stillbirths and fetal deformation (Janerich, Piper and Glebatis; Yalom, Green and Fisk; Silverman and Lee, p. 65). New drug-resistant bacterial strains, especially the "gram-negative" variety, have emerged and are estimated to cause an additional unnecessary 100,000 deaths a year, particularly in hospitals (Kagan, p. 306; Kunin, Tupasi, and Craig, p. 556).

One particularly tragic example concerns the development of cervical cancer in the daughters of women who had received diethylstilbestrol during their pregnancies twenty years earlier (Herbst, Ulfelder, and Poskanzer).

This leads to the question of the long-term impact of overprescribing and incorrect prescribing. Pierce Gardner and Leighton Cluff have noted that little attention is being paid to the "delayed untoward effects of drugs," in particular their role in the later development of cancer and degenerative diseases (Gardner and Cluff, p. 85). Since "adverse reactions" to drugs often take the form of heart and circulatory difficulties, arthritis, or cancer, it seems reasonable to associate the ever increasing incidence of these three chronic diseases with the observed overprescribing of today's medicine. The conversion of iatrogenic disease into chronic disease deserves greater attention than it has received.

The increasing drug dependency of American culture and, especially, the involvement in illicit forms of drug use are in part generated and maintained by the enthusiastic prescribing of the medical profession.

The physician himself is likely to be affected adversely, since these powerful substances with a general and diffuse impact on the organism lead him to neglect diagnosis (Lasagna). A. Dale Console testified, "Advertising and promotion efforts encourage the doctor to believe there is an easy way to practice medicine. They offer larger and larger shotguns which make pinpoint diagnosis, or for that matter any diagnosis at all, a pedantic exercise and a troublesome inconvenience that only the less well-informed academic bothers with" (U.S. Congress, Senate, Select Committee on Small Business, pt. 11, p. 4486). The "fixed-combination" antibiotics, which held forty percent of the drug market until banned by the FDA in 1968, were prime culprits in leading physicians to bypass diagnosis, but all the "broad spectrum" drugs available for use today have the same effect (ibid., pp. 4482–4483; Silverman and Lee, pp. 109–110).

U.S. Food and Drug Administration

Prior to the adoption of the Kefauver–Harris Drug Amendments of 1962 the U.S. Food and Drug Administration bore responsibility only for drug safety. The manufacturer reported the results of his tests, and the FDA reached a decision.

As a result of the thalidomide disaster (introduction into general use of a sedative and hypnotic drug that was found to cause malformation of infants born to mothers using it during pregnancy) and the Kefauver hearings, the Food, Drug, and Cosmetic Act was strengthened by three new provisions in 1962: (1) The FDA must be kept informed of the progress of tests of investigational drugs (i.e., drugs not yet approved by the FDA); clinical investigations must be preceded by sufficient preclinical studies to ensure reasonable safety; clinical investigations must be planned and executed by qualified investigators, and their names must be provided to the FDA. (2) On the basis of material submitted by the manufacturer the FDA decides whether or not the drug is effective in the way represented. (3) The FDA bears responsibility for ensuring that prescription drug advertising contains adequate information on effectiveness, side effects, and counterindications for the use of the drug advertised.

While observers credit this agency with improved regulatory performance since 1962, the ethical problems outlined above—relating to the pharmacological competence and human weakness of the prescribing physician, on one hand, and the commercial power of the drug industry on the other—are probably not amenable to substantial improvement as a result of any action that could be taken by the FDA.

Conclusion

The relationship between the medical profession and the drug industry is "unhealthy, and in many ways corrupt," as A. Dale Console expressed it, but at the same time economically beneficial to both parties (U.S. Congress, Senate, Select Committee on Small Business, pt. 11, pp. 4480–4481). The physician acts de facto as promoter and distributor of the products of industry, while these substances, in turn, enable him to process a large number of patients during the working day (the average length of an office visit in American practice is ten to twelve minutes).

To the extent that he acquiesces in this relationship the physician abandons his traditional role of master of his instruments of cure—and even, to some extent, that of diagnostician. Since his professional status is derived from the presumed possession of this knowledge, this status has been impaired by the medical profession's over-close relationship with the drug industry (Talalay, pp. 245–246).

The consequence for the American public is apparently about 200,000 avoidable deaths every year (the total number of deaths in the United States was 1.9 million in 1972), together with the observed increase in chronic disease. Some might ask if these factors do not actually outweigh the profession's contribution to the public health (Illich).

Two questions still await a satisfactory answer. Is a continuing stream of new pharmaceutical products beneficial to the public health, given the medical profession's demonstrated inability to understand and use them properly? Does the existing system of informing physicians about drugs promote adoption of the most scientific modes of treatment, or are the interests of a scientific medicine overwhelmed by those of the drug industry?

Undoubtedly public health would improve and many ethical problems would disappear if the volume of drugs available for use were vastly reduced, following the prescription given more than a century ago by Dr. Oliver Wendell Holmes. But the likelihood of this is minimal, as is that of a more effective governmental or social control over the manufacture and use of medicinal drugs. On the other hand, it is difficult to see why the public must remain passive while the medical profession and the drug industry work out their relationship at society's expense.

HARRIS L. COULTER

[*For further discussion of topics mentioned in this article, see the entries:* HUMAN EXPERIMENTATION; *and* PHARMACY. *Also directly related are the entries* PRISONERS, *article on* PRISONER EXPERIMENTATION; RESEARCH POLICY, BIOMEDICAL; *and* THERAPEUTIC RELATIONSHIP. *See also:* ADVERTISING BY MEDICAL PROFESSIONALS; *and* RISK. *See* APPENDIX, SECTION I, MEDICAL ETHICS: STATEMENTS OF POLICY DEFINITIONS AND RULES (BRITISH MEDICAL ASSOCIATION); *and* SECTION IV, AMERICAN PHARMACEUTICAL ASSOCIATION.]

BIBLIOGRAPHY

"Abbott Firm Charged in Drug Labeling Case." *Washington Post*, 8 May 1971, p. A3.

"Action of Gases on Bacteria." *Ohio State Medical Journal* 1 (1905): 82.

BEECHER, HENRY K. "Ethics and Clinical Research." *New England Journal of Medicine* 274 (1966): 1354–1360. The first critique by a physician of unethical practices in drug trials.

COULTER, HARRIS LIVERMORE. *Divided Legacy: A History of the Schism in Medical Thought.* Vol. 3: *Science and Ethics in American Medicine: 1800–1914.* Washington: Wehawken Book Co., 1973. A history of medical ethics and of the nineteenth-century relations between the medical profession and the drug industry.

Council on Economic Priorities. *In Whose Hands? Safety, Efficacy, and Research Productivity in the Pharmaceutical Industry.* Economic Priorities Report 4, no. 4–5 (1973). Critical examination of drug-industry claims by a consumer organization.

DAVIDSON, JOHN K. "The FDA and Hypoglycemic Drugs." *Journal of the American Medical Association* 232 (1976): 853–855. Criticizes the profession's use of tolbutamide in diabetes.

DOWLING, HENRY FILLMORE. *Medicines for Man: The Development, Regulation, and Use of Prescription Drugs.* New York: Knopf, 1970. A general discussion of the relations between the medical profession and the drug industry; less critical and insightful than the similar book by Silverman and Lee.

"FDA Criticizes Firm's Letter on Estrogen Drugs." *Washington Post*, 14 January 1976, p. A3.

GARDNER, PIERCE, and CLUFF, LEIGHTON E. "The Epidemiology of Adverse Drug Reactions: A Review and Perspective." *Johns Hopkins Medical Journal* 126 (1970): 77–87. One of the early warnings about the iatrogenic disease epidemic.

GODDARD, JAMES A. "The Drug Establishment." *Esquire*, March 1969, pp. 117–121, 152, 154. A former FDA Commissioner takes a skeptical look at the drug industry.

HARRIS, RICHARD. *The Real Voice.* New York: Macmillan Co., 1964. A scholarly account of the Kefauver hearings, but short, unannotated, and unindexed.

HERBST, ARTHUR L.; ULFELDER, HOWARD; and POSKANZER, DAVID C. "Adenocarcinoma of the Vagina: Association of Maternal Stilbestrol Therapy with Tumor Appearance in Young Women." *New England Journal of Medicine* 284 (1971): 878–881.

ILLICH, IVAN. *Medical Nemesis: The Expropriation of Health.* New York: Pantheon, 1976. A highly articulate and heavily documented polemic against the medical profession: "The medical establishment has become a major threat to health."

JANERICH, DWIGHT; PIPER, JOYCE M.; and GLEBATIS, DONNA M. "Oral Contraceptives and Congenital Limb-Reduction Defects." *New England Journal of Medicine* 291 (1974): 697–700.

KAGAN, BENJAMIN M.; FANNIN, SHIRLEY L.; and BARDIE, FELIX. "Spotlight on Antimicrobial Agents—1973." *Journal of the American Medical Association* 226 (1973): 306–310. A critique of the prescribing of antibiotics.

KUNIN, CALVIN M.; TUPASI, THELMA; and CRAIG, WILLIAM A. "Use of Antibiotics: A Brief Exposition of the Problem and Some Tentative Solutions." *Annals of Internal Medicine* 79 (1973): 555–560. Another critique of the prescribing of antibiotics.

LASAGNA, LOUIS. "The Pharmaceutical Revolution: Its Impact on Science and Society." *Science* 166 (1969): 1227–1233. Points out that medical care has suffered in consequence of the availability of too many new medicines.

LOWBURY, EDWARD JOSEPH LISTER, and AYLIFFE, G. A. J. *Drug Resistance in Antimicrobial Therapy.* Springfield, Ill.: Thomas, 1974. On the prescribing of antibiotics and the consequent emergence of new and resistant disease strains.

MAY, CHARLES D. "Selling Drugs by 'Educating' Physicians." *Journal of Medical Education* 36 (1961): 1–23. Also in United States, Congress, Senate, Select Committee, *Competitive Problems*, pt. 10, pp. 3938–3957. Calls attention to the misleading claims made by drug manufacturers.

MINTZ, MORTON. *By Prescription Only.* 2d ed. rev. Boston: Beacon Press, 1967. Issued originally in 1965 as *The Therapeutic Nightmare.* A classic study of the Kefauver hearings and the drug industry.

———. "'Misleading' Drug Promotion Letter?" *Washington Post*, 9 January 1976, p. B1.

———. "Reversing a Stand on the 'Pill'." *Washington Post*, 12 January 1976, p. B2.

Pharmaceutical Manufacturers Association. *Annual Survey Report, 1973–1974.* Washington: 1975. Gives sales and research statistics of PMA members.

SILVERMAN, MILTON, and LEE, PHILIP R. *Pills, Profits, and Politics.* Berkeley: University of California Press, 1974. A comprehensive account of the drug industry and medicine—scholarly, readable, and objective.

"Status of Problem of Usage of Tolbutamide—Preliminary Statements: FDA Statement, Friday, May 22, 1970." *Diabetes* 19 (1970): 467.

TALALAY, PAUL, ed. *Drugs in Our Society.* Baltimore: Johns Hopkins Press, 1964. One of the earlier examinations of the drug industry and medicine. Some of the contributions are very original, objective, and enlightening.

United States, Congress, Senate, Committee on Labor and Public Welfare, Subcommittee on Health. *Examination of the Pharmaceutical Industry, 1973–1974.* 93d Cong., 1st & 2d sess. Hearings on S. 3441 and S. 966. 6 pts. Washington: Government Printing Office, 1973–1974. The Kennedy hearings.

———, Congress, Senate, Committee on the Judiciary, Subcommittee on Antitrust and Monopoly. *Administered Prices in the Drug Industry.* 86th Cong., 2d sess. Pursuant to S. Res. 238. Washington: Government Printing Office, 1960–1961, pts. 14–26. *Drug Industry Antitrust Act.* 87th Cong., 1st & 2d sess. Pursuant to S. Res. 52 on S. 1552. 7 pts. Washington: Government Printing Office, 1961–1962. The Kefauver hearings, which first brought these issues to public attention and led to adoption of the Kefauver-Harris Amendments to the U.S. Food, Drug and Cosmetic Act.

———, Congress, Senate, Select Committee on Small Business, Subcommittee on Monopoly. *Competitive Problems in the Drug Industry.* 93d Cong., 2d sess. 28 pts. Washington: Government Printing Office, 1967–1975. The Nelson hearings.

———, Department of Health, Education, and Welfare, Task Force on Prescription Drugs. *Final Report: February 7, 1969.* Washington: Government Printing Office, 1969. A highly critical account of the prescribing practices of U.S. physicians and of drug-industry influence on the medical profession, compiled by a group of experts.

———, General Accounting Office; and United States, Department of Health, Education, and Welfare, Food and Drug Administration. *Supervision over Investigational Use of Selected Drugs.* Report to the Subcommittee on Reorganization, Research, and International Organizations, Committee on Government Operations, United States Senate. Washington: Comptroller General of the United States, 1973.

University Group Diabetes Program. "A Study of the Effects of Hypoglycemic Agents on Vascular Complications in Patients with Adult-onset Diabetes: I. Design, Methods, and Baseline Results." *Diabetes* 19 (1970): 747–783. "II. Mortality Results." *Diabetes* 19 (1970): 789–830. "IV. A Preliminary Report on Phenformin Results." *Journal of the American Medical Association* 217 (1971): 777–784. A ten-year, government-financed study conducted in 12 university medical schools and covering over 800 patients, most of whom were followed for eight years, this showed that tolbutamide—used in the treatment of about 1.5 million diabetics in the U.S. alone—is attended with a higher incidence of cardiovascular complications and mortality than when patients are treated with insulin or by dietary adjustment.

YALOM, IRVIN D.; GREEN, RICHARD; and FISK, NORMAN. "Prenatal Exposure to Female Hormones: Effects on Psychosexual Development in Boys." *Archives of General Psychiatry* 28 (1973): 554–561. Examination of the children of diabetic mothers who received estrogen and progesterone to prevent pregnancy complications suggests that prenatal sex hormone levels may influence some aspects of postnatal psychosexual adjustment in boys.

DRUG TRIALS

See HUMAN EXPERIMENTATION, *article on* BASIC ISSUES; DRUG INDUSTRY AND MEDICINE.

DRUG USE

I. DRUG USE, ABUSE, AND DEPENDENCE
Robert Neville

II. DRUG USE FOR PLEASURE AND TRANSCENDENT EXPERIENCE
Sidney Cohen

I
DRUG USE, ABUSE, AND DEPENDENCE

Drugs that affect thought, mood, or behavior have been used for a great many purposes. L. J. West lists ten: (1) to restore normal function, under conditions of disease, disorder, or discomfort, in either medical or nonmedical contexts; (2) to improve performance under various conditions; (3) to alter learning; (4) to reinforce behavior either positively or negatively; (5) for recreation; (6) to explore the inner self either of others or of oneself; (7) for the symbolic purposes of showing willingness to take risks, of demonstrating maturity, of defying prohibitions and controls, of belonging to peer groups, and of marking the transition from one culture to another; (8) to achieve mystical experience or partake of religious rites; (9) to manipulate others; and (10) to foster social or political policies by weakening the enemy or manipulating those dependent on drugs (West).

A wide range of drugs is available for one or several of these purposes, including sedatives and hypnotics, stimulants, psychedelics and hallucinogens, opiate narcotics, volatile solvents, nonnarcotic analgesics, clinical antidepressants, and the major tranquilizers. In light of West's list of purposes for drug use, alcohol, nicotine, and caffeine should be added, along with drugs used as "truth serum," those used for crowd control or to disorient individuals, and those drugs being developed to enhance or suppress sensory experience, including aphrodisiacs (Evans and Kline).

This article will deal with fundamental value questions associated with individual drug taking for whatever purpose, under four headings: (1) moral attitudes; (2) abuse and dependence; (3) criminality; and (4) drug abuse rehabilitation programs.

Moral attitudes

The range of attitudes toward drug taking extends from positive encouragement for easy access and use of drugs, whenever they do something someone wants and the costs to the user and others are acceptable, to the feeling that drugs should be used, if at all, only in desperate situations, because they involve an unnatural manipulation of one's mind and body and because they serve as unnecessary crutches. Whether one views drugs as problematic depends on the position occupied in the spectrum between those two extreme attitudes.

Indeed, those at the permissive end may look to drugs not only as remedies for abnormal functioning but as producing states that are better than normal. Let us put aside the instrumental use of drugs in religious or "self revelation" contexts, a use that Robert Veatch in 1974 called "neo-Protestant"; an interpretation of drug use reflecting this perspective was given by Weil in 1972. People indulge in some drugs, for instance alcohol and marijuana, because of the pleasant consciousness they bring and because they facilitate enjoyable social times. The permissive attitude sees drugs as avenues to enjoyable experiences and evaluates the costs of drug taking quite differently from the attitude that approves drugs, if at all, only in therapeutic contexts where the cost–benefit question would include the risks of using drugs to get well.

Permissive attitudes toward drugs are often tied to larger views of life. Life is difficult even without crises for most people, and drugs can be seen as helping them through life's difficulties. Many people take caffeine and nicotine for a lift and alcohol to blur sharp edges, and in recent years millions of people in America have come to count on mild tranquilizers. Yet, if there were no drugs, many conditions that the drugs are used to alleviate might be considered entirely normal—just the frustrations and pains of daily living.

Those with a restrictive attitude toward drugs, on the other hand, might say the problems should be faced directly for what they are, not with a drugged response. Those whom Klerman called "pharmacological Calvinists" emphasize the seriousness of life. Life is not something to be "gotten through" but rather a task in which to achieve. The truths of life are to be faced directly, even when they are unpleasant and when nothing can be done about them. The honesty, sincerity, and directness of one's own response is a great value, according to this perspective; indeed, it is an important goal to be able to respond to life with singleness of heart (Klerman). Psychoactive drugs interpose an externally conditioned level of response between the work and one's perceiving and willing heart so that true integrity is difficult if not impossible. Drugs, according to the restrictive attitude, should be used, if at all, only instrumentally, where a necessary condition of defining their benefit is their enabling one to live with honesty and seriousness. Robert Veatch, Associate for Medical Ethics, Institute of Society, Ethics, and the Life Sciences at Hastings-on-Hudson, has argued that the "Protestant drug ethic" disapproves of drugs that give pleasure but approves those that increase achievement, e.g. caffeine.

But why, ask more permissive people, should life be taken so seriously? Does not life itself mock our efforts? Why not take harmless pleasures where we can?

The value sets of those along the permissive–restrictive spectrum probably reflect personality types and social orientations that are systematically reinforced by more pervasive elements than the results of drug attitudes alone. The philosophic merits of the positions do not segregate neatly. The restrictive position is probably correct in holding that an honest and serious approach to life is a great value. The permissive position is equally correct in pointing out that pleasures and crutches are also great values. The restrictive position that taking drugs just to get through life is likely to lead to a falsification of experience and perhaps even a lessened ability to cope is probably right, although surely some situations are handled better, both experientially and in terms of objective outcome, with the aid of drugs. Those on the permissive end of the spectrum may point out rightly that drugs can enrich experience, which is both pleasant and beneficial.

Some issues dividing the permissive and restrictive approaches really have nothing to do with drugs. Is there an obligation to experience things directly and from the heart? Always? Under what conditions? Is there a moral or social right of private choice regarding what to experience and what pleasures to seek? (Veatch has a fine discussion of the problem of a choice.) Why do some people with crutches learn how to walk and others to love crutches?

Conflicts between attitudes, of course, are not the only moral issues involved in drug use, social commentators remind us. There are objective social situations that may be said to "drive" some susceptible people to drugs. Deprived of the pleasures of love, friendship, and self-satisfaction, a person looks more favorably on the pleasures of drug experiences. Should this be taken as warranting drug pleasures for the unfortunate, or rather as a call to action in remedying the lack of love, friendship, and self-satisfaction? Economic hardship and frustration, boredom, anomie, and countless other aspects of industrial society encourage people to take a mechanical approach to control of their responses to life. But perhaps those objective situations need amendment as much as the victims need balm.

It should be noted that, in discussing the morality of resorting to drugs in some situations, seemingly minor drugs such as caffeine and nicotine have been grouped with major ones such as alcohol and tranquilizers. This should not be taken to imply that there are no differences of seriousness among drugs. But regarding attitudes, there seems to be a carryover from one drug to another; if drugs are perceived to be valid ways out of troubles and into happiness, then which drug is used is a matter of context, expectations, availability, and costs.

Abuse and dependence

The language of drug use and abuse reflects the values and interests of the speakers. For instance, the World Health Organization Expert Committee on Drug Dependence defines drug abuse as "persistent or sporadic excessive drug use inconsistent with or unrelated to acceptable medical practice." Of course, one must determine what constitutes "excessive" use. But does the definition mean to suggest that some excessive drug use might be consistent with, or part of, acceptable medical practice? Does it suggest that using drugs for purposes or in dosages inconsistent with or unrelated to acceptable medical practice is what constitutes excessive use? At the very least, that definition supposes that medical goals such as treatment of mental illness or alleviation of physical pain determine legitimacy in drug use, and that using psychoactive drugs at dosage levels causing significant effects for nonmedical purposes—such as pleasure, consciousness alteration, or narcosis—is in fact drug abuse, or close to it. Questions are frequently raised about the standing the medical profession has to define use and abuse.

Sometimes drug abuse is loosely identified with drug "addiction." The tendency of medical professionals now, however, is to drop reference to addiction and speak of drug "dependency" instead. The World Health Organization distinguishes psychic from physical dependence and defines them in the following ways: *Psychic* dependence is

A compulsion that requires periodic or continuous administration of a drug to produce pleasure or avoid discomfort. This compulsion is the most powerful factor in chronic intoxication with psychotropic drugs, and with certain types of drugs may be the only factor involved in the perpetuation of abuse even in the case of the most intense craving. Psychic dependence, therefore, is the most universal characteristic of drug dependence. Operationally, it is recognized by the fact that the dependent continues to take the drug in spite of conscious admission that it is causing harm to his health and to his social and familial adjustment, and that he takes great risks to obtain and maintain his supply of the drug [Isbell and Cruściel].

Physical dependence is defined as:

A pathological state brought about by repeated administration of a drug and that leads to the appearance of a characteristic and specific group of symptoms, termed an abstinence syndrome, when the administration of the drug is discontinued or—in the case of certain drugs—significantly reduced. In order to prevent the appearance of an abstinence syndrome the continuous taking of the drug is required. Physical dependence is a powerful factor in reinforcing psychic dependence upon continuing drug use or in relapse to drug use after withdrawal [ibid.].

It is not at all necessary that people who are drug dependent thereby are drug abusers. Where the need for the drug is easily satisfied and where the effects of being on the drug do not impair activity, dependence may not dominate life. Many people dependent on barbiturates or opiates (morphine or heroin when easily obtained) can lead otherwise apparently successful lives; alcoholics, on the other hand, usually find it difficult under the influence of excessive drink to function normally even though the drug is easily available. If drug abuse is "bad use," drug dependency itself may not involve bad use. dependency itself may not involve bad use.

Some aspects of drug abuse have less to do with the effects of the drugs themselves than with the social factors necessary for their use. Illegal drugs are obtained in criminal ways, communities may be disrupted, economic productivity of the users and those who must care for them may be adversely affected, and the life of drug pursuit sometimes crowds out other important aspects of living.

The borderline between drug use and drug abuse often varies with diverse cultures. Acceptable alcohol consumption is a case in point. Diverse attitudes are held toward illegal activity, the importance of being "responsible" at all times, and many other factors. Many people would say that use is abuse when the user or others are definitely harmed by the use, although there is wide variability in considering what harm is in contrast to "acceptable cost."

To determine whether a certain drug use is abuse, and if so how bad, is a very difficult undertaking. Because drugs affect subjective experience, and people are different, even the same drugs in the same contexts have different meanings and effects. And then contexts differ too, as well as the histories of the individuals and groups involved. Moral judgments about drug use and abuse then must involve a complex weighing of a variety of factors, some of which will be discussed next.

Self. Drugs affect the person taking them. They have direct perceptible qualities, and except in the case of mild stimulants the usual case is that people take drugs to feel good. But the direct experience of drugs is also the culmination of previous strands of experience; drugs have a meaning depending on how ready the user is for them. "Trips" are good or bad; expectations and habits are reinforced or disoriented, depending on the drug, setting, and physical state. Taking drugs also has future consequences for the person's self, affecting his body, his future habits and expectations, his commitments and preparedness for life. It can be fairly said that no responsible and thorough judgment about a person's drug use can justifiably be given without a systematic interpretation of what that use means in the individual's direct experience, as a culmination of what he brings to the experience, and as a cause of its future effects. A moral evaluation of all of this entails direct reference to philosophical values concerning personal life-style of enjoying, achieving, and being free.

Interpersonal relationships. Drugs affect not only the user but also those with whom he is associated; his personal relationships mediate the effects of the drugs on himself. Psychoactive drugs are significant in interpersonal relations in many ways. They color the user's interpretation of those relationships, encouraging perhaps "good feeling," perhaps feelings of paranoia. They may affect the functions of relationships in personal growth; for instance, a classic psychoanalytic interpretation of heroin addiction views it as a way of fixing on passive oral dependence on one's mother (Radó). Drugs also affect the way others perceive the user. And, most of all, the activity of drug procurement and taking affect a person's health and also his ability to friends of drug users must accommodate themselves to the life of drug use, whether this means tolerating tobacco smoke, covering up for an alcoholic, or coping with a criminal junkie. For people whose lives are dominated by drug taking, this activity can be the most important determinant of personal relations, cutting them off from some people and putting them in touch with others. Drug use may require a superficiality in personal relationships that prevents serious ongoing emotional encounter.

Social career. Over and above one's personal experiences and relationships, a person generally has a career. This involves him in educational, familial, and economic institutions; in the long run a person's career is the shape of his life, his trajectory from birth to death. Drugs affect a person's health and also his ability to participate effectively in the institutions important for his career. Drugs can speed or impede education and productivity. They can make for tenser or more relaxed family structures. The wide differences in social, economic, and cultural classes affect participation in the institutions important for social development, and the moral meaning of drugs varies accordingly; for instance, amphetamines approved as helping middle-class hyperkinetic children to learn in school are sometimes viewed by lower classes as threats of social control.

Economic effects. Over and above the direct economic impact of drug use on the user's life, drug use affects the overall economy. Users are more or less productive, depending on the drugs, the circumstances of use, and the measure of productivity. Furthermore, the use of many psychoactive drugs is illegal, and this has enormous impact on the overall economy of societies. Users frequently must steal in order to pay for their drugs, and law enforcement agencies must be sustained to deal with this. In impoverished neighborhoods the social organization of drug use may be such as to prevent effective social and political organization; as in most things, the poorest people suffer most from the unfortunate consequences of drug use. This is true for legal drugs, such as alcohol, as well as for illegal ones, such as heroin and cocaine. It should be noted that economic effects are values for a larger society than that of the user and his personal contacts; therefore, the larger society has an interest in his drug taking that may be quite alien to his own real interests, and an issue of social versus individual good may be involved. Who should decide when a user's use is abuse so detrimental that something should be done about it?

Cultural implications. Any activity of sufficient magnitude to make an economic difference in society also makes a cultural difference. At the very least drug taking and the attitudes of various groups toward it partly define the values of a culture. A high frequency of alcohol use, for instance, or use of opiates may signify that people feel impotent to improve a condition from which they suffer. The tolerance of generally disapproved drug use is a measure of the permissiveness of a society. The acceptability of drug use of various sorts indicates where a society draws some significant lines distinguishing public from private spheres. Some people see drug abuse as a call for the treatment of a larger social pathology. Others see it as a justification for the regulation of drug use even when the users do not feel victimized by their use; this paternalism has been a dominant American attitude toward narcotic drug use.

Criminality

Drugs of various sorts are indeed abused by some, and frequently by those least able to help themselves—those suffering most from social inequalities and from psychological unreadiness to face the pressure of their social situation. For these reasons, as well as on moralistic grounds, societies have attempted to control drug use through taxation and prohibition. Actually, not all social policies regarding drugs have aimed at reducing drug use; the Opium Wars of the nineteenth century offer a counterexample. But most legislation has been against drugs; witness, in the United States, the experiments with prohibition of alcohol in the early part of this century and the attempt to control narcotics through the Harrison Act. Alfred R. Lindesmith has detailed the history of governmental policies in the United States to stop or contain the use of various drugs (1965).

One line of objection to the legal regulation of drug use is from a very individualistic libertarian, such as Thomas Szasz. What a person does in private is his own business, and the society has a right to intervene only when the private practice clearly has social harm. Now most of the social harm of drug use, the libertarian argument goes, derives from the fact that certain drugs are illegal. Whatever reasons a person might have for initiating the use of illegal drugs, the continued use involves him in a drug culture, associating with those who make their living dealing in illegal substances, and frequently paying for the drugs through illegal activity. As mentioned above, when an illegal drug culture becomes dominant in a particular segment of society, that segment finds it very difficult to organize itself or make use of the legitimate social institutions. Many people besides libertarians would agree with the following view: "It is clear that the *social and legal policies*, if ostensibly developed to control or prevent the use of some mind-altering drugs, are the cause of the main social problems that we find with those drugs" (Fort, p. 223).

But what about those who abuse legal drugs, for instance, alcohol? The libertarian position answers this, first by pointing out that Prohibition caused more problems than it solved and second by arguing that alcoholism is a private matter. Of course, some people will be hurt by alcoholism; certainly drunks should be prohibited from driving or otherwise endangering people's lives. But if a person chooses to ruin himself with drugs, that is his right (but it should be understood that society has no obligation to take care of him afterward).

The libertarian position usually is argued from two premises, one theoretical, one empirical. The theoretical premise is that individuals should be allowed to do what they want unless their exercise of liberty seriously endangers others. The empirical premise is that governmental attempts to regulate private affairs generally make matters worse; such policies are difficult to enforce and frequently create an opposition with greater powers of evil than the private matter under regulation.

There are several liberal (not libertarian) arguments against these premises. First, it is very hard to draw the line between private effects of drug use and those that might be harmful to others. Even when a drinker drinks alone and only alone, his drinking deprives others of his company and perhaps of his efforts on which they may have a claim. Tobacco smoke affects those nearby who are not smoking. At the very least, drug use affects the economy, and may do so harmfully.

Second, for a person to confine the effects of his drug use to a private sphere, he must have considerable resources to provide that cushion. For instance, he must relieve his family of the responsibility for caring for him when he suffers drug-induced sickness. Whereas this might be possible for rich people, it often is not for poor people. In conditions of poverty there is little privacy, and everyone suffers the burdens of all. So the conservative theoretical argument is class-biased.

Third, part of the impetus behind drug control legislation derives from the same motive as the control of food quality and of drugs for medical use. Basically the argument is that the public by and large is not in a position to protect itself from being abused by drugs. Just as consumers generally cannot judge the purity of the foods offered for sale and are at the mercy of physicians regarding the medicines offered to them, so people need protection against psychoactive drugs. People, particularly if they are young, do not know what they are getting into when they begin to smoke, drink, or use illegal drugs. And by the time they find out, the dependence on the drug and the involvement in the drug life-style may be so powerful that only heroes can get out.

Fourth, control of drugs is not as unsuccessful as critics complain. On the one hand, people are affected by what society legitimates and what it prohibits. Even if the laws are unenforceable, there are some who will not partake solely because of legal prohibitions, and there are some who will lose their resistance to harmful peer pressure if the legal sanctions are removed. On the other hand, *direct* prohibition may not be as effective as other forms of control. Legalizing drugs but making them available through specially controlled channels, as Great Britain currently does for narcotics, may be an effective way of reducing drug use.

Yet another perspective, neither conservative nor liberal, sees the whole focus on governmental regulation of drugs to be a mistake. The problem is not the chemicals people use and abuse, but the people who use them and the society in which they must live.

To consider the social critique first, it will be recalled that Marx called religion the "opiate of the masses." The interesting point is the supposition about opiates. Why do people need opiates? Because their lives are dehumanized and alienated from the activities providing real value and pleasure. The real problem is not the chemicals, which might well be used harmlessly if there were no pressures for abuse; the problem is the society. Consequently, the effort to prohibit or control psychoactive drugs both detracts attention from the real problem and serves to support the social status quo. If the means by which people suffer can be curtailed through repressive controls, then the roots of the suffering can be left as they are, to the benefit of those in power who profit by the social arrangements. According to this radical view, even the best-intentioned of liberals merely reinforce the interests of those in power to keep basic conditions as they are while dealing with only the more blatant symptoms of human suffering.

From this perspective both the conservative and liberal views are mistaken also on the nature of personal responsibility. The conservatives are mistaken in believing that moral norms apply only to relations among people; the liberals are wrong in saying merely that the public has an interest in those interpersonal relations. Rather, persons and societies are bound by obligations regarding the quality of life, whether or not they happen to accept the obligations and regardless of their utility or disutility. For instance, a person makes promises that bind his future self, and ought therefore to make himself able to fulfill those promises. Drug use should be thought of in the light of obligations to personal and social value. The fundamental problem for this radical position is: Who decides what the obligations are? A secondary problem is determining when people should be forced to fulfill their obligations.

In all of these arguments it is apparent that the social response to drug use and abuse raises basic issues of freedom, social control, and social and personal values. These are well focused in the problem of drug rehabilitation programs.

Rehabilitation

If a person perceives his drug use as abuse and seeks professional help, the ethical problems surrounding rehabilitation are those of behavior control therapies employed as extensions of self-control. But in most cases drug abuse is also perceived by society as a problem *for* society. Societies, then, intervene in their own interest, providing rehabilitation opportunities or rehabilitation programs, sometimes as legal alternatives to incarceration for criminal offenders. Most rehabilitation programs, besides aiming to be therapeutic for the abuser, are exercises in social control for the intended benefit of the society. Rehabilitation programs in the United States are aimed mainly at narcotic and alcohol abusers, who may use other drugs incidentally. Users of other illegal drugs such as marijuana and LSD are treated simply as criminal offenders if they do not also use narcotics. The following brief discussion considers treatment programs moving from "least change" in the abuser to "most change."

Heroin maintenance. Heroin maintenance policies allow heroin addicts to receive their drugs in prescribed dosages from licensed agents of the state, usually family physicians. In an account of British and American experiences with heroin maintenance, we find that American clinics flourished from 1912 to 1924; since 1924 British physicians have legally maintained addicted patients on opiates, although since 1966 the dispensing of heroin has been permitted only through special clinics; family physicians continued to supply other opiates and synthetic narcotics (Brecher, chap. 13). Heroin maintenance has the advantage of assuring chemically controlled preparations, greatly reducing the dangers of toxicity and overdosage. It also reduces or eliminates the criminal aspects of the addict's life having to do with procuring and paying for his drugs and removes the black market underground from the heroin business. Addicted individuals are then enabled to lead more stable and productive lives, other things being equal.

But on this policy individuals are not rehabilitated regarding their addiction itself, and probably not regarding the causes of their addiction. And so, critics point out, even though a legally approved and orderly life is possible, the very causes that led to addiction in the first place remain and are likely to continue to upset the addicts' lives. Furthermore, even though heroin may be a benign drug in controlled dosages, it still dulls the user's perception of life, according to the critics. Heroin maintenance is principally a policy of social control to solve a social problem. Although the user's drugs become safe and legal, making his life better in important ways, heroin maintenance is not therapeutic for the individual in the sense of improving his life in those areas that led to drug abuse.

Methadone maintenance. Methadone is a drug chemically similar to heroin, although different from heroin in that it does not, when taken orally in the usual prescribed dosages, produce the feelings of euphoria typical of initial heroin experiences. Addicts are maintained on methadone with the same motives as in heroin maintenance: providing a safe and legal drug, and allowing addicted persons to live a more stable and productive life. Similar objections regarding nonrehabilitation apply. Methadone is preferred to heroin mainly because it is not illegal.

In the United States, however, methadone maintenance programs typically are administered as programs for drug addicts, frequently related to courts, not as medicinal treatments provided by family physicians. This means that the "social organization" of the methadone programs is quite different from the heroin maintenance policy in Great Britain. Persons in the program are gathered and typed for their drug problems (whereas patients of a given physician have a common need of medical treatment, but not the same disease or need). Many methadone program centers have indeed become headquarters for the sale of illegal drugs, since they are common meeting grounds for interested personnel. Furthermore, methadone maintenance programs are blatant forms of social control, creating a physically dependent population subject to the policies of the social agencies.

Mixed treatments. Other rehabilitation programs aim to make users drug-free and to improve their psychological and economic capabilities. These involve either immediate detoxification or the use of methadone in progressively smaller doses until the user is no longer physically dependent. Alcohol and barbiturate detoxification are also handled in medically supervised ways. After detoxification various regimens of personal and group therapy are undertaken, and efforts may be made to secure the addicts some employment and to stabilize their home lives. How the approaches are mixed depends on the specific services and talents in each program, and also on what seems best for each drug user. Mixed treatment programs are highly variable in their effectiveness (*American Journal of Drug and Alcohol Abuse;* Mandell and Goldschmidt).

The degree to which mixed treatment programs are "rehabilitatively ambitious" depends upon both the program ideology and the kind of population served. Some programs count themselves successful if their clients, for a specified period of time, remain drug-free, commit no crimes, hold steady jobs, and maintain intact homes or living situations. "Drug-free" varies in meaning from merely not using the illegal or addictive drugs from which the person was detoxified all the way to avoiding all psychoactive drugs, including nicotine and alcohol. Programs that have comprehensive lists of drugs from which one should be free tend also to have high standards of psychological rehabilitation. Their treatments aim to penetrate to the reasons why people use drugs in the first place, and often focus on the concept of the "addictive personality." The degree to which a person's drug problem is a matter of personality or social condition probably depends on many independent factors.

Therapeutic communities. Therapeutic communities, deriving from the initial experiments with Alcoholics Anonymous and Synanon, are founded on the belief that rehabilitation requires the total immersion of the drug abuser in a treatment program (Yablonsky). Frequently staffed by former addicts or graduates of the program, therapeutic communities aim to break down psychological defenses and personality structures of their members on the grounds that those personalities are in need of thorough revision; the means for this are often group criticism sessions that appear brutal and sadistic to observers. Positive aspects of people's personalities are supposed to be reinforced, and the addicts' lives are reorganized in positive personal and social ways. Frequently therapeutic communities run businesses employing their own members, or place their members in local businesses. Some communities believe that members must remain in the community for life, having substituted dependence on the community for dependence on drugs. Other communities attempt to graduate their members to independent lives. Critics of therapeutic communities point out that they reinforce dependence so that a person becomes dependent or "psychologically addicted" to the community in nearly all aspects of his life, particularly regarding his feelings of self-worth. A person dependent on drugs alone at least has the possibility of compartmentalizing his dependence and perhaps treating that alone. Defenders of the communities answer that dependence on a community is still better than the life of crime and addiction, and that most drug-dependent people do not have strong enough personalities to live independently while compartmentalizing drug dependence.

Rehabilitation programs: conclusion. In the United States the development of drug rehabilitation programs has been affected by feelings in some quarters that drug abusers should be punished as well as rehabilitated, and also by ideological quarrels about goals and about who should be served. It is very difficult to determine scientifically the success rates of various treatment modalities, partly because most programs have not been set up so that accurate record keeping has a high priority and partly because reports by addicts about their drug use, criminal records, and social status cannot be trusted.

One of the greatest moral problems is that rehabilitation programs often promise what they are unlikely to fulfill, with two results. First, drug rehabilitation programs gain social and legal power over people that is often not justified by the programs' intent, i.e., mere rehabilitation; this also displaces responsibility from the user to the programs. Second, drug abusers frequently are not helped, and both they and society still have drug problems that need resolution. For these reasons, some critics of restrictive policies believe the illegal drugs should be decriminalized, that heroin and methadone maintenance programs should be instituted through regular channels of medical care, not special drug agencies, and that minimal goals of rehabilitation should be adopted. Then drug abusers can seek more ambitious kinds of rehabilitation on their own, encouraging better means of therapy and minimizing dependence on rehabilitation programs of controversial effectiveness.

ROBERT NEVILLE

[*Directly related are the entries* BEHAVIOR CONTROL; *and* PSYCHOPHARMACOLOGY. *For further discussion of topics mentioned in this article, see the entries:* ALCOHOL, USE OF; *and* PATERNALISM.]

BIBLIOGRAPHY

American Journal of Drug and Alcohol Abuse 2 (1975): 1–138. An issue devoted to the Drug Abuse Reporting Program (DARP).

BAZELON, DAVID L. "Drugs That Turn on the Law." *Journal of Social Issues* 27, no. 3 (1971), pp. 47–52.

BRECHER, EDWARD M., ed. *Licit and Illicit Drugs: The Consumers Union Report on Narcotics, Stimulants, Depressants, Inhalants, Hallucinogens, and Marijuana—Including Caffeine, Nicotine, and Alcohol.* Boston: Little, Brown & Co., 1972.

Drug Abuse Survey Project. *Dealing with Drug Abuse: A Report to the Ford Foundation.* Patricia Wald and Peter Hutt, co-chairpersons. Foreword by McGeorge Bundy. New York: Praeger Publishers, 1972.

EVANS, WAYNE O., and KLINE, NATHAN S., eds. *Psychotropic Drugs in the Year 2000: Use by Normal Humans.* Springfield, Ill.: Charles C. Thomas, 1971. Based on the meeting of the Study Group for the Effects of Psychotropic Drugs on Normal Humans held at the annual meeting of the American College of Neuropsychopharmacology, Puerto Rico, 1967.

FORT, JOEL. *The Pleasure Seekers: The Drug Crisis, Youth and Society.* New York: Bobbs-Merrill Co., 1969.

GOSHEN, CHARLES E. *Drinks, Drugs, and Do-Gooders.* New York: Free Press; London: Collier-Macmillan Publishers, 1973.

ISBELL, H., and CRUŚCIEL, T. L. "Dependence Liability of 'Non-Narcotic Drugs." *Bulletin of the World Health Organization* 43, supp. (1970), pp. 1–111.

KLERMAN, GERALD L. "Psychotropic Hedonism vs. Pharmacological Calvinism." *Hastings Center Report* 2, no. 4 (1972), pp. 1–3.

LENNARD, HENRY L. *Mystification and Drug Abuse. Hazards in Using Psychoactive Drugs.* Jossey-Bass Behavioral Science Series. San Francisco: Jossey-Bass, 1971.

LINDESMITH, ALFRED RAY. *The Addict and the Law.* Bloomington: Indiana University Press, 1965.

MANDELL, WALLACE, and GOLDSCHMIDT, PETER. *Interdrug: An Evaluation of Treatment Programs for Drug Abusers.* Baltimore: Johns Hopkins University School of Hygiene and Public Health, 1973.

PEELE, STANTON, with BRODSKY, ARCHIE. *Love and Addiction.* New York: Taplinger Publishing Co., 1975.

RADÓ, SÁNDOR. "The Psychoanalysis of Pharmacothymia." *Psychoanalytic Quarterly* 2 (1933): 1–23.

ROSENTHAL, MICHAEL P. "Legal Controls on Mind-and Mood-Altering Drugs." *Journal of Social Issues* 27, no. 3 (1971), pp. 53–72.

SZASZ, THOMAS S. "The Ethics of Addiction." *American Journal of Psychiatry* 128 (1971): 541–546.

VEATCH, ROBERT M. "Drugs & Competing Drug Ethics." *Hastings Center Studies* 2, no. 1 (1974), pp. 68–80.

WEIL, ANDREW. *The Natural Mind: A New Way of Looking at Drugs and the Higher Consciousness.* Boston: Houghton Mifflin, 1972.

WEST, LOUIS JOLYON. "Hallucinogenic Drugs: Perils & Possibilities." *Hastings Center Studies* 2, no. 1 (1974), pp. 103–112.

WITTERS, WELDON L., and JONES-WITTERS, PATRICIA. *Drugs and Sex.* New York: Macmillan Co., 1975.

World Health Organization, Expert Committee on Drug Dependence. *Sixteenth Report.* World Health Organization Technical Report Series, no. 407. Geneva: 1969.

YABLONSKY, LEWIS. *Synanon: The Tunnel Back.* Baltimore: Penguin, 1967. Originally published as *The Tunnel Back: Synanon.*

II

DRUG USE FOR PLEASURE AND TRANSCENDENT EXPERIENCE

Introduction

When youthful recreational drug users are asked why they consume their chosen chemical, the most frequent response is some variation of "to feel better." Their response has two implied meanings. One is the relief of some noxious physical or mental condition—relief from dysphoria. The other is the achievement of a euphoric feeling-tone—a "high." A surprising number of chronic drug abusers are actually treating themselves for some distressing psychological ailment. The disorder may be excessive shyness, boredom, depression, or frustration. Just as common, though, is the eternal search for euphoria, the willingness to explore a variety of chemical configurations for some transient feeling of gladness, joy, elation, or simply a relaxation of the tensions of the day. This article deals with the use of drugs for purposes of pleasure, self-transcendence, and religion, as conceived in a continuum, highlighting the principal ethical and social issues involved.

The value issues

The quest for chemically induced euphoria raises a number of ethical issues, including whether the achievement of gratuitous satisfaction is in itself morally objectionable, and whether the price paid for enjoyment enters into the transaction, affecting the quality of that enjoyment. Further, supposing that an inexpensive, nontoxic euphoriant could be produced, socioethical issues arise regarding its permissible use, its control and dispensation, and the general effect on society of the availability of "instant happiness" in pill form.

With the relaxation of the Western view that pleasure exacts a price and that gratification should be delayed, chemically induced pleasure per se is losing its pejorative connotations and becoming a positive value. Balanced against this burgeoning value are a number of disvalues, both to others and to oneself. In relation to others, indulgence in chemical pleasures can render the user a public charge, and the current system of illegalities involved can force the user into transgressions against others in order to support his own dependence on illegal chemicals. Personally, in addition to the loss of their freedom of choice in their exercise of "freedom of action," users incur neurophysiological penalties: repetitive stimulation of the nonspecific reward centers by chemical or electrical means results in a diminution of the pleasurable feeling-tone, so that more and more stimuli are needed to augment a waning emotional response. Ordinary pleasurable activities hardly register after vigorous electrochemical stimulation, and the biphasic quality of the central nervous system manifests itself—e.g., for every amphetamine "high" there is a subsequent postamphetamine depression. Learning may also be impaired, for learning involves the aversive and reward centers of the brain; where the reward is "for nothing," then nothing is learned. Pleasure is thus no longer a source of value in learning, for the learning and the pleasure have become dissociated.

The acceptable euphoriants

The detailed stories of alcohol, tobacco, and the caffeinated beverages need no retelling here

They are culturally accepted and widely used. The choice of ethanol and tobacco as pleasure-giving agents has been, in retrospect, not entirely wise. Their hazards have become international concerns, but like all items that have become ingrained into the cultural matrix they can hardly be eliminated. When culturally approved euphoriants are used, society adopts fairly lenient attitudes toward them. Unfortunately, the acceptable pleasure-giving drugs are neither invariably mood-elevating nor non-toxic.

Alcohol, although consumed for good cheer and fellowship, provokes aggression more than all psychochemicals. When consumed heavily, its psycho- and physiopathology involve every organ system. Many heavy drinkers drink to allay anxiety and tension; strangely enough, they are measurably more anxious while under the influence. Besides the relaxation that does occur at low doses, the well-known disinhibiting effects are sought after. They make the user more vivacious and gregarious and facilitate social interaction. However, a lingering reservation remains: Could those social graces have been better achieved by learning them?

The dilemma of alcohol is that, while lesser doses provide surcease and group well-being, chronic, higher intake levels can result in physical dependence and a multitude of personal, economic, and social disasters. About ten percent of all who drink for pleasure and good fellowship end up with addiction or an assortment of disabilities.

Tobacco is a puzzling substance. It is hardly a euphoriant, and it is clearly productive of varied disabilities. When cigarette users are asked why they persist in their smoking ritual, they are hard pressed to describe clearly the rewards of smoking. Some will mention mild feelings of relaxation, others slight stimulation, but most notice no mood alteration at all. Smoking seems like a minimally rewarding activity. The joys of inhaling the pyrolytic products of tobacco can hardly be a justification for its entrenched use. Instead, we should wonder whether it is the reinforcement of interminably repeated oral–manual behaviors that produces the reward, just as a fair part of the reward of a two percent bag of heroin is in the "fixing." Furthermore, nicotine has been shown to produce a withdrawal syndrome in laboratory animals when its administration is suddenly discontinued. This has also been amply confirmed in humans whose accounts of "kicking" cigarettes are commonplace.

The stimulants

Turning to those drugs used for pleasure under other than legal conditions, a number of oddities are encountered. "Highs" occur not only with stimulants but also with depressants. Attempts to stay "high" continuously are never successful. Although the initial period of use may be delightful, continued substantial usage can result in a transformation into greater dysphoria than ever. The individual may become locked into the habitual requirement that increasing amounts of the drug be procured and injected or swallowed. Most heroin addicts will say that they no longer use the drug to get "high"; rather, it is needed to avoid the withdrawal sickness. What was once a euphoriant experience has become a losing struggle to feel normal.

The stimulants are chosen for hyperphoric purposes because of their mood-elevating, energizing properties. Cocaine must be considered the supreme euphoriant. The "rush" comes within moments of "snorting" or injecting intravenously, and it is very intense. The effects dissipate quickly, and the process can be repeated dozens of times a day. The various amphetamines behave like a long-acting cocaine. They, and the amphetamine-related anorectics, provide a lift to the tired and a bit of brightness for the sad. A decade ago a wave of intravenous amphetamine users, the "speed freaks," demonstrated once again that indescribable pleasure infinitely prolonged was an impossible goal.

The hallucinogens are listed with the stimulants because they produce a state of sympathetic dominance with dilated pupils, hyperthermia, tachycardia, and a hyperaroused tracing on the electroencephalogram. The use of so potent a drug as LSD for pleasure needs to be explained (Cohen, pp. 45–63). In average doses (50–150 mcg) it induces an intensification of sensory awareness—particularly, but not exclusively, visual. Colors are more saturated. The perceived object has greater depth and luminosity and seems enormously meaningful. Immobile objects move. Time is slowed down. Synesthesias are noted: The scent of music can be detected and colors are tasted. Thought, emotion, and sensation seem to fuse into some primordial mental process. The critical function of the ego is obliterated along with the ego's boundaries. The LSD "trip," then, is an intense, novel esthetic experience made more so by one's inability to be critical of its effects.

Western societies are not the only ones that employ stimulants for conviviality. Caffeine-containing plants have a worldwide market, and there are others. Khat (*Catha edulis*) is a mildly euphoriant plant containing pseudo-norephedrine. It is chewed in a social setting and is a well-established stimulant in the Arabian peninsula, enjoyed by all classes and sold in the market place of every town. Originally it was imported from Ethiopia, where the authorities are quite permissive about khat use despite the interdiction against alcohol and drug use by the prevailing religion. The chewing of pituri leaves (*Duboisa hopwoodii*)—a mild stimulant that fostered animation and well-being—was a part of the culture of the Australian aborigines long ago.

The depressants

Chemical enjoyment is not merely a matter of procuring stimulation and exhilaration. Drugs that provide relaxation, withdrawal, stupor, and even coma are valued for those effects. For that reason, narcotics, hypnosedatives, and minor tranquilizers are also popular recreational agents. Anesthetics like ether, chloroform, and nitrous oxide have had their periods of popularity. These substances were used for ether frolics, chloroform jags, and laughing gas soirées even before their usefulness as anesthetics became known. Certain chemicals not ordinarily thought of as suitable for human consumption like gasoline, paint thinner, spray paint aerosols, and other volatile solvents are favored by some juveniles to induce a rapid, intoxicated "high."

Marijuana (*Cannabis sativa*) is second only to alcohol in worldwide popularity as a pleasure-producing drug. Typically, it has been the recourse of rural or tribal populations or of the urban poor, but with the recent indulgence of middle-class youths in drugs that alter consciousness, its use has become widespread. Cannabis, like LSD, is an example of a substance that has been used for both pleasure and self-transcendence. Yogis have, in the past, employed it as an aid in their meditations, and it has been used as a religious ceremonial drug in a number of cultures.

The antipathy toward marijuana in the United States is difficult to understand. As commonly used, it has a rather trivial effect upon consciousness and behavior. The stereotypes of sex-crazed dope fiends presented in the earlier popular literature can scarcely be confirmed by field studies and pharmacological investigations. Perhaps the excessive reaction to the introduction of cannabis occurred because it was a culture-alien drug introduced by strangers. Meanwhile, however, established attitudes and legislation toward cannabis are changing in a reasonable direction. Although a new intoxicant is far from needed, the public health hazards of marijuana require moderate, rather than draconian, controls.

The transcendental experience

The use of botanicals to achieve a transcendental experience may well have occurred early in man's history, even before plants were sought for their medicinal properties. It has been speculated that some primitive concepts of God could have been derived from the ingestion of natural hallucinogenic materials.

A definition is needed for the word "transcendental" because many transitional states between intense pleasure and the transcendent event exist. The transcendental experience (also called peak, mystical, religious, cosmic, Satori, etc.) is one in which most of the following phenomena are apperceived (adapted from Stace):

1. A feeling of oneness with the universe to the point where the boundaries between Me and Not Me have completely dissolved.

2. A feeling that this state is the real reality, much more valid than sober reality.

3. A sense of timelessness and of spatial non-existence (the "Now Moment" of Meister Eckhart) pervades the experience.

4. A sense of sacredness, awesomeness, and significance is felt by the person in the throes of such an experience.

5. The perceptual component may be a blazing white light or a vision of surpassing beauty.

6. A deeply felt positive mood, be it joy, ecstacy, bliss, peace, or love, is reported.

7. A common comment of those who have undergone such a state is that it is indescribable, ineffable.

8. In this condition paradoxes become resolved, and opposites are seen simply as two aspects of the same thing.

9. Sometimes (by no means invariably) abrupt, dramatic alterations of values and belief systems occur. The changes may last for a lifetime or for a day. If the changes are transient, the awesome experience may leave the person worse off than ever. The load of guilt and despair may be greater because he had "seen the Light" and failed.

A critical issue is whether the hallucinogen-induced transcendental state is the same as the spontaneously occurring one. All of the qualities described above have been encountered after the ingestion of large amounts of hallucinogens by some people under certain conditions. Nevertheless, there are differences. The nonchemical state is likely to be more highly valued and more impressive since it takes place without evident cause. Mystical states of this sort have been instrumental in the founding of religions, have changed people's lives dramatically, and have altered whole social systems. The chemical transcendental state that is purchased for a few dollars is less apt to transform the individual but has been known to do so in a few instances. Perhaps the greatest difference in the two conditions is that the spontaneous state comes forth following meditations, vigils, strenuous exercises, severe deprivation, suffering, or similar stressful events. The person involved is more likely to be prepared for the strange, out-of-the-body experience and is apt to use it as a constructive turning point in his existence. The LSD-induced experience often takes place in unprepared individuals, and they may undergo a disintegrating rather than an integrating outcome of such a cataclysmic event.

It should not be assumed that the transcendent experience cannot be induced by other pharmacologic classes of drugs. William James considered that "the sway of alcohol over mankind is unquestionably due to its power to stimulate the mystical faculties of human nature usually crushed to earth by cold facts and dry criticisms of the sober hour" (p. 72). DeQuincy's *Confessions of an English Opium Eater* contains passages indicating that he experienced transcendent experiences, on occasion, under laudanum. Nitrous oxide has evoked similar subjective responses. It may be that any drug that can produce a dissociation state is capable of evoking a transcendent experience under favorable personality and environmental conditions.

Cultural justification for peak experiences

The use of pleasure-giving chemicals has long been approved: Each society appears to have its own approved substances for this purpose. New euphoriants arouse concern, but if they become culturally approved hedonic materials, the moral issues are relegated to such questions as under what circumstances they should be sold, to whom, or what portion of the tax load they should bear. Since the condoned euphoriants in industrialized countries are at least as harmful as the illegal ones, the morality of interdicting certain substances while permitting others becomes obscure. Traditional usage and familiarity appear to be the major grounds of justification.

Can sociopolitical systems survive without some natural or synthetic euphoriant? Usually it is the religious sector of society that prohibits these items: among others, the Muslims, Latter Day Saints, and Seventh Day Adventists, for example. Clearly, the devout subsist quite well without chemicals that alter mood, but the popularity of mood-elevating materials indicates that most people are unwilling to relinquish them. Although no perceptible biological need for these substances exists, large numbers of humankind insist that they be available.

Religions have been based upon the use of theochemicals (usually as theobotanicals). These religions hold either that a religious experience can be induced by the use of chemicals or that chemicals are necessary for sacramental ritual. The Native American Church of North America uses the peyote cactus in the context of a communal, ritualized, direct experience of God. During the 1960s churches were formed by people who had religious experiences with LSD or mescaline: the Neo-American Church and the League for Spiritual Discovery, to name only two groups in the United States. Efforts to obtain recognition for the use of the latter-day psychedelics as sacramental agents were unsuccessful, and the advocates of an "acid" religion appear to have faded away. The residual mark upon the dominant culture is evident. It is likely that the interest in Eastern mystical religions and in meditative techniques in the Western world stems from the upsurge in the taking of LSD and related agents during the 1960s. Perhaps episodic recrudescences of psychedelic-induced sectarian fervor will occur in the future.

Our understanding of paranormal experience, including the transcendental state, requires revision. Traditionally, our culture has employed a sanity–insanity continuum. The phenomenology of the mystical state would then require that it be placed at the insanity end of the continuum since it was obviously not a sane event. In other societies, and increasingly in North America, the mystical state is becoming highly regarded as a superlative state of human awareness. Perhaps the time has come to consider that peak and cosmic experiences are forms of a new category: "unsanity." Unsanity would be a marked alteration of consciousness that brings about an integration of oneself and one's universe. Then the continuum would consist of insanity at one end—a disorganized altered state of consciousness in which the meaning, organization, and significance of existence have disintegrated. Sanity would occupy a middle position, and at the other end would be unsanity. It could not be a straight-line continuum; rather it would have to be bent into a circular alignment until only a gap separated the insane from the unsane state because of the ability of some people to slip between the two mental conditions, disintegration and integration. The reawakened interest in the direct experience of altered states of consciousness, however obtained, indicates that the simple insanity–unsanity polarity is being abandoned. Alterations of consciousness that can be called unsane are held to be of value either for themselves or as opportunities to learn from such unusual changes in mental functioning.

Conclusion

The hope for euphoria and relief from dysphoria are common motivations for nonmedical drug taking. Chemical hedonism is becoming more acceptable as groups endeavor to increase freedom of action by individuals. Whether drug-induced joy is qualitatively identical with the naturally occurring emotion is problematic because of the nature of the functions of the nervous system. The conflicts between personal freedom and social responsibility are apparent in the issues of illicit drug use.

The available culturally acceptable euphoriants vary from the relatively harmless (whether to the individual or to society) to the definitely harmful. The same can be said of the illegal euphoriants sold in the black market. It would seem reasonable to attempt to reduce the personal and social morbidity and mortality wherever possible.

The hallucinogens are the pharmacological class capable of evoking both the hedonic and transcendental states more precisely than other drug classes. The chemical transcendental experience is similar to, but not identical with, the spontaneous religious experience. Efforts to establish a psychedelic religion based upon LSD have failed because of the lack of a philosophy, an ethic, and an ability to integrate the psychedelic experience into daily life, as well as the assumption that the experience was something for everyone.

The chemical transcendental state has provided us with two opportunities: (1) to study the phenomenology of transcendent experience and (2) to revise our incorrect notions about altered states of awareness that cannot be understood by being subsumed under the word "insane."

SIDNEY COHEN

[*For further discussion of topics mentioned in this article, see the entries:* ALCOHOL, USE OF; PAIN AND SUFFERING; *and* SMOKING. *Also directly related are* HEALTH AS AN OBLIGATION; *and* MENTAL ILLNESS, *article on* CONCEPTIONS OF MENTAL ILLNESS.]

BIBLIOGRAPHY

CLARK, WALTER HOUSTON. *Chemical Ecstasy: Psychedelic Drugs and Religion.* New York: Sheed & Ward, 1969.

COHEN, SIDNEY. *The Beyond Within: The LSD Story.* 2d ed. Foreword by Gardner Murphy. New York: Atheneum, 1967.

HUXLEY, ALDOUS LEONARD. *The Doors of Perception.* New York: Harper, 1954.

JAMES, WILLIAM. *The Varieties of Religious Experience: A Study in Human Nature.* Gifford Lectures on Natural Religion, 1901–1902. New York: Modern Library, 1902.

MASLOW, ABRAHAM HAROLD. *Religions, Values and Peak-Experiences.* The Kappa Delta Pi Lecture series. Columbus: Ohio State University Press, 1964.

MASTERS, ROBERT E. L., and HOUSTON, JEAN. *The Varieties of Psychedelic Experience.* New York: Holt, Rinehart & Winston, 1966.

PAHNKE, WALTER N., and RICHARDS, WILLIAM A. "Implications of LSD and Experimental Mysticism." *Journal of Religion and Health* 5 (1966): 175–208.

SMITH, HUSTON. "Do Drugs Have Religious Import?" *LSD: The Consciousness-Expanding Drug.* Edited by David Solomon. Introduction by Timothy Leary. New York: G. P. Putnam's Sons, 1964, chap. 8, pp. 152–167.

STACE, WALTER TERENCE. *Mysticism and Philosophy.* Philadelphia: Lippincott, 1960.

WATTS, ALAN WILSON. *The Joyous Cosmology: Adventures in the Chemistry of Consciousness.* Foreword by Timothy Leary and Richard Alpert. New York: Pantheon Books, 1962.

ZAEHNER, ROBERT CHARLES. *Mysticism, Sacred and Profane: An Inquiry into Some Varieties of Praeter-Natural Experience.* Oxford: Clarendon Press, 1957.

DYNAMIC THERAPIES

Concepts and principles underlying the ethical issues discussed in this entry can be found in the entry BEHAVIOR CONTROL.

Introduction

The term "dynamic therapies" refers to both psychoanalysis and psychotherapy. In psychoanalysis, the analyst and patient undertake a general examination of the patient's personality, feelings, and social behavior for the purpose of changing disturbed or unsatisfactory patterns of feeling and action. Typically they meet several times a week for a fifty-minute session over a period of several years. The techniques of analysis, as of therapy, are verbal. Psychotherapy is a more general term than psychoanalysis. It sometimes refers to a course of treatment in which the therapist and patient meet over a relatively short period of time with specific goals, for example, the treatment of particular symptoms of distress. More generally, it refers to any course of treatment that employs psychoanalytical assumptions about personality and behavior even when the techniques of treatment vary. In this discussion, the term "therapist" is used for therapists and analysts alike, and the term "therapeutic relationship" refers as well to the analytic relationship (but not, of course, to medical relationships, which are also in one obvious sense therapeutic).

A neurotic condition is one in which a patient's feelings and behavior are disturbed by patterns of anxiety and internal conflict but in which he or she can respond relatively well to everyday demands. Recently, dynamic therapy has come to be used more and more effectively with psychotic patients, who are more seriously disturbed and incapacitated.

The patient who undertakes treatment with a psychotherapist or psychoanalyst commits himself to an extended interaction that involves characteristic goals, values, and responsibilities. The relationship between patient and therapist may be compared with the medical (doctor–patient) relationship and the teacher–student relationship, and it may be distinguished from both of them in the following ways.

The goal of psychotherapy, like the goal of medicine, is the health of the patient. Physical health is, however, a much less controversial notion than emotional or mental health. For medicine, the body can be seen as a mechanism, and the job of the physician is to intervene to the best of his limited ability to put the mechanism right when some part malfunctions. The patient's body is the more or less passive object of the doctor's ministrations, which may be medical or surgical. The disturbances treated by psychotherapy, on the other hand, are disturbances in the patient's relations to other persons, in his feelings about himself, and in his self-understanding—usually all three at once. Therefore, the aim of therapy is not restoration of bodily functions but a change in social behavior and in the patient's attitudes toward himself. At best, the therapist provides a context or situation in which the patient can act to achieve these aims. Thus, the goals of the therapeutic relationship and the roles of the participants stand in contrast to the goals and roles of the medical relationship. There are of course situations in which the two relationships intersect. A physician will have to be alert to the psychosomatic aspects of physical malfunction, as in hysteria, and a therapist must be prepared to find organic causes for emotional disturbance in some cases.

The therapeutic relationship can be compared with the mutual relationship of teacher and student. One model of the teaching relationship is the Socratic model set forth in Plato's writings. The teacher is like a "midwife" who leads students to use capacities and reveal knowledge that they possess latently. Paradoxically, the therapeutic relationship is more Socratic than teaching itself. In many teaching situations, the student cannot really be said to know "unconsciously" what he is learning, but in every therapeutic situation the patient is seen as trying to make conscious the unconscious attitudes, fears, and conflicts that shape his behavior.

Basic values

Self-examination. In psychoanalytical theory the patient and therapist are assumed to share certain values that are inherent in their mutual activity. The patient must share with the therapist the conviction that self-examination and self-discovery are the essential means of treatment. In this way, psychotherapy and psychoanalysis are distinguished from other ways of treating emotional and mental disturbance, e.g., behavioral therapy and drug treatment. Some theoreticians claim not only that self-examination is the chosen means of effecting change, but that the cultivation of self-knowledge is an end in itself. This view echoes the philosophical commitment, derived from Socrates, that the unexamined life is not worth living. Accordingly, self-examination not only brings about behavioral change and some resolution of the patient's problems but remains a continuing activity that enriches the patient's life. In this sense, the successfully treated patient is quite unlike the untreated "healthy" person, for he is trained to maintain special strategies of self-perception.

In treatment itself, the patient can achieve self-knowledge only indirectly. Because his anxiety blocks direct admission of his deepest fears and conflicts, he is encouraged to give the therapist clues to his emotional expectations and thought processes by saying whatever "comes into his head," the stratagem of free association. To the extent that the self-ascribed task of introspection involves structuring, editing, and thus censoring of his output by the patient, it is from a psychoanalytical point of view self-defeating.

Autonomy. Freud referred to the inherent value of self-examination in saying about the goals of psychoanalysis, "where id was, there ego shall be." This means in part that behavior which carries out understood and integrated purposes is to take the place of unconsciously motivated and therefore compulsive behavior. This is not to say that psychoanalysis values unspontaneous action. Rather, it recognizes that the spontaneous actions of a child are largely unavailable to adults and that the compulsive behavior of disturbed patients is not spontaneous at all but the product of inhibition and repressed conflict. Therefore, a second value in psychoanalytical theory, one also suggested in Freud's slogan, is autonomy.

According to the theory, all persons deal with childhood fears and crises by learning patterns of response. When such fears are particularly overwhelming and such crises particularly traumatic, the learned pattern is not so much one of coping as one of escaping. The person learns to shut off certain possibilities of action and achievement when the threat of trying and failing is, again unconsciously, felt to be too great. The disturbances produced in this way preempt some social and personal achievements: The patient, far from acting spontaneously, is the prisoner of inflexible patterns of feeling and action, which he uses undiscriminatingly and often inappropriately in his mature life. He is literally the prisoner of his past, paradoxically the prisoner of himself. The goal of therapy is invariably to free the patient, to allow him to build a more flexible set of emotional responses and actions to achieve self-worth and well-being. Like the value of self-examination, the value of autonomy is achieved only as an ongoing process, not as a stopping place.

Implications of the dominant values. The commitment of patient and therapist to the twin values of self-examination and autonomy has several implications. A comparison with medicine is again relevant. In medicine, it is often assumed that the patient submits himself voluntarily for treatment. This is important but relatively unproblematic: The patient must be informed of his disorder and of the planned treatment and its attendant risks, and must consent.

In psychotherapy, voluntariness is also important and much more problematic. First, the patient's reasons for seeking therapy and his conception of personal goals are to some extent shaped by his emotional and social expectations. The first difficult job of the analyst is to see that the choice of therapy is made as freely and realistically as possible. Second, the patient resists expressing his conflicts because to express them is to experience them, i.e., to do what he most fears, whether it is to stand up to a feared father, separate himself from an overprotective mother, or make some other move. The voluntary decision to proceed with therapy is one that must be remade continually; the relatively uncensored expression of conflict is itself an expression of autonomy.

There are other implications. The preeminent value given to self-examination and autonomy implies that all other values are to be seen in their light. For example, altruism and kindness have value not in themselves but only if they emanate from the actor's free choices and not from his fears and conflicts. Self-sacrificing generosity may be a symptom of unconscious patterns of self-destruction. This "transvaluation of values" is far-reaching; the determinant of value of action is not the traditional interpersonal one, whether the act benefits or harms those affected by it, but an intrapersonal one, whether the act is genuinely an autonomous response to the needs of others.

Certain other characteristic attitudes proceed from the choice of self-examination as the mode of therapy (in contrast, for example, to drugs). The therapeutic process involves repeated encounters with emotional pain and suffering—not the magnified (and therefore intolerable) pain that the patient unconsciously expects, but the real pain of suffering everyday frustrations and of accepting one's own limitations. The acceptance of pain and suffering is therefore a basic value in the psychoanalytical picture of the well-lived life. A related aspect of such a life is that it will involve gratification. The person will be able to delay gratification and control impulses, but also to achieve gratification in its time. In psychoanalytical theory and practice, disturbance is ordinarily traceable to fears about sexual activity and conflicts in attitudes toward sex; it follows that symptoms of disturbance will be found in adult sexual practices and fantasies. The capacity for vigorous heterosexual activity is traditionally held to be a mark of health. Some recent writers use a more subjective criterion for sexual health, the satisfaction of the individual and his capacity to give satisfaction to his sexual partners. Disagreement here turns on the question whether non-heterosexual preferences are clearly a symptom of neurotic conflict and of aborted emotional growth.

Responsibilities

Responsibilities of the therapist to the patient. The involvement of the therapist in the patient's life is intimate. Although they meet only at regularly scheduled times in a controlled setting, the patient is expected to trust the therapist with his seemingly most irrational fantasies and his most embarrassing fears. The more unguarded the patient is, the more the therapist can help. The patient, vulnerable to shame and disappointment, must trust the therapist, and the therapist must show that he deserves such trust.

Countertransference. The characteristic mode of response by the patient is called transference. This means that the patient acts out with the therapist his personal ways of dealing with other persons; he projects or transfers onto the therapist his expectations about others. The therapist will play many different roles as the patient relives and reconstructs different relationships. As that occurs, the therapist has two major responsibilities. He must allow the patient to proceed with transference by maintaining a basic situation in which the patient will not feel vulnerable or ashamed, and he must prod the patient into discussing and understanding the patterns revealed in the transference.

The demands on the therapist are very high. While an ordinary physician or teacher may allow his personal spontaneity to express itself in his role, the therapist's expressions of personality will be likely to impede transference. One responsibility of the therapist is to guard against such intrusions with self-awareness. Accordingly, therapists and analysts are usually required to be psychoanalyzed themselves to bring their own conflicts to some resolution and to become aware of their personal predilections for abusing the therapeutic relationship. For example, the therapist may unconsciously cultivate the patient's dependence for the satisfaction that this gives, and furthermore he may rationalize doing so as therapy even if it frustrates the long-term goals of the patient. In other words, the responsibility of the therapist is to minimize, if not avoid altogether, what is called "counter-transference."

Reality-testing. One way of describing the patient's problem is to say that he acts unrealistically in his disturbed everyday behavior. The paranoiac patient attributes malevolent purposes to persons who do not have them; a sexually unresponsive woman patient may fear the violence of men who in reality will not be violent. In helping the patient to see and understand such patterns, the therapist aids the patient to test his or her expectations against reality.

This responsibility of therapists becomes controversial when therapists are accused of imposing their own notion about reality on their patients. A way of examining the controversy is by asking whether the relevant sense of the term "reality" is one about which different therapists may have their own ideas. On one hand, it is clear in our examples that the absence of malevolent intentions or of the prospect of violence may be objective facts. That is not to say that a mistaken expectation of violence is always a symptom of neurotic conflict; there is a burden on the therapist to measure the patient's expectations by what is reasonable (even if mistaken) in the circumstances. This comment, however, reintroduces the controversy, since it will be said that therapists impose on patients their own idea of what is reasonable.

To this, there are two responses. The first is an admission that abuse *is* always possible. A therapist *may*, in contravention of his responsibilities, shape a patient's attitudes to fit his own political, ethical, or other beliefs. There are obvious cases of abuse as there are obvious cases of nonabuse in which the patient's fears and expectations, typically by his own admission, are "unreal." There is a gray area as well. Debates about whether all sexual inversion is derived from unreal perceptions and fears of the opposite sex, and is therefore neurotic, are one illustration of the controversial or gray area. The second response is that the therapist's responsibility is to help the patient to test his expectations against reality in those areas in which the nature of reality is to the highest degree uncontroversial.

"Arm's-length" aspects. For reasons which have been discussed, the therapist must be especially careful to minimize the patient's dependence on him. The therapist must allow the patient to bring about the transference and not collaborate, so that the patient may in time see the transference as the product of his expectations and fears. Moreover, dependence of any sort during therapy undercuts or makes more remote the goal of autonomy.

There are various opinions and practices with regard to implementing this responsibility. According to some theories the therapist is allowed or even encouraged to touch, advise, or otherwise guide the patient if this will serve the ultimate ends of therapy. Therapists of this kind are expected to exercise wide discretion in meeting patients' perceived needs. An extreme minority view sanctions sexual activity with patients in suitable situations. Such practitioners reject many of the strictures of orthodox Freudian practice as outlined above.

The traditional practice of therapists trained in methods set by Freud is to interfere as little as possible with patients' free associations and yet guide them to speak freely and reflect on the significance of those associations. This explains many formal constraints in psychoanalysis. The patient usually lies on a couch so that he will not be influenced by the therapist's facial expressions and gestures. Sessions are strictly limited to fifty minutes and are scheduled as regularly as possible. The therapist usually insists on making the patient's financial responsibilities as unambiguous as possible and on holding the patient financially liable for missed appointments. The patient is expected to make and adhere to a long-term commitment to therapy with brief vacations scheduled at the therapist's convenience. The rationale for these features of therapy/analysis is that they help the therapist maintain an arm's-length relationship within which the patient's own projections are most clearly distinguishable from the idiosyncratic contributions of the therapist. Even within these constraints, of course, there will be subtle differences among therapists and these will be reflected in their minute-by-minute decisions about whether and how to intervene in the patient's presentations.

The controversy between the relatively austere Freudian procedures and those of so-called revisionists can be seen as a controversy about the relation of methods to goals. On one side, it may be said that orthodox Freudian therapists set rigorous goals for patients and that the patient is so thoroughly "on his own" that the prospects of succeeding are minimized. The patient needs the active support of the therapist to change his conception of himself as fearful and rejected. On the other side, the argument is that such support achieves short-term gains at the cost of aggravating the patient's dependence and frustrating transference by changing a projected into an actual relationship. On one side, the goals are called too high and unreachable; on the other, they are called too transitory and limited. A therapist will shape his own conceptions of his responsibility in forming an opinion about this debate.

Responsibilities of the patient. Since the relationship of therapist and patient involves reciprocal roles, each has responsibilities defined by his role. The responsibility of the therapist is to the patient, but the patient's responsibilities are to himself, since his benefit is the object of the activity. Therefore, one tends to think of the therapist's responsibilities as ethical responsibilities; by comparison, the patient's responsibilities are not ethical ones unless a person can be said to have ethical responsibilities to oneself, e.g., to improve one's condition. Whether or not one calls them ethical, the patient's responsibilities reciprocate those of the therapist.

First of all, the patient has a responsibility to be honest. In ordinary cases it may be hard or easy to be honest; in therapy it is always hard, because one is asked to be honest in just those cases where one is most tempted to evade because the truth is inherently painful. As patient, one must question and reassess one's picture of oneself, one's goals, and the bases of one's most intimate relationships. One must also stand ready to entertain and tolerate paradoxical beliefs, for example the awareness of loving and hating the same person at the same time. Finally, one must be ready to entertain and admit irrational feelings of anger, aggression, and guilt even when this entails feeling altogether vulnerable.

A commitment of this kind to honesty is a commitment to spontaneity. One must say what "comes into one's mind," censoring it as little as possible, however nonsensical it seems and shameful it feels. The most important clues for a therapist will characteristically seem nonsensical or shameful to the patient: nonsensical when the patient as self-censor denies their possible importance, shameful when they intimate unconscious fantasies of danger.

The so-called arm's-length aspects of therapy impose discrete and unambiguous responsibilities on the patient. Part of the patient's early "work" is to examine his response to these constraints. He must assume a long-term schedule, appear at appointments, and take on what is typically a considerable financial burden.

Responsibilities of the therapist outside the therapeutic setting. In obvious ways the therapist plays a uniquely intimate role in the patient's life and has ethical responsibilities that extend beyond the work of the therapeutic appointments. However austere the therapist's way of proceeding may be, the patient at some point during treatment feels dependent on the therapist. This is not simply the result of the patient's transference; it is also a recognition of the fact that the therapist becomes an important and powerful person in the patient's life when the analysis is successful.

The patient's trust of the therapist rests on the assurance that his communications will be confidential. The therapist has an obvious duty to protect confidentiality as much as possible. While the therapeutic relationship shares this feature with the medical relationship, the need for confidentiality in the former is more pervasive since the patient in therapy typically sees his remarks as revelations about his own vulnerability and about what is most embarrassing to him. In this respect it is more closely comparable to the relationship of lawyer and client, where a breach of confidentiality is likely to endanger the client and frustrate the purposes for which the relationship exists.

The responsibility to protect patients' confidentiality is a negative one, viz., to refrain from certain actions. Does the therapist have a positive responsibility to be available to the patient outside fixed appointments? That would be like any physician's responsibility to respond in emergencies, but there are special complications. The long-term interests of the patient, autonomy and self-reliance, are undercut if the therapist encourages dependence and is seen as too readily available in periods of crisis. The job of the patient is to acquire his own strategies for crises. At the same time, the patient who is in genuine danger from himself (from suicidal wishes, for example) or from others may not survive to achieve these long-term goals. The therapist's role must be to guide him through such situations in such a way as to minimize both their recurrence *and* dependence.

So far, the discussion of the therapist's responsibilities has been about responsibilities to the patient. There are obvious situations in which these responsibilities will come into conflict with responsibilities to others. Suppose, for example, that the patient reveals plans to harm his family. (To simplify matters, imagine that the contemplated actions are legal.) Suppose also that the therapist believes the patient has the capacity to carry out his plans, intends to do so, and will do so. Does the therapist have a responsibility to warn intended victims or otherwise prevent harm? Theorists disagree. All emphasize that the therapist's primary obligation is to the patient, and intervention is justified only if the harm is particularly severe or irreversible. Beyond this, some say that the obligation to the patient is absolute and emphasize that the therapist must try to prevent such actions only through his work with the patient.

The most serious problems of this kind arise when the patient intends to do something harmful that is forbidden by law, in the most extreme kind of case when he intends to take a human life. In such cases one may say that even if the therapist's responsibilities *as therapist* extend only to the patient, he will usually believe on good grounds that the patient's interests cannot be advanced by letting him proceed. Decisions to intervene will be difficult, because the therapist must assess the likelihood that the patient will act and the impact of his intervention on the patient, both on continued treatment and on the patient's social and legal status. In any case, it cannot be said that such hard cases can be made easy because the obligation to protect confidentiality is absolute.

There are really two kinds of moral argument for intervention. The therapist *as therapist* has obligations to the patient, and this will precipitate moral dilemmas about the effects of intervention. But the therapist also has responsibilities as a person. A second kind of moral argument is that his responsibilities to potential victims will in serious cases override his responsibilities as therapist. There are in other words acute dilemmas in which the therapist must weigh his responsibilities as therapist and as a person, dilemmas that cannot be resolved by appealing to rules and general formulas.

Therapists are often asked to give so-called expert testimony about the competence of defendants to stand trial or face legal hearings. This does not involve a breach of confidentiality where the examinee is not the therapist's patient and where there has been no intimation of confidentiality. Even so, there is a moral dilemma since the therapist's role as therapist is in conflict with his legal role in two ways. First, he is accustomed by training to consider the needs of examinees (or patients) without having to balance those needs against the needs of other persons like potential victims. In testimony he is usually asked to make just such a judgment. Second, he is not ordinarily expected to decide between responsibility or nonresponsibility (sanity or insanity) but to see all persons as relatively autonomous in personally idiosyncratic ways. To be sure, his expertise allows him to predict with some confidence how certain persons will act in certain situations, but this is only one ingredient of the morally relevant determinations that the law asks of him.

There is another complicating feature of the therapist's role as expert witness. He is, by training, an expert on the effect on the subject (usually the defendant) of imprisoning him, of returning him to his home or the street, and so forth. As witness he is asked to leave these insights aside and make a judgment only about what the subject is likely to do. But at the same time he knows that his testimony is a determinant of what happens to the subject. He is therefore asked to do two things which are antithetical to his training: to disregard the welfare and interests of the subject and to intervene directly in the subject's life. The demand that he participate thus generates a moral dilemma.

Therapy in its social setting

The dilemmas of the therapist as expert witness are one kind of illustration of general conflicts between the values of dynamic therapy as an institution and other values. In most roles in daily life—as friend, parent, worker, teacher, student, political participant, etc.—the values of self-examination and autonomy are not treated as absolute but are qualified by other values. Ordinarily, the well-functioning individual is socially defined by such communal purposes as his contribution to the joint external enterprise rather than by such self-regarding goals as freedom from fear, conflict, and anxiety. To some extent, of course, there is no conflict insofar as psychological health allows one to be a full participant in other roles. At the same time, the pursuit of psychological health will also lead to the rejection of overdemanding friends or overprotective parents or to mitigation of an obsessive commitment to one's job. One may respond to this clash of values in several ways. One may employ the notion of autonomy to criticize those institutional practices that seem to impede self-determination. Or one may say that the claims of friendship, work, and political goals should in many instances override self-regarding claims.

The social role of psychotherapy and psychoanalysis comes under attack in another way. Because they involve great investments of time by the therapist and time and money by patients, they are available to few persons and involve, even for those few, an uncertain prognosis. These features provoke the criticism that dynamic therapy involves an inefficient and wasteful allocation of society's resources. The political questions here involve hard moral questions about what sorts of economic allocations are just and justifiable.

THOMAS H. MORAWETZ

[*Directly related are the entries* BEHAVIOR CONTROL; MENTAL HEALTH THERAPIES; MENTAL ILLNESS; SELF-REALIZATION THERAPIES; *and* THERAPEUTIC RELATIONSHIP. *Other relevant material may be found under* CONFIDENTIALITY; HEALTH AS AN OBLIGATION; INFORMED CONSENT IN MENTAL HEALTH; MENTAL HEALTH, *article on* MENTAL HEALTH IN COMPETITION WITH OTHER VALUES; *and* SEX THERAPY AND SEX RESEARCH, *article on* SCIENTIFIC AND CLINICAL PERSPECTIVES. *See* APPENDIX, SECTION IV, AMERICAN PSYCHIATRIC ASSOCIATION, *and* AMERICAN PSYCHOLOGICAL ASSOCIATION.]

BIBLIOGRAPHY

EDELSON, MARSHALL. *The Idea of a Mental Illness.* New Haven: Yale University Press, 1971.

ERIKSON, ERIK H. *Insight and Responsibility: Lectures on the Ethical Implications of Psychoanalytic Insight.* New York: W. W. Norton & Co., 1964.

FREUD, SIGMUND. *The Standard Edition of the Complete Psychological Works of Sigmund Freud.* 23 vols. Translated under the editorship of James Strachey. London: Hogarth Press, 1954–1964. See especially: "The Dynamics of Transference," vol. 12, pp. 99–108; "Remembering, Repeating, and Working-through," vol. 12, pp. 147–156; "Introductory Lectures on Psycho-Analysis," vol. 15, pp. 15–239 and vol. 16, pp. 243–463; "The Ego and the Id," vol. 19, pp. 13–66; "Civilization and Its Discontents," vol. 21, pp. 64–145; "New Introductory Lectures on Psycho-Analysis," vol. 22, pp. 7-182; and "Analysis Terminable and Interminable," vol. 23, pp. 216–253.

FROMM-REICHMANN, FRIEDA. *Principles of Intensive Psychotherapy.* Chicago: University of Chicago Press, 1950.

GOLDSTEIN, KURT. *Human Nature in the Light of Psychopathology.* The William James Lectures Delivered at Harvard University, 1937–1938. Cambridge: Harvard University Press, 1940.

Group for the Advancement of Psychiatry, Committee on Psychiatry and Law. *Criminal Responsibility and Psychiatric Expert Testimony.* Report, no. 26 (May 1954). Topeka, Kans.: 1954.

GUREVITZ, HOWARD. "Tarasoff: Protective Privilege versus Public Peril." *American Journal of Psychiatry* 134 (1977): 289–292.

HAMPSHIRE, STUART. *Thought and Action.* London: Chatto & Windus; New York: Viking Press, 1960.

HARTMANN, HEINZ. *Essays on Ego Psychology: Selected Problems in Psychoanalytic Theory.* New York: International Universities Press, 1964.

HORNEY, KAREN. *Neurosis and Human Growth: The Struggle toward Self-Realization.* New York: W. W. Norton & Co., 1950.

REDLICH, FREDERICK C., and FREEDMAN, DANIEL X. *The*

Theory and Practice of Psychiatry. New York: Basic Books, 1966.

RIEFF, PHILIP. *Freud: The Mind of the Moralist.* New York: Viking Press, 1959.

———. *The Triumph of the Therapeutic: Uses of Faith after Freud.* New York: Harper & Row, 1966.

ROGERS, CARL RANSON. *On Becoming a Person: A Therapist's View of Psychotherapy.* Boston: Houghton Mifflin, 1961.

SCHAFER, ROY. *A New Language for Psychoanalysis.* New Haven: Yale University Press, 1976.

SULLIVAN, HARRY STACK. *The Interpersonal Theory of Psychiatry.* Edited by Helen Swick Perry and Mary Ladd Gawel. New York: W. W. Norton & Co., 1953.

SZASZ, THOMAS STEPHEN. *The Ethics of Psychoanalysis: The Theory and Method of Autonomous Psychotherapy.* New York: Basic Books; Dell Publishing Co., 1965.

———. "Psychiatry, Ethics, and the Criminal Law." *Columbia Law Review* 58 (1958): 182–198.

WAELDER, ROBERT. "Psychiatry and the Problem of Criminal Responsibility." *University of Pennsylvania Law Review* 101 (1952): 378–390.

WINNICOTT, DONALD WOODS. *Collected Papers: Through Paediatrics to Psycho-Analysis.* New York: Basic Books, 1958.

EASTERN EUROPE

See MEDICAL ETHICS, HISTORY OF, *section on* EUROPE AND THE AMERICAS, *article on* EASTERN EUROPE IN THE TWENTIETH CENTURY.

EASTERN ORTHODOX CHRISTIANITY

Eastern Orthodox Christian ethics bases its ethical judgments on Holy Scripture and Holy Tradition. Holy Tradition consists of the "mind of the Church" and is discerned in the decisions of ecumenical and local councils, the writings of the Fathers of the Church, canon law, and the penitentials (guides for the administration of the sacrament of Penance).

Issues not directly treated in the ancient sources are dealt with by modern Orthodox ethicists by seeking to express ethical judgments that are in harmony with the "mind of the Church." Thus, their writings have a certain provisional character and are always subject to episcopal, synodical, or general ecclesial critique. There are occasionally differences of substance in the writings of modern Orthodox Christian ethicists. By and large, however, responsible Orthodox ethicists maintain a common ethical stance. Modern issues in bioethics often require of ethicists that they find parallels in the tradition and, with the help of reason, deduce new ethical applications from established doctrinal, historical, and pastoral positions.

Basic doctrine and ethical affirmations

The Eastern Christian doctrinal position tends to be cautious in defining positively the central affirmations of its faith. It prefers the *via negativa*, or "apophatic" method (i.e., saying what is not the case). In ethics, a practice may be proscribed as not in harmony with the ethos of the faith, but often no positive solution is offered other than the need for patience and acceptance of the situation.

Nevertheless, Eastern Orthodox Christianity does avail itself of positive or "kataphatic" doctrinal and ethical statements. These are taken seriously when they are normative in character, but not in a rigid, legalistic, or absolute fashion. All positive statements regarding divine revelation—the Tradition—are seen as limited and subject to mystery as a necessary dimension of all human understandings of the divine. In canon law and in ethics this has led to the practice of "economia," which authorizes exceptions to the rule without considering the exception a precedent or abrogating the rule. In most cases the justification for the application of "economia" is the avoidance of greater harm in the case of the strict application of the rule (Kotsonis). Several key doctrinal teachings have immediate ethical application with specific reference to bioethical issues.

Theological anthropology. The *humanum* of our existence is both a given and a potential. Some of the patristic authorities distinguish between the creation of human beings in the "image" of God, and in his "likeness." "Image" is the *donatum* of intellect, emotion, ethical judgment, and self-determination. In fallen humanity these remain part of human nature, albeit darkened, wounded, and weakened. The "likeness" is the human potential to become like

God, to achieve an ever expanding, never completed perfection. This fulfillment of our humanity is traditionally referred to as *theosis* or "divinization." Human beings are in fact "less than fully human." To achieve *theosis* means to realize our full human potential. Ethically, this teaching leads to the acceptance, on the one hand, of the existence of a "human nature," but, on the other, it clearly does not restrict our "humanum" to conformity to that nature. The "image" provides a firm foundation for ethical reasoning. The "likeness" prohibits the absolutizing of any rule, law, or formulation (Maloney).

Divine energies and human self-determination. Though God's essence is totally incomprehensible to the human mind, God's energies are present in every human experience. To speak of divine energies is to speak of God's actions in relation to the created world. The relationship of God's energies to human freedom and self-determination has obvious ethical implications. Orthodox Christianity teaches that, though God is Lord of history, he does not coerce or force obedience and conformity to his will. Coerced conformity is dehumanization, whereas fulfilled humanity—which is the divinization of human life—must be free, since God is free. This raises the question of Divine Providence and Human Responsibility. Orthodox Christianity holds these two in paradoxical tension: man is responsible and must act, but God accomplishes his will, either with or in spite of man's actions. Ideally, human actions are harmoniously integrated with divine purposes in a perfect synergy of divine and human wills. This belief is but an extension and application of the Orthodox doctrine of the divine and human natures in the one person of Jesus Christ. Ethically, this means that we are not permitted simply to wait upon God. Rather, we are committed to the exercise of self-determination and responsibility in conformity with both human reality and divine purpose (Florovsky, pp. 113–120).

Body–spirit. God is seen as the creator of both the material and the spiritual dimensions of reality. Eastern Orthodox Christianity sees these aspects of existence as closely bound together. The icon is an example of this belief. At first sight, the icon appears to be a stylized artistic representation of a holy figure. Yet the iconographer's purpose is to capture in form, line, color, and symbol both the spiritual and the physical reality of the figure. The sacramental use of material means (such as water, oil, bread, wine, etc.) for spiritual purposes also illustrates Eastern Orthodoxy's comprehension of the intimate relationship of matter and spirit. For bioethics, this key concept is important because it leads to a serious affirmation of the psychosomatic unity of human life. "Body" and "soul" are the constituents of human existence; the Orthodox emphasis on the Resurrection confirms its view that human life and human fulfillment are inextricably bound to both the physical and the spiritual dimensions of human existence. In more contemporary terms, body and personhood are essential for the fulfillment of human potential (Antoniades, 1: 204–208).

Law, motive, intent. Based on the above, ethical reasoning in Orthodoxy is a balanced combination of law, motive, and intent. Moral law is based in large part on the *donatum* of human nature. For Eastern Orthodoxy, natural law refers primarily to the elementary relationships that are necessary for the constitution and maintenance of human society. For the Fathers of the Church, the Decalogue is an excellent expression of the natural law common to all men (Harakas, 1964). In a similar yet more flexible pattern, there are modes of behavior that are either prescribed or proscribed for the lives of Christians growing in the image and likeness of God toward *theosis* or full humanity. These positive and negative injunctions are found in the Holy Scriptures, in the writings of the Fathers and in the canons of the Church. For the Orthodox these statements are normative in the sense that they embody the mind of the Church and reflect standards of behavior that are appropriate and fitting for the members of the Church and, potentially, for all human beings growing in the image and likeness of God for the full realization of personhood.

This first level of ethical direction is saved from legalism and rigid prescriptivism by the fundamental emphasis on love as a motive of action. Grounded thoroughly on a Trinitarian theology that understands the Holy Trinity first as a community of persons united in love, the Church teaches that being God-like means being loving. In general, the commandments of the moral law are embodiments of loving concern for the welfare of others. Consequently, in most situations the loving action is in conformity with the guidelines provided by the commandments (Harakas, 1970).

The possibility remains open, however, for the exception, i.e., for the exercise of "economia" when conformity to the prescribed action is perceived as detracting from the basic intent of all reasoning—the advancement of each person in community toward the fulfillment of the image and likeness of God. Thus, both order and compassion are harmonized in an approach to Christian ethics that seeks to avoid the extremes of legalism and relativism.

Bioethics

It is convenient to treat the Eastern Orthodox approach to bioethics under two major rubrics: the protection of life and the transmission of life. Implicit in the treatment of each of the bioethical issues are the affirmations implied in the doctrines of the image and likeness, *theosis*, human self-determination and responsibility, the intimate bond of body and personhood, and the interpenetrating relationship of commandment, love, and the realization of true human potential.

The protection of life. Orthodox Christian ethical thought universally holds that life is a gift of God and as such is the necessary prerequisite of all other physical, spiritual, and moral values. As a gift of God it is a moral good held by the individual and by societies in trust, and over which they do not have absolute control. Both the individual and societies, however, are charged with the moral responsibility of protecting, transmitting, and enhancing life. The concerns of bioethics relate primarily to the first two of these concerns. Generally speaking, human responsibility for the preservation of life means that we are not given the right to terminate human life. Even the exceptions to this rule are understood as arising when conflicting claims to life become mutually exclusive, and a choice must be made. The preservation and protection of life are thus seen as crucial in ethical decision making. Since life is the prerequisite of all other this-worldly goods such as education, intelligence, social worth, and service to humanity, it has an intrinsic value that may not be violated under normal circumstances.

Health care. It follows quite logically that the care of one's own health and societal concern for public health are moral imperatives (Androutsos, pp. 191–195, 250). Throughout its history, Eastern Orthodox Christianity has concerned itself sacramentally with the physical health of the faithful. The Sacrament of Holy Unction has not been conducted as a service of the "last rites." Rather, it is a healing service conducted both publicly and privately for the faithful. One of the constituents of the condition of original sin in which man actually finds himself is sickness. Total harmony of the creation with God would in fact eliminate sickness and ill health. The spiritual and physical dimensions of health are closely bound together in Orthodox thought. Thus, it was natural for the priest and the physician often to be one and the same person (Constantelos, 1967).

The issue of the allocation of scarce medical resources demands a general principle of distribution. Neither the ability to pay nor an aristocratic criterion of greater human value or worth is acceptable. Eastern Christianity has always distinguished between the essential value of human life and social worth. In spite of the enormous difficulties involved, the ethical imperative from the Orthodox perspective calls for the widest possible distribution of health care and life-protecting facilities and resources, rather than a concentration of such resources for the select few. The famous health care center established by Saint Basil in the fourth century in Cappadocia of Asia Minor was designed to reach as many people as possible. It and similar institutions embodied the Eastern Christian view on health distribution (Constantelos, 1968, chap. 11).

Rights of patients. The understanding that each person is created in the image and likeness of God with the personal destiny of achieving *theosis* implies that each patient has an essential and inviolate dignity as a person. The fact that individuals can achieve personhood only in community (*Unus Christianus, nullus Christianus*), requires the concern of the healthy for the ill. Those who deliver health care, therefore, do not morally discharge their responsibility by the mere mechanical application of healing methods and practices. Underlying every medical procedure ought to be a basic respect for the patient as God's image and likeness. The patient is never a thing. Consequently, medical practitioners are obligated, within reason and in the light of the patient's well-being, to maintain confidentiality and to obtain informed consent for procedures that entail excessive risk. Exceptions and restrictions on this obligation should be made in the light of the patient's welfare and whenever possible in consultation with those having immediate responsibility for the patient, e.g., his or her family.

Human experimentation. For the same reasons articulated in the previous section, the Eastern Orthodox Christians take a very hesitant stance vis-à-vis human experimentation.

Medical trial and error conducted for the well-being of the patient himself is often required and necessary. However, the submission of a patient to experimental procedures without significant regard for his or her direct personal benefit is wrong. There is no moral obligation of any person to be used by another for the benefit of a third party. Human self-determination requires that the patient decide. Such a decision must be based on adequate information regarding the procedures, ends to be achieved, and risks involved. The patient does not have the right to inflict harm upon himself unnecessarily. The researcher should use human experimentation procedures only after all other means of testing have been exhausted and there is every reasonable expectation of the avoidance of harm to the subject. In every case, experimenter and subject are morally obligated to exercise great caution. The hope of benefiting mankind in general does not outweigh the moral obligation of the protection of the individual life.

Abortion. Eastern Christianity has a long history of opposition to abortion. Its ethical teachings as embodied in canon law and in the penitential books, as well as in more formal ethical instruction, condemn abortion as a form of murder. Because our humanity is a psychosomatic unity and because Orthodox Christians see all of life as a continuous and never ending development of the image and likeness toward *theosis* and full humanity, the achievement of particular stages of development of the conceptus is not ethically relevant to the question of abortion.

In his second canon, St. Basil specifically rules out the artificial distinction between the "formed" and "unformed" conceptus (*The Rudder*, pp. 789–790). Thus, any abortion is seen as an evil. Since the physical and the personal aspects of human existence are understood as essential constitutive elements of our humanity, the conceptus—unfulfilled and incomplete as it may be—may not be destroyed under normal circumstances. Eastern Orthodox ethicists reject as unworthy those counterarguments which appeal to economic and social reasons and so hold life to be less valuable than money, pride, or convenience. Armed with modern genetic information, they also reject the argument that an abortion may be justified because a woman is entitled to control her own body. That basic affirmation of self-determination is not rejected; what is rejected is the claim that the conceptus is a part of the mother's tissue. It is not her body; it is the body and life of another human being entrusted to her for care and nurture.

Only in the case in which the life of the mother is endangered by the conceptus is it morally appropriate to consider the possibility of abortion. Yet, even here, the main operative value is the preservation of life. Numerous prudential considerations will be taken into account, though it is likely that the preservation of the mother's life will most often be chosen. In any case, it falls into the class of "involuntary sin" in which the evil of the event is recognized, while the personal guilt is mitigated (Papacostas, pp. 9–13, 83–105).

Organ transplants. In the case of organ transplants, the crucial ethical considerations are two: the potential harm inflicted upon the donor and the need of the recipient. Historically, the Orthodox Church has not objected to similar, though not identical, procedures, such as blood transfusions and skin grafts. In both cases, no radical threat to the life of the donor is perceived, and the lifesaving consequences for the recipient are substantial. Similar considerations affect the Orthodox Christian judgment of organ transplants. In no case should a person ignore or make light of the ethical implications of organ donation. Donating an organ whose loss will impair or threaten the life of the potential donor is never required and is never a moral obligation of any person. If the condition of health and the physical well-being of the donor permits, some transplants are not objectionable. Renal transplants are a case in point. A healthy person may consent to donate a kidney knowing that his or her health is not thereby impaired.

The recipient of an organ transplant ought to be in otherwise good health, and there should be a substantial expectation of restoration to normal living in order to warrant the risk to the donor.

Heart transplants present a special case. Objectively they are different from other sorts of organ transplants because they presuppose the death of the donor. Though some Orthodox hierarchs have objected to heart transplants because the "heart" is often designated in the devotional literature of the Church as the seat of the soul, most have not responded negatively to heart transplants in principle. However, caution has been expressed regarding the temptation to hasten the death of the donor for the sake of the recipient. Also, so long as this procedure does not yet have a high success rate, it is morally questionable to continue its practice until the phenomenon of tissue rejection is better understood.

Drug addiction. The *use* of stimulants, depressants, and hallucinogens for any purpose other than the restoration of health or the alleviation of abnormal pain, when properly and legitimately prescribed by a physician, is condemned; but Orthodox ethics, because of its teaching on "involuntary sin," is able to recognize the evil of the condition of drug addiction and yet also recognize that the essence of the evil is that personal self-determination has been lost, and with it a large measure of personal responsibility. Orthodox texts often refer to sinful conditions as "sickness" and "illness." In the case of drug addiction the cure is the restoration of self-determination. In the Orthodox view, the judgment that drug addiction and alcoholism are evil and sinful, on the one hand, and the judgment that they are illnesses, on the other hand, are not mutually exclusive. This is not to say, of course, that every sickness is the result of individual voluntary sins, a position specifically denied by the Orthodox doctrine of original sin.

Mental health: values, therapies, institutions. At the heart of the Eastern Orthodox Christian approach to mental health is the understanding of human wholesomeness in the doctrine of *theosis*. True and full human well-being is the consequence of our proper relationship with God (Demetropoulos, pp. 155–157). Mental health is one dimension of this total relationship. Since no individual human being perfectly achieves this relationship, it may be noted that, just as we are all in some measure "less than fully human," in the same manner we are all in some measure lacking in full mental health. The Orthodox concept of repentance or *metanoia* implies a change of mind, a transfiguration and transformation of the human mind. What is significant is that the teaching of the spiritual Fathers of the Eastern Church emphasizes the need for constant repentance on the part of every human being in the direction of his human goal and destiny.

Some recent studies have related traditional spiritual methodologies to standard psychotherapeutic theories, methods, and approaches (Faros). There are differences, of course, but there is also a remarkable number of parallels to be found between the ancient spiritual disciplines and modern schools of psychology.

Orthodox ethics sees the mentally ill as fellow human beings who need compassionate assistance. Therapies that degrade their essential humanity and attitudes that dehumanize the mentally ill in the eyes of society and deny assistance, relationship, and therapeutic support are in themselves immoral and dehumanizing.

Aging. In the ethical consciousness of the Church, respect and deference for the elderly, and especially for elderly parents, is an important moral responsibility. There is a strong feeling that children ought personally to care for their aged parents. It is only when circumstances are such that it is truly impossible for children to care for their aged parents that they may be placed in appropriate institutions for care. Such institutions have long been a part of the Eastern Orthodox Church's social mission (Constantelos, 1968, chap. 13).

Death, dying, and euthanasia. The traditional definition of physical death is "the separation of soul and body." Such a definition is not subject to objective observation. Thus it is not within the province of theology to determine the medical indications of death and the onset of the dying process. However, in reference to the terminally ill person, certain distinctions can be made. Physical life is generally understood to imply the ability of the person to sustain his or her vital activities. Physical death begins when interrelated systems of the body begin to break down. Death occurs when the systemic breakdown becomes irreversible. It may well be that physical life and death are events in a continuum in which it is impossible to discern when the dying process actually begins. Nevertheless, the bias of the Church and the traditional bias of the medical practitioner (cf. Oath of Hippocrates) is to do everything possible to maintain life and hinder the onset of dying and death. The medical use of drugs, surgical operations, and even artificial organs (mechanical kidneys, lungs, hearts, etc.) are considered legitimately used when there is a reasonable expectation that they will aid the return in due time to normal or close to normal functioning of the whole organic system.

The special case arises in that it is now medically possible to keep the body "alive" with a complex array of artificial organs, medications, transfusions, and the like. Under these conditions it may not be feasible to expect, with any degree of probability, the restoration of the organic functioning of the body. When, especially, there is no evidence of brain activity in conjunction with the systemic breakdown, we can safely say that the patient is no longer alive in any religiously significant way, and that, in fact, only certain organs are functioning. In such a case there is no moral responsibility to continue the use of artificial means. It is of interest that the Prayerbook of the Eastern Orthodox Church includes a whole service devoted to those in the process of dying. In the case of the individual whose death is prolonged and attended by much "struggling to die," the key sentence in the prayer calls upon God to separate the soul from the body, thus giving rest to the dying person. It asks God "to release your servant (name) from this unbearable suffering and this continuing bitter illness and grant rest to him" (*Mikron Euchologion*, p. 192).

However, it must be emphasized that this is a prayer directed to God, who, for the Orthodox, has ultimate dominion over life and death. Consequently, the preceding discussion in no way supports the practice of euthanasia. Euthanasia is held by some to be morally justified and/or morally required to terminate the life of an incurably sick person. To permit a dying person to die, when there is no real expectation that life can sustain itself, and even to pray to the Author of Life to take the life of one "struggling to die" is one thing; euthanasia is another, i.e., the active intervention to terminate the life of another. Orthodox Christian ethics rejects the alternative of the willful termination of dying patients, regarding it as a special case of murder if done without the knowledge and consent of the patient, and suicide if it is permitted by the patient (Antoniades, II, pp. 125–127). One of the most serious criticisms of euthanasia is the grave difficulty in drawing the line between "bearable suffering" and "unbearable suffering," especially from an Eastern Orthodox perspective, which has taken seriously the spiritual growth that may take place through suffering (Rom. 8:17–39).

Ethical decision making is never precise and absolute. The principles that govern it are in a measure fluid and subject to interpretation. But to elevate euthanasia to a right or an obligation would bring it into direct conflict with the fundamental ethical affirmation that as human beings we are custodians of life, which comes from a source other than ourselves. Furthermore, the immense possibilities, not only for error but also for decision making based on self-serving ends, which may disregard the fundamental principle of the sanctity of human life, argue against euthanasia.

Generally speaking, the Orthodox Church teaches that it is the duty of both physician and family to make the patient as comfortable as possible, to provide the opportunity for the exercise of patience, courage, repentance, and prayer. The Church has always rejected inflicted and unnecessary voluntary suffering and pain as immoral; but at the same time, the Church also has perceived in suffering a positive value that often goes unrecognized in the "logic of the world."

The only "eu-thanasia" (Greek for "a good death") recognized in Orthodox ethics is that death in which the human person accepts the end of his or her life in the spirit of moral and spiritual purity, in hope and trust in God, and as a member of his kingdom. True humanity may be achieved even on a deathbed.

The transmission of life. Orthodox Christian ethical thought considers that the transmission of human life is no less a fundamental responsibility of mankind than its protection. The Church sees this aspect of its concern as the divinely chosen means by which human beings contribute cooperatively in God's creative work. The transmission of human life is thus a holy and sacred moral responsibility. This responsibility is a generally human one and is taken up, sanctified, and made a part of the corporate life of the body of Christ in the Sacrament of Holy Matrimony. Though not the only purpose of marriage, the transmission of human life is an important duty and moral responsibility. This is readily seen in the fact that if each and every person now alive failed to contribute to the transmission of human life, it would be only a matter of time until human life would be extinguished from the face of the earth. The divine injunction "to be fruitful and multiply" (Genesis 1:28) is a fundamental moral imperative in the teaching of the Orthodox Christian Church. It is within this larger framework that we approach the specific issues of human sexuality, fertility control, population, artificial insemination, in vitro fertilization, and genetic screening and counseling.

Human sexuality. The Church teaches that human sexuality is a divinely given dimension of human life that finds its fulfillment in the marital relationship. This is also supported by empirical observation, for at their very biological basis, sexual differences clearly exist for reproductive purposes. Because of the fact that human reproduction requires a long period of time for the newly born child to achieve a level of development permitting physical self-care, and increasingly long periods for social, educational, emotional, and economic maturity, the human race long ago recognized the need for some kind of permanent relationship of the sexes for the purpose of serving the reproductive purpose. That permanent relationship is marriage.

However, the purpose of marriage is not limited or restricted to this aspect alone. The purposes of marriage and their ranking in importance are a point of difference among Orthodox authorities (both patristic and contemporary), but scriptural and patristic evidence argue for at least four purposes for marriage, without ranking them in order of primacy: (1) the birth and care of children, (2) the mutual aid of the couple, (3) the satisfaction of the sexual drive, (4) growth in mutuality and oneness, i.e., love. In the mixture of these purposes, the whole purpose of human sexuality is fulfilled and completed, ethically and humanly (Constantelos, 1975).

Ethical corollaries of this position are: (1) all the dimensions of human sexuality are properly fulfilled in marriage, and the married have the moral obligation to seek the enrichment and fulfillment of their marriage in all of its aspects, as indicated above; (2) premarital sexual relations between unmarried persons are sinful and as such are labeled fornication; (3) sexual relations between two persons, at least one of whom is married to a third person, are morally evil and as such are labeled adultery; (4) sexual relations between persons for payment is sinful and is labeled prostitution; (5) sexual relations between brothers and sisters, parents and children, and other close relatives are morally wrong and as such are labeled incestuous; (6) sexual relations between persons of the same sex are immoral and as such are labeled as acts of homosexuality in the case of males, and lesbianism in the case of females; (7) sexual relations between a human being and animals are condemned as immoral, being labeled acts of bestiality; (8) autoerotic activity is adjudged as an improper expression of human sexuality, and as such is labeled masturbation.

Fertility control. Fertility control, or contraception, is the practice by which mechanical, chemical, or other means are used, either before or after a sexual act, in order to prevent fertilization of the ovum by the sperm, thus circumventing the possible consequences of the sexual act—the conception and ultimate birth of a child.

General agreement exists among Orthodox writers on the following two points: (1) since at least one of the purposes of marriage is the birth of children, a couple acts immorally when it consistently uses contraceptive methods to avoid the birth of any children, if there are not extenuating circumstances; (2) contraception is also immoral when used to encourage the practice of fornication and adultery.

Less agreement exists among Eastern Orthodox authors on the issue of contraception within marriage for the spacing of children or for the limitation of the number of children. Some authors take a negative view and count any use of contraceptive methods within or outside of marriage as immoral (Papacostas, pp. 13–18; Gabriel Dionysiatou). These authors tend to emphasize as the primary and almost exclusive purpose of marriage the birth of children and their upbringing. They tend to consider any other exercise of the sexual function as the submission of this holy act to unworthy purposes, i.e., pleasure-seeking, passion, and bodily gratification, which are held to be inappropriate for the Christian growing in spiritual perfection. These teachers hold that the only alternative is sexual abstinence in marriage, which, though difficult, is both desirable and possible through the aid of the grace of God. It must be noted also that, for these writers, abortion and contraception are closely tied together, and often little or no distinction is made between the two. Further, it is hard to discern in their writings any difference in judgment between those who use contraceptive methods so as to have no children and those who use them to space and limit the number of children.

Other Orthodox writers have challenged this view by seriously questioning the Orthodoxy of the exclusive and all-controlling role of the procreative purpose of marriage (Zaphiris; Constantelos, 1975). Some note the inconsistency of the advocacy of sexual continence in marriage with the scriptural teaching that one of the purposes of marriage is to permit the ethical fulfillment of sexual drives, so as to avoid fornication and adultery (1 Cor. 7:1–7). Most authors, however, emphasize the sacramental nature of marriage and its place within the framework of Christian anthropology, seeing the sexual relationship of husband and wife as one aspect of the mutual growth of the couple in love and unity. This approach readily adapts itself to an ethical position that would not only permit but also enjoin sexual relationships of husband and wife for their own sake as expressions of mutual love. Such a view clearly would support the use of contraceptive practices for the purpose of spacing and limiting children so as to permit greater freedom of the couple in the expression of their mutual love.

Population. There would appear to be a direct contradiction between the ethical imperative to "be fruitful and multiply" and the need to respond ethically to the "population explosion."

Those few Orthodox writers who have addressed themselves to this question ask if the issue is not so much a question of population as it is one of the fair and just distribution of the world's resources. (Papacostas; Gabriel Dionysiatou; Evdokimov, pp. 163–174). However, in the light of strong evidence that food and mineral resources are limited, population control is, without question, of ethical significance. This is not necessarily in conflict with the Orthodox teaching on marriage. Of interest in this instance is a fourth-century quotation from St. John Chrysostom, made in reference to the purpose of marriage, which the saint considered to be primarily the satisfaction of the sexual drive:

> It was for two reasons that marriage was introduced; so that we may live in chastity [*sophrosyne*] and so that we might become parents. Of these the most important reason is chastity . . . especially today when the whole inhabited world [*he oikoumene*] is full of our race [John Chrysostom].

If overpopulation in the saint's eyes was a fact of the fourth century providing an argument to support his views on marriage, it implies that today the fact of overpopulation continues to have ethical significance. If it is true that humanity has in fact been obedient to the divine command and has been "fruitful" and has "multiplied" and has "filled the earth" (Gen. 1:28), then it would appear that this has ethical significance.

Thus, it seems valid to raise the question, within the context of Orthodox ethics, of the appropriate means of population control. Orthodox ethics disapproves of any means of population control that would violate and coerce the individual couple's choice regarding their obligation to procreate. It opposes the use of those means on a large scale that it opposes in individual cases, i.e., abortion. Those Orthodox teachers who oppose contraceptive practices of any nature, when faced with the facts of population pressures, are placed in the position of proposing widespread abstinence from sexual relations by huge numbers of people. Those who hold to the legitimacy of a reasonable use of contraceptives within marriages that have produced some offspring are prepared to accept the need and propriety of population control through educational methods, encouraging smaller families through contraceptive methods. All Orthodox ethicists, however, would hold that respect for the freedom of each couple to decide must be considered an important and significant factor of population control policy.

Artificial insemination. For obvious reasons, artificial insemination of unmarried women, or of married women without the consent and cooperation of the husband, is rejected by the Orthodox, in the first instance as a form of fornication, and in the second as duplicity and a form of adultery (Galanopoulos, pp. 455–456). What of the cases in which the husband gives his permission or urges the procedure upon his wife? In this situation, when a donor's semen is used, Orthodox ethicists readily view it as the intrusion of a third person into the sacred marital relationship and reject it as a form of adultery not ethically appropriate. In the instances in which the couple is not able to bear their own children, the other purposes of marriage remain in effect, and the marriage of the couple continues to be both valid and fulfilling. Such a couple may decide to adopt children.

In the case of insemination with the husband's sperm (AIH), there are differing opinions. Some ethicists hold that AIH is also improper because the child is not conceived as a result of natural sexual intercourse (Constantinides). This position, however, does not prohibit medical treatment of the husband for the correction of some medical defect that may be the cause of the failure to achieve conception. This view is countered by the consideration that the integrity of the marital relationship is not attacked by AIH. Rather, one of its main purposes is permitted to be fulfilled. It is questionable if the ethical argumentation connecting AIH with the requirement for the physical act of sexual intercourse is drawn from Eastern Orthodox sources.

Orthodox writers have not dealt with artificial inovulation and in vitro fertilization procedures.

It would seem consistent, though, to hold that, so long as the sperm and ovum are those of the husband and wife, and the wife carried the child to term, such procedures would not in themselves be objectionable. However, egg grafts from anonymous donors and the transplantation of a fertilized ovum to a foster mother who would then carry the conceptus to term would attack the integrity of the marriage and the mother-child relationship.

Another topic that has received little treatment from Orthodox writers is sterilization: vasectomy in the case of the male and tubal ligation in the case of the female. It would appear that the irreversible character of these procedures would cause most Orthodox to see them as a violation of one of the purposes of marriage, though it is conceivable that some cases involving serious threat to the life of the wife might justify the procedures. Obviously, the use of the operation to permit promiscuous sexual living would be rejected out of hand by Orthodox ethicists (Zozos).

Genetic counseling and genetic screening. At first glance it may appear that the Eastern Orthodox Church has little or nothing to say on genetic counseling and screening. Yet genetic counseling, which seeks to provide information to prospective parents before a child is conceived, simply makes more precise that which the Church has sought to do through its canon law, which prohibits marriages between closely related persons (*The Rudder*, pp. 977–999). This ancient compendium of prohibitions to inbreeding clearly has its historical antecedents in the observation that genetic defects tend to multiply when inbreeding takes place. Consequently, it would appear that genetic counseling most appropriately should take place before marriage. It seems equally clear that for the Orthodox the option of abortion is not ethically appropriate when amniocentesis indicates some genetic deformation.

Genetic screening of whole groups or populations to determine carriers of genetic disease would also be encouraged by Orthodox ethics, so as to provide as much information as possible to persons before marriage. Ethical prudence would cause two persons who are carriers of the same genetic disease not to marry, thus avoiding the high probability that deformed children would be born to them.

In this way, what is more or less crudely effected through the Church's rules regarding prohibited marriages because of consanguinity would be accomplished more accurately through scientific genetic screening. In the same spirit, it would be possible to support legislation prohibiting marriage between two carriers of the same genetic disease, especially in the case of a disease that is widespread and a threat to the total human genetic pool.

Conclusion

The common denominator of all the issues discussed is the high regard and concern of the Church for human life as a gift of God. Orthodoxy tends to take a conservative approach to these issues, seeing in them a dimension of the holy and relating them to transcendent values and concerns. An intense respect for human life is needed to hold the reins upon those who would attack it. The human person, from the very moment of conception, is dependent upon others for life and sustenance. It is in the community of the living, especially as it relates to the source of life, God in Trinity, that life is conceived, nurtured, developed, and fulfilled. The trust we have in others for the continued well-being of our own lives forms a basis for generalization. Eastern Orthodox ethics, consequently, functions with a pro-life bias that honors and respects the life of each person as a divine gift, which requires development and enhancement.

STANLEY S. HARAKAS

[*For further discussion of topics mentioned in this article, see the entries:* ABORTION; CONTRACEPTION; DEATH AND DYING: EUTHANASIA AND SUSTAINING LIFE; DOUBLE EFFECT; ETHICS, *article on* THEOLOGICAL ETHICS; GENETIC DIAGNOSIS AND COUNSELING; GENETIC SCREENING; MENTAL HEALTH, *article on* RELIGION AND MENTAL HEALTH; MENTAL ILLNESS, *article on* CONCEPTIONS OF MENTAL ILLNESS; ORGAN TRANSPLANTATION, *article on* ETHICAL PRINCIPLES; POPULATION ETHICS: RELIGIOUS TRADITIONS, *article on* EASTERN ORTHODOX CHRISTIAN PERSPECTIVES; *and* SEXUAL ETHICS. *Also directly related are the entries* DEATH; EMBODIMENT; HUMAN EXPERIMENTATION; INFORMED CONSENT IN HUMAN RESEARCH; LIFE; NATURAL LAW; *and* RATIONING OF MEDICAL TREATMENT. *See also:* REPRODUCTIVE TECHNOLOGIES, *article on* ETHICAL ISSUES; *and* STERILIZATION, *article on* ETHICAL ASPECTS.]

BIBLIOGRAPHY

ANDROUSTOS, CHRISTOS. *Systema ethikes* [Ethical system]. 2d ed. Thessalonike, Greece: Basil Regopoulos Publishing House, 1964.

ANTONIADES, VASILEIOS. *Encheiridion kata Christon ethikes* [Handbook of Christian ethics]. Constantinople: Fazilet Press, 1927.

BASIL, SAINT; RALLE, G. A.; and POTLE, M. *Syntagma ieron kanonon* [Compendium of sacred canons]. Athens: Hartophylakos Press, 1854, vol. 4, pp. 88–294.

CONSTANTELOS, DEMETRIOS J. *Byzantine Philanthropy and Social Welfare*. New Brunswick, N.J.: Rutgers University Press, 1968.

———. *Marriage, Sexuality and Celibacy: A Greek Orthodox Perspective*. Minneapolis, Minn: Light and Life Publishing Co., 1975.

———. "Physician-Priests in the Medieval Greek Church." *Greek Orthodox Theological Review*, Winter 1966–1967, pp. 141–153.

CONSTANTINIDES, CHRYSOSTOM. "Technike gonimopoiesis kai theologia." [Artificial insemination and theology]. *Orthodoxia* 33 (1958): 66–79; and 34 (1959): 36–52.

DEMETROPOULOS, PANAGIOTES. *Orthodoxos Christianike ethike* [Orthodox Christian ethics]. Athens: 1970.

EVDOKIMOV [EVDOKIMOFF], PAUL. *Sacrement de l'amour: Le Mystère conjugal à la lumière de la tradition orthodoxe*. Paris: Éditions de l'Épi, 1962. Greek ed. *Mysterion tes agapes* [Mystery of love]. Translated by Serapheim Orphanos. Athens: 1967.

FAROS, PHILOTHEOS. "Mental Patients and Verbal Communication of the Religious Message." *Theologia* 42 (1971): 602–606.

FLOROVSKY, GEORGES. *Bible, Church, Tradition: An Eastern Orthodox View*. Belmont, Mass.: Nordland Publishing Co., 1972.

GABRIEL DIONYSIATOU. *Malthousianismos: To englema tes genoktonias* [Malthusianism: The crime of genocide]. Volos: Holy Mountain Library, 1957.

GALANOPOULOS, MELETIOS. *Systema ieras exomologetikes* [A study of penitential practice]. Athens: Orthodox Source Books, 1954.

HARAKAS, STANLEY S. "The Natural Law Teaching of the Eastern Orthodox Church." *Greek Orthodox Theological Review*, Winter 1963–1964, pp. 215–224.

———. "An Orthodox Christian Approach to the 'New Morality'." *Greek Orthodox Theological Review*, Spring 1970, pp. 107–139.

JOHN CHRYSOSTOM. "Eis to apostolikon reton: Dia de tas porneias ekastos ten heautou gynaika echeto" [On the words of the apostle: Concerning the fornication each has with his own wife]. *Patrologiae cursus completus: Series graeca* [Patrologia Graeca]. 161 vols. Paris: Apud Garnier Fratres, editores & J.-P. Migne successors, 1857–1894, vol. 51, cols. 207–218, at col. 213.

KOTSONIS, JEROME. "Fundamental Principles of Orthodox Morality." *The Orthodox Ethos: Studies in Orthodoxy*. Edited by A. J. Philippou. Oxford: Holywell Press, 1964.

MALONEY, GEORGE A. *Man, The Divine Icon: The Patristic Doctrine of Man Made According to the Image of God*. Pecos, N.M.: Dove Publications, 1973.

MANTZARIDES, GEORGE. *Christianike Ethike: University Lectures*. Thessalonike, Greece, 1975.

Mikron euchologion, e aghiasmatarion [Shorter prayer book]. Athens: Apostolike diakonia tes ekklesias tes Ellados, 1956.

PAPAKOSTAS, SERAPHEIM. *To zetema tes teknogonias: To demographikon problema apo Christianikes apopseos* [Question of the procreation of children: The demographic problem from a Christian viewpoint]. Athens: Brotherhood of Theologians "Zoe," 1933, 1947.

The Rudder. Translated by D. Cummings. Chicago: Orthodox Christian Educational Society, 1957.

WARE, TIMOTHY. *The Orthodox Church*. Baltimore: Penguin Books, 1972.

ZAPHIRIS, CHRYSOSTOM. "The Morality of Contraception: An Eastern Orthodox Opinion." *Journal of Ecumenical Studies* 11 (1974): 677–690.

ZOZOS, CONSTANTINE. "The Medical, Legal, and Moral Aspects of Sterilization." *Ekfrasis*, Spring 1974, pp. 33–54.

EASTERN ORTHODOX ETHICS
See EASTERN ORTHODOX CHRISTIANITY.

ECOLOGY
See ENVIRONMENT AND MAN; ENVIRONMENTAL ETHICS.

EDUCATION
See MEDICAL EDUCATION; MEDICAL ETHICS EDUCATION; NURSING; POPULATION POLICY PROPOSALS, *article on* POPULATION EDUCATION.

EGYPT
See MEDICAL ETHICS, HISTORY OF, *section on* NEAR AND MIDDLE EAST AND AFRICA, *articles on* ANCIENT NEAR EAST, CONTEMPORARY ARAB WORLD, *and* CONTEMPORARY MUSLIM PERSPECTIVE. See also ISLAM.

ELDERLY
See AGING AND THE AGED.

ELECTRICAL STIMULATION OF THE BRAIN

Concepts and principles underlying the ethical issues discussed in this entry can be found in the entry BEHAVIOR CONTROL.

Electrical stimulation of the brain (ESB) was developed primarily as a research technique with therapeutic potential. Stimulation of areas deep within the brain can contribute parts of a functional "map" by making it possible to correlate observed behavioral responses with specific brain sites. ESB is therapeutically valuable both in its own right and as a diagnostic procedure. It can, for example, be used to locate the precise point at which the neurosurgeon will make a therapeutic lesion in order to relieve a variety of symptoms.

The technique typically involves the introduction within cerebral tissue of fine metallic conductors, which are insulated except at their tips, and with terminals located outside the scalp for making connections with instrumentation, including transmitting and receiving equipment and quite possibly a computer for processing incoming signals and selecting appropriate signals for transmission to the brain.

State of the art

The modern era of experimental work was initiated by the discovery of the motivational effects of electrical stimulation and of its effectiveness as a conditioned stimulus. This development was facilitated technically by the development of stereotaxic surgery, by means of which depth electrodes may be implanted deep in the brain with a very high degree of accuracy. The fine metallic conductors cause minimal and functionally undetectable damage to brain tissue. There is every indication that the conductors may remain in place indefinitely with no untoward effects. The problem of using cumbersome wiring between subject and instrumentation has been solved by development of miniaturized, multichannel radiostimulators, which may be implanted beneath the skin of the subject. ESB may now be administered in a variety of settings by remote control.

Clinical applications have been or are being developed for inhibition of epileptic seizures, sensory prostheses, and treatment of some types of psychopathology. ESB has already been used experimentally in primates to influence the following functions: motor activity, food intake, aggressive behavior, maternal relations, sexual functions, motivation, learning, anxiety, pleasure, and friendliness. ESB is clearly impressive as a behavior control technology, but it still leaves scientific problems unsolved and a large array of ethical issues to be considered.

Although ESB has established correlations between brain sites and behavioral functions, the details of the physiological response to ESB and the nature of the neural changes associated with the conditioning process are not well understood. ESB does not reveal the mode of action of a particular brain site, nor does it provide an integrated understanding of brain functions or of mediating processes in the brain at the time of stimulation.

Direct manipulation of the brain for purposes of controlling behavior raises serious questions for psychiatry. The nature of psychopathology, the relation between organic and functional concepts in the etiology of mental illness, and the choice of appropriate methods of treatment may all need revision in light of these new techniques. The relative simplicity and effectiveness of ESB may lead to its widespread use before its secondary and indirect effects on personality are well enough understood, and before social and ethical implications have been carefully considered.

Ethical issues

ESB presents ethical issues by virtue of its status as a major clinical procedure; because it must be considered an experimental rather than a standard therapy; because of its potential for placing the control of one's behavior in the hands of others who may or may not be of one's choosing, and for ends not necessarily one's own; and because it controls behavior by directly manipulating the brain.

So long as the procedure is requested or agreed to by a competent individual who understands the nature and consequences of the procedure and who elects the procedure for reasons that appear reasonable and medically justifiable, there need be only the assurance that the patient is well protected as a subject of human experimentation and that the patient is competent to give consent. In this ideal situation, however, all parties must still determine whether the risks are justified and whether the consent to those risks is both voluntary and informed.

It is in the nature of ESB that the ideal of the competent, rational patient may be difficult to fulfill, since it is precisely the organ of consent and reason which is "ill" or functioning improperly. Thus, the locus of decision and control is likely to shift radically to others, and with this shift the dangers of control by others must be taken into account. What criteria of normalcy are employed in diagnosing behavioral dysfunction; and what behavior is to be controlled? Who will make these decisions, and to what extent are the controllers accountable to the patient, his representatives, and the public?

One must weigh the costs and benefits of ESB relative to other modes of treatment or to no treatment at all. Should one alter the brain directly or the environment in order to change behavior? If treatment is imposed on the patient by others can this intrusion on the patient's freedom and dignity be balanced by the expansion of freedom and self-control to be gained as a result of treatment? If treatment were to be undertaken not in the individual's best interests but to protect society (as in the case of an assaultive prisoner), does the gain in protection of others warrant the denial of the freedom and dignity of the person who is subjected to the treatment? How best may these decisions regarding legitimacy and evaluation be made?

These questions take on special importance because they deal with the direct manipulation of the brain. No other organ is so intimately associated with our conceptions of personhood, identity, and self-control. The brain is the locus of mind and "agent" of behavior, and presumably modification of the brain will affect personality more than other kinds of manipulation. Thus the intrusiveness of ESB procedures may be a high cost to pay for the benefits. The very directness of ESB, the lack of mediation by consciousness, may have consequences for humans that ought to be avoided, if those consequences include a reductive, mechanistic, and manipulative image of man.

If ethical problems arise in the medical context, where relief of suffering is the object, and the expected benefits of ESB are therefore relatively easy to specify and widely perceived to be legitimate, how much more difficult it will be to decide whether and to what ends to use ESB to program the experiences and behavior of the well. The social history of psychotropic drugs is suggestive. The question remains of how best to weigh all considerations—the possible costs to personal dignity and individual liberty and the possible gains in happiness and knowledge.

JOEL MEISTER

[*Directly related are the entries* BEHAVIOR CONTROL; *and* MENTAL HEALTH THERAPIES. *See also:* INFORMED CONSENT IN MENTAL HEALTH; MENTAL HEALTH, *article on* MENTAL HEALTH IN COMPETITION WITH OTHER VALUES; PERSON; *and* RESEARCH, BEHAVIORAL. *For discussion of related ideas, see the entries:* ELECTROCONVULSIVE THERAPY; PSYCHOPHARMACOLOGY; *and* PSYCHOSURGERY.]

BIBLIOGRAPHY

COOPER, I. S.; CRIGHEL, E.; and AMIN, I. "Clinical and Physiological Effects of Stimulation of the Paleocerebellum in Humans." *Journal of the American Geriatrics Society* 21 (1973): 40–43.

DELGADO, JOSÉ M. R. *Evolution of Physical Control of the Brain.* New York: American Museum of Natural History, 1965.

———; MARK, V.; SWEET, W.; ERVIN, F.; WEISS, A.; BACH-Y-RITA, A.; and HAGIWARA, R. "Intracerebral Radio Stimulation and Recording in Completely Free Patients." *Journal of Nervous and Mental Disease* 147 (1968): 329–340.

———; ROBERTS, W. W.; and MILLER, N. E. "Learning Motivated by Electrical Stimulation of the Brain." *American Journal of Physiology* 179 (1954): 587–593.

ERVIN, FRANK R.; MARK, VERNON H.; and STEVENS, JANICE. "Behavioral and Affective Responses to Brain Stimulation in Man." *Neurobiological Aspects of Psychopathology.* Edited by Joseph Zubin and Charles Shagass. New York: Grune & Stratton, 1969, pp. 54–65.

GAYLIN, WILLARD; MEISTER, JOEL; and NEVILLE, ROBERT, eds. *Operating on the Mind: The Psychosurgery Conflict.* New York: Basic Books, 1975.

GILMAN, S.; AMIN, I.; and COOPER, I. S. "The Effect of Chronic Cerebellar Stimulation upon Epilepsy in Man." *Transactions of the American Neurological Association* 98 (1973): 192–196.

HEATH, R. G. "Electrical Self-Stimulation of the Brain in Man." *American Journal of Psychiatry* 120 (1963): 571–577.

INGRAHAM, BARTON L., and SMITH, GERALD W. "The Use of Electronics on the Observation and Control of Human Behavior and Its Possible Use in Rehabilitation and Parole." *Issues in Criminology* 7, no. 2 (1972), pp. 35–53.

LONG, DONLIN M. "Electrical Stimulation for Relief of Pain From Chronic Nerve Injury." *Journal of Neurosurgery* 38 (1973): 718–722.

MARK, VERNON H. "The Relief of Chronic Severe Pain by Stereotactic Surgery." *Pain and the Neurosurgeon: A Forty Years' Experience.* Edited by J. C. White and W. H. Sweet. Springfield, Ill.: Charles C Thomas, 1969, pp. 843–887.

———, and ERVIN, FRANK R. *Violence and the Brain.* New York: Harper & Row, 1970.

NEVILLE, ROBERT. "Zalmoxis, or The Morals of ESB and Psychosurgery." Gaylin, *Operating on the Mind,* pp. 87–116.

OLDS, JAMES, and MILNER, PETER. "Positive Reinforcement Produced by Electrical Stimulation of Septal Area and Other Regions of Rat Brain." *Journal of Comparative and Physiological Psychology* 47 (1954): 419–427.

"Physical Manipulation of the Brain." *Hastings Center Report,* special supp., May 1973.

SCHWITZGEBEL, ROBERT, and SCHWITZGEBEL, RALPH K. *Psychotechnology: Electronic Control of Mind and Behavior.* New York: Holt, Rinehart & Winston, 1973.

SHEER, DANIEL E., ed. *Electrical Stimulation of the Brain.* Austin: University of Texas Press, 1961.

VALENSTEIN, ELIOT S. *Brain Control: A Critical Examination of Brain Stimulation and Psychosurgery.* New York: Wiley-Interscience, 1973.

VAUGHAN, HERBERT G., JR. "Some Reflections on Stimulation of the Human Brain." *Neurobiological Aspects of Psychopathology.* Edited by Joseph Zubin and Charles Shagass. New York: Grune & Stratton, 1969, pp. 66–77.

ZUCKER, MITCHELL H. *Electronic Circuits for the Behavioral and Biomedical Sciences.* San Francisco: W. H. Freeman, 1969.

ELECTROCONVULSIVE THERAPY

Concepts and principles underlying the ethical issues discussed in this entry can be found in the entry BEHAVIOR CONTROL.

Nature and effect of the treatment

Electroconvulsive therapy (ECT) is a method of psychiatric treatment using an electric current to induce a convulsion. Much confusion has arisen over the years through the use of the misnomer "shock treatment" as synonymous with ECT. In fact, ECT has nothing to do with surgical or physiological shock (acute progressive circulatory failure), nor is it accompanied by "shock" in the usual psychological sense of the term.

The stimulus for the development of convulsive therapy was the ancient observation that some mental patients temporarily improved after a spontaneous seizure. Allied to this was the belief, now recognized as erroneous, that epilepsy and psychosis, particularly schizophrenia, never occurred together. In actuality, certain forms of epilepsy (temporal lobe) are often indistinguishable clinically from schizophrenia, while certain types of schizophrenic psychoses are thought to form a continuum with epilepsy, the so-called borderlands of epilepsy.

In 1938 two Italian psychiatrists, Drs. U. Cerletti and L. Bini, first induced a convulsion in a patient by a simple electrical apparatus using an alternating current (Cerletti and Bini). Up until that time convulsions had been induced by the intravenous injection of a mixture of camphor and oil, or subsequently by the use of inhalants such as aliphatic fluorinated ethers, which were administered through a vaporizer to a completely anesthetized patient. Although still in use, inhalation-induced convulsions have been largely replaced by ECT.

Despite many modifications in technique since Cerletti and Bini, the basic method of ECT has changed very little. Electrodes are applied to one side of the head (unilateral ECT) or to both (bilateral ECT) temples. The patient is prepared with a muscle relaxant and anesthetized with a short-acting barbiturate. The electric current is then applied for 0.1 to 0.5 seconds, resulting in a two-stage or tonic-clonic convulsion. (If the degree of muscle relaxation is sufficient the visible signs of convulsions—the tonic-clonic movements—may be entirely absent; an EEG recording, however, will demonstrate evidence of a cerebral seizure.) Afterward the patient regains consciousness in five to ten minutes but remains lethargic and mildly confused from anywhere from fifteen minutes to an hour (Kalinowsky and Hippius).

The number of treatments given during a course of ECT varies from patient to patient and from psychiatrist to psychiatrist, but in general two or three a week for a period of two to four weeks is standard, varying with diagnosis and age of patient. Although originally used for inpatients, it is now frequently given on an outpatient basis, with the patient returning home a few hours after the ECT.

The safety of ECT appears to be reasonably well established with only brain tumors serving as absolute contraindications. Relative contraindications include coronary artery disease or a recent heart attack. Nor does age seem to be a limiting factor in its use since elderly patients have undergone courses of ECT without adverse reactions. In all age groups, however, there is a transient loss of memory that varies with the age of the patient, the number of treatments received, and the method (unilateral ECT causes less memory impairment than does bilateral ECT) (Squire and Chace; Harper and Wiens). Innovations and techniques such as the administration of high concentrates of oxygen or the use of unilateral electrodes have lessened the memory loss considerably (Harper and Wiens).

Originally introduced as a treatment for schizophrenia, ECT is now generally recognized as more effective in the treatment of affect or mood disorders, particularly depression. The indications for its use are involutional depression, psychotic depression, and the depressive phase of manic–depressive psychosis. Although more controversial, its use has been advocated by some in the control of mania and certain forms of acute schizophrenia (acute catatonic excitement). Its value in behavior or personality disorders as well as neurosis has not been established. Its foremost indication remains the imminent probability of suicide. Data exist, however, to indicate that ECT can decrease the mortality of depression quite apart from any consideration of suicide. The total morbidity for depression is not limited to suicide but includes death from infection, malnutrition, and cardiovascular problems (Avery and Winokur).

Bioethical problems associated with ECT

The bioethical issues associated with ECT are similar to those involving psychosurgery and certain forms of drug treatment.

The consent issue. In most instances patients referred for ECT have severe restrictions on the consent process secondary to their illness, which is usually depression, often of a psychotic degree. Since the patient's decision-making powers are affected and other persons often decide on the use of ECT, the treatment demands little in the form of a therapeutic alliance. The patient consequently is "objectified" and "controlled" rather than persuaded to be a willing participant in the treatment. The same, however, could be said of any alternative treatment.

The question of brain damage. Most authorities state that ECT is not associated with permanent changes in brain tissue. In addition there are no autopsy findings that prove ECT causes "brain damage." Despite this, it is the impression of many clinicians that frequent and extended courses of ECT result in a blunted or "burnt out" personality. Since depression is a self-limiting illness whose natural history, at least for any acute episode, is toward improvement or suicide, the question of risk versus benefit assumes critical proportions. The resolution of this important therapeutic and ethical dilemma must await further studies of the basic pathophysiology of natural and induced seizure states, and of depressive illnesses.

Unpleasant psychological effects. It has been recognized for years that patients often demonstrate fears regarding ECT that seem out of proportion to the treatment risks (Kalinowsky and Hippius). These "psychopathological phenomena" can be quite disabling and difficult to manage although they usually terminate after the beginning of the treatment. When present, however, such phenomena greatly complicate the consent issue since the patient may, as a result of these fears, reject ECT on grounds that may appear "unreasonable" to the treating psychiatrist, thus leading to further manipulation and control.

The empiricism of ECT. Despite an impressive body of knowledge on the electrochemical and physiological accompaniments of ECT, the treatment remains largely an empirical one, similar to all treatments for illnesses of unknown etiology. Such a state of affairs is particularly unsatisfactory in behavior control technologies since the patient obviously cannot be adequately informed regarding all the implications of a treatment if the basis for the treatment's efficacy remains mysterious. It should be recalled, however, that ECT has been in use since 1938—a longer experience than we have had with either behavior or drug therapies—and that the patient can therefore be offered reasonable reassurance that treatment, when indicated, will be safe and free of known long-term permanent side effects. It remains to be seen whether or not more subtle consequences to nervous system functioning may be discovered.

Societal factors in the use of ECT. There are indications that in at least one state (Massachusetts) ECT is overutilized in private psychiatric hospitals, while the treatment is underutilized in public state mental hospitals (Dietz). Although the exact reason for the disparity in its use is controversial, it may be at least partly due to economic considerations: ECT is covered by many private health insurance policies. Most of the inmates of private psychiatric hospitals have some form of private health insurance, whereas public mental hospital patients rarely possess private insurance. There is also the question of the availability of anesthetists or anesthesiologists, who are more available for financial reasons at private rather than state hospitals. (There are no rigid requirements that ECT be administered with an anesthetist present, but such an arrangement is frequent enough to constitute "standard medical practice.")

An additional bioethical problem under the "societal factors" heading is the use of ECT to accelerate recovery from depression. If a depressed patient is protected from suicide he eventually will recover; with the use of selective antidepressant drugs he will recover more quickly; finally, improvement may be further accelerated in certain selected cases with the use of ECT. It is not unusual, therefore, in private psychiatric facilities for "productive" or "creative" individuals to receive ECT so that they may sooner resume busy professional careers. This poses an ethical problem of considerable magnitude if depression is considered to be at least partly the result of environmental circumstances, in addition to biological bases and perhaps even genetic predisposition. Is the ECT serving to return a patient to a professional or social situation that is responsible for the development of his depression in the first place?

Conclusion

The bioethical issues raised by ECT are of overriding importance in the light of the treatment's proven effectiveness when intelligently applied. The storm of controversy accompanying its use has led some to advocate its discontinuation altogether. This is surely an overreaction: ECT as a treatment procedure has great potential either for therapeutic benefit or for abuse. Specific challenges include a greater understanding of the physiological mechanisms underlying its effectiveness and increased integration with other treatment methods leading to greater cooperation between the patient and the treating physician.

RICHARD M. RESTAK

[*Directly related are the entries* BEHAVIOR CONTROL; *and* MENTAL HEALTH THERAPIES. *For discussion of related ideas, see the entries:* ELECTRICAL STIMULATION OF THE BRAIN; PSYCHOPHARMACOLOGY; *and* PSYCHOSURGERY. *Other relevant material may be found under* GENETIC ASPECTS OF HUMAN BEHAVIOR, *article on* GENETICS AND MENTAL DISORDERS; INFORMED CONSENT IN MENTAL HEALTH; *and* MENTAL HEALTH SERVICES, *article on* SOCIAL INSTITUTIONS OF MENTAL HEALTH.]

BIBLIOGRAPHY

ABRAMS, RICHARD. "Recent Clinical Studies of ECT." *Seminars in Psychiatry* 4 (1972): 3–12.
AVERY, DAVID, and WINOKUR, GEORGE. "Mortality in Depressed Patients Treated with Electroconvulsive Therapy and Anti-Depressants." *Archives of General Psychiatry* 33 (1976): 1029–1037.
BLACHLEY, P. H., and GOWING, D. "Multiple Monitored Electro-convulsive Treatment." *Comprehensive Psychiatry* 7 (1966): 100–109.
CERLETTI, U., and BINI, L. "L'Elettroshock." *Archives of General Neurology, Psychiatry and Psychoanalysis* 19 (1938): 266–268.
CRONHOLM, BORJE, and OTTOSSON, JAN-OTTO. "The Experience of Memory Function after Electroconvulsive Therapy." *British Journal of Psychiatry* 109 (1963): 251 ff. Special issue.
DIETZ, JEAN. "ECT Study Reveals Disparity between Public, Private Units." *Psychiatric News* 10, no. 15 (1975), pp. 1, 14–15.
ESSMAN, WALTER B. "Neurochemical Changes in ECS and ECT." *Seminars in Psychiatry* 4 (1972): 67–79.
FINK, MAX. "The Mode of Action of Convulsive Therapy: The Neurophysiologic-Adaptive View." *Journal of Neuropsychiatry* 3 (1962): 231–233.
———. "The Therapeutic Process in ECT." *Seminars in Psychiatry* 4 (1972): 39–46.
FURLONG, F. W. "The Mythology of Electroconvulsive Therapy." *Comprehensive Psychiatry* 13 (1972): 235–239.
GIAMARTINO, GARY A. "Electroconvulsive Therapy and the Illusion of Treatment." *Psychological Reports* 35 (1974): 1127–1131.
HARPER, ROBERT G., and WIENS, ARTHUR N. "Electroconvulsive Therapy and Memory." *Journal of Nervous and Mental Disease* 161, no. 4 (1975), pp. 245–254.
HURWITZ, THOMAS D. "Electroconvulsive Therapy: A Review." *Comprehensive Psychiatry* 15 (1974): 303–314.
KALINOWSKY, LOTHAR B. "The Convulsive Therapies." *Comprehensive Textbook of Psychiatry.* 2d ed. Edited by Alfred M. Freedman, Harold I. Kaplan, and Benjamin J. Sadock. Baltimore: Williams & Wilkins, 1975, vol. 2, pp. 1969–1976.
———, and HIPPIUS, HANS. *Pharmacological, Convulsive, and Other Somatic Treatments in Psychiatry.* New York: Grune & Stratton, 1969.
SQUIRE, LARRY R., and CHACE, PAUL M. "Memory Functions Six to Nine Months after Electroconvulsive Therapy." *Archives of General Psychiatry* 32 (1975): 1557–1564.
VOLAVKA, JAN. "Neurophysiology of ECT." *Seminars in Psychiatry* 4 (1972): 55–65.

EMBODIMENT

For a discussion of an important parallel topic basic to many ethical issues in bioethics, see the entry MIND–BODY PROBLEM.

"Embodiment" as used herein designates a specific phenomenon inherent to human life and experience which, although touched upon periodically in the history of thought, has been studied only in recent times as a focal issue. As thus used, embodiment refers to a distinctively different range of issues from those traditionally associated with the "mind–body" problem. One way to make this difference prominent would be to point out that, whether or not what has been called "mind" is or is not ultimately or metaphysically explainable by or reducible to "body," or vice versa, persons nevertheless experience their own bodies in specific ways which can be studied as such. Embodiment as a philosophical and bioethical phenomenon designates a fundamentally different range of issues from those entailed in any of the variety of positions adopted as regards the so-called mind–body problem.

Perhaps a better way of eliciting this phenomenon would be to note one striking peculiarity of Descartes's (1596–1650) efforts. On the one hand, he argued that mind (*res cogitans*) and matter (*res extensa*) are "substances": mutually exclusive, self-subsistent, and ontologically distinct entities, neither of which requires the other *to be* or to be *known*. This metaphysical bifurcation of reality (dualism) led Descartes to the view that, somehow, mind and body "interact," although specifying that "somehow" proved to be inordinately difficult if not impossible. However that may be, his reflections on the mind–body complex, *especially in his own case,* showed him that, as he put it, the mind is not "in" the body in the way a boatman is "in" a boat—that is, not contingently or accidentally. To the contrary, he discovered and reiterated often (Descartes, vol. 1, pp. 118, 255, 345, and *passim*), my mind is "intimately" connected to my body, and this "intimate union" itself constitutes a major problem. The difficulty he faced was, in one sense, that his metaphysical view could not provide the grounds for understanding that "intimacy": if everything must be *either* mind *or* body and nothing can be both, how can the one be "united" or "interact" with the other? In another sense, the difficulty is, as Gilbert Ryle pointed out (p. 22), that Descartes's efforts incorporate a "category mistake." Both "interactionists" and "reductivists" are guilty of the "dogma of the ghost in the machine": Since "the phrase 'there occur mental processes' does not mean the same sort of thing as 'there occur physical processes,' . . . it makes no sense to conjoin or disjoin the two" (ibid.).

Correct in its way though Ryle's point is, it raises very much the same problem, since he wants to say that "one and the same activity" (e.g., raising my arm) can be described and explained in more than one way (ibid., pp. 50–51). This, however, leaves obscure how logically *different* order statements can yet pertain to "one and the same thing."

More important, such an argument in effect obfuscates Descartes's insight—which is genuine, however much he and others tended to confuse it: *that one does in truth experience one's own body* as profoundly "intimate." The problem is that one must be able to account for this sense of "intimacy." Several philosophers immediately after Descartes saw this issue, and thereby pointed the way to the phenomenon of embodiment. Blaise Pascal (1623–1662) noted with marked irony that if one, like Descartes, composes all things of mind and body, surely that mixture would itself be intelligible. Yet, he insisted, not only do we not understand the body, and even less the mind, least of all do we know "how a body could be united to a mind. This is the consummation of [our] difficulties, and yet it is [our] very being" (Pascal, pp. 27–28). That is, whether or not there are two, or more, substances in reality, and however one may try to explain them, that we are *both* mind *and* body and do not comprehend their "intimate union," is thus "[our] very being." It is this "intimacy" which calls for attention, regardless of one's particular metaphysical point of view on mind and matter.

Benedict de Spinoza (1631–1677) saw, too, that Descartes's bifurcation created insuperable difficulties in understanding that union. While he focused his principal argument in metaphysical terms (arguing that neither mind nor matter could possibly be genuine "substances," but only the "attributes" of the one unitary substance which is reality itself), Spinoza nevertheless saw the importance of accounting for what Descartes had identified merely as a "union." Both mind and body are essential to one another, for Spinoza, and he came to conceive the one, body, as mirrored in the other, mind, as its "idea" (Spinoza, pp. 82–107). Although, to be sure, Spinoza's theory is far more complex, the point of major emphasis here is that by rejecting the Cartesian dualism he was enabled to come to a clearer understanding of that very "intimate union" which had impressed Descartes.

But it was not until relatively recently, in the early writings of Henri Bergson (1859–1941), that this sense of intimacy became a specifically focal issue on its own (although, it is true, Bergson did not fully explore this insight). The human body itself must be viewed as that whereby any person has a locus, a placement, in the world, and this is a unique and *sui generis* phenomenon (Bergson, pp. 11, 14–16, 57). What makes the body not simply an object like other objects in the world (that is, its "intimacy") is that it is fundamentally "*mine*." Thus, although it is indeed physical, it is not simply that—for it is the only one uniquely singled out as *mine*, and as a "center." The field of physical objects and events is spatially organized around *my* body as its center of reference. In the second place, when my body is viewed as such it is clear that its functioning as center is not at all limited to such matters as spatial location—it is as well that by means of which I am able to engage in activities of any sort. Thus spatial location and the familiar sense-qualities pertaining to physical things are *never* experienced independently of specific contexts of *action*. My body is an "actional center." For the perceiver, things are not first of all conglomerations of data, and only later taken up into contexts of bodily action; even being aware of such sensa, where possible, requires highly specialized and technical means and equipment—i.e., specific contexts of specialized action are present even here. Perception is thus not a matter of "data reception" ("input") followed by "internal neural translation" and then by "externalization" ("output"). *For the perceiver*, "things" are "menacing," "helpful," "handy," "obstacles,"—in short are, as Piaget also recognized, experientially organized as "poles of action" (Piaget, chap. 1) appearing only in and through specific activities directed toward them. As such, "things" of all sorts are *essentially* tied to these actional contexts.

My body is thus uniquely singled out for, and experienced by, me *as mine* (i.e., as "intimate"). It is as well a "center," *that by means of which I am in the world*, in the midst of objects, people, language, culture, etc., and is that by means of which the surrounding milieu is presented to me for my thought and action. These points, which had impressed Bergson, turn out to be fundamental to the phenomenon of embodiment, and are crucial for understanding the subsequent discussions of it.

As Edmund Husserl (1859–1938) showed in his many writings, the primary phenomenon is the experiential relation of consciousness to its own embodying organism (Husserl, 1950, pp. 130–131). Granted that *this* organism is uniquely singled out (Husserl, 1959, p. 97), the problem of embodiment is to determine in *what sense* this organism is "mine" or *how it becomes experienced* as "mine," and thus enables the experiencing of surrounding things (ibid., pp. 60–61).

What had so impressed and troubled Descartes—the "intimate union"—Husserl calls the experiential relation to the body as mine. But it was just this that Descartes and many others following him had obscured in the interest of trying to resolve the very different metaphysical question concerning the "mind–body" relation. It is to the embodiment phenomenon that Gabriel Marcel's analysis of the fundamental opacity at the heart of personal experience, my *body-qua-mine*, is addressed (Marcel, 1940, p. 40); it is here that Maurice Merleau-Ponty locates an essential ambiguity intrinsic to the body itself (Merleau-Ponty, 1962, pp. 237 ff., 269 ff., 350 f.). So "intimate" is this "union" that one is tempted to say, with Jean-Paul Sartre, "I *am* my body" (Sartre, pp. 368–427). So profound is the experiential connection to my animate organism that it is necessary to say that there *are* no other things in the world, no experienced world at all, except on condition of my having this body experienced by me as mine (Straus, 1966, pp. 38–58, 137–165). Thus, the sense in which things "belong" to me is a sense ultimately derived from the ways in which I experience my own body as "belonging" to me; the latter is the condition for the former (Marcel, 1935, pp. 223–225). This phenomenon becomes especially manifest in instances where mental disturbances occur, and the sense of "mineness" becomes severely compromised, or never develops (Bosch, pp. 61–112).

One central problem is this: In virtue of what is this one animate organism uniquely singled out for and in my experience? Which specific experiences are there, without which this organism would cease to be experienced by me as mine, or which give it its sense as mine (Straus, 1958)?

The issue is by no means a settled one (Zaner, 1964, pp. 198–261) but is exceedingly complex and subtle. In one way or another, however, it seems generally agreed that the animate organism becomes and remains an embodying organism solely to the extent that (1) it is not just a body (*Körper*) but an animate organism (*Leibkörper*), the sole object within which belong my own fields of sensation; (2) it is the only object "in" which I "rule and govern" immediately, within each of its "organs" and the total organism itself; (3) it enacts most immediately my "I can" (go, perceive, move, turn, grasp, and the like); (4) it is that "by means of which" I perceive and otherwise experience the field of objects in the world and thus is my access to the world and the focus of the world's actions on me; and (5) it is not only that whereby I experience other things, but is itself experienced by me at the same time as I experience other things (that is, the organism is essentially reflexively related to itself) (Husserl, 1959, p. 97).

In short, embodiment is fundamentally connected with various levels and modalities of bodily attitudes, stances, and movements (Buytendijk, pp. 238–275, 285–344, 401–488), personal striving or willing, and perceptual awareness of things. Solely to the extent that wishes, desires, advertences, movements, etc., are or can be actualized or embodied by means of corporeal kinaesthetic flow-patterns, which are functionally determinative of the several perceptual fields and of what appears therein, can one sensibly say that this organism is "uniquely singled out as 'mine.'" Processes of sensory "feeling" (coenesthetic, kinaesthetic, proprioceptive, etc.), elementary strivings (reaching, squinting, locomoting, etc.), and the consequent achievement of an actional/perceptual field of objects in the surrounding environs—all these, and doubtless still others, go to constitute the phenomenon of embodiment.

But it needs to be stressed that there is quite another dimension to this theme, to which surprisingly little attention has been devoted, but which turns out to be just as essential to embodiment. However tempting it is to say "I am my body" (as when, for example, someone strikes one in the face: "Don't hit *me!*" we say), many cases in psychopathology (Binswanger), literature, and even situations of daily life, show that the temptation is deceptive. The situation is far more complex, for the human self's relation to its own body is *not only one of "mineness" but also of radical "otherness."* However intimate and profound is the relation between me and my body, it is equally the case that this body is experienced as strange and alien, and in specific ways.

I *am* my body; but in another sense I am *not* my body—or *not just* that. Indeed, this otherness is so profound that we inevitably feel forced to qualify the "am": It is not identity, equality, or inclusion. It is "mine," but this means that I am in a way distanced from it, for otherwise there would be no sense to "belonging"; it would not be characterizable in any sense as "mine." So "close" is the union that the experience of my body's otherness can be psychically shattering (whether it be my body's happy obedience which I notice for the first time, or its hateful refusal to obey my desires to do something). So intimate is it that I have moments in which I genuinely feel "at one" with it. Yet, so other is it that there are times when I treat it as a mere thing *other than* me (for instance, obsessively stuffing it with food, or otherwise mistreating it; or as when I encounter it as "having a life of its own" to which I must willy-nilly attend—like it or not, "my" hair grows and must be trimmed for action in certain contexts, "my" hands cleaned, "my" bowels moved, "my" cold cured, and so on and on) (Zaner, 1966, pp. 85–87).

I find myself, in short, as a creature who is embodied by an animate organism whose connections to me (and I to it) *are themselves an experiential impasse*—an *aporia* in Plato's sense. Nothing so much as me my*self* is at once so utterly familiar and usual (Who else could I be?), yet so completely foreign and alien (Who, indeed, am I?). This unique complex, it seems, is not indicative of a mere inability to make up one's own mind, but is rather the disclosure of the essence of embodiment. And, what seems so distinctive is just this "mineness/otherness" dialectic which is the core of the human body-as-experienced (Engelhardt, pp. 28–60, 89–119, 130–168).

In these terms, to speak of embodiment is to speak of something *which I am*, and not something which can be placed over against me as a sort of object (*ob-jectum*). I am myself, as embodied, my own most fundamental problem, as Pascal had already perceived with remarkable insight. What is expressed by the "problem of embodiment" is precisely my *being as embodied*—i.e., the very context, setting, and meaning of human life itself. The self–body problem *is* thus *enacted* at every moment in the ongoing life of the human person.

Recognizing this embodiment/enactment contexture makes it possible to appreciate that the human body is essentially an *expressive* phenomenon. While the particular ways in which wishes, feelings, desires, strivings, and so on are expressed, doubtless vary from culture to culture, in the life of a particular individual to some extent, as well as historically, this does not deny but in fact goes to show that embodiment is expressive (Merleau-Ponty, pp. 174–199). Nor does the fact that people can and do dissemble and deceive themselves and others belie the expressiveness of the body; indeed, these are themselves expressive phenomena, however difficult it may be to discover and then to interpret them (Burrell and Hauerwas).

This expressiveness signifies that embodiment is essentially a *value phenomenon*. In simplest terms, precisely because this one specific organism is uniquely singled out as embodying me, and thus is most intimately "mine" (me-yet-not-me), whatever can happen to it happens to me. Though my body is that wherein I "rule and govern," I am as well subject to *its conditions*. Hence, it is fundamental to the experience of embodiment that "it matters" what can and does happen to my body. Its value character can also be made clear if one considers, quite apart from bioethical issues, what one thinks of someone who is "loose" with his or her body. Indeed, embodiment as "intimate union" seems clearly to lie at the heart of the prominent sense of "inviolability" of the person. Thus, too, is it more understandable that persons have a sense of their own "privacy" (i.e., with the basic desire to maintain one's *integrity*, as expressed in the traditional "principle of totality"), and that there are constraints felt as regards unwanted intrusions (as in psychosurgery) into the lived-body; that is, the life of the embodied person.

It is also clearer why current discussions of bioethical issues—abortion, psychosurgery, euthanasia, etc.—are so highly charged. On the one hand, medical practices (and much biomedical experimentation) are ways of intervening or intruding into that most intimate and integral of spheres—the embodied person. On the other hand, every person is embodied, enacts himself through that specific animate organism which is "his own" and is thus expressive of that very person. Bodily schemata, attitudes, movements, actions, as well as perceptual abilities are all value-modalities by which one articulates and expresses one's character, personality, habits, goals—in short, one's life as a totality.

From the perspective of embodiment, then, medical practice and biomedical research are planned or potential interventions into the sphere of intimacy—whether this sphere be only initial (as in infancy) or more developed. And, whether such interventions be mainly directed to the body or to the mind, all of them unavoidably have their impact on the other; that is, a person's *life as a whole* is necessarily affected. *Psyche* and *soma* are inextricably bound together as constituents of an integral context in all living humans (Zaner, 1975). So, because embodiment is necessarily expressive, and therefore a value phenomenon, *every medical intervention is by nature of the case ethical in character*—although this is obviously not in any way to settle the specific issues inherent to any specific intervention.

In these terms, attempting to settle such specific issues requires that the fundamentally ethical nature of any intervention be explicitly brought out and appreciated and, second, that the specific details of the particular case be determined as far as possible. That is to say, of course, that respect for the person—his/her values, goals, etc., in short, *integrity*—is not a value brought in extraneously, *but is inherent to the very nature of medical interchanges*. In different terms, an appreciation of embodiment as expressive and valuational *requires* respect for the perspective, concerns, values—integral, embodied life—of the patient himself. It might be added here that, in the event that adopting that point of view is not possible—as when one confronts an unconscious patient, e.g.,—the decision to intervene in ways which do not include the perspective of the patient requires *other* ethical grounds, and must thus be subject to critical assessment as such. Other exceptional cases can also be thought of—mental retardates, neonates, etc.—but these, too, do not escape the requirement to respect the patient's integrity, though they do require special ways of taking it into account (e.g., consulting relatives) as well as the additional ethical issues for decision making.

Accordingly, every medical issue is not only *ethical* but highly *context-specific. There is thus no way of settling any bioethical issue in the abstract* (e.g., "Should a life-threatening tumor be removed at the cost of memory?" Or "Is it morally good to cure emotional disorders by cutting the brain?"). Every medical practice, no matter how trivial, is *to begin with* value-laden, which means that each one either explicitly or (most often) implicitly is expressive of some vision of what is or is thought to be morally good. As Hauerwas effectively argues, furthermore, moral notions are such that they *construe the world* in a certain way as opposed to other ways of construing it (Hauerwas). In addition, it is suggested here that the primary issue for bioethics—given the perspective of an embodied person, and recognizing that moral notions are neither simply descriptive nor simply normative—is to educe, to make explicit, the "ways of construing the world" found in ongoing medical interventions and patient-responses. Only subsequently does it become possible to make informed judgments about the particular context-specific practices and issues facing medicine. How one can come to such truly informed judgments is an obvious problem, but it is not within the scope of this article; it is hopefully enough to have delineated the phenomenon of embodiment, its expressive and value character, and the ethical nature of medicine itself—its incorporation of ethical notions in each of its practices, and the necessity of having to assess these in their own terms as such.

RICHARD M. ZANER

[*Directly related are the entries* MAN, IMAGES OF; MIND–BODY PROBLEM; *and* PERSON.]

BIBLIOGRAPHY

BERGSON, HENRI. *Matière et mémoire.* 54th ed. Paris: Presses Universitaires de France, 1953. Translated as *Matter and Memory* by Nancy Margaret Paul and W. Scott Palmer. London: G. Allen Unwin; New York: Humanities Press, 1970.

BINSWANGER, LUDWIG. "The Case of Ellen West." *Existence: A New Dimension in Psychiatry and Psychology.* Edited by Rollo May, Ernest Angel, and Henri F. Ellenberger. New York: Basic Books, 1958, pp. 237–364.

BOSCH, GERHARD. *Infantile Autism.* Translated by D. Jordan and I. Jordan. New York: Springer-Verlag, 1970.

BURRELL, DAVID, and HAUERWAS, STANLEY. "Self-Deception and Autobiography: Theological and Ethical Reflections on Speer's *Inside the Third Reich.*" *Journal of Religious Ethics* 2 (1974): 99–117.

BUYTENDIJK, FREDERIK J. J. *Attitudes et mouvements: Étude fonctionelle du mouvement humain.* Paris: Desclée de Brouwer, 1957.

DESCARTES, RÉNE. *Philosophical Writings,* 2 vols. Translated by Elizabeth S. Haldane and G. R. T. Ross. New York: Dover Books, 1955. See especially "Discourse on Method," "Principles of Philosophy," and "Passions of the Soul" (all vol. 1).

ENGELHARDT, H. TRISTRAM, JR. *Mind–Body: A Categorial Relation.* The Hague: Martinus Nijhoff, 1973.

GURWITSCH, ARON. *Field of Consciousness.* Pittsburgh: Duquesne University Press, 1964.

HAUERWAS, STANLEY. "The Self as Story: Religion and Morality from the Agent's Perspective." *Journal of Religious Ethics* 1 (1973): 73–85.

HUSSERL, EDMUND. *Erste Philosophie (1923/24).* (Husserliana, vol. VIII.) Den Haag: Martinus Nijhoff, 1959. Pt. 2.

———. *Ideen zu einer reinen Phänomenologie und phänomenologischen Philosophie.* Vol. 1: *Allgemeine Einführung in die reine Phänomenologie* (Husserliana, vol. III). Vol. 2: *Phänomenologische Untersuchungen zur Konstitution* (Husserliana, vol. IV). Den Haag: Martinus Nijhoff, 1950, 1952. English paperback translation by W. R. Boyce Gibson as *Ideas.* New York: Macmillan, 1962.

MARCEL, GABRIEL. *Du refus à l'invocation.* Paris: Gallimard, 1940.

———. *Etre et avoir.* Paris: F. Aubier, 1935.

MERLEAU-PONTY, MAURICE. *Phénoménologie de la perception.* Paris: Gallimard, 1945. Translated by Colin Smith as *Phenomenology of Perception.* Atlantic Highlands, N.J.: Humanities Press; London: Routledge & Kegan Paul, 1962.

PASCAL, BLAISE. *Pensées.* New York: Modern Library, 1941.

PIAGET, JEAN. *The Origins of Intelligence in Children.* Translated by Margaret Cook. New York: International Universities Press, 1952.

RYLE, GILBERT. *The Concept of Mind.* New York: Barnes & Noble, 1949.

SARTRE, JEAN-PAUL. *L'Être et le néant.* Paris: Librairie Gallimard, 1943. Translated by Hazel E. Barnes as *Being and Nothingness.* Buffalo, N.Y.: Washington Square Press, 1966.

SPINOZA, BENEDICT DE. "Ethics." *Chief Works.* Translated by R. H. M. Elwes. New York: Dover Publications, 1951, vol. 2.

STRAUS, ERWIN. "Aesthesiology and Hallucinations." *Existence: A New Dimension in Psychiatry and Psychology.* Edited by Rollo May, Ernest Angel, and Henri F. Ellenberger. New York: Basic Books, 1958, pp. 139–169.

———. *Phenomenological Psychology.* New York: Basic Books, 1966.

ZANER, RICHARD M. "Context and Reflexivity: The Genealogy of Self." *Evaluation and Explanation in the Biomedical Sciences.* Edited by H. Tristram Engelhardt, Jr. and Stuart F. Spicker. Dordrecht, Holland: D. Reidel Publishing Co., 1975, pp. 153–174.

———. *The Problem of Embodiment.* Phaenomenologica, no. 17. The Hague: Martinus Nijhoff, 1964. 2d ed. 1971.

———. "The Radical Reality of the Human Body." *Humanitas* 2 (1966): 73–87.

EMBRYO TRANSFER

See REPRODUCTIVE TECHNOLOGIES, *articles on* IN VITRO FERTILIZATION, ETHICAL ISSUES, *and* LEGAL ASPECTS.

EMOTIVIST ETHICS

See ETHICS, *article on* NON-DESCRIPTIVISM.

ENVIRONMENT AND GENETICS

See GENETIC CONSTITUTION AND ENVIRONMENTAL CONDITIONING.

ENVIRONMENT AND MAN

This entry does not discuss concrete ethical dilemmas concerning the quality of human life in the biosphere. Rather, it examines those underlying concepts, attitudes, and ideals involved in the relationship between humans and their environment that tend to give shape to conflicting ethical views in many areas of bioethics, particularly those discussed in the next entry, ENVIRONMENTAL ETHICS.

I. WESTERN THOUGHT   Ian G. Barbour
II. EASTERN THOUGHT   Hajime Nakamura

I
WESTERN THOUGHT

In the history of Western thought, views of the relation of human life to the environment have been diverse. To simplify the representation of this diversity within a brief article, some of these views have been grouped under three headings: (1) man's dominion over nature; (2) man's participation in nature; and, intermediate between these extremes, (3) man's stewardship of nature. Each motif has characteristic implications for the way one treats the environment. A final section discusses the crucial ethical issues today. The generic terms "man" and "mankind" refer throughout to both men and women, and the term "nature" is an abbreviation for "nonhuman nature."

Man's dominion over nature

Biblical and classical roots. The first chapter of Genesis (1:28) includes the commission to "have dominion over the fish of the sea and over the birds of the air and over every living thing." Man alone is created "in the image of God" (Gen. 1:26) and set apart from all other creatures. Moreover, nature is desacralized in biblical religion. Ancient Israel believed that God had revealed himself primarily in historical events rather than in the sphere of nature. Such assumptions contributed to the Western cultural outlook within which science and technology could eventually arise; for if nature is orderly, but not divine or demonic, it can be understood and used by human beings.

A number of recent authors have claimed that this biblical idea of dominion was the main historical root of exploitative attitudes in the West. In a widely quoted article, Lynn White points to the separation of man and nature and the assertion of the rights of man over nature in biblical thought. Holding that ideas and attitudes are significant influences in history, White concludes that, because of its anthropocentrism and "arrogance toward nature," Christianity "bears a huge burden of guilt" for the environmental crisis (White, p. 1206). However, such an account seems to neglect the complexity of Western thought. To be sure, the Genesis passage about dominion was used in later centuries to justify exploitative practices. But we shall note that (1) several nonbiblical sources of the dominion theme can be identified in the West; (2) the stewardship theme in the Bible itself sets limits on dominion, which were subsequently ignored; (3) environmental destruction has been common in nonbiblical cultures of both East and West since antiquity; and (4) institutions as well as ideas must be examined as determinants of environmental behavior, even if institutions are themselves partly the product of ideas.

From the many strands of Greek thought, Plato, Aristotle, and the Stoics may be singled out for their influence on the West. Their writings portrayed a gulf between man and all other beings, based on the unique human capacity for reason. Aristotle stated that other creatures are devoid of the contemplative activity in which man is most akin to God; plant and animal life exist simply for the sake of human life. Cicero, drawing upon Stoic writings, insisted that we have no obligation to respect animals because they are not rational beings (Glacken, chap. 1). To the neo-Platonists of the Hellenistic era, the eternal forms are only imperfectly embodied in the world of nature. To the Gnostics and Manicheans, nature is the realm of evil from which the human soul seeks to escape. Greek and Roman views were indeed extraordinarily diverse, and we shall mention later some pantheistic authors who were more appreciative of the natural world. But the classical sources that were most influential on the early Church, the Middle Ages, and subsequent Western thought seem to have stressed the separation of man and nature.

The growth of science and technology. With the rise of the scientific world view in the seventeenth century, the dominion theme assumed increasing prominence. To Francis Bacon (1561–1626), the conquest of nature is the goal of science, for "knowledge is power." "Let the human race recover the right over nature which belongs to it by divine bequest." Bacon's *New Atlantis* called for a state-funded research establishment and a scientific elite through which man's rightful supremacy would be systematically extended. René Descartes (1596–1650) similarly extolled practical knowledge that would make us "the lords and masters of nature." Like the Stoics, he thought that man's unique rationality justified such sovereignty. Descartes elaborated a sharp dualism of matter and mind. Apart from human life, the world consists of particles in motion, and mathematics is the key to understanding it. Animals, he asserted, are machines without minds or feelings. The gap between mind and matter in Descartes's anthropocentric outlook posed the central problems for modern philosophy.

The mechanistic interpretation of nature was further developed by Isaac Newton (1642–1727). In earlier centuries the world had been viewed as a hierarchy of organisms, each with its place and purpose in an overall plan. For Newton and his followers, nature was taken to be constituted by impersonal masses and forces, operating according to deterministic laws. The measurable "primary" qualities, such as mass and velocity, were said to be objective characteristics of the real world; all other "secondary" qualities, such as color and temperature, were relegated to subjective responses in the observer's mind. Here was the "objectification" of nature as a realm essentially alien to man, and the detachment of the observer from the world he is observing. While Newton himself respected the cosmic watch as the product of the Divine Watchmaker, it is not surprising that his more secular successors had no scruples about exploiting it. If nature is a machine, it has no inherent rights, and man need not hesitate to manipulate and use it.

John Locke (1632–1704) provided the political philosophy within which such dominion could be justified and encouraged. Locke maintained that the political order is necessary primarily to protect the natural rights of the individual to "life, liberty, and property." Unfettered private ownership and use of resources, he said, would lead to economic and technological growth. By protecting the property rights of the rising middle class, the state would be encouraging industrial development and prosperity. Locke's influential writings thus endorsed individualism, the accumulation of wealth, technological development, and the subjugation of nature, all of which were to be promoted by the structures of government.

In the emerging industrial technology of the eighteenth century, man's dominion was increasingly achieved in practice as well as in theory. To the leaders of the industrial revolution, the environment was primarily a source of raw materials. In the new capitalism, the private ownership of resources encouraged treatment of the natural world as a source of commercial profit. Along with rising standards of living came increasing burdens on the environment. Since antiquity there had been deforestation, overgrazing, and soil erosion; but the technologies that developed in the last two centuries produced pollution and consumed natural resources at unprecedented rates. Here were combined the influences of biblical religion, dualistic philosophy, mechanistic science, and, above all, capitalist economics and industrial technology.

The American experience. For the New England Puritans, the new surroundings were strange and threatening, and their writings often refer to nature as an enemy to be subjugated. As the pioneers moved progressively westward, much that they encountered was hostile, a threat to survival, an obstacle to be overcome. Forests were cleared and wilderness destroyed to make way for civilization. The advancing frontier was interpreted in the light of the nation's "manifest destiny" to "conquer a continent." The New World offered apparently endless stretches of good land and seemingly unlimited natural resources. Air, water, and land appeared ample to absorb the waste products of a burgeoning civilization.

In the early days of the nation, there was considerable support for Thomas Jefferson's ideal of a nation of farmers. The agrarian, pastoral vision, which stressed the virtues of a rural society, contrasted with Alexander Hamilton's goal of an urban, mercantile nation in which commerce and manufacturing would thrive. The tensions between these two ideals continued, but it was clearly the industrial vision that prevailed. The steam locomotive, the "iron horse," had by the Civil War become a symbol of both technology and the conquest of nature in the growth of the nation. As industry thrived, urbanization accelerated, and the United States changed from a rural and small-town nation to a predominantly urban one. The new technologies and the private ownership of land and resources led to the concentration of economic power, the amassing of personal fortunes, and the rise of giant corporations for which the environment was a source of wealth.

To these geographic and economic factors may be added some characteristic American values that encouraged exploitative attitudes. Faith in technology and confidence in the expert have been pervasive. An optimistic belief in inevitable progress has until recently been shared by all levels of society. America has also been obsessed by the goal of growth, the assumption that "bigger is better." Impressive increases in productivity and gross national product have certainly occurred, but at social and environmental costs that have only slowly been recognized. It can also be argued that a male-dominated society has admired in public life the aggressive, active qualities that it calls "masculine," rather than the nurturing, conserving, sensitive qualities it has associated with women and family life. For a variety of reasons, then, the dominion theme has been even more prevalent in America than in the rest of the Western world.

Man's participation in nature

The opposing theme of man's participation in nature also has diverse religious and scientific roots.

Religious versions of participation. The religions of the ancient Near East sought the harmonious integration of man's life within the life of nature. Their rituals and festivals celebrated the annual cycle of the seasons and the fertility of nature rather than historical events. In nations around the Mediterranean, the earth-mother figure assumed differing forms: Athene, Artemis, Demeter, Isis, etc. Myths of dying and rising gods (e.g., Adonis, Osiris) were associated with the rebirth of life in the spring, and other gods and goddesses represented the power of a variety of natural forces.

Greek popular religion often made reference to the sacred in nature: the sacred grove or mountain, the spirit of the river or rock. Other versions of man's interconnectedness with nature is found in such Greek thinkers as the Epicureans. Lucretius, for example, held that the world was not designed for human use; in his nature poetry he reflected on the beauty and interdependence of the world as a natural process. But neither the nature religions of Asia Minor nor the naturalistic or pantheistic philosophies of Greece were as influential on later Western thought as the biblical and Greek themes previously mentioned. The theme of participation is of course strongly represented in Eastern thought, especially in Taoism in China and Zen Buddhism in Japan.

St. Francis's deep love of the natural world and his sense of union with it were far from typical of medieval thought, and yet they have continued to appeal to the popular imagination. He saw nature as a living whole and all creatures as objects of God's love, and hence as significant in their own right. He spoke of our sister the earth and greeted the birds as brothers, extending the family relationship and the circle of God's love to include all created beings sharing a common dignity and equality under God. Humanity is part of a wider community, and each creature has its own integrity, which must be respected.

Nineteenth-century Romanticism. Apart from several of the Christian mystics, the participation motif found few defenders in the early modern period. Only after the industrial revolution—and largely as a reaction to it—did this theme come into prominence. For such poets as Blake, Wordsworth, and Goethe, nature is not an impersonal machine but an organic process with which man is united. God is not the remote watchmaker but a vital force immanent in the natural world. Not rational analysis but feeling and imagination are the highest human capacities. Intuition grasps the unity of organic wholes, the interrelatedness of life. In natural settings a person can find a healing power, a sacramental presence, a bond with the human soul, an experience of peace and joy. Other Romantic authors extolled wild, sublime, untouched landscapes and forests. They idealized the "noble savage," uncorrupted by civilization, and exalted the "natural" and the "primitive."

The New England Transcendentalists referred in similar terms to the presence of the sacred in the realm of nature. Henry David Thoreau (1817–1862) held that nature is a source of inspiration, vitality, and spiritual renewal; it can teach us humility and simplicity. "In Wildness is the preservation of the World," he wrote. Thoreau criticized the frantic pursuit of progress and affluence, the growth of technological industrialism, and the pressures of an impersonal urban life. His year and a half of living alone at Walden Pond made him more aware of the interrelationships among creatures and the natural stability upset by humans; in solitude he found serenity and freedom. But he did not advocate giving up civilization. He sought rather a simplification of life and an alternation and balance between life with nature and life with civilization.

Starting in the 1870s, the writings of John Muir gave wide circulation to a philosophy of wilderness preservation. With the disappearance of the frontier, wilderness areas can be saved only by deliberate national policy. Muir, like Thoreau, found a divine harmony in nature, a freedom not possible in the artificial constraints of civilization, a source of serenity in a decadent society. He was founder of the Sierra Club (1892), and through his writing he campaigned tirelessly for legislation to protect wild areas of the American West.

George Perkins Marsh's *Man and Nature* (1864) was the first detailed study of the destructive influence of man on his environment. Marsh was an accomplished linguist familiar with ancient history and literature. As ambassador to Italy he traveled in Mediterranean countries and saw barren deserts where great cities and civilizations had once flourished. He traced the effects of deforestation and overgrazing on soil erosion, and the destruction of land by salination from excessive irrigation. He provided careful documentation of the fragility of the environment and the disruptive and often irreversible effects of civilization in disturbing the natural equilibrium.

The ecology movement. The scientific version of the participation theme received its first systematic expression in Darwin's *The Origin of Species* (1859) and *The Descent of Man* (1871), in which man was presented as a part of nature, in continuity with other forms of life. No sharp discontinuities separated human from animal life, either in evolutionary history or in present morphology and behavior. The theory of evolution seemed to undermine man's unique status, and close parallels to most human capacities could be found among lower forms. Moreover, Darwin's studies brought out the interconnectedness of the web of life and the complex balance of interactions in the biological world. Subsequent research in population dynamics has underscored the importance of the relation of organisms to their environment (habitat, food sources, predator–prey relationships, etc.). Recognition of the interdependence, diversity, and vulnerability of biological species prepared the way for ecology.

Within the twentieth-century science of ecology two concepts in particular have significant implications concerning man's relation to the environment: (1) The ecosystem concept emphasizes the interdependence of all forms of life in biotic communities and the far-reaching repercussions of thoughtless human intervention. Ecologists have traced the complex reciprocities and mutualities among organisms, the food chains linking diverse species, the interlocking cycles of elements and compounds, and the delicate balances that are easily upset. (2) There are limits to the growth of populations, and every environment has a finite carrying capacity. Human life, too, is dependent on limited resources —land, air, water, plant life, oil reserves, mineral ores, etc.—whose uses can be extended by technology, but not without new environmental costs. Writing in the 1930s, Aldo Leopold was one of the first ecologists to suggest that our ethical concern must be extended to apply to the land and nonhuman life as well as to human beings. Since Rachel Carson's *Silent Spring* (1962), the call for an ecological conscience has been repeated with increasing urgency, coupled often with a critique of the short-sightedness of most technological solutions.

In recent years, many biologists have focused on growth as the crucial problem. For Paul Ehrlich and Garrett Hardin, the control of population overshadows all other issues. Gains in agricultural and industrial production in developing countries cannot keep up with exponential population growth. In the Club of Rome study, *Limits to Growth* (1972), and the British *Blueprint for Survival* (1972), attention is directed also to the limits of industrial growth which are set by resource reserves and pollution levels. These studies advocate policies of "no growth" for industrial production as well as population. In opposing growth and in underscoring the finite carrying capacity of the environment, these authors adopt an ecological viewpoint, even if at times their outlook seems more anthropocentric than the views of other authors mentioned previously.

The youth counterculture since the late 1960s has been influenced by the ecology movement, but its approach has had more in common with Romanticism. Like the latter, it has valued feeling, imagination, and immediacy of experience more than intellect and rationality. The counterculture has been disenchanted with technology, holding that preoccupation with efficiency, productivity, and rational control has alienated industrial man from nature and from his fellow man. Some young people have dropped out of society to form communes in rural settings within which they hope to recover harmony with nature and with each other. Other youths have sought alternative life-styles that might encourage personal, social, and environmental harmony without a radical break from the prevailing social order.

Finally, process philosophy should be mentioned as an intellectual system influenced by evolutionary biology and consistent with an ecological outlook. Much of twentieth-century philosophy, especially in Europe, has perpetuated the Kantian separation of the human realm of freedom and history from the realm of nature. Existentialism, for example, emphasizes the distinction between the sphere of personal selfhood and the objective world of impersonal nature. In phenomenology this split is partially overcome in the analysis of "being in the world." But in the thought of Alfred North Whitehead (1861–1947) and his more recent followers the realms of selfhood and nature are more systematically brought together again. Nature is a creative process, a community of interacting organisms, not a deterministic machine. Every being is constituted by its relationships and dependent on its environment, but each is at the same time in its own way a center of experience. Process thought rejects matter–mind dualism (as in Descartes), the subordination of matter to mind (idealism), and the reduction of mind to matter (materialism). Mind and matter are not two opposing substances, but two aspects of events in systems having many levels of organization. Thus all beings are intrinsically valuable and worthy of respect as centers of at least rudimentary experience. In process thought, man is to be understood in the same categories as other beings. There are no metaphysical discontinuities, though the importance of any given category will vary widely among different levels of being. God transcends nature but is also immanent in the creative process; he does not intervene coercively from outside, but participates throughout cosmic history. Process thought thus avoids the separation of man and nature and of God and nature, which have in the past encouraged environmentally destructive attitudes.

Man's stewardship of nature

Stewardship in biblical thought. The third motif in Western thought, stewardship or responsible use, represents a middle ground between dominion and participation. In the story of Adam and Eve, man is put in the garden "to till it and keep it" (Gen. 2:15), and throughout the Bible, man does not have absolute and unlimited dominion, for he is responsible to God. "The earth is the Lord's" because he created it. The land belongs ultimately to God; man is only a trustee, caretaker, or steward, responsible for the welfare of that which is entrusted to him and accountable for his treatment of it. The biblical outlook is, in the last analysis, theocentric rather than either anthropocentric or biocentric.

Whereas dominion implies that the environment has no reason for existence except for our use, stewardship affirms the intrinsic value of nature. In the biblical view, the created world is valued in itself, not simply as an instrument of human purposes. In several of the Psalms (19, 89, 104, etc.) God is said to delight in the earth and the manifold variety of life quite apart from man. Even in the first chapter of Genesis, each form of life is pronounced good before mankind is on the scene. The Sabbath is a day of rest for the earth and living things as well as for man. Human life and nature stand together as jointly God's creation.

Many biblical passages express appreciation and wonder in response to nature. Job is overwhelmed by the majesty of natural phenomena. Jesus spoke of the lilies of the field and the sparrow's fall. Value pervades all life, not just human life. Furthermore, nature is part of the drama of redemption and will share in the ultimate harmony, as portrayed in the symbolic vision of the coming kingdom when "the wolf shall dwell with the lamb, and the leopard shall lie down with the kid" (Isa. 11:6). Paul imagines that "the whole creation has been groaning in travail together until now," but it will all take part in the final fulfillment (Rom. 8:22). Although the focus of attention was on man and human history, the world of nature was not neglected within the Bible as it was in much of later Christian thought.

St. Benedict might be taken as an early model of the stewardship perspective. Compared with St. Francis's deep feeling and sense of union with the natural world, St. Benedict's response was more practical, using nature but using it with care and respect. The Benedictine monasteries combined work and contemplation. They developed such sound agricultural practices as crop rotation and care for the soil, drained swamps, and husbanded timber all over Europe; they were creative in the practical arts related to nature.

There has also been a continuing literary tradition that has extolled the pastoral ideal of "cultivated nature," intermediate between the expanding city and the inhospitable wilderness. The pastoral poems of Theocritus, Virgil, and Horace celebrated the beauties of the country landscape and the virtues of the simple rural life in contrast to the growing urbanization of Greek and Roman life. The pastoral motif continued through the history of literature and was prominent in eighteenth-century Europe. Leo Marx has traced the importance of the "middle landscape," which combines the values of civilization and nature, in a number of nineteenth-century American novels.

The conservationist outlook. From more recent times Gifford Pinchot might serve as an example of a conservationist rather than preservationist viewpoint. Whereas John Muir campaigned for the preservation of untouched wilderness, Pinchot advocated "wise use" and "scientific management" of federal lands and eventually secured Theodore Roosevelt's support. In 1905 the U.S. Forest Service was established, with Pinchot at its head. Its policy was intermediate between unlimited exploitation and absolute protection: the maximum sustained timber yield that would conserve forests for the future. And if the Forest Service can be criticized for opening up federal lands too rapidly to mining, grazing, and lumber interests, it can also be appreciated for the dedicated service of the new professional foresters. A similar conservationist viewpoint prevailed in other government agencies.

The biologist René Dubos gives a contemporary rendition of a philosophy of "respectful use" and "creative intervention." He advocates managing and transforming the earth, but with awareness of the consequences and limitations of human activity. Man has unique capacities and potentialities, yet he is dependent on the biotic community, and there is no radical disjunction of man and nature. Dubos writes:

> We certainly must reject the attitude which asserts that man is the only value of importance and that the rest of nature can be sacrificed to his welfare and whims. But we cannot escape, I believe, an anthropocentric attitude which puts man at the summit of creation while still a part of it [Dubos in Barbour, p. 53].

Barry Commoner considers the relative roles of population, affluence, and technology in environmental pollution. He rejects the thesis of Ehrlich and others that population is the main problem, and instead lays the blame on heavily polluting technologies (especially plastics, detergents, fertilizers, and pesticides). He argues that continued growth is possible with ecologically sound agriculture and technology. But Commoner's rejection of a no-growth policy is based also on his concern for social justice. Improvement of living standards of the poor at home and abroad requires the growth of technologies that are not so environmentally destructive. While Commoner gives insufficient attention to the population question, he does look at issues of distributive justice and at the political and economic structures that perpetuate inequalities, which are ignored by many environmentalists (Commoner, chaps. 10–12).

Social justice and environmental preservation. Issues of justice and environment are intertwined, because exploitation of man and of nature arises from the same institutions and attitudes, such as the concentration of economic power, the pursuit of profit, and the manipulative mentality. In the words of C. S. Lewis, "What we call Man's power over Nature turns out to be a power exercised by some men over other men with Nature as its instrument." Karl Marx was keenly aware of the importance of economic institutions and technology in the exercise of power, but he shared the dominant Western view of "mastery over nature," though with some concern to conserve nature as an object of aesthetic satisfaction. The environmental records of the Soviet Union and the Chinese People's Republic seem to have been no better than those under capitalism.

Like the exponents of dominion, the environmental movement has in the past tended to neglect the issue of justice. From Thoreau to Muir to countercultural communes, the retreat to nature has often involved turning one's back on the city and "the dirty institutions of men." (To be sure, Thoreau advocated the abolition of slavery, and Muir campaigned for wilderness legislation, but the areas of their political activity were limited.) The political viewpoint of environmentalists has reflected their predominantly middle-class background and interests. Today, with widespread recognition of limited global resources, there is among environmentalists a greater concern for inequalities within and between nations. There is awareness that the United States, with six percent of the world's population, accounts for thirty to forty percent of the world's annual resource consumption. As the gap between rich and poor countries increases, questions of distribution, and hence of political and economic power, assume high priority.

Advocates of this third ("stewardship") position will thus often join those in the second ("participation") group in political action to protect the environment. Together they recognize that, by treating air and water as free commodities, the market economy has not charged industry for the true costs. Any "public goods" or unrestricted "commons" will be overused. These indirect costs ("externalities") can only be acknowledged by legislative action (e.g., effluent standards, taxes, or subsidies). Those in this third group differ from the second group in practice mainly by putting more emphasis on the human consequences of legislation. Who pays for and who benefits from a particular law? What happens to the men who lose their jobs in the rare cases when a factory closes because it cannot meet strict standards? Concrete decisions involve the agonizing balance of conflicting values: technological progress, environmental preservation, and social justice.

Continuing issues

The above classification of viewpoints into three groups is an oversimplification. On any issue there is a continuum of intermediate gradations. Moreover, a person might be closer to group two on some topics, and to group three, say, on other topics. As presented above the third view seems to be a compromise between two extreme positions, whereas its proponents claim that it can be defended in its own right. Let us summarize some of the underlying issues today:

1. God's relation to nature. The "dominion" theme has been associated historically with an emphasis on the transcendence of God. The religious versions of "participation," on the other hand, have stressed divine immanence, understood in terms of nature gods, or a pantheistic principle, or a mystical unity known by intuition. The "stewardship" view usually combines transcendence and immanence. Nature is to be neither exploited nor worshiped. Contemporary religious thought has explored this option in the liturgical celebration of nature and in sacramental, incarnational, or evolutionary theologies of nature.

2. **Man's status in nature.** Classical thought portrayed a radical disjunction, an absolute gulf, between man and nature. The ancient Near East, at the opposite extreme, incorporated man into nature, and Darwin in his day contended that any differences between humans and animals are minimal. Today it is more common to acknowledge the evolutionary continuity of human life with lower forms and the presence of structural and behavioral similarities, while insisting that there are significant differences in degree—which amount to qualitative distinctiveness—in human language, culture, and personal and interpersonal life. Priority can be given to human needs and the value of persons without ignoring the welfare of the nonhuman world. "Stewardship" implies the cooperation and partnership of man and nature for the fulfillment of the highest potentialities of both.

3. **Man's attitude to nature.** The attitudes of mastery, subjugation, control, and manipulation have been predominant in the modern West, especially in recent decades in the United States. Technological man since Bacon has often expressed in word and deed a Promethean or Faustian arrogance. In this cultural situation, and with economic and political pressures exerted by institutions having a vested interest in exploiting nature, the "stewardship" position can easily be co-opted or pushed in the direction of "dominion." This is a point at which the scientific version of the "participation" motif—the ecological recognition of finitude, limits, and interdependence—needs to be strongly stressed. Awareness of the far-reaching repercussions of human actions can encourage humility and caution in seeking to impose technological solutions.

4. **The role of technology.** If those in the first group have great confidence in technology, many of those in the second are ready to curtail or even abandon it because of its environmental and human costs. But the rejection of technology today would condemn most of the world to continued poverty, hunger, and disease. A middle position seeks through political processes the reduction of environmental degradation and a more equal distribution of the benefits of technology. One can be realistic about the exercise of economic power by affluent nations and by polluting industries and yet believe that legislation expressing new national priorities can effect change. There are also unexplored possibilities of intermediate technologies scaled to local needs and resources, which may offer a path between large-scale industrial technology and primitive agrarian conditions in both developed and developing countries.

5. **The problem of growth.** Between the options of "pro-growth" and "no growth" lies a policy that asks "Whose growth?" and "What kind of growth?" Global justice requires that in developed nations economic growth be channeled toward services that are not resource-intensive. Ecological wisdom requires research on technologies for recycling and waste reduction, in addition to effective birth control measures, which have not been discussed here. Both justice and ecology require that affluent nations practice restraint in consumption, for which changes in personal values and life-styles, and in national priorities and policies, will be necessary. But the "pro-growth" mentality is so deeply ingrained that those who favor "selective growth" will have to devote most of their efforts to limiting the inequitable demands that the industrial West places on the world's resources and environment. The path of responsible stewardship leads to individual action and institutional change to combine environmental preservation and social justice.

IAN G. BARBOUR

[*For further discussion of topics mentioned in this article, see the entries:* EVOLUTION; *and* LIFE. *Also directly related are the entries* ENVIRONMENTAL ETHICS; MAN, IMAGES OF; SCIENCE: ETHICAL IMPLICATIONS; *and* TECHNOLOGY. *See also:* FUTURE GENERATIONS, OBLIGATIONS TO; JUSTICE; MIND–BODY PROBLEM; *and* REDUCTIONISM.]

BIBLIOGRAPHY

BARBOUR, IAN G., ed. *Western Man and Environmental Ethics.* Reading, Mass.: Addison-Wesley Publishing Co., 1973.

BLACK, JOHN. *The Dominion of Man.* Edinburgh: Edinburgh University Press, 1970.

COMMONER, BARRY. *The Closing Circle.* New York: Alfred A. Knopf, 1971.

DUBOS, RENÉ. *A God Within.* New York: Charles Scribner's Sons, 1972.

Glacken, Clarence. *Traces on the Rhodian Shore.* Berkeley: University of California Press, 1967.
Leiss, William. *The Domination of Nature.* New York: George Braziller, 1972.
Marx, Leo. *The Machine in the Garden.* New York: Oxford University Press, 1964.
Nash, Roderick. *Wilderness and the American Mind.* New Haven: Yale University Press, 1967. Rev. ed. 1973.
Passmore, John. *Man's Responsibility for Nature.* New York: Charles Scribner's Sons, 1974.
Santmire, H. Paul. *Brother Earth.* New York: Thomas Nelson, 1970.
Tuan Yi-fu. *Topophilia: A Study of Environmental Perception, Attitudes, and Values.* Englewood Cliffs, N.J.: Prentice-Hall, 1974.
White, Lynn, Jr. "The Historical Roots of Our Ecologic Crisis." *Science* 155 (1967): 1203–1207. Reprinted in Barbour, *Western Man*, pp. 18–30.

II

EASTERN THOUGHT

Conformity to nature. Chinese thought traditionally tended to consider that all things could exist only insofar as they were in conformity with human nature. This gave rise to the attitude of esteem for the principle of nature that exists in the human mind. Since ancient times, the idea of Heaven (*T'ien*) was conceived by the Chinese in close relation with human life. According to a poem composed in the ancient period of the early Chou dynasty, Heaven created all human beings, and therefore Heaven, as the ancestor of man, handed down moral precepts which humans must observe. This idea was inherited by Confucius. He recommended acknowledging "the order of Heaven," which meant that "one should follow the morality given by Heaven." There was a general assumption that politics should follow laws based upon natural law.

The opinion that "man should follow his true nature" was also stated by other scholars in ancient China, but their interpretation was different from that of Confucius. Mo-tzu taught that the ruler should follow only what Heaven wished. Lao-tzu insisted that the correct way for man is to follow the way of Heaven; therefore, the basis of the correct way is *T'ien-tao* (the Way of Heaven). Yang-chu (who lived between the times of Mo-tzu and Mencius) stated: "Original human nature desires only sex and food. Therefore, it is better for one not to have relations with others but only to satisfy one's own desires. It is a natural law that one does what he wants." Mencius taught that "the true character of man is good; however, the evil mind arises by the temptation of material desires. Therefore, one should cultivate one's own mind and exhibit one's own true character."

An exception to Chinese thought was Hsü-tzu, who maintained that the true character of man tends to evil. Nevertheless, he recognized the possibilities in man of becoming good. Chuang-tzu taught that one should perfect one's true character, and his followers came to teach the theory that "one should return to his true character." In the San-kuo dynasty, Wang-pi (A.D. 226–249) also taught the doctrine of "return to one's true character." This thought was developed greatly in the Sung philosophy, where the central theme was the concept of a person's true character. The traditional current of thought in Chinese history is "to return to the true and natural human character." This idea was shared by the Japanese also, probably under the influence of Chinese thought.

Buddhism was also influenced by this current of thought. Buddhists did not look for truth in the phenomenal world but explored the inner thought as expressed in a peculiarly Chinese way: "If one realizes the truth that all existences are the same, one immediately returns to one's true nature." Both illusion and enlightenment of people were understood to be derived from the natural character of humanity:

The mind is the ground and nature is the king. Where there is nature there is the king, and where there is no nature, there is no king. Where there is nature, there are body and mind. Where no nature exists, there is neither body nor mind. Buddha is created by knowledge of self-nature; therefore, one must not look for the Buddha through the body. If self-nature comes by enlightenment, then the self-enlightened being is namely the Buddha.

There was, in India, no such idea of self-nature as the principle that maintains the body and mind as ignorant or enlightened. Some Chinese scholars have recognized a Taoist influence in this conception of self-nature. However, this concept could have appeared from the traditional ideas of the Chinese.

Medieval Confucians said that the whole of nature is to be found in any single item, which theory seems originally to have been due to the influence of Buddhist philosophy, especially the Buddhist Hua-yen school.

Nature as the absolute. Taoism taught that nature is the absolute. Chinese Buddhist Pure Land teachings adopted the ideas of Taoism. Chinese Buddhists had to pass through a process of complicated thought before they acknowledged a Chinese type of naturalism. In this connection, Chi-tsang reasoned as follows: Chinese philosophical thought, especially in Lao-tzu and Chuang-tzu, regarded existence as phenomena, and void (*Sunyata*) as a substance other than existence. Therefore, void was not identifiable with existence. Buddhism, on the contrary, taught that phenomena are actually the manifestation of the Absolute. Therefore, the absolute significance of the phenomenal world cannot be recognized in actual life in the philosophy of Lao-tzu and Chuang-tzu. In Buddhism, however, one can accept this phenomenal world as absolute states of existence, because actual life in this world is identical with absolute existence. Although this criticism by Chi-tsang may not be correct, at least he tried to recognize a significance in life in this world.

The T'ien-t'ai and Hua-yen sects further expanded on this thought. According to the T'ien-t'ai sect, appearance and actuality are not different kinds of substances because appearance is identical with reality. Therefore, they taught that "each existence in this world is the middle way." Each of the phenomenal forms of this world is a form of absolute existence. The Hua-yen sect developed this thought even further and taught the theory of "mutual penetration and identification of all things with one another." The supreme meaning emerges when all phenomena are perfectly identified by their harmonious interrelationships. Therefore, nothing exists outside of phenomena and their diverse forms of manifestation.

As a result, the actual natural world was acknowledged as absolute existence. In Zen Buddhism, for example, this naturalistic tendency led to the conclusion that each one of the existences of this world is, just as it appears, a manifestation of truth. Zen monks, of course, opposed and rejected any merely superficial naturalism. Nevertheless, the Chinese generally accepted the view that nature is the absolute. Finally, the T'ien-t'ai sect taught the theory that "all existences and even grass, trees, and earth can attain Buddhahood." That is to say, even physical matter existing in nature can realize enlightenment and become Buddha. Generally speaking, the tendency was to regard nature as the most beautiful and highest existence, on an equal plane with humans. Therefore, the Chinese Buddhists (especially Zen monks) tried to seek absolute significance in everyday life. Everyday life, just as it truly is, is identical with enlightenment. Zen masters also made much of natural beauty, as in the words of Master Ryōkan:

> In spring, the flowers;
> and in autumn the moon;
> In summer a refreshing breeze,
> and in winter the snow;
> What else do I need?

Zen Buddhists also endeavored to take the surroundings in the natural world as they are.

As the result of the tendency to regard nature or actuality as absolute existence, the Chinese came to adopt an attitude of optimism. Thus, they regarded this world as a good place in which to live; they finally came to believe that perfect existence must exist in this world. Here, the idea of the *Sheng-jen* (sage) was established. He was the perfect person such as the Chou King or Confucius. The sage is not a god but a man.

The Chinese identified nature with human life. The harmony of all existences is necessary in order to harmonize with nature and live in peace. Thus, they asserted the idea of "moderation." As the Chinese regarded man as a part of nature or the universe, they did not regard nature as opposed to man. Since they seldom thought nature needed to be transformed by experimental manipulation in order to master her ways or laws, natural science did not develop quickly in China. This fact is perhaps the chief reason why China lagged scientifically behind other countries in the modern world. Leaders of the People's Republic of China recognize this fact and are trying to improve and develop natural science, although this may lead to a conflict with traditional values maintained by Taoist-minded people.

Throughout all modern Eastern countries, Eastern thought has found an accommodation to technology and its manipulation of nature. In many Eastern countries studies of natural sciences and technology are strongly encouraged by governments and big businesses. Traditionalists also admit this line of development, which is unavoidable in modern civilization.

However, in some cases conflict with traditional values prevents the development of technology. This is most conspicuous with the Jains, who prohibit killing animals and the use of any insecticide, owing to the spirit of compassion towards all living creatures. They may not use animals for experiments; they may not exploit marshes, for the enterprise will kill many small living creatures living there. With the Sikhs and

Parsis there are no such prohibitions, and that is why they have been ahead of others in the modernization of India. For Chinese, Japanese, and Koreans who profess Mahāyāna Buddhism there is no problem of this kind.

Relationship between Heaven and man. The Chinese elaborated an organic theory of a "reciprocal relationship between Heaven and man." In the period of the Chan-kuo (Warring States period, 480–222 B.C.), "scholars of the positive and negative principle" advocated a kind of nature worship, which was carried into the Early Han period. According to this principle, natural phenomena and man-made institutions are mutually interrelated, and therefore if the King, who was the representative of man, governed the country well, then the phenomena of nature—such as weather, wind, and rain—would be favorable to man. If the reign of the King was poor, on the other hand, then natural calamities would arise. This idea was most strongly stressed by Tung Chung-shu (179?–104? B.C.) of the Early Han dynasty, who thought that disasters were sent from Heaven in order to admonish the King. The thought of *Ko-ming* (revolution) which means, literally, "to cut off (or take away) the mandate of Heaven from some particular ruler" played a role in checking or correcting the tyranny of aristocrats. This thought was influential in later periods in China. Some Indian Buddhists also held the theory that "disasters arise through poor government by the King."

Buddhism had sutras which stated the theory of disaster and which were highly regarded by the Chinese. A typical example of these sutras is the *Chin-kuang-ming* (*Suvarnaprabhasa*, Golden Splendor) sutra, which states in detail in the thirteenth chapter that, if the King does not protect the *dharma* (religion, law) well, a terrible calamity will arise. That is to say, as the result of maladministration on the part of the Emperor, falsehood and struggle will increase in his country, and the ministers and subjects will rise against him. Furthermore, the deities will become angry; wars will break out; the enemy will overrun the country; family members will fight each other; nothing will be pleasant or comfortable for man. Natural phenomena will at the same time become worse. Living beings will lack vigor, plagues will arise, and pestilence will sweep the land. Therefore, the Emperor should attempt his best in governing the country by the *dharma* (law).

Naturalized Buddhist monks from India propagated Buddhism in conformity with this organic way of Chinese thinking. Gunavarman, for example, taught Buddhism to Emperor Wen of the Sung dynasty in the following way:

The four seas are your land and all existences are your subjects. One pleasant word and all your subjects are happy. One act of a good ruler brings harmony to the people. If you only punish wrongdoers without killing and do not impose heavy taxes, then nature will harmonize with man, and fruits and crops will ripen well.

This organic form of ethical thought continued in China for a long time. Pure Land teachings of China also explained this ethical theory in terms of the relationship between the Buddha's grace and nature (Nakamura, p. 283).

Action and change as reality. The inclination to live contentedly in this given phenomenal world appears in ancient Shintoism as well as in modern sectarian Shintoism. The founder of the Konkō sect, for instance, teaches, "Whether alive or dead, you should regard the heaven and earth as your own habitation." The process of the phenomenal world is activity—mighty, self-maintaining, and procreative with the creative power and freedom of sublime wonder. Master Dōgen says, "Being is time, and time is being. Everything in the world is time at each moment." He said quite radically, "Birth and death is the life of Buddha."

In the long history of Japanese thought it was traditional to seek for the Absolute in the phenomenal (physical) world, and this way of thinking has played an effective role in the assimilation of Zen as well. For Master Dōgen, impermanence is itself the absolute state, and this impermanence is not to be rejected but to be valued. "Impermanence of grass, trees, and forests is verily the Buddhahood. The impermanence of the person's body and mind is verily the Buddhahood. The impermanence of the country and scenery is verily the Buddhahood."

Jinsai Itō (1627–1705), the Japanese Confucianist, regards heaven and earth as the evolution of great activity in which nothing but eternal development occurs, and hence he completely denies what is called death. According to Itō, the world of reality is nothing but change and action, and action is in itself good. All of the characteristically Japanese scholars believe in phenomena as the fundamental mode of existence. They unanimously reject the passivity of the neo-Confucianists of the Sung period.

Love of natural beauty. The way of thinking that recognizes absolute significance in the phenomenal world seems to be culturally associated with the Japanese traditional love of nature, which has been characteristic of the Japanese since ancient times. The Japanese in general love mountains, rivers, flowers, birds, grass, and trees, and represent them in the patterns of their kimono, and they are fond of the delicacies of the season, keeping their edibles in natural forms as much as possible in cooking. Within the house, flowers are arranged in a vase and dwarf trees placed in the alcove, flowers and birds are engraved in the transom, simple flowers and birds are also painted on the sliding screen, and in the garden miniature mountains are built and water is drawn. The literature is also closely tied up with warm affection for nature. *Makura no Sōshi* (Pillow Books) begins with general remarks about the four seasons and then goes into the description of the scenic beauties of the seasons and human affairs. There are many essays of this kind. If the poems on nature should be removed from the collections of Japanese poems, few poems would be left.

The love of nature, in the case of the Japanese, is tied up with their tendencies to cherish minute things and treasure delicate things. Contrast the Japanese love of individual flowers, birds, grass, and trees with the British enjoyment of the spacious view of the sea, the Dover Cliffs, and the countryside. Such aesthetic preferences of various nations are culturally significant traits of their respective peoples.

The Japanese enjoy nature as it is reflected in their compact range of vision, which is particularly evident in the following poem:

In my garden fall the plum blossoms—
   Are they indeed snowflakes
Whirling from the sky? [*Maunyōshū*, V, 822.]
The nightingale sings
Playing at the lower branches
Lamenting the fall of the plum blossoms [ibid., 824].

In this respect the Japanese love of nature differs somewhat from the Chinese attachment to the rivers and mountains. The point may be best illustrated by the comparison of the following two poems. Dōgen, the Japanese Zen master (thirteenth century A.D.), writes:

> Flowers are in Spring, Cuckoos in Summer,
> In Autumn is the moon, and in Winter,
> The pallid glimmer of snow.

The meaning of the above poem coincides with what is intended by the Chinese verse of *Wu-men-kuan* (Gateless Gate, by Wu-men Hui-k'ai).

> A hundred flowers are in Spring, in Autumn is the moon,
> In Summer is the cool wind, the snow is in Winter;
> If nothing is on the mind to afflict a man,
> That is the best season for the man.

The word "cuckoos" of the Japanese is replaced in the Chinese by the "cool wind," which gives an entirely different effect. The cool wind and cuckoos are both sensible objects, but while the former gives the sense of indefinite, remote boundlessness, the latter gives a limited and familiar, homely impression.

The Indians also love nature and construct gardens (*udyana, arama*)—where they plant grass and trees and lay out wells and springs—but they rarely try to imitate natural rivers and mountains on a small scale. The Indian ascetics also composed poems in praise of nature. They enjoy and extol nature as the sanctuary beyond worldly sensuous attachments, afflictions, and bondages. In their case nature is conceived to be something opposed to human vicissitudes, and at a distance from human feeling.

In the case of the Japanese, however, priests and laymen alike are attached to nature, which is at one with human beings. They esteem the sensible beauties of nature, in which they seek revelations of the absolute world.

> Cherry blossoms, falling in vain,
> Remind me of the Treasure plants,
> That adorn paradise [Emperor Kazan].

There is no inkling of a view that regards the natural world as cursed or gruesome.

Love of animals. The tender love of animals traditionally runs in the veins of the Japanese, but that love is concentrated on minute, lovable living things.

> A copper pheasant warbles out.
> Listening to its voice I thought,
> Could it be the father calling?
> Could it be the mother calling? [Master Gyōki.]

The image of the "copper pheasant" is very Japanese. In contrast, the people of India and the South Asiatic countries are often fond of a story such as willingly abandoning oneself to a hungry tiger who attacks one. Such a story is not quite congenial to the poetic sentiments of the Japanese, although both peoples wanted to express the idea of benevolence toward living creatures.

Surroundings are sentient. The conception was prevalent in medieval Japan that even grass and trees have spirits and consequently are eligible for salvation. The idea that even the things of "no-mind" (the objects of nature that have no spirits) can become Buddhas, based upon the

Tendai doctrines, was particularly emphasized in Japan. This constituted an important theme for study in the Japanese Tendai sect, and the idea was inherited also by the Nichiren sect. Nichiren (1222–1282) sought the superiority of the *Hokke* (Lotus) *Sūtra* in its recognition of the eligibility of the grass and the trees to become Buddhas. There appear time and again among the Japanese Buddhist writings the following lines: "When a Buddha, who has attained enlightenment, looks around the universe, the grass, trees, and lands, all become Buddhas." In "Noh" songs we often come across such an idea, which was taken for granted socially and religiously in those days. There is a story of a willow tree's becoming a Buddha, based upon the religious faith of the Jōdo-shin sect. The oral tradition of the medieval Tendai sect of Japan pushed the idea of grass and trees that become Buddhas so far as to preach "the non-becoming Buddhas of the grass and trees" (for they are already Buddhas). According to this theory, everything is by nature a Buddha—that is to say, to attain enlightenment through ascetic practice is one and the same thing as being a Buddha without recourse to ascetic practice. Not only the grass and trees but also rivers, mountains, and the earth are themselves Buddhahood, already possessed intact. There is no becoming a Buddha in the sense of coming to be something separate and different in nature. That is the reason why the nonbecoming of Buddhahood was preached. The logical conclusion of the idea of the acceptance of the given reality is here definitely and clearly crystallized.

Some Indian Buddhists also admit the spirituality of grass and trees, along with certain schools of Indian philosophy that also adopt such a view. However, most Indian philosophies maintain that all living things attain the state of deliverance through enlightened intelligence (*vidyā*), and that grass and trees do not become Buddhas in their actual state as they are. Such a tendency of thinking as discussed above still seems to be effective among the Japanese even in these days, when the knowledge of natural science prevails. In fact, there is probably no other nation on earth that uses an honorific expression prefixed to the names of everyday objects such as tea and water. This is not merely a question of an honorific expression, but a manifestation of a way of thinking that seeks a raison d'être and sacredness in everything that exists.

Exploitation of nature. Monks of Southern Asia did not want to be involved in productive work of any kind. They just practiced meditation without working physically. This attitude has been preserved throughout Asiatic countries except in China and Japan. In China, Zen priests began to cultivate fields attached to their own temples in the eighth century A.D. in order to secure a permanent supply of food. The reason for the transformation of the monks' activities is as follows. Originally, the monks lived on alms, begging food from lay believers; but because of the devastation caused by many wars, Zen monks escaped from the confusion in cities and towns and settled deep in the mountains, where they could not beg for alms. At this point they began cultivating the land and building their own lodgings by manual labor. Thus physical labor came to be cherished. The motto, "If one did not work a day, one should not eat on that day" became their favorite saying. This motto has also been greatly encouraged by the Communist government under Mao Tse-tung.

In Japan monks of many sects went so far as to engage in such economic activity as constructing roads, rest houses, hospitals, ponds, and harbors, and exploiting fields. Philanthropic works were encouraged in Japan: Rendering service to others was said to be the essence of Mahāyāna. For laymen all sorts of productive work except slaying animals and selling wines, weapons, and so on were encouraged.

Harmony with nature. One of the most prominent features of Buddhist thought is its zest for harmony with nature. The joy of enjoying natural beauty and of living comfortably in natural surroundings was expressed by monks and nuns in *The Poems of the Elders* (*Theragāthā* and *Therigāthās*). This attitude was inherited from the ancient Indian hermit-sages (*ṛṣis*) as described in the *Mahābhārata* and the *Rāmāyana*. This attitude is most clearly reflected in Japanese gardening. Japanese gardens, which were developed after those in China, differ greatly from the Western and Islamic ones, which concentrate on geometrical patterns and symmetry. Japanese gardens, although artificially made by adept gardeners, give us the impression of reproducing natural beauty as such, inducing us to imagine ourselves as living in real nature: Artifice and nature are not separate. Harmonious union with nature is the ideal of the Buddhists' view of the environment.

Contemporary implications. According to Buddhist philosophy, humans constitute just one class of living beings and have no right to an unlimited dissipation of natural resources. Also, humans have no right of unlimited exploitation of animal and plant life, which form part of nature. Hitherto, Westerners or moderns too often tended to think of themselves as quite separate and different from the natural world and to believe that they were qualified for the exclusive use of the natural world, for they were created by, and in the image of, God. But this assumption of dominion over nature by humans is ungrounded and unreasonable. The modern Western view has brought devastation to the environment. Now we are incurring nature's retaliation. In order to deal with the difficult situation in which we are now placed, we must find a solution. Solution can be approached in two ways: objectively and subjectively.

On the objective side, the views reported above would suggest that we should discard the arrogant attitude that we are entitled to exploit the natural world at our own will, without limit and regardless of the consequences. If nature is to be met with affection, we can scarcely afford to maintain the attitude of "conquerors of nature." The entire world of nature, rather than being monopolized by a limited number of countries, should be shared by all mankind. Just as individual egoism should be curtailed and placed under control, so national and ethnic egoism should be restrained in the name of justice and respect for nature.

On the subjective side, there is need for a corrective factor against the opinion of moderns that the progress of mankind consists in the unlimited satisfaction of human desires for material objects. Buddhism taught satisfaction with what is given to men. "To know what it is to be truly satisfied" was thought to mean knowing the way of spiritual happiness. The key to the relationship between man and environment may very well lie in the control of our desires according to this concept of satisfaction.

HAJIME NAKAMURA

[*For further discussion of topics mentioned in this article, see the entries:* BUDDHISM; CONFUCIANISM; DEATH, *article on* EASTERN THOUGHT; HINDUISM; TAOISM; *and* MEDICAL ETHICS, HISTORY OF, *section on* SOUTH AND EAST ASIA, *articles on* INDIA, PREREPUBLICAN CHINA, *and* JAPAN THROUGH THE NINETEENTH CENTURY. *See also:* ENVIRONMENTAL ETHICS; ISLAM; PURPOSE IN THE UNIVERSE; SCIENCE: ETHICAL IMPLICATIONS; *and* TECHNOLOGY.]

BIBLIOGRAPHY

BOSE, D. M.; SEN, S. N.; and SUBBARAYAPPA, B. V., eds. *A Concise History of Science in India.* New Delhi: Indian National Science Academy, 1971.

DE BARY, WILLIAM THEODORE; CHAN, WING-TSIT; and WATSON, BURTON, eds. *Sources of Chinese Tradition.* New York: Columbia University Press, 1960.

———; HAY, STEPHEN; WEILER, ROYAL; and YARROW, ANDREW, eds. *Sources of Indian Tradition.* New York: Columbia University Press, 1958.

JAGGI, OM PRAKASH. *Scientists of Ancient India and Their Achievements.* Delhi: Atma Ram & Sons, 1966.

LESLIE, CHARLES, ed. *Asian Medical Systems: A Comparative Study.* Berkeley: University of California Press, 1976.

NAKAMURA, HAJIME. *Ways of Thinking of Eastern Peoples: India, China, Tibet, Japan.* Rev. English trans. Edited by Philip P. Wiener. Honolulu: East–West Center Press, 1964.

RIEPE, DALE. *The Naturalistic Tradition in Indian Thought.* Seattle: University of Washington Press, 1961.

TSUNODA, RYŪSAKU; DE BARY, WILLIAM THEODORE; and KEENE, DONALD, eds. *Sources of Japanese Tradition.* New York: Columbia University Press, 1958.

ZIMMER, HEINRICH ROBERT. *Hindu Medicine.* Edited with a foreword and preface by Ludwig Edelstein. Publications of the Institute of the History of Medicine, The Johns Hopkins University. 3d ser.: The Hideyo Noguchi Lectures, vol. 6. Baltimore: Johns Hopkins Press, 1948.

ENVIRONMENTAL ETHICS

I. ENVIRONMENTAL HEALTH AND HUMAN DISEASE
　　　　　　　　　　　　　　　Samuel S. Epstein
II. QUESTIONS OF SOCIAL JUSTICE
　　　　　　　　　　　　　　Norman J. Faramelli
　　　　　　　　　　　　　and Charles W. Powers
III. THE PROBLEM OF GROWTH　　　*C. P. Wolf and*
　　　　　　　　　　　　　　Drew Christiansen

I
ENVIRONMENTAL HEALTH AND HUMAN DISEASE

This article will deal with the moral and ethical aspects of the relationship between environmental pollution and disease in the general and working populations. First, the scope and extent of environmentally induced disease is discussed. Second, the problem is analyzed with respect to the cost-benefit considerations of modern technology. Third, problems of biased data are described, including how these may impact on regulatory decisions. Finally, the need for a greater role of ethical consideration in decision-making processes is emphasized. Although the themes considered are universal, particular reference is made to problems in the United States.

Human disease

There is now growing evidence that much human disease hitherto regarded as spontaneous, including cancer, is caused by environmental pollutants. The contributory role of air pollution in chronic respiratory disease is also well recognized. Additionally, there is a growing body of evidence relating specific pollutants to psychobehavioral disorders and birth defects. Finally, emerging data incriminate specific pollutants with mutations and mental deficiency.

This realization of a relationship between pollution and disease is heightened by the exponential increase in human exposure to new synthetic chemicals and to their degradation and combustion products in air, water, and soil. In general, these new chemicals are not adequately characterized, either toxicologically or ecologically.

Chronic respiratory disease. Chronic respiratory disease, including bronchitis, emphysema, and asthma, is widely prevalent in the general population. Air pollution, especially particulates and sulfates from stationary sources, such as municipal incinerators and utility plants, and automobile emissions, are major causes of chronic respiratory disease. For example, it has been estimated that failure to meet current sulfur dioxide standards, which are exceeded in most U.S. cities, is responsible annually for six thousand premature deaths, six million to ten million avoidable asthma attacks, and twenty million to thirty million days of exacerbation of cardiovascular and respiratory disease. It is further estimated that automobile emissions contribute ten to fifteen percent of health hazards from air pollution. Additionally, many kinds of dust present in the work environment cause debilitating lung diseases, called pneumoconioses. About 125,000 U.S. coal miners suffer from coal workers' pneumoconiosis, and the disease is estimated to cause three thousand to four thousand deaths each year. Other pneumoconioses include asbestosis, as well as various cancers, from asbestos; byssinosis from cotton dust; bagassosis from sugar cane dust; and silicosis from various silica-containing dusts. Over all, nine thousand annual deaths in the United States are attributed to these occupational dust diseases.

Cancer. Cancer is now a major killing and disabling disease of epidemic proportions. More than 53 million of the 210 million U.S. population (twenty-five percent) will develop some form of cancer, and approximately twenty percent of Americans now die from cancer (Epstein, "Environmental Determinants"). It is estimated that 665,000 new cancer cases were diagnosed and that there were 365,000 cancer deaths in 1975. Thus, cancer deaths in 1975 alone were approximately five times higher than the total U.S. military deaths in the Vietnam and Korean war years combined.

Human misery apart, the total economic impact of cancer is massive. Estimates indicate that in 1969 the direct costs for hospitalization and medical care for cancer exceeded $500 million. The direct and indirect costs of cancer, including loss of earnings during illness and during the balance of normal life expectancy, were estimated in 1971 as $15 billion (ibid.); estimates for 1976 are about $18 billion.

The rate of recent increase of cancer deaths in the United States and other industrialized countries is more rapid than the rate of increase in population. It is of interest to note that besides cancer only two other major causes of death—homicide and cirrhosis of the liver—have significantly increased in the recent past. The increase in new cancer cases is real and over and above any increase due to age alone.

There is now a growing consensus that the majority of human cancers (some estimates go from seventy to ninety percent) are environmental in origin and that they are hence ultimately preventable. Prevention, however, demands vigorous legislative and regulatory initiatives, which have not yet been forthcoming. The basis for such estimates largely derives from epidemiological studies, in large community populations over extended periods, which have revealed wide geographic variations in the incidence of cancer of various organs. It should be noted, however, that the role of specific environmental carcinogens has been so far implicated or identified in only relatively few of the studies (ibid.).

It is of particular interest that a recent National Cancer Institute (NCI) atlas has demonstrated marked geographical clustering of high mortality rates from cancers of various organs in the United States among white male and female populations in heavily industrialized counties (Mason et al.). Such data are presumptive of associations between cancer rates in the general community and the proximity of residence to certain industries, in addition to exposure of workers in these industries. There are growing analytic data on the spillover of carcinogens from industry into the adjacent community by routes including air, water, and clothing of workers.

Apart from the importance of occupational factors in the incidence of "neighborhood" or community cancer in the population at large, specific occupational exposures are an important cause of cancer deaths, particularly in males. Various estimates have suggested that five to fifteen percent of all current cancer deaths in males are occupational in origin. These include lung and other cancers in insulation workers and others, such as construction workers, exposed to asbestos; bladder cancer in the aniline dye and rubber industry, induced by such chemicals as 2-naphthylamine, benzidine, 2-aminobiphenyl and 2-nitrobiphenyl; lung cancer in uranium miners of Colorado, in coke oven workers, and in workers even briefly exposed to bischloromethylether; skin cancer in shale oil workers; nasal sinus cancer in wood workers; cancer of the pancreas and lymphomas in organic chemists; and angiosarcoma of the liver, besides other cancers, in workers involved in the manufacture and fabrication of polyvinyl chloride (PVC).

The toll of cancer related to particular occupational exposures is overwhelming. For instance, it has been estimated that approximately twenty percent of all of the long-term asbestos workers die of lung cancer. Approximately thirty percent of all premature deaths in uranium miners are due to lung cancer. Many other occupational groups are at high cancer risk, including steelworkers, miners and smelters, rubber workers, and workers in a wide range of petrochemical industries (Epstein, 1976).

Birth defects. There is also growing recognition of the importance of environmental pollutants as causes of birth defects. The incidence of gross congenital malformations in the United States, although unknown in the absence of a comprehensive national registry, has been estimated at from three to four percent of total live births. Three major categories of human teratogens have so far been identified: viral infections, X-irradiation, and chemicals and drugs, such as mercurials, thalidomide, and diethylstilbestrol.

Thalidomide is a drug that was widely marketed in Europe in the late 1950s and early 1960s for "morning sickness." An alert medical practitioner associated the drug with the birth of thousands of babies with unusually malformed limbs. It should be noted that the teratogenicity of thalidomide was detected only because of the unusual birth defects it produced; had it induced a similar incidence of more common defects, in all likelihood, it would still be in use as a "safe" drug to this date. The phenoxy herbicide 2,4,5-T and related compounds, including their dioxin contaminants, have also aroused concern because of their widespread and increasing use and because of their teratogenicity in mice and rats; their incrimination in human birth defects is not yet clearly established.

Mutations. The first evidence that environmental pollutants may influence the genetic constitution of future populations resulted in the 1930s from the discovery that high-energy radiation induces mutations. The subsequent development of nuclear energy added a new dimension and enhanced awareness of the problem of genetic hazards. Once radiation-induced mutagenesis was discovered, there were reasons to suspect that some chemicals would act similarly, but proof of this was delayed until World War II, when mustard gas was shown to induce mutations in fruit flies. Many and varied types of chemicals have subsequently been shown to be mutagenic in various systems, including mammals. The likelihood that some highly mutagenic chemicals may come into wide use or indeed may already be in wide use is now of serious concern.

Psychobehavioral defects. Emerging data are establishing clear relationships between exposure to a wide range of environmental and occupational pollutants and acute and delayed psychobehavioral or neurotoxic effects, ranging from disturbances in fine judgment and discrimination and mild alterations in personality and behavior to advanced dementia. Illustrative of a common pollutant producing acute psychobehavioral effects is carbon monoxide, which, at concentrations found in common traffic conditions, produces disturbances in reflexes, judgment, and orientation. Further examples are the delayed effects of lead, which, at exposure levels and body burdens now considered to be relatively low, induce disturbances in personality and nerve conduction in workers, and mental deficiency and psychobehavioral disorders in young children.

Cost-benefit analysis

There is considerable controversy today over whether narrow "cost-benefit" language is truly adequate for arriving at sound environmental policies, since that kind of calculus does not in itself account for such ethical considerations as justice, human rights, and personal preferences. Furthermore, when a policy is determined by cost-benefit analysis, it is important to take into account what kind of benefit and what kind of risk are involved, who enjoys the benefit, who suffers the risk, and how those benefits and risks are evaluated and perceived by the parties involved. Thus, for example, it is sometimes difficult to know what conclusion to draw when the health risk is significant to a very small portion of the population and the benefits are minimal (perhaps esthetic) yet desirable to almost the entire nation.

Chemical products and processes. Since World War II there has been an exponential and largely unregulated increase in the numbers and quantities of synthetic organic chemicals manufactured and used in industrialized countries. The claimed need, generally in the absence of supportive documentation, to use increasing numbers of new synthetic chemicals makes it essential to recognize and evaluate carcinogenic and other human hazards with regard to the real or alleged matching benefits they confer; similar considerations extend to direct and indirect environmental degradations. Such costs must be balanced by factors including the persistence and environmental mobility of the chemical, the size of the general or occupational population exposed, and the reversibility of the adverse effects. The costs must be viewed realistically: The total annual U.S. costs of approximately $15 billion due to cancer have hitherto been largely discounted ("externalized," in the language of the economist). As the majority of human cancers, in both the general and the working population, are now considered to be due to environmental carcinogens, and hence preventable, there should be clear economic and other incentives to reduce this environmental and occupational burden. Yet, strangely enough, there has been a low fiscal allocation (less than ten percent of the $830 million budget in 1976) to environmental and chemical carcinogenesis in the budget of the National Cancer Institute (NCI).

The criterion of broad societal efficacy, once extended from therapeutic drugs to other synthetic chemicals, such as food and feed additives, pesticides, and industrial chemicals, may further simplify the cost-benefit calculus. As an illustration of this argument, food additives would have to be excluded from products unless they are both safe and either significantly improve the quality of the nutritive value of the food or lower its costs. Food dyes and other such "cosmetic" food additives do not meet the criterion of efficacy, let alone safety (further societal reflection is needed to determine what extent of risk if any is reasonable in achieving a purely cosmetic effect in foods).

Similarly, claims that occupational carcinogens serve industrially unique purposes must be examined openly by experts representing a broad range of viewpoints and interests, with particular recognition of the attendant and generally externalized or discounted human costs and of the lack of economic incentives to develop similarly efficacious and less hazardous product or process alternatives. In the absence of such alternatives, consideration should be directed to the possible banning of the manufacture and use of the carcinogen or to restricting its use to large industrial facilities willing and able to use closed systems that are continuously monitored with instrumentation of maximum sensitivity, with the results made accessible to exposed workers.

Inherent in toxicological and regulatory practice is acceptance, in principle, of the concept of balancing benefit—benefit to the general public rather than to industry—against cost to public health and the environment rather than cost or risk to industry. If this standard is to be taken seriously, one might ask: If the chemical product or process in question does not demonstrably serve a broad social and economically useful purpose for the general population, why introduce it and force the public at large to accept potential costs without general matching benefits? The concept of matching benefits against costs has generally been employed in practice to maximize short-term benefits to industry, even though this may entail minimal benefits and maximal costs to the general public. While such an approach is detrimental to the general public, it is also often detrimental to the long-term interests of industry, which may suffer major economic dislocation when hazardous products and processes, to which it has improperly developed premature major commitments, are belatedly banned from commerce.

Nuclear energy. It has been argued that conventional cost-benefit considerations cannot be applied to policy questions in the development of a national, large-scale, fission-based economy, since such questions are largely ethical rather than technological. These questions, relating to whether society should strike a Faustian bargain with atomic physicists and engineers, have been posed, for example, as the burden of requiring continuous monitoring in order to avoid unparalleled disaster (Kneese).

It appears that the immediate environmental

impact of the routine operation of the nuclear fuel cycle can be reduced to lower levels than is possible with fuel-fired plants. The attendant problems include strip-mine land reclamation, pneumoconioses, and community air pollution, especially from acid sulfates. However, such an immediate superiority of nuclear technologies does not necessarily exclude other potential alternatives, including solar and geothermal energy, on which very small research and development efforts have so far been expended, and on which the U.S. and other governments appear reluctant to increase their commitments. Furthermore, the technologies for controlling adverse environmental and public health impacts of fossil fuel utilization have been developed and are being used without major economic perturbations. Such alternatives apart, the advantages of fission are more easily stated in terms of cost-benefit analysis than are the hazards, particularly those associated with operation of the fuel cycle and with long-term waste storage. Estimates on the probability and consequences of major accidents at a nuclear facility vary. A not unfamiliar association of estimates of low consequence with direct or indirect economic interest, and of estimates of higher probability and higher consequences with the absence of such interest, appears to exist. Property damage apart, critiques by the Union of Concerned Scientists (1974) and the American Physical Society (1975) have indicated that a report contracted by the then Atomic Energy Commission (AEC), underestimated the health consequences of a power plant accident by a factor of sixteen to fifty. Furthermore, it is clear that the probability of accident rises with increasing numbers of nuclear facilities and is increased by sabotage, terrorism, and warfare. The continued insistence of the nuclear industry on subsidies, such as the limited liability conferred (in the United States) by the Price–Anderson Act, indicates that the industry itself is concerned about the probability and financial consequences of accidents.

Of equal concern is the problem of storage of high-level radioactive wastes. Unequivocal assurances by the then AEC that the Lyons, Kansas, salt formations were the best site for waste storage were shortly followed by disclosures that the site was a "leaky sieve" (Metzger, p. 158). Current emphasis on the design of surface storage facilities, intended to last a hundred years or so pending development of more permanent storage sites, will necessitate development of sophisticated automated monitoring and elaborate safeguards against accidents, terrorism, and theft.

The costs and hazards from nuclear technology are of utmost gravity and clearly extend to mankind many generations hence. It is clear that decisions on the benefits and costs of nuclear technologies must be shifted from the narrow and constrained arena of the private sector and technology to the wider domain of the public interest and representative government.

Burden of proof. The concept of the burden of proof in public policymaking and in cost-benefit analysis is somewhat unfamiliar to the scientific and technological communities. The absence of comprehensive and definitive data on a particular problem too often results in an unwillingness of scientists and technologists to express opinions or accept a role in the decision-making process. However, a lack of information or the presence of certain levels of uncertainty are key elements in the appropriate reallocation of the burden of proof from the public and the federal government to any party initiating the risk (Karstadt). In administrative conflicts and, increasingly, in judicial decisions, "the burden of proof rests on the party that *initiates* the risk, that *profits* from the risk, and that has the greatest resources to *do* something about the risk" (Nader). Thus, the absence of data on any particular adverse effect would indicate that policies should be developed on the basis of "worst possible case" assumptions about safety to public health and the environment. Subsequent development of such data would allow appropriate policy modifications.

Generation and interpretation of the data base

The data base. Decisions on costs and benefits can be only as good as their underlying data base. There is growing evidence to confirm and document the long-held suspicion that data generated by an interested party, whether by in-house industrial scientists, commercial testing organizations, or universities under contract, may exhibit constraints ranging from biased interpretation to frank manipulation; charges of biased interpretation have also been directed against public advocacy groups. Such constrained data can no longer be regarded as exceptional; for example, in common consumer products such as drugs, there have been the criminal conviction of a drug company's officials for fraudulent manipulation of data on the drug MER/29; the conviction for submission of false data by another company on the drug Dornwall; the nolo contendere plea of a laboratory for concealing information on Flexin; the withdrawal of Panalba from the market after its producer had been found to conceal data on its lack of efficacy ("Introduction of a Bill"); and the testimony by Food and Drug Administration (FDA) Commissioner Schmidt before Senator Kennedy on 20 January 1976 that a commercial testing laboratory under contract to a pharmaceutical company reported on nonexistent slides on animals under carcinogenicity tests with Aldactone. Deputy Administrator Quarles of the Environmental Protection Agency (EPA) also testified on 20 January 1976 on the manipulation of toxicological and carcinogenicity data on pesticides, including Chlordane and Heptachlor, tested by a commercial testing laboratory under contract to a pesticide manufacturer. A later report based on preliminary analysis of chronic toxicity data submitted before 1971 by industry to one of EPA's predecessor organizations (FDA) in support of tolerances for twenty-three pesticides and accepted for this purpose revealed that the data on only one of these pesticides were satisfactory; the twenty-three pesticides were selected for analysis since tolerances on a particularly large number of food commodities had been established for most of them. The other data were so flawed and biased as to invalidate any possible inferences as to carcinogenicity (Reuber). On the basis of such reports, the policy of the EPA is to have an extensive audit of data, in its files, previously submitted by industry in successful support of petitions for pesticide tolerances.

Turning to another class of consumer products, the automobile, the following examples provide further illustrations: In 1972 a well-known automobile company falsified emission control certification tests; however, it was able to ward off a criminal prosecution by paying a $7 million fine. During the first EPA suspension hearing in April 1972, the automobile manufacturers claimed that installation of emission controls in 1975 cars would induce a five to ten percent fuel penalty over 1972 cars. However, by April 1973, at the second EPA suspension hearing, the manufacturers admitted that there would be no fuel penalty. Finally, by June 1973 another automobile company publicly announced and then informed Congress that there would be a sales-weighted fuel economy of thirteen percent due to catalytic converters (Epstein, "The Public Interest").

Examples in the field of occupational disease abound. The seriousness and extent of the problem cannot be exaggerated. The chronology of the development of data, from 1972 to 1975, on the carcinogenicity of vinyl chloride, used in the production of PVC, has recently been reviewed in a report of the AAAS Committee on Scientific Freedom and Responsibility. The report documents the suppression of carcinogenicity data by the Manufacturing Chemists Association (MCA), allowing continued exposure, without warning, of tens of thousands of workers to high concentrations of vinyl chloride (Edsall).

Blatant examples of constrained data are also evident in recent analyses of the economic impact of proposed regulations. A now apparently standard response by certain sectors of industry to attempts by regulatory agencies to promulgate standards limiting environmental and occupational exposure to chemical carcinogens is to forecast major economic disruption and unemployment attendant on compliance. Apart from the questionable economic validity of such forecasts, they do not appear to address themselves to the unrecognized ("externalized") costs, economic and otherwise, of carcinogenic and other toxic effects due to human exposure to carcinogens (Epstein, 1976).

Inflated economic analyses have also been used by industry to discourage "Toxic Substances" legislation, which mandates the requirements for premarket notification to the EPA and in some cases toxicological testing of new chemical agents prior to their introduction to commerce and the workplace.

Policies for improving data base. It is now clear that there is a critical need to ensure impartial and competent testing of all products to which human exposure is anticipated. The present system of direct, closed-contract negotiations between manufacturing industries and commercial and other testing laboratories, including universities, is open to potential abuse and creates obvious mutual constraints. One possible remedy would be the introduction of a disinterested advisory group or agency as an intermediary between manufacturers and commercial or other testing laboratories. Proper legal and other safeguards could be developed to minimize potential abuses and conflicts of interest. Manufacturers could notify the intermediary group when safety evaluation was required for a particular product. The advisory group would then solicit contract bids on the open market. Bids would be awarded on the basis of economics, quality of protocols, and technical competence. The progress of testing would be monitored by periodic reports and site visits, as with federal contracts. At the conclusion of the studies, the advisory group would comment on the quality of the data, make appropriate recommendations, and forward these to the concerned regulatory agency for routine action.

This approach appears preferable to the secret award of unbid contracts. Additionally, quality checks during testing would ensure the high quality and reliability of data and minimize the need to repeat studies, thus reducing pressure on federal agencies to accept unsatisfactory data on an ex post facto basis. This approach not only would minimize constraints due to special interests but would also upgrade the quality of testing in commercial and other testing laboratories (Epstein, "Public Health Hazards").

Major constraints in the field of occupational health and safety would be reduced by the future employment of physicians and hygienists by joint management–labor groups or committees, rather than exclusively by management, as is the current practice.

Interpretation of the data base. The significant influence of economic and related constraints on expert advisory committees, both federal and nongovernmental, is being increasingly appreciated. In addition to constraints on the generation of objective data, constraints on the evaluation and interpretation of these data by regulatory agencies may also influence resulting policy decisions.

A major scandal in regulatory agencies is the personnel "revolving door." Agencies recruit directly from industry significant numbers of senior administrative staff, who often tend to perpetuate philosophies and practices consistent with interests of industry and not of the broader public. More seriously, a significant number of senior agency officials leave for industry positions during their active careers or after retirement. This creates a potential conflict of interest and an apparent inducement to agency officials to interpret data in a manner consistent with their future interests and not to take action possibly detrimental to future employers. As of January 1977, one-year restricted employment mobility clauses have been introduced in agency–employee contracts; such clauses are long-standing and commonplace in industry. Furthermore, a major mitigating factor would be development of an enhanced economic and professional status for regulatory employees and of third party testing and other measures that would restrict the discretionary power of senior agency officials, especially those eligible for early retirement.

In this connection, equally serious potential conflicts of interest and a source of pressure on agency scientists arise with the senior science advisers to agencies, generally university scientists, who may receive substantial contracts and grants from the agency and who may also be consultants to industries regulated by the agency and receive research contracts from these industries. These senior science advisers may play a key role, and one that is largely hidden from Congress, in major decision making in agencies (Epstein, 1977).

The public interest movement as an ethical renaissance

The "public interest" movement. The "public interest" movement, as a modern expression of social ethics and as an instrument of political change, is now less than one decade old. Public interest groups embrace a wide spectrum of heterogeneous styles and objectives, from conservationists and environmentalists, some of whom are traditionalist and conservative, to more activist consumer and citizen groups. Some contemporary concerns of a few public interest groups dealing with occupational health and safety issues present a potential for ad hoc alliances between labor and the public interest.

The public interest movement expresses the conviction that the "common good" is inadequately represented in the brokerage of decision making at the federal and local levels, where narrow and concentrated economic and political interests are joined and where considerations of social equity are largely precluded. The movement also expresses convictions that public health and environmental costs are poorly perceived and too readily discounted in regulatory decision making, and that the burden of proof for such costs is too readily accepted by government or inappropriately shifted from the private to the public sector. Specifically, these public interest convictions have been aroused because of considerations such as the following: restrictions in public access to regulatory data; the unwillingness of agencies to accept qualified citizen and consumer representatives in their decision-making processes; and the legislative failure to create formal mechanisms for representation of the public interest in the regulatory process.

Public access to data. Public interest groups are pressing for further legislation concerning open access to data. Formal and informal discussions among agencies, industry, and federal and nongovernment expert committees on all issues relating to human safety and environmental quality, and data relevant to such discussions, properly belong to the public domain. Indeed, under the 1967 Freedom of Information Act (U.S. Public Law 89-487, 80 Stat. 250), all federal records are intended to be open to the public except for specified and validated exemptions, such as trade secrets. Demands for access to internal data and working papers are being extended to agency memoranda and to quasi-governmental bodies in the United States, such as the National Academy of Sciences (NAS), which has successfully claimed exemption from the requirements of the Freedom of Information and Federal Advisory Committee Acts (Boffey).

Impetus to demands for routine access to internal agency memoranda is periodically reinforced by "leaks" of confidential agency documents, which make it clear that an agency is seriously derelict. In some instances, such "leaks" have resulted in congressional investigations and major policy shifts. There is no evidence to suggest that unrestricted access to agency data, excluding security and validated proprietary documentation, would in any way impede regulatory practice.

Public involvement in decision making. It is important that the citizen and consumer be adequately represented at the earliest stages of the decision-making process and of agency–industry discussions on the widest range of issues concerning the public interest. Decisions by agencies on setting standards, in areas including technological innovations and new products after closed discussions of data that have been treated confidentially, are no longer acceptable; similar considerations obtain for products already established in commerce with relation to data on safety and efficacy. Such methods of decision making are contrary to the long-term interests of industry, quite apart from the best interests of the public: The possibility of legal penalties and of third party suits and class actions further point to the need for citizen participation. While there is a growing acceptance by industry of the legitimacy of demands for representation by public interest groups, formal mechanisms for this purpose have not yet been adequately developed.

Federal advisory committees, as stipulated by the 1972 U.S. Federal Advisory Committee Act (FACA), are required to be a "means of furnishing expert advice, ideas and diverse opinions to the Federal government." While many committees are not particularly effective and meet only infrequently, others wield great power, and their recommendations provide important input into government actions. While the FACA made important strides in opening up deliberations of advisory committees to the public, there are still major problems in making the system responsive to broader interests. Most agencies exclude public interest representatives, whether technically qualified or not, and also qualified experts nominated by public interest groups, from their advisory committees. The lack of fair balance on committees hinges on the contention by government that industry members should be designated "public" representatives. Of equal concern is the increasingly common device, particularly in EPA, of appointing fact-finding "panels," in contrast to recommendation-making advisory committees, as the panels are alleged to be exempt from the requirements of the FACA for balance and openness (Epstein, 1977).

Need for an agency for consumer protection. While numerous agencies in the United States, particularly the Department of Commerce and the Small Business Administration, already support the positions and interest of industry, as of March 1977 no government agency was charged with the advocacy and protection of consumer and citizen interests. Whether these interests relate to concerns such as clean air, auto safety, meat inspection, or land use, there was no currently adequate mechanism for presenting the public interest.

It is freely recognized that business representatives outnumber consumer representatives by over one hundred to one in appearances before federal agencies (Epstein, "Statement"). The offices of consumer affairs of the various U.S. federal agencies are small and relatively ineffectual, reflecting as they do overall agency policies and pressure, and are clearly no substitute for an independent agency, uniquely and solely charged with representing the public interest. This is a critical deficiency in regulatory practice, especially as the health, safety, and other interests of the public have been and can be massively influenced by the decisions and actions of a wide range of federal agencies.

Summary

While problems of environmental pollution and human disease are long-standing, they have been greatly magnified by the accelerated utilization of natural resources and the progressive development of new technologies. The historical causes date from the industrial revolution, the petrochemical era of the 1930s, and the nuclear technologies since World War II.

Although "pollutant" is often pejoratively applied to synthetic industrial chemicals and to nuclear fission products, there is a wide range of other chemical pollutants comprised in four broad categories. The first group consists of natural chemicals in excess, such as nitrates and arsenic. Natural fungal or plant toxins in crops constitute the second group. The third group consists of complex organic and inorganic mixtures, such as community air and water pollutants and occupational pollutants, including coal tar pitch volatiles, which comprise a wide range of undefined as well as partially defined chemical components. Finally, there is the group of synthetic chemicals: agricultural chemicals, notably pesticides and fertilizers; food additives, which may be intentional, such as antioxidants and dyes, or accidental, such as pesticides, heavy metals, and plasticizers; fuel additives; household chemicals; and industrial chemicals. Most of these chemicals are derived from petroleum, which is now the basic stock for the synthesis of the majority of organic chemicals.

Pollutants may induce a wide range of adverse or toxic effects. Acute or chronic toxicity per se may appear in the growing embryo, infant, child, or adult with effects ranging from slight impairment of health and fitness to disabling disease and death. More specific manifestations of chronic toxicity include the induction of cancer (carcinogenicity), birth defects (teratogenicity) and genetic abnormalities (mutagenicity). That chronic toxicity may also impair immunity or produce psychobehavioral disorders has become increasingly evident. Some pollutants may induce more than one type of toxicity. Pollutants, or their precursors, may also interact outside (in vitro) or inside the body (in vivo) with other chemicals to produce otherwise unanticipated cumulative effects. Such effects can also result from interactions between otherwise harmless and common environmental chemicals, such as nitrites in vegetables and secondary amines in meat and fish, to produce a class of chemicals known as nitrosamines, which are usually potent carcinogens, mutagens, and teratogens.

The need to use increasing numbers of synthetic industrial chemicals makes it essential to recognize and estimate the human and ecological hazards they pose and their societal acceptability with regard to the real or alleged matching benefits they confer. Hazards from particular products or processes, or from particular synthetic chemicals, whether in consumer products or in the workplace, need not necessarily be accepted even when their matching benefits are, or are alleged to be, high. Equally efficacious but less hazardous alternative chemicals, processes, and products are generally available. The imposition of mandatory criteria such as broad societal utility, efficacy, or other characteristics prior to the introduction of synthetic chemicals into commerce may well simplify such equations (Epstein, "Public Health Hazards").

Rational judgments on the benefits of technological innovations and on the risks of environmental pollution are critically dependent on the availability of a comprehensive, unbiased, and valid data base to competent decision makers who represent a broad range of balancing interests. However, growing evidence has demonstrated that data on both risks and benefits, generated directly or indirectly by economically interested parties, frequently suffer from constraints ranging from biased interpretation to frank manipulation. Moreover, critical analysis of major current problem areas in environmental pollution demonstrates that, while these problems are frequently presented in a technological or scientific guise, their primary determinants are economic, political, and evaluative. As a corollary, the prevention of such problems requires sensitivity to social equity and justice—a calculus not sufficiently represented in contemporary policymaking, which is currently based, with rather excessive emphasis, on the narrower perspectives and interests of regulatory agencies, industry, and the technological or scientific communities.

SAMUEL S. EPSTEIN

[*For further discussion of topics mentioned in this article, see the entries:* DRUG INDUSTRY AND MEDICINE; JUSTICE; *and* RISK. *For discussion of related ideas, see the entries:* DECISION MAKING, MEDICAL; ENVIRONMENT AND MAN; FUTURE GENERATIONS, OBLIGATIONS TO; HEALTH AND DISEASE; HEALTH POLICY; PUBLIC HEALTH; SOCIAL MEDICINE; *and* TECHNOLOGY. *See also:* MEDICAL ETHICS, HISTORY OF, *section on* EUROPE AND THE AMERICAS, *article on* EASTERN EUROPE IN THE TWENTIETH CENTURY; *and* SMOKING.]

BIBLIOGRAPHY

American Physical Society. "Report on Light Water Reactor Safety." *Reviews of Modern Physics* 47, supp. 1 (1975), pp. S1–S123.

Boffey, Phillip M. *The Brain Bank of America: An Inquiry into the Politics of Science.* Introduction by Ralph Nader. New York: McGraw-Hill Book Co., 1975.

Edsall, John T. "Scientific Freedom and Responsibility: A Report of the AAAS Committee on Scientific Freedom and Responsibility." *Science* 188 (1975): 687–693.

Epstein, Samuel S. "Environmental Determinants of Human Cancer." *Cancer Research* 34 (1974): 2425–2435.

———. "The Political and Economic Basis of Cancer." *Technology Review* 78, no. 8 (1976), pp. 34–43.

———. "Public Health Hazards from Chemicals in Consumer Products." *Consumer Health and Product Hazards—Chemicals, Electronic Products, Radiation. The Legislation of Public Safety*, vol. 1. Edited by Samuel S. Epstein and Richard D. Grundy. Cambridge: MIT Press, 1974, pp. 45–99.

———. "The Public Interest: Overview." *Environmental Health Perspectives* 10 (1975): 173–179.

———. "Statement in Support of Legislation to Create the Agency for Consumer Protection." Committee on Government Operations, House of Representatives. *Establishing an Agency for Consumer Protection.* Hearings on H.R. 7575 before a subcommittee. 94th Cong., 1st sess., 17–20 June 1975. Washington: Government Printing Office, 1975, pp. 378–381.

———, and Gage, Kit. "The Federal Advisory Committee System: An Assessment." *Environmental Law Reporter* 7 (1977): 50001–50012.

"Introduction of a Bill Establishing a National Drug Testing Center." United States, Congress, Senate, *Congressional Record*, 91st Cong., 1st sess., 1969, 115, pt. 16: 21360–21370. Remarks and items supporting Senator Nelson's Bill S. 2729, dealing with problems and needed reforms in the pharmaceutical industry. No hearings held; the bill died in the Committee on Labor and Public Welfare.

Joint Review Committee of the Sierra Club and the Union of Concerned Scientists. "Preliminary Review of the AEC Reactor Safety Study." Washington, November 1974, p. 95. Unpublished study.

Karstadt, Myra L. "Protecting Public Health from Hazardous Substances: Federal Regulation of Environmental Contaminants." *Environmental Law Reporter* 5 (1975): 50165–50178.

Kneese, Allen V. "Benefit-Cost Analysis and Unscheduled Events in the Nuclear Fuel Cycle." *Resources* (A Newsletter of Resources for the Future, Washington), no. 44 (1973), pp. 1–5.

Mason, Thomas J.; McKay, Frank W.; Hoover, Robert; Blot, William J.; and Fraumeni, Joseph F. *Atlas of Cancer Mortality for U.S. Counties: 1950–1969.* DHEW Publication no. (NIH) 75–780. Bethesda, Md.: National Institutes of Health, 1975.

Metzger, H. Peter. *The Atomic Establishment.* New York: Simon & Schuster, 1972.

Nader, Ralph. "Professional Responsibility Revisited." Proceedings of the Conference on Science, Technology, and the Public Interest. The Brookings Institution, Washington, D.C., 8 October 1973. The proceedings are available from the Commission for the Advancement of Public Interest Organizations, Suite 1013, 1975 Connecticut Avenue, Washington, D.C. 20009.

Reuber, Melvin D. *Review of Toxicity Test Results Submitted in Support of Pesticide Tolerance Petitions.* Report for the Office of Pesticide Programs, United States Environmental Protection Agency, 9 April 1976. Available from the EPA in photoreproduced form.

II
QUESTIONS OF SOCIAL JUSTICE

An increasing concern with the natural environment has developed recently throughout the world. Many factors have contributed to this, for example, the growth in global population with its attendant demands upon the biosphere, the depletion of many natural resources and threat of extinction of many forms of plant and animal life, and the realization that human life and welfare are interrelated with the life and welfare of other organisms within the biosphere.

Concomitant with the victories of the environmental movement has been the apprehension of some social critics that the effort to protect the environment would place the "rights of nature" above the "rights of humanity," and especially the rights of poor people (Neuhaus; Faramelli). Many worry, for example, that in the name of conservation or of ecological balance justification would be found for exclusionary policies (e.g., zoning) that would ensure the preservation of unjust enclaves of privilege. There is also apprehension about the possibility that strong environmental legislation might cripple economic growth and increase unemployment. Those who are pessimistic about securing any redistribution of goods and services from the existing economic order frequently see in the environmental movement a threat to efforts to secure more benefits for the poor from increasing economic activity. These people have not been reassured by arguments that in a stationary-state economy redistribution would *have* to occur (Daly).

Although this view, which pits the environment against people, can be and has been seriously challenged by responsible critics, e.g., Leonard Woodcock and Congressmen Rangel, Dellums, and Conyers of the Black Congressional Caucus, it brings to the fore important theoretical questions that demand the attention of ethicists. Among the major issues that it raises are the following: (1) How do the rights of human beings and the rights of the environment come into conflict? (2) If, in justice, we are required to protect the environment, can we also adequately protect people and meet their just demands? (3) What assumptions about justice and the social order led to the perception that there is a conflict between the rights of persons and the rights of the environment?

Justice, the rights of human beings, and the rights of the environment

There are, of course, competing views of justice. Differing definitions of justice are generally differing proposals as to the means by which persons may legitimate and adjudicate among or reconcile the various *claims* made upon a society. Put differently, conceptions of justice represent different ways of evaluating or reconciling competing claims that someone or some group has a "right."

For people to make a rights claim is for them to voice "the conviction that what they are demanding is both legitimate and necessary" (Jenkins). The claims that people articulate as "rights" are incredibly variegated. They are, moreover, becoming more, not less, complicated as we move from a tradition wherein rights were seen primarily as negative ("freedom from" various intrusions) to a situation where rights lay claim "to" or "for" particular goods and services. The danger in this shift lies in a propensity to claim as a basic "right" that merits social recognition all kinds of desires of individuals and groups (ibid.), with the consequent difficulty, made ever more hazardous by the mere fact of proliferation, of determining when these claims are indeed grounded in justice. The adjudication between conflicting rights claims demands a theory of justice by which to distinguish genuine rights from arbitrary claims. But just such a theory of justice is precluded by the liberal tradition's inclination to identify rights and interests and, as we shall see, by its rejection of rational discrimination of rights and its belief in the identity of interests.

Rights for nonhuman beings and things?

Environmentalists have argued that it is erroneous to assume that only human beings have rights. These authors (Stone; Cobb) propose that nonhuman beings and things, indeed the natural environment as a whole, have rights that demand, in justice, recognition and protection by society. In developing this argument Stone, Cobb, and others have made explicit what has been implicit in much of the environmental movement for some time. So long as nonhuman natural things were considered simply "of value," it was unclear whether they were valuable in themselves or valuable simply because they were instrumental in achieving some *human* purpose or in meeting some *human* need. By claiming that nonhuman things have rights, environmentalists impute an intrinsic worth to natural systems independently of human purposes and needs. This eco-ethic is frequently accompanied by a rejection of the "homocentrism" of traditional Western thought.

There are two other aspects to the proposal that natural systems have rights. These aspects, if accepted, would have revolutionary implications for our concepts of rights and of justice. First, the rights at issue are less those of individuals than those of classes or species of things, and difficult to define natural units, such as landscapes and ecologies. Second, when it is species and natural systems that have rights, then one is concerned not just with adjudicating claims made in the present, but with the preservation and protection of those classes of things in the future. For instance, one of the issues in the controversy over the use of nuclear power is the legacy of radioactive wastes that are generated to serve present energy needs but at the same time pose problems for the future of the ecosystem.

All these propositions (that nonhuman entities have rights, that these rights pertain more to classes than to individuals, and that these rights belong to future and not just present entities) compel clarification of the concept of justice and its role in prioritizing rights claims.

Rights, duties, and the concern for social justice

Earlier it was noted that the proliferation of claims for "human rights" brings with it the great difficulty of judging which claims are authentic and which are not. Not only are more and more claims being made—and to the claims made in behalf of human subjects of rights must now be added claims made in behalf of the environment—there is also an uncertainty about our personal relationship to those claims. If we are responsible not only for ensuring the rights of those who make direct claims upon us but also for taking into consideration *future* claims by future persons and things, it becomes more and more difficult to determine just what our responsibilities are. In the resulting confusion, there is the tendency to give merely rhetorical allegiance to any one of the vast array of possible rights that could be affected by our choices and actions. When this happens, normative discourse can easily serve as the rationalization for any personal predilection. This indeed is to a large extent what is happening today.

The proliferation of rights claims results in a weakening of the moral binding power of all rights language by sheer reiteration. Moreover, the expansion of positive demands upon scarce social resources dilutes the imperative for social justice, with dire consequences for the welfare of the poor, who have relied on rights claims to make up for the failures of social policy. Let us see how this process takes place.

First, it has long been recognized that if rights are to be protected and ensured, there must be a duty correlative to each right. That is, it must be possible to specify an action that is definitely required by specifiable persons in situations in which rights are in jeopardy. Yet this is no easy task (Macklin), especially where there are competing rights claims.

Second, claims made on behalf of the environment are structurally similar to the more traditional protective or "natural rights" claims made on behalf of human subjects rather than to the more contemporary assertive "human rights" claims central to discourse concerning issues of social justice.

From this we can see how the claims of environmentalists can come to have a standing that would threaten the claims for action made in the name of social justice ("human rights" claims). In the social policy area, these concerns surface when we consider who pays for pollution control costs. For instance, will expenditures for environmental quality increase unemployment and thus foster poverty? In other words, do the claims made for clean streams take precedence over the claims made for human welfare?

Here we are either caught in a genuine cul de sac or need to rethink our notions of rights and of justice. The latter is, in our judgment, the case, and the *apparent* conflict pitting human welfare *against* the rights of the environment is rooted in an inadequate theory of justice.

The need for an adequate theory of justice

Central to the apparent conflict between human rights and environmental rights are two assumptions about the meaning of justice that have been bequeathed to us by eighteenth-century rationalism. The first assumption was that there is a body of self-evident and delimited "natural rights," which, if protected, will provide sufficient basis for the establishment of an acceptable human community. The second was that there is a "natural identity of interests," that is, that if all participants in politico-economic processes are given the chance to pursue their self-interest, the resultant distribution of goods, services, or political decisions will magically orchestrate themselves to the optimum welfare of all. This, for instance, is the "invisible hand" ordering economic life. The use of rights claims as a moral and legal tool of self-interest has led to the abandonment of the belief that rights are limited and self-evident. In the subsequent confusion, all that anyone knows in the presence of a rights claim is that its adherents feel that the claim is "legitimate and necessary." Similarly, the natural identity of interests assumption has created confusion in public policy, because it provides no rational ground for the adjudication of claims. The abandonment of the principle of limited and self-evident rights, the identification of rights with interests, and the assumed identity of interests have all contributed to the "misplaced debate" that pits the rights of human beings against the rights of the environment. It has also created anarchy in public policy and has led to the view that governmental intervention, whether to advance the "human rights" of poor people or to protect the environment, is seen as an illegitimate intrusion into the "natural order" of things.

If systematic judgments about the relative position of competing rights claims and duties can be made in any coherent and intelligent fashion, it is necessary to develop a coherent theory of justice (Macklin). Such a theory has either not yet evolved or has not yet been recognized, nor do we propose to develop one here. But it is possible to state what characteristics it would have if it were to help adjudicate concerns about the environment and the concerns of social justice:

1. It would provide persons and communities with the capacity to validate moral claims, including rights claims, and to order them hierarchically.
2. In order to have such a capability, the theory would reject the assertion that rights are "unanalyzable primitives" (Feinberg) and instead would disentangle instrumental from substantive rights (Cobb).
3. It would have the capacity to reformulate rights claims so as to protect and evaluate the concern expressed and to do so in such a way that potential reconciliation with other claims would be possible.

4. It would provide criteria for the identification of which agents are responsible for actions in which kinds of situations to ensure that validated concerns of claims are met. Among the kinds of criteria necessary would be, for example, those which enable an agent to know what degree of probability about future events would be required to establish an obligation from need-meeting action now in order to avoid an adverse consequence to future generations.

The kind of moral theory suggested would provide the missing link between general exhortations to rid the world of poverty or to reconceive the relationship between nature and history and the cacaphony of rights claims we have discussed.

People and the environment in public policy

Up to this point discussion has focused on characteristics of the state of normative discourse that help explain why the tension between the environment and social justice is so strongly felt. The proposed remedy for this situation is the development and legal implementation of a widely shared normative theory of justice. This cannot be expected to occur in the near future. Hence, for the foreseeable future the perception that concern for the environment and concern for people will be inimical will continue. In the meantime, public policy and its implementation must still go on. The question is, How shall this be carried out?

A partial answer may be found by extending a concept pioneered in the 1970 National Environmental Policy Act in the United States. As a result of that legislation, the responsible public official must "include in every recommendation or report on proposals for legislation and other major Federal actions" an environmental impact statement setting forth not only the expected environmental consequences of the proposed action but an assessment of the alternative to it. What this environmental legislation recognizes is that "at stake" in every federal action is its environmental impact. But if we are to make headway in developing policies relating synergistically the impacts they will have, *all* the impacts of those policies should be analyzed. Put differently, even if we do not yet have the conceptual tools or the normative consensus necessary to order our priorities among various claims, we can insist that a good faith effort is made both to determine who or what has a "stake" in what is proposed and to forecast what the consequences of those policies will be on those who will be affected. Viewed in this way, policy formation is an active and self-conscious reconciliation of the competing claims of the "stakeholders" in every policy. It would be possible to require that all parties, public or private, advocating legislation, administrative regulations, or public action, be put under the discipline of seriously attempting to assess all the consequences of thir proposals. Such a requirement would tend to discourage the advocacy of proposals that seriously ignore any major affected interest; it might even encourage creativity and innovation in policy development. And an indirect benefit might be slow inductive progress toward the ordering of normative priorities to correspond with the conceptual ordering of priorities through the more theoretical process suggested earlier. For example, social policies designed to foster both environmental equality and social justice are not impossible to conceive or even implement, if a broader range of alternatives are considered differently and a society moves beyond the dead-end of ecology versus human rights.

Conclusion

The contemporary debate about the environment and social justice is understandable but misplaced. It is understandable because, in the absence of a commonly shared theory of justice and in the midst of rhetorically inflated claims about policy effects, battle lines tend to form along the environment-versus-social-justice axes. As noted above, this polarization is the result of considering too few public policy alternatives and the implications of those alternatives. But it is misplaced because the debate, when posed this way, reinforces patterns of discussion that shut off more constructive efforts to discover precisely what most needs protection or support in the human community and in nature. And it is also misplaced because it tends to discourage a more imaginative search for policy approaches that will relate concern for people to concern for the environment. Until a consensus can be reached on a normative theory of justice, and in order to avoid anarchic and rhetorical approaches, we should place the participants in the political process under the pedagogy of assessing what the consequences for everyone and everything of "value"—that is, for all the "stakeholders"—will be as a result of every formal policy proposal. Such a consideration will result in a better understanding of "Who pays the costs?" and "Who receives the benefits?" of each policy proposal.

Until those questions are answered the misplaced debate will continue.

NORMAN J. FARAMELLI
AND CHARLES W. POWERS

[*See the following article:* THE PROBLEM OF GROWTH. *For further discussion of topics mentioned in this article, see the entries:* ANIMAL EXPERIMENTATION, *article on* PHILOSOPHICAL PERSPECTIVES; FOOD POLICY; FUTURE GENERATIONS, OBLIGATIONS TO; JUSTICE; POPULATION ETHICS: ELEMENTS OF THE FIELD, *articles on* DEFINITION OF POPULATION ETHICS *and* ETHICAL PERSPECTIVES ON POPULATION; *and* RIGHTS. *For discussion of related ideas, see the entries:* ENVIRONMENT AND MAN; *and* TECHNOLOGY, *article on* TECHNOLOGY AND THE LAW.]

BIBLIOGRAPHY

BURCH, WILLIAM R. *Daydreams and Nightmares: A Sociological Essay on the American Environment.* Harper & Row Monograph Series in Sociology. New York: 1971.

CALLAHAN, DANIEL J. *The Tyranny of Survival, and Other Pathologies of Civilized Life.* New York: Macmillan Co., 1973.

COBB, JOHN B., JR. "The Population Explosion and Rights of the Subhuman World." *IDOC International (North American Edition)*, no. 9 (12 September 1970), pp. 41–62. Published by IDOC (International Documentation on the Contemporary Church), 432 Park Ave. S., New York, N.Y. 10016. In a collection of documents on "A Theology of Survival."

DALY, HERMAN E. "Toward a Stationary-state Economy." *Patient Earth.* Edited by John Harte and Robert H. Socolow. New York: Holt, Rinehart & Winston, 1971, chap. 14, pp. 226–244.

Ethics 85, no. 1 (October 1974), pp. 1–66. Issue on John Rawls, *A Theory of Justice.*

FARAMELLI, NORMAN J. "Ecological Responsibility and Economic Justice: The Perilous Links between Ecology and Poverty." *Andover Newton Quarterly* 11, no. 1 (1970), pp. 81–93.

FEINBERG, JOEL. "The Nature and Value of Rights." *Journal of Value Inquiry* 4 (1970): 243–260.

JENKINS, IREDELL. "From Natural to Legal to Human Rights." *Human Rights: Initial Publication of the American Section of the International Association for the Philosophy of Law and Social Philosophy, Consisting of the Papers Prepared for Its Second Plenary Meeting.* AMINTAPHIL 1. Edited by Ervin Harold Pollack. Buffalo: Jay Stewart Publications, 1971, pp. 203–218.

KRIEGER, MARTIN H. "Six Propositions on the Poor and Pollution." *Policy Sciences* 1 (1970): 311–324.

———. "What's Wrong with Plastic Trees?" *Science* 179 (1973): 446–455.

LONGWOOD, MERLE. "The Common Good: An Ethical Framework for Evaluating Environmental Issues." *Theological Studies* 34 (1974): 468–480.

MACKLIN, RUTH. "Moral Concerns Appeals to Rights and Duties: Grounding Claims in a Theory of Justice." *Hastings Center Report* 5 (1976): 31–32, 34–38.

National Environmental Policy Act of 1969. Pub. L. No. 91-190. 83 Stat. 852. 42 U.S.C. sec. 4321 et seq. (1970).

NELL, ONORA. "Lifeboat Earth." *Philosophy and Public Affairs* 4 (1975): 273–292.

NEUHAUS, RICHARD JOHN. *In Defense of People: Ecology and the Seduction of Radicalism.* New York: Macmillan Co., 1971.

"The No-Growth Society." *Daedalus* 102, no. 4 (Fall 1973), pp. 1–245. Entire issue.

PASSMORE, JOHN ARTHUR. *Man's Responsibility for Nature: Ecological Problems and Western Traditions.* New York: Charles Scribner's Sons, 1974.

POWERS, CHARLES W. "Growth as an American Value: An Ethicist's Point of View." *Growth in America.* Edited by Chester L. Cooper. Woodrow Wilson International Center for Scholars, Contributions in American Studies, no. 21. Edited by Robert H. Walker. Westport, Conn.: Greenwood Press, 1976, chap. 3, pp. 26–38.

RAWLS, JOHN. *A Theory of Justice.* Cambridge: Harvard University Press, Belknap Press, 1971.

SILLS, DAVID L. "The Environmental Movement and Its Critics." *Human Ecology* 3 (1975): 1–41.

SMITH, JAMES N., ed. *Environmental Quality and Social Justice in Urban America.* Washington: Conservation Foundation, 1974.

STONE, CHRISTOPHER D. "Should Trees Have Standing? —Toward Legal Rights for Natural Objects." *Southern California Law Review* 45 (1972): 450–501. Reprint. Foreword by Garrett Hardin. Los Altos, Calif.: W. Kaufmann, 1974.

Yale Task Force on Population Ethics. "Moral Claims, Human Rights, and Population Policies." *Theological Studies* 35 (1974): 83–113.

III
THE PROBLEM OF GROWTH

Introduction

The term "environmental ethics" refers to a broad range of considerations about mankind's relationship to the physical environment. Among these are cultural attitudes toward the nonhuman world, proposals for resymbolizing humanity's relation to nature and to its own technology, and examination of normative guidelines for assessing human interventions in the physical environment. A special set of problems is raised by studies showing the projected consequences of exponential rates of economic growth and population expansion. This article will be concerned with those problems caused by economic growth and with the broader ethical questions that they raise. The problems can be summarized in the following way: (1) Anticipation of unprecedented natural and economic catastrophes through the depletion of natural resources, pollution of the environment, and the increasing demands of an expanding population raises serious doubts about the long-term survival of the human species and of the earth's ecosystem. (2) Proposals for the revision of economic life in the interest of global survival threaten to exacerbate difference over distribution of wealth and political power. (3) Related schemes for review of human attitudes toward the natural world with an eye to long-term survival question the appropriateness of Western values, as found in Judaism, Christianity, and secular, scientific rationalism for preventing global environmental catastrophes. (4) Questions arise as to whether an environmental ethics can succeed in providing social guidance without a radical change of ethos or a shift in religious outlook.

The growth problem

When human technologies were less extensive, humanity lived in relative balance with nature. Natural systems continued the rhythm of growth and decay in a process of dynamic compensation (Commoner). The expansion and transformation of technologies since the Second World War brought a break in this circle and introduced linear, noncompensating forces into the physical economy. Many products of late industrialization are not biodegradable and so become pure waste rather than recyclable matter. Other substances set up dangerous interactions with basic ecological processes. Phosphates, for example, lead to the eutrophication of water sources, and there is some evidence that fluorocarbons break down the earth's protective ozone layer.

Articulation of the problem. The potential in human technology for inflicting long-term and cumulative damage on the environment was first noted by George P. Marsh in his 1864 book *Man and Nature* (Glacken). In defining this problem, Marsh formulated the modern concept of natural equilibrium. He believed that even in the face of natural disasters nature is a self-compensating system. But human interventions in the environment, he thought, were uncontrolled disturbances with irreversible consequences. He even considered it possible that the earth would become uninhabitable due to human disruption of natural systems.

Such concerns have been voiced by the environmental movement, beginning with the 1962 publication of Rachel Carson's *Silent Spring*. The most notable of these cautionary studies was *The Limits to Growth* (Meadows et al.). Marsh had considered only the damage done the environment by simple people—shepherds, for example, working with primitive tools as they eked out a livelihood, very often at near-subsistence levels. The authors of *Limits*, however, studied the impact of advanced technology in meeting the growing demands of expanding populations for higher and higher material standards of living. They found that exponential growth in population, production, and consumption was accompanied by similar exponential increases in pollution and in depletion of natural resources, especially nonrenewable mineral reserves. They also noted that many of these trends were resistant to immediate change because of time lags. Certain pollutants, for example, would continue to accumulate at higher levels of the food chain, even after they were no longer produced and once fertility rates declined to replacement levels, populations would still continue to grow for some years before stabilizing. If these trends continued unchanged, the authors forecast sudden and uncontrollable collapses in ecological and economic systems.

Nature of the problem. The problem of growth, therefore, consists in the rapid increase in strains—in the form of pollution and resource depletion—which modern industrial economies place on the finite capacity of the globe. It emphasizes the cumulative and systemic effects of economic activity on the global environment which jeopardize the survival of the human species and of the biosphere which sustains it. The notion of limits to growth places in doubt the efficacy of material progress in both capitalist and socialist economies.

The debate among economists. Despite these problems, growth has many defenders. In conventional economic thinking it has been upheld as a cardinal economic value (Rostow; Barnett and Morse). It has a peculiar fascination in the United States, where national character and institutions have been shaped by material abundance (Potter). It is defended as the best—possibly the only way—to improve conditions of life for the poor (Rostow; Passell and Ross; Kahn et al.).

For a short time prior to the appearance of *Limits*, a few economists had questioned the ecological and social costs of economic growth. The standard economic measures of Gross National Product were criticized for neglecting the costs of production borne by third parties and future generations (Kapp). The assumed connection between social welfare and economic growth was also scrutinized in view of the disamenities (e.g., lack of privacy, quiet and clean air) and diseconomies (e.g., increased transportation time and costs, increased solid waste generation) associated with growth economics (Mishan). An economic policy based on the pursuit of growth was also thought to diminish consumer choice and social welfare, and the costs of growth were seen as regressive, diminishing the choices of the poor before those of the middle and upper classes (ibid.).

While some economists have begun to "internalize the externalities" of industrial production in their measurements (Schurr), the usefulness of these indices will depend on correct analysis of the social and environmental costs, the assignment of responsibility for them, and the effectiveness of the mechanisms employed to overcome them (Kapp). Mishan's work in challenging the assumption that growth increases choice and that social welfare is synonymous with choice has been extended by Hirsch (1976) to a general critique of the moral basis of growth economics.

Since *Limits*, debate has focused on three issues: whether economic policy should shift to a no-growth model, what the components of a no-growth economy should be, and what social mechanisms can be relied upon to produce the shift to a less hazardous economic system. The authors of *Limits* argued that an "equilibrium society" must supplant one based on exponential growth. In the sequel to *Limits*, entitled *Mankind at the Turning Point*, Mesarović and Pestel condemn not growth per se but "unbalanced and undifferentiated growth." They advocate an "organic growth" in which differentiated regional economies are kept in balance with the demands of basic human welfare and biological integrity. Thus, in certain developing areas of the world economic growth with attendant pollution and resource depletion might be permissible, while in industrial regions negative growth would be required to bring the economy within sustainable limits.

Among classical economists, J. S. Mill stands out for his perception of the potential benefits of a "stationary state" economy. For Mill, increased production was a desirable goal only in backward countries; in industrial countries the need was for better distribution of wealth for the sake of greater enjoyment of life. John Maynard Keynes also foresaw a point in economic development where essential needs would be satisfied and people would prefer to devote themselves to noneconomic purposes.

These speculations have their modern counterparts in the notions of "steady-state" economics and a shift from "quantitative" to "qualitative" growth. A frequent prescription is to reorient economic activity from capital- to labor-intensive production and to divert those energies from exploitation of nonrenewable resources toward expansion of the service sector—health, education, recreation, and the like.

Types of limits. Four main types of "limits" have been distinguished: physical, environmental or biological, social, and scientific or technological. Physical limits are said to be given in the First and Second Laws of Thermodynamics, the First in regard to the conservation of matter —waste or recycling must equal processed material resources (Olson, p. 3), the Second in respect to depletion of "low entropy" sources of materials and energy. Environmental limits are indicated in the concept of "carrying capacity" —the ability of a habitat to sustain particular numbers of particular species—and the biological limits of degradability, or "ecological resiliency." Social limits (Miles; Hirsch; Heilbroner) include political and economic limits as major subtypes. Political limits are expressed in terms of institutional arrangements and political "will." Economic limits are set at the point where rising marginal costs of physical growth exceed falling marginal benefits, thus reducing welfare rather than increasing it (Daly). Finally, scientific and technological limits concern the possible future knowledge and applications that may raise or remove the limits to growth. Space colonization as the ultimate evasion of the Malthusian trap is an example of what economist Charles Cicchetti terms the "Apollo syndrome"— the faith that technology, like God, will provide. Along with technological optimism (or pessimism), a second issue is the limits to substitutability. According to conventional economic wisdom, these appear indefinitely expansionable. "The reservation of particular resources for later use, therefore, may contribute little to the welfare of future generations" (Barnett and Morse, p. 11).

These types of limits interact and interpenetrate with one another. For example, physical limits are transferred and translated to social limits, so that physical density reveals itself symptomatically in forms of social stress such as crowding. In a manner reminiscent of Thomas Malthus's "positive" and "preventive checks,"

Forrester declares that the question of social choice regarding growth is one between physical and social limits, with the availability of choice resting largely on the social side. If those choices are not exercised wisely, he contends, the outcome will be imposed and enforced by physical necessity.

Criticisms of such "limits" emphasize the potency of both technological and value change. For example, the authors of Limits are criticized for taking an unduly pessimistic view of human values as resources for change (Cole et al.; Mesarović and Pestel). Values were "swept up" into the aggregation of economic trends and given no independent valence. Moreover, they were regarded as conservative and so yet another obstacle to establishing timely restraints on growth. Whatever the truth of these judgments, whether society can muster the political will to alter economies of growth remains one of the basic questions of the growth debate. It has also been charged that Limits underestimated the problem-solving potential of science and technology, and that technological solutions can be found to technologically created problems (Cole et al.; Kahn et al.).

The broader ethical issues

Global survival. The "limits to growth" debates raises the stakes for environmental ethics by widening the scope of concern from a local to a global context and by shifting the normative issue from quality of life to planetary survival. The nature of global survival, the characteristics of possible survivors, the allocation of costs exacted for the sake of survival, and moral limits on the irrational strength of claims made in the name of survival constitute one primary set of ethical issues raised by the growth debate.

The meaning of survival. From the time of the Greeks, philosophers extended moral credit to the notion that children represent a sort of personal continuity for mortal human beings. Only with the Enlightenment, however, did a generalized notion of posterity as a sort of symbolic immortality appear as an important motivating factor in the Western ethos. The fear of widespread ecological catastrophe has broadened the notion of survival to refer not only to the continuation of the human species but also to the maintenance of the basic biological conditions of terrestrial life as well.

For some thinkers, the preservation of the earth's biosphere alone—without humanity—would be sufficient to count as survival and even to make moral claims against human persons. This judgment is based on appeals to the beauty, order, and variety of forms of life other than humanity. More often, however, planetary survival is taken to include preservation of human life, with some admixture of other values defining the level of civilization thought to constitute a minimal threshold for a worthwhile human life beyond mere subsistence. The debate over survival as a value, therefore, centers on the definition of civilized life.

Searching for a standard. Definitions of survival are especially difficult to formulate in technological societies, where there is a tendency to confuse survival with material comfort (Callahan, 1973; Mishan). One approach to global survival, which tries to avoid the ideological deceptions inherent in allowing survival to be defined in terms of the preservation of a particular way of life, is the attempt to establish "basic human needs" as a baseline for international cooperation (Aspen Institute). This approach assumes that planetary survival depends on the sacrifices of many groups and that the costs and benefits of a global survival pact must seem reasonable to all parties.

An opposite line of argument contends that the provision of basic needs, in the form of relief and development assistance to the poor, is counterproductive, increasing suffering in the long term and hastening economic and environmental catastrophe. In this view, what is required is a system of incentives to stimulate conservation, and benign neglect of marginal groups who drain the economy without appreciably contributing to the social product (Forrester). Proposals for allocation of food assistance (triage) to starving peoples on the basis of the recipients' proven capacity to curb demand through population limitation frequently share the latter view. Recent economic studies of fertility decline, however, indicate that income distribution and the satisfaction of basic needs are preconditions of couples' limiting their family size (Rich).

At the extreme, survival is taken to exclude all other right-making considerations, so that a given group's claim to survival rests on its exclusive possession of a stock of resources, which it is obliged to manage only in its own interest and that of its own descendants (Hardin, 1974). Neither this "lifeboat" mentality nor the milder liberal economic view tries to set a standard of living compatible with global survival. Both demonstrate a preference for protection of future generations matched by a readiness to exclude marginal groups from use of the present store of resources.

Speculation about imaginable standards of survival is found in discussions of the "stationary-state economy" and the "equilibrium society" (Daly; Meadows et al.). There is some agreement among proponents of equilibrium that the needs of future generations should be taken into account so that long-term needs prevail over short-term ones (Meadows et al.; Forrester; Daly). But there is also considerable support for meeting basic human needs in the short run, especially in the area of food (Mesarović and Pestel; Aspen Institute). Other factors in setting a standard of civilized survival would include the ethics of production and distribution, the philosophy of community or human sociality, and notions of human fulfillment.

Survivors. Who should survive? Conceivably, species survival is consistent with the continued existence of a small remnant—a band of aborigines, say, living at a very low level of civilization. That prospect may seem unsatisfactory, because the meaningfulness of human survival is tied to "progress" and more generally to the values of civilized life. This objection gives rise to the notion that those entitled to survive are the bearers of civilization: long-term rational economizers, people who are scientifically inventive, latter-day ascetics, self-reliant organic gardeners, or nature mystics. Claims made for these groups are often nationalistic and sometimes racist and elitist (e.g., the presumption that procreation by the poor presents the greatest threat to survival). The difficulty in adjudicating the claims of competing groups to represent civilization and the apparent self-interest and prejudice in some such claims suggest that a more universal standard is necessary in deciding who shall survive. While not every group may expect to perdure indefinitely and, once comparable sacrifices are made on all sides, no group can demand exceptional protection for itself on historical grounds, one could reasonably argue that no group should be sacrificed for the sake of planetary survival under a standard that is less than universal.

Sacrifices. Because planetary survival will place different burdens on various groups, specification of any universal standard of survival implies at least rough comparability among the sacrifices exacted from different groups (Yale Task Force). Where large families are the only insurance parents have against neglect in their old age, for example, it would be unfair to see sterilization of poor couples in less developed countries as standing on parity with a shift to energy-efficient automobiles by affluent individuals in the industrialized countries. Distinctions are required between enjoyments and needs, between secondary (status) needs and the basic requirements of survival, and between limits on the exercise of fundamental rights and the blanket denial of such rights. Enjoyments should always be sacrificed before limits are imposed on the satisfaction of basic needs. Such limitations do not abrogate the rights they restrict, however; they only circumscribe their proper exercise. Hence an enjoyment such as eating steak twice a week would be surrendered before asking any group to ration its food; and meatless days might be imposed without infringing the right to food.

Moral limits. Survival is a strong value, and frequently its invocation signals a readiness to take exceptionally strong measures to protect the interest of one group against others. For the sake of survival men and women are willing to sacrifice many other values. The need for moral strictures on appeals to survival is motivated by the fear that this imperative may destroy "everything in human beings that makes them worth survival" (Callahan).

Building on what has already been said, there are at least five criteria for determining the prima facie moral acceptability of appeals to global survival: universality, just distribution of differential burdens, a common material standard of sufficiency, the constellation of values for civilized life that accompany the appeal and the disinterestedness of the proponents. (1) Universality requires that no human group be excluded from the class of potential survivors. (2) The norm of justice requires that the sacrifices asked of each group be truly comparable. (3) The norm of sufficiency asks whether the level of material well-being envisaged by the proponents is reasonable, distinguishing between surfeit and basic need, and defining a common standard of what is required to live a dignified human life. (4) A concern for civilized life demands clarity about the values that define human dignity and worthwhile survival. Are values we cherish threatened or only reasonably curbed? What is it we sacrifice for the sake of survival: freedom, justice, compassion, artistic expression, trust between persons? And what will replace them? (5) Disinterestedness is a common norm of all moral judgments. Thus, one can ask, does the person invoking survival, his nation, or class have anything to gain from adoption of proposed measures? Does the survivalist or his/her group sacrifice anything of comparable value to what is asked of others?

Obligations to future generations. Duties to posterity are postulated on a variety of grounds: justice as fairness, love implying care for the future of what we love, gratitude for life, responsibility to the God of all being. Though obligations to the future are widely recognized, the extent of the obligation of the present generation to provide for the future is much disputed. Where biological survival functions as the dominant value, there is a tendency to extend the obligation to remote generations (Forrester; Hardin, 1972; Callahan). Such temporalizing of universal moral standards has been criticized as unreasonable and unworkable. Ignorance of the future and lack of control over the long-term consequences of our action are seen as possible grounds for limiting the temporal scope of these obligations. Perhaps we are obliged to avoid worsening the world posterity receives from us and especially to avoid actions that have irreversibly harmful consequences (Callahan), but we cannot be expected to calculate the effect of our actions on very distant generations. The inclusion of the remote future in our moral calculus, moreover, may lead to avoidance of proximate obligations to our contemporaries and immediate descendants. Two proposed norms that limit obligations to proximate future generations are the extent to which persons can expect to share a common moral ideal with the future generations (Golding) and the extent to which parental love can imagine taking responsibility (Passmore).

A new ethics and a new ethos

Criticism of the traditional ethos and its ethics. In discussions of environmental problems, the ideas of ethos and ethics are frequently conflated. We can distinguish the two this way: An ethos is a set of beliefs characteristic of a particular culture and exemplified by its dominant institutions; an ethics is a system of moral action guides employed to adjudicate conflicts within a given ethos. Thus, the medieval Christian church's ban against usury regulated commercial transactions as long as ecclesiastical life gave shape to social institutions. But when the church was replaced by commerce as the dominant institution of European society the controls on usury were whittled away in keeping with the new ethos.

If a shift in ethos is indeed required to respond to the potential dangers of growth, then the elaboration of ecological ethics in the absence of a new world view may prove an empty exercise. An ethics developed in the context of the present technico-economic structure, for example, may already cede priority to the growth mentality in such a way that normative principles would provide only a neat casuistry for minor problems instead of the radical criticism and imaginative speculation which are needed to establish a stable-state economy.

Many environmentalists have questioned the adequacy of traditional Western moral ideas in treating ecological problems. At least three sorts of criticism are made: (1) The anthropocentrism of Judeo-Christian thought is incapable of dealing with ecological problems because it gives precedence to human needs rather than to the whole of nature of which humanity is only a part. (2) Ethical principles alone are inadequate to the crisis; only a change in world view on a par with religious conversion will guarantee survival. (3) Global survival is too grave a matter to be left to individual conscience and democratic decision-making processes; support must be given to knowledgeable elites to direct society's future.

Anthropocentrism. The charge that Judaism and Christianity are responsible for the evil consequences of the growth ethic rests on the claim that the biblical notion of man's dominion over the earth was the origin of the West's utilitarian indifference to nature. Whether or not the notion of dominion is to be interpreted as such critics suggest, the biblical tradition and the Western tradition more broadly contain many strands with quite distinct attitudes toward nature (Passmore). Monastic Christianity, for example, first in its Benedictine and later in its Cluniac form, showed considerable respect for nature and found in created things an avenue to contemplation of the Creator (Glacken; Dubos). It has been argued, moreover, that scientific rationalism, not religious anthropocentrism, is the true precursor of the present ecological crisis (Passmore). Another source of growth mania may be the anthropology of liberal economics, which excludes any common definition of essential human needs, opposes asceticism and self-examination, and encourages economic activity as a means of allaying psychological insecurity (Wolin).

Perhaps *homo economicus* is at fault; but that does not provide grounds for concluding that all person-centered ethical systems are incapable of constructive responses to problems of global ecology. Where essential human needs are distinguished from artificial ones and asceticism is understood as a beneficial form of behavior, a person-centered ethic might very well uphold ecological values. Similarly there may well be a convergence between protection of the biosphere and prevention of harm to human persons such that there is no need to resort to sacralization of nature to protect the environment.

Moral conversion. The fear that codes of moral rules, especially those built on traditional principles, will be insufficient to redirect the juggernaut of technological supergrowth has led to proposals to resacralize nature, especially on Oriental or Amerindian models. To form a culture that shows respect for nature, it is argued, changes are required in the deeper levels of society's affective response to the environment. This kind of shift, it is assumed, can be brought about only by a change in religious outlook.

A more moderate position argues that profound reverence for nature need not imply resacralization and that sound ecology is compatible even with uses of nature that alter the traditional ecology of a place (Dubos). The "land ethic" proposed by Aldo Leopold takes a stronger position, but one still short of mystification of the environment. "A thing is right," Leopold wrote, "when it tends to preserve the integrity, stability, and beauty of a biotic community." On the other side, the philosopher John Passmore considers that all schemes to deal with the environment as a religious problem are "rubbish," and that scientific rationalism and existing categories of ethical analysis are adequate to the task.

Taking a pragmatic view, Jay Forrester argues that religion is the domain of humanity's long-range intentions and so a primary instrument of social change. Organized religion should use its resources to support a future-oriented, meritocratic, and nature-oriented conservation ethic that will assure the preservation of scarce resources for future generations and should forgo a traditional commitment to altruism, egalitarianism, and humanitarianism, which will only create worse conditions in the future.

Political ethics. Like the authors of *Limits,* some environmentalists view the present crisis as refuting the false hope that economic growth alone will solve such obdurate problems as poverty and racial discrimination (Commoner), or that the wealth of a nation can replace the well-being of its people. Taken seriously, the admission of limits to growth may place in jeopardy the power structure which has encouraged unlimited productivity as the basis of political harmony and "progress" (Commoner).

Instead of the diffusion of power Commoner advocates, others argue for greater concentrations of power in the interest of timely response to unprecedented crises. The "systems dynamics" school (Forrester; Meadows et al.) contends that ordinary political judgments in a democratic society are counterproductive because of the counterintuitive behavior of complex social systems. Hence they would vest responsibility for societal guidance in an elite of computer experts. Forrester argues that in the interest of future generations power must be left to those who will use resources wisely. More crassly, Hardin proposes that those nations which have the resources for self-sufficient survival have no obligations to share with those who have less, and he supports coercive measures to encourage no-growth. A moderate solution, which may help to balance the need for control in the interest of future generations with respect for freedom and basic human rights, is to apply coercion, when it is needed, only on the macroeconomic level (Pirages and Ehrlich). Thus, by applying pressure to large economic institutions, freedom can be preserved for the individual human beings to whom it properly belongs.

DREW CHRISTIANSEN AND C. P. WOLF

[*Directly related is the entry* ENVIRONMENT AND MAN. *For further discussion of topics mentioned in this article, see the entries:* FOOD POLICY; FUTURE GENERATIONS, OBLIGATIONS TO; JUSTICE; *and* RATIONING OF MEDICAL TREATMENT. *Other relevant material may be found under* POPULATION ETHICS: ELEMENTS OF THE FIELD, *articles on* ETHICAL PERSPECTIVES ON POPULATION *and* NORMATIVE ASPECTS OF POPULATION POLICY.]

BIBLIOGRAPHY

Aspen Institute for Humanistic Studies. *The Planetary Bargain: Proposals for a New Economic Order to Meet Human Needs: Report of an International Workshop Convened in Aspen, Colorado, July 7–August 1, 1975.* Aspen Institute for Humanistic Studies, Program on International Affairs, Occasional Paper. Palo Alto, Calif.: 1976.

BARNETT, HAROLD J., and MORSE, CHANDLER. *Scarcity and Growth: The Economics of Natural Resource Availability.* Baltimore: Resources for the Future, Johns Hopkins Press, 1963.

CALLAHAN, DANIEL J. *The Tyranny of Survival and Other Pathologies of Civilized Life.* New York: Macmillan Publishing Co., 1973.

———. "What Obligations Do We Have to Future Generations?" *American Ecclesiastical Review* 164 (1971): 265–280.

CHEN, KAN, et al. *Growth Policy: Population, Environment, and Beyond.* Ann Arbor: University of Michigan Press, 1974.

COLE, H. S. D.; FREEMAN, CHRISTOPHER; JAHODA, MARIE; and PAVITT, K. L. R., eds. *Models of Doom: A Critique of "The Limits to Growth."* With a reply by the authors of *The Limits to Growth.* New York: Universe Books, 1973. Papers by members of the Science Policy Research Unit, University of Sussex, Brighton, England. Published in England as *Thinking about the Future: A Critique of "The Limits to Growth."* London: Chatto & Windus, Sussex University Press, 1973.

COMMONER, BARRY. *The Closing Circle: Nature, Man and Technology.* A Borzoi Book. New York: Alfred A. Knopf, 1971.

DALY, HERMAN E., ed. *Toward a Steady-State Economy.* San Francisco: W. H. Freeman & Co., 1973.

DUBOS, RENÉ. *A God Within.* New York: Charles Scribner's Sons, 1972.

FORRESTER, JAY W. "Churches at the Transition between Growth and World Equilibrium." *Zygon* 7 (1972): 145–167.

GLACKEN, CLARENCE J. "Man's Place in Nature in Recent Western Thought." *This Little Planet.* Edited by Michael Hamilton. Introduction by Edmund S. Muskie. New York: Charles Scribner's Sons, 1970, pp. 163–201. Cooperative project of the National Presbyterian Center and the Episcopal Cathedral in Washington.

GOLDING, M. P. "Obligations to Future Generations." *Monist* 56 (1972): 85–99.

HARDIN, GARRETT. *Exploring New Ethics for Survival: The Voyage of the Spaceship 'Beagle'.* New York: Viking Press, 1972.

———. "Lifeboat Ethics: The Case against Helping the Poor." *Psychology Today,* September 1974, pp. 38–43, 123–124, 126.

HEILBRONER, ROBERT L. *An Inquiry into the Human Prospect.* New York: W. W. Norton, 1974.

HENDERSON, HAZEL. "Redefining Economic Growth." *Environmental Quality and Social Justice in Urban America.* Edited by James Noel Smith. Washington: Conservation Foundation, 1974, chap. 8, pp. 123–145.

HIRSCH, FRED. *Social Limits to Growth.* Twentieth Century Fund Study. Cambridge: Harvard University Press, 1976.

ILLICH, IVAN. *Tools for Conviviality.* World Perspectives, vol. 47. Edited by Ruth Nanda Anshen. New York: Harper & Row, 1973.

JONAS, HANS. "Technology and Responsibility: Reflections on the New Task of Ethics." *Religion and the Humanizing of Man.* 2d rev. ed. Edited by James M. Robinson. Waterloo, Ont.: Council on the Study of Religion, 1973, pp. 3–20. Plenary Addresses, International Congress of Learned Societies in the Field of Religion, Los Angeles, 1–5 September 1972.

KAHN, HERMAN; BROWN, WILLIAM; MARTEL, LEON; and the Hudson Institute Staff. *The Next 200 Years: A Scenario for America and the World.* New York: Morrow, 1976.

KAPP, KARL WILLIAM. *The Social Costs of Private Enterprise.* New York: Schocken Books, 1971.

LEOPOLD, ALDO. *A Sand County Almanac.* New York: Oxford University Press, 1949. Enl. ed. 1966.

MEADOWS, DONELLA H.; MEADOWS, DENNIS L.; RANDERS, JØRGEN; and BEHRENS, WILLIAM W., III. *The Limits to Growth: A Report for the Club of Rome's Project on the Predicament of Mankind.* New York: Universe Books, 1972.

MESAROVIĆ, MIHAJLO, and PESTEL, EDUARD. *Mankind at the Turning Point: The Second Report to the Club of Rome.* New York: E. P. Dutton & Co./Reader's Digest Press, 1974.

MILES, RUFUS E., JR. *Awakening from the American Dream: The Social and Political Limits to Growth.* New York: Universe Books, 1976.

MISHAN, EZRA J. [EDWARD J.] *The Costs of Economic Growth.* New York: Praeger Publishers, 1967.

OLSON, MANCUR. "Introduction." *Daedalus* 102, no. 4 (1973), pp. 1–13. Introduction to an issue entitled "The No-Growth Society."

PASSELL, PETER, and ROSS, LEONARD. *The Retreat from Riches: Affluence and Its Enemies.* Foreword by Paul A. Samuelson. New York: Viking Press, 1973.

PASSMORE, JOHN. *Man's Responsibility for Nature: Ecological Problems and Western Traditions.* New York: Charles Scribner's Sons, 1974.

PIRAGES, DENNIS C., and EHRLICH, PAUL R. *Ark II: Social Response to Environmental Imperatives.* San Francisco: W. H. Freeman & Co., 1974.

POTTER, DAVID M. *People of Plenty: Economic Abundance and the American Character.* Charles R. Walgreen Foundation Lectures. Chicago: University of Chicago Press, 1954.

RICH, WILLIAM. *Smaller Families through Social and Economic Progress.* Overseas Development Council Monograph, no. 7. Washington: 1973.

ROSTOW, WALT WHITMAN. *The Stages of Economic Growth: A Non-Communist Manifesto.* Cambridge: University Press, 1960. 2d ed. 1971.

SCHUMACHER, ERNST FRIEDRICH. *Small is Beautiful: Economics as if People Mattered.* New York: Harper & Row; London: Blond & Briggs, 1973.

SCHURR, SAM H., ed. *Energy, Economic Growth, and the Environment.* Baltimore: Resources for the Future, Johns Hopkins University Press, 1972. Papers presented at a forum conducted by Resources for the Future, Washington, D.C., 20–21 April 1971.

TINBERGEN, JAN, coordinator. *RIO: Reshaping the International Order: A Report to the Club of Rome.* New York: E. P. Dutton & Co., 1976.

WOLIN, SHELDON S. *Politics and Vision: Continuity and Innovation in Western Political Thought.* Boston: Little, Brown & Co., 1960.

Yale Task Force on Population Ethics. "Moral Claims, Human Rights, and Population Policies." *Theological Studies* 35 (1974): 83–113.

ENVIRONMENTAL HAZARDS

See ENVIRONMENTAL ETHICS, *articles on* ENVIRONMENTAL HEALTH AND HUMAN DISEASE *and* QUESTIONS OF SOCIAL JUSTICE.

ETHICAL THEORY

See ETHICS.

ETHICS

In this twelve-part entry the first article, THE TASK OF ETHICS, *is intended as an introduction to the meanings, methods, and questions of ethics. It also serves as a guide to the remaining eleven articles, which explain in detail the varieties of ethical theories upon which bioethics is built.* RULES AND PRINCIPLES *describes various ways of distinguishing moral from nonmoral rules and some of the problems connected with the moral justification of rules and the appeal of moral principles.* DEONTOLOGICAL THEORIES *is concerned with theories founded on the notion of duty and contrasts them with consequentialist theories in general.* TELEOLOGICAL THEORIES *discusses theories founded on the notion of good. Some of the conceptions of good are described and some general objections to teleological theories are discussed.* SITUATION ETHICS *defends one type of teleological ethics, i.e., act-utilitarianism. The article on* UTILITARIANISM *sets forth the general features of utilitarianism as a consequentialist type of ethics, and discusses the distinction between act-utilitarianism and rule-utilitarianism.* THEOLOGICAL ETHICS *distinguishes three basic theistic orientations and then uses them to show the relationship of religion to the issues of normative ethics.* OBJECTIVISM IN ETHICS *deals with theories of ethical knowledge and other conceptions of ethical rationality.* NATURALISM *describes three types of naturalism in ethics, examines critically the "naturalistic fallacy," and explains some of the problems concerning the relation of science to ethics.* NON-DESCRIPTIVISM *distinguishes between two types of metaethical theory: descriptivism, which includes naturalism and intuitionism, and non-descriptivism, which includes emotivism and prescriptivism.* MORAL REASONING *discusses an important philosophical problem concerning the nature of moral reasoning, namely, the problem of showing how moral reasoning can be practical, by motivating and guiding action, and yet be universally applicable in the sense that it involves universally valid criteria of evidence. The final article,* RELATIVISM, *distinguishes among descriptive, normative, and metaethical relativism and relates each to the problem of medical ethics.*

I.	THE TASK OF ETHICS	*John Ladd*
II.	RULES AND PRINCIPLES	*Wm. David Solomon*
III.	DEONTOLOGICAL THEORIES	*Kurt Baier*
IV.	TELEOLOGICAL THEORIES	*Kurt Baier*
V.	SITUATION ETHICS	*Joseph Fletcher*
VI.	UTILITARIANISM	*R. M. Hare*
VII.	THEOLOGICAL ETHICS	*Frederick S. Carney*
VIII.	OBJECTIVISM IN ETHICS	*Bernard Gert*
IX.	NATURALISM	*Carl Wellman*
X.	NON-DESCRIPTIVISM	*R. M. Hare*
XI.	MORAL REASONING	*Philippa Foot*
XII.	RELATIVISM	*Carl Wellman*

I
THE TASK OF ETHICS

Remarks on terminology

In this article "ethics" will be taken to mean the philosophical inquiry into the principles of morality, of right and wrong conduct, of virtue and vice, and of good and evil as they relate to conduct. The term "ethics" has a number of other meanings. For example, "ethics" is often used to mean professional ethics as it is incorporated into ethical codes for professional conduct set forth, say, by medical associations. "Ethics" is also sometimes used to refer to the beliefs or practices of a particular sect or group, as in Christian or Jewish ethics, or to describe the values or conduct of a particular individual, as in Hitler's ethics. The subject matter of this article, however, is ethics in the traditional sense in which it has been a subject of investigation by Western philosophers since the time of Plato and Aristotle.

It should be noted that there is no standard distinction of meaning between "ethics" and "morality" (or "morals"). The most that can be said with certainty is that "ethics" comes from Greek, whereas "moral" comes from Latin. Individual philosophers often use the words to make a distinction that they feel is important; other philosophers use them synonymously.

Finally, a word of warning about *labels*. It is easy to overestimate the importance of labels. A label is merely a convenient device designed for a very limited purpose; making labels or learning about labels is not itself a philosophical activity. Again, there is no standard use among philosophers of labels like "naturalism," "utilitarianism," "descriptivism," "intuitionism," "subjectivism," "relativism," and "skepticism." Each individual writer employs such labels for his own particular purposes. Indeed, many philosophers refuse to provide labels for their own positions, although they are not at all reluctant to assign labels to positions with which they disagree.

One of the most mischievous consequences of

excessive preoccupation with labels is that it lends credence to the erroneous opinion that every theory comes in a hermetically sealed package that can be accepted or rejected only as a whole. It is also inconsistent with the idea that ethical theories can be and ought to be open-ended and capable of assimilation and differentiation as well as of change and development.

The distinctive nature of ethics

Ethics, law, custom, institutional practices, and positive morality. It is important to distinguish ethics from law, custom, institutional practices, and positive morality (i.e., the body of accepted popular beliefs of a society about morality). Ethics is logically prior to all of these in that it comprises principles and standards that can be used to criticize and evaluate those other norms. Thus, ethics is not directly concerned with what law is, or what custom is, or what institutions are, or what positive morality is; rather, it is concerned with what they *ought* to be. It is therefore fallacious to argue in any simple or direct way from the fact that certain laws, certain practices, or certain beliefs exist to the conclusion that a certain kind of conduct is right or wrong. Law, in particular, differs from ethics in that sanctions play a crucial, perhaps even an essential, role in law and legal decisions, and norms depend for their definition and determination on the actions of specific institutional organs such as the courts and the legislature. To argue directly from the illegality of a certain form of behavior, e.g., euthanasia, to its immorality is therefore to commit a kind of fallacy. Americans, as Tocqueville pointed out, tend to construe ethical (and political) questions as if they were legal questions. It is important constantly to be on one's guard against approaching medical ethics in that way. In this regard, considerable confusion results from the indiscriminate use of the concept of "right" to solve any and every moral problem in medical ethics, as in the "right to life," the "right to die," or the "right to medical treatment." When these supposed rights are discussed, hardly any effort is made to explain what kind of right is intended, e.g., whether or not it can be waived, who has the corresponding duty, and what kind of conduct it calls for.

Although ethics is logically independent of law, custom, practices, and positive morality, it does not follow that there are no ethically significant connections between them. We cannot avoid such moral questions as: Ought we always to obey the law, and ought we to conform to custom, to conventional medical practice, and to positive morality? And what are the limits to the obligation to obey and to conform?

The controversial character of ethics. In order to understand what philosophical ethics is all about it is necessary to begin by recognizing that ethics, by its very nature, is controversial. That is, like philosophy in general, it is essentially disputatious, while at the same time it must deal with issues that are both urgent and inescapable. In this regard, ethics is very much like politics and law, with which it is frequently compared.

The controversial character of ethics is best understood by noting three essential conditions of a genuine controversy as contrasted with, say, a simple clash of interests or tastes: (1) There must be a real inconsistency, rather than a merely verbal disagreement, between the two positions, that is, they must be logically incompatible in some sense. (2) Each of the opposing positions must be reasonable (i.e., nonarbitrary) and capable of being supported by argument. (3) A meeting of minds must, in principle at least, be possible; for an issue would not be controversial if it is taken for granted that arguing over it will inevitably and necessarily be futile.

It should be clear that in all three respects ethical disagreements are quite unlike mere differences of subjective preference or differences of taste. *De gustibus non est disputandum* —there is no disputing over tastes. Ethical issues, on the other hand, are eminently disputable in the three ways just mentioned; hence, it is a mistake to think that one can or ought to choose an ethical theory as one chooses foods from a menu.

In view of the controversial character of ethics, one should not expect moral philosophers to provide definitive answers to all of our moral problems, that is, answers of the kind that would be acceptable to any rational being at any possible time. But, although philosophers cannot in good conscience claim infallibility, it does not follow that what they say can be ignored. Why this is so should become clear presently.

Dogmatic skepticism and skeptical dogmatism. Broadly speaking an ethical theory is a theory about reasons—reasons for doing or refraining from doing something, reasons for approving or disapproving of something, or reasons for believing or asserting something about morality, virtuous or vicious conduct, good and evil rules, practices, institutions, policies and goals. The rightness or wrongness of an action may itself be regarded as a reason for performing or not performing that action. In this broad sense, then, the task of ethics may be described as the search for and establishment of reasons for various sorts of things connected with conduct, including such things as actions, motives, attitudes, judgments, rules, ideals, and goals. This general concern with reasons explains why ethics is commonly taken to be a rational activity and a matter of practical reason.

There is a prevailing opinion to the effect that there are no good reasons for any of the things just mentioned, that is, reasons that could in any serious way be regarded as objectively or intersubjectively valid. This position may be called *ethical skepticism* (like other philosophical labels, "ethical skepticism" is used differently by different philosophers). Ethical skepticism, in the sense intended here, entails the denial of the distinction between right and wrong; for, if there can be no good reason or at least no good moral reason for preferring one sort of conduct over another, then it follows that everything is permitted. The obverse of ethical skepticism is *ethical dogmatism*, the position that no reasons need be given for one's ethical views. Skepticism holds that no reason is possible, whereas dogmatism holds that no reason is necessary. It should be obvious that ethical skepticism, as depicted here, is itself a dogmatic position, and that ethical dogmatism, as depicted here, is essentially skeptical. Hence, the two positions may also be called "dogmatic skepticism" and "skeptical dogmatism," respectively. Persons who have not studied or thought very much about ethics as such often take for granted that those two positions are the only alternatives. Since it is not easy for anyone to remain completely morally neutral about *everything*, it is difficult, if not impossible, to find persons who are consistent skeptics in actual practice. On the other hand, practicing ethical dogmatists are unfortunately much too common—unfortunately, perhaps even tragically, because in practice a stubbornly dogmatic attitude on ethical issues inevitably results in a breakdown of communication between dissenting parties and, in the end, sanctions the use of coercive measures and violence. However, neither of the two doctrinaire positions, absolute skepticism and dogmatism, is accepted by any reputable philosopher, whatever else he may believe. It is easy to see why, for it is self-defeating to hold either that it is impossible or that it is unnecessary to give reasons for one's assertions: Anyone who openly advocates either position cannot, at risk of self-contradiction, attempt to give any reasons to anyone else for accepting what he says.

The methods of ethics

A brief discussion of the methods of argumentation that are likely to be encountered in ethical discussions will be helpful. The discussion presented here will be patterned after Peirce's article, "The Fixation of Belief" (Peirce), although there are some obvious differences between the view of ethics presented here and Peirce's view.

The first method is the *appeal to authority*. In the development of Western moral notions, the method of authority has had widespread use as the basis for morality. Much of Judaic as well as of Christian ethics illustrates this, although there are also authoritarian elements in Greek and even in Egyptian ethics. The appeal to authority seeks to establish a proposition on the ground that a particular authority declares it to be so. In ethics, for example, it is held that one ought or ought not to do something, to approve or disapprove of something, or to believe something to be right or wrong on the ground that the authority tells us to do so. The authority may be a human or a supernatural person, a group of persons, an institution (e.g., a church or a government), or a collection of writings: it may even be a fictitious person (e.g., society or the ideal observer). In order to validate the reported pronouncements of the authority, it is necessary to show (1) why the supposed authority is an authority, that is, it is necessary to establish its credentials, and (2) that the alleged pronouncements truly emanate from the authority, that is, it is necessary to establish the authenticity of the reports of its pronouncements. Various types of religious ethics amply illustrate how these two questions might be answered. With regard to ethics, the credentials of the authority are generally based on the authority's superior knowledge and moral superiority (e.g., the supreme love, benevolence, solicitude, or disinterestedness found in the authority). Thus, theologians usually ascribe these qualities to God and then use them to vindicate his ethical authority; the authenticity of the reports of God's commands depends on a variety of other assumptions that need not be detailed here. Insofar as religious systems depend on faith or revelation to establish the credentials of the authority, or the authenticity of what is attributed to it (e.g., in the Scriptures), this use of the method of authority has only a very limited appeal; it has cogency only for believers.

Philosophical ethics, in contrast to faith-based religious ethics, is directed to a universal audience rather than to a particular subgroup. As such, it does not require an act of faith. In this connection, it should be mentioned that not all versions of the natural law theory are fideistic or authoritarian. Aristotle's version, for example, is neither. Locke's version is authoritarian but not fideistic. By the same token not all theological ethics is fideistic or authoritarian [see article VII below on THEOLOGICAL ETHICS].

From what has been said about the method of authority it should not be concluded that it is never a valid form of argument. Of course it is, provided satisfactory answers are given to the two questions. We often use the method of authority to settle questions outside of ethics, for example, questions of law or of medicine.

A second method of argument often used in ethics is the appeal to *consensus hominum*, that is, the citing of the alleged agreement of people in general or of particular groups of people concerning an issue to establish a particular ethical contention. In arguments of this sort, essentially the same two kinds of question are relevant as were asked concerning the use of the appeal to authority, namely: (1) Is there any reason to believe that the people cited are especially wise and ethically competent? and (2) Do they, as a matter of fact, actually subscribe to the beliefs attributed to them?

A third method of argument is the appeal to "intuition," "self-evidence," or "gut feelings." The use of the concept of intuition by philosophers should be carefully distinguished from its more general use by nonphilosophers, say, physicians and patients. Philosophers, particularly those known as intuitionists (e.g., G. E. Moore), have developed theoretical arguments to the effect that ethical knowledge is ultimately founded on intuition; "intuition" is a technical concept in their system [see article VIII below on OBJECTIVISM IN ETHICS]. The use of the method of intuition in practice, however, has certain drawbacks; for one thing it carries no conviction for persons who have not already had the intuition, or what they think is an intuition. Thus it is very difficult, if not impossible, to persuade someone who is not already convinced that euthanasia is wrong on the grounds that it is self-evident that killing is wrong. More generally, something that may seem incontrovertibly self-evident to you today may not seem so tomorrow or to someone else. Particularly in the field of medical ethics, conditions and categories are changing so rapidly that a person's intuitions are apt to become out of date as quickly as the medical technologies that give rise to the ethical problems in the first place.

Finally, the method par excellence of the moral philosopher in both his theoretical and his practical investigations is what may be called the method of argumentation or the dialectical or Socratic method, that is, the method of asking questions and looking for answers backed up by good reasons. In philosophy, and ethics in particular, no position, theory, or answer is sacrosanct; it is acceptable only if it can be supported by argument. It is important, therefore, to pay close attention to the arguments and reasons that philosophers give for their various contentions, for without them what they say would simply be dogmatic and boring.

The subject matter of ethics: normative ethics and metaethics

The subject matter of ethics may be divided into two parts, which are generally called "normative ethics" and "metaethics." *Normative ethics* asks questions directly related to the criteria and standards of right and wrong action, what things are good and evil, and questions about moral conduct in general. *Metaethics*, on the other hand, is a second-order inquiry into the nature of ethical discourse itself. Normative ethics gives us ethical theories about what we ought to do, while metaethics gives us theories about what ethics is all about.

As a second-order type of inquiry, metaethics is related to normative ethics as philosophy of science is to the sciences, or philosophy of art is to the arts. It is often called "analytical ethics" or sometimes simply "moral philosophy." Historically, the distinction between normative ethics and metaethics is probably due to G. E. Moore, who differentiated between the questions: "What kinds of thing are good?" and "What is the definition (or meaning) of good?" (Moore). Accordingly, it is sometimes held that metaethics and normative ethics employ different methods; metaethics, for example, is said to use the method of conceptual, logical, or linguistic analysis; on an older view, that meant searching for definitions. At one time it was held (e.g., by Ayer) that metaethics is ethically neutral (i.e., non-normative), and that the moral philosopher ought to concern himself only with metaethics, leaving normative ethics to others. Today, this view of metaethics is repudiated by many moral philosophers. Perhaps the only essential distinction that remains between normative ethics and metaethics is a difference of levels or orders, the subject matter of normative ethics being moral conduct and the subject matter of metaethics being ethics itself (i.e., ethical discourse, judgments, concepts, and reasoning). Although the questions of normative ethics and metaethics often overlap or even merge, it will be convenient to discuss them separately.

The questions of normative ethics

As in other areas of ethics, what the basic questions of normative ethics are is a matter of dispute. It will become clear as we proceed that the particular questions selected for investigation already reflect a prior ethical commitment as to what is important and urgent. In order to understand what is at issue, therefore, it is extremely important that we take these biases into account.

In general terms, normative ethics may be said to concern itself with the problem: *What ought I (or we) to do?* This question may be very general—i.e., By what principle ought I (or we) to order my life (our lives)? Or, on the other hand, it may be very specific—i.e., What ought I to do here and now about this particular thing? In most ethical theories, the question is taken at the intermediate level: It is assumed to be about the rules, principles, or standards of right and wrong conduct.

One ramification that is of particular relevance in problems of bioethics is that the ought question is asked in a variety of different contexts. Consider the question: What ought to be done with an infant with a hopelessly incurable genetic disease, e.g., spina bifida? It is clear that this question might be asked by a particular physician with regard to a particular infant, or it might be asked by the governing board of a hospital trying to decide on a general policy, or it might be asked by the general public as a question of public policy. Considerations that are relevant to the ought question at one level may not be identical with those that are relevant at another level, although they are obviously interconnected.

Questions like, "What ought I (or we) to do?" reflect a number of different practical predicaments, which, in turn, generate several different categories of moral problems. Here we shall examine four of these categories.

1. Ought questions arising out of conflicts of interest. Let us begin with moral problems that arise out of conflicts of interests. Basically the same kinds of problem are found in connection with conflicts of wants, desires, needs, preferences, pleasures, or other types of "subjective value." The moral perplexity, the ought question with which we are concerned here, is due to the fact that in certain instances it is impossible to satisfy one interest (or "value") without sacrificing other interests. Consequently, we are forced to choose between interests. Which one ought to be preferred?

Philosophers who take conflict of interest problems to be the most basic category of moral problem usually claim that the chief function of ethics is to provide an objective procedure for adjudicating between conflicting interests (Rawls, 1951).

When conceived in this way, the task of ethics is to develop a way of ordering interests. The best known theory of ordering is utilitarianism, which uses the maximization principle to order interests [*see article* VI *below on* UTILITARIANISM]. Rawls's theory of justice provides another, rather sophisticated method of ordering interests (or goods) (Rawls, 1971). The kind of ought questions that are involved here are usually taken to be questions of distributive justice.

The basic assumption underlying this particular conception of ethics is that every interest (or "value") has, as such, a prima facie claim to be satisfied. That is, other things being equal (i.e., ideally), one ought to try to satisfy all the interests of everyone. Of course, almost all of the time other things are not equal, so that one interest has to give way to other more pressing interests. According to this approach it is the simple fact of being an interest that constitutes its claim to be satisfied rather than the quality of the interest itself, e.g., its object or its rationality. This assumption is rejected categorically by philosophers like Kant, who would not agree that it is the function of ethics to settle conflicts of interest (or "values").

2. Ought questions arising out of moral dilemmas. Another set of ought questions comprises those that arise from a conflict of moral demands, e.g., conflicts of duty. Such a conflict occurs when in a particular situation it is impossible for a person to do one of his duties without failing to do one of his other duties. Unlike conflicts of interests, conflicts of this kind presuppose the antecedent moral quality of the actions or activities that conflict and so should be called *moral dilemmas*.

Moral dilemmas are, of course, a common occurrence in the medical world; for example, doctors often have to choose between duties to different patients. Philosophers have generally adopted the (unfortunate) terminology of W. D. Ross and now refer to such conflicting duties as prima facie duties (Ross). The term "prima facie" is unfortunate because it suggests that the duty in question is only apparently a duty; for Ross and others who accept the distinction, on the other hand, a so-called prima facie duty is simply a real duty that is overridden by another duty. A prima facie duty is one that, other things being equal, is binding; however, if other things are not equal, one prima facie duty may be overridden by another prima facie duty. The overriding duty may be called "the absolute duty" [see article III below on DEONTOLOGICAL THEORIES]. Thus, a doctor's prima facie duty to save the baby might be overridden by his duty to save the mother. Depending on what the writer takes as his basic ethical framework, moral dilemmas may be regarded as conflicts of duties, of obligations, of loyalties, of rights, of moral rules, or of "morally relevant facts" [see article VIII below on OBJECTIVISM IN ETHICS].

The ought question in the case of moral dilemmas is: Which of my prima facie duties ought I to perform when it is impossible to perform all of them? Some philosophers have offered the principle of utility as a deciding principle in case of a conflict of duties. Others have offered a "sense of urgency (or importance)" as a deciding principle. Ross himself described the selection of an absolute duty from among a set of prima facie duties as a "fortunate" act. Many would agree with Gert that two impartial persons might reach different conclusions concerning a decision in a particular case [see article VIII below on OBJECTIVISM IN ETHICS].

Again, it should be observed that moral dilemmas come into existence only when all of the acts in question are in fact prima facie duties. Hence, ought questions arising out of dilemmas make sense only if the prima facie duties, rules, or other "morally relevant considerations" are not in dispute. This assumption about prima facie duties is not shared by some philosophers, e.g., by act-utilitarians and situationalists, who maintain that in principle there are no moral dilemmas and hence that ought questions of this type can be easily eliminated [see articles V and VI below on SITUATION ETHICS and UTILITARIANISM].

3. Ought questions arising out of ethical disagreements. In a pluralistic society such as ours and in a world divided by different ideologies, religions, and cultures, it is inevitable that people should disagree over the rightness or wrongness of various kinds of conduct. An act that is regarded as right in one group is regarded as wrong in another group and as morally neutral in still another. The differences in question here do not call for ordering or weighting as in the case of conflicts of interest or of moral dilemmas; here we are dealing with black and white, as it were—that is, positions that are incompatible in the sense that they are mutually exclusive logically. Solutions to problems of this type depend on the deployment of a complete ethical theory, the refutation of alternatives to it, and, in general, a theory of ethical reasoning.

It is unnecessary to point out that differences of opinion about right and wrong like those being considered here create moral problems for the medical practitioner. They cannot be avoided, however, by a simple resort to dogmatic skepticism or skeptical dogmatism. It should be obvious that the ought question can take different forms depending on how it is construed and what one's purposes are. If a doctor wants (or has the time) to try to change the other person's mind, then we are dealing with an ought question of the third type, i.e., an ethical disagreement; but if he has to act at once on the basis of what he thinks to be right, then he is faced with a moral dilemma: Does his duty to respect the ethical beliefs of others that he knows to be wrong override his duty to save the patient's life?

4. Ought questions turning on the distinction between duties and other oughts. Finally, there is a set of problems relating to the distinction between duties and other things that one ought to do for nonmoral reasons [see article II below on RULES AND PRINCIPLES]. It would be absurd to say that a doctor either has a duty to play golf on his day off or a duty not to do so, although it might be perfectly correct to say that he ought to play because he enjoys it. There are many issues relating to the scope of morality; some philosophers regard the scope as very large, so that almost everything that a person might do is either right or wrong. (This kind of moral rigorism is a consequence of certain versions of act-utilitarianism.) Most philosophers, however, narrow the scope; some do so very radically and hold, for example, that a person can have duties only to others and not to himself. This view raises interesting questions for medical ethics, for it automatically excludes the duty to oneself to take care of one's own body: Why should a patient take pills prescribed by a doctor if he has no duty to try to get well?

Metaethical questions

It should be pointed out at once that all of the preceding discussion concerning normative ethics properly comes under metaethics, since it is *about* normative ethics rather than being part of normative ethics. Five of the articles in this entry are directly concerned with metaethical issues. The issues of metaethics are, however, quite technical and so may not appear to be as directly relevant to medical ethics as are the issues of normative ethics. Nevertheless, it is impossible to deal with normative problems intelligently without entering into the field of metaethics, as should be evident from the foregoing discussion of ought questions.

Metaethics encompasses three broad categories of questions. The first group of questions concerns the connection between ethics (morality) and conduct. How can the belief that a certain action is right or wrong move a person to do or refrain from doing an action? On the other hand, how can a person ever fail to do his duty? (Mortimore). Traditionally, the problem has been formulated: How can ethics and moral reasoning be practical? [*See articles* x *and* xi *below on* NON-DESCRIPTIVISM *and* MORAL REASONING.]

The second group of questions concerns the connection between ethics (e.g., beliefs about right and wrong) and facts about the real world. What is the relationship of "value" to "fact," of "ought" to "is," or of "normative" to "descriptive"? (Hudson, 1969). This kind of issue might be important for medical ethics in establishing, for example, the connection between "normative" concepts like health and physiological conditions specifiable in purely scientific terms (Kovesi) [*see article* ix *below on* NATURALISM].

The third kind of question covered by metaethics concerns the logical relationship between ethical propositions of various degrees of generality, e.g., (1) all-embracing super-principles like the principle of utility or the categorical imperative, (2) moral rules and practices, and (3) individual, here-and-now moral decisions. Is one of these types logically more basic than the others? Are they linked deductively or inductively, conductively, constructively, or in some other way? [*See article* II *below on* RULES AND PRINCIPLES.]

One answer is that the rightness or wrongness of a particular action can be derived deductively from a super-principle, e.g., the principle of utility, with the aid of intermediate factual premises of an empirical nature. There are numerous objections to this kind of "mechanical" model of ethical reasoning. Another answer, which is adopted by the so-called rule-utilitarians, offers a two-tier model in terms of which the rightness or wrongness of particular actions is justified by appeal to a rule or practice, and the rules or practices are in turn justified by appeal to the super-principle, e.g., the principle of utility (Rawls, 1955). Yet another answer is to interpret the different kinds of ethical propositions as serving different functions in ethical discourse, that is, as providing answers to different categories of ought questions mentioned earlier. On this view, the super-principle may be taken as an answer to the question: How is a moral rule distinguished from a nonmoral rule?

In conclusion, it is important to realize that all of the questions outlined here are interconnected; one cannot satisfactorily answer one of them without addressing oneself to the others, and it should become clear as one probes into them that the problems of ethics are both complicated and subtle.

The following articles on ethics

In reading the articles on ethics that follow, it may be helpful to bear these three points in mind. First, while the articles as a group offer a comprehensive survey of the principal theories of ethics, one should not expect the articles to be entirely consistent with each other: each article reflects, to some extent, its author's own philosophical perspective. Second, philosophers prefer to organize their discussions around a problem or a set of problems; hence, the title of each article might be more accurate if it were prefaced by the words: "Problems concerning . . ." (e.g., rules, utilitarianism, etc.). Finally, it is important to recall the remarks made earlier about labels and to note that in almost every article the author takes pains to point out that the label in question is used in a number of different ways.

The first five articles (II–VI) are primarily concerned with questions of normative ethics. The next article (VII) describes some of the ways in which religious beliefs are connected with ethics. The last five articles (VIII–XII) deal with questions of metaethics; the last two of them discuss special problems that are particularly important for an understanding of the nature of ethics in general.

JOHN LADD

[*While all the articles in this entry are relevant, see especially the articles* RULES AND PRINCIPLES *and* MORAL REASONING. *For further discussion of topics mentioned in this article, see the entries:* HEALTH AS AN OBLIGATION; LAW AND MORALITY; NATURAL LAW; *and* OBLIGATION AND SUPEREROGATION. *This article will find application in the entries* ACTING AND REFRAINING; BIOETHICS; DECISION MAKING, MEDICAL; DOUBLE EFFECT; JUSTICE; *and* SEXUAL ETHICS. *For discussion of related ideas, see:* PRAGMATISM.]

BIBLIOGRAPHY

ARISTOTLE. *Nichomachean Ethics.*
AYER, ALFRED JULES. "On the Analysis of Moral Judgments." *Philosophical Essays.* New York: St. Martin's Press; London: Macmillan & Co., 1954, pp. 231–249, esp. p. 245.
HUDSON, WILLIAM D. *Modern Moral Philosophy.* Garden City, N.Y.: Anchor Books, Doubleday & Co., 1970. A lucid, critical exposition of contemporary metaethical theories.
———, ed. *The Is–Ought Question: A Collection of Papers on the Central Problem in Moral Philosophy.* Controversies in Philosophy. Edited by A. G. N. Flew. London: Macmillan & Co., New York: St. Martin's Press, 1969.
KOVESI, JULIUS. *Moral Notions.* Studies in Philosophical Psychology. Edited by R. F. Holland. New York: Humanities Press, 1967.
MCCLOSKEY, J. *Meta-ethics and Normative Ethics.* The Hague: Martinus Nijhoff, 1969. A defense of ethical intuitionism, including an explanation of the concept of prima facie duties.
MACINTYRE, ALASDAIR. *A Short History of Ethics.* Fields of Philosophy. Edited by Paul Edwards and John Hospers. New York: Macmillan Co., 1966.
MELDEN, ABRAHAM IRVING. *Ethical Theories.* 2d ed., rev. Englewood Cliffs, N.J.: Prentice-Hall, 1967. A standard collection.
MOORE, GEORGE EDWARD. *Principia Ethica.* Cambridge: At the University Press, 1903, pp. 3–6.
MORTIMORE, GEOFFREY, ed. *Weakness of Will.* London: Macmillan & Co.; New York: St. Martin's Press, 1971.
PEIRCE, CHARLES SANTIAGO SANDERS. "The Fixation of Belief." *Popular Science Monthly* 12 (1877): 1–15. Reprint. Charles S. Peirce. *Essays in the Philosophy of Science.* Edited by Vincent Tomas. New York: Liberal Arts Press, 1957, pp. 3–30. Reprint. *Collected Papers of Charles Sanders Peirce.* 8 vols. Vol. 5: *Pragmatism and Pragmaticism.* Edited by Charles Hartshorne and Paul Weiss. Cambridge: Harvard University Press, 1960, pp. 223–247.
PLATO. *Apology.*
———. *Protagoras.*
———. *Republic.*
RAWLS, JOHN. "Outline of a Decision Procedure for Ethics." *Philosophical Review* 60 (1951): 177–197, esp. 186–187.
———. *A Theory of Justice.* Cambridge: Belknap Press, Harvard University Press, 1971.
———. "Two Concepts of Rules." *Philosophical Review* 64 (1955): 3–32.
ROSS, W. D. *The Right and the Good.* Oxford: At the Clarendon Press, 1930, pp. 19–32.
SELLARS, WILFRED, and HOSPERS, JOHN, eds. *Readings in Ethical Theory.* 2d ed. Century Philosophy Series. Edited by Justus Buchler. New York: Appleton-Century-Crofts, 1970. A collection of important recent writings in metaethics.
TAYLOR, PAUL W. *Principles of Ethics: An Introduction.* Belmont, Calif.: Dickenson Publishing Co., 1975.

II
RULES AND PRINCIPLES

Questions about the role of rules and principles in ethics have a special relevance for many discussions in bioethics. First, it has been true historically that the ethical positions associated with the profession of medicine have almost always been articulated in sets of rules. Even when the form in which the ethical position is presented is an oath or a prayer, the substantive content is usually embodied in a set of more or less specific rules laid down for the conduct of those in the medical profession.

Second, it is clear that any adequate response to the ethical quandaries that beset the biomedical sciences today must take the form of a set of general action guides: rules or principles. Most of the prominent ethical dilemmas in modern biomedicine—e.g., the design of experiments involving human subjects, the distribution of scarce medical resources, and questions surrounding the issue of when exotic lifesaving technology is to be employed—arise in areas where medicine is most socially complex and highly organized. The difficulties involve crucial decisions to be made by individual agents; but primarily they demand policies that must be made at a high level of the organizational pyramid and that will affect different persons in different ways.

It is of course possible that the policies actually adopted will be overly rigid and insensitive to subtle differences among the various cases to which they apply. To acknowledge this danger, however, is hardly fatal to the argument in favor of rules. To suggest that the *possibility* of the promulgation of overly rigid rules in this area provides good reason for jettisoning rules and principles altogether—in favor, perhaps, of immersion in the particular aspects of individual cases—is surely mistaken. Persons whose lives are affected by the institutions and agencies within modern biomedicine are entitled to reliable expectations about the stance of medical professionals toward certain basic ethical issues. The only way to provide for such expectations is through the formulation of general policies, i.e., sets of rules and principles, that guide the conduct of those who work within the institutions. The policies need not, of course, be so rigid as to eliminate all discretionary decision making on the part of those to whom they apply.

General features of rules

In the context of moral discussions, both rules and principles are best conceived as general action guides specifying that some type of action is prohibited, required, or permitted in certain circumstances. The distinction between a moral rule and a moral principle is difficult to draw with precision, but principles are generally distinguished from rules by being both more general and more foundational.

The most obvious feature of rules is their diversity (Wittgenstein, pp. 56–88). Rules are important for a wide variety of human activities, and the features they have in one area may be completely lacking in another. To be convinced of this, one need only reflect on the roles played by rules in morality, games, language, scientific investigation, etiquette, legal procedure, religious ceremonies, and love affairs, not to mention the practice of medicine. In all these areas, rules surely exist in that they both guide an agent's own actions and provide a basis on which the action of others may be evaluated. It is difficult, however, to find a single model that will characterize the nature and function of rules in all those areas.

The force of rules. Consider first how the force of rules may vary in different areas. Some rules have the force of restrictions, while others have the force of regulations, permissions, or prescriptions. More important, there may be great variation in the point of having rules in an area at all. Sometimes rules are required to coordinate the activity of large groups of persons. In such cases it may be unimportant which particular rules are adopted as long as everyone follows the same ones. The rule of driving in the United States that all drivers must stay on the right except when passing is often cited as an example of a rule with this purpose. Rules by which different uniforms are assigned to a hospital staff in order to allow easy identification constitute another example. Other rules have as their point the efficient achievement of some particular goal. Rules that govern the monitoring procedures in an intensive care unit would be an example here. There are yet other rules the point of which is to create situations of stress where mental or physical dexterity can be demonstrated, such as the rules of games.

The sanctions of rules. Rules can also vary with regard to the sanctions attached to them. An almost limitless variety of rewards and punishments can be distributed in accord with conformity to or breach of rules. There is also significant diversity in the possible sources of such distributions. Sometimes the appropriate sanctions are applied by the actions of other persons, such as policemen. At other times, as in the case of rules like "Don't be exposed to too many X-rays," there are "natural" sanctions attached to a rule. At still other times it is alleged that the sanctions are in the hands of some supernatural agent to be distributed in either this world or the next.

The existence of rules. Conditions for the existence of rules also vary across different areas of their operation (Warnock, pp. 43–52). In some areas, a particular rule is said to exist only if it is the result of legislation by some "official" body. The official rules of such highly organized games as football and baseball exist only on this condition. Other rules are said to exist if a sufficiently large group of persons recognize them in their practical decisions and in their appraisal of the behavior of others. Linguistic rules and the rules of etiquette would be examples here. The rules of medical etiquette, which regulate referral, consultation, fee-splitting, etc., are in a somewhat ambiguous position with regard to the grounds for their existence. Such rules are partially grounded in traditional practices, partly in legislation by professional bodies (e.g., the American Medical Association), and partly in governmental regulation. The often acrimonious disagreements among health-care workers about these matters may very well be explained by a lack of agreement on the conditions for the existence of such rules.

The scope of rules. Rules also vary in their scope of application, with regard to both the persons to whom they apply and the locations within which they have application. Some rules apply to all persons in all places, e.g., moral rules, in some views. Some rules apply to some persons in all places, e.g., rules concerning physician advertising. Some rules apply to all persons in some places, e.g., rules forbidding smoking when oxygen is in use. Some rules apply to some persons in some places, e.g., rules for the maintenance of sterility in operating rooms. There is some dispute about the significance of these distinctions of scope, since by manipulation of the particular formulation of a rule it is possible to change its scope without changing its substantive content (Kovesi, pp. 73–85). There is nevertheless a difference in the generality of rules that is at least partially captured by these distinctions.

The concreteness of rules. One final way in which rules may vary is in concreteness. All rules associated with practical contexts will mention an action, but the description involved can vary in its concreteness. Thus, within traditional medical ethics one finds rules as abstract as "Do no harm," and as concrete as "Do not engage in fee-splitting." One would expect the relative concreteness of a rule to be reflected in the manner in which it regulates action. Thus, relatively abstract rules function primarily to set goals or parameters for action, while more concrete rules normally invoke prohibitions or requirements.

It is important to keep the diversity of rules in mind when discussing the nature of moral rules. Without such an awareness, one is likely to fasten upon some particular model for the function of rules drawn from a single area of human activity and to treat moral rules as if they must fit that model. If one supposes that moral rules must have the precision and concreteness of the rules of chess, or that they must be the result of explicit legislation like the rules of tournament golf, one is likely to despair of finding any moral rules at all. Once one recognizes, however, that rules function differently in the different areas of human life where they play significant roles, it is possible to appreciate the distinctive features of moral rules.

Moral rules in classical ethical theory

The diverse ways in which the nature and function of moral rules and principles can be understood is also illustrated in their treatment by classical moral philosophers. Most classical moral philosophers have agreed that rules and principles play a central role in moral deliberation and justification; there has also been agreement for the most part on the content of the correct moral rules. Moral philosophers as otherwise diverse as Kant and Mill or Plato and Hume have agreed that moral rules requiring, for example, fidelity, veracity, justice, and beneficence should be observed. Alongside this agreement, however, deep disagreement has existed about the nature and function of moral rules.

For Plato and Aristotle moral rules are requirements on action necessary for each person to achieve a satisfying and fulfilling life. The penalty for failure to abide by moral rules is the disintegration of the personality of the wrongdoer. In Immanuel Kant's ethical theory, moral rules play a quite different role. Kant argues that in all action by rational creatures a maxim, or general rule of action, is implied. Some maxims or rules of action are incumbent on an agent because of the particular projects he may be pursuing or because of his rational concern for his own happiness. Other maxims, however, which Kant called moral laws, are requirements on the actions of every rational agent no matter what his particular aims might be. The binding force of these rules, which give rise to what Kant called categorical imperatives, derives from what Kant took to be certain universal features of human rationality. Unlike Plato, Kant did not think there was any necessary connection between an agent's observance of moral rules and his achievement of happiness. Indeed, Kant felt that the parity between virtue and happiness, which is at the heart of Plato's doctrine, can be ensured only if there is some supernatural agent to distribute appropriate rewards and penalties in an afterlife.

J. S. Mill's conception of the role and nature of moral rules differs importantly from those of both Plato and Kant. Mill holds that moral rules are rules of thumb based on our experience of the general tendencies of actions to promote certain ends, which should be observed if the utilitarian goal of the greatest happiness is to be achieved (Cunningham, pp. 168–196). Rules are necessary since it is not usually possible, because of insufficient time or insufficient information, to plot the consequences of particular actions when an occasion for action arises. Moral rules for Mill, then, are necessary because of certain limitations inherent in human deliberation; they could be dispensed with in the practical deliberation of a creature with perfect knowledge.

Underneath the surface agreement on the part of classical moral philosophers about the importance of moral rules and about the content of these rules, there remained, then, large areas of disagreement. They disagreed about (1) the point of having moral rules at all; (2) the sanctions attached to the rules; (3) the techniques of argument required to support the particular rules they accepted; and (4) the conditions under which an agent might be required to make an exception to one of the rules.

All of these areas of controversy within classical ethical theory also make their appearance in recent discussions within bioethics. Thus, there are disputes about the point of having rules in bioethics at all. Some have argued that rules are necessary in order to ensure coordinated activity within biomedicine, which will lead to some overarching end, such as the greatest happiness of the greatest number. Others have argued that rules are required so that the practice of medicine will approximate some abstract model of justice or rational human conduct.

There are also disputed questions about the sanctions appropriate to rules in bioethics. Should the ultimate sanctions for the rules be located in the professional bodies to which health-care professionals belong, or in some governmental body more attentive, perhaps, to the desires of the general public? Or again, should external sanctions be rejected altogether in favor of making adherence to the rules a matter of individual integrity?

There are endless disputes within bioethics, of course, about the techniques of argument appropriate to establishing a specific moral rule. According to a familiar view, such arguments must rest on ultimate premises that direct the medical profession to promote some broadly teleological goal. Others have argued to the contrary that the premises should be of the sort that specify certain types of action as absolutely forbidden.

Finally, there are many questions within bioethics that turn on decisions about how to treat alleged exceptions to moral rules. On some views a rule that admits of exceptions is really no rule at all. According to other views, rules are primarily designed to handle "normal" cases and will always admit of exceptions in extraordinary cases. There clearly are extraordinary cases where a physician, for example, is justified in revealing confidential information gained in a medical encounter with a patient—where, say, prevention of some great disaster is in question. Such cases, some argue, demonstrate that it is misleading to speak of a rule enjoining confidentiality in the doctor–patient relationship. They further argue that such cases show that questions about when one is justified in breaching medical confidentiality are to be settled on a case-by-case basis. Others, taking a different view of rules and their exceptions, contend that such cases merely show that the applicability of any rule is called into question by extraordinary circumstances.

It is of some importance to recognize that many of the most recalcitrant problems within bioethics merely reflect disputes within classical ethical theory about the nature and function of moral rules and principles. To recognize this is to acknowledge the importance for bioethics of an adequate account of such rules and principles.

Moral rules and moral justification

Three explanations of moral justification. Rules have come into ethical theory primarily through a theory of justification. A common picture of the moral justification of actions is one that relies on the following principle:

> An action A is morally justified if it is in accord with the relevant moral rules, where these rules have been derived from a set of adequate moral principles.

According to the view of justification associated with this principle, questions about the moral justification of an action are typically raised at the level of rules. When the morality of a particular action is brought into question, the discussion is shifted to the justification of the relevant moral rules and eventually to the adequacy of those moral principles on which the rules rest. This model is usually taken to be thoroughly deductive, where the transitions from principle to rule and from rule to action are criticized from the standpoint of deductive logic. The moral rules would normally be seen as derivable from the moral principles together with some empirical data.

This scheme of justification is not unfamiliar in bioethics. Thus, in order to defend a particular experimental procedure against the charge that it unjustifiably puts experimental subjects at risk, someone might claim that informed consent has been obtained from the subjects. He could then appeal to the generally recognized rule that it is morally permissible to put experimental subjects at risk if their informed consent has been obtained. (Of course, other conditions must also be met.) If he is challenged further to support this rule, he may then turn to some higher-level principle from which it can be derived. For example, he might point out that the moral principle requiring us to treat others as ends, and not merely as means, supports the rule of informed consent.

Even though this model of justification may

be familiar, however, it has been criticized for a number of reasons. First, it has been argued that it can give no adequate account of justification, since it merely puts off the really difficult questions. One such question, how the adequacy of a particular principle or set of principles is to be determined, has seemed to many moral philosophers to admit of only two possible answers: Either the principles must be derived from yet more abstract principles, which themselves would then need to be justified, or the principles are in some sense self-evident. It is further pointed out that certain well-known difficulties confront those who adopt either of the two options.

Some recent existentialist moral philosophers, e.g., Sartre, and noncognitivist moral philosophers, e.g., Hare, have suggested, however, a third possibility: Ultimate principles are neither derivative nor self-evident, but are rather the objects of free and creative choice. According to this view, the moral principles that ultimately support the superstructure of an agent's morality are chosen by him. Any attempt to deny this essentially voluntaristic account of the foundations of morality is, it is claimed, bound to be shot through with self-deception or, in Sartre's terminology, "bad faith." These views, of course, do not answer questions about the justification of ultimate moral principles, but rather rule them irrelevant. Such questions are regarded as inappropriate at the level of such foundational practical principles.

Resistance to rules and principles. Some moral philosophers have found the difficulties of justification discussed above so formidable that they have suggested that the model that engenders them must itself be faulty. They have argued that justification, instead of flowing from principles to rules to action, rather moves in the opposite direction. According to their view, it is actions that are seen quite directly as justified or not, and rules and principles are adequate only insofar as they summarize correct moral perceptions. The function of rules and principles, thus, would not be to ground or support particular judgments about the moral character of individual actions, but rather to sum up in the manner of rules of thumb our particular perceptions about the moral character of particular actions. Rules and principles that play this role may nevertheless be important in the moral life, particularly in moral pedagogy, but they lack the foundational character that they have in the more traditional picture.

A second source for much of the resistance to rules and principles as playing an important role in morality is found in the feeling on the part of many that if rules and principles are given such a role, decision making will inevitably become rigid, legalistic, and insensitive to the nuances of particular cases. The rather ill-defined contemporary view, situation ethics, has attempted to develop this criticism (Fletcher, pp. 18–22). Proponents of situation ethics characteristically emphasize the complexity of particular moral dilemmas and conclude from this that any general action guide, whether rule or principle, will either be useless in coming to a reasoned decision or force the problem into a preconceived mold that may leave out the very features that demanded our moral attention. Situation ethics, then, proposes that we make moral decisions by immersing ourselves in the particular features of a problematic situation, and by freeing ourselves as much as possible from the influence of general action guides. Our decision making must, of course, be guided by some goals or ends, and situation ethics has opted here for a concern for the welfare of others. Ultimately, situation ethics seems to be coincident with act-utilitarianism, inheriting its weaknesses as well as its strengths.

The summary conception and the practice conception of rules. Both of these objections to the centrality of rules in morality can be illuminated by considering recent discussions of two distinct conceptions of how rules relate to the actions that they regulate (Rawls; Searle). First, according to what has been called *the summary conception*, rules function to summarize our perceptions about the permissibility or obligatoriness of a certain range of actions. In this view, we suppose that the possibility of performing a certain action exists prior to the adoption or recognition of a rule regulating that action. The rule arises out of an attempt to summarize our particular perceptions of when it is permissible, obligatory, or prohibited to perform the action. A rule that regulates smoking in certain areas—e.g., where oxygen is in use—is an example of such a rule. Smoking is an action possible prior to and independently of any rules regulating or specifying the activity. For various reasons, persons might come to recognize that smoking in certain places or situations, in unventilated rooms or around asthmatic children, should be forbidden. As a result of such recognition, rules prohibiting smoking in those situations could be formulated. Here the rule functions to regulate an activity that was possible independently of the rules, and can be said to summarize our perception of the inappropriateness of a number of individual acts of smoking.

This conception of the process by which rules come to be formulated and of the relation of such rules to the actions they regulate has been contrasted with a quite different view, the *practice conception of rules*. According to this view, rules do not function to regulate independently existing actions but rather serve to create the possibility of new forms of action. Actions within the context of games are often pointed to in this regard. Within baseball, for example, the actions of striking out, walking, and stealing a base are not possible prior to the rule-governed structure that gives a special significance to movements on the baseball field. The rules of baseball do not regulate actions that were possible prior to those rules, but rather create the possibility of performing those actions. The rules of baseball both make it possible to strike out and attach sanctions to striking out.

The distinction between the two conceptions has a special significance for any discussion of ethical rules within biomedicine. While the practice of medicine is not a game with health-care workers and patients playing according to well-defined rules, there is nevertheless a large part of medicine that is highly conventional. The standard doctor–patient encounter is not unlike a stylized performance where parties on both sides have a clear notion of the parts they are expected to play. A doctor is not just someone to whom we *happen* to turn for help, as we might turn to a friend or a neighbor. The doctor is rather certified as having a certain competence and the patient's encounter with him is guided by certain clear-cut conventions. The moral rule that enjoins confidentiality on the doctor–patient relationship, for example, is not based on our recognition that for the most part medical confidentiality has better overall results than its opposite. Rather, confidentiality is one of the constituents of the doctor–patient relationship. If we removed the requirement of confidentiality, we would not have the same relationship somewhat damaged, but another relationship altogether. In this sense, the rule of confidentiality seems to fit more closely the practice conception of rules rather than the summary conception. Other rules, such as that which enjoins the doctor to be truthful to the patient or that which invokes the notion of fairness in the allocation of scarce medical resources, also seem to approximate the practice conception. There are other rules in medicine, however, that would seem to fit more closely the summary conception. Consider, for example, a rule physicians might adopt specifying the conditions under which placebos would be prescribed.

These two conceptions of rules are not intended, of course, to be competing accounts of the nature of all rules. Clearly, some rules fit the practice conception and others fit the summary conception. The central question for ethical theorists has been which conception most nearly captures the nature of moral rules, and about the answer to this question there has been little agreement. If one focuses on moral rules dealing with such highly conventional activities as promise keeping and truth-telling, or such institutionally complex areas as punishment and social justice, one is inclined to understand moral rules in accord with the practice conception. Moral rules associated with beneficence, however, would seem to push one toward the summary conception. What is clear is that those who have objected most strenuously to giving rules a central role in the processes of practical deliberation and moral justification (e.g., act-utilitarians and situationists) have also held that moral rules fit the summary conception; their opponents have emphasized the similarity of moral rules to the rules of practices. The most plausible view would seem to be that some moral rules fit one conception and others another.

Conclusion

There is no doubt that many of the most difficult problems within bioethics resist solution partly at least because of disagreement about relatively abstract issues about the nature and function of moral rules and moral principles. Recent discussions within ethical theory have provided powerful analytical tools for approaching some of these issues, but a consensus on them does not appear to be at hand. The primary lesson to be learned from much of the recent work would seem to be that one should avoid applying some simple model of the nature and function of moral rules and principles to all the contexts in which they play a role. An ultimately satisfying account of these issues can surely be given only within the framework of a comprehensive ethical theory.

WM. DAVID SOLOMON

[*For further discussion of topics mentioned in this article, see the entries:* BIOETHICS; CODES OF MEDICAL ETHICS; DECISION MAKING, MEDICAL;

LAW AND MORALITY; *and* PRAGMATISM. *While all the articles in this entry are relevant, see especially the article* SITUATION ETHICS, *which offers a different perspective on the same topic, as well as the following:* THE TASK OF ETHICS, DEONTOLOGICAL THEORIES, TELEOLOGICAL THEORIES, UTILITARIANISM, NON-DESCRIPTIVISM, MORAL REASONING, *and* RELATIVISM. *This article will find application in the entries* DOUBLE EFFECT; HEALTH POLICY; THERAPEUTIC RELATIONSHIP; *and* TRUTH-TELLING. *For discussion of related ideas, see the entries:* ACTING AND REFRAINING; JUSTICE; NATURAL LAW; *and* OBLIGATION AND SUPEREROGATION.]

BIBLIOGRAPHY

CUNNINGHAM, ROBERT L. *Situationism and the New Morality.* Contemporary Problems in Philosophy. New York: Appleton-Century-Crofts, 1970. A collection of articles surveying issues raised by the contemporary attack on moral rules.

FLETCHER, JOSEPH FRANCIS. *Situation Ethics: The New Morality.* Philadelphia: Westminster Press, 1966. A popular defense of situation ethics.

KOVESI, JULIUS. *Moral Notions.* Studies in Philosophical Psychology. London: Routledge & Kegan Paul; New York: Humanities Press, 1967. Develops idea that moral rules are involved in the application of moral concepts.

RAWLS, JOHN. "Two Concepts of Rules." *Philosophical Review* 64 (1955): 3–32. The classic discussion of the distinction between the summary and practice conception of rules.

SEARLE, JOHN R. *Speech Acts: An Essay in the Philosophy of Language.* London: Cambridge University Press, 1969. A development of Rawls's earlier discussion of the two concepts of rules, with special application to linguistic rules.

WARNOCK, GEOFFREY JAMES. *The Object of Morality.* London: Methuen & Co., 1971. A defense of the moral centrality of virtue, as opposed to rules.

WITTGENSTEIN, LUDWIG. *Philosophical Investigations.* Translated by G. E. M. Anscombe. Oxford: B. Blackwell, 1953. Reprint. New York: Macmillan Co., 1958, 1973. An important philosophical work in which the notion of rule-governed activity is one of the major explanatory tools. It has had enormous influence on contemporary philosophical discussions of rules.

III
DEONTOLOGICAL THEORIES

The diversity and complexity of ethical opinions and theories invite classification. Dichotomies are especially popular, because they promise to organize a multitude of disagreements around a fundamental one from which they are derived. One of these dichotomies, teleology/deontology, introduced in 1930 by C. D. Broad, was an improvement on Sidgwick's confusing trichotomy (into intuitionism, egoism, and utilitarianism), and caught on during the subsequent decades. It is perhaps worth noting that deontology was not always contrasted with teleology. Ironically, deontology (from Greek, *deon*, duty, and *logos*, discourse) was first introduced, as the title of one of his books, by the archteleologist Jeremy Bentham. He used the word to suggest that his work had turned the art of determining what is one's duty, into a quantifiable discipline, based on the hedonic (pleasure) calculus: "the science of morality." The current use of "deontological ethics" strongly suggests a very different conception of ethical theory.

Even its current use as one-half of a classificatory pair has not been uniformly the same. Some use the two terms as a mutually exclusive and jointly exhaustive pair, others merely as coupled names for interestingly different types of theory. Some aim primarily at satisfying the canons of dichotomous classification, others at bringing to light the profoundly important difference between two groups of ethical theories, some of which are paradigms of deontological, others of teleological, ethics.

The Hebrew–Christian paradigm

The oldest of the four major paradigms of deontology, Hebrew–Christian ethics, conceives of the ideal life for man as obedience to the will of God, or to some positive law or rule believed to express that will, whatever may be the individual's own plans or desires. The central idea here is that, while the conception of the ideal life does indeed determine what is the right thing for one to do, such determination is independent of what one would oneself desire or choose to do, or what would maximize satisfaction all round. Of course, many ethical theorists, from Kant to the present day, also stress that what one actually desires may differ from what is the right thing for one to do but nevertheless maintain that the right thing is what one would choose to do if only one were in some way idealized, e.g., if one followed one's rational part, had full information about all the consequences of the alternatives, or did not suffer from weakness of will. By contrast, in the Hebrew–Christian ideal under discussion there is no necessary connection between the right thing for one to do and what even such idealized human dispositions would dispose one to do, because the right thing for one to do is to *submit* to someone else's will. And while that will is necessarily an ideal will, not all strands in the Hebrew–Christian tradition tie that ideal to rationality.

Another idea, closely related to the previous

one, is that there are certain absolute prohibitions which the morally good person will never disobey, whatever the consequences of obedience may be. In the Hebrew–Christian tradition these absolute prohibitions comprise the deliberate killing of the innocent (including abortion), suicide, adultery, and, a favorite of recent philosophical discussions, procuring the judicial punishment of the innocent. In this view, if one were moved by the contemplation of the disastrous consequences of obeying and of the favorable consequences of disobeying, one would not be following reason or morality but would necessarily be yielding to temptation or fear. A corollary sometimes extracted is that to advocate the systematic study of what a person should do if, in certain rather special circumstances, the consequences of obeying would be particularly disastrous, is to advocate systematically corroding people's moral fiber. For by such routine thinking about the "unthinkable" one is led to construe *disobedience* not as wickedness or even weakness, but as the rational thing to do, as when someone performs an act of euthanasia because he cannot endure the suffering of a terminally ill person or kills a deformed baby because he wants to spare it the physical and mental torments it will have to suffer.

Immanuel Kant

A second paradigm of deontology is Kant's ethics. At the cost of greatly oversimplifying it, its main theses can be summarized as follows: the requirements of morality always override all other kinds of reasons for doing something when they run counter to them. Therefore, these moral requirements, expressed by the imperatives of duty, cannot be explained as merely "hypothetical imperatives," i.e., as what we ought to do because it is the means to certain ends that our psychological makeup and the circumstances in which we find ourselves cause us to have. For, since men are only partly rational and partly desiring creatures, the merely hypothetical imperatives that are based on desire always can and often do come into conflict with the imperatives of duty. In order, therefore, that these imperatives of duty should always override incompatible nonmoral (hypothetical) imperatives, men would have to be—and as partly rational creatures they actually are—subject to a desire-independent or categorical imperative, an imperative based solely on the formal nature of the law, its universality; an imperative, therefore, which unconditionally requires of every rational being that it act only on maxims it can will to become universal laws of nature and that it follow these maxims whatever the consequences of doing so. Therefore a morally upright person cannot be one who assiduously and conscientiously does what, as reason tells him, is the best means to his ends whatever they happen to be but, rather, someone with a good will, i.e., someone always prepared to do *whatever* moral duty requires. And the good life for man cannot consist in the maximal satisfaction of desires and the attainment of merely desire-determined ends. It must, rather, be a life in which people form moral communities where everyone obeys the Categorical Imperative, and everyone therefore regards everyone else as "an end in himself," and so acts on the principle of never treating anyone merely as a means to his own desire-determined ends. It must, therefore, be a life in which individual rights, freedom, dignity, self-respect, and justice are maintained.

There are in Kant's ethics at least four importantly different emphases, the first two continuous with the Hebrew–Christian conception of the ideal life, the second two incompatible with it: (1) the insistence that the ideal life for man consists in submission to a certain will or command expressed in universal imperatives that hold for everyone and contain no exceptions; (2) the insistence that unlike hypothetical imperatives moral imperatives are unconditional, and so inescapable, containing no exceptions, and therefore absolute, i.e., binding whatever the consequences, and supreme, i.e., overriding all others with which they come into conflict; (3) the insistence that the will to which a morally good person submits is not the will of another, but his own will, insofar as he is rational and so respects law; and (4) the stress on certain liberal values, such as autonomy, freedom, dignity, self-respect, and the respect for individual rights.

Broad's definition of deontology attempts to do justice to these two paradigms. In his definition, deontological ethical theories are those maintaining that there are certain types of acts—and it is known what types these are—that are "absolutely wrong," i.e., such that every particular act of these types is wrong whatever else may be true of it, in particular, whatever its consequences or those of any of the alternative acts open to the agent. If a physician is such an "absolutist" deontologist then he may believe, for instance, that killing a fetus is wrong whatever its consequences, even if the mother were to die without the operation. He must, therefore, let the mother die rather than save her life by killing the fetus, for that would be using a human being as a means only. If he is a teleologist, then he will believe that performing an abortion is the right thing to do, if and only if in the circumstances it is the act with the best consequences.

The absolutist account has two serious weaknesses. One is that it identifies deontology with absolutism, a view that many philosophers, including some who want to call themselves deontologists, regard as irrational. But deontology should not be so defined, using the name for an extreme and widely unacceptable type of ethical theory. A second weakness is that it excludes a group of philosophers, namely, the Oxford Intuitionists of the 1930s and 1940s—H. A. Prichard and his followers, W. D. Ross, and E. F. Carritt—all of whom would want to be included among deontologists and are often treated as deontologists par excellence, but who reject absolutism as well as all forms of consequentialism.

The Oxford Intuitionists

The main theses of the Oxford Intuitionists are as follows. The determination of the moral rightness or wrongness of a particular act depends on its intrinsic nature. This intrinsic nature is composed of the many intrinsic properties of an act. Some of these are "right-making," others "wrong-making," and still others morally neutral. If an act involves killing a fetus, then it has a wrong-making property. If it is also an act of saving the life of a human being, then it has a right-making property. In that it is an act of performing a surgical operation, it is morally neutral. Determining the rightness or wrongness of an act must proceed through two stages. The first is the determination of all of its right/wrong-making properties. As soon as one such property is ascertained, the act is classified as "prima facie right," "right other things equal" (say, the act of saving the life of the mother), or "prima facie wrong," "wrong other things equal" (say, the act of killing a fetus). The second stage involves determining the relative "weights" of the various right/wrong-making properties, and the corresponding "stringencies" of the "prima facie duties" of a person who is contemplating performing an act with this set of right/wrong-making properties. Determining the relative stringency of these prima facie duties establishes the "final duty," "duty sans phrase," or "duty all things considered." In answer to how someone knows what are right-making and what are wrong-making properties of an act, and which are the more stringent, Prichard and his followers claim that he knows this by intuition, that is, in a manner comparable to that in which he knows the basic truths of mathematics. If a physician intuits the wrongness of killing a fetus as being weightier than the rightness (obligatoriness) of doing whatever will save the life of his patient (and if he does not detect any other right/wrong-making property in the act), he will judge killing the fetus to be wrong, all things considered, and will regard it as his final duty not to perform the operation. If he intuits the obligatoriness of saving his patient's life to be more stringent than the obligatoriness of not killing the fetus (and if he does not detect any other right/wrong-making property in the act) then he will think it his final duty to perform the abortion.

This theory of the Oxford Intuitionists shares with the two paradigms of deontological theories already discussed the acceptance of "intrinsicalism" (as it might be called), i.e., the view that acts are right/wrong on account of their intrinsic nature. However, this theory rejects absolutism on the ground that, however horrible the intrinsic nature of an act may be, the alternatives available in the circumstances in which it is performed may be still more horrible. The question cannot be answered a priori. The considerations must be weighed when *all* the procurable information about the nature of the act is in. For this reason, intrinsicalism can reject the absolutist's denial of the relevance of the consequences either of refusing to do or of doing an intrinsically (i.e., prima facie) wrong act. But while intrinsicalism can affirm the relevance of the consequences of any act to the determination of its rightness, it must do so in one of two ways, which distinguish it from the consequentialist and the teleologist in general. Consequentialists maintain that the rightness/wrongness of an act depends solely on its consequences. They need not thereby deny the intrinsicalist's claim that the rightness/wrongness of an act depends solely on its intrinsic nature, but must insist, against the intrinsicalists, that there is only one right-making property, "optimificity," i.e., being among the acts open to the agent the one with the best consequences, and only one wrong-making property, nonoptimificity. Now, the intrinsicalist may assert merely that in addition to optimificity there are other right-making properties, and that the consequences of an act bring such other properties to light. Suppose it is discovered that the consequences of killing a certain fetus include the preservation of the mother's life and that no other course of action has this consequence. Then this information about the consequences of this act may show, not, perhaps, that it is the optimific act, but rather something else, namely, that this act has the right- (or even obligatory-) making property of being the only way of discharging a physician's duty to look after the health and life of a patient. A physician may well regard this duty as his most stringent, more stringent even than his duty not to kill the fetus or the duty to do what has the best possible consequences.

An intrinsicalist could, however, deny that optimificity is a right-making property at all, on the grounds that optimificity is not an intrinsic property of an act, since whether an act is optimific depends not solely on *its* nature and consequences but also on the nature and consequences of the *other* acts open to the agent.

Lastly, the intrinsicalist differs from the consequentialist in a third way, in that he maintains that the intrinsic nature of an act is not determined solely by its consequences but also by its relation to the past. Unlike a consequentialist, an intrinsicalist can admit, as relevant to the question of whether the mother has a right to terminate her unwanted pregnancy, the fact that the pregnancy was the result of rape.

To avoid absolutism, the Oxford Intuitionists would have to reject Broad's definition, but could define as deontological any ethical theory which holds that an act is finally right if its right-making properties are weightier than its wrong-making ones, where right- and wrong-making properties are those on account of any single one of which an act would be prima facie right (or wrong), and if it were its only such property, also finally right (or wrong). This definition retains the intrinsicalism rejected by consequentialists but does not imply absolutism, since it allows that the consequences of an act are always relevant to its rightness, though not of course the only thing relevant. This still leaves open the question, on which deontologists might disagree, whether among the right/wrong-making properties to which the consequences of an act are relevant, its optimificity is or is not to be included. In this account of the distinction, an ethical theory can contain teleological and deontological elements and so can be classified as being both. Thus, rule-utilitarianism, the theory that an act is right if it conforms to rules that pass the test of utility, can be thought of as a utilitarian modification of intrinsicalism. For it retains the intrinsicalist claim that an act is right if and only if it conforms to universal principles or rules. But it replaces the deontological claim that these principles are known by intuition by the teleological claim that they are those which satisfy the criterion of utility. Thus a "pure" teleological theory, e.g., act-utilitarianism, might hold that a given act of euthanasia was right because it was the act, of all those open to the physician, that produced the greatest happiness or least unhappiness of all those affected. A pure deontologist, e.g., a Kantian, would reject this on the ground that this was to treat the patient merely as a means to the end of maximizing happiness. But a mixed theory, e.g., rule-utilitarianism, might reject the act if it is of a kind whose general performance or permissibility would not maximize the happiness of those affected by or engaging in the practice of euthanasia.

Recent contract theories

There is a fourth group of moral theorists who want to be and usually are thought of as typical deontologists. Their work, a revival of classical contract theory, has been much discussed recently, above all that of their most distinguished member, John Rawls. These contract theories differ from their prototypes, Hobbes, Locke, and Rousseau, in at least two important respects. They use the conceptual machinery of the classical contract philosophers, but not to justify political obligation, the claim that all those subject to a municipal legal order have a moral obligation to abide by it. They use it, rather, to derive with the help of this conceptual machinery whatever they take to be the principles that obligate us, above all the principles of justice. Nor do they rest their claims about what it is right for a person to do on the existence of a basic contractual obligation to do this thing, either because he is somehow bound by an actual original contract by which the contractors have bound themselves and their heirs or because he has himself actually consented to do this thing, although perhaps only tacitly, by accepting the benefits he received from his society.

Perhaps the best way to explain how modern contractarians use the idea of the social contract is to exhibit their close affinities with both the Oxford Intuitionists and the rule-utilitarians. Rawls's contract theory retains the intrinsicalism common to intuitionists and rule-utilitarians but, in answer to the question of how we know what are these principles or rules to which right acts must conform, offers an answer different from either of these, and closer to those in the contractarian tradition, especially Kant. In Rawls's view, these principles are those universal ones which would be chosen, for adoption in a society of which the chooser is himself a member, by any rational person with a certain, somewhat idealized, psychology (e.g., purely self-interested, nonenvious) and placed in a certain favored position (behind "a veil of ignorance," which precludes his having any advantage or disadvantage relative to others making a similar choice). Rawls conceives of the ideal life for man as necessarily lived in cooperation as well as in competition with others. Hence, though fully recognizing the importance of maximizing those "primary" good things in life whose value does not depend on the manner in which their cooperative production is organized, he places especially heavy emphasis on the values arising out of the satisfaction of the proper conditions of association, such as freedom, autonomy, human rights, dignity, self-respect, and just distribution of the jointly produced primary goods.

Since Rawls regards the emphasis on these values as the essence of the tradition in which he is writing, and since he finds these values overlooked or disparaged in the utilitarian and, more generally, the teleological tradition, he readily adopts the account of "teleology/deontology" first offered by William K. Frankena. Like him, Rawls takes teleology as the basic term; defines as teleological any theory that holds: (1) that the good can be defined independently of the right, and (2) that the right is defined as what maximizes the good thus independently determined; and then defines as deontological any theory that is not teleological. Although this definition has several weaknesses—for example, it turns into a deontological theory Moore's later ideal utilitarianism because it does not satisfy (1), and Christian salvationism because it does not satisfy (2)—it is considered by many moral philosophers in the United States and Great Britain the best now available.

KURT BAIER

[*Directly related are the other articles in this entry:* TELEOLOGICAL THEORIES, UTILITARIANISM, THEOLOGICAL ETHICS, *and* OBJECTIVISM IN ETHICS. *See also:* ABORTION, *article on* CONTEMPORARY DEBATE IN PHILOSOPHICAL AND RELIGIOUS ETHICS; DEATH AND DYING: EUTHANASIA AND SUSTAINING LIFE, *article on* ETHICAL VIEWS; INFANTS, *article on* ETHICAL PERSPECTIVES ON THE CARE OF INFANTS; *and* OBLIGATION AND SUPEREROGATION.]

BIBLIOGRAPHY

ANSCOMBE, G. E. M. "Modern Moral Philosophy." *Philosophy* 33 (1958): 1–19.
BENNETT, JONATHAN. "Whatever the Consequences." *Analysis* 26 (1966): 83–102.
BENTHAM, JEREMY. *Deontology.* London: Longman & Co.; Edinburgh: William Tait, 1834.
BRANDT, RICHARD B. *Ethical Theory.* Englewood Cliffs, N.J.: Prentice-Hall, 1959.
BROAD, CHARLES DUNBAR. *Five Types of Ethical Theory.* London: Kegan Paul, Trench, Trubner & Co.; New York: Harcourt, Brace & Co., 1930.
FOOT, PHILIPPA. "Morality as a System of Hypothetical Imperatives." *Philosophical Review* 81 (1972): 305–316.
FRANKENA, WILLIAM K. *Ethics.* 2d ed. Englewood Cliffs, N.J.: Prentice-Hall, 1963.
PRICHARD, HAROLD ARTHUR. *Moral Obligation.* Oxford: Clarendon Press, 1949, 1968.
RAWLS, JOHN. *A Theory of Justice.* Cambridge: Belknap Press of Harvard University Press, 1971.
ROSS, W. DAVID. *The Foundations of Ethics.* Oxford: Clarendon Press, 1939.
SMART, J. J. C., and WILLIAMS, BERNARD. *Utilitarianism: For and Against.* Cambridge: University Press, 1973.

IV
TELEOLOGICAL THEORIES

Teleological ethics (from Greek *telos*, end, and *logos*, discourse) is a term currently used mainly to refer to a group of ethical theories that are thought to share certain characteristics on account of which they are importantly different from those of another group, called "deontological." Theorists of both types agree that it is their task to provide us with guidelines for the good, or even the best possible, life for man, and they agree further that this can be done only by finding answers to the two fundamental ethical questions, "What is good?" and "What is to be done?" Teleologists then part company with deontologists on the issue of the role of the good in ethical theory. Teleologists regard "What is good?" not only as the logically prior of the two questions, which must therefore be answered first, but also attempt to ascertain what should be done by discovering the best way of attaining the good. Different teleologists then disagree with one another on three main problematic issues: (1) exactly what is good; (2) whose good is to be taken into account; and (3) exactly what sorts of guidance can ethical theory provide for the individual in his deliberations about what is to be done.

What is good? Teleologists give a great variety of answers to their primary question, "What is good?" We can distinguish between *monistic* answers, which contend that there is only one good, e.g., happiness, satisfaction of desire, self-realization, or perfection, and *pluralistic* answers, which include all or most of the goods embraced by monists and more, e.g., knowledge, friendship, self-respect, freedom, etc., and which maintain that life is the better the more of these goods it holds or the more nearly it holds them in the proper proportion.

We can also distinguish between *objective* (or intersubjective) answers, which claim that the good is the same for all (e.g., Aristotle's happiness in accordance with virtue) so that those who pursue other ends must be considered misguided, and *subjective* answers, which say that two people's conceptions of the good life *for them* may differ, one seeking, say, power, the other contentment, yet neither of them being mistaken.

We can further distinguish between those answers that construe the good as *nonquantifiable*, e.g., salvation, which one must either win completely or lose altogether, and those that make the good *quantifiable*; in some such answers, e.g., when the good is conceived as pleasure, happiness, or power, the good is with no ideal limit other than maximization; in other answers of this type, e.g., when the good is conceived as self-realization, the good has a limit, and then one can approximate and perhaps even reach it, but one cannot have indefinitely more of it.

The last difference to be noted among teleological answers to this question is that between versions of the good that are *forward-looking* or *consequentialist*, such as future happiness, and *nonconsequentialist* versions of the good, which give the future no greater weight than the past and allow consideration of the future in relation to the past. Such *nonconsequentialist* views could consider retribution a good, whereas a *consequentialist* view would tend to stress reform or deterrence as an objective of punishment. It would seem that *nonconsequentialist* teleological theories, especially if pluralistic and objective, could with equal justification be called deontological.

Whose good? Concerning the second main question, whose good is to be considered in determining what is the right thing for someone to do, we can distinguish two main types of answers, *person-relative* and *person-neutral*.

Person-relative answers conceive of the good as what is welcome from a given point of view, e.g., rain from the farmer's but not the vacationer's. One's own good is what is welcome from one's own point of view. How good one's own life is depends on how much in it is welcome, how much unwelcome from one's own point of view. We can distinguish four major classes of such answers, depending on whose good it is that is taken into account in determining what is the right thing to do.

The first, ethical egoism, says that only the agent's own good should be considered. In this class belong hedonistic and eudaimonistic egoism (in which the good is conceived as quantifiable and without ideal limit other than maximization), personal perfectionism (in which the good is conceived as quantifiable but with an ideal limit), and salvationism (in which the good is conceived as nonquantifiable).

The second type of such answers, which might be called "ethical elitism," says that only the good of the elite is to be considered. In another of its forms, perfectionism would be an instance of this type. Thus, in Nietzsche's view, the right thing to do is what maximizes human excellences. He argues that any individual's life can retain the deepest significance only if he lives for the good of the rarest and most valuable human specimens, that mankind must therefore strive continually to produce great men and make possible the realization of their excellences. Society must therefore set aside the claims of the less gifted to have their good taken care of.

The third, which might be called "ethical parochialism," says that only the good of the agent's appropriate in-group (e.g., class, sex, party, church, race, country) should be taken into consideration. This view can, of course, be combined with any type of answer to the teleologist's primary question, "What is good?"

The fourth, "ethical universalism," insists that the good of mankind as such must be taken into account. To this fourth type belongs utilitarianism, whose most popular versions combine an answer of this fourth type to the second main question ("Whose good?") with some version of a quantifiable conception of the good in answer to the first question ("What is good?").

It should be noted that though all these answers are *person-dependent*, ethical elitism and utilitarianism are *agent-neutral*, while ethical egoism and parochialism are *agent-relative* theories. That is to say, in the case of the former two, the class of persons whose good is to be taken into consideration is not, as with the latter two it is, defined by reference to the agent, i.e., the person for whom the morally right action is to be ascertained. Thus, in dealing with the problem of triage, a utilitarian physician must ignore questions of how his allocation of scarce lifesaving drugs will affect the physician's own or his in-group's good, but must allocate them so as to promote the greatest good of the greatest number of persons (whether they are valuable or not); an ethical elitist must promote the greatest good of the most valuable persons (however many or few they may be); an ethical parochialist must promote the greatest good of members of his in-group (whatever their number or merit); and an ethical egoist must promote his own greatest good. And all of them must do so regardless of the way in which this will affect the good of those who fall outside the purview of the definition.

Person-neutral answers, by contrast, conceive of the good as that whose existence, occurrence, or prevalence is an absolutely, i.e., *person-neutrally*, good thing. In this view, the question of *whose* good is the existence, occurrence, or prevalence of something is at best a secondary, at worst an illegitimate question. Such views can admit that developments favorable from a given point of view will tend to appear (absolutely) good to those who have adopted that point of view, but when the bias caused by the special perspective of that point of view is corrected, the appearance of good may then be revealed as deceptive. Writers such as Hegel and Marx, but oddly enough also G. E. Moore, belong in this category.

For illustration let us consider the case of A, a surgeon, who is a sadistic hedonist. He believes that pleasure and pleasure alone is good in itself, and he derives pleasure from, among other things, watching people suffer. In the person-relative answer, the amount of pleasure A would derive from watching B's pain during and after surgery would necessarily count in determining whether it is right for A to perform such surgery on B. If A derives very great pleasure, that could outweigh, in the calculation of whether A ought to operate on B, the misery inflicted on B by the surgery. By contrast, in the person-neutral view, the fact that A's pleasure is derived from watching B's pain may disqualify A's pleasure from being taken into account at all. It may even make it the case that the occurrence of the operation, if A enjoys it, is (absolutely) worse than if A were to suffer in sympathy with his patient.

The most widely discussed of these theories is undoubtedly utilitarianism. Utilitarians typically conceive of the good as some kind of (positive or negative) "payoff" attaching to things that one brings about or that befall one. These payoffs, sometimes called "utiles," are thought of as a common unit of measure, permitting an evaluative (hedonic, felicific, satisfaction, etc.) calculus, which would enable one to conduct one's moral life in a rational manner comparable to that of the efficient businessman. As the good businessman maximizes profits, so the moral person maximizes the net balance of good over bad (pleasure over pain, happiness over unhappiness). One of the major difficulties with this attractively simple idea has been to reduce the payoffs yielded by different courses of action to an empirically verifiable common denominator, both in the case of a single person and, even more so, in the case of a whole society. The main disagreements about how the good of mankind as a whole is to be conceived revolve around what is meant by the greatest happiness or pleasure of the greatest number.

One issue here is whether what is to be compared is the *sum total* of happiness or the *average* happiness resulting from alternative policies. The "sum-total" view would, for example, advocate an increase in population if that increased the total of the balance of pleasure over pain, even if that balance was decreased in every individual life, while the "average" view would advocate it only if it raised the average amount of pleasure or happiness. A second disagreement is about whether the principle of utility is a test of the moral acceptability of the acts of individual agents or of the institutions of a society. The classical utilitarians (Bentham, Mill, Sidgwick) were more concerned with improving the social order than with improving the morals of private individuals. Bentham, for example, divides ethics into private ethics and legislation, the former concerned with providing principles and rules for use by private individuals, the latter with providing those to be used by legislators. In one plausible interpretation of Bentham, both were to use the same principle (which Bentham calls the principle of utility), namely, of acting so as to maximize the happiness of those "under their direction." In other words, private individuals were to use the principle of maximizing the agent's happiness, legislators that of maximizing the happiness of those subject to the law. Bentham therefore was, as far as private ethics is concerned, an egoist rather than a utilitarian, as these terms are usually defined. The individual citizen is bound by the law only in the sense that the sanctions attached to it make it prudent but not morally obligatory for him to follow it. If the legislator follows the principle of maximizing the happiness of those subject to the law, then the law so weights the subject's choice that if they follow the principle of maximizing their own happiness, they will be maximizing the sum total of society's happiness. Sidgwick by contrast appears to have accepted the moral obligatoriness of the law, at least to the extent to which the laws themselves satisfied the principle of utility, i.e., if they were designed so as to achieve the greatest net balance of good over evil in the society they govern. Later utilitarians (e.g., G. E. Moore) concentrate on what Bentham called private ethics, and devote little attention to the problem of whether or not such social institutions as marriage, private property, and the criminal law impose a moral obligation to do what the rules of these institutions require.

Guidance for individual deliberation. Our third question concerns the extent to which ethical theory can provide guidance to the individual in his deliberation about what to do. This institutional or social problem of ethics arises in the recent controversy between so-called act-utilitarians and rule-utilitarians. Act-utilitarians maintain that the only right-making property is "optimificity," having consequences at least as good as those of any alternative acts open to the agent, and the only wrong-making property is "nonoptimificity." Utilitarians typically interpret optimificity in terms of the hedonic (pleasure) or felicific (happiness) calculus. They do not for instance admit that an act's being, say, an act of craniotomy (the cutting or crushing of the fetal skull to reduce its size for removal when normal delivery is not possible) makes it necessarily a wrong act, or its being required by the law necessarily makes it right. Thus, although Bentham recognized the importance of the function of compulsory social rules and of the difference between acts of legislators and of private individuals, he was nevertheless an act-utilitarian. That is, he held that the rightness of a particular act, whether of a legislator or of a private individual, was determined solely by its optimificity, and not at all by its conformity with certain moral rules.

A related issue concerns the kinds of rules a teleologist can admit in his system. The act-utilitarian can, like Bentham, allow that a society should have compulsory rules, such as laws, but he must insist that, as far as determining the rightness of particular acts is concerned, such compulsory rules are no more than rules of thumb, i.e., rough indicators of what it would be right for an individual to do. On this view, private individuals admittedly will, by and large, do better by following these rules than by working out the optimific act on their own. But it is conceivable, perhaps likely, that an individual will occasionally have adequate reason to think that the optimific act will involve breaking the law. If a physician has information that a particular act of euthanasia (or suicide, etc.) would have the best consequences, then he must regard these particular acts as right, even if they constitute a breach of a compulsory social rule or law whose existence is desirable from the point of view of utility. On this, contemporary act-utilitarians thus agree with Bentham. But they differ from him in insisting that the optimificity of a particular act is to be construed, not in Bentham's way as the greatest happiness of the agent, but that of the greatest number of persons.

Rule-utilitarians, by contrast, contend that the rightness of an act is determined by its conformity with all the compulsory social rules that pass the test of utility. Different rule-utilitarians differ from one another on precisely how this is to be interpreted; whether the rules are actual or ideal rules, whether their utility is the utility of everyone actually following them, or the utility of their being accepted as binding in the community. The debate is still in full swing, and no generally agreed results have so far emerged. It is not even generally agreed whether there is any plausible form of rule-utilitarianism that does not collapse into, that is, does not yield exactly the same moral precepts as, act-utilitarianism.

Among the many objections to teleological theories, one deserves special attention, namely, that they imply that the end justifies the means. It is hard to formulate this general objection clearly, but it appears to make at least two distinguishable accusations. One is that teleological ethics, by setting up one end as the good and therefore as so important that its attainment has an overriding claim on us, and by defining the right as whatever brings about the good, encourages us to ignore the rights of those who stand in the way of our attaining that end. If our end is the Nietzschean maximization of human greatness or the conversion of the whole world to our favorite ideology or the control of fertility, the means needed to accomplish such an aim may well run roughshod over many rights and needs whose recognition and satisfaction would stand in the way of that end. Plainly even the more modest end of maximizing pleasure or happiness is open to this objection, if that maximization is interpreted, in the manner of classical utilitarianism, as bringing about the greatest total balance of happiness or pleasure over unhappiness or pain. For that end may actually require the violation of the rights and the neglect of the needs of some members of a community. In the field of population policy, for instance, the maximization of happiness requires an increase in the number of people whose happiness is summed, even if the happiness of every member is thereby decreased, as long as the total is increased. Similarly, in the area of allocation of scarce resources, it may require (to maximize the sum total of happiness) that more of the available resources be allocated to those who already have plenty in order to eliminate the sufferers living at the minimal level of subsistence. This would hold equally in the international and the national sphere. On this point, too, the debate continues.

The second accusation against teleological theories is that they espouse "consequentialism," the principle that consequences alone determine the rightness of an act, i.e., that the only right-making property of an act is its optimificity. This involves the denial that any acts other than optimific ones are intrinsically right, and in particular the denial of absolutism, the contention that there are acts of certain types that are wrong whatever the consequences. Here we must distinguish between the weaker view, that the consequences of a particular act are never irrelevant to its rightness, and the stronger view that certain types of acts are absolutely wrong, i.e., wrong whatever the consequences. If rule-utilitarians can be classified as teleologists, then there are some teleologists who are not open to the milder version of this accusation, for rule-utilitarians do of course allow that acts other than optimific ones can be right. However, they are open to the stronger accusation, for they reject absolutism.

KURT BAIER

[*While all the articles in this entry are relevant, see especially* UTILITARIANISM, *which builds on this article,* DEONTOLOGICAL THEORIES, *and* NATURALISM. *Also directly related is* OBLIGATION AND SUPEREROGATION.]

BIBLIOGRAPHY

BENTHAM, JEREMY. *An Introduction to the Principles of Morals and Legislation.* Oxford: Clarendon Press, 1907.
BRANDT, RICHARD B. *Ethical Theory.* Englewood Cliffs, N.J.: Prentice-Hall, 1959.
LYONS, DAVID. *Forms and Limits of Utilitarianism.* Oxford: Clarendon Press, 1965.
MILL, JOHN STUART. *Utilitarianism, Liberty, and Representative Government.* New York: E. P. Dutton & Co., 1910.
MOORE, G. E. *Principia Ethica.* Cambridge: University Press, 1903.
SIDGWICK, HENRY. *The Methods of Ethics.* 7th ed. London: Macmillan & Co., 1907.

V
SITUATION ETHICS

This article offers a very brief account of situation ethics as a morality without rules and how it would apply to moral questions about biological and medical choices.

Situation (or "contextual") ethics is best classified as act-utilitarianism. Recent philosophical exponents include J. J. C. Smart in *Outline of a System of Utilitarian Ethics,* Kai Nielsen in *Ethics Without God,* E. F. Carritt in *A Theory of Morals,* and A. C. Garnett in *Ethics: A Critical Introduction.*

Act-utilitarians determine what is right by electing that course of action which offers the most beneficent consequences or greatest utility in each act, each particular situation. On the other hand, rule-utilitarians decide what is right by following preconceived rules (as advocated by R. B. Brandt, A. C. Ewing, Kurt Baier, and, in an earlier day, by Bishops Butler and Berkeley). It should be noted, however, that some rule-utilitarians adumbrate only the most general norms, such as "do good and avoid evil" and the Golden Rule, which are not rules of specified conduct. This was the case, for example, with Hastings Rashdall (*A Theory of Right and Wrong*) and Henry Sidgwick (*Methods of Ethics*).

Most religious ethicists elaborate a divine-command theory of morality, combined with a special or "general" revelation of the rules of God's imperative will. It should not be surprising, therefore, that situation ethics finds little favor in religious circles. Nevertheless, there are a few theological advocates in modern times, including Eberhard Griesbach, Paul Tillich, Dietrich Bonhoeffer, Joseph Fletcher, Paul Lehmann, J. A. T. Robinson, James Pike, and Helmut Thielecke.

Frankena ventured to call situation ethics "modified act agapism," because one exponent (the present writer) has chosen to set it within the framework of *agape*, the Greek word meaning "love" or concern for persons (Frankena, p. 3). By "modified" Frankena means (quite correctly) that although situationism is act rather than rule ethics it still has a place for rules of thumb, as long as they are not treated as always-obliging rules. Thus it is possible to say that ordinarily it is right to tell the truth, yet in some situations it would be wrong if telling the truth, on balance, had "unloving" (i.e., bad) consequences.

When norms are either universalized or absolutized they are perverted into what logicians call the material fallacy of faulty generalization. Cognitively, act ethicists take note of what is most frequently the right action in similar situations, as judged by the benefits gained, and this offers a meaningful rule of thumb. They might say, for instance, "We ought to get the patient's consent to do surgery, especially amputations," but this would be a guideline only—not a moral law or rule. If the patient lay in coma or was psychotic or a small child, they might act without consent.

Love being understood as good will or beneficence, the ethics of love is equivalent to utilitarian ethics. Philosophers commonly identify the two. Mill expounded this equivalence in his *Utilitarianism:* "To do as you would be done by, and to love your neighbour as yourself, constitute the ideal perfection of utilitarian morality" (p. 204). Fletcher has contended that love and justice are the same (1966, pp. 87–102). Mill showed that utility and justice are the same (pp. 226 ff.). In this way love and utility are commensurate. Utility, expressed in nonagapistic language, means the greatest possible preponderance of good over evil, calculated as to remote as well as immediate consequences; i.e., it aims at a net gain of *nonmoral* good. Human happiness is thus the ethical measure of right actions, not moralistic or legalistic rectitude. Agapistic modes of exposition, as in Christian ethics, are complementary but not necessary to situation ethics.

"Net gain" is a key concept here. Act ethics is relativistic; it appreciates the importance of coming to terms with the finite human condition —the need, that is, in all decision making to weigh and choose between competing values. Rarely if ever do we have a chance to select between purely good and evil alternatives. This "gray area" is an elementary parameter in ethical investigation.

Moral agents are in effect choosers in significant situations, trying to choose a course between conflicting values that yields a balance of benefit or, in a classical phrase, the "proportionate good" (Fletcher, 1970). This particular good is conceived qualitatively, not just quantitatively; some moral choices may favor the quality of a nonmoral good over its quantity; sometimes quantity can actually subvert quality —as in the case of resuscitating a patient whose cerebral function is lost irretrievably. So-called cost-benefit judgments are for this reason the main substance of moral decisions. This is the relativity of ethics.

The weight of tradition is on the side of making moral decisions or "forming conscience" according to rules. The conventional wisdom of the past, which is still at work in some parts of the world today, has just assumed that righteousness or rightness means being faithful to set norms. This is ethical legalism, basing good conduct on obedience to moral laws. Some of its exponents have actually held that many acts are morally wrong even though their consequences are good (Ramsey, p. 40).

Conventional rule ethics or legalism was sometimes linked with divine commands as an ultimate sanction for the rules. This is not, however, the case with rule-utilitarianism; its norms are formulated empirically and without dogmatic or revelational claims. We should all be prompt to acknowledge the great debt we owe to the idealism and ethical concern that have gone with the traditional morality, even though some hideous things were done in its name—and still are. Its defenders give ground reluctantly and painfully but meanwhile the maturing debate throws valuable light on the nature and function of ethics.

The deontological–teleological typology of ethical theories in the traditional literature is increasingly unacceptable. Conventional discourse in an earlier era understood "duty" (Greek *deontais*) as obedience to rules. But doing one's duty in some situations might require a moral agent to depart from rules sometimes, in order to realize the greatest good possible. In short, act ethics can be quite as dutiful, as responsive to obligation, as rule ethics. In act ethics one's duty is pragmatically to get the best possible results, and one's goal (*telos*) is to fulfill one's duty to realize beneficial results.

Thus duty means to do what is best in the

situation, which may mean either obeying or flouting any given rule. It is this flexibility that invites the rule-committed to charge act ethics with holding that "the end justifies the means." The only adequate retort is to bow to the charge, for nothing we do (no means) is morally justifiable except by the good end it seeks. (Random behavior is "meaningless.") Ethically, actions as such are not self-validating. What gives a "means" its *meaning* is the end to which it is directed. On this reasoning one could rightly engage in civil disobedience, could not be an absolute pacifist, might violate one or the other of the Ten Commandments, could fail to keep a promise or fulfill a contract, and so on.

Except for those who persist in thinking of duty as faithfully following rules, the traditional division between duty and goal ethics (deontology and teleology) is empty, because each kind of these systems presupposes the other. This makes it a distinction without a difference. All ethics of whatever kind predicates obligation or duty. The act–rule distinction, on the other hand, holds up much more adequately.

In sum, situation ethics is a utilitarian or consequentialist ethics, motivated by concern for human well-being, decisionally flexible in method, and guided in its judgments by the greatest good realizable rather than by adhering to prefabricated norms or moral rules.

How, then, does this mode of ethical analysis work out in bioethics, in biological and medical policy and decision making?

It should be obvious that situationists would not either approve or disapprove things like abortion or artificial insemination *in toto*, as a class or category of human acts. They would not deny patients their freedom to be choosers (to be moral agents) nor condemn them when they exercise it, although realistic counsel, including medical advice, might sometimes cut across a patient's wishes. Each case would be weighed on its own merits, clinically and consequentially. Situationists could not lay down such blanket opinions as "Sterilization is wrong." The closest they could come might be, "In this case sterilization would be wrong." Their judgments would hang on the foreseeable results medically, psychologically, and socially.

Some moralists believe that fetal life possesses human rights (e.g., a right to be born) but such a belief is not an ethical proposition; it is a religious teaching or metaphysical speculation with a moral entailment logically, viz., a taboo on abortion. No situationist could adopt universal negatives of this sort. (A few antiabortion moralists, however, have been consequential enough to use the principle of proportionate good to justify therapeutic abortion, to save patients' lives in certain kinds of cases—for example, tubal pregnancies and uterine cancers.)

Of the many ethical issues posed by the "life sciences" (bioethics) those having to do with life itself, as distinguished from health, are perhaps especially searching. Questions as to whether we may let a patient die in terminal illnesses without further medical intervention ("indirect euthanasia") and even "direct euthanasia" and suicide are appropriate to act ethics. Such solemn decisions might be made where the good to be gained outweighed counter considerations. Life itself in act ethics is a relative rather than absolute value—a value to be seen in relation to other values, at most only *primum inter pares* (first among equals)—and therefore it could be decided occasionally that the benefit balance appraisal favors termination. Situation ethics fits quality-of-life appraisals of human initiatives in death and dying, but not sanctity-of-life prohibitions (Fletcher, 1974, pp. 156–160).

This same situational decision making would be applied to fetal research and experimentation, *in utero* and *ex utero*. Even if it was believed that fetuses have human rights (which would be established by nonethical reasoning), it could still be decided that risk of damage or death in some circumstances is justified by proportionate good. Only if fetal life was declared categorically "untouchable" would fetal research be condemned. The same relativity applies to decisions about abortion and terminating defective newborns who are afflicted, for example, with a severe cerebrospinal defect (such as spina bifida with a myelomeningocele and hydrocephaly).

A wide spectrum of bioethical issues invites analysis. Transplant medicine poses some issues, as do proposals to devise chimeras (man–animal fusions) and cyborgs (man–machine prostheses)—each of which could have medical and social utility. Situation ethics asks what is beneficial rather than what is "natural" or any other criterion. In matters of experimentation, for example, any act ethicist would be open to the possibility of approval for such steps as fetal thymus for production of homologous antibodies (in cancer therapy or control of transplant rejection), or the maintenance of brain-dead cadavers as a source of rare hormones or viable organs for treatment of chronic diseases such as renal failure. But a case of *preponderant nonmoral benefit* would have to be made for each such procedure. Many other illustrative problems could be cited.

Physicians sometimes speak of situation ethics as "clinical" in the sense of looking at each case on its own merits. Through training and practice they are familiar with the importance of diagnosis and prognosis case by case. They know that the "laws" of medical science do not apply categorically to all patients with a given complaint; they are accustomed to judge what is best not so much by general principles as in their light. Being case-centered, they reject any notion that medicine should be practiced according to moral rules. Guidelines they welcome, yes, but not categorical moral rules.

Many emerging bioethical questions take shape in fields such as behavior and genetic control, but in all instances situation ethics would put aside moralistic–legalistic generalities and "class action" rules, asking instead the only serious ethical question, namely, "What is the best thing to do in *this* case, for this particular person or problem?" Bioethics directed to acts rather than rules makes good sense to biologists and physicians.

JOSEPH FLETCHER

[*While all the articles in this entry are relevant, see especially the articles* RULES AND PRINCIPLES, DEONTOLOGICAL THEORIES, TELEOLOGICAL THEORIES, *and* UTILITARIANISM. *For further discussion of topics mentioned in this article, see the entries:* ACTING AND REFRAINING; FREE WILL AND DETERMINISM; JUSTICE; LAW AND MORALITY; NATURAL LAW; OBLIGATION AND SUPEREROGATION; PRAGMATISM; *and* RIGHTS. *This article will find application in the entries* ABORTION; BIOETHICS; CARE; CIVIL DISOBEDIENCE IN HEALTH SERVICES; DECISION MAKING, MEDICAL; FETAL RESEARCH; HUMAN EXPERIMENTATION; LIFE; *and* TRUTH-TELLING.]

BIBLIOGRAPHY

COX, HARVEY. *The Situation Ethics Debate.* Philadelphia: Westminster Press, 1968.
FLETCHER, JOSEPH. *The Ethics of Genetic Control.* Garden City, N.Y.: Anchor Press, 1974.
———. *Situation Ethics.* Philadelphia: Westminster Press, 1966.
———. "Virtue Is a Predicate." *Monist* 54 (1970): 66–85.
FRANKENA, WILLIAM K. *Ethics.* 2d ed. Englewood Cliffs, N.J.: Prentice-Hall, 1973.
MILL, JOHN STUART. *Essential Works.* Edited by Max Lerner. New York: Bantam Books, 1961.
RAMSEY, PAUL. *Fabricated Man.* New Haven: Yale University Press, 1970.

VI
UTILITARIANISM

We have to distinguish between utilitarianism as (1) a *metaethical* view about the *meanings* of the moral words (e.g. that 'right' means 'utility-maximizing'), and (2) a *normative* view about what it is right to do (e.g. that it is always right to do what will maximize utility). Most of the famous utilitarians such as Bentham and J. S. Mill, though they did not always sufficiently distinguish the two questions, were more interested in the second. If interpreted in the first way, utilitarianism is a kind of Ethical Naturalism (provided that we make the somewhat questionable assumption that 'utility' itself is a descriptive, not a normative concept); if in the second, however, it may be embraced by thinkers holding widely differing views about the meanings of the moral words. G. E. Moore, the antinaturalist, was a utilitarian of a sort, and it is also possible for an adherent of Non-Descriptivist Ethics to be a utilitarian.

As a theory about what it is right to do, utilitarianism is a type of *consequentialism*, holding that the morality of actions is to be judged by examining their consequences. But it is important to understand that the word "consequences" here is to be taken in a broader sense than usual: the consequences of an act, in this sense, include all the differences made to the history of the world by the fact that the act is performed. They thus include the performance itself of the act; any attempt to draw a line between the act and its consequences, and to say that in judging morally we have to pay attention to the act in itself and ignore the consequences, is bound to confuse the issue by taking "consequences" in a narrower sense than a careful consequentialist would. If I give a patient an overdose of some barbiturate, a strict division between act and consequences might lead us to say that my giving him the dose is the act, and his death the consequence. A careful consequentialist, however, would include both under "the consequences". If it be objected that this is a misuse of the word, the consequentialists do not need to wince, because it was their opponents who invented the term "consequentialist" and thus caused the confusion.

It should be noted that it is not possible to draw a line between act and consequences in such a way as to separate what is morally relevant from what is not. In the example just given it is relevant that the patient dies, which is a consequence, even in the narrowest sense, of the giving of the dose. The fact that we commonly say, not "He gave him the dose, with the consequence that he died", but "He killed him by giving him the dose", shows that what on one description appears as a consequence of an act can on another description appear as an act.

There are, however, two other distinctions which *are* morally relevant, and which are often confused with that between act and consequences. The first is between different sorts of consequences, in the wide sense. It is usually thought that certain sorts of consequences are morally relevant, others not. In the above example, some have held that to bring about the patient's death (a consequence of giving him the dose) was morally wrong, and a thing that ought not to have been done, but that a further consequence, that the patient suffered no more pain thereafter, ought not to be considered relevant when judging the morality of the act. Whether it ought can be debated; but it is confusing to represent the debate as one between consequentialists and anti-consequentialists, since for both parties it is the relevance or irrelevance of these respective *consequences* that is in question. In general, nearly everything that anti-consequentialists put in terms of a distinction between act and consequences can be put more clearly in terms of one between different consequences (in the wide sense) of an act.

The second morally relevant distinction is that between (1) the narrower class of consequences intentionally brought about and (2) the more comprehensive class (which includes [1]), of all the consequences which flow from an act, whether they are intended or not. It is usually held that, in judging the morality of the agent, it is the intended consequences of his act that have to be considered, and that unintended consequences do not justify blame. The fact that we blame people for the unintended consequences of their acts when they were negligent is no real exception to this rule: what is blameworthy is the omission to take care to inform themselves of the probable consequences. It is necessary also to distinguish between *direct intention*, of which desire that the intended thing should happen is a necessary condition, and *oblique intention*, where we directly intend A, and in pursuit of this intention knowingly cause B, obliquely intending it, though not desiring it. The surgeon who terminates a pregnancy may directly intend to save the mother's life but only obliquely intend the death of the fetus. Both kinds of intention are relevant to the morality of the agent, and discussions of the principle of double effect become clearer if the distinction is observed.

In this connection it is important to distinguish, also, between judgments about the moral rightness of the act and those about the moral worth of the agent. The use, in theological discussions about the morality of acts, of expressions such as "sinful" obscures this distinction. This word applies most naturally to the agent. If so applied, the agent's intentions and motives will be highly relevant; the same act done with one intention or one motive may justify the imputation of sin to the agent, but done with another, not. Theologians are commonly interested in what will happen on Judgment Day; and for deciding the final disposal of souls it is important to know what state they are in.

Utilitarianism, however, has seldom been advocated primarily as a way of judging agents. If it is so used, it will have to pay as much attention to motives and intentions as any other theory. Good motives and intentions will be benevolent ones, arising from the desire to do what is best for the people affected by our acts. As we shall see (and contrary to what many anti-utilitarians think) this may engender a respect (because of the good which comes therefrom) for the moral principles which anti-utilitarians too revere. But utilitarians have generally been less interested in the question "Will this act put me in the category of good or of sinful men?" than in the question "Ought I now to do this act? Would it be the right thing to do?" and they have said that we must answer this question by examining the consequences.

Here, however, the further difficulty arises that we can never know for sure what the consequences of an act will be. We have to rely on merely probable predictions even for those consequences which we have time to consider; and we never have time even to consider the great majority of the consequences, stretching out to the end of history. To meet this difficulty, we have to make another distinction, different from but related to one already made, namely between a *post eventum* judgment which might be made, say, by the Archangel Gabriel at the end of history, in possession of all the facts, and that made by an agent when he is wondering, in a state of fairly deep ignorance of the future, whether he ought to do a certain act. One of these judgments is made in full knowledge of the act's consequences; the other is made in ignorance of them, and often in circumstances in which it is only too easy, by self-deception, to conclude that the consequences of a certain act would be for the best for all those affected taken as a whole, when what really moves us (only we do not realize it) is that they would be for the best for *ourselves*, never mind about the others.

A utilitarian who does not allow for this source of error may find himself thinking, on what seem to him good utilitarian grounds, that he morally ought to do something, which in the event proves not to have been for the best; or a non-utilitarian may do the opposite, sticking to some well-established moral principle without calculating the consequences, and it may turn out to have been for the best. Thus a utilitarian may give a lethal dose to a suffering terminal patient, convinced that he is acting for the patient's and for everybody else's good, when really it is his own convenience that moves him; and a non-utilitarian may refrain from a similar act because he does not think that innocent life ought to be taken; and it may turn out that the patient, in what life remains to him, gets (say through the extra time that he has with his wife and children, to reconcile them all to the loss of parting) some good that outweighs the pain which he suffers, and which in any case can be controlled by careful medication. Anti-utilitarians have argued, on these grounds, that there are utilitarian reasons for not being a utilitarian; but this is too hasty. It follows only that utilitarian reasoning ought not to be too briskly applied, and that a more careful utilitarian, recognizing the pitfalls mentioned above, may think it safest, even on utilitarian grounds, to stick to a well-tried moral principle, as being most likely to be for the best.

This line of thought enables us to bypass the well-canvassed distinction between act- and rule-utilitarianism. Rule-utilitarianism, of which there are many varieties, was invented in order to defend utilitarianism against the objection that it leads to conclusions which to ordinary moral opinion seem outrageous. *Act-utilitarianism* is the view that it is individual acts which are to be judged according to their utility: i.e., the act which ought to be done is that which will yield the most utility (ignoring cases where there are two acts which tie for this prize). *Rule-utilitarianism* (in one of its forms) is the view that the act which ought to be done is that which is prescribed by the set of principles which has the highest acceptance-utility (i.e., whose general acceptance in society will yield more utility than the general acceptance of any other set). Since, it is assumed, the general acceptance of a set of moral principles which includes most of the well-established ones would have better consequences than the acceptance of the principle that each individual act should be judged on grounds of its own utility without bringing in any other principles, this would be a way of giving a rule-utilitarian justification for conforming to the well-established principles, and thus reconciling utilitarianism (of this rule-utilitarian sort) with received opinion.

The maneuver depends on the assumption that the principles which are well-established in a given society are actually the ones with the highest acceptance-utility; otherwise dilemmas will arise with which there is no time to deal here. But the move to rule-utilitarianism is strictly speaking unnecessary. Given the uncertainties and pitfalls involved in moral decision, a careful *act*-utilitarian can argue, as G. E. Moore did, that the probability of acting for the best is maximized by sticking to the well-established principles (Moore, p. 162). Such a person will be both an act- and a rule-utilitarian, and the two positions will have coalesced. If, unlike Moore, we allow that there may be rare and unusual cases in which it is perfectly plain that it would be for the best to break one of these principles, and that in those cases we ought to do so, we can remain act-utilitarians for those cases, while agreeing that in the vast majority of cases it is rational, even for a consistent act-utilitarian, to stick to the rules.

The cleft between act- and rule-utilitarians can be made to seem wider by concentrating on unusual cases—which is why anti-utilitarians not only harp on them, but invent others more fantastic. Suppose, it is said, you were faced with a situation in which, by deliberately killing one innocent man, you could save many other innocent men from violent deaths at the hands of others: ought you not, as a utilitarian, to kill the one, and is not this at variance with received opinion? Such cases arose when the Nazis ordered the directors of mental institutions to kill off incurables, threatening, if not, to kill many more themselves. The utilitarian should reply that such unusual cases are a bad guide when selecting principles to regulate our lives in general. It is easy to cook up cases in which an act-utilitarian would have to say that some well-established principle ought to be departed from; and he should not be afraid to say this. But he should go on to say that in real as opposed to cooked-up cases there would nearly always be other considerations which would make it for the best to stick to the well-established principle (witness the actual history under the Nazis of the Bethel institution whose director, by a marvellously courageous stand on principle, coupled with brilliant diplomacy, got the authorities to change their policy).

Therefore, in selecting principles (necessarily fairly simple and general) to guide our lives by, to teach to our children, to encourage others to follow, etc., we ought to forget about these unusual cases and concentrate on those which are more likely to occur; for then we shall get a set of principles which will give the right answer in the great majority of cases. If we do get such a set of principles established in society, then we shall be not merely intellectually but emotionally committed to them; breaches of them by us will excite compunction; by others, indignation; and thus a particular morality will have got firmly and generally accepted in our society. The union of rule- with act-utilitarianism thus brought about will have engendered what has been called "motive-utilitarianism": the view that we ought to cultivate and be guided by those motives, traits of character, etc., whose general adoption would be for the best. Our moral intuitions will be in accord with these motives, and will prescribe acts which (given that the principles have been well chosen) are in nearly all cases the right ones. But intuitions are a reliable guide only when they have been schooled in accordance with a good set of moral principles; the intuitions of those who have had a poor moral education can be bad counselors (which is why intuitionism as a theory in moral epistemology has to be rejected).

There is also a theoretical reason why rule- and act-utilitarianism coalesce. To understand this, we have to distinguish between the expressions "general" and "universal" as applied to principles. "General" is the opposite of "specific": a principle has a high degree of generality if it places few and simple conditions on its own application. For example, the principle that the innocent ought never to be killed is more general than the principle that they ought not to be killed except when to kill them would save them from needless suffering, and then only at their own request. Both these principles are, however, equally universal, which means that neither contains a reference to individuals, but only to kinds of individuals. By contrast, the statement that *John* ought to be painlessly killed is not universal, because it contains a reference to the individual, John. It may be "universalizable", if its author is prepared to substitute for it the universal principle that in all cases of a certain sort (perhaps minutely specified) patients in a certain condition (likewise minutely specified) ought to be painlessly killed. The doctrine of the universalizability of moral judgments holds that all moral judgments have to be so universalizable.

The rule-utilitarian has to choose whether the rules whose acceptance-utility he is to judge are allowed to be highly specific, or whether they must be of above some stated degree of generality. In either case, if they are moral principles, and if the thesis of universalizability is accepted, they will have to be universal. If he allows them to be highly specific, then his view collapses into act-utilitarianism; for if the moral judgments of an act-utilitarian are universalizable, he will be able to substitute for them universal principles (albeit highly specific ones) which give the reasons for the judgments. But in that case the rule-utilitarian, too, can adopt the same universal principles as his rules, which will then also be highly specific. If this can be done in all cases, the two positions become for practical purposes equivalent.

It has been objected to utilitarianism that there are cases where the breach of some well-established principle by one person will be for the best, provided that others observe it, because he will get some advantage, and not enough people will be breaking the principle to do any harm. Suppose that a few people can without detection, and with some financial advantage to themselves, break the laws about exhaust emissions from cars, and that the resulting air pollution is not sufficient to endanger public health. It is said that in such cases a utilitarian should prescribe the breaking of the principle that beneficial laws should be observed, and that this is contrary to received opinion. Rule-utilitarians try to meet this case by saying that although the *act* would be for the best, the *rule* forbidding such acts (i.e., the well-established principle that is being broken) has a high acceptance-utility; and that thus on their view the act would be wrong, because it would be a breach of a principle having a high acceptance-utility. But to this it is objected that there is a principle with an even higher acceptance-utility, namely, the more specific principle that one should observe the general well-established principle except in cases where it is clear that everyone else will observe it, but that then one should break it (Lyons, ch. 3).

Such a move puts the rule-utilitarian and the act-utilitarian (whose views have again coalesced) back at variance with received opinion. But they can escape the ill consequences of this (1) by pointing out that cases in which one can know that all the others are going to observe the principle are very rare; and that (for reasons given above) one should not base one's selection of moral principles on such rare cases, and received opinion has not: it is therefore easy to explain why received opinion condemns the act, and why it is (in general) right to do so. (2) They can point out, also, that there is a hidden disutility involved, namely the fact that all the others are having something done to them (namely, to be taken advantage of) which they do not want to have done to them. They do not know that it is being done; but it is a disutility to me to have things done to me that I very much want not to be done, even if I do not know about it (suppose, for example, that my wife's analyst uses his position in order to seduce her without my discovering). This hidden disutility to so many others might outweigh the advantage gained by the offender.

There has not been space to deal with these difficulties adequately, nor to list all the distinctions between types of utilitarianism, nor all the objections that have been made to them; those that seem the most important have been selected. Four more difficulties should be mentioned briefly. First, utilitarian calculations involve judgments comparing the utility accruing to one person with that accruing to another; this presents theoretical difficulties, although we all do it in practice.

Secondly, in the important issue of population policy, and in other issues (e.g., abortion, contraception) where what is in question is the morality of bringing or not bringing a person into existence, a distinction has to be made between classical or total utilitarianism and average utilitarianism. According to *total utilitarianism*, the utility which has to be maximized is the total utility accruing to all existing beings: this would be increased by the addition of an extra member to the population, if the positive utility accruing to him, however small, was greater than that lost by the others owing to his advent. This view, therefore, requires a more liberal population policy than *average utilitarianism*, which requires us to maximize average utility; on this view, we are not required to increase the population beyond the point at which the average utility accruing to existing beings begins to fall.

Thirdly, it has been debated whether the beings whose utility is to be considered include all beings, including dumb animals, which are capable of suffering and enjoyment, or only humans; Jeremy Bentham took the former view, which seems most defensible.

Fourthly, the merits of *positive utilitarianism* and its rival *negative utilitarianism* have been canvassed. The first, in its calculation of utilities, reckons both benefits and harms and balances them against each other; the second reckons only harms. Since the deprivation of a benefit is a harm, it is difficult to sustain the distinction.

Lastly, it has often been disputed what is meant by "utility", and by such phrases as "for the best" which have been used above. Answers put in terms of pleasure and the absence of pain, as given by Bentham and Mill, are now generally thought to be too restrictive. The easiest short answer is to say that that act is for the best which is in the greatest interest of those affected, taken in sum; and that what is in a person's greatest interest (what maximizes *his* utility) is what he would choose to happen if he were fully informed and completely prudent. Morality, then, on the utilitarian view, emerges as a kind of universalized prudence (utilitarians should not be accused, as they sometimes are, of equating it with *selfish* prudence or expediency). We are to give as much weight to the interests of all as the prudent man gives to his own. The affinity between this maxim and the Christian Golden Rule is obvious.

R. M. Hare

[For the background and varieties of utilitarianism, see the fourth article in this entry, TELEOLOGICAL THEORIES. On act-utilitarianism and the use of rules, see the articles above on RULES AND PRINCIPLES and SITUATION ETHICS. On the universalizability of moral judgments, see the tenth article in this entry, NON-DESCRIPTIVISM. For an extensive discussion of G. E. Moore, see the ninth article in this entry, NATURALISM. For discussion of related ideas, see the entries: ANIMAL EXPERIMENTATION, article on PHILOSOPHICAL PERSPECTIVES; DOUBLE EFFECT; and RISK.]

BIBLIOGRAPHY

BAYLES, MICHAEL D., ed. *Contemporary Utilitarianism.* Garden City, N.Y.: Doubleday & Co., Anchor Books, 1968. Bibliography.

DIGGS, B. J. "Rules and Utilitarianism." *American Philosophical Quarterly* 1 (1964): 32–44.

HARE, R. M. "Medical Ethics: Can the Moral Philosopher Help?" *Philosophical Medical Ethics: Its Nature and*

Significance: Proceedings of the Third Trans-disciplinary Symposium on Philosophy and Medicine, Held at Farmington, Connecticut, December 11–13, 1975. Edited by Stuart F. Spicker and H. Tristram Engelhardt, Jr. Philosophy and Medicine, vol. 3. Edited by H. Tristram Engelhardt, Jr. and Stuart F. Spicker. Boston: D. Reidel Publishing Co., 1977, pp. 49–61.

HASLETT, D. W. *Moral Rightness.* The Hague: Martinus Nijhoff, 1974.

LYONS, DAVID. *Forms and Limits of Utilitarianism.* Oxford: Clarendon Press, 1965. Bibliography.

MOORE, G. E. *Principia Ethica.* Cambridge: At the University Press, 1903. Especially chapter 5.

QUINTON, ANTHONY. *Utilitarian Ethics.* New Studies in Ethics. Edited by W. D. Hudson. London: Macmillan & Co., New York: St. Martin's Press, 1973. Bibliography.

RESCHER, NICHOLAS. *Distributive Justice: A Constructive Critique of the Utilitarian Theory of Distribution.* Indianapolis: Bobbs-Merrill, 1966. Bibliography.

SMART, J. J. C., and WILLIAMS, B. A. O. *Utilitarianism: For and Against.* Cambridge: At the University Press, 1973. Bibliography.

VII

THEOLOGICAL ETHICS

Meaning of theological ethics

The term "theological ethics" (or "moral theology") in its inclusive sense is employed to designate the activity of critical reflection about the bearing of beliefs in God or gods on the understanding of the moral life. As such, it embraces, at least in principle, analysis of the moralities of Christianity, Islam, Buddhism, Hinduism, Judaism, and other religions, including what are sometimes called "primitive religions." In its restricted sense, however, theological ethics refers to discourse about a particular theistic morality embraced by the writer or by a significant part of the society to which the writer belongs. Thus in European-American society the most common expressions of theological ethics are Christian and/or Jewish.

This understanding of theological ethics may be employed even when the meaning of religion is extended to include such belief systems as Marxism and secular humanism, as has recently become fashionable in some circles. If such belief systems are to be considered "theologies" by virtue of their exhibiting the phenomena of faith (that is, trust and commitment) in valued objects or states of affairs (that is, "gods"), then theological ethics may also take the form either of examining the general bearing of such faiths on morality or of analyzing the particular relation of, for example, secular humanism to morality. Nevertheless, such an extension of the meaning of religion is both problematic and marginal to the enterprise of theological ethics as ordinarily conducted. Therefore, and because of limits of space, no further reference will be made to it in this article.

Three types of normative judgments are basic to theological ethics: judgments of obligation, virtue, and value. Judgments of obligation (or duty) respond to the question, "What morally ought to be done?" Thus a particular action, norm, or policy may be probed as to whether it is morally right, wrong, or permissible, such as whether a physician ought to lie to, or withhold information from, a dying patient, or what kind of policy a hospital morally ought to have in such matters. Judgments of virtue (or moral character) respond to the question, "What qualities or dispositions of a person, for which the person can be said to be accountable, are commendable or reprehensible?" Thus a physician may be considered worthy of praise for conscientiousness or of blame for uncharitableness, and a patient worthy of praise for courage or of blame for incivility. Judgments of value respond to the question, "What objects or states of affairs are good or bad?" or "In what sense and to what degree are particular objects or states of affairs good or bad?" Thus the health of a patient may be judged to be a good in itself, a particular national health program to be not as good as another, and the use of intensive care units to be bad as a means to accomplish certain ends. It is noteworthy that in medical and public policy circles these three types of ethical judgments (obligation, virtue, and value) are occasionally collapsed into one category and simply called "values" or "value judgments," with a resulting tendency to confusion in the identification and assessment of specific moral problems.

Theological ethics shares these three types of normative judgment with philosophical ethics (or moral philosophy). What distinguishes the two enterprises from each other, therefore, is not a difference about the need to employ judgments of obligation, virtue, and value, but rather the orientation that is brought to their employment. Philosophical ethics may, but need not, assess the bearing of theistic beliefs and attitudes on the moral life, and to a very considerable degree proceeds today without doing so in European-American society. It more characteristically finds its orientation, to the extent it acknowledges a need to set forth its general bearings or background beliefs, in secular views of human nature, human worth, or human good, sometimes accompanied by logical, epistemological, and/or cosmological doctrines. Theological ethics, on the other hand, is committed by its very nature to the examination of the moral life from the viewpoint of theistic beliefs and attitudes. It characteristically seeks its orientation in inquiries about the appropriate human response to whatever is held to be God's nature, will, or activity, and examines and advocates theories of obligation, virtue, and value associated with that response.

Sharing the same type of orientation with theological ethics are everyday religious moralities of one sort or another. Such moralities involve beliefs and attitudes about God or gods, and their judgments about obligation, virtue, and value are related in some manner to their theistic beliefs and attitudes. How then are they to be distinguished from theological ethics? The answer is that such moralities are a kind of routine practice that can be participated in with only a minimum of theoretical reflection and criticism, while theological ethics is essentially a theoretical activity. It is a highly developed science that examines theistic orientations, the moralities that are associated with such orientations, and the relations between theistic orientation and morality. This is not to say that two mutually exclusive sets of persons are implied by the distinction, those who practice a religious morality and those who engage in theological ethics. This is no more true than to say that the set of persons who engage in health-enhancing actions and the set of persons who are medical scientists are mutually exclusive. Most (but not all) theological ethicists are participants in one or another religious community, adhering to many (if not all) of its tenets on morality. On the other hand, the overwhelming majority of participants in religious moralities are not also involved in theological ethics, properly speaking.

Theistic orientation

A theistic orientation is one that contains two specifiable characteristics. First, there is a belief in one or more religious objects (that is, God or gods). These objects (or states of affairs) can take various forms in different belief systems, but common to all belief systems of this kind is the understanding that such objects transcend ordinary human experience of the sort natural science can describe (actually or potentially), and that the nature and/or activity of such objects is of very considerable, if not overriding, importance to human experience. A plurality of such objects in a belief system is an indication of polytheism. A single such object that is also held to be relevant to the experience of all persons and groups of persons points to monotheism. And a single such object whose relevance is affirmed only to a particular group of persons suggests henotheism (Niebuhr, pp. 24–31). The principal focus of this article will be on monotheism, the belief in one such object that is ordinarily referred to as God and considered to be universally relevant.

The second characteristic of a theistic orientation is the response that human beings are called upon to make to God. This response, or attitude of the believer, is dependent for its appropriateness upon how God is conceived. Emphasis may be placed on his being holy, gracious, powerful, loving, compassionate, just, fearsome, merciful, forgiving, etc.; his bringing persons out of bondage, revealing his will through a prophet, suffering for his people, etc.; his ordaining of general norms or duties that human beings should follow, establishing a way of life that if pursued will lead eventually to happiness, etc. Among the types of human response that may be considered appropriate to the believer are penitence, sorrow, joy, trust, hope, commitment, imitation, submission, acceptance of a new relationship, undertaking a pilgrimage, obedience to his law, or embarking on a way of life in pursuit of human excellence (Little and Twiss, p. 61).

There is obviously a wide variety of theistic orientations, not only among the different religions of the world, but even to a considerable degree within the same religion. The effect of this rich plurality of orientations is to make theological ethics, at least in its inclusive sense, a highly complex enterprise. For it must examine and clarify the bearing not only of one but of many theistic orientations on the moral notions of obligation, virtue, and value. This task is made even more intricate by the recognition that these moral notions are themselves subject to varying interpretations.

In order to reduce the variables in this article, most of the theistic orientations will be viewed as sundry expressions of one or more of the three basic orientations, and the remainder left to other writings for those readers wishing to pursue them. It is sufficient here to acknowledge that the three basic orientations do not exhaust the full range of possibilities. The first of these basic types of theistic orientation (TO_1) holds that God is the author of a universal ordering which has always been normative for human persons everywhere. Furthermore, the appropriate response to God is to perceive this normative ordering, by reason and/or by revelation, and freely to abide by it. To do so ordinarily implies the acceptance of a way of life that progressively leads to the fulfillment and happiness of the wayfarers. An example of TO_1 is the traditional Catholic teaching on natural law, in which God is perceived to have created human beings with certain ends discoverable by reason. The pursuit of these ends progressively leads, when aided by grace, to human excellence coupled with supreme happiness in fellowship with God (Thomas Aquinas I-II, 1–5; 94). Another example is the Hindu doctrine of karma, in which a person is and gets today what he deserves from his past. Moreover, he becomes and receives in the future what is fitting to his thoughts, attitudes, and actions today (Sarma, pp. 53–59).

The second type of theistic orientation (TO_2) assumes that God has intervened in history to establish a new relationship with humanity, and that a certain event associated with that intervention becomes revelatory or illuminative for the interpretation of other events in history. The appropriate response of one who is a believer is to make this event central to the understanding of one's own existence, assigning meaning to one's life in accord with the accepted meaning of the event and taking upon oneself the commitments associated with the event. Sometimes the event is seen not so much as establishing an entirely new divine–human relationship, but as renewing a relationship that has always existed though (because of human sin or finitude) has not previously been sufficiently perceived or honored. An example of TO_2 is the Exodus experience of Judaism in which God is understood to have delivered his people out of bondage in Egypt and to have established a covenant with them, specifying the kind of monotheism they are to embrace and the duties they are to perform. The appropriate Jewish response today to this event is to acknowledge it as constitutive of one's own history and to conduct one's life in keeping with the divine–human covenant that is understood to have come about through it (Kadushin, pp. 93–95; Hillers, pp. 46–71). Another example is the event in which God is understood by Muslims to have revealed his will to his prophet Muhammad, as is recorded in the Koran. The appropriate response of the believer is to submit to this will, and in so doing diligently to fulfill the five specified duties of profession of faith, prayer facing Mecca five times daily, alms-giving, fasting in the Arabic month of Ramadan, and pilgrimage to Mecca once in a lifetime (Donaldson, pp. 40, 111).

Theistic orientation of the third type (TO_3) places emphasis on characteristic attributes or qualities of God rather than on a normative order he has established in the world (TO_1) or an intervention he has made into history (TO_2). In some instances these attributes are to be imitated, as when believers affirm that they ought to be loving because God is loving. In other instances the response is not imitation (which may be held to be presumptuous), but the expression of attitudes or emotions that are considered fitting responses to a designated quality of God, as when believers respond with an overwhelming sense of awe to God's majesty and holiness. The ascetic response to Allah's (God's) mercy and providence in Sufism (a Muslim mystical religion) and the Christian response of faith, hope, and love to God's grace are further examples of TO_3 (Donaldson, pp. 194–212; Augustine, pp. 36–38, 132, 135–136). However, Gods' grace in the Christian religion can also be understood as decisively expressed in a particular event such as the life and death of Jesus, in which case the orientation may actually be a combination of TO_2 and TO_3. And when al-Ghazzālī reasserted in Sufi mysticism a strong belief in special revelation by Allah to Muhammad, a similar combination resulted (Ghazzālī, pp. 54–68).

Relation of theistic orientations to moral notions

How do theistic orientations bear on moral notions of obligation, virtue, and value? Most simple answers to this question are quite misleading, such as claims that morality generally depends on religion or, alternatively, has no important connection with it. Actually, a number of possible relations between theistic orientation and moral notions do exist and need to be considered, as recent writers on theological ethics have pointed out (Frankena, pp. 295–296; Graber, p. 54). Among these relations, the chief ones seem to be (1) historical, (2) logical, (3) psychological, (4) epistemological, (5) linguistic, and (6) ontological.

Historical relation. A historical relation is one pointing to the genesis of a moral notion in a particular orientation, as for example agapic love had its origin (or at least played its first major role) in early Christian religion from which it later achieved prominence in Western moral consciousness. However, to identify a moral notion with a particular orientation as its historical source is not to claim that without this orientation the notion would not have developed elsewhere, or that having developed historically in one orientation it cannot be adopted into another one (either religious or secular). Sometimes theistic orientations are recipients, not just generators, of moral notions. For example, the notions of generosity and vengeance in the Koran may have got there in part because of the influence of pre-Islamic nomadic morality on Muhammad. And some Aristotelian and Stoic virtues were adopted by the early Christian community, although considerably transformed by their function in that new orientation.

Logical relation. A logical relation pertains (1) to the explication of moral judgments entailed by theistic (and other) orientations. Thus Jewish belief and worship lead implicitly to the moral obligation to seek justice for those who are deprived of it. And Protestant claims about the acceptance of the grace of God through faith entail claims about love and service to the neighbor. A logical relation may also pertain (2) to the justification of moral judgments by theistic (and other) orientations, a role that historical relations, however important, cannot perform. Such justification can aim at showing that a certain theistic orientation is necessary (whether sufficient or not) to establish that one or more moral judgments are valid. For example, it may be suggested that only if there is a God who has commanded that we provide medical care for the indigent are we morally obligated to do so. If successful, this procedure would make at least one judgment in morality logically dependent on religion, and the idea of a secular morality that is fully autonomous would be, to this extent, invalid (or at least in need of qualification). On the other hand, justification may take the form of showing that a certain theistic orientation is sufficient (although not necessary) to establish the validity of a moral judgment. For example, it may be claimed that if there is a God who has commanded that we provide medical care for the indigent we are morally obligated to do so. But this procedure, if successful, would leave open the possibility that the justification for this moral judgment may also be provided by appeal to some other consideration, and to this extent a secular (or other theistic) orientation fully autonomous from this one would be possible. For the most part, justification in theological ethics in the modern world has concentrated on those logical relations between theism and morality that involve judgments of their sufficiency and not of their necessity, although passages in some important writings do point more in the direction of logically necessary relations (Barth, 1957, pp. 540–542).

Psychological relation. A psychological relation centers on the motivation that theistic orientation may provide for doing one's duty, for improving one's character, or for developing one's values in a more worthwhile direction. For example, there is obviously a significant incentive for right moral action for those who accept the Hindu doctrine of karma, since one's status in one's next existence is understood to be determined largely by one's moral performance and development in this existence. And the Christian notion that Christ died for the world has provided some persons with a strong motivation to open their lives to philanthropic service of others. The psychological function of theistic orientation, however, cannot also provide justification for particular judgments of morality. Nevertheless, it has often been sufficiently impressive in advancing the level of adherence to morality that some nontheistic philosophers (such as John Stuart Mill) have pondered at length the nature and possibility of a secular alternative to the motivational function of theism in improving moral performance.

Epistemological relation. An epistemological relation pertains to claims of knowledge that theistic orientation is held to contribute to morality. Appeals to this kind of relation generally take the form that there are basic truths of morality not generally (or sufficiently) known except through theistic orientation. An example is the claim that only if one truly loves God does one know what it really means to love one's fellow human beings, a claim made by a number of major religions. Another example is the position of early Calvinism that human sinfulness has so weakened the natural human knowledge of morality that there is need through theistic orientation to accept an extraordinary republication of basic morality, such as the Ten Commandments. Still again, it may be claimed that while some people may have a natural knowledge of morality, others (usually less informed) need the special help that theistic orientation provides. Or it may be believed that while all persons have a rational knowledge of basic duties of morality, none has knowledge of morality's higher duties (such as those practiced by the saint or hero— consider orders of nuns devoted to the medical care of the indigent) except by means of some theistic orientation, a belief that was widely current in medieval Catholicism. Finally, it should be acknowledged that the flow of ideas in the epistemological relation may also move in the opposite direction, such as when it is claimed that the knowledge or practice of morality contributes in some important manner to our knowledge of God.

Linguistic relation. The linguistic relation between theistic orientation and morality usually takes the form of asserting that the meaning of some (or all) moral terms requires appeal to religious terms. Of course, this can be declared true by definition or stipulation, as when all moral terms are held to be religious terms in some quite extended sense of religion. But this is not what advocates of the linguistic relation usually have in mind. Rather they argue for such a position as "good" means "valued by God," or "right" means "ordained by God." The linguistic relation may also take the form of affirming that understanding the meaning of God makes reference to moral terms necessary. It may be held, for example, both that one does not know the meaning of the term "God" except by relating it to the terms "good" and "just," and that "good" and "just" are moral terms.

Ontological relation. The last of these relations between theistic orientation and morality is ontological. Simply put, this holds that the existence of moral obligations, virtues, and values (or some part of them) is vitally connected with the existence or sustaining activity of God. This may take the form of saying that, since everything existing depends on the existence of God, morality could not exist without God. The thrust of this relation may also move in the opposite direction, such as when it is claimed that God's existence is enhanced by the moral lives of his human creatures, a claim especially espoused in the ethics of process theology.

All three types of theistic orientation (TO_1, TO_2, and TO_3) have employed in one or another of their expressions each of these six kinds of relations to morality. Nevertheless, some theistic orientations tend to stress certain relations with morality. Thus when God is conceived as the author of a universal ordering (TO_1), the ontological, psychological, and linguistic relations seem to be especially prevalent. When God is understood to have intervened in history through one or more decisive events (TO_2), historical, epistemological, logical, and psychological relations are often prominent. And when God is perceived as possessing attributes that appropriately lead to certain human responses (TO_3), logical, linguistic, and psychological relations to morality come to be emphasized.

Moral notions

The notion of obligation has played a more important role in theological ethics in the twentieth century, especially in Western Christianity and Judaism, than have the notions of virtue and value. This also seems to be true of the applied field of bioethics. For here such moral problems as those associated with death and dying, experiments on human persons, and distribution of health care have been formulated more often in terms of what morally ought to be done rather than what human characteristics are morally commendable or what states of affairs are morally to be preferred to others. Nevertheless, in theological ethics an increasing interest in considerations of virtue is now becoming evident (Hauerwas, pp. 48–89; Carney, "Virtue–Obligation"). Whether the notion of virtue will also come to play a larger role in the applied field of bioethics or not will have to await further developments.

Obligation theory. Within obligation theory a controversy about the role of moral rules in religious morality—sometimes called the "rule-situation" debate, at other times the dispute over "situation ethics"—has been especially prominent in recent years. To understand the issue at stake it is necessary to distinguish between two types of moral norms. First, there are those norms that are of very wide generality and do not specify some particular action that is required, forbidden, or permitted. Examples are that "we should love our fellow human beings" and that "we should obey the will of God." Norms of this sort will here be called "principles," although this term bears a different connotation in some other writings. It is distinctive of principles that actions on the basis thereof may be enormously diverse, since no concrete action pattern is included in the statement of these norms. Second, there are norms that do specify some particular action, such as "we should refrain from telling lies to patients" and "we should not use persons in experiments without their consent." Such norms will here be called "rules," although the meaning of this term also varies in some other writings. It is distinctive of rules that actions on the basis thereof center on some concrete action pattern (such as lying, or consent in experimentation), although it is not always clear how to relate such a pattern to a specific situation.

Both situationalists and nonsituationalists agree on the need for ethical principles in moral action. But situationalists hold that rules are advisory and not obligatory for human action, since they are only recommendations that arise out of accumulated human experience. The correct moral procedure is to apply the cherished principle (or principles) directly to the situation encountered and, after perhaps considering the counsel given by relevant rules, to do whatever seems indicated by the principle in the situation, whether or not this conforms with what such rules advise (e.g., is informing a particular dying patient of his situation the most loving thing to do in that situation?). Nonsituationalists, on the other hand, hold that moral rules, if valid, do have a binding force in what they require, prohibit, or permit, and that there are some moral rules which are valid. Decisions about validity come about by inquiring whether a rule under review is the best action-pattern expression of the cherished principle (or principles) for a given class of situations (e.g., would informing the dying of their state be the rule most in keeping with love?). Thus the correct moral procedure for nonsituationalists is to accept valid moral rules as obligatory for the classes of situations for which they are intended, and to review from time to time the validity of the moral rules they follow, especially when events or persons raise challenges about them.

The rule–situation controversy would be misleadingly characterized by claiming that one side in the dispute is not interested in obligatory norms and the other not interested in actual situations. For situationalists do hold to obligatory norms, although only principles and not rules. And nonsituationalists do address themselves to actual situations, both in applying rules and in validating them.

Among situationalists there are differences about the nature of the appropriate principle. Some hold that love is the highest principle or criterion of human action, e.g., Joseph Fletcher, others that the command of God is, e.g., Karl Barth. Among the former the criterion is often interchangeably expressed either as doing for the neighbor what love requires or as promoting the good of the neighbor. In either case the procedure is a teleological one requiring a full-fledged value theory about what is truly good or loving. Unfortunately such a theory is seldom supplied. Situationalists of the other sort employ a deontological procedure in their obligation theory by emphasizing a formal characteristic of human action (i.e., conformity with God's command) rather than the good to be promoted by human action. This kind of situationalism requires an epistemological theory that clarifies how we are to know God's particular commands, which involves an elaborate doctrine of revelation that is difficult to provide. Of course, one can combine these two types of situationalism in various ways. For example, one can claim that our ultimate principle is to love our neighbor or to promote his good, but the best way of doing this is simply to follow God's commands. Or it can be claimed that the ultimate principle is to obey God's commands and that what he repeatedly commands is that we love our neighbor or promote his good.

Among nonsituationalists there are also both teleologists and deontologists, but of course they also (unlike situationalists) employ obligatory rules as expressions of principles in specified classes of situations. In addition, another kind of difference can be observed among them regarding exceptions or qualifications to moral rules. Augustine of Hippo, for example, held that it is always wrong to tell a lie, whatever the circumstances or consequences thereof, even though the blameworthiness of some lies (e.g., benevolent ones) is not as great as others (e.g., malevolent ones). Thus for him and others who have followed in his footsteps there are no exceptions or qualifications to valid moral rules. Muhammad, on the other hand, claimed that one should refrain from telling lies when one is not involved in any of three types of circumstances, namely, in war, in love, and in attempting to reconcile friends who have become alienated. There are two ways of conceiving of such limits to a rule about lies or other practices in religious ethics, and both have been employed at one time or another by theological ethicists. On the one hand, one can say that there are exceptional circumstances or consequences that make the breaking of a rule morally permissible. On the other hand, it can be claimed that a valid rule contains certain qualifying conditions that, if present in a particular situation, make the rule inapplicable, and thus not to follow it is not to break it. Presumably the latter procedure was the recourse of Muhammad. It is also the recourse most often employed today in life-and-death matters by nonsituationalists in theological ethics who believe, as almost all do, that there are some limited circumstances in which it is morally right to let someone die.

Another issue in the obligation theory of theological ethics is whether there are two levels of duty in religious morality, one level that pertains to all persons and another level that is proper only to some. It is said that some obligations are such that it is justifiable to impose them on everyone whether or not they give their consent, such as telling the truth, keeping promises, and refraining from unprovoked assaults on human persons. Special obligations associated with different occupational roles are also usually included as first-level duties (e.g., the duty of physicians not to abandon patients in their care). On the other hand, it is said that there are some obligations that should not be imposed upon persons without their consent but may be freely adopted by those who covenant themselves to a religious–moral way of life requiring them. Such obligations may include assuming accountability for an act one could not have avoided, accepting responsibility for an act of someone else, and engaging under special circumstances in inordinate self-sacrifice (Carney, "Accountability," pp. 320–327). The performance of such "second-mile duties" is most often referred to as works of supererogation. Sometimes, however, they are called the deeds of "saints and heroes" and are considered to be motivated by personal and group "ideals" (in which case they are often treated more as expressions of virtue than as second-level obligations). They have been especially prominent throughout the history of Roman Catholic ethics but are receiving increased attention in Protestant and philosophical circles.

Nevertheless, there is controversy over the moral validity of dividing duties into first-level and second-level ones. For some theologians, such as Luther, see the employment of this distinction as a means by which some persons avoid the full demands of faith in God and others pervert their works of supererogation into vain claims to moral superiority over their fellow human beings. However one assesses the religious and ethical aspects of this controversy, it seems undeniable that the theory and practice of second-level duties have had a profound influence in history on the establishment of health-care facilities and the enlistment of health-care personnel.

Virtue theory. The moral notion of virtue performs two fundamental functions in theological ethics. First, it indicates the kind of person who is to be rightly considered good, just, faithful, loving, holy, and so forth. As such, it provides both a normative complement to a merely descriptive account of human nature and an ideal of human personhood at which to aim. Second, the notion of virtue offers an alternative to notions of obligation for the discernment of the morality of human acts. It does this not by asking what is required by some principle or rule of obligation, but by inquiring what characteristics or qualities (such as faithfulness, fairness, a loving disposition, and so forth) constitute the goodness of a person, and by designating those acts to be good (or bad) to the extent that they are appropriate to the model of a good (or bad) person.

In addition, the notion of virtue may also perform an important derivative function to that of obligation. This may occur in the following manner. If we first know what acts are right or wrong by appeal to one or another principle or rule of obligation, we may be concerned to develop the personal dispositions or traits of character that provide habituation and motivation to perform or resist those acts. In this derivative employment of virtue theory, we try to develop those virtues that support the general types of obligations we already acknowledge (but perhaps do not always fulfill because of moral weaknesses), and those virtues in turn enable us "to put our hearts into our duties," thus increasing the probability that our performance will more closely approximate our profession.

There are two general theories in religious ethics as to how one becomes virtuous or develops a good character. The first is said (somewhat incorrectly) to derive from the influence of Aristotle on Christianity and Islam, and (more correctly) to be exemplified in much of the medieval literature on cases of conscience. This is the theory that a person becomes good by doing good acts, much as surgeons become better surgeons by regular practice. It may be called the "acquiremental theory," inasmuch as it is largely achieved by the repetition of human acts. But it was vigorously repudiated by the Protestant Reformation, because it was believed to lead to a works-righteousness rather than, as Luther and Calvin advocated, a faith-righteousness. They in turn (drawing upon Augustine and others) set forth, as an alternative theory about how one becomes good, the acceptance in the depths of one's heart of the forgiveness and reconciliation offered by God. It was anticipated that this acceptance would lead to new dispositions toward God (trust) and the neighbor (love), much as a physician or patient might be judged to be a different (and better) person following changed dispositions toward those persons with whom he or she is involved. Since this manner of becoming virtuous or righteous focuses on the development of changed attitudinal relations to God and neighbor, it may be called the "relational theory." The dispute over the alternative theories has occasionally flared up with some intensity, although a few theologians, such as Thomas Aquinas and Jonathan Edwards, can be read as advocating a combination of them as closer to the truth (Thomas Aquinas I-II, 58–63; Edwards, pp. 1–41).

One of the topics in theological ethics that appears in both virtue and obligation conceptualities is that of moral default. When considering such default as an obligation notion, the basic category is one of guilt. The offender is considered by himself or others to have engaged in a violation of an important norm and feels (or should feel) guilty for doing so. On the other hand, the acknowledgement of serious default as a virtue consideration ordinarily takes the form of shame. One is judged by himself or others not so much as being in violation of something, but as having fallen short of some ideal and as needing to overcome his humiliation or to "find cover" for some nakedness of his moral personhood. It is of some note that in many religions the concept of sin is related to both guilt and shame. If one is guilty, the violation must be forgiven; if one is ashamed, the person must be restored. Next to the awakening and nurturing of a sense of the holy in human experience, perhaps nothing is as important for religion as its therapies for overcoming guilt and shame.

Another topic common to both virtue and obligation frameworks in theological ethics is that of rewards and punishments. There would seem to be three answers religion has provided to the problem of relating human character and action to good fortune and bad. The first is that persons actually do get what they deserve, if not in this life then in a later life. This answer is especially central in the Hindu doctrine of karma, in early Islam, and in some expressions of Christianity. The second answer is that the doctrine of rewards and punishments is not so much a description of what people actually do receive in relation to the moral quality and actions of their lives as what they deserve to receive. This is to say, rewards and punishments are symbols, not real states of affairs, that bear testimony to the worthiness or unworthiness of persons. This notion is more commonly found in abstract systems of theological ethics than in everyday religious moralities. The third answer is that rewards and punishments are not extrinsic states of affairs in persons' lives (now or hereafter), but intrinsic conditions of their characters. According to this doctrine, persons become what they truly worship, honor, and do. It is noteworthy that the first answer (i.e., that persons actually get what they deserve) can function within a pure obligation framework, but that the second answer (i.e., rewards and punishments are simply symbols of worthiness or unworthiness) and the third (i.e., rewards and punishments are intrinsic to one's character) require either a pure virtue framework or an obligation framework that is combined with a virtue one.

Natural law theory has also played a major role in some expressions of theological ethics. This too can be interpreted with primary emphasis either on obligation notions or virtue ones. Most commonly it has been given an obligation focus. Those who employ it in this manner seek an understanding of what constitutes the fullness or perfection of the functioning of human nature in order to discover and declare obligatory those norms of human action that seem most commensurate to human nature so conceived. When this endeavor presupposes that the ordering of human life is part of an ordering by God of his entire creation, and allows its determinations to be guided in part by this presupposition, then the natural law doctrine is a part of theological ethics. Natural law with an obligation focus has also been of interest to some secular thinkers, though of course with little or no theistic orientation to their inquiries and advocacies. On the other hand, there are theists (and nontheists) who have placed the emphasis in natural law doctrine on virtue considerations. They do this by focusing on the normative nature of human persons, and simply set forth an understanding of traits of character and actions appropriate thereto. They are less interested in the law of human nature as something external and obligatory to human nature (though derived therefrom) than they are in human nature itself and its appropriate functioning.

Value theory. The notion of value in theological ethics focuses on the question of what things or states of affairs are worth pursuing in life, or what things or states of affairs are more worth pursuing than others. The most usual answer to this question is that goods of the soul are more important than goods of the body (Augustine referred to them respectively as "eternal goods" and "temporal goods"), and that the primary goods of the soul pertain to the cherishing of God and the neighbor. To these primary goods one's own personal qualities (also goods of the soul) are usually considered to be secondary, chiefly in an instrumental sense. Furthermore, goods of the body are ordinarily to be seen as also worthwhile, and most especially if they are needed by the neighbor for his basic sustenance. It should also be observed that a teleological theory of obligation in theological ethics ideally requires a well-developed and hierarchically ordered theory of value.

The concept of love would seem to be the most central one in theological ethics, especially in the Western world (Outka; Ramsey, 1950, pp. 1–45, 92–132, 234–248). Sometimes this functions as a value notion, as when it is employed to specify the objects or states of affairs that are held to be important. At other times, it is a virtue notion, especially when it is acclaimed as "the form of all virtue" or the inclusive motive of good human action. Still again it is considered to be a general principle of obligation, not only teleologically ("do what love requires") but also deontologically ("conform your actions to the covenant-love of God"). This central relevance of love to so much of the territory that theological ethics covers may be a reason for claiming, as one of the major religions has, that "the greatest of these is love."

FREDERICK S. CARNEY

[*Directly related are the entries* BUDDHISM; CONFUCIANISM; HINDUISM; ISLAM; JUDAISM; PROTESTANTISM; *and* ROMAN CATHOLICISM. *While all the articles in this entry are relevant, see especially the articles* THE TASK OF ETHICS, RULES AND PRINCIPLES, DEONTOLOGICAL THEORIES, TELEOLOGICAL THEORIES, *and* SITUATION ETHICS.]

BIBLIOGRAPHY

AUGUSTINE. *The Enchiridion on Faith, Hope and Love.* Translated by J. F. Shaw. Edited by Henry Paolucci. Analysis by Adolph von Harnack. Chicago: Henry Regnery, 1961.

BARTH, KARL. "The Command of God." *Church Dogmatics.* 4 vols. Vol. 2: *The Doctrine of God.* 2 pts. Edited by G. W. Bromiley and T. F. Torrance. Edinburgh: T. & T. Clark, 1957, pt. 2, chap. 8, pp. 509–781.

———. "Freedom for Life." *Church Dogmatics.* 4 vols. Vol. 3: *The Doctrine of Creation.* 4 pts. Edited by G. W. Bromiley and T. F. Torrance. Edinburgh: T. & T. Clark, 1961, pt. 4, sec. 54, pp. 324–564.

CARNEY, FREDERICK S. "Accountability in Christian Morality." *Journal of Religion* 53 (1973): 309–329.

———. "The Virtue-Obligation Controversy." *Journal of Religious Ethics* 1, no. 1 (1973): pp. 5–19.

DONALDSON, DWIGHT M. *Studies in Muslim Ethics.* London: S.P.C.K., 1953.

EDWARDS, JONATHAN. *The Nature of True Virtue.* Foreword by William K. Frankena. Ann Arbor Paperbacks, AA37. Ann Arbor: University of Michigan Press, 1960, 1969.

FRANKENA, WILLIAM K. "Is Morality Logically Dependent on Religion?" *Religion and Morality: A Collection of Essays.* Edited by Gene Outka and John P. Reeder, Jr. Garden City, N.Y.: Anchor Press/Doubleday, 1973, pp. 295–317.

GHAZZĀLĪ, AL- [AL-GHAZĀLĪ]. *The Faith and Practice of al-Ghazālī.* Translated by W. Montgomery Watt. Ethical and Religious Classics of East and West, no. 8. London: George Allen & Unwin, 1953.

GRABER, GLENN C. "A Critical Bibliography of Recent Discussions of Religious Ethics by Philosophers." *Journal of Religious Ethics* 2, no. 2 (1974), pp. 53–80.

HAUERWAS, STANLEY. *Vision and Virtue: Essays in Christian Ethical Reflection.* Notre Dame, Ind.: Fides Publishers, 1974.

HILLERS, DELBERT R. *Covenant: The History of a Biblical Idea.* Seminars in the History of Ideas. Baltimore: Johns Hopkins Press, 1969.

KADUSHIN, MAX. *Worship and Ethics: A Study in Rabbinic Judaism.* Evanston: Northwestern University Press, 1964.

LITTLE, DAVID, and TWISS, SUMNER B., JR. "Basic Terms in the Study of Religious Ethics." *Religion and Morality: A Collection of Essays.* Edited by Gene Outka and John P. Reeder, Jr. Garden City, N.Y.: Anchor Press/Doubleday, 1973, pp. 35–77.

NIEBUHR, HELMUT RICHARD. *Radical Monotheism and Western Culture: With Supplementary Essays.* New York: Harper & Brothers, 1960.

OUTKA, GENE. *Agape: An Ethical Analysis.* Yale Publications in Religion, no. 17. New Haven: Yale University Press, 1972.

RAMSEY, PAUL. *Basic Christian Ethics.* New York: Charles Scribner's Sons, 1950.

———. *Deeds and Rules in Christian Ethics.* New York: Charles Scribner's Sons, 1967.

———. *The Patient as Person: Explorations in Medical Ethics.* The Lyman Beecher Lectures at Yale University. New Haven: Yale University Press, 1970.

SARMA, DITTAKAVI SUBRAHMANYA. "Hindu Ethics." *Essence of Hinduism.* Bhavan's Book University, no. 171. Bombay: Bharatiya Vidya Bhavan, 1971, chap. 4, pp. 37–59.

SATHAYE, SHRINIWAS G. *Moral Choice and Early Hindu Thought.* Jaico Books, no. J-353. Bombay: Jaico Publishing House, 1970.

THOMAS AQUINAS. *Summa Theologiae* I-II 1–5; 55–67; 90–108; II-II 64, 7. Translated by the Fathers of the English Dominican Province as "Treatise on the Last End," "Virtues," "Treatise on Law," and "Whether It Is Lawful to Kill a Man in Self-defense?" *Summa Theologica: First Complete American Edition.* 3 vols. Vol. 1: *First Part, QQ. 1–119 and First Part of the Second Part, QQ. 1–114, with Synoptical Charts.* New York: Benziger Brothers, 1947, questions 1–5, pp. 583–615; questions 55–67, pp. 819–877; questions 90–108, pp. 993–1119; vol. 2: *Containing Second Part of the Second Part, QQ. 1–189 and Third Part, QQ. 1–90, with Synoptical Charts,* question 64, art. 7, pp. 1471–1472.

VIII
OBJECTIVISM IN ETHICS

For those interested in bioethics, an ethical theory is primarily of value insofar as it provides some help in solving the perplexing moral problems that arise in the practice of medicine and related fields. Ethical theories that attempt to provide such help are called normative ethical theories.

In some sense all normative ethical theories should be regarded as objectivist theories, for a normative ethical theory is an attempt to provide a systematic account of morality such that one will be able, at least in principle, to determine correct answers to at least some moral problems. Most objectivist theories claim that there is a correct answer to every moral problem, and it is this more extreme claim that has given rise to the various forms of ethical skepticism, e.g., emotivism, relativism, and subjectivism, which deny that there is one and only one right answer to any moral problem. Some objectivist theories amount to little more than the claim that there is, at least in principle, a correct answer to all moral problems, their main claim to philosophical interest being that they are stated in philosophical terminology, e.g., the moral sense theory and intuitionism.

As a guide to conduct

The unique feature of the moral sense theory is its claim that the object of moral perception is the actual particular situation just as the object of our sense of beauty is the actual particular painting, etc. It is the real situation with all of its detail that each person is to perceive morally. This theory has the virtue of noting that moral agreement is, in fact, fairly widespread. It also makes clear that disagreement in moral judgments counts against the objectivity of morality no more than disagreement in perception counts against the objectivity of colors or sounds. However, it does not account for the many differences between moral perception and seeing and hearing, e.g., the absence of any specific sense organ and the absence of any scientifically determined objective correlation, such as with light waves and sound waves. This theory really is little more than an extended metaphor; it amounts simply to the claim that we often do know what the morally correct answer to a moral problem is, but it does not explain in any way how we know this.

Intuitionism is a slightly more sophisticated theory. The object of intuition is not a particular situation, but something more general. The different versions of intuitionism differ from one another in what they claim to be the object of intuition. In the most common form of intuitionism, we are said to intuit certain prima facie duties, or moral rules, and we then look at particular situations and apply the relevant rule. This form of intuitionism takes mathematics rather than sense perception as its model. Just as people see that five plus five is ten, or that twenty-eight divided by four is seven, so everyone sees that one morally ought not to kill or steal or lie. Moral disagreements are explained away in the following ways: (1) Just as some people are so mentally defective that they do not see the obvious mathematical truths, so some people are so morally defective that they do not see the obvious moral truths; (2) the situation may be unclear so that people disagree about what rules are applicable; and (3) emotional factors are sometimes distorting.

Intuitionism allows for a slightly more articulate account of moral thinking than the moral sense theory. An intuitionist who observes a doctor performing some medical experimentation on his patients by leading them to think it is part of their treatment says the doctor is doing wrong because he sees the situation as one involving deception and intuits that deception is morally wrong. Thus the intuitionist can say why what the doctor is doing is wrong, viz., because it involves deception, whereas the moral sense theorist is limited to saying that he simply sees that it is wrong. This makes it seem as if all the advantage is with the intuitionist. However, he has a problem that the moral sense theorist does not have. The intuitionist, because he gives deception as a reason for saying that what the doctor did was morally wrong, seems to be committed to holding that all deception is wrong. This does not seem to accord with the ordinary view, for almost all agree that deception may sometimes be morally justified, e.g., when its intent is to prevent significant suffering. It turns out that the various duties or rules of the intuitionist may conflict. Thus he needs a second intuition to tell what is morally right in a situation involving a conflict between two or more rules. The moral sense theorist deals with each situation on its own and so never runs into this kind of problem.

This problem has led some intuitionists, e.g., Sidgwick and Moore, to the view that we intuit not rules of conduct but good consequences. Rules of conduct or moral rules (the two are generally, though mistakenly, taken to be equivalent; Baier, chaps. 4 and 5; Gert, chaps. 1 and 4) are then treated as those rules which if acted upon result in more of the intuited good being produced than any other rules. There are many variations of this teleological form of intuitionism. The most common form is a version of utilitarianism in which pleasure or happiness is taken as good and moral rules are taken as those which result in the greatest amount of happiness for the greatest number. The problem with such a form of intuitionism is that it not only seems to allow any kind of deception practiced by a doctor as long as no harm comes to the patients but also allows involuntary experimentation on patients if such experimentation results in benefits to others that are greater than the harm caused the patient. And though some doctors seem to agree with this conclusion, it is not universally accepted (Gert, p. 99).

Teleological intuitionism differs, only in a technical sense, from another group of theories, which we can call "naturalist" theories. The naturalists are suspicious of the faculty of intuition required by intuitionism and thus, instead of saying that we intuit what things are good, make it a matter of meaning rather than intuition that such things are good (Bentham, chaps. 1 and 2). So they not only deny that we have any special faculty of intuition which enables us to say that pleasure is good but also deny that any such faculty is needed; for they claim that good simply means pleasure—or whatever else it is that they want to substitute for what the intuitionist intuits as good. But besides facing most of the objections that teleological intuitionists face, the naturalists have the additional problem that their proposed definitions of good do not seem very plausible (Moore, chap. 1).

All of the theories discussed so far may seem to be simply academic discussions of moral matters, that is, they may seem to be discussing moral problems as if they were merely problems in some academic subject. But it is generally recognized that the primary function of moral judgments is not to provide information; it is to guide conduct. Thus theories were developed in which correct moral judgments were explicitly put forward as the advice that would be given by an ideal observer about how to act in a moral situation. These "ideal observer theories" can be viewed as variants of moral sense theories, with the ideal observer looking at the actual situation in all of its detail, or as an intuitionist applying all of the relevant moral rules and ideals and balancing them in the appropriate fashion. The focus of attention shifted from determining how the ideal observer came to his moral judgments, e.g., by a moral sense, intuition, or by definition of terms, and to a consideration of the characteristics of the ideal observer. Among the characteristics discussed, three seemed most important: that the observer be informed, impartial, and sympathetic. Thus the correct moral answer to any moral problem was that which would be arrived at by the ideal observer. The very close connection between the ideal observer and God seems obvious.

All of the objectivist theories discussed so far assume that all rational men, if impartial and fully informed, would arrive at the same moral answers. It is this assumption—that there are unique moral answers to every moral question—that has been primarily responsible for the various skeptical theories that are currently so popular. For none of the standard objectivist theories is able to deal adequately with the fact of moral disagreement. The strength of objectivist theories is that they account for those clear cases in which there is complete moral agreement. Everyone agrees that a doctor should not perform an unnecessary operation simply in order to increase his income. But there is not complete agreement on whether or not to allow a terminally ill patient in severe pain to die. Most objectivists are committed to the view that, given all of the facts of the particular case, there is one and only one morally right course of action. But an objectivist need not hold with this. All that he needs to accept is that it is an objective matter as to what are the morally relevant considerations. Thus everyone agrees that in the absence of countervailing reasons doctors ought to try to keep their patients from dying. Everyone also agrees that the facts that the patient is terminally ill and that he is suffering great pain count as reasons for allowing the patient to die. The disagreement arises in determining when, if ever, these reasons outweigh the obligation to keep the patient alive.

Rationality and impartiality

Most contemporary ethical theorists agree that correct moral judgments are those that would be arrived at by impartial rational persons (sympathy is no longer regarded as necessary). Thus attention has focused on clarifying the concepts of rationality and impartiality. Of course, it is recognized that such persons must be as fully informed about the particular situation as possible, and that many, if not most, moral disagreements arise from a disagreement over facts or estimates of probability, but these problems are considered to be practical rather than philosophical. An objectivist theory is required only to account for moral agreement and disagreement in those (rare) situations where the facts are not in dispute.

For almost all theorists, the account of impartiality is such that it follows analytically that, if one rational person makes an impartial judgment, then other rational persons with the same information will make exactly the same judgment. It is easy to see that, with this account of impartiality, any moral question can have only one answer. But it does not seem to be an essential feature of impartiality that all impartial persons will agree. Judges on various courts, all of whom are considered impartial, often reach different decisions, even though they are all presented with the same set of facts, because they weigh the facts differently.

What is required by impartiality is that one base one's decision in a situation only on the relevant features of the case. In a moral decision that means that one uses only morally relevant considerations in making that decision. Objectivist ethics is distinguished from the various skeptical positions (i.e., subjectivism, relativism, emotivism, and prescriptivism) by its claim that what are the morally relevant considerations is an objective matter. Objectivists would hold that any moral judgment incompatible with the one favored by all of the morally relevant considerations is incorrect. And it is incorrect even if the judgment (1) accurately *describes* my feelings or attitudes (subjectivism), or (2) is accepted by my society (relativism), or (3) genuinely *expresses* my feelings or attitude (emotivism), or (4) is one that I would prescribe, even universally (prescriptivism). However, objectivism need not hold (though almost all objectivists have held) that all rational impartial observers will give the same weight to all of the morally relevant considerations.

What are the morally relevant facts? Here again most objectivists are in substantial agreement. The fact that someone will be harmed—e.g., killed; made to suffer pain, anxiety, sadness, or displeasure; disabled; or deprived of freedom, opportunity, or pleasure—is a morally relevant fact. Similarly, the fact that someone will be prevented from suffering one or more of these harms is morally relevant. Also morally relevant are facts concerning deception, the breaking of a promise, cheating, the breaking of a law, or the neglect of one's duty, e.g., as a doctor. What is in dispute is whether facts about a person's gaining greater abilities, opportunities, pleasure, or even the mere satisfaction of a desire are also morally relevant. An impartial person making a moral decision must take into account only these morally relevant facts; he cannot take into account any other facts, such as that it is he and his friends rather than people for whom he does not care who will suffer if this decision is taken rather than that one. One way of guaranteeing impartiality of this kind is to say that the person cannot use any fact about himself that distinguishes him from any other person (Gert, pp. 89 ff.; Rawls, pp. 136–142). He must make his moral decision as if he knew nothing about the identity of the various parties involved.

It may seem that, following this procedure, all impartial persons would always reach the same decision, but that is not the case. Two persons, completely impartial, may reach different decisions in a particular case because they may give different weights to the different harms involved. For example, one doctor may recommend compulsory genetic screening of all married couples because he holds that the harm preventable by such screening outweighs the deprivation of freedom that it involves, while another who believes that exactly the same amount of harm will be prevented by the screening does not think that it justifies the deprivation of freedom involved. An objectivist need not be dismayed by this conclusion, for he can hold that this disagreement occurs only within a very limited area and that both parties agree on the morally relevant facts. Thus, he knows that both doctors would hold it immoral to prevent genetic screening of those couples who desire it, and both would hold that such screening should be made available to all who want it. For in these cases harm can be prevented, and there is no countervailing deprivation of freedom.

Rationality and desire

The final issue that needs to be discussed is an explanation of agreement concerning the morally relevant considerations. This is where the concept of rationality becomes central, for it is the agreement of all impartial rational persons that determines the morally relevant considerations. Unfortunately the standard account of rationality (Rawls, p. 142 f.) in philosophy and in the social sciences, especially economics, puts no limit on the content of rational desires. On this account, to act rationally is simply to act in such a way as most efficiently to bring about the maximum satisfaction of one's desires, taking into account only their intensity, probability, etc. Since there is no limit to the content of rational desires, there is no limit to what counts as a morally relevant consideration. It may seem that this account of rationality is therefore incompatible with objectivity in ethics, but in fact it results in the most extreme form of objectivism. For objectivity can be obtained with this account of rationality only by having each desire count the same as every other desire of equal intensity and probability. Thus the standard account of rationality requires that impartiality, by ruling out all consideration of content, rule out all disagreement.

But there does seem to be a limit to the content of rational desires. Certain desires, viz., for death, pain, disability, or loss of freedom, opportunity, or pleasure, are taken by laymen as well as by psychiatrists as irrational, if one has no reason for so wanting. It is an essential feature of rationality to desire to avoid these harms or evils for oneself, unless one has some reason not to. This explains why there is universal agreement on the morally relevant considerations. All rational persons, if impartial, agree that these evils are to be avoided or prevented for everyone, unless there is an adequate reason for not doing so.

Once we have given this content to the concept of rationality, we are not threatened by anarchy if we allow impartial rational persons to disagree. The content we have given to reason limits the possible disagreement to the ranking of the various evils. This is not only a disagreement that an objectivist can accept, but one that he must accept if he is to provide an acceptable account of our actual moral experience. Moral discussions can then be explicitly focused on the relevant matters. When we have a moral disagreement due to a different ranking of the evils, then we see if we can come up with an alternative that lessens the evils on one side or the other. In this way, moral disputes need not result in each party's regarding the other as ill-informed, irrational, or partial, whereas if one accepts most objectivist theories one must regard all moral disagreement as due to ignorance, irrationality, or partiality. Nor, in reaction to these extreme objectivist theories, are we tempted to fall into skepticism and regard no moral problems as capable of a correct solution. Rather we can acknowledge the common ground, and thus have the opportunity to work together to see if we can lessen what all of us regard as evils, even though we may disagree on whether one evil is more or less important than some other evil.

Thus an objectivist theory that does not insist on unique answers to every moral problem, while it does not provide the kind of mechanical decision procedure that some doctors may have hoped for, does provide some help in solving the perplexing moral problems that arise in medicine and related fields. For it makes clear what factors are morally relevant, and thus guides the decision maker (physician, scientist, patient, subject, policymaker, etc.) toward that alternative that seems to be favored by most of the morally relevant considerations. But it also makes clear what these parties already knew: that there are some cases in which no single alternative is favored by most of the morally relevant considerations. But even in these cases, where each must decide which alternative he will adopt, the objectivist theory can be useful in helping him to rule out all of the morally unacceptable alternatives.

BERNARD GERT

[*For further discussion of topics mentioned in this article, see the entries:* NATURAL LAW; *and* OBLIGATION AND SUPEREROGATION. *While all the articles in this entry are relevant, see especially the articles* RULES AND PRINCIPLES, MORAL REASONING, *and* RELATIVISM. *See also:* DECISION MAKING, MEDICAL; DOUBLE EFFECT; JUSTICE; LAW AND MORALITY; *and* PRAGMATISM. *This article will find application in the entries* DEATH AND DYING: EUTHANASIA AND SUSTAINING LIFE, *article on* ETHICAL VIEWS; ENVIRONMENTAL ETHICS; HEALTH CARE, *articles on* RIGHT TO HEALTH-CARE SERVICES *and* THEORIES OF JUSTICE AND HEALTH CARE; INFORMED CONSENT IN THE THERAPEUTIC RELATIONSHIP, *article on* LEGAL AND ETHICAL ASPECTS; *and* TRUTH-TELLING, *article on* ETHICAL ASPECTS.]

BIBLIOGRAPHY

MORAL SENSE THEORIES

SHAFTESBURY, ANTHONY ASHLEY COOPER. *Characteristicks of Men, Manners, Opinions, Times* . . . London: 1711.

HUTCHESON, FRANCIS. *A System of Moral Philosophy.* London: 1755.

BONAR, JAMES. *Moral Sense.* Library of Philosophy.

Edited by J. H. Muirhead. New York: Macmillan Co., 1930.

DEONTOLOGICAL INTUITIONISM

REID, THOMAS. *Essays on the Active Powers of Man.* Edinburgh: 1788.
PRICHARD, HAROLD ARTHUR. *Moral Obligation: Essays and Lectures.* Oxford: Clarendon Press, 1949.
ROSS, WILLIAM DAVID. *The Right and the Good.* Oxford: Clarendon Press, 1930.

TELEOLOGICAL INTUITIONISM

SIDGWICK, HENRY. *The Methods of Ethics.* London: Macmillan & Co., 1874.
MOORE, GEORGE EDWARD. *Principia Ethica.* Cambridge: University Press, 1903.

NATURALISM

BENTHAM, JEREMY. *An Introduction to the Principles of Morals and Legislation.* London: 1789.
MILL, JOHN STUART. *Utilitarianism.* London: 1861.
PERRY, RALPH BARTON. *General Theory of Value.* New York: Longmans, Green & Co., 1926.

IDEAL OBSERVER THEORIES

SMITH, ADAM. *The Theory of Moral Sentiments.* London: 1759.
FIRTH, RODERICK. "Ethical Absolutism and the Ideal Observer." *Philosophy and Phenomenological Research* 12 (1952): 317–345.

THEOLOGICAL VIEWS

THOMAS AQUINAS. *Summa Theologiae,* I–II (1266–1274). Especially questions 90–97.
NIEBUHR, REINHOLD. *An Interpretation of Christian Ethics.* Living Age Books, no. 1. New York: Meridian Books, 1956.
RAMSEY, PAUL. *Deeds and Rules in Christian Ethics.* New York: Charles Scribner's Sons, 1967.

IMPARTIAL RATIONALITY

HOBBES, THOMAS. *Man and Citizen.* Edited with an introduction by Bernard Gert. Garden City, N.Y.: Anchor Books, 1972. Contains *De Cive* (1651), translated by Thomas Hobbes; and *De Homine* (1658), translated by Charles T. Wood, T. S. K. Scott-Craig, and Bernard Gert.
KANT, IMMANUEL. *Grundlegung zur Metaphysik der Sitten.* Riga: 1785.
BAIER, KURT. *The Moral Point of View: A Rational Basis of Ethics.* Ithaca: Cornell University Press, 1958.
GERT, BERNARD. *The Moral Rules: A New Rational Foundation for Morality.* New York: Harper & Row, 1970. Revised paperback. Torchbook, 1973.
RAWLS, JOHN. *A Theory of Justice.* Cambridge: Harvard University Press, Belknap Press, 1971.

IX
NATURALISM

The expression "ethical naturalism" is a common noun rather than a proper name. It is used as a label to classify ethical theories rather than as a name to refer to any single ethical thesis. It is commonly and properly used in at least three senses. Ethical naturalism is any ethical theory that (1) holds that the standard of value or obligation is provided only by nature or the natural; (2) explains the nature of value and obligation entirely within a naturalistic metaphysics; or (3) defines ethical words in terms of natural characteristics. The first definition provides a foundation for the moral rules and human rights so crucial to issues in medical ethics. The second claims to be the ideal philosophical basis for bioethics. The third explains how the life sciences can be logically relevant to judgments of value or obligation.

Main types of ethical naturalism

In the first sense, an ethical naturalism is any theory that takes nature or the natural as its standard of value or obligation. It stands opposed to any act or way of life that is artificial, unnatural, or a perversion of nature. One version takes natural purposes as its standard. Aristotle, for example, believed that each kind of thing has its specific end or goal determined for it by its very nature. The good of each individual is to realize its natural end. Another version takes natural law as determining right and wrong. Thus, Cicero claims that specific rules commanding certain kinds of conduct and forbidding others exist in the laws of nature independent of society and its laws. This moral law of nature dictates what humans ought or ought not to do. An important offshoot of the natural law theory is the view that all men possess certain natural rights. Locke defended the inalienable rights to life, liberty, and property. Still another version of ethical naturalism in the first sense takes presocial, primitive human nature as its standard. Rousseau claimed that innate human nature is fundamentally good but that it becomes corrupted and perverted by society with its artificial institutions. His ethical ideal was to live naturally, spontaneously acting upon innate human feelings and impulses. All versions of ethical naturalism in the first sense agree that the moral ideal is the natural life; they disagree as to the precise definition of "nature" and "the natural" (Lovejoy, Chinard, Boas, and Crane).

Ethical naturalism in the second sense is any theory that interprets value and obligation within the framework of a purely naturalistic metaphysics. It refuses to appeal to any supernatural entities like God, an immortal soul, or Platonic forms in explaining the good for man and the grounds of obligation. It may well be that Aristotle was an ethical naturalist in this sense as well as the first, but Cicero certainly was not, because he thought of the natural law as expressing the will of God and morally binding for just that reason. Some philosophers seem to be metaphysical naturalists first, in order to avoid any ontological commitment to supernatural entities, and then become ethical naturalists in order that their ethical theory may be consistent with their ontology. Thus, Samuel Alexander tries to show how he can extend his analysis of spatio-temporal nature to include an explanation of value, and John Dewey argues at length that both the existence of values and our knowledge of them can be understood purely in terms of the interests and intelligence of natural organisms. Others seem to reject all supernatural religion and morality and become ethical naturalists because they wish to place human life and human well-being at the center of ethics. Humanists, both non-Marxist and Marxist, are less concerned with ontological commitment than with the import of their ethical theories for individual self-realization and social reform.

Ethical naturalism in the third sense is any theory that defines ethical words in terms of natural characteristics. It rejects both ethical intuitionism, which holds that ethical words stand for nonnatural characteristics, and nondescriptive ethics, which maintains that ethical words do not refer to any kind of quality or relation. No very satisfactory definition of what counts as an ethical word has been given, but examples are "good," "bad," "right," "wrong," and "ought." Nor is it at all clear just what makes any characteristic "natural." The two most plausible suggestions are that natural characteristics are qualities or relations that exist in the spatio-temporal realm of nature *or* that they are qualities or relations that are given in experience or analyzable into such empirically observable characteristics. In any event, it is helpful to classify each version of ethical naturalism in the third sense according to the natural science from which it borrows its defining terms. To hold that "good" means either pleasant or desired is to hold a psychological version of ethical naturalism. To define "good" as that which serves physiological needs, or "right" as that which is conducive to the survival of the species, is to adopt a biological form of ethical naturalism. Finally, an anthropological or sociological version would be the proposition that "right" means to be in conformity with the mores of the agent's society. Diverse as these ethical theories are, they share a fundamental allegiance to empiricism, the view that experience is the source of all concepts and the ground of all knowledge.

Naturalistic fallacy

George Edward Moore charged all versions of ethical naturalism in the third sense with committing "the naturalistic fallacy." Although he repeated this charge frequently throughout the first three chapters of *Principia Ethica*, he never did make it clear which of several alleged errors he had in mind. Since the nature and fallaciousness of "the naturalistic fallacy" have been much debated for decades, the interpretation and evaluation of Moore's critical arguments are of considerable philosophical importance.

Moore often suggests that the fundamental error of ethical naturalism lies in defining, or trying to define, what is incapable of definition (Moore, p. 7). He argues that goodness, the primary ethical characteristic, is simple rather than complex and claims that the only philosophically illuminating sort of definition is one that analyzes some complex object into its parts. It follows, of course, that the word "good" cannot be defined in any philosophically relevant way. Since the fundamentally irreducible quality of goodness cannot be defined at all, it obviously cannot be defined in terms of natural characteristics. Ethical naturalists have typically tried to rebut this charge by arguing either that goodness is a complex property or that definitions that do not analyze a complex object are philosophically legitimate.

In other passages, Moore says (p. 13) that the error of ethical naturalism is in mistaking the proper subject matter of ethics. The crux of this charge is that the naturalists assign the wrong ontological status to values and obligations. By defining the vocabulary of ethics in terms of the concepts of the natural sciences, they locate the subject matter of ethics within the realm of nature. Actually, the ethical characteristics of goodness and oughtness are *non*natural properties that have a kind of being quite different from that of natural properties like yellowness, sweetness, or pleasantness. Moore agrees with the naturalists that values and obligations are facts, but he insists that they are nonnatural facts known only by intuition. Other critics of naturalism give a more radical interpretation of the ontological error by denying that values and obligations are facts of any kind. They maintain that values are ontologically different from facts and that what *ought to be* the case has a different metaphysical status from what *is* the case. Since the subject matter of ethics is values and oughts, it is a metaphysical muddle to define ethical words in terms of any kind of facts, either natural or nonnatural. Ethical naturalists have argued vigorously, on both metaphysical and epistemological grounds, against any ontological commitment to a nonnatural realm or to the being of "oughts." Even if their arguments are granted, however, it need not follow that the subject matter of ethics is the facts of nature. If either emotivism or prescriptivism is on the right track, ethical sentences do not describe any facts at all.

In still other places (e.g., p. 11), Moore says that the error of ethical naturalism consists in misrepresenting our ethical knowledge. This epistemological error treats our rational insight into values and obligations as though it were a species of empirical knowledge essentially like the knowledge provided by the natural sciences. Since the goodness of something is not a property that can be observed by external perception or felt by inner sensation, our knowledge of values must come from reason rather than experience. And since how people ought to act, as contrasted with how they do act, cannot be established by observation or experiment, our knowledge of obligations must also be a priori. Ethical naturalists have replied that, if values and obligations are properly interpreted through their naturalistic definitions, it is clear how scientific information can answer ethical questions. Much of the debate between the intuitionists and the naturalists from 1903 until about 1944 centered upon just this epistemological issue. The ethical naturalists were trying to show how empiricism can explain our knowledge in ethics just as it explains our scientific knowledge; the ethical intuitionists were insisting that only rationalism can give an adequate account of our ethical insight.

A fourth set of passages implies that Moore intended to charge the ethical naturalists with making a logical error. His complaint is that they try to establish their conclusions by using logically invalid arguments. One of his prime targets is Mill's notorious proof of utilitarianism. Mill argues that happiness is the one and only thing that is intrinsically desirable on the ground that it is the one and only thing desired for itself. Moore insists that from the psychological fact that happiness is desired it does not follow that it is desirable, worthy of being desired (p. 67). He also objects, on logical grounds, to Rousseau's argument that the simple life is best *because* it is natural (p. 42). From the fact that something is natural it does not follow that it is good, much less that it is the best. More contemporary examples of arguments that might be charged with this sort of logical fallacy are that genetic engineering is wrong because it is artificial and the argument that all men have a human right to medical care because all human beings need medical care. Roughly, the logical error can be said to consist in deducing an ethical conclusion from purely factual premises or, to use a slogan modeled on a passage in Hume, deducing an "ought" from an "is."

In spite of considerable disagreement about the precise nature of "the naturalistic fallacy," it used to be widely agreed that any argument from facts to values or from "is" to "ought" does commit some sort of logical error. More recently, however, serious attempts have been made to show that some such arguments are logically valid. Logicians, such as Prior, have constructed arguments that meet all the requirements of formal logic and by which one can infer an allegedly ethical conclusion from apparently factual premises (1960, pp. 201–202, 204). Cautious language is imperative here because it remains undecided whether such arguments show that ethical conclusions can be inferred from factual premises or merely that the distinction between factual and ethical sentences is vague. Other philosophers have advanced arguments that, they claim, move from purely factual premises to ethical conclusions and are valid not by virtue of their logical form, but by virtue of their content. Searle, for example, has defended the validity of an argument from the fact that someone has promised to do something to the conclusion that he ought to do it (p. 44). Although such arguments have been widely criticized, their plausibility leaves it an open question whether "the naturalistic fallacy" is genuinely a logical error at all.

Science and ethics

What is primarily at stake throughout the debate about the alleged fallaciousness of "the naturalistic fallacy" is the relevance of science to ethics. What bearing, if any, do the findings of the life sciences have upon bioethics? Suppose that medical science should establish the fact that in some cases of terminal cancer there is no way to relieve the intense pain while sparing the life of the patient. This factual premise might seem to some to support the ethical conclusion that in some cases euthanasia is right. Or suppose, to take a different sort of example, medical sociology should find that the medical resources required to save one life by organ transplant would, if allocated to chemotherapy, save two or even three lives. Many would conclude, in this instance, that the use of medical resources for organ transplantation is not the best allocation. The philosophical problem is to explain just how scientific facts like these can be relevant to ethical conclusions about what is right or good. Leaving aside the possible, but implausible, view that factual information has no bearing whatsoever upon bioethics, there are four solutions proposed to this problem.

1. The traditional solution is that factual information is relevant to ethical conclusions insofar, and only insofar, as it can be subsumed under some universal ethical principle. Thus, granted that the right act is always the act that is most conducive to pleasure or the absence of pain, the fact that in some cases euthanasia is the only effective way to remove great pain implies that euthanasia is sometimes right. Again, the factual minor premise that the use of medical resources for organ transplant does not save the largest number of lives implies that it is not the best allocation of medical resources only when coupled with the ethical major premise that the best allocation of medical resources is the one that saves the most lives. On this view, the findings of the life sciences are reasons for ethical conclusions only when combined with universal principles provided by a very different discipline, ethics.

2. The solution proposed by ethical naturalism in the third sense is that factual premises imply ethical conclusions by virtue of the defined meaning of ethical concepts. There is no need to add any moral principle to the factual premise because the naturalistic definition of the ethical word used in the conclusion shows that the conclusion merely reaffirms what has already been asserted in the premise. If the word "right" really does mean "most conducive to pleasure or the absence of pain," then to conclude that in some cases euthanasia is right is merely to reassert in other words what is asserted by the premise that in some cases euthanasia is the only means to the absence of pain. Again, the fact that the use of medical resources for organ transplant does not save as many lives as some alternative allocation clearly implies that it is not the best allocation if "the best allocation" is defined, at least for medical contexts, as "the allocation that saves the most lives."

3. A more recent suggestion is that factual information supports ethical conclusions by some special sort of nondeductive inference (Toulmin, pp. 160–161). Just as the scientific method reasons in terms of an inductive logic that cannot be reduced to any form of deduction, so ethical reasoning is governed by a special logic of ethics. What links the factual premise to the ethical conclusion is neither a tacit ethical principle nor a naturalistic definition, but a rule of ethical inference. Since these arguments do not follow the rules of standard deductive logic, they are not deductively valid; since they do claim to follow a different set of logical rules appropriate to ethics, they are still valid in some logical sense. Although this sort of solution has been advocated by several philosophers, they have not been able to agree on the rules for this special sort of logic or on what gives logical validity to it.

4. Finally, some emotivists argue that factual information is not *logically* relevant to ethical conclusions at all, but that it is relevant in some *psychological* sense. The fact that in some cases euthanasia is the only way of terminating intense pain does not logically imply that in some cases euthanasia is right; but since most people dislike pain and approve of whatever removes pain, the factual information does psychologically support the conclusion that euthanasia is sometimes right just because the belief in this factual premise tends to cause one to have a favorable attitude toward euthanasia. Similarly, the fact that the medical resources used to save one life by organ transplantation could be used in some other way to save two or more lives is psychologically relevant to the conclusion that it is not the best allocation of medical resources because, given our human desire to save as many lives as possible, belief in this fact tends to cause a less favorable attitude toward this allocation than to some alternative one.

Debate continues among philosophers as to just how factual statements can serve as reasons for or against ethical conclusions. Ethical naturalism remains one of the more plausible ways of explaining how the life sciences can make their contribution to the solution of the problems of bioethics.

Further implications

Ethical naturalism, in one or another of its various senses, is of considerable importance to bioethics in a number of additional ways. First, conclusions in medical ethics are frequently defended by an appeal to moral rules. The practice of taking organs for transplant from live donors is often criticized as a violation of the no harm rule, i.e., the general rule that the physician or surgeon ought never intentionally to injure a patient except for the greater medical benefit of that patient. Euthanasia and abortion pose moral problems for bioethics just because most people accept the rule that it is generally wrong to kill a human being. What status are we to ascribe to such moral rules? Unless one is willing to admit that moral rules are mere conventions—artificial and arbitrary—one is committed to the view that they are in some sense natural. And this, of course, is the view that the standard of right and wrong is the natural, where the natural is conceived in terms of the law of nature. Although the traditional language of natural law theory is generally rejected in contemporary bioethics, the substance of this version of ethical naturalism in the first sense remains in any view that takes moral rules seriously.

Second, many issues in bioethics hinge on the existence and content of certain human rights. Whether the routine harvesting of organs from cadavers without express consent is morally justified depends in part upon whether the human right to property implies that either the deceased individual or the next of kin "owns" the cadaver. It is widely taken for granted that certain types of experimentation with human subjects are morally right only if the subjects have given their free informed consent. But this presupposes the existence of some human right to privacy or personal security that such experimentation would violate were it not waived by consent. These human rights, implicitly affirmed or denied in medical ethics today, are simply the traditional natural rights under a new name. Once more ethical naturalism in the first sense, viz., the offshoot of natural law theory that upholds the existence of natural rights, turns out to be of unsuspected importance for bioethics.

Third, if bioethics is to reach any conclusions whatsoever about what is good or bad, desirable or undesirable, it must have some theory of value. While philosophy offers a wide array of value theories, not all of these are equally suitable for the purposes of bioethics. A supernatural or nonnatural theory of value will at least place ethics at a considerable distance from the life sciences and may sever the connection between them entirely. A naturalistic ethics, even in the second sense, will interpret values and disvalues in terms of categories such as pleasure, desire, human needs, life, or species survival. Ethical naturalism, therefore, claims to be the ideal philosophical basis for bioethics, since it offers an ethical theory that makes the life sciences directly and decisively relevant to the solution of ethical problems. This does not, however, prove that any version of ethical naturalism is true, but it does indicate both its theoretical and practical importance for bioethics.

CARL WELLMAN

[*Directly related are the entries* BIOLOGY, PHILOSOPHY OF; MEDICINE, ANTHROPOLOGY OF; NATURAL LAW; RIGHTS; *and* SCIENCE: ETHICAL IMPLICATIONS.]

BIBLIOGRAPHY

ADAMS, ELIE MAYNARD. *Ethical Naturalism and the Modern World-View.* Chapel Hill: University of North Carolina Press, 1960.

ALEXANDER, SAMUEL. *Space, Time and Deity.* London: Macmillan, 1920, vol. 2, pp. 236–314.

ARISTOTLE. *Nicomachean Ethics.* William David Ross, trans. *The Basic Works of Aristotle.* Edited by Richard McKeon. New York: Random House, 1941, pp. 927–1112. See pp. 935–952, bk. 1.

BRANDT, RICHARD B. *Ethical Theory.* Englewood Cliffs, N.J.: Prentice-Hall, 1959, pp. 151–182.

CICERO. *De Legibus.* Clinton Walker Keys, trans. *De Re Publica, De Legibus.* Loeb Classical Library. London: William Heinemann; New York: G. P. Putnam's Sons, 1928, pp. 287–519. See especially pp. 317–331.

DEWEY, JOHN. *Theory of Valuation.* Chicago: University of Chicago Press, 1939.

HUDSON, WILLIAM DONALD, ed. *The Is–Ought Question.* London: Macmillan, 1969.

HUME, DAVID. *Treatise of Human Nature* (1738). Edited by L. A. Selby-Bigge. Oxford: Clarendon Press, 1888, pp. 455–476.

LOCKE, JOHN. "An Essay Concerning the True Original Extent and End of Civil Government" (1690). *Treatise of Civil Government and A Letter Concerning Toleration.* Edited by Charles L. Sherman. New York: Appleton-Century Co., 1937, pp. 3–162.

LOVEJOY, ARTHUR ONCKEN; CHINARD, GILBERT; BOAS, GEORGE; and CRANE, RONALD S., eds. *A Documentary History of Primitivism and Related Ideas.* Contributions to the History of Primitivism. Baltimore: Johns Hopkins Press, 1935. Contains fifty-seven meanings of "Nature."

MILL, JOHN STUART. *Utilitarianism* (1863). *Utilitarianism, Liberty and Representative Government*. Edited by A. D. Lindsay. New York: E. P. Dutton & Co., 1950, pp. 1–80.

MOORE, GEORGE EDWARD. *Principia Ethica*. Cambridge: Cambridge University Press, 1903.

PRIOR, ARTHUR N. "The Autonomy of Ethics." *Australasian Journal of Philosophy* 38 (1960): 199–206.

———. *Logic and the Basis of Ethics*. Oxford: Clarendon Press, 1949.

ROUSSEAU, JEAN JACQUES. *A Discourse on the Origin of Inequality* (1755). G. D. H. Cole, trans. *The Social Contract and Discourses*. London: J. M. Dent & Sons, New York: E. P. Dutton & Co., 1913, pp. 155–246.

SEARLE, JOHN R. "How to Derive 'Ought' from 'Is'." *Philosophical Review* 73 (1964): 43–58.

STEVENSON, CHARLES LESLIE. *Ethics and Language*. New Haven: Yale University Press, 1944, pp. 152–173.

TOULMIN, STEPHEN EDELSTON. *An Examination of the Place of Reason in Ethics*. Cambridge: Cambridge University Press, 1950, pp. 130–165.

X
NON-DESCRIPTIVISM

When classifying ethical theories, it is essential to make clear the questions to which they are answers; for if different theories are answers to different questions, they do not necessarily conflict. The main division is between theories about the meanings of the moral words or the analyses of the moral concepts, and theories about the rules of valid reasoning on moral questions of substance. Non-descriptivism belongs to the former class, whereas utilitarianism, for example, belongs to the latter. Thus utilitarianism, which answers the question "How ought we to decide what we ought to do?" is not necessarily in conflict with non-descriptivism, which is an answer to the question "What does 'ought' mean?" The relation between the two types of question is that the rules of valid reasoning in any field depend upon the rules governing the uses of words (that is, upon the conceptual framework) in that field. Thus the fact that "all" and "some" have the meanings that they have makes the following an invalid inference: "Some abortions are wrong; therefore all abortions are wrong." We shall see that, so far from utilitarianism and non-descriptivism being incompatible, a certain form of the first is a consequence of a certain form of the second.

Within theories about the meanings of the moral words the main division is between descriptivist and non-descriptivist theories. A crude and unilluminating first attempt at stating the difference between them is to say that descriptivism regards moral judgments as factual statements *describing* people, acts, etc., or attributing properties to them, whereas non-descriptivism treats them as having an altogether different role in our language (see below). But the difference is best illustrated by examples. Examples of descriptivist theories are various forms of ethical naturalism according to which moral judgments are equivalent to certain statements of empirical, or at least not specifically moral, fact; when their analysis is understood, they can be verified or falsified by observation of these non-moral facts. For example, if it were true (as it would be on a very crude type of naturalistic theory) that "wrong" meant "forbidden by the Church", then we could determine that euthanasia was wrong simply by observing the verbal behavior of those in ecclesiastical authority (e.g., by reading papal encyclicals). This assumes that the word "authority" is a purely descriptive word, which in many uses it is not.

An important sub-class of naturalism is subjectivism (a view which goes back to the ancient Greek philosopher Protagoras). This is a descriptivist theory, in that it holds that moral judgments state facts of a sort. But whereas on some descriptivist theories (the objectivist ones) the facts in question will be facts about the external world, independent of the thoughts, dispositions, etc., of people, on others (the subjectivist ones) they will be facts about these latter. It is an interesting exercise to determine to which of these classes the "ecclesiastical" type of naturalism just mentioned belongs. Subjectivism is a type of naturalism because it equates moral facts with a certain kind of non-moral facts, namely, facts about people's thoughts, etc.

Another example of a descriptivist theory is ethical intuitionism. This rejects naturalism, but maintains that moral judgments are statements of a different sort: they attribute certain specifically moral (sometimes called non-natural) properties of rightness, wrongness, goodness, badness, etc., which are not the same as, nor analyzable in terms of, non-moral properties of any kind. Moral properties are, as was said, *sui generis* (of a sort all their own). The way we ascertain whether something has or lacks these properties is by intuition, i.e., by exercising a special moral faculty whereby we apprehend their presence or absence. The distinguishing mark of intuitionist theories, among descriptivist theories, is their irrationalism; that is to say, at certain crucial points in the process which is supposed to end with an answer to a moral question, they appeal, not to argument, but to our alleged ability to know, without argument, the truth of certain moral facts. Although intuitionism was generally thought to have been discredited by its essential irrationalism, and now has few overt defenders, a great number of present-day philosophers (perhaps the majority) become crypto-intuitionists when they descend to discuss practical issues; i.e., they think that the way to do this is to look for moral judgments which they and their readers can accept without argument. This often amounts to a mere appeal to received opinion (to what people, including the philosopher in question, think), and thus comes close to a form of ethical relativism or subjectivism, the dividing line between which and intuitionism is often invisibly narrow.

The terms "cognitivism" and "non-cognitivism" are commonly used to distinguish the same types of theory as are "descriptivism" and "non-descriptivism". The former pair make the division according to whether a theory does or does not allow of the existence of *knowledge* of some moral facts. Other ways of making the distinction are by asking whether or not a theory allows that moral judgments may be *true* or *false*, or whether or not it allows that moral adjectives may be the *"names"* of *properties*. All these alternative ways of marking the distinction are unreliable, because of the obscurity of the words "knowledge", "true", and "property". Some non-descriptivist theories may well be able to give a sense to the claim that we can know that a moral judgment is true, or that an act has the property of being wrong. But we may perhaps define descriptivism and non-descriptivism by saying that the former, unlike the latter, claims that the meanings of the moral words are a function solely of the truth-conditions of propositions containing them; that is to say, there is no further element in their meaning which could leave us the choice of asserting or denying them, once the facts of the case are given. Thus descriptivism, unlike non-descriptivism, is committed to a "verificationist" account of the meanings of the moral words.

Non-descriptivism arose because of dissatisfaction with all these kinds of descriptivism. It was thought that G. E. Moore had produced conclusive arguments against all forms of naturalism, including subjectivism; but his own and other attempts to produce a viable intuitionist theory had encountered insuperable difficulties over the nature of the *sui generis* moral properties and our means of discerning them. The first non-descriptivist theory to hold the field was of an irrationalist type, called "emotivism" (although non-descriptivism does not *have* to be irrationalist). Leaving aside certain premonitory remarks of Berkeley and Hume, and a longer and very Humean passage of J. S. Mill, the first full-scale exposition of non-descriptivism was by the Swedish philosopher Hägerström; this was succeeded by Ogden and Richards' work. Emotivism was embraced by some, though not all, of the Logical Positivists (the "Vienna Circle") and their followers. Its most trenchant expositions were by Carnap and Ayer; and its fullest and most considered defense by Stevenson.

Emotivists are non-descriptivists because they hold that there is a central element in the meaning of moral judgments in their typical uses which goes beyond the describing of the acts, people, etc., judged, or the ascribing to them of properties, whether objective or subjective. This element, called by Stevenson the emotive meaning, does not necessarily exhaust the meaning (there may be descriptive elements as well), but it prevents any analysis in wholly descriptive terms. The emotive meaning of a moral term consists in its tendency to express or evince feelings or attitudes (e.g., of approval) that the speaker has, or to produce or evoke such attitudes in a hearer. One of the commonest and most time-wasting confusions in moral philosophy is to equate these non-descriptivist views with the subjectivist view (a form of naturalism; see above) that moral judgments are *statements that* the speaker as a matter of psychological or behavioral fact has such feelings or attitudes. Though both views are opposed to the objectivist kind of descriptivism, and are therefore sometimes loosely grouped together under the name "subjectivism", this wide use of the term is highly misleading, because it obscures the main division of ethical theories into descriptivist and non-descriptivist, and because it suggests that all types of non-descriptivist theory are open to the same objections as are subjectivist descriptivist theories; this is not so.

The mistake becomes even more serious when made about the form of non-descriptivism which has largely superseded emotivism. This is sometimes called prescriptivism. It holds that the peculiar element in value-judgments (of which moral judgments are a sub-class) which prevents their being wholly descriptive, is their prescriptivity, i.e., their action-guiding function. There are prescriptivist elements clearly detectable in the moral philosophies of Socrates, Plato,

Aristotle, and Kant, though overlaid by their predominant descriptivism. Some of the earlier emotivists, by comparing moral judgments with imperatives, adopted a form of prescriptivism (Carnap; Ayer, p. 160; Stevenson, p. 21). More recent writers, however (e.g., Hare, 1952, 1963) have departed in two important respects from emotivist views. First, they have insisted that prescriptive judgments (including imperatives) can function as premises or as conclusions in logical inferences, the term "emotive", which suggests irrationalism, is therefore inappropriate in characterizing the meaning of moral judgments, even if it has a prescriptive element. Secondly, they have claimed, in agreement with many other ethical writers who are not prescriptivists, that moral judgments are universal or at least universalizable. This feature, which they share with descriptive judgments, can be explained as follows: it is self-contradictory to make dissimilar moral judgments about two cases which one admits to be similar in all non-moral respects.

On these two features of prescriptivity and universalizability, it has been the hope of recent writers to found a theory of moral reasoning which would enable moral questions, such as those in bioethics, to be rationally discussed, and thus escape the imputation of irrationalism which attached to the earlier emotivists. Such a theory owes a great deal to Kant: we are to seek prescriptions which we can universalize; i.e., the "maxims of our actions" (Kant's phrase) are to be such as we are prepared to accept for identical cases in which we are at the receiving end. Such cases can be hypothetical ones (it is not necessary for there to be any likelihood that we shall actually find ourselves in the situations of our victims); all that is necessary is that we should prescribe *as if* we were to be the sufferers from the consequences of the actions prescribed. In the hypothetical case, the sufferer is to be supposed to have the same desires as the present victim has, not as the present agent has; therefore objections of the sort "If it were done to me I would not mind" are ill-taken.

Such a theory of moral reasoning leads naturally to the requirement of impartiality in moral judgment, i.e., the requirement to treat the equal interests of every affected party as of equal weight. It thus leads, in spite of its Kantian connections, to a form of utilitarianism; for if one gives equal weight to the equal interests of each party, one will be led to maximize the satisfaction of those interests, taken as a whole. The theory also has affinities with the Christian (and indeed pre-Christian) Golden Rule that we should do to others as we wish them to do to us sc. in precisely similar situations (Hillel, Babylonian Talmud, *Sabbath*, 31a; Matthew 7:12; see also Tobias 4:16).

It also has links with the so-called rational-contractor theory, which holds that we should adopt those principles which would be agreed to by a set of self-interested parties who were without knowledge of their roles in the society to be governed by the principles (Rawls; Richards; both, however, attempt to draw non-utilitarian conclusions from their theories). It also has affinities with the so-called ideal-observer theory, which holds that we should adopt those principles which would be prescribed by an impartially benevolent spectator who knew all the facts (Haslett). In Christian morality God often serves as this ideal prescriber. Most of these theories are, however, propounded by descriptivists (for an interesting reference to the possibility of formal analogies between descriptivist and non-descriptivist theories in their approaches to actual moral issues, see Richards, p. 85).

All these theories have considerable bite when discussing questions in bioethics. For example, questions about the legitimacy of euthanasia can be handled as follows: if, in a particular case, we can determine that to kill the patient would be, in sum, most in the interests of all affected parties, including his own, then it is legitimate. However, if it could be shown that a law, or even a rule of medical ethics, permitting this would (as it well might) lead in practice to abuses (i.e., to the use of euthanasia in many cases where the conditions for its legitimacy were not in fact realized), then the law or rule ought not to be adopted. This would be especially important if cases in which euthanasia would be justified (i.e., in which there was no way of doing better for the interests of the parties) are rare, as is plausibly claimed. These two arguments (which, though contrary in tendency, are not actually inconsistent) lead together to the conclusion that euthanasia is sometimes morally right but that it would be morally wrong to make it legal (a combination of views actually held by many medical men). If such a conclusion can be shown to be in accord with the Golden Rule (as it probably can), it should be of interest to Christian moralists (Hare, 1975).

R. M. HARE

[*While all the articles in this entry are relevant, see especially the articles* UTILITARIANISM *and* NATURALISM.]

BIBLIOGRAPHY

Ayer, A. J. *Language, Truth and Logic.* London: V. Gollancz, 1936, chap. 6.

Berkeley, George. *A Treatise Concerning the Principles of Human Knowledge* (1710). Introduction, sec. 20.

Carnap, Rudolph. *Philosophy and Logical Syntax.* London: Kegan Paul, Trench, Trubner & Co., Psyche Miniatures, 1935. Partial reprint. *The Age of Analysis.* The Great Ages of Western Philosophy, no. 6. Edited by M. White. Boston: Houghton Mifflin, 1955, pp. 209–225.

Hägerström, Axel, "Kritiska Punkter i Värdepsykologien." [Critical Points in the Psychology of Value] *Festskrift tillägnad E. O. Burman.* Uppsala: K. W. Appelbergs Boktryckeri, 1910, pp. 16–75. For references and summary, see his *Philosophy and Religion.* Translated by Robert T. Sandin. London: G. Allen & Unwin, 1964, p. 315.

Hare, R. M. "Ethical Theory and Utilitarianism." *Contemporary British Philosophy.* 4th sers. Edited by H. D. Lewis. The Muirhead Library of Philosophy. London: George Allen & Unwin, 1976.

———. "Euthanasia: A Christian View." *Philosophic Exchange* 2, no. 1 (1975), pp. 43–52.

———. *Freedom and Reason.* Oxford: Clarendon Press, 1963.

———. *The Language of Morals.* Oxford: Clarendon Press, 1952.

———. *Practical Inferences.* London: Macmillan & Co., 1971. Bibliography.

Haslett, D. W. *Moral Rightness.* The Hague: Martinus Nijhoff, 1974.

Hume, David. *Treatise of Human Nature* (1738). Bk. 3 pt. 1, sec. 1.

Kant, Immanuel. *Grundlegung zur Metaphysik der Sitten* (1785). Translated by Lewis White Beck as *Foundations of the Metaphysics of Morals.* Indianapolis, Ind.: Bobbs-Merrill, 1956.

Mill, J. S. *A System of Logic* (1843). Bk. 6, chap. 12.

Moore, G. E. *Principia Ethica.* Cambridge: At the University Press, 1903.

Ogden, C. K., and Richards, I. A. *The Meaning of Meaning.* London: Kegan Paul, Trench, Trubner & Co.; New York: Harcourt, Brace & Co., 1923, chap. 6, p. 125.

Protagoras. "Protagoras." *Die Fragmente der Vorsokratiker, griechisch und deutsch.* Edited by Hermann Diels. Berlin: Weidmannsche Verlagsbuchhandlung, 1951–1952, vol. 2, pp. 253–271. Also in Philip Wheelwright. *The Presocratics.* New York: Odyssey Press, 1966, pp. 239–248.

Rawls, John. *A Theory of Justice.* Cambridge: Harvard University Press, 1971.

Reichenbach, Hans. *The Rise of Scientific Philosophy.* Berkeley: University of California Press, 1951.

Richards, D. A. J. *A Theory of Reasons for Action.* Oxford: Clarendon Press, 1971.

Stevenson, Charles L. *Ethics and Language.* New Haven: Yale University Press, 1944.

Urmson, J. O. *The Emotive Theory of Ethics.* London: Hutchinson, 1968. Bibliography.

XI
MORAL REASONING

What is moral reasoning? Reasoning is the process by which one arrives at a conclusion from premises, and in moral reasoning one arrives at a moral conclusion. It may contain the word "moral" or its cognates, or an expression such as "virtue" or "good action" that is normally understood in a moral sense. But no vocabulary peculiar to ethics is required; one may merely say that something should be, or ought to be done, implying that it is for moral reasons that this is so. A moral conclusion is a moral judgment, however expressed.

To describe moral reasoning we must find the premises from which moral conclusions can be derived, and one way of seeing the task is as that of finding criteria for the application of the concepts used in moral judgment. In moral reasoning we use the criteria; in moral philosophy we search them out. Thus, when Plato asked about the nature of justice, or courage, or temperance, he insisted that we needed this knowledge if we were to argue correctly about the morality of something that was done, and much of the history of ethics is about discussions of this kind. Kant, for instance, proposed that we test the moral permissibility of an action by considering whether the principle under which it fell could be willed as a universal law. And Bentham and his Utilitarian followers thought that an action could be shown to be right by showing that it tended to produce the greatest happiness of the greatest number. For reasons that will be mentioned later many philosophers of the present century have abandoned the search for such criteria of morality; nevertheless John Rawls's *Theory of Justice,* one of the most important and interesting of contemporary books on ethics, is precisely concerned with the criteria by which institutions and actions may be judged to be just.

So far nothing has been said to show why there should be any special difficulty about describing moral reasoning or why, indeed, moral reasoning should be any different from reasoning about any other subject. Obviously there are difficult concepts involved, but there are many elusive concepts outside of ethics, as, for instance, that of *metaphor* or *tragedy* or *intelligence* or *genius.* Why is the problem of moral reasoning different from the problem of reasoning elsewhere?

Undoubtedly there is a special difficulty about

moral reasoning, and it comes from the fact that in some sense or other (and the interpretation is a matter of controversy) moral reasoning is practical reasoning: it is reasoning about what to do. David Hume made the minimal claim when he said that morals excite passions and produce or prevent actions, and insisted that our theory of moral reasoning must take this into account. He himself went farther, demanding a quite general connection between moral judgment and the will; but the mere fact that men ever do something because they think it right to do so is enough to create problems for certain philosophers. For some have had theories of the psychology of human action by which the search for pleasure and the avoidance of pain is the only motive on which anyone can act. Bentham, for instance, believed this and had to reconcile it with the fact that moral action is possible. It was public interest that determined morality but private interest that determined what was done. Reasoning about what was right and good was reasoning about the general happiness; reasoning about what to do was about one's own pleasure and pain. No one could be influenced by moral considerations unless public and private interest seemed to coincide, and no one ever acted morally except for the sake of gaining pleasure or avoiding pain. Bentham accepted both conclusions. For Kant, however, the problem was more acute. He insisted that only action done for the sake of duty had true moral worth, while no action was done for the sake of duty if done purely to satisfy the agent's desire. How could there be actions not determined by the usual causality of desire? This is not the place to discuss Kant's solution; his problem is mentioned to show that even on the minimal interpretation of the dictum that moral reasoning is practical some philosophers find themselves in trouble. They ask, "How is moral action possible at all?"

For most, however, the problem arises only upon a different understanding of the practical nature of moral judgment and hence of moral reasoning; it is not just that men sometimes do what they think they ought to do, but that there is a general, or even universal, connection between moral judgment and conduct. How close a connection should we see? Some, following Plato, go so far as to deny that anyone ever sees that he ought to do one thing and then deliberately does something else. No one, they say, ever really sees the better and chooses the worse.

So if, in what looks like moral reasoning, he seems to draw the conclusion that he should do a thing, but does not choose to do it, the conclusion has not really been drawn. One may, however, insist that the phenomenon called (somewhat misleadingly) "weakness of will" does exist, that people may be fully aware that they should do a thing but have no intention of doing it. Taking this line, one will say that moral judgment and action are not quite as closely tied together as some have thought and will allow the possibility of reasoning to a moral conclusion without the accompanying action or intention.

Whether or not this is to be called "practical reasoning" is another matter. There is a tradition that follows Aristotle in insisting that the conclusion of a piece of practical reasoning is an action, and it would be possible to adopt this terminology whatever one said about the problem of "weakness of will." It will be more natural, however, to say that moral reasoning may be to a conclusion not accompanied by action (or intention) if one is satisfied that a moral judgment may on occasion be like this.

In what sense then is moral reasoning practical reasoning, if not in the sense that it leads to a conclusion that is necessarily accompanied by a choice? The answer seems to lie in the idea that certain considerations give reasons for acting, and that the premises from which a moral conclusion follows must be premises of this kind. In moral reasoning we argue that something should or should not be done, and if it should be done this implies that there are reasons for doing it. This account has the advantage of applying directly to reasoning about other people's action as well as one's own.

Practical reasoning that meets these conditions is something extremely familiar in everyday life. A doctor starts, for instance, with the objective of curing his patient and reasons that it will be necessary to get his blood pressure down. Further he calculates that the administration of a certain drug will have this effect. He reasons, in other words, about the necessary conditions and the sufficient conditions (the lowering of the blood pressure necessary for the cure; the taking of the drug sufficient for the lowering of the blood pressure) until he comes to something that is within his power, such as the administration of the drug. This is something he can do, and it is seen as the first link in a chain leading to the desired end. The starting point is, as Aristotle says, "*to orekton,*" something wanted, and the reasoning is with a view to achieving the desired thing, which may be, but is not necessarily, external to what is to be done. So the doctor might reason about how to cure the patient or about how to act in accordance with his wishes.

Similarly, it might be supposed, moral reasoning has as its starting point something wanted; either an end stated in terms that are already moral, such as to be just, to do morally good actions, or perhaps an end such as the relief of suffering or the preservation of the dignity of human life. The problem about moral reasoning, seen as a form of practical reasoning, lies in the status of these ends. For suppose that a certain individual does not have them? How, then, shall we be able to proceed by practical reasoning to the conclusion that he should, for example, refrain from killing or injuring others, or should help those in need? We have a very strong desire to say that we can always draw this conclusion, if only because "should" and "should not" are instruments of social pressure, and we want to bring pressure against *anyone* who acts in an antisocial way. What is more, we should like to be able to say that each and every man *has reason* to be moral whatever his particular ends may be.

It is not surprising, therefore, that there should have been many attempts to show that there are ends which everyone must have and to use these as the starting point for practical moral reasoning. Can it not be said, for instance, that we necessarily desire our own happiness, and may it not be possible to show that therefore we should live a virtuous life? Plato's *Republic* contains one of the most famous examples of a theory of this type. It is argued that, although just actions apparently benefit others and are to the disadvantage of the just man, nevertheless the way to be happy is to be just. Some have elaborated such theories, but others have rejected them out of hand. For one thing it is really very difficult to show that an unjust action could never be advantageous. And in any case it has seemed to many that no one who wants to be just in order to be happy is truly moral.

Kant put forward this latter point of view in an extreme form, insisting that a moral judgment is a categorical imperative, not one telling us how to achieve the end of happiness, or any other that we might happen to have. Kant's own solution was complex, and it is hard to interpret at certain points. But it consisted essentially in equating moral action with rational action, not because in acting morally a man did what was likely to get him what he wanted, but because moral action was action determined merely by the principle of universalizability and therefore (Kant said) by the faculty of reason.

Most philosophers agree that neither Plato nor Kant was successful in solving the problem. How can we represent moral reasoning as practical and also universally applicable?

At present the answer most often given is one that derives from Hume's moral philosophy. Hume's answer to his own question about how we can guarantee a connection between moral judgment and the will lay in his theory of moral sentiment. He distinguished, on the one hand, judgments of fact (which he called conclusions of reason) and, on the other, sentiments or feelings (which he called passions). Conclusions of reason were, in themselves, inert; only in conjunction with a passion could they produce or prevent action. Moral judgments were expressions corresponding to sentiments; to judge an action virtuous was (at least in one rather simple version of Hume's theory) to say that one felt a pleasing sentiment of approbation when it was contemplated; and there seemed to be no difficulty in seeing why we should do the actions that pleased us thus. The connection between moral judgment and the will had, let us suppose, been taken care of. What had become of reasoning to a conclusion about what was right and wrong?

The answer is that a gap had appeared between the facts from which moral reasoning may be supposed to proceed and the conclusion to which it led. On the one hand, there is the contemplation of the facts and reasoning about them, on the other, a moral sentiment, and between the two there is a gap: Hume's famous gap between "is" and "ought." There is some disagreement about the interpretation of Hume's theory at this point; some say that the gap only consists in this: that *deductive* reasoning cannot join the two. (In other words, one will never be forced on pain of contradiction to move from premises to conclusion.) But one may doubt whether any form of reasoning is possible from "is" to "ought" if there are not even rules of evidence linking the two. And Hume does not suggest that there are; he thought that as a matter of fact sentiments of approbation were felt toward what was "useful or agreeable to ourselves or others," but he did not suggest that any other connection linked the two.

An even more thoroughgoing version of this theory was worked out in the thirties and forties of the present century, under the influence of logical positivism, but having as its adherents many who would not count themselves as logical positivists. C. L. Stevenson, who did more work on this than anyone else, argued in *Ethics and Language* that moral language expresses attitudes and that there is no logical connection between a man's attitudes and his beliefs. The connection between beliefs and attitudes is merely causal, and there is no reason why a given set of "premises" may not bring one individual to one "conclusion" and another to another conclusion. In fact there is no limit—except that imposed by the requirement that attitudes be consistent—to the moral judgments that may be "derived" from a given set of facts. Stevenson talks about moral "arguments" and "rational methods" of trying to resolve moral disputes, but it is not clear that such descriptions are warranted. When A. J. Ayer, who had put forward a similar view, said that in his opinion there was argument about facts but not about moral values, he was perhaps representing the position more accurately.

A similar doubt as to whether argument from fact to value is left standing can be raised about the imperativist theory of R. M. Hare. He too denies that there are rules of evidence linking facts to moral judgment, except those requiring not only consistency but also that moral judgments shall ultimately concern kinds of facts, not individuals or individual occasions. Within these limits one may make any moral valuations one chooses, and in fact Hare stresses the element of decision in adopting principles after rehearsing the facts. He speaks of the decisions as being "based on" the facts, but again the description may be challenged. On this theory it does not seem to be here that one finds something properly called moral reasoning, but rather in the reasoning from one moral imperative to another. The latter is of course an important kind of reasoning, whose rules have been developed by Hare himself. But so far as the original problem of practical reasoning is concerned one may say that it has been banished rather than solved.

The correct theory of moral reasoning is thus a matter of current controversy. Will our actual moral reasoning necessarily be affected by these uncertainties? The answer seems to be that sometimes it will not be affected. So long as certain ends are taken for granted, we can proceed as if the theoretical difficulties about the foundations of moral reasoning did not exist. Aristotle points to this kind of case when he says that a doctor does not deliberate about whether to cure his patient; he takes the end for granted and asks how it may be achieved. But Aristotle's example raises for us the difficulties that can so often be put aside. Debating the moral issues of euthanasia we ask whether a doctor must always try to cure his patient and, if not, when he should give up this end, and whether he should ever actually seek the patient's death. Can the philosophers help us here? Obviously they can so far as what is wanted is a clarification of ideas such as that of passive and active euthanasia; but when the question touches problems about the ultimate starting point of moral reasoning the area of uncertainty and controversy is reached.

PHILIPPA FOOT

[*While all the articles in this entry are relevant, see especially the articles* THE TASK OF ETHICS, RULES AND PRINCIPLES, NON-DESCRIPTIVISM, *and* RELATIVISM. *See also:* DEATH AND DYING: EUTHANASIA AND SUSTAINING LIFE, *article on* ETHICAL VIEWS.]

BIBLIOGRAPHY

ANSCOMBE, G. E. M. *Intention.* 2d ed., Oxford: Basil Blackwell, 1963. See sections 33–44.

———. "Thought and Action in Aristotle." *New Essays on Plato and Aristotle.* Edited by R. Bambrough. London: Routledge & Kegal Paul, 1965.

ARISTOTLE. *Nicomachean Ethics.* Translated by Martin Ostwald. Indianapolis: Bobbs-Merrill Co., 1962.

AYER, A. J. *Language, Truth and Logic.* 2d ed. London: Victor Gollancz, 1947. See Chapter VI, pp. 102–120.

BENTHAM, JEREMY. *An Introduction to the Principles of Morals and Legislation* (1780). New York: Humanities Press, 1970.

GEACH, P. T. "Dr. Kenny on Practical Inference." *Analysis,* January 1966, pp. 76–79.

HARE, R. M. *Freedom and Reason.* Oxford: Oxford University Press, 1963.

———. *The Language of Morals.* Oxford: Oxford University Press, 1952.

———. *Practical Inferences.* London: Macmillan, 1971. See pp. 69–73.

HUME, DAVID. *A Treatise of Human Nature* (1739). Oxford: Oxford University Press, 1941.

KANT, IMMANUEL. *Critique of Practical Reason* (1788). Translated by Lewis White Beck. Indianapolis: Bobbs-Merrill, 1956.

———. *Foundations of the Metaphysic of Morals* (1785). Translated by Lewis White Beck. Indianapolis: Bobbs-Merrill, 1959.
KENNY, A. J. "Practical Inference." *Analysis* 26 (1966): 65–75.
MACINTYRE, ALASDAIR C. "Hume on 'Is' and 'Ought'." *Against the Self-Images of the Age*. New York: Schocken Books, 1971, pp. 109–124.
PLATO. *Republic*.
RAWLS, JOHN. *A Theory of Justice*. New York: Oxford University Press, 1972. Cambridge: Belknap Press of Harvard University Press, 1971.
STEVENSON, CHARLES L. *Ethics and Language*. New Haven: Yale University Press, 1944.
THOMSON, JUDITH J. and DWORKIN, GERALD, eds. *Ethics*. New York: Harper & Row, 1968.
WALLACE, G., and WALKER, A. D. *The Definition of Morality*. London: Methuen & Co., 1970.

XII
RELATIVISM

The expression "ethical relativism" refers to a number of diverse theories, each of which asserts that something ethical is relative to something else, for example that what one ought to do is relative to the mores of one's society or that one's values are relative to one's personality. There are many versions of ethical relativism, not all of equal importance for bioethics. Some, such as cultural relativism, threaten the truth of any objective bioethics and the legitimacy of any innovative medical practice. Others, such as situation ethics, maintain ethical absolutism in our changing world. One version differs from another either in what it asserts to be relative or in that to which it takes it to be relative. In all versions, however, to assert that one thing is relative to another is to assert that the former varies with and depends upon the latter.

It is helpful to distinguish between ethical relativism and ethical nihilism. While the relativist asserts that something ethical is relative, the nihilist denies its very existence or reality. Reflecting upon the fact that an act practiced with approval in one society is condemned in another, the relativist concludes that right and wrong are relative to culture, and the nihilist concludes that there really is no difference between right and wrong. Confronted with ethical disagreement among different sorts of individuals, the relativist asserts that the truth of any ethical statement is relative to the personality of the speaker, while the nihilist denies that ethical sentences have any truth-value at all.

Any version of ethical relativism can most fruitfully be classified according to the status of its central thesis. If the thesis is a factual statement, the theory is an instance of descriptive relativism. If the thesis is an ethical statement, a judgment of value or obligation, it is a normative relativism. If the thesis is an epistemological claim, i.e., a statement about our ethical knowledge, pertaining to the meaning of ethical statements or the validity of ethical arguments, it is a metaethical relativism.

Descriptive relativism

Any ethical relativism in which the central thesis simply asserts that as a matter of fact something ethical varies with and depends upon something else is a descriptive relativism. The two most important species of descriptive relativism are cultural and psychological. A cultural relativism asserts that something ethical (such as the mores, the institutions creating rights and duties, the human goals or ends, and the moral concepts or ethical judgments) is relative to culture. Psychological relativism asserts that something ethical (such as goals, evaluations, or conscience) is relative to the psychology of the individual.

In its Statement on Human Rights, the Executive Board of the American Anthropological Association asserted in 1947 that "standards and values are relative to the culture from which they derive" (Executive Board, p. 542). This assertion is far from unambiguous. If it means that one is morally bound only by the standards of one's own society and that what is good in one society may be evil in another, then it is a version of normative relativism. If, on the other hand, it means to assert that in point of fact the moral standards accepted by one society and the goals valued by any people vary with and are causally determined by the culture of that society or people, it is a version of descriptive relativism. Since the Executive Board clearly takes this assertion to be established by the sciences of anthropology and social psychology, the latter interpretation is the more plausible.

The Executive Board argues that this cultural relativism poses an awkward dilemma for any contemporary declaration of human rights (ibid., pp. 542–543). If the declaration affirms the moral standards and human aspirations of Western cultures, it will be inapplicable to the many non-Western societies with their very different standards and values. But given the cultural variation in standards and values, there is no universal code of human rights that could be asserted by all men and applied to all societies. What status, then, is one to assign to article 25 of the United Nations' Universal Declaration of Human Rights, which affirms the right of every human being to medical care, or to the Declaration of Helsinki, which proclaims the right of all persons to be subjected to potentially dangerous clinical research only after free informed consent? Are these declarations morally applicable to medical practice and public health programs throughout our world, or are they merely ethnocentric expressions of our provincial biases?

Another version of descriptive relativism is the theory that the individual's values, in the sense of goals and evaluations, are relative to his or her personality. A timid person will shun danger and consider it a great evil, but a more adventurous individual will discount risk and even value it as a source of excitement. Similarly, the importance an individual attaches to sexual satisfaction will vary with and depend upon the strength of his or her sex drive. Since this version of psychological relativism simply asserts an empirical fact about the individual, it is a descriptive relativism.

This descriptive relativism may, however, have important implications for bioethics, specifically for the moral justification of medical advice. When a physician or surgeon advises a patient, the advice typically does not limit itself to factual information about the available procedures and the probable outcomes of each. Medical advice is most helpful to the patient when it includes some judgment of what would be best for the patient in the light of all this complex technical information. But evaluations, including judgments of which course of action is best, are supposed to be relative to the personality of the individual. Presumably, then, a somewhat timid physician or surgeon will be less inclined to advise risky procedures and a more adventurous doctor will evaluate potentially dangerous treatments more favorably. The patient, however, will probably have a rather different personality and, accordingly, a different set of values from that of the doctor. Since it is the patient's life and health that are at stake, it is the patient's values and not the doctor's that determine the ethical propriety of the advice. How, then, is it morally justifiable for one individual, the physician or surgeon, to give medical advice to another individual, the patient?

Normative relativism

When the central thesis of some version of ethical relativism is an ethical statement, rather than a factual statement, it is a version of normative, rather than descriptive, relativism. Once more, the most important varieties are either cultural or psychological, but some interesting versions fall into neither of those two classes.

It is often claimed that right and wrong are relative to culture, that the very same kind of act is right in one society and wrong in another because of the different mores of the two societies. Since this claim goes beyond describing the facts about what people believe right and wrong in order to take a stand on which acts really are right and wrong, it is a normative relativism. If true, this version of cultural relativism is important for bioethics in at least two ways. First, many medical acts are of the kinds forbidden, permitted, or required by the mores of many societies. Some societies harshly condemn the mutilation of corpses, an act involved in the harvesting of cadaver organs for transplant, while other societies permit such actions. Similar prohibitions and permissions apply to the acts of abortion, infanticide, and euthanasia. It makes all the difference in the world to the moral status of such acts whether their rightness or wrongness is relative to culture or somehow independent of the mores of the agent's society. Second, the explosion of knowledge in the life sciences and the application of this new knowledge to medicine have inevitably led to great innovation in medical practice. But new practices, such as organ transplantation or artificial insemination, often deviate dramatically from traditional culture and established custom. Are we to conclude that new medical procedures are morally justifiable only so long as they remain within the bounds of custom, or is there some moral standard independent of the mores by which medical innovation can be morally justified?

A second version of normative relativism is the theory that values are relative to the desires, preferences, or interests of the individual (Perry, pp. 115–116). Notice that this theory differs from the previous one in two ways. What is declared to be relative is values rather than obligations, and that to which they are held to be relative is the psychology of the individual rather than culture. The word "values" in this theory means "things that have value or disvalue," not "goals or evaluations." This change in the meaning of the word "values" is crucial, for it distinguishes the normative relativism of this section from the descriptive relativism of the previous section. This theory denies that anything can be good or bad in itself and asserts that everything of value is good or bad *for* some individual. It goes on to assert that what makes anything good or bad for someone is that individual's desires, preferences, or interests. Bioethics often presupposes that medical treatment is morally justified only if it is beneficial for the patient. But if this version of ethical relativism is true, there can be no universal rule on such matters; one must ask for whom the treatment is to be beneficial. The act of taking out one kidney for transplant may be beneficial for the altruistic person who cares more deeply about the welfare of a loved one than about his or her own health, but harmful for the individual who prefers his or her own health to the life of another. Similarly, the relative value of length of life as compared with the quality of life may depend upon the desires of the individual patient. If so, there can be nothing like *the* moral solution to the ethical problems of organ transplantation or euthanasia; such problems must be solved anew in the case of each patient.

A third version of normative relativism, and one that is neither cultural nor psychological, is situation ethics. The central thesis of this theory is that obligation is relative to the situation in which the agent is acting, that what one ought to do varies with and depends upon the circumstances (Fletcher, p. 65). Thus the very same kind of act, for example an act of euthanasia, may be right in one situation (where a terminally ill patient is in pain too intense to be allayed and has begged for release) but wrong in another situation (where the pain can be reduced considerably by medication and the patient is psychologically prepared to go on living). Situation ethics stands opposed to ethical legalism, the view that there is a set of moral rules specifying which kinds of action are always right and which kinds are always wrong. Situation ethics tends to emphasize the differences between cases and to make it easier to justify untraditional conduct in changing circumstances; ethical legalism tends to emphasize the similarities between acts and to insist that exceptional circumstances do not justify exceptions to accepted moral standards. That these opposed approaches to bioethics do frequently lead to different conclusions is illustrated by the recent literature dealing with issues such as euthanasia, abortion, organ transplants, and genetic engineering.

Metaethical relativism

Any ethical relativism in which the central thesis is an epistemological claim about (i.e., a statement about our knowledge of) ethical statements or ethical arguments is metaethical. One version of metaethical relativism is the view that the meaning of the expressions "normal" and "abnormal" as used in clinical psychology is relative to culture. This is an ethical relativism because the expressions "normal" and "abnormal" are ethical words used to brand some person or action as socially acceptable or unacceptable; it is a *meta*ethical relativism because its central thesis is an epistemological claim about the meaning of these ethical words. Ruth Benedict has argued that every society has its distinctive pattern of culture in which certain values are selected and emphasized while other possible objects, acts, and personalities are devalued or even excluded. She gives many illustrations of the way in which a personality type or mode of conduct is judged normal or abnormal to the degree that it does or does not fit into the accepted pattern of culture. She concludes that "normality is culturally defined" (Benedict, p. 72). The relevance of this metaethical relativism to medical ethics is obvious. A diagnosis of mental illness is often made on the basis of a judgment of psychological abnormality, and the goal of psychotherapy is typically to restore the patient to normality. But is the diagnosis of mental illness more than an expression of the cultural bias of the psychologist? And what is the moral justification for attempting to get the patient to conform to a cultural pattern that may be alien to him or her? Both the way in which one conceives of mental illness and the stand one takes on the morality of psychotherapy will hinge upon whether one accepts or rejects this version of metaethical relativism.

Conclusion

It is of considerable importance to distinguish carefully among descriptive, normative, and metaethical relativisms, because the evidence for or against each is so very different. Since descriptive relativisms make purely factual assertions, they can presumably be shown to be true or false by scientific investigation. Since normative relativisms assert something about value or obligation, they can be established or refuted only by the very different methods of ethics. Since metaethical relativisms make claims about meaning, truth, or reasoning, they are to be accepted or rejected on the basis of epistemological considerations. Only by sorting out the various kinds of ethical relativism is there any hope of distinguishing the true versions from the false ones.

Although the three kinds of ethical relativism are very different, they are not unrelated. One version of ethical relativism often serves as a premise in an argument to support some other version. For example, the mores of any society vary with and are dependent upon the culture of that society (descriptive relativism); it is the mores of the agent's society that make an act right or wrong; therefore, right and wrong are relative to culture (normative relativism). The logic of the argument is cogent, but the truth of its premises can, and should, be questioned. Some anthropologists and social psychologists have challenged this descriptive relativism by presenting empirical evidence that there are some cultural and psychological universals, e.g., the values of loyalty or love and the taboo against incest. Many philosophers deny that the mores are the standard of right and wrong on the grounds that we condemn some mores as unjust or inhumane. The proper conclusion to draw after critically assessing the many arguments for and against the various kinds of ethical relativism is not The Truth about Ethical Relativism, but many truths about the several versions of ethical relativism.

CARL WELLMAN

[*While all the articles in this entry are relevant, see especially the articles* RULES AND PRINCIPLES, OBJECTIVISM IN ETHICS, *and* MORAL REASONING. *For further discussion of topics mentioned in this article, see the entries:* RIGHTS; *and* RISK. *Also directly related are the entries* HUMAN EXPERIMENTATION; MEDICINE, ANTHROPOLOGY OF; MENTAL HEALTH; MENTAL ILLNESS; *and* THERAPEUTIC RELATIONSHIP. *See also:* HEALTH CARE, *articles on* RIGHT TO HEALTH-CARE SERVICES *and* THEORIES OF JUSTICE AND HEALTH CARE; *and* LIFE.]

BIBLIOGRAPHY

BENEDICT, RUTH (FULTON). "Anthropology and the Abnormal." *Journal of General Psychology* 10 (1934): 59–82.

BRANDT, RICHARD B. *Ethical Theory.* Englewood Cliffs, N.J.: Prentice-Hall, 1959, pp. 83–150, 271–294.

Executive Board, American Anthropological Association. "Statement on Human Rights: Submitted to the Commission on Human Rights, United Nations." *American Anthropologist* n.s. 49 (1947): 539–543.

FLETCHER, JOSEPH FRANCIS. *Situation Ethics: The New Morality.* Philadelphia: Westminster Press, 1966.

HERSKOVITS, MELVILLE JEAN. *Man and His Works.* New York: Alfred Knopf, 1948, pp. 61–78.

LADD, JOHN. *Ethical Relativism.* Belmont, Calif.: Wadsworth, 1973.

MACBEATH, ALEXANDER. *Experiments in Living.* London: Macmillan, 1952.

MOSER, SHIA. *Absolutism and Relativism in Ethics.* Springfield, Ill.: Charles C. Thomas, 1968.

PERRY, RALPH BARTON. *General Theory of Value.* New York: Longmans, Green & Co., 1926; Cambridge: Harvard University Press, 1954, pp. 115–145.

WELLMAN, CARL. "The Ethical Implications of Cultural Relativity." *Journal of Philosophy* 60 (1963): 169–184.

WESTERMARCK, EDVARD ALEXANDER. *Ethical Relativity.* London: Kegan Paul, Trench, Trubner & Co., 1932.

ETHICS COMMITTEES

See DEATH AND DYING: EUTHANASIA AND SUSTAINING LIFE, *article on* PROFESSIONAL AND PUBLIC POLICIES; INFANTS, *articles on* PUBLIC POLICY AND PROCEDURAL QUESTIONS *and* ETHICAL PERSPECTIVES ON THE CARE OF INFANTS; RESEARCH POLICY, BIOMEDICAL; HUMAN EXPERIMENTATION, *article on* SOCIAL AND PROFESSIONAL CONTROL.

EUGENICS

I. HISTORY   Kenneth M. Ludmerer
II. ETHICAL ISSUES   Marc Lappé

I
HISTORY

"Eugenics" is a term with several connotations. Literally meaning "well-born," the term at different times has meant different things: a science that investigates methods to ameliorate the genetic composition of the human race; a program to foster such betterment; a social movement; and, in its perverted form, a pseudoscientific retreat for bigots and racists.

Background and origins

Although the idea of improving the hereditary quality of the race is at least as old as Plato's *Republic,* modern eugenic thought arose only in the nineteenth century. The emergence of interest in eugenics during that century had multiple roots. The most important was the theory of evolution, for Francis Galton's ideas on eugenics—and it was he who created the term "eugenics"—were a direct logical outgrowth of the scientific doctrine elaborated by his cousin Charles Darwin. Galton advanced his ideas, moreover, just when it was becoming current to apply the doctrine of evolution to nonbiological situations. He developed his views during the period when evolutionary theory was exerting its greatest impact upon intellectual thought, having flavored doctrines known as "naturalism" and "Social Darwinism." These doctrines encompassed a group of ideas, the most important of which were the analogy between biological organisms and social systems, the conviction that scientific research represented the most accurate approach to knowledge, and the appeal to scientific method for solutions to social problems. Social Darwinists also tended to equate a person's genetic "fitness" with his social position and popularized the catchwords "survival of the fittest" and "struggle for existence."

However, other factors were also important in the growing interest in eugenics. One in particular was the philosophical belief, inherited from certain eighteenth-century thinkers, in the notion of human perfectibility. Another was the rising fear, which coexisted with this faith in progress, that the quality of the American stock was deteriorating. According to this view, the "unfit" in civilized societies constituted a major menace; they were surviving in increasing numbers because of modern medicine and charity, and they were procreating far more rapidly than the "fit," thereby diluting the concentration of valuable hereditary characteristics in the population.

Despite those factors favoring an interest in eugenics, an organized eugenics movement did not occur in the United States or elsewhere before 1900. The principal reason was ignorance regarding the process of heredity. The physical basis of heredity was still unknown; chromosomes were not found in the nucleus until the 1880s, and only after 1900 were they shown to be carriers of genes. Furthermore, no rule was known to govern the transmission of traits from one generation to the next; the only formula biologists could follow was the broad theory used by breeders that "like produces like" and the assumption that all traits result from a "blend" of parental characteristics. Finally, popular belief in the inheritance of acquired characteristics vitiated the supposed need for eugenic proposals, since this theory contradicted the idea that improvements in the hereditary composition of man could occur only from breeding programs.

The birth of modern genetics in 1900 enabled earlier interest in eugenics to be mobilized into an organized movement. Mendel's law provided an explanatory framework in terms of which the transmission and distribution of traits from one generation to the next could be understood, thereby permitting eugenic proposals to be conceived on a heretofore impossible scale. Eugenic breeding programs now became widely appealing because at last they could be based upon biological theory rather than upon the imprecise rules used by breeders. Other early developments in genetics stimulated further interest in eugenics: the emergence of a belief that most traits are determined by single genes acting independently, and the popularization after 1900 of the work of August Weismann, who disproved the Lamarckian theory that acquired characteristics can be inherited.

The eugenics movement; positive and negative eugenics

The potential applicability of the science of genetics to social problems was recognized almost immediately. Organizations focusing on eugenics were created around the world: in England, the United States, Germany, Scandinavia, Italy, Austria, France, Japan, and South America. The eugenic societies in each country exhibited some national differences, but all were committed to the popularization of genetic science, demanding that social legislation be directed by what they judged to be biological wisdom. The center of this trend was the American eugenics movement, whose national headquarters was at the Eugenics Record Office at Cold Spring Harbor, Long Island, and whose acknowledged leader was the geneticist Charles B. Davenport. Unless otherwise specified, the remainder of this discussion will refer to eugenics in America.

From approximately 1905 to the early 1930s, eugenicists in the United States presented a two-part policy to improve the hereditary level of the American people. One part they called "negative eugenics," the elimination of unwanted characteristics from the nation by discouraging "unworthy" parenthood. Through specific devices—including marriage restriction, sterilization, and permanent custody of defectives—eugenicists hoped to stop the breeding of people whose physical disabilities or behavioral characteristics they thought to be genetically determined. Such hereditary "degenerates" included epileptics, criminals, alcoholics, prostitutes, paupers, the feebleminded, and the insane. The second part of eugenic policy was "positive eugenics," the striving to increase wanted traits in the population by urging "worthy" parenthood. Since eugenicists recognized that the social and technical difficulties facing positive eugenics were enormous, most of them recommended only that the public be educated about the "facts" of heredity to encourage "superior" couples to respond voluntarily to the plea to have more children.

At the turn of the century the movement attracted a diverse group of individuals. Eugenicists came in largest numbers from the native-born, Anglo-Saxon, Protestant, upper middle class and formed part of that generation's educators, scientists, scholars, journalists, physicians, lawyers, and clergy. They lived in all sections of the country and in both urban and rural areas. It is notable that until around 1915 many of the most enthusiastic members of the movement were geneticists. It is also notable that some eugenicists were racists who claimed a scientific sanctuary in eugenics for their bigotry. After the First World War leadership of the movement fell predominantly into the hands of such persons, the most famous of whom was the New York socialite Madison Grant.

Early in the century eugenic programs seemed feasible in light of the scientific knowledge of the day. By the First World War, however, additional developments in genetic science had cast doubts upon the scientific merit of eugenic proposals. A number of investigations showed that environment as well as heredity significantly influences human development; other studies demonstrated that many characteristics are produced by multiple genes interacting rather than by single genes acting independently as previously supposed. Moreover, early studies in statistical genetics showed that the process of reducing the frequency of a gene in a population by breeding programs would require a prohibitively long period, not the short time that eugenicists had assumed. Taken as a whole, these developments demonstrated that the process of inheritance was much more complex than originally had been thought, and they effectively invalidated the genetic assumptions underlying the movement.

In response to this new scientific information, many geneticists during the First World War years lost interest in the eugenics movement. Still, the majority of the eugenicists remained as captivated by the movement as they had been at the beginning. Although they debated the particulars of the programs they advocated, they did not, for the most part, examine the ethical and scientific assumptions underlying their proposals. Few of them ever bothered to examine such fundamental issues as whether their programs made scientific sense in view of the more recent findings in genetics or whether mankind really is wise enough to decide which characteristics are desirable.

In their continuing allegiance to the movement, most eugenicists were sincere and well-intentioned. However, the majority had no scientific training, and their claim that the eugenic program rested firmly upon a valid genetic foundation rested more on faith than on personal acquaintance with the experimental evidence. To most of them, eugenics assumed the proportions not of a science but of a social and moral crusade. Some so venerated science and the scientific method that they came to regard the acceptance of eugenic programs as a religious duty imposed by the theory of evolution. Like many Americans of the time, eugenicists were frightened by what they interpreted as a sharply increasing incidence of physical and mental degeneracy, and they were equally alarmed at civilization's supposed interference with the working of natural selection. Accordingly, they were interested in eugenics because it offered a "scientific" solution to pressing social "problems" compatible with the world view of the naturalistic mind. The intellectually stagnant qualities of the eugenics movement can best be understood in this light.

Impact of the eugenics movement

Ironically, it was at the time that the scientific justification of eugenics was weakening that the movement started to achieve its greatest political triumphs. As the First World War closed, the movement began a decade of intensive and successful campaigning for "eugenic" legislation. Eugenicists were active peripherally in a variety of issues, including prohibition, birth control, custodial care for "defectives," and antimiscegenation bills. More important, however, they made a significant impact in two areas of major social import: sterilization legislation and immigration restriction.

The first American state to pass a sterilization law was Indiana, which in 1907 enacted a bill requiring the compulsory sterilization of inmates of state institutions who were insane, idiotic, imbecilic, or feebleminded, or who were convicted rapists or criminals, upon the recommendation of a board of experts. By 1931 thirty states had passed compulsory sterilization measures, some of which applied to a very wide range of "hereditary defectives" including "sexual perverts," "drug fiends," "drunkards," and "diseased and degenerate persons." The laws were passed primarily in response to the lobbying in the various state legislatures of zealous eugenicists, who viewed the measures as important planks of "negative eugenics." The most influential person was Harry Laughlin, Davenport's assistant at the Eugenics Record Office, who pursued the sterilization campaign with missionary fervor and whose writings on the subject guided interested lawmakers and citizens for more than twenty years.

More important than the sterilization laws was the Immigration Restriction Act of 1924. The controversial feature of this law was not the restriction of immigration per se but the *selective* restriction clause. This provision drastically limited the entry of individuals from Southern and Eastern Europe on the grounds they were "biologically inferior," favoring instead the admission of "Nordics" from the northwest corner of the Continent. Following the First World War, at a time of intense nationalism and isolationism, of deep hostility and animosity toward anything foreign, the social climate was ripe for the enactment of a law restricting immigration. Nevertheless, the racist leaders of the eugenics movement, masking their hatreds under a veneer of scientific objectivity, seized this opportunity to argue that selective immigration restriction was a biological imperative. They stressed the Social Darwinistic supposition that economic and social status indicate hereditary value; they insisted that heredity is far more important than environment; and they claimed that any mixing between "American" and "foreign" blood would produce offspring inferior to both parental types. The leading popularizer of this view in Congress was Laughlin, who had been appointed "expert eugenics agent" of the House Committee on Immigration and Naturalization in 1920, and whose most important congressional testimony, "Analysis of America's Modern Melting Pot," greatly influenced congressmen and the public alike. By this time, as previously mentioned, eugenicists' view of biology had lost its scientific validity, but the public did not appreciate this, in part because few geneticists let their disapproval of the movement be generally known. During the course of the immigration debates, eugenicists' authority was effective and uncontested; their misappropriation of genetic theory gave their view a scientific sanction which opponents could not combat. Eugenicists considered the passage of this law to be their greatest triumph.

In the 1920s the racist side of eugenics was emerging in the German as well as in the American eugenics movement. With the rise to power of Hitler, who himself had long emphasized eugenic strictures for race improvement, the eugenics movement in Germany became irrevocably intertwined with the Nazi regime. Leading eugenicists became Nazi officials, and other government leaders also found eugenic concepts appealing. Eugenic ideas of race and of Aryan superiority were sanctioned by law on 14 July 1933, when Hitler decreed the Hereditary Health Law, or Eugenic Sterilization Law, which was created to make certain that "less worthy" members of the Third Reich did not transmit their genes. This law started the process that resulted in the experimentation with euthanasia in 1939 and ultimately to the mass murder of millions of other "undesirables."

Decline and rebirth of the movement

In the 1930s, in the aftermath of the eugenics movement's greatest political victories, the American people repudiated it. Several factors contributed to the movement's downfall. To a generation frightened by the Nazis' creed of Aryan superiority and alarmed by their morbid fascination with biological fitness and human breeding, the racist ideology of the American eugenics movement no longer carried its earlier appeal. Distrust of eugenics in the United States seemed well-founded, since throughout the 1930s some American eugenicists had publicly applauded the Nazi "eugenic" measures. The eugenics movement lost further ground as other social conditions changed. The Great Depression castigated economically, without distinction, both the Nordic and non-Nordic; people whose incomes had disappeared could no longer maintain with impunity that money and social prestige are indicators of genetic excellence. Finally, in the 1930s the scientific argument against eugenic proposals became fully developed. Increasingly, traits claimed by eugenicists to be the result of single genes acting alone were found to have complex etiologies—some were discovered to be the product of many genes interacting together, some of heredity interacting with environment, and many of environment alone. Research in psychology and anthropology also began to emphasize the influence of culture and environment in the development of an individual, race, or society, thereby providing further evidence against those eugenicists who held the Social Darwinist view that social position measures biological fitness. Perhaps the most devastating argument against eugenic schemes came in the 1920s and 1930s from the culmination of the earlier work in population (statistical) genetics. Research in this field showed that even allowing eugenicists their claim of the omnipotence of heredity and of the Mendelian recessive mode of inheritance of many traits, a eugenic breeding program would not cause the swift elevation of the genetic level of the population that eugenicists had predicted. As a result of all these events, eugenics in the 1930s fell into disrepute.

Though the old eugenics movement underwent an ignominious fall, the period from the 1950s to the 1970s witnessed a rekindling of interest in eugenics. The roots of this renewed interest began in the 1930s, when a new leadership, genuinely concerned with mankind's genetic future, rescued the eugenics movement from its ruins and assumed the task of rebuilding it. The new leaders cast off the class and race biases of the movement's founders, conceded the absurdity of earlier eugenicists' biological claims, and put together a new eugenics creed that was both scientifically and philosophically in keeping with a different America.

Since the Second World War, the revived eugenics program has developed along two main paths. The first and less dramatic is genetic counseling, the instructing of prospective parents about the likelihood that their children will be born with a genetic condition, a service which many medical centers are already offering. The second and more controversial part, sometimes itself called the "new eugenics," is the startling new view of man's future resulting from the postwar advances in molecular biology. With the composition of the gene and the principal molecular mechanisms of inheritance having been clarified, futurists foresee a day when physicians will distribute test-tube babies and do away with hereditary conditions by genetic surgery. The new eugenics is often referred to as "genetic engineering," though in fact it involves more than just genetic surgery as that name implies. In a narrow sense it may require environmental manipulation; nonetheless it utilizes genetic techniques. Although at present genetic engineering is still a branch of experimental biology, not yet ready for application to man, its overtones for eugenics are clear. The "new eugenics," like the old, is highly controversial; but the modern controversies center primarily on ethics rather than science, focusing on issues of means and not of ends.

The future of eugenic applications in the United States is not easy to determine, for American society today provides no clear signs, only ambiguous and contradictory clues. Modern medical research is increasingly interested in utilizing genetic techniques. For example, recent discoveries have made it possible to detect *in utero* fetuses afflicted with certain genetic diseases by examination of amniotic fluid. At the same time, however, contemporary medicine has not lost its traditional determination to conquer disease by utilizing the environmental procedures of surgery, diet, and drugs. If completely successful, the environmental approach theoretically might end the need for any further consideration of eugenics. Of course, there is no reason why the eugenic and environmental approaches cannot be pursued together, but so far there is much disagreement concerning their relative desirability and merit. Moreover, at a time when American society is struggling to determine how to foster individualism without encouraging lawlessness or civil disobedience, and how to contain crime and violence without resorting to repression, the ancient problem of clarifying the responsibility of the individual to the group and of the present generation to the future remains unresolved. On no other issue is it so important to reach a consensus before enacting far-reaching eugenic programs.

KENNETH M. LUDMERER

[*Further discussion of the history of the ethical issues in eugenics is found in the entry* EUGENICS AND RELIGIOUS LAW. *Also directly related are the entries* EVOLUTION; GENETIC ASPECTS OF HUMAN BEHAVIOR; GENETIC CONSTITUTION AND ENVIRONMENTAL CONDITIONING; GENETIC DIAGNOSIS AND COUNSELING, *article on* GENETIC COUNSELING; GENETIC SCREENING; REPRODUCTIVE TECHNOLOGIES; *and* STERILIZATION, *articles on* ETHICAL ASPECTS *and* LEGAL ASPECTS. *Other relevant material may be found under* BIOLOGY, PHILOSOPHY OF; EUPHENICS; GENETICS AND THE LAW; MEDICAL ETHICS UNDER NATIONAL SOCIALISM; POPULATION POLICY PROPOSALS, *article on* GENETIC IMPLICATIONS OF POPULATION CONTROL; PRENATAL DIAGNOSIS, *article on* ETHICAL ISSUES. *See also:* RACISM.]

BIBLIOGRAPHY

DAVENPORT, CHARLES BENEDICT. *Heredity in Relation to Eugenics.* New York: Henry Holt & Co., 1911.

DUNN, L. C. "Cross Currents in the History of Human Genetics." *American Journal of Human Genetics* 14 (1962): 1–13.

GALTON, FRANCIS. *Hereditary Genius: An Inquiry into Its Laws and Consequences.* London: Macmillan &

Co., 1869. Reprint. Gloucester, Mass.: Peter Smith Publisher, 1976.
GRANT, MADISON. *The Passing of the Great Race: Or, the Racial Basis of European History.* New York: Charles Scribner's Sons, 1916.
HALLER, MARK HUGHLIN. *Eugenics: Hereditarian Attitudes in American Thought.* New Brunswick: Rutgers University Press, 1963.
HOFSTADTER, RICHARD. *Social Darwinism in American Thought.* Rev. ed. Beacon Paperbacks, no. 16. Boston: Beacon Press, 1955.
LAUGHLIN, HARRY HAMILTON. *Eugenical Sterilization in the United States.* Chicago: Psychopathic Laboratory, Municipal Court of Chicago, 1922.
LUDMERER, KENNETH M. *Genetics and American Society: A Historical Appraisal.* Baltimore: Johns Hopkins University Press, 1972.
OSBORN, FREDERICK HENRY. *Preface to Eugenics.* Harper's Social Science series. Edited by F. S. Chapin. New York: Harper & Brothers, 1940.
PICKENS, DONALD K. *Eugenics and the Progressives.* Nashville: Vanderbilt University Press, 1968.

II
ETHICAL ISSUES

Objectives and means in eugenic practice

Attaining any eugenic objectives for improving the effective genetic quality of human populations over time will entail policy considerations that challenge contemporary values. The scope of a hypothetical eugenic policy would impinge on virtually every facet of human life. Eugenics, in Sir Francis Galton's words, deals with *"all the influences that improve the inborn qualities of a race"* (emphasis added).

In animal populations, the objectives of eugenics were attained through programs that selectively bred individuals whose gross physical characteristics corresponded to arbitrary ideals. Undesirable stock were prevented from procreation through sterilization or slaughter. In human populations, the early eugenicists sought to decrease the propagation of the physically or mentally handicapped and to encourage the procreation of those they deemed to embody the best features of the "race." A broad range of disabilities was considered to be genetically caused, and effective elimination of the affected individuals was sought by means remarkably similar to those used by animal husbandmen. Such steps were expected to reduce the incidence and ultimately the gene frequency for presumptively deleterious "genetically based" conditions like degeneracy, paraplegia, tuberculosis, promiscuity, and criminality, as well as those conditions which had a bona fide genetic basis, like albinism. Other geneticists, notably Hermann Muller, advocated *positive* eugenic ends through the selective use of semen from "more desirable" persons to propagate their kind.

The chief stumbling block in attaining any of these early objectives of human eugenics has been the absence of adequate data on which to formulate policies that would be scientifically and ethically sound. More recently, genetic policy has shifted attention from broad eugenic objectives to focus on specific disease entities or physical disabilities with straightforward genetic causes. Most of the objectives of positive eugenics remain unattainable because of fundamental uncertainties regarding the genetic contribution to "positive" attributes like stature and intelligence.

In animal populations the only measurable eugenic successes have been made by extraordinary winnowing of less desirable sires or dams in favor of their more desirable counterparts. Applied to humans, such mechanisms would seem inevitably to violate traditional societal norms and breach virtually all rules of justice. Indeed, extreme forms of genetic intervention such as were practiced in Nazi Germany are stark reminders of the extremes to which pseudoscience can be bent to meet ideological objectives.

Value questions in genetic intervention

Ethical issues arise even when extremely modest eugenic ends are sought or less coercive measures are contemplated to bring about genetic change in human societies than were used by the Nazis. For instance, a decision to embark on a marriage counseling program that includes genetic information raises important questions of the propriety of the state's interest in genetic aspects of the reproductive decision making of couples. Any policy to discourage the marriage or procreation of individuals raises the question of possible infringement of privacy—a concept recently expanded by U.S. Supreme Court decisions to encompass reproductive acts, e.g., in *Griswold* v. *Connecticut* and *Eisenstadt* v. *Baird*.

Whether or not the state should embark on eugenic programs at all hinges on unexamined features of the ethos of the body politic. Some eugenicists have maintained that there is a fundamental obligation to improve the human species, while theologians like Paul Ramsey have maintained that an obligation for improvement is a derivative notion bereft of fundamental justification (Ramsey). Other scholars, like Bentley Glass, maintain that there is an irreducible obligation, in the light of genetic knowledge, not knowingly to produce a child with the prospect of physical or mental defect.

Progress through genetic improvement is part of a natural law ethic held by several major scholars. Sperry maintains that the preeminent place of humans in the order of nature makes manifest the direction of evolution. He argues that being human confers an obligation to continue this pattern of improvement (Sperry). Some evolutionary and molecular biologists (Simpson; Stent) contest this viewpoint and argue that there is nothing in the evolutionary program that would suggest an innate tendency toward evolutionary progress, nor is there any reason to conceive of human destiny as having a divine imperative toward species improvement.

Another viewpoint is that the very fact that humans can be aware of their origins and of the forces that shape their future mandates active use of that knowledge; to be human means to change nature and the human estate, and human consciousness and rationality argue for building on and directing the processes of nature (Fletcher). On the other hand, it is contended that there is nothing in nature or in any doctrine of divine providence that would require that a person succeed in achieving eugenic ends (Ramsey).

Minimal ethical considerations

Recent definitions of eugenics now embrace the concept of a basic obligation to *maintain* as well as to improve the genetic potentialities of the species. According to the theologian James Gustafson, responsibility for preventing deterioration of human genetic systems becomes enlarged because of our increased contemporary knowledge of genetics. We now know that present generations *are* causally linked with the genetic health of future generations, and thus, he argues, we have a moral responsibility to them. He identifies the injunction of not harming others as a minimal guiding principle undergirding any genetic policy (Gustafson). The rabbinical tradition, as well as that of Theravedic Buddhism, emphasizes that obligation by stressing the moral requirement that one not knowingly multiply the individual or collective griefs of mankind through taking even the most "mild" genetic risks (Narot).

Several theologians go farther in arguing that the moral principle of *primum non nocere* (first, do no harm) should be assigned the greatest weight in policymaking, and hence would preclude any attempt at genetic intervention that might generate unintentional harms. In traditional medical ethics, the basic precept of not harming was considered more important than attempting to improve an individual's condition. Medical interventions before the twentieth century often led to more lives lost than saved. Even today, where surgical practice is involved, the physician is under a primary injunction not to *add* to the patient's injuries. Before intervening, the surgeon must establish that reasonable cause exists for breaching the integrity of the body; that the techniques or explorations under consideration are appropriate to the ends being considered; that a reasonable attempt has been made to enlist the cooperation of the patient in consenting to the risks entailed in the intervention; and, finally, that informed consent to proceed has been obtained.

Validating eugenic policy

Analogous arguments can be applied to the propriety of initiating genetic interventions. A minimum ethical requirement for eugenics might include answers to he following: (1) Has the human genetic condition been demonstrated to be sufficiently impoverished or endangered to justify intervention? (2) Are the ends sought desirable, and who makes that judgment? (3) Are the means to be used necessary and sufficient ethically to achieve those ends? (4) Are the individuals whose lives will be affected by the interventions given a voice in the decision making? (5) Can assurances be given that implementing eugenics programs will not generate more harm than good? This approach links policy decisions to both utilitarian (e.g., the greatest genetic good for the greatest number) and deontological (are the inherent moral duties acceptable on their face?) considerations.

Should those requirements be met, second order ethical issues enter. One is the question of the appropriate level at which eugenic policies ought to be applied. For instance, it might be possible to achieve strictly limited eugenic ends for a specific disease entity whose gene frequency *was* in fact endangering individual families, while difficult or impossible to achieve the same ends for a more complex disease. A second problem is the weight to be given to efficacy in reaching eugenic ends. Some policies for achieving eugenic objectives could focus on specific disease entities and embrace intrauterine diagnosis and abortion of carriers of deleterious genes, while others could center on more prevalent manifestations of complex gene–environment interactions, such as hypertension. Programs to minimize the transmission of the responsible genes for hypertension would nevertheless be likely to prove cumbersome and inefficient. But it is a basic ethical tenet that inefficiency or any other index of likely efficacy alone does not invalidate the ends being sought.

Defining the genetic status quo

All objectives of eugenic policy must ultimately be measured against the perceived necessity of change. And any rationale for changing the human gene pool requires an understanding of its present state. What is the genetic status quo? Is the genetic composition of society deteriorating in such a way that remedial actions are mandatory in order to keep the situation from getting worse? Geneticists (e.g., Sang) believe that the apparent increase in human heterozygosity for deleterious genes will inevitably lead to their spread through the population. J. H. Sang sees the present genetic condition as being roughly equivalent to the beginnings of environmental pollution a generation ago. Others challenge this view (Medawar). They argue that, while some genetic dangers confront mankind, most are not of great urgency or gravity.

The truth appears to lie closer to the moderate view. Most contemporary geneticists downgrade the weight to be assigned to presumptive genetic deterioration and focus on the consequences of dynamic trends in populations instead. They are unclear about the ultimate meaning of the vast amount of genetic heterogeneity revealed by study of human genetic polymorphisms (situations in which two or more forms of the same allele or gene pair are present in a frequency greater than that generated by mutation alone). For example, it is uncertain whether selective forces or purely neutral events like genetic drift have generated the degree of genetic diversity observed in human populations. The ethical weight one should give to sustaining, reducing, or increasing the current level of diversity depends basically on an as yet unobtainable understanding of the evolutionary significance of genetic heterogeneity or the complex sources for the load of mutations we carry (Neel).

Ultimately, the genetic status quo of a given population is determined by dynamic factors and its exposure to mutagens like ionizing radiation. The average fertility, differential mortality, and age characteristics of the breeding members of a population can themselves affect the genetic "load" that the population bears. Geneticists have identified many components that contribute to the load of deleterious genes a population carries, including radiation, mutation rate, and age at reproduction, each of which is an independent variable requiring its own analysis for weighting and policy decisions.

In addition to those factors, the intrinsic rate of increase of the population will have major impacts on genetic load. For example, if the population is growing rapidly it will tend to have a lower mutation rate and lower genetic load than will a stationary or declining population, because spouses will tend to reproduce when younger and hence prior to the accretion of excessive numbers of new mutations. Demographic effects of this sort mean that eugenic policies may conflict with other valued policies, like those for population limitation. In turn, "slow growth" policies in which reproductive age is protracted and/or delayed, will decrease the incidence of deleterious traits that are expressed with increasing probability with age (e.g., Huntington's chorea or multiple polyposis coli), but increase the risk of age-associated reproductive damage (e.g., Down's syndrome).

It is likely to prove crucial for eugenic considerations to recognize that the current population is not in genetic equilibrium and that acting to sustain or change demographic patterns will itself have potentially important effects on the frequency of specific genetic conditions. Slower-growing populations will have progressively less opportunity for sudden fluxes in the numbers of rare or novel genotypes, and the progressive trend toward equalizing family size seen in developing countries will minimize the differential fertility necessary for evolutionary change. While reduction of family size in itself does not mean that changes in gene frequency cannot still occur, in concert with population constraints it will reduce the variance in offspring and the effective number of breeding couples. Such a reduction means that the effective breeding population itself will shrink and that the opportunity for selection, so crucial to eugenic improvement, will also diminish.

Initially, that trend will give the appearance of eugenic effect since under the conditions of fertility limitation the frequency of recessive genes actually declines for the first few generations. But the long-term effect of reduction in population size, coupled with improved hygienic conditions and medical care, is to push frequencies for deleterious genes higher by "saving" affected and carrier individuals for procreation (Imaizumi, Nei, and Furusho). But what constitutes a tolerable load of harmful recessive genes, and how "deleteriousness" is to be defined, are still major value questions to be answered.

Defining "desirable" and "undesirable" genes

Like demographic shifts that affect gene frequencies, all eugenic policies require changes in the relative frequencies of human genes. But intentionally changing gene frequencies means determining which genes are to be assigned "favorable" and which "unfavorable" status. Definitions of "good" and "bad" genes are clouded in the ambiguities inherent in complex genetic causation. The classic example is the gene for hemoglobin S in which carriers of a single gene are afforded protection against the adverse health consequences of malaria, while those with two genes have sickle-cell anemia. Even as straightforwardly "harmful" a gene as that for retinoblastoma, an eye tumor, may be so closely coupled with one or more genes for intelligence as to be inseparable in any attempt at eugenic practice. Indeed, the gene for retinoblastoma itself may be so "pleiotropic" that one of its ancillary effects could affect factors bearing on intelligence.

Thus, significant questions of fact remain for all but the most egregiously lethal or semilethal genetic combinations. Even greater questions of value lie in determining how one should assign ethical weight to such genes when they are known. For instance, the gene involved in hyperbetalipoproteinemia, a familial condition characterized by high cholesterol levels and deficient receptors for low-density lipoproteins, is associated in white male carriers (heterozygotes) with over a fifty percent chance of a first heart attack before the age of fifty. Fully 0.1 to 0.5 percent of the population carry the responsible gene. Yet the policy one should implement in newborns or adults found to have this disorder is far from simple (Erbe). Moreover, for eugenic effects to be of consequence, aggressive genetic counseling must be exerted against procreation *prior* to the manifestation of the gene's full deleterious consequences, a difficult task both in practice and in principle.

In part, the dilemma of eugenic policy is that its effectiveness is conditioned on the long-term scope of its application, while the genetic decisions made by individual couples are most often predicated by more immediate life-style and personal decisions. Should some or most couples confronting the prospects of reproduction where Type II hyperbetalipoproteinemia is at stake be counseled to consider the more future-related, populationwide implications of their decision? By and large, genetic counselors have shied away from the inevitable eugenic "taint" of long-range goals. Emphasis on seeing "good" and "bad" genes in the context of genetic heterogeneity with wide variability in expression has become the rule rather than the exception.

Problems in changing the genetic status quo

It is widely recognized that the *effective* genetic makeup of any individual as expressed in that person's phenotype is the end result of a series of interactions between genes and the environment. Gene expression itself is affected by modifier genes, position effects, and the constellation of other genes in the person's genetic makeup. Most important, several key ethical considerations essential to any scientifically sound eugenics policy would focus on the nongenetic factors that might distort the expression of the genotypes in question.

Socioeconomic factors are strongly associated with increased incidences of birth defects like anencephaly (a lethal congenital condition in which the upper brain case and cerebrum fail to form) or reduced scores on intelligence tests. "Eugenic" policies could conceivably be implemented for these or other socioeconomically associated "defects" on the assumption that their statistical associations with class standing accurately reflect social differentials in gene frequency. This was, of course, the rationale behind extremely oppressive eugenics programs in the past. Class-related genetic policies are not only inappropriate where the conditions of life of the persons involved are in fact the major causative factors of disability, but politically freighted as well. Environmental influences that affect gene expression, such as diet (vitamin B deficiency), trace elements (lead contamination), and intrauterine existence, are known to lead to the appearance of "familial" patterns of defect or disability that are in fact largely or solely environmental rather than genetic in origin. In such instances, policies predicated on genetic models will almost always be socially injurious and unjust. Additionally, defects and tumors might be the result of sporadic somatic mutations rather than germinal ones, and hence not subject to the influence of procreative policies. Hemophilia and retinoblastoma are both examples. Hence, complexity in genetic causation calls for constraint in policy implementation if breaches of individual rights are to be avoided.

The state versus the individual

In addition to adequate consideration of the possible confounding effect of environmental influences, eugenic policymakers would need to consider the ethical issues around the weight to be given to personal autonomy. Theoretically, an operating principle for justifying eugenic policies could be derived form the demonstrated eugenic effect of population policies in which the reproductive behavior of individuals is constrained or otherwise influenced. Incentive programs in the form of tax credits, or disincentives in the form of penalties for those carrying presumptively deleterious genes, have been suggested at different times in the history of the eugenics movement. Such policy considerations place a greater weight on the public good to be obtained by compliance with a politically mandated solution than with autonomously derived decisions about reproduction that place the individual or his or her family above the state. The relative primacy of the state or the individual has traditionally underscored much of the eugenics debate.

Seen in modern terms, increased genetic knowledge and the enhancement of our ability to predict the consequences of reproductive behavior with an unprecedented degree of accuracy bring two ethical traditions into conflict: One assigns a high value to autonomy and self-determination in decision making, while the other imbues individuals with a sense of duty to act for the common good. When faced with these alternative orientations, ethicists have differed on where to place the greatest moral weight. Some ethicists believe that society does not have an unmitigated right to intervene in parenthood and reproductive behavior (Twiss). Others allow that right and accept that the state has the minimal moral obligation to educate and instruct in matters affecting reproduction. Such instruction is appropriate to raise the consciousness of persons with regard to the future, specifically the immediate social consequences of their reproductive behavior (Gustafson). The ethical arguments for the latter position are based on society's implicit obligation to future generations at least not to leave them worse off genetically than we are.

This obligation is reinforced by the recognition that our increasing sophistication in developing predictive techniques allows us to project the genetic consequences of individual human actions with unprecedented statistical accuracy. For instance, we now have a growing awareness of the legacy left by individuals who carried the gene for Huntington's chorea into this country in the eighteenth century, unknowingly spreading the deleterious gene they bore (as well as an indeterminate number of beneficial ones) to thousands of descendants. Under similar circumstances today, some ethicists believe that the range of moral accountability is greatly increased, and the notion of negligence becomes a more dominant reason for assessing moral liability.

Can eugenic policy be just?

Ethicists concur that any eugenic policy should incorporate some of the features of a just system. However, they disagree on what the achievement of "justice" requires. Those who believe that justice requires doing the greatest good for the greatest number might be expected to support those eugenic programs which achieved the "greatest genetic good." Others believe that the violation of one person's rights, especially those that embody basic liberties like procreation, for the benefit of others is only rarely, if ever, justified. Indeed, many follow John Rawls's suggestion that justice requires a fundamental kind of fairness in the allocation of goods, and they propose that justice requires that such goods and opportunities be distributed in a manner that benefits those who are least advantaged. While Rawls does not specifically apply this notion to eugenic policies, many if not most such policies seem to require sacrifices from those who are already (genetically if not otherwise) the less advantaged. Whether a specific program would be "fair" from a Rawlsian viewpoint depends on the priorities assigned to human goods and identification of those goods which are fundamental (Lappé).

Ultimately, we do not know which goods are considered sufficiently fundamental such that they *must* be distributed equally; nor do we know which ones are able to be distributed unequally without compromising the status of those most well off while aiding those who are least advantaged. Rawls does identify one human good, that of liberty, as being irreducible. In his view, liberty may be restricted only for the sake of liberty itself (Rawls).

At face value, the attainment of eugenic aims conflicts strongly with liberty. The implementation of eugenic programs has historically entailed selective, and often arbitrary and coercive, restrictions on some members of society and not others. The problem of justice is exacerbated by the fact that by definition the negative aspects of any eugenic policy on liberties would be directed at the least, rather than the most, fortunate members of society.

Some geneticists believe that any inequalities in social policies that flow from eugenic considerations could be rectified by allotting an individual's genetic contribution to the next generation on the basis of some assessment of his or her social worth. Two prominent geneticists have advocated assigning an Index of Social Value which apportions social value to some traits and not others (Gottesman and Erlenmeyer-Kimling). In practice, it would probably prove difficult to separate those traits that have little or no genetic contribution from those that have much, since ascertainment of the heritability of complex human traits is notoriously difficult. Any such exercise would ultimately require some determination of an individual's "genetic health," a concept that itself is fraught with ambiguities.

One manner of handling the problem is to allow a mechanism to produce a consensus in order to restrict liberty according to certain rules. Those who submit to restrictions, for instance, might then have a claim on those who have benefited from their actions. In the case of the mentally retarded, it might be considered ethically acceptable to have them or their proxies involved in any decisions to restrict childbearing in return for better conditions for their fewer offspring—conditions better than they themselves received.

But Rawls asserts that it is axiomatic for each person to wish to have greater natural assets and hence to wish such assets for one's own or another's children. Would we encourage donor insemination for the mentally retarded if the donor semen from high-IQ persons could be used? Certainly it is problematic to assume that actions should be taken that ensure for every person's descendants the best genetic endowment possible. While it is possible to argue that for a period of time a society should take steps at least to preserve the general level of natural abilities and to prevent the diffusion of serious defects, it is not clear how those aims would now be ethically realized.

Conclusions

The problem of reaching consensus on ethically acceptable approaches to eugenic policy is likely to remain knotty. Encouraging policies that already have public sanction—such as restricting consanguineous marriages or encouraging prospective genetic counseling, with amniocentesis and selective abortion—could effect a reduction in the incidence of rare genetic *diseases* but would not appreciably affect genetic load. Populationwide policies whose fundamental effect is to shift childbearing or other demographic characteristics through provision of public information, counseling, voluntary screening, and other noncoercive public devices would appear to be more consistent with equity and justice than are policies directed at compulsory control of individuals or groups; yet they are the very ones that would be most hampered by lack of compliance or participation.

At present we also lack a sufficient understanding of the critical demographic factors that already affect gene frequency, the incidence of chromosomal aneuploidy, as is found in Down's syndrome, and the opportunity for selection in human populations. Thus, we do not know with reasonable certainty that the current genetic status of the population (as it is affected by shifts in demographic variables) is in fact undergoing a dysgenic trend, or whether it is stabilizing in terms of the genetic load.

We have not systematically studied the socioeconomic factors that interact strongly with environment to produce disparate incidences of congenital malformations, many of which have appreciable genetic components. Furthermore, we have not sufficiently evaluated the desirability of utilizing existing methodologies for genetic intervention—such as population screening for carrier status, or genetic counseling and amniocentesis, which are now of value to *individuals*—as means of effecting populationwide change.

In our current state of ignorance of the benefits and consequences of the options, the consensus of most geneticists (Neel) is that our obligations to improve environmental components are prior to eugenic considerations. Chief among these environmental factors are those which increase the genetic load (mutagens) and those which undermine the expression of human genetic potential (and here poverty would rank highest). Stabilizing the quantity of the population so that equitable distribution of resources remains possible also has priority over eugenic considerations. Many geneticists would also concur that, where eugenic policies are to be put into effect, they should preserve the status quo of the gene pool, particularly by minimizing the contribution of novel mutations.

Ultimately, the fundamental ethical issues broached by eugenics embrace the problem of the relative primacy of the individual or the collective whole. Should the ends of medicine serve the race or individuals? Should any aspect of human procreation be placed under state control? How can desirable eugenic programs best be implemented to enhance individual responsibility, justice, and equity in the distribution of society's resources? These remain among the strongest unanswered ethical questions in eugenics.

MARC LAPPÉ

[*For further discussion of topics mentioned in this article, see the entries:* FUTURE GENERATIONS, OBLIGATIONS TO; JUSTICE; PATERNALISM; PRIVACY; *and* STERILIZATION. *Also directly related are the entries* GENE THERAPY; GENETIC DIAGNOSIS AND COUNSELING; GENETIC SCREENING; *and* GENETICS AND THE LAW. *See also:* POPULATION POLICY PROPOSALS, *article on* GENETIC IMPLICATIONS OF POPULATION CONTROL; *and* REPRODUCTIVE TECHNOLOGIES, *article on* ETHICAL ISSUES. *For discussion of related ideas, see the entry* MEDICAL ETHICS UNDER NATIONAL SOCIALISM.]

BIBLIOGRAPHY

CROW, JAMES F. "The Quality of People: Human Evolutionary Changes." *BioScience* 16 (1966): 863–867.

ERBE, RICHARD W. "Mass Screening and Genetic Counseling in Mendelian Disorders." *Ethical, Social and Legal Dimensions of Screening for Human Genetic Disease.* Edited by Daniel Bergsma. *Birth Defects: Original Article Series* 10, no. 6 (1974), pp. 85–99.

FLETCHER, JOSEPH F. *The Ethics of Genetic Control: Ending Reproductive Roulette.* Garden City, N.Y.: Anchor/Doubleday, 1974.

GOTTESMAN, IRVING I., and ERLENMEYER-KIMLING, L. "Prologue: A Foundation for Informed Eugenics." *Social Biology* 18, suppl. (1971), pp. S1–S8.

GUSTAFSON, JAMES M. "Genetic Engineering and the Normative View of the Human." *Ethical Issues in Biology and Medicine.* Edited by Preston N. Williams. New York: Schenkman Press, 1973, pp. 46–58. Symposium on the Identity and Dignity of Man, Boston University, 1969.

HUXLEY, JULIAN. "Eugenics in Evolutionary Perspective." *Perspectives in Biology and Medicine* 6 (1963): 155–187.

IMAIZUMI, YOKO; NEI, MASATOSHI; and FURUSHO, TOSHIYUKI. "Variability and Heritability of Human Fertility." *Annals of Human Genetics* 33 (1970): 251–259.

INGLE, DWIGHT JOYCE. *Who Should Have Children? An Environmental and Genetic Approach.* Indianapolis: Bobbs-Merrill, 1973.

JONES, ALUN, and BODMER, WALTER F. *Our Future Inheritance: Choice or Chance? A Study.* London: Oxford University Press, 1974.

LAPPÉ, MARC. "Can Eugenic Policy Be Just?" *The Prevention of Genetic Disease and Mental Retardation.* Edited by Aubrey Milunsky. Philadelphia: W. B. Saunders Co., 1975, chap. 21, pp. 456–475.

LIPKIN, MACK, JR., and ROWLEY, PETER T., eds. *Genetic Responsibility: On Choosing Our Children's Genes.* New York: Plenum Press, 1974. Proceedings of the Symposium "Genetics, Man, and Society" held at the American Association for the Advancement of Science meeting, 29 December 1972.

MEDAWAR, PETER. "The Genetical Impact of Medicine." *Annals of Internal Medicine* 67, no. 3, pt. 2 (1967), pp. 28–31.

MULLER, HERMANN J. *Man's Future Birthright: Essays on Science and Humanity.* Edited by Elof Axel Carlson. Albany: State University of New York Press, 1973.

NAROT, JOSEPH R. "The Moral and Ethical Implications of Human Sexuality as They Relate to the Retarded." *Human Sexuality and the Mentally Retarded.* Edited by Felix F. de la Cruz and Gerald D. La Veck. New York: Brunner/Mazel, 1973, chap. 15, pp. 195–205. Conference on Human Sexuality and the Mentally Retarded, Hot Springs, Arkansas, 1971.

NEEL, JAMES V. "Lessons from a 'Primitive' People." *Science* 170 (1970): 815–822.

OSBORN, FREDERICK HENRY. *The Future of Human Heredity: An Introduction to Eugenics in Modern Society.* New York: Weybright & Talley, 1968.

RAMSEY, PAUL. *Fabricated Man: The Ethics of Genetic Control.* A Yale Fastback, no. 6. New Haven: Yale University Press, 1970.

RAWLS, JOHN. *A Theory of Justice.* Cambridge: Harvard University Press, Belknap Press, 1971.

SANG, J. H. "Nature, Nurture, and Eugenics." *Postgraduate Medical Journal* 48 (1972): 227–230.

SIMPSON, GEORGE GAYLORD. "The Concept of Progress in Organic Evolution." *Social Research* 41 (1974): 28–51.

SPERRY, R. W. "Science and the Problem of Values." *Zygon* 9 (1974): 7–21.

STENT, GUNTHER. "Molecular Biology and Metaphysics." *Nature* 248 (1974): 779–781.

TURNER, JOHN R. G. "How Does Treating Congenital Diseases Affect the Genetic Load?" *Eugenics Quarterly* 15 (1968): 191–197.

TWISS, SUMNER B., JR. "Ethical Issues in Priority-Setting for the Utilization of Genetic Technologies." *Ethical and Scientific Issues Posed by Human Uses of Molecular Genetics.* Edited by Marc Lappé and Robert S. Morison. *Annals of the New York Academy of Sciences* 265 (1976): 22–45.

EUGENICS AND RELIGIOUS LAW

I. JEWISH RELIGIOUS LAWS   *David M. Feldman*
II. CHRISTIAN RELIGIOUS LAWS

                       William W. Bassett

I
JEWISH RELIGIOUS LAWS

The laws of incest and consanguinity in the Old Testament would seem to have a rationale in eugenics, though this is never specified in the biblical text. The traditional commentators, too, advert only to the natural repugnance against incest. In the talmudic discussion as well as in the legal codes, the subject is treated as a sexual offense, involving a breach of morality rather than a eugenic error. (The Talmud is the repository of rabbinic exposition of biblical law and teaching, spanning more than five centuries of discussion in the Academies. The legal codes are, in turn, based on the Talmud and on subsequent development of the law, such as in Responsa, or formal opinions in response to new case-law inquiries.)

Even bastardy is a moral rather than a eugenic category. The *mamzer* (in Jewish law, the product only of an adulterous or incestuous liaison, not of an "out-of-wedlock" relation between unmarried persons) is not physically illborn; his status is compromised only legally and socially, rendered so in punitive (or deterrent) judgment against parents not free to have entered the relationship. But no difference obtains between the *mamzer* born of adultery—even technical adultery, such as when the document of divorce for her previous marriage was impugned—and the *mamzer* born of incest. Hence, no eugenic motive can be assigned here.

Moreover, the one "maimed in his privy parts" bears the same legal disabilities as the *mamzer*. Thus, a man of "crushed testicles or severed member" is excluded from "the congregation of the Lord" (Deut. 23:2); this verse is interpreted to mean only that he may not enter into conjugal union with an Israelite woman. Hence, the castrated male is under the ban *because* the act of castration is forbidden. But one "maimed in his privy parts" as a natural result of birth defect or of disease, as opposed to one castrated by his or another's deliberate assault, is free of this disability. The legal situations were thus analogized: "Just as the *mamzer* is the result of human misdeeds, so only the castrated one who is such as a result of human misdeeds is to be banned." With that distinction made in both cases, and with the further fact that the banned *mamzer* and castrated are permitted to marry, for example, a fellow *mamzer* or proselyte, it must be concluded that moral outrage and punitive judgment rather than eugenic considerations are evidently operative.

Eugenics, in the sense of choosing a marriage partner with the well-being of the progeny in mind, is more clearly present in talmudic counsel and legislation. A person is counseled to choose a wife prudently, and guidance is offered in doing so in accordance with the intellectual and moral virtues of the prospective bride; and since, we are told, a child normally takes after its mother's brothers, one should regard the maternal uncles in making his decision (Bava Batra 110a). A hidden physical blemish in a bride is grounds for invalidating a marriage, unless the husband can be presumed to have known of it in advance.

Actually, heredity as a eugenic principle takes its legal model from rulings with respect to circumcision. An infant whose two brothers died as a possible result of this operation may not be circumcised. He is deemed to have inherited the illness (probably hemophilia) that proved fatal to his two brothers. The Talmud goes on to say that an infant whose two maternal cousins showed that weakness may also not be circumcised. That is, the statistical evidence yielded by two sons from the same mother can also be reflected in two sisters of that mother (Yevamot 64b). This is a remarkably early recognition that hemophilia is transmitted through maternal lineage—in itself a significant eugenic discovery.

The statistical evidence, or the presumption, of adverse hereditary factors in a third family member, when those factors are seen to exist in two of them, thus becomes the basis of talmudic laws of eugenics. With modern laboratory means to determine the presence of these factors, the principle would of course operate even sooner, without waiting for statistical evidence in two members. The Talmud rules that a man may not marry into a family of epileptics or lepers, or—by extension—a similar disease (ibid.). This may be the first eugenic edict in any social or religious system.

True, the pure hereditary nature of this recommendation is not unanimously agreed upon. While one view in the Talmud attributes the "heredity" to physical characteristics of the marriage partner, another view sees the basis as "bad luck." In a recent Responsum, where the questioner considered abortion because the mother was epileptic, the Rabbi responded that the latter of the two views above may be the right one, and that fear of bad luck is an inadequate warrant for abortion (Feldman, p. 292).

In an other than legal context, the Mishnah (the foundation layer of the Talmud), speaks of the faculties that a father bequeaths to his son: "looks, strength, riches, and length of years" (Eduyot II, 9). Here, too, the commentaries align themselves on both sides: one sees the bequeathing of faculties as a natural hereditary process, the other sees them as reward for the father's virtues.

Two other talmudic ideas with eugenic motifs have reflections in current practice. In the interests of fulfilling the injunction to "love one's wife as much as himself and honor her more than himself," a man is advised to seek his sister's daughter as a bride; his care for her will be the more tender due to his affection for his own sister. Yet in the thirteenth century Rabbi Judah the Pious left a testamentary charge to his children and grandchildren, which became a source of guidance to others on the level of precedent for subsequent Jewish law. In this famous testament, he advises against marriage with a niece as having adverse genetic results. Modern rabbinic authorities conclude that, if this be the case, "natures have changed" in this regard since the days of the Talmud; but generally they dismiss the fears as medically unjustified.

A second point is a talmudic notion that eugenic factors operate in the matter of intercourse during pregnancy. Conjugal relations, we are told, should be avoided during the first trimester as "injurious to the embryo"; but encouraged during the final trimester as desirable for both mother and embryo, that the child then comes forth "well-formed and of strong vitality" (Niddah 31a). A medieval Jewish authority makes the matter a point of pride in comparative culture: the Talmud recommends coitus during the final trimester, while the Greek and Arab scholars say it is harmful. Don't listen to them, he says (Responsa Bar Sheshet, No. 447). Nonetheless, the Talmud prohibits the marriage of a pregnant or nursing widow or divorcee; the second husband, it is suggested in Maimonides's formulation, may be less considerate of another man's fetus and may inadvertently damage it through abdominal pressure in intercourse (Yevamot 36a); and in the nursing situation, assuming that a pregnancy weakens the mother's milk, the new father may fail to take the necessary steps to supplement the diet of his stepchild. And a pregnant woman who feels either an urgent physical or psychological need for food during the Yom Kippur Fast is to be fed for the sake of her fetus's welfare as well as her own (Yoma, 82a).

More homiletic than legal is the notion that defective children can be the result of immoral or inconsiderate modes of intercourse—an idea expounded but ultimately rejected by the Talmud (Nedarim 20a). Yet in more modern times the Hasidim ("pietistic" Jewish groups with a mystic orientation) maintain that spiritual consequences of the act are indeed possible; that if a man has pure and lofty thoughts during or preparatory to cohabitation he can succeed in transmitting to the child an especially lofty soul. Hence dynastic succession of leadership, as opposed to democratic selection, obtains among Hasidic groups.

A study of biblical and talmudic sources written by Max Grunwald in 1930, cited by Jakobovits, discerns a broad eugenic motif. Grunwald writes that Judaism

. . . quite consciously strives for the promotion of the quality as well as quantity of the progeny by the compulsion of matrimony, the insistence on early marriage, the sexual purity of the marital partners and the harmony of their ages and characters, the dissolubility of unhappy unions, the regulation of conjugal intercourse, the high esteem of maternity, the stress on parental responsibility, the protection of the embryo, etc. To be sure, there can be no question here of a compulsory public control over the health conditions of the marriage candidates, but that would positively be in line with the principles of Jewish eugenics: the pursuit after the most numerous and physically, mentally, and morally sound natural increase of the people, without thinking of an exclusive race protection [Jakobovits, p. 154].

Though abortion is warranted primarily for maternal rather than fetal indications, screening of would-be parents for actual or potential defective genes, such as in Tay–Sachs disease, would, like premarital blood tests, be much in keeping with the Jewish traditional eugenic concern.

DAVID M. FELDMAN

[*For further discussion of topics mentioned in this article, see the entries:* ABORTION, *article on* JEWISH PERSPECTIVES; JUDAISM; *and* RELIGIOUS DIRECTIVES IN MEDICAL ETHICS, *article on* JEWISH CODES AND GUIDELINES. *Also directly related are the entries* FETAL–MATERNAL RELATIONSHIP; GENETIC DIAGNOSIS AND COUNSELING; GENETIC SCREENING; *and* PRENATAL DIAGNOSIS.]

BIBLIOGRAPHY

FELDMAN, DAVID M. *Birth Control in Jewish Law.* New York: New York University Press, 1968. Paperback ed. *Marital Relations, Birth Control, and Abortion in Jewish Law.* New York: Schocken Books, 1974.

JACOBS, LOUIS. *What Does Judaism Say About . . . ?* New York: Quadrangle/New York Times Book Co., 1974, pp. 165–166.

JAKOBOVITS, IMMANUEL. *Jewish Medical Ethics: A Comparative and Historical Study of the Jewish Religious Attitude to Medicine and Its Practice.* New ed. New York: Bloch, 1975.

ZEVIN, SHELOMOH YOSEF. *Le'Or Ha-Halakhah.* 2d ed. Tel Aviv: Abraham Zioni, 1957, pp. 147–158.

II
CHRISTIAN RELIGIOUS LAWS

Christian religious laws historically comprehend a large spectrum of rules to guide individual conduct and social relationships among the baptized. The laws most likely to have eugenic significance are the canons prohibiting the marriage of relatives. These regulations also formed the basis for the civil law prohibitions against the marriage of relatives in both the Continental legal systems and in the common law statutory scheme. Though the principal justification given for such prohibitions in Christian law has been ethical and social, there is substantial evidence that they also may reflect considerations classified as eugenic in contemporary scientific research.

The ecclesiastical regulations that forbid marriage between persons closely related by consanguinity are among the most ancient canons of the Christian tradition. Penalties attached to the violation of religious exogamic laws have varied historically in their severity, as, indeed, have the ways of measuring the degrees of kinship and defining within which degrees the crime of incest shall be punished. But the core of the tradition of canon law remains constant and reflects an extreme reluctance to accept the marriages of close relatives as humanly or religiously feasible.

For Roman Catholics all marriages within the direct line of blood relationship, i.e., between an ancestor and a descendant by parentage, and within the collateral line to the third degree, i.e., to second cousins, are forbidden (*Code of Canon Law*). In the Greek Orthodox tradition, marriage in the direct line and in the collateral line to the sixth or seventh degree by the Roman method of computation is prohibited in canon 54 of the Synod in Trullo, 691/2 (Hefele). All Oriental Christians forbid marriages in the direct line; Armenians, Jacobites, and Copts prohibit it in the collateral line to the fourth degree, Melkites to the sixth degree, Serbs and Chaldeans to the third degree, and Ethiopians without distinction. Among the Protestant reformers the restrictions of the medieval canon law were accepted by some, e.g., Melanchthon, Kemnitz, only the Old Testament regulations of Leviticus 18:6–13 by others, e.g., Bucer and, perhaps, Luther, and only the closest tie of direct parental relationship by still others, e.g., Wycliffe. In the Anglican community, *The Book of Common Prayer* contains the table drawn up by Archbishop Parker based on Leviticus in naming relatives incapable of marriage (Wheatly). Most Protestant churches today follow the prohibitions of civil law regarding incest and kinship marriage (Acte for Kynges Succession; Acte for Succession of Imperyall Crowne; Concerning Precontracte and Degrees).

The sources and commentaries upon the Christian laws record debate about the extent of the prohibition, the possibility of dispensation within certain close degrees of kinship, and about the related question of the divine or natural law origin of the laws. They reveal, however, only the most sketchy discussion of the foundations of the regulations themselves.

The classical reasons given for the prohibition of consanguineous marriages are ethical and social. The first reason was called the *respectus parentelae*, namely, that such marriages would undermine the respect due to parents and consequently to all those who are closely related (Thomas Aquinas, *Summa Theologiae* II-II, 154, 9). Secondly, they constitute a moral danger to family life arising from the possibility of early moral corruption of the young dwelling within the same household in which marriage could be allowed (ibid.; Sanchez, 7.52.12; 53). Thirdly, the prohibition of consanguineous marriages prevents the disruption of the family by sexual competition and forces the multiplication of friendships and the spread of charity (Augustine). These three reasons seem to have been sufficient to justify the laws, so that most scholars did not go beyond them to seek a further justification. Esmein, for example, said the laws arose simply out of an instinctive repulsion for incest and were not reflective of any known adverse physical consequences.

It is only in comparatively modern times that an explicitly eugenic reason for the prohibition has received scientific attention. Writing in 1673, Samuel Dugard noted: "There is a *judgment* which is said often to accompany these Marriages, and that is *Want of Children* and a *Barrennesse*" (p. 53). "The Children are weak, it may be; grow crooked, or, what is worse, do not prove well; presently, Sir, it shall be said what better could be expected? an unlawfull Wedlock must have an unprosperous successe" (p. 51). A. J. Stapf's *Theologia Moralis* in 1827 alluded to this possibility (p. 359). A fuller treatment is found in Le Noir's 1873 edition of St. Alphonsus's Moral Theology. Westermarck in 1889 and Laurent in 1895 spoke at length of a physiological justification of the canons to prevent indiscriminate inbreeding and the risk of a high incidence of deleterious genetic effects. Wernz, in 1911 (n. 416 [72]), writing from a comprehensive knowledge of the canonical tradition, said the ancient writers also knew of the undesirable effects of excessive inbreeding. He noted reasons derived from contemporary medical science in the writings of Gratian (early twelfth century) (c. 35 q. 283, c. 20), Pope Innocent III (1161–1216) (Schroeder), and Thomas Aquinas (*Commentum*). Since the late nineteenth century nearly all commentators on the canonical rules speak of eugenic objections to marriages of blood relatives.

It is possible to find in the ancient ecclesiastical commentators an awareness of a eugenic foundation to the prohibition expressed in primitive and undifferentiated modes of speech. For example, a persistent belief was kept alive among theologians and canonists that children of incestuous relationships will die or will be greatly debilitated, or that the familial line will be cursed with sterility. Benedict the Levite (850?) wrote of these marriages: "From these are usually born the blind, the deaf, hunchbacks, the mentally defective, and others afflicted with loathsome infirmities" (Benedict the Levite). Furthermore, in the explanations of the name of the impediment (i.e., the impediment of consanguinity), if one traces their origins through medieval glossography to the *Etymologies* of Isidore of Seville (560?–636), there appears an awareness of a physiological factor in the blood bond of close relatives that must be weakened before marriage safely can be contracted.

The antecedents of the Christian canons in the Mosaic law (Lev. 18:6–13) and the Roman law were taken as expressions of natural law by the canonists and were continued in the barbarian codes. In his *Ecclesiastical History* (I, 27) where the Venerable Bede (673–735) notes these laws, he records a quotation from a letter of Pope Gregory I to Augustine of Canterbury, written in 601. The reason given by Gregory for forbidding marriages of close relatives is, "We have learned from experience that from such a marriage offspring cannot grow up" (Bede). This letter and this reason not only are later picked up and cited by Gratian (c. 35, q. 5, c. 2) and Thomas Aquinas (*Summa Theologiae Suppl.* 54, 3), but may be found in virtually all the canonical collections of the early Middle Ages. Though comment on this passage is rare, comment was, perhaps, unnecessary. The passage from Gregory seems clearly to say that from forbidden consanguineous marriages experience teaches that children are affected or unable to grow up. There is thought to be a physiological consequence to incest. In the light of this it seems probable that the labored argumentation over the question of how close the relationship must be for marriage to be forbidden by natural law must have been conducted in some awareness of a popular belief in the biological consequences of such unions. The fear of genetic anomalies or biological debilitation from indiscriminate inbreeding may not be perfectly articulated. It is difficult to imagine, however, that some physiological dangers to offspring may not have been intended in the frequent citation of Pope Gregory to sustain the severity of the prohibition.

Thomas Sanchez (1625), who wrote the greatest of the canonical commentaries on marriage, says that the most suasive ground for forbidding incestuous unions is that there is a sharing of the blood among close relatives and that the physical image of a progenitor (*imago, complexio, effigies, mores, virtus paterna*) passes to offspring, so that the blood must be weakened through successive generations before marriage should be contracted (7.50; 7.51.1–2). Thus, preventing marriages of close relatives to protect the offspring by allowing several generations to pass before procreation can be called a measure of eugenic foresight, however simple the scientific awareness to support it may have been.

In summary, a eugenic foundation to Christian religious laws forbidding the marriage of close relatives is clearly articulated and commented upon by modern scholars from the late eighteenth and nineteenth centuries. Evidence of this kind of awareness may be discovered earlier in the canonical sources, however, going back at least to the seventh century. It would seem consistent with the eugenic connotation of those laws rooted in antiquity, together with a Christian sense of responsibility for offspring that partly motivated them, to consider further eugenic restrictions on marriage in Christian communities today, in light of contemporary knowledge of genetics.

WILLIAM W. BASSETT

[*For a further discussion of topics mentioned in this article, see:* EUGENICS. *See also:* FUTURE GENERATIONS, OBLIGATIONS TO; GENETICS AND THE LAW; NATURAL LAW; *and* STERILIZATION. *This article* will find application in the entries GENE THERAPY; GENETIC ASPECTS OF HUMAN BEHAVIOR; GENETIC CONSTITUTION AND ENVIRONMENTAL CONDITIONING; GENETIC DIAGNOSIS AND COUNSELING; GENETIC SCREENING; PRENATAL DIAGNOSIS; and REPRODUCTIVE TECHNOLOGIES.]

BIBLIOGRAPHY

An Acte for the Establishement of the Kynges Succession. 25 Hen. VIII, c. 22 (1533–1534). *The Statutes of the Realm*, vol. 3, pp. 471–474, especially pp. 472–473.

An Acte for the Establisshement of the Succession of the Imperyall Crowne of this Realme. 28 Hen. VIII, c. 7 (1536). *The Statutes of the Realm*, vol. 3, pp. 655–662, especially pp. 658–659.

AUGUSTINE. "Of Marriage between Blood-Relations, in Regard to Which the Present Law Could Not Bind the Men of Earliest Ages." *The City of God*. Translated by Marcus Dods. Introduction by Thomas Merton. New York: Modern Library, 1950, bk. 15, chap. 16, pp. 500–502. "De jure conjugiorum, quod dissimile a subsequentibus matrimoniis habuerint prima connubia." *Patrologia Latina*, vol. 41, cols. 457–460. Also. *Writings of Saint Augustine*. Vol. 7: *The City of God, Books VIII–XVI*. Translated by Gerald G. Walsh and Grace Monahan. The Fathers of the Church: A New Translation, vol. 14. Edited by Roy Joseph Deferrari. New York: 1952, pp. 450–454.

BEDE [VENERABLE BEDE]. *Bede's Ecclesiastical History of the English People*. Edited by Bertram Colgrave and R. A. B. Mynors. Oxford: At the Clarendon Press, 1969, bk. 1, chap. 27, pp. 78–102, esp. p. 85. Facing texts in Latin and English.

BENEDICT THE LEVITE [BENEDICTI DIACONI]. "Capitularium collectio: Pertz monitum." *Patrologia Latina*, vol. 97, bk. 3, sec. 179, col. 820.

Code of Canon Law: A Text and Commentary. Edited by Timothy Lincoln Bouscaren and Adam C. Ellis. Milwaukee: Bruce Publishing Co., 1946, canon 1076, secs. 1, 2; pp. 487–489. English with commentary.

Concerning Precontracte and Degrees of Consanguinite. 32 Hen. VIII, c. 38 (1540). *The Statutes of the Realm*, vol. 3, p. 792.

COUSSA, ACACIO. *Epitome Praelectionem de Iure Ecclesiastico Orientali*. 3 vols. Vol. 3: *De Matrimonio*. Rome: Typis Monasterii Exarchici Cryptoferratensis, 1948.

DAUVILLIER, J.-DECLERCQ C. *Le Mariage en droit canonique oriental*. Paris: Recueil Sirey, 1936.

DUGARD, S[AMUEL]. *The Marriages of Cousins German, Vindicated from the Censures of Unlawfullnesse, and Inexpediency, Being a Letter Written to His Much Honour'd T.. D.* Oxford: Printed by Hen: Hall for Thomas Bowman, 1673. Work attributed to Dugard although taken largely from Jeremy Taylor's *Ductor Dubitantium*. Attributed to Simon Dugard in the British Museum Catalog.

ESMEIN, ADHÉMAR. *Le Mariage en droit canonique* (1891). 2 vols. 2d ed. Paris: Recueil Sirey, 1929.

FLEURY, J. *Recherches historiques sur les empêchements de parenté dans le mariage canonique des origines aux fausses décrétales*. Paris: Recueil Sirey, 1933.

GRATIAN [GRATIANUS, THE CANONIST]. *Decretum divi Gratiani, universi iuris canonici pontificias constitutiones & canonicas brevi compendio complectens*. Lyons: I. Pideoius, 1554, pt. 2, causa 35, question 2, canon 20, p. 1217; question 5, canon 2, pp. 1218–1221.

GREGORY I. "Gregorius Augustino Episcopo." *Registrum Epistolarum*. 2 vols. 2d ed. Edited by Paulus Ewald and Ludovicus M. Hartman. Monumenta Germaniae Historica, Epistolarum, vols. 1, 2. Berlin: Apud Weidmannos, 1957, vol. 2, pp. 332–343, especially pp. 335–336.

HEFELE, CHARLES JOSEPH. "The Quinisext or Trullan Synod, A.D. 692." *A History of the Councils of the Church: From the Original Documents*. Translated and edited by William R. Clark. Vol. 5: A.D. 626 to the Close of the Second Council of Nicaea, A.D. 787. Edinburgh: T. & T. Clark, 1896, sec. 327, pp. 221–239, especially canon 54, p. 231.

ISIDORE OF SEVILLE. "De adfinitatibus et gradibus." "De agnatis et cognatis." "De conivgiis." *Etymologiarum sive originum, libri XX*. 2 vols. Edited by W. M. Lindsay. Oxford: E typographeo Clarendoniano, 1911, vol. 1, bk. 9, chaps. 5–7, unpaginated. *Patrologia Latina*, vol. 82, cols. 353–368.

LAURENT, E. *Mariages consanguins et dégénérescences*. Paris: A. Maloine, 1895.

LIGUORI, ALFONSO MARIA DE'. "De matrimonio." *Theologia moralis*. 4 vols. Edited by D. Le Noir. Paris: Ludovicum Vivès, 1872–1874, vol. 3, bk. 5, tractate 6, pp. 661–858, especially pp. 783–784.

MEYVAERT, PAUL. "Bede's Text of the *Libellus Responsionem* of Gregory the Great to Augustine of Canterbury." *England before the Conquest: Studies in Primary Sources Presented to Dorothy Whitelock*. Edited by Peter Clemoes and Kathleen Hughes. Cambridge: Cambridge University Press, 1971, pp. 15–33.

Patrologiae cursus completus: Series Latina [Patrologia Latina]. 221 vols. Paris: Apud Garnier Fratres, editores & J.-P. Migne successores, 1851–1894. *Supplementum*. Paris: Éditions Garnier frères, 1958– .

SANCHEZ, TOMÁS. "De impedimentis." *Disputationum de sancto matrimonio sacramento*. 3 vols. in 4. Vols. 1, 2, Genoa: Iosephum Pavonem, 1602. Vols. 3, 4, Madrid: Ludouici Sanchez, 1605, vol. 3, bk. 7, disputation 50, pp. 332–336; disputation 51, pars. 1–2; pp. 336–337; disputation 52, par. 12, p. 352; disputation 53, pp. 352–354.

SCHROEDER, HENRY JOSEPH. "The Twelfth General Council (1215): Fourth Lateran Council." *Disciplinary Decrees of the General Councils: Text, Translation, and Commentary*. St. Louis: B. Herder Book Co., 1937, pp. 236–296, in particular canon 50, pp. 279–280. Latin text. "Canones Concilii Lateranensis IV (Oecumen. XII): Anno 1215 Habiti." *Disciplinary Decrees*, pp. 560–584, in particular canon 50, p. 578.

STAPF, AMBROSIUS JOSEPH. *Theologiae Moralis*. 2 vols. Innsbruck: Typis & sumptibus Wagnerianis, 1827, vol. 2, sec. 312, p. 359.

The Statutes of the Realm, Printed by Command of His Majesty King George the Third, in Pursuance of an Address of the House of Commons of Great Britain, from the Original Records and Authentic Manuscripts (1810–1833). 11 vols. in 12. Reprint. London: Dawsons of Pall Mall, 1963.

THOMAS AQUINAS. *Opera omnia*. Vol. 30: *Commentum in Libros IV Sententiarum*. Edited by Stanislai Eduardi Fretté. Paris: Ludovicum Vivès, 1878, distinctions 40 & 41, question 1, article 4, pp. 770–771.

———. *Summa theologiae* II–II 154, 9. "Utrum incestus sit species determinata luxuriae." *Summa theologiae.* Vol. 3: *Pars secunda secundae.* Rome: Marietti, 1948, pp. 722–723. Translated by the Fathers of the English Dominican Province as "Whether Incest Is a Determinate Species of Lust?" *Summa Theologica: First Complete American Edition.* 3 vols. Vol. 2: *Second Part of the Second Part, QQ 1–189 and Third Part, QQ 1–90, with Synoptical Charts.* New York: Benziger Brothers, 1947, question 154, article 9, pp. 1823–1824.

———. *Summa theologiae suppl.* 54, 3. "Utrum consanguinitas de iure naturali impediat matrimonium." *Summa theologiae.* Vol. 4: *Pars tertia et supplementum.* Rome: Marietti, 1948, pp. 838–840, especially "Sed contra," p. 839. Translated by the Fathers of the English Dominican Province as "Whether Consanguinity Is an Impediment to Marriages by Virtue of Natural Law." *Summa Theologica: First Complete American Edition.* 3 vols. Vol. 3: *Supplement QQ 1–99, Appendices, Articles, Index.* New York: Benziger Brothers, 1948, question 54, article 3, pp. 2758–2760, especially "On the contrary," p. 2759.

WAHL, FRANCIS X. *The Matrimonial Impediments of Consanguinity and Affinity.* Washington: Catholic University of America Press, 1934.

WERNZ, FRANZ XAVER. *Ius decretalium.* 6 vols. Rev. ed. Vol. 4: *Ius matrimoniale ecclesiae catholicae.* Florence: Libraria Gaichetti, 1911.

WESTERMARCK, EDWARD A. *The History of Human Marriage* (1889). 3 vols. 5th ed. rev. London: Macmillan & Co., 1921; New York: Allerton Book Co., 1922. Reprint. New York: Johnson Reprint Corporation, 1971.

WHEATLY, CHARLES. "Of the Preface and Charge and the Several Impediments to Matrimony." *A Rational Illustration of the Book of Common Prayer of the Church of England.* 8th ed. London: C. Hitch et al., 1759, chap. 10, sec. 3, pp. 376–383. A commentary.

EUPHENICS

Euphenics (Greek *eu* = good or well, plus French or Greek *phen* = to show) was coined by Lederberg in 1963 as a noun to describe the engineering of human development. Currently it refers to the application of medical and surgical treatments to genetic disease (Lederberg, p. 521). Euphenics represents the realization of biological knowledge applied to human hereditary malfunctions, i.e., therapeutic medical genetics. It is an alternative to eugenics that has as its goal the reshaping of the gene pool of an entire population.

Nearly two thousand hereditary diseases are now known to afflict the human species; some such diseases have been medically cured symptomatically, but no genetic cure at the level of the gene itself has yet been effected. To date only about one hundred of these are understood at the level of primary gene products (enzyme or protein). For nongenetic diseases such as invasions by pathogenic microorganisms, a cure rids the afflicted of that which was the cause. However, in the case of hereditary diseases the source of affliction is a malfunctioning or nonfunctional gene, a component of the very biologic core of each individual. Treatment of such diseases relieves or even eliminates their symptoms, but the defective gene or genes persist and can be passed on to offspring.

Even though no effective method of altering or replacing abnormal genes in human patients is yet available, more and more genetic diseases are being treated by alleviating the malfunctions involved. Medical strategies currently in use as therapies for genetic diseases are enumerated with examples in Table 1.

Many genetic diseases, such as phenylketonuria, hemophilia, and cystic fibrosis, reduce the reproductive potential of the afflicted individual to nearly zero when untreated. These individuals are not sterile in the sense of lacking in functional gamete production; their low reproductive potential is a result of physical debilitation, social unacceptability, or prereproductive death caused by the genetic disorder they suffer. In most cases successful treatment of the disease symptoms also removes barriers to reproduction. Thus euphenics commonly raises the reproductive potential (Darwinian fitness in evolutionary terms) of the treated individual. It is a fundamental evolutionary precept and axiom of population genetics that whenever the Darwinian fitness of a particular genotype is raised relative to all others in a population the frequency of the genes characterizing the "selected" or more advantaged type will increase in the gene pool of subsequent generations. Herein lies a principal ethical dilemma of euphenics. As medical science is able to treat more genetic diseases, the frequency of each malfunctioning gene concerned will increase in subsequent generations. Indeed, it has been estimated that, at our current rate of medical advancement in the treatments of genetic defects, within five to ten generations the frequency of serious genetic defects could double from its current one in every twenty live births to one in ten (Augenstein, p. 32; see also Holloway and Smith).

Although each genetic disease may be rare in the populations surveyed, collectively these diseases already pose serious challenges to our medical resources and health-care delivery systems. It is estimated that currently twelve million Americans suffer from genetic diseases; at

TABLE 1

Medical Strategies Currently in Use as Euphenic Measures with Examples

1. Addition of a missing product
   Insulin for diabetes mellitus
   Antihemophilic globulin for hemophilia
   Gamma globulin for gamma globulin deficiency
   Thyroid hormone for familial goiter
   Human growth hormone for pituitary dwarfism
2. Restriction of a substance
   Dietary restriction of phenylalanine for phenylketonuria
   Dietary restriction of galactose for galactosemia
   Dietary restriction of fructose for fructose intolerances
   Dietary restriction of protein for arginosuccinicaciduria, citrullinemia, hyperammonemia, and hyperargininemia
   Avoidance of muscle relaxants for cholinesterase deficiency
3. Drug therapy
   Allopurinol for gout and Lesch-Nyhan syndrome
   Phenobarbital for hyperbilirubinemia and Crigler–Najjar syndrome
   D-penicillamine for Wilson's disease
   Cortisol for adrenogenital syndrome
   Pyridoxine for primary hyperoxaluria
4. Immunologic inhibition
   Rhogam for Rh incompatibility between mother and child
5. Surgery
   Remove lens for inherited cataracts
   Remove eye for retinoblastoma
   Remove spleen for spherocytosis
   Open valve for pyloric stenosis
   Close spinal column for spina bifida
   Transplant kidney for cystinosis, polycystic kidneys, and Fabry's disease
   Transplant bone marrow for Thalassemia and agammaglobulinemia

Source: Adapted from Howell, pp. 278–279, and Omenn, p. 56.

least forty percent of all infant mortality results from genetic factors; and every married couple stands a three percent risk of having a genetically defective child (United States, National Institutes of Health, p. 6). It is the compounding effects of all treatable genetic diseases that lead to concerns about the deterioration of the gene pool in future generations if the reproductive potential of large numbers of individuals is significantly raised by treating a multitude of different genetically defective conditions.

Conversely, it has been argued that genetic deterioration of the human species is a red herring (Lappé, p. 419), that the genetic consequences of the successful treatment of diseases caused by rare recessive genes are slight (Crow, p. 865), and that the rate of genetic deterioration is so slow that future generations will certainly find solutions to cope with any difficulties that arise (Medawar, p. 33). Positions such as these are usually based on single examples of rare genetic diseases, which show that more than a thousand years would be needed to double the frequency of any particular gene in question following successful treatment. It is also common to slip into "eugenic rationales" in discussions of the prevention of genetic deterioration. Thus, the possibility of sterilizing carriers (heterozygotes) becomes an issue, while this has nothing to do with euphenic practices.

The relationship of euphenics to genetic deterioration of the gene pool must be clearly known (1) before considering the collective impact on society of all current and potential genetic ameliorations and (2) in order to clearly separate eugenic goals or objectives from the problems created by euphenics. Euphenics may be an alternative to eugenics, and the two practices can, and we think should, be mutually exclusive. We suggest that this distinction must permeate all discussions of the euphenics–genetic deterioration dilemma. The solution to any such dilemma will eventually have to be an ethical compromise, and we should not unduly complicate our task by superimposing eugenic issues on an already difficult situation.

Several important factors that are generally not included in cursory views of the euphenics–genetic deterioration dilemma must be enumerated before moving on to remaining ethical problems. First, although the incidence of any particular treated genetic defect increases only slightly in a population as a whole, the effects within any family expressing the defect are much more pronounced. For example, pyloric stenosis (constriction of the passageway between the stomach and small intestine) produces obstructions in about 3 per 1,000 newborn infants. Before corrective surgery (to release and open the pyloric sphincter between the stomach and the small intestines) was available most individuals with this polygenic condition (produced by the action of many genes) died as children without making any contribution to the gene pool. When the condition became surgically treatable, afflicted children were then enabled to reach adulthood and reproduce. It was found that among their children 70 out of each 1,000 needed surgical treatment for pyloric stenosis. This is more than a twentyfold increase in one generation within families harboring this genetic defect (Stern, p. 628). Because inherited diseases are familial, the impact of even a slight populational increase becomes localized and hence intensified in particular families. Such situations may result in heavy burdens within certain families, while any effects go virtually unnoticed in the rest of society.

Second, because each genetic disease is rare and is held at low frequency by the low or zero reproductive fitness of those afflicted, the human mating system has not routinely generated recombinants (new genetic combinations) bearing two or more serious conditions within the same genotype. We have never observed the interaction of two or more genetic diseases in one person, e.g., someone with hemophilia and sickle-cell anemia or someone with phenylketonuria (inability to metabolize an essential amino acid) and galactosemia (inability to digest milk sugar). We are totally ignorant of the genetic interactions and phenotypic expressions of any such recombinants. We do know that in other organisms where double and triple mutants can be derived experimentally (they are never found in nature), the interaction of two or more mutations effecting lower viability or fertility typically exceeds the detrimental effects of either mutant alone. This problem is well known to animal and plant breeders and to anyone who has conducted Mendelian transmission exercises with the geneticist's favorite, *Drosophila melanogaster* (fruit fly). The treatment and subsequent reproduction of human beings with viability and/or fertility debilitations not only raise the frequency of each condition separately, but they progressively increase the probability of producing genotypes with multiple disorders. This potential needs to be thoroughly evaluated before the incidences of too many genetic diseases are allowed to increase unchecked.

Third, besides being ignorant about phenotypes due to genetic interactions involving multiple afflictions, we also know nothing of treatment interactions that might result. The array of genetic interactions possible from the collective accumulation of ten, or a hundred, or a thousand treatable genetic diseases makes even speculation on treatment interactions impossible.

Euphenics creates another set of ethical problems with regard to the distribution of medical care and the priorities given to medical services. It has been suggested that, once we accept the premise of equal right and access to medical care along with the present reality that we simply cannot, physically cannot, treat all who are in need, it seems more just to discriminate by virtue of categories of illnesses, for example, rather than between the rich ill and the poor ill (Outka, p. 24). Thus, with our assuredly less-than-optimal conditions for supplying medical services, it can be argued that persons with rare, noncommunicable disease should receive low priority, especially when costs are high and resources are scarce. Acceptance of this stance would place virtually all euphenic measures at a very low priority, because each measure is bound to treat a rare, noncommunicable (in the usual public health sense of the word) disease. Yet collectively, as we have noted, genetic diseases constitute a major medical problem. Again we urge that genetic diseases be considered collectively as a class of medical problems for which we are ethically obligated to provide medical services. It must also be recognized, however, that, if this view served to accelerate medical treatments of numerous genetic diseases, it would at the same time intensify the euphenics–genetic deterioration dilemma.

It has been pointed out that, in cases where normally functioning carriers (individuals with one dose of a detrimental recessive gene and one of the normal functioning gene) actually display "hybrid vigor," i.e., higher reproductive ability than individuals with two doses of the normal gene (homozygotes), public health might be better improved by treating "normals" than by treating sufferers (Turner, p. 196). Thus, for diseases such as sickle-cell anemia, cystic fibrosis, and Tay-Sachs, where carriers may have a higher reproductive potential than normal homozygotes, it may be argued that in the interest of future generations normal homozygotes also need, and are entitled to, some sort of treatment to compensate for their lower fitness. Do normal homozygotes in these situations have any claim to medical care? If so, what kinds of treatment would be appropriate, how would they be distributed, and what priority should they receive?

Yet another set of ethical issues concerning euphenics is that associated with partially treatable genetic defects. The concerns here regard situations where the rescuing of individuals from certain death seriously compromises their quality of life (Motulsky, p. 656). For example, it is now possible to isolate children born with immunodeficiency diseases inside a germ-free bubble to protect their defenseless systems from contracting any infections. They would prove fatal. This rare class of inherited diseases has little treatment other than the artificial bubble environment. Unless or until an efficient treatment is found, some kind of bubble is the only environment afflicted individuals can look forward to. Such partial treatments place "quality of life" and "sanctity of life" in an ethically difficult juxtaposition. Whether or not children might have the right to sue their parents for wrongful life in view of the psychological stresses experienced as a result of parental decisions to produce them and/or sustain them in spite of known genetic malfunctions and prognoses has already been questioned (Stern, p. 627; Ghent).

We must also ask how far any "right to reproduce" extends to such partially treated individuals. For example, suppose a boy with the form of immunodeficiency disease known as X-linked infantile agammaglobulinemia (also known as Bruton agammaglobulinemia) grew up in a germ-free bubble to adulthood, fell in love with a young nurse, and married her. Does this couple have a right to children, possibly by collecting sperm from the young man and using it to artificially inseminate his wife to produce their child? We know that all daughters will be carriers of their father's defective gene and all sons will be normal in this respect. Should there be any restriction on reproduction or on the production of female carriers? Should there be concern about increasing the frequency of a gene that produces a disease that has such a currently restrictive medical prognosis? Should we supply all the bubbles needed in anticipation of a cure? There are many proposed answers to these questions, but no unanimity. The difficulties exist entirely at the euphenic level; none needs to be extended into the realm of eugenics to conjure up ethical dilemmas.

Are we at an impasse—damned if we do and damned if we do not? Golding (p. 463) has said that the tragedy of the euphenic situation may be that we will have to reckon with the fact that the amelioration of short-term evils and the promotion of good (avoidance of gene pool deterioration) for the remote future are mutually exclusive alternatives. Is this so? Is there any way out of this "euphenic fix?" Callahan (p. 7) says that there are only two ways to bridge the middle ground between unlimited individualism (individual rights) and unlimited regard for the community (societal rights): by establishing upper limits to what individuals can demand and by establishing internal restraints on the harm communities may do to individuals. The task is to arrive at an ethical compromise that properly balances the rights of individuals and those of society.

There seems to be a divergence of views regarding the individual–societal rights balance that ranges from abstention from reproduction by persons receiving euphenic aid (Wallace, p. 44) to claims of no responsibility to society's gene pool on the part of families receiving treatment for genetic disorders (Kushnick, p. 624).

As a "working" solution to the myriad problems raised by euphenics we propose the following concept as a humanitarian "stopgap," given the complexities and perplexities inherent in current euphenic implications.

It would seem most efficacious, given all the problems discussed above, that a "holding stance" be taken with regard to any deliberate change whatsoever in the human gene pool for the immediate future. At the same time it seems imperative that we continue to relieve to the best of our abilities the pain and suffering of individuals afflicted with genetic diseases and anomalies. We therefore suggest that the ethical compromises that must accompany euphenic manipulations be formulated about the objective of producing no change in the frequency of any genes whose function is altered by euphenic measures. We term this concept *genetic conservation* and define it as the maintenance of the genetic composition of the human gene pool in its present proportions. This must be understood to include active measures such as various degrees of control over reproductive potentials of individuals in whom euphenic measures would raise fitness and thereby produce future changes in gene frequencies. Genetic conservation includes acceptance of neither dysgenic nor eugenic gene frequency alterations given our current lack of knowledge regarding the compounding and interaction effects of increasing genes involved with genetic diseases and the pleiotropic (multiple, secondary) and evolutionary effects of decreasing or eliminating these same genes. We recommend genetic conservation because we do not currently have the requisite armamentaria with which to interfere either profitably or wisely with the structure of the human gene pool.

Genetic conservation focuses our problems squarely on the balance between individual and societal rights and responsibilities. It is not an answer but a way of approaching an answer that incorporates what we believe to be honest and just parameters. Within this domain we still have to decide upon the uses and abuses of technologies such as sterilization, artificial insemination, fetal monitoring, therapeutic abortions, and carrier determinations as they counterbalance euphenic measures in genetic conservation.

If genetic conservation is not needed now, when we can still effectively treat only a handful of genetic diseases, when do we reach the critical point at which we should be concerned about the additive and interaction effects of euphenic measures? Do we have the responsibility to act before even small changes occur and thereby set a precedent for future euphenic developments, or should we treat the "euphenic fix" as a negligible societal problem fraught with academic dilemmas too far removed from the real problems of contemporary society to warrant premeditated actions? The task of incorporating euphenic measures into the medical practices of our society is clearly upon us. It would be wise to consider all costs and benefits judiciously and proceed with the utmost care. The gene pool is, after all, our own.

LEE EHRMAN AND JAMES J. NAGLE

[*Directly related are the entries* EUGENICS; *and* GENETIC CONSTITUTION AND ENVIRONMENTAL CONDITIONING. *For further discussion of topics mentioned in this article, see the entries:* GENE THERAPY; GENETIC DIAGNOSIS AND COUNSELING; GENETIC SCREENING; *and* RATIONING OF MEDICAL TREATMENT. *See also:* ABORTION; PRENATAL DIAGNOSIS; REPRODUCTIVE TECHNOLOGIES, *article on* ARTIFICIAL INSEMINATION; RIGHTS; *and* STERILIZATION.]

BIBLIOGRAPHY

AUGENSTEIN, LEROY. *Come, Let Us Play God.* New York: Harper & Row, 1969.
CALLAHAN, DANIEL. "Science: Limits and Prohibitions." *Hastings Center Report* 3, no. 5 (1973), pp. 5–7.
CROW, JAMES F. "The Quality of People: Human Evolutionary Changes." *BioScience* 16 (1966): 863–867.
GHENT, J. F. "Tort Liability for Wrongfully Causing One to Be Born: Annotation." *American Law Reports* 3d ser. 22 (1974): 1441–1448.
GOLDING, MARTIN P. "Ethical Issues in Biological Engineering." *UCLA Law Review* 15 (1968): 443–479. Symposium: Reflections on the New Biology.
HOLLOWAY, SUSAN M., and SMITH, CHARLES. "Effects of Various Medical and Social Practices on the Frequency of Genetic Disorders." *American Journal of Human Genetics* 27 (1975): 614–627.
HOWELL, R. RODNEY. "Genetic Disease: The Present Status of Treatment." *Medical Genetics.* Edited by Victor A. McKusick and Robert Claiborne. New York: Hospital Practice Publishing Co., 1973, pp. 271–280.
KUSHNICK, THEODORE. "When to Refer to the Geneticist." *Journal of the American Medical Association* 235 (1976): 623–625.
LAPPÉ, MARC. "Moral Obligations and the Fallacies of 'Genetic Control'." *Theological Studies* 33 (1972): 411–427.
LEDERBERG, JOSHUA. "Experimental Genetics and Human Evolution." *American Naturalist* 100 (1966): 519–531.
MEDAWAR, PETER BRIAN. "Do Advances in Medicine Lead to Genetic Deterioration?" *Mayo Clinic Proceedings* 40 (1965): 23–33.
MOTULSKY, ARNO G. "Brave New World?" *Science* 185 (1974): 653–663.
OMENN, GILBERT S. "Genetic Engineering: Present and Future." *To Live and to Die: When, Why, and How.* Edited by Robert H. Williams. New York: Springer-Verlag, 1973, pp. 48–63.
OUTKA, GENE. "Social Justice and Equal Access to Health Care." *Journal of Religious Ethics* 2, no. 1 (1974), pp. 11–32.
STERN, CURT. "The Place of Genetics in Medicine." *Annals of Internal Medicine* 75 (1971): 623–629.
TURNER, JOHN R. G. "How Does Treating Congenital Diseases Affect the Genetic Load?" *Eugenics Quarterly* 15 (1968): 191–197.
United States, National Institutes of Health, National Institute of General Medical Sciences. *What Are The Facts about Genetic Disease? Most Ubiquitous of All Human Maladies.* DHEW Publication no. (NIH) 76–370. Bethesda, Md.: Department of Health, Education, and Welfare, 1976.
WALLACE, BRUCE. "Genetic Engineering: The Promise of Things to Come." *Essays in Social Biology.* Edited by Bruce Wallace. Vol. 2: *Genetics, Evolution, Race, Radiation Biology.* Englewood Cliffs, N.J.: Prentice-Hall, 1972, pp. 41–44.

EUROPE

See MEDICAL ETHICS, HISTORY OF, *section on* EUROPE AND THE AMERICAS.

EUTHANASIA

See DEATH AND DYING: EUTHANASIA AND SUSTAINING LIFE; INFANTS; ACTING AND REFRAINING; LIFE-SUPPORT SYSTEMS.

EVOLUTION

Essentials of Darwin's theory

Although Charles Darwin (1809–1882) did, of course, have both precursors and partial anticipators, and although there have been considerable rival accounts, the public is quite right to link the idea of biological evolution indissolubly with his name. The epoch-making landmark is, therefore, the publication in 1859 of *The Origin of Species by Means of Natural Selection* This, the full title, is significant and apt. There was also a subtitle, which has since acquired a sinister ring: *or the Preservation of Favoured Races in the Struggle for Life.* Darwin's theory is evolutionary inasmuch as it asserts, and provides an account of, the origin of species. Evolution is thus contrasted with the idea of the fixity of species, all of which were presumably specially created; as is assumed, for instance, in the picturesque creation stories of Genesis. Natural selection is here the key term in an account of how evolution, in this sense, has occurred and still is occurring.

None of these various notions was by itself new. For instance, the general hypothesis of the derivation of all present species from a small number, or perhaps even a single pair, of original ancestors was propounded by Maupertuis, as the President of the Berlin Academy of Sciences, in 1745 and 1751, and by Diderot, the organizer of the great French *Encyclopédie,* in 1749 and 1754. Again, there is a garish account of struggles for existence and of natural selection—combined with a notion of natural kinds, or fixed species, emphatically detached from any idea of divine creation—in the philosophical epic *Concerning the Nature of Things* by the Epicurean Roman poet Lucretius in the first century B.C. The actual phrases "a struggle for existence" and "the survival of the fittest" occurred in relevant contexts in, respectively, *An Essay on the Principle of Population* (1798)— the *First Essay*—by T. R. Malthus, and in an article by Herbert Spencer for the *Westminster Review* of 1852. So Darwin's claims to theoretical originality must rest upon the matter and the manner of his putting it all together.

He himself in his *Autobiography* described *The Origin* as "one long argument from the beginning to the end" (Darwin, 1958, p. 140), massively illustrated and supported by his vast knowledge of natural history. It runs:

If . . . organic beings present individual differences in almost every part of their structure, and this cannot be disputed; if there be, owing to their geometrical rate of increase, a severe struggle for existence at some age, season, or year, and this cannot be disputed; then . . . it would be an extraordinary fact if no variations had ever occurred useful to each being's own welfare, in the same manner as variations have occurred useful to man. But if variations useful to any organic being do occur, assuredly individuals thus characterized will have the best chance of being preserved in the struggle for life; and from the strong principle of inheritance they will tend to produce offspring similarly characterized. This principle of preservation, or the survival of the fittest, I have called natural selection [Darwin, 1859, pp. 169–170].

Certainly this core argument is valid, and the premises from which it proceeds can scarcely be contested. The Malthusian power to multiply and the tendency of like to produce like are two of the most familiar and fundamental facts of both the animal and the plant worlds. But this core argument is very far from proving what Darwin never thought it did prove: that all known species are descended, like varieties, from other species. Big questions remain. Has there, for one thing, been enough time for the often enormously elaborate and subtly arranged sorts of known life to have evolved in this way; or do we have to postulate some further, faster, much more directed mechanism? How, for a second thing, do hereditable differences or—for that matter—hereditable similarities appear, and are there sufficient of these, and of the right kinds?

In 1865 William Thomson (Lord Kelvin) argued that the physics of the cooling of the earth could not admit the vast ages which the geologists thought that the record of the rocks had revealed; and it was only well after Darwin's death that Kelvin could be shown to have been in fact mistaken. Darwin himself was convinced that natural selection has been the most important, but not the exclusive, means of modification. In a later work on *The Descent of Man* he protested that, whereas "my critics frequently assume that I attribute all changes of corporeal structure and mental power exclusively to the natural selection of such variations as are often called spontaneous," in fact he had from the beginning always "stated that great weight must be attributed to the inherited effects of use and disuse, with respect both to the body and the mind" (1874, p. viii). This was in fact Darwin's view, although such passages must come as a shock to those familiar with the ferocity of the neo-Darwinian rejection of any such Lamarckian suggestion of the inheritance of acquired characteristics.

On the second count Darwin was fully aware that he just did not know: "Our ignorance of the laws of variation is profound" (Darwin, 1859, p. 202). The work done in the early 1860s by Gregor Mendel—digging to lay the foundations of genetics—became known to his successors only after it had been independently repeated at the turn of the century by Hugo de Vries.

Darwinism and religion

The impact of Darwin's work has been wide-ranging and profound: It was altogether fitting that his compatriots buried him in Westminster Abbey, hard by the tomb of Sir Isaac Newton. It was Darwin and not any of his evolutionist predecessors who persuaded the scientific world that all the variety of life today has been the outcome of a gradual process of descent with modifications from a few original very simple forms —perhaps only one. Darwin succeeded where others had failed because he indicated a mechanism that he could prove must be operating, and because he deployed a mass of detailed evidence and argument to show how the development could have occurred.

Outside of science the first upset was to established religious ideas. For, although Darwinism was no more incompatible with a literal reading of Genesis than earlier evolutionary accounts of the origin of the solar system, or of the present condition of the earth's crust, it was nevertheless felt to be far more obnoxious. It seriously threatened, as they did not, the assumed special status of man. The epitome of this unease was the scandalized and scandalous question put by Bishop Wilberforce to the physiologist and publicist T. H. Huxley at an extraordinary meeting of the British Association for the Advancement of Science: "Is it on your father's or your mother's side that you are descended from a monkey?"

At least equally important was the threat to what should be known as the Argument or Arguments to Design. The most powerful expression of these in popular form was found in William Paley's *Natural Theology: or Evidences of the Existence and Attributes of the Deity Collected from the Appearances of Nature* (1802).

How could things so complex and so subtly integrated as organisms, and organs such as the human eye, come about, save by divine design? To this the Darwinian now replied that they could be, and were, the products of natural selection operating on spontaneous variation. His reply leaves the more sophisticated Argument to Design, which some find in the "Fifth Way" of St. Thomas Aquinas, untouched. The popular version appeals to divine design to fill lacunae in the scientific account of nature: "the God of the gaps." The more sophisticated alternative appeals first to whatever may be found to be the most fundamental regularities in the universe, including perhaps those from which evolution follows as a consequence. It then urges that this order cannot be intrinsic, but must be imposed by a divine orderer. The classical response to this is to be found in the posthumous masterpiece of the Scottish philosopher David Hume, *Dialogues concerning Natural Religion* (1779).

Social Darwinism

As the prestige of Darwin's theory grew, people began to see in it an endorsement for very diverse social and political policies. In America, William Graham Sumner (1840–1910) and, in Britain, Herbert Spencer (1820–1903) saw it as authorizing economic laissez-faire and no-holds-barred competition. J. D. Rockefeller declared in one of his Sunday school addresses:

The growth of a large business is merely a survival of the fittest. . . . The American Beauty rose can be produced in the splendour and the fragrance which bring cheer to its beholder only by sacrificing the early buds which grow up around it. This is not an evil tendency in business. It is merely the working out of a law of nature and a law of God [Hofstadter, p. 31].

At the opposite end of the spectrum of political economy Karl Marx wanted to dedicate the first volume of *Capital* to Darwin (Darwin declined), and in a speech at Marx's graveside Friedrich Engels claimed, "Just as Darwin discovered the law of development of organic nature, so Marx discovered the law of development of human history." Later socialists contrived to see Darwin's ideas as positively endorsing their own panacea, state monopoly. Thus in 1905 the future British Labour Prime Minister Ramsay MacDonald wrote that

. . . the Conservative and aristocratic interests . . . have armed themselves . . . with the law of the struggle for existence, and its corollary, the survival of the fittest. . . . Darwinism is not only not in intellectual opposition to Socialism, but is its scientific foundation. . . . Socialism is naught but Darwinism economized, made definite, become an intellectual policy, applied to the conditions of human society [Ferri, p. v].

A little later still, in the pre–World War I Vienna of his youth, the future Führer of German National Socialism wrote:

By means of the struggle the élites are continually renewed. The law of selection justifies this incessant struggle by allowing the survival of the fittest. Christianity is a rebellion against natural law, a protest against nature. Taken to its logical extreme Christianity would mean the systematic cult of human failure [Bullock, p. 693].

All such attempts to derive social, political, or bioethical norms directly from the findings of natural science must be unsound. Insofar as the survival of the fittest, and the so-called law of selection, really are laws of nature there cannot be any question of choosing either to obey or to disobey them. For where any descriptive law of nature holds it follows necessarily, by definition, that any occurrence incompatible with that law must be physically impossible. If there is to be room for a prescriptive law that lays down not what is but what ought to be, then it has to be physically possible either to do or to refrain from doing what the prescriptive law says ought to be done.

Many are inclined to believe that what is natural is by the same token good. Given this assumption it is almost impossible not to infer that a process of selection which is both natural and guaranteed to ensure the survival of the fittest must be for the best. The inference is, nevertheless, fallacious. The crux is the different criteria for what is fittest and what is best. Darwin's theory guarantees the survival of the fittest only and precisely insofar as actual survival is the criterion of fitness to survive. If some alternative and independent criterion is introduced, then Darwin's core argument ceases to be valid. The actual conclusion of Darwin's argument is entirely consistent with the familiar facts: that, for instance, men who are in every way wretched specimens exterminate superb species of animals; or that genius has often been laid low by the activities of unicellular creatures having no wits at all. The survivors, simply by surviving, show that they had what it took to survive. But those who seek in Darwinism an assurance that the good will in the end prevail surely measure merit by some quite other standard. For to accept what the theory actually has to offer as if this were necessarily good would be to make time-serving your supreme principle: Whatever in fact wins will be as such in your eyes best.

Darwinism and progress

Darwin himself never misconstrued the expressions "struggle for existence" or "natural selection" in his theory as normative. Yet he was perhaps a little inclined to see in his own picture of biological evolution some promise of what he could value as progress. Thus in the penultimate paragraph of *The Origin* he wrote:

As all the living forms of life are the lineal descendents of those which lived long before the Cambrian epoch, we may feel certain that . . . no cataclysm has desolated the whole world. Hence we may look with some confidence to a secure future of great length. And as natural selection works solely by and for the good of each being, all corporeal and mental endowments will tend to progress towards perfection.

Before Hiroshima the second of these statements was reasonable enough. But the third was not warranted by what went before: No reason had been given for assuming that it is all, for the losers also, for the best; and it is perfectly possible for a species with great "corporeal and mental endowments" to be extinguished as a result of some competitive weakness elsewhere. Many of Darwin's successors, including several most distinguished biologists, have tried to develop his characteristically cautious hint of a promise of progress. After all, it certainly is the case that those animals which all would rate as, by other standards, higher, first appear in the most recent and, in that quite different sense, higher geological strata. (Darwin pinned into his copy of that once hugely popular book, Chambers' *Vestiges of the Natural History of Creation*, the memorandum slip: "Never use the words 'higher' and 'lower.'")

In our century Julian Huxley, one of the grandsons of T. H. Huxley, made several attempts to find in "the facts of evolutionary biology . . . a verifiable doctrine of progress" (1923, p. 19). It is, he said, a fundamental need "to discover something, some being or power, some force or tendency . . . moulding the destinies of the world—something not himself, greater than himself, with which [he can] harmonize his nature . . . achieve confidence and hope" (ibid., p. 17). Joseph Needham, another contemporary biological Fellow of the Royal Society, who maintains both a Christian and a Marxist–Leninist allegiance, reviews the same materials to find that "the new world-order . . . is no wild idealistic dream, but a logical extrapolation from the whole course of evolution, having no less authority than that behind it, and therefore of all faiths the most rational" (1943, p. 41). Later, in *History Is on Our Side*, he concluded: "Whatever force hinders the coming of the world cooperative commonwealth . . . is ultimately doomed. Against the world-process no force can in the end succeed" (1946, pp. 209–210).

The basic fault of every kind of social Darwinism is in logic. Albert Einstein expressed the heart of the matter with memorable clarity: "As long as we remain within the realm of science proper, we can never meet with a sentence of the type 'Thou shalt not kill.' . . . Scientific statements of facts and relations . . . cannot produce ethical directives." On the other hand, the trouble with the present sort of appeal to evolution as a guarantee of progress is that it ignores the decisive matter that members of our species already possess, or very soon will possess, the capacity to destroy all life on this planet. The point is made with salutary brutality by A. M. Quinton, writing on "Ethics and the Theory of Evolution." Man, he insists, "has certainly won the contest between animal species in that it is only on his sufferance that any other species exist at all, amongst species large enough to be seen at any rate" (p. 120).

Evolutionary biology cannot, therefore, provide either social and political norms or a promise of future progress. But what it can offer is a perspective. Huxley and Needham both wanted more, and sometimes believed that they had found it. What are available are possibilities of seeing our human activities in an evolutionary perspective. Thus, thirty years after he began his quest in *Essays of a Biologist*, Julian Huxley wrote:

In the light of evolutionary biology man can now see himself as the sole agent for further evolutionary advance on this planet, and one of the few possible instruments of progress in the universe at large. He finds himself in the unexpected position of business manager for the cosmic process of evolution [1953, p. 132].

Darwin, Lamarck, and the social engineers

Suppose that we do think of the life of mankind in this way, and suppose too that we are by temperament or conviction social engineers, eager to remake other people and human society nearer to our heart's desires. Then we are likely to be depressed by the implications, or the apparent implications, of Darwinism, especially when this is supplemented and reinforced by the findings of the geneticists. For this system of ideas defines certain limits to our ambitions. It suggests that our basic raw material—people, and especially the youngest—is in various genetically determined ways intractable; and that whatever shaping and molding we contrive to effect will have to be redone to all members of all succeeding generations.

Suppose again that we believe, conformably to our role as members of a quasi-Platonic elite of social engineers, that a very small group of ruthless and determined men can by seizing or creating a mighty state machine utterly transform a people and a society. Then we may well think that the neo-Darwinians present too passive a picture of the mechanisms of biological change. Surely there must be something we can do, other than just wait for natural (or artificial) selection to operate on spontaneously occurring variations?

These suppositions describe a ready market for Lamarckian notions of the inheritance of acquired characteristics. The Chevalier de Lamarck (1744–1829) developed a theory of evolution well before Darwin. Among the supposed laws of this theory are: (1) The production of a new organ in an animal body results from a new need which continues to make itself felt; (2) the development and effectiveness of organs are proportional to the use of these organs; and (3) everything acquired or changed during an individual's lifetime is preserved by heredity and transmitted to that individual's progeny. Almost all contemporary biologists believe that the vital third of these Lamarckian laws has been definitely refuted by experiment.

During the 1930s and 1940s similar ideas were put forward in the USSR by T. D. Lysenko. He succeeded in persuading the Soviet authorities both that his own researches were leading to results that would make possible a great leap forward in agricultural production and that orthodox Mendelian genetic theory was un-Marxist, and therefore to be eradicated in and by the party and the state. No such great leap forward in fact resulted. But it is indeed the case—though this has, of course, no bearing on the scientific question of which is the more true and tested—that Darwin's theory is more passive and less voluntaristic than that of Lamarck;

that it is, in Marxist terminology, metaphysical rather than dialectical. In the persecutions consequent upon the rise of Lysenko Russia's most distinguished geneticist, academician N. I. Vavilov, died, a martyr for science, in a concentration camp.

A less obscurantist response to the scientific picture of biological evolution is to make proposals for improving the human gene pool by encouraging the multiplication of approved genes and discouraging that of the disfavored. It was Darwin's cousin Francis Galton who introduced the label "eugenic" to describe such proposals, though the idea of policies for improving the human breed is at least as old as Plato's *Republic*. Eugenicists are, typically, concerned because modern medicine has enabled the victims of many formerly destructive genetic defects to survive to reproduce; and because those who rise or who stay at the top in socially mobile societies—and who are, therefore, presumed to have some of the best genes—tend to have fewer children than those who stay at or fall to the bottom.

More or less modest eugenic proposals, and attempts to draw moral and political conclusions from evolutionary biology, still are being made and will surely continue to be made; although it may sometimes seem that today we have rather more violent denunciation of putative proposals and alleged attempts than actual proposals and actual attempts. Consider, for instance, the controversies touched off in the last few years by the publication of Arthur Jensen's "How Much Can We Boost IQ and Scholastic Achievement?" and of Edward Wilson's *Sociobiology*.

Conclusion

Certainly evolutionary biology can provide a perspective for our thinking about human life, and certainly our policies must never fail to take account of stubborn biological and genetic facts. But policies are about possible choices; and no facts can by themselves determine in what sense choices ought to be made. Which is one, although only one, of the reasons why it must be wrong, for instance, to assume that admitting that any genetically determined average differences in talent or temperament between different racial and social groups would be to concede the propriety of advantaging or disadvantaging individuals for no other or better reason than that they happen to belong to one such group or another. The facts will remain whatever they are, whether we choose to face them or not, whether we like them or not. But it is always for us to decide what we ought to do.

ANTONY G. N. FLEW

[*Directly related are the entries* BIOLOGY, PHILOSOPHY OF; MEDICINE, PHILOSOPHY OF; NATURAL LAW; PURPOSE IN THE UNIVERSE; *and* SCIENCE: ETHICAL IMPLICATIONS. *For discussion of related ideas, see the entries:* EUGENICS, *article on* ETHICAL ISSUES; GENETIC ASPECTS OF HUMAN BEHAVIOR, *article on* PHILOSOPHICAL AND ETHICAL ISSUES; *and* RACISM.]

BIBLIOGRAPHY

BARNETT, SAMUEL ANTHONY, ed. *A Century of Darwin*. Cambridge: Harvard University Press; London: Heinemann, 1958. Essays on the impact of Darwin in various areas.

BULLOCK, ALAN LOUIS CHARLES. *Hitler: A Study in Tyranny*. Completely rev. ed. Pelican Books, A564. Harmondsworth, England: Penguin Books, 1962.

CANNON, HERBERT GRAHAM. *Lamarck and Modern Genetics*. Manchester: Manchester University, 1959. A short polemical study by one of the few modern biologists with any time for Lamarckian ideas.

CHAMBERS, ROBERT. *Vestiges of the Natural History of Creation*. London: J. Churchill, 1844; New York: Wiley & Putnam, 1845. A dozen editions, 1844–1860.

DARWIN, CHARLES ROBERT. *The Autobiography of Charles Darwin*. Edited, with original omissions restored, by Nora Barlow. London: Collins, 1958. The first complete edition; all earlier, and some later, versions omit Darwin's various remarks about religion.

———. *The Descent of Man, And Selection in Relation to Sex*. 2 vols. London: John Murray, 1871. 2d ed. 1874.

———. *The Origin of Species by Means of Natural Selection, Or the Preservation of Favoured Races in the Struggle for Life*. London: John Murray, 1859. There is a modern edition of the first version by J. W. Burrow in the Pelican Classics series, Harmondsworth, England, and Baltimore: Penguin, 1968. A variorum text for the successive editions of Darwin's lifetime is edited by Morse Peckham, Philadelphia: University of Pennsylvania Press, 1959. References are to the modern edition.

DEWEY, JOHN. "The Influence of Darwinism on Philosophy." *The Influence of Darwinism in Philosophy and Other Essays in Contemporary Thought*. New York: Henry Holt & Co., 1910. Dewey always claimed that Darwinism was one of the continuing and formative influences on his own thought.

EISELEY, LOREN. *Darwin's Century: Evolution and the Men Who Discovered It*. Garden City, N.Y.: Doubleday Anchor Books, 1958. An exhilarating account of evolution and the men who discovered it.

FERRI, ENRICO. *Socialism and Positive Science: (Darwin-Spencer-Marx)*. Translated by Edith C. Harvey from the French ed. of 1896. 4th ed. The Socialist Library, no. 1. London: Independent Labour Party, 1906.

FLEW, ANTONY GARRARD NEWTON. *Evolutionary Ethics*. New Studies in Ethics. London: Macmillan; New York: St. Martin's Press, 1967. Reprint. *New Studies in Ethics*. 2 vols. Edited by William Donald

Hudson. London: Macmillan & Co.; New York: St. Martin's Press, 1974, vol. 2, pp. 217–268. A short philosophical study, aspiring to be definitive.
HOFSTADTER, RICHARD. *Social Darwinism in American Thought, 1860–1915.* Philadelphia: University of Pennsylvania Press; London: Oxford University Press, 1944. Hofstadter achieves what he undertakes admirably.
HUXLEY, JULIAN SORELL. *Essays of a Biologist* (1923). Pelican Books A51. Harmondsworth, England: Penguin Books, 1939.
———. *Essays of a Humanist.* London: Chatto & Windus; New York: Harper & Row, 1964.
———. *Evolution in Action.* London: Chatto & Windus; New York: Harper & Row, 1953. The Huxley works are landmarks in the search described in this article.
HUXLEY, THOMAS HENRY, and HUXLEY, JULIAN SORELL. *Evolution and Ethics: 1893–1943.* London: Pilot Press, 1947. A piquant confrontation between mighty grandfather and rather weaker grandson, hosted by the latter.
JENSEN, ARTHUR R. "How Much Can We Boost IQ and Scholastic Achievement?" *Harvard Educational Review* 39 (1969): 1–123.
KROPOTKIN, PETER ALEKSEEVICH. *Mutual Aid: A Factor of Evolution.* Rev. and cheaper ed. London: Heinemann, 1907. A counterblast to the traditional emphasis on competition, written with an eye to human applications.
NEEDHAM, JOSEPH. *History Is on Our Side: A Contribution to Political Religion and Scientific Faith.* London: G. Allen & Unwin, 1946.
———. *Time, The Refreshing River: Essays and Addresses, 1932–1942.* London: G. Allen & Unwin, 1943. Needham has an unusual width of interest, and a quite extraordinary combination of commitments.
QUINTON, ANTHONY M. "Ethics and the Theory of Evolution." *Biology and Personality: Frontier Problems in Science, Philosophy, and Religion.* Edited by Ian T. Ramsey. Oxford: Blackwell; New York: Barnes & Noble, 1965, pp. 107–131.
SPENCER, HERBERT. *The Principles of Ethics.* 2 vols. A System of Synthetic Philosophy, vols. 9, 10. London: Williams & Norgate, 1892, 1893. Very weary stuff, by a producer of evolutionary ethics more highly regarded by his contemporaries than by successors.
WADDINGTON, CONRAD HAL, ed. *Science and Ethics.* London: G. Allen & Unwin, 1942. A wartime discussion perhaps of antiquarian interest only.
WIENER, PHILIP PAUL. *Evolution and the Founders of Pragmatism.* Foreword by John Dewey. Cambridge: Harvard University Press, 1949. A serviceable study of a subject that may be of somewhat similar antiquarian interest.
WILSON, EDWARD OSBORNE. *Sociobiology: The New Synthesis.* Cambridge: Harvard University Press, Belknap Press, 1975.

EXPERIMENTATION, ANIMAL
See ANIMAL EXPERIMENTATION.

EXPERIMENTATION WITH HUMAN SUBJECTS
See HUMAN EXPERIMENTATION; PRISONERS, *article on* PRISONER EXPERIMENTATION; INFORMED CONSENT IN HUMAN RESEARCH; FETAL RESEARCH; CHILDREN AND BIOMEDICINE; AGING AND THE AGED, *article on* HEALTH CARE AND RESEARCH IN THE AGED; WOMEN AND BIOMEDICINE, *article on* WOMEN AS PATIENTS AND EXPERIMENTAL SUBJECTS.

EXTRAORDINARY TREATMENT
See DEATH AND DYING: EUTHANASIA AND SUSTAINING LIFE, *articles on* ETHICAL VIEWS *and* HISTORICAL PERSPECTIVES; INFANTS, *article on* ETHICAL PERSPECTIVES ON THE CARE OF INFANTS; LIFE-SUPPORT SYSTEMS; RIGHT TO REFUSE MEDICAL CARE.

Encyclopedia of BIOETHICS

WARREN T. REICH, *Editor in Chief*
GEORGETOWN UNIVERSITY

VOLUME 2

THE FREE PRESS
A Division of Macmillan Publishing Co., Inc.
NEW YORK

Collier Macmillan Publishers
LONDON

Copyright © 1978 by Georgetown University

All rights reserved. No part of this book may be reproduced or transmitted in any form or by any means, electronic or mechanical, including photocopying, recording, or by any information storage and retrieval system, without permission in writing from the Publisher.

THE FREE PRESS
A Division of Macmillan Publishing Co., Inc.
866 Third Avenue, New York, N.Y. 10022

Collier Macmillan Canada, Ltd.

Library of Congress Catalog Card Number: 78–8821

Printed in the United States of America

Library of Congress Cataloging in Publication Data
Main entry under title:

Encyclopedia of bioethics.

   Includes bibliographies and index.
   1. Bioethics—Dictionaries. 2. Medical ethics—Dictionaries. I. Reich, Warren T. [DNLM: 1. Bioethics—Encyclopedias. QH302.5 E56]
QH332.E52   174'.2   78-8821
ISBN 0-02-926060-4
ISBN 0-02-925910-X (this edition)

The *Encyclopedia of Bioethics* project was made possible by a grant from the National Endowment for the Humanities. In funding the project the Endowment matched gifts from non-federal sources including The Joseph P. Kennedy, Jr. Foundation, the Raskob Foundation, The Commonwealth Fund, The Loyola Foundation, Inc., and The David J. Greene Foundation, Inc. The views expressed in this encyclopedia are not necessarily those of the Endowment or of these foundations.

Complete and unabridged edition   1982

FAITH HEALING

See MIRACLE AND FAITH HEALING.

FEES

See SURGERY; HEALTH CARE, *article on* HEALTH-CARE SYSTEM; HEALTH INSURANCE; MEDICAL ETHICS, HISTORY OF, *section on* EUROPE AND THE AMERICAS, *articles on* MEDIEVAL EUROPE: FOURTH TO SIXTEENTH CENTURY *and* WESTERN EUROPE IN THE SEVENTEENTH CENTURY.

FERTILITY AND FERTILITY CONTROL

See POPULATION ETHICS: ELEMENTS OF THE FIELD; POPULATION ETHICS: RELIGIOUS TRADITIONS; POPULATION POLICY PROPOSALS; CONTRACEPTION; STERILIZATION; ABORTION; PRENATAL DIAGNOSIS.

FETAL–MATERNAL RELATIONSHIP

The human fetus commands our consciousness almost as a new stage of life. At least a newly acknowledged stage, Fletcher concludes, as he assesses the increased interest in the prenatal human being and witnesses an increasing willingness to intervene in fetal life and in the fetal environment (Fletcher, pp. 326–327). The deliberations of a national commission established by the U.S. Congress to study human research also emphasize the growing concern for the fetus, both as a vulnerable individual and, in the medical view, as a patient with health-care needs (National Commission, pp. 61–72). The field is quite new, however, and an analysis of the fetal–maternal relationship must admit to only a small body of confirmed knowledge.

Maternal psychology

Pregnancy begins under diverse circumstances. Some mothers plan for pregnancy, others subconsciously want to be pregnant, still others are disinterested or unwilling. Pregnancy may commence in an atmosphere of love, fear, hate, or indifference. Virtually nothing is known about any influence of these early behaviors and emotions on the development of later ties between mother and fetus. But simple observation convinces us that ties do develop and increase as pregnancy progresses.

Bibring has described psychological stages that begin with planning for the pregnancy, move through a period of questioning, then accepting, the existence of the pregnancy, and arrive at an important step, the perception of fetal movement (Bibring). Before quickening, or the feeling of the baby moving within the womb, the mother perceives the fetus more as a part of herself than as a separate individual. After quickening, the unborn baby is more real as an independent being and will often be called by a name and given personality traits. Quickening heightens attachment considerably, and even unwanted babies may begin to be accepted.

The "feeling of life" also has a central position in the traditional disciplines of law and theology. The English common law and the early law of the United States held that abortion was permissible before quickening but was a misdemeanor or otherwise prejudiced afterward. Similarly, the moment of quickening has been considered by various religious groups to be the time the fetus becomes truly human or the time the soul joins the body. These basic definitions, evolved in law, medicine, and religion, attest to the psychological importance of feeling life and lend credence to the concept that maternal–fetal bonding accelerates after the perception of fetal movements.

The phenomenon of maternal attachment to the fetus is presumably vulnerable, and failure of attachment might be expected to correlate with increased morbidity for the unborn infant or the postnatal child. Data to support this hypothesis are very few, in part because of the difficulties in designing a prospective study. One investigator does report an association between a specific type of stress during pregnancy and childhood morbidity in the first four years of life. During the first month after delivery Stott recorded retrospective information about a series of pregnancies and then prospectively collected data about the children. Maternal illness during pregnancy and temporary situational stresses were not associated with increased childhood morbidity, but chronic situational stresses during the pregnancy, characterized by persistent interpersonal tensions, did show an association. The children whose mothers experienced chronic tension suffered more physical illness, more behavior disturbances, and a higher incidence of delayed development. The postnatal environment was also inadequate for these children, however, and a clear separation of prenatal and postnatal influences seems impossible. Stott claims that the relationship with prenatal stress was closer and more direct and suggests that a parallel may exist between his findings in human pregnancy and the well-established effects of chronic stress which limit successful pregnancies in several animal species (Stott).

Although maternal–fetal attachment seems to intensify, especially after quickening, the bond may still be fragile until certain events occur after delivery (Klaus and Kennell). In many animal species a sensitive period exists right after birth. If mother and offspring are not allowed to be close together, or if certain maternal behaviors such as touching or licking are prevented, rejection rather than acceptance of the baby occurs. Human mothers do not show such clear behavior patterns, but there is evidence that differences do exist between mothers who have touch contact with their premature infants by five days of life and thereafter and those who see but do not touch their babies until three to five weeks of age. Attachment behaviors, such as looking and smiling at the baby and cuddling the baby, are more evident, and feeding skills are better developed, in the mothers who have not had a long separation from their infants (ibid., pp. 1023–1025). It is as if the close and constant contact of mother and fetus establishes a relationship that will continue naturally after birth, presumably to the well-being of the infant, unless the contact is artificially interrupted.

Grief reactions in mothers give further evidence of the strong maternal–fetal attachments that develop during pregnancy. When a baby dies at birth, mourning is easily identifiable in the mother. It occurs whether the mother had intimate contact with her baby or did not see or touch her baby (ibid., p. 1021). Mourning also occurs with the birth of a retarded or defective child in recognition of loss of the desired normal child who had been held in fantasy through the last months of pregnancy (Solnit and Stark). Parents who elect abortion of a genetically abnormal fetus after prenatal diagnosis also show grief reactions, which they interpret as loss of a desired healthy child; mothers experience their losses midway through pregnancy, when the fetus is about twenty weeks old.

Another common phenomenon in women who seek prenatal diagnosis is a conscious postponing of the usual psychological attachment to the fetus until diagnostic results are available. If many days go by after quickening, postponing affection for the fetus becomes more and more difficult, and a mother often recognizes increasing difficulty with the option of abortion.

Fetal well-being

As our perception of the fetal–maternal relationship changes focus from mother to fetus, the information available changes from psychological to physiological. The fetus is growing, developing, and changing continuously in ways more drastic than at any other stage of life. While all of this is happening, the fetus is utterly dependent on the mother to fill all of its needs. The newborn infant appears very dependent but at least can obtain his or her own oxygen and, when hungry, can initiate adult behavior that will provide food. The fetus can do neither; but interestingly, although totally dependent on the mother, the fetus has its needs met automatically and has no need to induce specific plans or thoughts in the mother.

Many changes are necessary for a woman to carry a pregnancy successfully, and it is likely we understand only a few of them. Whether a change is induced by the fetus or fetal tissues or by control systems elsewhere in the woman's body is often unknown. One of the most basic of changes requires the mother's immune system to tolerate a mass of growing, foreign tissue in her body (Beer and Billingham). The usual reaction, when not pregnant, would be to reject foreign cells and tissue, but this reaction is blocked as the fetus grows and thrives.

Another phenomenon of pregnancy, which is common to many women and may also be protective of the fetus, is the appearance of nausea and vomiting in the early months (Hook). Mothers who show these symptoms, except in the extreme, are less likely to miscarry and are more likely to have healthy babies. The first part of pregnancy is concerned with complex changes in the embryo and the genesis of the body organs; this is a time that is highly vulnerable to noxious chemical agents. Later pregnancy sees major growth and weight gain by the fetus and is a time when the mother requires extra food. Thus the nausea and vomiting of early pregnancy could result in minimizing fetal exposure to exogenous toxins at an especially crucial time; and such nuisance symptoms may actually represent protective efforts initiated by the fetus or fetal-placental unit.

Optimal conditions for growth and maturation within the fetal environment are not fully known, although the conditions existing in most healthy pregnancies are obviously satisfactory and perhaps approach the optimum. Exactly what chemical conditions are best for cell division and growth or what levels of sensory input are best for brain development are questions for the future. We do know a number of conditions that are not optimal for the embryo or fetus, however. As suggested earlier, constant interpersonal tension in the mother may be harmful to the fetal central nervous system. Several drugs and chemicals and a few viruses and other infectious agents are known to alter fetal development and cause severe birth defects. Thalidomide and the rubella virus are well-known examples.

Two common habits in our population are also known to yield suboptimal conditions for fetal development. These are smoking and alcohol ingestion. Smoking has been shown to cause lower birth weights in several studies and may increase newborn mortality (Rantakallio, pp. 411–413). High alcohol consumption during pregnancy may cause mental retardation, growth failure, and minor birth defects (Jones et al.). In these examples the harm of drug abuse is felt by both mother and fetus, although more severely by the latter. It was once thought that under deprivation conditions, like calorie or iron deficiency, fetal well-being was preserved at the expense of the mother's tissues. In such situations the mother may suffer more than the fetus, but it has become clear that fetal development is also significantly impaired (Winick). The reasonable conclusion is that mother and fetus are a unit and that the unit is vulnerable to various insults.

Interests and responsibilities

The existential duality of two individuals living very closely together and, simultaneously, of a single fetal–maternal unit, which thrives or suffers as a unit, creates complexities for ethical analysis. If one concentrates on two separate albeit intricately associated individuals, separate interests can be conceived, which may be in conflict. Since either individual may survive at the expense of the other, including the other's life, conflicting interests would seem to be self-evident. On the other hand, emphasis on the unitary nature of mother and fetus would not admit to conflicting interests. In Western society, despite an emphasis on concomitant health of mother and fetus, most problems seem to arise from perceived conflicting interests of two separate individuals.

Prenatal human life is very fragile. Incubation within the uterus has evolved as a semiautomatic process, which will be successful often enough to make possible survival of our species. How should a mother's responsibilities to her growing fetus be viewed? Must she always avoid behaviors or environments that might pose a risk to fetal well-being? Should she always choose actions that are thought to maximize the potential for fetal development? Always is too strong a word for the real world, but it could be suggested that a mother should and does feel responsibility for her unborn child and should especially avoid anything that has a significant chance of harming the fetus.

Insofar as the act of initiating or accepting a pregnancy is voluntary, it would seem that a mother should simultaneously assume responsibility for the well-being of her fetus. The initiation of pregnancy may already add complicating factors, however. Swigar reports that during acute grief, as occurs in loss of a parent, a pregnancy may be started as a natural part of coping with the loss (Swigar, Bowers, and Fleck). A short time later the pregnant woman may view the pregnancy as an added burden that makes coping impossible. Also, throughout the pregnancy other circumstances may change, leading a mother to deny her first acceptance of the fetus. If a society condones abortion, choice remains with the mother for a period of time, and she may reaffirm her commitment and responsibility to her unborn baby or may divest herself of any responsibility. When a society forbids abortion, and thereby places high regard with the very young fetus, it seems proper that the society share responsibility for the fetus with the mother and give aid when needed.

Psychological attachment of mother to fetus increases during pregnancy, and physiological adjustments to the pregnancy are continually occurring in the mother. At the same time the fetus is becoming more complex, resembling children or adults more closely, and evolving complicated behaviors. These changes seem naturally to encourage and reinforce maternal commitment to the fetus. This may suggest that responsibility to the fetus should increase the farther a pregnancy proceeds, and proscription of certain maternal behaviors should become stronger. From one point of view this is true, since both society and parents desire the birth of healthy babies rather than damaged ones, and the likelihood of live birth and survival increases as the pregnancy progresses. Responsibility must be present in early pregnancy as well, however, if damaged embryos are to be prevented, for it is in the early weeks that the fetus is most vulnerable to environmentally induced defects.

The thrust of the fetal–maternal relationship, the explicit goals of most parents, and the needs of society all unite in a supposition of life with maximum potential for the fetus. As long as there is choice available in starting and continuing an early pregnancy, it seems justified to choose in favor of fetal well-being when privacy interests of the mother and health interests of the fetus seem to conflict. At present our understanding of a healthy pregnancy places some constraints on a mother's behavior, but most women find such constraints quite acceptable. It would be well for society to encourage responsible behavior toward the fetus and to provide the means for mothers to understand and carry out those responsibilities.

MAURICE J. MAHONEY

[*For discussion of related ideas, see the entries:* HEALTH AS AN OBLIGATION; PRENATAL DIAGNOSIS; *and* PRIVACY. *Other relevant material may be found under* ABORTION, *articles on* MEDICAL ASPECTS, *and* CONTEMPORARY DEBATE IN PHILOSOPHICAL AND RELIGIOUS ETHICS; ALCOHOL, USE OF; FETAL RESEARCH; SMOKING; *and* WOMEN AND BIOMEDICINE, *article on* WOMEN AS PATIENTS AND EXPERIMENTAL SUBJECTS.]

BIBLIOGRAPHY

BEER, ALAN E., and BILLINGHAM, R. E. "Immunobiology of Mammalian Reproduction." *Advances in Immunology* 14 (1971): 1–84.

BIBRING, GRETE L. "Some Considerations of the Psychological Processes in Pregnancy." *Psychoanalytic Study of the Child* 14 (1959): 113–121.

FLETCHER, JOHN. "Parents in Genetic Counseling: The Moral Shape of Decision-Making." *Ethical Issues in Human Genetics: Genetic Counseling and the Use of Genetic Knowledge.* Edited by Bruce Hilton, Daniel Callahan, Maureen Harris, Peter Condliffe, and Burton Berkley. New York: Plenum Press, 1973, pp. 301–327.

HOOK, ERNEST B. "Changes in Tobacco Smoking and Ingestion of Alcohol and Caffeinated Beverages during Early Pregnancy: Are These Consequences, in Part, of Feto-Protective Mechanisms Diminishing Maternal Exposure to Embryotoxins?" *Birth Defects: Risks and Consequences.* Edited by Sally Kelly, Ernest B. Hook, Dwight T. Janerich, and Ian H. Porter. New York: Academic Press, 1976, pp. 173–183. Proceedings of a Symposium on Birth Defects: Risks and Consequences. Sponsored by the Birth Defects Institute of the New York State Department of Health, held in Albany, N.Y., 7–8 November 1974.

JONES, KENNETH L.; SMITH, DAVID W.; ULLELAND, CHRISTY N.; and STREISSGUTH, ANN PYTKOWICZ. "Patterns of Malformation in Offspring of Chronic Alcoholic Mothers." *Lancet* 1 (1973): 1267–1271.

KLAUS, MARSHALL H., and KENNELL, JOHN H. "Mothers Separated from Their Newborn Infants." *Pediatric Clinics of North America* 17 (1970): 1015–1037.

National Commission for the Protection of Human Subjects of Biomedical and Behavioral Research. *Research on the Fetus: Report and Recommendations.* DHEW Publication no. (OS) 76-127. Washington: U.S. Department of Health, Education, and Welfare, 1975.

RANTAKALLIO, PAULA. "Predictive Indices of Neonatal Morbidity and Mortality." *Clinical Perinatology.* Edited by Silvio Aladjem and Audrey K. Brown. St. Louis: C. V. Mosby Co., 1974, pp. 409–421.

SOLNIT, ALBERT J., and STARK, MARY S. "Mourning and the Birth of a Defective Child." *Psychoanalytic Study of the Child* 16 (1961): 523–537.

STOTT, D. H. "Follow-up Study from Birth of the Effects of Prenatal Stresses." *Developmental Medicine and Child Neurology* 15 (1973): 770–787.

SWIGAR, MARY E.; BOWERS, MALCOLM B.; and FLECK,

STEPHEN. "Grieving and Unplanned Pregnancy." *Psychiatry* 39 (1976): 72–80.

WINICK, MYRON. *Malnutrition and Brain Development.* New York: Oxford University Press, 1976, pp. 25–31, 107–110.

FETAL RESEARCH

Historical background

An interest in how a fetus can survive inside a mother has existed for millennia. Hippocrates and Aristotle, recognizing the importance of respiration in adults, speculated that it must occur in fetuses. For many centuries, however, knowledge of the fetus as a semi-independent entity was obscured by the prevalent notion that blood vessels of the mother ran right through the placenta and supplied the fetus with maternal blood, as if it were just another organ of the mother. Such thinking was espoused by prominent physicians as far back in time as Galenus and Vesalius.

The proposition that maternal and fetal vessels were separate was not advanced until 1564, when Arantius espoused it without adducing any evidence for his thesis. The core difficulty was that the circulation of blood was not understood before 1628, when William Harvey published his major work on it. In a later work, published in 1651, Harvey accurately concluded that, if maternal and fetal vessels were continuous, maternal arteries would have to anastomose with fetal veins, a most unlikely proposition. Yet as late as 1687, as prestigious a scientist as Fabricius ab Aquapendente would still proclaim that maternal and fetal vessels were, indeed, continuous and that the fetal heart, which had been shown by Nymman in 1628 to beat at a rate different from the maternal heart rate, was beating to no independent purpose (*ex sui generis*).

Other scientists reflected upon the problem of how a fetus, totally bathed in amniotic fluid, could use its lungs. Charleton, in 1659, suggested that the fetus breathed from a bubble atop the amniotic fluid. As late as 1798, more than a century after Harvey described the fetal circulation, Scheel, finding meconium, the fetal bowel excretory product, in fetal lungs, concluded that the fetus inspired amniotic fluid and that it was at least a partial source of oxygen. He was not aware that this was a phenomenon of severe fetal distress only, as was to be proved in the twentieth century.

In the midst of these debates among physiologists, obstetricians interjected observations of great importance, based on experience rather than theory. In 1683 the French obstetrician Mauriceau stated that, when a fetus died as a result of the prolapse of its umbilical cord, it was because circulation through the cord ceased and fetal blood was not "revivified" in the placenta. He denied fetal pulmonary respiration. Later, in 1794, after Scheele had discovered oxygen in 1772, Girtanner correctly updated Mauriceau's theory by saying that it was a lack of oxygen that caused the death of fetuses with prolapsed cords. Roederer, perhaps the most perceptive of all, made two crucial observations. He noted that, if an umbilical cord was torn apart during delivery, no mother ever exsanguinated through the cord, as one could have expected if maternal and umbilical vessels were continuous. Conversely, he observed that, when the placenta was delivered intact and attached to the fetus, no fetus ever exsanguinated through the placenta, as again would be expected if the fetal vessels and maternal vessels were continuous through the placenta.

Suffice it to say, then, that by the end of the eighteenth century it was realized that the fetus was not simply another maternal organ, directly supplied by the mother's arteries (Hellegers). Fetal research became an enterprise of investigating a vascularly separate entity within the mother. In large measure, initial studies involved the supply of oxygen to the fetus, since it was realized that oxygen was the most "vital spirit" upon which fetal survival depended.

Methods: preliminary distinctions

Structure versus function. Over the centuries, fetal research has involved studies of structure and studies of function. In simplified form, these methods may be called embryological or anatomical and physiological or biochemical, respectively. Purely anatomic structural research presents many fewer ethical problems than functional research. Structural research can be done on fetuses exteriorized from their mothers and therefore need not involve any risk to the mothers themselves. Work of the last two decades has repeatedly shown, however, that exteriorizing the fetus radically alters its functions from those *in utero*. Optimal studies of fetal function therefore require research in situ and hence may involve risk to the mother also. For this reason, most functional fetal research is done on animals. While in general qualitative functioning of mammalian, and especially of primate, fetuses bears great resemblance to the human, quantitative similarities are less clear and extrapolation of data from one species to another presumes too much. Sooner or later, then, animal findings must be confirmed in the human, to result in clinically useful applications of animal findings.

Beneficial versus nonbeneficial research. As in all human research, so also in fetal research, ethicists have differentiated between beneficial (or therapeutic) and nonbeneficial (or nontherapeutic) research. In the former case the research subject stands to gain as much from the research, whether it be a procedure or a drug, as to lose from it. In the latter case, the subject cannot possibly benefit himself, and any benefits can therefore only accrue to others.

Fetal research, developed especially in Anglo-Saxon, Scandinavian, and northwest European countries, as well as in Uruguay, has led to an increasing ability to monitor the fetal status *in utero*. Three major forms of monitoring have been developed. The first, and numerically least important, is intrauterine diagnosis of congenital abnormalities by culturing fetal cells. The second, resulting from the analysis of fetal products in amniotic fluid, permits analysis of fetal age and allows optimal timing of delivery to avoid those cases of prematurity previously caused by too early induction of labor or Caesarean sections. Therefore, if medical conditions dictate that a fetus might be better off delivered than *in utero*, more accurate analysis of the potential danger of prematurity became possible. The third form of monitoring permits acute intrauterine diagnosis of the state of well-being of the fetus. These techniques involve measurement of fetal heart action (EKG), as a reflection of fetal oxygenation, measurement of fetal scalp blood in labor (again as a reflection of its oxygenation), ultrasonographic measurements of its rate of growth, and hormonal measurements reflecting both fetal and placental metabolic functions.

The first measurement of the fetal EKG clearly is an example of beneficial research. It could in no way harm the fetus, yet it provides a more accurate method of measuring the fetal heart rate than was possible with a regular stethoscope. Most research on fetal monitoring is similarly of the beneficial variety in that it is designed, after much animal work, to determine the status of the particular fetus being monitored. Where, for instance, it is necessary to obtain amniotic fluid, as in the first experiments to determine fetal maturity, the calculus is more problematical. There is some risk in the procedure (amniocentesis), which must be balanced against the usefulness of the information to be gained on the fetal status. Suffice it to say that it is potentially beneficial. Clearly nonbeneficial research is any procedure done on an extrauterine previable fetus since such fetuses cannot survive and cannot themselves derive any benefit from the information gained, even though future fetuses may.

If the injunction to pursue the beneficial research is followed rigidly, the first application of animal data to humans tends to be carried out in desperate circumstances, where the fetus may still benefit. The desperate situation may itself obscure the usefulness of a new procedure, which might well be beneficial if applied under less desperate circumstances. The "beneficial research" rule, therefore, not only restricts the number of cases in which a new procedure can be tried but may also lead to confusion as to the circumstances under which it might be most usefully employed.

Benefit and risk. While the risks of a nonbeneficial procedure fall solely on the research subject (both mother and fetus if performed in situ), the benefits may extend beyond the research subject to the population as a whole. Since researchers often feel an obligation toward mankind as a whole rather than toward the research subject alone, there is a tendency for the scientific community, in calculating these risks and benefits, to focus on the potential benefits for all rather than for the research subject alone. In its extreme form, such risk-benefit calculation can lead to a "scientific imperative." In such a view, all knowledge will ultimately benefit someone, or all of us; and the benefits weighed against the risks to the individual research subject warrant much greater risk taking than if the potential benefit is to accrue to the research subject alone. At issue is the extent to which one individual may be used experimentally for the benefit of others than himself.

Consent

It is generally agreed that fully informed and willing adults may submit themselves to certain risks for the benefit of others than themselves. Such is the risk undertaken by adults who donate blood to a blood bank. A key question in the ethics of fetal research is how the researcher can justify the presumption of such consent by a fetus, which cannot express such consent directly. Where it is held that others than the fetus (or similarly incompetent individuals, such as children or the mentally retarded) can give consent on their behalf to becoming the subject of research, the ethical question still remains: By what reasoning can it be justified that one individual give such consent on behalf of another?

Moral status of the fetus

Where it is held that a fetus is no more than another maternal organ, such as an appendix or a tonsil, it will be clear that the ethics of fetal research involves no more than the donation of a nonessential organ to the scientific community. It presents only a problem of the ethical procurement of such an organ. However, it is clear that many see a difference between a fetus and just another maternal organ. Most individuals assign greater value to the fetus than to a single organ and therefore see greater ethical difficulty in its procurement than in that of other organs. The greater the value attached to the fetus, therefore, the greater the difficulties seen in the ethics of fetal research. In general, those who value the fetus most highly will most seriously question both its procurement for, and its use in, research.

Sometimes this dilemma is oversimplified. It is expressed in the notion that, if induced abortion is an evil, so must research on the fetus be, at least if the fetus is alive. However, the following paradox becomes apparent: Those who most oppose abortion do so precisely because they view the fetus as already a child in the moral sense. So stated, the ethical issue could be reformulated as follows: For what reasons and under what conditions may child research be done? And who may give the consent for a fetus or child to be enrolled in an experiment?

Various ethical positions

Several schools of thought have been expressed on the ethics of fetal research on live previable fetuses. The most permissive holds that, if fetuses may be killed in abortion, surely to do research on them is no more harmful (Gaylin and Lappé). A variant holds that this opinion is incorrect at least while the fetus is still *in utero*, for a mother could always change her mind about undergoing the abortion. Once the abortion has been done, however, this view would also permit any research, since the fetus's death is inevitable (Wasserstrom).

The opinion that the inevitability of death is sufficient warrant for incorporating an individual in research protocol has been put forward also in the case of those about to be executed under capital punishment. It has not, however, found favor in practice. Neither has it in the past been found acceptable to inject cancer cells into terminally ill patients for purposes of research.

The most "restrictive" view of the ethics of fetal research holds that no one can give consent on behalf of another unless it was well known that the other would indeed give such consent (Ramsey). Since that knowledge cannot be obtained from fetuses or other incompetents, it can never be presumed that one can give proxy consent for such nonbeneficial research on behalf of an incompetent. Of course, if a fetus, child, or other incompetent could benefit from the research, proxy consent could be given, since it is presumed that the research subject would indeed wish to receive such direct benefits.

Yet a different view has also been expressed. It holds that the most "restrictive" view, described above, is too personalized or individualistic. Human beings are social beings and do not live in isolation. Observation of man shows his instinctive interest in saving the life of his fellow man when it can be done at little or no risk. Such instincts and acts are part of the nature of man, and man ought to act this way precisely because of his innate human nature. Extrapolating from this nature of man, it may therefore be held that proxy consent can be given in those cases where it can reasonably be postulated that a human being would give consent precisely because he ought to do so from this observable nature of human beings (McCormick).

Obviously, under the latter construct, the question remains how the term "little or no risk" is, in practice, to be construed. The occurrence of pain, while not an issue of risk, might be considered a separate warrant for not presuming consent. Another element of this position would hold that the research should neither shorten nor prolong the life of the inevitably dying fetus. In sum, this view holds that fetal research (or its presumed moral analogy, child research) can indeed be done, but under restricted circumstances. Like the most restrictive position, described above, it does not view impending death as a sufficient ethical warrant for the research, nor does it recognize the scientific imperative that even assured benefits for others than the research subject warrant the subject's use.

Regulations

In the face of these varied considerations, and in the light of much publicly expressed concern, both the United States and Britain have issued regulations to govern fetal experimentation.

The first governmental commission to study the ethics of fetal research was British and consisted of an advisory group, chaired by the obstetrician-gynecologist Sir John Peel. Its report was published in 1972 (Great Britain). After receiving evidence from many sources the group recommended as follows:

1. Viable fetuses should not be subjected to nonbeneficial research.

2. The point of viability should be set at twenty weeks, corresponding to a weight of 400–500 grams.

3. Dead fetuses or tissues may be used in accordance with the provisions of the Human Tissue Act, which governs the postmortem use of human tissue.

4. Where the provisions of the Human Tissue Act do not apply, and with parental approval (the sex of the parent being left unspecified), research is permitted on the whole dead fetus or its organs, providing it is not done in the operating room or the delivery room, providing no money is exchanged, and providing full records are kept.

5. Whole alive previable fetuses may be used if they weigh less than 300 grams, providing clause 4 above is observed, and providing it is done in departments directly related to a hospital, with sanction of its ethics committee. The committee must assure itself of the validity of the research, the impossibility of performing it in any other way, and the competence of the investigator.

6. No possibly harmful agent may be administered to a fetus *in utero* to determine the effects of the agent.

In the United States, a commission called the National Commission for the Protection of Human Subjects in Biomedical and Behavioral Research was established by law in July 1974 and was given the specific charge to study the ethics of fetal research. After holding extensive hearings and commissioning a series of position papers from scientists, ethicists, and lawyers, the Commission issued its recommendations.

In summary form they may be stated as follows: In general, fetal research may be done if informed consent is obtained from the mother, with the father not dissenting. When specifying recommendations for different types of research, the Commission, like its British counterpart, attempted to classify forms of fetal research by the status of the fetus. Its major guidelines were as follows:

1. Therapeutic (or beneficial) research may be performed on the pregnant woman providing the fetus is placed at minimum risk consistent with meeting the health needs of the mother.

2. Nontherapeutic (or nonbeneficial) research may be done on pregnant women, providing minimal or no risk accrues to the well-being of the fetus.

3. Nontherapeutic research on the fetus *in utero* who is not to be aborted may be done, providing it is for important reasons and has been preceded by adequate work on pregnant animals and nonpregnant humans, and providing there is minimal or no risk to the well-being of the fetus.

4. Nontherapeutic research on the fetus *in utero* who is to be aborted should follow the same rules as under clause 3 above, but if there is difficulty in interpreting the rule, the Secretary of the Department of Health, Education, and Welfare (DHEW) may approve the research if it is approved by a National Ethical Review Board.

5. Nontherapeutic research on the fetus during the abortion procedure or on the live nonviable fetus outside the womb may be performed if (a) it is important and the information cannot otherwise be obtained; (b) animal research and research on nonpregnant women has preceded it; (c) the fetus is less than twenty weeks old; (d) the abortion procedure is not changed for purposes of the research; and (e) the duration of fetal life is not altered by the research. Cases of difficulty in interpretation are again to be submitted to the national review board and the Secretary of HEW.

6. Nontherapeutic research on the possibly viable fetus may be performed if no additional risk is imposed on the well-being of the fetus and the general rules regarding importance, previous research, and consent, as described above, are followed.

7. Research on dead fetuses or their tissues may be done if consistent with local laws and with commonly held respect for dead babies.

8. The advisability, timing, and method of abortion should not be altered for research purposes.

9. There should be no monetary transactions in fetal research.

10. Fetal research funded by the U.S. government outside of the United States should apply the same principles as within the country.

11. Research on abortion techniques should be allowed to continue in accordance with existing regulations.

Most of the recommendations of the Commission were promulgated in the *Federal Register* of 8 August 1975. There was, however, one glaring difference of opinion between the Commission and the Department of Health, Education, and Welfare. The Commission (clause 5 above) held that the duration of life of a live nonviable fetus outside of the uterus could not be altered by the research. As acknowledged in the *Federal Register* of 13 January 1977, the DHEW ruled instead that "vital functions of the fetus will not be artificially maintained except when the purpose of the activity is to develop new methods for enabling fetuses to survive to the point of viability."

The difference in points of view was glaring. The Commission had held that research prolonging the life of a fetus was permissible only if *that* fetus stood to survive thereby. The DHEW suggested it could be done if *other* fetuses could benefit from the knowledge even though the research subject fetus could not. In the proposed regulations of January 1977, the Commission's view was agreed to.

In summary, the British and U.S. regulations have certain things in common but differ on other grounds. They have in common that dead fetuses and their tissues are to be afforded the respect of other dead human bodies and tissues. Fetuses with a chance of survival are to be treated like children. Willful damage to the fetus *in utero* may not be caused, presumably lest a mother change her mind about abortion. Significant differences are that in the U.S. regulations fathers can veto the research, while in the Peel Report there is no such specific provision. In Britain it is proposed that no nonbeneficial research is to be done on the fetus *in utero* or the viable fetus. In the United States, it may be done if there is minimal or no risk. Finally, the British proposals set viability at twenty weeks or at 300 grams when research is done on the whole previable fetus, whereas the U.S. proposals set it at twenty weeks or 500 grams, but allowing for the fact that this point may be reconsidered from time to time. Finally, the U.S. proposals suggest the establishment of a national review board for difficult cases.

It may be expected that, as experience is gained with such regulatory processes, the proposals will evolve from time to time.

ANDRÉ E. HELLEGERS

[*Directly related is* FETAL–MATERNAL RELATIONSHIP. *For further discussion of topics mentioned in this article, see the entries:* ABORTION; CHILDREN AND BIOMEDICINE; INFORMED CONSENT IN HUMAN RESEARCH; *and* RISK. *See also:* CADAVERS; HUMAN EXPERIMENTATION, articles on BASIC ISSUES, PHILOSOPHICAL ASPECTS, *and* SOCIAL AND PROFESSIONAL CONTROL; ORGAN DONATION; PRENATAL DIAGNOSIS; *and* RESEARCH, BIOMEDICAL.]

BIBLIOGRAPHY

GAYLIN, WILLARD, and LAPPÉ, MARC. "Fetal Politics: The Debate on Experimenting with the Unborn." *Atlantic Monthly*, May 1975, pp. 66–71.

Great Britain, Department of Health and Social Security, Scottish Home and Health Department, Welsh Office. *The Use of Fetuses and Fetal Material for Research: Report of the Advisory Group.* London: Her Majesty's Stationery Office, 1972. The Peel Commission Report.

HELLEGERS, ANDRÉ E. "Some Developments in Opinions about the Placenta as a Barrier to Oxygen." *Yale Journal of Biology and Medicine* 42 (1970): 180–190.

MCCORMICK, RICHARD A. "Experimentation on the Fetus: Policy Proposals." National Commission, *Research on the Fetus: Appendix*, chap. 5, pp. (5-1)–(5-11).

National Commission for the Protection of Human Subjects of Biomedical and Behavioral Research. *Research on the Fetus: Report and Recommendations.* DHEW Publication no. (OS) 76-127. *Appendix.* DHEW Publication no. (OS) 76-128. Bethesda, Md.: Department of Health, Education, and Welfare, 1975.

RAMSEY, PAUL. *The Ethics of Fetal Research.* New Haven: Yale University Press, 1975.

United States, Department of Health, Education, and Welfare, Office of the Secretary. "Protection of Human Subjects: Fetuses, Pregnant Women, and In Vitro Fertilization." *Federal Register* 40, no. 154 (8 August 1975), pp. 33526–33552.

———, Department of Health, Education, and Welfare, Office of the Secretary. "Protection of Human Subjects: Proposed Amendments Concerning Fetuses, Pregnant Women, and In Vitro Fertilization." *Federal Register* 42, no. 9 (13 January 1977), pp. 2792–2793.

WASSERSTROM, RICHARD. "Ethical Issues Involved in Experimentation on the Nonviable Human Fetus." National Commission, *Research on the Fetus: Appendix*, chap. 9, pp. (9-1)–(9-10).

FOOD POLICY

Introduction

The phenomenon of widespread hunger and malnutrition is not new in the history of humankind. What is new is the growing sense that today famine and starvation are by no means inevitable. The technical ability to feed the human family exists. The political will, translated into national or international policy, is nonetheless frequently absent. Therein lies the central ethical issue in planning and implementing a food policy which meets contemporary needs.

The ethical issue—and consequent challenge especially to those nations and groups that could be of assistance—was expressed clearly in the final resolution of the United Nations World Food Conference in November 1974:

Every man, woman and child has the inalienable right to be free from hunger and malnutrition in order to develop fully and maintain his physical and mental faculties. Society today already possesses sufficient resources, organizational ability and technology, and hence the competence, to achieve this objective. Accordingly, the eradication of hunger is a common objective of all the countries of the international community, especially of the developed countries and others in a position to help [World Food Conference].

In examining the ethics of food policy, this article includes (1) a survey of the world situation of hunger and food supplies; (2) a study of various policy responses; and (3) an analysis of the frameworks within which ethical questions might be addressed.

World food situation

The dimensions of the world food problem are unfortunately easy to describe in terms of human suffering. The United Nations Food and Agriculture Organization (FAO) reported in 1974 that the incidence of hunger and malnutrition around the world had not appreciably declined in the past twenty years despite significant progress in food production, as well as progress in economic development (Food and Agriculture Organization). Hunger, or undernutrition, refers to inadequacy in the quantity of food available for a person, with resultant loss of weight, disease, and lowering of overall human energy levels. Malnutrition refers to inadequacy in the quality of food, occurring when the diet lacks sufficient protein necessary for ordinary health as well as sufficient vitamins and minerals. Protein deficiency is especially harmful to the physical and mental development of the young (in early years of life protein shortages impair the growth of the brain and the central nervous system). Vitamin A deficiency, a common factor in malnutrition, is a major cause of blindness in many countries, afflicting an estimated total of 100,000 children in the Far East alone each year. It is possible for a malnourished person not to be undernourished, but it is likely that a person who is undernourished will also be malnourished.

In its World Food Survey prepared in 1963 the FAO estimated that twenty percent of the population in the developing countries were undernourished, and sixty percent were malnourished. These figures have been contested by some as being based upon a faulty reckoning of what is necessary for healthy survival (Poleman). A more refined assessment prepared by FAO prior to the World Food Conference in 1974 stated that malnutrition affects approximately 460 million people and is especially severe among young children. The areas most seriously afflicted with food shortages are in the Far East (excluding the Asian centrally planned economies, for which insufficient information is available). Here the FAO estimates that more than thirty percent of the population—more than 300 million people—suffer from an insufficient protein-energy supply.

Central to hunger problems in any part of the world is the fact that the poorer segments of the population face the fullest impact of scarce food supply. Whatever may be the ecological, technological, or other reasons for poor distribution of available food among different geographical areas in the world, a basic feature of the current food situation is the extreme inequality in distribution among different socioeconomic groups (Schertz). Urban poor in the developing countries suffer most, being deprived of access to subsistence of food and without adequate income to purchase food.

Supply and demand. Global availability of food is impacted by both "supply" and "demand" factors. In terms of supply, the most disturbing fact in recent years has been the fluctuating production of basic feed grains. In 1972 the world output of wheat, coarse grains, and rice declined for the first time in more than twenty years, bringing a reduction of about thirty-five million tons; this occurred again in 1974, with a drop of between forty and fifty million tons. Since world grain production (in 1974, approximately 1.2 billion tons per year) must increase twenty-five million tons per year simply to keep pace with increasing demand, this shortfall in production meant that the two major world reserves of food had to be tapped: stocks of grain held by the principal exporting countries and cropland idled under farm programs in the United States. The combination of grain stockpiles and cropland reserves amounted in 1961 to the equivalent of 105 days of world grain consumption; by 1972 this was 69 days. The total reserves fell sharply to 55 days in 1973 and 33 days in 1974. Preliminary estimates are 31 days in 1976 (Brown, 1975, p. 8).

Fluctuating weather patterns have been a major element in grain production shortfalls. Rainfall was poor in the Soviet Union and in parts of Asia and Africa in 1972, with deleterious effects on harvests. There appears to be evidence that the monsoon belt is gradually shifting southward, affecting the Indian subcontinent and sub-Saharan Africa. And some meteorologists are suggesting that projections for North American grain production are highly unstable because of increased weather variability in the midwestern regions of Canada and the United States (Thompson).

Also affecting production has been the shortage and rising price of fertilizer. Nitrogen-based fertilizer, a product of the petrochemical industry, is critical in increasing crop yields—especially of the hybrid strains of seed used in the so-called Green Revolution. But fertilizer production lagged in the late 1960s, and then the rising price of oil caused costs to increase more than six times between 1972 and 1974.

Other major production constraints include arable land availabilities and water supply shortages (Eckholm). Bringing new land into production is extremely costly; not allowing cultivated land to lie fallow and be regenerated undercuts future production. Fresh water resources are being increasingly tapped for extensive irrigation, but salinity and siltation threaten expansion of these projects.

Another supply factor to be considered is that the world's fish catch has increasingly been a source of high quality protein, both for direct consumption (Japan and the Soviet Union are large consumers) and for animal feed (especially for the United States and Europe). Between 1950 and 1970 there was an annual increase of five percent in the fish catch. But the catch peaked in 1970, at seventy million tons, and has declined several million tons since then. Overfishing, competition among nations, and phenomena such as erratic ocean currents off the Peruvian coast have meant both a decline in available fish and a rise in prices. Some sort of international agreement limiting the annual intake of fish on a species-by-species, region-by-region basis is necessary if the ocean is to continue to be a source of food supply (Brown and Eckholm, pp. 145–163).

The other side of global availability of food, the demand, is influenced by both rising population and rising affluence. The world's population of four billion people is now growing at a rate of nearly 2 percent annually, which means a doubling every thirty-seven years. The growth rate in the developing countries, however, is around 2.4 percent annually. World production of food increases approximately 2.8 percent annually; developing country production increases 2.7 percent annually. Serious questions have been raised—especially during the United Nations World Population Year of 1974—as to whether food production could keep up with the demand created by population growth (World Population Conference).

The added factor of rising affluence puts more pressure on food supplies. Effective demand for food is increasing now around 3.5 percent, outstripping population increases. This reflects a notable shift in dietary standards and patterns in developing countries as well as developed countries. The FAO projects for the developing countries that rising affluence—most particularly among the newly rich oil-producing nations—will mean annual rates of growth in demand of 3.3 percent for cereals, 4.8 percent for fish, and 4.6 percent for meat (Food and Agriculture Organization, pp. 24–27). In particular the greater demand for grain-fed livestock products will put pressures on world grain availability.

Nutrition and rich nations. The problem of poor nutrition also exists in many developed countries, under two aspects. First, there are many hungry people in rich countries, hungry because they are poor and unable to afford good food. The U.S. Department of Agriculture estimates ten to twelve million people go to bed hungry every night in the richest nation in the world. Nutrient deficiency is also a serious problem that affects rich as well as poor in developed countries. Poor-quality diets—substandard intakes of vitamins and minerals—are alarmingly common in the United States among persons below and above poverty-level incomes. Explanations include the inadequacy of good nutrition education and the widespread availability of low quality convenience food ("junk food"). A second aspect is the presence of nutritional diseases of abundance (Food and Agriculture Organization, p. 13). The consumption of excess calories and dietary patterns marked by high content of cholesterol and saturated fats lead to a high prevalence of obesity and arteriosclerosis. Diabetes, hypertension, and heart disease are consequences of this diet and are responsible for the largest number of deaths and permanent disabilities in the developed nations.

Perspectives and policies

What is the nature of the world food problem? The perspective that one takes in answering this question determines in large part the policy response that will be promoted. At least four perspectives, and consequent policy responses, are possible.

First, the world food problem can be seen as caused by climatic vagaries, a temporary but recurring phenomenon brought about by poor weather conditions. The policy response called for, therefore, is a system of world grain reserves. After the example of Joseph's management in ancient Egypt (Genesis 41), grain is stored in times of plenty to be distributed in times—or regions—of scarcity. Second, the food problem can be seen primarily as the outcome of a losing race with population-growth rates. As Thomas Malthus noted at the beginning of the nineteenth century, the increase in population tends to outstrip the increase in available food resources, and famine and disease automatically cut back on the population (Malthus). A more humane policy response would be a large-scale effort to control population growth. Third, the lack of adequate production capabilities can be seen as the cause of the world food problem. According to this position, an increased output of grain is dependent upon more intensive use of the technology associated with the Green Revolution—high yield grains, nitrogen fertilizer, sophisticated machinery, and so forth. The policy response, then, is further commitment to Green Revolution techniques.

A fourth perspective on the world's food problem views it primarily as indicative of inefficient and inequitable economic systems, national and international. While grain reserves, population control, and technological improvements are in fact called for, more serious attention must be paid, this position argues, to corporate business policies, development models, trade patterns, consumption patterns, pricing systems, monetary relationships, investment practices, and so forth. The food problem is seen as symptomatic of the larger socioeconomic-political crisis of national and global development. Policy responses must therefore primarily address this larger issue, e.g., through negotiations for a new international economic order such as that debated by the United Nations (United Nations, General Assembly).

International efforts. The United Nations World Food Conference has recommended three approaches to meeting global hunger needs: (1) short-term—increase and improve food aid; (2) middle-term—establish a food reserve; and (3) long-term—improve agricultural production in food-deficit nations. Each approach involves significant economic, political, and ethical issues.

Food aid from the United States, for example, has operated since 1954 under Public Law (PL) 480, the so-called "Food for Peace" program. Originally established to provide an outlet for huge farm surpluses, PL 480 has in recent years declined in gross tonnage actually shipped overseas (either through sale programs on easy terms or through relief grants dispersed by voluntary agencies). It has also been used as a support for nations with military and strategic importance to the United States. Moreover, patterns of dependency are frequently created by aid programs, particularly if they extend beyond response made during critical emergency periods (DeMarco and Sechler, pp. 36–52).

Global food reserve systems face political and management difficulties more serious than technical challenges. Control of market operations, sharing of agricultural data, and expense of stockpiling either in producing or in consuming nations are a few of the political questions that must be faced if any system of international reserves is to be instituted.

The challenge of increasing productivity in the poor countries themselves is linked to the wider issue of alternative development models. During the 1950s and 1960s "development" tended only to mean efforts to increase industrialization, with the agricultural sector largely bypassed. It is now evident that an increase in food production will require more attention to farmers—especially to small farmers who make up the bulk of the agricultural sector in the developing nations (McNamara). Land reform, irrigation, seed and fertilizer supplies, credit and marketing support, and local cooperative efforts must be included in development approaches aimed directly at assistance for the poor majorities in those nations. Such structural changes will promote more self-reliant food-growing strategies among the poor, without the need for significant capital outlay (United Nations, Department of Economic and Social Affairs).

Ethical frameworks

As the quantitative and qualitative dimensions of the world food problem have become evident, serious debate has been stirred over the ethical aspects of possible solutions. A humanitarian ethic has traditionally supported feeding the hungry. Indeed, the example frequently given of altruism is sharing one's bread with the poor—a universal sign of generosity. Likewise, religious bodies have responded to the needs of the hungry, motivated by their sacred teachings. Mosaic law, for example, mandated that an extra row of grain was to be left standing in the field at harvest time, so that the poor might claim it as their portion (Leviticus 19: 10–11). Biblical prophets reminded the people of Israel that observance of their covenant with God required a sharing with the hungry (Isaiah 58:1–10). Jesus identified himself with the hungry and told his followers that in feeding—or not feeding—the hungry, they were relating to him in ways that determined their entry into his kingdom (Matthew 25:31–46).

A spirit of aiding the hungry, whether religious or humanistic, has provided the motivation for countless acts of sharing. But in recent years this spirit of generous sharing has faced a serious challenge, itself based on the ethical grounds that sharing does more harm than good. In the mid-1960s, William and Paul Paddock argued that global food scarcity would soon become so severe as to necessitate a policy of discriminating as to who should receive food aid (Paddock and Paddock). They suggested the "triage" analogy, originally a medical principle for classifying the wounded into those who would survive only with help, those who would survive without any help, and those who would not survive despite help. Applying this analogy to the nations of the world, the authors urged that food aid should go only to those who fell within the first category; the other nations should be ignored since sharing with them would be either unnecessary or hopeless.

An even starker analogy has been suggested by Garrett Hardin in arguing for a "lifeboat ethic" (Hardin). The rich nations of the world are in lifeboats with limited capacities and are surrounded by poor, hungry nations struggling in the water. The rich should not yield to the humanitarian impulse to allow the poor into the lifeboats or else all will sink. Feeding the poor, Hardin argues, simply encourages continued increase in their population and will lead to an even greater catastrophe at a later date. The more ethical thing to do, therefore, is to let millions starve today rather than threaten billions tomorrow.

Counterarguments to triage and lifeboat ethics point out that (1) the world food supply is not yet so desperate as to warrant such drastic measures; (2) inequitable distribution patterns and wasteful consumption in affluent countries should be corrected before allowing people to starve to death; (3) widespread political, economic, and social turmoil would be the consequences of policies that dismiss major populations of the world; and (4) food aid should not be considered in isolation but in context of development programs that will improve food production, provide employment opportunities, and promote population stabilization.

In the ethical debate prompted by the need for food policy decisions, it is possible to discern at least six key and interrelated questions which must be addressed.

1. *Should the policy of food allocation be governed primarily by market operations?* In the operation of the market, the law of supply and demand dictates the allocation of goods. Effective demand is measured in terms of money available to meet the price asked. When applied to food distribution, particularly in times of scarcity when prices are high, market operations may not necessarily—indeed, do not usually—serve *need*. Need is measured not in terms of money available but of protein requirements for a particular population. Thus the Soviet Union could enter the international market in the summer of 1973 and purchase twenty million tons of grain, affecting the poor nations by (1) markedly driving up the world price of grain, and (2) limiting available reserves to be used as relief aid (Schertz).

2. *Should food aid policies be used as instruments of a nation's foreign policy?* Programs of general foreign aid have traditionally had political as well as humanitarian motivations. The question arises whether food assistance should be placed in the same category as other types of aid and therefore be subjected to the politics of strategic and economic considerations. Or should food assistance be given primarily on the basis of the severity of hunger and malnutrition experienced by a particular nation? Immediately prior to the Rome Food Conference in 1974, for example, most PL 480 food assistance from the United States was going not to the poorest coun-

tries, which the United Nations had classified as the "most seriously affected," but to a few Southeast Asian countries with military importance to the United States (Rothschild).

3. *Should food relief today be limited because of more serious problems tomorrow?* Those who support triage or lifeboat ethics argue that in the long run their stance is more humanitarian than food relief programs today. Immediate relief, according to this view, is counterproductive since (1) it encourages the poorer nations to postpone building their own agricultural productivity and (2) it allows the present generation to live and breed more people to be fed in the future—an increasingly impossible task. Whether or not one agrees with this view, it is certainly true that future consequences of present policies must be weighed in any ethical evaluation.

4. *Can some nations be classified as "expendable"?* If hard choices must be made in food policy decisions, what guidelines are used to determine who shall live and who shall die? This question is phrased in a harsh fashion, but as a matter of fact decisions that have a "life-or-death" character about them are made daily. For example, at the close of the World Food Conference the United States declined to provide additional emergency relief assistance for several famine-stricken nations such as those in the Sahel but instead sold additional foodstuffs to Syria, a key nation in the politics of the Middle East. In effect this meant that some Africans died and some Syrians lived. Guidelines for such decisions, some argue, should not weigh political factors more heavily than human needs. Racial, ethnic, cultural, and religious factors which might be considered in the decision-making process also should be ethically evaluated.

5. *Does the state have an obligation to feed the hungry?* This question raises the much larger issue of the ethical character of overall public social policy. The extent of the obligation of the community to provide for the general welfare of all citizens is the background for any discussion of the specific provision of food for those who, for one reason or another, are hungry. One ethical approach is exemplified by Japanese social policy: All children are provided with nutritionally balanced lunches, and pregnant mothers and young children are provided with special food-supplement programs. This is a national policy designed to promote a healthy, strong population. Another approach is taken by the United States, where food stamps, school lunches, and maternal and child nutrition programs are "relief-oriented." Used as poverty assistance programs, these efforts are not integrated specifically into national health and welfare goals (Simon, pp. 82–89).

6. *Is there any obligation to reduce consumption in more affluent societies?* The wide—and in many instances, widening—gap between rich and poor nations presents a strong ethical challenge: What is the morality of affluent consumption patterns in a world of so much hunger? Grain-fed beef and pork, wastage of food, profligacy in energy use, and widespread use of fertilizers for lawns all tell of consumption patterns that have an effect upon both availability and price of foodstuffs in the world. Some would argue that simpler life-styles in the rich nations have become a moral imperative in order to facilitate more equitable distribution of the world's resources; others counter that changes in consumption patterns would have little or no effect on distribution and that no one has a right to what another has legitimately earned.

Alternative frameworks. The general social policy framework within which food allocation is viewed will determine the kind of answers given to the ethical questions raised above. Three general policy frameworks can be suggested here, and their implications for food policy briefly outlined.

An *economic* framework provides rationalization for the operation of the market as the most efficient means of securing more food for more people. A shortage of food, it is contended, will cause food prices to rise and thereby stimulate greater production. Nothing should be done to interfere with the market mechanism—such as interfering in trade patterns, limiting profits, regulating prices—or production will suffer and less food will be available. Within this framework, food is automatically allocated to those who can afford it, within nations or among nations. Food relief is only possible once market demand is met; otherwise an artificial scarcity will be created and the benefits of market-stimulated production will not be felt. Moreover, this framework legitimates the policies of triage or lifeboat ethics in order to create a stable environment within which the market can operate. As for domestic policy, food should be provided by the state only on an emergency relief basis; within the economic framework general welfare policies are viewed with suspicion. Finally, it is meaningless to encourage a curtailment of consumption in the affluent countries since this action would cause a lessening of demand, a lowering of prices, and a curtailment of production.

A *charity* framework posits the sharing of food with the hungry as a responsibility to be met primarily through relief assistance programs. Collections of money for purchasing food for starving people is a typical response within this framework. The operation of the market is not challenged head-on, as long as sufficient surplus food is produced that can be shared (as was the case in the 1950s and 1960s in the United States). A strong rejection of the use of food aid for political purposes, as well as a reluctance to classify any nation as "expendable," characterizes this framework. But this position flows more from a humanitarian sense of solidarity than from a stringent ethical critique or social analysis of the proposals. Similarly, triage and lifeboat ethics are ruled out as "unthinkable," and "morally obscene." State obligations to feed the hungry are, within this framework, problematic since the emphasis tends to be placed more on personal response than on political arrangements. Reduction of consumption and the living of simpler life styles is promoted only in order to provide a surplus to share; if that surplus can be found without such curtailment, the charity framework would not emphasize any obligation for a change.

Central to a *justice* framework is the statement of a fundamental human "right to food." The statement by the United Nations (cited in the introduction to this article) that every human possesses a fundamental "right" to have basic food requirements met can be seen to rest upon both religious and humanistic grounds. A doctrine of creation, common at least within the Judeo-Christian tradition, affirms that the goods of the earth are for all creatures and not simply for a privileged few. This right is grounded in a view of human nature (whether inspired by a religious or a nonreligious perspective) which sees every man and woman to be of equal dignity and equal destiny. Just as life is inviolable, so the requisites of life are inalienable. Food, the necessary sustenance of life, therefore cannot be treated simply as another commodity dependent in its allocation on the operation of the market. It is ethically unacceptable, according to this view, to permit a situation in which a person's ability to live is dependent upon his or her ability to pay.

While the justice framework does not exclude the charitable sharing of food, it goes much farther by providing a systemic approach not only to the issue of food policy but also to other key issues within national and international economics and politics (Goulet). Hence it links food relief policies to agricultural development policies; emphasizes the connection between population stabilization efforts and general nutritional levels; relates trade, investment, and income distribution patterns to their impact on availability of basic human needs; and questions national budget priorities, e.g., defense spending over social welfare programs. Triage and lifeboat ethics are rejected not only because they are not universalistic in their interpretation of rights but also because of their narrow focus on relief assistance unrelated to other developmental factors. The "right to food" should be promoted through the state's guaranteeing every citizen a nutritionally adequate diet and the social policies (e.g., full employment) necessary to achieve that end. Because gross disparities in consumption levels are seen to contribute to unequal distribution of food, the justice framework would emphasize the obligation of simpler life-styles in affluent societies. The "right to food" for a basic level of survival should take precedence over any right of superfluities.

Conclusion

The crisis of the world food situation is likely to remain in the forefront of contemporary problems for some time. Hence the ethical issues raised by food policy decisions will continue to confront policy makers and ordinary citizens. As should be evident from the analysis presented in this article, it is impossible to discuss national or international food policy without discussing other public policy topics such as foreign assistance programs, population programs, socio-economic development plans, corporate business practices, social values and priorities, and so forth. Similarly, the stance one takes toward the problem of feeding the hungry will be part of one's overall social ethical stance and hence relates to a more general understanding of the obligation of the rich to the poor, the welfare role of the state, the role of economic factors in human development policies, and so forth.

Given the complex character of this topic of the ethics of food policy, then, at least three projections can be made regarding issues that will figure in all future debates and decisions: (1) The stark challenge of world hunger will probably increase in the immediate future, posing even more difficult ethical dilemmas to be faced;

(2) pressures on the rich, well-fed nations by the poor, hungry nations will present intensifying threats to world peace; and (3) the quality of response made to the problem of hunger will not only represent but also affect the general ethical standards that are essential to the maintenance and growth of human community.

PETER J. HENRIOT

[*Directly related are the entries* POPULATION ETHICS: ELEMENTS OF THE FIELD, *articles on* DEFINITION OF POPULATION ETHICS, THE POPULATION PROBLEM IN DEMOGRAPHIC PERSPECTIVE, ETHICAL PERSPECTIVES ON POPULATION, *and* NORMATIVE ASPECTS OF POPULATION POLICY; *and* POPULATION POLICY PROPOSALS, *article on* CONTEMPORARY INTERNATIONAL ISSUES. *For further discussion of topics mentioned in this article, see the entries:* JUSTICE; POVERTY AND HEALTH; *and* RIGHTS.]

BIBLIOGRAPHY

BERG, ALAN, and MUSCAT, ROBERT J. *The Nutrition Factor: Its Role in National Development.* Washington: Brookings Institution, 1973.
BROWN, LESTER RUSSELL. *The Politics and Responsibility of the North American Breadbasket.* Worldwatch Paper, no. 2. Washington: Worldwatch Institute, 1975.
———. *Seeds of Change: The Green Revolution and Development in the 1970's.* Foreword by Eugene R. Black. New York: Overseas Development Council, Praeger Publishers, 1970.
———, and ECKHOLM, ERIK P. *By Bread Alone.* New York: Overseas Development Council, Praeger Publishers, 1974.
CALLAHAN, DANIEL. "Doing Well by Doing Good: Garrett Hardin's 'Lifeboat Ethics'." *Hastings Center Report* 4, no. 6 (1974), pp. 1–4.
DEMARCO, SUSAN, and SECHLER, SUSAN. *The Fields Have Turned Brown: Four Essays on World Hunger.* Washington: Agribusiness Accountability Project, 1975.
ECKHOLM, ERIK P. *Losing Ground: Environmental Stress and World Food Prospects.* New York: W. W. Norton & Co., 1976.
Food and Agriculture Organization. *Population, Food Supply and Agricultural Development.* Rome: Food and Agricultural Organization of the United Nations, 1975. Available from UNIPUB Inc. in New York.
GOULET, DENNIS. "World Hunger: Putting Development Ethics to the Test." *Christianity and Crisis* 35 (1975): 125–132.
HARDIN, GARRETT. "Living on a Lifeboat." *BioScience* 24 (1974): 561–568.
KOTZ, NICK. *Let Them Eat Promises: The Politics of Hunger in America.* Introduction by George S. McGovern. Englewood Cliffs, N.J.: Prentice-Hall, 1969.
MCNAMARA, ROBERT S. *One Hundred Countries, Two Billion People: The Dimensions of Development.* New York: Praeger Publishers, 1973.
MALTHUS, THOMAS ROBERT. *On Population* (1798). Edited and introduced by Gertrude Himmelfarb. New York: Modern Library, 1960. First published as *An Essay on the Principle of Population.*
MAYER, JEAN. "Coping with Famine." *Foreign Affairs* 53 (1974): 98–120.
PADDOCK, WILLIAM, and PADDOCK, PAUL. *Famine, 1975! America's Decision: Who Will Survive?* Boston: Little, Brown & Co., 1967.
POLEMAN, THOMAS T. "World Food: A Perspective." *Science* 188 (1975): 510–518.
ROTHSCHILD, EMMA. "Food Politics." *Foreign Affairs* 54 (1976): 285–307.
SCHERTZ, LYLE P. "World Food: Prices and the Poor." *Foreign Affairs* 52 (1974): 511–537.
SIMON, ARTHUR. *Bread for the World.* Grand Rapids, Mich.: Wm. B. Eerdmans Publishing Co., 1975.
SINGER, PETER. "Famine, Affluence, and Morality." *Philosophy and Public Affairs* 1 (1972): 229–243.
THOMPSON, LOUIS M. "Weather Variability, Climatic Change, and Grain Production." *Science* 188 (1975): 535–541.
United Nations, Department of Economic and Social Affairs. *Poverty, Unemployment and Development Policy: A Case Study of Selected Issues with Reference to Kerala.* ST/ESA/29. New York: United Nations, 1975.
———, General Assembly, Ad Hoc Committee: Chairman. *Declaration on the Establishment of a New International Economic Order.* A/AC.166/L.50. Draft Resolution, no. 1, adopted without vote as resolution 3201 (S-VI) at the 2229th meeting, 6th special sess., U.N. General Assembly, 1 May 1974. Also A/9559 (GAOR, 6th special sess., supp. no. 1), pt. 2 (to be issued) and A/RES/3201, 3202 (mimeographed).
World Food Conference, Rome, 5–16 November 1974. *Report of the World Food Conference.* E/CONF.65/20. New York: United Nations, 1975.
World Population Conference, Bucharest, 19–20 August 1974. *Report of the United Nations World Population Conference, 1974.* E/CONF.60/19. New York: United Nations, 1975.

FRANCE

See MEDICAL ETHICS, HISTORY OF, *section on* EUROPE AND THE AMERICAS, *articles on* MEDIEVAL EUROPE: FOURTH TO SIXTEENTH CENTURY, INTRODUCTION TO THE MODERN PERIOD IN EUROPE AND THE AMERICAS, FRANCE IN THE NINETEENTH CENTURY, *and* WESTERN EUROPE IN THE TWENTIETH CENTURY.

FREE WILL AND DETERMINISM

The topics of free will and free action are of the highest importance to bioethics, because there is a maxim in ethics reflecting an intuition of common sense that "ought" implies "can." In other words, it would make no sense to claim that a physician ought to cure a cancer patient if the means to effect that cure did not exist; nor would it make sense to say that a dying man ought to refrain from dying unless he were able to do so. Responsibility and praiseworthiness can be meaningful only if the recipients of praise and blame can deliberate, choose, and control their actions. Thus, free will is crucial to bioethics in that, if physicians and other health professionals could not initiate action or refrain from it, the consequences of ethical reasoning would have no claim upon them. Justifying an act, claiming to have acted in order to conform with an ethical or practical norm, would have no meaning; one would merely be able to report the causes that produced an action, i.e., just explain it. Equally the status of the freedom and free will of the recipients of health care is an essential focus of bioethical dispute as exemplified by questions about whether it is proper to force treatment upon patients, or whether it is proper to deceive a patient so that he accepts a course of treatment or submits to an experiment.

In addition to such questions about the actions of professionals and patients, questions about freedom bear on the nature of our technologies. It is a matter of concern, for example, whether psychosurgery and other techniques create or hamper freedom, and whether such techniques provide or limit opportunities to affect and possibly control the actions and/or ideals of individuals—including their entire orientation to the world. Another set of questions arises about whether patients are free with regard to their behavioral repertoire. If not, a crucial result is that it may be necessary and proper to treat as illnesses what we now consider to be crimes. Thus, punishment and cure are rivals that can be understood in terms of whether one believes in the freedom and agency of persons. All of these questions, however, become moot if determinism is to be believed.

Determinism is the general view that all events, including human actions, are produced by prior conditions, which make those events and actions inevitable. The characterizations of the prior conditions take a variety of forms, but they all seem motivated by the idea that, if there were no specifiable laws governing the order of natural events, events would be thoroughly random and unintelligible, and science would be impossible. But science is highly successful at predicting and often controlling events. Thus, man must construe nature as following a rule in order for coherent experience to be possible.

Before we accept this assumption of science and the determinism in it, it should be first pointed out that the sets of specific rules that people devise in system-building vary widely. Some of these perspectives, which are called upon to support determinism, draw on the nature of the cosmos; some draw on the nature of society; and some draw on the nature of the individual. All of them are deterministic in that they place individual human action into a tightly structured explanatory nexus that specifies the world order. Thus, for the determinist, all action conforms necessarily to or with the rules of the nexus. Further, although the rules are primarily conceived as generalizations, which are descriptive of the natural course of events only on an abstract level, they are also thought to be applicable to specific situations.

Alternative positions on the freedom of the will can by and large be discussed under three paradigms: (1) Determinism is correct and there is no free will; the only world is the world of science. This view is represented by, for example, Holbach. (2) There are at least two separate worlds, more correctly spheres of existence or universes of discourse, one subject to deterministic theories and one subject to moral law. James and Kant exemplify this version. (3) Free will and determinism are compatible in one realm, because there are gaps in the causal nexus. This would accord with the ideas of Hume and Eccles. The three options recur and evolve in the history of philosophy and science.

The physical sciences

For Hobbes, Newton, Laplace, and Baron d'Holbach, human action was seen on a model of mechanical determinism. Hobbes, rejecting Descartes's dualism of mind and body, held that mind is matter in motion and explained will in terms of appetite—i.e., as a reaction to a strong desire. Hobbes did not consider that there was a separate faculty called the will which could be either free or caused. Rather, he understood observable physical force or coercive control to be a necessary condition for freedom. Thus, for Hobbes, freedom is akin to political or social liberty under strictly enforced laws.

Hobbes justified responsibility and sanction on utilitarian grounds. Pain, punishment, and the like may go into a person's deliberation as possible consequences of a proposed action. Such deliberation can influence the appeal of the proposed act and thus produce the determined outcome of the deliberation. But this justification for institutional sanctions is at best paradoxical, because punishment, in its full sense of retribution, presupposes the guilt of the agent while the very notion of guilt is in tension with a physicalistic theory of mind. Hobbes's physicalistic orientation makes it difficult to understand how an agent could weigh and choose alternatives, and without alternatives "guilt" is inappropriate, since "ought" implies "can." Hobbes is thus a compatibilist, a "soft determinist" (as James called such theories).

After Hobbes, the Newtonian influence on Laplace and the moral inferences drawn by Holbach resulted in a thoroughly deterministic position apparently free of paradoxical traces. Laplace believed that from a complete description of the entire universe at any one moment the complete description of it at any other moment (earlier or later) could be derived by way of the laws of mechanics. Holbach conceived of the will as a modification of the brain and thus brought the will under the rubric of physicalistic determinism. He likened ideas to physical events influencing action. As a result, he claimed to see no moral difference between one's jumping or one's pushing another person out a window.

Of course, Newtonian determinism has been replaced by the uncertainty principle of Werner Heisenberg, and thus physics appears again to have a structure that would accommodate compatibilism. However, this kind of gap in the determined nexus has nothing to do with the freedom required for morality. It merely makes a place for randomness rather than strict causal determinism (unless one were to hold that subatomic particles were choosing freely but within general constraints upon their choices).

Although the Newtonian model of mechanical causation is often taken to be prototypical, other deterministic systems have been developed. For example, Spinoza, a determinist par excellence, held that there is only one substance, self-conceived and self-sufficient, constituting all of nature. He expanded his deterministic view on the model of a mathematical deductive system: All events follow in their exact logical order like the lines in the demonstration of a geometric theorem. Spinoza held that mind and body are like the distinct but inseparable sides of a coin, and what appears to happen to each follows a fixed order of parallel sequences. Mental phenomena (ideas, decisions) and physical phenomena (movement, behavior) can then be said to occur only within their respective attributes—those of thought and extension—although a correlative order holds between them. It follows, for Spinoza, that common conceptions of free will are ill-drawn, because people fail to realize that they exist in a tightly structured causal system. Thus, he concluded that there is no free will in the sense that one's desires or actions would be uncaused. There is, however, free will in understanding and loving nature and working within a knowledge of its laws. Freedom, then, is inner satisfaction derived from intellectual love of God, i.e., understanding and appreciating the divine order of nature.

The biological sciences

Many important perspectives on determinism stand between Holbach and current thought, not the least of which is that of Franz Joseph Gall's reductionist neurophysiology. Further, even Sir John Eccles has pondered the relation between a free will and the brain, arguing that consciousness can change neurological states, thus trying, through a soft determinism, to give free will a place in the biologists' world. His thought is reductionistic yet dualistic. He suggests that an ego, a will, interacts with the brain. His position in this regard is not fully specified, but he admits that the current state of science is too weak for solving "the problem." In spite of a sense of urgency present in the strivings of Eccles and other great biologists (see, e.g., Monod), winning a place for free will within reductionistic thinking (which attempts to derive the laws of one science from those of another and gives rise to biology's threat to the concept of freedom) is surely far from accomplished, as John Findlay shows in *Psyche and Cerebrum*. For, although a physical substratum may be necessary for human personal existence, it is surely not sufficient. As Marjorie Grene argues in many of her essays about the mind–body problem, the mediation of socialization is clearly required as well, and this requires a psychology.

Psychology

Two powerful and prominent schools of psychology, the psychoanalytic and the behavioristic, have deterministic features. The early years of an individual's life determine the personality in Freudian thought, and patterns of behavior in the Skinnerian view. In Freudian thought, the instinctive reflex-nature of human beings, the so-called primary process, works through mechanisms of defense to avoid anxiety. Almost all of anxiety depends on the personality as it is determined in the first five years of life. This dynamic eventuates in repressed ideas that work in a causal fashion to produce determined behavior. In Skinnerian thought, behaviors that occur randomly are followed (randomly or otherwise) by so-called reinforcers, which produce alterations in the frequency of similar behavior. In

Freudian thought, willing is a conscious activity, but it is determined by an unconscious dynamic as relentless or inviolate as Newton's laws. In Skinnerian thought the notion of will (free or not) falls away both a priori (given the logic of the methodology) and a posteriori (given the failure to predict, control, or even discover inner mental occurrences). Thus, Skinner urges that we drop such concepts and go "beyond freedom and dignity."

The threats to free will from each of these psychologies is blunted by their inherent conceptual difficulties. Chomsky, in his reviews of Skinner's *Verbal Behavior* and *Beyond Freedom and Dignity*, has provided a philosophical critique of behaviorism that demonstrates how its crucial concepts (stimulus, response, and reinforcement) are often used in a question-begging way and how the methodology of behaviorism is violated when the questions are not begged. A. C. MacIntyre, in his book *The Unconscious*, has criticized Freud's thought especially where the theory of the personality is modeled on a nineteenth-century physical dynamics. The notions and images of defense mechanisms, repression, and censorship are fraught with paradox. The use of libidinal energy and conservation of energy, though deterministic if taken literally, seem most intelligible if taken metaphorically and heuristically.

Furthermore, a satisfactory psychology would have to have a satisfactory learning theory. Such a learning theory would have to accommodate a philosophical account of knowledge, including an account of observation. There are compelling arguments in the philosophical literature for the logical priority of knowledge to observation: In order to see, one first has to know. This Platonic view has been well articulated by Hanson, who also displays how language-dependent the making of observations is. Thus, the physical, social, and linguistic contexts are ingredient in observing a sunrise or seeing what the hour is (which depends, for example, on knowing about time zones and the effect of seasons on daylight).

The contextual nature of human knowledge and thinking, in addition to raising severe doubts concerning deterministic psychologies, also suggests a distinction of importance to the status of a belief in determinism. It can be argued that the distinction between cause and reason would be obliterated by "hard" determinisms. If one held that mental processes are wholly determined by physical events, then there could be no sense of "reason" as the employment of an argument to convince people of the truth of determinism or any other doctrine. The conception of a valid argument is lost, and one is caused only to utter sounds or nod one's head (O'Connor).

Moreover, in making observations, asking questions, and conducting an inquiry, people operate under a set of assumptions which orient them in such a way that there is an organization to their seeing, answering, and knowing when the inquiry is complete. Thus, for example, if a coroner is asked how a person died, he gives a physiological account; if a policeman is asked in the same case, he gives an account of action; if a widow is asked, she may give a psychological account, telling how the victim was angry and reckless. All this is meant to suggest how in our thinking we can change focus, can conduct our thinking from varying points of view. Thus, even if our sciences were compellingly deterministic, we, like Kant and William James, might be led to consider, as a separate purchase on the world, a nondetermined realm of personal existence.

The realm of freedom

Deterministic science frames its explanations in terms of the composition and structure of objects, together with descriptions of antecedent conditions. In the eighteenth century David Hume gave the classical empiricists' description of the scientific conception of cause. Hume said that humans come to believe in a causal connection simply through observing a regular succession of events of one kind (e.g., releasing objects) in relation to events of another kind (they fall). These come to be construed as representing a definite, unalterable series of events linked so as to constitute a chain of events attributable to nature as a pattern.

Reacting in part to Hume's skepticism about the necessity with which effect follows cause, Immanuel Kant specified a dilemma, or set of antinomies, about determinism (Kant, *Critique of Pure Reason*, A426, B454 sq.). Philosophically, some idea of freedom is needed to explain how the autonomy of practical reason (legislation by rational good will, i.e., planning action within the bounds of the logic of the concept "ought") can be efficacious in the world order (ibid., A547, B575). The trouble is that this autonomy seems to violate the most fundamental experience of nature as determined by laws rather than randomly. In this regard Kant undertook to provide an account of ethics in a realm of freedom, just as he had earlier undertaken to provide an account of science in a deterministic realm.

Kant displays how freedom and free will come into ethics in opening his *Groundwork of the Metaphysic of Morals* thus: "It is impossible to conceive anything at all in the world, or even out of it, which can be taken as good without qualification, except a *good will*" (p. 61). He unpacks the notion of a good will in terms of the willingness of the individual to endorse the absolute moral rule of always treating persons as ends in themselves, never as means merely. This notion of good will has little or nothing to do with the notion of action, free or caused (Frankfurt). Good willing need not be efficacious or successful in the determined sphere.

If Kant defines "good will" as a willingness to affirm a universal moral rule (the "categorical imperative"), he is not merely initiating or legislating a sense of "will." His definition connects with the tradition that made rational will an attitudinal concept and made moral rationality an essential requirement of good willing. This suggests two senses of "will": one about thinking, affirming, etc., and one about acting. And Kant did specify two senses of "will"—*Wille* and *Wilkür*—which correspond to these. In this schema, *Wille* is not freedom to act, but it is freedom to reason at a very abstract level. Such *Wille* is constrained by logic alone. The logic works this way: If universalizing the rule that describes one's moral behavior would result in a self-destructive social context, the action is considered incoherent, and the universalizing is self-contradictory (Silber). *Wilkür*, on the other hand, is the will that produces action.

The importance of these two kinds of will depends upon a third dimension that refers to ideals in terms of which one acts or refrains, strives to change the world or oneself for the sake of some highly esteemed yet unrealized value. Here notions of personal dignity, privacy, and self are most at home, most applicable to those who have forged a hierarchical view of themselves (Frankfurt). Specifically such a person is aware of the distance between what he wants and does on the one hand, and what he wants to become, the ideals he wants to realize, on the other. In this context two sorts of freedom can be distinguished. One is the freedom of self-determination, the freedom of the individual to choose for himself the story of the life he would like to strive to live. The other is autonomy, which is freedom from contradiction in one's moral logic. Morality consists in part in respecting the striving of every other rational agent to control what he becomes even though his goals and self-control may be placed under stresses, for instance, by impulses from within that crave immediate gratification. Thus, under the logic of autonomy, it is irrational to protect an individual from himself. In these two senses, then, the issues about freedom center upon whether one is allowed to be or to become oneself or allowed to choose one's goods and goals, and even whether one is allowed to fail as an independent rational adult. It is a consequence of this distinction that one can only destroy one's own autonomy and can do so only by willing to violate the self-determination of another.

Alternative philosophical reactions

If one appreciates the importance of the second universe of discourse, where freedom reigns, and is also aware of the power of the first realm, where determinism pervades the conceptualizing, one will be likely to seek a way to unite the two. "Compatibilism" or "soft determinism" names the kind of theory that attempts to loosen the determinations of the natural order so as to introduce a place for human control and responsibility. This, if accomplished, not only makes morality intelligible but also vindicates our intuitions of freedom.

Although a soft determinism is desirable because both hard determinism and freedom have their conceptual power, and although it is plausible because without consistency or regularity in the natural order, one would not know how to use his freedom, there are severe difficulties in soft determinism (O'Connor).

One difficulty is that the sense of "freedom" which soft determinists admit is freedom from coercion, freedom from external cause. The distinction between compulsion and self-control is thus lost. Were the soft determinists to offer an analysis of uncaused, self-controlled action, it would likely be uncaused in the sense of random. Such a position cannot distinguish in moral or explanatory terms the difference between acting on an unknown posthypnotic suggestion and acting deliberately on one's own intention. The idea of being able to choose an alternative course of action is not acknowledged. Equally the inevitability of hard determinism is lost.

Other philosophical work in this century aims at showing the inapplicability of deterministic language to typical human action. There is the powerful philosophical criticism of construing will as a faculty or entity. Ryle, under Wittgenstein's influence, shows that the word "will" is not referential, i.e., does not name an entity. Melden shows that "will" does not relate to the causal sphere, does not explain how bodily actions get initiated. Austin shows that the ways in which such words as "if" and "could" work do not require us to consider choices as determined.

Focus on bioethics

Biomedical technology relates to the topic "free will and determinism" in the general way that any aspect of human interaction and control does, and in a way particular to it. The particular relationship gives rise to the special peculiarities of *bio*ethics. Concerning liberty in general, a draft for military services is analogous to a draft for experimental subjects. Much ethical thought in biomedical literature is concerned with how prisoners or the poor are recruited or exploited for experimental purposes. This reflects a concern that the will of such individuals not be coerced. Biomedical technology is also uniquely involved in the issue of free will and determinism because it apparently has the scientific capacity to affect the willing of the individual. The law might require someone to do something against his own will (compliance against one's will), but psychotherapy, psychosurgery, or psychopharmacology might bring someone to change his ideals and life-style (form or reform the individual's will), thus undermining his self-determination. Law may define rights, while technology remakes the individual, changes his willing.

This point broaches the relation between rights and the will through what is in effect a denial of determinism. For the recognition of freedom, wants, and rights culminates in the doctrine of informed consent that has acquired enormous importance in biomedical ethics. In sum, the requirement that consent be informed stems from the connection between rationality and personhood; the requirement that the informed person grant consent stems from the connection between freedom and personhood. All this is as it should be in order to secure respect for the essential aspects of personhood: that aspect of will which reasons (autonomy) and that aspect which idealizes (self-determination).

But lest one think that technology only compromises freedom, an important reminder is in order. Without technology certain aspects of life and will would be greatly impoverished, e.g., a woman might become or remain pregnant against her will. The point is that procedures create options—new courses of action, which, in being efficacious, make the general affirmation of certain moral judgments or choices more apt. With the presence of a means, willing an end for oneself or being willing to see it be available to others becomes importantly distinguishable from merely wishing that that end were the case (Thomson).

Concepts of will bear on medicine in a variety of other ways. Consider, first, whether within the strictures of maintaining respect for persons as ends in themselves there is any limit to be placed on use of the technology in the service of human desires. For example, pregnancy cannot ordinarily be called a medical condition in that it is not an illness, sickness, disease, or injury. Should medicine restrict itself to aiding in the recovery from these negative states and/or preventing them? If so, perhaps it has no business effecting abortions. Here desire may fail to make a compelling claim on what an institution will sanction. Second, the concept "sick role" involves the fact that the patient cannot recover merely by willing (desiring) to recover. Third, sickness or injury may undermine the exercise of rational free choice: If one loses consciousness or becomes delirious, one cannot think straight. This kind of contingency may sometimes provide the warrant for overriding what an individual may sometimes claim as right, e.g., to refuse treatment or to die. Circumstances may lead us to think that an individual's rationality has been undermined, and in such cases we might be inclined to say that it is the illness, not the person, that is claiming the right. If illness undermines reason, power, and hope, and if thereby it compromises personhood, then force—physical or otherwise—seems to be warranted to restore freedom. That is, we might treat individuals as objects rather than as persons in order to restore them to personhood. This might be described as freeing the will—disencumbering it, as the major tranquilizers, such as phenothiazine derivatives, may be said to do.

Current technologies that manipulate the embodied person seem to amount to tampering with the capacity to will (in any sense) in a way that violates free will (in any sense). But the concepts do not allow us to say that educating an individual in order to create a rational agent violates the freedom of his or her will. Analogously, correcting a disease condition in the brain may be providing the person with an efficacious will, which he controls in accord with free choice and personal ideals.

The possibility of serious moral abuse of this technology is part of Thomas Szasz's concern with the way we treat the so-called mentally ill. The abuses of potential and actual enforced treatment, especially in the areas of behavior control, psychosurgery, psychopharmacology, etc., may be seen as assaults on the will itself, the core of personhood, since these procedures seem to affect thought and spontaneous behavior at their root. Justification of procedures under the rubric of "cure" can be a powerful rhetorical move.

A quagmire of questions arises in pondering whether we make or remake persons through behavior modification, psychosurgery, etc. Most theories about determinism, however, focus especially on whether agency rather than reason is determined. In kleptomania, for example, the individual is caused to steal rather than caused to think stealing is permissible. The nagging puzzle here is how thought relates to action. Some doctrine of embodiment seems needed to account for how events outside the individual affect thought and the performance of action.

Another aspect of the way the concept of will bears on bioethics has to do with the apparent trend to medicalize moral problems. A deterministic view of man and behavior is one in which the concepts of responsibility and punishment lose out to the concepts of illness and cure, respectively. Personhood and free will are thereby lost. In response to this, Morris has argued that there is a right to be punished rather than cured or rehabilitated.

Finally, the formal ethical principle of rational will, i.e., autonomy, when employed by the medical practitioner, is a source of respect for individuality and idiosyncrasy. If understood and acted upon, this sense of "will" would suffice to guarantee the morality of the health professional.

Conclusion

As a natural entity each human person participates in the world at three levels. As a material and biological creature, one has a relationship with the cosmic order. As a social being, one has a place in a social order and social interaction with other persons. As a psychological being, one has an intrapersonal dynamic—a relationship with oneself. These are three levels on which the human agent as creature may be contemplated. Determinism is the position that claims that all events are the necessary result of prior causal conditions—supernatural, natural, social, psychological—which were themselves the necessary results of earlier conditions apparently ad infinitum.

There is, however, also a fourth level, which describes an aspect of human personhood—the realm of freedom. It has to do with understanding the logical possibilities of ethical thought. The notion of freedom is equivalent to understanding the limit that rationality places on certain kinds of interpersonal actions. The understanding of the human person requires respect for each individual as an end rather than a means, as free rather than as an object merely subject to the laws of nature. The notions of agent and goal-seeker are then essential to ethics. The freedoms from contradiction, restraint, and dogma are the freedoms of reason, agency, and self-determination. They are all important to an ethic for biomedicine.

EDMUND L. ERDE

[*For a further discussion of topics mentioned in this article, see:* BEHAVIORISM, *article on* PHILOSOPHICAL ANALYSIS. *Also directly related are the entries* MENTAL HEALTH THERAPIES; MIND–BODY PROBLEM; *and* REDUCTIONISM. *This article will find application in the entries* BEHAVIOR CONTROL; GENETIC ASPECTS OF HUMAN BEHAVIOR; POVERTY AND HEALTH; PRISONERS, *article on* PRISONER EXPERIMENTATION; PSYCHOPHARMACOLOGY; PSYCHOSURGERY; *and* RIGHT TO REFUSE MEDICAL CARE. *For discussion of related ideas, see the entries:* EMBODIMENT; *and* PATERNALISM.]

BIBLIOGRAPHY

ARENDT, HANNAH. *The Human Condition.* Charles R. Walgreen Foundation Lectures. Chicago: University of Chicago Press, 1958.

AUSTIN, JOHN LANGSHAW. *Philosophical Papers.* 2d ed. Edited by J. O. Urmson and G. J. Marnock. Oxford: Clarendon Press, 1970.

CHOMSKY, NOAM. "Review: Beyond Freedom and Dignity." *New York Review of Books,* 30 December 1971, pp. 18–23.

———. "Reviews: Verbal Behavior." *Language* 35 (1959): 26–58.

ECCLES, JOHN C. "The Physiology and Physics of the Free Will Problem." *Progress in the Neurosciences and Related Fields.* Edited by Stephan L. Mintz and Susan M. Widmayer. Studies in the Natural Sciences, vol. 6. New York: Plenum Press, 1974, pp. 1–40. Proceedings of Orbis Scientiae held by the Center for Theoretical Studies, University of Miami.

EDWARDS, PAUL, and PAP, ARTHUR, eds. *A Modern Introduction to Philosophy: Readings from Classical and Contemporary Sources.* 3d ed. New York: Free Press, 1973. Selections from some sources referred to in this essay, e.g., Holbach and James, as well as a comprehensive bibliography on "free will and determinism."

ENGELHARDT, H. TRISTRAM, JR., and SPICKER, STUART F., eds. *Mental Health: Philosophical Perspectives.* Philosophy and Medicine, vol. 4. Dordrecht, Holland: D. Reidel Publishing Co., 1977.

FINDLAY, JOHN N. *Psyche and Cerebrum.* The Aquinas

Lecture, 1972. Milwaukee: Marquette University Press, 1972.
FRANKFURT, HARRY G. "Freedom of the Will and the Concept of a Person." *Journal of Philosophy* 68 (1971): 5–20.
HANSON, NORWOOD RUSSELL. *Patterns of Discovery: An Inquiry into the Conceptual Foundations of Science.* Cambridge: Syndics of the Cambridge Univeristy Press, 1958.
KANT, IMMANUEL. *Critique of Pure Reason.* Translated by Norman Kemp Smith. London: Macmillan & Co., 1964.
―――. *Groundwork of the Metaphysic of Morals.* Harper Torchbooks, The Academic Library. Translated by H. J. Paton. New York: Harper & Row, 1964.
MELDEN, ABRAHAM I. *Free Action.* Studies in Philosophical Psychology. New York: Humanities Press; London: Routledge & Kegan Paul, 1961.
MONOD, JACQUES. *Chance and Necessity: An Essay on the Natural Philosophy of Modern Biology.* Translated by Austryn Wainhouse. New York: Knopf, 1971.
MORRIS, HERBERT. "Persons and Punishment." *Monist* 52 (1968): 475–501.
O'CONNOR, DANIEL J. *Free Will.* Problems in Philosophy Series. Garden City, N.Y.: Doubleday, Anchor Books, 1971.
RYLE, GILBERT. *The Concept of Mind.* New York: Barnes & Noble; London: Hutchinson & Co., 1949.
SILBER, JOHN R. "Procedural Formalism in Kant's Ethics." *Review of Metaphysics* 28 (1974): 197–236.
SKINNER, B. F. *Beyond Freedom and Dignity.* New York: Bantam Books, 1972.
SZASZ, THOMAS S. *The Myth of Mental Illness: Foundations of a Theory of Personal Conduct.* Rev. ed. New York: Harper & Row, 1974.
THOMSON, JUDITH JARVIS. "A Defense of Abortion." *Philosophy and Public Affairs* 1 (1971): 47–66.

FUTURE GENERATIONS, OBLIGATIONS TO

Discussion of the question of man's obligations to *distant* future generations is rarely found in the writings of ethical theorists prior to the contemporary period. Current interest in the problem derives from such issues as environmental pollution, the use of natural resources, population control, eugenics, and, generally, the effects of our actions or inaction on the health and well-being of future generations. In connection with these matters proposals have been advanced on the grounds of an obligation that the present generation is alleged to owe to future mankind. We shall consider the ethical basis of such obligations and also, briefly, the justifiability of certain eugenic proposals.

The concept of obligations to future generations

Although this concept is widely used, only a few writers have attempted to clarify its meaning (Golding, "Ethical Issues," pp. 451–457; Golding, 1972; Feinberg). The fact that a concept has a coherent meaning does not imply that it is ethically sound: The question "What do we mean when we speak of obligations to future generations?" should not be confused with the question "Do we have obligations to future generations?"

The first step in elucidating this concept is to distinguish obligations to future generations from any other obligations or duties that we may have. Failure to make this distinction can be an obstacle to clear moral thinking. Merely from the fact that restraint in the use of natural resources would be good or "the right thing to do" for some reason, it does not follow that we owe that restraint to future generations. The latter requires a special kind of moral justification. As Golding points out, obligations to future generations are derivative from the (single) obligation to promote what is good *for* future generations, assuming we have such an obligation (Golding, 1972). The practical importance of stressing this is that it will often be very difficult to show that some proposal will promote what is good for future generations, since we do not know what their conditions of life will be.

Second, obligations to future generations should be distinguished from obligations to our immediate posterity, even though the content of some of the obligations may be the same. That is to say, obligations to future generations are obligations to individuals with whom the present generation cannot reasonably expect to share a common life, i.e., individuals beyond the next three or four generations. The problem of the ethical basis of obligations to such distant future generations (which are as far away as one hundred generations and more in some eugenic schemes) will be discussed in the next section.

Because the alleged obligations are owed *to* what are at present nonexisting individuals, and because those nonexisting individuals are alleged to have rights against the present generation, the concept of obligations to future generations appears to be logically odd. The statement "A has a right against B" seems to imply that A can claim something from B. But future generations are not in a position to claim anything from us, nor, unlike immediate posterity, will they ever be.

A few philosophers have dealt with this problem (Golding, 1972; Baier; Feinberg). Feinberg maintains that rights involve interests and claims. He draws attention to the fact that the law protects the rights of a fetus under the doctrine of "future interests." The rights of a fetus are protected on the assumption that it will be born, i.e., that its potential interests will be actualized; and its claims can be made by a guardian appointed for this purpose. The rights of future generations are dealt with along similar lines: Future generations have rights against us if future generations will exist, and claims for the protection of their interests can be made by currently living individuals. Golding emphasizes that the oddness of the idea of the rights of future generations arises only on theories of rights that virtually identify rights with occurrent acts of claiming or acts of will. There is no difficulty once the distinction between "making a claim" and "having a claim" is acknowledged. He also points out that there are different types of rights (Golding, "Towards a Theory of Human Rights," pp. 541–547). There is no conceptual difficulty in conceding that future generations can have "welfare rights" against us, that is, entitlements to certain of the goods of life, such as good health and health care.

Ethical basis of obligations to future generations

If the concept is not logically odd, it still remains to show whether or not we have any obligations to future generations. It will readily be seen that two of the standard sources of obligations (and rights) are inapplicable to this case; namely, contract (agreement) and mutual benefit. We cannot enter into an agreement with distant future generations, nor are we participants with them in a mutually beneficial social arrangement. Four recent approaches to the ethical basis of obligations to future generations will be considered here.

1. Utilitarianism. Utilitarian theories of ethics judge the rightness or wrongness of an act (or of a type of act) by reference to whether its consequences are, on balance, good or bad. According to classical utilitarianism, this means that an act is right if it produces a greater amount of happiness than unhappiness. Given this forward-looking emphasis, utilitarianism appears to be especially pertinent to our problem, but it is not clear that the maximization of the welfare of future generations would be recognized by utilitarians as an obligation owed *to* posterity. The utilitarian principle, according to Narveson, implies a duty to maximize the average happiness of a community of existing individuals. Only under this condition can the promotion of utility be conceived as an obligation to persons. Narveson's discussion, however, is not undertaken within the context of the future-oriented programs and ecological concerns that have motivated the question of obligations to future generations. He is primarily interested in whether utilitarianism entails a duty to produce as many children as possible so long as their happiness exceeds their misery; and for the reason given his answer is negative (Narveson).

Nevertheless, it may still be asked whether or not the effects of an act on the welfare of future generations must be taken into account in the calculation by which the utilitarian determines its rightness, in addition to its effects on the welfare of the currently living. It seems, though it is not certain, that Jeremy Bentham (1748–1832), the founder of modern utilitarianism, would reject the assertion that we have a duty to take futurity into account. "Can it be conceived," asks Bentham, "that there are men so absurd as to . . . prefer the man who is not to him who is; to torment the living under pretence of promoting the happiness of them who are not born, and who may never be born?" (Cited in Berlin, p. 57.)

In contrast, the position that probably prevails among utilitarians, however, is expressed by Sidgwick: "It seems clear that the time at which a man exists cannot affect the value of his happiness from a Universal point of view; and that the interests of posterity must concern a Utilitarian" (Sidgwick, p. 381). Still, the extent to which the welfare of future generations should be taken into account in a utilitarian calculation is problematic, for Sidgwick concedes that the effects of one's actions on posterity are necessarily more uncertain than their effects on the present. A utilitarian framework does not justify very much planning for the sake of futurity, especially when it entails severe sacrifices on the part of the present generation, as Passmore points out (p. 85). This conclusion also holds for other theories, too.

A major difficulty with a utilitarian approach to our problem is that the calculation of average utility presupposes a definite population for whom the calculation is made, but the numerical composition of future generations is indefinite and unknown. Stearns therefore suggests that obligations to future generations derive from a duty of benevolence to maximize "total utility" rather than average utility: We ought to see to it that human existence continues at a high level of intrinsic value (Stearns). Stearns admits, however, that the duty to maximize average utility for existent persons may come into conflict with the duty to maximize total utility, and he provides no way to resolve the conflict.

2. **Rawls.** In contrast to the utilitarian approach, John Rawls takes justice rather than happiness or welfare as his starting point. His discussion is more relevant to such issues as the use of natural resources than it is to eugenic planning, for example. Rawls is concerned with saving for posterity; we ought to save for posterity because we ought to act as justly to posterity as to our contemporaries. A principle of "just savings" requires that real capital accumulation ought to be set aside for the next generation, though the amount cannot be stated precisely.

The life of a people is conceived as a scheme of cooperation spread out in historical time. It is to be governed by the same conception of justice that regulates the cooperation of contemporaries . . . thus imagining themselves to be fathers, they are to ascertain how much they should set aside for their sons by noting what they would feel entitled to claim from their fathers [Rawls, p. 289].

Rawls's argument for just savings rests upon a number of hypothetical suppositions, which are adopted for theoretical purposes. In a significant respect Rawls reinstitutes the two sources of obligation that were rejected earlier, contract and mutual benefit. He presumes that all past, present, and future mankind meet together to establish the most fundamental principles of social order for themselves as rational and self-interested individuals. These persons are also conceived as being under a "veil of ignorance." They lack knowledge of the particulars of their life, e.g., the social stratum into which they will be born; they also do not know the generation into which they will be born. Rawls presents an elaborate argument to the effect that these "contractors" will adopt certain principles of justice as a basis for their future cooperation, among which is included the principle of just savings. This is a sort of insurance policy that would be adopted by self-interested individuals who are seeking to minimize their risks, given that they do not know their position in the order of generations. The just savings principle, it should be noted, does not require any generation to make heroic sacrifices for the sake of posterity, and Rawls would oppose savings schemes whose enforcement necessitates the abandonment of democratic political procedures.

Aside from the debatability of Rawls's other hypothetical suppositions, the conception of society as a "scheme of cooperation" spread out in historical time does not appear convincing, because earlier generations do not benefit from the cooperative behavior of later generations. Rawls in fact views the contractors as making reciprocal exchanges, each one giving up something for the other. In the case of those who will be born into future generations, these exchanges are "virtual" rather than real, according to Rawls. For "reciprocation" here means that each generation gives up something for other generations, including previous generations. This kind of exchange is, of course, a fiction; and in contrast to Rawls's other hypothetical presumptions it is not a fiction that we have any reason to accept. To the extent that Rawls's argument requires this notion of reciprocal exchange, therefore, the argument for the just savings principle does not appear successful. In any case, the implications of Rawls's approach for the problem of obligations to distant future generations are limited, though important enough, perhaps. The ethical basis of proposals for enhancing the genetic quality of mankind, for example, must be sought elsewhere.

3. **Passmore's approach.** While the utilitarians start from happiness and Rawls from justice, Passmore's key concept is that of love. We love our children and show a loving concern for their well-being. It follows, Passmore maintains, that we should show a loving concern for *their* children's well-being too, because our children's well-being is bound up with their loving concern for their children's well-being. And so on, down the generations, a "chain of loves" is set up, from us into the future (Passmore, pp. 87–90). In a sense the obligation to promote the good of future generations is derived from our obligation to promote the good of immediate posterity.

But is it really possible to love one's remote descendants, say, one's descendants in the tenth degree? Passmore admits that this is not possible, for we cannot love people about whom we know virtually nothing. Yet, if we cannot love them, we can at least be concerned about them and take some interest in their welfare, because of our love for immediate posterity. This, then, is what constitutes the basis of obligations to future generations, in Passmore's approach, and one may question whether it takes us very far, for the obligation it would justify seems very attenuated. In practical terms, in fact, Passmore believes that the present generation should concentrate its efforts on promoting good conditions of living for the next few generations rather than focus upon remote futurity. He also agrees with Rawls that we should not be required to give up democratic institutions for the sake of future generations.

4. Golding. Golding substantially agrees with the practical conclusions arrived at by Passmore, though his theoretical approach is somewhat different. Two concepts are central to his position: the concept of "moral community" and the concept of the "social ideal" (Golding, 1972). A moral community is a group of persons who stand in moral relations to each other, the most important of which are obligations and rights. These obligations and rights may be very specialized or very general, depending upon how, or for what purposes, the moral community is generated.

One of the varied ways in which a moral community may be generated is that the members of a group share a common view of the good, a good that is social in its scope and not idiosyncratic. In essence, this is a view of the good life—the good life for a society. Golding calls a conception of the good of this sort a "social ideal." As a result of a common commitment to a social ideal, he maintains, the members of the group have obligations and rights toward each other. Second, a moral community can be generated, according to Golding, because the members share in, have applicable to them, a social ideal in consequence of certain characteristics they have in common. Human rights have an ethical basis if there is a common human nature and an ideal of the good life applicable to men in virtue of this nature.

As far as the basis of obligations to future generations is concerned, Golding maintains that the crucial question is whether the present generation and future generations constitute a single moral community in either of the two respects just described, whether they are united under an ideal of the good life. If there is such an ideal, it will determine what it is that the present generation ought to desire for future generations. But, though the basis of an obligation to promote a good life for future generations is secured in this way, according to Golding, he nevertheless argues that in practice it is extremely difficult to determine what the obligations are, concretely speaking. For in practice the present generation can be guided only by current ideals of the good life, and it is far from certain that these ideals will be shared by future generations or, what is probably more important, that these current ideals will be applicable to future generations. We do not know what the conditions of life of future generations will be, especially in the case of very distant generations, so we do not know what we ought to desire for them. Against this conclusion, Callahan argues that we do at least know that future generations will need certain basic goods such as air, water, and sources of nourishment, and that future generations are entitled to these goods as matters of human rights. The ethical imperative, according to him, is not so much the promotion of a good life for future generations as it is not harming future generations in any fundamental way (Callahan).

It should be noted that one problem that confronts every theory of the basis of obligations to future generations is that of potential conflicts between these obligations and obligations to nearby generations. To what extent may the good (and the rights) of nearby generations, or even the present generation, be overridden for the sake of far-off generations? This question goes to the heart of policy proposals currently advanced by some conservationists, eugenicists, and proponents of population limitation. Some of these proposals raise many problems regarding conflicts of values or rights (Golding, "Ethical Issues"; Golding and Golding; Passmore, Part Two). In general, it may be argued that conflicts between the promotion of the good of nearby generations and of remote generations should be resolved in favor of the former. As Golding points out, our obligations to our children and grandchildren are much clearer than our obligations to remote generations, for our ideal of the good life is more applicable to the former, and our knowledge of their conditions of life is more secure (Golding, 1972).

Eugenics and obligations to future generations

For purposes of this discussion "eugenics" will be taken to refer to long-range proposals for improving the genetic constitution of the human race. Eugenics is concerned with influencing the direction of human evolution and is therefore to be distinguished from genetic counseling, which aims to avert personal or family tragedy. This section is a very brief discussion of some aspects of certain eugenic proposals in relation to the issue of obligations to future generations. (For a detailed treatment, see Golding, "Ethical Issues.") It is difficult to see any justification for such proposals other than the existence of an obligation to promote the good of distant futurity. It should be noted that these proposals involve many disputed empirical questions, as well as ethical and valuative considerations.

Positive eugenics. Positive eugenics, according to Sir Julian Huxley, "has a far larger scope and importance than negative. It is not concerned merely to prevent genetic deterioration, but aims to raise human capacity and performance to a new level" (Huxley, p. 123). Any proposal for positive eugenics must, first, specify the traits that are to be enhanced. Second, it must show that the enhancement of these traits is likely to be advantageous for the community of the future. This is required if our obligation in respect to future generations is the promotion of what is good *for* them, i.e., the social ideal as applied to their conditions of life, as Golding would put it.

The Nobel laureate geneticist Hermann J. Muller maintains that the most important genetic objectives for the good of mankind at large are health, intelligence, and temperamental qualities that favor fellow-feeling and social behavior (Muller, pp. 114 f.). Muller proposes that the sperm of selected donors who exhibit these traits should be preserved for future use, so as to maximize their incidence in the population. This proposal, first of all, raises conceptual and empirical questions. What do such terms as "health," "intelligence," "spirit of cooperation," "sense of duty," and "appreciation of nature" (in Muller's writings on the subject) mean? Are intellectual and moral traits genetically transmitted? There may be conceptual difficulties in maintaining that moral traits, for example, are heritable. Whatever it is that cooperative parents transmit to their children, it certainly need not manifest itself as a spirit of cooperativeness. The social meaning of any such trait will depend upon the conditions in which it is manifested.

Assuming for the sake of argument, however, that such traits are genetically transmissible, should we seek to maximize their incidence in future generations? Consider intelligence and the sense of duty, two traits that are generally valued. It is far from clear that we should want to increase their incidence, or their strength, under every condition whatsoever. It is doubtful that we should welcome an increase in the relative numbers of intelligent men unless altruists could be increased in the same proportions; for otherwise the result could be a disadvantageously large number of clever and crafty mean men. On the other hand, the survival of civilization may depend on a delicate balance between fellow-feeling and self-interest that could be upset by the kind of eugenic planning that Muller has in mind. In the absence of knowledge of the conditions of life for future generations, the applicability of our social ideals to future generations is highly uncertain, and the existence of obligations regarding the above matters is problematic.

Negative eugenics. Negative eugenics seems to be in a much better case and may be treated even more briefly. It has been argued that medicine has had a "dysgenic" effect, because it prolongs the life of carriers of harmful genes, thus enabling their transmission to future generations. Huxley states:

Without selection, bad mutations inevitably tend to accumulate; in the long run, perhaps 5,000 to 10,000 years from now, we shall certainly have to do something about it. . . . Most mutations are deleterious, but we now keep many of them going that would otherwise have died out [cited in Golding, "Ethical Issues," p. 453].

Thus, the "load of mutations" increases in every generation, and some kind of artificial selection for survival is needed, lest mankind commit biological suicide. This, in sum, is one of the arguments for schemes of negative eugenics: prevention of a weakening of man's genetic capacity.

Though there are conceptual problems in designating some genes as "bad," it is certainly the case that there are genetic diseases which we would like to see eliminated, e.g., Huntington's chorea and retinoblastoma. It may become feasible—at present it is not—to identify with certainty individuals who have inherited the gene for Huntington's chorea, a highly debilitating nervous disease that eventuates in death, before they reach the age of reproduction and perhaps sterilize them. (Techniques such as sterilization, especially when compulsory, raise ethical issues in their own right.) It is hard to imagine that certain genetic diseases could ever be anything but disadvantageous for future generations, and it could therefore be argued that we have an obligation to do what we can, consistent with reason and morality, to eliminate them. But there still would be problems. The gene for sickle-cell anemia, whose elimination is now encouraged, is advantageous under certain conditions, as it provides protection against malaria. For all we know, we may not be doing future generations a favor by attempting to eliminate it. It should also be borne in mind that all of us are carriers of some deleterious genes. Any program of negative eugenics that succeeds in eliminating all these genes would do so at the price of killing off the human race.

MARTIN P. GOLDING

[*For further discussion of topics mentioned in this article, see the entries:* ENVIRONMENTAL ETHICS, *articles on* QUESTIONS OF SOCIAL JUSTICE *and* THE PROBLEM OF GROWTH; *and* EUGENICS, *article on* ETHICAL ISSUES. *Also directly related are the entries* JUSTICE; POPULATION ETHICS: ELEMENTS OF THE FIELD, *article on* NORMATIVE ASPECTS OF POPULATION POLICY; POPULATION POLICY PROPOSALS, *articles on* CONTEMPORARY INTERNATIONAL ISSUES, SOCIAL CHANGE PROPOSALS, COMPULSORY POPULATION CONTROL PROGRAMS, POPULATION DISTRIBUTION, *and* GENETIC IMPLICATIONS OF POPULATION CONTROL; *and* RIGHTS. *For discussion of related ideas, see the entries:* ETHICS, *article on* UTILITARIANISM; GENE THERAPY, *article on* ETHICAL ISSUES; *and* REPRODUCTIVE TECHNOLOGIES, *article on* ETHICAL ISSUES.]

BIBLIOGRAPHY

BAIER, ANNETTE C. "Can Future Generations Correctly Be Said to Have Rights?" American Philosophical Association Symposium, Atlanta, Georgia, 28 December 1973. Unpublished.

BERLIN, ISAIAH. *Two Concepts of Liberty.* An Inaugural Lecture delivered before the University of Oxford on 31 October 1958. Oxford: At the Clarendon Press, 1958.

CALLAHAN, DANIEL. "What Obligations Do We Have to Future Generations?" *American Ecclesiastical Review* 164 (1971): 265–280.

FEINBERG, JOEL. "The Rights of Animals and Unborn Generations." *Philosophy and Environmental Crisis.* Edited by William T. Blackstone. Athens: University of Georgia Press, 1974, pp. 43–68.

GOLDING, MARTIN P. "Ethical Issues in Biological Engineering." *UCLA Law Review* 15 (1968): 443–479.

———. "Obligations to Future Generations." *Monist* 56 (1972): 85–99.

———. "Towards a Theory of Human Rights." *Monist* 52 (1968): 521–549.

———, and GOLDING, NAOMI HOLTZMAN. "Ethical and Value Issues in Population Limitation and Distribution in the United States." *Vanderbilt Law Review* 24 (1971): 495–523.

GUSTAFSON, JAMES. "Basic Ethical Issues in the Bio-Medical Fields." *Soundings* 53 (1970): 151–180.

HUXLEY, JULIAN. "Eugenics in Evolutionary Perspective." *Eugenics Review* 54 (1962): 123–141.

JONES, ALUN, and BODMER, WALTER F. *Our Future Inheritance: Choice or Chance? A Study by a British Association Working Party.* Oxford: Oxford University Press, 1974.

MULLER, HERMANN JOSEPH. *Man's Future Birthright: Essays on Science and Humanity.* Edited by Elof Axel Carlson. Albany: State University of New York Press, 1973.

NARVESON, JAN. "Utilitarianism and New Generations." *Mind* 76 (1967): 62–72.

PASSMORE, JOHN ARTHUR. *Man's Responsibility for Nature: Ecological Problems and Western Traditions.* New York: Charles Scribner's Sons, 1974.

RAMSEY, PAUL. *Fabricated Man: The Ethics of Genetic Control.* New Haven: Yale University Press, 1970.

RAWLS, JOHN. *A Theory of Justice.* Cambridge: Harvard University Press, 1971.

SIDGWICK, HENRY. *The Methods of Ethics.* 7th ed. London: Macmillan & Co., 1907. New York: Dover Publications, 1966.

STEARNS, J. BRENTON. "Ecology and the Indefinite Unborn." *Monist* 56 (1972): 612–625.

G

GENE THERAPY

*The articles in this entry, unlike the articles in some of the other composite entries, were planned as an interdependent unit. Thus, the first five articles—*ENZYME REPLACEMENT, GENE THERAPY VIA TRANSFORMATION, GENE THERAPY VIA TRANSDUCTION, CELL FUSION AND HYBRIDIZATION, *and* PRODUCTION OF FOUR-PARENT INDIVIDUALS—*present the various techniques for altering genetic structure and then simply introduce the ethical issues that affect the use, control, and continuation of research in these fields. The sixth article,* ETHICAL ISSUES, *serves as a unifying article that brings together the analysis and the ethical principles affecting all of the above developments.*

   I. ENZYME REPLACEMENT   *Elizabeth F. Neufeld*
  II. GENE THERAPY VIA TRANSFORMATION
                                          Richard O. Roblin
 III. GENE THERAPY VIA TRANSDUCTION
                                          Richard O. Roblin
 IV. CELL FUSION AND HYBRIDIZATION
                                              George Poste
  V. PRODUCTION OF FOUR-PARENT INDIVIDUALS
                                              Beatrice Mintz
 VI. ETHICAL ISSUES   *Roger L. Shinn*

I

ENZYME REPLACEMENT

Most genetic diseases are due to a mutation in a gene that encodes the chemical structure of some enzyme: the defect in the gene is translated into a defect in the enzyme molecule, so that the enzyme functions poorly, if at all. Since enzymes are the catalysts governing all chemical reactions in the body, the loss of activity of even one enzyme can have far-reaching and serious repercussions. Essential metabolic reactions may fail to take place; toxic products may accumulate; control mechanisms may be disrupted. In some cases, the effects of the enzyme deficiency can be modified by altering the environment. More often, the consequences of the enzyme deficiency seem inescapable, unless some way can be found to supply the enzyme itself.

There is no great difficulty in replacing an enzyme that normally functions in a body fluid. The administration of certain concentrated blood fractions to hemophilia patients is a commonly practiced form of enzyme replacement (though not usually so designated), because the active components in the concentrates are the enzymes required to cause blood to coagulate.

Replacement of enzymes that function within cells is another matter. All enzymes are proteins, too large to penetrate through the plasma membrane at the cell surface. Replacement of an enzyme that would normally function in the nucleus or the cytoplasm is, to present knowledge, impossible. Enzyme replacement is fundamentally different from hormone replacement, such as administration of insulin to diabetic patients, because protein hormones like insulin function at the cell surface and need not penetrate into cells.

However, enzyme replacement seems feasible, at least in theory, in the case of enzyme constituents of the cellular organelle known as the lysosome, since this organelle is in communication with the exterior of the cell. By a process known as pinocytosis, a bit of the cell surface membrane invaginates and forms a pocket, which pinches off into a rounded vesicle and moves to the interior of the cell. The vesicle then fuses with a lysosome, bringing to it any enzyme that may have been trapped in the original pocket.

Lysosomes are involved in a recycling process, whereby large and complicated molecules are broken down to small pieces, which are returned to the general metabolism. If one of the many lysosomal enzymes is deficient, the breakdown process is interrupted and partially degraded large molecules accumulate; the lysosomes become very large and numerous, preempting much of the cellular space and eventually causing clinical disease. There are more than two dozen known diseases attributable to a lysosomal enzyme deficiency. The inevitably fatal outcome of most lysosomal enzyme deficiency diseases creates understandable pressure to test the feasibility of enzyme replacement.

Experience since the first attempt in 1964 has clarified the precautions necessary to avoid harm to the patient. The enzyme preparation must be free of microorganisms and of fever-causing impurities. In order to minimize immunological reactions, the enzyme should be of human origin and as pure as possible. Trials must begin with small doses to monitor biochemical changes and to record side effects. As of early 1975 encouraging biochemical changes resulting from the injection of highly purified enzymes have been reported in a few instances, but the state of the art has not progressed sufficiently to test large doses, over prolonged periods, for possible therapeutic benefit.

The same ethical problems arise as in the testing of any experimental drug. The potential benefit must outweigh the risks; the severity of the disease should not foster an attitude that any experiment is permissible simply because the patient is already doomed. It is unfortunate that there are no analogous animal diseases on which the benefit of the treatment could be objectively evaluated.

The human materials from which the enzymes can be prepared (blood, urine, placenta) are plentiful and require no sacrifice or consent on the part of specific donors. However, the minute quantities of pure lysosomal enzymes that can be extracted from these sources may render the cost of treatment exorbitant. Until the technological problems of production are overcome, there will surely be a question of who is to receive the treatment if it is found beneficial.

In summary, enzyme replacement for lysosomal deficiency diseases is possible in theory and presents no novel ethical dilemmas. However, the simplicity of the procedure is only apparent, and much more effort will have to be invested before its therapeutic value can be assessed. In order to prevent false hopes or premature discouragement, it is important that enzyme replacement be known as a potential but not an actual mode of treatment for that special group of inherited diseases.

ELIZABETH F. NEUFELD

[*While all the articles in this entry are relevant, see especially the articles* GENE THERAPY VIA TRANSDUCTION *and* ETHICAL ISSUES.]

BIBLIOGRAPHY

"Enzyme Therapy in Genetic Diseases." Edited by Daniel Bergsma, Robert J. Desnick, Robert W. Bernlohr, and William Krivit. *Birth Defects: Original Article Series* 9, no. 2 (1973).

Enzyme Therapy in Lysosomal Storage Diseases. Proceedings of the Workshop on Cell, Biological and Enzymological Aspects of the Therapy of Lysosomal Storage Diseases. Edited by J. M. Tager, Gerrit Josephus Maria Hooghwinkel, and W. Th. Daems. Amsterdam: North-Holland Publishing Co.; New York: American Elsevier Publishing Co., 1974.

II
GENE THERAPY VIA TRANSFORMATION

Gene therapy refers to the future possibility of introducing new, functional genetic information contained in molecules of DNA into human cells with the intention of treating human genetic disease. Gene therapy via genetic transformation would envision using isolated fragments of purified human DNA to accomplish this end. This approach is suggested by the observed alteration of the genetic properties of certain types of bacteria (Hotchkiss and Gabor) and of the fruit fly *Drosophila* (Fox, Yoon, and Gelbart) by exposing their cells to isolated DNA molecules in vitro. Since mammalian cells (including human cells from patients with genetic diseases) can now be grown in vitro, it is possible to expose such cells to isolated pieces of purified DNA in order to determine whether such treatments can permanently alter the genetic characteristics of the cells. Although this type of experiment has been tried many times with mammalian cells, and some successful results have been reported, to date none of the apparently successful experiments have proved to be reproducible upon further investigation. In contrast to the inconsistent results with isolated pieces of purified DNA, transfer of at least one particular human gene to mouse cells growing in vitro by addition of purified whole human chromosomes has recently been reproducibly achieved (Burch and McBride; Willecke and Ruddle).

Permanent genetic modification of human cells by means of added DNA would probably require the following processes. First, the added DNA would have to be taken up by the cells and transported to the cell nucleus. Second, the newly introduced DNA would have to become permanently associated with the cellular chromosomal DNA. This could take place through physical integration of the added DNA into the preexisting DNA of the chromosomes of human cells. Finally, the newly introduced genetic information encoded in DNA must be correctly transcribed into another molecule, an RNA "message," and this RNA message must be correctly translated into a functional protein in order to express the new genetic information. Since there are molecular "start" and "stop" signals on DNA, which control transcription into the RNA message and translation into protein, the added DNA must include these signals and must be integrated in such a way that the signals function correctly.

Some of the technical problems that have so far prevented permanent genetic alteration of mammalian cells in vitro by isolated DNA are (1) the tendency of isolated DNA to be degraded by intra- and extracellular enzymes, which abolishes the capacity of the DNA to carry genetic information; (2) the low frequency (approximately one part in ten million) of the DNA for any specific gene in the total DNA extracted from an organism, which means that adding the specifically required DNA segment to genetically deficient cells will be a very rare event; (3) the low probability of integration of random fragments of added DNA into chromosomal DNA of the cell.

Current approaches to overcoming these problems include (1) enrichment of the added DNA for specific DNA molecules, so that the DNA preparation consists entirely of the desired specific DNA, and (2) in vitro attachment of the human DNA sequence which one wants to integrate to a DNA molecule from a virus like SV40, since such virus DNA molecules have the capacity to integrate themselves into mammalian cell chromosomal DNA. If human DNA sequences were attached to SV40 DNA, it is expected that the SV40 DNA could serve as a carrier to help integrate the human DNA sequences into cellular chromosomal DNA. [See GENE THERAPY, article on GENE THERAPY VIA TRANSDUCTION.]

Ethical issues raised by proposals for genetic therapies include assessment of efficacy (Will the technique work at all, or will it work in enough of the patient's cells to improve the patient's condition?) and side effects in the patient and his or her offspring (Will the technique cause acute illness or a more slowly developing illness like cancer? Will it alter the patients' eggs or sperm beneficially to repair the genetic lesion, or detrimentally?). Since there are currently almost no animal experimental models of human genetic disease, questions of efficacy and side effects will end up being evaluated in the first treated human patients, unless animal models are developed.

The following preliminary ethico-scientific criteria have been proposed for first attempts at genetic therapies (Friedmann and Roblin). (1) There should be adequate biochemical characterization of the prospective patient's biochemical disorder. (2) There should be prior experience with untreated cases of what appears to be the same genetic defect so that the natural history of the disease and the efficacy of alternative therapies can be assessed. (3) There must be adequate characterization of the quality of the exogenous DNA vector used in the genetic therapy attempt. (4) There should be extensive studies in experimental animals to evaluate the therapeutic benefits and adverse side effects of the prospective gene therapy techniques. (5) For some genetic diseases, the patients' skin cells grown in vitro exhibit the genetic disorder. In such cases, it would be possible to determine whether the proposed gene therapy technique could restore the missing function in the skin cells of the prospective patient. This could be done first in vitro, without any of the risks of treating the whole patient. Some side effects, such as chromosome damage and cellular changes suggesting malignancy, could be assessed by such in vitro experiments. Only after a potential gene therapy technique had satisfied all these safety and efficacy criteria would it be considered for use in patients.

RICHARD O. ROBLIN

[While all the articles in this entry are relevant, see especially the articles GENE THERAPY VIA TRANSDUCTION and ETHICAL ISSUES.]

BIBLIOGRAPHY

Burch, John W., and McBride, O. Wesley. "Human Gene Expression in Rodent Cells after Uptake of Isolated Metaphase Chromosomes." *Proceedings of the National Academy of Sciences of the United States of America* 72 (1975): 1797–1801.

Fox, Allen S.; Yoon, Sei Byung; and Gelbart, William M. "DNA-Induced Transformation in *Drosophila*: Genetic Analysis of Transformed Stocks." *Proceedings of the National Academy of Sciences of the United States of America* 68 (1971): 342–346.

Friedmann, Theodore, and Roblin, Richard. "Gene Therapy for Human Genetic Disease?" *Science* 175 (1972): 949–955.

Hotchkiss, Rollin D., and Gabor, Magda. "Bacterial Transformation, with Special Reference to Recombination Process." *Annual Review of Genetics* 4 (1970): 193–224.

Willecke, Klaus, and Ruddle, Frank H. "Transfer of the Human Gene for Hypoxanthine-Guanine Phosphoribosyltransferase via Isolated Human Metaphase Chromosomes into Mouse L-Cells." *Proceedings of the National Academy of Sciences of the United States of America* 72 (1975): 1792–1796.

III
GENE THERAPY VIA TRANSDUCTION

Gene therapy via transduction envisions the future possibility of using viruses or viral DNA as "carriers" for the introduction of new specific, foreign DNA sequences into cells for the purpose of ameliorating human genetic disease. It has already been shown that DNA of a monkey virus called SV40, when added to cultures of human cells growing in vitro, can apparently be integrated into the chromosomal DNA of the human cells (Aaronson and Todaro). Biochemical techniques exist for physically attaching foreign pieces of DNA to the SV40 DNA (Jackson, Symons, and Berg). It seems likely that infection of cells with a foreign SV40 hybrid DNA molecule would result in the integration of the linked foreign DNA and the SV40 DNA into the cellular chromosomal DNA. In this way the SV40 viral DNA would be used as a "carrier" to overcome the barrier to integration of foreign DNA in human cells.

In addition, new genetic techniques, using enzymes which fragment DNA into gene-sized pieces, appear likely to facilitate the isolation of specific human DNA segments (Cohen; Kedes et al.). Once isolated, such human DNA segments could, in theory, then be linked to SV40 DNA to produce a recombinant DNA molecule which might be capable of low-efficiency integration into the DNA of human cells. This could be a quite general approach to facilitating the integration of specific exogenous human DNA sequences into chromosomal DNA of recipient human cells.

A fundamental limitation of currently envisioned approaches to gene therapy via transduction is the presence of virus genetic information in the "carrier" used to facilitate the integration of foreign DNA into the chromosomal DNA of the recipient cell. Expression of virus genes in the recipient cell may be deleterious for the cell and the organism as a whole. For example, infection of human cells with SV40 DNA in vitro causes the infected cells to resemble cancer cells (Aaronson and Todaro). In addition, the newly introduced genetic information, even if correctly translated into the appropriate protein, may not have the desired effect of curing the patient, since the protein might be overproduced or be functionally inadequate in its new cellular environment. Thus, problems of possible deleterious side effects are raised by proposals for gene therapy via transduction, in addition to the questions mentioned previously [see GENE THERAPY, *article on* GENE THERAPY VIA TRANSFORMATION.]

One attempt at gene therapy via transduction has already been made in humans. This situation involved three sisters who have abnormally high levels of the amino acid arginine in their blood, apparently because they suffer from an inherited inability to produce the enzyme arginase, which normally breaks down arginine (Terheggen et al.). All three sisters received injections of the Shope papilloma virus of rabbits, on the theory that the virus DNA contains a gene coding for the synthesis of a virus-specific arginase enzyme that might reduce the high blood arginine levels (Rogers et al.). It is, however, still debatable whether the virus does indeed code for a virus-specific arginase, and this is a crucial point in the scientific rationale for the therapy (Friedmann and Roblin).

This first attempt at gene therapy in man appears to have been unsuccessful but not actively harmful (Brody). It has been criticized as "somewhat premature" (Raivio and Seegmiller) and defended as "justified because of the tragic circumstances involved" (Anderson). This controversy highlights the ethical issues [see GENE THERAPY, *article on* GENE THERAPY VIA TRANSFORMATION] which will continue to surround attempts to alter human genes in order to treat human genetic disease.

Richard O. Roblin

[*While all the articles in this entry are relevant, see especially the articles* GENE THERAPY VIA TRANSFORMATION *and* ETHICAL ISSUES.]

BIBLIOGRAPHY

AARONSON, STUART A., and TODARO, GEORGE J. "Human Diploid Cell Transformation by DNA Extracted from Tumor Virus SV40." *Science* 166 (1969): 390–391.

ANDERSON, W. FRENCH. "Genetic Therapy." *The New Genetics and the Future of Man.* Edited by Michael Pollock Hamilton. Grand Rapids, Mich.: William B. Eerdmans Publishing Co., 1972, pp. 109–124.

BRODY, JANE E. "Genetic Aid Fails in Test on Humans." *New York Times*, 1 March 1975, p. 30.

COHEN, STANLEY N. "The Manipulation of Genes." *Scientific American* 233, no. 1 (1975), pp. 24–33.

FRIEDMANN, THEODORE, and ROBLIN, RICHARD. "Gene Therapy for Human Genetic Disease?" *Science* 175 (1972): 949–955.

JACKSON, DAVID A.; SYMONS, ROBERT H.; and BERG, PAUL. "Biochemical Method for Inserting New Genetic Information into DNA of Simian Virus 40: Circular SV40 DNA Molecules Containing Lambda Phage Genes and the Galactose Operon of *Escherichia coli.*" *Proceedings of the National Academy of Sciences of the United States of America* 69 (1972): 2904–2909.

KEDES, LAWRENCE; CHANG, ANNIE C. Y.; HOUSEMAN, DAVID; and COHEN, STANLEY N. "Isolation of Histone Genes from Unfractionated Sea Urchin DNA by Subculture Cloning in *E. coli.*" *Nature* 255 (1975): 533–538.

RAIVIO, KARI O., and SEEGMILLER, J. EDWIN. "Genetic Diseases of Metabolism." *Annual Review of Biochemistry* 41 (1972): 543–576.

ROGERS, STANFIELD; LOWENTHAL, A.; TERHEGGEN, H. G.; and COLUMBO, J. P. "Induction of Arginase Activity with the Shope Papilloma Virus in Tissue Culture Cells from an Argininemic Patient." *Journal of Experimental Medicine* 137 (1973): 1091–1096.

TERHEGGEN, H. G.; LAVINHA, F.; COLUMBO, J. P.; VAN SANDE, M.; and LOWENTHAL, A. "Familial Hyperargininemia." *Journal de génétique humaine* 20 (1972): 69–84.

IV

CELL FUSION AND HYBRIDIZATION

Fusion of different cells and combination of their genetic material into a single nucleus is a highly efficient method for the transfer of genetic information between cells. This principle is adequately illustrated by the fusion of the male and female gametes during fertilization. However, apart from the gametes and a few other specialized cells in which fusion occurs in response to highly regulated physiological stimuli, *spontaneous* fusion between cells in the body is a rare event, and the resistance to fusion is also displayed by cells when cultured in vitro. However, in the 1960s it was found that certain viruses could be used to induce cells to fuse; and that discovery, together with the realization that highly different cells could be fused together to yield viable hybrid cells, has dictated the emergence of experimentally induced cell fusion and hybridization as a potent investigative technique in many diverse areas of biology and medicine (Harris; Ephrussi; Poste; Davidson and de la Cruz).

The range of cell types and species over which experimentally induced cell fusion and hybridization can be achieved has yet to be fully explored. Even so, viable hybrid cells have already been produced between human cells and cells from a numerous animal species and between cells from equally diverse animals. Recently, entirely new cells have been constructed by fusing nuclei from one cell type into enucleated cytoplasms from another (again not necessarily from the same species) (Ringertz and Bolund). The main conclusion from these studies is that cells from different species are completely compatible when fused together and are capable of remarkable metabolic integration, and they also retain the ability to proliferate for considerable periods. These findings suggest that cells do not in general possess intracellular mechanisms for the recognition and expression of incompatibility similar to those responsible for recognition and rejection of organs, tissues, and intact cells exchanged between different individuals.

Cell fusion and hybridization methods have already made an important contribution to efforts to map the human genome, that is, the assignment of specific genes to individual chromosomes and the analysis of the linear order of genes on single chromosomes. Analysis of human gene expression in hybrid cells offers considerable advantages over the classical methods involving pedigree analysis. The rapid growth rate and short generation times of cultured cells means that information on expression of particular genes can be obtained over a period of a few weeks rather than the decades necessary for comparable pedigree studies in human populations. Furthermore, fusion and hybridization of human cells carrying particular genes of interest can be engineered to obtain information that would be impossible to obtain in human populations other than by controlled matings. The overall stimulus for the analysis of the human gene map is that such knowledge will contribute to a better understanding of the genetic defects associated with chromosome abnormalities and also allow more accurate diagnosis and counseling of those with genetically determined diseases.

In addition to their value as an experimental tool for studying human genetics in vitro, hybrid cells produced by fusion of two or more cell types with defined properties also provide a powerful system for analyzing the general organization of cells and the factors regulating specific cell functions. To date, this widespread use of cell fusion and hybridization techniques has been limited exclusively to cells grown in vitro, and no serious ethical problems have been created by this work. Of greater concern from an ethical standpoint, however, is the possibility of using cell fusion as a vehicle for introducing new genetic information into cells in the body. In practical terms, however, this approach would require a considerably more sophisticated methodology than is currently available. As with any technique of gene transfer it is necessary to ensure that the transferred gene(s) be introduced at the correct chromosomal loci in the recipient cell and that unwanted genetic material be excluded or at least not expressed. Those represent formidable technical problems. Even if methods were available to fulfill the requirements, an important limiting factor still remains in the difficulty of achieving fusion of donor cells carrying the required genetic information with the desired recipient cells. The wide range of cell types found in the tissues of the body dictates that fusion of the introduced donor cells with a particular class of recipient cells would be extraordinarily difficult to achieve. However, cell types that can be isolated temporarily from the body for manipulation and then returned to the host could be subjected to such treatment. The list of cells in this category is at present restricted to blood cells and the male and female gametes. The latter are of interest in the context of this article since cell fusion techniques could be used to modify the genetic composition of gametes, including those from humans.

Of most immediate concern is the possibility of using cell fusion techniques to introduce nuclei into newly fertilized eggs from which the nucleus has been removed. The transfer of nuclei into enucleated eggs can be used to create embryos of defined genetic constitution. Because the embryos developing from such eggs possess a genetic makeup identical to the individual that donated the transplanted nucleus, transfer of nuclei from a single donor into a number of eggs could be used to create genetically identical progeny, that is, a clone. Since the nuclei transplanted into the egg are taken from diploid somatic cells, rather than the haploid germ cells, the ultimate development of these methods would mean that nuclei (either within intact cells or as isolated structures) could be transferred from easily obtained cells such as those in the blood or the skin and then introduced into large numbers of enucleated eggs by fusion to create multiple identical progeny.

Cloning by surgical transfer of nuclei into enucleated eggs has been demonstrated already with frog embryos (Gurdon) but the use of cell fusion techniques to achieve similar transfer of nuclei into enucleated eggs has not yet been successful with either amphibian or mammalian eggs. However, few such cell fusion experiments have been attempted (or at least published) and there is no a priori reason why this type of cell fusion should not be possible. Indeed, in view of the rapid growth in our understanding of the factors that control the ability of cells to fuse, it is considered that the fusion of somatic cells with mammalian gametes could probably be engineered at present with little difficulty if the problem were to receive detailed attention as a combined effort between researchers experienced in handling gametes and others familiar with cell fusion methods. A greater technical limitation in nuclear transfer experiments of this kind may well prove to be the difficulty of prior removal of nuclei from mammalian eggs since these are much smaller than amphibian eggs and thus less amenable to surgical enucleation. In addition, researchers may be reluctant to undertake cell fusion and nuclear transfer experiments using mammalian eggs knowing that the development of a successful experimental protocol for cloning experimental animals could probably be used without too much modification for cloning experiments with humans.

The ethical issues raised by cell fusion are futuristic, in the sense that no work is now possible in humans. However, this technique holds out hope for the following possibilities, which raise value questions: increased knowledge of genetic defects; more accurate genetic diagnosis and counseling; the introduction of new genetic information into body cells (for therapeutic purposes); and the creation of embryos of defined genetic constitution and the production of genetically identical progeny (cloning). Eventually the technique could involve extensive control of genetic health and human reproduction—and a consequent impact on human freedom.

GEORGE POSTE

[*While all the articles in this entry are relevant, see especially the articles* PRODUCTION OF FOUR-PARENT INDIVIDUALS *and* ETHICAL ISSUES. *Also directly related is the entry* REPRODUCTIVE TECH-

NOLOGIES, articles on ASEXUAL HUMAN REPRODUCTION, ETHICAL ISSUES, and LEGAL ASPECTS.]

BIBLIOGRAPHY

DAVIDSON, RICHARD L., and DE LA CRUZ, FELIX F. *Somatic Cell Hydridization*. New York: Raven Press, 1974.

EPHRUSSI, B. *Hybridization of Somatic Cells*. Princeton: Princeton University Press, 1972.

GURDON, J. B. *The Control of Gene Expression in Animal Development*. Cambridge: Harvard University Press, 1974.

HARRIS, H. *Cell Fusion*. Oxford: Clarendon Press, 1970.

POSTE, GEORGE. "Mechanisms of Virus-Induced Cell Fusion." *International Review of Cytology* 33 (1972): 157–252.

RINGERTZ, N. R., and BOLUND, L. "Reactivation of Chick Erythrocyte Nuclei by Somatic Cell Hybridization." *International Review of Experimental Pathology* 13 (1974): 83–116.

V
PRODUCTION OF FOUR-PARENT INDIVIDUALS

Conventionally, each individual has only two natural parents. From the union of the maternal egg and paternal sperm, a single-cell fertilized egg, or zygote, is formed. As it divides, ultimately producing many cells, the genetic material is repeatedly replicated. Thus, despite emergence of diverse cell specializations during development, the cells from one zygote all retain a common biparental hereditary endowment.

With recent advances in methods of handling very young mammalian embryos outside the mother have come ways of modifying the embryos. As a result of some of these new techniques, individuals with four rather than two natural parents became a laboratory reality a decade ago (Mintz, 1965), at first in mice, then in rabbits, rats, and even sheep. To accomplish this, undifferentiated embryos, with only a few cells, are first removed from pregnant females and placed in culture; cells of two genetically unrelated embryos are then brought into contact so as to adhere and form an artificial composite. Though each contributing embryo had its own mother and father and would ordinarily have become a complete animal, the composite becomes organized into a single unified embryo in which cells of disparate genetic composition coexist without fusing (except, during later stages, in skeletal muscle tissue). For further development, the composite is surgically transferred to the uterus of a surrogate mother. Thousands of experimental animals have been produced in this way and have been born fully viable. But they are literally cellular *mosaics*: each tissue can comprise two genetically different kinds of cells, harmoniously integrated. Although the conjoined origin of such individuals recalls the multispecies chimera of Greek mythology (an animal with a goat's head, a lion's body, and a serpent's tail), the parts of the chimera remained agglomerated rather than becoming developmentally integrated into the various organs; hence neither the analogy nor the term chimera is apt.

Mosaic four-parent laboratory animals have provided a great many unique and fruitful ways of analyzing, in the framework of the intact organism, some of the most complex and heretofore obscure events in mammalian development (Mintz, 1974). Moreover, animals produced by conjoining normal embryo cells and embryo cells bearing specific genetic diseases, or heritable disease susceptibilities, have enabled investigators to define the tissue site of the initial lesion or susceptibility in complex disease syndromes (Mintz, Custer, and Donnelly). Some of these diseases, such as blindness or anemias, have been clinically mitigated or averted when normal cells were made to coexist with genetically defective cells or were allowed to replace the latter during development. An even more striking result has recently been obtained by combining normal mouse embryo cells with cells from a certain malignant tumor known as a teratocarcinoma: Once in the normal embryo environment, the cancer cells became normal and developed into normal functional tissues of all kinds (Mintz and Illmensee).

The technology required to create mosaic four-parent individuals varies from species to species, for example, in the composition of the culture medium in which the embryos are handled. A medium suitable for consistently normal development of comparable early-stage human embryos in vitro has not yet been perfected and would necessitate an unpredictable number of trials with embryos from volunteer subjects. Hence, the experimental production of such human mosaics is not imminent. There is, however, a good probability that the appropriate technology could be devised if there were adequate incentive.

Would this, in fact, be a worthwhile objective? One scientist has seriously considered the proposition that it might be (Edwards) on the ground that a genetically defective embryo could be "rescued" or an inherited disease averted by an early infusion of genetically normal cells into a prospectively abnormal embryo. Another scientist has even informally proposed that a novel form of racial "integration" could be achieved through the intermingling of skin cells of different colors within individuals.

Among the moral and ethical questions that might be debated are: What would happen, psychologically and legally, to the rights of the "individual," to the conventional identities of self, sibling, parent, family? Should one, instead, take a more casual view, in effect regarding genetically mosaic individuals as a mere extension of what occurs whenever blood is transfused, a cornea grafted, or a kidney transplanted, other than between identical twins?

If the debate were to focus solely on such relatively abstract questions, it would bypass the fact that—in this matter as in some others in the realm of the "new biology" and "genetic engineering"—there is substantial laboratory and clinical evidence on the dangers, as well as the benefits, of the procedure, and this knowledge should provide an indispensable basis for the evaluation.

Most deleterious genes are recessive, so that the embryos destined to be defective are produced, along with some normal ones, from apparently normal parents who carry the gene. But most genes are not yet expressed at the early stages during which embryo "rescue" by normal cells is still practicable. Therefore, embryos at risk could not be identified. (Because of this problem, the normal cells in experiments in mice must be introduced into all candidate embryos from such parents, in order to rescue the putative defectives. In the handling, some normal embryos inevitably fail to survive.) As a much more desirable alternative, some human defects can be diagnosed *in utero* at a later stage, by amniocentesis, which is relatively harmless and allows the option of aborting a seriously impaired fetus.

The notion of using "normal" human embryo cells for rescue purposes is also fraught with other, unknown, hazards stemming from the fact that—unlike the situation in laboratory mice—we do not have genetically uniform, or inbred, strains of people; nor do we aspire to produce them. Therefore, when embryo cells from two sources are combined, undetected lethal genes may be included in one or both contributors. Moreover, one cell strain may bring along some undesirable disease-causing virus, or may introduce undetected cancerous cells.

One of the more bizarre possible complications stems from the fact that the sex of an individual is chromosomally determined in its cells from the time of fertilization, long before the reproductive tract develops. Thus, genetically female and genetically male cells may be combined by chance and may result in sterility or even hermaphroditism in the mosaic individual. To avoid this, a sample of cells from each donor embryo would first have to be taken and analyzed chromosomally. Another bizarre and even more common complication would result if the two contributing cell strains bore genes for different skin color, even if the differences were slight. Far from achieving an "integrated" blend, we know from observations on mice (Mintz, 1974) that the resultant four-parent individual would in fact be transversely striped as the result of transverse migration of pigment cell clones during development.

In the human clinical literature, there appear to be rare individuals who, judging from their cellular sex chromosome mosaicism, their striping, or other clues, have probably arisen from an accidental conjoining of fraternal-twin embryos. Such individuals have only two parents, albeit two populations of cells of distinctive hereditary composition. But the disparity in cellular genotypes would usually be much greater if completely unrelated embryos from separate pairs of parents were artificially conjoined.

The facts thus furnish a sound basis for the ethical evaluation of such experimentation as being undesirable in humans, even if it is highly useful and informative in laboratory animals. In the human population, medical progress may be expected increasingly to supply more reasonable options of early diagnosis and treatment.

BEATRICE MINTZ

[*While all the articles in this entry are relevant, see especially the article* ETHICAL ISSUES. *For further discussion of topics mentioned in this article, see the entries:* ABORTION; HUMAN EXPERIMENTATION; PRENATAL DIAGNOSIS; REPRODUCTIVE TECHNOLOGIES; *and* RISK.]

BIBLIOGRAPHY

EDWARDS, R. G. "Fertilization of Human Eggs in Vitro: Morals, Ethics and the Law." *Quarterly Review of Biology* 49 (1974): 3–26.

MINTZ, BEATRICE. "Gene Control of Mammalian Differentiation." *Annual Review of Genetics* 8 (1974): 411–470.

———. "Genetic Mosaicism in Adult Mice of Quadriparental Lineage." *Science* 148 (1965): 1232–1233.

———; CUSTER, R. PHILIP; and DONNELLY, ANDREW J. "Genetic Diseases and Developmental Defects Analyzed in Allophenic Mice." *International Review of Experimental Pathology* 10 (1971): 143–179.

———, and ILLMENSEE, KARL. "Normal Genetically

Mosaic Mice Produced from Malignant Teratocarcinoma Cells." *Proceedings of the National Academy of Sciences of the United States of America* 72 (1975): 3585–3589.

VI
ETHICAL ISSUES

Genetic therapy is so new a possibility that there are no established ethical codes to guide it. Ethical discussion of genetics and eugenics is at least as old as Plato, but the traditional debates do not touch on proposals for genetic therapy. Hence the possibilities now opening up, like some others in bioethics, require new thinking. But thinking never starts entirely anew. In this case new questions require attention to some of the fundamental ethical insights of the past in order to see what meaning they may have for present and future thinking. It is also possible that new visions and opportunities may call forth new ethical insights.

A starting point is the almost universal belief that therapy—healing—is an ethical good. Only rarely is this conviction challenged by beliefs that illness is an illusion, not to be treated medically, or is retribution for wrongs committed by a person in this or an earlier incarnation. But even such belief systems normally make some place for therapy. More generally, the art of healing, which is the primary aim of medicine, has also been a major concern of religion and ethics. The Western tradition demonstrates the case in both its major sources: Among the Greeks, the Pythagoreans maintained an intimate relationship among ethics, religion, and health; the Hebrews frequently conjoined salvation and healing.

Thus for most ethics, religious or secular, therapy is a good. To be able to heal a disease and to refuse to do so is a moral wrong. The burden of proof is not on the advocacy of therapy but on its rejection.

But not every proposed therapy is good. Some purported therapies may not be genuine therapies. Other therapies may not be worth the costs or the unintended consequences. Genetic therapy magnifies some ethical issues and raises some of its own.

A comparison helps to define some of the issues at stake. At present, prenatal diagnosis (usually by amniocentesis) followed by abortion of defective fetuses is an established, although fairly recent, practice. It is ethically controversial, but some guidelines are clear. Its advantages are that it enables some parents, who are worried about genetic hazards, to produce infants with the assurance that they will not suffer certain specified liabilities. The procedures are well tested. The risks are definable and in most respects slight. The disadvantage is that abortion does not heal, but destroys the fetus.

Genetic therapy reverses the gains and losses. Its advantage is that its outcome, if it proves effective, is the healing of diseases, not the destruction of fetuses. Its disadvantage is that its potentialities are largely unknown. It is not an established practice. The limited literature about it consists mostly of reports about experimentation on bacteria, plants, insects, more rarely on animals, and still more rarely on human beings. There is no record of demonstrated effectiveness. Conjectures about its promises, its risks, and its threats vary widely.

For such reasons it is important to resist the technological enthusiasm that insists that whatever is possible should be tried. Geneticists, like other people fascinated by new powers, have sometimes assumed that everything possible must be done (Muller, p. 521). An ethical inquiry must assume a human ability to decide that, among many possibilities, some are morally obligatory, some are optional, and some are pernicious. Moral judgment requires not only a general framework of valuation but also detailed investigation of particular issues.

Starting with the assumption that therapy is a good, this article will examine a series of five problems in genetic therapy.

Problem 1: What is a genetic liability?

The purpose of therapy is to heal illness. Sometimes the reality of illness is obvious, definable, and beyond controversy. It is usually easier to identify ailments than to define ideal human types. There is immense disagreement on what an ideal genetic inheritance might be; there is far less disagreement on the undesirability of specifiable illnesses.

But even illness is not self-defining. People and societies define illness. The meaning of the language of illness and health is determined as much socially as medically. This is the most obvious in the case of mental illness and health, where societies often deny freedom to dissenters and the abnormal by hospitalizing them rather than imprisoning them. If we broaden the concept of genetic illness to include genetic liabilities, it becomes still more obvious that some definitions may be arbitrary, socially conditioned, and prejudiced.

Skin color, for example, is genetically determined to a considerable degree. Skin color in many societies has been a basis for segregation, assignment to a social status or even an ineradicable caste, and slavery. Skin color may therefore be a genetic liability in some social situations. The ethical answer, however, is not a genetic therapy to remove the liability of a given skin color; it is a social change that makes the color no longer a liability. Such social changes do take place. Some families have, in a few years, shifted from efforts to lighten skin color to a conviction that "black is beautiful."

Similarly sex has been regarded as a genetically determined liability. A long history of prejudice and even of sophisticated metaphysics has nourished notions of male superiority in many societies. A shorter history of amniocentesis followed by elective abortion has shown a small number of women who wish to abort a fetus of the undesired sex (Etzioni, p. 119). Sex determination by parents may or may not increase, by abortion or by other methods; but it cannot be called genetic therapy except on the absurd assumption that sex is an illness. The issue belongs in this discussion only as one evidence that the definition of genetic liabilities is as much a social as a medical judgment.

Quite apart from human prejudice, the definition and seriousness of human ailments depend partly on social situations. Astigmatism and myopia, which might be life-threatening liabilities in a jungle society, are relatively trivial in scientific–industrial societies. It is, at least at present, far easier to supply people with eyeglasses than to undertake a program of genetic therapy that is exceedingly costly, that involves risky experimentation, and that may never succeed. The example suggests that genetic therapy is, for the foreseeable future, a process of last resort to be undertaken for exceptionally serious illnesses that cannot be effectively treated by other methods.

A different issue arises in the case of longevity. Many diseases shorten human life, and the advance of medicine has brought increases in life expectancy. At present research is under way on the genetic causes of the process of aging. There are serious conjectures about the possibility of discovering the cause and cure of aging, with the result that life expectancy might be greatly increased and death might normally come only from accidents rather than from natural causes. The possibility, although only speculative thus far, raises fundamental ethical questions. Is it good that birth and death are part of the normal rhythm of human life? Would it be better or worse if the human species should become a "permanent" population, with death and birth both rare events?

There is still another way in which the definition of genetic liabilities is problematic. Liabilities may be genetically linked to assets. For example, it appears that the sickle-cell anemia trait is linked to a capacity to resist malaria. Thus the combination of qualities has apparently been an advantage for survival in some situations and has been preserved by natural selection. In a social situation where malaria is not a serious threat, a genetic therapy for sickle-cell anemia appears desirable. It would require a strained argument to maintain that sickle-cell anemia should be deliberately perpetuated because of some possible advantage in a changed social situation that may or may not come some day. But the example suggests that there may be an ethical preference for conventional medical treatment of many ailments rather than for genetic therapy.

The mention of sickle-cell anemia raises a last issue in the definition of genetic liabilities. It is the stigmatization attached to certain illnesses and to identifiable groups in which those illnesses are prevalent. It happens to be the case that sickle-cell anemia affects principally people of African origin. It is also the case that Tay-Sachs disease occurs almost exclusively among people of Ashkenazi Jewish ancestry. Likewise thalassemia major (Cooley's anemia) is most common among people of Mediterranean origin. And cystic fibrosis is most common among people of North European descent. The biomedical fact is that every identifiable group of people has in its gene pool certain problematic characteristics. The social and ethical fact is that human prejudice sometimes adds stigmatization to those problems—perhaps by overemphasizing them in social categorization, by underemphasizing them in medical research and treatment, or by doing both at once. So common a human practice does not of itself mean that genetic therapy is either good or bad. But it offers further evidence of the fact that scientific investigation and medical practice always operate in a social context. That social context is often more determinative of the ethical meaning of medical acts than are the acts themselves in their abstract and unreal isolation.

Problem 2: Risk and consent

If the good of therapy is almost beyond ethical controversy, the methods of genetic therapy are more problematic, because they involve some of the most radical contemporary scientific experiments and projections. More than most undemonstrated medical possibilities, they involve risks.

Earlier speculation on genetic therapy represented it as a form of microsurgery, in which the scientist or surgeon might remove a deleterious gene from a chromosome and replace it with another gene. Biologists sometimes replied that chromosomes are so minute that such procedures are unlikely to become possible soon (Muller in Wolstenholme, p. 255).

Current inquiry focuses on such methods as transformation, transduction, and cell fusion. In all these cases the aim is to treat genetic ailments by introducing into human cells new genetic materials through biochemical processes. The possibilities of such action have been demonstrated in some forms of life, but the processes are exceedingly complicated. They require getting precisely the right genetic material to precisely the right place and integrating it with the defective material in such a way as to remedy the difficulty. The intricacy of the process and the possibility of unintended combinations of genetic material are so great as to raise serious ethical problems.

These problems are related to those encountered in all medical experimentation. An experiment by definition has unpredictable results. Hence experimentation presents an unavoidable ethical dilemma. On the one hand, medical progress depends upon experimentation, which has produced and continues to produce immense gains. On the other hand, experiments always involve some risk of harm to patients.

Several standard practices moderate this dilemma. The first is to test new procedures on nonhuman life before trying them on people. There have been ethical protests against cruelty in experimentation on animals (Roberts, pp. 33–51), and indiscriminate experimentation on animals raises serious ethical problems. However, the dominant ethical conviction (or, at least, the dominant human prejudice) considers experimentation on human beings, without prior testing on animals, to be irresponsible. Even so, the shift from animals to human beings always involves some unpredictability and some risk. Furthermore, some specialists point out that, because there are currently only one or two animal models of recessive genetic diseases, some experimental genetic therapies can be tested only on human patients.

A second practice is to ask the informed consent of the patient. The patient may give such consent in the hope that the benefits of treatment will outweigh the dangers. Such research is usually called beneficial research. Or the patient may volunteer to participate in an experiment not aimed to benefit the patient (nonbeneficial research) for the sake of contributing to the progress of medicine. There are problems in the doctrine of informed consent: Is the patient's information accurate, and is the consent authentic, especially considering the subtle forms of duress that rise out of the patient's dependence on the physician? Despite such problems informed consent remains a fundamental ethical principle in medical research.

Experimentation in genetic therapy has the special problem, as does experimentation on infants, that informed consent is normally impossible. The benefits and hazards of such research, when it crosses the boundary between the nonhuman and the human, bear not upon persons who may give consent but upon genetic material which may develop into persons either helped or harmed by the experiment.

Given the complexity of genetic therapy, human experiments are highly risky. Failures may produce human monstrosities. Even successes may have side effects that are harmful or disastrous. The ethical questions are momentous: What risks are justifiable for the sake of possibly great gains?

It is not surprising that researchers are challenging the old medical principle, "Do no harm." Some are substituting the principle, "Be willing to do some harm that greater good may come." Yet that new principle conflicts with another widely recognized ethical principle, "Never use persons solely as means to the ends of others." If that principle is accepted, researchers may still ask whether DNA or fetal material, which is unquestionably human but not yet fully personal, is entitled to some of the ethical respect due to persons.

A utilitarian ethic can meet such problems with a cost-benefit analysis. It justifies harm to a few—or even a few hundred—persons or potential persons for the sake of discovering therapy that could eventually help thousands. However, the utilitarian ethic and the cost-benefit analysis, which are applied to so many problems in contemporary life, run into difficulties when persons are concerned. A speculative example shows why. A researcher might find no ethical problems in irradiating seeds to produce mutations, with the expectation that most mutations would be undesirable but .01 percent might be beneficial. The cost of damaged seeds might be negligible compared with the value of the knowledge gained or the commercial value of the rare success leading to a desirable new species or variety. The most enthusiastic experimenter would have some ethical reluctance to do the same with human sperm and ova. Human failures cannot be discarded like plants. A sensitivity to human dignity intrudes upon the cost-benefit analysis and the utilitarian ethic.

Obviously there is no human existence without risk, so it may seem totally unreasonable to impose the demand of riskless experimentation upon genetic research. But there is a difference between accepting risk in voluntary informed consent and imposing risk upon persons still to be conceived and born. The warnings against risk are the more impressive because of (1) the high probability of error in such complex experimentation and (2) the very grave nature of the errors that may be expected.

The scientific community has pointed out the risk that experimentation may produce mutant bacteria designed for genetic therapy but actually hazardous to life and subject to no known treatment. Research is directed toward two kinds of precautions: first, the physical containment within laboratories of the bacterial material; second, the development of "safe bugs" that cannot survive and reproduce except under laboratory conditions. For two years scientists observed a self-imposed worldwide moratorium on hazardous experimentation, pending adoption of guidelines for research. In July 1976 the U.S. National Institutes of Health, after a long consultative process, issued guidelines. The British did the same soon after, and scientists in most of the world accepted the restraints. The situation is an example of the way in which ethical responsibility depends upon complex scientific judgments. The guidelines have been criticized from two opposite viewpoints: Some scientists think they are too restrictive and will delay valuable scientific progress; others think they are not stringent enough and may permit grave dangers.

Continuing ethical discussion of such issues may be expected. It is unlikely that unanimity among researchers, ethicists, or the public will come soon. But it can be assumed that the issue of risk will weigh heavily in all efforts at genetic therapy.

Problem 3: Advantages to individuals and to the species

The advantages to the individual and to the human species do not always coincide. Some conflicts are evident even in ordinary medical practice. Modern medicine, by treating successfully some age-old illnesses of genetic origin, enables people to survive and reproduce, when they could not have done so in the past. It thus interferes with the processes of natural selection, which operate cruelly toward individuals but strengthen the human species. The increase in the genetic load upon future generations is a matter of concern to many geneticists.

In debates on this issue physicians and genetic counselors usually do their work with a primary loyalty to the patient or client. Physicians may warn patients of the possibility of procreating defective offspring, but they do not withhold treatment because of the risk that patients may pass on a genetic liability. Genetic counselors may discuss with clients the risks in procreation, but they usually seek to help clients make their own decisions. Geneticists, by contrast, are likely to put greater emphasis on the species. They tend to regard the human race or large populations of the human race as their patients or clients. This is not to divide the two groups into embattled camps (physicians and genetic counselors vs. geneticists); it is simply to distinguish characteristic accents and emphases, with the awareness that ethical insight often comes out of the interplay of various concerns.

Genetic therapy, if it proves successful, may soften the conflicts between individual good and species good. If, for example, therapy moves from the point of healing diabetes to the point of correcting the genetic causes of diabetes, the result benefits individuals and the human race. In this way genetic therapy may become the best of all possible therapies.

But at this point genetic therapy meets another issue, already mentioned (Problem 1). Definitions of ailments are, in part, socially determined. A genetic liability in a particular social situation may not be a liability in all situations, and it may be related to genetic strength in some situations. Thus Robert S. Morison has observed: "It is very hard to identify a bad gene. It is even harder to identify good ones" (in Hoagland and Burhoe, p. 63). And Theodosius Dobzhansky has written: "A load-free mankind may turn out to be a dull stereotype, with no particular physical or mental vigor. . . . No heredity is 'good' regardless of the environment" (Dobzhansky, pp. 411, 412; cf. Glass, p. 63).

If genetic therapy should eliminate some present genetic ailments at the cost of severely narrowing the human gene pool, it may make the human race less capable of meeting future situations in which the physical and social environment may be greatly changed. To ascribe an illusory permanence to existing conditions is a common human error. So there is reason to take seriously warnings against any rash introduction of genetic changes that may be irrevocable. Even so (see Problem 1 above), only a cruel ethic would preserve genetic illness and suffering on the chance that it might someday prove valuable in a now unforeseeable situation. Likewise it is possible that the scientific advances that may make genetic therapy practical may also enable the human race to cope better with unforeseeable futures.

Problem 4: Human manipulation

Discussions of bioethics usually come, sooner or later, to the issue of manipulation of human beings. But precise definition of manipulation and its ethical meaning is difficult.

Modern ethical consideration of manipulation shows a convergence of two lines of thought, so different in most respects that their considerable agreement is remarkable.

The first of these is the heritage of many traditional religions, including notably Judaism and Christianity. Here life is seen as a divine gift, mysterious and wonderful. It is not a human accomplishment. Reverence for the gift of life includes a warning against human *hubris*, which would rashly tamper with a gift so wonderful and so far beyond human origin. People impressed by this gift respond with awe and gratitude. When they consider any acts so radically innovative as genetic therapy, they warn against "playing God."

However, this warning does not translate immediately into specific ethical prohibitions or permissions. The same religious heritage emphasizes that the divine gifts include the human freedom and rationality that make possible the enhancing of life. All civilization—from such acts as the cultivation and cooking of food to brain surgery and genetic practice—involves the use of distinctively human gifts to modify the original gifts of creation. Any medical treatment can be considered a human intervention in the given order of things. One might construct a spectrum of activities, ranging from such modest steps as the improvement of diet and the calculated use of vitamins to such ambitious measures as brain surgery and genetic therapy. All can be called manipulation. Genetic therapy is a far more radical possibility than the production of orange juice so that people in cold climates may ingest Vitamin C, and the ethical cautions surrounding it should therefore be the greater. But there should not be any absolute, a priori prohibition of either.

The second line of thought comes out of modern evolutionary theory and ecology. Curiously, although evolutionists and traditional religionists have sometimes quarreled, they frequently develop comparable attitudes toward human interventions in nature. Barry Commoner has formulated one of the major laws of ecology: "Nature knows best." He does not intend that quite literally and does not, in fact, really argue that nature *knows* anything. But he explains his law to mean, "Any major man-made change in a natural system is likely to be *detrimental* to that system" (Commoner, p. 41). Although he does not specifically apply his "law" to genetic therapy, its relevance is plain. Well-intended interventions in natural processes often have unintended effects, sometimes disastrous.

Again, replies are possible. Human beings, like microbes and beasts, are always modifying their environments. All civilization and all medical practice are human interventions into natural processes. The twentieth century has become more aware than many past centuries of the harm done to natural systems and to human welfare by calculated interventions that do not take into account the remarkable complexity and fragility of the ecosystem. Geneticists will be wise to take account of the wisdom of ecologists. But to refuse any action that might be called intervention or manipulation would not only mean the prohibition of genetic therapy; it would mean the reversal of the whole tradition of scientific therapy.

If there is, therefore, no ethical basis for the a priori rejection of genetic therapy, there are ethical cautions. One such caution urges appreciation of the present genetic heritage of mankind, which is not the result of human manipulation. The same caution warns against likening human reproduction to industrial production with its calculated quality controls. It may be desirable to put inspectors at the end of assembly lines with the duty of detecting faulty products, which can be melted down and reprocessed. Human therapy works differently.

A second caution is against the depreciation of everything imperfect or in some way substandard. That can lead to a loss of compassion. It can also lead to undesirable guilt in the parents of genetically defective children. At this point there is some help in recognizing that everybody is genetically imperfect. Moral wisdom involves the acceptance of some imperfections and the effort to overcome others. If genetic therapy should succeed in overcoming some imperfections, it will nevertheless need to accept some others.

Problem 5: The social context of decision

The possibility of genetic therapy represents a major enhancement of human powers. As in the case of any increase of power, ethical questions arise as to who shall use this power and for what purposes. It is logically possible to separate the ethics of therapy, which is itself a good, from the ethics of particular therapists, who may use it beneficially or unjustly. But programmatically the separation is not so easy. The ethical value of genetic therapy will depend largely upon the social context in which it is used, upon the ethical sensitivity of those with the power to use it, and upon the values for which it is used.

Every generation of mankind tries to perpetuate its values, including its prejudices, through indoctrination of the following generation. Genetic therapy may be seen as a new ability to add genetic indoctrination to the psychological, educational, and social indoctrination already practiced. Even so ambitious and hopeful a geneticist as Herman Muller worried that the proposals he advocated might be badly used. Even while advocating programs of genetic improvement, he worried about their use by a generation with values so corrupt that it might use genetics to produce the wrong qualities (Muller, p. 537).

So long as programs are limited to therapy, the risks of abuse are minimal—even though the very definition of ailments, to repeat a point, is partly social. The problem increases with the realization that therapy need not stop with therapy. That is, the techniques for genetic therapy are identical with the techniques for more ambitious programs of reconstituting the human genetic heritage, producing a super race, or even the breeding of a new race of chimeras (perhaps combining genetic material from people and apes, so as to form subhuman beings to do the world's monotonous work). If it is said that all this is fantasizing, the answer is that the whole effort at genetic therapy at present verges on fantasy, and that much of modern history is the story of fantasies that become realities.

Some geneticists have warned about the problem of use and supervision of genetic accomplishments (Roberts, p. 24; Dobzhansky, p. 413; Morison in Hoagland and Burhoe, p. 41). Herman Kahn, not a geneticist but a "futurologist" much interested in technical solutions to human problems, has gone so far as to propose an "index of forbidden knowledge," including some knowledge of genetics. "Genetic engineering has in it the makings of a totalitarianism the like of which the world has never seen" (Kahn, p. 24). His proposal, of course, simply relocates the problem, transferring power from the practitioners of genetics to the political authorities who draw up and enforce the prohibitions upon knowledge.

All these examples reinforce the point that the ethical issues are due not simply to the nature of genetics but to the uses to which it can be put; and those uses depend very largely upon the social context of research and practice. The answer may include some cautions about genetics; it obviously includes some warnings about social processes that have no necessary relation to genetics.

Another issue relating to the social context of decision has to do with priorities. Is it more important, at this point in history, to seek to overcome abnormally low intelligence by genetic devices or by relieving protein deficiencies in the diets of millions of children? The scientific answer may be that both are important and that there is no conflict between them. But the social context may mean that, in a world of limited resources, choices are necessary—choices made by the governmental agencies or foundations that fund research and therapy.

In international discussions of genetics, participants from the "third world" frequently think of research in genetics as a luxury possible only for rich societies and irrelevant to most of the world (Paul Verghese, in Birch and Abrecht, p. 39). They may resent expenditures in this area, when the same funds might be used to relieve immediate suffering. It may be answered that genetic research might result in elimination of diseases that are now an immense economic burden on mankind, especially on the poor societies.

The issues of the social context of decision entail many ethical problems. They are not confined to bioethics or to the ethics of genetic therapy. Yet they deserve mention here because they may prove to be the most important of all ethical issues in this as in other areas.

Conclusions

The foregoing discussion has proceeded on the basis that therapy is a good, but that all therapies involve a variety of ethical issues related to the definitions of illness, the factors of risk and consent, the relative advantage of the individual and the species, the hazards of human manipulation, and the social context of medical practice and research. Such factors pose ethical warnings.

Because therapy is a good, there can be no a priori ethical judgment against it. To the contrary, the burden of proof rests on those who resist it. But because of the ethical warnings it is important to proceed with restraint and sensitivity on so innovative a possibility as genetic therapy.

It is particularly important to resist the temptation to think that everything possible is desirable. Scientific investigators, therapists, and society at large will be wise to face consciously the ethical issues in contemporary decisions, rather than let the ethical questions be settled by default.

ROGER L. SHINN

[Directly related are the entries GENETIC ASPECTS OF HUMAN BEHAVIOR; GENETIC CONSTITUTION AND ENVIRONMENTAL CONDITIONING; GENETIC DIAGNOSIS AND COUNSELING; GENETIC SCREENING; and GENETICS AND THE LAW. Other relevant material may be found under HEALTH AND DISEASE; INFORMED CONSENT IN HUMAN RESEARCH; and PURPOSE IN THE UNIVERSE. See also: HUMAN EXPERIMENTATION, articles on BASIC ISSUES and PHILOSOPHICAL ASPECTS; and RISK.]

BIBLIOGRAPHY

BIRCH, CHARLES, and ABRECHT, PAUL, eds. *Genetics and the Quality of Life.* Elmsford, N.Y.: Pergamon Press, 1976.
COMMONER, BARRY. *The Closing Circle: Nature, Man, and Technology.* New York: Alfred A. Knopf, 1971.
DOBZHANSKY, THEODOSIUS. "Changing Man." *Science* 155 (1967): 409–415, especially 411–413.
EDWARDS, R. G., and FOWLER, RUTH E. "Human Embryos in the Laboratory." *Scientific American* 223, no. 6 (1970), pp. 44–54, especially pp. 44–45.
ETZIONI, AMITAI. *Genetic Fix.* New York: Macmillan Publishing Co., 1973.
FRIEDMANN, THEODORE, and ROBLIN, RICHARD. "Gene Therapy for Human Genetic Disease?" *Science* 175 (1972): 949–955.
GLASS, HIRAM BENTLEY. *Science and Ethical Values.* Chapel Hill: University of North Carolina Press, 1965.
HAMILTON, MICHAEL POLLACK, ed. *The New Genetics and the Future of Man.* Grand Rapids, Mich.: William B. Eerdmans, 1972.
HARRIS, MAUREEN, ed. *Early Diagnosis of Human Genetic Defects: Scientific and Ethical Consideration.* Fogarty International Center Proceedings, no. 6. H.E.W. Publication, no. (NIH) 72-25. Bethesda, Md.: National Institutes of Health; Washington: Government Printing Office, 1971.
HILTON, BRUCE; CALLAHAN, D.; HARRIS, M.; CONDLIFFE, P.; and BERKLEY, B., eds. *Ethical Issues in Human Genetics: Genetic Counseling and the Use of Genetic Knowledge.* Fogarty International Center Proceedings, no. 13. New York: Plenum Press, 1973.
HOAGLAND, HUDSON, and BURHOE, RALPH W., eds. *Evolution and Man's Progress.* New York: Columbia Press, 1962. First published in *Daedalus* (Summer 1961).
KAHN, HERMAN, interviewed by URBAN, G. R. "Herman Kahn Thinks about the Thinkable: 'Most of the Traditional Causes of War Have Disappeared'." *New York Times Magazine,* 20 June 1971, p. 12 ff., especially p. 24.
KASS, LEON R. "Making Babies—The New Biology and the 'Old' Morality." *Public Interest,* no. 26, Winter 1972, pp. 18–56.
LAPPÉ, MARC, and MORISON, ROBERT F., eds. *Ethical and Scientific Issues Posed by Human Uses of Molecular Genetics. Annals of the New York Academy of Sciences* 265. New York: 1976.
LIPKIN, MACK, JR., and ROWLEY, PETER T., eds. *Genetic Responsibility: On Choosing Our Children's Genes.* New York: Plenum Press, 1974.
MULLER, HERMANN J. "What Genetic Course Will Man Steer?" *Proceedings of the Third International Congress of Human Genetics: The University of Chicago, Chicago, Illinois, U.S.A., September 5–10, 1966.* Edited by James Franklin Crow and James V. Neel. Baltimore: Johns Hopkins Press, 1967, pp. 521–543.
ROBERTS, CATHERINE. *The Scientific Conscience: Reflections on the Modern Biologist and Humanism.* New York: George Braziller, 1967.
ROSLANSKY, JOHN, ed. *Genetics and the Future of Man: A Discussion.* 1st Nobel Conference, Gustavus Adolphus College, 1965. New York: Appleton-Century-Crofts, 1966.
WOLSTENHOLME, GORDON E. W., ed. *Man and His Future: A Ciba Foundation Volume.* 1st American ed. Boston: Little, Brown & Co., 1963.

GENETIC ASPECTS OF HUMAN BEHAVIOR

The articles in this entry, unlike the articles in some of the other composite entries, were planned as an interdependent unit. The first article, STATE OF THE ART, *provides an introduction to the concepts of behavioral genetics that are applicable to bioethics. Each of the following three articles—*MALES WITH SEX CHROMOSOME ABNORMALITIES (XYY AND XXY GENOTYPES), RACE DIFFERENCES IN INTELLIGENCE, *and* GENETICS AND MENTAL DISORDERS*—discusses the role that genetic factors play in those behavioral areas and introduces some ethical issues involved in the use and control of the results of research in each of these areas. The fifth article,* PHILOSOPHICAL AND ETHICAL ISSUES, *brings together the major conceptual issues and ethical principles that apply to the preceding four articles and analyzes such questions as the relationship between de-*

terministic scientific theories and individuality and the extent to which genetic scientists are accountable for unforeseen consequences of their research.

   I. STATE OF THE ART   *Irving I. Gottesman*
  II. MALES WITH SEX CHROMOSOME
      ABNORMALITIES (XYY AND XXY GENOTYPES)
                         Ernest B. Hook
 III. RACE DIFFERENCES IN INTELLIGENCE
                         John C. Loehlin
 IV. GENETICS AND MENTAL DISORDERS
                        Irving I. Gottesman
  V. PHILOSOPHICAL AND ETHICAL ISSUES
                         Arthur L. Caplan

I
STATE OF THE ART

The purpose of this article is to provide an introduction to the domain and concepts of behavioral genetics in a context that may be useful to bioethics. The ethical issues embedded in the field, which will be discussed in this and the following articles, include the modifiability of behavior with important genetic underpinnings, the equating of determinism with heredity, race differences in intelligence and alleged racial inferiority, chromosome abnormalities and antisocial behavior and responsibility, screening for, and labeling of, deviant behavior potentials on genetic grounds, considerations of the rights to marriage and procreation of the mentally disordered and their relatives, and an overview of philosophical-ethical issues related to human behavioral genetics.

The field of behavioral genetics. Behavioral genetics will be defined below; it could as well have been called the genetic aspects of human psychological traits to avoid the awkwardness of having to define "behavior," with the term genetic being understood to refer to genes, the units of heredity, rather than to development (as in ontogenetic). The aspects of behavior that lend themselves to measurement and analyses as "units" are called traits and are assumed to be biophysical "somethings." Traits acquire their names by three implicit steps: (1) We observe that people differ in what they do and in how they do it, infer a quality common to their actions drawing on the richness of ordinary language, and attach an adverb (X behaves aggressively). (2) The quality is applied to the actor instead of just to his behavior by means of an adjective (X is aggressive). (3) Finally, we abstract the property as a thing and give it a noun form (X has the trait of aggression). Traits are inferred for human behavior, and it is useful to construe them in a hierarchical organization, with narrowly defined traits (e.g., aggression) being clustered to form higher-order traits (e.g., authoritarianism).

Social, political, and ethical conclusions will be drawn from the data and theories generated by behavioral geneticists, with or without their consent. The virginity and neutrality of data are beyond recall once they are married to the printed page. Too often *misapplied* human genetics is the result. Behavioral genetics of human characteristics is controversial because it implies intentional tampering with the gene pool of our species—acts directed toward changing, in one fashion or another, that which uniquely specifies Homo sapiens, our mental behavior.

The battle that has raged for the past hundred years over the nature and origin of the behavioral differences observed among human beings was known as the "nature–nurture" controversy or as the "heredity–environment" debate. Most of the arguments have coalesced over time as part of the field now known as behavioral genetics (Fuller and Thompson; Hirsch; McClearn and DeFries; Ehrman and Parsons). The field of human behavioral genetics can be defined approximately as the interface between differential psychology (Anastasi; Tyler) and human genetics (Cavalli-Sforza and Bodmer; Lerner and Libby; Stern).

Complexity and uncertainty are inherent in the domain of behavioral genetics since we almost always deal with weak inferences about the chain of events in hypothesized gene-to-behavior pathways; the biology of complex behaviors is not known. Such necessary ambiguities lend themselves to exploitation by both legitimate and illegitimate contenders for recognition as experts; the consumer must therefore be cautioned about the politicization and simplification of the data base for behavioral genetics. Confusion in the public arena is furthered by pronouncements of specialists in compartments of psychological or genetic knowledge other than behavioral genetics. It is too easy to forget that eminence and expertise in one specialty do not transfer to other specialties. There does not appear to be a conspiracy to suppress "forbidden knowledge" about man's nature. However, there is obstinacy in recognizing the technological barriers that limit the so-called genetic comparisons between populations of our species or between two persons who differ in the measured level of some trait of interest (Thoday). Dobzhansky has forcefully made the point that the dogma of tabula rasa and the dogma of genetic predestination in regard to human characteristics both deserve to be labeled pejoratively as myths (1976).

Historical development of behavioral genetics. A brief intellectual pedigree of the field of behavioral genetics could provide a perspective on its merits and flaws. Both differential psychology and eugenics were launched by Francis Galton, an English gentleman-scholar of independent means and liberal thought, with the publication of his book *Hereditary Genius* in 1869. Galton was very much influenced by Charles Darwin, his second cousin and contemporary. The ideas of Darwin's *Origin of Species* (1859) with its emphases on chance variation, pre-Mendelian inheritance of somatic-physical characteristics, and the principles of natural selection were quickly assimilated by Galton and applied to the transmission of mental, especially intellectual, traits. Galton's scientific ideas provided the foundation for today's behavioral genetics and must be seen in their historical context in regard to the social, economic, and political conditions of Victorian Britain (Buss). Social Darwinism (the misapplication of the principle of natural selection—survival of the fittest—to social and economic institutions), however, cannot be blamed on either Darwin or Galton, but rather on Herbert Spencer and his disciples. Galton went astray when his enthusiasm for perfecting the human species led him to make a total ideology out of eugenics (Blacker); it was not enough for it to be a branch of science, it had to be both a religion and a basis for social policy. It should be noted that the eugenics societies of both the United Kingdom and the United States have, in their efforts to salvage the humane core of eugenics and to protect it from abuse, changed the names of their journals from *Eugenics Review* and *Eugenics Quarterly* to *Journal of Biosocial Science* and *Social Biology*, respectively.

At the turn of this century, Mendel's ideas about the genetic basis for the transmission of physical resemblances between parents and offspring were rediscovered. The simple beauty and explanatory power of the Mendelian laws fused with the Darwinian theory of organic evolution, Galtonian ideas about the perfectibility of the species, and admiration for the natural sciences' contributions to problem solving. With further additions from the growing awareness of individual differences and of class stratification in urban-industrialized society, and from the meteoric rise of the mental testing movement in education, the key elements were in place for what was to become human behavioral genetics. Scientists eagerly embraced, applied, and misapplied these ideas to such ills of society as feeblemindedness, insanity, pauperism, and alcoholism; contemporary research on the same problems (in modern clothing) must struggle with both the stigma of past distortions and the complexities of the issues involved in making advances.

Genetic aspect. There is no denying the genetic individuality of the four billion or so humans now living on this planet, since there are some seventy trillion potential human genotypes. The Declaration of Independence was too succinct with its phrase "all men are created equal." "Equal" in that context continues to mean deserving equal protection of the law and the right to equal opportunity for developing talents to the fullest extent compatible with the rights of others. Jefferson could not have meant that the capacities of each human were identical; the distinction between equality and identity permits a science of human behavioral genetics to call itself moral—equality is accepted, identity is rejected.

The bibliographies to this and the following articles adequately document the masses of empirical data that have been gathered about the environmental and genetic factors alleged to contribute to the observed variation in human psychological characteristics. Such characteristics or traits do not lend themselves readily to classical Mendelian genetic research because, unlike such traits as flower color or pea smoothness, behavior is continuous (rather than discrete), complex, and fluid (time and experience dependent); hence, the definition of units of behavior is difficult if not arbitrary. A few rare behavioral traits, such as Huntington's disease and some two hundred kinds of mental retardation, do lend themselves to simple Mendelian analyses in terms of major gene effects. However, the vast majority of behaviorally relevant traits, if they have an appreciable genetic component to their variation, must be analyzed by the methods of quantitative genetics. The variation is assumed to arise from the simultaneous occurrence of many small, discontinuous polygenic effects that result in smooth, bell-shaped distributions of such traits as height, blood pressure, and IQ test scores when combined with environmental sources of variation. The partitioning of the sources of variation in biometrical behavioral genetics is a means to an end: the elucidation of the molecular bases of trait variation and of how the genotype interacts with environment to produce a trait. The variation referred to in the foregoing is variation in the phenotype—the observable, measured characteristics of a human being—as opposed to the genotype, which is inferred to underlie the trait but many, many steps removed from it at the other end of the gene-to-behavior pathway. A useful way to conceptualize the contribution of heredity to a trait is in terms of heredity's determining a reaction range. Within this framework a genotype determines an indefinite but nonetheless circumscribed assortment of phenotypes, each a function of one of the possible environmental arrays (complexes) to which the genotype may be exposed. It helps to remember that different phenotypes may have the same genotype, and different genotypes may have the same phenotype.

Up to the present, behavioral genetics has been oriented toward structural gene (those which produce proteins) effects; an entirely new world of complexity will open when provision is made for the advances in molecular biology that incorporate the concepts of gene regulation by other kinds of genes (regulators that turn structural genes on and off) and by cybernetic elements in the cytoplasm and membrane of cells that permit quantitative variation in gene products so as to meet the changing needs of the organism.

Conclusion. A specific set of ethical considerations for the field of behavioral genetics has yet to be formulated. The principles enunciated for psychology, human genetics, and medicine will have to be reworked and restated. Few completely unique principles seem to be needed. The professional-ethical issues of doing research in this area are somewhat different from their parent disciplines, as pointed out in the following articles, dealing with males with sex chromosome abnormalities (XYY and XXY genotypes), race differences in intelligence, and genetics and mental disorders. An overview and synthesis of philosophical and ethical issues is offered in the final article in this entry.

IRVING I. GOTTESMAN

[*While all the articles in this entry are relevant, see especially the article* PHILOSOPHICAL AND ETHICAL ISSUES. *For further discussion of topics mentioned in this article, see the entries:* EUGENICS; EUPHENICS; *and* EVOLUTION. *Other relevant material may be found under* MENTALLY HANDICAPPED; *and* POPULATION POLICY PROPOSALS, *article on* GENETIC IMPLICATIONS OF POPULATION CONTROL.]

BIBLIOGRAPHY

ANASTASI, ANNE. *Differential Psychology: Individual and Group Differences in Behavior.* 3d ed. New York: Macmillan Co., 1958.

BLACKER, CHARLES P. *Eugenics: Galton and After.* London: Duckworth, 1952.

BUSS, ALLAN R. "Galton and the Birth of Differential Psychology and Eugenics: Social, Political, and Economic Forces." *Journal of the History of the Behavioral Sciences* 12 (1976): 47–58.

CAVALLI-SFORZA, LUIGI L., and BODMER, W. F. *The Genetics of Human Populations.* A Series of Books in Biology. San Francisco: W. H. Freeman & Co., 1971.

DOBZHANSKY, THEODOSIUS. "The Myths of Genetic Predestination and of Tabula Rasa." *Perspectives in Biology and Medicine* 19 (1976): 156–170.

EHRMAN, LEE, and PARSONS, PETER A. *The Genetics of Behavior.* Sunderland, Mass.: Sinauer Associates, 1976.

———; OMENN, GILBERT S.; and CASPARI, ERNST, eds. *Genetics, Environment, and Behavior: Implications for Educational Policy.* New York: Academic Press, 1972.

FELDMAN, M. W., and LEWONTIN, R. C. "The Heritability Hang-up." *Science* 190 (1975): 1163–1168.

FIEVE, RONALD R.; ROSENTHAL, DAVID; and BRILL, HENRY, eds. *Genetic Research in Psychiatry.* Proceedings of the Sixty-Third Annual Meeting of the American Psychopathological Association. Baltimore: Johns Hopkins University Press, 1975.

FULLER, JOHN L., and THOMPSON, W. ROBERT. *Behavior Genetics.* New York: Wiley, 1960.

GOTTESMAN, IRVING I. "Biogenetics of Race and Class." *Social Class, Race, and Psychological Development.* Edited by Martin Deutsch, Irwin Katz, and Arthur R. Jensen. New York: Holt, Rinehart & Winston, 1968, chap. 1, pp. 11–51.

———. "Developmental Genetics and Ontogenetic Psychology: Overdue Detente and Propositions from a Matchmaker." *Minnesota Symposia on Child Psychology.* Edited by Anne D. Pick. Minneapolis: University of Minnesota Press, 1974, vol. 8, pp. 55–80.

———, and ERLENMEYER-KIMLING, L. "Prologue: A Foundation for Informed Eugenics." *Social Biology* 18, supp. 1 (1971), pp. S1–S8.

HIRSCH, JERRY, ed. *Behavior-Genetic Analysis.* McGraw-Hill Series in Psychology. New York: McGraw-Hill, 1967.

LERNER, ISADORE MICHAEL, and LIBBY, WILLIAM J. *Heredity, Evolution, and Society.* 2d ed. San Francisco: W. H. Freeman & Co., 1976.

LOEHLIN, JOHN C., and NICHOLS, ROBERT C. *Heredity, Environment, and Personality: A Study of 850 Sets of Twins.* Austin: University of Texas Press, 1976.

MCCLEARN, G. E., and DEFRIES, J. C. *Introduction to Behavioral Genetics.* A Series of Books in Psychology. San Francisco: W. H. Freeman & Co., 1973.

SHIELDS, JAMES, and GOTTESMAN, IRVING I., eds. *Man, Mind, and Heredity: Selected Papers of Eliot Slater on Psychiatry and Genetics.* Baltimore: Johns Hopkins Press, 1971.

STERN, CURT. *Principles of Human Genetics.* 3d ed. A Series of Books in Biology. San Francisco: W. H. Freeman & Co., 1973.

THODAY, J. M. "Limitations to Genetic Comparison of Populations." *Journal of Biosocial Science,* supp. 1

(1969), pp. 3–14. Biosocial Aspects of Race: Fifth Symposium of the Eugenics Society.
TYLER, LEONA E. *The Psychology of Human Differences.* 3d ed. The Century Psychology Series. New York: Appleton-Century-Crofts, 1965.
URBACH, PETER. "Progress and Degeneration in the 'IQ Debate' (I)." *British Journal for the Philosophy of Science* 25 (1974): 99–135. Pt. 2, pp. 235–259.
WITKIN, HERMAN A., et al. "Criminality in XYY and XXY Men." *Science* 193 (1976): 547–555.

II
MALES WITH SEX CHROMOSOME ABNORMALITIES (XYY AND XXY GENOTYPES)

The reported association of criminality with specific chromosome genotypes in males is clearly pertinent to the field of bioethics.

The evidence for this association, first documented in 1965 (Jacobs, Brunton, and Melville) comes from observations that men with XYY, XXY, or other genotypes involving extra sex chromosomes are found in prisons and other security settings significantly more often than expected by chance. In studies reported by 1973 the rates of the XYY and XXY genotypes in newborn males and samples of older "normal populations" were about 0.1 percent for each, whereas the rates in security settings for those with some mental disorder or retardation (mental-penal settings) were 2 percent for XYY and about 1 percent for XXY, i.e., twentyfold and ten times greater than background respectively (Hook, 1973). The rates in exclusively penal settings were 0.4 percent and 0.3 percent for the XYY and XXY genotypes respectively, i.e., four and three times greater than expected respectively. The rate of increase of the much rarer XXYY genotype over background in security settings is about one order of magnitude greater than that for the XXY genotype.

The precise reasons for these increases over the background rates are still uncertain, however. Moreover, many adult males with these genotypes apparently have not come into conflict with the law. It appears likely that adverse socioeconomic factors contribute to the institutionalization of males with extra sex chromosomes (as they do for XY males) but there is no evidence that such adverse factors are exclusively responsible for the differential in "deviance." Certainly the XYY genotype is not reported more frequently in male infants born to parents of lower socioeconomic background.

An early popularized view that XYY men in prison were particularly likely to be aggressive and violent offenders (which was given unfortunate emphasis by false press reports that the mass murderer Richard Speck was an XYY) has not been borne out by comparison of the crimes of XYY and XY prisoners in the same settings.

There have been several studies of males with sex chromosome abnormalities detected in cross-sectional studies of adult populations (Noel et al.; Zeuthen et al.; Witkin et al.). The most extensive of them revealed an increased history of "criminality" in XYYs compared to XYs detected in the same community sample which remained after adjustment for parental socioeconomic status, educational index of the proband, height, and score on a rough test of cognitive function (Witkin et al.). While there was evidence for a tendency to intellectual dysfunction in XYYs, the relationship of this to the association with "criminality," which involved trivial offenses in many cases, is still obscure and under further investigation.

Specific issues in bioethics pertinent to questions posed by these genotypes are those related to "informed consent" for studies of individuals, confidentiality of diagnoses, "complete disclosure" of diagnostic information, and screening for genetic disorders. (See more extensive discussion of such themes under the appropriate headings in this encyclopedia.) A more general issue concerns the social implications of discoveries in human behavior genetics and inferences that may be drawn from data in this field.

Specific ethical questions arise in counseling, because an association of the genotypes with deviance is not inevitable, and the precise nature and extent of associated behavioral difficulties are still undefined. Many workers believe that, once the diagnosis is made, complete disclosure of the findings is necessary, as well as discussion of what is known and not known concerning the genotype. Four clinical situations may be distinguished for consideration of the effects such a course may have upon the individuals involved.

1. *When the diagnosis is made prenatally, in cells cultured from the amniotic fluid of a woman in early pregnancy who might wish to abort her fetus (usually one concerned about having an infant with trisomy 21—mongolism or Down's syndrome).* In the United States the decision to terminate a pregnancy is, legally, that of the woman involved, and there appear to be no grounds for withholding knowledge of any genotype that may have significant implications for her decision.

2. *When the diagnosis is made in an infant or*

very young child. Discussing the genotype with the parents may induce anxieties about the prognosis. Critics of one study contend that the parents' reaction to the knowledge of the diagnosis will lead to the type of deviant behavior feared in such children (i.e., it will result in a "self-fulfilling prophecy"). Others believe that the opposite effect will result, as parents will be more likely to seek both definitive evaluation of their child and appropriate intervention for such difficulties as may emerge. Studies on newborn infants to date have reported some anxieties in parents who have been informed of the diagnosis, but no adverse behavioral consequences have been correlated with these per se. (See Beckwith and King; Hook, 1976, for further discussion and references.)

3. *When the diagnosis is made in an older child or an adult in whom there are no apparent behavioral problems.* There may be no apparent clinical relevance of the genotype, at least to behavior. In this case, knowledge of the genotype will probably give rise to less concern than in the situation described above.

4. *When the diagnosis is made in an older child or adult in whom there are significant behavioral problems or psychopathology.* Discussion of the diagnosis in this situation inevitably raises the question of the relationship of the genotype to the observed behavior. Parents with guilt feelings about their role in contributing to their child's difficulties may find knowledge of the genotype important because they may be reassured by learning there is another hypothesis to account for the problem. (One hopes that such knowledge will not lead to a fatalistic attitude or diminish attempts to provide appropriate intervention and amelioration of contributory environmental difficulties.) Affected individuals accused of crimes or those already in security settings may wish to introduce such information into their defense, although the likelihood of its being useful to them in this context is moot. Despite several attempts, the existence of an abnormal sex chromosome complement has not been recognized as evidence pertinent to an "insanity" defense in the United States.

In only one instance (in Australia) does it appear even to have been considered by a court in reaching a verdict of not guilty by virtue of insanity, and the decision was based primarily on other evidence (Bartholomew and Sutherland). This, of course, does not preclude the possibility that the presence of the genotype may be admissible evidence in some jurisdictions in the United States in the future.

As any study of the chromosomes of a male individual for clinical reasons, e.g., a search for autosomal abnormality such as mongolism, may result in discovery of an extra Y chromosome, some of the circumstances discussed above may arise occasionally in the practice of any clinical geneticist.

The greatest controversy to date, however, has concerned a systematic screening program for the XYY genotype in newborn populations in the United States (see Culliton, 1974, 1975; Beckwith and King; Hook, 1976). The main criticisms are those summarized above in situation 2. Since individuals studied in such a screening program are almost all ostensibly normal at birth, the contention is that revealing diagnostic information to the parents concerning a genetic condition for which there is no specific therapy (unlike, for example, phenylketonuria) is an unwarranted intrusion, which will have adverse consequences upon the affected infant. Some newborn screening programs in Europe avoid this criticism by not informing parents of diagnoses of sex chromosome abnormalities in male newborns (Nielsen), but this is contrary to the precept, generally accepted in the United States, of "complete disclosure" of diagnostic information. It has been claimed that because of these factors "informed consent" for participation in such a screening study in the United States is impossible to obtain. The main disagreement between opponents and defenders of such a study concerns the probability of adverse or beneficial outcome to those actually diagnosed as newborns.

Another disagreement concerns the possibility of social harm to others from such a screening study for newborns. The critics contend that such investigations and, by implication, other studies of the XYY genotype or genetic aspects of misbehavior tend to emphasize in the public mind the possibility of genetic contributions to deviance and diminish the force of arguments for ameliorating social deprivation. Others have countered that such studies contribute to our understanding of how environmental intervention may ultimately be made more specific to human needs. (Of course, similar differences in viewpoint may arise concerning genetic studies of aspects of human behavior other than deviance.) It seems unlikely that complete agreement on these issues will ever be reached, since fundamentally different concepts of social priorities, the uses of science, and the implications of discoveries in human genetics are involved.

ERNEST B. HOOK

[*While all the articles in this entry are relevant, see especially the article* PHILOSOPHICAL AND ETHICAL ISSUES. *For further discussion of topics mentioned in this article, see the entries:* CONFIDENTIALITY; INFORMED CONSENT IN HUMAN RESEARCH; *and* TRUTH-TELLING, *article on* ETHICAL ASPECTS. *Directly related are* GENETIC SCREENING; *and* GENETIC DIAGNOSIS AND COUNSELING, *article on* GENETIC COUNSELING. *For more on the insanity defense, see* MENTAL ILLNESS, *article on* LABELING IN MENTAL ILLNESS: LEGAL ASPECTS.]

BIBLIOGRAPHY

BARTHOLOMEW, ALLEN A., and SUTHERLAND, G. "A Defense of Insanity and the Extra Y Chromosome: R v. Hannell." *Australian and New Zealand Journal of Criminology* 2 (1969): 29–37.
BECKWITH, JON, and KING, JONATHAN. "The XYY Syndrome: A Dangerous Myth." *New Scientist* 64 (1974): 474–476.
CULLITON, BARBARA J. "Patient's Rights: Harvard Is Site of Battle over X and Y Chromosomes." *Science* 186 (1974): 715–717.
———. "XYY: Harvard Researcher under Fire Stops Newborn Screening." *Science* 188 (1975): 1284–1285.
HOOK, ERNEST B. "Behavioral Implications of the Human XYY Genotype." *Science* 179 (1973): 139–150.
———. "Extra Sex Chromosomes and Human Behavior: The Nature of the Evidence Regarding XYY, XXY, XXXY, and XXX Genotypes." *Genetic Mechanisms of Sexual Development.* Edited by H. L. Vallet and I. H. Porter. New York: Academic Press, forthcoming in 1978. Summarizes all data available by 31 December 1976.
———. "Geneticophobia and the Implications of Screening for the XYY Genotype in Newborn Infants." *Genetics and the Law.* Edited by Aubrey Milunsky and George J. Annas. New York: Plenum Press, 1976, pp. 73–86. A discussion of this and another paper follows on pp. 87–101.
JACOBS, PATRICIA A.; BRUNTON, MURIEL; and MELVILLE, MARIE M. "Aggressive Behavior, Mental Sub-normality and the XYY Male." *Nature* 208 (1965): 1351–1352.
NIELSEN, JOHANNES. "Chromosome Examination of Newborn Children." *Humangenetik* 26 (1975): 215–222.
NOEL, B.; DUPORT, J. P.; REVIL, D.; DUSSUYER, I.; and QUACK, B. "The XYY Syndrome: Reality or Myth?" *Clinical Genetics* 5 (1974): 387–394.
WITKIN, HERMAN A., et al. "Criminality in XYY and XXY Men." *Science* 193 (1976): 547–555.
ZEUTHEN, E.; HANSEN, M.; CHRISTENSEN, A.-L.; and NIELSEN, J. "A Psychiatric-Psychological Study of XYY Males Found in a General Male Population." *Acta Psychiatrica Scandinavica* 51 (1975): 3–18.

III

RACE DIFFERENCES IN INTELLIGENCE

When the individuals constituting subgroups within a society show average differences in some trait with which social rewards are associated, an ethical question arises: Is this differential reward just? Subsidiary to this central question others arise: What are the origins of the trait differences? Can these differences be modified—and if so, by what means and at what cost? What are the historical and economic bases of the association with differential rewards? Can and should this association be eliminated?

All of the above questions have been raised about observed differences in intellectual achievements among groups with different racial or ethnic origins within modern societies. The nature of the discussion, and some of the social and ethical issues arising, are illustrated by the continuing controversy concerning the average difference in intelligence test performance between U.S. blacks and whites.

Elements of the question

The U.S. black and white populations need be only loosely defined for the purposes of this discussion. U.S. whites are a population with largely European ancestry. U.S. blacks are a population with predominantly African ancestry but a substantial European admixture—estimates in the range of ten to thirty percent have most often been obtained, based on blood groups (Reed). Most of the African ancestors were brought to America as slaves in the seventeenth and eighteenth centuries. Virtually all of the evidence on differences in the intellectual performance of U.S. blacks and whites has been based on social rather than biological definitions of these groups. Thus it will occasionally be the case that an individual who is identified as "black" will actually have less African ancestry than another who is identified as "white." Nevertheless, it is clear that on the average, when representative groups of U.S. individuals identified as "blacks" and "whites" are compared, they can be expected to differ substantially on biological variables associated with different ancestral gene pools, as well as on social variables associated with different historical backgrounds, education, and contemporary environmental circumstances. If the two groups then differ in some behavioral characteristic, such as average performance in intelligence tests, the question arises as to whether this difference is primarily associated with the difference between the groups in biological ancestry or the difference in social and cultural history.

This is not an easy question to deal with empirically, since existing social groups almost inevitably differ in a variety of biological and historical factors, and thus comparisons are inherently ambiguous. The situation is not made any simpler by the fact that a considerable degree of political and emotional involvement tends to be characteristic of investigators in this area, leading them frequently to overinterpret their data and occasionally, one suspects, to bias its gathering in perhaps subtle and unconscious ways to fit the preconceptions of the researcher. However, a good deal of evidence of one sort or another about differences in intelligence test performance between U.S. blacks and whites has been accumulated over the years.

This evidence is reviewed and discussed from various points of view in several books or edited collections (e.g., Jensen; Loehlin, Lindzey, and Spuhler; Miller and Dreger; Richardson and Spears; Shuey). Some of the authors (Jensen; Shuey) conclude that genetic differences are probably implicated to a considerable degree in U.S. black–white differences in performance on intelligence tests; others tend to emphasize environmental interpretations (Miller and Dreger; Richardson and Spears); still others take an agnostic position (Loehlin, Lindzey, and Spuhler). There is general consensus that at present unselected groups of U.S. blacks and whites can be expected to differ in average performance on intelligence (IQ) tests by ten to twenty IQ points, in favor of the whites. That is by no means a negligible difference—for example, the average difference between two white individuals picked at random is about seventeen IQ points. However, the total range of individual differences in IQ is something like ten times as large as the average between-group difference (measurement at the extremes of the distribution is somewhat uncertain). It follows, therefore, that an individual's race is at best a weak predictor of his intellectual performance—a fact of some social importance.

Additional conclusions for which there is a fair amount of evidence (ibid.) are that the average differences in IQ test performance by blacks and whites exist prior to school entry, and that ordinary educational manipulations have not been notably successful in changing them; that average black–white IQ differences have been fairly constant over the period of fifty years or so that they have been measured in the United States; that the degree of European admixture is not strongly predictive of IQ test performance within the U.S. black population; and that attempts to match groups of blacks and whites on socioeconomic criteria tend to reduce but not to eliminate altogether the IQ differences between them. It should be noted that such matching requires favorable selection within the black group or unfavorable selection within the white group, or both, and this selection may involve genetic as well as environmental factors.

Finally, it is widely agreed that a variety of environmental conditions ranging from nutrition to education can affect IQ test scores at least to some degree, mostly to the disadvantage of U.S. blacks. Thus it is reasonable to conclude that at least part of the observed black–white IQ difference is due to environmental factors, and the debate concerns whether all of it has arisen in this way, or whether some may reflect differences in the genetic makeup of the two populations. Of course, if a genetic comparison between the populations should turn out to favor the blacks, the observed IQ disadvantage would a fortiori be environmental in origin.

At a second level removed from the question of race differences, there have been questions raised concerning the IQ test itself (for a sampling, see Block and Dworkin): Do the tests in fact measure intelligence or merely conformity to middle-class academic norms? To what extent do differences in test score among individuals (rather than groups) reflect hereditary as opposed to environmental differences among them? It should be emphasized that the latter question is only loosely linked with the questions about racial differences. It is quite possible for variation among individuals on some trait to be largely a function of differences among their genotypes, while differences between two groups of these individuals are entirely due to environmental differences between the groups. To see that this can be so, suppose that individual variation on some trait in some population is mostly genetic in origin, but susceptible to a modest degree of environmental influence. If the population is divided into two genetically comparable groups, and one is subjected only to favorable environmental influences while the other is subjected only to unfavorable ones, an average difference will be created between the groups that will be purely environmental in its origin. The individual variation within each group will, however, remain largely genetic.

Although there has been some recent controversy on the matter, the evidence still seems to support a considerable genetic component in individual differences in IQ test scores, although the exact proportions of heredity and environmental influence remain in some doubt. But, as noted, the question of whether there is any appreciable genetic component in U.S. racial differences in IQ test performance is an independent one, and in 1976 it remains quite open.

Given that empirical uncertainty on these matters persists and is likely to do so for some time, it still is possible to explore some of the ethical implications of one or the other answer to the question whether average racial differences in intelligence test scores are to an appreciable extent genetic in origin, as well as some of the ethical issues raised by the agreed-on facts about U.S. black–white differences in IQ test performance.

Ethical implications

Origins of differences. Does it make a practical difference if the average black–white IQ discrepancy has a genetic component or not? In many situations—for example, to an employer considering a job applicant—it should make essentially no difference. The employer is concerned, or should be concerned, with the intellectual capacity of the applicant, not with the average intellectual capacity of some group to which the applicant may belong—and even less with the relative contribution of heredity and environment to that average. As noted earlier, race is a very weak predictor of IQ. If the employer wants an estimate of the applicant's general intellectual capacity, he should not look at his skin color, he should give him an IQ test. He would probably be even better advised to give the applicant a test measuring the specific ability requirements of the job—general intelligence tests tend not to be very predictive of success on most jobs, if candidates are already selected on educational background. Similarly, a person's race is not a sensible basis for selection in other contexts where intellectual capacity or educability may be at issue, such as admission to college, the armed forces, or the like—quite irrespective of whether the difference between races might or might not be partially genetic in origin.

When, then, might the origins of the difference be relevant? They could be relevant when formulating social policy intended to modify the difference, or in assessing the outcome of existing social practices. For example, if a genetic difference existed, one would not assume that a simple equalization of environmental opportunities should eliminate all differences in performance between groups. However, knowledge that a genetic difference existed might be expected to encourage research on the nature of the particular biological mechanisms involved, which should improve prospects of effective action to deal with the difference. Likewise, firm knowledge that the difference was *not* genetic should help focus attention on the relevant environmental mechanisms underlying it. And if both genetic and environmental factors are involved, there may be complex interactions between them to be explored.

This is not meant to imply that the only socially appropriate response to a difference between groups is to try to eliminate it. Other alternatives exist, such as diminishing the extent to which the difference is associated with social rewards (Jencks et al.) or accepting the status quo, on a principled defense of inequality or on the grounds that the social costs of change are likely to exceed the benefits. However if social policy does dictate an attempt to change a group difference, or in making the decision to elect such a policy over other possible alternatives, accurate knowledge concerning the causal mechanisms underlying the difference is presumably highly desirable.

Labeling. Another class of ethical questions involves the psychological impact on the groups concerned of the imputation of biological differences between them. Surely if it should prove to be the case that a portion of the average difference in IQ test scores between U.S. blacks and whites were due to genetic differences between them, it would be extremely important that this not be overinterpreted by the members of either group. Each individual should realize that his intellectual capacities are what they are, and are not affected in the slightest by whether someone classifies him as "black" or "white" or "rich" or "poor" or "Jewish" or "Chinese" or anything else. A clear appreciation of the tremendous range of individual variation within any racial group is probably a critical factor in achieving this understanding.

It has sometimes been argued that, because of the risk of falsely labeling blacks as biologically inferior, research and publication on the issue of black–white IQ differences should be suspended (Block and Dworkin). But surely this is at least an equally hazardous course. First, it is by no means self-evident that a genetic origin of the differences is more psychologically devastating than a cultural origin. Would it really be better for morale in the black community, for example, to attribute a difference in average IQ to black mothering rather than to a different distribution of genes? Second, is social action based on ignorance and public stereotypes ever really preferable to social action based on knowledge? A moratorium on research in this area or on intelligence testing would seem to come perilously close to a choice of ignorance. Third, to restrict research and discussion on these (or any other) issues imperils the fundamental scientific and academic freedom to follow the evidence wherever it may lead.

On the other hand, it is obviously desirable that research involving racial and ethnic differences be carried out and reported with some sensitivity to the possibilities of misinterpretation that abound in this area. Researchers in the social and biological sciences are sometimes prone to report effects as "statistically significant" without an indication of the proportion of the total variation in the phenomenon that these effects account for—a practice no doubt encouraged by the fact that this proportion is often embarrassingly small. In a study of racial differences in a trait of social importance, such a practice is clearly indefensible. To report a difference between groups without due regard to the variation within the groups is socially irresponsible. Likewise it is irresponsible to report a fallible finding (as all empirical findings are) without some indication of the probable range of uncertainty involved. Violations of the ordinary canons of good scientific practice and honesty of discourse are serious matters where social policies affecting thousands or millions of lives are involved. It is probably unwise for a technical specialist to try to make social policy—it is better that he simply report the facts clearly and honestly as he sees them. But it is surely not too much to ask that he consider some of the more likely public misinterpretations of his findings and explicitly forestall them (for a history of public controversy in this area, see Cronbach).

Use of intelligence tests. Still other ethical questions relate to the use of intelligence tests themselves. In the first place, is it ethically proper to use an imperfect instrument or uncertain research for making socially important decisions? The answer critically depends on the nature of available alternatives. Defenders of intelligence tests point out that such tests, while admittedly imperfect, have consistently demonstrated predictive power superior to that of interviews, letters of recommendation, teachers' ratings, and other frequently advocated alternatives. Also in their favor may be counted the fact that the tests are not biased by such superficial characteristics of the respondent as his skin color or how he dresses. They doubtless reflect more subtle kinds of bias in favor of the kinds of experience familiar to the test constructors. Still, the good empirical consistency between tests based on widely different content (including tests with content explicitly intended to minimize cultural differences) suggests that the impact of these sorts of bias may be overstated by some of the more vocal critics of intelligence testing.

There is, however, an aspect of decision making based on imperfect measures that has deservedly received considerable attention in recent years, since it bears directly on the fairness of intelligence and other ability tests to members of disadvantaged groups, including racial minorities (Schmidt and Hunter). Consider the following situation: two groups differ in average score on a fallible test; applicants scoring above a certain level on that test are selected for a job (or for admission to a selective college, or whatever); the test is strictly unbiased for individuals—that is, it does not systematically overestimate or underestimate the chances of job success for individuals at any score level in either group. Paradoxically, the result of using this test may still be bias at the group level, in the sense that the test will tend to select a smaller proportion of the lower-scoring group than is actually capable of performing effectively on the job. One can compensate for this in various ways, such as lowering the score required for selecting members of the lower scoring group—but this will reintroduce bias at the level of the individual. Less qualified people will be hired over better qualified people, a poorer group overall will be chosen, more individual errors of prediction will be made, and so forth. This unwelcome paradox rests squarely on the fallibility of the test. The greater the predictive accuracy of the test, the less will be the discrepancy between fairness at the individual level and fairness at the group level. A shift away from tests to less accurate methods of prediction, such as interviews, will only make matters worse, although possibly concealing the discrepancies more effectively from public scrutiny.

There is another hazard in the use of intelligence tests (or any other fallible and standardized selection procedure for access to jobs). That is, that errors of prediction are likely to be nonrandom. Individuals or groups who systematically perform in a way that leads to low scores on the tests will tend to be excluded. Assuming the selection procedure is generally valid, most of these individuals will be properly eliminated, that is, their actual chances of success on the job will be poor. But no selection procedure is perfectly valid, so some people who could in fact perform well on the job will also be rejected. So long as there exist a variety of routes of entry to the job, probably no irremediable harm is done, since these people can reach the job some other way. But if a really effective predictor of job success is developed, a test so good that everyone starts using it, a small but permanently discriminated-against group of individuals is created: the class of people that could do well on the job but happen to do badly on some quite irrelevant aspect of the test. From the point of view of the test users, the test is satisfactory—it provides them with a larger proportion of candidates that turn out well than would some other way of selection. But from the point of view of the discriminated-against group things are not so attractive. Thus social justice may require the maintenance of some less than optimally efficient procedures for entry to education and jobs, as well as continued scrutiny of the most successful procedures to minimize their undesirable side effects.

Intelligence tests and differential social rewards. Finally, it is appropriate to ask how closely intelligence test performance is in fact related to status and income in the particular case under consideration, U.S. blacks and whites. At least one investigator has reviewed the evidence on this question at some length (Jencks et al.). He and his colleagues conclude that the cognitive skills measured by IQ tests do not in fact play a very large role in accounting for differences in status and income among individuals in the United States, nor in accounting for such differences between blacks and whites. Some of the individual differences in success may be due to personality attributes like energy, ambition, and social skills; many may simply be due to luck, i.e., to factors such as happening to be in the right place at the opportune time. Tests of cognitive skills, by virtue of their ability to predict educational success, have a moderate ability to predict occupational level. (This does not, by the way, appear to be an artifact of the use of such tests in admission to colleges and professional schools—the correlations between IQ and occupation were as high in 1917 as in more recent years when standardized tests have widely been employed in educational selection.) Within occupations, however, tests show only a very modest ability to predict income. To put it more concretely, it takes a higher IQ, on the average, to meet the educational requirements for becoming a doctor than a plumber, and a doctor, on the average, makes more money. But the difference between a doctor who makes a great deal of money and one who does not is more likely to be a matter of entrepreneurial skills, social connections, or drive than of intellectual gifts per se.

A good deal of the difference in average income between U.S. blacks and whites seems also to be unrelated to differences in IQ test performance. Jencks cites figures suggesting that in the early 1960s equating blacks and whites for intelligence test scores would only have reduced the income gap between them by about one-third. Blacks and whites of the same IQ level received approximately the same amount of education—indeed, at low IQ levels, blacks got more education than whites of comparable IQs. Jencks concludes that decreasing the income differential between blacks and whites in a given occupation who have equal educational qualifications and ability test scores would be a far more effective step toward equality than attempting to alter average differences in IQ. Employment and income patterns of U.S. blacks have been undergoing some changes in the last decade, so these conclusions might need some qualification today. Nonetheless, some of the emotions surrounding the issue of black–white IQ differences might be lessened if the fairly modest role of intellectual capacities in achieving differential social rewards were kept in mind.

Conclusion

The present discussion has focused on U.S. blacks and whites. The broader question of differences in intellectual capacities across the spectrum of human races has occasionally been addressed by anthropologists and psychologists in the general context of the study of human evolution. In recent years, however, it has received only a fraction of the attention devoted to the black–white issue. The broad question is a very difficult one to tackle empirically because of the confounding of cultural and biological variation. A number of attempts have been made to develop means of assessing cognitive abilities that are equivalent across cultures, but the success of such efforts has so far been debatable. Nevertheless, as a scientific question, race differences in intelligence are much more interesting in a broad evolutionary perspective than in the narrow focus of the black–white problem—even though many of the social and ethical issues are most sharply posed by the particular case of black–white IQ differences.

JOHN C. LOEHLIN

[*While all the articles in this entry are relevant, see especially the article* PHILOSOPHICAL AND ETHICAL ISSUES. *Other relevant material may be found under* RESEARCH, BEHAVIORAL. *For an explanation of the meaning of IQ test scores, see* MENTALLY HANDICAPPED. *See also:* RACISM.]

BIBLIOGRAPHY

BLOCK, N. J., and DWORKIN, GERALD, eds. *The IQ Controversy: Critical Readings.* New York: Pantheon, 1976.

CRONBACH, LEE J. "Five Decades of Public Controversy over Mental Testing." *American Psychologist* 30 (1975): 1–14.

JENCKS, CHRISTOPHER; SMITH, MARSHALL; ACLAND, HENRY; BANE, MARY JO; COHEN, DAVID; GINTIS, HERBERT; HEYNS, BARBARA; and MICHELSON, STEPHAN. *Inequality: A Reassessment of the Effect of Family and Schooling in America.* New York: Basic Books. 1972.

JENSEN, ARTHUR ROBERT. *Educability and Group Differences.* New York: Harper & Row, 1973.

LOEHLIN, JOHN C.; LINDZEY, G.; and SPUHLER, J. N. *Race Differences in Intelligence.* San Francisco: W. H. Freeman, 1975.

MILLER, KENT S., and DREGER, RALPH MASON, eds. *Comparative Studies of Blacks and Whites in the United States.* Quantitative Studies in Social Relations. New York: Seminar Press, 1973.

REED, T. EDWARD. "Caucasian Genes in American Negroes." *Science* 165 (1969): 762–768.

RICHARDSON, KEN, and SPEARS, DAVID, eds. *Race and Intelligence: The Fallacies Behind the Race–IQ Controversy.* Baltimore: Penguin Books, 1972.

SCHMIDT, FRANK L., and HUNTER, JOHN E. "Racial and Ethnic Bias in Psychological Tests: Divergent Implications of Two Definitions of Test Bias." *American Psychologist* 29 (1974): 1–8.

SHUEY, AUDREY MARY. *The Testing of Negro Intelligence.* 2d ed. New York: Social Science Press, 1966.

IV

GENETICS AND MENTAL DISORDERS

Our contemporary knowledge, however incomplete, about the contribution of genetic factors to the etiologies of mental disorders raises the following bioethical issues: the age-old question of determinism of moral behavior (given the supposed equivalence of genes with determinism); the rights to procreate, to marry, to get an abortion, to commit suicide; the rights to employment and to health insurance after an illness; the stigma of having a history of mental illness or of being the relative of a psychiatric patient; the risks incurred by parents who would adopt the offspring of some kinds of patients; and the effectiveness, currently and in the future, of forms of eugenic intervention.

Despite the uncertainty that surrounds the causes of serious mental disorders, the weight of evidence available in 1976 favors the conclusion that unspecifiable genetic factors play an important role in the development of most cases. Consideration of disorders less disruptive than the psychoses, such as the neuroses, alcoholism, or personality disorders, calls for even greater caution before assigning a causal role to genetic factors. Empirical studies that show the risks of developing the same disorder in the relatives of index cases provide our basic data in this area of psychiatric genetics. The mere fact that a condition of psychiatric interest is familial is hardly sufficient evidence that genetic factors are importantly involved; such factors can be invoked when studies of twins and other family members conform consistently to some reasonable genetic model of transmission, even when environmental factors are controlled by the use of adoption strategies. At present (1976) we have no corpus delicti for demonstrating the chromosomal, biochemical, or biophysical basis of any of the common psychiatric disorders, each of which affects more than 1 in 100 people.

Once some rare forms of mental retardation were shown to fit a simple Mendelian form of recessive inheritance and found to be associated with inborn errors of metabolism at the beginning of this century, a neuropsychiatry dominated by the natural sciences eagerly embraced, applied, and misapplied the neat Mendelian laws to its burdensome "diseases"—insanity, feeblemindedness, and alcoholism. Regrettably, unwarranted eugenic fervor was stimulated by the political distortion of some of these biological ideas so that efforts to apply genetic information in an enlightened manner to the problems of the human condition are too often impugned. It was not until 1916 that E. Rüdin in Munich published the first sound family study of schizophrenia in the siblings of schizophrenics; he found that only some 8.5 percent of the siblings could be expected to develop a psychosis in their lifetimes, far fewer than the twenty five percent required to fit a simplistic recessive gene theory.

No simple theories fit any of the data in the genetics of mental disorders. This state of affairs has led to espousing a diathesis-stressor framework for evaluating causes—a constitutional, often hereditary, predisposition of the body to a disease, group of diseases, or structural or metabolic abnormality. Within this framework the genes provide the predisposition toward mental disorder, and the environment provides the stressors that combine in some fashion to produce a patient. Both the genes and the environment are necessary to develop a serious mental illness; neither by itself is sufficient as a cause.

In addition to the slightly more than 200 Mendelian or single locus diseases associated with mental retardation (such as phenylketonuria with an incidence of 4 per 100,000 infants), some of the other problems associated with childhood have a more obscure but noteworthy genetic component. Reading disability, sleepwalking, and stuttering are among such traits. Most often the evidence comes from the much higher concordance rate for identical twins (who have all their genes in common) than for same-sex fraternal twins (who share fifty percent of their genes in common on average). The twin method has shown that juvenile delinquency resembles the pattern for measles with very high concordance rates in both identical (MZ) and fraternal (DZ) twin pairs: Thus "infectious" agents in the environment should be sought since genetic contributors have been excluded. So-called childhood psychoses (infantile autism and very early "schizophrenia") are so rare that they might be explained as rare spontaneous mutations (1 in 100,000); their etiology is unknown. Conditions resembling adult schizophrenia but having their onset around the time of the first signs of puberty may be genuine early onset cases of schizophrenia with a heavy loading of genetic contributors to liability.

Twin and family studies of personality traits suggest that some reminiscent of "temperament" (as contrasted with superficial traits) have appreciable genetic components; normal social introversion is the best example. When it comes to the extremes of personality functioning, the personality disorders diagnosable in a clinic, one careful twin study showed an MZ concordance rate of thirty-three percent and a DZ rate of six percent. A study of the biological and adoptive fathers of adoptees who grew up to have diagnosable psychopathic personalities in a hospital setting reported that nine percent of the former compared to two percent of the latter were also psychopaths. Both studies suggest that genetic predispositions play an appreciable role in some cases. Alcohol addiction has many causes, some mainly cultural and some apparently with a genetic component. A Swedish study of chronic alcoholism reported concordance rates of seventy-one percent and thirty-one percent respectively in MZ and DZ twin pairs. The results of a Danish study of the adopted-away offspring of alcoholics was compatible with a diathesis-stressor explanation also.

When one has observed high-strung dogs or horses and then been told the trait was familial, it raises the suspicion that some forms of psychoneurosis may have significant genetic components. However, twin studies of neuroses without further subdivision by types are not impressive to a geneticist. When a specific diagnosis of anxiety state is the subject of research, however, and care is taken to make the psychiatrist unaware of the co-twin's diagnosis, concordance rates of forty-one percent and four percent are obtained in MZ and DZ twin pairs respectively. On an a priori basis one would expect important genetic contributors to the liability to the very common milder depressions called neurotic or reactive. The lifetime risk of developing some kind of mild or serious mood disorder, based on firsthand knowledge of everyone in a defined area, runs as high as 8.5 percent for males and 17.7 percent for females. Neither twin nor family studies of the neurotic depressions can identify a significant genetic contribution to the individual differences in the liability to respond to stressors with these kinds of reactions.

Affective psychoses, also known as manic–

depressive disorders and more recently subdivided into bipolar and recurrent unipolar affective disorders, are highly heritable. The lifetime risk for being hospitalized with an affective psychosis is about 3.5 percent for men and 5.8 percent for women in the United Kingdom data. Based on many studies we would expect about fourteen percent of the parents, siblings, or offspring of manic–depressive index cases to develop some kind of affective disorder or to commit suicide. If the index case has a bipolar psychosis (with swings from severe depression to mania), the risks rise to about twenty percent. Twin studies confirm the high genetic loading in affective psychoses.

The bioethical issues raised by these data must be tempered by other facts: The disorders are usually well controlled by chemotherapy or electroconvulsive therapy; the first attack often comes after the end of the reproductive period; most persons have only a few bouts during their lifetimes; and between attacks the patients are difficult, if at all, to distinguish from unaffected persons.

More research has been done on schizophrenia from a genetic point of view than on any of the other conditions described above. It is a tragic condition, because it often begins in late adolescence or early adulthood and seldom results in a complete recovery even with modern chemotherapy. The evidence for the existence of important and specific genetic factors in schizophrenia rests on the low observed lifetime risk in the general population of one percent contrasted with the high risks in relatives of index cases, even when separated by adoption and reared in schizophrenia-free environments. The rate in identical co-twins of schizophrenics is about fifty percent while that in fraternal co-twins, siblings, and offspring is about ten percent. If both parents are schizophrenic, their children run a risk of about forty-six percent of developing the same condition. Further, no environmental agent has been identified that will reliably produce schizophrenia in a person unrelated to a schizophrenic. Children of schizophrenics adopted away from them early in life still go on to develop schizophrenia in their adoptive homes with normal adoptive parents; in one study five of forty-seven such offspring with severely sick biological mothers grew up to be schizophrenics.

Pessimism and fatalism should not be generated by this recounting of facts. The actual risk to the child of a schizophrenic depends both on the severity of illness in the parent and the status of the other parent; if the affected parent has a mild schizophrenia and the co-parent is normal, the risk falls to near the general population risk of one percent. The vast majority of offspring are remarkably normal, and the small subset who will become schizophrenic are not at present identifiable in advance. It is uncertain whether adoption into a "good" home lowers the risk for adopted-away children. The mode of transmission is uncertain, so that we must rely on the empirical facts in taking any socially significant action.

It is a fact that twenty states currently have laws prohibiting the marriage of schizophrenics; the laws are not enforced. Fears are often raised that, unless the childbearing of schizophrenics is curtailed, a dysgenic trend will be encouraged. However, the fertility of schizophrenics is greatly reduced for various reasons and fears are unwarranted; most babies born to patients arrive before the first onset. The question of how a disadvantageous condition like this can be maintained despite the greatly reduced fitness of male (0.5 children per patient) and female (0.9 children per patient) schizophrenics can be answered in part by a polygenic theory wherein "carriers" of subthreshold numbers of genes are not selected out of the breeding population, together with the late onset of the condition in many patients. The fecundity of *married* schizophrenics is quite close to that of controls; only fifty-four percent of females and thirty-three percent of males marry in recent times. Such facts, together with the fact that about ninety percent of schizophrenic patients have nonschizophrenic parents, would make eugenic intervention quite ineffective in reducing the population incidence of new cases. At the level of the individual family, however, suffering would be relieved by contraceptive protection of schizophrenic patients, not because of the tolerable risk for genetic reasons to the offspring, but because schizophrenics find parenthood an added burden to their coping skills, and their children are denied adequate parenting.

Further research is required to understand how genetic factors interact with which environmental ones over the life span to produce mental disorders. Knowing that genetic factors are involved cannot be equated with a rigid determinism. The efficacy of special treatments and preventive measures are not ruled out by the very high heritability of the liabilities to the psychoses. To the extent that environmental events are idiosyncratic stressors that interact with various of the predisposed genotypes for the psychoses, modification of our society for a specific stressor will have little effect on the overall incidence or prevalence of the psychoses. Although a few individuals may be so predisposed genetically that almost any environment would be sufficiently stressful for them, the vast majority with a genetic high risk for breakdown will depend on encounters with the environmental stressors for potentiation of their predispositions. The estimates of the degree of genetic determination of liabilities depend on extant ecological conditions; novel rearing or living arrangements might reduce the values dramatically.

The poisonous mixture of genetic conclusions and political ideology that led to the Nazi sterilization and marriage laws regarding psychiatric cases, retardates, epileptics, the blind, and the deaf must never be forgotten; the laws were carried out by special eugenic courts composed of a lawyer, a physician, and a geneticist. Our firm knowledge about the etiologies of psychological disorders is so fallible that caution, tempered by humanitarian concerns for individuals, must govern efforts at eugenic intervention in psychiatric genetics.

IRVING I. GOTTESMAN

[*While all the articles in this entry are relevant, see especially the article* PHILOSOPHICAL AND ETHICAL ISSUES. *Other relevant material may be found under* ALCOHOL, USE OF; MEDICAL ETHICS UNDER NATIONAL SOCIALISM; MENTALLY HANDICAPPED; *and* STERILIZATION, *article on* ETHICAL ASPECTS. *See also:* FREE WILL AND DETERMINISM; *and* PSYCHOPHARMACOLOGY.]

BIBLIOGRAPHY

BOK, SISSELA. "Ethical Problems of Abortion." *Hastings Center Studies* 2, no. 1 (1974), pp. 33–52.
CHEZ, RONALD A. "Mental Disability as a Basis for Contraception and Sterilization." *Social Biology* 18, supp. 1, September 1971, pp. S120–S126.
CURRAN, WILLIAM J. "Ethical and Legal Considerations in High Risk Studies of Schizophrenia." *Schizophrenia Bulletin*, no. 10, Fall 1974, pp. 74–92.
GOTTESMAN, IRVING I., and ERLENMEYER-KIMLING, L., eds. "Differential Reproduction in Individuals with Physical and Mental Disorders." *Social Biology* 18, supp. 1, September 1971.
——, and SHIELDS, JAMES, with MEEHL, PAUL E. *Schizophrenia and Genetics: A Twin Study Vantage Point.* Personality and Psychopathology, no. 13. New York: Academic Press, 1972.
HESTON, LEONARD L., and GOTTESMAN, IRVING I. "Genetic Counseling in Psychiatry." *Modern Medicine*, 29 May 1972, pp. 48–52.
HIRSCHHORN, KURT. "Practical and Ethical Problems in Human Genetics." *Birth Defects: Original Article Series* 8, no. 4 (1972), pp. 17–30.
HULSE, FREDERICK S. "Scientific Ethics and Physical Anthropology." *American Journal of Physical Anthropology* 31 (1969): 245–248. Presidential Address.
ROSENTHAL, DAVID. *Genetic Theory and Abnormal Behavior.* McGraw-Hill Series in Psychology. New York: 1970.
SHIELDS, JAMES. "Schizophrenia, Genetics and Adoption." *Child Adoption: The Journal of the Association of British Adoption Agencies*, no. 80, no. 2 of 1975, pp. 19–24.
SLATER, ELIOT. "German Eugenics in Practice." *Man, Mind, and Heredity: Selected Papers of Eliot Slater on Psychiatry and Genetics.* Edited by James Shields and Irving I. Gottesman. Baltimore: Johns Hopkins Press, 1971, pp. 281–292. Original appearance. *Eugenics Review* 27 (1936): 285–295.
——, and COWIE, VALERIE. *The Genetics of Mental Disorders.* Oxford Monographs on Medical Genetics. London: Oxford University Press, 1971.

V
PHILOSOPHICAL AND ETHICAL ISSUES

History

Most historians of science would locate the origin of modern scientific genetics with the rediscovery of the work of Gregor Mendel concerning inheritance in pea plants carried out simultaneously by Bateson, Correns, and Tschermak around 1900. Behavioral genetics, as a recognized subdiscipline of genetic science, is even more recent, dating from the research of scientists such as R. C. Tryon, Konrad Lorenz, J. P. Scott, and J. L. Fuller after the Second World War.

Despite the relative infancy of the disciplines of genetics and behavioral genetics, ethicists have long been concerned with explicating the relationship between deterministic scientific theories of human and animal nature and behavior, and philosophical conceptions of agency, individuality, and morality.

The importance of incorporating scientific opinions on persons' malleability and educability and the causal sources of their behavior into philosophical conceptions of ethics and theories of value is reflected in the writings of numerous philosophers. Plato, in his efforts to construct the morally ideal state in the *Republic*, took careful note of the fact that human beings differ greatly as to their innate natural abilities and capacities, as well as in their willingness to utilize and fulfill these constitutional abilities and capacities. His conception of the just state was one in which there is an exact correlation between natural human abilities and the specialized functions or roles he thought necessary for harmonious political and moral life.

Plato's commitment to a rudimentary form of behavioral genetic science can be seen in his recommendations concerning the state's role in regulating reproduction and in educating the young. He advocated strict segregationist policies of marriage and social class based on previously established hereditary tendencies and behavioral abilities among potential parents.

With the rise of modern science, the philosophical efforts of Hobbes, Machiavelli, Butler, and Rousseau in constructing ethical and political systems and prescriptions were greatly influenced by their respective views of human nature, heredity, and the causal or motive sources of human behavior and social activity. The persistent efforts of political philosophers throughout the seventeenth and eighteenth centuries to ascertain what man would be like in a primitive state of nature were aimed at cutting through the veneers of culture and civilization to find the essential causal sources of human behavior. The philosophical essays written during these periods reveal a powerful and persistent belief that an empirical understanding of human nature and its biological makeup is a necessary prerequisite for constructing political and ethical systems. It was thought that the only plausible social and ethical systems were those that were consistent with scientific facts about human biological nature.

The same belief—that man's moral and political views should be constructed in light of and not in spite of man's biological nature—is reflected in Mill's attempt to provide a proof for the principle of utility as an overarching moral standard in his classic work *Utilitarianism*. Mill argued that the only thing persons *can* ultimately desire and do in fact desire is happiness. Despite the fact that Mill recognized that people differ as to what sorts of objects and activities make them happy, the reality that human beings were "biologically programmed" to seek happiness was the only sort of factual proof that, he argued, could be provided for demonstrating the validity of utility as an acceptable moral standard.

Many theologians, ethicists, and political theorists have long been interested in elucidating the key elements that constitute and determine human nature and behavior. Thus it is somewhat puzzling that the rise of modern scientific genetics and behavioral genetics has not occasioned more philosophical interest in the relevance of empirical scientific findings concerning the individual and social behavior of persons for contemporary theories of moral value and political organization. While many twentieth-century biologists such as Julian Huxley, G. G. Simpson, C. H. Waddington, Theodosius Dobzhansky, and Konrad Lorenz have tried to erect moral systems based upon scientific findings concerning the genetics and evolution of human behavior and culture, most twentieth-century philosophers have remained oblivious to the normative efforts of scientists such as these (Wilson).

This curious state of affairs can best be explained by analyzing the views of twentieth-century positivist critics of previous attempts at moral philosophizing, such as Moritz Schlick and A. J. Ayer, as well as the views of the influential school of British moral philosophers known as the intuitionists led by G. E. Moore, H. A. Prichard, and David Ross (Hancock; Warnock). The early positivists argued that prescriptive ethical knowledge of any sort was incapable of verification or test by scientific empirical methods and thus was incapable of obtaining rational assent or of being critically evaluated and discussed. They considered ethics to be merely a matter of subjective opinion and irrational or emotional beliefs and, consequently, beyond the realm of scientific investigation concerning human nature, human behavior, heredity, or anything else.

Moore and the intuitionists argued that the demonstration or discovery of empirical truths concerning man or human behavior, of the sort offered by Mill or Plato, allowed no inferences to be drawn about appropriate standards of morality and conduct. Moore in particular argued that it was fallacious to think that it was possible to move, logically, from a set of descriptive empirical facts concerning what *is* the case regarding human nature, conduct, and causation to a set of prescriptive or normative conclusions concerning what *ought* to be the case concerning human morality and social arrangements. Any attempt to cross the logical gulf that Moore saw between facts and values would necessarily founder on a logical fallacy. Those scientists and philosophers who attempted to build ethical and moral norms on biological or genetic facts were guilty of committing what Moore called the "naturalistic fallacy."

The arguments of the positivists and the intuitionists exerted a powerful hold over much of twentieth-century ethical theorizing (Hancock).

For the most part, the efforts of biologists to provide scientific and factual underpinnings for theories of ethics and normative judgments were dismissed as logically fallacious or simply beside the point of moral and political argument. However, recent advances in our scientific understanding of the genetics of human behavior, of human social and political activity, and of the biology of human nature are compelling a renewed interest in the relevance of findings in genetics for ethics. Moreover, questions are now being raised about the adequacy of the criticisms by positivists and intuitionists of a scientifically based ethics. In addition, research into the genetic underpinnings of human behavior has raised a host of ethical questions that run the gamut from the morality of conducting such research to questions of the liability, culpability, and responsibility of those individuals who manifest disorders or behavior that society considers criminal, deviant, unhealthy, or symptomatic of illness or disease.

Distributive justice

Some recent work concerning the genetic basis of human behavior has focused on overt and complex behavioral traits such as IQ, criminality, and mental illness. Attempts have been made to correlate certain types of genotypes with certain varieties of behavior that frequently result in the institutionalization of persons in prisons, hospitals, nursing homes, and insane asylums. Particularly prominent in behavior genetic research in this area have been the attempts to correlate criminal behavior and other deviant behaviors in males with chromosomal abnormalities such as the presence of one or more extra sex chromosomes—XYY, XXY, XXYY [see article II above, MALES WITH SEX CHROMOSOME ABNORMALITIES (XYY AND XXY GENOTYPES)]. Such research has raised anew some traditional philosophical puzzles concerning free will, determinism, equity, desert, and equality.

Normative theories of justice can be divided into two types: distributive theories, which are concerned with establishing the proper division of the goods, benefits, evils, and resources existing in a society among its members, and retributive theories, which are concerned with the establishment of appropriate rewards, compensations, and punishments for specific human actions.

Many scientists have worried that the demonstration of genetic sources for aberrant or deficient behaviors in individuals could or should alter present conceptions of justice and equality. For example, if it were shown that some biological subgroup of human beings had a mean IQ score substantially below that of the mean of some other group, then some behavioral geneticists argue that such a fact would cause us to reevaluate the goods, duties, and services provided to these various groups [see article III above, RACE DIFFERENCES IN INTELLIGENCE]. Such factual demonstrations concerning behavioral traits would apply to populations of individuals and not to the individuals themselves, since great amounts of variation regarding any behavioral trait exist among members of any given biological group. The claim is frequently made that social policies, if they are to be just and nonarbitrary, must attend to genetically caused behavioral variations among biological groups. Thus, given individual biological differences in behavior or abilities, the equalization of opportunity for a whole group of disadvantaged people or the minimization of the functional importance of a given behavioral trait may not result in a just or egalitarian distribution of social goods and evils (Jencks et al.; Herrnstein).

Philosophers concerned with formulating theories of distributive justice have long grappled with the question of the import and relevance of individual variations in human aptitudes, skills, abilities, and behaviors. Some philosophers, such as the utilitarians Bentham and Mill, have argued that the only standard relevant to the determination of just distributions of social goods are the benefits and costs to persons of the performance of a particular action or social policy. Other philosophers—social-contract theorists such as Rousseau or John Rawls—have argued that theories of justice must also take cognizance of the social positions, social arrangements, capacities, aspirations, and attitudes of the human beings involved in any political or social distribution (Rawls). While it has long been realized by all parties in debates about justice that human beings differ as to their biological endowments, backgrounds, and social and cultural environments, behavioral genetics has raised for the first time the empirical claim that such differences may be correlated with, and perhaps exclusively caused by, genetic differences that vary systematically among biologically defined groups. While the validity of such claims would not settle the basic philosophical dilemma about the relevance of human differences for considerations of justice, it raises difficulties for those views of justice that are committed to the basic equality of human beings as an initial premise for determining valid rules or standards of distributive justice. However, since it is possible to formulate theories of justice that do not presuppose a merit criterion such as abilities, social usefulness, wealth, or intelligence, but rest instead upon considerations such as overall, aggregate social utility or individual needs, there is no reason to suppose that the empirical demonstration of systematic biological differences resulting from genetic sources must necessarily affect social policy or ethical assessments of justice or equity.

It should be added that discussions of justice and social policy frequently conflate and confound concepts such as equity, social equality, and individual interest. Ethical decisions about how to count individual interests in arriving at just and fair social policies are distinct from questions of constructing a proper standard or standards of social distribution (Benn; Williams). The fact that persons differ genetically and behaviorally must be shown to be relevant to the assessment of both who should count or be counted in ethical policy matters and how much and to whom goods and evils ought to be given. It is quite consistent to maintain either that all men are equal in their unique moral value as individuals but that not all men are entitled to the same amounts of wealth, honor, property, duties, and obligations, or the converse, that all men ought to receive equal shares of the available goods and evils despite the fact that all men are not equal with regard to morally relevant personal properties. The establishment of human equality and criteria for social equity and justice are projects that have been and will continue to be independent of biological differences among individuals (Olafson; Blackstone).

Retributive justice

The question of biological determinism. A central issue surrounding the study of the genetic determinants of human behavior is that of determinism. A number of researchers in the broad area of the genetics of human differences (Jensen; Herrnstein; Block and Dworkin) have argued that variations in the abilities, powers, and capacities of human beings of various biologically defined groups can be directly attributed to variations in the genetic factors possessed by members of these various groups. Arguments have been made on the basis of very tenuous evidence to the effect that some behavioral differences such as variations in degrees of intelligence, criminality, mental illness, aggression, and emotion present in the individuals of a particular biologically defined subpopulation of human beings (i.e., women, blacks, XYY males) are the end results of innate genetic differences between such biologically specified populations.

Claims of this sort have a number of important implications for understanding and analyzing human nature, human conduct, and moral beliefs and practices. Some scientists on the basis of the available evidence have advocated various versions of genetic hereditarianism or determinism. Such views maintain that key aspects of human behavior are heritable, innate, instinctual, and programmed from birth in each individual person. Complex behaviors are held to have their ultimate causal source in inherited genetic factors that are not easily amenable to modification, amelioration, or improvement. Moreover, such variations in the distributions of abilities and skills are held to be highly correlated with the distributions of genetic factors among various groups, races, and populations of human beings. Inspired by recent work in human behavioral genetics, genetic determinists argue that human nature is in many respects fixed unalterably at birth, is highly resistant to developmental and environmental modification, and is a heritable trait or feature analogous to other phenotypic features possessed by human beings.

Genetic determinism has inspired a set of strident objections from those who view such theories as politically or ideologically inspired, methodologically unsound, morally pernicious, and far too biased in favor of the power of innate, inherited, genetic factors to control human behavior, action, and abilities (Block and Dworkin; Kamin). Critics of genetic determinism have tried to argue that, while it is true that human behavior does have certain limits imposed upon it by genetic inheritance, there is far more room for malleability and change than is admitted on hereditarian views. They argue that genetic inheritance acts as a kind of open program in individuals allowing various environmental and developmental factors to greatly influence the range of outcomes to be expected from a given complement of genes in terms of observed behaviors and traits. Such criticisms attempt to challenge the rigid determinism of genetic hereditarians by showing that behavior in human beings is subject to alteration, modification, change, and improvement by means of the proper environmental adjustments. The notion that certain biological groups differ inherently in their capacities and abilities is held to be directly traceable to differences in the environmental circumstances encountered by such groups.

It is certainly true that some of the work conducted in the area of human behavioral genetics has reawakened debate about the implications of deterministic theories for morality and ethics. Many would argue that social policies ought to reflect the manner and degree to which human behavior can be affected by environmental manipulation and social engineering. However, it is interesting to note that both proponents and critics of strong versions of genetic determinism are forced to acknowledge the difficulties posed by scientific theories that indicate the causal sources of human behavior consist of factors beyond human control and choice. Even those who are strongly committed to the claim that human behavior is malleable and that environmental forces play a powerful role in determining human behavior are, despite their rejection of genetic hereditarianism, often still committed to a strong version of scientific determinism. For to say it is the environment rather than our genes that determines who we are, what we do, and what we are likely to do in the future still leaves relatively little room for the traditional interpretations of human action and behavior as being caused by free will, intentional choice, and rational decision. Thus, the issues raised by recent investigations and debates concerning the influence of genetic factors upon human behavior lead directly to a consideration of an age-old philosophical conundrum—how to reconcile free will, moral responsibility, and human agency with deterministic scientific theories of human behavior. The notion that we can punish or reward persons for their conduct and behavior—the basic assumption of retributive justice, which in turn assumes that persons can be held liable and responsible for their actions—seems incompatible with scientific views of human behavior that posit genetic or environmental factors as the sufficient causes of such behavior.

Philosophical positions on responsibility and retributive justice. Much philosophical attention has historically been devoted to the question of when persons can be held responsible or liable for their actions. Aristotle, in his *Nichomachean Ethics*, puzzled over the question of whether a sea captain could be held liable for tossing the cargo of his ship overboard during a storm since the circumstances requiring such behavior were beyond the captain's control. Irresistible compulsion and coercion are frequently offered as exculpatory reasons for mollifying the consequences of what would usually be considered prima facie examples of criminal or antisocial behavior. Genetic determinism seems to raise significant new problems for theories of retributive justice since claims are frequently made that certain behaviors—e.g., criminality, aggression, alcoholism, and sexual aberrations—in some individuals are the causal end products of genetic factors beyond the will and control of the individuals who possess them (Szasz).

Some philosophers (e.g., Dworkin) have argued that deterministic scientific theories of human behavior, including theories of genetic determinism of the sort cited above, are incompatible with the notion of moral responsibility or liability. If deterministic theories accurately describe the sources of human behavior, then free will and conscious choice have no role in guiding human action. Therefore, such hard determinists would claim, there is no sense in holding persons liable, culpable, or responsible for actions they did not cause or mediate.

Other philosophers have argued that determinism and responsibility are in fact compatible. A deterministic account that men with extra sex chromosomes will tend to commit aggressive crimes does not show that such men cannot be held responsible or accountable for their acts. Such accounts, frequently termed "soft" determinism insofar as they reconcile moral responsibility with determinism, argue that deterministic causal accounts, far from obviating considerations of moral responsibility, are prerequisites for such concepts to be meaningful. If we are to hold persons morally responsible for what they do, it is necessary to show a causal connection of a deterministic sort between the intentions, motives, plans, and volitions that individuals have and the actions they perform. Moreover, the fact that a given deterministic account of human behavior is true does not, in itself, show that under other circumstances a different behavior could not have occurred. Scientific determinism, on the compatabilist account, is necessary as a prerequisite for both moral responsibility and liability for punishment or reward, since it is necessary to show that if a person had chosen to act in a way other than he or she did choose to act, that act would have eventuated as the causal, deterministic result of this alternative choice.

Yet a third philosophical position regarding retributive justice and moral responsibility might be labeled human self-determinism. This view holds that, even if deterministic accounts of human behavior or activity are scientifically unassailable, they are so blatantly at odds with our conscious awareness of the fact that we do choose to perform certain actions freely and voluntarily that human freedom and, in turn, human moral responsibility reside in a distinctive form of human causation that stands outside the domain of scientific theorizing. This view explicitly denies the possibility of reducing all accounts of human volitional behavior and choice to scientific deterministic theories of the sort represented by behavioral genetic accounts of human activity. It is a species of holism or emergentism that denies the possibility of reducing all human behavior to causal empirical interactions given the prominence of notions such as free will, choice, and intention in our personal awareness of the behaviors we initiate (Chisholm).

All three of these positions concerning responsibility and the feasibility of retributive justice have implications for the assessment of empirical findings in behavioral genetics. Those scientists who adopt a belief in the ultimate reducibility of human behavior to complex, causally deterministic accounts of genetic/environmental interactions provide grist for the mill of hard deterministic views of the futility of retributive justice. Philosophical compatibilists may try to accommodate such reductionistic accounts with the notion of moral responsibility by arguing for the causal efficacy of human intentions and volitions that are reducible to genetic/environmental factors. However, the assessment of behaviors not impelled by such conscious mental states, such as XYY-induced aggression or genetically programmed altruistic behavior, remains unclear. The same problem arises on self-deterministic accounts of responsibility and liability since presumably such self-determination would occasionally be beyond the reach of certain individuals who were mentally ill, insane, retarded, or genetically predisposed toward certain behaviors [see article IV above, GENETICS AND MENTAL DISORDERS]. Since retributive justice presupposes the meaningfulness of concepts such as moral responsibility, free will, culpability, choice, and intention, one thesis of genetic determinism poses yet another in a long historical series of deterministic scientific challenges to man's view of his conscious causal role in nature and society (Edel).

Rights

Considerations of justice lead directly to questions concerning the implications of behavioral genetics for the language of rights. Rights questions can be divided in three ways—the question of what sorts of things can have rights, what rights should be or are possessed, and how are conflicting claims to rights adjudicated? Much furor has surrounded the question of the rights of the mentally ill, mentally retarded, and criminally insane, particularly when these conditions are primarily genetic in origin. Conflicts arise between the rights of the patient to remain autonomous, informed, and free and the perceived needs of society and the medical profession to intervene therapeutically in the life of the individual. Different theories of ethics assign differing weights to the relevance of biological differences among individuals for moral and ethical decisions. Under certain moral theories, the moral status of the mentally ill or mentally retarded individual as a bearer of rights must be determined independently of genetic considerations concerning behavioral differences given the *human* status of the ill or the retarded.

It should be noted that the determination that an individual has moral standing or can be the subject of rights and entitlements does not, in itself, show what specific rights are possessed by the individual, or how questions of conflicting claims to rights among persons are to be equitably resolved. It is sometimes thought that if an individual can be shown to be capable of having rights, then the individual possesses any and all rights accorded any individual in a similar society or social setting. However, the mere demonstration that an individual is capable of having rights does not settle the issue of which, if any, rights are possessed by the individual, or what priorities and weights will be assigned to the individual's rights when in conflict with the rights of others. In analyzing questions of conflicting claims, as in cases of truth-telling, confidentiality, or paternalism, in a medical context, the demonstration that a person can have rights is a necessary but not a sufficient condition for resolving conflicting rights claims (Melden).

Nor does the establishment of certain specific rights for persons end moral debate since conflicts between individuals concerning rights are bound to occur given the varied interests of persons involved in social, medical, and institutional interactions. In many instances questions as to how to adjudicate among conflicting claims often devolve upon specific beliefs in complex theories of distributive or retributive justice. Behavioral genetics introduces questions of assessing and predicting social consequences, future behaviors, and the efficacy of therapeutic techniques that bring a great degree of uncertainty into moral calculations. Given the uncertainty surrounding behavioral genetic knowledge and techniques for the management and modification of human behavior, the onus seems to be on those who propose to infringe, violate, ignore, or override the acknowledged rights of individuals to justify rights-constraining activity on scientific grounds (Block and Dworkin).

Methodological issues

Since behavioral genetics as practiced by psychologists, geneticists, ethologists, and physicians is aimed at elucidating the variables mediating vast portions of individual and social behavior, criticism has surrounded the theories and hypotheses advanced by researchers in this nascent science. Because the potential for the abuse and misuse of genetic knowledge of any sort is so large, it is of vital importance to recognize the tenuousness of current scientific understanding of the genetics of human behavior (Ehrman and Parsons) [*see article* I *above*, STATE OF THE ART]. Many factors, including genetic, environmental, cultural, and sociopolitical variables, enter into the expression of human behaviors. The complexity of the causal chains and etiologies involved in moving from a gene to a specific behavior is enormous. Much of the criticism devoted to behavioral genetic research efforts in the areas of IQ, criminality, mental illness, mental retardation, aggression, and sexuality has resulted from attempts by some scientists to fit complex causal etiologies and behaviors into overly simplistic crude genetic models of a "one gene–one behavior" variety (Kamin; Lewontin).

First, it should be noted that human behavioral genetics is a new science, and its hypotheses are, for the most part, tentative and speculative. Second, relatively little is known about the causal effects of developmental environments, regulatory genes, genetic environments, and the ecological and selectional histories upon structural genes that code for specific traits or behaviors. Third, all genes and genotypes can determine only a range of behaviors of a type, rather than a specific or particular behavior, given the number of factors entering into the ultimate expression of any given type of behavior (Dobzhansky). Fourth, it is almost always fallacious to attempt to divide or classify the causes of human behavior or any human feature as caused solely by genetic or environmental factors. Nature and nurture work together in a reciprocal harmony to produce phenotypic or behavioral traits. Finally, it would be logically fallacious to assume that, because a given trait or behavior has its primary source or cause in a specific set of genetic factors, the behavior is no more than the interactions of these factors, or the most efficacious way of understanding human behavior is through genetic analysis (Hull). Reductionism and methodological genetic individualism require more than the existence of a gene/environment story to be shown valid.

While these points are generally accepted by both proponents and critics of behavioral genetics, they are often ignored in the heat of ethical debate. It is especially incumbent on those who hold an ethical theory that accepts empirical scientific findings as relevant to ethical beliefs concerning morality, rights, and justice to take these methodological limitations into account in promulgating their views.

Ethics of behavioral genetics research

As work in the domain of behavioral genetics continues, the question of the social responsibility of such scientists and their liberty to inquire freely into the genetics of human behavior will undoubtedly come more and more to the fore. The question of the liability of genetic scientists for the foreseen and unforeseen consequences of their research on social groups and upon social policies is particularly pressing, given the possibilities that governments, institutions, professions, and businesses will utilize and, perhaps, abuse genetic information. It seems fair to point out that behavioral geneticists qua scientists have no more expertise for resolving questions concerning freedom of inquiry and social responsibility than do lay persons. Moreover, since the data base for resolving ethical and policy questions about scientific research involves information from so many areas of technical expertise, it is imperative that input from all areas of society be obtained before attempts at moral solutions are made. Given the fact that science is a social enterprise and that behavioral geneticists are fallible human beings, the widest possible participation should be sought in trying to lay out the limits of scientific expertise, responsibility, and freedom.

ARTHUR L. CAPLAN

[*Directly related are the entries* ETHICS, *articles on* UTILITARIANISM, OBJECTIVISM IN ETHICS, NATURALISM, *and* NON-DESCRIPTIVISM; FREE WILL

AND DETERMINISM; JUSTICE; MENTAL ILLNESS; REDUCTIONISM, article on ETHICAL IMPLICATIONS OF PSYCHOPHYSICAL REDUCTIONISM; RIGHTS; and SCIENCE: ETHICAL IMPLICATIONS. *For discussion of related ideas, see the entries:* EUGENICS; EUPHENICS; GENE THERAPY; GENETIC CONSTITUTION AND ENVIRONMENTAL CONDITIONING; GENETIC DIAGNOSIS AND COUNSELING; GENETIC SCREENING; GENETICS AND THE LAW; MENTAL HEALTH SERVICES, article on SOCIAL INSTITUTIONS OF MENTAL HEALTH; *and* POPULATION POLICY PROPOSALS, article on GENETIC IMPLICATIONS OF POPULATION CONTROL.]

BIBLIOGRAPHY

BENN, STANLEY I. "Egalitarianism and the Equal Consideration of Interests." *Equality.* Edited by J. Roland Pennock and John W. Chapman. Nomos: Yearbook of the American Society for Political and Legal Philosophy, no. 9. New York: Atherton Press, 1967, chap. 4, pp. 61–78.

BLACKSTONE, WILLIAM T., ed. *The Concept of Equality.* Minneapolis: Burgess Publishing Co., 1969.

BLOCK, NED JOEL, and DWORKIN, GERALD, eds. *The I.Q. Controversy: Critical Readings.* New York: Pantheon Books, 1976.

CHISHOLM, RODERICK M. "Human Freedom and the Self." *Introductory Readings in Ethics.* Edited by William K. Frankena and John T. Granrose. Englewood Cliffs, N.J.: Prentice-Hall, 1974, pp. 289–294.

DOBZHANSKY, THEODOSIUS G. *Genetics of the Evolutionary Process.* New York: Columbia University Press, 1970.

DWORKIN, GERALD, ed. *Determinism, Free Will, and Moral Responsibility.* Central Issues in Philosophy Series. Englewood Cliffs, N.J.: Prentice-Hall, 1970.

EDEL, ABRAHAM. *Ethical Judgment: The Use of Science in Ethics.* Glencoe, Ill.: Free Press, 1955.

EHRMAN, LEE, and PARSONS, PETER A. *The Genetics of Behavior.* Sunderland, Mass.: Sinauer Associates, 1976.

HANCOCK, ROGER N. *Twentieth Century Ethics.* New York: Columbia University Press, 1974.

HERRNSTEIN, RICHARD J. *I.Q. in the Meritocracy.* An Atlantic Monthly Press Book. Boston: Little, Brown & Co., 1973.

HULL, DAVID L. *Philosophy of Biological Science.* Foundations of Philosophy Series. Englewood Cliffs, N.J.: Prentice-Hall, 1974.

JENCKS, CHRISTOPHER; SMITH, MARSHALL; ACLAND, HENRY; BANE, MARY JO; COHEN, DAVID; GINTIS, HERBERT; HEYNS, BARBARA; and MICHELSON, STEPHAN. *Inequality: A Reassessment of the Effect of Family and Schooling in America.* New York: Basic Books, 1972.

JENSEN, ARTHUR ROBERT. *Genetics and Education.* New York: Harper & Row, 1972.

KAMIN, LEON J. *The Science and Politics of I.Q.* Complex Human Behavior. Edited by Leon Festinger and Stanley Schachter. Potomac, Md.: Lawrence Erlbaum Associates; New York: John Wiley, Halsted Press, 1974.

LEWONTIN, R. C. "The Analysis of Variance and the Analysis of Causes." *American Journal of Human Genetics* 26 (1974): 400–411.

LOEHLIN, JOHN C.; LINDZEY, GARDNER; and SPUHLER, J. N. *Race Differences in Intelligence.* A Series of Books in Psychology. San Francisco: W. H. Freeman, 1975.

MELDEN, ABRAHAM IRVING, ed. *Human Rights.* Basic Problems in Philosophy series. Belmont, Calif.: Wadsworth Publishing Co., 1970.

OLAFSON, FREDERICK A., ed. *Justice and Social Policy: A Collection of Essays.* A Spectrum Book, no. S-13. Englewood Cliffs, N.J.: Prentice-Hall, 1961.

RAWLS, JOHN. *A Theory of Justice.* Cambridge: Harvard University Press, Belknap Press, 1971.

SZASZ, THOMAS STEPHEN. *Law, Liberty, and Psychiatry: An Inquiry into the Social Uses of Mental Health Practices.* New York: Macmillan Co., 1963.

WARNOCK, MARY. *Ethics since 1900.* 2d ed. Oxford Paperbacks University series, no. 1. New York: Oxford University Press, 1966.

WILLIAMS, BERNARD. "The Idea of Equality." *Moral Concepts.* Edited by Joel Feinberg. Oxford Readings in Philosophy. New York: Oxford University Press, 1969, chap. 12, pp. 153–171.

WILSON, EDWARD OSBORNE. *Sociobiology: The New Synthesis.* Cambridge: Harvard University Press, Belknap Press, 1975.

GENETIC CONSTITUTION AND ENVIRONMENTAL CONDITIONING

The determinants of individuality

Each person is unique, unprecedented, unrepeatable. This belief, long implicitly accepted in Western civilization as the fundamental credo of humanistic philosophy, can now be scientifically supported by knowledge of the interplay between genetic constitution and environmental forces.

Except for identical twins, no two persons have the same array of genes; the chance that a particular array has occurred in the past or will occur again in the future is practically nil. Furthermore, genetic uniqueness is compounded by environmental uniqueness, because the phenotypic expression is conditioned at every step in development by environmental forces that are forever changing; even identical twins become different persons because, during intrauterine life and increasingly after birth, they are exposed to different environmental conditions.

Each social group also has an individuality, which is determined by genetic and environmental factors. The effects of the genetic determinants are usually obvious in some distinctive ethnic characteristics of the group; the effects of environmental determinants are manifested in the group's conventions and habits.

The field of bioethics includes issues involving the various rights and duties both of individual persons and of social groups; among other things, it must take into consideration their desire to preserve their uniqueness. However, the relative importance of the person and of the group differs from one situation to another. Under primitive conditions, ethical systems probably give less weight to the rights of the individual person than to the behavior patterns that influence the survival of the group. In contrast, modern societies tend to emphasize the rights of the individual person.

Until recent times, the phrase "human rights" referred chiefly to political, social, and economic issues. The new kind of concern symbolized by the word "bioethics" has gained prominence only during recent decades. One of the reasons for its emergence is the fear that scientific knowledge will be increasingly used to manipulate human nature and thus to achieve greater and greater social control over the individual person. Before examining the validity of this fear, I shall illustrate with a few examples that human societies have always influenced the biological and behavioral characteristics of human beings—even though the influence was not necessarily premeditated.

A casual walk through a museum provides evidence of the striking changes in human characteristics that have occurred during the past few centuries. For example, the armor of medieval knights reveals that, even in this aristocratic class, people were then much smaller than the average American or European today; similarly, most eighteenth-century belles were dwarfs when compared with today's average college girl. Behavioral characteristics also have changed. Conflicts of attitudes between generations have always been so prevalent that they have been a perennial subject of discussion for moralists and novelists, beginning with the descriptions of father–son relationships recorded on Sumerian tablets and Egyptian papyri several thousand years ago.

Biological and behavioral changes have thus been a constant feature of human life, and most of them have been the consequence of social influences. Curiously, however, there is now much anguish about the possibility of further change. Paradoxically we are more afraid of the changes that might be brought about consciously by scientific knowledge than of the changes resulting from the operation of blind social forces. I do not share this sense of alarm, and furthermore I do not believe that our means for altering at will man's nature are more effective than those which have been used unconsciously in the past.

Manipulations of the genetic endowment

The biological developments that are most publicized in newspapers, magazines, and other mass media are those commonly listed under the catchy phrase "genetic engineering." They include different approaches to genetic manipulation, such as (1) asexual, clonal reproduction achieved by cultivating cells from a donor whose genetic characteristics it is desired to perpetuate; (2) introduction into an organism of genetic material from another species, so as to produce a new hybrid organism—a chimera; (3) modification of the genes by chemical treatment or by other laboratory procedures.

Theoretical achievements along these approaches are indeed spectacular and not to be dismissed as mere science fiction; but they are still rather far out. They have been developed with primitive organisms and may eventually result in techniques that will alter chemical properties of specialized cells, with possible practical applications. But this does not justify the claim repeatedly made that they will soon make it possible to modify the human genetic endowment.

Even granted that methods of genetic engineering will continue to be developed, there will always be enormous theoretical and practical limitations to their use in human beings. The human organism is so complex and so dependent on social support that any significant alteration of its genetic structure would upset the subtle integration of its attributes, probably with disastrous results. Furthermore, there is as yet no understanding of the relationships on the one hand between the chemical structure of a gene or its position on the chromosome, and on the other hand its functional effect. In consequence, there is no known way to design, select, or transfer a gene capable of doing a desired job.

Another unsolved problem of genetic engineering is how to deal with the fact that each of the important features of the human character is controlled by several genes. There seems to be no chance of finding a gene responsible for, say, intelligence, or honesty, or ability in mathematics, or aptitude for long distance running, or any other attributes that might appear desirable. Since excellence in any physical or mental activity depends upon the combined action of several different genes, significant genetic improvements of human nature would require modifications of the whole genetic apparatus rather than of isolated parts of it—an enterprise that cannot even be visualized at the present time.

At first sight, eugenic approaches to genetic manipulation seem to be of more immediate practical applicability than genetic engineering, especially in view of the recent advances in artificial insemination. However, the eugenic approach presents considerable technical difficulties that are inherent in the essential mechanisms of sexual reproduction. Genes that may be together in the diploid cells of an organism and whose interaction endows that organism with certain qualities commonly become separated from one another in the gametes. The process of segregation makes it impossible to predict from the appearance and performance of a person during life what will be the characteristics of the offspring.

Because of the inescapable uncertainties in the transmission of genetic attributes, there has been a trend in the eugenic movement to play safe and to concentrate efforts on the elimination of genetic defects, but programs of negative eugenics also run into great difficulties. The majority of obviously harmful hereditary conditions are determined by recessive genes and become manifest only in the homozygous state (in which both members of a gene pair are the same, or identical). As a consequence, removing the homozygous carriers from the reproductive pool would have little effect on the frequency of the gene in the population, since the far more numerous heterozygous carriers usually appear normal and remain undetected (heterozygous meaning the members of a gene pair are different, usually one normal and one abnormal, i.e., potentially disease-causing). The method of amniocentesis admittedly makes it possible to detect a genetic defect in the fetus before birth and to induce abortion at an early stage; but the range of applications of this technique is rather narrow.

Assuming that techniques may eventually be developed to detect the heterozygous carriers of many deleterious genes, this would probably create other dilemmas for eugenic policy. In many cases, heterozygotes for "bad" genes tend to have some advantages over homozygotes for supposedly "good" genes (this heterozygote "advantage" is referred to as "polymorphism"). For example, heterozygotes for sickle-cell hemoglobin possess high resistance to malaria, and it has even been claimed that artistic ability tends to be high in schizophrenic pedigrees. In any case, human beings are heterozygous for such an enormous number of genes that many of the most gifted persons are unquestionably carriers of one or more genes that might be considered unfavorable in the homozygous state. The situation is so complex that, in the present state of ignorance, there is no way to formulate an overall policy for the carriers of potentially harmful recessives.

In addition to the scientific difficulties that stand in the way of genetic engineering and of eugenics, other difficulties of a social nature are bound to interfere with all programs of genetic manipulation.

First is the lack of consensus concerning the characteristics to foster or to reject. The definition of the ideal person differs with time, place, and social group, and furthermore it changes rapidly; the heroes of one generation commonly become the villains of the next. All human societies, on the other hand, have a stubborn resistance to change with regard to certain of the conventions and ways of life that they consider normal. Before any program of genetic engineering or eugenics can be implemented on a population scale, it must therefore be made compatible with the traditions of the group. In the final analysis, the greatest stumbling block to genetic improvement may be the difficulty of modifying parent–child bonds, husband–wife relationships, sexual urges and habits, and other aspects of ordinary social life that are extremely resistant to any form of social manipulation.

Environmental conditioning

Whereas it is unlikely that genetic engineering or eugenics can significantly alter the human genetic endowment in any predictable future, environmental factors can achieve a fairly rapid genetic shift within the spectrum of genetic polymorphism of a given population. The classical example of such a shift is sickle-cell anemia, a case in which the presence or absence of malaria is the environmental factor responsible for selection pressure. When a shift occurs within the spectrum of genetic polymorphism, it is the population that adapts, rather than individual persons; the environment determines which form of a particular polymorphic gene comes to predominate under a given set of conditions.

A more direct environmental approach to the genetic problem is to modify the phenotypic (i.e., physical, biochemical, and/or physiological) expression of defective genes; the disorders resulting from genetic defects can be alleviated by treating the affected person, preferably early in development. Classical examples of this approach are the replacement of missing gene products such as gamma globulin (a serum protein essential in the immune response) or insulin, control over production of toxic metabolites in phenylketonuria (a biochemical genetic disease in which the amino acid phenylalanine cannot be reduced to its metabolites and thus builds up in the cells causing damage) by reducing the dietary intake of phenylalanine, and perhaps the use of placental glucocerebrosidase in Gaucher's disease (a neurological disease caused by an excess of unmetabolized fat [lipid] because of a lack of the effective enzyme, glucocerebrosidase).

The influence of environmental factors on the expression of the genetic endowment is much broader than indicated by the few examples just mentioned; it is reflected not only in the control of pathological situations, but in practically all the traits of a population. Throughout history, in fact, humankind has shaped itself by unconscious or conscious manipulations of the environment without undergoing significant changes in its genetic constitution. A few cases will suffice to illustrate how political and social forces are now altering the phenotypic expressions of human nature in several parts of the world.

Most people in Oriental countries used to be smaller than Caucasians, but Asiatic teenagers in Hawaii, Japan, or Taiwan now grow as fast and as tall as their Caucasian counterparts. Patterns of disease also are changing in these countries. Instead of falling prey to infections as in the past, Asiatic people who have adopted the Western ways of life increasingly suffer from vascular diseases, just as Caucasian people do. The biological changes that are now occurring in Oriental countries thus reflect those that occurred in Western countries a few decades after the Industrial Revolution as a result of economic prosperity and environmental changes. Irrespective of racial background, furthermore, behavioral patterns also change along with biological characteristics.

The rapidity with which these changes have occurred in certain national groups excludes the possibility that they are of genetic nature. In fact, experience shows that the human phenotype can be profoundly altered by environmental forces within one single generation without any change in the genotype. In the state of Israel, for example, the boys and girls raised in the kibbutzim tower over their parents who came from the ghettos of Europe and also differ from them in behavior.

The control of behavior

Let us now consider whether control of population behavior can be achieved by new procedures based on modern science. Individual behavior can be modified by different techniques, such as surgery and electric stimulation of certain parts of the brain; the administration of drugs having more or less specific effects; the induction of anxiety states; operant conditioning; and various advertising practices. While each one of these techniques has proved effective under special conditions, there would be great difficulties in applying them continuously on a large scale, because any one of them would soon elicit neutralizing counterforces. Students of advertising have discovered that most advertising campaigns eventually become counterproductive and "self-liquidating." If one cannot fool all the people all the time, the reason is not because people discover the truth, but rather because most human beings have a tendency to doubt what they are instructed to believe.

In any case, there is no evidence that the new techniques of behavior control based on science are more effective than those developed empirically in the past. Modern synthetic drugs have not yet produced any alteration of mob psychology comparable to the peyote religion of American Indians; such drugs have not released greater military fanaticism than that of the Scandinavians gone berserk after consuming the holy mushroom, or of the Arabian cult of Assassins after they had visualized paradise under the influence of hashish. Modern psychological theories have not yielded results more spectacular than those obtained by religious reformers such as Wesley or by political demagogues such as Mussolini or Hitler. Behavioral control has been practiced on a huge scale in all societies since the beginning of time, and its most powerful methods have always been some form of environmental conditioning.

Social factors in environmental control

Biologically as well as politically, man is a social animal. No human being can exist alone. The social group is as significant a unit of human life as the individual person, and both must be given equal weight in any discussion of bioethics.

The many situations in which conflicts now exist between the rights of the individual person and the rights of the social group are the most widely discussed bioethical problems of our times—for example, whether to prolong or to terminate the life of a person, young or old, who is so defective or incapacitated as to be a heavy charge on the family or on society. Rules of conduct in these matters are not clearly defined because, as illustrated by the following examples, a given practice may be thought ethically wrong in one situation yet be thought justifiable under other circumstances.

In 1973 a commission of American physicians spent a few weeks investigating health problems in the People's Republic of China. They were greatly impressed by the improvements that had taken place in the state of health of the Chinese people during the previous two decades, especially among children. They noticed, however, that these improvements had been achieved in part by the use of medical techniques that are potentially dangerous and that would be regarded as medically unethical in the United States and Western Europe. For example, the Chinese "barefoot" doctors observed by the American physicians routinely administered the antibiotic chloramphenicol even for trivial illnesses, without checking for the occurrence of blood abnormalities in patients receiving the drug.

As pointed out in the American report, the widespread and indiscriminate use of chloramphenicol probably had a wide range of beneficial effects such as minimizing the incidence of hematogenous osteomyelitis (an infectious bone disease), preventing the development of rheumatic fever following streptococcal pharyngitis (inflammation of the pharynx), and rapidly arresting meningococcal or typhoid infections. On the other hand, while much grave disease was thus avoided by the routine use of chloramphenicol, there is great likelihood that this policy was responsible for many cases of blood marrow aplasia. Uncontrolled use of a potentially dangerous drug would be considered malpractice in the United States where the sociomedical benefits to be derived from the practice would be small. In contrast, sacrificing a number of people to chloramphenicol in order to improve the health of millions of persons, especially children, may have been judged a socially justifiable policy under the conditions that prevailed in continental China from the 1950s to the 1970s.

Ethical attitudes concerning abortions also have complex and variable social determinants. From scientific, religious, philosophical, and emotional points of view, there are reasonable grounds for opposing abortion. The fetus who is being destroyed is a human being, at least potentially. Furthermore, official tolerance of abortion may progressively lead to a casual attitude toward infanticide first at the moment of birth and eventually at later stages of development. On the other hand, it is a fact that abortion has long been practiced on a huge scale all over the world, usually under deplorable conditions. The ethical justification for legalizing abortion is that the new policy makes it possible to practice it under strict medical control, thus practically eliminating mortality among the aborted women and greatly decreasing the number of unwanted children.

These two examples illustrate that the phrase "reverence for life," appealing as it is, has an ambiguous meaning and can lead to opposite attitudes under different conditions. From the biological point of view, the fate of the group as a whole is more important than the welfare of any member of it; from the humanistic point of view, it is the individual person that counts; from the social point of view, attitudes on this score differ according to circumstances. One of the most interesting and appealing paradoxes of human life is that practically all civilizations have regarded the sacrifice of a person, real or symbolic, as contributing to the furtherance of the group. Christians uphold both self-sacrifice ("the Cross principle") and the sanctity of individual life.

The person and the social group

A community is more than a congeries of people, places, and buildings; it consists of an association of persons who willingly function as part of an integrated whole because they share certain customs and values. Belonging to a community therefore implies limitation of individual rights and acceptance of certain behavioral patterns.

The exercise of free will is now regarded as one of the most fundamental human rights, but in practice all human beings are conditioned by their total environment. Taboos, customs, parental training, and other educational practices have been used traditionally and most often unconsciously to shape attitudes and tastes; all such forms of conditioning interfere in some degree with the manifestations of free will. Social forces have always shaped human mentality even under conditions of absolute political freedom.

The fact that the development of all traits is irreversibly conditioned by environmental stimuli, especially those experienced early in life, makes it theoretically impossible to design a social structure that would provide absolute freedom. The most and best that society can do is to provide environmental conditions as diversified as possible so that each person has a wide range of options from which to select in order to develop his particular attributes and for the pursuit of his goals.

One of the criticisms most commonly leveled against modern societies is that they "manipulate" human beings, especially through the mass media, and thus foster social conformity. In fact, however, the modern world is at least as favorable to the exercise of individual freedom and to the development of individual characteristics as any period in history. All over the world and at all times, socialization has meant conditioning of tastes, attitudes, and goals, in other words "manipulation" of people. Human beings have always been "manipulated" by the particular social group in which they are born and develop. The very existence of so many different cultures is evidence that the potentialities common to all human beings have been molded into different existential shapes by social constraints and by educational processes.

Primitive societies have a fairly narrow repertoire of conditioning practices, whereas modern technological societies expose their members to a multiplicity of simultaneous conditioning stimuli. The traditions of such societies are complex, diversified, and fruitfully discordant. They nurture within themselves the seeds of innumerable revolutions. They are inventive and flexible. By increasing the range of stimuli and of options, they make possible a wider range of social phenotypes, or in other words of social diversity.

The greatest threats to the modern world, in fact, come not from social homogenization, as is commonly stated, but rather from social instability. While diversity of social influences and of environmental conditions is an asset because it facilitates the unfolding of human potentialities, there are limitations to the use that can be made of this asset. Freedom involves not only what to do but also what not to do. Complete permissiveness is unbiologic and would inevitably result in the disintegration not only of the social order but also of individual lives.

Individual vs. "statistical morality"

The Hippocratic Oath deals with the rights, duties, and personal relationships of physician and patient. The rules of medical morality which it codifies have retained much of their value, but this does not mean that they are universal or immutable. For example, attitudes toward the death of very young and very old people differ profoundly from one population to another. Among contemporary Westerners attitudes toward venereal diseases, mental disorders, contraception, abortion, or euthanasia are rapidly changing. It can be anticipated that the increase in scientific knowledge and in the cost of medical care will bring about still further changes in attitudes with regard to pathological conditions that create biologic or economic dangers for society.

The Hippocratic code was formulated at a time when medical action affected only the individual patient and his immediate family. In contrast, some of the modern biomedical technologies affect the population as a whole and even the following generations. The very effectiveness of modern medicine therefore makes it imperative that individual morality be supplemented by "statistical morality"—an attitude concerned not so much with the welfare of the individual patient and his immediate family as with the population at large now and in the future.

Present discussions concerning the advisability of continuing mass programs of smallpox vaccination illustrate the statistical nature of certain medical problems. Post vaccinial encephalitis (inflammation of the brain) causes every year a significant number of fatalities, but on the other hand smallpox has all but disappeared from many parts of the world. Continuing mass vaccination is therefore tantamount to sacrificing every year a certain number of persons for the sake of protecting society against any recurrence of smallpox epidemics. Such problems are now appearing in many medical specialities. Renal dialysis, organ transplantation, and the use of organ substitutes present difficult problems that are not purely medical but refer to selection of patients and donors, the shortage of competent personnel, the cost to the community, and the social disturbances created by the accumulation of patients requiring exacting care for medicated survival. These problems of medical ethics demand a kind of social evaluation rarely needed in the traditional physician–patient relationship. They must be considered from the point of view of their social consequences as well as from the point of view of the patient's welfare.

The formulation of medical programs for undeveloped countries presents still other types of conflict between the individual and the population at large. From the humanitarian point of view, it is imperative to improve the health of old people wherever they are. From the social point of view, however, it is more useful to focus scientific research and medical action on children and young adults, because they are the creative forces of the present and the future. Wherever resources are limited, choices thus have to be made between sympathy for the individual person and concern for society as a whole—choices that are bound to be painful because the demands of societywide morality are usually in conflict with other humanitarian impulses.

Many medical problems do not require soul searching. All over the world, much disease can be traced to a variety of social defects that are well understood, for example, shortage of decent food or of clean water and exposure to toxic substances such as lead in slum dwellings or chemical fumes in factories. It is intellectually stimulating to discuss the bioethical problems posed by esoteric medical technologies that affect small numbers of peope. From the standpoint of social morality, however, it would be more important to focus public and scientific attention on issues that are simpler but affect immense numbers of people, in prosperous as well as in developing countries.

Since the financial and personnel resources that can be devoted to medical problems are limited even in the most prosperous nations, matters of priorities will inevitably generate conflicts also in the field of scientific research. From the social point of view, for example, it might be more profitable to focus research on the prevention of cancer and heart disease, rather than on the treatment of these conditions, even though matters of therapy appear of greater urgency for the stricken person and for the family.

Thus, many bioethical problems arise more from the social than from the scientific aspect of the question. As the cumulative cost of medical research and of medical care increases, both in money and in personnel, medical ethics must be reconsidered in the harsh light of social economics.

Summary

There is hardly any chance that the genetic constitution and the fundamental behavioral patterns of the human species can be significantly altered by scientific interventions in the near future. Limiting our vision of bioethics to such improbable occurrences is tantamount to escapism from many other pressing ethical problems of our present societies. The real issue is to design physical and social environments in which the human animal can function best, by getting the most "mileage" out of the genes that govern its nature and behavior.

The preceding paragraph does not mean that "you cannot change human nature," because the word "nature" implies more than genetic constitution. Profound changes in the biological and behavioral characteristics of humankind—its phenotypic nature—have continuously occurred in the past and continue to occur now. Moreover such changes can take place rapidly, and they affect whole populations. However, they are not the consequence of alterations or manipulations of the genes or of the brain, but of the fact that all phenotypic expressions can be rapidly shaped by environmental forces. The most acute and immediate problems of bioethics have their origin in the influence that surroundings and events have on biological and mental personality—and hence on the future.

The greatest bioethical problem of our time is how best to provide conditions that enable all human beings to actualize their potentialities in a socially acceptable and desirable way, while interfering as little as possible with the exercise of their free will. The characteristics of the physical and social environment, rather than the manipulation of the genes and the brain, present the problems of greatest significance for the welfare of our contemporaries and for the future of humankind.

RENÉ DUBOS

[*For further discussion of topics mentioned in this article, see the entries:* BEHAVIOR CONTROL; ENVIRONMENTAL ETHICS; EUGENICS; GENE THERAPY; LIFE; *and* PRENATAL DIAGNOSIS. *Also directly related are the entries* EUPHENICS; EVOLUTION; FREE WILL AND DETERMINISM; GENETIC ASPECTS OF HUMAN BEHAVIOR; *and* PUBLIC HEALTH. *Other relevant material may be found under* GENETIC DIAGNOSIS AND COUNSELING; GENETIC SCREENING; HEALTH CARE; *and* MASS HEALTH SCREENING. See APPENDIX, SECTION I, OATH OF HIPPOCRATES.]

BIBLIOGRAPHY

BLOOM, BENJAMIN S. *Stability and Change in Human Characteristics.* New York: John Wiley, 1964.

DOBZHANSKY, THEODOSIUS. *Heredity and the Nature of Man.* New York: Harcourt, Brace & World, 1964.

———. *Mankind Evolving.* New Haven: Yale University Press, 1962.

DUBOS, RENÉ. *Beast or Angel? Choices That Make Us Human.* New York: Charles Scribner's Sons, 1968.

———. *Man, Medicine and Environment.* New York: Praeger, 1968.

———. *Man Adapting.* New Haven: Yale University Press, 1965.

———. *The Mirage of Health: Utopias, Progress, and Biological Change.* New York: Harper & Row, 1959.

———. *So Human an Animal.* New York: Charles Scribner's Sons, 1968.

MEADE, J. E., and PARKES, A. S., eds. *Biological Aspects of Social Problems.* New York: Plenum Press, 1965.

NEUBAUER, PETER, ed. *Children in Collectives: Childrearing Aims and Practices in the Kibbutz.* Springfield, Ill.: Charles C. Thomas, 1965.

ROSLANSKY, JOHN D., ed. *Genetics and the Future of Man.* Amsterdam: North Holland Publishing Co., 1966.

SONNEBORN, T. M., ed. *The Control of Human Heredity and Evolution.* New York: Macmillan Co., 1965.

WADDINGTON, CONRAD H. *The Ethical Animal.* London: George Allen & Unwin, 1960; Chicago: University of Chicago Press, 1967.

WOLSTENHOLME, GORDON E., ed. *Man and His Future.* Boston: Little, Brown Co., 1963.

GENETIC COUNSELING

See GENETIC DIAGNOSIS AND COUNSELING, *article on* GENETIC COUNSELING; PRENATAL DIAGNOSIS.

GENETIC DIAGNOSIS AND COUNSELING

I. GENETIC DIAGNOSIS — *Robert F. Murray, Jr.*
II. GENETIC COUNSELING — *Robert F. Murray, Jr.*

I
GENETIC DIAGNOSIS

Heterozygote detection to prevent disease

One important goal of genetic diagnosis and counseling is to prevent genetic disease in individuals or families. The usual approach to this problem has been to identify moderate- to high-risk families (recurrence risk of twenty-five percent or greater) through the birth of an affected child and then provide comprehensive genetic counseling to such families. Even when direct advice is not given, counselors expect many of the parents in the moderate- to high-risk group to decide to have fewer or no further offspring after they learn of the increased recurrence risk for affected children or, whenever possible, to have prenatal diagnosis followed by selective abortion to avoid the birth of a second or third affected child.

This approach to prevention, in which reproductive control is instituted only after the birth of the first affected child, will only slightly lower the number of affected children born (Motulsky, Fraser, and Felsenstein). It follows that, if any significant reduction in the case load of a particular disorder is expected, couples or individuals at risk to have affected children will have to be identified before the first affected child is born. This is one reason why detecting the healthy or clinically unaffected gene carrier is so important.

The generally accepted definition of a gene carrier is a clinically healthy person who is heterozygous for a mutant gene determining an autosomal or X-linked recessive trait (King). The heterozygous gene carrier has two genes at the same place or locus on a specific pair of chromosomes. Both genes code for the same function, but they differ in structure, although usually only very slightly. One gene of the pair is usually the more common or "normal" one and the other is an atypical or abnormal one. If the mutant or abnormal gene must be present in double dose to produce clinical disease, it is termed recessive. If only one gene is required for the full clinical effect, it is termed dominant.

Some heterozygous carriers of genes that determine autosomal dominant traits of late onset, e.g., Huntington's disease, go through most or all of their reproductive lives before there is clinical evidence that the inherited disorder is present. From a clinical perspective and for pragmatic reasons, then, the definition of the heterozygous gene carrier can be justifiably broadened to include any persons who carry in their genes or genetic material a genetic abnormality that may possibly produce disease in them at some time during their lives and can be inherited by their offspring in whom it may or may not produce a disorder of clinical significance. In certain respects it is just as important, if not more important, for the carrier of a gene determining an autosomal dominant trait such as myotonic dystrophy or Huntington's chorea to be aware of it as it is for a woman who carries the gene that causes the Duchenne type of muscular dystrophy (an X-linked recessive trait), or an individual who is found to carry a gene that causes Tay-Sachs disease (an autosomal recessive trait), to be informed that they are carriers. Detecting the heterozygous carriers of genes determining autosomal dominant traits prior to childbearing and preventing them from having offspring could, if imposed absolutely, in one generation dramatically reduce the frequency of the gene in the population to the level where essentially all cases of the disease that occurred would be due to spontaneous mutation. If all female carriers of X-linked traits were detected and prevented from having affected male offspring by prenatal diagnosis and selective abortion, and if affected male offspring were prevented from having daughters, the frequency of a given X-linked trait could be dramatically reduced in one or two generations. Even if this approach to prevention can be ethically justified, it is essential that the methods of heterozygote detection in X-linked conditions meet the criteria of sensitivity and/or reliability for modern methods of heterozygote detection. As these tests are developed to detect the heterozygous carrier and to distinguish affected fetuses, parents who wish to will be able to choose selective abortion to avoid having affected offspring.

The goal of preventing or eliminating inherited disease can be most efficiently achieved, in a medical sense, by detecting heterozygotes for autosomal dominant and X-linked recessive traits. The technology for detecting such autosomal dominant traits in carriers is only in its infancy and can be applied primarily to metabolic disorders such as certain forms of porphyria and hypercholesterolemia. The emphasis in heterozygote detection will probably continue to be on the heterozygous carriers of disorders inherited as autosomal recessive traits where tests are more reliable. In these instances genetic counseling is much more accurate, and prenatal diagnosis is more often possible.

From a medical standpoint, heterozygote detection appears to be clearly justifiable when the following criteria have been met:

1. A defined high-risk subgroup exists in the population.
2. A relatively simple and reliable test exists for the heterozygote.
3. Couples where both are heterozygous for an autosomal recessive trait are able to have unaffected children through amniocentesis and/or low-risk prenatal diagnosis.
4. Effective medical therapy is available for the affected individuals and/or offspring of matings at risk (Kolodny, p. 197).

A relatively frequent inherited disorder that meets a majority of these criteria is Tay-Sachs disease. As knowledge of the etiology of genetic disorders grows, and with further technical development, many more disorders will be counted among the conditions in which heterozygote detection is clearly medically justified.

Ethical conflicts in heterozygote detection

Whenever programs designed to prevent disease are launched, there is a determined attempt by public health agencies to get most if not all persons at risk involved in the testing program, so as to eliminate the disease if at all possible. Efforts to eradicate disease are expensive, and public officials want the programs to be as cost-effective as possible in order to reduce the economic burden of disease on society. The value to be emphasized in their view is the general welfare. From a public health perspective, therefore, it would probably be most desirable to force everyone in the population at risk to undergo heterozygote testing and, whenever couples at risk are found, to require them to be counseled. Couples so identified would also have to have all pregnancies monitored by amniocentesis in order to detect any or all affected or potentially affected offspring so that, when it is appropriate for the disorder in question and there are no moral or religious prohibitions, couples would be encouraged to choose abortion to avoid the birth of an affected child.

A program of this kind, despite its goal of relieving suffering in persons with genetic disease and in their families, would infringe upon the important value of freedom. Patients have always had the right to choose their own medical destiny after being fully informed of the alternative courses of action. In this instance they would also have to consider the value of potential human life in spite of the presence of serious disease and perhaps early death. It is important to weigh the loss of an individual's desire or even right to remain ignorant of his or her genetic makeup against the risk of the person's someday having a child or children with an inherited disorder. In a society where it might become illegal *not* to know or deliberately to avoid knowing one's heterozygote carrier status, the person who opts to remain ignorant would be a criminal. Although their rights are threatened, couples who can accept selective abortion as a way of terminating genetically diseased children, avoiding the suffering in the family, and enhancing the chance of having a healthy child might find the net effect of a mandatory heterozygote detection program to be positive from an ethical perspective. The hard question to be answered might be whether selective abortion could be considered an acceptable action to take for all genetic diseases detectable prenatally.

There are many cases where heterozygotes can be readily detected but where effective therapy is not available and prenatal diagnosis is not possible or has not yet been established as safe. Such is the case with the hemoglobinopathies, e.g., sickle-cell disease or thalassemia major, for which reliable treatment is unavailable and prenatal diagnosis, although possible, is quite risky. Further developments of this procedure or new diagnostic methods may significantly reduce this risk. Mandatory heterozygote testing in that latter situation is only weakly justified from an ethical perspective since the options available for the heterozygous couples at risk who wish to avoid having affected children are primarily negative ones. These include (1) to avoid having any more children; (2) to adopt children; (3) artificial insemination of mother by a noncarrier donor; (4) artificial insemination of a noncarrier surrogate mother by the father's sperm; or (5) to wait for a medical breakthrough that provides effective treatment.

Persons coerced into heterozygote testing programs may feel that being given genetic information without also being offered ways to avoid having children with serious inherited disease is not justified on ethical grounds. Not only are they uncompensated for the loss of their free and uninhibited reproduction, but the genetic information may provoke anxiety in the heterozygote. This is often because the information is usually unexpected and also because this knowledge, if passed on to others, may significantly influence the desirability of the individual in the eyes of others as a marriage partner. The knowledge of how much heterozygotes for genetically determined disorders are stigmatized by that information is essential to a balanced consideration of the ethical conflicts involved.

These ethical problems are magnified and complicated further if one considers the options for acting on heterozygote detection in X-linked recessive disorders. The method of preventing or eliminating genetic disease in prenatal diagnosis of many X-linked disorders is the selective abortion of male fetuses wherever methods of distinguishing affected from unaffected males are not yet available. Parents and society must balance the destruction of a healthy fetus against the risk of a diseased one. Selective abortion of female offspring of affected male heterozygotes is one way of attempting to eliminate X-linked disease. To try to eliminate genetic disease by destroying otherwise healthy gene carriers, however, simply cannot be ethically justified under the traditional value systems we follow. One ought not to destroy healthy persons to control disease. On the other hand, in families where boys are born with severe X-linked disorders such as the Duchenne form of muscular dystrophy and one or more affected children are already present, the severe emotional and financial burden of rearing such children in the family could conceivably tip the ethical balance of values in the minds of the parents so that taking the fifty percent risk of aborting a potentially unaffected male fetus might appear to be justifiable when considered against the chance of yet another child affected by a chronic progressive illness.

The capabilities for heterozygote detection have expanded rapidly and will continue to do so. But it is clear that there are so many factors involved in the possible outcome of testing that, when all ethical factors are considered, it appears better to continue such programs only on a noncoercive basis, whereby persons enter detection programs only after they have been thoroughly educated about (1) their risk of being heterozygous and (2) the possible outcomes of the testing program in a medical, psychological, and social sense, prior to having the test performed (Lappé, Gustafson, and Roblin). There should also be leeway in the operation of the program for the expression of ethically different points of view.

Special problems of inherited diseases of late onset in heterozygotes

In some instances the clinical diagnosis of a genetically determined condition carries with it not only the usual medical problems but also special ethical problems. These special problems differ from those in common medical disorders because the condition diagnosed may have grave implications not only for the patient but also for a large percentage of his or her offspring.

There are several autosomal dominant conditions of late onset that result in either chronic debilitation or premature death but where the affected individual may already have completed his or her family by the time the signs or symptoms of the disease have appeared or been detected. The diagnostician, whether a family practitioner or medical subspecialist, must come to grips with the conflict between the ethical principle of truth-telling and the patient's right, and in some cases even the patient's desire, to remain ignorant of information about which he can do nothing but worry. But the need of many to remain as free as possible of unnecessary anxiety stands in opposition to the desire of society to reduce its burden of illness in the interest of the general welfare and to protect future generations as much as possible from the burden of disease caused by deleterious genes.

An example of this kind of disorder is Huntington's disease, a progressive, degenerative neurological condition that will inevitably result in death. When diagnosis is made very early, i.e., at a time when the patient has no idea he or she is affected, the counselor is faced with a critical decision. Should the patient be told about an incurable condition like this one, with such a devastating prognosis that a significant proportion of those who know about the outcome or have seen its end stages commit suicide (Reed and Chandler)? Or should the information be withheld until the time when the symptoms become obvious to the patient, thereby postponing at least for a while the long period of anxiety and depression that such patients so frequently experience when first told of the diagnosis? The second course of action may avoid severe emotional upset and perhaps postpone suicide, but in the interval between the diagnosis and the appearance of definite symptoms the patient and spouse may have one or more children, each of whom has a fifty percent risk of being affected with the disorder.

There may be considerable reluctance, and the potential heterozygote may even resist being told of his or her potential for developing a chronic, often debilitating, and perhaps lethal genetic disease even when there is a potentially reliable method of detecting the heterozygote (Klawans et al.). One study of a possible method for detecting heterozygous carriers of a gene for Huntington's disease revealed that twenty-three percent of the persons at risk to inherit the gene stated they would not take advantage of the test because they felt they were better off remaining ignorant of their carrier status (Eldrige and Stern).

There are clearly conflicting ethical issues here: (1) the physician's or counselor's obligation to tell the truth and (2) the patient's or client's right to full disclosure stand in conflict with (3) the client's desire for well-being and (4) the client's occasional desire to remain ignorant of events that he or she is powerless to influence and that might even lead to reproductive isolation and/or self-destruction. (5) The value or the quality of a life destined to end in the tragedy of Huntington's disease may also be questioned, but it should be remembered that no one is guaranteed an average life expectancy at birth no matter how healthy one's childhood. What good things they accomplish in the years prior to the onset of this kind of disease may more than balance the suffering of the patients and their families. (6) There are also economic costs for society and the families of the afflicted individuals, because their last years are invariably spent in a mental institution or hospital for chronic disease. (7) Finally, there are concerns for the potential health of future offspring and the quality of the gene pool. By implementing vigorous programs of sterilization or rigid restriction of reproduction of those persons at risk to develop this disease, it is possible drastically to reduce it or nearly eliminate it. But the value of eliminating this source of personal suffering and economic burden must be weighed against the pain and loss of freedom of the thousands of people involved in such a program.

Everyone at risk to carry the gene and therefore pass it on to their offspring must be involved in this kind of eugenics program, but, in the absence of a reliable method of detecting a gene carrier, half of those at risk will be unnecessarily penalized since they will not by chance be gene carriers. From their standpoint a program like that is clearly unjust.

No course of action can be totally satisfying under these circumstances. The clinician or counselor must carefully consider the patients' needs and their capacity for tolerating emotional pain and weigh this against the risk of harm to future generations and the right of persons who quite accidentally inherit mutant genes to as full and complete a life as possible. The ethical equation will probably always contain too many unknowns to provide a satisfactory solution.

ROBERT F. MURRAY, JR.

[*Directly related are the entries* GENE THERAPY; GENETIC SCREENING; GENETICS AND THE LAW; *and* PRENATAL DIAGNOSIS. *For further discussion of topics mentioned in this article, see the entries:* ABORTION; EUGENICS; STERILIZATION; *and* TRUTH-TELLING.]

BIBLIOGRAPHY

ELDRIDGE, R., and STERN, R. "Huntington's Disease—Attitudes toward Genetic Counseling." *American Journal of Human Genetics* 25 (1973): 26A.

KING, ROBERT C. *A Dictionary of Genetics.* New York: Oxford University Press, 1968, p. 35.

Klawans, Harold L., Jr.; Paulson, George W.; Ringel, Stephen P.; and Barbeau, André. "Use of L-DOPA in the Detection of Presymptomatic Huntington's Chorea." *New England Journal of Medicine* 287 (1972): 1332–1334.

Kolodny, Edwin H. "Heterozygote Detection in the Lipidoses." *The Prevention of Genetic Disease and Mental Retardation*. Edited by Aubrey Milunsky. Philadelphia: W. B. Saunders Co., 1975, chap. 7, pp. 182–203.

Lappé, Marc; Gustafson, James M.; and Roblin, Richard. "Ethical and Social Issues in Screening for Genetic Disease: A Report from the Research Group on Ethical, Social and Legal Issues in Genetic Counseling and Genetic Engineering of the Institute of Society, Ethics and the Life Sciences." *New England Journal of Medicine* 287 (1972): 1129–1132.

Motulsky, Arno G.; Fraser, George R.; and Felsenstein, Joseph. "Public Health and Long-Term Genetic Implications of Intrauterine Diagnosis and Selective Abortion." *Birth Defects: Original Article Series* 7, no. 5 (1971), pp. 22–32. Symposium on intrauterine diagnosis held at Indianapolis, Indiana, 13–14 October 1970. Edited by Daniel Bergsma.

Myrianthopoulous, Ntinos C. "Review Article: Huntington's Chorea." *Journal of Medical Genetics* 3 (1966): 298–314. An excellent article that reviews the state of knowledge of Huntington's chorea with the emphasis on the epidemiology and age of onset of clinical symptoms.

Reed, E. Edward, and Chandler, Joseph H. "Huntington's Chorea in Michigan: I. Demography and Genetics." *American Journal of Human Genetics* 10 (1958): 201–225.

II
GENETIC COUNSELING

Historical aspects

Genetic counseling can be considered a special form of medical advising, differing from the conventional doctor–patient situation in the following respects: (1) the disorders or diseases involved are primarily a consequence of abnormalities of the genes or the genetic material. (2) The focus of decisions to be made is usually on future children, i.e., with the chance of a condition's occurring or recurring in the offspring of an individual couple. (3) The family or couple, rather than an individual patient, is often of primary concern.

Genetic counseling developed as an outgrowth of the eugenics movement during the late 1920s and early 1930s. A symposium on genetic counseling was held as early as 1934. Early counseling was carried out by geneticists at the Eugenics Society Record Office.

The first university-based heredity clinic in the United States was established in 1940 at the University of Michigan. The Dight Institute was established at the University of Minnesota in 1941 and was one of the first agencies publicly to offer genetic counseling services in America, although similar services had been previously offered by the Eugenics Education Society on a smaller scale (Reed).

Although in the earliest days most counselors were Ph.D. geneticists motivated in large part by the principles of negative eugenics, physicians began to play a more significant role in counseling after medical genetics was recognized as a medical discipline, with the establishment of divisions of medical genetics at the University of Washington in 1956 and the Johns Hopkins School of Medicine in 1957, and the Department of Medical Genetics at the University of Wisconsin in 1958. The publication of a small book entitled *Counseling in Medical Genetics* by Dr. S. C. Reed of the Dight Institute in 1955, along with the rapid expansion of interest and research in medical genetics, stimulated greater interest in counseling among physicians.

Definition of genetic counseling

The definition and purview of genetic counseling have changed considerably since it was introduced. Its focus has broadened from merely establishing and interpreting the recurrence risk of genetically determined disorders to the generally accepted, more comprehensive definition of the American Society of Human Genetics (Fraser), which describes the purposes and scope of modern genetic counseling.

In genetic counseling, one or more specially skilled and/or trained individuals attempts to communicate to a counselee and/or his family the diagnosis, genetic mechanism, prognosis, and alternative courses of action available to manage a genetically determined disorder. Where necessary, the counselee should be able to decide on a course of action that will be consistent with the medical, economic, and psychosocial burden of the disorder, the immediate and long-range goals, and ethical and religious values that will lead to the best possible adjustment for all involved.

This broad definition makes the counselor responsible for communicating accurate information, for being sure that the diagnosis is correct, and for helping the family make decisions about the financial and psychological adjustments required to meet the needs of the affected person or persons in the family.

The number of disorders determined by single genes of large effect is approaching 2,000 and continues to increase annually. This number does not include conditions determined by two or more genes. Genetic diseases can and do involve all organs and organ systems. Although they are diagnosed with greatest frequency at birth or in the first years of life, certain ones are more characteristic of middle age. Because no single medical practitioner can master all the skills needed to provide the kind of genetic counseling required by the comprehensive definition of counseling now accepted, most workers in the field agree that comprehensive counseling can best be delivered by a team of properly trained, highly motivated professionals. Since genetically caused conditions involve all organ systems and all age groups, comprehensive counseling is best pursued in a medical center under the supervision of a physician—primarily because an accurate diagnosis is the cornerstone of reliable, comprehensive genetic counseling (Murphy and Chase, pp. 11–15).

Why people seek counseling

Most people who need or want genetic counseling have either had a child with a defect presumed to be of genetic origin, need information, or are seeking a special diagnostic test or procedure. The most frequently encountered reasons for seeking genetic counseling are summarized in the following seven situations.

1. The client may have just learned that one parent, sibling, or close relative has a hereditary condition that he or she is at risk to inherit and wants to know what that risk is and, perhaps, what can be done to avoid it.

2. A couple has given birth to a child with a genetically determined abnormality and wants to know what the chance of having another such child might be and what they can do about it.

3. A couple who have had two or three spontaneous abortions or have been unable to conceive wish to know if their problem has any genetic basis.

4. A mother or father with one or more abnormal offspring may be contemplating sterilization, if the cause is genetic, to prevent the birth of future affected offspring.

5. A couple who have had a previously affected child may want to have special diagnostic tests or to undergo prenatal diagnosis for a genetically determined or inherited condition.

6. Parents of a child with a genetically determined abnormality would like to know whether *that* child or their unaffected children are at risk to have affected offspring.

7. An individual involved in a modern genetic screening program may have been identified as the carrier of a mutant gene which is responsible for an inherited disorder when two such genes are inherited by the same person.

Although these are the most common reasons why families seek counseling, there are a wide variety of other important but less common reasons why counselees may seek or be referred for genetic counseling:

1. Sometimes there is a conflict in the family over which member of the couple or which side of the family is responsible for the genetic defect.

2. The mother and father may wish to know whether anything they did, e.g., drug ingestion, was directly responsible for causing the abnormality.

3. An unmarried person who is a carrier may wish to know what risk there is of having a child with the disease caused by the mutant gene.

4. A couple has learned through a genetic screening program that they are both carriers of the same mutant gene and are, therefore, at a twenty-five percent risk to have a child with an inherited disease determined by that gene.

5. A couple has learned that they are distant relatives and wish to know what effect this will have on their chance of having an abnormal child.

The genetic counseling process

The correct diagnosis of a genetically determined disorder rests on a comprehensive medical evaluation, which should include: (1) a good history and physical examination of the proband (the affected person through whom the family is identified), the parents, and at least one unaffected brother or sister (occasionally an autopsy may provide physical information); (2) an accurate family pedigree of at least three generations; and (3) all indicated laboratory studies. Once the correct diagnosis is made and the pattern of occurrence in the family has been established, the genetic counseling process can begin.

In all medical situations the person seeking help is, theoretically, free to make his or her own decision about the therapeutic course of action; but, in fact, most persons usually follow the advice they receive from their physician. A similar relationship exists between most physician-counselors and the clients who have to deal with a genetically determined disease in themselves or their offspring, since persons with genetic disorders must be cared for and treated just like anyone else who is ill. The counselor-counselee relationship takes on an added dimension, because the clients, their parents, or other genetically related family members may have to make decisions that affect their unborn children and that may affect future generations of offspring, and because of the statistical nature of genetically determined events. Meaningful counseling also requires effective communication, probably a minimum level of education in the counselee, and a high degree of motivation in both the counselor and counselee.

The complete counseling process consists of four distinct parts. First, there is the process of client characterization, in which the intellectual, psychological, and socioeconomic makeup of the client is determined by the counselor. Second, there is the educational aspect of the counseling process in which the counselor not only provides information about the disease but also tries to teach the counselee how genes are inherited and the way that the simple laws of chance operate. The counselee not only should be able to understand what reproductive options are available but also must appreciate what the consequences of following each option might be.

As a third part of the process, the counselor must evaluate the educative process to be certain that the client has received and understood the facts transmitted and is mentally and emotionally able to make any necessary decisions.

Finally, there is a continuing need for followup, both immediate and long-term, to be certain that the most important facts and concepts transmitted have not been forgotten, and to determine what effect counseling might have had on whatever decision or decisions the family has made.

Decision making in genetic counseling is fraught with deeply felt emotional tensions within and between individuals, and sometimes has life-and-death consequences. The counselor's purpose in followup is to provide emotional support and be a source of additional information needed to help the counselees take action appropriate for them.

These components of counseling—namely, characterization, education, evaluation, and followup—go on continuously and simultaneously. Even as counseling proceeds, new information about the pattern of occurrence of the disorder in the family or new information from other families with a similar problem may make it necessary to reevaluate and/or reeducate the family or present new recurrence risks. Periodic review of the family records in the genetics clinic may show that new information should be added. Patients with some genetic disorders should be reexamined and the working diagnosis reconfirmed periodically. In this way, errors in diagnosis can be rectified since the consequent course of action can have such a profound impact on future decisions made by the family.

Qualifications for genetic counseling

Although there appears to be agreement on a working definition of genetic counseling, there is considerable difference of opinion among those in the field about who is best qualified to do counseling. There are conflicting opinions about what the qualifications of counselors ought to be. There is no licensing procedure for genetic counselors, nor is there any generally accepted standard describing minimum training and/or experience for counselors. The largest group (eighty percent) are counselors with the M.D. degree, and the next largest (eleven percent) are the Ph.D. geneticists (Sorenson and Culbert). The so-called medical paraprofessionals —nurses, psychologists, and social workers— now constitute the smallest group (nine percent) of genetic counselors, but their participation is growing rapidly. These health professionals, called genetic assistants, genetic associates, or genetic advisers, are trained in special two-year programs and receive a Master of Science degree in counseling; as counselors or counselor assistants, they may eventually do most of the counseling in simple, uncomplicated cases.

There are very good reasons why the physician might be best suited for genetic counselor, and equally good reasons have been presented that favor the non-M.D. health professional as genetic counselor. Furthermore, in spite of considerable discussion and debate on the topic, there is no consensus on the criteria for certification of genetic counselors. Such criteria might develop when the understanding of the counseling process and the critical ingredients in counseling are more clearly defined.

The needs in comprehensive counseling have made it necessary to develop a counseling team. It consists of specially trained professionals led by the physician-counselor, who maintains overall responsibility for the counseling process including verification of the diagnosis, provision of accurate and comprehensive counseling, and adequate followup, as well as coordination of such special diagnostic procedures as amniocentesis. A paramedical professional might deliver and interpret the medical and genetic information. He or she would work with the family under the direct supervision of the physician and collaborate with the physician to coordinate any additional special services required by the clients.

Since there is no clear consensus on the qualifications of the ideal counselor, it has been proposed that there be certification of genetic counseling clinics based on the presence of a counseling team consisting of qualified physicians and nonphysician counselors who have available or have access to all the supportive expertise and diagnostic capabilities necessary to deliver comprehensive genetic counseling (McKusick). Until the time when some type of uniform quality control or certification exists in genetic counseling, the counselee will have to approach an unknown counseling clinic with caution.

Psychological and social considerations in genetic counseling

When a child or adult is found to have a genetically determined disorder, there are almost always emotional reactions in the affected individual, the parents, and other members of the family. The emotional reactions vary according to the nature and severity of the condition, as well as the expectations of the family.

The nature of the communications process would indicate that strong emotions or unresolved emotional conflicts will almost certainly interfere with the educational aspect of counseling. In the opinion of some, the focus of effective counseling should be on the emotions of the counselees and not on transmitting factual material about the genetic disorder itself. Franz Kallmann, a psychiatrist and one of the early genetic counselors, characterized genetic counseling as "explanatory or manipulative forms of short term psycho-therapy aimed at reducing anxiety and tension." The therapist must deal not only with guilt, hostility, and anxiety but also with the client's subjective interpretation of genetic odds (Kallmann).

With a high genetic risk of fifty percent (generally accepted by geneticists), or one in two, optimistic parents who have one affected child and who thought that all their children would be affected by the condition may be happy because they have an even chance to have a normal child. By the same token, parents with a one in four or twenty-five percent recurrence risk may choose to emphasize the three in four odds of having an unaffected or healthy child. In contrast to these attitudes is that of the parents who see a low genetic recurrence risk of one in twenty or less as unacceptable because, in their depressed and sometimes paranoid state of mind, they feel that the second affected child will be born to *them* rather than to some other couple at risk. Such couples feel singled out for punishment and may tend to think that "everything happens to me." Sometimes feelings of hostility are mixed up with guilt feelings and are expressed in questions like "Why did this happen to me?" or "What did I do to deserve this?" Parents who do have a second affected child after genetic counseling were found by Fletcher, in his work with mothers undergoing amniocentesis after having one affected child, to develop unusual guilt feelings, which he called "cosmic guilt." These mothers felt that some external force ("God" in some cases, some undefined force in others) has singled them out for punishment. Mothers without professed religious belief also expressed these feelings (Fletcher). The genetic counselor makes every effort to help such parents cope with and dispel their guilt feelings. For, unless these feelings are brought into perspective, parents will not be able to make realistic, rational decisions in their own best interests.

Not only the client's subjective state of mind but also the way that risk figures are presented can influence their interpretation. Furthermore, there is much evidence from studies of decision making in nongenetic counseling situations that discussing a risk situation ahead of time will tend to influence persons in such a way that they are willing to take a greater risk. Probably, this phenomenon, called the "risky-shift" phenomenon, operates in genetic counseling as well (Pearn).

In addition to strong feelings of hostility and guilt, parents may also use denial to cope with the birth of an abnormal child. This mechanism is dangerous, because parents may have a second affected child if they delay seeking genetic counseling before realizing that the recurrence risk of the condition is moderate to high. As many as forty percent of parents may wait years after an affected child has been born before seeking genetic counseling. Fifteen percent of this group gave birth to a second affected child during their period of denial. Counselors must recognize and help clients deal with denial where it exists, if parents wish to plan more children (Shore).

Parents will invariably experience a period of grief and mourning for the loss of the perfect child they expected. The severity and length of the mourning period will vary according to the perceived burden of the condition. This grief is sometimes severe and repetitive. But it must be worked through. During this period the family needs to be supported by counselors and other specially trained health professionals. If the counselor is to facilitate the grief process, he must be trusted implicitly by the parents and at the same time be supportive and understanding.

The emotional makeup of the counselor is as important as that of the consultant. It is essential that counselors with particular biases or prejudices about certain counselees or genetic diseases be conscious of this. In most cases it is probably best to have someone else work with the family or to be certain that under these circumstances the counselor's role in the counseling process is a minor one. Where the psychological situation in a family is especially complex or severe, it may be necessary for the parents to work with a psychiatrist or psychotherapist. It is not likely that every genetic counseling unit will eventually have a genetics-trained psychotherapist available, as has been recommended by one experienced psychologist-counselor.

What is the outcome of counseling?

The counselor is interested in knowing at least three things about the outcome of counseling: (1) What information has the counselee received and retained? (2) Has the counselee been able to cope effectively with the presence of a handicapped child or individual in the family? (3) What reproductive decisions will the counselees make?

The client may learn about and understand the genetic and medical information conveyed by the counselor but might still have a negative response to counseling because of persistent feelings of denial, anxiety, hostility, or depression. These feelings, which are often seen in genetically determined disorders, are part of the so-called coping response, which parents must go through while attempting to come to grips with personal tragedy (Falek and Britton). Residual or unresolved feeling will lead some persons to block or confuse information presented to them during counseling, even though they appear to follow and understand the counseling information they receive.

Until recently, the evaluation of counseling effectiveness dealt almost solely with making some estimate of the reproductive choices of those counseled since this was the only real alternative open to them if they wished to avoid future affected offspring. Judged by this criterion, genetic counseling appeared to be fairly effective. In one series of couples studied, however, it was found that only twenty-four percent of low-risk (lower than one in ten) couples had fewer offspring as a consequence of counseling. Forty-three percent of moderate- to high-risk couples (a risk of one in two or greater) with a less severe genetic condition had fewer children than originally planned while sixty-five percent of high-risk couples with more severe conditions were deterred (Carter et al.). There have been similar findings in other studies. Results like these have been interpreted as indicating that genetic counseling has been successful in influencing reproductive behavior.

Most counselors do not believe that counseling should be used as a negative eugenic tool. They also feel that, even if an attempt were made to use it as such, it would fail. Some evidence supporting this position came from a study of an attempt to use genetic counseling to reduce the incidence of sickle-cell disease in a small Greek village near Athens. The program did not reduce the incidence of sickle-cell disease, but it did result in the social stigmatization of the female carriers of the sickle-cell gene in the village, so that many of them felt obligated to seek a spouse in another village (Stamatoyannopoulos).

In addition to unexpected social stigmatization, genetic programs may fail because family planning is unsuccessful. One study of the offspring of couples at risk to have children with cystic fibrosis revealed that forty percent of the pregnancies that occurred after counseling were accidental. Often families who say they do not wish further children use ineffective or no contraceptive methods. In other cases there are religious beliefs or cultural practices that prevent counselees from limiting their families.

Another way to evaluate genetic counseling is to try to measure what information clients have learned from counseling and what they understand, since it is presumed that they will make their decision based on what they know about the condition in question.

In some studies of information received and retained by counselees, learning seemed to be poor, with fewer than half having an accurate understanding of the disorder or the recurrence risk (Murray et al.). Most studies have indicated that genetic mechanisms and genetic recurrence risks are uniformly poorly understood and recalled. It appears that many, if not most, families may be making genetic decisions based on a poor or inadequate understanding of the real recurrence risks.

The genetic counselor, although he or she may be committed to not directly influencing the decision of the counselee, will almost always have some personal goals in counseling—either conscious or subconscious. Since most are health professionals, they do want to help reduce the pain and suffering that accompany the birth of a severely defective child. In spite of their best intentions there are very likely to be attempts, usually indirect, to be somewhat directive in counseling where the risk of such tragedy is high. Counselees can be made aware of this tendency if the health-related purposes of counseling are set forth at the outset in clearly understood terms.

Ethical problems in genetic counseling

As the capability for intervening in the process of mating and reproduction has increased and genetic diagnostic tools have become more effective, the number and complexity of moral and ethical problems encountered by the genetic counselor have increased.

Confidentiality. One particular ethical issue has confronted the counselor since the early days of genetic counseling. Who owns genetic information? The answer to this question will determine whether or not the genetic counselor requires the counselee's permission to contact other genetically related family members who might be at risk to be a carrier of or be affected by a mutant gene. If information about a genetic carrier state or inherited disease is considered the same as all other medical information, the answer to the question is that genetic information is owned by the person identified with the condition, and, as with any other medical information, that person must give his or her permission before the information can be transmitted to anyone else. This means that a family member might, for personal reasons, prevent the counselor from informing other family members to let them know that they are at risk. If, on the other hand, genetic information, because of its wider implications, is "owned" by all genetic relatives, the counselor can feel free to identify other family members at risk on his own initiative. It has been argued that genetic information should be considered public health information, because, although genetic disease is not "contagious" or "communicable" in the usual sense of the words, it can be considered contagious or communicable in a genetic sense, i.e., vertically transmissible, since genetic relatives have a predictable probability of getting the condition. This, it is argued, is sufficient reason for removing genetically determined conditions from the private to the public area of medical information (Shaw, "Survey").

Freedom and manipulation. The modern definition of genetic counseling clearly states that the counselee will make his or her own decision. This decision should be made as free of coercion and/or influence of the counselor as is humanly possible. The concept of counseling is termed nondirective. It is obvious that "true" nondirective counseling is really impossible. But counselors should, and most do, try to reach that ideal. There are certain situations in which it might be justifiable to either tell or direct counselees in their decision making. Such situations might occur when: (1) the genetic risk is moderate to high as judged by medical geneticists (twenty-five percent or higher) and the condition is severe (the burden of disease is high); (2) the condition is severe, and the affected individual will eventually become an economic and social burden on society (e.g., severe mental retardation or severe congenital malformation); (3) the family has neither the psychological nor the economic resources to cope with the condition, particularly if they already have one affected child; (4) the genetic risk is 100 percent, and the burden of the condition is high.

It has been suggested that just as the physician may intervene to direct the decisions of patients when their behavior is essentially harmful, i.e., either self-destructive or harmful to society, the counselor is justified in "directing" the decisions of clients when on balance those decisions would appear to lead to significantly more harm than good to the family, to the potential child, or to society (Birch). Counselors who are strongly opposed to "directive" counseling under any circumstances argue that, once counselors begin to tell families what to do for any reason, they will find increasing numbers of situations in which they will impose their personal beliefs or values on others. The freedom of choice of the counselee will gradually be eroded. The only way a counselor might possibly be ethically justified in being directive is to inform the counselee before counseling begins of the fact that he may sometimes recommend a course of action in the interest of society or in what *he* perceives to be in the best interest of the counselee. When there is a conflict between the interests of the individual and the interests of society, the burden of justification is on the counselor who supports what he or she perceives to be the interests of society.

Truth-telling. A recurrent ethical theme debated among genetic counselors and physicians alike is related to truth-telling (Lubs). The general principle followed in medical practice is that the patient should be told the whole truth except when, in the physician's judgment, it will do more harm than good. A difficulty with this position lies in the fact that the physician cannot really know with confidence what is harmful to an individual or family, except possibly in certain unusual situations. What might be unacceptable or harmful to one individual may be quite acceptable and even beneficial to another. What might seem to be harmful or injurious now might not be later. What might be harmful in one family might not be in another. Since only those involved can really judge what is harmful to them, the purist would say that no counselor has the right to withhold information from a family or individual under any circumstances.

There are, however, at least two situations when the counselor might be justified in not telling the truth or withholding information from a family or person being counseled. The most frequently encountered is the instance where genetic information revealed in the course of a diagnostic workup is suggestive of or consistent with nonpaternity, especially when this was previously unsuspected. With no certainty that the husband–wife relationship is reasonably secure, there is a definite potential for disrupting an otherwise stable family if this information is revealed. On the other hand, where the bond between husband and wife is especially strong, this unwelcome information may only cause temporary pain and upheaval, which can eventually be resolved with the help of the counselor. Either way, there is a potential risk of serious harm. Those who favor truth-telling point out that the husband has a right to know about cases of potential nonpaternity and that he is being wronged when this information is withheld. On the other hand, since this situation is so often the result of a brief and perhaps long-forgotten episode in the mother's life, it might be better not to dredge it up. A major consideration in this situation is the fact that the innocent child involved may be the person who is most seriously injured when the counselor tells the whole truth.

Another case where the truth may justifiably be withheld is the rare case of the female who looks like a female but whose sex chromosomes are XY instead of XX. This condition, known as testicular feminization, is seen in females who are often quite sexually attractive and who might marry but who are infertile because their internal reproductive organs do not develop. Instead of ovaries, they have testicles, usually located in the lower abdomen, which may be associated with a hernia.

The experience of one counselor who has counseled several of these patients is that, if they are given the information that they carry a Y chromosome, this knowledge can have a psychologically devastating effect on them or their parents from which they may not recover (Macintyre). Most husbands of these women, when confronted with such information, are unable to cope with it, and it has usually resulted in separation and divorce. Since experience has shown that in this situation the harm of giving such information outweighs the benefit, many counselors feel that it is best to provide only enough information to enable the patient with the condition to understand her reproductive limitations.

An essential, if not *the* essential, ingredient in genetic counseling is providing information. All pertinent information should be transmitted to or provided for the counselee. Where there are exceptions to this rule, the burden of proof must be placed on the counselor to justify the reason or reasons for withholding information from the person being counseled. In other words, there may be a right not to know that should be respected in counseling as much as the right to know the truth.

Conclusion

The counselor must constantly maintain an awareness of the conflicts of rights and values that arise in the course of genetic counseling. It is probably wisest to maintain a flexible rather than a dogmatic position with respect to these conflicts, since new and more potent methods of diagnosis and treatment of genetic disease will no doubt make it necessary to shift the emphasis and direction in counseling for particular genetically determined disorders to meet the growing needs of counselees and their families.

ROBERT F. MURRAY, JR.

[*Directly related are the entries* GENE THERAPY; GENETIC SCREENING; GENETICS AND THE LAW; *and* PRENATAL DIAGNOSIS. *For further discussion of topics mentioned in this article, see the entries:* ABORTION; CONFIDENTIALITY; EUGENICS; *and* TRUTH-TELLING.]

BIBLIOGRAPHY

BIRCH, CHARLES. "Genetics and Moral Responsibility." *Genetics and the Quality of Life.* Edited by Charles Birch and Paul Abrecht. Elmsford, N.Y.: Pergamon Press, 1975, chap. 1, pp. 6–19.

CARTER, C. O.; EVANS, K. A.; FRASER ROBERTS, J. A.; and BUCK, A. R. "Genetic Clinic: A Follow-Up." *Lancet* 1 (1971): 281–285. Long-term follow-up on the reproductive activity of patients who have received genetic counseling in a well-established clinic.

FALEK, ARTHUR, and BRITTON, SHARON. "Phases in Coping: The Hypothesis and Its Implications." *Social Biology* 21 (1974): 1–7. Discusses the stages of psychological stress that parents of handicapped children or persons affected with chronic and/or debilitating conditions must go through.

FLETCHER, JOHN. "The Brink: The Parent–Child Bond in the Genetic Revolution." *Theological Studies* 33 (1972): 457–485. Report of a study of the ethical and moral conflicts in parents undergoing amniocentesis to diagnose a genetically determined condition after having already had one affected child.

FRASER, F. C. "Genetic Counseling." *American Journal of Human Genetics* 26 (1974): 636–659. Reviews the generally accepted modern process of genetic counseling, highlighting some of the problems that arise and including discussion of the extent and availability of counseling services.

HILTON, BRUCE; CALLAHAN, DANIEL; HARRIS, MAUREEN; CONDLIFFE, PETER; and BERKLEY, BURTON, eds. *Ethical Issues in Human Genetics: Genetic Counseling and the Use of Genetic Knowledge.* Fogarty International Proceedings, no. 13. New York: Plenum Press, 1973. Proceedings of a symposium sponsored by the John E. Fogarty International Center for Advanced Study in the Health Sciences and the Institute of Society, Ethics and the Life Sciences, 10–14 October 1971.

KALLMANN, FRANZ J. "Some Aspects of Genetic Counseling." *Genetics and the Epidemiology of Chronic Diseases: Symposium 17–19 June 1963, Ann Arbor, Michigan.* Edited by James V. Neel, Margery W. Shaw, and William J. Schull. Public Health Service Publication, no. 1163. Washington: Department of Health, Education, and Welfare, Public Health Service, Division of Chronic Diseases, 1965, pp. 385–395.

LEONARD, CLAIRE O.; CHASE, GARY A.; and CHILDS, BARTON. "Genetic Counseling: A Consumer's View." *New England Journal of Medicine* 287 (1972): 433–439.

LUBS, HERBERT A. "Privacy and Genetic Information." Hilton, *Ethical Issues in Human Genetics,* pp. 267–275.

———, and DE LA CRUZ, FELIX, eds. *Genetic Counseling: A Monograph of the National Institute of Child Health and Human Development.* New York: Raven Press, 1977.

MACINTYRE, M. NEIL. "Discussion" [of "Privacy and Genetic Information" by James L. Sorenson]. Hilton, *Ethical Issues in Human Genetics,* pp. 280–281.

McKUSICK, VICTOR A. "The Growth and Development of Human Genetics as a Clincal Discipline." *American Journal of Human Genetics* 27 (1975): 261–273.

MURPHY, EDMOND A., and CHASE, GARY A. *Principles of Genetic Counseling.* Chicago: Year Book Medical Publishers, 1975. Gives a detailed definition of the term "consultand" (pp. 86–88). A book about the theory of genetic counseling.

MURRAY, R. F., JR.; BOLDEN, R.; HEADINGS, V. E.; QUINTON, B. A.; and SURANA, R. B. "Information Transfer in Counseling for Sickle Cell Trait." *American Journal of Human Genetics* 26 (1974): 63a. Abstract of paper presented at 26th Annual Program of the American Society of Human Genetics, 16–19 October 1974, Portland, Oregon.

PEARN, J. H. "Patients' Subjective Interpretation of Risks Offered in Genetic Counseling." *Journal of Medical Genetics* 10 (1973): 129–134. Discusses the ways that counselees may interpret recurrence risk odds presented to them in genetic counseling and the factors that might influence that interpretation.

REED, SHELDON C. "A Short History of Genetic Counseling." *Social Biology* 21 (1974): 332–339. Brief discussion of the origins of genetic counseling and its evolution to its present status.

SHAW, MARGERY W. "Review of Published Studies of Genetic Counseling: A Critique." Lubs, *Genetic Counseling,* pp. 35–52.

———. "Survey of Counseling Practices: Discussion." Hilton, *Ethical Issues in Human Genetics,* pp. 13–17. Principal discussant.

SHORE, MILES F. "Psychological Issues in Counseling the Genetically Handicapped." *Genetics and the Quality of Life.* Edited by Charles Birch and Paul Abrecht. Elmsford, N.Y.: Pergamon Press, 1975, chap. 15, pp. 160–172. Presents some of the psychological traumas that parents of children with handicaps due to genetic factors have to cope with.

SORENSON, JAMES R., and CULBERT, ALBERT J. "Genetic Counselors and Counseling Orientation—Unexamined Topics in Evaluation." Lubs, *Genetic Counseling,* pp. 131–154.

STAMATOYANNOPOULOS, GEORGE. "Problems of Screening and Counseling in the Hemoglobinopathies." *Birth Defects: Proceedings of the Fourth International Conference, Vienna, Austria, 2–8 September 1973.* Amsterdam: Excerpta Medica; New York: American Elsevier Publishing Co., 1974, pp. 268–276. Report of a seven-year study of the effect of screening and counseling for sickle-cell disease in a Greek village.

GENETIC IMPLICATIONS OF POPULATION CONTROL

See POPULATION POLICY PROPOSALS, *article on* GENETIC IMPLICATIONS OF POPULATION CONTROL.

GENETIC INTERVENTION or GENETIC ENGINEERING

See GENE THERAPY; REPRODUCTIVE TECHNOLOGIES; EUGENICS; EUPHENICS; GENETIC DIAGNOSIS AND COUNSELING; GENETICS AND THE LAW; GENETIC CONSTITUTION AND ENVIRON-

MENTAL CONDITIONING; RESEARCH, BIOMEDICAL; RESEARCH POLICY, BIOMEDICAL.

GENETIC SCREENING

"Genetic screening" is an umbrella term covering a heterogeneous collection of diagnostic procedures aimed at abnormalities known to be in some fashion inherited. In the minds of both professionals and public, it usually refers to conditions of relatively simple and straightforward inheritance, such as single gene disorders or abnormalities of chromosome number. Thus screening for sickle-cell disease, involving a single gene defect, is thought of as genetic screening, while screening for high blood pressure, which is probably multifactorial, with both genetic and environmental components, is not so regarded.

Newborn metabolic screening

Genetic screening got its start in the early part of the 1960s with the development of a relatively simple and inexpensive test, suitable for use in newborns, for phenylketonuria (PKU), a rare autosomal recessive in which the child inherits one defective gene from each parent. PKU leads to irreversible mental retardation and is usually found only in populations of northern European descent. People with this condition are deficient in phenylalanine hydroxylase, an enzyme critical to the metabolism of phenylalanine, which is a common—indeed, almost unavoidable—amino acid found in many foods. Experimental work in the 1950s had disclosed that a synthetic diet almost free of phenylalanine and begun shortly after birth seemed to prevent the worst consequences of PKU and to lead to children with IQs in the normal range. When early diagnosis became possible, it seemed logical to institute newborn screening programs in an effort to begin dietary treatment of those affected as early as possible. In the United States the avenue chosen to achieve this end was the law. Representatives of what has been called the mental health establishment—researchers, doctors, relatives of the affected, and government child health experts—combined in a lobby that achieved the legislated institution of compulsory PKU screening of newborns in most states by about the middle of the decade (Bessman and Swazey, pp. 54–56).

For a while, PKU screening was regarded as a great achievement in public health medicine. But by the end of the decade a certain amount of disenchantment had begun to set in. There were arguments about which tests were best to use, and when they should be done (Holtzman, Meek, and Mellits). In the United States babies are screened in the hospital on about the third day after birth, but more reliable results are apparently achieved in Britain, where visiting nurses collect the sample at about three weeks. Such home visits are, of course, more expensive than in-hospital procedures. The almost universal screening had also resulted in many false positives, not only because it turns out that there are several variants on classic PKU, but also because transiently raised blood levels of phenylalanine of no particular clinical consequence are not uncommon. Retesting became a logistical problem but needs to be done because the synthetic diet can be harmful to a normal child. Even in classic PKU, the diet is less than ideal: unappetizing, disruptive to family life, and very restricted. Furthermore, it has now become clear that the diet probably does not restore intelligence totally; while children with treated PKU usually do have IQs in the normal range, they are generally significantly lower than those of their nonaffected sibs. The diet is restorative enough, however, for these children to lead much more nearly normal lives. Among other things, this means that they remain in the community, marry, and have children, rather than live out their lives in institutions, childless. Because PKU is a recessive, this change in reproductive capacity will not appreciably increase the gene frequency for many generations, but a serious reproduction problem has nevertheless arisen. PKU women give birth to children who are retarded no matter what their genotype because of a toxic uterine environment, and it is by no means certain that a temporary maternal return to the special diet (which is usually terminated around the age of six) will be completely effective in preventing such a sad outcome. Several economic studies of PKU screening have concluded that it is cost-beneficial (based usually on simple calculations of screening costs vs. institutionalization costs), but they frequently have taken no account of all these additional problems. Nor is it possible to calculate the social harm done by compelling such a medical procedure by law, as is common in the United States.

PKU screening in many ways is a type case of the kinds of difficulties encountered by all mass screening procedures, which frequently turn out to be less perfect in practice than had been hoped when they were instituted. It is important to emphasize that PKU screening also grew out of the same tradition as previous forms of mass screening, in that it was an attempt to find and treat a discernible clinical entity (McKeown, p. 2). This model of newborn screening is being extended, with many states in the United States moving to test the blood sample drawn for PKU for a number of other (usually rare metabolic) disorders, not all of which are treatable and for some of which the treatment is disputed.

Chromosome screening

Genetic screening quickly began to move away from the traditional case-finding-and-treatment model in a number of ways. One departure was the institution of screening newborns for abnormalities of chromosome number. In 1965 Patricia Jacobs and her colleagues published a study revealing that there was a larger than expected number of men with an extra Y chromosome to be found incarcerated in mental–penal institutions. This led to much speculation, particularly in the popular press, that the XYY man was, at the very least, born antisocial and might even be born criminal. By contrast, it was known even then from other population studies that most XYY men led apparently normal lives and did not get into trouble with the law. But the controversy grew, and it was felt that the argument could be settled if large numbers of newborns were screened for all chromosome anomalies. The first purpose of such screening was to ascertain the true newborn incidence of such abnormalities, which was then largely unknown. Second, if those who were found to have such abnormalities (particularly XYY) were then followed throughout childhood and into adult life, it could perhaps be determined if behavior problems were in fact a common occurrence in this group. Such studies resulted in a dilemma regarding issues such as disclosure and truth-telling. Full disclosure to the parents was felt by some researchers to be unwise, because it might alter the child's upbringing and cause him harm, but failure to disclose also carried a moral (and perhaps legal) price. In addition, at least one study, in Boston, began to move beyond mere follow-up and decided to attempt treatment in areas like family counseling and speech therapy (Culliton).

This program came under sharp attack from critics who charged that its consent procedures were deficient. The critics also argued that the attempted therapeutic intervention and the partial disclosure of the nature of the abnormality rendered the project both scientifically worthless for purposes of studying the true nature of any behavior difficulties associated with the XYY chromosome complement, and also affected the child's upbringing adversely (Beckwith and King). This form of screening then attempted to follow the traditional model of case-finding followed by treatment. But the "case" consisted of vague, unquantifiable, and in at least some cases probably nonexistent behavior patterns rather than a well-defined metabolic sequence; also, the "treatment" is amorphous and unfocused.

Carrier screening

Simultaneous with these programs a completely different kind of screening developed. The object of the search shifted from a disease to a carrier, a person who is clinically well but risks having a child with a disease. These programs do not involve case-finding and treatment in the conventional sense, but rather represent an attempt to identify the person at risk and to intervene in his or her reproductive life, an approach not taken by any previous screening program. In the United States, these programs were directed mainly at two conditions, sickle-cell anemia and Tay–Sachs disease. The two share some similarities. Both are autosomal recessives, for instance, meaning that both parents must be carriers (heterozygotes) before it is possible to give birth to a child with the disease (homozygote), and there is a one-in-four chance of such a child with each pregnancy. But the conditions also have some important and instructive differences.

Tay–Sachs disease is very rare but is found most frequently in Ashkenazi Jews, those of eastern European descent. A child with Tay–Sachs disease appears to be developing normally for the first few months of life and then begins inexplicably to regress. The clinical course includes blindness and paralysis; the child soon becomes bedridden and totally unresponsive and is generally dead before the age of four. There is no adequate treatment for the disease, but prenatal diagnosis makes possible identification of an affected fetus early enough for a legal abortion. This means that a heterozygote couple having no moral objection to such an abortion can be virtually assured that they never need give birth to a child with Tay–Sachs disease. Development of a relatively simple and inexpensive test for heterozygosity made mass screening technically possible late in the 1960s, and the voluntary screening programs among Ashkenazi Jews have met with varying degrees of success; participation has been greater in the United States, for instance, than in Canada (Beck et al.).

By contrast, however, sickle-cell screening differs on almost all counts. For one thing, this hereditary defect of red blood cells is relatively common for a genetic disease, particularly among blacks, although it is also found with appreciable frequency among whites of Mediterranean descent and occasionally in other groups. The clinical course is fairly heterogeneous; while it is true that the lives of many people with sickle-cell disease are painful and short, it is also true that many of those with the disease are not so severely affected, and it is possible for them to live close to normally. No truly satisfactory treatment yet exists, but there are hopes that one will develop in the not too distant future. Perhaps more important, no suitable method of prenatal diagnosis for the disease is widely available, so it is not possible to assure a heterozygote couple that they need not have an affected child. Furthermore, even when such prenatal diagnosis is possible, the dilemma over the necessity of abortion in sickle-cell disease would still be acute, since the clinical course is so various, and there is no way of telling in advance how badly affected the child will be. Several heterozygote mass screening tests have been developed, but most have some technical deficiencies that have resulted in misdiagnosis. They have been widely applied among American blacks, sometimes in a very disorganized and disruptive fashion. And even in a well-run program, with careful counseling and follow-up (historically a rarity), the couple found to be heterozygous has only a few discouraging and difficult options: adoption, artificial insemination with a noncarrier donor, or simply taking the one-in-four risk with each pregnancy (Schmidt).

Perhaps most important is that, like PKU, sickle cell has ended up in the legislative arena. By the early 1970s, at least twelve states had passed laws concerning sickle-cell screening, and in about half those states the laws were compulsory, requiring screening at school entry or prior to obtaining a marriage license. In retrospect, it has become clear that, because of a confusion of language in many of the laws, legislators thought they were promulgating disease screening (in the PKU mold) rather than carrier screening (in the Tay–Sachs mold). As a result of a belated outcry in the black community (particularly the black medical community), the laws have frequently not been funded or enforced. The fact that such testing could be so quickly mandated, however, given the lack of sound medical rationale for such urgent compulsory programs, is an interesting and perhaps disturbing sidelight on the confused interaction between the medical and political processes in the United States (Powledge).

Prenatal diagnosis

Another area of genetic screening with potentially wide application is prenatal diagnosis. By and large, prenatal diagnosis has been applied in high-risk pregnancies, such as those where the mother has already given birth to an affected child. Advanced maternal age is the other high-risk situation where the prenatal diagnostic technique known as amniocentesis (in which fluid from the womb is withdrawn by hypodermic and examined) has often been employed. Women over thirty-five, and particularly women over forty, are at significantly elevated risk of bearing a child with Down's syndrome (mongolism). This abnormality of chromosome number, the single most important cause of severe mental retardation, can be diagnosed early by amniocentesis. In some major medical centers an early amniotic tap on such women is routine. A mass application of amniocentesis for Down's syndrome has been proposed, beginning with older women but eventually reaching all pregnancies; such a program, it is argued, would virtually eliminate Down's syndrome from the population altogether. The program would also —almost as a side effect—turn up all the other gross chromosome aberrations, thus forcing the abortion issue for conditions where the degree of deficit is not disabling (such as XO, known as Turner's syndrome, where sterility appears to be the chief problem) or not established (the aforementioned XYY). It might also turn out to be cost-beneficial to run other kinds of tests on the amniotic cells grown for chromosome study. We could thus gradually institute a full-blown prenatal screening program, probably automated, whose limits would be set, in theory at least, only by our failure to keep devising new diagnostic tests for additional abnormalities, most of which would be individually very uncommon.

Susceptibility screening

The final—and potentially broadest—area of expansion is what might be called "susceptibility" screening. Mass screening for high blood pressure can be thought of in this way. It involves searching for a condition that is at least in part genetically determined and is not in itself a disease, but which has the potential for causing lethal damage at some point in the future (Sackett). Another example is the search in children and young people for elevated blood cholesterol and other lipids, which may indicate that they risk heart disease perhaps thirty or forty years later (Chase, O'Quin, and O'Brien). Perhaps the most promising area of this type of research, from a scientific standpoint at least, is the reported associations between a wide variety of histocompatibility (HL-A) antigens (heretofore used to assess the degree of tissue match between transplant donor and recipient) and many different kinds of neoplastic, auto-immune, and infectious diseases (McDevitt and Bodmer).

Although the nature of the relationship between the antigens and subsequent disease development is not yet clear, the antigens may turn out to be handy markers with great predictive value. It may become possible to identify, perhaps even at birth, the kinds of disorders to which each of us is most genetically vulnerable, and, the argument goes, to take steps to alter our environment in such a way as to render us less liable to disease. Some such attempts are already under way. It is now common practice in the United States to screen for high blood pressure and to treat even those people found to have only moderately elevated blood pressure with antihypertensives and advice about obesity, smoking, and salt intake. Small children are being placed on low-cholesterol diets when they are found to have raised blood lipid levels, thought to be often genetically determined, although many years of research will be necessary before it is conclusively known whether such early diet alteration is in fact effective in preventing later cardiovascular disease. Nevertheless, it is clear that susceptibility to diseases of many kinds—including infectious disease—often has some genetic basis, and there will be continued and persistent attempts to identify that basis. The stated aim of such work will be the same general aim that screening has always had: early identification and mitigation of disease.

Ethical issues

One would somehow expect that such vast explorations of our genotypes would raise a set of brand new ethical issues, and yet, curiously, the issues are often the familiar ones that plague many areas of medicine. The difference is one of scale: we are talking here of situations that potentially affect huge segments of the population, perhaps even our entire species. Space permits discussion of only a few, and the discussion does not pretend to be exhaustive.

Informed consent. Virtually all genetic screening programs have been research (as opposed to service) programs at the outset, even if their organizers did not recognize that fact publicly or privately. Most of the conditions sought have been quite rare, and thus small pilot projects in advance of mass programs have been logistically impractical. Legal consent requirements for research programs, particularly ones involving children, are more stringent than those for service programs, and because frequently the consent procedures used have been minimal or even nonexistent, ethical and legal issues arise.

Formal consent procedures should certainly be observed in genetic screening. In many places such procedures are legally required, but even when they are not, observing them will serve to strengthen a positive sense of joint participation that ought to exist between medicine and its clients. Furthermore, when the screening program is experimental, those who agree to participate should, at a minimum, be apprised of that fact. The tissue samples taken should be used only for the tests specified in the consent form and then disposed of; if the organizers wish to make additional research use of the sample, for running other kinds of assays for instance, the consent form should say so. The nature of available treatment and the extent of follow-up should also be clearly specified. In problematic studies—exemplified, for instance, by newborn screening for XYYs—the controversial nature of the abnormality and the vagueness of the treatment possibilities ought to be forthrightly discussed with potential screenees or their proxies beforehand. In short, the screeners ought to place more emphasis on "informed" than on "consent," no matter how difficult that goal may sometimes seem.

Confidentiality. Confidentiality is a critical and much-ignored issue in genetic screening.

Mass screening takes place under the aegis of public health, and confidentiality is traditionally not strongly protected in public health situations. However, given that genetic screening is frequently aimed at research more than service, and given the potential for abuse inherent in easy access to others' genotype information, steps should be taken to assure that information gathered in screening programs is carefully protected, probably by law. Additional attention will have to be devoted to consideration of who should have access to this information. Insurance companies will surely want it if they can get it, particularly when it is likely to predict future susceptibility to chronic disease. Should prospective employers perhaps have access to that information? Should the press, in cases, for instance, where the person under scrutiny is a candidate for public office? These and many other questions related to confidentiality simply have not been widely considered. But it can, as a beginning, probably be reasonably argued that the information obtained in genetic screening programs ought to be more carefully protected than it is.

Some initial steps toward legal protection of such records have already been taken. Legislation setting up New York State's Birth Defects Institute, a statewide repository for many kinds of information, contains language requiring that the information be protected, as does the legislation establishing the new Maryland Commission on Hereditary Disorders. Legal requirements, however, are largely pointless unless it is technically possible to protect such information, particularly once it is stored in a computer. Britain's computerized genetic register system, known by the acronym RAPID, has a number of built-in protection procedures that may serve as a prototype (Emery, Elliott, Moores, and Smith). On the other hand, we have apparently not yet devised any foolproof ways of making data banks impregnable (Turn and Ware). This, of course, is an increasingly serious problem for nonmedical records too.

Truth-telling. This is a perennial dilemma in medicine: is it ever right to withhold the truth from a patient, even when the doctor believes the truth may be harmful? XYY screening provides a textbook example: can truth-telling here be justified for its own sake if the researcher believes that the child's upbringing will be altered for the worse and a controversial diagnosis thus turned into a self-fulfilling prophecy? Another example is nonpaternity. Genetics cannot establish beyond question that a man *is* a child's father, but such studies can prove when a man *is not*, and such a finding is not uncommon in heterozygote screening programs. Unless one dissembles, there is no way to tell a man that on the one hand he runs no risk of siring a child with such-and-such disease without also telling him that the child with such-and-such disease he already has is not really his. Many researchers simply lie (one sickle cell screener, for instance, tells the noncarrier "father" of an affected child that he runs no further parental risk because his child's condition is the result of a fresh mutation—a possible, although highly unlikely, event). That may appear to be the compassionate solution in the short run, but there are those who argue that the truth-telling obligation transcends such short-term considerations and that, as a practical matter, medicine pays a price in loss of patient confidence when it lies.

Allocation of scarce resources. The important issue of resource-allocation has frequently been considered in relation to genetic screening only by implication, such as the previously cited cost-benefit studies that have always concluded that screening is indeed cost-beneficial. It has already been pointed out that each such study has failed to consider some economic factor that might legitimately be included, and therefore it is possible to make the figures come out any way one wants. Leaving aside that question, however, one may still argue that even a truly cost-beneficial screening program may not represent the best use of the limited resources available to any society. The program to wipe out Down's syndrome is a dramatic and eye-catching proposal, yet the same resources might be better used on programs (prenatal nutrition, for instance, or better delivery room procedures) to prevent mild mental retardation, which is much more common and may be a greater public and private burden. On the other hand, some would argue that there is an ethical obligation to screen when we can, even if it is *not* cost-beneficial. Ethnic considerations also contribute to allocation dilemmas, particularly in the United States. Recessive disorders are almost by definition ethnic, and it is possible to argue that, historically, the amount of attention devoted to such rare conditions is not unrelated to the affected group's political clout. Tugs of war by contending groups in the legislative arena may be the traditional way of getting things done in the United States, but it may also be true that such contention is neither the most ethical nor the most rational way to decide medical priorities. And in developing countries, preoccupied as they are with the crushing problems of basic nutrition and serious infectious diseases that their wealthier neighbors have largely overcome, any expenditure on exotica like uncommon recessives would probably be clearly out of the question. Perhaps one might also argue that the wealthier nations have an obligation first to help the poorer ones solve their basic problems before they, the rich, set off in pursuit of the rare.

Conclusion

Certainly some of the difficulties with genetic screening arise from our inability to decide whether it belongs in the public health area along with other mass screening procedures, or whether it represents something new and different in medicine. There are arguments on both sides. Biologically speaking, it is becoming increasingly clear that most (perhaps almost all) disease, even infectious disease, has some genetic basis, if only in the relative strengths of our individual abilities to resist invading toxic agents. In this view, mass screening for high blood pressure is just as "genetic" as newborn screening for PKU, even though the exact mode of inheritance in the former is not yet understood, and even though the contribution of environment to its development is probably strong. Thus there are no valid distinctions between genetic screening and other kinds of screening.

The opposite argument is based on the notion that, whatever the biological reality, there is a public conception of genetic disease that renders it qualitatively different in the minds of most people; it is specially awesome and irrevocable, and specially mysterious. That makes genetic screening a uniquely sensitive and delicate area of health care, one that ought to be bound by special ethical restrictions.

TABITHA M. POWLEDGE

[Directly related are the entries EUGENICS; GENETIC DIAGNOSIS AND COUNSELING; MASS HEALTH SCREENING; and PRENATAL DIAGNOSIS. For further discussion of topics mentioned in this article, see the entries: CONFIDENTIALITY; GENETIC ASPECTS OF HUMAN BEHAVIOR, articles on MALES WITH SEX CHROMOSOME ABNORMALITIES (XYY AND XXY GENOTYPES) and GENETICS AND MENTAL DISORDERS; INFORMED CONSENT IN THE THERAPEUTIC RELATIONSHIP; RATIONING OF MEDICAL TREATMENT; and TRUTH-TELLING. See also: POPULATION POLICY PROPOSALS, article on GENETIC IMPLICATIONS OF POPULATION CONTROL.]

BIBLIOGRAPHY

BECK, E.; BLAICHMAN, S.; SCRIVER, C. R.; and CLOW, C. L. "Advocacy and Compliance in Genetic Screening." *New England Journal of Medicine* 291 (1974): 1166–1170.
BECKWITH, JON, and KING, JONATHAN. "The XYY Syndrome: A Dangerous Myth." *New Scientist* 64 (1974): 474–476.
BESSMAN, SAMUEL P., and SWAZEY, JUDITH P. "Phenylketonuria: A Study of Biomedical Legislation." *Human Aspects of Biomedical Innovation.* Edited by Everett Mendelsohn, Judith P. Swazey, and Irene Taviss. Cambridge: Harvard University Press, 1971, pp. 49–76.
CHASE, H. P.; O'QUIN, R. J.; and O'BRIEN, D. "Screening for Hyperlipidemia in Childhood." *Journal of the American Medical Association* 230 (1974): 1535–1537.
CULLITON, BARBARA J. "Patients' Rights: Harvard Is Site of Battle over X and Y Chromosomes." *Science* 186 (1974): 715–717.
EMERY, ALAN E. H.; ELLIOTT, DOROTHY; MOORES, MICHAEL; and SMITH, CHARLES. "A Genetic Register System (RAPID)." *Journal of Medical Genetics* 11 (1974): 145–151.
"Ethical and Social Issues in Screening for Genetic Disease." Institute of Society, Ethics, and the Life Sciences. Research Group on Ethical, Social, and Legal Issues in Genetic Counseling and Genetic Engineering. *New England Journal of Medicine* 286 (1972): 1129–1132.
Ethical, Social and Legal Dimensions of Screening for Human Genetic Disease. Edited by Daniel Bergsma, Marc Lappé, Richard O. Roblin, and James M. Gustafson. Birth Defects: Original Article Series, vol. 10, no. 6. National Foundation, March of Dimes. New York and London: Stratton Intercontinental Medical Book Corp., 1974.
HARRIS, HARRY. *Prenatal Diagnosis and Selective Abortion.* London: Nuffield Provincial Hospitals Trust, 1974.
HOLTZMAN, N. A.; MEEK, A. G.; and MELLITS, E. D. "Neonatal Screening for Phenylketonuria. I. Effectiveness." *Journal of the American Medical Association* 229 (1974): 667–670.
JACOBS, PATRICIA, et al. "Aggressive Behaviour, Mental Sub-Normality and the XYY Male." *Nature* 208 (1965): 1351–1352.
KOMROWER, G. M. "The Philosophy and Practice of Screening for Inherited Diseases." *Pediatrics* 53 (1974): 182–188.
LESSLER, KEN. "Screening, Screening Programs and the Pediatrician." *Pediatrics* 54 (1974): 608–611.
McDEVITT, HUGH, and BODMER, WALTER. "HL-A, Immune-Response Genes, and Disease." *Lancet* 1 (1974): 1269–1275.
McKEOWN, THOMAS. "Validation of Screening Procedures." *Screening in Medical Care.* Edited by Lord Cohen, E. T. Williams, and G. McLachlan. London: Oxford University Press, 1968, pp. 1–13.
POWLEDGE, TABITHA M. "The New Ghetto Hustle." *Saturday Review of the Sciences*, 27 January 1973, pp. 38–40; 45–47.

Raine, D. N. "Screening for Disease: Inherited Metabolic Disease." *Lancet* 2 (1974): 996–998.

Sackett, David L. "Screening for Disease: Cardiovascular Disease." *Lancet* 2 (1974): 1189–1191.

Schmidt, R. M. "Hemoglobinopathy Screening: Approaches to Diagnosis, Education, and Counseling." *American Journal of Public Health* 64 (1974): 799–804.

Turn, Rein, and Ware, W. H. "Privacy and Security in Computer Systems." *American Scientist*, March–April 1975, pp. 196–203.

GENETICS AND THE LAW

Introduction

Advances in reproductive technologies and the prevention and treatment of hereditary diseases raise questions basic to societal values, thus creating a nexus between law and genetics. Since law in other countries on this subject is scanty and difficult of access, this discussion will be limited to American law, which is derived in part from the English common law system.

The confrontation between law and applications of genetic knowledge is inevitable, especially in the United States, where the legal system empowers the state to protect and preserve public health and at the same time guarantees, by the Constitution, certain fundamental rights to each citizen. The tension between the interests of the state in ensuring a healthy, productive population and the rights of the individual to be left alone, free from interference by the state, is the main arena where legal questions on genetics may arise.

Several areas of the law that have felt the impact of advances in genetic technology will be discussed in this article: (1) the constitutional right of privacy in reproductive decisions; (2) state action to prevent genetic disease by genetic screening laws; (3) licensure of genetic counselors; (4) the confidential relationship between doctor and patient or counselor and counselee; (5) common law involving prenatal torts; (6) the informed consent doctrine as applied to treatment and experimentation; and (7) federal activity to regulate health hazards to which citizens are exposed, such as recombinant DNA and environmental hazards that may cause human mutations.

Procreation and privacy

Two landmark decisions by the U.S. Supreme Court have increased the genetic counselor's flexibility in providing options to a family faced with a high risk of having a genetically defective child. These involve contraception and abortion.

In 1965 the Court struck down a state statute that prohibited the use of contraceptives, finding it to be an unconstitutional invasion of marital privacy (*Griswold* v. *Connecticut*). Eight years later the Court, relying on the *Griswold* precedent, found many state abortion statutes to be unconstitutional. In that case, the Court said that a state may not invade the zone of privacy surrounding the woman and her doctor concerning a decision to abort during the early months of pregnancy (*Roe* v. *Wade*).

How did the Supreme Court find a constitutional "right of privacy" in these cases, when privacy is not mentioned in the Constitution? It did so by defining privacy as a "fundamental" right emanating from a penumbra of rights embodied in the First, Third, Fourth, Fifth, and Fourteenth Amendments. Normally a state statute must bear a rational relationship to a legitimate state objective. The statute's content and operation must not be arbitrary, capricious, or too broad to withstand a challenge to its constitutionality. However, when an individual's fundamental right, such as the right to privacy, is curtailed or limited by a state action, or when a suspect classification—e.g., race, sex, alienage, religion—is inherent in the action or statute, the Court looks with strict scrutiny at the purpose and means employed and will sustain the questioned statute only if the state has demonstrated a *compelling* interest in its goal and proposes the least restrictive method of achieving it.

Since privacy in the use of contraceptives and in the decision to abort is now constitutionally protected, the genetic counselor may offer these alternatives to a couple who may wish to restrict their reproduction because of a high genetic risk. The development of amniocentesis and prenatal diagnosis from amniotic fluid cells now offer many parents an opportunity to have normal children without risk of bearing abnormal children. This freedom of choice may not be disturbed by the state unless a compelling interest can be shown.

A thornier issue concerns the state's interest in preventing the birth of genetically diseased children who often become an economic burden and sap the resources of the state. This issue is complicated because it touches on the fundamental right of procreational privacy. According to the *Roe* rationale, the state cannot intervene to prevent births, because it does not have a compelling interest in the fetus until the period of viability. Thus, at present, mandatory abortion is outside the purview of state intervention, and voluntary abortion remains as the mechanism for decreasing the incidence of genetic disease that can be diagnosed antenatally. Perhaps at some future time a state will test its power to prevent the birth of genetically abnormal children, but as yet no state has done so.

Another method of preventing genetic disease is by compulsory sterilization. While involuntary sterilization of mental defectives has been upheld by the courts (*Buck* v. *Bell; In re Moore*), it is not clear that such actions will remain valid in light of the subsequent judicial development of the right of procreational privacy. No state has yet proposed to sterilize carriers of genetic disease, i.e., individuals who are otherwise healthy, or to deny marriage licenses to high-risk couples.

Genetic screening laws

The legislative response to the rise of mass genetic screening technology has been phenomenal. Within six years after the perfection of an automated method with which to test infants for phenylketonuria (PKU)—an inherited mental deficiency—compulsory screening laws had been enacted in forty-three states. Neonatal PKU screening is fast, simple, inexpensive, and relatively noninvasive, and PKU can be controlled by a special diet which prevents mental retardation. No legal challenges have been brought against the compulsory aspect of these laws.

The second wave of genetic screening laws raised more complicated ethical and legal questions. Between 1970 and 1973 programs designed to identify parents who carried the gene for sickle-cell anemia, a hereditary defect of red blood cells, were written into law in thirteen states. The purpose of these laws was not to screen for disease; it was to warn healthy persons about their procreative risks. Many of the laws were compulsory, requiring testing at birth, at entry into school, or at the time of application for a marriage license. Since these programs were aimed at a particular racial group, blacks, the classification was inherently suspect, and serious doubts were raised as to the constitutionality of the statutes in question. By mid-1972 the U.S. Congress enacted a law that did much to correct the poorly operated state programs. Substantial federal funds were made available to establish state screening programs that were premised on voluntary participation. The law demanded that testing data be kept confidential in order to decrease the probability of stigmatization by genotype, and it required that subjects be offered adequate genetic counseling services in addition to the testing. In 1976 sickle-cell screening laws were enacted in seventeen states, most of which comply with federal guidelines.

A third wave of screening laws began to sweep through the United States in the mid-1970s. Essentially, these laws are an extension of the earlier PKU legislation. Simple tests have been devised for a number of other inborn errors of metabolism, such as galactosemia, homocystinuria, tyrosinemia, and histidinemia. All of these diseases are accompanied by mental retardation except tyrosinemia, and all can be treated by dietary manipulation. Liver disease and cataracts are found in galactosemia; displaced lenses and a tendency to form blood clots are features of homocystinuria; tyrosinemia is characterized by vomiting, diarrhea, shortness of breath, water retention, and enlarged liver and spleen, often resulting in cirrhosis and early death; histidinemia shows a variable clinical pattern but often growth retardation and speech and hearing defects are present. Tests for these diseases have now been incorporated into the neonatal screening programs in several states.

In 1973 the Maryland legislature created a Commission on Hereditary Disorders to make recommendations concerning state-supported genetic screening. The Maryland approach embodies many principles from the federal law; particularly, it abjures compulsion and guarantees the privacy of data derived from screening.

May states require that people submit to compulsory genetic screening? To answer this question, the well-recognized interest of the state in protecting the public health must be balanced against the constitutionally protected rights of the individual—freedom of religion, equal protection under law, and the right to privacy. The sickle-cell laws were targeted at a racial minority in a way that created a substantial risk of discrimination. Some school officials, employers, and insurance carriers tended to stigmatize carriers of the sickle-cell gene who were otherwise healthy; this could be argued to be a violation of the equal protection clause of the Fourteenth Amendment. Mandatory neonatal screening laws are probably a valid exercise of the public health power so long as they are applied fairly to all racial, religious, and ethnic groups.

For the near future the patterns seem set. Mass screening to detect treatable genetic disease in infants will be compelled by law in most states; testing programs aimed at warning people about procreative risks will be voluntary. Government programs to provide voluntary prenatal screening for chromosomal disease could have an important influence in reducing the incidence of such disorders. If tests are devised to demonstrate genetic susceptibility to certain environmental hazards, then the government may offer genetic screening to alert those persons at risk.

Licensure of genetic counselors

Because of the rapid expansion of knowledge of genetic disease, there is an increasing need for trained genetic counselors. Traditionally, the medical profession has been regulated by the state through medical licensure laws and by professional organizations through board certification. At present there are no formal or informal mechanisms for regulating the activities and proficiencies of genetic counselors.

The impetus for regulation is being felt from two sources. First, several genetic screening laws now encourage quality control of counseling performed by employees of the states. Second, there is widespread debate over the definition of genetic counseling and what kind of training is sufficient for a person to offer counseling services. Because it is likely that our society will need only a few thousand counselors in the near future, it may be preferable to look to professional organizations to develop certification programs rather than to argue for separate legislation for licensing in each of the fifty states. Eventually, however, state legislation will probably be needed for many paramedical professionals, including genetic counselors, in order to ensure quality delivery of health care.

Confidentiality in the genetic counseling setting

The confidentiality of the doctor–patient relationship has long been recognized in tort law, and breach of that duty to the patient that causes damage can result in an action for invasion of privacy, for violation of a contractual agreement, or for destruction of a fiduciary relationship.

The principle of medical confidentiality has a long historical background. The Oath of Hippocrates (fifth century B.C.) states: "What I may see or hear in the course of the treatment . . . which on no account one must spread abroad, I will keep to myself holding such things shameful to be spoken about." Similarly, the Principles of Medical Ethics of the American Medical Association warn: "A physician may not reveal the confidences entrusted to him in the course of medical attendance, or the deficiencies he may observe in the character of patients, unless he is required to do so by law or unless it becomes necessary in order to protect the welfare of the individual or of the community."

Patients have relied on the confidential character of their communications with physicians. This relationship has carried over to the counselor–counselee setting, and it is here that special problems arise. Does the counselor have a right, or even a duty, to warn relatives at risk when, by doing so, he breaches the confidential relationship with his patient?

Although no case law exists in the genetics context, there is a fairly strong judicial precedent that validates disclosure of confidential information by the doctor to a third party if it can be demonstrated that the third party has an overriding interest in the information. Thus, precedents exist for a justified breach when the spouse or prospective spouse has a vital need for the information, when an unrelated person is put in jeopardy without the information, or when government requires disclosure.

Perhaps the most difficult legal problem in genetic counseling is the determination of the appropriate uses of genetic information. The practice of unauthorized disclosure should not be encouraged. Instead, the counselor is advised to ascertain whether the patient agrees to limited disclosure of genetic data during the first counseling session. Naturally, disclosure to third parties who are not relatives—such as school teachers, employers—without patient consent carries a much greater risk of liability than disclosure to relatives or spouses who may suffer injury if not forewarned.

Prenatal torts

With the growing use of prenatal diagnosis and the availability of legal abortions, a physician can screen his vulnerable patients for a variety of genetic risks such as carrier status before conception and genetic fetal abnormality after conception. Failure to offer such procedures may constitute a tort if the doctor should have known of the danger and if an infant is born later with severe genetic disease.

Wrongful birth and wrongful life actions have arisen in the courts when the plaintiff complains that a child should not have been conceived or born. In the former tort, wrongful birth, physicians have been held liable for failure of sterilization operations. In the latter, wrongful life, liability may ensue if a doctor fails in his duty to inform a pregnant woman of an increased risk of birth defects in her fetus in time for her to obtain an abortion. In some cases the child has sued, but these actions have thus far been unsuccessful because of the logical and legal absurdity that results when a person complains that he or she should not have been born.

The history of wrongful birth and wrongful life actions is grounded in prenatal tort law. Injury to a fetus after viability has long been recognized in the United States, but a cause of action matures only after the live birth of the damaged fetus. Since 1946 an injury received by the fetus before viability is cognizable at law.

In the past, courts have generally refused to recognize that a tort was committed when a physician failed to apprise the pregnant patient that she had a risk of carrying a severely defective fetus and neglected to present the alternative of abortion. During the 1960s two lawsuits involving congenital effects of rubella (German measles) explored this problem, which has special relevance to genetic counseling: whether a doctor who has failed to warn his patient of a substantial risk that she will give birth to a child with congenital anomalies should be held liable for such a birth (*Gleitman* v. *Cosgrove; Stewart* v. *Long Island Hospital*). Physicians who were sued for failure to inform their patients of such a risk offered two defenses. First, because abortion was illegal, the patient could derive no benefit from the information. Second, since the only alternatives open to the patient were to destroy the fetus or bear the child, the court had no measure with which to assess damages. It was not possible to compare the value of a defective life to nonexistence. Until 1975 courts consistently agreed with such a defense by defendant physicians. In that year the Supreme Court of Texas held that a doctor could be held liable for the economic burden caused by his failure to inform a patient that her fetus might be seriously deformed (*Jacobs* v. *Theimer*). This decision opened the door to malpractice actions based on the failure to diagnose a particular genetic disease in the fetus.

In 1976 two other cases created a further wedge in wrongful life actions. In one case, two obstetricians were found liable for failing to inform the parents of their risk of having a child with Tay-Sachs disease and for failing to offer the proper diagnostic tests (*Howard* v. *Lecher*). In the other case, the parents of two children who died of polycystic kidney disease were allowed to sue the physician who falsely reassured them, after the first child died but before the second child was conceived, that they need not worry about a recurrence when, in fact, the risk was twenty-five percent. In this case complaints were allowed not only for economic loss but also for emotional pain and suffering (*Park* v. *Chessin*).

Another aspect of wrongful life actions is the possibility that such a suit may be brought by the infant against his parents who, with knowledge of a positive prenatal diagnosis, refuse abortion and knowingly bring the defective fetus to term. Such a wrong by a parent has a slender precedent in a suit by a bastard against his father (*Zepeda* v. *Zepeda*). While the court admitted that a tort had been committed, it refused to fix damages, stating that it was for the legislature to create such a remedy if it so desired.

Informed consent in genetic counseling

The doctrine of informed consent is a development in law that may have a broad impact on genetic counseling. Informed consent essentially requires that the patient be told the nature and consequences of the disease and treatment and the alternatives available to him. It may be broadly interpreted to mean that the counselee should be informed as to the nature and scope of the genetic information being gathered, the significance of positive test results, the genetic risks involved in procreation, the burdensomeness of the disease in question, and the alternatives available to him to avoid serious consequences. The last might include contraception, sterilization, adoption, artificial insemination, amniocentesis, and abortion. Thus the law places a duty on the physician to make certain disclosures that are necessary in order for the patient to make an informed, rational decision. Failure to do so may be negligence.

Such a legal obligation places a tremendous burden on the genetic counselor. A careful family history and physical examination must be recorded, and sophisticated laboratory tests must be ordered to diagnose properly any one of many rare genetic disorders. More than 2,500 genetic diseases are now catalogued (McKusick), and approximately 100 new ones are being discovered each year.

In addition to keeping abreast of this relatively new medical specialty, the counselor must adequately inform the counselee. An understanding of Mendelian inheritance is often difficult to communicate to the average adult who has little knowledge of reproductive biology and probability. With patients who are minors, who are mentally defective, or who do not speak the native language of the counselor, the problem of comprehension may be insurmountable. In these cases an adult relative, guardian, or ombudsman may be needed for proxy consent or for help in relaying the message.

There are many new procedures being developed in medical genetics that require human experimentation. In addition to amniocentesis, other approaches to prenatal diagnosis are being tested. Ultrasound images of the fetus aid in determining size, shape, growth patterns, and fetal movements. Fetoscopy is a procedure for directly visualizing the fetus, but the danger of induced abortion is not known. Fetal biopsies and fetal blood samples have met with some success. Some genetic diseases are being investigated to determine whether enzyme replacement or gene replacement is feasible. In these areas of research informed consent of the patient-subject is vital to an understanding of the potential risks and benefits that the experiment entails. Research on fetuses that are destined to be aborted is under particular scrutiny.

Genetic hazards in the environment

Recently experiments have been devised to piece together DNA—the genetic material—from organisms of different species. This endeavor results in recombinant DNA with new genetic messages. Widespread concern has been voiced as to the human heath hazards of such experiments. National and international meetings have been held to assess the risks and benefits in order to prevent the conduct of experiments considered to be unreasonably dangerous. The National Institutes of Health in the United States has published guidelines for this research. International cooperation is fervently desired.

Genetic disease originates by a process called "gene mutation." In the 1920s it was discovered that X-rays cause mutations in fruit flies. Since that time overwhelming evidence has shown that the mutation process is similar in viruses, bacteria, plants, animals, and man. Furthermore, mutations can be induced by various physical and chemical agents. An agent that causes a mutation is called a *mutagen*.

One of the greatest concerns today is how to reap the benefits of scientific advances in atomic energy and in the chemical and drug industries without polluting the gene pool for future generations by increasing the mutation rate. In addition to genetic insults, certain chemicals and radiation also increase the incidence of cancer (carcinogens) and nongenetic birth defects (teratogens). Since there is considerable overlap among mutagens, carcinogens, and teratogens, many scientists believe that a substance that causes cancer is also likely to cause an increase in birth defects and genetic diseases.

A flurry of legislative activity has addressed this problem. The U.S. government has taken an active interest in protecting its citizens from environmental hazards. There has been a proliferation of governmental agencies and statutory law bearing on this question. The Council on Environmental Quality, the Environmental Protection Agency, the National Institute of Environmental Health Sciences, the National Institute of Occupational Safety and Health, the Food and Drug Administration, and the Occupational Safety and Health Administration are examples of federal institutions charged with monitoring and regulating environmental hazards including mutagens. Congress has also passed statutes requiring testing and control of hazardous and noxious substances. These include the Clean Air Act, the Water Pollution Control Act, the Federal Insecticide Act, and the Atomic Energy Act. The Toxic Substances Control Act was passed in 1976, requiring registration of all new chemicals ninety days before marketing. It also requires the manufacturer to test, or to pay for tests of, each substance that has been determined by the Environmental Protection Agency to be likely to present an unreasonable risk to health or the environment. Tests will be made for toxicity, mutagenicity, carcinogenicity, and teratogenicity. If these legal requirements are effective, then man-made genetic diseases should be reduced.

MARGERY W. SHAW

[*Directly related are the entries* CONFIDENTIALITY; ENVIRONMENTAL ETHICS, *article on* ENVIRONMENTAL HEALTH AND HUMAN DISEASE; GENETIC DIAGNOSIS AND COUNSELING; GENETIC SCREENING; MEDICAL MALPRACTICE; *and* PRENATAL DI-

AGNOSIS. *Other relevant material may be found under* ABORTION, *article on* LEGAL ASPECTS; CONTRACEPTION; MENTALLY HANDICAPPED; *and* STERILIZATION, *article on* LEGAL ASPECTS. *For discussion of related ideas see the entries:* FUTURE GENERATIONS, OBLIGATIONS TO; *and* PRIVACY. *See also:* GENE THERAPY; GENETIC ASPECTS OF HUMAN BEHAVIOR; *and* REPRODUCTIVE TECHNOLOGIES, *article on* LEGAL ASPECTS.]

BIBLIOGRAPHY

CHILDS, BARTON. *Genetic Screening: Programs, Principles, and Research.* Washington: Committee for the Study of Inborn Errors of Metabolism, National Academy of Sciences, 1975.

EHRMAN, LEE; OMENN, GILBERT S.; and CASPARI, ERNST., eds. *Genetics, Environment, and Behavior: Implications for Educational Policy.* New York: Academic Press, 1972.

HALLER, MARK H. *Eugenics: Hereditarian Attitudes in American Thought.* New Brunswick: Rutgers University Press, 1963.

HOLDER, ANGELA RODDEY. *Medical Malpractice Law.* Foreword by Milton Helpern. New York: John Wiley & Sons, 1975.

HOOK, ERNEST B.; JANERICH, DWIGHT T.; and PORTER, IAN H., eds. *Monitoring, Birth Defects and Environment: The Problem of Surveillance.* Proceedings from a Symposium on Monitoring, Birth Defects and Environment, Albany, N.Y., 1970. New York: Academic Press, 1971.

KABACK, MICHAEL M., and VALENTI, CARLO, eds. *Intrauterine Fetal Visualization: A Multidisciplinary Approach.* From a fetoscopy conference held at Newport Beach, California. Amsterdam: Excerpta Medica; New York: American Elsevier Publications Co., 1976.

MILUNSKY, AUBREY, and ANNAS, GEORGE J., eds. *Genetics and the Law.* New York: Plenum Press, 1975.

MCKUSICK, VICTOR A. *Mendelian Inheritance in Man: Catalogs of Autosomal Dominant, Autosomal Recessive, and X-Linked Phenotypes.* Baltimore: Johns Hopkins Press, 1975.

SPERBER, MICHAEL A., and JARVIK, LISSY, eds. *Psychiatry and Genetics: Psychosocial, Ethical and Legal Considerations.* New York: Basic Books, 1976.

COURT OPINIONS

Buck v. Bell, Superintendent. 274 U.S. 200. 71 L. Ed. 1000. 47 S. Ct. 584 (1927).

Gleitman v. Cosgrove. 49 N.J. 22. 227 A.2d 689 (1967).

Griswold v. Connecticut. 381 U.S. 479. 14 L. Ed. 2d 510. 85 S. Ct. 1678 (1965).

Howard v. Lecher. 386 N.Y.S.2d 460 (1965).

Jacobs v. Theimer. 519 S.W.2d 846 (Tex. 1975).

Park v. Chessin. 387 N.Y.S.2d 204 (1976).

Roe v. Wade. 410 U.S. 113. 35 L. Ed. 2d 147. 93 S. Ct. 705 (1973).

In re Sterilization of Joseph Lee Moore. 289 N.C. 95. 221 S.E.2d 307 (1976).

Stewart v. Long Island College Hospital. 58 Misc. 2d 432. 35 App. Div. 2d 531 (N.Y. 1971).

Zepeda v. Zepeda. 41 Ill. App. 2d 240. 190 N.E.2d 849 (1963).

GERMANY

See MEDICAL ETHICS, HISTORY OF, *section on* EUROPE AND THE AMERICAS, *articles on* INTRODUCTION TO THE MODERN PERIOD IN EUROPE AND THE AMERICAS; WESTERN EUROPE IN THE SEVENTEENTH CENTURY; CENTRAL EUROPE IN THE NINETEENTH CENTURY; WESTERN EUROPE IN THE TWENTIETH CENTURY; *and* EASTERN EUROPE IN THE TWENTIETH CENTURY. *See also* MEDICAL ETHICS UNDER NATIONAL SOCIALISM.

GREECE

See MEDICAL ETHICS, HISTORY OF, *section on* EUROPE AND THE AMERICAS, *article on* ANCIENT GREECE AND ROME.

HANDICAPPED PERSONS

See Chronic Care; Mentally Handicapped; Infants.

HEALTH AND DISEASE

I. HISTORY OF THE CONCEPTS   *Guenter B. Risse*
II. RELIGIOUS CONCEPTS   *Frank De Graeve*
III. A SOCIOLOGICAL AND ACTION PERSPECTIVE
   Talcott Parsons
IV. PHILOSOPHICAL PERSPECTIVES
   H. Tristram Engelhardt, Jr.

I
HISTORY OF THE CONCEPTS

Introduction

Health and disease are common but ambiguous notions. As has been recently stressed, this vagueness is based on their normative as well as their descriptive nature. Since models of explanation and evaluation are intimately connected with biological knowledge, social structure, and cultural values, health and disease concepts have varied greatly throughout history.

Health, in a sense, is defined by what does not constitute health, namely disease. That term, too, reflects a series of value judgments regarding psychophysiological characteristics considered to be deviant or pathological. Disease has been defined as person-centered, discontinuous, and undesirable—the so-called organismic view of disease. Such a notion reflects three independent explanatory frameworks that have been used for disease: (1) as a biological abnormality; (2) as a behavioral discontinuity; and (3) as a phenomenological occurrence (Fabrega).

The first category is rooted in the idea that distress and disability are based on abnormal processes and structural alterations of the human organism. These biological deviations have had, for the most part, a universal validity, in contrast with variable social and behavioral factors. Disease as a behavioral discontinuity comprises the full range of behavioral responses to pain and dysfunction as determined by social, psychological, and cultural factors. Finally, disease as a phenomenological occurrence is perceived as a completely altered state of being for the affected person, often as a result of supernatural possession or invasion. All three categories, which can coexist in any given culture, stress the discontinuity of health and disease states. They consider the former highly desirable and the latter evil, and attempt to assign responsibility to individualized extraneous agents for the occurrence of disease.

Two concepts of disease have been alternatively dominant throughout history. One of them, termed *ontological*, viewed disease as something alien and external to the healthy person. As a specific and objective reality distinct from the patient harboring it, disease ran its own course and exhibited pathognomonic (specifically characteristic) symptomatology and pathological lesions. In such a view, the stress was on the common character of symptoms and lesions shared by sick individuals thus allowing for distinctive patterns and a clear separation between health and disease. In accepting such a concept, physicians and patients put considerable stress on diagnosing those entities through the assignment of specific labels while also searching for specific therapies to eradicate each disease. Finally, the ontological concept encouraged the search for specific etiologies (disease causations), fostered the notion of the immutability of disease species and deemphasized the seemingly endless variations of symptomatic sequences and lesions observable in individual patients.

The other concept has been called *physiological* since it viewed disease merely as a consequence of disturbed functions operating within individual human beings. Such a viewpoint blurred the distinctions between healthy and diseased states, stressed the uniqueness of each patient's illness, and internalized the pathological process by placing it in the organism instead of ascribing the disturbances to outside causes. Under those circumstances, the stress was on individual prognosis rather than diagnosis, and the institution of general treatments aimed at the disturbed bodily functions.

Ancient cultural conceptions

If we may extrapolate from contemporary primitive societies certain essential characteristics already present in prehistoric times, disease was essentially seen as a behavioral and phenomenological discontinuity with strong social and religious implications. Thus, concepts of disease such as disease–object intrusion and soul loss were probably genuine paleolithic traits of almost universal distribution (Clements). More recent concepts such as spirit intrusion and breach of taboo reveal a gradual social and religious sophistication and organization which used disease as a social sanction and moral offense.

Cosmological schemes and religious values largely determined the causal agents responsible for disease. Hence, certain symptoms and signs were often lumped together as more or less coherent units and were believed to have been inflicted by individualized agents: a witch, god, or demon. By contrast, other syndromes perhaps just as dramatic in their appearance, distress, and dysfunction were viewed in naturalistic causal terms. Most commonly these manifestations constituted traumatic diseases acquired during hunting, warfare, and agricultural pursuits.

Whether viewed in magico-religious or naturalistic terms, many symptom sequences were lifted out of the bewildering array of discomforts and given a specific name and cause. There was a strong belief that disease is often substantial and caused by harmful foreign agents either spiritual or material, such as in the concepts of disease–object intrusion and spirit possession. In the breach of taboo category, varying syndromes were contextually associated with punishment by a specific supernatural agent. Thus, attempts to reify distress and dysfunction occurred in practically all primitive societies. These crude ontologies were and still are highly functional; they provide the necessary focus for specific therapies, while assigning meaning to disagreeable sensations through the establishment of diagnoses. Such procedures were bound to lessen fear and anxiety in both the affected individual and his circle of family and friends.

A higher level of sophistication in the discernment of patterns concerned with distress and dysfunction occurred in the ancient Near East. As was true in prehistoric settings, magico-religious interpretations of such syndromes as well as naturalistic ones coexisted side by side. Texts obtained from Mesopotamia, for example, reveal selected clusters of distress used by the *āšipu* or magical expert to arrive at a diagnosis, often equated with the supernatural cause: "the hand," "seizure," or "grasp" of a certain deity (Labat, 1951). This is likewise true for the symptoms treated by the lay healer or *asû*. The assumption was that certain specific chains of events, namely, disease symptoms, were initiated by supernatural agents and then proceeded on predetermined courses to predictable outcomes. Most of the diagnostic sequences were assembled according to the bodily parts in which the prominent symptoms could be observed—"if his face and eyes are red, if he suffers in his middle and his belly"—thereby closely relating discomfort and loss of function to certain bodily regions. Moreover, the sequence of observed bodily parts was organized from the top of the head to the feet, a systematization which, like the Egyptian Edwin Smith Papyrus, may merely reflect the didactic, textbook character of the Mesopotamian writings.

Another important characteristic shared by both ancient Mesopotamians and Egyptians was that disease was not a static cluster of symptoms but a process with pathognomonic or specifically characteristic manifestations, critical days, exacerbations, remissions, and recurrences. Such developments could be understood as omens, valuable for prognosis, as well as being useful criteria for instituting appropriate therapies. The desire for specific diagnosis of disease was shared by healer and patient alike. The former could then project a prognosis and institute therapy, and attach meaning to the patient's sufferings while responding to the imperative to relieve it. The stress on prognosis may have arisen both from a need for the lay practitioner to limit his activities in order to avoid later reproaches or sanctions, and from a basic urge of the afflicted to know their fate.

Ancient Egyptian medical texts similarly used anatomical frameworks to link their symptom sequences together. The association was perhaps reinforced by a religion that assigned protective deities to certain human organs and a funerary practice confirming their existence. In the case of the Edwin Smith Papyrus, primarily dealing with trauma, one can see a careful systematization of symptom sequences for diagnostic and prognostic purposes. Possible pathways of development in certain situations are well described and critical stages and symptoms clearly identified. Since the inception of the papyrus coincided with Egypt's period of monumental funerary constructions, it is tempting to speculate that the text reflects a massive compilation of cases involving thousands of workers. Under those conditions, trauma was clearly seen in naturalistic terms as an occupational hazard to be reckoned with.

Ancient Egypt also reveals a subsequent step in the biological disease conceptualization: the appearance of a general operative principle called WHDW. A comparison of various identifiable disease entities led to the establishment of a physiopathological mechanism or agent—in the case of WHDW a peccant or corrupt matter whose presence in the body spread putrefaction and therefore had to be eliminated. Perhaps the late Egyptian custom of taking monthly purgatives, reported by Herodotus, was an application of such a theoretical principle. Indirect evidence regarding this and other practices suggests a hygienic concept of health to which Egyptians seemed to have adhered seriously.

Ancient physiological conceptions

The attempt to conceptualize disease blurred the distinguishing lines between several symptom sequences that had attained ontological status. Efforts to understand the individuality and physiological unity of the human organism led to the establishment of physiological concepts of disease. These ideas, prevalent in ancient India and China, were mainly derived from comprehensive philosophies of nature. Both civilizations postulated the existence of general operative principles, which maintained proper harmony among the bodily parts while preserving their overall specific composition.

Such was the nature of the ancient Indian *dosas* (humors)—air, bile, and phlegm—viewed as bodily secretions and principles supporting the activity of the human organism. Likewise, equilibrium of the Chinese yin and yang forces assured health and prevented disease. In each case human physiology constituted a microcosm closely related to the speculative macrocosmic processes. Stress on individual psychophysical characteristics—the so-called constitution—provided the basis for detailed and prescriptive efforts to achieve and maintain health. As defined in India's old text, the *Caraka Samhita*, health —"the supreme foundation of virtue, wealth, enjoyment, and salvation"—was more a moral than a biological concept. Only through good health were persons capable of realizing their transcendental goal of divine freedom, and *Ayurveda*, the science of health and longevity, was the appropriate vehicle to achieve it.

About the same time, ancient Greece engaged in similar conceptualizations and value judgments concerning health and disease. The treatises of the Corpus Hippocraticum ascribed to the Cnidian School seemed to have shared many of the characteristics of the Mesopotamian and Egyptian syndromes by associating the symptom clusters with anatomical structures. Although the Cnidians demonstrated an awareness of bodily fluids, disease units were classified according to the organs and bodily parts affected beginning from head to toe, as in the Edwin Smith Papyrus. Such a focus on organ-related symptom sequences allowed for specific therapeutic recommendations while often pointing the way toward specific causes responsible for distress. A fundamental difference from its Near Eastern antecedents was the absence of magicoreligious explanations.

The Coan School clearly adopted a physiological conceptualization of disease by borrowing heavily from the contemporary philosophies of nature and employing the micro-macrocosm analogy. In this scheme, bodily humors (phlegm, blood, yellow and black biles) played essential roles in guaranteeing functional integrity. Such Greek societal values as harmony, symmetry, beauty, and balance decisively shaped the ideal of "perfect" health and the humoral theory of disease.

Through speculative explanations, Hippocratic medicine viewed each individual as possessing a particular humoral mix responsible for somatic appearance and mental attitude. Such "constitution" and "temperament" provided "relative" health and predisposed the individual to certain imbalances. His discomforts and biological struggle with the environment were deemed unique and the result of life-style, geographic location, and climate. The focus of disease as a particular form of human experience was the diseased individual. As a corollary, health could be preserved through a dietary regime aimed at maintaining a proper internal humoral balance and through avoidance of noxious environmental factors. The individual was thus held accountable for his own health.

There were notable exceptions to this predominantly physiological approach to disease. One was epilepsy, the "sacred disease." In certain instances symptoms or signs appeared representing rather unique organic responses. Some, like the epileptic convulsion and the gouty toe, were dramatic and disabling enough to achieve early recognition and forced healers to distinguish them as enduring reified types.

Galen's synthesis retained the humoral theory of disease but also attempted to indicate the anatomical and functional disturbances accompanying distress. His teleologically oriented system strengthened the concept of health with biological arguments; indeed, all parts of the human organism were built to function harmoniously in the preservation of health according to divine nature. Thus, living a healthy life, in Galen's view, was a moral obligation, and disease a phenomenon reflecting ignorance, intemperance, or both.

The advent of Christianity reinstated the Old Testament concept of disease as reflecting the commission of sins. Confession and prayer directed toward the cleansing of patients' souls were the proper treatment. On the other hand, the New Testament included ideas and events that prompted some seemingly conflicting concepts of disease—disease is a consequence of sin yet subject to the healing of the spirit (Mark 5:21–43), a weakness which can be converted to spiritual strength (1 Cor. 12:7–10). Under the influence of these concepts, care of the sick became a very important work of charity. It was a service that brought divine rewards to those who practiced it. These apparently conflicting Christian concepts of disease were fused in various ways during medieval times.

From ancient and medieval to modern theories

The new biological conditions generating widespread epidemics in the Roman Empire during A.D. 165 and 251, Byzantium (A.D. 542), and medieval Europe (1348), contributed to a gradual shift in the concept of disease. Thousands of human beings involuntarily afflicted by similar symptom clusters and physical signs made suffering appear to be a thing, an *ens* or independent reality, and again popularized ontological conceptualizations of disease. This was the case with the Islamic physician al-Razi or Rhazes, who in the ninth century described a disease entity very similar to contemporary smallpox.

Such epidemics as the notorious "Black Death" displayed characteristic sequences of suffering, reenacted thousands of times over a widely scattered territory. These occurrences tended to confirm the independent existence of disease, its episodic and destructive character, and the presumed existence of outside agents responsible for its transmission, as well as the apparent helplessness of those affected by the disaster. Though the humoral theory of disease remained the traditional rational explanation for biological deviancy, medieval and later Renaissance authors increasingly sought to distinguish separate disease entities and provided them with specific names. The appearance of two seemingly new forms of distress, syphilis and typhus, during the Italian wars of 1494–1559, only accelerated the trend. The repetitious patterns stimulated the search for specific etiologies and effective therapies, but such a classification of diseases also tended to provide a real identity to limited patterns of distress at the expense of others.

One of the early authors to examine the question of specific causes was the Italian physician Girolamo Fracastoro (1483–1553), who had experienced an outbreak of bubonic plague at Verona in 1510 and another of typhus in 1528. Fracastoro proposed a new concept—the contagion, a material composed of imperceptible particles called *seminaria*, which gave rise to specific diseases. Depending on their density and power of penetration, these contagious particles caused skin or deep-seated organ afflictions. They propagated through direct contact, through an object, or, over a short distance, through the air.

Famous authors such as Paracelsus, a contemporary of Fracastoro, and van Helmont in the seventeenth century, reinforced the ontological concept of disease with their own theories. Paracelsus (1493–1541) spoke of particular *seeds of disease* endowed with characteristic spiritual powers, a veritable *ens morbi* sown by God after the fall of man. These seeds were alien and extraneous to the affected human organism and were capable of causing particular diseases. Van Helmont (1577–1644), expanding on these ideas, conceived of disease as a specific, invisible idea, which became flesh in a host through a successful battle with the organism's local or central principle of vitality. Finally, William Harvey (1578–1657) formulated the idea that diseases were parasites endowed with a vital principle of their own. Extrapolating from his embryological studies, Harvey equated contagium with the emission of semen. Once impregnated locally in the affected organism, it remained alive and was disseminated by virtue of its own vitality.

The ontological concept of disease received new impetus through the works of the physician Thomas Sydenham (1624–1689), who lived through the time of the last plague epidemic affecting London in 1676. Sydenham differentiated diseases as substantial forms, specific expressions of nature's struggle to balance the humors and recover health. Thus, intensive analyses of clinical cases could help to delineate essential syndromes, discover specifically characteristic symptoms, and trace their temporal development. Each disease represented a distinguishable entity while affected individuals were merely embodiments of the disease.

Sydenham's ontological viewpoint allowed diseases to be classified into orders, families, genera, and species as if they were realities like plants and animals. Numerous physicians established complex classifications of disease—the science of nosology—such as the schemes of the Swedish botanist Linnaeus (Carl von Linné, 1707–1778), Boissier de Sauvages (1706–1767), William Cullen (1710–1790), and Philippe Pinel (1745–1826).

While most physicians chose to see diseases as specific discontinuities, others sought to discover in medicine some basic underlying principles similar to Newton's universal law of gravity. John Brown (1735–1788) placed special emphasis on one biological property, "excitability," considered to be the basic quality inherent in living matter and the ultimate cause of life. Brown's excitability was distributed equally throughout the organism across organ and tissue boundaries. It was sustained by the steady influx of environmental and internal stimuli, with health a rather precarious central balance between adequate stimulation and inherent excitability. Thus, Brown developed a scheme depicting successive states of nonhealth on both sides of the equilibrium on a sliding scale. He allowed even mathematical calculation for the varying degrees of excessive or inadequate excitability in lieu of establishing fixed and independent disease entities. Health and disease were merely phases of a continuously changing balance between stimuli and excitability, a notion close to the modern ecological viewpoint.

The last two centuries

In the early nineteenth century, the advent of large hospitals capable of housing thousands of patients recreated an artificial epidemic scenario. Systematic autopsies firmly linked clinically defined illnesses with specific gross pathological changes. This fact is reflected in the writings of prominent physicians of the Paris Medical School, who described with the utmost precision numerous diseases, including tuberculosis, rheumatic heart disease, typhoid fever, and many others.

Pinel called for analysis of distress sequences, making them rallying points for clinicopathological research, while simultaneously deemphasizing individual biological responses. Pierre F. Bretonneau (1771–1862) emphatically declared that diagnosis, prognosis, and treatment depended upon the specific character of a disease.

The ontologists, of course, received further confirmations of their ideas through epidemics of typhoid fever, diphtheria, and later cholera and yellow fever. These problems greatly stimulated the search for specific causes and eventually led to the establishment of the germ theory of disease in spite of persistent opposition. The efforts made by Louis Pasteur (1822–1895), Robert Koch (1843–1910), and others demonstrated that specific microorganisms were responsible for specific diseases. Bacteriology tended to identify the invasion of specific microbes closely with the entire disease, relegating environmental and social factors to the background.

Among the early nineteenth-century proponents of a shift in emphasis to the functional factors operative in the organism, was François J. V. Broussais (1772–1838). While accepting the specificity of disease entities, Broussais accused the representatives of the Paris School of setting up factitious disease entities based solely on static anatomical changes.

The subsequent rise of experimental physiology, in both France and Germany, spurred by such notable scientists as François Magendie (1783–1855), Claude Bernard (1813–1878), Carl Ludwig (1816–1895), Emil DuBois Reymond (1818–1896), and Hermann von Helmholz (1821–1894), exposed the existence of complex levels of organic function. Concepts such as Bernard's "internal environment" and Cannon's "homeostasis," the subsequent discovery of neurohormonal biofeedback, and immunological and genetic mechanisms, seemed to indicate that disease was a general notion and that individual diseases responded to general laws of physiology.

Moreover, the new biological understanding of the human organism gradually restored the primacy of the psychophysiological individuality of each patient in terms of susceptibility or resistance to disease, as well as the manifestations and actual course of the various diseases. As has been pointed out, those authors arguing for a physiological concept of disease felt that illnesses were more contextual than substantial, the result of complex physiological interactions between the individual organism and the environment.

The spectacular therapeutic triumphs in the first half of the twentieth century can be attributed largely to previous efforts at establishing specific disease entities. Whether viewed as convenient abstractions or real objects, these discrete units allowed the search for etiology or disease causation as well as the establishment of diagnostic, prognostic, and curative criteria. Finally, individual diseases acquired social meanings and were manipulated easily to receive public support for research aimed ostensibly at their eradication or prevention.

Conclusion

This brief historical perspective has depicted primarily the growth of a biological concept of disease in close association with Western values and scientific progress. The total development reflects a gradual fine-tuning of distress sequences, established by a series of value judgments in response to the prevailing biological conditions. Within this category, both the ontological and physiological conceptualizations have been operative for millennia, merely reflecting a repeated change in emphasis.

Whether anatomical, physiological, or biochemical explanations are used, the idea that disease is an entity that can be detached from man and linked to deleterious causes has had great appeal. It has been suggested that such an outside projection and effort to particularize distress constitutes a protective psychological mechanism designed to shield the healer from some of the emotional implications of suffering and disability (Engel).

As mentioned before, the organismic concept implies that disease is a distinct and undesirable or "bad" event in both time and space for which the affected individual is not responsible. This state can be "conquered" and "eliminated" through resolute action by medical professionals, thereby halting or avoiding suffering and disability.

Hence, it is easy to see how viewing disease as a discontinuous "thing" has profoundly shaped the organization and functioning of health-care systems. As the view is now established, people believe themselves to be generally "healthy" until, on the basis of certain sensations, they believe they are "diseased," and therefore in need of medical services specifically to remove the temporary disability. Medical intervention is viewed as an episodic contact, often reluctantly initiated by those who judge themselves to be "diseased." Having entered into a contractual agreement with the professional, both healer and patient strive anxiously to define and label the disability in order to recommend specific measures aimed at the "removal" of the disease. Perceptions deemed deviant, which cannot be fitted into this reductionist and largely biological framework, encounter serious difficulties in diagnosis and care within the established medical system. The same is true for so-called preventive measures instituted in the "absence" of disease.

Efforts have been made recently to formulate a broader concept of health and disease in which both conditions, rather than static and discrete, are seen as shifting phases in a complex, multilevel set of activities involving humans and their environment (Dubos). The ecological perspective is concerned with the relationships of whole populations with their physical environments, in which disease is a natural and constant phenomenon among the multiplicity of adaptive mechanisms (Engel). One can go even farther and state that certain levels of "dis-ease" are actually necessary for the integrity of human adaptive and defense mechanisms—witness the failure of the human immunological system in the absence of continuous viral challenges.

If disease is thus seen as a natural event in man's relationship with the physical as well as the social environment, it has been suggested that the general systems theory is better suited to describe and analyze the various levels of this rather complex association. This approach would tend drastically to blur the present boundaries of health and disease and have far-reaching implications for patterns of health care (Fabrega). Whether a unified or "systems" view of disease becomes prevalent will depend on the willingness of both healers and patients to abandon a psychologically more palatable system operative since the dawn of man.

GUENTER B. RISSE

[*Directly related are the other articles in this entry. For discussion of related ideas, see the entries:* HEALTH AS AN OBLIGATION; HINDUISM; MIRACLE AND FAITH HEALING, *article on* CONCEPTUAL AND HISTORICAL PERSPECTIVES; PROTESTANTISM, *article on* DOMINANT HEALTH CONCERNS IN PROTESTANTISM; *and* THERAPEUTIC RELATIONSHIP, *article on* HISTORY OF THE RELATIONSHIP. *Other relevant material may be found under* MEDICAL ETHICS, HISTORY OF, *section on* NEAR AND MIDDLE EAST AND AFRICA, *article on* ANCIENT NEAR EAST; *section on* SOUTH AND EAST ASIA, *articles on* PRE-REPUBLICAN CHINA *and* CONTEMPORARY CHINA; *and section on* EUROPE AND THE AMERICAS, *article on* MEDIEVAL EUROPE: FOURTH TO SIXTEENTH CENTURY.]

BIBLIOGRAPHY

ORIGINAL SOURCES

Charaka Samhitā. 6 vols. Translated into Hindi, Gujarati, and English by the Shree Gulabkunverba Ayurvedic Society. Jamnagar, India: 1949.

FRACASTORO, GIROLAMO. *De contagione et contagiosis morbis et eorum curatione.* Translated with notes by William Cave Wright. New York: G. P. Putnam's Sons, 1930.

GALEN. *A Translation of Galen's Hygiene (De sanitate tuenda).* Translated by Robert Montraville Green. Springfield, Ill.: Charles C. Thomas, 1951.

HIPPOCRATES. *Hippocrates.* 4 vols. Translated by William H. S. Jones. Loeb Classical Library. Cambridge: Harvard University Press, 1948–1953.

Huang-ti nei-ching su-wen. The Yellow Emperor's Classic of Internal Medicine. New ed. Translated by Ilza Veith. Berkeley: University of California Press, 1972. See especially bk. 1–2, pp. 97–131.

LABAT, RENÉ. *Traité akkadien de diagnostics et prognostics médicaux.* 2 vols. Vol. 1: *Transcription et traduction.* Collection des travaux de l'Académie internationale des sciences, no. 7. Leiden: Brill; Paris: 1951.

NEEDHAM, JOSEPH. *Science and Civilisation in China.* Vol. 2: *History of Scientific Thought.* London: Cambridge University Press, 1956.

PINEL, PHILIPPE. "Analyse appliquée à la médecine." *Dictionnaire des sciences médicales.* Paris: Panckoucke, 1812, vol. 2, pp. 23–30. See also Pinel, *La Médecine clinique,* for a more accessible source.

———. *La Médecine clinique rendue plus précise et plus exacte par l'application de l'analyse, ou recueil et résultat d'observations sur les maladies aiguës, faites à Salpêtrière.* Paris: Brosson, Gabon & Cie, 1802. Numerous editions.

SYDENHAM, THOMAS. Preface to "Medical Observations Concerning the History and Cure of Acute Diseases." *The Works of Thomas Sydenham.* Translated by R. G. Latham. London: Sydenham Society, 1848–1850, vol. 1, pp. 11–24.

SECONDARY LITERATURE

CLEMENTS, FOREST EDWARD. "Primitive Concepts of Disease." *American Archeology and Ethnology* 32 (1932): 185–243.

DUBOS, RENÉ. "Determinants of Health and Disease." *Man, Medicine and Environment.* New York: Praeger Publishers, 1968, pp. 67–88.

ENGEL, G. L. "An Unified Concept of Health and Disease." *Perspectives in Biology and Medicine* 3 (1960): 459–485.

FABREGA, HORACIO, JR. "Concepts of Disease: Logical Features and Social Implications." *Perspectives in Biology and Medicine* 15 (1972): 583–616.

KING, LESTER S. "Nosology." *The Medical World of the Eighteenth Century.* Chicago: University of Chicago Press, 1958, pp. 193–226.

KUDLIEN, FRIDOLF. "The Old Greek Concept of 'Relative Health'." *Journal of the History of the Behavioral Sciences* 9 (1973): 53–59.

PAGEL, WALTER. "Van Helmont's Concept of Disease—To Be or Not to Be? The Influence of Paracelsus." *Bulletin of the History of Medicine* 46 (1972): 419–454.

RIESE, WALTHER. *The Conception of Disease: Its History, Its Versions, and Its Nature.* New York: Philosophical Library, 1953.

RISSE, GÜNTER B. "The Brownian System of Medicine: Its Theoretical and Practical Implications." *Clio Medica* 5 (1970): 45–51.

———. "Rational Egyptian Surgery: A Cranial Injury Discussed in the Edwin Smith Papyrus." *Bulletin of the New York Academy of Medicine* 48 (1972): 912–919.

SIGERIST, HENRY E. *Civilization and Disease.* Ithaca: Cornell University Press, 1943.

TEMKIN, OWSEI. "Health and Disease." *Dictionary of the History of Ideas.* 4 vols. Edited by Philip P. Wiener. New York: Scribners, 1973, vol. 2, pp. 395–407.

II

RELIGIOUS CONCEPTS

Throughout history religious man has interpreted the vicissitudes of his mental and physical adequacy, as well as his efforts to maintain or to restore it, within the framework of his religious attitude and behavior. To him, health is a powerful symbol of what it etymologically means, i.e., the "wholeness" of man, often conceived as his original state. Disease, then, is viewed as symbol and symptom of his "fall" into historical existence, and healing as the symbol of his often metahistorical redemption or salvation, in fact the restoration of wholeness. This is realized by human or divine healers and by the instrumentality of religious or magicoreligious practices. Religion (subjection to the ultimately transcendent) and magic (manipulation of powers present in the finite), known to blend in concrete attitudes and expressions, will be jointly considered. This article is written from the point of view of the comparative study of religions, i.e., on the basis of a nonnormative, value-neutral study of the phenomena in a representative and meaningful variety of religious contexts.

Disease

Disease is a privation or withholding of something good, an undue negative situation, inflicted by an evil or, as punishment, by a good and just power, or incurred by an evil deed. Exceptionally it is a meaningful manifestation of the will of a god, and even a special privilege. A particular illness may be related to known or demonstrable physical causes, but there is always a spiritual dimension to that causality. Infliction by an evil power may be "black magic" (a curse or spell by a living person), interference of a dead person, or arbitrary machinations of malevolent supernatural agencies. An evil deed by which disease is incurred may be the profanation of something sacred, nonobservance of commands or prohibitions, negligence of the necessary precaution at certain critical moments of "passage," breach of a spiritual relationship, or any other way of offending a god, spirit, or ancestor. The offended supernatural being or the deity representing the cosmic or moral order violated may then inflict disease as a punishment on the trespasser and restore the right order of things. Thus, in the Vedic tradition, disease is usually regarded as the punishment of sin; *Varuṇa*, the moral governor par excellence on the threefold level of cosmic, social, and ritual order, punishes and restores order upon due satisfaction.

Different views may successively or simultaneously occur in the same religious context. The same Vedic tradition does not exclude black magic or demonic malevolence, and stresses the "law of karma": Every act, good, bad or indifferent, produces a morally qualified "fruit" (*phala*). Bad karma shows in a specific disease (pulmonary consumption for the killer of a brahmin, black teeth for a drunkard, bad breath for a malignant informer, etc.) or in an inferior existence. The maturation of deeds (*karmavipāka*) is not limited to the individual life cycle and may manifest itself in a subsequent birth. Elsewhere, too, sin and disease transcend the individual dimension: "The fathers have eaten sour grapes and the children's teeth are set on edge" was obviously a common proverb, which Jeremiah and Ezekiel felt necessary to reject in favor of personal punishment for personal guilt. Collective guilt and punishment are typically expressed in the myths of the fall and related to the breach of a spiritual relationship between original man and his Creator. This idea is not just a generalization of personal responsibility: Because the good order of creation has been disturbed objectively, the whole of creation is suffering the consequences.

Disease can be interpreted as a temporary withdrawal of the soul from the body, as the introduction into the body of noxious substances causing pain or absorbing vitality, as the invasion of an organism by an alien power, or as a possession of body or mind by the power of evil.

Healing

The religious or magicoreligious interpretation of disease logically results in analogous patterns of healing. Here again, even on the primitive level, empirical knowledge of cause and effect is obvious; many interventions have a properly medical character. Still, the religious dimension remains of primordial importance. In the threefold division of healing in the *Avesta*—*kereta* (the knife, i.e., surgery), *urvara* (herbs, i.e., pharmacology), and *manthra* (sacred formula, i.e., prayer, incantation)—the last is preferred, and the *manthro-baeshaza* is called "the physician of physicians."

The form of the healing interventions will follow the interpretation of the disease. If the illness is the result of black magic, it will be neutralized by countermagic. If interference of a dead person is the cause, the reason for his resentment will be carefully diagnosed and removed, or a propitiation ritual may be in order. If the disease is viewed as inflicted by a malignant supernatural agency, exorcism might be necessary, or an appeal to a superior agency is made. One may also try to bribe or mislead the malevolent power, e.g., by assuming another identity, or by transferring the disease to another being. If the cause is identifiable as an evil deed, confession and penance will be required and whatever is necessary to have the right order restored. Restitution will be offered and mercy implored to effect reconciliation.

The interpretation of the disease will dictate the technique employed: recovery of the soul wandering about in the invisible world, or held by a spirit; removal of the pathogenic substance in the body; homology of the afflicted part in the body with a corresponding factor in the inviolable world of the sacred; ritual identification of the patient with a Savior-God who suffered and overcame the same affliction, etc.

Human healers: medicine men, shamans

A healer is one who knows the hidden powers in nature and can apply them in cures. The concrete erudition in his diagnosis and the empirically acquired knowledge of healing properties are very much a part of the primitive healer's profession, but they are not divorced from his familiarity with the supernatural powers. Furthermore, the practitioners would consider their magicoreligious knowledge of these powers equally empirical, i.e., corroborated by the success of their application. Even in more advanced civilizations of the pretechnological era, a practice of medicine free from the occult was unknown. In Egypt the masters of medicine were high ecclesiastical dignitaries, performing their art in the temples as interpreters of the gods. Pindar, when enumerating what Asclepius learned from Cheiron, the divine inventor of pharmacology, coordinates exorcism with external and internal medicine.

Healers do come in families, but even then the profession is handed on as a religious mystery. More often healers are called or chosen because of their innate ability for religious healing. Their call may be recognized in many ways: dreams, significant events, physical or mental peculiarities that manifest their otherness, fascination with the supernatural. Their own miraculous recovery from severe afflictions can prove a divine confirmation of their call. Once chosen, healers are often subjected to the ordeals of initiatory death and resurrection, unless their own disease is interpreted that way.

The Ngaanga Mbwoolo of the Yaka in Central Africa, for example, is usually a frail, sickly person, who dreams he is rescued from the water. He faints and loses consciousness repeatedly. A specialist initiates him, exposing him to the power of *mbwoolo* (healing magic), submerging him in water, giving him a new name when emerging, and instructing him in the severe tabus he will have to observe as a medicine man.

Perhaps the most typical healer in the so-called primitive societies is the shaman, someone acting under the power of inspiration rather than a technician using religious lore. His psychic transformation is essential as a prerequisite for his healing power, because he is to be the intermediary between the world of gods and demons and the world of men. This profound transformation is often related to a dream or vision during severe illness, hallucination, near-starvation, or exhaustion, bringing him to the brink of life and death. In an ecstatic trance, usually self-induced, he ascends and descends mystically to the worlds above and beneath, where he challenges or questions or is willingly possessed by the spirit causing the disease, and ultimately overcomes him in a heroic struggle. Shamans are especially competent in the cure of mental disorders or strongly psychosomatic illnesses.

Divine healers: gods of healing and culture heroes

Human healers derive their power and their medicine often from the gods or from a culture hero, the mythical first physician. In Egypt the god Thoth invented the formulae for giving remedies their power. The pharaoh, leading priest and divine representative of the gods, was well versed in the art of healing. In the Hellenistic-Egyptian cult Sarapis and Isis were divine healers. Likewise Marduk and the goddess Ninkarrak in the Mesopotamian tradition, Apollo, Cheiron, Asclepius with his retinue and the nymphs of medicinal springs in the Greco-Hellenistic tradition. At the Upanāyana ceremony in India, Agni is invoked as physician and maker of remedies. Tutelary deities of famous sanctuaries have an important healing function. Especially in the case of afflictions that require miraculous healing powers or a massive restoration of the good order of creation clearly theurgic medicine is required, and the intervention of healing gods is requested.

Culture heroes are often shamanic prototypes. In Hawaii Koreamoku obtained the knowledge of medicinal herbs "from the gods." According to the Vendīdād, Thrita was the first physician in Iran, and among the Mayas Itzama is said to have been the originator of medicine.

Instruments of healing

The main instruments of healing are charms or incantations, amulets, and pilgrimages. A charm (Latin: *carmen*) is a magical formula sung or recited to bring about a beneficial result, e.g., to banish a disease or to confer magical power on an object used for a cure. The Greek word *epode* is used by Pindar when he describes Asclepius healing people with magic melodies. Often sacred texts are used in charms, as myths in Egypt and Vedic mantras in India.

Amulets are objects, usually worn or carried on the body to keep harmful influences at a distance, to cure diseases, or to confer strength and luck. They are often "charmed" or reinforced by bringing them in contact with something of superior efficacy, by being applied at a powerful time or in a sacred context. In India investiture with an amulet was included in the new or full moon sacrifice, or in the initiation ceremony. The amulet was steeped for three days in a mixture of curds and honey, and blessed with a hymn. Quotations of sacred writings and written charms are favorite amulets. The African fetishes, usually statuettes, are a sort of amulet. The Ngaanga Mbwoolo invariably uses them for protection and healing. Together with his panacea—ashes or scrapings, or a mixture of both reduced to powder—he exposes them to the cosmic power of the new moon, all the time softly singing incantations to them.

A special sort of amulets are the mementos of holy persons or ancestors, especially relics of their bodies. They may become the object of a cult, and specifically of pilgrimages. Seeking cures for oneself or vicariously for others has always been one of the main incentives for pilgrimages. Already in Ancient Egypt faraway temples were visited for healing purposes and sanatoria were organized in the temple area, where patients could spend the night and receive medical advice from the healing gods in prophetic dreams. This "incubation" is probably derived from the dream-oracles, with interpretation by qualified specialists and a therapeutic practice based on it, that we find in the archaic cultures.

Healing in the monotheistic biblical tradition

Monotheism transforms the basic religious views of disease and healing profoundly but tolerates survivals; on the popular level it is not impervious to magic. In Islam, for example, charms and amulets are very much in use. Handwritten verses of the Qur'ān are burned with medicinal substances to give additional power to the ashes used in medication. The phylacteries, tassels, and doorpost symbols of the Jewish tradition are of the amulet type. Although recourse to occult powers is in contradiction to the religious mentality of the Bible, its possibility and efficacy are taken for granted. White magic with therapeutic intent is acceptable in rabbinical writings, and charms have survived in esoteric as well as in popular Judaism. In the incantations, the name of the demon inflicting the disease allows the healer to address him personally and to exert power over him, thus neutralizing his influence. It is lawful to "cheat" the powers of evil by a substitution of victims or by changing the name of the victim. One can try to buy the spirits off with suitable bribes.

Chapters 13 and 14 of Leviticus place disease and healing in a cultic-religious context. Disease as well as healing has a divine cause: "Salve he only can bring that wounded us; hand that smote us shall heal" (Hosea 6:2). The ills men suffer are retribution for their wickedness. Leprosy, for example, can be a punishment for slander (Miriam: Num. 12:8–11), avarice (Gehazi: 2 Kings 5:26–27) or presumption (Uzziah: 2 Chron. 26:16–21).

Cure for sickness was sought through prayers, sacrifices, or the blessing of priests and prophets. Asa is blamed because in his sickness "he did not have recourse to the Lord, trusting rather in the skill of physicians" (2 Chron. 16:12).

Similarly, in the New Testament disease is closely related to sin and the devil, and healing to Jesus, curing "all those who were under the devil's tyranny" (Acts 10:38) and saying to the cripple, "Behold thou hast recovered thy strength; do not sin any more, for fear that worse should befall thee" (John 5:14). Healing by Jesus is closely related to forgiveness and faith: "Son, thy sins are forgiven" (Mark 2:5), or "Go, thy faith has made thee whole" (Mark 10:52). Paul calls the power of healing a charism of the Holy Spirit (1 Cor. 12:9).

According to traditional Christian theology, the original state of health was a "preternatural gift" dependent upon man's perseverance. However, the recovery of grace by fallen man did not restore this gift here and now, but as an eschatological transformation by God. Man, even when redeemed, remains vulnerable and beset by the "sequels of sin," disease and death. However, as in other religions, his faith, his reliance on the intervention of the saving God, directly or through healing agents, did not leave him helpless. The practice of praying and of blessings, preventive as well as curative, was always sanctioned by the Church.

Whereas the Catholic and Eastern Orthodox traditions integrated much of the healing techniques found in the non-Christian tradition, the reaction against ritualism in the Reformed churches resulted in distrust of any but a purely spiritual or a soundly scientific healing. In fundamentalist churches, especially in Pentecostalism, the gift of healing remains in vogue. The Church of Christ, Scientist, founded in 1879 by Mary Baker Eddy, interprets sickness as the failure of the human mind to understand and to obey God perceptively. She maintained that it is imperative to return to the "cures produced in primitive healing by holy, uplifting faith."

The "Charismatic Renewal Movement" in the Catholic Church reintroduces faith healing, insisting on a strict observance of the evangelical pattern—a spiritual approach centered on faith, prayer, and the forgiveness of sins, the laying on of hands, and the confession of Jesus as the one who actually heals. The mainline Catholic theology, however, would interpret the charism of healing in the tradition of the Church's care for the sick, with words and gestures of material and spiritual comfort, charity, and compassion. Catholicism and Orthodoxy have traditionally appealed to the Mystery of the Incarnation for a strongly sacramental ritual in the healing ministry, especially through the Anointing of the Sick, the confession of sins, and communion. In the anointing the symbolism of oil, which "nourishes, softens, lubrifies, illuminates, impregnates, and penetrates," is that of medication. Confession and penance counteract the cause of disease. The Eucharist is, in the words of Ignatius of Antioch, *"pharmakon athanasias"* (medicine of immortality).

Other sacred signs, called sacramentals, figure prominently in Catholic healing ritual: e.g., benedictions against rabies, poisonous snakes, sore throats, etc., and the Catholic equivalent of amulets—scapulars, medals, icons, and statues, especially of miraculous origin, all imparting protective or healing powers. The *encolpia* of the dignitaries of the Orthodox Church used to be called *phylacteria* and, as Nicephorus says in his Antirrethicus III, served "for the protection of life, for health of soul and body, for healing in sickness, and for the averting of attacks by unclean spirits."

Most prominent as amulets were the relics. A cult of relics and of pilgrimages to the sanctuaries containing them came about, based on the text of Acts 19:11–12: "God did miracles through Paul's hands . . . so much so, that, when handkerchiefs and aprons which had touched his body were taken to the sick, they got rid of their diseases." Other pilgrimages were made to places believed to be hallowed by apparitions of the Blessed Virgin. There the sick lie alongside the processional route of the Blessed Sacrament or reliquaries of the saints and are immersed in the water of a miraculous spring. The Church, while trying to keep these practices free of superstition and of "pious (or, on occasion, impious) fraud," has staunchly defended the principle itself of sacramentality on which they are based.

Conclusion

Quite generally we find in the religions of mankind the belief that disease and healing are connected with mysterious forces that influence man and his world. Religion brings these forces into play for healing. The conviction that the healer is able to neutralize the evil spirits causing disease is certainly in no small measure responsible for the success of magicoreligious therapeutics. The confidence in supernatural healing agents, the security resulting from a coherent vision of creation and from faith in a benevolent and provident God, the peace of mind resulting from forgiveness, the certainty of ultimate salvation, the social dimension of common worship—all seem to be helpful for patients to find relief and reassurance. The psychotherapeutic or psychohygienic effects of the religious experience seem to create conditions favorable for healing: freedom from remorse, acceptance by others, willingness to put up with a certain measure of discomfort as religiously meritorious, readiness to accept one's own limitations, the decision to change one's ways and to make the most of concrete possibilities are attitudes, inspired and fostered by religion, that are truly healthy and healing.

FRANK DE GRAEVE

[*For further discussion of topics mentioned in this article, see the entries:* DEATH; *and* MIRACLE AND FAITH HEALING. *Also directly related are the entries* BUDDHISM; EASTERN ORTHODOX CHRISTIANITY; HINDUISM; ISLAM; JUDAISM; PROTESTANTISM, *article on* DOMINANT HEALTH CONCERNS IN PROTESTANTISM; ROMAN CATHOLICISM; *and* TAOISM. *See also:* MENTAL HEALTH, *article on* RELIGION AND MENTAL HEALTH; PAIN AND SUFFERING, *article on* RELIGIOUS PERSPECTIVES. *The articles under the entry* MEDICAL ETHICS, HISTORY OF *provide further information for various historical periods and geographical regions.*]

BIBLIOGRAPHY

DAWSON, GEORGE GORDON. *Healing: Pagan and Christian.* London: Society for Promoting Christian Knowledge; New York: Macmillan Co., 1935.

DE BEIR, LEO. *Religion et magie des Bayaka.* St. Augustin bei Bonn, Germany: Anthropos-Institut, 1975.

HASTINGS, JAMES, ed. *Encyclopaedia of Religion and Ethics.* 12 vols. Edinburgh: T. & T. Clark, 1908–1922. See especially the series of articles "Disease and Medicine," vol. 4, pp. 723–772.

HILTNER, SEWARD. *Religion and Health.* New York: Macmillan Co., 1943.

KÖTTING, BERNHARD. *Der frühchristliche Reliquienkult und die Bestattung im Kirchengebäude.* Arbeitsgemeinschaft für Forschung des Landes Nordrhein-Westfalen, Geisteswissenschaften, no. 123. Cologne: Westdeutscher Verlag, 1965.

———. *Peregrinatio religiosa: Wallfahrten in der Antike und das Pilgerwesen in der alten Kirche.* Forschungen zur Volkskunde, vols. 33–35. Münster: Regensburg, 1950.

LEWIS, I. M. *Ecstatic Religion: An Anthropological Study of Spirit Possession and Shamanism.* Pelican Anthropology Library. Hammondsworth, England: Penguin Books, 1971.

Le Monde du sorcier: Égypte, Babylone, Hittites, Israël, Islam, Asie Centrale, Inde, Népal, Cambodge, Viet-Nam, Japon. Sources orientales, no. 7. Paris: Éditions du Seuil, 1966.

Les Pèlerinages: Égypte ancienne, Israël, Islam, Perse, Inde, Tibet, Indonésie, Madagascar, Chine, Japon. Sources orientales, no. 3. Paris: Éditions du Seuil, 1960.

TILLICH, PAUL. "The Relation of Religion and Health: Historical Considerations and Theoretical Questions." *Review of Religion* 10 (1946): 348–384.

TRACHTENBERG, JOSHUA. *Jewish Magic and Superstition: A Study in Folk Religion.* New York: Behrman's Jewish Book House, 1939. Reprint. Cleveland: World Publishing Co., 1961.

III

A SOCIOLOGICAL AND ACTION PERSPECTIVE

Introduction

Health, and its negative, illness, have probably been major human subjects of preoccupation and concern as long as anything like human society has existed. Some kind of illness has been very prevalent in many different nonliterate societies, such as those of Central Africa (De Craemer and Fox) or the Navaho of the American Southwest (Kluckhohn and Leighton). The main tradition of the conception of and concern with health that has dominated Western civilization, however, has its roots more in Greek culture than in any other single source. The earliest connected literary remains of this tradition are usually called the body of the Hippocratic writings, which date from the fifth century B.C.

The Hippocratic writings contained an exceedingly important formula that has survived ever since, mainly in its Latin form: *vis medicatrix naturae* (the healing power of nature).

This formula seems appropriate to provide a major point of reference for the present analysis. It will be somewhat interpreted in terms of possible relevance to the social and action contexts that are stressed in the title of the present article. *Vis medicatrix* may be thought of as a property of living systems, in the first instance individual organisms, by virtue of which such systems have the capacity to cope, often without outside intervention, with disturbances to health or cases of illness, that is, cases not so severe as to exceed certain limits. It was one of the primary tenets of Hippocratic medicine that the physician should take account of the inbuilt capacity of his patients to recover from illness and work along with such capacities, rather than to intervene arbitrarily without reference to such capacities (Henderson, 1970).

I share the view, which has been prevalent almost throughout the history of Western medical thinking, that the *primary* focus of the problems of health and illness lies in the state of the organism as a living biological system. The task of the present article, however, is to put this biological aspect of the problem in a somewhat larger setting. Of all living species, man is predominantly characterized by being, to paraphrase Aristotle, in the first instance a "social animal," and in the second instance a bearer of symbolically organized cultural traditions. In the usual senses of the history of science, then, man—though very obviously an organism in the biological sense—is more than an organism. He is a behaving system, a personality, a member of structured social systems, and a participant in cultural systems and patterns of meaning of what is sometimes called "the human condition." Though there are obviously many continuities between the problems of health and illness at nonhuman levels and the human problems, our primary concern is with certain features of the health–illness complex that are most specifically human.

Health, as a central feature of the state, in the first instance, of the living organism, is quite clearly one of the most important features of the human condition for problems of bioethics. I shall argue in the next section that problems of health should not be regarded as only organic, even though they are "rooted" in the state of the organism, but this does not negate the importance of the organic reference. Furthermore, I wish to stress that I regard health as a state of the human individual, not of collectivities such as populations or species or, in more concretely human terms, groups or societies.

In empirical human experience, after all, the life of the individual is bounded by birth—whatever the interpretation of states between conception and birth—and by death—however organic death be defined. In any case, problems of health are intimately intertwined with those of both birth and death, perhaps especially the latter. It is clearly no accident that birth and death have constituted primary foci of meaning for every known human religion. Such problems seem to be at the very center of the ethical problems of the life condition.

If this be true of the beginning and the end of the life cycle of the human individual, it is clearly also true of what happens to such individuals in between. If we must recognize the existence and importance of mental as well as organic health, as I think we must, the focal religious category of suffering can surely be organic—though presumably not in the strict sense physical; a stone presumably does not feel pain when it is hit with a sledgehammer—but there seems to be no reason to limit the concept of suffering to the organic level. Thus mental or spiritual pain or anguish seems very real and not a mere epiphenomenon of organic events. Insofar as there is mental health or illness it is clearly intertwined with these.

Finally, another classic field of religioethical concern, the "problem of evil," seems to stand on a different level, relatively dissociated from the state of the organism. It articulates with the problem of justice, precisely in the sense of what happens to individuals. Since injustices, objectively suffered or subjectively felt, affect the whole person, they can scarcely fail to interact with the problems of health.

Definition of health

In seeking the focus of a definition of health, I should like to rely on the work of an eminent contemporary biologist, Ernst Mayr, who has introduced, at least for the American biological world, the concept of *teleonymy* (Mayr, 1974). Teleonymy, a term Mayr uses to avoid the metaphysical connotations of the old teleology, may be defined as the capacity of an organism, or its propensity, to undertake successful goal-oriented courses of functioning, especially behavior. In the common classification of aspects of organic systems, behavior has a special place, which should be distinguished from structure in its anatomical references and from nonbehavioral processes such as those involved in underlying metabolism. In one sense, behavior involves the mobilization of the resources and capacities of the organism as a whole in the regulation of its relation to its environment. A prototypical category of behavioral orientation concerns food seeking and securing activities on the part of organisms, which activities become more elaborate and organized as a function of advance on the evolutionary scale. Similar contexts are protection of the organism's integrity from such threats as those of predators or infectious disease.

Mayr's concept of teleonymy is meant explicitly to link the behavioral category of goal-orientedness with the levels of anatomical structure and physiological process. It is, in a certain fundamental sense, a *functional* concept in that it has to do with understanding the conditions under which certain properties of the requisite types of organisms are possible. For purposes of the present analysis, however, I should like to extend the conditions, the frame of reference, beyond the internal anatomical–physiological conditions of the organism itself and its physical and organic environments, to include also the action level, in the technical sense of the theory of action, within the framework of the human condition. This notably involves considerations of the personality of the individual as differentiated from both organism and its behavioral system, and also includes the levels of social interaction and cultural symbolization and their background in problems of meaning in Max Weber's sense of that concept.

Against this background, then, I should like to venture to put forward a tentative definition of health. It may, in this broadened sense, be conceived as the teleonomic capacity of an individual living system. The teleonomic capacity that we wish to call health is a capacity to maintain a favorable, self-regulated state, which is a prerequisite of the effective performance of an indefinitely wide range of functions both within the system and in relation to its environments. The very generalization of the concept is one of its most fundamental properties. One centrally important reference of the concept is the capacity to cope with disturbances of such a state which come either from the internal operations of the living system itself or from interaction with one or more of its environments. Again, let it be said that, although I accept the focusing of the concept of health at the organic level of the individual, I wish to extend its relevance, on the one hand, into the physical environment and, on the other, into the action environment and the telic system, i.e., its psychological and sociocultural environments.

From this point of view, the illness of an individual should be seen as an impairment of its teleonomic capacity. It may be grounded in any one factor or in a complex combination of factors, the effect of which is to interfere with satisfactory and effective functioning. Thus, a state of illness may result from some failure to cope with the exigencies of the external environment, as in the case of invasion by agents of bacterial infection. It may equally arise from malorganization of the relations of organic and action level subsystems of the more general human condition, or from internal pathological processes such as a malignant tumor.

Though a special point has been made of centering or "anchoring" the phenomena we denote by the concepts of health and illness in the individual organism, for humans at least I wish quite explicitly not to confine them to the organic level but to speak also of mental health and illness. Making allowance for complex states of interdependence, as in the so-called psychosomatic field, I think of mental health and illness as anchored in the personality of the individual, as part of the action system which must be analytically distinguished from the organism. I think, however, of organism and personality as involved in complex relations of interdependence and interchange that cannot be analyzed here.

Another problem, besides that of the status of the concept of mental health, concerns the respects in which health is to be differentiated from other states of individual well-being, and illness from other varieties of human "ills," as this problem for instance has been posed by the well-known definition of health adopted by the World Health Organization.

In my opinion the most important criterion for drawing this line concerns the individual's capacity to *control* the state referred to and some of the conditions on which it depends, especially conditions internal to the individual of reference. Thus, though many persons are "ignorant" in various respects, given opportunity and adequate intelligence, they can, if "motivated," diminish their ignorance by learning. If faced by temptation to commit a crime they can mobilize their moral standards and abstain, but if afflicted with a bacterial infection they cannot simply "decide" not to be sick. This criterion has the advantage of applying equally to organic and to mental illness.

The essential point concerns the capacities for "voluntary action" on the part of the individual. It is, by consensual definition, undesirable to be ill, both for the individual himself and for others associated with him. But if "really" ill and not malingering, he cannot be expected simply to "pull himself together" and proceed to get well. He must rather seek to gain control of certain conditions normally beyond the range of voluntary action through the help of an outside therapeutic agency, through self-care, or simply through "letting" the *vis medicatrix* operate. Put in terms of Freud's psychology, therapy is in general beyond the range of direct "ego-function" capacities but requires some kind of management of id-functions.

It is above all failure to discriminate these two categories of well-being of the human individual which, in this direction at least, vitiates the famous World Health Organization definition of health. It is simply not acceptable to identify health with well-being in general. Of course it does not follow that the health–illness context is the only one in which the mechanisms of voluntary control prove inadequate. The other principal one historically has always been closely associated with that of health but should be clearly differentiated from it, namely, to put it in Christian terms, the religious "state of grace." In that area it has repeatedly been asserted that the human individual cannot attain grace by his own efforts alone but only, again in Christian terms, by divine intervention. This, however, is clearly very different from the boundary between illness and failure or incompetence in the areas normally subject to voluntary control.

Within this latter area, however, it is also important to make another distinction related to the problem of individual control. Of course in the spheres of ego function, of voluntary action, the degree of control open to any particular individual is always limited by conditions beyond his control, such as those altogether outside human control like proverbially the weather, and those in the control of other human agencies, as in the much-discussed sphere of power relations. This sphere of uncontrolled factors very generally includes exposure to risk of the genesis of illness. To identify "powerlessness" in this sense with illness is, however, highly confusing. They are quite different aspects of the human predicament.

I should finally like to put forward the concept

of health under a rubric that has proved to be of substantial importance in the theory of action. This way of looking at it, however, will certainly prove to be unfamiliar to most people who have been considering the problem of health. The rubric I have in mind is that health should be considered to be a "generalized symbolic medium of interchange" (Parsons, 1969). The prototype of such a medium is money. In a somewhat different context, language may be treated as such a medium. Finally, a recent attempt has been made to deal similarly with the concept of intelligence (Parsons and Platt), treated as a circulating symbolic medium rather than as a trait of the individual.

It seems to me to be reasonable at least to experiment with the conception that health can feasibly be fitted into this theoretical context. The attempt to do so, however, raises exceedingly complicated theoretical issues that can only be suggested within the limits of the present brief article. It is hoped that future work will develop these matters substantially further. A very brief outline will be presented below.

Illness and life-expectancy

In whatever senses health and illness may be considered predominantly problems focusing at the organic level, there can be little doubt that problems of coping with illness or its threat, as well as of maintaining the levels of health of a population, are paramount problems at the level we have called that of action. It is at this level that the primary concern formulated in the title of this article comes to focus.

The essential background of the problem lies in the fact that there is a certain underlying expectation that a normally participating member of a human community will manifest certain variant, but probably determinable, levels of health defined as teleonomic capacity. There is a long history of the bearing of these problems on the human condition. Perhaps the most dramatic aspect of it has lain in the history of vital statistics. The most significant single figure in this complex has been the expectancy of human life at birth. There are, of course, many different aspects of this story, and, in historical perspective, there is serious incompleteness of evidence. The general impression, however, seems to be that it was not until well into the nineteenth century that really great generalized increases in the expectancy of life began to occur. The figures are uncertain, but we might possibly fasten on thirty years as the norm for a "good" society of earlier periods. In the United States in 1975 the figure had reached seventy-two years, and the American record in this respect is by no means the best among the advanced countries. This enormous advance in the expectancy of life has of course occurred in spite of the twentieth having been a century of great wars and many other vicissitudes.

A substantial part of this effect has been attributable to the control of life-threatening diseases by measures of medical care and public health. Another major component has been contributed by improvements in the standard of living such as better nutrition, housing, and the like. This is not the place to attempt to assess the relative importance of these contributions.

To be sure, many of the individuals of reference have been, as it were, kept alive by "artificial" means. These have covered a wide range, including protection against specific diseases by immunization measures such as vaccination for smallpox and many others, all the way to control of diabetes through the use of insulin or other therapeutic measures to improve regulation of the sugar levels of the human blood. It would, however, be difficult to interpret the term "artificial," which was stated in quotation marks in the preceding sentence, in a simple pejorative sense. We underwent severe controversies over the problem of the justification of "interference with natural selection" in an earlier era, and there would today be very few defenders of the rigoristic point of view that the "unfit" should be allowed to die without medical intervention or public health measures. There is, it seems, no generalized basis for believing that susceptibility to particular diseases under so-called natural conditions is an index of general teleonomic incapacity that would merit such drastic selective policies.

By and large, the evidence seems to indicate, however, that the maintenance of the recently acquired high levels of expectancy of life, and the concomitant more general improvements of the state of health of modern human populations, would not have been possible without certain kinds of institutionalized intervention in the situations that define the conditions of health and counteract the incidence of a wide variety of illnesses, including the many historic cases where the consequences of illness have included substantially higher death rates than would otherwise obtain.

Social role of sickness

Perhaps at this point the question of illness as a social role category and the social dynamics of its relation to the problems of health can appropriately be introduced (Parsons, 1951, 1958, 1975). Surely, in previous stages of both organic and sociocultural evolution, there have evolved numerous improvements in the capacity of organisms and individuals to cope with disturbances in their health, that is, their teleonomic capacity as I have defined it. Equally clearly, however, the modern age introduced a new set of conditions in this respect which may very broadly be characterized as the application of scientific methods and procedures to this problem. Medicine and public health, as professional complexes in modern society, do not have meaning unless they do constitute innovations relative to the primordial process of natural selection, though they of course have had many forerunners.

It is in this sense that such a concept as that of the "sick role" and its counterpart, the therapeutic role, become meaningful in modern society. This clearly includes not only the social status and treatment of individuals who actually are undergoing impairment of the normally expected levels of health but also various possibilities of, and threats to, current states of health, such as those involved in the spread of an epidemic disease (Parsons, 1975). The essential point is that the actuality or threat of illness is treated as an explicit problem for more or less rational action on the part of selected elements of the population concerned, including those whose health is impaired or who stand under threats of such impairment. This population includes, on the one hand, the laity (in a medical frame of reference) and, on the other hand, the most competent specialists with expertise in coping with complicated and specific problems of illness or its possibility.

Sociologically, the problem is the nature of institutionalization of the relations between actually or potentially sick people, on the one hand, and participants in health-care agencies, on the other. Very generally, in the modern world the latter have come to center on members of the medical profession so far as the treatment of sick individuals is concerned. The practice of medicine, in this sense, should be distinguished from the administration of public health measures, such as the attempt to eliminate infectious diseases by mass programs of immunization or of control of sources of infection, which, however, do not operate primarily on individuals.

With due allowance for the complexity of the social organizations involved in therapeutic functions in the first sense above, the relation traditionally called that between physician and patient may be taken as central and prototypical, for certain broad purposes. The two role types are institutionally defined as complementary to each other in such ways that the meaning of either assumes involvement with the other.

In discussing the problem of institutionalization of both the sick role and that of therapeutic agencies, particularly physicians, a certain background is essential to keep in mind. I should like to stress two sets of background considerations. The first concerns the centrality and virtual universality of the positive valuation of health and the correlative, negative valuation of states of illness. It is surely fundamental that it is considered by and large to be a good thing to be healthy and a bad or unfortunate thing to be sick. Of course, assuming that such valuations concern the state of individuals, I would hold that it is only on certain romantic fringes of modern thinking that exceptions to these statements could be made, exceptions that have something to do with systematic cults of pessimism and antiworldly values. Valuations of health, however, cannot be fully institutionalized without a number of other contributions. For instance, there must be some criteria for the selection of means that are feasible for controlling states of illness and threats to health, taking into account also their usefulness for larger populations. Additionally, there must be mobilization of the resources essential for coping with the states that do come to be treated as feasibly controllable. Thus, one of the most important chapters in the institutionalization of modern medicine and public health consisted in the achievement of control over a number of infectious diseases. Vaccinations against smallpox and the contributions of Pasteur and Koch are legendary historical aspects of this story. The mobilization of facilities for effective control generally involves the availability of trained personnel, to say nothing of the grounding of the knowledge of what to do in the requisite branches of science. It also involves the institutionalization of relevant modes of contact between the professional personnel involved and the people in need of some kind of professional help. This is the context in which I would like to consider the state of illness and that of therapeutic agents defining, as social roles, the sick role and the therapeutic role respectively.

One may start by suggesting that the core type of role of members of the therapeutic agency, particularly physicians, is an example of what has come to be known as the professional type of role. The physician may be presumed to be committed to the valuation of health in a sense parallel to that in which Gerald Platt and I have maintained that the academic professional is typically committed to the valuation of cognitive rationality (Parsons and Platt). The two values, of course, overlap substantially in that scientific knowledge has become a key factor in the role of the control of illness and the promotion of health.

Like other professional roles, that of physician is categorizable in pattern variable terms. It has a high incidence of universalistic standards, for example, of the generalizability of propositions about diagnosis and probable therapeutic consequences of medical measures. It is functionally specific in that the relations of physicians to their patients are focused on problems of the patients' health rather than other ranges of personal problems. It is performance-oriented in that the task of the physician is to intervene actively to cope with actual cases of illness and its threats, not to sit passively by and "let nature take its course." It is also predominantly affectively neutral, though with the kinds of qualification that Renée Fox has proposed under her concept of "detached concern" (Fox). Finally, by contrast with certain other occupational roles —for example, with that of the business entrepreneur or executive—the professional role generally, including that of physician, is governed by an orientation toward collective values. The most central manifestation of this is the professional ideology that puts the welfare of the patient ahead of the self-interest of therapeutic agents, physicians in particular.

Complementary to the role of physician, or more generally of the therapeutic agent, the social role of the actually or potentially sick person should be considered. In this connection, social roles should be clearly distinguished from other aspects of the patient's condition, most notably those that figure prominently in processes of medical diagnosis and plans for therapeutic intervention. The social role is a framework of expected social relationships normatively defined, conformity with which is not part of the technical discipline of medicine or other health services. We may organize the description of the sick role's four main properties as follows.

First, the exemption from normal social responsibilities on account of illness, where it occurs, is distinguished from their willful and deliberate avoidance. Where simulating a state of illness can be ascribed to deliberate motivation, medical people speak of "malingering" (Field). Apart from the phenomena of malingering, however, to be sick is to be in a state that is institutionally defined as not the sufferer's personal fault. The question of how far personal responsibility may have been involved in coming into such a situation, as through unnecessary exposure to sources of infection, should be distinguished from the sense in which the person may be held personally responsible for "pulling himself together" by an act of will and simply behaving as though he were not sick. If the sick person is genuinely sick, it is presumed that this is not an efficacious mechanism of social control.

Second, a closely related feature of the role of illness, depending on the character and acuteness of the state of illness, is that the impairment of teleonomic capacity and the risk of future impairments are held to justify exemption from some performance expectations which are normally applied to altogether healthy persons. The simplest type of case would have to do with abstention from the performance of both occupational and family obligations and the sanctioning of this abstention by medical authority, such as the medical injunction to stay home in bed for a certain period. Of course, many patients afflicted with "chronic" illnesses may have to accept certain handicaps with respect to the fulfillment of performance expectations or undergo some kind of a regimen of management of their condition with perhaps minimal actual abstention from performance.

Third, there is a sharing of the positive valuation of health and the negative valuation of illness between therapeutic agent and sick person. For the sick person, of course, such valuation by other persons closely bound up with his daily life, such as members of his family, can be very important. This sharing involves a commitment to the attempt to recover a state of health or, in the case of chronic illnesses or threats of illness, to accept regimens of management that will minimize the current impairment of teleonomic capacity and future risks that the actual or presumptive illness may entail. In view of certain recent discussions (Parsons, 1975), it is important to emphasize that this commitment is not confined to recovery from states of acute illness but definitely extends to medically acceptable management of chronic states and of threats to health.

Fourth, the sick role is characterized by a commitment to cooperate with relevant therapeutic agencies—most notably, for our purposes, with physicians. It should again be made clear that this cooperation is not definable as a one-way relationship, namely, the patient's simply putting himself under the control of the therapeutic agent. It is a far more complex, reciprocal interaction between these role carriers.

Like every other type of professional role and social function of organizations in which professional roles are prominent, those of the physician or other therapeutic agency have important functions that may be characterized as functions of "social control" in a well-established sociological sense of that expression. It is this aspect both of the role of illness and of the physician–patient relation that in earlier publications I had predominantly in mind in connecting illness with the idea of deviant behavior. This categorization applies with particular salience to cases where the state of illness approximates being "purely mental"; but the range is far broader than this, extending throughout the field of so-called psychosomatic illness. Deviant behavior and illness are not coextensive; but the fact that there is considerable overlap seems scarcely open to doubt.

To complete this particular part of the analysis, I would suggest that a certain relatively definite patterning is characteristic of agencies of social control in their orientation to the objects of such control. In this connection I should treat, as a limiting case, social control alone or predominantly through rational, bureaucratic-type authority. The kinds of social control I have in mind involve resort to much subtler mechanisms. As, however, was originally worked out in connection with a paradigm of the nature of psychotherapeutic processes, I think this is a very widespread organizational phenomenon and that it has an important bearing on health and illness. That it should not be construed to be confined to this context, however, should be made evident by the way in which comparable conceptualization has been employed for the analysis of the socialization process, most recently on my own part in work with the socialization component of higher education, in collaboration with Platt (Parsons and Platt).

The roles that have prominent functions of social control can be characterized by four primary features. The first is that the agent of control tends to treat his objects or subjects, whichever way the relationship is formulated, with a certain permissiveness. I would link this with the conception, in the institutionalized version of the sick role, that being ill cannot ordinarily be conceived to be the fault of the sick person and that illness can justify certain exemptions from "normal" expectations of performance. Permissiveness is an important consequence of the affective neutrality of the therapeutic role in that the typical physician is not supposed to react to many manifestations of his patient's sentiments and behavior with the usual value-controlled patterns of reaction.

A second component of the social control paradigm is that the agent of control must, at a certain level, be supportive of the sick person in the latter's endeavor to control. This involves a certain acceptance of the state of the individual "in trouble," whether that be illness with its disabilities and incapacities or ignorance on the part of the uneducated. Perhaps the prototype of the supportive relation is parental concern for the welfare of young children who have not yet attained the capacities for autonomous performance in many areas. In the medical context the supportive relation pertains to an aspect of what Renée Fox has called detached concern (Fox). It is perhaps most conspicuous where patients are involved in painful and dangerous states and where it is of great importance to them to believe that their physicians deeply "care" about their welfare and are doing everything they can to help.

Linked with permissiveness and support are two other characterizations of the role of agent of social control. One is the maintenance of a certain aloofness from the more exigent and imperative preoccupations of the persons to be controlled. In the psychotherapeutic context, this above all takes the form of refusing to reciprocate various kinds of overtures in which patients engage toward their therapist, whether they be overtures defining the therapist in an unrealistic type of cohesive, nonprofessional relationship, for example, as love object, or with inappropriate hostility and aggression.

Finally, the last of the features of the paradigm of social control concerns the manipulation of rewards, the occasions for actually implementing such rewards, and the occasions for withholding them. In psychotherapy by far the commonest mode of such implementation is to reinforce the patient's own attainment of insight into his condition and the motivational background manifested in it. Particularly in psychoanalytic practice, the most powerful instrument of such manipulation of rewards is what has often been called "the psychoanalytic interpretation" (Edelson).

The present section of this article should be understood as an attempt to delineate in three different contexts, namely the professional role, the sick role, and the more general paradigm of social control, the principal structural components of the role complex in which, on the one hand, the handling of sick people and their illnesses, and, on the other hand, the orientations and activities of therapeutic agents, notably physicians, take place.

Illness as impairment of the state of health and the possibilities and threats of its occurrence to someone in the state of health stand at the very center of the problems of bioethics. In common with other organic species, humans are capable of exposure to possibilities of illness, and in fact to the certainty that the misfortune of illness will be inflicted on many in any given generation. This is an essential part of "man's fate." Moreover, it is an essential part of this aspect of the human condition that there is no direct correspondence between the incidence of illness—including its severity, its accompanying suffering, or the risk of premature death it entails—and those fundamental ethical concerns in terms of which an individual might be considered morally responsible for the consequences of his voluntary actions.

The "dialectic" of the relation between health and illness is bracketed within the still deeper set of dilemmas of the human condition concerning life—which comes to the individual through human organic birth, and through that channel alone—and the inevitable, though in timing very uncertain, fate of individual death. It is surely not without significance that, though pregnancy and the imminence of childbirth can scarcely be called in the usual sense illnesses from which a woman suffers, in the modern world they have, especially through the role of physicians, come to be drawn into the "health complex," and similarly, though eventual death is as "normal" as any human phenomenon, a very large proportion of expected or potential death situations are also brought into the health complex.

Clearly birth and death of individuals constitute a principal focus of the problems of bioethics which have become so salient (though not for the first time) in modern societies. Surely, then, the intimate connection between birth and death on the one hand, and the problems of evil and suffering on the other, constitute the very center of the problems of bioethics and, perhaps even more basically, of the inevitable problematics of religious meaning.

As a conclusion of this discussion, it may be suggested that in addition to the problems of suffering and evil there is another basic dilemma in the life course between birth and death. It may indeed be a particularly important consequence of the emergence of what we here call the health complex in modern society, that a third irresolvable focus of human dilemmas has again become salient in human history, namely, what may be called the problem of *capacity*. If the problem of suffering comes to focus in human exposure to the impact of a deprivation that is independent of the individual's agency, and if the problem of evil emerges more generally in the exposure to the impact of consequences which are independent of active intentions, the problem of capacity focuses on the fact that, however much we may *want* to do something, we may be prevented by incapacity from actually doing it. Of course illness is far from being the only source of human incapacity, but it is a focally prominent and symbolic one, especially in a society with an activistic orientation.

Health as interaction medium

At the beginning of this article it was suggested that health might be treated as a symbolic circulating medium regulating human action and other life processes. Since that is an unfamiliar concept, a brief further elucidation seems to be in order. The concept has been worked out with reference to such phenomena as money and political power at the level of human social interaction, and in certain respects language may be viewed as such a medium. There seem also clearly to be important parallels at the organic level. The best known are, at the macrophysiological level of the individual organism, the hormones such as insulin, adrenalin, or cortisone (Cannon), and at the microphysiological level the enzymes in their role in the synthesis of proteins in the cell (Luria).

As is most familiarly illustrated by the case of money, a unit in an interacting system—such as an individual or an organ or tissue—must receive an "income" of the medium in question with which to acquire essential means of its functioning, and in the latter process must "expend" this resource. This is what is meant by the "circulation" of the medium. I shall not take the space to attempt to elucidate these complicated processes further here.

From this standpoint health may be conceived as circulating, within the organism, within the personality, and between personality and organism. Thus good health is an "endowment" of the individual that can be used to mobilize and acquire essential resources for satisfactory functioning as organism and personality. Health, in this meaning, would only function if it is "used" and not "hoarded."

When conceived as such a medium, health stands midway between the action-level media such as money, power, and language, and intra-organic media such as hormones and enzymes. If the general idea of continuity as well as interdependence between such levels is tenable, and it has underlain the whole pattern of analysis of this article, the further development of the implications of this approach to the concept of health should prove to be of far-reaching significance for the future. Its working out, however, will require a great amount of analytical effort.

Conclusion: health and illness in the human condition

This article has endeavored to approach the problem of understanding health and illness by treating them as central aspects of the human condition. Just as man himself is both living organism and human actor, who is personality and social and cultural being at the same time, so health and illness are conceived, as human phenomena, to be both organic and sociocultural. At the organic level, health is conceived to be one highly generalized underlying capacity, to be distinguished from strength, agility, or intelligence. Similarly, at the action level, the meaning of health is to be carefully distinguished from the relevant aspect of intelligence, from knowledge, from ethical integrity, and from other qualities of the individual. To return to the World Health Organization's conception yet again, health definitely cannot be identified with human welfare in general, but must be further specified.

At the same time it cannot be confined to the organic level, to say nothing of the still more narrowly conceived "physical." It must be conceived as bridging both organic and "social" or, more generally still, "action" in the sense of symbolic involvements. It is by virtue of this dual involvement, and not simply its organic relevance, that health is a focal center of the problem area of bioethical concerns. It concerns the underlying conditions of the organic life of human beings, their biological births, and their ultimate deaths, and levels of functioning in between, but at the same time it concerns the problem of the meaning of this life and its vicissitudes. To squeeze out either aspect would be to vitiate the significance of the concept as a whole.

TALCOTT PARSONS

[*Directly related are the other articles in this entry. For further discussion of topics mentioned in this article, see the entries:* DYNAMIC THERAPIES; PUBLIC HEALTH; *and* THERAPEUTIC RELATIONSHIP. *This article will find application in the entries* HEALTH CARE; *and* LIFE-SUPPORT SYSTEMS. *See also:* BIOLOGY, PHILOSOPHY OF; MEDICINE, PHILOSOPHY OF; MENTAL ILLNESS; *and* PAIN AND SUFFERING.]

BIBLIOGRAPHY

CANNON, WALTER BRADFORD. *The Wisdom of the Body.* New York: W. W. Norton & Co., 1963.

"The Concept of Health." *Hasting Center Studies* 1, no. 3 (1973), pp. 2–88. Issue contains five essays on health and illness.

DE CRAEMER, WILLY, and FOX, RENÉE CLAIRE. *The Emerging Physician: A Sociological Approach to the Development of a Congolese Medical Profession.* Hoover Institution Studies, no. 19. Stanford: Hoover Institution on War, Revolution, and Peace, Stanford University, 1968.

EDELSON, MARSHALL. *Language and Interpretation in Psychoanalysis.* New Haven: Yale University Press, 1975.

FIELD, MARK G. *Soviet Socialized Medicine: An Introduction.* New York: Free Press, 1967.

FOX, RENÉE CLAIRE. *Experiment Perilous: Physicians and Patients Facing the Unknown.* Glencoe, Ill.: Free Press, 1959.

HENDERSON, LAWRENCE JOSEPH. *The Fitness of the Environment: An Inquiry into the Biological Significance of the Properties of Matter.* New York: Macmillan Co., 1913.

———. *On the Social System: Selected Writings.* Edited by Bernard Barber. The Heritage of Sociology. Edited by Morris Janowitz. Chicago: University of Chicago Press, 1970.

———. *The Order of Nature: An Essay.* Cambridge: Harvard University Press, 1917.

KLUCKHOHN, CLYDE, and LEIGHTON, DOROTHEA. *The Navaho.* Cambridge: Harvard University Press, 1946.

LOUBSER, JAN J.; BAUM, RAINER C.; EFFRAT, ANDREW; and LIDZ, VICTOR MEYER, eds. *Explorations in General Theory in Social Science: Essays in Honor of Talcott Parsons.* 2 vols. New York: Free Press, Macmillan Publishing Co., 1976.

LURIA, SALVADOR EDWARD. *Life: The Unfinished Experiment.* New York: Scribner's, 1973.

MAYR, ERNST. *Populations, Species and Evolution: An Abridgement of "Animal Species and Evolution."* Cambridge: Harvard University Press, Belknap Press, 1970.

———. "Teleological and Teleonomic: A New Analysis." *Methodological and Historical Essays in the Natural and Social Sciences.* Edited by Robert S. Cohen and Marx W. Wartofsky. Boston Studies in the Philosophy of Science 14, Synthese Library 60. Dordrecht, Holland: D. Reidel Publishing Co., 1974, pp. 97–117.

PARSONS, TALCOTT. "The Definitions of Health and Illness in the Light of American Values and Social Structure." *Patients, Physicians, and Illness: Sourcebook in Behavioral Science and Medicine.* Edited by E. Gartly Jaco. Glencoe, Ill.: Free Press, 1958, pp. 165–187.

———. *Politics and Social Structure*. New York: Free Press, 1969. See Chapter 14, "On the Concept of Political Power"; Chapter 15, "On the Concept of Influence"; and Chapter 16, "On the Concept of Value-Commitments."

———. "The Sick Role and the Role of the Physician Reconsidered." *Milbank Memorial Fund Quarterly* 53 (1975): 257–278.

———. *Social Structure and Personality*. New York: Macmillan Co., Free Press of Glencoe, 1964.

———. *The Social System*. Glencoe, Ill.: Free Press, 1951. See especially Chapter 10.

———, and Fox, Renée Claire. "Illness, Therapy and the Modern Urban American Family." *Journal of Social Issues* 8, no. 4 (1952), pp. 31–44.

———; Fox, Renée Claire; and Lidz, Victor M. "The 'Gift of Life' and Its Reciprocation." *Social Research* 39 (1972): 367–415. Reprint. *Death in American Experience*. Edited by Arien Mack. New York: Schocken Books, 1973, pp. 1–49.

———, and Platt, Gerald. "The Cognitive Complex: Knowledge, Rationality, Learning, Competence, Intelligence" and "The Core Sector of the University: Graduate Training and Research." *The American University*. Cambridge: Harvard University Press, 1973, pp. 33–162.

IV
PHILOSOPHICAL PERSPECTIVES

Health and disease are cardinal concepts in medicine, for they inform medical care. That which is wholesome is to be pursued, that which leads to diseases is to be avoided. The concepts of health and disease seem both to describe reality and to place a value upon it. Decisions regarding the meaning and scope of concepts of health and disease profoundly influence the character of health care (e.g., if alcoholism, homosexuality, menopause, and aging are considered diseases, then medical treatment and research will be focused upon them). One might, then, suspect that concepts of health and disease conceal value judgments that would otherwise be more explicitly treated as bioethical issues. For example, should alcoholics be held responsible for drinking? Are homosexual acts immoral because they are unnatural? What are the needs of the elderly to which we should respond?

This article will examine philosophical questions that bear upon the concepts of health and disease, focusing on the following five areas:

1. What is the meaning of such concepts as disease, sickness, illness, and health? In particular, are some of these concepts more descriptive and others more explanatory?

2. Is the meaning of concepts such as health and disease to be discovered or invented? And to what extent do value judgments, whether based on cultural decisions or on the recognition of essential biological functions, structure concepts of health and disease?

3. Is the definition of mental disease different from that of somatic disease?

4. Do concepts of animal diseases function the same way as our concepts of human diseases?

5. To what extent are bioethical issues concealed in concepts of health and disease?

Many other philosophical issues could be addressed here. At the very beginning, one should note that drawing distinctions among concepts such as sickness, deformity, disability, dysfunction, and disfigurement is a philosophical enterprise. In the course of this article, attention will be restricted primarily to concepts of disease, illness, health, and well-being, and these terms will receive somewhat stipulative boundaries. This is in particular the case with regard to the concept of illness, which will be taken to include sicknesses, dysfunctions, deformities, etc.—it will signify medical complaints.

Illness and well-being, disease and health

Terms such as "illness," "disease," "sickness," "fever," "measles," "rash," "bronchogenic carcinoma," "a viral infection," are meant to identify particular states of affairs. Some such terms function simply to identify a constellation of phenomena, e.g., "He has a fever," "She has a rash," "John is sick," "I feel ill." Others explain what is observed, e.g., "He has a reaction to a flu shot," "She has contracted a case of measles from John," "John has bronchogenic carcinoma," "I have diarrhea due to ECHO-18 virus." The first (e.g., identifying someone as having a rash or a fever) is a primarily descriptive endeavor. Still, insofar as such descriptions involve collecting a number of signs and symptoms in order to identify a pattern of phenomena, a first level of explanation is attempted (i.e., there may be no descriptions that are not also explanatory). A constellation of phenomena is identified and held to be recurrent so that if such a constellation of phenomena is encountered again in the future it can be identified, e.g., "That is a case of Cushing's syndrome," "That is a case of vitiligo," "That is a case of phthisis." Such diagnoses of syndromes, of recurrent constellations or patterns of signs and symptoms, allow predictions to be made (prognoses) and allow control of phenomena (therapy), e.g., "That is only vitiligo, there is nothing to worry about," "You have phthisis—it is best that you retire to a sanitarium." Such predictions and attempts at therapy can succeed even in the absence of any causal explanations of the phenomena.

In fact, a great deal of medicine has historically been concerned simply with classifying patterns of signs and symptoms so that they can with ease be recognized in the future. Thomas Sydenham's (1624–1689) classic *Observationes Medicae* (1676) suggested classifying diseases in definite species following the methods of botanists in erecting their classifications of plants. His work was followed by that of François Boissier de Sauvages de la Croix (1706–1767), *Nosologia Methodica sistens Morborum Classes Juxta Sydenhami mentem et Botanicorum ordinem* (1768), and William Cullen's (1712–1790) *Synopsis Nosologiae Methodicae* (1785). In these works, illnesses received systematized descriptions. Sauvages, for example, placed diseases in ten classes, subdivided those into 44 orders and 315 genera, and finally placed 2,400 species of diseases under those genera. He distinguished such general classes as fevers (e.g., intermittent fevers), debilities (e.g., fainting), pains (e.g., heartburn), and fluxes (e.g., diarrhea). These classifications and their explanatory power ("This is another case of X disease") functioned without a causal account (though such was also given) of the occurrence of particular phenomena. Such phenomenological levels of medical descriptions and explanation are still employed today whenever a new illness is identified for which a causal explanation is not yet forthcoming. For example, no causal explanation was available for the mysterious Legionnaires' disease when it occurred in Philadelphia in 1976.

Medicine is also capable of enriching explanatory enterprise when it begins to put forth accounts of the constellations of observed phenomena by relating what is observed via general laws of physiology and psychology to other phenomena. The result is a two-tier account of illnesses. The first tier is that of the observed constellations of phenomena, e.g., a description of yellow fever. The second tier is that of a model advanced to account for the originally observed phenomena, e.g., an explanation that what is observed in yellow fever is due to an acute viral disease caused by a group B arborvirus transmitted by either Aedes aegypti (i.e., urban yellow fever) or by forest mosquitoes (i.e., sylvan yellow fever), and which produces scattered necrosis and necrobiosis of parenchymal cells in the liver and other changes. The second tier is that of nomological, lawlike regularities that are imposed upon phenomena in order to account for their genesis and character. Once a model is available, one is no longer constrained, for example, to say only that "yellow fever usually progresses in the following way." One can add "because of _____," filling in the blank with a pathophysiological explanation.

The laws of pathophysiology and pathopsychology thus come to relate new phenomena to the originally described constellation of signs and symptoms. The new phenomena are often identified as causes of the illness. Moreover, disease models enable one to search for inferred entities. Consider, for example, the treatise *Syphilis, Sivi morbi Gallici* (1530), by Girolamo Fracastoro (1484–1553), in which it is argued that the French disease or syphilis was contagious. In fact, in a subsequent work, *De Contagionibus et contagionis Morbis et eorum Curatione* (1546), he proposed three different forms of contagion: (1) by contact, (2) by contact and fomites (e.g., scabies, phthisis, and leprosy), and (3) by contact, fomites, and action over a distance (e.g., pestilential fevers and variola). Moreover, he proposed an entity that caused the infection—a seed of contagion.

Later talk about such inferred entities as "seeds of contagion" was strengthened by demonstrating intervening variables such as pathological anatomical findings. Giovanni Battista Morgagni (1682–1771) in his *De Sedibus et Causis Morborum per Anatomen Indagatis* (1761) correlated clinical observations with postmortem findings, and Philippe Pinel (1755–1826) incorporated anatomical considerations into his *Nosographie philosophique* (1798), producing nosological classifications that incorporated not only clinical observations but anatomical considerations as well. A turning point was reached when Xavier Bichat (1771–1802) argued that constellations of symptoms and signs could be explained in terms of underlying pathological processes; the key to medical advance was, according to Bichat, to be found through autopsies. This shift to the study of pathological findings as a way of accounting for clinical observations began to give content to the processes thought to be underlying illnesses. When such models were supplemented by theories of microbiology, biochemistry, and genetics, they resulted in well-developed causal explanations of illnesses.

In the process of moving from mostly descriptive to explanatory accounts of illnesses, the meanings of illnesses were radically altered. For example, individuals who once were thought to die of acute indigestion are today found to die of a myocardial infarction (heart attack). Yellow fever, which was once feared as a disease transmitted by direct personal contact, is now understood to be transmitted by an insect vector (carrier). Also, illnesses that were once lumped together can be distinguished, e.g., typhoid and typhus fevers. Further, illnesses once thought discrete can be seen to be manifestations of one disease, e.g., consumption and scrofula (the King's evil), can both be explained in terms of the pathological effects of Mycobacterium tuberculosis. Moreover, findings not immediately associated with illnesses can, given a disease model, be used to indicate the presence of a disease prior to an individual's becoming ill—for example, the finding of a lesion indicative of bronchogenic carcinoma (lung cancer) on a routine X-ray examination. Such hidden maladies have been termed lanthanic diseases (Feinstein, p. 145). Thus, on the basis of a disease model one can identify a set of observables, a disease state, which is much broader than the immediately confronted constellations of signs and symptoms that constitute an illness.

In short, the meaning of the phenomena observed (i.e., the illnesses with their clinical signs and symptoms) is reinterpreted in terms of disease models. One might stipulatively distinguish illnesses from diseases where "illness" identifies constellations of signs and symptoms, and "disease" identifies illnesses joined to disease models or explanations where the content of the illness is augmented by the phenomena found on the basis of a disease model. It should be noted that disease explanations are only one of the possible kinds of explanations for illnesses (others would include immediate divine punishment or demonic possession). Finally, being healthy can take on a technical status once a feeling of well-being is certified by proper medical examination. The availability of models of disease (and thus of health) provides a theoretical vantage point from which phenomena can be explained, predicted, and controlled.

Disease and illnesses—discoveries or inventions?

A great deal of contemporary philosophical discussion concerning concepts of health and disease is focused upon whether their features are discovered or invented. The question turns upon whether certain physiological and psychological functions can be identified as natural to humans so that their absence would define disease states. If the functions that are proper to humans can be defined without reference to particular social or individual expectations, then the definitions of health and disease are not relative. In that case, absolute answers should be available with regard to whether menopause, alcoholism, drug addiction, homosexuality, or aging should count as diseases. The answers would then be discovered, not invented.

Leon Kass and Christopher Boorse have argued that one can, indeed, specify those functions which are integral to being human, and thus secure nonarbitrary conceptions of disease. These arguments can also be used to sort out essential from nonessential, if not proper from improper, applications of medicine to human life. Kass and others contend that the recognition of such nonarbitrary concepts of health and disease can help prevent medicine from being recruited into solving social, not truly medical, problems.

In contrast, Joseph Margolis, Engelhardt, and others have argued that definitions of disease and health depend upon value judgments and, as a result, can be understood only in terms of particular cultures and their ideologies (Margolis). Partisans of this viewpoint stress that an evolutionary appreciation of human functions makes functions dependent upon particular environments. Thus, since there are no standard environments, there can be no standard definition of human functions. Moreover, intraspecies variations increase the potential adaptability of species. In addition, traits may, in very obscure ways, contribute to the survival of the species. For example, it has been contended that homosexuality may be genetic and may have evolved because homosexual members of primitive societies functioned as helpers and thus increased the overall survival of the species. In fact, the value of species survival can itself be brought into question. For, even though the sickle-cell trait contributes to the general adaptability of the human race (i.e., allowing better survival and reproductive efficiency in the presence of falciparum malaria and in the absence of antimalarial drugs), one may not accept the sickle-cell trait as a state of health (i.e., one may wish to have that range of individual adaptability that would allow one to be a test pilot, for example, but might be precluded by sickle-cell trait) or accept sickle-cell disease as the price one pays for species adaptability. Unlike evolution, which is a group- or species-centered concept, definitions of human health and disease tend to be individually oriented.

Such definitions may not, therefore, include certain individual problems as states of health, even if those problems contribute to the survival advantage of the species. In addition, even very widespread phenomena can in the case of humans be treated as diseases, e.g., menopause and presbyopia.

If concepts or notions of human illness, disease, and health are in part social constructions, there will be marked differences between the ways in which diseases are identified for humans and the ways they are identified for other animals. That is, social or cultural criteria are not available for the animals involved as they are for humans, but are imposed by another species. As a result, in the case of other animals, whether or not a state of affairs contributes to the survival advantage of that species, as well as whether an absence of a particular capacity is very widespread in the species or not (e.g., the inability of bats to see versus presbyopia) will be decisive with regard to the definition of animal diseases in a way that such considerations do not obtain for human diseases. Unless one has a special interest in the animals concerned (i.e., they are pets or domesticated animals), one is likely, for example, to view infections in such species as elements of natural ecological balances in which overpopulation in particular niches is avoided by increased death rates of the members of the species. Some transference, though, from judgments concerning human diseases to judgments concerning animal diseases is likely. As one comes to recognize certain states of affairs as being the pathophysiological substrata of human illnesses, one will have had the advantage of recognizing these states of affairs as pathological on the basis of their underlying states of human suffering, thus enabling them to be identified as states of disease in animals where criteria for identifying cases of illnesses are not as clearly available.

In any event, conditions of humans can be identified as illnesses on the basis of:

1. Their dysteleology—their preventing a function either (a) discovered to be essential to humans (i.e., according to the position of Boorse and Kass)—or (b) defined by a culture to be necessary to humans (according to the position of Margolis and Engelhardt).

2. Their causing pain, which is a concomitant of a process either (a) discovered not to be essential to humans or (b) defined by a culture not to be necessary to humans (compare the pain of teething, which is not held to be a state of illness since it is integral to a normal human function, with the pain of a migraine headache, which is not clearly an element of a process making such a contribution) (King).

3. Their being constituted by a deformity identified on the basis of aesthetic considerations— consider vitiligo.

Insofar as the grounds for such teleological and aesthetic judgments can be identified in the nature of man, illnesses are discovered. Insofar as such grounds are not available, illnesses are invented within particular cultures. Moreover, if it is easier to discover that a function is proper to humans by noting the result of its absence, or if it is easier to define a function as integral to health by seeing some of the consequences of its absence, it will be easier to define diseases than the concept of health. In fact, health may take on a somewhat global sense of the ability to achieve human purposes and goals. Health in that global sense will then identify the physical and psychological substrata of autonomous action. Any more concrete specification of health is likely, though, to be made in terms of healths —absences of particular diseases (Burns, p. 45).

The above issues concern the role of values in the identification of illnesses. The question of the role of values in the development of disease models (the second tier) must also be faced, as well as questions of the extent to which such models are discovered or invented. Philosophical discussions within the traditions of medicine concerning the status of disease models have a long history. In general, disease models have been characterized as either ontological or physiological (functional). Ontological theories have held that the term "disease" names things in the world such as the word "cat" names cats. It is within this tradition that talk of disease entities, *entia morbi*, arose. Such ontological accounts can themselves be distinguished into two types: (1) those which identify the disease entity, the *ens morbi*, with a particular thing (an infectious agent) and (2) those which identify diseases with enduring patterns of signs and symptoms—a conceptual realist position. Ontological views of the first sort can be found in the works of Paracelsus (1493–1541), and those of the second sort in the long tradition of classifying types of diseases inspired in modern times by Thomas Sydenham. This second view understood diseases as enduring types, almost as Platonic ideas. Sydenham argued, "Nature in the production of disease, is uniform and consistent; so much so, that for the same disease in different persons the symptoms are for the most part the same; and the self same phenomena that you would observe in the sickness of a Socrates you would observe in the sickness of a simpleton" (Sydenham, p. 15). It is within such a view of diseases that one can talk of "Jones being (having) a typical case of typhoid." Also, within that context one can talk of typical cases as rare, "One rarely sees a typical case of secondary syphilis any more." Patients embody the eternal disease types, where the typical is the full, complete expression of a disease, not necessarily its usual expression. The type is an ideal type.

In contrast, individuals such as F. J. V. Broussais (1772–1838), Carl Wunderlich (1815–1877), and Ernst Romberg (1865–1933) argued in a somewhat nominalistic fashion that only individuals, patients, were real. Definitions of diseases represented arbitrary identifications of certain physiological deviations. For such physiologists or functionalists of disease, individuals were real (and, in a somewhat nonnominalist fashion, so were the basic laws of physiology). But classifications of disease were arbitrary. Within such traditions one can find conceptual roots for critiques against misguided views of disease that seek to find unique causes. Also, such functional views allow consideration of disease entities, but now as observational-theoretical units ranging from theoretically well understood and delineated states of affairs (i.e., diseases proper) to syndromes (i.e., recognized recurring constellations of signs and symptoms). Current uses of "disease" in standard nomenclatures and nosologies have this character. Moreover, such functional concepts of disease would allow one to choose on pragmatic grounds whether one wishes, for example, to treat tuberculosis as an infectious, genetic, or environmental disease (knowing that all three sorts of factors contribute to the development of tuberculosis) on the basis of which variables are most easily manipulated. One may decide that, because little is known about the exact inheritance of resistance against tuberculosis, and because any eugenic programs to eliminate that disease would be very slow in taking effect, it would be best to treat tuberculosis as an infectious disease (as well as an environmental disease, one depending upon socioeconomic conditions such as housing, food, etc.).

One is confronted then with a view of illness and disease that has value judgments playing roles at all levels. Judgments concerning human teleology and proper human form act to identify certain constellations of phenomena as being pathological, as being illnesses. Insofar as ontological accounts of the status of diseases fail, value judgments play a role in structuring disease explanations. Thus, depending upon realist or nominalist convictions concerning the status of disease models, and depending upon whether the identification of functions as proper to humans is dependent upon cultural biases or not, concepts of disease and health will be held to be either discovered or invented.

Finally, it should be remarked that, while definitions of diseases may be dependent upon value judgments, certain physiological and psychological functions are likely to be esteemed by persons no matter what their particular culture. Thus, basic cardiovascular, gastrointestinal, and respiratory functions are likely to be valued in all cultures, because they are conditions for the possibility of achieving any values. On the other hand, the inability to discriminate colors, to taste phenylthiocarbimate, or to roll the sides of one's tongue inward may or may not be counted as diseases, depending upon the contribution that these functions make within a particular context or environment (including the cultural ambience). Definitions of illnesses and diseases will (insofar as essential human functions cannot be identified and disease models are arbitrary) be dependent upon social conventions. If this is the case, it will readily follow that definitions of illnesses and diseases have deep social roots as well as broad social consequences.

Models of disease: mental, somatic, and otherwise

Controversy has also arisen concerning the status of nonsomatic models of disease. Individuals such as Thomas Szasz have argued that only somatic diseases are legitimately diseases, while those states of affairs usually termed mental diseases are, in fact, simply problems with living (Szasz). Such arguments are based upon a Cartesian assumption that disease predicates are somatic predicates and can thus only be predicated of bodies, while problems with living, states of affairs similar to moral vices, can be predicated of minds. Following similar lines of argument, individuals have contended that enterprises such as psychotherapy are tantamount to applied ethics (Breggin, p. 60).

Against such views, one might respond that such stark dichotomies or dualisms usually fail to offer satisfactory accounts of reality—it is not at all clear within such explanations what the role of the brain is with respect to mental life. If mental life is dependent upon brain function, then mental diseases can at least in some senses be reduced to physical diseases (e.g., depression can be presumed to be dependent upon a neurophysiological substrate and thus in principle open to pharmacological treatment). If such correlation and consequent partial reduction are not possible, then the problems of relating mental and physical life become acute (e.g., how can one explain that pharmacological treatment does influence states of mental health?). But, if one views diseases as the result of applying explanatory models to the organization of signs and symptoms, it then does not matter whether the signs and symptoms identify physical states of affairs (e.g., "He has a rash"), or psychological states of affairs (e.g., "I feel bad," or "I feel depressed"). Nor does it really matter whether the models produced to correlate these phenomena are pathophysiological or psychopathological (cf., "The type of stress associated with A type individuals leads to an increased instance of coronary artery disease," "Schizophrenia is due to an immunological defect," "Depression is caused by retroflexed rage," "Coronary artery disease is a function of diet and genetic background"). Most accounts of diseases will in fact mingle physical and psychological observables as well as mixing pathophysiological and pathopsychological constructs in the explanation of the patterns of phenomena identified as illnesses.

As a consequence, one may come to view distinctions among somatic, psychological, or social models of disease as based primarily upon pragmatic needs to accent the usefulness of particular modes of therapeutic intervention. One may, for example, advance sociological models of disease, construing diseases primarily in terms of social variables (e.g., general economic conditions, the state of water and sewage services, or problems associated with particular exercise and dietary habits), while prescinding from the pathophysiological or psychopathological states of particular individuals.

Finally, a distinction is made between medical and nonmedical models of therapy in a way different from the distinctions among somatic, psychological, and sociological models of disease. This distinction is often meant to contrast the autonomy of clients in nonmedical therapeutic models with the dependence of patients upon health-care practitioners in medical models. Medical models tend to be more closely bound to what Talcott Parsons has called the "sick role," which (1) excuses ill individuals from some or all of their usual responsibilities, (2) holds them not responsible for being ill, (3) holds that they should want to become well (a therapeutic imperative), and (4) holds that they should seek out experts to treat their illness (Siegler and Osmond, p. 41). Medical models thus tend to support paternalistic intervention by health-care practitioners and to relieve patients of responsibility for their health or for directing their own care and cure. Nonmedical models, in contrast, tend to accent individual, client responsibility.

Somatic models of disease may employ medical or nonmedical models of therapy (compare treating hypertension with antihypertensive agents versus enjoining the afflicted individual to find ways to change his or her life-style with regard to stress and eating patterns). The same is true of psychological models (compare treating a depression chemotherapeutically versus approaching that state of affairs à la Szasz as a problem in living) and of sociological models (compare controlling the spread of typhoid by vaccination versus eliciting changes in habits of washing hands after defecation, etc.).

One should note that as disease explanations identify more variables as predisposing towards particular diseases, individuals will become increasingly responsible for becoming ill, even though they will remain nonresponsible for being ill (e.g., one is not responsible for *being* a person with bronchogenic carcinoma as one is responsible for being a malingerer—one cannot be told to stop it, but one can be held responsible for *having developed* bronchogenic carcinoma because of one's smoking habits). As the impact of life-styles upon the development of diseases becomes clearer, the responsibility of individuals for their health will increase the possible scope of nonmedical models of therapy (i.e., one will be able to enjoin certain life-styles as wholesome).

The ethical force of nosologies

The history of medicine offers numerous instances of the use of concepts of health and disease in order to carry out covert ethical and political judgments (i.e., not simply reflecting general discoveries of the proper goals of humans). For example, in the United States prior to the War Between the States, blacks were seen to have specific diseases peculiar to them, such as the disease of running away to the North (drapetomania), or the disease of an absence of wholesome inclination to do effective plantation work (dysaesthesia aethiopis or hebetude of mind and obtuse sensibility of body), for which explanatory accounts were advanced (Cartwright, pp. 707–711). Masturbation, considered in the nineteenth century a serious moral fault, was also appreciated as a serious disease for which castration, excision of the clitoris, and other invasive therapies were employed. Individuals were even determined to have died of masturbation, and postmortem findings "substantiated" this cause of death (Engelhardt, "Disease of Masturbation," p. 244). The disease of masturbation illustrates in part the effect of ethical norms on the psychology of discovery. Moreover, the political use of the diagnoses of mental illnesses in the Soviet Union, as well as at least some isolated evidence of such use in the United States (Engelhardt, "Fear of Flying," p. 20), indicates the continued use of these concepts as social means of enforcing value judgments. (Accounts such as those by Boorse and Kass hope to avoid such difficulties by an appeal to natural functions and norms.)

But not all such social employment of disease definitions is necessarily malevolent. One may find, for example, individuals advocating a view of alcoholism or drug addiction as diseases in order to recruit the forces of medicine in the control of these phenomena. Moreover, such conditions may be termed diseases in order to relieve alcoholics and drug addicts of the social opprobria that would otherwise attend them. It is often important to remind individuals of the function that disease concepts have in excusing individuals and soliciting sympathetic and helpful attention.

The concepts of health and disease, of illness and well-being, serve various explanatory, evaluative, and social purposes. As a result, how one views concepts of health and disease will have important consequences for health care. If one views health in the expansive fashion that the World Health Organization (WHO) suggests—namely, as not merely the absence of disease or infirmity but as "a state of complete physical, mental, and social well-being" (Constitution of the World Health Organization)—then the role of health care is expanded dramatically. Yet without a more restricted understanding of well-being, medicine would be focused on ameliorating such states of non-well-being as poverty and ignorance. A more restrained definition of health would propose health to be that state of physiological and psychological function which allowed the achievement of those goals and states held to be proper to humans. Because of the generality of such a definition, it may be necessary to make reference to diseases, that is, we may best decide which functions are proper to humans through recognizing the consequences of their absence. Concepts of health may be specifiable in detail only through reference to diseases. Finally, one should note that health is not equivalent to the appropriate goals of therapy in that the goals of treatment may fall far short of the restoration of perfect health. One might compare, for example, the goals of therapy in the case of a patient with terminal cancer and the goals of therapy in the case of a patient who has been cured of cancer but only after removal of major portions of the intestines and the establishment of a colostomy.

H. TRISTRAM ENGELHARDT, JR.

[*This article will find application in the entries* AGING AND THE AGED, *article on* HEALTH CARE AND RESEARCH IN THE AGED; ALCOHOL, USE OF; DRUG USE, *article on* DRUG USE, ABUSE, AND DEPENDENCE; HEALTH AS AN OBLIGATION; HOMOSEXUALITY; *and* MENTAL ILLNESS. *For discussion of related ideas, see the entries:* MIND–BODY PROBLEM; *and* RACISM, *article on* RACISM AND MEDICINE. *Other relevant material may be found under* PATERNALISM; PUBLIC HEALTH; SMOKING; *and* THERAPEUTIC RELATIONSHIP, *article on* CONTEMPORARY MEDICAL PERSPECTIVE.]

BIBLIOGRAPHY

BOORSE, CHRISTOPHER. "On the Distinction between Disease and Illness." *Philosophy and Public Affairs* 5 (1975): 49–68.

——. "What a Theory of Mental Health Should Be." *Journal for the Theory of Social Behavior* 6 (1976): 61–84.

BREGGIN, PETER ROGER. "Psychotherapy as Applied Ethics." *Psychiatry* 34 (1971): 59–74.

BURNS, CHESTER R. "Diseases versus Healths: Some Legacies in the Philosophies of Modern Medical Science." Engelhardt, *Evaluation and Explanation,* pp. 29–47.

CARTWRIGHT, SAMUEL A. "Report on the Diseases and Physical Peculiarities of the Negro Race." *New Orleans Medical and Surgical Journal* 7 (1851): 691–715.

"Constitution of the World Health Organization." *The First Ten Years of the World Health Organization.* Geneva: 1958, p. 459. Preamble.

ENGEL, GEORGE L. "A Unified Concept of Health and Disease." *Perspectives in Biology and Medicine* 3 (1960): 459–485.

ENGELHARDT, H. TRISTRAM, JR. "The Concepts of Health and Disease." Engelhardt, *Evaluation and Explanation*, pp. 125–141.

———. "The Disease of Masturbation: Values and the Concept of Disease." *Bulletin of the History of Medicine* 48 (1974): 234–248.

———. "Fear of Flying: The Psychiatrist's Role in War." *Hastings Center Report* 6, no. 1 (1976), p. 21. One of three short essays under this title, pp. 20–22.

———. "Ideology and Etiology." *Journal of Medicine and Philosophy* 1 (1976): 256–268.

———, and SPICKER, STUART F., eds. *Evaluation and Explanation in the Biomedical Sciences*. Philosophy and Medicine, vol. 1. Dordrecht and Boston: D. Reidel Publishing Co., 1975.

FABREGA, HORACIO, JR. "Concepts of Disease: Logical Features and Social Implications." *Perspectives in Biology and Medicine* 15 (1972): 583–616.

FEINSTEIN, ALVAN R. *Clinical Judgment*. Huntington, N.Y.: Robert E. Krieger Publishing Co.; Baltimore: Williams & Wilkins Co., 1967.

FOUCAULT, MICHEL. *The Birth of the Clinic: An Archeology of Medical Perception*. New York: Pantheon Books, 1973.

KASS, LEON R. "Regarding the End of Medicine and the Pursuit of Health." *Public Interest* no. 40 (1975), pp. 11–42.

KING, LESTER S. "What Is Disease?" *Philosophy of Science* 21 (1954): 193–203.

MACKLIN, RUTH. "The Medical Model in Psychoanalysis and Psychotherapy." *Comprehensive Psychiatry* 14 (1973): 49–69.

———. "Mental Health and Mental Illness: Some Problems of Definition and Concept Formation." *Philosophy of Science* 39 (1972): 341–365.

MARGOLIS, JOSEPH. "The Concept of Disease." *Journal of Medicine and Philosophy* 1 (1976): 238–255.

SIEGLER, MIRIAN, and OSMOND, HUMPHRY. "The 'Sick Role' Revisited." *Hastings Center Studies* 1, no. 3 (1973), pp. 41–58.

SYDENHAM, THOMAS. "Preface to the Third Edition." *The Works of Thomas Sydenham*. 2 vols. Vol. 1: *Medical Observations Concerning the History and Cure of Acute Diseases*. Translated by Dr. Greenhill. London: Printed for the Sydenham Society, 1848–1850, pp. 11–27.

SZASZ, THOMAS. *The Myth of Mental Illness: Foundations of a Theory of Personal Conduct*. Rev. ed. New York: Harper & Row, 1974.

TEMKIN, OWSEI. "The Scientific Approach to Disease: Specific Entity and Individual Sickness." *Scientific Change: Historical Studies in the Intellectual, Social, and Technical Conditions for Scientific Discovery and Technical Invention, from Antiquity to the Present*. Edited by A. C. Crombie. University of Oxford, 1961. London: Heinemann; New York: Basic Books, 1963, pp. 629–647.

TOULMIN, STEPHEN. "Concepts of Function and Mechanism in Medicine and Medical Science (Homage à Claude Bernard)." Engelhardt, *Evaluation and Explanation*, pp. 51–66.

WARTOFSKY, MARX W. "Organs, Organisms and Disease: Human Ontology and Medical Practice." Engelhardt, *Evaluation and Explanation*, pp. 67–83.

HEALTH AS AN OBLIGATION

Does it make sense to speak, as people often do, of health as something one has an obligation to pursue or maintain? Do people who act in ways detrimental to their own health not only act imprudently, but in violation of what in the moral sense they *ought* to do as well? If there is such an obligation in regard to health, to whom is it an obligation and what is its origin?

In order to answer these questions, it will be necessary to consider first some of the general characteristics of obligations or duties. Every obligation can be understood as an obligation to act in some specifiable way. Even obligations that do not seem to be of this sort can be redescribed as being obligations in regard to action. Thus, an obligation to be a person of honest or charitable character is an obligation to *act* in honest or charitable ways, and an obligation never to be insulting is an obligation to *act* always in a noninsulting way. Thus every obligation is fundamentally an obligation in regard to action.

To be under an obligation to perform some action A is to be in circumstances such that one is prima facie morally culpable if one does not do A. That is not to say that it can never be justified to act in violation of one's obligations; rather, it is to say that, if one does so, one bears the burden of proof to show why doing so is justified, the obligation notwithstanding. Thus, if I promise to meet you promptly at five, I am thereby under an obligation to do so. But violation of the prima facie obligation can be justified, as, for example, when I am late because I stopped to save a drowning child. But even then, the obligation retains some force; I owe you the justification, since it is to you that I am obligated in virtue of the promise. Had I been able to telephone you upon my discovery of the drowning child, you would, of course, have been able to obviate the need for later justification by releasing me from my obligation. That obligation gave you certain rights in regard to my behavior. But you were under no obligation not to waive those rights.

Obligations can arise from promises, contracts, agreements, and even implicit understandings. Such obligations may be called special obligations. Perhaps obligations can arise from other sources as well; we shall return to this point below. Each special obligation involves an agent who is under the obligation, someone to whom the agent is obligated and who thereby has rights that follow from the obligation, and some action in respect to which the agent is obligated. Descriptions of special obligations thus have the general form: person X is obligated to person Y to perform action A. If he wishes, Y may release X from the obligation, but if he does not do so, then X owes Y either the action or an acceptable justification for nonperformance—Y has a right to these things.

The action may be one the performance of which is to the benefit of Y, as when X has an obligation to pay Y an agreed amount of money for the rental of Y's house. The action may be to the benefit of some third party, in whose well-being Y perhaps takes some interest, as when X is under an obligation to care for Y's child because X is under contract with Y to do so. Or the action may be to X's own benefit, as when X promises Y that X will take a needed vacation. Actions that benefit the agent may be called *self-regarding* actions. There can be various kinds of reasons for performing them; among such reasons may be that one is under an obligation to someone else to do so. When that is the case, the fact that X is the person who stands to benefit from X's performance of action does not entail that X is under no obligation to perform it.

Does it make sense to speak of obligations to oneself, as in, "You owe it to yourself to stop overeating?" One plausible interpretation of such admonitions is that if you do not change your ways, then you will be acting contrary to your own interests, for example, whatever your aspirations and objectives may be they will be impeded by your imprudence. By this interpretation, the admonition is sound advice, but it is not *moral* advice, i.e., advice based on considerations of what constitutes right action as opposed to merely useful action. (On the difference between prudential and moral judgment, see Frankena.) It would be imprudent of you not to reform, but the claim does not have the moral force of the claim, for example, that you owe it to another to pay your debt to him.

Is there a moral sense in which you can be said to owe it to yourself to adopt prudent behavior? Kant held that duties to oneself, e.g., to preserve one's life, maintain one's health, and develop one's talents, were the most central of obligations. But there is a problem in making sense of the notion of an obligation to oneself. For an obligation from which one can release oneself at will has no binding force, and hence is not an obligation at all in any significant sense. Yet, if X is the person to whom X is purportedly obligated, it would seem that X, in his capacity as holder of the rights that are entailed by the obligation, has it within his authority to waive those rights, and thereby to release himself from the obligation. Thus, if one has an obligation to preserve or enhance one's health, it cannot sensibly be said to be a special obligation *to* oneself—even though such obligations would be self-regarding.

It might still be possible to argue that one has an obligation to oneself in regard to one's health if it could be shown that there are obligations that, unlike the special obligations we have been discussing, do not arise from any sort of promise, contract, or agreement and are not subject to nullification at the discretion of those who have rights under the obligations. This was Kant's position; he argued that pure reason, independently of any actions performed by people or relationships among them, was the source of moral obligations including obligations in regard to one's own health and life. Others have argued that statements about rights and obligations can be supported by appeal to natural law, by which is meant not the laws of the natural sciences such as physics or biology, but moral laws that are claimed to be inherent in nature. Such a view might include the belief that the preservation of life and health are intrinsic goods that one is obligated to oneself to protect in virtue of natural law. However, unless one is willing to adopt some such position, in spite of the difficulties they invite, it does not seem likely that one can coherently construe self-regarding obligations in respect to health as being obligations to oneself. (For a discussion of Kantian ethics, see Rawls and Sidgwick; for a discussion of the concept of natural law, see Melden.)

Still, the statement that one owes it to oneself to be healthy need not be taken as simple nonsense. Rather, it can be taken as a prudential maxim of broad applicability. For, whatever one's aspirations and objectives, whatever one's conception of the good life, it is likely that one's interests will be advanced by health and impeded by illness. One owes it to oneself to be healthy, then, at least in the sense that one is usually foolish from the point of view of one's own present and future desires if one acts so as to undermine one's health. But this is not by itself a moral issue, nor is there any literal sense of obligation involved. It may be unfortunate in many ways for one to be foolish, but it is not necessarily a violation of any obligation.

The fact that one cannot literally be obligated to oneself to maintain or enhance one's health does not entail that one has no such obligation to anyone. On the contrary, one may have explicit obligations to others that require that one protect and defend one's health. Athletes, astronauts, and others for whom excellence of physical condition is a prerequisite for professional success could be contractually bound to act prudently with respect to their health. In such cases the obligation to be healthy would itself be an explicit duty, voluntarily undertaken, self-regarding, requiring the individual under threat of legal and moral culpability to act prudently in regard to health. But the obligation to be healthy can have other origins as well, on the basis of which the obligation to health is not explicit, but is derivative from other obligations.

If Smith is obligated to Jones to do A, it seems reasonable to hold that Smith is obligated as a consequence to do whatever is necessary for the doing of A. Thus, if Smith, for whatever reason, is obligated to Jones to provide Jones with a home-cooked dinner, Smith is also obligated to obtain the necessary food. And if Smith defaults, he cannot escape culpability on the grounds that he simply did not obtain that food. He was under an obligation to obtain it because doing so was necessary to the fulfillment of his obligation to cook and serve it to Jones. Are there perhaps obligations ostensibly unrelated to health the honoring of which requires a regard for one's health?

Surely some obligations do not require that one maintain one's health. For example, one who freely undertakes to provide financially for the education of another can fulfill his obligation to do so by the appropriate arrangement of trusts or insurance, without thereby becoming obligated to perform any self-regarding actions. But, equally surely, other obligations do require the maintenance of one's health. If Jones promises to be Smith's traveling companion for a year, Smith has a right to expect Jones to seek to maintain a state of health compatible with travel. If Jones defaults, in virtue of illness induced by imprudent behavior against sound medical advice, Smith can justly claim to have been let down in a morally culpable way.

Certain special relationships involve the explicit undertaking of special obligations, such as are incurred in a marriage, which, whatever else may be true of it, binds the participants to one another in certain legally specified ways. But beyond these obligations both voluntarily undertaken and legally imposed, it is reasonable to claim that further obligations exist, implicit in that special relationship, obligations that reflect the individual characteristics of the particular pattern of expectations and interdependencies that have nurtured the relationship. To enter into a marriage, at least of the traditional sort, is typically to undertake voluntarily to pursue and sustain a pattern of interaction that typically will require the maintenance of one's health. The fact that illness frequently does—and is more frequently presumed to—befall its victims independently of culpable behavior on their part, in no way contradicts the fact that illness is sometimes invited or induced by imprudent behavior. In such cases, it can make sense to *blame* someone for being ill—precisely because that illness results from the violation not only of prudential maxims but of moral obligations as well.

Similar arguments apply to the maintenance of life itself. No one has a special moral obligation to himself, literally speaking, to refrain from suicide. But one may have obligations to others the fulfillment of which is incompatible with one's death. Thus the survivors of a suicide feel anger and resentment, along with their remorse and guilt, not only commonly but justifiably, when the suicide leaves obligations unfulfilled.

As we have seen, some would argue that obligations that require the maintenance of one's health can arise from one's moral circumstances independently of the special relationships one has to others. The Kantian position is of this type, as are other positions also. One need not embrace Kantian rationalism to believe that there are obligations that transcend the specific actions and special relationships of mankind. For example, one who accepts a theologically based conception of moral obligations might argue that obligations in regard to one's health and life have a divine origin. Thus, the empiricist John Locke held that one's life is not property over which one has jurisdiction; rather it is held, so to speak, in trust, in virtue of being of divine ownership in consequence of which one has obligations to preserve and protect it. Much of the discussion of rights within the religious tradition has taken place in terms of the natural law mentioned above; Locke's views are a case in point.

Or it might be argued that one's status as a member of the human community itself imparts obligations of a variety of sorts—obligations to one's fellow man—such as the obligation to be charitable to those in need or the obligation to refrain from gratuitous assault. If such obligations do exist, some of them may have as consequences the existence of further obligations that are self-regarding in respect to health. However, from the point of view of secular humanism (the view that moral values arise solely out of human wants and needs but are nonetheless of fundamental importance in the choice and justification of behavior), the case for such self-regarding obligations has not been made in any adequate way, and thus the burden of proof rests with one who claims that such obligations do exist to explain more clearly what they are and how they come into being.

But even short of such entirely general and possibly illusory obligations, one can be in moral circumstances that make plausible the presence of obligations that arise independently of one's contracts, agreements, or familial relationships. Thus, in an isolated society with just one physician—of substantial age—one might well argue that the physician is under an obligation to provide for future medical care; that he *ought* to take an apprentice and pass on his skills for the sake of his survivors and future generations. And one might well mean not merely that it would be desirable to others for the physician to do so, but that he is morally culpable should he refuse. Nor does there seem to be anyone in a position to release him from such an obligation. His acceptance of the apprentice, moreover, might entail further self-regarding obligations, at least until the apprenticeship is completed.

There are thus five principal ways in which a self-regarding obligation in respect to health may be said to arise: (1) as a special obligation to oneself, (2) from an explicit contract or agreement pertaining to one's health, (3) as a consequence of other obligations the fulfillment of which requires the maintenance of one's health, (4) in virtue of special relationships to others, for example, one's family, and (5) in consequence of one's more general moral circumstances. The first way is illusory. The remaining taxonomy is overlapping; an obligation might well fall under more than one of the descriptions (2) through (5).

Of course, to speak of an obligation to be healthy is elliptical: even when a self-regarding obligation in respect to health does exist, that obligation cannot strictly speaking be to *be* healthy, for health is often beyond one's capacity to achieve or maintain. Rather the obligation can only be to take available and reasonable actions conducive to the state of health that is required by the logically prior obligations in virtue of which one has the obligation to be healthy.

Without such prior obligations, one would have no obligation to be healthy. People who are alone—devoid of the relationships with others that bind them to others—may also be devoid of the aspirations that bind them to the future. Such tragic figures may lack the desire to maintain their health, their bodily integrity, or even their lives. There seems to be no persuasive basis for arguing that under such circumstances they have any obligation to do so. Further, we can imagine circumstances under which the pursuit of health might be less tragically shunned. For example, a physician might wish to endure a temporary serious illness in order to experience directly the plight of the seriously ill—with a view toward enhancing his future ability to deal sensitively with his patients. Or a writer, aspiring to produce the most compelling portrayal of illness, might willingly embrace a chronic ailment in the service of his art. It is hard to see how either could reasonably be accused of thereby violating any moral obligation, in the absence of other considerations.

Health, then, is an almost universally valuable state. Yet many are imprudent in the extreme in regard to their own health. It would be hard to argue convincingly that one literally owes it to oneself to be healthy, no matter how desirable health is acknowledged to be. But most owe it to others, for a variety of reasons, to seek to maintain good health. Those who lack any such obligation are free to abandon their health, their bodily integrity, and even their lives without fear of thereby violating a moral obligation. But that freedom exists only in consequence of the chilling impoverishment of their lives.

SAMUEL GOROVITZ

[*Directly related are the entries* CARE; ETHICS, *articles on* RULES AND PRINCIPLES *and* DEONTOLOGICAL THEORIES; HEALTH AND DISEASE; NATURAL LAW; OBLIGATION AND SUPEREROGATION; *and* RIGHTS. *This article will find application in the entries* HEALTH CARE; LIFE; MENTAL HEALTH; POVERTY AND HEALTH; PUBLIC HEALTH; RIGHT TO REFUSE MEDICAL CARE; SMOKING; SUICIDE; *and* THERAPEUTIC RELATIONSHIP.]

BIBLIOGRAPHY

Dworkin, Gerald. "Paternalism." *Monist* 56 (1972): 64–84.
Frankena, William K. *Ethics.* 2d ed. Englewood Cliffs, N.J.: Prentice-Hall, 1973. See especially pp. 114–116.
Kant, Immanuel. *Lectures on Ethics.* Translated by Louis Infield. London: Methuen & Co., 1930, pp. 117–118.
Locke, John. *The Second Treatise of Government.* New York: Liberal Arts Press, 1952. See chap. I, 6, pp. 5–6.
Melden, Abraham I. *Human Rights.* Belmont, Calif.: Wadsworth Publishing Co., 1970, pp. 40–75.
Mill, John Stuart. *On Liberty.* New York: Liberal Arts Press, 1956, chap. 4, pp. 91–113.
Rawls, John. *A Theory of Justice.* Cambridge: Belknap Press of Harvard University Press, 1971, pp. 251–257.
Sidgwick, Henry. *The Methods of Ethics.* London: Macmillan Co., 1901. See especially pp. 511–516.
Singer, Marcus. "On Duties to Oneself." *Ethics* 69 (1959): 202–205.

HEALTH CARE

I. HEALTH-CARE SYSTEM — *Philip R. Lee and Carol Emmott*
II. HUMANIZATION AND DEHUMANIZATION OF HEALTH CARE — *Jan Howard*
III. RIGHT TO HEALTH-CARE SERVICES — *Albert R. Jonsen*
IV. THEORIES OF JUSTICE AND HEALTH CARE — *Roy Branson*

I
HEALTH-CARE SYSTEM

The concepts of respect for persons and rights to health care provide a useful perspective for ethical evaluation of health-care systems. Means for the provision of health care vary widely both among and within countries. Wide differences in societal values and ethical notions are graphically expressed in the allocation of health-care resources and the manner in which services are provided. National and even regional patterns of treatment and organization defy simple comparisons. Western developed countries, however, share sufficient common elements to provide illustrative examples.

It is indicative of the limitations of the Western medical model that this discussion of health-care systems concentrates on provisions for the treatment of existent disease. The term "disease care system" or "medical care system" might be more accurate, for a true health-care system should include not only the diagnosis, treatment, and rehabilitation of diseased individuals but also a variety of preventive approaches, including changes in individual behavior and manipulations of the environment that may decrease the burden of disease for individuals and populations.

The ultimate purpose of a health-care system is to sustain the health of a population. In the modern health-care systems of developed countries this purpose is addressed not only by the actual mechanisms for the provision of care, but also by complex financing and regulatory systems. In the past, medical care consisted of a simple exchange between a patient and a practitioner. The payment for care was determined by informal contract between the parties. Although the exchange was occasionally regulated by states or professional associations, it was for the most part a matter between the one who sought the care and the one who provided it. More recently social, economic, scientific, and technical changes have transformed this simple relationship into one between the patient and a host of professional health-care providers in a complex, multifaceted health-care system.

Analysis of mechanisms for the provision of health-care services should be considered in this broader context, since the provision of care, the financing mechanisms, and the regulatory systems are mutually interdependent. Methods of financing constrain delivery options. Regulatory power derives from fiscal leverage and impacts on the access to care, the kinds of care provided, and the quality of care.

This article will discuss: (1) financing systems; (2) regulatory practices; (3) health-care delivery systems, described in terms of the different modalities of care as characterized by (a) levels of health care and (b) stages of treatment in the health–disease spectrum; (4) health-care resources—both manpower and facilities; (5) a survey of approaches to the organization of services; and finally, (6) some comments on health-care policy choices.

Financing health care

Financing mechanisms of health-care systems raise problems of distributive justice and the rights of persons to medical care. Typically, these are problems about who bears the obligation to provide and to pay for care, how persons can be selected for care in a fair way, what kinds of care (e.g., primary care or open-heart surgery) are available and about whether it is just to devote a certain portion of national assets to health care at the expense of other public needs.

The emergence of these ethical questions of

health-care systems has followed the evolution of collective mechanisms for the financing, regulation, and delivery of health care. The first major impetus for collective action in health care was the need to protect populations against communicable diseases and to provide hospital care for the poor. Initiative for collective efforts to assure hospital care to the general public came from the private sector in the form of voluntary and church-supported hospitals. Collective financing mechanisms for large segments of the population began first in Germany and then spread throughout most of the developed world. France followed the German example through development of a compulsory social insurance system to meet much of the public's health-care costs; other countries, such as Sweden, Britain, and Canada, pay health-care costs out of general revenues.

Despite increasing public responsibility for health-care financing in recent decades, the health-care systems of most developed countries are characterized by many combinations of private and public financial responsibility. The United States, Italy, Spain, and Portugal have fragmented, pluralistic systems in which services are paid for by the patient, by private insurance, or by government for certain classes of persons or services. Relying more on public finance and control are the delivery systems of the Netherlands and France, followed by those of Germany and Switzerland. In the Scandinavian countries the responsibility for financing and ownership of facilities is largely assumed by local or national government. In Britain, financing is almost entirely derived from national taxation. The Soviet Union operates a centralized national system for financing and planning, in which administration is decentralized and health care is provided by physicians and other health-care personnel who are government employees.

Most health-care systems, however, are composed of a mix of publicly and privately rendered services. Even systems collectivized as completely as the British system typically allow some private services. Private financing generally consists of out-of-pocket expenditures and insurance coverage. Public financing consists of either provision of services directly by government or through government subsidy for care purchased in the private system.

The United States, however, has lagged behind most European nations in the development of public financing systems. It differs from most developed countries in a heavy reliance on insurance sold by the private sector in preference to government financing. These private health insurance plans began during the Great Depression and grew slowly until the Second World War when they experienced rapid expansion as fringe benefits in labor–management negotiations. Health insurance coverage has since broadened, until almost all employed Americans are covered by some form of insurance. Still, in this broad coverage there are gaps both in individuals and in services covered. Children and low-income families are often inadequately covered by private insurance, as are many basic, nonhospital primary care services, prescription drugs, and dental care. As a result, private insurance pays less than a third of the total U.S. health-care bill (Saward).

Almost since its founding, the American government has directly provided health care to certain beneficiaries. Merchant seamen were the first to receive federal health care. They continue to receive it through the U.S. Public Health Service. Members of the Army and Navy were also provided limited medical services. Later benefits were extended to veterans with service-connected disabilities. Health-care programs originally established for veterans with service-connected disabilities have been significantly expanded to offer a broad range of services to all veterans. American Indians are also provided with medical services directly by the federal government. State governments traditionally provided hospital care for the mentally ill and contagious disease hospitals for those with communicable diseases. County and city governments operated general hospitals for the poor.

In the Social Security Act of 1935, the U.S. government initiated grants to states for the health-care expenses of special groups such as crippled children. Those programs were followed by programs for the care of the dependents of servicemen during the Second World War. In the 1960s the federal government initiated limited financial support for state-financed medical care programs for the indigent aged.

In 1965 the federal role dramatically changed with enactment of Medicare and Medicaid legislation as amendments to the Social Security Act. Medicare provided federal subsidy for the medical care costs of approximately twenty million Americans over the age of sixty-five. The Medicaid program is a joint federal–state program that has gradually assumed the burden of paying the medical bills of some twenty-five million poor persons.

Health-care providers are compensated for their services in a variety of ways. Nationalized health systems typically place all providers on salaries or on contract, as are the consultants and general practitioners in Great Britain. Physicians and dentists in the United States and in other mixed financing systems are usually compensated on a fee-for-service basis. Many physicians are salaried, however, in programs for federal beneficiaries, in large prepaid group practices, in public hospitals, and on medical school faculties. Despite the predominance of the fee-for-service model in the United States, there is a strong trend toward group practice prepayment programs. These prepayment programs offer a full range of services for a set annual fee per person. Legislation encouraging the further development of such "Health Maintenance Organizations" (HMOs) has been enacted and an increasing number are being established.

Financing and payment mechanisms have considerable influence on service patterns. In the United States private insurance policies encourage expensive hospital care because inpatient services are usually broadly covered, while the cost of ambulatory services is limited. Fee-for-service payment of providers has been criticized because it provides an incentive to deliver covered services and accords low priority to preventive services. HMOs based on set annual per capita fees are believed to reverse these incentives. Efficiency is thus enhanced, but safeguards must be set up to counter the possibility that less care will be provided than the patient requires.

Total health-care costs of a nation are usually measured as a percentage of its Gross National Product (GNP). All countries have experienced significant increases in the share of the GNP devoted to health and health care. For example, in 1950 the United States devoted 4.6 percent of its GNP to health and health-care costs. By 1973 this figure had risen to 7.7 percent, and in 1975 it reached 8.3 percent, or $118 billion. The relative investment in health care as compared to other expenditures is closely related to a nation's economic development and per capita income. Typically, wealthier countries expend a greater percentage of their GNP on health care than less prosperous ones.

There are only limited opportunities to control total spending in health-care systems that have multiple sources of financing and that pay physicians on a fee-for-service basis. Costs can be more tightly controlled in systems that are more centrally administered. The United Kingdom places a tight lid on health-care expenditures, devoting a smaller proportion of national wealth to health care than those publicly financed systems which pay on a fee-for-service basis, such as the Canadian health-care system.

The provision of health care has become so complex and expensive that it has required some degree of government intervention to correct obvious inequities. No nation enjoys unlimited resources. Choices must be made. At the base of most health-care decisions lies the moral problem of attempting to achieve efficiency in the use of available resources while at the same time recognizing the needs of individuals in an equitable manner.

Regulating health care

The trend toward governmental financing of health care has been accompanied in all countries by a growing government role in health-care planning and regulation. This direct relationship between increased public financing and control can be seen in all health-care systems, though no two systems are exactly the same. In England, for example, there is little attempt to regulate the decisions of individual physicians, but the whole system is regulated by the annual appropriation for health-care services by Parliament. The total health-care appropriation ultimately determines the scope of services that can be provided. In the United States, in contrast, fee-for-service payment for providers remains the predominant model, but there is increasing government intrusion into the autonomy of the medical profession. Professional Standards Review Organizations (PSROs) are mandated to review professional decisions about the type of care provided and federal reimbursement policies significantly affecting the fees charged by physicians.

The regulation of health care by government has met a great deal of resistance. Medicine has a long tradition of professional autonomy and self-regulation through voluntary adherence of the individual practitioner to a code of professional ethics. Even within the medical profession, collective efforts to enforce quality standards have arisen only in recent decades.

Regulation of health care deals with determinations of the quality of care; distribution of health manpower, facilities, and services; the regulation of hospital charges and physicians' fees; and the regulation of the drug industry to assure the safety and efficacy of drugs and other products used in medical care. Early regulatory efforts were designed to assure the quality and safety of hospitals and other facilities from gross environmental hazards, such as fire. Standards for the education and licensure of health professionals were also an early development designed to protect the public from unqualified practitioners. Quality standards, promulgated through hospital accreditation procedures, attempted to assure the quality of facilities some years before attempts were made to monitor the distribution of hospital beds.

In recent years, in the United States, regulatory actions by government have increased. Some of them, such as air and water purity control, the screening of therapeutic drugs, the regulation of medical devices, and occupational safety requirements result from public demand stimulated by alleged abuses. Others arise from budgetary considerations that appear to demand certain restrictions. These have increased significantly as the federal government has assumed the responsibility of providing or paying for medical care through welfare and insurance programs. Among these are limitations of reimbursements to hospitals for care provided under government programs, mandated review of appropriateness and quality of services provided by physicians under these programs, regulation of rates charged by hospitals and physicians, efforts to reduce restraints on competition in the medical market place, and review by public agencies of proposals to construct health-care facilities. These regulatory efforts are based upon legislation by the Congress and are developed and administered by a variety of federal agencies and other public or private bodies under contract to the government. Extensive regulatory activities in the practice of medicine are justified by an appeal to the public good. Nevertheless, they pose ethical problems such as intrusion into the discretion of physicians about manner and extent of care, rationing of medical resources, and bureaucratic delays in meeting urgent needs. However, governmental regulation is now an established feature of health-care services in many nations.

Change in the ethical premises regarding rights to care or the proper distribution of health resources have accompanied growing public involvement in health-care finance and regulation and added a significant new facet to the ethical bases for the delivery of health care. Most modern nations are based on at least a presumption of equality. As governments assume a growing posture in the financing and regulation of health care, they are, for the first time, able to affect the distribution of health-care resources. With that leverage, most have responded to a political or ideological climate that cannot abide gross inequality in the distribution of resources. Greater public responsibility for distribution of manpower, facilities, and services encourages the creation of standards of equity in access to care. Cost controls also become necessary because runaway costs and strained budgets can severely curtail both access and quality.

Delivery of health-care services

Financing and regulatory mechanisms have a profound effect on the kind of care provided by different combinations of health personnel in different delivery settings. Recognition of these interdependences allows a more comprehensive assessment of the health-care modalities, resources, and organizational arrangements that characterize health-care delivery systems.

Modern health-care services are characterized by complex modalities of health care. The diversity of these services arises from the levels of sophistication of services and stages of intervention in the health–disease spectrum.

Levels of health care. One of the most useful typologies divides health-care services into levels of care, characterized by degrees in specialization of manpower and in the elaborateness of facilities and equipment required to provide the service. The most basic aspect of a health-care system is the way in which it provides the opportunity for first and continuing contact with a provider of care. This is called primary care. Primary care services include initial diagnosis, treatment of general medical problems, and referral to specialists when necessary. Health education and preventive services are also included in this level. Providers of primary care are best able to integrate and coordinate all types of needed services, in order to provide continuity as patients move through other levels of care.

In the United States general practitioners traditionally provided primary care until the rapid development of specialization after the Second World War. Since then internists, pediatricians, and family medicine specialists have become increasingly important providers of primary care.

In addition, when the supply of medical specialists, surgeons, and obstetrician-gynecologists exceeds the demand for their specialized services, many of these physicians also provide primary care. Such care is usually extended in physicians' offices and group practice facilities, although hospital outpatient clinics and emergency rooms and neighborhood health centers supply primary care for many low-income urban dwellers.

Variations among countries in the provision of this crucial facet of medical care highlight different approaches to care based on varying values. Countries like the Soviet Union and the People's Republic of China, confronted by limited resources, rank access to primary care for all above optimum quality or high technology medicine for smaller numbers. The "barefoot doctors" of China and the "feldshers" or physician assistants of the Soviet Union provide basic medical services in the most isolated settings. Both countries place limitations on care that requires high cost and complex technology in urban hospitals.

The United States and Sweden represent the opposite end of the spectrum. Concentration on the more sophisticated levels of care has compromised broad primary care coverage. Provision of all levels of care at optimum levels is conceivable, but limitations on the resources of even the wealthiest of countries demand some trade-offs. The United States has thus tolerated wide disparity in the availability of basic medical services. Patients with certain kinds of medical problems or patients of higher socioeconomic status have been allowed to consume disproportionate amounts of health-care resources while others go with only minimal levels of care. This difference in American and Chinese policy trade-offs between quality and access clearly reflects the wide diversity in criteria for distributive justice.

America's poor record in equalizing access to primary care is closely related to its achievements in providing sophisticated secondary and tertiary care to some segments of the population. Secondary care refers to those services provided by consulting specialists and subspecialists. Though patterns vary among countries, in the United States these services are provided both in office and in hospital settings. Insurance plans and government subsidies have made expensive specialty care available to most socioeconomic groups, but the concentration of secondary care centers in cities or regional medical centers often denies rural Americans, particularly the poor, equal access to those specialized services.

Tertiary care represents the most specialized form of care, delivered by subspecialists in large regional centers, university medical centers, or, in some cases, only in national centers. This sophisticated level of care includes complex technology, high-cost procedures such as open-heart surgery, organ transplants, and renal dialysis. Such services are unavailable in many parts of the world because of limited medical expertise. Some countries that have the technical acumen have limited the availability of these services by allocation policies that favor the more equal distribution of basic services. Great Britain, for instance, has a highly sophisticated medical system, but many areas of Britain provide no renal dialysis because its high cost would restrict funds for provision of quality primary care.

Though external to the traditional medical model, self-care is coming to be acknowledged as another significant level in a total health-care system. Individual assumption of responsibility for health-promoting behavior and the treatment of minor health problems with traditional remedies or nonprescription drugs can have a great impact on the stresses placed on the other levels of care, particularly primary care. Great variation in self-care among and within health-care systems can be attributed to individual differences but also to cultural and ethnic norms, economic considerations, and geographic availability of health services.

The health–disease spectrum. Health care consists of more than the diagnosis and treatment of patients with existent disease. Quite distinct health resources, skills, and treatments are required at different stages in the developmental stream of disease processes. First and often overlooked are preventive measures. Traditionally, these include immunization and environmental sanitation measures, but broader preventive measures such as early screening and environmental and life-style improvements are receiving increasing attention.

Just as individuals are taking increasing interest in the development of health-promoting life-styles, so governments are beginning to ask how they can contribute to disease prevention and health promotion. Most countries have taken some action to deter the alcohol consumption of automobile drivers. Policies vary widely concerning tobacco consumption, nutrition programs, the use of automobile seat belts, gun control, and a wide variety of other preventive measures that might significantly affect health status and the demands made on the health-care system. Such efforts, particularly those aimed at improving public knowledge about particular hazards to health, have not been particularly effective. Among countries surveyed on this point, annual expenditures on health education varied from only four to ten cents per person (Maxwell).

Diagnosis and treatment of disease have unquestionably consumed the majority of attention and resources of American medicine. The decline in infectious diseases has greatly lengthened life-expectancy, increasing in turn the problems of chronic disease among the growing aged population. These health problems—many related to environmental and social stresses and the life-styles of many Americans—require approaches quite different from those that contributed to the decline of infectious disease rates (Fuchs).

This changing burden of disease, reflected in growing chronic disease rates in the more developed world, has revealed deficiencies in the final two stages in the provision of health care: rehabilitation and long-term support. The American emphasis on the treatment of acute disease has forced many patients who could be rehabilitated to self-sufficiency with appropriate social services into the medical care system. Financing mechanisms favor acute hospital care and discourage the development of more efficient and suitable community-based alternatives for long-term care.

Health-care resources

Manpower. The delivery of different modalities of care hinges on the availability of trained personnel and specialized facilities. Physicians are central to the manpower resources of health-care systems in the developed world. The availability of physicians, however, varies greatly between countries. For example, in 1970 the United States had 158 physicians per 100,000 population, while Portugal had only 87 and the USSR had 238. Although there is some evidence that increases in health resources reach a point of diminishing returns, variation of this magnitude can be assumed to affect seriously the level and quality of care.

Absolute numbers, however, tell only part of the story. For example, significant increases in practicing physicians in Germany in the 1950s did nothing to improve the delivery of primary care, because most of the physicians became hospital-based specialists. In the United States an almost fifty percent increase of physicians in the 1960s has done little to alleviate the shortage of primary-care physicians or the geographic maldistribution of physicians. The United States in 1975 enjoyed the services of 2,000 fewer primary-care physicians than it had in 1931. Growing population has exacerbated the problems, decreasing the proportion of primary care physicians from 94 to 55 per 100,000 population between 1931 and 1972.

The American medical profession has made some strides in correcting the exaggerated trend toward specialization by creating a new specialty of Family Practice in 1969 and by expanding the training programs in general medicine and pediatrics.

Although most health-care systems suffer from specialty maldistribution, countries that exercise more rigid central planning are better able to correct inappropriate distribution. In the United States twenty-five percent of all doctors specialize in surgical fields. In Sweden, however, only fifteen percent of physicians are surgeons, and in England and Wales the figure is thirteen percent. The effects of such overstaffing on the amount of expensive and sometimes unnecessary surgical procedures can be profound.

Physicians in many countries are as poorly distributed geographically as by specialty. This problem is, again, particularly pronounced in the United States. Some states have almost three times the proportion of physicians to population as other states. Within states, the urban areas have, on the average, twice as many physicians in relation to the population as rural areas. Within metropolitan areas, low-income inner-city residents are often denied access to physicians and dentists because of the overconcentration of practitioners in the wealthy urban and suburban areas. Maldistribution can be better controlled in centrally organized health systems such as England's by limiting the allotment of specialty training and practice positions open to physicians.

Specialization within the medical profession has been accompanied by even greater specialization within the new health-care professionals, technicians, and other workers. In the United States the health labor force has become the nation's third largest, in 1975 employing 4.5

million people in at least 375 different job categories. Technological innovations that require highly skilled technicians create the need for many new health occupations. Even more revolutionary are the new roles of physician assistants and nurse practitioners, who contribute to the American system a mid-level professional similar to the Russian feldshers.

The changing role of the physician in a pluralistic, highly specialized system is reflected in the declining ratio of physicians to other health personnel. The distribution of power and delegation of responsibilities vary widely among systems. For example, German doctors delegate less work to nurses and other professionals than their English counterparts. Reason and economy dictate better use of other professionals, particularly the mid-level health practitioners. Appropriate distribution of power and responsibilities among U.S. health personnel, however, require changing entrenched behavior and reforming systems of legal liability so as not to place almost exclusive responsibility on physicians.

Facilities. Increasing technology and specialization have dramatically changed the role of the hospital. The modern hospital has become the focus of most surgical and medical specialty care for the severely ill and seriously injured. Economies of scale have made such centers the site for many expensive diagnostic and therapeutic devices. In the United States enormous investment in hospital facilities was initiated by a federal construction program (Hill–Burton) beginning in 1946. Most industrialized countries provide 90 to 120 general and psychiatric hospital beds per 10,000 population; and although the United States falls below these norms with only 81 beds per 10,000 population, occupancy rates suggest a costly oversupply of general hospital beds. Duplication of special facilities places additional financial stress on hospitals. Inadequate planning, uncontrolled expansion of services, rising demand, improved technology, and price inflation have resulted in a rise in hospital costs to forty to fifty percent of total health-care expenditures in many more developed countries.

Organization of health-care services

Impediments to effective coordination of services persist in most health-care systems. Not even totally nationalized health-care systems approach an appreciable degree of coherence and integration of services. The American system, however, is surpassed by none in its heterogeneity and fragmentation.

Fragmentation exists between levels of care when those needing a diversity of services are not provided with continuity of care, coordinated by a single primary care practitioner. Institutions offering different types of care are even more fragmented. Public health agencies oversee traditional preventive measures, a variety of institutions perform diagnosis and treatment of acute physical diseases, another system cares for the mentally ill, still another for the mentally retarded, a multiplicity of programs serve the chronic alcoholic or drug addicted, and an array of proprietary and public institutions operate nursing homes and other long-term care services.

Although not mutually exclusive, four different delivery subsystems for traditional disease care can be identified within the United States. They are characterized by different geographic locations and financing mechanisms. They are subsystems serving: (1) inner-city poor; (2) urban and suburban middle and upper classes; (3) rural dwellers of all income levels; and (4) beneficiaries of various federal programs. The contours of these four broad subsystems highlight major variations in the types of care and sources of financing.

1. *The inner-city poor*, including a disproportionate number of aged and minorities, rely largely on publicly supported hospitals for inpatient, outpatient, and emergency care. In many of the states, inadequately funded Medicaid programs for the disadvantaged have not allowed adequate access to private physicians, few of whom choose to work in the often depressed economic and living conditions of the inner city. As a result, outpatient clinics and emergency rooms in public or teaching hospitals provide most primary care. Without personal physicians, the urban poor must contend with fragmented and often impersonal care.

2. *Urban and suburban middle and upper classes* enjoy what is often termed "mainstream" medical care. Through private health insurance, Medicare, and personal payment, patients have access to office-based generalists and specialists. Community hospitals are used and patients are able to choose among a variety of physicians, since there is a greater proportion of physicians to population, particularly in suburban settings. Patient satisfaction with medical services is very high within this subsystem.

3. *Rural medical care* is the task of the general practitioner and the small community hospital. It offers limited access to specialists. A shortage of professionals imposes long working hours on rural practitioners and limits their opportunities for continuing education. Patients in rural areas seem to make less demands for care and the physicians do not generate laboratory, X-ray, and hospital care to the same extent as urban practitioners.

4. *The federal subsystem* comprises several independent systems, which serve members of the armed forces and their dependents, veterans, merchant seamen, and American Indians. Primary care is provided in dispensaries, often by paramedical personnel; while specialists provide hospital-based services. In some delivery systems, such as the Veterans Administration, the provision of ambulatory care is often very limited in contrast to inpatient services.

The two variables of financing and geography, which differentiate these subsystems of American health care, represent two of the most important determinants of a population's access to care. Inability to pay has long been considered the central barrier to receiving medical care. But geographic proximity and accessible transportation determine to a great extent the availability of health manpower and facilities. The inverse relationship between health-care needs and resources is apparent in that, while wealthy areas attract the highest concentration of health facilities and physicians, illness is most prevalent among the poor.

Fragmentation presents a third barrier to care in most health-care systems. For example, increases in the capabilities of the U.S. health-care system have created enlarged expectations and an amorphous, disjointed array of means of meeting them. The complexity of health-care delivery and financing systems acts as a deterrent to acquiring needed care.

Most developed countries have attempted to correct differences in access to care through a combination of governmental financing and reorganization. The slow American commitment to public financing has been chronicled. Organizational innovations have been even slower to evolve. After a strong repudiation of group practice in the 1930s, American physicians have gradually moved away from the solo practitioner model toward single or multispecialty groups. Multispecialty groups, using either prepayment or fee-for-service plans, can contribute significantly to the quality of and access to care. Close association of professionals provides a quality incentive and control mechanism rarely found in solo practice. Continuity of care is improved through internal referral patterns.

The pluralistic U.S. approach encourages myriad federal health programs serving the special needs of population subgroups. One of the most ambitious was the development of Neighborhood Health Centers during the administration of President Lyndon Johnson. These centers attempted to bridge traditional barriers to care for the urban poor. Other programs have been created for migrant workers, Alaskan natives, crippled children, drug addicts, and a number of other groups. Still other efforts have been devoted to the sufferers of specific diseases, such as mental retardation and kidney disease.

This incremental, categorical extension of public support for health care reflects an ambivalence about the concept of the right to health care for all persons and the freedom of individual practitioners and institutions. If care is to be available broadly and equitably, limits must be put on the autonomy and free enterprise of providers. Neither alternative has been wholly acceptable. As a consequence, complexity and fragmentation within the system have been compounded.

Health-care policy choices

The variety of health-care needs, services, and personnel in developed nations prevents simplicity of organization. Reorganization and refinement of health delivery systems present monumental problems to all who would attempt their rationalization. Effective implementation of financing, regulatory, and delivery mechanisms awaits the emergence of a consensus regarding appropriate trade-offs among acceptable standards of health-care quality, access, and costs.

These three notions—quality, cost, and access—are central to current attempts to develop effective policies regarding provision of health-care services. Quality refers to assessment of the efficacy and efficiency of medical care provided at the various levels. Cost refers to the aggregate cost to those who consume and pay for care. Access refers to the way in which resources are distributed so that persons needing care can obtain it. Those three key categories of health policy draw together, through complex relationships, the modalities, resources, and organizational patterns that are the elements of any modern health-care system.

In the United States, the standard of high quality medical care for individual patients has long been a central goal of the medical profession and therefore the standard to which the system as a whole aspired. Emphasis is placed on skilled technical services, such as those provided by specialists using increasingly complex and costly facilities, equipment, and technical personnel.

Commitment to equality of access has been slow to evolve. Traditionally, economic status of a patient determined access to high quality care, while charity and local governments dispensed inferior care to those less fortunate and implicitly less deserving. Recent years have brought a broadening of commitment to equitable access for all. Yet concrete policy and implementation lag far behind. As a result the United States stands alone among the wealthy, industrialized nations of the world as the last to institute a form of social insurance or national health service in order to equalize access to health care.

In the past decade the need for health cost control has plagued virtually every Western European country, Canada, and the United States, despite the marked differences in the way the services are provided and financed, the way physicians and institutions are paid for their services, and the mix of primary and highly technical, specialized services.

Policymakers are only beginning to recognize —largely through the growing pressure of budgetary constraints—the reciprocal relationships and necessary trade-offs among these policy standards of quality, access, and cost. Rigid cost controls can result in unequal distribution of care and compromised access for many persons. High quality standards and continued growth of expensive medical innovations can appreciably escalate costs and perhaps limit the access of large portions of the population to basic medical care.

A direct relationship exists between the policy goals—quality, access, and cost—and the ethical notions of distributive justice and respect for persons. The quality standards employed in development of health-care systems reflect clear notions of respect for persons, balanced with notions of distributive justice or rights to health care. The policy demands of reasonable, affordable cost greatly affect the distribution of care and the notion of fairness and equity.

Analysis of the ethical bases of any health-care system must derive from a broad understanding of the institutional arrangements that finance, regulate, and deliver health-care services. Serious ethical reflection on health-care systems has only begun but promises to provide valuable insights that can enlighten the difficult health-care choices of the future.

PHILIP R. LEE AND CAROL EMMOTT

[*For a further discussion of topics mentioned in this article, see:* JUSTICE. *Also directly related are the entries* HEALTH INSURANCE; HEALTH, INTERNATIONAL; HEALTH POLICY; MEDICINE, SOCIOLOGY OF; POVERTY AND HEALTH; PUBLIC HEALTH; *and* SOCIAL MEDICINE. *This article will find application in the entries* AGING AND THE AGED, *articles on* ETHICAL IMPLICATIONS IN AGING, *and* HEALTH CARE AND RESEARCH IN THE AGED; CHRONIC CARE; ENVIRONMENTAL ETHICS, *article on* ENVIRONMENTAL HEALTH AND HUMAN DISEASE; HOSPITALS; INSTITUTIONALIZATION; MASS HEALTH SCREENING; *and* MENTAL HEALTH SERVICES, *articles on* SOCIAL INSTITUTIONS OF MENTAL HEALTH, *and* EVALUATION OF MENTAL HEALTH PROGRAMS. *See also:* DRUG INDUSTRY AND MEDICINE; FOOD POLICY; MEDICAL ETHICS, HISTORY OF, *section on* EUROPE AND THE AMERICAS, *article on* EASTERN EUROPE IN THE TWENTIETH CENTURY; MEDICAL SOCIAL WORK; *and* POPULATION POLICY PROPOSALS, *article on* CONTEMPORARY INTERNATIONAL ISSUES.]

BIBLIOGRAPHY

BEAUCHAMP, DAN E. "Public Health as Social Justice." *Inquiry: A Journal of Health Care Organization, Provision and Financing* 13 (1976): 3–14.

DOUGLAS-WILSON, IAN, and MCLACHLAN, GORDON, eds. *Health Service Prospects—An International Survey Published on "The Lancet's" 150th Anniversary in October 1973.* Boston: Little, Brown & Co., 1973; London: Lancet, Nuffield Provincial Hospitals Trust, 1973.

FRY, JOHN, and FARNDALE, WILLIAM ARTHUR JAMES, eds. *International Medical Care: A Comparison and Evaluation of Medical Care Services throughout the World.* Oxford: Medical & Technical Publishing, 1972.

FUCHS, VICTOR R. *Who Shall Live? Health, Economics, and Social Choice.* New York: Basic Books, 1974.

GLAZIER, WILLAM H. "The Task of Medicine." *Scientific American* 228, no. 4 (1973), pp. 13–17.

MAXWELL, ROBERT. *Health Care: The Growing Dilemma.* A McKinsey Survey Report. New York: McKinsey & Co. Available only from this firm of management consultants.

MECHANIC, DAVID. *Medical Sociology: A Selective View.* New York: Free Press, 1968.

OUTKA, GENE. "Social Justice and Equal Access to Health Care." *Journal of Religious Ethics* 2, no. 2 (1974), pp. 11–32.

Parker, Alberta W. "The Dimensions of Primary Care: Blueprints for Change." *Primary Care: Where Medicine Fails.* A Wiley Biomedical-Health Publication. Edited by Spyros Andreopoulos. Foreword by Merlin K. DuVal. New York: John Wiley & Sons, 1974, pp. 15–80.

Saward, Ernest W. "The Organization of Medical Care." *Scientific American* 229, no. 3 (1973), pp. 169–175.

United States, Department of Health, Education, and Welfare, Public Health Service. *Forward Plan for Health—FY 1977–81: June 1975.* DHEW Publication no. (OS) 76-50024. Washington: 1975.

White, Kerr L. "International Comparisons of Medical Care." *Scientific American* 233, no. 2 (1975), pp. 17–25.

II

HUMANIZATION AND DEHUMANIZATION OF HEALTH CARE

Numerous critics of twentieth-century America contend that relationships among people in various walks of life have become increasingly dehumanized. In health care the contrast between humanitarian ideals and the reality of antihumanitarian practices has magnified moral concern about dehumanization and intensified the quest for alternatives to existing styles of care.

Advocates of change have highlighted a variety of problems: inequities in the delivery of services, practitioner indifference to the emotional needs of patients, and conflicting value priorities in professional schools, which subvert commitments to humanism (Becker and Geer; Kendall and Jones).

Depersonalization of the health professional is another concern. As the locus of diagnosis and treatment shifts from the offices of independent practitioners to large-scale institutions, providers of health care themselves suffer loss of autonomy, individuality, and personal responsibility (Halberstam; Mills). Their degrees of freedom are also restricted by the intrusive hands of third parties such as insurance companies, government agencies, and peer review committees.

Connotations in the literature

Although dehumanization is a widely used concept, specific definitions of it are rare. Among the exceptions are the views of dehumanization as a "loss of human attributes" (Vail) and as a "loss of dignity" (Lewis).

In most writings "dehumanization" and "depersonalization" are used interchangeably, but some scholars distinguish between them. Thus, Leventhal defines *depersonalization* as the splitting of the psychological and physical self and *dehumanization* as the feeling that one is isolated from others and regarded as a thing rather than a person (Leventhal, p. 120).

According to Hayes-Bautista, personalization–depersonalization refers to the relative number of human providers in attendance; whereas humanization–dehumanization refers to the relative degree of humanism practiced by those present. Therefore, services can be relatively depersonalized but humanized, or relatively personalized but dehumanized. In the latter case, the participation of many providers may heighten the dehumanization.

A review of relevant literature suggests that the terms "dehumanization" and "depersonalization" have a wide variety of ethical connotations, which are usually implicit rather than explicit in these writings (Howard). The concepts refer to the process of objectification in human perception, to man's exploitation of fellow man, to indifference or coldness in human interaction, to the repression and constriction of human freedom, and to social ostracism and alienation.

When dehumanization connotes objectification, it means that people are being perceived and treated as *things*—as unfeeling quantifiable objects with standardized parts and wholes. More specifically it may mean that people are being viewed as *machines*. Out of necessity patients may be forced to "interact" with hardware instead of human providers and may themselves be seen as extensions of tubes, respirators, and monitors or, in the case of transplant patients, as mere receptacles for foreign and artificial organs.

Another form of objectification occurs when patients are perceived as *pathologies*. As a result of the professionalization process and specialization, health workers tend to stereotype patients as organs or diseases (as "gall bladders" or "hypertensives") rather than whole people. The providers are focusing on their own priorities, not on their patients' perspectives and needs.

When dehumanization refers to exploitation in medical settings, it signifies that people are being used instrumentally without regard for their pain and suffering, as guinea pigs are used. The prototype here is the Nazi experience, wherein the "doctors of infamy" conducted medical research in concentration camps (Mitscherlich and Mielke). Analogies have been drawn in the United States when subjects have been coerced into participation and have not been adequately informed of the nature of the experiment, the risks involved, or alternative treatments that could save their lives.

A more subtle mode of exploitation takes place when patients and providers are degraded and humiliated as *nonpersons* or *lesser persons*. Lesser people are branded by an ascribed or acquired characteristic (such as skin color, sex, or mental illness), which becomes the rationale for discrimination. They are considered undeserving of the rewards, perquisites, and standard of care to which full-status persons are entitled, and accordingly lower priorities and reduced benefits are allocated to them.

"Depersonalization" commonly implies an *absence of warmth* in human interaction. Emotional distance is alleged to be functional for professionals who deal with sick patients. But the lack of visible feeling among health workers may be perceived by patients as cold indifference, not detached concern, and a denial of the humanness of server and served alike.

At the core of most descriptions of dehumanized patients and personnel is powerlessness and *loss of autonomy*. Individuals are not captains of their fate. They are coerced, coopted, and manipulated into conformity. Architects and planners use the term "dehumanized" to refer to *static and sterile medical institutions* that force human beings to behave in unnatural or unusual ways, thereby constricting their freedom, growth, and completeness. Thus, patients may be restricted to their sleeping quarters or nearby halls and be compelled to visit with intimates under circumstances that preclude privacy. In contrast to this regimentation, humanizing environments respond to the comprehensive and changing needs of patients. Structures and routines are flexible and accommodating because designers have anticipated human variability and the full range of social and asocial behaviors.

Related to the idea of dehumanization as a loss of options is its reference to human beings who are *denied the ultimate option*—the opportunity for life or the opportunity for death. The development of technologies to extend "life" has fostered debate concerning the moral wisdom of preserving lives where the capacity for social functioning is seriously impaired. The concept of dehumanization is used by both sides. One group says it is dehumanizing to let death occur or to terminate a life that can be maintained. Their critics say it is dehumanizing to maintain lives that are not worth living.

Last, the term "dehumanization" connotes *isolation* or *abandonment* of patients, especially when confinement is involuntary, unanticipated, long-term, and associated with a hopeless prognosis. In these situations caretakers tend to be undertrained, overworked, and drained of whatever emotions they permit themselves to feel. The "turning off" process feeds on itself, isolating and alienating providers as well as patients.

Ingredients of humanized care

The various meanings of dehumanization and depersonalization are suggestive by implication of the meanings of humanization. To some extent at least, the ingredients of humanism vary according to time and culture. For present-day American society the following eight conditions appear to be necessary and sufficient for humanized care (Howard).

Three "images of man" are essential in a humanized health-care relationship. Patients must be viewed as *unique* and *irreplaceable whole persons, inherently worthy* of the caretaker's concern. A humanistic relationship also involves three aspects germane to the social structure of the interaction. Within the limits imposed by physical and cultural constraints, patients must be *autonomous* persons who have significant control over their destinies; they should *share in decisions* affecting their care; and their relations with providers should be *equalitarian* or *reciprocal*, not deferential or patronizing. Finally, two emotional factors are required: practitioners must feel and express *empathy* and *warmth* (positive affect) for their patients.

These eight conditions are derived from ideas in the literature, interviews with patients, and theoretical speculation aimed at limiting and integrating salient concepts. Yet the model is tentative and open to criticism on certain grounds discussed below.

In regard to the condition of "inherent worth," a relevant counterargument is that some people are more valuable to society than others and should accordingly be given higher medical priority and preferential treatment (Feingold). Such discrimination against persons of lesser worth could prolong their suffering and further undermine their value to society. Thus, the dehumanization would perpetuate itself, especially if those discriminated against condoned their low priority.

The condition of patient uniqueness may be regarded as somewhat contrary to the universalistic ethic that "a patient is a patient is a patient" from the formal standpoint of provider obligation (Parsons, 1964). But the two mandates are not contradictory. The universalistic ethic says that all patients are equally worthy of being defined as patients; whereas the ethic of uniqueness says that everyone entitled to care also deserves to be treated as an individual with unique attributes, needs, and desires.

Even when they appreciate the "wholeness" of patients, health workers are forced by limited skills and resources to narrow their focus of concern. At some point their responsibility ends, and society's duty begins. Moreover, a humanistic approach to patients recognizes their right to privacy, which comprehensive care may infringe upon. Knowledge of another's secrets is a form of latent power that can be dehumanizing on its face if in sharing secrets with caretakers the patient experiences a violation of self (American Psychiatric Association).

In postulating freedom as a condition of humanized care, one must allow for limitations. If there are no restrictions on the sovereignty of patients, their demands may dehumanize providers by imposing on their freedom. And where caretakers and resources are scarce, one patient's freedom to fulfill his needs may nullify the options of others, as in a zero-sum game. It would be false to assume, however, that freedoms are always in conflict. The humanizing effects of autonomy can be additive. Behavioral choices among in-patients can lessen their dependency, alienation, and loss of identity, thereby facilitating a functional return to the outside world for everyone involved. Patient independence can also enhance the freedom of health professionals who are absolved of the burden of providing total care.

The participation of patients in decisions affecting their care is a humanitarian ideal that may have dehumanizing overtones. Empirical evidence suggests that sick people cannot psychologically absorb all the information necessary to help make rational decisions (Ley and Spelman; Howard and Tyler). And when patients do understand therapeutic options and prognoses, their anxiety and fear may so immobilize them that they merely rubber-stamp their practitioners' recommendations. A significant number of health workers believe that in the name of humanism terminal patients should be spared the whole truth even if this denies them the opportunity to plan for death.

Equalitarianism in practitioner–patient relations may be impossible to achieve because the expertise of professionals may categorically give them superior social status (Freidson, 1967, 1970). Patient vulnerability and dependency can strengthen this latent power of professionals. But there are equalizing forces such as fiduciary exchange (fee for service), consumer collective bargaining, the high economic and social prestige of many patients, and mandates that health professionals serve humanity without exploitation. These factors lead to mutual dependency and reciprocity, thereby increasing the probability of humanistic relationships between providers and recipients of care.

The concepts of empathy and affect are perhaps the most debatable ingredients of humanized care. If health workers really put themselves in their patients' shoes, they might totally expend their emotional energy and be ill-equipped to handle the demands of practice. Affective involvement with patients can threaten objective decision making and the authoritative posture that may be necessary to assure therapeutic compliance. Thus, practitioners are admonished to be affectively neutral in feelings and actions (Parsons, 1953, 1964), a prescription that may depersonalize the providers as well as their patients. Some compromise is called for, permitting genuine but restrained emotional commitments to patients.

Causes of dehumanization

Dehumanization in health care is linked to dehumanization in the larger society, because the institutions that deliver care are not closed systems. Their goals, rules, rewards, and authority structures resemble those of other institutions to which they are formally and informally related. They also share common ideologies concerning the appropriate allocation of resources and the just distribution of goods and services. Thus, the same social forces that contribute to dehumanization in economic and political milieus lead to dehumanization in health care. These include five interdependent processes: aggregation of services, bureaucratization of services, secularization of values, professionalization of skills, and proliferation of technologies.

The pressure to centralize facilities and manpower is partially the result of the technological revolution and the escalating cost of medical hardware. It is also fostered by rapid and massive increases in medical knowledge and the specialization necessary to master that knowledge. When practitioners with diverse skills combine under one roof, they can more readily consult and divide the cost of diagnostic and therapeutic tools. But health systems of skyscraper size have impersonal faces and interiors (Lindheim). They are pressed in the name of efficiency to standardize the process of care, to convert individuals into ciphers, and to develop secondary forms of communication to span the vast distances. Anonymity and "anomie" (Durkheim's term for social disengagement) are frequent by-products.

Bureaucracy goes hand in hand with aggregation. Work tasks are rationalized and routinized so workers can be rotated and replaced. Rules are formalized and procedures ritualized to guarantee conformity to written words. Paper work competes with sick people for the attention of disinterested professionals who compile and copy reams of information about patients. The computer steps in to process these data for a variety of relevant parties (U.S. DHEW). Responsibility for patient care is diffused and may be abdicated if someone points the finger of blame. Selves are lost in the system and stalemated by inertia.

The secular trend in health care can be seen in the transfer of control from voluntary to proprietary institutions where absentee landlords rule the roost and the primary goal is profit (Howard and Tyler). As the health industry commands a larger and larger share of the gross national product, competition for the dollar supersedes humanitarian considerations unless humanism is a route to monetary rewards.

Professionalization is a humanizing force when it improves the quality of care, but it also has dehumanizing consequences. Most important is the monopoly over the healing market granted to select professions. This restricts the number of legitimate care-givers and the scope of their activities, forcing patients to choose from the chosen. The "pecking order" among the professions confers rank in accord with technical ability rather than human skills. And, paradoxically, those caretakers who have the most contact with patients as aides and gatekeepers are usually people with low professional status and little if any training in human relations.

A multitude of dehumanizing ideologies permeate health systems and are often internalized by the victims themselves. These include the following beliefs: that health care is a privilege rather than a right; that certain illnesses are morally reprehensible; that terminal illness justifies pariah status; that physicians are omnipotent; and that the ability of health workers to provide tender loving care is inversely correlated with their technological competence.

The thrust toward dehumanization in health services is not without its counteroffensive. The mere recognition of the need for change is a first step. More concrete catalysts can be found in the movements for national health insurance, consumer power, women's liberation, and decentralized medicine. National health insurance could elevate the "lesser" people to full entitlement. Consumer participation in health enterprises, beyond token representation, could elevate the priority of humanism. The rising tide of sex consciousness among female providers and patients could free them from the shackles of stereotyped status and make them more sovereign shareholders in the process of care.

Geographical decentralization could ease access, facilitate pluralistic responses to community needs, and humanize the physical context of care. The proliferation of new health roles typified by nurse practitioners and physician assistants is another form of decentralization. Segments of physician expertise and power are diffused to satellite professionals.

Even the shift from voluntarism to a profit orientation could be a humanizing influence if dissatisfied patients had the option of going elsewhere. In a competitive market consumers and providers would seek environments which enhanced their dignity. Dehumanizing systems would have to change or offer enticing tradeoffs such as lower costs or better care. And over the long run experience might very well show that humanistic approaches to care are actually less costly, by any standard, than dehumanizing alternatives.

JAN HOWARD

[*For further discussion of topics mentioned in this article, see the entries:* HOSPITALS; MEDICAL PROFESSION; PATERNALISM; PATIENTS' RIGHTS MOVEMENT; RACISM; *and* THERAPEUTIC RELATIONSHIP. *Also directly related is* CARE. *Other relevant material may be found under* MENTAL ILLNESS, *article on* LABELING IN MENTAL ILLNESS: LEGAL ASPECTS; MENTAL HEALTH SERVICES, *article on* SOCIAL INSTITUTIONS OF MENTAL HEALTH; *and* WOMEN AND BIOMEDICINE, *article on* WOMEN AS PATIENTS AND EXPERIMENTAL SUBJECTS. *See also:* CONFIDENTIALITY; DEATH AND DYING: EUTHANASIA AND SUSTAINING LIFE; PRIVACY; RATIONING OF MEDICAL TREATMENT; *and* TRUTH-TELLING.]

BIBLIOGRAPHY

American Psychiatric Association. Conference on the Confidentiality of Health Records, Key Biscayne, Florida, 6–9 November 1974. Working Papers. Official report. Natalie Davis Spingarn. *Confidentiality: A Re-*

port of the 1974 Conference on Confidentiality of Health Records. Washington: 1975.

BECKER, HOWARD S., and GEER, BLANCHE. "The Fate of Idealism in Medical School." *American Sociological Review* 23 (1958): 50–56.

FEINGOLD, EUGENE. "Strategies for Research: Commentary on Jan Howard's Conceptual View." Howard, *Humanizing Health Care*, pp. 109–113.

FREIDSON, ELIOT. *Professional Dominance: The Social Structure of Medical Care*. New York: Aldine-Atherton Publishing Co., 1970.

———. "Review Essay: Health Factories, the New Industrial Sociology." *Social Problems* 14 (1967): 493–500.

HALBERSTAM, MICHAEL J. "Liberal Thought, Radical Theory and Medical Practice." *New England Journal of Medicine* 284 (1971): 1180–1185.

HAYES-BAUTISTA, DAVID. Personal comments at Symposium on Humanizing Health Care, San Francisco, 1–2 December 1972. Sponsored by DHEW, National Center for Health Services Research and Development.

HOWARD, JAN. "Humanization and Dehumanization of Health Care: A Conceptual View." Howard, *Humanizing Health Care*, pp. 57–102. Bibliography.

———, and STRAUSS, ANSELM, eds. *Humanizing Health Care*. New York: Wiley-Interscience, 1975.

———, and TYLER, CAROLE C. "Comments on Dehumanization: Caveats, Dilemmas, and Remedies." Howard, *Humanizing Health Care*, pp. 231–257.

KENDALL, PATRICIA L., and JONES, JAMES A. "General Patient Care: Learning Aspects." *Comprehensive Medical Care and Teaching: A Report on the New York Hospital–Cornell Medical Center Program*. Edited by George G. Reader and Mary W. Goss. Ithaca: Cornell University Press, 1967, pp. 73–120. See especially pp. 84–86.

LEVENTHAL, HOWARD. "The Consequences of Depersonalization during Illness and Treatment: An Information-Processing Model." Howard, *Humanizing Health Care*, pp. 119–161. Excellent bibliography.

LEWIS, CHARLES E. "A Physician's Perspective." Howard, *Humanizing Health Care*, pp. 263–268.

LEY, P., and SPELMAN, M. S. "Communications in an Out-Patient Setting." *British Journal of Social and Clinical Psychology* 4 (1965): 114–116.

LINDHEIM, ROSLYN. "An Architect's Perspective." Howard, *Humanizing Health Care*, pp. 293–304.

MILLS, C. WRIGHT. "Old Professions and New Skills: The Medical World." *White Collar*. New York: Oxford University Press, 1951, pp. 115–121.

MITSCHERLICH, ALEXANDER, and MIELKE, FRED. *Doctors of Infamy: The Story of the Nazi Medical Crimes*. Translated by Heinz Norden. New York: Henry Schuman, 1949.

PARSONS, TALCOTT. "Illness and the Role of the Physician: A Sociological Perspective." *Personality in Nature, Society, and Culture*. 2d rev. ed. Edited by Clyde Kluckhohn and Henry A. Murray. New York: A. Knopf, 1953, pp. 609–617.

———. *The Social System*. New York: Free Press, 1964.

United States; Department of Health, Education, and Welfare. *Records, Computers, and the Rights of Citizens*. Report of the Secretary's Advisory Committee on Automated Personal Data Systems. DHEW Publication no. (OS) 73–94. Washington: Government Printing Office, 1973.

VAIL, DAVID. *Dehumanization and the Institutional Career*. Springfield, Ill.: Charles C. Thomas, 1966.

III
RIGHT TO HEALTH-CARE SERVICES

Humanization is the first important ethical concern associated with health-care services: It centers on respect for persons and on the provision of services which reflect that respect. The second important ethical consideration concerns the distribution of services in an equitable manner throughout a society; ethical analysis here will center on the concept of distributive justice: Can health services be provided in such a way that the benefits and burdens of these services fall fairly on persons and classes in a society? Since the concept of distributive justice is treated more fully elsewhere, this article will discuss the right to health care as a bioethical example of distributive justice: Is it reasonable to assert that persons have a "right" to health care such that others, either individuals or communities, have an obligation to make such care available?

The Universal Declaration of Human Rights, Article 25, 1, adopted in 1948 by the United Nations, declares "everyone has a right to a standard of living adequate for the health and well-being of himself and his family, including food, clothing, housing, medical care and necessary social services." This statement was expanded in the International Covenant on Economic, Social, and Cultural Rights (Article 12), adopted by the General Assembly of the United Nations in 1966:

(1) The States Parties to the Present Covenant recognize the right of everyone to the enjoyment of the highest attainable standard of physical and mental health.
(2) The steps to be taken by the States Parties to the Present Covenant to achieve the full realization of this right shall include those necessary for:
 (a) the provision for the reduction of the stillbirth rate and of infant mortality and for the health development of the child;
 (b) the improvement of all aspects of environmental and industrial hygiene;
 (c) the prevention, treatment and control of epidemic, endemic, occupational and other diseases;
 (d) the creation of conditions which would assure to all medical services and medical attention in the event of sickness [Robertson].

Similar statements appear in other international declarations and in the pronouncements of various public and private groups. For example, the American Medical Association has resolved that "it is the basic right of every citizen to have available to him adequate medical care" (American Medical Association).

The present article will (1) set forth the concept of rights in general; (2) review the arguments concerning the right to health care in particular; and, (3) describe some attempts to translate the affirmation of such a right into public policy.

Rights

The word "rights" has a consecrated place in the vocabulary of democratic governments, but not without controversy. During three centuries of widespread use in political and philosophical discussion, it has represented both the demands of the dispossessed for access to possession and the claim of the possessors to continued tenure. Debate has raged over the nature and justification of these claims and demands. Disagreement and confusion reign about the object of rights, which range from the concrete right of property to the abstract pursuit of happiness. Between the extremes of the debate, a wide area of consensus about human and civil rights has made possible the creation of relatively equitable laws and structures for the administration of justice in democratic nations. Even there, however, the application of the concepts of rights to particular situations has stirred controversy. Opinions about the nature, origin, and extent of rights are often so confused as to merit Bentham's scoff, "natural rights is simple nonsense . . . rhetorical nonsense, nonsense upon stilts."

Although the notion of rights was not unknown previously, its history begins significantly in the seventeenth century (Strauss; Burns). The history can be described broadly as moving through four phases, which are alternately philosophical and political. The first phase, inspired by John Locke (1632–1704), consisted in setting forth arguments for the independence of the individual, in certain essential aspects, as against ruler and government. Those essential aspects were, in Locke's formulation, life, liberty, and property. This first, more theoretical phase moved into a second, more practical one when political systems came into existence with the explicit intent of establishing social institutions which might give actual realization to the "unalienable rights to life, liberty and the pursuit of happiness" (American Declaration of Independence, 1776) of the "natural, imprescriptible and inalienable rights . . . liberty, property, security and resistance of oppression" (French Declaration of the Rights of Man and of Citizens, 1789).

Other rights were spelled out, and civil and social arrangements devised to implement them: the rights of freedom of religion, expression, assembly, due process before the courts, representative government, etc. The first two philosophical and political phases evolved as part of the complex movements in Western civilization away from religious and political absolutism toward more participatory forms of ecclesiastical and civil life. The formulation of rights reflected these movements. Rights define spheres of personal autonomy that authorities are forbidden to enter and, indeed, are obliged to protect. Rights are conceived as titles of an individual to protection of life, liberty, and possessions. Hugo Grotius provided a definition that became classical: "A right is a moral quality of a person entitling him justly to possess or to perform something" (Grotius).

The third and fourth phases of the history of the notion of rights begins at the end of the eighteenth century. Even as the great declarations of rights were being proclaimed, several voices called for a wider conception. Tom Paine, in *Rights of Man* (1791–1792) and *Agrarian Justice* (1797), proposed a broad social security system as a matter of right, not of charity. The *Manifeste des Egaux* of the Babeuf Conspiracy (1796) proposed the right to work and the right to education. Nineteenth-century socialists developed these themes, proposing social reforms that would not only protect the rights of privacy and possessions but also provide the means of a decent life to those in need. In the fourth phase these proposals, coming from a variety of ideological positions, began to be realized as many countries gradually initiated programs of social security, unemployment insurance, and welfare programs. These rights and their social implementation were acknowledged in the Universal Declaration of Human Rights of the United Nations (1948). The International Covenant on Economic, Social, and Cultural Rights (1966) distinguished such rights from the more traditional civil and political rights. The third and fourth phases of the history of the notion of rights add to the protection of personal autonomy the additional positive obligation of a government to provide for the enhancement of that autonomy. The so-called economic, social, and cultural rights include the right to work, to organize unions, to fair remuneration, to social security, to free education, to adequate housing, to nutrition and medical care. It is common to speak of the rights of the first and second phases of this history as "natural rights," which, in their institutional realization, are approximated by "civil and political rights." The rights of the third and fourth phases are frequently called "human rights," which are recognized in "economic, social, and cultural rights" (UNESCO; Cranston; Raphael).

Philosophers and jurists have discussed extensively the nature of legal and moral rights. This complex discussion cannot be rehearsed here (Muirhead; Benn). However, out of the discussion certain distinctions and generalizations pertinent to the right to health can be abstracted. It is usual to distinguish between special rights, arising from particular transactions or relationships, such as promises, contracts, or parenthood, and general rights, pertaining to all persons prior to any particular transactions or relationships. One who enters a contract has a special right to fulfillment of its terms by the other party; both parties have a general right to honest and full disclosure prior to any contractual engagement. Another distinction is between legal rights and moral rights. The former are created and sanctioned by enactments of governmental authority; the latter are asserted as valid apart from, and morally prior to, any legal rule. Moral rights are sometimes invoked to demand some legal enactment and are often invoked in criticism of existing law. This article maintains that, insofar as there is a right to health care, it would be a general moral right, although in many countries a legal right has been enacted in recognition of the moral right.

Certain generalizations can be made about the concept of rights as it appears in the philosophical discussions and political debates. First, the concept of rights refers not only to the single individual, as do many other moral notions, such as duty, responsibility, or virtue. It rather refers to the obligation incumbent upon other persons in respect to the individual claiming the right. This is expressed in the maxim: Rights and duties are correlative. A classical definition underlines this feature: A right is an active moral power of a person to receive something from another as a matter of moral necessity (Pufendorf). A modern author suggests that "a man has a right whenever other men ought not to prevent him from doing what he wants or refuse him some service he asks for or needs" (Plamenatz, Lamont, and Acton, p. 75).

A second generalization is that a right is usually asserted as a strong moral justification for possessing, doing, or receiving something of considerable importance in human life. The philosophical theories that have been offered to justify rights stress this point. In the natural law doctrine, espoused by Locke, Grotius, and Pufendorf, long the most congenial home for the doctrine of rights, rights were conceived as those spheres of autonomy and freedom of action within which the natural and divinely posited ends of human persons could be attained (Maritain). Many contemporary theories of rights are inspired by Kant's notion of the autonomy of the person. In these theories, autonomy means that each person is a unique center of reflection and affection and an original source of action. In this, all persons are equal. Respect for this autonomy is the origin of all moral judgment, insofar as respect recognizes the other as equally a person as oneself.

Certain rights can be seen to follow closely upon the affirmation of autonomy: the right to equal protection under law, to due process, to equal consideration of interests, and the so-called negative rights, not to be treated cruelly or to be exploited as chattel. "Rights in this category," writes Feinberg, "are probably the only ones that are human rights in the strongest sense: unalterable, 'absolute', exceptionless and non-conflictable, and necessarily and peculiarly human" (1973, p. 97).

Rights, understood primarily as manifestations of the autonomy of persons reserved from interference by government, were formulated largely during the first and second phases of the history related above. However, in recent years it has become obvious that the complexity of industrial society has made many of these traditional rights problematic. The autonomy of persons which they affirm is often frustrated by conditions of social and economic life. It is often inimical to adequate social cooperation. The expanded conception of rights that developed during the third and fourth phases of the history emphasized the development and enhancement of personal autonomy, as well as its recognition and protection. Thus, the theoretical justification begins to turn on human needs, understood as those possibilities for autonomy which can be realized only under certain social and material conditions.

This shift in emphasis from autonomy to need as the basis of certain moral rights requires some reformulation of traditional arguments. First, need is an elastic term, varying greatly with different persons and situations. Needs may be capricious, transient, trivial, and highly idiosyncratic. It seems clear that a need as such does not generate the right to its satisfaction. Also, needs are often satisfied by material goods and services, which may be plentiful, adequate, scarce, or nonexistent. It seems peculiar to claim a right to the plentiful or to the nonexistent, while claims on the adequate or scarce seem inevitably to collapse into arguments over priorities of need. In addition, the aggregate of personal needs may require satisfaction that would be socially inefficient or even damaging. Finally, material goods and services are produced by persons who themselves make claims to their own products and their allocation. Great complexity is introduced into the theory of rights by these features of human need.

Human rights

Arguments in favor of natural rights center on the notion of autonomy. Those in favor of human rights add the notion of need, with the complexity noted above. Is it possible, despite this complexity, to formulate arguments that make the affirmation of economic and social rights reasonable and morally compelling?

There are several prerequisites to any reasonable argument in favor of social and economic rights. First, need must be defined in a way that reduces its elasticity. Proponents of such rights have attempted to do so by identifying certain human needs as fundamental, such as nutrition, housing, and health. While these attempts are not altogether satisfactory, they do identify areas of critical importance for human living. Second, limitation of resources, efficiency of allocation, and levels of priority are realistically recognized by asserting that no one person or group has an unlimited claim to satisfaction of needs but only to "equal opportunities" within fairly constituted social institutions. Finally, the rights that producers of goods and services claim over their products can be acknowledged by systems which allow both for fair compensation and for adequate and equitable supply. In democratic states these variations are effected by regulated free markets, in communist countries by centrally planned social and economic systems.

Each of these approaches is conceptually and practically problematic. Unlike the first four phases of the history of rights, in which theory anticipated practice, current social and economic practices almost universally precede their speculative justification. Various institutions have been established that realize, at least partially, social and economic rights. Oddly enough, in some nations these rights are more fully acknowledged than the traditional civil and political rights.

These approaches have several consequences for the notion of rights. Social and economic rights are much less closely attached to the individual than civil and political ones. They are more susceptible of balancing and rationing in their realization. They engender obligations, not in all others, as do the civil and political rights, but only in those responsible for production and allocation of goods and services. They can be realized by gradual implementation. Thus, a legal right such as due process implements a moral claim that is owed to every individual, as such, in every formal and public consideration of his or her case. This right is not to be balanced or rationed or gradually implemented but is an absolute, full, and immediate moral recognition of one's autonomy as a person. On the other hand, the right to education is conditioned by the potentials of individuals, by availability of teachers and materials, by cultural differences, and by other social needs.

It might be said, then, that rights known as economic and social are moral claims, made by persons or on behalf of persons, that social institutions be so arranged that material goods and services can be obtained with reasonable effort, without detriment to dignity and self-respect and in such a way as to promote individual freedom and social cooperation. The correlative obligation is incumbent upon those who command the skills for producing these goods and services and, more important, on those who hold responsibility for the policymaking and administration of social institutions. Social and economic rights are mediate and more remote implications of the basic right to be respected as an autonomous person—mediate because it is necessary to show, by certain empirical arguments, how the particular right, such as health care or education, does relate to personal autonomy; more remote because, unlike civil and political rights, they may be left unrealized for individuals because of material conditions, without necessarily disdaining their autonomy as persons.

One author writes of rights based upon need:

A natural need for some good as such . . . is always in support of a claim to that good. A person in need is always "in a position" to make a claim, even when there is no one in the corresponding position to do anything about it. Such claims, based on need alone, are the "permanent possibilities of rights," the natural seed from which rights grow. When manifesto writers speak of them as if already actual rights, they are easily forgiven, for this is but a powerful way of expressing the conviction that they ought to be recognized by the state here and now as potential rights and consequently as determinants of *present* aspirations and guides to *present* policies [Feinberg, 1970, p. 255].

Whether social and economic rights can be called "human" rights or only moral rights, whether they be called "possibilities of rights" or conditional rights, their proclamation does manifest heightened sensitivity to the moral problem of being human and living humanly in the contemporary world.

Right to health

John Locke, in some passages, added to "life, liberty, and property" two other rights: "integrity of body" and "freedom from bodily pain." He was maintaining, of course, only that all persons should be made secure against assaults that could damage their health or destroy their lives (Locke).

However, as industrialization and urbanization made social living more complex, threats to health other than physical assault were posed by human agency, although often unknowingly. Pollution, crowding, increased mobility, and working conditions constituted health dangers. Johann Peter Frank wrote his *Complete System of Medical Policy* (1779–1827), in which comprehensive measures for public health were detailed. Governments everywhere instituted programs of water control and sanitation, immunization and quarantine, food and drug inspection, and industrial safety regulations, as it became clear (or was suspected) that conditions in those areas threatened health and that single individuals as such could not protect themselves against them. Such measures, with the addition of fertility control, remain the most important recognition of a right to health in its most basic sense: a recognition of the autonomy of persons within a community of persons (Symposium on Human Rights). In England, Sir Edwin Chadwick (1800–1890) stimulated major reforms in sanitation, housing, and labor conditions. In the United States a federal quarantine law was enacted in 1796 after several epidemics of yellow fever. Congressman Lyman noted its passage with the remark that "the right to the preservation of health is inalienable" (Chapman and Talmadge). The United States soon afterward established the Marine Hospital Service to care for sick sailors. Through the nineteenth century private and public efforts were made to improve social conditions that affected health (Rosen).

The "right to health," then, belongs to the natural moral rights as an implication of the right to preservation of life. It consists of measures taken by authorities to protect a populace from recognized dangers to health that cannot be warded off by individual effort. The affirmation of a right in this sense does not detract from the importance, sometimes termed a moral duty, of the individual's care for his own health, when it can be effected by personal behavior. Neither does a right to health in this sense prove that an individual has a right to medical care, if medical care is taken to mean diagnosis and treatment, by scientific means, of the illness of an individual.

Right to medical care

In the tradition of Western medicine, medical care has been provided to individuals either on the basis of a contract between physician and a patient, or as a matter of religious charity, or occasionally as a public welfare benefit. Recently medical care for personal illness and disability has been asserted to be the subject of a right. This implies that neither the monetary considerations of a contract, nor charitable benevolence, nor the stigma of poverty is the moral basis for medical care. The right to medical care affirms that no person should lack medical care because of inability to pay for it. It implies that there should not be two systems of care, differing in quality and quantity, for those who can pay and those who cannot. Further, no person should lack care because the social, cultural, and geographical organization of medical care makes it difficult to gain access to providers.

In the late nineteenth century nations began to recognize these problems. Beginning in 1883 Germany, under Chancellor Bismarck, enacted broad programs of sickness, accident, and old-age insurance. Certain nations, such as the Soviet Union and the People's Republic of China, have dramatically rearranged the structure of medical care to overcome these obstacles. Other nations, such as Great Britain and Sweden, have modified this structure in a variety of ways. The United

States is moving toward a reduction of barriers to care principally by a gradual extension of the financing of care by private and public insurance (Anderson). In many countries the impetus for these social changes has come from labor parties and movements where the importance of health for employment is clearly recognized.

The philosophical argument for such a right arises from the intuitive proposition that illness is the sole reason for medical treatment. It seems unreasonable and unfair to add ability to pay as a necessary condition of treatment, for there is no rational justification for distinguishing between rich ill and poor ill for this purpose. All who are ill, then, deserve to be treated equally (Outka).

A number of arguments might be proposed against this line of reasoning. Providers of care may contend that medical skills are personal accomplishments and that those who acquire them have the right to determine their allocation (Nozick; Sade). The differing seriousness of illness and the differing effect on individuals and upon the society may require grounds of discrimination other than illness itself. The uncertain effect of medical care in curing illness, and even more in preserving health, casts doubt on whether illness should engender a moral obligation to be treated (Kass). The fact that many illnesses are due to voluntary persistence in deleterious habits clouds the title of the patient to publicly provided care.

All these counterarguments have merit. Any reasonable argument for a right to medical treatment must take account of both the intuitive position and the counterarguments. Such an argument might proceed as follows:

1. The health of individuals, defined as maximal attainable psychosomatic freedom of action, is an important ingredient in their self-regard as autonomous persons and a significant, if not inevitable, basis for respect of that autonomy by others.
2. The health of individuals is an important ingredient of the well-being of a society, contributing to its cultural and political vitality and supporting its economy. In particular, differences in health status among social classes can reinforce other inequalities which seriously undermine social well-being.
3. To the extent that medical care, on an assessed basis, can contribute to the health of individuals and thus improve the possibilities of self-respect and of equality, there is a moral obligation, arising from the basic moral obligations of respect and justice, to arrange social institutions so that care is available to all in equal need and so that such care can be provided in a manner conducive to self-respect. It can be said, then, that all persons have a moral right to medical care.

The phrase "on an assessed basis" is the key to any social policy that would implement this right. It means that an assessment of several empirical factors must be made. First, questions of the effectiveness of medical treatment must be answered by continual studies (Cochrane). Second, estimates of the impact of various illnesses on individuals and upon the society as a whole must be made. Third, costs of certain treatment forms and the possibilities and costs of alternatives, medical or nonmedical, must be estimated. Finally, the impact of various degrees of regulation and control upon the efficacy of medical care, upon the morale of providers, and upon the humaneness of the provision of care must be thoughtfully considered. This assessment will allow a reasonable, if not perfect, definition of the conditions under which the right to medical care can be implemented. Since the right to health care and to medical care has been classified as an economic and social right, it is, as we have suggested, properly susceptible to such assessments and discrimination. It is an assertion that it is profoundly immoral to withhold the means requisite to relief of pain and cure of illness from those in need. It is, at the same time, a reminder that these means are not unlimited, are relatively costly, and compete with other social goods.

Even when a right to health care and to medical care is acknowledged, their implementation will differ greatly in different nations. Cultural attitudes toward health and medicine, economic resources, availability of manpower, the form and authority of government will have significant influence on the implementation of health rights. Also, the needs of the population, its geographical environment, food and water resources, and style of life are important indicators for policies which implement health rights.

The implementation of health rights requires consideration of all the elements of health services. Their sociological, technological, political, and economic organization must be understood. To the extent that these are amenable to policy decisions, judgments about priorities and trade-offs must be made and effective patterns of organization must be planned. Thus, the quality of care, access to care, and the cost of care are recognized as variables; primary, secondary, and tertiary care are recognized as correlative but separable modalities of medicine; preventive approaches are acknowledged as more influential on health status than high-technology therapies. The objective of all these considerations is to design or redesign the institutions providing health and medical care in such a way that major health needs of a society are met to the extent that medical science and technology can be most effective.

In developed countries these goals will be met quite differently from the way they are met in developing and underdeveloped nations. The problems in developed countries will concern the authority of health professionals, concentration on technological medicine providing acute care, and disproportionate costs resulting from both. In developing nations, problems will center on the institution of preventive measures and the provision of primary care, as well as upon the availability of financing and manpower (King, 1966, 1974).

In all nations, however, health rights can be implemented in limited ways. These rights do not, as we have noted, pertain to the individual as intimately as civic and political rights. Thus, each person cannot demand as a right that his or her disease, no matter how trivial or how exotic, be medically treated. Persons can demand as a right that there be sufficient trained professionals, with adequate and available resources, so that preventable illness can for the most part be prevented, treatable diseases for the most part be treated, and palliation of incurable disease be provided. Thus, while not all health needs of each person will be met, provision is made for meeting a range of needs and for dealing with many of the most debilitating and most traumatic problems, as these are viewed in a particular culture.

Conclusion

The right to health or medical care, then, is a moral, general, and conditional right. It is moral because it rests upon the assertion that personal autonomy and the conditions for its security and enhancement should be respected. It is general because it does not arise from a particular transaction or special relationship but is prior to these and lends them their moral quality. It is conditional because its implementation in social institutions depends upon the existing states of these institutions, upon existing material resources, and upon the continuing reassessment of the effectiveness of medical interventions.

The difficulty of defining and implementing a right to health care should not disguise its moral importance. Feinberg writes, "having rights enables us 'to stand up like men,' to look others in the eye and to feel in some fundamental way the equal of anyone" (1970, p. 252). Illness and disability severely challenge a person's physical and emotional autonomy. If the social and economic organization of health care makes the sick person not only a seeker of care and support but a mendicant, illness is a challenge to moral autonomy as well. The debility of illness, its concomitant social and economic disadvantages, and the humiliation of being dependent are the factual origins of a claim to the right to care, an affirmation of one's personal moral dignity in the crisis of illness. In addition, modern scientific medicine, which has a tendency to neglect the less dramatic but more socially useful fields of prevention, primary care, rehabilitation, and chronic care, needs the moral reminder that the health of all persons, not only the exotically ill, is its obligation. Right to health care may not be among the strong rights of the civil and political order. Still, it serves the valuable moral and political purpose of introducing respect and justice into social and economic institutions that manifest a growing complexity and indifference to personal need.

ALBERT R. JONSEN

[*While all the articles in this entry are relevant, see especially the following article on* THEORIES OF JUSTICE AND HEALTH CARE. *For further discussion of topics mentioned in this article, see the entries:* HEALTH AND DISEASE; JUSTICE; NATURAL LAW; *and* RIGHTS. *Also directly related are the entries* HEALTH INSURANCE; HEALTH, INTERNATIONAL; HEALTH POLICY; PUBLIC HEALTH; *and* SOCIAL MEDICINE. *This article will find application in the entries* ADOLESCENTS; AGING AND THE AGED, *article on* HEALTH CARE AND RESEARCH IN THE AGED; CHILDREN AND BIOMEDICINE; ENVIRONMENTAL ETHICS, *article on* ENVIRONMENTAL HEALTH AND HUMAN DISEASE; FOOD POLICY; GENETIC SCREENING; HOSPITALS; MASS HEALTH SCREENING; POVERTY AND HEALTH; PRISONERS, *article on* MEDICAL CARE OF PRISONERS; *and* RATIONING OF MEDICAL TREATMENT. *For discussion of related ideas, see the entries:* HEALTH AS AN OBLIGATION; MEDICAL ETHICS, HISTORY OF, *section on* EUROPE AND THE AMERICAS, *article on* EASTERN EUROPE IN THE TWENTIETH CENTURY; *and* PATIENTS' RIGHTS MOVEMENT.]

BIBLIOGRAPHY

American Medical Association, House of Delegates. *AMA House of Delegates Proceedings*, 23rd Clinical

Convention, 30 November–3 December 1969, p. 183. Resolution 62.

ANDERSON, ODIN WALDEMAR. *Health Care: Can There Be Equity? The United States, Sweden and England.* New York: Wiley, 1972.

BENN, STANLEY I. "Rights." *Encyclopedia of Philosophy.* 8 vols. New York: Macmillan Publishing Co., Free Press, 1967, vol. 7, pp. 195–199.

BROWN, STUART M., JR. "Inalienable Rights." *Philosophical Review* 64 (1955): 192–211.

BURNS, J. H. "The Rights of Man since the Reformation." *An Introduction to the Study of Human Rights.* Edited by Francis Aimé Vallat. London: Europa Publications, 1972, pp. 16–30.

CHAPMAN, CARLETON B., and TALMADGE, JOHN M. "The Evolution of the Right to Health Concept in the United States." *Pharos of Alpha Omega Alpha: Honor Medical Society* 34 (1971): 30–51.

COCHRANE, A. L. *Effectiveness and Efficiency: Random Reflections on Health Services.* London: Nuffield Provincial Hospitals Trust, 1972.

CRANSTON, MAURICE WILLIAM. *What Are Human Rights?* New York: Basic Books, 1963.

FEINBERG, JOEL. "The Nature and Value of Rights." *Journal of Value Inquiry* 4 (1970): 243–257. Commentaries by Carl Wellman and Jan Narveson, pp. 257–260.

———. *Social Philosophy.* Englewood Cliffs, N.J.: Prentice Hall, 1973.

FRANKENA, WILLIAM K. "Natural and Inalienable Rights." *Philosophical Review* 64 (1955): 212–232.

FRIED, CHARLES. "Rights and Health Care—Beyond Equity and Efficiency." *New England Journal of Medicine* 293 (1975): 241–245.

GROTIUS, HUGO. *De jure belli ac pacis libri tres* (1670). I, i, iv.

HART, H. L. A. "Are There Any Natural Rights?" *Philosophical Review* 64 (1955): 175–191.

KASS, LEON. "Regarding the End of Medicine and the Pursuit of Health." *Public Interest,* no. 40, Summer 1975, pp. 11–42.

KING, MAURICE. "Personal Health Care: The Quest for a Human Right." Symposium, *Human Rights in Health,* pp. 227–243.

———, ed. *Medical Care in Developing Countries.* London: Oxford University Press, 1966.

LOCKE, JOHN. *A Letter on Toleration* (1620). *Epistola de Tolerantia: A Letter on Toleration.* Latin text edited with a preface by Raymond Klibansky. Translated with introduction and notes by J. W. Gough. Oxford: At the Clarendon Press, 1968.

———. "The Second Treatise of Government: An Essay Concerning the True Origin, Extent, and End of Civil Government" (1614). *Two Treatises of Government.* Edited by Peter Laslett. 2d ed. Cambridge: At the University Press, 1967, chap. 2, sec. 6, pp. 288–289.

MARITAIN, JACQUES. *The Rights of Man and the Natural Law.* New York: C. Scribner's Sons, 1943; London: G. Bles, 1944.

MELDEN, ABRAHAM IRVING. *Rights and Right Conduct.* Oxford: B. Blackwell, 1950.

———, ed. *Human Rights.* Basic Problems in Philosophy Series. Edited by A. I. Melden and Stanley Munsat. Belmont, Calif.: Wadsworth Publishing Co., 1970.

MUIRHEAD, J. H. "Rights." *Hastings' Encyclopedia of Religion and Ethics.* 13 vols. Edited by James Hastings. New York: Charles Scribner's Sons, 1919, vol. 10, pp. 770–777.

NOZICK, ROBERT. *Anarchy, State and Utopia.* New York: Basic Books, 1974.

OUTKA, GENE. "Social Justice and Equal Access to Health Care." *Journal of Religious Ethics* 2, no. 2 (Spring 1974), pp. 11–32.

PLAMENATZ, JOHN; LAMONT, W. D.; and ACTON, H. B. "Symposium: Rights." *Psychical Research, Ethics and Logic.* Aristotelian Society, supp., vol. 24. London: Harrison & Sons, 1950, pp. 75–110.

PUFENDORF, SAMUEL. "Jus est potentia moralis activa, personae competens ad aliquid ab altero necessario habendum: Vocabuli ambiguitas." *Elementorum jurisprudentia universalis libri duo.* 2 vols. New and rev. ed. (1672). Def. 8. sec. 1. Modern ed. 2 vols. Photographic reproduction. Vol. 1: *The Photographic Reproduction of the Edition of 1672.* Introduction by Hans Wehberg. Classics of International Law, no. 15. Edited by James Brown Scott. Oxford: At the Clarendon Press, 1931, vol. 1, p. 66. Translation. Vol. 2: *The Translation.* Translated by William Abbott Oldfather. Vol. 2, p. 58.

RAPHAEL, DAVID DAICHES, ed. *Political Theory and the Rights of Man.* Bloomington: Indiana University Press, 1967.

RITCHIE, DAVID GEORGE. *Natural Rights: A Criticism of Some Political and Ethical Conceptions.* Muirhead Library of Philosophy. London: George Allen & Unwin, 1894.

ROBERTSON, ARTHUR HENRY. *Human Rights in the World.* Manchester: Manchester University Press, 1972.

ROSEN, GEORGE. *History of Public Health.* Foreword by Felix Marti-Ibañez. MD Monographs on Medical History, no. 1. New York: MD Publications, 1958.

SADE, ROBERT M. "Medical Care as a Right: A Refutation." *New England Journal of Medicine* 285 (1971): 1288–1292.

STRAUSS, LEO. *Natural Right and History.* Chicago: University of Chicago Press, 1953.

Symposium on Human Rights in Health, London, 1973. *Human Rights in Health.* Introduction by G. E. W. Wolstenholme and Katherine Elliott. Ciba Foundation Symposium, n.s. no. 23. New York: Associated Scientific Publishers, 1974.

UNESCO, ed. *Human Rights, Comments and Interpretations: A Symposium.* Introduction by Jacques Maritain. New York: Columbia University Press; London: Allen Wingate, 1949. Reprint. Westport, Conn.: Greenwood Press, 1973.

IV

THEORIES OF JUSTICE AND HEALTH CARE

This article addresses the problem of how society's finite health-care resources should be allocated. Specifically, it presents and analyzes the principal theories of justice that have been developed in regard to the distribution of health resources within society. The article is concerned, therefore, with macroallocation of resources.

It is helpful to begin by noting the broadest definition of justice: Justice is rendering to each his due. Distributive justice is the form of justice concerned with distributing among persons the benefits and burdens that are due to them. Abstractly stated, distributive justice requires persons to be treated alike unless there are relevant differences among them. Persons have been described as relevantly similar or different according to their utility to society, their deserts because of efforts or achievements, and their equal humanity. This discussion will look first at a position based on utility, then will examine one based on desert, and finally will analyze several that rely in different ways on equality. Each position will be examined as to how it applies distributive justice to a concrete macroallocation question: whether or not to develop and make available one highly sophisticated medical device, the totally implantable artificial heart.

What the article does not directly confront is how groups distribute scarce medical resources among a restricted group of individuals needing them, the issue of microallocation. Hence the article will not focus on a discussion of such problems as the allocation of scarce kidney dialysis machines among a small number of eligible patients [see RATIONING OF MEDICAL TREATMENT].

Utility of health care

For some analysts a health-care system is just only if it improves the health of the entire society. The goal is to achieve for the majority the highest level attainable of infant survival, the most years of life expectancy, and the most total work days free from hospitalization. Joseph Fletcher, for example, is convinced that the ethics of health-care delivery must be utilitarian. Fletcher's understanding of utilitarianism is seeking the greatest good, in this case health, for the greatest number possible (Fletcher, pp. 107–108).

Others who have argued that ethical considerations demand that health care maximize the majority's level of health are more careful about how the few who are not included in the utilitarian principle's "greatest number" should be treated. Clark C. Havighurst assumes the utilitarian calculus in his comments on the ethics of government funding of an artificial heart program. If the potential gains of public or private projects exceed potential losses, funds should be approved to support them even if society cannot afford to provide compensation to losers. However, if practical limits allow, justice may indicate that those who are harmed by the actions benefiting the majority receive special compensation (Havighurst, p. 249).

Another utilitarian proposes that society design the most comprehensive, discerning, and impartial cost-benefit analysis possible. Tom Beauchamp, like Havighurst, contends that health-care programs such as development and distribution of the artificial heart would be defensible only if justified by a cost-benefit analysis. However, he concedes that sometimes it would not be permissible to follow the dictates of such a utilitarian calculus. Indeed, he would invoke broader considerations of justice, not merely in terms of compensation but as a "threshold" consideration. He believes that rule-utilitarianism—judging what is a right action or policy on the basis of rules that promote the greatest general good—is capable of resolving conflicts between justice and utility (Beauchamp, pp. 20, 27–29).

Criticisms of this position arise from general criticisms of the theory of utilitarianism. In its most obvious form, utilitarianism would require that a health-care system provide the best care possible for the greatest number of people. Critics ask: If a health-care system improved the health of the majority of society, would it be a just system if it slighted the care of such minorities as the aged, children, the disabled, and the dependent poor? It is not obvious that minority populations should be allowed to die early in order to improve the society's success in decreasing average morbidity, or even mortality, rates. Arrangements to compensate those minority populations who are harmed in the pursuit of optimal health for the majority are not an adequate substitute for the demands of distributive justice. Justice must take better account of all groups in governing the initial distribution of medical and health-care benefits.

Some critics declare that distributive justice cannot be accommodated to utilitarianism since, whatever sophisticated form utilitarianism might take, it depends on computing the sum of good produced, rather than the justice of how the total is distributed to individuals (Miller, p. 39). If utility is a moral demand independent of justice, the utilitarian position could well be judged not to be an application at all of the principle of distributive justice.

Entitlement to health care

While the utilitarian position discussed above has been dismissed by some critics as not founded on distributive justice, another position, which can be called the entitlement position, has been attacked for being so concerned with the rights of individuals that it also ignores crucial issues of distributive justice.

Robert Nozick is the leading advocate of the entitlement position. He argues that his views do satisfy the requirements of distributive justice, but he makes it clear that his theory begins with the individual. He starts his major work on the subject by saying that individuals have rights, and there are things no person or group may do to them without violating their rights (Nozick, p. ix). He later cites the physician as an example of a person whose rights to offer his skill for whatever personal reasons should not be violated by society's idea of how medical care can most appropriately be distributed to meet its needs. "Just because he has this skill, why should *he* bear the costs of the desired allocation, why is he less entitled to pursue his own goals, within the special circumstances of practicing medicine, than everyone else?" (ibid., p. 234).

The entitlement position can just as stoutly defend the rights of consumers of medical care. Provided they acquire their wealth justly, whatever level of health care patients purchase would be just, and society would be acting unjustly in regard to patients if it interfered by imposing its schemes for allocating health care. The position includes the proviso that an individual cannot consume an irreplaceable resource, such as the single culture necessary for the only vaccine that can eradicate a plague, if the disappearance of that resource worsens the position of others to the point that they cannot be compensated (ibid., p. 178).

Nozick brushes aside traditional forms of distributive justice as a threat to the individual because they take possessions from individuals and allocate them to others according to some overarching social pattern. He stresses the worth of past actions and conditions of individuals, rather than the utility of future results or present patterns of equality. He sometimes even refers to his position as the "historical-entitlement" view of justice, because it is on the basis of one's present and past that one's due is determined. The entitlement position would say that if physicians, indeed a large number of physicians, wished to devote their entire resources, time, and energy to developing and marketing an artificial heart to be sold at prices that would make them available for only a very few, primarily the wealthy, the physicians would be justified in doing so, however little it contributed to improving the overall health of the society or however much it violated some pattern of equality.

To understand criticisms of the entitlement to health care position it is necessary to look at Nozick's outline of how he moves from his assumptions to his conclusions:

1. People are entitled to their natural assets.
2. If people are entitled to something, they are entitled to whatever flows from it (via specified types of processes).
3. People's holdings flow from their natural assets.

Therefore,

4. People are entitled to their holdings.
5. If people are entitled to something, they ought to have it (and this overrides any presumption of equality there may be about holdings) [ibid., pp. 225–226].

Nozick claims not to be advocating any of the major alternative ways of specifying distributive justice, including desert (determining distribution of goods, including health care, according to persons' merits or efforts or achievements), but when Nozick connects the language of entitlement to natural assets, he seems to be slipping into reliance on desert. Entitlements are usually considered to be rights arising from promises among individuals or rules agreed upon by a society, yet such promises and rules do not rely on the degree of personal merit, effort, or achievement possessed by the individuals claiming the entitlement. Nozick's use of the word entitlement borrows some of its force from its hint that a person deserves the advantage he enjoys from his natural assets. Yet this clashes with a common intuition shared by many and articulated by John Rawls: No person deserves his place in the distribution of native endowments any more than he deserves his initial starting place in society: One of the most important tasks of human justice is to overcome the results of the "natural lottery" in native assets (Rawls, p. 104). In sum, one critique of Nozick's first premise, which is concerned with natural assets, is to ask if people are entitled to (let alone deserving of) their natural liabilities, including physical handicaps and congenital diseases.

Another challenge to Nozick is directed at his second premise, particularly that people are entitled to *whatever* flows from their natural talents and abilities. His critics think he is idiosyncratic in his basic intuition that each person is entitled to do whatever he wants with what he receives or acquires. In the real world, rights of individuals and groups clash. Thomas Nagel insists that no person possesses absolute entitlements like those Nozick believes in (Nagel, p. 147). Even if a defender of the entitlement position were to believe that clashes among individuals claiming their rights to acquisition and transfer of holdings could generally be solved through contracts made in the marketplace of supply and demand, the question would remain as to whether health care is desired in the way consumers desire other goods. Are not the urgency and level of demand, especially in acute medical care, often the result of accident or forces well beyond the consumer's control and free choice? A proponent of a free market system of distribution of goods might still believe that the specific characteristics of medical care dictate the employment of other more relevant modes of distribution.

Nozick's critics also attack his third premise, arguing that he does not sufficiently recognize that neither a patient's nor a physician's wealth flows simply from his or her natural assets. Society can claim much of the responsibility for both the advantages and disadvantages with which the individual physician and patient approach the training of physicians as well as the transfer and acquisitions of health care. Therefore, more than Nozick acknowledges, society is justified in intervening in the distribution of health-care delivery.

Rejection of one or more of these premises prevents critics from agreeing with Nozick's conclusion that a person's entitlements override presumptions of equality. Critics cannot agree that physicians are entitled to offer services in whatever manner and for whatever price they choose, and that consumers are entitled to acquire health care at whatever level their wealth and income allow.

By contrast to the first two positions explored so far, the next three positions relating justice to health care make equality a necessary part of distributive justice, although it is the last position that attempts to be strictly egalitarian.

Decent minimum of health care

Charles Fried has relied on the concepts of both desert and equality to specify how distributive justice requires that society provide a decent minimum of health care (and only a minimum) to its citizens. On the one hand, Fried thinks that patients who have the means to obtain a higher level of health care than others deserve the freedom to purchase it. At least in the United States, a social system has been instituted where those who want more individualized services can get them if they are willing to pay more. The system that allows variation in wealth and income is not itself morally suspect. He does not see why a sector like health care should be carved out and governed by different moral principles (Fried, 1976, pp. 32, 33).

On the other hand, in all areas of society a notion of a decent, fair standard obtains. Fried argues that the decent minimum in respect to health should be distributed equally to all members of the society. He is not entirely clear on the crucial point of whether the equality is in terms of a guarantee to all citizens of an equal decent minimum *level of health* or in terms of making available a *fixed amount of health care*. Fried says that the concept of a decent minimum is always relative to what is available over all and what the best available might be. He suggests that at present maternal and child care and humane surroundings of all health care are essential elements in the decent minimum that ought to be provided directly by the government (ibid., p. 32). However, he also says that he prefers assuring each person a fixed amount of money to purchase medical services as the person chooses, realizing that this would continue a system where the poor would be unable to get the level of care available to the rest of society. The advantaged, after all, could add the government allotment for health care to their already existing resources and purchase much more than the decent minimum available to the poor. A decent minimum would probably not provide every citizen with equal access to an artificial heart. It would allow development and purchase of artificial hearts by those who could afford it, and the wealthy could justifiably include their government health subsidy in their payment for an artificial heart.

As is the case with all mixed positions, more thoroughgoing advocates of desert or equality can criticize the confusion created by compromise. A decent minimum of health care, through taxation, forcibly removes some financial assets justly acquired by persons, while failing to achieve for all citizens equality in levels of health care, let alone of health. For example, Fried is not clear about what should be done for the poor who spend their fixed health allotment to meet nonhealth expenses, then discover they have a condition that is certainly and imminently fatal but can be treated and probably cured through standard medical procedures. Should society actually refuse to provide medical treatment to such persons?

On the other hand, the decent minimum position can be praised for its sensitivity to the moral intuition that society should not entirely neglect any of its citizens' basic needs. It also recognizes limitations by restricting assistance to services that lie below a level that would disrupt the entire political and economic system of a society.

Maximin level of health care

The position that advocates a "maximin level of health care" attempts to be more generous than Fried's decent minimum in distributing health care in a way that does not increase social inequality. The maximin position has been based on John Rawls's theory of justice. Rawls asks us to imagine persons in an "original position" of ignorance about variations in natural abilities or respective places in the social order, but who know general facts about the human condition and natural and social laws. He says that, if persons in such a position were asked to select principles governing society, they would prudently defend their interests by choosing rules whose effect on each of them would be the least damaging possible. In other words, they would choose a maximum minimum, or "maximin."

Not knowing their chances of increasing or decreasing their share of primary goods, such rational agents would follow a maximum-minimum approach to justice and would adopt the principle that each member is entitled to "an equal right to the most extensive total system of equal basic liberties compatible with a similar system of liberty for all." Aware, however, that incentives for the more talented and productive individuals lead to benefits for every person, they would adopt a second, "difference," principle, which permits social and economic inequalities. But in a continuation of the maximin approach, inequalities would be permitted only so long as they lead "to the greatest benefit of the least advantaged," and only if the inequalities were "attached to offices and positions open to all under conditions of fair equality of opportunity" (Rawls, p. 302).

Ronald Green applies Rawls's theory of justice to health care, arguing that rational agents in the original position, confronted with how to relate justice to health care, would secure the highest minimal level of health care for themselves and their loved ones. He believes that such rational agents could consider health care a primary social good comparable to civil rights and liberties. If they did, health care would be governed by Rawls's egalitarian first principle: Equal access, irrespective of income, to the most extensive health services the society allows, would be approved as the position conforming to the demands of distributive justice (Green, 1976, p. 117).

But how would maximin reasoning understand justice in circumstances where the amount of health care to be distributed was limited? Green says that a "lexical ordering" of health care is the single most important implication of contract theory for the macroallocation issue (Green, 1977, p. 14). By lexical ordering he means addressing first the urgent health-care needs of the worst-off group, and only when substantial progress with them had been achieved would the special health needs of the next least-advantaged group be treated, and so on until medical programs for the advantaged were undertaken.

One difficulty with this position is identifying the worst-off group. Are they the medically or the economically least-advantaged? Green sometimes says that they are the group with the worst health; those with conditions that cause the greatest physical and mental suffering. At other times he assumes that the medically worst off are the same as the economically least advantaged, and says that a just health-care system would first attack diseases that he assumes to be found disproportionately among the poor: arthritis, hypertension, malnutrition, and work-related injuries (ibid., pp. 31–33).

A second difficulty is the definition of health. When Green says health is a primary social good and the measure of the disadvantaged, does he understand health in physical terms as a well-working physical organism, functioning as most other physical organisms do? Or does he assume that the fundamental social good of health is a more inclusive state of complete physical, mental, and social well-being, as the World Health Organization defines health? And are the least advantaged to be defined by their distance from such an inclusive understanding of health?

There does not seem to be any inherent reason why the maximin position could not consider those needing an artificial heart among the medically least advantaged, and therefore place a high priority on developing it. As it happens, Green calls for a moratorium on development of the artificial heart. He holds the opinion that rational agents would consider those with childhood diseases or conditions of high morbidity (great physical or mental suffering) as less medically advantaged and therefore to be favored in the allocation of health resources over the likely users of the artificial heart, that is, those beyond a "normative age" who have heart disease. In fact, if those whose lives are prolonged suffer from strokes or cancer, the artificial heart might contribute to increasing morbidity in society. If the least advantaged are to be equated with the poorest, Green is certain that the artificial heart does not selectively benefit lower-income individuals (ibid., pp. 11, 12, 17).

Interpretations of Rawls other than Green's might regard health care as a material good comparable to wealth and therefore to be distributed according to Rawls's "difference principle." According to such an interpretation, a government that paid for the health care of the affluent might still serve justice if by doing so a single and more effective health-care system improved most the health of the least advantaged. In this fashion utility would be allowed to modify equality. As we shall see, thoroughgoing egalitarians object to such reasoning.

Equal access to equal levels of health

The first part of the position requiring equal access to equal levels of health stresses the first part of the formal definition of distributive justice: Persons are to be treated alike. All persons should have equal access to health care regardless of their financial, geographic, or other differences. Persons with similar medical cases should receive similar treatment. Except for the sickness itself, no differences among the sick are relevant. As grounds for his position of equal access to health care, Gene Outka points to the moral perspective of equal regard for all persons and the fact that in general humans are equally vulnerable to accidents and illnesses for which they are not responsible.

The second part of the position (compatible with Outka's) focuses on the second part of the formal principle of justice: Persons are to be treated alike *unless there are relevant differences*. Level of sickness is identified by this position as the most significant relevant difference. It will not be enough for persons to receive the same amount of health care. Uniquely among the views surveyed in this article, according to this position distributive justice requires that greater and lesser degrees of medical need must be met in different individuals for persons to be treated as equals. As Robert Veatch states the position, "Justice requires everyone has a claim to health care needed to provide an opportunity for a level of health equal, as far as possible, to other persons' health" (Veatch, 1976, p. 134).

Some egalitarians might find Veatch's formulation, limiting care of individuals to other persons' level of health, as too weak a view of equality. They would concede that the position may be adequate for guaranteeing care of medical needs, but what of wants or desires concerning health? Should these be limited by a single level of health? David Miller contends that "because people have varied needs and *wants*, physical resources such as food, medicine, and education should not be assigned in equal quantities to each man, but in different proportions to different people, according to their peculiar characteristics" (p. 149) He bases his position on the notion that "every man should enjoy an equal level of well-being." This level is not met by only satisfying needs, but by providing "as large a proportion of each person's further desires as resources will allow." Each person's different level of desires is measured on an individual "scale of well-being," indicating the allocation of resources that would give him the least and greatest well-being. Justice requires that each person should enjoy as high a position on his own scale of desire for well-being as every other person enjoys (Miller, p. 144).

The position advocating equal access to equal levels of health attempts to meet attacks from nonegalitarians rather than responds to objections from egalitarians. Veatch's saying that justice requires that health care "provide an opportunity" for equal levels of health points to the voluntarism behind many desert interpretations of distributive justice. Individuals who repeatedly refuse to take advantage of treatments provided should not be required to improve their health. Nevertheless, those who develop health needs because they voluntarily engage in activities clearly detrimental to health, such as heavy drinking or smoking or recreational stunt flying, ought to contribute more than others to the cost of their restoration to an equal level of health (Veatch, 1976, pp. 13, 14).

When Veatch says that health care is to provide health equal, "as far as possible," to other persons' health, he accepts utility as a limitation. For example, treatments are not necessary for the incurably sick who cannot find any use for the resources allocated to them. Also a distinction must be observed between need and desire. Outka suggests cosmetic surgery as an example of a technique to which society would not be required to provide equal access. Veatch adds treatment for baldness, prenatal sex selection, attending physicians accompanying travelers, and electrical stimulation of pleasure centers of the brain (Outka, p. 92; Veatch, 1976, p. 141).

It is not surprising that the position advocating equal access to equal levels of health has difficulty making an unequivocal judgment regarding the artificial heart. Surely those who seek an artificial heart to prolong life cannot be dismissed as desiring a trifle. They could make a plausible case that they need the artificial heart if their level of health is to be equal to others'. Veatch, however, argues that the high costs of an artificial heart might make it impossible to restore those who are in even more serious medical need to a level of health equal to that of others. He says that his position could justify a legislature's not voting to expend public money on such a project (Veatch, 1977, pp. 16, 17).

Nonegalitarians remain unconvinced by the position favoring equal access to equal levels of health. Those emphasizing entitlement or desert believe that persons' rights are being violated if their wealth is taxed in order to provide equal levels of health to all citizens. Utilitarians accept the point made by strict egalitarians about the difficulty of separating needs from desires, but conclude that even genuine health needs can never be met by finite human resources, and therefore that need is a practical impossibility as a criterion for just distribution of health care. Even utilitarians who accept the desirability of maximizing equality as a social good do not believe that justice requires the impossible, and therefore reject the position that distributive justice demands that society provide its citizens with equal levels of health (Beauchamp, p. 20).

Conclusion

Convergence among the positions appears at certain points. Even those positions that select a single concept such as utility, desert, or equality as the essence of distributive justice do not ignore the importance of the others. For example, Havighurst and Beauchamp reject a narrow view of utility, arguing that other considerations must be taken into account. Equality, as developed by Outka and Veatch, is complemented by desert and utility. On a more practical level, all the positions examined here that hold equality to be significantly relevant to health care also say that justice requires that society provide at least a minimum level of care. They continue to debate how justice might further specify such a minimum, but none argues that the minimum should include providing artificial hearts to those whom doctors declare need them.

Several issues remain unresolved. Some philosophers believe that taking the moral point of view requires selecting a single understanding of distributive justice as the most adequate and then applying it to the facts of health-care delivery. Others hold that the varying nature of particular problems dictates whether utility or desert or equality should control the just distribution of health care. Even more difficult to resolve is how to relate justice to other ethical principles such as respect for persons and utility. Even those who subscribe to a nonutilitarian definition of distributive justice usually grant that utility is a valid independent ethical principle. Ethical decisions often involve more than simply considering what justice requires. They would admit that sometimes justice must be overruled. The just act may not, all things considered, be the right act.

The discussion among philosophers and ethicists is marred by a lack of conceptual clarity. For example, all the positions outlined here are affected by how health is defined, yet none carefully restricts the meaning of the term. Furthermore, there is no consistent distinction made between medical care and health care. The latter term is usually employed in this literature, although the discussion usually focuses on what many writers mean by medical care. If health care, particularly health in its more inclusive definitions, were actually meant, the task of determining what distributive justice requires would prove to be even more complex.

Roy Branson

[*While all the articles in this entry are relevant, see especially the previous article,* RIGHT TO HEALTH-CARE SERVICES. *Also directly related are the entries* JUSTICE; *and* ETHICS, *article on* UTILITARIANISM. *This article will find application in the entries* AGING AND THE AGED, *article on* HEALTH CARE AND RESEARCH; FOOD POLICY; HEALTH, INTERNATIONAL; HEART TRANSPLANTATION; HOSPITALS; KIDNEY DIALYSIS AND TRANSPLANTATION; MENTAL HEALTH SERVICES, *article on* SOCIAL INSTITUTIONS OF MENTAL HEALTH; POVERTY AND HEALTH; *and* RATIONING OF MEDICAL TREATMENT. *See also:* HEALTH AS AN OBLIGATION; *and* PUBLIC HEALTH.]

BIBLIOGRAPHY

BARRY, BRIAN M. *The Liberal Theory of Justice: A Critical Examination of the Principal Doctrines in "A Theory of Justice" by John Rawls.* Oxford: Clarendon Press, 1973.

——. "On Social Justice." *Oxford Review*, no. 5 (1967), pp. 29–52.

BEAUCHAMP, TOM L. "Macroallocation and Policy Decisions." Unpublished paper for the Conference on Policy Making in Health Resource Allocation: Concepts, Values, Methods, sponsored by the National Legal Center for Bioethics, Washington, D.C., 1977. 32 pp. Proceedings to appear in 1978.

BEDAU, HUGO A. "Radical Egalitarianism." Bedau, *Justice and Equality*, pp. 168–180.

———, ed. *Justice and Equality*. Englewood Cliffs, N.J.: Prentice-Hall, 1971.

FEINBERG, JOEL. *Social Philosophy*. Englewood Cliffs, N.J.: Prentice-Hall, 1973.

FLETCHER, JOSEPH. "Ethics and Health Care Delivery: Computers and Distributive Justice." Veatch, *Ethics and Health Policy*, chap. 6, pp. 99–109.

FRIED, CHARLES. "Equality and Rights in Medical Care." *Hastings Center Report* 6, no. 1 (1976), pp. 29–34.

———. "Rights and Health Care—Beyond Equity and Efficiency." *New England Journal of Medicine* 293 (1975): 241–245.

GREEN, RONALD M. "Health Care and Justice in Contract Theory Perspective." Veatch, *Ethics and Health Policy*, chap. 7, pp. 111–126.

———. "The Nuclear-Powered Totally Implantable Artificial Heart—A Rawlsian Reassessment." Unpublished paper for the Conference on Policy Making in Health Resource Allocation: Concepts, Values, Methods, sponsored by the National Legal Center for Bioethics, Washington, D.C., 1977. 32 pp. Proceedings to appear in 1978.

HAVIGHURST, CLARK C. "Separate Views on the Artificial Heart." Veatch, *Ethics and Health Policy*, chap. 15, pp. 247–255.

MILLER, DAVID. *Social Justice*. Oxford: Clarendon Press, 1976.

NAGEL, THOMAS. "Libertarianism without Foundations." *Yale Law Journal* 85 (1975): 136–149. Review of *Anarchy, State, and Utopia* by Robert Nozick.

NOZICK, ROBERT. *Anarchy, State, and Utopia*. New York: Basic Books, 1974.

OUTKA, GENE. "Social Justice and Equal Access to Health Care." Veatch, *Ethics and Health Policy*, chap. 5, pp. 79–98.

RAWLS, JOHN. *A Theory of Justice*. Cambridge: Harvard University Press, Belknap Press, 1971.

VEATCH, ROBERT M. "What Is a 'Just' Health Care Delivery?" Veatch, *Ethics and Health Policy*, chap. 8, pp. 127–153.

———. "Legislative Prerogatives and the Right to Health." Unpublished paper for the conference on Policy Making in Health Resource Allocation: Concepts, Values, Methods, sponsored by the National Legal Center for Bioethics, Washington, D.C., 1977. 21 pp. Proceedings to appear in 1978.

———, and BRANSON, ROY, eds. *Ethics and Health Policy*. Cambridge, Mass.: Ballinger Publishing Co., 1976.

WILLIAMS, BERNARD A. O. "The Idea of Equality." Bedau, *Justice and Equality*, pp. 116–137.

HEALTH-CARE REGULATION
See HEALTH CARE, *article on* HEALTH-CARE SYSTEM; HEALTH POLICY; MEDICAL PROFESSION.

HEALTH HAZARDS
See ENVIRONMENTAL ETHICS, *article on* ENVIRONMENTAL HEALTH AND HUMAN DISEASE.

HEALTH INSURANCE

Purposes, types, and early history

In modern society every individual is exposed to the risk of sickness, which may result in heavy expenditures for hospital, surgical, general medical, and other health services as well as in losses of income. Health insurance has the purpose of distributing the financial impact of sickness among the members of a group, the insured. Such insurance may operate either on the principle of reimbursement or on the principle of provision of services. In the first type of arrangement the insurer charges each insured a fixed sum (premium) and reimburses each insured for his expenditures for the covered health services as well as income losses, should that be part of the agreement. In the second type of arrangement, which is frequently called a prepayment health plan, the insurer makes the necessary arrangements for rendering the health services covered by the plan with hospitals, surgeons, other doctors, and pharmacists and charges the subscriber a fixed periodic amount payable in advance, often designated as fee, for the provision of the needed care.

Health insurance may be either left to voluntary arrangements with private insurance organizations of the profit or nonprofit type or made mandatory pursuant to legislation establishing compulsory coverage of workers and other population groups. Compulsory health insurance, like private health insurance, may operate either on the reimbursement principle or on the provision of services principle.

Compulsory national health insurance with public insurance organizations originated in Germany in 1883. In that year Germany enacted a statute that instituted mandatory health insurance of workers against the costs of health care and income losses through local or regional public entities created for that purpose. The covered employers had to enroll their workers and bear one-third of the contributions, which were fixed by law. Other countries on the European continent followed the German example, especially Austria in 1888 and Norway in 1909/1911. A great impetus to the expansion of compulsory health insurance was given by the inclusion of this type of coverage in the British National Insurance Act of 1911. Japan joined the ranks in 1922, France in 1928, New Zealand in 1938, and Italy in 1943. The United Kingdom changed from the reimbursement system to the provision of services with the enactment of the National Health Service Act of 1946. Some other nations followed that example (see below).

Private health insurance based on individual policies underwent very slow growth. Because of the inherent frequency of the insured event, the uncertainties in developing actuarial data for the frequency and severity of the losses covered, the difficulties in policing losses, and the amount of administrative work involved, commercial insurance carriers were not attracted to the health field. The few companies that entered the field before the First World War limited coverage to the loss of earnings resulting from sickness and filtered out short-term cases by means of waiting periods. Such policies were written by commercial carriers in the United States, beginning in 1898. Clauses providing for limited health-care benefits were gradually added. Surgical benefits appeared in 1903, and medical benefits in 1910. At that time limited amounts for hospital expenses were likewise added as an incidental benefit (Dickerson, pp. 152, 169, 311). The business of private carriers expanded with the advent of group policies. Commercial group insurance against hospital expenses was offered as early as 1910, and against surgical expenses since 1928. Group policies providing medical expense benefits came into use in the early 1940s (ibid., pp. 152, 169). A large number of these group policies were procured by employers pursuant to collective bargaining agreements, because fringe benefits were not affected by the wage-stop during the Second World War. The main impetus to expansion of commercial health insurance in the United States came from the success of the prepayment health plans in the 1930s (see below).

The development of private health insurance in Germany paralleled in many respects that in the United States. Prior to the great inflation in 1924 few commercial insurance companies provided health insurance, and those that did restricted their policies to indemnities for loss of earnings caused by sickness (Manes, 1922, vol. 2, p. 76). Insurance protection against health-care expenditures was furnished either by the governmental health insurance system for employees, established in 1883, or by local mutual or fraternal associations. As a result of the inflation in 1924, however, the demand for health insurance grew rapidly and commercial insurance against health-care expenditures became an important branch of the insurance industry (Manes, 1932, vol. 3, p. 120; Apelbaum, p. 912). Although the governmental health insurance system has been extended continuously since the Second World War to include new occupational groups and now covers nearly ninety percent of the population, private health insurance in Germany still plays an important role in providing full or supplementary protection for a substantial number of citizens.

Health insurance in the United States

Present extent of mandatory and voluntary health insurance. The United States is the largest industrial nation in the world that has not introduced a general comprehensive national health insurance or health services program. Provisions to that effect were not included in the Social Security Act of 1935 because of the opposition of the medical profession.

This, of course, does not mean that in the United States there are no compulsory health insurance programs either for particular population groups or for health-care expenditures resulting from specified events. Nor does it mean that there are no massive public programs providing health care for needy persons. It does, however, signify that more than half of the personal health-care expenditures of the population are met either by direct payments of the consumers or under private insurance or prepayment plans.

The chief compulsory insurance programs covering health-care expenditures are the state or federal workers' compensation systems and Medicare. Since the beginning of workers' compensation legislation in 1911 the governing statutes provided protection against medical costs of industrial accidents or occupational diseases. In a number of states the coverage of such medical expenditures is unlimited, while others still apply ceilings. During the fiscal year 1975 the payments for medical expenses under workers' compensation programs totaled $1.99 billion. Medicare, which was established in 1965 to provide health insurance for persons over sixty-five and was extended in 1972 to permanently and to totally disabled recipients, covers a much larger share of the total annual health-care bill of the U.S. population. Medicare reimburses hospital expenses of the aged and permanently and totally disabled and, on a supplementary, voluntary, and contributory basis, their medical expenditures. In 1975 the cost of Medicare totaled $15.6 billion, comprising $11.3 billion for hospital insurance expenditures and $4.3 billion for medical services. In addition, governmental expenditures for individual health care under medical assistance, military dependents', veterans' and general hospital programs totaled in excess of $25.7 billion.

The aggregate expenses for individual health care during the fiscal year 1976 amounted to $120.4 billion, of which $48.4 billion was defrayed by governmental expenditures. In other words, $72.0 billion (or 59.8 percent of the total) had to be supplied by private funds, consisting primarily of direct consumer payments, and monies defrayed by commercial insurers or prepayment plan operators. Direct consumer payments amounted to $39.1 billion (or 54.3 percent of the total sum covered by private funds), payments by commercial insurance companies or prepayment plan operators to $31.4 billion (43.6 percent of that total), and payments for industrial facilities or by private charity to $1.56 billion (2.2 percent of the total). These figures, however, comprise all types of health services. Broken down into various types of health-care expenditures, the figures for 1976 show that individual health-care expenditures for hospital care and surgical and other physicians' services (amounting to $81.8 billion out of the total of $120.4 billion) were largely covered by public provisions or private insurance and that the direct consumer payments for these types of services constituted only $15.11 billion or 18.5 percent of the total expenditures for that type of service, while direct consumer payments for all other health services amounted to $24.0 billion or 62.0 percent of the total of $38.68 billion expended by public and private sources for such services (Gibson and Mueller, Tables Nos. 2 and 3).

The foregoing data reveal the modern importance in the United States of private health insurance by commercial carriers or prepayment arrangements as supplementary protection in addition to compulsory insurance and public assistance. This position was attained by a development that occurred between the depression in the late 1930s and the present.

Growth of voluntary health insurance in the United States. Although some industries provided hospital and medical services for their workers long before the beginning of the present century, and although some life insurance carriers wrote policies providing income maintenance during sickness or reimbursement of medical expenses in lieu thereof as early as at the turn of the century, modern voluntary health insurance in the United States started with the organization of prepaid hospital care plans prompted by the depression of the 1930s. The Baylor University Hospital plan for teachers, established in 1929, is commonly referred to as the pioneer example. Other such single hospital plans followed, but it was quickly realized that better results could be reached if several hospitals in an area combined and developed a prepayment plan open to all members of the community. Sacramento, California, was the first city to start such a scheme. Seven hospitals formed a nonprofit organization, which contracted with the subscribers for the provision of hospital services. Similar plans were formed in other cities, and in 1933 the American Hospital Association (AHA) endorsed that approach. In 1937 it began to give formal approval of such plans if they complied with the AHA's standards. The sign of approbation was a blue cross, and the plans established under the auspices of the Blue Cross Commission of the AHA became known as Blue Cross Plans. Subsequently the Blue Cross Commission was replaced by the Blue Cross Association. The growth of such plans was spectacular, growing from thirty-eight in 1937 to seventy-six within the span of eight years. The development of similar prepayment plans for the delivery of physicians' services was at first stunted by American Medical Association (AMA) opposition to agreements between doctors and an outside organization for the rendering of services to patients who had contracted for such care with the intermediary. Nevertheless, some local and even statewide medical societies ignored the disapproval of the AMA and established prepayment plans for physicians' services. Statewide plans were organized in California and Michigan. The California plan was incorporated in 1939 as a nonprofit corporation called California Physicians' Service and was an instant success. Because of legal doubts plans of that type were specifically sanctioned by legislation passed in 1941 and upheld against attacks by the State Insurance Commission in a widely noted decision of the State Supreme Court rendered in 1946. In 1942 the AMA relinquished its opposition to plans sponsored by the medical profession and, through its Council on Medical Service, issued standards for their approval. Approved plans were entitled to exhibit a blue shield as a sign of approbation and thus became known as Blue Shield Plans. In 1946 the various local plans formed a national organization, the National Association Blue Shield Plans, which currently counts seventy-two members. In addition, a number of so-called independent plans sprang up through either community efforts or particular individual initiative. The most important plans in this category are the Kaiser Foundation plans, originating in 1942 in Richmond, California, and now operating in several West Coast cities as well as in Colorado, Hawaii, and Ohio, and the Health Insurance Plan of Greater New York, launched in 1947. The need for, and wide acceptance of, prepayment plans prompted the commercial carriers likewise to develop individual and group policies covering hospital, surgical, and other medical expenses. Hospital expense insurance was offered commercially in 1934 and surgical insurance in 1938. While at first coverage was rather limited, in the course of time the insurance industry broadened and diversified its policies.

The extent of private health insurance protection in the United States under commercial policies, Blue Cross/Blue Shield plans, and independent plans grew rapidly between 1940 and the present. The gross enrollment under private plans providing hospital insurance existing in 1940 is estimated at slightly above 12 million individuals. The gross coverage of that type of insurance at the end of 1974 is reported at 208 million individuals, of whom 115 million had coverage under commercial policies, 84 million under Blue Cross/Blue Shield plans, and 9 million under independent plans. The share of the commercial carriers in the hospital insurance field between the years 1940 and 1974 grew from 31 percent to 55 percent, while the share of the Blue Cross/Blue Shield plans, despite their growth in the number of subscribers, dropped from 50.5 percent to 40.3 percent. The 208 million figure represents gross enrollment, without correction for duplicate coverage. Adjustment of these figures to net coverage (i.e., coverage of different individuals) is subject to divergent estimates. It is, however, certain that more than three-quarters of the U.S. population is protected by private hospital insurance and that nearly the same proportion is covered for surgical services. Of the 115 million persons having hospital insurance written by commercial carriers in 1974, 75 percent were covered under group policies.

Private health insurance plans vary greatly in the scope and type of expenditures covered. While a large segment of the population under sixty-five is protected by voluntary insurance against basic medical expenditures for hospital and surgical services, the number of persons having additional protection in the form of major medical or comprehensive coverage under commercial policies or Blue Cross/Blue Shield plans is considerably less, but rapidly growing. It reached the 132 million mark (not adjusted for duplication) in 1974 (Mueller and Piro, Table 8). One of the chief obstacles to the development of comprehensive coverage for a fixed and reasonable premium was the traditional delivery system of medical care by sole practitioners on a fee-for-service basis. In order to facilitate the organization of group practices operating on a prepayment arrangement as typified by many so-called independent plans, Congress in 1973 enacted the Health Maintenance Organization Act of 1973 (PL 93-222). It provides federal assistance for the establishment of new HMOs complying with federal requirements and exemption from restrictive state laws and practices.

As has been mentioned before, the principal mandatory or optional health insurance protection created by legislation consists of the Medicare provisions for compulsory hospital insurance and optional medical and other health-care insurance (now covering 22 million persons above age sixty-five and 2 million disabled persons), which were inserted in the Social Security Act by amendments of 1965, and the medical benefits prescribed by state and federal workers' compensation laws. Mandatory health insurance against nonoccupational sickness for persons under sixty-five was established by legislation in two states: Hawaii enacted the first comprehensive Mandatory Prepaid Health Care Law in 1974, while Rhode Island in the same year passed a Catastrophic Health Insurance Plan Act. In 1976 Minnesota enacted a comprehensive health insurance act, which grants tax benefits to employers who provide the prescribed health coverage. The numerous proposals for comprehensive or catastrophic federal health insurance of able-bodied people under sixty-five have so far failed to result in legislation, although efforts to that effect date back to 1915 ("Chronology").

Ethical issues and protection against abuses. Health insurance, like any other type of insurance, consists in an actuarial determination of the probable magnitude of the aggregate risk to which a group as a whole is exposed and the allocation of the respective costs to the members of the group in equal or otherwise equitable shares. In other words, insurance is a risk-sharing or risk-spreading device. The actuarial basis and risk-distributing features of the system distinguish it from immoral aleatory or gambling agreements. The insurer bears the risk of unforeseen events or miscalculations, but insurance activities are not unethical if the premiums or fees charged are actuarially sound and fair. Expressed in technical language the rates charged to policyholders or subscribers must be adequate and not excessive. Within this framework health insurance, whether private or public, performs a desirable social function. When insurance is a private business activity, i.e., is for profit, the built-in profit factor should assure a fair return from underwriting but should not be disproportionate. The proper basis for measuring reasonable profit is a much-debated issue. In public insurance a crucial issue of fairness is raised by provisions that charge unequal rates for equal benefits. In public health insurance, for example, health-care benefits that are not wage-related are often provided in return for a percentage levy on earnings. This, for instance, is the case with respect to the hospital benefits under Medicare in the United States or the medical benefits under Australia's Medibank system. Within limits, however, such intergroup transfers are considered desirable.

Because of the interrelation between risks covered and premiums or fees charged, health insurance is particularly vulnerable to various types of abuses on the part of the insurers or on the part of the insured or on the part of the suppliers of the stipulated health-care services.

The most flagrant abuses by private insurers consist in coverage gaps not properly disclosed and excessive charges for administration and profits. Ordinarily commercial health insurance coverage is not comprehensive and full but accords reimbursement only for specified types of care and is subject to deductions, co-insurance, exclusions, or limitations on amounts. As a result there is danger that policies are marketed that expose the insured to gaps in the protection that are not clearly realized by policyholders and may be medically unsound. Health insurance should focus on adequate coverage of basic care requirements unless only major or catastrophic expenditures are clearly intended to be covered.

The most flagrant abuses by the insured consist in overutilization of the benefits provided. Since health insurance is based on actuarially predicted normal needs, overuse on the part of some endangers the viability of the whole structure and at any rate increases the costs for members of the group who practice restraint. It is established, however, that the idea of prepaid medical care has affected the whole perception of medical needs (Berkanovic et al.).

Finally, the most dangerous abuses of health insurance are produced by suppliers of medical services who engage in useless or unnecessary procedures, make charges that are exorbitant in view of the quality of care given, or make outright fraudulent claims.

In response to persistent complaints about abuses of the types described, various control mechanisms have been established. In particular, a number of states have extended the supervision by insurance commissioners beyond the mere assurance of solvency of commercial carriers to policy terms and rates charged and also brought other types of insurers under the commissioners' jurisdiction. In addition, the federal Retirement Income Security Act (PL 93-406) has imposed detailed reporting and disclosure duties on the administrators of employee welfare benefit plans, including plans that provide medical, surgical, or hospital care or benefits for employees through the purchase of insurance or otherwise.

Health insurance outside the United States

A number of nations have followed the example of the United Kingdom and established national health service systems covering the whole population, e.g., New Zealand, Sweden, and the USSR (Simanis, pp. 69–102), but a larger group of countries operate national health insurance systems modeled after the original German system. Most of them are now quite comprehensive with respect to the categories of persons and expenditures covered. Hence there is little room for private health insurance. A notable exception is the Federal Republic of Germany, which maintains substantial though steadily diminishing exclusions with respect to persons and services covered. For that reason private health insurance by private corporations still plays a considerable role, although it is dwarfed by the statutory system. Private health insurance in Germany in 1974 covered 16.3 million insured, of whom 10.7 million had supplementary coverage, while 4.5 million had comprehensive coverage. Premiums earned during 1974 amounted to DM 5.8 billion, of which 81.2 percent were reported by the ten largest commercial carriers. Medical care and income loss payments to the insured for the same period totaled 71.8 percent of the premiums. The premium income of the private health insurance amounts to about 12 percent of that of the statutory health insurance system, which at the beginning of 1974 covered 33.2 million members (Germany, pp. 376, 382).

In Australia the Health Insurance Act 1974, the Medibank Act, as amended, establishes a novel mixed system of public and private insurance, operated by a federal institution called Medibank. The basic coverage, consisting of free accommodation and treatment in public hospitals and at least 85 percent of doctors' fees listed in a schedule is available free of charge to persons with low incomes. Persons receiving a higher income pay a levy of 2.5 percent on their earnings, not to exceed specified maximums fixed at different amounts for single persons and families. Extra hospital benefits falling into three classes (including stay in private hospitals) can be secured for additional charges. Finally, wage earners above certain income levels may obtain Medibank private insurance for a fixed charge payable in advance rather than the graduated levy. A person who prefers insurance with a private insurer may refrain from joining Medibank. Private insurers thus remain competitive with Medibank.

New Zealand likewise provides free care in public hospitals and extensive general medical service benefits, though not amounting to comprehensive free medical care. As a result voluntary private health insurance covering stay in private hospitals or additional medical service expenditures payable by the patients are still filling a legitimate need (New Zealand, Royal Commission, chap. 43, secs. 13–22).

STEFAN A. RIESENFELD

[*Directly related are the entries* HEALTH CARE; HEALTH POLICY; *and* POVERTY AND HEALTH, *article on* POVERTY AND HEALTH IN THE UNITED STATES. *Other relevant material may be found under* AGING AND THE AGED, *articles on* SOCIAL IMPLICATIONS OF AGING *and* HEALTH CARE AND RESEARCH IN THE AGED; *and* MEDICAL PROFESSION, *article on* ORGANIZED MEDICINE. *For discussion of related ideas, see the entries:* HEALTH, INTERNATIONAL; HOSPITALS; *and* PUBLIC HEALTH.]

BIBLIOGRAPHY

APELBAUM, JOHANNES. "Krankenversicherung: A. Individualversicherung." Manes, *Versicherungslexikon*, pp. 911–918.
BERKANOVIC, EMIL; REEDER, LEO G.; MARCUS, ALFRED C.; and SCHWARTZ, SUSAN. *Perceptions of Medical Care: The Impact of Prepayment*. Lexington, Mass.: Lexington Books, D. C. Heath & Co., 1974.
BURNS, WILLIAM J. "The Michigan Enabling Act for Non-Profit Medical Care Plans." *Law and Contemporary Problems* 6 (1939): 559–564.
"Chronology of Health Insurance Proposals, 1915–1976." *Social Security Bulletin* 39, no. 7 (1976), pp. 35–39.
DICKERSON, OLIVER DONALD. *Health Insurance*. The Irwin Series in Insurance. Homewood, Ill.: Richard D. Irwin, 1959. Rev. ed. 1963.
DUPEYROUX, JEAN-JACQUES. *Sécurité Sociale*. 5th ed. Précis Dalloz: Droit. Paris: Librairie Dalloz, 1973, pp. 140–171, 341–387.
EVANG, KARL. "Prepaid Medical Care in Norway." *Medical Care and Family Security*, pt. 1, pp. 1–80.
Germany, Federal Republic, Statistisches Bundesamt. *Statistisches Jahrbuch für die Bundesrepublik Deutschland: 1975*. Stuttgart: W. Kohlhammer, 1975.
GIBSON, ROBERT M., and MUELLER, MARJORIE SMITH. "National Health Expenditures, Fiscal Year 1976." *Social Security Bulletin* 40, no. 4 (1977), pp. 3–22.
Health Insurance Institute. *Source Book of Health Insurance Data: 1975–1976*. New York: 1976.
LEAR, WALTER J. "Medical Care in the United States." *Medical Care and Family Security*, pt. 3, pp. 195–344.
MANES, ALFRED. *Versicherungswesen: System der Versicherungswirtschaft*. 2 vols. 3d rev. and enl. ed. Leipzig: B. G. Teubner, 1922. 5th ed., rev. 3 vols., 1932.
———, ed. *Versicherungslexikon: Ein Nachschlagewerk für alle Wissensgebiete der gesamten Individual-und Sozial-versicherung*. 3d ed., rev. and enl. Berlin: E. S. Mittler & Sohn, 1930.
Medical Care and Family Security: Norway: Karl Evang, M.D., England: D. Stark Murray, M.D., USA: Walter J. Lear, M.D. Preface by David Abrahamsen. Foreword by Caldwell B. Esselstyn. Englewood Cliffs, N.J.: Prentice-Hall, 1963.
MUELLER, MARJORIE SMITH, and PIRO, PAULA A. "Private Health Insurance in 1974: A Review of Coverage, Enrollment, and Financial Experience." *Social Security Bulletin* 39, no. 3 (1976), pp. 3–20.
New Zealand, Royal Commission of Inquiry. "Health." *Social Security in New Zealand: Report of the Royal Commission of Inquiry, March 1972. Journal of the House of Representatives of New Zealand* (1972), Appendix, Report no. H53. Wellington: A. R. Shearer, Government Printer, 1972, pt. 8, pp. 392–480.
PEART, HARTLEY F., and HASSARD, HOWARD. "The Organization of California Physicians' Service." *Law and Contemporary Problems* 6 (1939): 565–582.
RIESENFELD, STEFAN ALBRECHT. *Prepaid Health Care in Hawaii*. Legislative Reference Bureau, Report no. 2. Honolulu: University of Hawaii, 1971.
SIMANIS, JOSEPH G. *National Health Systems in Eight Countries*. Health Insurance Statistics: HI Series.

DHEW Publication no. (SSA) 75-11924. Baltimore: Social Security Administration, Office of Research and Statistics, 1975. Bibliography.

STEIN, OSWALD. "Krankenversicherung: B. Socialversicherung." Manes, *Versicherungslexikon*, pp. 918–926.

SURMINSKI, ARNO. "Die PKV-Unternehmen 1974." *Zeitschrift für Versicherungswesen* 26 (1975): 548–551.

United States; Department of Health, Education and Welfare; Social Security Administration. *Social Security Bulletin: Annual Statistical Supplement, 1974.* DHEW Publication no. (SSA) 76-11700. Washington: Government Printing Office, n.d.

———; Department of Health, Education and Welfare; Social Security Administration, Office of Research and Statistics. *Social Security Programs throughout the World, 1973.* Research Report, no. 44. DHEW Publication no. (SSA) 74-11801. Washington: Government Printing Office, 1974.

HEALTH, INTERNATIONAL

International health care comprises those activities for the prevention, diagnosis, and treatment of disease and the underlying causes of disease that require the coordinated consideration and action of individuals, groups, and agencies from more than one country. International health care is the aggregate of those efforts that are directed through collaborative actions toward improving and maintaining the health of the world's population.

Understanding the problem

A distinction should be made between health care and medical care. Health care embraces the latter, with its disease orientation, but entails in addition the identification and eradication of all factors, social as well as personal, that contribute to ill health. Thus, health depends not only on the provision of minimal medical care, but even more on the availability of safe drinking water, sufficient food, protection against communicable diseases and from environmental pollution, and means of fertility control. Many international agreements have spoken of these benefits as basic inalienable rights of every person. Such services, however, are the very ones that are desperately lacking in all poor countries (Gish). It is estimated, for example, that more than one-half of the world's people have no access to modern means of health care (Bryant, p. ix), while more than seventy percent have no access to safe drinking water (*Human Rights in Health*, p. 21).

Standards for safe water, adequate nutrition, and vector control are relatively free of cultural differences. Nonetheless, how these health-care benefits can be provided and how delivery systems are to be financed, constructed, maintained, and managed are dependent on local circumstances. In much the same way the determination must be made in each local situation as to what constitutes the minimal medical care to which every individual has a right. What an individual in New York City may claim as his right in terms of minimal medical care is far greater than that expected by the inhabitant of the New Guinea highlands.

One priority need in international health activities is for purely descriptive information, perhaps through developing methods of measurement to allow for comparisons in economic, social, medical, and technological terms of what health services can be delivered in a given area for an acceptable cost, at specified levels of complexity, and by health personnel trained according to local resources and requirements. Such information would provide the eventual possibility of defining more exactly the limits to the implementation of the rights to minimal medical care in a variety of situations and areas. The difficulties of developing such scales arise from the problems of balancing the impact of competing economic and social priorities within and between participating nations, of ameliorating the influence of political rivalries among countries, of exploring their different perceptions of real health needs in a given region, and of enlisting cooperation. This is obviously more difficult where people feel that they do not have complete control of programs and their outcomes.

Motivations for international health work

Motivations for international health work have been and are particularly complex. Among the motivations are self-interest, political considerations, and seeing health work as constituting reparations for past exploitation.

Self-interest. Past successful international health ventures have all been characterized by recognition on the part of the participating countries of the benefits that would accrue to themselves through their cooperation. This is true of quarantine efforts to control smallpox, cholera, and plague; of agreements regarding biological standardization of pharmaceuticals; of the coordinated treatment of venereal disease among seamen; and of the current work to eradicate malaria and smallpox. When the impact of cooperation on a given country is more remote or difficult to assess, the involvement of that nation in an international health enterprise is more difficult to obtain. Some development of a widespread sense of supra- and internationalism among the community of nations is required if mankind is ever to scale the barriers imposed by nationalism (*Symposium*).

Political considerations. Political considerations have played a dominant role in international health activities. The USSR and the other Cominform countries withdrew from the World Health Organization (WHO) in 1949–1950 in a dispute over the payment of dues and remained out of the organization for seven years. In 1964 South Africa was deprived of voting privileges by the Seventeenth World Health Assembly, and in 1966 Portugal was suspended from regional activities in the African region. Southern Rhodesia was considered to be in suspension as of 1975, while Israel has not been a member of the non-Arab Committee of WHO's Eastern Mediterranean region since 1968 (Goodman, pp. 203–204). The People's Republic of China joined WHO only in 1974. The question arises whether it is proper to exclude countries from WHO on grounds that have little to do with health. It should be possible to gain recognition of the fact that health activities around the world, more than almost any other activity, should be related to, yet privileged and separated from, other political considerations governing international relations.

Reparations for past exploitation. Another recurring issue in international health is the role and responsibility of economically more privileged nations for the health needs of less developed countries. Some argue that part of the present wealth and power of the richer nations derives from past economic exploitation of poorer nations. This position is pressed with increasing persistence in the meetings and agencies of the United Nations. Partly in response to this pressure, the richer nations have somewhat reluctantly been contributing to the development of the poorer nations through such agencies as the United Nations and through such mechanisms as bilateral assistance.

Toward a basis for international responsibility

There is little agreement as to the root of the responsibilities of one nation for another, the range of implied commitments, or the manner in which they might best be fulfilled. A consensus seems to be emerging, however, in some way related to the concept of distributive justice.

Distributive justice. Distributive justice might be considered that species of justice whereby a given social unit, ranging from an individual to a nation, is enabled by virtue of its own effort and responsibility, helped by other social units, to secure its proportionate share in the common good (Messner, p. 293). The "common good," in turn, can be taken as the sum of those conditions of social life by which individuals, families, and groups can achieve their own fulfillment in a relatively thorough and ready way.

In the area of health the common good includes among other things access to safe air and water, adequate nutrition and shelter, and competent medical personnel and facilities. If mental health is included, the common good extends to the provision of the means for education, employment, and recreation.

Both the due to which each social unit has a claim and the social cooperation essential to the actualization of that due lie at the heart of any consideration of distributive justice. Cooperation presupposes a community in which it operates. The existence of such a community is often the clearest when it is a case of individuals constituting a nation. It is less clear, but no less true, in the case of nations relating to one another. On the basis of propinquity, common interests, and the mutual interdependence needed for a true society, the world today is truly a community of nations. As such, considerations of distributive justice apply to the world community (Cronin, p. 316).

The practical realization of the demands of distributive justice on the international level raises conflicts pertaining both to the determination of what is due to each nation and to the manner in which the cooperation necessary for its sharing is best achieved. Mankind is just now approaching the question of how much any one country can legitimately expect other nations to assist it in fulfilling its responsibilities to its own people.

It is easier to say what international distributive justice forbids than what it demands in matters of health and medical care. It requires that no nation engage in policies or practices that work against the basic health rights of other nations. In this regard, one very evident failure of justice occurs when clearly definable obstacles are placed in the way of a nation's fulfilling its responsibilities to ensure and enhance the health of its citizens. Some of the more obvious examples of such obstacles are encouragement of physician migration, atmospheric and oceanic pollution through nuclear testing and industrial practices, and nationalistic practices and policies designed to neutralize the effectiveness of cooperative international health efforts. Less clear examples are all those international exchanges that serve to undermine economic and social development on which any health progress is dependent. These include the Cold War, extreme forms of nationalism, and trends toward isolationism.

Continuing problems in international health

Among the most serious dilemmas in health on a worldwide scale concern the interaction between food supplies, population growth, environmental sanitation, economic poverty, and the availability of medical services. Health is so dependent on each of these factors that the lack of any one can be considered a proximate danger to the safety of a community. These are issues with which each country struggles, over which no one has yet sufficiently triumphed, and which place extremely heavy demands on the individual nation and the community of nations for their solution.

Extragovernmental bodies. Complicating efforts at international cooperation in the health field is the fact that national governments have only a partial role in resolving or aggravating problems in international health. Much of the work in international health is done by extragovernmental bodies such as missionary organizations, private foundations, and business organizations. Even more pervasively, professional organizations or private commercial interests in one nation may create situations that interfere with the health needs of individuals in other countries.

For example, private pharmaceutical companies may carry out research in other countries that would not be sanctioned in their parent country. The subjects of such research may be inhabitants of poorer nations or less privileged classes of any country. Thus foreign drug firms have done research on high-vitamin diets in Taiwan and on antischistosomiasis preparations, which were later found to be carcinogenic, in Brazil. Research on contraceptives has become particularly sensitive because of inflated fears about genocidal intent. The questions of exploitation and informed consent that such projects raise lead inevitably to demands for international regulation. Similar issues are raised by international commerce in biologically active products, such as trade in poorly standardized pharmaceuticals or the purchase of blood by private American concerns from Haitian nationals. The government of the country whose people may be subject to exploitation has the primary responsibility for controlling such programs. Because their capacity for scientific regulation is often limited, it may sometimes be easier for such nations simply to ban all foreign research within their borders.

Migration of medical personnel. International migration of medical personnel provides a significant example of the complexity of international health problems. Between 1961 and 1971 almost seventy-six thousand physicians entered the United States as immigrants or exchange visitors. This migration raises concerns about international exploitation because of the disparity in medical resources between the United States and the nations from which the doctors come. Studies have shown that countries such as Turkey have given more foreign aid to the United States through investment in training doctors than they have ever received in health assistance from America (Taylor, Dirican, and Deuschle).

The migration of physicians has been facilitated by a continuing surplus of house-staff training positions in American hospitals and the many appealing professional opportunities for medical practice in the United States. American hospitals have an understandable interest in encouraging this migration in order to have sufficient staff to provide the best care possible to the populations they serve. Yet, if they deplete the medical personnel of more needy countries, these hospitals incur the serious obligation to ensure that the health of their communities is not purchased through the irresponsible enticement of foreign-trained personnel. This obligation must be exercised by setting up controls in cooperation with concerned organizations such as the American Hospital Association, the American Medical Association, the Association of American Medical Colleges, and the individuals and institutions they represent. To the extent that the United States with all its affluence has thus far failed to establish a medical care system based on its own production of personnel, it is failing in its international responsibilities. For the long term, specific policies are needed to begin to redress the world imbalance of doctors and other health personnel. Official government agencies have control over the immigration and visa apparatus under which foreign-trained physicians enter and remain in a country; in addition, such government agencies alone enjoy a direct relationship with the governments of those countries that lose their physicians.

Yet the question arises whether restrictive national policies should be developed that encroach on individual rights to travel and migrate. The United States government, for instance, faces the dilemma of whether to reintroduce some of the restrictive policies that were changed in the immigration reforms of 1965. Currently physicians are granted immigration preference in the United States because there is considered to be a doctor shortage in that country. The determination of this shortage is based more on demand for services than on actual need, and the demand for physician services continues high. National attention should be directed, therefore, not only to immigration laws but also to increased investment in the training of physicians and on delegating responsibilities to other members of the health team so as to provide more efficient and economical health care.

Responsibility for correcting the injustice implicit in the international migration of physicians remains also with the countries losing their physicians. The common tendency to train doctors according to urbanized and hospital-based standards rather than meet the needs of large rural populations, and a medical education that emphasizes the theoretical and the specialized rather than the real needs arising from rural poverty, present the physician with the choice between emigration and either unemployment or employment under very difficult circumstances.

Conclusion

To achieve distributive justice in international health activities, then, each country must assume the responsibility, take the initiative, and make the investments needed to solve its own health-care problems, while at the same time shouldering its responsibility as part of the world community. In particular, the more affluent countries must be careful to curtail exploitative practices, showing great restraint over any impulse to "sell" Western technology, medical care patterns, and forms of financing to peoples who can benefit more from their own established social systems or from a new synthesis of health services more appropriate to local conditions. This will require the exercise of international humility and a reversal of any paternalistic domination of the health-care systems of developing countries.

JAMES R. MISSETT AND CARL E. TAYLOR

[*Directly related are the entries* ENVIRONMENTAL ETHICS; FOOD POLICY; HEALTH POLICY; MASS HEALTH SCREENING; *and* PUBLIC HEALTH. *Other relevant material may be found under* POVERTY AND HEALTH, *article on* POVERTY AND HEALTH IN INTERNATIONAL PERSPECTIVE. *For discussion of related ideas, see:* DRUG INDUSTRY AND MEDICINE; *and* FUTURE GENERATIONS, OBLIGATIONS TO.]

BIBLIOGRAPHY

BRYANT, JOHN. *Health and the Developing World.* Ithaca, N.Y.: Cornell University Press, 1969.

CAHILL, KEVIN M., ed. *The Untapped Resource: Medicine and Diplomacy.* Maryknoll, N.Y.: Orbis Books, 1971.

CRONIN, JOHN F. "Our International Obligations: Thoughts on the International Common Good." *World Justice* 1 (1960): 298–317.

FRY, JOHN, and FARNDALE, W. A. J., eds. *International Medical Care.* Oxford: Medical & Technical Publishing Co., 1972.

GISH, OSCAR. *Doctor Migration and World Health.* London: G. Bell, 1971.

GOODMAN, NEVILLE M. *International Health Organizations and Their Work.* 2d ed. Baltimore: Williams & Wilkins Co.; Edinburgh: Churchill Livingstone, 1971.

Human Rights in Health. Ciba Foundation Symposium, n.s. no. 23. Amsterdam, London, and New York: Associated Scientific Publishers, 1974.

MESSNER, JOHANNES. "International Social Justice." *World Justice* 3 (1962): 293–309.

NOZICK, ROBERT. "Distributive Justice." *Philosophy and Public Affairs* 3 (1973): 45–126. Edited reprint in *Anarchy, State, and Utopia.* New York: Basic Books, 1974, pp. 149–231.

Symposium on Health of Mankind. Ciba Foundation 100th Symposium. Edited by Gordon E. W. Wolstenholme and Maeve O'Connor. Boston: Little, Brown & Co., 1967.

TAYLOR, CARL E., DIRICAN, RAHMI; and DEUSCHLE, K. W. *Health Manpower Planning in Turkey.* Baltimore: Johns Hopkins University Press, 1968.

Teamwork for World Health. Ciba Foundation Symposium. Edited by Gordon E. W. Wolstenholme and Maeve O'Connor. London: J. & A. Churchill, 1971.

WARWICK, DONALD. "Contraceptives in the Third World." *Hastings Center Report* 5, no. 4 (1975), pp. 9–12.

HEALTH POLICY

I. EVOLUTION OF HEALTH POLICY
    Stephen P. Strickland

II. HEALTH POLICY IN INTERNATIONAL
    PERSPECTIVE   *Odin W. Anderson*

I
EVOLUTION OF HEALTH POLICY

History of the concept

Public policies on health matters are probably as old as civilization's first governments, but health policy as a discrete area of decision making and a distinct field of study has emerged only recently. In Babylon, about two thousand years before Christ, the Code of Hammurabi regulated the practice of medicine in part by specifying a system of rewards and punishments for surgeons, depending on whether the operations they performed were successful or not. A health policy of vital, enduring consequence was made by the Consuls of Rome in the first century A.D., when they encouraged the construction of aqueduct and sewage systems for the young city-state's burgeoning population.

Social policies regarding health, including those formulated and followed by health professionals vis-à-vis the public, may not always require governmental action. Hippocrates' explanations of medicine as science, though they did not go unchallenged, evolved into a set of principles that governed the physician's role for more than two thousand years. In many societies, that role was long dominant in health matters. Since, despite the Babylonian example, few societies regulated the practice of physicians through their government, the Hippocratic Oath, though taken as a private, professional pledge, was in effect the backbone of a social health code. Nonetheless, policymaking ordinarily implies governmental decisions. One definition of policy is, simply, a strategy adopted by governments for the solution of problems (Sundquist). Yet, as some of the illustrations that follow suggest, a public policy decision may restrict or exclude an active role for government.

History of health policy in the United States

The parameters of health policy have broadened gradually through the centuries. In the United States, substantial enlargement of the federal government's role in health affairs since World War II has given health policy a clearer meaning, although there are still those who assert that there is no U.S. national health policy. What such commentators probably mean is that the various governmental strategies for enhancing health status do not add up to a comprehensive, or even a coherent, national health policy. Be that as it may, it is important to understand that policymaking may involve decisions that are direct or indirect, comprehensive or incremental, positive or negative (Strickland, *Politics*, pp. x–xi).

In the early days of the American Republic, most statesmen believed that, by constitutional divisions, the states had primary responsibility for ensuring, insofar as any government could, the health status of the people. Isolated departures from this perceived barrier seemed to prove its practical good sense as well as its philosophical soundness. In 1813 the Congress, with the endorsement of President Thomas Jefferson as well as that of President Madison, authorized the purchase and free distribution of cowpox vaccine to physicians throughout the states to fight off a growing epidemic of smallpox, following Dr. Edward Jenner's novel and successful experiment in England. Unfortunately, in an early caricature of the bureaucratic tradition, the wrong vaccine got shipped to North Carolina, and several deaths resulted. For many decades thereafter, the control of epidemics, through quarantine and other measures, was left to the states.

In the course of arguing federal versus state responsibility in the matter of quarantine laws in 1796, Representative Samuel Lyman of Massachusetts maintained that the states' "right to preserve health and life was inalienable" (Chapman and Talmadge, pp. 334–335). Retrospectively, one might find in that argument an early inference of a corresponding right to health for the people, but no such right was actively asserted in other U.S. legislative assemblies for another century and a half. In fact, while state and municipal governments were occasionally looked to for responsive actions in situations of social emergency, and more consistently for provision of standards of hygiene and later for licensure of physicians, self-reliance was the personal credo and the de facto health policy of U.S. society from colonial days until recent years. There were exceptions to the rule. The federal government was thought to have responsibility for providing direct medical care for those in its military and naval forces and for merchant seamen. Later, the responsibility was extended to other special groups: lepers, Indians, and, after the First World War, veterans.

Theodore Roosevelt had been much interested in Bismarck's pioneering social legislation and was especially keen on Germany's Sickness Insurance Act of 1883, which guaranteed medical treatment and cash benefits for sick or injured wage-earners. As Progressive Party candidate for President in 1912, Roosevelt made national health insurance a plank in his new party's platform. Another half-century would pass before the idea would be accepted that it was appropriate for the federal government to underwrite medical insurance for large groups of citizens, other than the special wards.

In the intervening years, most of the national health legislation in the United States had been in the form of federal grants to the states. As part of Franklin Roosevelt's New Deal, grants were transmitted through the U.S. Public Health Service to assist state and local public health agencies fight tuberculosis, pellagra, and venereal diseases, and even to train health professionals in the process. In 1946 the Hill–Burton Hospital Act, while following the traditional approach of federal grants to states, nonetheless enlarged the federal health role in a major way. From the time of the bill's enactment to 1975, four billion, one hundred million federal dollars had helped construct 6,584 hospitals. Meanwhile, wartime successes in biomedical research and development (such as the first widespread production and use of penicillin) bolstered support for expanding the federal government's role in another area: support for biomedical research, in this case mainly through grants to researchers housed in academic institutions.

In most instances of expansion of the federal role, even indirectly through grants to states or local institutions, the justification has been the existence of a crisis. The New Deal grants followed upon the results of the first U.S. national health survey in the early 1930s. Those results showed that what had been believed to be a generally healthy citizenry contained population pockets distinguished by chronic health problems, including some statistically doomed to short lives. The 1941–42 military draft produced an alarming figure: fully one-third of the young men examined were disqualified for service for physical or psychological reasons (Strickland, *Politics*, p. 19). After the war, the acute shortage of hospital beds provided an unassailable justification for the very large federally supported, state-run hospital construction program that was to follow. Gradually, by an aggregation of incremental, separate actions, a kind of national health policy evolved in the United States. President Harry Truman tried to promote a more consistent, unified policy, calling in 1947 for a balanced program of (1) construction of health facilities, (2) increased support for biomedical research, (3) federal aid to medical education, and (4) a national health insurance plan that would make certain that no one went without medical care because of inability to pay. He was successful in securing congressional approval of the first two components and beaten badly on the third and fourth. The practical reason for the mixed result was obvious at the time; the American Medical Association (AMA) fought national health insurance and aid to medical schools as "socialized medicine," supported hospital construction, and was neutral on federal support of medical research. Constituting the most prominent and powerful advisers to most policymakers in that era, the AMA spokesmen carried their case. The division also illuminated an unwritten, unenunciated rule of U.S. national health policy: governments, whether federal, state, or local, should espouse no policies, nor launch any programs, that would impinge upon the traditional manner of the practice of medicine.

The defeat of the proposal, in 1949, to provide federal support for medical schools, despite their chronic financial condition and the widespread recognition of a national shortage of physicians, was a somewhat more complicated affair than the simultaneous defeat of the Murray–Wagner–Dingell national health insurance bill. But that defeat (like the temporary demise of the possibility of national health insurance) was just as clearly a policy decision as was, for example, the legislative victory represented by the enactment of the Hill–Burton Act. In short, from 1949 to 1963 it was national policy *not* to provide direct federal funds for the education and training of physicians.

Those organizations that lined up to support or to oppose one or more features of the Truman administration's health policy proposals, together with the respective outcomes of those proposals, tell us much about the political context of health policymaking in the United States in the 1940s and 1950s. The AMA was clearly the dominant interest group of that period, although organizations representing the other segments of what became identified as the "health care industry" increased in number and strength throughout from the war's end. Principal among them were the Association of American Medical Schools, American Hospital Association, American Nurses's Association, the Health Insurance Association of America, and the American Public Health Association (APHA).

In the 1960s other health interest groups

came to the fore. These included the Committee of 100, supported by, among others, the United Auto Workers and other labor unions, which has the passage of a comprehensive health insurance plan as its only goal. Consumer advocate Ralph Nader established a Health Research Group, which represents no identifiable industrial or professional forces and takes an active interest in many national health issues in the name of the average consumer. One group of organizations especially successful in influencing health policy in the last quarter-century consists of those promoting the cause of research on specific diseases, including the American Cancer Society and the American Heart Association. Support also came from within the policy establishment. Specifically, from the early 1950s through the middle 1960s Congress consistently provided more funds for biomedical research (through appropriations for the categorical disease programs—e.g., cancer, heart, arthritis, infectious diseases—of the National Institutes of Health) than whatever the administration in power proposed.

In 1965 over one-third of the health budget of the U.S. Department of Health, Education, and Welfare (DHEW) was spent on medical research. The other major item in that budget was for hospital construction. In that year the scope and the philosophy of U.S. national health policy was radically altered. Under the prodding of President Lyndon Johnson, Congress enacted Medicare—providing medical insurance for the elderly—and Medicaid—providing federal subsidies for state programs that paid for medical care for the poor. For the first time in its history, the federal government thereby asserted responsibility for financing medical care for segments of the general population other than its historical wards.

The adoption of those programs caused related practical changes in the policy arena. First, the concept of national health insurance was freed from entrapment in debates over socialized medicine; now particular alternative approaches could be discussed on their merits. Second, where funds for financing medical care for citizens (nonmilitary, nonveteran) previously represented a small portion of the federal health budget, by 1976 Medicare and Medicaid had raised that component to sixty-eight percent of the $37.7 billion budget (*Special Analyses*, p. 169).

Meanwhile, relationships among the health interest organizations have changed. Earlier the AMA could often command the support of the major organizations representing nurses, hospitals, and the health insurance industry on issues of common concern. This is no longer the case. By 1975 each such organization was promoting its own national health insurance measure, each asserting the importance of a federal role but beyond that diverging widely in its delineation of that role, the populations to be covered, the financing and administrative mechanisms, and the scope of benefits to be provided.

Another phenomenon has occurred in the last decade that, combined with the others, seems to signal a new era in health policy in the United States. Groups of analysts outside the health lobby organizations have developed who are primarily concerned about health policy issues. Health policy centers or programs, some free-standing and some affiliated with educational institutions, usually involving professionals from various disciplines, are now in operation across the country. In addition to performing nonpartisan, cross-disciplinary analysis, and in some cases training students in health policy analysis, these new organizations are already supplementing in a very important way the expertise that is sometimes available and always proffered to policymakers by organizations representing particular segments of the health-care industry or special advocacy groups.

Responsive and responsible health policy

In individual and broad social terms, good health historically has been accorded high value. Benjamin Disraeli said: "The health of the people is really the foundation upon which all their happiness and all their powers as a state depend." President Truman's Commission on the Health Needs of the Nation asserted: "For the State, health is the wellspring of a nation's strength, its provision and protection one of the first obligations" (*Building*, p. 1). Thus the health policies of nations, from ancient days to the present, have had an implicit ethical base. Governmental interventions in the practice of medicine, the eradication of disease sources, the development of new biomedical knowledge, and the financing or provision of health services have traditionally been justified by reference to "the common good."

An important new element in health policy is the emergence of a growing number of ethicists, and others with different training but similar concerns, who have encouraged the systematic examination of values and assumptions underlying or prompting particular social attitudes and actions. They have helped elevate and illuminate the ethical issues that are almost always involved in health policy decisions but often in the past have been treated only implicitly, if at all. Thus today the matter of "distributive justice" is sure to be considered, alongside questions of technical efficacy and cost, when policymakers at several levels next address the issues of providing public funds for kidney transplants. And ethicists and sociologists as well as physicians, medical researchers, and lawyers have served on a national commission to propose guidelines for the protection of human subjects in biomedical and behavioral science research projects. In current public deliberations about a whole range of current health policy issues—from health resource planning and coordination, to review systems for ensuring medical care quality, to strategies for encouraging young physicians to serve in areas where doctors are sorely needed—considerations of equity loom as large as economic factors, and social responsibility and personal liberty are being weighed specifically, if with difficulty. The presence of professional ethicists in policy deliberations will not inevitably produce more ethical outcomes, but the systematic examinations of ethical concerns should be welcomed.

The burgeoning of interest in health policy, in the United States and other countries, is an inevitable result of an enlarged governmental role in health. In most Western nations, an increasing percentage of the gross national product is being spent for health purposes. In 1974 Americans spent, directly or indirectly, more than one hundred billion dollars for health-related purposes, almost eight percent of the gross national product. In Sweden, the Netherlands, West Germany, France, and Austria the percentage has been slightly lower but comparable (Maxwell, p. 18). Meanwhile, improvements in health status are not widely discernible. The principal, overarching health policy issue facing these and other countries today is whether the large investments being made are bringing appropriate dividends. Supplementary questions include whether the health resources being relied upon are in an appropriate mix and whether those resources are being managed and coordinated in optimal ways.

An increasing number of citizens are manifesting a desire to participate, at least indirectly, in shaping the answers to some of these questions and in helping resolve some of these issues. In the United States the persistent public interest, the heightened concern of health providers about their future roles, the proliferation of health policy study centers, and the more consistent attention to ethical issues mean that the public policy process as it concerns health is going to be livelier than ever in the years ahead. This could mean that decisions will be more difficult to arrive at than before; it will not necessarily guarantee that they will be sounder. Still, wherever in the world the combination of forces described here is at work, they can only make policymaking itself a more open, hence a healthier, activity. And it is just possible that the consequent decisions may produce healthier results for the people.

STEPHEN P. STRICKLAND

[*Directly related is the other article in this entry:* HEALTH POLICY IN INTERNATIONAL PERSPECTIVE. *Also directly related are the entries* MASS HEALTH SCREENING; POVERTY AND HEALTH; PUBLIC HEALTH; *and* RESEARCH POLICY, BIOMEDICAL. *Other relevant material may be found under* CHRONIC CARE; ENVIRONMENTAL ETHICS, *articles on* ENVIRONMENTAL HEALTH AND HUMAN DISEASE, *and* QUESTIONS OF SOCIAL JUSTICE; GENETIC SCREENING; HEALTH CARE, *articles on* RIGHT TO HEALTH-CARE SERVICES *and* THEORIES OF JUSTICE AND HEALTH CARE; *and* MENTAL HEALTH, *article on* MENTAL HEALTH IN COMPETITION WITH OTHER VALUES. *For discussion of related ideas, see the entries:* AGING AND THE AGED, *article on* HEALTH CARE AND RESEARCH IN THE AGED; DEATH AND DYING: EUTHANASIA AND SUSTAINING LIFE, *article on* PROFESSIONAL AND PUBLIC POLICIES; *and* INFANTS, *article on* PUBLIC POLICY AND PROCEDURAL QUESTIONS.]

BIBLIOGRAPHY

The Budget of the United States Government, Fiscal Year 1976. Washington: U.S. Government Printing Office, 1975, pp. 129–134.

Building America's Health. Report to the President's Commission on Health Needs of the Nation. Vol. 1: *Findings and Recommendations.* Washington: Government Printing Office, 1952.

CHAPMAN, CARLETON B., and TALMADGE, JOHN M. "Historical and Political Background of Federal Health Care Legislation." *Law and Contemporary Problems* 35 (1970): 334–347.

HARRIS, RICHARD. *A Sacred Trust.* New York: New American Library, 1966.

MAXWELL, ROBERT. *Health Care: The Growing Dilemma.* A McKinsey Survey Report. New York: McKinsey & Co., 1974.

RUSSELL, LOUISE B.; BOURQUE, BLAIR B.; BOURQUE, DANIEL P.; and BURKE, CAROL S. *Federal Health Spending, 1969–74.* Washington: Center for Health Policy Studies, National Planning Association, 1974.

SIGERIST, HENRY E. *Civilization and Disease.* Chicago: University of Chicago Press, 1943.

Special Analyses, Budget of the United States Government, Fiscal Year 1976. Washington: U.S. Government Printing Office, 1975, pp. 169–195.

STRICKLAND, STEPHEN P. *Politics, Science, and Dread Disease.* Cambridge: Harvard University Press, 1972.

———. *U.S. Health Care: What's Wrong and What's Right.* New York: Universe Books, 1972.

SUNDQUIST, JAMES L. *Politics and Policy: The Eisenhower, Kennedy, and Johnson Years.* Washington: Brookings Institution, 1968.

TORREY, E. FULLER, ed. *Ethical Issues in Medicine: The Role of the Physician in Today's Society.* Boston: Little, Brown & Co., 1968.

II
HEALTH POLICY IN INTERNATIONAL PERSPECTIVE

All private and public enterprises are immersed in ethical and public policy issues. This is particularly true of health services because of their lifesaving and pain-relieving functions. This article will examine the various types of health service systems that have emerged in industrialized countries and their bearing on the rights and responsibilities of the individual and the state.

Types of health services

The activities classified under the health sector of a society can be subclassified under public health and personal health services. The public health activities are directed toward whole populations and treat the individual only incidentally. The personal health services activities are also directed to whole populations, but the central concern is the individual patient in a relationship to a professional trained in individual health services. In public health activities the state takes the initiative to protect the public from communicable diseases transmitted by impure water and food, uncollected garbage, and population congestion. The public policy objective is to protect a population from its own effluents, and this aim can be carried out only by collective action. In personal health services activities the initiative for gaining attention must be taken by the individual adult or by a parent for a child. Collective action through government can take the form of assuring an adequate supply of health services facilities and personnel and financial mechanisms through private or public means, so a family is not ruined financially in the event of a costly medical care episode.

Hence health maintenance of the public, as described under public health activities, must by its very nature be a collective and governmental responsibility; historically this has been the case. Health-care services for illnesses and injuries do not inherently need to be a collective and governmental responsibility; in fact, in Western countries that have undergone the industrial revolution, the rise of the middle class, parliamentary government, and the emergence of the free market, there has been and continues to be a mix of private and public financing and responsibility. The trend, however, has been toward increasing public responsibility and funding, so that the private sector is diminishing in importance. There are some problems the private sector is inherently incapable of solving, e.g., distributing services to population groups that do not have the capacity to purchase. The rationing of the marketplace has been replaced by the rationing of free services funded by taxation. The latter is assumed to be more equitable.

Ethical issues and public policy

Who is my brother? What is our obligation to share with strangers? To what extent should some individuals be asked to sacrifice for others? To what extent is collective action mutually sustaining among individuals so that all gain by being able to control external forces better by collective action rather than by purely individual efforts? It is obvious that classical public health measures (e.g., pure water) are mutually sustaining for the individual and the group. It would seem that ethical considerations intrude here less than in personal health services. In personal health services, however, which entail the pain and anxieties of individual sufferers seeking succor from another individual—a health services professional—a host of ethical problems emerge involving both the individual and the collective.

In this regard, it would seem to be significant that no country has experienced a purely market-oriented type of personal health services delivery system. This would be a system in which the hospitals would be operating for profit, providing services for those who could afford to buy them as they would a car or a refrigerator. A patient who suffered a heart attack or a serious injury would be denied admission to the hospital unless he could prove he had the money. Physicians likewise would be under no obligation to provide charity care, as was traditional in pre–national health insurance arrangements in

European hospitals. At best, government would buy services from the profit-oriented private sector for people in low-income groups, assuming the state felt so obliged.

It seems reasonable to surmise that personal health services have been left out of the pure market and profit sector of society because they are an extension of the caring functions of the family. Fundamental human needs for nurture, affection, and care are regarded as beyond calculable price, and their satisfaction should not depend on purchasing power in the market place. This is not to say, of course, that there is no price and that costs can be ignored. The family unit needs an economic base of some kind, but it does not operate on a strict profit and loss basis with regard to members within it.

Modern, scientifically oriented health services did not emerge until the later nineteenth century. The infrastructure of facilities and personnel, as we know it today, was not really shaped until the 1920s, and the rapid cost acceleration did not occur until after World War II. In Western Europe and North America the personal health services were then shaped on the liberal-democratic economic and political model, which divided economic and social functions according to private and public sector responsibilities. In the richer countries, like France, Germany, Great Britain, and the United States, the modern hospital continued for the most part to be privately owned by churches or non-profit, self-perpetuating community boards. Physicians were usually private practitioners selling their services to whoever could afford to pay, and also normally providing free care in the hospitals to charity patients in return for experience and access to private patients. Health services for the poor (classified as paupers and low-income working class) were paid by philanthropy or public funds. In other Western European countries, such as the Scandinavian countries, the hospitals were normally owned by the local governments, and hospital-based physicians were salaried. The middle-class private sector was too small to sustain a hospital system. Enough purchasing power emerged, however, to sustain a growing class of physicians in private practice.

Availability of personal health services was regarded as a collective responsiblity no matter how grudgingly implemented, although free care was limited to those who were defined as destitute. Individuals at other income levels were expected to pay for their services as used. Health services were clearly regarded as a community good not to be denied merely because of inability to purchase.

In personal health services systems in highly developed industrial countries the service units, such as short-term and long-term hospitals, physicians and their variety of specialties, nurses, pharmacists, laboratory technicians, and related personnel are quite similar types in the various countries. This means that medical science itself has created the same service unit types in response to the technology of medicine. What differs widely is how these service units are organized and funded and the underlying rationale for their creation.

In time, the financing of personal health services turned increasingly to obligatory taxation of some sort, from payroll deduction to personal income tax. There was a shift from a noblesse oblige and charity concept directed to paupers and the working class to a general system for everybody. As personal health services became more costly, particularly hospital-based services, the middle classes also began to support some sort of action through government. Voluntary health insurance plans were regarded as inadequate and inequitable. During the time before and after World War I when the relative merits of voluntarism versus compulsion were debated, issues of individual self-reliance versus the collective were raised. Issues of regressive versus progressive tax methods were hotly discussed because they meant taking from those who had and giving to those who had not. Underneath it all was a rationale of protecting family financial solvency across all income levels, because all levels were threatened. It was an ethic of middle-class prudence. The insurance or contingency rationale was a natural one in a liberal-democratic and largely capitalistic type of economic and political system.

Great Britain and the USSR

Eventually, the personal health systems moved more toward a health services concept, particularly in Great Britain in 1946, after twenty-five years of experience with a limited health insurance scheme covering the services of general practitioners for workers below certain incomes. Characteristic of this concept is the provision of complete services rather than the indemnification of costly episodes of illness. There was a shift from buying and selling to providing—essentially, a shift from self-reliance to collective responsibility. Increasingly, a concept of maintaining the general health of the people both through public health measures and through personal health services, including early diagnosis and rehabilitation, came into prominence. It was envisioned in a democratic framework as being for the benefit of the individual. The benefit to society in the form of less absence from work, greater productivity on the job, better learning ability in school, and so on were alluded to but not emphasized.

The health services system in the USSR, which was formulated after Lenin came to power in 1922 and took its current shape by the 1930s was fashioned in a very different economic and political system. The USSR system is the extreme of collectivistic concepts and organizations. The USSR was able to start from almost nothing, because at the time of the Revolution there was a weak infrastructure of hospitals and physicians. Such private practitioners as were left in the country either were liquidated or else joined in the development of an exceedingly rational public health service system—that is, rational as judged from formal organizational norms of planning, coordination, centralized funding, and centralized control. There were no vested interests to contend with.

Central to the USSR concept is collective party responsibility for the health of the population in order to increase production among workers and to assure a future healthy population through extensive provision for maternal and child health programs. By Western standards the USSR system is generously conceived, judged by the number of hospital beds, physicians, and auxiliary personnel, and the great use of services. It appears that the USSR places a high priority on health services in relation to other consumer goods and services.

The USSR health service appears to be the acme of collectivistic concern for the peoples' illnesses, leaving exceedingly little to individual initiative. There is a great deal of health surveillance of the population in well-baby health centers and compulsory annual physical examinations at places of work.

Equalizing access to health services

A value common to all these systems is the drive to equalize access in some way, first by reducing or eliminating the cost to the patient at time of service, and second, by distributing the services so that access is not discouraged by distance or inadequate beds and personnel.

The elimination of cost at time of service is accomplished by making a health service universal and compulsory. The assurance of equal distribution of services units as well as uniform and high quality are much more difficult to implement. All systems so far are only approximating these objectives. No system seems to be willing to allocate the amount of resources necessary for relatively equal distribution. There are other priorities.

The most readily measurable indication of equality is access to services comparing income groups, age, sex, and residential areas. Since illnesses are more common in low-income and upper age groups, the presumption is that low-income groups should, in fact, have more services than upper-income groups, because they need more. Hence, an average across-the-board indicator of use would still imply inequality. A perverse indicator would be one in which all income groups and classes die of the same causes in equal proportions, since all must die eventually.

In all developed countries the trend is toward the USSR and British models (although Western countries are likely to stop far short of the relatively lavish number of services units and the centralized control in the USSR model). The convergence takes shape in the strong tendency to create health services delivery regions and structures for easier monitoring of costs through strict budget controls, putting more and more physicians on salaries, regulating the costs of hospitals, and long-range planning. The objective is to close the seeming open-endedness of health services demands and costs. Still, there is no necessary relationship between how much a country spends on its health services and the organizational structure in which services are provided. Basically, each country provides services as determined by public policy rather than by its administrative and cost control mechanisms. No country has yet dared (with the possible exception of the USSR) to find out what the saturation level of public demand may be. Other priorities have intervened.

To counteract the seeming inexhaustible demands for health services there are governmental attempts to influence changes in lifestyles deleterious to health, such as overeating, lack of exercise, and smoking, in order to reduce the demands on and the cost of the health services. Whether a more closed system will be more egalitarian than an open system characterizing most Western countries can by no means be easily answered. It will depend in large part on how generously conceived the closed system will be in keeping queuing at a minimum and other methods of rationing services. As heroic lifesaving measures proliferate (e.g., renal dialysis), ethical problems of rationing will intensify. So far it seems that all systems have provided more or less what the public has asked for. Now, however, the still increasing volume of this demand is worrying politicians. They are looking for methods to curb the costs.

Accusations against the public by public officials are increasing, viz., that the public is requesting a great many unnecessary services. Further, physicians are accused of overprescribing. It seems that the proportions of gross national product that countries are willing to allocate to their health services are approaching limits within their taxing tolerances. The range currently is from five to eight percent, Great Britain being on the low side and Sweden, Canada, and the United States on the high side. (It is higher in the USSR.) In view of the aging populations it is reasonable to assume that expenditures as a percentage of gross national product will push toward ten to twelve percent. Eventually, there may be centrally established norms to give the people what is regarded as need by certain arbitrary standards rather than what the public wants. If the public does not get what it wants, the possibility arises that it will then create a private sector, unless a private sector is proscribed, an increasing possibility.

This convergence of trends will be likely to stabilize at various levels of private–public sector relationships, concepts of adequate quality, and costs, depending on the individual–collective balance in the continuum of health services systems from the United States to the USSR. Other countries will range within this continuum. The tension between individual and collective responsibilities and public debates on equality will continue. The tension will continue because it seems to be politically difficult for the body politic to arrive at an acceptable concept of distributive justice. Some people will continue to have more than others. The best that can be done is to narrow the range between the poor, the better off, and the rich.

ODIN W. ANDERSON

[*Directly related is* HEALTH CARE, *articles on* HEALTH-CARE SYSTEM, RIGHT TO HEALTH-CARE SERVICES, *and* THEORIES OF JUSTICE AND HEALTH CARE. *For discussion of related ideas, see the entries:* JUSTICE; POVERTY AND HEALTH; PUBLIC HEALTH; *and* RATIONING OF MEDICAL TREATMENT. *Other relevant material may be found under* CARE; HEALTH INSURANCE; *and* HEALTH, INTERNATIONAL. *See also:* MEDICAL ETHICS, HISTORY OF, *section on* EUROPE AND THE AMERICAS, *articles on* BRITAIN IN THE TWENTIETH CENTURY *and* EASTERN EUROPE IN THE TWENTIETH CENTURY.]

BIBLIOGRAPHY

ANDERSON, ODIN W. *Health Care: Can There Be Equity? The United States, Sweden, and England.* New York: John Wiley & Sons, 1972.

———. *Health Services in the USSR.* Selected Papers, no. 42. Chicago: Graduate School of Business, University of Chicago, 1973. Available from the University of Chicago, Graduate School of Business.

DOUGLAS-WILSON, IAN, and MCLACHLAN, GORDON, eds. *Health Service Prospects: An International Survey.* Lancet and Nuffield Provincial Hospital Trust. Boston: Little, Brown & Co., 1973.

FRY, JOHN. *Medicine in Three Societies: A Comparison of Medical Care in the USSR, USA, and UK.* New York: American Elsevier Publishing Co., 1970.

GLASER, WILLIAM A. *Paying the Doctor: Systems of Remuneration and Their Effects.* Baltimore: Johns Hopkins Press, 1970.

———. *Social Settings and Medical Organization: A Cross-National Study of the Hospital.* New York: Atherton Press, 1970.

MECHANIC, DAVID. "Ideology, Medical Technology, and Health Care Organization in Modern Nations." *American Journal of Public Health* 65 (1975): 241–247.

WILENSKY, HAROLD L. *The Welfare State and Equality: Structural and Ideological Roots of Public Expenditures.* Berkeley: University of California Press, 1975.

HEALTH SCREENING
See MASS HEALTH SCREENING.

HEART TRANSPLANTATION

Historical development

The idea of transplanting human organs and tissues is not novel—the *Iliad* described a "chimera" and later, in the third century B.C., a Chinese physician is reported to have drugged two soldiers, opened their chests, and exchanged their hearts.

The modern origins of cardiac grafting date from 1905, when Alexis Carrel and C. C. Guthrie attached the heart of a small dog into the neck of a larger dog. Thereafter the work of F. C. Mann and others (1933), H. Sinitsyn (1948), and E. Marcus and others (1953) continued with heterotopic (outside the pericardial cavity) animal experiments. In 1958 M. Goldberg, E. F. Berman, and L. C. Adman performed a series of orthotopic (within the pericardial cavity) canine heart transplantations, with animal survival time ranging from 21 to 117 minutes. Two years later, R. R. Lower and N. E. Shumway reported that their dogs survived for twenty-one days posttransplantation.

Almost two decades earlier, P. Medawar and his associates had described and demonstrated the immune-response mechanism, together with the importance of inducing an "actively acquired tolerance" in recipients of homografts. This phenomenon remains a major problem in heart transplantation. In 1962 K. Reemtsma, W. E. Williamson, and F. Iglesias initiated immunologic suppression posttransplantation. On 23 January 1964 J. D. Hardy performed the first attempt at heart heterotransplantation (from one species to another). The first human heart homograft (within the same species) was performed by C. N. Barnard on 3 December 1967; and three days later, on 6 December, A. Kantrowitz performed the first cardiac homograft in the United States when he transplanted the heart of a two-day-old infant into a nineteen-day-old infant. Barnard's patient survived for eighteen days; Kantrowitz's patient survived for six and one-half hours.

The mass media described 1968 as the "Year of the Transplant"; indeed, 105 hearts were transplanted in that year, 26 in November alone. The number of hearts transplanted in 1968 is more than thirty-seven percent of all cardiac transplants done between December 1967 and December 1975. Initial enthusiasm diminished in the mid-1970s, when it became generally acknowledged that a moratorium on cardiac transplantation had occurred. One way to measure the effect of that reduced activity is to note that about fifty-eight percent of all the heart allografts done in the ninety-two months after Barnard's initial operation were done in the twenty-three-month period from January 1968 to November 1970.

By December 1975, 286 hearts had been transplanted into 277 recipients by sixty-four teams in twenty-two countries. In the meantime, while the procedure markedly diminished in frequency and was abandoned in most settings, N. E. Shumway maintained an even pace with transplantations at Stanford Medical Center, averaging one per month. Of twenty-seven transplants performed in 1974, for example, fifteen were at Stanford.

Intercultural aspects

Long before Aristotle located the seat of soul, emotions, and intellect in the heart, Ezekiel had prophesied that the Lord God would give a new heart to the house of Israel: "I will take out of your flesh the heart of stone and give you a heart of flesh" (11:19, 36:26). This notion of heart, which became a central tenet of the cultural and religious sensibilities of the Judeo-Christian West, was of more than an efficient pump; this heart would enable the recipient to keep the Lord's statutes and laws. So tenaciously did this purpose of the heart persist that later Greek and Roman attempts to locate soul in the liver promptly and quietly failed. Meanwhile, celebrated in scripture, myth, poetry, and song, the heart continues to be regarded as the shrine of love, thus influencing attitudes regarding surgical interventions in the human heart.

Opposition to cardiac homotransplantation on expressly religious grounds is atypical in the United States and Europe, although there are certain groups, e.g., Jehovah's Witnesses, who do specifically object. Beyond those cultures whose religious and philosophical heritage is rooted in the Judeo-Christian tradition, but within those countries with the technical capacity to perform the procedure, the moral legitimacy of cardiac transplantation varies. Confucianism objects to all surgeries on a certain view of filial piety: The body is given by parents, and intentionally to injure or mutilate it is an act of disrespect toward them. The dominant traditional religion in China, Confucianism, is now officially condemned, and there is almost certainly the technical capability to perform heart allografts; but this procedure has not yet been reported from China, perhaps because ancient and modern methods are intermixed, scientific data are not regularly exchanged with other countries, and the procedure is not considered to be cost-effective. India, where Hinduism is the most prominent religion, has reported cardiac transplantation although traditional Indian medicine has been homeopathic rather than allopathic. Medical traditions in Islam contain no prohibition against surgeries of any sort so long as they are designed for and intend the benefit of the patient; and heart transplantation has been reported from Turkey.

Medical regulation

The ethics of human experimentation, which continues to provoke controversy, was an initial problem with cardiac transplantation. Indeed, early assessment of this procedure raised serious questions about whether both donors and recipients were sufficiently protected, and particularly whether the consent mechanisms were adequate. Almost from the beginning of cardiac transplantation, therefore, "due regard for the welfare and safety of each individual patient" came to be emphasized by both national and international medical associations. The American Medical Association (AMA) had succinctly listed three ethical guidelines for human experimentation in 1946; the Nuremberg Code was propounded by the War Crimes Tribunal in 1947; and the Declaration of Helsinki was adopted by the World Medical Association in 1964. All of these sought the protection of human subjects of experimentation through adequate consent, clinical design (including bench and animal studies), risk-benefit ratio, and the like.

The American Medical Association in 1966 and The British Medical Association in 1967 followed with their own reinforcement of these regulations; and in December 1968, a year after Barnard's initial operation, the AMA adopted a "Statement on Heart Transplantation," which included among five major points the provision that "the determination of death in organ donors must be made by no less than two physicians not associated with the surgical team performing the transplant" (American Medical Association, 1969, pp. 1704–1705). The other major points dealt with research design, patient welfare and safety, procurement, and public information. While the letter of that law has probably been followed, complications resulting from medical versus legal criteria for death have sometimes compromised its spirit. Within the United States there is, for example, no uniform statutory definition of death and the subsequent debate between "good medical practice" and "law and order" remains unresolved while a new consensus emerges. The discussion itself is replete with ethical implications for private and professional conduct within an ordered society that is ostensibly governed by law.

Ethical issues

Procurement. The initial availability of donor hearts for a given recipient was largely a matter of sheer coincidence rather than careful planning. Indeed, procurement remains one of the two most critical problems as viewed by cardiac surgeons (rejection and infection together constitute the other); and this in turn has raised related questions concerning the criteria and definition of death. Several countries have encountered this problem; and in the United States it is complicated by the fact that different states employ different criteria—ranging from silence (which implies current medical practice) to heart and lung function to total and permanent absence of brain function—with the result that procurement encounters certain legal obstacles in some states. The advent and widespread use of the pump oxygenator, which can maintain metabolizing processes almost indefinitely in some cases, has further compounded the issues. Two developments, however, portend a lessening of the problems associated with viable donor heart availability and procurement in the United States: the trend among state legislatures either to be silent or to enact brain-death-oriented criteria, and the adoption by each of the states of a uniform anatomical gift act, which facilitates the donation of organs by overcoming the confusion and medicolegal problems generated by conflicting state laws.

Legal questions have been raised in renal transplantation regarding ownership of a grafted organ, but in every instance, and without litigation, legal possession has devolved to the recipient. In the nature of the case of unpaired organs like the heart, similar questions have arisen academically but not for litigation. Compensation for the donor heart has ordinarily been expressed by recipient or transplant program responsibility for all hospital costs incidental to transplantation. Indeed, some surgeons engaged in cardiac transplantation have argued that there should be no charge whatsoever to the recipient. With presumption of recipient ownership of the transplanted organ, no further serious questions of the legal obligation of a recipient to either donor or donor-family have been raised.

During the fifteen months following the initial human heart transplant, 118 allografts were performed on 116 patients in eighteen different countries by some forty different surgeons. In the same period about fifty liver transplants were performed. The experience with liver grafts is important to note in consideration of the ethics of cardiac transplantation because, like the heart, the liver is an unpaired organ, there is no artificial or prosthetic liver available, the donor–recipient pool is unequal, the surgical risk is high (eighty-two percent in the case of liver), and prognosis is poor (only five recipients of liver allografts have survived beyond six months). Yet there has been virtually no lay or scientific discussion of the ethical, economic, or technical implications of this type of human transplantation.

Efficacy. Toward the end of James Hardy's published report of the first human heart transplant, this one a heterograft involving a chimpanzee's heart, the operation was defended as the patient's only chance for survival. Almost three years later, when Christiaan Barnard performed the first human heart allograft, it was stated that the patient's disease was "incurable by any known treatment other than cardiac transplantation." Now, after seven years' experience and certain advances that allow the claim that this procedure has been demonstrated to be both feasible and capable of prolonging useful life, the rationale given at its inception—that this is an innovative therapy of last resort for otherwise prompt and certain death—remains more or less constant.

There is surely a large ethical question entailed in the rhetoric of those who proclaim the efficacy of cardiac transplantation. Minimally, there is an almost infinite variety of meanings along the spectrum between "survival" and "a useful life." And especially in Western philosophical and theological traditions, the assessment and determination of what is good for the self by the self has been highly prized. The anxiety, sometimes latent and sometimes unconcealed, that an individual's good-for-himself might be subjected to the tyranny of another, that a subject could become objectified and a "thou" rendered into an "it," is clearly reflected in the language of the several codes and declarations cited earlier.

There may still be room, however, to acknowledge more explicitly the therapeutic calculus and how it is reckoned by patients, surgeons, families, and other publics who have a reasonable but not restrictive interest in both costs and benefits. To speak of "survival" or "useful life" is to employ a genre of categories that is not disinterested but value-preferenced from the start; and that is why particular questions about the adequacy of consents from patients who are candidates for heart transplantation, together with the more general issues of value judgments, are raised in this context.

Cost-benefit. Cognate questions have arisen with special reference to the conflict between patient perspective and social cost-benefit. The average monetary cost of cardiac transplantation in the United States in 1975 was $36,130 per patient, inclusive of operation and early postoperative care (Rider et al.). Among early cases costs ranged upward to $160,000. Some within both medical and lay communities have doubted that, all things considered, there is proportional cost-benefit to either individual patients or the public at large.

One controlled study (1971) showed that the mean survival time of potential recipients was seventy-four days, and that of transplant patients eighty-nine days—a comparison that led one analyst to conclude that the statistical difference is insignificant and does not justify wide clinical application. A later (1973) assessment of human cardiac transplantation concluded, however, that almost all of the long-term survivors had achieved impressive degrees of physical, social, and vocational rehabilitation. The criteria for the more recent appraisal include histocompatibility, diagnosis of rejection, immunosuppressive therapy, cardiac output, and the like; but nowhere are nonquantifiable norms, or those particular idiosyncrasies which mark us as individual persons, either identified or interpreted. We probably need more autobiographical comment from the recipients themselves before concluding more than that the procedure is technically feasible; in the absence of such comment, measuring the values, tangible or intangible, at stake in this procedure continues to evoke considerable discussion. To this point, in a democratic society, there has been a formal reluctance to jeopardize the supreme value attributed to the individual by articulating general benefits to be derived from a collectivist-oriented policy, but the potential conflict of interest between technological success and humane service signifies a critical and pressing ethical issue: Ought we to do everything we are technically able to do?

Risk-benefit. Histocompatibility remains a serious and urgent problem in cardiac transplantation. In Shumway's experience, which is the most extensive and comprehensive to date, there is no consistent correlation between patient survival and the degree of histocompatibility as determined by present matching techniques. Nevertheless, what does appear to be significant is that the most serious complication of immunosuppression is increased susceptibility to infection.

Because Shumway's patients comprise one-third of all human heart allografts, the data from his series are significant for ethical assessment: the overall patient survival rates are forty-eight, thirty-seven, and twenty-five percent at one, two, and three years respectively; and for those patients who survive the first three postoperative months, the rates are seventy-seven, sixty, and forty-two percent respectively; fifty-six percent of all postoperative deaths are attributed to complications from infections, most frequently (sixty percent) sited in the lungs; allograft rejection, in both acute and chronic forms, remains the single most challenging problem; hospitalization for all patients has averaged sixty days, and in-house costs have averaged $36,130 per patient. Of ninety-five patients transplanted since Shumway's initial attempt on 6 January 1968 with Michael Kasperak, who was dying from chronic viral myocarditis, thirty-three were alive in December 1975. Five of Shumway's recipients have survived for more than five years, the longest living survivor being more than six years posttransplant (Rider et al.).

The success rate of an innovative therapy is an important datum for ethical evaluation; but among the 286 operations done by sixty-four teams in twenty-two countries, the variables, both moral and technical, are enormous, and determination of "success" is therefore more or less arbitrary. Shumway's experience suggests that rejection and infection can be controlled well enough to offer dying patients a fifty percent chance of one-year survival and a forty percent chance of two-year survival; but, as has been already noted, survival alone is inadequate evidence of efficacy. Just now the data suggest that cardiac transplantation is more than promising but less than accepted therapy and that it will remain investigational until its survivor rate approximates that of kidney recipients or, as is the case with chronic renal failure, there is an available prosthesis.

Psychological aspects. Since the advent of open-heart surgery in 1955, psychiatrists have studied attendant emotional problems in patients undergoing cardiac surgery. The incidence of postoperative psychoses in some groups has run as high as twenty-five percent of all patients. In addition, with transplantation, massive doses of steroids for immunosuppression tend to produce an initial euphoria, which is followed by depression and, for brief periods, psychosis. Some patients have totally withdrawn, lying mute and catatonic; others have been confused and bewildered by the most common utensils (e.g., a toothbrush); still others have manifested hysterical crying. Patient disorientation, memory lapses, and confusion seldom last longer than a few days; but these symptoms have been noted to recur with altered doses of immunosuppressive drugs. Several male patients have reported an awakened feeling of virility.

These data, while inconclusive because they are neither complete nor yet amassed in sufficient quantity to generalize, nevertheless suggest that each patient should have preoperative psychiatric and psychological evaluations. Because it is widely held that an occasional psychosis is not as important as the principal and urgent need for surgery, there has been neither sufficient time nor sufficient interest to accommodate this aspect of cardiac surgery to date.

Histocompatibility. Of related interest is the matter of tissue typing and matching. In 1923–1924 Holman's work suggested that skin graft recipients resisted "foreign matter" in the form of donor-grafts and that an immune response of some sort was at work in graft rejection. In the late 1920s J. B. Brown and E. C. Padgett demonstrated that skin grafts between monozygotic twins could be successfully performed, thereby indicating that identical twins shared the same immune-response mechanism. Two decades later Medawar and his associates showed that the destruction of foreign transplanted tissue occurs through an active immunization mechanism in the recipient.

Since the early phases of cardiac transplantation, tissue typing and matching has been done on a preferred scale from A to D. In addition, ABO blood groups are assessed for compatibility. Among many surgeons, however, tissue typing and matching was definitely subordinate to fascination with surgical innovation and an articulated obligation to their in extremis patients; indeed, some even stated publicly that they would not want immunologists to stay the scalpel simply because there was not an adequate tissue match. In some sophisticated centers, moreover, it was said that rejection would not occur with the human heart; but, in fact, rejection occurs in human and dog heart grafts in about the same ratio as in human and dog kidney transplants.

At least part of the ethical assessment of any innovative procedure hinges on the adequacy of scientific and clinical design. The "need" of patients is surely important, but that need also includes the best possible circumstances for success of the proposed therapy. This is emphatically the case with an innovative therapy of this sort, and therefore to ignore or neglect histocompatibility seems plainly incongruent with that obligation. The typical heart transplant candidate, indeed, may already offer the poorest chance of success for a variety of reasons, not least among them being that this is a desperate measure (which, coincidentally, is also the reason that it would be ethically repugnant to use healthy patients in order to demonstrate or establish the risk-benefit ratio of cardiac transplantation). Thus there is arguably both a moral and a scientific duty to employ scarce therapies (in this case, organs available for transplantation) to the best possible advantage; and that, in turn, suggests that the operational priorities of experimental cardiac transplantation should be altered by making patient need subordinate to allograft viability predicated on immunologic compatibility. Meanwhile, the early priorities of human heart homografting have prompted some critics to suggest that the advancement of science has predominated over the well-being of patients.

Publicity. Some ethical problems result from developments that are quite tangential, if at all related, to the principal issues of cardiac transplantation. Mass media, for example, have played an important role in this operation since Barnard's initial procedure, which the press celebrated as "heart transplant success" and "the ultimate operation." Indeed, television crews were admitted to Barnard's operating theater and almost instantaneously he, and later other prominent cardiovascular surgeons, was lionized on TV screens throughout the world. One result of this immediate notoriety was an enormous pressure exerted on surgeons by a desperate but hopeful public that was itself sometimes misled by euphoric and incorrect reporting, with the consequence that some transplantation undoubtedly was done under less than optimal circumstances by less than adequately staffed and trained teams on patients who were at inordinate risk. Moreover, certain breaches occurred in the conventional code of professional conduct as surgeons seized the potential value of media exposure for building and research funds, for potential donors and recipients, as well as for gaining the upper hand in some intramural professional quarrels.

The absence of such readily available exposure, by comparison, has consigned to virtual anonymity the surgeon who performed the first surgery inside the heart—an operation, in many ways, more daring and dramatic and important than transplantation. On 13 July 1912 Théodore Tuffier, who was professor of experimental surgery at the Sorbonne, invaginated the wall of the aorta a short distance above the aortic valve with his index finger and dilated the narrow ring of the valve to relieve progressive and severe stenosis. Eight years later Tuffier reported to the Fifth International Surgical Society that his patient was still alive and temporarily improved. But his achievement aroused little enthusiasm among either his colleagues or the general public. The role of mass and instantaneous media, as a part of ethical assessment in surgical innovations, has been too long and too much neglected.

A related avenue for ethical assessment lies in asking: Who are the real heroes in medical and surgical innovation? The names that recur time and again in the literature are those of the surgeons who head the transplant teams; when the names of patients occur, it is chiefly for purposes of identifying a particular incident. Although it is a very subtle affair, the value questions of such an innovative therapy as cardiac transplantation are very much associated with a reward system; and it may merely be observed here that this system in modern medicine strongly favors the one who operates rather than the one operated upon. Publicity associated with the space programs, in both the United States and USSR, poses a rather novel alternative: The persons celebrated in these technical triumphs are the astronauts, whose encapsulated relationship with earthbound support groups may not be very different from that of an anesthetized patient on an operating table. Basically at consideration here is the question of proportionality between risk and reward.

Artificial heart implant

On 4 April 1969 D. A. Cooley implanted a total prosthetic heart replacement into Haskell Karp. Three days later, Karp's artificial heart was replaced by a human heart; but the patient died thirty-two hours later of pneumonia and renal failure. After M. E. DeBakey's left ventricular bypass on 21 April 1966 and Barnard's transplant of 3 December 1967, Cooley's use of the prosthetic heart is widely noted in the literature as the third major advance in cardiac transplantation. Although unique in experience with cardiac replacement to date, the total artificial heart has definite ethical significance.

Availability of a viable totally implantable artificial heart would, of course, resolve all those ethical issues associated with procurement; as it would also overcome many, if not all, of the problems related to tissue matching and timing of surgery. In this process of research and development, the artificial heart will also doubtless raise some new ethical questions, but we do not yet possess such an instrument. A left ventricular assist device (LVAD)—which consists of an implanted pump, connected by tubing with a power and servo-control console at the patient's bedside—is currently developed, and animal studies show that circulation can be maintained for many days to weeks without obvious evidence of discomfort, loss of alertness, or other adverse effects. Nevertheless, biological or engineering failure of the device does occur and is inevitable after a time. Withal, the LVAD remains an assist device, not a totally implantable heart.

In view of some comments that intend to associate cardiac with renal transplantation, it should be noted that the analogy is not altogether suitable, in part precisely because of the present lack of a prosthesis for circulation that is comparable to the one available for dialysis, and in part because of currently insurmountable physiologic and biochemical differences between the two organs and systems. Eventually, however, it is hoped that patients with terminal cardiovascular disease may be managed by implantation of a nuclear-powered mechanical device rather than by transplantation; and when that is possible, the problems of donor supply and rejection/immunosuppression will be eliminated.

HARMON L. SMITH

[*Directly related are the entries* DEATH, DEFINITION AND DETERMINATION OF, *article on* CRITERIA FOR DEATH; DEATH AND DYING: EUTHANASIA AND SUSTAINING LIFE, *article on* PROFESSIONAL AND PUBLIC POLICIES; HUMAN EXPERIMENTATION; ORGAN DONATION; *and* ORGAN TRANSPLANTATION. *For further discussion of topics mentioned in this article, see the entries:* CONFUCIANISM; HINDUISM; INFORMED CONSENT IN HUMAN RESEARCH; ISLAM; JUDAISM; *and* TECHNOLOGY, *article on* TECHNOLOGY AND THE LAW. *For discussion of related ideas, see the entries:* KIDNEY DIALYSIS AND TRANSPLANTATION; RATIONING OF MEDICAL TREATMENT; RESEARCH, BIOMEDICAL; *and* RISK.]

BIBLIOGRAPHY

American College of Surgeons/National Institutes of Health Organ Transplant Registry, Chicago. "ACS/NIH Organ Transplant Registry: First Scientific Report." *Journal of the American Medical Association* 217 (1971): 1520–1529.

———. "ACS/NIH Organ Transplant Registry: Second Scientific Report." *Journal of the American Medical Association* 221 (1972): 1486–1491.

———, Advisory Committee to the Registry. "ACS/NIH Organ Transplant Registry: Third Scientific Report." *Journal of the American Medical Association* 226 (1973): 1211–1216. Reprint requests to 55 E. Erie St., Chicago, Ill. 60611 (John J. Bergan, MD).

American Medical Association, Judicial Council. "Ethical Guidelines for Organ Transplantation." *Journal of the American Medical Association* 205 (1968): 341–342.

———, House of Delegates. "Statement on Heart Transplantation." *Journal of the American Medical Association* 207 (1969): 1704–1705.

BERGAN, JOHN J. "A Review of Human Solid Organ Transplantation." *Bulletin of the American College of Surgeons* 60, no. 3 (1975), pp. 24–26.

CASTELNUOVO-TEDESCO, PIETRO, ed. "Psychiatric Aspects of Organ Transplantation." *Seminars in Psychiatry* 3 (1971): 1–168. A collection of fifteen articles.

FADALI, A. MONEIM A., and SOLOFF, LOUIS A. "An Assessment of Cardiac Transplantation." *American Heart Journal* 86 (1973): 721–732.

FOX, RENÉE CLAIRE, and SWAZEY, JUDITH P. *The Courage to Fail: A Social View of Organ Transplants and Dialysis.* Chicago: University of Chicago Press, 1974.

GRIEPP, RANDALL B.; STINSON, EDWARD B.; DONG, EUGENE; CLARK, DAVID A.; and SHUMWAY, NORMAN E. "Acute Rejection of the Allografted Human Heart: Diagnosis and Treatment." *Annals of Thoracic Surgery* 12 (1971): 113–126.

———. "Determinants of Operative Risk in Human Heart Transplantation." *American Journal of Surgery* 122 (1971): 192–197.

HAZÁN, S. J. "Psychiatric Complications Following Cardiovascular Surgery. Part I. A Review Article." *Journal of Thoracic and Cardiovascular Surgery* 51 (1966): 307–319.

JONSEN, ALBERT R. "The Totally Implantable Artificial Heart." *Hastings Center Report* 3, no. 5 (1973), pp. 1–4.

MOORE, FRANCIS C. *Transplant: The Give and Take Issues of Transplantation.* New York: Simon & Schuster, 1972.

NAJARIAN, JOHN S., and SIMMONS, RICHARD L., eds. *Transplantation.* Philadelphia: Lea & Febiger, 1972.

RICHARDSON, ROBERT G. *The Scalpel and the Heart.* New York: Charles Scribner's Sons, 1970. First published in 1969 as *The Surgeon's Heart.*

RIDER, A. K., et al. "The Status of Cardiac Transplantation, 1975." *Circulation* 52 (1975): 531–539.

STINSON, EDWARD B.; PAYNE, ROSE; GRIEPP, RANDALL B.; DONG, EUGENE; and SHUMWAY, NORMAN E. "Correlation of Histocompatibility Matching with Graft Rejection and Survival after Cardiac Transplantation in Man." *Lancet* 2 (1971): 459–462.

THOMPSON, THOMAS. *Hearts: Of Surgeons and Transplants, Miracles and Disasters along the Cardiac Frontier.* New York: McCall Publishing Co., 1971.

United States; Department of Health, Education, and Welfare. *The Totally Implantable Artificial Heart: Economic, Ethical, Legal, Psychiatric, Social Implications.* A Report of the Artificial Heart Assessment Panel of the National Heart and Lung Institute. DHEW Publication, no. (NIH) 74–191. Bethesda, Md.: National Institutes of Health, 1973.

WOLSTENHOLME, GORDAN ETHELBERT WARD, and O'CONNOR, MAEVE, eds. *Law and Ethics of Transplantation.* Ciba Foundation Blueprint. London: Churchill, 1968.

YAO, J. K. Y. "Human Heart Transplantation." *Canadian Journal of Surgery* 16 (1973): 67–76.

HEMODIALYSIS
See KIDNEY DIALYSIS AND TRANSPLANTATION.

HINDUISM

Hinduism is a religious system which has grown organically over at least four millennia. It contains, even today, many primitive survivals and folk superstitions which are significant for the study of Indian bioethics. The formalized religiosocial system of Hinduism took something like its final shape in the Gupta Period (c. A.D. 300–550), which can with some justification be called "the Classical Age" of Hindu India. This article attempts to describe briefly the basic ideas that have influenced the development of bioethical concepts and attitudes in Hindu India and to survey those aspects of the Hindu system of medicine which are specially connected with bioethics.

Presuppositions

One of the most important conditioning factors in Hinduism is the doctrine of transmigration. In its final form this postulates that there is an inmost self (*ātman*) in each being (ranging from the highest god to the meanest insect), which is essentially immutable. This self becomes involved in matter (which in some philosophical systems is considered to be fundamentally illusory) and hence takes incarnate form. According to the conduct of the embodied being, the soul or self is carried at death to another body, which will enjoy bliss or suffer in accordance with its previous behavior. This process is called *saṃsāra*, and the force that produces it is known as *karma* (action).

Transmigration links all living beings in a single system. Unlike the religious systems based on Judaism, Hinduism makes no sharp distinction between human and animal. The doctrines of *karma* and *saṃsāra* have been largely responsible for the development of principles of nonviolence (*ahiṃsā*) and vegetarianism. In classical Hinduism, however, nonviolence is considerably modified for the layman, as distinct from the ascetic, to permit righteous warfare, the punishment of criminals, and self-defense.

The process of transmigration is looked on as painful, and the main quest of classical Hinduism has been "release" (*mokṣa*) from the cycle of birth and death in a state of timeless bliss. This condition is differently interpreted. The *Sāṅkhya* school, once very influential, and the still surviving heterodox sect of Jainism define release as the complete separation of the individual soul from matter. The *Advaita Vedānta* system, most influential in intellectual Hinduism, interprets it as a full realization of the illusory character of the material world and of the specious personality, and the recognition of the soul's identity with an underlying impersonal world spirit, often called Brahman. Theistic Hinduism of the *Viśiṣṭādvaita* school, which has had the greatest influence in the formation of popular ideas, interprets release as union with the personal God (whether called Viṣṇu or Śiva) who is the ultimate fact of the universe and out of whom the cosmos emerged.

In theory release should be the aim of all striving, but there are many other aims that for the layman are fully legitimate. In classical Hinduism there is a sharp theoretical distinction between the norms of the ordinary person and those of the ascetic (*sannyāsī*), who has "given up the world." The latter should aim directly at release. The ordinary man approaches the goal by slower stages, over many lives. For him there are three legitimate aims: *dharma*, adherence to religious and ethical norms in order to ensure a happier rebirth; *artha*, amassing wealth for the benefit of oneself and one's family; and *kāma*, the satisfaction of personal desires and drives. The three aims, which may be paraphrased as Piety, Profit, and Pleasure, are in descending order of priority; but, provided they are pursued in due order, all three are legitimate.

Hinduism is essentially monotheistic but teaches that the one primeval being, or God, produced at the beginning of this great cycle of time innumerable supernatural beings, all of whom are endowed with individual volition; some generally carry out the will of the High God; others have turned against him and are constantly working to undo the work of creation. The theme of the constant battle between gods and demons, light and darkness, good and evil is not so strongly emphasized as it was in ancient Iran, but it is an important feature of the mythology of the earliest traceable phase of Indian religious thought. It remained an important feature of Hindu mythology, with much influence on popular beliefs: The world is full of demons, which are normally at war with the gods, and which can be potent factors in causing disease.

The condition of the cosmos is steadily declining. Within the great cycle of time in which the universe exists, there are many lesser cycles, at the beginning of which the world's pristine virtue and bliss are temporarily restored, only to decline until a rock bottom of evil and misery is reached. At present we are in the fourth and most evil phase of the cycle, the *Kali yuga,* but the end is hundreds of thousands of years away. Thus traditional Hinduism is backward looking. The eternal norms of human conduct were followed much more closely in former periods, and the best that can be done nowadays is to imitate the example of the past as far as possible.

Neither the doctrine of *karma* nor that of cosmic decline implies fatalism. Within the framework of his objective condition a man can and should mold his own destiny; and, though cosmic decline is ultimately inevitable, the process can be slowed down and even temporarily reversed by human effort and by the intervention of the gods. The texts contain many passages stressing the virtue of human effort (*puruṣakāra*), as against the mere acceptance of bad conditions resulting from destiny (*niyati*) or chance (*yadṛcchā*).

Social norms

The hierarchical social order was believed to be eternal. The four great classes (*varṇa*) emerged at the beginning of time from the body of the Creator, and they were the fundamental basis of society. The *Brahman* (priest), the *Kṣatriya* (warrior and ruler), the *Vaiśya* (merchant), and the *Śūdra* (worker) formed four eternal classes or estates, members of which had differing functions and status and were often looked on as different species, analogous to horses and cattle. For a member of a lower class to attempt to fulfill the functions of a higher one was an affront to nature and to the gods. Below the four great classes were the untouchables, theoretically outside the social order altogether, who fulfilled impure functions such as removing garbage, cremating corpses, working in leather, etc. These were more or less isolated from all close contact with the better classes.

The four classes formed, to the modern mind, a largely artificial division of society, and the objective social condition of any time or place rarely corresponded to the norms of the textbooks. More important than the class in everyday life was the caste (*jāti*), a group of families generally following the same profession and theoretically contained within one of the four classes. Castes were also hierarchically graded and were normally endogamous. They were controlled by local councils of elders, which exerted great power on their members.

The family

In the social ethics of India the fundamental unit was the family rather than the individual. The extended family (*kula*) was a corporate entity, and the individual was subordinate to it. It was patrilinear, patriarchal, and patrilocal, but the authority of the patriarch was limited by traditional law. He did not have the right to dispose of the family property arbitrarily, nor did he have complete power over the lives of the family members. The ritual of *śrāddha*, whereby the dead ancestors communicated with and received sustenance from the living, was a very powerful force in cementing the family together. Without male descendants to perform *śrāddha* a man and his ancestors suffered in the afterlife, and hence the Indian was very philoprogenitive. A male heir was essential for his spiritual welfare as well as for the survival of his line. In view of heavy child mortality, it was incumbent upon him to produce as many children as possible, in the hope that at least one would be a son who would carry on the line.

The wife became integrated into her husband's family, theoretically (though by no means always in practice) completely subordinate to him. It was considered indecent to leave a girl unmarried after her first menstruation, and marriage normally involved the payment of a heavy dowry. Thus the birth of a daughter was often looked on as a misfortune. In some medieval communities female infanticide was practiced, but this social custom has no basis in the sacred texts, which look on abortion and infanticide as very grave forms of murder.

Individual conduct

Within the framework of the three aims, the life of the individual of good class was circumscribed by ritual observances and taboos. Many sacraments, from before birth to after death, marked the progress of his life. The brahman was expected to devote a considerable amount of time each day to prayer and ritual, and members of other castes were encouraged to imitate him.

Much emphasis was placed on ritual purity, and many of the taboos of Hinduism were directed at its maintenance. While their conscious purpose may have been otherwise, many of the prescriptions of purity are hygienic, and they seem to have evolved in the group subconscious in order to maintain health in a subtropical climate. Among such rules we may mention the insistence on a daily bath, the custom of eating with the right hand and washing the anus and the sexual organs with the left, the ban on eating cooked food left overnight, and the very strict taboo against contact with corpses, whether human or animal. The latter is most important from the medical point of view. Also significant in this respect is what in modern north Indian languages is called *jūṭhā*, impurity brought about by contact with the body fluids of others, especially saliva and mucus. *Jūṭhā* is still a very potent factor in the life of the ordinary Indian. From the traditional point of view the Western custom of blowing the nose on a handkerchief and placing the latter in the pocket is positively filthy and repulsive—far better to hawk and spit on the ground, where the offensive matter is unlikely to come in contact with more than the soles of the feet and shoes of others.

In regard to sex, Hinduism displays an ambivalent attitude, based on conflicting texts. In general the ascetic norms discourage all sexual activity, which not only distracts the individual in his quest for release but also results in a great loss of power, both physical and spiritual. On the other hand sex is the greatest source of pleasure, which, in the framework of the three aims, is a perfectly legitimate object to pursue. The ethics of classical Hinduism would confine sexual activity to one stage of a man's life. An initiation ceremony (*upanayana*) was prescribed for boys of the upper classes; it was not an initiation into sexual life, but preceded a long period of celibate studentship. After this the young man was married, normally to a bride chosen by his parents, and raised a family. Ideally he was expected in late middle age to give up family cares and devote the rest of his life to religion. The moral code involved the scrupulous performance of ritual and the observance of taboos, but much emphasis was also placed on more strictly ethical principles. The virtues of honesty, hospitality, and generosity were encouraged, and the basic duty of nonviolence was often interpreted in a positive sense, as actively benefiting others.

Hindu medicine

A complex medical system, known as *āyurveda*, "the science of (living to a ripe old) age," evolved in India. This developed over the first millennium B.C., and the oldest text of the elaborated system, the *Caraka Saṃhitā*, appears to go back to approximately A.D. 100. But it contains much older material and has certainly been added to in later times. There are numerous later medical texts that show progress in respect of practical medical measures and introduce new drugs, but the basic medical theory was virtually unchanged. Noteworthy among later medical texts is the *Suśruta Saṃhitā*, probably of the sixth century A.D., which contains the only detailed account of ancient Indian surgery.

The knowledge of physiology among the Hindus was not very advanced, and the functions of the internal organs were not well understood. The controlling factors in the organism were believed to be the three primary fluids or humors (*doṣa*) of the body—wind (*vāta*), gall (*pitta*), and mucus (*kapha*). In order that the body might function properly and health be maintained, the three had to be well balanced. A preponderance of one fluid over another led to disease, and the imbalance of the three was the immediate cause of almost every illness, both physical and mental, though the ultimate cause might be *karma*, demons, or some of many other factors. It was the function of the doctor to reestablish a harmonious balance of the fluids and to remove if possible the causes that led to the disharmony.

On this unpromising basis it is surprising that a medical system was elaborated which was quite as successful as any other known before the days of Harvey, and probably more successful in several respects. The success of *āyurveda* was achieved through centuries of careful observation and trial-and-error, and with the aid of an exceptionally wide pharmacopoeia.

The ethics of health and disease

Moral factors, linked with the all-embracing doctrine of *karma*, were believed to be very significant in promoting health or causing disease. Illnesses might be caused by the sins or shortcomings of a previous existence, and longevity was also believed to be largely due to *karma*. The doctrine encouraged inner acceptance of disease and gave a ready-made explanation of its cause, but nowhere is a man advised to submit to illness without attempting its cure. Since the evil brought about by *karma* cannot be estimated with certainty and the bad effects of sins can be offset by the merit gained by good deeds, there was every reason why a sick man should seek all the medical help at his command in order to achieve health.

Karma was generally thought to be the source of congenital defects. A man born with a deformed hand, for example, was believed to have incurred this misfortune as a result of an evil deed (for instance striking a brahman) committed by the same hand in a previous life. This again did not prevent attempts at improving his condition by surgery, since the duration of the punishment through *karma* was not known, and the trouble might be only temporary.

As well as *karma*, other factors were believed to promote health or disease. Devotion (*bhakti*) to God, who might set aside the law of *karma* for those who had faith in him, encouraged longevity and health. Neglect of religious duties and lack of faith, on the other hand, might lead to the withdrawal of divine protection, with the result that demons might take possession, leading to disease or madness.

Mental and spiritual training in concentration and meditation, commonly known as yoga, was also believed to promote health and longevity, and this is still believed by its practitioners both in India and elsewhere. The complicated and difficult practices associated with the system of yoga based on physical exercise (*haṭha-yoga*) were thought to produce not merely longevity but, for the most advanced adepts, actual immortality.

More closely linked with ethics was the general view that equanimity and kindness are therapeutic in their effects. Excess in every respect is looked on with disfavor by the medical texts, which in general encourage a via media fully consistent with the doctrine of the three aims of life. The most impressive emphasis on the value of altruism and love in the promotion of health and longevity occurs in the seventh-century text of the Buddhist physician Vāgbhaṭa (i. 2).

The ethics of medicine

The activities of the physician (*vaidya*) were declared by Caraka (i. 30.29) to be closely linked with the doctrine of the three aims. By relieving suffering and adding to the sum of human happiness he fulfills the first aim, carrying out his religious duty; from the generous fees of his wealthy patients he achieves the second aim, riches; while the third aim, pleasure, is achieved by the satisfaction he obtains, first from a high reputation as a healer, and second from the knowledge that he has cured many people whom he loves and respects.

The two latter aims were not to be disparaged. The few famous physicians occurring in story and tradition were not selfless servants of humanity, but very wealthy men, like the fashionable practitioners of modern times. Moreover there was no ban on the *vaidya*'s advertising his skill, and we read of a young doctor, recently qualified and in search of patients, walking through the streets of an ancient Indian city and shouting "Who is ill? Who wants to be cured?" (*Mahāvagga* viii. i. 8–13). Charlatanry was rampant in ancient India, and quacks would come canvassing as soon as they heard that a well-to-do person was sick (Caraka i. 29).

Nevertheless the first aim, *dharma* or religious duty, overrode the other two, and a very high moral standard was expected of the physician. At the outset of his training he went through a very solemn initiation, at which his teacher (*guru*) instructed him that he was to live a frugal and ascetic life, celibate and vegetarian, while undergoing training. He must obey his teacher implicitly "unless instructed to commit a mortal sin." Then the teacher continues:

When you have finished your studies, if you want to have a successful, wealthy, and famous practice, and to go to heaven when you die, you must pray every day, when you get up and go to sleep, for the welfare of all beings, especially cattle and brahmans, and you must strive with all your power to heal the sick. You must not betray your patients, even at the risk of your own life. . . . You must always be pleasant of speech . . . and always strive to improve your knowledge. In the home of a patient your words, mind, thoughts, and senses should be directed on your patient and his treatment. . . . Nothing that takes place in his house must be told outside, and you must not tell anyone of his condition, if by that knowledge the patient or others might be harmed [Caraka iii. 8.7].

This well-known passage has frequently been compared with the Hippocratic Oath. The same chapter of the *Caraka Saṃhitā* contains, however, certain counsels which may not be quite consistent with the ethics of Western medicine. The physician is advised to refuse treatment to the king's enemies, miscreants (a vague term that may include not only criminals but also partisans of heretical antibrahmanic systems such as Buddhism), and loose women. He should not accept as a patient a case that is obviously terminal (except, presumably, to alleviate pain). In the latter case, the question, "Should a doctor tell?" is left largely to the discretion of the doctor. Caraka lays down that if a physician concludes that the condition of the patient is hopeless he should keep this knowledge to himself if he believes that it would harm the patient or others.

Though India was generally conservative in its outlook, believing in the continued efficacy of traditional norms and prescriptions, Hindu medicine shows a surprising openness to new ideas, at least in respect of practical treatment. The theoretical basis in the doctrine of the three fluids was never abandoned and underwent little modification in the course of time; but the *vaidya* was advised to be constantly on the lookout for new drugs and treatment methods, and the texts, if compared chronologically, show a steady increase in the number of items in the pharmacopoeia. Even after his long apprenticeship was over, it was recommended that the physician continue to improve his knowledge by studying his patients and even obtain fresh knowledge of unusual remedies from hermits, cowherds, and hillmen (Suśruta i. 36.10). Regular professional gatherings of physicians formed a means for the exchange of ideas and knowledge. The descriptions of these colloquiums show that many physicians had knowledge not obtained from textbooks, which they tried to keep to themselves (Caraka iii. 8.15–18; ii. 32.72–75).

Though he might be wealthy, not bound by any rules of an ascetic character, the physician was apparently encouraged by the texts to consider himself a sort of secular priest, with a special, almost supernatural charisma bestowed on him at the initiation ceremony at the beginning of his studies. The upper-class man who had undergone the normal initiation or *upanayana* was "twice-born" (*dvija*), and thus superior to the *śūdra* or woman, who had had only one birth. The *vaidya*, in contrast, was "thrice born" (*trija*). As the words of his teacher show, this exalted status involved a high standard of conduct. He should always be "of a calm mind, pleasant of speech, . . . the friend of all beings" (Suśruta i. 10.3). He was to some extent relieved of the burden of caste taboos. He could enter the homes of people of a lower caste than his, handle their bodies, and even taste their urine when making a diagnosis. He was not averse to using animal products in compounding his drugs, and prescribed these quite freely. Yet the taboo on handling a corpse seems to have applied to a physician as strongly as to an ordinary man. No medical text prescribes the actual dissection of a corpse, much less any form of vivisection of humans or animals, which would not only have brought ritual pollution but would also have been repugnant to the principle of nonviolence.

Social service medicine

There is evidence that in some Hindu kingdoms the provision of medicine was controlled by the state. Suśruta (i. 10.3) refers to a system of licensing of qualified medical practitioners to avoid quackery. Some texts on law and statecraft give prescriptions concerning the punishment of a doctor giving inefficient treatment, resulting in injury or death to the patient (*Kauṭilīya* iv. 1; Kane, iii. 19). How far this advice was followed in practice is uncertain, since the medical texts and general literature indicate that quackery was rampant in every Indian city.

The provision of free medical care to the poor was looked on as part of a king's duty to provide protection for his subjects, which was generally interpreted in a positive sense (Caraka i. 30.29, with Mehta's notes). From the days of the benevolent Buddhist emperor Aśoka (third century B.C.) onward, the better rulers of India did at least something in this respect. Dispensaries manned by professional doctors, providing free services to the poor, existed in many cities. These might be provided by the state but were also often financed by private charity. In South India especially, hospitals and dispensaries were often attached to the great temples. Free medical services might be provided by doctors themselves, for they were encouraged to treat the poor, learned brahmans, and ascetics without charge (Suśruta i. 2.8; vi. 11.12–13). The most outstanding examples of the provision of free medical services in this part of Asia, however, are to be found in Buddhist Sri Lanka and Cambodia, and are thus outside the scope of this article.

Mental health

The rigidly controlled social and familial system of Hindu India produced many misfits. The "dropout" in Hindu society could often find a socially acceptable outlet in one of the many ascetic sects or as an independent ascetic; but this did not prevent the incidence of insanity, either temporary or permanent, which is dealt with in medical texts in some detail. The most frequent factor in serious insanity was the attacks of "demons," which often found access to the patient through his own sins. Magic and exorcism were therefore resorted to in serious cases of insanity and were prescribed by the medical texts, which in general were disinclined to have recourse to the supernatural except in extreme cases.

While the texts prescribe gentle treatment for the simpler and less intractable types of mental illness such as melancholia, the attitude to the more violent and aggressive forms of insanity was typical of the same period in the West. The seriously insane patient was confined, chained, starved, beaten, branded, and deliberately frightened. The treatment no doubt reflects a degree of social scorn and disapproval of a patient who was believed to have brought his troubles on himself by his own sins, and who was in any case repulsive, intractable, and dangerous, but it may have been effective in some cases, being a crude form of the shock therapy of modern psychiatry.

Conclusion

The Hindu system of medicine was one of the most effective to have been devised in antiquity. It was more or less a closed system as far as its theory went but was always open to new ideas in respect of practical therapeutic methods. Thus *ayurveda* has survived to the present day and still plays a significant role in Indian medicine. With the support of the state āyurvedic physicians are now trained in special colleges, where they learn the elements of scientific physiology as well as the traditional medical craft. Thus they provide an alternative medical service which is effective in the curing of many diseases and is appreciably cheaper than "Western" medicine. In the process of adapting itself to new conditions, however, the system is losing its personality and is on the way to becoming a poor relation of the modern international system of medicine. Meanwhile serious research is going on in India on the effectiveness of āyurvedic therapy. The system still has usefulness, and some of its prescriptions have already led to the discovery of new drugs in Western medicine.

A. L. BASHAM

[*Directly related are the entries* MEDICAL ETHICS, HISTORY OF, *section on* SOUTH AND EAST ASIA, *articles on* GENERAL HISTORICAL SURVEY *and* INDIA; *and* POPULATION ETHICS: RELIGIOUS TRADITIONS, *article on* A HINDU PERSPECTIVE. *For further discussion of topics mentioned in this article, see the entries:* ABORTION; BUDDHISM; CADAVERS, *article on* GENERAL ETHICAL CONCERNS; CONFIDENTIALITY; DEATH, *article on* EASTERN THOUGHT; ENVIRONMENT AND MAN, *article on* EASTERN THOUGHT; HOSPITALS; INFANTS, *article on* INFANTICIDE: A PHILOSOPHICAL PERSPECTIVE; MENTAL ILLNESS, *article on* CONCEPTIONS OF MENTAL ILLNESS; *and* TRUTH-TELLING. *See also:* MIRACLE AND FAITH HEALING. *See* APPENDIX, SECTION I, OATH OF HIPPOCRATES, *and* OATH OF INITIATION (CARAKA SAMHITA).]

BIBLIOGRAPHY
ANCIENT INDIAN MEDICAL TEXTS

CARAKA. *Caraka Saṃhitā*, 6 vols. Edited by P. M. Mehta with translations in Hindi, Gujarati and English by the Shree Gulabkunverba Ayurvedic Society. Jamnagar: Shree Gulabkunverba Ayurvedic Society, 1949.

SUŚRUTA. *Suśruta Samhitā*. Edited by N. R. Acharya. Bombay: Nirnayasagara Press, 1947. 3 vols. Translated by Kaviraj Kunja Lal Bhishagratna. Varanasi, India: Chowkhamba Sanskrit Series, 1973.

VĀGBHATA. *Astāngahrdayasaṃhitā. The First Five Chapters of Its Tibetan Version*. Edited and translated, with original Sanskrit, by Claus Vogel. Introduction and Commentary on Tibetan translating-technique. Abhandlungen für die Kunde des Morgenlandes, vol. 37, no. 2. Wiesbaden: Franz Steiner, 1965.

OTHER ANCIENT INDIAN TEXTS

KAUTILĪYA. *The Kautilīya Arthaśāstra*. 3 vols. Edited by R. P. Kangle. University of Bombay Studies: Sanskrit, Prakrit and Pali, nos. 1–3. Bombay: University of Bombay, 1960–1965. Text and translation.

Mahāvagga [The Great Division]. "The Mahāvagga, I–IV." "The Mahāvagga, V–X." *Sacred Books of the East*. Edited by F. Max Muller. Vol. 13: *Vinaya Texts: Part I*. Vol. 17: *Vinaya Texts: Part II*. Translated by T. W. Rhys Davids and Hermann Oldenberg. Oxford: Clarendon Press, 1881, 1882. Reprint. Delhi: Motilal Banarsidass, 1968, vol. 13, pp. 73–355; vol. 17, pp. 2–325; esp. vol. 17, pp. 176–179.

WORKS ON INDIAN MEDICINE

CHAKRAVORTY, RANES. "The Duties and Training of Physicians in Ancient India as Described in the Suśruta Saṃhitā." *Surgery, Gynecology, and Obstetrics* 120 (1965): 1067–1070.

FILLIOZAT, JEAN. *The Classical Doctrine of Indian Medicine: Its Origins and Its Greek Parallels*. Translated by Dev Raj Chanana. Delhi: Munshiram Manoharlal, 1964. Original version. *La Doctrine classique de la médecine indienne: Ses origines et ses parallèles grecs*. Paris: P. Geuthner & Imprimerie nationale, 1949. Detailed treatment of origins; valuable comparisons with Greco-Roman medicine.

JOLLY, JULIUS. *Medicin*. Grundriss der indo-arischen Philologie und Altertumskunde, vol. 3, no. 10. Strassburg: K. J. Trubner, 1901. Translated and supplemented with notes by C. G. Kashikar as *Indian Medicine*. Foreword by J. Filliozat. Poona: C. G. Kashikar, 1951. Bibliography. Translation not seen by author.

KUTUMBIAH, P. *Ancient Indian Medicine*. Foreword by S. Radhakrishnan. Bombay: Orient Longmans, 1962.

MENON, I. A., and HABERMAN, H. F. "The Medical Students' Oath of Ancient India." *Medical History* 14 (1970): 295–299.

MULLER, REINHOLD F. G. *Grundsätze altindische Medizin*. Acta historica scientiarum naturalium et medicinalium, vol. 8. Copenhagen: Einar Munksgaard, 1951.

REDDY, D. V. SUBBA. "Medical Ethics in Ancient India." *Journal of the Indian Medical Association* 9 (1941): 385–400.

———. "Medical Relief in Medieval South India: Centres of Medical Aid and Types of Medical Institutions." *Bulletin of the History of Medicine* 37 (1961): 287–288.

SHARMA, PRIYA VRAT. *Indian Medicine in the Classical Age*. Foreword by V. Raghavan. Chowkhamba Sanskrit Studies, vol. 85. Varanasi: Chowkhamba Sanskrit Series Office, 1972. Interesting and important as the work of a fully qualified āyurvedic physician.

ZIMMER, HEINRICH ROBERT. *Hindu Medicine*. Edited with a foreword and preface by Ludwig Edelstein. Publications of the Institute of the History of Medicine, The Johns Hopkins University. 3d ser.: The Hideyo Noguchi Lectures, vol. 6. Baltimore: Johns Hopkins Press, 1948.

WORKS ON PERIPHERAL TOPICS

BASHAM, ARTHUR LLEWELLYN. *The Wonder That Was India*. 3d rev. ed. London: Sidgwick & Jackson, 1967; New York: Taplinger Publishing Co., 1968. A general survey of Indian culture, useful for background reading.

KANE, P. V. *A History of Dharmaśāstra: Ancient and Mediaeval Religious and Civil Law*. 5 vols. in 7. Government Oriental Series, Class B, no. 6. Poona: Bhandarkar Oriental Research Institute, 1936–1962. This monumental survey of ancient Indian law, custom, and ritual, one of the most remarkable achievements of single-handed Indian scholarship, contains much of relevance to the introductory sections of this article.

WALKER, GEORGE BENJAMIN. *Hindu World: An Encyclopedic Survey of Hinduism*. 2 vols. London: Allen & Unwin; New York: Praeger, 1968. Concise articles on many aspects of Hindu religion and culture.

HIPPOCRATIC OATH

See CODES OF MEDICAL ETHICS. See also APPENDIX, SECTION I, OATH OF HIPPOCRATES.

HISTORY OF MEDICAL ETHICS

See MEDICAL ETHICS, HISTORY OF.

HOMEOPATHY

See MEDICAL PROFESSION, *article on* MEDICAL PROFESSIONALISM; ORTHODOXY IN MEDICINE.

HOMOSEXUALITY

I. CLINICAL AND BEHAVIORAL ASPECTS
    D. L. Creson

II. ETHICAL ASPECTS
    George A. Kanoti and Anthony R. Kosnik

I
CLINICAL AND BEHAVIORAL ASPECTS

A number of issues must be considered in any discussion of the clinical and behavioral aspects of homosexuality. First, there is a problem with definition. Second, only with an awareness of the difficulties involved in reaching a consensus as to the nature and causes of homosexuality is it possible productively to consider the bewildering and often contradictory literature that attempts to describe, explore, and explain homosexuality. Third, distinctions, both popular and scholarly, that have been drawn between sexual normality and sexual abnormality must be considered. Finally the possibility and the desirability of changing homosexual behavior require attention.

Problems in definition and diagnosis

The term "homosexuality" is frequently used as if its meaning were self-evident, but even a superficial review of the available literature suggests that such an assumption is unwarranted. In place of a general consensus, diverse definitions of homosexuality—many of them operational—reflect the scientific interest and frequently the personal biases of individual clinicians and scientists.

The most prevalent and generally accepted view of homosexuality is that it represents an acquired condition that is both psychological and pathological. Authors who do not consider the homosexual condition to be pathological tend to view it as a variation in normal development so extreme as to deserve special consideration. The focus of much contemporary research remains on the etiology and pathogenesis of homosexuality or the homosexual condition. Even among those who accept such a model there are marked disagreements as to the nature of homosexuality.

It might be assumed that at least the presence of homosexual behavior would serve as a common theme in the definitions of homosexuality proposed by those who view homosexuality as a condition or affliction of the individual. There are, however, problems in defining homosexual behavior that are as difficult and complicated as defining homosexuality itself. Many researchers —Irving Bieber (p. 248), for example—do insist on "repetitive homosexual behavior in adulthood" as a prerequisite for a "diagnosis" of homosexuality. Other scholars exchange the dilemmas inherent in defining homosexual behavior for the dilemma of homosexuality without overt homosexual behavior. Judd Marmor (p. 4) represents one such point of view when he emphasizes the importance of psychological motivation in any definition of homosexuality. Marmor stresses the importance of erotic desire for members of the same sex, and he explicitly states his conviction that homosexuality need not be accompanied by overt homosexual behavior. In Marmor's view, a view shared by many, homosexual behavior is incidental to an underlying psychological condition—a condition or affliction that may manifest itself in ways other than participation in overt homosexual acts.

Despite the widespread acceptance of homosexuality as a psychological condition there is still contemporary interest in clinical and scientific circles in the possibility that homosexuality represents a purely physiological condition or at least a physiological predisposition to a psychological homosexuality. This view, more prevalent in the past than at present but recently regaining favor, continues to draw support from a limited amount of research data. In 1952 F. J. Kallman reported a study of homosexuality among twins. In reporting his research Kallman described a marked difference in the rate of concordance in overt homosexual behavior between dizygotic (nonidentical) and monozygotic (identical) twins. The concordance in overt homosexual behavior among dizygotic twins was slightly higher than would be expected, on the basis of data available for the general population. The concordance reported for monozygotic twins was, however, complete (Kallman). Although Kallman's work has been criticized and replication has not been consistent, interest in homosexuality as a physiological condition has received a new impetus with the recent controversy around research findings that suggest endocrinological anomalies may occur in "exclusively" homosexual males (Kolodny et al.; Tourney, Petrilli, and Hatfield).

Despite arguments that homosexuality represents a specific condition of the individual, much of what is known about homosexual behavior does not fit easily into such a model. Early authors took cognizance of ambisexuality and bisexuality and in many cases showed some awareness of the wide diversity that characterized human sexual behavior, but it was the early surveys of Kinsey and his co-workers (Kinsey, Pomeroy and Martin; Kinsey et al.) that first confronted clinicians, scientists, and the public at large with the profound implications of that sexual diversity. These surveys suggested an incidence of overt homosexual behavior in the population of the United States that was much higher than had previously been estimated. Kinsey and his group reported that thirty-seven percent of the male population studied and nineteen percent of the female population studied engaged in overt homosexual behavior at some time following the onset of their adolescence. In the case of male subjects the homosexual behavior was to the point of orgasm. The Kinsey percentages have been criticized on a number of methodological grounds (Hunt) and probably slightly exceed the actual incidence of homosexual behavior in the United States.

Despite doubts about the accuracy of his percentages, the diversity in sexual object choice described by Kinsey has been further documented since the surveys were published. Overt homosexual behavior is not limited to a small, identifiable group in Western societies that can easily be described on the basis of a constellation of shared psychological, social, and physiological traits. The prevalence and diversity of homosexual behaviors complicate all efforts to define homosexuality and raise questions as to the scientific merit of such concepts as homosexuality and heterosexuality.

Nature and causes of homosexuality

Because of the problems in defining homosexuality any discussion of its nature and its causes is made difficult. Whether homosexuality is understood as a physiological, psychological, or sociological phenomenon will determine where one looks for causal relationships.

The traditional view that homosexuality is an acquired psychological condition reflecting a divergence from normal development has led to more than sixty postulated etiologies for homosexuality (Hatterer). Most of them emphasize the importance of early relationships with parental figures. Because most of the data on which such theories are based were collected in clinical settings and were strongly influenced by Freudian theory, little attention has been given to the role of later life experience in the development of a homosexual orientation.

A male homosexual's relationship with his mother is most frequently posed as a cause for his sexual object choice. Both strong, domineering, exploitive mothers and ambivalent, dependent, insecure mothers have been described as playing an etiologic role in the development of homosexuality. The male homosexual's father and the father–son relationship have also been considered causal in the development of homosexuality, as has the pattern of the relationship between the two parents during the early personality development of the homosexual. Such theoretical explanations of homosexuality are best understood in terms of family dynamics that prevent or hinder the child from developing important sequential identification with female and male parent figures necessary for a heterosexual orientation.

What work has been done in attempting to relate female homosexuality to early developmental influences has resulted in causal explanations that are variations on the ideas of identification with parental figures suggested for male homosexuality (Bacon). The importance of considering the developmental causes of female homosexuality as distinct from developmental causes for male homosexuality has become clearer in recent years. Judith Bardwick (p. 140) has described the sexual identification of middle-class American girls as essentially a bisexual identification, one in which "traditional feminine needs" and "achievement needs" are in conflict. The work of looking at female sexual object choice as developmentally distinct from that of males is just beginning but is receiving more attention than it received in the past.

Although few scientists would advance a causal explanation of homosexuality that was purely physiological, the time has passed when physiological variables can be ignored in any discussion of sexual behavior. Various areas of the brain are intimately tied to sexual behavior (MacLean), and the brains of males and females develop differently as a result of endocrine influences. Differences in the central nervous systems of males and females correlate with differences in behavior (Paintin). The fact that such differences do not explain homosexuality does not mean they need not be considered in any comprehensive explanation of the phenomenon.

Anthropological and sociological descriptions of homosexual behavior in diverse social cultural systems (Tripp, pp. 68–94) and descriptions of homosexual life-styles in Western society (Hoffman)—along with investigations of homosexual liaisons in such restrictive institutions as prisons and other correctional facilities (Ward and Kassebaum)—raise additional questions about the adequacy of developmental models to explain homosexuality. Studies by ethologists of sexual behavior in other animal species and particularly those dealing with nonhuman primates suggest the possibility that nonspecific sexual responses serve social functions other than recreation and procreation. This appears to be just as true of homosexual behaviors as it is of heterosexual behaviors.

Research into the origin and development of gender identity (Money and Ehrhardt) points to important questions about the meaning of maleness and femaleness, questions that have implications for attempts to understand homosexuality. In their extensive study of male and female homosexuality, Marcel Saghir and Eli Robins suggest a significant relationship between early differentiation of gender identity and subsequent involvement in homosexual behavior (Saghir and Robins). The relationship between gender identity and homosexual behavior, however, appears complex and remains poorly understood and poorly defined.

A satisfactory explanation of homosexuality is not now possible. The bewildering array of psychological, physiological, and sociological variables that appear relevant to any such explanation have yet to be drawn together into an internally consistent model that explains homosexuality. It may be that homosexuality represents one or more definable conditions or that what we have called homosexuality is descriptive of behavior with diverse antecedents, behaviors that have meaning only in the rich and varied contexts of human life.

Medical and social distinctions between normal and abnormal

The question of whether homosexual behavior should be considered normal or abnormal is not only a medical question but also a social question that relates to the prevailing norms of the sociocultural system. A practical issue in considering such a question is the extent to which social ideals and cultural norms influence clinical and scientific views of human sexuality and in particular definitions of homosexuality. An article in the *Journal of Homosexuality* by Vern L. Bullough traces aspects of the historical relationships between sexual behavior and social deviancy. Bullough points out difficulties that result from considering sexual variations and social deviancy as synonymous (Bullough).

Clinicians and scientists share a social heritage with a general population that equates homosexuality and deviancy. To the extent that that heritage provides a priori assumptions about homosexuality, it adds to the difficulties in objectively reaching an understanding of the normality or abnormality of homosexual behavior.

The rise of gay militancy as chronicled by Teal and the tenuous acceptance by both the courts and some sectors of the public of the "right to be gay" has had an effect on the clinical view of homosexuality (Teal). This effect is evident in the continuing debate in psychiatric circles on whether homosexuality per se should have been dropped from the official diagnostic nomenclature of the American Psychiatric Association.

It is probable that in the future clinicians and scientists will refer not to homosexuality and heterosexuality but rather to descriptions of sexual behavior. In that event efforts to understand homosexual behavior will be only a part of a much broader effort to understand sexual preference. The focus of scientific interest will be the complex variables that mold and perpetuate sexual desire, and clinical efforts will be directed at problems and difficulties that come into being as a consequence of erotic desire in particular contexts.

Changing homosexual behavior

A medical assumption that homosexual behavior is symptomatic of a pathological condition has the inevitable consequence of therapeutic attempts to reverse or change the behavior in a direction considered normal. The traditional tools of medicine have been employed in such efforts with meager success. Dr. D. J. West, in discussing the failure of medicine to develop a widely applicable means for changing homosexual behavior, writes: "Vast numbers of men and women feel that a homosexual orientation is their natural bent, and no amount of pressure from the criminal law, from moral condemnation, or from medical advice is likely to make any fundamental difference to their way of life" (p. 161).

Where homosexuality is part of a well-developed life-style much of medicine and clinical psychology has come to terms with the refractive nature of homosexual behavior. There is an increasing clinical emphasis on support and problem solving that seeks to help the homosexual accept himself as he or she is and assist the individual in developing ways of adapting to the heterosexual society.

It should be remembered when discussing efforts to change homosexual behavior that individual sexual expression is seldom clearly either homosexual or heterosexual. Rather, the clinician and the behavioral scientist must deal with idiosyncratic variables relating not only to variations in sexual object choice but also to such factors as gender identity, social expectations, and self-esteem. The decision to attempt to change homosexual behavior must, with the present state of knowledge, be made on a case-by-case basis and then without assurance of outcome.

D. L. CRESON

[*Directly related are* SEXUAL BEHAVIOR; SEXUAL DEVELOPMENT; SEXUAL ETHICS; *and* SEXUAL IDENTITY. *See also:* MENTAL ILLNESS, *article on* CONCEPTIONS OF MENTAL ILLNESS; RESEARCH, BEHAVIORAL; *and* SEX THERAPY AND SEX RESEARCH.]

BIBLIOGRAPHY

BACON, CATHERINE LILLIE. "A Developmental Theory of Female Homosexuality." *Perversions, Psychodynamics and Therapy.* Edited by Sándor Lorand and Michael Balint. New York: Random House, 1956, pp. 131–159.

BARDWICK, JUDITH M. *Psychology of Women: A Study of Bio-Cultural Conflicts.* New York: Harper & Row, 1971.

BIEBER, IRVING. "Clinical Aspects of Male Homosexuality. Marmor, *Sexual Inversion,* pp. 248–267.

BULLOUGH, VERN L. "Heresy, Witchcraft, and Sexuality." *Journal of Homosexuality* 1 (1974/75): 183–201.

HATTERER, LAWRENCE J. *Changing Homosexuality in the Male: Treatment for Men Troubled by Homosexuality.* New York: McGraw-Hill, 1970, pp. 34–41.

HOFFMAN, MARTIN. *The Gay World: Male Homosexuality and the Social Creation of Evil.* New York: Basic Books, 1968.

HUNT, MORTON M. *Sexual Behavior in the 1970s.* Chicago: Playboy Press; New York: Dell, 1974, pp. 303–319.

KALLMANN, FRANZ J. "Comparative Twin Study on the Genetic Aspects of Male Homosexuality." *Journal of Nervous and Mental Disease* 115 (1952): 283–298.

KINSEY, ALFRED CHARLES; POMEROY, WARDELL B.; and MARTIN, CLYDE E. *Sexual Behavior in the Human Male.* Philadelphia: W. B. Saunders Co., 1948, p. 623.

———; POMEROY, WARDELL B.; MARTIN, CLYDE E.; and GEBHARD, P. H. *Sexual Behavior in the Human Female.* Philadelphia: W. B. Saunders, 1953, p. 453.

KOLODNY, ROBERT C.; MASTERS, WILLIAM H.; HENDRYX,

JULIE; and TORO, GELSON. "Plasma Testosterone and Semen Analysis in Male Homosexuals." *New England Journal of Medicine* 285 (1971): 1170–1174.

MACLEAN, PAUL D. "Brain Mechanisms of Elemental Sexual Functions." *The Sexual Experience*. Edited by Benjamin J. Sadock, Harold Kaplan, and Alfred Freedman. Baltimore: Williams & Wilkins Co., 1976, pp. 119–127.

MARMOR, JUDD. "Introduction." *Sexual Inversion: The Multiple Roots of Homosexuality*. Edited by Judd Marmor. New York: Basic Books, 1965, pp. 1–24.

MILNE, HUGO, and HARDY, SHIRLEY J., eds. *Psycho-Sexual Problems: Proceedings of the Congress Held at the University of Bradford, 1974*. Baltimore: University Park Press, 1975.

MONEY, JOHN, and EHRHARDT, ANKE A. *Man and Woman, Boy and Girl: The Differentiation and Dimorphism of Gender Identity from Conception to Maturity*. Baltimore: Johns Hopkins University Press, 1972.

PAINTIN, D. B. "The Physiology of Sex." Milne, *Psycho-Sexual Problems*, pp. 3–19.

SAGHIR, MARCEL T., and ROBINS, ELI. *Male and Female Homosexuality: A Comprehensive Investigation*. Baltimore: Williams & Wilkins Co., 1973, pp. 18–31, 192–203.

TEAL, DONN. *The Gay Militants*. New York: Stein & Day, 1971.

TOURNEY, GARFIELD; PETRILLI, ANTHONY; and HATFIELD, LON. "Hormonal Relationships in Homosexual Men." *American Journal of Psychiatry* 132 (1975): 288–290.

TRIPP, C. A. *The Homosexual Matrix*. New York: McGraw-Hill Book Co., 1975.

WARD, DAVID A., and KASSEBAUM, GENE G. "Lesbian Liaisons." *The Sexual Scene*. Edited by John H. Gagnon and William Simon. Trans-action Books. Chicago: Aldine Publishing Co., 1970, pp. 125–136.

WEST, D. J. "Homosexuality." Milne, *Psycho-Sexual Problems*, pp. 157–165.

II
ETHICAL ASPECTS

An analysis of the ethics of homosexuality is hampered by a lack of agreement among ethicists on several important issues. Among those can be included the definition of homosexuality, the role of cultural and empirical factors, ethical positions regarding homosexual behavior, and an understanding of the meaning of human sexuality that can serve as a context for offering an assessment of homosexuality. The issue is complicated by the diverse emotional responses to homosexuality.

The task of ethical analysis is not made easier by the fact that a variety of technical terms and colloquial expressions used to describe homosexuals or homosexuality contain strong overtones of moral attitudes as well as emotional reactions and prejudices. For example, the terms "invert" or "pervert" suggest negative attitudes; "queer" and "fag" are derisive; and the term most commonly employed by homosexuals to describe both male and female homosexuals is "gay"—which has a positive connotation.

The most emotionally neutral and nonjudgmental definition of homosexuality is: a predominant, persistent, and exclusive psychosexual attraction toward members of the same sex. A homosexual person is one who feels sexual desire for and a sexual responsiveness to persons of the same sex and who seeks or would like to seek actual sexual fulfillment of this desire by sexual acts with a person of the same sex. A distinction is drawn by a majority of authors on the subject between the homosexual *condition* and the homosexual *act*. Homosexual *condition* refers to the person who is sexually attracted exclusively to persons of his or her own sex or who wishes to express sexually deep feelings exclusively with persons of his or her own sex. Homosexual *act*, on the other hand, refers to sexual contacts between persons of the same sex. Such acts may be performed by persons in a homosexual condition or by persons in a heterosexual condition. Thus, many heterosexuals engage in homosexual *acts* without being in a homosexual *condition*.

Cultural and religious attitudes toward homosexuality

There is scarcely a culture or society, ancient or modern, that does not reveal some evidence of homosexual behavior and practice. In some cultures (including some North American Indian tribes) homosexuals were regarded as representatives of the deity and were credited with powers of divination and of magic. In other cultures, homosexual and heterosexual practices formed part of the religious cult by which the divinity was worshiped. Aside from those examples of homosexuality in a religious context, the lack of accurate and complete historical data, especially for pre-Christian cultures, makes it difficult to assess whether the general attitude of a given society toward homosexuality as a form of sexual expression was one of approval, toleration, or condemnation. The long-held assumption, for instance, that in ancient Greece homosexuality was widely and generally held in favorable esteem is today seriously challenged (Karlen). It would seem more accurate to infer from the available data that the prevailing attitude in most ancient societies was one of permissiveness rather than approval. In no culture was homosexuality the dominant form of sexual expression.

In Judaism and Christianity homosexual behavior has been condemned explicitly and consistently with utmost severity. This attitude in the Judeo-Christian world is best explained by the fact that in the Bible homosexual behavior is viewed as a crime worthy of death (Lev. 18:22; 20:13) and a sin "against nature" (Rom. 1:26), which excludes one from entry into the kingdom of God (1 Cor. 6:10). Even more threatening was the punishment visited by God upon Sodom for the assumed sin that was named after that city (Gen. 19:1–29).

The Sodom story has been particularly influential in forming not only religious attitudes but civil legislation as well. The Code of Theodosius and the Code of Justinian, for instance, prohibited all sodomistic practices under pain of death by fire. The explicit motivation behind the prohibitions was to protect the state from the wrath of divine judgment such as befell Sodom. The sixth-century Code of Justinian had considerable impact on both the ecclesiastical and the civil laws of the Middle Ages and indirectly on civil law in the West well into our own time. The prevailing conviction drawn from the Stoics that pleasure was evil and that procreation was the primary justification for sexual expression provided a philosophical rationale to support the view that homosexual behavior was unnatural and immoral.

Biblical scholarship in recent years has raised some serious questions about the legitimacy of the conclusions drawn from the above-mentioned scriptural sources. Some contend that many of the Old Testament condemnations of homosexuality are for cultic reasons (such practices constituted part of the Canaanite worship of Baal) and not because they were sexually improper. Others insist that the homosexual interpretation given to the Sodom incident originated in a first-century A.D. reinterpretation of Genesis 19 as a reaction to the basest features of Greek sexual immorality. Then, too, it must be acknowledged that none of the biblical writers who mentioned homosexuality as a deliberately chosen option knew anything of inversion as an almost irreversible condition fixed in childhood. These developments, together with insights from the behavioral and social sciences, have paved the way for serious reconsideration among most religious groups of their former attitudes both to homosexuals and to homosexual behavior.

Homosexuality and society

Incidence of homosexuality. Scientific research has revealed homosexuality to be a phenomenon of bewildering complexity, defying classification according to rigid, predetermined notions that would categorize all people simply as either heterosexual or homosexual. Empirical evidence suggests that only in a few individuals is sexual orientation totally exclusive. In those individuals in whom heterosexual orientation is dominant, there seems to be a latent potentiality for homosexual interest, which may or may not rise to consciousness. It has been suggested that Western cultural pressures mask an awareness of this interest and create a deceptive picture of rigid polarity between heterosexual and homosexual orientations. It is estimated that some four percent of American men are exclusively homosexual all their sexual lives, as are a smaller percentage of women.

The homosexual's self-awareness. One of the most important aspects of homosexuality is the awareness of being different from the majority of people. The consciousness of being "different," of belonging to a minority, leads homosexuals to suffer from the same problems of discrimination and oppression as all minority groups. The added factor that their "differentness" is morally reprehensible and must be kept secret leads to an even more profound alienation for the homosexual. Societal attitudes and legislation that condemn homosexual expression force homosexuals to live in two worlds. In the work world they share the cultural values and norms of whatever sector they participate in. Their homosexual world is largely one of leisure time or recreational activities that frequently center around the "gay" bar. For most, these bars are social institutions where friends can be met, the news of the homosexual world heard, gossip exchanged, invitations to parties issued, and warnings about current danger spots and attitudes of the police given.

Myths about the homosexual. Prejudices resulting from biased education or societal attitudes have created several popular myths regarding the homosexual that further compound the problem, namely:

1. The myth that male homosexuals are easily identifiable as effeminate or female homosexuals as masculine; that homosexuals recognize each other and form a veritable secret society; that homosexuals invariably tend toward particular professions. On the contrary, most homosexuals

are not recognizable in mannerism or in social and professional life.

2. The myth that every homosexual is attracted to children and adolescents and seeks to have physical contact with them. The incidence of homosexual offenders in this regard is proportionately no greater than that of the heterosexuals.

3. The myth that all homosexuals are unstable or promiscuous and unable to form enduring relationships. Studies of carefully matched samples of heterosexuals and homosexuals reveal no appreciable differences in traits that would manifest personality dysfunction. Lasting friendships are formed in spite of the strong social pressures that militate against them.

4. The myth that homosexuals simply require will-power or an experience of heterosexual intercourse or heterosexual marriage to correct their condition. True sexual inversion is for all practical purposes irreversible.

Social attitudes toward homosexuals. Authorities agree that no one can live in an atmosphere of universal rejection, widespread pretense, and derision, or in a society that outlaws and banishes one's activities and desires without experiencing a fundamental influence on one's personality. A 1970 study reports that eighty-six percent of the general American population disapproved of homosexuality, indicating the extent of the rejection of the homosexual in American society.

There appears to be a general consensus among experts that the extreme opprobrium that society attaches to homosexual behavior, particularly by way of criminal statutes and restrictive employment practices, has done more social harm than good and goes beyond what is necessary for the maintenance of public order and human decency. This consensus is paving the way to a growing movement toward legal reform that would remove legal penalties against acts in private among consenting adults and eliminate discrimination against the homosexual in employment and housing policies. Among the stimuli that have contributed notably to this movement are: the Westminster Report (1956); the Wolfenden Report (1957); the American Law Institute Model Penal Code (1962); the statement of a group of English Quakers (*Towards a Quaker View of Sex*); the U.S.-founded Gay Liberation movement, 1969; and the Final Report of the Task Force on Homosexuality, U.S. Department of Health, Education, and Welfare (1969). Full equality in employment and housing, full security, and full acceptance by society will not be achieved by changes in the law alone, but such changes may help to facilitate the recasting of public attitudes that is ultimately needed.

Ethics and homosexuality

The question of normative ethical positions on homosexuality revolves around the question of the meaning of human sexuality. There are diverse ethical opinions on the meaning of human sexuality, ranging from the integrist to the relational to the recreational.

The integrist understanding of human sexuality sees an inherent relationship between the procreative or life-giving and the unitive or love-giving aspects of human sexuality. Although the authors who advocate this understanding of human sexuality do not require that every expression of human sexuality be procreative, and although many who advocate it accept contraception for married persons as an ethically justifiable way of carrying out parental obligations, they maintain that both the procreative and unitive aspects of human sexuality are humanly significant and are intended to be united in marital intercourse. Among authors advocating this understanding are Paul Ramsey, George Gilder, and the majority of Roman Catholic writers.

The relational and recreational positions both hold that the procreative or reproductive aspect of human sexuality is not so terribly significant and that the *human* meaning of sexuality consists in the role that sexuality and sexual acts have to play in relating one human person to another and thereby fulfilling spiritual, emotional, intellectual, and physiological needs (the relational understanding) or in providing human persons with entertainment, pleasure, or escape from the humdrum of living (the recreational understanding). Among authors advocating a relational view are John McNeill, the authors of the Quaker statement, Norman Pittenger, and others. Advocates of a recreational understanding of human sexuality are numerous in contemporary society.

Apart from the question of the meaning of sexuality, the ethical and moral questions raised by homosexuality do not differ substantially, if at all, from the ethical and moral questions raised by heterosexuality, since the question of the normative aspects of interpersonal and intersocial behavior are the same for homosexuality as for heterosexuality.

The current ethical positions on homosexuality can be grouped under the three categories used to describe the ethical analysis of the meaning of human sexuality: the integrist understanding of human sexuality, which sees homosexual acts as unnatural; the recreational understanding, which regards homosexual behavior as natural and therefore justifiable; the relational understanding, which evaluates homosexual acts according to their ability to contribute to the growth and development of the person.

Integrist understanding and homosexuality. The integrist understanding is directly opposite to the recreational in its analysis of homosexuality. The difference rests on basically irreconcilable positions. Where the recreational view sees homosexuality as morally good (an ethically normative form of sexual expression between humans), the integrist view sees homosexuality as objectively immoral, i.e., an aberration or violation of natural order (natural law) or the order of divine creation. Thielicke and Barth both argue in this fashion on the basis of the creation of the sexes. Furthermore, the integrist view, while cognizant of empirical evidence, does not accept factual evidence as ethically normative. Its normative sources are either religious teaching, tradition, or philosophical concepts of nature, order, purpose of sexual organs, marriage, etc.

Harvey's analysis of the ethics of homosexuality is an excellent example of this position. He argues that, since the homosexual act by its very essence cannot result in the transmission of life, it frustrates the very purpose of the sexual organs and must be judged unnatural, i.e., a serious disruption of the order of nature and/or the will of the divine creator. Harvey, however, shows great compassion for the plight of the homosexual person. He recognizes that homosexuality has not been chosen by the person and suggests various counseling techniques and practices to assist the homosexual's psychological and spiritual adjustment to his homosexuality. With other authors, such as A. Hyatt Williams and the early Hettlinger (1966), Harvey also emphasizes the incompleteness of homosexual acts, the rejection of the natural complementariness of the opposite sex, and the impossibility of parenthood as evidence of the unnaturalness of homosexuality. Furthermore, these incompletenesses create a consistent frustration to the homosexual in his or her search for an "ideal love." Thus, homosexual relations face severe obstacles to being durable, stable, and unexploiting.

Recreational understanding and homosexuality. The chief arguments of the recreational view are founded on at least two considerations: (1) a rejection of the integrist's assumptions concerning homosexuality, and (2) incorporation of empirical evidence as the normative factor in the ethical analysis. Proponents of this viewpoint argue that, contrary to the integrist's understanding of sex, homosexuality is natural. Moreover, empirical evidence shows that all infrahuman mammals from the lowest form of mammal to the brightest subhuman primate exhibit homosexual behavior. Furthermore, statistically significant numbers of persons engage in homosexual behavior, and every culture, from the most primitive to the most sophisticated, has some form of homosexuality. Authors such as Churchill report that, although homosexuality has aroused anxiety and disapproval within any number of cultures, the majority of cultures provide an outlet for some type of socially approved expression of the homosexual needs of males and females and have attempted to regulate rather than suppress homosexual behavior. Furthermore, homosexual behavior is found in all cultures and groups that do not make a pronounced effort to suppress it and, more often than not, is noted even in groups that do actively condemn it. But, as Opler notes, in practically all cultures homosexuality is regarded as a deviation from the majority values and norms of behavior.

These authors argue that the extent of homosexual behavior in all levels of mammalian creatures and the nondeleterious effects of such behavior on the lives of the participants constitute a new source of information which challenges the assumptions upon which the integrist view builds its analysis of homosexuality. Furthermore, proponents of the recreational view argue that the integrist view is incorrect in making the nonprocreative aspect of homosexuality the normative basis for its judgment.

Relational understanding and homosexuality. The relational understanding represents an analysis of homosexuality that avoids the conceptual frameworks of both the integrist and the recreational views. In some sense the relational understanding is a compromise or moderating one. But its analysis rests not so much on a tempering of the integrist and recreational positions as on the relational interpretation of the meaning of human sexuality.

Some of the important influences on the development of the relational position on homosexuality are the personal experiences of homosexuals and the growth of sociological and psychological empirical information on the extent and effects of homosexuality. The relational view regards homosexuality as normative in the same respect as it regards heterosexuality as normative, i.e., the quality of the relationship between persons is the criterion. Relationships that foster self-giving and mutuality, that stimulate growth in humanness and aid personal fulfillment, are normative. Authors such as the framers of the Quaker statement, Pittenger, and McNeill see the unethical aspect of homosexuality not in the action of sexual contact between members of the same sex but in the quality of the relationship between those persons. Selfishness, manipulation, exploitation, etc., are regarded as the source of evil, not the sexual actions themselves. This view reflects a personalistic approach and rejects an approach based on a concept of "natural" rooted in the action itself.

There are variations of the integrist view that incorporate part of the relational ethics without rejecting the normative value of heterosexuality. Those representing this position, such as Charles E. Curran, are acutely aware of the existential plight of the homosexual person, especially in some areas of Western society. They acknowledge the inhumanity of many social institutions and practices (e.g., promotion criteria), toward the homosexual person. They also recognize the basic psychological drive shared by the heterosexual and homosexual for companionship, affection, security, intimacy, and love. Consequently, their approach attempts to be responsive to the homosexual *person* living in contemporary society, with all its existential contingencies and pressures, and does not ground its analysis on an abstract, impersonal base.

In various ways, this view attempts to avoid recognizing homosexuality as an ethically normative expression of sexuality and, at the same time, to be sensitive to the homosexual *person*. Some who hold this view attempt to reconcile those two seemingly irreconcilable positions by disclaiming any moral responsibility on the part of the homosexual person for his homosexuality, yet condemning homosexual behavior. Others attempt to reconcile the positions by recognizing that in certain existential circumstances homosexual behavior is not condemned because the person is in intolerable circumstances that prevent him or her from pursuing the norm of heterosexual contact. Thus, they are forced to compromise the norm of heterosexuality in face of the loss of great human values, such as love, companionship, affection, etc.

Conclusion. Briefly, then, a wide spectrum of views on homosexuality is to be found. Each variant reveals a normative judgment toward homosexuality that favors the integrist, the recreational, or the relational understanding of human sexuality. Careful reading of authors is necessary to discover which view they favor. However, all these views reveal elements of the relational ethics and therefore resist judging homosexuality in a black-and-white fashion. No one seems seriously to judge the ethics of homosexual behavior exclusively on the basis of a procreational meaning of human sexuality. The tempering of traditional ethical-moral positions such as that of the Roman Catholic position on homosexuality as evidenced in the Vatican's 1975 *Declaration on Certain Questions Concerning Sexual Ethics* (Sacred Congregation) seems to suggest that a move toward moderation is afoot (Kosnik et al.).

All are sensitive to the humanness of homosexuals and do not view homosexuality in the abstract. They view homosexuals as persons seeking to find wholeness within the context of a predominantly heterosexual society and tradition.

The ethical debate hinges on several still unsettled questions surrounding homosexuality. Among these are: the etiology of the homosexual condition and acts; the dilemma of the homosexual condition; the impact of homosexual acts upon society; the relationship between private individual behavior and public policy; the meaning of the differentiation of the sexes; the meaning and institutional significances of marriage; the role of empirical evidence in normative thought; the concept of evil; and the role of religious, philosophical, and social tradition in ethical analysis and evaluation. A comprehensive ethics of homosexuality is evolving as these major questions begin to be unraveled. The integrist position is beginning to lose support, especially in influencing the direction of law and public opinion concerning homosexuals. Today, no one view seems to have the ascendency in ethical literature.

GEORGE A. KANOTI AND
ANTHONY R. KOSNIK

[*Directly related are the entries* SEXUAL BEHAVIOR; SEXUAL DEVELOPMENT; SEXUAL ETHICS; *and* SEXUAL IDENTITY. *Other relevant material may be found under* NATURAL LAW; *and* SEX THERAPY AND SEX RESEARCH.]

HOMOSEXUALITY: Ethical Aspects

BIBLIOGRAPHY

American Law Institute. "Deviate Sexual Intercourse by Force or Imposition" and "Loitering to Solicit Deviate Sexual Relations." *Model Penal Code: Proposed Official Draft.* Philadelphia: 1962, sec. 213.2, pp. 543–544; sec. 251.3, p. 591. Recommends that private homosexual acts between consenting adults be excluded from the criminal law.

American Psychiatric Association, Task Force on Nomenclature and Statistics. "Position Statement on Homosexuality and Civil Rights." *American Journal of Psychiatry* 131 (1974): 497. The American Psychiatric Association's decision to list homosexuality under a new psychiatric category.

BAILEY, DERRICK SHERWIN. "Homosexuality and Christian Morals." *They Stand Apart: A Critical Survey of the Problems of Homosexuality.* Edited by John Tudor Rees and Harvey V. Usill. London: William Heinemann, 1955, chap. 3, pp. 36–63. Homosexuality is viewed as part of a broader problem of sexual immorality. A critical theological analysis of the historical Christian and biblical interpretations that have contributed to the development of negative Western attitudes toward homosexuality.

———. *Homosexuality and the Western Christian Tradition.* London: Longmans, Green & Co., 1955. Traditional homosexual interpretation of the biblical story of Sodom and Gomorrah is challenged. Homosexuality's status in pre-Justinian law is discussed.

BARNHOUSE, RUTH TIFFANY, "Homosexuality." *Anglican Theological Review,* supp. ser. no. 6 (1976), pp. 107–134.

BARTH, KARL. *Church Dogmatics: A Selection.* Translated and edited by G. W. Bromiley. Introduction by Helmut Gollwitzer. New York: Harper & Row, 1962, pp. 194–229. Traditional ethical argument concerning the evil of homosexuality.

BUCKLEY, MICHAEL J. *Morality and the Homosexual: A Catholic Approach to a Moral Problem.* Foreword by John C. Heenan. Westminster, Md.: Newman Press, 1959. Thoroughly traditional moral view in response to the Wolfenden Report.

CHURCHILL, WAINWRIGHT. *Homosexual Behavior among Males: A Cross-Cultural and Cross-Species Investigation.* New York: Hawthorne, 1967; Englewood Cliffs, N.J.: Prentice-Hall, 1971. Cross-cultural sources are quoted and employed to support the normality of homosexual behavior in mammalian sexuality.

CURRAN, CHARLES E. "Homosexuality and Moral Theology: Methodological and Substantive Considerations." *Thomist* 35 (1971): 447–481. Succinct ethical analysis of current ethical positions held by religious ethicists. Excellent review of the ethical methodological problems in the analysis of the ethics of homosexuality.

GILDER, GEORGE F. *Sexual Suicide.* New York: Quadrangle/New York Times Book Co., 1973.

Great Britain, Committee on Homosexual Offenses and Prostitution. *The Wolfenden Report: Report of the Committee on Homosexual Offenses and Prostitution.* London: Her Majesty's Stationery Office, 1957; Authorized American ed. Introduction by Karl Menninger. New York: Stein & Day, 1963. Citations to American edition. Takes a position on legality of homosexual behavior vis-à-vis public order and private morality.

HARVEY, JOHN F. "Homosexuality." *New Catholic Encyclopedia.* 15 vols. New York: McGraw-Hill, 1967, vol. 7, pp. 116–119. Concise, clear example of an integrist analysis of homosexuality written by a Roman Catholic theologian.

———. "Pastoral Responses to Gay World Questions." Oberholtzer, *Is Gay Good?,* chap. 5, pp. 123–139. A sensitive application of the integrist (predominately procreational) position to the human problems of homosexual existence in a heterosexual world.

HETTLINGER, RICHARD FREDERICK. *Living With Sex: The Students' Dilemma.* New York: Seabury Press; London: SCM Press, 1966, pp. 94–112. Student handbook that treats traditional arguments and attitudes towards homosexuality.

———. *Sex Isn't That Simple.* New York: Seabury Press, 1974, pp. 138–154. 2d ed. *Human Sexuality: A Psychosocial Perspective.* Belmont, Calif.: Wadsworth Publishing Co., 1975. A reversal of the author's previous moral and emotional judgment of homosexuality. The relational position is taken.

JONES, H. KIMBALL. "Homosexuality—A Provisional Christian Stance." Oberholtzer, *Is Gay Good?,* chap. 6, pp. 140–162. The author responds to a more traditional position by arguing strongly for incorporation of personal and empirical evidence into the ethical analysis of homosexuality.

KARLEN, ARNO. *Sexuality and Homosexuality: A New View.* New York: W. W. Norton & Co., 1971. Contains many interviews with homosexuals and descriptions of the homosexual community in contemporary society. Extensive bibliography.

KOSNIK, ANTHONY; CARROLL, WILLIAM; CUNNINGHAM, AGNES; MODRAS, RONALD; and SCHULTE, JAMES. *Human Sexuality: New Directions in American Catholic Thought: A Study Commissioned by the Catholic Theological Society of America.* New York: Paulist Press, 1977. A final report of a committee established by the Board of Directors of the Catholic Theological Society of America in 1972.

MCNEILL, JOHN J. *The Church and the Homosexual.* New ed. Mission, Kans.: Sheed Andrews & McMeel, 1976.

MARMOR, JUDD, ed. *Sexual Inversion: The Multiple Roots of Homosexuality.* New York: Basic Books, 1965.

MARNEY, CARLYLE. "The Christian Community and the Homosexual." *Moral Issues and Christian Responses.* Edited by Paul T. Jersild and Dale A. Johnson. New York: Holt, Rinehart & Winston, 1971, pp. 181–191. A recreational view that attacks arguments of a procreational position. Somewhat inaccurate in its interpretation of traditional biblical sources.

National Institutes of Mental Health. *Final Report of Task Force on Homosexuality: The Hooker Report.* Edited by John M. Livingood. Chevy Chase, Md.: National Institutes of Mental Health, 1969. DHEW Publication no. (HSM) 72-9116. Washington: Government Printing Office, 1972. Concludes that legal sanctions against consenting adult homosexual should be abolished. Takes issue with studies that treat homosexuality as a clinical entity.

OBERHOLTZER, W. DWIGHT, ed. *Is Gay Good? Ethics, Theology, and Homosexuality.* Philadelphia: Westminster Press, 1971.

OPLER, MARVIN K. "Anthropological and Cross-Cultural Aspects of Homosexuality." Marmor, *Sexual Inversion,* pp. 108–123. Attempts to synthesize biological, psychological, social, and cultural factors in its analysis of homosexuality.

PITTENGER, WILLIAM NORMAN. *A Time for Consent: A Christian's Approach to Homosexuality.* 2d ed. rev. and enl. London: SCM Press, 1970. Argument for the acceptability of homosexual acts between persons intending a permanent union and who are in love.

RAMSAY, PAUL. *Fabricated Man: The Ethics of Genetic Control.* New Haven: Yale University Press, 1970.

Sacred Congregation for the Doctrine of the Faith. "Declaration on Certain Questions Concerning Sexual Ethics." *The Pope Speaks* 21 (1976): 60–73. Official Roman Catholic analysis of the morality of sexual behavior. The document recognizes the possibility of a nonresponsible homosexual condition but condemns homosexual acts.

SALZMAN, LEON. "Latent Homosexuality." Marmor, *Sexual Inversion*, pp. 234–247. Challenge to the validity of the Freudian theory that all humans are latently homosexual to a greater or lesser degree.

SHINN, ROGER L. "Homosexuality: Christian Conviction and Inquiry." *The Same Sex: An Appraisal of Homosexuality.* Edited by Ralph W. Weltge. Philadelphia: Pilgrim Press, 1969, pp. 43–54. A brief but thorough example of a moderating ethics.

THEILICKE, HELMUT. *The Ethics of Sex.* Translated by John W. Doberstein. New York: Harper & Row, 1964. Analyzes homosexuality within the context of Christian (Lutheran) theology of sexuality and marriage.

Towards a Quaker View of Sex: An Essay by a Group of Friends. Edited by Alastair Heron. London: Friends Home Service Committee, 1963. Theological review of homosexuality which supports the positive aspects of homosexual affection.

VON ROHR, JOHN. "Toward a Theology of Homosexuality" and "A Response to the Responses." Oberholtzer, *Is Gay Good?*, chap. 1. pp. 75–97; chap. 13, 239–252. Survey of the traditional Christian ethical stance and argument concerning homosexuality.

WARWICK, DONALD P. "Tearoom Trade: Means and Ends in Social Research." *Hastings Studies* 1, no. 1 (1973), pp. 27–38. Critique of a sociological study of homosexuality that reveals some of the ethical questions raised in such studies and indicates the benefits and risks of empirical research.

WILLIAMS, A. HYATT. "Problems of Homosexuality." *British Medical Journal* 3 (1975): 426–428. Medical-psychiatric appraisal of homosexuality in a society that has become more permissive and less punitive toward homosexuals and homosexual acts.

HOSPITALS

This article deals with the hospital in contemporary society. It also discusses the hospital's recent transition from a philanthropic institution essentially free from public accountability to its present status as an institution subject to public scrutiny like other social institutions.

While most bioethical issues in the health field eventually occur in the hospital, this article will focus on the major ethical problems that derive from the hospital's broadening institutional responsibilities. No specific differentiation will be made here among hospitals based on religious sponsorship, proprietary (for profit), university teaching hospitals, or publicly supported ones. The hospital model for this discussion is the voluntary, not-for-profit, short-term, acute-care, community hospital. The major problems confronting this model are essentially the same for all hospitals, and certainly the moral obligations of all hospitals are the same. In terms of the source of their funds, both operating and capital, by far the majority of hospitals today are at least quasi-public institutions.

The hospital in contemporary society

Historical perspective. The advent of Christianity saw the early development of the hospital. Even the well-organized early hospitals, *valetudinaria*, of the Romans, as well as those of the Arabians, came into existence after the time of Christ. The hospitals of Europe had their beginning in the hospices, almshouses, and *infirmaria* or sick bays of monastic institutions in the fourth and fifth centuries (Garrison; Ives). The establishment of Lazar houses or *leprosaria* for leprosy victims in the early centuries was a further extension and application of the teachings of Christ through the efforts of religious orders. With the coming of the Black Death in the fourteenth century, the leprosarium was replaced by the plague or pest house (Major). Up to and throughout the Middle Ages, the sick in hospitals were ministered to by monks and nuns, not by physicians (Castiglioni).

The doctor actually was not central to, nor did he begin to dominate, the hospital scene until the eighteenth century, when pathology was founded as a science and newer discoveries in physics and chemistry advanced clinical medicine. Professional dominance in the hospital became the accepted order in the nineteenth century with the acquisition of surgical knowledge, the advent of general anesthesia, and advances in antisepsis. Assuming the physician's commitment to be first and foremost the interest and safety of his patients, society fostered the concept that the hospital was a milieu in which the physician should have readily available to him anything and everything that could further his ministering to the sick and injured.

Until the mid-twentieth century, the hospital was the doctor's workshop, characterized by a distinct one-on-one physician–patient relationship, uncluttered by third-party intervention. The hospital was regarded as a philanthropic institution. Aggrieved patients did not have access to the courts to seek redress for alleged negligence by the hospital, which was protected by the legal doctrine of charitable immunity. The corporate concerns of its governing authority, the board of trustees, rarely extended beyond fiscal matters, hospital census, and building programs. The practicing physicians constituted the hospital medical staff organization, functioning largely as an island unto themselves, with the mystique of the physician impenetrable, to the extent that rarely ever was the medical staff called to account for the quality of its work or held responsible for the influence it exercised on hospital functions.

The role of the administrator was tenuous; his effectiveness usually was measured in terms of a balanced budget and the extent to which harmony was maintained with the medical staff. Nursing and other supporting personnel, while in the employ of the institution, nonetheless also came under the professional dominance of the medical staff. The role and rule of the physician were pervasive, and for the most part accepted; seldom did he have to assume responsibility for the influence he wielded.

The hospital today. Since the mid-twentieth century the hospital has been subject to societal forces which have had a major impact on it. Included in those forces have been the persistent concern over escalating costs; demands for improved effectiveness, i.e., the application of sound principles of management and organization to the entire hospital structure; and the insistence on public accountability as manifested in the consumer and patients' rights movements, in the disappearance of protective charitable immunity, in statutory and common law pinpointing responsibility, in the growing and continuing concern regarding quality controls, and in the imposition of regulatory measures from both the private and public sectors.

No longer regarded as a philanthropic institution or just the physicians' private workshop of the past, the hospital today is a social institution, a place where the multiple skills and services of many disciplines, along with costly facilities and equipment, can be coordinated and managed in the best interests of the patients it is supposed to serve. The institution's governing authority, the board of trustees, must accept the responsibility for effectively coordinating and managing all the resources provided to it by society. To that end, society has vested authority in the hospital's board of trustees and in turn holds it both morally and legally responsible for everything that goes on within the institution, including the activities of all professionals privileged to work in it.

Carrying out sound fiscal planning, promulgating proper personnel and other nonmedical administrative policies, and seeing that those policies are effectively implemented are functions not new to hospital boards. However, a hospital board's having to accept corporate responsibility for the quality of the actual medical care rendered within the institution is indeed new. It dates from the 1965 landmark *Darling* decision (Southwick, 1973) and has been reinforced by many subsequent similar decisions (Southwick, 1976; Blaes).

Throughout history society has found it necessary periodically to reappraise the relationships of the medical profession in society (Williams, 1972), to evaluate the effectiveness of controls on its members, and to exact accountability from them. The medical profession is currently in one of those reappraisal periods. It has been shown that very little, if any, controls have been exercised by organized medicine or state licensing boards (Derbyshire; Holman) regarding surveillance and continuing competence. It is also realized that, except for a few closed panel prepaid health-care plans and the developing Professional Standards Review Organizations, whose effectiveness in this area has not yet been shown, the only real quality control mechanism for maintaining surveillance of clinical practice in the United States is the properly organized and effectively functioning hospital medical staff organization. Thus it comes about that the hospital is used by society as a prime means of exacting accountability from the medical profession. The Joint Commission on Accreditation of Hospitals, the main voluntary agency for accrediting hospitals, has similarly stressed the medical staff's accountability to hospital boards of trustees (Joint Commission). Legally speaking, the medical staff today is very much a part of the total hospital organization. As the contemporary title implies, the hospital administrator is the institution's chief executive officer. He is directly accountable to the hospital's board of trustees and receives his authority from them. He and the members of his administrative team, through the application of the functions of planning, organizing, directing, supervising, and controlling, must coordinate and direct *all* the institution's available resources in the most effective manner in consonance with institutional goals.

The physician in the hospital setting today has two roles. His primary role is to apply his professional skills to the care of the sick and injured. The other is to assist, as a member of the organized hospital medical staff, in carrying out functions assigned to the medical staff organization. Those functions deal with aspects of quality control, such as carrying out surveillance of clinical practice, determining the extent to which professional standards are complied with, and making recommendations to the board of trustees for disciplinary or other corrective action in accordance with their findings. In addition, the medical staff, through formal organizational mechanisms, also gives advice and makes recommendations on nonclinical matters such as planning, community relations, setting objectives, building, and budgeting.

One of the important developments of the societal forces impacting on hospitals in the past four decades has been the defining of the physicians' position and relationships in the hospital setting. While not particularly welcomed by the medical profession, the effect has been to centralize controls in the institution and to reduce greatly the previously dichotomized administrative controls and long-standing organizational schizophrenia. The stage has been clearly set for the medical staff today to be integrated into the institution-wide management process.

Some major problems facing hospitals today. Today's hospital is still very much in transition, continually having to face challenges that very often greatly influence its relationships with society. Some of the impending challenges in the United States include coping with a formally established national health policy; adjusting to the inception of national health insurance; adapting to more stringent fiscal controls, such as prospective rate setting; participating in meaningful areawide planning; being subject to more centralized decision making through regionalized hospital systems or multiple hospital complexes with central control; assuming a much more aggressive role in imposing and maintaining more effective controls on the actual quality of medical care rendered within them; and reappraising its present costly commitment to acute illness and high technology care to the exclusion of other factors that affect health (Illich). Those challenges are in the offing. Some problems of a more immediate nature facing hospitals today are as follows:

Escalating costs and numerous problems emanating from this dimension continue to be a top-priority concern for hospitals, influencing all their decisions and actions. While the hospital's governing authority has both a moral and a legal responsibility to manage its resources and to provide to the community high-quality medical care in the most effective manner, the fact of the matter is that the hospital does not have control over those resources. Patients are the source of hospital revenue, and the hospital's physicians admit patients and discharge them as they see fit. Nor does the hospital have direct control over its costs or purchases. New facilities, equipment, services, programs, and employee staffing patterns are often dictated or greatly influenced by physicians, who, if they do not have their requests met, can withhold their admissions or take their patients elsewhere. Questions arise as to whether cost and other controls can be effectively applied in the hospital setting alongside the entrepreneurial fee-for-service mode of practice.

The administrator of the hospital is in an exceedingly tenuous position in attempting to apply the management process on an institution-wide basis, whether it is with respect to meeting community needs, implementing certain board policies, or identifying or dealing with problems of competence or conduct of various professional groups in the hospital setting (Williams, 1976).

Hospital trusteeship, in its traditional mold, is no longer adequate to meet the challenges facing hospitals. The need for change here is not fully comprehended by hospitals. Dealing with hospital problems, many of which often approach crisis dimensions, requires more than a hurried overnight acquaintance with them. The chief executive officer, who is supposed to apply or cause to be applied the principles and practices of management throughout the entire institution, cannot do so today without the informed support and involvement of the corporation's board of trustees. Trustees, per se, do not administer the hospital, but they have to understand what those principles of management are, because they are both morally and legally responsible for managing the affairs of the institution.

Differing ideologies and concepts of the hospital frequently preclude coordination of institutional resources. Not all segments of organized medicine are willing to accept some of the premises that hospital boards have been forced to lay down, particularly with respect to the organization and management of the institution, to physicians rendering accountability to boards composed mostly of nonmedical persons, or to the establishment of newer methods of delivering health care involving hospital-based physicians. It is not uncommon to see those situations result in conflict, confrontation, and economic hardship for the institution and the community it serves.

The hospital today is inundated by regulatory measures, to the point of counterproductivity. The inception of the Medicare and Medicaid programs in the United States was soon followed by widespread abuses and spiraling costs necessitating constraints of federal health expenditures, and hence the regulations. In preparation for national health insurance, hospitals can expect even more regulatory measures.

Ethical considerations

The hospital is a principal site for the delivery of health care. Practically all the ethical problems of the health-care delivery system can be found in the hospital to varying degrees. Because the force of public accountability has begun only recently to make the hospital see that it has social responsibilities and that it will come under increasing public scrutiny in the future, many hospitals have not yet recognized that they will have consciously to become moral agents, as Dr. Edmund D. Pellegrino points out in his Harvey Weiss Lecture of January 1976.

The hospital's purpose. In order for those who have both the moral and legal responsibility for the conduct of the hospital to understand the extent of that responsibility and the ethical obligations that derive from it, the hospital's purposes must be clearly and meaningfully defined. The establishment of corporate objectives, the structuring of the necessary organization, the institution-wide application of administrative functions for controlling institutional resources, attaining predetermined objectives, ascertaining the success and effectiveness of the institution, and justifying its continuing existence—those functions can be fulfilled only when the purposes of the institution are clearly defined, understood, and accepted.

It is generally assumed that the hospital's purposes are quite well defined, that it is first and foremost for the provision of high-quality care for the sick and injured, and that the safety and interest of patients is supposed to take precedence over all other concerns. The roots of that assumption are in the origins of the hospital, and they have survived through the centuries. But is that assumption valid today? Are hospitals really first and foremost for patients and the public?

The existence of certain situations in contemporary hospitals raises sensitive questions: hospital administrators dismissed for having incurred the displeasure and anger of their medical staffs as they stood firm on issues of medical competence, patient safety, and board policy; institutions allowing themselves to be pressured into closing existing facilities and moving to the suburbs, leaving deprived areas without adequate facilities; reluctance on the part of hospitals to introduce or become associated with newer and needed methods of delivering health care for fear of being viewed as competing with traditional modes of practice; hospitals that back down from commitments to affiliate with a medical school because of opposition from its medical staff. These situations raise serious questions about the validity of the assumption that hospitals are first and foremost for the patients.

Has the hospital's commitment to technology brought about a change in its prime purpose, or has the prime purpose just been temporarily lost sight of? The hospital's prime purpose of assuring that the interest and safety of the patient take precedence over all other concerns is still the accepted concept of society, and it is an ethical obligation of the highest order for the hospital's governing authority to understand, accept, and fulfill the expectations society has for its hospitals. Only with a full understanding of the institution's purposes can the extent of institutional morality be determined.

Institutional responsibility. Institutional responsibility is a synthesis of all the ethical obligations that follow as a result of moral commitment to the institution's purposes. It is a relatively new concept in hospitals, having first received attention in the 1960s under the rubric of corporate responsibility. Unfortunately that term carries with it a negative connotation because of its association with medical horror stories involving gross incompetence, the establishment of liability, and the necessary meting out of punishment for violation of the hospital's purposes. As noted by Pellegrino in the lecture cited above, for the courts to have to define institutional responsibility with such emphasis is a severe criticism of the failure of hospitals to be alert to their ethical obligations. Many of the disciplines associated with the hospital have had little, if any, orientation to the hospital's purposes or ethical objectives. The governing authority of the hospital, as the institution's conscience, has an obligation to see that the hospital's purposes are conveyed to, understood by, and accepted by all concerned. That is not always an easy task. It must be consciously and assiduously worked at. Failure to achieve this will perpetuate the present imbalance of patient advocacy and run the grave dangers of intervention by law or government.

It is incumbent on the institution to see that individual professions in the hospital honor their own principles of ethics and the commitments derived from them. It must be alert to the possibility that those professions at times might allow economics and etiquette to obscure the real intent behind their codes of ethics. Institutional responsibility must transcend both personal and professional goals. This can be a difficult responsibility to discharge, because members of the medical profession rely heavily on their centuries-old tradition of critical self-analysis and have not fully conceptualized the hospital and the expectations society has of it. The governing authority's discharging of its overall responsibility can be difficult, because, in the eyes of many physicians, the governing authority of the hospital and its administration are cast in adversary roles rather than as guardians of an institution whose purpose is subscribed to by all concerned.

Proper discharge of the hospital's responsibility and attainment of its purposes will not result unless those in leadership positions realize that the hospital's value structure, i.e., institutional ethics, is an indispensable factor in the decision-making process. Failure fully to appreciate this factor is reflected in the increasing dependency on legal expertise and welcoming of legal reasons as a basis for decision making. Such legal resort can be used as a means of dodging one's responsibility. There must, of course, be respect for and compliance with the law. At the same time it must also be remembered that ethical obligations frequently extend considerably beyond what is required by the law.

Governance and management. The understanding and acceptance of the institution's ethics by those responsible for the operation of the institution is not enough. There must be a system of effective governance and competent administration to see that the time-tested administrative functions of planning, organizing, supervising, and controlling of institutional resources are implemented and continually supervised. With the appropriate interaction of people, the acceptance by them of the institution's ethics, and the assurance of professional input, coordination of resources can occur and lead to the fulfillment of the institution's ethical obligations.

Proposals, as have occurred in some states, that legislatures set standards for trusteeship should suggest to those responsible for hospitals that they have overlooked their own ethical obligation to apply introspection to existing systems of governance. Review and improvement of time-honored methods of selection, orientation, and continuing education of trustees is long overdue (Greenleaf, 1974). If trustees were appointed to boards only after first advising them of the institution's moral commitment and ethical values, obtaining their commitment to the institution's purposes and its values, and ascertaining their sense of social responsibility and willingness to spend time and become involved, they would still be subjected to considerably less screening than is required by law of the professionals who, in effect, come under the control of those trustees. If institutions are to be public servants, then top priority has to be given to critical review and improvement of board organization and function (Greenleaf, 1972, 1974). It is not only experienced observers in the field and students of organizational and managerial behavior who are critical of hospitals' systems of governance today. Existing deficiencies are quite obvious to many professions working in the hospital. If institutional goals are to transcend personal and professional goals, then it must be readily apparent to all associated with the hospital that standards for the performance and personal conduct of those doing the governing are at least as high as, or higher than, for those being governed.

Administration–board relationships. To insure maximum fulfillment of its ethical obligations, trustees must reappraise traditional administration–board relationships (Greenleaf, 1974). The administrator is supposed to be the servant of the board; yet circumstances have thrust him into situations that tend to obscure this relationship. The limited amount of time invested by trustees in their task, their reluctance to challenge established methods, and other factors have brought about a dependency of trustees on the administrator, which leads to an ineffective board (Law). Traditionally, the trustee has had to tread a narrow line between setting policy and not getting into administration. This has contributed to his passive role in governance and his undue dependency on the administrator—in effect, a transference of trustee responsibility to administration. This dependency can also result by design. Not all administrators actually want an effective board (Drucker). If trustees are effectively to discharge their public trust, then the traditional concept of limited trustee involvement in hospital affairs must be reassessed (Greenleaf, 1974); trustees will have to accept more personal responsibility and involvement. This can be, and has been, done (Law).

Serious ethical problems existing in the area of contractual arrangements between hospitals and hospital-based physicians can be traced to the trustees' lack of involvement and their undue dependency on the administrator. When open-ended contracts are written without proper fiscal controls, exorbitant amounts of money can accrue to a physician, upwards of $400,000 annual income being reported. Contractual arrangements conducive to such situations are common. Many physicians in such instances obviously become more oriented to economics than to institutional ethics, but they are by no means solely to blame for exploiting the resources of the hospital in that manner. Those contractual arrangements do not exist without the approval of administration and the board of trustees. Trustees who have not been involved or developed a sense of personal responsibility about their public trust are ill prepared to stand firm as a cohesive board and give support to an administrator under pressure from professional societies. To apply sound fiscal principles to the rest of the institution and to exempt one sensitive area suggests an abdication of moral responsibility.

Patients' rights. Many of the issues pertaining to patients' rights have ethical implications for the hospital. Some of the issues, such as confidentiality, informed consent, and other patients' rights are discussed elsewhere in this encyclopedia. The hospital has an obligation to assure the patient, in keeping with the purpose of the hospital, that his or her rights will be protected at all times while a patient in the institution. An important step in this direction was taken with the development of the American Hospital Association's Patient's Bill of Rights. To assure the protection of patients' rights, the administrator must be prepared to serve as patient advocate. Explicit policies must exist that will assure the patient that he is kept informed of whatever procedures are performed on him, that he is provided with appropriate education on health matters, and that his options for treatment will be made clear to him. No longer is the decision-making power in the medical encounter a medical monopoly. The patient wants to be and is entitled to be part of that process. If appropriate policies designed to assure patients' rights are ignored or defied, the administrator of the institution must see that the rights accorded the patient are no less than he would insist upon for himself, his family, or any members of the institution's board of trustees. The institution has a moral obligation to inform the patient if any misadventure has befallen him, even though it has seemingly been corrected, or even if any professional in the institution is opposed to the patient's being so informed.

Quality control. The area of quality control and maintenance of acceptable professional standards traditionally has been left to physicians. The hospital today has a moral obligation, as demanded by society, to reject the long-standing informal policy of "hands off the hospital medical staff." Ideally, all members of the medical staff will accept the obligations inherent in their own code of ethics and also those that flow from the ethics of the institution. In the event that such obligations are ignored, the obligation of the institutions' officials to take corrective action should be clear. Regardless of whether all individuals do or do not follow their professional code of ethics, it is incumbent upon the institution to maintain ongoing systems for determining standards, for maintaining surveillance, and for initiating corrective action as indicated (Joint Commission on Accreditation). By the institution's insisting upon the existence of such systems and the application of sound management principles in this area, as elsewhere in the hospital, it is fulfilling its own responsibilities and also assisting the medical staff leaders to fulfill theirs. This involves more than the obligation to correct faults: such systems assure that early professional obsolescence can be detected and programs implemented to prevent incompetence. There are a number of contentious issues in this area dealing with organization, controls, full-time paid officials of the medical staff, to whom and to what extent is the physician accountable, etc. The fabric of these issues is frequently heavily interwoven with politics and emotion. In dealing with them, the officials of the institution must be able to relate the issues and the implications of them to the hospital's purposes and ethical obligations.

The area of institutional planning has important ethical implications for those who direct and influence its development. For example, the accessibility of facilities is not to be denied to an individual because of race, religion, or other similar discriminatory factors; yet this can happen when health facilities serving a whole urban core community are moved to the suburbs. Another aspect of institutional planning with serious ethical implications is the institution's need to examine critically its continuing involvement in duplication of facilities and equipment and its apparent continuing commitment to new technology and participation in what has been called the medical arms race.

Conclusion

At a time of ferment and change, when traditional values are being questioned and persons and groups are experiencing new-found freedoms, it is essential that society's expectations for its hospitals are not lost sight of. The difficulty of managing hospital affairs today is not made easier by the political and professional pressures that are attendant upon so many of its problems. As those responsible for the direction of hospitals cope with those pressures and problems, their decisions and actions must be rooted in the hospital's prime ethical purpose of making certain that the interest and safety of the patient takes precedence over all other concerns.

KENNETH J. WILLIAMS

[*For further discussion of topics mentioned in this article, see the entries:* HEALTH CARE; MEDICAL PROFESSION; NURSING; *and* WOMEN AND BIOMEDICINE, *article on* WOMEN AS HEALTH PROFESSIONALS. *Also directly related are the entries* CARE; HEALTH POLICY; MEDICAL EDUCATION; MEDICINE, SOCIOLOGY OF; PATERNALISM; PATIENTS' RIGHTS MOVEMENT; *and* RELIGIOUS DIRECTIVES IN MEDICAL ETHICS. *See also:* INFORMED CONSENT IN THE THERAPEUTIC RELATIONSHIP; MEDICAL MALPRACTICE; *and* MEDICAL ETHICS, HISTORY OF, *section on* EUROPE AND THE AMERICAS, *article on* MEDIEVAL EUROPE: FOURTH TO SIXTEENTH CENTURY. *For a* PATIENT'S BILL OF RIGHTS, *see* APPENDIX, SECTION III, *which also contains other documents on patients' rights, and* SECTION I, ETHICAL AND RELIGIOUS DIRECTIVES FOR CATHOLIC HEALTH FACILITIES.]

BIBLIOGRAPHY

ANNAS, GEORGE J. *The Rights of Patients.* New York: Avon Books, 1975.
BLAES, STEPHEN M. "The Legal Perspective: The Darling Case Shows a Silver Lining." *Trustee* 29, no. 5 (1976), pp. 9–11.
CASTIGLIONI, ARTURO. *A History of Medicine.* Translated and edited by E. B. Krumbhaar. New York: Alfred A. Knopf, 1941.
DERBYSHIRE, ROBERT C. "Medical Licensure and Professional Discipline 1976." *Annals of Internal Medicine* 85 (1976): 384–385.
DRUCKER, PETER FERDINAND. *Management: Tasks, Responsibilities, Practices.* New York: Harper & Row, 1974, chap. 52, pp. 626–636.
GARRISON, FIELDING HUDSON. *An Introduction to the History of Medicine, with Medical Chronology, Suggestions for Study and Bibliographic Data.* 4th ed. Philadelphia: W. B. Saunders Co., 1966, pp. 176–189.
GREENLEAF, ROBERT K. *The Institution as Servant.* Cambridge: Center for Applied Studies, 1972.
———. *Trustees as Servants.* Cambridge: Center for Applied Studies, 1974.
HOLMAN, EDWIN J. "Medical Discipline: Everybody's Business." *Hospital Practice* 11, no. 11 (1976), pp. 11–15.
ILLICH, IVAN D. *Medical Nemesis: The Expropriation of Health.* London: Calder & Boyars, 1975.
IVES, A. G. L. *British Hospitals.* Britain in Pictures. The British People in Pictures. London: Collins, 1948, pp. 8–30.
Joint Commission on Accreditation of Hospitals. *Accreditation Manual for Hospitals.* Chicago: Hospital Accreditation Program, 1971. 1973. 1976.
LAW, JOHN T. "The Process for Organizational Change at the Hospital for Sick Children." *The Role of the Voluntary Trustee.* Edited by W. Jack McDougall. London, Ont.: Research and Publication Division, School of Business Administration, University of Western Ontario, 1976, pp. 111–117.
MAJOR, RALPH H. *A History of Medicine.* 2 vols. Springfield, Ill.: Charles C Thomas, 1954, vol. 1, pp. 335–346.
SOUTHWICK, ARTHUR F. "The Hospital as an Institution—Expanding Responsibilities Change Its Relationship with the Staff Physician." *California Western Law Review* 9 (1973): 429–467.
———. "The Legal Perspective: The Medical Staff Privileges—A Matter of Fairness." *Trustee* 29, no. 8 (1976), pp. 7–12.
WILLIAMS, KENNETH J. "Medical Issues—Past and Present." *Hospital Medical Staff* 1, no. 1 (1972), pp. 2–13.
———. "The Quandary of the Hospital Administrator in Dealing with the Medical Malpractice Problem." *Nebraska Law Review* 55 (1976): 401–416.

HUMAN EXPERIMENTATION

The first article presents the HISTORY *of the practice of human experimentation. This is followed by an article on the* BASIC ISSUES, *which offers an introduction to and overview of all the important issues in human experimentation that bear on bioethics. Then, there is a discussion of the* PHILOSOPHICAL ASPECTS *of the subject and a concluding article on* SOCIAL AND PROFESSIONAL CONTROL, *which focuses on the major alternatives available in establishing rules governing human experimentation.*

I. HISTORY *Gert H. Brieger*
II. BASIC ISSUES *Alexander Morgan Capron*
III. PHILOSOPHICAL ASPECTS *Charles Fried*
IV. SOCIAL AND PROFESSIONAL CONTROL
 Mark S. Frankel

I
HISTORY

The history of experimentation on humans is as old as that of medicine itself. Every society has had its healers, and in all times some of those entrusted with the health of their fellow men have, by one means or another, attempted to devise better ways to prevent illness, to restore the sick to health, and to rehabilitate those who have turned to the healer's care. A set of core values has emerged from the great tradition of a continuing search for better ways of coping with the problems of illness. Each new generation of students of the healing arts is in turn expected to acquire a set of values along with the technical matters of medicine. Although the values pertaining to the use of human subjects in medical experimentation have not always been made explicit, it is of interest to trace their historical development. Experimentation on human beings is nothing new (Guttentag). It continues to take place whenever physicians are in the process of introducing new diagnostic and therapeutic procedures or agents. Man is an inveterate experimenter, and man has been the chief experimental animal both as experimenter and as subject (Hill).

Everyone who has thought about the problems posed by human experimentation agrees to its antiquity, but its history has yet to be written. This gap in our historical literature may be partly attributed to the fact that such a history would virtually be a history of medicine itself; furthermore, it would require a tremendous breadth of knowledge not only of medical practice and research but of political, cultural, economic, and particularly philosophical conditions as well. A mere catalog of human experiments, while interesting and perhaps even instructive, is not sufficient.

A history of the subject should also encompass changes in definitions and types of experiments, changes in social attitudes toward experiments and experimenters, and the evolution of responses to the problems posed by them. Whether these responses be societal and peer pressures or religious and legal, their development is part of the complex patterns of all societal regulations. They obviously cannot be discussed in a vacuum, in a context outside of the host of other moral and social forces in any given time. Only a few questions can be raised here, even fewer well answered. One of the most obvious questions concerns the attitudes and values of medical scientists of earlier times: Was there a discussion of human experimentation, indeed of any current ethical issues similar to those confronting us today? Has there been a change in values that the historians of medicine can describe?

For convenience of discussion the types of human experimentation may be readily divided into at least four groups: (1) natural experiments; (2) experimentation incidental to medical treatment; (3) deliberate experiments involving second parties; and (4) autoexperiments.

Natural experiments

When we think of the various kinds of scientific experiments we usually visualize the scientist and his subjects, whether volunteers or not, paid or unpaid, whether rats or college students. Yet the idea of human experimentation should not be confined merely to the laboratory setting. Nature presents us with bolder experiments than we would ever dare to perform ourselves (Beecher). Nature, however, is not always a good experimenter. She makes her experiments without mercy and performs them casually. One often cannot reason accurately from them at all (Hill). So it is small wonder that for millennia medical knowledge made relatively slow progress.

Wars, famines, and subsequent pestilences have dogged mankind throughout history and have in their wake left untold numbers of experiments. The food shortages of World War I, for instance, led to the reduction of the incidence of deaths from coronary artery disease, thus leading to many nutritional experiments in an attempt to link the arteriosclerotic process to dietary intake of fats, or of carbohydrates, trace metals, and a host of other possible causal factors. Others described a reduction in the severity of diabetes during times of starvation.

Experimentation incidental to medical treatment

Ancient Greek concerns with the welfare of patients are amply demonstrated throughout the Hippocratic writings. In the Oath, inspired by the Pythagoreans, the doctor swore to apply his knowledge for the benefit of the sick. In the *Epidemics* occurs the famous stricture, "To help or at least to do no harm." Similarly, the first aphorism of Hippocrates warns the physician that trying out unknown remedies may be deleterious to the patient. As usually translated the aphorism reads, "Life is short, the art long, opportunity fleeting, experience treacherous, judgment difficult." The Greek words allow "experience treacherous" to be rendered as "experiment is perilous," thus perhaps intended as a warning about experimenting with new and unknown therapeutic measures, whether diet, drugs, or surgery. The Hippocratic physicians were not experimental scientists in the modern sense. Their strictures merely serve to emphasize that the treatment of any patient is an experiment, sometimes a perilous one.

The twentieth century has witnessed not only a phenomenal growth of biomedical science and research but also numerous changes in the patterns of medical care. One of the most prominent has been the increasing use of hospitals for treatment of the sick, so that the hospital has become the site for incidental experimentation of a kind we have long since come to approve. Until fairly recently many people have regarded hospitals only as a place to die, as a last resort, and as a place for medical care for the poor. Many were apprehensive that hospitals existed for experimenting upon hapless and hopeless patients, where students, while learning about medicine, made use of human "teaching material" in some way deleterious to the patients. This is not merely a late nineteenth- or twentieth-century idea; for example, we find it in one of the epigrams of Martial of the first century A.D.: "I was ailing: But you at once attended me, Symmachus, with a train of a hundred apprentices. A hundred hands frosted by the North winds have pawed me: I had no fever before, Symmachus; now I have" (v. 9). It is difficult to describe exactly how prevalent and extensive the process of learning from patients has been, but certainly it has to be viewed as one type of human experimentation.

It is of interest, and was a point not lost upon the public, that from the 1870s on the leaders of experimental medical science were also frequently the leading medical teachers and hospital practitioners. One group of late nineteenth-century writers who repeatedly drew attention to the hospitals as places for ruthless experiments and unbounded medical enthusiasms were the British and American antivivisectionists.

Deliberate experimentation involving second parties

Our main sources for the charge that they disples of deliberate human experimentation was attributed to the Alexandrian physicians Herophilus and Erasistratus of the third century B.C. Our main sources for the charges that they dissected living persons are Celsus, the Roman encyclopedist of the first century A.D., and the later Roman writer Tertullian. The Hippocratic authors of the fourth century B.C. probably did not dissect human cadavers and relied on other sources for their knowledge, including chance findings of skeletons, observations of injuries, and animal dissections. Those Alexandrian physicians of the next century who followed the teaching of Hippocrates believed that a knowledge of anatomy was necessary, for without such knowledge treatment of disease was not deemed possible. Those physicians, Celsus said, believed that four requisites existed for a theory of medicine: "a knowledge of hidden causes involving diseases, next, of evident causes, after these of natural actions also, and lastly of the internal parts" (Celsus, p. 9). Thus these physicians, known as the Dogmatists, stressed the need for anatomical study. Following Aristotle they also believed that the living body was important to understand, that dissection of the dead would lead to different results, that dissection taught only morphology, and that death changed man. Vivisection taught about functions, that which was truly worth knowing.

In an authoritarian state such as that of the Ptolemies of Egypt, who were known both for their ruthlessness and for their support of the arts, scholarship, and science, the climate for such vivisections was set. The Alexandrians condoned vivisection on several counts—it was practiced only on condemned criminals, it was essential for the advance of knowledge, and the torture of the few would lead to great benefits for the many—all arguments that would be echoed in later times. The history of anatomy, and particularly among the Alexandrians, can be understood only in terms of philosophical arguments and assumptions. Vivisection was thus demanded by the consideration that it was the only road to true knowledge (Edelstein, *Ancient Medicine*). It should not be viewed as merely the sport of cruel men.

Celsus, in whose own time apparently no dissection of any kind was practiced, left no doubt as to his feelings on the matters he reported:

"[W]hat remains, cruel as well, to cut into the belly and chest of men whilst still alive, and to impose upon the Art which presides over human safety someone's death, and that too in the most atrocious way" (Celsus, p. 23). He called those who performed these arts medical murderers, and ever since his day historians have continued to discuss, and to believe or disbelieve, these alleged Alexandrian practices.

The Alexandrians certainly did not look upon human vivisection in the same way as have subsequent generations. They performed it, as Celsus indicated, for the benefit of future generations, and one must not assume that they performed it endlessly. One of the few comments of Erasistratus that has been preserved reveals at least a concern for ethical consideration by the physician: "Most fortunate indeed wherever it happens that the physician is both, perfect in his art and most excellent in his moral conduct. But if one of the two should have to be missing, then it is better to be a good man devoid of learning than to be a perfect practitioner of bad moral conduct, and an untrustworthy man" (Edelstein, "Professional Ethics," p. 334).

The list of possible examples of deliberate human experiments since the time of the Alexandrians is long and cannot be cataloged here. It includes many therapeutic trials such as those of James Lind on a small scale proving that citrus fruits are preventive for scurvy (1753) to more recent large-scale experiments such as the Tuskegee syphilis studies comparing the fate of treated and untreated victims of the disease (1932–1972). In the history of therapeutic methods other than drugs, such as surgery, an equal list might be compiled. The story of transfusion, for instance, includes numerous examples (Zimmerman and Howell). With the advent of microbiology in the last decades of the nineteenth century, many physicians undertook experiments to prove or disprove certain bacteria as causes of specific diseases or various immunizing methods as effective preventives. The list of published experiments in the world's medical literature is vast if one stops to search for it carefully. Most probably an equally vast amount of all types of human experiments have never reached the stage of publication and are therefore beyond the reach of the historian.

Of the many examples, three deserve fuller discussion. One was a well-controlled experiment carried out in Boston in 1721. The second, the work of William Beaumont, was that of a lonely frontier physician seizing upon a rare chance to study the function of digestion in a sometimes unwilling subject. The third was a carefully conceived experiment conducted in Cuba in 1900, which, like the Boston experiment, concerned the understanding and control of a much-dreaded epidemic disease.

The first example illustrates nicely the need to look for, and identify if possible, in each age the predominant nonscientific cultural currents if the procedures of science are to be properly understood. Nowhere is this better illustrated than in the discussion surrounding the smallpox inoculation experiment of the Reverend Cotton Mather and Dr. Zabdiel Boylston in early eighteenth-century Boston. Not to take into account the theological concerns of the day is merely to chronicle another series of experiments.

Between 1630 and 1702, one biblical life span, Boston suffered five major outbreaks of smallpox. Then came a nineteen-year respite, followed by the major epidemic of 1721–1722. Because the disease was not endemic, some children escaped exposure during their earlier years. Consequently, in America, in contrast to Europe, the adult mortality was higher. Cotton Mather had read about the process of inoculation brought from the Near East to England and wished to apply the technique in Boston in an attempt to cut the fearful toll of the disease. He convinced only one of the medical practitioners of the town that inoculation of small doses of infectious material would cause only mild disease and then subsequent immunity. Boylston agreed to the idea but first tried the procedure on his son and two of his slaves. It is generally assumed that he did not use himself as a subject because he had successfully recovered from smallpox during the 1702 outbreak and hence assumed he was now immune. Mather and Boylston then inoculated about 250 Bostonians, with striking results. Their group suffered only a two percent mortality from the disease. The general population meanwhile counted more than eight hundred deaths for a mortality rate of fifteen percent.

Despite the remarkable results, the outcry against Mather and Boylston was fierce. One of the most vociferous opponents of inoculation was Dr. William Douglass, an English physician of considerable accomplishments and far better training than Boylston. That Douglass should lead such vehement opposition may in part be explained by professional jealousy. But Douglass probably sincerely believed that Mather and Boylston were undertaking a reckless trial, impermissible behavior for the profession that was to help and to do no harm. Also of great importance in evaluating this professional controversy is to keep in mind a theological issue of far greater importance to the eighteenth-century participants than the question of scientific validity of the experiment. It was an age when men feared and trembled before their God. To interfere with his handiwork, to change his will by tampering with disease inoculation was thus a dangerous activity indeed. These fears must be understood in the context of the theological climate of Puritan New England. Simply to assess this incident in the history of human experimentation on the basis of the ethics involved is not sufficient.

The second example deals with William Beaumont, an army surgeon stationed on Michigan's Upper Peninsula in the early 1820s, who seized upon a chance to study the gastric digestion, something that prior to his time would have required vivisection. He tended the shotgun wound of his experimental subject, Alexis St. Martin, for three years and then began to observe the digestive function through the opening that remained when all else had healed. Beaumont then also deliberately experimented with different kinds of foods and learned much about the digestive function and the gastric juices. "I consider myself but a humble inquirer after truth—a simple experimenter," Beaumont wrote in his classic book of 1833 (p. 31). "I had opportunities for the examination of the interior of the stomach and its secretions. . . . I have availed myself of the opportunity afforded by a concurrence of circumstances which probably can never again occur, with a zeal and perseverence proceeding from motives which my conscience approves" (p. 6). Beaumont's contemporaries lauded him for his zeal and for his results. Historians have celebrated his achievements as one of the few true nineteenth-century American contributions to experimental medicine.

The unfortunate Alexis had on several occasions literally to run away from Beaumont to put a stop to the tedious experiments. There are very few contemporary comments about the ethics of the situation. A popular health journal, *The Family Oracle of Health,* published in London, noted that Beaumont regretfully reported that his subject had absconded ("Curious Experiments," p. 433). "We can only say," wrote the editors in 1825, "that however anxious we may be for the promotion of science, we should certainly have acted in precisely the same manner" (ibid.). When one considers that the physician in this case virtually hounded his subject, meanwhile also providing for his medical and bodily well-being, it is not surprising that the impatient Alexis ran off. What is perhaps surprising is the apparent lack of comment by reviewers upon the experimental situation.

One of the most dramatic and thus also one of the best-known episodes in the annals of human experimentation occurred in Cuba in 1900 when Major Walter Reed and his co-workers proved beyond a shadow of a doubt that yellow fever required the mosquito as an intermediate host to spread the infection. Despite the death of Dr. Jesse W. Lazear, who was a member of the Reed board that had been appointed by the U.S. government, and despite the dangers to which the more than a dozen volunteers were exposed, Reed was hailed as a hero, his results looked upon as a true scientific triumph. One of the things Reed showed clearly was that yellow fever was not caused by the *Bacillus icteroides*, as suggested by the South American physician Guiseppi Sanarelli. The antivivisectionists frequently mentioned Sanarelli's experiments as examples of human vivisection, but they rarely pointed a finger at Reed.

This brings out certain features of Reed's work that bear stressing in the context of this discussion. The first three cases of experimental yellow fever included two members of the board, Lazear and James Carroll. Carroll, after a prolonged illness, finally recovered, but the unfortunate Lazear succumbed. There has been debate whether he allowed himself to be bitten by a mosquito known to have been infected with yellow fever or was accidentally exposed. Reed, in his preliminary report, mentioned that Lazear had become accidentally infected on a hospital ward, because he knew that Lazear was concerned about his life insurance. Reading the documents leads one to the conclusion that Lazear was indeed a knowing self-experimenter.

When Reed arrived in Cuba he was greatly saddened by the news of his colleague's death. Yet the question of the mosquito spread of the disease was not yet fully settled, because by then only three cases were at hand, and they could conceivably have been exposed to the disease outside of the confines of the camp in which they were working. A group of men kept in quarantine and then exposed to infected mosquitoes and another group kept away from mosquitoes but in close contact with the soiled clothing and bedding of known yellow fever cases would settle the issue, but Reed needed the experimental subjects. The Governor General of Cuba, Leonard Wood, himself a former captain in the Medical Corps, gave permission and provided facilities. That he was a medical man and could readily see both the importance and the dangers in the work is not always appreciated. Wood also provided money to pay the volunteers. Thus these were true volunteers who were carefully chosen and well informed. The volunteers had to be over the age of twenty-four, the age of freedom from parental control in Spain. The same rule did not apply to the U.S. soldier-volunteers.

Some of the Havana papers did raise a hue and cry about the supposed barbarities of the Reed board. The Spanish consul supported Reed so long as he did not use minors and had written consent from each participant. Fortunately, none of the additional thirteen volunteers, nine of whom developed yellow fever, died from the illness. The intriguing question remains, then, why there was so little outcry against these experiments. Several answers come to mind, not the least of which is the startling nature of the result. As Reed wrote to his wife in December 1900, "aside from the antitoxin of diphtheria and Koch's discovery of the tubercle bacillus it will be regarded as the most important piece of work, scientifically, during the nineteenth century" (Truby, p. 159). Reed's careful planning and the comprehensive nature of the experiments that once and for all explained the mechanism of spread of a dreaded and dreadful disease, and the careful medical attention given volunteers also helped to defuse criticism. The immediate practical results seen in the clean-up of Havana and the safe and successful building of the Panama Canal were further proof of the magnitude of the Reed board's findings.

We have here, then, a short-term, truly experimental situation that was safely concluded and had almost immediate widespread implications, an example of risk to a few for the benefit of the many. Small wonder that few voices were raised in question. Certainly this was not the case with numerous late nineteenth-century experiments with other microbial diseases. Syphilis, tuberculosis, cholera, and many more all were used in either autoexperiments or in groups of voluntary or involuntary subjects. The outcry against the "inoculation of these loathsome diseases" was in the decades around 1900 at times very loud, and antiscientific groups such as the antivivisectionists especially prized the criticism coming from within the medical profession itself.

Autoexperimentation

Autoexperiments doubtless are as old as all human experimentation; that is, they began with man's search for new sources of foods and healing herbs. In the modern sense of experimental science, self-experimentation is usually said to have begun with the work of Santorio Santorio in the early seventeenth century. Self-experimentation became especially prominent in the later nineteenth century, when a number of investigators exposed themselves to a variety of infectious organisms they were studying. John Hunter more than a hundred years previously had deliberately inoculated himself with venereal disease, thus contracting both syphilis and gonorrhea and confusing the two markedly different diseases for several subsequent generations of patients and their physicians. Few autoexperimenters died as a result of their studies, though Dr. Jesse W. Lazear, who succumbed to yellow fever in 1900, was a notable exception.

Autoexperimentation continues to this day and, in one case at least, has led to a Nobel Prize. Werner Forssmann, in 1929, wished to study ways of approaching the chambers of the heart avoiding the dangers resulting from external intracardiac injections. He was aware of the work of French physiologists of the 1860s who had recorded intraventricular pressures by means of a catheter passing from the jugular vein to the heart in a horse. As a surgical house officer Forssmann requested permission to try a similar experiment on himself. Though he was denied official sanction he performed several such experiments, walking up staircases and down long corridors from the operating room suite to the X-ray department in order to record the event for posterity. His chief, after initial chagrin that his orders had been disobeyed, urged Forssmann to publish his results. "The technique is so new and unusual," Forssmann quotes Dr. Richard Schneider, "that people are bound to try to put you down. And they'll use ethical arguments. The most effective argument against them will be that you're working to improve treatment" (Forssmann, p. 86). Subsequent events have certainly borne out the prediction of usefulness. The ethical question involved has not been so clearly resolved.

Many additional examples of autoexperimentation could be described (Altman). What has been rarely discussed, however, are the ethical implications involved. It has been traditional among clinical investigators to subject themselves to any experiment to which they propose to subject others. Many articles in the medical literature uphold this standard. One of the dangers inherent in this standard, however, is that a high level of risk may be imposed on other subjects: "Sometimes the experimenter rationalizes in a peculiar way the risk to which he puts his subjects. He first carries out the experiment upon himself and then subjects others to the same procedures. The viewpoint that it is all right to risk someone else because the experimenter has undergone the same hazard seems extraordinary. It is as if it were all right to commit mayhem or murder if one at the same time practices self-mutilation or committed suicide" (Ratnoff and Ratnoff, p. 85).

Criticisms of human experimentation

While it is not our purpose here to trace the legal aspects of human experimentation, it should suffice merely to point out that English cases, at least, go back to the mid-eighteenth century, to *Slater* v. *Baker*, 1767. This case established the rule that the physician experimented at his peril. Mr. Baker, the defendant surgeon, tried to straighten an improperly healed leg fracture by use of a new instrument that caused the leg to extend or lengthen. The method of the time was to do the reverse, to cause healing by compression. Of interest in this case is that neither side raised the question of experimentation, and the court, moreover, seems to indicate that many physicians and surgeons have used uncommon methods "for the sake of trying experiments" (Ladimer and Newman, p. 182).

Before closing with a word about the influence of Claude Bernard, we should return briefly to one rich source of literature about human experimentation. Between 1875 and 1910 both the British and the American antivivisection societies published numerous exposés of the supposed excesses of the medical experimenters. While primarily interested in animal welfare, the antivivisectionists also concerned themselves with the use of human subjects, especially indigent hospital patients. A good many people tolerated animal vivisection, George Bernard Shaw, an active antivivisectionist himself, quipped, because they feared if experiments were not made on animals they would instead be carried out on themselves. The man who vivisected in the morning, many opponents claimed, too often appeared in the operating theaters of the hospitals in the afternoon. The leaders of the antivivisection movement claimed that the ambitions of the medical men led them to look upon patients, especially charity patients in hospitals, as so much experimental material.

Those antivivisectionists who were not mainly against medical science but were genuinely interested in the moral dilemmas posed by the use of animals to help gain understanding of disease and thereby to improve therapy, frequently pointed out that, if it was ethically defensible to experiment upon dumb and unsuspecting animals, it would merely be the next step to justify it for humans. Albert Leffingwell, an American physician who wrote much against vivisection around the turn of the century, carried the argument to the next step, fearing that, once we admit that patients in hospials have no rights superior to scientific demands, then there is hardly a limit to which such experimentation may not be carried on the poor, the ignorant, the feeble-minded, and the defenseless (Leffingwell).

Claude Bernard

One of the nineteenth century's most renowned physicians, Claude Bernard, was himself the frequent target of the antivivisectionists. He speaks of this in his *Introduction to Experimental Medicine*, published in Paris in 1865. Though now more than a century old, this book was written in such clear French prose that it is still read by students and scholars the world over for its beautiful exposition of the scientific method. Claude Bernard, about whom his illustrious student Paul Bert said, "He discovered as others breathed" (Bernard, p. xvii), believed that the observer listens to nature; the experimenter questions nature and forces her to reveal herself. Observation shows, experiment teaches, Bernard was fond of saying. Both are necessary to the advancement of understanding.

The *Introduction* consists of three parts. The middle part, being devoted to vivisection, is of most interest here. To understand nature we must get at the inner parts, to learn laws and properties of living matter. Men in all ages, Bernard claimed, have felt this need, and so in all ages they have experimented. Have we a right to perform experiments and vivisections on man, Bernard asked.

Physicians make therapeutic experiments daily on their patients, and surgeons perform vivisections daily on their subjects. Experiments, then, may be performed on man, but within what limits? It is our duty and our right to perform and experiment on man whenever it can save his life, cure him or gain him some personal benefit. The principle of medical and surgical morality, therefore, consists in never performing on man an experiment which might be harmful to him to any extent, even though the result might be highly advantageous to science, i.e., to the health of others [p. 101].

Bernard agreed with those whose morals condemned experiments such as dangerous operations on condemned criminals, though he did regard it as permissible to make useful to science the tissues of those decapitated immediately after their execution. We must not deceive ourselves, he wrote, for "morals do not forbid making experiments on one's neighbor or on one's self; in everyday life men do nothing but experiments on one another. Christian morals forbid only one thing, doing ill to one's neighbor. So, among the experiments that may be tried on man, those that can only harm are forbidden, those that are innocent are permissible, and those that may do good are obligatory" (p. 102).

Claude Bernard's contributions to the principles of investigation are clear and unmistakable. He and his work must be seen in the tradition of nineteenth-century materialistic science, keeping in mind that he opposed the mechanistic tendency to reduce physiological functions to simple physical and chemical terms. Physiology was the road to the future of medicine, "the noblest of professions and most depressing of trades." His views about experimental science, including vivisection, were motivated by the philosophy of all good teachers: to show how knowledge is derived and how new knowledge may be acquired.

Conclusion

A major question raised by looking at the history of human experimentation relates to an apparent change in ethical sensibilities over time. There have been instances prior to our own time when there was an outcry against the use of human subjects, but they were relatively rare, at least as reflected by the literature checked so far. Objections were raised by Celsus, by the antivivisectionists, and in our own day an intense debate has been carried on over questions of informed consent, prisoner-volunteers, the use of children as subjects, and other related topics. The atrocities in the name of human experimentation committed by the National Socialists are discussed elsewhere in this encyclopedia, but most certainly when their extent became fully known in the course of the trials in Nuremberg, the world was aghast and medical scientists became more conscious of their ethical responsibilities.

Not only Nuremberg and the Nazi atrocities but subsequent worldwide concern with conservation and with civil rights have all served to raise awareness. And certainly ethical sensibilities do change. No one today treats servants in the way our forebears did two and three generations ago, and yet they were not evil, thoughtless, or without their ethical sensibilities. Children no longer work in factories, and their mothers and fathers no longer work fourteen to sixteen hours, six days a week. Standards do change.

An additional important factor in raising to high levels of discussion and leading, in fact, to the need for an encyclopedia of bioethics has been the rapid rise of experimental medicine. The clinical investigator, while born of the sciences of medicine following the decades of Claude Bernard and Louis Pasteur, has gained increasing stature and position in today's biomedical establishment. His reward system is based primarily on research productivity. These "new men of power" (Barber) have a whole new magnitude of power to do good, and also a whole new magnitude of power to do harm.

The world of the nineteenth century has changed to one that not only has given rise to a huge increase in resources allocated to medical science but has also seen the development of a far more extensive accountability and fairly close peer review. These procedures have led to much closer scrutiny and fuller discussion of what goes on behind the doors of the many scientific laboratories. As research has become increasingly sophisticated, so have the controls both scientists and society have sought to exercise over the conduct of experimentation. These controls, however, have evolved more slowly, and only a few voices have been raised to bring them about. Medical science, just as science as a whole, is not simply a set of technical operations based on rational principles. Medical research should also be seen as a set of activities subject to clear ethical standards.

GERT H. BRIEGER

[*For discussion of related ideas, see the entries:* ANIMAL EXPERIMENTATION; *and* RISK. See APPENDIX, SECTION I, OATH OF HIPPOCRATES; *and* SECTION II: DIRECTIVES FOR HUMAN EXPERIMENTATION. *See also:* CHILDREN AND BIOMEDICINE; INFORMED CONSENT IN HUMAN RESEARCH; MEDICAL EDUCATION; MEDICAL ETHICS

[UNDER NATIONAL SOCIALISM; and PRISONERS, article on PRISONER EXPERIMENTATION.]

BIBLIOGRAPHY

ALTMAN, LAWRENCE K. "Autoexperimentation: An Unappreciated Tradition in Medical Science." *Hippocrates Revisited: A Search for Meaning.* Edited by Roger J. Bulger. New York: Medcom, 1973, chap. 19, pp. 193–210.

BARBER, BERNARD. "Some 'New Men of Power': The Case of Biomedical Research Scientists." *Annals of the New York Academy of Sciences* 169 (1970): 519–522.

BEAUMONT, WILLIAM. *Experiments and Observations on the Gastric Juice and the Physiology of Digestion.* Plattsburgh: F. P. Allen, 1833. Facsimile reprint. New York: Dover Publications, 1959.

BEECHER, HENRY K. "Experimentation in Man." *Journal of the American Medical Association* 169 (1959): 461–478.

BERNARD, CLAUDE. *An Introduction to the Study of Experimental Medicine* (1865). Translated by Henry C. Green. New York: Dover Publications, 1957. This paperback edition is readily available and is a classic in the history of medicine.

BLAKE, JOHN B. "Smallpox Inoculation in Colonial Boston." *Journal of the History of Medicine and Allied Sciences* 8 (1953): 284–300.

BRIEGER, GERT H. "Some Aspects of Human Experimentation in the History of Nutrition." *Use of Human Subjects in Safety Evaluation of Food Chemicals.* National Academy of Sciences, National Research Council Publication no. 1491. Washington: 1967, pp. 207–215. Proceedings of a conference held at the National Academy of Sciences, Washington, D.C., 29–30 November 1966.

CELSUS, AULUS CORNELIUS. *De medicina.* 3 vols. Translated by W. G. Spencer. Loeb Classical Library. Cambridge: Harvard University Press; London: W. Heinemann, 1935–1938. The "Prooemium," vol. 1, pp. 2–41, is particularly useful as a historical source for Alexandrian medicine.

"Curious Experiments on the Process of Digestion, Made on the Living Subject. By Dr. Beaumont, of Fort Niagara." *Family Oracle of Health, Economy, Medicine, and Good Living: Adapted to All Ranks of Society, from the Palace to the Cottage.* Edited by A. F. Crell and W. M. Wallace. London: Printed by C. Smith, publisher varies, 1825, vol. 2, pp. 431–433. Volume numbering varies.

DOWLING, HARRY F. "Human Experimentation in Infectious Diseases." *Journal of the American Medical Association* 198 (1966): 997–999.

EDELSTEIN, LUDWIG. *Ancient Medicine: Selected Papers of Ludwig Edelstein.* Edited by Owsei Temkin and C. Lilian Temkin. Baltimore: Johns Hopkins Press, 1967.

———. "The Hippocratic Oath." Edelstein, *Ancient Medicine,* pp. 3–64.

———. "The History of Anatomy in Antiquity." Edelstein, *Ancient Medicine,* pp. 247–302.

———. "The Professional Ethics of the Greek Physician." Edelstein, *Ancient Medicine,* pp. 319–348.

FORSSMANN, WERNER. *Experiments on Myself: Memoirs of a Surgeon in Germany.* Translated by Hilary Davies. New York: St. Martin's Press, 1974.

FOX, RENÉE C. "Ethical and Existential Developments in Contemporary American Medicine: Their Implications for Culture and Society." *Milbank Memorial Fund Quarterly* 52 (1974): 445–483. Bibliography.

FRENCH, RICHARD. *Antivivisection and Medical Science in Victorian Society.* Princeton: Princeton University Press, 1975. The best source for the British antivivisection movement and medicine's responses.

FREUND, PAUL A., ed. *Experimentation with Human Subjects.* New York: George Braziller, 1970. A series of articles, philosophical, legal, but not primarily historical. Some historical examples.

———. "Introduction to the Issue 'Ethical Aspects of Experimentation with Human Subjects'." *Daedalus* 98 (1969): viii–xiv.

GUTTENTAG, OTTO E. "The Problem of Experimentation on Human Beings: II. The Physician's Point of View." *Science* 117 (1953): 207–210.

HILL, ARCHIBALD V. "Experiments on Frogs and Men." *The Ethical Dilemma of Science and Other Writings.* New York: Rockefeller Institute Press, Oxford University Press, 1960, pp. 24–38.

Human Vivisection: A Statement and an Inquiry. 3d rev. ed. American Humane Association, 1900.

IVY, A. C. "The History and Ethics of the Use of Human Subjects in Medical Experiments." *Science* 108 (1948): 1–5.

KATZ, JAY, ed. *Experimentation with Human Beings: The Authority of the Investigator, Subject, Professions, and State in the Human Experimentation Process.* New York: Russell Sage Foundation, 1972. Many useful historical sources included.

LADIMER, IRVING, and NEWMAN, ROGER W., eds. "Legal Review and Analysis." *Clinical Investigation in Medicine: Legal, Ethical, and Moral Aspects: An Anthology and Bibliography.* Boston: Law-Medicine Research Institute, Boston University, 1963, chap. 3, pp. 169–264. Source for *Slater* v. *Baker* (1767). This anthology reprints many useful articles and is the best single source to use for a beginning study of the history of human experimentation.

LEFFINGWELL, ALBERT. "The Final Phase: Experimentation on Man." *An Ethical Problem: Or, Sidelights Upon Scientific Experimentation on Man and Animals.* New York: C. P. Farrell; London: G. Bell & Sons, 1914, chap. 18, pp. 289–325. Written by a physician who was a leading antivivisection spokesman. Full of examples of human vivisection or experimentation.

PAPWORTH, MAURICE H. *Human Guinea Pigs: Experimentation on Man.* Boston: Beacon Press, 1968. Contains many examples, mostly recent. Good source for the subject of human experimentation but not for its history.

PEARCE, RICHARD M. "The Charge of 'Human Vivisection' as Presented in Antivivisection Literature." *Journal of the American Medical Association* 62 (1914): 659–668. Bibliography.

RATNOFF, OSCAR D., and RATNOFF, MARIAN F. "Ethical Responsibilities in Clinical Investigation." *Perspectives in Biology and Medicine* 11 (1967): 82–90.

SHAW, GEORGE BERNARD. *The Doctor's Dilemma* (1906). Penguin Plays. New York: Penguin Books, 1954, 1975. The eighty-page preface is a stinging and very witty attack on medical science.

TERTULLIAN. *Quinti Septimi Florentis Tertulliani De Anima.* Edited with text in Latin and translated into English with English introduction and commentary by Jan Hendrick Waszink. Amsterdam: J. M. Meulenhoff, 1947, esp. pp. 185–186. Commentary on chap. 10, para. 4 with reference to chap. 25, para. 5. Translated by Edwin A. Quain as "Tertullian on the Soul." *Tertullian: Apologetical Works and Minucius Felix Octavius*. The Fathers of the Church: A New Translation, vol. 10. Edited by Roy Joseph Deferrari. New York: 1950, pp. 179–309, esp. chap. 10, para. 4, p. 200; chap. 25, para. 5, pp. 238–239. Also translated by Peter Holmes as "A Treatise on the Soul." *The Ante-Nicene Fathers: Translations of the Writings of the Fathers down to A.D. 325*. Edited by Alexander Roberts and James Donaldson. American reprint of Edinburgh ed., revised by A. Cleveland Coxe. Vol. 3: *Latin Christianity: Its Founder, Tertullian*. Grand Rapids, Mich.: Wm. B. Eerdmans Publishing Co., 1957, pt. 1, sec. 9, pp. 181–235, esp. chap. 10, p. 189; chap. 25, p. 206. Also "Liber de Anima." *Patrologiae cursus completus: Series latina*. 221 vols. Edited by J.-P. Migne. Paris: Apud Garnier Fratres, editores & J.-P. Migne successores, 1851–1894, vol. 2, cols. 681–798, esp. chap. 10, cols. 703–704; chap. 25, col. 734. Tertullian argues that vivisection is both cruel and useless because the cutting itself alters the body parts. Waszink says other physicians argued that knowledge may be gained accidentally from wounds.

TRUBY, ALBERT E. *Memoir of Walter Reed: The Yellow Fever Episode*. New York: P. B. Hoeber, 1943. The most useful of many Reed biographies. Reprints Reed's letters to his wife and to Surgeon General George Sternberg in which Reed expresses his concerns about the human experiments he is conducting.

VERESSAYEV, VIKENTY [Smidovich, Vikentii Vikentevich]. "Experiments on Living Men and Women." *The Memoirs of a Physician* (1901). Translated by Simeon Linden. New York: Alfred A. Knopf, 1916, app. B, pp. 332–366.

ZIMMERMAN, LEO M., and HOWELL, KATHARINE M. "History of Blood Transfusion." *Annals of Medical History* 4 (1932): 415–433.

II
BASIC ISSUES

Human experimentation is the central pillar of modern biomedicine. Knowledge gained from research with animals would be irrelevant to medicine if it were not tested in human beings, and the sophisticated and often expensive forms of diagnosis and therapy that for many people have come to characterize contemporary medicine are merely the fruits of human research. Standing, then, in the pivotal position between basic research and animal laboratories, on the one hand, and routine clinical application in the health-care system, on the other, is the complex social practice known as human experimentation.

Human experimentation as a social practice

Research employing human subjects is undeniably as old as medicine itself. But systematic experimentation became an integral part of medical practice only in modern times, with the development of scientific medicine, which can be traced to Harvey in the seventeenth century, Jenner in the eighteenth, and Barnard, Pasteur, and their medical contemporaries in the nineteenth (Beecher, 1970, pp. 5–12). The issues raised by the use of human beings in research are, however, not new. The central fact of human experimentation remains that one person is intentionally being used as a means to benefit others, thus posing an ethical dilemma as old as society itself. The increasing demands for experimentation to establish the efficacy and safety of biomedical procedures not only magnify the absolute numbers of people involved but also make clear that the practice must be viewed in social terms and is not simply a matter for the conscience of an individual investigator (Jaffe, pp. 200–205).

Human experimentation can be placed in its social context in two ways. First, it is a human activity conducted within social institutions, according to norms developed and fostered by organized, professional bodies, and frequently with the sanction and even the sponsorship and financial support of governmental agencies. Second, the social context of human research emerges from an examination of its justification, for the legitimacy of any particular experiment will depend basically upon one's view of society. Those who regard individuals as part of a collective whole, which allocates rights and protects them, can justify a wide scope for experimentation that serves to advance the interests of the collectivity. Conversely, for those who emphasize the inviolability of human persons and the moral obligation to respect their dignity and to refrain from treating them simply as parts related to a social whole, any human experimentation whose purpose is to benefit persons other than the subjects of the experimentation requires very strong justifying reasons. Similarly, social contractarians may deny that the rights surrendered to society create any obligation to participate in risk-laden activities not for one's own benefit but solely so that others may gain from the sacrifice. Because fundamental premises about the nature of human society and values are seldom made explicit by writers on experimentation, widely differing views on the permissible nature and scope of human research have been voiced (Jonas, pp. 2–15).

Conceptual clarifications

The term "research" is a broad one which includes both investigations that require intentional manipulation of the objects being studied and those that involve the gathering of data about naturally occurring events. Many scientists would limit the meaning of "experimentation" to the first group and call the second "observation" (McCance, pp. 189–190). There is an undeniable difference between them, not the least because it is unlikely that a mere observer will increase the direct, physical risks to his subjects, as a person who deliberately alters their conditions may. Nonetheless, both observation and direct manipulation can provide the basis for an experiment that tests a hypothesis or provides new information (Beecher, 1970, pp. 1–2). Moreover, if the purpose of clarifying the concept of "experimentation" is to set the ambit of social regulation, there may be no reason to distinguish between these two means of conducting human research. For the investigator who observes the behavior of human beings and their reactions to events of which he is not the cause is, no less than one who actively manipulates his subjects' environment, making use of them for scientific purposes. The data collected could prove to be embarrassing to them, and the very fact of being observed may harm their sense of personal integrity and privacy.

Just as the absence of palpable risk may thus for some purposes not be a good ground for excluding an activity from the definition of experimentation, the fact that an activity gives rise to a risk, even one of unknown dimensions, is by itself not enough to bring it within the meaning of human experimentation. Risk and progress are inseparable, for patients as for the society as a whole. In a particular case even a well-accepted and supposedly innocuous technique may cause an unexpected disaster. If risk-taking is synonymous with experimentation, medical experimentation would have to be said to take place continually during routine care in every hospital and doctor's office (Ivy, p. 1; Shimkin, p. 205).

Pointing out that therapy has much in common with experimentation may have the great benefit of reminding physicians to treat their patients like collaborators in a shared adventure and to observe patients' progress with great care (Katz, "Education," pp. 294–295). But the attempted equation of all therapy with experimentation is misguided if it is used to argue that the direction of experimentation should be left solely to the judgment of a physician, as patient care traditionally has been. The similarity between experimentation and treatment on this ground ought not to obscure other differences that may be material to distinguishing between them for most purposes.

A prime source of confusion on this point arises from the fact that physicians in the course of regular care and management sometimes try new and unproved procedures on their patients, especially in desperate situations when existing modalities have been ineffectual. It seems proper to term this experimentation, even though it may occur as an isolated innovation rather than as part of a well-regulated clinical trial; yet the intent of the physician is plainly different from that prevailing in other types of research. Consequently, a distinction may be drawn between such *beneficial* experimentation and experimentation in which the sole purpose is *nonbeneficial* —to acquire knowledge rather than to improve the subject's condition. Beneficial experimentation is sometimes called "therapeutic experimentation" (National Commission, p. 6), but that term seems erroneously to exclude trials of diagnostic or preventive techniques which like new therapies are intended to benefit the persons on whom they are used. It must be remembered that the term "beneficial" refers merely to the intent with which an experiment is performed, not to its outcome; also, "nonbeneficial" should not be equated with "useless," although the intention is information of use to science and not any direct benefits to the subject.

In the actual conduct of medicine and the other life sciences, beneficial and nonbeneficial experimentation are often intermingled with each other and with nonexperimental procedures. The difficulty in any particular case of separating these elements does not mean that such distinctions ought not to be drawn, merely that care is needed in the undertaking. An experimental step, which might include calculated inaction, taken to benefit a particular subject (in such circumstances, properly termed a patient-subject), will frequently yield information of general applicability. Medical scientists would, indeed, regard it as improper not to exploit the experience gained from a trial in one patient-subject to aid in the care of others. Complementarily, nonbeneficial research may in fact have unintended helpful consequences for its subjects, or some of them may at some future date be ill with the same condition again and be aided by the knowledge gained from the experiments; yet this does not place the research, if it was undertaken for scientific purposes, into the category of intended beneficial experimentation.

As useful as these definitions may be for analytic purposes, their application in practice—particularly for regulatory purposes—can be problematic. Research on prophylaxis using healthy subjects illustrates the difficulty of deciding whether an experiment should be classed as beneficial or nonbeneficial. The immunity potentially created for subjects in a vaccine trial, for example, makes it seem like a beneficial experiment; it is not just the *knowledge* gained from the study that may be helpful to them if they are exposed to the relevant disease in the future, but an actual change in their *condition*. Nevertheless, such a trial, aimed at discovering the safe and acceptable dosage of the vaccine that will produce adequate immunity in the population, is directed at gaining knowledge, and any immunity that *may* be produced in the sample population is only a by-product.

The pivotal role played by the intention of the investigator in drawing the distinction between beneficial and nonbeneficial experimentation makes it especially difficult to apply. The existence of research protocols susceptible of prior review by scientific peers and others provides a possible way of establishing an investigator's intent, but attempts to divine it after the fact are more likely to give rise to dispute, as is well illustrated by the experimentally induced hepatitis in children at the Willowbrook State School in New York in the 1950s and 1960s (Goldby; Krugman, pp. 966–967). A further difficulty arises when the purpose of the procedure is differently perceived by the participants, as when a physician wishes to perform a study unrelated to the care of a patient who nonetheless believes that since it has been proposed by his physician it must be beneficial.

Basic ethical issues

The basic ethical problem in human experimentation—as well as the central social issue—is that experimentation is the use of one person by another to gather knowledge or other benefits that may be only partly good, if at all, for the first person; that is to say, the experimental subject is not simply a means but is in danger of being treated as a mere token to be manipulated for research purposes (Jonas, p. 3). In social terms, the propriety of such an action and the rules for its conduct if permitted are largely influenced by one's presuppositions concerning the relationship of individual human persons to the social whole. Within an ethical framework, propriety and rules are dependent upon one's premises about the value and dignity of the human person and about the personal and social values that medical practice is intended to promote.

In the context of experimentation, two areas of potential disagreement about the purpose and value of modern medicine are most relevant. The first concerns the practitioner's obligations, if any, to the science of medicine. Most commentators hold that the improvement in health care over the past century can be traced to the movement of medicine from art to science. (The accuracy of this position is not important. Whether medicine really is more scientific and whether the increasingly scientific cast to medical practice has made it "better" or "more successful" is not at issue, so long as the developments are believed in by most physicians and lay persons.) The scientific aspects of medicine rest in turn on deliberate and well-planned experimentation. From this it may be argued that those who have been taught medicine ought to participate in the development of scientific knowledge by approaching each patient as an intentional object of study for the benefit of future patients and of scientific knowledge. Such a duty on the part of physicians could be seen as the modern equivalent of the promise made by novices under the first portion of the Hippocratic Oath, to pass on instruction free of charge to the children of their teachers. The modern physician is an investigator whose teacher-father is science itself. Others argue, however, that in becoming a physician one covenants only to provide care for patients and that in applying or receiving the fruits of past experiments neither physicians nor patients obligate themselves to contribute to the care of others in the future.

The second important disagreement about the nature of medical practice, as it affects human experimentation, is in the interpretation of what is generally regarded as the basic canon of medical ethics: Do no harm. Although this point of disagreement is brought into sharpest focus in research that employs unconsenting subjects (such as the unborn or mentally incompetent adults) it is not so limited. All human experimentation points to an unresolved tension at the heart of medicine, the contours of which derive from one's interpretation of "do no harm."

This commandment cannot be read literally, of course, since every medical intervention contains the potential for some harm. Yet a strict reading would suggest that the phrase means that the harm being risked must be the minimum necessary for the benefit of the patient. Although it is through experimentation that medicine may be able to yield better (that is, more effective and less harmful) means of prevention, diagnosis, and therapy, that does not overcome the violation of the strict reading of the "do no harm" rule when a patient is made the subject of an experiment. Even the use of a consenting volunteer in an experiment with minimal risk places the physician-investigator in a dilemma between loyalty to the medical canon and service to medical science.

Those who read the commandment less rigorously find no contradiction in a physician-investigator's dual loyalties. In their view, since the prohibition on harm is only relative, it is not violated when a physician conscientiously undertakes an experiment involving a risk of harm (but not the intent to harm) that is reasonably related to the benefit to be derived for humankind. Emphasis is placed instead on the fact that, by advancing medical knowledge, the physician-investigator best serves the true meaning of the exhortation to avoid harm, which is to improve the quality of care provided to *all* patients and reduce the total harm associated with medical practice.

The design of experiments

Experiments with human beings that are so poorly designed as to yield no relevant observations are unethical since they expose subjects to risk and inconvenience to no useful end whatsoever. Moreover, the methods used should expose the minimum number of people to risk and permit the strongest possible conclusions to be drawn from the data. Within these limits, a wide range of scientific approaches has been legitimately employed. These extend from the trial of a new technique by a solitary doctor in a single patient-subject or small series to elaborate double-blind studies with thousands of randomly assigned subjects at a number of cooperating medical centers. Research at the former end of the spectrum is sometimes criticized because "uncontrolled" variables may render its results inconclusive or even erroneous. Nevertheless, it has long been the source of much clinical knowledge and continues to be widely employed in some branches of medicine, particularly in the initial stages of discovery, as corroboration for controlled experimentation, and where alternative methods would be difficult to apply or are felt to raise too many ethical problems (Beecher, 1970, pp. 109–112).

Because of the difficulty in eliminating extraneous variables, rigorous design is more readily achieved through prospective, randomized trials. This methodology was initially championed by Sir Austin Bradford Hill and has found increasing use as medical scientists become more conversant with the statistical methods which underlie it and as instances come to light which demonstrate the flaws in experiments that lacked proper statistical design (Hill). Even when used in this technical sense, "experimentation" has varied meanings. For example, one or more existing techniques may be tested against a new one, or the technique under study may be compared with nonintervention or with an inert substance, called a placebo. Or retrospective data may be employed—for example, information already available on the experience with alternative modalities in patients "matched" on all relevant criteria to the study subjects, in which case the design is termed quasi-experimental. The subjects and investigators may know which alternative is being employed; the subjects may be kept in ignorance, in which case the procedure is called a blind experiment; or both subjects and investigators may not know, in which case the study is said to be double-blind. Blind procedures, which are designed to compensate for bias and expectations, are most widely applied in pharmaceutical trials, but they have been used even in such areas as surgery (Beecher, 1961).

Despite its greater statistical validity, randomized experimental design presents an ethical problem when it is employed in research on patients rather than on volunteer subjects, particularly where the experiment is designed to compare an existing, accepted therapy with a new one. To deprive a patient of an approved treatment not because of any peculiarities in his or her own condition but because of the dictates of random assignment would seem to contradict the physician's primary obligation, on which the patient relies, to promote and protect the patient's well-being above all else. Indeed, the intrusion of scientific ends into the treatment setting may so disrupt the single-minded care of the patient that the procedure, although intended to benefit the patient, is better denominated *nonbeneficial* experimentation. Conversely, since randomized clinical trials are properly employed only in those cases in which well-informed medical opinion is undecided about which alternative is the better one for the patients in question, it may be more ethical to randomize patients and thus give each an equal *ex ante* chance of receiving the treatment that will eventually be proved preferable (Shaw and Chalmers, pp. 487–494).

As with all polar arguments, the facts of a particular case typically make it difficult to apply either position in its pure form to determine the ethicality of randomization. At the very least, one must doubt that the balance between alternative ways of proceeding is ever truly in perfect equipoise. Furthermore, if a physician-investigator makes fine distinctions among his patients in deciding which ones to admit to the trial he may bias the results and render them useless; however, if he fails to exercise careful clinical judgment, he may wrongly expose some patients to an undue risk.

Fortunately, statisticians have developed techniques that can reduce these dilemmas, if not eliminate them. A "crossover" design permits patient-subjects to be rotated from one procedure to the others; grouping of patient-subjects by factors known to be relevant and then randomizing within the group allows a smaller total sample to be used; and careful statistical analysis as the trial proceeds permits the researchers to cease should one alternative be shown to involve unjustified levels of harm or to alter the ratio of subjects assigned to each group (Weinstein, pp. 1281–1284; Zeisel, pp. 477–478). Statistical and decision theory cannot, however, provide the ultimate conclusion on which experimental design should be used, since that turns on one's judgment about what level of validity is necessary to protect future patients at what level of risk to present subjects.

There is no legal obstacle to a randomized clinical trial, provided that there has been candid disclosure of the procedure to the subjects (Fried, pp. 14–43). Some researchers seek to justify nondisclosure on the grounds that patients may be unnecessarily harmed because they will be upset by the idea of randomizing, that their withdrawal may bias the results, and that it is at any event irrelevant to them since their physician is still doing what he believes is best for them (Lasagna, p. 506). But this leads patients to distrust physicians and to conclude that they have been given placebos without warning when in fact they have not (Beecher, 1970, p. 271).

Requisites for the conduct of experiments

Before an experiment may properly be conducted, there are certain generally agreed upon requirements that must be met. Some of these relate to the scientific method employed in the research and others to ethical or social considerations. Although there are minor differences in the specifics, most of these requirements appear in the principal codes (United States, Defense Department; World Medical Association) and governmental statements (United States, Department of Health, Education, and Welfare).

There are three basic scientific requisites for experimentation. First, an experiment should be based on laboratory and animal experimentation and other scientifically established facts so that the point under inquiry is well-focused and has been advanced as far as possible with nonhuman means. The use of other living beings in research is subject to its own set of moral dictates, and those strictures on human studies which arise from our sense of autonomy and awareness of pain may be equally applicable in studies involving other beings that are capable of these same sensations (Singer, pp. 11–26).

Second, an experiment should follow proper scientific and medical procedures. The means employed to perform tests and observe phenomena should be capable of yielding the data that the research intends to provide. In most instances, this will mean that the experiment should be built around a testable hypothesis and conducted according to the criteria for validity already described. In such a case, the experiment will usually either prove or disprove the hypothesis (the latter being the preferred approach in the basic sciences but often unacceptable in medical science); sometimes an equivocal result will point to necessary refinements for an experiment to yield a definitive answer. Some human research, however, does not proceed on the basis of a testable hypothesis. Epidemiological studies or tests designed to yield profiles of the "normal" range for various characteristics in a population are acceptable forms of human research although not directed at proof of a particular point.

Finally, an experiment must be carried out by persons who are scientifically qualified, and the devices used must be those contemplated in the research design. Although proper execution of experiments is so obvious as hardly to require mention, it bears emphasis since it is surprising how often this requisite of experimentation is not fulfilled in actual practice. The concern here is less with a momentary slip—which, as in ordinary practice, may amount to professional negligence despite the generally high standards of the practitioner—than with the failure to limit experiments to research settings that possess personnel and equipment fully equal to the difficulty of the task at hand. Much research involves large teams of biomedical scientists, so that adequate qualifications of paraprofessional and technical staff are as much an issue as the scientific and medical qualifications of the principal investigator. Moreover, since much research takes place within medical schools, many projects involve persons still in training, who by definition are not yet fully qualified and certified. The investigator and the institution are ethically and legally responsible to assure that the people and procedures employed in carrying out an experiment are suitable. Ironically, the high level of expertise that characterizes most clinical research conducted at major institutions probably means that the results obtained in initial trials overstate the efficacy and safety that will actually be associated with new procedures once they are adopted in general medical practice.

In addition to these scientific requisites, for an experiment to go forward certain ethical requirements should also be met. First, all foreseeable risks and reasonably probable benefits, to the subject of the investigation and to science or more broadly to society, must be carefully assessed, and second, the comparison of these projected risks and benefits must indicate that the latter clearly outweigh the former. Moreover, the probable benefits must not be obtainable through other less risky means [see RISK]. The ethical calculation will also be affected by the type, severity, and reversibility of the harm. Under the Nuremberg Code, for example, all experiments in which death or disabling injury could occur are disapproved, except, perhaps, in those cases in which the investigators also serve as subjects.

To be consistent with Western notions of bodily integrity and personal autonomy, experimentation may go forward only with the informed and voluntary consent of the subject (Capron, pp. 403–423; Ramsey, p. 5). The consistent insistence in all codes and ethical discussions on researchers' obtaining of consent indicates that utilitarian justifications have not thus far been found persuasive in establishing a right in society to insist on involuntary participation in research. This requirement has posed special problems for research with subjects who are incapable of comprehending what is being proposed, such as the unborn, the very young and very old, and the mentally deficient, and with subjects whose captive or dependent status casts doubt on their freedom of choice, such as students and residents of institutions (Katz, pp. 955–1108)

Finally, participation by a subject in an experiment should be halted immediately if the subject finds continued participation undesirable or a prudent investigator has cause to believe that the experiment is likely to result in injury, disability, or death to the subject.

Violations of the principles of licit experimentation have repeatedly come to light, most dramatically in the horrendous abuses of the Nazi concentration camps but in civilian settings since the war as well (Beecher, 1966; Pappworth). As a result, traditional reliance on informal means of control, relying largely on biomedical professionals themselves (Parsons, pp. 118–120), has been supplemented by formal mechanisms involving lay persons, often with explicit sanctioning power of governmental funding and prosecuting authorities. Regulation has taken the form of prior review of protocols and more rigorous attempts to employ informed consent as a means for making subjects active collaborators whose understanding of the research and its purposes enables them to make valuable contributions to its success (Fox, pp. 419–420; Mead, pp. 162–165). The direct involvement of these bodies serves to emphasize that human experimentation is no longer a matter of occasional interest to a few physicians but is a major social activity in which large numbers of people are involved under the aegis of profit-making as well as academic institutions, and on which future progress in pharmacology, genetics, and all areas of the life sciences is dependent. While supervision by formal bodies may reduce instances of unwarranted or unconsented research, it seems certain also to increase the impetus toward governmental compensation of subjects harmed by negligence and the inevitable maloccurrences of biomedical exploration.

ALEXANDER MORGAN CAPRON

[*Directly related are the other articles in this entry as well as the following entries:* INFORMED CONSENT IN HUMAN RESEARCH; RESEARCH, BEHAVIORAL; RESEARCH, BIOMEDICAL; *and* RISK. *Other relevant material may be found under* CHILDREN AND BIOMEDICINE; FETAL RESEARCH; GENE THERAPY, *article on* ETHICAL ISSUES; HEART TRANSPLANTATION; PRISONERS, *article on* PRISONER EXPERIMENTATION; REPRODUCTIVE TECHNOLOGIES, *article on* ETHICAL ISSUES; RESEARCH POLICY, BIOMEDICAL; *and* SURGERY. *For discussion of related ideas see the entries:* ANIMAL EXPERIMEN-

TATION; DRUG INDUSTRY AND MEDICINE; and MEDICAL ETHICS IN LITERATURE. See APPENDIX, SECTION II: DIRECTIVES FOR HUMAN EXPERIMENTATION (*entire section*); and SECTION IV, AMERICAN PSYCHOLOGICAL ASSOCIATION.]

BIBLIOGRAPHY

BEECHER, HENRY K. "Ethics and Clinical Research." *New England Journal of Medicine* 274 (1966): 1354–1360. Description of recent unethical experiments, and a critique.

———. *Research and the Individual: Human Studies*. Boston: Little, Brown & Co., 1970. The major work of one of the medical pioneers through whose efforts public and professional attention has recently been focused on the field of human research.

———. "Surgery as Placebo." *Journal of the American Medical Association* 176 (1961): 1102–1107. Examines instances of accepted surgery in which the placebo effect appears to have been the sole functioning agent.

CAPRON, ALEXANDER MORGAN. "Informed Consent in Catastrophic Disease Research and Treatment." *University of Pennsylvania Law Review* 123 (1974): 340–438. Analysis of the functions and limitations of informed consent and of the appropriate development of the law on this doctrine.

Fox, RENÉE C. 'A Sociological Perspective on Organ Transplantation and Hemodialysis." Ladimer, "New Dimensions," pp. 406–428.

FREUND, PAUL A., ed. *Experimentation with Human Subjects*. New York: George Braziller, 1970. Expanded and revised version of the Spring 1969 issue of *Daedalus*. Contains an excellent collection of articles.

FRIED, CHARLES. *Medical Experimentation: Personal Integrity and Social Policy*. Amsterdam: North-Holland Publishing Co., 1974. Through the glass of the randomized clinical trial, examines legal and philosophical aspects of the conflict between obligations to society and obligations to individuals.

GOLDBY, STEPHEN. "Experiments at the Willowbrook State School." *Lancet* 1 (1971): 749. A forceful accusation, which provoked an apology from the journal's editors, of the use at Willowbrook of mentally defective children in nonbeneficial experimentation.

HILL, A. BRADFORD. "Principles of Medical Statistics." *Lancet* 1 (1937): 41–43. Initial presentation on the value of randomization in clinical research, which is reiterated in his *Principles of Medical Statistics*, now in its ninth edition (New York: Oxford University Press, 1971).

IVY, A. C. "The History and Ethics of the Use of Human Subjects in Medical Experiments." *Science* 108 (1948): 1–5. Traces the history of medical experimentation from the Early Ages to Nazi Germany and examines the use of animals, students, lay persons, prisoners, and mental incompetents as subjects.

JAFFE, LOUIS. "Law as a System of Control." Freund, *Experimentation with Human Subjects*, pp. 197–217.

JONAS, HANS. "Philosophical Reflections on Experimenting with Human Subjects." Freund, *Experimentation with Human Subjects*, pp. 1–31.

KATZ, JAY. "The Education of the Physician-Investigator." Freund, *Experimentation with Human Subjects*, pp. 293–314.

———, ed. *Experimentation with Human Beings: The Authority of the Investigator, Subject, Professions, and State in the Human Experimentation Process*. New York: Russell Sage Foundation, 1972. A comprehensive selection of historical and current materials on human experimentation, including many illustrative examples and complete case studies, arranged according to a two-level analytical framework.

KRUGMAN, SAUL. "Experiments at the Willowbrook State School." *Lancet* 1 (1971): 966–967. Defends hepatitis experiments as indirectly benefiting the children, their peers, and their families.

LADIMER, IRVING, ed. "New Dimensions in Legal and Ethical Concepts for Human Research." *Annals of the New York Academy of Sciences* 169 (1970): 293–593. The result of a conference held by the New York Academy of Sciences, 19–21 May 1969.

LASAGNA, LOUIS. "Drug Evaluation Problems in Academic and Other Contexts." Ladimer, "New Dimensions," pp. 503–508.

MCCANCE, R. A. "The Practice of Experimental Medicine." *Proceedings of the Royal Society of Medicine* 44 (1951): 189–194. A basic, early statement of the elements of clinical investigation.

MEAD, MARGARET. "Research with Human Beings: A Model Derived from Anthropological Field Practice." Freund, *Experimentation with Human Subjects*, pp. 152–177.

National Commission for the Protection of Human Subjects of Biomedical and Behavioral Research. *Research on the Fetus: Report and Recommendations*. DHEW Publication No. (OS) 76-127. Bethesda, Md.: Department of Health, Education, and Welfare, 1975.

PAPPWORTH, M. H. *Human Guinea Pigs: Experimentation on Man*. London: Routledge & Kegan Paul, 1967.

PARSONS, TALCOTT. "Research with Human Subjects and the 'Professional Complex'." Freund, *Experimentation with Human Subjects*, pp. 116–151.

RAMSEY, PAUL. *The Patient as Person: Explorations in Medical Ethics*. The Lyman Beecher Lectures at Yale University, 1969. New Haven: Yale University Press, 1970. An excellent presentation of a well-developed moral structure for analyzing many facets of the doctor–patient relationship, based upon Christian theology.

SHAW, LAWRENCE W., and CHALMERS, THOMAS C. "Ethics in Cooperative Clinical Trials." Ladimer, "New Dimensions," pp. 487–495.

SHIMKIN, MICHAEL B. "The Problem of Experimentation on Human Beings: I. The Research Worker's Point of View." *Science* 117 (1953): 205–207. Cogently outlines the broad guiding principles under which the author believes the use of human subjects in medical research should be approached.

SINGER, PETER. *Animal Liberation: A New Ethics for Our Treatment of Animals*. A New York Review Book. New York: Random House, 1975. An unstinting critique of the ways in which animals are abused. Argues for a radical redefinition of animals' rights.

United States, Defense Department. "Nuremberg Code." U.S. v. Karl Brandt. *Trials of War Criminals before Nuremberg Military Tribunals under Control Law No. 10*. Washington: Government Printing Office, 1947, vol. 2, pp. 181–183. See also Appendix to this encyclopedia.

———, Department of Health, Education, and Welfare. "Protection of Human Subjects." 45 C.F.R. §46 (1976).

WEINSTEIN, MILTON C. "Allocations of Subjects in Medical Experiments." *New England Journal of Medicine* 291 (1974): 1278–1285. Provides alternatives and additions to randomized trials, such as adaptive designs.

World Medical Association. "Declaration of Helsinki." *New England Journal of Medicine* 271 (1964): 473–474. An international document to be used as a moral guide for doctors in clinical research. Distinguishes clinical research in which the aim is therapeutic and clinical research for science's sake alone.

ZEISEL, HANS. "Reducing the Hazards of Human Experiments through Modifications in Research Design." Ladimer, "New Dimensions," pp. 475–486.

III
PHILOSOPHICAL ASPECTS

The concept of experimentation is wide and ill-defined. The intuitive notion, at least as it applies to human experimentation in the biomedical field, is of an action taken toward solving a problem in respect to a subject or subjects in order to observe the effects of the action and thus to suggest the formation or establish the invalidity, modification, or confirmation of a hypothesis offered as a possible solution to the problem.

The need for definition arises first because hypotheses may also be formed, rejected, modified, or confirmed on the basis of observations alone, the investigator doing no more than studying epidemiological data regarding subjects who remain entirely statistical and anonymous items. Such impersonal statistical studies may be considered outside any reasonable limit required for the concept of experimentation. It should be noted, however, that to the extent that knowledge is gleaned in such observational studies, future persons might be benefited; to the extent that resources are diverted for the purpose of such investigations, present persons might be prejudiced. Nevertheless, a conception of human experimentation that includes such studies is obviously too wide. The central notion and concern must relate to intentional impositions upon the subjects of experimentation. Such impositions may, of course, include withholding or varying treatment that would otherwise be accorded for the purpose of acquiring information.

A further problem of definition is raised by the meaning that the term experimentation has acquired in technical statistical theory. An experiment in technical literature is defined as the testing of a hypothesis by the prospective assignment on a random basis of instances to two or more categories so that the subsequent observation of these categories may provide confirming or disconfirming evidence for the hypothesis under consideration. If the data studied relate to processes already completed (in technical language, if it is retrospective), then the inquiry is said to be quasi-experimental in design. And the intervention in or observation of uncontrolled, isolated instances, because providing no statistically validated basis for inference, does not qualify as experimental at all in the technical sense. In spite of the importance of and growing interest in this limited, technical concept of experimentation, no such narrow definition will be adopted here. Not only have many important observations and interventions been made—and they continue to be made—on statistically uncontrolled and episodic bases, but a sounder and less doctrinaire concept of scientific truth would not exclude conclusions based on such observations, although their inaccuracies and ambiguities must be recognized.

Finally, considerable confusion in discussions of human experimentation arises from the failure to keep in mind a further crucial distinction, that between therapeutic and nontherapeutic experimentation. In therapeutic experimentation a course of action (or studied inaction) is undertaken in respect to the subject for the purpose of determining how best to procure a medical benefit to that subject. In nontherapeutic experimentation, by contrast, the sole end in view is the acquisition of new information. The presence of a purpose to help the subject in therapeutic experimentation may tend to obscure a divergence of interest between experiment and subject, and perhaps for this reason considerable ingenuity has been expended to show that the two categories of experimentation are really hardly distinct. On one hand, it is pointed out that often action undertaken for the benefit of a particular person will add to the general store of knowledge. Indeed, it would seem irresponsible from almost any point of view deliberately not to take advantage of the experience in a particular case when it can benefit other persons. Thus therapeutic experimentation is said to have an ineluctably nontherapeutic component. Correspondingly, the subject of what appears to be nontherapeutic experimentation may himself on some future occasion be the beneficiary of the information acquired. Two examples illustrate the difficulty of maintaining the integrity of the two analytical categories.

1. For a particular illness two alternative therapies are available, and medical opinion is

split on which alternative is better. Either alternative appears to some reputable practitioners to be the better one to pursue in the interests of an individual patient. In order to resolve the controversy and thus to benefit future patients, for a group of patients the choice between the two best therapies is determined on a wholly random basis. Every patient receives a therapy that in the opinion of some doctors is the best available. By studying the outcomes a definitive evaluation of the two alternatives might be possible.

2. Inmates at a state home for severely retarded children receive an experimental vaccine against infectious hepatitis and then are deliberately infected with hepatitis in order to determine the efficacy of the vaccine. Although only healthy subjects are used, it is argued that infectious hepatitis is endemic in institutions such as those to which the subjects are confined, that they are highly likely to contract the disease eventually in any case, and thus that the experiment may be beneficial to them and is clearly intended to be beneficial to other inmates of such institutions.

The first case looks like a case of therapeutic experimentation, and yet the design of the experiment is such that the experiment itself is not intended to, nor likely to, benefit the particular subjects except insofar as the disease is one that may recur. In the second case the subjects would plainly not be deliberately infected if their benefit was the sole end in view. Yet the subjects may benefit from the research, and, even if they do not, others similarly situated may. The two cases illustrate the fact that in the conduct of actual experimentation therapeutic and nontherapeutic elements will often be intermingled. It is not correct, however, to conclude from this that the distinction is without meaning and cannot be maintained. Rather, no more is shown than that a particular course of action may contain both elements. The fact that nontherapeutic elements are present in much ordinary treatment may also show that nontherapeutic experimentation is important and perhaps indispensable. But, of course, the necessity of nontherapeutic experimentation has not been denied, nor have we yet explored the conditions under which ethically it may go forward. The confusion of the two categories, however, impedes pursuit of the central ethical questions involved.

That there are truths regarding his own nature or the nature of the universe as a whole that man should not seek to plumb is an ethical position of no serious interest. And the question of what if any risks one may be prevented from encountering freely and knowingly in the pursuit of knowledge is properly treated under the general heading of paternalism. This leaves as the most crucial ethical issue in human experimentation the extent to which the interests of the individual may be compromised for the benefit of a larger group, even if he is also a member of that group. To the extent that experimentation is clearly therapeutic this issue is not raised, and exactly to the extent that the experimentation has nontherapeutic elements the issue must be answered. One may suspect that the interest in presenting the distinction between therapeutic and nontherapeutic experimentation as difficult and confused arises from an unwillingness to confront this ethical issue. The distinction and its difficulties are represented in the dual role of the physician, when the experimenters are physicians. A long tradition in Western society assumes a special relationship between the physician and his patient. In this sense the physician might be seen as simply the surrogate of the patient's own autonomy. On the other hand, the physician has often sought to view himself both as a scientist, pursuing the objective value of truth in a dispassionate and impersonal way, and as a general benefactor of mankind caring for the health of the species. These two roles—of physician and of scientist—correspond precisely to the concepts of therapeutic and nontherapeutic experimentation. This correspondence shows that the central philosophical issue relates to the rights and obligations of the individual insofar as he is patient or subject, and that the question of the proper role of the physician may be seen as derivative. These issues, however, until recently—perhaps until the shocking revelations of the brutal experiments of the Nazi doctors—have tended to remain obscured and confused behind vague generalities concerning the physician's devotion to science, to humanity, and to his patients—all at the same time. The historical record shows that these platitudes concealed laxness and even brutality, particularly where the subjects were the poor, the ignorant, or the helpless.

The argument might plausibly be made that an individual has an obligation to participate in nontherapeutic experimentation insofar as he is the beneficiary of experimentation undertaken in the past. For, by receiving only the benefits while sharing none of the burdens of advancing biological knowledge, the individual would be taking unfair advantage of the sacrifices of others. It is this argument—rather than the frequent and perhaps unavoidable alloy of the therapeutic and the nontherapeutic in experimentation—that provides the strongest basis for curtailing the absolute autonomy of individual patients, subjecting them sometimes to impositions that care for their own health would not dictate. The argument, however, is too general to be conclusive. First, there is no unfairness if the knowledge gained from prior nontherapeutic experimentation was donated freely by the subjects of those experiments. The beneficiary of such gifts may be under an obligation of gratitude to give similar gifts to others, but this is not an obligation that he can be compelled to fulfill, and therefore it is not one that should be exacted from him without his full consent. Further, where previous subjects have participated in nontherapeutic experimentation for fair compensation, no obligation arises for a subsequent individual, to whom is not offered or who does not choose to accept such compensation, to allow himself to be subjected to nontherapeutic experimentation. Also to the extent that present sufferers are the beneficiaries of knowledge gained not from nontherapeutic experimentation but from observational studies and therapeutic experimentation, there would of course follow from the receipt of such benefits no obligation to make any sacrifice in order to render a benefit to others.

Finally, and more generally, to the extent that the ideal subjects for experimentation are persons afflicted with particular diseases, the argument from fairness does not go through. The group in which benefits and burdens should be fairly shared should be the group of all actual and potential patients, that is, society as a whole. There is no reason why a person should be singled out to make this particular sacrifice simply because he is a convenient subject for experimentation. If there is an obligation to participate in experimentation, it can exist only in the context of a reasonably well understood formal structure intended to even out the burdens and benefits of experimentation over the society as a whole. In the absence of such a scheme, the argument of obligation can be made only on the basis that everyone always has an obligation to do what under the circumstances appears to render a net benefit to the society or the human group as a whole. If so wide a notion of obligation is rejected—as it would seem to be in all but the most thoroughgoing utilitarian moralities—then it would seem there is no general obligation to participate in experimentation.

If the individual has no obligation to participate in experimentation, it follows that the physician has no right to procure such participation by deception, overbearing insistence, or compulsion. The physician may request participation, presenting to his patient an opportunity not only to further the patient's own interests, but also to render substantial benefits to others. If illness and treatment were conceived in these terms often enough, then gradually there might come into existence the kind of socially accepted and pervasive understanding of the need for experimentation from which obligations might arise. That would particularly be the case if not only the ill but the healthy too were regularly called on as experimental subjects.

Undoubtedly respect for individual autonomy, which leads us to enroll persons in experimentation pursuant only to their free choice or as part of a fair scheme of participation, does to some degree deny society the benefits of sound experimentation. And thus some questionable and perhaps harmful therapies will continue to be used, while the introduction of beneficial new therapies may be delayed. But this cannot be a conclusive argument for an obligation to participate, for, if we adhere to the Kantian precept that we may never use another human being as a means alone, no matter how exalted our ends, then we must be prepared on occasion to forgo certain net social advantages that an imposition on some individual or group of individuals might procure. On the other hand, to the extent one embraces the utilitarian ideal of maximizing the greatest good of the greatest number—which is perhaps the most modern and clearest expression of the vague notion of the "common good" —then the prospect of available net benefits becomes a sufficient ground for an obligation to participate in experimentation and indeed for permitting deception and compulsion in obtaining such participation.

The most difficult question of all relates to the nature of fair experimental schemes pursuant to which an obligation to participate might be urged. If the scheme is fair, is it conceivable for healthy and sick individuals to be required to participate in experimentation even against their will? The picture of persons being subjected to a harmful experimental procedure by force is so repellent that such a formulation seemed unacceptable from the start. And a system that subjects patients to experimentation not by force but by deception and nondisclosure is equally unacceptable, equally destructive of relationships of trust and goodwill. But an obligation to participate might be implemented without recourse to such objectionable means. The participation pursuant to some fair scheme might be made a condition of receiving medical benefits. If no force or fraud is used, would the person seeking medical help have grounds to complain of nontherapeutic experimental elements in his care, provided the very care he was seeking was made possible by the acquiescence of others in such experimentation? On the other hand, it might be argued that, even where the scheme is fair and all-inclusive, society may not condition medical care on accepting the hazards of nontherapeutic experimentation. Any such argument would have to assert the primacy of a right to medical care even in the face of the claims of fairness. This assertion is far from implausible, as it may proceed from the premises that medical care is care for one of the essential attributes of human personality, bodily integrity, and an individual may never be required to subordinate his personal integrity, even to the otherwise just claims of his fellow men. On this view the physician is the agent and servant of the patient's concern for his personal integrity.

Finally, it must be stressed that experimentation is justified not only on the instrumental claim that it will improve future medical care but also because the very aim of adding to the system of human knowledge about nature and about human nature is one of the highest purposes of intellectual activity. This is the concept of experimentation as intellectual adventure. This perhaps deepest justification, however, carries with it the greatest need for obtaining the willing and intelligent participation of experimental subjects. In this way the spiritual adventure of experimentation becomes a joint venture characterized not by domination and arrogance but respectful cooperation.

CHARLES FRIED

[*While all the articles in this entry are relevant, see especially the previous article,* BASIC ISSUES. *For discussion of related ideas, see the entries:* INFORMED CONSENT IN HUMAN RESEARCH; JUSTICE; OBLIGATION AND SUPEREROGATION; RESEARCH, BIOMEDICAL; *and* RIGHTS, *article on* RIGHTS IN BIOETHICS. *Other relevant material may be found under* HEALTH CARE, *articles on* RIGHT TO HEALTH-CARE SERVICES *and* THEORIES OF JUSTICE AND HEALTH CARE; *and* TRUTH-TELLING, *article on* ETHICAL ASPECTS. *See also:* ETHICS, *articles on* DEONTOLOGICAL THEORIES *and* UTILITARIANISM; *and* MEDICAL ETHICS UNDER NATIONAL SOCIALISM.]

BIBLIOGRAPHY

BEECHER, HENRY K. *Research and the Individual.* Boston: Little, Brown & Co., 1970. This and Pappworth's *Human Guinea Pigs* awakened concern for the ethics of medical experimentation.

"Experimental Design." *International Encyclopedia of the Social Sciences.* Edited by David L. Sills. New York: Macmillan, Free Press, 1968, vol. 5, pp. 245–263. An excellent introduction to the statistical aspects of the subject.

FOX, RENÉE C. *Experiment Perilous.* Glencoe, Ill.: Free Press, 1959. Considers the notion of experimentation as a cooperative adventure.

FREUND, PAUL A., ed. *Experimentation With Human Subjects.* New York: George Braziller, 1970. This and "New Dimensions in Legal and Ethical Concepts for Human Research" are recent collections containing materials of great interest.

FRIED, CHARLES. *An Anatomy of Values.* Cambridge: Harvard University Press, 1970, chaps. 11, 12.

———. *Medical Experimentation: Personal Integrity and Social Policy.* Amsterdam: North-Holland; New York: American Elsevier, 1974. The author's views are developed at length.

KATZ, JAY, with the assistance of CAPRON, ALEXANDER, and GLASS, ELEANOR S. *Experimentation with Human Beings.* New York: Russell Sage Foundation, 1972. An extraordinarily useful collection of bibliographic material and of primary sources, both historical and contemporary, including the texts of several of the codes governing experimentation, such as the Nuremberg Code and the Declaration of Helsinki.

"New Dimensions in Legal and Ethical Concepts for Human Research." *Annals of the New York Academy of Sciences* 169 (1970): 297–593.

PAPPWORTH, MAURICE H. *Human Guinea Pigs.* Boston: Beacon Press, 1967.

RAMSEY, PAUL. *The Patient as Person.* New Haven: Yale University Press, 1970. An excellent presentation of one moral-theological view.

RAWLS, JOHN. *A Theory of Justice.* Cambridge: Harvard University Press, Belknap Press, 1971. A recent distinguished treatise on the concepts of liberty, justice, and fairness from a generally Kantian perspective.

IV
SOCIAL AND PROFESSIONAL CONTROL

This article focuses on the principal alternative strategies and mechanisms for the professional and social control of human experimentation. Ideally, controls perform two broad functions. First, they establish rules governing the relationship between the investigator and his subjects in the research process; it is of prime importance to create a check on improper behavior by the researcher in order to protect subjects from abuse. Second, such controls can also help to generate a climate receptive to the rigors of scientific research—an atmosphere in which "good" science and the humane treatment of subjects flourish, so that the full benefits promised by science can be realized without the sacrifice of moral values. Too often this second function goes unappreciated by those skeptical of the processes of social control.

Two basic types of controls are discussed here: professional and public/quasi-public. The assignment of controls to either one of the two categories is based on their source of implementation rather than on their origin. In the final analysis, what is most important for protecting human subjects and instilling confidence in medical research is how the controls actually affect behavior within the research setting.

Professional controls

An important test of professional status is autonomy. As a group, researchers resent what they consider to be outside surveillance, believing that their status as responsible, independent investigators is thereby questioned or threatened. They prefer to be evaluated in terms of what their colleagues would do rather than by rigid standards developed by sources external to the profession. As a consequence, the research community has been involved in the development of several control mechanisms, which, in addition to whatever protection they have afforded research subjects, have had the effect of reinforcing professional autonomy.

Medical education and training. During the extended period of professional training, the medical profession has its best opportunity to influence the behavior of its future members (Katz, 1969). Robert K. Merton (pp. 76–77) observes that the medical schools are socially defined as the guardians of the profession's values and norms with the responsibility of transmitting to students both the technical skills and the moral standards of the profession. However, a recent study concluded that medical schools in the United States appear to be much more successful in imparting technical knowledge to their students than they are in inculcating the values associated with the ethics of human research (Barber et al., p. 174). Education is a long-term process and, while essential for any fundamental change in professional ethics and behavior, must be reinforced by other, more direct mechanisms for regulating the ethical conduct of human research.

Professional codes, guidelines, and moratoria. A characteristic of most professional groups is a self-administered code of ethics designed to govern the behavior of their membership. The code as a "regulatory" device is based on the dual premise that the profession is the most competent judge of deviant behavior of its members and is capable of enforcing its standards. Ideally, then, by articulating the values involved in human experimentation, codes of ethics can increase professional awareness of the ethics of research and provide a yardstick by which to judge professional behavior.

However, as a mechanism for protecting human research subjects, ethical codes suffer from several shortcomings. Their highly generalized and hence ambiguous language may invite conscious evasion or neglect by individual investigators (Barber et al., pp. 59–92; Katz, 1969). Moreover, there is little incentive for researchers to make the codes more precise, for their vagueness gives the experimenter wide latitude in which to claim that any self-serving behavior on his part is consonant with the collective goals of the profession. Besides, taking precautions to protect subjects is costly in terms of time and energy, and an obstacle to those caught up in the competitiveness of scientific research. Lax enforcement of the codes is another problem. Since professional associations exist primarily to serve their members, it is not surprising that the medical profession has rarely spoken out against members accused of improper behavior or incompetence (Jaffe, p. 408).

Professional groups may also contribute to the development of standards through the promulgation of guidelines applicable to a specific issue, such as the statement on heart transplantation by the Board of Medicine of the National Academy of Sciences (1968). Such statements are typically more detailed than the broader ethical codes. However, when the issuing groups are not the professional associations, or are convened for the single purpose of promulgating a particular set of guidelines, they have no real authority over those engaged in the activity in question. Furthermore, in the case of ad hoc groups, it remains unclear how their composition or the timing of their convening should be determined.

Clinical moratoria involve the suspension of the use of an experimental procedure or drug for an undetermined period (Fox and Swazey, p. 122). During this time, the activity in question is tested and reevaluated in the laboratory with the hope of solving the problems that precipitated the suspension of its use on humans.

A moratorium may result from pressures invoked formally, such as by withdrawing hospital operating privileges from individual surgeons, or informally, as when colleagues argue that to continue human trials would damage the public image of the profession. The impact of the moratorium on clinical investigation will, in large part, be determined by the procedures used to initiate it. But whatever methods are used, proponents of a moratorium must battle against the internalized values of the medical profession, which require that its members do all that is possible to advance medicine and to relieve human suffering. What often happens, then, is that only after substantial and repeated failure —during which time patient-subjects may be subjected to serious injury or to death—will a moratorium actually be implemented. This is well illustrated by the heart transplant moratorium (ibid., pp. 122–148), which was initiated only after many posttransplantation deaths had already occurred.

Licensure. Professional licensure is a means used to grant status and recognition to a body of specialized knowledge. It is aimed more at practitioners of medicine than at researchers, but since both functions are frequently performed by a single individual, licensure may serve as a check on the conduct of clinical research. Its working assumption is that fear of having one's license to practice medicine revoked will prevent the investigator from abusing the rights and well-being of research subjects. But while licensure regulations are codified by statute, what has occurred, at least in the United States, is that a "state protected environment" has emerged in which the medical profession has maintained its authority to determine the criteria for licensure and to sit in judgment of colleagues accused of improper behavior (Cohen, p. 73). Since state licensure boards in the United States have seldom initiated disciplinary action against physicians for poor medical practice (Derbyshire), licensure appears not to promise much protection for research participants.

Recently, there have been a number of proposals to expand the composition of licensing boards to include public members with interests extraneous to the field being licensed (Cohen). In Oregon, for example, a state statute establishes a nine-member Psychosurgery Review Board with the power to suspend or revoke licenses to practice medicine and requires that two of its members be from the general public while none of the seven physician members may be directly involved in conducting psychosurgery or intracranial brain stimulation (Dow, Grimm, and Rushmer). Although the appointment of one or two public members is generally regarded as a healthy step forward, some view it as only a token structural change that will ultimately make little difference with the actual behavior of licensing boards, as the few laymen will inevitably be co-opted by the large physician-scientist majority (Cohen, p. 82).

Journals. Another possible control strategy that has received considerable attention is the use of the editorial power of professional journals. For example, both the Council of Biology Editors (Woodford) and the European Association of Editors of Biological Periodicals ("Ethics in Human Experimentation") have issued policy statements that call for editors to apply ethical criteria when assessing manuscripts for publication. Specifically, the issue is how, if at all, research data obtained unethically should be presented to the journal readership.

Some have argued that when research procedures violate ethical standards the manuscript should not be published (DeBakey, p. 116). The practical rationale for this position is that failure to publish would discourage researchers from engaging in unethical procedures. Fear is also expressed that, if such research were to be published, science would be adversely affected in the arena of public opinion. To implement this view, proponents have suggested that authors include in their articles descriptions of how the informed consent of subjects was obtained and that a statement of approval of the protocol from the institution's committee on research ethics accompany the manuscript (Woodford). A contrasting position is held by some who believe that, since the experiment cannot be undone and the data may be valuable, the material should be published along with critical comment by the editor or invited participants in order to bring the ethical issues to public attention (Katz, 1974). If it is true, as Justice Brandeis observed, that sunlight is the most effective of all disinfectants, then publication would avoid the low visibility of unethical research and subject it to close professional and public scrutiny so that steps can be taken to remedy the problem.

Whichever side of the argument one accepts, there remain several unresolved issues with respect to implementing strategy. Unless all journal editors adhere to similar ethical standards and procedures, authors may simply seek out those with the least stringent requirements.

Moreover, given the fact that few institutional review committees perform any type of effective continuing review (Barber et al., pp. 155–156), of what real consequence is the stipulation that the committee at the author's institution indicate its approval of the initial protocol?

Subject advocates. It is generally recognized that there is a distinction between the relationship of a physician to his patients and that of a researcher to his subjects. In the case of the former, the physician is primarily concerned with the patient's welfare; in the latter instance, the investigator is mainly dedicated to the solution of some scientific problem. Because of this distinction, it has been proposed that a "physician friend," or subject advocate, with no direct interest in the proposed research, be appointed to protect the welfare of the subject, while the investigator would be responsible for conducting the experiment (Guttentag). In the United States, a number of local institutions have adopted a procedure giving the subject advocate authority to halt a study or withdraw a subject if he or she felt it necessary (Cowan).

The implementation of the subject advocate concept has not been without its problems. In many institutions, it has been chiefly applied to nontherapeutic research. Cases have occasionally arisen where an investigator has been uncooperative with proposals to have a third party intercede between him and his patient-subjects (ibid.). Further, as Katz and Capron (p. 231) argue, the close ties between the physician friend and patient may make the task of informing the subject and obtaining his consent even more difficult. As an alternative, they suggest that a less involved physician or a lay person act as the subject's advocate. In the latter case, however, it is questionable how well suited the layman will be to understand and communicate medical information to the subject.

Peer review. Peer review is both a process and a set of structures that have been initiated to produce normatively desirable behavior on the part of investigators using human research subjects. An unpublished 1975 report by the Canada Council revealed that at least thirty-five Canadian Universities had created either ad hoc or standing committees to review research protocols. In most instances the committees were mandated by government agencies funding medical research. Peer review committees have also been an integral part of Great Britain's medical research program, with prime impetus for their creation coming from England's Medical Research Council (Great Britain, Medical Research Council). In the United States, peer review had its formal beginnings in guidelines issued in 1966 by the U.S. Public Health Service (Frankel, p. 52). Whatever the originating source of these peer review mechanisms, they have all had their content and implementing procedures so strongly influenced by the medical profession that it seems appropriate to include them in a discussion of professional controls.

The typical peer review mechanism is a committee located within the institution at which the research is conducted. Underlying such a decentralized system of control is the premise that committees made up of local people familiar with local practices and mores will be in an ideal position to assess the ethical dimensions of research protocols. Moreover, review performed by knowledgeable peers may uncover technical deficiencies in the research design that, when communicated to the investigator, will eventually result in more adequate safeguards for subjects as well as better science.

Unfortunately, in practice the review committee system has not always functioned well. In both Canada and the United States considerable diversity exists in the composition and operating procedures of committees in different institutional settings. Even within a single institution, policies and procedures often tend to reflect the size, commitment, and diversity of the committees' members. What this often means, therefore, is that the ultimate effect of the proposed investigation on the well-being of subjects is determined in a haphazard way, depending on the dispositions of those serving on a particular review committee at the time the proposal is submitted. In fact, many Canadian institutions operate without either ethical codes or formal review procedures to guide committee deliberations. Further, in both Canada and the United States, there is no formal means of communicating decisions among the committees of various institutions so that they all could profit from similar cases decided at an earlier date. Surely, a method for sharing decisions would ultimately improve the capability of these committees to evaluate and resolve the complex and sensitive problems of human research (Calabresi).

It is also clear from the United States experience that members of the medical profession have been reluctant to police their colleagues in the context of institutional committee review (Gray, pp. 30–55). Perhaps because the reviewers and the researcher have been molded with the same set of professional values, many committees have acted more as a clearinghouse for proposals than as rigorous critics (Barber et al., p. 165). These revelations have been a prime impetus behind rising demands that more laymen be added to the review committees. Regulations issued in 1974 by the U.S. Department of Health, Education, and Welfare require that institutions receiving funds from that agency include nonscientists on their committees. How effective these committees will be remains to be seen. Very little consideration has thus far been given to how the lay members should be recruited or how they will be compensated for assuming a responsibility that could make heavy demands on their time. Finally, if the system of peer review is to be effective as a control mechanism, committees must be given an explicit and realistic mandate with workable procedures for achieving its objectives (Veatch).

Public and quasi-public controls

It is generally recognized that some degree of autonomy is essential for scientists to perform effectively in pursuing the benefits of medical research. But the possibility of arbitrary and capricious action by scientists, which may create unacceptable burdens for human subjects, has made it necessary for society to assume greater control over the conduct of human research. In some instances this has meant the creation of mechanisms at the national and subnational level that seek to give public authorities sufficient control over the activities of researchers without unduly hampering their creative function. On other occasions, less formal sources of authority, such as quasi-public bodies whose activities have a pronounced public effect, may function as a check on the actions of clinical investigators.

Legislative action. The concern of legislatures with the rights of medical research subjects is typically expressed through hearings and in proposed or enacted legislation. For example, a series of dramatic hearings in the U.S. Congress (United States, Senate) was instrumental in bringing public attention to the ethical and legal dilemmas involved in experimentation on humans and led to the passage of important legislation (National Research Act). Individual state legislatures have also enacted statutes, either regulating research on certain categories of subjects, such as fetuses (Reback), or on the use of specific experimental techniques (Dow, Grimm, and Rushmer). Other countries as well have enacted legislation regulating the testing of drugs on humans prior to their approval for general use (e.g., Gesetz zur Neuordnung).

Hearings perform a number of useful control functions. In addition to expanding the scope of legislative and public attention given to a problem, they permit divergent voices to bring their concerns before a public policymaking body and create a record of public support or disapproval for pending legislation. Too often, however, hearings are one-sided, as the committee chairman carefully selects those witnesses who support his position. Furthermore, hearings are usually enveloped in an aura of sensationalism, which limits their contribution to rational decision making.

Legislation does have the advantage of imposing uniformity on research conducted in different institutions within the same political jurisdiction. One group of subjects should not be exposed to substantially more risky procedures than another simply because of where they live in the community or the nation. Legislation also establishes parameters that researchers, judges, and administrators can use to guide their decision making. But legislation also has its shortcomings. There is always the danger that the statute will be badly drafted. A poorly phrased law may confuse those whose actions it seeks to control and, consequently, perhaps endanger those whom it was intended to protect. Moreover, once enacted, a statute is not easily amended or repealed, leaving the clarification of its intent to the time-consuming and sometimes costly process of litigation or to the discretionary power of a small group of administrators.

Administrative and regulatory bodies. Regulatory agencies establish specific standards, pursuant to broader enabling legislation, which are intended to delimit the activities of specific clientele groups. Administrative agencies also possess the authority to issue rules regulating behavior, but they function primarily to promote some designated activity, such as the support of health research. Examples of agencies concerned with human experimentation include the Canadian Department of Health and Welfare and the U.S. Department of Health, Education, and Welfare. In the latter case, two branch organizations are involved: the Food and Drug Administration and the National Institutes of Health. Since the mid-1960s, both have issued several interpretive rulings, derived from national legislation, regarding human research (Curran). Other federal agencies in the United States that support or conduct medical research have generally used the policies and regulations of these agencies as a model.

Administrative rules and regulations have the advantage of being relatively easy to change within a short time span. Moreover, a regulatory agency created for the single purpose of regulating a specific activity can concentrate its institutional resources on a particular problem area. But even precisely drawn regulations have not automatically resulted in effective agency control over the conduct of human research. In Canada, the Department of Health and Welfare has issued safety and ethical guidelines for human research on drug abuse, but places *sole* responsibility for maintaining the safeguards on the investigator and his employing institution (Canada). Review procedures at the national level, then, are inconsequential in guarding against abuse. Experience in the United States also raises questions about the efficacy of administrative remedies. Lax enforcement by authorities of the Food and Drug Administration and the National Institutes of Health has diluted the impact of past administrative actions. Insufficient manpower and dependence for information on the conduct of human investigations by those whom it regulates—the drug industry—have been some of the difficulties hampering the Food and Drug Administration's enforcement efforts (United States, Comptroller General). The National Institutes of Health's main mission of promoting a successful health research program works against whatever responsibility it has assumed for protecting human subjects. Because the values of agency officials are so closely aligned with those of the research community, there is a tendency for the former to bend to the wishes of the latter. The result is that the protection of human subjects is frequently subordinated to the demands of science.

Judicial remedies. Another strategy for developing controls for human experimentation is reliance upon the evolution of judicial decisions to establish gradually the limits of permissible research. But litigation concerning human experimentation is rare, leaving little case law to guide peer review committees or public policymaking bodies. Although several cases decided in England and the United States from the eighteenth through the twentieth century addressed the issue of human experimentation, they were concerned with the use of novel treatments by medical practitioners rather than with experimental procedures used by physician-investigators (Curran, pp. 542–545). Two recent cases, however, one Canadian (*Halushka*, 1965) and one in the United States (*Kaimowitz*, 1973), have dealt directly with the issue of informed consent in the context of experimentation. Of course, the courts have on a number of occasions considered the issue of informed consent in the context of medical malpractice. But whether the decisions regarding patient treatment will be applied similarly to nontherapeutic research remains uncertain.

While litigation does hold people accountable, its appropriateness and efficacy as a means for controlling the conduct of human research are questionable. The paucity of lawsuits in this volatile area raises the question of how well research subjects are prepared to recognize grounds for legal action and subsequently to take action on them. Recourse to litigation can be so cumbersome and expensive that subjects may be discouraged from taking that route. Moreover, judicial policymaking is limited by the inability of the courts to shape disputes or to select freely from among innovative policy alternatives, judges' lack of expertise in matters of medicine, and the absence of sufficient, knowledgeable staff to investigate the issues. Nevertheless, the judiciary does have a policymaking role to play through its power to rule on the legality and reasonableness of the procedures and decisions of other control mechanisms as well as of the actions of those participating in the research process.

Commissions and consultative bodies. Public commissions or advisory bodies are sometimes convened to study existing and anticipated social problems in order to provide policymakers with data and recommendations for their consideration. The appointment of such bodies removes the issue from the arena of polarized debate into an atmosphere where calmer deliberations can prevail. Two prominent examples pertinent to human research policymaking are the British Advisory Group established to study the ethical, medical, social, and legal implications of using fetuses and fetal material for research (Great Britain, Department of Health and Social Security) and the U.S. National Commission for the Protection of Human Subjects of Biomedical and Behavioral Research (National Research Act).

Whether national advisory mechanisms can be successful in influencing the direction of controls for human experimentation is an open question. Their typically short life span and a membership that operates on a part-time basis clearly reduce their usefulness. Too often the appointment of a commission is a political strategy for delaying action on a controversial issue. Since it must compete for power with other institutions in the political system, its recommendations are frequently ignored or criticized by public officials with the authority to make policy decisions. Yet, in the case of the National Commission for the Protection of Human Subjects, the requirement that the Department of Health, Education, and Welfare publicly respond to its recommendations may overcome the tendency toward obfuscation that has frequently characterized past handling of commission reports.

Compensation schemes. Despite all precautions that may be taken to protect research subjects from injury resulting from an experiment, it is inevitable that, in situations of uncertain but real risk, some harm will occur. Insurance schemes, which compensate subjects who suffer injury during the course of an experiment, may exert some controlling influence over the conduct of research. Reasons for initiating a compensatory scheme may include a desire to fulfill some moral obligation to subjects who assume the risks or to encourage persons to participate in research. But the control features of such a mechanism derive from its potential for causing those involved in conducting, reviewing, or sponsoring the research to analyze more carefully the risks and benefits of the proposed investigation and to give full consideration to the availability of alternative procedures for obtaining approximately the same results (Calabresi, p. 398). In this sense, insurance mechanisms are considered to be more than mere retroactive remedies for subjects injured by experiments; they are also viewed as a means of deterrence.

When the compensation of subjects has occurred in the past, as with the case of the survivors of the Tuskegee, Alabama, syphilis experiments, awards have been granted to subjects who have been willing to go through the arduous and uncertain process of litigation. In addition to the difficulties that this poses for subjects, litigation may also create an environment hostile to a productive health research program. If researchers are confronted with the threat of a lawsuit and thereby deterred from exploring innovative medical techniques, then society may be ultimately forced to forgo new advances in health care because both investigators and their institutions are unwilling to accept the risk of liability.

There is at least one existing law that incorporates an insurance scheme into its provisions relating to clinical drug testing. The German Drug Act of 1976 (Gesetz zur Neuordnung) requires that, if subjects are not already covered by private insurance, the sponsor of the research, usually a drug manufacturer, must establish insurance for the subject, which in the case of death or permanent employment disability must amount to at least 500,000 German marks (approximately $143,000). Perhaps the most frequently suggested compensation proposal is a "no-fault" insurance plan whereby compensation for injury would not be contingent upon assignment of fault, but would be granted merely upon demonstrating that harm resulted from the experimental procedure (Ladimer).

A number of difficulties will be associated with implementing compensation schemes (United States, DHEW, Office of the Secretary). Should the mechanism for financing the fund be privately or publicly financed, or should the two approaches be combined? In research directed at treating the ill patient, the task of distinguishing between harm induced by the experiment and that caused by the progression of the illness will surely be problematic. Finally, as a deterrent to unethical research a no-fault plan would be of little value if it became a license for investigators to engage in high-risk procedures. Recognizing this potential dilemma, some commentators have recommended that, while the decision to award compensation should not be contingent upon demonstrating fault, the possibility that fault was involved should not be ignored. Investigators who have been negligent would be subject to legal action by either the injured participant or the compensation fund (Adams and Shea-Stonum).

News media. The news media, most notably the press, can mobilize opinion for or against a proposed course of action without involving the inducements for individual compliance or the coercive sanction generally associated with other control mechanisms. By reinforcing or altering public attitudes toward medical innovations, the media can exert considerable influence on their development. At the initial stages of the heart transplant phenomena, for example, positive media accounts were instrumental in encouraging physicians to continue the experimental procedure (Fox and Swazey, p. 145). But the media's portrayal of the heart surgeon eagerly waiting for the next eligible donor to be declared legally dead aroused public skepticism and indignation. As public support waned, so did the number of physicians willing to proceed with heart transplant operations.

The ability of the news media to bring issues to public attention cannot be denied. In doing so, they perform an important function of educating the public about both the successes and the failures of medical research. But news accounts may also oversimplify or distort the issues; headlines tend to favor the dramatic. As a consequence, the public may become confused or be given a false impression about the "facts." In the end, such effects are a disservice to medicine as well as to the public it seeks to serve.

Conclusions

The planning and assessment of control strategies and mechanisms for human experimentation must be done within the context of an often inevitable trade-off between the desire to protect human subjects and the quest for improved medical techniques. One value cannot be vigorously pursued without limiting opportunities for achieving the other; hence, some degree of balance between the two is preferred. How well controls contribute to that end should be the criterion by which existing mechanisms and strategies are judged and new ones planned. Clearly, a mixture of professional and social controls is needed to produce accountability in human research. Standards of professional competence and performance must be judged, in large part, by the professional community. Moreover, the efficacy of social controls is often limited by the costs of time, manpower, and commitment required to implement them. Yet the need for public/quasi-public controls emerges from the recognition that human experimentation encompasses much more than technical considerations and that no one group is the sole "expert" with regard to questions concerning human values.

Whatever form they ultimately take, controls will, in the end, be no better than those responsible for implementing and enforcing them. Thus, an important benefit of the mechanisms and strategies outlined above is the guidance they provide to those involved in the research process. It would be shortsighted to view them merely as devices that place constraints on the clinical investigator. Rather, by clarifying areas of uncertainty they can help the researcher to resolve difficult ethical dilemmas and, as a result, generate renewed public confidence in and support for medical research.

MARK S. FRANKEL

[*Directly related are the entries* CODES OF MEDICAL ETHICS; INFORMED CONSENT IN HUMAN RESEARCH; MEDICAL EDUCATION; MEDICAL PROFESSION; RESEARCH POLICY, BIOMEDICAL; *and* RIGHTS. *For further discussion of topics mentioned in this article, see the entries:* COMMUNICATION, BIOMEDICAL; FETAL RESEARCH; HEART TRANSPLANTATION; *and* MEDICAL ETHICS, HISTORY OF, *section on* EUROPE AND THE AMERICAS, *articles on* BRITAIN IN THE TWENTIETH CENTURY *and* NORTH AMERICA IN THE TWENTIETH CENTURY. *Other relevant material may be found under* MEDICAL MALPRACTICE; *and* PRISONERS, *article on* PRISONER EXPERIMENTATION. *See* APPENDIX, SECTION II: DIRECTIVES FOR HUMAN EXPERIMENTATION (*entire section*); *and* SECTION IV, AMERICAN PSYCHOLOGICAL ASSOCIATION.]

BIBLIOGRAPHY

ADAMS, BERNARD R., and SHEA-STONUM, MARILYN. "Toward a Theory of Control of Medical Experimentation with Human Subjects: The Role of Compensation." *Case Western Reserve Law Review* 25 (1975): 604–648.

BARBER, BERNARD; LALLY, JOHN J.; MAKARUSHKA, JULIA LOUGHLIN; and SULLIVAN, DANIEL. *Research on Human Subjects: Problems of Social Control in Medical Experimentation.* New York: Russell Sage Foundation, 1973.

Board of Medicine, National Academy of Sciences. "Cardiac Transplantation in Man." *National Academy of Sciences: News Report* 18, no. 3 (1968), pp. 1–3. Reprint. *Journal of the American Medical Association* 204 (1968): 147–148.

CALABRESI, GUIDO. "Reflections on Medical Experimentation in Humans." *Daedalus* 98 (1969): 387–405.

Canada, Health and Welfare, Non-Medical Use of Drugs Directorate. "Health, Safety and Ethical Guidelines in Human Research on Drug Abuse." Ottawa: 1973. Insert in kit on research grants.

COHEN, HARRIS S. "Professional Licensure, Organizational Behavior, and the Public Interest." *Milbank Memorial Fund Quarterly* 51 (1973): 73–88.

COWAN, DALE H. "Human Experimentation: The Review Process in Practice." *Case Western Reserve Law Review* 25 (1975): 533–564.

CURRAN, WILLIAM J. "Governmental Regulation of the Use of Human Subjects in Medical Research: The Approach of Two Federal Agencies." *Daedalus* 98 (1969): 542–594.

DEBAKEY, LOIS. "Ethically Questionable Data: Publish or Reject?" *Clinical Research* 22 (1974): 113–121.

DERBYSHIRE, ROBERT C. "Medical Ethics and Discipline." *Journal of the American Medical Association* 228 (1974): 59–62.

DOW, ROBERT S.; GRIMM, ROBERT J.; and RUSHMER, DONALD S. "Psychosurgery and Brain Stimulation: The Legislative Experience in Oregon in 1973." *The Cerebellum, Epilepsy, and Behavior.* Edited by Irving S. Cooper, Manuel Riklan, and Ray S. Snider. New York: Plenum Press, 1974, pp. 367–389.

"Ethics in Human Experimentation." *British Journal of Nutrition* 29 (1973): 149. Report of the Working Group, European Association of Editors of Biological Periodicals.

Fox, Renée, and Swazey, Judith P. *The Courage to Fail: A Social View of Organ Transplants and Dialysis.* Chicago: University of Chicago Press, 1974.

Frankel, Mark S. "The Development of Policy Guidelines Governing Human Experimentation in the United States: A Case Study of Public Policy-Making for Science and Technology." *Ethics in Science and Medicine* 2 (1975): 43–59.

Gesetz zur Neuordnung des Arzneimittelrechts, 24 August 1976. *Bundesgesetzblatt*, issue 110, 1 September 1976, p. 2445. New German Drug Act.

Gray, Bradford H. *Human Subjects in Medical Experimentation: A Sociological Study of the Conduct and Regulation of Clinical Research.* Health, Medicine and Society; Wiley-Interscience Series. New York: John Wiley & Sons, 1975.

Great Britain, Department of Health and Social Security, Scottish Home and Health Department, Welsh Office. *The Use of Fetuses and Fetal Material for Research: Report of the Advisory Group.* London: Her Majesty's Stationery Office, 1972. The Peel Commission Report.

———, Medical Research Council. "Responsibility in Investigations on Human Subjects." *British Medical Journal* 2 (1964): 178–180.

Guttentag, Otto E. "The Problem of Experimentation on Human Beings: II. The Physician's Point of View." *Science* 117 (1953): 207–210.

Halushka v. University of Saskatchewan. 53 D.L.R.2d 436. (C.A., Sask., 1965).

Jaffe, Louis L. "Law as a System of Control." *Daedalus* 98 (1969): 406–426.

Kaimowitz v. Department of Mental Health for the State of Michigan. Civil Action No. 73-19434-AW. (Cir. Ct., Wayne Co., Mich., July 10, 1973).

Katz, Jay. "Correspondence: Editorial Rewritten." *Clinical Research* 22 (1974): 10–11.

———. "The Education of the Physician-Investigator." *Daedalus* 98 (Spring 1969): 480–501.

———, and Capron, Alexander Morgan. *Catastrophic Diseases: Who Decides What? A Psychosocial and Legal Analysis of the Problems Posed by Hemodialysis and Organ Transplantation.* New York: Russell Sage Foundation, 1975.

Ladimer, Irving. "Clinical Research Insurance." *Journal of Chronic Diseases* 16 (1963): 1229–1235. Editorial.

Merton, Robert K. "Some Preliminaries to a Sociology of Medical Education." *The Student-Physician: Introductory Studies in the Sociology of Medical Education.* Edited by Robert K. Merton, George G. Reader, and Patricia L. Kendall. Cambridge: Commonwealth Fund, Harvard University Press, 1957, pp. 3–79.

National Research Act. Pub. L. No. 93–348. 88 Stat. 342. 42 U.S.C. 289l-1.

Reback, Gary L. "Fetal Experimentation: Moral, Legal, and Medical Implications." *Stanford Law Review* 26 (1974): 1191–1207.

United States, Comptroller General. *Federal Control of New Drug Testing Is Not Adequately Protecting Human Test Subjects and the Public: Food and Drug Administration, Department of Health, Education, and Welfare: Report to the Congress.* Washington: General Accounting Office, 1976.

———; Department of Health, Education, and Welfare; Office of the Secretary. "Protection of Human Subjects." *Federal Register* 39 (1974): 18914.

———; Department of Health, Education, and Welfare; Office of the Secretary. *HEW Secretary's Task Force on the Compensation of Injured Research Subjects.* Publication no. OS–77–003. *Appendix A: Materials Cited in the Principal Report.* Publication no. OS–77–004. *Appendix B: Materials and Correspondence Not Cited in the Principal Report, But Relevant to the Work of the Task Force.* Publication no. OS–77–005. Bethesda, Md.: National Institutes of Health, 1977.

———; Senate, Committee on Labor and Public Welfare, Subcommittee on Health. *Quality of Health Care—Human Experimentation, 1973: Hearings Before the Subcommittee on Health of the Committee on Labor and Public Welfare.* 93d Cong., 1st sess., on S. 974, S. 878, and S.J. Res. 71, 1–4. Washington: Government Printing Office, 1973.

Veatch, Robert M. "Human Experimentation Committees: Professional or Representative?" *Hastings Center Report* 5, no. 5 (1975), pp. 31–40.

Woodford, F. Peter. "Ethical Experimentation and the Editor." *New England Journal of Medicine* 286 (1972): 892.

HUMAN NATURE, THEORIES OF

See Man, Images of; Environment and Man; Person; Mind–Body Problem; Free Will and Determinism; Mental Health Therapies.

HUNGER

See Food Policy.

HYPNOSIS

Concepts and principles underlying the ethical issues discussed in this entry can be found in the entry Behavior Control.

What hypnosis is

Hypnosis is a process by which one person helps another withdraw attention from reality and guides the other through fantasies that are experienced as real. These experiences may be so convincing as to be temporary delusions. When hypnotized, subjects tend more readily to accept without criticism ideas and plans suggested to them. Hypnosis is properly used when a competent psychologist, doctor, or dentist employs it to assist in such tasks as the recovery of disturbing forgotten memories, the alleviation of pain, and the reduction of anxiety.

Historical origins of misunderstandings

There is widespread misunderstanding of hypnosis, originating with its early history in Western Europe. That began with Mesmer (1734–1815), a physician qualified by the standards of his time, who claimed that cures he achieved by practices such as stroking were manifestations of a biomagnetic force. His claims were investigated, his theory discredited, his successes overlooked, and he was adjudged a charlatan. Because of official disapproval, the development of techniques for inducing trance and the exploitation of hypnotic phenomena were left to "quacks" and itinerant entertainers. Early explanations of the phenomena were couched sometimes in terms of demon-possession and transcendence of the natural, and at other times in terms of pretense and fraud. Late in the nineteenth century doctors again began to view hypnosis as a natural phenomenon and to make use of it, e.g., Bernheim, Janet, and Freud, but it was not until 1933 that there appeared the first book about it from an experimental laboratory, the one directed by Clark Hull at Yale. Something of the aura of disrepute surrounds it even today. Teaching it is still sadly neglected by universities; this has helped perpetuate misunderstanding of hypnosis even among professional practitioners.

The ethical concerns about hypnosis

Four kinds of concern about hypnosis may be distinguished. The first is the concern of society in general about whether hypnosis is potentially dangerous and should be regulated. In particular, there is concern about its power to compel behavior and distort experience. The second is that of professionals contemplating its use. For example, a marriage counselor wonders whether it is "right" to promote by hypnosis the particular decision he deems the wise one for his client. The third is the concern of those who contemplate being hypnotized. For example, a prospective patient doubts the morality of "surrendering" initiative and rationality in seeking hypnosis to stop smoking and asks advice from his spiritual adviser, who in turn has to face the issue. Finally, the fourth kind of concern is that of lawyers, judges, law enforcement officers, and politicians who ask whether hypnosis is always, never, or sometimes a proper procedure for obtaining confessions and evidence.

The concern about hypnotic powers

The power to compel behavior. The general social anxiety about hypnosis springs from the myth that the hypnotist wields such power that, with a few words, he can compel his subject to carry out almost any suggested action and accept almost any suggested idea. This myth has been fostered by unrealistic fiction such as *Trilby* and *The Manchurian Candidate*. The victim is pictured as changed, by a deceptively innocuous ritual, into a hapless puppet—still oddly conscious, but stripped of capacity to weigh alternatives rationally and of ability to give or withhold consent. If hypnosis did actually confer such power, society would be faced with a management issue of such magnitude as might require reconsideration of some basic ethical views. However, the myth is untrue.

The myth arose, no doubt, from early anecdotes and demonstrations that were intended to impress. Still, even competent experimenters, running well-controlled laboratory studies, were also mistaken. Rowland and, separately, Young thought hypnosis caused their subjects to comply with requests to pick up poisonous snakes barehanded and to throw fuming nitric acid at a laboratory assistant (Kuhn and Russo). However, Orne and Evans, with greater sophistication, replicated their work but also showed that nonhypnotizable simulators (subjects who had been strongly motivated by a coexperimenter to try to deceive the hypnotist into thinking they had been hypnotized) also carried out the same acts. Clearly, the simulators were confident that no responsible investigator would knowingly put someone else in jeopardy, and it is probable that the genuinely hypnotized thought so too.

It has become clear that hypnotic subjects, just as others, understand when an act is "only an experiment." To escape the artificiality of the laboratory context, some scholars have closely examined retrospective accounts of the use of hypnosis in real life to compel innocent, reluctant victims to behave in criminal or immoral ways. However, even when the accounts seem honest, they are ambiguous, since compliance can be interpreted as the result of personal relationships which, without hypnosis, might well have elicited compliance, or even as the result of inclinations already part of the person that required only the "excuse" of hypnosis to become manifest. Furthermore, there are equally convincing reports of subjects who, when pushed to behave in ways repugnant to them, come out of trance.

Of course, responsible investigators *do* have scruples. Their unwillingness to take a risk conceivably implies some uncertainty on their part, and it may fairly be said that the compelling power of hypnosis has not been finally tested—perhaps never can be. It can be said that no one has yet shown that hypnosis has ever induced anyone to do something where no other method could. Moreover, when the controls in experiments have been more thorough, then it has more clearly indicated that hypnotic suggestion is limited in its coercive power.

That is not to say that hypnosis has no power. Subjects are moved to carry out, posthypnotically, behaviors about which they are somewhat reluctant, sometimes indeed wondering about the origins of the impulses. Hilgard has shown that it is possible to overcome the resistances of some highly suggestible subjects who have undertaken beforehand to strive to resist suggestions to carry out some trivial act. Hypnosis can be used either to add some strength to, or subtract some strength from, a subject's impulses—both impulses of which the subject approves and also ones of which he disapproves. That is equally true of other forms of interpersonal influence, such as persuasion, eliciting promises, and offering rewards. These techniques also involve ethical problems, and the only point being made is that hypnosis is not a special case on that score.

The power to distort experience. It is quite clear that hypnosis may produce distortions. A procedure that can elicit hostility to a friend, lead to identification of an enemy as an ally, and falsify memories of the preceding hour cannot be dismissed offhand as ethically negligible. Those who report such phenomena are not pretending or lying. Good subjects really see what is not there, feel stuck to their seats, are deluded about where they live, and forget what they know.

These happenings are not exactly as they seem to a naive observer. Human beings every now and then, and sometimes willingly and wittingly, abandon themselves to temporary fantasies, as in watching movies or playing sports. On occasion, fantasies are experienced as if real and may have considerable impact on behavior and personality. The two modes or levels of mental functioning—one concerned with reality, the other with fantasy—usually alternate, but dominance of one does not imply total suspension of the other, and material from one percolates into the other. For example, a dreamer may realistically think: "This is only a dream." Similarly a hypnotized subject is neither continuously nor totally oblivious to the real world. For instance, he may genuinely have become blind, as suggested, to a particular piece of furniture in the room, and yet somehow blandly avoid bumping into it. Paradoxically, he must be seeing it, yet does not know his own seeing. Some measure of contact with reality is being retained.

Altogether then the belief that hypnosis confers on the hypnotist unlimited power to compel others to behave in any suggested way and to distort perceptions to the point of unchecked irrationality is a myth. The concern felt by some about any use at all of hypnosis is based on false ideas about it.

The ethical concerns of professional practitioners and their patients. Trance induction procedures are many and varied. Mostly they involve an initial voluntary abandonment of transactions with current reality, and the directing of attention to a monotonous sensory stimulus such as a blinking light or the sensations in one's own hand. Obviously these procedures are in themselves ethically neutral; the ethical problems arise in the uses made of established trance. Trance utilization includes giving short-term and long-term suggestions for the occurrence of various experiences and behaviors—ones that may in turn alter perceptions and attitudes, influence actions, and modify intentions; and it also includes eliciting memories, fantasies, and creative thinking from the patient in a context of diminished critical self-scrutiny. The ethical issues involved are those dealing with diminished rationality, with deception, and with the exerting of influence on others.

The problems of diminished rationality. Those who hold that the good life involves unrelenting struggle for the maintenance of maximum voluntary self-direction based on rational judgments would be bound to reject hypnosis on moral grounds. Few contemporary theorists take such a simplistic and uncompromising stand. Gormley, for instance, writing from a Roman Catholic viewpoint, argues that there are no compelling ethical objections to the serious use of hypnosis, as in therapy or research, given that the hypnotist is "known for his probity." He notes that unethical hypnotists may use the techniques to further evil ends (Gormley).

Many believe that fantasy not only is inevitable but also has uses and intrinsic merit. They hold that the goodness or badness of a particular fantasy depends on its content, on the circumstances of its occurrence, and on its effects. Thus hypnotically instigated fantasy, designed and used to undermine conscientious objections to marital infidelity would be deemed bad, but sexual fantasies planned to reduce neurotic anxiety over morally acceptable sexual behavior would be good.

The problems of deception. Some theorists believe further that the positive effects of imagination are strengthened when supported by *con-*

viction, and in practice try to win the belief of their patients. So, when a doctor prescribes a placebo, he does so in such a way as to have his patient believe it to be an active drug. The administration of hypnotic suggestions also contains both inadvertent and planned placebo components. A typical ethical defense of such intentional deception is that the end justifies the means.

Despite use of placebos, ethical conduct requires truth and frankness whenever possible without detriment to the patient. Prospective hypnosis patients should be fully informed about what is involved and should have the option of declining the treatment. They should be helped to understand that the process is a collaborative effort to become so involved in an experience that, for the time being, the question of its reality will probably not arise. They should be told that suggestions may be given to them and that hypnosis will increase the likelihood of positive responses, especially to suggestions of the hypnotist.

The increased suggestibility ought to make the hypnotist especially careful in framing suggestions and on guard about subtle cues that he may be communicating. He should also follow up his work, for one can never be quite sure how a suggestion has been interpreted. It is also desirable that both patients and their therapists understand that hypnosis does not make it easy to achieve long-lasting changes in behavior. Actually, hypnosis is a subtle instrument which can increase the effectiveness of interpersonal helping *only* when it is used with skill and sensitivity; and even then it has its failures.

The problems about influencing. The main use of hypnosis is psychotherapeutic. For example, Frankel outlined its value in treating phobias, and Hilgard and Hilgard in pain alleviation. There are ethical requirements in all patient–helper relationships: those of fair dealing, humane concern, responsibility for outcomes, consent, and confidentiality. They are extremely important in hypnosis. Additionally, in psychotherapy, there sometimes arise problems concerned with how the goals of treatment shall be decided. Because hypnosis appears to be a directive procedure, its use in treatment may seem to deny the patient the privileges and responsibilities of making decisions. The broad ethical problem is that of how far and by what methods it is right to influence the lives of others.

On these issues opinions are divided. One division is between those who pragmatically support a managerial and manipulative approach to the control of overt behavior, and those who value for themselves and others experiences of independence, critical inquiry, understanding, and spontaneity. Another division, related but not the same, is between those who are by intellectual conviction determinists and those who accept the doctrine of free will. Adherents of the different positions practice different kinds of psychological treatment. Some members of each position at times make use of hypnosis, but they do so in different ways. For instance, psychoanalysts, who are mostly determinists but nevertheless promote understanding, self-determination, and spontaneity, make use of hypnosis to facilitate the emergence into consciousness of emotional content which, behind the scenes, has been exerting a constricting and damaging effect on behavior and experience. Quite differently, a therapist convinced of the efficacy of convictions may use direct forceful suggestions, conveyed in trance, to inculcate views he thinks will alter his patient's neurotic life-style. Again differently, a behavior therapist may use hypnosis to guide his patient through experiences of successfully and calmly coping with situations about which he was previously phobic. Thus the ethical dilemmas in hypnotherapy are not particularly related to hypnosis itself, but rather arise out of basic issues that divide psychological healers into "schools."

Ethical concerns of legal personnel. There are uses of hypnosis in law. The main ones are in obtaining confessions or denials by suspects and in helping witnesses recollect evidence. The limited coercive power of hypnosis makes it unlikely that an uncooperative suspect will be hypnotized and, if he is, that he will directly divulge his guilt. Possibly, a suspect could be tricked into hypnosis and into experiences in which he might inadvertently give himself away. Such a procedure would be an unethical infringement of normal civil rights.

Hypnosis does, on occasions, facilitate the recovery of forgotten memories with cooperative suspects and witnesses. The impressive instances are where emotionally traumatic events, repressed from consciousness because of their horror, are revived, but forgotten minutiae also are sometimes recaptured. However, the notion of complete and perfect recall is another myth—it is simply not true that a person can be taken back in time to relive, with complete accuracy, the experience of a former occasion. More important, there is no guarantee that what is recalled in hypnosis is true. Even when a subject is sure a memory is correct, it may be false. Hypnosis does increase the amount of remembered material, but the increase consists of a complex mixture of fact and fantasy. Independent corroboration is essential in respect of all information gathered through hypnosis whenever its accuracy matters.

Other effects

It remains to consider "side effects" and "after effects." There are very few unanticipated consequences of hypnosis per se. Immobility may cause some muscle stiffness and, occasionally, transient headaches and dizziness occur. Very infrequently there may be reluctance to waken immediately, but there is no danger of staying in trance for a long time. A few subjects continue to carry on unnecessary posthypnotic behaviors or feel conflicted about them because suggestions have not been cancelled; an ethical hypnotist is careful about this possibility, and also follows up his cases. Subjects do not become progressively addicted to hypnosis, and it does not foster increasing dependency as a personality trait.

A hypnotized person may spontaneously produce vivid and emotionally laden memories—rarely in the laboratory, but more often where trance is used for purposes relevant to real-life concerns. The hypnotist should be professionally equipped to deal helpfully with such contingencies. In a few instances, hypnotic therapy has been followed by significant psychological disorder, such as severe anxiety and even paranoidal delusion. One can never be sure that these disorders would not have emerged in any case. What is indicated is the need for patients to be selected with professional competence and care. Three useful rules are that even minor signs of psychopathology, especially of psychosis, should be taken seriously; that a therapist should not attempt to treat with hypnosis a patient he could not cope with in some other way; and that particular caution should be exercised when the relationship is at all likely to be other than brief and relatively superficial.

Although anyone with a little training may learn to hypnotize, it is too serious a matter for stage entertainment and parlor tricks. Harmless in itself, it may be used injudiciously to bring about unpleasant and psychologically harmful experiences; for instance, direct removal of symptoms that meet psychological needs may have unfortunate results. For instance, the complete and unconditional abolition of a physical pain which happens also to be serving as atonement for guilt may precipitate a more severe expression of remorse. Therefore, it is desirable that the use of hypnosis be limited to trained persons such as psychologists, psychiatrists, physicians, dentists, and social workers. Short courses of training in hypnosis that are open to persons without basic training in the above or like professions, and which offer so-called diplomas, are to be deplored.

Conclusion

The induction of hypnosis is a rather neutral affair. Trance causes no harm, and likewise has no therapeutic efficacy in and of itself. There is no illness that requires it for treatment. There are many persons in whom it cannot be successfully induced. It is ethically neutral.

There are ethical problems in respect of how, and for what purposes, a hypnotic trance is utilized. For instance, an unscrupulous person may be able deceptively to use hypnosis as part of a strategy to achieve disclosure of a secret or a sexual seduction. Further, while hypnosis is not antitherapeutic, when misused it is capable of increasing damage caused by therapeutic ineptitude.

Its main applications are in the alleviation of pain; the treatment of psychosomatic disorders such as asthma and dermatitis; the control of habits such as nail-biting and smoking; counteracting destructive attitudes such as anxiety and inferiority feelings; securing increased application and effort as in studying and sport; and in the psychotherapy of emotional disorders. In these tasks it may be used in a number of quite different ways. There are considerable differences in ethical opinions about the relative acceptability of the different ways of using it. These differences reflect more general differences of opinion about how far and in what ways it is right to exert influence on others.

A. G. HAMMER

[*Directly related is the entry* BEHAVIOR CONTROL. *Other relevant material may be found under* CONFIDENTIALITY; INFORMED CONSENT IN THE THERAPEUTIC RELATIONSHIP; MENTAL HEALTH THERAPIES; *and* TRUTH-TELLING.]

BIBLIOGRAPHY

American Psychological Association, Ad Hoc Committee on Ethical Standards in Psychological Research. *Ethical Principles in the Conduct of Research with Human Participants.* Washington: 1973.

"Antisocial Behavior and Hypnosis." *International Journal of Clinical and Experimental Hypnosis* 20 (1972): 61–130. Special issue.

DU MAURIER, GEORGE LOUIS P. B. *Trilby.* New York: Harper & Brothers, 1895.

FRANKEL, FRED H. *Hypnosis: Trance as a Coping Mechanism.* Topics in General Psychiatry. New York: Plenum Medical Book Co., 1976.

FROMM, ERIKA, and SHOR, RONALD E., eds. *Hypnosis: Research Developments and Perspectives.* Modern Applications of Psychology. Chicago: Aldine-Atherton, 1972. See especially "Individual Researches within Specific Areas," pt. 4, pp. 481–571.

GORDON, JESSE E., ed. *Handbook of Clinical and Experimental Hypnosis.* New York: Macmillan, 1967. See especially Perry London, "Ethics in Hypnosis," chap. 18, pp. 591–612.

GORMLEY, WILLIAM JAMES. *Medical Hypnosis: Historical Introduction to Its Morality in the Light of Papal, Theological, and Medical Teaching.* Catholic University of America, Studies in Sacred Theology, 2d ser., no. 126. Washington: 1961.

HILGARD, ERNEST ROPIEQUET. *The Experience of Hypnosis.* A Harbinger Book. New York: Harcourt, Brace & World, 1968.

———, and HILGARD, JOSEPHINE R. *Hypnosis in the Relief of Pain.* Los Altos, Calif.: William Kaufmann, 1975.

HULL, CLARK LEONARD. *Hypnosis and Suggestibility: An Experimental Approach.* The Century Psychology series. New York: Appleton-Century Co., 1933.

KUHN, LESLEY, and RUSSO, SALVATORE, eds. *Modern Hypnosis.* New York: Psychological Library, 1947. Reprint. Hollywood: Wilshire Book Co., 1958. See especially Lloyd W. Rowland, "Will Hypnotized Persons Try to Harm Themselves or Others?" pp. 39–44; Paul Campbell Young, "Hypnotic Regression—Fact or Artifact?" pp. 56–63; Wesley Raymond Wells, "Experiments in Waking Hypnosis," pp. 45–55, and "Ability to Resist Artificially Induced Dissociation," pp. 75–87; and Milton H. Erickson, "Concerning the Nature and Character of Post-Hypnotic Behavior," pp. 105–142.

MARCUSE, FREDERICK L. *Hypnosis: Facts and Fiction.* Pelican Books, A446. Baltimore: Penguin Books, 1959.

ORNE, MARTIN T. "Hypnosis, Motivation, and the Ecological Validity of the Psychological Experiment." *Nebraska Symposium on Motivation: Current Theory and Research in Motivation Series* 18 (1970): 187–265.

———, and EVANS, FREDERICK J. "Social Control in the Psychological Experiment: Antisocial Behavior and Hypnosis." *Journal of Personality and Social Psychology* 1 (1965): 189–200.

———, and HAMMER, A. GORDON. "Hypnosis." *The New Encyclopaedia Britannica.* 30 vols. Chicago: 1976, Macropaedia, vol. 9, pp. 133–140.

SHOR, RONALD E., and ORNE, MARTIN T., eds. *The Nature of Hypnosis: Selected Basic Readings.* New York: Holt, Rinehart & Winston, 1965.

I

IMMUNIZATION

See PUBLIC HEALTH; HUMAN EXPERIMENTATION, *article on* HISTORY; MEDICAL ETHICS, HISTORY OF, *section on* EUROPE AND THE AMERICAS, *article on* NORTH AMERICA: SEVENTEENTH TO NINETEENTH CENTURY.

IN VITRO FERTILIZATION

See REPRODUCTIVE TECHNOLOGIES, *articles on* IN VITRO FERTILIZATION, ETHICAL ISSUES, *and* LEGAL ASPECTS.

INCARCERATION

See INSTITUTIONALIZATION; PRISONERS.

INDIA

See MEDICAL ETHICS, HISTORY OF, *section on* SOUTH AND EAST ASIA, *articles on* GENERAL HISTORICAL SURVEY *and* INDIA; HINDUISM; POPULATION ETHICS: RELIGIOUS TRADITIONS, *article on* A HINDU PERSPECTIVE.

INDUSTRIAL MEDICINE or INDUSTRIAL HEALTH

See ENVIRONMENTAL ETHICS, *article on* ENVIRONMENTAL HEALTH AND HUMAN DISEASE; PUBLIC HEALTH.

INFANTS

The first article, MEDICAL ASPECTS AND ETHICAL DILEMMAS, *offers an overview of medical problems in infancy, with a systematic presentation of the ethical issues and the ethical implications of medical decision making. The second article,* ETHICAL PERSPECTIVES ON THE CARE OF INFANTS, *offers a summary and critique of the major ethical positions taken in response to the ethical issues described in the first article. Public policy on whether medical care should be given to or withheld from infants is examined in the third article,* PUBLIC POLICY AND PROCEDURAL QUESTIONS. *The fourth article,* INFANTICIDE: PHILOSOPHICAL PERSPECTIVE, *explores the ethical question of infanticide—both historically and in the context of modern medicine—in more detail than is given in the second article.*

I. MEDICAL ASPECTS AND ETHICAL DILEMMAS
    John Michael Hemphill
    and John M. Freeman
II. ETHICAL PERSPECTIVES ON THE CARE OF INFANTS
    Warren T. Reich and
    David E. Ost
III. PUBLIC POLICY AND PROCEDURAL QUESTIONS
    Warren T. Reich and
    David E. Ost
IV. INFANTICIDE: A PHILOSOPHICAL PERSPECTIVE
    Michael Tooley

I
MEDICAL ASPECTS AND ETHICAL DILEMMAS

The ethical dilemmas in the medical care of infants are primarily the consequence of several developments: an increased ability to save the lives of infants who formerly would have died (Avery), directing concern toward their future quality of life; improved capabilities in diagnosing disease and disability in advance of their manifestations; and an attitude which takes seriously the dilemma of decision making by others on behalf of the infant.

This article will delineate the highly complex issues involved in decision making in infant care as seen from a medical perspective, including birth defects; problems when the outcome is unpredictable; experimentation with infants; and the management of infants selected for nontreatment. It will confine itself to an explanation of the medical context and ethical dilemmas in infant care and a description of the options, methods, and difficulties involved in resolving these dilemmas, as perceived from a clinical perspective. The article following this one will offer a more formal ethical inquiry into the issues raised in this article.

Therapy in the newborn with a birth defect

One may define four categories of birth defects, each of which raises different ethical considerations about quality of life: (1) static conditions associated with mental retardation; (2) progressive conditions associated with mental retardation; (3) static conditions associated with physical disability but normal intelligence; (4) progressive conditions causing physical disability but with normal intelligence.

These categories highlight two major variables which interact in the ethical considerations: the quality of life with mental and physical disabilities and the expected duration of life (Hemphill and Freeman). In this discussion static disease refers to mental or physical deficits that are already present and are unlikely to get worse. Some of these diseases will shorten life span to variable degrees. Progressive disease implies that mental or physical problems will get worse and usually that life span will be shortened. In progressive disease there is the possibility (current or future) of intervening and reversing or retarding further deterioration.

1. Static conditions associated with mental retardation. Intelligence cannot be determined in the neonatal period. However, many conditions can be identified at that period which are universally associated with retardation of variable degree; other conditions have a known probability of mental retardation.

Example A. A newborn with Down's syndrome (mongolism) is born with duodenal atresia (blockage of the upper intestine) and therefore cannot be fed. The operation to repair the intestine is relatively simple; if performed, the child will survive and will be retarded, with an IQ in the thirty-to-sixty range. He will never function independently but will relate to his peers and to his environment. If the operation is not performed, the child cannot be fed and will starve to death. Should the child be operated upon (Shaw; Gustafson)?

Example B. A premature infant has evidence of bleeding into the brain. After several days of vigorous treatment, she shows signs of severe brain damage; in all likelihood she will never be able to ambulate or communicate. When first fed at one week of age, she is found to have an intestinal obstruction like that in the baby in Example A. Should this baby be operated upon despite the prospect of profound retardation much more severe than that in Example A?

Issues. If either child did not have predictable retardation, one could reasonably anticipate that he or she would receive the best available medical care for the intestinal obstruction. The primary ethical question, then, is whether the future degree of retardation should enter into the decision, and if so, at what level of retardation one should "draw the line." A second issue pertains to the degree of certitude which is required concerning the future retardation. A third question is: If it is justified to withhold ordinary medical therapy (intestinal surgery) because of the retardation, would it not also be consistent to withhold normal care from a Down's syndrome baby who is physically well and requires no surgery? A fourth question, constant throughout this discussion, is: Who should decide?

2. Progressive conditions associated with mental retardation. A number of diseases are detectable in infancy which are either untreatable or, if treated, cause progressive mental impairment. Some metabolic diseases, such as phenylketonuria and maple syrup urine disease, are treatable with diet. If they are diagnosed and treated early, the child may have normal or nearly normal mentation. If these diseases are diagnosed late, varying degrees of mental retardation will result.

Other conditions, such as Tay-Sachs disease or metachromatic leukodystrophy (MLD), are currently untreatable and result in death in childhood. The duration of survival depends on the quality of the nursing care and the vigor of medical therapy when complications ensue. Other diseases such as Huntington's disease may allow years of normal life prior to the onset of progressive symptoms (Hemphill).

Issues. In addition to the questions raised in category (1) above, there are three major issues involved in these diseases with progressive symptoms. First, decisions in cases such as Tay-Sachs disease or in metachromatic leukodystrophy involve *the vigor of the struggle* against the inevitable early death. Questions of embarking on therapy are directed at intervening illnesses or complications of the basic disease processes. Should the child who can no longer swallow be tube fed, or should a gastrostomy (feeding by a tube into the stomach) be attempted to ease the patient's care? Are there "ordinary" procedures that should be considered heroic when the child has lost contact with the environment?

Second, as technology advances, *new forms of therapy* will be directed at the disease process itself. When new therapy becomes available, one option would be to treat advanced cases first, so that, if the therapy is unsuccessful or harmful, the harm will be done to those children with the poorest prognosis. This option could create further dilemmas, since the intervention might merely halt progression of the disease, preserving the most handicapped. An alternative option would be to direct therapy at the least affected, where halting the process will do the most benefit, but where complications of new therapy would do the most harm.

A third issue involves *early diagnosis*. If an infant is diagnosed as having Tay-Sachs disease on the basis of a screening test and develops an intervening illness, should the intervening illness be treated, if one knows that he will die of his disease in early childhood? A serious dilemma is caused by the fact that it is common medical practice to offer early detection and abortion of the fetus with these conditions, yet to vigorously treat the infant with the same conditions.

3. **Static conditions with physical disability but normal intelligence.** Birth defects take many forms. Many that are associated with physical disability and the prospect of normal intellect require medical or surgical intervention. Examples of such defects may include spina bifida, congenital limb amputations, craniofacial anomalies, the short gut syndrome secondary to infarction of the bowel, and exstrophy of the bladder. Other more common, but less severe, anomalies, include cleft lip and palate, major facial nevi (birthmarks), hypospadias (malformation of the urethra), and ambiguous genitalia. Still others, such as dwarfism, deafness, or blindness, require less medical intervention but may require major accommodations of life-style. Most children with these anomalies will have normal intelligence. Each will have a greater or lesser degree of physical impairment, and each in a different fashion. The question, then, is whether a physical defect, such as blindness, facial deformity, or the inability to walk is sufficient reason for not sustaining an infant's life.

Example C. A newborn has an omphalocele (intestines exposed outside the abdomen). The surgeon operates but finds that most of the intestine has died, and that after repair only ten inches of intestine are left, not enough to absorb adequate nutrition. The infant will have to be fed special nutrients by vein to survive, but she is normal in all other respects. There is a one percent, or less, chance that the remaining intestine will ever be sufficient to allow the child to be fed. The child can be maintained indefinitely, i.e., for months or years, by intravenous feeding, but this requires that she be in the hospital. There is a high risk of infection and other complications from the intravenous feeding.

Issues. In this case, life can be sustained only by artificial means. However with artificial feeding, the person is otherwise normal. The necessity of artificial feeding will, however, severely restrict the quality of that life. One issue is whether the physician should begin to sustain life when there is little likelihood that that life will ever become self-sustaining. A second issue is whether such a life should be sustained until the person is capable of making his own decision. A third issue is whether there are minimal abilities (i.e., breathing, circulation, feeding) which are required before further medical intervention is warranted.

Is there a difference in these types of cases between the infant and the adult? Are the ethical considerations involved in embarking on such therapy different for an infant who has had no conscious past, and who has no knowledge of his future, than for the adult who has a past, who can evaluate his future, and who can participate in the decision?

Example D. A newborn with a myelomeningocele (spina bifida) is found to have no movement of the legs and loss of sensation below the level of the umbilicus. The sac protruding from the back, containing the displaced spinal cord, is intact, the head is normal size, and there is not a major curvature of the back. No other anomalies are obvious. Should the child's back be surgically closed and full therapy undertaken (Lorber, 1971; Freeman, 1973; Shurtleff et al.)?

Expected outcome. If *vigorous* treatment is undertaken, the child will remain completely paralyzed in the legs and, while able to "ambulate" on crutches and braces up to the back, will be wheelchair-bound by the teenage period (Freeman, "To Treat or Not to Treat"). He has a ninety percent chance of developing hydrocephalus requiring a shunt which may need to be revised on multiple occasions. He will never achieve voluntary control of bowel and bladder, although devices and training may provide cleanliness. He is unlikely to have normal sexual function. He may require orthopedic operations on back, hips, and legs. He has an eighty percent chance of having an IQ above eighty. The cost of the medical care will be $50,000–$100,000 for his projected lifetime.

If *no* surgical treatment is undertaken and the child is provided with ordinary nursing care, the child has approximately a forty percent chance of being alive at three months and a ten to twenty percent chance of surviving one year or more (ibid.). The usual cause of early death is the development of meningitis through the open sac on his back. If the child survives sufficiently long and if the hydrocephalus goes untreated, he will become retarded and blind. Major deformities of spine and limbs will develop and interfere with care and sitting. Impairment of kidney function from incomplete bladder emptying will often lead to death in the teenage period.

Issues. The issue is whether it is ethical to withhold treatment of a nonprogressive physical disability because it will affect the future quality of life. If so, at what level of disability and what quality of life? Another issue is whether a decision to make the initial surgical intervention to prevent meningitis commits one to the total comprehensive treatment (shunting, further neurosurgery, orthopedic work, etc.), or whether the decision to support the life may be reconsidered with each new problem requiring additional surgical or drug intervention.

4. Progressive conditions with physical disabilities but normal intelligence. There are many examples of these conditions, including cystic fibrosis, some types of congenital heart disease, sickle-cell disease, polycystic kidney disease, muscular dystrophy, and Werdnig-Hoffman's disease.

Example E. Werdnig-Hoffman's disease is a disease of unknown cause, resulting in progressive degeneration of the motor nerve cells of the spinal cord. The affected children are of normal intelligence, but over the first year or two of life have increasing paralysis and frequent pneumonias which are the ultimate cause of death. The disease is inherited in an autosomal recessive fashion and is always fatal, usually by age two.

When such a child, age six months, presents with his first, second, or third episode of pneumonia, should it be treated? How vigorously? Should a tracheostomy be done so that the child can breathe better, even if it will mean that he spends the rest of his days in a hospital? Should one cure this episode knowing that the patient will have another episode of pneumonia in the near future?

Issues. The issues in these progressive physical conditions are similar to those considered under the progressive mental conditions but made more poignant by the fact that the intelligence of the child will be normal, and he or she will, therefore, be more aware of his or her increasing limitations. A further issue is whether extension of life for the infant is as important as extension of life for the adult.

Further complicating factors

Unpredictability of outcome. In the cases discussed above, the medical decisions are primarily concerned with future quality of life; the article following this one presents a variety of ethical methods explaining how decisions affecting infants ought or ought not be based on such qualitative considerations. Both assume that prognoses with or without therapy can be predicted. In fact, however, for many sick newborns without defined birth defects one cannot predict the outcome (Pildes et al.). Rapid advances in the care of the premature and the newborn have resulted in an increase in the number of infants who survive, but without advances in our ability to assess the quality of that survival. Controlled studies can easily measure the effects on mortality; studies of the effects on morbidity may take years. Sequelae or side effects of even short-term therapy may not be present until normal maturational processes make analysis of specific impairments possible. For example, therapy which interferes with visual or auditory processing may not become evident until the child is seven or eight years old and reading and cortical processing become testable. Furthermore, during the years before testing of the long-term effects of one treatment is complete, the therapy often will have changed, and the results of the long-term study may be of little consequence.

How vigorous should therapy be for the sick newborn who has an indeterminate prognosis? For the infant with difficulty breathing at birth the physician does not know if the child will be normal or intellectually impaired. Over time, and with appropriate long-term studies, a statistical prediction of outcome may be possible (Fitzhardinge et al.; Fisch et al.). How should statistical predictions affect decisions about therapy for the individual baby? If the chances are fifty percent that he will be profoundly retarded and neurologically impaired, should therapy be begun? At what statistical point should the line be drawn, and why?

In the newborn where time is critical, decisions are often made on "gut reaction" with inadequate information; some infants are, therefore, never treated. Would more children be saved if all such children were initially treated with vigor, and therapy terminated or active euthanasia allowed if after a defined period of time the poor prognosis became clear?

Experimental therapy in the newborn. One does not have a full appreciation for the dilemmas of therapy in infants unless one takes account of the fact that much of therapy in the newborn is experimental, although rarely conducted as controlled research. The physiology and metabolism of the newborn and premature are different from those of the older child and adult. Animal models and experience with the adult do not provide strict analogies to newborn disease or predictable responses to specific therapies. Therefore, the assessment of the efficacy and toxicity of therapy involves the administration to the newborn himself of new drugs or new procedures. Critical evaluation of such therapy involves controls in which some children do not receive the new therapy, and in some cases may not receive what has become established therapy.

However, unless such experimentation is carried out in a controlled fashion, each newborn receives randomized therapy based on the whim or prejudice of the physician, without the information about benefits that controlled studies could provide. Who can make decisions about such studies? Who can evaluate risks and benefits of specific therapy when both are unknown? Is it ethical to do such studies in the newborn? Is it ethical not to?

For example, a newborn with a bacterial infection may not develop any symptoms until the infection is widespread and rampant. At one time it was believed that when the mother's membranes had been ruptured for a prolonged period of time prior to delivery of the child, the risks of infection of that newborn, and subsequent death, were high. Chloramphenicol, a new antibiotic, had been shown to be safe and effective in older children and adults for combating infection with organisms that were common causes of death in the newborn. It became common practice to administer this antibiotic *prophylactically* to prematures and newborns born after prematurely ruptured membranes. A higher than expected death rate was noted in this group. Was it due to the drug?

A study was proposed in which some infants would receive this drug prophylactically, in a standard fashion, and others would receive no therapy until infection was apparent. The results of the study indicated that the children who did *not* receive any therapy were more likely to survive than those who received the drug. It was found that the dose of chloramphenicol which had been appropriate for the older child or adult was not appropriate for the newborn, because the newborn was unable to metabolize the drug properly, resulting in high levels of the drug and in death (Burns, Dodgman, and Cass).

While the ethics of medical experimentation with children is discussed in another entry [see CHILDREN AND BIOMEDICINE], several observations can be made about the question itself in the *treatment* of newborns. The chloramphenicol case raises the question whether a parent, physician, or committee should allow a control group of infants to be deprived of therapy generally considered to be appropriate. The irony of the experiment was that the infants who received less than "standard" therapy did better than those who received standard therapy.

Management of the untreated infant

A special dilemma affects any case in which the decision is made not to treat. If such a decision is the most appropriate, what would constitute the most ethical management of the untreated infant? Options for management of these infants in examples A and B who have an intestinal obstruction include:

1. To provide "usual" nursing care, but not to feed, and allow the child to starve and dehydrate. The expected duration of life will be seven to twenty-one days.
2. Not to feed but to sedate the child so that he or she will not cry too much.
3. To give intravenous fluids which will prolong the existence but only postpone death from starvation.
4. To give intravenous fats and proteins (hyperalimentation) which will maintain the nutrition and life for an indefinite period of time.
5. To provide "usual" nursing care, including feeding, knowing that the child is likely to aspirate and die.
6. To give intravenous potassium to terminate

the child's life quickly and avoid the period of starvation.

In deciding the appropriate management of the untreated infant, two dominant ethical issues arise. One question is whether distress and suffering experienced or anticipated during this period justifies active euthanasia; and if so, under what circumstances (Freeman, 1972; Cooke). A second concern is whether the decision not to treat is morally equivalent to euthanasia through active means (Freeman, 1973; Duff and Campbell, 1976) and whether *any* effort to maintain life is reasonable after the selection for nontreatment has been made. If a patient dies, is it ethically relevant whether the physician wanted him to die or just didn't want him to have to live with his handicap?

The quest for ethical guidelines

We have presented the dilemmas of decision making regarding therapy in the newborn. Distinct categories of defects, selected cases, and commonly experienced complicating factors have been presented to emphasize that each child is an individual and that there are many subtleties to be considered in decisions about initiating therapy, choosing among alternative therapies, and withholding therapies. The attempt to find ethically acceptable medical criteria for resolving dilemmas in the treatment of defective infants necessitates a reexamination of principles and claims as they are perceived in the practice of medicine.

Conflicting principles. In the contemporary setting of pediatric medicine, two ethical principles guiding the medical profession seem to come into conflict. First, physicians are dedicated to the principle of preserving life and overcoming disease, and second, they are taught the dictum *primum non nocere* (first, do not harm). At the time of Hippocrates when few therapies were available, harm could often be directly attributable to a specific therapy. Today, with advancing technology, harm may be far more subtle, a choice between potential outcomes. In general, physicians take "do no harm" as a given principle, and it does not enter the decision process except in a medico-legal sense. Most physicians concentrate, rather, on "doing what is best," maximizing benefit to the patient, the family, and at times to society. The subtleties and conflicts involve knowing "what is best" and for whom. Not infrequently, the question arises whether preservation of an individual's life is "best" (e.g., Freeman, "Psychosocial Aspects of Meningomyeloceles"; Duff and Campbell, 1973, 1976).

Conflicting interests. Considering the consequences of medical decisions and attempting to do what is "best," the physician must ask, "Best for whom?" In dealing with children, the competing interests may be even more striking than when dealing with adults. When these interests are in conflict, which of these interests should be decisive in the medical judgment? Those interests and some of the difficulties in determining them are described below.

Interests of the infant. Preservation of life and improvement of health have generally been identified with "the best interests of the patient." Many are now utilizing the concept of the future quality of life in judging the best interests of the infant (Duff and Campbell, 1973, 1976; Engelhardt, "Ethical Issues"; McCormick; Jonsen et al.).

This quality of life includes the degree of physical well-being, intellectual ability, and psychosocial adjustment and adaptation of the handicapped person.

Two observations can be made about the interests of the patient, one dealing with the infant's right (or claim) to life, and the other pertaining to the difficulty in judging what the quality of the infant's life will be.

Given the requirement to act "in the interests of the patient," are the interests of the infant any less than the interests of the older child or adult? For many there is a difference in perception of and attitude toward the fetus, the infant, and the older child (Tooley; Engelhardt, "Bioethics"; Langer; Fletcher, 1975). This is not a fixed, age-related difference, but a gradient from conception increasing with age after birth to some undefined point. The embryo unperceived and unfelt appears to have less worth than the fetus felt, but not yet seen. The newborn seen and perceived, but lacking established relationships, presents greater dilemmas than the fetus, but less than older children. Yet, within the first year, or perhaps even within the first six months of life the infant has achieved a standing equal to the older child. Something has apparently occurred during this period to account for this change in perception and attitude and its effects on medical decisions.

Others perceive a fundamental equality of value among human beings, regardless of their stage of development. This attitude would favor criteria that do not permit systematic age-related discrimination. Recognition of these two attitudes enables understanding of two different decision-making standards "in the interest of the patient."

Second, health-care professionals increasingly feel that, while life per se has value, the value of an individual life is related to its quality—i.e., the value of life for the individual is related to his or her own perception of the quality of that life. On the one hand, it is considered reasonable for an adult to accept or refuse treatment, on the basis of his or her own perception of the quality of life. Even when the adult is incapable of making his own evaluation and decision, others have some basis on which to project themselves into the patient's situation and to decide on his or her behalf. However, in the case of the infant who has never had, and will never have, a different quality of life, do we even have the *ability* to project? The *loss* of abilities or the *loss* of quality is quite different from never having had those abilities. Growing up paraplegic, blind, or deaf may be very different from *losing* motor ability, sight, or hearing at a later age. Does the retarded youngster or adult "suffer" over the skills and abilities he never had (Engelhardt, "Ethical Issues")? When we attempt to project ourselves into the newborn's situation are we really practicing self-deception? There is the very real possibility that physicians may be projecting themselves into the position of the parents who are grieving over the loss of the normal child they did not have (Fletcher, 1974).

Interests of the family. In general, most families put the interests of the child above their own interests. Occasionally families will consider the effect of the child on their economic situation, on their marriage, and on the other children and decide that these considerations outweigh the "interests" of the newborn. How the family perceives the complexities of the medical situation and its effects on the family is strongly affected by the information and prejudices of the person discussing the problem with them. Physicians in general have acted as the child's advocate but have "gone along" with the families' decision about treatment when that decision seemed reasonable. When the physician deemed that interests of the child were clearly different from the interests of the family, and when the family has opted for nontreatment, an ethical dilemma has occurred which has required involvement of committees or courts to achieve resolution.

Interests of the health personnel. Intensive and long-term care of the defective infant affects physicians, nurses, and hospital staff. The impact on the health professional can range from a feeling of fulfillment to minor inconvenience, to serious distress, to profound and demoralizing suffering. Should these potential feelings affect the decision to treat the defective infant or management decisions about the untreated infant?

Interests of society. The interests of society determine the attitudes of both physician and parents toward the defective child. In fact, the extent of the affected child's disability will, to a large extent, be determined by the accommodations and opportunities society offers. Choices, then, are limited by the constraints of public attitudes and resources. Conversely, because physicians and other decision makers make their judgments within the context of the (societal) resources that are available to them, public policy must go beyond the accumulation of individual medical decisions.

A medical decision to treat a retarded infant, for example, may well depend on the conditions in the local chronic care facility, or even to provide other living arrangements within the community will depend on how the society orders its priorities. If changes in living arrangements occur, the medical decision may well be altered. Thus, the medical decision both directly and indirectly reflects society's assessment of an individual's worth, whether determined by his economic productivity, by the joy he may bring to a few individuals, or by his enhancement of society's pluralistic attitudes simply by being cared for.

Decision making. Because the infant cannot make his or her own decisions, the judgment is left to a third party: parents, physicians, or some other individual or group. There is currently a struggle as to who among these should dominate. The following observations are presuppositions and cautions related to the question of agency in decision making.

The physician's decision can be laden with values, personal and professional biases, and a tendency to dominate decision making because of control of new information. Parents are not only prejudiced by their own value system and their desire for perfection in the child, but are usually deeply affected by feelings of shock, guilt, shame, and anger, and strongly influenced by the authority of the medical staff and the extent of their own contact with the baby. The first observation, then, is that an awareness of one's own system of values and prejudices is critical for all those involved in the decision-making process.

Second, it is often suggested that decisions should be made by parents and physician together, though not to the exclusion of other relevant parties (Duff and Campbell, 1976). It is important to face squarely the issue of how these decisions should be made when the interests of the decision makers themselves are in conflict—whether through court review or simply through establishing a priority of responsibilities.

Third, there is the question of responding to the decisions of others, especially in deciding whether to challenge decisions in cases of apparent neglect. The physician and, to some extent, also the hospital administrator, the nurse, and others must decide whether to question the decision of parents (for example) who are obstructing treatment, not for the good of the child, but because they do not want to raise such a child, even when other solutions may be available.

Finally, decisions made on behalf of infants must be scrutinized even more carefully than for other patients. As we have noted, the infant's perceived value to others may be less, and his preferences may be incorrectly assumed to be identical to those of parents and physicians.

Summary

We have presented the ethical dilemmas of therapy in the newborn in the framework of selected cases to emphasize that each child is an individual, and that there are many subtleties to be considered in the decision to embark on therapy and in the choice of therapies. These subtleties include independent variables of life span, physical disability, predicted intelligence, age of onset, and, perhaps most significantly, the often tragic nature of all the available options. Decisions must be ethical, but the most ethical choice may be unclear.

We must further realize that the therapy we give may be experimental and unpredictable and, therefore, scrutinize its effects not only on survival but on quality of survival.

To a large extent ethical dilemmas in the newborn are the result of advancing technology. We are now able to save infants who formerly would have died, but in some cases the survivors will have questionable quality of life (Avery, 1975). Now that we can save them, we must face the decision: Should we? As technology advances we will be able to save the twenty-week fetus, nurture the ten-week embryo, culture the blastocyst and fertilize the ovum. Ethically, how will we decide which, if any, of these will receive the "benefits" of technology?

JOHN MICHAEL HEMPHILL
AND JOHN M. FREEMAN

[*Directly related are the other articles in this entry;* CHILDREN AND BIOMEDICINE; *and* PAIN AND SUFFERING. *For further discussion of topics mentioned in this article, see the entries:* DEATH AND DYING: EUTHANASIA AND SUSTAINING LIFE, *article on* ETHICAL VIEWS; DECISION MAKING, MEDICAL; *and* LIFE. *For discussion of related ideas, see the entries:* ABORTION, *articles on* MEDICAL ASPECTS *and* CONTEMPORARY DEBATE IN PHILOSOPHICAL AND RELIGIOUS ETHICS; ACTING AND REFRAINING; FETAL RESEARCH; MENTALLY HANDICAPPED; PATERNALISM; *and* PRENATAL DIAGNOSIS.]

BIBLIOGRAPHY

[The bibliography for this article is combined with those for the following two articles and can be found under article III below, PUBLIC POLICY AND PROCEDURAL QUESTIONS.]

II
ETHICAL PERSPECTIVES ON THE CARE OF INFANTS

Introduction

This article presents and evaluates contemporary ethical views of physicians and ethicists regarding moral dilemmas in the care of infants. Although the ethical discussions do not respond to all the subtle clinical dilemmas presented in the previous article, they present a variety of conceptual and ethical frameworks for dealing with many of these questions. Some aspects of these ethical problems are unique to the situation of infants, but many pertain to general ethical issues on the protection and care of human life.

The article has two parts: (1) ethical positions in the criteria for treatment found in the medical literature and (2) ethical theories applied by philosophers and theologians to the care of infants. The latter part is subdivided into the following sections: (a) traditional deontological positions; (b) contemporary positions emphasizing a rethinking of the concept of "person"; (c) consequentialist positions; and (d) approaches rejecting the "humanhood" standard and proposing either an ethic of care or an ethic of avoiding harm (a negative formulation of the consequentialist position).

Ethical criteria in medical practice

For a survey of ethical views among physicians in the 1970s regarding the care of defective infants, one can look to the rather intense controversy in the medical literature concerning criteria for treating, not treating, or terminating babies afflicted with spina bifida with meningomyelocele (see *Example D* in the previous article). There is a growing body of medical writings that is distinctive insofar as the authors suggest moral values or principles as reasons for determining the "medical criteria." The three positions summarized below, taken from the medical literature, though understandably less concerned about ethical theory, represent the same range of views one finds among ethicists and can be evaluated on the basis of general ethical theory.

One position, that of Zachary, relies on the religiously and philosophically based ethical principle that the direct and deliberate killing of a human being is wrong—a principle which also excludes encouraging a baby to die, either by giving no treatment at all, e.g., no feeding, or by not treating complications such as infections. Zachary adopts the positive principle that the baby should be "encouraged to live," implying that this support of life is to be by "ordinary" care. This strong inclination to early treatment for all spina bifida patients does not, however, lead to a policy of universal application of all available means of treatment. An important medical presupposition is Zachary's view that the purpose of early operation in meningomyelocele cases is not to save the child's life, as though the child would die if the operation were not undertaken: many survive without any spinal operation, and some die as a result of it. The ultimate criterion, then, is "whether there are advantages of early operation which outweigh any possible extra risks that such operation might have for the life of the child" (Zachary, p. 274).

Citing some undesirable long-term results in patients receiving full and vigorous treatment—severe mental and physical disability and questionable benefit in terms of survival rate and other risks to life—John Lorber advocates a second perspective: selectivity for treatment (Lorber, 1971). Five "adverse criteria" indicative of the future disability of the infant, particularly the site of the spinal lesion and the degree of paralysis, should be applied at birth with "the utmost strictness" (Lorber, 1973, p. 204); infants with any one or any combination of the five criteria should not be given active treatment (Lorber, 1974). Lorber intends his criteria to exclude from any special treatment beyond custodial nursing care all patients whose prognosis is for an unacceptable quality of life, e.g., those who will have severe physical handicaps such as might require long braces, crutches, or wheel chair for locomotion (1972, p. 856). Lorber points to various value considerations as possible justification for the use of his criteria: negatively, to avoid suffering and hardships for the patient, e.g., repeated operations and hospitalization, and excessive cost (1971, p. 228; 1973, p. 202). Positively stated, the quality of life should be consistent with self-respect, happiness, and marriage; an important social criterion is the ability to "earn their own living in competitive employment and be self-supporting with a secure, independent place in society" (Lorber, 1972, p. 867).

Other experts (cited by Freeman, 1973) find a high percentage of children who would not have qualified for treatment by Lorber's standards, who had an IQ of 80 or above and who had moderate to mild disabilities. Thus, variations on the selectivity position appear, depending partly on clinical success but also on values. For example, Shurtleff is more tolerant of patients Lorber would have excluded because of *physical abnormalities* as known from neuromuscular evaluation; instead, he regards *brain function* at a socially acceptable level as the most important human criterion. Medical criteria for maximum therapy include assessment for brain mass and function and absence of a major malformation that would preclude self-care as an adult. A principle is cited: "The emotional and physical pain we cause child patients to suffer must be justifiable in terms of some reasonable hope for a future happy social adjustment" (Shurtleff et al., p. 1010).

A third position is represented by Freeman, who is more optimistic than many about both the quantity of survivors and the improved quality of their lives following vigorous treatment, though at great social and financial cost (Freeman, "To Treat or Not to Treat"). His concern is for those selected for nontreatment. Since passive euthanasia, with its slow death and suffering for patient, family, and professional personnel, is not humane, and since active euthanasia is illegal, vigorous treatment in virtually all cases of meningomyelocele is the best practical policy. But active euthanasia would sometimes be the most humane policy for the most severely affected infants who are "selected for nontreatment." It can be morally justified for the reasons that would have justified abortion of the same infant as well as those that would justify nontreatment (Freeman, 1972, 1973, "To Treat or Not to Treat"). Duff and Campbell (1976) go further, suggesting that killing the severely afflicted meningomyelocele infant can be an obligatory part of the physician's task. They cite person-oriented philosophical concerns but offer no medical criteria.

Several observations can be made about these views (Reich, "Quality of Life," pp. 495–501; Veatch, 1977). First, determining the medical criteria for treatment and nontreatment is not just a technical question of diagnosis and prognosis: it is a value question, and the medical literature increasingly includes an ethical dimension. Second, there is a need to offer more precise ethical justification for the medical criteria. For example, desired results are *asserted* by both Lorber (ability to be self-respecting and self-supporting) and Shurtleff (ability to care for self), but clear theoretical reasons are not offered as to *why* the value of life should hinge on a certain level of physical ability, in the one view, or mental ability, in the other.

Third, Zachary's position can be criticized for not giving sufficient attention to suffering as a factor in establishing criteria. On the other hand, the objection has been brought against Freeman's view that the physician's or nurse's interest in being free from distress or suffering should not be regarded as a greater claim than the infant's acknowledged right to life (Cooke). These difficulties are "hazards of the profession": caring for the sick and dying will always produce strain (Robertson, 1975, p. 260).

Fourth, there seems to be strong indication in much of the medical literature (Lorber, 1973) and even in an ethical report ("Ethics of Selective Treatment") that the values determining criteria for treatment are the values perceived by the medical expert. Concern has been expressed about the need to include the values of parents, society, and other more universally grounded values (Veatch, 1977).

Fifth, there is need for more frankness about medical practices designed directly to cause the early death of the infant. The medical literature seems to indicate that the debates on survival ability of untreated infants may be strongly influenced by whether those infants are sedated so they do not cry and then are fed only on demand (Lorber, 1972, p. 871; 1973, p. 203; Freeman, 1973, pp. 136–138; "To Treat or Not to Treat," pp. 16–17).

Ethical theories and the care of infants

Traditional deontological positions. Here we attempt a synthesis of those views that regard all human life as a sacred trust, to be protected by two complementary rules: one must not directly and deliberately kill an innocent human being; and there is an obligation to sustain human life, particularly incumbent on those who have the responsibility of caring relationships. Characteristic of the Judeo-Christian and some philosophical traditions, this approach is deontological in that it holds that there are obligatory limits on what may be done because of moral claims that are not simply dependent on the anticipated consequences of such actions. Our obligations are not determined principally by the wants and needs of parents and physicians or, principally, by the consequences of the child's living out its life (Gustafson).

Moral value of the infant's life. The two duties of not killing and of sustaining life are rooted in the concept of the inherent worth of the individual and the principle of the essential equality of all human life, which rejects discrimination as regards protection and care of humans of any age on the basis of personal assets, status, or class.

These two rules are principally supported by the arguments that (1) every human has an inalienable moral *right* to life and hence to receive protection and minimal care, and (2) the *good* of life gives rise to an obligation to respect life in these ways. Defenders of this position also offer auxiliary arguments based on considerations of *good consequences*—that such rules produce the most beneficial consequences for mankind (justice, peace, and human dignity)—and on the need to avoid the *bad consequences* of infant euthanasia: the danger to all weak or defective classes of humans if society should permit the taking of lives regarded as expendable, the danger that the health-care professions might abandon their traditional protective attitude toward life if they adopt the practice of taking human lives of whatever age, and the destructive consequences for an ambience of trust so necessary at the beginning of life for the healthy development of the child if a policy of thorough caring for life were abandoned (cf. Bok, 1974, p. 33; John Fletcher, 1975, p. 77; Weber, p. 74).

Common to this group, for purposes of our interests, is the belief that this twofold obligation clearly applies to all infants, even those suffering crippling defects, for beginning at least at birth every human individual has a right to live. In other words, for this group every infant is a person in the sense of being a bearer of rights that are to be recognized by others and protected by society. (Most in this group would place the moment at which an individual deserves protection and care much earlier. Views within the group range from conception to quickening or viability [see ABORTION].)

Limits and exceptions to the rules. In the traditions being considered here, no serious moral theory has held that existing human life is the absolute value; life is not the greatest good nor death the greatest evil. While one ought always to respect human life, the rules to protect and sustain life have limits and exceptions because of other real human goods and conflicting obligations.

For example, in Jewish law the obligation to preserve human life is suspended if it conflicts with *higher duties*—the prohibitions against idolatry, murder, and certain sexual offenses. Further, since Jewish religious law does not regard the quality of life preserved a relevant moral factor, it is inclined to favor aggressive treatment of defective newborn children regardless of the extent of their impairment (Bleich).

According to some authors, there are *rare exceptions to personhood*. Some infants may not be subjects of rights deserving protection and care, even though they have passed the point in life when humans ought to be regarded as such: They are so severely malformed that they may not qualify as minimally human for those purposes. Examples given are the anencephalic infant (Smith) and genetic "monstrosities" (Gustafson). Active or passive euthanasia of these nonpersonal infants would be morally permissible.

Another basis for restricting our obligations in this area is found where the *conditions necessary for the duty are absent*. For Ramsey, the duties to protect and care for human life are seen in the larger framework of a duty never to abandon care. If sick persons of any age cannot receive our care, either because they are permanently unconscious or are experiencing intense, non-relievable pain, positive euthanasia would be morally permissible (Ramsey, "On (Only) Caring for the Dying").

The most widely discussed factor limiting the positive duty to sustain life is the *extraordinary means* principle, which finds relevance in the distinctions between killing and allowing to die and between prolonging life and prolonging dying. Some authors view this principle as providing a helpful ethical framework for dealing with obligations to the infant (Ramsey, ibid., pp. 113–164; Reich, "Testimony," "Quality of Life," pp. 505–509; Weber).

While the *term* "extraordinary means" is used with a great number of meanings, we restrict this discussion to the classic extraordinary means position, with its moral presuppositions, as it has been discussed most commonly among Catholic authors (e.g., Kelly, pp. 128–141) and contemporary Protestant thinkers (Ramsey, "Cn (Only) Caring for the Dying"; Gustafson, pp. 546–557).

One presupposition to this distinction involves the significance of *omission*. An omission leading to death is not regarded as in itself blameless while commission is wrong; not giving or taking what is necessary to preserve life is as wrong as active euthanasia when the omission is not justified (Kelly, p. 128). In traditional Catholic ethics and for some contemporary philosophers, the justification of omission is seen as different from that of active killing because of two key factors. (1) A *positive* duty of doing good obliges only when certain circumstances are present while a *prohibition* of evil by its very nature obliges always unless exceptions can be justified. (2) It is much more likely that there be a proportionate reason for *not* performing a life-sustaining act, e.g., in achieving other goods and in avoiding the unreasonableness or harm that may be involved in the life-sustaining effort. A second presupposition is the principle of double effect, particularly its requirement that the death of the patient may only be indirectly intended as the by-product of a deliberate omission whose direct intention is some other goal justified by a proportionate reason.

The extraordinary means principle states that there is an obligation to use ordinary means to preserve life, but no strict obligation to use extraordinary means. From an ethical perspective, usually no significant distinction is made between discontinuing and not starting a treatment that is judged nonobligatory. Extraordinary means are those that cannot be obtained or used without excessive expense, pain, or other inconvenience, or which do not offer a reasonable hope of success (Kelly, p. 129; Veatch, *Death, Dying*, pp. 108–109). The principle focuses on two elements—no reasonable hope of success in *medical* terms and excessive hardship caused by the *treatment*—thereby avoiding judgments on the worth of a particular *kind* of life in itself. Because of the prudential nature of the judgment, it is presumed that the *patient* normally judges what is extraordinary and nonobligatory. Because the baby cannot make his own decisions about treatment, objective standards must be cautiously applied. This principle can be applied to the care of the infant in the following way (cf. Weber, pp. 89–98; Reich, "Quality of Life," pp. 505–509):

1. *Reasonable hope of success for the infant.* If, e.g., surgery can "bring the baby through the crisis" to a relatively self-sustaining situation and if the baby can be kept alive for a not inconsequential period of time without excessive burden, there is an obligation to treat. On the other hand, the dying child may or even should be permitted to die as treatment becomes futile or harmful or does not reasonably enhance the prospects for life. Most would hold that death must be imminent—when dying has taken "irreversible control" (Ramsey, "Reference Points in Deciding," p. 95) or if only a few days or hours of life would result from treatment (Robertson, 1975, p. 237). Some would extend this time period to a few months or even a year (Weber, pp. 90–91). An infant who is not expected to survive childhood is not by that fact alone to be considered already dying.

2. *Excessive hardship for the infant.* First, if a single procedure or a more prolonged comprehensive treatment involves excessive pain or risk, it is not obligatory. Second, if the battle to survive at the same time suppresses higher values—e.g., peace, personal communication, a growing atmosphere of love—for a long period of time (perhaps a year), that would not be an obligatory burden. These burdens might include some combination of repeated surgery, lengthy dependence on life-support devices, prolonged hospitalization and absence from home, severe mental or physical handicap anticipated as a result of lifesaving surgery, etc. Within the framework of values and moral rules characterizing the traditional "extraordinary means" ethic, infants would not be excluded from treatment on the basis of the *kind of defect* they have, whether physical, mental, or relational, but only insofar as the comprehensive treatment causes *excessive hardship*. And a treatment that sustains life in a handicapped condition does not necessarily by that fact cause excessive hardship. Some of these hardship factors are present in more extreme cases of the meningomyelocele disorder described in *Example D* in the previous article. On the other hand, as regards the affected infant, treatment would seem obligatory for the Down's syndrome baby with duodenal atresia (*Example A* in the previous article), since surgery offers reasonable hope of success, the surgery itself cannot normally be regarded as extraordinary (Gustafson, p. 546), and the hardship for the infant would not normally be excessive.

3. *Excessive hardship for the family.* The extraordinary means ethic is strongly inclined to judge excusing factors of hardship principally in terms of the child's well-being. It assumes that some unavoidable distress or suffering is appropriately embraced in fulfilling one's duties to care for the life of one's children (ibid., p. 555). Still, if the prospect "overtaxes the ordinary powers of man" or of this family in particular, the treatment can be extraordinary (nonobligatory) for them; but other persons and resources should be sought out to care for the child. Authors have mentioned excessive cost—e.g., jeopardizing the education of the other children for the benefit of preserving the afflicted child's life by a few years—or even excessive emotional hardship as excusing factors.

4. *Excessive hardship for society.* The extraordinary means ethic has traditionally viewed the limits of the duty to sustain life in terms of burdens on specific individuals. While "burden on society" created by treatment of the child is a slippery slope, which some feel does not excuse from the duty to treat (Weber, p. 97), it seems that norms of distributive justice would have to be applied in determining whether society must assume the burden of providing *every* possible form of lifesaving treatment, even the most expensive, for all diseased and defective infants.

The extraordinary means standard is relative. The attempts among Catholic moralists, particularly in the 1950s, to identify *specific* medical procedures, indications, and costs as extraordinary have been largely abandoned because of rapid advances in medical science, changes in the economy of health, and especially the inadequacy of predetermined casuistic answers to what must essentially be a prudential judgment.

Objections to these views. Objections against the positions reported in this section include the following: (1) It is questionable whether moral rights can be attributed to very young children, for rights seem to require that one be able to claim value for oneself (Hauerwas, 1975, p. 226; Engelhardt, "Bioethics and the Process of Embodiment," p. 495). (2) The right to life includes the right not to be killed but not the positive right to treatment. (3) Whether life is good in a way that ought to affect our behavior depends not on some universal notion of life as a good, but on whether an individual life is good on balance. (4) The difference between killing and allowing to die is of little moral consequence: Both have the same goal or purpose—the best death under the circumstances—and both practices have the same effect, the death of the patient. Indeed, it is more difficult morally to justify allowing someone to die a slow and painful death than to justify causing a quick death in the same circumstances (Joseph Fletcher, "Ethics and Euthanasia"). (5) Those who would allow few or no exceptions to the obligation to sustain life do not take adequate account of the implications of human suffering and the wants and desires of other parties. (6) The uncertain way of regarding anencephalic infants and "monstrosities" indicates that many in this tradition do not have a satisfactory theory of embodiment as it applies to the human person. (7) The extraordinary means ethic is not very helpful, for it is surrounded by ambiguities and based on relative assessments. What is "excessive hardship" for infant or family, and according to what concepts of suffering is it to be judged? According to what norms of efficaciousness does one judge that treatment is successful or futile?

Rethinking the meaning of person. The term "person," as it has traditionally been used in ethics, is equivalent to "subject of rights." In more recent bioethical reflection the concept of "person" is being rethought by some contemporary philosophers in response to such issues as abortion, criteria for determining death, and euthanasia for defective infants. While the following two theories contain elements of various ethical systems, one distinguishing feature they have is a starting-point which draws distinctions between "human" as a biological category and "person" as a moral status, such that mere membership in the species Homo sapiens is not sufficient to secure the moral status of "person."

A strict definition. One modern account (Tooley) holds that, however else one may characterize "person," the concept entails "having a serious right to life." Thus, with respect to defective infants, this view focuses on the question whether denying life-sustaining treatment to such entities (or actively killing them) constitutes the violation of the right to life of a person.

Having a right to something means (1) that one be capable of desiring that thing; and (2) that merely having this desire places other individuals under a prima facie obligation to refrain from denying its fulfillment. Both conditions are necessary, for, e.g., if I do not desire to be free or am incapable of having that desire, then confining me cannot logically constitute a violation of my "right" to freedom.

The key element in this formulation is, of course, the element of desire. Although we often speak casually of desiring something—an apple, a car, a certain status in life—what we are fundamentally doing, on analysis, is desiring that some particular proposition be true (ibid., p. 70). But to desire that a proposition be true I must first understand the concepts involved in it. In other words, to have a right to something entails having the minimal conceptual equipment to be capable of desiring that thing. A normal adult person can certainly conceive of himself as a self-conscious subject of experiences and is capable of desiring that he should continue to exist as such a subject in the future. An infant, however, is incapable of having such a desire, for the infant lacks the requisite concepts. Consequently, this argument would conclude, an infant is not a person having a right to life that would be violated by active or passive euthanasia. Once the infant has developed the necessary concepts, particularly the concept of himself as an enduring, self-conscious subject of experience, however, he would possess such a right.

Some critical reflections can be offered on the Tooley thesis. First, the denial of personhood to the infant that renders a nontreatment decision morally permissible carries with it a number of difficulties. It must be stressed that the accounts presented apply equally as well to normal, healthy infants as they do to defective infants: It is morally permissible to kill healthy infants if it is desirable on other grounds to do so. Such an act is no more morally reprehensible than to kill a newborn kitten and requires no more serious justifying reasons. On these grounds, it might be more seriously wrong to draw blood from an infant for nontherapeutic purposes than it would be actively to kill the infant, for Tooley's argument suggests that, while an infant may not have a serious right to life, the infant may well—along with animals—have a right not to be subjected to unnecessary pain. There is something counter-intuitive about a theory which requires a serious justifying reason for subjecting an infant to the prick of a hypodermic needle, yet permits us to take that infant's life on less serious grounds.

Second, this analysis entails that, while animals do not have a serious right to life, they may well have other rights that must be respected, just as persons do (Tooley, p. 91). Such a formulation, by equating the criterion for personhood only with having a serious right to life (rather than with having rights generally), not only drives a wedge between personhood and rights but also raises serious questions about adjudication between animal rights and human personal rights where the two are in conflict. On this basis, a good deal of animal experimentation would constitute a violation of rights.

Third, it has also been suggested that the conclusion, "Infanticide is morally permissible," does not follow from the proposition that infanticide does not constitute the violation of a person's right to life, for some actions may still be morally wrong without violating anyone's rights (Benn, p. 97).

A broader concept of person. Even if the infant is not a person in the stringent sense outlined above, i.e., as a bearer of rights (in particular as bearer of a serious prima facie right to life), there may be other, less stringent conceptions of personhood the conditions for which an infant might well fulfill. Engelhardt ("Bioethics and Process of Embodiment") offers an analysis that does not deny value, even very high value, to entities that are not self-conscious. We value a human fetus, on the whole, much more highly than we value the fetuses of other species; and we do so on the basis of the human fetus's potential to become a person. While it does not have rights as such, the fetus may possess high value. Alternatively, depending upon circumstance (e.g., a fetus with a serious genetic defect), the fetus may have considerable disvalue.

Infants, however, while they do not have rights in the strict sense, are qualitatively different from fetuses in that they are capable of assuming and sustaining a social role in human relationships: the role of "child" (Engelhardt, 1974, pp. 230–232). In virtue of this social role, rights are imputed to and obligations assumed as regards infants. In brief, infants are treated *as if* they were persons, although in the strict sense they are not.

With regard to the decision whether to treat defective infants, this view suggests (1) that such infants do not have a right, in the strict sense, to their lives, and consequently the right to life of a person is not violated in the decision not to treat; (2) that, because of the special social status and high value of infants, the decision not to treat is a decision requiring serious justifying reasons, though those reasons may be wholly consequentialist in nature; and (3) that a significant factor in deciding to treat should be the potentiality of these infants to attain human personal life.

From the perspective of the infant (or the person that the infant might become), one might further argue that there can be, under some circumstances, an obligation *not* to treat the infant. Such an argument would rest on the claim that there are some conditions of life—e.g., an existence characterized by severe and continuous pain—that transform continued existence from a blessing to an injury (Engelhardt, "Ethical Issues," pp. 185–186). Claims like this, under the rubric "tort for wrongful life," have been advanced in law and have raised the moral question whether there may be in some circumstances a duty not to give existence to, or prolong the existence of, another person: "life can be of a negative value such that the medical maxim *primum non nocere* ("first do no harm") would require not sustaining life" (ibid., p. 187). In cases where death is inevitable and treatment would serve merely to prolong suffering, active euthanasia could be regarded as permissible—perhaps even mandatory—provided that effective procedural safeguards could be established and that the practice did not result in an erosion of the social role of "child" generally—conditions which may in fact never obtain (ibid., p. 188).

Engelhardt's social personhood theory, with its attendant distinction between rights and values, does not fall prey to the problem in Tooley's position—of seeming to restrict the immorality of infanticide to the absence of conditions required for personhood. It provides rather more protection for the healthy infant in that it demands at least very serious reasons for the disposal of a "person in the non-strict sense." Yet a problem with Engelhardt's social personhood is that it is only socially normative, not morally normative; it is a description of the way in which we do, in fact, treat infants in one social context. In other social contexts, and at other times, infanticide has been prevalent; simply being a female infant, for instance, has been in some social contexts a sufficient disvalue to warrant infanticide. And social attitudes do not constitute moral principles. Further, Engelhardt presumes that allowing infanticide will not support the social interests which we have—e.g., in the development of moral values of care and compassion. This is, however, only an empirical assumption, not a necessary consequence of permitting infanticide.

A second objection is that the establishment of birth as the condition for social personhood appears to be arbitrary; Engelhardt has suggested that it is reasonable to distinguish, in terms of personhood, between an infant and an aborted fetus solely on the grounds of whether or not the aborted fetus is capable of living and interacting as a child, "given *only the usual aids* offered after a *normal* term birth" (Engelhardt, 1976, p. 53; emphasis added). But on these grounds "social personhood" can be withheld from premature infants or term infants requiring extraordinary care for the preservation of life. The moral relevance of "a normal term birth" or "the ability to survive without extraordinary care" needs still to be established.

Third, there are also problems of adjudication between the rights of a person in the "non-strict" sense and those of a person in the strict sense, given a conflict situation. For example, are the imputed rights of infants any less obliging simply in virtue of being imputed?

Consequentialist positions. Several positions on the care of infants have been developed according to consequentialist methods, which hold that moral claims concerning value of life are based primarily on predictable qualitative consequences for the individual or for others whose interests are involved in the case.

An act-utilitarian perspective. Joseph Fletcher argues for pediatric euthanasia—whether active or passive, for there is, in his view, no effective difference between the two—as an appropriate response to a great variety of neonatal deficiencies. He employs an ethic of act-utilitarianism, which holds that the greatest good should be done for the greatest number, the good to be judged according to human need in every situation.

In his position on euthanasia (Joseph Fletcher, 1973), not *any* means can be justified by *any* end, but the killing of a human is good whenever the desirable consequences—such as the avoidance of misery and dehumanization—outweigh any disvalue in the action, according to a cost-benefit anaylsis. Neonatal euthanasia is warranted, in Fletcher's "quality-of-life ethic," whenever death is judged to be a desirable goal representing a proportionate good in pursuit of the highest good, which is human happiness.

Specific criteria for judging which children should be supported or terminated does not depend on a biological standard: such a vitalistic approach to life is idolatrous (Joseph Fletcher, 1968). Fletcher suggests twenty criteria of "humanhood" or "personhood" which can serve as a basis for deciding our obligations concerning the continuation or termination of descriptively human lives (Joseph Fletcher, 1974; "Medicine and Nature of Man"). The *minimal* criteria seem to include cerebration, self-awareness, intelligence, self-control, control of existence, and communication. In terms of "intelligence," below a 40 IQ score an individual is questionably a person; below the 20 mark he or she is not a person. Cerebration or neocortical function is the precondition for all the other criteria. The remaining indicators are *optimal* criteria.

Without these minimal criteria the individual is a subpersonal object which can be terminated without moral offense. The Down's syndrome baby, for example, is not a person. Direct, involuntary euthanasia is the preferred way of dealing with such a baby. Indeed, it is morally wrong to fail to commit euthanasia when keeping the infant alive is not warranted by consequences. Letting the baby die without rational control or human decision has no moral value: *Only* direct euthanasia can give positive moral meaning to the life of such an afflicted child (Joseph Fletcher, 1968).

However, if the minimal criteria from cerebration to communication ability are present, on closer analysis it is clear that for Fletcher their presence does not give value to a human, for human life is a value only extrinsically, because of circumstances of need. The presence of the criteria serves the function, in Fletcher's ethic, of identifying that group of individuals for whom a utilitarian calculus is to be applied for determining whether the life of the individual should be preserved or terminated—depending on the situational needs of society, of other individuals, or of the infant himself. What is most decisive in Fletcher's situation ethics is not a concept of the rights of babies but the freedom of other people to terminate "reproductive failures" before or after birth according as these agents determine what is the most convincing need (see ibid.).

One objection brought against Fletcher's thesis —and one applicable to the theories of Tooley and Engelhardt as well—is that his very attempt to define who is human (or a person) and who is not human and on that basis to determine who is to be excluded from human care and protection is a misguided project in terms of social and moral consequences: Discrimination on the basis of "humanity" has been used to justify too many injustices and atrocities throughout history (Bok, pp. 40–41; Hauerwas, 1973, pp. 220–221; Weber, pp. 75–78; Reich, "Quality of Life," pp. 492–493). Second, intelligence and rationality are basic values but are not criteria of the human separated from the values and community for which they exist: The more important moral question is whether we are becoming more human by providing the retarded with respect and care (Hauerwas, 1973, p. 221). Third, there are other objections that focus on the utilitarian method: The calculus on the future meaningfulness of the child's life and the future burdens on others places too much power in the hands of the mature and employs arbitrary criteria for comparing these consequences with others—such criteria as IQ level and the mere presence of Down's syndrome (cf. Reich, "Quality of Life").

A theory of religious consequentialism. Richard McCormick presents a theory that discusses only the question, What infants should be saved? omitting any discussion of active neonatal euthanasia. He takes the position that a substantive standard is needed to make appropriate decisions regarding severely deformed and handicapped infants and suggests as a criterion the potential for human relationships associated with the infant's condition (McCormick).

McCormick's explanation begins with the extraordinary means principle but goes beyond it. He believes the term "extraordinary" is so relativized to the condition of the patient that it is the condition or the quality of life of the patient that is decisive. His substantive argument begins with the notion that life is a value to be preserved *insofar as* the higher, spiritual purposes of life are attainable. Not only the limits of the duty to preserve life but the very duty itself is based on the possibility of attaining these higher values. Because these higher spiritual goods are the inseparable love of God and neighbor, life finds its meaning in human relationships, and life is a value to be preserved only insofar as it contains some potentiality for human relationships. If this potentiality is totally absent or if it would be totally undeveloped or utterly subordinated to the mere effort for survival, no treatment is obligatory: The baby may be allowed to die.

It is the task of physicians to provide concrete categories or presumptive biological symptoms for the judgment about the baby's relational potential. Such biological symptoms are Down's syndrome, anencephaly, and other categories in between. According to McCormick, most would probably agree that the anencephalic infant lacks relational potential, but that the same could not be said of the mongoloid infant.

McCormick concludes by offering some cautions: If we must err it should be on the side of life; all human lives are valuable regardless of condition; parents and physicians should not make their proxy judgment in utilitarian fashion on the basis of valued functions of the child, but only in terms of the child's good.

Among the objections raised against McCormick's position (cf. Reich, "Quality of Life," pp. 501–505; Weber, pp. 82–86), the first pertains to his normative theory on value of life in theological perspective. There is an ambiguity about the value of life in the moral sphere, for while he holds that all human lives have inherent value, this apparently has no *normative* ethical significance about the care we ought to give those lives, unless they have a *particular* value, specifically the quality of relatedness. Though McCormick suggests that his view is based on "the Judeo-Christian perspective," those traditions characteristically have not based the value of life or the rules expressing duties to protect and sustain life on isolated qualities such as the relational ability to love, to "experience our caring and love" (McCormick, p. 176), or other conditions for personal achievement. They have valued life and supported the duty to care for life because all human life has a transcendent value, either inherently or extrinsically because of God's relationship to it.

Second, making quality-of-life judgments on a consequentialist assessment of a certain *kind* of life raises the objection mentioned above against "humanhood" standards. The relational criterion—a vague, relative standard—offends against the principle of equality which has characterized rules governing the just conservation of life in the natural rights tradition. (Limits and exceptions to the duty by their nature produce a certain inequality, but a justified one.) Third, focusing on the capacities of the *isolated* individual neglects the approach that life is known as a value to be preserved also in terms of the kind of people (humane, merciful) that *we* ought to be. Fourth, regarding McCormick's suggestion that diagnostic categories such as Down's syndrome be used as medical indications for treatment/nontreatment, they represent a misleading quantification of an elusive quality, for each of them embraces great variations in mental and physical handicaps.

Approaches rejecting the humanhood standard. A final group of ethical approaches to the protection and care of infants begins with the premise that the foregoing theories that ask whether infants are *persons* or are *normatively human*, and on the basis of characteristics required for personhood or for the truly human determine which infants shall be excluded from the protection and care of the human community, are dangerous and should be abandoned, for these distinctions have been misused to justify discrimination and atrocities against people in the past and are currently the basis for the inhumane care and institutional cruelty provided for the retarded (Bok, p. 41; Hauerwas, 1973, p. 218; 1975, p. 226).

An ethic of care. Both the deontological and the consequentialist positions are misleading, according to Hauerwas. He argues that ethical reflection should not focus on the decision making called for by the question, What should we do? but on the prior question, How should the agent see and understand what the moral situation itself is? It is our character and roles that determine the descriptions we think appropriate about the moral situations we confront. The roles constituting our lives are complex patterns of expectations that provide the context for attribution of responsibility, praise, and blame (Hauerwas, 1975, p. 229).

It is the role of parents that reveals that they have a positive obligation to care for their children. Physicians have become the extension of this parental commitment by providing, among other forms of care, the ability to keep many babies alive. The dilemmas about keeping a defective child alive are forced on us because, on the one hand, there is the moral commitment of parents to care for and to protect their children and, on the other hand, modern technology has expanded our care far beyond the original intention in seeking new means of care. Among the expanded forms of care available, one should distinguish those that "buy time" for making a reasonable diagnosis—e.g., respirating certain infants at birth—and treatment done to initiate a curative process, for the more care we find ourselves committed to give, the more difficult it is to terminate medical forms of care.

A basis for principles in the care of defective children is found by asking what kind of care we should give them in order for us to be humane. Infants have claims upon rational beings because they are capable of *suffering;* and we are liable as rational moral agents because of our capability of alleviating the ills of the predicament. We should have "moral care and respect" for children insofar as they suffer; and as suffering members of the human species, they should be given life-sustaining treatment or not in a manner equal to other sufferers. There is no reason why the *defective* condition of some children should disqualify them from receiving the care we should give any children insofar as they suffer, "unless their defect is so severe there is little possibility that they will ever be able to respond to care" (ibid., p. 228).

The actual extent and kind of care we should give newborns must be seen in conjunction with the responsibilities that are found in the roles of parents and physicians. Yet many defective children are being allowed to die because parents often have a distorted notion of parental responsibility; they seem to assume that they are responsible for assuring total well-being, perfect physical health, and protection from suffering and death for their children, and that without these elements the children might better be dead. On the contrary, we are responsible to see that the basic physical, emotional, and moral needs of our children are supplied. There are limits to the positive aid parents are expected to give to keep their children alive; but it may be preferable to describe these limits as "putting these children to death" rather than merely "allowing them to die," so that we may learn what we take the role of parenting to be. Certain ways of explaining the withholding of treatment from defective children and letting them die as distinct from putting them to death are deceptive except where one is describing the decision not to prolong death.

The physician's task is not to make a family happy or to protect parents from the suffering occasioned by a retarded child, but to secure the health of these children, even if it may mean tragedy for the family. The doctor need not use every technique available to keep every child alive; he can limit his care, but only if he has the assurance that there are other kinds of care present in the community, which is questionable in today's society.

Several observations can be made about this attempt to resolve dilemmas in the care of infants on the basis of an ethic of character and roles. First, it does not offer a very specific account of the actual limits of human care in these situations—a shortcoming the author acknowledges. Second, there is some ambiguity in "the inability to respond to care" used as a criterion for exempting some children from the care offered to other human beings. If this refers to therapeutic response to medical treatment, Hauerwas has said little that others have not said without relying on his system, for few would regard it as obligatory to offer futile treatment. If the capacity referred to is some other quality of the infant, such as his ability to give interpersonal response to someone else's company keeping—also a form of care according to the author—the criterion would be functionally similar to the "personhood" approach, which Hauerwas rejects. Third, one reason Hauerwas's ethic of care cannot provide more specific norms on when it is wrong not to preserve life is because it relies so heavily on the role of parenthood as a *moral* basis of duty—precisely at a time when culture, technology, and individual wants are altering the meaning and expectations of parenthood. His argument requires appeal beyond the commitments actually found in parenthood, perhaps to an elaboration of the claims of the child and the moral needs of society.

Avoiding harm. Some attempts have been made to avoid criteria of humanhood by applying to infants the principle of avoiding harm. This principle involves a negative form of consequentialism, which necessitates weighing the harms attendant upon the adoption of a particular policy.

Sissela Bok offers important principles for the ethics of infanticide in the context of a discussion of abortion. Bok determines the nature and extent of the protection that should be offered to infants by considering the harm that comes from taking life (Bok). Underlying the elemental sense of the sacredness of life and the elemental fear of its extinction are certain reasons for protecting life: (1) Killing is the greatest of all dangers *for the victim.* Awareness of a threat to life causes intense anguish; the taking of life can cause great suffering; and since the experience of life is so valuable and absorbing and is the precondition of other values, no one should be unjustly deprived of it. (2) Killing is brutalizing and criminalizing *for the killer.* (3) Killing often causes *the family of the victim and others* to experience grief and loss. (4) *All of society* has a stake in the protection of life; permitting killing sets patterns for victims, killers, and survivors that are threatening and ultimately harmful to all (ibid., p. 42).

Society's reasons for protecting life are present with respect to infanticide, especially "the brutalization of those participating in the act and the resultant danger for all who are felt to be undesirable by their families or by others" (ibid.). The reasons for protecting life gain in strength during pregnancy—viability is a decisive cut-off point—and they prohibit both late abortions and infanticide. Although infanticide is practiced in some primitive societies as a means of family limitation, its public acceptance in all other societies is unthinkable, considering the availability of contraception, early abortion, and means of caring for children in institutions and through adoption.

The overriding consideration is "the threat which would be felt by all if infanticide as a parental option were thought to be possible" (ibid., p. 50). Still, there is a difference between the active killing of infants and not undertaking the battle for life or not carrying it as far as one would otherwise "in those rare cases where an infant is born with a severe malformation, such as the absence of a brain" (ibid.).

This theory raises certain questions. How does one establish that deliberate termination of an infant's life is more harmful to him or the parents than its continuation? How does one justify the claim that harm to society from a practice of infanticide should count for more than short-term benefit claimed by physicians and parents in some cases of infanticide?

Although Bok's approach does not offer ethical solutions to many of the neonatal dilemmas—because the author did not present her position in the context of those questions—Jonsen and Garland apply the ethic of avoiding harm in more detail. They say the principle "Do no harm" is the most suitable guide for the complex decisions regarding neonatal survival, provided there be compensating benefit for the patient. As a negatively formulated principle it has the advantage of not requiring exceptions; the difficulty in its use is defining harm (Jonsen and Garland; Jonsen et al.).

Life-preserving interventions do harm to infants if the infants (1) are unable to survive infancy, for such infants "are already in the dying state" (cf. Jonsen et al., p. 762); (2) will live in intractable pain that can be alleviated neither by immediate nor by long-term treatment; or (3) cannot participate even minimally in human experience—meaning the infant must have "some inherent capability to respond affectively and cognitively to human attention and to develop toward initiation of communication with others" (ibid.). A baby with Down's syndrome would not fulfill the criteria; one with a trisomy 18 would.

The approach taken by Jonsen and Garland, based on an ancient medical-ethical principle, raises crucial questions about identifying the "harm" that would justify abandoning the infant. First, it is not clear why it would be "doing harm," as a general rule, if one supports the life of an infant who cannot survive infancy. Such a life can often be precious to the parents and presumably also to the infant. Furthermore, a case could be made that it does harm not to survive childhood or adolescence. Would that prospect justify abandoning an infant? Second, no convincing reason is given why it would "do harm" to sustain an infant who lacks the ability to respond cognitively and affectively to others. The child unable to respond may not be able to experience any "harm," and the parents and others in a caring relationship may find their humane support of the child rewarding and not harmful. The "do no harm" principle is helpful for some medical decisions, but it is insufficient in itself for resolving many of the dilemmas of neonatal care, especially those involving the worth of lives afflicted with serious defects.

Conclusion

The rapidly developing debate on the care of infants shows a strong similarity to the more thoroughly developed abortion debate, at the same time involving many issues from the discussions of euthanasia. It raises again in rather poignant fashion the crucial question of the value of human life and when the protection and support of that life should begin and end; and it offers a variety of interpretations of the claims and responsibilites of various parties, the ethical principles to be used, and the means for resolving conflict situations.

Two comments can conclude this survey. First, it would be unfortunate if the discussion were obfuscated by the implications of commonly used terminology. For example, "defective infants" could imply that there is a class of infants who do not measure up to normality as a value concept, thus prejudicing the discussion. Second, while physicians and ethicists are each beginning to employ the others' tools in arriving at reasonable standards and decisions, a comparison of this article with the one preceding indicates that the discourse of the two groups is far from integrated. Further communication between these two professions is highly desirable as ethical inquiries continue.

WARREN T. REICH AND DAVID E. OST

[*Directly related are the other articles in this entry. For further discussion of topics mentioned in this article, see the entries:* CARE; DECISION MAKING, MEDICAL; LIFE; *and* PERSON. *Also directly related are the entries* ABORTION, *articles on* MEDICAL ASPECTS *and* CONTEMPORARY DEBATE IN PHILOSOPHICAL AND RELIGIOUS ETHICS; CHILDREN AND BIOMEDICINE; DEATH AND DYING: EUTHANASIA AND SUSTAINING LIFE, *article on* ETHICAL VIEWS; ETHICS, *articles on* DEONTOLOGICAL THEORIES *and* UTILITARIANISM; *and* PAIN AND SUFFERING. *Other relevant material may be found under* ACTING AND REFRAINING; HEALTH CARE, *articles on* RIGHT TO HEALTH-CARE SERVICES *and* THEORIES OF JUSTICE AND HEALTH CARE; LIFE-SUPPORT SYSTEMS; MAN, IMAGES OF; MENTALLY HANDICAPPED; NATURAL LAW; RIGHT TO REFUSE MEDICAL CARE; *and* SOCIALITY. *See* APPENDIX, SECTION III, PEDIATRIC BILL OF RIGHTS.]

BIBLIOGRAPHY

[The bibliography for this article can be found following the next article, PUBLIC POLICY AND PROCEDURAL QUESTIONS.]

III
PUBLIC POLICY AND PROCEDURAL QUESTIONS

As regards policy decisions that govern the treatment of defective and diseased infants one may distinguish among (1) substantive policies dealing with euthanasia and sustaining life established both by law and by professional groups; (2) procedural policies that determine where the authority and responsibility for decision making should be fixed—with parents, physicians, other health personnel, or committees; and (3) the socioeconomic policies that deal, at a societal level, with a just distribution of resources and burdens in the case of highly defective children.

The unique aspect of all these policies is that they deal with infants who cannot withhold or grant consent to treatment, to euthanasia, or to the broader socioeconomic policies affecting them, and yet these policies have the possible effect of not continuing the lives of those who could live and who have not had an opportunity for life experiences (cf. Gustafson, p. 540). This makes the question of substantive and procedural policies a particularly weighty ethical question.

Policies on euthanasia and sustaining life

Medical organizations such as the American Medical Association have issued formal policies opposed to active euthanasia. Medical policies governing nontreatment of infants develop in an informal fashion as policies of "ordinary and reasonable care" for specific disease conditions and can be traced in the medical literature, as noted in the previous article.

U.S. law generally presumes that personhood exists following a live birth and thus that all infants are entitled to the usual legal protections, regardless of physical or mental characteristics (Robertson and Frost, p. 884). Contemporary laws in virtually every nation oppose active euthanasia of infants even with merciful motive, although in cases of infanticide of defective infants by parents penalties are sometimes mitigated and parents are frequently acquitted (Maguire, pp. 22–54). Views on legal policies dealing specifically with infants have taken the form of interpretations of existing law and recommendations of appropriate legal protections.

One interpretation of existing U.S. law is that parents, physicians, nurses, and hospital administrators who permit withholding of a medical or surgical procedure necessary to maintain the life of a defective infant with resultant death of the infant may be criminally liable on charges of homicide by omission, child neglect, failure to report child abuse, and/or conspiracy (Robertson, 1975). A different interpretation is that the physician's liability is limited to cases in which "normal and reasonable care," as opposed to extraordinary efforts, is withheld (Heymann and Holtz, p. 386). Despite possible legal sanctions, selective withholding of medical treatment from infants is widely practiced in the United States and elsewhere (Duff and Campbell, 1973; Lorber, 1973). Furthermore, as of 1976, neither a parent nor a physician had ever been prosecuted for withholding care from a defective newborn infant and allowing it to die (Robertson and Fost, p. 886), though there have been civil cases in which treatment has been ordered or the court has not permitted nontreatment to occur.

Further discussion focuses on what the legal policy *ought* to be. Some support the present prohibition of active pediatric euthanasia, for if it were legalized it would be "a most dangerous weapon in the hands of the State or ignorant or unscrupulous individuals" (Lorber, 1973, p. 204). Those supporting the present law's prohibition of withholding treatment argue that allocating the right to life—i.e., determining who shall be treated—on the basis of physical or mental characteristics or social contribution requires an arbitrary choice inconsistent with a democratic society and sets one on a slippery slope of neglecting the value of the person (Robertson and Fost, p. 887). Robert Burt has held that the current system of formal illegality of withholding treatment, coupled with lack of prosecution except for occasional exemplary cases, is the best policy for ensuring careful decisions, for the possibility of criminal liability is a powerful safeguard (Burt, pp. 445–446). Robertson, on the other hand, has pointed out that such a policy drives these decisions underground, resulting in disrespect for the law, arbitrary and abusive nontreatment decisions (because they cannot be aired openly), and unequal treatment before the law because of the possibility of selective prosecutions (Robertson, "Discretionary Non-Treatment," p. 462). Duff and Campbell are so committed to a change in current law that they have taken a position of civil disobedience against laws that may prohibit allowing infants that do not promise meaningful life to die and against conspiracy laws that may prohibit consultation on euthanasia (Duff and Campbell, 1973, p. 894; 1976, p. 489).

Two alternative policies have been proposed as changes in the law: (1) establishment of *criteria* for the class of infants or the limited circumstances that would justify either withholding treatment or actively terminating the lives of infants, or (2) establishment of an acceptable *process* of decision making for governing either of these actions.

In the first category, some commentators favor criteria describing *circumstances* justifying *nontreatment*. Robertson favors identifying the class of infants from whom treatment may lawfully be withheld by specifying defective characteristics and the familial or institutional situations justifying withholding treatment. For example, if there were a consensus that profoundly retarded, nonambulatory hydrocephalics who are blind and deaf or infants who are unlikely to survive beyond a year are not owed ordinary treatment, this may be a basis for developing criteria. Further protection could be gained by requiring certification of clinical findings by two nonattending physicians (Robertson, 1975, pp. 266–267).

Criteria of this type would reduce the risk of arbitrary decision making but would be difficult to establish and revise, due to the complexity of cases and changing medical, social, and moral factors (Robertson and Fost, p. 888). Furthermore, it is not clear whether these criteria would be established by an institutional committee, a professional group, a national commission of mixed membership, a legislature, or the courts. Much depends on what kind of consensus can and should be obtained.

In spite of these problems, one policy recommendation favors the legal fiction of "nonpersonhood" for a broadly defined class of high-risk infants whose lives could be supported, abandoned, *or actively terminated;* these decisions would be entrusted to parents for the same reasons that women are permitted to decide about abortion: quality-of-life considerations for the infant and the fact that parents bear the principal risks, burdens, and benefits (Marks with Salkovitz).

The *procedural* approach to improved public policy can be viewed in the light of due process, which requires that all relevant information be taken into account and calls for an impartial decision maker. Robertson and Fost (p. 888) take the procedural approach for legitimizing *nontreatment;* in their view, due process is best accomplished by a disinterested party—one or a group of physicians, a judge, or a mixed lay and medical committee. Further procedural safeguards could be provided for other decision makers (Robertson, 1975, p. 267): by prohibiting the withholding of treatment for, say, seven days, to diminish the adverse influence of postpartum anxieties, by requiring that parents be counseled and thoroughly informed, and by requiring physicians to obtain a consultation and to state in writing their reasons for withholding treatment.

Duff and Campbell also favor the *procedural* approach but with different conclusions: Parents and physicians should be granted discretion to decide for the death of a defective infant, whether by *active or passive* means, governed only by general guidelines formulated by the health professions; in cases of disagreement between physician and family the courts should help decide when death may be chosen (Duff and Campbell, 1973, p. 894; 1976, p. 492).

Who should decide?

Apart from the procedural question involved in legal policy, there is a further ethical question: Given a policy permitting some nontreatment decisions, who ought to make these decisions? According to the informed consent doctrine, it is the patient himself who has the responsibility for deciding whether to accept or reject treatment; but with the infant, the question is, Who should make the proxy decision?

A variety of ethical considerations are appealed to in answering the question of responsibility in decision making: the *values* likely to be represented by an agent, *qualifications* for making a reasonable decision, *role* expectations, the *right* to decide, and where such *power* or *authority* should reside. A further ethical question about agency has until now received little attention: whether the *authority* for accepting and refusing lifesaving treatment for an incompetent infant becomes a wedge for assuming an autonomous *right* to determine the qualities which shall be normative for such judgments (cf. Gustafson, p. 553; Reich, "Testimony," p. 56).

Parents. Several reasons have been offered why parents should be the effective decision makers for their infants' treatment: (1) There is a principle that in society decisions should normally be left to the most intimate and smallest unit involved (cf. Gustafson, p. 542); and the child is a member of a family in a more complete sense than he is a member of any other group (Weber, p. 105). (2) As in other medical situations, parents are presumed to be the best qualified decision makers, for by their very role they are most likely to be morally committed to the welfare and best care of their child (cf. Hauerwas, 1975, p. 225; Veatch, 1976, pp. 202–203). (3) Moreover, parents will enjoy the greatest benefits and bear the greatest emotional and economic cost whether the infant survives or not (Smith, p. 40; Engelhardt, "Bioethics and Process of Embodiment," pp. 498–499; Duff and Campbell, 1976, p. 490). (4) The importance of parents as decision makers is underscored by the fact that some who believe that only physicians should make the decision recognize that when parents are strongly in favor of therapy, there is no question as to what must be done (Freeman, "To Treat or Not to Treat," p. 14).

There are, however, objections to this policy. Fost has argued that there is no compelling reason to assume that parents are either competent to decide or will decide in the infant's best interest, and that, in fact, many adults fail to make competent and rational decisions concerning their own treatment (Fost, p. 5). Furthermore, parents are subject to a variety of emotional pressures—factors that may impair their capacity to make objective decisions ("Ethics of Selective Treatment," pp. 87–88; John Fletcher, 1974). Medical reaction to this situation varies. While Freeman believes the parents' decision is primarily emotional and can hardly be an informed decision (1973, p. 141), others argue that parents often *can* make an informed decision if "they are afforded access to a range of resources beyond the expertise and bias of a single doctor and . . . sufficient time for contemplation of the alternatives" (Shaw, p. 886; cf. Shurtleff et al., pp. 1009–1010). The danger that the family may act out of self-serving motives (Maguire, p. 189) or make ill-considered decisions for whatever reason is offset by the possibility, even the duty, of the physician to report the case to public or judicial authorities who could intervene to save the child (Robertson and Fost, p. 884).

Physicians. In terms of competence for understanding the character of the defect and the prognosis for the infant possessing it, the physician is the best qualified decision maker and in this respect possesses a kind of objectivity that the parents may not. Such considerations have led to the claim that, in spite of the physician's own emotional pressures and while the parents should play a role in the decision not to treat, it is in the final anaylsis a physician's decision (Freeman, "To Treat or Not to Treat," p. 14; "Ethics of Selective Treatment," p. 88). Other reasons are given why only the physician should choose life or death for his patient: the child has been put in his care (Ingelfinger); and it would be dishonest to ask parents to decide, because in the vast majority of cases parents are influenced by the doctor's opinion (Rickham). Yet there appears to be a policy among physicians recognizing the moral autonomy of parents in deciding whether a defective infant should live if the child will be *mentally* deficient, e.g., if afflicted by Down's syndrome (see Gustafson, pp. 539–540).

A policy favoring the physician's authority, however, runs the risk of fostering paternalism in the medical profession and encouraging physicians' manipulation of information (Duff and Campbell, 1973, p. 894; Fost, p. 16), thus rendering proxy "informed consent" a mere fiction. One may also question the objectivity of the physician's decision: Prognoses are always probabilistic, and physicians frequently disagree among themselves about therapeutic decisions. Even if this were not the case, medical qualifications are not moral qualifications, and they do not guarantee that a medically sound decision will ipso facto be a morally sound one (Robertson, "Discretionary Non-Treatment," p. 463; Smith, p. 39). In addition, some studies have shown that physicians' judgments on such issues depend less upon their medical training than upon their moral and religious commitments; consequently, a policy locating the decision in the hands of the physician means "adopting his particular religious or philosophical bias as determinative of wisdom and morality in this case" (Maguire, p. 181). And physicians have no authority to impose their moral values as such on the patients or the patients' parents (see Gustafson, pp. 539–546).

From elements in the medical, ethical, and legal literature, one can construct a mediating position in the current standoff between parents and physicians as decision makers, provided distinctions are made among various *kinds* of decisions, judgments, and roles. There are values and competencies in the physician's role whose loss in a struggle over rights would be detrimental. Since medicine is not merely technique but an art as well, and since the physician has special responsibilities to his patient, he or she can be regarded as the leader of the entire decision-making effort, for the physician is in a unique position to determine the condition and prognosis of the patient; to obtain proper medical consultation; to make a judgment on what *ought* to be done on the basis of his own values; to give the parents a clear, thorough, and sympathetic explanation; if the physician's own judgment is recommended, to explain its medical and value dimensions, with alternatives; to offer assistance in obtaining additional medical advice and moral counsel; to involve other health-care personnel depending on role and competence; when parents prove incapable of weighing options and taking responsibility, to exercise a restricted but legitimate paternalism in guiding them to a decision; and through this process to offer encouragement and consolation. On the part of parents, their informed *consent* should be regarded as a true *decision* and not a mere concurrence in the judgment of another. Their granting or withholding of permission is the ultimate and effective decision on what *shall* be done (unless further appeal, e.g., to a court, seems required). Their consent can be *sufficiently* informed, free, and voluntary even without total medical knowledge and under pressures of strong emotional and professional influences.

Nurses. It is felt by some that nurses in neonatal intensive care should be involved in the decision-making process for several reasons. Not only are they affected by the decision; their responsibility and liability—which in some cases are independent of the physician's and which have increased as nurses have assumed some functions formerly performed by physicians—also point in this direction (Harris; Robertson, 1975, p. 228; Robertson and Fost, p. 885).

Committees. Four kinds of committees can be offered as candidates for decision maker: (1) a review body that gives or withholds ratification in cases where both physician and parents have agreed that nontreatment is the desirable course; (2) a consulting committee, designed to advise and assist parents and physician in arriving at a decision about treatment; (3) a committee that would act as a proxy decision maker when the patient is incompetent, either by attempting to decide what the infant would want done if he or she could judge or by deciding what would be socially and ethically acceptable in the interests of the infant; and (4) an institutional review committee that would review decisions and practices after the clinical decision had been acted upon.

The advantages of such committees are manifold: they can (1) provide an important safeguard against ill-considered judgments; (2) achieve greater objectivity about value-of-life questions since they are neither personally nor professionally involved in the treatment of the infant; (3) lighten the burden of a decision that is always at best a difficult one; (4) provide a wider airing of viewpoints and reasons, including a voice for the infant through appointment of an advocate, and (5) bring to bear greater ethical expertise than can be provided by parents and physician alone.

There are costs attendant upon this policy as well: (1) The very objectivity of such committees makes them remote from the personal reality of the infant and the family and insulated from the consequences of the decision (cf. Ingelfinger, p. 914). (2) Committees are vulnerable to pressures to accept the decisions of attending physicians (Robertson, 1975, p. 265). (3) The machinery of group decision making does not guarantee consensus, and in the absence of uniform standards of judgment there is no assurance that relevantly similar cases will be decided in similar ways by any two committees, thus denying equal treatment to the two infants involved. (4) If publicly sanctioned committees mete out life or death, they risk losing society's pervasive symbolic commitment to the importance of saving identifiable human lives (ibid.). (5) Finally, the formal designation of a committee as decision maker may have the effect of involving the state in "relative assessments of life" in contradistinction to its primary commitment to the sanctity of all persons (Robertson, "Discretionary Non-Treatment," pp. 461–462). This effect is realized when the state merely accords validity to a reviewing body developed by the medical profession itself.

Macroallocational policies

To a large extent, social policies determine the sorts of medical treatment available for the defective infant; yet with few exceptions the ethical dimensions of these policies have received little scrutiny.

On the assumption that we should enhance the maximum development of every child, some believe society should place a high priority on medical, psychological, educational, and social services to the disadvantaged child (Zachary) Others are concerned that life-support treatment may perpetuate an undue proportion of retardates or marginally functional survivors (Shurtleff et al., p. 1010). Some offer ethical considerations for the thesis that the proportion of defective persons in society may be controlled by public policy decisions of resource allocation (Smith).

Other concerns pertain to cost. Some argue that the life of an infant cannot be measured in terms of costs. Others point out that there is, or should be, a limit to costly and scarce neonatal intensive care and follow-up treatment (Kramer, pp. 75–93). Furthermore, the diversion of health personnel to highly defective infants may jeopardize the lives of other would-be patients who might fare better with treatment—especially in developing nations—and may restrict the availability of other valued services (ibid., p. 78). Still others are convinced that social and economic costs of maintaining defective newborn infants are a minute and manageable portion of health expenditures (Robertson and Fost, p. 887); that taxpayers as a group rather than individual families should bear the cost of care; that precise data showing the trade-offs with health and other expenditures do not exist; and that ceasing to care for the defective would not necessarily lead to a reallocation within the health budget with a net savings in suffering or life (Robertson, 1975, pp. 259, 261).

The question, then, is how to discriminate justly, at the macroallocational level, in the distribution of lifesaving resources to defective infants. Some believe that an economic framework, especially when combined with cost-benefit analysis—though not divorced from ethical principles—provides a useful tool for establishing criteria for treatment on a societal as opposed to individual level (Kramer, pp. 79–83). Solutions to these questions depend on principles such as distributive justice that are discussed elsewhere in this encyclopedia.

Although serious examination of public policy in this area is only beginning, allocational considerations are having an impact at the local institutional level, as, e.g., when treatment is not offered unless financial resources are assured (Shurtleff et al., p. 1005) and when criteria for treatment of newborn infants are determined by the need to avoid excessive cost and so as to give better attention to those more likely to benefit from total care (Lorber, 1971, p. 288). In any case, socioeconomic issues can be expected to have an increasing influence on the ethics of infant care.

Conclusion

The deliberate nontreatment of defective newborn children—or involuntary passive euthanasia of infants—has clearly become a common practice. Active pediatric euthanasia, whether commonly practiced or not, is increasingly being advocated. The primary debate at the policy level seems to be whether the governing principle should be one of equal, nondiscriminatory protection; whether lack of cognitive or other abilities in the patient should justify different treatment from that accorded to other patients; or whether psychosocial and economic benefit and harm to a variety of parties should be the principal criterion. A secondary debate, inextricably bound up with the first, deals with the limits of parental and professional autonomy in decisions affecting the lives of infants.

WARREN T. REICH AND DAVID E. OST

[*For further discussion of topics mentioned in this article, see the entries:* ACTING AND REFRAINING; CIVIL DISOBEDIENCE IN HEALTH SERVICES; DEATH AND DYING: EUTHANASIA AND SUSTAINING LIFE, *article on* PROFESSIONAL AND PUBLIC POLICIES; DECISION MAKING, MEDICAL; HEALTH CARE, *articles on* RIGHT TO HEALTH-CARE SERVICES *and* JUSTICE AND HEALTH CARE; INFORMED CONSENT IN THE THERAPEUTIC RELATIONSHIP; LIFE; MEDICAL MALPRACTICE; PATERNALISM; *and* THERAPEUTIC RELATIONSHIP. *For discussion of related ideas, see the entries:* ADOLESCENTS; CHILDREN AND BIOMEDICINE; *and* RIGHTS. See APPENDIX, SECTION III, PEDIATRIC BILL OF RIGHTS.]

BIBLIOGRAPHY

AVERY, MARY ELLEN. "Considerations on the Definition of Viability." *New England Journal of Medicine* 292 (1975): 206–207.

BENN, S. I. "Abortion, Infanticide, and Respect for Persons." *The Problem of Abortion*. Edited by Joel Feinberg. Belmont, Calif.: Wadsworth Publishing Co., 1973, pp. 92–104.

BLEICH, J. DAVID. "Theological Considerations in the Care of Defective Newborns." Swinyard, *Decision Making*, chap. 25, pp. 512–586.

BOK, SISSELA. "Ethical Problems of Abortion." *Hastings Center Studies* 2, no. 1 (1974), pp. 33–52.

BURNS, LAFAYETT E.; DODGMAN, JOAN E.; and CASS, ALONZO B. "Fatal Circulatory Collapse in Premature Infants Receiving Chloramphenicol." *New England Journal of Medicine* 261 (1959): 1318–1321.

BURT, ROBERT A. "Authorizing Death for Anomalous Newborns." Milunsky, *Genetics and the Law*, pp. 435–450.

COOKE, ROBERT E. "Whose Suffering?" *The Journal of Pediatrics* 80 (1972): 906–907.

DUFF, RAYMOND S., and CAMPBELL, A. G. M. "Moral and Ethical Dilemmas in the Special-Care Nursery." *New England Journal of Medicine* 289 (1973): 890–894.

———, and CAMPBELL, A. G. M. "On Deciding the Care of Severely Handicapped or Dying Persons: With Particular Reference to Infants." *Pediatrics* 57 (1976): 487–493.

ENGELHARDT, H. TRISTRAM, JR. "Bioethics and the Process of Embodiment." *Perspectives in Biology and Medicine* 18 (1975): 486–500.

———. "Ethical Issues in Aiding the Death of Young Children." *Beneficent Euthanasia*. Edited by Marvin Kohl. Buffalo, N.Y.: Prometheus Books, 1975, pp. 180–192.

———. "On the Bounds of Freedom: From the Treatment of Fetuses to Euthanasia." *Connecticut Medicine* 40 (1976): 51–55.

———. "The Ontology of Abortion." *Ethics* 84 (1974): 217–234.

"Ethics of Selective Treatment of Spina Bifida: Report by a Working-Party." *Lancet* 1 (1975): 85–88.

FISCH, ROBERT O.; BILEK, MARY K.; MILLER, LYNN D.; and ENGEL, ROLF R. "Physical and Mental Status at Four Years of Age of Survivors of the Respiratory Distress Syndrome: Follow-up Report from the Collaborative Study." *Journal of Pediatrics* 86 (1975): 497–503.

FITZHARDINGE, P. M.; PAPE, K.; ARSTIKAITIS, M.; BOYLE, J.; ASHBY, S.; ROWLEY, A.; NETLEY, C.; and SWYER, P. R. "Mechanical Ventilation of Infants of Less than 1,501 Grm. Birth Weight: Health, Growth, and Neurologic Sequelae." *Journal of Pediatrics* 88 (1976): 530–541.

FLETCHER, JOHN. "Abortion, Euthanasia, and Care of Defective Newborns." *New England Journal of Medicine* 292 (1975): 75–78.

———. "Attitudes toward Defective Newborns." *Hastings Center Studies* 2, no. 1 (1974), pp. 21–32.

FLETCHER, JOSEPH. "Ethics and Euthanasia." *American Journal of Nursing* 73 (1973): 670–675. Reprint. *To Live and to Die: When, Why and How*. Edited by Robert H. Williams. New York: Springer Verlag, 1973, chap. 9, pp. 113–122.

———. "Four Indicators of Humanhood—The Enquiry Matures." *Hastings Center Report* 4, no. 6 (1974), pp. 4–7.

———. "Medicine and the Nature of Man." *The Teaching of Medical Ethics*. Edited by Robert M. Veatch, Willard Gaylin, and Councilman Morgan. A Hastings Center Publication. Proceedings of a Conference Sponsored by the Institute of Society, Ethics, and the Life Sciences and Columbia University College of Physicians and Surgeons, 1–3 June 1972. Hastings-on-Hudson, N.Y.: Institute of Society, Ethics, and the Life Sciences, 1973, pp. 47–58.

———. "The Right to Die: A Theologian Comments." *Atlantic Monthly*, April 1968, pp. 62–64.

FOST, NORMAN. *Ethical Problems in Pediatrics. Current Problems in Pediatrics*, vol. 6, no. 12. Edited by Louis Gluck. Chicago: Year Book Medical Publishers, 1976.

FREEMAN, JOHN M. "Is There a Right to Die—Quickly?" *Journal of Pediatrics* 80 (1972): 904–905.

———. "Psychosocial Aspects of Meningomyeloceles." *Practical Management of Meningomyelocele*. Edited by John M. Freeman. Baltimore: University Park Press, 1974, pp. 218–223.

———. "To Treat or Not to Treat." *Practical Management of Meningomyelocele*. Edited by John M. Freeman. Baltimore: University Park Press, 1974, pp. 13–22.

———. "To Treat or Not to Treat: Ethical Dilemmas of Treating the Infant with a Myelomeningocele." *Clinical Neurosurgery* 20 (1973): 134–146.

GUSTAFSON, JAMES M. "Mongolism, Parental Desires, and the Right to Life." *Perspectives in Biology and Medicine* 16 (1973): 529–557.

HARRIS, CHERYL HALL. "Some Ethical and Legal Considerations in Neonatal Intensive Care." *Nursing Clinics of North America* 8 (1973): 520–531.

HAUERWAS, STANLEY. "The Demands and Limits of Care—Ethical Reflections on the Moral Dilemma of Neonatal Intensive Care." *American Journal of the Medical Sciences* 269 (1975): 222–236. Reprint. Hauerwas, *Truthfulness and Tragedy*, chap. 13, pp. 169–183.

———. "The Retarded and the Criteria for the Human." *Linacre Quarterly* 40 (1973): 217–222. Reprint. Hauerwas, *Truthfulness and Tragedy*, chap. 11, pp. 157–163.

———; Bondi, Richard; and Burrell, David B. *Truthfulness and Tragedy: Further Investigations in Christian Ethics*. Notre Dame: University of Notre Dame Press, 1977.

HEMPHILL, MICHAEL. "Pretesting for Huntington's Disease: An Overview." *Hastings Center Report* 3, no. 3 (1973), pp. 12–13.

———. and FREEMAN, JOHN M. "Ethical Aspects of Care of the Newborn with Serious Neurologic Disease." *Clinics in Perinatology* 4 (1977): 201–209. "Symposium on Neonatal Neurology." Edited by Joseph J. Volpe.

HEYMANN, PHILIP B., and HOLTZ, SARA. "The Severely Defective Newborn: The Dilemma and the Decision Process." *Public Policy* 23 (1975): 381–416.

INGELFINGER, F. J. "Bedside Ethics for the Hopeless Case." *New England Journal of Medicine* 289 (1973): 914–915.

JAMES, L. STANLEY, and LANMAN, JONATHAN T., eds. "History of Oxygen Toxicity and Retolental Fibroplasia." *Pediatrics* 57 (1976): i–ii, 591–642. Supplement.

JONSEN, ALBERT R., and GARLAND, MICHAEL J. "A Moral Policy for Life/Death Decisions in the Intensive Care Nursery." Jonsen, *Ethics of Newborn Intensive Care*, pp. 142–155.

———, and GARLAND, MICHAEL J., eds. *Ethics of Newborn Intensive Care*. San Francisco, Calif.: Health Policy Program, School of Medicine, University of California; and Berkeley, Calif.: Institute of Governmental Studies, University of California, 1976.

———; PHIBBS, R. H.; TOOLEY, W. H.; and GARLAND, M. J. "Critical Issues in Newborn Intensive Care: A Conference Report and Policy Proposal." *Pediatrics* 55 (1975): 756–768.

KELLY, GERALD. *Medico-Moral Problems*. St. Louis: Catholic Hospital Association of the United States and Canada, 1958.

KRAMER, MARCIA J. "Ethical Issues in Neonatal Intensive Care: An Economic Perspective." Jonsen, *Ethics of Newborn Intensive Care*, pp. 75–93.

LANGER, WILLIAM L. "Infanticide: A Historical Survey." *History of Childhood Quarterly* 1 (1974): 353–365.

LORBER, JOHN. "Early Results of Selective Treatment of Spina Bifida Cystica." *British Medical Journal* 4 (1973): 201–204.

———. "Results of Treatment of Myelomeningocele: An Analysis of 524 Unselected Cases, with Special Reference to Possible Selection for Treatment." *Developmental Medicine and Child Neurology* 13 (1971): 279–303.

———. "Selective Treatment of Myelomeningocele: To Treat or Not to Treat?" *Pediatrics* 53 (1974): 307–308.

———. "Spina Bifida Cystica: Results of Treatment of 270 Consecutive Cases with Criteria for Selection for the Future." *Archives of Disease in Childhood* 47 (1972): 854–873.

McCORMICK, RICHARD A. "To Save or Let Die." *Journal of the American Medical Association* 229 (1974): 172–176. Also published in *America* 131 (1974): 6–10. Subsequent discussion and rejoinders are found in *America* 131 (1974): 169–173.

MAGUIRE, DANIEL C. *Death by Choice*. Garden City, N.Y.: Doubleday, 1974.

MARKS, F. RAYMOND, with SALKOVITZ, LISA. "The Defective Newborn: Analytic Framework for a Policy Dialog." Jonsen, *Ethics of Newborn Intensive Care*, pp. 97–125.

MILUNSKY, AUBREY, and ANNAS, GEORGE J., eds. *Genetics and the Law*. New York: Plenum Press, 1976.

PILDES, ROSITA S.; CORNBLATH, MARVIN; WARREN, IRVINA; PAGE-EL, EDWARD; DIMENZA, SALVATORE; MERRITT, DORIS M.; and PEEVA, ANTONIA. "A Prospective Controlled Study of Neonatal Hypoglycemia." *Pediatrics* 54 (1974): 5–14.

RAMSEY, PAUL. *Ethics on the Edges of Life: Medical and Legal Intersections*. New Haven: Yale University Press, 1978.

———. "On (Only) Caring for the Dying." *The Patient as Person: Explorations in Medical Ethics*. New Haven: Yale University Press, 1970, pp. 113–164.

———. "Reference Points in Deciding about Abortion." *The Morality of Abortion: Legal and Historical Perspectives*. Edited by John T. Noonan, Jr. Cambridge: Harvard University Press, 1970, pp. 60–100.

REICH, WARREN T. "Quality of Life and Defective Newborn Children: An Ethical Analysis." Swinyard, *Decision Making*, chap. 24, pp. 489–511.

———. "Testimony." *Medical Ethics: The Right to Survival, 1974: Hearings before the Subcommittee on Health of the Committee on Labor and Public Welfare, United States Senate, on Examination of the Moral and Ethical Problems Faced with the Agonizing Decisions of Life and Death*. 93rd Cong., 2d sess., 11 June 1974, pp. 53–62. Committee print.

RICKHAM, PETER PAUL. "The Ethics of Surgery in Newborn Infants." *Clinical Pediatrics* 8 (1969): 251–253.

A Commentary. Reprint. *Neonatal Surgery* by Peter Paul Rickham and J. H. Johnson. New York: Appleton-Century-Crofts, 1969, chap. 10, pp. 128–131.
ROBERTSON, JOHN A. "Discretionary Non-Treatment of Defective Newborns." Milunsky, *Genetics and the Law*, pp. 451–465.
———. "Involuntary Euthanasia of Defective Newborns: A Legal Analysis." *Stanford Law Review* 27 (1975): 213–269.
———, and FOST, NORMAN. "Passive Euthanasia of Defective Newborn Infants: Legal Considerations." *Journal of Pediatrics* 88 (1976): 883–889.
SHAW, ANTHONY. "Dilemmas of 'Informed Consent' in Children." *New England Journal of Medicine* 289 (1973): 885–890.
SHURTLEFF, DAVID B.; HAYDEN, PATRICIA W.; LOESER, JOHN D.; and KRONMAL, RICHARD A. "Myelodysplasia: Decision for Death or Disability." *New England Journal of Medicine* 291 (1974): 1005–1011.
SMITH, DAVID H. "On Letting Some Babies Die." *Hastings Center Studies* 2, no. 2 (1974), pp. 37–46.
SWINYARD, CHESTER A., ed. *Decision Making and the Defective Newborn: Proceedings of a Conference on Spina Bifida and Ethics.* Foreword by Robert E. Cooke. Springfield, Ill.: Charles C Thomas, 1978.
TOOLEY, MICHAEL. "A Defense of Abortion and Infanticide." *The Problem of Abortion*. Edited by Joel Feinberg. Belmont, Calif.: Wadsworth Publishing Co., 1973, pp. 51–91.
VEATCH, ROBERT M. *Death, Dying, and the Biological Revolution: Our Last Quest for Responsibility.* New Haven: Yale University Press, 1976.
———. "The Technical Criteria Fallacy: The Case of Spina Bifida." *Hastings Center Report* 7, no. 4 (1977), pp. 15–16.
WEBER, LEONARD J. *Who Shall Live? The Dilemma of Severely Handicapped Children and Its Meaning for Other Moral Questions.* New York: Paulist Press, 1976.
ZACHARY, R. B. "Ethical and Social Aspects of Treatment of Spina Bifida." *Lancet* 2 (1968): 274–276.

IV
INFANTICIDE: A PHILOSOPHICAL PERSPECTIVE

It is commonly believed that infanticide is in itself morally wrong, and seriously so. This belief is rarely questioned, particularly because the issue is so deeply emotional. Despite this, some have argued that there are strong reasons, both ethical and historical, for re-examining the question of the morality of infanticide. The following article is written from a perspective that takes seriously such challenges to the belief that infanticide is very wrong. The article begins by reviewing some pertinent historical evidence, and then examines critically the underlying ethical issues.

Historical considerations: attitudes and practices of other cultures

In less advanced societies, infanticide was widely accepted, into the present century. It was very common to destroy infants that were deformed, diseased, illegitimate, or regarded as ill omens. But the practice was not restricted to such cases. In many societies, custom determined how many children a family should have, and infanticide was enjoined as a means of achieving the desired family size. In such societies the practice of infanticide was prevalent (Westermarck, pp. 394–396).

This was the case, for example, in most of the South Sea Islands. It was very common in Melanesia, and even more so in Polynesia. It was almost universally practiced in Australia among the Aborigines, where a woman might be punished for rearing too many children. The practice was not as general in New Zealand, but it was still quite frequent and was not regarded as wrong. It was rejected by some tribes in North and South America but very common in others. It was practiced less in Africa than in the South Sea Islands, but in some cultures, such as the Swahili, infanticide was very common and not seriously disapproved (ibid., pp. 396–405).

Infanticide was also widely accepted in more highly developed cultures. It was a common practice among the ancient Arabs, who sometimes regarded it as not merely permissible but a duty. Female infanticide was common in the poorer districts of China and, although prohibited by both Buddhism and Taoism, was not regarded as wrong by most people (ibid., pp. 407–408).

Infanticide was also accepted and practiced by the two most advanced cultures of ancient Europe, Greece and Rome. In Greece, exposure of weak and deformed infants was an ancient custom and, in at least one state, Sparta, was required by law. Exposure of healthy infants was not generally approved, but it was practiced very widely and not regarded as a grave offense.

This attitude toward infanticide was shared by the greatest of the Greek philosophers, Plato and Aristotle. Aristotle, in the ideal legislation proposed in his *Politics*, holds that deformed infants should not be allowed to live. Plato, in the *Republic*, goes farther and advocates the destruction not only of defective children but of those who are the product of inferior parents or of individuals past the ideal childbearing ages.

The status of infanticide in Rome was similar. The exposure of healthy infants was probably less common; however, this apparently reflects not a difference in moral outlook, but the need for a population sufficient to maintain a large army. So while the killing of healthy infants was disapproved, it was not viewed as an especially serious crime. And the destruction of weak or deformed infants was not merely accepted but required by custom, and possibly even by law (ibid., pp. 408–411).

What account can be given of the vast difference between these societies and societies, such as our own, that emphatically reject infanticide? Some have suggested that cultures that accept infanticide do so because of less parental love for their offspring. But this explanation appears not to be borne out by the evidence. The testimony of anthropologists is that, in most societies where infanticide was prevalent, once a decision had been made to allow a child to live the care and concern shown by its parents compared favorably with that typical of contemporary Western society (ibid., pages 404–405).

If it were merely a question of explaining the prevalence of infanticide, this could perhaps be done, in many cases, by reference to the difficult conditions of life. The crux, however, is a difference in *moral attitude*, since the anthropological evidence strongly supports the view that people in societies that practiced infanticide did not regard the killing of an infant as a situation of great moral conflict. Exposure of infants was apparently carried out without any deep sense of regret and, particularly in the case of weak and deformed children, was thought of as a perfectly natural way of behaving (ibid., pp. 396–397; 402–403).

The fact that these societies viewed infanticide as a perfectly normal practice does not, of course, show that the practice is ethically acceptable. Many societies, for example, have viewed slavery and the torture of enemies as perfectly natural ways of behaving. Nonetheless, the fact remains that in these societies the prevailing moral attitude was that infanticide as such posed no great moral problem.

What accounts for the great change in moral attitude toward infanticide? By far the most important factor was the emergence of Christianity as the dominant religion in Europe, and its subsequent influence upon law and morality. But in spite of Christianity's vigorous opposition to infanticide, attitudes changed slowly; infanticide was an important factor in limiting population growth in Europe until the nineteenth century, despite the fact that infanticide was considered a crime deserving punishment by death (McKeown, pp. 143–161; Langer).

The reasons for the Christian church's vigorous opposition to infanticide are to some extent a matter of debate. Some hold that its opposition rested primarily upon concepts such as those of stewardship, of man's sociospiritual solidarity, of life as a gift from God, and of man as God's creation. Other writers, including historians of ethics such as Westermarck and Lecky, maintain that the main source of the opposition to infanticide was the belief that man was immortal, together with the belief that unbaptized infants would endure eternal torment in hell (Lecky, pp. 22–23; cf. Westermarck, p. 411).

All of the above considerations are theological in nature. It is possible to argue, however, that Christian opposition to infanticide rested primarily upon purely philosophical claims, such as the belief that human beings differ radically in kind from other living animals and, as a result, enjoy a moral status not enjoyed by other living things (Noonan; Adler).

Which of these conflicting historical interpretations is correct cannot be considered here. What is important in the present context is the fact that the Christian church, even if it held that infanticide could be shown by philosophical argument to be morally objectionable, did appeal very heavily to arguments that incorporate Christian theological assumptions, and these theological arguments surely played a very important role in altering the moral outlook of people toward infanticide. The question, then, is whether there is in fact a convincing argument against infanticide if such theological assumptions are set aside and the case is argued on purely philosophical and scientific grounds.

The ethical question: Is infanticide morally wrong?

The three most important issues involved in the question of the moral acceptability of infanticide are, first, whether there is an important ethical distinction between actively terminating the life of an infant, and merely allowing it to die; second, whether the destruction of human infants suffering from severe and irreparable mental impairment is seriously wrong; and finally, whether the destruction of normal human infants is seriously wrong. The discussion here will be confined to the last of these three questions.

A common response to the question of the morality of infanticide is that infanticide is morally wrong because it involves the destruction of an innocent human being. Properly understood, this might turn out to be a defensible position. But as formulated it is potentially misleading, since it is easy to construe it as asserting that it is membership in a particular biological species that is *in itself* the property that makes it seriously wrong to destroy something, and it is clear that this cannot be right. If one were to discover nonhuman animals, on earth or elsewhere, who spoke languages, formed societies with highly developed social and political institutions, were advanced in science and culture, and so on, the fact that they belonged to a different biological species would not be grounds for viewing it as any less wrong to kill them than to kill normal adult human beings. Species membership, defined in purely biological terms, is not *in itself* a morally relevant characteristic.

Though not morally relevant in itself, membership in a particular biological species might still turn out to be morally relevant by being connected with some other property that is in itself morally significant.

What types of properties might be held to be morally significant in themselves with respect to the question of when it is morally permissible to destroy something? In order to answer this question, it is critical to distinguish between potential and nonpotential properties. An individual's potential properties are a matter of the properties it will later come to have if it is not interfered with. An individual's nonpotential properties are a matter of characteristics it currently possesses, independently of any future development.

It is clear that there are some nonpotential properties that make it seriously wrong to destroy something. It may be difficult to say *precisely* what those properties are, but it seems plausible, if one reflects upon the conditions under which one would deem it wrong to kill individuals belonging to some nonhuman species, that it is a matter of enjoying a certain sort of mental life or, perhaps, of being capable of interacting in certain ways with others possessing the same capacity.

It will be convenient to introduce the term "person" to refer to any being that possesses the relevant nonpotential properties. That is to say, a person is by definition any individual that possesses those nonpotential properties, whatever they may turn out to be, and regardless of whether they are empirical properties, theological ones, or metaphysical ones, that make it seriously wrong to destroy something. We can then say that one of the things that someone who asserts that killing innocent human beings is wrong presumably wants to affirm is that killing innocent persons, human or otherwise, is wrong. But usually he will be asserting more than this. He will probably also be asserting that the destruction of potential persons is wrong, where a potential person is something that, in view of its biological or ontological nature will, if not interfered with, come to possess the nonpotential properties that make something a person.

There is a third thing that an individual who asserts that killing innocent human beings is wrong may want to affirm, namely, that it is also wrong to kill any individual that, even if neither a person nor a potential person, is of such a nature that it would develop into a person were it not for the presence of certain defects, or negative factors, which inhibit the normal process of development.

There are, then, three main reasons that might be advanced for thinking that the destruction of healthy human infants is seriously wrong in itself. First, it might be claimed that human infants have developed to the point where they are persons. Second, normal human infants, even if not persons, certainly will become persons if not interfered with, and it has been maintained by some that the destruction of such potential persons is also seriously wrong in itself. Finally, all human infants, whether normal or not, would seem to possess a positive biological or ontological nature that would, were it not for certain defects, result in the individual's developing into a person, and some have maintained that the destruction of anything with such a nature is seriously wrong in itself.

There is one other important line of argument that needs to be considered. This is the view that infanticide, even if not seriously wrong in itself, may be wrong because of its consequences.

Are human infants persons?

The first argument against infanticide runs as follows. The reason that it is seriously wrong to destroy adult human beings is that they are persons, where the term "person" refers to any being possessing those nonpotential properties, whatever they may be, that make it seriously wrong to destroy something. It is true, as a matter of fact, that normal human infants also possess those person-making characteristics. Hence the killing of normal human infants is seriously wrong.

Evaluation of this argument requires answers to two questions. First, what nonpotential properties must an entity possess to be a person? Second, what state of development have human infants in fact achieved with respect to these person-making characteristics?

Both questions raise difficult issues. With

respect to the first, the natural starting point is with the view that something that has no mental life at all cannot be a person. But while this is surely a necessary condition, it does not seem sufficient. Nonhuman animals enjoy at least a rudimentary mental life, including sensory experiences and desires, and yet it is generally believed, at least in Western society, that it is not seriously wrong to kill such beings. It is important to emphasize that this view may be mistaken and that those who maintain that it is wrong to destroy any sentient being may be right. But for the sake of the present discussion it will be assumed that this is not the case. The issue, then, is what more, beyond sensation and rudimentary desires, is needed in order for an organism to be a person.

Several answers have been proposed. Among the more plausible suggestions are the following: (1) A person consists of mental states, existing at different times, and unified by memory; (2) a person is an entity possessing both consciousness and the capacity for thought; (3) a person is an entity that enjoys not only consciousness, but self-consciousness, that is, is aware of itself as a conscious being existing through time; (4) a person is an entity that is aware of itself not only as a possessor of states of consciousness but as a potential agent; (5) a person is an entity possessing consciousness and capable both of envisaging a future for itself and of having desires about such future states of itself; (6) a person is an entity capable of interacting socially and morally with other beings with similar capacities.

How can it be determined what the correct answer is? One might appeal to intuitions, but this seems very unsatisfactory, since the intuitions of people differ on this matter even more than on most moral issues. A somewhat better approach is to attempt to relate the question to other moral issues. This could be done by finding a set of general moral principles that would lead to plausible results over a wide range of moral issues, and from which one could derive an answer to the question what nonpotential properties make it seriously wrong to destroy something. But while this approach has the virtue of encouraging the formulation of a coherent moral outlook based on principles of wide applicability, it will not resolve disagreements among people, since the intuitions of people about what general moral outlook is best also differ greatly.

What is really needed is a convincing account of the foundations of morality, an account that would enable one to determine what the correct moral principles are. Whether such an account is even possible is a matter of great controversy. Some philosophers have attempted to show that the notion of justified moral principles is, in the final analysis, conceptually incoherent. Other philosophers, however, believe that it is possible to provide a justification.

The issues involved in this question are among the most difficult in metaethics, and they cannot be considered here. There is, however, one idea that deserves mention, since it suggests one approach that may help one to evaluate alternative answers to the question, What makes something a person? The basic idea is this: Even if moral principles turn out not to have truth-values, so that it is impossible to justify them in the sense of showing that they are epistemically reasonable, it may be possible to justify them in a different way, namely, by showing that it is rational for a society to bring it about that all its members act upon those principles. To do this would involve showing that it was in the interest of society to inculcate those moral principles. If, then, it could be argued that the interest of society is ultimately a function of the interests of its members, and that the interests of individual members are ultimately a function of their needs and desires, it would follow that for a moral principle to be capable of being justified in this way it would have to be somehow related to the satisfaction of the needs and desires of individuals in society.

How would this help to resolve the issue? It would do so in that it suggests that, in considering the alternative answers advanced as to the morally relevant concept of a person, the question to ask is what nonpotential properties are such that the destruction of an individual with those properties would entail the nonsatisfaction of some important desire or need. The first three answers do not seem to pick out properties that meet this requirement. The alternative that satisfies the requirement in the most straightforward way is the fifth: If a person is an entity capable of envisaging a future for itself and of having desires about future states of itself, those desires will go unsatisfied if such an entity is destroyed. Moreover, since the desire that there be future states of oneself will normally be very strong, one can see why it is that we regard the destruction of persons as not merely wrong but seriously so.

It must, however, be stressed that other answers may also satisfy the suggested requirement. Moreover, one might challenge the requirement itself, since the force of the argument in support of it is not easy to evaluate. The issues are difficult ones, and much more metaethical investigation will probably be needed to arrive at an answer that can be justified in a satisfactory fashion.

The second important issue is whether human infants have, as a matter of fact, developed to the point where they possess properties that suffice to make them persons. It might be thought that, in the absence of a compelling answer to the first question, there could be no possibility of dealing fruitfully with this second question. But that need not be so. The number of properties that might plausibly be viewed as person-making characteristics is, after all, quite limited. Normal adult human beings possess all of those properties, so that even if one does not know which of those properties really are the ones that make something a person, one is still justified in holding that normal adult human beings are persons. On the other hand, it is clear that human infants lack many of those properties. If it turned out that they lacked *all* of them, then one would be justified in concluding that human infants are not persons, even if one did not know precisely which of those properties were the ones that make something a person.

But is there good reason for believing that newborn humans lack all such properties? A number of interrelated considerations have been advanced in support of the claim that they do. First, it has been argued that there does not seem to be anything in the behavior of human infants that suggests the presence of a capacity for thought, or any awareness of oneself either as a conscious being existing through time or as a potential agent. In response, it has been contended that there is at least some evidence that an infant remembers its experiences, and hence is a person in the sense of an entity that enjoys states of consciousness existing at different times, unified by memory. The problem with this reply, however, is that, if memory is interpreted in a weak sense which makes the claim that human infants remember their experiences uncontroversial, namely, in a sense in which memory does not involve any capacity for thought, then it is equally true that nonhuman animals remember their experiences, and yet most people refuse to classify such animals as persons.

A second line of argument is that studies of the development of the human brain strongly support the conclusion that newborn humans have not developed to the point where their central nervous systems provide a basis for the higher mental functions. The electrical activity in the brain of a human infant, as recorded by an electroencephalograph, is neither as intense nor of the same wavelengths as that characteristic of fully conscious adults. The electroencephalograms of newborn humans closely resemble those of other species, such as frogs, and it is a long time before they are as complex as those of the great apes. Even more important is the evidence provided by direct anatomical studies. The brains of newborn humans are not fully developed. The complex functional systems of conjointly working cortical zones, which are the basis of the higher mental functions, are not present in humans at birth, as is the case with the respiratory and other systems. Many neurophysiologists even believe that the brain states that are the basis of higher mental functions, rather than maturing independently, do so only as a result of the child's interaction with its environment.

A third consideration that seems relevant to the question whether newborn humans are persons turns about the relative capabilities of human infants and animals of other species. Other animals are not in general inferior to human infants with respect to those behavioral capacities that are indicative of level of mental functioning. On the contrary, even quite young animals of many other species exhibit behavior of which human infants are incapable, and which provides some grounds for believing that those other animals enjoy a level of mental life qualitatively superior to that enjoyed by newborn humans. So it would seem that, unless one is prepared to grant that many nonhuman animals are also persons, one cannot consistently appeal to the behavioral capacities of newborn infants to support the claim that they are persons.

Is it seriously wrong to destroy potential persons?

The second argument against infanticide is as follows: It is wrong to kill persons. But it is also wrong, and seriously so, to destroy potential persons, that is, to destroy things that, in view of their biological or ontological nature, are such as will, if not interfered with, become persons. Since normal human infants, even if not persons, are certainly potential persons, their destruction is seriously wrong.

Perhaps the first thing to notice about this argument is its bearing upon the question of the morality of abortion. If it is sound, it appears to be equally an argument against abortion, at least in cases where the woman's life is not in danger. A fertilized human egg cell will, if not interfered with, normally develop into a person. So if infanticide were wrong because it involved the destruction of a potential person, it would seem that the same should be true of abortion. Moreover, it would seem not to make any difference how early in the development of the embryo the abortion was carried out. This is a conclusion that many people find morally unacceptable. This in itself does not, of course, show that the destruction of potential persons is not morally wrong. But it does suggest that it is difficult to accept the present argument against infanticide while rejecting the extreme conservative position on abortion.

The claim that it is seriously wrong to destroy potential persons has, however, been directly challenged as well. There are at least two important objections to it that deserve careful consideration. One runs as follows: First, if it is seriously wrong to destroy potential persons, then it must also be seriously wrong to prevent an organism that is a potential person from realizing that potential. For if it were not, one could first damage the brain of a human fetus or infant, so that it was no longer of such a nature as to develop into a person, and then destroy what was now something less than a potential person, and thereby avoid doing anything seriously wrong. Second, consider a system of objects whose structure and causal interrelationships are such that the system will, if not interfered with, give rise to a person. If it is wrong to prevent a unified organism from developing into a person, surely it is also wrong to prevent a system of interrelated objects that would normally develop into a person from doing so. For how can it make a moral difference whether the potentiality and its ground reside in a single organism or in a structured system of causally interrelated objects? What matters, it would seem, is simply the potentiality whose realization is prevented.

If sound, this argument shows that it is seriously wrong to destroy a potential person only if it is also seriously wrong to prevent the coming into being of a person, the potentiality of which is grounded in the nature and structure of some system of causally related objects. This implies that the destruction of potential persons is seriously wrong only if certain methods of contraception are also seriously wrong, namely, those involving interference that prevents a system of objects that includes a human sperm and ovum, causally so related that the normal outcome will be a fertilized human egg cell, from giving rise to that outcome.

If, then, one believes that contraception cannot be wrong in itself, this argument seems to show that one cannot maintain that the fact that infanticide involves the destruction of a potential person makes it seriously wrong. Moreover, even if one thought that some methods of contraception were morally objectionable, this argument might still raise difficulties, since very few who think that contraception and infanticide are wrong hold that they are wrong to the same degree, as would seem to be implied if the reason they were wrong were the same: prevention of the coming into being of a potential person.

The second objection to the view that infanticide is wrong because it involves the destruction of a potential person attempts to establish an even stronger conclusion, namely, that the destruction of potential persons is seriously wrong only if intentionally refraining from reproduction is also seriously wrong.

The argument turns upon the following critical principle: If C is some type of causal process that normally gives rise to a state of affairs of type S, and if processes of type C would not have moral value were it not for the fact that they do result in states of affairs of type S, then if other factors such as motive, risk to the agent, etc., are equivalent, there is no serious moral difference between intentionally refraining from bringing about the existence of a process of type C and intentionally destroying a process of type C *before* any state of affairs of type S has been produced by it. Some philosophers find this principle plausible; others do not. Some object to it because they think it entails consequentialism, the view that the rightness or wrongness of an action is a function of the goodness or badness of its consequences. This objection is mistaken; the principle in question is perfectly compatible with a thoroughgoing rejection of consequentialism. Another objection is to the effect that there is a moral difference between acts and intentional omissions. In reply it can be said, first, that the claim that there is *any* moral difference between acts and intentional omissions is at present being deeply questioned by some contemporary moral philosophers (Rachels). Second, it is not enough to claim that there is *some* moral difference, for example, between killing a person and intentionally letting a person die, in cases where the motives, etc., are the same. If the above principle is to be rejected, it must be claimed that there is a *serious* moral difference, and this is a highly controversial contention.

Given the above principle, the argument proceeds as follows: If the world were different, so that normal members of the species Homo sapiens did not develop into persons, human organisms would not possess the moral significance they now do. It then follows that, *if* newborn humans are only potential persons, human organisms at that point in their development would not possess any moral significance were it not for the fact that they later become persons. Application of the above principle then leads to the conclusion that there cannot be any serious moral difference between destroying a human infant before it has become a person, and intentionally refraining from bringing about the existence of a human infant. Hence there can be no serious moral difference between infanticide, on the one hand, and, on the other, intentionally refraining from activity that would initiate a causal process that would give rise to a person, provided that motivation and other factors are equivalent in the two cases. So if intentionally refraining from intercourse is not seriously wrong in itself, neither is infanticide.

To sum up, the view that infanticide in the case of normal human infants is seriously wrong because it involves the destruction of potential persons has, either by itself or in conjunction with other assumptions, three important consequences. The first is that abortion appears to be seriously wrong, no matter how early it is performed. The second is that certain methods of contraception appear to be equally wrong. The third is that, if intentionally refraining from bringing into existence certain types of causal processes—those that would not have moral significance were it not for the state of affairs to which they give rise—is morally comparable to destroying such processes before they bring about the morally significant result, then intentionally refraining from conception is morally comparable to infanticide.

Is it seriously wrong to destroy latent persons?

The third argument against infanticide is this: It is wrong to kill persons. But it is also wrong, and seriously so, to destroy latent persons, that is, to destroy things whose positive biological or metaphysical nature is such that they would, were it not for the presence of certain defects, develop into persons. Since human infants, both normal and abnormal, are latent persons, their destruction is seriously wrong.

The nature, or ontological constitution in question, may be conceived of in radically different ways. It may be construed naturalistically, as a matter of a certain sort of biological makeup. Or it may be construed metaphysically, in terms of an ontological structure that somehow transcends the physical realm.

The latter view appears to have been adopted by some contemporary writers (cf. Adler; May). And it is certainly possible that there is a difference between the human species and all other species that is not explicable in biological or neurophysiological terms. It is, however, a view that it is difficult to offer much evidence in support of.

There is, however, a more important objection to this third line of argument, and it is one that applies regardless of whether one views the ontological constitution of a thing as something biological or as something metaphysical. The basic problem with this line of argument arises from the fact that, while not all latent persons are potential persons, all potential persons are latent persons. This suggests that if it is prima facie seriously wrong to destroy latent persons, it is also prima facie seriously wrong to destroy potential persons. And this in turn means that, if the arguments, outlined in the previous section, in support of the view that it is not seriously wrong to kill potential persons are sound, they also support the conclusion that the destruction of latent persons is not seriously wrong in itself. So unless those arguments can be rebutted, it is not possible to appeal *either* to the fact that infanticide involves the destruction of potential persons *or* to the fact that infanticide involves the destruction of latent persons as grounds for holding that infanticide is morally wrong.

Does the practice of infanticide have undesirable consequences?

The first three arguments against infanticide were directed to showing that infanticide is seriously wrong in itself, independently of its consequences. This final argument, in contrast, attempts to show that infanticide, even if not intrinsically wrong, is morally unacceptable because it is likely to have quite undesirable consequences.

Different suggestions have been advanced as to the undesirable consequences of admitting the practice of infanticide. The two most important are these:

First, it has been contended that acceptance of infanticide would lead to a weakening of respect for the lives of other human beings. A society that had no qualms about destroying human infants when it was socially useful to do so would rapidly come to accept the view that it was morally permissible to destroy other human beings, such as the weak, the handicapped, and the elderly, when that was socially useful.

Second, it has been suggested that acceptance of infanticide would result in a weakening of parental feeling, so that people would come to have less concern for those offspring that they did not destroy and treat them more harshly than they would in a society where infanticide was unacceptable. Besides being undesirable in itself, such a reduction in tenderness and consideration could not but have an adverse effect upon the personality development of the children so treated.

If acceptance of infanticide did have either of these consequences, there would be good reason for prohibiting it, even if it is not wrong in itself. The crucial issue, then, is whether there is good evidence in support of the claim that acceptance of the practice of infanticide will, as a matter of empirical fact, have either of these consequences.

In the case of the claim that acceptance of infanticide would lead to a weakening of respect for the lives of other human beings, the underlying train of thought seems to be this: If one accepts infanticide, one is willing in some cases to accept violations of the principle that one ought not to kill innocent human beings, and doing this will surely make one more willing to countenance violations of that principle in other cases.

Those who reject this line of argument claim that it rests upon a confusion. They argue that the basic moral principle involved is not that one ought not to kill innocent human beings but that one ought not to kill innocent *persons*. And then, if human infants are not persons, acceptance of infanticide does not involve accepting violations of the relevant basic moral principle, and so there is no reason to conclude that one will thereby become more prone to violate moral principles when it is socially useful to do so.

In reply to this, it might be argued that, even if it is true that the basic moral principle involved here is that one ought not to kill innocent persons, many people still think in terms of human beings, and acceptance of infanticide will tend to lessen the sense that such people have of the wrongness of killing human beings who are persons.

But in response to this it has been argued that, even if this is to some extent the case, the consequence in question can be avoided. What will be needed is a program of education. The willingness of most people to accept the termination of human life involved in cases of irreparable brain damage and abortion performed early in pregnancy suggests that such people recognize, even if not in a very clear or explicit fashion, that what matters is not membership in a particular biological species but whether one is a person—an individual enjoying a certain sort of mental life. Exposure to ethical discussions of the conditions under which it is seriously wrong to destroy something would enable people to formulate a clear and explicit moral perspective that was congruent with their underlying moral intuitions. As a result, it is not clear that there is good reason to believe that acceptance of the practice of infanticide need result in any weakening of respect for the lives of persons.

The other important consequence that it has been argued would follow from acceptance of the practice of infanticide is that parents would behave with less tenderness toward their children. Those who reject this contention have responded as follows: First, even though pets are not thought of as persons, it is not thought morally acceptable to treat them cruelly. Why would the case be different for human infants, if people viewed them as sentient beings that were only potential persons? Second, parents are deeply attached to their children simply because they are their own. Third, there is the direct evidence provided by societies where infanticide was an accepted practice. As was noted above, the testimony of anthropologists indicates that the tenderness with which parents treated those children who were allowed to live, and the concern that they exhibited for the well-being of those children, were often superior to that which is typical in our own society.

The attempt to argue that infanticide is wrong because it has undesirable consequences is thus no less debatable than the attempts to show that infanticide is intrinsically wrong. All of the major arguments against infanticide are open to objections that some consider convincing. The claim that newborn humans are persons seems problematic. The contention that the destruction of potential persons is seriously wrong seems to have implications that many cannot accept. And finally, it is far from clear that there are good grounds for holding that acceptance of the practice of infanticide would have undesirable consequences.

Even if it did turn out that none of these arguments were sound, might not one still be justified in holding that infanticide is seriously wrong? For while it is true that in the case of some moral issues one can offer an argument, in the case of others one can only fall back upon one's fundamental moral intuitions. And why may infanticide not be a case of the latter sort, where argument is neither possible nor necessary?

This line of thought raises some deep metaethical issues, which cannot be considered here. However, it can be argued that even if those metaethical problems are set aside, this view is exposed to an apparently serious objection, in view of the historical facts mentioned earlier. First, the moral intuitions of people with regard to infanticide disagree very dramatically across different cultures. Second, the ethical outlook of Western society on the question of infanticide appears to have derived mainly from a specific theological outlook, and one that would now be rejected by many. These facts suggest that it cannot be justifiable to regard the moral intuitions of our own society as adequate grounds for the view that infanticide is seriously wrong.

The question of the morality of infanticide is an extraordinarily difficult one. It would seem that any final resolution of the issue must await further research. A central part of such research will be philosophical reflection upon the crucial moral questions of, first, what makes something a person, that is, what nonpotential properties make it seriously wrong to destroy something; and second, whether the destruction of potential persons is seriously wrong. And it may be, for reasons indicated earlier, that a satisfactory answer to these questions cannot be arrived at without a solution to the central problem of metaethics, that of the ultimate foundations of morality. In addition, it may turn out that the morally relevant concept of a person is such that further scientific investigation of the psychological and neurophysiological constitution of human infants is needed before it can be determined whether they have in fact developed to the point where they possess properties that suffice to make them persons.

MICHAEL TOOLEY

[*Directly related are the other articles in this entry, especially* ETHICAL PERSPECTIVES ON THE CARE OF INFANTS, *and the following entries:* ABORTION, *article on* CONTEMPORARY DEBATE IN PHILOSOPHICAL AND RELIGIOUS ETHICS; ACTING AND REFRAINING; LIFE; PERSON; *and* RIGHTS. *For discussion of related ideas, see the entries:* CHILDREN AND BIOMEDICINE; *and* ETHICS, *articles on* DEONTOLOGICAL THEORIES, TELEOLOGICAL THEORIES, UTILITARIANISM, *and* THEOLOGICAL ETHICS. *Other relevant material may be found under* ANIMAL EXPERIMENTATION, *article on* PHILOSOPHICAL PERSPECTIVES; *and* CONTRACEPTION.]

BIBLIOGRAPHY

ADLER, MORTIMER. *The Difference of Man and the Difference It Makes.* Introduction by Theodore T. Puck. Based on the Encyclopaedia Britannica Lectures delivered at the University of Chicago, 1966. New York: Holt, Rinehart & Winston, 1967.

ARRISTOTLE. *Politics.* 7, 16 1334B–1336A. Brief statement on infanticide and abortion.

BAIER, KURT. *The Moral Point of View: A Rational Basis of Ethics.* Contemporary Philosophy. Ithaca: Cornell University Press, 1958. Interesting discussion of fundamental metaethical issues, centering on the question of the foundations of morality.

BENN, S. I. "Abortion, Infanticide, and Respect for Persons." Feinberg, *The Problem of Abortion,* pp. 92–104. Account of what it is to be a person, supporting the view that infants are not persons, and then arguing that infanticide, even if not intrinsically wrong, may be open to consequentialist objections.

BENNETT, JONATHAN. "Whatever the Consequences." *Analysis* 26 (1966): 83–102. Incisive criticisms of a commonly held view of the moral significance of the distinction between an action and its consequences.

BENTHAM, JEREMY. *Bentham's Theory of Legislation.* 2 vols. Translated by Charles Milner Atkinson. London: Oxford University Press, 1914, vol. 2, pp. 38–39. Brief statement of views on infanticide.

BRANDT, R. B. "The Morality of Abortion." *Monist* 56 (1972): 503–526. Discusses the selection and justification of moral principles dealing with termination of life. Unusually incisive and interesting.

COHEN, MARSHALL; NAGEL, THOMAS; and SCANLON, THOMAS, eds. *The Rights and Wrongs of Abortion.* A Philosophy and Public Affairs Reader. Princeton: Princeton University Press, 1974. Raises issues that are relevant to the question of the morality of infanticide.

DUFF, RAYMOND S., and CAMPBELL, A. G. M. "Moral and Ethical Dilemmas in the Special-Care Nursery." *New England Journal of Medicine* 289 (1973): 890–894. Ethical dilemmas that confront doctors concerned with the care of defective infants.

FEINBERG, JOEL, ed. *The Problem of Abortion.* Basic Problems in Philosophy Series. Belmont, Calif.: Wadsworth Publishing Co., 1973. Most of the articles in this excellent anthology deal with issues that are relevant to infanticide as well as to abortion, and some of the articles contain discussions of the question of the morality of infanticide.

FLANDRIN, JEAN-LOUIS. "L'Attitude à l'égard du petit enfant et les conduites sexuelles dans la civilisation occidentale." *Annales de démographie historique* (1973): 143–205.

FLETCHER, JOSEPH. "Indicators of Humanhood: A Tentative Profile of Man." *Hastings Center Report* 2, no. 5 (1972), pp. 1–4. Brief but useful discussion of the properties that make something a person.

GOODRICH, T. "The Morality of Killing." *Philosophy* 44 (1969): 127–139. Raises some important questions about the fundamental moral principles relevant to the question when it is morally wrong to kill something.

Hocart, A. M. "Infanticide." *Encyclopaedia of the Social Sciences.* 15 vols. Edited by Edwin R. A. Seligman. New York: Macmillan Co., 1932, vol. 8, pp. 27–28. A brief summary of cultural differences with respect to acceptance of infanticide.

Langer, William L. "Infanticide: A Historical Survey." *History of Childhood Quarterly: The Journal of Psychohistory* 1 (1974): 353–365.

Lecky, William Edward Hartpole. *History of European Morals from Augustus to Charlemagne.* 2 vols. 3d ed. London: Longmans, Green & Co., 1910. Contains an account of the differences between pagan and Christian attitudes toward infanticide that, though useful, is both rather emotional and philosophically naive.

Luriia, Aleksandr Romanovich. *Higher Cortical Functions in Man.* Translated by Basil Haigh. London: Tavistock Publications, 1966.

McKeown, Thomas. *The Modern Rise of Population.* London: Edward Arnold, 1976. Contains a good discussion of the role of infanticide in limiting population growth in Europe.

May, William E. "What Makes a Human Being to Be a Being of Moral Worth?" *Thomist* 40 (1976): 416–443.

Minkowski, Alexandre, ed. *Regional Development of the Brain in Early Life.* Oxford: Blackwell Scientific, 1967. Symposium organized by the Council for International Organizations of Medical Sciences and by the Délégation à la Recherche Scientifique et Technique.

Noonan, John Thomas, Jr. "An Almost Absolute Value in History." *The Morality of Abortion: Legal and Historical Perspectives.* Edited by John T. Noonan, Jr. Cambridge: Harvard University Press, 1970, pp. 1–59.

Plato. *Republic.* Bk. 5, 460. Plato's views on infanticide.

Rachels, James. "Active and Passive Euthanasia." *New England Journal of Medicine* 292 (1975): 78–80.

Robinson, Roger James, ed. *Brain and Early Behaviour Development in the Fetus and Infant.* New York: Academic Press, 1969. Proceedings of a C.A.S.D.S. Study Group on "Brain Mechanisms of Early Behavioural Development" held jointly with the Ciba Foundation, London, February 1968.

Sterman, M. B.; McGinty, Dennis J.; and Adinolfi, Anthony M., eds. *Brain Development and Behaviour.* New York: Academic Press, 1971.

Westermarck, Edvard Alexander. "The Killing of Parents, Sick Persons, Children—Feticide." *The Origin and Development of the Moral Ideas.* 2 vols. London: Macmillan & Co., 1906–1908, vol. 1, chap. 17, pp. 393–413. A reasonably unemotional description of differences in attitudes toward infanticide in various cultures.

Williams, Glanville Llewelyn. "The Protection of Human Life." *The Sanctity of Life and the Criminal Law.* Foreword by William Warren. James D. Carpentier series, 1956. New York: Alfred A. Knopf, 1957, chap. 1, pp. 3–33. A useful discussion of infanticide from both a moral and a legal point of view.

INFORMED CONSENT IN HUMAN RESEARCH

I. SOCIAL ASPECTS   *Bradford H. Gray*
II. ETHICAL AND LEGAL ASPECTS   *Karen Lebacqz and Robert J. Levine*

I
SOCIAL ASPECTS

An analysis of the social aspects of informed consent in human experimentation can focus on several distinct problems. In this article, we shall briefly consider the origin of concern about informed consent, the problem of making an abstract concept operational, factors that affect the quality of informed consent in transactions between professionals and clients, and the problem of assuring that accepted standards of informed consent are consistently met.

Origin of concern

One can begin by attempting to account for the concept itself. That is, why has informed consent come to be regarded as an essential element of certain types of human relationships, particularly in researcher–subject and doctor–patient relationships? An answer can be sought in the several social functions served by informed consent (Capron, pp. 364–376) or in broad social values that underlie the concept of informed consent and help explain the strong societal response to revelations of instances in which the standard of consent seems to have been grossly violated. Or one may seek the explanation for the growth of concern about informed consent in broad historical changes in the nature of disease and related changes in doctor–patient relationships (Freidson) or in the development of human experimentation as an activity distinct from patient care and the rapid growth in its support by publicly accountable government agencies, particularly since World War II. An understanding of the social nature of the concept of informed consent has only begun (Veatch, p. 22). Much historical and comparative work remains to be done.

Making informed consent operational

A second sense in which informed consent is social is in the way it is given operational meaning: What must be disclosed for consent to be informed, and what constitutes sufficient pressure or constraint to render consent involuntary? Relatively little is known about systematic variations among researchers and others in how these questions are answered (Barber et al., pp. 29–57). It is clear, however, that there is disagreement regarding both broad categories of research (e.g., research on prisoners) and particular examples. Total disclosure often seems to be a practical impossibility, and who is to say whether the prisoner, the patient, or the student is "really" free to give voluntary consent? Though general guidance is provided by statutory and case law, the routine procedural answer to such questions comes from human subjects review committees (*Institutional Guide to DHEW*), which, among other functions, review investigators' judgments about what must be disclosed to subjects. Evidence to date indicates that these committees have not given equal attention to the less concrete matter of avoiding duress when consent is obtained (Cowan; Gray, chap. 3; Melmon, Grossman, and Morris). A systematic study of variations among institutional review committees is being conducted in the United States by the National Commission for the Protection of Human Subjects of Biomedical and Behavioral Research. This study approaches decision making by review committees as a social process, the outcome of which is influenced by the composition and procedures of the committee, by their institutional setting, and by the characteristics of the research under review.

No matter how informed consent is made operational by an investigator or review committee, whether it actually takes place depends upon another set of social factors. We shall now turn to the factors that complicate the informed consent transaction, whether between doctors and patients or between researchers and subjects.

Quality of informed consent

Although a small but growing empirical literature exists regarding factors that affect the quality of informed consent in professional–client interactions, little is known of differences among disciplines or specialties in norms that support the disclosure or control of information. However, such differences undoubtedly exist. Healing ideologies vary in emphasis on the importance of a patient's faith in the healer, and research methodologies vary in dependence on the naiveté of subjects. While human subjects review committees may reduce the effect of such variations among disciplines or specialties, it is unlikely that they can be eliminated quickly by such committees, even should this be deemed desirable. As norms, their existence may not only be hard to detect, but they may also provide justification to investigators for noncompliance with review committee decisions requiring certain disclosures. Nevertheless, there are some sources of pressure today for greater standards of disclosure by professionals. A growing body of malpractice case law reflects increased expectations of disclosure by professionals in their relations with clients (Capron). There are also indications that patients commonly desire more information than physicians are willing to supply (McIntosh). While there is some evidence that there is exaggeration in the fears of some physicians that certain disclosures might cause patients to make irrational decisions (Alfidi), the effects of new levels of disclosure by professionals are not well documented at present.

The other half of the subject–researcher relationship also contributes to variations in the quality of consent. Client expectations of disclosure by professionals are related to social class (e.g., Duff and Hollingshead). Patient expectations regarding control over the professional–client relationship are similarly class-related. The fact that these factors vary within the population is important, because there is some evidence that the expectations of subjects regarding disclosure by professionals and the extent to which subjects are willing to leave the making of decisions to professionals are factors that affect the adequacy of informed consent (Gray).

A related aspect of the researcher–subject relationship is the amount of difference or social distance—in education, social class, or race—between investigators and subjects. This social distance translates into a differential in power: The greater the discrepancy between researcher and subject, generally the less likely is informed consent to occur (ibid., p. 138). The impact of social distance on consent is particularly serious in view of the evidence that clinic patients who are low in education and in power may be differentially selected for research that has an unfavorable ratio of risks to benefits (Barber et al., pp. 53–57).

A further complication of discrepancies in background between subjects and researchers is that even nontechnical words may not transmit intended meaning. For example, the words "new drug" do not necessarily convey "experimental drug." Or a subject may assume that being a "study patient" means only that students will be present. While words like "experiment" may create exaggerated fears, the frequently used euphemisms may fail to communicate the essential aspects of the situation.

The setting in which consent is obtained and the way consent forms are presented to subjects may also affect the quality of consent (Gray, pp. 202–234). A key point is that signed consent forms cannot be equated with informed consent. The danger is always that the form rather than the substance of consent will become the end sought. This is not to suggest that consent forms serve no purpose. The only existing data show that the quality of consent is improved by the introduction of consent forms (McCollum and Schwartz). However, the way in which consent forms are used can affect the quality of consent. The use of a consent form unaccompanied by an oral explanation is not a reliable way of transmitting information from researcher to subject. In addition, the delegation to subordinates of soliciting subjects' consent can present hazards to meaningful consent in some circumstances (Gray, pp. 215–221).

Doctor–patient relationship as a barrier to consent

The traditional doctor–patient relationship can itself be a barrier to informed consent when a physician-investigator is dealing with a patient-subject, since surrender of control to the physician is part of the socially expected attitude of the sick individual (Parsons, pp. 433–465). The individual who is accustomed to deferring to physicians and who does not recognize the differences between the doctor–patient relationship and the researcher–subject relationship may not expect disclosures of the sort required by informed consent. Thus, the obtaining of informed consent may necessitate a degree of education of the subject, at least to the basic points that participation is voluntary and that a decision by the subject is required.

Informed consent as a social control problem

Let us now change perspectives and briefly consider informed consent as a problem of social control—that is, how can there be assurance that investigators will obtain informed consent from subjects? Disclosure and control must be considered together in this article because an important base of power or control in a human relationship is the possession of valuable knowledge by one of the parties. Disclosure requirements go to the heart of such power and would lead the social theorist to expect some degree of resistance by the party in control. (Consider, for example, the debate in some medical journals over proposals to share with patients their own medical records.)

The social control problem is intensified by the fact that in many instances it is in investigators' immediate interests to do the minimum possible in obtaining consent. Even where investigators' personal values recognize the importance of informed consent, other factors may impinge. Some evidence (and reason) suggests that, in research that carries a recognizable degree of risk, well-informed subjects are more reluctant to participate than are poorly informed subjects (Gray, pp. 140–154; Martin et al.). The investigator is located in a social system that rewards success in research. Furthermore, the conduct of research is complex and demanding; the recruitment of subjects is but one of many problematic elements in the process. Although the researcher is expected to do research and to obtain informed consent, the latter may make the former—to which most rewards are attached—more difficult. There are some indications that investigators' particular location in the scientific reward structure affects their ethical standards (Barber et al., pp. 59–92).

A further complication in assuring that informed consent is obtained is that it is often not a highly visible activity. The negotiations between the doctor and patient, and between the researcher and subject, typically occur in private. The tangible aspect of informed consent—the consent form—can be highly misleading as an indicator of informed consent. In one study of women who had signed consent forms and participated in a double-blind investigation of drugs for inducing labor, thirty-nine percent understood only later that they were subjects of research (Gray, pp. 128–140).

So long as the rewards for successful recruitment of subjects are high and the probability is low that failure to obtain consent will be brought to light and penalized, it is safe to predict that shortcomings in informed consent will be a common feature of human experimentation. Thus, there have been calls for providing investigators with special training to increase their sensitivity to ethical and legal issues involved in research on human subjects and for increasing the monitoring by human subjects review committees of the conduct of the research they have approved. Among the more specific suggestions that have been made are the "two-stage consent form" in which the subject's understanding of the research is documented in a brief quiz (Miller and Willner), the presence of witnesses or "consent committees" when consent is obtained (U.S. DHEW), and postparticipation interviews with samples of subjects to assure that no pattern of consent shortcomings has developed (Gray, p. 251).

BRADFORD H. GRAY

[*Directly related is* HUMAN EXPERIMENTATION, *articles on* BASIC ISSUES, PHILOSOPHICAL ASPECTS, *and* SOCIAL AND PROFESSIONAL CONTROL. *Other relevant material may be found under* PATIENTS' RIGHTS MOVEMENT; RESEARCH, BIOMEDICAL; RIGHTS, *article on* RIGHTS IN BIOETHICS; RISK; THERAPEUTIC RELATIONSHIP, *articles on* SOCIO-HISTORICAL PERSPECTIVES *and* CONTEMPORARY MEDICAL PERSPECTIVE; *and* TRUTH-TELLING.]

BIBLIOGRAPHY

ALFIDI, RALPH J. "Informed Consent: A Study of Patient Reaction." *Journal of the American Medical Association* 216 (1971): 1325–1329.

BARBER, BERNARD; LALLY, JOHN J.; MAKARUSHKA, JULIA LOUGHLIN; and SULLIVAN, DANIEL. *Research on Human Subjects: Problems of Social Control in Medical Experimentation.* New York: Russell Sage Foundation, 1973.

CAPRON, ALEXANDER MORGAN. "Informed Consent in Catastrophic Disease Research and Treatment." *University of Pennsylvania Law Review* 123 (1974–1975): 340–438.

COWAN, DALE H. "Human Experimentation: The Review Process in Practice." *Case Western Reserve Law Review* 25 (1974–1975): 533–564.

DUFF, RAYMOND S., and HOLLINGSHEAD, AUGUST B. *Sickness and Society.* New York: Harper & Row, 1968.

FREIDSON, ELIOT. *Profession of Medicine: A Study of the Sociology of Applied Knowledge.* Edited by Robert Bierstedt. New York: Dodd, Mead & Co., 1970.

GRAY, BRADFORD H. *Human Subjects in Medical Experimentation: A Sociological Study of the Conduct and Regulation of Clinical Research.* Health, Medicine, and Society, A Wiley-Interscience Series. New York: John Wiley & Sons, 1975.

Institutional Guide to DHEW Policy on Protection of Human Subjects. DHEW Publication no. (NIH) 72-102. Bethesda, Md.: National Institutes of Health, 1971.

McCOLLUM, AUDREY T., and SCHWARTZ, A. HERBERT. "Pediatric Research Hospitalization: Its Meaning to Parents." *Pediatric Research* 3 (1969): 199–204.

McINTOSH, JIM. "Process of Communication, Information Seeking and Control Associated with Cancer: A Selective Review of the Literature." *Social Science and Medicine* 8 (1974): 167–187.

MARTIN, DANIEL C.; ARNOLD, JOHN D.; ZIMMERMAN, T. F.; and RICHART, ROBERT H. "Human Subjects in Clinical Research—A Report of Three Studies." *New England Journal of Medicine* 279 (1968): 1426–1431.

MELMON, KENNETH L.; GROSSMAN, MICHAEL; and MORRIS, R. CURTIS, JR. "Emerging Assets and Liabilities of a Committee on Human Welfare and Experimentation." *New England Journal of Medicine* 282 (1970): 427–431.

MILLER, ROBERT, and WILLNER, HENRY S. "The Two-Part Consent Form: A Suggestion for Promoting Free and Informed Consent." *New England Journal of Medicine* 290 (1974): 964–966.

PARSONS, TALCOTT. *The Social System.* New York: Free Press, 1964.

United States; Department of Health, Education, and Welfare. "Protection of Human Subjects: Proposed Policy." *Federal Register* 39 (1974): 30648–30657.

VEATCH, ROBERT M. "Ethical Principles in Medical Experimentation." *Ethical and Legal Issues of Social Experimentation.* Edited by Alice M. Rivlin and P. Michael Timpane. Washington: Brookings Institution, 1975, pp. 21–59.

II
ETHICAL AND LEGAL ASPECTS

"The voluntary consent of the human subject is absolutely essential." This, the first sentence of the 1947 Nuremberg Code, signals the centrality of the consent requirement in research using human subjects (United States, Defense Department). Prior to Nuremberg, statements of medical and other professional organizations apparently made no mention of the necessity of consent. Subsequently, the tendency to focus on "informed consent" has been reinforced by public outcry over the inadequacy of consent in certain landmark cases, e.g., Willowbrook, Jewish Chronic Disease Hospital, Tea Room Trade (Katz, 1972, pp. 1007, 9, 325), and Tuskegee (United States, Tuskegee Syphilis Study). Indeed, the issue of "informed consent" has so dominated recent discussion of the ethics of research that one might be led to think erroneously that other issues (e.g., research design, selection of subjects) are either less important or more satisfactorily resolved.

Grounding of "informed consent"

Philosophical basis. The philosophical foundations of the requirement for "informed consent" may be found in several lines of reasoning (Veatch, 1976, pp. 2–21; Engelhardt, p. 6). Based upon the Hippocratic admonition "to help, or at least, to do no harm," one can justify seeking consent for the benefit of the patient; to do so provides a mechanism for ascertaining what the patient would consider a "benefit." However, a focus solely on patient benefit would allow physicians not to seek consent when they judge that doing so might harm patients (see the discussion of incomplete disclosure below). Thus this justification alone does not suffice to establish a requirement to seek consent.

The requirement can also be justified on utilitarian grounds: The practice of seeking consent may produce the "greatest good for the greatest number" by forestalling suspicion about research, thus ensuring a subject population and increasing the efficiency of the research enterprise. Again, however, the justification fails to stand alone, since it can be used to justify *not* seeking consent: The social good might be better served by avoiding the inefficient and frequently time-consuming consent process. Carried to its extreme, the social benefit argument might support the use of unwilling subjects, as in Nazi Germany.

The firmest grounding for the requirement to seek consent is the ethical principle "respect for persons." As stated by Justice Cardozo, this principle asserts that "every human being of adult years and sound mind has a right to determine what shall be done with his own body" (Veatch, 1976, p. 16). In Kantian terms, this principle ensures that the research subject will be treated as an "end" and not merely as a means to another's end (Macklin and Sherwin, p. 443). Thus the purpose of the consent requirement is not to minimize risk, but to give persons the right to choose.

Religious basis. Several fundamental tenets of the Judeo-Christian tradition also provide grounding for the requirement to seek consent. This tradition affirms that each human life is a gift from God and is of infinite and immeasurable worth (the "sanctity of life"). The infinite worth of the individual requires that persons treat each other with respect and not interfere in each other's lives without consent.

The consent requirement can also be grounded explicitly in the notion of covenant. Seeking consent is an affirmation of the basic faithfulness or care required by the fundamental covenantal nature of human existence (Ramsey, 1970, pp. xi–xiii).

Legal basis. The legal grounding for the requirement for consent to research (Annas, Glantz, and Katz, pp. 29–37; Fried, pp. 18–25) is based on the outcome of litigation of disputes arising almost exclusively in the context of the practice of medicine. There is virtually no case law on the basis of which legal standards for consent to research, as distinguished from practice, might be defined (there is one Canadian case, *Halushka* v. *University of Saskatchewan*). The law defines, in general, the circumstances under which a patient, or by extension a subject, may recover damages for having been wronged or harmed as a consequence of failure to negotiate adequate consent.

The legal bases for the consent requirement —which also shed light on the ethical dimensions of consent—are twofold (Annas, Glantz, and Katz).

1. Traditionally, failure to obtain proper consent was treated as a "battery" action. Closely related to the principle of respect for persons and self-determination, the law of "battery" makes it wrong a priori to touch, treat, or do research upon a person without the person's consent. Whether or not harm befalls the patient/subject is irrelevant: It is the "unconsented-to touching" that is wrong.

2. The modern trend in malpractice litigation is to treat cases based upon failure to obtain proper consent as "negligence" rather than "battery" actions. The negligence doctrine combines elements of "patient benefit" and self-determination. To bring a negligence action, a patient/subject must prove that the physician had a *duty* toward the patient; that the duty was *breached;* that *damage* occurred to the patient; and that the damage was *caused* by the breach. In contrast to battery actions, negligence actions remove as a basis for the requirement for consent the simple notion that "unconsented-to touching" is a wrong—rather, such touching is wrong (actionable) only if it is negligent and results in harm; otherwise, the patient/subject cannot recover damages.

Under both battery and negligence doctrines, consent is invalid if any information is withheld that might be considered material to the decision to give consent.

Functions of "informed consent"

Katz and Capron have identified the following functions of "informed consent" (pp. 82–90): to promote individual autonomy, encourage rational decision making, avoid fraud and duress, involve the public, encourage self-scrutiny by the physician-investigator, and reduce the civil and/or criminal liability of the investigator and his or her institution.

In general, the *negotiations* for informed consent are designed to safeguard the rights and welfare of the subject while *documentation* that the negotiations have been conducted properly safeguards the investigator and institution (Levine, 1975, pp. 52–65). The net effect of the documentation may, in fact, be harmful to the interests of the subject. The retention of a signed consent form tends to give the advantage to the investigator in any adversary proceeding; moreover, the availability of such documents in institutional records may lead to violations of privacy and confidentiality (Reiss, pp. 74–86). Where investigators now retain consent forms signed by subjects, Reiss proposes that the interests of subjects would be better served if they retained statements of responsibility signed by investigators.

Whether or not negotiations for "informed consent" to research should be conducted according to different standards than consent to practice is controversial. Feinstein observes that it is our custom to adhere to a "double standard": "An act that receives no special concern when performed as part of clinical practice may become a major ethical or legal issue if done as part of a formally designed investigation" (Feinstein). In his view there is less need for formality in the negotiations for informed consent to a relationship where the interests of research and practice are conjoined—e.g., as in research conducted by a physician-investigator having the aim of demonstrating the safety and/or efficacy of a nonvalidated therapeutic maneuver—than when the only purpose of the investigator–subject relationship is to perform research. Capron, on the other hand, asserts: "Higher requirements for informed consent should be imposed in therapy than in investigation, particularly when an element of honest experimentation is joined with therapy" (Katz, 1972, p. 574). Levine (1975, pp. 41–42) concludes that patients are entitled to the same degree of thoroughness of negotiations for informed consent as are subjects of research. However, patients may be offered the opportunity to delegate decision-making authority to a physician, while subjects should rarely be offered this option. The most important distinction is that the prospective subject should be informed that in research, as contrasted with practice, the subject will be at least in part a means and perhaps only a means to another end.

Two interpretations of the consent requirement

Interpretations of the meaning and application of "informed consent" reflect a tension between respecting the autonomy of persons and protecting them from harm. Hans Jonas and Paul Ramsey develop a covenantal model in which subjects are respected and protected by ensuring that they give truly *informed* consent. Benjamin Freedman stresses the individual's *freedom of choice*, whether or not the choice is informed (Jonas; Ramsey, 1970; Freedman).

For Jonas and Ramsey, the consent requirement is derived from the duty to treat persons as ends, not merely as means. In research, subjects are "used" as means to the end of acquiring knowledge (in Jonas's terms, they are "sacrificed" for the collective good). Such "use" of persons is justified only if the subjects so identify with the purposes of the research that they will those purposes as their own ends. Only then are they not being "used," but instead they have become "co-adventurers" (Ramsey's term). The consent requirement thus affirms a basic covenantal bond between the researcher and the subject and ensures respect for the subject as an end, not merely a means.

To establish a true covenant, the subject's consent must be *informed*. Only subjects who genuinely know the purposes and appreciate the risks of research can assume those risks and adopt those purposes as their own ends. Ideal subjects, therefore, would be researchers themselves (Jonas, p. 17). The less one understands the risks and identifies with the purposes of research, the less valid is one's consent. Jonas therefore establishes a "descending order of permissibility" for the recruitment ("conscription") of volunteers. Both Ramsey and Jonas restrict the use of subjects unable to consent or to understand what is involved, permitting the use of such subjects only in research directly related to their own condition (Jonas) or their own survival and well-being (Ramsey).

This interpretation reflects certain assumptions that can be challenged. First, while neither Jonas nor Ramsey focuses exclusively on patients as subjects, their approach appears to be influenced largely by the "medical practice model." That approach may not be adequate to deal with research not based on the medical practice model—e.g., social science research.

Second, while Ramsey argues that it is wrong to use a person in research without consent irrespective of risk (because one can be "wronged" without being "harmed"), he nonetheless appears to share with Jonas the assumption that most research is risky and involves "sacrifice" on the part of the subject. In fact, most research does not present risk of physical or psychological harm; rather, it presents inconvenience (e.g., of urine collection) and discomforts (e.g., of needle sticks) (Levine, 1976, pp. 8–11). Even Phase I drug testing, traditionally assumed to be risky, has been estimated to present subjects with "risks" slightly greater than those involved in secretarial work, and substantially less than those assumed by window washers and miners (Arnold, p. 18).

But the most important challenge is Freedman's alternative interpretation and use of the basic principles (1975). Like Jonas and Ramsey, Freedman derives the consent requirement "from the duty which all of us have, to have respect for persons." Unlike Jonas and Ramsey, however, he interprets the requirement of respect for persons to allow the possibility of a "valid but ignorant" consent.

Freedman proposes that striving for "fully informed consent" is generally undesirable, and that what is required is "valid consent," not necessarily "informed consent." To be valid, consent must be responsible and voluntary. Thus valid consent "entails only the imparting of that information which the patient/subject requires in order to make a responsible decision" (Freedman, p. 34). A choice based upon less or other information than another responsible person might consider essential is not necessarily a sign of irresponsibility. Overprotection is a form of dehumanization and lack of respect; for example, to classify persons as incompetent in order to "protect" them from their own judgment is the worst form of abuse.

This approach also has several weaknesses. Much hinges on what is taken to be a "responsible" choice. Freedman suggests that "responsibility" is a dispositional characteristic and is to be judged in terms of the person, not in terms of a particular choice. However, there is still an element of paternalism introduced in judging another to be a "responsible" person. Moreover, this approach may not provide sufficient protection for those subjects who tend too readily to abdicate responsibility for choice.

It is clear that debates over the interpretation of "informed consent" depend on interpretations of the basic ethical principle of respect for persons and the extent to which that principle requires protection from harm or respect for autonomy.

"Informed consent": conditions and exceptions

According to the Nuremberg Code, to consent to participate in research one must: (1) be "so situated as to be able to exercise free power of choice"; (2) have the "legal capacity" to give consent; (3) have "sufficient . . . comprehension" to make an "enlightened" decision; and (4) have "sufficient knowledge" on which to decide. Recent discussion emphasizes the "knowledge" or information component of consent—hence the term "informed consent." Cook argues that the term violates the original intention of the Code, which focused on freedom of choice rather than the quantity or quality of information transmitted and which used the term "voluntary consent," not "informed consent" (Cook, pp. [13-11] ff.).

Most commentators agree that compromise of any one of the four conditions specified by the Nuremberg Code jeopardizes the ethical acceptability of the research.

"*Free power of choice.*" The Nuremberg Code proscribes "any element of force . . . or other ulterior form of constraint or coercion" in obtaining consent. Any flagrant coercion—e.g., when competent, comprehending persons are forced to submit to research against their expressed wills—clearly renders consent invalid. Yet more subtle or indirect "constraints" or "coercions" may obtain when prospective subjects are highly dependent, impoverished, or needy. Some commentators argue that consent obtained from such persons must be considered to have violated the intent of the Code. This argument has been posed most sharply with respect to prisoners and other institutionalized populations, since institutionalization often involves both dependency and impoverishment. Some argue that consent to participate in research is not valid when it is given (1) to procure financial reward in situations offering few alternatives for remuneration; (2) to seek release from an institution either by evidencing "good behavior" or by ameliorating the condition for which one was confined; or (3) to please physicians or authorities upon whom one's continued welfare depends (Myers; Branson).

West argues, however, that such indirect forms of constraint do not constitute coercion in a strict sense and thus do not render consent involuntary (p. [2-5]). Drawing upon Nozick's philosophical analysis of coercion, West argues that "coercion" consists in a threat to render one's circumstances worse if one does not do something. Hence, a threat to withdraw basic necessities of existence, or in some other way to render an inmate's situation worse if he or she declines to participate in research would constitute coercion and render consent invalid. Similarly, to condition release upon participation would constitute coercion, since it would make the inmate's situation worse by removing normal alternatives for seeking release. But the provision of better living conditions in exchange for participation in research does not constitute a threat to make conditions worse; rather, it is an enticement to make conditions better. Thus, it is not coercion. While enticements can invalidate consent by undermining the rational *grounds* for choice, they do not undermine the *voluntariness* of the choice. Similarly, a desire to "get well" or to influence favorably institutional authorities is not an "ulterior" constraint in the strict sense of the Nuremberg Code, though it may be a very real psychological constraint.

A crucial question is whether consent is invalidated if participation in research may be considered possible grounds for early parole. Until recently, such consideration was widely condoned, indeed, so widely that the American Medical Association took the position that persons convicted of "vicious crimes" should not qualify for participation in research, lest they be paroled (Katz, 1972, p. 1025). Today, the practice of reporting participation in research to parole boards has been stopped in response to arguments that it renders participation "coercive" and consent invalid. Logically, of course, consent to participate in other activities that are used as grounds for parole consideration (e.g., work, educational programs) would also then be invalidated as "coercive" unless there is some ground for distinguishing those activities from research. It is a matter of dispute whether participation in research differs from other activities because, to the extent it involves "selling one's body" as opposed to "selling one's labor," it might be viewed as akin to prostitution (Wartofsky).

In any case, the practice of reporting participation in research to parole boards is not "coercion" in the strict sense, but simply an "inducement." Many inmates protest the practice of forbidding such reporting. Most commentators agree that such inducements must be minimized in order to ensure that subjects participate in research under conditions that approximate the "free living state," and thus that freedom is maximized as much as possible (Jonsen et al.).

Competence and comprehension. The Nuremberg Code requires both "legal capacity" to consent and "sufficient understanding" to reach an "enlightened" decision. Definitions of competence often include elements of comprehension, e.g., to evaluate relevant information (Shuman, p. 55), to understand the consequences of action (Katz, 1974, p. 183), and to reach a decision for rational reasons (Annas, Glantz, and Katz, pp. 203–216; Macklin and Sherwin, p. 445).

Goldstein charges that linking determinations of competence to assessments of comprehension is "pernicious" since refusal to participate in research might be judged "irrational" by the investigator and then used as grounds for declaring the person incompetent (Goldstein, p. 15). He argues that the purpose of the "informed consent" requirement is to guarantee the exercise of free choice, not to judge its rationality. Therefore, he proposes that competence should be presumed and that ordinarily only a "showing that the patient is comatose" should be accepted as proof of incompetence (ibid., p. 26).

Assessments of incompetence. While there is disagreement as to the grounds for assessing incompetence, most commentators agree that such assessments are limited in several ways. First, a judgment of "incompetence" applies only to certain areas of decision making, e.g., to one's legal but not to one's personal affairs (Ennis and Siegel, p. 74). Second, confinement to a mental institution is not per se equivalent to a determination of incompetence (Murphy, 1975, p. 44). Third, some who are legally competent are functionally incompetent, while some who are legally incompetent are functionally competent (Annas, Glantz, and Katz, p. 215). Fourth, those whose incompetence is intermittent (e.g., due to temporary inebriation or insanity) rather than permanent might, while competent, make binding decisions to participate in research when incompetent (Levine, 1976, p. 31; Veatch, 1976, p. 39). And finally, with the exception of children where the *class* by definition is legally incompetent, judgments about incompetence should be made on an individual basis (Shuman, p. 63).

The famous *Kaimowitz* psychosurgery case appears to negate several of these limitations. The court found that the effects of long-term institutionalization were so destructive of the person's decision-making ability that by implication all persons incarcerated for long periods of time would be "incompetent" to make certain choices, especially those related to participation in irreversible, experimental interventions having unknown risk. The decision has been criticized severely both because it appears to make a class judgment regarding incompetence and because it links the determination of incompetence to the very fact of institutionalization (Annas, Glantz, and Katz, p. 212; Mason, p. 321; "Medical Treatment," p. 562).

The Nuremberg Code makes no provision for the use of subjects lacking legal capacity and/or comprehension. Most subsequent codes and discussions have allowed their use with certain restrictions: that mentally competent adults are not suitable subjects (Levine, 1976, p. 33; also discussed in 1969 guidelines of the

American Medical Association); that the veto of a legally incompetent but minimally comprehending subject is binding (Annas, Glantz, and Katz; Veatch, 1976, p. 3); and that consent of the legal guardian must be obtained; see the 1964 Declaration of Helsinki.

Proxy consent. Debate has arisen whether a guardian may give "proxy consent" to the participation of an incompetent person in research. Adopting the "battery" argument, Ramsey claims that the use of a nonconsenting subject is wrong whether or not there is risk, simply because it involves an "unconsented touching." Unconsented touching is not wrongful, however, when it is for the good of the individual. Hence, proxy consent may be given for the use of nonconsenting subjects in research only when the research includes therapeutic interventions related to the subject's own recovery (Ramsey, 1970, p. 11; May, p. 247; Fried, p. 23).

However, Ramsey acknowledges that benefit does not always justify unconsented touching; such touching of a competent adult is wrong even if it benefits that person. Why, then, can benefit be presumed to justify such touching for a child (or other subject unable to give consent)? McCormick proposes that the validity of such interventions rests on the presumption that the child, if capable, would consent to therapy. This presumption in turn derives from a child's *obligation* to seek therapy, an obligation that the child possesses simply as a human being (McCormick, 1974, p. 9). Because children have an obligation to seek their own well-being, we presume they would consent if they could, and thus presume also that "proxy consent" on their behalf would not violate respect for them as persons.

By analogy, McCormick suggests that, as members of a moral community, children have other obligations to which one would presume their consent and give "proxy consent" on their behalf. One such obligation is to contribute to the general welfare when such contribution requires little or no sacrifice. Hence, nonconsenting subjects may be used in research not directly related to their own benefit so long as the research fulfills an important social need and involves no discernible risk. Ramsey counters this argument with respect to children, claiming that McCormick's position fails to recognize that a child is not an adult with a full range of duties and obligations. Instead, they have rights to be protected (Ramsey, 1976).

Adopting this premise about the nature of the child as a moral being, Freedman draws different conclusions (p. 38). Since a child is not a moral being in the same sense as an adult, he argues, the concept of "wrongful touching" does not apply. The child has no right to be "let alone," but only a right to be protected. Hence, Freedman concludes that the only relevant moral issue is the risk involved in the research, and, like McCormick, that children could be used in research unrelated to their therapy provided it presents them no discernible risk. Thus, the debate centers on the status of the child (a paradigmatic incompetent) as a moral being and interpretations of the requirements of respect for persons.

The inevitable conflict between the goals of promoting autonomy and providing added protection are demonstrated vividly in disagreements both over standards of competence and the use of incompetent subjects. This conflict is developed further in the current controversy over the assignment to various individuals or groups (e.g., "consent committees") of the responsibility to monitor negotiations with persons having "limited capacities to consent" (Levine, 1975, pp. 43–71).

Disclosure of information. The Nuremberg Code requires that the subject be told "the nature, duration, and purpose of the experiment; the method and means by which it is to be conducted; all inconveniences and hazards reasonably to be expected; and the effects upon his health or person which may possibly come." Subsequent codes have modified those requirements: Disclosure of the *purpose* of research is often ignored, and other stipulations such as disclosure of the availability of alternative modes of treatment are added. U.S. federal regulations require (1) a fair explanation of procedures, (2) disclosure of risks, (3) explanation of benefits, (4) description of alternatives, (5) an offer to answer questions, and (6) the statement that the subject may withdraw at any time (*Federal Regulation*, p. 32). In addition, they forbid the use of exculpatory language.

While these requirements have the force of law, they are by no means exhaustive of possible standards for disclosure. To them one might add the following: a statement of overall purpose; a clear invitation to participate in research, distinguishing maneuvers required for research purposes from those necessary for therapy; an explanation of why that particular person is invited (selected); a suggestion that the prospective subject might wish to discuss the research with another person; and an explanation, when appropriate, of the fact that the risks are unknown—since much biomedical research is designed to determine the nature, probability, and magnitude of harms that might be produced by a therapeutic maneuver (Levine, 1976, pp. 11–16). Veatch would add the names of members of any review boards that had approved the research, an explanation of who is responsible for harm done, and an explanation of the right, if any, to continue receiving treatments found useful (Veatch, 1976). In short, there is no universal agreement on standards for disclosure of information or on what it takes for a person to have "sufficient knowledge" to give "informed" consent.

Those who agree on the need for disclosure of information in a particular category—e.g., the "risks"—often disagree on the nature of the information that must be made known. The Nuremberg Code requires explication of hazards "reasonably" to be expected. Does this include an infinitesimal chance of a substantial harm, or a substantial chance of an infinitesimal harm? Neither the quality nor the probability of the risks to be disclosed have been clearly determined by judicial decision or law.

Disagreements over particulars arise in part from disagreements about underlying standards: Is disclosure to be determined by (1) general medical practice or opinion, (2) the requirements of a "reasonable person," or (3) the idiosyncratic judgment of the individual? While the legal trend may be shifting from the first to the second (Curran, p. 25), it may be argued that only the third is truly compatible with the requirement of respect for the autonomy of the individual person (Veatch, 1976, pp. 25–28; Annas, Glantz, and Katz, pp. 31–37).

Yet even those who adopt the third standard disagree as to its implications. As noted above, Freedman holds that the idiosyncratic judgment of the individual is overriding, to the point that the prospective subject can choose to have less information than a "reasonable" person might require. Veatch, however, argues that anyone refusing to accept as much information as would be expected of a "reasonable person" should not be accepted as a subject (1976, p. 29).

Professional codes (e.g., the code of the American Medical Association) and federal regulations reflect the traditional medical maxim of "do no harm" and the legal doctrine of *therapeutic privilege* according to which a physician may withhold information when in his or her judgment disclosure is either infeasible or potentially harmful (Annas, Glantz, and Katz, p. 35). Recent critics have argued that invoking the doctrine of therapeutic privilege to justify withholding of information from a prospective subject in order to assure cooperation in a research project is almost never appropriate; it gives the investigator entirely too much license to serve vested interests by withholding information that might be material to the decision of a prospective subject (ibid.). In fact, courts are according physicians less and less latitude in this regard in the context of practice.

Similarly, in research activities contingent upon subjects' lack of awareness of purposes or procedures it has been thought permissible either to withhold information or to practice deliberate deception provided harms are minimized and subjects "debriefed" (given a full explanation) afterward. Baumrind opposes deceptive practices, arguing not only that they violate the principle of respect for persons but also that in the long run they will invalidate research on scientific grounds. Various proposals have been made to minimize the need for and harmful effects of deceptive practices: Subjects might be invited to consent to incomplete disclosure with a promise of full disclosure at the termination of the research (Levine, 1975, pp. 30–31); subjects might be told as much as possible and asked to consent for specified limits of time and risk (Ashley, p. 19); or consent based on full disclosure might be negotiated with mock or "surrogate" populations (Veatch, 1975, p. 52; Levine, 1976, pp. 26–29).

Conclusions

The use of a person as a research subject can be justified only if that person, or one authorized to speak on his or her behalf, consents to such use. The legal and ethical requirement for consent is grounded in fundamental tenets of the Judeo-Christian religious tradition as well as in basic ethical principles which create the obligation that all of us have to treat persons as ends and not merely as means to another's end. Most major disagreements over the form and substance of the consent requirement derive from conflicting interpretations of the basic principles.

Those who are interested in making operational the requirement for consent have a tendency to focus nearly all of their attention on the consent form. Federal regulations prescribe what information must be included in and excluded from these forms. Members of institutional review boards and researchers collaborate in a struggle to create reproachless forms. This seems to reflect an assumption that the consent form is an appropriate instrumentality through which we might fulfill our obligation not to treat persons merely as means. As we have indicated, there are important reasons to question this assumption.

Moreover, preoccupation with consent forms reflects an emphasis on "informed consent," a concept that stresses rational decision making. As we focus on "informing" and on "rational" decision making, we may become inattentive to other equally important attributes of consent—e.g., voluntariness. Insistence upon rational decision making at the expense of idiosyncratic modes of self-determination is not necessarily respectful of persons.

KAREN LEBACQZ AND
ROBERT J. LEVINE

[*For further discussion of topics mentioned in this article, see the entries:* CHILDREN AND BIOMEDICINE; HEALTH CARE, *article on* HUMANIZATION AND DEHUMANIZATION OF HEALTH CARE; INFORMED CONSENT IN THE THERAPEUTIC RELATIONSHIP; INFANTS; *and* PRISONERS. *Also directly related are the entries* PATERNALISM; PSYCHOSURGERY; *and* TRUTH-TELLING. *See* APPENDIX, SECTION II, NUREMBERG CODE, DECLARATION OF HELSINKI, *and* ETHICAL GUIDELINES FOR CLINICAL INVESTIGATION (AMERICAN MEDICAL ASSOCIATION).]

BIBLIOGRAPHY

ANNAS, GEORGE; GLANTZ, LEONARD; and KATZ, BARBARA. "The Law of Informed Consent in Human Experimentation." Available from The National Commission for the Protection of Human Subjects of Biomedical and Behavioral Research, Bethesda, Md. June 1976.

ARNOLD, JOHN D. "Alternatives to the Use of Prisoners in Research in the United States." National Commission, *Research Involving Prisoners: Appendix,* pp. (8-1)–(8-18).

ASHLEY, BENEDICT M. "Ethics of Experimenting with Persons." *Research and the Psychiatric Patient.* Edited by Joseph C. Schoolar and Charles M. Gaitz. New York: Brunner/Mazel Publishers, 1975, chap. 2, pp. 15–27. Proceedings of the Eighth Annual Symposium, 16–18 October 1974, Texas Research Institute of Mental Sciences.

BAUMRIND, DIANA. "Nature and Definition of Informed Consent in Research Involving Deception." Available from The National Commission for the Protection of Human Subjects of Biomedical and Behavioral Research, Bethesda, Md. January 1976.

BRANSON, ROY. "Philosophical Perspectives on Experimentation with Prisoners." National Commission, *Research Involving Prisoners: Appendix,* pp. (1-1)–(1-46).

COOK, JOYCE MITCHELL. "The Problems of Informed Consent Focussing on Prisons." National Commission, *Research Involving Prisoners: Appendix,* pp. (13-1)–(13-61).

CURRAN, WILLIAM J. "Ethical Issues in Short Term and Long Term Psychiatric Research." *Medical, Moral and Legal Issues in Mental Health Care.* Sixth Annual Taylor Manor Hospital Scientific Symposium, Baltimore, April 1974. Edited by Frank J. Ayd, Jr. Baltimore: Williams & Wilkins Co., 1974, chap. 3, pp. 18–27.

ENGELHARDT, H. TRISTRAM, JR. "Basic Ethical Principles in the Conduct of Biomedical and Behavioral Research Involving Human Subjects." Available from The National Commission for the Protection of Human Subjects of Biomedical and Behavioral Research, Bethesda, Md. December 1975.

ENNIS, BRUCE J., and SIEGEL, LOREN. *The Rights of Mental Patients: The Basic ACLU Guide to a Mental Patient's Rights.* An American Civil Liberties Union Handbook. New York: Avon Books, 1973.

Federal Regulation of Human Experimentation, 1975. Prepared by Freeman H. Quimby, Susan R. McKenzie, and Cynthia B. Chapman. Subcommittee on Health of the Committee on Labor and Public Welfare, United States Senate, 94th Cong., 1st sess. Washington: Government Printing Office, 1975.

FEINSTEIN, ALVAN R. "Clinical Biostatistics. XXVI: Medical Ethics and the Architecture of Clinical Research." *Clinical Pharmacology and Therapeutics* 15 (1974): 316–334.

FREEDMAN, BENJAMIN. "A Moral Theory of Informed Consent." *Hastings Center Report* 5, no. 4 (1975), pp. 32–39.

FRIED, CHARLES. *Medical Experimentation: Personal Integrity and Social Policy.* Amsterdam: North-Holland Publishing Co.; New York: American Elsevier Publishing Co., 1974.

GOLDSTEIN, JOSEPH. "On the Right of the 'Institutionalized Mentally Infirm' to Consent to or Refuse to Participate as Subjects in Biomedical Research." National Commission for the Protection of Human Subjects of Biomedical and Behavioral Research. *Research Involving Those Institutionalized as Mentally Infirm. Appendix.* Bethesda, Md.: 1978.

INGELFINGER, FRANZ J. "Informed (But Uneducated) Consent." *New England Journal of Medicine* 287 (1972): 465–466.

JONAS, HANS. "Philosophical Reflections on Experimenting with Human Subjects." *Experimentation with Human Subjects.* Edited by Paul A. Freund. New York: George Braziller, 1970, pp. 1–31.

JONSEN, ALBERT R.; PARKER, MICHAEL L.; CARLSON, RICK J.; and EMMOTT, CAROL B. *Biomedical Experimentation on Prisoners: Review of Practices and Problems and Proposal of a New Regulatory Approach.* Health Policy Program, Discussion Paper. San Francisco: Health Policy Program, University of California, 1975.

KATZ, JAY. "Human Rights and Human Experimentation." *Protection of Human Rights in the Light of Scientific and Technological Progress in Biology and Medicine.* Proceedings of a Round Table Conference of the Council for International Organizations of Medical Sciences, Geneva, 14–16 November 1973.

Geneva: World Health Organization, 1974, pp. 181–217.

———, ed. *Experimentation with Human Beings: The Authority of the Investigator, Subject, Professions, and State in the Human Experimentation Process.* New York: Russell Sage Foundation, 1972.

———, and Capron, Alexander Morgan. *Catastrophic Diseases: Who Decides What? A Psychosocial and Legal Analysis of the Problems Posed by Hemodialysis and Organ Transplantation.* New York: Russell Sage Foundation, 1975.

LEVINE, ROBERT J. "Appropriate Guidelines for the Selection of Human Subjects for Participation in Biomedical and Behavioral Research." Available from The National Commission for the Protection of Human Subjects of Biomedical and Behavioral Research, Bethesda, Md. December 1975.

———. "The Nature and Definition of Informed Consent in Various Research Settings." Available from The National Commission for the Protection of Human Subjects of Biomedical and Behavioral Research, Bethesda, Md. December 1975. Published as "Informed Consent to Participate in Research." *Bioethics Digest* 1, no. 11 (1977), pp. 1–13, no. 12 (1977), pp. 1–16.

McCORMICK, RICHARD A. "Proxy Consent in the Experimentation Situation." *Perspectives in Biology and Medicine* 18 (1974): 2–20.

———. "A Reply to Paul Ramsey: Experimentation in Children: Sharing in Sociality." *Hastings Center Report* 6, no. 6 (1976), pp. 41–46.

MACKLIN, RUTH, and SHERWIN, SUSAN. "Experimenting on Human Subjects: Philosophical Perspectives." *Case Western Reserve Law Review* 25 (1975): 434–471.

MASON, JOHN R. "Kaimowitz v. Department of Mental Health: A Right to Be Free from Experimental Psychosurgery?" *Boston University Law Review* 54 (1974): 301–339.

MAY, WILLIAM E. "Experimenting on Human Subjects." *Linacre Quarterly* 41 (1974): 238–252.

"Medical Treatment and Human Experimentation: Introducing Illegality, Fraud, Duress and Incapacity to the Doctrine of Informed Consent." *Rutgers-Camden Law Journal* 6 (1975): 538–564.

MURPHY, JEFFRIE G. "Incompetence and Paternalism." *Archiv für Rechts-und Sozialphilosophie* 60 (1974): 465–486.

———. "Total Institutions and the Possibility of Consent to Organic Therapies." *Human Rights* 5, no. 1 (1975), pp. 25–45.

MYERS, MATTHEW. "Testimony." Transcript of the Meeting Proceedings (14th) Held at Bethesda, Maryland on January 9–10, 1976: National Commission for the Protection of Human Subjects of Biomedical and Behavioral Research. PB-250 822. Springfield, Va.: National Technical Information Service, 1976, pp. 30–69. Statement concerning the National Prison Project, American Civil Liberties Union Foundation.

National Commission for the Protection of Human Subjects of Biomedical and Behavioral Research. *Research Involving Prisoners: Appendix to Report and Recommendations.* DHEW Publication No. (OS) 76-132. Bethesda, Md.: Department of Health, Education, and Welfare, 1976.

RAMSEY, PAUL. *The Patient as Person: Explorations in Medical Ethics.* The Lyman Beecher Lectures at Yale University. New Haven: Yale University Press, 1970.

———. "A Reply to Richard McCormick: The Enforcement of Morals: Nontherapeutic Research on Children." *Hastings Center Report* 6, no. 4 (1976), pp. 21–30.

REISS, ALBERT, JR. "Selected Issues in Informed Consent and Confidentiality with Special Reference to Behavioral Social Science Research Inquiry." Available from The National Commission for the Protection of Human Subjects of Biomedical and Behavioral Research, Bethesda, Md. February 1976.

SHUMAN, SAMUEL I. "The Emotional, Medical and Legal Reasons for the Special Concern about Psychosurgery." *Medical, Moral, and Legal Issues in Mental Health Care.* Sixth Annual Taylor Manor Hospital Scientific Symposium, Baltimore, April 1974. Edited by Frank J. Ayd, Jr. Baltimore: Williams & Wilkins Co., 1974, chap. 6, pp. 48–80.

United States, Defense Department. "Nuremberg Code." U.S. v. Karl Brandt. *Trials of War Criminals before Nuremberg Military Tribunals under Control Law No. 10.* Washington: Government Printing Office, 1947, vol. 2, pp. 181–183.

———, Tuskegee Syphilis Study Ad Hoc Advisory Panel. *Final Report.* Washington: Public Health Service, 1973.

VEATCH, ROBERT M. "Ethical Principles in Medical Experimentation." *Ethical and Legal Issues of Social Experimentation.* Edited by Alice M. Rivlin and P. Michael Timpane. Brookings Studies in Social Experimentation. Washington: Brookings Institution, 1975, pp. 21–59.

———. "Three Theories of Informed Consent: Philosophical Foundations and Policy Implications." Available from The National Commission for the Protection of Human Subjects of Biomedical and Behavioral Research, Bethesda, Md. February 1976.

WARTOFSKY, MAX W. "On Doing It for Money." National Commission, *Research Involving Prisoners: Appendix*, pp. (3-1)–(3-24).

WEST, CORNEL RONALD. "Philosophical Perspective on the Participation of Prisoners in Experimental Research." National Commission, *Research Involving Prisoners: Appendix*, pp. (2-1)–(2-22).

INFORMED CONSENT IN MENTAL HEALTH

Social concern about the role of patient consent in mental health care has been apparent at least during the past century, when civil commitment statutes were first subjected to widespread criticism and reform. Concern about patient consent in all medical practice has been a more recent phenomenon and seems not—or at least not yet—to have attained the same intensity as for mental health care. The thesis underlying this article is that central questions regarding the practical possibilities for attaining consensual relations between patient and professional, and the ethical imperatives that press toward such a goal, are essentially similar for both mental health and all physical health care. This thesis will be suggested first by addressing the popular perception of substantial differences between the roles of consent in the two fields; second, by considering common arguments for abolition or reformation of civil commitment statutes for "mentally ill persons" and linking those arguments to considerations common to all health-care practice; and third, by exploring arguments for dispensing with consent for one group of potential mental health patients—convicted criminals—and identifying some parallels that might be drawn from these arguments for all health care.

Comparing the roles of informed consent in mental and physical health care

Questions about the ethical imperatives and practical difficulties in obtaining informed consent for medical interventions appear in both physical and mental health care. The questions are framed in the same terms for both, but the issues of informed consent present a more troublesome appearance for mental health care. In fact, the barriers to rationality may be generally no greater in such care than for most physical health patients. A prospective mental patient is nonetheless more likely to perceive himself and to be perceived by the professional practitioner as "less than fully rational." This evaluation is likely to be a critical reason why such a prospective patient considers himself, and is accepted by the professional as, a candidate for mental health care. Furthermore, the norms of "cure" for mental health are much more obviously open to dispute than norms for physical health cure. Thus mental health practice gives at least apparently greater latitude for the imposition by professionals of their norms (or the socially accepted norms) of "mental health" in disregard of the patient's autonomous will. (This is not to say, however, that all conceptions of "health" in physical medicine are "objective" and value-free.) Finally, coerced care of persons considered "mentally ill" is openly and universally approved by legal rule (particularly through civil commitment statutes)—more so than for physical medicine. Legal rules also do sanction some kinds of care in physical medicine without patients' consent, e.g., compulsory vaccination (Capron, pp. 404–418).

All of these apparent differences in the role of informed consent between physical and mental health care are matters of degree. Though the magnitude of those differences as perceived by practitioners and patients (among others) is clearly important, it is not clear whether the perceived measure of difference necessarily leads to lesser or greater adherence to norms of informed consent in mental as opposed to physical health care. Precisely because questions about informed consent appear so pronounced in mental health care, both patients and practitioners are more likely than in physical medicine to be consciously aware of these questions. Thus, if it is true, for example, that many psychiatric patients are no less rational than other patients and no more eager to defer to their doctors' treatment, the heightened mutual awareness of irrational components in mental health treatment decisions increases the likelihood that these aspects will be addressed rather than ignored. Because of this increased awareness generally, it may be that the underlying intent of the informed consent ethos—that the professional practitioner must actively solicit and assiduously respect the patient's autonomous will—might be more likely of achievement, for many patients at least, in mental health than in physical health care.

Justifications for nonconsensual health care of the mentally ill

As noted, civil commitment laws in every state justify some form of coerced health care for persons found "mentally ill." The laws are not limited to provision of mental health care to the "mentally ill," but are regularly invoked to justify nonconsensual physical health care as well to persons who refuse such care as a result of their "mental illness." This broader coverage of the statutes further indicates the important links between mental and physical health care regarding issues of informed consent. Current controversy regarding civil commitment laws revolve essentially around one central theme: whether it is ever ethical and/or practical to provide health care without regard to the recipient's consent. Examination of this controversy will thus illuminate important issues regarding the role of informed consent in health care generally. Criticism of these laws is directed at two different aspects: (1) the class of persons who might properly be subjected to nonconsensual care, and (2) the sorts of treatment that might properly be imposed.

To whom might nonconsensual care properly be provided? It can plausibly be argued that many people who refuse health care are in fact ambivalent in their desires. For example, consider a person who unsuccessfully attempts suicide but who subsequently refuses any life-saving treatment (whether physical or mental health care). This person might be showing ambivalence between word and deed, the words refusing treatment but the deed (in the failure and public discovery of the attempt) calling attention to the person's need for treatment. The suicide attempt might stem from irrational self-perceptions of worthlessness and extreme need. Thus this person's refusal of care might also be symptomatic of an irrational belief that there is no possible help or deserved assistance, rather than reflecting an "informed" rejection of care (Katz, p. 771).

Those who argue for civil commitment of such persons, on grounds that they unconsciously desire care but cannot consciously admit that wish, attempt to sidestep the arguments regarding the improprieties of coerced care by denying that such care is coerced. But unless this argument is limited in some way, it obviously proves too much. Unless some reliable indicators can be found to distinguish between people who refuse care "ambivalently" and those who refuse "unambivalently," this justification for civil commitment wholly confounds any notions of individual autonomy and self-determination, which form the ethical underpinnings of the informed consent requirement.

Civil commitment laws attempt to accommodate this problem of principle by authorizing coerced care only for those found "mentally ill." But this definitional limit raises the same conceptual problems as the attempt in the criminal law to square the free will/determinism circle by holding all persons responsible for their antisocial acts except those persons found "insane" (Stone, pp. 218–228). The criteria for defining "exculpating insanity" can ultimately be used to include any mental process that leads a person to commit an aberrant antisocial act, which thus collapses "insanity" and "violating the criminal law" into one concept (Morris, 1968, pp. 516–528). Similarly, "mental illness" can include any refusal of treatment that a professional practitioner would prescribe for a "mentally ill person" (Livermore, Malmquist, and Meehl, pp. 78–89). Some civil commitment statutes in fact provide a standard suspiciously close to this: that care may be imposed on a person who, "because of mental illness, fails to understand his need for treatment" (Brakel and Rock, pp. 39–40).

Many commitment statutes offer additional apparent limitations, restricting coerced care to persons found "mentally ill *and* dangerous to self or others." Several courts have constitutionally mandated such added limitations (Stone, pp. 47–60), but the limitations do not adequately solve the ethical or practical problems raised by nonconsensual care. The criteria for unacceptably "self-dangerous" behavior in particular are open to intense normative dispute. Consider, for example, whether "suicide" by overwork, overweight, or enlistment in the Marines might be viewed as inappropriately aberrational behavior (Greenberg, p. 248). Acts considered wrongly "dangerous to others" are less subject to idiosyncratic personal value perspectives, but such acts are already proscribed by ordinary criminal laws. If preventive detention to anticipate criminal acts is considered appropriate, and if "mental illness" is reliably predictive of such acts, then a case might be made for the added state power of such commitments. But psychiatric predictions of dangerous acts—toward others or, indeed, toward self—are notoriously unreliable, and the ethical propriety of applying social constraints to anticipate criminal behavior generally is seriously questionable (Dershowitz; Diamond).

Some critics have argued that the absence of any clearly principled demarcations for "civilly committable conduct" demonstrates that such commitment statutes should be wholly abolished (Szasz). But it may be that this argument makes too much of the conceptual weaknesses underlying civil commitment statutes. As noted in this discussion, demarcations between consensual and nonconsensual mental health care generally are available more as a principled hope than as a reliable practical reality in relations between patients and mental health practitioners. While this proposition may not justify the stark validation of nonconsensual care expressed by civil commitment statutes, it could justify accepting some explicit coercion so long as that coercion works toward the ultimate values of truly consensual mental health care. On this ground, some have proposed retaining civil commitments but imposing rigid time limits on them (Katz, p. 773). Proponents of this position argue that any scheme purporting to respect the informed consent of a prospective mental health patient must in some way acknowledge the reality that conscious and unconscious wishes may diverge, and that always withholding care in response to one's expressed conscious wish may give inadequate credence to and respect for his unconscious wishes.

The notion that truly informed consent can be achieved after the passage of time following initially coerced contact between practitioner and patient may be best attuned to the practical realities of obtaining such consent in all of health-care practice. Because all patients will typically be subject to numerous coercions (from intrapsychic pain, for example, whether due to psychological or physical causes) in their initial dealings with professional practitioners, the psychological dominance of these coercions can best be abated by repeated efforts of patient and practitioner consciously directed toward that goal. The process necessarily takes time. This attribute of the realization of truly informed consent in all health care suggests the suitability of using time passage to demarcate appropriately from inappropriately coerced care in civil commitment statutes. From this perspective, it may be questioned whether "mental illness" is a necessary or proper legal label for designating persons subject to nonconsensual health care. Such limitation may reflect a practical conviction that coerced care, while an inescapable aspect of all health care, should have visible and stringent limitations, to emphasize the normative preeminence of informed consent values, more than a principled view that coerced care is uniquely justified for "mentally ill" people.

Imposing a time limit on commitment, invoked in response to a professional's belief that a protesting patient "truly" wishes care, is intended to resolve the practical problems created by any regime of forced care (Burt, 1974, pp. 277–280). Such a position cannot resolve the questions of principle concerning whether coerced care is ever justified normatively. Insofar as such questions rest on competing models of psychological functioning—whether "free will" or "psychic determinism" best explains choices of conduct, whether "conscious" or "unconscious" wishes are truer depictions of a person's intent—it may be that such questions admit no resolution in principle.

What modes of treatment might properly be coerced? If the propriety of coerced care can properly be assessed only after the passage of time, it would appear that any mode of treatment with irreversible effects must not be imposed until the requisite time has passed. But just as no rigid line of principle separates coerced from noncoerced care, or ambivalent from unambivalent refusals of care, similarly no line clearly separates reversible from irreversible modes of treatment. Residence in a psychiatric hospital, for example, itself has many irreversible consequences. The time spent, of course, can never be recaptured, and beyond this the social and self-labeling attendance in such residence may have significant reinforcing and thus irreversible implications (Rosenhan).

Many people consider these consequences less drastic and less clearly irreversible than interventions intended to operate without enlisting the subject's conscious collaboration—such as electroconvulsive therapy (ECT), psychosurgery, or even administration of psychotropic drugs. Some courts and legislatures, reflecting this view, have ruled that civil commitment status as such does not justify nonconsensual administration of such therapies. One court has held that psychosurgery is not permitted under any circumstances for involuntarily hospitalized persons (*Kaimowitz*). Other courts and legislatures have ruled that an involuntarily committed person must be permitted to refuse such therapies as psychosurgery and ECT; they have provided special review procedures to regulate availability of these procedures to involuntarily committed persons who consent to them (California Code; *Wyatt* v. *Stickney*, p. 380).

Nonconsensual mental health care in the criminal justice system

Because mental health care is intended, much more directly than physical medicine, to facilitate behavior change, arguments for coerced mental health care have special force in the criminal justice system. Mental health professionals have, more frequently and with more assurance in the past than now, offered their services to that system. This offer, epitomized in the currently disfavored term "rehabilitative ideal," presents more persuasive arguments for the ethical propriety of forgoing "informed consent" as a prerequisite for care than arguments available under civil commitment laws. Unlike such laws, the immediate basis in the criminal laws for nonconsensual state intervention is not a person's perceived need for treatment. That basis is the person's violation of social norms for conduct. If nonconsensual mental health care can reliably change that person's proclivity to engage in criminal conduct, a strong case appears for its imposition (Burt, 1975, pp. 31–32). It can be argued to the contrary, however, that a convicted criminal should be free to reject "cure" and to opt instead for noncurative "punishment," such as incarceration (von Hirsch, pp.

124–131). By this argument, coerced mental health care is seen as more violative of an individual's integrity and dignity than coerced physical incarceration.

This dispute currently has only future application, since, notwithstanding the grandiose claims for the "rehabilitative ideal," none of the psychotherapies deployed within the criminal justice system in the past decades has had any demonstrable success in reforming criminals (Morris, 1974, pp. 1–27). But if, in the future, some type of therapy, e.g., psychosurgery, is shown to have reliable effectiveness, the issue will be sharply posed: Should the recipient's informed consent, as a prerequisite for all mental health care, be an overarching ethical imperative or simply reflective of a practical (but by then outdated) proposition that nonconsensual care is ineffective?

Consideration of this future possibility for mental health care places in sharp relief one further difference between such care and physical medicine—that the latter is widely viewed as more reliably effective in accomplishing its intended purposes than the former. This greater effectiveness comes in part from beliefs, for example, that a person's appendix may "successfully" be removed without his conscious collaboration, whereas his oedipal complex cannot thus readily be resolved (Katz, pp. 772–773). But, as noted, many modes of psychotherapy purport to operate without enlisting the patient's conscious collaboration. Further, the patient's psychological status is increasingly viewed as a critical element for the curative prospects of physical health-care interventions (Jenkins).

The widely shared view that all current modes of mental health care are less effective than many techniques of physical medicine points toward perhaps a critical difference regarding the role of informed consent in mental as opposed to physical health care. Although there are few clear demarcations in principle between the roles of informed consent norms in physical and mental health care, nonetheless differences are widely felt to exist in practice. It may be that widespread doubts regarding the demonstrable efficacy of any modes of psychotherapy—among both lay and professional publics—have some direct impact in highlighting the importance of consent to such therapy. By contrast, because physical medicine is widely seen as effective (an appendectomy typically accomplishes its proclaimed goal), both the lay and the professional publics may be more comfortable with the notion that any "physically ill" person (who exhibits, for example, the medically accepted symptoms of appendicitis) should be willing, in fact and in principle, to accept the treatment offered by physicians for that condition. Thus strict attention to the niceties of informed consent may appear less compelling for the practitioners and recipients of physical health care than for mental health.

If this hypothesis is correct, it would further suggest that, as the psychotherapies are viewed to be more demonstrably effective, widespread public concern with the need for consent to such therapies might diminish. Nonconsensual deployment of mental health care in the criminal justice system will be one gauge of this trend. Such a result would be paradoxical. Since concern for patient consent purports to be derived from social commitment to protect individual self-determination, it would be ironic if this concern should diminish regarding the psychotherapies precisely when and because they become more effective in overriding patients' capacity for self-determination.

ROBERT A. BURT

[*Directly related are the entries* INFORMED CONSENT IN THE THERAPEUTIC RELATIONSHIP; INSTITUTIONALIZATION; MENTAL ILLNESS; *and* PATERNALISM. *Other relevant material may be found under* BEHAVIORAL THERAPIES; ELECTRICAL STIMULATION OF THE BRAIN; ELECTROCONVULSIVE THERAPY; HYPNOSIS; MENTAL HEALTH SERVICES, article on SOCIAL INSTITUTIONS OF MENTAL HEALTH; *and* PSYCHOSURGERY. *For a discussion of informed consent in human experimentation related to mental health, see* RESEARCH, BEHAVIORAL. *See also* APPENDIX, SECTION III, DECLARATION ON THE RIGHTS OF MENTALLY RETARDED PERSONS; *and* SECTION IV, AMERICAN PSYCHIATRIC ASSOCIATION *and* AMERICAN PSYCHOLOGICAL ASSOCIATION.]

BIBLIOGRAPHY

AYD, FRANK J., JR., ed. *Medical, Moral and Legal Issues in Mental Health Care.* 6th Annual Taylor Manor Hospital Scientific Symposium. Baltimore: Williams & Wilkins Co., 1974.

BRAKEL, SAMUEL J., and ROCK, RONALD S., eds. *The Mentally Disabled and the Law.* Rev. ed. Chicago: University of Chicago Press, 1971.

BURT, ROBERT A. "Of Mad Dogs and Scientists: The Perils of the 'Criminal-Insane'." *University of Pennsylvania Law Review* 123 (1974): 258–296.

———. "Why We Should Keep Prisoners from the Doctors: Reflections on the Detroit Psychosurgery Case." *Hastings Center Report* 5, no. 1 (1975), pp. 25–34.

California Welfare & Institutions Code, secs. 5325–5326 (West 1972). See also supplement.

CAPRON, ALEXANDER MORGAN. "Informed Consent in Catastrophic Disease Research and Treatment." *Uni-*

versity of Pennsylvania Law Review 123 (1974): 340–438.
DERSHOWITZ, ALAN M. "Preventive Confinement: A Suggested Framework for Constitutional Analysis." Texas Law Review 51 (1973): 1277–1324.
DIAMOND, BERNARD L. "The Psychiatric Prediction of Dangerousness." University of Pennsylvania Law Review 123 (1974): 439–452.
GREENBERG, DAVID F. "Involuntary Psychiatric Commitments to Prevent Suicide." New York University Law Review 49 (1974): 227–269.
JENKINS, C. DAVID. "Recent Evidence Supporting Psychologic and Social Risk Factors for Coronary Disease." New England Journal of Medicine 294 (1976): 987–994, 1033–1038.
Kaimowitz v. Michigan Department of Mental Health. Civil No. 73-19434-AW. (Cir. Ct. Wayne Co., Mich., 10 July 1973). Reprint. Brooks, Alexander D. Law, Psychiatry and the Mental Health System. Boston: Little, Brown & Co., 1974, pp. 902–921.
KATZ, JAY. "The Right to Treatment—An Enchanting Legal Fiction?" University of Chicago Law Review 36 (1969): 755–783.
KINDRED, MICHAEL; COHEN, JULIUS; PENROD, DAVID; and SHAFFER, THOMAS, eds. The Mentally Retarded Citizen and the Law. New York: Free Press, 1976. Sponsored by The President's Committee on Mental Retardation.
LIVERMORE, JOSEPH M.; MALMQUIST, CARL P.; and MEEHL, PAUL E. "On the Justifications for Civil Commitment." University of Pennsylvania Law Review 117 (1968): 75–96.
MORRIS, NORVAL. The Future of Imprisonment. Studies in Crime and Justice. Chicago: University of Chicago Press, 1974.
———. "Psychiatry and the Dangerous Criminal." Southern California Law Review 41 (1968): 514–547.
ROSENHAN, DAVID L. "On Being Sane in Insane Places." Santa Clara Lawyer 13 (1973): 379–399.
STONE, ALAN A. Mental Health and Law: A System in Transition. Crime and Delinquency Issues. DHEW Publication no. (ADM) 75-176. Rockville, Md.: National Institute of Mental Health, Center for Studies of Crime and Delinquency. Washington: Government Printing Office, 1975.
"Symposium: Mental Disability and the Law." California Law Review 62, no. 3 (May 1974). Special Issue.
SZASZ, THOMAS. Law, Liberty and Psychiatry: An Inquiry into the Social Uses of Mental Health Practices. New York: Macmillan Co., 1963.
VON HIRSCH, ANDREW. Doing Justice: The Choice of Punishments. New York: Hill & Wang, 1976.
Wyatt v. Stickney, 344 F. Supp. 373 (5th D. Ala. 1972).

INFORMED CONSENT IN THE THERAPEUTIC RELATIONSHIP

I. CLINICAL ASPECTS   Eric J. Cassell
II. LEGAL AND ETHICAL ASPECTS   Jay Katz

I
CLINICAL ASPECTS

Purpose and nature of informed consent. This article is intended as a brief introduction to the clinical setting of informed consent in the therapeutic relationship. Apart from legal requirements, informed consent acts to improve the care of patients. It does so by increasing the bond of trust, by facilitating autonomy through the provision of choice, and by increasing the patient's participation in his own care.

Statements by the physician concerning the patient's situation not only should be understandable explanations of the disease state but should emphasize what things mean in terms of the patient's expected ability to function, his mode of life, and what is important to him. Discussions of alternatives should be couched in the same individual terms and must include, in addition to an understanding of competing risks and benefits, a specification of the probable durations of time involved in the courses contemplated.

The discussion of risks and hazards of the diagnostic or therapeutic options, as well as information about anticipated pain or suffering, are, in theory and practice, the most troublesome aspects of informed consent. The functions of informed consent are not promoted either by completely hiding the painful truth of risks or suffering or by the provision of an encyclopedic and graphic list of all possible known risks or hazards. To facilitate freedom of choice, the patient must know only those risks or hazards that would influence a rational person in the same position, and with the same alternatives (Mills; Alfidi; Mitchell and Cragin). However, to facilitate trust, patients should receive an honest answer to any specific question they ask about anticipated benefits, risks, or suffering. To increase their participation in their own care, patients must also know about any hazards where their preparation in advance would make them endure it more easily. The informed patient is one who knows what his alternatives are. The physician's presentation of alternatives may contain biases that must be guarded against. Presumably, the doctor has already weighed the risks and benefits of the option for which he or she is obtaining consent. Since the physician may be embarking on something dangerous to the patient, he may have to overconvince himself and his patient of its correctness. In his presentation to the patient, he is liable to assign undue weight to those factors that support his decision and less weight to those factors that detract from it. It may be difficult for him to review in full for the patient the thinking that led to his choice. Nevertheless, the patient is entitled to a close approximation of the thought process.

Obstacles to informed consent. The most frequently mentioned barrier to informed consent is the patient's lack of medical knowledge (Lieberman). The barrier exists not only on a factual level, but at the level of basic understanding of medicine. Disclosing the facts of a contemplated procedure—its risks, hazards, and benefits—even if presented in an understandable manner, may not be sufficient to produce a truly educated consent. For example, if one of the possible consequences of a course of action is an allergic reaction, the mere disclosure of that fact, or even its probability, does not and cannot make clear the wide range of phenomena contained within the word "allergic." Nor does an even more complete discussion reach the hazards that may follow the treatment of an allergic reaction, or the wide and varying times from minutes to months over which allergy may manifest itself. Such a presentation of facts will almost always be inadequate if "informed" is meant to mean "educated" (Ingelfinger). However, the primary object of information is to facilitate the patient's care rather than avoid legal action. Factual knowledge is used, therefore, to extend the patient's understanding and meet the needs and fears that are unique to each patient—not only those common to all patients. On occasion, the decision may be so difficult, the hazard so great, or the patient's indecision or lack of trust so compelling that factual knowledge must be sufficiently extensive to provide the most comprehensive background to decision.

One of the things least understood by patients is that decision making in medicine—indeed, the practice of medicine itself—is an enterprise of competing probabilities. Almost nothing is sure. The facts being presented to the patient and used by the physician in his choice of action may be undermined by a finding of tomorrow or a new interpretation of yesterday. Further, physicians are well aware that the probabilities of a successful outcome or a risk are based on the findings of a population of similar cases but cannot predict what will happen in an individual instance. Because they do not understand this, patients are prone to believe that, if a rare complication occurs, someone must be at fault. In addition, certain statistical laws, especially those that cover rare events, are counterintuitive and may further mislead a layman.

A major block to informed consent may be the emotional state of the patient or his ability to think or reason clearly. While the courts are quite demanding with regard to the evidence required to prove incompetence, relative inability to participate rationally in the process of informed consent is a common occurrence in the seriously ill. A symptom such as pain may speak so loudly to the patient that he will say or do almost anything to be free of the distress. In later moments he may not remember the extent of the suffering, or may project onto others the reasons for a behavior of which he may be ashamed. Similarly, the patient's previous experience with a situation, or its occurrence in a family member, may so color his perceptions that he is unable properly to assign priority to the information he is receiving. Fear is particularly destructive in this regard. While many fears, such as the fear of choking, are shared by all, certain risks or consequences will evoke apparently inordinately great fear in a particular patient because of the patient's personality or life history. These factors do not militate against the principle of informed consent, but rather place a special burden on the physician that the information he or she conveys contain not only important facts of the situation but take into account the patient's personality and the alteration in cognition imposed by the illness. In this regard, it should be noted that the cognition of persons who are seriously ill may be qualitatively different from the thinking of the same persons when they are well. Their thinking resembles, in certain respects, the thought of preoperational children as described by Piaget (Flavell; Cassell). However, neither the patient nor the attendants may be aware of the change in thought, despite the serious consequences it may have for autonomous decision making. It has been repeatedly demonstrated that patients may not remember even the most important information (Golden and Johnston). Memory and perception in the illness situation, as in other areas, tend to be selective and variable. In addition to these blocks to truly informed consent, patients who are unable to trust require special effort by the physician to enable their understanding and participation, since this process deals at the same time with the need of the patient to be informed and the problems of care created by the inability to trust.

Ambiguous nature of informed consent. Informed consent, in addition to its benefits, can also be dangerous if poorly handled. Although it is meant to provide the information necessary for a rational person to make choices, present or contemplated sickness is the enemy of clear reason. The relative importance of the risks and benefits, no matter how clearly presented, may not be evaluated adequately by the patient. Rather, the patient's fears or hopes, or those of his family (Korsch), may frighten the patient away from necessary therapy or encourage a choice based on false expectations. An overly explicit or detailed presentation by a physician in the hope of avoiding legal difficulties may do great harm. So, too, might a detailed explanation to someone who has clearly indicated he does not wish to know more.

The harm may arise when the patient, informed of a risk of small statistical probability, acts as if the risk were a large and impending threat. For example, a patient was informed that following replacement of a heart valve small clots could form and cause a stroke. Surgery was successful and relieved the previously incapacitating heart failure. The patient was almost disabled by the fear of stroke and interpreted virtually every headache or tingling of fingers or toes as evidence that stroke was occurring. Another patient misunderstood the meaning of the word sterility which, he was told, might follow chemotherapy for Hodgkins disease. He became impotent after the treatment, and it was months before the physician became aware of the problem and many more months before it was solved.

Informed consent may be heard not as a statement of possible risks but rather as a dire prophecy. To make the point, patients who receive necessary medications for serious illness and then read the adverse reactions statement for the drugs in the *Physicians' Desk Reference* may feel more threatened by the drug than by the disease. The truth is not meant to be used as a weapon in this or any other clinical situation.

Conclusion. Anyone who becomes acquainted with these obstacles and problems may wonder, with some physicians, whether the difficulties negate the possibility of informed consent. As noted above, informed consent can never be educated consent in the sense that the patient will have all the knowledge and trained judgment that the physician brings to the decision (Ingelfinger). However, the patient brings something to the process that the physician cannot know, and that is an understanding of his or her individual priorities, needs, concerns, beliefs, and fears. If nothing else, the need for informed consent is an acknowledgement that a medical procedure or act is meant to be done for (not to) a person. An overly legalistic interpretation of the need for informed consent may as much deny the personhood and the patient's need to exercise choice as would neglect of the need to inform simply because difficulties exist. That patients forget things or may deny unpleasant realities in no way detracts from their right to exercise choice or participate in their own care to the degree that they can and wish or do not wish. It should always be remembered that informed consent is a concept whose force should come not primarily from its legal necessity, but rather from its enhancement of good medical care. The needs of the patient should dictate its form and content.

ERIC J. CASSELL

[*Directly related are the entries* THERAPEUTIC RELATIONSHIP, *articles on* CONTEMPORARY SOCIOLOGICAL ANALYSIS *and* CONTEMPORARY MEDICAL PERSPECTIVE; *and* TRUTH-TELLING. *For discussion of related ideas, see the entries:* DECISION MAKING, MEDICAL; PAIN AND SUFFERING; *and* RISK. *Other relevant material may be found under* HEALTH CARE, *article on* HUMANIZATION AND DEHUMANIZATION OF HEALTH CARE; INFORMED CONSENT IN HUMAN RESEARCH; *and* INFORMED CONSENT IN MENTAL HEALTH.]

BIBLIOGRAPHY

ALFIDI, RALPH J. "Informed Consent and Special Procedures." *Cleveland Clinic Quarterly* 40 (1973): 21–25.

CARPENTER, WILLIAM T. "A New Setting for Informed Consent." *Lancet* 1 (1974): 500–501. Comment.

CASSELL, ERIC J. *The Healer's Art: A New Approach to the Doctor–Patient Relationship.* Philadelphia: Lippincott, 1976.

FLAVELL, JOHN H. *The Developmental Psychology of Jean Piaget.* The University Series in Psychology. Edited by David C. McClelland. Princeton: Van Nostrand Reinhold Co., 1963, pp. 156 ff.

GOLDEN, JOSHUA A., and JOHNSTON, GEORGE D. "Problems of Distortion in Doctor–Patient Communications." *Psychiatry in Medicine* 1 (1970): 127–149.

INGELFINGER, F. J. "Informed (But Uneducated) Consent." *New England Journal of Medicine* 287 (1971): 465–466. Editorial.

KORSCH, BARBARA M. "Physicians, Patients, and Decisions." *American Journal of Diseases of Children* 127 (1974): 328–332.

LIEBERMAN, MARVIN. "The Physician's Duty to Disclose Risks of Treatment." *Bulletin of the New York Academy of Medicine* 50 (1974): 943–948.

MAYERSON, EVELYN W. *Putting the Ill at Ease.* Harper Medical. New York: Harper & Row, 1976.

MILLS, DON HARPER. "Whither Informed Consent?" *Journal of the American Medical Association* 229 (1974): 305–310.

MITCHELL, JOHN A., and CRAGIN, CHARLES L., III. "Informed Consent—A Doctor's Dilemma." *Journal of the Maine Medical Association* 64 (1973): 94–97, 101.

ROMANO, JOHN. "Reflections on Informed Consent." *Archives of General Psychiatry* 30 (1974): 129–135.

II
LEGAL AND ETHICAL ASPECTS

The doctrine of informed consent, introduced into U.S. case law in 1957, represents judges' groping efforts to delineate physicians' duties to inform patients of the benefits and risks of diagnostic and treatment alternatives, including the consequences of no treatment, as well as to obtain patients' consent (*Salgo v. Stanford University*). The doctrine's avowed purpose was to protect patients' right to "thoroughgoing self-determination" (*Natanson v. Kline*). The legal implications of informed consent, however, remain unclear. The doctrine is in fact more of a slogan, which judges have been too timid or too wise to translate into law, at least as yet. It has been employed with little care but great passion to voice a dream of personal freedom and individual dignity. Though its legal impact in protecting patients' right to self-decision making has been scant, the threat of informed consent has opened profound issues for the traditional practice of medicine.

The medical framework

It has been insufficiently recognized, particularly by judges, that disclosure and consent, except in the most rudimentary fashion, are obligations alien to medical practice. Hippocrates' admonitions to physicians are still followed today: "Perform [these duties] calmly and adroitly, concealing most things from the patient while you are attending to him. Give necessary orders with cheerfulness and serenity, turning his attention away from what is being done to him; sometimes reprove sharply and emphatically, and sometimes comfort with solicitude and attention, revealing nothing of the patient's future or present condition" (Hippocrates). Thus it is not surprising that the Hippocratic Oath is silent on the duty of physicians to inform, or even converse with, patients. Similarly Dr. Thomas Percival, whose 1803 book *Medical Ethics* influenced profoundly the subsequent codifications of medical ethics in England and the United States, commented only once on the discourse between physicians and patients, restricting his remarks to "gloomy prognostications." Even in that context he advised that "friends of the patient" be primarily informed, though he added that the patient may be told "if absolutely necessary" (Percival, p. 91). The Code of Ethics of the American Medical Association, adopted in 1847, and the Principles of Medical Ethics of the American Medical Association, adopted in 1903 and 1912, repeat, in almost the same words, Percival's statement. The AMA Principles of Medical Ethics, endorsed in 1957, delete Percival's wording entirely and substitute the vague admonition that "physicians . . . should make available to their patients . . . the benefits of their professional attainments." The pertinent sections of the *Opinions of the Judicial Council of the AMA*, interpreting the Principles, note only the surgeon's obligation to disclose "all facts relevant to the need and performance of the operation" and the experimenter's obligation, when using new drugs and procedures, to obtain "the voluntary consent of the person" (American Medical Association, Judicial Council, "Principles"). Nine years later, the AMA House of Delegates in endorsing, with modifications, the Declaration of Helsinki, asked that investigators, when engaged "in clinical [research] primarily for treatment," make relevant disclosures to and obtain the voluntary consent of patients or their legally authorized representative.

Thus in the context of therapy no authoritative statement encouraging disclosure and consent has ever been promulgated by the medical profession. The AMA's tersely worded surgical exception was compelled by the law of malpractice. Its experimental exception represented primarily an acquiescence to United States Public Health Service and the U.S. Department of Health, Education, and Welfare requirements, which in turn were formulated in response to congressional concerns about research practices. When disclosure and consent prior to the conduct of therapeutic research were endorsed by the AMA, it did not extend those requirements to *all* patient care but limited the exception to "clinical [research] primarily for treatment."

Two significant conclusions can be drawn: (1) "Informed consent" is a creature of law and not a medical prescription. A duty to inform patients has never been promulgated by the medical profession, though individual physicians have made interesting, but as a rule unsystematic, comments on this topic. Judges have been insufficiently aware of the deeply ingrained Hippocratic tradition against disclosure and, instead, seem to have assumed that individual physicians' lack of disclosure was aberrant with respect to standard medical practice, and hence "negligent," in the sense of "forgetful" or "inadvertent," conduct. (2) When judges were confronted with claims of lack of informed consent, no medical precedent, no medical position papers, and no analytic medical thinking existed on this subject. Thus physicians were ill prepared to shape judges' notions on informed consent with thoughtful and systematic positions of their own.

The legal framework

With the historical movement from feudalism to individualism, consent, respect for the dignity of human beings, and the right of individuals to shape their own lives became important principles of English common law and, in turn, of American common law. Yet, as these principles gained greater acceptance, questions arose in many areas of law about the capacity of human beings to make their own decisions and about the need to protect them from their "own folly." The tug of war between advocates of thoroughgoing self-determination and those of paternalism has continued unabated. The informed consent doctrine manifests this struggle. While in physician–patient interactions the legal trend during the past two decades has been to increase somewhat the right of patients to greater freedom of choice, the informed consent doctrine has not had as far-reaching an impact on patients' self-determination as many commentators have assumed. This fact has been insufficiently appreciated and has led to confusion, further compounded by the courts' rhetoric that seemed to promise more than it delivered.

Consent to medical and surgical interventions is an ancient legal requirement. Historically an intentional touching without consent was adjudicated in battery. The law has not changed at all in this regard, and a surgeon who operates on a patient without permission is legally liable, even if the operation is successful. In such instances any inquiry into medical need or negligent conduct becomes irrelevant, for what is at issue is the disregard of the person's right to exercise control over his body. The jurisprudential basis of these claims is personal freedom:

> ... under a free government at least, the free citizen's first and greatest right, which underlies all others—the right to himself—is the subject of universal acquiescence, and this right necessarily forbids a physician or surgeon, however skillful or eminent ... to violate without permission the bodily integrity of his patient by ... operating on him without his consent ... [Pratt v. Davis].

But what does consent mean? In battery cases it means only that the physician must inform the patient what he proposes to do and that the patient must agree. Medical emergencies and patients' incompetence are the only exceptions to this requirement.

In mid-twentieth century, judges gradually confronted the question whether patients are entitled not only to know what a doctor proposes to do but also to decide whether the intervention is advisable in the light of its risks and benefits and the available alternatives, including no treatment. Such awareness of patients' informational needs is a modern phenomenon, influenced by the simultaneous growth of product liability and consumer law.

The law of fraud and deceit has always protected patients from doctors' flagrant misrepresentations, and in theory patients have always been entitled to ask whatever questions they pleased. What the doctrine of informed consent sought to add is the proposition that physicians are now under an affirmative duty to offer to acquaint patients with the important risks and plausible alternatives to the proposed procedure. The underlying rationale for that duty was stated in *Natanson* v. *Kline*:

> Anglo-American law starts with the premise of thorough-going self-determination. It follows that each man is considered to be master of his own body, and he may, if he be of sound mind, expressly prohibit the performance of life-saving surgery, or other medical treatment. A doctor might well believe that an operation or form of treatment is desirable or necessary but the law does not permit him to substitute his own judgment for that of the patient by any form of artifice or deception [*Natanson* v. *Kline*].

The language employed by the *Natanson* court in support of an affirmative duty to disclose derives from the language of the law of battery, which clearly makes the patient the ultimate decision maker with respect to his body. Thus the courts reasoned, with battery principles very much in mind, that significant protection of patients' right to decide their medical fate required not merely perfunctory assent but a truly "informed consent," based on an adequate understanding of the medical and surgical options available to them.

Yet in the same breath judges also attempted to intrude as little as possible on traditional medical practices. In doing so their impulse to protect the right of individual self-determination collided with their equally strong desire to maintain the authority and practices of the professions. Law has always respected the arcane expertise of physicians and has never held them liable if they practiced "good medicine." The law of consent in battery represented no aberration from this principle since most physicians agree that patients at least deserve to know the nature of the proposed procedure. However, the new duty of disclosure that the law, in the name of self-determination, threatened to impose upon physicians was something quite different. For the vast majority of physicians significant disclosure is not at all part of standard medical practice. Most doctors believe that patients are neither emotionally nor intellectually equipped to be medical decision makers, that they must be guided past childish fears into "rational" therapy, and that disclosures of uncertainty, gloomy prognosis, and dire risks often seriously undermine cure. Physicians began to wonder whether law was now asking them to practice "bad" medicine.

In the early informed consent cases, judges simply did not resolve the conflict between self-determination and professional practices and authority. The result was distressing confusion. In obeisance to the venerable ideal of self-determination, courts purported to establish, as a matter of law, the physician's

... obligation ... to disclose and explain to the patient in language as simple as necessary the nature of the ailment, the nature of the proposed treatment, the probability of success or of alternatives, and perhaps the risks of unfortunate results and unforeseen conditions within the body [*Natanson* v. *Kline*].

The threat of such an obligation greatly disturbed the medical profession. It recognized that serious implementation of such a standard would significantly alter medical practice. Physicians argued that in order fully to serve patients' best interests, they must have the authority to exercise medical judgment in managing patients. Courts likewise bowed to this judgment. In the very sentence that introduced the ambiguous but exuberant new phrase, "informed consent," the court showed its deference to medical judgment and its hesitancy to disturb traditional practice:

... in discussing the element of risk a certain amount of discretion must be employed consistent with the full disclosure of facts necessary to an informed consent [*Salgo* v. *Stanford University*].

Thus the extent to which evolving case law, under the banner of individualism, was challenging traditional medical practice—which for millennia has treated patients paternally as children—remained confusing. In those earlier cases (*Salgo* v. *Stanford University*, *Natanson* v. *Kline*) judges were profoundly allegiant to both points of view, but the balance was soon tipped decisively in favor of protecting medical practices.

Battery or negligence. The striking ambivalence of judges toward the doctrine of informed consent manifested itself in the competition between battery and negligence doctrines as a means of analyzing and deciding the claims of lack of informed consent. Battery offered a more rigorous protection of patients' right to self-determination. The inquiry into disclosure and consent would not be governed by professional practices but instead would rest on the question: Has the physician met his expanded informational responsibility so that the patient is able to exercise a choice among treatment options? A negative answer to this question would show that the physician's actions constitute trespass, rendering him liable for an unauthorized and "offensive" contact (*Dow* v. *Kaiser Foundation*).

However, in virtually every jurisdiction judges resolved the competition in favor of negligence law. In doing so, judges were able to defer to medical judgment by evaluating the adequacy of disclosure against the medical professional standard of care, asserting that this standard will govern those duties as it does other medical obligations. As a consequence, physicians remain free to exercise the wisdom of their profession and are liable only for failure to disclose what a reasonable doctor would have revealed. Furthermore, negligence theory does not redress mere dignitary injuries, irrespective of physical injuries, and requires proof that the patient, fully informed, would have refused the proposed treatment. Interferences with self-determination, standing alone, are not compensated.

In rejecting battery, judges made much of the fact that such an action required "intent," while negligence involved "inadvertence"; it was the latter, they believed, that accounted for the lack of disclosure. They overlooked that the withholding of information on the part of physicians is generally quite intentional, dictated by the very exercise of medical judgment that the law of negligence seeks to respect. In stating that the nondisclosures were "collateral" to the central information about the nature of the proposed procedure and hence not required for a valid consent, judges discarded the very idea of informed consent—namely, that absence of expanded disclosure vitiates consent. They refused to extend the inquiry to the total informational needs of patients, without which patients' capacity for self-decision making remains incomplete. At bottom, the rejection of an expanded battery theory and of its proposed requirement of informed consent followed from the threat they posed to the authority of doctors and traditional medical practice.

Thus informed consent, based on patients' thoroughgoing self-determination, was a misnomer from the time the phrase was born. To be sure, a new cause of action has emerged for failure to inform of the risks of, and in most jurisdictions alternatives to, treatment. Some duty to disclose risks and alternatives, the courts were willing to say, exists; the extent of that duty is defined by the disclosure practice of a reasonable physician in the circumstances of the case. The new claim is firmly rooted in the law of negligent malpractice, in that plaintiffs are still required to prove the professional standard of care by means of medical expert witnesses. In these, the majority of jurisdictions, traditional medical practice—which generally opposes disclosure—has scarcely been threatened at all in legal reality. The legal life of informed consent, except for dicta about self-determination and the hybrid negligence law promulgated in a handful of jurisdictions, was almost over as soon as it began. Judges had briefly toyed with the idea of patients' self-determination and then largely cast it aside. Good medicine, as defined by doctors, remains good law almost everywhere.

Modifications in professional standard of care. In a few jurisdictions, beginning in 1972 in the District of Columbia with the decision in *Canterbury* v. *Spence,* the new cause of action for failure to inform combined elements of battery with negligence, creating a legal hybrid. The court purported to abandon the professional standard of care with respect to disclosure, asserting that

> . . . respect for the patient's right of self-determination on particular therapy demands a standard set by law for physicians rather than one which physicians may or may not impose upon themselves [*Canterbury* v. *Spence*].

Thus the court laid down a judge-made rule of disclosure of risks and alternatives, which for all practical purposes resembled an expanded battery standard of disclosure.

The preoccupation with risk disclosure, however, continued unabated. From the very beginning, despite all the talk about "informed consent," judges did not lay down any rules for a careful inquiry into the nature and quality of consent, which on its face any meaningful implementation of the doctrine required. Instead major emphasis was placed on risk disclosures. Since in the cases before courts plaintiff-patients only complained of the injurious results of treatment, this emphasis is understandable. Yet to focus solely on risks is to bypass the principal issue of self-determination—namely, whether the physician kept the patient from arriving at his own decision. The *Canterbury* court, too, restricted its concerns largely to risk disclosures; and added the requirement that

> an unrevealed risk that should have been made known must materialize for otherwise the omission, however, unpardonable, is legally without consequence [ibid.].

Thus the court foreclosed legal redress for the patient who, fully informed of the potential effects of, for example, a maiming operation, would have chosen an alternative medical course, even though some of the risks did not materialize.

But to the extent these jurisdictions have abandoned the professional standard of disclosure, traditional medical practice has been challenged; "good medicine," in the eyes of the profession, may no longer be a sufficient defense. Seemingly, in these jurisdictions self-determination has begun to encroach upon the province of medical paternalism. That encroachment, however, may be substantially an illusion, for the touted abandonment of the professional standard of disclosure in *Canterbury* was far from complete. Medical judgment to truncate full disclosure must be "given its due," the court said, when "it enters the picture." The court left ambiguous when the plaintiff must establish the appropriate standard of disclosure by an expert witness, or when he must produce such a witness in order to rebut a defendant-physician's claim that good medical judgment was exercised.

What is clear is that the physician has a "therapeutic privilege" not to disclose information where such disclosure would pose a threat to the "well-being" of the patient. But the ambit of this privilege as well as the relationship of its invocation to a directed verdict is not clear, and this for "good" reasons: Even in these most liberal jurisdictions with respect to patients' rights, courts still cannot face squarely the question of how much they are willing to challenge the traditional medical wisdom of nondisclosure. The law remains ambiguous with respect to this, the core issue of informed consent.

Tensions between self-determination and paternalism. Beyond its allegiance to medical paternalism, noted above, the *Canterbury* court showed its preference for paternalism in another way. Under negligence law, the courts have stated that lack of disclosure cannot be said to have caused the patient's injury unless the patient, if adequately informed, would have declined the procedure; this is the crucial problem of causation in informed consent cases. Such an approach to causation is quite appropriate where law seeks not to compensate interference with self-determination, but only physical injuries resulting from inadequate disclosure. Yet the *Canterbury* court, and every court that has considered the matter subsequently, held that the decision whether or not to undertake therapy must be examined not from the point of view of the patient-plaintiff but from that of a "prudent person in the patient's position," limiting the inquiry to whether a "reasonable patient" would have agreed to the procedure. This substitution of a community standard of a "reasonable" person cuts the heart out of the courts' purported respect for individual self-determination. Questions of the influence of hindsight and bitterness are familiar to juries, as is the problem of self-serving testimony generally. While those are delicate problems, they do not justify abrogating the very right at issue in cases of informed consent: the right of individual choice, which may be precisely the right to be an "unreasonable" person.

Epilogue on law. Thus law has proceeded feebly toward the objective of patients' self-determination. While a new cause of action, occasionally hybridized with battery, has emerged for the negligent failure to disclose risks and alternative treatments, it remains a far cry from the avowed purpose of the informed consent doctrine, namely, to secure patients' autonomy and right to self-determination. In not tampering significantly with the medical wisdom of nondisclosure, yet creating a new cause of action based on traditional disclosure requirements, courts may have accomplished a different result, very much in line with other purposes of tort law—namely, to provide physically injured patients with greater opportunities for seeking compensation whenever it can be argued that disclosure might have avoided such injuries. In doing so judges may have hoped, through the anticipatory tremors of dicta, to urge doctors to consider modifying their traditional disclosure practices. But judges have been unwilling, at least as yet, to implement earnestly patients' right to self-determination.

Whither informed consent?

The disquiet that the doctrine of informed consent has created among physicians cannot be fully explained by the small incremental step courts have taken to assure greater patient participation in medical decision making. More likely it was aroused by the uncertainty over the scope of the doctrine and by an appreciation that medical practice, indeed all professional practice, would be radically changed if fidelity to thoroughgoing self-determination were to prevail. In what follows, some of the issues raised by the idea of an informed consent doctrine, based on a premise of self-determination, will be discussed.

Patients. Traditionally patients have been viewed as ignorant about medical matters, fearful about being sick, childlike by virtue of their illness, ill-equipped to sort out what is in their best medical interest, and prone to make decisions detrimental to their welfare (Parsons). Thus physicians have asserted that it makes little sense to consult patients on treatment options; far better to interact with them as beloved children and decide for them. In the light of such deeply held convictions, many physicians are genuinely puzzled by any informed consent requirement. Moreover, its possible detrimental impact on compassion, reassurance, and hope—ancient prescriptions for patient care—has raised grave ethical questions for the medical profession.

Those concerns should not be dismissed lightly. What may be at issue, however, is not an intrinsic incapacity of patients to participate in medical decision making. For not all patients, and probably not even most, are too uneducated, too frightened, or too regressed to understand the benefits and risks of treatment options available to them. Moreover, their capacities for decision making are affected to varying degrees, for example, by the nature of the disease process, its prognosis, acuteness, painfulness, etc., as well as by the personality of patients. The medical literature is largely silent on the question of who—under what circumstances and with what conditions—should or should not be allowed to participate fully in medical decision making.

But why has not the sorting-out process, distinguishing between those patients who do and those who do not have the capacity for decision making, been undertaken long ago? One answer suggests itself: Once those patients have been identified who, in principle, can make decisions on their own behalf, physicians would be compelled to confront the questions of whether to interact with them on a level of greater equality; whether to share with them the uncertainties and unknowns of medical diagnosis, treatment, and prognosis; and whether to communicate to them their professional limitations as well as the lack of expert consensus about treatment alternatives. Such an open dialogue would expose the uncertainties inherent in most medical interventions; and to the extent medicine's helpful and curative power depends on the faith and confidence which the physician projects, patients may be harmed by disclosure and consent.

Physicians' objections to informed consent, therefore, may have less to do with the incompetence of patients as such than with an unrecognized concern of the doctrine's impact on the dynamics of cure. Put another way, the all too sweeping traditional view of patients has misled doctors into believing that medicine's opposition to informed consent is largely based on patients' incompetence, rather than on an apprehension, however dimly perceived, that disclosure would bring into view much about the practice of medicine that physicians seek to hide from themselves and their patients; for example, the uncertainties and disagreements about the treatments employed; the curative impact of physicians' and patients' beliefs in the unquestioned effectiveness of their prescriptions rather than the prescriptions themselves; the difficulty in sorting out the contributions that *vis medicatrix naturae* ("the healing power of nature") makes to the healing process; the impact of patients' suggestibility to cure, etc. Thus the question: When does informed consent interfere with physicians' effectiveness and with the dynamics of cure?

Little attention has been paid to the fact that the practice of Hippocratic medicine makes patients more incompetent than they need be. Indeed patients' incompetence can become a self-fulfilling prophecy as a consequence of medical practices. That the stress of illness leads to psychological regression, to chronologically earlier modes of functioning, has been recognized for a long time. Precious little, however, is known about the contributions that physicians' attitudes toward and interactions with their patients make to the regressive pull. Also, little is known about the extent to which regression can be avoided by not keeping patients in the dark, by inviting them to participate in decision making, and by addressing and nurturing the intact, mature parts of their functioning. This uncharted territory requires exploration in order to determine what strains will be imposed on physicians and patients alike, if Anna Freud's admonition to students of the Western Reserve Medical School is heeded:

. . . you must not be tempted to treat [the patient] as a child. You must be tolerant toward him as you would be toward a child and as respectful as you would be towards a fellow adult because he has only gone back to childhood as far as he's ill. He also has another part of his personality which has remained intact and that part of him will resent it deeply, if you make too much use of your authority [quoted in Katz, p. 637].

Physicians. Traditionally physicians have asserted that their integrity, training, professional dedication to patients' best medical interests, and commitment to "doing no harm" are sufficient safeguards for patients. The complexities inherent in medical decision making, physicians maintain, require that trust be patients' guiding principle. The idea of informed consent does not question the integrity, training, or dedication of doctors. Without them, informed consent would be of little value. What the idea of informed consent does question is the necessity and appropriateness of physicians' making all decisions for their patients; it calls for a careful scrutiny of which decisions belong to the doctor and which to the patient.

Physicians have preferences about treatment options that may not necessarily be shared by patients. For example, no professional consensus exists about the treatment of breast cancer. The advantages and disadvantages of lumpectomy, simple mastectomy, radical mastectomy, radiation therapy, chemotherapy, and various combinations among these are subject to much controversy. Dr. Bernard Fisher, chairman of the National Surgical Adjuvant Breast Cancer Project, has said that we simply do not know which method is best (Fisher). Thus the question must be answered: How extensive an opportunity must patients be given to select which alternative? Informed consent challenges the stereotypical notion that physicians should assume the entire burden of deciding what treatment *all* patients, *whatever* their condition, should undergo. Indeed, can the assumption of this burden be defended purely on medical grounds in the first place? Is not the decision in favor of one treatment for breast cancer over another, like many other treatment decisions, a combination of medical, emotional, aesthetic, religious, philosophical, social, interpersonal, and personal judgments? Which of these component judgments belong to the physician and which to the patient?

Much needs to be investigated in order to learn the practical human limits of any new obligations to disclose and to obtain consent:

1. Informing patients for purposes of decision making requires learning new ways of interacting and communicating with patients. Such questions as the following will have to be answered: What background information must patients receive in order to help them formulate their questions? How should physicians respond to "precipitous" consents or refusals? How deeply should doctors probe for understanding? What constitutes irrelevant information that only tends to confuse? What words and explanations facilitate comprehension? Physicians have not been in the habit of posing such questions.

2. Underlying informed consent is the assumption that physicians have considerable knowledge about their particular specialties, keep abreast of new developments, and are aware of what is happening in other fields of medicine that impinge on their area of professional interest. This is not so; indeed, it may be asking too much. Moreover, since physicians have their preferences for particular modes of treatment, can they be expected to present an unbiased picture of alternative treatments?

3. Physicians have consistently asserted that informed consent interferes with compassion (Silk). Doctors believe that, in order to maintain hope or to avoid the imposition of unnecessary suffering, patients in the throes of a terminal illness, and other patients as well, should not be dealt with honestly. But the evidence for such allegations is lacking. When physicians are asked to support them with clinical data, they are largely unable to do so (Oken). Indeed, the few studies that have been conducted suggest that most patients do not seem to yearn for hope based on deception, but for hope based on a reassurance that they will not be abandoned, that everything possible will be done for them, and that physicians will deal truthfully with them. Moreover, evidence is accumulating that informed patients become more cooperative, more capable of dealing with discomfort and pain, and more responsible. Whether the often alleged conflict between "compassionate" silence and "cruel" disclosure is myth or reality remains to be seen. Disclosure may turn out to be a greater burden to those who have to interact with patients than to the patients themselves.

4. Informed consent confronts the role of faith in the cure of disease and the complex problems created by the uncertainties inherent in medical practice. To some extent the two issues are intertwined. The effectiveness of a therapeutic program, it has often been said, depends on three variables: the "feeling of trust or faith the patient has in his doctor and therefore in his therapy . . . , the faith or confidence the physician has in himself and in the line of therapy he proposes to use . . . , and the therapy [itself]" (Hoffer, p. 124). Informed consent could interfere with the first two variables and thus undermine the effectiveness of treatment. Precisely because of the uncertainties in medical decision making, the physician, to begin with, defends himself against those uncertainties by being more certain about what he is doing than he realistically can be. There is perhaps some unconscious wisdom in what he has been doing since Hippocrates' days, for the unquestioned faith the doctor has in his own therapy is also therapeutic in its own right. Thus, to be a more effective healer, a physician may need to defend himself against his uncertainties by believing himself to be more powerful than he is. That defense will be threatened by informed consent, for it would now require him to be more aware of what he does not know, and therapeutic effectiveness in turn might suffer. Finally, patients' response to treatment also depends on faith in the physician and his medicines. Knowing of the "ifs" and "buts" may shake patients' faith and undermine the therapeutic impact of suggestibility, which contributes so much to recovery from illness.

Physicians' traditional counterphobic reaction

to uncertainty, adopting a sense of conviction that what seems right to them is the only correct thing to do, has other consequences as well. Defensive reactions against uncertainty have led to overenthusiasm for particular treatments that have been applied much more widely than an unbiased evaluation would dictate. The ubiquitous tonsillectomies performed to the psychological detriment of untold children is a classical example. Moreover, by not acknowledging uncertainty to themselves, doctors cannot acknowledge it to their patients. Thus consciously and unconsciously physicians avoid the terrifying confrontation of uncertainty, particularly when associated with poor prognosis. As a result, communications with patients take the form of an evasive monologue. The dialogue that might reveal these uncertainties is discouraged (Davis).

While disclosure of information would reduce patients' ignorance, it would also diminish doctors' power within the physician–patient relationship. As Waitzkin and Stoeckle have observed, the "physician enhances his power to the extent that he can maintain the patient's uncertainty about the course of illness, efficacy of therapy, or specific future actions of the physician himself" (p. 187). Thus new questions arise: What consequences would a diminution of authority have on physicians' effectiveness as healers? How would patients react to less powerful doctors? Would they accept them or turn to new faith healers?

Limits of self-determination. Patients' capacity for self-determination has been challenged on the grounds that neither total understanding nor total freedom of choice is possible (Ingelfinger). This of course is true. Any informed consent doctrine, to be realistic, must take into account the biological, psychological, intellectual, and social constraints imposed upon thought and action. But those inherent constraints, which affect all human beings, do not necessarily justify treating patients as incompetents. Competence does not imply total understanding or total freedom of choice.

What needs to be explored is the extent to which medicine, like law, should presume competence rather than incompetence, in interactions with patients. Neither presumption comports fully with the psychobiology of human beings; both of them express value judgments on how best to interact with human beings. Once the value judgment is made, one can decide on the additional safeguards needed to avoid the harm that any fiction about human behavior introduces.

The idea of informed consent asks for a presumption in favor of competence. If that is accepted, it may also follow that human beings should be allowed to strike their own bargains, however improvident. The then Circuit Judge Warren E. Burger, in commenting on a judicial decision to order a blood transfusion for a Jehovah's Witness had this to say: "Nothing in [Justice Brandeis' 'right to be let alone' philosophy, suggests that he] thought an individual possessed these rights only as to *sensible* beliefs, *valid* thought, *reasonable* emotions or *well-founded* sensations. I suggest he intended to include a great many foolish, unreasonable and even absurd ideas which do not conform such as refusing medical treatment even at great risk" (*Application of President of Georgetown College*). A physician may wish, and even should try, to persuade his patients to agree to what he believes would serve their medical interests best; but ultimately he may have to bow to his patients' decision, however "senseless" or "unreasonable," or withdraw from further participation. The alternatives, deception or coercion, may be worse, for either would victimize not only patients but physicians as well.

Conclusion

The narrow scope that courts have given to the informed consent doctrine may reflect a deeply held belief that the exercise of self-determination by patients is often against the best interests of otherwise responsible adults and that those interests deserve greater protection than personal freedom. It may also reflect a judicial recognition of law's limited capacity to regulate effectively the physician–patient relationship. Therefore, once having suggested that patients deserve at least a little openness in communication, courts may have concluded that they had gone as far as they could. Judges, at least for the time being, have largely left it up to the medical profession to confront the question of patients' greater participation in medical decision making.

Despite their snail's pace, the courts' approach may have merit. Implementing a right of self-determination has tremendous consequences for medical practice. Many difficult problems, each with vast ethical implications, need to be considered by the medical profession. Thus introspection and education, responsive to the legal and professional problems that new patterns of physician–patient interaction will create, may ultimately provide firmer foundations for new patterns of physician–patient interactions than forced change through outside regulation. The latter, however, may increase if the profession does not rise to the challenge of addressing these long-neglected problems.

JAY KATZ

[*For further discussion of topics mentioned in this article, see the entries:* CHILDREN AND BIOMEDICINE; INFANTS; PATERNALISM; PRISONERS; PSYCHOSURGERY; THERAPEUTIC RELATIONSHIP; *and* TRUTH-TELLING. *Also directly related are the entries* DECISION MAKING, MEDICAL; INFORMED CONSENT IN HUMAN RESEARCH; *and* INFORMED CONSENT IN MENTAL HEALTH. *See* APPENDIX, SECTION I, PRINCIPLES OF MEDICAL ETHICS (AMERICAN MEDICAL ASSOCIATION); SECTION II, DECLARATION OF HELSINKI; SECTION III, A PATIENT'S BILL OF RIGHTS, *and* PATIENTS' RIGHTS: REGULATIONS FOR SKILLED NURSING FACILITIES.]

BIBLIOGRAPHY

American Medical Association, Judicial Council. *Opinions and Reports of the Judicial Council.* Chicago: 1969.

———, Judicial Council. "Principles of Medical Ethics." *Opinions and Reports of the Judicial Council,* pp. vi–vii.

BURGER, WARREN E. "Reflections on Law and Experimental Medicine." *UCLA Law Review* 15 (1968): 436–442.

DAVIS, FRED. "Uncertainty in Medical Prognosis: Clinical and Functional." *American Journal of Sociology* 66 (1960): 41–47.

FISHER, BERNARD. "The Surgical Dilemma in the Primary Therapy of Invasive Breast Cancer: A Critical Appraisal." *Current Problems in Surgery,* October 1970, pp. 1–53.

GLASS, ELEANOR S. "Restructuring Informed Consent: Legal Therapy for the Doctor–Patient Relationship." *Yale Law Journal* 79 (1970): 1533–1576.

HENDERSON, L. J. "Physician and Patient as a Social System." *New England Journal of Medicine* 212 (1935): 819–823.

HIPPOCRATES. "Decorum." *Hippocrates.* 4 vols. Translated by W. H. S. Jones. The Loeb Classical Library. London: William Heinemann; New York: G. P. Putnam's Sons, 1923, vol. 2, pp. 278–301, especially par. 16, pp. 296–299.

HOFFER, A. "A Theoretical Examination of Double-Blind Design." *Canadian Medical Association Journal* 97 (1967): 123–127.

INGELFINGER, F. J. "Informed (But Uneducated) Consent." *New England Journal of Medicine* 287 (1972): 465–466.

KATZ, JAY, ed. *Experimentation with Human Beings: The Authority of the Investigator, Subject, Professions, and State in the Human Experimentation Process.* New York: Russell Sage Foundation, 1972.

———, and Capron, Alexander, Morgan. *Catastrophic Diseases: Who Decides What? A Psychosocial and Legal Analysis of the Problems Posed by Hemodialysis and Organ Transplantation.* New York: Russell Sage Foundation, 1975.

MCCOID, ALLAN H. "A Reappraisal of Liability for Unauthorized Medical Treatment." *Minnesota Law Review* 41 (1957): 381–434.

OKEN, DONALD. "What to Tell Cancer Patients: A Study of Medical Attitudes." *Journal of the American Medical Association* 175 (1961): 1120–1128.

PARSONS, TALCOTT. *The Social System.* Glencoe, Ill.: Free Press, 1951.

PERCIVAL, THOMAS. *Medical Ethics.* Edited by Chauncey D. Leake. Baltimore: Williams & Wilkins Co., 1927.

PLANTE, MARCUS L. "An Analysis of 'Informed Consent'." *Fordham Law Review* 36 (1968): 639–672.

SILK, ARTHUR D. "A Physician's Plea: Recognize Limitations of Informed Consent." *American Medical News,* 12 April 1976, p. 19.

WAITZKIN, H., and STOECKLE, J. D. "The Communication of Information about Illness: Clinical, Sociological, and Methodological Considerations." *Advances in Psychosomatic Medicine* 8 (1972): 180–215.

COURT DECISIONS

Application of President of Georgetown College. 331 F. 2d 1010 (D.C. Cir. 1964). Certiorari denied. 377 U.S. 978. 12 L. Ed. 2d 746. 84 S. Ct. 1883 (1964).

Canterbury v. Spence. 464 F.2d 772 (D.C.C.A. 1972).

Cobbs v. Grant. 104 Cal. Rptr. 505. 502 P.2d 1 (1972).

Dow v. Kaiser Foundation. 12 Cal. App.3d 488. 90 Cal. Rptr. 747 (Ct. App. 1970).

In re Estate of Brooks. 32 Ill. 2d 361. 205 N.E.2d 435 (1965).

Mohr v. Williams. 104 N.W. 12 (Minn. 1905).

Natanson v. Kline. 186 Kan. 393. 350 P.2d 1093 (1960). 187 Kan. 186. 354 P.2d 670 (1960).

Pratt v. Davis. 118 Ill. App. 161 (1905). Aff. 224 Ill. 300. 79 N.E. 562 (1906).

Salgo v. Stanford University. 317 P.2d 170 (Cal. 1st Dist. Ct. App. 1957).

Schloendorff v. New York Hospital. 105 N.E. 92 (N.Y. 1914).

Wilkinson v. Vesey. 295 A.2d 676 (R.I. 1972).

INSEMINATION, ARTIFICIAL

See REPRODUCTIVE TECHNOLOGIES, *articles on* ARTIFICIAL INSEMINATION, ETHICAL ISSUES, *and* LEGAL ASPECTS.

INSTITUTIONAL GUIDELINES FOR HUMAN RESEARCH

See APPENDIX, SECTION II, *for the texts of guidelines and codes. See also* HUMAN EXPERIMENTATION, *article on* SOCIAL AND PROFESSIONAL CONTROL.

INSTITUTIONAL REVIEW BOARDS

See HUMAN EXPERIMENTATION, *article on* SOCIAL AND PROFESSIONAL CONTROL.

INSTITUTIONALIZATION

Introduction

Persons may be institutionalized as a result of being convicted of a crime or as a result of mental commitment. Although the present article will, for illustrative purposes, examine some issues in criminal confinement, it will center on mental institutionalization, for it is there that bioethical concerns are the strongest.

Grounds for institutionalization. Mental institutionalization can itself occur either through the process of "criminal" commitment or through the process of "civil" commitment. "Criminal" commitment is a term referring loosely to the mental hospitalization of disordered persons who have been involved in the criminal system. Thus, it would include persons committed as mentally incompetent to stand criminal trial, persons committed after being found not guilty by reason of insanity, and persons committed as sexual psychopaths.

"Civil" commitment refers to the mental hospitalization of disordered persons when such hospitalization is not triggered by criminal charges. There are two separate bases of the commitment power—the paternalistic (*parens patriae*) power and the police power.

In the exercise of its paternalistic power, the state is presumably authorized to hospitalize and treat those persons who, because of mental illness, are in need of treatment and are unable to make appropriate personal decisions about hospitalization and treatment. In contrast, when the state commits an individual pursuant to its police power, it does so not necessarily for the good of that individual, but rather to protect the public from an individual who is mentally ill and dangerous to others.

Rights of the institutionalized. Regardless of the basis of their commitment, institutionalized individuals possess certain rights even while incarcerated. At one time the courts took a "hands-off" approach to the rights of the institutionalized and refused to limit the discretion of institutional administrators (Wexler, 1973, p. 91). The "hands-off" approach has now given way to an approach of increased judicial involvement, particularly in the United States. Increasingly, courts are addressing ethical issues of institutionalization and are converting moral rights to legal rights. Rights relating to communication, visitation, and religious freedom have, for example, been rather well developed (Ennis and Siegel). Other "rights"—such as the right to sexual expression—are less developed but are beginning to receive the attention of commentators and advocates (Friedman, p. 85).

At the least, institutionalized persons should be entitled to a humane environment. The landmark *Wyatt* v. *Stickney* decision recognized such a right as part of an overall right to treatment. *Wyatt*, for example, guaranteed patients a right to a bed, to nutritious meals, to a semblance of privacy, and to other conditions necessary to combat what would otherwise be an inhumane and antitherapeutic environment.

The movement toward decarceration. In any discussion of institutionalization, it is also important to note that, in part because of the inhumane and antitherapeutic nature of institutional environments, there is now a clear-cut social movement toward deinstitutionalization or decarceration. For a variety of reasons, the emerging presumptions—social, psychological, and now even legal—disapprove confinement of patients and prisoners if it is avoidable, reject lengthy periods of confinement if shorter periods might suffice, and reject confinement in secure facilities if confinement in less secure facilities might be suitable.

The decarceration trend has been spurred on principally by the vastly expanding literature documenting the adverse effects of incarceration, by the increased awareness that alternatives to institutionalization are often satisfactory, and by the moral view that we ought not to deprive persons of any more liberty than is necessary to achieve legitimate governmental goals.

Despite the decarceration trend, institutionalization will remain a reality for a large number of persons. Perhaps the main concerns facing institutionalized persons are (1) their rights during confinement, particularly with regard to treatment issues, and (2) the length of time they may be involuntarily confined. It is, then, to these topics of "treatment issues" and "durational limits" to which this article will now turn (Wexler, 1976).

Treatment issues

Right to treatment. The relevant treatment questions include the constitutional or legal right *to* treatment, and the emerging right of prisoners, patients, and "hybrids" to refuse treatment. While both issues are of rather recent legal origin, the right-to-treatment area is considerably more developed than the area of the right to refuse treatment. Yet, in terms of U.S. Supreme Court pronouncements, the question whether there is a constitutionally prescribed right to treatment remains unresolved. Many observers had expected the question to be resolved by the Court in the case of *O'Connor* v. *Donaldson*, but the Court decided the case on narrow grounds and left open the right to treatment question. The Court ruled simply that it is inappropriate to confine without treatment nondangerous persons capable of adequate community adjustment. But the Court did not consider whether such persons could be confined if treatment were forthcoming or whether persons confined because of dangerousness have a right to treatment.

Regardless of the Court's ultimate verdict on the existence of a constitutionally grounded right to treatment and its contours, the right—or at least some semblance of it—is now so firmly embedded in lower court decisions, modern statutory enactments, and legal commentary that its continued existence, with or without a constitutional imprimatur, is almost assured. The right may have a different theoretical basis and scope, however, when applied to patients committed pursuant to the state's paternalistic power (*parens patriae* patients) and when applied to patients committed pursuant to the police power of the state (police power patients).

The "core" recipients of a right to treatment are presumably those *parens patriae* patients who are committed because they are mentally ill, legally incompetent to make hospitalization and treatment decisions, and in need of treatment. To the extent that a need for treatment is part of the rationale for commitment, confinement without treatment would be legally unwarranted.

Police power patients are also generally thought to enjoy a right to treatment, but the scope and theoretical base for this assumption is far shakier than in the case of *parens patriae* patients. Courts have often accorded police power patients a right to treatment on the theory that detention is ordinarily appropriate only for a finite period, following a trial with many procedural protections and after a finding that the subject has committed a specifically defined offense. Since police power commitments deviate substantially from that criminal model, the quid pro quo or "trade-off" for the departure ought to, according to the theory, result in a right to treatment even for police power patients. The theory has, however, been sharply criticized on a number of grounds (Wexler, 1976, pp. 77–78, n. 51).

Right to refuse treatment. Even if treatment must be made available to all types of patients, including police power patients, it by no means follows, as some psychiatrists and psychologists mistakenly believe, that therapy can be forced on unwilling patients and that those patients' rights to treatment would be infringed if therapy were not thrust upon them. The legal system—through case law, legislation, and administrative regulations—is steadily defining and refining a right to resist treatment.

In a handful of recent cases, the courts have addressed the right-to-resist issue. One such case is *Mackey* v. *Procunier*. Mackey, a California prisoner, was transferred to a secure mental hospital in order to receive electroconvulsive therapy. The transfer for that purpose was apparently agreed to by Mackey. Once at the hospital, however, Mackey was apparently subjected, without his consent, to a very different procedure, "anectine therapy." Anectine is a muscle relaxant that induces paralysis and respiratory arrest. Its standard use, with an anesthetic, is as an adjunct to electroconvulsive therapy in order to prevent bone fracture. Mackey claimed, however, that the drug was administered to him while he was awake and as part of a program of "aversive therapy." That is, he received anectine injections contingent upon his engaging in inappropriate behavior. When administered to conscious patients, anectine has been described as creating a sensation of drowning, dying, and suffocating. When Mackey's case reached the U.S. Court of Appeals for the Ninth Circuit, that court ruled that proof of his allegations could raise serious questions under the Eighth Amendment cruel and unusual punishment clause and the emerging First Amendment protection of mental privacy.

A similar problem arose with security patients at the Iowa Security Medical Facility. There, nurses administered injections of apomorphine, a drug that induces uncontrollable vomiting, to misbehaving patients. The U.S. Court of Appeals for the Eighth Circuit, in the case of *Knecht* v. *Gillman*, held that the administration of apomorphine without the informed consent of the patient ran afoul of the constitutional proscription against cruel and unusual punishment.

Clearly, then, aversive and punitive programs of behavior control can no longer be resorted to with complete freedom. Indeed, forcibly subjecting patients even to schemes of positive reinforcement—"reward therapy"—is a process also headed for legal difficulty. Programs of "token economies" in which tokens, earned for appropriate behavior, can later be cashed in to purchase desired items and events, may pose difficulties, because patients may begin in such programs at legally unwarranted stages of deprivation. Similar difficulties may be encountered in the "tier systems" in which privileges are dependent upon one's place in a hierarchy and in which one's place, in turn, is dependent upon appropriate behavior.

In many token economies and tier systems, food, beds, privacy, and ground privileges are used as "reinforcers," available only if earned by engaging in appropriate behavior. Yet decisions defining rights of patients, such as *Wyatt v. Stickney,* are increasingly suggesting that patients are constitutionally entitled to such items and events as part and parcel of a humane psychological environment. Considerable doubt is thus cast on the legality of the continued use of such contingently available reinforcers (Wexler, 1973).

As in the case of the right to treatment, the right to resist treatment operates differently with respect to *parens patriae* and police power patients. According to the emerging view, if a patient or prisoner is mentally competent to decide about matters of therapy and gives informed consent, the therapy can go forward; but if the competent person refuses consent, the state lacks a sufficiently compelling interest to thrust the therapy upon the patient. If, on the other hand, the patient is incompetent instead of competent, his or her acquiescence or refusal is not determinative, and in certain instances a surrogate decision maker (guardian, Human Rights Committee, court, etc.) can consent even to intrusive therapies if less restrictive techniques seem unsuitable and if the proposed therapy seems, in a cost-benefit sense, to be in the best interest of the incompetent patient (Spece; Shapiro; Wexler, 1975).

Since incompetency is regarded as part and parcel of the *parens patriae* commitment power, *parens patriae* patients will presumably be eligible for forced treatment if the other important tests—best interest and least restrictive alternative—are also met. Police power commitments, on the other hand, are based on potential dangerousness but do not necessarily require a level of mental disability amounting to incompetency. Police power patients, therefore, may be in a position to refuse intrusive treatment.

Informed consent. In order to preserve the delicate balance between according police power patients a right to refuse therapy and allowing them at the same time to submit to therapy to which they give their informed consent, it will be necessary for the courts to come properly to grips with the concept of consent.

One impediment to approaching the consent notion carefully has been the conceptual confusion generated by the case of *Kaimowitz v. Department of Mental Health,* a decision that barred the performance of experimental psychosurgery on involuntarily confined patients. The *Kaimowitz* court held that psychosurgery could not be performed without the informed consent of the subject, and held further that even apparently *acquiescing* patients could not submit to the procedure because such persons would be unable to give legally adequate informed consent.

Informed consent, according to *Kaimowitz* and other authorities, can be broken down into the three constituent elements of competence, knowledge, and voluntariness. With respect to confined patients submitting to psychosurgery, the *Kaimowitz* court found each of the required elements unsatisfied. Competence was absent because the court viewed the process of institutionalization as creating a dependence among patients that rendered them cognitively incapable of making decisions as serious and complex as whether to submit to psychosurgery. Knowledge was found wanting because the risks of psychosurgery were deemed so uncertain that consent to psychosurgery could not be regarded as truly "informed." And voluntariness was absent largely because the desire for release was regarded as so overpowering that it would coerce patients into submitting to psychosurgery in order to improve their prospects for discharge.

The *Kaimowitz* court has been criticized for its loose use of the three informed consent elements, which renders *Kaimowitz* difficult to distinguish analytically from instances where informed consent *ought* to be found (Wexler, 1974; Murphy). It is clear that the *Kaimowitz* court intended its holding to be confined to the special facts involved in the case—that is, to procedures that are experimental, highly intrusive, dangerous, and irreversible. But such a limitation is not particularly satisfactory in a conceptual sense. In any event, there is a tendency at least among certain advocates to expand the *Kaimowitz* rationale to cover a host of therapeutic situations other than psychosurgery.

Of particular concern is the notion of voluntary consent. If the legal system wishes to preserve the delicate balance mentioned earlier between allowing consensual therapy and disallowing nonconsensual therapy, it must pierce through the rhetoric, fueled by the *Kaimowitz* case or at least by sloppy readings of that case, that institutions are inherently coercive and that, because the lure of release is so overpowering, voluntary consent is unobtainable in an institutional setting. Should the inherent coercion formula be accepted and applied to a broad spectrum of intrusive therapies, the effect would be to vitiate the right *to* treatment, because the following reasoning would apply. Patients have a right to treatment. They also have a right to resist treatment in the absence of informed consent. But informed consent cannot be given by institutionalized patients, because any such consent would be inherently coerced rather than voluntary. Thus, patients cannot be forcibly subjected to therapy, nor can they voluntarily submit to it, for their submission will be equated with forcible subjection.

By such reasoning, our institutions—for security-status patients, for prisoners, and for involuntarily confined civil patients—would be converted, by the force of law, to humane holding facilities and nothing more. And under the label of paternalism, patients would be *deprived* of a treatment option, and some would continue to be confined because of their untreated dangerous behavior. That result seems as unacceptable as the opposite problem—the traditional position from which we are rapidly moving—of a therapeutic free-for-all, where therapists are allowed to determine the appropriateness of treatment procedures for particular patients and to subject even competent protesting patients to those procedures.

The solution seems to lie in recognizing that pressure to select a particular option, even if the pressure is generated by a desire to avoid or reduce incarceration, should not itself be deemed the legal equivalent of coercion. In other areas of the law, that recognition is readily apparent. Plea bargains are upheld as voluntary even though motivated by a desire to avoid or reduce incarceration. Reasonable conditions of probation and parole are regarded as voluntarily agreed to even though, once again, their acceptance by prospective probationers and parolees is often motivated by a desire to avoid or reduce incarceration. Coercion, therefore, should not be viewed as a doctrine that condemns pressure as such, but rather as a doctrine that guards against unfair or unreasonable choices. Coercion should more readily be found if benefits are promised to a patient for the *mere act* of participating in a program, or if a patient is threatened with additional adverse consequences for not participating, than if merely an opportunity is offered to participate (with no benefits or detriments flowing from the participation decision as such) in a program that, should it actually alter the patient's behavior and undercut the reason for his detention, may lead to his release.

Durational limits

Legal commentators and recent cases have urged the creation of durational limits on confinement with respect to both the criminal system and the system of civil commitment. In terms of the criminal law, for example, Norval Morris, concerned about therapeutic excesses and about our total inability to predict dangerousness, has urged that criminal punishment should be justified by retribution and deterrence, should not have an independent goal of preventive detention, and ought not to have a reformative goal other than to make rehabilitative programs available to willing participants who are confined for periods set solely by retributive and deterrent considerations. Morris suggests intricate interplays between the deterrent and retributive variables to determine whether incarceration is justified, but in determining the maximal length of permissible incarceration—the question of key concern for present purposes—considerations of deterrence fall out of Morris's scheme, leaving retribution to reign supreme.

According to Morris, deterrence should drop out of the maximum length of incarceration determination because of the following process of reasoning:

To use the innocent as a vehicle for general deterrence would be seen by all as unjust. . . . Punishment in excess of what the community feels is the maximum suffering justly related to the harm the criminal has inflicted is, *to the extent of the excess*, a punishment of the innocent, notwithstanding its effectiveness for a variety of purposes [Morris, p. 1173].

With deterrence thereby removed from the picture, Morris proposes that maximum lengths of incarceration be determined by the principle of retribution or "desert": "no sanction greater than that 'deserved' by the last crime or bout of crimes for which the offender is being sentenced should be imposed." Morris, therefore, views retribution as a protective principle, a principle necessitating that a "retributive lid" be clamped on the length of permissible incarceration.

Case law in the United States accepts the principle that a period of criminal confinement ought to be confined by a retributive lid—or at least by a "rough" retributive lid—and that considerations of reformation, deterrence, and preventive detention ought not to play an important role in creating constitutionally permissible maximum periods of incarceration. Actually, the statement of the principle, as opposed to its accepted application, goes back as far as 1910, when the Supreme Court in *Weems* v. *United States* announced a constitutionally grounded mandate to the effect that "it is a precept of justice that punishment for crime should be graduated and proportioned to offense." In recent years, the courts have begun to invoke the *Weems* "proportionality" rule, which is considered to be part and parcel of the Eighth Amendment proscription against cruel and unusual punishments, to set aside at least those dispositions and sentences which are grossly disproportionate to the moral blameworthiness or seriousness of the triggering offense.

Probably the most pertinent of the proportionality decisions involved an explicitly indeterminate sentence that was given to a criminal sex offender of a type clinically indistinguishable from sexual psychopaths who often find themselves under indeterminate commitment to security mental hospitals. The case in question is the 1972 California case In re *Lynch*. Lynch, upon his second conviction for indecent exposure, was given a wholly indeterminate sentence (which might therefore theoretically entail lifetime confinement). Invoking the proportionality language of *Weems*, coupled with the California constitutional proscription against cruel and unusual punishments, the *Lynch* court found the indeterminate sentence imposed by the trial court to be without satisfactory legal support.

Like certain other courts that have recently grappled with proportionality questions, the *Lynch* court was unimpressed with state assertions that the interest of general deterrence and the need for sex offender isolation were sufficient to sustain the heavy penalty. Moreover, the state's assertion of a need for isolation did not carry the day even though the court recognized that, with respect to the sexual conduct at issue, the prospect for recidivism is very real. The prospect of recidivism in the context of indecent exposure, real as it may be, simply does not, in the words of the *Lynch* court, "justify the greatly enhanced punishment" of indeterminate confinement.

Although *Lynch* was decided in the context of the criminally *convicted*, carryover to the category of special offenders who are criminally *committed* (and, by a somewhat different line of reasoning, to the category of the civilly committed) is rather compelling. If, for example, one accepts the *Lynch* principle that indeterminate confinement of at least certain convicted sex offenders must be legally replaced by a rough retributive lid, and if one accepts the research findings that convicted sex offenders are clinically comparable to committed sexual psychopaths, it is difficult to justify on due process and equal protection grounds—and surely on grounds of sound social policy—the propriety of wholly indeterminate confinement for the category of sexual psychopaths. From a constitutional and public policy standpoint, it is therefore important to recognize that, if two sexually deviate groups are in fact virtually indistinguishable, the decision whether a sexual deviate will serve a determinate term as a convicted criminal or an indeterminate term as a committed patient must in actuality rest on prosecutive, psychiatric, or judicial whim. The constitutional and policy objection can be reduced, of course, if the emerging requirement of a ceiling on convicted sex offender confinement is carried over to the category of committed sexual psychopaths.

The argument for constitutional limits on the length of confinement in therapeutic and noncriminal contexts (where retribution is inappropriate and where it is inappropriate to use retributive lids as guides to setting ceilings on confinement) received a boost by important and much quoted language in the U.S. Supreme Court case of *Jackson* v. *Indiana*. In the course of setting a constitutional clamp on the period that a defendant may be committed as incompetent to stand trial, the *Jackson* Court stated broadly that, "at the least, due process requires that the nature *and duration* of commitment bear some reasonable relation to the purpose for which the individual is committed."

The *Jackson* due process durational limit language, coupled with similar language in the later *O'Connor* v. *Donaldson* civil commitment case, indicates that, even for civil commitments, a substantive due process rationale will require that incarcerative ceilings be set.

Thus, durational limits on confinements seem to be emerging with respect to all types of institutionalized persons: criminally convicted, criminally committed, and civilly committed. The setting of such limits provides further evidence of the social and legal trend away from the notion of institutionalization.

DAVID B. WEXLER

[*Directly related are the entries* HOSPITALS; MENTAL HEALTH SERVICES, *article on* SOCIAL INSTITUTIONS OF MENTAL HEALTH; PATERNALISM; PATIENTS' RIGHTS MOVEMENT; PRISONERS; RIGHT TO REFUSE MEDICAL CARE; *and* RIGHTS. *Other relevant material may be found under* HEALTH CARE, *articles on* HUMANIZATION AND DEHUMANIZATION OF HEALTH CARE, RIGHT TO HEALTH-CARE SERVICES, *and* THEORIES OF JUSTICE AND HEALTH CARE; *and* MEDICAL ETHICS, HISTORY OF, *section on* EUROPE AND THE AMERICAS, *article on* EASTERN EUROPE IN THE TWENTIETH CENTURY. *For discussion of related ideas, see the entries:* BEHAVIOR CONTROL; INFORMED CONSENT IN HUMAN RESEARCH; INFORMED CONSENT IN MENTAL HEALTH; INFORMED CONSENT IN THE THERAPEUTIC RELATIONSHIP; MENTALLY HANDICAPPED; MENTAL ILLNESS; *and* PSYCHOSURGERY.]

BIBLIOGRAPHY

BROOKS, ALEXANDER D. *Law, Psychiatry, and the Mental Health System.* Boston: Little, Brown & Co., 1974.
ENNIS, BRUCE J., and SIEGEL, LOREN. *The Rights of Mental Patients: The Basic ACLU Guide to a Mental Patient's Rights.* An American Civil Liberties Union Handbook. New York: Avon Books, 1973.
FRIEDMAN, PAUL R. *The Rights of Mentally Retarded Persons: The Basic ACLU Guide for the Mentally Retarded Persons' Rights.* An American Civil Liberties Union Handbook. New York: Avon Books, 1976.
In re Lynch. 105 Cal. Rptr. 217. 503 P.2d 921 (1972).
Jackson v. Indiana. 406 U.S. 715. 32 L. Ed.2d 435. 92 S. Ct. 1845 (1972).
Kaimowitz v. Department of Mental Health for the State of Michigan. 1 Mental Disability Law Reporter 147 (1976) (Mich. Cir. Ct., Wayne Co. 1973).
Knecht v. Gillman. 488 F.2d 1136 (8th Cir. 1973).
LIVERMORE, JOSEPH M.; MALMQUIST, CARL P.; and MEEHL, PAUL E. "On the Justifications for Civil Commitment." *University of Pennsylvania Law Review* 117 (1968): 75–96.
Mackey v. Procunier. 477 F.2d 877 (9th Cir. 1973).
MILLER, FRANK W.; DAWSON, R.; DIX, G.; and PARNAS, R. *The Mental Health Process.* 2d ed. University Casebook Series. Mineola, N.Y.: Foundation Press, 1976.
MORRIS, NORVAL. "The Future of Imprisonment: Toward a Punitive Philosophy." *Michigan Law Review* 72 (1974): 1161–1180.
MURPHY, JEFFRIE G. "Total Institutions and the Possibility of Consent to Organic Therapies." *Human Rights* 5 (1975): 25–45.
O'Connor v. Donaldson. 422 U.S. 563. 45 L. Ed. 2d 396. 95 S. Ct. 2486 (1975).
SHAPIRO, MICHAEL H. "Legislating the Control of Behavior Control: Autonomy and the Coercive Use of Organic Therapies." *Southern California Law Review* 47 (1974): 237–356.
SINGER, RICHARD G., and STATSKY, WILLIAM P. *Rights of the Imprisoned: Cases, Materials, and Directions.* Contemporary Legal Education series. Indianapolis: Bobbs-Merrill, 1974.
SPECE, ROY G. "Conditioning and Other Technologies Used to 'Treat?' 'Rehabilitate?' 'Demolish?' Prisoners and Mental Patients." *Southern California Law Review* 45 (1972): 616–681.
Weems v. United States. 217 U.S. 349. 30 S. Ct. 544 (1910).
WEXLER, DAVID B. "Behavior Modification and Other Behavior Change Procedures: The Emerging Law and the Proposed Florida Guidelines." *Criminal Law Bulletin* 11 (1975): 600–616.
———. *Criminal Commitments and Dangerous Mental Patients: Legal Issues of Confinement, Treatment, and Release.* DHEW Publication no. (ADM) 76–331. Rockville, Md.: Department of Health, Education, and Welfare; Public Health Service; Alcohol, Drug Abuse, and Mental Health Administration; National Institute of Mental Health, 1976.
———. "Foreword: Mental Health Law and the Movement toward Voluntary Treatment." *California Law Review* 62 (1974): 671–692.
———. "Token and Taboo: Behavior Modification, Token Economies, and the Law." *California Law Review* 61 (1973): 81–109.
Wyatt v. Stickney. 344 F. Supp. 373 (M.D. Ala. 1972).

INSURANCE, HEALTH
See HEALTH INSURANCE.

INTENSIVE CARE
See LIFE-SUPPORT SYSTEMS.

INTERNATIONAL CODE OF MEDICAL ETHICS
See APPENDIX, SECTION I, INTERNATIONAL CODE OF MEDICAL ETHICS.

INTERNATIONAL HEALTH
See HEALTH, INTERNATIONAL.

INTUITIONISM
See ETHICS, *articles on* DEONTOLOGICAL THEORIES *and* NON-DESCRIPTIVISM.

INVOLUNTARY COMMITMENT
See INSTITUTIONALIZATION; MENTAL HEALTH SERVICES, *article on* SOCIAL INSTITUTIONS OF MENTAL HEALTH; INFORMED CONSENT IN MENTAL HEALTH.

IRAN
See MEDICAL ETHICS, HISTORY OF, *section on* NEAR AND MIDDLE EAST AND AFRICA, *articles on* ANCIENT NEAR EAST *and* PERSIA.

ISLAM

The word "Islam" comprises two principal meanings: (1) the religion founded by the Prophet Muhammad and based upon the revelations in the Koran and the traditions of his sayings and doings gathered in the so-called *hadith*, and (2) the culture and the civilization of the countries conquered by Muslim armies or islamicized peacefully. All of these countries had, of course, their pre-Islamic traditions, and some of them possessed high cultures, as the Persians and the Greeks. These cultural traditions were partially eliminated and partially islamicized, though sometimes by a rather superficial varnish, as, e.g., the ceremonies of the pilgrimage to Mecca. In the following article issues of non-Islamic origin will be discussed along with authentically Islamic factors. The article concentrates on the past, when Islamic civilization was still completely self-contained and consistent with its principles. A short remark will be made at the end about the very complex present situation, which is marked by a vehement conflict between the influences of Western technology, culture, and thought, on the one hand, and the Islamic tradition, on the other.

Main sources of Islamic bioethics

Islam as a religiocultural system has been molded by a variety of sources and influences, among which the Greek legacy has played a very important part. Thus, Greek medicine has been taken over in toto by translating the classical texts of Hippocrates, Galen, Rufus, Dioscorides, and others into Arabic. Philosophy has also been widely shaped after the Greek model: practically the whole of Aristotle's writings were handed down to the Arabs.

Ethical ideas were found in philosophical and medical books alike, e.g., Plato's *Phaedo*, Aristotle's famous *Nicomachean Ethics*, Galen's "On that the best physician must also be a philosopher," and so on. As for practical ethics, the Persian element must also be mentioned. It manifested itself mainly in the so-called *Fürstenspiegel* (ethical code books for princes) as well as in the didactic poems written in Persian. A famous early specimen of the first category is the *Qābūsnāme* (tenth century), where physical training, such as riding, swimming, and so on, plays a considerable role. But as this Persian tradition mainly refers to courtly manners, the art of government, politics, conducting wars, and practical questions such as buying land, a slave, or a horse, it seems neither necessary nor possible to discuss it here in any more detail.

The concept of symmetry. The magic word of the Galenic system was the *eukrasia* or, more comprehensively, the symmetry (Arabic: *i'tidāl*). It was by conserving symmetry in the different spheres of one's life that an individual protected his health, and it was by teaching his patients how to conserve or restore it that a physician made himself indispensable to them.

In Greek medicine, symmetry meant an equilibrium among the four humors—blood, phlegm, yellow bile, and black bile—and the four qualities—hot, cold, moist, and dry. According to the Galenic system the blood was damp and hot, the mucous damp and cold, the yellow bile dry and hot, and the black bile damp and cold. If the four humors and the qualities combined with them were in a state of mutual equilibrium (Greek: *eukrasia*, literally "the state of being well mixed" or "well-tempered"), one was healthy. If one of the humors, and therewith one of the qualities, became so dominant that the balance was considerably disturbed, the *eukrasia* was displaced by a *dyskrasia* ("state of being ill mixed" or "ill-tempered"), which meant that one was sick. Whereas the Hippocratic school had believed in the existence of an ideal equilibrium of the four humors, Galen modified the theory by teaching that, in fact, no such ideal state existed. The influence of such exterior factors as climate, age, profession, customs, and so on caused a dominance of one of the four humors to be observed in every human body. This gave an individual his own special habit and complexion, his "temperament," which according to the four humors may be sanguine, phlegmatic, choleric, or melancholic.

The Galenic system of therapy rested on the principle of *contraria contrariis*. In other words, the Galenic system was an allopathy par excellence. "Hot" diseases were cured by "cold" remedies, "moist" by "dry," and vice versa. Hearing these terms, a modern listener not familiar with Galenic medicine might think of cold and hot drinks, of baths and other hydropathic treatments, but these were only a small part of what Galen and his disciples meant when they talked of hot, moist, etc. These qualities were ascribed to every part of nature: Musk was hot and dry, cucumber was cold and damp in the second degree, costus was hot and dry in the second degree, and so on. No drug taken from one of the three realms of nature could fail to be categorized in this scheme.

Symmetry, habits, and environment. The idea of symmetry was extended with respect to the so-called six things not regulated by nature itself (in Latin: *sex res non naturales*), namely, "the air," eating and drinking, sleeping and being awake, rest and motion, "evacuation and retention" (referring to bodily secretions liable to man's will, mainly sexual intercourse), and the emotions. In connection with the atmosphere, the question of climate and of the dwelling place was usually discussed. Sometimes other issues such as habits, professions, garments, and even plants, pictures, and wallpapers were taken into consideration. Thus, the Galenic system not only implied a strong ethical element, it taught people to care for a healthy environment or "ecological equilibrium," to put it in a modern expression. We must be aware, though, that the realization of many of these wise exhortations was possible only for people with the necessary financial resources and a certain personal freedom and mobility.

More or less the same rules are to be found in the great medical handbooks of medieval Islam, as the famous "Royal Book" (*Liber Regius*) by ᶜAlī ibn ᶜAbbās al-Madjūsī (tenth century) or the even more famous *Canon* of Ibn Sīnā (Avicenna, died 1037), and more particularly in books concerned exclusively with "the conservation of health" (Arabic: *ḥifẓ aṣ-ṣiḥḥa*), which formed a special and important division of medical care in the Greek and Roman as well as Islamic culture.

Regarding cultural influences of this kind of hygienic or prophylactic methods, the most important issue seems to be balneology or the habits of bathing. Public bathhouses (*ḥammām*) formed an integral part of every Islamic town, and their number in large towns such as Baghdad and Damascus, according to reliable sources, often surpassed a hundred during the golden age of Islamic culture.

Islamic background and influence. The juncture of Greek, Persian, and Islamic bioethics resides (with modifications to be treated later on) in the concept of moderation, symmetry, or harmony. Harmony is, in fact, one of the leading concepts of Islamic anthropology. Islam molded a type of person whose main characteristics are harmony and contentment—a type opposite to the dynamic, ever studious, never reposing man so typical of the inhabitants of the Western world since the Renaissance. In molding this type of man, Islam was helped a good deal by Sufism or Muslim mysticism, which, at the latest, from the thirteenth century onward, took a leading part in the shaping of the Muslim way of life.

However, the daily life of the average Muslim was, and still is, molded mainly by the Koran and the example of the Prophet's life and sayings, handed down in the so-called *hadith* or records. On the whole, the outlook of these sacred scriptures on life is positive. The Koran is full of praise of the creator, whose wonderful power manifests itself in creation. Man is exhorted to "eat and drink but without being immoderate" (7, 29) and to marry a wife or, under the condition of treating them equally (with equal justice), two, three, or four (4, 3). He is, in addition to this, allowed to have sexual intercourse with "whom his right hand possesses" (meaning his slave girls).

It must, however, be said that both Sufism and its counterparts, orthodox Sunnism and Shiism, gradually moved away from the dynamic activism of the first centuries to either a contemplative quietism, the ideal of self-annihilation, or narrow-minded orthodox legalism. The excessive "trust in God" or belief in predestination led to a strong indifference toward the vicissitudes of man's life. Superstitious beliefs in demons, djinns, the Evil Eye, and so on gradually superseded rational thinking. The development of the sciences was more and more paralyzed by a dogmatic belief in the authorities.

Those medical writers of medieval Islam who fought decay in medicine usually said that there was only one remedy against it: returning and closely clinging to the foundations laid by "the Ancients," meaning Hippocrates, Galen, etc. Despite a very small number of minor corrections, the Galenic system as a whole was never questioned by any of the great physicians of medieval Islam until new medical ideas came from Europe. On the other hand, a so-called prophetic medicine was developed by a number of religious scholars. On the basis of a few Koranic verses (sometimes interpreted most sophistically, since the Holy Book is almost totally devoid of references to medical things) and some medical advice (not surpassing bedouin medical lore) that the Prophet was reported to have given now and then, they developed a framework that was gradually filled up with Galenic material. By this process the authority of Galen was replaced by that of the Prophet. What that meant—or, at least, could mean, is shown by the following story: A doctor who opposed the reported prophetic prohibition of bleeding on Saturdays fell seriously ill and did not recover until he repented. It was the "prophetic medicine" that sanctified the use of amulets and talismans, potions from the rinsed ink of Koranic verses, and other magical lore in a religious guise. With all the rich heritage taken over from the Ancients, it was the Prophet who won the victory over them after a struggle of about half a millennium. This is also true for the principle of symmetry: It was followed only where it did not come into conflict with the Islamic Law, its rituals, its manners and customs, its prohibitions, and its punishments.

Particular issues in bioethics

Fasting, prayer, and pilgrimage. Two authentic contributions of Islam to bioethics are ritual prayer and fasting. Ritual prayer, which must be performed five times a day, consists of two main parts: ritual ablution and a number of movements such as kneeling, bowing down, and getting up, accompanied by the recital of prayer verses, praise of God, and the confession of faith.

Fasting during the month of Ramadan (ninth month of the Islamic calendar) forbids eating, drinking, smoking, and sexual intercourse from dawn until dusk. Apart from the pious meanings of fasting and prayer, the Muslims regard the performance of both as being very favorable to the health of the believers. As for fasting, its alleged effect may, however, be questioned—at least under the conditions of modern life. On the one hand, many of those who fast overstrain themselves by working during the day without eating and by spending a large part of the night eating and chatting in the company of family, friends, and neighbors. On the other hand, it is statistically proven, according to some critics, that overall consumption increases during the fasting month.

The rules of *iḥrām* or state of consecration during the ḥajj (pilgrimage to Mecca) include the interdiction of sexual intercourse as well as hunting, cutting the hair, and trimming the nails. Whether or not these requirements, which do have a bearing on personal health, are also relevant to the values associated with the ethics of health care need not be argued here. Suffice it to say that because of modern conveniences the state of *iḥrām* no longer lasts as long as it used to, but may in fact be shortened to a few hours if the pilgrim travels by plane.

Sexuality. The Islamic attitude toward sex is opposite to that of the traditional Christian church. One of the patriarchs of Islam, Ḥasan ibn ʿAlī, grandson of Muhammad, is reported to have married successively (not more than four at a time) two hundred women, without damage to his status as a very pious man. The sexual act is regarded as meritorious (on the part of the male)—equivalent to giving alms if exercised within the wide limits set by the Koranic revelation, as was mentioned above. No wonder, therefore, that the pharmaceutical literature of medieval Islam was extremely rich in aphrodisiacs. Circumcision is prescribed not in the Koran but only in the *hadith*. Nevertheless, it has to be undergone by each Muslim boy, usually between the ages of five and ten, and is regarded as being not only of hygienic importance but also highly favorable to man's sexual faculty and one of the indispensable marks of a Muslim man.

In marriage, the traditional role of the wife was far inferior to that of the husband—she being obliged to absolute obedience to his orders, he having the right to divorce her for the most arbitrary reasons. These conditions have been changed, at least legally, in the jurisdiction of most Islamic countries today.

Abortion, though strictly interdicted by the Hippocratic Oath, was not altogether forbidden by Islamic law. Abortifacients were therefore always to be found in medieval Arabic pharmacopoeias. Furthermore, certain methods of contraception are discussed in the legal books, particularly coitus interruptus.

Wine and music. Contrary to the liberal (for men only) regulation of sex life, Islam has made of wine drinking one of the great sins. In fact, the intention of the Koran, where the condemnation occurs in various degrees, seems to have been to banish drunkenness but not the pleasure of wine drinking in general. The great poets of Islam have not ceased praising wine again and again—sometimes with mystical connotations, but more often with real wine in mind. Physicians would recommend different kinds of wine in many cases, though this was sometimes subjected to orthodox criticism.

Approximately the same is true of music, which was used from time immemorial as a psychotherapeutic treatment for melancholics but fell under the verdicts of orthodox jurists. Thus, the famous Jewish physician, theologian, and philosopher Maimonides makes an official excuse in a treatise where he recommends wine and music to one of his patients, a Muslim prince of the Ayyubid dynasty in Egypt, contending that the physician is responsible only for the patient's bodily welfare.

Mutilation. Two issues merit a brief remark, though one of them is now completely obsolete and the other has disappeared in most Islamic countries. One is the sterilization of men. It was widely practiced—not for the good of anything like modern family planning, as is sometimes alleged by modern Muslims, but for supplying the vast demand for eunuchs as guardians of the harems. The other is the cutting off of the right hand—and in case of recidivism the left, and then the feet—as a Koranic penalty for thieves. Under modern medical supervision this punishment is still imposed in Saudi Arabia.

Medical care systems and methods. Medical care was on a high level during the halcyon era of Islamic civilization (ca. A.D. 750–1250). Medical institutions centered upon the courts. Most of the great physicians were in the service of a court, but many of them would keep a private "shop" (*ḥānūt*) for patients from the town. In the classical period, even smaller towns used to have their physician or practitioner, as Goitein has been able to prove for Egypt from the Geniza sources. In the towns the medical men would generally choose a certain specialization. The specialty medical practices mentioned in the *Adab aṭ-ṭabīb* (see below) and in other Arabic sources were: physiologist (*ṭabā'i'ī*, literally specialist of the "natures," meaning the humors), oculist, orthopedist (*mujabbir*, literally bone setter), surgeon, phlebotomist, and cupper. Many of these physicians were Christians or Jews.

It was, however, the hospitals that played the most important role in the system of medical care. The three most famous hospitals of the medieval Islamic world were that of the Buwayhid prince ʿAḍud ad-Daula in Baghdad, founded in 981, that of Nūr ad-Dīn in Damascus, founded in 1154, and that of the Mamluk al-Manṣūr Kala'ūn in Cairo, founded in 1284. These hospitals were equipped with dispensaries, teaching rooms, and a number of departments or special halls that corresponded more or less with the aforesaid specializations. The only noteworthy additions were rooms for cholerics or melancholics (*mamrūr*, literally sufferers from a disease in the *mirra*, the bile). We are also told that maniacs were treated in these rooms, but that should not be confounded with a modern, standard treatment of mental disease as is sometimes done. Recourse to chains and whipping in the treatment of the insane was a normal procedure of the time. The list of specialists engaged by ʿAḍud ad-Daula (tenth century) for his hospital in Baghdad included physiologists, oculists, surgeons, and orthopedists. One mention is made in the sources of a dispensary sent into the countryside. Military expeditions were of course usually accompanied by at least one physician.

In the application of medicines and therapies, a hierarchy of three degrees had been instituted previously by the Greeks, and the doctor was bound to abide by it. This involved proceeding from the slightest to the most serious intervention—using at first only simple drugs, then compound drugs, and only as an *ultima ratio* having recourse to "cutting and burning." Bleeding and cupping were, however, very normal treatments, but the bleeder and cupper were on a lower social level than the physician, often playing the role of mere underlings to the latter. Cauterization, though stigmatized by a *hadith* (saying) of the Prophet, was applied in many cases from cancer to hydropsy, and sometimes certain surgical operations were performed, such as amputation and lithotomy. In a semimythical way, a Caesarean section is described in the famous *Shahnameh* of Firdausi; but no case of this operation seems to be known from Islamic times prior to modern Western influence.

The code of practical ethics would allow the physician to see the woman he treated. He would consult her, feel her pulse, etc., through a curtain. Operations on her genital organs would generally be left to a midwife or a "sage-woman." Even lithotomy on a female patient is reported to have been performed at times by a midwife under the direction of a surgeon.

Surgery. Surgery in general did not rate very highly, and only a small number of Arabic physicians wrote books on it. That the physicians of the Islamic Middle Ages preferred to do without surgery is easily understandable if one considers the lack of aseptic methods, the knowledge of which was beyond thought before the emergence of microbiology. Under such conditions surgical operations must have led to wound fever and death in many cases. Additional insurmountable barriers to progress in the medical art were a dogmatic clinging to the Galenic system and the fact that anatomical dissection was strictly forbidden by religious law. No indications exist that any doctor of the medieval Islamic world dared to disobey this prohibition.

Physician ethics. The basic ethical text for the physician of medieval Islam, regardless of his religious belief, was the Hippocratic Oath, forbidding the doctor to kill, to help in killing, to abort, or to divulge secrets confided in him by his patients, and obliging him to do everything in the best interests of the sick person.

Other important Greek sources, the influence of which can be traced in the pertinent Arabic literature through the centuries, were Galen's probably spurious commentary on the Oath, Galen's "On that the best physician must also be a philosopher," where the physician is warned against being inveigled by worldly goods and pleasures, and the same author's "On examining the physician," which is in fact a guidebook for laymen instructing them how to find a good doctor, laying the main stress again on the behavioral side. A good physician should not be a courtier or a lackey to rich people. He should not indulge the cravings of his patients as long as any possible damage to their health is involved. And he should watch over the preparation of the drugs he prescribes, letting no one interfere with his prescriptions and treatments except responsible and reliable persons. His main concern should not be how to become rich and famous, but how to cure men's illnesses, and he should be ready to treat the poor without fees.

These ideals were handed on by Arabic physicians. The number of pertinent sources is very limited, one of the most important being the book *Adab aṭ-ṭabīb* (practical ethics of the physician) by a Jewish author of the tenth century A.D. Here, as always where the history of social ideas is concerned, it is difficult to differentiate between ideal and reality. We may, however, be sure that if not the majority at least some adhered to these ideals, as for example the great philosopher-physician Rāzī (died 925), who is expressly reported to have cured poor people gratis.

Already in ancient Greece it had been disputed among physicians and philosophers whether the doctor should also be a spiritual guide. There is, in fact, no doubt that the office of court physician, where the doctor was expected to say yes or no to the manifold intentions of his protégé, had ethical implications. Furthermore, the sources indicate that many of the great physicians were clearly aware of the interdependence of physical and spiritual activities. Ibn Sīnā (Avicenna) highlights it in his famous didactic poem on medicine in the following lines:

Plenty of joy increases the body's prosperity;
In certain cases, however, obesity thus produced may become harmful.
Grief, on the other hand, may be pernicious for weak ones,
Useful, however, for those who want to lose weight.

Here again, moderation or harmony is at stake. Typically enough, a famous Muslim physician of the thirteenth century, who served several princes of the Aiyubid dynasty in Syria, wrote a guidebook of psychohygiene, which is based on the belief in this interdependence and in which the watchword is again the Aristotelian medium between extremes, otherwise labeled symmetry.

The waning and resurgence of traditional values in medicine. By the introduction of magic under religious disguise, the rational Galenic system was undermined and finally destroyed. It may be that this development was at least partly due to the constant shortage of physicians after the golden age of the first centuries. A Muslim author of the thirteenth century complains that the practice of medicine, though among the so-called collective duties (as was also the Holy War), was almost completely neglected by Muslims, who would prefer to become judges or *faqihs* (doctors of religous law), because in these professions fame and wealth were much easier to attain than in a medical practice.

Talismans and ink-therapy (or letter-sorcery) were available for everyone. They only needed to be religiously sanctified to become widespread in medical use, as in fact they did. It needs hardly be said that, with this downfall of Greek medical science, its ethical standards also ceased to exist.

It is, therefore, a remarkable and highly felicitous fact that Greek-Arabic medicine, the so-called *Unani* medicine, has been and is still being revived in Pakistan and Muslim India, mainly under the auspices of the National Hamdard Foundation. In the colleges run by the Foundation *ḥakīms* (physicians) are trained and educated in the spirit of that historic medicine which saw man as a union of body and spirit and taught him that moderation, harmony, and equilibrium were a peerless panacea for fragile human beings.

J. C. BÜRGEL

[*Directly related are the entries* MEDICAL ETHICS, HISTORY OF, *section on* NEAR AND MIDDLE EAST AND AFRICA, *articles on* PERSIA, CONTEMPORARY ARAB WORLD, *and* CONTEMPORARY MUSLIM PERSPECTIVE; *and* POPULATION ETHICS: RELIGIOUS TRADITIONS, *article on* ISLAMIC PERSPECTIVES.]

BIBLIOGRAPHY

BÜRGEL, J. CHRISTOPH. "Psychosomatic Methods of Cures in the Islamic Middle Ages." *Humaniora Islamica* 1 (1973): 157–172.

———. "Secular and Religious Features of Medieval Arabic Medicine." *Asian Medical Systems: A Comparative Study.* Edited by Charles Leslie. Berkeley: University of California Press, 1976, pp. 44–62.

CAMPBELL, DONALD EDWARD H. *Arabian Medicine and Its Influence on the Middle Ages.* 2 vols. Trubner's Oriental Series. London: K. Paul, Trench, Trubner & Co., 1926.

ELGOOD, CYRIL. *A Medical History of Persia and the Eastern Caliphate, from the Earliest Times until the Year A.D. 1932.* Cambridge: University Press, 1951.

———. "Tibb-ul-Nabbi or Medicine of the Prophet. Being a Translation of Two Works of the Same Name. I. The Tibb-ul-Nabbi of Al-Suyútí. II. The Tibb-ul-Nabbi of Mahmud bin Mohamed al-Chaghayni: Together with Introduction, Notes and a Glossary." *Osiris* 14 (1962): 33–192.

HAMARNEH, SAMI. "Arabic Historiography as Related to the Health Professions in Medieval Islam." *Sudhoffs Archiv: Vierteljahrsschrift für Geschichte der Medizin und der Naturwissenschaften* 50 (1966): 2–24.

———. "Development of Hospitals in Islam." *Journal of the History of Medicine* 17 (1962): 366–387.

MEYERHOF, MAX. "Science and Medicine." *The Legacy of Islam.* Edited by Thomas W. Arnold and Alfred Guillaume. Oxford: Clarendon Press, 1931, pp. 311–355.

ROSENTHAL, FRANZ. "The Defense of Medicine in the Medieval Muslim World." *Bulletin of the History of Medicine* 43 (1969): 519–532.

SCHACHT, JOSEF, and BOSWORTH, C. E., eds. *The Legacy of Islam.* 2d ed. Oxford: Clarendon Press, 1974. Extensively revised version of 1931 ed.

———, and MEYERHOF, MAX. *The Medico-Philosophical Controversy between Ibn Buṭlān of Baghdad and Ibn Riḍwān of Cairo: A Contribution to the History of Greek Learning among the Arabs.* Egyptian University, Faculty of Arts, Publication no. 13. Cairo: 1937.

ISRAEL

See MEDICAL ETHICS, HISTORY OF, *section on* NEAR AND MIDDLE EAST AND AFRICA, *article on* CONTEMPORARY ISRAEL. See also JUDAISM.

ITALY

See MEDICAL ETHICS, HISTORY OF, *section on* EUROPE AND THE AMERICAS, *articles on* MEDIEVAL EUROPE: FOURTH TO SIXTEENTH CENTURY, WESTERN EUROPE IN THE SEVENTEENTH CENTURY, *and* WESTERN EUROPE IN THE TWENTIETH CENTURY.

J K

JAPAN

See MEDICAL ETHICS, HISTORY OF, *section on* SOUTH AND EAST ASIA, *articles on* GENERAL HISTORICAL SURVEY; JAPAN THROUGH THE NINETEENTH CENTURY; TRADITIONAL PROFESSIONAL ETHICS IN JAPANESE MEDICINE; *and* CONTEMPORARY JAPAN: MEDICAL ETHICS AND LEGAL MEDICINE. See also BUDDHISM; CONFUCIANISM; TAOISM.

JEHOVAH'S WITNESSES

See PROTESTANTISM, *article on* DOMINANT HEALTH CONCERNS IN PROTESTANTISM; RIGHT TO REFUSE MEDICAL CARE.

JEWISH ETHICS

See JUDAISM.

JEWISH HOSPITALS

See RELIGIOUS DIRECTIVES IN MEDICAL ETHICS, *article on* JEWISH CODES AND GUIDELINES; JUDAISM; HOSPITALS.

JOURNALISM AND BIOMEDICINE

See COMMUNICATION, BIOMEDICAL, *article on* MEDIA AND MEDICINE.

JUDAISM

Introduction

Historical antecedents. In medical ethics two of the Jewish people's most notable contributions to human civilization converge: medicine and ethics. From the beginning Jews have shown a special concern with the healing art. Already the Hebrew Bible includes in its religious legislation some revolutionary concepts of preventive medicine and public health. In the Talmud (the authoritative body of Jewish law and lore composed between ca. 300 B.C.E. and 500 C.E.), among numerous medical references, are found the earliest mention of such innovations as artificial limbs, some of artificial insemination, oral contraceptives, and Caesarean operations on living mothers. Several of the Talmud's authors practiced medicine. They were succeeded by what became in the Middle Ages the common phenomenon of the rabbi–physician. Indeed, it has been estimated that over one-half of the best-known rabbinical scholars and authors in medieval times were physicians by occupation, including such notables as Maimonides (1135–1204), Nachmanides (1195–1270), the Ibn Ezras (1092–1167), the Ibn Tibbons (twelfth and thirteenth centuries), and Immanuel ben Solomon of Rome (ca. 1265–1330). The disproportionate predominance of Jews in medieval medicine was no doubt due in part to their exclusion from other professions, especially in Muslim countries.

These historical antecedents no doubt contributed significantly to the extraordinary predilection among Jews for a medical career in modern times. Only aptitudes conditioned by centuries of nurture could have produced such a disproportionate preoccupation with medicine as enabled Jews to receive some twenty percent of all Nobel prizes for medicine—a proportion more than forty times the ratio of Jews in the world. Likewise many leading medical historians were Jews, ranking authorities like Max Neuburger, Arturo Castiglioni, and Charles Singer.

Paramount throughout this long association between Judaism and medicine has been the emphasis on ethics as their common denominator. Building materials for the ever rising edifice of Jewish medical ethics came from all strata of Jewish religious literature, while the architects were legal experts who, as already noted, themselves often combined rabbinical and medical experience. The Bible provided the foundations: the sanctity and dignity of human life, the duty to preserve health, an uncompromising opposition to superstition and irrational cures—including faith healing—a rigid code of dietary restraints and sexual morality, and many basic definitions of moral imperatives in medical practice, including the rights of the dead. The significance of the biblical contribution to the development of medical ethics in Western society may be appreciated when it is contrasted with the Code of Hammurabi and other ancient legislations, which provided, for instance, for the amputation of a doctor's arm if he proved unsuccessful in an operation on his patient.

On these biblical foundations, the Talmud established the legal framework in virtually all fields of medical ethics, setting forth the main principles applicable to a wide variety of medicomoral issues, such as abortion, sterilization, contraception, euthanasia, malpractice claims, etc. The Talmud, followed by the great medieval codes of Jewish law, even enacted certain eugenic recommendations against marriages suspected to result in physically or morally defective children. Anyone who has read the famous Oath of Asaph Harofé (sixth century) or the medical writings of other early Jewish contributors to the history of medicine will recognize at once how profoundly the spirit of the Hebrew Bible and the Talmud suffused their ethical outlook. Little wonder that there was never a Jewish form of the Hippocratic Oath, though there were Christian and Muslim as well as pagan versions of it widely in use for centuries. Jews simply fell back on their own ethical heritage enshrined in their religious law.

It was left to the voluminous rabbinical "responsa," accumulated over the past thousand years, to interpret these principles in the light of contemporary conditions and in response to new problems created by the advance of medical knowledge and techniques. The Responsa serve as case-law in the evolution of Jewish law. Now being published annually, notably in Israel and America, are hundreds of such rabbinical Responsa or decisions, usually given in reply to personal inquiries by individual questioners. The latest of these often massive works ranges over the entire gamut of ethical problems in medicine, from transplants to artificial insemination, and from experimentation on humans or animals to autopsies.

Recent sources and developments. Despite the rich profusion of literary material scattered over a period in excess of three thousand years of continuous productivity, Jewish medical ethics featured neither in the vocabulary of Jewish literature nor as a distinct rubric within it until quite recently. (The very term "Jewish medical ethics" was first used as a title of a doctoral thesis submitted in 1955 and subsequently of a book first published in 1959 by the present writer.) Previously the large number of religious rulings on medical practice found in the sources of Jewish law were accessible only to Hebrew scholars, and only a few monographs on specialized subjects existed in various periodicals or occasionally as separate publications, written in the vernacular.

In the past two decades this field, seeded by these diverse grains and roots and fertilized by the constantly increasing output of new Responsa, has begun to yield an ever growing harvest. Stimulated partly by the rising worldwide concern with medicomoral problems and partly by the contemporary search for relevance in the presentation of Jewish values, Jewish medical ethics is now an established and popular discipline of Judaism that begins to command its own experts and literature. The subject is now featured in several books as well as numerous articles dispersed in a wide range of rabbinical, scientific, professional, and popular journals. Striking testimony to the belated literary "recognition" of Jewish medical ethics is furnished by a comparison between the old *Jewish Encyclopedia*, published in 1901, which contained no entries on the subject at all, and the new *Encyclopaedia Judaica* of 1972, which has fairly extensive entries under abortion, artificial insemination, autopsies, birth control, euthanasia, homosexuality, and transplants. In Jerusalem two institutes are exclusively devoted to various projects of research, publications, and lectures on medicine in Jewish law.

General principles

Jewish law operates and develops on the basis of ancient authority and precedent, founded primarily on the Bible and the Talmud, and ultimately codified as authentic law in such legal codes as Maimonides's twelfth-century *Yad Ha-Hazakah* (The Strong Hand) and Joseph Karo's sixteenth-century *Shulhan Arukh* (The Prepared Table). Its subsequent evolution is entirely determined by the opinions expressed in the commentaries on these works and especially in the Responsa issued to the present day by the leading masters of Jewish law, usually communal rabbis or deans of rabbinical academies of outstanding scholarship. In dealing with new situations or problems, they must invariably relate their decisions to the rulings of earlier authorities, or arrive at them by a process of deduction and logical argumentation from broad principles previously established.

This process is dynamic enough to allow for a certain diversity of interpretation, particularly since the original sources themselves, notably the Talmud, embrace many divergent views that may, within some well defined limits, all be regarded as authentic. Consequently, on many modern medicomoral issues some more stringent and some more permissive opinions continue to divide leading rabbis, though the basic attitudes are fairly uniform. To give but two examples in anticipation of the details to be presented later: on euthanasia all rulings are agreed in condemning any active hastening of death as homicide; but opinions differ on whether, under certain circumstances, medical treatment may be withdrawn and resuscitation efforts be suspended. Similarly, on abortion, while there is general agreement on requiring the operation to save the mother's life and on forbidding it for purely social or economic reasons, its legality to prevent the birth of defective children is still subject to conflicting verdicts.

Beyond this limited diversity of interpretation, there are differences between Traditional and Progressive Judaism. These differences concern not so much the definition of Jewish law as the degree to which one is to regard its provisions as binding. (This article, while written from an Orthodox point of view, broadly reflects the diverse views found in the traditional sources of Jewish law and ethics. On some specific issues Reform and Conservative Jewish opinions may be less definitive and/or more permissive.)

Virtually the entire corpus of Jewish medical ethics traces its countless rulings to certain fundamental principles. The most important among these may be listed as the religious obligation to protect life and health, the absolute sanctity of human life from birth to the definitive establishment of death, the reliance on competent medical opinion and skills, and the dignity of man in death as in life.

The duty to preserve life and health. Judaism, like other religious faiths, recognizes the inner conflict between the essentially divine (and therefore providential) character of disease on the one hand, and on the other hand human efforts to control disease. The Talmud (*Berakhot* 60a) resolved this conflict by resorting to the "Divine sanction": God expressly permitted recourse to doctors in the biblical provision, among the liabilities for the infliction of injuries, that "he shall surely cause him to be healed" (Exodus 21:19). The problem of "flying in the face of Providence" therefore never troubled the rabbis as it for long troubled Christian theologians, whether in opposing the flight from plagues in the Middle Ages or in objecting to the use of chloroform to ease the pangs of childbirth in the nineteenth century. Jewish thinkers regarded the conquest of pain and disease as no less mandatory than the application of water to the thirsty throat or of the plow to the uncultivated soil in man's striving for survival and prosperity, to use an analogy first mentioned by Maimonides.

Indeed, in the words of the *Shulhan Arukh*: "The Torah [the law in the first five books of the Bible] gave permission to the physician to heal; moreover, this is a religious precept, and it is included in the category of saving life." Accordingly, a doctor is considered to be fulfilling a religious obligation, and if he withholds his services, "he is deemed as if he were shedding blood," unless a more competent doctor is available. The duty to protect life and health devolves upon others, too. Based on the law "You shall not stand upon your neighbour's blood" (Leviticus 19:16), anyone in a position to rescue another person from any danger to life or limb who refuses to do so is guilty of a serious offense. Even in respect of one's own body, the law rules emphatically: "It is forbidden to rely on miracles or to endanger one's life." Likewise, any mutilation of the body is strictly prohibited, and the codes list several detailed instructions for the protection of one's life and health. In fact, regulations concerning health must be observed more stringently than ritual laws.

From these principles is derived the general rule that all religious laws are automatically suspended in the face of any risk of life, however remote, the only exceptions being the three cardinal transgressions of idolatry, immorality (i.e., incest and adultery), and murder. Thus the circumstances under which the Sabbath and other laws, including circumcision, are completely suspended in cases of the slightest danger to life or modified for reasons of health, are stipulated in great detail in the codes and commentaries. They insist with every emphasis: "It is a religious precept to desecrate the Sabbath for any person afflicted with an illness which may prove dangerous; he who is zealous [in such desecration] is praiseworthy, whilst he who asks [questions] sheds blood" (*Shulhan Arukh*).

The sanctity of human life. The sanctity of life has a very specific connotation in Jewish law. It regards every human life as absolute and infinite in value. By mathematical definition, infinity can no more be increased by multiplication than it can be reduced by division. Hence, a physically or mentally handicapped life, in whatever state of debility, is worth no less than a full and healthy life. And, by the same token, one person has exactly the same value as a million people.

This reasoning is governed above all by moral considerations; for it is the indispensable foundation for the sanctity and equality of all human life and thus for the moral order itself. The argument is as simple as it is compelling. If a person who has only another hour to live would lose his absolute title to life—presumably because it is all but worthless—it would follow that a patient expected to live for two more hours would enjoy twice this infinitesimal value. As the expectancy of life increases to, say, a week, or a year, or five years, so would its worth appreciate correspondingly. Consequently, no two human beings would have the same value. The worth of all would become relative, relative to their expectancy of life, or their state of health, or their usefulness to society, or any other arbitrary criterion.

This would be the thin end of the wedge dividing mankind into people of superior and inferior worth, into those who have a greater and others who have a smaller claim to life. In the Jewish view, there can be no defensible line drawn between the Nazi horror of liquidating so-called inferior members of society by the million and the advocates of euthanasia for individuals who have become "worthless."

Jewish law expresses this cardinal principle in both legal and moral terms. Implicitly following the Talmud (*B. Kamma* 26b), a person who killed a child while falling from a high roof would in principle be guilty of a capital offense like any ordinary murderer, even though he hastened the child's certain death by only a few moments. Or, in the more general formulation of the law by Maimonides: "He who kills, whether [the victim be] a healthy person or a sick person approaching death, or even a patient already in his death-throes, is treated as a capital criminal."

Nowhere is the supreme value of human life more impressively presented than in the stern warning that, according to the Talmud (*Sanhedrin* 4:5), had to be given to witnesses in a capital trial. Admonishing them that they were "answerable for the blood of him [who might be wrongfully condemned if their testimony was unreliable] and the blood of his [potential] offspring to the end of the world," the courts solemnly declared, inter alia: "Therefore was but a single person [originally] created in the world, to teach that if any man has caused a single soul to perish, Scripture imputes it to him as if he caused the whole world to perish; and if any man saves a single life, Scripture imputes it to him as if he saved the whole world."

However much Judaism cares about the mitigation of human pain, often even at the expense of modifying its own most sacred observances, what it cannot do is to purchase relief from suffering at the cost of life itself. Here, as in other cases affecting priceless public stakes such as national defense or freedom, the individual may be required to subordinate his interests to those of society at large. For any sanction of euthanasia could not but cheapen life generally by making its preservation contingent upon considerations of expediency or relative merit.

By the same token of life's infinite value, the deliberate sacrifice of one human being can never be sanctioned for the purpose of saving another, or any number of others. The reasoning given in the Talmud (*Sanhedrin* 74a) is significant. While the exclusion of idolatry and immorality from the usual rule of suspending all laws when they might conflict with life requires scriptural proof, the unconditional prohibition of bloodshed (except in self-defense or war) is based simply on "the logical argument" of "How do you know that your blood is redder than his?" —that is, what makes your life worth preserving more than another's? Later authorities apply this reasoning also to deny the right of voluntary self-sacrifice for the sake of saving other lives, by reversing the argument: "How do you know that his blood is redder than yours?"

On the strength of the analogy of infinity's being unaffected by multiplication, this principle is then extended to the rescue of any number of lives at the deliberate expense of a single life. To use the Talmud's illustration: if a hundred innocent people were taken hostage and their lives would be spared only if they delivered one of them to the oppressor for execution, they must not surrender that one life even if the others would thereby be saved (*Tosefta, Terumot,* 7:23).

These principles clearly affect the rulings on euthanasia, on organ transplants where these might hasten the donor's demise, and on hazardous experiments on humans, whether with or without the subject's consent.

The limits of life. The inviolability of life, founded on the aforementioned considerations, extends from birth to the establishment of death. It is modified in respect to prenatal life.

This exclusion derives from the exegesis of the verse "He that smites *a man,* so that he dies, shall surely be put to death" (Exodus 21:12), which the rabbis construed to mean "a man, but not a fetus." The same conclusion is also indicated in the only (albeit somewhat indirect) reference to abortion in the Bible: "And if men strive together, and hurt a woman with child, so that her fruit depart, and yet no harm follow, he shall be surely fined . . . but if any harm follow, then shall you give life for life" (Exodus 21:22–23). This passage, by a curious twist of literary fortunes, eventually marked the parting of the ways between the Jewish and Christian rulings on abortion. The Christian tradition, following the *Septuagint* mistranslation of the Hebrew for "no harm follow" by the Greek for "[the child be born] imperfectly formed," distinguished between a formed and an unformed fetus, branding the killing of the former as murder ("life for life"). Since 1588 (by a decree of Pope Sixtus V) the distinction was finally eliminated and the killing of any human fruit from the moment of conception was considered murder. According to the Jewish interpretation, however, if "no harm follow" refers to the mother's survival following her miscarriage; there is then no capital guilt, and the attacker is merely liable to pay compensation for the loss of her fruit. "But if any harm follow," i.e., if the woman is fatally injured, the man responsible for her death has to "give life for life," and the capital charge of murder exempts him from any monetary liability.

Accordingly, in Jewish law the infinite value of life sets in only from birth, legally defined as the moment when the head or the greater part of the child's body emerges from the birth canal. Until that moment the child enjoys some very sacred rights that may not be violated except on the most urgent medical grounds. But its title to life is distinctly inferior to that of the mother, so that, in any mortal conflict between them, it is mandatory to destroy the unborn child if there is no other way to save the mother. But it should be added that Judaism recognizes the Catholic position as being compatible with the obligations imposed upon non-Jews by the Noachide code (a Jewish talmudic designation for the seven biblical laws given to Adam and to Noah and binding on all mankind).

The generation of life. The Jewish attitude toward moral problems raised in the generation of life stems above all from the positive precept of procreation. In the absence of any valid medical indications, procedures like abortion, sterilization, and contraception do, to be sure, constitute intrinsically immoral practices. But the objection to them is rooted less in their destruction of potential life per se, or in their defiance of the "natural law" (as held in Christian thought)—which can be quite legitimate, e.g., in acts of healing—than in the deliberate refusal to participate with the Creator in the propagation of the race.

Judaism lists the duty to "be fruitful and multiply" as the first of its 613 biblical commandments (Genesis 1:28). The Jewish code of law, *Shulhan Arukh,* opens the code on marital legislation: "Every man is obliged to marry a woman in order to be fruitful and to multiply, and whoever does not engage in procreation is as if he shed blood, he reduces the Divine image and causes the Presence of God to depart from Israel." Technically, the commandment is fulfilled by having at least a son and a daughter, i.e., by perpetuating life through the reproduction of those responsible for one's own birth. In the Jewish view, therefore, it is an offense (or a transgression of a "positive precept") to abstain from marriage, or from conjugal intercourse within marriage, no less than to prevent or interrupt pregnancies.

This approach is reflected in the peculiar slants and omissions of the relevant biblical legislation, too. While the injunction to propagate the race is specific and explicit, references to interfering with the generation of life are only of the vaguest kind. Abortion, as already noted, is not expressly mentioned at all, though it was widely practiced in ancient times. Birth control, also not unknown in antiquity, occurs only incidentally in the condemnation of Onan for refusing to consummate the levirate bond in succession to his deceased brother by resorting to *coitus interruptus* (Genesis 38:9–10). Somewhat more explicit, but still rather indirect, is the prohibition of sterilization, or rather castration; but even this is forbidden only in the context of banning animals so maimed from serving as sacrifices on the altar (Leviticus 22:24).

Except in carefully defined cases, the Jewish tradition has always regarded these acts as strictly proscribed by biblical law. On sterilization in particular, Judaism pioneered a religious attitude otherwise unknown until millennia later. The Hebrews were the only nation of antiquity to ban the emasculation of men and animals alike—a ban not endorsed by Buddha, Confucius, Jesus, or Muhammad. One recalls the widespread use of eunuchs in ancient courts and the early Christian self-emasculation "for the sake of holiness," followed until the late Middle Ages by the practice of castrating children in order to preserve their soprano voices as church choristers. By contrast, the Law of Moses excluded a man who "has his stones crushed" from the Temple service (Leviticus 21:20) and him "that is crushed or maimed in his privy parts" from "entering into the congregation of the Lord" (Deuteronomy 23:2), that is, from Jewish marriage.

According to later sources, these operations may be performed only for pressing health reasons. On birth control, for instance, the Talmud (*Yevamot* 12b) permits (or requires, according to one view) the use of a contraceptive tampon by minor, pregnant, or lactating women to prevent any danger to their own or their offspring's life that might result from a conception under those circumstances.

The law proscribes sexual relations among spouses in times of famine, but not on economic grounds. The restriction—from which childless couples are in any case exempted—is meant simply to curb pleasures and comforts equally forbidden on days of national or private mourning. More characteristic of the spirit, if not the letter, of Jewish law is the story related in a famous thirteenth-century moralistic work: A poor person complained that he could not afford to support any more children and asked a sage for permission to prevent his wife from becoming pregnant again. Said the sage: "When a child is born, the Holy One, blessed be He, provides the milk beforehand in the mother's breast; therefore do not worry!" But the man continued to fret. Then a son was born to him. After a while the child became ill and the father turned to the sage: "Pray for my son that he shall live!" "To you applies the verse," exclaimed the sage. " 'Suffer not thy mouth to bring thy flesh into guilt' " (Ecclesiastes 5:5).

Even less relevant in a Jewish context is the fear of "population explosion." For many overpopulated and underdeveloped countries this fear may be realistic enough. Where human lives are in jeopardy, the moral sensibilities of Judaism, applying the general rule of placing life above law, would in any case not be unsympathetic to certain population control measures designed to counter any threat of widespread starvation. Moreover, most rabbinic authorities doubt if the precept to "be fruitful and multiply" is altogether incumbent on "the Sons of Noah," i.e., on non-Jews. The precept, it should also be noted, followed as it is by the command "and fill the earth and conquer it" (Genesis 1:28), is expressly linked with man's conquest of nature; only by harnessing the world's resources can man fulfill the first part of the verse.

But the specter of "population explosion" would hardly haunt a people that for most of its history has been threatened with annihilation and is currently agitated by acute "population shrinkage." To Jews, suffering from a gross imbalance between a disproportionately low natural increase and an alarmingly high artificial decrease through drift, assimilation, and intermarriage, a restrictive attitude to birth control and abortion is of vital demographic as well as moral concern to assure continued physical no less than spiritual survival.

A further cardinal principle governing the rules of Jewish law on the generation of life is the strict limitation of the right and duty of procreation, and indeed of sexual intimacies, to personal relations sanctified by marriage. Any pre- or extramarital sexual adventures are utterly abhorrent to Judaism, as a betrayal of a divine trust, a debasement of human dignity, a surrender to self-indulgence, and an erosion of the family bond as the basic unit of society. The free availability of contraceptives and abortions, particularly before or outside marriage, would also remove the most powerful deterrent to promiscuity.

A similar distinction between the purely legal and wider moral concerns is applied to artificial insemination. Anticipating by about seventeen centuries the first baby so conceived, born in 1866 ("Symposium," p. 227), the Talmud (*Hagigah* 15a) considered the feasibility of an artificial (albeit accidental) impregnation, concluding that a virgin would not lose her status if she had become pregnant by bathing in water previously fertilized by a male. Based on this passage, most modern Responsa hold that a married woman's conception *sine concubito*, even if the sperm donor were a close relative, does not constitute adultery or incest. Yet, despite this leniency, and however great the benefits in individual cases, the numerous rabbinical verdicts of recent years are virtually unanimous in utterly condemning the practice as incompatible with the sanctity of marriage and undermining the moral fabric of society. Judaism, they aver, recoils from reducing human generation to stud-farming methods, leading to a mechanization of life whereby a father is replaced by test tubes and syringes, and his responsibility to care for his natural offspring by a payment for the sample he supplied, and whereby the decision on who is to sire a mother's child is left to the arbitrary whims of a doctor, a laboratory technician, or a computer. Equally repugnant is the essentially clandestine character of the operation, without any proper records or other safeguards, which conceals the donor's identity and deceives the public by a fraudulent entry in the child's birth certificate. This in turn deprives such children of the inalienable right to a father and other relatives who can be identified, and exposes them to the risk of incestuous marriages among siblings whose paternity, unknown to each other, is identical. None of these objections apply, of course, to insemination from the husband.

Underlying the often highly technical discussions on the subject is the fundamental recognition of man's incomparable dignity. As the delicately balanced fusion of body, mind, and soul, created in the image of their Creator, humans must be generated out of the intimate love joining husband and wife together, out of identifiable parents who care for their progeny, and out of a home that provides affectionate warmth and compassion.

The conclusion of life. The inalienable rights conferred on man by his Creator extend to and beyond his death. Indeed, in the Jewish scale of values, rights figure in death even more prominently than in life. While a person lives, the emphasis is on the obligations he owes to others rather than on the privileges he may claim. The moral imperatives of the Bible are enshrined in Ten Commandments—or, for the conscientious Jew, in 613 positive and negative precepts—not in a Bill of Rights. The dead, on the other hand, "are free from any religious duties" (*Niddah* 61b). In life, man's ineluctable assignment is work and struggle: "for man is born unto toil" (Job 5:7). In death, the key word is peace: "for the latter end of man is peace" (Psalms 37:37).

Jewish ethics therefore ranks as the noblest form of charity—"loving-kindness of truth" in the language of the rabbis—services rendered to those who can no longer fend for themselves, including the utmost consideration for the dignity of the dying.

Their ordeal usually poses an ethical dilemma to which Judaism is particularly sensible. All too often there is a tragic conflict between the sanctity of life and the relief of human suffering, especially in the terminal stage of life. While, as has been explained, in any direct confrontation between the two life must prevail, because of its supreme and absolute value, the anxiety to mitigate pain and misery nevertheless invokes the most far-reaching concessions, short only of an actual attack on life itself.

The leading principles that determine the relevant rules of Jewish law can be illustrated by two examples.

Although the term "euthanasia" to denote "the action of inducing gentle and easy death" was first used by the British moral historian William E. H. Lecky only in 1869 and is now applied to a practice severely condemned in Jewish law, the term itself is by no means alien to students of the Talmud. In fact, the precise Hebrew equivalent ("*mithah yafah*") for the Greek "*eu thanatos*" is used to demand efforts to ensure an "easy [literally, pleasant] death" even for capital criminals. Thus, in the hypothetical case of a death sentence (in practice capital punishment had been abolished), the Talmud interprets the precept "love your neighbor as yourself" (Leviticus 19:18) to mean "choose an easy death for him," for instance, by resorting to the swiftest possible form of execution, and by drugging the convict with an intoxicating drink beforehand (*Sanhedrin* 43a, 45a, 52b). This extreme concern to remove all avoidable pain from the agony of death even when judicially imposed on the most heinous offenders applies, of course, a fortiori to innocent sufferers as they approach the end.

A medical illustration of this principle occurs in the story of the Syrian King Ben Hadad, whose messenger Hazael, sent to inquire from the Prophet Elisha whether the king would survive his sickness, was told: "Go, say unto him 'You shall surely recover'; howbeit the Lord has shown me that he shall surely die" (2 Kings 8:10). Jewish ethics in some respects regards peace as an even greater virtue than truth; hence it rates the patient's peace of mind higher than the doctor's truthfulness if this might undermine the patient's hope or mental tranquility.

Once death has set in, the mortal remains must be "returned to the earth as it was" (Ecclesiastes 12:7) by interment with the minimum delay. Significantly, the rights of the dead to speedy burial and to respectful treatment generally are again explicitly stated only in regard to capital criminals; even the most depraved humans are created in the divine image and deserving of reverence in death (Deuteronomy 21:23). It is only from this biblical reference that the Talmud derives the strict laws against "disgracing the dead" in general, by undue exposure, unwarranted exhumation, and especially incisions or disfigurement of any kind.

Another category of bodies enjoying the special protection of the law, again because they are more defenseless than others, is that of persons who die without leaving any family to care for their funerals. The obligation of burial then devolves on any Jew, including the high priest, who otherwise must not defile himself by contact with the dead even for his own father or mother (Leviticus 21:11).

These provisions are in striking contrast to the attitude of modern civil legislation (such as the English Warburton Anatomy Act of 1832) which remove any legal protection against dissection from the bodies of executed prisoners and of "unclaimed persons," treating such bodies as *res nullius* (i.e., an impersonal, ownerless object).

Altogether, Jewish law regards the human body as divine property and therefore as inviolable. Its custody and care are the responsibility of the person himself in his lifetime, and after death of his next of kin or, in their absence, of the community at large. Neither the state nor any other agency can ever claim the right of possession of a human body.

These considerations explain the restrictive Jewish attitude to autopsies and anatomical dissection as well as to the utilization of cadaver tissue for organ transplants. The Talmud mentions some anatomical experiments, though never in a medical context. The problem of using bodies for teaching and research purposes was first raised by a Jewish medical student at the University of Göttingen (Germany) in 1737, but it was not until several decades later that Rabbi Ezekiel Landau of Prague, the leading rabbinical scholar of his time, in a celebrated Responsum, laid down the guidelines that inform the Jewish religious view on autopsies to the present day. While affirming the strict law against "disgracing the dead" by any violation of the corpse's integrity, he held that this law, like any other, was superseded by the saving of life. Hence, a postmortem could be sanctioned only if another patient with similar symptoms was "at hand" who might directly benefit from the findings of the operation.

Several later authorities extended the sanction to the bodies of persons who had freely given their consent in their lifetime. Some other limited extensions were also gradually added. But the hiatus between the doctors' ever growing demand for bodies to meet research needs and the rabbis' fierce determination to defend the rights of the dead remained wide enough to engender much bitterness. In Israel, in the continued absence of any legal requirement of family consent for autopsies, this conflict occasionally led to street riots and cabinet crises, as it prevented the Hebrew University, founded in 1925, from establishing a medical school until 1948.

Summary of ethical directives

General

Sick visitation. Judaism ranks the visitation of the sick among the finest expressions of true charity and the *imitatio Dei* ideal, as God himself, in the imagery of the rabbis, visited Abraham following his circumcision (Genesis 18:1). The *Shulhan Arukh* devotes an entire chapter to detailed instructions on this precept. The principal object includes the caller's being moved to pray for the patient's recovery and rendering any service required for his comfort. The duty devolves on all; even the great must visit their juniors. Hence, the functions of the local sick-visitation societies, which were very active in Jewish communities since the Middle Ages, were performed mainly by lay people. Rabbis merely contributed their share like others.

The sick enjoy a special claim to private and public assistance. Contributions toward their needs take precedence even over the erection of a synagogue, and it is morally indefensible to refuse such aid.

The physician. The physician's duties, rights, and liabilities are also regulated in considerable detail. Since he is engaged in the supreme service of the life and health of others, Jewish law treats the physician as a quasi-religious official. For instance, the *Shulhan Arukh* bars doctors—like rabbis before the professionalization of the rabbinate in the Middle Ages—from accepting payments except "for loss of time and for the trouble taken," since religious duties are to be performed gratuitously. Yet the patient is liable to honor any prior agreement even if it stipulated an excessive fee, since the physician "sold him his expert knowledge which is beyond price." On the other hand, the charges of apothecaries are more strictly controlled; the price of drugs must not be raised above their proper level, and an agreement exacted in an emergency to pay an extortionate charge need not afterward be kept.

Also like rabbis, physicians are exempt from liability for damages caused by errors of judgment, provided they were duly licensed (in talmudic times, by the ecclesiastical courts). This was enacted "for the social order"; for although ordinarily a person is always liable to payment of damages even if caused unwittingly, an exception is made in this case "so that doctors will be found to heal."

In Jewish sources professional secrecy was never enjoined on physicians with the zeal with which it is now guarded by the medical fraternity under the influence of the Hippocratic Oath. It was covered simply by the general injunctions against any kind of gossip: "You shall not go . . . as a talebearer" (Leviticus 19:16) and "reveal not the secret of another" (Proverbs 25:9). In fact, confidentiality should be disregarded when the public or some other overriding interest is at stake, such as in cases of medical evidence in court actions to secure justice for another party or to prevent the deliberate concealment of a hereditary defect in matrimonial arrangements.

Jewish law raises no objection to the treatment of women by males. But it urges the utmost chastity to avoid any lewdness in thought or deed. Parents, however, should not be subjected by their own children to any act that causes a loss of blood (including injections) except in emergencies when no one else is available.

Consent, risky operations, and experiments. Since one may endanger one's own life no more than anyone else's, Jewish law requires no consent for operations deemed essential by competent medical opinion.

Any chance to save life, however remote, must be pursued at all costs. Hence, untried cures and even possibly fatal operations may be sanctioned (with the informed consent of the patient) in a desperate gamble to save or prolong life, if no safe treatment is available. But since one may never sacrifice one's own or any other person's life however many lives might thereby be saved, possibly hazardous experiments must not be carried out except on sick subjects who would themselves be the beneficiaries if the experiment succeeded. Some authorities, however, permit a person to expose himself to an uncertain risk of life in an effort to save another from certain death; but this is an act of charity, not of obligation. On the other hand, the duty to volunteer for experiments that involve no risks to life or health devolves on anyone who may thereby directly promote the health interests of others, and it may not be unethical to perform such tests even without the subject's consent, provided they are in fact completely harmless.

Tests of new medications and surgical procedures should first be performed on animals, but every care must be taken to protect them from any avoidable pain.

Surgery

Organ transplants. There is no firm objection to organ transplants and to the utilization of cadaver tissue, provided (1) this is done for life-saving purposes, (2) the death of the donor has been definitely established, and (3) the anticipated benefit to the recipient is substantially greater than the risk.

Included in this sanction, according to most authorities, are corneal grafts, since the restoration or preservation of eyesight is to be regarded as a life-saving act (as blind people may more easily be liable to fatal accidents). Hence, some require that a donor should stipulate in his bequest that his eyes be used only for patients suffering from, or threatened with, complete blindness; but others waive this condition. In any event, the prior consent of the donor or his family should be obtained, and the unused part of the eye after the removal of the cornea should not be disposed of except by burial.

Similarly, kidney transplants, because of their relatively high success rate, have generally been permitted. The prevailing rabbinic view on kidney donations from living donors, too, is permissive, though such donations are to be regarded as acts of supreme charity, not as an obligation.

But the fairly numerous Responsa on heart transplants show a good deal more hesitation and disquiet. In view of the very high rate of failures leading to the death of most recipients, compounded by the grave doubts as to whether the donors were in fact truly (not just "clinically") dead when the vital organ was excised, several rabbis have condemned the operation as "double murder." The whole procedure is clearly still in an early experimental stage, and the risks involved cannot be justified so long as it is carried out for general research purposes rather than for the primary objective of saving the life of the patient at hand. On the definition of death and the right to stop resuscitation efforts, see below.

Cosmetic surgery. There are four possible counterindications to plastic surgery for cosmetic ends: the theological implications of "defying Providence"; the possible risks involved in any operation; the objection to any mutilation of the body; and the ethical censure of human vanity, especially among males. In the sparse rabbinic writings on the subject, these reservations could be discounted, provided the danger is minimal; and especially (1) if the operation is medically indicated, e.g., following an accident or for grave psychological reasons; (2) if the correction of the deformity is designed to facilitate or maintain a happy marriage, or (3) if it will enable a person to play a constructive role in society and to earn a decent livelihood.

The beginnings of life

Artificial insemination. Artificial insemination from the husband is generally permitted, as is semen sampling for fertility tests, if deemed necessary to promote the prospects of a conception within marriage. But the semen should preferably be obtained through normal intercourse.

But for the reasons already given, Jewish jurists object strongly to donor insemination, counseling adoption instead. Nevertheless, most authorities do not regard artificial insemination by donor (AID) as constituting an adulterous relationship or a child so conceived as suffering the serious disabilities of bastardy, though some marriage restrictions may apply, since the true father is unknown. The mutual legal bonds and obligations between him and his child—whether in matters of care, incest, or inheritance—cannot be revoked or transferred.

Contraception and sterilization. The voluminous literature on contraception and sterilization can hardly be summarized in a few lines. All the sources insist that in these capital judgments, literally determining whether a life is "to be or not to be," every case must be judged individually on its merits by rabbinical experts based on the most competent medical opinion.

While the current views among leading rabbis are by no means uniform, they broadly agree that contraceptive precautions should be taken only on medical grounds. Where the danger resulting from a pregnancy is not acute, sanction for certain types of contraceptive devices will more readily be given for a limited period, usually of two years, and if the precept of procreation (i.e., by the birth of a son and a daughter) has already been fulfilled.

In a recent volume of Responsa, the various methods of birth control have been graded, from the most to the least favored, according to their degree of direct interference with the generative act and organs, as follows: (1) oral contraceptives ("the pill") and intra-uterine devices (IUD), provided they do not have an abortifacient effect, to be used so long as there is some sound medical indication; (2) female sterilization, if the danger of a pregnancy is permanent; (3) postcoital removal of semen by tampon or douche; (4) cervical rubber cap to close the mouth of the uterus; (5) spermicides; (6) tampon or diaphragm inserted before coitus; (7) IUD, if the effect is to abort the impregnated egg; and (8) male condom, to be used only in extreme cases of acute danger and if other means are unavailable or unacceptable. The "safe period" method is always legitimate, but Jewish law would counsel a divorce rather than permanent abstention, since it regards the regular payment of the conjugal dues as mandatory in every marriage.

The sterilization of males is a more serious offense than of females; it is never permitted except if urgently necessary as a therapeutic measure. According to some authorities, therefore, prostatectomies should avoid severing the seminal ducts, if at all possible.

Abortion. Generally the view prevails limiting the sanction to cases involving a hazard to the life of the mother, whether physical or psychological (e.g., by violence or suicide). While some rabbis brand the destruction of a fetus under any other circumstances as "an appurtenance of murder," others have recently extended the sanction to (1) mothers whose health is seriously at risk, (2) substantial fears that the child may be born with grave defects, and (3) cases of rape or incest. But such pregnancies should be terminated preferably within the first forty days or at least within the first three months.

The end of life

Preparation for death. The predominantly "this-worldly" character of Judaism is reflected in the relative sparsity of regulations on the inevitable passage from life to death and in the absence of any sacramental rites. The concern for the patient's physical and mental welfare remains supreme to the end, and every caution is urged to ensure that nothing shall aggravate his condition or compromise his will or ability to live. He should never be informed that his affliction is terminal, unless one can reasonably assume that such information will come as a relief rather than as a shock to him.

If he is aware of his condition, he should be encouraged to order his temporal affairs and to recite his confession before God; but even then he should be reassured: "Words can cause neither life nor death." However, if he suffers undue pain, analgesics may be administered, even if by inducing sleep or unconsciousness they deprive him of the opportunity to make these preparations and even if there is some risk of thereby unwittingly hastening the end, so long as this is done genuinely for the sole purpose of rendering him insensitive to acute pain.

Once death is believed to be imminent, any movement of the body or parts of it should be avoided. For "the matter can be compared with a flickering flame; as soon as one touches it, the light is extinguished." But one is permitted to "remove an impediment" to death, e.g., by stopping a clattering noise near the patient that delays his demise. No preparations for burial or mourning should be made before death has set in; indeed until then the patient is deemed "a living person in all respects."

Resuscitation and the definition of death. The classic talmudic definition, establishing death by the definite cessation of breathing and pulsation, remains essentially valid for most practical purposes. Such criteria as "clinical death," irreversible brain damage, or a flat EEG reading cannot be recognized in Jewish law, not merely because the diagnosis may be mistaken but because any fraction of life, in whatever state of animation, is still infinite in value. Even after all signs of life have disappeared, especially in cases of sudden death, a period of twenty to thirty minutes should elapse before treating the patient as dead, "lest he was merely in a swoon."

However, in view of modern resuscitation methods, which can restore respiration to eighty percent of unconscious patients, this definition now requires some modification, and death must not be presumed, even after the heart and lung action have stopped, so long as there is any hope of revival. Hence, resuscitation efforts must not be left untried, nor may any incision be made on the body (for autopsies or the removal of an organ) until both criteria are satisfied: the complete stoppage of all spontaneous life functions, and the certainty that artificial revival will prove unavailing. There may be conditions (see below) under which artificial respiration may be suspended in hopeless cases. But whether and when to apply or withdraw a respirator must be determined by the interest of the patient himself, not by the concern to obtain a viable organ for transplant purposes, since it would be wrong to manipulate one life for the sake of another.

Euthanasia. Euthanasia proper is opposed without qualification in Jewish law. It condemns any active and deliberate hastening of death as sheer murder, whether the physician acts with or without the patient's consent. Judaism rates suicide as even worse than murder in some respects, since there can be no atonement by repentance for self-destruction.

On "passive euthanasia" in the final phase of lingering life, however, contemporary rabbinic views diverge. Some will not allow any relaxation of efforts, however artificial and ultimately hopeless, to prolong life. But others do not require the physician to resort to "heroic methods"; they sanction the removal of medications or machines that serve only to draw out the dying patient's agony, provided no natural means of subsistence (e.g., food) are withdrawn.

Autopsies and dissection. Upon death, the bodies of Jews, following purification by ritual washing, must be buried with the minimum delay. Their integrity must not be violated in any way, whether by incisions, embalming, or cremation. Some authorities permit the use of the body for anatomical research and study if the deceased expressly so directed in his lifetime, thus renouncing the honor due to him. But most make an exception only for autopsies required to save life, with the additional safeguard of family consent and rabbinical sanction being obtained. While no blanket permission for indiscriminate postmortem examinations can therefore be given, it is usually granted when there is some reasonable prospect that the findings may help to save another existing patient's life, especially if the cause of death was completely obscure or due to suspected hereditary traits. Recently it has been suggested to broaden the sanction to include controlled tests of new medications and cases of suspicion that mistakes occurred in the diagnosis or treatment. But any permission would be subject to reducing the operation to a minimum, performing it with the greatest dispatch and utmost reverence, and returning all parts of the body for burial.

IMMANUEL JAKOBOVITS

[*For further discussion of topics mentioned in this article, see the entries:* ABORTION, *article on* JEWISH PERSPECTIVES; CADAVERS, *article on* JEWISH PERSPECTIVES; CONTRACEPTION; DEATH AND DYING: EUTHANASIA AND SUSTAINING LIFE, *article on* HISTORICAL PERSPECTIVES; EUGENICS AND RELIGIOUS LAW, *article on* JEWISH RELIGIOUS LAWS; LIFE, *article on* VALUE OF LIFE; ORGAN DONATION, *article on* ETHICAL ISSUES; ORGAN TRANSPLANTATION, *article on* ETHICAL PRINCIPLES; POPULATION ETHICS: RELIGIOUS TRADITIONS, *article on* JEWISH PERSPECTIVES; RELIGIOUS DIRECTIVES IN MEDICAL ETHICS, *article on* JEWISH CODES AND GUIDELINES; REPRODUCTIVE TECHNOLOGIES, *article on* ETHICAL ISSUES; SEXUAL ETHICS; *and* STERILIZATION, *article on* ETHICAL ASPECTS. *Also directly related are the entries* DEATH, *articles on* WESTERN RELIGIOUS THOUGHT: DEATH IN BIBLICAL THOUGHT *and* POST-BIBLICAL JEWISH TRADITION; DEATH, DEFINITION AND DETERMINATION OF, *article on* PHILOSOPHICAL AND THEOLOGICAL FOUNDATIONS; OBLIGATION AND SUPEREROGATION; PAIN AND SUFFERING, *article on* RELIGIOUS PERSPECTIVES; SUICIDE; *and* TRUTH-TELLING, *article on* ETHICAL ASPECTS. *See* APPENDIX, SECTION I, OATH OF ASAPH *and* DAILY PRAYER OF A PHYSICIAN ("PRAYER OF MOSES MAIMONIDES").]

BIBLIOGRAPHY

[Biblical and talmudic sources are given in the text. Other works containing the principal sources of Jewish law include codes and Responsa.]

CODES

KARO, JOSEPH. *Shulhan Arukh.*
MAIMONIDES. *Mishneh Torah.* Also called *Yad Ha-Hazakah.*

MODERN RESPONSA

FEINSTEIN, MOSES. *Igrot Moshe.* 5 vols. New York: Author, 1959–1973.
WALDENBERG, ELIEZER JUDAH. *Tzits Eli'ezer.* 11 vols. Jerusalem: 1945–1970.
WEINBERG, JECHIEL. *Seridei Esh.* 4 vols. Jerusalem: Mosad HaRav Kook, 1961–1969.
WEISZ, ISAAC JACOB. *Minhat Yitshak.* 5 vols. 1st ed. London, Jerusalem: Author, 1955–1972.

GENERAL

FELDMAN, DAVID M. *Birth Control in Jewish Law.* New York: New York University Press; London: University of London Press, 1968. Paperback ed. *Marital Relations, Birth Control, and Abortion in Jewish Law.* New York: Schocken Books, 1974.
FRIEDENWALD, HARRY. *The Jews and Medicine.* 2 vols. Baltimore: Johns Hopkins Press, 1944.
JAKOBOVITS, IMMANUEL. *Jewish Medical Ethics: A Comparative and Historical Study of the Jewish Religious Attitude to Medicine and its Practice.* New ed. New York: Bloch Publishing Co., 1975.
———. "Medicine and Judaism." *Journal of a Rabbi.* New York: Living Books, 1966, pp. 137–220.
Medical Ethics: A Compendium of Jewish Moral, Ethical and Religious Principles in Medical Practice. 5th ed. Edited by M. D. Tendler. New York: Federation of Jewish Philanthropies, Committee on Religious Affairs, 1975.
PREUSS, JULIUS. *Biblisch-Talmudische Medizin.* 3d ed. Berlin: S. Karger, 1921. Edited translation by Samuel Paley. New York: Ktav Publishing House, 1971.
ROSNER, FRED. *Modern Medicine and Jewish Law.* New York: Yeshiva University, 1972.
"Symposium: Artificial Insemination, Medicolegal Implications." *American Practitioner* 1 (1947): 227–234.
ZIMMELS, HIRSCH JACOB. *Magicians, Theologicians and Doctors.* London: Goldston, 1952.

JUSTICE

The concept of justice

Actions that are wrong in the special way called "unjust," and likely to arouse that special kind of indignation associated with perceived injustice, are of three basic types: first, invidious *discrimination* or arbitrarily unequal treatment in legislating, administering, or enforcing rules, or in distributing burdens or benefits; second, *exploitation,* that is, taking advantage of another's trust or natural handicaps to gain unfairly at his expense, or placing another at an unfair disadvantage in competitive or cooperative undertakings; and third, *judgmental injustice,* which consists in the making of false derogatory judgments about persons or their works, statements that "aren't fair" to the persons they are about. Of these, the third has only minor application to bioethics. The second would apply to methods of purchasing and selling blood and biological "spare parts," and to experiments that take advantage of poor people. The category of deepest philosophical interest, however, is the first, and especially those actions in it which consist in allocating medical services, medicines, and equipment in short supply.

The distribution of benefits is an occasion for justice, according to David Hume, only when wanted or needed goods or services are moderately scarce relative to the demand for them. If such goods existed in profuse abundance, he argued, then the very concept of justice in their distribution would never have occurred to anyone, for "what purpose make a partition of goods where every one has already more than enough?" (Hume, sec. III, pt. I). But, on the other hand, if the scarcity of needed goods or services were so extreme that there is no way to prevent the majority of the population from perishing and the remainder from suffering extreme deprivation and misery, then "the strict laws of justice are suspended in such a pressing emergency, and give place to the stronger motives of necessity and self-preservation" (ibid.). And, wherever we can imagine such conditions to have always obtained, there the concept of distributive justice could not have arisen.

The basic principle of distributive justice, and one upon which most writers can agree as a starting place for analysis, is that like cases are to be treated alike and different cases treated differently in direct proportion to the differences between them. If two unemployed persons in line for a food dole appear to be quite alike, and yet one is given twice as much as the other, then our sense of justice will be offended, unless we learn that one is twice as heavy as the other, or has twice as many dependents, or has some nutritional deficiency requiring a large portion immediately. In the absence of some such special explanation, the inequality in the treatment accorded individuals who are essentially alike is completely arbitrary, and *arbitrary inequality* is thought by most philosophers to be the essence of distributive injustice. Similarly, if two individuals who appear quite different in some conspicuous way are treated alike, injustice has been done unless there is some important similarity underlying their differences and justifying their common treatment.

It will always be possible, however, to point to *some* differences. A defender of a discriminatory law banning Negroes, say, from national parks might argue that the statute does indeed treat like cases alike and different cases differently. All persons who are alike in being white are treated alike; all persons alike in being black are treated alike; people who are different in respect to skin color are treated differently. The natural reply, of course, to this sophism is that skin color is not a *relevant* similarity or difference for the purposes of justice. Injustice is done when persons who are alike in every *relevant* respect (not in absolutely every respect) are treated differently, or when individuals who are different in some *relevant* respect are treated alike, or in a way quantitatively disproportionate to the *relevant* difference between them.

Our initial principle of distributive justice is quite vacuous, then, until supplemented by criteria for determining the relevance of similarities and differences among persons. It is impossible to formulate such criteria in a general way. Which characteristics of persons are relevant will depend on such factors as the purposes of legislators in making rules; whether we are awarding prizes, conferring rewards, assigning grades, or making compensation, and in each case, for what kind of victories, achievements, services, skills, excellences, or losses; whether we are imposing burdens, like taxation and military conscription, or allocating benefits, like economic income or medical services; and the nature of the burdens or benefits themselves. In some of these contexts the determination of relevance is straightforward and beyond controversy. That an ambulance rushing an accident victim to the hospital is different from other vehicles is relevant to justifying the imposition of less rigid speed limits on it, whereas a vehicle's character as a delivery truck or a taxicab is quite irrelevant to evasion of speed laws. These are judgments about which reasonable persons do not differ. Similarly, there are virtually no contexts in which skin color is relevant, and hardly any in which sex is relevant. In some contexts, however, where judgments of relevance are not narrowly determined by the rules of a contest, the purpose of a rule, or the like, appeal must be made to basic attitudes that are inherently controversial. The most notorious controversies of this kind have been over the relevance of such disparate factors as merit and need in the distribution of wealth and health services throughout a community.

Most of the Greek city-states during the Hellenic period had democratic and antidemocratic factions in constant political struggle. The former demanded equalitarian distributions of wealth and privilege (except for slaves); the latter insisted that distributions be based on such factors as merit, service, and skill. The latter group suffered somewhat from a rhetorical disadvantage in the struggle, for the Greek word "just" (*dikē*) was very closely connected, if not identical in meaning, with the word for "equal" (Vlastos, pp. 39 ff.). Plato and Aristotle both aligned with the antiequalitarian party, since both were convinced that there are many inequalities that are perfectly just (even though that opinion when voiced must have had the sound of paradox, even strict contradiction). Aristotle's way of reconciling the antiequalitarian position with common sense was especially ingenious. He analyzed distributive justice as proportionate rather than absolute equality. By this he meant that the equality required by justice is an equality of ratios: one citizen's portion must be to his own relevant characteristics (or "merits" in a broad sense) as any other citizen's portion is to *his* relevant characteristics. Distributive justice between any two citizens A and B is achieved then when:

$$\frac{A\text{'s portion}}{B\text{'s portion}} = \frac{A\text{'s possession of relevant characteristic } X}{B\text{'s possession of relevant characteristic } X}$$

This turned out to be a formula that even democrats could accept, since they disagreed with Aristotle only about how to fill in the variable X. Strict "meritarians" would replace X with skill or virtue; others would insert hereditary class, labor, or past contribution; equalitarians would substitute simply "need." Many writers today would defend different answers in different contexts, varying with the particular benefit or burden to be distributed.

Formulating criteria for the relevance of characteristics, however, hardly exhausts the moralist's concern with distributive justice, for even if we can agree about what organization of society justice ideally prescribes, we are left with the problems of evaluating transactions and allocations in our actual imperfect society, and of effecting transitions to more equitable arrangements. The nineteenth-century British moral philosopher Henry Sidgwick rested much of his theory of justice on the distinction between "ideal" and "conservative justice." The latter term refers to the requirements that contracts be honored and that "natural and normal expectation" not be disappointed. The dictates of ideal and conservative justice conflict, however, whenever reform is affected through abrupt change of rules and procedures in a society, and very often when this happens the conflicting principles are not subject to satisfactory reconciliation. Any reform of an imperfect practice is likely to be unfair in its effects on some people, for to change the rules "in the middle of the game," even when those rules were not altogether fair, will disappoint the honest expectations of those whose prior commitments and life plans were made in genuine reliance on the continuance of the old rules. The justice (on balance) of changing the rules in such circumstances will depend on the degree of their unfairness and also on the extent and degree of the reliance placed on them. In virtue of the latter factor, justice requires, other things being equal, that reforms be made in a piecemeal and gradual fashion and that, wherever possible, persons should be given ample advance notice. In hard cases, however, where relevant factors balance equally, reformers must weigh quite legitimate incompatible claims against one another in such circumstances that, whichever decision is made, it will be unfair to someone or other.

One of the more important systematic works on justice to appear in recent years is John Rawls's *A Theory of Justice* (1971). Not unmindful of Sidgwick's distinction between ideal justice and problems of justice in the nonideal world, Rawls restricts his book, for the most part, to what he calls "ideal theory" (as opposed to "nonideal theory"). His work is addressed primarily to the task of deriving the ultimate principles of justice that would regulate an ideal "well-ordered society," an artificial model in which it is presumed that all laws are just and universally complied with. He concedes that the more urgent problems about justice belong to nonideal theory (or "partial-compliance theory" as he also calls it), for example, questions of justice against a background of unjust laws (the problem of civil disobedience), or questions that arise only after a just law has been broken (problems in the theory of criminal punishment), but he makes the deliberate assumption that ideal theory is more fundamental since it can provide "the only basis for the systematic grasp" of the more pressing problems of nonideal theory. In this respect his work is strikingly reminiscent of one feature of Plato's *Republic*.

Rawls's method of deriving the ultimate principles of justice, and the basic structure of society that they prescribe, falls in the tradition of social-contract theory. A proposed principle of justice is valid provided it would be chosen over any alternative principle that could be proposed to a gathering of normally self-interested, rational persons, each concerned to advance and protect his own interests. These hypothetical persons can be imagined to have gathered together voluntarily in an "original position" of equal power, for the purpose of designing the institutions that will regulate their future lives. We are also to suppose that each wears a "veil of ignorance," which prevents him from knowing facts about his own future condition that could tempt him to base his choice on the desire to promote his own interest to the disadvantage of others. If these rational choosers are neither rich nor poor, white nor black, male nor female, old nor young—so far as they know—then they cannot be lobbyists for any particular class interests and must choose their principles from a more disinterested vantage point. Formulating the correct principles of justice, then, is simply determining which principles such hypothetical rational persons, in such hypothetical circumstances, would choose to govern the design of their political and economic institutions.

The principle Rawls derives by this "contractarian method" for determining the justice of distributions of wealth is a further articulation of the traditional doctrine that persons are to be treated equally unless there are relevant differences among them. His basic thesis is that the only sufficient reason for departing from equality in the distribution of economic goods is that an unequal distribution would be to *everybody's* advantage, even that of the worst-off persons. The mode of distribution itself can have effects on the amount of goods produced, and in some cases, a particular sort of unequal distribution might be an essential component of a system that vastly increases the amount of goods subsequently available for distribution. Surely any reasonable person, Rawls argues, would prefer a system that gives him a larger amount of goods, though a smaller portion than his neighbor's, to a system that gives him a smaller absolute amount, though one that is an equal share. But unless unequal distribution benefits those with smaller shares, Rawls insists, it cannot be just.

Justice and social utility

A distinction is very commonly made between two classes of reasons that can have bearing on difficult decisions made by authorities, judges, political leaders, and others who must impose burdens or allocate goods and services in scarce supply. One class of reasons appeals to "utilitarian considerations," the other to considerations of individual justice, rights, and deserts. Conflicts between reasons in these mutually exclusive but not altogether homogeneous categories generate some of the most difficult dilemmas of social decision making. "Utilitarianism" is the name of an ethical theory that has taken many different forms, but common to them all is the view that social decision makers should consider *only* reasons of the utilitarian kind. The leading rivals of utilitarianism are the dualistic theory that grants cogency to *both* classes of reasons and requires us, when they conflict, to balance them against each other on the particular occasion of their conflict, and the contractarian theory of John Rawls, which gives absolute priority in cases of conflict to considerations of individual justice.

According to utilitarianism, social decision making is largely a matter of mathematical calculation. When faced with a decision involving the interests or fates of numerous individuals, the utilitarian would consider the various options open to him and predict as carefully as he can the effects of each possible choice on those persons whose interests are involved. Every such person is to "count as one, and no one as more than one" in the famous formula attributed to Bentham by Mill (Mill, chap. V, par. 36). The right decision, then, will be the one that will cause the largest balance of benefit over harm all-round, or in the unhappy case where every possible decision will cause more harm than good, the right decision is the one that will cause the smallest balance of harm over benefit. The utilitarian will discount predictions of remote or speculative consequences and give greater weight to consequences that are immediate and relatively certain. Utilitarians also attach great weight in their calculations to "the public interest," though most of them, following Bentham, think of "the public interest" as a mere shorthand way of referring to the sum total of individual interests affected. Some utilitarians concede that there is a difference in kind between utilitarian considerations and appeals to justice, but argue that in cases of conflict utilitarian considerations must always win out, since it appears to them to be contrary to reason ever to prefer one alternative to another that does more good or less harm all told. Other utilitarians analyze justice itself in utilitarian terms, so that conflicts between justice and social utility are logically impossible. Mill, for example, wrote that "justice is a name for certain moral requirements, which, regarded collectively, stand higher in the scale of social utility and are therefore of more paramount obligation than any others" (ibid., final par.).

The utilitarian *analysis* of justice seems implausible, at least at first sight, because of the large number of actual and hypothetical examples of apparent conflict between justice and social utility. Slavery in ancient Greece, for example, surely exploited the slaves unfairly for the good of their masters, but it may have made possible the great flowering of Hellenic culture from which we are still all benefiting. In time of war, brave soldiers are commonly left to perish for the sake of the greater number of their comrades who live to fight another day, and innocent civilians are ruthlessly destroyed for the sake of a victory that will benefit both the victors and the world community as a whole. Dictators in backward countries have sacrificed whole generations in order to achieve the economic growth deemed necessary to the elimination of widespread suffering, misery, poverty, and ignorance. Even the unrivaled affluence of Western nations has been said to be a historical vindication of the atrocious conditions imposed unfairly on the working masses in the earlier periods of industrial capitalism. All of these seem to be examples of actions or policies that promote social utility effectively but at the expense of individual justice.

No one would deny that social utility normally reinforces justice, or that conflicts between the two are the result of a much too hasty and inexact use of the utilitarian calculus, a failure to measure costs accurately, or to consider unwanted side effects. But it seems dogmatic to insist that apparent conflicts between utility and justice are always and necessarily based on such errors. The case against utilitarianism is derived in powerful measure from the spontaneous repugnance most of us feel when individuals are ruthlessly sacrificed for the sake of a social good they cannot share. The antiutilitarian, therefore, will insist that each individual has certain inviolable rights that limit what can be done to him without his consent even for the welfare of society as a whole, and that these rights are not properly subject to utilitarian bargaining and calculating. The antiutilitarian of Rawlsean bent will argue further that no rational, normally self-interested person in the "original position," ignorant of whether he is to be a slave or a master, sickly or healthy, worker, manager, or owner, would opt to be governed by the utilitarian moral principle, for the risk of his eventual exploitation or sacrifice, as required by that principle, would seem unreasonable. Those arguments of Rawls, designed to reinforce the argument from natural repugnance, are highly controversial, especially insofar as they put the stamp of rationality on a very conservative attitude toward the making of choices under conditions of uncertainty. That attitude is expressed in what is called the "maximin rule" in game theory, which directs us, when we are uncertain of outcomes, to play it safe by selecting the alternative whose worst possible outcome would be superior to the worst possible outcomes of the other alternatives. Thus, Rawls's rational chooser in the original position would select principles "for the design of a society in which his enemy is to assign him his place" (Rawls, p. 152). Those principles, Rawls insists, would not be utilitarian.

The problem of allocating scarce medical resources

The provision of medical services raises various problems that call for the application of principles of justice. There is first of all the question of how medical resources—for example, manpower, skills, and technology—on all levels of care should be distributed among the citizens of a democracy. Should these matters be determined by the operation of a free market place, by some form of utilitarian calculation, or by direct reference to a principle of justice proclaiming an equal right to health care for every person irrespective of income, geographical location, or social services rendered? A second set of problems, while less common, is even more vexatious: those requiring decisions about who should receive medical assistance in those special circumstances where there are not enough resources to go around. It will never be possible to guarantee that adequate lifesaving therapy will be made promptly available to all who need it. Surgeons at battlefield hospitals in wartime, for example, or during sudden unexpected natural calamities may be so overwhelmed by the numbers of wounded that they cannot possibly save them all. In cases of that kind, working criteria of a wholly medical kind can be formulated in advance. Patients whose wounds are so minor that they are likely to survive even without immediate medication or surgery and those whose wounds are so serious that they are likely to die in any case are put at the bottom of a priority list, whose only purpose is to permit the saving of as many lives as possible, where each life is presumed to be as valuable as every other, and all claims to life are of equal weight. If resources are inadequate even to help all of those in the top priority category, then harried surgeons, too rushed to ponder philosophical conundrums, are wholly justified in treating patients in the top category on a "first come, first served" basis.

Other examples of unavoidable scarcity derive from the necessary lag between the discovery and the mass production of a lifesaving drug. In the period of temporary shortage, authorities may have to decide in a very deliberate way which segments of the population should have prior claims, and such decisions are not always made on wholly medical grounds. Early in the Second World War, for example, American authorities decided to reserve penicillin, then new and in short supply, for military casualties. Paul Freund describes the options faced by American officers in a military hospital in North Africa in 1943. There was enough penicillin to treat either but not both of two groups of soldiers: those who had been wounded in combat and those who had contracted venereal disease while on leave. The decision was made to give priority to those with venereal disease on the ground that they could be cured quicker, and thus made ready for combat faster, and also that while infected they constituted a menace to others (Freund, 1971, p. 280). The doctors' dilemma in this case appears, at first, to be a straightforward conflict between individual justice and social utility resolved in favor of the latter. On closer examination, however, it becomes more difficult to separate the interwoven strands that neatly. All utilitarian considerations, properly so called, are forward looking; they concern only the consequences of the alternative actions or policies open to us. On the other hand, considerations of justice are both of a backward looking and a forward looking kind. When we look to the past and discover that a promise was made or an expectation created, or that the past services of some soldiers make them more deserving of assistance than soldiers whose past behavior had a different character, we are considering matters highly relevant to a weighing of moral claims on the scales of justice. But we can also find considerations relevant to justice in our projections of future consequences. Perhaps a failure to provide quick and effective cure to the venereal disease patients would deprive soldiers still in battle of the reinforcements necessary for victory, and that many more of them are killed or maimed as a result than would otherwise be the case. In that event, one might plausibly claim that a decision to withhold medication from their potential replacements is unfair to *them*. In many complex moral problems of this kind, considerations of justice, as well as utilitarian considerations, are found on both sides of the case.

Some of the most pressing problems of justice for the medical profession today are also of a kind that seem likely always to be with us, namely, problems of allocating natural or artificial organs for transplant or insertion in the bodies of patients who will otherwise die. It either is now possible or will soon be possible for medical technology to transplant livers, kidneys, and hearts; to replace parts of these organs with synthetic devices; and to create artificial kidneys, lungs, intestines, and other bodily parts. But it seems unlikely that the supply of these items will ever be adequate to the need. Consequently, responsible parties must decide how these scarce items can be justly allocated.

Questions of justice arise both about procedures for contributing organs and about procedures for allocating those on hand. Questions of the former kind are restricted to the provision of live organs for transplant. In the case of paired organs, for example kidneys, there seems to be no injustice in permitting a person, when fully and precisely informed of the risk, voluntarily to donate an organ to a patient in need. The ancient maxim of law *volenti non fit injuria* (one is not wronged by that to which he fully consents) is a maxim of justice too. One is not treated unfairly when his own voluntary choices are implemented for him, and there is no reason to believe that highly altruistic and compassionate choices cannot be fully voluntary. The question of whether it can be fair to *compel* persons to donate one of their paired organs in time of depleted supply and sudden dire need is more difficult. The law has always been reluctant to force people to come to the aid of others, in contrast to refraining from hurting others. In recent years, however, more and more legal commentators have come to the view that one person can deliberately hurt another by omission as well as commission, and that when there is little risk to oneself, one has a duty to come to the aid of those in danger (Ratcliffe, pp. 225 ff.). In almost any imaginable situation in which, say, a kidney is required, more than one person would be in a position to contribute, and the fairest method of choosing between them, as in choosing among those eligible for jury service, would probably be a random lottery. The practical difficulty, however, of designing and implementing a fair lottery under conditions of emergency might very well count against any proposal for legal coercion. Even discounting that factor, there are moral considerations that should make us hesitate. The loss of a kidney is surely a greater sacrifice than that required of a taxpayer, of a champion swimmer who must get wet in order to save the life of a drowning child only a few yards away, or of an unwilling donor of blood, a substance that is naturally self-replenishing. The greater the risk to the donor, the weaker the case for compulsion. The problem becomes morally difficult as the risk to the donor approaches an even balance with the expected benefits, duly discounted for uncertainty, to the recipient. Fortunately, we can probably avoid the circumstances that generate this problem, and hence also the need for coercion, by reliance on the compassionate good will of friends and relations of patients, and of other volunteers.

Another controversial question is whether justice condones permitting or encouraging persons to *sell* their organs to hospitals for the purpose of transplantation in others. On the one hand, some argue that where the price offered is determined by a free and open market place, and both the seller and the buyer are free to decide whether the price is worth what the one is to lose and the other to gain, then the transaction between them is a fully voluntary one, and by the *volenti* principle, no one is treated unjustly by an agreement to which he fully consents. On the other hand, some argue that such a system will be inevitably exploitative if the seller is induced to incur a serious risk of physical harm in exchange for money. As Paul Freund has pointed out, the debate is similar to that which rages over the moral propriety of a volunteer army, where the soldiers are recruited from the poorer classes and persuaded by the inducement of higher pay to shoulder the potential hazards of military combat. Freund notes how widespread and natural is the repugnance evoked whenever society appears to maintain "an impoverished class of citizens to serve as risk-takers for others" (Freund, 1971, p. 279).

Somewhat different problems of justice arise when the organ to be donated is an unpaired one, for example, a heart or a liver, without which the donor cannot live at all. There is no difficulty, again because of the *volenti* principle, in patients agreeing well in advance to the use of such organs after their death; the problem arises where use of the organ requires fatal surgery on a consenting patient before he is legally dead. At first sight, a patient's consent to such "altruistic euthanasia" seems on the same moral footing with cases of "altruistic suicide," which are generally applauded in our society, a soldier in training falling on a hand grenade to save the lives of his buddies, a pedestrian throwing himself in the path of onrushing vehicles in order to push an imperiled child out of danger, and so on. If there is any difference between the two kinds of cases, morally speaking, it is that the voluntary donor, being typically a terminal patient with brief life expectancy, contributes something of less value to himself. His contribution, in a way, is more reasonable, since a relatively small loss is suffered for the sake of a great gain to another, and by the same token less heroic than the sacrifice of the spontaneously altruistic suicide who dies while healthy and young. In any case, it can be argued that, where the rights of third parties are not involved, how a person disposes of his life is his own business, and that he cannot be treated unfairly by a transaction to which he freely consents. But even if this antipaternalistic judgment represents ideal justice, especially as concerned with the claims of needful recipients and willing donors, it does not yet speak for a clear moral consensus in our society. To permit nonsuffering terminal patients the option of premature death by surgical intervention, as Freund notes, may be to trouble their consciences and impair their peace of mind in their last days, and in the present, perhaps transitional, stage of moral uncertainty, it could place undue strains on physicians who must either deliberately kill an innocent nonsuffering human being or else let another human being suffer a preventable death.

Procedures for allocating artificial organs have raised even more troublesome problems of justice than procedures for acquiring organs. The most dramatic example, widely discussed in the late 1960s and early 1970s, is the artificial kidney (renal dialysis equipment). Technical developments are expected to reduce the cost of hemodialysis so that one day the scarcity that generated the problem may disappear, but it seems clear that problems of the same general kind will be with us far into the future.

Medical administrators must formulate criteria to govern the fair allocation of scarce organs, and in order to do this they must decide which characteristics of the needy are relevant for the purposes of justice, and the relative weights these characteristics have when compared with each other. After an initial screening phase, only those whose claims are worthy of serious consideration remain, but since the number in this group may exceed the number of kidney machines available, further selection will have to be made, possibly using more sensitive criteria applied in a less sweeping way. When all medical "likelihood of success" tests have been applied, general agreement on the requirements of justice ceases, and writers are divided into two contending groups. One faction would count as relevant at the stage of final selection still another set of considerations which, although they are somewhat heterogeneous, we can call "social worth factors." The other group is adamant in confining the remaining selection procedure to a random lottery. The controversial social worth factors include such matters as the number of dependents the applicant has (Rescher would give priority to the mother of four young children over a middle-aged bachelor), the probable value of future contributions by the applicant to the public good, (a forward looking factor), and the value of past services rendered to the public good (a backward looking, nonutilitarian factor). Writers who urge consideration of these factors argue that odious as such comparisons among sick people may be, selections do have to be made among them, otherwise they shall all perish, and since that is so, every nonarbitrary factor will be considered as carefully as possible by any decision maker who takes justice at all seriously. Social worth factors, they continue, are not arbitrary in any usual sense, in the way that skin color or sex, for example, would be. They are not "arrived at through will or caprice"; they are not despotic, unreasonable, unsupported, or capricious. A capricious decision is one "proceeding from some whim or fancy," and there is nothing whimsical or fanciful in the consideration of dependent children, future contributions, and gratitude for past services. Indeed, to ignore the reasons one does have and decide instead on random grounds, or no grounds at all, is itself to be arbitrary, and hence unfair. In supporting the "future contribution factor," Rescher argues that a hospital is a trustee for the public interest, and in allocating its lifesaving resources in that role it must look to a "return on its investment" for all society. As for the "past services rendered factor," he argues, "It would be morally indefensible of society in effect to say: 'Never mind about services you rendered yesterday: it's only the services to be rendered tomorrow that count with us today'" (Rescher, 1969, p. 179).

On the other side, it is often argued that appraisals of contributions, past or present, inevitably involve subjective preferences; they require comparisons of incommensurable factors; they do not lend themselves to quantification or exact measurement. Moreover, they are likely to embody majority value judgments and discriminate against nonconformists and dissidents. Perhaps bohemians and hobos are not as deserving of material rewards as are more industrious and productive folk, but it is quite another thing to say that they have a weaker claim to remain alive. Indeed, at the Renal Dialysis Center in Seattle, predominantly lay committees evaluated the social worth of applicants by considering among other things such factors as scout leadership, civic activities, and even church attendance, which shows how "social worth" comes to be interpreted as middle-class respectability.

The opponents of social worth criteria point out that, in every other area of social affairs where innocent human life must be deliberately sacrificed, "the governing standard is not the merit, or need, or value of the victim, but equality of worth as a human being" (Freund, 1971, p. 281). Thus, whenever we can save some but not all, the only morally tolerable method of decision is to choose among them on a random basis. Only by constructing some sort of lottery can we avoid making the judgment that some human lives are more worth saving than others, that the "natural and inalienable right to life" is subject to differences of degree. The rationale of the lottery of life and death in emergency situations is that it is the only way of assuring from the start that there will be equal opportunity for survival among the imperiled, with absolutely no favoritism for any of them. Randomness is not to be confused with arbitrariness.

To be arbitrary is to act for no reason, or no relevant reason. To resort to randomness when there is no reason to do so is to be arbitrary. But to choose randomly for a reason, namely, to avoid giving preference to some individuals when others have presumably equal claims, is to perform an action for which there is a clear and compelling rationale, and that is the very opposite of arbitrariness.

Nicholas Rescher has proposed a compromise between the advocates of social worth evaluations and the adherents of the principle of the equality of human life. Rescher maintains that there is no precise way of assigning relative weights to the various factors he deems relevant. Within rather wide limits, any number of weight assignments are equally reasonable. He then proposes the following point-rating system as one out of a number of equally acceptable possibilities. Five factors are allowed relevance: (1) the relative likelihood of success, (2) life expectancy, (3) family dependence, (4) future contributions, (5) past services rendered. The purely medical factors (1) and (2) are together assigned a weight equal to the nonmedical factors (3), (4), and (5). At this point, Rescher concedes the desirability of introducing an element of chance. The five-criteria point system is used to reduce the pool of plausible applicants to a number somewhat larger than the number of machines available for allocation. Then the final selection is made by means of a lottery. Among the advantages Rescher cites for introducing randomness at the final stage of selection is that it would be psychologically easier for the rejected patient and those who will miss him to know that his rejection was due to bad luck purely, rather than deliberate human choice, and that it "relieves administrators of the awesome burden of ultimate and absolute responsibility" (Rescher, 1969, p. 184).

JOEL FEINBERG

[*For further discussion of topics mentioned in this article, see the entries:* ETHICS, *article on* UTILITARIANISM; *and* RIGHTS. *This article will find application in the entries* BLOOD TRANSFUSION; DEATH AND DYING: EUTHANASIA AND SUSTAINING LIFE; HEALTH CARE, *articles on* RIGHT TO HEALTH-CARE SERVICES *and* THEORIES OF JUSTICE AND HEALTH CARE; HEALTH POLICY, *article on* HEALTH POLICY IN INTERNATIONAL PERSPECTIVE; HEART TRANSPLANTATION; KIDNEY DIALYSIS AND TRANSPLANTATION; LIFE-SUPPORT SYSTEMS; ORGAN DONATION, *article on* ETHICAL ISSUES; ORGAN TRANSPLANTATION; POPULATION ETHICS: ELEMENTS OF THE FIELD, *article on* ETHICAL PERSPECTIVES ON POPULATION; POPULATION POLICY PROPOSALS; POVERTY AND HEALTH, *article on* POVERTY AND HEALTH IN INTERNATIONAL PERSPECTIVE; RACISM; RATIONING OF MEDICAL TREATMENT; *and* SUICIDE. *Also directly related are the entries* FOOD POLICY; FUTURE GENERATIONS, OBLIGATIONS TO; GENETIC ASPECTS OF HUMAN BEHAVIOR, *article on* PHILOSOPHICAL AND ETHICAL ISSUES; INFANTS, *article on* INFANTICIDE: A PHILOSOPHICAL PERSPECTIVE; LAW AND MORALITY; *and* RISK.]

BIBLIOGRAPHY

ARISTOTLE. *Nicomachean Ethics*. Translated by Martin Ostwald. Indianapolis: Bobbs-Merrill, 1962, bk. 5.

BEECHER, HENRY K. "Scarce Resources and Medical Advancement." *Experimentation with Human Subjects*. Edited by Paul A. Freund. New York: George Braziller, 1970, pp. 66–104.

BRANDT, RICHARD B. "The Concept of Rights." *Ethical Theory*. Englewood Cliffs, N.J.: Prentice-Hall, 1959, pp. 434–441.

CHILDRESS, JAMES F. "Who Shall Live When Not All Can Live?" *Soundings* 53 (1970): 339–355.

DANIELS, NORMAN, ed. *Reading Rawls: Critical Essays on "A Theory of Justice."* New York: Basic Books, 1975.

DEL VECCHIO, GIORGIO. *Justice: An Historical and Philosophical Essay*. Translated by Lady Guthrie. Edited by A. H. Campbell. Edinburgh: University Press, 1952.

FEINBERG, JOEL. *Social Philosophy*. Englewood Cliffs, N.J.: Prentice-Hall, 1973, pp. 55–119.

FREUND, PAUL A. "Ethical Problems in Human Experimentation." *New England Journal of Medicine* 273 (1965): 687–692.

———. "Organ Transplants: Ethical and Legal Problems." *Proceedings of the American Philosophical Society* 115 (1971): 276–281.

FUCHS, VICTOR R. *Who Shall Live?* New York: Basic Books, 1975.

FULLER, LON L. "The Case of the Speluncean Explorers." *Harvard Law Review* 62 (1949): 616–645. Reprint. *The Problems of Jurisprudence*. Edited by Lon L. Fuller. Brooklyn, N.Y.: Foundations Press, 1949, pp. 2–27.

HUME, DAVID. "Of Justice." *An Inquiry Concerning the Principles of Morals* (1777). LaSalle, Ill.: Open Court, 1960, sec. 3.

JONSEN, ALBERT R., and HELLEGERS, ANDRÉ E. "Conceptual Foundations for an Ethics of Medical Care." *Ethics of Health Care*. Edited by Laurence R. Tancredi. Washington: National Academy of Sciences, 1974, pp. 3–20.

MILL, JOHN STUART. "On the Connection Between Justice and Utility." *Utilitarianism* (1863). Indianapolis: Bobbs-Merrill, 1971, chap. 5.

MORRIS, HERBERT. "Persons and Punishment." *Monist* 52 (1968): 475–501.

NOZICK, ROBERT. *Anarchy, State, and Utopia*. New York: Basic Books, 1974.

OUTKA, GENE. "Social Justice and Equal Access to Health Care." *Perspectives in Biology and Medicine* 18 (1975): 185–203; *Journal of Religious Ethics*, no. 1 (1974), pp. 11–32.

"Patient Selection for Artificial and Transplanted Organs." *Harvard Law Review* 82 (1969): 1322–1342.

PERELMAN, CHAIM. *The Idea of Justice and the Problem of Argument*. Translated by John Petrie. New York: Humanities Press, 1963.

Pieper, Josef. *Justice.* Translated by Lawrence E. Lynch. London: Faber & Faber, 1957.
Ramsey, Paul. *The Patient as Person.* New Haven: Yale University Press, 1970.
Ratcliffe, James M., ed. *The Good Samaritan and the Law.* Garden City, N.Y.: Doubleday & Co., 1966. See especially "Law, Morals, and Rescue," by Anthony M. Honoré, pp. 225–242.
Rawls, John. *A Theory of Justice.* Cambridge: Harvard University Press, Belknap Press, 1971.
Rescher, Nicholas. "The Allocation of Exotic Medical Lifesaving Therapy." *Ethics* 79 (1969): 173–186.
———. *Distributive Justice: A Constructive Critique of the Utilitarian Theory of Distribution.* New York: Bobbs-Merrill, 1966.
Sidgwick, Henry. "Justice." *The Methods of Ethics* (1893). 7th ed. New York: Dover, 1907, bk. 3, chap. 5.
Thomas Aquinas. *Summa Theologiae* II–II 57–122. Translated by the Fathers of the English Dominican Province as "Treatise on Prudence and Justice." *Summa Theologica: First Complete American Edition.* 3 vols. Vol. 2: *Second Part of the Second Part, QQ 1–189 and Third Part, QQ 1–90, with Synoptical Charts.* New York: Benziger Brothers, 1947, II–II, questions 57–122, pp. 1431–1704.
Vlastos, Gregory. "Justice and Equality." *Social Justice.* Edited by Richard B. Brandt. Englewood Cliffs, N.J.: Prentice-Hall, 1962, pp. 31–72.
Williams, Bernard. "The Idea of Equality." *Philosophy, Politics and Society.* 2d series. Edited by Peter Laslett and W. G. Runciman. Oxford: Blackwell; New York: Barnes & Noble, 1962, pp. 110–131.

KIDNEY DIALYSIS AND TRANSPLANTATION

When all other therapies have failed to control the course of their disease, patients in the end-stage of chronic renal failure have two modes of treatment available to them that may prolong their lives: hemodialysis (the use of a machine to perform the blood-purifying functions of the kidney), and a kidney transplant. The clinical use of renal dialysis began in the early 1940s when a pioneering version of this machine was first tried on a few patients who were in acute kidney failure. The first series of human kidney transplants, in nine patients, was carried out from 1951 through 1953 without achieving long-term survival. Then, in 1954, a transplant involving identical twins marked the first successful clinical trial. In 1960 the invention of a cannula-shunt apparatus made possible long-term or chronic dialysis for patients with end-stage kidney disease. And in the early 1960s the development of immunosuppressive chemotherapy began enabling physicians to grapple with what is still the major medical problem in transplantation, the body's rejection of "foreign" transplanted tissue.

According to statistical data gathered in 1975, in the United States alone more than 8,800 kidney transplants were performed from 1951 through June 1973, and by July 1974 approximately 12,000 patients were being maintained on chronic dialysis. Thus, over the course of the past thirty years, dialysis and transplantation have developed in therapeutic efficacy, and they have been used to sustain the lives of thousands of people. Nevertheless, these entwined forms of treatment are still fraught with medical ambiguities and ethical dilemmas for medical scientists, health professionals, renal patients and their families, the lay public, and the body politic.

Many of the medical and ethical issues associated with renal transplantation and dialysis are common to most areas of therapeutic innovation, particularly those which involve critically or terminally ill patients. Among these issues are: the often stressful and conflicting double roles played both by the physician who is also a clinical investigator, and the patient who is also a research subject; problems of uncertainty that arise from current limitations in medical knowledge and technique; ambiguities about the stage of development of a relatively new treatment—that is, how "experimental" or "therapeutic" it is; concerns about the ethical and legal requirements of human experimentation, notably, the difficulties of obtaining truly informed, voluntary consent from a patient-subject, of striking a proper balance between the potential benefits and risks to such a subject, and of protecting his or her integrity and privacy; decisions about the allocation of scarce material and nonmaterial resources (including human organs) that must be made at individual, institutional, and larger societal levels; and finally, the personal and social implications of the medical commitment to use extraordinary means to prolong the lives of terminally ill patients.

In addition to the problems shared with other forms of therapeutic innovation, dialysis and renal transplantation have contributed to the development of certain more distinctive issues. These include questions about the relative "quality of life" that dialysis (in a hospital, at a proprietary center, or at home) and transplantation (of a live or a cadaveric kidney) have to offer; the "gift of life" significance of these modes of treatment—the transcendent meaning that givers and receivers can experience both in the dialysis and transplant situations, and the mutual tyranny that these ways of exchanging life for death can also impose on those who participate in them; the justification for inflicting a major surgical injury on a live donor in order to help a dying recipient; and the operational definition of death that is proper as well as necessary to use when cadaveric organs are transplanted. Paradoxically, now that in the United States most financial costs for dialysis and renal transplantation are borne by the government (U.S. Public Law 92-603, 1972), these treatments still present equality and equity problems, which are forerunners of difficulties that will probably become more salient and widespread in the American health-care system of the future.

This article will focus on some of the socioethical phenomena that accompany dialysis and renal transplantation, particularly in the United States, with special emphasis on the experiment–therapy dilemmas they entail, their gift-exchange aspects, and both the quality of life and allocation of scarce resources questions that they pose.

Experiment–therapy dilemmas

Despite the thousands of patients who have undergone dialysis and/or transplantation since these procedures were introduced, the question of where they fall on the experiment–therapy spectrum remains a core ambiguity, of import to both physicians and patients. For the question of how experimental or how routinely therapeutic a given procedure is judged to be has far more than semantic meaning. The definition of a treatment's status plays a major role in physicians' decisions about the circumstances under which that treatment may justifiably be used, and about the types of patients on whom it may properly be tried. Physicians find it difficult to define the overall experimental–therapeutic status of procedures like dialysis and renal transplantation. The types of criteria that are employed, sometimes implicitly, include quantitatively expressed probability estimates of the patient's medical prognosis with a transplant or dialysis, and qualitative judgments about the type of life that one or the other treatment may offer the patient, however long or short his period of survival.

A host of uncertainties about how well or poorly a given patient may do with a transplant or with dialysis, medically and sociopsychologically, contribute to the difficulty of establishing quantitative and qualitative criteria by which to locate these treatments on the experiment–therapy continuum. A cardinal example of the uncertainties that confront transplant physicians and their patients is whether the kidney recipient will be able to "walk the therapeutic tightrope": to prevent his body's rejection of the transplanted kidney, the recipient must continually take doses of immunosuppressive drugs that vastly increase his susceptibility to severe, massive, and often lethal kinds of infection. The course of the chronic dialysis patient also is difficult to predict with accuracy. For example, there are few explicit medical criteria for predicting how a patient will respond to dialysis, physiologically and chemically, or how he will cope with the psychological and social stresses that accompany life-dependency on an artificial kidney machine.

Because of such uncertainties, dialysis and transplantation continue to be utilized only for patients in the terminal stages of chronic renal failure, when all other treatments have been exhausted. Thus, although numerous physicians contend that transplantation and dialysis have become "accepted" clinical procedures, they remain less than "routine" treatments in the "ordinary practice of medicine." As the Advisory Committee to the Renal Transplantation Registry acknowledged in its 11th Registry Report, 1972, for example, "the exact place of transplantation in the care of end-stage renal failure has yet to be defined."

Gift-exchange aspects

A particularly complex and pivotal set of issues inherent to transplantation are those associated with the fact that it constitutes a gift of such magnitude and moment. A vital organ is donated by one individual in order to "give life" to another person who is terminally ill. Like other forms of gift exchange, organ donation is implicitly structured and regulated by a set of norms. These are the same "symmetrical" and "reciprocal" norms that Marcel Mauss identified in his classical monograph *The Gift:* the obligations to give, to receive, and to repay. Failure to live up to any of these expectations produces social strains that affect the giver, the receiver, and those associated with them.

The dynamics of organ gift-exchange can be especially intricate in the case of renal transplantation. Because the kidneys are paired vital organs, in contrast to the heart or liver, the loss of one kidney is not immediately life-threatening. Because one kidney can safely do the work of two, the medical profession, and the larger society of which it is part, permit live as well as cadaveric renal transplants to take place.

In situations where a live renal transplant is contemplated, the members of the prospective recipient's family face strong normative pressure to offer a gift of a kidney. This is partly because a live transplant from a donor who is genetically related to the recipient (his parents, siblings, or children), and who is a "good tissue match," has a better medical prognosis than a cadaveric transplant. The rejection reaction engendered by all transplants except those between identical twins is less likely to occur rapidly, and with a severity that jeopardizes the functioning of the implanted organ. Furthermore, the honor, intimacy, and generosity of the family and of each of its members are symbolically involved in the individual and collective willingness of these close kin to give of themselves to their dying relative in this supreme, life-sustaining way. Because the biomedical and psychosocial pressures on prospective live family donors are so compelling, medical teams involved in the renal transplantation situation have felt obliged to devise "gatekeeping" mechanisms that protect relatives against an excessively coerced or self-coerced gift of a kidney.

The terminally ill patient who is offered such a gift is subject to complementary normative pressures to accept the live kidney from the candidate-donor. To refuse the organ transplant implies a rejection of a gift of life, of the person who proffered it, and of the family unit to which the donor and recipient jointly belong.

If the live transplant does take place, then, in keeping with the paradigm of Mauss, there is a sense in which the donor–recipient relationship takes on a debtor–creditor dimension. The donor has made the utmost gift; the recipient has received something that is inherently unrepayable. Under some circumstances, the extraordinary meaning of what has been interchanged may reinforce the solidarity of the donor and recipient so that their mutual self-esteem is enhanced without jeopardizing their autonomy. But it is also possible that the essentially unreciprocal gift that links them may bind them one to the other in a mutually tyrannical way. The donor may hover over the person and life of the individual who has received his kidney, and the recipient may feel that, because he can never hope to repay the gift of life that he has received, he must allow or even encourage this.

As the foregoing implies, the complex impact of the triple norms of gift-exchange makes it difficult for transplant teams, patients, and their families to predict and evaluate the psychological, social, and moral consequences that a live kidney transplant will have for the donor, the recipient, and their kin.

Although dialysis does not involve offering and accepting a vital bodily part in the same literal sense that a kidney transplant does, its life-giving implications are as momentous. Whether dialysis is conducted in a medical center or at home, it entails the continuous exchange of life for death through the donation of time, energy, skill, and concern by the persons who help run the kidney machine and attend the dialysis patient. In the case of home dialysis, this exchange is all the more remarkable, because it requires and permits a lay person (usually the patient's spouse) to assume an unprecedented amount of medical responsibility for operating a complex life-support system. As a consequence, especially home dialysis confronts patients with the problem of receiving and reciprocating a recurrent gift of life.

Quality of life

As already suggested, the quality of life that dialysis and transplantation offer to persons with end-stage kidney disease is a matter of deep and constant concern to the medical professionals, patients, and families involved. As with many other medical interventions, there are few reliable predictive measures that accurately forecast how well a patient will respond to these modes of treatment. Moreover, there is no ready set of criteria to define and measure a "good" or "bad" quality of life: different individuals and social groups have their own indices by which they judge how "successfully" a person's life is being extended by a procedure like transplantation or dialysis.

Numerous transplant and dialysis patients "do well," leading what they and others consider to be active, rewarding lives, particularly when contrasted with their pretreatment existence. Other patients fare less well. But, irrespective of the ostensible differences in adjustment that may characterize them, all transplant and dialysis patients face certain stresses that may significantly undermine the quality of their prolonged lives.

The potential "gift of life" burdens that a kidney transplant can impose are preeminent among these stresses. In addition, the renal transplant recipient may find it difficult to tolerate the medical uncertainties surrounding the constant threat that his body will reject the transplanted kidney, and his increased vulnerability to grave infections, as a consequence of the immunosuppressive drugs that he is taking to forestall such a rejection reaction. The unpredictability of an eventual rejection, and the ever-present possibility that, for unknown reasons, his original renal disease may redevelop in his transplanted kidney, confront the recipient with the stress of knowing that someday he probably will have to return to life on the dialysis machine and to the prospect of being a candidate for another transplantation. He must also deal with the psychosocial as well as the biomedical side effects of the immunosuppressive drugs he is receiving: For example, cosmetic changes in his appearance, due to high doses of steroids, may alter his self-image and his presentation of self.

Many patients undergoing long-term dialysis seem to experience even greater strain than the recipients of transplants. Among the difficulties they report are the severe dietary restrictions that they are required to observe, the general state of enervation to which they are prone, the sexual impotence that some patients develop, and, above all, the degree to which they feel that they are "unnaturally tied to a machine." For some patients and their families, the kidney machine becomes a major, anthropomorphic presence in their existence, a presence that they characteristically feel is as "monstrous" as it is "miraculous." Perhaps the best indicator of how oppressed by their life with the machine many dialysis patients and their families feel is the almost millenarian expectation with which they look forward to their "deliverance" through the medium of a kidney transplant, though they have enough expert knowledge to recognize that even the best-matched live kidney transplant will not provide "salvation."

The quality of life questions that chronic hemodialysis raises are further complicated by the unresolved debate about the advantages and disadvantages of in-center versus at-home dialysis that is still occurring among different groups of physicians, nurses, and technicians administering the procedure. All agree that home dialysis is far less costly, in the strictly financial sense. The strongest proponents of home dialysis contend that, in addition to its economic assets, it frees patients and their families from the vise of triweekly, six-to-eight-hour visits to a hospital or center for treatment. And they insist that virtually all patients can be taught successfully to manage dialysis at home. Other medical professionals are more skeptical about the general applicability of home dialysis. Some medical teams conducting dialysis in large cities, with populations who are ethnically and socioeconomically heterogeneous, claim that home dialysis is too risky or difficult for many patients whom they treat, because of their family situations, cultural traditions, personal life-styles, and/or the physical layout and facilities of their residences. Some of the same physicians, nurses, and technicians are also impressed by the degree of stress that the responsibility for conducting dialysis at home creates for the patients, their spouses, and other close kin, even when they are comparatively well suited and equipped to attempt it. What makes it all the more difficult to appraise the assets and liabilities of different types of dialysis is the element of self-fulfilling prophecy that seems to influence the outcome of this procedure. Evidence suggests that those medical teams that are the most convinced that home dialysis is an efficacious and desirable mode of treatment are also the ones that obtain the best results with it.

From a certain point of view, no matter how well a patient seems to be adjusting to life with dialysis or a transplant, the fact remains that he is chronically dying. For the treatment that the patient is receiving is at once extraordinary and a "half-way technology," superimposed on an end-stage kidney disease that cannot be cured or even ameliorated at the present stage in the development of medicine. Because this is so, when a transplant patient who has lost one or more donor kidneys through rejection, or a dialysis patient who is struggling with his machine-dependent life, decides that dying is a lesser evil than the treatments available to him, the medical team is faced with a peculiar set of moral and metaphysical questions. The most common ways that patients on dialysis signal their desire to die are by requesting their physicians to stop treatment or by engaging in overtly or covertly suicidal behavior, such as pulling out their cannula-shunt or going on dietary binges. In this situation, if the medical team decides to accede to the patient's wish to discontinue his life by tapering off or not initiating further treatment, are they legitimizing and facilitating suicide? Are they performing euthanasia? Or are they merely allowing a terminally ill patient to die a "natural" death?

Allocation of scarce resources

The experiment–therapy dilemmas, the gift-exchange phenomena and the quality of life issues related to dialysis and renal transplantation all bear upon the allocation of scarce resources problems that these treatments also present. The most fundamental resource allocation questions concerning these procedures are: In the light of their present phase of development, the prevalence of end-stage kidney disease that they are designed to treat, and the net balance of suffering, reprieve, and fulfillment with which they imbue the lives of dialysis and transplant patients and their families, to what extent ought monies, equipment, technical competence, personnel, time, hospital space, vital organs, concern, anguish, and hope to be invested in these particular ways of keeping terminally ill persons alive? And what proportion of such inherently scarce, precious resources should be allotted to dialysis and transplantation as compared to life-prolonging therapies for other catastrophic diseases; the treatment of less grave, more "ordinary" medical conditions; the implementation of preventive medicine procedures of various sorts; the furtherance of medical knowledge through medical research; and the education of new generations of health professionals? On the macroscopic level, the issue becomes: How much should our society invest in the health–illness–medicine sector taken as a whole, as compared with other needed and valued activities?

The case of hemodialysis and renal transplantation is a particularly illuminating one in these regards. For, with the passage of U.S. Public Law 92-603 in 1972, financial coverage under Medicare was extended to include most of the treatment costs of dialysis and kidney transplants for most patients. Ostensibly, this constituted a felicitous solution to the overwhelming financial problems with which end-stage renal patients and their families were formerly confronted. But, in fact, the law has not resolved many of the scarce resources questions associated with these procedures and has helped to create some new ones. For example, passage of the law has intensified uncertainty and debate over the projected long-range costs of treating large numbers of patients with terminal kidney disease, and whether our society can afford to defray them. It has created a problem of equity by singling out persons suffering from renal disease for special financial coverage while many other equally expensive catastrophic illnesses have been ignored. Furthermore, the act of making financial help equally available to persons with end-stage kidney disease has increased the pressure on physicians not to deny any such patients access to dialysis and transplantation, no matter what the medical, psychological, or social contraindications may be. Is the most just and proper system of allocating scarce medical resources, then, one that entails no patient selection or triage whatsoever? This is an issue that has become more acute and unsettling for medical teams caring for renal patients since U.S. Public Law 92-603 was passed. It converges with the problems of quality of life and of intervention in the human condition already confronting medical professionals involved in the dialysis–transplantation enterprise. Finally, although the law provides economic resources for patients with terminal kidney disease, it does not eliminate other types of scarcity. The sheer availability of monies does not command the services of the many highly trained persons, the specialized facilities and equipment, particularly dialysis machines and beds, and the abundant supply of donor kidneys that are requisite for hemodialysis and renal transplantation.

In our opinion, dialysis and kidney transplantation have received so much public and professional attention not only because they represent dramatic medical advances, but also because of the crucial ethical and existential issues that they have evoked. Together, they constitute a paradigm of some of the basic questions of value and belief with which modern Western society is currently wrestling.

RENÉE C. FOX AND JUDITH P. SWAZEY

[*Directly related are the entries* HEART TRANSPLANTATION; ORGAN DONATION; ORGAN TRANSPLANTATION; *and* RATIONING OF MEDICAL TREATMENT. *For a description of the dialysis process and further discussion of the ethical issues, see* LIFE-SUPPORT SYSTEMS. *See also:* CHRONIC CARE; DEATH AND DYING: EUTHANASIA AND SUSTAINING LIFE, *article on* ETHICAL VIEWS; JUSTICE; LIFE, *article on* QUALITY OF LIFE; MEDICAL ETHICS, HISTORY OF, *section on* EUROPE AND THE AMERICAS, *article on* EASTERN EUROPE IN THE TWENTIETH CENTURY.]

BIBLIOGRAPHY

ABRAM, HARRY S. "The Psychiatrist, the Treatment of Chronic Renal Failure, and the Prolongation of Life." *American Journal of Psychiatry* 124 (1968): 1351–1358.

———; MOORE, GORDON L.; and WESTERVELT, FREDERIC B. "Suicidal Behavior in Chronic Dialysis Patients." *American Journal of Psychiatry* 127 (1971): 1119–1204.

BLAGG, C. R.; HICKMAN, R. O.; ESCHBACK, J. W.; and SCRIBNER, B. H. "Home Dialysis: Six Years' Experience." *New England Journal of Medicine* 283 (1970): 1126–1131.

CHILDRESS, JAMES F. "Who Shall Live When Not All Can Live?" *Soundings* 53 (1970): 339–355. Reprint. *Readings on Ethical and Social Issues in Biomedicine.* Edited by Richard W. Wertz. Englewood Cliffs, N.J.: Prentice-Hall, 1973, pp. 143–153.

CRAMOND, WILLIAM A. "Renal Transplantation: Experiences with Recipients and Donors." *Seminars in Psychiatry* 3 (1971): 116–132.

DUKEMINIER, JESSE, JR. "Supplying Organs for Transplantation." *Michigan Law Review* 68 (1970): 811–866.

FELLNER, CARL H. "Selection of Living Kidney Donors and the Problem of Informed Consent." *Seminars in Psychiatry* 3 (1971): 79–85.

———, and SCHWARTZ, SHALOM H. "Altruism in Disrepute: Medical versus Public Attitudes toward the Living Organ Donor." *New England Journal of Medicine* 284 (1971): 582–585.

FOX, RENÉE C., and SWAZEY, JUDITH P. *The Courage to Fail: A Social View of Organ Transplants and Dialysis.* Chicago: University of Chicago Press, 1974.

FRIEDMAN, ELI A., and KOUNTZ, SAMUEL L. "Impact of HR-1 on the Therapy of End-Stage Uremia." *New England Journal of Medicine* 288 (1973): 1286–1288.

KAPLAN DE-NOUR, A., and CZACZKES, J. W. "Emotional Problems and Reactions of the Medical Team in a Chronic Hemodialysis Unit." *Lancet* 2 (1968): 987–991.

KEMPH, JOHN P. "Psychotherapy with Donors and Recipients of Kidney Transplants." *Seminars in Psychiatry* 3 (1971): 145–158.

MCKEGNEY, F. PATRICK, and LANGE, PAUL. "The Decision to No Longer Live on Chronic Hemodialysis." *American Journal of Psychiatry* 128 (1971): 267–274.

MAUSS, MARCEL. *The Gift: Forms and Functions of Exchange in Archaic Societies.* Translated by Ian Cunnison. Glencoe, Ill.: Free Press, 1954.

MOORE, FRANCIS D. *Transplant: The Give and Take of Tissue Transplantation.* New York: Simon & Schuster, 1972.

Renal Transplant Registry, Advisory Committee. "The 11th Report of the Human Renal Transplant Registry." *Journal of the American Medical Association* 226 (1973): 1197–1204. Reports by the Human Renal Transplant Registry and the American College of Surgeons/National Institutes of Health Organ Transplant Registry are published yearly in the *Journal*.

SIMMONS, ROBERTA G.; HICKEY, KATHLEEN; KJELLSTRAND, CARL M.; and SIMMONS, RICHARD L. "Donors and Non-Donors: The Role of the Family and the Physician in Kidney Transplantation." *Seminars in Psychiatry* 3 (1971): 102–115.

STARZL, THOMAS E. *Experience in Renal Transplantation.* Philadelphia: W. B. Saunders, 1964.

SWAZEY, JUDITH P. "The Scribner Dialysis Shunt: Ramifications of a Clinical-Technological Innovation." *The Management of Health Care.* Edited by William J. Abernathy, Allan Sheldon, and Coimbatore Prahalad. Cambridge, Mass.: Ballinger, 1975, pp. 229–252.

THORWALD, JURGEN. *The Patients.* Translated by Richard Winston and Clara Winston. New York: Harcourt Brace Jovanovich, 1972.

United States, Department of Health, Education, and Welfare. *Treatment of Chronic Kidney Failure: Dialysis, Transplantation, Costs, and Need for More Vigorous Efforts.* Report to Congress by Comptroller General of United States. Washington: General Accounting Office, 1975.

WOLSTENHOLME, GORDON E. W., and O'CONNOR, MAEVE, eds. *Ethics in Medical Progress.* Ciba Foundation Symposium. Boston: Little, Brown & Co., 1966. Retitled *Law and Ethics of Transplantation.* Ciba Foundation Blueprint. London: J. & A. Churchill, 1968.

L

LABELING IN MENTAL ILLNESS
See MENTAL ILLNESS, *article on* LABELING IN MENTAL ILLNESS: LEGAL ASPECTS.

LATIN AMERICA
See MEDICAL ETHICS, HISTORY OF, *section on* EUROPE AND THE AMERICAS, *article on* LATIN AMERICA IN THE TWENTIETH CENTURY.

LAW AND MEDICINE
See LAW AND MORALITY; TECHNOLOGY, *article on* TECHNOLOGY AND THE LAW; ABORTION, *article on* LEGAL ASPECTS; DEATH, DEFINITION AND DETERMINATION OF, *article on* LEGAL ASPECTS OF PRONOUNCING DEATH; DEATH AND DYING: EUTHANASIA AND SUSTAINING LIFE, *article on* PROFESSIONAL AND PUBLIC POLICIES; GENETICS AND THE LAW; HUMAN EXPERIMENTATION, *article on* SOCIAL AND PROFESSIONAL CONTROL; INFANTS, *article on* PUBLIC POLICY AND PROCEDURAL QUESTIONS; INFORMED CONSENT IN HUMAN RESEARCH, *article on* ETHICAL AND LEGAL ASPECTS; INFORMED CONSENT IN THE THERAPEUTIC RELATIONSHIP, *article on* LEGAL AND ETHICAL ASPECTS; INSTITUTIONALIZATION; MEDICAL MALPRACTICE; MENTAL ILLNESS, *article on* LABELING IN MENTAL ILLNESS: LEGAL ASPECTS; ORGAN DONATION, *article on* LEGAL ASPECTS; PRIVACY; REPRODUCTIVE TECHNOLOGIES, *article on* LEGAL ASPECTS; RESEARCH POLICY, BIOMEDICAL; STERILIZATION, *article on* LEGAL ASPECTS.

LAW AND MORALITY

Bioethical problems are often discussed in legal as well as in moral contexts. Lawyers, as well as ethicists, are concerned with such questions as abortion, euthanasia, experimentation upon human beings, among others. This is not surprising; the law is seriously concerned with protecting such basic rights as the right to life, bodily integrity, and privacy, and these are the rights involved in these ethical questions.

The overlap between law and morality has been a source of the substantial debate about the relation between law and morality, which is not confined to the bioethical context. The debate is best divided into two main issues, although the discussion of these issues often overlap: (1) What, if any, bearing does the moral status of a rule have on its status as a law? (2) To what extent, if any, should the legal system be used to enforce moral perspectives?

Moral status and legal status

Natural law traditions. Western legal thought has been dominated by a natural law tradition. There are many variants of this tradition, and the differences among them will be discussed below; what they have in common is a belief in a body of laws governing all people at all times and in a source for those laws other than the customs and institutions of a given society. Such beliefs are frequently accompanied by the additional beliefs that no societies are authorized to create laws that conflict directly with natural laws, and that any such conflicting laws may therefore be invalid. In short, the natural law tradition asserts the existence of a set of laws whose status as laws is based upon their moral status.

The beginning of this tradition is in the ancient world. Aristotle (384–322 B.C.) drew a distinction between that part of justice which is natural and should have the same force everywhere and that part which is legal and has its force only in those places where it has been adopted by the people who live there. That distinction was developed extensively by the Stoics; they emphasized two further points about natural justice, viz., that it is based upon right reason and that it is in agreement with nature. Cicero (106–43 B.C.), whose legal writings are based upon the Stoic tradition, emphasized the claim that no legislation can alter the validity of natural laws, which remain binding on all people. Some of these ideas were incorporated into Roman law, and the later Roman lawyers probably identified *jus naturale* (the philosophical notion of natural law) with *jus gentium* (a system of laws that had developed in the Roman world and governed the relations among free men independently of their nationality). This identification strengthened the idea of natural law as universal law.

These classical ideas gave rise to a number of different natural law traditions, the two most important of which are the religious tradition—which culminated in the writings of St. Thomas Aquinas (1224–1274)—and the secular tradition—which is exemplified in such authors as Hugo Grotius (1583–1645) and John Locke (1632–1704).

St. Thomas Aquinas defined a law as an ordinance of reason for the common good, promulgated by him who has the care of the community. He then distinguished four types of laws: eternal laws, natural laws, human laws, and divine laws. The eternal laws are laws promulgated by God on the basis of divine reason. The natural laws are the eternal laws in man—man's natural inclinations toward the proper acts and ends. In short, St. Thomas postulated an eternal, unchanging set of laws implanted by God in man and knowable from reason on the basis of man's natural inclinations. Human laws are valid only insofar as they do not conflict with those divinely promulgated, unchanging laws. Valid human laws either are conclusions drawn from the basic natural laws or are determinations of details left undetermined by the natural laws.

The natural law theories of Hugo Grotius and John Locke also contain theological references, and St. Thomas does emphasize the rational basis of natural law. Nevertheless, Grotius and Locke represent a different tradition of natural law, one that puts more emphasis on natural law as rationally derivable rather than on natural law as divinely ordained law. In addition, their tradition—especially in the writings of Locke—puts great emphasis on the natural law's protection of natural rights, rights that all human beings have independently of the state and its laws. Locke explicitly drew the conclusion that a state loses its legitimacy insofar as its laws are in violation of natural rights, such as the right to life or liberty.

These natural law traditions continue to influence discussions about the relation between the law and bioethics. Writers influenced by the theological version of the natural law tradition continue to argue that any valid laws must be in conformity with the divinely ordained natural law. Thus, many Roman Catholic writers argue that there must be civil laws prohibiting abortion, euthanasia, and the mutilation (including sterilization) of innocent persons, because those procedures are in conflict with the natural law. To those who would object that this is an illegitimate use of the law to enforce morality, these writers reply that it is of the very nature of legitimate law that it prohibits such activities. Writers influenced by the ideas of natural rights thinkers like Locke continue to argue that no purported law is legitimate if it allows the violation of the basic rights of human beings. This type of argumentation is particularly prevalent in countries such as the United States where the courts possess the ability to declare laws unconstitutional when they infringe upon basic human rights. Recent decisions (e.g., *Griswold v. Connecticut,* in which the U.S. Supreme Court ruled that a Connecticut law prohibiting the use of contraceptives is unconstitutional) have suggested that jurists are prepared to extend those rights to include ones not explicitly mentioned in the written Constitution. This type of argumentation has also been used by German jurists as the basis for denying the efficacy of various decrees and laws promulgated during the Nazi regime.

Legal positivism. The natural law tradition has not been universally accepted. There has also been a long tradition of thinkers, dating back to antiquity, who have insisted that the only laws that exist are those adopted by a given society, that there is no necessary connection between the legal status of a law and its moral status. Defenders of this position, the position of legal positivism, are not opposed to the moral criticism of individual laws and of whole legal institutions; positivists often advocate changes in the law on the basis of moral considerations. All that the positivists insisted on was that an immoral law, however much it should be changed, remains valid as a law until repealed by the appropriate social mechanisms.

Jeremy Bentham (1748–1832) and John Austin (1790–1859) were the two most influential proponents of this view, although earlier figures like Jean Bodin (1530–1596) and Thomas Hobbes (1588–1679) should also be mentioned. The basic thesis of positivism has often been conflated with another of Austin's theories, the imperative theory of law (law is the command of the sovereign). Since this latter theory has not survived critical examination, it is crucial to distinguish it from the basic theme of positivism: that what the law is is a separate question from what the law ought to be.

Some legal positivists have taken their view to mean that laws must be obeyed no matter how immoral they are. But the most important positivists, Bentham and Austin, have clearly argued that there are circumstances in which an immoral law should be violated despite its status as a law; this weakens, of course, the force of the claim that a law retains its status as a law despite its immorality.

In any case, legal positivists insist that questions about the relation between law and morality must be settled independently of questions about what the law is. The legal status of a rule is independent of its moral status. This leads us, therefore, to the second of our questions: When ought the law to be used to enforce certain moral positions?

Use of the legal system to enforce morality

The law is clearly used on some occasions to enforce moral viewpoints. We believe that murder is wrong and that the coercive mechanism of the law should be used to prevent murders. However, even if we believe that euthanasia is wrong or that one should come to the aid of others in distress, should the law be used to enforce these beliefs?

The liberal approach. John Stuart Mill (1800–1873), in his classic, *On Liberty*, advocated the liberal answer to that question—that society should use the coercive mechanisms of the law only to prevent actions that harm someone other than the performer or another who has consented to the performance of the action. In other words, Mill argued that the social enforcement of morality was inappropriate when only the agent or others who had consented would be harmed.

Mill's followers have therefore opposed the existence of laws creating "victimless crimes," among which must be included laws against suicide and voluntary euthanasia. They have approved of court decisions that allow rational adults to refuse medical treatment on religious grounds, even though the refusal would result in their dying. Finally, those who have denied fetal humanity have been advocates of legislation permitting abortion upon request.

A number of points must be kept in mind about the liberal position. (1) It does not require legislation prohibiting all actions that harm others. Whether there should be legislation will depend upon such factors as the existence of harmful side-consequences, the possibility of enforcement, etc. All that the liberal position entails is that such actions, because they harm others, are candidates for appropriate legal prohibition. (2) Actions that harm others may be prohibited legally even when the others consent if their consent is not efficacious (because, for example, the consenting party is a minor, is insane, or is coerced). This point is extremely important in connection with legislation governing medical experimentation. Consider, for example, the problem of experiments upon children, where the experiments are not primarily intended to aid in their therapy and where there are potential hazards. In such a case, consent of the children does not count, and the relevance of parental consent is unclear. Mill's principles would therefore allow for legislation enforcing some socially determined moral standards in this area. Similar difficulties arise for experimentation upon the insane and upon imprisoned criminals subject to subtle forms of coercion. (3) This liberal position is not identical either with the English common law tradition or with American constitutional law. Both allow for legal prohibitions that are unacceptable in the liberal framework. For example, the consent of the person killed in an act of voluntary euthanasia is, at least in theory, no defense against a charge of murder in either legal system.

Adherents of the liberal approach have in recent years expanded upon it and modified it in a number of ways. One question that has received considerable attention is that of determining whose consent is efficacious. Our current understanding of mental illness makes it very difficult for us to accept a sharp dichotomy of those competent to consent and those incompetent, since there are many degrees of mental disturbance. What moral standards should society, therefore, impose through legislation in the treatment of the mentally ill, and to what extent can those decisions be left to the therapist and the consenting patient? Another question that has received considerable attention is the extent to which society can legitimately use the law to temporarily prevent an individual from carrying out certain decisions in the hope that the individual will change his mind. If, for example, we were morally opposed to voluntary euthanasia, could we within the liberal framework legally require a period between a request for voluntary euthanasia and the actual implementation of that request? Still a third question that has received considerable attention is the legitimacy of legally imposing certain positive moral duties. Mill was primarily concerned with challenging the legitimacy of laws prohibiting immoral actions; it is unclear how he would have dealt with good samaritan laws—laws that would require, for example, trained medical personnel to come to the aid of accident victims. Would such laws that require positive actions and not mere forbearances be a legitimate legal enforcement of morality? A final question that has received considerable attention is whether society can pass laws designed to prevent harm to fetuses, animals, or future generations. If it could, this would markedly change the liberal attitude toward laws governing abortion, experimentation upon animals, and conservationist and ecological measures.

Arguments for legislating morality. From its very beginning, the liberal tradition has had its critics. Writers in the natural law tradition objected, of course, to the liberal presupposition that the moral and legal status of rules could be separated. But even some of those who agreed with positivism have argued that there is a wider scope for legislating morality than the scope allowed for by Mill.

James Fitzjames Stephen (1829–1894), in his influential *Liberty, Equality, and Fraternity* (1873), argued that one of the purposes of both the criminal and civil law is to promote and encourage virtue while discouraging vice. Stephen conceded that certain areas of morality could not be dealt with by the law because the relevant laws could not be enforced without destroying privacy and individual rights; he claimed, however, that there are many areas of morality that should be treated by the law despite Mill's strictures. This point of view has been extended by Lord Devlin, the distinguished English jurist. Devlin contends that the continued existence and strength of a society require a common moral code. There is, therefore, a social interest in the preservation of such a code, and it is at least sometimes appropriate to enforce part of the code through the use of the law. Devlin limits his conclusions to cases where this enforcement of morality will not violate human rights.

Devlin applied this approach to English abortion legislation in the 1960s. He argued that the severe punishment of the illegal abortionist cannot be justified on the grounds that such an abortionist poses a threat to the health of the mother, since that threat exists primarily because the abortionist's activities are illegal. Instead, such laws can only be explained and justified as an attempt by society to protect its fundamental views on sexuality and on human life.

Additional considerations on enforcing morality. There are then a number of differing systematic approaches to the question of which aspects of morality should be enforced legally. In addition to those systematic approaches, various authors and courts have suggested additional considerations that must be weighed in deciding whether or not to legally enforce moral standards. These considerations are usually, although not always, advanced as reasons for not enforcing the standards in question. Among the most prominent of the considerations are the following.

1. *Respect for differing views in a pluralistic society.* In the 1973 U.S. Supreme Court's discussion of abortion statutes in *Roe* v. *Wade*, the Court suggested that legislation enforcing a moral viewpoint is inappropriate when those who are experts in the relevant area disagree as to the legitimacy of that viewpoint. This principle is in keeping with a wider movement against legislating disputed moral positions. A number of important considerations support this mode of thought. To begin with, people seem to have a right to follow their own conscience rather than to be compelled to follow the conscience of the rest of society. Moreover, there are tremendous detrimental consequences for a society when many of its citizens feel that the law is being

used to coerce them into following the moral views of others. Such considerations are even more important in societies where there are substantial moral disagreements among the citizens.

Such weighty considerations have to be balanced, of course, against considerations in favor of legally enforcing the moral position in question. Even if a pluralistic society should forgo passing many laws out of deference to the views of those who think that the actions that would thereby be illegal are not wrong, there remain some cases in which the force of the law should be applied because of the evil consequences of the actions in question. If such actions produce very harmful results and infringe the rights of a sufficiently large number of individuals, then the possible benefits that may be derived from passing and enforcing a law preventing those actions may well override the considerations against passing such a law. It is, of course, extremely difficult to balance these considerations.

2. *Respect for privacy.* There are laws that cannot be enforced without infringing the privacy of the citizens involved. It has been suggested (by the U.S. Supreme Court in *Griswold v. Connecticut*) that such laws are illegitimate just because of the inability to enforce them in an acceptable fashion. For that reason, the Court declared unconstitutional a Connecticut law prohibiting the use (and not merely the production) of contraceptive devices. It has also been argued that laws regulating the patient–doctor relation are inappropriate because they can be enforced only by the state's entering into and examining a relation that must be private.

Once more, of course, these considerations have to be balanced against legitimate state needs and interests. We do have many legitimate laws that invade the privacy of the doctor–patient relation. Doctors must, for example, report gunshot wounds to the police. In short, privacy is not an absolute right, and there may be laws that ought to be passed in the interest of public safety, for example, even if enforcing them requires an invasion of privacy. All that can properly be said is that difficulty in enforcing these laws without invading the privacy of others is a weighty consideration against passing such laws.

3. *The side-consequences of passing such a law.* It is sometimes argued that certain moral positions ought not to be enforced legally just because the laws that codify them will be violated anyway and their surreptitious violation will lead to many tragic results. Thus, it has been argued that laws prohibiting abortion only result in women's seeking unsafe, illegal, and very dangerous abortions. Again, it has been argued that laws prohibiting voluntary euthanasia only result in surreptitious acts of voluntary euthanasia and in informal decisions to "let the patient die," acts and decisions that can be abused. It goes without saying that these considerations have to be balanced against the benefits, if any, of the laws in question.

Considerations 1 to 3 are reasons why certain actions should not be illegal even if they are immoral. There are, however, considerations for making actions illegal even if they are not immoral. Two deserve special notice.

4. *The difficulty of distinguishing between fraudulent and legitimate cases.* Suppose that there are no moral objections to voluntary euthanasia. Some have argued that it would still be wise legally to prohibit such killings because it is difficult to distinguish cases of honest requests from cases of consent obtained by subtle fraud or duress. Again, some have argued that, despite the moral permissibility of experimenting upon consenting adults, there should be laws prohibiting experiments conducted upon prison inmates, because one cannot really tell when the consent of such inmates is really voluntary.

5. *Slippery-slope arguments.* It is often argued that legalizing certain morally acceptable actions leads at a later date to irresistible pressures for legalizing immoral actions, and that the only way to avoid sliding down this slippery slope is to prohibit even the acceptable action. Thus, it has been argued that voluntary euthanasia should be illegal, even if morally acceptable, as a way of ensuring against the later legalization of involuntary euthanasia. Naturally, both of these factors must be weighed against the desirable results that might result from legalizing the morally acceptable actions.

Conclusion

It is clear, then, that there are no easy answers to questions about the relation between law and morality. There are strong considerations favoring legal positivism, but there are other considerations favoring a natural law doctrine. And even if one is a legal positivist, there are conflicting considerations that one has to weigh in deciding about the appropriate relation between one's moral code and society's legal code.

BARUCH A. BRODY

[*Directly related are the entries* NATURAL LAW; PRIVACY; *and* RIGHTS. *This article will find application in the entries* ABORTION, *article on* CONTEMPORARY DEBATE IN PHILOSOPHICAL AND RELIGIOUS ETHICS; ANIMAL EXPERIMENTATION, *article on* PHILOSOPHICAL PERSPECTIVES; DEATH AND DYING: EUTHANASIA AND SUSTAINING LIFE, *articles on* ETHICAL VIEWS *and* PROFESSIONAL AND PUBLIC POLICIES; FUTURE GENERATIONS, OBLIGATIONS TO; RIGHT TO REFUSE MEDICAL CARE; *and* SUICIDE. *For discussion of related ideas, see the entries:* CHILDREN AND BIOMEDICINE; INFORMED CONSENT IN MENTAL HEALTH; *and* ROMAN CATHOLICISM.]

BIBLIOGRAPHY

AUSTIN, JOHN. *The Province of Jurisprudence Determined* (1832).

DEVLIN, PATRICK. *The Enforcement of Morals.* London: Oxford University Press, 1965.

ENTREVES, ALESSANDRO PASSERIN D'. *Natural Law: An Introduction to Legal Philosophy.* London: Hutchinson's University Library, 1951.

FEINBERG, JOEL. "Legal Paternalism." *Canadian Journal of Philosophy* 1 (1971): 105–124.

FRIEDRICH, CARL JOACHIM. *The Philosophy of Law in Historical Perspective.* Chicago: University of Chicago Press, 1958.

FULLER, LON L. *The Morality of Law.* Storrs Lectures on Jurisprudence, 1963. New Haven: Yale University Press, 1964.

GIERKE, OTTO FRIEDRICH VON. *Natural Law and the Theory of Society, 1500 to 1800: With a Lecture on the Ideas of Natural Law and Humanity by Ernest Troeltsch.* Translated by Ernest Barker. Beacon Paperback, no. 50. Boston: Beacon Press, 1957.

GROTIUS, HUGO. *De jure belli ac pacis.* Vol. 2: *The Translation.* Translated by Francis W. Kelsey. Oxford: At the Clarendon Press, 1925.

HART, H. L. A. *Law, Liberty, and Morality.* The Harry Camp Lectures. Stanford: Stanford University Press, 1963.

———. "Positivism and the Separation of Law and Morals." *Harvard Law Review* 71 (1958): 593–629.

LOCKE, JOHN. *Two Treatises of Government.* Edited by Peter Laslett. Cambridge: At the University Press, 1960.

MILL, JOHN STUART. *On Liberty.* London: J. W. Parker, 1858.

PACKER, HERBERT L. *The Limits of the Criminal Sanction.* Stanford: Stanford University Press, 1968.

STEPHEN, JAMES FITZJAMES. *Liberty, Equality, Fraternity.* London: Smith, Elder & Co., 1873.

THOMAS AQUINAS. *Summa Theologiae.* I–II 90–95. *Basic Writings of Saint Thomas Aquinas.* 2 vols. Edited by Anton C. Pegis. New York: Random House, 1945, vol. 2, pp. 742–789. "Law." *Introduction to Saint Thomas Aquinas.* Edited by Anton C. Pegis. New York: Random House, Modern Library, 1948, pp. 607–650.

WILLIAMS, GLANVILLE. *The Sanctity of Life and the Criminal Law.* New York: Alfred A. Knopf, A Borzoi Book, 1957.

LEBANON

See MEDICAL ETHICS, HISTORY OF, *section on* NEAR AND MIDDLE EAST AND AFRICA, *articles on* CONTEMPORARY ARAB WORLD *and* CONTEMPORARY MUSLIM PERSPECTIVE.

LIFE

I. VALUE OF LIFE   *Peter Singer*
II. QUALITY OF LIFE   *Warren T. Reich*

I
VALUE OF LIFE

Introduction

The issue of the value of life lies behind many contentious issues in bioethics, from population control to euthanasia, and from the use of expensive lifesaving machines to the use of animals in research. A rational decision about whether to take strenuous measures to save the life of a seriously handicapped infant necessarily involves an estimate of the value of that infant's life. The same issue is raised by the economic question of how much a society ought to invest in new, extremely expensive medical techniques, as well as by experiments on animals that can be justified only if animal life is of decisively lower value than human life. Population policy must face the issue of the value of future, possible lives, while the value of embryonic-fetal life is relevant to discussions of abortion. In a more subtle way, some conception of the value of human life is implicit in almost every ethical discussion of life-and-death questions.

Despite its importance, the issue remains one of the most baffling in philosophical ethics, and fundamental disagreements are common. The present article does not attempt to set forth and explain all the major positions on this topic—for that would be too demanding a task—but to explore the implications of a variety of positions that can be taken on this question.

In all societies known to us, there have been some prohibitions on the taking of life (Westermarck). Presumably, no society can survive if it allows its members to kill one another without restriction. Just which beings are included in the circle of protection, however, is a matter on which societies have differed. Among many tribal societies, the only serious offense is to kill an innocent member of the tribe itself. In more sophisticated nation-states protection is generally accorded to all within the state's boundaries, although in some cases this protection has been denied to minorities. Nowadays most people grant, in theory if not in practice, that it is in ordinary circumstances wrong to kill any person. Today few attempt to defend, on moral grounds, the restriction of the circle of protection to a particular tribe, race, class, sex, or nation. The moral inadequacy of these forms of group egoism does not require detailed demonstration. We can therefore begin with the view that the life of every person has value, and consider whether this view needs to be supplemented in taking into account all human beings; all potential or possible people; all sentient beings; all living beings.

The value of a person's life

The word "person" is sometimes used, in popular discourse, as equivalent to "human being"; but for the purposes of ethics this is not strictly correct without further justification. "Person" had its origin in the Latin word for a mask used by an actor, and subsequently came to mean one who plays a role, or is an agent. Its current philosophical sense is, according to the *Oxford English Dictionary*, "a self-conscious or rational being"; and it is in this sense that the word will be used in this article.

In ordinary circumstances it is wrong to take a person's life; but *why* is it wrong? Is it because a person has a right to life? Is it because human life is sacred? Is it a matter of neither rights nor sanctity, but of utilitarian considerations? All these views deserve examination. The differences among them pose a problem that is not merely theoretical, since different positions have different practical outcomes.

The right to life. As long as people have talked of "rights" they have talked of a "right to life." For Thomas Hobbes, the right to preserve one's own life was the only right one had in the state of nature. John Locke added rights to liberty and property but retained the right to life in first place; and Thomas Jefferson gave it the same priority when he proclaimed the right to life, liberty, and the pursuit of happiness in the Declaration of Independence of the United States of America. If we recognize a doctrine of natural rights at all, it seems difficult to deny that there is a right to life, since without this right other rights would be pointless.

Additional support for the idea of a right to life may be found in the value we place on individual freedom and autonomy. There can be no greater violation of an individual's autonomy than the taking of his life without his consent.

This additional consideration leads to an important possible consequence of the idea of a right to life which has not always been recognized by its supporters. Generally, to say that we have a right means that anyone who infringes our right without our consent does wrong; but we are not obliged to insist on our rights. If I have a right to payment for services I have rendered you, I can, if I wish, waive my right and tell you to keep your money. By this account, it seems, paradoxically, possible to suggest that on the "right to life" view neither suicide nor voluntary euthanasia is a violation of this right. On the other hand, many hold that this right is "inalienable," meaning that it cannot, or ought not, be surrendered or transferred. Thus, there is much debate today on whether respect for individual freedom requires us to permit suicide or voluntary euthanasia and, if so, what the implications would be for the "right to life."

The sanctity of human life. Those who regard human life as sacred deny that we have the right to end our own life or that of others. This position can be found in Plato's *Phaedo* (62 b-d) and in doctrines attributed to Pythagoras; but it is today associated mainly with Christian writers, especially those of Roman Catholic persuasion. Thus Thomas Aquinas asserts that whoever takes his own life sins against God, in the same way that whoever kills a slave sins against the master to whom the slave belonged (*Summa Theologiae* II-II, 64, 5). Hence he considered suicide as evil as murder. This view—that God alone has the right to decide when we shall live and when we shall die—can still be seen in the frequent charge that advocates of euthanasia are attempting to "play God." Even if we accept the theological background of this charge, however, it appears difficult to sustain the one-sided view that we are playing God when we terminate life but not playing God when we try to keep alive someone who would have died without our intervention.

Since the idea of the sanctity of human life is really a doctrine about all humans, rather than persons as we have defined these terms, I shall postpone more detailed discussion until the appropriate section of this article.

Utilitarian considerations. Utilitarian philosophers have, until recently, given less attention to questions involving life and death than to other problems, to which their principle of maximizing the surplus of happiness over misery is more easily applied. It is not really clear, however, why, on a utilitarian view, it should be wrong to take an innocent person's life. On the utilitarian view, life itself has value only insofar as it is a prerequisite of happiness. Certainly, if we kill people at random, we will cause great unhappiness to the families and friends of those we kill. This may seem to be a mere side effect of the real wrongness of killing, but its importance should not be underestimated.

The utilitarian must also take into account another important side effect: the fact that the knowledge that some people have been killed will put others in terror of the same thing happening to them; but if we painlessly kill a hermit who has no contact with anyone else, and we do it in such a manner that no one can possibly find out about our deed, then the utilitarian considerations based on the effects of homicide on the living do not apply. What does the utilitarian say then?

The standard utilitarian answer would seem to be that, since a person's existence, in normal circumstances, contains a positive balance of happiness or pleasure over misery or pain, killing someone reduces the surplus of happiness in the universe.

There are, however, some objections to this answer. First, if we happened to know that the hermit was leading a miserable life, and that this state of affairs could not be expected to improve, it would not merely be allowable to kill him: It would be obligatory. The utilitarian principle requires involuntary euthanasia in these circumstances. The tough-minded utilitarian might accept this conclusion, adding only that it concerns a hypothetical situation and that, so far as actual cases are concerned, if a person desires to continue living, this must be taken as a good indication that his life contains a positive balance of happiness, however miserable he may seem to us.

A more fundamental objection to the standard utilitarian view is that it appears to put the decision not to procreate on a par with the decision to kill, since the outcome of both is one less happy person in the universe. Since the problem of the value of potential persons or possible people is not limited to utilitarian theory, however, I shall postpone discussion of this objection.

A curious variant of this standard version of utilitarianism is pessimistic utilitarianism. Pessimists maintain that life has, on the whole, more misery than joy. If we couple this belief with utilitarian ideas, we reach the conclusion that if everyone could be painlessly and instantly killed, this would be the right thing to do. It says a good deal for our optimistic temper that this position has not been taken very seriously in the present century.

The cost-benefit approach. No subject is so sacred that economists shrink from the attempt to measure it; and economists have tried to assess the value of life in monetary terms, in order to ascertain what level of resources it is rational for a society to put into such areas as exotic lifesaving therapy, public health, and road safety. Distasteful as it may be to mention dollars and human lives in the same sentence, in practice we cannot avoid relating the two. People may be fond of saying that a human life is "beyond price," but they would be less enthusiastic about a government that took this saying literally and devoted every dollar it had to keeping people alive, at the expense of education, art galleries, and national parks. On the other hand, governments may also spend too little, in purely economic terms, on such lifesaving measures as road signs and preventive medicine.

It must be admitted, though, that the cost-benefit approach sometimes gives an absurdly precise figure. What, for instance, are we to make of a report that tells us that the value of a human life in Britain in 1973 was £17,000? (Cooper and Culyer, p. 293). Not surprisingly, when we look behind the figures, the appearance of mathematical precision breaks down. It may be possible to reach a tolerable estimate of the costs of some of the side effects of death, like loss of productive output, but the more central question of the value of life to the individual whose life it is defies normal accounting procedures.

One approach is to use the price system and find out what people are prepared to pay to avoid death (Schelling). Of course, people will probably give all they have or can obtain to avoid certain death, since money is valueless to a dead person; but the question becomes more manageable if we consider expenditure that reduces the risk of death. Assume that the probability of dying inside one's own car as a result of an accident is 1 in 1,000, and that a newly designed car is available that eliminates this risk entirely. In order to afford it, however, we will, for the next ten years, have to give up such luxuries as taking our vacations in faraway places and dining at good restaurants. Would we buy the safe car? By asking a variety of questions of this type, it is possible to get some idea of how much value a person places on his own life.

When we deal with dreaded but very unlikely events, it is difficult to assess how much it is rational for us to spend to make the event still less probable, and there is evidence to suggest that people do not behave rationally when faced with these situations anyway. Certainly society as a whole is prepared to spend enormous sums in an effort to save the lives of particular, known individuals—miners trapped by a rockfall, for example—although it refuses to spend much smaller sums that would be statistically certain to save lives (though we could not tell whose life was saved), for example, by providing free routine screening of women for cancer of the cervix. Since each statistic is in fact an individual person, this attitude suggests that our society has an incoherent set of priorities so far as saving life is concerned. This incoherence becomes even more blatant if we take into consideration the millions of lives that affluent nations could save by increasing aid to poorer nations. If we were to judge by the amount most people contribute to famine relief or other forms of overseas aid, the average citizen of an affluent nation values the life of an Asian or African at rather less than a ticket to a football game or the theater.

The value of human life

With rare exceptions—among them Fletcher (1972), Tooley (1973), and Warren (1977)—writers on such topics as abortion and euthanasia do not distinguish the value of the life of a person from the value of the life of a human being; but, as we saw earlier, a common meaning of "person" today is that of a rational or self-conscious being, while some human beings (members of the species Homo sapiens) are neither actually rational nor self-conscious. Examples are human fetuses, newborn babies, and humans with severe brain damage. The non-equivalence of meanings of "person" and "human being" is further indicated by the fact that there are nonhumans, like chimpanzees, dolphins, and other mammals, who are more rational and self-conscious than very young or mentally deficient members of our own species. The question arises, then, whether a human life has some special value over and above that of the life of a nonhuman, even when the human being in question is not a person in the sense accepted here.

The issue has important practical consequences in two respects. First, when we are trying to decide whether to keep alive a human being who is not a person, should we place the same value on that life as we would on the life of a self-conscious, rational being, or should we regard it as more like one of the nonhuman animals whom in some respects it more nearly resembles? Second, should we treat chimpanzees and other animals in ways that we do not treat mentally defective members of our own species; for instance, perform medical experiments on them that we know will lead to a slow, unpleasant form of death? Our actual behavior in both these cases implies that the life of a human being is always more valuable than the life of a member of another species. But can this view be justified?

In those parts of the world influenced by Judeo-Christian thought the life of a human being is regarded as having unique value, irrespective of whether the human being is a person as I have defined the term in this article. True, there have been disputes among Christians about the status of the fetus; but from the moment of birth, at the very latest, the life of any being born of human parents has been held to be as sacrosanct as the life of a normal person.

The Judeo-Christian foundations for this position are clear. Human beings, alone among animals, are made in the image of God and have immortal souls. By possessing a spiritual nature, humans are linked with God and the angels, far above the highest nonhuman beings, which are merely corporeal, and which cease to exist when their bodies decay. The Christian doctrine is that all human beings have souls, and not merely those with self-awareness or other mental capacities. In their writings about abortion, it was the act of killing a being with an immortal soul that the Christian Fathers were especially concerned to condemn—not surprisingly, when one recalls that this act was thought to consign the victim to an eternity of either heaven or hell. For this reason alone they considered the distinction between human and nonhuman crucial for determining when life is sacred.

We have already noted the view of some that to kill a human is to usurp the role of God. In the first chapter of Genesis, however, it is said that

God granted man dominion over the other animals. This was taken to mean that God had given man the right to kill them and otherwise make use of them for his own ends (Passmore; Singer, *Animal Liberation*, chap. 5).

The legitimacy of drawing a sharp distinction between the value of the life of *any* human being and of any nonhuman has now become part of secular as well as religious ethics in Western nations and other countries influenced by Western ideas. But if the religious doctrines on which the Judeo-Christian position rests are rejected, it is hard to see on what basis the distinction can stand. The mere fact that a being is a member of our own species does not in itself seem to be a relevant moral consideration, any more than membership of our own tribe, race, or nation could be, in itself, a justifiable ground for discrimination within our species. If we are to hold that the life of a human deserves greater protection than that of a nonhuman, we must support our claim by reference to morally relevant characteristics that only humans have, but there do not appear to be any such characteristics, since, as we have seen, a chimpanzee, or even a pig, surpasses a month-old infant or a severely retarded human in respect of any characteristic that could reasonably be said to give the lives of normal adult humans special value. It therefore appears that the attribution of unique value, above that possessed by any nonhuman, to the lives of all human beings cannot be justified in nonreligious terms (Singer, *Animal Liberation*, chap. 1). One or two writers have attempted to argue that membership of the human species is a morally significant fact (Benn), but in general philosophers have not faced the difficulty of defending a moral distinction based on species alone.

The value of potential and possible persons

On the basis of the above considerations, it may be granted that the life of a human being with severe brain damage, lacking any degree of self-awareness or ability to communicate with others, is of no greater value than the life of a dog or monkey. Still, it may be argued, normal infants are in a quite different situation, since they are persons, potentially if not actually. So we must now confront the question postponed earlier: What is the value of a potential person?

The value of potential persons. In discussing the value of potential persons, it is helpful to separate the idea of a potential person from the idea of a possible person. I shall draw this distinction in the following manner. A normal newborn baby is not a person, but it has the potential to become one; we may say that it is, potentially, a person. If, on the other hand, a childless couple are discussing whether they shall conceive a child, then the person who will eventually exist if they decide in favor of so doing (and all goes according to plan) is a possible person. It is possible for that person to exist in the future, but there is at the time of their discussion no entity that can be said to have the potential to become that person.

The idea that a potential person should be treated with the same respect as an actual person does seem to explain some of our strong objections to infanticide. Many people, however, have much weaker objections—or no objections at all—to abortion; yet the potentials of the newborn infant and the fetus are very similar. Many would hold that neither is a person, and that both will become persons in the course of normal development. That is, both have the potential to become persons. To allow the potential person the same rights as the actual person therefore seems to rule out abortion, except, perhaps, when necessary to save the life of the mother. Perhaps one could argue on other grounds—for instance, a woman's right to control her reproductive processes—that abortion is less serious than infanticide; but the grounds for allowing abortion would have to be very serious if the fetus is considered to have the same status as a potential person that a newborn baby has.

It is not possible to explore this topic fully in the present article. It is worth noting, however, that the horror with which Western societies regard infanticide is by no means universally shared by other societies. The pioneering anthropologist Edvard Westermarck takes twenty pages, in his *Origin and Development of the Moral Ideas*, to list the different societies known to practice and approve of infanticide. The list covers every continent and a variety of cultures, from nomadic tribes to sophisticated city dwellers. The practice was common in Ancient Greece, and compulsory infanticide of deformed infants had the approval of Plato and Aristotle; under the Roman Empire, too, it had the support of leading thinkers, including the notably humanitarian Seneca. All this suggests that respect for potential persons is not a universal value; like respect for human life, it is strongest in countries that have been under Christian influence.

The value of possible persons. If we do value the lives of *potential* persons, we must also ask ourselves what value we place on the lives of *possible* persons. If we value potential persons highly because they will become actual persons, then it would seem that we ought to value the lives of possible persons highly too. On this view, to refrain from having a child would be to commit a moral offense comparable with infanticide—neither act kills an actual person, but both prevent the existence of an actual person at some future date (Hare).

Hardly anyone accepts the conclusion that we have an obligation to bring as many people as possible into the world. On the other hand, if we do not respect the lives of potential persons because we want to have as many actual persons as possible, why is a being's potential important at all? Why should the value of a being's life not depend on what it actually is, rather than what it potentially is? Michael Tooley has argued strongly against the belief that a being's potential gives it a right to life.

It is generally true of the views about the value of life that we have been considering that they refer only to people already in existence. They do not tell us whether the creation of additional life is also to be valued. Among the classical utilitarians, for instance, only Henry Sidgwick noticed that there is an issue here.

We saw earlier that the standard utilitarian reason why killing a person is wrong, independently of the pain the act may cause, is that it diminishes the amount of happiness in the universe. This implies that the ultimate utilitarian objective is the greatest total amount of happiness achievable; and one way of increasing the total is to increase the number of happy beings in existence, until we reach the point at which the decrease in the happiness of existing beings that would result from further additions to an already crowded world outweigh the increases to the total sum of happiness brought about by the creation of more beings. This implication is contrary to the common assumption that population policy should be aimed at producing the best life for a given nation, or for everyone alive, rather than the largest total amount of happiness that the nation or planet can sustain.

Some utilitarians have attempted to show that Sidgwick's view is not the only possible interpretation of utilitarianism on this issue; but perhaps the most notable work on this topic has been done by the Oxford philosopher Derek Parfit. Parfit, while recognizing the alarming implications of Sidgwick's view, has found serious difficulties with alternative proposals too (Narveson; Parfit).

Other theories give no better guidance in this area. To say that a person has a right to life, or that a person's life is sacred, seems to have little bearing on whether it is good to create additional life. These are deontological theories; they prohibit certain acts, acts of taking life, but they do not tell us whether life is in general a good thing, and if so, why. In this sense, these theories reduce to rules about when life may be taken and tell us little or nothing about the value of engendering new life.

The value of animal life

We have seen that, religious considerations apart, it is difficult to justify valuing the lives of human beings who lack such characteristics as rationality and self-awareness above the lives of animals. But what value does the life of an animal have?

In Western societies, nonhuman lives are very cheap. We kill animals for food, although we could feed ourselves more economically without so doing (Singer, *Animal Liberation,* chap. 4); we kill animals by the million in our laboratories, often for trivial purposes [see ANIMAL EXPERIMENTATION]; we kill millions more for their furs and skins, or just for the fun of killing.

Some Western writers have objected to these practices. From Pythagoras and Plutarch through Leonardo da Vinci to Tolstoy, Schweitzer, and George Bernard Shaw, vegetarians and antivivisectionists have struggled against the dominant "human chauvinist" ideology of the West; but they have always been a small minority.

In the East, animal life has generally been given greater consideration. Hinduism condemns the killing of animals, and strict Hindus are vegetarians. Buddhist doctrine takes a similar position, although Buddhists are permitted to eat the flesh of animals that others have killed, and this enables them to circumvent the vegetarianism that appears to be implied by their respect for animals. Jainism, on the other hand, is even stricter than Hinduism. For the Jain, to avoid killing anything at all is the highest ideal. Moreover, the Jain's regard for animal life extends to the positive act of assisting sick and injured animals; Jains have gone so far as to establish hospitals for animals.

In the face of such sharply conflicting traditions, it is difficult to decide how to value animal life. If, as suggested earlier, we cannot justify refusing to value the lives of severely retarded humans along with the lives of animals of similar capacities, we may have to question our practice of killing animals for food and in scientific experiments, since it is difficult to accept that we are morally entitled to kill retarded humans for these purposes.

When we compare the lives of animals (and severely retarded humans) with those of normal humans, or persons, it does seem that there could be reasonable grounds for preferring the latter, if a choice must be made. A utilitarian might base this preference on considerations like the more protracted anticipations of death that a being with greater foresight and self-awareness would have; the greater grief of friends and relations, who would be more sharply aware of what had happened, and would remember it longer, than the mate, mother, or child of a dead animal, would be another factor; so would the greater capacity for happiness of the human being, at least if we agree with John Stuart Mill that Socrates, even when dissatisfied, was happier than a pig satisfied. From other ethical standpoints, it could be argued that the great autonomy and capacity for moral personality of the normal human gives rise to a stronger claim to life.

These grounds for valuing the life of a person more highly than that of an animal are not, however, grounds for ignoring or discounting the value of an animal's life when we are not forced to choose between a person and an animal. Most of the characteristics on the basis of which we value the lives of persons are present to some degree in many animals: Animals can communicate with each other, recognize and form relationships with other members of their species, act in ways that reveal some ability to reason, suffer pain, and experience pleasure, and some of them even show the sentiments of sympathy for others that underlie the moral capacities in human beings. Some have, of course, argued that the animal behavior in question here can be adequately explained without inferring that animals possess, as humans do, the power of conceptual thought; but only the most thoroughgoing skeptic is likely to deny that animals can feel pain (Regan and Singer).

The value of life as such

So far we have been considering the value of specific forms of life, all of which have had important characteristics—like consciousness—in addition to the basic characteristic of life itself. We must now consider the question: Does life without these important characteristics have any value? Is there value in the life of something that is not conscious, and so cannot experience pleasure or pain or be aware of its own existence in any way? Plants, presumably, come into this category. (It has been claimed that plants are conscious, but these claims have not been substantiated under controlled laboratory conditions.) Is it wrong idly to pluck a blade of grass?

It is sometimes said that we cannot respect all life, since if we did, we would have to starve. This charge is valid only against an absolutist ethic that prohibits all taking of life for any reason. It is quite possible to hold, without risk of starvation, that all life has some value, and we ought never to take life except when necessary to preserve life. This appears to be the position taken by some varieties of Buddhism and, in the West, by the "reverence for life" ethic of Albert Schweitzer.

Nevertheless, the view that life itself, without even the capacity for consciousness, has intrinsic value is by no means self-evident, and it is not easy to see on what basis it can be defended. Schweitzer offered no real arguments for his ethical position, and Buddhists and Jains who hold this type of view have generally defended it on religious or quasi-religious grounds that carry little conviction for those who do not share their fundamental outlook.

A more commonsense view is that, while it may be wrong to kill a plant, the wrongness of so doing depends on the value of the plant for humans or other animals, rather than on any wrong done to the plant itself. It is difficult to argue that beings that cannot experience anything can have rights. For utilitarians, too, the only things of value are conscious experiences; thus, without consciousness, life would have no intrinsic value.

PETER SINGER

[*Directly related is the other article in this entry,* QUALITY OF LIFE. *This article will find application in the entries* ABORTION, *article on* CONTEMPORARY DEBATE IN PHILOSOPHICAL AND RELIGIOUS ETHICS; ANIMAL EXPERIMENTATION, *article on* PHILOSOPHICAL PERSPECTIVES; DEATH AND DYING: EUTHANASIA AND SUSTAINING LIFE, *article on* ETHICAL VIEWS; INFANTS, *articles on* ETHICAL PERSPECTIVES ON THE CARE OF INFANTS *and* IN-

FANTICIDE: A PHILOSOPHICAL PERSPECTIVE; LIFE-SUPPORT SYSTEMS; and RATIONING OF MEDICAL TREATMENT. For further discussion of topics mentioned in this article, see the entries: ETHICS, articles on DEONTOLOGICAL THEORIES and UTILITARIANISM; MENTALLY HANDICAPPED; PERSON; RIGHTS; RISK; and SUICIDE. For discussion of related ideas, see the entries: BIOETHICS; ENVIRONMENTAL ETHICS, article on QUESTIONS OF SOCIAL JUSTICE; and OBLIGATION AND SUPEREROGATION.]

BIBLIOGRAPHY

ADLER, MORTIMER JEROME. *The Difference of Man and the Difference It Makes.* Introduction by Theodore T. Puck. Based on the Encyclopaedia Britannica Lectures delivered at the University of Chicago, 1966. New York: Holt, Rinehart & Winston, 1967.

BARTH, KARL. *Church Dogmatics.* 4 vols. Vol. 3: *The Doctrine of Creation.* 4 pts. Edited by G. W. Bromiley and T. F. Torrance. Translated by A. T. Mackay, T. H. L. Parker, Harold Knight, Henry A. Kennedy, and John Marks. Edinburgh: T. & T. Clark, 1961, pt. 4.

BENN, STANLEY I. "Egalitarianism and the Equal Consideration of Interests." *Equality.* Edited by James Roland Pennock and John W. Chapman. Nomos: Yearbook of the American Society for Political and Legal Philosophy, no. 9. New York: Atherton Press, 1967, chap. 4, pp. 61–78. Papers of a meeting of the American Society for Political and Legal Philosophy and the American Society for International Law, Washington, April 1965.

COOPER, M. H., and CULYER, A. J., eds. *Health Economics.* Harmondsworth, England: Penguin Books, 1973.

DAWSON, R. F. F. "A Practioner's Estimate of the Value of Life." Cooper, *Health Economics,* pp. 336–356.

FLETCHER, JOSEPH. "Indicators of Humanhood: A Tentative Profile of Man." *Hastings Center Report* 2, no. 5 (1972), pp. 1–4.

———. *Morals and Medicine.* Boston: Beacon Press, 1960.

FRIED, CHARLES. *An Anatomy of Values.* Cambridge: Harvard University Press, 1970, chap. 12.

GLOVER, JONATHAN. *Causing Death and Saving Lives.* New York: Penguin Books, 1977.

GOODRICH, T. "The Morality of Killing." *Philosophy* 44 (1969): 127–139.

HARE, R. M. "Abortion and the Golden Rule." *Philosophy and Public Affairs* 4 (1975): 201–222.

HAYZELDEN, J. "The Value of Human Life." *Public Administration* 46 (1969): 427–441.

HENSON, RICHARD G. "Utilitarianism and the Wrongness of Killing." *Philosophical Review* 80 (1971): 320–337.

KOHL, MARVIN. *The Morality of Killing.* New York: Humanities Press, 1974.

LEACH, GERALD. *The Biocrats.* Rev. ed. Harmondsworth, England: Penguin Books, 1972. New York: McGraw-Hill, 1970.

LECKY, WILLIAM EDWARD HARTPOLE. *History of European Morals from Augustus to Charlemagne.* 2 vols. London: Longmans, 1869. Reprint. 2 vols. in 1. New York: George Braziller, 1955, chap. 4.

McCLOSKEY, H. J. "The Right to Life." *Mind* 84 (1975): 403–425.

NAGEL, THOMAS. "Death." *Nous* 4 (1970): 73–80. Reprint. *Moral Problems.* 2d ed. Edited by James Rachels. New York: Harper & Row, 1975, pp. 360–370.

NARVESON, JAN. "Moral Problems of Population." *Monist* 57 (1973): 62–86.

PARFIT, DEREK. "On Doing the Best for Our Children." *Ethics and Population.* Edited by Michael Bayles. Cambridge: Schenkman, 1975, pp. 100–117.

PASSMORE, JOHN. "The Treatment of Animals." *Journal of the History of Ideas* 36 (1975): 195–218.

RAMSEY, PAUL. "The Morality of Abortion." *Life or Death: Ethics and Options.* By Edward Shils, Norman St. John-Stevas, Paul Ramsey, P. B. Medawar, Henry K. Beecher, and Abraham Kaplan. Introduction by Daniel H. Labby. Seattle: University of Washington Press, 1968, pp. 60–93.

REGAN, THOMAS, and SINGER, PETER, eds. *Animal Rights and Human Obligations.* Englewood Cliffs, N.J.: Prentice-Hall, 1976, pt. 1.

ST. JOHN-STEVAS, NORMAN. *Life, Death and the Law.* London: Eyre & Spottiswoode, 1961, chaps. 6, 7.

———. *The Right to Life.* London: Hodder & Stoughton, 1963.

SCHELLING, T. C. "The Value of Preventing Death." Cooper, *Health Economics,* pp. 295–321.

SCHWEITZER, ALBERT. *The Philosophy of Civilization.* 2d rev. ed. Vol. 2: *Civilization and Ethics.* Translated by C. T. Campion. London: A. & A. Black, 1946.

SIDGWICK, HENRY. *The Methods of Ethics.* 7th ed. London: Macmillan, 1907, pp. 415–416.

SINGER, PETER. *Animal Liberation.* New York: New York Review of Books, Random House, 1975.

———. "A Utilitarian Population Principle." *Ethics and Population.* Edited by Michael Bayles. Cambridge: Schenkman, 1975, pp. 81–99.

TOOLEY, MICHAEL. "Abortion and Infanticide." *Philosophy and Public Affairs* 2 (1972): 37–65. Revised reprint. "A Defense of Abortion and Infanticide." *The Problem of Abortion.* Edited by Joel Feinberg. Belmont, Calif.: Wadsworth, 1973, pp. 51–91.

WARREN, MARY ANNE. "Do Potential People Have Moral Rights?" *Canadian Journal of Philosophy* 7 (1977): 275–289.

———. "On the Moral and Legal Status of Abortion." *Monist* 57 (1973): 43–61.

WESTERMARCK, EDVARD ALEXANDER. *The Origin and Development of the Moral Ideas.* 2 vols. 2d ed. London, New York: Macmillan, 1912–1917, vol. 1, chaps. 14–21; vol. 2, chaps. 35, 44.

WILLIAMS, GLANVILLE LLEWELYN. *The Sanctity of Life and the Criminal Law.* Foreword by William Warren. James D. Carpenter series, 1956. New York: Alfred A. Knopf, 1957.

II

QUALITY OF LIFE

Quality of life is a term that has gained widespread use in both common parlance and ethical discourse only since the 1950s.

In bioethics there are four major questions for which quality-of-life considerations are relevant:

(1) To what extent should quality-of-life considerations have a bearing on biomedical decisions regarding the sustenance, termination, or shortening of human life, e.g., in questions dealing with abortion and euthanasia? (2) In what way should quality-of-life factors be used to justify the prevention of human or possibly human bodily life, e.g., through contraception, sterilization, refraining from sexual intercourse, or abortion? (3) Do quality-of-life considerations justify interventions which might alter the nature of man, e.g., through selective breeding utilizing artificial insemination, cloning, etc.? (4) What behaviors or policies do quality-of-life considerations require or permit in dilemmas of environmental health, public health, population priorities, and so forth? This last question overlaps into another major area characterized by a search for quality-of-life indicators for developing prudent public policy as regards the overall quality of life—biomedical, psychosocial, ecological, economic, cultural, etc.—for human beings in society (Gitter and Mostofsky; U.S., Environmental Protection Agency, p. iii; Rescher, pp. 60, 72–73). Some of the literature in this enormous area dealing with human environments offers an examination of conceptual problems, values, and principles (Michalos; Bunge; Baier; Rescher; McCall; Wingo; Gerson) that provides a helpful analytic framework for understanding quality-of-life arguments in biomedical decision making and policymaking.

The four issues mentioned above raise questions about different kinds of behavior, and the debates surrounding them tend to appeal to distinctive clusters of values and principles. This article will deal with the first question—pertaining to the continuation or termination of human lives in a biomedical context—because this question has been the most prominent quality-of-life issue in bioethics and provides models for other issues of its kind.

The quality-of-life problem to be discussed below is usually understood in the following way. Modern biomedicine can frequently sustain a human life only in an impaired or defective condition without the ability to cure or substantially improve the basic defect. Furthermore, it is possible to diagnose defects, e.g., *in utero*, to an extent not previously possible. With more attention now focused on the unremedied condition of the individual patient, it is increasingly being argued that life need not be sustained or may be directly terminated if the quality of life is not satisfactory. The ethical question is: What principles should govern the use of a quality-of-life concept in a possible justification for the termination or continuation of human life, especially the life of the defective, diseased, comatose, seriously ill, or dying patient of any age?

The following four sections will (1) offer conceptual tools for understanding quality-of-life arguments and clarify some of the meanings and uses of quality-of-life principles in discussions of the value of life; (2) identify the distinctive characteristics of a quality-of-life ethic in terms of moral theory and give examples of typical quality-of-life arguments in the area of abortion and euthanasia; (3) offer a critique of the quality-of-life ethic together with some alternate theories for taking account of human qualities in biomedical ethics; and (4) make some concluding remarks.

Quality of life: preliminary clarifications

Although in popular presentations the quality-of-life ethic is sometimes based on slogans, imprecise theory, and false dichotomies, it represents an increasingly significant insight and therefore deserves careful analysis. Yet because it does not enjoy a long history of precise and critical use, any analysis of it must be somewhat tentative.

Proponents of a quality-of-life ethic claim that modern biomedicine necessitates a radical shift in ethics. A quality-of-life ethic must replace a sanctity-of-life ethic, it is argued, principally because a sanctity-of-life ethic takes an absolutist position on preserving physical life, is more concerned with quantity of life or mere survival of bodily life, and fails to take seriously into account the quality of that life. On the other hand, many claim that the quality-of-life ethic denies that human life is a sacred value ("A New Ethic"; Fletcher, "Ethics and Euthanasia," "The 'Right' to Live"). These claims will be examined in the following paragraphs.

Quality of life as morally normative. Quality or qualities of life may be used in a *descriptive*, an *evaluative*, or a *morally normative* statement.

A *descriptive* statement makes an observation about the presence or nature of a quality, characteristic, or property. Since every characteristic can be called a quality, every human individual has qualities of life or a quality of life. The term quality of life is used descriptively in a medical context to refer to a broad set of medical concerns (American Medical Association). When used to describe a patient's present or future condition, the term quality of life is morally neutral.

An *evaluative* statement about quality of life indicates that some value or worth is attached to the characteristics of a given individual or to a kind of human life. This means that a quality of life is good or desirable or valuable. When a quality is valued in this way, the life may be appreciated, desired, or judged worth living as something sacred, beautiful, or beneficial, but this does not necessarily imply that actions terminating or supporting life are right or wrong. Some call this a nonmoral value judgment (Frankena, pp. 28–32). For example, one may value a physically mobile life; but the question remains unanswered as to what actions are permissible in reference to a life that is less than normally mobile.

A *morally normative* (or prescriptive) statement on quality of life entails a moral judgment on valued qualities of life, which involves saying that certain norms indicate which attitudes toward or ways of treating human bodily life are morally good or bad, right or wrong. Quality of life has its real significance for bioethics, then, only when it is used in a morally normative judgment that states whether one ought to support and protect life on the basis of a perception of human qualities.

Much of the medical literature shows an ethical preference for quality-of-life standards in deciding which lives should be saved (e.g., Shaw); sociological studies offer descriptive accounts of how certain qualities actually enter into medical decisions in life-and-death cases (Crane); and some physicians have developed specific schemes for drawing the line between handicaps or defects that do not meet quality-of-life standards and those that do (cf. Reich). But from an ethical perspective the quality-of-life question is not adequately treated until one gives moral reasons why a certain quality or qualities *should* be decisive in a morally normative statement (rule, norm, or maxim) indicating what sort of action is permissible or impermissible for making life-and-death decisions in conflict situations.

Some problems in the quality-of-life and sanctity-of-life distinction

The sacredness of human life. The broad issue is the endeavor to explain and account for a moral respect-for-life principle which states that it is wrong, whether absolutely or prima facie, to terminate or shorten or not to sustain a human life. (Sanctity of life is a term that can be used in a generic sense interchangeably with respect for life. In this article, however, it is used more narrowly to refer to a distinctive ethical argument.) A quality-of-life ethic, like a sanctity-of-life ethic, is simply an attempt to articulate a respect-for-life principle, or a judgment on the moral value of life. Differences between the two approaches arise in accounting for the basis of the morally normative judgment and the limits and exceptions to the rules.

Furthermore, a quality-of-life ethic may actually hold that a human life is sacred, in the sense that it is respected evaluatively and normatively, provided certain qualities are present. Yet this is essentially what the sanctity-of-life position commonly affirms, namely, that a human life is respected and valued because of certain important qualities, such as the transcendent or spiritual qualities of human nature (cf. Clouser). Consequently, the claim that a sanctity-of-life ethic affirms the sacredness of human life whereas a quality-of-life ethic does not is a misleading dichotomy: The sacredness of human life does not adequately distinguish the two approaches; yet they differ significantly insofar as the sanctity-of-life ethic is inclined to hold that, as a rule, all or virtually all human lives are sacred, i.e., are morally deserving of protection and support.

The sacredness of human biological life. Some have claimed that the sanctity-of-life ethic respects bodily or physical life, whereas a quality-of-life ethic respects the quality of human life (Fletcher, "The 'Right' to Live," p. 46). In fact, however, most theories that support the moral judgment of the sanctity of human life do not place morally normative value on mere biological life in isolation from other characteristics or qualities. If we ask what makes acts of terminating or shortening human bodily life wrong, many ethical systems supporting a sanctity-of-life principle, whether on grounds of moral philosophy or of theology, would appeal to other reasons which entail some further human quality or qualities (cf. Frankena). For example, the ethical systems of Judaism and Christianity may appeal to the extrinsically oriented quality of being related to the God who created humans in his image, or to the intrinsic characteristic of being endowed with a spiritual and immortal soul (ibid.) or simply to physical life as the indispensable condition for human values and valuing (Gustafson, 1971, p. 140).

Certainly it is possible to hold a sanctity-of-life principle simply because of what human bodily life is without relying on the value of other aspects or qualities of human life. This principle has been maintained, for example, on grounds that it is self-evidently right on nonreligious premises, as well as on the basis of the sanctity of all life—plant, animal, and human—a position held in Hinduism and Jainism in the East and in the ethics of Albert Schweitzer in the West (Frankena, pp. 47, 55–57).

But a sanctity-of-life ethic and a quality-of-life ethic are falsely dichotomized if one fails to see that, generally speaking, the sanctity-of-life principle depends on moral judgments concerning qualities of human life that go beyond mere biological life to embrace mental, social, religious, and other qualities. Thus, the dichotomy between the value of biological life versus qualitative life is seriously misleading. More accurately, the quality-of-life and sanctity-of-life approaches differ, in the major ethical theories, on the moral significance of one quality compared with another.

The absoluteness of the sanctity-of-life norm. The distinctiveness of the quality-of-life ethic is said to inhere in its rejection of the *absoluteness* of the rules that characterize the sanctity-of-life ethic ("A New Ethic"; Fletcher, "The 'Right' to Live," pp. 46–47). In fact, however, a belief in the sanctity of life can take either an absolute or a nonabsolute form. Certainly the majority of ethical systems that maintain a sanctity-of-life ethic hold that it is only generally or presumptively wrong, or wrong prima facie, or wrong *ceteris paribus* not to preserve and protect human life, "but not always actually or finally wrong, since other moral considerations may still make it right in certain circumstances" (Frankena, p. 33). For example, in many sanctity-of-life perspectives, a patient may be allowed to die when higher obligations or values of greater significance intervene; and abortion is permitted for a variety of reasons, e.g., when good and proportionate reasons are the ones intended, as by the application of a principle of double effect. Consequently, the distinctiveness of the quality-of-life ethic is not found in the nonabsoluteness of its principles.

Quality of life versus quantity of life. The terms quantity of life and quality of life sometimes are contrasted in the notion that it is not quantity of life that is to be valued but quality (Fletcher, "The 'Right' to Live," p. 48). The "quality, not quantity" slogan contains both misleading and helpful connotations. When quantity means mere bodily life without regard for any qualities other than bodiliness, the contrast is misleading, for, as we have seen, it is a judgment on valued qualities, not quantitative bodiliness, that lies at the root of both quality-of-life and sanctity-of-life positions.

Concern for quantity of life could also mean concern about the length of an individual's life, or about the number of individual lives saved or lost (Frankena, p. 27). Those whose respect for life is expressed in sanctity-of-life considerations may frequently (though not always) feel obliged to struggle to extend the length of a human's life and to save more rather than fewer lives. But their moral reason will usually revolve about values deeper than temporal duration and numbers of lives. However, there is a way in which the quantity–quality contrast helps to explain the distinctiveness of a quality-of-life ethic. Quality of life can be contrasted with mere quantity in the sense of population growth (Wingo, p. 3). When the quantity of lives to be sustained in society surpasses the resources required to preserve them, a quality-of-life ethic would be more inclined to claim that those individuals who enjoy a higher quality of life have a greater claim on the means for supporting life (cf. Ehrlich and Ehrlich). It should be noted, however, that this quantity–quality dilemma puts a strain on both the quality-of-life and the sanctity-of-life ethic. In resolving the question of allocation of resources, even some who hold that all human beings have a right to life (a typical sanctity-of-life position) will acknowledge that the quality of those contending for sustenance may be a relevant factor in deciding which lives shall be saved.

Quality of life versus qualities of life. A quality-of-life ethic places morally normative significance on the satisfaction or nonsatisfaction experienced through some more or less clearly defined quality of life. In this context, quality of life tends to be used as an aggregate term, referring to the overall condition of life (cf. Bunge, p. 151). The global quality may be that of either an individual person or of society as a whole: Both concepts are used in value-of-life arguments discussed in biomedical ethics.

Overall quality of life is sometimes identified with a single, normatively valued quality, while for others the quality of life embodies or depends on the presence of a cluster of particular qualities (Baier, pp. 64–65). In the former case, a single-quality criterion is used, and in the latter a multiple-quality criterion, in judging whether life should be protected and preserved.

Furthermore, the aggregate quality of life does not necessarily refer to either a fixed amount of quality or to a high or excellent quality of life. As a global term, it refers to a rank on some evaluative scale, and "it may be a rank that deserves to be called a fault rather than an excellence" (Baier, p. 63). Since there are degrees of quality, the notion of "losing the quality of life" may simply mean losing *some* of life's quality or qualities.

Quality-of-life arguments

The preceding section indicated that a variety of perceptions and uses of quality-of-life concepts are relevant to judgments on the value of human life and explained some difficulties in identifying the uniqueness of a quality-of-life ethic. The present section will pinpoint the distinctive character of a quality-of-life ethic in the larger framework of ethical theory and give examples of a quality-of-life ethic in arguments concerning selective abortion and euthanasia.

Quality-of-life ethic: a question of moral theory. We have seen that quality (or qualities) of life can be used in opposing moral arguments. What is distinctive about a quality-of-life ethic is the moral theory on which it depends. A quality-of-life ethic can be defined as one that (1) depends on an ethical theory of consequentialism; (2) assigns relative and unequal value to human lives on the basis of the possible consequences of variable qualities; and (3) espouses the norm that the conservation and protection of human life are not required or do not carry an overriding obligation, unless the directly experienced qualities or the qualities expected to be experienced actually invest that life with sufficient value.

In contrast, a sanctity-of-life ethic (1) typically depends on a deontological theory of ethics (though other moral theories can be used to support its norms); (2) assigns equal value to human lives regardless of their condition, usually on the basis of inherent values; (3) frequently presupposes the general moral orientation favoring a strong moral belief that human life should be treasured and respected; and (4) espouses the presumptive norm that human life ought to be sustained and protected and that life ought not to be taken without a very serious justification (Clouser; Callahan, 1969).

The basic debate is one of ethical methodologies, principally in terms of consequentialist versus deontological ethics (Frankena, pp. 46–47).

According to the *consequentialist* method—on which quality-of-life arguments depend—moral claims concerning value of life are based solely or principally on predictable qualitative consequences for the individual and/or for others whose interests are involved in the case. For example, actions related to the termination or sustenance of the life of a defective individual would be justified by the achievement of desirable consequences ("benefits" such as happiness) and the avoidance or minimization of undesirable consequences ("harms" such as suffering).

Among a variety of consequentialist views, two major types can be singled out. In a *personalistic* perspective, a consequentialist method can be used to justify actions by reference to the anticipated benefits to an individual. Stated positively, the quality-of-life norm might be: "Preserve the patient's life if it is expected to be a meaningful life." A negative formulation is the maxim of medical ethics: "Do no harm [e.g., by terminating a patient's life or maintaining it in a seriously defective condition] unless there is corresponding benefit to the patient." In a *social* perspective, a consequentialist method judges what is right by the principle of utilitarianism, a classic formulation of which is: the greatest good for the greatest number. Whether the patient's life should be maintained or terminated would depend on a calculus of the impact that quantified, predicted consequences of the patient's life and condition would have on the quality of life of all concerned parties and of society itself.

Another method of moral reasoning is found in *deontological* theories, which hold that there are obligatory limits on what can be done to an individual based solely or principally on moral claims that are demonstrably or self-evidently right or wrong on a priori grounds. These claims are usually expressed in terms of the inherent and inalienable moral rights of individuals and are not dependent on the possible desirable or undesirable consequences of the human being living out the course of his or her life.

Examples of the typically consequentialist quality-of-life ethic are given under the next two headings. Most of the ethical implications indicated in each of these sections—on selective abortion and "euthanasia"—are applicable to the other.

Selective abortion. We are concerned here with quality-of-life arguments only in reference to selective abortion following prenatal genetic diagnosis. Prenatal diagnosis may indicate the presence of such abnormalities as a chromosomal anomaly, e.g., Down's syndrome; a sex-linked disorder, e.g., hemophilia; or a metabolic disorder, e.g., Tay-Sachs disease (Lebacqz, pp. 18 ff.). Abortion, undertaken selectively after prenatal diagnosis to prevent the birth of an infant with a genetic disease by terminating the life of the fetus, is sometimes justified by quality-of-life arguments (ibid., pp. 102–111, 146–160). According to these arguments, life should be maintained or terminated on the basis of what would preserve or enhance the quality of life, that is, "human well-being" (Fletcher, 1971, pp. 778, 781).

Whose quality of life is appealed to? Utilitarian quality-of-life arguments appeal to the quality of—or the consequences to—the "greatest number." This would include the human race (e.g., its biological quality), civil society (e.g., in terms of social and economic costs), the parents and family (e.g., the suffering or rewards of caring for the patient, strain on the marriage, or economic stress), and the fetus/infant itself (its own well-being) (Gustafson, "Genetic Counseling").

Quality-of-life arguments dealing with selective abortion usually justify termination of the life of the individual fetus or, systematically, of entire categories of affected fetuses, for the good of society at large. The quality of life of society is measured principally in terms of eugenic, economic, and demographic benefits (Lebacqz, pp. 146–147).

Selective abortion of those affected by a genetic disorder as well as those who are carriers of "deleterious" genes is sometimes performed for purposes of improving the quality of the gene pool, if it is judged that carriers (heterozygotes) actually have a negative impact on the genetic condition of society. Whether the fetus's life should be terminated for eugenic reasons depends on whether that action will improve the genetic health of society and prevent or eradicate disease (ibid., pp. 124, 379)

On an economic basis, abortions are justified if comparative cost analyses indicate that the cost of a prenatal screening program with selective abortion—say, for Tay-Sachs disease or Down's syndrome—is less than the cost for medical care, including possible institutionalization of the nonaborted children (ibid., p. 104).

Furthermore, on the assumption that both society and individual families have a legitimate interest in restricting the number of children to be born, selective abortion is a justified means of quality control to assure that the children who are born will be optimal in health and other human qualities such as the ability to lead independent and useful lives (Morison, pp. 208–211).

In addition to social reasons, consequentialist arguments are also based on the quality of life of the fetus itself. In this argument, abortion for the sake of benefit to the fetus is sometimes supported by the claim that there is a right to be well-born. If one cannot be born without a preventable serious genetic defect, it is better not to be born at all (Lebacqz, pp. 109, 192–193).

The implications of consequentialist arguments on selective abortion depend in part on whether there is a presumption that at some point in time the prohibition of killing applies to the fetus except when reasonable exceptions can be made. Particularly when such a rule does apply, quality-of-life arguments are affected by a conflict between benefits to the individual versus benefits to society.

Terminating life in other situations. The following paragraphs will examine the quality-of-life arguments for terminating or not sustaining the lives of defective infants, the dying, the terminally ill, and the seriously ill. In most cases the same principles also apply to the fetus.

"Humanhood" versus "accessory" qualities. A distinction can be drawn between what may be called humanhood qualities and accessory qualities. While these terms may be used in either a descriptive, an evaluative, or a morally normative sense, they are defined here as used in morally normative judgments.

Humanhood qualities are those distinctively human qualities that are so valued that they are seen as necessary conditions for extending to the humans possessing them the benefit of those norms that the human moral community offers for the preservation of life. Those are the norms that prohibit killing and that require the sustenance of life, the latter requirement being the duty at least of those who have the responsibility of a caring relationship, such as physicians and parents. Accessory qualities, on the other hand, are secondary or subordinate qualities that are regarded as criteria for either intensifying or diminishing the obligation to preserve life.

The humanhood–accessory distinction can be applied to human life at any stage; but it is used less frequently in quality-of-life arguments concerning abortion, in which the moral status of the fetus plays a less definitive role than eugenic and other social consequences.

Humanhood qualities become criteria for moral obligations in the context of a moral theory; they may be used in either a deontological or a consequentialist ("quality-of-life") mode of moral reasoning.

A typical quality-of-life argument employs a consequentialist perspective in which the morally normative value of a human life depends fundamentally on whether a humanhood quality is expected to lead to some higher quality such as happiness or well-being. Given the absence of this quality, it is not wrong actively to terminate a human life or to abandon to death an individual for whom one has responsibility (Gustafson, "Mongolism," p. 537). In a deontological perspective, on the other hand, there is a strong tendency to embrace all humans equally in the basic right to life and then to allow for exceptions to obligations arising from those rights. However, some discrimination does occur by definition, especially when "humanhood" qualities are those that account for personhood. In this approach, an individual would have a right to the preservation of life if, e.g., he had the quality of rationality (Tooley) or of membership in a rational species (May), or whatever quality one takes to be essential to personhood.

Accessory qualities are used in a supplementary way to specify or restrict the basic duties associated with humanhood qualities. Accessory qualities may be referred to variously in arguments for *intensifying* the basic duty to sustain the life, e.g., of a person who has optimal qualities of intellect or personality, is socially useful, or is in dire medical need. Arguments for *diminishing* the basic respect-for-life principles sometimes hinge on accessory qualities such as the condition of being a dying person.

Typical quality-of-life arguments. Although there is no general agreement on the use of humanhood and accessory qualities of life, these categories can be employed in analyzing two quality-of-life arguments in the area of "euthanasia."

Position number one, described above as consequentialism in a social perspective, is a classic utilitarian position as proposed, for example, by Joseph Fletcher.

Human life has no value unless it produces personal well-being, here interpreted as human happiness for the largest number of people. The direct or indirect killing of a human is good whenever it is justified by this highest value or whenever the desirable consequences outweigh any disvalue in the action, according to a cost-benefit analysis (Fletcher, "Ethics and Euthanasia"; *The Ethics of Genetic Control*, p. 185).

Fletcher's minimal "humanhood" criteria are associated with rational function; they include neocortical function (which is a prerequisite to the others), intelligence (humans scoring below 40 IQ are questionably persons, and below the 20 mark they are not persons), communication, and other qualities. Accessory qualities (he calls them "optimal criteria"), which contribute to an individual's better well-being, include curiosity, balance of rationality and feeling, and idiosyncrasy (Fletcher, "Medicine," "Four Indicators"). If the humanhood qualities are lacking, there is no moral offense in terminating a human life; if they are present, the moral value of preserving or terminating a life is known only by a utilitarian assessment of the needs of all those affected in each situation ("Medicine," p. 55).

Position number two is a typical (consequentialist) quality-of-life argument, described above as one that focuses on a criterion of personal rather than social well-being.

An example here is the view of Richard McCormick, who speaks of infants but whose principles are applicable to humans of any age. It is the quality or kind of life that is critical in knowing which humans should be saved. The argument is based on the conviction that life is a value to be preserved insofar as the higher, spiritual purposes of life are attainable—the love of God and the love of neighbor, which is its expression. Because of the supremacy of these inseparable loves, life finds its meaning in human relationships. The humanhood quality or key criterion for determining whether there is a duty to preserve life is thus the *relational potential* of the individual. If this potentiality is absent or completely subordinated to the mere effort for survival, the individual may be allowed to die. Concrete standards of relational potential must be developed by physicians in the form of "biological symptoms," that is, diagnostic categories such as Down's syndrome, anencephaly, and other categories. By these quality-of-life criteria, the Down's syndrome baby would very likely meet the humanhood standard of relational potential, whereas the anencephalic infant would not (McCormick, 1974).

Many other quality-of-life arguments have been offered, though not always on the basis of a thoroughly elaborated theory. A partial quality-of-life theory has been suggested on the basis of "wrongful life" or "wrongful birth," a concept that has appeared in legal cases in the United States (Brantley) and in many other countries

(Tedeschi). These cases are tort actions for "wrongfully" causing a child to be born, e.g., with birth defects or impaired ability to enjoy life. While some of these suits have succeeded in their initial stages, as of 1977 all such suits had failed in U.S. courts (Brantley, pp. 147–149). Nonetheless, a philosophical argument has been developed on the basis of a "wrongful life" concept regarding responsibility for those who are not competent to decide for or against treatment on the basis of their own projected quality of life. In this argument, "wrongful life" implies that an individual should never have existed; and a child's life, when characterized by severe pain and deprivation, can be an "injury of continued existence." On utilitarian grounds, and because of a positive duty to avoid harm, this argument provides for justifying euthanasia of small children for the sake of the children themselves (Engelhardt).

Levels of qualities, and who decides. An analysis of the qualities used in consequentialist arguments for determining the moral factor of "humanhood" indicates that there are three levels of qualities serving as criteria. The *principal* quality (e.g., rational abilities, relational potential) receives its moral significance from an *ultimate* quality (e.g., happiness, love) constituting the goal and meaning of life, which can be achieved only through the principal quality. *Operational* qualities (e.g., IQ, diagnostic categories) are used as more substantive criteria for knowing whether the principal qualities are present. The interdependency of the three levels of qualities, all of which can be called normative, is frequently not brought to the surface in moral arguments on quality of life.

Another significant distinction emerges regarding who decides on quality-of-life criteria. *First-person* criteria depend on the subject's own judgment of what qualities—physical, mental, social, spiritual, etc.—are required for continued existence, together with his appraisal of the extent to which he possesses them. *Proxy* criteria are used in the decisions of those responsible for judging on behalf of an individual who is incompetent to decide for himself. *Third-party* criteria are standards on quality of life employed by those who influence the subject's decision—physicians, relatives, and others who introduce either their personal preference or consensus-type standards on what quality of life should be. While first-person criteria have special relevance for suicide (Hume) and the right to refuse treatment (Kelly), proxy and third-party criteria appear most influential in determining legal and other policy standards for decisions regarding the lives of fetuses, infants, and other incompetent patients.

Critique and alternate theories

Critique of the quality-of-life ethic

General objections against the consequentialist calculus. A number of objections commonly brought against consequentialist modes of reasoning constitute the most basic objections against a quality-of-life ethic. First, when the overall quality of life or categorical condition of life is the standard for the moral judgment on value of life, this relative criterion is at odds with the concept of the moral equality of all human beings and the fundamental principles of justice which undergird the ethical and legal protection of life in Western society (Lebacqz, p. 384; Kass, p. 187; Ramsey, 1978). Second, the consequentialist ethic relies on the prediction of good and bad results, and those results depend on contingent factors such as emotional and social support for the handicapped. Hence this moral theory is vulnerable to arbitrary and static assessments. To minimize the problem of arbitrariness, consequentialist theories frequently use operational criteria; but these standards raise the objection of excessive quantification. For example, use of IQ scores as a quality-of-life criterion suffers from an excessive and elusive reductionism of human worth; and categorization of moral value of life according to syndrome—Down's syndrome, for instance—overlooks the fact that individuals afflicted with many such defects manifest a wide range of physical and mental capabilities and handicaps. Finally, in consequentialist systems, judgments on a great variety of complex questions related to disease, health, life, and death are dominated by a single value, such as personal or human well-being or whatever constitutes human growth; and this single value tends to encourage an oversimplification of complex ethical questions.

The slippery slope argument. The terms "slippery slope" argument and "wedge" argument are frequently used interchangeably. Whatever term is used, the argument can take two distinct forms.

The first slippery slope argument states that, if a certain kind of behavior, once permitted as a general line of conduct, would predictably lead to other behavior harmful to humanity, this is a good reason against admitting the first practice or reform. In reference to the quality-of-life debate, the first slippery slope argument does not seek to refute the inherent moral reasons given in support of the practice initially proposed. Rather, it objects that, if the quality-of-life ethic is adopted in one area, e.g., to justify selective abortion, this practice or reform will inexorably lead to acquiescing in other objectionable practices.

Parents' fear of defect and expectation of normalcy in offspring have intensified since prenatal diagnosis and genetic counseling have become available (cf. Lebacqz, p. 199). Furthermore, a general willingness in society to abort a defective fetus strengthens the attitude favoring selectivity in accepting or rejecting nascent life. These fears, expectations, and attitudes of selectivity— which are enhanced by quality-of-life arguments and which serve as concrete criteria for the utilitarian benefit–harm calculus—will encourage and increasingly lead to abortion for lesser reasons and to infanticide in situations where parents have had no opportunity for prenatal diagnosis.

The second slippery slope argument looks to the moral reasons offered for one practice or reform and concludes that they logically entail other practices that are harmful to humanity. Accordingly it is claimed that quality-of-life arguments intended to justify only selective abortion are also arguments for infanticide, or those formulated only in support of infanticide also logically threaten the protection of other humans, and that consequently the entire line of argument fails to offer proper protection to human life. Since a quality-of-life ethic extends the respect-for-life principle only to those humans whose good or bad quality of life will produce good results or at least a minimum of bad results, this establishes a logical precedent for a widespread destruction of life and for the withdrawal of care from human relationships (Lebacqz, pp. 195–218; Kass). For example, the unacceptable quality of life used to justify selective abortion and the termination of defective infants in the arguments reported above are largely determined by assessments of the social acceptability of certain physical diseases, mental abnormalities, or personal relationships before any rule for preserving life would apply. But a criterion that depends on social acceptance is an evaluative standard open to such variance and reinterpretation that it logically threatens the protection of many humans in those conditions and in other circumstances of life.

Beneficiaries of quality-of-life decisions. Quality-of-life arguments for selective abortion based on the notion of benefit to the fetus (or infanticide for the benefit of the infant) raises the logical problem of eliminating abnormalcy or bad health by eliminating the individual—i.e., the problem of whether an individual can really be the beneficiary of an action that terminates its very existence (Camenisch). Similarly, the "wrongful life" argument, in both philosophical and legal discussions, raises several objections: how and under what conditions existence itself can be an injury, and the impossibility of measuring damages equitably against intangible benefits associated with life (Brantley, pp. 140, 154–155).

Alternate theories on quality of life. In contrast to a quality-of-life ethic, an alternate mode of moral reasoning might, for example, adopt a theory of justice that provides a different starting point for coping with an undesirable quality of life. Whereas many quality-of-life theories depend on a *merit* theory of justice, meaning that the right to life is earned by the accomplishments or the potential contribution of the individual, some claim that a preferred theory is an *equality* theory of justice that is characterized by impartiality in acknowledging the right or claim to the preservation of life. Fundamental to this theory is that all humans are included in the respect-for-life principle, even if some justifiable exceptions are made in conflict situations. This position —which acknowledges the basic claim to protection and care on the part of even those humans that lack physical, mental, or relational qualities —also affirms the importance of other qualities essential to the human condition as well as to individual lives: trust, security, generosity, and the like (Lebacqz, pp. 220–237).

Whether this or some other system is chosen as one's starting point in accounting for a respect-for-life principle, there are many ethical theories that take account of the adverse conditions or qualities of life in developing an ethic of medicine. There are a number of distinctions and rules useful for resolving conflict situations arising in the treatment of diseased, abnormal, or defective individuals that do not share all the elements of a quality-of-life ethic.

For example, some sanctity-of-life writers, acknowledging an important distinction between dying and nondying patients, make exceptions to the rule about the preservation of life in certain circumstances. One sanctity-of-life ethicist permits active or passive euthanasia when certain key qualities (probably "accessory" qualities) are present, but only in the case of the dying patient: the inability to receive care and the presence of excessive pain (Ramsey, 1970).

Attempts are also made to resolve quality-of-life conflicts by drawing the distinction between using ordinary means to preserve life (which are affirmatively obliging) and extraordinary means (which are not affirmatively obliging). The principle states that there is no strict obligation to accept medical treatment that would involve excessive pain, expense, or hardship or that does not offer a reasonable hope of benefit (Kelly, p. 129). Qualities and even, on occasion, overall quality of life are clearly elements in this principle—e.g., the pain or excessive hardship (such as frequent, repeated surgery offering little hope of benefit) which may accompany or follow upon the procedure. There are many debates on the meaning and usefulness of this principle—some holding that the terms ordinary and extraordinary should be abandoned, others maintaining that they are simply equivalent to or should be reduced to what the "reasonable person" would accept or refuse, what is or is not medically indicated, etc. (McCormick, 1978; Ramsey, 1978).

There is, however, a fundamental methodological debate as to whether the "extraordinary means" rule should be interpreted as part of a deontological or a consequentialist ethic. One position holds that the admittedly consequentialist judgment that a procedure is an extraordinary means is a restricted, situational judgment on the unreasonable hardship or futility associated with treatment and must be justified as an exception to the rule requiring the preservation of life—a respect-for-life rule that does not systematically exclude certain "kinds" of life from protection on qualitative grounds (Reich). The other position holds that the "extraordinary means" judgment is frequently a decision on the acceptability of this or that treatment on grounds of usefulness, hardship, or personal acceptability; but judgments of this type imply and lead to a quality-of-life ethic, which must determine substantive standards for judging which quality or kind of life represents a value that ought to be preserved (McCormick, 1974, 1978).

Conclusion

The ways in which quality of life enters into moral judgments on the preservation and protection of life are extremely complex. They depend on all the elements of ethics—conceptual presuppositions, basic moral theory, relevant principles, and modes of resolving conflicts.

Many of the difficult decisions on preserving life in biomedical situations depend on conceptual presuppositions—the meaning, value, and disvalue of death, dying, life, health, disease, suffering, pain, and freedom—and on an understanding of the roles of parents and health-care providers.

Qualities of life are most often appealed to normatively to resolve conflict situations. Yet certain conflicts dominate in the struggle to identify the most morally significant quality of life. Material and measurable qualities contrast and conflict with the intangible and immeasurable qualities of life. Qualities associated with individuals (e.g., the ability and freedom to act in typically human ways) contrast with qualities associated with the human community (such as caring, loyalty, and justice). Whether quality of life should be judged subjectively by the person experiencing it (or expected to experience it in the future), objectively according to some theory of knowledge or social consensus, or by a third party such as physician or guardian is a question causing further conflict.

A distinctive quality-of-life ethic typically depends on a consequentialist moral theory for determining the relevant moral criteria for the preservation of life; but it is not the only method that is based on and takes account of valued qualities. Deontological positions are based on the value of a fundamental quality of humanhood and on such community qualities as interpersonal fidelity. In a fundamental sense, then, one can say that the respect-for-life debate, insofar as it focuses on quality of life, is a debate about the normative implications of two or more kinds of moral theories which emphasize different valued qualities. The outcome of the debate depends on what one thinks of the implications of those theories.

WARREN T. REICH

[*Directly related is the other article in this entry,* VALUE OF LIFE. *For further discussion of topics mentioned in this article, see the entries:* DOUBLE EFFECT; *and* ETHICS, *articles on* DEONTOLOGICAL THEORIES, UTILITARIANISM, *and* TELEOLOGICAL THEORIES. *This article will find application in* ABORTION, *article on* CONTEMPORARY DEBATE IN PHILOSOPHICAL AND RELIGIOUS ETHICS; AGING AND THE AGED, *article on* ETHICAL IMPLICATIONS IN AGING; DEATH AND DYING: EUTHANASIA AND SUSTAINING LIFE, *article on* ETHICAL VIEWS; DECISION MAKING, MEDICAL; ENVIRONMENTAL ETHics, *articles on* QUESTIONS OF SOCIAL JUSTICE *and* THE PROBLEM OF GROWTH; INFANTS, *articles on* ETHICAL PERSPECTIVES ON THE CARE OF INFANTS *and* INFANTICIDE: A PHILOSOPHICAL PERSPECTIVE; LIFE-SUPPORT SYSTEMS; PAIN AND SUFFERING; *articles on* PHILOSOPHICAL PERSPECTIVES *and* RELIGIOUS PERSPECTIVES; POPULATION ETHICS: ELEMENTS OF THE FIELD, *article on* NORMATIVE ASPECTS OF POPULATION POLICY; PRENATAL DIAGNOSIS, *article on* ETHICAL ISSUES; RATIONING OF MEDICAL TREATMENT; RIGHT TO REFUSE MEDICAL CARE; *and* SUICIDE.]

BIBLIOGRAPHY

QUALITY OF LIFE IN BIOMEDICAL DECISION MAKING

AMERICAN MEDICAL ASSOCIATION. *Quality of Life.* 3 vols. First National Congress on the Quality of Life. Acton, Mass.: Publishing Sciences Group, 1974–1975.

BRANTLEY, JOHN R. "Wrongful Birth: The Emerging Status of a New Tort." *St. Mary's Law Journal* 8 (1976): 140–159.

CALLAHAN, DANIEL J. *Abortion: Law, Choice, and Morality.* New York: Macmillan Co., 1970.

———. "The Sanctity of Life." *Updating Life and Death: Essays in Ethics and Medicine.* Edited by Donald R. Cutler. Boston: Beacon Press, 1969, chap. 8, pp. 181–223. Followed by commentaries by Julian R. Pleasants, James M. Gustafson, and Henry K. Beecher, with a final response from Callahan, pp. 223–250.

CAMENISCH, PAUL F. "Abortion: For the Fetus's Own Sake?" *Hastings Center Report* 6, no. 2 (1976), pp. 38–41.

CLOUSER, K. DANNER. " 'The Sanctity of Life': An Analysis of a Concept." *Annals of Internal Medicine* 78 (1973): 119–125.

CRANE, DIANA. *The Sanctity of Social Life: Physicians' Treatment of Critically Ill Patients.* New York: Russell Sage Foundation, 1975.

ENGELHARDT, H. TRISTRAM, JR. "Ethical Issues in Aiding the Death of Young Children." Kohl, *Beneficent Euthanasia,* pp. 180–192.

FLETCHER, JOSEPH. "Ethical Aspects of Genetic Control: Designed Genetic Changes in Man." *New England Journal of Medicine* 285 (1971): 776–783. Reprint. Reiser, *Ethics in Medicine,* chap. 65, pp. 387–393.

———. "Ethics and Euthanasia." *To Live and to Die: When, Why, and How.* Edited by Robert H. Williams. New York: Springer-Verlag, 1973, chap. 9, pp. 113–122.

———. *The Ethics of Genetic Control: Ending Reproductive Roulette.* Garden City, N.Y.: Anchor Press/Doubleday, 1974.

———. "Four Indicators of Humanhood—The Enquiry Matures." *Hastings Center Report* 4, no. 6 (1974), pp. 4–7.

———. "Medicine and the Nature of Man." *The Teaching of Medical Ethics.* Edited by Robert M. Veatch, Willard Gaylin, and Councilman Morgan. A Hastings Center Publication. Proceedings of a conference sponsored by the Institute of Society, Ethics, and the Life Sciences and Columbia University, College of Physicians and Surgeons, 1–3 June 1972. Hastings-on-Hudson, N.Y.: Institute of Society, Ethics, and the Life Sciences, 1973, pp. 47–58.

———. "The 'Right' to Live and the 'Right' to Die." Kohl, *Beneficent Euthanasia,* pp. 44–53.

FRANKENA, WILLIAM K. "The Ethics of Respect for Life." *Respect for Life in Medicine, Philosophy, and the Law.* By Owsei Temkin, William K. Frankena, and Sanford H. Kadish. The Alvin and Fanny Blaustein Thalheimer Lectures, 1975. Edited by Stephen F. Barker. Baltimore: Johns Hopkins University Press, 1977, chap. 2, pp. 24–62.

GUSTAFSON, JAMES M. "Genetic Counseling and the Uses of Genetic Knowledge—An Ethical Overview." Hilton, *Ethical Issues in Human Genetics,* pp. 101–119. Includes discussion.

———. "God's Transcendence and the Value of Human Life." *Christian Ethics and the Community.* Philadelphia: Pilgrim Press, 1971, chap. 5, pp. 139–149. Original appearance. *Proceedings of the Catholic Theological Society of America* 23 (1968): 96–108.

———. "Mongolism, Parental Desires, and the Right to Life." *Perspectives in Biology and Medicine* 16 (1973): 529–557.

HILTON, BRUCE; CALLAHAN, DANIEL; HARRIS, MAUREEN; CONDLIFFE, PETER; and BERKLEY, BURTON, eds. *Ethical Issues in Human Genetics: Genetic Counseling and the Use of Genetic Knowledge.* Proceedings of a symposium sponsored by the John E. Fogarty International Center for Advanced Study in the Health Sciences and the Institute of Society, Ethics, and the Life Sciences, 10–14 October 1971. Fogarty International Proceedings, no. 13. New York: Plenum Press, 1973. Papers followed by discussion.

HUME, DAVID. "On Suicide." *Of the Standard of Taste and Other Essays.* Edited by John W. Lenz. The Library of Liberal Arts. New York: Bobbs-Merrill Co. 1965, pp. 151–160.

KASS, LEON R. "Implication of Prenatal Diagnosis for the Human Right to Life." Hilton, *Ethical Issues in Human Genetics,* pp. 185–199.

KELLY, GERALD. "Preserving Life." *Medico-Moral Problems.* St. Louis: Catholic Hospital Association of the United States and Canada, 1958, chap. 17, pp. 128–141.

KOHL, MARVIN. *The Morality of Killing: Sanctity of Life, Abortion, and Euthanasia.* New York: Humanities Press, 1974.

———, ed. *Beneficent Euthanasia.* Buffalo: Prometheus Books, 1975.

LEBACQZ, KAREN A. "Prenatal Diagnosis: Distributive Justice and the Quality of Life." Ph.D. dissertation, Harvard University, 1974. Parts of this work condensed as "Prenatal Diagnosis and Selective Abortion." Reiser, *Ethics in Medicine,* chap. 64, pp. 376–387.

McCORMICK, RICHARD A. "A Proposal for 'Quality of Life' Criteria for Sustaining Life." *Hospital Progress* 59, no. 9 (1975), pp. 76–79.

———. "The Quality of Life, the Sanctity of Life: A Theological Perspective." *Hastings Center Report* 8, no. 1 (1978), pp. 30–36.

———. "To Save or Let Die: The Dilemma of Modern Medicine." *Journal of the American Medical Association* 229 (1974): 172–176.

MAY, WILLIAM E. "What Makes a Human Being to Be a Being of Moral Worth?" *Thomist* 40 (1976): 416–443.

"Medicine and the Quality of Life." *Western Journal of Medicine* 125 (1976): 1–16. Seven views.

MORISON, ROBERT S. "Implications of Prenatal Diagnosis for the Quality of, and Right to, Human Life: Society as a Standard." Hilton, *Ethical Issues in Human Genetics,* pp. 201–220. Includes discussion.

"A New Ethic for Medicine and Society." *California Medicine: The Western Journal of Medicine* 113, no. 3 (1970), pp. 67–68. Editorial.

RAMSEY, PAUL. *Ethics on the Edges of Life: Medical and Legal Intersections.* The Bampton Lectures in America. New Haven: Yale University Press, 1978.

———. *The Patient as Person: Explorations in Medical Ethics.* The Lyman Beecher Lectures at Yale University, 1969. New Haven: Yale University Press, 1970.

REICH, WARREN T. "Quality of Life and Defective Newborn Children: An Ethical Analysis." *Decision Making and the Defective Newborn: Proceedings of a Conference on Spina Bifida and Ethics.* Edited by Chester A. Swinyard. Foreword by Robert E. Cooke. Springfield, Ill.: Charles C Thomas, 1978, chap. 24, pp. 489–511.

REISER, STANLEY JOEL; DYCK, ARTHUR J.; and CURRAN, WILLIAM J., eds. *Ethics in Medicine: Historical Perspectives and Contemporary Concerns.* Cambridge: MIT Press, 1977.

SHAW, ANTHONY. "Dilemmas of 'Informed Consent' in Children." *New England Journal of Medicine* 289 (1973): 885–890.

TEDESCHI, G. "On Tort Liability for 'Wrongful Life'." *Israel Law Review* 1 (1966): 513–538.

TOOLEY, MICHAEL. "A Defense of Abortion and Infanticide." *The Problem of Abortion.* Edited by Joel Feinberg. Basic Problems in Philosophy series. Belmont, Calif.: Wadsworth Publishing Co., 1973, pp. 51–91.

WILLIAMS, GLANVILLE. *The Sanctity of Life and the Criminal Law.* James S. Carpentier series, 1965. New York: Knopf, 1957.

QUALITY OF LIFE IN HUMAN ENVIRONMENTS

BAIER, KURT. "Towards a Definition of 'Quality of Life'." *Environmental Spectrum: Social and Economic Views on the Quality of Life.* Edited by Ronald O. Clarke and Peter C. List. New York: D. Van Nostrand Publishing Co., 1974, chap. 5, pp. 58–81.

BUNGE, MARIO. "'Indicators': Judging the Quality of Life." Shea, *Values and the Quality of Life*, pp. 142–156.

EHRLICH, PAUL R., and EHRLICH, ANNE H. *Population, Resources, Environment: Issues in Human Ecology.* A Series of Books in Biology. San Francisco: W. H. Freeman, 1970.

GERSON, ELIHU M. "On 'Quality of Life'." *American Sociological Review* 41 (1976): 793–806.

GITTER, GEORGE A., and MOSTOFSKY, DAVID I. "The Social Indicator: An Index of the Quality of Life." *Social Biology* 20 (1973): 289–297.

MCCALL, STORRS. "Quality of Life." *Social Indicators Research: An International and Interdisciplinary Journal for Quality-of-Life Measurement* 2 (1975): 229–248.

MICHALOS, ALEX C. "Measuring the Quality of Life." Shea, *Values and the Quality of Life*, pp. 24–37.

RESCHER, NICHOLAS. "Quality of Life and Social Indicators." *Welfare: The Social Issues in Philosophical Perspective.* Pittsburgh: University of Pittsburgh Press, 1972, chap. 4, pp. 60–77.

SHEA, WILLIAM R., and KING-FARLOW, JOHN, eds. *Values and the Quality of Life.* Canadian Contemporary Philosophy series. Edited by John King-Farlow and William R. Shea. New York: Science History Publications, 1976.

United States, Environmental Protection Agency, Office of Research and Monitoring, Environmental Studies Division. *The Quality of Life Concept: A Potential New Tool for Decision-Makers.* Washington: 1973. Based on a symposium held at Airlie House, Warrenton, Virginia, 29–31 August 1972.

WINGO, LOWDON. "The Quality of Life: Toward a Microeconomic Definition." *Urban Studies* 10 (1973): 3–18.

LIFE-SUPPORT SYSTEMS

The intensive care unit has become the site of modern medicine's most elaborate methods and technologies. In these special wards of hospitals, critically ill patients are surviving by means of what are loosely called "life-support systems." From the term life-support systems we often infer that the most advanced technology of medical science is employed. Some of these techniques, however, are extremely basic and simple. The technical sophistication of these systems ranges from mouth-to-mouth respiration and external cardiac massage to extracorporeal membrane oxygenators and artificial pacemakers. Hence, life-support systems are a heterogeneous collection of administered synthetic, semisynthetic, and natural agents which by manual or automated means support or substitute for certain vital functions of a patient during the critical, life-threatening phases of an illness or injury. This article describes the function and purpose of these systems and, in a second part, catalogues the ethical problems that their existence and use occasion.

Function of life-support systems

All life-support systems, from simple to sophisticated, have a common purpose: to aid, support, or supplant a vital function that has been seriously impaired. Almost always, they are intended as temporary aids or substitutes while the body has the opportunity to recuperate from some major damage. Although occasionally they may contribute to the healing process, this is not essential to the service they provide. What is essential is that the support system is in lieu of a vital function, thus its premature removal will lead to the demise of the patient. These systems are essentially temporary adjuncts to vital processes and will be withdrawn when healing has progressed. For example, it is expected that a patient with a drug overdose who needs artificial respiration will breathe spontaneously when the effects of the drug wear off. Although a machine which is breathing for the patient does not promote healing, it provides support for a vital function until the patient recovers.

The vital processes, to which these systems are an adjunct, are definable in a very basic manner: they are the delivery of minimal nutrients to the cells of the living organism and the removal of waste products from the organism. Nutrition requires a supply of (1) oxygen and (2) substrate, carried by the blood and delivered to the tissue to provide sufficient energy for the processes that maintain tissue integrity. The next sections will provide a schema for understanding the components of the basic life processes and will describe how temporary alternatives may substitute for natural functions.

Supply of oxygen. The normal mechanism for transport of oxygen calls for (1) a pump to bring into the lungs oxygen from the surrounding air, (2) a place for exchange of oxygen from the air to the blood, (3) a means by which oxygen can be carried by the blood and relinquished to the tissues, and (4) a pump to drive the blood from the lungs to the tissues. When any or all of these biological mechanisms fail, mechanical systems have been devised to provide alternatives.

Oxygen delivery to the lungs. The pump that brings oxygen into the lungs consists of the chest cage, its muscles, the diaphragm, and the central neurological drive to breathe. During inspiration, the chest cavity expands, causing lung volume to increase, thus drawing air into the lungs. The timing and coordination of this process are guided by the central nervous system. The process may fail either because of the inability of the brain to regulate respiration, a loss of the nervous connections between the brain and the respiratory muscles, or weakness or deformity of the chest wall. The central nervous system may have a depressed regulatory ability because of a narcotic overdose; trauma to the spinal cord may effectively sever the nerves between the brain and the respiratory muscles; and such defects, as severe curvature of the spine or injury to the chest wall, may render someone incapable of inspiring a sufficient amount of oxygen with each breath.

Failure of the respiratory system can usually be treated by mechanical support. In the 1940s, the "iron lung" was invented for polio patients whose respiratory muscles were paralyzed. The patient was placed into a tanklike device which expanded the lungs by creating a negative pressure between the chest wall and the metal chamber surrounding the patient's chest, thus drawing air into the lungs. Today, although no longer needed for polio patients, similar machines, called negative pressure ventilators, are occasionally used for infants. More commonly, a positive pressure ventilator is employed. This is essentially a bellows which blows oxygen into the lungs, usually by way of a tube passed through the nose or mouth into the trachea. The timing and volume of respiration are tuned to the patient's needs. The feedback loop, normally present between the central nervous system and the lungs, which regulates respiration is completed by a physician who monitors oxygen and carbon dioxide in the blood and adjusts the machine accordingly. It is conceivable that the adjustment of the ventilator could be performed by computer.

Gas exchange in the lungs. The site for exchange of gases between the air and the blood is the terminal portions of the airway, called the alveoli. Here a thin membrane separating the air from the blood in the capillaries permits rapid diffusion of gases across it while maintaining an intact interface. Oxygen passes from the alveoli into the blood and carbon dioxide from the blood to the alveoli, both moving from a region of high to lower concentration. If these alveoli are not expanded and filled with air during inspiration, or if the membrane becomes so damaged as not to permit exchange of gases, or if the blood bypasses the alveoli, adequate oxygenation cannot occur. Respiratory distress syndrome of the newborn, smoke inhalation, and a clot in the vessel of the lung are examples of each of such problems. Support may be provided by increasing the concentration of oxygen inspired and attempting to keep the lungs expanded. Initially, this can be done by enriching the oxygen in the surrounding air, but use of a ventilator may be required to deliver oxygen under sufficient pressure to maintain expansion of the lungs. In some conditions of severe impairment of the lungs, adequate oxygenation cannot be accomplished by ventilator alone. In recent years, an extracorporeal membrane oxygenator has been used. This device consists of a large thin membrane, external to the patient, which allows exchange of gas across it. Oxygen-depleted blood, which is removed from the patient, flows continuously through the oxygenator and thence back into the body. The blood is separated from a reservoir of oxygen by the membrane, thus allowing diffusion of oxygen into, and carbon dioxide out of, the blood. The extracorporeal membrane oxygenator not only substitutes for the process of gas exchange but replaces the entire process of pumping air into the lungs, which is no longer necessary when adequate oxygen is provided by the machine.

Oxygen transport by the blood. Oxygen is transported from the lungs to the tissue by means of hemoglobin, a protein in the red blood cell. Without hemoglobin, only a trivial amount of oxygen could be carried by the blood while a patient breathes surrounding air. When there is excessive blood loss, transfusion from a human donor is the only way to restore hemoglobin and, hence, the oxygen-carrying capacity of the blood. Although synthetic oxygen carriers may someday become feasible, at present there is no alternative to transfusion. Of course, the supply of blood as well as certain factors in the patient's condition place limits on this technique.

Delivery of oxygen to the tissues. In the next step of supplying oxygen to the tissues, a pump is required to move the blood from the lungs to the tissues and back to the lungs to be replenished with oxygen. Our biologic pump, the heart, may fail for many reasons: disease of the heart muscle, failure of its conducting system, which initiates and coordinates the beating of the heart, or inability to cope with abnormal demands placed on the heart by a disease state. In an emergency, a failing heart can be compensated for by one of three methods. First, the mechanical pumping action of the organ may be augmented by other mechanical means, either external massage of the heart by hand or by counterpulsation. This is a technique in which a balloon placed in the aorta, the artery that supplies blood to the body, is inflated and deflated with the heartbeat and thus improves circulation to the body and to the heart itself. Second, the contractal action of the heart can be augmented by drugs such as digitalis and adrenalin. Finally, an artificial pacemaker, a transistorized device that electrically stimulates the heart, causing it to contract at regular intervals, can be connected to the heart itself by means of a wire passed through the veins into the heart. This will regulate heart rhythm when irregularities (arhythmias) are causing inadequate cardiac output. The pacemaker is one of the few life-support systems that can easily become a permanent replacement for a bodily structure. A patient may take oral medication to augment cardiac function; but, when cardiac function is so poor that counterpulsation is required, there is no portable substitute.

Supply of nutrients. In addition to oxygen, tissues require substrate in the form of foodstuffs. Unlike the hibernating animal, most humans do not build up large fat stores; humans can live from stored fat and sugar for only a few days. Patients whose illness makes oral feeding impossible or ineffective become rapidly undernourished. From the end of the nineteenth century, water and sugar were administered beneath the skin and later directly into the veins. However, even with such intravenous feeding, a patient's tissues can break down without being adequately replaced by new tissue. This can severely retard the healing process. Since the 1960s, total parenteral nutrition has helped to solve this problem. An intravenous administration of a mixture of nutrients—fats, protein and sugar, vitamins and minerals—will not only prevent the net loss of tissue but can also promote growth. This "intravenous soup" is believed sufficient to provide all necessary nutrients, although so little is known about elements provided by ordinary diet in minuscule amounts that deficiencies may not be readily recognized. Total parenteral alimentation is in widespread use; doubtless there will soon be patients receiving it for long periods, even as outpatients. If this happens, it can no longer be considered a temporary life support.

Removal of wastes. An adequate supply of nutrients coupled with suitable oxygen supply is effective in maintaining metabolic stability only if there is means for removal of wastes generated by metabolic processes. Exchange transfusion of fresh blood for the blood of a patient who has built up inappropriate wastes may be employed in some situations. For example, the high levels of the bilirubin pigment, which causes jaundice in some newborns, can be decreased by exchange transfusion, a process in which most of the patient's blood is removed and simultaneously replaced with donor blood. Similarly, exchange resins may enhance the removal of certain substances from the blood; as examples, the use of the drug sodium polystrene sulfonate aids in exchanging potassium for sodium, and the drug calcium EDTA hastens the removal of lead. These substances have only limited uses. For longer-term management of waste removal, the process of dialysis has been devised.

The principle of dialysis is simple. Blood is exposed to the dialysate, a solution containing the needed amount of plasma constituents. Interposed between the blood and the dialysate is a thin membrane, which provides a barrier so that the substances will move in and out of the plasma according to their concentration gradients. Dialysate can be placed into the peritoneal cavity—the thin vascular lining of the abdominal organs—through a catheter, permitted to equilibrate for a time, and then removed, bringing with it the waste substances. Although dialysis is still widely used, there is now a more efficient method called hemodialysis.

Hemodialysis is performed in a machine with a large coil covered with permeable membrane. The patient's blood flows through the coil, permitting the exchange of substances with the dialysate through the membrane. In this system, blood must flow out of the body, through the machine, and back into the body. For this purpose, a needle or catheter is placed into an artery and a vein providing exit and return routes. The invention of a permanently implantable arteriovenous connection, called a "shunt," to allow regular access to these vessels (1952) made possible chronic dialysis, a biweekly cleansing of the blood for patients whose kidneys have ceased, temporarily or permanently, to perform this necessary function. Similar methods for patients suffering from inadequate liver function are not yet as highly developed, although exchange transfusions and exchange resins are used to help in the detoxification and excretion of metabolic products in hepatic failure. However, even though the enzyme systems of the liver are very complex, its function may at some time be mimicked by machine as successfully as dialysis serves as an "artificial kidney."

Implanted organ systems. The aforementioned schema of providing oxygen, nutrients, and waste removal outlines some of the means by which vital processes can be supplanted when the body's organ systems are incapable of providing adequate function. In general, the devices and techniques mentioned require machinery or manipulation external to the body; these are not generally portable. There are, however, more portable substitutes for many of these vital substances, such as the artificial pacemaker and the filtration system of a donated human kidney. Apparently, the only difference between these portable systems and the more bulky mechanical ones is size and convenience. However, a qualitative difference actually separates the portable from the nonportable devices. Once the pacemaker or the kidney is implanted, there is resignation to the fact that it is no longer a temporary replacement. It has been resigned that the vital tissues will not recuperate. As witness to this attitude, it is common practice to remove the nonfunctioning kidney from a recipient of a donated kidney. The transplanted kidney is not a temporary means of life support but is incorporated into the body of the patient. Some envision the day when many vital organs and systems can be permanently substituted, just as artificial limbs have been used for many centuries. With this in mind, researchers have proposed an artificial heart—a synthetic blood pump—driven by an attached power source, which would be totally implanted in the chest cavity to take over the functions of a defective natural heart. If such a device is developed, a life-support system will permanently and totally replace a vital function. Clearly, the ethical questions that surround the temporary use of an artificial system may not be applicable when a permanent and total replacement is available.

Summary. The foregoing has been a brief outline of the means currently available (or conceived) to support the essential functions of the body during a period of critical illness. For the most part, these systems are applied to bridge the gap during which the natural course of the disease would lead to the death of the patient and to provide the opportunity for healing, usually with medical or surgical assistance.

It is important to recognize, however, that although alternatives have been devised for many vital functions, a patient can be overwhelmed by multiple organ failures which will lead to death despite the adequacy of each component of a multiple life-support system. This description of the medical perspectives provides the background for a discussion of the ethical issues attendant upon the existence and use of life-support systems.

Ethical issues

The second part of this article will survey the ethical issues surrounding the ability to support life by means of refined technical devices and procedures which can substitute for vital functions of the body. While many dramatic debates are stimulated by problems popularly termed "pulling the plug," the question of termination of life support is by no means the only ethical issue associated with the use of these systems. This article divides these issues into three major topics: (1) questions of medical practice, (2) questions of policy, and (3) philosophical questions.

Ethical questions in medical practice concerning life-support systems. Life-support systems are employed by physicians after determining that the condition of a patient requires one or more of the technical devices or procedures described in the first section of this article. It might be said that such determinations have four critical ethical moments: (1) decisions to initiate life-supporting care, (2) decisions to wean the patient from the system, (3) decisions to continue life-support systems, and (4) decisions to terminate the care when death is likely to ensue. Each of these "moments" raises different ethical problems and appears rather differently with different sorts of technology. Many of these problems are treated more fully elsewhere in this encyclopedia, in articles to which reference will be made.

Initiating care. Occasionally, the clinical decision to assist a patient's vital functions with a life-support system can be made after careful evaluation of the patient's condition and even in consultation with the patient. This is usually the case when hemodialysis is being considered as substitute for seriously impaired kidney function. In this sort of care, time is available for evaluation and consultation; an alternative, namely transplantation of a kidney, is under certain circumstances available; and in many cases, a long history of renal disease has readied the patient for this decision. In addition, there are sometimes medical problems that might incline physicians not to recommend dialysis, since its use would complicate the illness rather than manage it. It is also conceivable that patients, reflecting on the style of life demanded by dialysis, might refuse it [see ORGAN DONATION; ORGAN TRANSPLANTATION]. Inherent in these decisions must be a careful assessment of the hazards or untoward effects of a support system, as the risks may be as serious as the underlying disability.

However, decisions to initiate life-support systems of other sorts seldom enjoy either careful evaluation or extensive consultation with the patient. Ventilators and oxygenators are often applied in emergency situations when patients become suddenly and critically ill, as in acute heart failure, or when they are in a comatose condition. In the true emergency situation, it is difficult to make an accurate assessment of the patient's condition and prospects. Most commonly, it is considered imperative and ethical to take whatever measures are available to save life and to attempt an evaluation afterward. Although this practice has been challenged, it seems ethically defensible. However, it sometimes results in patients' being sustained by medical means but with little or no hope of recovery (Tait and Winslow).

In such emergency situations, the informed consent of the patient can rarely be obtained; often others who might give proxy consent are unavailable. All the ethical problems surrounding informed consent are attached to decisions to initiate life-support systems. However, a peculiar problem arises when family or friends urge physicians to use all available means to save patients whom the physicians consider unsalvageable. Physicians sometimes accede to these urgings against their better judgment, out of sympathy for the family or fear of liability to malpractice. It seems reasonable, however, to maintain that physicians have the right to refuse to employ treatments which, after careful assessment and consultation, they judge to be of no avail. The customary ethics of medical practice calls upon physicians who make such a judgment to alleviate whatever suffering they can and, if possible, to assist patient and family to find other professional assistance.

Another ethical problem attending the initiation of treatment is that of cautious evaluation of the patient when time and circumstances allow. As a patient's condition deteriorates, more and more measures may be employed, even though there is little likelihood that they will contribute to recovery or even survival. Many hospitals have considered the resuscitation of a dying patient as similar to an emergency case, requiring employment of immediate and vigorous measures. The results of such resuscitations may be a few more days or hours of dying life. However, increasingly, suggestions are being made that careful, daily evaluation of seriously ill patients in intensive care units should lead to more discriminating orders about the resuscitation of such persons ("Optimum Care").

It is also important, from an ethical point of view, to consider that the intensive care that employs life-support systems has certain adverse side effects. The condition of the patient is very closely monitored, and medical interventions of various sorts, which entail risks, are carried out on the basis of the information provided by the monitoring. It is possible that some of these may be quite unnecessary or irrelevant to the patient's actual needs. Some information derived from techniques that are still quite experimental may be ambiguous in import; still, there is often strong temptation to use it in management of the patient. Increased information, then, while it contributes to improved diagnosis and therapy, may also increase risks to the patient. Furthermore, much of the monitoring depends on technical procedures and devices which are fallible. Misinformation may also increase the risks to the patient. Another adverse side effect of intensive care is psychological disturbances caused by deprivation of familiar surroundings and by the "sensory overload" occasioned by the noise, light, and activity of the intensive care ward. Patients sometimes become disoriented and even show psychotic symptoms of hallucination and destructive behavior. While this is temporary and can be prevented by good care, sometimes significant decisions may be made about patients and by patients when they lack the capacity for rational judgment.

A final ethical problem that might attend the initiation of life-supporting care is selecting which patient among many should be the recipient of care that is in short supply, sometimes called medical triage. This problem traditionally arises at a time of great catastrophe, such as earthquake or war. However, it also attended the introduction of penicillin and of hemodialysis. It may sometimes arise when a critically ill patient is brought to intensive care, only to find all beds filled. The problem of rationing of care will usually be a policy problem, made at a level other than clinical decisions [see RATIONING OF MEDICAL TREATMENT].

Weaning the patient from life-support systems. Most life-support systems are intended to be temporary substitutes for a vital function. Those systems designed to substitute for or support cardiopulmonary functions are applied to avert a life-threatening crisis and will be withdrawn when the patient's own body can again carry out the pumping of blood and oxygen exchange necessary for life. Even dialysis, now used for irreversible renal failure, also serves as a substitute for temporary loss of function. Whenever this is the case, decisions must be made about weaning the patient from the machine. This is often a perilous process. Momentary withdrawal of support or diminution of levels of oxygen can result either in the patient's restored capabilities once again taking over or in serious damage and even death. There is a risk-benefit calculation which is often extremely difficult and imposes the heaviest burden on the conscience of those who have the technical skills to make it [see RISK; DECISION MAKING, MEDICAL].

Continuing life-support systems. If the patient is weaned successfully, he or she returns to self-sustaining functions and, it is hoped, to normal life. If weaning is unsuccessful, the patient dies or remains dependent on the machine. Further, certain sorts of life-support systems, such as chronic dialysis for irreversible kidney failure, are intended as continuous substitutes for natural function (if transplantation is not possible). When patients appear to be permanently machine-dependent, questions about continuing life support arise. Certain of these questions may be asked by the patients themselves. Some patients may find the strict regimen of dialysis, as well as its physical and psychological effects, intolerable. They may ask for discontinuance of dialysis, aware that the certain consequence is death [see RIGHT TO REFUSE MEDICAL CARE]. "Dialysis suicide," in which patients fail to appear for treatment, do not follow their diet, or pull out their shunts, is not unknown.

The life-support system may support life for a person suffering from ailments that would, in the absence of the device, be fatal. The totally implantable heart, now in the early stages of development, might sustain strong cardiopulmonary functions in a patient ravaged by cancer (United States, National Heart and Lung Institute; Jonsen, 1975). In such cases, the quality of life supported by the artificial system becomes a matter for ethical reflection [see LIFE, *article on* QUALITY OF LIFE]. Here, unlike most decisions to initiate support or wean patients from it, the decisions belong primarily to patients. Physicians may be the best judges of the efficacy of treatment; patients alone are the best judges of its desirability.

Recognition of patients' right to make decisions about continuation of life-support systems does not absolve physicians from responsibility. They must consider whether they have the ethical obligation to argue against a patient's request to terminate care should they consider it misguided, the result of temporary discouragement, or influenced by organic brain dysfunction. At times, patients may not express overtly the wish to terminate treatment, but seem to reject treatment by lack of cooperation. In such cases, physicians must determine whether such lack of cooperation is equivalent to a decision not to continue. Even if they do so determine, they must ask whether they can ethically or legally accede to such a covert signal from their patient. The decision to discontinue treatment, which belongs to the patient, thrusts the physician into a turmoil of ethical problems. At the very least, it might be said that the physician's duty is to inform the patient honestly about his or her condition and its physical, psychological, and social prognosis; to discern accurately the patient's wishes; and, if necessary, to assist in the formulation of the patient's decision. While efforts to educate or to encourage are commendable, adamant resistance to a clearly formulated decision by the patient or easy acquiescence in relatives' wishes to continue, would be ethically reprehensible.

Quite another set of problems arises when patients are unable to make decisions about continuation of support. Patients in deep coma, with severe damage to vital organs, may be kept breathing; profoundly regressed psychotics may be kept alive by dialysis. The question is raised whether the quality of life of these persons justifies continuation of these treatments. "Good" or "poor" quality of life is a judgment that a person may make about his or her own life; but, when made by others, it is highly suspect. Grounds for such judgment are unquestionably extremely subjective, colored by the class, status, education, and values of the one making the judgment. Further, in the Western legal tradition (with some horrendous exceptions such as the acquiescence of German law in the Nazi "elimination" of the senile and retarded), poor or low quality has never been thought to justify the extinction of human life.

Terminating care when death is likely. The question is most crucial when a patient has suffered severe and apparently irreversible damage to a vital organ system and is in deep coma from which recovery is unlikely. Since heart action and breathing may be artificially sustained, such patients are not legally dead according to the legal definition of death current in most jurisdictions: "the total stoppage of the circulation of the blood, and a cessation of the animal and vital functions consequent thereupon, such as respiration, pulsation, etc." (*Black's Law Dictionary*, 4th ed. 1951). Is it legally and ethically right to withdraw artificial systems from such a person, thereby resulting, in high probability, in his or her legal death? Recently certain efforts have been made to "redefine" death. The most prominent effort was first enunciated in the 1968 report of the Harvard Medical School Ad Hoc Committee to Examine the Definition of Brain Death, "A Definition of Irreversible Coma" (Harvard). This report has been widely interpreted as proposing that coma, diagnosed as irreversible by certain medical means, counts as a criterion for death in addition to the traditional criterion of cessation of respiration and pulsation. Several states have adopted legislation in accord with the recommendations of the report [*see* DEATH, DEFINITION AND DETERMINATION OF].

The basis of the so-called brain death criterion is that discernible absence of electrical activity in the cortex of the brain is a definitive sign of the death of the cerebral cortex, supposedly the center of "higher" human activities. When this is the case, it can be said that artificial life systems are merely supporting the "animal" activities of a corpse. However, persons in profound coma may show some brain activity. Should their "animal" functions be artificially continued, even when there is very little hope of return to consciousness or of recovery? This is an ethical rather than a legal question.

The ethical issues have been posed by applying certain distinctions, long familiar to ethics and to medicine, to this new problem. These are the distinctions between "ordinary and extraordinary," "usual and unusual," "useful and useless," and "imperative and elective." The discussions surrounding these distinctions and their applicability are extensive and subtle (Veatch). Several points must be noted about the most commonly used (and perhaps most controversial) of these distinctions, "ordinary vs. extraordinary" [*see* DEATH AND DYING: EUTHANASIA AND SUSTAINING LIFE, *article on* ETHICAL VIEWS]. First, many ethicists have remarked that physicians and ethicists use these terms quite differently. Medically, "extraordinary" tends to be identified with the new, experimental, rare or complex; morally, "extraordinary" is defined not by reference to the mode of treatment but to its effects upon the patient (and upon the patient's family). Thus, a definition fashioned originally in Catholic moral theology has been widely quoted with approval, "*extraordinary* means of preserving life [means] all medicines, treatments, and operations, which cannot be obtained or used without excessive expense, pain, or other inconvenience, or which, if used, would not offer a reasonable hope of benefit" (Kelly; Ramsey).

Commentators have stressed the final words, "[without] reasonable hope of benefit." They point out that the key to the proper use of this distinction, when the patient is incapable of participating in the decision, is the physician's judgment that the life-support system has little promise of "tiding the patient over" to a state of self-support. If this is the case, it is ethical to discontinue the artificial support. However, the difficulty of applying this judgment lies in the meaning given to "reasonable hope" and "benefit." Reasonable hope refers to a judicious assessment of the patient's state in the light of current medical science and the experience of the physician—an assessment filled with many variables. Benefit refers to some state of recovery or health. It is extremely difficult to find any consensus on the meaning of these terms, which are "value" words rather than references to some empirical state. At the very least, proponents of the ethical distinction between "ordinary" and "extraordinary" are generally agreed it is unreasonable to describe as "benefit" the perpetuation of a vegetative state, in which a perfused cadaver is maintained by artificial mechanical systems.

A final ethical problem arises after it has been decided that the person has indeed died. It may then be asked, Should the "dead" body continue to be perfused with oxygen in order to maintain fresh organs for transplantation, for experimentation, or for anatomical and physiological demonstrations? [See ORGAN DONATION.] This practice is fraught with legal hazards in any jurisdiction where "brain death" is not the sole criterion of legal death. However, its ethical propriety is still under discussion. The proposed benefits, which may be rationally listed and defended, come into conflict with profound cultural and religious feelings about the proper treatment of the recently dead (Gaylin).

Questions of policy. There are several policy questions about life-support systems that involve ethical questions. Most of these systems are, in the words of one influential author, "halfway technology"—that is, they do not themselves heal the disease or disorder, but rather bypass it, substituting temporarily or permanently for the damaged function (Thomas). As such, they often create chronic states of modified illness rather than curing. This has considerable consequences for the persons, for the medical care system, and for society. Persons may find themselves living compromised lives at great expense to themselves or their families. Physicians enticed by the brilliance of the technology may concentrate great effort and resources on these systems to the detriment of other, possibly more effective, forms of care. The costs of the sophisticated machinery, the need for highly trained technicians, the temptation to develop more intensive care units than are necessary, all contribute to the escalating costs of medical care borne by individuals and by society. In addition there is suspicion, seemingly confirmed by some studies, that much of the technology employed in intensive care units results in little or no improvement in care or cure of the patients (United States, Office of Technology Assessment). It is ethically as well as economically imperative to carry out cost-effectiveness studies, so that medical technologies are introduced into general use on the basis of solid evidence of their efficacy and superiority over other forms of treatment. Finally, the existence of this technology can create in society false hopes about the capabilities of medical care and, in addition, create fears in many persons of being captive to these machines, of "having one's death prolonged."

These are issues that can be resolved only by the adoption of social policies. Many advanced nations have attempted to develop rational policies about health care, the use of medical resources, and the pursuit of biomedical research. These policies stress efficient allocation and application of public resources to medical care. However, the goal of efficiency creates ethical dilemmas. They are usually best posed as problems of justice: What are fair criteria for inclusion and exclusion of categories of persons or of single persons in the medical care system? Is it just to exclude some without providing alternatives? How can a fair distribution of resources be made between different public needs? Who has the right to make allocation decisions? (Fuchs.)

In addition to problems of justice in allocation of resources, life-support systems create the need for policies about death and dying. How is death to be defined legally? How can persons retain control over the manner of their own dying? How is the society to care for those who are beyond cure? These questions, which all societies have faced, are made more urgent by medical technology which supplants "natural" forces and postpones "natural death." Specific forms of legislation about the definition of death and the authority of dying persons to refuse life-prolonging medical care have been enacted to meet these needs in many jurisdictions. The ethical assumptions underlying these needs and the legislation are increasingly discussed (Veatch).

Philosophical issues. Current life-support systems and the promise of even more sophisticated technology occasion a reflection deeper than the ethical considerations hitherto discussed. This reflection can be called philosophical, since it centers not so much on what the right course of action is as on how essential features of human life and of human nature are understood. Indeed, although the ethical questions come quickly to mind, they will be dealt with only in the most superficial way without some grounding in the deeper philosophical reflection. In recent years, a new interest in the philosophy of biology and the philosophy of medicine has drawn thinkers to these questions (Jonas).

Life-support systems substitute a machine or a machine-made product for a vital function. This leads to the philosophical question about the relationship between human persons and the technological artifact. This question has its origins in the reflections of Greek sophists about "nature" (*physis*) and "art" (*techne*). There are definitional matters to be settled: How does one define the natural or the artificial? Living and nonliving? Human and technical? But there are deeper problems: What is so essential about the human that its replacement by machine would abolish humanness? Can humans, who are by their own nature makers of machines, become themselves mechanized? Are there "natural" limits to the incursion of artifacts into human life? Is there "natural" death whose time can be recognized and should be respected?

These are profound questions. They are not stimulated by each kind of life-support system in itself, but rather by the general existence and possibilities of technological substitutes for human functions. Certain devices, such as the dialysis machine and the artificial heart, make it possible to pose the questions in a particularly dramatic way: Can there be a "man–machine symbiosis"? Ought we create "bionic persons"? However, the general question of the relationship between persons and machines as psychosomatic adjuncts must be posed in a careful and discriminating manner. The realistic possibilities of the technology must be distinguished from more imaginative potentialities; the actual purposes of its application must be clarified and distinguished from the intrinsic capabilities of the devices. Goals of restoration of function or recovery of health must be stated and distinguished from other uses less related to the traditional purposes of medical care. In sum, while life-support systems can serve laudable objectives, those objectives must become better defined and thoughtfully evaluated from medical, ethical, and philosophical perspectives.

ALBERT R. JONSEN AND
GEORGE LISTER

[*Directly related cross references are mentioned in the text. Other relevant material may be found under* INFANTS; INFORMED CONSENT IN HUMAN RESEARCH; INFORMED CONSENT IN THE THERAPEUTIC RELATIONSHIP; *and* KIDNEY DIALYSIS AND TRANSPLANTATION. *See also:* BIOLOGY, PHILOSOPHY OF; HUMAN EXPERIMENTATION, *article on* BASIC ISSUES; JUSTICE; *and* MEDICINE, PHILOSOPHY OF.]

BIBLIOGRAPHY

FOX, RENÉE C., and SWAZEY, JUDITH P. *The Courage to Fail: A Social View of Organ Transplants.* Chicago: University of Chicago Press, 1974.

FUCHS, VICTOR ROBERT. *Who Shall Live? Health, Economics, and Social Choice.* New York: Basic Books, 1974.

GAYLIN, WILLARD. "Harvesting the Dead: The Potential for Recycling Human Bodies." *Harper's,* September 1974, pp. 23–30.

Harvard Medical School, Ad Hoc Committee of the Harvard Medical School to Examine the Definition of Brain Death. "A Definition of Irreversible Coma." *Journal of the American Medical Association* 205 (1968): 337–340.

JONAS, HANS. "Technology and Responsibility: Reflections on the New Tasks of Ethics." *Social Research* 40, no. 1 (1973), pp. 31–54.

JONSEN, ALBERT R. "Scientific Medicine and Therapeutic Choice." *New England Journal of Medicine* 292 (1975): 1126–1127.

————. "The Totally Implantable Artificial Heart." *Hastings Center Report* 3, no. 5 (1973), pp. 1–4.

KELLY, GERALD A. *Medico-Moral Problems.* St. Louis: Catholic Hospital Association, 1958.

"Optimum Care for Hopelessly Ill Patients." *New England Journal of Medicine* 295 (1976): 362–364.

RAMSEY, PAUL. *The Patient As Person: Explorations in Medical Ethics.* The Lyman Beecher Lectures, Yale University. New Haven: Yale University Press, 1970.

TAIT, KAREN M., and WINSLOW, GERALD. "Beyond Consent—The Ethics of Decision Making in Emergency Medicine." *Western Journal of Medicine* 126 (1977): 156–159.

THOMAS, LEWIS. "Notes of a Biology Watcher: The Technology of Medicine." *New England Journal of Medicine* 285 (1971): 1366–1368.

United States, National Heart and Lung Institute. *The Totally Implantable Artificial Heart: Economic, Ethical, Legal, Medical, Psychiatric, Social Implications.* Report by the Artificial Heart Assessment Panel. DHEW Publication no. (NIH) 74–191. Bethesda, Md.: National Institutes of Health, 1973.

————, Office of Technology Assessment. *Development of Medical Technology: Opportunities for Assessment.* Washington: Government Printing Office, 1976.

VEATCH, ROBERT M. *Death, Dying, and the Biological Revolution: Our Last Quest for Responsibility.* New Haven: Yale University Press, 1976.

WÄHLIN, AAKE; WESTERMARK, LARS; and VLIET, ANSJE VAN DER. *Intensive Care.* Wiley-Interscience Publication. New York: J. Wiley & Sons, 1974.

LITERATURE AND MEDICINE
See MEDICAL ETHICS IN LITERATURE.

LIVING WILLS

See DEATH AND DYING: EUTHANASIA AND SUSTAINING LIFE, *article on* PROFESSIONAL AND PUBLIC POLICIES; RIGHT TO REFUSE MEDICAL CARE.

LONGEVITY

See AGING AND THE AGED, *articles on* THEORIES OF AGING AND ANTI-AGING TECHNIQUES *and* ETHICAL IMPLICATIONS IN AGING.

MAIMONIDES, PRAYER OF

See CODES OF MEDICAL ETHICS. See also APPENDIX, SECTION I, DAILY PRAYER OF A PHYSICIAN.

MALPRACTICE

See MEDICAL MALPRACTICE.

MAN, IMAGES OF

Images are vivid representations of human life and its place in the cosmos. They function as instruments for conveying various ways in which the meaning of human life is grasped, expressed, and dealt with. Images are thus intimately involved in justifications of ways in which human beings are treated. The clinician who treats patients with "absolute objectivity" may well have an image of himself as a thinking machine equipped with an X-ray eye. The research specialist may treat human subjects as though they appeared to him to be high-grade organisms whose personal histories, as distinguished from the chronicle of their pathologies, are irrelevant. Medical scientists who are animated by other and more traditional images, such as the ideal physician or loving father, may feel that they have violated the canons of science. In more general terms the norms of policy and practice affecting human well-being are likely to be more directly related to compelling images than to logically articulated doctrines or rationally contrived theories.

The history of Western man is a story of master images competing for sovereign power in the human spirit. These images have had different names and different effects in diverse cultures and philosophical systems. The names I have used for the images have been purposively chosen so they will not be associated with any one conceptual scheme. It is at once the virtue and the difficulty of images that one conceptual scheme may embody several images that when systematically related are incompatible. The images I have selected are not meant, therefore, to classify types but rather to elicit presuppositions, particularly in respect to issues in bioethics, that are all the more powerful because they are seldom explicitly recognized. The first pair of images to be considered helps illustrate this as their conflict with each other sets the stage for the strife of images in Western culture.

The singular creature versus the odd specimen

All of the major religions of the West and much of its philosophy employ the image of *the singular creature*. Man is thus represented as being a creature endowed with a value not derived from his life in the natural world. The laws of nature govern part of his life but not the essence of it. Essential humanity is manifested only in relationship to a being, or an order of being, transcending the natural order in power and dignity.

The biblical religions, which have largely shaped Western culture, represent *the singular creature* as the child of God. God has created man after his own likeness, in his own image. Man has thus a destiny altogether different from that of any other creature in the cosmos. The biblical religions do not agree among themselves on the time and place of the consummation of that unique destiny. They do agree that human life cannot be explained in purely naturalistic categories. They agree further that this uniqueness of being levies unique obligations upon human beings. An important element of his singularity is manifested in man's capacity for responding to commandments and solicitations. So, while he may largely behave like a reactive organism and thus seem to be highly predictable, man is under divine orders to achieve moral character. No other creature is so charged; no other is less predictable.

Much of systematic philosophy from classical antiquity to the present has laid great weight upon man's uniqueness. Philosophers generally have not relied upon images to express this doctrine. Plato is the grand exception. He says that man is the plaything of the gods. Not that man is a mere puppet in the hands of capricious deities; but that they enjoy his expression of godlike capabilities: rationality and creativity.

The odd specimen represents man as the product altogether of purely natural forces that function according to invariant mechanical laws. So man is an accident of nature rather than the planned offspring of higher powers. Thus in both ancient and modern forms *the odd specimen* strips human life of every illusion of uniqueness.

Not unique, then, but odd; an accident, but a curious one. Man is afflicted with strange hungers. For instance, he craves to know his world rather than merely react to it. Some of what he wants to know adds nothing to his ability to survive and prosper as a biological organism. Indeed his craving for knowledge may lead him to jeopardize his future on this planet.

Man is an odd specimen in another striking respect. He can create imaginary worlds, all-inclusive illusions. These he can treasure more than any of his physical possessions, more than any of his social perquisites. He can lose himself and his environment in visions; and count them well lost. This is an odd way for nature to behave.

When *the odd specimen* rules the spirit there is only one intelligible explanation for this freakish behavior of nature: Man's oddity is simply a subjective assessment of his being; he is not so perceived by any other being. Odd, then, as he may seem to himself, he is nonetheless a product of unintending natural forces wholly obedient to physical law.

So for *the odd specimen* the attribution of unique value to any particular human being simply expresses a subjective preference springing from subrational feeling or impulse. Systems of social conventions are forthcoming to reward some preferences and penalize others. But there is no really rational way to justify preferring a B. F. Skinner over a Charles Manson, a St. Francis over a Hitler.

There are many situations in the biomedical sciences in which specialists agree upon investigatory and restorative procedures though they do not share the same image worlds. Sooner or later fundamental disagreements over policy are bound to arise. As much at stake as anything else is the right of one image rather than any of its foes to preside over conflict in policy and procedure. If man is perceived and construed as *the singular creature*, if persons are thus endowed with unique value, then whatever policies and practices conflict with those truths are immoral, regardless of whether or not they are legal. But if *the odd specimen* has a superior grasp of the realities, then moral principles are really conventionalized rules concocted by *the odd specimen* to conceal from himself the brute nature of his organic drives. Then prudence rather than conscience ought to decide when the rules can be ignored either for the advancement of science or for private gain.

A remarkable variety of related conflicts develops from the generic strife of *the singular creature* with *the odd specimen*.

The sinner versus the victim

All of the biblical religions of the West and many strands of philosophical thought represent man as a creature guilty of transgression against moral laws grounded in ultimate reality. Such aberrations spring from conditions of the soul, the control center of one's being. The adult person is responsible for such conditions as his grossly inflated self-love, his capriciousness, his distorted perceptions of the reality and needs of other selves. This self-responsible creature is *the sinner*.

According to certain religious myths angelic beings preceded man in sin. But these demonic powers do not force man to sin. They may be accessories or accomplices, but man cannot legitimately blame them for his sin and guilt. Nor can he blame any lower creature or any aspect of the natural order. He may live in a society given over to wickedness, but that state of affairs is not the prime mover of his corruption.

The victim represents deviance as a product

of forces over which the individual deviant has no control. Ancient myths portray this creature as the victim of inexorable fate. Modern versions of this image systematically depersonalize the cosmic machinery. But the product remains the same: a creature flawed by a random misfiring of purely natural forces—genes, impersonal social institutions, universal subrational biological impulses. The modern *victim* has a great range of options from which to choose the prime cause of his unfortunate condition.

So the *odd specimen* comes across the horizon again. Now his oddity is his inability to relate appropriately and efficiently to his social environment. He is compulsively out of step—not because he sees and wills a value beyond his society or perversely clings to a value his society no longer credits; he is simply a creature whose impulses have not been adequately socialized. So the fundamental policy question about *the victim* is whether he can be so adjusted, reprogrammed, as to render him harmless if not productive. Thanks to biomedical science ways of doing this are now so sophisticated that the subject of such treatment, along with his loved ones, may feel nothing but gratitude for it. *The victim* is trained to esteem contentment above individualistic values.

Important ambiguities infest the conflict between *the sinner* and *the victim*. Many of them ride on the metaphor of *disease*. Thus, from the side of *the sinner:* Moral corruption is an infection that permeates the whole being of individual persons and the entire body of mankind. To be human is to be a sinner. That is both the grandeur and the misery of the human condition.

The metaphor of *disease* is also at home with *the victim*, except that he does not use it as a metaphor. For him drunkenness is not a sin or moral fault. It is a sickness and must be perceived and treated as a sickness. So also for sexual promiscuity, homosexuality, and addictive gambling, so far as such behavior is still presumed to be socially unacceptable if not actually reprehensible.

With the indispensable assistance of psychoanalytic doctrines, *the victim* has generalized the metaphor of *disease* to a level comparable to that achieved in Christian ages by *the sinner*. Society as such is sick, is systematically neurotic; incurably so, in all likelihood. So *the victim* sees deviance not as transgression but as behavior that happens to conflict with arbitrary social conventions.

The ambiguities of *disease* do not succeed in concealing the sharpness of the conflict between *the sinner* and *the victim*. *The sinner* is a being endowed with a limited but sovereign will and with real though impaired rationality. In his life moral corruption is not an effect of antecedent nonpersonal causes. His alienation from the true good is brought on by a fundamental fault in his own intentions. So above all else he needs justification and rectification.

The victim on the other hand is a triumph of causal explanation. In his world nothing and no one is to be morally faulted. The law may hold him responsible for destructive behavior, but he cannot help doing what he does. He is a product, not a self-mover.

Wide generalization of *the victim* provides a license for treating maladjusted and dysfunctional people as so many defective units produced by the biosocial system. Residual moral sentiments might yet dictate that such flawed products be treated with compassion and forbearance, except when the social cost of doing so is prohibitive. But if *the victim* is right, the long-range health of society depends on making the requisite changes in the biosocial system. It remains to be seen whether *the victim* has an objective understanding of health and a sufficiently rational capacity for calculating its proper instrumentation.

The soul-self versus the plastic man

The *soul-self* is a being endowed by the Creator with a unique packet of high capabilities. Human beings are obviously alike in outer aspects, but no person is a replica of any other person. So, while individuals have a common inclusive destiny, the place each person shall occupy in that ultimate community must be realized or attained by each. That ultimate perfection is intimately related to and may be a consequence of the attainment of a stable moral character in the present world.

Plastic man is able and ready to become whatever is demanded by external social forces. This creature is a function of his social environment; his reality is whatever role or compound of roles he is called upon to play in a given context. Accordingly, well-developed antennae are absolutely indispensable for his well-being; appeal to singular powers of mind, heart, or will would be at best profoundly perplexing. *Plastic man* is made to be manipulated. When he is not being manipulated he feels neglected, unloved, unreal. The aim of his existence is infinite adaptability rather than stable pattern or the exercise of either personal or social power. In a presumptively humane society he does not need to fear that the instruments of terror will be used to give him the shape required, except in dire crises.

The conflict between *soul-self* and *plastic man* rages in bioethics. Appeals to an absolute right to life in the abortion controversy are animated more by the image of the fetus as already a soul-self, literally in embryo, than by a general principle of reverence for all forms of life. The chief opposing view exploits the image of an organism that cannot have any compelling value, any enforceable rights, because real human beings are products of a social environment, and their rights are so many conventionalized social commitments. So appeal to the unique potentialities of a fetus is speculative. The right of the host female to her health and happiness must override all speculative ascription of right and value to something not yet really human.

Soul-self and *plastic man* pursue each other in the burgeoning field of genetic engineering. Until now the genetic stock of human life has been subjected to the random play of such forces as famine, plague, climatic conditions, warfare, and irrational distributions of both natural resources and other forms of wealth. Now science has made it possible to produce desirable mutations in man's genetic stock. Scientific thought has replaced that ancient abstraction, "human nature," so dear to the heart of *soul-self;* the substitute, so necessary for *plastic man*, is a complex genetic spread rendered conceptually manageable by arbitrary parameters. So there is no fixed limit on what can be made of man by human beings equipped with the technical prowess.

Similar considerations apply to scientific experimentation with live human subjects. *Plastic man* gives every encouragement to extend experimentation far beyond its present modest endeavors, envisioning a humanity far healthier than any the earth has ever seen. To this, *soul-self* protests that such a vision confounds the power of scientific man with the prerogatives of Deity. God has endowed humanity with a fixed nature, and to that he has given a sacred value. Human efforts to modify that nature, thus violating that sacred value, will surely draw down upon the age of *plastic man* a terrible reprisal.

But why should *plastic man* be frightened? He is not burdened with an immortal soul; his psyche has already been patterned out of those ancient terrors of everlasting punishment. He is altogether in favor of new patterns for his psyche—and new ones beyond those. He is altogether opposed to bondage to the delusive images of *soul-self*.

The pilgrim versus the stranger

The pilgrim is man searching in faithful perseverance across an alien world for his true homeland. *The stranger* has no real home here or hereafter. Alienation is his natural and unrelievable condition. *The pilgrim* may or may not have a true home on earth; the biblical religions do not agree on this. They do agree that eventually alienation is overcome once and for all. Then fitly prepared souls are united in a sublime and everlasting community. But *the stranger* has no such destiny. He knows that he is a cosmic orphan. He grants that it is possible to synthesize artificially an ephemeral sense of belonging. Yet he doubts the value of the effort and self-deception required for that short-term gain.

Pilgrim and *stranger* make common cause against everything making for complete domestication of the human spirit within any humanly contrived environment, present or future. They would choose implacable anxiety and unappeasable guilt rather than total acquiescence to the ordinances of any man-made and man-controlled society, no matter how humane its advertisements and rationalizations.

Theoreticians and policymakers who want to banish alienation from human life have rugged foes in *the pilgrim* and *the stranger*. But that does not mean that these two images are good companions on the road to a blessed community at the end of history. *The pilgrim* believes that he has been created for life in community, but he must constantly be on guard against the seductions of false community. For *the stranger* "community" is merely a name for a society to which people have voluntarily surrendered their individuality and relinquished their freedom.

The worker versus the player

The economic actualities of job, productivity, usefulness, efficiency, and buying power define the meaning of life for *the worker*. In his world, "What do you do for a living?" is an ultimately serious question. A person is not someone who happens to work; personhood is defined by work. A person is not someone who happens to acquire things; acquisitiveness is the prime instinct in real human beings.

The player puts the capacities for the creative enjoyment of life far ahead of its economic aims and instruments. For him the world itself is a game. The really meaningful life consists in mastering its rules and thereupon shaping the expenditures of one's energies as creatively as possible.

In ancient civilization *the player* was largely a creature of an affluent leisure class. Indeed, well into modern times he was carried on the back of *the worker*, and *the worker* was enslaved by the system of production. Today the "work ethic" is being widely and passionately rejected. In the midst of ambiguities embracing these negations the conflict between *the worker* and *the player* persists. This can be felt in the severe difficulties both socialist and capitalist societies are having with leisure time. The problem is to give it a value other than reward for productive work time—which leaves work time as the critical index to *worker's* worth. In American society the systematic exploitation of leisure time is a very large business, requiring the dedication of myriad workers and the manipulation of hosts of plastic people.

Worker and *player* collide in bioethics. Should costly medical instruments be used to prolong the life of a person no longer productive or useful—except for purposes of experimentation? Should scarce medical instruments be reserved either for people who can pay for them or for people who are more productive than others? Or, on the other hand, should the decisive criterion be this double-jointed one: whether a person is, or might again become, capable of enjoying his or her distinctive life; and of expressing that enjoyment in a unique way.

Worker and *player* pursue each other into genetic engineering. To what end is cloning to be dedicated? *Worker* reminds us that healthy and happy people make better workers, better producers and consumers, than sick and unhappy people do. *Player* asks: But suppose the link between the highest kinds of creativity and certain neuroses were not accidental? Freud was suspicious of art. Why was he so naive about science?

The steward versus the exploiter

The steward capitalizes the sense of life as a gift for which the giver exacts accountability and levies appropriate rewards and penalties. To *the steward* the resources of nature and human nature are given to have and to hold in trust, not in fee simple or absolute possession. *The exploiter* sees the resources of nature and of human nature as so many challenges to enterprise. Life and its good things are not gifts. Life is a fact; the good things are the fruit of one's forcible and cunning extraction. And *the exploiter's* accountability is purely positivistic, that is, only to human juristic rules backed by the sanctions of proved power. So daring coupled with prudence is as close as *the exploiter* means to get to the moral order.

Thanks to scientific technology *the exploiter* has moved up into a range of enterprise undreamed of earlier. One bite of a machine now used in strip mining contains more coal than a horde of miners with manual tools could extract in a week. And if positive law rather than moral sentiment demands it, the devastated terrain can be swiftly returned to a reasonable resemblance of its pristine order.

So the contemporary *exploiter* has high confidence in his ability to live with the consequences of his ever expanding enterprise; or at least with the more immediate consequences. Sufficient unto the evils of the day are the medicaments of scientific technology.

Steward and *exploiter* have rendered bioethics contested territory. *The steward* argues the sanctity of human life as a charge sufficiently grave to be divine. He is likely to appeal to some intolerable consequences if this charge is not fully honored. In medical technology *the exploiter* sees new challenges to enterprise, but not necessarily for his own aggrandizement above all else. As a man of science he envisions advancement of knowledge. Knowledge for what? asks *the steward*. *The exploiter* answers: Eventually for the amplification of man's power to control nature; thus to reduce the play of chance to the vanishing point.

The provincial versus the boundary-breaker

This contest may prove to be the most fateful of all. *The provincial* is man committed to living within everlastingly fixed boundaries; it is possible to achieve a truly human existence only within them. This image has often carried with it a religious-philosophical conviction that the ultimate source or giver of life has legislated these boundaries, and he will not tolerate any violation of them. *The provincial* may believe that the boundaries are far-flung; he may be a cosmopolite rather than a rustic. But, however generous the divine definition of the limits, in any case determining an ample space for every legitimate human interest, the limits are there and they are everlasting. Accordingly *the provincial* is predisposed to see every great devastation of human life as a divine requital for transgression of the boundaries of the human province.

Man as *boundary-breaker* will have none of this. For him all boundaries are provisional; those legislated by every human society, with the sole exception of the scientific community, are arbitrary. No doubt some limits are necessary, but there is nothing metaphysical or mystical about that. So the scientific exploration alike of the structure of the cosmos as a physical system and of the inner life of man is rule-oriented. The rules are methodological; nothing is gained, much is lost, by viewing them as divine ordinances. As for the "laws of nature," history proves conclusively that they are open to modification; modern science is a story of conceptual revolutions. True, nature itself is orderly. But the principles defining that orderliness are human achievements, and there is nothing absolutely fixed in the human province.

Scientific technology has immensely amplified the power of *the exploiter* to make both nature and large segments of mankind into prey. Investigative science, sometimes called pure, has given *the boundary-breaker* fantastic powers for dissolving ancient mysteries of mind and matter —first by conceptual revolutions, thereafter by actual probes into what heretofore was known as the inviolable soul and out into the far reaches of the cosmos.

So it comes about that *the boundary-breaker* now views all rules as human "how-to" fiats, thus as revisable or disposable as forward-pressing interests dictate. *The provincial* persists in seeing the prime rules as moral prescriptions to which all "how-to" fiats must give an accounting.

The clash of these images reverberates throughout bioethics. Scientific research is committed to breaking through arbitrarily stipulated boundaries of knowledge. What the boundaries define, in *the boundary-breaker's* eyes, is like a bivouac in which science momentarily rests before making the next advance. That rather than an encounter with an absolute "Thou Shalt Not Pass."

So as science clears away thickets and tangles in man's knowledge new avenues are opened for controlling all of the forces of nature that prey on human life. But *the boundary-breaker* is not content with this. He intends to bring down the ancient boundary between man and nature, that barricade of archaic images alienating the human province from the animal kingdom. Once that boundary is breached human beings can be treated as organisms differing from "lower" creatures only in degree of structural complexity. For this purpose a range of "how-to" fiats comes into view: rules for inducing from the organism behavior patterns that yield optimal satisfaction; better yet, "how-to" rules for producing human organisms free of destructive proclivities from the start. *The provincial* holds the boundaries between the human province and the animal kingdom to be inviolable. Human beings can in fact be treated as though they were cattle or vermin, but not with impunity. The penalty for this ultimate transgression is the dehumanization of the dehumanizers. If the program of dehumanization is sufficiently systematic and adequately rationalized it may carry a whole civilization down into a bestial darkness.

So *the provincial* insists that investigative and therapeutic "how-to" rules, techniques, and policies must answer to rules irreducibly moral. A practical application: he says it is not right that medical review boards should be entirely staffed with medical personnel. On every such jury there should be someone who holds a watching brief for the indefeasibly human, a champion of the integrity of the human province.

The struggle among images

The strife among the great perennial images has revolved around the question whether man's being and value are unique. In the contemporary scene this issue achieves a unique expression in the conflict between the images of *prophet* and *forecaster*. Thus some of the ancient conflicts between biblical religion and magic are renewed.

In the biblical religions *the prophet* is a person (or a community) ordained to reveal the religious-ethical demands of God. This revelation is tensed: It is grounded in history; it contains the indispensable directives for faithful existence in the present; it sketches the shape of that community in which man's being is ultimately perfected, the Kingdom of God. *The prophet's* references to the near future are all provisional: The people of God shall surely suffer if they fail to keep covenant. But *the prophet's* allusions to the ultimate future are all absolute: The perfection of human community in God's good time is altogether certain; but only God is the determiner of that time.

The forecaster has grasped cosmic law and thus is able to deduce the shape and substance of things to come. Discovery of those laws is a historical process. The structure thus apprehended is timelessly perfect. But *the forecaster's* interest in this realm of law is not contemplative, it is practical. He wants so to amplify and clarify this supreme knowledge that human behavior may be predicted, both in general and in particular. As a realist *the forecaster* admits that this ideal is imperfectly realizable for the time being. Yet he believes that it contains the seed of the future. *The prophet* does not know when a generation will appear that will properly heed the transcendent demands of the ethical order. He is sure that only in so doing is man's well-being assured.

Both *the prophet* and *the forecaster* have visions of a human community abounding in health, power, and happiness. *The prophet* believes that the shape of that blessed community has been given in history, and the route to it is contained in certain ethical absolutes backed by divine sanctions. As man of science *the forecaster* has an attitude toward moral rules that is prudential. In *the prophet's* view prudential considerations pertain to the order of "how-to," whereas moral rules always embrace something uniquely human. Respect for that is the beginning of wisdom.

The charter of this article does not include a warrant for predicting the outcome of any of the contests among the root images of man in Western life.

JULIAN N. HARTT

[*For further discussion of topics mentioned in this article, see the entries:* ABORTION; BEHAVIOR CONTROL; ENVIRONMENT AND MAN; ENVIRONMENTAL ETHICS; ETHICS, *articles on* SITUATION ETHICS, OBJECTIVISM IN ETHICS, *and* RELATIVISM; FREE WILL AND DETERMINISM; FUTURE GENERATIONS, OBLIGATIONS TO; NATURAL LAW; PERSON; PURPOSE IN THE UNIVERSE; RESEARCH, BEHAVIORAL; *and* RESEARCH, BIOMEDICAL. *Other relevant material may be found under* BIOETHICS; BIOLOGY, PHILOSOPHY OF; EVOLUTION; JUSTICE; MEDICAL ETHICS IN LITERATURE; MEDICINE, PHILOSOPHY OF; *and* REDUCTIONISM. *See also:* DRUG USE, *article on* DRUG USE FOR PLEASURE AND TRANSCENDENT EXPERIENCE; MEDICAL ETHICS, HISTORY OF, *section on* EUROPE AND THE AMERICAS, *article on* EASTERN EUROPE IN THE TWENTIETH CENTURY.]

BIBLIOGRAPHY

ARENDT, HANNAH. *The Human Condition.* Chicago: University of Chicago Press, 1958.
BLACK, MAX. *Models and Metaphors.* Ithaca: Cornell University Press, 1962.
BOULDING, KENNETH. *The Image.* Ann Arbor: University of Michigan Press, 1956.
CAMUS, ALBERT. *The Stranger.* Translated by Stuart Gilbert. New York: Vintage Books, 1946.
DRUCKER, PETER F. *The End of Economic Man.* New York: John Day Co., 1939.
EMMET, DOROTHY M. *The Nature of Metaphysical Thinking.* London: Macmillan & Co., 1946.
FARRER, A. M. *The Glass of Vision.* London: Dacre Press, 1948.
FREUD, SIGMUND. *Civilization and its Discontents.* Translated by James Strachey. New York: W. W. Norton, 1961.
HARNED, DAVID BAILY. "The Deviant Self: Everyman as Vandal." *Virginia Quarterly Review* 51 (1975): 329–346.
HART, R. *Unfinished Man and the Imagination.* New York: Herder & Herder, 1968.
HARTT, JULIAN N. *The Lost Image of Man.* Baton Rouge: Louisiana State University Press, 1963.
HUIZINGA, JOHAN. *Homo Ludens.* Boston: Beacon Press, 1955.
LEWIS, C. S. *Out of the Silent Planet.* New York: Macmillan Co., 1943.
MACPHERSON, C. B. *Democratic Theory: Essays in Retrieval.* Oxford: Clarendon Press, 1973.
MARGENAU, HENRY. *The Nature of Physical Reality.* New York: McGraw-Hill, 1950.
PEPPER, STEPHEN. *World Hypotheses.* Berkeley: University of California Press, 1942.
PLATO. "Timaeus," "Republic," "Laws." *The Dialogues of Plato.* Translated by Benjamin Jowett. New York: Random House, 1937.
RICOEUR, PAUL. *Fallible Man.* Translated by Charles Kelbley. Chicago: Henry Regnery, 1965.
RIEFF, PHILIP. *The Triumph of the Therapeutic.* New York: Harper & Row, 1966.
SARTRE, JEAN-PAUL. *Being and Nothingness.* Translated by Hazel E. Barnes. New York: Philosophical Library, 1956.

MANIPULATION

See BEHAVIOR CONTROL; BEHAVIORAL THERAPIES; TECHNOLOGY; GENE THERAPY; REPRODUCTIVE TECHNOLOGIES.

MARXISM

See MEDICAL ETHICS, HISTORY OF, *section on* EUROPE AND THE AMERICAS, *article on* EASTERN EUROPE IN THE TWENTIETH CENTURY.

MASS HEALTH SCREENING

Mass health screening involves the rapid and economical testing of a large number of apparently well individuals to sort out those who probably have a disease from those who probably do not. The actual test employed in a "screen" may vary from blood sampling (as in the VDRL exam used to suggest the presence of a venereal disease), to blood pressure tests (used to suggest the presence of hypertension), to chest X-rays (when evidence of diseases of the lung is being sought), to microscopic smears of cells from the cervix (looking for cervical tissue abnormalities), to a combination of medical history, physical exam, and a variety of physiological studies all administered within a brief period of time and testing several different functions (multiphasic screening).

Screening differs from diagnosis in a number of ways. A screening test is not intended to be definitive; hence, persons with positive (or suspicious) findings must be referred to their physicians for a more conclusive examination (i.e., diagnosis). Built into screening (from the Middle English *skrene* or sieve) is the realization that the test results for some individuals will erroneously indicate that they have the disease (false positives), while the results for others will erroneously indicate that they do not have it (false negatives). A screening test is "sensitive" to the extent it is able to give a positive finding when the person tested truly has the disease under study; it is "specific" to the extent that it is able to give a negative finding when the person tested is free of the disease under study. Diagnosis (from the Greek word meaning "to distinguish") implies that a final (and presumably correct) decision is made as to the presence or absence of the disease in a given individual. Diagnosis goes beyond the single test or battery of tests employed in screening to a synthesis of the most complete and accurate information available about the particular individual. Because they involve the testing of a specific individual for the presence of probable disease, diagnostic examinations are both more expensive and more accurate than screening tests.

Screening is employed for either of two purposes: (1) to prevent the onset of new disease (primary preventive screening) or, more commonly, (2) to detect the probable presence of disease at an early stage so that the individual may then receive definitive diagnosis and therapy (secondary preventive screening). Only the former is properly mass "health" screening. Both types deal with presumably presymptomatic persons; but in primary preventive screening even those testing positive will still be considered healthy, although at special risk of becoming diseased.

History of mass health screening

Early detection programs have been advocated at least since the middle of the nineteenth century. Mass screening was first applied between 1910 and 1930 as a response to a number of community needs, such as keeping individuals with certain physical disabilities or communicable diseases out of military service, controlling the spread of communicable diseases like tuberculosis and syphilis, and protecting industry from the subsequent loss of their workers' services through preemployment examinations. In the 1930s began the examination of marital blood specimens to halt congenital syphilis and the mass screening of school children for deafness so that enlarged adenoids might be treated with radium. The majority of such early screening projects concentrated on the detection of presymptomatic cases, i.e., on secondary preventive screening.

The growth of primary preventive screening is a somewhat later phenomenon that might be considered to have really begun with the expansion of the insurance industry in the 1930s and the accompanying need to assess accurately the risk of providing coverage to certain classes of individuals. It was clearly present, however, when, in the early 1950s, automated testing and multiphasic health screening for conditions expected to pose a disease risk were initiated. Prepaid health-care programs such as Kaiser–Permanente played a significant role in the development of this automated testing and primary preventive screening. Finally, advances in the detection of genetically transmitted diseases such as sickle-cell anemia and Tay Sachs disease opened the possibility of testing for people clinically well but at risk of having a child with the disease.

Criteria for an adequate screening program

There are many requirements for an adequate screening program. It is important, for ethical assessment, that all of them be fulfilled or, at least, attended to with an explanation of why they could not be fulfilled. There must be: (1) a definition before screening begins of the purpose of each screening procedure and of the attainable objectives of the screening program as a whole; (2) the availability of a useful and reliable battery of devices and/or procedures capable of detecting the disease prior to its usual time of diagnosis in the study population; (3) a known effective therapy for the disease being sought when the purpose of the screen is not only for education or for epidemiological information; (4) one or more critical points in the history of the disease (after diagnosis first becomes possible and before it is ordinarily made in the community) such that therapy instituted before one of these points is more effective in interfering with the course of the disease than is therapy undertaken afterward; (5) the availability of well or of asymptomatic persons who are motivated to undergo a series of examinations that may be expensive, time-consuming, and sometimes unpleasant; (6) a definition of the target population with some knowledge of its demographic and social characteristics; (7) a determination of what volume of patients is required for a valid screen and then to keep the program viable; (8) financial resources to support the screening preparation, execution, and follow-up; (9) a more comprehensive communicable, chronic, or genetic disease program of which the screening is but one integral part; (10) a sense of responsibility on the part of the screeners to ensure that those screened are informed of the physical, financial, and social implications of the test results; (11) where possible, educated community involvement in the formulation of the program design and objectives, the administration of the program, and the evaluation of its results; (12) updating of the original program design and objectives in the light of program experience and new medical developments; (13) absence of compulsion to participate in the screen; (14) confidentiality of the screening results; and (15) open evaluation of the results of the screen itself and of the program.

Problems in mass health screening

Problems in mass health screening include effectiveness, evaluation, costs, benefits, follow-up, the "normal range" of test results, and who should be screened. Other issues, which will not be considered here, are who should screen, how it should be organized, and how it should be carried out.

Effectiveness and evaluation of mass health screens. The effectiveness of mass health screening is often uncertain. For most known diseases, it is still unclear whether primary prevention controls disease any more effectively than secondary preventive screening, and further whether such discovery of presymptomatic disease through secondary preventive screening is any more effective than the diagnosis and treatment of early symptomatic disease.

One source of ethical problems in both primary and secondary preventive screening is the making of screening process decisions that lead to a known incidence of expense and inconvenience on the part of both screeners and screened, based on an estimate rather than a knowledge of the risks of the disease. In the case of primary preventive screening there is the additional element of the screen dealing principally with risks themselves, i.e., with the known fact that only a portion of those examined and found to have the predisposing factors will ever become diseased. The detection of these factors will, however, expose all individuals found to have them to the fears, concerns, inconveniences, and possible prejudices endured by those actually having the disease itself.

Evaluation of the effect of mass health screening in terms of decreasing morbidity and mortality and of its cost-benefits depends on the actual incidence, prevalence, and natural history of the disease. These are often very hard to determine. Comparisons of the results of morbidity surveys with clinical examinations commonly lead to large discrepancies. In addition, almost no mass screening program at the present time is equipped to detect the presymptomatic or asymptomatic stages of a large portion of those selected chronic and disabling conditions which are most significant in terms of frequency of occurrence, or chronic disability, or amenability to public health prevention and/or control.

At the same time there is a lack of thoroughly validated and appropriately sensitive tests for the majority of the socially and financially most debilitating chronic illnesses. In spite of this, however, there is little doubt that mass screens have resulted in the referral of many individuals for subsequent definitive diagnosis and treatment of such diseases as tuberculosis, glaucoma, hearing defects, syphilis, diabetes, cancer, and hypertension.

As of 1977 there is insufficient proof that prognosis is improved by mass health screening that utilizes periodic health exams. The preventive effect of such examinations seems to occur principally when a person has already developed a significant disease. In other words, periodic health examinations are chiefly beneficial in secondary prevention, i.e., the early detection of already present disease. In periodic health screening primary prevention usually occurs through the diminished incidence of those diseases which commonly accompany the discovered disease.

Most screening surveys must currently be considered experiments, and consideration should be given to applying to such examinations the same discretionary controls that are employed in other, more obvious, examples of experimentation with human subjects. As has been noted above, this would entail minimally some simple explanation to those screened about the possible hazards or inconveniences of the test, i.e., an X-ray exam, about the possible personal implications of a positive test, about facilities for follow-up if the test is positive, and about the presumed relationship between the screened-for condition and its possible health consequences. In the case of primary preventive screening, it should also include an explanation as to whether the relationship between the risk factor and the disease is solely an observed one or in addition is considered a causative one. For instance, the observed relationship between a risk factor and a disease is sometimes questionable, as in the proposed association of an XYY chromosome pattern with subsequent criminal behavior and in the presumed relationship of certain early personality profile tests with subsequent antisocial activity. At other times the existence of the relationship is already established, but there is a question as to whether the relationship is causative, as in the hypothesized causative effect of certain habits (e.g., smoking) on the eventual development of certain diseases, and in the postulated causative relationship between elevated serum cholesterol values and the onset of coronary artery disease.

Costs and benefits of mass health screening. Mass health screening projects can be very expensive. Screening demands much time, money, effort, and personnel before it can even begin; it then needs a steady stream of examinees if it is to remain in operation. Often the entire exam or one of its components must be repeated at specified intervals if the test is to be most effective. The selection of the screening test is not dependent, however, solely on its effectiveness and dependability; minimum cost and maximum benefits are also issues.

These terms "costs" and "benefits" always raise the question of the yardstick by which they should be determined. While the actual process of such determination is important for adequately evaluating the program, the chief ethical requirement is that there be full disclosure of the way in which costs and benefits are determined, defined, and then applied to the screen in question.

The sources of costs in mass health screening projects are multiple. There are the costs (in terms of equipment, materials, and manpower) of the screening tests themselves, of contacting the desired public and then insuring that they attend, of the number of false positives who will require further investigation if the screening test is of low specificity, of repeated screening as determined by the incidence and duration of the disease under scrutiny, and of following the true positives to ensure adequate treatment.

This is a complicated process, because the real costs (money, personnel, and equipment) are not always the same as the monetary expenditures. The real costs of a disease, for instance, include the personal fears, the social prejudices, the psychological reactions, the cultural superstitions, and the other variables associated with illness and death. Thus, even when a society expends no money directly on the control or treatment of a disease, it may still be subject to the costs of that disease.

Because the determination of the real costs of a screening program is so difficult, it imposes on the advocates of such a program the obligation not to belittle this complexity or use it to conceal other possible relationships between costs and benefits as they might appear to affected individuals. In short, complexity must be presented as something complicated, but done so in a simple manner.

The benefits of mass health screening depend on the psychological and physical impact of the disease for the community and/or individual, the prevalence and/or severity of the disease, the sensitivity of the tests, and the extent to which the disease can be affected by treatment after it is detected. Like the consequences bred by any piece of knowledge, there is no benefit to be derived from health screening that will not have its attendant cost. The determination of the presence of any treatable disease will bring demands for the application of that treatment. The determination of the prevalence of any currently nontreatable disease will lead to new insistence on research into possible cures. Some screens may lead to monetary expenditures that will eventually result in decreased costs from the disease. For instance, tuberculosis screening did finally bring about overall savings to the community, but largely because it is a disease that afflicted mostly young adults. Current screening for heart disease, on the other hand, may only increase health-care costs without any attendant societal savings because of the older age of those so afflicted. In the end, it may be concluded that cost-benefit calculations alone are useful more as an affirmative argument in support of a screening program than in opposition to one that is advocated on humanitarian or other grounds.

Abnormal test results. The issue of laboratory test results ("values") outside the accepted range of "normal" is another issue accented by the increasing use of mass health screens. It is already known that there is some variation in published data on many "abnormal" laboratory values, that there is a wide range in physician interpretation of "abnormal" results, and that there is often an ignoring of "abnormal" results in arriving at a diagnosis. Commonly there is no general agreement among physicians that a given value or constellation of values constitutes clear evidence of the presence or absence of a particular disease. The manner in which "normal" values are defined and determined must be examined anew with special attention to the variations in "normal" values by age, sex, and race. Even then, a generation of testing of groups of differing age, sex, etc., then following them up, may be required to determine whether in individuals with marginally abnormal values significant disease develops to a greater extent or sooner than in those individuals with "normal" test results. This is an especially important issue because of the life-style changes, together with the adverse insurance and employment effects, resulting from "abnormal" laboratory values such as fasting blood sugar, uric acid, and serum cholesterol.

Who should be screened. Finally there is the important question of who should be screened. Mass health screening is directed not only at the detection of disease in the general population; it is also applied (as selective screening) quite frequently to patients admitted to a hospital, to school children in a search for sensory deficits which may impede their learning process, and episodically to selected groups of such children in the hopes of identifying either future behavioral difficulties or the presence of genetically transmitted characteristics like the sickle-cell trait. Different problems are posed by the screening of each of these groups: the general population, hospital patients, and school children.

The general population has the right to expect dual protection against undetected carriers of communicable disease and against itself being subjected to unnecessary health screening. It has the additional right to expect that its privacy will not be invaded, no matter how necessarily, by screening procedures that involve no candid explanation of the implications of the test, especially where no effective therapy exists for the sought condition. Those who agree to the screen have the right at its conclusion to reassurance as to the state of their own health. Notification only of those with positive test results is not enough; those with negative results are entitled at least to information as to this fact. Funds for such notification should be built into every screening budget. In this way the screening procedure can serve as an important vehicle for health education.

In a very few present instances is there any justification for publicly mandating screening. Exceptions would be screening for those communicable diseases which may asymptomatically exist in a transmissible state in certain individuals at peculiar risk to disseminate them widely to a susceptible population (i.e., typhoid in food handlers and tuberculosis in hospital personnel), and also screening for sensory deficits (seeing, hearing, speaking, or reading difficulties) in school children which would pose grave obstacles to their progress in learning. In the former case the screening is of a selected population who might be a particularly vicious menace to other individuals; in the latter, it is part of the overall responsibility of the state to determine that its children possess the minimal physical skills required for education.

One reason that there should be considerable hesitation in legislating screening is the present uncertain state of the art. All screening without exception involves errors in misidentifying some who have and who do not have a condition. Legislated screening would thus expose a certain proportion of those screened to misidentification and its attendant inconveniences, loss of privacy, and possible loss of privileges (i.e., insurability) without either free individual consent on their part or any overriding benefit to society.

This applies in particular to screening for conditions, such as sickle-cell and Tay Sachs diseases, that are most prevalent in certain ethnic groups. The sole justifications for such screenings are (1) genetic counseling of those able to transmit the disease to their offspring, (2) epidemiological surveys of the prevalence of the disease, and (3) patient education of those having the disease or the ability to transmit it. None of these reasons is sufficient to compel individuals legislatively to submit to screening examinations. In addition, even social pressure on susceptible ethnic groups to submit to such screening exposes them to the possibility of future discriminatory actions that may work to their disadvantage, not because each person is individually found to have the condition in question (i.e., high blood pressure) but because the mere fact that they belong to a certain ethnic group (e.g., blacks) is mistakenly taken to mean that they have a high *individual* (as contrasted to group) risk of having the condition.

Hospital and clinic populations provide a particularly profitable yield for screening procedures. This is because such patients do not require time-consuming and expensive promotional campaigns, because laboratory facilities are readily available, and because the results of the screenings can be quickly and easily followed up.

The performance of health screens on hospital and clinic populations affords both advantages and its problems. Various studies report that the profile of laboratory determinations performed on individuals admitted to hospitals indicates in a significant proportion of patients the existence of abnormal test results that have not been sought by the physician and are often of direct help to him in caring for the patient. It can thus be argued that, appropriately used, laboratory screens strengthen the physician's diagnostic precision and improve his management of disease.

The proliferation of such tests has at the same time, however, increased the opportunity for their excessive and unnecessary misuse. In addition, there is some evidence that the incidence of false positives in such screens is high and that some of the "cheaper" tests have little effect on diagnosis. Sigmoidoscopic (colon flexure) examination, tests for glucosuria, and two-hour postprandial blood sugar, the EKG, and anterior–posterior chest X-rays are considered to be the screens leading to the highest number of true positives.

The screening of school children for conditions other than communicable diseases or sensory deficits which may directly interfere with their learning process raises the issues of informed consent, or coercion, and of access to the results. If the parents are provided with an adequate explanation of the reasons for the screen and of the implications of both positive and negative results, the demands of informed consent would seem to be largely satisfied. Coercion is another matter, as in any situation where givers and withholders of rewards are intimately involved in an examination process. For there to be no coercion, the fact of participation or nonparticipation and the results of any school-based screen for conditions not directly and immediately relevant to the education process should be unavailable to the school authorities without the written permission of the parents. In addition, parents should have access to any and all physical and psychological testing results concerning their children.

In any event, everyone who consents to becoming involved in any physical or psychological screening process is entitled to know its results and their implications. Active follow-up is required, therefore, for both positives and negatives. Follow-up of positives is particularly crucial when the disease is one that can be successfully treated. When there is little that can be done to stem the course of the disease or ameliorate its effects, careful attention should be devoted beforehand to whether the screen should be conducted at all. Two justifications for screening for currently irreversible diseases would be medical experimentation and the collection of evidence about exactly when a disease does become irreversible. With any shortage of funds for mass health screening, priority should be given to the early detection of those diseases for which early treatment has been shown to be effective. In any event, screeners should always guard against producing anxiety and invalidism in otherwise healthy people. Reassurance to those screening negatively is but the corollary to accurate and balanced information for those screening positively.

JAMES R. MISSETT AND CARL E. TAYLOR

[*For further discussion of topics mentioned in this article, see the entries:* CHILDREN AND BIOMEDICINE; GENETIC ASPECTS OF HUMAN BEHAVIOR; GENETIC SCREENING; HOSPITALS; HUMAN EXPERIMENTATION; INFORMED CONSENT IN HUMAN RESEARCH; PRIVACY; *and* RISK. *Also directly related are the entries* GENETIC DIAGNOSIS AND COUNSELING; HEALTH POLICY; HEALTH INSURANCE; PRENATAL DIAGNOSIS; *and* PUBLIC HEALTH. *See also:* HEALTH CARE; *and* MENTAL ILLNESS, *article on* LABELING IN MENTAL ILLNESS: LEGAL ASPECTS.]

BIBLIOGRAPHY

American Medical Association, Council on Medical Service. *A Study of Multiple Screening.* Chicago: American Medical Association, 1955.

BERGSMA, DANIEL, ed.; LAPPE, MARC; ROBLIN, RICHARD O.; and GUSTAFSON, JAMES M., eds. *Ethical, Social, and Legal Dimensions of Screening for Human Genetic Disease.* Birth Defects Original Articles series, vol. 10, no. 6. New York: Stratton Intercontinental Medical Book Corporation, 1974.

Commission on Chronic Illness. *Chronic Illness in the United States.* 4 vols. Cambridge: Harvard University Press, 1956–1959.
GELMAN, ANNA C. *Multiphasic Health Testing Systems: Reviews and Annotations.* HSRD 73-70-42 PB 196651. Rockville, Md.: U.S. Health, Education & Welfare Department; National Center for Health Services Research & Development, 1971.
HUTCHISON, GEORGE B. "Evaluation of Preventive Services." *Journal of Chronic Diseases* 11 (1960): 497–508.
Institute of Society, Ethics, and the Life Sciences; Research Group on Ethical, Social, and Legal Issues in Genetic Counseling and Genetic Engineering. "Ethical and Social Issues in Screening for Genetic Disease:" *New England Journal of Medicine* 286 (1972): 1129–1132.
National Research Council, Committee for the Study of Inborn Errors of Metabolism. *Genetic Screening: Programs, Principles, and Research.* Washington: National Academy of Sciences, 1975.
"Screening for Disease." *Lancet* 2 (1974): 818ff. A series of articles by eighteen authors running from *Lancet* no. 7884 (5 October) to no. 7895 (21 December).
TEELING-SMITH, G. "Economic Aspects of Screening Programs." *Proceedings of the Royal Society of Medicine* 61 (1968): 767–768.
WARSHAW, LEON J. "Principles of Screening and Types of Programs." *Bulletin of the New York Academy of Medicine* 45 (1969): 1259–1268.

MEDIA AND MEDICINE

See COMMUNICATION, BIOMEDICAL, *article on* MEDIA AND MEDICINE.

MEDICAL EDUCATION

This entry deals with ethical issues in medical education itself. For a discussion of the teaching of medical ethics in medical schools, see MEDICAL ETHICS EDUCATION.

Medical schools stand in a particular relationship to society, a relationship which confronts them with special ethical problems and obligations. They are the sole portals of entry into the profession and the principal guardians of medical knowledge. Their educational philosophy determines what is a "good" physician, and this ideal in turn shapes the whole of medical practice—who shall study medicine, how medical needs are defined and met, and how resources are allocated (Freidson).

This article maps out in a preliminary way the obligations that derive from the special situation of medical schools in modern society and the dominant values and presuppositions of medical education upon which they depend.

Despite the recent intense interest in the teaching of bioethics, curiously little analytic and formal assessment of the ethics *of* medical education has taken place (Veatch and Sollitto; Veatch, Gaylin, and Morgan). What moral obligations do medical schools owe to the communities they claim to serve—students, patients, and society? How are the conflicting needs of each to be resolved? To what extent are the actual values of medical schools congruent with their obligations? In whom are the obligations vested?

Questions of this kind are addressed only obliquely in the vast literature of medical education. They are answered usually by exhortations to what is "good" for students, patients, society, or medicine. The "good" is grounded more in ideological assertion than in critical definition. No genuinely ethical and philosophic foundation exists against which to examine current practices.

Philosophy and ethos of medical education

Any future formal consideration of the ethics of medical schools and medical education must take into account the values on which they are justified today. These values constitute more an ethos than a philosophy of education.

The traditional value system. The dominant contemporary value system holds that medicine is largely an enterprise of science and technology. The physician is considered the guardian of this arcane knowledge, and he is ordained to apply it to society's needs (Kornberg; Campbell, pp. 134–135; Thomas). To do this, he must subscribe to certain values—faith in the rational solution of medical problems, disinterested concern for patient and society, dedication to competence in practice and to the community of science, which transcends personal interest (Fox, 1976).

The philosophical corollaries of the dominant ethos are positivism and reductionism. Medical faculties, as a result, are disinclined to give serious attention to the social, behavioral, and personal dimensions of illness. Subjects like medical sociology or family, community, and preventive medicine are intellectually peripheral. Ethical issues are seen as matters of opinion not susceptible to rational discourse.

This ethos has shaped medical education profoundly in the last half-century. It is expressed in the Flexnerian doxology: Medicine is a scientific endeavor to be taught in the university by full-time academicians who are specialists, and in university controlled hospitals (Flexner, especially pp. 25, 106). On this view, it is reasonable to select students primarily for scientific and quantitative capability. Teaching must inculcate the scientific method of problem solving. The physician is to be trained as an applied biological scientist if he is to keep abreast of the continuous flow of biomedical information.

On this view, the prime social mission of medicine is to develop and apply "high technology"—that which eradicates disease through apprehension of its causes. Palliation or containment of disease processes has a lesser priority. The social determinants of illness are acknowledged, but are not subject to "scientific" solution, and therefore not intrinsic to medicine. Social sciences and humanities are useful adjuncts, but not to the intellectual enterprise of medicine.

An emerging value system. In the last several decades, varying degrees of modification of the dominant scientific value system have been offered (Hiatt, 1976; Canada; Evans; Gregg; Carnegie Commission). Community and patient needs have been advanced as the ordering principles for medical education. On this view, medicine is designed primarily to alleviate the major health needs of a country, not just those defined as "medical." Medicine should, therefore, be shaped by epidemiology and ecology, and matched to demographic, socioeconomic, and cultural sources of ill health. Advanced technologic medicine is most useful when it has wide community benefit; decisions on the use of medical resources rest with both the public and the professional. Primary care, prevention, and health maintenance should receive heavy emphasis (Alpert and Charney; Rutstein).

Medical education modeled on this value system would differ from the dominant pattern. The social and behavioral sciences and even the humanities are as pertinent as the biological sciences; students are selected on the basis of social concern and interest in "people" and their mundane problems; emphasis is on caring as much as curing; education and health-care planning are to be closely articulated at every stage. The community, not the university hospital, is the proper locus for medical education. Many of the functions now assigned to physicians are to be delegated to nurses, so that fewer rather than more physicians will be needed (Bryant; Ramalingaswami).

Humanistic medical education. In addition to these competing concepts of medical education, there has been in the last decade a growing perception of the need for a more "humanistic" physician. The values subsumed under this title are still rather broadly defined—ranging from humanistic psychology, on the one extreme, to philosophy and history, on the other (Pellegrino, 1974; Banks and Vastyan).

Formal programs teaching human values have appeared as part of the curriculum in a number of schools (McElhinney). They comprise a wide selection of disciplines in the humanities and the "humanistic" end of the social sciences spectrum. They promise to have a leavening effect on the dominant value system. This growing attention to the teaching of ethical and value issues involved in medical care should prepare educators and students for a more formal consideration of the ethical obligations of medical schools and the educational process.

The newer value systems have not yet had an important effect on any medical school. They are intermingled to varying degrees with the dominant ethos. Even in newer schools established specifically to advance community or humanistic values, the scientific ethos remains dominant. Without denying the unequivocal benefits conferred on mankind by the scientific value system, its justification for the educational process will be increasingly scrutinized in the years ahead. The ethical positions taken by medical educators will be strongly flavored by the degree of their commitment to each of these ordering value systems.

Source and nature of ethical obligations

The ethical justification of an institution rests in the degree to which it matches performance with the purposes for which society establishes and supports it. Its ethical obligations are, in turn, grounded in the nature of the institution, the expectations it engenders, and the social essentiality of the services it provides.

In this respect, there are at least four facts in the social situation of medical schools that give them unusual power and correspondingly increase the weight of their ethical obligations: (1) Their triple functions of teaching, advancing knowledge, and care of patients are all essential to the social welfare. (2) They are the only means of access to the knowledge of medicine, its practice and licensure. (3) They accept, voluntarily, public funds ostensibly to serve the public interest. (4) They are accorded exceptionally wide discretionary latitude in performing their social function.

Medical schools are accorded a privileged

position simply because mankind is subject to illness that can be cured, contained, or prevented by special knowledge not in everyone's possession. Medical schools generate and transmit such knowledge and the skills needed to apply it, and in doing so they enter moral relationships with three communities: society in general, students and faculty, and patients. An overall ethics of medical education must delineate the obligations owed to each group and must attend to resolution of the conflicts of obligations that complicate any attempt to serve the interests of disparate groups simultaneously.

Obligations to society in general

The social obligations of medical schools impose at least four moral obligations upon them: (1) A continuous supply of medical personnel must be assured and adjusted in number and kind to perceived and anticipated social need. (2) Medical graduates must be safe and competent practitioners; those who lack integrity must be denied a degree. (3) Equity of access to the profession must be assured for all segments of society. (4) Medical knowledge must be preserved, constantly validated, transmitted, and extended by research.

Supply of medical personnel. The numbers, kinds, and distribution of physicians are of grave social concern. Traditionally, they have been construed in terms of economics and personal choice, to be regulated by the usual forces of supply and demand. Medical schools have sedulously avoided explicit measures that might consciously modulate these forces. Rather, they have produced physicians in accord with dominant educational values, with the result that a disparity between the supply of specialists and generalists has occurred.

If medical schools, however, are the sole source of physicians and accept public funds to produce them, matching need and supply becomes an ethical obligation. Is the medical educator's claim to serve public interest morally defensible when the spectrum of physicians is out of phase with the epidemiology and ecology of ill health? Should the public be forced to seek legislation or other public policy to redress the imbalance? Is this not a commentary on the need for a heightened ethical sensitivity on the part of all whom society has charged with so crucial a position in the physician supply line?

These questions do not imply that medical schools must oscillate with every shift in public preference or that their only function is to produce primary-care physicians. That kind of *reductio ad absurdum* is as morally suspect as neglect of the issue. It does mean that medical educators must reflect more seriously on the ethical implications of their role in society. Only an active, ethically responsive interaction with the public interest will resolve the knotty distinctions between actual needs, wants, demands, and mere whims.

A further question pertains to the obligation to provide for geographic distribution of medical graduates. Clearly, medical schools do not have legal or regulatory authority to determine where physicians practice. Such authority would confer powers over some of the most fundamental rights of the citizen, and they should remain with the appropriate agencies of government like legislatures and licensing bodies. At issue, for example, is the constitutional question of whether the established right of individuals to live and work where they please takes precedence over the still debatable right to equity of access to medical care.

There is, however, a moral obligation that medical schools take all measures that would improve the probabilities of optimum distribution to underserved areas. They can accomplish this by selecting students motivated to primary care, providing appropriate curricula to those who wish to enter that field, making scholarships available, demonstrating how good care can be provided efficiently at remote sites and with satisfaction to patient and physician, extending the medical school regionally in continuing education, offering back-up consultation, and establishing integrated family practice residencies (Ginzberg and Yohalem; Prywes; Pellegrino, 1973).

Most effective of all would be the recognition by the majority of medical schools that congruence is necessary between the education physicians receive and the epidemiology of needs they are intended to serve. Specifically preparing eighty-five percent of students to enter primary-care specialties could, on this view, be an ethical obligation. The allocation of resources, the design of curricula, and the reorganization of priorities for this purpose would be in direct conflict with the scientific value system of medical education. Yet if medical schools are to act morally, they must examine this conflict. The traditional value system would usually cast issues of this kind in terms of ideological or emotional commitments to "quality." The moral implications of the term "quality," together with the dominant ethic of medical education it usually signifies, should be critically assessed.

In recent years medical educators and legislators have come into increasing conflict on the issues of supply and need of physicians in the United States. Departments of family practice and even details of curricula have been imposed by law in state schools. Legislation is being sought in others to make commitment to practice in remote areas a condition of admission to medical school. Educators protest the violation of academic freedoms; legislators contend that access to medical care supersedes those freedoms. The resolution of these conflicts in obligations is usually attempted in economic or political terms. But since the differences are about goals and priorities the fundamental decisions are ethical in nature. Neither legislators nor educators seem as yet prepared for ethical discourse, but this is the only rational way to order one set of obligations to another.

Medical education cannot confront such dilemmas without a more intensive and conscious assumption of the moral obligation it incurs in the matter of manpower supply. These obligations must be clearly set forth and rationally justified and implemented, if the public is to be credibly informed about what is indeed at stake in the argument between government and medical schools.

Assuring competency and integrity. Society rightly expects that the medical school will not confer the degree on an incompetent or dishonest student. This obligation is especially stringent since admission to an American medical school is tantamount to graduation, and graduation is tantamount to licensure and the right to practice.

How well does the current pattern of clinical instruction lend itself to the close surveillance required to assure competence? The student's closest and most continuous contact is with the house staff; supervision by senior clinicians is too often brief and intermittent; attention is focused on cognitive information rather than actual performance in decision making and patient care. Few students are failed in the clinical years, and the further advanced the training the greater is the reluctance to do so. Yet it is precisely at the later stages of clinical education that crucial defects may appear. Are these practices sufficient to meet the ethical obligation to society?

If the obligations to assure competence are to be seriously observed, perhaps we should change present modes of clinical instruction to include more frequent observation of history and physical examination, more direct involvement by senior clinicians, and assessments of judgment, humaneness, and capacity to fulfill patients' needs. Attention to methods and criteria of performance examinations to complement cognitive information will enhance the possibility of assuring competence in medical school graduates (National Board of Medical Examiners). Even more effective would be a decision to make graduation contingent on satisfactory performance of clearly defined clinical tasks, and not on the passage of time or completion of prescribed courses.

Much more difficult is the assessment of integrity. Only the grossest violations are detectable in the admission process; but defects of character do become evident in the critical circumstances of clinical medicine. There is need for greater knowledge as to (1) whether present methods of supervision adequately detect defects of character; (2) how general the tendency is to overlook certain questionable acts out of motives of fear of legal challenge or simple desuetude or laxity; and (3) what the relationship is between lapses of integrity in medical school and later behavior in practice.

The educator must balance his obligation in fairness to help the student learn through error with his obligation to assure integrity in the graduate. It is questionable whether a medical school can excuse itself on the grounds that lack of integrity is a matter for licensing bodies, professional societies, and hospital credentials committees.

The want of general agreement on what constitutes disqualifying dishonesty and the subjective character of the judgments makes this a hazardous enterprise. How do we balance and reconcile the civil rights of students and the ethical obligation of the medical school to society?

Equity of access for all applicants. There can be little debate in a democratic society that every applicant for entry into a profession should have equal consideration. The grosser violations of this principle, like the ethnic quotas of years past, have all but disappeared, though more subtle forms of discrimination still persist sporadically.

The central issues today center on the preferential admission of minority students to remedy the serious disproportion between their representation in the medical profession and their representation in society. Minority physicians have a higher probability of settling in minority communities. Increasing their numbers in medical schools has the potential social benefit of improving accessibility to care for neglected populations, whose needs exceed those of the general population.

How does the medical school resolve the conflicts in its obligations to redress a long history of social injustice and its obligations to provide equal opportunity for admission for all? The serious question of reverse discrimination calls for some leadership in establishing priorities in obligations on the part of medical educators. Many have assumed the responsibility for minority admission and have allocated the resources to assist disadvantaged students—although often half-heartedly—but this conviction seems to be weakening in the face of counterpressures. How firmly grounded in its special ethical obligations to society is the obligation of medical schools to improve minority representation?

Some medical educators have favored the admission of students with humanistic attitudes in hopes of producing more compassionate physicians and more who are interested in primary care. A variety of attitudinal standards have been proposed to complement or even substitute for some of the more objective criteria of academic performance. There are some serious ethical questions in this move. Is it proper to deprive students with high academic scores on the basis of attitudinal scales as yet untested for validity, or even more subjective opinions about who is and who is not humanistic? Is the bright person by definition less humane, and the less academically accomplished more so?

There will always remain the possibility of inequity in medical school admission, but it must be reduced to the absolute minimum. Some have suggested that a lottery would accomplish this. But then, is it more just for individual and society to trust to chance or to a combination of criteria responsibly administered by admission committees sensitive to their ethical obligations?

Guardianship of medical knowledge. It is in the medical schools that technical and scientific knowledge necessary to modern medical care is concentrated most effectively on problems of disease. Medical schools are also the indispensable loci for human investigations and for training clinical investigators. They can function, moreover, as the channels whereby university disciplines in the social sciences and humanities can most immediately be focused on medical problems (Hiatt, 1975, pp. 259–264; Pellegrino, 1970).

Medical schools are therefore society's main guardians of medical knowledge, charged with its preservation, validation, transmission, and extension. For that reason, scholarship and research are moral obligations, not luxuries or refinements. Without them, medical care must become static in quality and efficacy. A firmer grasp of this fact would mitigate public misapprehension as well as the academic hubris about research.

Ethical dilemmas arise when a school sets the research obligation against its other obligations to students and patients. The eager promise of medical schools to fulfill all its obligations fully is too facile. Since resources are necessarily limited, each school is obliged to select a research mission consistent with quality performance in teaching and patient care. These cannot be compromised at the expense of overly ambitious investigative efforts.

It is the total research effort of all the medical schools of a country that must be preserved, not the presumed right of each school to pursue the amount of research its faculty deems requisite to some ill-defined notion of "quality" or prestige.

These obligations of medical schools to society as a whole are illustrative of the range of questions that constitute one part of the domain of the ethics of medical education. Another part of that domain pertains to the obligations to students, faculty, and patients.

Obligations to faculty, students, and patients

The ethical obligations of medical schools to their faculties are not particularly different from those in other educational institutions. Moreover, faculty rights are increasingly being propounded by professional associations and faculty unions and bargaining units. The corollary need for a code of faculty ethics has, however, escaped equally assiduous attention.

The medical school has many of the same ethical obligations to students as any other educational institution—competent teaching, access to the knowledge and skills needed for competence, fair methods of evaluation, rights of due process, and participation in decisions that affect personal well-being. In consequence of its claim to be an educational instrument, the school is morally compelled to give the most serious attention to the quality of its instruction.

This responsibility—which is still insufficiently attended to—must extend far enough to detect inadequate or disinterested teachers and to rehabilitate, reassign, or terminate them.

To these ordinary obligations are added those unique to clinical education. Medical schools dedicate large segments of the curriculum to supervised experience in the care of patients. The clinical clerkship and residency training are important to society as well as to the student, because they assure a long period of supervised experience before independent practice is permitted. Students need a sufficiently broad spectrum of clinical experiences and increasing opportunities to make some independent decisions. Residents and fellows need wider latitudes of decision making to advance their clinical skills.

To attain these educational objectives, very delicate conflicts of obligation must be resolved constantly: Where does pedagogy end and independent responsibility begin, and when does student or resident training seriously impair patient safety? Two extremes are equally detrimental to society—unsupervised dominance of house staff and overrestrictive supervision, which impedes acquisition of clinical skills.

The way a clinical teacher resolves such conflicts will subtly shape the student's developing value system. The image of the "good" physician is sculpted by influential teachers with effects often more lasting and significant than the knowledge imparted. The teacher, therefore, incurs a stringent obligation to regard lapses in the ethical and human quality of his behavior before students and patients as among the most serious. The "image" of the "good" physician is under much scrutiny today; one of the most important moral obligations of medical school faculties is to take some responsibility for that image.

A corollary obligation is to assist the student in coping with the unique emotional challenges in medical education—the first encounter with a cadaver, with death and dying, and with the incurable, the recalcitrant, and the self-punishing victims of alcohol or drug excess. The values a student sees exhibited and the way he develops his own coping mechanisms condition his behavior as a practitioner. Assistance in developing mature and humane personal development is part of a medical school's obligation to students and their future patients.

Questions of value and purpose now enter almost every phase of medical decision. Specific instruction in dealing with conflicts of values is now a requirement of a sound medical education. The medical school must recognize its responsibility to prepare students to care for patients who may reject their values or their medical treatments. Moral and technical authority are no longer vested entirely in the profession. Their separation may be more consistent with the rights of self-determination afforded in a democratic and morally pluralistic society.

Patients cared for by the faculty of medical schools in their teaching hospitals manifestly have a moral claim first to the rights owed to any patient—competent, personalized, considerate care, protection of confidentiality, and right to informed consent. Fulfillment of these ordinary obligations, however, is vastly complicated when the patient is also a teaching subject.

The simple fact of being ill places any patient in a state of special vulnerability in which his humanity is diminished. The sick person cannot command his body; he experiences pain and anxiety; he lacks knowledge of what is awry, and how to repair it; he is heavily dependent upon the good graces of others. In a teaching hospital, this vulnerability is further heightened by a complex skein of human relationships of which the patient is the center. Students, fellow residents, faculty, investigators, and other health professionals all enter the highly sensitive arena of his care. The possibility of serious and multiple conflicts of purpose, interest, and moral obligations is painfully present.

No matter how just the claim of students to a proper clinical education, their participation in the care of patients is a privilege and not a right. If serious conflicts occur, the obligations to the patient must take precedence. The medical school must assume the primacy of the patient, even when it may contravene its other obligations to teach students or residents.

Necessarily, society permits some intrusion into the traditional privacy of the patient's care to assure a continuing supply of clinically competent physicians. Every student must perform enough history and physical examinations; every surgical resident must perform his first operation; every new medication, or diagnostic or surgical procedure, must be tried for the first time on someone. There is no other way to advance knowledge or assure the competence of clinicians. We are all beneficiaries of the teaching subject's sacrifice of some of his own comfort and safety.

These permitted invasions must be recognized as such and sedulously guarded so that no patient ever becomes mere teaching "material" or a mere experimental "subject." Certain special obligations bear upon medical schools and teaching hospitals, in addition to those incurred in the care of any patient.

The patient has a right to know who is taking care of him. Students must be introduced as such, and not as "doctor"; house staff too must be clearly identified; the responsible attending physician must be known and his part in the procedure as operator or supervisor spelled out. All the sensitive issues of disclosure and consent must somehow be guaranteed while permitting graded clinical experience for students and house officers.

It is inconceivable that society will withdraw the privileges of students and residents to learn clinical skills by actual practice. The deleterious social consequences would outweigh the abuses that exist in certain situations today. To produce only didactically trained physicians would frustrate another obligation of medical schools —to assure a supply of competent practitioners. Nonetheless, the most careful supervision by experienced clinicians is mandatory if the privilege is not to be severely limited in the future.

Institutional ethics

The ethical obligations briefly sketched in this article are corporate obligations of a medical school in modern society. They bind specific members of the faculty, administration, and staff acting as individuals. But these individuals also act as members of the faculty body of committees, administration, and boards of trustees. These collective and corporate actions have moral consequences, so that the institution itself becomes a moral agent. Its collective acts, like those of individuals, are therefore legitimate objects of moral scrutiny.

An institution, in short, makes a "profession" like an individual: That "profession" is to serve certain social needs. Performance must match that declaration as closely as possible or it is ethically inauthentic. Academic, fiscal, and managerial decisions cannot be isolated from their impact on the moral obligations discussed here. The value presuppositions, the ethos underlying the ethics of medical education, must also be clearly identified and morally justified. Setting the value rheostat at a point that accommodates the obligations owed to society, faculty, students, and patients, and that equitably resolves the conflicts between these obligations is the central task of institutional ethics.

The contemporary social climate portends a genuine renewal of interest in the values and purposes of contemporary life and all its institutions. Medical schools deserve the same critical scrutiny because of the special position they enjoy in society (Fox, 1974). Their value presuppositions, policies, and actual operation will be subject to more explicit and stringent tests of ethical justification.

This article has drawn a preliminary map of the topography of issues and obligations that make up the domain of the ethics of medical education. This domain deserves rigorous and urgent examination by critical methods of ethics —descriptive, normative, and metaethical.

EDMUND D. PELLEGRINO

[*Directly related are the entries* HEALTH CARE, *articles on* HEALTH-CARE SYSTEM *and* HUMANIZATION AND DEHUMANIZATION OF HEALTH CARE; *and* MEDICAL ETHICS EDUCATION. *For discussion of related ideas, see the entries:* CARE; MEDICAL PROFESSION, *article on* MEDICAL PROFESSIONALISM; MEDICINE, SOCIOLOGY OF; PRIVACY; RACISM, *article on* RACISM AND MEDICINE; SOCIAL MEDICINE; *and* THERAPEUTIC RELATIONSHIP, *articles on* CONTEMPORARY SOCIOLOGICAL ANALYSIS *and* CONTEMPORARY MEDICAL PERSPECTIVE. *Other relevant material may be found under* JUSTICE; *and* REDUCTIONISM, *article on* PHILOSOPHICAL ANALYSIS.]

BIBLIOGRAPHY

ALPERT, JOEL J., and CHARNEY, EVAN. *The Education of Physicians for Primary Care.* DHEW Publication no. (HRA) 74-3133. Bethesda, Md.: Bureau of Health Services Research, 1973.

BANKS, SAM A., and VASTYAN, E. A. "Humanistic Studies in Medical Education." *Journal of Medical Education* 48 (1973): 248–257.

BRYANT, JOHN H. "One World or Many: America's Role in Medical Education." Miller, *Medical Education*, pp. 171–184.

CAMPBELL, E. J. MORAN. "Basic Science, Science, and Medical Education." *Lancet* 1 (1976): 134–136.

Canada, Department of National Health and Welfare, Marc Lalonde, Minister. *A New Perspective on the Health of Canadians: A Working Document.* Ottawa: 1974.

Carnegie Commission on Higher Education. *Higher Education and the Nation's Health: Policies for Medical and Dental Education.* New York: McGraw-Hill, 1970.

EVANS, LESTER J. *The Crisis in Medical Education.* Ann Arbor Science Library. Ann Arbor: University of Michigan, 1964.

FLEXNER, ABRAHAM. *Medical Education in the United States and Canada: A Report to the Carnegie Foundation for the Advancement of Teaching.* Introduction by Henry S. Pritchett. The Carnegie Foundation for the Advancement of Teaching, Bulletin no. 4. New York: 1910.

Fox, Renée C. "Ethical and Existential Development in Contemporaneous American Medicine: Their Implications for Culture and Society." *Milbank Memorial Fund Quarterly* 52 (1974): 445–483. Bibliography.

———. "The Sociology of Modern Medical Research." *Asian Medical Systems: A Comparative Study*. Edited by Charles Leslie. Berkeley: University of California Press, 1976, pp. 102–114.

Freidson, Eliot. *Professional Dominance: The Social Structure of Medical Care*. New York: Atherton Press, 1970.

Ginzberg, Eli, and Yohalem, Alice M., eds. *The University Medical Center and the Metropolis*. New York: Josiah Macy Jr. Foundation, 1974.

Gregg, Alan. "The Golden Gate of Medicine." *Annals of Internal Medicine* 30 (1949): 810–822.

Hiatt, Howard H. "The Need for University Involvement in Medical Education." *Daedalus* 104, no. 1 (Winter 1975), pp. 259–272.

———. "The Responsibilities of the Physician as a Member of Society: The Invisible Line." *Journal of Medical Education* 51 (1976): 30–38.

Kornberg, Arthur. "Research: The Lifeline of Medicine." *New England Journal of Medicine* 294 (1976): 1212–1216.

McElhinney, Thomas K., ed. *Human Values Teaching Programs for Health Professionals: Self-Descriptive Reports from Twenty-nine Schools*. 3d ed. Institute on Human Values in Medicine, Report no. 7. Philadelphia: Society for Health and Human Values, 1976.

Miller, George E., ed. *Medical Education and the Contemporary World: Proceedings of a Symposium Conducted by the University of Illinois College of Medicine, September 13–14, 1976, Chicago, Illinois*. A Publication of the John E. Fogarty International Center for Advanced Study in the Health Sciences. DHEW Publication no. (NIH) 77–1232. Bethesda, Md.: U.S. Department of Health, Education, and Welfare; Public Health Service; National Institutes of Health, 1977.

National Board of Medical Examiners, Committee on Goals and Priorities. *Evaluation in the Continuum of Medical Education: Report*. Philadelphia: 1973.

Pellegrino, Edmund D. "Educating the Humanist Physician: An Ancient Ideal Reconsidered." *Journal of the American Medical Association* 227 (1974): 1288–1294.

———. "The Most Humane Science: Some Notes on Liberal Education in Medicine and the University." *Bulletin of the Medical College of Virginia* 47, no. 4 (1970), pp. 11–39. The Sixth Sanger Lecture.

———. "The Regionalization of Academic Medicine: The Metamorphosis of a Concept." *Journal of Medical Education* 48 (1973): 119–133.

Prywes, Moshe. "The Balance of Research, Teaching, and Service in Medical Education." *Minerva* 9 (1971): 451–471.

Ramalingaswami, V. "One World or Many: America's Role in Medical Education." Miller, *Medical Education*, pp. 185–193.

Rutstein, David D. "Medical Education for the Future." *Blueprint for Medical Care*. Cambridge: MIT Press, 1974, pp. 139–160.

Thomas, Lewis. "The Future Place of Science in the Art of Healing." *Journal of Medical Education* 51 (1976): 23–29.

Veatch, Robert M.; Gaylin, Willard; and Morgan, Councilman, eds. *The Teaching of Medical Ethics*. Hastings Center Publication. Hastings-on-Hudson, N.Y.: Institute of Society, Ethics and the Life Sciences, 1973.

———, and Sollitto, Sharman. "Medical Ethics Teaching: Report of a National Medical School Survey." *Journal of the American Medical Association* 235 (1976): 1030–1033.

MEDICAL ETHICS

See Bioethics; Medical Ethics, History of; Medical Ethics Education; Medical Ethics in Literature; Medical Ethics under National Socialism.

MEDICAL ETHICS EDUCATION

Medical ethics teaching, in one form or another, is as old as medicine itself. Plato's Academy, at least if we take the content of the dialogues to reflect the subject matter considered there, was the scene of numerous exchanges on the ethics of actions taken in a medical context. Socrates, for example, in book three of the *Republic* (III, 406D-E) explores with Glaucon the appropriateness of a carpenter's deciding to turn his back on a physician's prescribed lengthy regimen so that Glaucon can resume his daily work and normal life-style. Decision making in a medical context has always been appropriate for philosophical and sometimes theological reflection, at least in the Western tradition.

Likewise, training of physicians and medical professionals has generally included some form of instruction in medical ethical norms from the perspective of the physician. It is clear that the classical codes of medical ethics served an important educational function. The Hippocratic Oath, the single most important document in the history of Western medical ethics, explicitly functions as part of the instruction of the young physician in the Western world. The first part of the Oath is a covenant between the student and his teacher. The student pledges to hold his teacher "as equal to my parents," to teach his teacher's offspring the art of medicine without fee, but otherwise "to no one else."

The ethical precepts of the second part of the Oath we can presume provided the framework for the apprentice to learn the ethical practices and philosophical premises of the particular Pythagorean school of medicine reflected in the Oath (Edelstein). Similarly the ancient Hindu oath from the *Caraka Samhita* was an oath of initiation phrased in terms of the precepts in which a teacher should instruct the disciple. Ethical admonitions were also contained in the legends of the training of Hua Tu, the great Chinese surgeon of the second century B.C.

Even today much of the instruction in physician ethics takes place in the apprenticeship mode. The student, spending years in contact with his instructors, gradually absorbs a set of ethical precepts, which he recognizes as norms whether or not they are put into practice. This apprenticeship mode of teaching is often defended as the only effective way to socialize the student into a particularly professional ethic. However, more systematic efforts at medical ethics education have emerged in the modern period, to some extent in the training of the medical professional and, especially, in the teaching of medical ethics outside of the special professional and clinical context. Often these medical ethics teaching traditions have evolved within an explicitly theological context.

The Roman Catholic tradition of moral theology has produced a body of literature used in the teaching of courses in medical ethics not only in Catholic medical and nursing institutions but also, indeed primarily, in seminaries and undergraduate university institutions. From at least the early seventeenth century there is a continuous tradition of volumes on "pastoral medicine" and "medical deontology" culminating with Carl Capellmann's *Pastoral-Medicin* (1877) and Giuseppi Antonelli's *Medicina pastoralis* (1891). That tradition continues with the major American twentieth-century volumes in Catholic medical ethics, including M. P. Bourke's *Some Medical Ethical Problems* (1921), Gerald Kelly's *Medico-Moral Problems* (1949), Charles J. McFadden's *Medical Ethics for Nurses* (1949), and Thomas J. O'Donnell's *Morals in Medicine* (1956), all used as course texts and all standing explicitly in the Catholic tradition of moral theology. The early volumes began as extracts of medically relevant material from the standard textbooks of moral theology rather than as offshoots of the discipline and tradition of physician ethics.

Similarly, a body of Jewish medical ethical literature has grown from the tradition of talmudic scholarship. The two most important surveys of the halakhic tradition, Jakobovits's *Jewish Medical Ethics* (1959, rev. ed. 1975) and Rosner's *Modern Medicine and Jewish Law* (1972), both stand in the tradition familiar to the talmudic student rather than to the secular Western medical professional. They form the basis of medical ethics education for the rabbi and other students of the halakhic tradition.

Protestant theological thought has been less explicit in medical ethics teaching. Two volumes from the 1950s formed the basis for much medical ethics instruction in Protestant seminaries and religion departments in the recent past: *The Ethical Basis of Medical Care* by Willard Sperry (former dean of Harvard Divinity School) and *Morals and Medicine* by Joseph Fletcher. These volumes have recently been supplemented by a vast literature in medical ethics coming largely from an ecumenical theological perspective, but including the writings of a number of the more conservative leading Protestant theologians.

By comparison, relatively little medical ethical literature survives from earlier secular physician education. One medieval text by Arnold of Villanova (1235?–1315?), a leading physician of his day, entitled *On the Precautions That the Physician Must Observe*, addresses practical problems of physician behavior but avoids the pretense of idealistic ethical generalization. It is not clear to what extent this was used explicitly as an educational document. Friedrich Hoffman, professor of medicine at Halle, published a series of lectures to medical students in 1738 under the title *Medicus Politicus*. Its subtitle was "rules of prudence whereby the young physician should direct his studies and his life if he wishes to acquire and maintain both reputation and successful practice." Published in the latter part of the eighteenth century, John Gregory's *Lectures on the Duties and Qualifications of a Physician* appears more explicitly designed for the educational context. It originated as a series of lectures to medical students and provided the context for debates about the role of the physician, which led to the most important modern document on medical ethics from the physician community, Percival's *Medical Ethics* (1803).

Contemporary medical ethics teaching

While there are scattered reports of medical ethics teaching throughout the world, no precise statistics are available. Consequently, the remainder of this entry emphasizes the medical ethics education that is taking place in Europe and the United States. In Western Europe informal reports indicate that in most countries little formal curriculum in medical ethics remains. There are scattered reports of teaching in basic university and theological training. In medical education institutions there is often informal instruction in the apprenticeship mode. In some institutions preclinical and clinical instruction may include sessions dealing more or less explicitly with medical ethics. Formal offerings, however, are apparently the exception rather than the rule. In the Netherlands, at Nijmegen and Leiden, options in ethics are available for medical students at both the preclinical and clinical stages, and a Professor of Medical Ethics has been appointed at the newly established faculty in Maastricht. In Sweden, medical ethics is not part of the formal curriculum, but teachers sometimes include it on their own initiative. At Stockholm's Karolinska Institute one formal appointment in medical ethics has been made. The Nottingham Medical School in England has a teaching program with many structured contacts with students (Jones and Metcalfe). It is reported that, from the end of the Civil War in Spain (1939), Spanish medical schools taught medical deontology from a Roman Catholic perspective as a compulsory discipline in the final year of medical school. Resentment against the government and against the Catholic hierarchy's support for the government, however, has modified this policy; currently, medical ethics courses are not compulsory, although every school can offer such a program. Other West European countries for which information is available report no formal and systematic instruction in medical ethics in the medical education process.

Similar reports come from other parts of the world. Australia's University of Melbourne reports a series of five lectures on medical ethics in the fifth year of the medical student's course of study. The University of St. Thomas in the Philippines has a Section of Medical Ethics in its Faculty of Medicine and Surgery which offers a series of courses in the first three years of the curriculum. Browne reveals an explicit awareness of the development of medical ethics teaching in Africa. He reports the emerging tension between the European medical ethical traditions introduced by Great Britain and France and Africa's indigenous systems of ethics, which may lead to quite different resolutions of ethical problems. Although it is difficult to get exact data about teaching in Eastern Europe, reports indicate the conscious development of medical ethics instruction in that area as well. In Hungary there is an integrated teaching endeavor focusing on a roundtable discussion at the Medical University of Pécs, Institute of Marxism-Leninism. There is substantial interest in the teaching of deontology (ethics) in medical schools in the Soviet Union (Gol'dshtein; Karlsen and Kosarev; Sokolov; Priduvalov, Chernega, and Abrosimov). At least four medical schools there have active teaching efforts. While we have no information on teaching in the People's Republic of China specifically devoted to medical ethics, there is record of great emphasis on the transformation of medical education to give an impetus to the ideological remolding of medical workers. A report from the Revolutionary Committee of Hua Shan Hospital, Shanghai, concludes that "the revolution in medical education is closely related to the revolution in medical and health work. . . . The participation of worker–peasant–soldier students in the struggle to study in college, run it and transform it with Marxism–Leninism–Mao Tsetung Thought has greatly enhanced the proletarian political atmosphere of our hospital" (p. 319). At least in this report of one Chinese teaching institution, ideological transformation appears to be the dominant value question.

In the United States contemporary medical ethics has evolved rapidly. Extensive teaching of medical ethics in the United States today takes place in undergraduate colleges and universities, but formal medical ethics teaching is known to exist at any level from the primary school to the professional and graduate academic levels. While no decisive data are available except for medical school teaching, according to the Report of the Commission on the Teaching of Bioethics the number of courses explicitly oriented to medical ethics at the college and university level is growing. These courses are offered in departments of philosophy, theology, religion, biology, psychology, and sociology. Courses are designed to serve not only those who are majoring in premedical studies, but others professionally oriented to fields related to medical ethics (such as divinity, law, and public administration), as well as the student who is interested in the field either as a way of studying the field of ethics or simply as a way of becoming an informed consumer of health care. Courses at the undergraduate level have one of two characteristic foci. Some are concerned with global issues such as population or ecology. Others concentrate more narrowly on the moral problems of the physician–patient relationship. The majority of courses survey widely the issues of medical ethics, but more and more courses are emerging that focus more narrowly on a specific issue area such as the ethics of death and dying, human experimentation, the control of human behavior, or genetics.

Probably the next largest number of courses in medical ethics taught in the United States is in the theological seminaries. Some theological traditions have traditionally taught medical ethics. Others have, within the last decade, found medical ethics an appropriate way of teaching the discipline of ethics to their students. Law schools have traditionally offered instruction in legal medicine, but often this has been oriented primarily toward medical malpractice law. A number of American law schools have added specific courses on ethical and legal issues in medicine to their curricula. These courses not only include discussion of the current laws related to the issues of medical ethics, but introduce the law student to analysis of the philosophical issues underlying legal questions in medicine. In short they ask what the law ought to be as well as what the law is.

High school teaching of medical ethics is a recent and rapidly growing development. Normally offered as an experimental elective or as an area of concentration within a course in biology or civics, medical ethics courses at the high school level are perhaps the most rapidly growing dimension of American medical ethics teaching. Adult education in community education programs and in religious educational settings is another rapidly growing, but poorly documented, phenomenon. More and more adults are being offered opportunities to study ethical problems they face in obtaining health care and in making decisions about their care.

The one area of American medical ethics teaching that is well documented is teaching in medical schools. Two national studies, in 1972 and 1974 (Veatch, pp. 97–102; Veatch and Sollitto), surveyed all American medical schools. In 1974, 112 American medical schools were surveyed; of 107 schools responding, 97 indicated some kind of medical ethics teaching. But of those 97, 19 reported that teaching is in the form of discussion in courses not specifically identified as "ethics" courses (courses such as introduction to the patient, social medicine, psychiatry, or pharmacology). Forty-seven schools reported elective courses, up from thirty-seven schools in 1972. Fifty-six schools reported special programs such as conferences or lectures on issues in medical ethics, up from seventeen two years earlier. In comparing the data of the survey from two years previously it is clear that the special program or conference is frequently the way a school begins to develop medical ethics teaching. After interest in the special program is aroused at the school, an experimental course is developed. If that is successful, then regular course offerings emerge sometime later. The fact that there has been particularly rapid growth of the special programs in the two years may indicate that a growth of elective courses can be anticipated in the future.

The required course in medical ethics is a rarity in the American medical school. Six schools have some kind of required exposure to medical ethics. The resistance to requiring exposure has been great and has been made even greater by the shrinking of the medical school curriculum time because of pressures to increase clinical training and simultaneously to reduce the total number of teaching hours and years.

The survey also found that in 1974 there were thirty-one faculty members in American medical schools devoting at least half their time to teaching medical ethics. This is an increase from nineteen two years previously. While no comparable figures are available for the number of people whose primary area of specialization is teaching medical ethics outside of medical school, it has been estimated that it is at least twice as many as within medical school facilities.

The concepts and goals of medical ethics education

Underlying this rapid evolution of medical ethics education are a number of conceptual issues about how medical ethics ought to be taught and why. The Commission on the Teaching of Bioethics, a group that met in the United States from 1972 to 1976 to examine the goals, problems, and future of bioethics teaching, identified four goals, some or all of which may be sought in a particular teaching effort. These include (1) helping the student identify and define moral issues in a biomedical context, (2) developing strategies for analyzing moral problems in medicine, (3) relating moral principles to specific issues and cases, and (4) training a small group for careers in bioethics. The first three goals are educational objectives appropriate for any lay person who must make ethically informed choices about his or her own health care. The fourth is much more specialized. It is the objective of doctoral-level programs designed to train a small number of specialists who themselves will go on to teach medical and biological ethics in universities and professional schools. Of the four objectives the first two—especially the first—will also be appropriate for training those who will be not primarily decision makers but counselors, advisers, and consultants. Thus the physician, if he is seen as one who should be not a primary decision maker but rather one who assists the patient in making decisions about his own care, must still learn to recognize an ethical issue when he encounters one. He must learn how to distinguish technical questions of medicine, which are more appropriately his own area of competence, from value dimensions, for which he may or may not have a special competence. Only if he learns to recognize and reflect on the ethical and other value dimensions will he be able to articulate to the patient the value considerations that are inherent to the decision-making process. Similarly, a counselor or consultant (even when not a primary decision maker) should know and understand the different theories or strategies for resolving ethical conflicts within medicine. Only if the consultant recognizes that those standing in different traditions may reach different resolutions of ethical conflicts will he be prepared to present the problems to the patient in a manner that will permit appropriate resolution by the patient.

The third goal, relating moral principles to specific cases and issues, will, for the most part, be the task of the primary decision maker. While this would normally be the patient or the group consuming the health care, there are certain specific ethical choices that logically must be made by health care providers. The physician must, for instance, decide how much information is to be transmitted to the patient about a particular diagnosis. The nurse must decide when to object to a particular kind of care on ethical grounds. Only with a knowledge of the systems of relating principles to cases will the decision maker be able to carry out that task effectively.

Beyond the question of the goals of bioethics and medical ethics teaching there are questions of the patterns of teaching. Regarding the question of the most appropriate format for teaching, one approach often used—especially in science and medicine departments—is the issues method of organization. The course presents a series of issues such as death and dying, experimentation, genetics, behavior control, abortion, etc. While this approach permits a precise focus on the problem areas as they arise in the medical context, it has been criticized because it does not lend itself to systematic analysis of the basic ethical and philosophical problems underlying these issues. Freedom and self-determination are basic themes underlying not only behavior control but the doctrine of informed consent as well. This objection has led to proposals to organize medical ethics teaching thematically (McCormick). Thus the structure of the course might include such headings as freedom, informed consent, benefits to the patient, benefits to society, autonomy and self-determination, the sanctity of life, social justice, truth-telling, and contract keeping. The price of this approach, however, may be the loss of specificity and clinical vividness, which make the issue organization so attractive. One resolution is to structure the course simultaneously by themes and issues. Thus a course might be organized according to basic themes of a course in ethics but subtitled with an issue area in medical ethics that raises in a particularly vivid way the problems posed by that theme.

Occasionally other methods of organization will be appropriate. A course for doctoral candidates in bioethics may use a historical approach. A course in the clinical years in a medical school might be organized so that several sessions would be held in each of the clinical rotations. The price of that structure, however, may be great redundancy unless the sessions are carefully coordinated.

Another question pertains to the mode or method of teaching bioethics. At undergraduate institutions the presumption has generally been that the teaching is done in the traditional course format. This may take the form of a seminar or a lecture, although there is generally a preference for some vehicle which encourages student participation in class discussion. At the high school level development of "modules" has been increasingly popular. These are smaller units designed to be used within a course. They can be taught as self-contained units for special study projects or for an entire class within the normal curriculum. For graduate academic training in medical ethics and for training of health-care professionals the concept of the traditional course format has been increasingly questioned. In medical schools a number of modes of teaching have been attempted including not only lectures and seminars but also intensive weekend workshops, clinical "rounds" where cases are discussed in the traditional manner of the particular service, evening seminars, internships in which the student works full time in medical ethics for periods of a month or more, and grand rounds that offer educational opportunities not only for medical students but for house staff and teaching faculty as well. Some of the more innovative programs combine all of these formats into an integrated program in medical ethics teaching. Departments or programs of medical ethics or of humanities that include medical ethics have been created at several American medical schools including the Pennsylvania State University, the State University of New York at Stony Brook, the University of Kansas, the University of Texas Medical Branch at Galveston, and Georgia Medical College. Interfaculty programs oriented to the teaching of medical ethics which combine teaching in medical faculties with graduate departments and other professional schools have been pioneered at Harvard, Case Western Reserve University, and Georgetown University.

In the Soviet Union there has been considerable public debate about the appropriate locus of teaching within the medical school. Proponents of teaching medical deontology in the context of history of medicine (Gol'dshtein) have debated with proponents of placing it within the context of social hygiene and public health (Karlsen and Kosarev). The literature also emphasizes the need for more integrated teaching involving both clinical and theoretical disciplines (Priduvalov, Chernega, and Abrosimov). A debate in the Soviet literature about teaching medical students relatively early or late in their training period is one that parallels a Western controversy concerning the teaching of medical ethics.

Opportunities for graduate-level study in medical ethics are beginning to emerge in a small number of American universities. These normally combine the systematic training in an academic discipline, such as philosophy, with more advanced training in medical ethics. A clinical training program is included in the curriculum to give the student exposure to the specific medical problems that emerge in the hospital context. Such programs have been pioneered at Georgetown University, the University of Tennessee, the University of California Medical Center at San Francisco in conjunction with the Pacific School of Religion at Berkeley, and Harvard.

A third question in the teaching of medical ethics is: Who ought to do the teaching? The teaching of bioethics requires skills in the biological and medical sciences as well as in the discipline of ethics and in systematic analysis. The Commission on the Teaching of Bioethics has recommended that, especially for teaching done at the university and professional schools, the minimum standard for teaching ought to be that the teacher have professional competence in at least one of these areas and what the commission called "competent amateur" standing in the other. By competent amateur standing was meant serious postgraduate training, preferably at least a year of full-time study, a master's degree, or its equivalent. One common pattern has been team teaching in which the scientific and the more philosophical disciplines can each be represented by someone with full professional standing. The members of the team would also have competent amateur standing in the other perspective. The assumption is that one should no more presume that ethical analysis can be done on an ad hoc amateur basis than that analysis of the scientific aspects of bioethical issues can be similarly conducted.

It is clear that medical ethics education is a rapidly evolving field. More courses are offered today, at least in the United States, than ever before, and more explicit training of medical ethics educators is taking place. Whether this is a passing phenomenon or the development of a major educational specialty remains to be seen.

ROBERT M. VEATCH

[*For discussion of related ideas, see the entries:* JUDAISM; MEDICAL EDUCATION; *and* ROMAN CATHOLICISM. *See* APPENDIX, SECTION I, OATH OF HIPPOCRATES *and* OATH OF INITIATION (CARAKA SAMHITA). *Other relevant material may be found under* BIOETHICS; *and* MEDICAL ETHICS, HISTORY OF, *section on* SOUTH AND EAST ASIA, *article on* INDIA, *section on* EUROPE AND THE AMERICAS, *articles on* ANCIENT GREECE AND ROME, BRITIAN IN THE NINETEENTH CENTURY, *and* NORTH AMERICA IN THE TWENTIETH CENTURY.]

BIBLIOGRAPHY

BLOMQUIST, CLARENCE. "The Teaching of Medical Ethics in Sweden." *Journal of Medical Ethics* 1 (1975): 96–98.

BROWNE, S. G. "Teaching Medical Ethics in Africa." *Medical Journal of Australia* 7, supp. 1 (1972), pp. 63–65.

The Commission on the Teaching of Bioethics. *The Teaching of Bioethics*. Hastings-on-Hudson, N.Y.: Institute of Society, Ethics and the Life Sciences, 1976.

EDELSTEIN, LUDWIG. "The Hippocratic Oath: Text, Translation and Interpretation." *Ancient Medicine*. Edited by Owsei Temkin and Lilian C. Temkin. Baltimore: Johns Hopkins Press, 1967, pp. 3–64.

GOL'DSHTEIN, L. M. "Rol' deontologii v meditsinskom obrazovanii" [The Role of Deontology in Medical Education]. *Sovetskoe zdravookhranenie* 30, no. 4 (1971), pp. 49–50.

GOROVITZ, SAMUEL. *Teaching Medical Ethics: A Report on One Approach*. Cleveland: Moral Problems in

Medicine, Department of Philosophy, Case Western Reserve University, 1973. Not easily available.

JONES, J. S. P., and METCALFE, D. H. H. "The Teaching of Medical Ethics in the Nottingham Medical School." *Journal of Medical Ethics* 2 (1976): 83–86.

KARLSEN, N. G., and KOSAREV, I. I. "O prepodavanii deontologii v meditsinskom vuze" [Instruction in Deontology in a Medical School]. *Sovetskoe zdravookhranenie* 30, no. 4 (1971), pp. 46–49.

MCCORMICK, RICHARD A. "Issue Areas for a Medical Ethics Program." Veatch, *The Teaching of Medical Ethics*, pp. 103–114.

PRIDUVALOV, F. M.; CHERNEGA, R. P.; and ABROSIMOV, V. A. "Deontologiia v sisteme meditsinskogo obrazovaniia" [Deontology in the Medical Education System]. *Sovetskoe zdravookhranenie* 31, no. 9 (1972), pp. 66–67.

The Revolutionary Committee, Hua Shan Hospital, Shanghai First Medical College, Shanghai. "Hospital-run Medical Colleges Are Fine." *Chinese Medical Journal* n. s. 1 (1975): 315–324.

SOKOLOV, D. K. "Opyt prepodavaniia deontologii v meditsinskom institute" [Experience in the Teaching of Deontology in a Medical Institute]. *Sovetskoe zdravookhranenie* 31, no. 9 (1972), pp. 64–65.

VEATCH, ROBERT M. "National Survey of the Teaching of Medical Ethics in Medical Schools." Veatch, *The Teaching of Medical Ethics*, pp. 97–102.

———; GAYLIN, WILLARD; and MORGAN, COUNCILMAN, eds. *The Teaching of Medical Ethics.* Proceedings of a conference sponsored by the Institute of Society, Ethics and the Life Sciences and Columbia University College of Physicians and Surgeons, 1–3 June 1972. Hastings-on-Hudson, N.Y.: Institute of Society, Ethics and the Life Sciences, 1973.

———, and SOLLITTO, SHARMON. "Medical Ethics Teaching: Report of a National Medical School Survey." *Journal of the American Medical Association* 235 (1976): 1030–1033.

MEDICAL ETHICS, HISTORY OF

A central concern of bioethics is the entire history of medical ethics, which deals with ethical problems in the physician–patient relationship. The present entry, which surveys the history in some detail, is divided into four sections, PRIMITIVE SOCIETIES, NEAR AND MIDDLE EAST AND AFRICA, SOUTH AND EAST ASIA, *and* EUROPE AND THE AMERICAS. *Within each section, the articles are arranged chronologically. Other articles, describing the history of specific topics or fields within bioethics, such as the history of behavioral psychology, the history of human experimentation, and the history of the therapeutic relationship, are situated topically within the encyclopedia.*

I. PRIMITIVE SOCIETIES   Lucille F. Newman
II. NEAR AND MIDDLE EAST AND AFRICA
  1. ANCIENT NEAR EAST   Darrel W. Amundsen
  2. PERSIA   Rahmatollah Eshraghi
  3. CONTEMPORARY ARAB WORLD
       Samuel P. Asper and Fuad Sami Haddad
  4. CONTEMPORARY MUSLIM PERSPECTIVE
       Muhammad Abdul-Rauf
  5. CONTEMPORARY ISRAEL   Seymour Siegel
  6. SUB-SAHARAN AFRICA   Frederick T. Sai
III. SOUTH AND EAST ASIA
  1. GENERAL HISTORICAL SURVEY
       Paul U. Unschuld
  2. INDIA   O. P. Jaggi
  3. PREREPUBLICAN CHINA   Paul U. Unschuld
  4. CONTEMPORARY CHINA   John J. Kao and Frederick F. Kao
  5. JAPAN THROUGH THE NINETEENTH CENTURY   Joseph M. Kitagawa
  6. TRADITIONAL PROFESSIONAL ETHICS IN JAPANESE MEDICINE   Taro Takemi
  7. CONTEMPORARY JAPAN: MEDICAL ETHICS AND LEGAL MEDICINE   Rikuo Ninomiya
IV. EUROPE AND THE AMERICAS
  A. ANCIENT AND MEDIEVAL PERIODS
    1. ANCIENT GREECE AND ROME
        Darrel W. Amundsen
    2. MEDIEVAL EUROPE: FOURTH TO SIXTEENTH CENTURY
        Darrel W. Amundsen
  B. MODERN PERIOD: SEVENTEENTH TO NINETEENTH CENTURY
    1. INTRODUCTION TO THE MODERN PERIOD IN EUROPE AND THE AMERICAS
        Laurence B. McCullough
    2. WESTERN EUROPE IN THE SEVENTEENTH CENTURY   Albert R. Jonsen
    3. BRITAIN AND THE UNITED STATES IN THE EIGHTEENTH CENTURY
        Laurence B. McCullough
    4. NORTH AMERICA: SEVENTEENTH TO NINETEENTH CENTURY   Chester R. Burns
    5. CENTRAL EUROPE IN THE NINETEENTH CENTURY   Guenter B. Risse
    6. FRANCE IN THE NINETEENTH CENTURY
        Dora B. Weiner
    7. BRITAIN IN THE NINETEENTH CENTURY
        M. Jeanne Peterson
  C. CONTEMPORARY PERIOD: THE TWENTIETH CENTURY
    1. INTRODUCTION TO THE CONTEMPORARY PERIOD IN EUROPE AND THE AMERICAS
        Laurence B. McCullough
    2. EASTERN EUROPE IN THE TWENTIETH CENTURY   Benjamin Page
    3. WESTERN EUROPE IN THE TWENTIETH CENTURY   Clarence Blomquist
    4. BRITAIN IN THE TWENTIETH CENTURY
        Gordon Wolstenholme

5. NORTH AMERICA IN THE TWENTIETH CENTURY　*Albert R. Jonsen, Andrew L. Jameton, and Abbyann Lynch*
6. LATIN AMERICA IN THE TWENTIETH CENTURY　*Augusto León C.*

I
PRIMITIVE SOCIETIES

Both alongside of and prior to the highly articulated ethics of the health-care professions of literate societies, values exist which guide the practice of medicine and its analogues in non-literate or primitive societies. Within such societies, various more or less articulated sets of values direct the provision of health care. Such "medical ethical" concepts are derived in part from common understandings of how things are, and in part from collective moral judgment about how things should be. While it is a fact common to humanity that we must confront human suffering, concern ourselves with it, and somehow account for its unequal distribution, the ways of doing so are as diverse as are the cultures of mankind. Far from postulating the naivete of noble savages, it is the purpose here to suggest that medical systems are part of social systems, that definitions of health, illness, therapy, and the rules of behavior relating to them are made in a cultural context. What will be presented here are thus rules for ethical behavior in health care in a number of primitive societies at various times rather than sets of highly articulated maxims for the conduct of medicine.

In an anthropological discussion of non-Western medical ethics, it is important to differentiate between the constructs "morals," "values," and "ethics." *Morals* may be defined as judgments on individual activity; *values* as stated expressions of the cultural framework within which these judgments are made; and *ethics* as socially derived generalizations induced from individual morality. The terms "morals," "values," and "ethics" are descriptive of elements of the everyday world of individuals. They do not indicate those concepts which are used by philosophers and ethicists in arguments concerning the justifiability of particular kinds of conduct. "Primitive" medicine refers to curing that takes place outside the systematic practice of disease-oriented medicine. It is by definition nonsystematic, more often a residual category of multifarious practices relating to many kinds of healing, some religious in origin, some secular. In common with sufferers everywhere, however, the sick person perceives himself and is perceived by others to be temporarily in a position of vulnerability, of powerlessness and uncertainty, and in need therefore of strength and help from outside. It is a common societal response to require the healer to acknowledge and not to exploit vulnerability. This unequal relationship may be said to create the ethical setting.

The separation of disease and illness from life processes is a product of Western scientific medicine, as is the development of practitioners who deal solely with somatic disease. Primitive healing usually refers to coping with suffering in all realms—including those in the natural, supernatural, and social worlds. Human suffering may be perceived in different societies as deriving from forces or circumstances either internal or external to the person, his family, or the entire community. Those differences in perceived etiology require different kinds of practitioners. Although not generally recognized, many forms of primitive medicine are characterized, as is modern medicine, by specialization, and thus by differing definitions of the nature and origin of disease. Specialists are extremely varied. They may include diagnosticians, such as diviners and Navaho hand-tremblers; healers whose power is perceived as primarily supernatural, such as sorcerers, shamans, exorcists, and priest-healers; primarily secular curers, such as midwives or bonesetters; and technical specialists such as Chinese acupuncturists, Malay herbalists, Pomo "sucking doctors," and Thai "injection doctors."

Primitive medicine, then, refers to a diverse and nonunified set of perspectives and practices that do not easily bear comparison. However, certain ethical precepts emerge as characteristic of some examples, even if they are not applicable to all. Ethical principles common to much of primitive medical practice are described below under the general rubrics of (1) personal example, (2) charisma and reciprocity, and (3) authenticity.

Personal example

The ethical precepts do not differ from the values held in many nonindustrial societies, but the shaman or doctor often holds a special position—particularly if his power is seen to have divine origins—in which the standard values of society become concentrated. The demonstration of goodness, purity, even perfection, by personal example is a prime ethical principle of primitive medicine. The moral component of primitive medicine is provided by the social group who observe and support—in much the same way that a modern medical system depends on popular acceptance and shared definitions for success. As Lévi-Strauss suggests, "the efficacy of magic implies a belief in magic. The latter has three complementary aspects: first, the sorcerer's belief in the effectiveness of his techniques; second, the patient's . . . belief in the sorcerer's power; and finally, the faith and expectations of the group, which constantly act as a sort of gravitational field within which the relationship between sorcerer and bewitched is located and defined" (Lévi-Strauss, p. 24). The group is sometimes the personal community of the sick person, such as those participating in a Navaho "sing." Sometimes it is the entire community—as is often the case with an exorcism of malevolent spirits—depending on whether the misfortune is seen to have occurred to the sick person and his family alone or whether the existence of a sick person is symptomatic of a misfortune incurred by the entire community. In either case the practitioner takes upon himself or herself responsibility as an intermediary between supernatural or unseen unknowable forces and those suffering from their ill effects. In order to do this effectively, the practitioner must demonstrate his own moral leadership.

Charisma and reciprocity

Curing in the primitive sense cannot be separated from the curer. It is a personalistic endeavor and its ethics are personalistic. The curer acts not as representative of a medical institution but in his own right. The following ethical concerns derive from the concept of *charisma* with its definition of healing ability as a gift that must be acquired by extraordinary means and must be shared freely and with all those who need help, to be effective. Reciprocity to mankind for a divinely awarded gift is the central ethical theme.

Romano, in describing the charismatic healer and folk saint Don Pedrito Jaramillo, defines the ethics of the healer role in a Mexican-American community in South Texas:

In order to perform his or her role "properly," the healer should be "at the service of the public," and respond to each request for assistance. In this capacity, he or she should not impose a fixed charge for services, but rather operate on the basis of relative means and reciprocal exchange. In addition, the healing role should be performed without overt display of fear, doubt, or diffidence. The requirement of a relatively selfless personality is closely associated with the belief that it is basically "un-Christian" to profit from the suffering and ills of humanity. The healer should also acknowledge the omniscient presence of God as the ultimate source of health and well-being. If requested, services rendered may also include efforts toward the solution of personal and social problems [Romano, p. 1153].

In another description of this folk saint, Dodson notes:

He made no charges for the prescriptions; the patients gave whatever they cared to, or nothing at all; nevertheless, what the people gave voluntarily was sufficient for Don Pedro, who himself often gave to the poor and the sick who came to him. For himself he wanted only enough for his very frugal living. He thought that since God had bestowed on him the power to help humanity, he could also take that power away from him if he used it for his own benefit [Dodson, p. 12].

This concern for his patients extended to leaving provisions for those who might arrive while he was away from home. Charisma, perceived in life, may continue after death, as in the case of Don Pedrito Jaramillo, who became a folk saint and whose shrine was still a pilgrimage place noted for healing power seventy years after his death.

The description of a Himalayan shaman by Berreman stresses also the characteristic of diffidence in relation to payment. Things appear, are brought, but are never asked for. "God provides." The "gift" in this instance is the actual presence of the deity, which acts through the person of the *devta* or holy man, while he himself takes no credit for healing or helping those who present themselves. Charisma is a quality that is communicable, and in India the concept of *darshan* means that one acquires good by being in the presence of a good person. The gift must not be withheld (Berreman).

Authenticity

Authenticity of the folk practitioner is often established through recruitment by supernatural election—a "calling" or notification by some sign that the person indeed has special powers of healing. The "sign" may range from peculiar circumstances of birth that set the future healer apart, occurrences during early life, a seemingly miraculous cure from disease, or special abilities or powers observed during his or her lifetime. Zinacanteco shamans may be chosen by "divine election" before birth, or their vocation may become manifest during the course of an illness or as a result of seizures or epilepsy (Fabrega and Silver). In *Black Elk Speaks*, the life story of a holy man of the Oglala Sioux, Black Elk describes the vision he experienced at the age of nine during a grave illness:

I am sure now that I was then too young to understand it all, and that I only felt it. It was the pictures I remembered and the words that went with them; for nothing I have ever seen with my own eyes was so clear and bright as what my vision showed me; and no words that I have ever heard with my ears were like to words I heard. I did not have to remember these things; they have remembered themselves all these years [Neihardt, p. 41].

An ethical consideration for the person so elected is reflected in the responsibility to heed the call.

Training may be by ordeal also, or may be by extended apprenticeship. The vision as sought among American Plains Indians constituted verification by ordeal including altered states of consciousness and heightened susceptibility to suggestion induced by privation of food and sleep. During this time the novice would "receive" his song and his strengths and would learn of his sacred animal and the contents of his medicine bundle, those material things that would help in his healing practice. Lifetime respect for the animal species or totem, or in the case of healing through spirit possession respect for the familiar spirit, was an important obligation. The Sioux holy man Black Elk stated that, after his vision, contrary to common practice in his community, he no longer enjoyed the hunt and could no longer kill the animals that had become his helpers. A deep knowledge of the culture, its traditions, and the secrets of previous medicine men were all a part of the requirements. In addition, practice with an experienced practitioner assured that irresponsible or ignorant healing practices would not take place. Verification of ability was generally by indications of success. An interesting discussion of concerns for authenticity appears in Lévi-Strauss's article "The Sorcerer and His Magic," in which the sorcerer first questions and then joins the shamanistic brotherhood. A consistent ethical concern, despite extreme diversity of traditions of healing, is in authenticating the calling and the training of healers. The importance of the authenticity of healers is not only the protection of individual patients and families but protection of the community itself, in light of their moral responsibilities.

Conclusion

The ethical setting of primitive medicine, made up of suppliant, healer, and surrounding social group, and a mutually held belief in certain common values, provides the roots of what in literate societies is found in articulated systems and codes of medical ethics. These include judgments concerning the need for medical practitioners to be of special moral integrity and to demonstrate their goodness by personal example; the obligation of the healer to be of assistance to others without profit to himself; and the authentication of medical practitioners or ways of identifying the true or efficacious healer. These concerns are leitmotifs of interest found in many, if not most, primitive societies. They may be viewed as preliminary forms from which discussions about the integrity of health-care providers, rights to health care, and the certification of health-care professionals spring.

LUCILE F. NEWMAN

[*For further discussion of topics mentioned in this article, see the entries:* HEALTH AND DISEASE, *articles on* HISTORY OF THE CONCEPTS *and* A SOCIOLOGICAL AND ACTION PERSPECTIVE; PAIN AND SUFFERING; *and* THERAPEUTIC RELATIONSHIP, *article on* HISTORY OF THE RELATIONSHIP. *Also directly related is* MIRACLE AND FAITH HEALING.]

BIBLIOGRAPHY

BERREMAN, GERALD D. *Hindus of the Himalayas: Ethnography and Change.* 2d rev. ed. Berkeley: University of California Press, 1972.

DODSON, RUTH. "Don Pedrito Jaramillo: The Curandero of Los Olmos." *The Healer of Los Olmos, and Other Mexican Lore.* Edited by Wilson M. Hudson. Texas Folklore Society Publications, no. 24. Dallas: Southern Methodist University Press, 1951, pp. 9–70. Reprint. 1966.

FABREGA, HORACIO, JR., and SILVER, DANIEL B. *Illness and Shamanistic Curing in Zinacantan: An Ethnomedical Analysis.* Stanford: Stanford University Press, 1973.

HARNER, MICHAEL J., ed. *Hallucinogens and Shamanism.* New York: Oxford University Press, 1973.

KLUCKHOHN, CLYDE, and LEIGHTON, DOROTHEA. *The Navaho.* Rev. ed. Edited by Lucy H. Wales and Richard Kluckhohn. Natural History Library, no. N28. Garden City, N.Y.: Doubleday-Anchor Books, 1962.

LABARRE, WESTON. *The Peyote Cult.* Enl. ed. New preface by author. New York: Schocken Books, 1969.

LÉVI-STRAUSS, CLAUDE. "The Sorcerer and His Magic." Middleton, *Magic, Witchcraft, and Curing*, pp. 23–

41. Reprinted from Lévi-Strauss, Claude. *Structural Anthropology.* New York: Basic Books, 1963.
MIDDLETON, JOHN, ed. *Magic, Witchcraft, and Curing.* American Museum Sourcebooks in Anthropology. Garden City, N.Y.: Natural History Press, 1967.
NEIHARDT, JOHN G. *Black Elk Speaks: Being the Life Story of a Holy Man of the Oglala Sioux as Told through John G. Neihardt (Flaming Rainbow): Illustrated by Standing Bear.* A Bison Book, no. 119. Lincoln: University of Nebraska Press, 1961.
ROMANO V., OCTAVIO IGNACIO. "Charismatic Medicine, Folk-Healing, and Folk-Sainthood." *American Anthropologist* 67 (1965): 1151–1173.
TURNER, VICTOR W. *The Drums of Affliction: A Study of Religious Processes among the Ndembu of Zambia.* Oxford: Clarendon Press; London: International African Institute, 1968.
VOGEL, VIRGIL J. *American Indian Medicine.* The Civilization of the American Indian Series, no. 95. Norman: University of Oklahoma Press, 1970.

II
NEAR AND MIDDLE EAST AND AFRICA

1. ANCIENT NEAR EAST — Darrel W. Amundsen
2. PERSIA — Rahmatollah Eshraghi
3. CONTEMPORARY ARAB WORLD — Samuel P. Asper and Fuad Sami Haddad
4. CONTEMPORARY MUSLIM PERSPECTIVE — Muhammad Abdul-Rauf
5. CONTEMPORARY ISRAEL — Seymour Siegel
6. SUB-SAHARAN AFRICA — Frederick T. Sai

1
ANCIENT NEAR EAST

Diverse civilizations are subsumed under the rubric "Ancient Near East." The documents used in this article relevant to the medical ethics of the Ancient Near East, however, concern Egypt, Assyria–Babylonia, and Persia. The primary sources employed for Egypt range from the first half of the third millennium B.C. to the fourth century B.C., those for Assyria–Babylonia from the eighteenth to the seventh century B.C., and those for Persia from the fourth century B.C. to the sixth century A.D.

No literature specifically devoted to any aspect of medical ethics appears to be extant from the Ancient Near East. If any physician ever saw fit to express his views on the moral and ethical bases of health care or the more mundane and pragmatic rules of medical etiquette, his writings are not known. What information is available must be gleaned from medical, legal, and other literature and, consequently, is quite sketchy, leaving many questions unanswered.

The magico-religious element

There has been a tendency sometimes to overemphasize and at other times to underemphasize the magico-religious basis of Ancient Near Eastern medicine. This is not the place to discuss the merits of the various assessments of the degree to which Ancient Near Eastern medicine should be considered rational or superstitious. It is, however, important to consider the question as it impinges upon medical ethics. It is safe to say that, with certain qualifications, the Ancient Near Eastern cultures were permeated with religion and what we may today call magic. Disease was often viewed as inflicted by the gods, perhaps as a punishment for sin, or by various demonic forces, perhaps arbitrarily. The supernatural agency could be viewed as being active within the individual or perhaps even as the disease itself. On the other hand, direct, external disabilities (broken bones, dislocations, contusions, and so forth) were usually viewed without reference to the divine, and the treatment employed was generally devoid of magic. Some scholars seem especially refreshed when encountering in the ancient texts "rational" methods of medicinal treatment of internal diseases. It should be noted, however, that the efficacious medicinal treatment of diseases does not necessarily conflict with a belief in a supernatural etiology. Additionally, the psychological effect of incantations or ritualistic utterances on the patient who had faith in their efficacy should not be too readily discounted. This is not to say that there were no instances where the physician did attempt to attribute a particular disease specifically to what we would call natural causes. What is more remarkable is that physicians did not seek to excuse themselves from a responsibility to their patients on the ground of the common belief in the supernatural etiology of disease. Regardless of their ailments, those suffering from internal diseases as opposed to injuries were not considered untouchable, demon-possessed creatures to be shunned by the physicians. Even in Persia the fact that the ill were regarded as unclean and were segregated did not detract from the physician's overall responsibility to attempt to cure to the extent of his ability and knowledge.

There are some semantic problems. The word physician has thus far been used in a general sense to apply to anyone whose professional role was that of a healer. Conventionally the word sometimes applies only to a specific category of healers, as in Egypt and Assyria–Babylonia. Except when a specialized use is indicated, "physician" will refer here to any practitioner of the healing arts. In Egypt there were three categories: the physician (as *swnw* is usually translated), the priest of Sekhmet, and the magician (Edwin Smith Papyrus [a sixteenth-century B.C. copy of an original probably written between 3000 and 2500 B.C.], case 1; compare *Papyrus Ebers* [from roughly the same period], col. 99). Although each of these was a profession distinct unto itself, any two or even all three might be combined in the same man. The healers in Assyria–Babylonia were the magical expert and the physician (Ritter). Of the two, only the former was a priest, and the two professions were both functionally and ideologically distinct. In Persia medical practice was divided among those who healed by the knife, those who healed by herbs, and those who healed by the holy word (*Zend-Avesta: Vendîdâd* 7, 44; compare *Ardibehist Yast* 2, 7 [both were written probably before the end of the fourth century B.C.]). These also appear to have been essentially separate professions, although all were priests. In spite of the diverse cultural traditions of Egypt, Assyria–Babylonia, and Persia, they shared a common feature in the practice of the healing arts, as magical and rational methods were both used. Sometimes one individual would employ both approaches while at other times the magical expert and the physician would treat the same patient either together or separately.

The Egyptian physician, as revealed by the medical papyri, made a prognosis before undertaking treatment. Where the prognosis was favorable, the physician's comment was "an ailment that I shall treat"; if it was uncertain, "an ailment that I shall combat"; and if the prognosis was unfavorable, "an ailment not to be treated." The Edwin Smith Papyrus contains the record of fifty-eight examinations, each followed by either treatment or a decision not to treat. The author recommends treatment in forty-two cases and leaves sixteen untreated. In three of the hopeless cases (6, 8a., and 20), some alleviative treatment is indicated. In the *Papyrus Ebers* a small number of cases are regarded as untreatable (e.g., cols. 108–110), and in one hopeless case there is an attempt to relieve the patient. That specific alleviatory instructions were given only in a minority of these hopeless cases does not necessarily indicate a lack of compassion. Incidental remarks in the papyri show physicians carefully and gently treating their patients, showing marked kindness to the ill, injured, and maimed.

In Assyria–Babylonia the magical expert (*āšipu*), whose medical repertoire consisted mostly of incantations and charms (although he occasionally employed ointments and purgatives), was a prognosticator who did not hesitate to withdraw from hopeless cases. His colleague, the physician (*asû*), who administered medicines, performed some surgery, and only seldom used incantations, seems very rarely to have abandoned his patients but continued with treatment to the end. This difference may be due in part to the fact that the *āšipu* treated primarily chronic illnesses while the *asû* dealt usually with acute diseases and injuries (Ritter).

Persia presents a unique case. Medicine, as represented in the sacred literature, occupied a place of central importance in Zoroastrianism. The physician was an essential element in the category of purification, for prophylaxis, hygiene, and treatment of disease. In the Zoroastrian struggle between the forces of good and evil, the physician was clearly aligned with the good, and his patients, when polluted by disease, turned to him, being compelled not only by the motivation shared by the ill in any culture but also by religious necessity. And, significantly, Persian physicians do not appear to have refused to treat hopeless cases.

Euthanasia and abortion

On the question of euthanasia there is no direct evidence pro or contra. In Persia it seems extremely unlikely that the physicians, being Zoroastrian priests, fighters of disease and auxiliaries of the forces of good, entrusted with the healing of patients who were, in their illnesses, victims of the forces of evil, would have hastened the end of those whom they would not have abandoned even when their condition was hopeless. But this is only conjecture on a point for which there does not appear to be any evidence. Both in Assyria–Babylonia and in Egypt anyone who committed suicide was regarded as having cut himself off from the gods. There survives from Egypt (end of third millennium B.C.) a touching dialogue between a man contemplating suicide and his soul (in *Ancient Near Eastern Texts*, pp. 405 ff.). Although the man is not considering suicide owing to illness, the psychological struggle that is portrayed reveals a culture in which suicide was not accepted simply as a personal option without strong moral and religious compunctions. Very likely the physician would have viewed active euthanasia, if he would even have considered it at all, as opprobrious and immoral.

Although prescriptions for induced abortion are found in the Egyptian medical papyri, there is doubt concerning its legality. Induced abortion was prohibited, however, in Assyria by the Middle Assyrian Laws (fifteenth century B.C.?) which stipulated that, if a woman had an abortion by her own act, whether or not she survived the ordeal, she was to be impaled on a stake and left unburied (A, 53). There is no mention, however, of the involvement of physicians. In Persia the practice was prohibited by the *Vendîdâd*. The woman, the nurse who performed the abortion, and the woman's father or the father of the fetus, if the abortion was sought at their instigation, were all held guilty of willful murder (*Zend-Avesta: Vendîdâd* 15, 2, 12–14).

Regulation of medical care; experimental medicine

Little is known about the regulation of the medical profession in ancient Egypt. It has been commonly thought that medical care was free, based on the erroneous assumption that all medical practitioners were priests and that sacerdotal medicine would have been motivated by charity. The sources do not bear this out, but they also supply no information on fees. Although there appears to have been no system of medical licensure, medical procedure seems to have become rigidly prescribed over the centuries. Diodorus Siculus (first century B.C.; his material from Egypt was based on Hecataeus, sixth century B.C.) writes that physicians gave treatment in accordance with ancient written procedures, and if their patients died the physicians were absolved from any charge. If they deviated from traditional methods in any way they were subject to the death penalty on the assumption that few physicians could ever be wiser than the physicians of old (1, 82, 3). Aristotle, however, describes a slightly more flexible situation in Egypt in which physicians could alter their prescriptions after four days, although if one of them altered it earlier he did so at his own risk (Aristotle, 1286a, 4).

The first recorded attempt to protect the patient from the incompetent physician is from Babylonia and appears in the Code of Hammurabi (1727 B.C.?), where it is specified that, if a physician performed a major operation with a bronze lancet on a nobleman and caused the patient's death or operated with a bronze lancet on his eye, resulting in the loss of that member, the physician's hand would be cut off (#218). If an operation with a bronze lancet resulted in the death of a commoner's slave, the physician was required to replace the slave (#219), and if the operation caused the loss of the slave's eye the physician was to pay one-half the slave's value in silver (#220). No punitive regulations are extant governing medical procedures other than surgery. This is understandable particularly in a culture permeated by magical beliefs. The unsuccessful use of incantations or sympathetic magic (and the administration of medicinal herbs may be included in this category), where the healing role of the practitioner is nearly passive because of the supernatural agents at play, stands in marked contrast to the active immediacy of surgery. Also included in the Code of Hammurabi is the fixing of fees for surgery, where the amount is determined by the social status of the patient (#215–217, 221–223). While this reveals the absence at least of free surgical care in Babylonia at that time, it also indicates a desire on the part of the legislator to peg medical expense to the economic means of the patient.

As in Assyro-Babylonian civilization, so also in Persia there was a gradation of fees for the healer (here not limited to surgery) proportional to the social standing of the patient (*Zend-Avesta: Vendîdâd* 7, 41–43). As indicated above, in Persia there were three types of healers: those healing with the knife, those healing with herbs, and those healing with the holy word. Although the last was to be preferred above the others (ibid., 7, 44), it was to the qualifications of the surgeon that attention was given. If a worshiper of Ahuramazda (god of light and goodness) wished to practice the art of healing, he was first to perform surgery three times on worshipers of the Daevas (i.e., followers of any religion other than Zoroastrianism). If none of these patients should die, he was held qualified to practice on the faithful. If he should presume to attend any worshiper of Ahuramazda and "wound him with the knife" before he was qualified or when he had failed the test, he would be punished as for willful murder (ibid., 7, 36–40). Commentary on *Zend-Avesta* (sixth century A.D.?) specifies, however, that the culprit should be branded.

This example of *experimentum in corpore vili* has parallel elsewhere in the Ancient Near East.

A physician suggested in a letter to the Assyrian king (seventh century B.C.?) that a particular prescription be tested on members of the domestic staff before being administered to a member of the royal family. While Caesarean sections are known to have been performed in Babylonia in the second millennium B.C. as a last resort to save the infants of dying women, the evidence suggests that the procedure was used only on slaves. Save for such occasional references, little is known of the ethics of medical experimentation in the Ancient Near East. It should be noted that all three cultures under consideration developed a written tradition of medical knowledge and procedures. Especially when such traditions were allied ideologically with religion, opportunities for medical experimentation that deviated from accepted practice became extremely limited.

The ideal physician

If there was ever an attempt in ancient Egyptian or Assyro-Babylonian literature to articulate precisely the qualities that ought ideally to be possessed by the physician, it does not appear to be extant. It is, however, in the Pahlavi literature of the Sassanian Persians that at last a description of the qualities of the ideal physician appears, from book three of the encyclopedic *Dinkard*. The section in question probably was written during the sixth century A.D., although there is a distinct possibility that it was based directly on a lost portion of the much older *Zend-Avesta* of which only approximately one-third survives. The perfect physician

. . . should know the limbs of the body, their articulations; remedies for disease; . . . should be amiable without jealousy, gentle in word, free from haughtiness; an enemy to disease, but the friend of the sick, respecting modesty, free from crime, from injury, from violence; expeditious; . . . noble in action; protecting good reputation; not acting for gain, but for a spiritual reward; ready to listen; . . . possessed of authority and philanthropy; skilled to prepare health-giving plants medically, in order to deliver the body from disease, to expel corruption and impurity; to further peace and multiply the delights of life [Bk. III-157, 19].

Relevance of Ancient Near East medical ethics

The degree of influence of Egyptian medicine on Hippocratic medicine is a debatable question. Even the extent to which one can speak of the survival of distinctly Egyptian medicine in Ptolemaic and Roman Egypt and its possible interaction with Greek medicine is also in doubt. While Zoroastrian-Persian medicine probably continued earlier Assyro-Babylonian traditions and was in turn affected by Greek medicine, the extent of its direct influence on and relationship with Islamic and Indian medicine remains still uncertain. Medical ethics, however, is even less apt to be borrowed by members of one society from another culturally alien to it than are its medical theory and concomitant technique. Thus the relevance of Ancient Near Eastern medical ethics to the broader spectrum of the development of Western medical ethics is, at the most, only peripheral.

DARREL W. AMUNDSEN

[*Compare related topics from this article with:* ABORTION, *articles on* CONTEMPORARY DEBATE IN PHILOSOPHICAL AND RELIGIOUS ETHICS *and* LEGAL ASPECTS; HEALTH AND DISEASE; *and* MIRACLE AND FAITH HEALING. *Additional relevant material may be found in the following three articles dealing with Persia and the Arab countries.*]

BIBLIOGRAPHY

PRIMARY SOURCES CITED ON ANCIENT NEAR EAST

Ancient Near Eastern Texts Relating to the Old Testament. 3d ed. with supp. Edited by James Bennett Pritchard. Princeton, N.J.: Princeton University Press, 1969.

BISHOP, DALE LAWRENCE. "Form and Content in the Videvdad: A Study of Change and Continuity in the Zoroastrian Tradition." Ph.D. dissertation, Columbia University, 1974. English translations of *The Dinkard* by Kohiyár and of the *Avesta* by Darmesteter are somewhat unreliable. Bishop presents a translation and discussion of the medical sections of these two works and of a previously untranslated commentary on the *Avesta*.

"The Code of Hammurabi." Translated by Theophile J. Meek. *Ancient Near Eastern Texts*, pp. 163–180.

The Dinkard: The Original Pahlavi Text; The Same Translated into Zend Characters; Translations of the Text in the Gújráti and English Languages; A Commentary and a Glossary of Select Terms. Edited by Peshotan dastur Behramjee Sanjana. English translation from the Gújráti by Ratansháh Erachsháh Kohiyár. Bombay: Duftur Ashkara Press, 1874–1891.

"A Dispute over Suicide." Translated by John A. Wilson. *Ancient Near Eastern Texts*, pp. 405–410.

The Edwin Smith Surgical Papyrus: Published in Facsimile and Hieroglyphic Transliteration with Translation and Commentary. 2 vols. Translated by James Henry Breasted. Oriental Institute Publications, vols. 3, 4. Chicago: University of Chicago Press, 1930.

"The Middle Assyrian Laws." *Ancient Near Eastern Texts*, pp. 180–188.

The Papyrus Ebers: The Greatest Egyptian Medical Document. Translated by B. Ebbell. Copenhagen: Levin & Munksgaard; London: H. Milford, Oxford University Press, 1937.

The Zend-Avesta. 3 vols. Edited and translated by James

Darmesteter. Vol. 1: *The Vendîdâd.* Vol. 2: *The Sîrôzahs, Yasts, and Nyâyis.* The Sacred Books of the East, vols. 4, 23. Oxford: Clarendon Press, 1880–1883. Reprint, Westport, Conn.: Greenwood Press, 1972.

CLASSICS AVAILABLE IN A VARIETY OF TRANSLATIONS

ARISTOTLE. *Politics.*
DIODORUS SICULUS. *Library of History.*

SELECT LIST OF EASILY ACCESSIBLE SECONDARY LITERATURE

BIGGS, ROBERT. "Medicine in Ancient Mesopotamia." *History of Science* 8 (1969): 94–105.
BRANDENBURG, DIETRICH. "Avesta und Medizin: Ein literaturgeschichtlicher Beitrag zur Heilkunde im alten Persien." *Janus* 59 (1972): 269–307.
GHALIOUNGUI, PAUL. *Magic and Medical Science in Ancient Egypt.* London: Hodder & Stoughton, 1963.
GRAPOW, HERMANN. *Kranker, Krankheiten und Arzt: Vom gesunden und kranken Ägypter, von den Krankheiten, vom Arzt und von der ärztlichen Tätigkeit.* Grundriss der Medizin der alten Ägypter, vol. 3. Berlin: Akademie-Verlag, 1956.
OPPENHEIM, A. LEO. "Mesopotamian Medicine." *Bulletin of the History of Medicine* 36 (1962): 97–108.
REINER, ERICA. "Medicine in Ancient Mesopotamia." *Journal of the International College of Surgeons* 41 (1964): 544–550.
RITTER, EDITH K. "Magical-Expert (=Āšipu) and Physician (=Asû): Notes on Two Complementary Professions in Babylonian Medicine." *Studies in Honor of Benno Landsberger on His Seventy-Fifth Birthday, April 25, 1965.* The Oriental Institute of the University of Chicago Assyriological Studies, no. 16. Chicago: University of Chicago Press, 1965, pp. 299–321.
SIGERIST, HENRY ERNEST. "Ancient Egypt." *A History of Medicine.* 2 vols. Vol. 1: *Primitive and Archaic Medicine.* Historical Library, Yale Medical Library, Publication no. 27. New York: Oxford University Press, 1951, pp. 215–373.
———. "Medicine in Ancient Persia." *A History of Medicine,* vol. 2: *Greek, Hindu, and Persian Medicine,* Historical Library, Yale Medical Library, Publication no. 38, 1961, pp. 195–209.
———. "Mesopotamia." *A History of Medicine,* vol. 1, pp. 375–497.
WILSON, JOHN A. "Medicine in Ancient Egypt." *Bulletin of the History of Medicine* 36 (1962): 114–123.

2
PERSIA

Persia is the name that has been used in European countries for the ancient kingdom in southwestern Asia. The people of that country have always called their country Iran. Today both names are used.

Persia (or Iran) has a long history and has been the center of a number of ancient empires extending beyond the border of the present kingdom. As a highway for the movement of people and ideas, from the prehistoric period onward, Iran has held an important position between East and West. What it received from both sides it never ceased to adapt, to absorb, to synthesize, and to transmit.

Persian medicine had great fame in antiquity. Historians mentioned extreme specialization as one of the outstanding characteristics of Persian medicine in the seventh century B.C., approximately three centuries before Hippocrates.

Within the limits of this article, the general background of Persian medicine cannot be discussed; only the history of medical ethics in Persia can be considered.

Prehistoric period (paleomedicine)

Traditional Iranian legends and tales, embodied in mythological sources, tell us something about early medicine in Iran. The epic poem *Shahmameh* (The Book of Kings), written by the great poet Abulghasem Ferdowsi around the beginning of the eleventh century B.C., is the best of the mythological sources. Although it is a narrative tale, it represents the prehistoric Persian culture with regard to medicine.

Ferdowsi pointed to the high degree of medical knowledge that the Iranians had attained. He explained how Rostam, a prehistoric Persian hero, had been born by Caesarean section while the mother was kept bemused during the operation by the use of wine as an anesthetic. Further, although criminal abortion was a severely punishable offense, many abortifacients were known to Iranian physicians.

Protohistory period

The period beginning in Persia with the Elamite civilization (3000 B.C.?) and ending roughly at the start of the first millennium B.C. would represent the medical practice of the most ancient civilization.

Elamite civilization began in Susa (Shoosh) in southwestern Iran as early as it did in neighboring Mesopotamia, and it was closely tied culturally to Sumer, Babylonia, and Assyria. When the Elamite king ruled Mesopotamia in the thirteenth century B.C., he possessed the stone pillar on which was inscribed the code of Hammurabi, and early King of Babylonia (1955–1913 B.C.?). This stela, discovered at Susa in Iran in 1902 by the French orientalist Jean Vincent Schiel, is preserved in the Louvre in Paris. A notable feature of the code is that it contains a code of ethics, which comprised the standards of medical men, their remunerations, and punishments. Although this code belongs to Babylonia, it also shows the Elamite contributions to medical ethics.

Aryan periods (ninth to fourth century B.C.)

Aryans were a branch of Indo-European people who were originally centered at the Volga River, north of the Caspian Sea. Iran was occupied by groups of Aryan tribes as early as the Protohistoric period. They were the dominant force of the plateau in the ninth century B.C. In 550 B.C., Cyrus the Great established the Achamenid dynasty, the first Persian empire. He conquered Babylonia and proclaimed the first Declaration of Human Rights in history. He then extended his dominion and ruled over a vast realm extending from Anatolia to the border of India.

Another important event of the Aryan periods was the appearance of the great Iranian prophet, Zoroaster (more probably in the seventh century in the northwest of Iran, Azarbaijan). He was an ethical prophet of highest rank, stressing constantly the need for man to speak cleanly, to act cleanly, and to think cleanly. Some parts of the *Avesta*, the holy book of Zoroaster, were known as the first medical text.

According to the Avestan principles of medicine, in most diseases the physical and natural causes (such as cold, heat, stench, dirt, hunger, thirst, anxiety and/or old age) played an important part; others were said to have been caused by magic or evil. Infection was recognized as a natural cause of diseases, so physicians were warned that they should not move carelessly from case to case.

With a divorce taking place between theology and medicine, medical matters came into the hands of physicians who were drawn from the Iranian noble families. There were also hospitals for practical training.

Three kinds of practitioners issued from the schools: healers with herbs, healers with knives, and healers with holy words. The last were the most highly trained. There were rules for testing the skill of healers. Surgeons had to undergo a very severe trial before they were allowed to operate.

Among the healers there were a certain number of specialties. One of these was called *Durustpat* (master of health); his aim was to remove the cause of diseases. Purification of earth, air, water, and food was the duty of *Durustpats* and rulers. In the realms of both health and ethics, Zoroastrian medicine reached a high level. According to the Avestan rules, to destroy life was to destroy the highest form of the Ahura Mazda (God) Creation. Abortion was forbidden; punishment for procuring abortion was the same punishment as for willful murder. In the case of abortion, the man and the woman were both guilty.

There were also rules for the care of pregnant females, whether two-footed (human) or four-footed (animal). In the case of sick or pregnant animals, it was the duty of the animal's master to provide veterinary service and to take care of the young of animals for six months, just as they cared for children for seven years.

Physicians were the privileged class of the people. It was desirable that the prestige of the physician should be untarnished by professional misconduct. Elgood (1951, p. 13) quotes this ideal as summed up by Maneckji N. Dehala:

The first indispensable qualification of a physician was that he should have studied well the science of medicine ... who hears the case of his patient with calmness ... who is sweet-tongued, gentle, friendly, zealous of honour of his profession, averse to protracting the disease for greed of money and who is God-fearing. An ideal healer heals for the sake of healing.

Then he mentions the other ranks who practice for reward and renown, or for the sake of merit and money: "The lowest in the scale is the greedy and heartless physician, who dishonours his noble profession." He says that the duty of physicians is "to watch carefully the effect of medicine that he prescribes ... to visit the invalid daily at a fixed hour, to labour zealously to cure him, and to combat the disease of the patient, as if it were his own enemy" (Elgood, 1951, p. 13).

In Zoroastrian medicine, physicians could receive a suitable fee for services rendered. The amount of the fee depended upon the status of the patient. The *Durustpat*, or master of health, who carried great responsibility, was promised a reward in heaven in return for his labor on earth.

Hellenistic period (330 B.C. to A.D. 224)

The Hellenistic period of Greek medicine began with the conquest of Iran by Alexander in 330 B.C.

In the five centuries that lay between the Achamenid and the rise of the Sassanians, historical fact is very meager, and we know almost nothing about medical and ethical points of view. We know that Hippocrates (460?–377 B.C.) was an important figure in a tradition that elevated Greek superstitious medicine into science, his ethical outlook is reflected in the Hippocratic Oath.

During this period Greek civilization was so pervasive that neither the successor of Alexander nor the Parthians could remain unaffected by Greek civilization.

Sassanian period (A.D. 224 to 632)

The Sassanian period is the period of Jundishapur medicine, which lasted from the foundation of the Sassanian dynasty (A.D. 224) to the Arab conquest of Iran (A.D. 632).

The Sassanian dynasty ruled for more than four centuries, renewed Persian tradition, reformed Zoroastrian religion, which had become polluted with foreign ideologies, and founded the Jundishapur University.

Shapur I (A.D. 241–272) captured many Roman cities. He banished the inhabitants of Antioch and settled the physicians and scientists in Gundishpur (Jundishapur). Shapur II (A.D. 309–379) enlarged the city and founded a university. When Nestorius, a patriarch of Constantinople, was excommunicated by the Byzantines, he and his followers were invited to transfer their rich, great library and school of learning to Jundishapur.

Khosrow Noshisran (or Anushirvan the Just, A.D. 531–579) enriched the university. He dispatched his prime minister, Burzuya or Buzurgmehr, to India to procure the different arts and sciences of the Indian culture. He returned with various Indian books, arts, and drugs, and Indian physicians. By Noshisran's order, a gigantic work (thirty volumes) on poison was composed, and many Greek and Indian books were translated into Pahlavi (ancient Persian). Around A.D. 555, a general meeting of physicians was held at Jundishapur to discuss various medical problems.

When the Neoplatonist School in Athens was closed, the fame of Noshisran and his university enticed the Greek exiles to come to Persia. The arrival of refugees from Athens strengthened the Greek influence in Jundishapur even though it remained markedly cosmopolitan. The teaching of the university was not in the Greek language but in Pahlavi, Syriac, and a dialect specific to the university. The system of medicine also was not purely Greek but a combination of Persian, Greek, and Indian medicine, with the result that the university developed an eclectic method of its own.

From an ethical point of view, although the school was under the influence of the Greeks, the rise of Zoroastrians under the Sassanian dynasty brought forward again Zoroastrian moral teachings.

Writing in this period, Burzuya says:

> I read in medical books that the best physician is the one who gives himself over to his profession. . . . So I determined to follow this and to aim at no earthly gain . . . I exerted myself in the treatment of patients . . . in those cases where I could not hope to effect a cure I tried to make their suffering more bearable. Whenever I could I used to attend in person to my cases . . . For no one whom I treated did I demand any fee or any sort of reward [Elgood, 1951, p. 52].

Islamic period

The Islamic period began with the Arab conquest of Iran in A.D. 652. In describing the medical system of that period, it is customary to speak of Arabian medicine. It must be remembered that what crystallized during the Islamic era comprised the heritage of Assyrians, Babylonians, Persians, Greeks, and Indians. A great amount was transferred into Arabic through Jundishapur, developed and cultivated mostly by Iranians, and transferred to Europe by the means of translation of Persian physicians' books.

When the Arab invasion of Iran occurred, the Jundishapur Medical School was at the height of its glory. It surrendered to the Muslims in A.D. 636 and was left undisturbed. It continued to exist for another century and a half, and became the greatest center of medical education throughout the Islamic world. The last important act of the school was the publication in A.D. 860 of the first official pharmacopoeia, which was used throughout the Islamic world.

When Al Mansure, the Abbasid Caliphate, enlarged the city of Baghdad, he built a *Bimaristan*, which became the Metropolitan Hospital and the cradle of the Medical School of Baghdad. This idea was suggested to him by the Jundishapur court physician; *Bimaristan* is a Persian word meaning "the place of sick people."

For many years after the entry of the Muslims into the realm of medicine, Jundishapur provided the staff for the new Arab foundation. The Persian physicians played a conspicuous part in the translation of books from Pahlavi and other languages into Arabic, and when the Muslims began to compile books in Arabic and to establish Islamic medicine, the famous leading authors (e.g., Tabari, Rhazes, Haly Abbas, Avicenna) all were Iranian.

During the first four centuries after Islam, there were many great writers who wrote mainly in Arabic and occasionally in Persian. Akhavaynī (tenth century A.D.) and Jurjānī (eleventh century A.D.) are the first physicians who began to write exclusively in Persian. This tradition continued after them and did not suffer much change until the mid-nineteenth century.

The works of these great men in Arabic and Persian became the leading medical textbooks of the East and West for many centuries. Some of these principal medical works (e.g., *Avicenna's Canon of Medicine*) became classic books and were used at many medical schools in Europe as late as 1650.

Medical ethics in Persia after Islam

Medicine in the Islamic era during the first two centuries after Islam was more or less the same as in Jundishapur, but from the point of view of medical ethics the appearance of Islam was a very important event.

Jundishapur medical ethics was a combination of Persian, Greek, and Indian moral principles. The conjunction of the new ethico-cultural system of Islam with Jundishapur medical ethics secured the moral aspect of medicine. The physician of the Islamic era, at the same time, was accustomed to the ethical principles of Persia and acquainted with the Hippocratic Oath as well as the new Islamic moral doctrine. They were acquainted with philosophy and had a good knowledge of theology. Nizami Aruzi (twelfth century A.D.) detailed the qualifications of a physician in *Chahar Maghaleh* as follows:

A physician should be of tender disposition and wise nature, excelling in acumen, this being a nimbleness of mind in forming correct views, that is to say, a rapid transition to the unknown from the known, and no physician can be of tender disposition if he fails to recognize the nobility of the human soul; nor of wise nature unless he is acquainted with Logic, nor can be excel in acumen unless he be strengthened by God's aid and he who is not acute in conjecture will not arrive at a correct understanding of any ailment [Elgood, 1951, p. 234].

In this period, medicine was reckoned an honorable and useful profession, and the books of the leading Persian figures of the time are full of advice for physicians. Avicenna (A.D. 980–1063) placed the moral points under a different subject of his splendid work, *Canon of Medicine*. Rhazes (Razi) (A.D. 860–932) compiled a special booklet on the duty of physicians, and Haly Abbas (Ahwazi) (died in 994) allocated a chapter in his principle work, *Liber Regius*, to medical ethics [see translation of this text in the APPENDIX, SECTION I].

Present period

For many centuries after Islam, medicine in Iran was more or less the same as described. The foundation of Dar-ul-Fanun or École Polytechnique in Tehran in 1852 changed the situation. From the beginning, it was a military academy, but the college began shortly to develop into a university. The foundation of the Faculty of Medicine was laid by a number of excellent European and Iranian teachers. The school curriculum at first was a combination of Iranian and Western medicine, and the ethical point of view was under the influence of Iranian tradition.

From the year 1858 the university sent a number of fellows to Europe each year. With the return of these scholars and the establishment of a modern hospital in Tehran in 1868, the program of the school and the practice of medicine became gradually westernized.

Since the period of Reza Shah, 1923–1941, the program of the Medical School of the modern University of Tehran—and after that the other medical school—has been based completely on modern medicine; medical ethics and the history of medical tradition are both taught. Graduates of the Tehran Medical School are asked to take an oath, an excerpt from which follows:

Now that I . . . have been found eligible to practice medicine, in the presence of you, the board of judgment of my thesis and others here present, I swear by God and the Holy Book of Koran and call to witness my conscience that in my profession I will always be abstemious, chaste, and honest and, as compared with the glory of the art of medicine, I will hold in contempt all else—silver, gold, status, and dignity. I promise to help the afflicted and needy patient and never divulge patients' secrets. I will never undertake dishonest work such as producing abortion and recommending a fatal drug.

What I do, I will try always to be approved by God and be known for my uprightness.

RAHMATOLLAH ESHRAGHI

[*Other relevant material may be found under* ISLAM; *in the previous article on* ANCIENT NEAR EAST; *and in the article* ANCIENT GREECE AND ROME *in this entry's subsequent section on* EUROPE AND THE AMERICAS.]

BIBLIOGRAPHY

ACKERKNECHT, ERWIN HEINZ. *A Short History of Medicine.* New York: Ronald Press Co., 1968.

AQILI ALAVĪ, M. H. *Khulāsat al-Ḥikmah* (1770). Tehran: 1870.

AKHAVAYNĪ AL-BUKHĀRĪ, ABŪ BAKR RABĪ ᶜIBN AHMAD AL-. *Hidāyat al-muta'allimīn fī al-ṭibb* [Direction of teachers, in medicine]. Edited by Jalāl Matīnī. Dānishgāh Intishārāt [Meshed University Publications], no. 9. Meshed, Iran: Meshed University Press, A.H. 1344 [1965]. Added title page in English: *Hidāyat-al-muta' allimīn fi l-tibb,* by Abu Bakr Rabī'b. Ahmad al-Akhawainī al-Bukhārī.

AVICENNA [HUSAYN BIN ᶜABD-ULLĀH HASAN BIN ᶜALĪ BIN SĪNĀ] [IBN SINA]. *al-Qānūn fi al-Ṭibb* (11th Century). Edited and translated by Mazhar H. Shah as *The General Principle of Avicenna's Canon of Medicine.* Karachi: Naveed Clinic, 1966.

BROWNE, EDWARD GRANVILLE. *Arabian Medicine: Being the Fitzpatrick Lectures Delivered at the College of Physicians in November 1919 and November 1920.* Cambridge: University Press, 1921.

CLENDENING, LOGAN. *Source Book of Medical History, Compiled with Notes.* New York: Paul B. Hoeber, 1942. Reprint. Dover Publications, 1960.

DAVIS, NATHAN SMITH. *History of Medicine, with the Code of Medical Ethics.* Chicago: Cleveland Press, 1903, 1907.

ELGOOD, CYRIL LLOYD. *A Medical History of Persia and the Eastern Caliphate, from the Earliest Times until the Year A.D. 1932.* Cambridge: University Press, 1951.

———. *Safavid Medical Practice; or, The Practice of Medicine, Surgery, and Gynaecology in Persia between 1500 A.D. and 1750 A.D.* London: Luzac, 1976.

ESHRAGHI, R. *Akhlāq-i-Pizishki* [Medical ethics]. Meshed, Iran: Meshed Medical School, 1969.

HALY ABBAS [ᶜALĪ IBN AL-ᶜABBĀS, AL-MAJŪSĪ AL-ARRAJĀNĪ]. *Kāmil-ul-Sanāᶜat* or *Kitāb al-Malikī* [The perfect practitioner, or, the royal book] [*Liber Regius*] (110th century). Translated into Latin by Stephanus Antiochenus in 1127 as *Liber regalis disposito nominatus ex arabico.* Edited by Antonius Vitalis Pyrranensis. Venice: Bernardinus Ricius impensa Joannis Dominici de Nigro, 1492. See also Constantine the African [Constantinus Africanus] for a Latin version.

JURJĀNĪ, ISMĀᶜĪL IBN HASAN AL-. *Zakhīrah-i-khvārazmshāhī* [The treasury, or thesaurus, of Khawārazmshah] (11th Century). 5 vols. Anjuman-i Āṣār-i Millī Silsilah-i intishārāt [Institution for publishing national literary treasures]. Tehran: Tehran University Press, 1960.

MOKHTAR, AHMED MOHAMMED. *Rhases contre Galenum: Die Galenkritik in den ersten zwanzig Büchern des "Continens" von Ibn ar-Razi.* Inaugural-Dissertation zur Erlangung der Doktorwürde der Hohen Medizinischen Fakultät der Rheinischen Friedrich-Wilhelms-Universität zu Bonn. Bonn: 1969. Text in Arabic and German.

NAṢĪR AL-DĪN ṬŪSĪ, MUHAMMAD IBN MUHAMMAD. *Akhlāq-i-Nāṣirī* [Nasir's ethics] (13th Century). Translated by G. M. Wickens as *The Nasirean Ethics, by Nasīr al-Dīn Ṭūsī.* UNESCO Collection of Representative Works: Persian series. London: Allen & Unwin, 1964.

NIẒĀMĪ-I-ᶜARŪẒI-I-SAMARQUNDĪ, AHMED IBN UMAR IBN ᶜALI. *Chahār Maqālā* [Four discourses] (12th century). Tehran: Elmi Bookstore, 1936. Translated by Edward Glanville Browne as *Revised Translation of the 'Chahár Maqála' ("Four Discourses") of Niẓámi-i ᶜArúdi of Samarqand, Followed by an Abridged Translation of Mírzá Muḥammad's Notes to the Persian Text.* E. J. W. Gibb memorial series, vol. 11, no. 2. London: Cambridge University Press, Trustees of the E. J. W. Gibb Memorial, Luzac & Co., 1921.

RHAZES [ABŪ BAKR MUHAMMAD IBN ZAKARĪYĀ AL-RĀZĪ]. *al-Ḥāwī fi al-ṭibb* [Âlhavi] (9th Century). *Kitābul ḥawī fi't-ṭibb, (Continens of Rhazes), (An Encyclopedia of Medicine).* 25 vols. Edited by the Osmania Oriental Publications Bureau from the unique Escurial and other manuscripts. Hyderbad-Deccan: Dāiratu'l-Ma'ārif-il-Osmānia, 1955 [1374 A.H.]. In Arabic, added title pages in English.

SAID, HAKIM MOHAMMED. *Al-Tibb al-Islam: A Brief Survey of the development of Tibb (Medicine) during the Days of the Holy Prophet Mohammad (p.b.u.h.) and in the Islamic Age.* Presented on the Occasion of the World of Islam Festival, London, April–June 1976. Karachi: Hamdard National Foundation, 1976. See especially "Medicine and Iran," pp. 80–82.

———. "Traditional Greco-Arabic and Modern Western Medicine: Conflict or Symbiosis." *Hamdard: The Organ of the Institute of Health and Tibi Research* 18, nos. 1–6 (1975), pp. 1–76. Six numbers issued as one. Paper presented in the "Islam and Occident" series at the University of Berne, Switzerland, 12 February 1975. Appendix C, "Ethical Basis of Medicine," pp. 71–76.

SIDDĪQĪ, MUHAMMAD ZUBAYR. *Studies in Arabic and Persian Medical Literature.* Foreword by Bidhan Chandra Roy. Calcutta: Calcutta University Press, 1959.

WILCOCKS, CHARLES. *Medical Advance, Public Health, and Social Evolution.* Commonwealth and International Library, Liberal Studies Division. Oxford: Pergamon Press, 1965.

3

CONTEMPORARY ARAB WORLD

The first recorded code of medical ethics, that of Hammurabi in 2250 B.C., originated in Babylon in what is now known as the Middle East. In the ninth century A.D., Al-Rahawi, an Arab scholar, wrote a 223-page book on medical ethics. Up to the end of the thirteenth century, the subject received further attention from other Arab authors, such as Al-Tabari, Al-Majoussi, Ibn Radwan, and Al-Baghdadi (Haddad and Bitar). Their writings were profoundly influenced by Hippocrates, whose precepts had been translated by Ibn Abi Usaybia in his book "On the Origins of the History of Categories of

Physicians." The Middle East then slumbered only to awaken in the beginning of this century.

The Arab Middle East is a huge geographic area extending 5,000 miles from Muscat and Oman on the Arabian Sea westward to Morocco on the Atlantic Ocean. Population density varies widely from one nation to another, related chiefly to climatic conditions and availability of water, food, and natural resources. Cultures differ from region to region. Some peoples are nomadic, others city dwellers. Colloquial Arabic also differs widely from one region to another. Chiefly Arab, these nations also have many minority groups, such as Armenians in Lebanon and Syria, Kurds in Iraq, Indians and Pakistanis in the Gulf states, Italians and Greeks in Egypt, and Frenchmen and Spaniards in North Africa. Male dominance is generally the rule, but women are rapidly becoming more "liberated." The wearing of the veil by women is fast disappearing, and educational facilities for girls are increasing. Arab dress among both men and women is moving toward Western styles despite the comfort of Arab clothing in desert regions. The Middle Easterner in general is very religious and emotional, characteristics intimately reflected in the ways medical ethics is understood and applied. In the Moslem faith, the Koran, as well as the Hadith Ash Sharif, is the basis of the code of ethics in general and medical ethics in particular, as is the Bible in the Christian faith.

The ease of international communication, especially the great impact that television has had on the population, is bringing about a profound and rapid change in the Arab world, now shifting steadily toward conformity with Western and international standards. Whatever may be true today of medical ethics in the Middle East may not be so in the near future.

Teaching of medical ethics

At present ethics is taught in well over three-quarters of the thirty-eight medical schools in the area as part of Forensic Medicine or History of Medicine. In a few schools it is taught as a separate course. The basic teachings are those of Hippocrates, richly supplemented by Koranic law, local habits, traditions, and conventions (Abu Zikri, p. 191). The University of Damascus uses a textbook on medical ethics written by Dr. Ziad Darwish of the Faculty of Medicine.

In Lebanon there are two official codes of ethics, one for the Lebanese Society of Neurology, Neurosurgery, and Psychiatry, adopted in 1960 (Drooby) and one for the Lebanese Order of Physicians, ratified as law by the Lebanese Government in 1969 (Ajlani). Similar codes are published and enforced in other countries, such as Syria and Egypt (Abdel Naby, pp. 182–200).

The number and quality of postgraduate courses offered in medical centers are rapidly increasing, as is attendance by specialists, practitioners, and health-care workers. Genetic screening and counseling, recognition of occupational diseases, and a host of other important subjects are being brought to the attention of health professionals, and these advances, together with their ethical aspects, are being introduced in all Arab nations.

Ethical issues

Relationship between doctor and patient. Doctors are held in very high esteem by the public, and medicine is considered a noble profession, especially in those nations with free enterprise. When the doctor has the confidence of his patient his word is final and his decision rarely questioned.

Suits for malpractice are seldom entered, and open complaints against doctors are not often voiced.

Advertising oneself publicly is not acceptable. It is considered poor taste for a physician to publish articles or to appear on television with the intent of self-promotion. It is the accepted rule that in news media "medical men speaking on medical subjects should remain anonymous" (Ayoub, p. 207).

Some patients, especially among the affluent in urban centers, show a lack of confidence in their doctors by seeking examination by other physicians. Only occasionally do they reveal, even if asked, that they have been examined elsewhere. This often results in confusion on the part of the patient regarding his illness, and in repetitious laboratory and X-ray examinations and dangerous multiplicity of medications. Moreover, when a physician learns that his patient has been examined elsewhere, he seldom makes contact with the other physicians. This is not to say, however, that there is not good communication among doctors when patients are specifically referred for consultation.

In those countries in which free medical services are offered to all citizens, emphasis has shifted from the individualistic medical approach to the more impersonal care provided by group practice or hospital teams. The new social morality in medicine, based on the belief that medical care is a basic human right, has modified and in some instances replaced the traditional individualistic approach to health services (Sagherian, pp. 8, 20–21).

Relationship between doctor and his colleagues. This relationship is of two kinds, depending on whether medicine is a free or a national enterprise. In the latter, the relationship among colleagues is becoming steadily governed by the unwritten code of ethics of government employees.

Where medicine is free enterprise the relationship among physicians is influenced by two major factors: the accepted code of ethics and competition. The former has the greater impact. The latter is almost completely lacking in rural areas because of the scarcity of physicians, but competition becomes an important element in large cities, where doctors are found in great concentration. Often the two factors function in opposite directions. When medical ethics may be compromised by professional competition, differences are commonly settled at the level of the physicians involved and seldom result in court action.

Epidemics and communicable diseases. When epidemics or alarming diseases such as rabies occur, governments often deny their existence, but later, after the hazard of spread is less or the disease is controlled, officials may admit and even detail the occurrence. This technique is used to prevent panic and fear among the population while public health workers and others seek to contain the disease. The reporting of communicable disease is officially required of physicians, yet penalties for failure to do so are seldom exacted.

Health care. Free medical care has been introduced in Kuwait, Bahrain, Qatar, the United Arab Emirates, Saudi Arabia, and some other Arab states. Physicians and surgeons are full-time government employees receiving no professional fees and working in government dispensaries and hospitals. Patients are offered full medical facilities, including hospitalization, medication, and appliances. In spite of these facilities and to keep personal enterprise alive, a limited number of nongovernment doctors are allowed to practice privately. Previously established private hospitals are permitted to continue operating. They are kept quite busy by patients who prefer to pay for services rendered rather than go to free government facilities and by foreigners living in the region.

Other states, such as Syria, Lebanon, and Jordan, provide their poor with free medical care and medicines, administered in government dispensaries and hospitals by part-time doctors. In addition, through a social security system with preset fee schedules, the cost of medical treatment to the employee is shared in different proportions by the government, employer, and employee. In these states part-time physicians often have a private practice, which they conduct in late afternoon and early evening for those who can afford to pay.

In some Arab nations, well-to-do members of large families accept responsibility for the welfare of even distant relatives, thus forming a kind of family social security system. In states that have become affluent from oil revenues but still lack adequate medical facilities, the indigent ill may be sent abroad at government expense. Indeed, even the wealthy who require sophisticated treatment may be sent at government expense to hospitals in the developed Arab nations or to European or American clinics. Occasionally, eminent consultants are flown into these nations to examine distinguished patients.

Family planning. Only recently have Arab nations begun to recognize the importance of family planning. Saudi Arabia, sparsely populated, understandably does not advocate contraception, yet through television and other media emphasizes the value of spacing children to protect the health of the mother. In Lebanon and Syria contraceptive devices, sold for hygienic purposes, are readily available, and clinics accept for sterilization those with adequately sized families who request it. Yet both contraception and sterilization are poorly utilized, and families with six to ten children are common.

Although Moslem law permits a man four wives, polygamy is rapidly decreasing, and monogamy is becoming customary. Infertility of a wife is an accepted cause for divorce. Modern medical centers have facilities for investigating and treating infertility, but they are not fully utilized at present. Artificial insemination is rarely practiced.

Abhorrence of the defective is common. The body of a baby stillborn in the hospital is often unclaimed by its parents. A child who suffers permanent disability or deformity may be much loved by family but may be kept secluded from public view.

Aging, death, autopsy, and transplantation. Grandparents are as much a part of family life as are children, and several generations of a family are often housed together. The decrepit and senile are kept within the family and, indeed, the elderly member who is found in hospital to have an incurable disease is usually taken home to die. Burial of the dead occurs normally within one day, a custom necessitated by the rarity of embalming and rapid decomposition of the corpse in hot weather. This custom is common among all ethnic and religious groups living in Middle East countries, including Christians, Moslems, Jews, Hindus, and other sects.

Autopsies are rarely requested by either attending physicians or families of the deceased because of the widespread belief among many Moslems and Christians in the bodily resurrection of the dead. Removal of organs or disfigurement of the cadaver accordingly is not commonly acceptable. Yet despite this belief permission for autopsy is being increasingly requested and granted, especially in centers with the necessary facilities.

Transplantation of cadaver organs is rarely practiced because of the same doctrinal beliefs limiting autopsies. In Lebanon, where transplantation is forbidden by law, the Parliament was considering a bill to permit use of cadaver organs when civil strife erupted in 1975. Donation of organs from healthy subjects is permissible. Yet friends and relatives of anephric patients, for example, seldom agree to give up a kidney, and renal dialysis remains the commonest treatment for chronic uremia. This technique is well advanced in Lebanon and is rapidly improving and increasingly available elsewhere.

Drugs, addiction, and experimentation. Among Arab nations there is no uniform practice regarding the availability and sale of drugs to the public. In some countries anyone can purchase from a pharmacist highly sophisticated drugs without a doctor's prescription. Dispensing of addictive drugs, however, is regulated.

In some countries, for example Saudi Arabia and Kuwait, the importation, sale, and consumption of alcohol are forbidden. In other countries, such as Lebanon, there is open sale of all kinds of alcoholic beverages, but despite the abundance, availability, and low cost of locally produced arak and wine and imported whiskies, open drunkenness is rarely observed, acute alcoholism seldom seen in hospital emergency rooms, and alcoholic cirrhosis infrequent.

Cigarette smoking is prevalent among the populace of Arab nations. Imported cigarettes do not carry a warning on the health hazard of smoking required on all packages sold in the United States and Great Britain. This is a striking incongruity. If the habit is dangerous, then Ministries of Public Health of the Arab nations should insist that packages of both imported and locally produced cigarettes be labeled appropriately.

The sale of marijuana is forbidden, but it is said to be readily available in several countries. Its use, however, is uncommon even among university students (Nassar, Melikian, and Der-Karabetian, p. 219).

In medical centers the rules regarding human experimentation are similar to those in Western nations. It is known, however, that occasionally pharmaceutical companies from countries with strict regulations have conducted drug trials in some Arab nations with less rigid laws. Such studies appear to have been done with careful regard to informed consent and observation of subjects during the experiment.

Conclusion

Despite the varied geographic, social, financial, and (to a lesser extent) ethnic background of inhabitants of the different Arab countries, there is a definite and steady trend toward uniformity in medical ethics, as well as toward conformity with Western understanding and application of such ethics.

SAMUEL P. ASPER AND
FUAD SAMI HADDAD

[*See the following article for a Muslim perspective on medical ethics in the Arab world. See also* APPENDIX, SECTION I, OATH OF HIPPOCRATES.]

BIBLIOGRAPHY

ABDEL NABY, ALY. *Forensic Medicine and Toxicology.* Cairo: Sherif's Bookshop, n.d.
ABU ZIKRI, A. "Ethics in Obstetric Practice." *Journal of the Kuwait Medical Association* 10 (1976): 191–196.
AJLANI, RIAD, ed. "Code de Deontologie, Decret No. 13187." *Bulletin de l'Ordre des médecins du Liban* 4 (1970): 1–15.
AYOUB, IBRAHIM S. "Code of Medical Ethics." *Lebanese Medical Journal* 24 (1971): 205–208.
DROOBY, A. S. *Code of Medical Ethics.* Beirut: Lebanese Society of Neurology, Neurosurgery and Psychiatry, 1959, pp. 1–5.
HADDAD, FARID SAMI, and BITAR, EMILE W. "Les Principes déontologiques dans la médecine arabe." *International Congress of the History of Medicine, 22d, Bucarest, 1970. Comptes Rendus.* Bucharest: n.d., pp. 67–68.
NASSAR, NABIL T.; MELIKIAN, LEVON H.; and DER-KARABETIAN, AGOP. "Studies in the Non Medical Use of Drugs in Lebanon: I. The Non Medical Use of

Marijuana, LSD, and Amphetamine by Students at the American University of Beirut." *Lebanese Medical Journal* 26 (1973): 215–232.

SAGHERIAN, ARTIN A. "Implications of Ethics for Medical Practice." M.A. thesis, Department of Philosophy, American University of Beirut, 1973.

TIKRĪTĪ, RĀJĪ ABBĀS AL– [RAJY ABBAS AL–TAKRITY]. *Professional Medical Ethics*. Baghdad: Al-Ani Printing Press, 1970.

4
CONTEMPORARY MUSLIM PERSPECTIVE

The countries of the Middle East, from Morocco to Pakistan and from Turkey to the Sudan, along with lands in other areas of the world, are largely dominated by the Islamic religion. The moral values prevailing in those countries are inherently related to Islamic teachings that stress the presence of God, his ever watchful eye, and the recording of all an individual's deeds, words, and thoughts. Therefore, in any study of the bioethical values and attitudes prevailing among those peoples one must take into account Islamic moral judgments and divine law known as the *sharī'a*, which are based upon the sacred text—the Qur'an, which is believed to be the Word of God—and upon the model of the Prophet, represented in his recorded words and known as Hadith.

Although it is true that Western thought and secular pressures have recently influenced Islamic lands to varying degrees, the debate over ethical issues arising from recent medical and technological progress—much lagging in the geographical area under discussion—has not been as great in the Arab nations as it has been in the West. Moreover, the strong belief of the Muslim people in God, their reverence for him, and their belief in divine sanctions in life after death considerably color their attitude toward material affairs. Their reverential fear of God is more profound in proportion to the gravity of the action involved. And since what is at stake in the practice of medicine is human life itself, it will be seen that the Muslim physician's religious conscience has been and still is the basic determinant of his attitude toward his patients, no matter what material benefit may be involved. This article will therefore attempt to relate the ethical standards of Islam to certain controversial issues debated in medical circles recently as a result of remarkable advances in the sciences of biology and medicine.

Principles of Islamic medical ethics

Islamic medical ethics can be summarized as follows:

Whoever kills a human being *not* in lieu of another human life nor because of mischief on earth, is as if he has killed all mankind. And whoever saves a human life is as if he has saved all mankind [Qur'an, V, 32].

God has not created a disease but that He has created a cure for it—except old age [Bukhari vol. 4, p. 6].

The worth of human life and the value of efforts to save it are eloquently conveyed in the foregoing Qur'anic verse. The wisdom saying following it expresses the Muslim physician's awareness of God's hand not only in creating and sustaining life but also in preordaining its illnesses and providing their cures. Those key ideas are reflected in the following "oath" taken by contemporary Muslim physicians, gathered from the introductions of some classical medical works dating from the early centuries of the Islamic era:

Hereby we take this oath in Thy Name, Our Lord, Creator of all the heavens and the earth, to follow Thy counsel as Thou hast revealed it unto the Prophet Muhammad, peace and blessings be upon him.

Give us the strength to be truthful, honest, modest, and merciful.

Give us the understanding that ours is a sacred profession that deals with Your most precious gifts of life and intellect.

Make us worthy of this favored station with honor, dignity, and piety so that we may devote our lives to serving mankind, poor and rich, wise and illiterate, Muslim and non-Muslim, black and white, with patience and tolerance, with virtue and reverence, with knowledge and vigilance, with Thy love in our hearts and compassion for Thy servants, Thy most precious creation.

A Muslim physician's attitude toward his patients derives from his belief in God, who commands him to aid his fellow men by delivering them from their pains and restoring them to health. In treating his patients, he should be conscious of God's rewards and of his retribution, not of possible lawsuits demanding exorbitant monetary compensation should an error occur. Both patient and physician believe that God is the ultimate cause of illness and health and that human efforts, including medical treatment, are provided by him. It is recognized that these can fail if restoration to health is contrary to God's preordained decree. Quite apart from the outcome of treatment, the physician's effort is appreciated. If the patient recovers, all believe that he was cured by God. In case of failure, no mortal is to blame unless there was obvious negligence; in any case, all resign themselves to God's will. Therefore a Muslim physician does not and should not reveal to his patient the details of his illness if he feels that the patient would be unduly disturbed by the revelation. Yet, if an uneasy patient is unduly frightened and the truth might reduce his fear, a full disclosure in this case would be helpful. The physician is to use his own discretion and heed the advice of the Prophet, who said: "Convey good tidings, and give not distressing news. Seek to make things easy and impose not burdensome views."

Muslim medical experts from the first century of Islam (seventh century A.D.) did not neglect to write about the ethical code to be upheld by the physician. The ethical principles can be summed up as follows:

1. The physician should be willing to forgive the wrongs of others.
2. He should be willing to counsel all.
3. He should be truthful, honest, and objective.
4. He should be merciful and considerate of the needs of others.
5. He should strive to lead an upright, morally clean life (Hamarneh, 1971, pp. 1100–1109).

Particular issues

Health care. In spite of the great heritage of medical science developed in the Arab world during the early centuries of Islam, through the contributions of such scholars as al-Razi (d. 926), Avicenna (979–1037), and Averroes (1126–1198), popular practices and the ministrations of the "medicine man" persisted in some areas until recent years, bolstered by local superstitions and misguided beliefs. Thus, to be examined or treated by a "modern" doctor has been regarded as taboo; but with wider education and greater appreciation of scientific achievements, a change in attitude toward health care has recently become more discernible.

Today the medical profession in Middle Eastern countries is well organized along modern lines, equipped in most cases with up-to-date research facilities, although the level of attainment still lags behind that of the West because of economic and political factors. The ratio of available physicians, hospital beds, and expensive equipment is far below that in the West. On the other hand, treatment in government hospitals in many of those countries is free for all segments of society; justice prevails and favoritism is rare, although the utilization of available services depends, of course, upon physical proximity and the educational level. Urban dwellers, for example, have greater access to medical facilities than those living in rural areas.

Research and experimentation. Islamic belief in the sanctity of human remains has given rise to an aversion to the use of human cadavers for medical research. In addition, the prohibition in Islam of the torture of animals has caused protests to be raised about animal experimentation. Nonetheless, both animal experimentation and research on human cadavers have been tolerated on the basis that they presumably will lead to bettering human health. The use of drugs containing alcohol is the source of debate insofar as Islamic law prohibits the consumption of alcohol. Some hold that such drugs ought not to be given Muslim patients, whereas others tolerate their use in view of the small amount of alcohol consumed.

The practice of postmortem dissection was also greeted with loathing, resentment, and vehement protest. Then a *fatwa* (religious ruling) was issued, according to which such dissection is permitted under Islamic law, provided it is legally required and limited to the parts absolutely necessary for purposes of autopsy (Dijwi, "Hukm," "Tashrih"). After all, in the case of a deceased person who happened to swallow a valuable object such as a jewel or a gold coin, the law of *shari'a* permits the opening of the deceased's stomach for the purpose of retrieving it (Ibn Hazm, p. 166).

Organ transplantation. Organ transplantation has caused considerable shock and elicited widespread aversion in the Arab nations. A Muslim finds it difficult to imagine a vital part being separated from the body of one person and grafted onto the body of another. The practice of organ transplantation also seemed to raise problems with the belief that people will be raised after death and go to paradise or to hell. If the donor of an organ is among those condemned to hell, and the recipient destined for paradise, what will happen to the organ in the afterlife?

Although this medical intervention has been greeted with repugnance, as were blood transfusions when they were first introduced, the view has emerged that it is justified because of necessity. Nonetheless, organ transplantation requires the consent of the donor if alive and the approval of his relatives if dead.

Mental health care. The care of victims of mental disorders has been known from early times. Special wards and hospitals, known as *Birmaristans*, were assigned to those sufferers. However, in a society where a knowledge of psychopathology is virtually absent, the beginning of erratic behavior is usually resented and despised. When such behavior becomes habitual, it is interpreted as the work of subterranean evil spirits, which possess the body of the victim, and aversion turns to deep sympathy. Resort is then made to various traditional remedial treatments to exorcise the evil spirit, including a practice known in some areas as *al-Zar*, a kind of shamanistic exercise. Sometimes the spirit is benevolent, and the patient may perform miracles, in which case he is endowed with the status of sainthood. Should the victim's behavior become dangerously violent, the appropriate governmental agencies may then be called upon to admit him to an asylum.

Abortion. The practice of abortion is prohibited by Muslim law, and no one disputes its prohibition, especially when the fetus is four months old or more, unless the life of the mother is endangered. The prohibition of abortion is not based simply on the sanctity of human life or on the denial that a woman has the right to control her own body. It is based upon the divine law, which makes the destruction of innocent human life, in particular the killing of one's own child, second in gravity only to the sin of disbelief in God. The fetus is regarded as a separate, living or potentially living, entity and not as a part of the woman's body. Even if it were admitted that the fetus is a part of the woman's body, an abortion would be regarded as seriously wrong insofar as a person under Islamic law is forbidden to mutilate or mistreat himself.

Some jurists hold that abortion during the first four months of pregnancy is merely undesirable. They base their argument on a *hadith* which describes the stages of development of a fetus in the womb of its mother: from a drop into a clot of congealed blood after forty days; thence to a morsel of flesh after a similar period, and then the *ruh*, i.e., the soul, is breathed into it after another forty days. (Bukhārī vol. 2, p. 132; Ibn Mājah). These liberal jurists believe that during the initial period the fetus is not as yet a fully human being; therefore its destruction is less than a grave sin. The majority of jurists, however, reject this view, as the *hadith* in their judgment was in no way intended to offer a scientific medical explanation of the beginning of human life but was rather intended to call attention to the wonders of divine creation.

Birth control. The use of modern birth control methods has, unlike abortion, caused much controversy in Muslim society. At an early time authors argued over the practice of *'azl*—i.e., withdrawal by the male at the moment of ejaculation in order to prevent conception. When newer methods of birth control were introduced and widely advocated, opinion was sharply divided. Some regarded the practice as infanticide, whereas others saw nothing wrong with it since it did not involve the killing of a fertilized cell. Conflicting *fatwas* were issued, with supporters and critics on both sides. Recently opposition to contraceptive practices has subsided, partially because of the specter of overpopulation with its dangers and partially because of a growing awareness of scientific realities. Currently there is a consensus that birth control is permissible under the following conditions: (1) The method is harmless to both partners and does not involve permanent sterilization (unless this is medically necessary); (2) the practice is agreed to by both parties; (3) it is not evilly motivated such as by fear of the burden of child care or hatred of bearing children; (4) it does not result in a severe drop in Muslim population or endanger its security (Ghazzāli, vol. 2, p. 47; Abdul-Rauf, 1972, p. 68).

Dying and euthanasia. Administering any medical treatment to save the life of an apparently dying person whose recovery might be possible would not offend Islamic sensibilities; but the artificial prolongation of the life of a terminal patient who has permanently lost consciousness is morally unjustifiable. Life is human by virtue of the individual's consciousness, which makes a person a responsible agent. When consciousness is irretrievably lost, there is no hope of recovery, and the use of medical technology for the mere survival of a vegetable-like being is (quite apart from the objectionable material waste) a violation of human dignity and appears as a challenge to the divine will. On the other hand, it is tantamount to murder to hasten the death of a patient even when he suffers from endemic pains. In such cases, a pain-killing remedy may and should be administered. It should not be forgotten that God has determined the span of each life and that the moment of its termination is in his hands alone. To endeavor to save a threatened life where there is some hope of recovery does not violate this principle. What is attempted here is to remove a threat posed to a full human life. It is quite another matter to intervene in hopeless cases by prolonging a degrading "vegetable" existence, or arbitrarily to terminate the life of a sufferer in the name of "mercy killing."

MUHAMMAD ABDUL-RAUF

[*Directly related is the entry* ISLAM. *For a further discussion of topics mentioned in this article, see:* POPULATION ETHICS: RELIGIOUS TRADITIONS, *article on* ISLAMIC PERSPECTIVES. *See also the preceding article on* CONTEMPORARY ARAB WORLD.]

BIBLIOGRAPHY

ABDUL-RAUF, MUHAMMAD. *Islam: Creed and Worship.* 2d ed. Washington: Islamic Center, 1974.
———. *The Islamic View of Women and the Family.* New York: Robert Speller & Sons, 1977.
———. *Marriage in Islam: A Manual.* An Exposition Banner Book. New York: Exposition Press, 1972.
ALWAYE, MOHIADDIN. "The Conception of Life in Islam." Majallat al-Azhar [Journal of al-Azhar University] 45 (March 1973), pp. 1–5.
AMMAR, SLEÏM. "Rhazès (850–923 J.C.): Psychologie médicale-psychosomatique et déontologie." *Tunisie médicale* 47 (1969): 5–13.
BUHAYRI, MUHAMMAD ʿABD AL-WAHHAB. "Fatwa al-marhum al-Shaykh Bakhit fi tashrih al-mayyit" [Fatwa of the late Sheikh Bakhit on autopsy]. Majallat al-Azhar [Journal of al-Azhar University] 6 (1935): 627–633.
BUKHĀRĪ, MUHAMMAD IBN ISMĀʿĪL AL-. *Sahih.* Cairo: Halabi Press, A.H. 1349 [1930], 4 vols.
DIJWI, YUSUF AL-. "Hukm tashrih al-mayyit fi al-shariʿah al-Islamiyyah" [Autopsy from the Islamic point of view]. Majallat al-Azhar [Journal of al-Azhar University] 6 (1935): 472–473.
———. "Tashrih al-amwat" [Autopsy]. Majallat al-Azhar [Journal of al-Azhar University] 6 (1935): 577–578.
EL-QADI, AHMAD. "Professional Ethics: Ethics in the Medical Profession." *Journal of the Islamic Medical Association of the United States and Canada* 7, no. 2 (1976), pp. 27–30.
The Encyclopaedia of Islam. New ed. 5 vols. Leyden: E. J. Brill; London: Luzac & Co., 1960. See articles on Ibn Rushd, vol. 2, pp. 410–413; Ibn Sīnā, vol. 2, pp. 419–420; and al-Rāzī, vol. 3, pp. 1134–1136.
GHAZZĀLĪ, AL- [AL-GHAZĀLI]. *Ihyā ʾulūm al-dīn.* 4 vols. Cairo: 1965. Translated by Bankey Behari as *The Revival of the Religious Sciences: A Translation of the Arabic Work Ihyaʾ ʿUnum al-Din.* Farnham, England: Sufi Publishing Co., 1972. Abridged trans.
GURAYA, MUHAMMAD YUSUF. "The Importance of Health in Islam and in Islamic Countries." *Hamdard* 14 (1971): 103–122.
HAMARNEH, SAMI KHALAF. "A Brief Survey of Islamic Medicine during the Middle Ages." *Journal of the Islamic Medical Association of the United States and Canada* 7, no. 1 (1976), pp. 21–25.
———. *Catalogue of Arabic Manuscripts on Medicine and Pharmacy at the British Library.* History of Arabic Medicine and Pharmacy, vol. 3. Cairo: Hamarneh, 1975. Supported by a grant from the Smithsonian Institution.
———. "The Physician and the Health Professions in Medieval Islam." *Bulletin of the New York Academy of Medicine* 47 (1971): 1088–1110. Based upon two seventh- (thirteenth-) century works, *Classes of Physicians* by Ibn Abi Usaybiʿah and *The Comprehensive Compendium of Medicine* by Ibn al-Quff, his disciple.
HUSAINI, I. M. "Islam and Modern Problems." Majallat al-Azhar [Journal of al-Azhar University] 45 (February 1973), pp. 7–14.
IBN HAZM, ʾALI IBN AHMAD. *Al-Muhalla.* Cairo: Muniriyyah Press, A.H. 1349 [1930], vol. 5.
IBN MĀJAH, MUHAMMAD IBN YAZID. *Sunnan ibn-i mājah sharīf kāmil.* Cairo: Fajjalah Press, 1965, vol. 1.
PASHA, HAKEEM M. AZEEZ. "Establishment of Unani Hospitals in Islamic Countries." *Bulletin of the Institute of History of Medicine* (Hyderabad) 3 (1973): 68–70.
PEARSON, J. D., ed. *Index Islamicus, 1906–1955: A Catalogue of Articles on Islamic Subjects in Periodical and Other Collective Publications.* Cambridge, England: W. Heffer & Sons, 1958. Supplements 1956–1960, 1961–1965, 1966–1970, and yearly.
TIRMIDHĪ, MUHAMMAD IBN ISA AL-. *Sahih.* 6 vols. Cairo: Fajjalah Press, 1965, vol. 4. Includes commentary *Tuhfat al-Ahwadhi.*

5
CONTEMPORARY ISRAEL

The State of Israel, being the first political structure in nearly two thousand years to be governed by Jews, attempts to reflect in its laws and mores the ethics and values of Jewish tradition. Though the majority of the Jewish population does not consider itself orthodox and thereby bound by rabbinic law, there is a politically strong orthodox minority whose platform calls for making the law of the land congruent with traditional norms. Because all of the governments that have governed Israel since its founding in 1948 have been coalition governments, no one party having a clear majority, the influence of the orthodox parties, frequently needed to achieve a majority in the Israel Knesset (parliament), has been great. There is also a growing counterpressure in Israeli society for more contemporary values, which frequently conflict with ancient ordinances. The legal system of Israel, partially inherited from the Turks and the British Mandate authority, leaves matters of personal status such as marriage, divorce, conversion to another religion, and inheritance in the control of religious courts. This is true in the Jewish community as well as among the Christian and Muslim minorities (Baker, pp. 159–182, 207). Because the imposition of halakhic (Jewish legal) norms depends frequently on the changing political situation, the viewpoints of various religious authorities are sometimes accommodated to make them more acceptable (Birnbaum).

Particular areas of tension in Israeli society have been centered on the problems of autopsies and abortion. Jewish tradition is restrictive in its permission for autopsies. In 1953 the Knesset enacted a bill entitled the Anatomy and Pathology Law (Rosner, pp. 146 ff.). It provided that an autopsy could be performed (1) if the autopsy is required by law (such as in the matter of foul play), (2) if the cause of death cannot be established without an autopsy, (3) to save a life, and (4) in cases of genetic or inherited disease where the family may be guided or counseled concerning future children. Further provisions were that, if a person agreed in writing that his body be used for science, it is permitted to dissect the body for medical instruction and research. It is also permitted to remove organs of the deceased for purposes of transplant (ibid.). It was also directed that the deceased must be buried in accordance with Jewish law, and all organs removed for examination must be returned for burial.

There have been widespread complaints of abuses in the matter of autopsies. Claims were made that autopsies were performed without the family's permission. Orthodox Jews staged demonstrations in the cities of Israel; and the Chief Rabbi of Israel, on 15 October 1966, issued an opinion which stated that "autopsy in any form whatsoever is prohibited by the law of the Torah . . . there is no way to allow it except in a matter of immediate danger to life, and then only with the approval of an eminent rabbi." The statement was signed also by 356 rabbis. Various commissions were appointed to recommend changes in the Anatomy and Pathology Law, but the law remained as it was enacted in 1953.

The question of abortion has been another source of controversy. Abortions were legal from the beginning of the history of the State of Israel only for therapeutic reasons. On 11 January 1977 a new law was enacted by the Knesset to take effect one year after passage (*Sefer hachukim*). It provided for abortions after approval by a three-member committee consisting of two physicians (one of them a gynecology-obstetrics specialist) and a social worker. Some of the reasons for a legal abortion are: if the mother is under sixteen or over forty; if the pregnancy resulted from rape, incest, or was out of wedlock; if it can be determined that the child would be born physically or mentally defective; or if the birth would damage the physical or emotional life of the mother. This is, of course, more liberal than the dominant orthodox view, which generally limits abortions to narrow therapeutic grounds where the mother's life is in danger. There have been tendencies among orthodox authorities to broaden the grounds for abortion (Feldman; Maier and Waldenberg; Bleich, pp. 325–372). The tense political situation has tended to drive these opinions into private rather than public expression. On the record, the orthodox rabbinate opposes the abortion statute. In this controversy they are supported by more secularly inclined groups who view abortion as diminishing the Jewish population of Israel in a time when the Arab minority is multiplying at a higher rate than the Jewish sector.

An important source of guidance on bioethical issues in Israel are the Responsa (answers to questions) of the leading rabbinic authorities. These Responsa guide the decisions of traditional Jews in Israel and in the Diaspora. They are collected in volumes that have wide circulation and are in Hebrew. The most important writers of the Responsa are the late Sephardic Chief Rabbi, Ben Zion Uziel; Rabbi Zevi Waldenberg; Rabbi Yehuda Untermann, the former Ashkenazic Chief Rabbi; and Rabbi Shlomo Goren, the present Ashkenazic Chief Rabbi.

The Shaare Zedek Hospital in Jerusalem, which is under orthodox auspices, founded the Falk-Schlesinger Institute for Medical Halachic Research. The Institute publishes the journal *Assia* (Physician). In each of the issues of *Assia* problems of bioethics are discussed. These include such questions as human experimentation and genetic engineering (no. 15, 1976); castration, artificial insemination, and contraception (no. 13, 1976); and transplants, animal experimentation, and euthanasia (no. 16, 1977).

The issues of *Assia* also contain codes of behavior for nurses and physicians, especially as regards the treatment of the sick on Sabbaths and festivals, when many types of activities are forbidden. It also outlines the ways in which to obtain the consent of patients for medical treatment and to participate in experiments.

The State of Israel is an example of a newly emerging society that is attempting to harmonize traditional values embedded in an ancient religious tradition with the contemporary outlook in confronting the concerns and dilemmas of bioethics.

SEYMOUR SIEGEL

[*For further discussion of topics mentioned in this article, see the entries:* ABORTION, *article on* JEWISH PERSPECTIVES; CADAVERS, *article on* JEWISH PERSPECTIVES; DEATH, *article on* WESTERN RELIGIOUS THOUGHT: POST-BIBLICAL JEWISH TRA-

DITION. *Also directly related are the entries* JUDAISM; *and* RELIGIOUS DIRECTIVES IN MEDICAL ETHICS, *article on* JEWISH CODES AND GUIDELINES. *See* APPENDIX, SECTION I, OATH OF ASAPH.]

BIBLIOGRAPHY

"Animal Experimentation." *Assia*, no. 16 (Sh'vat 5737 [January 1977]), p. 18. Text in Hebrew. Hebrew title: "Tsar baale chayim litzorech nisyonot refuiim" [Suffering of animals for the sake of medical experiments].

Assia: Original Articles, Abstracts and Reports on Matters of Halacha and Medicine. Edited by Avraham Steinberg. Jerusalem: The Falk Schlesinger Institute for Medical Halachic Research at Shaare Zedek Hospital, 1972–, irregular. Text in Hebrew, English table of contents.

BAKER, HENRY E. *The Legal System of Israel.* Rev. ed. Jerusalem: Israel Universities Press, 1968.

BIRNBAUM, ERVIN. *The Politics of Compromise: State and Religion in Israel.* Rutherford, N.J.: Fairleigh Dickinson University Press, 1970.

BLEICH, DAVID. *Contemporary Halachic Problems.* The Library of Jewish Law and Ethics, vol. 4. Edited by Norman Lamm. New York: Ktav Publishing House, Yeshiva University Press, 1977.

"Euthanasia." *Assia*, no. 16 (Sh'vat 5737 [January 1977]), p. 21. Text in Hebrew. Hebrew title: "Harigat adam mityasser" [Killing of a person who is in agony].

FELDMAN, DAVID. *Birth Control in Jewish Law.* New York: New York University Press, 1968.

"Genetic Engineering." *Assia*, no. 15 (Marchesvuon 5737 [October 1976]), pp. 30–31. Text in Hebrew. Hebrew title: "Handasah genetit." A translation of "Genetic Engineering." *Journal of the American Medical Association* 220 (1972): 1356–1357. Editorial.

"Human Experimentation—Helsinki Declaration, 1964." *Assia*, no. 15 (Marchesvuon 5737 [October 1976]), pp. 29–30. Text in Hebrew. Hebrew title: "Nisyonot bevney adam."

"Israel, State of (Health, Welfare, and Social Security)." *Encyclopaedia Judaica.* 16 vols. New York: Macmillan Co.; Jerusalem: Keter Publishing House, 1971–1972, vol. 9, pp. 982–994.

MAIER, D., and WALDENBERG, E. Y. "Medico-Halachic Aspects of Tay-Sachs Disease." *Assia*, no. 13 (Adar 5736 [February 1976]), pp. 6–10. Text in Hebrew. Hebrew title: "Hebetim hilchtyim urefuiyim bitsmonet tay-sachs."

"Priorities in the Saving of Life." *Assia*, no. 16 (Sh'vat 5737 [January 1977]), p. 19. Text in Hebrew. Hebrew title: "Kedimiyot behatzacat nefashot."

ROSNER, FRED. *Modern Medicine and Jewish Law.* Studies in Torah Judaism, no. 13. Edited by Leon D. Stitkin. New York: Yeshiva University, Department of Special Publication, 1972.

Sefer hachukim shel Medinat Israel [Laws of the State of Israel]. Jerusalem: Government Printer, 1977, no. 842 (9 February 1977).

UZIEL, B. "Responsa of Chief Rabbi B. Uziel in Medico-Halachic Problems: Autopsies; Reliance on Physician's Opinion; Castration; Artificial Insemination, The Concept of 'Trefah' as Applied to a Human Being; Examination of Semen; The Use of I.U.D. Contraceptives." *Assia*, no. 13 (Adar 5736 [February 1976]), pp. 27–34. Hebrew title: "Sikumeyhatchuvot binyeney refuah vihalachah mitoch sheelot uteshuvot Mishptei Uziel."

6
SUB-SAHARAN AFRICA

Medical ethics in sub-Saharan Africa is extremely complicated and cannot be considered in any sense homogeneous. This is because the geographic area is vast—22,950,124 square kilometers, or about nine million square miles—and is made up of thirty-nine independent countries with innumerable sociocultural groupings. Many of the countries are nation-states only superficially, since within their borders there are ethnic groups that have little in common with their co-nationals while being more closely affiliated with groups in other countries. Quite apart from indigenous cultures, these countries, until the late 1950s and early 1960s, were under the domination of more powerful metropolitan countries that impressed their cultures upon local cultures. The interaction between an external impressed culture and a local one is more complicated in the field of medicine than in any other.

Traditional and scientific methods

Some of the countries have had contact with scientifically based medicine for less than fifty years and others for little more than one hundred years. The development of medical ethics in all the African countries has therefore tended to follow the existing European ethical values, principally those of France and Great Britain. European medical values and practices, faced with traditional African medical practice, took the position that all traditional medical practices and their practitioners were bad. The healers were considered no more than quacks and deceivers, and therefore were either ignored or actively suppressed. Even the traditional midwives or "birth attendants," as they are now known, who from time immemorial have provided some help to a woman at a most difficult time, were looked upon with disfavor. To a certain extent this attitude was supported by the beliefs and practices of the religions that came with the colonizers. Since many of the approaches to traditional healing included the intervention of gods and spirits, which the Christian religion found abhorrent, the total practice was eventually suppressed. Another important reason for opposition was that European medical practice imposed upon its doctors an ethical code which required that they not associate with practitioners whose training and beliefs differed from their own.

After acquiring independence, however, and with the need for national self-assertion in all spheres, these countries are witnessing a gradual disappearance of some of these negative attitudes. The rise of black consciousness and the acceptance of the notion that "blackness" is no longer a sign of inferiority have meant that African nations and individuals have begun to assimilate the acquired medical knowledge with the roles and potentials of their own traditional medicine and medical practice. In some countries the requisite political courage has been expressed in decisions that have been taken, including the promulgation of laws recognizing traditional medical practice. Financial grants have been made for research into the methods and the preparations used in traditional medicine. In a few instances medical scientists are actively involved. This new collaboration between traditional and imported medical practice is likely to be helped further by the indigenization of the churches and the improvement of the quality of their leadership. Whereas previously priests and ministers from the majority of churches had been very simply trained and had assumed a dogmatic approach to their subjects, a growing number can now be considered learned men and theologians, who are able to help formulate the churches' views on subjects of such crucial importance as the conflict between traditional and modern medical practice.

The medical profession itself, in the majority of countries, now feels relatively free to develop new ways of practice and to attempt collaboration with traditional birth attendants, herbalists, and other healers without fear of losing either the respect or the comradeship of their European colleagues.

Much of traditional medicine touches on the realm of psychiatry. It has been shown that involvement of traditional practitioners in psychiatric treatment makes for a more humane treatment and much better integration of patients into society (Lambo). Other efforts that may be cited are the involvement of the University of Ghana Medical School in training programs for traditional birth attendants. In many countries the newer medical schools (such as Makerere in Uganda, Nairobi in Kenya, and Yaoundé in Cameroun) are striving to gain much more credibility and relevance within their own societies. They are, therefore, embarking on programs that identify and preserve traditional practices considered valuable enough to be preserved (Jelliffe and Bennett). Traditional practices that can be considered harmless are therefore to be left alone, and practices that can be considered truly harmful are to be eliminated.

Standards for medical practice

Practically every English-speaking country has a general medical council or board responsible for registration, accreditation, and supervision of medical practice. In most of these countries the responsibility rests with the Ministries of Health. In a few, medical practice is completely autonomous. In French-speaking countries the boards of control are generally quite distinct from the Ministries of Health. Many of these medical councils or boards, however, have been fashioned along lines more responsive to West European norms and needs than to African ones. They have had little time to devote to the development of ethical guidelines relevant to social and cultural conditions peculiar to life within African countries. Some agreed principles remain fundamental, however. Privacy of the patient is respected, and so is confidentiality, although here there is a question of disclosure required by the government before payment for medical service or by employers for acceptance of sick leave.

Health-care service

There are very few scientifically trained medical personnel in Africa. There is a doctor-to-population ratio ranging from 1:3,000 in such better-off cities as Dakar, Accra, and Nairobi to 1:200,000 in some poorer dry rural areas, such as most of Northern Nigeria and all of the Sahel region. There are countries within which there may not be a single specialist in any recognized field of medicine. This immediately raises the issue of the kind of medicine that is most suitable in such conditions. European medicine has developed and gained the reputation of "face-to-face" medicine and has also concentrated more on curative than on preventive medicine. In Africa, on the other hand, the practice of face-to-face medicine, if it is accepted as the ideal, will mean the exclusion of eighty to ninety percent or more of the population from medical coverage. It will also place an inhuman load on the few medical practitioners and quickly reduce them to no more than purveyors of drugs and injections. Fendall sees this as posing the "quantity versus quality" dilemma, although not all agree with his view.

The doctor in Africa is now being asked to view his role in light of certain priorities—his first being promotive and preventive health services, the second being curative, in terms of individual patient treatment in offices or sophisticated hospitals. In attempting the first, many have pointed out that not much can be done until medical practice is so arranged that the community is both the consumer and the provider of its own health care. This can be done only if delegation of health care to nonphysician personnel is done on a basis of genuine needs and of how they can best be handled rather than on a basis of locality or categories of persons. The debate will continue, but almost all the new medical schools have agreed that the doctors' training should be responsive to the needs of the communities and relevant to the organization and priorities set by Ministries of Health.

In medical education itself, the issue has been one of "excellence versus quantity" in the total numbers of doctors to be trained. The majority of African medical schools have felt it necessary to enroll students of the highest possible scientific caliber and to train them to so-called internationally accepted standards. In the past this has meant that very few doctors would be turned out in any given year; but much more, that in many countries the best and only available scientific skills are diverted into medicine, depriving other areas of national development of potential manpower. An ethical issue of considerable importance raised by the lack of trained personnel is the fact that this skimming off of needed personnel to medicine, together with their costly training, is disastrous for many national needs. For in the end the doctors produced choose to become specialists who can practice medicine only when given reasonably sophisticated support facilities and services, frequently relieving existing hospital needs rather than those of preventive medicine. The frustration this leads to, as well as the wastefulness in terms of national need, is one of the major ethical issues on the African medical scene.

Another issue is charging fees to people who are very poor. If the practitioner is not supported by the national exchequer, he must charge fees to maintain his practice. However, a form of conduct that appears far from ethical exists in most of the West African countries, both French-speaking (Senegal, Ivory Coast, Togo), and English-speaking (Sierra Leone, Ghana, Nigeria, Kenya). Doctors who are paid a full salary with various amenities by the government still choose to charge the ordinary citizenry a fee for services. In Ghana, in the mid-1970s, doctors were paid an allowance specifically in lieu of such a practice. Yet there are some who take the government allowance and also continue with a fee-based practice. Such flagrantly unethical behavior may lead to a situation where confidence in medical leadership cannot be maintained.

Population, family planning, and abortion

Population control as it is advocated in the Western world has unfortunately helped to blur the issues of family planning and led to a debate that should have been completely unnecessary. Many African countries rightly consider themselves to be underpopulated. Some, such as Gabon, Cameroun, Central African Republic, maintain that they want much larger populations. All feel that they need development for the benefit of their people, but with very few exceptions they refuse to admit that curbing population growth is relevant to the needs for increased development. Unfortunately some doctors have hidden behind this national and international debate and have failed to recognize the doctor's role in family planning. Many doctors seem not to understand that postponing pregnancies until the woman is biologically most prepared and helping to stop reproduction when biological factors are no longer in her favor constitute an important medical consideration. They also fail to recognize that the spacing of the family, which used to be practiced in Africa based on either sexual abstinence or a geographic separation of man and wife, is itself very necessary for ensuring adequate health of both mother and child. The excessive mortality that simple childbirth puts on women fourteen to forty-five years of age has not been fully recognized by the majority of the medical profession (World Health Organization). Where it is recognized, adherence to inappropriate laws and practices imposed from Europe means that family planning services are withheld involuntarily from the majority of the population in need. The Catholic Church, through its influence on the French-speaking countries, has ensured that satisfactory medical leadership in family planning is not strong. This is enforced by the French contraception laws of 1920, which are still on the statute books of many French-speaking countries, despite their repeal by France and by Mali in 1972 (Wolf).

In the field of contraception, the major ethical

question posed to the doctor is, therefore, whether he should encourage free supplies of contraceptives, knowing full well that Europe and America, which are the sources of these supplies, require that they be handled almost exclusively by the doctor. Would he not be advocating a dual standard? The doctor needs, however, to balance against this possibility the results of withholding such supplies from populations that have nothing else to resort to.

Other serious ethical questions are raised in providing contraception to women who are either not married, according to the traditional norms prevailing in their locality, or want to practice contraception without the knowledge of their regular partner. Yet so tenuous are some of the marital relationships, so difficult is it to get some husbands into a hospital or family planning clinic when they are well, that insistence on consent by both man and woman might, in the end, do an injustice to the woman. Physicians must resolve this ethical dilemma within their own national frontiers.

The question of abortion is debated seriously. Many of the abortion laws in Africa are based on those of England and France, which repealed them in 1967 and 1974 respectively. However, in the majority of former British and French possessions the old laws were still on the statute books in 1975, and the increasing number of illegal abortions with their consequences of mortality, morbidity, and sterility have still not prompted the medical conscience to have the laws reviewed. Zambia did review its laws and alter them, but stipulations within the new law (1973), particularly the stipulation that the approval of two medical practitioners is required, make it unlikely that it will serve the majority of those in need. The Africa Regional Conference on Abortion held in Accra in 1973 agreed to call for a review of the laws, yet nothing was done. It is true that African societies generally do not accept abortions because they value highly the continuity of lineage; the unborn child, for example, may be a reincarnation of an ancestor. However, it would be untrue to say that abortions were not known in Africa before the arrival of the white man. In many cultures, pregnancies resulting from taboo relationships or from some types of adultery are interfered with—generally by women—and the men are kept in the dark. Once this is accepted, it means that there may be a need to review other grounds that might be considered suitable for abortion.

The doctor's dilemma in this field is twofold. He knows that, despite the law and his adherence to it, increasing numbers of women are risking their lives by recourse to backstreet abortionists. He also knows that there are so few doctors in relation to other needs of society that to make abortion laws more liberal may mean increasing the load on doctors still further. Given these problems, it is difficult to understand the view of some doctors in African countries that education, information, and services for fertility regulation should not be more widespread than they are now.

Throughout sub-Saharan Africa, the medical profession is seeking ways and means to understand better the kind of medical education needed to encourage training practices that would produce doctors better equipped to work within the African context.

FREDERICK T. SAI

[*Directly related are the entries* HEALTH, INTERNATIONAL; HEALTH POLICY, *article on* HEALTH POLICY IN INTERNATIONAL PERSPECTIVE; *and* POVERTY AND HEALTH, *article on* POVERTY AND HEALTH IN INTERNATIONAL PERSPECTIVE. *For a further discussion of topics mentioned in this article, see:* ABORTION, *article on* LEGAL ASPECTS. *Other relevant material may be found under* POPULATION ETHICS: RELIGIOUS TRADITIONS; *and* POPULATION POLICY PROPOSALS. *For discussion of related ideas, see:* MEDICINE, ANTHROPOLOGY OF.]

BIBLIOGRAPHY

BRYANT, JOHN H. *Health and the Developing World.* Ithaca, N.Y.: Cornell University Press, 1969.

FENDALL, N. R. E. *Auxiliaries in Health Care: Programs in Developing Countries.* Baltimore: Johns Hopkins Press, 1972.

JELLIFFE, D. B., and BENNETT, F. J. "Indigenous Medical Systems and Child Health." *Journal of Pediatrics* 57 (1960): 248–261.

LAMBO, T. ADEOYE. "The African Mind in Contemporary Conflict." *WHO Chronicle* 25 (1971): 343–353. Jacques Parisot Foundation Lecture, 1971.

SNYDER, FRANCIS G. "Health Policy and the Law in Senegal." *Social Science and Medicine* 8 (1974): 11–28.

United Nations. *Demographic Yearbook, 1973.* New York: United Nations Publishing Service, 1974.

WOLF, BERNARD. *Anti-Contraception Laws in Sub-Saharan Francophone Africa: Sources and Ramifications.* Laws and Population Monograph Series, no. 15. Medford, Mass.: Fletcher School of Law and Diplomacy, Law and Population Program, 1973.

World Health Organization. *World Health Statistics Annual, 1972.* Vol. 1: *Vital Statistics and Causes of Death.* Geneva: 1975.

Remaining articles for MEDICAL ETHICS, HISTORY OF appear in Volume 3.